P9-CFK-347

THE OXFORD COMPANION TO LAW

THE OXFORD COMPANION TO LAW

BY

DAVID M. WALKER

M.A., Ph.D., LL.D., F.B.A.,
One of Her Majesty's Counsel in Scotland,
Of the Middle Temple, Barrister,
Regius Professor of Law in the
University of Glasgow

CLARENDON PRESS · OXFORD

1980

Oxford University Press, Walton Street, Oxford OX2 6DP

OXFORD LONDON GLASGOW
NEW YORK TORONTO MELBOURNE WELLINGTON
IBADAN NAIROBI DAR ES SALAAM LUSAKA CAPE TOWN
KUALA LUMPUR SINGAPORE JAKARTA HONG KONG TOKYO
DELHI BOMBAY CALCUTTA MADRAS KARACHI

Published in the United States of America by Oxford University Press, New York

British Library Cataloguing in Publication Data

Walker, David Maxwell
The Oxford companion to law.
1. Law—Dictionaries
I. Title
340'.09181'2 K48 79-40846

ISBN 0-19-866110-X

Typeset by CCC in Great Britain by William Clowes (Beccles) Limited, Beccles and London. Printed in the United States of America.

PREFACE

My object in compiling this book has been to make available as concisely as possible information about some of the principal legal institutions, courts, judges and jurists, systems of law, branches of law, legal ideas and concepts, important doctrines and principles of law, and other legal matters which not only a reader of legal literature but readers in other disciplines and indeed any person whose work or reading in any way touches on legal matters may come across. As the book has been compiled and published in Britain it naturally has regard mainly to legal subjects as these are found in the legal systems of the United Kingdom, but I have tried also to take account of the major topics of legal history generally, of jurisprudence and legal theory, comparative law, international law, European Communities Law, and of the major topics, on the one hand, of those legal systems which have developed in other English-speaking parts of the world and which have been substantially influenced by and have strong links with United Kingdom systems of law, and, on the other hand, of the legal systems of the Western European countries with which the United Kingdom has long had, and now increasingly has, links. I have included entries on a few matters outside these limits where they have been frequently referred to in Western legal literature.

I have tried to complement rather than to duplicate existing legal works of reference, such as general textbooks, legal dictionaries, collections of legal maxims, and the treatises, commentaries, and textbooks on particular branches of legal science and on the law of particular jurisdictions on particular matters. Practically every entry could, with advantage, have been greatly expanded and none can be taken to deal fully with its subject. To have done so would have been quite impossible. I have necessarily on many matters excluded points of detail and qualifications of and exceptions to the broad statements. The references at the end of some articles and the Bibliographical Note (Appendix II) may give a lead to the major sources of further and detailed information. In particular I have not sought to compile a concise legal encyclopaedia of the law of any country, and readers desiring detailed information on rules of law must look in the textbooks of the particular system with which they are concerned. Still less is this book intended as a Layman's Home Lawyer. I have deliberately not given references to authority for many statements, as this would have taken up much space and been subject to frequent alteration. It is necessary also to emphasize that not only details but major institutions and doctrines of law differ in every jurisdiction and everywhere are constantly developing and being changed so that any statement may be superseded within quite a short time.

In compiling this book I have been greatly indebted for assistance and information to the Deputy Keeper of the Public Records, the Keeper of the Records of Scotland and the staff of the Scottish Record Office, the Librarian and staff of the National Library of Scotland, the Keeper and staff of the Advocates' Library, the Librarian and staff of the Institute of Advanced Legal Studies, the Librarian and staff of the Middle Temple, and the Librarian and staff of Glasgow University Library. I am grateful to several scholars who commented on various articles in draft but, above all, to Professor O. Hood Phillips, Q.C., D.C.L., who read the entire text in typescript and made many very valuable suggestions for improvements.

I also owe a large debt of thanks to the secretaries of the Faculty of Law, Glasgow University, namely Mrs. M. Buchanan, Mrs. D. Cameron, Mrs. D. Campbell, Mrs. D.

Hunt, Mrs. V. Jack and Mrs. A. Simpson, who over the years typed and retyped manuscript entries.

I am deeply indebted to the Delegates of the Clarendon Press for having encouraged me to proceed with compiling the book, and to the staff of the Press, and particularly Mr. P. H. Sutcliffe, for their interest in the work and their constant attention to all the problems of publication.

It is beyond possibility that I have included or dealt adequately with every point on which some enquirer might seek information or that I have avoided all errors or inaccuracies. For these faults I accept sole responsibility.

Department of Private Law,
University of Glasgow,
Glasgow, G12 8QQ.

D.M.W.

CONTENTS

ABBREVIATIONS

A.C.	Appeal Cases (House of Lords and Privy Council), 1865–
A.G.	Attorney-General
Adm.	Admiralty
All E.R.	All England Reports, England and Wales, 1936–
B.	Baron (judge), Court of Exchequer, England and Wales
C.A.	Court of Appeal, England and Wales, 1875–
C.B.	Chief Baron, Court of Exchequer, England and Wales; also Court of Common Bench (Common Pleas)
Ch.	Chancery Division, High Court, England and Wales
C.J.	Chief Justice
C.J.E.C.	Court of Justice of the European Communities
C.L.J.	Cambridge Law Journal, 1921–
C.P.	Court of Common Pleas; also Common Pleas Division, High Court, 1875–80
D.P. or Dom. Proc.	*Domus procerum* (the House of Lords)
D.P.P.	Director of Public Prosecutions, England and Wales
E.C.S.C.	European Coal and Steel Community
E.E.C.	European Economic Community
Exch.	Court of Exchequer; also Exchequer Division, High Court, 1875–80
Fam.	Family Division, High Court, England and Wales, 1972–
Foss	E. Foss, *The Judges of England*, 9 vols., 1848–64, or *Biographia Juridica*, 1 vol., 1870.
G.A.T.T.	General Agreement on Tariffs and Trade
High Court	The High Court of Justice, England and Wales, 1875–
H.L.	House of Lords
Holdsworth	Sir W. S. Holdsworth, *History of English Law*, 17 vols., 1922–66
I.C.J.	International Court of Justice
I.L.O.	International Labour Organization
J., JJ.	Justice(s), i.e. judges of the High Court, England and Wales; also Justice of the Supreme Court of the U.S.A.
J.P.	Justice of the Peace
J.R.	Juridical Review, 1889–
K.B.	Court of King's Bench; also King's Bench Division of the High Court, 1901–52
K.C.	King's Counsel
L.A.	Lord Advocate of Scotland
L.C.	Lord High Chancellor of Great Britain
L.C.J.	Lord Chief Justice of England
L.J., L.JJ.	Lord(s) Justice(s), i.e. judge(s) of the Court of Appeal, England and Wales, 1875–
L.J.C.	Lord Justice-Clerk of Scotland
L.J.G.	Lord Justice-General of Scotland
L.K.	Lord Keeper of the Great Seal
L.P.	Lord President of the Court of Session, Scotland

ABBREVIATIONS

L.Q.R.	Law Quarterly Review, 1885–
L.R.	Law Reports, 1865–
M.L.R.	Modern Law Review, 1937–
M.P.	Member of Parliament, United Kingdom
M.R.	Master of the Rolls, England and Wales
N.A.T.O.	North Atlantic Treaty Organization
N.I.	Northern Ireland
O.E.C.D.	Organization for Economic Co-operation and Development
O.E.E.C.	Organization for European Economic Co-operation
P.	Probate, Divorce, and Admiralty Division, High Court, England and Wales, 1875–1971; also President of that Division or now of the Family Division, High Court.
P. & M.	Sir F. Pollock and F. W. Maitland, *History of English Law before the Time of Edward I*, 1898
P.C.	Privy Council or Privy Councillor
P.C.I.J.	Permanent Court of International Justice
P.D.A.	Probate, Divorce, and Admiralty Division, High Court, England and Wales, 1875–1971
Q.B.	Court of Queen's Bench; also Queen's Bench Division, High Court, England and Wales, 1875–1901 and 1952–
Q.C.	Queen's Counsel
q.v.	*quod vide*, which see
qq.v.	*quae vide*, which (pl.) see
R.	*Rex* (the King) or *Regina.* (the Queen)
R.S.C.	Rules of the Supreme Court, England and Wales
S.C.	Session Cases, Court of Session, Scotland
S.G.	Solicitor-General
S.I.	Statutory Instrument
S.L.R.	Statute Law Revision
S.R. & O.	Statutory Rules and Orders
Supreme Court	The Supreme Court of the U.S.A.
s.v.	*sub voce* (under the heading)
U.K.	United Kingdom of Great Britain and Northern Ireland
U.N.	United Nations
U.N.E.S.C.O.	United Nations Educational, Social, and Cultural Organization
U.N.O.	United Nations Organization
U.S.	United States; United States Reports (U.S. Supreme Court)
U.S.A.	United States of America
v.	*versus*, against
V-C	Vice-Chancellor
W.S.	Writer to the Signet
Y.B.	Year Book; Y.BB. Year Books

UNIVERSITATIS GLASGUENSIS
BIBLIOTHECAE
D. D. D.

A

A and B lists. In British company law, lists made up when a company is being wound up, of contributories, that is of persons liable, in so far as their shares were unpaid or in accordance with their guarantees, to contribute to the discharge of the company's liabilities. The A list includes those holding shares at the date of winding up, and the B list those who were members within the past year but have ceased to be so; the latter are not liable to contribute in respect of debts and liabilities incurred after they ceased to be members.

***A fortiori*, argument** (by the stronger, so much the more). The argument that if rule A applies to case X, the present case is a stronger case for its application than case X, and accordingly rule A should apply to the present case also. It is not a logically compelling argument because whether the present case is truly stronger than case X may be open to argument, and there may be reasons why one rule should apply to one case and a different rule to a stronger case.

A mensa et thoro (from board and bed). A term adopted from canon law indicating the scope of a decree of divorce in the ecclesiastical courts which enjoined the separation of the spouses and the discontinuance of matrimonial cohabitation, but did not dissolve the marriage, nor permit the remarriage of either spouse. It is now represented by judicial separation, in which context the phrase is still used.

A posteriori (from the logically subsequent). A term used of reasoning from experience and not from axioms or postulates, from effects to causes, empirically or inductively. See also *a priori.*

A priori (from the logically prior). A term used of reasoning from abstract ideas to their consequences, from assumed axioms or postulates rather than from experience or particular instances, deductively.

***A priori* theories of law and justice.** A general descriptive term for those theories of law and justice, sometimes called 'metaphysical', developed by deduction from one or more assumed fundamental postulates. Thus Kant (q.v.) took as his basic postulate the free human will, because morality and justice are meaningless unless it be supposed that the subject is free to direct his will to one choice rather than another. Any principles derived from the categories are universally valid, independently of person, time, and place. From this pure categorical imperative, Act, according to a maxim which can be adopted at the same time as a universal law, Kant derived what he termed 'natural laws' drawing their validity from the category of free will, whereas positive law draws its validity from the expression of an authoritative will. The principles of justice are all derived from the basic idea of freedom. Fichte (q.v.) and Ahrens (q.v.) adopted a generally similar standpoint. Such theories stand in sharp contrast to empirical theories of law and justice which claim to impose obligation by virtue of derivation from the social facts and also to relativist theories, such as of Bentham (q.v.) or Pound (q.v.) which emphasize the unique nature of the judgment as to justice in every particular situation.

A vinculo matrimonii (from the bond of matrimony). A term derived from canon law indicating the effect of a decree for total dissolution of marriage. Divorce *a vinculo matrimonii* was obtainable from ecclesiastical courts prior to the Reformation only if the marriage was void by reason of a pre-existing canonical impediment, but after the Reformation such a divorce was obtainable in England on the ground of adultery only by private Act of Parliament, until divorce by judicial process was introduced in 1857. In Scotland after the Reformation the courts obtained powers to grant such divorce on the ground of adultery (at common law) or of desertion (under a statute of 1573).

Abandonment. The relinquishment of an interest or claim or of possession of property with intent to terminate proprietary interests therein, as by throwing away, or leaving and not seeking to retrieve it. In marine insurance the term is used of surrendering a ship or cargo to the underwriters on their paying its value as a constructive total loss; it then becomes their property for any scrap value obtainable.

Abandonment of action. The discontinuance of an action in court, normally because the plaintiff sees no chance of success.

Abandonment of domicile. The departure of a person from the country in which he has his domicile (q.v.) with the intention of acquiring a domicile in another country. Whether abandonment should be inferred has normally to be determined by reference to intention apparently disclosed by the circumstances of the case.

Abatement. Interruption of legal proceedings on a defendant's plea to a matter which prevented the plaintiff proceeding at that time or in that form, such as objections to the place, mode, or time of the plaintiff's claim. The modern rule is that an action does not abate by reason of marriage, death, or bankruptcy of a party if the cause of the action survives or continues. Criminal proceedings are not abated by the death of the sovereign or of the prosecutor, but are by the death of the accused.

Abatement of debts. The proportionate reduction of payments where a fund is insufficient to meet all debts in full.

Abatement of freehold. This arose where, on the death of a person seised of land, a stranger entered before the entry of the heir or devisee and kept the latter out. It is now obsolete since real estate vests in the deceased's administrator, not his heir, and pending a grant of administration, in the President of the Family Division of the High Court.

Abatement of legacies. Where the fund for the payment of legacies (q.v.) bequeathed by a will is insufficient to pay them all in full, they must abate, i.e. be cut down. Residuary legacies abate first, then general legacies, unless there is an apparent intention that any particular legacy is to be paid in full, or an order of priority is prescribed, then specific legacies, which abate only if the assets are insufficient to pay the deceased's debts. Demonstrative legacies do not abate, unless the assets are insufficient to pay the deceased's debts, until the funds from which payment is directed to be made is exhausted, when they abate along with general legacies. Within each class all legacies abate in the same proportion.

Abatement of nuisances. In England a public nuisance (q.v.) may be abated, i.e. stopped, destroyed, or removed, by anyone to whom it does a special injury, but to the extent necessary to prevent that injury only, and provided that the abator does not commit a breach of the peace in so doing.

A private nuisance may be abated by any person aggrieved on giving reasonable notice, but without causing a breach of the peace and without causing any greater damage than is essential. Notice is not necessary if the nuisance can be abated without entry on the wrongdoer's land, nor in case of emergency. In Scotland a common law nuisance may be abated under similar conditions to an English private nuisance.

Abbas modernus (or *abbas siculus*). The mediaeval title given to Nicolaus de Tudeschis (q.v.).

Abbott, Benjamin Vaughan (1830–90). U.S. lawyer, secretary of the New York Code Commission 1864 and draftsman of its penal code, a commissioner to revise the statutes of the U.S., 1870–72, and author with his brother Austin of many volumes of reports, forms of pleadings, and of the *United States Digest* (1879) and other Digests.

Abbott, Charles, Lord Tenterden (1762–1832). Practised first as a special pleader before being called to the bar by the Inner Temple in 1796 and built up a large practice. In 1802 he published his book on *Law relative to Merchant Ships and Seamen*, a long-neglected subject, which won him briefs and high praise. He never took silk but became a puisne judge of the Common Pleas in 1816, but was shortly thereafter transferred to the King's Bench and in 1818 succeeded Ellenborough as Chief Justice. As such he presided at the trial of the Cato Street Conspirators. His judgments were distinguished by their clarity of reasoning and perspicacity. In 1827 he was made a peer as Lord Tenterden, but took little part in politics in the House of Lords. Though not a great or learned judge, he was sensible and reasonable, and a master of legal principles, and his textbook has survived in successive editions to the present time as a work of great value. Some of his judgments are important, and in his last years he contributed to reforming statutes.

Abbreviate. In Scottish bankruptcy proceedings, a brief notice in statutory form, of the petition for sequestration of the bankrupt and of the court's deliverance thereon, which must be recorded in a statutory register within two days and prevents the bankrupt disposing of his estate.

Abbreviatio Placitorum. A volume consisting of a number of pleas abstracted from the rolls of proceedings in the *Curia regis*, King's Council, Parliament, and common law courts, in the period Richard I to Edward II, said to have been made by Arthur Agarde, Deputy Chamberlain of the Exchequer, and others in the times of Elizabeth I and James I, but in part made rather later, and printed by the Record Commissioners in 1811. Though only an abridgement and containing many errors, it is a useful source as containing the only first-hand information available of the working of the *Curia Regis* in its early days, and the earliest illustrations of the working of and earliest authoritative statements of the common law.

Abdication. The voluntary renunciation or abandonment of his office by a king or superior magistrate, as distinct from resignation, whereby an inferior restored his office into the hands of him from whom he received it. On the flight of James II in 1688 it was declared in England by the Bill of

Rights that he had abdicated the government and that the throne was thereby vacant. It is now thought that the sovereign's abdication can be effected by statute only and this was the course adopted when Edward VIII abdicated in 1936: see His Majesty's Declaration of Abdication Act, 1936. Abdication may be purely voluntary, as by reason of age, as in the case of Queen Wilhelmina of the Netherlands in 1948, or forced by circumstances, as in the cases of Nicholas II of Russia in 1917 and William II of Germany in 1918.

A'Beckett, Sir William (1806–69). English lawyer who migrated to New South Wales, became solicitor-general, a judge of the Supreme Court, and in 1853 Chief Justice of Victoria. He wrote various legal biographies and other works.

Abduction. The crime in English law of taking a girl under 16 from the possession of her parent or guardian, or a girl under 18 or a defective woman of any age from such possession for the purpose of unlawful sexual intercourse, or a girl under 21 with property or expectations of property from such possession to marry or have unlawful sexual intercourse, or of taking away and detaining any woman with the intention that she shall marry or have unlawful sexual intercourse with a person, by force or for the sake of her property or expectations of property. Abduction or kidnapping of any child is also an offence. Abduction of voters is also criminal.

In international law it is an infringement of the sovereignty of one state for agents of another state to abduct a person from the former state to the latter for trial and punishment there.

Abercromby, James, First Lord Dunfermline (1776–1858). Third son of General Sir Ralph Abercromby, was called to the English Bar in 1800 and later became a commissioner in bankruptcy and, in 1827, Judge-advocate-general. He was an M.P. from 1807 to 1830, in which year he became Chief Baron of Exchequer in Scotland. When the office was abolished in 1832 he re-entered Parliament and was elected Speaker in 1835. He retired in 1839, being then created Baron Dunfermline. Thereafter till his death he interested himself in public affairs in Edinburgh.

Abershaw or Avershawe, Louis Jeremiah (?1773–95). A notorious highwayman who terrorized the roads between London, Kingston, and Wimbledon until captured and hanged.

Abet. To encourage, incite, or assist another, especially to commit crime, usually found in the phrase 'aid and abet'. The act is called abetment or abetting and a person who instigates an abettor. An abettor, unlike the accessory, must be present at the commission of the crime, encouraging or assisting the principal. Assistance before the offence is counselling and procuring, and after the offence, e.g. to escape, is not abetting. Mere presence at the scene of a crime and failure to prevent it do not amount to abetting. An abettor was formerly, in the case of felonies, liable as a principal in the second degree, and now in all cases he is liable as a principal offender.

Abeyance. The situation where a right is not presently vested in anyone. It was formerly a fundamental rule of English real property that there must be no abeyance of seisin (q.v.) and any disposition of land contravening this rule was void, but this rule has been superseded since 1925. In certain exceptional cases abeyance persists, as in the principle that the fee simple of the glebe of a church is in perpetual abeyance, and that the freehold of the glebe is in abeyance between the death of one incumbent and the admission of the next one. The commonest modern case is where a peerage is in abeyance when there is no person presently entitled to it, though it has not lapsed.

Abingdon law. A phrase for summary justice, it being said that during the Commonwealth, Maj. Gen. Browne hanged his prisoners and then tried them. Cf. Jeddart justice.

Abinger, Lord Chief Baron. See SCARLETT, J.

Abjuration. Retracting or renouncing by oath. At common law in England it was the oath, taken by one who had claimed sanctuary, to leave the realm for ever. This disappeared with the abolition of sanctuary in 1624. Later statutes imposed abjuration as one of the penalties on Popish recusants. By the Abjuration Act, 1701, every person entering on public life had to be tendered an oath, known as the oath of abjuration, to the effect that he renounced the title of the pretended Prince of Wales (the son of James II) to the English throne, and recognized the rights of the dynasty established under the Act of Settlement; but all oaths of abjuration were abolished by the Promissory Oaths Act, 1871. It has been superseded by the oath of allegiance.

Ableman* v. *Booth (1859), 21 Howard 506. The Wisconsin courts freed Booth, an abolitionist, who had been convicted of violating one of the fugitive slave laws, in reliance on a state personal liberty law directed against the enforcement of the fugitive slave laws. The Supreme Court affirmed federal supremacy and denied the right of a state judiciary to interfere in federal cases, and the federal government rearrested and imprisoned Booth. The decision led the Wisconsin legislature to defend state sovereignty by adopting resolutions similar to the Kentucky and Virginia resolves.

Abolition Acts. Acts passed 1807–38 by the British Parliament on overseas possessions dealing with slavery. The Act of 1807 made it illegal for British subjects to engage in the slave-trade, that of 1811 made slave trading a felony, that of 1824 made it a capital offence, and that of 1833 abolished slavery and emancipated slaves after an apprenticeship period of four to six years. Amendment of the Act in 1838 improved the conditions of apprentices.

Abominable crime. In older legislation the euphemism for the crimes of buggery and bestiality (qq.v.).

Abortion. Expulsion of the human foetus from the womb before it has reached a state of development sufficient to permit it to survive independently, a state reached between the 21st and 28th weeks of pregnancy. It may occur accidentally or spontaneously (when it is usually called miscarriage) or be induced. Induced or artificial abortion has extensive legal implications. Most religions have long condemned it, but in the nineteenth century it was commonly made subject to heavy criminal penalties with exceptions occasionally recognized for therapeutic reasons, particularly the preservation of the life of the mother. The general illegality of abortion, however, gives rise to unauthorized criminal abortions, frequently performed by unskilled persons, with heavy mortality and risk. In 1920 the government of the U.S.S.R. authorized abortion on demand, and since 1945 moves to legalize abortion have been widespread. At common law in England it was criminal to procure abortion, unless done in good faith to save the life of the mother: *R.* v. *Bourne*, [1939] 1 K.B. 687. Apart from that it is criminal for a pregnant woman to take any drug or use any instrument with intent to procure her own miscarriage, or for any person to cause a woman to take anything or use any means on her with that intent. Alternatively a conviction may be secured for killing a child capable of being born alive. In Scotland abortion is criminal at common law, unless, probably, if carried out for therapeutic purposes; attempted abortion is criminal, if the woman is pregnant.

The Abortion Act, 1967 (Eng. and Scot.) permits the termination of a pregnancy by a medical practitioner if two medical practitioners are of the opinion, formed in good faith, that continuance of the pregnancy would involve risk to the life of the pregnant woman, or injury to the physical or mental health of her or any existing children of her family, greater than if the pregnancy were terminated, or that there is a substantial risk that if the child were born it would suffer from such physical or mental abnormalities as to be seriously handicapped. The operation must, save in emergency, be carried out in a hospital managed under the National Health Service Acts. This Act gave rise to a steep rise in the number of abortions in the U.K. and to acute controversy on social, ethical, and medical grounds, but an inquiry in 1973–74 recommended no change in the law.

Abridgement of All Sea Laws. *See* Welwod.

Abridgement of the Book of Assises. An anonymous alphabetical abridgement of matter from the Year Books, printed by Pynson in 1509 or 1510 and by Tottel in 1555. It is sometimes called *Liber Assisarum*, which confuses it with that work (q.v.). About a quarter of the 1,000 cases abridged comes from the *Liber Assisarum*, the rest from all reigns between Edward I and Edward IV, and they are arranged under 76 alphabetical titles, the chief topics being criminal law, land law, and procedure. It draws heavily on Statham's *Abridgement*.

Abridgements of the Statutes. The great quantity of statutes passed by the mediaeval English Parliaments gave rise to abridgements of the statutes. The earliest was a *Vieux Abridgment des Statutes*, including statutes down to 1455 in alphabetical order and published in French by Letton and Machlinia in 1481. There were later editions by Pynson. This was followed by Guillaume Owein's *Le Bregement de Toutes les Estatutes* of 1521, *The Statutes: The Abbreviation of Statutes translated out of French into English by John Rastall*, published in 1519 and 1527, in which the statutes are distributed under alphabetical titles and given both in the original language and in English, and F. Pulton's *Kalender or Table, comprehending the effect of all the Statutes, Magna Carta untill 7 Jacobi with an abridgment of all the statutes whereof the whole or any part is generall, in force and use*, published in 1560 and in later editions. There were also some less important works.

Later similar works include Edmund Wingate's *Exact Abridgement of all Statutes in force and use from the beginning of Magna Carta untill 1641*, published first in 1642 and later continued several times to later dates; T. Manby's *Exact Abridgment of all the Statutes made in the reigns of Charles I and Charles II to 1673* (1674); and Keble's and J. Cay's *Abridgment of the Publick Statutes in force from Magna Charta* (1739), continued in later editions to 1760 (1766). Later still were T. W. Williams' *Compendious Digest of the Statute Law* (1787) continued in later editions, and J. Gabbett's *Digested Abridgment and Comparative*

ABRIDGEMENTS OF THE YEAR BOOKS

View of the Statute Law of England and Ireland to 1811 (1812).

In Scotland Lord Kames (q.v.) published a *Statute Law of Scotland Abridged with Historical Notes* in 1757, and John Swinton published an *Abridgment of the Public Statutes in Force relative to Scotland from 1707 to 1754* (1755) and an *Abridgment of Statutes in Force relative to Scotland from 1707 to 1787* (1788) which was continued by W. Forsyth to 1827. Forsyth also compiled a *Dictionary of the Statute Laws of Scotland*, 1424–1707, and the whole of the *British Statutes in force relative to Scotland from 1707 to 1839*. There were also Abridgements by Alexander and Bruce.

Abridgements of the Year Books. Shortly after the first abridgement of the statutes came abridgements of the Year Books (q.v.). The first extant, attributed to Nicholas Statham in *Epitome Annalium Librorum tempore Henrici Sexti*, published about 1495, contains an abridgement in Norman French under titles in roughly alphabetical order of cases from Edward I to the end of Henry VI, and includes some cases not found in the Year Books. It is an excellent mirror of fifteenth-century law.

The next abridgement is the anonymous *Abridgement of the Book of Assizes* published by Pynson in 1509 or 1510 and reprinted by Tottell in 1555. It is an alphabetical arrangement of matter taken principally from the Year Books on the same plan as Statham's *Abridgement.*

Much more important is Sir Anthony Fitzherbert's *La Grande Abridgement*, first printed in 1514, and remarkable not only for accuracy but for its research. It contains extracts from Bracton's Note Book and from many unprinted Year Books, and was a model for future compilers of abridgements. It was extensively used by Staunford for his treatise on the prerogative.

The last was Sir Robert Brooke's *La Grande Abridgement* published in 1568; though based on Fitzherbert's, it contains much new material and draws on sources other than the Year Books.

These abridgements are almost entirely digests of case-law and ancestors of the modern Digests.

The later abridgements

In the seventeenth century there began to appear the later abridgements in which topics are divided under headings and the sources drawn on are Parliamentary records and statutes as well as cases, and which accordingly foreshadow modern legal encyclopaedias and collections of treatises on branches of the law.

The first of these is William Hughes' *Abridgement of the common law with the cases thereof drawn out of the old and new books of law* (1657) and *Grand Abridgement of the Law continued: or Collection of the Principal Cases and Points of the Common Law of England in the Reports, from the First of Elizabeth to the present time* (1660–63).

Much more important is Sir Henry Rolle's *Abridgement des Plusieurs Cases et Resolutions del Common Ley*, in law French, published, with an introduction by Sir Matthew Hale, in 1668, which contains many cases not elsewhere discoverable and seeks to make some arrangement of them within each title.

These were followed by Knightley D'Anvers's *General Abridgement of the Common Law* (1705–37), which was a translation of Rolle down to the title 'Extinguishment' with additional cases, by W. Nelson's *Abridgement of the Common Law; The Principal Cases argued and adjudged in the Courts of Westminster Hall* (1725–26), a poor work, by *A General Abridgement of Cases in Equity. By A Gentleman of the Middle Temple* (1732) which collects equity cases and adds some from manuscript sources, and finally Charles Viner's *General Abridgement of Law and Equity: Alphabetically digested under Proper Titles with Notes and References* in 23 volumes (1741–53). There was a second edition in 1791 and a supplement in 1799. This huge work was based on Rolle, but was built up from all other accessible materials, and is a most valuable compilation.

The transition to the modern encyclopaedia

The transition to the modern encyclopaedia, to the alphabetically arranged collection of systematic treatises and articles on all the branches of the law, is marked first by William Sheppard's *Epitome of all the Common and Statute Laws of this nation in force* (1656) and his *Grand Abridgement of the Common and Statute Law of England* (1675) which was based on *Coke upon Littleton* and *Coke's Institute.* But a great advance is seen in Sir John Comyn's *Digest of the Laws of England* (1762–67) frequently re-edited and extended down to 1882, in which the logical character of the work is more apparent, and the transition was completed in Matthew Bacon's *New Abridgement of the Law* (1736–66) which expanded and went through editions down to 1832. It combines a digest of the law with exposition of it and remains of value. William Cruise's *Digest of the Laws of England respecting Real Property* (1804) expounds the theory of the rules digested and also adopted a scientific classification, largely based on Blackstone, instead of alphabetical arrangement. Finally there came Charles Petersdorff's *Practical and Elementary Abridgement of Cases argued in the Courts of King's Bench, Common Pleas, Exchequer and at Nisi Prius; and of the Rules of Court from 1660* in 15 volumes (1825) with supplements down to 1870, which closely approaches a modern encyclopaedia in being arranged both alphabetically and analytically.

J. D. Cowley, *Bibliography of Abridgements, Digests,*

Indexes and Dictionaries of English Law to 1800; P. H. Winfield, *Chief Sources of English Legal History*.

Abrogation. The annulling of a rule by usage or contrary legislation. In the case of statute it is normally called repeal (q.v.). See also DESUETUDE.

Absence of accused. In cases of treason, an accused must be present at the preliminary enquiry and at the trial. In other indictable offences his presence at the trial is not essential. In summary offences the court may accept a plea of guilty in writing.

Absolute discharge. In Britain, a mode of dealing with a person convicted of an offence where the court, having regard to the circumstances, the nature of the offence, and the character of the offender, regards punishment as inexpedient and discharges him unconditionally.

Absolute dispositions, *ex facie*. In Scots law, an almost obsolete mode of creating security over land by conveyance of land which is *ex facie* outright but which is truly a transfer in security only, the fact that it is so qualified, the terms of the conveyance, and the undertaking to reconvey on the debt being discharged being contained in a separate deed, a back-letter or back-bond.

Absolute duties. Duties to which, according to Austin (q.v.), no rights corresponded. The category included duties not to persons but to God or to animals, duties to persons indefinitely or to the community, self-regarding duties, and duties owed to the sovereign. The first are not legal but religious or moral duties, the second are the aggregate of duties to many individuals, the third are probably impossible, and the fourth depend on Austin's view that there can be no legal relationship between subject and sovereign. These are no doubt of a special nature and the relations between subject and sovereign depend on the view adopted of the legal nature of the state.

Absolute, fee simple. An estate in land inheritable by heirs general, not defeasible nor determinable nor subject to condition. A fee simple absolute in possession is, since 1925, the only freehold interest capable of existing as a legal estate and is the largest interest in land known to the law.

Absolute interest. Full and complete ownership of property, liable to be determined only by the failure of successors in title.

Absolute liability. In the law of tort and delict, liability in damages, imposed under certain statutory provisions, and incurred by reason of the mere occurrence of an accident of a kind deemed prohibited, without regard to care or precautions taken and without need for proof of negligence or fault. It is sometimes confused with strict liability (q.v.), which is a slightly lower standard of liability. Similarly in criminal law absolute liability is liability for specified conduct or results independently of intention or other mental factor.

Absolute owner. A phrase in contracts for the sale of land meaning that the vendor will convey as beneficial owner.

Absolute privilege. In the law of defamation (q.v.) the privilege or legal entitlement to make a statement defamatory of another with complete immunity from liability to be sued, whatever the circumstances, as contrasted with qualified privilege (q.v.) where the freedom is conditional on the statement complained of not having been made maliciously. Absolute privilege attaches to statements made in parliamentary proceedings, in reports published by order of either House of Parliament, statements made in judicial proceedings, or documents made in such proceedings, statements made by an officer of state to another in the course of his duty and in a few other cases. The justification is the need for complete frankness of speech, even though it may be defamatory.

Absolute responsibility. The doctrine in international law whereby state responsibility may arise even though no intention or negligence can be imputed to the state.

Absolute, rule or order. A rule or order which is immediately of full effect and complete, as distinct from a rule or order *nisi*, which is made *ex parte*, on the application of one party only without notice to the other and which does not become absolute until the other party has had an opportunity of being heard and of showing cause why it should not be made absolute. If he shows cause the rule or order *nisi* is discharged, but if he does not it becomes absolute.

Absolute, term of years. A leasehold for a term which is to last for a certain fixed period, though liable to come to an end before the expiry of that period by re-entry by the landlord or on certain other events. It includes a term for less than a year, for a year, for years and for fractions of a year or from year to year. If a leasehold satisfies the requirements of a term of years absolute, it is a legal estate; if not, it is an equitable interest.

Absolute title. Registration of title under the Land Registration Act, 1925 as first proprietor of a freehold estate in land with absolute title vests in the person so registered an estate in fee simple in possession in the land, together with all rights, privileges, and appurtenances belonging or appurtenant thereto, but (a) subject to the incumbrances and other entries, if any, appearing on the register, (b) unless the contrary is expressed on the register, subject to such overriding interests, if any, as affect the registered land, and (c) where the first proprietor is not entitled for his own benefit to the registered land, subject as between himself and the persons entitled to minor interests, to any minor interests of such persons of which he has notice. Registration as first proprietor of a leasehold interest has a similar effect but is also subject to all implied and express covenants, obligations, and liabilities incident to the registered land.

Absolute warrandice. In Scots law a warranty that the grantee's title to land will not be reduced by anyone at all, implied in and normally also expressed in a conveyance of land for a full price.

Absolvitor. Decree in Scots law, absolving the defender from the claims made against him in a civil action. See *Assoilzie*.

Abstention. The deliberate refraining from doing, or the not-doing, some act, as distinct from inadvertent omission or failure to do it. Legal liability attaches in general for abstentions only where there was a positive legal duty to do the act which has not been done.

Abstract of title. In English real property law, a statement in chronological order of all the instruments and events under which a person is entitled to particular unregistered land, showing the history of the title and all incumbrances to which it is subject, prepared by the vendor's solicitor and delivered to the purchaser's solicitor, examined by him and checked by reference to the original deeds. The latter may make requisitions to ascertain any relevant fact undisclosed, to dispel doubts, or to remedy defects. The object is to enable the purchaser or mortgagee to judge the evidence deducing the vendor's title and any incumbrances affecting it. A contract for the purchase of land implies a right to an abstract unless this is expressly renounced. An abstract is also usually given in the case of a mortgage. In the absence of contrary stipulation a vendor must deduce his title from a good root of title at least fifteen (formerly thirty) years old, though in certain cases a longer title may be demanded.

In the case of registered land the vendor must supply the purchaser with authority to inspect the register and, if required, with a copy of the subsisting entries and of the plan, and also with copies, abstracts, and evidence in respect of any subsisting rights and interests relative to the registered land as to which the register is not conclusive and of any matters excepted from the effect of registration as the purchaser would have been entitled to if the land had not been registered, but beyond these requirements the vendor is not bound to supply written evidence of title.

Aburnius Valens. A Roman jurist of the Sabinian school, named by Hadrian to be *praefectus urbi feriarum Latinarum* and a pontiff and author of a *De fideicommissis libri vii.*

Abuse of civil proceedings, or of process. The tort or delict of initiating, and carrying through, civil proceedings against another, involving interference with liberty, or property, or liable to affect reputation, not only unsuccessfully or unjustifiably, but maliciously and without reasonable or probable cause. The plaintiff must show that the action was unsuccessful and that he sustained damage, and that the proceedings were taken in bad faith or in circumstances yielding an inference that they were taken maliciously and without probable cause. There is no liability merely because proceedings fail, or a judgment is reversed on appeal, nor because stringent measures are adopted unnecessarily, if they were legally justifiable. An action clearly frivolous or contrary to good faith may be stayed as an abuse of the process of the court, and a court has an inherent jurisdiction to protect itself from abuse of its own process.

Abuse of rights. The principle that in certain circumstances a person will not be permitted by law to exercise a right which he has by that law, on the basis that the right is conferred for certain purposes and may not be exercised for purposes foreign to these. The question arises particularly where a right is exercised with the sole or predominant purpose of harming another, e.g. A sinking a well on his land to abstract underground water beneath B's land. The doctrine of abuse of rights is recognized in most European legal systems, but without any very definite content or settled formulation, but is very inadequately recognized in common law systems. There is some support for it in some decisions of international tribunals. The doctrine has dangers because hardly any exercise of a power could not be challenged on this ground.

***Ac etiam* clause.** A clause, introduced by *ac etiam* (and also) stating the real cause of action in old cases where a fictitious cause was first stated in order to give the court jurisdiction. In the King's Bench in the Bill of Middlesex an allegation of trespass in Middlesex was necessary for jurisdiction, and this was followed by (*ac etiam*), a statement of the true cause of action, which might not have arisen in Middlesex. In the Common Pleas in the Writ of trespass *quare clausum fregit* a clause commencing *ac etiam* similarly gave the true cause of action and allowed that court to assume jurisdiction and thereby to capture business from the King's Bench. In both cases the initial allegation was fictitious and only the *ac etiam* clause disclosed the real cause of action. These forms were abolished in 1832. See *Latitat*; Bill of Middlesex.

Academic jurists. See JURISTS.

Acceleration. Where an interest in property in remainder or in reversion falls into possession earlier than it would otherwise have done, because the preceding interest is void or extinguished, as by surrender, merger, or lapse. Where a peerage has been disclaimed under the Peerage Act, 1963, a writ of acceleration, passing the peerage to the next heir, may not be issued.

Acceptance. In general, an act whereby a party agrees to a proposal, terms, or offer made to him, or undertakes a trust, office, or duty. In contract it is the indication by a person to whom an offer has been made by mode or conduct that he assents to the offer made to him. If unqualified, a binding contract is thereupon concluded.

Acceptance of bill. Acceptance of a bill of exchange is an indication that the drawee of the bill assents to the order of the drawer to pay the bill according to the tenor of his acceptance. It must be written on the bill and signed by the drawee, his signature alone being sufficient, and must not express that the drawee will perform his promise by any other means than the payment of money. It may be general, or qualified, i.e. conditional, partial, local, qualified as to time, or having the acceptances of some one or more of the drawees, but not of all.

Acceptance of goods. Where goods are delivered to a buyer which he has not previously examined, he is not deemed to have accepted them in performance of the contract of sale unless and until he has had a reasonable opportunity of examining them to ascertain whether they are in conformity with the contract. A buyer is deemed to have accepted goods tendered in performance of a contract of sale when he intimates to the seller that he has accepted them, or when the goods have been delivered to him and he does any act in relation to them which is inconsistent with the ownership of the seller, such as using or reselling them, or (except where otherwise provided) when, after the lapse of a reasonable time, he retains the goods without intimating to the seller that he has rejected them.

Acceptance of service. Where a solicitor, acting for a defendant or defender and with his authority, accepts a writ or summons on the latter's behalf, it has the same effect as if served on the defendant or defender personally. It is effected by writing on the principal writ or summons words clearly indicating acceptance of service on behalf of the defendant or defender.

Acceptilatio. In the civil and Scots laws, the verbal extinction of a verbal contract but without payment or performance, or with the acceptance of merely nominal satisfaction of the contractual undertaking.

Access. Approach or means of approach. In matrimonial law, it is the possibility of having intercourse with a spouse. Hence the presumption that a woman's husband was the father of her child may be rebutted by clear evidence of non-access by the husband at or about the time of conception.

Where spouses have separated and custody of a child has been awarded to one, the other is normally allowed access, i.e. opportunity to see and visit the child.

In land law an owner of land adjoining a highway has a right of access to it for any traffic required for the reasonable use of his property, and an owner of land fronting on water has a right of access to and from the water.

Access to the countryside for purposes of recreation is now secured by legislation.

Accessio. A mode of acquiring ownership of property. An owner of real or heritable property becomes owner of what is attached to it, as when a building is erected on land, or fixtures attached to a building: *quicquid plantatur solo, solo cedit.* An owner of personal or moveable property is owner of accessions to it, such as of accessories to his car, or what naturally accrues to it, as by the birth of young to his animals: *accessorium sequitur principale.* Artificial or industrial accession is the addition to the value of a subject by human art or labour exercised thereon, as by trees planted or buildings erected, as distinct from natural accession.

Accession. In international law the unconditional acceptance by one state of a treaty concluded between others.

Accession council. As soon as conveniently possible after the death of a monarch of the United Kingdom, an Accession Council meets, composed of the Lords Spiritual and Temporal, members of the late monarch's Privy Council, the Lord Mayor and aldermen of London and, in recent instances, the High Commissioners in London of Commonwealth countries. The new monarch then takes the oath for the security of the Church of Scotland prescribed by the Act incorporated in the Act of Union with England, 1706.

Accession declaration. A new monarch of the United Kingdom must before the first opening of Parliament or at his coronation declare that he is a Protestant and promise to uphold the statutes securing the Protestant succession to the throne: Accession Declaration Act, 1910.

Accession, deed of. A deed by a bankrupt's creditors approving of a trust secured by the debtor for the behoof of creditors generally.

Accession of the sovereign. On the death of a sovereign of the United Kingdom, his or her heir immediately and automatically becomes sovereign, and neither proclamation nor the making of the declaration required by the Accession Declaration Act, 1910, nor coronation, is necessary to his becoming sovereign for all purposes.

Accessorium sequitur principale. The accessory thing goes with the principal thing, the rule that fixtures or crops go with land, the young of animals with their mother.

Accession Treaty. The Treaty of Brussels of 1972 providing for the accession of the United Kingdom, Ireland, and Denmark to the European Communities. (Norway was also a party but did not ratify the treaty.) Under it the new members acceded to the existing Communities and accepted all their rules and became entitled to full membership on 1 January 1973 conditionally on incorporation of the Community law into their municipal laws. There was annexed to the Treaty an Act of Accession laying down the conditions of accession and making necessary modifications to earlier treaties and subsidiary instruments. It provided for a transitional period, expiring in 1977, for tariff reductions on trade between the U.K. and the original members.

Accessory. In English criminal law a distinction was formerly drawn in cases of felony (q.v.) between principals and accessories, the latter being divided into accessories before the fact, any who directly or indirectly by any means counselled or procured the commission of the felony but who were not actually present at or aiding in the commission of the felony, who were liable to the same punishment as the principal felon, and accessories after the fact, any who with knowledge that a felony had been committed, and, not being the wife of the felon, in any way secured or attempted to secure the escape of the felon by harbouring him or otherwise, who were themselves guilty of felony. In treasons (q.v.) all participants in any way are deemed principals. In misdemeanours (q.v.) all who would have been deemed accessories before the fact if the crime had been felony were deemed principals and punishable as such, and all who would have been deemed accessories after the fact were not guilty of any offence, unless the facts disclosed a substantive offence such as conspiracy to defeat the ends of justice. The distinction between felonies and misdemeanours was abolished by the Criminal Law Act, 1967, and all offences are governed since then by the rules relating to misdemeanours, so that the distinction between principal and accessory has virtually vanished. All who aid, abet, counsel, or procure the commission of any offence are liable as principal offenders.

As to Scotland see ACTOR OR ART AND PART.

Accessory obligation. In Scots law, an obligation additional and subordinate to another obligation and undertaken as security for the fulfilment of the primary obligation.

Accident. Colloquially, any undesired and unintended happening, especially involving harm or injury. In legal contexts such a happening entails no legal liability only if it was an inevitable accident, such an event as no normal human foresight or reasonable precautions could have avoided or prevented, e.g. damage by an earthquake. Legal liability, however, attaches if the accident was one of the kinds of happenings which, though unintended, should reasonably have been foreseen and by the exercise of reasonable precautions not allowed to happen, e.g. a vehicle collision, while in particular circumstances there may be strict liability (q.v.) if such a thing happens, or even absolute liability (q.v.) for the happening, whatever care and precautions may have been taken. In equity (q.v.) accident means an unforeseen event, misfortune, loss, act, or omission which is not the result of negligence or misconduct.

Accioly, Hildebrando Pompeo Pinto (1888–). Brazilian jurist and diplomat, author of *Tratado de Direito internacional público* (1933–35) and other works on international law and relations.

Accolade. The ceremony traditional in creating a knight, formerly by the king putting his hand on the knight's neck or shoulder or even striking him with his fist, and latterly by the king touching him on the shoulder with the flat of a sword.

Accolti, Francesco (Franciscus de Accoltis) (*c.* 1416–84). An Italian commentator, professor at Ferrara and Siena, author of *Commentaria super Lib. II Decretalium* (1481) and *Consilia seu Responsa* (1481).

Accommodation Bill. A bill of exchange signed by a party, as drawer, acceptor, or indorser, without receiving value therefor and for the purpose of accommodating or lending his name to another person. The accommodating party is liable on the bill to a holder for value, whether or not he was known to be such, but is entitled to be indemnified by the party to whom he has lent his name. The acceptor of an accommodation bill is a surety for another person who may or may not be a party thereto.

Accommodation works. Works such as bridges or fences, which a statutory undertaker, such as a railway or canal company, is required to make and maintain for the convenience of the owners or occupiers of lands adjoining the railway or canal.

Accomplice. One associated with another or others in the commission of a crime. The evidence of an accomplice is admissible against the other participant, but the jury must be warned to treat his evidence with reserve and that it is dangerous to convict without corroboration.

Accord and satisfaction. In English law, an agreement between two persons, one of whom has a right of action against the other, that the latter will give or pay or do something to or for the other and that the other will accept it in satisfaction of his claim. The accord is the agreement, the satisfaction is what is given. Accord and satisfaction bars the right of action, but neither element without the other has this effect.

Accorso, Francisco (Accursius) (*c.* 1182–*c.* 1260). Italian jurist, a pupil of Azo and later professor at Bologna and a leading figure in the renaissance of legal studies. He arranged the vast body of notes and comments on Justinian's Institutes, Digest, and Code into more methodical shape than ever before, so that it superseded all previous glosses and summaries and became known as the *Glossa ordinaria* or *magistralis* or the Great Gloss. His eldest son Franciscus (1225–93) also professor at Bologna, lectured on law at Oxford in about 1275, but returned to Bologna about 1280.

Account, action of. At common law in England an action lay against a receiver or bailiff, or by one merchant against another in respect of dealings between them in their mercantile capacities, for not rendering a proper account of profits. Persons entitled to an account were allowed to appoint auditors and if the accountant was in arrear, the auditors could commit the accountant to prison until the account was discharged. A plaintiff in either Queen's Bench or Chancery division may today endorse his writ with a claim for an account.

Account payee only. Words sometimes written across the face of a cheque to indicate that the sum in the cheque is to be credited by the collecting banker to the named payee only. The words have no statutory warrant but are commonly given effect to in banking practice and disregard of them may be negligence.

Account settled. A written statement of the account between a person bound to account to another and that other, showing what balance is due and to whom, which they have agreed and accept as correct. It is a good plea in defence to an action for an account. It may be set aside for fraud or serious errors.

Account stated. An admission by one person that money is due to another, from which law implies a promise to pay. An I.O.U. is evidence of an account stated. There need be no duty to account to another, so that an action for an account is not competent.

Accountant in Bankruptcy. An official who formerly had the management of the proceeds of bankrupts' estates, now replaced by the Chief Registrar in Bankruptcy.

Accountant-General or Accomptant-General. The Accountant-General of the Court of Chancery and the Accountant-General of the Court of Exchequer were officers formerly responsible for receiving and banking all moneys received into their respective courts. The Paymaster-General was substituted for both in 1872. In 1883 the Supreme Court Pay Office was established to act for all the courts and in 1925 the Accountant-General of the Supreme Court replaced the Paymaster-General.

Accountant of Court. An officer of the Court of Session who since 1889 has exercised supervision over trustees and commissioners in bankruptcies, over tutors, curators, and judicial factors, and has charge of all moneys consigned under orders of the court.

Accountant to the Crown. Any person who has received money for the Crown and is accountable therefor.

Accounting, action of. In Scots law an action of accounting, or of count, reckoning, and payment, is proper when one person is under a duty to render an account to another for his intromissions with the latter's property, as in cases of agent and principal, or trustee and beneficiary.

Accredit. To furnish an ambassador or other diplomatic agent with documents and authority which will ensure that he is received and given the credit properly due to him.

Accretion. The gradual addition of one thing to another, as where land is extended by accumulation of soil washed up, or a fund is increased by interest. It also occurs where one of several joint owners dies and his share of the property passes or accresces to the others.

In Scots law accretion occurs where an imperfect title is conveyed to a party and later is perfected, whereupon the validation attaches to that party also: *jus superveniens auctori accrescit successori.*

Accrual. Addition to a fund or property, as by the birth of young to animals, or the increase of land by alluvion. A right is said to accrue when it vests in a person, particularly if it does so gradually or without act on his part, or by the termination of a prior right.

Accumulation. The adding of income to capital indefinitely so as to produce a constantly growing fund. Prior to 1800 there was no rule in English law to prevent a settlement directing accumulation without distribution of the estate for duration of all lives in being plus 21 years (see Perpetuity), but, in consequence of *Thellusson* v. *Woodford* (1798) 4 Ves. 227,343; (1805) 11 Ves. 112,151, the Accumulations Act, 1800 (the "Thellusson Act"), set limits on the periods for which accumulations might be directed. It is now represented by the Law of Property Act, 1925, s. 164, as amended.

In Scotland there was at common law no rule against perpetuities but the Accumulations Act, 1800, was extended to Scotland and is now represented by the Trusts (Sc.) Act, 1961, s. 5, as amended.

Accursius. See ACCORSO.

Accusatory system. The system of criminal procedure in which a person, or a representative of the community, levels an accusation of crime against a person, and a magistrate, or a group of individuals drawn by chance from society, have to determine whether the accusation is substantiated or not. The system is said to symbolize and regularize the primitive combat. The alternative is the inquisitorial system (q.v.).

Accused, statement of. A statement sometimes made by a person brought before examining justices in answer to the charge after due caution that it will be taken down; if made, it is taken down, read over to the accused, signed by the magistrates and, if he chooses, by the accused.

Accused, rights of. The rights granted to a person accused of crime designed to ensure that he has a fair trial. The major rights are: not to be detained unnecessarily long without being tried, to be informed of the charge against him, to have fair opportunity to consult legal advisers and to prepare a defence, to be represented by counsel or solicitor, to have fair opportunity to cross-examine prosecution witnesses, to be allowed to give evidence and lead witnesses to counter the evidence given for the prosecution, to be heard in mitigation of sentence and, in certain cases and within limits, to appeal.

Different legal systems secure these rights to different extents and in different ways; what is very important is how far the rights accorded on paper by the relevant code or statutes are actually effective; and difficult questions arise as to self-incrimination, and protection from illegally-obtained evidence.

Achenwall, Gottfried (1719–72). Professor of philosophy, and later of law, at Gottingen, author of an outline of contemporary politics in European countries, *Constitutions of the Present leading European States*, works on economics, a *Jus Naturae* (1755), and other works. He was also a leader in the study of statistics.

Achéry, Jean Luc d' (1609–85). French Benedictine scholar, who published the *Spicilegium* (*Veterum aliquot scriptorum ... Spicilegium*) or *Dacheriana*, a collection of canonical literature comprising the *Dionysiana* and *Hadriana* (qq.v.) developed into a systematic whole, and numerous religious works.

Acknowledgment. An admission that a debt is due or a claim or liability exists, having the effect of cancelling the period of limitation which has run.

In relation to wills, if a will is not actually signed in the presence of witnesses, the testator must later acknowledge his signature in their presence, expressly or by conduct implying acknowledgment.

Acquiescence. Failure to object to an infringement of right, yielding an inference of assent and excluding a claim to equitable relief, arising where a person, in the full knowledge of his rights and of the conduct infringing them, allows the person to infringe, does not object, and leads the person infringing to believe that he has waived or abandoned his rights. The plea is frequently combined with laches (q.v.).

Acquired rights. Any rights in corporeal or incorporeal objects and having value, vested under a system of municipal law in a legal person, particularly those of aliens living in another state, which they are entitled to have protected by international law. Whether particular claims are 'rights' for this purpose is a matter of difficulty; the category includes property and economic concessions. Accordingly a state may nationalize or otherwise seize the property of its own nationals but may not do so in the case of acquired rights of aliens without provoking demands from the aliens' states for compensation. Interference with acquired rights may also take the forms of obstruction in the exercise of the rights, failure to pay rents or interest in respect of them, and other interferences. The cases where interference with the acquired rights of aliens are the rights to sequestrate (not confiscate) private property in belligerent-occupied territory and to confiscate enemy property taken at sea, the right of states to sequestrate alien enemy property in their own territory, and the rights of requisition and angary.

Acquirenda. In Scots law, property which may subsequently be acquired, particularly by a bankrupt or by a married woman after her marriage.

Acquisition of title. Title to property may be acquired originally or derivatively. Original acquisition comprises making things, taking possession of ownerless things such as fish and fowl, and the accession of land to existing land by natural forces. Derivative acquisition comprises acquisition of title from a previous owner, by gift, purchase, succession, or operation of law.

Acquittal. The setting free or deliverance of a person from a criminal charge, on a verdict of not guilty, or under a successful plea of pardon, of *autrefois convict* (q.v.) or *autrefois acquit* (q.v.). It bars a subsequent prosecution for the same offence, but whether it bars a subsequent prosecution for another offence depends on whether the evidence sufficient to support a conviction on the first charge would have been necessary for a conviction on the later charge.

Act. Any happening controlled by the human will, as distinct from an involuntary movement or from an event which happens naturally, or any voluntary muscular or glandular change together with its circumstances and immediate consequences. An involuntary movement, e.g. a sneeze, is not properly called an act.

Acts are sometimes distinguished into mental acts, comprising thinking, planning, and devising; these are generally of no legal significance, at least until disclosed, and frequently also until and unless resulting in and evidenced by overt physical actings; and physical acts, comprising doing or not doing various things. These acts may be positive, doing something, or negative, abstaining from doing or deliberately omitting to do something. Total inadvertence cannot properly be called a mental act, nor can bare omission or failure to act, as by forgetfulness, properly be called a physical act.

Law in general assumes that mental and physical activity can generally be controlled by a mental faculty usually called the will, and that a man can turn his mind to planning an activity, and can control his muscles to do or not to do particular actings. But weakness or disease of the mind may render him incapable of planning, and these factors, or drugs or rage, may overcome his self-control so that his will is, at least temporarily, not in control of his physical actings. In general a man can be held legally responsible for only those actings, mental and physical, which he could have willed and controlled, not for involuntary actings, such as sneezing or sleep-walking or things done while subject to an epileptic seizure.

The word act is, however, commonly confined to physical acts, to external manifestations of the actor's will. Voluntary physical acts may involve several distinct factors, particularly the motive (q.v.), which prompts the person to make, or refrain from making, the muscular change, the intention (q.v.), or other mental state which leads him to make or not make the change, the change or act itself, the circumstances in which the act was done or not done, and the immediate consequences which follow from the act or abstention. The muscular movement or act itself, such as swinging a stick or pressing the trigger of a pistol, is neutral by itself; the categorization of the act as legally innocent or culpable will depend on the other factors, the circumstances and immediate consequences.

Problems of difficulty arise in relation to consequences and may raise either or both of the issues of causation (q.v.) and remoteness (q.v.), i.e. did the act bring about what followed in time? and, is the consequence sufficiently closely related to, or on the other hand so distantly connected with, the act that its force must be deemed still effective, or spent and a limit set to the actor's liability for the act?

Act and Warrant. The document issued to the trustee in a Scottish sequestration, which confirms his appointment and confers on him title to, and powers to ingather, the bankrupt's assets.

Act in the Law (juristic act). An expression of the will or intention of a legal person directed to the creation, transfer, or extinction of a legal right and effective for that purpose, such as the making of a contract or will, or the execution of a conveyance. The main elements of an act in the law are the legal power of the actor to bring about the desired legal result, the direction by the actor of his will to the end in question, the expression or manifestation of that will, sometimes in a prescribed manner, and the absence of any factor, such as its being illegal or contrary to public policy, which vitiates or affects the essential validity of the transaction in question. An act in the law may be unilateral, such as making a will, or bilateral, where the wills of at least two parties are involved, each reacting on the other, as in the making of an agreement between two parties. See also ACT OF THE LAW.

Act of Adjournal. A regulation passed by the High Court of Justiciary in Scotland (q.v.), under powers originally inherent in it as the Court of the King's Justiciar and affirmed by the Act of 1672 instituting the High Court of Justiciary, or conferred by later statutes, regulating matters of procedure in the Scottish criminal courts. They are passed by a quorum of judges, recorded in the High Court Minute Books and in the Books of Adjournal, are now deemed statutory instruments and are cited, e.g. A.A. (Sittings of the High Court of Justiciary, Glasgow) 1967. They may be repealed, expressly or by implication, or fall into desuetude.

Act of Assembly. An item of legislation passed by the General Assembly of the Church of Scotland under powers partly assumed and partly founded on the General Assembly Act (of the Scottish Parliament), 1592, confirmed by the Confession of Faith Ratification Act (of the Scottish Parliament), 1690 and by the Protestant Religion and Presbyterian Church Act (of the Scottish Parliament), 1707.

Before the annual General Assembly passes any Act which is to be a binding rule and constitution of the Church, it must first, under Act IX of Assembly 1697, commonly known as the Barrier Act, be proposed as an overture from a synod or presbytery to the Assembly and, having been passed by the Assembly as such, be remitted to the consideration of the several presbyteries and their opinions be reported to the next General Assembly, who may then pass it, save that if a proposal has been sent down thrice as an overture, and presbyteries have failed to return their opinions, the Assembly may pass it without again sending it down. Besides Acts passed under the Barrier Act procedure the annual Acts of Assembly include some enactments on matters on which presbyteries have not been consulted. Other Acts deal with procedure in the church courts. The Acts of Assembly are found in various older collections and are now printed annually.

Act of Assembly (Scottish). See SCOTTISH ASSEMBLY ACT.

Act of Attainder. See ATTAINDER.

Act of Bankruptcy. An act indicative of insolvency, the commission of which by a debtor makes him liable to have a receiving order made against him in England, if within three months thereafter a creditor or creditors to the amount of £200 or over petitions therefor: see Bankruptcy Act, 1914, ss. 1–4. Bankruptcy commences on the day and at the time at which an act of bankruptcy is committed.

Act of Congress. An expression of the will of the President and congress of the United States, usually stating or altering the federal law of those States.

Mode of Legislation

Proposed legislation originates as a bill, which must be introduced by members of each House. There is no distinction, as in Britain, between government bills and private member's bills and as the President and members of the Cabinet are not members of the Congress, they cannot introduce bills directly, but must act through members of each House. Each House has an Office of Legislative Counsel to aid in the drafting of proposed legislation.

Bills are introduced by being filed with the secretary or clerk of the particular House. Each is given a number and referred by the presiding officer of the House (in practice by his clerk) to the appropriate committee. Each committee has regular meeting days at which proposed measures are considered. A bill may be referred by the chairman to a subcommittee for study and in cases of importance public hearings may be held, at which witnesses may present arguments for or against the bill. Bills are normally reported to the House only if approved and many do not survive this stage. If and when reported to the House, it is put on the calendar of that House and will normally come up for consideration in turn.

In proceedings on the floor of the House the debate is led by a member designated to represent it by the committee, usually the chairman. Debate in the House is limited but in the Senate unlimited, which permits filibustering. The bill is then voted on.

When one House has passed a bill it is sent to the other House for its consideration. If passed without amendment, it is submitted to the President for approval. If the second chamber makes amendments, the bill must be sent back to the first House for its concurrence. If it refuses to concur and the second chamber insists on its amendment, the bill is sent to a conference committee, consisting of members appointed by each House, which usually reaches agreement and whose report is usually confirmed by the vote of the two branches.

It is then submitted to the President for approval. He may veto it and frequently has done so; he must do so in whole and cannot do so in part. If he does not sign the bill within 10 days, it becomes law automatically. If he vetoes the bill, he returns it with a veto message giving his reasons. His veto may be overruled by a two-thirds vote of each house, and this occasionally happens.

The Parts of an Act

An Act commences with its number as a law of a numbered Congress, then the chapter and session number, which gives the order of approval in that session, the bill number and whether of the Senate or the House of Representatives, all merely for identification purposes, and whether it is an Act or a Joint Resolution. Then follow the preamble stating the purpose of the Act, and the enacting clause, which normally reads, Be it enacted by the Senate and House of Representatives of the United States of America in Congress assembled.

Then follow the sections of the Act, numbered consecutively, save that the first section is unnumbered.

Classification

There is no settled distinction between Public and Private Laws, the decision being one for the Editor of Laws of the Department of State, and the assignment of chapter numbers to new laws does not distinguish between the two. Public laws in general involve public policy and the national interest, whereas private laws are passed for the benefit of an individual or group; in case of doubt laws are classified as public.

Publication

The first officially published form of federal legislation is of each enactment as a separate sheet or pamphlet, known as a 'slip law', in two series Public and Private Laws, containing together both the Acts and, from 1940, joint resolutions of both Houses. In the same form Acts are later bound and indexed for the year as the Statutes at Large. Unofficial editions are also issued.

In practice reference is made to the United States Code, an official compilation of federal legislation from 1789 of a public character, rearranged by subjects under 50 'titles', and to a limited extent rewritten, under the supervision of a committee of the House of Representatives. The title and section numbers of the Code differ from the official titles and section numbers of the Act from which the material is derived. Some of the titles of the Code have been re-enacted as positive law. There are several private annotated editions of the U.S. Code.

Citation

In the U.S. Statutes at Large public and private laws are contained in separate volumes, each containing the laws in chronological order. Chapter numbers begin afresh with each volume, whereas law numbers are consecutive through an entire Congress.

Acts are commonly cited by volume and page of the *Statutes at Large*, e.g. 66 Stat. 282, but sometimes by public or private, Congress, session and chapter number, e.g. Public Law 415, 82nd Congress, 2nd session, chapter 478.

Many Acts are also known by popular name, e.g. the Taft-Hartley Law.

Until 1941 Acts and joint resolutions of both Houses of Congress were numbered chronologically and separately, but since then both have been listed without distinction as Public Laws.

Commencement

An Act becomes effective from the date of the President's approval unless another date is specified in the Act.

Interpretation

The principal basis for interpretation is the search for the ordinary meaning of the words used, but increasingly search is made for the intention of the legislature. Courts have frequently used extrinsic aids to establish legislative intention, as a guide to the interpretation, such as the legislative history of the Act. Failing assistance from such sources the court may seek to determine the purpose of the Act, or rely on some of the traditional maxims of interpretation, or on presumptions. Previous judicial interpretations are of great weight.

Interpretation is of great importance and may result in the statute being greatly restricted.

Act of Court. Legal memoranda of the nature of pleas, especially in the admiralty court.

Act of God. An event happening independently of human volition, which human foresight and care could not reasonably anticipate or at least could not prevent or avoid, such as earthquake or quite extraordinary storm. In torts act of God is a defence in cases of liability for intention or negligence and of strict liability. An insurer or a common carrier is impliedly, and frequently also expressly, not liable for loss caused by such an event.

In Scots law the corresponding concept is *damnum fatale* (q.v.).

Act of Grace. In England an Act of Parliament giving a general and free pardon was formerly called an Act of Grace. In Scotland by the Act of Grace, 1696, c. 32 a creditor who had had his debtor imprisoned was required to aliment his debtor if the latter were unable to do so himself; failing his making provision as required by the Act within ten days after being ordered to do so, the debtor was to be liberated. The Act applied in all cases of imprisonment by a creditor for a civil debt and substantially mitigated the law of imprisonment for debt. A prisoner claiming the benefit of the Act was bound, when desired, to execute a disposition *omnium bonorum* for behoof of all his creditors, and if he refused was not entitled to aliment so long as he persisted in his refusal. It became obsolete with the virtual abolition of imprisonment for debt in 1880 and was repealed as obsolete in 1906.

Act of Indemnity. An Act of Parliament passed to exempt a person from the legal consequences of something he has done or omitted to do, in breach of law, particularly if done in good faith. At the Restoration (q.v.) the Convention Parliament (q.v.) passed an Act of Indemnity to exonerate all who had committed acts against the monarchy, with certain exceptions. For long before the abolition of tests for municipal offices an annual Indemnity Act was passed to relieve dissenters who had accepted office. The Indemnity Act, 1920, was passed to restrict legal proceedings in respect of acts done in good faith for the defence of the realm during the First World War. Other instances have been to excuse the acts of military authorities in the exercise of martial law, or to relieve Members of Parliament from sitting while disqualified, e.g. Rev. J. G. MacManaway's Indemnity Act, 1951, or to relieve Ministers from the consequences of failing to lay before Parliament documents which should have been laid, e.g. Price Control and other Orders (Indemnity) Act, 1951.

Act of Oblivion. The statute 1660, 12 Car. 2, c. 11, which, except for certain named persons, gave a general indemnity for all illegal acts done during the interruption of government lasting from the beginning, in 1645, of the rebellion against Charles I to the restoration of Charles II in 1660.

Act of Parliament (or Statute). An expression of the will of Parliament, usually stating or altering the law in some respect.

Proposed legislation is initiated in the form of a Bill, which may be proposed by a Minister on behalf of the Government, or by an ordinary member (Private Member's Bill), or by a local authority, person, or body seeking Parliamentary powers or authority for some proposed action (Private Bill). Either of the first two kinds of Bill relates to a matter of public policy and has wide general application and is deemed a Public Bill; the third kind affects the rights and interest of the petitioners only. As an alternative to a Private Bill a body seeking Parliamentary powers may obtain them by obtaining a Provisional Order (q.v.) from a Minister of the Crown, which Order is subsequently confirmed by a Provisional Order Confirmation Bill, or by a Special Procedure Order (q.v.) under the Statutory Orders (Special Procedure) Acts, 1945–65, or in the case of Scotland by a Provisional Order (q.v.) from the Secretary of State for Scotland under the Private Legislation Procedure (Scotland) Act, 1936, which is subsequently confirmed by a Provisional Order Confirmation Bill. For the procedure of dealing with the various kinds of Bills see PARLIAMENT.

When a Bill has passed through all the requisite stages in both Houses it is presented for Royal Assent. See ROYAL ASSENT. As soon as this has happened a print of the Bill in final form is prepared in the Public Bill Office of the House of Lords, and assigned a chapter number. A proof of the Act is certified by the Clerk of Public Bills, sent to the Queen's Printer and the Controller of H.M. Stationery Office is requested to instruct the publication of the Act.

Two prints of the Act are prepared on vellum, endorsed with the words of the Royal Assent and signed by the Clerk of the Parliaments, and become the official copies of the Act. One is preserved in the House of Lords and one sent to the Public Record Office.

Classification

Statutes have been variously classified at different times.

The first distinction is between public Acts which will be judicially noticed without further proof, and private Acts, the existence and terms of which have to be proved in any court in which it is cited. But since 1850, and now under the Interpretation Act, 1978, every Act passed since 1850 is deemed and taken to be a public act and judicially noticed as such, unless the contrary is expressly provided and declared in the Act. With few exceptions all modern Acts are public in this sense. This distinction does not correspond to the classification of bills in Parliament into public and private.

The second distinction is that between Acts of general application to all persons in the country, local Acts which are limited in respect of area, and personal acts which are limited in respect of application to persons.

Accordingly the modern Queen's Printer's classification of statutes is into—

(a) Public general statutes (of general application and judicially noticed);
(b) Local, which may be public (i.e. judicially noticed) or declared to be not public; and
(c) Private, including personal, which may be printed by the Queen's Printer, or not so printed.

Since 1868 separate series of annual volumes of Public General and of Local and Personal and Private Acts have been published.

Statutes are also sometimes designated unofficially by such descriptions as *Declaratory Acts*, intended to remove doubts as to the existing law on some matter, *Consolidating Acts*, bringing together in one Act all the existing statute law on the topic,

Codifying Acts restating the whole law on the topic and superseding both existing common law and statutory rules on the subject, *Enabling Acts*, empowering and making it lawful to do what would not otherwise be lawful, *Remedial Acts*, intended to remedy some defect in the law, *Adoptive Acts* which apply in an area only if the local authority adopts them for that area, *Clauses Acts* supplying common form clauses which may be adopted in subsequent private bills, and *Statute Law Revision Acts* and *Statute Law (Repeals) Acts* which repeal large numbers of spent, obsolete, and unnecessary provisions.

Parts of Act

An Act comprises various parts:

The preamble is intended to state the reasons for and intended effect of the legislation. It is not a necessary part of an Act and is frequently now omitted. There may be a table entitled Arrangement of Sections showing what topic is dealt with in each numbered section. The long title sets out in general terms the purpose of the Act, e.g. An Act for Codifying the Law Relating to the Sale of Goods.

The enacting formula now runs: Be it enacted by the Queen's Most Excellent Majesty, by and with the advice and consent of the Lords Spiritual and Temporal, and Commons, in this present Parliament Assembled, and by the authority of the same, as follows:—. But Consolidated Fund Acts, Appropriation Acts and Finance Acts have a modified formula defining the sole responsibility of the Commons for the grant of money.

The substance of the Act is stated in a series of numbered sections, each with a note descriptive of the subject-matter thereof printed in the margin, and divided into subsections, paragraphs, and subparagraphs. Groups of consecutive sections may be called Parts, with a Roman numeral and a title, and in very large Acts the Parts are sometimes divided into Chapters—Parts and Chapters dealing with distinct topics. Without division into Parts a heading is sometimes prefixed to a group of related sections. The practice is to state the leading principles in the opening section, and then to deal with qualifications, exceptions, and details. Formal sections, such as those dealing with definitions, extent of application, and short title are now normally placed at the end.

Provision is frequently made empowering a Minister to make regulations under the Act prescribing matters of detail or machinery, and stating how this power is to be exercised, normally by statutory instrument (q.v.).

Interpretation and extent

Near the end of the Act, or possibly at the end of a Part or Parts, there may appear an interpretation section defining the meaning of particular words and phrases used in the Act for the purposes of that Act only, in terms which may be wider or narrower than the ordinary meanings of these words. The meaning may be defined by reference to the meaning assigned to the word in previous legislation on the same or a related topic.

There is frequently also a section or sections defining the extent of application of the Act, whether to the United Kingdom or to England and Wales only, or to Scotland only, or in parts to each country, and whether or not any sections extend to Northern Ireland, the Isle of Man, or the Channel Islands. Failing any such provision the presumption is that the statute applies to the United Kingdom.

Short title, citation and commencement

There is also a section stating the short title, by which the Act will normally be known and cited, e.g. the Sale of Goods Act, 1893. It may be provided that the Act or part of it, may be cited along with certain earlier legislation as, e.g. the Merchant Shipping Acts, 1894 to 1978.

The practice of conferring short titles on Acts began in 1845 and Short Title Acts of 1892 and 1896 christened many earlier statutes. The Statute Law Revision Act, 1948, gave short titles to other statutes. Some important older statutes were and are called from the place where the Parliament met, e.g. the Statute of Merton, 1235, others, such as the statute of Elizabeth (The Charitable Gifts Act, 1601, 43 Eliz. c. 4) from the name of the monarch, and some statutes are commonly and unofficially known from the name of the sponsor, e.g. Lord Campbell's Act (the Fatal Accidents Act, 1846).

Another section states the date when the Act is to come into force. If no express provision is made, an Act comes into force when it receives the Royal Assent, but it is common to provide that an Act or parts of it are not to come into force till a later stated date, or a stated period of time after the passing of the Act, or to empower a Minister by statutory instrument (q.v.) to bring the Act or sections of it into operation on a date or dates to be appointed by him ('appointed days').

Schedules

There frequently follow the text of an Act one or more schedules, frequently containing matters of detail, forms, and lists of prior Acts amended or repealed, in whole or in part, by the new Act, stating in what respects they have been affected.

Citation

Prior to 1963 United Kingdom statutes were cited by the serial number which the particular Act took as a chapter in the legislation of that session of Parliament, the session being numbered by the regnal year or years of the monarch over parts of which the session extended. (The regnal year extended from one anniversary of that monarch's accession to the next, and consequently normally extended over parts of two calendar years.) A session might all fall within one regnal year, or extend over parts of two, or even run into a third regnal year. Thus there were, e.g. the Patents Act, 1949 (12, 13 & 14 Geo. 6, c. 87), the Shops Act, 1950 (14 Geo. 6, c. 28), and the Income Tax Act, 1952 (15 & 16 Geo. 6 and 1 Eliz. 2, c. 10).

By the Acts of Parliament Numbering and Citation Act, 1962, all Acts passed after 1962 are cited simply by the calendar year and chapter number of the Act, e.g. The Television Act, 1963 (1963, c. 50).

The chapter numbers of Public General Statutes are printed in arabic numbers, of Local Acts in Roman numbers, and of Private or Personal Acts in arabic numbers in italic type.

Commencement

Till 1793, if no date were fixed for the commencement of a statute, it came into force on the first day of the session in which it was passed. Thereafter the date of receiving the Royal Assent was taken to be the date of commencement, and the phrase 'the passing of this Act' still means that date and not any date which may be fixed for the commencement of the Act.

Duration

Once passed an Act continues in force until amended or repealed, whether or not it or any part of it is brought into force, unless it is expressly provided that the Act is to expire on a stated date or occasion, or to expire unless continued in force by resolution of Parliament. But Acts due to expire are frequently continued by an Expiring Laws Continuance Act.

The doctrine that statutes may be abrogated by desuetude, by long disregard and non-enforcement, though applicable to pre-1707 Scots Acts, has no application to English, British, or United Kingdom legislation.

Promulgation

No formal promulgation is deemed necessary to make an Act binding or to acquaint persons with its provisions. Ignorance of the law is no excuse, and every person must acquaint himself with the statute law relevant to all his activities. Promulgation is in effect made by printing and publication of statutes.

See also CONSOLIDATION OF ENACTMENTS; INTERPRETATION; LEGISLATION; STATUTE LAW; STATUTE LAW REVISION.

See generally W. F. Craies, *Statute Law*; T. Erskine May, *Parliamentary Practice*; C. P. Ilbert, *Legislative Methods and Forms*; P. B. Maxwell on *Interpretation of Statutes*; H. Thring, *Practical Legislation.*

Act of Party. An expression of the will or intention of a person, directed to the creation, transfer, or extinction of some right and having legal effect for that purpose, e.g. the execution of a conveyance.

Act of Satisfaction, 1653. An Act whereby Cromwell divided Ireland into two parts: Clare and Connaught, which was left to Irish gentry and landowners; and the rest confiscated to meet the claims of the adventurers in Ireland and of the English army.

Act of Security. An Act passed by the Scottish Parliament in 1704 (A.P.S. XI, 69) providing that on the death of Queen Anne her successor should be nominated by Parliament, and should be of the royal line of Scotland and a Protestant, but, unless the Sovereignty of the kingdom were secured and guaranteed, that successor was not to be the person designated to succeed to the crown of England. The purpose was to define the constitutional position and vest authority in Parliament rather than the Queen's ministers. To this the English reply was the Alien Act, whereby, unless the Act of Security was repealed and the Hanoverian succession assured or a treaty for union in prospect by the end of 1705, Scots were to be treated as aliens in England. The Act was repealed when the Union of 1707 was agreed.

Another Act of the same name was passed in 1707 (A.P.S. XI, 402) in view of the apprehensions for the safety of the Presbyterian Church under the Union with England, then under discussion. It was provided that the Church of Scotland as then established was to continue without alteration in all succeeding generations, and the four universities were also to remain within the kingdom for ever. Though not embodied in the Articles of Union, this Act of Security was to be an indivisible part of it and the successors of the reigning sovereign were to subscribe and swear to it on their accession.

Act of Sederunt. An ordinance or rule made by the Court of Session (q.v.) under its inherent right as a supreme court or under the statutory powers of the court (Scots Act, 1540, c. 93) to make acts and ordinances for the ordering of process and the expedition of justice. Some early Acts of Sederunt were declaratory of law, others changed the law and were put before the Scottish parliament for ratification and had the force of statutes.

Since 1756 the Court does not appear to have exceeded the bounds of regulating procedure, and the power, now exercised under numerous statutes, is confined to the regulation of court business and procedure. They are recorded in the Books of Sederunt, there are many older collections of the Acts of Sederunt and a consolidating Act of

Sederunt of 1913 codified most of those still in force, being replaced by the Rules of Court in 1936. They are cited, formerly by date, e.g. A.S. 14 Dec. 1756, and now, e.g. A.S. (Rules of Court Amendment No. 3) 1967. Acts of Sederunt are now among the statutory instruments.

Act of Settlement. The English Act 1701, 12 & 13 Will 3, c. 2, which enacted that after the death of William III and his sister-in-law Princess Anne (later Queen Anne), and failing issue of either of them, the Crown of England should descend to Sophia, Electress of Hanover (daughter of Elizabeth, Queen of Bohemia, and grand-daughter of King James I of England and VI of Scotland) and the heirs of her body, being Protestants. It was a fundamental element in the Revolution Settlement in the years following 1688. By the Treaty of Union, 1707, it was provided that the succession to the Crown of Great Britain should be that provided for the succession to the Crown of England by the Act of Settlement.

It also provided that the sovereign must be a member of the Church of England by law established, and shall vacate the throne on becoming or marrying a Roman Catholic, that no person who has an office or place of profit under the King or receives a pension from the crown is capable of serving as a member of the House of Commons, that judges' commissions be made *quamdiu se bene gesserint*, and their salaries ascertained and established, but upon the address of both houses of parliament, it may be lawful to remove them, and that no pardon under the Great Seal of England is to be pleadable to an impeachment by the commons in Parliament.

The provision as to offices of profit under the crown depends on the Ministers of the Crown Act, 1937 and 1957, and the House of Commons Disqualification Act, 1975, subsequently amended by various Acts, and that as to judges' tenure of office in England by the Appellate Jurisdiction Act, 1876, the Judicature Act, 1925, s. 12, and the Courts Act, 1971, save that judges of all superior courts of the United Kingdom appointed after 1959 must retire on attaining 75: Judicial Pensions Act, 1959, s. 2. The Lord Chancellor is not affected by the rules as to tenure.

Act of Settlement, Irish. An Act of 1662 which substantially confirmed the Cromwellian settlement of Ireland in favour of the adventurers who supported the parliamentary cause. It began to reduce the land held by Irish Catholics from over 60 per cent in 1641 to 22 per cent in 1688 and 15 per cent in 1703 and established the Protestant ascendancy in Ireland which lasted till after Catholic Emancipation in 1829.

Act of State. An act of the executive as a matter of policy performed in the course of its relations with another state, including its relations with the subjects of that state, unless they are temporarily within the allegiance of the Crown. It is essentially an exercise of sovereign power done by virtue of the prerogative and hence cannot be challenged or controlled in municipal courts, so that the courts cannot question its validity and if an act done by the Crown or one of its servants with Crown authority is an act of state, the courts cannot enquire whether it is tortious or not. There can be no act of state against anyone who owes allegiance to the Crown, and it is not a defence to an action in tort brought by a British subject or by a friendly alien resident within British territory.

The official acts of every state or sovereign whose independence has been recognized by the Crown are acts of state in British law. Even if the actor is a British subject, in carrying out the act of state he commits no offence against British law, but British courts may inquire into the validity of decrees of a foreign sovereign government temporarily in Britain in their application to one of its own nationals resident here.

Act of Supremacy. The statute 1559, 1 Eliz. I, c. 1, which after the reversion to Catholicism under Mary Tudor, revived the legislation of Henry VIII abolishing the papal authority in England, repealed the heresy laws revived by Mary Tudor, and enacted that no foreign person was to have any power or jurisdiction in England, that all ecclesiastical power and authority should be annexed to the Crown of England, and that all ecclesiastical and lay officers should take an oath accepting the Queen as the only supreme governor in spiritual as well as temporal things. It was extended and its provisions made more stringent in 1563. The oath was replaced by one under the Promissory Oaths Act, 1868, which every clergyman of the Church of England must take before ordination or institution, and every archbishop or bishop before confirmation: Clerical Subscription Act, 1865.

Act of the Law. The creation, transfer, or extinction of a right by the operation of some rule of law itself without the intervention or consent of a party concerned. Thus when a man writes a book the law confers copyright on him; if he becomes bankrupt, it transfers his assets to another; if time elapses, it extinguishes certain rights. Similarly the law itself imposes on a man duties not to injure fellow-men, and not to infringe the criminal law. See also ACT IN THE LAW.

Act of Uniformity. Name given to various statutes regulating worship in the reformed Church of England, viz. 1549, 2 & 3 Edw. VI, c. 1, obliging clergy to use the Book of Common Prayer; 1552, 5 & 6 Edw. VI, c. 1, reinforcing the previous Act; 1559, 1 Eliz. I, c. 2, reviving the authority of the Book of Common Prayer of Edward VI with slight alterations, after the Marian reaction, and requiring it to be used, under penalties; it reserved also the jurisdiction of the ecclesiastical courts; and 1662, 14 Car. 2, c. 4, requiring the use of the Prayer Book scheduled thereto in all churches and applying the penalties of the earlier Acts to those contravening. All have now been partly repealed.

Act of Union. A term sometimes applied to the statute 1536, 27 Hen. VIII, c. 26, by which Wales was incorporated with and united into England and made subject to English law. It is more often applied to the Union with Scotland Act (of the Parliament of England) 1706, 6 Ann, c. 11, and the Union with Scotland (Amendment) Act (of the same Parliament) 1707, 6 Ann, c. 40, and to the Union with Ireland Act (of the Parliament of Great Britain) 1800 (39 & 40 Geo. 3, c. 67).

Acta. A general term for the enactments of Roman emperors, latterly covering *edicta*, *decreta*, and *rescripta*. After the end of the Republic emperors and magistrates took an oath to observe the *Acta* of previous emperors, except those whose *Acta* were rescinded on their deaths, or excluded from the oath. But good enactments of bad emperors might be preserved.

Acta Apostolicae Sedis. The official publication of the Holy See, issued since 1909, and the exclusive means of promulgating the laws of the Holy See, unless the See itself provides otherwise. It appears at least once a month, is in Latin, but usually prints documents in other languages. It includes encyclical and decretal letters, writings, and addresses of the Pope; decrees and decisions of the Congregations, tribunals, and commissions, and other matters.

Acta Diurna (or *Acta populi*, *Acta Publica* or simply *Acta*). A daily gazette in imperial Rome containing an official narrative of noteworthy events. It was started by Julius Caesar and the *Acta* were displayed daily on a white board and later taken down and preserved in the archives.

Acta Senatus (or *commentarii Senatus*). The official record of the proceedings in the Roman Senate under the Empire, made by a senator selected by the emperor.

Actio. In Roman law originally the activity of the plaintiff, but latterly the whole proceedings of a litigation, particularly the first stage, *in iure*, as distinct from *iudicium*, the second stage, before the *iudex*. The term is often combined with a word descriptive of the kind of claim, e.g. *actio mandati*, and sometimes has a material sense, as a claim or right.

Actio arbitraria. A Roman law action in which the formula (q.v.) required the judge, if he found the plaintiff's case to be valid, to make an order requiring the defendant to make over a thing to the plaintiff and fixing a sum which, in the event of failure to do so, the defendant should pay to the plaintiff.

Actio bonae fidei. An action in the Roman law available to a party in a *negotium bonae fidei*, in which the judge was allowed to take equitable considerations and pleas into account, as contrasted with *actio stricti juris*. All actions instituted by praetorian law were *actiones bonae fidei*.

Actio directa vel contraria. Contracts and obligations in Roman law gave rise to two actions, the *actio directa* to enforce implement of the main obligation and the *actio contraria* to enforce the counter obligation. Thus the *actio tutelae directa* was competent to the ward against his tutor to compel him to account, the *actio contraria* to the tutor against the ward for reimbursement of the expenses incurred in managing the ward's affairs.

Actio injuriarum. In the Roman law, an action for outrage, insult, or wanton interference with rights, for any intentional act which showed contempt for the personality of the victim and was such as to lower him in the estimation of others or outrage his feelings. *Injuria* was, according to Justinian's Institutes, one of the four recognized classes of harms, covering insult, contumely, or dishonour.

In Scots law the term is also used and has sometimes been misused as the term for the loss caused by unjustifiably causing the death of a blood-relative of the claimant.

Actio in personam. A personal action, in which the claimant claims by virtue of a right of obligation that the defendant should do, give, or make good something to or for him.

Actio in rem. A real action, in which the plaintiff claimed some thing as his against everyone else by virtue of a right of property. In particular, in courts exercising Admiralty jurisdiction an action *in rem* is one in which the plaintiff seeks to make good a claim to or against property, e.g. a ship or cargo, in respect of which he has a claim. In certain cases he can have the *res* arrested and detained.

Actio legis Aquiliae. In Roman law, the action founded on the *lex Aquilia*, which gave a remedy for wilful or negligent damage to property, originally for unlawfully killing another's slave or beast or for unlawfully damaging another's property, by burning, breaking, or destroying it, but later generalized into a remedy for any unlawful damage to property.

Actio per quod servitium amisit. A right of action given at common law in England to a master for loss to him resulting from injury to his servant done by a third party, whereby he lost the services of the latter, extended to a husband for loss of services resulting from injury to his wife. Similarly a father had until 1970 an action on this ground against a third party who by rape or seduction of his child capable of service to the parent caused him at least nominal loss; in effect it gave the parent a remedy for seduction of the child.

Actio personalis moritur cum persona. The rule that a personal claim perished on the death of a party. In Roman law the general rule was that a cause of action transmitted both to and against the representatives of the claimant or the wrongdoer. But there were important exceptions to this. In English law the maxim has been largely abrogated by the Law Reform (Miscellaneous Provisions) Act, 1934, and most causes of action survive against or for the benefit of the deceased's estate.

In Scots law the maxim survives in relation to claims for injury to feelings, but has never been applied to claims for patrimonial (i.e. financial) loss, which may be continued by the deceased's executor, while all claims transmit against the wrongdoer's representatives.

Actio quanti minoris. An action in Roman law whereby the purchaser of goods found to be defective claimed an abatement from the price in proportion to the defect. Under the Sale of Goods Act, 1893, a claim of this nature is competent. In relation to the sale of land such a claim is incompetent in Scotland.

Actio redhibitoria. An action in Roman law brought to avoid a sale and recover the price on account of vice or defect in the thing sold, which made it useless or so imperfect and inconvenient that the buyer would not have purchased it if he had known.

Actio spolii. In the canon law, an action, founded on the Roman law interdict *unde vi*, for the recovery of things stolen. It is thought to have suggested the assize of novel disseisin (q.v.).

Actio stricti juris. In Roman law an action in which the judge had to have regard to the strict rules of law only, as opposed to an *actio bonae fidei*.

Actio utilis. In Roman law, an action granted by the praetor under his judicial authority by way of extension of a recognized form of action to a category of persons or of cases not within the scope of the action as known in the civil law.

Action for annulment. In European Community law, proceedings to have the European Court review an act of the Council or Commission, other than a recommendation or opinion, and find the act illegal and declare it null. Parties may attack a Community act on any of the grounds of lack of competence, infringement of essential procedural requirement, infringement of the Treaty, or of any rule of law relating to its application, or misuse of powers (*détournement de pouvoir*). In general only member states or Community institutions may bring actions, but certain articles of the E.E.C. Treaty allow natural or legal persons to bring similar actions.

Action. A legal proceeding whereby one person seeks, in a civil court of justice by civil procedure, to enforce a right against, restrain the commission of a wrong by, or obtain a legal remedy from, another party. In particular contexts, synonyms may be used, such as 'suit' (the former equity term) or 'cause', particularly in matrimonial contexts. The general rule is that a person has a cause or ground of action if another person infringes any private right vested in him, though if the right is founded on statute it may be that a statutory remedy only may be pursued. If the right is a public one, the appropriate remedy for infringement is normally by proceedings of a public nature, such as prosecution rather than by action.

Under the old English system of forms of action (q.v.), actions were divided into real actions, claiming the right to recover lands; personal actions, claiming a debt, a chattel, or damages in lieu thereof, or damages for injury to him or his property; and mixed actions, partaking of the natures of both other classes.

Actions are today distinguished into action *in personam*, where the judgment of the court is an order or prohibition against the defendant, and actions *in rem*, found particularly in Admiralty practice, in which the plaintiff seeks to make good a claim to or against property, e.g. a ship or cargo, in respect of which, or of damage done by which, he makes an actionable demand.

Older writers frequently used the word in a wider sense, including both civil and criminal proceedings, but the latter are now usually distinguished from actions and called prosecutions.

Action, Forms of. A form of action in older English law was the technical mode of framing the writ and pleadings appropriate to the particular injury which the action was intended to redress. Originally all actions (other than suits in Chancery) were commenced by purchasing an original writ from the Chancery and each form of action was founded on a particular original writ. The writs early became limited in number and stereotyped in form, and the clerks in the Chancery writ office considered that they could not issue an original writ for which there was no precedent, and accordingly the forms of action, and the injuries for which remedies were competent, were restricted.

Actions were either real actions, whereby the plaintiff claimed the right to recover lands, or hereditaments; personal actions, whereby he claimed a debt, a chattel, or damages in lieu, or damages for injury to himself in person or property; or mixed actions, partaking of the nature of both real and personal actions, demanding both some real property and damages for a wrong done.

The principal real actions were actions of right proper, founding on a title in fee simple and seisin by the plaintiffs or by an ancestor under whom he claimed; actions of right in their nature, founding on gift or dower; actions of entry, when a tenant had come into possession of land; actions as to interests in land, to recover incorporeal hereditaments and *profits à prendre*; and actions of nuisance and of waste.

The forms of personal actions recognized prior to 1875 were: (1) On contract: (a) covenant, founding on a deed alone; (b) assumpsit, founding on a simple contract only; (c) debt, based on either a deed or a simple contract; (d) *scire facias*, on a judgment; (e) account; and (f) annuity. (2) On tort, where the writs were: (a) trespass *quare clausum fregit* (to real property); (b) trespass *de bonis asportatis* (to goods); (c) case; (d) trover; (e) detinue; (f) replevin; (g) *audita querela*; (h) conspiracy; (j) champerty; (k) attaint; (l) *decies tantum*; (m) forcible entry.

Of the mixed actions the most important was ejectment, whereby originally a lessee for a term of years, if ousted from possession, could have a remedy *quare ejecit infra terminum*, which came to be regarded as applicable only against purchasers from the lessor, and *ejectione firmae*, an action of the nature of trespass, applicable in all cases where the termor was dispossessed. Ejectment (q.v.) developed subsequently into a mode of trying the title to land and gradually usurped the place of all the other real and mixed actions for the recovery of corporeal hereditaments.

All the forms of action were abolished by statute in the mid-nineteenth century, finally by the Judicature Act, 1875. But while they lasted the forms of action profoundly influenced the development of the substantive law and even yet have not wholly lost their importance.

F. W. Maitland, *Forms of Action at Common Law*.

Action *in personam*. A civil claim resulting in a judgment of the nature of a command or prohibition directed against the unsuccessful party and only consequentially against his property in that he may suffer his property being sold to satisfy the judgment. It is distinct from a 'personal action' under the former system of forms of action (q.v.) and also from an *actio personalis* within the meaning of the maxim *actio personalis moritur cum persona* (q.v.) which was an action so connected with an individual that it fell on his death.

Action *in rem*. An action, particularly one brought in a court exercising Admiralty jurisdiction, in which the claimant seeks to make good his claim to or against certain property, such as a ship or cargo, in respect of which, or in respect of damage done by which, he makes his claim. It is quite different from a 'real action' under the former system of forms of action (q.v.). Originally a suit in Admiralty was commenced by arrest of the defendant or of his goods so as to make him put up bail for satisfying any judgment against him. Though this became obsolete the Admiralty court established the right to arrest property which was in dispute and to enforce its judgment against the property arrested on the basis that a maritime lien attached to the property from the time the claim was made to the extent of that claim. The right to enforce a maritime lien by action *in rem* was confined to the property by which the damage had been caused or in respect of which the claim had arisen. Modern law extends the right based on the maritime lien to many claims which do not give rise to a maritime lien.

Action on the case. In early English law an action lay only if a writ was obtainable from the Chancery, and writs were available only for recognized causes of action. See ACTION, FORMS OF.

Latterly however the clerks in Chancery began to issue writs for cases similar to those for which there were precedents. The development of the action on the case was long attributed to the Statute of Westminster II, 1285 (q.v.) which authorized the clerks in Chancery to issue writs in similar cases to those for which precedents existed, but the modern view is that it was developed independently by the courts. The writ was worded similarly to that for trespass but omitted the words *vi et armis*, and an action on the case could be had for every circumstance where a man had suffered loss or damage. It was no objection that the case was new in substance, so long as not new in principle. By the sixteenth century case had become a distinct form of action.

Actions on the case were either trespass on the case, covering wrongs similar to those covered by the writ of trespass, but without immediate violence; or general actions on the case, giving a remedy for all other wrongs, particularly for waste, deceit, and nuisance. The main distinction between trespass and case was that trespass lay for violence actual or implied, to tangible matter, and the plaintiff's interest was immediate, whereas case lay where there was no violence, or the matter affected was intangible, or the injury consequential only or the plaintiff's interest was in reversion. The one essential feature of case what that the plaintiff should set out in his writ the facts on which he claimed redress. Special, i.e. actual, damage had to be pleaded and found. The generality of this remedy opened the way for the development of such modern torts as defamation and negligence.

See C. H. S. Fifoot, *History and Sources of the Common Law*, Ch. 4; A. K. R. Kiralfy, *Action on the Case.*

Actions, real and personal. Synonyms for actions *in rem* and *in personam* respectively.

Acton, John (d. 1350). Canon of Lincoln and writer on canon law, wrote a learned commentary on the ecclesiastical constitutions of Otho and Ottobone, papal legates in England in the thirteenth century. First printed in 1496 in Lyndwood's *Provinciale*, his notes have long been held valuable for the interpretation of the canon law in England.

Acton Burnell, Statute of (or *Statutum de Mercatoribus*, 1283, 4 Edw. I). Was ordained by the king and his council at an assembly known as the Parliament of Shrewsbury (or Acton Burnell). Representatives of the commons participated in discussion of it, but not in its actual passing. The statute was intended to give merchants a speedy remedy for the recovery of debts, but this was not effective until the enactment of the Statute of Westminster II, 1285, c. 18, known as the *Statutum Mercatorum*. It was repealed in 1863.

Actor. A person who acts or does, the term borrowed from the Roman law, for the claimant, or plaintiff, in judicial proceedings, and also for the proctor or advocate in civil causes. The abbreviation Act. is still used in the Calling Lists of the Court of Session as indicating the pursuer's counsel.

Actor or Art and Part. In Scottish criminal law the principal participant in a crime is the actor, those who assist being responsible art and part. By common law and statute every person accused of a criminal offence is impliedly charged as guilty 'actor or art and part', i.e. as principal or accessory, no distinction in liability being drawn between primary and merely subsidiary participation.

Actor sequitur forum rei (a plaintiff must follow the defender to the jurisdiction to which he is subject). A general rule, subject to many qualifications, as to the forum in which the plaintiff should bring his action, if of a personal nature. The rule is founded on the principle that a plaintiff must seek judgment from a court which can, if necessary, make its order against the defendant effective.

Actus non facit reum, nisi mens sit rea (an act does not make a man guilty, unless his mind were guilty). The maxim often referred to as embodying the common principle of criminal law that guilt requires both wrongful conduct and the concurrence of the requisite mental element, whether intention or recklessness or even, exceptionally, negligence. But modern legislation has introduced numerous exceptions to the principle, where a person is made criminally liable if he does the forbidden conduct, irrespective of his state of mind, or, sometimes, unless he can exculpate himself on defined and limited grounds. In common law crimes the maxim still, in general, represents the law; in statutory offences it depends in every case on the terms of the Act or instrument creating the offence.

Actus reus. One of the essential elements of a completed crime (the other being *mens rea*), namely the human conduct which if done with *mens rea* is contrary to law. It is part of the definition of each crime to lay down what constitutes the *actus reus* thereof.

In some cases the *actus reus* lies in the conduct itself, e.g. perjury (the giving of false evidence on oath), or driving while under the influence of drink, where the conduct is criminal though no harm results to anyone, but in others the *actus reus* lies in the result, or in the conduct together with its direct result, e.g. murder, where the essence of the crime is the killing, the causing of death, the act or method being only part of the crime.

Actus reus is commonly the commission of some act, e.g. assault, but may be an omission, e.g. failure to exhibit lights on a vehicle, or a commission by omission, e.g. refraining from feeding an animal. An omission of either kind can be *actus reus* and criminal only where there was a legal duty to act and not to omit.

There is no *actus reus* and no crime if circumstances exist which amount to a lawful justification or excuse.

Ad colligenda bona, administrator. A person given a limited or temporary grant of administration (q.v.) to collect the property of a deceased where it is of perishable nature and a regular grant of administration cannot at once be made.

Ad factum praestandum (for the performance of an act). A term used in Scots Law of an obligation specifically to perform some duty, such as to deliver property or execute a deed, or a decree of court to such an effect, as contrasted with an obligation or decree to pay money.

Ad idem (at the same point). A phrase frequently used to signify that negotiating parties have reached agreement creative of a binding contract between them.

See also *Consensus in idem.*

Ad litem (with a view to a lawsuit). A term used of a guardian appointed to advise a child in relation to particular legal proceedings only.

Ad medium filum (to the middle line or thread). The line to which it is presumed that the lands of owners on opposite sides of a road or non-tidal river extend.

Ad valorem (according to value). A term used of duties or customs charged on certain articles, as contrasted with a flat rate duty without regard to value.

Ad vitam aut culpam (for life or till misconduct). A term in Scots law descriptive of the conditions on which certain categories of officers, e.g. judges or professors, formerly held their offices.

Adam, William (1751–1839). A nephew of the famous architects, Robert and James Adam, was called to the Scottish bar in 1773 but never practised there and in 1774 was elected M.P. for a Surrey borough. Later he represented various Scottish constituencies. He fought a duel with Charles James Fox in 1779 but later took part in negotiating the Fox-North coalition. He was called to the English Bar in 1782. In 1788 he was one of the managers of the impeachment of Warren Hastings, and in 1794 unsuccessfully moved an address to the Crown praying for clemency for Thomas Muir (q.v.) and Palmer.

In 1796 he took silk, in 1815 he was made a P.C., and in 1816 Lord Chief Commissioner of the Jury Court in Scotland (q.v.). He presided in that court till it was abolished by the Court of Session Act, 1830, but remained a judge of that latter court till his retirement in 1833. He was an intimate friend of Sir Walter Scott, edited *The Ragman's Rolls* for the Bannatyne Club and published a book on the Scottish jury system, a *Practical Treatise and Observations on Trial by Jury in Civil Causes* (1836) and other shorter works on the same subject.

Adat law. A complex collection of Indonesian customary laws and practices, mainly pre-Islamic and accordingly sometimes at variance with prescriptions of formal Islamic law. Nineteen *adat* law areas or circles were distinguished in the Dutch East Indies as cultural and geographic units, each having general internal similarity of *adat* systems.

C. Van Vollenhoven, *Het Adatrecht van Nederlandsch-Indie* (1906–33).

Additional Petition and Advice. A supplement to the Humble Petition and Advice (q.v.), prepared by the second Protectorate Parliament in 1654. It ended the period of attempts at a written constitution aiming at limited monarchy or republicanism.

Addled Parliament. A Parliament summoned by James I (his second English Parliament) on 5 April 1614 in the hope of raising money. The name was given because the Parliament was dissolved on 7 June without having passed any Act or granted any supply to the Crown. When it assembled the Commons demanded the abolition of impositions (q.v.), a demand which did not receive the support of the Lords, and the restoration of ejected clergy to their livings, and heated but futile discussions and eventual deadlock ensued. There were allegations that attempts had been made by a group of counties at the instance of the King to influence the elections, to obtain the return of members sympathetic to his views, but the attempt failed and the King strenuously denied the allegations.

Address. In the British Parliament, the reply of the Houses, particularly of the House of Commons, to the Queen's speech (q.v.) delivered at the opening of a new Parliament or a new session of Parliament. Two members in each House selected for the occasion move and second an address of thanks to the Sovereign. The debate on the address is an opportunity for discussing the whole policy of the Government and its proposals, and the Opposition criticizes the proposals and regrets omissions. After agreement the address, which takes the form of a single resolution thanking the Queen for her most gracious speech, is presented to Her Majesty who conveys her thanks for the address to the Lords through the Lord Steward of the Household and to the Commons through the Comptroller of the Household.

Either House may present an address to the Crown on almost any subject connected with the well-being or administration of the country or to express congratulation or condolence. An address from the House of Lords is agreed to by the House on motion, and an address from the House of Commons is agreed to by the House on motion made in the House and ordered to be presented to the Sovereign by the House or by members being Privy Councillors or members specially nominated. A joint address from both Houses originates in a resolution made in either House, communicated to the other, and an order is made by each House as to the manner in which the address is to be presented.

Ademption. The complete or partial withdrawal of a legacy by an act of the testator during his life, other than by the revocation of his will or the cancellation of the legacy. A specific legacy is held to be adeemed if that specific thing has ceased to belong to the testator before his death. A demonstrative legacy is not adeemed merely because the fund from which payment was directed is no longer in the testator's estate at his death, unless it is stated to be the only source of payment of the legacy. If it is given to a child, it is probably liable to be adeemed in so far as a portion has subsequently been paid to that child. If paid to a stranger in discharge of a moral obligation, it is liable to ademption if and so far as that obligation has been discharged before the testator's death. General legacies are subject to the same principles as demonstrative legacies.

Adherence. In international law, the action of a third state, with the consent of the original parties thereto, in accepting a treaty.

In the British law of treason, being adherent to the Queen's enemies in the realm, giving them aid or comfort in the realm or elsewhere, is treason: Treason Act, 1351; *R.* v. *Casement*, [1917] 1 K.B. 98,137; *Joyce* v. *D.P.P.*, [1946] A.C. 347.

In Scots law, it is the obligation of spouses to remain with and be faithful to each other, until or unless judicially separated or divorced. The word is also used of the action of a superior court when it agrees with and affirms the judgment of a lower court, e.g. the Court adhered.

Adhesion. In international law, accession (q.v.) by a state to a treaty with respect only to certain principles contained in it or to certain portions of it. The line between adhesion and accession with reservations is very uncertain.

Adhesion, contracts of. Contracts where the exercise of the will of one of the parties is in effect limited to adhering to or accepting the contract offered by the other or rejecting it. This arises where inequality of bargaining power exists and the terms of the contract are entirely dictated by one party, not negotiated, or where all the traders in a particular field offer the same or virtually the same terms, so that the customer has to take the terms offered or leave them.

Aditio hereditatis. In Roman law, acceptance of an inheritance, without which the bequest fell. It might be made by a formal act of acceptance (*cretio*), by acting as heir, or by mere expression of intent.

Adjective law. A generic term, invented by Bentham, for the bodies of principles and rules of law, other than the substantive law (q.v.), those dealing not with the rights and duties of persons, but with the means whereby those rights and duties may be declared, vindicated or enforced, or remedies for their infraction secured. It comprises, accordingly, the principles and rules relative to the jurisdiction of particular courts, court procedure, pleading and practice, evidence, appeal, execution of judgments, representation and legal aid, costs, conveyancing and registration, and procedure in administrative and non-contentious applications.

Adjourned summons. In the High Court in England a summons heard before a Master which is adjourned for further hearing before a judge, but also a summons in any court adjourned for further hearing.

Adjournment. The putting off of a hearing or meeting which is unfinished to another day, which may be fixed or *sine die*, i.e. deferred indefinitely.

In the procedure of the United Kingdom House of Commons the motion that the House do now adjourn is moved by a member of the Government at the conclusion of every sitting and provides the occasion for a short debate in which private members may raise matters which they wish to ventilate. A similar but longer opportunity arises on the days when the House adjourns for one of its seasonal recesses (Easter, Whitsun, Summer, Christmas).

Adjournment motions are also moved as an expression of respect on the death of a distinguished statesman, for the purpose of discussing 'a definite matter of urgent public importance', or to discuss a matter without necessarily reaching any decision on it.

Adjournment may be moved in the course of a debate, but the Speaker may refuse the motion if he considers it an abuse of the rules of the House, or put the question without debate. Such a motion is used to delay or prevent the House reaching a decision on the matter under discussion.

Adjournment may also take place automatically, as when the Speaker has to adjourn the House for lack of a quorum.

In High Court procedure adjournment is used of transferring applications made in chambers into court for public hearing, or vice versa.

Adjudication. In English and American law, the judgment or decision of a court, particularly the order which declares a debtor bankrupt and his property vested in a trustee for realization and division among his creditors.

In Scots law, it is the form of diligence (q.v.) by which a creditor attaches land. It may be adjudication in security, or adjudication for debt, whereby the creditor has the debtor's land transferred to him in settlement, initially redeemably but eventually outright. There are also recognized adjudication in implement, whereby a purchaser of land can have the title to lands purchased, the seller of which refuses to transfer the title, judicially transferred to him, and adjudication *contra haereditatem jacentem* whereby, when a debtor's heir renounced his succession to the lands, the creditor might obtain a decree to constitute the debt with a view to an adjudication of the land.

The term is also used of the determination of the Inland Revenue what, if any, stamp duty is chargeable on a particular instrument. The fact that a deed has been so considered is indicated by an adjudication stamp.

Adjustment of average. In maritime law the process of determining the values which the different interests and items thereof are to contribute to making up a general average (q.v.) loss. Under the York–Antwerp Rules, 1950, general average falls to be adjusted both as regards loss and contribution on the basis of values at the time and place where the maritime adventure ends.

Adjustment of loss. The settling and fixing of the amount of indemnity due to the assured under a marine insurance policy, after making necessary deductions and allowances, and the fixing of the amount which each underwriter is liable to pay.

Adjustment of pleadings. In Scottish civil procedure, the alteration of pleadings by counsel before the record (q.v.) is closed, so as to focus better the issues between the parties.

Adkins* v. *Children's Hospital (1923), 261 U.S. 525. An Act of Congress for the District of Columbia fixed minimum wages. A hospital which employed workers at less than the minimum fixed wage challenged the statute as in violation of the due process clause of the Fifth Amendment. Notwithstanding previous decisions permitting wage and price regulation legislation, the Supreme Court overruled the statute in question. The decision was a classic case of identification of freedom of contract and *laissez-faire* economics with constitutional rights and stood until overruled by *West Coast Hotel Co.* v. *Parrish* (1937) 300 U.S. 379.

Admeasurement of dower. At common law a writ by which an heir could obtain redress if, during the heir's infancy, the predecessor's widow had been assigned more land than she was entitled to as her dower.

Adminicle. In the Scottish action of proving of the tenor (q.v.) of a lost deed, a deed, copy, draft, or other evidence which tends to establish the existence and terms of the deed sought to be re-established.

Administration. A term formerly used of the body of persons appointed to carry on the government of the country, now more usually called 'the Government'. Also used in English law of the process of winding-up a deceased's affairs, ingathering his assets, paying debts, and distributing the balance to those entitled by law on intestacy or under the deceased's will, and of the obtaining from the Family Division of letters of administration to an administrator. An ancillary administration, subordinate to the original administration, must be obtained, in the case of foreigners, in the country where the foreigner's assets are situated.

The term is also used of the management of an estate in cases where difficulties in administration have arisen under orders for administration made by a judge of the Chancery Division.

Administration action or matter. An action or matter initiated in the Chancery Division by a creditor, or a trustee, executor, or administrator, who requires the direction of the court on the matter, to secure the proper administration of the estate of a deceased person. If the court makes an order for the general administration of the estate, the personal representatives cannot exercise their powers without the sanction of the court.

Administration bond. A bond with one or more sureties, unless sureties are dispensed with, for double the amount at which the deceased's assets are sworn, which must be given by any person to whom letters of administration of the estate of a deceased are granted in England. It is not required of an executor. The conditions of the bond are to make up and exhibit an inventory, to administer the estate, according to law, and to exhibit any later will which may be found and then to deliver up the letters of administration.

Administration, letters of. The authority which a person receives from the Family Division, to administer the estates of a deceased person. Administration may be granted *cum testamento annexo* where the deceased left a will but appointed no executor, or the executor has died or declines to act, or *de bonis non administratis*, where the executor dies intestate or the administrator dies, or *durante minore aetate*, where the executor nominated is under age, or *pendente lite*, where there is dispute as to the person entitled.

Administration, oath of. An oath which must be taken by persons applying for a grant of administration as to his right or title to the grant and that he will make due administration of the estate.

Administration of estates. The general term in English law for the whole process of dealing with the property of a person deceased. It covers the offices of personal representatives of the deceased, executors, and administrators, the grant of probate or letters of administration, the duties of the representatives to get in the assets, pay debts and legacies, and distribute the residue to persons entitled.

Administration, public. The science of exercising the function of departments of state, public corporations, departments of local government authorities, administrative agencies, and public utilities, and thereby giving effect to and applying in concrete cases the principles and policy determined by the national executive and, within the limits of the authority delegated to them, the corporations, local government authorities, and the directing bodies of the various public utilities. The powers and authority of the various agencies of public administration are defined by administrative law, but these agencies also operate very largely within the spheres of ordinary private law, e.g. when employing staff and purchasing stores.

Administration, right of (or *jus administrationis*). The right under Scots law, now abolished, of a husband, by virtue of marriage, to manage the heritable property of his wife, by virtue of which his consent to all her acts affecting the property was essential.

Administrative agency. A body established to exercise on behalf of the government and in the public interest, but so far as possible independently of political considerations, a regulative or controlling function over the conduct of particular professions, trades, industries, or services, which can also rely more on professional skills and experience than if the regulation were by politicians or civil servants.

The earliest of such agencies in Britain arose from the need for regulation arising from the rapid development of railways, the abuses attendant thereon, such as unfair preferences to some traders, and the inadequacy of the common law procedure applied to them. The Railway and Canal Traffic Regulation Act, 1854, was wholly inadequate and the Regulation of Railways Act, 1873, created a board of railway commissioners which gave way to the Railway and Canal Commission in 1888. The Railways Rates Tribunal of 1921 was more judicial than regulatory.

Since 1930 Area Traffic Commissioners have controlled the provision of bus services and road transport by a licensing system.

Since 1945 regulation by administrative agencies has in some cases, e.g. electricity generation and supply, been superseded by the nationalization of the industry concerned. In other cases, as under the National Health Service, new agencies such as Area Health Boards have been created to manage and provide a public service.

In the United States, administrative agencies have been in the main of the commission type, largely independent of the executive, and not directly accountable to it. The first such agency was the Interstate Commerce Commission of 1887, established to regulate interstate commerce particularly by controlling the rates and facilities of railroads. It, however, initially lacked effective power, though under the Transportation Act of 1920 it was given the duty of building up the national railroad system and came to have delegated legislative, administrative, and quasi-judicial functions.

Among later agencies on the same general lines are the Federal Trade Commission (1914), the Federal Power Commission (1920), the Federal Communications Commission (1934), the Securities and Exchange Commission (1934), the National Labor Relations Board (1935), the U.S. Maritime Commission (1936), and the Civil Aeronautics Board (1940).

Apart from these bodies the executive departments of the federal government all possess significant administrative and regulatory powers, and the tendency has been for the internal organization of these departments to approximate to the regulatory commission in form.

A leading trait of American administrative agencies in respect of enforcement has been the combination in them of the functions of legislator, prosecutor, and judge which in the view of many, violates a fundamental principle of natural justice. Judicial review is rarely available to compel effective enforcement of the law by administrators.

Administrative business. The management of affairs, including the taking of decisions, in circumstances where the persons concerned have complete freedom to act as they think right, as distinct from judicial business, which is conducted subject to defined rules, particularly as to hearing parties and evidence.

In the Chancery Division the term is applied to the part of the court's business concerned with executing trusts and wills, and deciding issues incidental thereto, as distinguished from contentious business.

Administrative county. In England and Wales the area for which a separate county council was elected prior to 1973. Though there were 40 counties in England and 12 in Wales, there were 61 administrative counties.

Administrative functions. One of the major groups of functions of government, involving the application of general policy in particular cases and particular situations, including some exercise of discretion in particular cases. 'Ministerial' functions are similar, but permit little or no discretion in their exercise, as in the execution of a warrant. They are distinct from legislative functions and judicial functions and are exercised by a large number and range of officials from Ministers of the Crown downwards, the permissible extent of discretion diminishing with lower grades of officials.

Administrative law. The body of principles and rules stating and governing the functions and powers of all the agencies of government concerned with the application, working out, and practical administration of government policy. These agencies include Ministers and departments of state, local authorities, public corporations, agencies, boards, commissions, and other bodies created to perform particular functions. The number and variety of administrative bodies and the range of their functions have developed enormously in the twentieth century. To a considerable extent disputes arising in the course of the administration of policy are determined by administrative tribunals (*q.v.*). Administrative law differs from constitutional law in being concerned with the subordinate agencies of government rather than with the supreme executive and legislature, and with the operation and function of government and the application of rules in practice rather than with the formulation of the general policy of internal government.

The functions entrusted to administrative bodies are multifarious and their powers great, both being normally derived from statute. Their functions include the regulation of town and country planning, the operation of the public education system, the provision of a system of social security, the direction of an industry or public services, the regulation of social evils such as gambling, and many other similar activities. Their powers are frequently conferred in wide terms, conferring a wide measure of discretion on the persons administering the function in question. Their powers may include: legislative power delegated by Parliament, to make rules and regulations for the detailed working out of the general principles laid down by Parliament; directive powers, whereby they may require private persons to do or refrain from doing conduct of stated kinds; investigating powers, to require information, statistics, returns and so on from individuals; licensing powers, to grant or refuse permission for the carrying on of particular kinds of business or trades or particular uses of premises; taxing powers, to impose and exact compulsory payments of rates, levies, or dues in certain circumstances; and quasi-judicial powers, to investigate objections and decide disputes arising incidentally out of the administration of the service in question.

A major defect of administrative law in Britain has been the totally unplanned and piecemeal development of such bodies and their powers, which frequently vary markedly from one to another. A general principle is that the body having a power must exercise it within the limits of the authority conferred and comply with the substantive, formal, and procedural conditions laid down for its exercise. Statutory powers must be exercised in good faith.

One of the crucial problems of administrative law is how far, in what circumstances, and by what means the ordinary courts will intervene at the instance of an aggrieved individual to question, control, and overrule some acting of an administrative authority. In short, how far is the administrator unfettered in the exercise of his discretion in the furtherance of his functions, and how far is he subject to control by law? In France administrative authorities are exempt from control by the ordinary courts but are subject to strict control by administrative courts and the *Conseil d'Etat*, or supreme administrative court, has great power to investigate and give redress for excess of official action. In Anglo-American systems no separate administrative courts exist and the individual's only redress against alleged misuse of power is by invoking legal remedies originally devised for other purposes. Some advance was, however, marked by the creation, in the U.K. in 1959 of the Council on Tribunals, which is not a court of appeal but has a supervisory function over tribunals and must approve the codes of procedure of many of them.

Delegated legislative power is controlled to some extent by Parliament, and to a small extent by the courts which may declare null an exercise of a power which is *ultra vires* or, probably, unreasonable, but have no control over the content of the delegated legislation.

In England control by the ordinary courts of the exercise of administrative power has been effected mainly by the use of the prerogative writs and orders. The writ of *habeas corpus* (q.v.) requires a person exercising restraint of the person of another to produce the person detained and show legal cause for the detention. If this cannot be done, the court will release the person. This writ is mainly used in alien and extradition cases, or where military authorities assert jurisdiction over a person. *Quo warranto* (abolished in 1938) was used to try the legal title of one presuming to exercise a public office. By an order of *mandamus* (q.v.) the court orders a public officer to perform some function which he should have performed and which the petitioner is entitled to have performed. It may be used to require an officer to exercise a discretion, but cannot be invoked to direct him how to exercise it. By an order of *certiorari* (q.v.) the court may order an inferior judicial body to submit the record of its proceedings so that it may consider whether they must be quashed on the ground of error in law. *Prohibition* (q.v.) lies to an inferior court or quasi-judicial tribunal to restrain it from proceeding in excess of its jurisdiction. An *injunction* (q.v.) may also be used to compel a defendant to refrain from some course of conduct. In 1978 an 'application for judicial review' was introduced as a means of claiming judicial review. In many cases, however, the statute governing the administrative action in question severely restricts or wholly excludes review by the courts.

In Scotland none of these writs or orders is used and control is effected by ordinary actions of declarator, reduction, or suspension and interdict (qq.v.) again subject to there being no statutory exclusion of challenge or review.

In the United States, delegated legislative power, to be valid, must be limited by prescribed standards, and grants of power by Congress are subject to scrutiny by the Supreme Court to see whether or not they are inordinate, but this has generally been exercised realistically and not unduly narrowly. Delegated legislation is of two main classes, contingent legislation where the application of the act is contingent on certain conditions, it being for the administrative agency to determine whether or not the conditions exist, and supplementary or subordinate legislation, or the authority to make administrative rules which shall have the force of law. All U.S. administrative agencies have such rule-making authority, but power to impose penalties cannot be thus conferred. Administrative agencies also have power to issue rulings to clarify the statutes they apply. Such rule-making powers have latterly tended to be exercised in consultation with the interests affected, or after a hearing, under the American Administrative Procedure Act, 1946, of objections to proposed rules.

Judicial control of delegated legislation is exercised mainly by reliance on the *ultra vires* doctrine (q.v.).

Both executive departments and administrative agencies possess quasi-judicial powers, e.g. to deal with applications for air transportation permits. Persons aggrieved by refusal of an application can normally secure judicial review of the decision if they can allege that due process of law was not observed. Also details of formal adjudicatory procedure have been laid down by Congress in the federal Administrative Procedure Act of 1946, which imposes procedural standards on the administrative process far exceeding those required under the due process clause.

In America the tendency has been for statute to regulate judicial review of administrative tribunals' decisions. The two main methods are: where the administrative order is not self-operative, to require the agency to bring a suit for its enforcement, in the course of which the court can review the agency action; and to provide for a party aggrieved petitioning for review in the appropriate court. But non-statutory review may also be obtained through the prerogative writs and very commonly the grant of an injunction. Non-statutory review has been invoked in many cases where no remedy is available under statute, or the statute does not cover the situation which has arisen. Judicial control exists where a challenge can be based on *ultra vires*, disregard of natural justice, or absence of substantial evidence for the determination, though the trend has been to narrow the scope of judicial review.

H. W. R. Wade, *Administrative Law*; J. F. Garner, *Administrative Law*; S. A. de Smith, *Judicial Review of Administrative Action*; H. Street, *Governmental Liability*; B. Schwartz, *French Administrative Law and the Common Law World*; J. Bennett Miller, *Outline of Administrative and Local Government Law in Scotland*; K. Davis, *Administrative Law* (U.S.).

Administrative procedure. The course of procedure followed by administrative departments and authorities in the exercise of quasi-judicial functions, in deciding claims or controversies brought before them, or the appropriate tribunal or person, for adjudication. Administrative procedure lacks the formality, and carefully formulated course of happenings, laid down for civil and criminal courts. The only requisites appear to be that any rules of procedure laid down for the particular tribunal or process of adjudication should be observed, and that the rules of natural justice (q.v.) be observed. But to a large extent controversies have been entrusted to administrative tribunals to avoid the formalism of court procedure and the development of formalism is contrary to the spirit of the legislation. Appeal is frequently limited, and frequently lies to a special appellate tribunal, or to a Minister, rather than to an ordinary court.

Administrative tribunals. Bodies and persons outwith the hierarchy of courts, strictly so-called, set up under numerous statutes and entrusted with the investigation and decision of matters in controversy and disputes arising out of the functioning of public administration. They exist commonly to deal with disputes between a citizen and a department of state such as to claims for social security benefit, but also to deal with disputes where specialized knowledge or experience is necessary, such as assessment of compensation for land acquired compulsorily, or which are thought unsuitable for the courts, such as the fair rent for premises let. Increasingly in the twentieth century disputes arising not out of public administration but out of private law relationships, e.g. as to fair rent, have been entrusted to tribunals rather than to ordinary courts, and such are preferably called statutory tribunals or simply tribunals. In the United Kingdom there is no consistent pattern as to size, composition, procedure, rights of appeal, and so on, and every kind of tribunal in *sui generis*.

Since the Tribunals and Inquiries Act, 1958, steps have been taken to keep many of these tribunals more under review by the establishment of the Council on Tribunals (q.v.) which has a general oversight over tribunals and enquiries and must be consulted before any new procedural rules are made, and a right of appeal to the courts on a point of law has been conferred on more tribunals.

Administrator. A person appointed by the Family Division of the High Court to ingather the assets of a person who has died intestate, or without executors appointed, surviving, or accepting, to pay his debts, and to distribute the balance to those entitled on intestacy or under the deceased's will. Until administration is granted the deceased's personal estate vests in the President of the Family Division. Administration is normally granted to one or more of the persons interested in the residuary estate of the deceased or to a trust corporation, but a creditor may obtain a grant if none of the persons entitled to the estate is willing to take it out. If the estate is, or is believed to be, insolvent, the Public Trustee (q.v.) can, on certain conditions, obtain a grant. In certain cases a limited grant of administration may be made. The office of administrator, unlike that of executor, is not transmissible, and if an administrator dies a fresh grant *de bonis non* (i.e. *non administratis*) must be made.

Administrator in law. A title given in Scots law to a father as tutor (or mother as tutrix) of a child in pupillarity (q.v.), or curator, or curatrix of a child in minority. The term is exegetic of the designation of the parent as tutor or curator and does not confer any further powers nor does it constitute a distinct office.

Admiral, Lord High. The title of Admiral signifies the commander of a fleet of ships of war. The first instance of an appointment under that title by an English king appears to have been in 1295. The function of the early admirals was the command of fleets only and they do not appear as judicial figures till about 1360. After 1323 the practice seems to have been to divide the coastline of England into two admiralties, one north and one south and west of the Thames. An admiral of all the fleets was appointed in 1360 and in 1412 the title Admiral of England, Ireland and Aquitaine first appears, while in 1540 one man was designated Lord Admiral. In 1619 Buckingham was appointed Lord Admiral and became responsible for the Navy Board, which was charged with the care and maintenance of the royal ships, a responsibility never later relinquished. The title Lord High Admiral is found in 1638, and also the title of First Lord. In 1628 the office was first put into commission and from 1709 was continuously in commission except when the Duke of Clarence (later William IV) was Lord High Admiral in 1827–28. Queen Elizabeth II has revived and assumed the title of Lord High Admiral. The office of First Lord of the Admiralty succeeded that of the first commissioner of the Boards of Admiralty of the seventeenth century. The office of First Lord was abolished in 1964, being superseded by that of Secretary of State for Defence.

As a commander of forces at sea distinctions were established between admirals of the white, blue, and red, and later the distinct grades of Admiral of the Fleet, Admiral, Vice-Admiral, and Rear-Admiral. In modern times senior administrative appointments are held by various grades of admirals and the ranks also extend to officers of the engineering, paymaster, and other branches of the service. In the U.S. the rank was unknown until David Farragut was appointed successively Rear-Admiral (1862), Vice-Admiral (1864), and Admiral (1866). The grade of Admiral of the Fleet was created for George Dewey in 1899.

For bibliography and lists of office-holders see *Handbook of British Chronology*, p. 125.

Admiral, Lord High, of Scotland. This Office of State commanded the royal ships and sailors and had supervisory duties over all seaports, harbours, and seacoasts. He had sole jurisdiction in maritime and seafaring causes, civil and criminal. He originally presided in person in the Admiralty Court, the jurisdiction of which was defined by statute in 1554 and 1567, but later delegated these duties to a Deputy Judge-Admiral. The court sat in Edinburgh, dealt with all crimes committed at sea, as well as actions concerning crimes, faults and trespasses committed upon the sea or in the ports or creeks thereof or in navigable waters so far as the sea ebbs and flows. The office was hereditary in the Dukes of Lennox but was in abeyance during the Commonwealth and Protectorate but was revived in 1660. The court resumed and was held by a Judge-Admiral rather than an Admiral-Depute. In 1673 the King conferred the office of Lord High Admiral of Scotland on James, Duke of York (later James VII and II), and the heritable office on the Duke of Richmond and Lennox, who in 1702 appointed a Lord High Admiral.

After the union Vice-Admirals were appointed by the Crown to be judges in civil affairs relating to naval and commercial concerns in Scotland, and there was also a Judge-Admiral, appointed in 1746 by the Vice-Admiral but thereafter by the Crown. The office seems now to have lapsed.

Admiralty action. Any action which may be tried in the Admiralty court in the exercise of its customary or statutory jurisdiction. The most distinctive feature of Admiralty process is the procedure *in rem*, whereby a ship or other property which has given rise to the action may be arrested and, if need be, sold and the proceeds thereof used to satisfy the plaintiff's claim.

Admiralty court. A court constituted in 1970 as part of the Queen's Bench Division of the High Court to take causes and matters involving the exercise of the High Court's Admiralty jurisdiction or its jurisdiction as a prize court, and previously dealt with in the Probate, Divorce, and Admiralty Division. It is staffed by such judges of the High Court as the Lord Chancellor may from time to time nominate to be Admiralty judges. The jurisdiction is derived partly from the customary jurisdiction of the High Court of Admiralty, but now largely from statute, and includes jurisdiction in wartime as a prize court (q.v.). Certain county courts have limited admiralty jurisdiction.

Admiralty, Court of, in Scotland. The origins of this court are uncertain, and it seems probable that the earliest Scots admirals were given jurisdiction similar to that conferred by other European States at that time and that they absorbed the jurisdiction of the coastal burghs in maritime matters. The office of High Admiral was certainly known in the fifteenth century and became hereditary in the Bothwell family from 1488. A Crown charter of the office to Lord Francis Stewart in 1587 gave him as wide jurisdiction as enjoyed by the Admirals of France, Spain, England, and other countries.

The Lord High Admiral of Scotland and the Isles thereof exercised his jurisdiction through Vice-Admirals and by Admirals-Depute, later called Judge-Admirals, holding courts in the principal seaport towns. Inferior grants of jurisdiction of an Admiralty nature were also made to Edinburgh and Glasgow over the Firths of Forth and Clyde, with appeal to the Court of Admiralty. It extended to maritime causes, civil and criminal, over the whole of Scotland and the court became the only court of first instance in maritime disputes and questions of crime committed at sea. It seems to have administered the customary law of the sea, and in claims involving foreigners it adopted, and applied the law merchant earlier and more fully than did the common law courts, and hence it became popular as a commercial tribunal.

The Court of Admiralty was in constant conflict with the common law courts on matters of jurisdiction. The Court of Session advocated to itself maritime and non-maritime causes, though only in the latter had it an admittedly superior jurisdiction. In 1609 the Admiralty court was declared by statute to be a sovereign court, but this did not imply its independence of the Court of Session. Under the Commonwealth a new Admiralty Court was established subordinate to the Court of Admiralty in England, with five judges. The separate court revived in 1661. An Act of 1681 similarly failed to ensure the independence of the Admiralty Court by failure to define 'maritime causes' in which the Admiral's jurisdiction was declared privative. The same conflict of jurisdiction arose with the High Court of Justiciary in criminal causes. By the Act of 1681 the exclusive jurisdiction of the Admiralty Court in criminal maritime causes at first instance was affirmed and in 1828 the jurisdiction of the High Court of Justiciary was declared cumulative with that of the Admiralty, in all crimes which might be tried in the latter court.

The prize jurisdiction of the court differed from that of the English Court in that it was inherent in the Admiral and did not require any special warrant to enable it to be exercised.

By the Articles of Union of 1707, Art. 19, the Admiralty jurisdiction in Scotland passed to the Lord High Admiral or Commissioners of Admiralty of Great Britain, but the court continued its judicial functions. The Commissioners appointed Vice-Admirals for Scotland. Its prize jurisdiction was transferred to the English Admiralty court by the Court of Session Act, 1825, its criminal jurisdiction to the High Court of Justiciary and the sheriff courts in 1828 and its remaining functions transferred to the Court of Session and Sheriff courts by the Court of Session Act, 1830.

The Admiralty jurisdiction (except in prize) is accordingly now exercised by the Court of Session and Sheriff courts along with their ordinary jurisdictions with only minor differences in procedure in certain cases.

Admiralty, Court of, in Ireland. The High Court of Admiralty of Ireland modelled on the corresponding court in England sat in Dublin and exercised an admiralty jurisdiction but not, though it was claimed in 1793, jurisdiction in prize.

After the Irish Judicature Act of 1877 it was amalgamated with the Irish High Court of Justice.

Admiralty Courts (U.S.). These were first established by the Constitutional Convention of 1787 which vested admiralty jurisdiction exclusively in federal courts. They had previously been exercised by courts of vice admiralty commissioned by royal patents or by colonial governors, or by common law courts till 1776, and then by state admiralty courts till 1787. The Constitution provides that the judicial power extends to all cases of admiralty and maritime jurisdiction. The jurisdiction is now vested in the U.S. district courts subject to review by the Circuit Court of Appeals and by the Supreme Court. It extends to all maritime matters, civil and criminal, within the three mile limit, including shipping on the Great Lakes and navigable rivers.

Admiralty Division. A name for the former (1875–1970) Probate, Divorce, and Admiralty Division of the High Court (q.v.), so far as exercising admiralty jurisdiction.

Admiralty, Droits of. The Lord High Admiral formerly had right by royal grant to custody of all ships found derelict on the high seas, ships or goods of the enemy found in English ports or captured by uncommissioned vessels, and of all flotsam, jetsam, and lagan (qq.v.), subject to the rights of the Lord Warden and Admiral of the Cinque Ports, of lords of manors, and of persons who had Admiralty rights in certain parts of the coast by royal grants. If not claimed by owners, these things were condemned by the Court of Admiralty as *droits* of Admiralty. In the modern Civil List it is provided that rights of Admiralty accruing to the Lord High Admiral are to be paid into the Exchequer. The property formerly liable to be condemned is now dealt with under the Merchant Shipping Act, 1894, ss. 523–525 and, if unclaimed by an owner or the holder of a Crown grant or of manorial rights, must be sold and the proceeds paid to the Crown. The right to droits carried jurisdiction with it; inquisitions into droits were held at ports and Vice-Admirals or droit gatherers reported them to the Admiral.

Admiralty, High Court of, in England. The High Court of Admiralty of England was originally the court of the deputy of the Admiral. It may have been established about 1340 and at first there were separate admirals of the north, south, and west, each with deputies and courts. These became merged into a single court in the fifteenth century. From the early fifteenth century there is one Lord High Admiral and one High Court of Admiralty.

The courts seem to have been instituted originally to prevent or punish piracy, and to deal with prize questions, but civil jurisdiction early developed and Acts of 1389, 1391, and 1401 sought to restrain the admiralty courts from dealing with non-maritime cases. The Offences at Sea Act, 1536, conferred jurisdiction in criminal cases on the admiral or his deputy and three or four other substantial persons appointed by the Lord Chancellor. An indirect result of this Act came to be the transfer of the criminal jurisdiction to the judges of the common law courts. The civil jurisdiction extended to all mercantile and shipping cases. The early jurisdiction seems to have been exercised through procedure similar to that of common law courts, though international relations resulted in the introduction of procedure based on Roman law and resembling that being adopted at that time on the European continent.

The court became important in the fifteenth and sixteenth centuries absorbing the jurisdiction of local courts, and developing a general jurisdiction in commercial matters. In the early seventeenth century there were chronic disputes between the courts of common law and of Admiralty over jurisdiction, with the result that the Admiralty court sank into comparative insignificance. A compromise was attempted in 1632 and it conceded some jurisdiction to the Admiralty court, but after the Great Rebellion the common law got the upper hand. In 1691 the jurisdiction and powers vested in the Lord High Admiral of England were transferred to the Lords Commissioners of the Admiralty.

The court became important again by reason of the prize cases arising during the wars of the eighteenth and early nineteenth centuries, above all during the time that Lord Stowell was judge. In 1834 the criminal jurisdiction was transferred to the Central Criminal Court and in 1844 it was extended to judges of assize. Statutes of 1840 and 1861 enlarged the jurisdiction of the court and it dealt with the steadily growing volume of litigation in shipping, collision, and salvage cases. Though modern legislation has restored much of the lost jurisdiction, Admiralty law has largely lost the international character it once possessed.

In modern times its jurisdiction has included, firstly an ordinary jurisdiction, comprising an exclusive criminal jurisdiction over British subjects, the crew of a British ship, and in cases of piracy at common law, in practice since the sixteenth century exercised by judges of the common law courts and exercised from 1834 by the Central Criminal Court, and from 1844 by the justices of Oyer and Terminer and Gaol Delivery; a civil jurisdiction, covering mercantile and shipping cases, collision, salvage, and similar cases, much of which was stolen by the Court of King's Bench in the seventeenth century; and Admiralty droits (q.v.). Secondly, it included the prize jurisdiction, adjudicating on the rights of the captors of ships and goods taken at sea to the things captured. This began to emerge in the sixteenth century and became a distinct jurisdiction in the following century. It was greatly developed and its principles settled by the decisions of Lord Stowell (vide SCOTT, WILLIAM) during the Napoleonic wars. The prize jurisdiction has developed into a quite distinct body of law.

Till 1859 the practitioners in the Admiralty Court were the same as those in the ecclesiastical courts, advocates and proctors, but in 1859 practice was opened to barristers and solicitors. So too, in place of the Attorney-General and Treasury Solicitor, there were the Queen's Advocate-General and the Queen's Proctor. There were also an Admiralty advocate and an Admiralty proctor. The offices of Queen's Advocate-General and of Admiralty advocate were allowed to fall vacant in the late nineteenth century but never abolished.

Under the Judicature Acts, 1873 and 1875, the High Court of Admiralty became part of the High Court of Justice, being linked with the Court of Probate and the Court of Divorce and Matrimonial Causes in the Probate, Divorce, and Admiralty Division. This division continued to exercise the admiralty jurisdiction of the High Court till 1970 when that Division was renamed the Family Division, Admiralty business being transferred to the Queen's Bench Division, and an Admiralty court constituted as part of the latter Division. Appeals from the court in the exercise of its prize jurisdiction have, however, since 1833, lain to the Judicial Committee of the Privy Council.

W. S. Holdsworth, *History of English Law*, Vol. I; W. Senior, *Doctors' Commons and the Old Court of Admiralty*; R. G. Marsden (ed.) *Cases determined in the High Court of Admiralty.*

Admiralty jurisdiction. In England, this extends at common law, to ship collisions, claims for salvage services, droits of Admiralty, possession of ships, bottomry and respondentia (qq.v.) bonds, claims for seamen's wages, and certain other kinds of cases. Under statute it also extends to towage claims, claims by seamen for wages and by shipmasters for wages and disbursements, claims for damage to cargo, actions for limitation of liability for loss caused by maritime casualty, rights over property recaptured from pirates, power to fine for improper use of national colours on a ship belonging to a British subject.

There is also the jurisdiction to condemn ships taken as prize, and over booty of war.

In Scotland Admiralty jurisdiction is exercised by the Court of Session and Sheriff Courts concurrently with common law civil jurisdiction and by the same judges. It falls to be exercised where a maritime case has to be determined, where proceedings are taken against a ship or goods on board ship, whether the cause of action be maritime or non-maritime, and where jurisdiction is expressly assigned to a court of Admiralty jurisdiction. In such causes these courts are bound to administer the general maritime law, which is common to Scotland and England.

Admiralty, Local courts of. Many seaport boroughs in England had by their charters grants of a court of Admiralty but these, save for the courts of the Cinque Ports, were revoked in 1835. The courts of Jersey and Guernsey have Admiralty jurisdiction, as have the courts of the Isle of Man.

The coast of England and Wales is divided into nineteen districts for each of which a Vice-Admiral of the coast may be appointed. These officials are appointed by letters patent under the Great Seal and represent the Lord High Admiral or the Commissioners for executing their office. The judge, registrar, and marshal of each Vice-Admiralty court is appointed by letters patent, but no appointments have been made for a long time.

Abroad, the place of Vice-Admiralty courts is supplied by Colonial Courts of Admiralty. They have jurisdiction similar to the High Court of Admiralty and appeal lies to the local court of appeal and ultimately to the Privy Council. A secretary of state may be empowered by commission under the Great Seal to establish a Vice-Admiralty court in any British possession and to transfer thereto the jurisdiction of a Colonial Court of Admiralty.

Admissibility. In relation to evidence, the question whether particular items of evidence may properly be adduced. It depends partly on rules of law or practice, e.g. that hearsay evidence is generally inadmissible, and partly on the logical relevancy of the particular evidence, i.e. its connection with the matter under investigation and its likeliness to convince the court as to the state of some fact which is part of the matter under investigation. Admissible evidence is accordingly evidence not excluded and also relevant to the matter under inquiry. Evidence may be admissible for one purpose and not for another. The wrongful admission or rejection of evidence may be a ground of appeal.

Admission. In civil proceedings, any facts admitted, or taken to be admitted, by a party in his pleadings, including all allegations made by the other party which are not specifically denied. One party may call on the other to admit certain facts and a judge may direct a defendant to admit matters not really in dispute. Admissions may also be made before or during proceedings.

In criminal proceedings an admission is any statement by an accused admitting either the offence charged, or any facts relevant to establish his guilt thereof. A statement made extra-judicially by an accused, e.g. to the police, amounting to an admission is not admissible in evidence if made in consequence of any threat, promise, or inducement, but a genuinely free and voluntary statement is admissible, including one made after caution. Silence, or non-denial of a charge or statement made in the accused's presence, is not an admission, unless very exceptionally.

Admittance. In English real property the act of the lord of a manor in accepting a person as the tenant of copyhold lands. It might be voluntary, when a new grant was made, or to a new tenant when the lands had escheated or reverted to the lord or the previous tenant had surrendered or devised the lands or had died intestate, or be unavoidable, where a lord admitted a new tenant whose claim was in accordance with the custom of the manor. Admittance might be express, where the steward of the manor delivered a symbol of seisin (q.v.) of the land to the new copyholder, a fact which was enrolled in the court rolls, and indeed prior to 1894 admittances could only be made in the customary court of the manor, or by implication, where the lord did something, such as accepting rent, showing an intention to accept the new tenant. Land formerly copyhold is now freehold and vests in the person having the best right to be admitted.

Admonition. A censure on the conduct of an offender against the honour or privileges of either House of Parliament, whether a Member or not, voiced by the Lord Chancellor in the Lords or the Speaker in the Commons, in either case on behalf of the House. It is recorded in the Journals of the House. It is also the least serious form of ecclesiastical censure, of the nature of a warning. But disobedience is punishable as contempt or contumacy. In military law and in Scotland, an offender may be dismissed with an admonition, which is a mere censure on his conduct.

Adolescence. In the Roman law the period between the age of puberty (14 for males, 12 for females) and majority (25). This corresponds to the period of minority (till 18, formerly 21) in Scots law, but the term has no special legal significance.

Adolphus, John Leycester (1795–1862), English barrister and author. In 1821 published *Letters to Richard Heber, Esq.* in which he discussed the authorship of the Waverley novels and attributed them to Sir Walter Scott. His *Circuiteers, an Eclogue* parodies the style of two colleagues on the northern circuit. He became a judge of Marylebone County Court in 1852 and worked to complete his father's *History of England.* In collaboration he edited several volumes of reports.

Adoption. The act of a person taking as his lawful child a person who is not in fact his child. In Roman law adoption (*adoptio* or *adrogatio*) was recognized as a means of continuing a family but was largely used as a means of liberating a son from *patria potestas.* At common law in England and Scotland adoption was not recognized, but was introduced by statute in 1926. The legislation provides for the adoption of persons under 18 who have not been married. Only adoption societies, which must be registered with and are subject to the control of local authorities, or local authorities, may make arrangements for adoption. They must ensure that parents understand the consequences of adoption of their child, and make enquiries into prospective adopters. Advertisements relative to adoption are restricted.

Application for an adoption order is made to a court which must appoint a person or body to safeguard the infant's interests. A natural parent may adopt his or her own child, alone or jointly with his or her spouse. The child must have been in the care and possession of the applicant for at least three consecutive months preceding the order.

Before making an order the court must be satisfied that persons whose consents are necessary have consented, that the order will be for the welfare of the infant, and that no unauthorized payment has been made or agreed to be made. The court may impose terms and conditions and dispense with any consent necessary for adoption in particular circumstances. Adoption orders must be registered by the Registrar-General of Births, Deaths and Marriages.

The effect of an adoption order is to transfer all rights, duties, obligations, and liabilities of parents or guardians in respect of future custody, maintenance, and education, including rights to appoint a guardian, from the natural parents to the adoptive parents, as though the adopted child were one born legitimate. For the purposes of marriage, family allowance, and national insurance, claims of damage on death, and intestate succession, an adopted child is deemed a child born in wedlock to the adopter.

The term adoption is also used of a person accepting as binding a contractual obligation which is void, or at least voidable, such as a bill on which his name has been forged, or a contract made by an agent on his behalf but in excess of the agent's authority.

Adoption doctrine. In international law, the doctrine that while international law and municipal law are distinct, municipal law presumes a mandate from its sovereign to incorporate international law. On this view a judge is entitled to resort to a rule of international law without demanding that it have been promulgated as a rule of municipal law, but he is bound by any superior rule of municipal law, such as statute or binding precedent of his own system, as that manifests the will of the sovereign in his municipal system to exclude contradictory international rules. In Anglo-American law it is presumed that the legislature did not intend to derogate from international law, and there is considerable historical support for this doctrine in U.K. law though there are some contrary cases.

Adoptive Act. An Act of Parliament which comes into operation only if a public body or local authority, entitled under the Act to do so, resolves to adopt and apply the Act within its area.

Adrogation. In the Roman law, a form of adoption, applicable to a person *sui iuris* and accordingly merging his family in that of the adopter. It required an investigation of its desirability by the pontiffs and the approval of the *comitia curiata* under the presidency of a pontiff. Later the comitial approval became a formality and the process was effected by imperial rescript.

Adscripti (or *adscriptitii glebae*). In Roman law a class of slaves who were attached to and transferred along with the land on which they lived. The category to which the term was applied in England merged into the later villeins (q.v.).

Adult. A person of full age (18, formerly 21).

Adultery. Consensual sexual intercourse between a married person and a person of the other sex, married or not, not being his or her spouse, during the subsistence of the marriage. Sexual intimacy short of physical coitus is not adultery, and it is questionable whether artificial insemination of a woman by another than her husband is adultery.

In Greek law an adulterer caught in the act could be killed and various states applied severe penalties to adultery.

Adultery was forbidden by the Mosaic law and prohibited by the Ten Commandments. There are many prohibitions of it in the Old Testament and mention of capital penalties for it. In the New Testament adulterers were classed with murderers, and this influenced modern legal systems to treat it as criminal or, at least, grave fault justifying divorce or legal separation.

In Roman law death seems to have been the penalty for adulterers caught in the act, but it was less severely punished during the Republic. The *Lex Julia* of 18 B.C. restored death as the penalty in certain cases, but in other cases banishment of both parties to different islands was the penalty. These penalties were made more severe by the Christian emperors; Constantine restored the death penalty and Justinian confirmed it. In later societies it was commonly criminal and even capital, but this has generally long been abandoned.

Adultery has long been a ground for judicial separation and for divorce; in England it is now evidence of the breakdown of a marriage for the purposes of divorce. It was formerly also a tort actionable by writ of trespass in an action for criminal conversation, but later damages might be recoverable only in the Divorce Court; the claim for damages has now been abolished. In Scotland adultery is evidence of the breakdown of a marriage and was formerly also the basis for a claim of damages.

Advance. Money paid before it is due. An advance note is a note given by the master of a ship to a seaman who has signed on, promising to pay to his order money not exceeding a month's wages if he goes to sea pursuant to his contract.

Advance freight. Freight (q.v.) paid in advance, which enables a shipper to endorse a bill of lading with a freight release and a consignee to take delivery immediately on arrival.

Advancement. The making of a capital payment to a person, particularly a young one, for a definite purpose for the person's benefit, before the payment would otherwise have been made. In the division of an intestate parent's estate any advancement made by portion during the parent's life has to be accounted towards that child's share of the estate.

Trustees had formerly no power, unless authorized by express provision of the settlement or by order of the court, to make a capital payment for the benefit of a person under age, but the Trustee Act, 1925, authorized this, subject to conditions.

The equitable doctrine of advancement is to the effect that if a purchase or investment is made by a father or person *in loco parentis* in name of a child or by a husband in name of his wife, it is presumed to be intended as an advancement for the child's or wife's benefit and to that extent satisfies any legacy later bequeathed to that person. The presumption arises in certain other cases also, depending on the relationship, but it may be rebutted by parole evidence.

Adventure, joint. A relationship in Scots law akin to partnership, but constituted normally for a single adventure or expedition or the attainment of a particular object rather than for the carrying on of a continuing business.

Adverse possession. In English real property, occupation of real property in manner inconsistent with the right of the true owner. Possession may be adverse *ab initio* as where a squatter seizes and retains possession, or adverse only from a later date, as where a tenant at first pays but later refuses to pay rent. Possession is not deemed adverse if referable to a lawful title.

Adverse possession entitles the possessor to be protected in his possession against anyone who cannot show a better title and after 12 years (30 years in the case of the Crown or land owned by a spiritual or eleemosynary corporation) the true owner's title is excluded and the possessor becomes owner, even though the true owner had been ignorant of the possessor's occupation.

Adverse witness. One whose conduct discloses bias against or hostility to the party who has called him, not merely one who gives honest evidence contrary to that party's case. That party's counsel may, with the leave of the judge, be allowed to lead or cross-examine the witness, or contradict him by other evidence.

Advertisement. A general intimation or announcement, particularly with a view to encouraging offers to contract. In relation to certain kinds of contract legislation controls the manner and content of advertisement. Statements made in an advertisement may be deemed merely designed to attract interest and not intended to be obligatory, but may be held to have been intended as undertakings or warranties (q.v.) and consequently to be obligatory.

Advertisements of Queen Elizabeth. Ordinances prepared by bishops in 1564–65 on the instructions of Queen Elizabeth I, intended to enforce order and uniformity in the ritual of the Church of England and in the vestments of the clergy, to counter the practice of many clergy in using unauthorized forms and containing strict regulations for their discipline. The Privy Council refused, however, to confirm them, and they were enforceable by episcopal authority only.

Advice of Pierre de Fontaines (or Advice to a Friend). A compilation of the usages and customs of Vermandois, written *c.* 1255 by Pierre de Fontaines, a councillor of St. Louis of France, dealing chiefly with procedure and including translated passages from Justinian's *Digest* and *Code.*

Advice on evidence. In England advice given by counsel, after the pleadings have been closed, on the evidence which will have to be led to support his client's case at the trial of the action.

Advising, in the Inner House of the Court of Session, a sitting at which the judges advise, or deliver opinions in, a case which they have had 'at *avizandum*', i.e. continued for consideration of their judgment. See also *Avizandum.*

Advisory, Conciliation and Arbitration Service. A body established in 1975 to promote collective bargaining and develop collective bargaining machinery. Its functions include conciliation, arbitration, advice, and inquiry and questions of the recognition of a union for purposes of collective bargaining. It may issue Codes of Practice to guide industrial relations.

Advisory opinion. An opinion given by the International Court of Justice, conveying guidance to the body requesting the advice. The Security Council and the General Assembly of the U.N. and other U.N. organs and special agencies authorized by the General Assembly may request an advisory opinion. Such opinions are not binding and cannot formally affect the legal position of subjects of international law, and have sometimes been disregarded. The Court has been unwilling to give an advisory opinion if to do so would in effect decide an issue between states which they have not submitted to the Court.

The Judicial Committee of the Privy Council may also be requested to give an advisory opinion.

Advocacy. The art and science of pleading cases on behalf of parties, particularly orally, before courts and juries. It requires a thorough appreciation of the relevant facts, a good knowledge of the relevant law, persuasive presentation, and argumentative powers.

Advocate. In the Roman law a person who assisted another's case by giving him advice on law or procedure or by speaking on his behalf. Advocates were a distinct class from the jurisconsults and possessed only a moderate knowledge of law, not enough to entitle them to give consultations. They developed from the second century B.C. onwards as specialists in forensic oratory; a leading representation of the class is Cicero, as is, later, the younger Pliny. Under the principate the distinction continued; the orator needed just enough law to understand the legal advice obtained from a jurisconsult. Under the later Empire the standing of advocates improved considerably. They studied law, formed a special corporation, were limited in numbers and were attached to particular courts where they acted generally as representatives of their clients. Justinian limited the honoraria which advocates might receive. In the Eastern Empire the higher offices were increasingly filled by advocates.

The term advocate is today the title of counsel in Scotland (see Advocates, Faculty of) and South Africa and its counterparts in other languages, e.g. *avocat, avvocato,* are the title of counsel in certain other countries. In the English ecclesiastical courts and the Admiralty court prior to 1857 advocates, i.e. members of the College of Advocates (q.v.) had exclusive right of audience. In the U.S. the term advocate is sometimes used synonymously with attorney, counsel, or lawyer.

Advocate, Crown (or Admiralty Advocate). A member of the College of Advocates (q.v.) who was special adviser of the Lord High Admiral and, when that office was put into commission, of the Lords of Admiralty. In the Court of Admiralty he was second law officer of the Crown; taking precedence after the Advocate-General (q.v.).

Advocate-Depute. A member of the Scottish bar appointed by the Lord Advocate (q.v.) to assist him in the discharge of his duties, particularly in the prosecution of crimes. The appointment does not preclude practice at the bar in civil cases. There are now customarily some eight or ten advocates-depute, the senior of whom (usually a Q.C.) acts for the Home (Edinburgh) circuit, the others taking in turn, for periods of six months at a time, the North, South, and West circuits, while one acts in Sheriff Court cases in which counsel's presence seems necessary. The appointments are personal to the Lord Advocate, from whom they hold their commissions, by whom they are revocable, and who is responsible for their actions. On the resignation of a Lord Advocate they formerly went out of office, though remaining till successors were appointed. For long appointments were also political, but this tradition was departed from in 1970. Advocates-depute consider the papers in cases investigated by local procurators-fiscal (q.v.) and decide whether prosecution is warranted and, if so, whether on indictment in the High Court or Sheriff Court, or summarily in the Sheriff Court. They conduct the prosecution case in all trials in the High Court of Justiciary, save where the Lord Advocate or Solicitor-General appears in person, and in doing so have all the powers which the Lord Advocate would personally have.

Advocate, Queen's. See ADVOCATE-GENERAL.

Advocate-General. (1) Her Majesty's (or the Queen's) Advocate-General, commonly called the Queen's Advocate, was the principal law officer of the Crown drawn from the College of Advocates (q.v.) and in the Admiralty and ecclesiastical courts, and was from about 1600 the Crown's chief adviser on matters of civil, canon, ecclesiastical, and maritime law, and from the time of the establishment of the Foreign Office in 1782 its standing legal adviser on international law. He was appointed by letters patent, was included in the expression 'The Law Officers of the Crown' and for long took precedence of both the Attorney General and the Solicitor General, but from 1862 he was postponed to them. Since the retirement of the last holder in 1872 the office has been in abeyance. Thereafter a person was appointed specially to assist the law officers on points of international law until 1886 when the post of Legal Adviser to the Foreign Office was created.

J. L. Edwards, *The Law Officers of the Crown.* List of holders of office in A. D. McNair, *International Law Opinions,* III, Appx. I.

(2) In the Court of Justice of the European Communities, four officers possessing the same qualifications as the judges, having the duty of acting with complete impartiality and independence and making, in open court, reasoned submissions on cases brought before the court, in order to assist the court in the performance of its task of ensuring that in the interpretation and application of the treaties the law is observed. It is not his function to advise on law or procedure. He sits on and speaks from the Bench like the judges, and tries in his submissions to draw together all the facts, arguments, and authorities for the use of the judges. He makes his submissions in public before the judges retire to consider their judgment, and he does not retire with them. His opinion is reported, like those of the judges, and may be subsequently cited as authority. In some ways his function is like that of a judge of first instance, expressing a view on the relevant law and how the case should be decided before the judges give their judgments, but it differs in that he hears the case with, not before, the judges and his view is never decisive and cannot be commented on by the parties to the case. In cases where examination of witnesses is necessary it is done by a chamber of three judges and an advocate-general.

Advocate-General, Admiralty. An office, distinct from that of King's Advocate-General, who was formerly the special legal adviser of the Lord High Admiral and subsequently of the Lords of the Admiralty.

Advocates, College of. A society of persons, doctors of civil law of Oxford or Cambridge, who associated for the practice of the civil and canon law about 1490 and were incorporated by royal charter in 1768. Members had in 1567 purchased the site later known as Doctors' Commons whereon were built premises for the Admiralty and ecclesiastical courts and houses for judges. Latterly the college consisted of the Dean of the Arches, who was always president, and the fellows, who were elected from those admitted as advocates under a rescript of the Archbishop of Canterbury. To become an advocate one had to be a D.C.L. of Oxford or LL.D. of Cambridge. Down to 1857 members of the college had the exclusive right to practise in the Admiralty and ecclesiastical courts. Thereafter advocates were permitted to practise in all courts and barristers were permitted to practise in the Probate and Divorce Courts created in that year and, after 1859, also in the Admiralty Court. The college was abolished by the Court of Probate Act, 1857.

G. D. Squibb, *Doctors' Commons.*

Advocates, Faculty of. The body of advocates which comprises the Scottish bar. Advocates are mentioned in Scottish statutes from 1424 but the Faculty of Advocates was instituted by the Act which established the College of Justice in Scotland in 1532, and advocates are members of the College of Justice. By custom the Faculty is deemed a corporation. It has its home in the Parliament House in Edinburgh, where the Parliaments of Scotland sat till 1707, and where the Court of Session and

High Court of Justiciary sit. Unlike the English bar, it is a small body and the great majority of the members practise, or have practised, in the courts.

The Faculty regulates its own admission. By Act of Sederunt of 1619 an examination and thesis in civil law were prescribed. The qualifications now are examination in the main branches of law, exemption from examination being given to holders of degrees in law of Scottish universities, and professional training partly in a solicitor's office, partly as a pupil to an advocate in practice.

The Faculty consists of Queen's Counsel (i.e. senior counsel), advocates or junior counsel, and Intrants, who are persons preparing for admission to the Faculty and call to the Scottish bar. When an Intrant is called he is admitted at a private ceremony and then takes the oath of allegiance before a Lord Ordinary in open court and signs the roll of advocates. The Faculty is headed by the Dean of Faculty; in court he leads all other members of the Bar, except the Lord Advocate. He is assisted by a Vice-Dean, Treasurer, and Clerk. All are elected at the anniversary meeting of Faculty each year, and in practice normally re-elected so long as they are willing to serve. In matters of etiquette and professional standards the Dean is assisted by a Council of senior members of Faculty, and the Dean and Council may investigate a matter of professional conduct, censure, or even disbar a member of Faculty.

The status of Queen's Counsel (q.v.) was not conferred on senior members of the Scottish Bar, other than Law Officers and Dean of Faculty, until Queen Victoria, on petition by the Faculty of Advocates, created a roll of Queen's Counsel in Scotland in 1897. Recommendations are made by the Lord Justice-General (q.v.) through the Secretary of State for Scotland. In Scotland only the Lord Advocate and Solicitor General sit within the bar, all other Q.C.s at the bar. The general practice was for long that Q.C.s might not draw or revise pleadings, nor appear in court nor before Provisional Order Tribunals, Royal or Parliamentary Commissions, or departmental inquiries, unless accompanied by a junior counsel. This is not now obligatory. In criminal cases a Q.C. may appear alone.

The Faculty established a library in 1681 and thereafter built up a magnificent collection of books. In 1925 all except the legal books, manuscripts, and papers were gifted to the National Library of Scotland which was created in that year.

An advocate is the only person who may appear on behalf of a party in any court in Scotland, civil, criminal, ecclesiastical, or administrative, save where legal representation is statutorily forbidden. In the House of Lords, Privy Council, and before Parliamentary committees, advocates may appear equally with members of the English Bar. In the Sheriff courts (q.v.) and other lower courts solicitors also have the right of audience. By custom an advocate must be instructed by an enrolled solicitor, Parliamentary agent, clerk to local authority, or lawyer furth of Scotland.

An advocate's fees are deemed to be honoraria and accordingly cannot be sued for. By custom he cannot be sued for professional negligence, ignorance, or defective advice.

Advocates, Society of, in Aberdeen. A society of legal practitioners practising in the courts in the city of Aberdeen, who formed a society in 1685 and obtained a Royal Charter in 1774. This society is a body of solicitors and unconnected with the Faculty of Advocates (q.v.).

Advocation. In Scottish civil procedure, a procedure whereby a case was carried from an inferior to a superior court. It was replaced by appeal in 1868.

In Scottish criminal procedure it is a mode of appeal from an inferior criminal court to the High Court of Justiciary. It is today rarely employed and usually by the prosecution, and particularly to review irregularities in the preliminary stages of a case.

Advocatus diaboli (or devil's advocate). The common name of the promoter of the faith, an office of the Sacred Congregation of Rites of the Roman church, whose function is to prepare and submit all available arguments against the promotion of a person by beatification or canonization.

Advowson. In English law, the right of patronage or of presentation to a church or benefice. It is an incorporeal hereditament (q.v.), but partakes of the nature of a spiritual trust.

The advowson of a rectory is normally appendant to a manor, but may be appendant to a house or land alienated from the manor. The advowson of a vicarage is normally appendant to the rectory from which it is extracted, but it may be appendant to a manor. If severed from the manor or other hereditament, it is an advowson in gross.

An advowson may also be presentative, where the patron has the right of presentation to the bishop or ordinary and also to require him to institute his clerk if he finds him qualified, or collative, where bishop and patron are the same person and the bishop by collation performs what is otherwise done by presentation and institution. Collation is completed by induction of the collatee. Or an advowson might be donative, where the sovereign, or a subject under licence from the Crown, has founded a church and by donation in writing had put a clerk in possession, without presentation, institution, or induction. Since 1898 donative advowsons have been made presentative.

Aediles. Roman magistrates, two elected from an early date and from 367 B.C. four in number, as assistants to the tribunes with duties in relation to roads, public places, temples, the public games, water supply and cleanliness. They had oversight of the archives, both *plebiscita* and *senatus consulta*, minor criminal jurisdiction and some civil jurisdiction in relation to the market place, in connection with which they issued edicts (q.v.) in the same way as the praetors, which were of great importance in developing the law of sale. Under Augustus the legal functions were transferred to the prætors. The office disappeared in the third century A.D.

Aegean civilization, law in. In the Aegean civilization, centred on the Peloponnese and Crete and covering roughly the period 4,000 B.C. to 1,000 B.C. there is evidence of a highly civilized and well-regulated life, politically and socially organized, and with a substantial volume of commerce, which all suggest a considerable body of custom and possible law but of this no details have yet been discovered, though law was highly developed in contemporary Babylon.

Aedric, laws of. See ANGLO-SAXON LAW.

Aelius Paetus Catus, Sextus (*fl.* 200 B.C.). A Roman jurist, author of the *Tripertita* (*Scil. Commentaria*), which contained the XII Tables, an account of its development by legal interpretation and the forms of *legis actiones*. It was the first commentary on the Twelve Tables. He is also said to have made a collection of forms under the name of *Ius Aelianum*, which may be the same as the third part of the *Tripertita*.

Aelius Tubero, Quintus (first century B.C.). Roman jurist and annalist, reputed an expert on public and private law and author of several legal works which failed to achieve great popularity.

Aemulatio vicini. In Scots law a proprietor of land may make any lawful use of his land, however offensive or harmful to a neighbour, but not if he does so *in aemulationem vicini*, mainly for pure spite or other oblique motives.

Aequitas. In Roman law, equity or fairness, a principle sometimes appealed to as a corrective of the rigour of strict law in particular contexts. It did not have the breadth of meaning or of application or detailed substance acquired by equity (q.v.) in English law but appears to have been little more than a principle of interpretation.

Aequo et bono, ex. From the fair and good angle, or on the basis of equity. In Roman law certain *iudicia bonae fidei* might be decided on this basis. With the consent of parties the International Court may decide an issue on this basis, if necessary despite the strict law.

Aes et libra (copper and scales). In early Roman law several kinds of transactions were affected *per aes et libram.* This required the presence of the parties to the transaction, five Roman citizens as witnesses, a citizen holding a pair of scales (*libripens*) and a piece of copper (*aes*). The transferee grasped the thing to be transferred and uttered formal words asserting that it was his and bought with the piece of copper, struck the scales with the copper and gave it to the transferor. The ceremony was a relic of a time when money was weighed not counted. This symbolic sale was utilized to effect various legal transactions, notably mancipation, i.e. the acquisition of ownership of a *res mancipi* (q.v.), to effect the release of a debtor from liability, to making a debtor the bondman of his creditor (*nexum*), to marriage by the form known as *coemptio*, a modified form of bride purchase, to a form of will mancipating the property to be left to a *familiae emptor*, to adoption, where the sale had to be done thrice to free the adoptee from his father's power, and to emancipation to release a son from his father's power. Transactions effected by copper and scales seems to have lasted, in some cases, to Justinian's time but in others to have disappeared earlier.

Aethelbert, laws of. See ANGLO-SAXON CODES.

Affidavit. In English law a written statement in the name of a person, the deponent, who makes it and signs and swears (or affirms) to its truth before a Commissioner for Oaths. Evidence may be given by affidavit by agreement, if the court or a judge so directs and is the customary mode in many kinds of judicial proceedings, particularly in the Chancery Division. Save under particular statutes affidavits are unknown in Scotland. In the U.S. the general uses of affidavits are the same as in England but are not used to the same extent as substitutes for oral evidence.

Affiliation. In English law, the process whereby a single woman, or a married woman living apart from her husband, obtains from a magistrates' court an order that the man judged by the court to be the father of her bastard child (putative father) shall pay her a weekly sum for the maintenance and education of the child. Payments continue till 13 or, if the court thinks fit, 16, or 21.

Affiliation and aliment. In Scots law, the action whereby a woman, single or married, obtains from the sheriff court a decree against the man (other than her husband, if any) deemed to be the father of her bastard child for the aliment and education of the child. If the woman is married she must overcome the presumption *pater est quem nuptiae demonstrant* by clear and cogent evidence, and in all cases she must satisfy the court that the man called as defender was the father of her child.

Affinity. Relationship not by blood (consanguinity) but by marriage only, as between a husband and his wife's relatives and conversely. By reason of the theory that marriage makes one flesh, the blood relations of each spouse become relations by affinity of the other spouse in the same degree, and this relationship in some cases is deemed an impediment to marriage. Secondary affinity is the relationship of one spouse's blood relations with the other spouse's blood relations. Collateral affinity is the relationship between the husband and the relations of his wife's relations. Affinity was an impediment to marriage in Mosaic law and in Roman law it was an impediment in the direct line infinitely but not in the collateral line. The early Christian church founded on these sources but the ambit of affinity gradually extended particularly among collaterals. The classical definition of affinity from Gratian became relationship of person emanating from carnal intercourse. By the tenth century the similarity between impediments based on affinity and on consanguinity became complete. After many developments the modern Code of Canon Law changed the basis of affinity from sexual intercourse to valid marriage and while accepting the impediment from affinity as extending infinitely in the direct line, it limited it to the second degree of affinity in the collateral line. Since affinity was an impediment of ecclesiastical law only, the Church exercised the right to dispense with it.

In modern legal systems the definition of the prohibited degrees of relationship is a matter of positive law. In England a table is provided in the Book of Common Prayer and the Marriage Act, 1949, as amended; in Scotland in the Marriage (Scot.) Act, 1977.

Affirm. (1) An appellate court affirms a judgment appealed against when it upholds and repeats it.

(2) When, in any circumstances where the taking of an oath is required, a person declares that the taking of an oath is contrary to his religious belief or that he has no religious belief, he is entitled to make an affirmation instead, under the Oaths Act, 1888.

(3) Where a contract is voidable (but not void) and the party entitled to challenge the contract, in the full knowledge of the circumstances, elects to treat the contract as good and to perform it, he is said to affirm it.

Affirmation. A solemn declaration that the declarant will tell the truth. By the Oaths Act, 1888, any person objecting to taking an oath, on the ground of his religious belief or on the ground that he has no religious belief, may affirm to the same effect as if he had taken the oath.

A person becoming a member of the House of Commons or of the House of Lords may affirm, instead of taking the oath of allegiance.

Afflictis, Matthias de (Matteo d'Afflitto) (*c.* 1430–*c.* 1523). Italian jurist and royal official, author of *De Iure Protimeseos* (1544), *Commentarius in constitutiones Siciliae et Neapolis, Commentarius super tres libros feudorum* (1598), *Lecturae super consuetudinibus Neapolitani Siciliaeque regni* (1550), and other works.

Afflictive punishment. A category of punishment known in France between the 1500s and the Revolution. The category including maiming, which included slitting or piercing the tongue, cutting off the lips or nose, cutting or burning off the hand, non-maiming corporal punishments, including branding, flogging, the pillory, and the carcan (an iron collar round the offender's neck with a chain by which he was attached to a wall), and non-corporal afflictive punishments, including consignment to the galleys for years, imprisonment for years, exile, servile labour, and *amende honorable* (q.v.).

Affray. In English criminal law, an unlawful skirmish or fight by at least one person, in which a weapon is drawn or a stroke given or offered, or a display of force by one or more persons without actual violence, to the terror of the Queen's subjects. A single attacker may be guilty of affray, his victim acting only in self-defence. The offence may be committed though no violence be actually done, provided terror is caused. There must be a real disturbance, but abusive or threatening words alone will not constitute an affray. In American law the same generally applies, with modifications in certain states.

Affreightment. The contract whereby one person undertakes for reward, called freight (q.v.) to carry goods for another in his ship. It is normally contained in and evidenced by a charter-party (q.v.) or bill of lading (q.v.).

Afonsine Ordinances (Ordenaçôes Afonsinae). A codification of Portuguese law completed in 1446 by the regents of King Afonso V. It was probably inspired by the Siete Partidas (q.v.) and dealt with royal jurisdiction, privileges of the nobility, succession, and a revised penal code. It included also restrictive rules of behaviour for Portuguese Jews and is thought to have contributed to the conversions to Christianity in subsequent years. These ordinances were replaced by the Manueline Ordinances in 1521.

Africanus, Sextus Caecilius (second century A.D.). A Roman jurist, author of *Quaestiones* in nine books, mainly containing decisions of Julianus. He is believed also to have written a commentary on the *Lex Julia de adulteriis*.

Agard or Agarde, Arthur (1540–1615). A distinguished antiquary and latterly deputy-chamberlain of the Exchequer, devoted much energy to cataloguing state papers. Some of his work was not published till long after his death. He was a friend of Camden, Stow, Selden, Spelman, and Cotton. He first discovered the true authorship of the *Dialogus de Scaccario*, published *The Repertoire of Records* (1585) and compiled an *Abbreviatio Placitorum in Banco Regis*, 1272–1367.

Age. The period of time which has elapsed since a person's birth, relevant particularly to marriage, legal capacity to contract, to liability to care and protection and to liability to special correctional regimes.

In the Roman law a distinction was drawn between infants (children under seven), children under puberty (fixed at 12 for girls and later at 14 for boys) who were known as pupils and were subject to guardianship by a tutor, and adolescents, children over that age: boys over puberty but under 25 were known as minors and were subject to curatory, and their transactions were voidable if advantage had been taken of their inexperience; girls over puberty remained subject to lifelong tutelage. Young men became major at 25.

In Scots law the same distinction is drawn between pupils (children from birth to 14 for boys, 12 for girls) and minors but minors of both sexes may either have or not have curators, and emerge from curatory at 18 (formerly 21), though their contracts remain voidable till they attain 22 (formerly 25), on proof of minority at the time of contract and lesion, i.e. substantial prejudice from the contract, this being more readily established if the minor did not have a curator.

In English law and common law systems children are deemed infants or minors from birth till attaining majority on the day of the 18th (formerly the day before the 21st) anniversary of birth, when they attain full adult capacities.

Many English and Scottish statutes have introduced variations on these rules; thus young persons may since 1929 marry at 16. In criminal law a child under seven was at common law deemed incapable of committing any crime, but this has been raised by statute to 10 in England and eight in Scotland. Many kinds of special correctional measures, e.g. Borstal training, are made applicable only to persons within certain limits of age.

In the United States the age of majority is generally 21, but lower in some states for females. There are again, however, exceptions such as that a senator must be 30, and the president 35.

At the other end of the age scale retirement from work is compulsory under various statutes at stated ages; thus judges of the Supreme Court retire at 75, of the inferior courts at 72 and in various employments at ages fixed by the regulations of the service concerned. The courts have never fixed any age as that beyond which a male is presumed incapable of begetting children but in the case of females the court may presume a woman incapable of childbearing when her age and other relevant facts clearly suggest that conclusion; no age is conclusive.

Agency. The legal relationship between one person, the agent, having authority to act, and having consented to act, on behalf of another, the principal, particularly to place the principal in contractual relations with a third party. The term is also sometimes used more widely of one acting in the interest of another.

An agent is not, as such, an employee, but an employee may be an agent. Agency was never developed by Roman law but in the early middle ages Roman principles were extended and developed, and the canon law also made developments. Anglo-Norman law created the positions of the *ballivus* and *attornatus*, but it was only in the seventeenth century in Europe during the ascendency of natural law that the principle of agency achieved recognition.

In general, whatever a person could do himself he may do by means of an agent. Agents are distinguished into various classes, firstly into special agents, who have authority to act for a special purpose only, and general agents, who have authority arising from their ordinary business or profession to act for their principals, or authority to act on behalf of their principals generally. Among particular classes of agents there are recognized mercantile agents, a class comprising factors and brokers, auctioneers, stockbrokers, solicitors, insurance brokers, shipbrokers, house and estate agents, while, at least in certain circumstances, carriers, shipmasters and others may be held to be agents.

Agency may be created expressly, by deed, writing or oral appointment, or impliedly, from the conduct or situation of the parties, or by ratification by the principal of an agent's actings on his behalf, or by statute. Agency of necessity arises where a person acts in the interest of another to preserve his property, or carries out his duties in his necessary absence, and agency by estoppel where one person has acted so as to lead another to believe that a third party was his agent and that other has transacted with the third party subject to that belief.

The extent of the agent's authority depends on the authority expressly conferred on him, or implied by the nature of his duties, the scope of his profession, or the custom of particular trades or professions. The general rule is that an agent cannot delegate his powers or duties to another, without express or statutory authorization: *delegatus non potest delegare*, but implied authority or custom or emergency or the principal's acquiescence may allow exceptions to this principle.

The principal is entitled to expect that his agent will carry out personally the business he has undertaken, use ordinary care and skill in doing so, use his judgment and discretion honestly and in the interest of the principal, keep accounts of monies received and paid on the principal's behalf, maintain confidence as to all that comes to his notice in the course of the agency, and not permit his own interest to conflict with his duty to his principal. The agent may not accept any bribe, secret commission or benefit not known to the principal.

The agent, on the other hand, is entitled to remuneration in accordance with the contract, by way of commission, fees or other payment, to be reimbursed his outlays, and be indemnified against liabilities incurred, and has a lien over the principal's property for all his claims.

In relations with third parties the principal is bound by acts done by his agent within the scope of his express authority, or his general authority, or the authority he has represented the agent, or enabled him to represent himself, as having, but not by contracts made outwith the agent's authority. The principal can enforce contracts made by his agent within the scope of his authority. He is liable to third parties for loss or damage to them arising from acts expressly authorized, or acts done by the agent within the course of his employment as agent, or the ostensible scope of his authority.

In relations with third parties an agent is personally liable if he does not disclose the existence or name of his principal, but in general not if he discloses at least the existence of a principal. A person purporting to act as agent is deemed to warrant that he has authority to do the act in question.

The relationship is terminated by act of either party, or by operation of law, on the completion of the undertaking, by lapse of the period agreed on, or by death or bankruptcy of either party.

See also FACTOR.

Agent of necessity. A person who is deemed in circumstances of necessity and urgency to have implied authority to act as agent. Thus, carriers by land are deemed authorized in case of necessity to sell perishable goods, and masters of ships are deemed to have implied authority to borrow money to prosecute the voyage or to sell cargo which is deteriorating.

Agent, Law. This was the customary designation of solicitors in Scotland prior to 1933, and is still used by some firms and persons.

Agent provocateur. A person who commits an offence or seeks to encourage another to commit an offence in order to detect or obtain evidence of the commission of the offence by that other and to secure his conviction.

Agistment. The contract whereby one man takes another's cattle, horses or other animals to graze on his land for reward, and to redeliver them on demand. The agister must take reasonable care of the animals and not expose them to avoidable risks and he is liable for injury which could have been prevented by the exercise of reasonable care.

Aggravation. That which increases the seriousness of a wrong or crime. A crime may be aggravated by the circumstances in which it was committed, such as theft by a police officer, or by the mode of commission, such as assault with a weapon, or by the intent, such as assault with intent to rape, or by the seriousness of the crime, such as assault with great violence or causing actual bodily harm.

Aggression. In international law, an act of expansion by one state at the expense of another by means of unprovoked military attack. This is commonly regarded as contrary to customary international law. Treaties and official declarations, including the Covenant of the League of Nations and the Charter of the U.N. have sought to outlaw aggression. The Security Council has the function of deciding whether aggression has taken place and of recommending what should be done about it. The practice of the U.N. has usually been to order a ceasefire and to denounce a state as an aggressor only if it fails to obey that order. Thus the Soviet Union was held to be an aggressor in Hungary in 1956. Argument has frequently arisen whether particular instances of military intervention were aggression or justifiable defence or justified by the consent of the state in which force was used. It is even more doubtful whether economic or propaganda activities directed against another state can be considered to constitute aggression.

Agnates. In the Roman law blood relations related to a person through males only, but extended in Scots law to blood relations on the father's side as distinct from cognates, relations on the mother's side. Agnatic relationship was based on the idea of the family held together by the *patria potestas* of the senior living male.

Ago, Roberto (1907–). Italian jurist, professor at Milan and Rome, delegate to many international conferences, member of the Institute of International Law, author of many books, notably on international law.

Agoranomi. Athenian magistrates with duties similar to the Roman aediles (q.v.). They maintained order in the markets, tested weights and measures, collected harbour dues and enforced shipping regulations.

Agrarian laws. Laws dealing with the disposal of the public land of ancient Rome. The disposition of the public land from the early Republic was an issue in dispute between the privileged and the poor. As no gratuitous disposal of public land could be made without the consent of the people, many ordinances affecting public land were proper laws. Roman history contains many instances of agrarian legislation latterly frequently proposed to placate or reward supporters or soldiers. The expression is still used by analogy in some modern countries for distributing land to tenants.

Agraeus, Claude Jean (seventeenth century). Swedish jurist, author of *Leges Sudromanicae et Wesmanicae* (1666).

Agreement (or *consensus in idem*). The concurrence of the wills of two or more persons on some common matter, evidenced by acts apparent to or communicated to, and understood by, each other. Agreement undisclosed is ineffective. Agreement has by itself no legal effect but is a prerequisite of any valid contract, payment, compromise, variation or discharge of contract, or conveyance.

The discovery that apparent agreement has concealed actual disagreement gives rise to problems of mistake or error (qq.v.) and the securing of agreement by misrepresentations or other improper means to problems of fraud and misrepresentation (qq.v.).

The term is also used as a synonym for a contract (q.v.), or sometimes for a distinguishable element of a contract, particularly where formalities are necessary for the legal validity or enforceability of the agreement as a legal contract.

Agreement of the People for a firm and present Peace upon Grounds of Common Right. A statement by the Levellers, or extreme faction in the army, embodying their constitutional theories, presented to the Council of the Army in 1647. It proposed that Parliament should be dissolved at the end of September 1648, that thenceforth biennial parliaments should be chosen, and that representatives of the nation should have sovereign power, but reserving to persons' own decisions religion, impressment for military service, punishment for participation in the late differences and equality before the laws. The agreement was discussed in the council of the Army without any conclusion being reached. A modified version was presented to the House of Commons early in 1649. It was based on the Heads of Proposals (q.v.) but adopted the proposal of the first Agreement that there should be a single elected House.

Agricultural Adjustment Acts. Two Acts (1933 and 1938) part of the U.S. New Deal programme to alleviate distress in the agricultural community by restricting output and raising prices. The 1933 Act was declared unconstitutional by the Supreme Court in 1936. The 1938 Act replaced it, establishing marketing quota criteria to collect surplus commodities until market prices rose.

Agricultural credits. Money advanced for agricultural purposes and usually secured by a mortgage of agricultural land.

Agricultural lands tribunals. Tribunals consisting of a lawyer, a landowner and a farmer, established to give or withhold consent to the operation of notices to tenants to quit agricultural holdings, and where necessary to grant certificates of bad husbandry.

Agricultural law. A generic term for those principles and rules of law particularly relevant to persons engaged in agriculture including, e.g. land law, agricultural holdings, nuisance, liability for animals, agricultural production, farming subsidies, and related topics.

Agricultural Policy, Common. A facet of the economic unity being promoted by the European Economic Community. The object is to establish one policy for the Community, not to co-ordinate national policies, by increasing productivity, ensuring a fair standard of living for agricultural workers, stabilizing markets, providing certainty of supplies and ensuring supplies to customers at reasonable prices. To achieve this member states must adopt a common organization of agricultural markets, by common rules of competition or compulsory co-ordination of national marketing organizations, or a European organization of the market. In pursuance of the objects the Community has adopted regulations in respect of various sectors of agricultural and horticultural produce, whereby markets and prices are controlled. It has also planned a structural re-organization of agriculture and a modernization of the farming industry. The financing of the common agricultural policy is part of the Community budget.

Aguesseau, Henri Francois d' (1668–1751). French jurist, studied under Domat, became a master of forensic eloquence and was thrice chancellor of France (1717–18, 1720–22 and 1737–50). He was a prolific writer, mainly on law, his *Oeuvres complètes* extending to many volumes, and attempted in vain a codification of French law but succeeded in having adopted considerable reforms, particularly important ordinances on donations, testaments and succession. He also achieved greater uniformity in the execution of law.

F. Monnier, *Le Chancelier d'Aguesseau*, 1863.

Agustin, Antonio. See AUGUSTIN, ANTONIO.

Agylaeus, Henri (Agylée) (*c.* 1533–95). French jurist, author of an edition of Justinian's *Novels* and *Justiniani Edicta: Justini, Tiberii, Leonis philosophi constitutiones et Zenonis una* and other works.

Ahmad ibn Hanbal (780–855). Muslim theologian and jurist, whose teachings are the basis of the Hanbali school, one of the four orthodox schools of law of the larger of Islam's two branches, the Sunni. His chief work is his collection of the Traditions of the Prophet Muhammad, considered by him to be a sound basis for argument in law and religion.

Ahrens, Heinrich (1808–74). German jurist, professor of legal philosophy at Graz, 1850, and of political science at Leipzig, 1860. He was author of *Cours de droit naturel ou de philosophie du droit* (1838); *Die Philosophie des Rechts und des Staates* (1850), and *Juristische Encyclopaedie* (1855). The former two were very popular and influential books.

Aid prayer. A plea in a real action at common law that if a particular tenant was about to lose land by default the reversioner might come before judgment and pray to be received to defend his right in the land.

Aiding and abetting. A term of no precise significance in criminal charges, signifying guilt by accession. Aiding and abetting was formerly the activity of a principal in the second degree to a felony, as contrasted with the principal in the first degree who was the principal offender. Aiding is assisting; abetting is being present assisting or encouraging the principal offender at the time of the commission of the offence. In all offences one who aids and abets is since 1967 liable to be tried and punished as a principal offender.

Aids. Originally voluntary payments made by feudal vassals to assist their lord on occasions of need, such as to ransom the lord if he were taken prisoner, but latterly demanded as a matter of right. The occasions when they might be demanded and the amount were frequently matters of dispute. According to Glanvill (q.v.) a lord was entitled to aids for paying the relief due to his own overlord, for the knighting of his eldest son, and for the provision of portion to his eldest daughter on marriage. According to Magna Carta, ransom of the lord from captivity was due instead of relief on his succession to the land. Magna Carta specified the aids payable to the King as supreme feudal overlord as being those three only, and even then, only a reasonable sum, unless the common counsel of the realm agreed otherwise. The barons in turn agreed to claim from their tenants only these three customary aids. The last exaction by the Crown was on the knighting of James I's eldest son in 1609 and on the marriage of his daughter in 1613. All feudal aids were abolished in 1660 but parliamentary taxes were called aids until 1688.

The term aid was also applied to the special contributions of boroughs to the king's revenue, and to a payment in lieu of the military service due from the Crown's knights. The borough aids were at first fixed but later assessed by Crown officers on a population basis.

Aiel, Aile or Ayle. A writ of aiel (*de avo*) was the real action, a variant of the writ of right, proper for the recovery of land where a man's grandfather (*aiel*) or great-grandfather (*besaiel*) died seised of land in fee simple and the heir was dispossessed of the inheritance by a stranger on the day of the ancestor's death. See also BESAIEL; COSINAGE.

Air law. The body of law concerned with civil aviation, regulating aerial navigation and the use of airspace. It is largely international or, by agreement, uniform in most countries.

The basic principle is that every state has sovereignty over the airspace above its territory and territorial sea. Each state can accordingly regulate the entry of foreign aircraft into its airspace but normally permits foreign private aircraft to fly through its territory.

A Chicago Conference of 1944 formulated 'five freedoms of the air', viz. the privileges of free passage through air-space, of non-commercial landing, e.g. to re-fuel, of commercial discharge from the aircraft's country, of commercial embarkation to the aircraft's country, and of conveying passengers, mail and cargo between any two contracting states, but they have not been ratified. The right of commercial landing is usually granted by bilateral agreements and these may embody any of the last three 'freedoms'.

Aircraft have the nationality of the state in which they are registered and must satisfy the authorities of that state in respect of airworthiness, crew and documents. Each state is responsible for the establishment and enforcement of air navigation rules and air traffic control, as well as the operation of ground facilities. Over the high seas the Rules of the Air established by the International Civil Aviation Organization apply.

The criminal law of the state of registration applies to an aircraft in flight save when the aircraft is within the territorial airspace of another state. A 1958 Geneva Convention extends the concept of piracy to crimes against aircraft.

The liability of air carriers to passengers and consignees of goods is mostly determined by a Warsaw Convention of 1929, later amended, and adopted as uniform law by many countries. In many cases these rules are also applied to non-international carriage.

By most legal systems liability for damage done by an aircraft to persons or things on the ground is absolute, and this was adopted by a Rome Convention of 1952.

Almost all countries are members of the International Civil Aviation Organization, a specialized agency affiliated to the United Nations, with headquarters in Montreal, a major function of which is to prepare and keep up to date international standards and practices relating to civil aviation.

C. N. Shawcross and K. M. Beaumont on *Air Law* (U.K.); J. Fixel, *Law of Aviation* (U.S.); B. Cheng, *Law of International Air Transport.*

Airspace. The atmosphere outside and surrounding the earth, particularly the portion of it directly above the territory and the territorial sea of a particular state. International law has adopted the common law maxim *cuius est solum eius est usque ad coelum*, so that airspace above a state territory is subject to that state's sovereignty. In consequence all aircraft, except those on unscheduled civil flights, require permission to fly over foreign territory, which is more readily presumed in the case of private than of public aircraft.

No delimitation has been agreed between 'air space' and 'outer space' but state sovereignty does not extend outwards indefinitely and satellites in orbit round the earth are not treated as infringing any state's airspace.

In 1961 the U.N. commended two principles to states, that international law applied to outer space and celestial bodies and that these were free for exploration but not subject to appropriation by any state.

D. Johnson, *Rights in Air Space.*

Aitchison, Craigie Mason, Lord (1882–1941). Was called to the Scottish bar in 1907 and took silk in 1923. He specialized in criminal work and was the leading advocate of his time in that sphere. He sat in Parliament, 1929–33, and was Lord Advocate in 1929, and held the office of Lord Justice Clerk from 1933 till his death. He was a just and sound rather than a learned judge, and like some of the eighteenth-century judges in his conviviality.

Aix-la-Chapelle, Treaties of. (1) A treaty of 1668 between France and the Triple Alliance of England, Sweden and the Dutch Republic, which ended the War of Devolution (1667–68). It did not settle the question of the right to devolution and was only a truce in Louis XIV's expansionist plans.

(2) A treaty of 1748 between Britain and France, which ended the War of the Austrian Succession (1740–48). It confirmed the right of succession of the House of Hanover in Britain and Hanover and effected mutual restoration of conquests in Canada and India. It guaranteed the right of Maria Theresa to the Austrian lands, but conceded to Prussia its conquest of Silesia. It was only a stage in the struggle between Britain and France for Canada and India which was resumed in 1756.

Aktenversendung. The practice known in Germany between about 1550 and 1879 of local courts remitting the papers in difficult cases to a local university law faculty for the collective decision of the professors. The latter heard no evidence or argument but decided the cases on the basis of the papers submitted only. The practice developed because the law professors were free from political influence and could be trusted to adhere to the rules. But they were frequently ignorant of local conditions, local customs and statutes. It arose also from the multiplicity of lower-level courts with untrained judges and the absence of effective appellate courts.

Though some rulers forbade the seeking of legal advice from outwith their states it was common to do so. In the eighteenth century some rulers forbade resort to outside sources and then later some abolished the practice altogether.

By custom law faculty decrees after Aktenversendung were binding though the inquiring court had to issue the decree and make it enforceable by its own process.

Legislation in the early nineteenth century abolished the practice in criminal cases and in 1879 general imperial legislation abolished it altogether. The practice contributed to the high standing of professors in Germany and kept their knowledge in touch with practical problems, but in many cases diverted their attention from their tasks of instructing and writing. It also was a major factor in introducing Roman law and Italian legal scholarship into Germany and in developing that body of knowledge and adjusting it to the needs of contemporary society.

Aktiengesellschaft. In German law, a corporation limited by shares, one of the two main types of corporations which may be formed for all lawful (not necessarily business) purposes. (The other is the *Gesellschaft mit beschränkter Haftung*, q.v.) It may be formed by five persons obtaining registration in the Commercial Register of the court of its domicile. The board of directors and the management must be separate and there is extensive regulation of accounts and meetings.

Alabama case. The *Alabama* was a warship built in Liverpool during the American Civil War by the Confederate states. She left British waters unarmed in June 1862, her guns being fitted at the Azores. She then acted as a commerce raider and destroyed various merchant ships belonging to the United States before being sunk in 1864. The U.S. Government claimed reparation from the United Kingdom for the damage done by the *Alabama* and other warships, particularly the *Florida*, and ultimately, under the Treaty of Washington of 8 May, 1871, the claim was submitted to an international arbitration tribunal comprising representatives of Brazil, Italy, Switzerland and the parties, which met at Geneva on 17 December, 1871. In the treaty of submission, the parties agreed on, as principles regulating the deliberations of the tribunal, certain rules which they agreed to observe between themselves in future, to bring to the notice of other maritime powers, and to invite the latter to accede to them. They were to the effect that a neutral state should take care to prevent the fitting out in its ports of a vessel believed intended for use against a power with which it is at peace, and to prevent the departure of such a ship from its jurisdiction, should not permit either belligerent to use its ports as a base against the other, nor to use them for obtaining supplies or men, and should exercise diligence to prevent violations of these duties. These rules have subsequently come to be regarded as internationally accepted. The first principle was adopted by Art. 8 of Hague Convention XIII, 1907.

At the arbitration the U.S. claimed also for indirect losses, but these claims were later departed from. On 15 September 1872, the award was issued finding the United Kingdom liable for the direct losses only and assessing damages at $15,500,000 in gold (£3,229,166).

Alamannic Code. This code, belonging to Suabia in southern Germany, the territory occupied by the Alamanni, consists of an earlier part, the *Pactus Alamannorum* dating from about 600, and a later part, the *Lex Alamannorum*, the date of which is about 718. The author of the later part is probably Duke Lanfred (709–30). The code comprises three parts, dealing with ecclesiastical causes, public law and private law, with some final additions. This body of folk-law appears to have remained in observance in south Germany and Switzerland down to A.D. 1000 or later.

Alanus Anglicus (early thirteenth century). A canonist of Welsh origin and a leading teacher at Bologna, author of a *Collectio Decretalium* (1206), a systematic collection of decretals (q.v.), *Apparatus* to *Compilatio prima antiqua* (after 1207), glosses on *Compilatio secunda antiqua* (after 1210) and *Apparatus Decretorum* (*Ius Naturale*), the last being one of the four great apparatuses on Gratian's *Decretum* which appeared about 1200.

Alaric's Breviary. See BREVIARIUM ALARICIANUM.

Alaska Boundary Dispute. A conflict between the U.S. and Canada about the extreme southern portion of Alaska. The U.S. purchased Alaska from Russia in 1867, the boundaries being those defined by an Anglo-Russian Treaty of 1825, and the Klondike Gold Rush made definition of the boundaries important. A joint High Commission having failed to settle the dispute, it was referred in 1903 to the arbitration of six impartial jurists who found in favour of the U.S. Though resented in Canada at the time the award was later accepted as a reasonable compromise.

Albani, Giovanni Gerolamo (1509–91). Italian canonist and cardinal, author of *De Donatione Constantini Magni* (1584), *De Cardinalatu* (1541), *De Potestate Papae et Concilii* (1544) and other works.

Albany Plan of Union (1754). A proposal formulated at a meeting of delegates from most of the American colonies, held at Albany in 1754, called for the purpose of formulating a concerted Indian policy, that the colonies federate for their greater security. The plan, drafted by Benjamin Franklin, provided for a representative governing body or Grand Council to be chosen triennially by the colonial legislatures with power to nominate executives, impose taxes, take charge of military affairs, and regulate Indian affairs. There was to be a President-General and the acts of the Grand Council were to be valid throughout the colonies unless vetoed by the Crown. The congress approved the plan, but the colonies refused to ratify it and the plan was never presented to Parliament. But it showed the value of such congresses and the plan was an ancestor of the Articles of Confederation and of the U.S. Constitution.

Alberico de Porta Ravennate (twelfth century). Italian glossator, author of glosses to parts of the *Corpus Juris* and a collection of *Distinctiones*.

Alberico di Rosciate (*c.* 1290–1354). A leading Italian jurist, a commentator and author of a *Dictionarium Juris* (1498), *Commentaria in Digestum* and *Tractatus super Statutis* and other works. His

Quaestiones statutorum was perhaps the first doctrinal treatise on private international law.

Alberto Gandino (*c.* 1245–*c.* 1310). Italian jurist, author of treatises *Quaestiones statutorum* and *Tractatus de Maleficiis* (1494).

Alberto Longobardista (twelfth century). Italian jurist, author of a commentary on the Lombard law.

***Albinatus, ius* or *droit d'aubaine*.** The right in old French law of the king to take all the property of an alien on his death, unless he had exemption therefrom. It was abolished in 1791.

Albisson, Jean (1732–1810). French lawyer and counsellor of state, took an active part in the drafting of the Napoleonic codes and was author of *Lois municipales et économiques de Languedoc* (1780) and many reports and proposals.

Albornoz, Egidio (or Gil Alvarez Carillo) de, Cardinal (?1290–1367). A Spanish cardinal, influential in papal policy, Vicar-General in the Papal States and restorer of papal authority there, and promulgator of the Italian code *Constitutiones de la Marca de Ancona* (*Liber constitutionum sanctae matris Ecclesiae*) commonly called the *Costituzioni egidiane*, 1357, which was later extended to all the papal states and lasted till 1816. It carefully regulated all government in the Italian republics so as to prevent local officials becoming too powerful.

Album. At Rome a whitened board on which were displayed public notices, including the *acta diurna* (q.v.) the annual edicts of the praetors, the lists of senators and jurors and other notices.

Alcala, Ordinance of (1348). Made by Alfonso XI of Castile which gave the force of law to the *Sieté Partidas* (q.v.) and formulated a hierarchy of sources of Castilian law, viz. statutes of the Cortes, the Fuero Real (q.v.), municipal fueros, and the *Siete Partidas.* It also confirmed the fueros or privileges of the nobility.

Alcalde. The Spanish title for mayor of a town or village, formerly for various kinds of officials having judicial functions and distinguished as *alcaldes de corte*, judges of the palace court, *alcaldes de crime*, judges in criminal cases, and others. Many kinds of alcaldes were found in the Spanish colonies in the Americas.

Alciati, Andrea (Alciatus) (1492–1550). An Italian humanist jurist, one of the first to interpret the civil law in the light of the history, languages, and literature of antiquity and to do original research rather than merely to copy earlier glosses of the texts. In 1518 he became professor at Avignon and his reputation grew rapidly; later he taught at Milan, Bourges, Pavia, and Bologna. He wrote *Annotationes in tres Libros Codicis* (the last three books), *Dispunctiones, Praetermissa* and *Paradoxa* and numerous miscellaneous treatises, as *De Verborum Significatione* (1530), but no systematic treatise. His *Responsa* range over the whole field of legal science. He also wrote *Emblematum Liber*, a collection of moral sayings in Latin verse and translated from Greek and Latin.

Alcohol. A constituent of many beverages, particularly wines, beers and spirits. The drinking of beverages containing alcohol has well-known effects on the central nervous system, initially as a stimulant but in larger concentrations as increasingly depressant, resulting in slurred speech, unsteady gait, and inability to make fine discriminations, and ultimately in sleep, coma and death. It affects the higher functions of the brain, thinking, remembering and making judgments. These physical consequences have important legal results. The individual's capacity to contract or transact legally is impaired and may be nullified, his capability to drive a vehicle is diminished, possibly to a dangerous degree, and his self-control is affected so that he may act in a way in which he would not when sober, e.g. quarrelling, fighting, amorous behaviour. Consequently to drive a vehicle when under the influence of alcohol is commonly an offence. In relation to a crime committed when under the influence of alcohol the general rule is that to have consumed alcohol is not a defence save in the extreme cases where the individual is no longer capable of forming the intent necessary for the particular crime or of controlling his conduct.

Alderman. From Anglo-Saxon ealdorman, originally synonymous with elder, and a term used of various high officials, and even of an earl or of the king; but latterly confined to a senior category of member of municipal corporations and county councils in England and Wales, and of municipal corporations in Ireland and parts of the United States. Outside London aldermen were abolished in 1972.

In Scotland the corresponding office in burghs was that of bailie (q.v.).

In the U.S. aldermen sat in municipal councils during the colonial period and after the Revolution acquired greater power and jurisdiction, but during the nineteenth century the power of the mayor increased and that of the aldermen declined, and the tendency has been for their numbers to decline.

Aldobrandini, Silvestro (1499–1558). Italian jurist, author of a commentary on Justinian's

Institutions, Addizioni ai commentarii di Filippo Decio sulla Decretali and *Trattato dell'usura.* His son became Pope Clement VIII.

Aleatory contract. A bargain, such as a wager, the outcome of which to some or all of the parties depends on an uncertain event.

Alessandri, Alessandro (Alexander ab Alexandro) (1461–1523). Italian jurist, author of a popular *Dies Geniales* (1522) modelled on Gellius' *Noctes Atticae* and Macrobius' *Saturnalia* and containing a confused mass of materials about law, antiquities and other matters.

Alexander III, Pope (Rolando Bandinelli) (*c.* 1105–81). Teacher of canon law at Bologna and author of the *Stoma* or *Summa Magistri Rolandi*, one of the earliest commentaries on the *Decretum Gratiani*, and of *Sentences* based on a work of Abelard. He was pope 1159–81 and as such held the Third Lateran council (1179) and was the beginning of the line of great lawyer-popes which culminated in Innocent III.

Alexander of Hales (1185–1245). English Franciscan theologian, author of a *Summa Theologica*, part of which dealt with law and justice, which was extensively used by later theologians.

Alexander, Sir William (?1761–1842). Was called to the bar in 1782, practised in Chancery and became a Chancery master in 1809 and Chief Baron of Exchequer in 1824, where he showed himself an able judge; he resigned in 1831.

Alfenus Varus, Publius (first century B.C.) a Roman jurist, consul 29 B.C., author of *Digesta* in 40 books, the first work to which that name was given, epitomes of which were used in Justinian's *Digest.*

Alfred the Great (?848–900). King of England 871–900, remembered for his long struggle against the Danes, which ended in 896 when they withdrew to Northumbria, East Anglia or the Continent. Alfred worked hard at civil reorganization in areas ravaged by the Danes and paid attention to the administration of justice. He issued a code of law largely founded on and incorporating rules contained in the codes of several of his predecessors. While avoiding unnecessary changes in custom, he sought to limit the bloodfeud and imposed heavy penalties for breaches of oaths or pledges. He also did much to encourage scholarship and himself translated Latin books.

M. Turk, *The Legal Code of Alfred the Great* (1893); F. Attenborough, *Laws of the Earliest English Kings* (1922).

Alias (otherwise (called)). A term used, particularly in criminal contexts, of the false names under which a person has passed at different times.

Alibi (elsewhere). A plea in defence to a criminal prosecution to the effect that at the time of the crime charged the accused was not at the place of the crime but was at another specified place, and therefore could not have committed the crime charged. The word is often incorrectly used for any kind of excuse.

Alien. A person, who, by the municipal law of one state, is not a national or citizen of that state though he may be resident there. No state can demand that its citizens be allowed to enter or reside on the territory of another state. Whether to admit aliens or not, and, if so, subject to what conditions, is a domestic matter for each state. Expulsion is also within the discretion of the state, probably subject to limitations by principles of humanity. If admitted an alien probably cannot be required to serve in the host country's armed forces but may be required to serve in police or defence forces. He is not necessarily entitled to political rights. In early times an alien was looked upon as akin to a criminal or an outlaw but the Roman law, by the idea of the *ius gentium*, accepted the idea that aliens had rights.

Since the late Middle Ages countries have been held entitled to exercise diplomatic protection over their nationals, resident in or visiting foreign countries. In some cases, under the regime of capitulations (q.v.), treaties secured immunity of aliens from local jurisdiction coupled with the privilege of their state in maintaining a special system of courts for its nationals on the territory of the host state. Apart from such cases an alien has to submit himself to local law and jurisdiction and should in ordinary cases seek redress of grievances against the host state in the local courts; but if he has unsuccessfully invoked that jurisdiction he may call for diplomatic intervention by his own state. In general, the alien cannot expect better treatment than the local national enjoys. In the twentieth century, however, support has grown for the recognition of an international minimum standard, recognizing fundamental human rights, but it is difficult to find an acceptable content for this standard. Among the major grounds of state responsibility to aliens are denial of justice as by gross deficiency in the administration of justice, and expropriation or nationalization of an alien's property which, it is generally accepted, is lawful if adequate, effective and prompt compensation is made, though some jurists hold the view that an alien cannot complain if he receives the same treatment in this respect as nationals, and expropriation for defence or police purposes is lawful even without compensation being payable.

By English common law, an alien is a subject of a foreign state, not born within the Queen's Dominions and consequently not owing allegiance to the Queen, or a person having no nationality of origin, or having lost his British nationality. Now under statute it means a person who is not a citizen of the United Kingdom and colonies, or any specified Commonwealth country, or the Republic of Ireland, nor a British subject without citizenship, nor a British protected person.

An alien friend is one whose sovereign or state is at peace with the United Kingdom and an alien enemy one whose sovereign or state is at war with the United Kingdom or who, whatever his nationality, is voluntarily resident or carrying on business in enemy, or enemy-occupied territory.

An alien friend has no title to be admitted to Britain, but while in the country owes temporary allegiance to the Crown and is entitled to protection, may sue and be sued, may hold property, and is criminally liable. But he may not become a member of the Privy Council or of either House of Parliament or hold any responsible office in the civil or military services and may not vote or sit on a jury. Landing in the United Kingdom is restricted and aliens arriving may be detained or refused entry in certain cases. Aliens deemed undesirable may be deported. An alien may, in the discretion of the Secretary of State, be granted 'political asylum' if to deport him would endanger his life or liberty or expose him to persecution on racial, religious, national or political grounds.

An alien may acquire British citizenship and become a British subject by naturalization (q.v.).

An alien enemy present in the United Kingdom in wartime may be allowed to leave the country but is liable to be interned. His private rights are, however, unaffected and he may sue in the courts.

An alien enemy resident in wartime in an enemy country cannot sue in any court in the U.K. and contracts with such a person are avoided, except those involving a property element which are merely suspended. Save by royal licence they cannot do business with or in the United Kingdom.

Alien and Sedition Acts. Four Acts passed by the U.S. Congress in 1798, restricting aliens and curtailing civil liberties in an attempt to silence criticism of the Federalist party. The three alien laws had the effect of lengthening the period of residence before naturalization, empowering the President to order out of the country aliens thought to be dangerous, and legalizing the apprehension or deportation of resident aliens when war was declared against the United States. The Sedition Act outlawed the publication of false and malicious writings against the government and the raising of opposition to any Act of Congress or by the President. These Acts gave rise to Kentucky and Virginia legislative resolutions for nullification of the Acts. The Acts expired or were repealed between 1800 and 1802.

Alienation. Any voluntary disposal of property, during life or, by will, on death. Questions arise whether law restricts freedom of alienation. In English feudal land law the Statute *Quia Emptores* in 1285 recognized alienation of a fee simple by substitution of a new tenant for an old but forbade future subinfeudation, the creation of a new fee subordinate to that of the tenant. In early English law there is evidence of restrictions on alienation of land on death but after Glanvill's time with the development of primogeniture these disappeared.

In Scots law as in continental systems, there have long been recognized restraints on alienation at death so as to ensure certain minimal provisions for certain members of the family and restrictions were introduced in England in 1938. See FAMILY PROVISIONS; LEGAL RIGHTS.

Alienation in mortmain. The alienation of lands to any corporation, lay or ecclesiastical.

Alieni juris. In the Roman law, a term descriptive of a person who was not *sui juris* (q.v.) but was under the tutelage or curatory of another, a father, husband or master.

Aliens Act (1705). An Act passed by the English Parliament to put pressure on Scotland to agree to the union of the Parliaments, to the effect of prohibiting all Scottish exports to England if the Scots did not accept the Hanoverian succession on the death of Queen Anne.

Aliment. In Scots law, the maintenance or support, in kind or in money, which certain persons are bound to afford to certain others connected with them by blood or marriage. A husband is bound to aliment his wife, though she has been guilty of adultery, and a wife her husband if he be indigent. A parent is liable to aliment his children till they can maintain themselves and a child his parents. A child is not bound to aliment his siblings unless as representing their father. The obligation is to supply food, clothing and other necessaries to an extent which preserves freedom from want; it may be enforced by action.

Alimentary provision. In Scots law, a provision made for the aliment of a beneficiary, the capital from which it is payable being held by trustees. If declared alimentary the beneficiary may not anticipate, assign or otherwise burden his interest at least to the extent of a reasonable provision.

Alimony. In English law, the financial provision to a wife from her husband while she was obliged to live apart from him. It might be awarded temporarily *pendente lite*, pending the outcome of a matrimonial cause between the spouses. The amount was in the court's discretion. After decree of judicial separation, alimony became a permanent allowance which continued for the whole period of the separation. A wife's allowance is now called maintenance (q.v.) and permanent financial provision.

In the U.S. alimony is awarded in similar circumstances but in some cases it may be awarded without separation or divorce proceedings.

Alison, Sir Archibald (1792–1867). Scottish judge and author, was called to the Scottish Bar in 1814, became an advocate-depute in 1822 and was Sheriff of Lanarkshire from 1834 till his death. He was elected Rector of Marischal College, Aberdeen (1845) and of Glasgow University (1857) and was made a baronet in 1852. He wrote a *Principles of the Criminal Law of Scotland* (1832), and a *Practice of the Criminal Law of Scotland* (1833), which are both still highly regarded, and a very successful *History of Europe During the French Revolution* (1833–42) with a continuation down to the accession of Napoleon III (1852–59), a *Principles of Population* (1840), *Life of Marlborough* (1867 and 1872), *Lives of Lord Castlereagh and Sir C. Stewart* (1861), and various essays.

All fours. A case is said to be on all fours with another, particularly a precedent, if it is indistinguishable from it in any respect material to the principle of law applicable.

Allegation. Generally any statement of fact made in a pleading or affidavit. In ecclesiastical law, it is any pleading after the libel, the first pleading, and may be distinguished into responsive allegations, counter allegations and exceptive allegations.

Allegiance. The obedience which every subject owes to the sovereign in return for the protection which that sovereign affords to his subjects. It was commonly held to be of four kinds, natural allegiance, due from all persons born within the British dominions, from the time of and by reason of their birth there; acquired allegiance, obtained by naturalization or denization; local allegiance, due from aliens so long as they remained within British dominions or retained a British passport; and legal allegiance, created by the oath of allegiance (q.v.) or due by children born abroad of natural-born British subjects which they could divest themselves of by declaration of alienage. At common law a natural-born subject could not divest himself of his allegiance, but this has been possible since the Naturalization Act, 1870. The older concept of allegiance has disappeared by reason of the British Nationality Act, 1948. Allegiance is now not a source but a consequence of nationality, and allegiance is owed by British subjects and British protected persons, by friendly aliens within the jurisdiction of the Crown, and by friendly aliens outside the jurisdiction who still have a call on the protection of the Crown e.g. by having a British passport. Alien enemies within the realm with the Crown's licence owe allegiance. Violation of allegiance by levying war on the Queen or adhering to the Queen's enemies is high treason. A person enjoying the protection of the Crown by virtue of allegiance is entitled to protection by the Crown against attack within Her Majesty's dominions and is entitled to diplomatic protection by the Crown.

Allegiance, Oath of. An oath, or affirmation to the same effect, which must be taken by various officers of state, judges, justices of the peace, Members of Parliament on taking their seats, and by English clergymen before ordination, to the effect that the taker will be faithful and bear true allegiance to the Queen and her successors according to law. A similar oath must be taken by a person wishing to become a British protected person or an alien on becoming naturalized.

Allen, Sir Carleton Kemp (1887–1966). Born in Australia, he became Stowell Civil Law Fellow at Oxford and Professor of Jurisprudence there, 1929–31, and then Warden of Rhodes House. Among his books the best known are *Law in the Making* (1927), *Bureaucracy Triumphant* (1931) and *Law and Orders* (1945). The latter show his vigorous criticism of administration and administrative law.

***Allen* v. *Flood*,** [1898] A.C. 1. Ironworkers objected to shipwrights working with them as they had previously and improperly worked on ironwork. The ironworkers' representative told their employer that if he did not dismiss the shipwrights the ironworkers would be called out on strike. It was argued that the threat was a malicious inducement to the employer to dismiss the shipwrights, who claimed damages. It was held that they were not entitled to damages because the ironworkers' representative's acts were not unlawful however malicious his motives and even though causing deliberate harm to the plaintiff.

Allgemeines bürgerliches Gesetzbuch. The general civil code for the Austrian Empire published in 1811 and founded on the Emperor Joseph II's partial codification of 1787. It comprised three parts and 1,502 sections and resembles, but improves on, the Prussian *Allgemeines Landrecht* (q.v.).

Allgemeines Landrecht. The Prussian General or Territorial Code of 1794, projected by Frederick the Great in 1749, drafted by Samuel von Cocceij, and intended to be a *Corpus Juris Fredericianum*, of which the first book on civil procedure was enacted in 1781, and revised 1780–93. It was divided into two parts and dealt with private law, church law, criminal law, commercial law and public and administrative law. It superseded the common Pandect and Saxon laws and claimed authority subsidiary only to provincial statutes and legal systems.

Alliance. In international law, a league between independent states for combined action, offensive or defensive. They are usually made by treaties for purposes defined therein, and numerous notable alliances are recorded in history.

Allocutus. The demand made of a prisoner convicted on indictment of treason what he has to say why the court should not proceed to sentence. The prisoner may move the court in arrest of judgment and show cause why sentence should not be passed but this is now rare.

Allodial land. Land not held of and under any superior but owned outright, as distinct from feudal land in which any landowner below the Crown can hold only an estate in the land of and under his feudal superior in consideration of certain services or payments. In France and Germany much land was originally allodial, but most of it was brought into feudal holding by the thirteenth century, though some land in Saxony remained allodial. After the French Revolution all land became allodial. In England bookland prior to 1066 may have been allodial, and the highest estate in land recognized in modern law, the estate in fee simple absolute in possession, is for practical purposes equivalent to outright ownership. In Scotland though most land is held feudally, some land is allodial, including the property of the Crown and of the Prince of Scotland, parish churches, churchyards and manses of the Church of Scotland, and land held udally (q.v.) in Orkney and Shetland.

Allotment. In company law the allocation of a certain number of shares to an applicant; in land law a portion of land assigned under an inclosure award, or a portion of land let by a local authority to a person for cultivation by him for the production of vegetables or fruit for himself.

Alluvion. In Roman law a variety of accession, the addition to existing land made by earth or sand washed up by the sea or a river. In English law if the increase is imperceptible it belongs to the owner of the existing land, or of the seashore, but if the increase is sudden and substantial to the seashore, the new land belongs to the Crown.

Almanac. A publication setting out the days of the weeks, months, and years, distinguishing fast days, feast days, term days and the like from common days. In England the almanac annexed to the Book of Common Prayer is part of the law and the courts must take notice of it. The term is also applied to various annual publications of a reference character and containing useful information.

Almon, John (1737–1805). Friend and confidant of John Wilkes, political writer, and compiler of the *Parliamentary Register*, the first monthly record of the proceedings of the United Kingdom Parliament, from 1774 to 1781, and of a summary of the debates from 1742 to 1774. His work is one of the main sources of Cobbett and Wright's *Parliamentary History* (q.v.). He also wrote a *History of the Parliament of Great Britain from the death of Queen Anne to the Death of George II* (1764), *Biographical Literary and Political Anecdotes* (1797) and various political works.

Almoner. In England a Lord Almoner is nominally charged with distributing royal alms, notably on Maundy Thursday. There is also an hereditary Grand Almoner. It also signified an officer of State of Scotland of uncertain duties probably connected with distribution of charity and the relief of poverty. The term was also formerly used of an officer of hospitals with similar functions.

Alnager or aulnager. A public sworn officer of the king who inquired into the assize of cloth, fixed seals to it, and collected a subsidiary or aulnage duty on cloth sold, which was abolished in 1689 and, in Ireland, in 1817.

Alsatias. The colloquial name (which first appears in Shadwell's plays in the time of Charles II) for recognized areas of sanctuary for criminals, survivals of the mediaeval sanctuaries (q.v.), which lasted till the end of the seventeenth century in London. The one which gave its name to all the others was Alsatia or Whitefriars, between Fleet Street and the Thames, but the Southwark Mint, the Minories and other places were other convenient refuges for thieves. The privilege of Whitefriars originated in the exemption from jurisdiction of an establishment of Carmelite friars in that area and was confirmed by James I in a charter of 1608. Statutes of 1696 and 1722 made it an offence to resist the execution of legal process in any of the 'pretended privileged places' named therein, while an Act of 1724 made it a transportable offence for any three persons to assemble for protection against the collection of debt. See Shadwell's *The Squire of Alsatia* and Scott's *The Fortunes of Nigel.*

Althaus, Johannes (Althusius) (1557–1638). Professor of law at Herborn, 1590, and author of

Jurisprudentiae Romanae methodice digestae libro duo (1586), *Politica methodice digesta* (1603), the first detailed political system devised in Germany and *Dicaeologicae libri tres* (1617). In the *Dicaeologica* he developed a complete legal system with particular reference to private law. He is important in the history of political theory for his advocacy of doctrines of popular sovereignty and his theory of federalism as a means of achieving national unity.

Althing. The national assembly of freemen in mediaeval Iceland and also the name of the modern Icelandic parliament.

Altius non tollendi. In Roman and Scots law, a servitude (q.v.) right attaching to one tenement of land restricting the liberty of the owner of an adjacent tenement to build higher on his own property so as to block the light to the windows of the dominant tenement.

Alvanley, Lord. See PEPPER ARDEN, R.

Alverstone, Lord. See WEBSTER, R. E.

Alveus. The bed of a river. In tidal water *prima facie* it belongs to the Crown. In non-tidal waters the bed belongs to the owner of the land through which the river flows; or, where the river divides two estates, the bed belongs in equal shares to the riparian proprietors, each owning to the middle line.

Amalfi, laws of (or Tabula Amalfitana, or Capitula et Ordinationes Curiae Maritimae nobilis civitatis Amalphae). A code of maritime law, formulated at Amalfi near Naples, probably about the end of the eleventh century or in the twelfth and recognized throughout the Mediterranean to the end of the sixteenth century.

Ambassador. A person appointed by the sovereign, or other head of state, to represent that person at a foreign court and to act as the recognized channel for communications with the foreign government, negotiations, complaints and representations. An ambassador is the highest rank of diplomatic representative and they were originally accredited only for monarchies, but later were sent out to republics, and now ambassadors are normally sent to all countries with which the sending country has diplomatic relations. By international law ambassadors, their families and staff are entitled to protection by the host country, to the respect due to the representatives of the head of another state, and to immunity from taxation and legal process, civil and criminal.

See also DIPLOMATIC PRIVILEGES.

Ambiguity (double meaning, doubt or uncertainty of meaning). In written instruments law recognizes two kinds of ambiguity, the patent ambiguity, which is obvious on the face of the instrument, such as a blank in a material respect, and the latent ambiguity, which is not apparent but becomes apparent only when the surrounding circumstances are known, e.g. a bequest to A.B., there being more than one person of that name. The general rule is that to resolve a patent ambiguity evidence extrinsic to the instrument, such as of facts known to the testator or other granter of the instrument, is admissible to enable the court to ascertain the meaning of words used, but not to give a meaning to a word or phrase capable of being given an ordinary interpretation. Extrinsic evidence is admissible to explain a latent ambiguity.

Ameer Ali, Syed (1849–1928). Born in Orissa, graduated at Calcutta and was called to the English bar in 1873. He entered politics and in 1883 became one of the Indian members of the Governor-General's council. In 1890 he became a judge of the High Court at Calcutta but settled in England in 1904 and in 1909 became a member of the Judicial Committee of the Privy Council. He wrote a treatise on *Mohammedan Law* (1880–84) and collaborated in many other legal works, and was a leader of Islamic thought and life.

Amende honorable. A punishment known in France from the twelfth century to the Revolution, used in case of an offence against the honour and authority of God, of the King, of the public weal or of a private person. It was later imposed in cases of public scandal and took two forms; simple amende honorable required the offender's presence in the Council Chamber where, kneeling and with bared head, he craved pardon of the persons injured by the act. Amende honorable *in figuris* took place in public; the offender, clad only in a shirt with a torch or taper in his hand and frequently with a halter round his neck, appeared before the door of the court house or church or both and on bended knees admitted his offence and craved pardon of God, of the King, of the officers of law and of the offended person. The latter was seldom imposed alone, but was frequently required before execution. Women as well as men were subject to it. The phrase is now used for a public apology and reparation.

Amendment. An alteration to some text, by substitution, alteration or addition, or any combination of these methods, intended to improve it in some respect.

A Bill put before Parliament may be sought to be, and frequently is, extensively amended, particularly at the Committee Stage.

In the statute law of the United Kingdom any provision of any statute may be amended in any way expressly or impliedly, by any subsequent statute, and in no case is there need for any special formality or any special majority. A large proportion of statute law is indeed intended to amend rules of common law or existing rules of statute law rather than to create wholly new rules.

Similarly procedural law contains provisions for the amendment of pleadings and records in civil and criminal proceedings, amendment of indictments, and substantive law provisions for amendment of charters, trusts, contracts and other kinds of documents.

Amendments to U.S. Constitution. Those changes which have been made since the Constitution was first ratified in 1789. Amendments proposed must be passed by Congress and submitted to the States for ratification, which requires the assent of three-fourths of the States. Only 30 have been submitted to the States and five were never ratified. The first 10 amendments constitute the Bill of Rights and became effective in 1791. Of the remaining fifteen the most important are the 'Civil War Amendments', the 13th (1865) which abolished slavery, the 14th (1868) which affirmed individual immunities and privileges, limited State action and recognized racial equality, and the 15th (1870) which gave the franchise to negroes. One, the 21st (1933), repealed the 18th (1919) which had established prohibition of liquor.

Amends, tender of. An offer to pay money in satisfaction of the harm done by an alleged wrong. Various statutes make a tender of amends a defence to an action for a particular kind of wrong, particularly where the wrong has been done innocently, as in the case of accidental defamation.

Amerciament or amercement. In English law a pecuniary penalty for an offence, when the offender was brought before the court of his lord, whether king or subject-superior, and was at the mercy of the lord. The amount seems originally to have been arbitrary but later came to be settled by custom and after Magna Carta the law was that, save where the amount was fixed, a freeman should be amerced only according to his means and be assessed by his peers. The court was said to amerce and its decision was recorded in the roll of the court. *Misericordia* was the liability, amercement its assessment in money terms.

Amercements were also levied on counties, hundreds and towns by the superior courts.

Amercements applied to judicial proceedings; thus a party who lost his case was 'in mercy', as was a jury which made a false presentment.

American Bar Association. A voluntary association of lawyers formed in 1878 to advance the science of jurisprudence, promote the administration of justice, uniformity of legislation and decision, uphold the honour of the legal profession and promote social intercourse among members. The association's work is supervised by a House of Delegates acting through a board of governors and various committees. There are numerous sections each devoted primarily to particular fields of law or branches of practice. It devotes much attention to maintaining standards of professional conduct, improving standards of legal education, supporting measures to improve the administration of justice and the uniformity of legislation. It publishes a monthly *American Bar Association Journal* and a *Yearbook*.

American law. See UNITED STATES LAW.

American Law Institute. A body, based on an idea contained in an address by W. N. Hohfeld (q.v.) in 1914, formed in 1923, to be a permanent organization for the improvement of the law.

It has prepared 'Restatements' of many of the major branches of the substantive law, based on the most generally accepted and most reasonable views of what the law should be. But these differ from codifications in having no authority beyond that of the Institute and the compilers of the volume, and from textbooks in not attempting to cite authority for their propositions nor seeking to reconcile divergent cases or opinions. Nevertheless the eminence of the scholars who have prepared the volumes and the care in their preparation has given them an authority similar to that of standard treatises.

American Law Review. A journal started by John Chipman Gray and John Goodman Ropes in 1866. It rapidly became important, and was later edited by Oliver Wendell Holmes, Jnr., and Arthur G. Sedgwick.

Ames, James Barr (1846–1910). Educated at Harvard he was in the first class to use Langdell's *Cases on Contracts.* He became a professor in the Harvard Law School in 1877, Dean in 1895, and Dane Professor in 1903. He was a man of great learning and a powerful teacher and had great influence on his students. During his time standards at Harvard steadily became more severe. For many years he was the unquestioned leader of American legal scholarship. He published case-books on many subjects and an important collection of his papers, *Essays in Legal History,* was published posthumously.

Ames, Samuel (1806–65). Was admitted to the Rhode Island Bar in 1826 and became the leader thereof. In 1854 he became chairman of the commission which produced the first complete restatement of the statute law of Rhode Island. In

1856 he became Chief Justice; his judgments were always treated with high respect.

Amicus curiae. Occasionally counsel is permitted or requested to argue for an interest not a party to the case before the court, such as the general public interest, or the interest of an authority or profession concerned indirectly, so as to enable the court to have in mind when deciding the case the view of that interest as well as of the parties. Thus counsel for the Law Society was heard as *amicus curiae* in *Allen* v. *McAlpine*, [1968] 2 Q.B. 229, in which solicitors were accused of professional neglect, where it was said that the duty of *amici curiae* was to help the court by expounding the law impartially, or if one of the parties were unrepresented, by advancing the legal arguments on his behalf. An *amicus curiae* may not normally appear save by leave of the court. The U.S. Supreme Court permits private persons to appear as *amici curiae* with the consent of both parties or the permission of the court but permits federal, state, and local governments to submit their views in a case concerning them without any consents.

Amiens, Treaty of (1802). An agreement between Britain, France, Spain, and the Batavian Republic effecting a short intermission in the Napoleonic Wars.

Amistad case. In 1839 some African natives brought from Africa to be enslaved and being shipped from Havana to Guanaja on the schooner *L'Amistad*, captured the vessel and were themselves captured off Connecticut. The question arose whether they were free or slave; the U.S. court eventually found them free and they were returned to Africa. The case aroused excitement and fierce passions, and exhibited all the worst features of deception and dishonesty to keep the negroes as slaves, and is one of the legal landmarks of the slavery question.

Amnesia (loss of memory, forgetfulness). Loss of memory of the events at the time of an alleged crime, even if proved to be genuine, the onus of proof being on the accused, may not be a good plea in bar of trial, on the basis that the accused is unfit to instruct his defence. See *R.* v. *Podola* (1960), 43 C.A.R. 220.

Amnesty. An act of grace or oblivion on the part of the Crown or Parliament obliterating some conduct, usually of a rebellious or criminal character, so that it cannot thereafter be founded on as meriting punishment. It is wider than pardon in that it obliterates all legal remembrance of the offence whereas pardon merely relieves from punishment. It may be granted subject to exceptions; thus the amnesty granted on the restoration of Charles II excluded those who had participated in the execution of Charles I. The last general act of amnesty in Britain was in 1747, which pardoned those who had participated in the second Jacobite rebellion. President Johnson granted an amnesty in 1868 after the Civil War. Peace treaties after wars frequently include an amnesty clause pardoning acts committed contrary to the laws of war by the belligerents and their forces, but named persons and persons guilty of war crimes are frequently excepted, as in the Treaty of Vereeniging in 1902, which terminated the South African War and in the Peace Treaties after the World Wars.

Amos, Andrew (1791–1860). A member of the first criminal law commission and first professor of English law in the University of London (now University College, London), 1829–37. Thereafter he served as legal member of the governor-general's council in India in succession to Macaulay, 1837–43, and, after a period as a county court judge, was Downing Professor of the Laws of England at Cambridge, 1848–60. He published an edition of Fortescue's *De Laudibus Legum Angliae* (1825), a book on *Fixtures* (1827), and books on topics of constitutional law.

Amos, Sheldon (1835–86). Son of the foregoing, was professor of jurisprudence at University College, London, 1869–79, and author of a *Systematic View of the Science of Jurisprudence* (1872), *The Science of Law* (1874), *Lectures on International Law* (1874) and other works. Later he was judge of a court of appeal in Egypt.

Amoveas manus. A writ commanding the return to a person of property belonging to him now in the possession of the Crown. Thus under the pre-1660 practice on petitions of right judgment, if against the Crown, was called judgment of *amoveas manus* since it directed that the King's hands be lifted from the property and possession restored to the petitioner. Both writ and judgment were abolished in 1947.

Ampliatio. In Roman criminal procedure, if the jury was undecided as to guilt they might by calling *amplius* demand repetition of the evidence. The practice fell into disuse in the principate.

Analogy (similarity). Reasoning by analogy is reasoning that if principle A applies to case X and the present case is similar in all material respects to case X, principle A should apply to the present case also. It has been much utilized in Anglo-American case-law. No analogy is logically compelling or leading to a necessary conclusion, but it is a common and useful way of arriving at a decision. It may however be countered by contending that the present case is as, or more, similar to case Y, decided on principle B, than to case X or by other arguments such as convenience, fairness or other. Analogy truly depends on the implicit recognition or contention that case X, or case Y, and the present case are both instances of a more general principle.

Analysis of legal problems. The resolution of problems into their component elements to facilitate the discovery of the principles of law appropriate to provide the answer to each problem. There is no set or infallible way of doing this. In the first place all the facts of the situation have to be ascertained and irrelevant facts discarded, though what is relevant and what is irrelevant depends on the problem being examined, on the legal system involved, and can usually only be decided by experience. Some facts may initially have to be assumed as a hypothesis, or have later to be investigated or verified. Secondly each legal difficulty raised by the facts in relation to the remedy desired must be distinguished, though many difficulties will be apparent only to the experienced lawyer, e.g. which court has jurisdiction, whether particular facts can be proved and, if so, how, and each legal difficulty must be individually investigated, e.g. what is the rule where goods have been sold and delivered and are found to be defective? This involves knowledge of the sources of the law (q.v.) and the literature, the weight to be attached to different statements or interpretations of the relevant rule, and an evaluation of the likelihood or otherwise of being able to bring the facts within the ambit of the rule in question, e.g. of being able to prove that the defendant acted without taking reasonable care. Lastly, the conclusions reached have to be formulated in the way appropriate, as advice to a client, an opinion, a draft of a writ of summons or other pleading or a decision.

Analytical jurisprudence. The general name for the approach to jurisprudence (q.v.) which concerns itself mainly with the classification of legal principles and rules and with the analysis of the concepts, relationships, words, and ideas used in legal systems, such as Person, Obligation, Right, Duty, Act and many more. It is mainly associated with positivism, the approach to law which concerns itself with positive law, i.e. legal systems and rules actually in force as distinct from ideal systems or law which should be. Analytical jurisprudence, though foreshadowed by Hobbes, is chiefly associated with Bentham and Austin (qq.v.), the latter of whom in *The Province of Jurisprudence Determined* and *Lectures on Jurisprudence* defined jurisprudence as 'the science concerned with the exposition of the principles, notions and distinctions which are common to ... the ampler and maturer systems of law' and set out to elucidate the notion of law properly so-called (to him, the commands of a superior) as distinguished from law improperly so-called, which included moral maxims and international law.

Analytical jurisprudence has been extensively cultivated and developed by English jurists, notably by Markby, Holland, Salmond (qq.v.), and others, and is almost necessarily particular rather than general, concerned with the terms and ideas used by a particular system of positive law since these vary from one system to another.

Analytical jurists consciously or unconsciously look on logical harmony and completeness as the main merit of a legal system, but no legal system is without anomalies and gaps, and so their logically perfect systems are in fact legal systems that should be, rather than that are. Nevertheless analytical jurisprudence has done much to clarify the concept of law and to elucidate the language and concepts of particular legal systems.

J. Austin, *Lectures on Jurisprudence*; J. Bentham, *Limits of Jurisprudence Defined*; *Of Laws in General*; W. W. Buckland, *Reflections on Jurisprudence*; H. L. A. Hart, *Concept of Law*.

Ananie, Jean d' (Joannes de Anania) (? – 1458). French jurist, professor at Bologna, wrote commentaries on the *Decretals*, *Consultations* and a *De Revocatione Feudi Alienati* (1546).

Anathema. A mediaeval ecclesiastical penalty. It was the great curse of the Church excommunicating a person or denouncing a doctrine or practice as damnable, normally pronounced by a bishop.

Anatolius (Sixth century A.D.). A professor of law at Constantinople, member of the commission which produced Justinian's *Digest* and author of a fragmentary dialogue with a student, *Dialogus Anatolii*.

Ancarano (or Ancharano), Petrus de (1333–1416). An Italian jurist and commentator on the canon and civil laws, author of a great *Commentaria in Decretales* (1535), *Lectura super Sexto* (1517), *Lectura super Clementinas* (1483) and *Consilia sive Responsa* (1496). He also wrote *Repetitiones*, studies on particular problems of law.

Ancestor. A lineal ascendant in a family or progenitor. In real property law prior to 1926, however, it meant any person from whom real property was inherited, though he might not be a progenitor.

Ancienne Coûtume de Normandie. The body of ancient laws and customs of the duchy of Normandie as they existed down to 1205 when

King John lost the duchy (except for the Channel Islands). It has been printed in the *Grand Coûtumier de Normandie*, which is the basis of the law of Jersey and, to a lesser extent, Guernsey.

Anciens Usages d'Artois (or *Coutumier d'Artois*). A compilation made *c.* 1300 by an unknown lawyer, probably of Arras, borrowing extensively from the Advice of Pierre de Fontaines (q.v.) and possibly from the Etablissements de Saint Louis (q.v.).

Ancient demesne. Manors of ancient demesne were manors in the actual possession of the Crown in the reigns of Edward the Confessor and William the Conqueror and designated Terrae Regis in Domesday Book. Tenants in socage (q.v.) of these lands were frequently known as tenants in ancient demesne, a peculiarity of their tenure being that prior to 1833 they could be sued in matters affecting their lands only in the Court of Common Pleas or the court of ancient demesne of the manor, constituted by the tenant in ancient demesne strictly so-called. Such land passed by common law conveyance. There were also customary leaseholders where tenure had characteristics of both freehold and copyhold, and copyholders of lease tenure who held at the will of the lord by copy of court roll, but neither of these groups were properly described as tenants in ancient demesne. The entry in Domesday Book is conclusive whether a particular manor was or was not ancient demesne. The tenure was abolished in 1922.

Ancient Law. A famous book by Sir Henry Maine (q.v.).

Ancient law. Ancient law can be defined somewhat arbitrarily as the bodies of legal institutions, concepts, ideas, and principles found in the societies which evolved prior to the fifth century A.D. Ancient law is quite distinct from primitive law (q.v.). Much ancient law is not at all primitive but mature and sophisticated; some primitive law is not at all ancient. Customary practices which may have been accepted as binding or enforceable doubtless existed in many societies from very early times but knowledge of the existence of social controls of kinds which are recognizably legal depends on the development of writing and the making of records, and their subsequent discovery and interpretation. Knowledge of law in pre-literate societies is only inference of probabilities, and the forms of law and social regulation inferred can only have been taboos, custom, and common acceptance.

It is a matter of extreme difficulty to construct a connected account of government, courts and law in ancient society, because evidence is frequently lacking or fragmentary and its interpretation, in the light of concomitant social conditions, very difficult. Attention must be concentrated on the collections of legal materials which have survived.

Civilizations first emerged in Mesopotamia and Egypt in the fourth millennium B.C. and developed later round the eastern Mediterranean, in Anatolia, Palestine, the Aegean area, Greece, and Italy and from all these areas records exist of so-called codes or bodies of written law. But in all cases they presuppose long periods of evolution of customary practices and represent only partial restatements or reductions based on ancient practices; the law-giver is not publishing a civil code but issuing a list of amendments to previous practice, or of instructions to judges how to deal with problems in changed circumstances.

In Sumeria and Babylon the administration of justice was a function of the priesthood and a department of the divine government of the city-state's patron god.

When Sumer was unified under the third dynasty of Ur the judges remained members of the priestly order but a specialized branch of it. Cases were heard in or at the gate of a temple. But later provincial courts with lay judges were introduced and Hammurabi developed civil courts with secular judges. Thereafter the regular court was that of the local governor and senior officials. There were appeals to the judges of Babylon and to the king. There is not, as there is in Egypt, evidence of widespread corruption, but the penalties prescribed were savage.

Mesopotamia

Babylonia, between the lower Euphrates and the lower Tigris, was a notable early centre of civilization. Written collections of law seem to have been not uncommon in the third millennium B.C. and it may be that every city-state had a collection.

The earliest text known is the Code of Ur-Nammu of Ur (*c.* 2100 B.C.). They seek to rectify various economic abuses and deal with compensation for injury; it is noteworthy that compensation has already superseded retaliation, an indication which makes these laws more mature than later codes of the less civilized Semites.

The incomplete Code of Lipit-Ishtar of Isin (*c.* 1930 B.C.) in Sumeria deals with a miscellany of subjects, slaves, gardens, inheritance, bigamy, division of the paternal estate, and fines for damage to a hired ox.

The earliest laws in Akkadian are the Laws of Eshnunna (*c.* 1900 B.C.) which deal with penalties for trespass, business transactions, betrothal and marriage, loss of property, sales and purchases, assaults and injuries, responsibility for damage, and other matters.

The most important legal material from Mesopotamia is the Code of Hammurabi, king of Babylon (*c.* 1780 B.C.) who welded the many city-states of

Sumer and Akkad into one kingdom and gave it one language and one legal system. His code has more orderly arrangement than any earlier one and is probably not so much borrowed from the earlier surviving codes as derived from a common origin. It deals with administration of justice, offences against property, land and houses, merchants and agents, marriage, family property and inheritance, assault and personal injury, professional responsibilities, agriculture and the ownership of slaves. But it is far from being a complete system of law; it is a body of amendments, revisions and recapitulations of principles already settled. These principles were observed in Babylonia for at least 1,000 years.

The body of laws known as the Assyrian Laws, from Assyria in the upper Tigris, probably belong to about 1350 B.C.; they are Middle Assyrian, and some fragments of Old Assyrian Laws are known. The noteworthy feature is that they always introduce modifications, amendments to the existing law which was basically the common law underlying the Code of Hammurabi. The surviving tablets deal with women, inheritance and sale of land, encroachment on a neighbour's property, and rights to irrigation water, debt and theft of cattle.

It is difficult to discuss the administration of justice in Babylonia and Assyria. In theory this was a duty of the king but civil cases seem to have been dealt with in the assembly at Babylon, but by a single royal officer in Assyria.

Anatolia

About 1700 B.C. a Hittite kingdom was created with its capital at Hattusas and its forces later raided and plundered Babylon.

Various tables and other collections of laws have been found, which evidence growth and development. They deal with various crimes, land tenure, marriage and other matters, and clearly supplement or modify an existing body of customary law; many deal with specific cases, and the tables may have been compiled from decisions of courts.

Little is known about Hittite courts, but there would appear to have been local courts composed of city governors and senior citizens, and in more important cases the commander of the local garrison presided. The Hittite law is comparatively advanced in that capital offences are few and compensation and restitution are remedies rather than retribution, which is uncommon. Traces of the blood-feud survived and guilt attached to the whole family of a wrongdoer.

Egypt

It is probable that an Egyptian law code existed long before it is first mentioned by Greek authors, and vague references suggest that something in the nature of a corpus of law went back to the earlier part of the third millennium B.C. In early times the Pharaoh was king, high priest, chief judge and commander-in-chief. By the time of the Old Kingdom (*c.* 2700–2100 B.C.) a hierarchy of administrators had developed; the country was divided into provinces under governors with local courts.

Under the Middle Kingdom (*c.* 2000–1700 B.C.) the law courts were in the capital, Ithet-Tawy, and all cases other than petty ones were heard there. Viziers as deputies of the Pharaohs acquired more power including judicial powers.

From the eighteenth dynasty (*c.* 1500 B.C.) there is a good deal of information about legal procedure. The viziers were chief administrators and judges. Death, mutilation and hard labour were penalties for serious crimes, fines and beating for lesser ones. Civil cases were mostly concerned with land and inheritance.

Under the New Kingdom, the Pharaoh, both human and divine, god and king, had complete legislative, executive and judicial powers, and was assisted by a huge bureaucracy.

In Ptolemaic Egypt (332–30 B.C.) there were parallel administrative and legal systems and parallel bureaucracies for Greek and Egyptian subjects. Egyptian custom was given the status of law. Egyptians and Greeks had separate courts with their own judges and rules. In general the language of the transaction determined the court of resort in civil mattters, but much law consisted of administrative decision and many private disputes were decided by bureaucrats rather than by courts.

Alexandria became the centre of intellectual life of the Eastern Mediterranean, and the great library and museum attracted most of the scholars of the known world to study there. Greek culture flourished but native institutions in custom, law and religion continued.

Roman Egypt (30 B.C. to 642 A.D.), though a special province of the Roman Empire important for the sake of the grain supply, continued to be governed by its Ptolemaic bureaucracy under a prefect appointed by the Emperor, who was chief judge as well as head of the army and administration. He exercised jurisdiction assisted by a *iuridicus*. Egypt was divided into three judicial districts and the Ptolemaic courts were eliminated or radically changed. Edicts were issued to establish particular practices which were cited as precedents. Roman law applied to Roman citizens and in Alexandria, though a Greek city, Roman government superseded the Ptolemaic courts. The major Greek cities were endowed with *boulai*. Roman citizenship could be earned until the Edict of Caracalla (212 A.D.) gave it to practically all groups. Under Diocletian Egypt was divided into three provinces and Latin replaced Greek as the bureaucratic language. After 395 Egypt was administered from

Constantinople and after 642 Egypt passed under the Arabs.

Palestine and Syria

Though the presence of man at Jericho is attested from *c.* 9000 B.C. city-states did not emerge until the third and second millennia B.C. In the second millennium Palestine was torn by war between the Egyptian and Hittite powers, but sometime in that period the Israelites settled in Palestine. The early books of the Old Testament reflect the memory of tribal groups in Palestine whom the latter Israelites, though invaders, considered to be their true ancestors.

In the thirteenth century B.C. Moses came to the fore as a leader and promulgated a body of rules of conduct which, according to tradition, he received from God on Mount Sinai and which, in reality, he appreciated were necessary for a stable and just community. These with supplemental interpretations became the basis of the body of Jewish law and tradition.

Like Hammurabi, Moses did not create a code but codified a list of traditional rules, drawn from various sources, which were to be relied on in future. Some were peculiar to the patriarchial class from which the Hebrews claimed descent, some borrowed from the Sumerians and other people with whom the Hebrews had been in contact and some were original. The Hebrews were a nomadic people and Moses not a ruler but a moral leader. The only sanction for the Mosaic Code was the sanction of God, and in the Mosaic Code there are purely religious provisions mingled with social and penal prescriptions, and all represented as a divine law issued by Jehovah in person. The Hebrew law is uniquè among ancient systems in making religion the basis of law and establishing a moral code. But it is less systematized or coherent than the Mesopotamian and Hittite codes and gives little evidence of development resulting from long judicial experience. Nevertheless Mosaic law was assimilated by the Hebrew tribesmen and its moral basis recognized and accepted, and this acceptance of the moral basis of law has been of great permanent importance, through Christianity, to Western civilization.

The Hebrews had no structure of courts, or state regulation. The so-called Judges were chieftains having control over limited areas and expected to maintain traditional rules of justice. They had no power to enforce apart from force. The Mosaic Code was kept in existence by the heads of families, and councils of elders. That the law was maintained was because it had been promulgated as the revealed word of God, embodied the essentials of tribal and family custom and breach involved denial of tribal faith.

By 1200 B.C. the Hebrews were organized in tribes largely linked by their common faith in Jehovah. By *c.* 1020 B.C. Saul had been made king and *c.* 1000 B.C. David was accepted as being king over all Israel. He based the governmental machinery and organization of the kingdom on that of the Canaanites and built up Israel into a powerful empire. His successor Solomon developed trade and culture, but after his death the state split into Israel (in the north) and Judah (in the south). The eighth century prophets criticize the Israelites for social, moral and religious shortcoming, for idleness and arrogance and by the end of that century most of Palestine and Syria had fallen to Assyria, which in turn fell to Babylonia and that in turn to the Persians in 539 B.C. During the Babylonian exile many of the historical books of the Old Testament were composed. The Persians allowed the Jews to return to Palestine. Alexander the Great incorporated Palestine, and Syria into the Greek and thereafter the Ptolemaic empire, but in 198 B.C. Palestine was annexed by the Seleucids of Syria. Greek culture developed in Palestine until the country was taken over by Rome. Palestine and Syria were on the whole peaceful and prosperous under the Romans. Hellenization continued. The adoption of Christianity as the established religion of the empire made Palestine an area of particular interest. In the sixth century there were renewed Persian inroads and then the Arabs conquered the country in A.D. 634–36.

Through Christianity the basic principles of Moses' rule have been powerfully influential throughout the Western world and more widely.

Greece

The Minoan civilization which flourished in Crete in the second millennium B.C. and the rather later Mycenaean civilization both attained considerable maturity but neither has left, so far as known, any evidence of a legal system. The Homeric poems, dealing with the Trojan war of *c.* 1200 B.C. reveal little of the customary unwritten laws of the early Greek communities. The kings seem to have been judges in peace as well as leaders in war, but there is also mention of the elders of the people sitting in judgment, though of principles on which they acted there is no indication.

In the seventh century B.C. however, most of the Greek city-states, apart from Sparta, seem to have progressed to written laws. In each case the legislation was made by, or attributed to, a distinguished person who was commissioned to frame a code which was then accepted by the community. The earliest code of this kind was traditionally said to have been that framed by Zuleucus for Locri and he and Charondas of Catana, who made laws also for other Greek cities of Sicily and Italy, were highly esteemed. Other legislators were Aristeides of Ceos, Pheidon of Corinth,

Philolaus of Corinth who made laws for Thebes, and Androdamas of Rhegium. Lycurgus of Sparta is probably legendary. At Athens in 621 B.C. Draco (q.v.) is credited with a harsh code which may not have done more than put into order and written form existing practice though it was regarded as advantageous to have a known code with fixed penalties in place of a mysterious body of customs applied by the nobility. Solon (q.v.), however, who reformulated the law in 594 B.C., as part of his reforms of the constitution, laws, and economy, was always afterwards regarded as the founder of the Athenian legal system.

The inscriptions known as the laws of Gortyn (q.v.) in Crete give some information about early Greek law in the fourth century. The laws appear to be a testament of a earlier code and refer to an existing body of written law, sometimes making innovations therein. Trial is by a single judge and procedure simple, and judgment is generally for an award of damages.

In mainland Greece there never developed a generally accepted body of law. Each city-state had its own system, though there were common features. Most is known about the system at Athens. The Athenian legal system had its origin in the legislation of Solon. Various kinds of magistrates had mainly judicial functions, the Thesmothetae who dealt mainly with public actions and criminal trials, the Forty who dealt with minor disputes and remitted more important cases to public arbitrators, and the Introducers, who dealt with certain actions summarily. The major courts were large, up to 1,000 and even more in great political trials, but in them appeals to emotion and prejudice, sophistry and misrepresentation of law seem to have been common. Special courts sat to deal with cases of homicide.

The distinction between public and private wrongs was recognized from Solon's time. In public actions the penalties were death, deprivation of civil rights, fines and confiscation. Imprisonment was an additional penalty in a few cases. In private actions penalties were pecuniary only.

More important in subsequent influence than substantive law have been the writings of the Greek philosophers. Much of the writing of Plato and Aristotle about the state, law, justice, and related subjects is not only the earliest substantial thinking about these subjects but embodies much of continuing importance and value, and much that has been profoundly influential on later thinking about law. Conversely, and possibly surprisingly, a material factor in Greek law was the absence of a body of thought about legal system, analysing and systematizing positive law, examining the difficulties and probing the implications.

In the Hellenistic period (323–30 B.C.) Greek culture and ideas of every kind spread rapidly to Egypt, Palestine, Asia Minor and elsewhere. Attic law was widely adopted outside Greece in forms modified to suit local conditions. In the first century B.C. the Romans conquered Greece and in 30 B.C. the whole Hellenistic world became part of the Roman Empire. Thereafter the influx of Roman law began though it came slowly and Hellenistic law continued, subject to Roman influence, at least until A.D. 212 when Roman citizenship was extended to nearly all inhabitants of the Roman Empire. For this period no code or contemporary treatise exists.

Rome

The people of Rome steadily developed in ascendancy over all Italy and built an empire which came to take in all the lands surrounding the Mediterranean. In their republican period (753–31 B.C.) they developed a *ius civile* based on custom and legislation applicable to Roman citizens only and also, from the mid-third century B.C., a *ius gentium* applicable to foreigners also. In 451–450 B.C. they adopted a body of legislation known as the Twelve Tables (q.v.) as a body of written law which upper-class magistrates could not alter to the detriment of ordinary litigants. Thereafter there was much legislation, *leges* or enactments of one or other of the assemblies of the Roman people, *plebiscita* or resolutions of the plebeian assembly, and *edicta* or proclamations of superior magistrates as to their intended approaches to the grant or refusal of judicial remedies. These became an important instrument of legal change, often materially modifying the statutory law. *Senatus consulta*, or resolutions of the senate had no legislative force at this period. By the end of the republic the Romans had developed an extensive systematized and mature body of law.

Under the principate (31 B.C.–A.D. 235) the power of the emperor increased; *leges* ceased to be enacted, *senatus consulta* became resolutions endorsing the proposals of the emperor, the *edicta* were in A.D. 131 restated and consolidated and made unalterable save by the emperor, and *constitutiones principum* in the form of edicts, proclamations, instructions, answers and decisions of the emperors became the chief form of legislation. In the second and third centuries A.D. Roman law attained its classical stage, one marked not only by maturity of the law but by the flourishing of a host of jurists who commented on the law, wrote treatises, analysed and discussed the problems of the positive law. In 212 A.D. Roman citizenship was extended throughout the empire, the distinction between the *ius civile* and *ius gentium* finally disappeared and Roman law was extended to all the lands round the Mediterranean, increasingly influencing the indigenous law.

From the later third century onwards, the dominate, the emperors were increasingly autocratic rulers and their legislation was the sole form of lawmaking. Legal science was in decline, as were all culture, government and the economy of the failing empire. In A.D. 324 Constantine founded Constantinople as the new administrative capital of the empire and the empire divided into the Latin West and the Greek East. In the fifth century the Visigoths sacked Rome in A.D. 410 while many of the provinces were lost and finally the Western Empire disappeared in A.D. 476.

The Eastern Empire, however, survived and in the sixth century the Emperor Justinian appointed commissions to simplify and digest the surviving writings of the jurists of three centuries earlier and the later legislation. This took the form of the *Digest* or Pandects, the *Institutions*, an outline of the elements of the law, the *Codex* of constitutions (534 A.D.) and a series of new legislative measures (*Novellae constitutiones*). This collection of material, later collectively known as the *Corpus Iuris Civilis*, is the main source of knowledge of the earlier and classical law, and continued to be the chief lawbook of the Roman world. At Constantinople itself it was a major source of the law of the later Eastern Empire.

The Roman law of the classical period, particularly as preserved and altered in Justinian's compilations, is the greatest legal achievement of antiquity in that for the first time rules of positive law are there found systematically arranged, analysed and commented on and with substantial steps made to discover general ideas such as obligation and property and to find the general principles and the problems. The Roman jurists were, however, practised lawyers and did not make any headway beyond the stage reached by the Greek philosophers in discussing the problems of justice, duty and others. Moreover, the Roman law more than any other system of ancient times has had great and continuing influence on mediaeval and modern systems of law. Most of the legal systems of Western Europe, and through them some of the legal systems of countries outside Europe, have bodies of law which in structure and arrangement, fundamental categories, concepts and terminology, methods of thought and actual rules, owe something, and frequently a great deal, to Roman law.

The Celtic countries

The Irish Law Tracts were not put into writing until the sixth or seventh century A.D., but they go back to a much earlier period of tribal, rural society, organized hierarchically in tribal areas under chieftains, having been preserved for centuries by oral tradition before being reduced to writing and representing a very old inheritance of learning. They go back to old Indo-European tradition and are probably in principle the oldest in Europe, wholly unaffected by Roman law or law from any Mediterranean country. In ancient times the Welsh laws had a similar basis to the Irish ones.

The Germanic countries

Roman writers give some information about the customs and laws as well as the other facets of the culture of Germanic peoples. Written statements of the law of Germanic tribes did not appear till the early mediaeval period but it was based on centuries of earlier customary practices.

Ancient international law

Peaceful relations, by way of trade, and war, for conquest or arising out of quarrels, between different tribes, city-states and different peoples, are known from remotest antiquity. Victories and defeats were alike ascribed to the gods. Independent states were regularly either bound to others by treaties of alliance or friendship or at war with others. Treaties negotiated by ambassadors, subsequently ratified, were well known. Such are known from all the areas where civilization had developed. Relationships of sovereignty and vassalage between states were well-known. Extradition of political offenders, fugitives and criminals was common. Rules existed about capture and return of runaway slaves. Alliances were often sealed by dynastic intermarriage. Resident ambassadors and frequent passage of envoys were regular. There was of course no formal international law but there was a body of recognized conventions, mostly founded on the internal law of Babylon, and the Akkadian cuneiform language was a diplomatic language known and used even in Egypt. There is much in the diplomatic relations of the years B.C. which anticipates the practice of the mediaeval and even the modern world.

Summary

Law and legal institutions emerged distinct from custom in the parts of the world which early became civilized and at a very early stage in that process of evolution. Development, however, took place in different ways at different places in the ancient world and reached different stages of maturity.

In Sumeria the theory of theocratic government did develop an idea of administering justice as a condition of social life but the priesthood never seems to have got much beyond retribution as a means of justice. The Hittites seem to have evolved a system unique in the ancient world, conscious of individual interests and rights. In Egypt law and courts existed but progress seems to have been stifled by corruption. The Hebrews took a great step forward in founding law on religious belief and morality. Greece marked a great step forward, not so much in legal institutions and rules, as in deep and serious thinking about the concepts which underlie law, the state, justice, right and wrong, while the Romans not only advanced far further than any previous people in the range of their legal rules, but began to examine, question, analyse, speculate, systematize and reduce to order a process which by the third century A.D. had produced a highly developed, refined and mature body of legal writing and thinking, as well as highly developed bodies of law and procedure.

See also ASSYRIAN LAW; BABYLONIAN LAW; BYZANTINE LAW; CELTIC LAW; EGYPTIAN LAW; GREEK LAW; HELLENISTIC LAW; HITTITE LAW; JEWISH LAW; MEDIAEVAL LAW; ROMAN LAW.

Cambridge Ancient History, passim; (Mesopotamia) G. Goetze, *The Laws of Eshnunna*; G. R Driver and J. Miles, *The Babylonian Laws*; *The Assyrian Laws*; (Anatolia) E. Neufeld, *The Hittite Laws*; J. Cuq, *Études sur le droit babylonien, les lois assyriennes et les lois hittites*; (Egypt) J. Pirenne, *Histoire des Institutions et du Droit Privé de l'ancienne Egypte*; E. Seidl, *Einführung in die ägyptische Rechtsgeschichte bis zum Ende des Neuen Reiches*; *Ägyptische Rechtsgeschichte der Saitenunds Persenzeit*; *Ptolemaische Rechtsgeschichte*; E. Taubenschlag, *Law of Graeco-Roman Egypt in the light of the Papyri*; (Palestine) J. Kent, *Israel's Laws and Legal Precedents*; J. Smith, *The Origin and History of Hebrew Law*; (Greece) J. Lipsius, *Das attische Recht und Rechtsverfahren*; L. Beauchet, *Histoire du Droit Privé de la République Athenienne*; G. Calhoun, *Growth of Criminal Law in Ancient Greece*; *Introduction to Greek Legal Science*; P. Vinogradoff, *Outlines of Historical Jurisprudence*, Vol. II; J. Jones, *Law and Legal Theory of the Greeks*; A. Harrison, *The Law of Athens*; D. MacDowell, *Law in Classical Athens*; (Rome) H. Jolowicz and J. Nicholas, *Historical Introduction to Roman Law*; C. Westrup, *Introduction to Early Roman Law*; A. Watson, *Law of Obligations, Persons, Property, Succession in the Later Roman Republic* (4 vols.); F. Schulz, *Classical Roman Law*; W. Buckland, *Textbook of Roman Law from Augustus to Justinian.*

Ancient lights. In English law, windows through which light has been enjoyed, otherwise than by consent or permission, originally for so long that the memory of man ran not to the contrary and, since 1832, for upwards of 20 years without interruption. Thereafter the right becomes absolute and indefeasible and entitles the owner to prevent an adjacent owner from obscuring the light. The windows may be enlarged or multiplied but the right to light, so far as increased, must be acquired separately, and the owner of adjoining land may, before the prescriptive period had run, erect a screen or other obstacle preventing the acquisition on increase of any right to light. There is no right to a view.

Scots law knows no doctrine of ancient lights but a servitude (q.v.) right to light and prospect may be acquired. The doctrine does not prevail generally in the United States either.

Ancient messuages. In England, houses or buildings erected before the time of legal memory, i.e. the accession of Richard I, though now any building is an ancient messuage if it was erected before the time of living memory and it cannot be proved to be modern. Rights of common frequently attached to ancient messuages and houses replacing them on the same sites.

Ancient monuments. Archaeological and historical sites and buildings which, in many cases, have by statute been placed under the ownership or guardianship of the Department of the Environment, or a local authority, subject to provisions for their protection, and access thereto by the public.

Ancient Pleas of the Crown. The name given to Kelham's translation of Britton (q.v.) of 1762.

Ancient Serjeant. The eldest of the Queen's Serjeants. See SERJEANT.

Ancients. A term applied to certain members of the Inns of Court and of Chancery. In the Middle Temple and the Inns of Chancery all members except students seem at one time to have been called ancients, and in the Middle Temple certain senior barristers are called ancients though for dining purposes only. In Gray's Inn ancients were formerly a distinct grade of senior barristers.

Ancillary relief. A remedy given incidental to the main remedy.

***Anderson* v. *Gorrie*,** [1895] 1 Q.B. 668. The plaintiff sued three judges of the Supreme Court of Trinidad and Tobago for damages for acts done, allegedly maliciously and without jurisdiction, in judicial proceedings. The Court of Appeal held that no action lies against a judge of a court of record for doing something within his jurisdiction, even though maliciously and contrary to good faith, as to permit an action would prejudice the independence of the judiciary. The Court has subsequently minimized any distinction between superior and inferior courts in respect of judicial immunity, in *Sirros* v. *Moore*, [1975] Q.B. 118.

Anderson, David (1863–1934). Called to the Scottish bar in 1891 and was Chairman of the Scottish Land Court, as Lord St. Vigeans, 1918–34.

Anderson, Sir Edmund (1530–1605). Took part in all the leading state trials in the latter years of Queen Elizabeth, became sergeant-at-law to the Queen in 1579, and was one of the commissioners appointed to try Mary Queen of Scots, in 1586. He became Chief Justice of the Common Pleas in 1582. He had a reputation for impartiality in civil cases and for harshness when dealing with Catholics and Nonconformists.

Anderson, James (1662–1728). Scottish antiquary and genealogist, became a Writer to the Signet. He came to the fore in 1705 with a vindication of the independence of Scotland, *An Historical Essay, showing that the Crown and Kingdom of Scotland is Imperial and Independent*, rebutting the pamphlet by William Attwood, *The Superiority and Direct Dominion of the Imperial Crown and Kingdom of England over the Crown and Kingdom of Scotland*, reviving the claim of Edward I to the crown of Scotland. For this Anderson received the thanks of the Parliament of Scotland, while Attwood's book was ordered to be burned by the common hangman. In 1727 he published a valuable collection of historical documents, *Collections relating to the History of Mary, Queen of Scotland.* He then embarked, under the patronage of the Scottish estates, on the great *Selectus Diplomatum et Numismatum Scotiae Thesaurus*, finally published by Thomas Ruddiman in 1739.

André, Jean d' (Johannes Andreae or Giovanni Andrea) (1275–1348). A French commentator-jurist, professor of canon law at Bologna, said to be the most distinguished authority on procedure in the Middle Ages and author of a *Glossa ad Decretales* and *Additions to the Speculum Iuris of Durandus* (1475).

Andrea di Isernia (?1250–1316). Italian jurist, author of *Riti della Magna Curia, Lectura, Consilia* (1575), *Commentarium in Codicem et Authenticum* (1581), and a vast *Commentaria in Usus Feudorum* (1598).

Andrews, Sir James, Bart. (1877–1951). Called to the Irish Bar in 1900, he became a Lord Justice of Appeal for Northern Ireland 1921–37, and Lord Chief Justice of Northern Ireland, 1937–51.

Andrews, James de Witt (1856–1928). American jurist, professor at Northwestern and Chicago, remembered for his commentary on the American legal system, Andrews' *American Law* (1900) and an edition of Blackstone.

Angary (or *ius angariae*, from *angaria*). A post-system, hence compulsory right of transport, was formerly the right of belligerents under international law to lay embargo on and seize neutral ships in their harbours and compel them and their crews to transport troops, munitions and provisions on payment of freight. This right was sometimes excluded by treaty and fell into disuse during the eighteenth century. The modern right of angary is the right of belligerents to destroy or take up and use in case of absolute necessity, for the purpose of offence or defence, neutral property on their territory or on enemy territory. There can be no requisition on the open sea. It is usually justified as an exercise of the power of a sovereign state. It does not extend to compelling neutral individuals to render services. There is a duty to pay compensation for damage done. It is comparable to the power of states to requisition property of their own nationals, or to seize enemy property on their territory.

Angelus Carletus de Clavasio (1411–95). Italian canonist, later beatified, author of a famous *Summa Angelica de Casibus Conscientiae* (1487).

Angelus de Ubaldis (Angelo degli Ubaldi) (1328–1407). Brother of Baldus (q.v.) Italian jurist and commentator, author of *In Digestum Commentaria* (1477) treatises on obligations and war, and *Responsa* (1582).

Anglo-American Commission. A joint international commission appointed by the United Kingdom and the U.S. in 1898 to negotiate a plan for the settlement of matters in controversy between the U.S. and Canada. The subjects for consideration were the Behring Sea sealing question, the Great Lakes, the boundary question, and certain other matters. The commission met but failed to achieve any results.

Anglo-American law. Since United States law (q.v.) has been strongly influenced by its origins in English law (q.v.) there is still a substantial family resemblance, particularly in matters still largely regulated by common law principles, e.g. contract, tort, though much less so in matters dependent largely or entirely on statute. The name Anglo-American law is sometimes given to this common core of the two systems, particularly as contrasted with civil law systems, such as French and German law, fundamentally based on civil (Roman) law. But there is no actual system of Anglo-American law.

Anglo-Indian codes. The Charter Act 1833 provided that laws generally applicable to all classes of persons in India should be enacted and appointed a Law Commission, comprising four members of whom the most distinguished was Macaulay. They prepared an Indian Penal Code, the draft being mainly Macaulay's work, which was enacted and proved a great success and the model for the later Anglo-Indian Codes. It was followed by Codes of Civil and Criminal Procedure in 1859 and 1861, a Succession Act in 1865, an Evidence Act and a

Contract Act in 1872, the Specific Relief Act of 1877, Negotiable Instruments Act, 1881, Indian Trusts Act, 1882, Transfer of Property Act, 1882, and Indian Easements Act, 1882. Revision and replacement of the Codes was also effected. In general these codes were very beneficial and successful and did much to pave the way for modern Indian and Pakistan law, except in matters determined by Hindu or Mohammedan religious laws, and they influenced the law of the Sudan and Nigeria.

C. P. Ilbert, *Legislative Methods and Forms*; Whitley Stokes, *The Anglo-Indian Codes.*

Anglo-Saxon codes. The term for written compilations of law dating from the period A.D. 449–1066 in England. They are all in English, a fact which itself evidences their native and Teutonic, rather than Roman, origins. They seem to have enacted the customary law of the people and to take for granted a great body of unwritten custom. They are not codes (q.v.) in the modern sense and are not the sole sources of Anglo-Saxon law; there are also charters, chronicles and other materials.

The earliest are the laws of King Aethelberht of Kent dating from the end of the sixth century, and promulgated about A.D. 600. They have no known model and nothing to suggest that they were intended for a Christian community. They were issued with the advice of the elders of the people. They are not general legislation so much as settling the scale of pecuniary penalties for specific offences. The concept of the King's peace and breach thereof is known, but nobles, ceorls and others also have their peace, for the breach of which penalties must be paid, as well as compensation for death, injury, or loss of property. The legislation is concerned almost entirely with criminal law.

Later there were laws of Hlothere and his nephew Eadric (685–86).

The last of the Kentish codes is the laws of Wihtraed (*c.* 695), which are laws agreed by the witan and for the most part devised to give a privileged position to the clergy. The Kentish legislation is concerned almost exclusively with criminal law, wrongs to be redeemed by money. There is little civil law, but a good deal of procedural law, particularly as to oaths. There is nothing about land or succession.

The West Saxons were the next people to promulgate written law, in the dooms of Ine (*c.* 694), which evidence borrowing from Wihtraed. Many are more elaborate and detailed than the Kentish laws and generally more mature. The proportion of criminal law is less and there are provisions about agriculture and trade.

Offa of Mercia (757–96) is known to have legislated and his laws probably generally resembled the laws of Kent and Wessex. But this apart there was a gap in legislation between Ine and Alfred.

Alfred the Great (871–99) seems to have intended to replace the still-existing laws of Aethelberht and Ine and probably also of Offa by a body of law of general application throughout his kingdom, taking the best from the older legislation and adding only a little of his own. His code (*c.* 890) is confused but significant as beginning a continuous period of legislation. It deals with a great mixture of topics but the latter half of the code is filled with rates of compensation for private wrongs and personal injuries. It is still legislation for a rude country where slaves were kept and roughly treated.

It was followed by legislation of Edward the Elder, Athelstan, Edmund and Edgar. There are two brief sets of laws of Edward (899–924), dealing respectively with the regulation of trade, seeking to confine it to towns in the presence of the portreeve, with disseisin and perjury, and the others with procedure, requiring the holding of a court every four weeks, and directed against cattle-theft. Athelstan's legislation (924–38) includes provisions for the relief of the poor and two pieces of essentially delegated legislation, one intended to repress cattle-stealing. The volume of legislation is considerable but it is confused and repetitive. Penalties are savage. Edmund (939–46) has left three short series of laws, dealing with ecclesiastical matters, theft and the repression of bloodfeud. Edgar (959–75) was responsible for four surviving series of laws, requiring the hundred-moot to assemble every four weeks, requiring the shire-moot to meet twice a year and be attended by bishops and ealdormen and other matters; and the main theme is the repression of cattle-stealing and trade in stolen beasts. Edgar's laws were later held in high esteem and Cnut undertook that Edgar's laws should be observed.

Ethelred (978–1016) legislated extensively, his laws dealing with keeping the peace, with a court in each wapentake of the Danelaw, religious observance and many other topics. Many are pious exhortation rather than enforceable rules.

Cnut (1017–35), on becoming King of England, accepted Edgar's laws but also issued a code of his own dealing partly with ecclesiastical, partly secular, matters and much merely declaratory of existing law; it summed up all past laws and enjoyed great respect and influence in the twelfth century. Two versions of his law, the *Consiliatio Cnuti* and the *Instituta Cnuti* were made after the Norman Conquest and the *Quadripartitus* (q.v.) gave prominence to Cnut's law. Looking at this legislation as a whole there is apparent a transition from declaration of existing custom to the making of new law, but always assuming the existence of much customary law. The laws are not codes in the sense of substantially complete bodies of law. In terms of juristic sophistication they are centuries behind the

Roman law of 1,000 years earlier; they are the work of legally less competent and mature advisers of less powerful rulers.

F. Schmid, *Die Gesetze der Angelsachsen*; F. Liebermann, *Die Gesetze der Angelsachsen*; B. Thorpe, *Ancient Laws and Institutes of England*; H. Richardson and G. O. Sayles, *Law and Legislation from Aethelberht to Magna Carta* (1966).

Anglo-Saxon law. The Anglo-Saxon period of English history extends from A.D. 449 to 1066. It covers three epochs. The first was from 449 to 800 when the country was settled by bands of Angles, Saxons, and Jutes, who founded small states, which gave way to the three larger states of Wessex, Mercia, and Northumbria. In 800 Wessex became supreme. The second epoch, 800-1017 witnessed the invasion of the Danes who for a time seemed likely to conquer England until Alfred pushed them back and by treaty with Guthrum kept them to the north and east of a line from London to Bedford and the north. In the tenth century England was organized under Alfred's successors. But the Danes came and in 1017 Cnut became King of England. In the epoch 1017–66 rule was divided between Danes and the dynasty of Wessex.

During this period society was evolving and law is found in the forms of customary rules developed by various communities, some of which are later recognized and included in collections promulgated by kings, in the form of rules enacted by kings, a form of law which becomes increasingly important from the codes of Wihtraed of Kent and Alfred of Wessex onwards and private legal transactions. Grouped by method of publication the Anglo-Saxon laws comprise laws and collections promulgated by public authority, including the laws of the Kentish kings, Aethelberht, Hlothere, Eadric and Wihtraed, of Ine, Alfred, Edward the Elder, Aethelstan, Edmund, Edgar, Aethelred, and Canute, and treaties, such as that between Alfred and Guthrum; statements of custom, including the law of the Northumbrian priests, and various fragments of local custumals found in Domesday Book; later private compilations including the post-Conquest collections, the laws of Edward the Confessor, the *Leges Henrici Primi*, the *Quadripartitus*, the *Rectitudines Singularum Personarum*, the *Gerefa* (qq.v.), and certain others and charters.

The older Anglo-Saxon codes, particularly those of Kent and Wessex, evidence a close relationship to the Barbarian laws of Lower Germany, notably the laws of the Saxons, Frisians, and Thuringians. Indeed, along with the Scandinavian laws, the Anglo-Saxon laws are the truest expression of Teutonic legal thought, not least in that they are written in English, whereas the *leges barbarorum*, of the Continent were materially influenced by Roman ideas and are all in Latin. There are no traces of Celtic influence or of any legal survival from the Roman occupation. Though the legislation of Alfred, Edward, Aethelstan, and Edgar shows some resemblance to the capitularies of Charlemagne and his successors, the similarity is not due to borrowing so much as to like circumstances. The Scandinavian invasion brought in some northern legal customs. Roman law had little direct influence but indirectly, through the Church, which was heavily influenced by Roman culture and ideas, it had an influence. The church influence added sanctity to the person of the King, introduced the custom of conveying land by written deeds and the concept of the will, and by its stress on motive and intention modified the older ideas as to liability.

The legal precepts known from this period are partly customary rules, partly enactments by kings, and partly private transactions. Much of the matter in the early codes is customary but adopted and promulgated by the kings, though as time goes on the proportion of enacted rules increases. About one half of the surviving materials deal with criminal law and procedure and of these at least half deal with tariffs of fines, others with punishment and others again with procedure. Matters of private law are relatively unimportant. Public law and administration also occupy a substantial part of the materials, principally the power and interests of the King, but many also relate to the church. In the early codes the tariffs of fines occupy most of the code, whereas in the later ones this is being superseded by a system based on confiscation, outlawry, and corporal and capital punishment.

One significant feature of the law is the opposition between folk-right, or the body of principles, formulated or not, which is an expression of the legal consciousness of the community in question, and privilege. The folk-right was of tribal origin and existed for East and West Saxons, East Angles, Northumbrians, Danes, Welsh, and others and continued even after the disappearance of tribal kingdoms. The formulation and application of the folk-right was in the tenth and eleventh centuries centred in the shire-moots, and the old law of land, contracts, succession, and the customary tariffs of fines were mainly regulated by folk-right, while king's and lords' reeves were supposed to manage local affairs according to folk-right. Folk-right was declared by the people themselves in their communities through the leading men, such as the eldest thanes. But folk-right could be modified or infringed by special law or royal grant flowing from an exercise of kingly power. Such special rules or grants were often suggested by the parties interested, particularly the church. In this way concessions of testamentary power and confirmations of grants and wills overruled the folk-right rules of succession; a privileged form of land-tenure—bookland—was created; and in time the kinds of right based on royal grants of privileges swamped the earlier folk-right.

Another material feature is the struggle to maintain the peace of society. Peace was conceived as the rule of a master in a certain area, house, estate, or kingdom, and so there were recognized the peaces of authorities of various grades, the father, the master, the lord, but particularly the King, so that there was a gradual development of rules about the King's peace. Fines were imposed for breaches of the peace of persons of different ranks, but particularly of the King.

Kinship groups were very important and not only marriage, wardship, and succession but personal protection and revenge and regulation of good behaviour were regulated by the law of kinship. All the kindred shared responsibility for a man's actions. This principle was later used to enforce responsibility and to keep lawless individuals in order. The collective responsibility of the kindred seems to have failed quite early and been replaced by that of guilds or voluntary associations of townships and hundreds. As time went on society came to be organized more on the basis of wealth and of official employment in the service of the king; a man's position in relation to land tended to become the chief factor in fixing his position in society.

Slavery was recognized and accepted but the church encouraged manumission, but the slave class tended to be merged with the predial serfs who cultivated the land and were sold with the farm.

Marriage seems to have consisted in a sale of rights of protection of a woman by her parents to her husband. It required agreement on the price and transfer of the bride. Later it came to be an arrangement between parents, bride, and bridegroom, to which the church added the requirement of the presence of a priest. The wife was under the control of her husband but had proprietary capacity. Divorce seems to have been recognized, by consent or for infidelity or desertion.

Succession on death was probably regulated by the custom of the district and there were no settled rules. The basis of a will was known in Anglo-Saxon law, but there is no law of executors; friends or the priest saw to the fulfilment of a man's last wishes.

Land law was more important than movable property. Land was cultivated in strips within open fields, each landholder having several discontiguous strips and rights in common fields. Land was held as folkland, according to folk or customary law, or as bookland, held according to the terms of the book in which the grant was recorded, or as laenland, as temporary gift or loan for one or more—frequently three—lives, frequently by the church to great men. Many of the Anglo-Saxon charters constitute bookland.

Movable property meant, above all, cattle and cattle-theft was common. The laws repeatedly tried to ensure that all sales were effected before witnesses.

The sphere of contract was unimportant, an appendage to the law of property. The earliest contracts were agreements for the payment of compensation for injury or death in which sureties for performance were important. Sureties were also employed in sale, but witnesses were also important. The church later enforced undertakings by penance and began to extend the idea of an enforceable agreement by assuming wide jurisdiction over breaches of faith, and also gave new applications to the old form of sureties.

In the undifferentiated sphere of tort and criminal law the natural mode of redress was force, which led to blood-feuds between kindred-groups. Some of the provisions of Anglo-Saxon codes sought to regulate the mode in which the feud might be prosecuted, and others sought to define the occasions on which physical force might be used. The acceptance of compensation in lieu of prosecuting the blood feud was doubtless at first voluntary but had become obligatory by the time of Aethelberht's laws which are largely occupied by a tariff of compensation payable for various wrongs and varying with the rank of the injured man. There are traces also of noxal surrender and of the *lex talionis*. But from an early period it was recognized that wrong must be paid for by *bot* or compensation to the injured man, or *wer* to his relatives if he were killed, but also by *wite* to the king, because the wrong was a breach of the King's peace.

A number of offences existed which could not be compensated in money, and this category increased and it was accepted that such offences should be dealt with by the king. The influence of the church and the developing organization of the state contributed to this idea. Thus there is visible the genesis of a system of criminal law.

Liability depended on the doing of the deed, not on the wrongdoer's state of mind; a man acted at his peril. In time under church influence attention was paid to the authority of the delinquent and the strict rules of liability began to be regarded as archaic.

In the time of Cnut the first explicit list of royal pleas is found, with different lists for Wessex, Mercia, and the Danelaw, and this developed into the list of pleas of the Crown known to Glanvill.

In respect of procedure the main effort of early law was to limit the conditions under which persons might resort to self-help in aid of the law rather than in lieu of it. The law of procedure was unformed and still lacked all the later rigidity. It was the task of the plaintiff to summon the defendant to court. If he came the plaintiff had to make an oath that his claim or accusation was made in good faith and make his claim formally, give pledges that he would duly prosecute his claim, and support his claim with sufficient evidence. The defendant might deny the charge on oath and had to give pledge that he would abide by the judgment of the court. The court decided which party must go to the proof and how the proof was to be made; the successful party was the one who successfully made his proof. There were various procedures in different kinds of cases.

The law of Anglo-Saxon times accordingly in many respects contained the seeds of later developments. Over the six centuries during which it developed there was continuing growth and maturing, but 1066 brought a new influence into the growth of English law and radically changed the course of development.

It is hardly possible to speak of a legal profession in the Anglo-Saxon period and what little legal literature there is is a few short handbooks, stating a few rules only.

F. M. Stenton, *Anglo-Saxon England* (1947); R. Schmid; *Die Gesetze der Angelsachsen* (1832); F. Liebermann, *Die Gesetze der Angelsachsen* (1907–16); G. B Adams and others, *Essays on Anglo-Saxon Law* (1876); F. Pollock and F. Maitland, *History of English Law* (1897); W. Holdsworth, *History of English Law*, vol. II; F. Seebohm, *Tribal Custom in Anglo-Saxon Law* (1911), *The English Village Community* (1905); H. Chadwick, *Studies on Anglo-Saxon Institutions* (1905); F. Maitland, *Domesday Book and Beyond* (1897); F. L. Attenborough, *Laws of the Earliest English Kings* (1920); A. J. Robertson, *Laws of the Kings of England from Edmund to Henry I* (1925); H. Richardson and G. O. Sayles, *Law and Legislation from Aethelberht to Magna Carta* (1966).

Anien, or Anianus (fifth cent.). A principal officer of Alaric II, king of the Visigoths, charged with making law for the Visigothic kingdom. He made a version of Gaius' *Institutes* in two books, excluding rules inappropriate to Visigothic society, an abridgement of the *Codex Theodosianus*, and saved Paul's *Receptarum Sententiarum libri quinque* from being lost.

Animals. A generic term including beasts, birds, fish, and other living creatures, other than humans. Tame and domestic animals are the property of their owners like other goods. Wild animals, a class comprising all animals not tame or domestic, are not the property of anyone until reduced to possession, when ownership is acquired by the captor and retained so long as he retains control of the animals. But the game laws provide for the poacher forfeiting animals and fish taken illegally by him. Ownership of wild animals is lost if they regain their liberty. At common law liability for harm done by domesticated animals e.g. a dog, existed if there were negligence in controlling them, or they did harm, having previously exhibited vice of the same general kind, and thereby attached to the owner knowledge of the animal's dangerous propensity, and there was liability for any harmful act done by an animal of a species deemed wild, e.g. a lion, even if trained. The Animals Act, 1971, imposes strict liability in England on the keeper for damage done by an animal of a dangerous species, and for damage done by an animal not of a dangerous species if likely to be severe or the animal were abnormally likely to do the damage. The owner of domesticated animals is liable if they escape, trespass on another's land and do harm such as is ordinarily in their nature to commit, and at common law trespassing animals could be seized and impounded to secure compensation for the damage done. At common law there was no general obligation to fence land or keep domestic animals off the highway but the owner is now liable if he was negligent in allowing them to be there, and a person who brings an animal on to the highway must use all reasonable care to prevent it doing damage to other persons there. There are numerous statutory provisions penalizing cruelty to animals, protecting various species and intended to prevent disease.

Animus (Lat. the mind). A word generally used as meaning intention or purpose, and a component of many phrases, e.g. *animus cancellandi*, intention to cancel or destroy; *animus domini*, intention of holding as owner; *animus furandi*, intention to steal; *animus manendi*, intention to remain; *animus possidendi*, intention to possess; *animus revocandi*, intention to revoke. *Animus et factum*, intention and fact, are the mental and physical elements of many kinds of conduct.

Ann. In Scots law, a half-year's stipend due after a minister's death to his widow, child or next of kin.

Annals of Congress. An unofficial record of speeches and statements in Congress, 1789–1824.

Annapolis Convention. A gathering of delegates from adjacent states which met at Annapolis, Md., in 1786, to resolve disputes over the navigation of the Potomac and to work out a system of interstate commerce. In view of the complexity of the issues at stake Hamilton proposed that all the states should convene to render the constitution of the federal government adequate to the exigencies of the Union, and he and Madison used the meeting as a way to secure revision of the Articles of Confederation. In 1787 the convention invited all states to a convention at Philadelphia in May 1787, for the sole purpose of such revision, and this became the Federal Constitutional Convention.

Annates. From the thirteenth century there was a recognized custom in the Roman Church for a bishop to claim the first year's profits of the living from a newly inducted incumbent. In time the popes, though at first only temporarily and under financial stress, claimed the privilege for themselves; sometimes they claimed them only from those benefices the patronage of which they had reserved to themselves but in some cases they claimed the first-fruits of all benefices in a country or in Christendom. From these claims developed the custom that in all countries subject to the ecclesiastical jurisdiction of Rome the first year's profits of the benefice were paid to the papal treasury. The clergy never willingly submitted to this and the system never operated completely or uniformly. The bishops and barons of England protested at the Council of Lyons in 1245. In 1534 Henry VIII claimed annates for the Crown.

Annexation. In international law, the unilateral act of one state in adding to its territory other territory hitherto independent, unclaimed, under protectorate or belonging to another state. This has frequently followed victory in war. It is also the way in which a state sometimes terminates a protectorate and makes the area a colony, as when Great Britain annexed the protectorate of Cyprus in 1914. Proclamation of annexation is not enough; there must also be subjugation or defeat of the former owning state or renunciation by it, and a normal exercise of state authority by the annexing state and acquiescence or recognition by other states to give a good title. No particular formalities are demanded for annexation; it may be by royal decree, as when Italy annexed Ethiopia in 1936, or joint resolution of Congress as when the U.S. annexed Texas in 1845 and Hawaii in 1898. But defeat and occupation of a country by another does not always or necessarily lead to annexation; thus the Allies occupied Germany in 1945 but disclaimed any intention to annex. Under the U.N. Charter annexation based on illegal use of force is ineffective.

Anniversary. The same date in a later year as that on which some event happened. Certain anniversaries have, by law or custom, been observed and celebrated in various ways. Thus both the anniversaries of the discovery of the Gunpowder Plot (5 November, 1605) and the restoration of Charles I (29 May, 1660) were appointed to be observed. Good Friday and Christmas Day were traditional Christian anniversaries now observed by statute as holidays.

Annual Practice ('The white book'). The Practice of the Supreme Court of Judicature, comprising the Rules of the Supreme Court, extensively annotated, and revised and republished annually.

Annual Register. A periodical commenced by R. Dodsley in 1758, with Edmund Burke as editor, 1758–88, to give a view of the history, politics and literature of the year. In the early volumes the historical section was largely an abridgement of parliamentary debates, but later this was replaced by more comprehensive narrative. It is still published.

Annualrent, bond of. A former form of security over land in Scots law, whereby the lender purchased from the borrower an annual rent or interest in return for the sum borrowed. Latterly at least there was a clause of redemption in favour of the borrower. The annualrent right did not carry the property in the lands themselves, but was a mere burden on the rents to secure the interest, and was redeemable on repayment of the purchase money, i.e. the sum lent.

Annuity. A right created by statute, deed, or will to receive a definite annual sum of money, not charged on land, or charged on personal property or a mixed fund. If charged on real property or leasehold it is normally called a rentcharge. It is personal property. It may be perpetual, or for the annuitant's life or for some other period, and may be conditional, e.g. on not remarrying.

Annulment. If a judicial proceeding is annulled it is deprived of effect and rendered inoperative, either retrospectively or prospectively. Annulment of a marriage is legislative or judicial invalidation of it, as in law never having existed, as distinct from dissolution, which terminates a valid marriage. To justify annulment there must be a radical defect, such as pre-existing marriage of one party, insanity at the time, or sexual impotence.

See also DIVORCE; NULLITY.

Anomie. A term introduced by Durkheim (q.v.) to refer to an individual or a social structure in which norms or standards of behaviour are absent, useless, or inconsistent. Among individuals anomie arises when ends are more important than means and illegitimate or anti-social means are used, leading often to delinquency, crime, and suicide. In societies anomie exists where common values and standards have ceased to be understood or accepted and yet have not been replaced by others, so that there is an absence or a breakdown of the social standards required to regulate human behaviour.

Ansegisus of Fontanelle (*c.* 770–833). A noble of the Carolingian court, abbé of Luxeuil, and diplomat, who made a collection of Charlemagne's and Louis's capitularies in four books, arranged by dates, published in 827 and continued by later scholars. It enjoyed great authority and some chapters passed into Gradian's *Decretum.*

Anselmo, Antoine (1589–1668). Belgian jurist, author of *Codex belgicus*, *Tribonianus belgicus* (1662), *Commentaria ad perpetuum edictum* (1665) and other works.

Anselmo Dedicata, Collectio. A tenth century collection of canon law materials by an unknown collector. The materials are systematically distributed in 12 books, dealing with 1. ecclesiastical hierarchy; 2. bishops; 3. councils; 4. priests and deacons; 5. minor clergy; 6. regulars and widows; 7. laity; 8. practice of virtues; 9. baptism; 10. worship and benefices; 11. feasts; 12. heretics and schismatics. It was widely used and very influential and it governed Italian and German canonical life until the appearance of Burchard's *Decretum* (q.v.).

Anson, Sir William Reynell (1843–1914). Became a Fellow of All Souls in 1867, practised at the bar for a short time and became Vinerian Reader in Common Law at Oxford in 1874 and Warden of All Souls in 1881. He wrote a *Principles of The English Law of Contract* (1879) which has remained a standard text, and a *Law and Custom of the Constitution* (1886–94) which has not lasted so well though useful for reference. He sat in Parliament and was Parliamentary Secretary to the Board of Education, 1902–5, in which office he was only moderately successful, and was made a Privy Councillor.

Answer. In judicial proceedings, a reply or counter-statement. In ordinary actions it is an affidavit in reply to interrogatories; in divorce proceedings it is the respondent's reply to the petition; in ecclesiastical courts it is called an allegation, but personal answers may also be demanded.

Anthropology. See JURISPRUDENCE AND ANTHROPOLOGY.

Antichresis. In Roman law, an agreement whereby a debtor transferred to his creditor land or goods to be used and occupied by the latter in lieu of interest on the money lent. It was later called mortgage, as distinct from vif-gage, where the fruits of the grounds were not transferred.

Anticipation. The doing of something before a particular time or date. In property law anticipation consists of assigning or dealing with income before it is actually receivable. It was common, in settlements of property on a woman, to introduce a clause restraining anticipation, to prevent her losing the benefit of future income, as by assigning it to her husband, but all restraints on anticipation were abolished in 1949.

In patent law a patent is bad for anticipation if the invention patented has previously been known in the United Kingdom.

Anticipatory breach of contract. The refusal by a party to a contract, before the due date for performance, to perform, or his disabling himself from being able then to perform. Such has the effect of breach entitling the other party to accept the breach and at once sue for damages or to reject the breach and await the date for performance, suing then if performance is not then made.

Anti-trust laws. The general name for the U.S. legislation seeking to control large business groupings whose actings unreasonably restrict freedom of competition and promote monopolies or practices contrary to the public interest. Down to the 1880s the courts favoured a *laissez-faire* policy, upholding the liberty of corporations to fix prices and promote monopolies. The matter was brought into politics in the next few years.

The main legislation is the Sherman Antitrust Act (1890) which authorized the federal government to take action against any combination in the form of trusts or otherwise, or conspiracy in restraint of trade. The Act was of limited success as the Supreme Court held that many reasonable manufacturing bodies fell outwith the terms of the Act. In 1914 the Clayton Antitrust Act forbade practices which substantially tended to lessen competition by outlawing price fixing, interlocking directorates, and acquisitions by one company in competing companies.

The creation of the Federal Trade Commission (1914) also tended to the same end. This commission is empowered to require annual reports from corporations and to investigate business practices and monopolistic practices, to prevent the establishment of combinations for maintenance of resale prices and the making of false claims for patents. Its investigations have led to better regulation of public utilities.

For the U.K. see MONOPOLIES; RESTRICTIVE TRADE PRACTICES.

Antiqua Statuta (or *Vetera Statuta*). The collective name of all the statutes, including the Statutes of Uncertain Date, passed before the reign of Edward III and printed by Pynson in 1508.

Antient Entries. The name given by Rolle (q.v.) to Rastell's (q.v.) *Collection of Entries.*

Antoine or Antonius, Godefroi (?–1618). A founder and chancellor of the University of Giessen, author of *Disputationes feudales* (1604), *De Camerae Imperialis Jurisdictione, Disputatio apologetica de potestate Imperatoris legibus soluta* (1609), and other works.

Antonio di Butrio (*c.* 1338–1408). An Italian commentator and canonist, professor at Bologna and Florence, author of *Commentaria in quinque libros Decretalium* (1473), *Commentaria in Sextum* (1479), *Consilia* (1492), and other works.

Anzilotti, Dionisio (1867–1950). Italian jurist, a founder of the positive school of international law and founder of the *Rivista di diritto internazionale* (1906–) and author of *Corso di diritto internazionale* (1912). He was a judge of the Permanent Court of International Justice, 1921–30, and President 1928–30.

Apology of the Commons (1604). A statement and defence of British parliamentary proceedings and privileges. The Commons proposed another and larger revenue to replace that obtained from the court of wards and was reprimanded by James I. In its *Apology* the Commons asserted the privileges of free elections, freedom of members from arrest during parliamentary sessions, and freedom of speech. They refrained however from formally presenting the *Apology* to the King.

Apostolic Constitutions (or Ordinances of the Holy Apostles through Clement). The largest collection of ecclesiastical law surviving from early Christianity. They comprise eight books, six being an adaptation from the *Didascalia Apostolorum* written in Syria about 250 A.D., the seventh a paraphrase and enlargement of the *Didache* and a Jewish collection of prayers, and the eighth a miscellaneous collection of matters, including a series of canons and the so-called Apostolic Canons, 85 canons derived from preceding constitutions and from canons of the councils of Antioch and Laodicaea. It is thought that the constitutions were compiled in Syria about 380 A.D. They were a source of the later canon law.

Appanage (or *Apanage*). Originally the means of subsistence given to younger children as distinct from the estate falling by primogeniture to the eldest. The system was common on the continent, notably in France where lands and lordships were conferred on the King's younger children. They were finally abolished in 1832. In England the system was never widespread and the term is confined to the appanages of the Crown, the duchy of Cornwall assigned to the King's eldest son on his birth or his father's accession, and the duchy of Lancaster.

Appeal. An application to a person, body, court, or tribunal superior to one which has decided an issue, to reconsider that decision and, if thought fit, to alter it. The major functions of appeal are to satisfy litigants that a case has been reconsidered, and secure rulings of general future application. Appeal may be limited or excluded in some cases. The most important questions in relation to an appeal are whether appeal is permitted at all or excluded, to what person or court it lies, and whether appeal lies on questions of fact, or on questions of law, or on both grounds. An appeal on a question of fact involves reconsideration of what facts must be held to have been established, of evidence relative thereto, and of what inferences should be drawn from them. An appeal on a question of law involves accepting facts found by the court of trial and considering only whether those facts justify the conclusion in law reached by the trial court, e.g. whether the facts found amount to legal negligence, or to a particular statutory contravention. The result of an appeal may be to affirm, modify, or reverse the decision of the court below. Sometimes a further appeal lies to a yet higher court, but at some point there must be finality and a court beyond which there is no appeal. Whether and where appeal lies and to what person or court is a matter for each legal system.

Particular forms of appeal have included the appeal by writ of error, which was the basis of the appellate jurisdiction of the House of Lords down to 1875; the basis of the appeal was the existence of an error in the record of proceedings in the lower court, matters not disclosed by the record being unimpeachable; the appeal by rehearing, which is a re-examination of the transcripts of evidence heard in the lower court, but not a rehearing of the witnesses; the appeal by case stated or stated case, in which the inferior tribunal sets down on paper the facts it has found admitted or proved and formulates a question raising a point of law to be answered by the appellate court; appeal by way of prerogative order, particularly for an order of *certiorari*, in which the appellate court is asked to consider the legality of the proceedings in the inferior court and to quash its decision as wrong in law. See also CASSATION.

Appeal Committee. A committee appointed each session by the House of Lords to consider matters relating to petitions to appeal to that House in its judicial capacity. It is deemed to consist of all the peers present during the session, three being a quorum, but it is normally attended only by the

Lords of Appeal in Ordinary. See also APPELLATE COMMITTEE.

Appeal, Court of. See COURT OF APPEAL; SUPREME COURT OF JUDICATURE.

Appeal of felony. In case of death by murder or manslaughter the feudal lord of the deceased, the widow, or the heir male might bring an appeal, in substance an accusation or challenge or claim for loss to himself rather than for harm to the public. An appeal might be brought even after the appellee had been tried on indictment and acquitted. The defendant had the right to trial by battle. The parties had to fight personally, save that a woman, a priest, an infant, a person over 60, or lame or blind might hire a champion. The battle took place before the judges of the King's Bench or Common pleas, and the parties were each armed with a staff an ell long and a leather shield, and battered each other from sunrise to star-rise or until one cried 'Craven'. The defendant could clear himself by the ordeal or, after this was abolished, by jury trial *per patriam*. If beaten in combat or found guilty, the defendant suffered the same judgment as if convicted on indictment, and the Crown had no power to pardon because the appeal was a private suit. It became obsolete but was not abolished and in 1817 Ashford brought a writ of appeal in the King's Bench against Thornton for the alleged rape and murder of Mary Ashford. Thornton had already been tried and acquitted of the charge at assizes; he demanded trial by battle against Ashford who declined to accept the challenge and Thornton was discharged: see *Ashford* v. *Thornton* (1818), 1 B. & Ald. 405; in the following year appeals of felony and trial by battle were abolished by statute.

Appeals, Statute of (1533). A major Act of the English Reformation, restraining appeals to Rome. It extended the principle of Richard II's statute of *praemunire* (q.v.), prohibiting the transfer of appeals in matrimonial cases from the courts of the English archbishops to the Roman curia. It enabled Archbishop Cranmer to declare void Henry's marriage to Catherine of Aragon and his marriage of Anne Boleyn to be recognized as lawful. The Act also contains the declaration that England was governed by one Supreme Head and King, thus asserting that the King was head of the State in both ecclesiastical and secular matters.

Appearance. The formal act of a defendant in a civil action intimating his intention to defend, failing which the plaintiff may proceed to take judgment against him. The term is also used of parties being present before a court, personally or by counsel and solicitor, when a proceeding is heard. In criminal cases the appearance of the accused, voluntarily or secured by his arrest, is normally a prerequisite of his being tried, but in summary trials the presence of the accused can in certain cases be dispensed with.

Appellate Committee. A committee appointed every session since 1945, by the House of Lords, consisting of the Law Lords, to which are referred for hearing all appeals taken to the House of Lords. The committee reports its decision to the House. See also APPEAL COMMITTEE.

Appellate Jurisdiction Acts. A series of statutes (1876, 1887, 1913, 1929 and 1947) which mainly regulate the House of Lords as a final court of appeal in the United Kingdom. See HOUSE OF LORDS.

Appellatio. The means whereby a litigant in Roman law disputed a judgment and had the case brought before a higher magistrate, normally the one who appointed the magistrate who heard the case at first instance. It is not found in Republican times and appeared only in the *cognitio extra ordinem* in the principate. It was effected orally or in writing, and this obliged the trial judge to transmit the documents to the higher magistrate with a written report. A new trial was held at which fresh evidence might be adduced. Justinian reformed the practice and settled the principle that all judgments except those of the praetorian prefect were appealable. In later usage the term was used as covering also *provocatio* (q.v.) which had previously been an appeal in criminal procedure, and also in relation to administrative decisions of magistrates.

Appendant and appurtenant. Words used to describe a hereditament (q.v.) annexed to another hereditament, the former being the adjunct or accessory and the latter the principal, such as a common of pasture and a house with land. A grant of the principal will carry the adjunct also. Things appendant can be claimed only by prescription whereas things appurtenant can be claimed by prescription or by express grant. Principal and adjunct must be such that they can properly be enjoyed together.

Appin Murder. After the Jacobite Rebellion of 1745–46 the estates of many landowners in the Highlands of Scotland were forfeited and persons appointed by the government as factors (estate agents) to manage them. Among these was Colin Campbell of Glenure who in 1749 became factor of Ardshiel, Callart, and Mamore in the Lochiel country. He employed an assistant factor and adviser James Stewart (James of the Glens) who had fought with the Stewarts in the Jacobite army but after the rebellion returned to farm Acharn in

Duror. On the Ardshiel estates, as elsewhere on the forfeited estates, the tenants paid their rent to the government factor, but paid a further rent which was remitted to the clan chief, in exile in France. An intermediary between the Appin people and their chief was Allan Breck Stewart, who also had fought in the Rebellion.

In 1752 Glenure removed a number of tenants on the Ardshiel estate and on 14 May was riding from Ballachulish towards Kentallan with three companions when, at the wood of Lettermore, he was shot. The murderer escaped. Suspicion fell on Allan Breck Stewart and he was searched for widely. James of the Glens and others were arrested, and James was charged as abettor of Allan Breck, named as the actual murderer.

After some delay, involving the most complete denials of justice, James was brought to trial on 21 September at Inveraray before the Duke of Argyll and Lords Elchies and Kilkerran. Argyll was hereditary Lord Justice-General of Scotland, and also head of the Clan Campbell, long enemies of the Stewarts.

In truth it was a trial of a Stewart and a Jacobite by a jury of Hanoverians, 11 of the 15 jurors being Campbells, for the murder of a Campbell. The jurors were chosen by the presiding judge. There was no evidence against James of knowledge or complicity in the murder, but only some suspicious circumstances. James Stewart was found guilty and hanged at Ballachulish on 8 November, and his body was hung in chains and remained till 1755. It was a judicial murder, the continuance of a clan feud under judicial forms. The story has ever since fascinated writers.

R. L. Stevenson, *Kidnapped* and *Catriona*; D. Mackay, *Trial of James Stewart.*

Appius Claudius (fifth century B.C.). Chairman of the commissions of ten appointed at Rome in 451 and 450 B.C. which formulated the Twelve Tables (q.v.).

Appius Claudius Caecus (*fl. c.* 300 B.C.). Traditionally said to be a famous jurisconsult who got his secretary to publish the pontifical calendar and a book, later called the *Ius Civile Flavianum* giving the formularies of the civil actions, and thus publishing the secret of litigation which the pontiffs had hitherto kept. The tradition is unreliable.

Application for judicial review. A form of procedure whereby a person seeking to challenge an administrative act or opinion may obtain from the High Court one of the prerogative orders (q.v.) or a declaration, injunction, and damages (qq.v.).

Apply, liberty to. A direction by a judge or master entitling parties to come back to the court without having to take out another summons.

Appointed day. A day fixed by a statute for some purpose, such as the coming into force of certain provisions of the Act. It may be fixed by the Act itself, or power may be conferred on a minister by order to fix the appointed day. There may be several appointed days for different provisions of the Act. The device accordingly enables the operation of the Act or parts of it to be postponed until arrangements have been made for putting the relevant provision into force.

Appointment, power of. A power conferred on a person by deed or will to appoint a person or persons to take a particular share or interest in property, real or personal. A true power is discretionary and not obligatory. It commonly arises where a testator, instead of giving his estate to certain persons, confers a power on his trustees to appoint those whom they choose to take the estate.

A power may be general, where the donee of the power may appoint anyone whom he chooses, or special, where he may appoint only one or more of a limited group of persons.

A power must be exercised in the way indicated, if any, e.g. by deed or by will. See also FRAUD ON A POWER.

Apportionment. An allotment or distribution in shares, used particularly of the rule under the Apportionment Act, 1870, that rents, annuities, dividends, and other periodical payments in the nature of income are deemed to accrue from day to day, so that they fall to be apportioned on, e.g. the death of a tenant for life.

Apportionment Act. An Act passed by the U.S. Congress after each census to determine the number of representatives each State may send to the House of Representatives. The same name is given to Acts passed by state legislatures for similar purposes.

Apprehension. In Scots law, the formal term for taking a person into custody on a criminal charge, synonymous with arrest. It may always be done under a written warrant granted by a magistrate, normally a sheriff, may under various statutes be done without warrant, and may at common law be done without warrant where a constable sees a person committing or attempting a serious crime, or is credibly informed that a person has committed a crime, or sees a person acting in such a way as to be liable to injure himself or others.

Apprentice. A person who binds himself contractually to serve with a master for a definite term to learn the latter's trade or skill, the latter being bound to instruct the apprentice. It is a very ancient form of training in professional or craft skills. Formerly used in connection with professions, e.g.

barristers (see *Apprenticii ad legem*) and surgeons, and still so used in Scotland in relation to solicitors and accountants, the term is now mainly confined to tradesmen.

In England apprenticeship has been regulated by law since the fourteenth century. In mediaeval times there were recognized gilds of craftsmen, comprising masters and apprentices. The gilds imposed restrictive conditions on numbers of apprentices. The Statute of Labourers and Apprentices of 1536 prescribed an apprenticeship of at least seven years as a prerequisite to exercising any craft; the justices of the peace had to administer the Act. Gild apprenticeship declined rapidly in the eighteenth century and a statute of 1814 repealed most of the 1536 Act; this marked the end of compulsory and of domestic apprenticeship. Henceforth apprentices worked under contract, lived at home, and received some wages. In the twentieth century formal apprenticeship has declined considerably and is increasingly being supplemented or replaced by technical training in educational establishments. There have accordingly developed in many countries a wide variety of colleges and training establishments covering not only trades and crafts but also administrative and professional skills, management and executive functions, many at post-graduate level.

Apprenticii ad legem (or *apprentices en la lay*). The term in use down to about 1455 for those who had been called to the bar but were not yet serjeants at law (*servientes ad legem*), a class who about that time began to be called utter barristers because, it is said, when arguing moots they sat uttermost on the forms. They have latterly come to be called barristers (q.v.).

Apprising. A form of diligence (q.v.) in Scots law, now replaced by adjudication (q.v.), under which a debtor's lands might be sold by the Sheriff to the extent necessary to pay his debts and the creditor paid out of the proceeds. If a purchaser could not be found, a portion of the land was made over to the creditor. The debtor might redeem the lands within seven years.

Approbate and reprobate. The Scottish counterpart of the doctrine of election (q.v.), to the effect that a person cannot simultaneously accept or take advantage of a deed and reject it; he must elect which course to follow.

Approbation des Loix. A work of which the full title is *Approbation des Loix, Coustume et Usages de l'Isle de Guernezey, differentes du Coustumier de Normandie, d'Anciennete observés en ladite Isle*, published first in 1715, and purporting to describe the laws and customs of Guernsey as they existed in 1580 and were confirmed by Queen Elizabeth in Council in 1581. Whether or not this be correct, the work, which deals with both civil and criminal law and points to the differences from the custom of Normandy, is the main source of knowledge of the law of Guernsey.

Appropriation. Generally, taking something as property, or setting property aside for a particular purpose.

In Parliamentary usage appropriation is the setting apart of particular sums of public money for particular heads of expenditure, such as justice, defence, etc., and the annual Appropriation Act fixes the sums thus appropriated and also authorizes the Treasury to borrow money in anticipation of revenue up to the total sum appropriated.

In English ecclesiastical law, appropriation arises when a benefice is perpetually annexed to a spiritual corporation which is the patron of the living, and which entrusts the cure of souls to a vicar as being the deputy (*vicarius*) of the patron, as contrasted with impropriation where the benefice is annexed to the use of a lay person or corporation. The living may be disappropriated if the patron presents a clerk who is instituted and inducted, or if a corporate patron should be dissolved.

In private law goods or money are appropriated when they are set apart for a specific purpose, such as payment of a particular debt. In respect of payments a debtor owing several sums to the same creditor may, when he makes a payment, appropriate or attribute that sum to a particular debt. If he does not do so, the creditor may do so, and if neither does so, the law appropriates it to satisfying the debt earliest in date, a rule important where some debts have become unenforceable by reason of lapse of time.

In criminal law appropriation or taking for oneself what one is not entitled to do is criminal.

Approval, sale on. The sale of goods to a buyer subject to the condition that he may return the goods within a stated time. If he does so, the sale lapses, but if he does not do so within the specified time or within a reasonable time, or on the other hand adopts the contract, as by using or reselling the goods, the contract is perfected. Approval may be to let the buyer ascertain if he wishes to keep the goods, or to enable him to resell if he can.

Approved schools. A term introduced in 1932 instead of certified reformatory schools or certified industrial schools for schools provided by local authorities or by voluntary bodies and approved by the Home Secretary or the Secretary of State for Scotland for the education and training of children and young persons committed to them under various statutory powers, particularly those found guilty of various kinds of juvenile delinquency. In 1968 the terms community home and residential establishment were introduced instead.

Approvement of waste. The right of enclosing superfluous waste lands granted or rather confirmed to the owners of the soil of waste lands by the Commons Act, 1235. This statute was confined to cases between lords of the manor and their tenants. It was extended by the Commons Act, 1285, to cover enclosure or approvement against tenants of a neighbouring manor. The powers under both Acts are not confined to lords of manors but may be exercised by any owner of waste, but sufficient pasture must be left for commoners. The right is today rarely exercised.

Appurtenances. Things belonging to another thing, such as yards and gardens to a building.

Approximation of laws. The E.E.C. Treaty provides that the approximation of laws of the member states is necessary for the proper functioning of the Common Market. This involves a greater measure of integration than the harmonization of national laws affecting trade necessary to implement the economic freedoms of the E.E.C. The object is not the creation of a body of European law but rather a system of a federal type which will contribute towards political and economic integration. In pursuance of this end the Council may issue directives to member states to enact laws affecting the establishment or functioning of the Common Market, and member states must refrain from passing laws or adopting administrative measures contrary to the E.E.C. Treaty. The Treaty also contains specific directions for the approximation of national laws in the field of customs, movement of workers, establishment, liberal professions, taxation, and export aids. The Council and the Commission have taken many initiatives for legislation, on such subjects as European patents, company law, and many other topics. Approximation of laws may also be achieved by multilateral international conventions, such as the Convention on Jurisdiction and the Enforcement of Civil and Commercial Judgments of 1969, and these may require significant modifications of the law of the member states and of acceding states.

Apsley, Lord. See BATHURST, HENRY.

Aquaeductus. In civil law and Scots law the servitude (q.v.) right to lead water by pipe or channel across or over the servient tenement of land for the use of the dominant tenement. It is frequently combined with the right to draw water from the servient tenement.

Aquaehaustus. In civil law and Scots law, the servitude (q.v.) right of drawing and taking water from one tenement of land for the use of the dominant tenement, for primary purposes, the use of humans and animals, but possibly for other purposes too, but not to the exclusion of the right of the servient tenement also to take water.

Aquilius Gallus, Gaius (first cent. B.C.). Roman jurist, credited with inventing the *stipulatio Aquiliana* and the *iudicium de dolo.*

Aquinas, Thomas (1226–74). Born near Naples, entered the Dominican order, studied at Naples and Paris and taught theology in Paris, Rome, and Naples. He was canonized in 1323, and in 1879 Pope Leo XIII directed that his teachings should be taken as the basis of theology in the Catholic Church. No theologian, apart from Augustine, has had greater influence on the theological thought of the Western Church. He is the greatest scholastic philosopher and produced an outstanding synthesis of philosophic thought down to his time.

His writings include a *Commentary on the Sentences of Petrus Lombardus*, commentaries on some of the Scriptures, on Boethius, and on certain works of Aristotle, but all are less important than his *Summa contra Gentiles* (1259–64), which is the chief work of the middle ages on natural theology, and his great *Summa Theologica* (1265–74), intended to be the sum of all known learning, of which Part II, dealing with morality in general, founded on Aristotle's *Ethics*, includes discussion of the basics of law and politics. This basis he found in natural law, which gives the pattern for all positive law, and which is a part of the eternal law of God. Natural law is the pattern for all positive law and the basis of political allegiance. The rule and measure of human action is reason, and the object of the law is the rational ordering of things which concern the common good. Law he defined as an ordinance of reason for the common good made and promulgated by him who had care of the community.

His legal writings are also a summary of the major legal principles which had come down to him tested by scholasticism. Since law was ultimately derived from God it was supreme in the State. Since man is a rational being he participates in the divine reason which he is under a duty to discover and obey, so that the human legal order originates in the efforts of man as a rational being to follow the dictates of Divine Reason. There were four kinds of law, eternal law, natural law, divine law, and human law; eternal law was the Divine Reason's conception of things; the natural law was man's participation in the eternal law; divine law was the precepts promulgated by God for mankind and revealed in the Scriptures; and human laws were the determinations drawn by human reason from the precepts of natural law to suit particular conditions. He recognized that some of the applications of natural law precepts were changeable. He also made an elaborate analysis of justice and sought to synthesize Aristotle's thoughts on the subject and what appeared to him valuable from the intervening centuries.

His work can indeed be called the first systematically complete philosophy of law in the history of jurisprudence, combining the philosophical traditions of Plato, Aristotle, and Stoicism, the law of the Bible, the principles of Roman law, and of Gratian's *Decretum*, and other elements. Moreover, he related his philosophy to his theory of morals and in turn to his view of the nature of man and of the Universe. It is important as emphasizing that law is not merely an arbitrary set of rules but depending ultimately on reason and right.

A. P. D'Entreves (ed.), *Aquinas: Selected Political Writings.*

Arabin, William St. Julian (?–1841). A noted eccentric, serjeant-at-law in 1825, he became one of the commissioners of the central criminal court in 1827 and is noted for a book of dicta and charges to juries called *Arabiniana* (1843).

R. E. Megarry. *Arabinesque-at-law.*

Arangio-Ruiz, Vincenzo (1884–). Italian jurist, Roman law scholar, author of many works including: *Istituzioni di diritto romano* and a *Storia di diritto romano.*

Arbitration (or the submission of a dispute to the decision of a person, other than a court of competent jurisdiction). The practice was well known among the Greeks and there is evidence for the existence of public arbitrators in many states. In Athens private arbitrators were frequently appointed to settle claims on an equitable basis and so relieve the pressure on the courts. Arbitration had its greatest influence in inter-state relations.

When Rome became a great power references were made to it by cities and states and were usually referred to the Senate. With the formation of provinces arbitration lost its international character, though senatorial decision of disputes between provincial communities continued till the third century A.D.

In modern English law reference to arbitration arises from agreement by the parties, or in certain cases is prescribed by statute. References by agreement may be made orally, but are normally by written agreement, regulated in England by the Arbitration Act, 1950. Any civil dispute may be referred, except those involving personal status, or transactions illegal or void. A question of law may be referred to arbitration, but is unsuitable therefor. Parties may in general appoint whom they please as arbitrator or, in Scotland, arbiter. If each may appoint an arbitrator, they must appoint an umpire (in Scotland, oversman) to decide if they disagree. Certain trade associations provide panels of arbitrators who operate under published rules of procedure. In certain cases the court may appoint an arbitrator or umpire. The hearing must conform to the principles of natural justice.

The award must determine all the differences referred to arbitration but no other matters. It may be made in any form. It may, by leave of the court, be enforced like a judgment, or be enforceable by action in court.

An arbitrator is not liable for lack of knowledge, skill or care, but may be removed for misconduct or unreasonable delay.

An arbitrator in England may, and must if the agreement requires, or he is directed by the High Court, state a case for the decision of the High Court on any question of law arising, or any award or part thereof.

The court may remit matters back for the arbitrator's reconsideration, or set aside the award as improperly procured, or because the arbitrator misconducted himself or the proceedings.

In Scotland arbitrations are conducted by a single arbiter, or by two arbiters with power to nominate an oversman. There is no power, save under statute, to state a case for the opinion of a court, and an award cannot be challenged save on the grounds of corruption, bribery, or falsehood, of going beyond the scope of matters referred or misconduct by the arbiter.

Arbitration, international. The practice of referring disputes between rulers and cities to impartial persons of high authority was common in the thirteenth century and cases continued common in succeeding centuries. Peace treaties came frequently to contain provisions for reference of matters in dispute to a neutral sovereign or body of persons. Thus the treaty between Cromwell's England and France in 1655 provided for arbitration by Hamburg in the calculation of damage suffered by either side since 1640. The Treaty of Utrecht provided for commissioners to determine the boundaries between Hudson's Bay and French territories.

Modern arbitration is usually dated from the proceedings undertaken by virtue of Jay's Treaty between Britain and the U.S. in 1794 which provided for the adjudication of various legal issues by mixed commissions. In 1822 Tsar Alexander was arbiter in a dispute between Britain and the U.S. The *Alabama* claims resulted in agreement to refer a number of points to a joint High Commission before which the U.S. claimed the reference of the question of liability to an agreed tribunal. The Washington Treaty of 1871 gave a basis for the arbitration which was held at Geneva and included representatives of both sides, the Emperor of Brazil, the King of Italy, and the Swiss Confederation. Other notable instances were the Behring Sea fisheries case (1893) and the Venezuela boundary dispute (1899).

Treaty may provide for settling a particular dispute, or provide for arbitration to resolve differences arising from the primary concern of the treaty, or may provide that certain or all kinds of differences between the parties are to be settled by arbitration. Unless the treaty stipulates to the contrary, an arbitral award is final and binding, and the party favoured by the award may enforce it by such means as are open to it under international law.

In earlier cases arbitral tribunals were often invited to resort to principles of justice and equity and to propose extra-legal compromises, but by the end of the nineteenth century arbitration had come increasingly to be a means of decision according to law.

The Hague Conference of 1899 thus found wide international use of arbitration, and erected at the Hague a Permanent Court of Arbitration which is not in fact a tribunal but a machinery to furnish a body from which countries which had agreed on arbitration might readily draw suitable arbiters. Parties to the Hague Convention may nominate up to four persons to the panel of arbiters, and when parties agree to submit a dispute to arbitration each appoint two from the panel and the four select an umpire. This was followed by various treaties of general arbitration between European powers and, after 1908, between the U.S. and European powers.

At the Hague Conference of 1907 proposals for a general treaty requiring compulsory arbitration were discussed but not adopted. The Hague Conventions of 1899 and 1907 contain a code of arbitral procedure, and Model Rules formulated by the International Law Commission were adopted by the U.N. General Assembly in 1958.

The Permanent Court had a moderately useful existence, but has been little invoked since 1919.

The Covenant of the League of Nations bound all signatory states not to go to war unless they first submitted the matter in dispute to arbitration or judicial settlement or to an enquiry held by the Council, and it provided a variety of treatments for disputes.

C. Phillipson, *International Law and Custom of Ancient Greece and Rome*; M. Tod, *International Arbitration among the Greeks*; J. L. Simpson and H. Fox, *International Arbitration*; J. B. Moore, *History and Digest of International Arbitrations*; A. de la Pradelle and N. Politis, *Recueil des Arbitrages internationaux*.

Arbroath, Declaration of. A letter addressed to Pope John XXII by the barons of Scotland, dated at the monastery of Arbroath in April 1320, containing the protest of Scotland against English aggression and asserting the rightfulness of King Robert the Bruce's claim to the throne. In reply Pope John exhorted both sides to make peace.

Archeion. An historical commentary on the central courts of justice in England published by Lambarde (q.v.) in 1635.

Archaeonomia. A collection of Anglo-Saxon laws published in 1568 with a Latin version by Lambarde (q.v.) which restored the Anglo-Saxon laws to the knowledge of lawyers.

Archbishop. The chief bishop and head of the clergy within his province. In the Church of England there are two archbishops: Canterbury, styled Primate and Metropolitan of All England, and York, styled Primate of England, each of whom in his own province superintends the ordinary diocesan bishops—and within his own diocese exercises episcopal jurisdiction.

The Archbishop of Canterbury has a seat in the House of Lords, takes precedence next after the Royal Family and before the Lord Chancellor, is *ex officio* chairman of the General Synod of the Church. By statute he may grant throughout both provinces all the licences, dispensations, and faculties within the Pope's jurisdiction prior to 1533, under which powers he grants special marriage licences and degrees (Lambeth degrees).

The Archbishop of York has a seat in the House of Lords, and takes precedence immediately after the Lord Chancellor.

Within his province each archbishop has, next and immediately under the sovereign, supreme power, authority, and jurisdiction in all matters ecclesiastical. He is elected by the dean and chapter of the cathedral in pursuance of a licence by the sovereign under the Great Seal (called a *congé d'élire*) and letter missive containing the name of the person to be elected. For holders of the office see *Handbook of British Chronology*.

The archbishop has authority to visit and inspect the bishops and clergy of his province, and in serious cases may deprive an inferior bishop.

Each archbishop (and bishop) has a court, the consistory court, held by his chancellor as his official principal in the cathedral church or elsewhere for the trial of ecclesiastical causes within the diocese. See CONSISTORY COURTS. Each archbishop also has a provincial court, which is a court of first instance in certain cases, but is mainly a court of appeal from the diocesan courts within the province. That of Canterbury is the Court of Arches, that of York the Chancery Court of York. See PROVINCIAL COURTS.

The Court of Faculties of the Archbishop of Canterbury has jurisdiction over the appointment of notaries public and the issuing of such faculties and licences as that Archbishop may grant in both provinces. See COURT OF FACULTIES.

In the Scottish Episcopal Church, which is not an established church in Scotland but a voluntary association, there is no archbishop, one bishop being elected as Primus by the group of diocesan bishops.

In the Church of Ireland there are two archbishops, Armagh, who is Primate of All Ireland, and Dublin, who is Primate of Ireland. The archbishoprics of Cashel and Tuam were reduced to ordinary bishoprics in 1833–34.

In the Roman Catholic hierarchy there are archbishops in England, Scotland, and Ireland, but these have no legal standing or powers, and their authority depends on the law and practice of their own church only.

Archbold, John Frederick (1785–1870). Called to the bar in 1814 and devoted himself to legal writing. He produced an annotated edition of Blackstone in 1811 and about 1824 a *Summary of the Law relative to Pleading and Evidence in Criminal Cases*, which, extensively revised by editors, is still a standard work. He also wrote various books on civil practice, and a *Justice of the Peace and Parish Officer* (3 vols., 1840) a practical guide for country magistrates. The third volume developed into a separate treatise on the Poor Law.

Archdeacon. In the Church of England a clergyman having jurisdiction under the Crown and next after the bishop of the diocese over a portion of the diocese, called an archdeaconry. He is appointed by the bishop, or presented by a patron to the bishop who then institutes him. His duties are to hold visitations of parochial clergy, to inspect and reform abuses among the clergy, to examine and present candidates for ordination, to induct to benefices, and various other functions. He formerly had a court, in which he might hear certain ecclesiastical causes, subject to appeal to the bishop. See ARCHIDIACONAL COURT. In the Roman Catholic Church the office of archdeacon is titular only.

Arches, court of the. The common name for the court of the Official Principal of the Archbishop of Canterbury, the provincial court of the Archbishop. The courts of both the Official Principal and the Dean of the Arches were at St. Mary-le-Bow which was built on arches. The two offices became merged in 1875. It was a court of first instance in all ecclesiastical causes, a jurisdiction acquired by encroachment on diocesan jurisdiction, and also a court of appeal from all diocesan courts within the province. Appeal lay formerly to the Queen in chancery, but latterly to the Queen in Council, and was heard by the Judicial Committee of the Privy Council.

Since 1963 the Court of Arches is the provincial court of the Archbishop of Canterbury with jurisdiction in appeals from judgments of consistory courts (q.v.) of the province. The judges since 1963 are the holder of the twin offices of Dean of the Arches and Auditor of the Chancery Court of York, two persons in holy orders appointed by the prolocutor of the Lower House of Convocation of the province, and two laymen appointed by the Chairman of the House of Laity after consultation with the Lord Chancellor and having the judicial experience thought appropriate by him. Certain proceedings may be heard by the Dean alone, but others must be heard by all the judges. The Queen in Council may hear appeals from the Court of Arches.

B. Woodcock, *Mediaeval Ecclesiastical Courts in the Diocese of Canterbury*. *Holdworth, H.E.L.*

Arches, Dean of the. Originally the official who had jurisdiction over thirteen 'peculiars', i.e. districts exempted from the jurisdiction of the bishop of the diocese (now abolished), of the Archbishop of Canterbury in London. In 1874 it was provided that the judge of the Court of the Arches should become Official Principal when a vacancy occurred in the latter office, which happened in 1875, so that it coalesced with that of the Official Principal who exercised the Archbishop's metropolitan jurisdiction. Now the judge of the provincial court of the Archbishop of Canterbury, the Arches Court of Canterbury or the Court of Arches (q.v.), is officially described as Official Principal but commonly styled the Dean of the Arches. Since 1874 he must be a barrister of ten years' standing in actual practice or a person who has been a judge of the Supreme Court, and a member of the Church of England. Since 1874 the same person is appointed by the two Archbishops, with the approval of the Queen, to be Dean of the Arches and Auditor of the Chancery Court of York. The Dean of the Arches and Auditor is now *ex officiis* Official Principal of the two Archbishops and also Master of the Faculties to the Archbishop of Canterbury.

See ARCHES, COURT OF; DOCTORS' COMMONS. For holders of the office see APPENDIX.

Archidiaconal courts. Courts held by archdeacons of the Church of England, presided over by a person called the archdeacon's official or by the archdeacon himself, having concurrent jurisdiction with the consistory court (q.v.) of the diocese, to which appeal lay. They were abolished in 1963.

Archidiaconus (Guido de Baisio) (?–1313). Italian canonist and cardinal, so-called from his having been archdeacon of Bologna, author of *Lectura super Sexta Decretalium* (1472).

Archon. The title of the holder of the highest office in many states of ancient Greece. At Athens there were at first three archons, a number raised to nine in about 680 B.C. From 487 they were chosen by lot and thereafter they had no political importance. The archonship imposed executive and judicial duties. In earlier times the archons tried cases alone, save those involving life or rights of citizenship where there was trial by the Areopagus (q.v.), and had jurisdiction in civil cases also. Later the archon had merely to decide if there was a case to try, in what court, and to preside at the trial. The archon eponymus was nominal head of the state and had special responsibilities for the protection of property and in his judicial capacity had responsibility for all cases involving family rights and inheritance. The archon basileus was charged with the religious duties of the former kings and accordingly dealt with lawsuits arising from these matters, and those between claimants to priesthood, charges of impiety, and homicide. The archon polemarchus was originally in charge of the armed forces, but his main duties were judicial, dealing with private suits involving foreign residents or between them and citizens.

Archpresbyter. The dean of a cathedral is sometimes called the archpresbyter of the diocese and sometimes assigned a dignity next to the bishop, though this is sometimes assigned to the chancellor of the diocese or to the archdeacon.

Arden, Richard Pepper. See PEPPER ARDEN, RICHARD.

Ardizzone, Jacobus de (thirteenth century). Italian jurist, who rearranged the materials comprising the *Libri Feudorum* and was author of a *Summa in usus Feudorum* (1518).

Arena, Jacobus de (fourteenth century). Italian jurist, author of notes on the Code and the Digest, *Compendium Moralium Notabilium*, and other works.

Areopagus. The oldest council of ancient Athens, which met on the hill of the Areopagus. From the earliest times it had special jurisdiction in cases of homicide and wounding with intent to kill. It was originally composed of nobles and was advisory to the king and from the introduction of the archons in the eighth century controlled Athens until Solon limited its power in favour of the Assembly (*ecclesia*). It nevertheless retained guardianship of the laws and tried prosecutions for unconstitutional acts, as well as cases of homicide. The council was further weakened by Cleisthenes, who strengthened the *ecclesia* and established a new council, the *boule*. From 487 B.C. archons were elected by lot and this further diminished the prestige of the Areopagus and it finally lost all political power in 462, and thereafter retained only limited religious jurisdiction, particularly in cases of murder.

Argentré, Bertrand d' (1519–1590). French jurist and historian, author of *Commentaires sur les quatre premiers livres de l'ancienne coutume* (1568), *Commentarii ad praecipuos juris Britanniae titulos* (1605), *Traité de l'ancien état de la petite Bretagne* and other works on Breton history.

Argou(x), Gabriel (1640–1703). French feudalist and jurist, author of *Institution au droit français* (1692), a concise exposition of the customary legislation of France which attained a high reputation.

Argument. Statements which seek to persuade, support, or justify a conclusion.

Argumentum ab inconvenienti. An agreement from the inconvenience of an alternative explanation.

Argumentum a contrario. An argument for opposite treatment from the contrary or converse case.

Argumentum ad baculum. An argument containing a threat, appealing to fear or timidity.

Argumentum ad hominem. An argument attacking the opponent's personality, characteristics, or peculiarities.

Argumentum a fortiori. An argument from another case weaker than the one being presented.

Argumentum ad misericordiam. An argument making an appeal to pity.

Argumentum ad populum. An argument which appeals to the beliefs of the general body of people.

Argumentum ad verecundiam. An argument appealing to respect for great men to establish the point.

Aristo, Titius (first century A.D.). A Roman jurist, a member of Trajan's council, and author of notes on Labeo's *Posteriores*, Sabinus' *ad Vitellium*, and possibly on Sabinus' *ius civile*.

Aristocracy. As a political system, the rule of the best people, the intellectually and morally superior, as distinct from the rule of one (monarchy), of the few (oligarchy), or of the people (democracy). But it is difficult to determine who are best or of superior qualities and aristocracy tends to come to mean the uppermost social stratum of a society or, in Britain, the peerage. Some aristocracies, Brahmans, mediaeval nobility, and others, have been hereditary, others, such as many later Roman emperors, dignitaries of the Catholic church, and leaders in some republics have been non-hereditary, though in many such groups there is a tendency for the aristocracy to become hereditary. Many English noble families, e.g. the Cecils, have produced national leaders in generation after generation.

In modern thought there is a tendency to replace the concept of the aristocracy by the concept of meritocracy, the leadership of those who have shown their merit and ability to lead.

Aristotle (384–322 B.C.). Greek philosopher, first studied under and worked with Plato at Athens (367–347 B.C.), then moved from place to place for some years (347–335 B.C.), and finally spent a dozen years in Athens again (335–322 B.C.) as head of the Peripatetic School in the Lyceum. Most of his extant writings represent lectures given by him at Athens in his later years. The range of his studies and writings was wide, covering not only philosophy and psychology, logic, morals, and politics, but literary criticism, physics, and biology. He had great respect for facts, and a passion for observing and classifying them systematically.

Among his contributions to thought most relevant to law and government are an essay on *Monarchy*, the *Alexander, or on Colonies*, the accounts of *158 Constitutions*, compilations of *The Customs of Barbarians* and of *Cases on Constitutional Law*, the *Eudemian Ethics*, *Nicomachean Ethics*, and the *Politics*. Of these the last two have exercised the greatest effect on subsequent thought. In the *Ethics* the question is: what makes a man a good citizen? and it discusses moral virtues and vices, and treats of justice. In the *Politics* the questions are: by what organizations, institutions and laws can the greatest amount of good character be secured? He lays down that the state exists for the good life, that law is the sovereign of states, and governments are servants of the law. He distinguishes lawful monarchy and arbitrary tyranny.

Aristotle's legal philosophy is incompletely known, but his extant works cast much light on legal matters; his systematization of valid inference has had great influence in legal reasoning. His passion for classification drew many of the fundamental distinctions in legal thinking; he distinguished constitution and laws, or public and private law, strict law and equity, and divided justice into universal justice and particular justice, and the latter into distributive and corrective justice, of which the former is concerned with public law, with the distribution of honour, wealth, and other assets among members of the community according to their merits, whereas corrective justice underlies punishment and reparation in private law. He distinguished acts having harmful consequences according to the subjective attitudes accompanying them. Much of his thinking about law and justice was developed from examination of the Athenian law of his time, but his analyses have permanent value and interest.

In the middle ages the study of Aristotle was, along with theology, the major part of the higher educational system. His *Organon* (the treatises on logic) was for centuries a major instrument of education. Scholasticism sought to reconcile the Aristotelian doctrines with the revelation of the Bible and the teachings of the Fathers of the Church. His *Ethics* and *Politics* became known in the West only in the thirteenth century, but ever since have been influential.

A. E. Taylor, *Aristotle*; D. Ross, *Aristotle*; W. Jaeger, *Aristotle*; E. Barker, *Political Thought of Plato and Aristotle.*

Armed forces. Bodies of persons raised and maintained for the defence of the state from external aggression and for the maintenance of peace and order internally. In the United Kingdom the existence of the forces is authorized by royal prerogative and statute. The armed forces comprise the Royal Navy, the Army, and the Royal Air Force, intended for operations at sea, on land, and by air respectively. In addition, there are the Royal Marines and the territorial and reserve forces. All are controlled by the Ministry of Defence. Each force has a body of administrative law, rules of discipline, a system of courts (courts-martial (q.v.)) and procedure, and scale of punishments, but members of the forces also have the rights and liabilities of ordinary citizens under the ordinary law of the land, but have in certain respects privileges and in others disabilities thereunder. Members of the forces are not exempt from the jurisdiction of the ordinary civil and criminal courts.

The Navy as a permanent institution dates from the fifteenth century and the only authority for its existence is the royal prerogative. Originally the Lord High Admiral issued instructions placing persons serving under military law and in time a practice and collection of precedents of offences and punishments grew up which, under Charles II, was embodied in Articles of War, which were the foundation of later Naval Discipline Acts. The Royal Marines date from 1755 and are controlled by the Admiralty.

The maintenance of a standing army in time of peace without the consent of Parliament was declared illegal by the Bill of Rights, 1688, and since then the army has continued to exist only by virtue of Parliamentary authority.

Originally Articles of War were issued by the Crown or the commander-in-chief. In 1689 the first Mutiny Act was passed and this was repassed annually till 1879 when replaced by the Army Discipline and Regulation Act, now replaced by the Army Act.

The Royal Air Force was established in 1918 and was governed originally by adaptation of the Army Act and now by the Air Force Act, which closely resembles the Army Act. The existence of the Air Force is also dependent on Parliamentary sanction.

The women's services were raised and maintained under temporary legislation during the First and Second World Wars. They were authorized permanently by the Army and Air Force (Women's Service) Act, 1948; the Women's Royal Navy Service is not authorized by statute, but is treated as part of the armed forces of the Crown for the purpose of certain enactments.

Historically the armed forces of the Crown were governed by Imperial legislation, but during the First World War the Dominions were recognized as having competence to control their own forces abroad. Now each Commonwealth country controls its own forces.

In the United States, Congress has power under the Constitution to make rules for the government and regulation of the land and naval forces. It adopted in 1789 and later revised Articles of War for the Army and in 1800 a similar code for the Navy. In 1951 these codes, as amended, were replaced by the Uniform Code of Military Justice. This provides for discipline and offences, the convening of courts-martial as required. Findings and sentence may be reviewed by boards of review and the Courts of Military Appeals, the determinations of which may not be further revised by the Supreme Court though the federal courts, including the Supreme Court, may examine the legality of detention by the *habeas corpus* procedure.

Armistice or truce. An agreement between belligerent forces for temporary cessation of hostilities. The state of war continues and at sea the rights of visit and search, to capture neutral vessels breaking blockade, and to seize contraband are unaffected. An armistice may be partial, concluded for substantial parts of the opposing forces and theatre of operations, or general, concluded for the whole, or substantially the whole, of the belligerents' forces and the whole region of war. Both, but particularly general armistices, are of political importance, and can be concluded only by commanders-in-chief; in the case of general armistices ratification by governments is often considered necessary. It is normally concluded in writing. The one essential condition is the cessation of hostilities. Violation of armistices is a wrong by international law. Where an armistice has been concluded for an indefinite period, either party may at any time give notice of termination and recommence hostilities. An armistice is frequently the prelude to negotiations for peace. The best known is that agreed between the Allies and Germany, effective on 11 November 1918, which ended hostilities in the First World War.

Armorial bearings, or arms. Devices, badges, crests, supporters, and other heraldic insignia formerly borne by armed knights on shields and surcoats and now retained as distinctive personal badges. The bearing of arms is regulated by the law of arms, as developed in each of the main European countries. In modern times arms have been commonly granted to public and local authorities and other corporate bodies as well as to individuals.

In England or Ireland arms may be borne only by virtue of ancestral rights, normally proved from the records of the College of Arms, or by virtue of grant. From early times the right to grant and regulate armorial bearings has been inherent in the office of a King of Arms within his province as representative of the sovereign. See ARMS, KINGS OF. Since the seventeenth century the right has been subject to the Earl Marshal's warrant to enable them to grant new arms, which warrant is granted in response to a memorial from the proposed grantee.

Complaints as to the unwarrantable use of arms are dealt with not by the common law courts but by the Court of Chivalry, which is a civil law court whose procedure is regulated in accordance with civil law forms, and from which appeal lies to the Judicial Committee of the Privy Council: see *Manchester Corporation* v. *Manchester Palace of Varieties, Ltd.*, [1955] P. 133.

In Scotland a coat of arms is a form of incorporeal heritable property held of the Crown and protected by law. The power to grant arms is part of the royal prerogative and long exercised on behalf of the Crown by the Lord Lyon King of Arms (q.v.). Since 1540 the practice has been for the sovereign to grant warrant to Lyon to grant the arms or augmentation requested. In 1672 was established the Public Register of All Arms and Bearings in Scotland and by statute no person may possess or claim to use arms unless the arms have been recorded in that register. Persons domiciled in Scotland having arms granted outside Scotland matriculate them in Lyon's Register.

Lyon Court has jurisdiction to punish contraventions of the law of arms and Lyon has power to erase or order the removal of unwarrantable arms. Lyon may also decide in a competition as to a right to arms. Appeal lies to the Court of Session and then to the House of Lords.

Arms, Kings of. These are officers of arms, appointed by the sovereign by letters patent under the Great Seal.

In England and Ireland the Kings of Arms are: Garter, who is sovereign over all other officers of Arms in England and who has jurisdiction over all England and over persons not domiciled in the United Kingdom; Clarenceux, whose province is England south of the Trent; and Norroy and Ulster, whose provinces are England north of the Trent and Northern Ireland. They had formerly to visit their provinces to record the arms and pedigrees of all who could establish right to the title of esquire or gentleman. They are under the jurisdiction of the Earl Marshal, and as his officers their acts cannot be questioned in courts of law.

In Scotland the Lyon King of Arms has jurisdiction to exercise the royal prerogative of granting arms, and of matriculating, i.e. re-registering, arms on the succession of an heir of the grantee.

Army. The branch of the armed forces operating generally and mainly on land.

From 1689 to 1879 the army was governed by annually-passed Mutiny Acts, which had the constitutional consequence of requiring Parliament to meet annually, and by Articles of War made by the Crown under the authority of these Acts. The Acts were consolidated by the Army Discipline Act, 1879, replaced by the Army Act, 1881, which likewise had to be renewed annually and which continued to recite the illegality of maintaining a standing army in peacetime without the consent of Parliament, a formula declared by the Bill of Rights, 1688. It was in turn replaced by the Army Act, 1955, which is renewable annually by Order in Council but not for more than five years unless Parliament otherwise determines.

The disciplinary code of the Army, to which its personnel are subject as well as to the ordinary law of the land, is contained in the Army Act and is based on the ordinary criminal law of England.

See C. Clode, *The Military Forces of the Crown*; G. Omond, *Parliament and the Army; Manual of Military Law.*

Arnisaeus, Henningus (?–1636). Teacher of medicine, philosophy, and politics in Germany and author of *Doctrina politica* (1606); *De jure majestatis libri tres* (1610); *De auctoritate principum* (1611); *De Republica* (1615), and other works.

Arnot, Hugo (1749–86). Passed advocate in 1772, but having poor health turned to history and literature. He published a *History of Edinburgh* in 1779 and in 1785 *A Collection and Abridgment of Celebrated Criminal Trials in Scotland, 1536 to 1784, with Historical and Critical Remarks*, which contains accounts of many trials not otherwise readily discoverable.

Arntzenius, Hendrik Johan (1734–97). A Dutch jurist, author of a learned *Institutiones iuris Belgici de conditione Nominum* (1783–98) and many other works.

Arraign. In English criminal procedure, to call a prisoner to the bar of the court to answer the charge laid against him in the indictment (q.v.). The prisoner is called to the bar by name, the indictment is read to him, he is asked whether he pleads guilty or not guilty, and his plea is recorded. It is a matter for the discretion of the presiding judge whether a plea may be withdrawn after arraignment or not. The plea may be one of guilty of the crime charged or of any other crime of which he may lawfully be convicted on that indictment, in which case his confession is recorded and he is sentenced; or he may stand mute, in which case a jury must be empanelled to determine whether he is mute of malice or mute by the visitation of God; or he may plead that the court has no lawful jurisdiction over him for the crime charged, but that another court has jurisdiction; or he may demur to the indictment, referring to the court the question whether, even admitting the facts alleged against him to be true, they constitute him guilty of the crime charged; or he may tender one or other of the special pleas in bar of an indictment, viz. *autrefois acquit* (q.v.), *autrefois convict* (q.v.), pardon, or special liability to repair a road, bridge, etc.; or he may plead the general issue, i.e. not guilty, in which case he puts himself in the hands of the jury. Where the general issue is pleaded the onus is on the prosecution to prove every fact or circumstance constituting the offence charged, and the prisoner may give in evidence every matter negating the allegations and also all matters of excuse and justification, but the onus of proof thereof lies on him.

Arrangement, Deed of. An agreement between a debtor and certain of his creditors, entered into by written instrument, for the benefit of his creditors and the discharge of his liabilities.

Arrangement, Scheme of. In company law, an agreement between a company and its creditors or any class of them or its members or any class of them, subject to statutory procedure and requiring the approval of the court, to compromise claims, alter the rights of members or otherwise resolve difficulties.

Array. In English criminal procedure, the list of jurors empanelled to try a prisoner. The prisoner may challenge the array, i.e. the whole number empanelled, but only on the ground of the partiality or default of the sheriff who summoned the jurors. A challenge to the array ought to be in writing, so that it may be put on record, and the other party may demur to it. If the ground of objection is established, the court will quash the array and direct a new panel of jurors to be summoned. Even if a challenge to the array is determined against the party making it, he may afterwards still have his challenge to the polls, i.e. object to each juryman separately as he is about to be sworn.

Array, Commission of. From the latter twelfth to the end of the sixteenth century writs, known as commissions of array, were periodically directed to persons in England under the authority of the Assize of Arms, 1181, and various other statutes concerned with the Defence of the Realm, authorizing them to muster and array or put into military order the inhabitants of every district and select from them men for military service. These writs purported to be issued to raise forces for the defence of the country from foreign invasion, for which purpose they were legal, but they were really used to impress men for wars in Scotland or France, for which they were technically illegal. The form of the commission was fixed by statute in 1403, so as to prevent the inclusion of any new penal clauses. In the mid-sixteenth century they were superseded by commissions of lieutenancy.

Arrest. The actual restraint of a person, committing, or suspected of having committed, a crime, with a view to his detention. At common law any private person may arrest where treason or serious crime has been actually committed or attempted, or there is immediate danger of such being committed, or a breach of the peace has been or is actually being committed or is reasonably apprehended. Constables and other persons employed for the preservation of the peace have these powers of arrest, and also powers to arrest on reasonable suspicion of serious crime, whether it has been committed or not, or where a reasonable charge of serious crime is made to them by a third party. In addition, many statutes give authority to constables, or in some cases other specified persons, or even to anyone, to arrest without warrant a person actually committing, or reasonably suspected of having committed, various specified crimes. An arrest without warrant is generally unlawful unless the person arrested knows or is informed of the grounds for his arrest.

In any case where there is power to arrest without warrant, a warrant for arrest may be granted on sworn information by, in England, any justice of the peace, or, in Scotland, by any judge, sheriff, magistrate, or justice of the peace. A warrant may be executed anywhere in England and Wales and backed in Scotland, Northern Ireland, the Isle of Man, or the Channel Islands, and conversely. The warrant may be executed at any time and the constable need not have the warrant in his possession but, if it is demanded, he must show it to the person arrested as soon as possible.

In arresting reasonable force may be used and a prisoner may be handcuffed if necessary. Doors may be forced to effect arrest and a person arrested may be searched, and property in the offender's possession may be taken if likely to be material evidence in the case.

In English law since 1967 an important distinction exists between arrestable offences, for which the sentence is fixed by law or for which a person may be sentenced to five years' imprisonment, where arrest is competent without warrant, and non-arrestable offences, where arrest requires a warrant and certain offences can be committed in relation to the former class only, such as assisting a person guilty of an arrestable offence. This distinction has replaced the former distinction between felonies and misdemeanours (q.v.).

Arrest in civil proceedings is now uncommon, and must be effected by virtue of a precept or writ issued out of some court.

In Admiralty actions *in rem* a ship or its cargo can be arrested by affixing the warrant to the mainmast and later substituting a copy thereof. Release can be obtained by giving bail (q.v.) for the sum claimed and costs.

Arrest of judgment. In a criminal case an accused may between conviction and sentence move the court in arrest of judgment, i.e. that judgment be not pronounced, because of a defect in the indictment or other irregularity in the record. In view of the modern simplified form of indictment the motion is rarely likely to succeed. The court can also arrest judgment if of the opinion that the indictment discloses no offence known to the law. Arrest of judgment does not preclude a fresh indictment.

Arrestment. In Scots law, the process whereby a creditor detains moveable goods or assets of his debtor in the hands of a third party. It may be arrestment on the dependence or in security, which merely prevents the third party disposing of the debtor's assets, pending the decision of the creditor's claim, or arrestment in execution, which follows on a decree of court and freezes the assets in the third party's hands and which may, if need be, be followed by an action of furthcoming, which makes the assets available to the creditor in satisfaction of his decree. By custom and statute certain subjects are privileged from arrestment. Arrestment to found jurisdiction, a practice borrowed from Holland, is the arrestment in Scotland of goods or assets of a foreigner not otherwise subject to the jurisdiction of the Scottish courts and suffices to give the Scottish courts jurisdiction in a claim against the foreigner.

Arrêt. In old French law, the official verdict of one or other of the various bodies and individuals which rendered official verdicts on legal, administrative, or political matters.

Arrêt de prince. A kind of embargo or detention by a state of foreign ships to prevent the spread of news of political importance.

Arson. At common law in England, the felony of wilfully and maliciously burning the dwelling-house of another. By statute setting fire to various kinds of places was also criminal. The crime was replaced by an offence under the Criminal Damage Act, 1971.

Art and part. In Scottish criminal law no distinction is drawn between the liability of primary participants in crime and accessories or accomplices, all who participate in any degree being equally guilty. The phrase may be related to *artifex et particeps* and have distinguished between participation as principal and as accessory. Formerly criminal libels had to charge a person 'art and part' so that he could be convicted, whatever his share in its commission, but every charge now impliedly charges the accused 'actor or art and part', so that he can be convicted whether he participated as principal or as accessory, and the phrase has come to mean 'by accession'.

Articled clerk. A person serving under articles, i.e. a written agreement with a practising solicitor or accountant, under which the latter undertakes to instruct in the practical skills and knowledge of the profession.

Articles. Clauses of a document, or sometimes the document itself, e.g. articles of clerkship, or of partnership.

Articles, Lords of the. A committee of the Scottish Parliament to which the real business of legislation was delegated. It originated as a device for improving the conduct of business, preparing business for the full Parliament. When Parliament met, certain persons were chosen to decide upon the articles presented by the King, leave being given to the other members to go away, returning only to pass the legislation *en bloc*. This device is found from the fourteenth century. In 1535 the Lords of the Articles were constituted a commission with full power to make statutes, and this practice was regularly followed thereafter. The numbers of Lords varied widely before the Reformation, but averaged about thirty. Between 1568 and 1639 Lords of the Articles were appointed in every Parliament, sat daily, and their recommendations were passed *en bloc* on the last day of the session. In 1606 James VI nominated the committee and thereafter royal influence was dominant in their selection. The Committee of the Articles was revived in 1661, membership being closely coincident with membership of the Council. The device of the Articles was condemned by the Convention Parliament of 1689 and there was bitter opposition to attempts to revive it, and it was abolished by statute in 1690.

Articles of Association. See MEMORANDUM AND ARTICLES OF ASSOCIATION.

Articles of Confederation. After the outbreak of the American War of Independence the Continental Congress appointed a committee of one representative from each state to draft an instrument of confederation. The committee produced its Articles of Confederation and Perpetual Union in 1776. The following year Congress adopted a draft and submitted it to the states as being the proposals most likely to obtain general acceptance, and by early 1779 most of the states had adopted the Articles. The Articles were in force as the supreme law of the land from March 1781 to March 1789. Under the Articles states might send two to seven delegates to Congress annually, but each state had only one vote. Congress was to manage foreign and Indian affairs, settle disputes between states, regulate coinage, borrow money, and establish a postal system. All rights not expressly ceded to Congress were retained by the states and this included taxation and the regulation of commerce.

The Articles had serious defects. There was neither federal executive nor judiciary, and Congress had no means of enforcing its measures. States exercised rights nominally ceded and defaulted on their undertakings. The delegation of authority in taxation and commerce was also a mistake. By 1786 the Articles were discredited. Nevertheless the underlying idea was right and served till experience had shown the way to a more permanent system which was secured by the Constitution. Without the Articles it is doubtful whether a union of the states would have survived.

Articles of Religion. See ARTICLES, THE THIRTY-NINE.

Articles of roup. In Scots law, the written statement of conditions on which subjects, particularly land or buildings, are exposed for sale by auction. The highest bidder executes a Minute of Preference and Enactment and the Articles and Minute constitute the contract.

Articles of the eyre. The headings under which justices in eyre visiting a county in the thirteenth–

fourteenth centuries made inquiries and on which they required an answer. The Articles ranged over the whole field of government and their object was to extract from juries information on every subject where a possible answer might afford ground for extracting an amercement from somebody or driving someone to make fine with the King. They comprised inquiries into the duties of counties, hundreds, townships, and boroughs, the misdeeds of officials, the rights of the Crown, such as escheats and other sources of revenue. On the answers the later proceedings of the eyre were founded against persons presented by the juries and against the juries themselves if they made a false presentment. As time went on the Articles lengthened. The Articles of inquiry contained in the hundred rolls were added to the Articles of the eyre, and known as the New Articles.

Articles of the peace. A person exhibits articles of the peace if he complains on oath to a court of summary jurisdiction that he has reasonable cause to fear that another person will do him, his wife, or child, bodily harm or burn his house. If the court is satisfied that the fear is founded on reasonable grounds, it is bound to require sureties to keep the peace; the court's jurisdiction so to order is founded on immemorial usage only, and is distinct from the court's power to bind to the peace on their own motion or as part of the punishment authorized by various statutes. If the defendant fails to do so, he may be imprisoned for not more than six months.

Articles of War. The first Articles were composed in 1527 by Ferdinand I, King of Bohemia and later Emperor and constituted an important code for conduct in wartime, seeking to enforce discipline on soldiers. They forbade plundering by soldiery without permission and the looting of cities and laid down that artillery, ammunition, and military stores were reserved for the Emperor.

In England ordinances for the government of troops raised for warfare were issued by the Crown with the advice of the Constable, or were enacted by the commander-in-chief in pursuance of authority for that purpose given in his commission from the Crown. They were in force only during the service of the troops for whom they were issued and ceased to operate on the conclusion of peace. Regulation of the army in peacetime did not come into existence until the Mutiny Act of 1689. The earliest complete code seems to have been the Statutes, Ordinances, and Customs of Richard II of 1385. During the Civil War the King and the Parliamentary leaders alike governed their troops by Articles of War, of 1639 for the royalist forces and of 1642 for the Parliamentary forces. Articles of War were issued by Charles II in 1666 for the first Dutch War, in 1672 for the second Dutch War and in 1685, by James II, on the outbreak of Monmouth's rebellion.

The early Articles were of great severity, imposing loss of a limb or death for almost every crime. Gradually they were modified and the Ordinances or Articles of War of 1672 formed the basis of the Articles of War of 1878 which were consolidated with the annual Mutiny Act in the Army Discipline and Regulation Act, 1879, now replaced by the Army Act.

An Act of 1754 empowered the Crown to make Articles of War for the government of the Honourable East India Company's troops; the language used indicates that they were intended for Europeans only, but they seem to have been applied to native troops also. In 1813 statute gave power to the governments of Fort William, Fort Saint George, and Bombay to make laws, regulations, and Articles of War for the government of all officers and soldiers in their services who were natives of the East Indies. Each presidency in India accordingly framed its own code. In 1845 the Governor General in Council provided a common code for native forces in India by enacting Articles of War. The Indian Articles of War were amended by later Acts of the Governor General in Council and replaced by the Indian Army Act of 1911.

Articles, The Thirty-Nine. These are the great and fundamental confession of faith of the Church of England, based on the Ten Articles of Henry VIII of 1536, the Thirteen Articles of 1538 and the Forty-Two Articles drawn up under Edward VI (1553) abolished under Mary but restored and revised under Elizabeth. They were adopted by the clergy of England in united convocation in 1562 and promulgated thereafter by authority of Queen Elizabeth. After their ratification by the Convocation of Canterbury in 1571, a statute of that year provided that all ecclesiastical persons should subscribe to them and those who maintained doctrines contravening them should be deprived. They have stood unaltered since 1571 though divergent interpretations have developed on various points.

Subscription to the articles is no longer obligatory in many circumstances where it formerly was so, but it is still obligatory on the clergy, on chancellors, and officials of ecclesiastical courts, and on certain lay persons.

Articuli ad Novas Narrationes. A tract, in Latin, dating from about 1450, probably a commentary on the *Old Natura Brevium*, though the name suggests that it is a commentary on *Novae narrationes* (q.v.). It follows the latter too closely, however, to be a commentary thereon.

It begins with a classification of pleas in the King's court, then gives a description of the courts and the pleas within their jurisdiction, a division of the writs into two categories, and finally a series of notes on specific writs. It was printed many times in the sixteenth century.

Articuli cleri. The name of a statute of 1315, made at Lincoln, relating to the liberties and licences of the clergy. The name was also used by Coke of the petitions presented by Archbishop Bancroft to the Star Chamber in 1605.

Articuli Inquisitione super statum Wintoniae. Articles of inquiry as to observance of the statute of Winchester, a statute of uncertain date, ascribed to 1285, 13 Edw. I, or to 1306, 34 Edw. I, Stat. 2.

Articuli super Cartas. The statute of 1300, 28 Edw. I, c. 1, confirming Magna Carta and the Carta de Foresta, but without the saving clauses contained in the Confirmation of Liberties of 1297 (25 Edw. I, c. 1).

Artificers, Statute of (1563). An attempt to stabilize society and to deal with pauperism and unemployment. The Act sought to restrict the movement of labour by insisting that men live and work in the place where they were born, and ordered the regulations governing apprenticeship in all trades to be enforced and the justices of the peace to fix wages. The Act, though having good intentions, was unsuccessful in combating the evils.

Artificial insemination. The introduction of semen into the female genital tract, not by sexual intercourse but by an instrument such as a syringe. The practice, well known in application to animals, has been applied to humans and raises difficult moral and legal problems, particularly where insemination results in childbirth. The semen may be that of the woman's husband (AIH) or that of a donor (AID).

The case of AIH does not raise difficulties, but AID does. In *Orford* (1921), 58 D.L.R. 251 a Canadian court held that AID was adultery and this view was strengthened by *Russell*, [1924] A.C. 687 which held that the essence of adultery was fecundation *ab extra*. But in *MacLennan*, 1958 S.C. 105, a Scottish court held that adultery was *conjunctio corporum* only and that AID was not adultery. In *Doornbos* (unrep.) a U.S. court held AID to be adultery and the child illegitimate. This was followed in *Gursky* (1963), 242 N.Y.S. 20. 406 and *Anon.* (1964), N.Y.S. 2d. 835. See also *R.E.L.*, [1949] P. 211; *Slater*, [1953] P. 235.

The Catholic church has strongly rejected AID as immoral.

Artificial person. Any entity other than a human being, such as a corporate body (q.v.), which is recognized in law as a legal person capable of having rights and duties.

Arumaeus (van Arum), Dominicus (1579–1673). Professor of law at Jena from 1602 and said to be founder of the study of public law in Germany, and collector of *Discursus academici de jure publico* (5 vols., 1617–23). He also wrote extensively on Roman law.

Ashbourne, Lord. See GIBSON, EDWARD.

Ashburton, Lord. See DUNNING, J.

Aschaffenburg, Gustav (1866–1944). Psychiatrist and criminologist, author of an important *Das Verbrechen und seine Bekämpfung* (Crime and its Repression) (1903), and founder of the *Monatschrift für Kriminalpsychologie*.

Ashbury Railway Carriage and Iron Co. Ltd.* v. *Riche (1875), L.R. 7 H.L. 653. The A Co.'s Memorandum of Association stated the purpose of the company as being to manufacture and sell and hire railway plant, fittings, and rolling stock. The company contracted to construct a railway in Belgium. It was held that the company was not liable for breach of that contract as the construction was outside the scope of the objects clause of the Memorandum, accordingly *ultra vires* of the company, and incapable of ratification by the members. See also *Ultra vires*.

Ashby* v. *White (1702), 2 Ld. Raym. 938; 3 Ld. Raym. 320. White, mayor of Aylesbury, prevented Ashby, a voter, from casting his vote. Ashby recovered damages at assizes, but in the Queen's Bench three judges held that no action lay since election disputes were for the House of Commons. Ashby appealed to the Lords, who upheld the dissenting judgment of Holt, C. J., holding that while the Houses of Parliament may judge of the exercise of their established privileges, whether a privilege exists or not is a matter of law for the courts. The Commons then resolved that they had the sole right of deciding all matters relating to elections, but the Lords denied this. In view of the Lords' attitude Paty and others brought actions against the constables of Aylesbury (See PATY'S CASE).

Ashe, Thomas (?–1618). Was called to the bar in 1582 and became a Bencher in 1597. He published abridgments of several of the early reporters, including one of Coke's reports with indexes of the cases and subject matter. His major work was his *Promptuary* (1614), which gives under alphabetical heads references to where the law can be found, a kind of mixture of abridgment and index.

Ashmole, Elias (1617–92). English antiquary; he qualified in law in 1638 but continued to study mathematics, physical sciences, and astrology, but these interests were superseded by history and antiquities. In 1652 he published a *Theatrum Chymicum* and in 1672 his major work, the *Institutions, Laws and Ceremonies of the Order of the Garter.* After the Restoration he devoted himself mainly to antiquarian and heraldic studies and was friendly with Selden and Dugdale (qq.v.). He was in high favour with Charles II and held various offices, but declined that of Garter King-of-Arms in favour of Dugdale. In 1682 he presented to Oxford University a fine collection of curios which formed the nucleus of the Ashmolean Museum, and subsequently presented his library to the university.

Asportation. The actual carrying away which, in English criminal law, was an essential of the common law crime of larceny. Before 1733 it was essential that an indictment for larceny should contain the allegation that the prisoner *cepit et asportavit.* To establish asportation it had to be shown that a thing attached had been completely detached, and any other thing removed from the place where it had been.

Asquith, Cyril, Lord Asquith of Bishopstone (1890–1954). Son of Lord Oxford and Asquith (q.v.), was called to the bar in 1920, became a judge of the King's Bench Division in 1938, a Lord Justice of Appeal, 1946, and a Lord of Appeal in Ordinary in 1951. In 1951 he is believed to have refused the office of Lord Chancellor. He was an excellent scholar, a member of the Law Revision and Law Reform Committees, chairman of several Royal Commissions, and wrote, with J. A. Spender, the life of his father.

Asquith, Herbert Henry, 1st Earl of Oxford and Asquith (1852–1928). Became a fine classical scholar, a fellow of Balliol and was called to the bar in 1876. In 1886 he entered Parliament and at once made a mark. In 1888 he appeared for Parnell before the Parnell Commission. In 1892–95 he was a successful Home Secretary, and then Chancellor of the Exchequer, 1906-08, and Prime Minister, 1908-16, in which capacity he had to deal with the constitutional crisis of 1910, being responsible for the Parliament Act, 1911, and the Irish Home Rule controversies. In 1916 he was driven from office and lost his seat in 1918 but returned to the House in 1920. He was defeated again in 1924 and accepted a peerage and became a member of the Judicial Committee of the Privy Council. He published several volumes of memoirs. A son became Lord Asquith of Bishopstone (q.v.).

J. A. Spender and C. Asquith, *Life of Lord Oxford and Asquith.*

Assart (or Essart). In the old English forest laws (q.v.) assart was the act of digging up trees or bushes in a forest so that they could not grow again; the term was also applied to land which had been asserted. Assart differed from waste in that assart was rendered usable for arable or pasture land, whereas in waste the bushes were liable to grow again. A man might obtain a licence to assart, usually for an annual rent; if he did so without licence, the land assarted reverted to the King until redeemed by a fine.

Assassination. The murder of a person by lying in wait for him and then killing him, particularly the murder of prominent people from political motives, e.g. the assassination of President Kennedy. It is punishable as treason or murder.

Assault. In Scots law, an attack on the person of another. It is both a crime and a delict. The slightest amount of force is enough and it need not take effect and there need be no actual injury. To aim a blow or a missile at a person is an assault. It may be done directly, or indirectly as by pulling away a chair. Assault is aggravated by the intention such as to rape, by the mode such as the use of a weapon, by serious injury, or by the fact that it is against an officer of law; but is justifiable if done in self-defence or in the course of duty so long as not excessive in the circumstances.

Assault and battery. Distinct kinds of conduct amounting in English law both to crimes and to torts, though it is common to use the term 'assault' for both kinds of conduct. An assault is any act done intentionally or, possibly, recklessly which causes another person to apprehend immediate and unlawful personal violence, e.g. presenting a weapon at another. A battery is any act done intentionally or, possibly, recklessly by which one inflicts unlawful personal violence on another, e.g. striking him. A battery accordingly frequently implies a prior assault.

Aggravated assaults are those made more serious by intent, e.g. intent to rob, or result, e.g. causing grievous bodily harm, or nature, e.g. indecent assault.

Assembly, liberty of. One of the liberties of the subject, importing the liberty of any two or more persons to meet and remain together in a public or, with the occupier's permission, private place for any purpose, provided that they do not intend to cause a breach of the peace or other crime or tort and even though they have reason to believe that their meeting may be opposed. Accordingly a meeting to be addressed by a political speaker or trade union leader is prima facie as lawful as a meeting to play cards or listen to music.

In England it was held in *Thomas* v. *Sawkins*, [1935] 2 K.B. 249, that police may enter a public meeting held in private premises if they have reasonable grounds for believing that otherwise seditious speeches would be made or breaches of the peace take place.

The liberty to hold an assembly or public meeting on public premises is restricted by the need to preserve the liberty of other persons to use the streets, squares, parks, and other places commonly used for public meetings, and by the need to preserve public order. There is no common law right to hold public meetings in Trafalgar Square, Hyde Park, on a common, or on the foreshore. Closely connected with liberty of assembly is the liberty to organize a procession or march along a public highway. See also ASSEMBLY, UNLAWFUL.

Assembly, unlawful. The common law crime committed if three or more persons assemble to commit, or when assembled do commit, a breach of the peace, or assemble to commit a crime by open force, or assemble for any common purpose, lawful or unlawful, in such a manner as to give firm and courageous people in the neighbourhood reasonable cause to fear that a breach of the peace will occur. An assembly is not unlawful merely because the participants realize that somebody may seek to break up their gathering and thus give rise to a breach of the peace.

Assembly, Consultative. The institution of the Council of Europe consisting of representatives drawn from the member states roughly in proportion to their populations. Representatives to the Assembly are elected or appointed by their national governments. Each representative may have a substitute who may sit, speak, and vote in his place. Representatives do not sit or vote in national blocs but according to political views and political groups, transcending national frontiers, have developed in the Assembly. They are not spokesmen of their governments (which are officially represented in the Committee of Ministers). The Assembly is the Council's deliberative body and may debate matters within the aim and scope of the Council and present recommendations to the Committee of Ministers and thus to member states; it is also a voice of European public opinion. It may also deliver opinions on matters referred to it by the Committee for opinion. A large number of international organizations make reports to the Assembly on their activities. The Assembly elects its own President and holds at least one session each year in Strasbourg. It has established a substantial number of general and particular committees. The official languages are English and French.

Assembly, European Parliamentary. One of the institutions common to the three European Communities from the beginning. In 1962 it redesignated itself European Parliament (q.v.).

Assembly of Notables. Consultative bodies of nobles and officials summoned at irregular intervals by the French kings. They developed from meetings of the royal council and membership rarely included persons of the Third Estate. The meetings were informal and merely discussed certain items of royal policy and their powers were less than those of states-general. Meetings were held periodically between 1470 and 1787 at the last of which preparations were made for recalling the states-general.

Assent. Agreement with or consent to some matter.

Assent of personal representatives. In English law, an act on the part of the personal representatives of a deceased for giving effect to the gift to a devisee or legatee, or the succession of next of kin, without which such a person has a merely inchoate right. Since 1926 an assent to the vesting of a legal estate in land must be in writing, signed by the personal representative, and naming the person in whose favour it is made. In respect of chattels it may take the form of express assignment or transfer of the subject-matter to the person entitled, or be implied by conduct justifying the inference that he has completed the deceased's disposition of the subject of bequest, as by authorizing the legatee to take delivery from a third party. After the assent a legatee's title to the legacy is complete. If assent is unreasonably withheld, the Chancery Division will compel a personal representative to give his assent.

Assent, Royal. See ROYAL ASSENT.

Asser, Tobias Michael Carel (1838–1913). Dutch jurist, professor at Amsterdam, started the *Revue de Droit International et de Legislation Comparé* in 1869 and was a founder of the Institute of International Law in 1873. He prevailed on the Dutch government to call the Hague Conference for the Unification of International Private Law in 1893 and received the Nobel Peace Prize in 1911, jointly, for his work in creating the Permanent Court of Arbitration at the Hague Peace Conference of 1899.

Assessment. Fixing the amount in money of some claim. Thus assessment of damages may be made by the court, or effected by a writ of enquiry or by reference to a master. Assessment of hereditaments or fixing their rateable values for rating purposes is done by officers of the Inland Revenue. Assessment of income tax is made by inspectors of taxes and General or Special Comissioners of Income Tax.

Assessor. In Roman law, an experienced lawyer who sat beside the governor of a province or other magistrate to assist and advise in the administration of the law. In modern law it usually means a lay person called to sit with a court trying a question requiring scientific or technical knowledge and to give the judge the benefit of his expert knowledge. In England the High Court or Court of Appeal may call in assessors in any action or matter. This is most common in cases involving matters of seamanship and navigation, where nautical assessors, who are always Elder Brethren of Trinity House, are called in. In Scotland the Court of Session may summon to its assistance a specially qualified assessor, particularly in admiralty or patent actions. The House of Lords may call in one or more assessors in appeals in Admiralty actions. The Judicial Committee of the Privy Council may call for the attendance of the archbishops and bishops of the Church of England as assessors in ecclesiastical appeals.

An assessor has no voice or power in deciding any issue before the court, his function being to assist the court's deliberations by enlightening it on the technical aspects of the matter being inquired into.

The term is also used of a person instructed by insurers to investigate an alleged loss, to determine whether the claim is genuine and admissible, and as to the amount of the loss sustained.

In Scotland the title is also used of a qualified lawyer who sits with a lay magistrate and advises him on law and procedure, and also of the official of a local authority who sets value on property for the purposes of rating. In the U.S. the term is also used in the last sense.

Assign. To transfer property, particularly personal or moveable property. In England an assignment is a total alienation of chattels personal (q.v.), a chattel interest in real property, or an equitable interest in real property. It may be voluntary or for consideration.

In Scotland an assignation is the mode of transferring outright or in security, incorporeal rights such as debts, shares, copyright, and insurance policies. Not all kinds of rights are assignable and certain kinds of rights require particular formalities.

Assisa continuanda, de. A writ, addressed to the judge of assize for the continuation of a cause when, unless this were done, material facts could not be proved.

Assisa de Ponderibus et Mensuris. An enactment of uncertain date, but ascribed to 1303, 31 Edw. I, setting out the weights and measures to be used.

Assisa de utrum. An ancient writ abolished in 1833, available to a cleric against a layman or conversely, to have determined whether particular lands were held by lay or ecclesiastical tenure.

Assise or Assize. This term, literally a 'session', was first used of a legislative enactment, such as the Assise of Clarendon or the Assise of the Forest.

An assise of the time of Henry II, the 'Grand Assize', provided that questions of seisin, and title to land should be tried by an inquiry of sixteen men sworn to speak the truth, and such an investigation, and the jurors, became known as assises, and the word came to be applied to the kinds of investigations which such bodies conducted, such as the assises of novel disseisin (q.v.) and mort d'ancestor (q.v.).

Magna Carta provided that assises of novel disseisin and mort d'ancestor should be held only in the shires where the lands in question lay, and for this purpose justices were sent into the county annually, and these became known as justices of assize. The Statute of Nisi Prius, 1285, enacted that justices of assize should try the issues in ordinary actions and return the verdict to the court at Westminster and accordingly the trial of civil actions took place at Assizes (q.v.) before the judge on circuit.

Assise of Arms. An ordinance issued in 1181 by Henry II which revived the ancient fyrd or national defence force.

Assise of Bread (or Assisa panis). An ordinance of 1266, for fixing the price of bread and ale.

Assise of Clarendon. The enactment in 1166 which established the assize of novel disseisin (q.v.).

Assise of Darrein Presentment (*de ultima presentatione*). A mediaeval real action which lay where a person or his ancestor has last presented a clerk to a living and had him instituted but had had his right of presentation challenged by another on the next occurrence of a vacancy. The writ called on the sheriff to summon an assise who were required to find whether the plaintiff was the last person who in time of peace as patron presented a clerk to the living in dispute. If so, he was seised of the advowson and was entitled to present again. It came to be superseded in practice by the action of *quare impedit* and was abolished in 1833.

Assise of Fresh Force. Fresh force was force recently done and if alleged to amount to a disseisin, the party disseised might within forty days bring an assise or bill of fresh force to recover his land.

Assise of Mort d'Ancestor. After Henry II had introduced inquisition by assize it was provided in 1176 that a writ could be obtained where it was alleged that a person had been deprived of his lands on the death of an ancestor by the intrusion of a stranger, to have an assize summoned to discover whether the plaintiff's ancestor died seised in his demesne as of fee of the tenement of land in dispute, whether he died since the period of limitation, and whether the plaintiff was his next heir. The action was abolished in 1833.

Assise of Northampton. An ordinance made in 1176 at the Council of Northampton by which Henry II divided England into six circuits and assigned three itinerant justices to each. It also introduced the assize of mort d'ancestor (q.v.).

Assise of Novel disseisin. In 1166 Henry II, acting on the principle that a person claiming to have been disseised, or ejected from, his land, must be entitled to recover possession before his challenger's claim to the lands could be considered, gave a person complaining of his recent disseisin a right to have a royal writ to the sheriff, requiring him to summon a jury of twelve to view the lands and dispute and give their verdict to the king's justices whether the defendant had justly and without a legal judgment disseised the plaintiff of his free tenement within the period of limitation. It was abolished in 1833.

Assise of Nuisance. A writ which lay against a man requiring him to redress or remove a nuisance which he had created in the freehold of another.

Assistance, writ of. A writ summoning judges and other legal persons to the House of Lords. The Attorney-General and Solicitor-General are understood to receive such writs, though they do not obey them.

The name is also given to a writ originally used by the Court of Chancery to enforce an order for the possession of land, a function now performed by a writ of possession, and also now sometimes used to put a receiver or other person in possession of specific chattels, where other procedure would be futile.

Assistants. The persons who, under the charters of the early New England colonies, formed the governing body. Thus the Massachusetts Bay Company charter of 1629 provided for eighteen assistants, to be elected annually by the stockholders, who were advisers to the governor. The Court of Assistants had legislative, executive, and judicial powers. It consisted of the Governor and any six assistants. When the Company became the Commonwealth, the assistants became magistrates and the upper house of the General Court, a bicameral legislature. Similarly the Connecticut charter of 1662 provided for twelve assistants.

Assisted person. A person litigating with financial assistance from the Legal Aid fund.

Assize, Commission of. A commission directed to certain persons authorizing them to try the possessory assises (q.v.) in countries named in the commission. Though this conferred a limited civil jurisdiction, difficult questions of law were often involved and royal judges were regularly among those commissioned, and commissioners of assize were regularly given the powers conferred by other commissions, such as oyer and terminer, also, and sometimes also given special statutory powers. Hence judges of assize came to exercise power little inferior to that of courts of common law.

Assize, Grand. A special variety of jury created probably in 1179 to determine which of two parties to a writ of right had the better right to lands held by free tenure, entitling a tenant of land to defend his right either by battle or by the Grand Assize. If the tenant chose the latter, the claimant had to obtain a writ that 12 legal knights from the same neighbourhood be chosen by four knights from the country and neighbourhood, to say on oath which of the two parties had the better right in the land which was in issue. Their verdict was final and was returned to the justices in eyre or to the King's court at Westminster. If the claimant objected to the tenant's choice of the Grand Assize, he must support his objection with reasons. Battle remained a competent alternative and incidental issues, such as whether it was a case for the Grand Assize, could be settled by battle or comburgation. The Grand Assize fell into disuse in later times but was not abolished till 1833. The last case was *Davies* v. *Lowndes*, (1835) 1 Bing. N.C. 599.

Assizes. England and Wales, except London and Middlesex, was divided until 1972 into seven circuits (q.v.) for the purposes of Assizes when persons holding commissions as judges of assize visited the main towns to hold the assizes. At Manchester and Liverpool Crown Courts (q.v.) took the place of criminal assizes while in London the Central Criminal Court (q.v.) exercised the criminal jurisdiction of assizes. Commissions were granted by letters patent to judges of the High Court and sometimes to judges of the Court of Appeal, county court judges, and to Queen's Counsel. They sat not by virtue of their judicial appointments but by virtue of the commissions granted, namely, of oyer and terminer, by which they were empowered to try serious crimes, of gaol delivery, empowering them to try all prisoners committed for any crime whatever and thus clear the gaols, of *nisi prius* (q.v.), under which civil causes, in which issue had been joined in one of the Divisions of the High Court, were tried by a jury of 12 of the county in which the venue is laid, and of the peace, by which the sheriffs and all justices were bound to be present at the county assizes. A commissioner of assize might also hear certain kinds of matrimonial causes.

Assizes were held four times a year. At summer and winter assizes both civil and criminal business was dealt with on all circuits. At Easter and autumn circuits criminal business only was disposed of, except at certain towns.

Though in theory commissioners of assize had all the jurisdiction of the High Court, in practice the civil jurisdiction exercised was in common law actions and matrimonial causes only. The criminal jurisdiction was greater in volume and importance than the civil. In 1972 the civil jurisdiction of assizes was transferred to the High Court (q.v.) and the criminal jurisdiction to the Crown Court (q.v.). See also CIRCUITS; CROWN COURT; HIGH COURT.

J. S. Cockburn, *History of English Assizes*, 1558–1714.

Assizes, Petty. The generic name for the forms of action introduced in England by Henry II whereby a plaintiff might have a right of possession declared and protected by a verdict of an assize of 12 knights. This form of action was applied successively to one who complained of having recently been ejected from his land (assize of novel disseisin, 1166), of having been ejected from his ancestor's land (mort d'ancestor, 1176), of having been dispossessed of his right of presentation to a church (darrein presentment) and of whether land was held by lay tenure or in free alms (by church tenure) (*utrum*) (qq.v.).

Assizes, possessory. The assizes of novel disseisin, mort d'ancestor, and darrein presentment (qq.v.).

Assizes de Jerusalem (or Letters of the Holy Sepulchre). A collection of laws and legal treatises compiled, according to tradition, by Godfrey of Bouillon for the government of the Latin Kingdom of the Orient after the conquest of Palestine by the Crusaders in 1909. It was based on the home customs of various learned persons, and essentially reflects law common to all Europe. The text was lost when Jerusalem was lost in 1187 but new compilations were later made, the best being that of John of Ghibelin, Count of Jaffa and Ascalon, of 1266. The elements of the collection all date from and reflect conditions of the thirteenth century. They emphasize the rights and privileges of the high court of the barons and represent an ideal feudalism. In Cyprus, the Venetians, then holding power, recognized the Assizes at law, and they were translated into Italian in 1531. The Assizes as used in Euboea were in force until the Turkish conquest there, and a version, known as the Assizes of Antioch, was carried to Armenia.

Beugnot, *Recueil des historiens des croisades, Lois*; Grandelande, *Étude critique sur les Livres des Assizes de Jérusalem*.

Associate justices. The name given in the Federal and many of the state courts of the U.S. to the judges other than the chief justice.

Associates. Formerly officers of the English common law courts appointed by and holding office at the pleasure of the Chief Justices or Chief Baron, and paid by fees. In 1852 they were made salaried and their tenure to be during good behaviour. They are now represented by clerks of the Associates Department of the Crown Office Department of the Central Office of the Supreme Court. Their functions are clerical and administrative.

Association (or unincorporated association). A group of persons formed to promote a common lawful interest or purpose, where the group they form is not, and sometimes cannot be, incorporated as a corporation or legal entity distinct from the members. In English law unincorporated associations include churches other than the Church of England, the Inns of Court, many clubs and societies, partnerships, friendly societies, and trade unions. The general characteristics of such bodies are that relations between members are contractual, their property is vested in trustees for the group, and they sue and are sued by representatives, save where practice or statute allows the group to sue and be sued under its group name.

The name Association was also taken by bodies the founding of which was urged by English parliamentary reformers at the end of the eighteenth century, such as the Society for Constitutional Information. By the 1830's they were a regular means of putting the pressure of public opinion on Parliament.

Association agreements. Agreements between the European Communities and countries which are not members. There are two types of association, the automatic association of those overseas countries and territories which were dependencies of or maintained particularly close relations with a member state, mostly former French possessions in Africa, and links with outside countries, sometimes as a stage towards full membership, usually involving the formation of a free trade area, possibly a customs union. Each of the parties to the agreements must independently ensure that the provisions of the agreement are observed; the agreements are administered by bilateral joint committees. The most far-reaching agreements are those with former overseas dependencies of member-states and other developing countries, which culminated in the Lomé Convention of 1975 between the Communities and 46 developing countries.

Association, liberty of. The liberty of any two or more individuals to form a body for the furtherance of any objects. In the United Kingdom this liberty is recognized provided the objects are not illegal or their promotion such as to involve crime or illegality. An association with criminal objects, e.g. to commit murders or thefts, would be illegal; there is no legal liberty to form such a body; its existence would not be recognized and no rights of members would be enforced. The cases which have given rise to problems in the United Kingdom are associations of employees, commonly called trade unions. At common law a trade union was an illegal association in so far as its objects included restrictions on the conduct of any trade or business by its members. The Combination Acts of 1799–1800 treated combinations of workmen to regulate the conditions of their work as illegal; these Acts were repealed in 1824 but in 1825 the repealing Act was repealed and the common law revived save that combinations for stated limited purposes were not to be illegal.

The Trade Union Act, 1871, provided, however, that the purposes of a trade union should not merely by reason that they were in restraint of trade, be deemed unlawful as to render a member liable to prosecution for conspiracy or otherwise, and that they should not be unlawful so as to render void any agreement or trust merely because the purposes were in restraint of trade. The Trade Union and Labour Relations Act, 1974, replacing the 1871 Act, subsequently repeated this principle, so that in general it is only by statute that trade unions are legal.

In other countries questions have arisen as to the right to form and belong to certain political parties.

Association, Memorandum and Articles of. See MEMORANDUM AND ARTICLES OF ASSOCIATION.

Association, The. A plan adopted by the First Continental Congress in 1774, whereby members pledged themselves not to import, export, or consume British goods if their grievance continued unredressed. It was enforced through vigilance committees and was a major factor in dividing loyalists from patriots, and was also used in recruiting troops.

Association of American Law Schools. A body formed in 1900, on the initiative of the Section on Legal Education of the American Bar Association, which prescribes minimum standards for law schools in the U.S. in respect of staff, curricula, and library facilities and may inspect schools and check their compliance with the standards set. It separated from the A.B.A. in 1914. A number of law schools, however, remain outside the A.A.L.S. Since 1948 it has published the *Journal of Legal Education.*

Association, unlawful. An association of persons having purposes which are unlawful, or which acts in an unlawful way. The category of 'unlawful' clearly includes purposes to commit crime or tort, but probably includes any purpose deemed contrary to public policy. Thus at common law combinations of workmen seeking to impose restrictions on their own terms of employment were held to be unlawful as being in restraint of trade and they have been legalized only by statute. Even if the association is lawful in intention, two or more members may effect a criminal conspiracy if at common law they agreed to do any unlawful act or to do a lawful act by unlawful means, or now under statute in England if they agree to pursue a course of conduct which will necessarily amount to or involve the commission of any offence by one or more if the agreement is carried out. Similarly, two or more members may commit the tort of conspiracy if they agree to effect an unlawful purpose, i.e. one with the predominant purpose of injuring another person, and damage thereby results to the complainer.

An association may also be unlawful in such cases as a professional partnership of, or including, unqualified persons.

Association, voluntary. An organization or body formed by a number of people for the promotion of any lawful end. Membership by a qualified person is, at least nominally, voluntary but in some cases there are strong pressures to join. According to the purposes for which the association is formed it may be possible to have it incorporated by Royal Charter, or under the Companies Acts, or Industrial and Provident Societies Acts, or to have it recognized as a friendly society or trade union, or to leave it, or have to leave it, as an unincorporated association (q.v.).

Assoilzie (from Lat. *absolvere*, O.Fr. *absoiller*, *assoiler*). In Scots law, to find a defender not liable. Such a finding is also called 'granting absolvitor'.

Assumpsit, (i.e., he undertook). This was the name of the old English common law form of action for damages for breach of a simple contract. It was a development from trespass on the case (q.v.), which gave a remedy for harm done. The basis was an allegation that the defendant had undertaken to do something for the plaintiff and had harmed the plaintiff in person or property by doing it badly or wrongly. Originally *assumpsit* lay only if there was an express agreement to do or pay. But in 1602 it was decided in *Slade's case* (1602), 4 Co. Rep. 926, that proof of an independent agreement to pay was unnecessary, it being implied where a debt existed, and this became known as *indebitatus assumpsit*. This made *indebitatus assumpsit* an equivalent to the action of debt. Later this action was extended to the enforcement of not only express but implied contracts, and to actions on quasi-contracts (q.v.). It was abolished by the Judicature Acts, 1873–75.

Assumption of risk. The principle that a party has taken on himself the risk of loss, injury, or damage befalling him and consequently cannot claim against the party who has caused the loss. The assumption may be made by express contract or by implication.

Assurance. A synonym for insurance, assurance being commonly used of lives and insurance of marine, property, and other insurances. Another, older, usage is as a synonym for conveyance.

Assythment. The indemnification due under Scots law to the surviving relatives of a person killed by criminal conduct, due by the person responsible.

A claim for assythment was certainly competent down to the late nineteenth century, provided the act founded on was punishable as a crime, and was possibly still competent within narrow limits until abolished in 1976.

Assyrian Law. Assyria is the northern half of the lands watered by the Euphrates and Tigris, known later as Mesopotamia and now as Iraq. In Assyria remains have been found of several collections of laws and legal materials; there are Old Assyrian Laws found in Cappadocia, dating from about 1000 B.C. or, according to others, *c.* 2200 B.C., apparently relating to a court dealing with commercial disputes. More important are tablets of about 1400 B.C., known as Middle Assyrian Laws, which cannot be a code but seem to be a collection of enactments about married women, possibly a series of amendments of the existing law. The topics dealt with are crimes, rights, and duties of women, pledge for debt, control of corporal punishment, land, and miscellaneous points. Though very incomplete they yield some information about procedure and penalties.

G. R. Driver and J. Miles, *Assyrian Laws.*

Astle, Thomas (1735–1803). Abandoned the law for the career of an antiquary. He continued the *Rotuli Parliamentorum*, 1278–1503, published in 1767–77, and in 1783 became Keeper of the Records. He collected a great collection of historical documents and papers and published in 1792 an *Account of the Seals of the Kings, Royal Boroughs and Magnates of Scotland.*

Aston, Richard (1717–78). Was called to the English bar in 1740 and went to Ireland as Chief Justice of the Common Pleas 1761. He became unpopular and was transferred to the English King's Bench in 1765, where he was accused of corruption in connection with the trial of Wilkes.

Astreinte. In French law, compulsion on a defendant to perform some act he should have done, effected by the imposition of a penalty for each day on which it remains undone. It effects some of the functions of the English injunction and some of those of an order of specific performance.

Asylum. A place of refuge or sanctuary. In Greece temples and altars were inviolate and a person or thing under the protection of the deity could not be removed by force. Under the Roman Empire the statues of the emperors and the eagles of the legions were made refuges against acts of violence, particularly for ill-treated slaves and escaped criminals. Under Christianity the custom of asylum or sanctuary attached to a church or churchyard. In modern times the word has often been used of an institution giving shelter and treatment to afflicted persons, particularly the mentally ill, now called a mental hospital.

In international law many countries grant asylum to their own nationals. Neutral powers may grant asylum to belligerent land forces, or to belligerent warships for limited periods. The principle that each state exercises authority over persons on its territory and that other states have no authority there, even over their own nationals, implies that an individual who is allowed to enter and remain in another state's territory, including its embassies and public ships, gains asylum there against the claims of his own state. Asylum may be sought by persons persecuted for political or religious or other reasons. States have always insisted on their liberty to grant asylum if they choose to do so and, apart from treaty, no state is bound to refuse admission to any fugitive nor to return him to his own state. But a fugitive alien is not entitled to demand admission by another state.

Atalaric, Edict of. An edict issued by Atalaric, Ostrogothic King of Italy about A.D. 530, dealing in 12 articles with wrongful disseisin, concubinage, and other abuses.

At arm's length. Parties may be said to negotiate at arm's length when they are wholly independent, having opposing interests and are regarding their own interests.

Athelstan, Laws of. See ANGLO-SAXON CODES.

Atimia. In ancient Greece, the loss of some or all civil rights. Deprivation of rights might be temporary, as until a debt was paid, or permanent, as in the case of punishment for treason, cowardice in the face of the enemy, or it might be permanent deprivation of some rights only, as for certain moral offences.

Atkin, James Richard, Baron, of Aberdovey (1867–1944). Was called in 1891 and read in the chambers of Scrutton (q.v.). His early years at the bar were very lean, but he developed a good practice in commercial law, took silk in 1906, and was made a judge of the King's Bench Division in 1913, where he was a great success. In 1919 he was promoted to the Court of Appeal and in 1928 he became a Lord of Appeal. He had a great knowledge of law and an insistence on principles, and his judgments are highly esteemed; many are classics. He was deeply concerned for the rights of the ordinary man. He was also deeply interested in legal education and did much to bridge the gap between practising and academic lawyers.

Atkinson, John, Baron of Glenwilliam (1844–1932). Was called to the Irish bar in 1865 and to the English bar in 1890. In 1880 he took silk in Ireland, became Solicitor-General for Ireland in 1889 and in 1892 Attorney-General for Ireland. He resumed that office in 1895 and entered Parliament. In 1905 he was appointed a Lord of Appeal in Ordinary. He was not a profound lawyer but very competent; he resigned in 1928.

Atkyns, Sir Edward (?1630–98). Called to the bar in 1653, he became a Baron of Exchequer in 1680 and Chief Baron in 1686 but was not reappointed at the Revolution.

Atkyns, Sir Robert (1621–1709). Brother of Sir Edward Atkyns (q.v.), was called to the bar in 1645 and was successful at the bar. He was a judge of the Common Pleas, 1672–80, and was respected for his fairness and moderation, but was dismissed for his political views, and became Chief Baron of Exchequer in 1689–94. He was deemed a learned lawyer and wrote various pamphlets on the constitutional issues of the day. After Hale, it was said, there was no more learned lawyer of his time, and none more honest. He wrote various works on the *Jurisdiction of the Chancery* (1695) and the *Jurisdiction of the House of Peers* (1699), as well as *Parliamentary and Political Tracts* (1734).

Atlantic Charter. A joint declaration drawn up and signed by Mr. Winston Churchill and President Franklin D. Roosevelt at a meeting at sea off Newfoundland in August 1941. It was aimed to reassure isolationist opinion in America and neutral public opinion throughout the world as to the war aims of the United Kingdom and its allies and of friendly neutrals such as the United States. The statesmen affirmed that their countries did not desire territorial gains and refused to recognize territorial changes made without the desires of the peoples concerned. They affirmed also the right of self-government and their desire to restore such right to those at present subject to military domination, and asserted as national policy claims to economic security, access to trade, and improved standards of labour. These aims were later embodied in the Charter of the United Nations.

Attach. In English law to arrest a person under a writ of attachment, or to seize property and place it under the control of the court.

Attachment of the person is either one mode of enforcing obedience to the orders of the High Court, such as injunctions, under a writ issued by leave of the court or a judge directing the sheriff to arrest the person, or a mode of punishing disobedience to writs, contempt of court, disobedience to the awards of a court, abuse of its process, and the like.

Attachment of debts is a proceeding employed in High Court actions where a judgment for money has been obtained against a person who is himself the creditor of a third party. The judgment creditor may obtain an order *nisi* that sums due by the third party shall be attached to satisfy the judgment debt, which has the effect of preventing the third party paying his creditor until the court has disposed of an application that the third party should satisfy the judgment creditor's claim.

Attachment of the Forest. The lowest of the three courts formerly held in the King's forests, which had authority to inquire, but not to convict, in cases where the value of the trespass was less than four pence.

Attachment of privilege. Where a man by virtue of his privilege to be sued only in a particular court, such as was enjoyed by attorneys of the Court of Common Pleas, summoned another to answer in that court. It has been obsolete since the early nineteenth century.

Attainder. The extinction of civil rights and powers which formerly applied when judgment of death or of outlawry was recorded against a person convicted of treason or felony. Its principal consequences were the forfeiture and escheat of the person's lands, and the corruption of his blood which debarred him from inheriting or holding land or of transmitting title by descent to anyone. The consequences were all abolished in 1870, save in case of outlawry which status survived till 1938.

Attainder could also be effected by statute. Bills of Attainder were normally initiated in the House of Lords, the proceedings being as for other bills, but the party challenged might appear by counsel and lead witnesses. They were much used under Henry VIII to punish persons who had displeased the King. The offences charged were usually called treason, but they did not have to satisfy the definition of that or any other crime. The first Bill of Attainder is said to date from 1459 and the most famous examples are those introduced against the Earl of Strafford in 1641 and against Laud. An Act of Attainder convicted a person of an offence and inflicted a punishment. It was not necessarily preceded by any trial, but if there was a trial, it was before the Houses of Parliament and not, as in the case of impeachment, before the Lords on the accusation of the Commons. In later cases attainder was a means of imposing Parliamentary control on one of the King's ministers. To be effective the Bill required the King's assent. Though commonly used from the fifteenth till the seventeenth century it has not been used since the case of Lord Edward Fitzgerald, an Irish rebel leader of 1798, and is now obsolete.

A bill for reversing attainder is first signed by the Crown and presented by a peer to the House of Lords by command of the Crown. Such a bill has been necessary as a step to claim a peerage where the ancestor was attainted.

In America Acts of Attainder or of pains and penalties were passed by colonial legislatures in some cases prior to 1789.

The Constitution of the U.S. forbids Congress to pass any Bill of Attainder, and many state constitutions do the same. In 1964 the Supreme Court held invalid, as being a Bill of Attainder, a section of a bill forbidding payment of salaries to named persons accused of being subversive.

Attaint. A writ of attaint was a proceeding to enquire whether a jury had given a false verdict. For this purpose a grand jury of 24 was summoned by writ of attaint to try the validity of the verdict of the first (petty) jury. If the verdict of the grand jury was contrary to that of the first jury, their verdict was reversed and the members lost all civil rights and became liable to punishment. It originated at the time when jurors were regarded as witnesses. In criminal cases a writ of attaint was issued at the suit of the King, but became obsolete by the end of the fifteenth century. In civil cases it was issued at the instance of either party but gave way to the practice of granting new trials and became obsolete after *Bushell's case* in 1670. Proceedings by attaint were abolished in 1825.

Attempt. An endeavour to perform some act. In criminal law an attempt to effect some criminal purpose has to be distinguished from mere wish or intention to effect that purpose, and from preparations to effect that purpose. It involves intended and partial, but incomplete, or frustrated, perpetration of the criminal conduct purposed, or at least acts not reasonably explicable as other than an endeavour to effect the crime. The difficulty is commonly to determine whether conduct has gone so far as to amount to attempt, or as to be distinguishable from innocent conduct. An attempt may be made though completion of the crime was physically impossible.

At common law in England it was a misdemeanour to attempt to commit treason, any felony or misdemeanour. Under many statutes attempts are themselves substantive crimes and it is a general rule that a person charged with a crime may be convicted of an attempt to commit that crime.

Attendance Centre. A place at which a court may require a young person to attend for a prescribed number of hours, as a minor penalty. It is not competent for young persons who have previously been sent to approved school, borstal training, detention centre, or imprisonment. If the young person fails to attend he may be dealt with in any other way competent to the court.

Attestation. The practice of having deeds signed in the presence of witnesses, who also sign, as evidence of the authenticity of execution of the deed, their addresses and designations being stated in an attestation clause at the end of the writing. The number of witnesses is commonly two, but in certain cases one suffices.

In Scotland, a granter's execution of a deed is normally attested by the signature of two witnesses whose addresses and designations are contained in a testing clause at the end of the deed narrating the execution in his presence, or their addresses and designations are appended to their signatures. In certain cases one witness suffices.

Attestation clause. The clauses appended to a document stating that the persons subscribing it did witness the subscription of the document by the party or parties thereto. In Scotland it is called a testing clause.

Atto, Collection of. A compendium of canon law, *Breviarium*, in 500 chapters, made in Rome *c.* 1075 by Atto of Milan, aimed at presenting in brief the norms of law and morality of the Roman church. It had limited influence but was used in later collections.

Attorney. In English law a person appointed by another to act on his behalf and having authority to act for him. A private attorney is a person appointed by a power of attorney to act for another or to represent him in a particular matter.

Attorneys at law were formerly persons appointed by a party in court, with power to appear for and to bind him, particularly in relation to the procedural steps of litigation. From 1292 the attorneys were a distinct kind of lawyer and in 1574 they were excluded from the Inns of Court and became associated with the Inns of Chancery, which came to be reserved for attorneys and solicitors only. Mediaeval statutes gave the judges power to admit and to control them. They were officers of court and subject to the discipline of the court. Their training was largely by apprenticeship, learning the uses of the forms and processes of the legal machinery. In the seventeenth century the attorneys grew rapidly in numbers, came in contact with clients and did all the ordinary drafting of pleadings and conveyances, consulting barristers only in difficult cases.

In 1739 a Society of Gentleman Practitioners in the Courts of Law and Equity was established and in the nineteenth century persons increasingly came to practise both at attorneys and as solicitors. From the time the Law Society was established in 1831 persons seeking admission as attorneys or solicitors had to pass an examination. In 1873 the term was abolished and the title 'solicitor of the Supreme Court of Judicature' substituted and all solicitors, attorneys, and proctors became solicitors of the Supreme Court.

In the U.S. an attorney, or, in full, an attorney and counsellor-at-law, may exercise all the functions of the English barrister, attorney, and solicitor, though in some states when acting in equity matters he is styled solicitor, and when acting in admiralty he is styled proctor or advocate. See also ATTORNEY-GENERAL; BARRISTER; SOLICITOR.

Attorney-General. In England, the member of the Bar and of the government who is the Crown's principal legal adviser. He is appointed by letters patent, is *ex officio* leader of the English Bar, and has precedence in all courts. The office developed from the mediaeval offices of King's Attorney and King's Serjeant and the title first appeared in 1461. From about 1570 he superseded the King's Serjeants as the chief legal adviser of the Crown. He is normally knighted, almost invariably a Member of Parliament and of the Privy Council, and has sometimes been a Member of the Cabinet, and as such he answers for the government on legal matters and takes charge of bills concerned with the administration of justice and other matters of technical law, and is expected generally to assist the House of Commons on legal matters.

As principal legal adviser of the Crown he advises particularly on matters of international, public, and constitutional law, and advises the Committee for Privileges of the House of Lords. He is also usually consulted by the Lord Chancellor over the appointment of lower judicial officers, and sometimes over appointments to the High Court.

As senior counsel for the Crown he is *ex officio* leader of the Bar, presides at general meetings of the Bar, and is an *ex officio* member of various bodies, but yet remains subject to the discipline of the Benchers of his Inn. He conducts major revenue, criminal, and other cases on behalf of the Crown. Formerly he was regularly involved in State trials and important prosecutions. He may file criminal informations *ex officio* in the Queen's Bench Division, and formerly could have a case tried at bar in the Divisional Court. He is still responsible for the decision whether or not to prosecute in important cases and in many cases acts through the Director of Public Prosecutions (q.v.). By entering a *nolle prosequi* he may terminate criminal proceedings on indictment.

His fiat or consent is required by statute before various kinds of prosecutions or proceedings are commenced and was formerly required for criminal appeals to the House of Lords, and he acts on behalf of the public generally at public inquiries and in cases of public wrongs. He exercises a general supervision over charities as representative of the Crown as *parens patriae.*

Prior to 1895 he might continue private practice at the Bar, but since then he may not do so and in any event could not, in view of the volume of government legal business. He is now remunerated by salary, not by fees.

He is commonly promoted to the bench, sometimes direct to the office of Lord Chancellor, or of Lord Chief Justice, on a vacancy occurring, but recent practice has dispelled the view that he had a right to judicial preferment as a reward for his political services, and some holders of the office have not been appointed to the bench at all.

There is under him a small Law Officer's Department headed by a Legal Secretary.

There are also an Attorney-General of the Palatine Court of the County Palatine of Lancaster, and an Attorney-General to the Prince of Wales. Prior to 1921 there was an Attorney-General for Ireland. There was a separate Attorney-General for Northern Ireland until 1972 and in Scotland the Lord Advocate (q.v.) occupies a generally corresponding position.

Most parts of the Commonwealth have adopted the title and assigned similar functions to their chief law-officers of the government, though in some the governmental functions are performed by a Minister of Justice.

J. Edwards, *The Law Officers of the Crown.* For list of holders of the office see APPENDIX.

Attorney-General of the United States. The chief law officer of the federal government and head of the department of justice. The office was created by the Judiciary Act of 1789 but did not become a cabinet office till 1814. He is appointed by the President. Initially the Attorney-General advised the President and heads of department on legal matters and conducted government cases before the Supreme Court, but still had time for private practice. Latterly his responsibilities have been greatly increased. From 1861 he has had control over U.S. attorneys and marshals, and since 1870 he has been head of the Department of Justice, which has departments dealing with Anti-trust, Civil Rights, tax, prisons, and the Federal Bureau of Investigation. He is assisted by the Solicitor-General, who has charge of the government's litigation before the Supreme Court; and several assistant Attorneys-General and a large staff. Each state also has an officer usually called Attorney-General with generally similar functions. Official opinions of the Attorneys-General of the U.S. have been published and constitute valuable state papers.

For list of holders of the office see APPENDIX.

***Attorney-General* v. *de Keyser's Royal Hotel, Ltd.*,** [1920] A.C. 508. During World War I the Crown requisitioned a hotel for military purposes. It acted under the Defence of the Realm Regulations, which provided for payment of compensation to the owner. The owners were held entitled to compensation, it being observed that even if prerogative powers had entitled the Crown to take possession of the premises without compensation the prerogative powers had been superseded by the statutory power and the Crown could not have relied on the prerogative.

***Attorney-General for New South Wales* v. *Trethowan*,** [1932] A.C. 526. Both houses of the New South Wales legislature passed bills, one repealing the section of the Constitution Act, 1902, which provided that no bill to abolish the Legislative Council, nor any bill to amend or repeal that section, should be presented to the Governor for assent unless approved by a majority of the electors voting in a referendum, and the other abolishing the Legislative Council. *Held* by the Privy Council that neither bill could be presented for assent until approved by a majority of the electors voting as they had not been passed in the manner required by statute in force in the State.

Attornment. In English law, the agreement of the owner of an estate in land to become the tenant of one who has acquired the estate next in reversion or remainder.

It may be effected expressly by deed, or be implied in law. Formerly attornment was necessary in most cases to complete the grant of a reversion or remainder, but since 1709 such grants are effectual without the attornment of any tenant.

Atwood, William (d. ?1705). Barrister and Chief Justice of New York, wrote numerous works on constitutional theory and legal history, including *Jus Anglorum ab Antiquo* (1681), the *Fundamental Constitution of the English Government* (1690), *The Superiority and Direct Dominion of the Imperial Crown of England over the Crown and Kingdom of Scotland* (1704) and *The Scotch Patriot Unmask'd*, which were ordered in Scotland to be burned by the public hangman.

Auburn system. So called from the state prison built at Auburn, N.Y., where the regime was first adopted in the 1820's, was a system of solitary confinement of prisoners by night, and of work in association but in silence during the day, as opposed to the Pennsylvania system of solitary confinement by day and night. An experiment of strict solitary confinement of heinous offenders adopted at Auburn in 1821 was discontinued in 1824 as a failure. Later innovations at Auburn were the lockstep, i.e. marching in single file, facing towards the guard, with the right hand on the shoulder of the man in front, the striped suit, extensions of the walls between each cell and special seating arrangements at meals, all designed to ensure silence and prevent communication between prisoners.

For some 40 years after 1816 there was heated controversy in criminological literature as to the merits of the Auburn and Pennsylvanian systems. The latter system was tried in a number of states but generally abandoned in favour of the Auburn system, but European Commissions of enquiry usually favoured the latter system, and many European countries adopted it in modified form. The controversy was diverted by the development of the Irish system started by Maconochie (q.v.).

Auction. A manner of sale or letting at which persons present compete to purchase or hire by making offers of increasing sums. It is normally conducted in public, after advertisement, by a licensed auctioneer, who signifies acceptance of the final offer or bid by a tap with a hammer. Property put up for sale may be withdrawn before the fall of the hammer. A sale by auction may be notified as being subject to a reserve price, and the seller may reserve liberty himself to bid. A bid may be withdrawn before the fall of the hammer. In Scotland a sale of land or buildings by auction is frequently called a 'roup'.

It may be conducted by setting up an inch of lighted candle, the maker of the last bid before it goes out being the purchaser. A 'Dutch auction' is an offer of property at a certain price and then successively at lower prices until a bidder makes an offer.

Auctor in rem suam. Agent for his own advantage, a phrase descriptive of the case where an agent or trustee, in breach of his fiduciary duty, takes advantage of his position for his own benefit.

Audi alteram partem (or *audiatur et altera pars*), hear the other party. An expression of the principle, frequently referred to as one of the essential elements of natural justice, that any person or body adjudicating on any matter in dispute, or judging the conduct of a person, should hear not only the complainer's side but also the other party's contentions and not condemn him unheard.

Audience. The right of access to and hearing by the sovereign enjoyed by the peers of the realm, individually, and by the House of Commons collectively, but today more commonly the interview between the sovereign and a minister, an ambassador when presenting his credentials, or an important official on taking up or relinquishing some appointment. The term is also used of an interview with the Pope.

Audience, court of. Archbishops of Canterbury delegated much of their jurisdiction to their Official Principal, but not all, and exercised this reserved jurisdiction, concurrently with that of the Court of the Arches, in the court of Audience. The jurisdiction was later exercised by the judge of the court of Audience. It seems to have vanished at the Reformation and was expressly abolished in 1963. Archbishops of York also had a court of the Audience.

Audience, right of. The right of persons, other than parties to the cause, to be heard by a court or tribunal. Barristers in England and Ireland and advocates in Scotland have right of audience in all courts, tribunals, committees, inquiries, and other proceedings at which persons may present submissions, unless that right is in any particular case excluded by statute. In inferior courts solicitors commonly have concurrent right of audience, and before some tribunals persons not legally qualified also have right.

Audita querela. Was a writ issued, in English common law pleading, to give the defendant a remedy where a matter of defence had arisen since judgment and was the basis of the defendant's plea to the court. It is now replaced by application for a stay of execution.

Audit and auditor. Audit is the examination of the accounts kept by financial officers of state, departments, public authorities, companies, and other bodies, to determine their accuracy, whether payments have been authorized, and whether the accounts exhibit a true and fair view of the financial state of the body in question. Extensive provision is made by many statutes for the audit of the accounts of various bodies.

The function of auditing is generally a function of an accountant. A matter of legal importance is to what extent an auditor should content himself with ensuring the accuracy of the accounts and how far he should look for financial imprudence or malpractices and warn the public concerned thereof.

Auditor. (1) the judge of the Chancery Court of York (q.v.). The office has since 1874 been held by the same person as holds the office of Dean of the Arches (q.v.).

(2) One of the judges of the court of the Rota (q.v.). The *auditores Rotae* were originally ecclesiastics, appointed by the Pope to hear questions in dispute and report to him. By the time the court of the Rota appears under this name in 1422 the auditors had become a permanent tribunal to which the final decision of certain matters was assigned. The Rota still ranks as the supreme papal court of justice and the auditors have special ecclesiastical privileges. They must be in priest's or deacon's orders, number 12, including four non-Italians, and form an intellectual elite from which bishops, nuncios, and cardinals are frequently drawn. The *auditor camerae* is an official with important executive functions in relation to sentence passed by the curia, and to papal bulls and briefs.

Auditor of Court of Session. An official of the court appointed to tax accounts of expenses incurred in litigation or other legal business. There are similar officials in the sheriffs courts.

Augmentation, process of. A process, now obsolete, in the Court of Teinds (q.v.) in Scotland brought by the minister of a parish against the titular and the heritors of the parish to obtain an increase in his stipend.

Augmentations, Court of. In 1536 Henry VIII established the Court of the Augmentations of the Revenues of the King's Crown, with a chancellor and treasurer, King's attorney of the court and King's solicitor of the court, and the auditors of the revenues of the Augmentations. It was a common law rather than a prerogative court. The court was to have the order, survey, and governance of all the monasteries, priories, and other religious houses which were dissolved or might come to the King under the Act dissolving the monasteries, and all the manors and hereditaments which the King had or might purchase. The machinery of the court was modelled on that of the duchy of Lancaster and it became both a court and a revenue department.

In 1546 it was amalgamated with the Court of General Surveyors, and in 1554 dissolved by letters patent and united with the Court of Exchequer.

Augustin, Antonio (1516–86). Distinguished Spanish jurist, archbishop of Tarragona, a noted commentator on the Roman law and also a noted canonist, editor of an amended text of the Pandects, *Emendationum et Opinionum juris civilis libri quatuor* (1541), author of *De Legibus et Senatusconsultis* (1583), *Constitutionum Codicis Justinianae Collectio* (1567), *Antiquae Collectiones Decretalium* (1567), *De Emendatione Gratiani Dialogorum libri II* (1587), and other works.

Augustine, Saint (Aurelius Augustinus) (353–430 A.D.). Bishop of Hippo in Africa, and philosopher. Converted to Christianity in 387, he became a bishop in 395. His writings included *Confessions*, *De Civitate Dei*, dialogues, letters, and treatises. From the legal standpoint importance attaches to his doctrine that the Church has sovereignty over the State, which is justifiable only in so far as it keeps the peace and favours the service of the Church. The Church was in a position to decide whether a particular government was acting justly or not. Positive or State law was not true law. Important in his philosophy is the concept of Pax, the sacred ordering of things in society and life which brought peace, the regulating principle.

Aula regia (or *regis*). A court established by William the Conqueror, consisting of the great officers of State resident in the palace and held in the great hall, but moving with the royal household. It was also known as the *Curia Regis* and from it evolved the Privy Council, Parliament, and all the common law courts of justice.

See also COMMON PLEAS; EXCHEQUER; KING'S BENCH; KING'S COUNCIL; PARLIAMENT; PRIVY COUNCIL.

Auld Alliance. A longstanding alliance between Scotland and France probably first made in 1168 and renewed many times thereafter, springing from common enmity to England. It was undermined by the union of the Crowns in 1603 and vanished after the union of the Parliaments in 1707 though traces of it survived in the French support for the exiled Stuart kings down to at least 1745.

Aulic Council (Reichshofrat). An organ of the Holy Roman Empire. Though based on the older council of advisers of the emperor, the Aulic Council was founded by Maximilian I in 1497 as a supreme executive and judicial council. It later became an organ for the empire alone and not for the hereditary principalities and as such, its organization and powers were fixed by the treaty of Westphalia in 1648. It comprised about 20 members nominated by the emperor for his lifetime and, at first itinerant, later became fixed in Vienna. Its executive powers were small and gradually lost all except a few formal powers, but it exercised the emperor's judicial powers, with exclusive cognizance of cases relating to imperial fiefs, criminal charges against immediate vassals of the emperor, and cases reserved to him. In other judicial matters it was in competition with the Imperial Chamber (*Reichskammergericht*) (q.v.). In the eighteenth century as the latter declined the importance of the Aulic Council increased. It came to an end when the Holy Roman Empire was dissolved in 1806.

Austin, John (1790–1859). English jurist. He early entered the army and served till 1812, after which he read law and was called to the bar in 1818. In 1826 he was appointed professor of jurisprudence in the newly founded University of London, now University College, but before commencing to lecture he visited Germany and made contact with Savigny, Niebuhr, and others. The subject was not recognized as a necessary branch of legal studies and in 1835 he resigned his chair in despair, though in the short time he had had some outstanding persons in his classes. In 1833 he was made a member of the first Criminal Law Commission but soon resigned. In 1836 he was one of the commissioners to enquire into the government and administration of Malta.

In 1832 he published *The Province of Jurisprudence Determined*, the first ten of his lectures compressed into six. In 1833–34 he was a member of the royal commission on the criminal law and procedure. Thereafter he lived much abroad, returning to England only in 1848.

Interest in his lectures was revived when Sir Henry Maine (q.v.) lectured on jurisprudence and emphasized the value of Austin's examination of the meanings and uses of legal terms. In consequence Austin's widow in 1861 published a new edition of the *Province* and his *Lectures on Jurisprudence; or the Philosophy of Positive Law*, with a biographical sketch by herself.

The *Province* and the *Lectures* subsequently exercised enormous influence on jurisprudence in England, though it is now known that Austin had been greatly influenced by Bentham, with whom he was very friendly. The importance of his work was the strict delimitation of the sphere of law and its distinction from that of morality, elaboration of the idea of law as a kind of command, and the close examination of the connotations of such common legal terms and ideas as right, duty, liberty, injury, punishment, rights *in rem*, and rights *in personam*. His works are marred by a heavy, tedious, and repetitive style, and by excessive reliance on Roman and English law. Nevertheless, he was the founder of English analytical jurisprudence and from his day until recently the predominant approach of English jurisprudence has been that of analysis of legal terms and concepts, so that he has had a permanent influence on jurisprudential thinking.

J. Austin, *Lectures on Jurisprudence* (ed. R. Campbell); *The Province of Jurisprudence Determined* (ed. H. L. A. Hart); J. Brown, *The Austinian Theory of Law*.

Australian Law. Europeans first landed in Australia in 1606 but not till 1770 did Cook take possession for the British Crown. The first British Colony in Australia was established at Botany Bay, New South Wales, in 1788, when Port Jackson was founded as a penal station for criminals. Tasmania was an auxiliary penal station, but became a separate province in 1825. From the start, in default of anything else, English law was applied. The first British governors were officers in command of the garrison, the convicts, and the few free settlers. In the 1830s New South Wales acquired a legislative council, and by 1860 all except Western Australia had attained responsible government.

The Australian Courts Act, 1828, provided that all laws and statutes in force in England on 25 July 1828, should be applied to the administration of justice in New South Wales and Van Diemen's Land so far as they could be applied there. Though Victoria was split off from New South Wales in 1851, and Queensland in 1859, 1828 was the date for the reception of English law there too. Tasmania was separated from New South Wales in 1825.

South Australia was founded in 1836 and Western Australia first settled in 1829 and in these states the date of foundation was the date of the reception of English law. Subsequent to the dates of reception no English decisions have any binding authority but great efforts have been made to keep the common law decisions uniform with those of England, though this has involved acceptance of some unfortunate doctrines. Hence even the High Court has regularly followed the House of Lords, and there has been much copying of English legislation.

The six States were six separate British colonies till 1900, but all had acquired responsible parliamentary government before that date, New South Wales and Victoria deriving their constitutions from statutes of 1855, South Australia and Tasmania from Orders in Council of 1856, Queensland in 1859 and Western Australia under statute in 1889. These original grants were clarified by the Colonial Laws Validity Act, 1865, and the Australian States Constitutions Act, 1907, of the United Kingdom Parliament.

The Commonwealth was established in 1901 by the Commonwealth of Australia Constitution Act, 1900, in which the distribution of powers and some of the major principles governing their relations were adopted from the Constitution of the United States. It created a Federal Parliament comprising the Queen, a Senate and a House of Representatives, a Federal Executive and Judiciary but preserved the State constitutions, except so far as inconsistent therewith. The substantive sections of the Commonwealth constitution can be amended only by referendum of the people initiated by the Australian Parliament, the proposal requiring a majority of electors in each of a majority of States, and a majority of the whole.

The Constitution has been extensively interpreted by the High Court, and in some matters by the Privy Council. In some respects the High Court has adopted principles of interpretation from the U.S. Supreme Court's interpretations of the U.S. Constitution. In 1942 the Federal Parliament adopted the Statute of Westminster 1931, which repealed the Colonial Laws Validity Act, 1865, in relation to it (but not to the States).

Commonwealth Government

The executive consists of the Queen, the power being exercised by the Governor-General appointed by the Queen on the advice of the Commonwealth Government. There is a federal Executive Council corresponding to the Privy Council. The Government consists of Ministers of State, who are members *ex officio* of the Federal Executive Council and must be senators or members of the House of Representatives.

The Commonwealth Parliament consists of the Governor-General representing the Queen, the Senate, comprising 10 senators for each State, directly elected, and the House of Representatives, consisting of members elected by the electorate in the States in proportion to their populations. The Houses have equal powers save that the Senate may not originate or amend money bills.

The Federal Parliament is mainly concerned with external affairs, taxation, defence, customs and excise, social services. The Constitution does not seek to protect individual rights and liberties by constitutional guarantees, but prohibits certain kinds of action, sometimes by both Commonwealth and States, more usually by States, such as State taxation of Commonwealth property.

The Governor-General may assent to or withold assent from bills or reserve them for the royal pleasure, or return bills to the House from which they originated with suggestions for their amendment. The maximum life of a Parliament is three years.

The Commonwealth Constitution Act vests specific or enumerated powers in the Commonwealth, leaving the residue of governmental power with the States. The Commonwealth Parliament has exclusive power to legislate in relation to *inter alia*, defence, foreign affairs, trade, banking, insurance, copyright, patents, negotiable instruments, taxation, marriage, and divorce. Many Commonwealth powers are concurrent with State powers.

Parliamentary procedure is in principle similar to that of the United Kingdom Parliament, but has many differences in detail. The Senate may neither initiate nor amend a money bill. In the event of disagreement between the Houses, there is provision for a conference of managers appointed by the two Houses, and for joint sittings. In the event of deadlock, both Houses may be dissolved.

Executive

Cabinet Government operates at federal level, with, in addition, certain ministers who are outside the Cabinet.

The Australian Capital Territory is administered by the Minister of the Interior and the Northern Territory by an Administrator advised by a Council.

State governments

The States have independent and responsible self-government. Each of the six States has a Governor appointed by the Crown, a Cabinet controlling the executive, an elected Legislative Council (except Queensland) and an elected Legislative Assembly, a structure of local government, and a judicial and legal system. The States retain all the legislative powers they had when colonies, except those withdrawn from them or exclusively vested in the Commonwealth Parliament.

Each of the States has power to amend, and has in fact amended extensively, the State constitution originally conferred on it as a colony. The State Parliaments are not, however, sovereign. It is provided that in the event of inconsistency between federal and State laws, the federal law prevails. The Governor-General cannot disallow State laws.

Local government legislation is a State matter but the system of municipal government is broadly the same throughout the Commonwealth, each State being subdivided into small areas under various names.

Judicial organization

The High Court of Australia established by the Commonwealth Constitution has original jurisdiction in a number of cases, particularly Commonwealth and inter-State issues, and may also entertain appeals from State courts. Its role in interpreting the Constitution is very like that of the U.S. Supreme Court, but this is only a small part of its work because it is also a general court of appeal from State courts on State law. It also has an important function in developing a uniform common law rule for Australia, in that a decision on a matter of common law on appeal from any State binds all State courts. It has reserved liberty to overrule its own decisions. Apart from the High Court there are no federal courts of general jurisdiction, but federal jurisdiction is vested in State courts. There are federal courts in the federal territories.

The constitution does not expressly establish judicial review, but the High Court has jurisdiction to interpret the constitution. That this is less frequently invoked than the similar power of the Supreme Court of the U.S. is because the Australian constitution makes a much larger and more detailed grant of powers to the Commonwealth government.

Appeal lay from the High Court in federal matters to the Privy Council till 1968 and its decisions are not binding but influential on all Australian courts. Theoretically no House of Lords decision is binding, but the tendency has been to follow the House of Lords to avoid divergent views on matters of common law, and there is deference to the views of the Court of Appeal in England though in recent years the High Court has acted more independently.

Each of the States has a Supreme Court, and there are also Supreme Courts for the Northern Territory and the Australian Capital Territory. Within each State there are also intermediate civil courts, intermediate criminal courts, and courts composed of magistrates and justices of the peace handling minor cases; the names vary in the different States. New South Wales maintains the distinction between law and equity, but elsewhere courts have jurisdiction in both spheres. If a State court is exercising federal jurisdiction, judges are bound by decisions of the full Supreme Court of another State, but if exercising State jurisdiction, a Supreme Court is bound to follow decisions of the High Court of Australia, but decisions of the Supreme Court of another State have persuasive authority only.

Notable features are the existence of a federal Commonwealth Industrial Court and a Conciliation and Arbitration Commission, concerned with settling industrial disputes, Industrial Courts or Commissions in the States, with the functions of registering and supervising industrial associations, and the making and enforcement of industrial awards. There is also a federal Court of Bankruptcy and State Supreme Courts have been given federal jurisdiction in bankruptcy.

Public law

The legal position of the Crown is based on British constitutional and legal theory; thus Crown prerogatives and immunities remain, though in most States the Crown is now, in general, liable in contract and tort. The traditional English prerogative writs of mandamus, prohibition, and *certiorari* are, as in England, the main means of the control of administrative tribunals by the courts. There is a great range and variety of administrative tribunals.

Criminal law

Letters Patent of 1787 authorized the governor to try and punish conduct which would be treason, felony, or misdemeanour according to the law of England, and this effected a wholesale reception of English criminal law as at that date. Most criminal law is State law, but there is a Commonwealth Crimes Act, 1914–41, dealing with offences against the whole community. A Queensland Criminal Code came into force in 1901; Western Australia adopted largely the same code in 1902, and a revised version in 1913, and Tasmania a code in 1924. The other States have remained under English criminal law as modified by statute, but have consolidated the statute law.

Private law

The common law of England was adopted in 1828 in New South Wales and Van Diemen's Land, so far as it could be adopted. It is also the basic law of all the States which were part of New South Wales in 1828 or later. In South Australia and Western Australia English law was adopted from the outset. Since then great regard has been had in Australia to its development in England, so that, in general, subject to any Commonwealth or State legislation to the contrary, it still applies in the Commonwealth and in each State. But differences of detail and in statutory modifications of the common law have developed in all the States.

Family law is a matter for each State though there is a Commonwealth Matrimonial Causes Act.

Common law, including the law merchant, determines the law of contract and the mercantile contracts; most of the latter are largely covered by statute, some United Kingdom, some Australian, some State. Australian legislation has generally followed the English pattern. In respect of employment there is a large body of legislation relative to industrial arbitration and conciliation.

In land law the most notable divergence from the English pattern is the existence of the Torrens (q.v.) system of registered title to land. In succession intestacy depends on the Statute of Distributions, modified in certain respects; realty and personalty have been assimilated for succession. There is a Public Trustee in whom property vests pending administration.

Procedure

Pleading and procedure everywhere, other than in New South Wales, is based on the English system, with variations, but in New South Wales the pre-Judicature Acts system obtains. In New South Wales, save by consent, every issue of fact must be found by a jury of four. Elsewhere jury trials are not nearly so common.

Legal profession

The profession is organized on a State basis. The earliest judges in Australia had been trained in England and their approach determined the development of the profession in New South Wales where, as also in Queensland, the two branches of the profession were and are separate. In Victoria the two branches were formally merged in 1891, but nearly all practitioners practise in one branch only. In the other States fusion actually applies. In all 'fused' States the different functions of barrister and solicitor exist, but lawyers may qualify as both and usually do so.

In all States the University law schools are the main mode of preparation for legal practice, the degree course being followed by a period of practical training in chambers.

By reason of the heavy reliance on the common law, there has been heavy reliance on English textbooks and Australia has produced only a limited number of textbooks for local conditions and, as yet, few legal classics.

Harrison Moore, *The Constitution of the Commonwealth of Australia*; Nicholas, *The Australian Constitution*; Lane, *The Australian Federal System*; Sawer, *Australian Government Today* (1967); G. W. Paton, *The Commonwealth of Australia*; Wood, *Constitutional Development of Australia*; Benjefield and Whitmore, *Principles of Australian Administrative Law* (1966); Baalman, *Outline of Law in Australia* (1969); Howard, *Australian Criminal Law* (1965); J. Fleming, *Law of Torts* (1965).

Austrian Law. The size and European importance of Austria have varied over the centuries. In Roman times the Danube was the northern frontier of the Empire. Later Charlemagne pushed the frontier of his empire to the east and in 811 organized the territories known as Austria as the Marks of Ostland and Friuli, north and south of the Drave respectively. The true birth of Austria was in 976 when the Babenbergs became rulers; it became a duchy separate from Bavaria in 1156 and the first Hapsburg became German king in 1273; thereafter Austria expanded and became the main bastion of western Europe against the Turks. Though the Hapsburgs effected a union of territories, these still preserved their individualities and legal codes. The development of Hapsburg powers was greatly advanced by Ferdinand I (1526–64) who established the privy council, chancellery, council of war, and court treasury, and between the end of the Thirty Years' War (1648) and 1740 successive rulers of Austria sought to develop central power over the collection of provinces, including Bohemia, Moravia, and Hungary, which constituted the Empire.

Roman law penetrated Austria as it did Germany. Maria Theresa (1740–80) further strengthened and centralized administration. She restricted the patrimonial jurisdiction of the lords on their estates. She planned a unified legal system in place of the separate systems of the provinces, and as early as 1713 projected a code. A civil code (*Codex Theresianus*) was completed in 1767 but not approved, and in 1768 a unified criminal code and procedure (*Theresianische Halsgerichtsordnung—Constitutio Criminalis Theresiana*) was introduced for the German and Bohemian lands. Torture was abolished in 1776. Under Joseph II (1780–90) an attempt was made to organize the court system and to concentrate judicial authority. Civil procedure was organized in local courts, courts of appeal, and supreme court. Serfdom was abolished in 1781. In 1787 part of the code of private law was introduced for the German and Bohemian lands (*Josephanisches Gesetzbuch*) and in 1788 criminal procedure was codified. Only in 1811 was codification completed by the publication of a General Civil Code for the Empire of Austria (*Allgemeines bürgerliches gesetzbuch*), similar to the portions of the Prussian Code dealing with private law. This was a work of much originality and had great influence in Eastern Europe.

In 1804 Francis I made himself Emperor of Austria but was defeated by Napoleon and in 1806 the Holy Roman Empire was brought to an end by Napoleon. Austria, however, participated in the defeat of Napoleon and remained after 1815 the leading German state, received accessions of territory, became presiding power in the German confederation after 1815 and one of the great powers of Europe. But thereafter Prussia gradually acquired economic and political supremacy among German states. The revolutions of 1848 saw the establishment of a moderately democratic constitution, but neo-absolutism returned in 1849 under Francis Joseph.

In 1866, as a result of war, Austria lost the leadership of Germany to Prussia and fell further behind as a power when the German Empire was founded in 1871. From 1867 to 1918 Austria was linked with Hungary in a joint monarchy and recovered something of the rank of a great power, being linked with Germany and Russia in the league of the Three Emperors in 1873, with Germany in the Dual Alliance in 1879, and with Germany and Italy in the Triple Alliance in 1882. After 1918 the Austro-Hungarian empire was dismembered, Czechoslovakia and Jugoslavia being created while territory became separate. Absorbed into Germany in 1938 Austria was liberated in 1945 and became independent again in 1955.

The constitution is of 1929, restored, slightly modified, in 1945. It is headed by the federal President who is directly elected and who appoints the Chancellor and, on his proposal, other members of the federal government. There is a sharp separation of powers between legislature, executive and judiciary, and adequate guarantees of civil liberties and rights of minorities, many founded on laws going back to 1862 and 1867.

The legislature comprises two houses, the Bundesrat representing the federal provinces, and the Nationalrat representing the people. Bills must go through both houses, the Nationalrat first; they may be held up but not vetoed by the Bundesrat. Many matters are reserved to the State, but law-making and administration are the responsibility of the province in all matters not expressly assigned to the State, and in certain spheres the State lays down the principles but the provinces make and administer the rules.

The country is divided into nine *länder*, or federal States, each with a provincial governor, elected by the provincial diet (*landtag*). The main police force is federal. Every commune has a burgomaster and council.

The legal system distinguishes judicial and administrative power; there exist only federal courts. There is a Supreme Constitutional Court (*Verfassungsgerichtshof*), a Supreme Administrative Court (*Verwaltungsgerichtshof*), and a Supreme High Court (*Oberster Gerichtshof*). There are also four courts of appeal, a number of courts of first instance, district courts, and numerous local courts, industrial courts and other special courts. The ordinary courts in general, have both civil and criminal jurisdiction. Judges are appointed by the cabinet on nomination by judicial panels. There is a system of public prosecution headed by the Attorney-General. Official reports are published of important decisions of the Supreme Court, but no decision is a binding precedent for the future.

The law is contained for the most part in the *Allgemeines burgerliches Gesetzbuch* (ABGH) (Civil Code) of 1811, *Handelsgetsetzbuch* (HGB) (Commercial code), and *Zivilprozessordnung* (ZPO) (Code of Civil Procedure), but all supplemented by many laws on particular topics.

O. Langer, *Bibliographia juridica austriaca*; K. Hautsch, *Geschichte Oesterreichs*; L. Adamovich, *Grundriss des österreichischen Verfassungsrechts*; *Handbuch des österreichischen Verfassungsrechts*; J. Kimmel, *Lehrbuch des österreichischen Strafrechts*; A. Ehrenzweig, *System des österreichischen allgemeinen Privatrechts.*

Authentication. Certification by a person that a record is in due legal form and that the person certifying it is the person authorized to do so.

Authenticum. A collection of Justinian's constitutions, Latin ones in the original, Greek ones in a literal translation into Latin. It contains 134 constitutions dating from between A.D. 535 and 556 and arranged down to no. 127, substantially in chronological order. Its name is said to have arisen from the fact that Irnerius (q.v.) originally doubted its authenticity but later changed his mind; he thought that it was an official collection made for promulgation in Italy, but this is questioned by modern authorities. It may, however, have been semi-official and probably dates from the sixth century.

Author. In copyright law the person, including a translator, who creates a literary work by selecting the language used, or the person who actually executes a design or makes a picture on the paper or canvas. In the case of a photograph it is the owner of the material on which it is taken.

Authorities. The general name for the statutes, judicial decisions, and standard (or authoritative) books, which give authoritative direction or guidance on the principle or rule of law relevant to a particular case, i.e. direction or guidance which must be accepted. Also, in the singular, denoting a particular statutory provision, decision, or statement in a book of authority which justifies a statement or decision on a point of law; thus, e.g. *Donoghue* v. *Stevenson* (q.v.) is the authority for the rule imposing liability on a manufacturer for harm to the consumer of his products by defects therein.

Authority, or the authorities, are sometimes contrasted with principles, distinguishing rules laid down by a statute or binding precedent from general precepts discoverable from decisions, equity, morality, and other sources.

Authority. First, legal power to do acts of a particular kind; in this sense used particularly of the powers conferred by a principal or his agent who may have express authority, implied authority, ostensible authority, or no authority. Second, a person or body having legal powers in a particular sphere, e.g. licensing authority, local education authority, or planning authority.

Authority, legal. The power conferred by law to do something, backed by an implied threat of some legal sanction if the exercise of the power is impeded. Thus, a constable is by certain statutes given authority to arrest without warrant; the person liable to be arrested cannot resist save on pain of being held to have resisted the police.

Authority, political. Is concerned with the ethical, legal, and practical bases on which power within a state rests and with the organization of that power. Ethically the question is on what grounds a state may claim allegiance from subjects and obedience to its laws. These grounds include power, the need for public order, divine will, the consent of the governed, and the benefits experienced in practice. Legally the questions are of the identity and recognition of the State and of the system of institutions through which government is exercised, limited, and controlled. Politically the concept of political authority is concerned with the forces actually determining policy. The term is also used of the power and influence of leaders in public life.

Authority of law. A system or body of law and each individual principle or rule thereof can be said to have authority if it exists by virtue of and is drawn from one or more of the sources of law recognized by the constitution of the State in question and by the legal system itself as providing authoritative statements of principles and rules, which must be applied and followed. Thus, in the United Kingdom a rule of law laid down by statute or in a decision of a superior court has authority; a statement in a student's textbook has no authority, even though it is correct. The determination of what kinds of sources yield authoritative precepts of law could be changed.

Furthermore, law has authority if and so far as principles and rules are accepted and do regulate and control the conduct of the majority of the people, and they accept that these principles and rules must be obeyed and followed, and that obedience to them is not a matter of choice. Disobedience is flouting the authority of the law and, if sufficiently widespread, destroys the authority of the law. Thus a revolution destroys the authority of the law until some person or body enforces a new body of rules of law.

An alternative view is that the authority of law depends on its conformity or otherwise to an ideal external standard such as morality, natural law, or law of God. But uncertainty as to what the standard is or requires renders this view of little practical utility.

Auto-da-fe (or *auto-de-fe*). The Portuguese and Spanish terms for the public ceremonies attending the proclamation of sentences that terminated Inquisition trials; the first was in 1481, the last, in Mexico, in 1850. They were very lengthy and elaborate, show trials, designed to instruct and impress the general public. The victims were apostate former Jews or Muslims, Protestants, and sometimes bigamists or sorcerers. The maximum penalty which the inquisitor could impose was life imprisonment; the death penalty was not carried out at an *auto-da-fe*, but was imposed and carried out by the civil authorities later.

Autocracy. The form of government which is vested in a single person, absolute, responsible to no one, and beyond legal control. Even the most autocratic Roman emperors claimed that their power was delegated by the Roman senate and people. In the middle ages it was generally held that rulers were subject to divine law, natural law, and the customs of the realm. But Machiavelli defended the arbitrary rule of one man as justified by expediency and the revival of Roman ideas at the Renaissance contributed to the belief that a great ruler should be as omnipotent as the greatest Roman emperor.

The divine right of kings theory in the sixteenth and seventeenth centuries contributed to the belief that a king was God's representative and he, his office, and his authority were sacred and resistance was not only treason but impious. The strongest defence of autocracy was by Hobbes who maintained that man required an absolute ruler to keep him in check. Modern totalitarians support an absolutist position, but rather the autocracy of the State or the party, albeit personified in one man, than that of a monarch.

Automatism. A state of mind and body in which an individual may act without being conscious of what he is doing, or without controlling, or being able to control, his conduct, so that it is involuntary. If this state be proved to have existed at the material time, the individual is not legally responsible for the consequences of his acts. This defence has covered acts done while sleepwalking or under concussion, but may not cover reflex actions, spasms, and the like, and may not cover acts done while under the influence of drink or drugs voluntarily taken.

Autonomic law. A body of law created by a body of persons within the community on which has been conferred subordinate and restricted legislative powers. Such powers are conferred by charter or statute on universities, the governing bodies of certain professions, public corporations and authorities, and within their powers they may validly make subordinate legislation binding on their members. Such legislation will be recognized by the ordinary courts and enforced so long as validly made and not contrary to any rule of the general law of the land.

Autre vie, estate pur. In English law, a tenancy of land during the life of another, the *cestui que vie*. Prior to 1926 this was the lowest estate of freehold recognized; since 1925 it is an equitable interest in land only, and devolves on the tenant's personal representations.

Autrefois acquit (formerly acquitted). A plea in bar of a prosecution, that the prisoner has previously been tried for and acquitted of the same crime before a court of competent jurisdiction. Whether the plea is good or not depends on whether the evidence needed to obtain a conviction on the later charge would have been sufficient to do so on the former charge.

Autrefois attaint (formerly attainted). A plea in bar of prosecution, on the basis that, if a man had been attainted of treason or felony, he, being deemed legally dead, could not be indicted for another treason or felony so long as the attainder remained in force. Attainder having been abolished, the plea is now obsolete.

Autrefois convict (formerly convicted). A plea in bar of prosecution, on the basis that the prisoner had previously been tried for and convicted for the same offence by a court of competent jurisdiction.

Autun commentary. A commentary on Gaius' *Institutes*, of the fifth or sixth century by a law-teacher at Autun and found there in 1898.

Auxiliary jurisdiction. The former jurisdiction of courts of equity in aid of the common law, principally by assisting a plaintiff to establish his case by compelling discovery (q.v.). It has been obsolete since the Judicature Act, 1873.

Auzanet, Barthelemy (1591–1673). French jurist, author of notes on the *Coutume de Paris* and *Observations et Mémoires sur l'etude de la jurisprudence* (1708).

Avail of marriage. Under feudalism the lord had the right of disposing of a ward in marriage and was entitled to an avail or payment on the ward becoming marriageable. It seems that, originally, it was due only from a ward who declined to marry at the lord's request, but later also if he died without having married or been requested to marry. A double avail was due if the lord offered a suitable lady willing to marry, but the ward declined to marry her. The exaction was abolished in 1747.

Aver and averment. To aver is to allege in a pleading; an averment is an allegation or statement in a pleading, e.g. that the defendant failed to perform his contract.

Average. Damage or loss from the perils of the sea. Simple, petty, or particular average exists where any damage or loss is caused to ship or cargo, whether by accident or otherwise, such as heavy weather damage or deterioration of cargo. Such a loss must be borne by the owner of the ship or cargo concerned, or by his insurers.

General or gross average arises where any extraordinary sacrifice or expenditure is voluntarily and reasonably made or incurred in time of peril for the purpose of preserving the property imperilled in the common adventure, such as the jettison of cargo to lighten the ship. In such a case the loss must be borne proportionately by each of the interests involved in the adventure, ship, cargo, and freight. The principle is an ancient maritime one, found originally in the maritime law of Rhodes, developed in Roman law, embodied in later maritime laws and accepted in the modern custom of the sea. Though frequently embodied in contracts of affreightment, the rule of contribution can be based equally on an implied term of the contract. The sacrifice must have been voluntary, there must have been substantial peril, and the sacrifice must have contributed to the saving of the vessel.

The extent of the contributions are settled by average adjusters who have developed a substantial body of rules of practice on the matter. Charterparties and bills of lading frequently incorporate the York–Antwerp Rules of General Average, which are a body of rules formulated at a series of international congresses on the subject, the last at York in 1864 and adopted in Antwerp in 1877 and subsequently several times revised.

An average bond may be granted by the consignees of goods in favour of the shipowner binding them to pay the shipowner when their due proportion had been ascertained.

D. Maclachlan, *Merchant Shipping*; J. Arnould, *Marine Insurance*; R. Lowndes and G. R. Rudolf, *General Average*.

Averani, Guiseppe (1662–1738). A Italian jurist, deemed the most illustrious follower of Cujas, author of *Interpretationes Iuris* (1746), *Disputatio de iure belli ac pacis* (1708), and other works.

Averments. Statements of fact in pleadings in civil claims.

Aviation. See AIR LAW.

Avizandum. In Scottish practice a judge says 'Avizandum' or is said to make *avizandum* when, having heard argument, he desires time to consider his judgment. Similarly in England when a court reserves judgment the report bears *c.a.v.* (i.e. *curia advisari vult*).

Avocat. A category of lawyer in France, whose functions were generally legal advice, but in the *Cour de Cassation* included all formal stages also. This branch of the profession has now been fused with the *avoués* (q.v.).

Avoid. To set aside or make void, especially a contract or deed. In pleading, avoidance is meeting one's opponent's pleadings by new matter.

Avoué. A category of lawyer in France, prior to the Revolution called *procureur*, whose function was formulating claims in litigation and representations in court. This branch of the profession has now been fused with the *avocats* (q.v.).

Avulsion. The tearing away of a portion of land by the operation of flood-water, change in the course of a river or encroachment by the sea, and its deposit against the land of another. The separated land continues to belong to the original owner. See also ALLUVION.

Avvocato. In theory the more scholarly lawyer in Italy who prepares a party's claim and defences in litigation; in practice enrolment as an *avvocato* is open to anyone who has six years practice as a *procuratore*, and *avvocati* usually retain their qualifications as *procuratori* and may act in both capacities in a case, while both classes of lawyers may give legal advice.

Award. Any words containing a decision, especially of an arbitrator or arbiter.

Axones. In the Prytaneum at Athens, tablets, originally of wood, later of stone, revolving on a vertical axis containing Draco's and Solon's laws, which were accordingly quoted by the number of the *axon*. The highest known number is 16.

Ayala, Balthazar (*c.* 1548–84). Jurist of the Spanish Netherlands, auditor (judge) of the Spanish forces in Netherlands (1580), later a member of the great council and master of requests, author of *De jure et officiis bellicis et disciplina militari libri III* (1582) and as such one of the forerunners of Grotius in establishing international law.

Ayola, Manuel José de (1728–1805). Spanish jurist, author of *Notas a la Recopilacion de Indias, Diccionario del gobierno y legislacion de Indias y España*, and other works, part of a plan for the modification of legislation dealing with the Indies.

Ayliffe, John (1676–1732). Practised as a proctor in the chancellor's court till 1710. His main titles to fame are the *Parergon Juris Canonici Anglicani* (1726) and the *New Pandect of the Civil Law* (Vol. I, 1734, never finished), an elaborate treatise on modern Roman law. The former is still a work of high authority, and the latter is a useful treatise designed as much for the politician and the diplomat as for the lawyer.

Ayloffe, Sir Joseph (1709–81). An eminent antiquary whose main work is the *Calendars of the Antient Charters and of the Welsh and Scottish records, now remaining in the Tower* (1772). He also compiled a *Collections relative to Saxon and English Laws and Antiquities* which is still unpublished.

Ayrault, Pierre (Aerodius) (1536–1601). French jurist, one of the first to criticize criminal procedure for the advantages it gave to the prosecution. He wrote extensively on legal subjects, including *De la nature variété et mutation des lois* (1564) and *Decretorum rerumve apud diversos populos libri duo* (1567).

Ayres. Judicial sessions held in mediaeval Scotland, usually twice yearly, by justiciars travelling from sheriffdom to sheriffdom, at the head burgh of the sheriffdom, the ancestor of the later criminal circuits. Cf. English 'eyre' (q.v.).

Azo (*c.* 1150–1230). Italian jurist. He became professor of civil law at Bologna and stood in the first rank of the glossators (q.v.). Among his pupils were Accursius and Jacobus Balduinus. His *Summa Codicis* (1537) and *Apparatus ad Codicem* (1596) collected by a pupil and completed by additions of Hugolinus and Odofredus formed a methodical exposition of the Roman law and carried great weight with courts, so much so that knowledge of his work was considered indispensable to aspirants to judicial office; it was said '*chi non ha Azo, non vada a palazzo*'. His work was known to and influenced Bracton.

He is not to be confused with Azo Lambertaccius, a thirteenth century canonist, professor of canon law at Bologna and author of *Quaestiones in jus canonicum*; nor with Azo de Ramenghis, a fourteenth century canonist, professor of canon law at Bologna and author of *Repetitiones super libros Decretorum.*

F. W. Maitland, *Bracton and Azo.*

Azpilcueta, Martin (1493–1586). Spanish theologian and canonist, whose works included a *Manuale sive Encheiridion confessariorum et paenitentium*, long deemed a classic, *De Alienatione rerum ecclesiasticarum, De Furto,* and *De homicidio casuali.*

Azuni, Domenico Alberto (1749–1827). Italian jurist, author of a *Dizionario Universale Ragionato della Giurisprudenza Mercantile* (1786–88), *Sistema Universale dei Principi del Diritto Maritimo dell'Europa* (1795–96) translated into other languages, and *Origine du droit et de la legislation maritime* (1810). He later became a French citizen and president of the appeal court at Genoa and judge of the commercial court at Cagliari.

B

B. A baron, or judge, of the Court of Exchequer.

Babylonian Law. Babylonia was in ancient times the southern half of the lands watered by the Euphrates and the Tigris, later known as Mesopotamia and now as Iraq. It was one of the earliest centres of civilization and extensive records survive on clay tablets of business transactions, and in the third millennium B.C. the kings of various city-states are known to have promulgated bodies of law. Written collections of laws seem to have been not uncommon in the third millennium and probably every city-state had such a collection written in Accadian or Sumerian. One Urukagina of Lagash (*c.* 2750 B.C.) asserted that he established the ordinances of earlier times; there survive records of administrative reforms but no general code. Sargon of Accad was said to have spoken justice. At Ur Ur-Engur (*c.* 2400 B.C.) asserted that he made justice prevail and Ur-Nammu, founder of the Sumerian third dynasty of Ur (*c.* 2100 B.C.), enacted a code of law of which fragments survive. There are also numerous clay tablets containing judgments, contracts, and other legal documents. There is a code in 60 clauses possibly of Bilalama, King of Eshnunna (*c.* 1900 B.C.) in Accadian and laws in Sumerian by Lipit-Ishtar, King of Isin (*c.* 1900 B.C.) dealing with a variety of topics, mainly family and property. All Babylonia was later united under a dynasty of which one of the kings, Hammurabi, about 1750 B.C. issued a code in 282 sections. It reproduces many of Bilalama's laws and shows ordered arrangement. (Some authorities put Hammurabi earlier, as early as 2050 B.C. and the other kings earlier still.) This, and the numerous contract tablets, evidence a developed state of law in which the state and its organization are more powerful, the law is secular, and to a considerable extent, written. It is known that the laws of Eshnunna and of Hammurabi sought to prescribe and regulate prices.

Hammurabi's code is not a complete system or statement of Babylonian law; many matters are excluded. It was compiled from, and presupposes, a great mass of Sumerian customary law under which peoples had lived in civilized fashion for centuries and restates and modifies some of these customary rules. Though only a Semitic version survives, it may have been issued also in a Sumerian version. It exhibits an ordered arrangement and comprises a prologue, text, and epilogue. The prologue and epilogue are poetical, laudatory of justice. Apart from a few survivals of primitive customs it has developed beyond tribal customs. The King is the foundation of justice and the judges are under his supervision and an appeal lies to the King. There is no blood feud or private retribution.

The population was divided into three classes, nobles subject to heavier penalties for crimes, the general body of freemen, and slaves. Slaves, though unfree and chattels, could marry, own property, and had rights.

Marriage was by purchase, completed by delivery of a bride-price on the one hand and dowry on the other. A lawful marriage required a contract which stated the consequences of repudiation of it. Neither party was responsible for the other's ante-nuptial debts, but both were responsible for the post-nuptial debts. Divorce was recognized, subject to provision for the wife and children. Concubinage was recognized, subject to various limitations, as was the legitimation of children by such a mother. Parental rights, adoption, and succession on death were all regulated.

Private ownership of land was recognized subject to fixed annual charges due to the King, which included the duties of providing men for the army and statute labour. Much land was owned by the King and cultivated by officials, soldiers and craftsmen, who held their holdings in return for service. There was extensive regulation of agriculture which might be done by the owner, an employee or a tenant.

Many kinds of contracts were known, which had to be evidenced by documents or the oath of witnesses. Debt was secured on the debtor's person, but he could nominate a wife, child or slave to pay off the debt. Personal guarantees were often given, the guarantor being liable on default. Payment through a banker or by a written draft against deposit was familiar; bonds to pay were treated as negotiable. Interest was commonly charged on loans. There was extensive rules controlling caravan traffic; caravans undertook five-fold responsibility for loss, and ships were hired at a fixed tariff.

In criminal law the principle of the *lex talionis* ruled. Death was a regular penalty for theft, robbery and criminal negligence and specified forms of the death penalty were prescribed for particular kinds of crimes. Exile, disinheritance and branding were also penalties recognized in certain cases. Imprisonment was not recognized, but the commonest penalty was a fine, imposed for breach of contract, injuries or damage to property. The mental element in wrongs was recognized. Carelessness and neglect were severely punished but an accident was not deemed an offence.

The usual procedure seems to have been a preliminary hearing before one or two judges, who took depositions and referred the case to a larger tribunal of priests or of civil officials. In time, the latter tribunal became more important. Though the ordeal by water was known, reference to the defendant's oath was regular and much stress was laid on written evidence. A cause might be heard by the King himself and his decision was final and binding. In other cases a judgment was not binding until recorded in an agreement which embodied the parties' agreements to accept it.

It is apparent that Babylonian law had reached a stage of considerable maturity. The extent of its relations with and influence on Hebrew law is uncertain; both may have developed from the common customary laws of the Near East. It is certain that it was widely applied and was studied for centuries after it was first collected. See also ASSYRIAN LAW; HEBREW LAW; HITTITE LAW.

G. Goetze, *The Laws of Eshnunna*; G. R. Driver and J. Miles, *The Babylonian Laws*, 2 vols.

Bach, Johann August (1721–58). German jurist and historical scholar, author of *Commentarius de divo Trajano sive de legibus Trajani* (1747) and a classic *Historia jurisprudentiae romanae* as well as work on classical subjects and editions of the writings of other jurists.

Bachelor. A word of several distinct applications, all implying junior standing and, usually, youth. Thus it is used of unmarried men, of those belonging to the lowest grade of knighthood (Knights bachelor), formerly of ecclesiastics of an inferior grade, such as young monks or recently appointed canons, and of those holding the lowest grade of degree of a university, a preliminary qualification, enabling them to proceed to the degree of master which alone formerly entitled them to teach. The word *baccalaureus* in this sense first appears in the university of Paris in the thirteenth century, under the system of degrees established by Pope Gregory IX.

Unmarried men have in many cases been penalized by law, usually for not taking responsibility for children. In Sparta and Athens this was so and at Rome the *Lex Julia et Papia Poppaea* penalized unmarried persons. Isolated instances of such penalties are known in the middle ages and modern tax legislation commonly differentiates between bachelors and married men.

Bachofen, Johann Jakob (1815–87). Jewish jurist and anthropologist, professor of the history of Roman law at Basel and a judge of the Basel criminal court. He wrote works on Roman civil law, but is famous for *Das Mutterrecht* (1861), the first attempt to produce a history of the family as a social institution and a major contribution to modern social anthropology.

Bachoff ab Echt, Reinhard (Bachovius Echtius) (1575–*c.* 1640). A humanist, professor at Heidelberg, author of *Notae et animadversiones ad disputationes Treutleri* (1617), *In institutionum libros Quattuor commentarii* (1628), *Commentarii in primam* *partem Pandectarum* (1630), *Tractatus de actionibus* (1657) and other works.

Back-bencher. A member of the United Kingdom House of Commons who holds no ministerial office and accordingly sits on a seat behind the front benches, which are reserved for Ministers and Parliamentary Secretaries on the government side and for chief members of the opposition on the other side.

Back-bond (or back-letter). A bond undertaking to indemnify a surety; in Scots law, a deed qualifying the terms of another, apparently absolute, deed, explaining that the latter is truly a disposition in security.

Backing a Warrant. Where a justice of one district or jurisdiction issued a warrant it was formerly necessary, for it to be executed elsewhere, that it be endorsed by a justice of the latter area. Now a warrant issued by a justice may be executed anywhere in England and Wales without backing. Corresponding provisions apply in Scotland. But a warrant issued in England must be backed for execution in Scotland, Northern Ireland, or the Channel Islands, by a magistrate having jurisdiction there, and conversely. The backing is in effect an endorsement authorizing the execution of a foreign warrant.

Bacon, Francis (Baron Verulam and Viscount St. Albans) (1561–1626). Son of Lord Keeper Sir Nicholas Bacon, was called to the Bar in 1582 and elected to Parliament in 1584. In the House he soon came to the fore though he early offended Queen Elizabeth by urging greater tolerance in religion and this lost him the chance of becoming Solicitor-General. He played a considerable part in investigating Essex's revolt in 1601 and in securing his conviction for treason.

On the accession of James I Bacon was knighted, and dedicated to the King a paper on the mode of carrying through the union between Scotland and England, and another on church reform. He was one of the commissioners appointed by the King to discuss with Scottish commissioners the question of union (1604), but the project made no headway in Parliament.

In 1607 he became Solicitor-General and more an adviser to the King. In 1613 he became Attorney-General and shortly thereafter came into collision with Coke, whose transfer he secured from the Common Pleas to the King's Bench. Their dispute came to a head over the proceedings against a clergyman, Peacham, for alleged treason, and continued as a dispute between common law and chancery for superiority. Finally in 1616 Coke was dismissed for refusing to accept that it was the office of the Crown to settle questions of jurisdiction between courts.

Bacon was also concerned in the prosecution of Sir Walter Raleigh and others. In 1617 he succeeded Ellesmere as Lord Keeper and in 1618 became Lord Chancellor and Baron Verulam. As chancellor he rapidly cleared off arrears of cases and his justice was exemplary. In 1621 he became Viscount St. Albans, but in the same year he was charged with bribery. He had probably accepted gifts from suitors, which was common at the time, though he had probably not been influenced thereby. The great seal was taken from him, he was fined £40,000, imprisoned during the King's pleasure (in fact, briefly) and banned from court. He was never given the full pardon he sought.

All his life he was a prolific writer, of literary works, notably the *Essays* (1597, later enlarged), philosophical works, notably the *Advancement of Learning* (1605) and *Novum Organum* (1620), historical works, especially the *History of Henry the Seventh* (1622), and legal works. The last include *De Augmentis* and *Maxims of the Law* and the *Use of the Law*, published under the common title of *The Elements of the Common Lawes of England* (1630) and *Reading on the Statute of Uses* (1642).

J. Spedding, *Life and Letters of Lord Bacon* (1861); *Life and Times of Francis Bacon* (1878); J. Nichol, *Francis Bacon, his Life and Philosophy* (1888–9).

Bacon, Sir James (1798–1895). Called to the Bar in 1827 he latterly limited himself to chancery business. He became Chief Judge in Bankruptcy under the Act of 1869, and held the office till 1883, concurrently, from 1870, with a post as Vice-Chancellor which office he held till 1886, being the last surviving holder of that office. On his retiral he became P.C. He was a sound judge though latterly he was out of touch with legal developments.

Bacon, Matthew (*fl.* 1730). Author of *The Compleat Arbitrator*, and a *New Abridgement of Law and Equity* (1736 and later editions), largely based on materials written by L. C. B. Gilbert and accordingly, under each heading, rather a dissertation than merely a collection of cases. It includes statutes also, and particularly when Gilbert's monographs are reproduced, the cases are presented in an intelligible pattern. There were numerous editions and it remained in use till the mid-nineteenth century.

Bacon, Nathaniel (1593–1660). A half-brother of Francis Bacon, barrister and master of requests to the Lord Protector. He wrote a *Historical and Political Discourse of the Laws and Government of England*, which professed to be collected from notes of Selden's, a constitutional history with a strong bias against monarchical and ecclesiastical pretensions (1647–51 and later editions).

Bacon, Sir Nicholas (1509–79). Was called to the Bar in 1533 and in 1546 became Attorney of the Court of Wards and Liveries. In 1558, possibly by Cecil's influence, he became Lord Keeper of the Great Seal and was knighted. In the following year he was authorized to exercise the jurisdiction of the Lord Chancellor and hear chancery cases. He developed in stature as a statesman, and strove to strengthen the position of the church in England. He has the reputation of a full and sound knowledge of law, and upheld the rights of the court of Chancery in writings. He was anxious to simplify the arrangement of the Statutes and to make them more generally accessible. One of his sons was Francis Bacon (q.v.).

Bacquet, Jean (*c.* 1520–97). French jurist, author of *Traité des droits du domaine royale* and other works.

Bad Parliament. The Parliament of 1377, which was dominated by John of Gaunt and which annulled reforms effected by the Good Parliament of the previous year.

Bagehot, Walter (1826–1877). Banker, political analyst, economist, and essayist, editor of *The Economist*, 1860–77, and author of *The English Constitution* (1867), a valuable descriptive study, distinguishing the real and formal centres of power in the British system of Government. His other writings included *Lombard Street* (1873), *Literary Studies* (1879), and *Economic Studies* (1880).

Bail. In civil proceedings, security that a person concerned in civil proceedings will obey the requirements of the court, resulting in immunity from arrest in respect of those proceedings. Bail was formerly important in every action. The word is also used of the persons who give the security. In Admiralty proceedings the owner of a ship or other property arrested may generally secure its release by giving bail to the extent of the value of the property.

In criminal proceedings a person admitted to bail is released from police custody to the custody of sureties who are bound to produce him to answer the charge against him or forfeit the sum fixed when bail is granted. A person arrested may be released on bail by the police, a magistrate, or judge, but have a discretion to refuse release if the accused is of bad character or the charge serious and in certain other circumstances. By the Bill of Rights, 1688, excessive bail may not be demanded. A person convicted may be admitted to bail pending his appeal being heard.

Bail-bond. A bond executed by a person arrested and other persons as sureties, providing for forfeiture of money if the person arrested does not satisfy any judgment against him.

Bail court. The common name of a court established in 1830, a sitting of the Court of Queen's Bench at Westminster, for the determination of points concerned with pleading and practice. Its chief function was the justification of special bail and it was also known as the Practice Court. It was abolished in 1854.

Bailee. A person to whom goods are entrusted for a particular purpose, e.g. custody, carriage or repair, without intention of transferring ownership to him.

Bailey* v. *Drexal (1922), 259 U.S. 20. A majority of the Supreme Court invalidated a child labour act, which imposed prohibitive taxes on the products of child labour in interstate commerce, as being a regulation of local labour conditions and accordingly a violation of states' rights under the Tenth Amendment.

Bailie. In Scotland, an official appointed to exercise a local delegated jurisdiction civil or criminal. Feudal bailies were appointed by lords of regality or of barony lands to exercise their jurisdiction. Municipal bailies, the counterparts of alderman, were elected by the members of town councils from among themselves and had judicial and administrative powers. In particular they sat in the burgh police courts until 1975.

Bailiff. A legal officer to whom some authority is entrusted. In England the term was formerly applied to the King's officers generally, and particularly to the chief officer of a hundred. In manors, the bailiff managed the property of the manor and superintended cultivation. The term still applies to the chief magistrates of certain towns, and the keepers of certain royal castles. In the Channel Islands the bailiff is the chief civil officer in each island, who presides in the states and in the royal court, and represents the Crown in all civil matters. A bailiff or sheriff's officer, is a person appointed by a high sheriff in England to summon juries, collect fines and execute writs and processes. They are usually paid by salary and give a bond for the due execution of their office and the accounting for fees received. In Scotland a bailiff is an inferior officer concerned with guarding, e.g. a water bailiff who watches for poachers.

Bailli. A mediaeval French salaried royal official, known from the twelfth century. He was the chief royal representative in a district, standing between a prévôt and the central royal court, administering royal justice, commanding the local armed forces and with oversight of revenues. He had important juridical responsibilities holding court at local assizes composed of royal officials and prominent citizens, and later of judicial officers and lawyers, who gave opinions on the local customary law to be applied. The court of the *bailli* had original jurisdiction over nobility and appellate jurisdiction over cases originally heard by prévôts and some seignorial courts and also jurisdiction over cases which concerned the royal rights and domains. In time as duties increased, other officials were appointed to assist the *bailli* and eventually they took over most of his powers. Thus by the thirteenth century lieutenants were appointed to serve under the *baillis* and by the sixteenth century they had completely superseded the *baillis* in judicial functions. By the later middle ages the *bailli* was usually a noble and his military and police functions survived but his judicial and revenue ones had disappeared. In the seventeenth century *intendants* took over the *baillis*' administrative responsibilities and they became mere figureheads.

Bailment. From Old French bailler, to deliver or put in the hands of, a delivery of personal chattels to a person under an express or implied contract, gratuitous or for reward, to deal with the property in a particular way. The English law has by and since *Coggs* v. *Bernard* (1703), 2 Ld. Raym. 909, been heavily influenced by Roman law. Bailment comprehends the deposit of goods to be kept gratuitously (*depositum*), mandate, to do something to or with the goods bailed (*mandatum*), loan of goods for temporary use (*commodatum*), pledge of goods in security (*pignus*), and letting on hire (*locatio conductio*). The last class includes the lodging of goods for safe keeping for reward (*locatio custodiae*), the letting of goods on hire for use (*locatio rei*), the hiring of services for or in relation to certain goods (*locatio operis faciendi*) and the hiring of the carriage of goods (*locatio operis mercium vehendarum*). Bailments may also be distinguished as gratuitous (comprising the first three classes) or for reward (the last two classes). To constitute a bailment there must be a transfer of actual or constructive possession of a specific chattel by its owner or possessor (bailor) to another (bailee), so that the latter may deal with it in the way required, thereafter returning it in its original or altered form. Among the most important matters between bailor and bailee is the degree of care and diligence required of the bailee in each kind of bailment. In every case the bailee is required to take the degree of care which may reasonably be looked for, having regard to all the circumstances.

G. W. Paton, *Bailment in the Common Law*.

Bairns' part. In Scots law, the fraction of a deceased's estate to which by law the bairns, or children, are entitled. See LEGITIM.

Balduinus. See BAUDOUIN, FRANÇOIS.

Balduinus, Jacobus (?–1235). Famous Italian jurist, pupil of Azo, Professor of Civil Law at Bologna, and author of the earliest treatises on procedure.

Baldus de Ubaldis, Petrus (1327–1406). Italian jurist, a pupil of Bartolus, a prominent commentator, who taught at Perugia and Pavia, and often acted in judicial and diplomatic capacities, author of *Commentaria super Decretalibus* (1547), *Consilia* (1548), *Commentaria in Libros Feudorum* and *Commentaria in Digestum* (1616). His brothers Angelus (1328–1407) and Petrus (1335–1406) were also eminent jurists.

Baldwin, Simeon Eben (1840–1927). Was admitted to the Bar in 1863 but joined the Yale Law Faculty in 1869 and did much to revivify that school. He later served 17 years as associate justice and chief justice of the Connecticut Supreme Court. He also took part in politics and was governor of Connecticut for two terms. He wrote extensively, was one of the founders of the American Bar Association, and a member of numerous learned societies and took a leading part in the movement to raise the standard of legal education and admission to the Bar in the U.S.

Balfour, Sir James, of Pittendreich (?1525–83). He may have studied divinity and law, but seems not to have been a member of the Scottish Bar. He participated in the assassination of Cardinal Beaton in 1547 and served as a French galley-slave along with John Knox.

He became Official of Lothian about 1554, in 1561 an extraordinary and in 1563 an ordinary Lord of Session and in 1564 one of the four Commissaries also. By 1565 he was a Privy Councillor and seems to have been a chief adviser of Mary, Queen of Scots. He later (1566) became Clerk-Register, was knighted and made one of the commissioners for revising, correcting and publishing the ancient laws and statutes of Scotland. He counselled and probably participated in the murder of Darnley in 1567, but nevertheless managed to become Lord President of the Court of Session. In 1573 he had to take refuge in France but returned in 1580.

He changed his political and religious allegiance several times and played a part, usually discreditable, in all the disturbances of his time in Scotland; he was reputed the most corrupt man of his age. He was nevertheless a very able man. To him is attributed *Practicks: or a System of the More Ancient Law of Scotland* (*c.* 1574–83, but first published in 1754) which represents the manuscript collection of a lawyer over a very long period, and remains a valuable repertory of ancient statutes and decisions, for which, in many cases, it is the sole authority.

Balfour, Sir James, of Denmiln and Kinnaird (1600–57). Devoted himself to Scottish history and antiquities. He was friendly with all contemporary Scottish scholars in that field, and acquainted with Dodsworth and Dugdale, and contributed an account of the Scottish religious houses to the latter's *Monasticon Anglicanum.* In 1630 he was knighted and made Lyon King of Arms, but was deprived of the latter office by Cromwell in 1654. He left numerous manuscripts, chiefly on Scottish heraldry, and was a diligent collector of cartularies and other documents of the Scottish bishoprics and monastic houses.

Balfour, John Blair, Lord Kinross (1837–1905). Called to the Scottish Bar in 1861, he rose rapidly, entered Parliament, became Solicitor-General for Scotland in 1880 and Lord Advocate 1881–85, in 1886, and 1892–95, and was also Dean of the Faculty of Advocates, 1885–86 and 1889–92. He was the last of the old style of Lord Advocates, who were in substance ministers for Scotland, before the office of Secretary for Scotland was created in 1885. In 1899 he became Lord President of the Court of Session and in 1902 Baron Kinross of Glasclune. His health was poor latterly, and though his judicial reputation is good he did not show the brilliance he had done while at the Bar.

Balfour Declaration. A statement contained in the Report of the Inter-Imperial Relations Committee of the Imperial Conference of 1926, so called from having been drafted by a committee under the chairmanship of A. J. Balfour, describing the then position and mutual relations of the United Kingdom and its Dominions. It stated: 'They are autonomous Communities within the British Empire, equal in status, in no way subordinate one to another in any aspect of their domestic or external affairs, though united by a common allegiance to the Crown, and freely associated as members of the British Commonwealth of Nations.' It marked the full recognition of the autonomy of what were then called the Dominions and since then further developments have taken place, notable features of which have been that many colonies and dependencies granted independence have been willing to remain in the Commonwealth, that certain countries have left the Commonwealth and the recognition of republics within the Commonwealth.

The term is also used of a letter by Balfour to Lord Rothschild in 1917 pledging British support for the establishment in Palestine of a national home for the Jewish people. The declaration greatly raised the hopes of Zionists. This was endorsed by the principal Allied powers, accepted by the Conference of San Remo in 1920, and became an instrument of British and international policy. It was included in the British mandate over Palestine approved by the League of Nations in 1922 and led ultimately to the creation of the State of Israel.

Ball, John Thomas (1815–98). Called to the Irish Bar in 1840, he became Solicitor-General, 1868, and later Attorney-General for Ireland, 1868 and 1874, and, in 1875 Lord Chancellor of Ireland. He retired in 1880. He wrote an *Historical Review of Legislative Systems operative in Ireland*, 1888.

Balladore-Pallieri, Giorgio (1905–). Italian jurist, judge from 1960 and President from 1974 of the European Court of Human Rights, author of *Diritto internazionale pubblico* (1937), *Diritto costituzionale* and other works.

Ballot. A small ball used for secret voting and hence the method of secret voting, originally by depositing small balls in an urn or box, but now usually by marking a prepared piece of paper.

Ballot Act, 1872. Substituted voting by secret ballot for open voting at British Parliamentary elections. Demand for the ballot had been made by some eighteenth century and many nineteenth century reformers and it was one of the six points of the Peoples' Charter (1836). The idea gained increasing support in Parliament thereafter but it was not adopted as government policy until after inquiries into corruption at the general election of 1868, which caused Gladstone to favour the change. This change was at least as important as widening the franchise as the right to vote could not be effective so long as workmen and tenants had to vote under the eyes of their employers or landlords.

Balsamon, Theodore (twelfth century). A Byzantine jurist, Patriarch of Antioch, author of learned commentaries on the canon law, a collection of ecclesiastical constitutions and a commentary on Photius's *Nomocanon.*

Balthasar, Josef Anton Felix (1737–1810). Swiss jurist, author of a leading work on Swiss ecclesiastical law, *De Helvetiorum Juribus circa Sacra* (1768).

Baltic and International Maritime Conference. An international organization founded in 1905 to end the cut-throat competition among shipowners sailing to the Baltic and White seas.

Baltic Mercantile and Shipping Exchange. An organization handling much of the world's tramp ship chartering, so named because it originated in the Baltic, a seventeenth century London coffee house and tavern. It is not now confined to trade with the Baltic but deals with worldwide trade.

Baluze, Etienne (1630–1718). French historian and jurist, librarian to Colbert, author of the *De concordia sive de libertatibus Ecclesiae gallicanae* (1663), *Capitularia regum Francorum* (1677), and other works.

Bambergische Halsgerichtsordnung (or *Bambergensis*). A German work of the early sixteenth century, primarily a statute dealing with criminal procedure, with substantive law interpolated, but partly a popular textbook. It was composed by Johann von Schwarzenberg (1463–1528) president of the High Court of Bamberg. It shows Roman law influence, was far superior to previous works on the same subject, was accepted as authoritative elsewhere and formed the foundation of the Carolina (q.v.) and thus the basis of German criminal law for three centuries.

Ban. A word used in the laws of the Franks for a proclamation, the fine incurred for disobeying such a proclamation, and the district over which the proclamation was issued. The *bannum dominicum* was employed by all feudal lords against offenders, usually taking the form of an order to make amends for wrongdoing.

Later, in France the word was used for the summons calling out the feudal host, and hence of the vassals summoned, while the sub-vassals summoned by them were known as the *arrière-ban.* They were last summoned in 1758.

The word was also used in a primitive sense in the ban of the Holy Roman Empire, and in France of the penalty of exile. The execution of the ban of the empire was usually entrusted by the emperor to a prince or noble. At first the sentence was an act of the Emperor himself but later it was entrusted to the Aulic Council (*Reichshofrat*) and to the Imperial Chamber (*Reichskammergericht*) but these courts lost their power in 1711. In modern English it is sometimes used of excommunication and implies exclusion and moral reprobation. In its original sense of a proclamation it survives in the phrase 'banns of marriage' (q.v.).

Banc, in (or *in banco*). Sittings of one of the pre-1875 courts of common law (Queen's Bench, Common Pleas or Exchequer) as a full court, as distinct from the sittings of single judges at *nisi prius* (q.v.) or on circuit (q.v.) for the purpose of determining questions of law. Originally they were confined to term time, but from 1838 they could be held at the time of holding sittings for *nisi prius* in London and Middlesex and after 1845 could be held at any time save during the long vacation. Three judges usually sat when sitting in banc. The Judicature Act, 1873, transferred much of the business *in banc* to Divisional Courts (q.v.), and the Judicature Act, 1876 transferred some of it to single judges.

Bancus publicus* and *bancus superior. The court previously and again, after 1660, known as the King's or Queen's Bench, was during the Protectorate known as the Public Bench or Upper Bench.

Banishment. Exile or exclusion from the realm. It might arise from quitting the realm following on abjuration (q.v.) or be imposed by statute. Thus the Roman Catholic Relief Act, 1829 provided for the banishment of Jesuits. It differs from transportation (q.v.) which was a punishment and for a period only, and from deportation (q.v.) which applies to non-patrials.

Bank (or banker). A person or company which carries on the business of banking. Banks can be distinguished into the Bank of England, ordinary or commercial banks, concerned with receiving money on deposit, lending it to customers, paying money in response to cheques drawn, collecting cheques for customers, and general financial advice and services. Merchant banks are concerned more with promoting and financing companies, arranging amalgamations and reconstructions and with major commercial and industrial enterprises than with individual accounts. Trustee Saving banks are institutions established for the receipt of money from depositors without benefit to the trustees or organizers, and the National (formerly Post Office) Savings Bank, operates similarly at post offices. The category of bank also extends to certain finance houses, largely concerned with financing hire-purchase transactions.

Bank holiday. Days on which, by statute, banks are closed. The days are, in England, Easter Monday, the last Monday in May, the last Monday in August, and Boxing Day, if not a Sunday and 27th December if the 25th or 26th is a Sunday; in Scotland, New Year's Day, 2nd January, Good Friday, the first Monday in May, the first Monday in August and Christmas Day. If Christmas and New Year fall on a Sunday the following day is the holiday. If 2nd January be a Sunday, 3rd January is the holiday. In Northern Ireland the holidays are 17th March, or if a Sunday, 18th March, Easter Monday, the last Monday in May, the last Monday in August, Boxing Day, if not a Sunday, and 27th December, if the 25th or 26th be a Sunday.

Bank notes. Promissory notes for sums of £1, £5, £10 and some larger sums, payable to bearer, issued by a banker and payable on demand, without interest. They circulate as readily as coins. In England notes may be issued by the Bank of England only and were formerly payable in gold, but the Scottish banks have power to issue their own notes to statutorily limited extents. The former are legal tender, the latter are not.

Banker's draft. An order to pay money on demand drawn by a branch office of a bank on its head office or vice versa, or by one bank on another. The former two kinds are not, but the last is, a cheque.

Banking. The business of receiving money from customers on deposit, lending them money, paying money on account of cheques drawn by the customer on his banker, and collecting bills, cheques and similar documents handed in by the customer for the credit of his account. Bankers also perform many services for customers such as undertaking safe custody of valuables, opening credits to facilitate international trade, acting as executor or trustee, and advising and assisting customers on financial matters.

The law applicable to banking and banker's transactions is in large measure the law generally applicable to the contracts of loan, surety, securities and similar transactions. In certain particulars, however, particularly in relation to cheques (q.v.) bankers are given special protection.

J. Paget, *Law of Banking*.

Bankruptcy. The proceeding whereby the state, acting through an officer appointed for the purpose, takes possession of the property of a debtor so that it may be realized and, subject to certain preferable claims and priorities, distributed rateably among his creditors. It has the purposes of protecting the debtor from undue pressure by certain creditors, of preserving fairness among the creditors, and finally, of discharging the debtor from his liabilities and enabling him to start afresh. Though it is not now a crime, becoming bankrupt involves modification of status, resulting in certain civil disqualifications and quasi-penal consequences. It is regulated by separate bodies of legislation in each country of the United Kingdom. The law is entirely statutory but it has long been a subject of legislation, the first English Act dating from 1542. Early legislation was mainly directed against fraudulent debtors.

In England and Wales a person may be adjudged bankrupt, if he has committed an 'act of bankruptcy', on the petition of a creditor, or of the debtor himself. The court then makes a receiving order placing the debtor's estate under the control of the court through its officer, the official receiver. The latter calls a meeting of creditors and obtains from the debtor a statement of affairs, the debtor is examined publicly as to the reasons for his insolvency and an order is made adjudicating him bankrupt. On adjudication the official receiver becomes trustee of the bankrupt's property, unless the creditors appoint another trustee. A committee of inspection may also be appointed by the creditors from among themselves. The official receiver or trustee must take possession of and realize all the bankrupt's assets. Certain debts must be paid in full in priority to other claims, and certain others are postponed to all ordinary creditors. Creditors must prove their debts and the official receiver or trustee must admit or reject every proof lodged, or demand further evidence. Secured creditors may surrender their securities and prove for the full debt, realize the security and prove for the balance, or assess the value of the security and prove for the balance. The realized assets are distributed by way of dividends to the creditors. Any ultimate surplus belongs to the debtor.

In Scotland the term bankruptcy does not have a fixed technical meaning. It may connote simple insolvency, or notour bankruptcy, or the process of sequestration (q.v.). A person who is notour bankrupt, i.e. notoriously insolvent, a state evidenced by statutory indicia, may be sequestrated on his own or his creditors' petition, in which case a person is appointed trustee in bankruptcy and commissioners appointed to assist him. He realizes the bankrupt's assets, adjudicates on claims and distributes the realized estate to creditors in dividends.

The bankrupt may at any time, on application, be discharged by the Court, which in general releases him from all debts provable in the bankruptcy. The application is heard in open court and the courts' discretion to grant discharge is restricted and qualified by statute. Discharge may be conditional or unconditional, or suspended, or conditional and suspended.

A bankrupt is disqualified from being elected to or sitting in either House of Parliament, from acting as a J.P. or member of a local authority and from certain other offices.

Independently of bankruptcy, a debtor may make an arrangement with his creditors or some of them for his release from their claims, usually by making a composition with them, or assigning his property to a trustee for them.

In the U.S. bankruptcy is divided between federal and state governments. The first federal Act was passed in 1800 and since then there has been constant expansion of federal powers in this sphere. See also CRIMINAL BANKRUPTCY.

R. V. Williams, *Law and Practice of Bankruptcy*; H. Goudy, *Law of Bankruptcy in Scotland.*

Bankton Lord. See McDouall, Andrew.

Bannelier, Jean (1683–1766). French jurist, author of an *Introduction à l'etude du Digeste* (1750) and *Observations sur la coutume de Bourgogne* (1766).

Banns. Public legal notice of intended marriage. The practice was early adopted in the Christian church of giving advance notice. The practice was known in France in the ninth century and in England was ordered by a synod of Westminster in 1200. The Lateran Council of 1215 made the publication of banns compulsory throughout Christendom. In England and Scotland by church law and statute banns long had to be published in church as a normal preliminary to marriage, but alternatives came to be allowed, normally publication by notice by a registrar. The matter is regulated by statute. In 1977 banns were abolished in Scotland, notice to the registrar being substituted in all cases. In the U.S. the practice of banns was general in the colonial period but is now much restricted.

Bar. (1) A partition across a court or other place. In the Houses of Lords and of Commons the Bar is the boundary of the House and persons not members of the relevant House who are summoned thereto, e.g. to be admonished, or permitted to address the House, appear at the Bar to do so. When arguing Appeals before the House of Lords, sitting in the Chamber, counsel stood at the Bar.

In ordinary English courts the Bar is an imaginary barrier separating the bench and the front row of counsel's seats from the rest of the court. Only Queen's Counsel, barristers holding patents of precedence, solicitors (who are officers of the court) and parties appearing in person, are allowed within the Bar; utter (i.e. outer) barristers and the public remain without the Bar. Hence to be 'called to the Bar' signifies being admitted to practise as an utter barrister before the court. In Scotland both Queen's Counsel and advocates sit in the front row of seats at the Bar, only the Lord Advocate and Solicitor General sitting within the bar, in the well of the court.

The word 'bar' appears also in such phrases as the 'case at bar', meaning the 'case in court'.

(2) In countries of the United Kingdom the Bar is the collective term in each country for counsel, or persons who have been 'called to the Bar', i.e. admitted by one of the Inns of Court, or the Inn of Court of Northern Ireland, as a barrister, or by the Faculty of Advocates in Scotland as an advocate. It accordingly excludes persons admitted as solicitors. The name is derived from the facts that in the mediaeval Inns of Court the utter barristers, when arguing at moots, sat on forms called 'The Barr', or that in courts barristers stand behind a bar or barrier in court when they plead. Members of the Bar are subject to the professional discipline of their Inn, or the Faculty, as the case may be. In this sense the Bar consists of two grades, Queen's Counsel (q.v.) and barristers (or junior counsel or juniors) (q.v.) or, in Scotland, advocates (q.v.).

In countries, such as states of the U.S., where the legal profession is undivided, 'the Bar' is commonly used as a synonym for the whole legal profession. Thus in the U.S. there are State Bar Associations and the American Bar Association (q.v.).

Bar Committee. A body of 48 barristers elected in 1883 to collect and express the opinions of the members of the Bar on matters affecting the profession. It laid the foundations of a code of professional etiquette. It was replaced by the Bar Council in 1894.

Bar Council. The common name for the General Council of the Bar of England and Wales which in 1894 replaced the Bar Committee. It comprised the Law Officers, the Chairman and Vice-chairman, 48 elected practising barristers, additional members appointed on account of their position or to represent any section of the Bar not adequately represented, and persons co-opted *honoris causa*. It was supported by contributions from the Inns of Court and subscriptions. It issued rulings and gave guidance on professional etiquette and conduct but had no disciplinary power. It was concerned generally with the independence, honour and integrity of the Bar and the improvement of the administration of justice. In 1974 it, and the Senate of the Four Inns of Court, was replaced by the Senate of the Inns of Court and the Bar (q.v.) and a new Bar Council came into existence exercising through its Bar Committee, exclusive jurisdiction in matters of professional conduct and etiquette, falling short of disciplinary proceedings.

Bar, Plea in. A plea seeking wholly to bar or defeat an action or prosecution. At common law a plea in Bar challenged the right of action altogether and comprised pleas by way of traverse which denied all or the essential parts of the averments of fact in the declaration, and also pleas by confession and avoidance which admitted the facts averred to be true but alleged new facts which repelled their legal effect. In equity a plea in bar was resorted to when there was no defect in the face of the plaintiff's bill, it sought to displace the equity in favour of the plaintiff by reducing the case to a particular point. In Scottish criminal procedure a plea in bar of trial is a plea, such as of present insanity, or absence of jurisdiction which prevents the court proceeding to trial at all.

Bar, trial at. The trial of a cause or indictment before a full court of three or more judges instead of the normal single judge. It was the original mode of civil trial. Trial at bar in civil cases is nearly obsolete. In criminal cases it was competent though rare until 1971. The court has a discretion to grant or refuse it, save that where the Crown is interested the Attorney-General may demand it as of right. Important trials at bar have been the Tichborne case (*R.* v. *Castro* (1874), L.R. 9 Q.B. 350); *A. G.* v. *Bradlaugh* (1885), 14 Q.B.D. 667; and *R.* v. *Jameson*, [1896] 2 Q.B. 425.

Bar of the House of Commons. The place, marked by a movable bar, beyond which non-members may not enter the Chamber when the House is sitting. When an offender, not being a member, is to be reprimanded or admonished he is brought to the bar by the Serjeant-at-Arms and there reprimanded by the Speaker in name and by authority of the House. Since 1772 offenders are no longer required to kneel at the Bar. When a witness is examined by the House or a committee of the whole House he attends at the Bar and is there questioned.

Bar of the House of Lords. The place, marked by a movable bar, beyond which non-members may not enter the Chamber when the House is sitting. Witnesses who are to be examined by the House or a committee of the whole House are sworn and examined at the Bar. When appeals to the House of Lords were heard in the Lords' Chamber (prior to 1948) counsel stood at the bar in a little pen which had little room for them or their books and addressed their Lordships from there. Though argument is now heard in a committee room the Lords' decisions are voted on and their speeches handed down in the Chamber to counsel at the Bar.

Barbarian laws. See GERMANIC LAW.

Barbatia, Andreas (*c.* 1400–69). Canonist, author of commentaries on parts of canon law, *Consilia*, and annotations on some of the writings of Baldus.

Barbeyrac, Jean (1674–1744). French jurist, professor at Lausanne and later at Groningen and distinguished for his translations of and commentaries on Noodt, Grotius, and Puffendorf. He reduced the principles of international law to those of the law of nature and opposed many of the views of Grotius. Of his original works *Le Traité du Jeu* (1709) is the most notable.

Barbosa, Augustin (1590–1649). Portuguese jurist and humanist, author of *Tractatus de clausulis usufrequentioribus* (1631), *De Officio et Potestate episcopi*, *De Officio et potestate parochi*, and many other works.

Barbosa Pietro (?–1606). Portuguese jurist and judge, author of commentaries on titles of the Digest.

Barclay, William (1546 or 47–1608). Studied law in France at Paris and Bourges under Cujacius, Donellus, and Contius and became professor of civil law at Pontamousson in Lorraine about 1580. In 1600 he published *De Regno et Regali Potestate* refuting the arguments of George Buchanan in his *De Jure Regni apud Scotos*, of Junius Brutus in the *Vindiciae contra Tyrannos* and of Jean Boucher in his *De Justa Henrici III Abdicatione et Francorum Regno*. In 1604 he became professor of civil law at Angers and later published the *In Titulos Pandectarum de Rebus Creditis et de Jure jurando*, dedicated to James I of England, and a treatise *De Potestae Papae* (1609) which made a great impression throughout Europe.

Bare trust (or simple trust). Where property is vested in one person upon trust for another and the nature of the trust is not prescribed by the settler but left to the construction of law. A bare trustee is a mere passive depositary and can neither take any part of the profits nor exercise any control over the corpus of the trust property save at the instance of the beneficiary.

Barebones' Parliament. Also known as the Little Parliament or Nominated Parliament, an assembly of nominees summoned by Cromwell in 1653, and named from a Puritan member Praise-God Barebone. It assumed the title of the Parliament of the Commonwealth of England and was ridiculed for its ill-judged reforms.

Bargain. An agreement or arrangement, particularly one reached by haggling or after argument. The word is usually found in particular phrases or contexts such as 'unconscionable bargain', or 'collective bargaining', or 'bargains between trustee and beneficiary'.

Bargain and sale. A former mode of conveying land in England, intended to evade publicity. Based on the equitable principle that a contract to sell a legal estate raised a use in favour of the purchaser immediately on payment of the price, which was virtually as good as legal ownership, the vendor bargained with and sold the land as a fee simple estate and received payment, leaving himself impliedly seised to the use of the purchaser. The Statute of Uses (1535) however, provided that where A was seised to the use of B, B should have a legal estate. As this would have allowed a bargain and sale to be a secret method of conveying legal estates, the Statute of Enrolments enacted that no estate of inheritance or freehold should pass unless the bargain and sale was made by deed and enrolled in one of the King's Courts of Record. But a bargain and sale of a leasehold might be made privately so that lawyers adopted the method of conveyance by lease and release (q.v.).

Barmote Courts of High Peak. The Great Barmote and Small Barmote Courts of High Peak are ancient courts, with jurisdiction as to lead mining rights and civil pleas relative thereto in the King's field of High Peak, Derbyshire and other parts of the hundred of High Peak, where the Crown has rights to mineral duties. Two Great Barmote courts are to be held in the year at Moniash, and Small Barmote courts as required at places to be fixed by the Steward. The courts are courts of record and are regulated by statute. The Small courts have jurisdiction in actions of title, trespass and debt. The judge is the steward, appointed by the Crown under seal of the Duchy of Lancaster, or his deputy. The steward and grand jury may make new rules and customs for regulating mining in the district. The officers of the court are the barmaster and deputy barmasters.

Barmote Courts of Wirksworth and adjacent liberties. Courts similar to the Barmote Courts of High Peak (q.v.). They are courts of record regulated by statute and held by the steward. The steward for the Wirksworth Barmote Court is appointed by the Queen under the seal of the Duchy of Lancaster and the stewards of the other five manors or liberties in private ownership by the persons entitled to the first estate of freehold in the mineral duties.

Barnard's Inn. One of the Inns of Chancery (q.v.).

Baron, Éguinaire-Francois (1495–1550). A leader of the French school of humanist jurists, teacher at Angers, Poitiers, and Bourges, author of numerous works on Roman law, notably *Pandectarum juris civilis Oeconomia* (1555) and *Commentaria in quatuor Institutionum libros* (1574).

Baron. In Norman French, a man, probably a free man or a King's man holding land directly from the Crown, and originally not a title of dignity. The Normans applied the term to holders of Anglo-Saxon thanages and latterly the term connoted a man of distinction and importance. Later the greater barons usually were given other and superior titles. In England hereditary barons from the fourteenth century received individual summons to Parliament. The style of baron as a title of dignity seems to have been first used in 1387 and such creations became common under Henry VI. It is now the lowest (fifth) grade of the British peerage, those addressed as 'The Lord X'. Since the fifteenth century a baron has been created by letters patent and has been a hereditary right. Since 1958 a baron may also be created for life only. The barons of the Cinque Ports were originally the whole body of their freemen but the style was later restricted to the mayors, jurats and M.P.s for the Cinque Ports. They claimed the right to hold a canopy over the sovereign at his coronation. The title was also used for a *Freiherr* or freelord of the Holy Roman Empire. Baronies were granted by imperial letter in Germany and France from the sixteenth century, but the title is now given by courtesy only.

From about the twelfth century the judges of the Court of Exchequer were the Lord Chief Baron of the Exchequer and other Barons of the Exchequer.

See BARON, CURSITOR; BARON OF EXCHEQUER.

Baron court. An inferior franchise court in mediaeval Scotland. A baron had a right to hold a court for his tenants only if he had an express royal grant of the jurisdictional right. The court was usually held at the *caput*, or principal seat of the baron, by the baron himself or by one or more bailies, the judges being originally those who owed suit of court, but from the late sixteenth century the baron or his bailie was sole judge. It had civil jurisdiction in matters of debt, possession, and disturbances of the peace, criminal jurisdiction in cases of theft and slaughter where the accused had been caught in the act, and a quasi-administrative jurisdiction over neighbourly relations between tenants. Appeal could be taken to the sheriff or regality court. In early feudal times the baron court was important as being local, held frequently, and held before suitors familiar with the facts. Baron courts declined from the fifteenth century and the civil and criminal work was attracted to royal courts. These courts have now long been obsolete.

Baron, cursitor. When the Barons of Exchequer came to be chosen from lawyers rather than Exchequer officials it became necessary to have a person with the technical knowledge of the course of Exchequer business formerly possessed by the Barons. This person was the cursitor Baron, known from the fourteenth century. The office probably grew in importance and became a separate office as the status of the other Barons rose. He had to keep the other Barons informed as the course of the Exchequer, and to audit the accounts of the sheriffs. The latter function was transferred in 1834 to the Commissioners for auditing the public accounts and the office was abolished in 1857.

Baron of Exchequer. Originally officers of the *Curia Regis*, theoretically composed of the King's barons, to whom was entrusted supervision of the royal revenues and the decision of cases relative thereto. This group developed into a distinct court and the title Baron was borne down to 1881 when the court and its successor, the Exchequer Division of the High Court, was abolished, by the Chief Baron and ordinary judges (Barons) of the Court of Exchequer. The abbreviations were, e.g. Pollock, C. B. and Parke, B. These titles were also given to judges of the Courts of Exchequer in Scotland and Ireland.

For list of Chief Barons see Appendix.

Baronetcy. An hereditary dignity (not an order of knighthood nor a degree of the peerage) instituted by James I in 1611, to be granted to the persons who contributed to the expense of the Plantation of Ulster, and again in 1625 to those who contributed to the Plantation of Nova Scotia. There are five classes of baronets, viz. of England, created 1611–1707; of Ireland, 1618–1801; of Nova Scotia, 1625–1707; of Great Britain, 1707–1801; and of the United Kingdom, since 1801. The Crown may create baronetcies to an unlimited number; creation is by letters patent under the Great Seal. It is an incorporeal heritable dignity and descends to the heirs of the body. The privileges of a baronet are set out in the letters patent creating the dignity; they are the title Sir X. Y., Bart., and precedence above all knights except Knights of the Garter. Among themselves baronets rank by date of creation. Claims to a baronetcy are made by petition to the Queen in Council and heard by a committee of the Privy Council. There is an official roll of baronets.

Barony. In mediaeval Scotland, the holding of a tenant-in-chief who held of the Crown 'in free barony' and also the tract of land over which the right extended. A barony was a unit for tenure, though including discontiguous lands, and always had a *caput*, and normally defined rights of jurisdiction were conferred on a baron, exercisable in his baron court, personally or by his bailie, and inseparable from the barony. The civil jurisdiction was concerned with debt, possession and assythment, the criminal with theft and slaughter. Appeal might be taken to the sheriff court. The lands could be sold but the jurisdiction and the *caput* retained.

By the thirteenth century a barony was an important administrative unit within the sheriffdom, and in the thirteenth and fourteenth centuries many ordinary fees and thanages were granted new charters converting them into baronies. By that time also the extent of jurisdictional rights regarded as normally attaching to a grant of barony were regarded as fairly settled. The jurisdiction was cut down in 1747 and is now long disused.

English law recognized no tenure by barony; apart from the baron paying a higher relief (q.v.), there was nothing to distinguish a baron's tenure from tenure by knight-service.

Barratry. The common law offence in England of exciting and maintaining suits and quarrels in the courts or in the country, as by disturbance of the peace or spreading false reports and rumours which provoke discord between neighbours. This crime was abolished, as obsolete, in 1967. The word is also used of a fraud committed by the master or seamen of a ship committed on the owner or insurers, which is also criminal, but indicted as fraud, piracy or otherwise.

Barrier Act. An act of the General Assembly of the Church of Scotland of 1697, which provides that a proposal for a new act or the repeal of an old one, if sent up to the Assembly, must be remitted by it to the several presbyteries for their consideration and their opinion to be returned to the next General Assembly.

Barrier Treaty (1709). Made between Great Britain and the United Netherlands by which the latter agreed to guarantee the Protestant succession in Britain to the House of Hanover and the former undertook to secure a barrier for the Dutch in the form of a number of fortified towns.

Barrington, Daines (1727–1800). Was called to the Bar, and became successively Marshal of the Court of Admiralty, a justice in North Wales, and Recorder of Bristol. In 1785 he resigned nearly all his offices to devote himself to antiquarian interests. His main work is the *Observations on the Statutes* (1766), reviewing many of the chief Statutes from Magna Carta to the seventeenth century and illustrating them with legal, historical, and other notes. He appreciated the value of the early statutes for legal and general history and his book is still of interest. He wrote numerous other works on all sorts of subjects, revealing a wide ranging interest rather than deep learning.

Barraud, Jacques (1555–1626). French jurist, at one time compared to Cujas, author of *Coustumes du comté et pays de Poictou* (1625).

Barrister. The term used in England and Wales since the fifteenth century and in countries wish legal systems modelled on that of England for a person who has been called to the Bar, i.e. admitted to practise in the superior courts and entitled to represent a party to a cause in court. The term 'barrister-at-law' though found in some statutes is unknown in the Inns of Court and the modern preference is for the simple term 'barrister'. In

England a barrister may not simultaneously be a solicitor nor belong to certain other professions, nor be engaged in trade. Collectively Queen's Counsel (q.v.) and barristers are called 'counsel'.

In England the power to call to the Bar has long been exercised by the four Inns of Court (q.v.). A person desiring to be called to the Bar must satisfy certain educational standards, be accepted as a proper person to be called, keep a specified number of terms by dining in Hall a stated number of times, pass the examinations of the Council of Legal Education, and undertake a prescribed period of pupillage in a practising barrister's chambers or attend a Practical Training Course. After call he may join a group of barristers practising in a set of chambers in London or a major provincial city. He may also join one of the circuits (q.v.) and attend at the towns on the circuit. No form of partnership is allowed between counsel. The Benchers of the Inn to which a barrister belongs may censure, suspend or even, after a hearing by the Senate of the Four Inns of Court disbar him for misconduct. The courts may punish him for contempt of court. The work within the ordinary scope of a barrister's practice is advising on questions of law, drafting pleadings, conveyances and other legal documents, and advocacy in court. Barristers have exclusive rights of audience in the superior courts and, in inferior courts, concurrent rights of audience with solicitors. When appearing in court counsel wear wig, gown, and bands.

A barrister may not normally see or advise a client or accept a brief to appear on behalf of a client without the intervention of a solicitor. There are limited exceptions to this principle. It is contrary to etiquette for a barrister to do legal work of kinds ordinarily done by a solicitor, such as routine conveyancing.

A barrister owes the duty to his client to raise every matter, advance every argument and ask every question which may assist his client's case. He owes a duty to the court not to mislead it, not to invite it to condone or assist any illegality, and not to conceal from it any authority (q.v.) unfavourable to his case. He must accept any brief in the courts in which he practises if a proper fee is offered, and may not discriminate between clients on racial or ethnic grounds.

Counsel's fees are not legally recoverable. They must be fees for each piece of work done and not a salary or lump sum for work over a period. Brief fees are arranged between counsel's clerk and the solicitor.

All superior judicial offices in England and Wales are reserved to barristers and most lower ones are also filled by them. Certain other offices have for long generally been filled by barristers. Conversely certain offices imply prohibition on practice at the Bar.

After attaining substantial experience and practice a barrister may be appointed by the Queen on the recommendation of the Lord Chancellor to be a Queen's Counsel (q.v.).

The Irish Bar, comprising members of King's Inns, Dublin was divided in 1921 into the Irish Bar, which continues to centre on King's Inns and the Four Courts, and the Northern Ireland Bar, comprising members of the Inn of Court of Northern Ireland, Belfast. The Scottish Bar is the Faculty of Advocates; its members are designated advocates (q.v.) and not barristers. A person may, by qualifying in the legal systems of each of the countries, become a member of the Bars of any two or more of the national Bars.

In Commonwealth countries with legal systems founded on English law but which have departed from the distinction between counsel and solicitors, legal practitioners are sometimes designated 'barristers and solicitors'.

Barrister, Inner and Utter or Outer. In the mediaeval Inns of Court the younger members of the Inn who argued at the moots (q.v.) before the benches were called masters of the Utter or Outer Bar because they sat uttermost on the forms which constituted the Bar in the hall or library where the moots were held. The other members who sat inside on the same form and recited the pleadings were called masters of the Inner Bar or inner barristers and, later, students. When a man had kept commons and performed satisfactorily in moots and was of seven years' standing he might be called to the Utter Bar and the phrase still in use on Call Night is that each student is called to the degree of the Utter Bar. The term 'inert barrister' has been replaced by 'student' and the simple term 'barrister' has replaced 'utter barrister'. The terms 'Inner Bar' and 'Outer Bar' are sometimes used today for the distinction between Queen's Counsel (q.v.) and barristers, because the former sit within the Bar.

Barron* v. *Baltimore (1833). The issue was whether the Bill of Rights, and particularly the Fifth Amendment, limited state as well as federal action. The Supreme Court held that the prohibitions in the first eight amendments were protection against only federal and not state infringements. More than a century later the ruling was altered.

Barry, James, Baron Santry (1603–73). Was called to the English (1628) and the Irish Bars and became a Baron of the Irish Exchequer (1634) and Chief Justice of the Irish King's Bench in 1660. He was said to be a very able man and an excellent judge.

Barry, Redmond John (1866–1913). Was called to the Irish Bar in 1888 and became Solicitor-

General 1905, Attorney-General 1909 and Lord Chancellor of Ireland 1911–13.

Barter. The direct exchange of goods for other goods, without use of money. If even part of the consideration is money, as when 'trading in' a car, the transaction is sale rather than barter.

Bartholomew of Exeter (*c.* 1110–84). English bishop and canonist, supporter of Becket. He frequently served as papal judge delegate and promoted the development of decretal law and codification; he was a scholar of wide reputation and wrote extensively.

Bartolo di Sassoferrato (Bartolus) (1313–1357). A commentator, studied law at Perugia and Bologna and was judicial assessor at Todi and Pisa before in 1343 settling as a law teacher at Perugia. He acquired immense authority as an expositor of Roman law and his reputation endured for centuries. He was a practical lawyer and sought to derive principles suitable for the time in which he lived from the medley of customs, feudal laws, and statutes then prevailing. He developed the theory of statutes, as exceptions from the generality of application of the Roman law, a study which contains the seeds of modern principles of international private law. His writings, lectures later amplified from notes and comments, were published in collected editions. His main works were *Commentarius in tria Digesta*; *Commentarius in libros IX Codicis priores*; *Commentarius super libris III posterioribus Codicis*; and *Lectura super Authenticis*. His *Opera* appeared in many editions.

C. N. S. Woolf, *Bartolus of Saxoferrato*.

Bartolomeo da Brescia (?–1258). Italian canonist author of *Glossa ordinaria Decreti*, a revision of the whole of Joannes Teutonicus (*c.* 1240), which was appended to most manuscripts of the *Decretum*. *Historiae super libro Decretorum Procarda*, and other works, mostly revisions of the works of others.

Bartolomeo da Capua (1248–1328). Italian jurist, author of *Commentaria in Constitutiones Regni Utriusque Siciliae* (1568).

Bartolomeo Saliceto (1330–1412). A famous Italian jurist of the school of Commentators, author of important commentaries on the Code and the *Digestum Vetus* and other works on Roman law.

Barwick, Sir Garfield Edward John (1903–). Called to the New South Wales Bar in 1927, he became Attorney-General of Australia, 1958–63 and Minister for External affairs, and then Chief Justice of the High Court of Australia in 1964. He represented Australia at many international conferences.

Basdevant, Jules (1877–). French jurist, judge of the Permanent Court of International Justice and of the International Court 1946–64 (President 1949–52).

Base fee. In real property, originally an estate held not by free or military service, but by base service, at the will of the Lord. It is now a fee which is limited in duration and admits of an absolute fee existing in remainder upon it. It arises from a disposition by a tenant in tail which, though purporting to create an absolute fee, is ineffectual to bar either the remainders only, or both the issue in tail and the remainders. In certain circumstances a base fee can be enlarged into a fee simple absolute.

Base tenure. In feudal land law, holding by villeinage or other customary services as distinct from tenure from the Crown or by military service.

Baselines. In international law, the lines on the map of the coastline of a maritime state from which its territorial waters are measured and which limit its internal or national waters. The method of demarcating this line has given rise to much difficulty. The general basis for this line is the low water mark along the coast following all its sinuosities. Difficulties arise in deeply indented coasts and Norway has claimed to draw the baselines joining the extremities of promontories and offshore islands. This was upheld by the International Court in 1951 which held that the baseline must follow the general direction of the coast.

Basic Law. The constitution of the German Federal Republic, adopted in 1949.

Basic Norm. In Kelsen's (q.v.) pure theory of law (q.v.) the basic norm (*Grundnorm*) is the fundamental proposition of a particular legal system, which is the source of the validity of every legal norm or proposition of that legal system. The differences between legal systems depend on their different basic norms. Thus the basic norm of modern English and Scots law is probably: what the Queen in Parliament has promulgated or authorized or permitted to be stated as law by the judges ought to be obeyed. This norm is the ultimate source of the validity (but not the substance) of every rule contained in statute, or in subordinate legislation authorized by statute, and of every rule laid down by a court or judge acting within the jurisdiction conferred on it or him. The nature and origin of the basic norm itself is however a non-legal question, not determinable by the pure theory, but determined by the political fact of acceptance, and if a revolution caused a different basic norm to be adopted it would result in a different legal system. In countries having written constitutions the constitution in force for the time being will normally be the basic norm of the country's legal system. Indeed the whole jurisprudence of the U.S. Supreme Court could be said to be derived from the proposition: what the Constitution has prescribed or permitted to be stated as law ought to be obeyed. The Constitution is the supreme law of the land.

Basilica. The most important legal compilation made between the time of Justinian and the Middle Ages, compiled about A.D. 900 under the instructions of the Eastern Emperor Basil I, who seems to have abrogated some of Justinian's laws which were in disuse and caused revision of the others. It was completed under his son Leo VI The Wise. It comprises 60 books, divided into titles, each title containing the relevant parts of Justinian's *Digest*, *Code* and *Novels* and sometimes of the *Institutes* also, and in some titles parts of the *Procheiron* (q.v.) of Basil I. It is in Greek, Latin originals being in the form of late Greek versions. In some cases changes were made and Greek terms were substituted for Latin technical terms. It is as much a collection of canon as of civil law and is more systematically arranged than the *Digest* or *Code*. It became the foundation of Byzantine jurisprudence. Some of the surviving manuscripts have marginal scholia, some being extracts from sixth- and seventh-century writers and others introduced in a thirteenth-century revision. Various synopses and hand-books were based on it and not many copies of the full text were made. An index was made in the twelfth century. It was forgotten until the sixteenth century and not published in full till the seventeenth century.

Basilikon Doron. A work by James VI of Scotland (later James I of England) published in 1598 setting out his view of the position of the King in the State. The King is God's lieutenant and must be implicitly obeyed and never resisted. He is author and maker of the law and has power to mitigate, suspend and interpret it. He is subject to the control neither of the church nor of his subjects, is above the law and answerable only to God. These views he elaborated later in the same year in the *Trew Law of Free Monarchies*. The views expressed are in fact those associated with the theory of the Divine Right of Kings.

Basnage, Henri (1615–95). French Jurist, author of an important *La coutûme reformée du pays et duché de Normandie* (1678), *Traité des hypothèques* (1681), and other works.

Basoche, Basoque, Basogue, Bazoche or Bazouges. A French guild of clerks from whom procureurs (legal representatives) were selected. It seems to have grown up in each legal centre as a group of men skilled in law who would act as legal representatives. In the fourteenth century it broke up into advocates, *procureurs*, and clerks, but retained its own offices and was organized in a military manner. To the last the Basoche retained the privileges that to be a qualified procureur one must have been 10 years on its register, and it had disciplinary powers over and judicial powers in cases between members. There were distinct Basoches for the Châtelet de Paris, the Palais de Justice, the *cour des comptes*, the Parlement of Paris, and in various localities in France.

Bassiano, Giovanni (Bassianus) (twelfth century). Italian glossator, teacher at Bologna, pupil of Bulgarus and master of Azo, author of *Summa ad Authenticas*, *Summa quicunque vult*, *Summa Codicis*, and *Summa Institutionum*.

Bastard (or illegitimate child). A child born to a woman not married, or begotten of a married woman by a man other than her husband, or born to a widow so long after her husband's death that it cannot have been begotten by her late husband. Legal systems have normally distinguished between legitimate and illegitimate children. The child of a married woman is presumed legitimate, but this presumption is rebuttable, not merely by evidence of adultery but only by strong evidence that her husband could not have been the father, and the presumption is the other way if the spouses are living apart under a decree of judicial separation. The question may be determined in proceedings for declaration of legitimacy. The child of a void marriage is treated as legitimate if at the time of the intercourse or the marriage either parent reasonably believed that the marriage was valid.

The mother is under the primary onus to maintain her bastard but contribution by the father may be enforced by an affiliation order.

A bastard's natural relationships with his parents are recognized for certain purposes, but apart from that he has no ascendant or collateral relatives. Modern legislation has increasingly assimilated his legal position to that of a legitimate child, particularly in relation to rights of succession.

A bastard may be legitimated by statute, or by the subsequent marriage of his natural parents, from the date of the marriage. See LEGITIMATION.

Bastille. A mediaeval fortress in Paris, originally the castle at the gate, Saint Antoine, which in the seventeenth and eighteenth centuries became a state prison and place of detention for important prisoners. The number of prisoners was small but they were detained by virtue of *lettres de cachet*, orders from the King against which there was no appeal. Later it became a place of judicial detention and persons being tried by the *parlement* were also held there. It became a symbol of Bourbon despotism.

On 14 July 1789 the Paris mob stormed the prison and this symbolized the commencement of the French Revolution. The building was later dismantled.

Bate's Case (or the Case of Impositions) (1606), 2 St. Tr. 371. James I by letters patent under the great seal imposed an additional poundage on imported currants. Bate paid the statutory rate, but refused to pay the additional duty, as not having been authorized by Parliament. The Barons of Exchequer found for the King, holding that the Crown revenues were an essential attribute of the Crown, and that the King might impose what duties he pleased so long as only for the purpose of regulating trade and not of raising revenue. Parliament acquiesced in the decision, notable for the breadth of the political doctrines enunciated by the judges. It was probably correct, the regulation of foreign trade being part of the royal prerogative and such an imposition became illegal only when the king sought to justify taxation for revenue purposes rather than for regulation of trade. But the court buttressed its decision by statements based on political theory of a king which greatly enlarged the royal power. The Bill of Rights, 1688, declared that it was illegal for the Crown to seek to raise money without Parliamentary approval.

Bath, Order of the. In the Middle Ages bathing was sometimes part of the ceremony of creating a knight. Knights of the Bath are said to have been created between 1399 and 1661. The Order of the Bath was created in 1725 and purports to be a revival of a former Order. It comprises military and, since 1847, civil divisions and ranks fourth among the Orders of Knighthood. There are three grades in the Order, Knights Grand Cross, Knights, and Companions.

Bathurst, Henry, 1st Lord Apsley and 2nd Earl Bathurst (1714–94). He was called to the Bar in 1736 and in 1754 became a judge of the Common Pleas. In 1770 he was one of the commissioners for executing the office of Chancellor, and, to general surprise, was made Lord Chancellor as Lord Apsley in 1771 and held office till 1778. He succeeded to the earldom in 1775. He was Lord President of the Council 1779–82. He was reputed honest and straight but quite inadequate to the office and contributed nothing to the development of equity, though he had the sense to be guided by Sewell, M. R.

Battery. See ASSAULT AND BATTERY.

Battle, trial by. This institution seems to have been common among the barbarian tribes of Europe and to have been based not so much on an appeal to force as on the belief that God would give victory to the right. It was introduced into England by the Normans and was used to decide a variety of disputed questions.

It was used in civil cases, in real actions, on a writ of right, possibly in claims of debt.

In criminal cases, battle was used where a private person brought an accusation against another by an appeal of murder, treason or felony.

In civil cases infants, women and persons over 60 might decline battle and employ champions. Sometimes churches, communities, and great landowners retained permanent champions, and in 1275 statute provided that champions need not swear as to belief in the cause they were seeking to uphold. But in criminal cases the parties themselves fought. The battle took place in the presence of the robed judges of the Queen's Bench or Common Pleas. The combatants were each armed with a staff an ell long and a leather shield, and the fight continued until one cried 'Craven' or the stars appeared, in which case the defendant was deemed to have won. In a criminal case if the party challenged were defeated he suffered the same penalty as if convicted on indictment and the Crown had no power to pardon him as it was a private suit.

Battle was never popular and the charters of some towns granted exemption from it. It began to fall into disfavour in the twelfth century and it was condemned at the Fourth Lateran Council in 1215 and, though practically obsolete by the end of the thirteenth century, it survived as a possible alternative to the Grand Assize in real actions down to the nineteenth century, and as a means of proof in appeals of murder. It was abolished in 1819 in both real and criminal actions after the last attempt to bring an appeal of felony (*Ashford* v. *Thornton* (1819), 1 B. & Ald. 405).

G. Neilson, *Trial by Combat.*

See also APPEAL OF FELONY; ASSIZE, GRAND; COMPURGATION; ORDEAL.

Baudouin, François (Balduinus) (1520–1573). An eminent French humanist jurist and theologian. He emphasized the importance of history and tried to trace the origins of Roman law. His works included *Scaevola seu Jurisprudentia Muciana* (1558) and *De institutione historiae universalis et ejus cum iurisprudentia conjunctione* (1561).

Baumes Laws. Several statutes of New York State criminal code enacted in 1926, named for the chairman of the New York State Crime Commission. They were based on the principle of repression, a notable example of this being the Habitual Crimes Act, which provided for increasingly heavy sentences for repeated crimes, and the statute which imposed mandatory life imprisonment for persons convicted of a fourth felony.

Bavarian Codes. Codes enacted in Bavaria in the eighteenth century, prepared by Von Kreittmayer. They comprised the *Codex Iuris Bavarici Criminalis* (1751), the *Codex Iuris Bavarici Judiciarii* (1753), and the Bavarian *Landrecht*, the *Codex Maximilianeus Bavaricus Civilis* (1756) which however retained the law of the Pandects as of subsidiary force.

Bavarian Law. The Bavarian Code dates from about A.D. 750 and contains native rules of Bavarian tribes, Alamannic rules, Salic, Frankish, and Visigothic rules, rules from the Lombard Edict, and rules of church councils for the clergy.

Bayne, Alexander, of Rires (?–1737). Passed advocate in 1714, became curator of the Advocates' Library and in 1722 first Professor of Scots Law in Edinburgh University. He published an edition of Hope's *Minor Practicks*, a small *Institutions of the Criminal Law of Scotland* (1730), and *Notes on the Municipal Law, being a Supplement to the Institutes of Sir George Mackenzie* (1731).

Bays. Indentations of a coastline with a wide opening, deeper than mere curvatures of the coast, where tracts of open sea are partly enclosed by land. Bays include some gulfs, fjords, firths, and similar large inlets. Whether an indentation is a 'bay' is a question of fact. They are of importance in international law in relation to the boundaries of the territorial sea, particularly the question whether the boundary is to follow the coastline or be drawn straight across the mouth of the bay. Some authorities have asserted that a straight line is permissible, if not exceeding a stated length and six, 10, 12 and 20 miles have been proposed. The Geneva Convention on the Territorial Sea of 1958 adopted a closing line of 24 miles so that only water on the inland side of a line of that length across a bay is internal water.

Where a bay is bounded by the territory of two or more states the general principle is that the bay is divided down the median line between the baselines from which the territorial seas of the two states are measured.

By general acquiescence some bays are deemed part of the internal waters of a state though the closing line exceeds the limits permitted by general international law. Many states claim historic bays in this way. Thus the U.S. claims Delaware and Chesapeake Bays, Canada claims Hudson's Bay.

Beale, Joseph Henry (1861–1943). Practised law in Boston before joining the faculty of the Harvard Law School in 1890, where he was latterly Carter Professor of General Jurisprudence, 1908–1912, and Royall Professor of Law, 1912–37. He was author of many books, notably *A Treatise on the Conflict of Laws*, and reported on Conflict of Laws, for the American Law Institute Restatement.

Bearer. In relation of bills of exchange, cheques, and promissory notes, is the person in possession of a bill or note. A bill or note may be payable 'to bearer' in which case the person in possession is entitled to demand payment, or 'to X or bearer' in which case, if it is endorsed in blank by X, by his signature, the bearer is entitled to payment. Such a bill or note may be transferred by mere delivery to the effect of entitling any person in possession of it to present it for payment.

Beatty* v. *Gillbanks (1882), 9 Q.B.D. 308. Members of the Salvation Army sometimes marched in procession through a town; a rival organization systematically obstructed and opposed them and disturbances took place. Despite a notice from the magistrates directing persons not to assemble to the disturbance of the peace, the Salvation Army held another procession and the police called on them to disperse and on their refusal arrested the leaders. The local magistrates found them guilty of unlawful assembly and ordered them to find sureties to keep the peace. The Queen's Bench Divisional Court allowed their appeal holding that, if people assemble for a lawful object without intending to cause or provoke a breach of the peace they do not constitute an unlawful assembly, even though they know that their meeting may be opposed and a breach of the peace may take place in consequence.

Beaumanoir, Philippe de Remi, Sire de (?1250–96). French jurist, bailli at Senlis and Clermont, seneschal of Poitou, and author of the famous *Coutûmes de Beauvaisis* (1283), one of the most important mediaeval works on old French customary law, which sets out not only the law of Beauvaisis but the fundamental principles of contemporary private law.

Beaumont, Sir John William Fisher (1877–1974). Was called to the Bar in 1901, took silk in 1930 and was Chief Justice of Bombay 1930–1943 and from 1944 a member of the Judicial Committee of the Privy Council.

Beawes, Wyndham (*fl.* 1775). Consul at Seville and St. Lucar, published in 1751 *Lex Mercatoria Rediviva, or the Merchants' Directory*, a guide to men in business and not a law book, but including the relevant law under all the heads such as bills of lading, charter-parties, insurance and the like, and dealing also with some relevant matters of international law and foreign law. The book proved useful and successful and passed through many editions.

Bebenburg, Lupold von (1297–1363). German canonist, author of an important *Tractatus de juribus regni et imperii* and other works.

Beccaria, Cesare Bonesana, Marchese de (1738–94). An official of the Austrian Government in Italy, and later a councillor of state and a member of a commission to reform the law of Lombardy. His work on economics was important but he is now remembered for his treatise *Dei Delitti e delle Pene* (On Crimes and Punishments) (1764), the first systematic study of the principle of punishment, which had immediate success and enormous influence; many of the ideas were familiar but nevertheless it represented a great advance in criminological thought; it advocated prevention of crime rather than punishment and condemned capital punishment and torture. It had great influence in Russia, Prussia, Austria, in all of which it provoked great changes in criminal law, and not least in revolutionary France and in England on Bentham and Romilly. His other works were on literary style and on economic problems.

Becket, Thomas (1118–70). Studied civil and canon law at Bologna and Auxerre and became Archdeacon of Canterbury and Henry II's chancellor in 1154. He lived in luxury, conducted embassies, raised and led troops, and was the trusted adviser of the King. In 1162 he was made Archbishop of Canterbury, became devout and austere and a supporter of the Gregorian reforms of the church, which included clerical independence from lay control. He resigned the chancellorship and began to oppose the King. In 1164 Henry, in the Constitutions of Clarendon, asserted the King's right to punish criminous clerks, forbade excommunication of royal officials and appeals to Rome and claimed for the King the revenues of vacant sees and the power to influence episcopal elections. A breach took place and Becket took refuge in France where he remained from 1164 to 1170. A kind of reconciliation was then effected, but Becket refused to lift the excommunication of certain royal servants and four knights, taking literally some angry words of Henry, killed Becket in Canterbury Cathedral. His tomb became a place of pilgrimage and he was canonized in 1173. He may have been excessively obstinate but sincerely opposed what he believed, with some justification, to be inroads by the state into the spiritual sphere.

Bedchamber Question (1839). In 1839 Lord Melbourne's government resigned and Sir Robert Peel, who was invited to form a government, appreciating that nearly all the ladies of Queen Victoria's household were related to members of Melbourne's cabinet or to their political adherents, intimated to the Queen that he could not undertake to form a ministry unless permitted to make some changes in the higher offices of the court, including the ladies of her bedchamber. The Queen refused to make changes and Peel declined to accept office on these terms. The Melbourne ministry continued but fell in 1841 and Peel assumed office, meeting with no further difficulty on the bedchamber question. Since then the principle for which he contended has been admitted to be constitutionally correct.

Bedingfield, Sir Henry (1633–87). He was called to the Bar in 1657, became a judge of the Common Pleas in 1686 and Chief Justice shortly thereafter. He is said to have been a grave but rather heavy lawyer, who owed his promotion to Jeffreys, and made no mark in the law.

Bedlam. See BETHLEM.

Beier, Adrien (1634–1712). German jurist, one of the first to study labour law, author of *Tractatus de iure prohibendi* (1683), *De collegiis opificum* (1688), and other works.

Beheading. A mode of inflicting capital punishment. Among the Greeks and Romans it was deemed the most honourable form of death. In Britain it is said to have been introduced into England by William the Conqueror and was usually reserved for persons of high rank. It was commonly used in Tudor and Stuart times, as in the cases of Mary, Queen of Scots and Charles I. The instrument was usually an axe, but a sword was often used and an early form of guillotine, 'the maiden' was used in Scotland in the sixteenth century. It is thought that Lord Lovat was the last person beheaded in England, in 1747. The guillotine was extensively used in France during the Terror.

Beheading was also part of the common law method of punishing traitors, along with hanging, disembowelling and quartering. In 1814 the King was empowered to substitute hanging as the ordinary mode of execution but till 1820, even after traitors were hanged, their heads were cut off. Under the Nazi regime in Germany beheading was the prescribed method but in modern times it has almost completely disappeared.

Beirut. See BEYROUT.

Belgian law. Belgium as a distinct state dates only from 1830 though proposals had earlier been made that it be independent. But previously

Belgium had been part of Charlemagne's empire, and then part of the United Netherlands under Spanish rule till 1713, and thereafter under Austrian rule till 1795 when Belgium was annexed by France. Earlier legal history accordingly falls under that of the Low Countries. After 1793 under French administration the French system and French law was introduced. In 1815 the kingdom of the Netherlands was established but in 1830 Belgium and Holland separated. It became an independent kingdom in 1831.

It is under a constitutional hereditary monarchy, the constitution having been promulgated in 1831.

Executive power is exercised by the king and ministers chosen from the majority party in the legislature. The nine provinces and nearly 2,400 communes have a large measure of autonomous government.

The legislature comprises the King, the Senate and the Chamber of Deputies. The Senate is partly directly elected, partly elected by provincial councils and partly co-opted. The Chamber is directly elected by universal suffrage. The King may dissolve the Chambers together or separately.

Judges are appointed by the king for life (unless removed by judicial sentence) and are independent of the executive and the legislature. The country is divided into nine provinces subdivided into judicial regions.

There is a Supreme Court (Court of Cassation) which sits in three chambers. Five regional courts of appeal, three regional labour courts and assize courts in each province sit for political and criminal cases and in each province there are courts of first instance (Tribunals of First Instance, Tribunals of Commerce and Labour Tribunals). Local magistrates' courts deal with minor criminal and commercial cases and police tribunals decide criminal cases. There are also various special tribunals and a military court. In administrative matters there are a Conseil d'Etat, a Cour des Comptes and lower administrative courts.

A large part of Belgian law is codified, in the Civil Code of 1804, based on the Code Napoleon, the commercial code and codes of civil procedure, penal law and criminal procedure.

Law is studied in the Universities of Brussels, Ghent, Liège, and Louvain.

P. Graulich, *Guide to Foreign Legal Materials—Belgium, Luxembourg, Netherlands*; P. L. Wigny, *Droit constitutionnel*; J. Constant, *Manuel de Droit pénal*; R. Dekkers, *Précis de droit civil belge.*

Bell, Alexander Montgomerie (1808–66). Became a Writer to the Signet in 1835 and Professor of Conveyancing in Edinburgh University in 1856. His published *Lectures on Conveyancing* (1867, third edition 1882) were long a standard treatise on the subject.

Bell, George Joseph (1770–1843). Passed advocate of the Scottish Bar in 1791. In 1816–18 he was professor of conveyancing to the W. S. Society. He was appointed Professor of Scots Law in the University of Edinburgh in 1822 and was made a Principal Clerk of Session in 1832. Lord Advocate Jeffrey would have made Bell a judge had there been a vacancy at the time. Bell was a member of several Royal Commissions investigating various matters of Scots law and took part in promoting improvements in procedure and land law. He continued to hold his chair till his death.

In 1800 he published a *Treatise on the Law of Bankruptcy in Scotland*, reissued in 1804 as *Commentaries on the Municipal and Mercantile Law of Scotland* and later developed into *Commentaries on the Laws of Scotland and on the Principles of Mercantile Jurisprudence*, and in 1829 a *Principles of the Law of Scotland* (enlarged edition, 1830), which, though intended as a handbook for students, rapidly became a standard authority. Both passed through many editions before the end of the nineteenth century. He published also *Illustrations from Adjudged Cases of the Principles of the Law of Scotland* (3 vols., 1836–1938) a useful collection of synopses of cases, designed as a companion to his *Principles*, a book on the sale of goods, and several other less important works.

The clarity of statement in the *Commentaries* and *Principles* caused them to be repeatedly edited after his death, to be cited and followed in innumerable cases, and to guide both the studies of students and the decisions of judges. Both rank among the half-dozen classics of Scottish legal literature.

His elder brother, Robert, was an advocate, Professor of Conveyancing to the Society of Writers to the Signet and author of the *Dictionary of the Law of Scotland* (1807) and of numerous works on Scots law, especially leases and conveyancing. An elder brother, John, was an eminent surgeon, and a younger brother, Sir Charles Bell, was a distinguished anatomist and physiologist.

Belleperche, Pierre de (Petrus de Bellapertica (?1250–1308). A French commentator-jurist, professor at Toulouse and Orleans, Chancellor of France, responsible for a reduction of the *ordonnance* of reform of 1302, and author of *Lectura super librum Institutionum* (1513), *Commentaria in Digestum novum* (1571) a commentary on the Code, one on the Digest and *Repetitiones.*

Bellewe, Richard (*fl.* 1575). Little is known of Bellewe, but he is said to have been of Irish origin and to have belonged to Lincoln's Inn. In 1578 he published the collection, variously known as *Brooke's New Cases*, *Little Brooke* or *Bellewe's Cases temp. Hen. VIII, Ed. VI et La Roygne Mary*, of cases selected from Brooke's *Abridgement* and put in chronological order. A later edition of 1651 rearranged the matter under alphabetical heads. In 1585 he published *Les Ans du Roy Richard le Second*, cases of Richard II's reign culled from the Abridgements of Statham, Fitzherbert and Brooke.

Belli, Bello or Bellinus, Pierino (1502–1575). Counsellor of State and legal adviser to the Duke of Savoy, an early writer on international law, author of *De re militari et de bello* (1563), *Consilia sive Responsa*, and other works.

Belligerency. The state of being in fact engaged in a war, whether war has been declared or not and whether or not the belligerent state has resorted to aggressive war in violation of international law. If a state recognizes the belligerency of parties to a civil war it recognizes that they have belligerent rights and duties similar to those accepted in international law. Recognition of insurgency does not go so far. Non-belligerency or benevolent neutrality is the conduct of a state in departing from a neutral's duties of impartiality by giving assistance to one of the sides in a war. In certain circumstances the U.N. Charter authorizes or even requires such a departure.

Belligerents. In international law, those states which are legally qualified to make war and are doing so. This power is confined to full sovereign states, and part-sovereign states, e.g. vassal states, states within a federation, are prohibited from offensive or defensive warfare though they may in fact engage in hostilities. Insurgent states may become belligerent states through recognition by third states, though this is not binding on the parent state. Thus the U.K. recognized the belligerency of the Confederate States of America in 1861 by making a declaration of neutrality. Recognition is only justifiable if there exists a civil war, in which the insurgents have established a government which controls a substantial part of national territory and maintains an orderly administration, and maintains armed forces under responsible control which are observing the rules of warfare. But recognition is sometimes granted by a state to cover illicit interference with the affairs of the state in question and to justify aid to the insurgents.

A distinction may also be drawn between principal belligerents who wage war by virtue of a treaty of alliance or having otherwise been involved in the war, and accessory belligerents who provide assistance to a principal by rendering help only.

Belvisio, Jacobus di (1270–1355). A notable Bolognese jurist, author of *Practica Judiciaria in Materiis Criminalibus*, *Lectura Authenticorum*, commentaries on the *Libri Feudorum*, and other works.

Bemis, George (1816–78). Was admitted to the Boston Bar in 1839 but turned to international law and published a number of pamphlets on the rights and duties of neutrals, particularly in relation to the *Alabama* (q.v.) claims.

Bench. A long seat, freestanding or placed against a wall. Doubtless from the fact that judges originally sat on a bench, 'The bench' became the term for the judges collectively, or the judges of a particular court, such as The King's Bench or The Common Bench (i.e. Common Pleas), or a group of Justices of the Peace, as in the phrase 'bench of magistrates'. The phrase 'bench and bar' denotes the judges and the barristers collectively. In the Inns of Court (q.v.) the benchers (q.v.) are the senior members who control the Inn. A person made a judge is said to 'go to the bench' or be 'raised to the bench'. In the House of Commons the front benches are reserved for the leaders of the party in power and of the Opposition respectively. Members not entitled to sit there are known as 'backbenchers'. The bishops of Church of England are also sometimes called the 'bench of bishops'.

Benchers (or Masters of the Bench). The senior members of each of the Inns of Court, recruited by co-option by the existing benchers, normally from among the judges, Q.C.s and senior members of their Inn. The Benchers govern the Inn, admit persons as students, call students to the Bar and have disciplinary power over members of the Inn (though this is delegated to the Senate of the Four Inns). They may elect a few Honorary Benchers, distinguished persons not members of the Inn, and each year elect one of themselves to be Treasurer of the Inn for the year.

Benefice. The term for a kind of land tenure adopted in the Frankish kingdom in the eighth century whereby a seigneur leased land to a freeman *in beneficium*, for his benefit. The lease was normally for the tenant's life but was sometimes hereditary. As a term for land tenure the term died out in the twelfth century, but came to designate an ecclesiastical office to which the church attached the perpetual right of receiving income. Bishops and seigneurs began to treat each church and its endowments as property to be leased and they appointed a priest by leasing to him the church and endowments in return for the spiritual duties.

In the Church of England a benefice is a freehold office of the cure of souls, held by a cleric styled, according to the nature of his benefice, rector or parson, vicar, or perpetual curate, who has a freehold interest in the emoluments of his benefice until death or he vacates the benefice. The term benefice covers rectories or parsonages, vicarages, perpetual curacies, chapelries, independent churches, or chapels without districts and sinecure benefices.

A vacancy in a benefice is filled by the patron presenting a fit clerk to the bishop to be instituted or collated. Except so far as permitted by law no incumbent can hold more than one benefice at the same time, i.e. in plurality. All parochial duties are committed to and imposed upon the incumbent of the benefice and all fees and emoluments therefrom belong to him.

A benefice is avoided by attainment of the age of 70, death, resignation, exchange, cession, declaration of avoidance by the bishop, and deprivation, in which case the interest of the incumbent in the profits and emoluments and property of the benefice ceases on the day when he vacates it.

Beneficiary, a or the person for whose benefit a trust is created. In the case of private trusts they must be identifiable persons or bodies or classes of persons. In the case of charitable trusts the category of beneficiaries must be adequately identified.

Benefit of Clergy. In the early Middle Ages in England, the privilege of the ordained cleric accused of felony to evade trial by the royal courts, and to be subject to trial only by ecclesiastical courts, an immunity which developed after and because of Becket's (q.v.) murder. The practice came to be to claim the benefit of clergy when arraigned before the secular criminal court, or to claim it after conviction to prevent judgment. The ecclesiastical courts were supposed to try him again but in most cases he was acquitted and discharged. It could be claimed originally only by ordained clerks, monks and nuns. A married man could invoke it, the lowest clergy not being celibate, but bigamists could not, and a bigamist was one who had married twice, or married a widow. By 1350 it was extended to secular as well as religious clerks and later to all who could read. To satisfy themselves of the validity of a claim to clergy the judges introduced the 'neck verse', (Psalm 51:1) which the accused was required to read, or recite. In 1547 a peer could claim it for a first offence. In 1692 women were allowed to claim it. In 1705 the necessity for reading was abolished and the privilege became available to all.

Before the end of the thirteenth century there began the process of excluding certain offences from benefit of clergy. In the fourteenth century it was settled that it did not apply to persons charged with high treason. Henry IV decided that only those actually in orders could claim clergy a second time. Under Henry VII it was provided that those not actually in orders were, if convicted, to be branded and disabled from claiming it a second time. In the sixteenth century statutes took away the privilege in many cases from persons not actually in orders; but a reaction against Henry VIII restored it in some cases. In 1717 it was enacted that persons not in orders convicted of a clergyable offence were to be transported for seven years. The list of felonies 'without benefit of clergy' was greatly extended by the eighteenth century. The 'neck verse' test was abolished in 1705 and the branding in 1779. The plea was wholly abolished in 1827.

Most of the American colonies adopted benefit of clergy with the common law but it was generally abolished at the Revolution though it survived in the Carolinas until the mid-nineteenth century.

Benelux Economic Union. An economic union established by treaty effective in 1960 between Belgium, the Netherlands, and Luxembourg, seeking to promote the free movement of persons, goods, capital, and services, the co-ordination of the economic, financial, and social policy of the three states and the development of a common economic policy, as to other countries. It is not affected by these countries' membership of the European Communities. The executive authority is a quarterly meeting of the Committee of Ministers; ordinary business is managed by a Secretary-General and staff.

Benevolence. Generally a gift or act of kindness but also the term for gifts extorted from subjects without Parliamentary consent by various English kings. Earlier there had been forced loans, but Edward IV discarded any pretence of repayment and the word 'benevolence' was first used in 1473. In 1484 Parliament abolished benevolence as 'new and unlawful inventions' but Henry VII nevertheless used the device extensively and even persuaded Parliament to make those who had promised gifts liable for unpaid arrears. Henry VIII demanded benevolences in 1528 and 1545 and James I received gifts in 1614, but further attempts to exact money in 1615 and 1620 aroused such opposition that the practice was discontinued. The Petition of Right, 1627, and the Bill of Rights, 1688, reaffirmed their illegality.

Benjamin, Judah Philip (1811–84). Born in the West Indies, became a lawyer and statesman in Louisiana and entered the U.S. Senate in 1852. He was offered a seat in the Supreme Court. When the Civil War broke out he sided with the South and served the Confederate States as Attorney-General, Secretary of War (1861–62) and Secretary of State (1862–65), and towards the end of the war urged the arming of slaves with the promise of emancipation. After the defeat of the South he escaped and after many adventures reached England, where he was called to the Bar in 1866. He later received a patent of precedence and became a Q.C. and enjoyed a large appellate practice. While in Louisiana he had produced a book, a codification of a branch of the law of that State, which had established his standing. In England in 1868 he published a textbook on the law of *Sale of Personal Property* which at once established his reputation and has been a standard book ever since.

R. D. Meade, *Judah P. Benjamin, Confederate Statesman.*

Bentham, Jeremy (1748–1832). Philosopher, economist, and legal theorist, graduated from Oxford in 1763 and was called to the Bar, becoming a bencher in 1817. He had no success at the Bar and early developed a critical attitude to the law. In 1776 he published a *Fragment on Government, or a Comment on the Commentaries*, acutely criticizing Blackstone's (q.v.) antipathy to reform and attacking the prevalent views on politics and jurisprudence.

Thereafter he worked on *The Introduction to the Principles of Morals and Legislation* which contended that the greatest happiness of the greatest number should be the object of legislation. It appeared in 1789. He visited Russia and wrote extensively on the principle of the panopticon, a system of prison discipline under which prisoners would be under constant surveillance from a central point in a circular building. In letters and tracts and in the *Principles of Penal Law* he enunciated many principles which have been the origins of reform in penal policy, but though his ideas found favour with the government, his schemes did not work in practice. In *Truth* v. *Ashhurst* he criticized the constitutional views of Mr. Justice Ashhurst laid down to a Middlesex grand jury and in 1807 he developed views on legal procedure.

He gave an impetus to philosophical radicalism by participating in the founding of the *Westminster Review* in 1823, though he wrote little for it.

He left a vast quantity of material in manuscript and his works were not adequately edited nor printed until the late twentieth century, when a complete edition began to appear, notably the *Introduction to the Principles of Morals and Legislation* and *Of Laws in General.* His works cover jurisprudence, ethics, logic and, to a lesser extent, economics. On the first two of these his influence has been immense, resulting in the substitution of reason and utility for other justifications in law and morals. It is now recognized that much of the credit hitherto given to Austin (q.v.) for developing analytical jurisprudence should belong to Bentham. His *Rationale of Judicial Evidence* (1827) contains valuable criticism of the existing law. His works have had great practical influence in subsequent legislation, and gave great impulse to such projects as codification, consolidation of statutes, and the abolition of antiquated and arbitrary rules. Evidence and criminal law in particular have been influenced, and such later reformers as Brougham, Romilly, Horner, and Mackintosh (qq.v.) were very commonly merely giving effect to his ideas. At the same time he overrated the ease of codification, the stupidity and corruption of lawyers, and the possibilities of reform. But the impetus which he gave to reform on the bases of reason and utility is his greatest service.

Life by J. Bowring in Bowring's edition of *The Works of Jeremy Bentham*; Leslie Stephen, *The English Utilitarians* (1900); M. Mack, *Jeremy Bentham* (1962).

Bequest (or legacy). Properly a gift of personal property made by will to a legatee, but sometimes used of real property also (for which the proper term is 'devise').

Bérault, Josias (1563–1633). French jurist, author of *Coutûme reformée du pays et duché de Normandie* (1612).

Berber, Friedrich Joseph (1898–). German jurist, author of *Lehrbuch des Volkerrechts* (1960–64) and other works on international law and relations.

Berckringer, Daniel (?–1667). Professor of Philosophy and later of Rhetoric at Utrecht, 1640, and author of *Institutiones politicae sive de Republica* (1662), *Exercitationes ethicae, economicae, politicae* (1664).

Berger, Johann Heinrich (1657–1732). German jurist, author of *Electa Disceptationum Forensium* (1706), *Electa Jurisprudentiae Criminalis* (1706), *Dissertationes Juris Selectae* (1707), and *Oeconomia juris ad usum hodiernum accommodati* (1712), and other works.

Bering Sea Dispute. A dispute between the U.S. and Great Britain and Canada over the international status of the Bering Sea. In 1881 the U.S. in an attempt to control seal hunting off the coast of Alaska, claimed jurisdiction over all Bering Sea waters, but Britain refused to recognize this. In 1886 and subsequent years the U.S. seized vessels mostly Canadian, found sealing there, and claimed that the Bering Sea was, as it had been when Russia owned Alaska, an internal sea. In 1891 an agreement was made that both U.S. and British ships would police the area and a treaty was signed referring the dispute to arbitration. This was held at Paris in 1893 and found against the U.S. seizures and assessed damages. In 1911 and subsequently sealing conventions have sought to control sealing in the area.

Berkeley, William, Lord (?–1741). Was called to the English Bar in 1695 and appointed Master of the Rolls in Ireland in 1696 but discharged the duties by deputy and resigned in 1731.

Berlich, Mathaeus (1586–1638). German jurist, author of *Conclusiones practicabiles* (1615) and *Decisiones* (1625).

Berlin, Congress and Treaty of (1878). Settled the Eastern Question of the 1870s. Russia, aroused by Turkey's suppression of revolts in Bosnia and Bulgaria had intervened and forced Turkey to give way and, by the Treaty of San Stefano, to agree to the creation of a large independent Bulgaria. Austria and Britain sought revision of this treaty by a European Congress and Bismark persuaded the Russians to agree. The conference reduced Bulgaria in size and made Roumania yield Bessarabia to Russia. Austria was allowed to occupy Bosnia and Herzegovina and Britain to occupy Cyprus, though all three were to remain under Russian sovereignty. The congress effected revision of the Treaty of San Stefano without war but left Russo-German relations strained and the Ottoman empire practically excluded from Europe.

Bernard, Montague (1820–82). Called to the Bar in 1846, became first Professor of International Law and Diplomacy at Oxford in 1859 and was an original member of the Institute of International Law. He participated in negotiating the Treaty of Washington in 1871 and in the Alabama arbitration. He had in his time a high reputation in the field of international law and published thereon. In 1871 he became a member of the Judicial Committee of the Privy Council, and resigned his chair in 1874.

Bernardo di Compostella. There were two jurists of this name, one author of *Breviarium decretalium Innocentii III* (1208), also known as the *Compilatio romana.* The other was author of various works on canon law, an *Apparatus ad Decretales Gregorii IX, Casus Decretalium* and *Apparatus ad constitutiones Innocentii IV.*

Bernardo di Pavia (?–1213). Italian canonist, author of a *Breviarium Extravagantium* (*c.* 1190), *Summa decretalium, Summa de matrimonio* and other works.

Berne Convention. A convention framed at a conference at Berne in 1886 whereby the contracting states (which included the U.K.) were constituted into a union for the protection of the rights of authors over their literary and artistic works. It was replaced by the Revised Berne or Berlin Convention of 1908 further revised at Rome in 1928 and at Brussels in 1948. The main principle is that authors who are nationals of any Union country enjoy in countries other than the country of origin of their work, the rights granted to natives, as well as the rights specially granted by the convention.

Bertillon, Alphonse (1853–1914). French anthropologist, while in the prefecture of police invented a system of identification of criminals by anthropometry, bodily measurements, description, and photographs. The merits of Bertillonage were recognized in the U.S. and he later elaborated his system. Though not the discoverer of fingerprinting he participated in developing it. He published *L'anthropologie métrique* (1909).

H. T. F. Rhodes, *Alphonse Bertillon.*

Bertrand, Pierre (1280–1347). French cardinal and canonist, author of *Libellus super jurisdictione, Tractatus de origine jurisdictionum* (1557), and of *Apparatus* on the Sext.

Berwick upon Tweed. This town was captured and recaptured repeatedly during the Anglo-Scottish wars of the thirteenth and later centuries. It was represented in the Scottish Parliament from 1476 to 1479 but in the English, British, and U.K. Parliaments from 1482 to 1885. The Wales and Berwick Act, 1746 provides that England includes Berwick-upon-Tweed, and for Parliamentary and local government purposes it is now treated as in England, though not part of the county of Northumberland.

Berytus, Beyrout or Beirut. An ancient Phoenician city and from 15 B.C. a Roman colony, now capital of the Lebanon.

In the third to sixth centuries it was famous for its school of Roman law, the most famous of all in its time. Its greatest days were in the fifth century, the time of teachers, many of whose names are known, whom jurists of Justinian's time revered as 'the oecumenical masters'. Justinian closed all the schools except Constantinople and Berytus and scholars from both were on the commissions which compiled the *Corpus Juris.* The period of study was five years and particular books were dealt with in each year. The teaching was long in Latin, Greek being substituted about A.D. 400, and seems to have been mainly by comment on a classical text, induction of generalizations, examination of difficulties, and illustrations from practice. In A.D. 557 the city was destroyed by an earthquake and the school was transferred to Sidon.

P. Collinet, *Histoire de l'école de droit de Beyrouth.*

Besaiel. One of the old English real actions, a variant of the writ of right (q.v.) providing for a claim for the recovery of lands based on the seisin of a great-grandfather.

Besetting. Without legal excuse, watching and waiting about the place where a person lives or works, to trouble him or compel him to act or abstain from acting in any way in which he is entitled to act or abstain. It may be tortious or, in connection with trade disputes, criminal.

Besold, Christopher (1577–1638). German jurist, Professor of Law at Tubingen, 1610, and Ingolstadt, 1635, and a voluminous writer on legal and ecclesiastical subjects, notably *Opus politicium* (1641), *Politicorum libri duo* (1618), and *Consultationes de iuris publici quaestionibus* (1628).

Best, Sir William Draper, Lord Wynford (1767–1845). Was called to the Bar in 1789, built up an extensive practice, and in 1804 participated in the impeachment of Lord Melville. He became a judge of the King's Bench in 1818 and Chief Justice of the Common Pleas in 1824. He resigned from ill-health in 1829 but was ennobled as Lord Wynford and made a deputy-speaker of the House of Lords. He had a sound grasp of legal principles but is said to have displayed temper and bias on the bench and had no great reputation as a lawyer.

Best evidence rule. A general principle of evidence that a party seeking to prove a fact must adduce the best form of evidence that the circumstances allow. Thus direct evidence is to be preferred to hearsay, an original document to a copy. But many exceptions and qualifications are recognized where secondary evidence is admitted.

Bestiality. A species of buggery, the offence of a human of either sex having unnatural sexual relations with an animal *per anum* or *per vaginam*.

Beth-din or *bet-din*. A Jewish tribunal, known from Old Testament times, empowered to handle cases of civil, criminal, or religious law. In modern times the functions of *bette-din* have varied according to the conditions of the Jewish communities. They still operate all over the world under the direction of rabbinic scholars and adjudicate on matters affecting their own communities. In countries where divorces must be obtained from civil courts orthodox Jews must still obtain a religious divorce from a *beth-din* before they may remarry. In Israel all questions of personal status are determined by religious courts.

Bethell, Richard Bethell, 1st Lord Westbury (1800–73), won the Vinerian scholarship at Oxford and was called to the Bar in 1823. He quickly made his way and was a leader of the Chancery bar from about 1840. He had great abilities, as classical scholar, orator, and debater but was bitterly sarcastic in speech, haughty in manner, and unpopular among other lawyers. He entered Parliament in 1851 and became Solicitor-General in 1852, succeeding as Attorney-General in 1856. While in the House he promoted some material legal reforms, including the establishment of the Probate Court. In 1861 he became Chancellor, but was not so successful in the Lords as he had been in the Commons. He resigned in 1865 after certain abuses in the administration of bankruptcy came to light, though his personal fault was slight, but thereafter sat in the House of Lords and Privy Council. All his life he was dissatisfied with the state of the law and with legal studies. He urged the need for a Ministry of Justice and secured the passing of the first and second Statute Law Revision Acts. He played a great part in persuading the Inns of Court to develop an adequate educational system and urged the government to convert the Inns of Court into a legal university and after his retiral presided over a commission examining one of his great projects, a digest of English law. It effected nothing and very little of his ambitious plans for the reform of the law and of legal education were realized. As a judge he over-simplified, ignored relevant authority and misstated law though in some cases his judgments give useful surveys of the law.

T. A. Nash, *Life of Lord Westbury*.

Bethlem. Originally the Hospital of St. Mary of Bethlem in Bishopsgate, London, founded as a priory in 1247 and used for the confinement of the acutely insane from the fourteenth century. After the dissolution of the monasteries management was by the City of London. Persons convicted of crime were sometimes sent there. In 1547 it was incorporated as a royal foundation for the reception of lunatics. It was moved to Moorfields in 1673 and its capacity was then said to be 250. In 1808 a Select Committee recommended the building of a special establishment for criminal lunatics but this was not then done. In 1815 Bethlem was rebuilt at St. George's Fields with accommodation for 60. There were considerable improvements in standards of care in the nineteenth century and the premises were extended several times. It was moved in 1930 to Shirley near Croydon. Many criminal lunatics were, however, held in country asylums or private licensed houses. In 1845 the Lunacy Commissioners revived the idea of a state asylum and in 1863 Broadmoor (q.v.) was opened. Bethlem was later transferred to Beckenham and is called the Bethlem Royal Hospital and Maudsley Hospital. The word Bedlam, and the expressions Tom O'Bedlam and Bess O'Bedlam, are derived from a corrupt form of the name.

Betrothal (or engagement). An agreement between a man and a woman that they will marry. The importance and formalities of the agreement depend on the customs of the society but it is frequently marked by gifts and celebrations. It was formerly and in some countries still is a formal ceremony. Canon law distinguished two types *sponsalia de praesenti*, a true but irregular marriage, abolished as immoral by the Council of Trent, and *sponsalia de futuro*, a promise to marry later. Betrothal validly contracted, could be dissolved only by consent or by failure to fulfil one of the conditions of the contract.

In the United Kingdom it partook at common law of the nature of a commercial contract, so much so that damages could be recovered if the engagement were unjustifiably broken off or not fulfilled, but in England damages for breach of promise have now been abolished, it being thought better that parties should be free to change their minds without penalty.

Betting, Gaming, Wagering, and Lotteries. Betting takes place where one or more persons hazard a sum of money or other stake on the result of a uncertain event, e.g. on which football matches result in a particular way. Gaming takes place where several persons play a game of skill or chance for stakes hazarded by the players, e.g. roulette. Wagering exists where two persons agree each to hazard a sum or other stake on the result of some unknown fact or uncertain event; there can only be two parties or sides, and one party must win and the other lose, e.g. on the sex of an expected child. A lottery is a scheme for awarding prizes by lot or chance; it is not a lottery if there is an element of skill, even if chance also plays a part.

In Britain betting is lawful at licensed betting shops or with bookmakers holding permits at racecourses. Gaming is lawful commercially only at premises licensed by the Gaming Board for Great Britain, and gaming elsewhere is restricted by statute. Lotteries are in general illegal, but there are exceptions for small lotteries incidental to social, charitable, sporting or athletic activities.

In English law all contracts by way of gaming or wagering are void, and money lent is not recoverable: securities given for betting or gaming are deemed given for an illegal consideration.

In Scotland contracts by way of gaming or wagering are not void but unenforceable, being deemed *sponsiones ludicrae*, and lotteries are illegal by statute only.

Beveridge Report. A report on social insurance and allied services prepared at the request of the British Government in 1942 by Lord Beveridge. It recommended the establishment of an integrated system of social security which would secure freedom from want. It advocated the establishment of a national health service and a system of family allowances and assumed the implementation of a policy of full employment. It was warmly welcomed and took on something of the character of a charter of human rights. Most of its recommendations were implemented in the years 1945–50.

Beyer, Georg (1665–1714). German jurist, professor at Wittenberg, the first to lecture on German law, with whom began the academic study and literature of German private law. His lectures were published as *Delineatio Iuris Germanici ad Fundamenta sua Revocati* (1718) and aimed to present German institutions as set out in the older sources, to deduce their historical development and procure systematic treatment for German law, separate from Roman.

Beyma, Julius van (1540–98). Dutch jurist and humanist, author of *De justitia et jure* (1596), *Commentaria in varios titulos juris* (1645), and *Commentarius in titulum de Verborum Significatione* (1649).

Beyrout. See BERYTUS.

Bias. An inclination to decide an issue influenced by any consideration other than its merits. The avoidance of actual or apparent bias is an essential of all exercise of judicial power. Accordingly, no judge may act in a case in which he has any pecuniary interest, however small or indirect, and any judicial expression indicative of bias would vitiate a decision. Legislation frequently displays bias, such as against landlords or employers.

Bicameral system. System of government in which the legislature consists of two houses, commonly modelled on the British House of Lords and House of Commons or the U.S. Senate and House of Representatives. In theory each house acts as a check on the other, prevents hasty or harsh legislation and enforces deliberation and sometimes reconsideration. In federal states one house commonly represents the states, and the other the people. Since 1945 there has been a move to unicameralism, particularly in non-federal states, and in bicameral legislatures the power of the lower house has steadily been gaining at the expense of the upper house.

Bickersteth, Henry, Lord Langdale (1738–1851). Studied medicine and mathematics before turning to law, and became a friend and admirer of Burdett and a disciple of Bentham. He was called in 1811, was successful at the Bar and was offered various judicial posts before, in 1836, becoming Master of the Rolls and a peer. In the House of Lords he took small part in politics but devoted attention to the promotion of legal reforms. He was mainly responsible for the Act of 1838 establishing the Public Record Office and initiated its work. He declined the Chancellorship in 1850. He was not so good a judge as he had been a counsel, but some of his decisions have been important on points of equity.

T. D. Hardy, *Memoirs of Lord Langdale.*

Bigamy. The offence of purporting to marry another during the subsistence of a valid prior marriage. Originally an ecclesiastical offence it was made a capital felony in 1603. No offence is committed if the first marriage is legally void, but it is even if the second 'marriage' is invalid, quite apart from the fact that it is bigamous, because the essence of the crime is going through the ceremony and appearing to contract a second marriage. Nor is an offence committed by mere cohabitation with one other than one's lawful spouse, even if the cohabitation had gained the reputation of being marriage, nor if a former marriage has been dissolved by law. In English law it is a defence to show that the accused's spouse had been continuously absent for seven years and not have been known to be living within that time, but this does not affect the subsistence of the first marriage.

The offence can probably not be committed by a person whose personal law permits polygamy and who had married more than once in the country in which he was domiciled.

Bigelow, Melville Madison (1846–1921). Was admitted to the Tennessee Bar in 1868 but removed to Boston in 1870 and became a member of faculty when Boston University Law School was opened. His main writings were *The Law of Estoppel* (1872), *Leading Cases on the Law of Torts* (1875), *Elements of the Law of Torts* (1878), *Elements of Equity* (1879), *The Law of Bills, Notes and cheques* (1880), and the *Law of Fraud*, (1888–90); and above all, his *Placita Anglo-Normannica* (1879), law cases from William I to Richard I, and the *History of Procedure in England from the Norman Conquest 1066–1304* (1880) are both works of great and permanent value on English legal history.

Bigham, John Charles, Viscount Mersey (1840–1929). Was called to the Bar in 1870 and developed a large practice in commercial cases. He took silk in 1883 and became an M.P. in 1895, but made no figure in the House. In 1897 he became a judge, and regularly sat in the commercial court. He also presided over the Railway and Canal Commission. In 1909 he was made President of the P.D.A. Division but ill-health forced his retirement in 1910, when he became a peer. Thereafter, however, he did much public and judicial work, on committees of inquiry, on the Privy Council and as president of the board hearing appeals in prize in the years 1914–16. He was made a viscount in 1916. He was an able judge though tending to short-circuit proceedings, and preferred a common sense approach to tedious and over-refined arguments.

Bigot de Préameneu, Felix-Julien-Jean (1747–1825). Advocate at the *Parlement* of Paris, later a moderate member of the legislative assembly of 1791 and Counsellor of State. With Tronchet and Portalis he was designated to draft the French Civil Code. Later he was Minister of the Cults and, during the Hundred Days, Minister of State.

Bill. (1) A word probably derived from Fr. *Libelle*, and a generic term in English law for many forms of complaint or petition, including a written information as to the commission of a crime which became an indictment if found to be true by a grand jury, a complaint against the King, the initial writ for an action against a court official, the request sent up by the House of Commons to the King, which developed into the first draft of proposed legislation, and a petition sent up to the chancellor and council, which came to be the method of initiating proceedings in the Court of Chancery. There are also numerous other uses of the word.

In Scots law the word was formerly applied to various kinds of applications to a court.

(2) A draft or proposed piece of legislation put before Parliament for its consideration, and, if thought fit, for enactment with or without alterations.

Bills are or three kinds:

Public bills, which relate to matters of public policy, are introduced by a Minister in pursuance of government policy, and normally have general application to the United Kingdom, or to England, or to Scotland or Northern Ireland, or to a major area such as London. The presumption is that they apply to the whole United Kingdom. Public bills originated from petitions from the Commons on behalf of the nation, and from 1414 it became the custom to send the King not a petition but a draft which could be accepted or rejected by him but not amended. Public bills are drafted by Parliamentary Counsel to the Treasury, the principles having been laid down in instructions from the department concerned.

Private members' bills, which relate to matters of public policy, but are introduced by a private member to effect what he regards as a desirable reform in the law, and normally have general application to the United Kingdom, or to England, or to Scotland. At the commencement of each session private members ballot for the opportunity to bring in a bill in the limited time available for this purpose. A private member must draft his own bill, unless the Government takes it up, when it may lend the assistance of a Parliamentary draftsman, and may also make government time available for it. Only a few private members' bills become law each session, but these sometimes deal with matters of considerable value and importance, e.g. the Matrimonial Causes Act, 1937, and the Murder (Abolition of Death Penalty) Act, 1965, and many lesser but worthwhile pieces of legislation, especially on protection of animals, superannuation, and matters of social welfare originated as private members' bills. Procedurally a private member's bill proceeds as a public bill.

Private bills confer powers or benefits on particular persons, local authorities, statutory companies, and private corporations. Such a bill is solicited by the parties interested themselves, is founded on a petition and is regulated by a special procedure.

If it is found that certain of the Standing Orders relative to private business are applicable to a public bill, it is then treated as a hybrid bill and its passage governed by special procedure. A hybrid bill is a public bill which affects a particular interest in a manner different from the private interests of other persons or bodies in the same category or class. Thus the London Passenger Transport Bill 1930 was treated as a hybrid bill.

See also PARLIAMENTARY PROCEDURE; PRIVATE BILL.

Bill Chamber. A court which originated in 1532 as a vacation court of the Court of Session (q.v.) and continued to have this function till abolished in 1933. The name was derived from the Bills or petitions presented to the court. The court was held before the junior Lord Ordinary of the Court of Session (the Lord Ordinary on the Bills) and was a distinct sub-court with its own staff and procedure. In time it acquired a special jurisdiction, not confined to vacations, relating to privileged summonses, certain forms of diligence, the granting of suspension or suspension and interdict, and the administration of powers under miscellaneous statutes. It had a large volume of special procedure. In 1933 it was merged in the Court of Session.

Bill of Advocation. In Scottish criminal procedure, the initial writ addressed to the High Court of Justiciary requesting it to bring up a case from an inferior court for review, particularly of errors in the course of the trial.

Bill of Attainder. A bill brought in Parliament introduced in the House of Lords to declare a person attainted, that is, subject to the stain or corruption of blood formerly incurred by a criminal conviction for treason or felony. It proceeds as an ordinary public bill but the accused may be heard by counsel and call witnesses before both Houses. The most notable use of this form of parliamentary judicature was against Wentworth, Earl of Strafford in 1641. It has not been used since the eighteenth century. See also ATTAINDER.

Bill of complaint (or bill in Chancery, or bill in equity). In the Court of Chancery (q.v.) prior to 1875 an ordinary suit was begun by filing or exhibiting a bill of complaint in the form of a petition to the Lord Chancellor, asking for certain relief. The term 'bill' was accordingly frequently used as equivalent to 'suit' or 'claim in the Chancery court'. There were many kinds of bills, varying according to the relief sought.

Bill of costs. A solicitor's account of his fees, charges, and disbursements. Delivery of a bill is a prerequisite to an action for recovery of the sum claimed.

Bill of Exceptions. An old mode of appeal, whereby a party to an action could except or object to the trial judge's ruling on a particular matter though it did not appear on the record of the trial, by recording the ruling objected to, having the judge seal it and then arguing the rightness or wrongness of the ruling before the full court. Such trials were introduced in 1285 by the Statute of Westminster II and abolished by the Common Law Procedure Act in 1852. The procedure survives in Scotland as a means of challenging a trial judge's direction on some points to a jury.

Bill of Exchange. An unconditional order in writing addressed by one person to another signed by the drawer, requiring the addressee to pay on demand, or at a fixed or determinable future time, a sum certain in money to, or to the order of, a specified person, or to the bearer. Bills originated in the practices and usages of merchants, which were settled by the decisions of courts, and have become incorporated in common law. In the United Kingdom the case-law was largely codified in the Bills of Exchange Act, 1882, and this has been adopted or adapted in most Commonwealth countries and in the U.S.A. A bill is a means whereby if A owes money to B but is owed money by C he requests C to pay B direct by drawing a bill on C payable to B. Bills of exchange are the principal examples of the category of negotiable instruments which are written orders or promises to pay definite sums of money which may be transferred from one person to another by indorsement or by delivery, so as to enable the transferee to sue thereon in his own name, for which valuable consideration is presumed and which conveys a good title to the transferee who takes in good faith and for value, in spite of any defect of title in the transferor. Promissory notes (q.v.) and cheques (q.v.) are special forms of bills of exchange.

A bill is drawn by writing an order 'Pay X the sum of £Y' addressing it to the debtor (drawee) and dating and signing it. Bills and notes are payable on demand or at a fixed or determinable future time. Cheques are always payable on demand.

A bill, once drawn, is normally presented to the drawee for acceptance, which signifies his assent to the order to him in the bill; this is done by his writing on the bill words signifying acceptance, signed by the drawee. An acceptance may be general or qualified. Acceptance is excused in certain cases.

Bills are an important class of negotiable instruments (q.v.) but the negotiability of a particular bill may be restricted. A bill is negotiated, if payable to order, by indorsement by the payee and delivery to the new holder, and if payable to bearer, by delivery alone. A bill may thus pass from hand to hand just as does money. A cheque drawn on a banker is a special kind of bill of exchange, to which certain special rules are applicable.

The person primarily liable on a bill is the acceptor to whom the order to pay was addressed. If he cannot pay the payee, the drawer and indorsers are jointly and severally liable to the payee. Each indorser by indorsing undertakes that, if the bill be dishonoured, he will compensate the holder or a subsequent indorser who is required to pay, provided the requisite proceedings on dishonour are taken.

When a bill falls due for payment the holder (either the original payee or another to whom it has been negotiated) presents it to the acceptor and, if paid, delivers up the bill. If the bill is dishonoured by non-payment the holder protests the bill, by handing the bill to a notary public who again presents it and if it is again dishonoured notes the fact on the bill and extends a formal document recording the dishonour and gives notice of dishonour to the drawer and each indorser whose name appears on the bill.

A bill is discharged by payment, renunciation or cancellation by the holder, merger, material alteration, renewal by giving a new instrument in place of an existing one, and by any other form of discharge applicable to a simple contract to pay money.

J. B. Byles, *Bills of Exchange*.

Bill of rights or Declaration of rights. The general term for general statements declaratory of the human or civil rights which the proposers assert individuals should enjoy against states and governments. By themselves such general statements may generate public opinion and influence governments but have no legal force or validity. They can have such force only if accepted by a government as a limitation of its own powers and if the rights declared are enforceable against the government and actually recognized by it. Early declarations of this character, though directed against particular grievances rather than asserting rights comprehensively, were the English Petition of Right, 1627, and the Bill of Rights, 1688 (qq.v.), but the earliest general declarations proclaiming that stated natural rights of man must form part of the fundamental law of the state and be protected by it were the Constitution of Virginia of 1776, the American Declaration of Independence, 1776, and the French Declaration of the Rights of Man and of the Citizen, 1789, prefixed to the Constitution of 1791 and in varying words to later Constitutions. In the nineteenth and twentieth centuries declarations of rights were incorporated in the constitutions of nearly all European states, and the practice has been widely followed all over the world. Britain and most Commonwealth countries have as yet failed to adopt the practice, though Canada accepted a Bill of Rights in 1960.

In Britain, in the absence of a Bill of Rights and machinery for its enforcement, no rights are safe against Parliament, which is controlled and managed by the government of the day, which in turn may be subservient to outside influences, such as trade unions or international political organizations. The need for a general Bill of Rights in Britain seems clear to many people.

No Bill or Declaration of Rights is of any value unless the rights can be actually vindicated against the government. Russia and East European countries have substantial declarations of rights in their constitutions but these are wholly ineffectual. In Britain the statutory duty on education authorities to provide education for children in accordance with the wishes of their parents is blatantly disregarded by education authorities, while the liberty of individuals not to join, or to leave, a trade union has been in many cases excluded.

In the twentieth century progress was made at the international level. The promotion of respect for human rights and fundamental freedoms is an object of the U.N.O. In 1948 the U.N. adopted a Universal Declaration of Human Rights and in 1950 the Council of Europe adopted the European Convention for the Protection of Human Rights and Fundamental Freedoms, enforceable through the European Commission of Human Rights and the European Court of Human Rights. U.N.E.S.C.O. maintains a special commission on Human Rights and the Status of Women and the I.L.O. has done much in a practical way to secure rights in the employment field.

Bill of Rights (U.K.). A statute passed by the Convention Parliament of England in December 1689 as part of the Revolution Settlement (q.v.) for declaring the rights and liberties of the subject and settling the succession to the Crown. It was based on the Declaration of Right of 1688 accepted by William of Orange on accepting the Crown of England, and was declaratory of the law rather than an introduction of new principles, and may be said to state the constitutional terms on which William and Mary were accepted as king and queen.

It narrates the unconstitutional acts of James II, declares his actions illegal, mentions the summoning of the Convention Parliament and declares (1) that the pretended powers of suspending laws or execution of laws by royal authority without consent of Parliament, and (2) of dispensing with laws or the execution of laws by regal authority, are illegal; (3) that the commission for erecting the late Court of Commissioners for Ecclesiastical Causes, and all other commissions and courts of like nature, are illegal; (4) that levying money for the use of the Crown by pretence of prerogative without grant of Parliament is illegal; (5) that it is the right of the subjects to petition the king; (6) that the raising or keeping a standing army in time of peace, unless with the consent of Parliament, is illegal; (7) that the subjects which are Protestants may have arms for their defence as allowed by law; (8) that election of Members of Parliament ought to be free; (9) that the freedom of speech and debates or proceedings in Parliament ought not to be impeached or questioned in any court or place out of Parliament; (10) that excessive bail ought not to be required, nor excessive fines imposed, nor cruel and unusual punishments inflicted; (11) that jurors ought to be duly impanelled and returned, and jurors which pass upon men in trials for high treason ought to be freeholders; (12) that all grants and promises of fines and forfeitures of particular persons before conviction are illegal and void; (13) and that for redress of grievances and for the amending, strengthening, and preserving of the laws, Parliament ought to be held frequently.

It then confirms the succession of William and Mary and provides for the succession on their deaths, provides a new form of oath of allegiance and supremacy, narrates the acceptance by William and Mary of the Crown, settles the succession, and provides for the exclusion from the succession to the Crown of persons who profess the popish religion or marry a papist and requires every King and Queen of the realm to make the declaration against popery. Finally, it abolished the dispensing power.

The Bill of Rights is one of the great constitutional documents of English history and one of the great charters of the rights and liberties of the subjects under English law.

Bill of Rights (U.S.). The first 10 amendments to the U.S. Constitution, proposed by Congress in 1789 at the instigation of Madison, and ratified in 1791. It was modelled on the Virginia Declaration of Rights of 1776 (q.v.) and formulates as the main rights entitled to legal protection the main liberties usually considered inherent in societies derived from the United Kingdom. Many of the framers of the Federal Constitution believed that enumeration of individual rights was unnecessary and unwise, since the specification might imply extension of the sphere of governmental activity. Others believed that a statement of rights would have a restraining influence on governmental encroachments on individual rights. Many states, when ratifying the Constitution, recommended that rights be written into the Constitution without delay.

The rights guaranteed are: (1) freedom of religion, speech, the press, peaceable assembly and petition for redress of grievances; (2) right to keep and bear arms; (3) no quartering of soldiers in peacetime; (4) security against unreasonable search and seizure; (5) capital charges answerable only on indictment of a grand jury; no double jeopardy for the same offence; no requirement of self-incrimination, nor deprivation of life, liberty or property without due process of law, nor taking of private property without compensation; (6) in criminal cases, right to trial by jury, to know the charge and be confronted by the witnesses; (7) in common law actions, jury trial preserved; (8) no excessive bail, excessive fines, nor cruel and unusual punishment; (9) enumeration in the Constitution of certain rights not to be construed to deny or disparage others retained by the people; (10) powers not delegated to the United States, nor prohibited by it to the States, reserved to the States respectively, or to the people.

Bill of Sale. In English law a written instrument whereby one transfers to another his property in goods or chattels, growing crops, fixtures, and trade machinery, absolutely or in security, giving the other a title without delivery to him of the goods. They are largely regulated by Act of 1878 and 1882, and whether a document is within the Acts or not depends on the substance of the transaction.

A bill must be registered at the Central Office of the Courts in London within seven days, by presenting to the registrar the bill, a true copy thereof, and an affidavit verifying execution and attestation. Bills have priority according to the date of registration and after five years re-registration is necessary. There are penalties for non-registration. A bill may be transferred or assigned without need for re-registration.

Under an absolute bill both property and right of possession pass to the grantee of the bill on execution thereof.

A bill by way of security in statutory form vests the chattels in the grantee, leaving the right of possession in the grantor until the occurrence of a statutory 'cause of seizure'. When the grantee exercises his right of seizure the grantor's legal interest in them ceases, though he may apply to the court for relief or take proceedings for redemption.

Bill of Suspension. In Scots law, a mode of challenging the justice of a conviction in the inferior criminal courts, on such grounds as lack of jurisdiction, incompetence of the proceedings, or departure from the rules of natural justice.

Bills in Chancery. The initial pleadings put before the mediaeval chancellors praying for equitable relief, usually praying that a subpoena (q.v.) should be issued to secure the appearance and examination of the defendant.

Bills in eyre. Complaints directed by aggrieved persons to the justices in eyre (q.v.) in the period 1280–1330, requiring various reliefs for very varied grievances in non-technical form. They were not the ancestors of bills in equity addressed to the chancellors but sought equity from common law judges. But this jurisdiction declined and equity became available from chancery only.

W. C. Bolland, *Select Bills in Eyre.*

Binding over to keep the peace. In England a magistrates' court has power, possibly at common law, possibly based on the Act of 1361, which created the office of Justice of the Peace, to bind a person over to keep the peace, i.e. to order him to enter into a recognizance (q.v.), with or without sureties to keep the peace and be of good behaviour for a reasonable time. If he refuses or fails to do so he may be imprisoned for not more than six months. Magistrates' courts also have statutory powers on receiving a complaint, to bind over to keep the peace or be of good behaviour. Such order may be made without a conviction, by way of preventive justice.

Binder, Julius (1870–1929). German jurist, wrote on Roman law, German civil law and legal philosophy; originally a neo-Kantian he later became a neo-Hegelian. His chief works are *Philosophie des Rechts* (1925) and *System der Philosophie des Rechts* (1938).

Birkenhead, Lord. See SMITH, F. E.

Birth. The extrusion of a child from its mother's womb and its commencement of an existence separate from her. There is some uncertainty as to when a child has a separate existence, and by what this is tested, by independent circulation or by being observed to have breathed, or to have cried. Prior to its birth, a child is a living being and, unless the birth is a natural miscarriage or an abortion (q.v.) in circumstances legally authorized, causing its death by abortion is criminal and a wilful act causing a child capable of being born alive, to die before it has an existence independent of its mother is in English law a statutory crime (child destruction). It is not murder to kill a child in the womb or while in the process of being born. At common law the unborn child does not have legal personality and cannot claim for ante-natal pain, suffering or deformity, though in some jurisdictions an action has been allowed for having negligently caused the child to be injured before its birth. In some cases, however, particularly of succession, courts have treated an unborn child as if born for the purposes of a benefit (*nasciturus pro iam nato habetur*) and to that extent have conceded a measure of personality provided that the child is later born alive. But no action would lie in favour of a still-born foetus.

A birth must be registered with the local registrar.

Birth, concealment of. A crime created originally in 1623 because of the difficulty of proving that a mother had murdered her child, to make criminal the attempt by anyone, by secret disposition of the dead body of a child, to endeavour to conceal its birth.

Birth control. The general name for methods of preventing conception, frequently used to comprehend all fertility control including abortion. It involves difficult religious, moral, psychological and social and economic as well as legal problems.

In the nineteenth century advocacy of birth control and information on contraceptive methods was attacked in England as the selling of pornographic literature. In the U.S. the Comstock Act penalizing contraception was passed in 1873. But in the twentieth century birth control has increasingly been recognized and governmentally approved.

Today in the U.K. birth control information and advice is legal and increasingly widely available.

In British law the use of any method of birth control other than total abstinence from sexual relations does not prevent consummation of a marriage, nor is insistence on the use of methods of birth control by one spouse against the will of the other a ground for divorce but it may be a factor of disharmony leading to the breakdown of the marriage.

See also ABORTION; INFANTICIDE.

Bishop. In some Christian churches, the chief pastor and overseer of a diocese or area containing a number of congregations. In the pagan Hellenistic world the word *episcopus* was also applied to various officials, stewards and managers. From early in the Christian church the three grades of clergy, bishops, presbyters or priests, and deacons seems to have been well established.

In the Church of England a bishop is nominated by the Crown, by granting to the dean and chapter of the cathedral of the vacant diocese a licence under the Great Seal (a *congé d'élire*) to elect a bishop with a letter missive containing the name of the person to be elected, or by letters patent. Apart from purely ecclesiastical powers and functions a bishop may have, or by seniority attain to, a seat in the House of Lords and has a seat in the House of Bishops of the General Synod of the Church of England and in the Upper House of the Convocation of the province of Canterbury or of York. He is chairman of the diocesan synod. He must appoint a chancellor of the diocese with authority to execute for him the office of ecclesiastical judge in the consistory court of the diocese, but may reserve the right to exercise the ordinary jurisdiction by himself. He may be called on to sit as one of the clerical members of the Court of Ecclesiastical Causes Reserved (q.v.).

In other Protestant Episcopal churches, election is made in other ways, as by a diocesan convention. In the Catholic Church appointment of bishops is made in various ways but final appointment is subject to the approval of the pope. The bishop has right and duty to govern the diocese and possesses legislative, judicial, and coercive power to be exercised by the Code of Canon Law. He may make laws in or out of synod and these bind all his subjects unless specially exempted, so long as they do not conflict with common or particular laws of the Church or of a council. He may act as judge in any ecclesiastical cause not reserved to a higher judge. Normally this power is exercised by the official of the diocese, a judge appointed by the bishop and acting in his name and with his authority.

Bishop, Joel Prentiss (1814–1901). Practised law in Boston, but is remembered for his legal writings, especially his *Commentaries on the Law of Marriage and Divorce* (1852) and *Commentaries on the Criminal Law* (1856–58). His numerous other works included *Commentaries on the Law of Criminal Procedure* (1866), *on the Law of Contracts* (1887), and *on the Non-contract law or Torts* (1889).

Bishops, Case of the Seven. The trial, in 1688, of Archbishop Sancroft and six bishops for having petitioned James II against his ordering the Declaration of Indulgence (q.v.) to be read in all churches. They were acquitted and this reinforced the claim that juries in prosecutions for libel could return general verdicts (q.v.).

Bitsch, Konrad (fifteenth century). German theologian and jurist, author of *De vita conjugali* (1432) and *Commentarius in consuetudines feudorum* (1673).

Black Acts. The name given to an edition, printed in black letter, of Acts of the Scottish Parliament for the period 1535–94.

Black Book of the Admiralty. A work probably compiled by an official of the Admiralty in the reign of Henry VI, containing regulations for the Admiral and the fleet about wages, lights, prizes, visit and search, compensation for collisions, and other matters, and including a copy of the laws of Oléron. It included also a tract on the *ordo judiciorum*, showing that the Admiralty court modelled itself on civil law procedure. It was edited by Sir Travers Twiss for the Rolls Series in 1871–76.

Black Book of the Exchequer (or *Liber Niger Parvus*). A book incorporated by Alexander de Swereford in the *Red Book of the Exchequer* (q.v.) and possibly for that reason attributed to him. It dates from the early thirteenth century. Its contents are very mixed, treaties, Papal bulls, *cartae* and an account of the royal household in Henry II's reign, and it illustrates points of legal interest about the feudal government of England. It was edited by Hearne and printed in 1728, 1771 and 1774.

Black Book of the Tower. Also called *Liber Irrotulamentorum de Parliamentis* and *Vetus Codex*, a book in the Public Record office consisting of extracts from the Rolls of Parliament of the time of Edward I and II. It was used as a source in constitutional disputes in the sixteenth and seventeenth centuries and printed in 1661 with other materials under the name *Placita Parliamentaria.*

Black cap. A square cap worn by judges of the High Court on various State and solemn occasions on top of their wigs. It was formerly worn when imposing sentence of death.

Black Code. French laws put into effect in 1685 concerning the treatment of slaves.

Black Codes (U.S.). The bodies of state laws developed during the seventeenth to nineteenth centuries determining the jurisdiction over rules of conduct and punishment of slaves. In the seventeenth century a stringent set of rules was in effect in Virginia and elsewhere, but the codes underwent continuous adaptation. Though they varied there were substantial common provisions. The basic principle was that slaves were chattels not persons and that law must protect the right of property and secure the slave owners from possible violence. Neither state nor church sought to protect the slaves.

Any amount of negro blood determined the race of a person, free or slave. The status of the child was determined by that of his mother. Slaves could not contract, own property, strike a white person even if attacked or give evidence against a white, could not assemble in groups, be absent from their owner's premises without permission, own firearms or be taught to read or write.

Obedience to slave codes were enforced by state tribunals, by harsh punishments (though seldom by death) and by white patrols which kept watch over slaves. If there was danger that slaves might become ungovernable the laws were more strictly enforced.

After the Civil War southern states passed new Black Codes to control and suppress freed slaves and to maintain white supremacy, and to ensure a continued supply of cheap coloured labour. The Black Codes continued to assume the inferiority of freed slaves by vagrancy laws which imposed heavy penalties designed to compel all blacks to work, and apprentice laws providing for the hiring of young dependants to former owners. Some states forbade negroes to own property or restricted them to menial occupations. Marriage was provided for but intermarriage prohibited and many activities were subjected to restriction. The right to vote was completely denied. These codes convinced many northern leaders that the South had not truly accepted emancipation and this led to the era of Military Reconstruction (1867–77) during which there was heavy reliance on the Freedmen's Bureau to defend the interests of former slaves.

Black List. A common term for a list of persons with whom the addressees of the list are not to deal. Thus trade associations sometimes issue lists of traders who are not complying with the rules of the association. The issue of such a list is permissible if made in pursuit of a legitimate trade object and without ill-will towards the person injured but it is actionable if issued for the purpose and with the effect of injuring the person in his business, or to intimidate or coerce, or is defamatory. The term is also used of persons listed as defaulters on the Stock Exchange.

Black Market. The dealing in commodities or services in a prohibited way, e.g. in excess of a controlled price.

Black Parliament. A Scottish Parliament which met at Scone in 1320, thus named because of the savage punishments which it imposed on those concerned in the conspiracy of Sir William de Soulis against King Robert I (the Bruce). The name is sometimes also given to the Reformation Parliament of England (1529–36) by those who opposed its measures.

Black Rod (properly, the Gentleman Usher of the Black Rod). An officer of the House of Lords, traditionally a retired naval or military officer of high rank. The appointment originated in 1361 with the foundation of the Order of the Garter and Black Rod still attends Garter ceremonies, and the parliamentary appointment dates from the reign of Henry VIII. He is appointed by letters patent and carries an ebony stick surmounted with a gold lion. In the House he controls the admission of strangers, assists at the introduction of peers, and commits parties guilty of breach of privilege or contempt. He summons the Commons to the Lords at the opening and prorogation of Parliament, and formerly also to hear the Royal Assent signified to Acts. His deputy is the Yeoman Usher of the Black Rod, an office now combined with that of Serjeant-at-Arms in the House of Lords.

Blackburn, Colin Blackburn, Lord Blackburn of Killearn (1813–96). Was called to the Bar in 1838 but made slow progress. He wrote a very valuable book on the law of sale of goods, and collaborated in producing several volumes of reports (1852–58). In 1859, though he had not taken silk, he was made a judge of the Queen's Bench, and soon proved himself the most learned member of the court. In 1876 he became one of the first Lords of Appeal in Ordinary and was sworn of the Privy Council. He resigned as Lord of Appeal in 1886. He served on the commission on the courts in 1867 and was chairman of the commission on the draft commercial code in 1878. His greatest judicial work was as Lord of Appeal and his judgments illuminated many branches of the law and furthered their development.

E. Manson, *Builders of our Law.*

Blackburne, Francis (1782–1867). Was called to the Irish Bar in 1805, became Attorney-General for Ireland (1830), Master of the Rolls in Ireland (1842), Chief Justice of the Queen's Bench in Ireland in 1846, Lord Justice of Appeal (1856), and Lord Chancellor of Ireland in 1852 and in 1866–67.

Blackleg. The term applied to a person who refuses to join an appropriate trade union, or who participates in strike-breaking by working for an employer whose regular workmen are on strike.

Blackmail (originally black, i.e. unlawful, and mail, i.e. rent). The tribute exacted from landowners by Scottish robbers as the price of immunity from raids i.e. 'protection money'. Now it is the common name of the crime of, with a view to gain for himself or another or with intent to cause loss to another, making an unwarranted demand for money or other benefit with menaces. A demand is unwarranted unless the person making it does so in the belief that he has reasonable grounds for making the demand and that the use of the menaces is a proper means of reinforcing the demand. The nature of the act or omission demanded is immaterial. It is a form of extortion.

Blackstone, Sir William (1723–80). Read classics and mathematics at Oxford, became a Fellow of All Souls, and was called to the Bar in

1746. He served as assessor of the Chancellor's court at Oxford, a delegate of the University Press, and took a large part in settling the scheme for the Vinerian Professorship of English law in 1756–58. From 1753 to 1758 he lectured privately at Oxford on English law, the first lectures on English law ever given in a university. In 1758 he became Vinerian Professor and lectured on English law there till 1766, and his annual lectures became the basis of his *Commentaries.* His lectures spread his fame and his progress at the bar was rapid. He entered Parliament and took silk in 1761, became Solicitor-General to the Queen in 1763 and in 1766 he gave up his chair. In 1770 he became a judge of the King's Bench and later in the same year he transferred to the Common Pleas where he proved himself an able judge and sat till his death. His second son, James, was Vinerian Professor 1793–1824.

His first work was on *Elements of Architecture* (1743) but his reputation rests on his legal works. These are an *Essay on Collateral Consanguinity* (1750), *Analysis of the Laws of England* (1754) a synopsis for those who attended his lectures, an essay on whether tenants in ancient demesne were freeholders (1758) and an edition of the Charters of John, Henry III and Edward I (1759). The last is a work of the highest importance for the historian and constitutional lawyer and shows that Blackstone was a very competent and learned historical scholar. His major work was the *Commentaries on the Laws of England* (1765–70) in four books (each occupying a volume) dealing respectively with the Rights of Persons, the Rights of Things, Private Wrongs, and Public Wrongs. It was an immediate success and eight editions appeared in his lifetime and at least 15 more after his death, while in 1848–49 Serjeant Stephen produced new *Commentaries* founded on Blackstone but adapted to the changed state of the law. The *Commentaries* were equally popular in the U.S.A. and passed through many editions there, being the chief source of knowledge of English law for many years and it inspired Kent's *Commentaries on American Law.*

Though Blackstone was not at all a scientific jurist the book was at once recognized as a classic by reason of the breadth and depth of learning displayed, the systematic and logical structure of the book, the accuracy of statement, and the literary grace with which the matter is presented. The book was also the best historical account of English law which had yet appeared, and the first exposition of that law as a system, in a connected narrative, and it had considerable influence on the subsequent development of the law. The work was criticized for its acceptance of the existing state of the law, notably by Bentham (q.v.) in his *A Fragment on Government* (1776), but this was no more than the common attitude of his day, and he does criticize many of the rules of law. His account of the law of contract is undeveloped and he treats equity (q.v.) very lightly. It remains among the half-dozen classics of English law, and has been cited and relied on repeatedly in courts.

Clitherow's *Life*, prefixed to Blackstone's Reports, vol. I; H. G. Hanbury, *The Vinerian Chair*; Holdsworth, XII, 702.

Blair, Robert, of Avontoun (1741–1811). Passed advocate in 1764 and acquired a considerable practice; he became Solicitor-General for Scotland in 1789–1806 and Dean of the Faculty of Advocates in 1801. He twice refused promotion to Lord Advocate, and also promotion to the bench. In 1808 he became Lord President of the Court of Session but died suddenly in 1811. He has a good grasp of legal principles and a strong love of justice, but did not hold office long enough to make a great mark.

Blanesburgh, Lord. See YOUNGER, ROBERT.

Blasphemy. The crime, of religious origin, with counterparts known in the Jewish and Roman law, of orally or in writing, scoffingly, or irreverently ridiculing or impugning the doctrines of Christianity, uttering or publishing contumelious reproaches of Jesus Christ, or profanely scoffing at Holy Scriptures or exposing any part of them to contempt or ridicule. It may be confined to Christianity and not other religions, and to the Church of England. The basis of the crime is said to be the tendency to shake the fabric of society generally. It is not now committed by serious and restrained argument that the doctrines of Christianity, or the Scriptures, in whole or in part, are unfounded or mistaken: *Bowman* v. *Secular Society*, [1917] A.C. 406.

G. D. Nokes, *History of the Crime of Blasphemy.*

Blastares, Matthew (fourteenth century). Byzantine monk and canonist, compiler of the *Syntagma Kata Stoicheion* (1335) an encyclopaedic compilation of civil and ecclesiastical laws, with commentaries by himself and others. It is arranged in Greek alphabetical order of titles, and emphasizes ecclesiastical law but includes more extensive excerpts from related civil law than any former collection. It was extensively translated and influenced late Byzantine legal codes and those of other nations. It was translated into Slavonic, appeared in a Bulgarian version in the fifteenth century and a Russian version in the sixteenth. By the eighteenth century it was recognized as the standard exposition of Eastern Orthodox Canon law. It helped to establish Slavonic customs as to legal procedure and laws regulating state protection of the poor. It also transmitted to later generations the concept of a political realm ruled by a sovereign and subjected to his law, and transcending the interests of individuals and classes. By basing the authority of the emperor and of the church hierarchy in a single, divine source, his work communicated to Eastern Europe the ideal of a harmonious relationship between church and state.

Blockade. The right of a belligerent who holds his enemy besieged to forbid access by sea to and egress from enemy territory by the enemy and by neutrals. It must be applied impartially to vessels of all nations. The customary rule was that ports had to be invested by sea and by land. In 1584 the Dutch declared blockaded all the ports of Flanders still in Spanish hands, though they could not be blockaded effectively. In the eighteenth century treaties sought to define the measures of force which a belligerent had to possess in front of a port before it could be deemed blockaded. The Armed Neutralities permitted belligerents blockading rights over a port only where the attacking power had vessels before the port and sufficiently near to make access clearly dangerous. In the Napoleonic War the British, in 1806, proclaimed a blockade of the French coast from the Elbe to Brest. Napoleon made a counter-declaration of blockade of the whole British Isles and excluded British goods from the continent. The British blockade was extended in 1807 to all French ports and those of every country at war with Britain. In 1856 the Declaration of Paris laid down that to be binding a blockade must be effective, but in the American Civil War the North declared the whole of the Confederate States' coasts blockaded and extended the concept of blockade to ships proceeding to an intermediate port from which the goods might be taken on through the blockade. The Declaration of London, 1908, severely restricted rights of blockade but was not ratified. In the First World War blockade of Germany was never legally proclaimed and seizure and condemnation of cargoes was done under the rules of contraband. An Order in Council of 1915 made goods of presumed enemy destination or origin liable to be taken into port and to be confiscated if contraband.

Though blockade is universal, particular vessels may be licensed to enter or leave for particular purposes and warships of all neutral nations may be allowed to pass in and out.

Blockade does not come into being automatically by reason of war and if not notified generally, a neutral vessel could probably not be condemned for breach of blockade and it is customary to allow neutral vessels time in which to leave blockaded ports.

Breach of blockade is constituted by an attempted or actual passage through the area of a blockade to or from a blockaded port, unless by a vessel needing repairs, stores, or otherwise in need. Vessels which breach blockade may be captured and sent into port to be brought before a Prize Court.

Apart from termination of hostilities blockade may be terminated by being restricted or raised, or if the defending fleet drive off the blockading one, or if it is ineffective. To be effective a blockade must create a real danger of capture for blockade-runners.

The development of large submarines and aircraft has now made blockade much more difficult to enforce and of lesser importance.

Blockade, pacific. A development of the right of blockade in wartime made in the early nineteenth century, and amounting to a compulsive measure of settling an international dispute. Cases of pacific blockade are truly cases of intervention or of reprisals. Thus, Britain and France blockaded the coast of Holland in 1833 to make Holland consent to the independence of revolting Belgium; this was intervention. In 1916 the allied powers, as reprisal, blockaded the coast of Greece, which was not then a belligerent. It is questionable whether pacific blockade is permitted by modern international law. If it is, it is justifiable today only after the failure of negotiations to settle the matter in dispute. It may, however, be used by or on behalf of the United Nations as a means of collective action for enforcing the principles and obligations of the U.N. Charter. In 1967 and subsequent years Rhodesia was nominally blockaded in pursuance of a U.N. resolution.

Blodewell, Richard (?–*c.* 1513). Dean of the Arches, who about 1494 formed and became first President of the Association of Doctors of Law and of the advocates of the churches of Christ at Canterbury, which became known as Doctors' Commons (q.v.), the body of practitioners in the ecclesiastical courts.

G. D. Squibb, *Doctors' Commons.*

Blois, Ordinance of (1579). One of several great reforming ordinances promulgated by French Kings, it dealt with reorganization of justice in the kingdom, rules governing the royal domain, the reformation of French universities, rights of parents over children, and a statement of the main disciplinary decisions of the Council of Trent.

Blood, corruption of. In mediaeval English law a conviction for treason or felony entailed, *inter alia*, the escheat to the lord of the felon's lands because, it was held, his blood was attainted or corrupted. This prevented any descent from being traced through the convicted felon. It was abolished in 1834.

Blood feud. In early law an injured person or the relatives of one killed could exact similar vengeance from the wrongdoer and his kin provided that no more was sought than was justly due. Later it was accepted that compensation could be paid in lieu of pursuing the blood feud, though initially the injured person or the relatives could either accept or reject the payment, and certain offences, e.g. treason, were *botless* or irredeemable, and were punishable by death or mutilation and the forfeiture of the offender's property to the King. The money value set on a man according to his rank was *wer* and the compensation *wergeld* or *bot*. In addition there was *wite*, a penal fine payable to the King or other public authority as a penalty for having broken the King's peace.

Blood groups. Classifications of blood based on the properties of red cells as determined by antigens. There are at least nine major blood group systems in man. Apart from medical uses determination of individuals' blood groups is of legal evidential importance. Since blood groups are inherited determination of the blood group of persons concerned can provide unequivocal evidence that a man is not the father of a particular child, but cannot prove more than merely possible paternity. In England a blood test may be ordered by the court where paternity is in question. A blood sample may be ordered to be taken from a person held on a charge of rape.

Blood money. The money penalty paid by a murderer to the relatives of his victim. Early legal systems commonly move from allowing blood feud (q.v.) to allowing and then requiring payment of blood money, and commonly specify in some detail the amounts payable for causing the deaths of or injuries to victims of various degrees. It was not confined to causing death but extended to all crimes of violence.

See also WERGELD.

Blood relationship. The relationship between two persons having at least one common ancestor. It is of two kinds, the whole or full blood, where the two have the same pair of ancestors, e.g. father and mother, and the half blood consanguinean where they have a common male ancestor and the half blood uterine where they have a common female ancestor. The relationship is important in succession, descendants of the half blood commonly being postponed to those of the full blood.

Bloody Assizes (1685). The name given to the judicial proceedings in the West of England following the defeat of Monmouth's rebellion at Sedgemoor. Chief Justice Jeffreys (q.v.) and four other judges were appointed to hear cases against those accused of involvement in the rising. They sat at various towns in the West and are believed to have sentenced about 300 prisoners to death and about 800 to transportation while very many more were fined, imprisoned, or flogged. The trials were conducted brutally and the sentences were very harsh, and after the Revolution of 1688 the whole proceedings were severely condemned.

Blue Books. The customary name for the many reports, volumes of statistics and other governmental publications, so-called from being published in blue paper covers.

Blue Laws. The term of uncertain origin given in the U.S. to laws intended to enforce moral standards. The name may have been derived from an account purporting to list the Sabbath regulations of New Haven, Conn., printed on blue paper and published in 1781. Sunday observance acts and sumptuary regulations were transplanted from or modelled on English legislation in the seventeenth century and some survived into the eighteenth century. They were most common and most strict in Puritan, Bible-oriented communities, and usually forbade work or sport on Sundays. In most States there are some survivals restricting particular activities on Sundays. National prohibition (1919–33) has been the most notable twentieth century manifestation of the spirit of the blue laws.

Blue-ribbon jury. U.S. term for a jury chosen for special qualifications to try a complex or difficult case, also called a special jury or struck jury. In certain cases statutes provide for a special jury; sometimes it is available as of right, sometimes on the motion of a party, sometimes in civil cases, sometimes in criminal cases. In modern practice such juries are not commonly used.

Blue sky laws. The U.S. term for legislation designed to protect purchasers of stocks and bonds from fraud.

Bluntschli, Johann Caspar (1808–81). Swiss jurist, president of the Great Council of Zurich, and later Professor of Constitutional Law at Munich. He was author of *Public and Legal History of the City and District of Zurich* (1838–39), works on German public law, German private law, and a code for the canton of Zurich, part of which was the model for the Swiss code. Later at Heidelberg he wrote a *History of Public Law and Politics* (1864), *Modern Law of War* (1866), *Modern International Law* (1868), *Lessons of the Modern State* (1875), and other works. He was one of the founders of the Institute of International Law.

Board of Admiralty. See ADMIRALTY.

Board of Green Cloth. A board, part of the Lord Steward of the Household's Department, consisting of the Lord Steward and the Treasurer, Comptroller and Master of the Royal Household with the functions of receiving accounts for all the royal household expenses and paying them and the wages of royal servants, and seeing to the good government of the household and servants. It took its name from the green cloth on the table at which its officers formerly sat. The Board seems to have disappeared.

Board of Trade and Plantations. A British government department established as a council in 1674 and reorganized after the Revolution of 1688 as a board of eight with nominally, the great officers of State, and a staff. It was reconstituted in 1786 as the Board of Trade, comprising the lords of the committee for the time being of the Privy Council appointed for the consideration of matters relating to trade and foreign plantations, i.e. colonies. It consisted of the President of the Board and the holders of various offices, but the President was the responsible political head. The Board of Trade was converted into the Department of Trade and Industry and then into the Department of Trade in 1974.

Boarding-house. Premises in which the occupier allows one or, usually, more persons to live for a time, providing food and services and allowing the use of a bed and providing other accommodation in common, in return for agreed payments. It differs from an inn (q.v.) in that the keeper is under no obligation to receive anyone, whether traveller or not, is not subject to the strict liability of the innkeeper for property lost, and has no lien on the lodger's property for money due.

Bocland (or Bookland). Lands in England held, during the pre-Norman period, under charter, book or other deed as opposed to folkland or *laenland*. It was a privileged form of ownership, held only by churches and great men, which often carried freedom from the customary burdens on land other than liability to military service and to contribute to the repair of fortresses and bridges. It was readily alienable and devisable by will and could be entailed, in which event the owner lost his power to alienate. It might also be leased to free tenants on such terms as the owner thought fit and it was then known as *Laenland* (loan land).

Bodily harm. A statutory term not precisely defined but covering injury to the person which is trivial or technical as distinct from 'actual bodily harm' or the more serious 'grievous bodily harm'.

Bodin, Jean (1530–96). French political philosopher. He studied and lectured on law and settled in Paris as an advocate. His *Six Livres de la République* (1576) was intended not only to describe the ideal State but to promote reforms in France. Sovereignty in his view arises from human needs and the family is the corner-stone of the State, which is an association of families which recognize a law-making power in a person or group. The whole duty of a subject is obedience. He also held that the types of humanity and their laws and institutions corresponded to climatic zones. He regarded a strong monarchy in France then as essential, but appreciated that it could not last for ever. He took part in politics and was a confidant of the King, and wrote various other works not now so important, including a *Methodus ad facilem historiarum cognitionem* (1566).

Body. In relation to written instruments, the main part of a deed, as distinct from the recitals (q.v.) and attestation clause (q.v.), or of an affidavit as distinct from the title and jurat (q.v.).

Body-snatching. The secret disinterring from graveyards of bodies recently buried in order to sell them for purpose of dissection. Prior to the Anatomy Act, 1832, there was no provision for supplying bodies to medical students for anatomical study and disinterment was profitable, though criminal. In order to defeat the practice, iron coffins were sometimes used, or mortsafes, frameworks of iron bars over the graves, or relatives maintained watch over graves for a week or two.

Bohic, Henri (1310-51). French canonist, author of *Distinctiones* and a commentary on the Decretals, analysing the fourteenth-century conflicts on the canon law.

Bohier, Nicolas de (Boerius) (1469–1539). French jurist, author of *Tractatus de officio et potestate legati a latere in regno Franciae* (1509), *Commentaria in consuetudines Bituricenses* (1543), *Consilia* (1574), and his most esteemed work, *Decisiones in senatu Burdigalensium discussae ac promulgatae* (1547).

Böhmer, or Boehmer, Georg Ludwig (1715–97). German jurist, teacher at Gottingen, author of *Principia Juris Canonici* (1762), *Principia Juris Feudalis* (1765), *Observationes juris feudalis* (1764), *Observationes juris canonici* (1766), *Systematis juris civilis fragmenta* (1799), and other works.

Böhmer, or Boehmer, Johan Samuel (1704–72). German jurist, teacher at Frankfurt on Oder, author of *Observationes in Carpzovii practicam novum rerum criminalium* (1759) and *Elementa jurisprudentiae criminalis* (1732).

Böhmer or Boehmer, Juste Henning (1674–1749). Father of the two foregoing, one of the foremost scholars and jurists of his day, professor at Halle, author of *Exercitationes ad Pandectas, Introductio in Jus Digestorum* (1704), *Jus ecclesiasticum protestantium* (1710), *Jus parochiale* (1701), *Introductio in jus publicum universale* (1710), *Institutiones juris canonici* (1738), and *Corpus Juris Canonici cum notis* (1747).

Boiling to death. A punishment once common in England and in Europe, for killing by poisoning. It was used several times in the sixteenth century, notably in the case of the bishop of Rochester's cook in 1531 who was publicly boiled at Smithfield without benefit of clergy. Statute in that year made poisoning treason. It was repealed in 1547 and poisoning made punishable in the same mode as other murders.

Bologna, University of. One of the oldest and most famous in Europe. In the early Middle Ages Bologna was the greatest centre of legal studies in Europe, and thousands of students flocked there from all over Europe to study civil and canon law. Its famous teachers included Irnerius, Accursius and many more.

It was a student university, in which the student organizations chose the rector and the masters, and this had an influence on the development of many later universities, as distinct from the masters' university, typified by Paris. Today it continues as a huge co-educational state university.

H. Rashdall, *Universities of Europe in the Middle Ages.*

Bolts (or boltings). The argument of cases for educational purposes in the halls of the Inns of Court or of Chancery by barristers and students before courts consisting of an ancient or bencher and two barristers. Unlike moots (q.v.) they were held in private. They have long been obsolete.

Bona fide possessor. In Roman law a person possessing property but not owning it, because the person from whom he acquired the thing was not owner. If such a person took possession before time for usucapion had run, the praetor allowed him to recover it by an *actio Publiciana* on the fictional basis that the time for usucapion had run.

Bona fide purchaser. A person whose rights are in many circumstances not defeated by the fact that the seller to him did not have good title to what he has sold, provided the purchaser has given value and taken without notice of the defect in the title of the seller. This equitable principle is applied in many branches of law, e.g. in relation to equitable interests attaching to the property, and in relation to the recovery of property sold by a trustee in breach of trust.

Bona fides (good faith). A person acts *in bona fide* when he acts honestly, not knowing nor having reason to believe that his claim is unjustified. A *bona fide* possessor holds on a title which he honestly believes to be good. A *bona fide* purchaser purchases, believing that he is entitled to buy and the seller to sell. *Bona fides* ends when the person becomes aware, or should have become aware, of facts which indicate the lack of legal justification for his claim.

Bona vacantia. Things which no person can claim as property, including wreck, treasure trove, waifs and estrays (qq.v.), the property of dissolved corporations, and the residuary estate of persons who have died intestate and without relatives entitled to succeed. It does not cover property lost or abandoned. The Crown has right to all *bona vacantia.*

Bona waviata. Goods waived or thrown away in flight by a thief. They are held to belong to the Crown as a penalty on the owner for not having himself pursued the thief and recovered his goods.

Bond. A formal deed whereby one person binds himself to do some thing to or for another, normally to pay a specified sum of money immediately or at a future date. In England the main kinds of bond are common money bonds, to secure the payment of money; *post-obit* bonds, to pay money after the death of a specified person, usually in respect of a lesser sum now advanced to the granter; administration bonds, granted by an administrator for the due performance of his duties of administration of an estate; Admiralty bonds, granted to obtain the release of a ship arrested; and Lloyd's bonds, which are securities by companies for payment of money.

In Scotland a personal bond undertakes payment of a sum of money, with interest till paid, and was frequently conjoined with a disposition of lands or an assignation of incorporeal property in security of repayment, thereby becoming a bond and disposition (or assignation) in security. The bond and disposition in security has now been replaced by the standard security (q.v.). There are also recognized bonds for cash credit, bonds of caution, bonds of corroboration, and bonds of relief.

Bond and disposition in security. The standard form in Scotland from the eighteenth century till 1971 of giving security over land for money borrowed. It combined a bond, acknowledging a loan and undertaking repayment and, in the meantime, interest, and a disposition of land in security, redeemable, but irredeemable in the event of sale or foreclosure under the powers contained therein or added thereto by statute, under which the creditor took sasine of the lands but in security only. When the money was repaid the bond was discharged. Bond and discharge had to be recorded in the Register of Sasines (q.v.). Since 1971 this form has been replaced by the standard security (q.v.).

Bonded warehouse. A warehouse licensed by the Commissioners of Customs and Excise for the storage of excisable goods, on which duty is not paid until they are cleared, i.e. taken out for sale. The name is derived from the bond which must be entered into to secure that the Crown does not lose duty by the goods being removed without payment. In the U.K. the system of bonded warehouses goes back to 1700, the system being gradually extended to different kinds of goods. It began in the U.S. in 1846.

Bonet, Honore (*c.* 1340–*c.* 1410). French canonist, author of *L'arbre des batailles* (*c.* 1386) a codification of the laws of war founded on Legnano (q.v.), and works on theology.

Bonfils, Henry-Joseph François Xavier (1835–97). French jurist, author of a famous *Manuel de droit international public* (1894) and other works.

Bonham's case (1610), 8 Co. Rep. 114. A case reported by Coke, containing a dictum asserting the supremacy of the common law of England and its power to control and adjudge to be void an act of Parliament which was against common right and reason. The principle, however, has not been subsequently accepted and the supremacy of statute over any rule of the common law has now been long established.

Boniface VIII, Pope (Benedict Caetani, *c.* 1235–1303) (Pope 1294–1303). A distinguished canonist and papal administrator. As Pope he was involved in a fierce struggle to maintain papal authority against the powerful rising monarchies of western Europe. In 1302 he issued the bull *Unam Sanctam*, restating the supremacy of the spiritual order over the temporal power. He also published the *Liber Sextus*, the third part of the *Corpus Juris Canonici*.

Bonnefoi, Ennemond (Bonfidius) (1536–1574). French humanist jurist, professor at Geneva, editor of *Juris orientalis libri III* (1573).

Bonorum possessio. In Roman law, the right which praetors granted to persons to use certain remedies which would enable them to get possession of the goods of a deceased though they were not the heirs. He would give *bonorum possessio secundum tabules* to a person nominated heir in tablets sealed with the requisite number of seals. A person not disinherited would claim possession *contra tabulas*. In the case of intestate succession the praetor would grant *bonorum possessio ab intestato* to children, blood relations, and spouses. *Bonorum possessio* had always to be sought from the praetor and within a limited time, and the grant from the praetor merely meant that he could use certain remedies, notably the interdict *quorum bonorum*. The importance of *bonorum possessio* lay in its use as an agency for the reform of the old law of succession, though it was originally a means of giving heirs by civil law additional remedies to ensure that an estate did not fall into the hands of persons with no entitlement thereto.

Bonser, Sir John Winfield (1847–1914). Was successively Chief Justice of the Straits Settlement, 1893, and of Ceylon 1893–1902. From 1902 he was a member of the Judicial Commission of the Privy Council.

Book hand. A variety of script based on Carolingian minuscule of the ninth and tenth centuries, and becoming distinct about the twelfth century, so called because it was the type of writing used in copying manuscripts.

Book of Assizes (or *Liber Assisarum*). The fifth part of the 1679 standard edition of the Year-Books, containing the cases of the years 1 to 50 Edward III (1327–77) arranged by regnal years. It had previously been printed in 1516 and differs from other Year-Books in containing some cases heard in the King's Bench and a few cases in Chancery. It should not be confused with the book properly called the Abridgement of the Book of Assizes.

Book of Entries. A register of the proceedings of the Privy Council, begun at the beginning of Henry VII's reign and containing mostly judicial business.

Book of the Consulate of the Sea (or *Consolato del Mare*). A book of customary sea laws in force in the Mediterranean, probably of Catalan origin, probably drawn up for the use of the consuls of the sea at Barcelona, and later translated into several languages.

Book of the Council. A record of the English Privy Council running continuously from 1421–35 and 1540 onwards.

Book of the Covenant. The collection of laws contained in Exodus 20, 22 to 23.19, the oldest code after the Decalogue.

Booking. In Scots law a special form of burgage tenure (q.v.) by which lands and buildings in the burgh of Paisley are held.

Bookland. A mode of landownership in Anglo-Saxon law, by which ownership was granted to be held in accordance with the book. It was a mode almost confined to grants by the Crown to great and powerful persons, and frequently conveyed also many of the powers of the State to the grantee. Conveyance may have been by signing and delivery or transfer of the book, and from this developed grants by charter.

Bookmaking. The business of receiving or negotiating bets on the outcome of some event, e.g. which horse will win a particular race, or on the outcomes of several events, e.g. which football matches played on a given day will result in draws. Bookmakers and members of the public as makers of bets at racecourses on the result of horse-races are subject to the discipline of the Committee of Tattersall's which may exclude any person from any racecourse under the jurisdiction of the Jockey Club if he has not settled his betting losses.

Books, censorship of. See CENSORSHIP.

Books of Adjournal. The records of the High Court of Justiciary in Scotland.

Books of authority. The authority of treatises, commentaries, text-books, and similar expository writing as statements of law varies in different legal systems. In general it is less in common law jurisdictions than in civil law jurisdictions; in the former greater weight attaches to judicial decisions than to treatises; conversely in the latter case.

In English law the works of Glanville, Bracton, Littleton, Coke, and Blackstone (qq.v.) are undoubted authorities as to the law of their respective times as much as any judicial decision, and the rank of being books of authority attaches also to the works of Hale, Hawkins, and Foster. Other books, however, are not books of authority nor, properly speaking, sources of law, but only legal literature, of persuasive influence and sources of knowledge of the law, of materials which may be adopted or utilized in judgment. Nevertheless other books, in particular recognized standard treatises on various branches of the law, are important for their collection, synthesis, and critical evaluation of the materials of the law; judges frequently accept their formulations and hesitate to differ from them, though undoubtedly they may do so. Students' books are of no authority, though some are also used by practitioners and may be more systematic and constructive than some practitioners' books. There was formerly a rule of practice that a text-book could not be cited in court in the lifetime of its author, but this is not now strictly followed.

Probably too little respect is accorded to standard text-books, because only in them can one find the jumble of statutes and cases relevant to a particular topic put into some order and set out systematically, explained and criticized. The standard text-books alone make order and sense of English law.

Broadly the same principle applies in Scotland where authority attaches only to the 'institutional writings' (q.v.) and only persuasive influence to certain standard text-books.

In U.S.A. the works of Blackstone, Kent, and Story were authoritative and very influential in shaping American law, and in modern times the great treatises of e.g. Wigmore, Corbin, Williston, and Scott have persuasive value. Particular importance attaches to the Restatement of the Law by the American Law Institute, a multi-volume unofficial codification of what the general consensus of expert academic and professional opinion regards as the best and normally the preponderant, view.

In all these, and other common-law jurisdictions the influence of the books of authority is now largely spent, as later statutes and cases have modified or restated the principles.

In the twentieth century persuasive influence has sometimes been exerted by articles in legal periodicals, criticising decisions, suggesting different approaches for development, investigating points more thoroughly than can be done in books of more extensive coverage.

In civil law jurisdictions on the other hand greater weight attaches to treatises. In the Roman law the opinions of jurists were highly respected and many wrote treatises. By the third century A.D. the writings of the great jurists of an earlier period were deemed authoritative and the Valentinian Law of Citations (426) provided that the writings of Papinian, Paul, Gaius, Ulpian, and Modestinus should be pre-eminently so. Justinian gave statutory authority to his *Digest* which is largely composed of edited excerpts from their writings. From early modern times in continental Europe the chaotic state of the law, enacted, customary, and derived from Roman and Canon law, has enhanced the importance of the systematic treatise.

The enactment of the French Codes by Napoleon gave rise to commentaries on the Codes. In Germany the text-writers of the nineteenth century created a systematic structure of law based on Roman law which paved the way for the Civil Code of 1900, which in turn generated a body of text-writing. Accordingly in these countries great weight attaches to the extensive treatises, such as by Planiol, Aubry et Rau, and others on the French civil law, by Enneccerus, Kipp and Wolff on German law.

Furthermore the Roman and continental tradition is to write general treatises on the civil, i.e. non-criminal law, the commercial law, and so on, whereas in common law countries the practice, since Blackstone and Kent, has been for individual authors to write books on branches and topics of the law, on contracts, torts, real property, etc. This obscures the unity of the law and particularly of the private, or civil and commercial, law.

In civil law systems considerable weight may also attach to notes on, and criticisms of, decisions by academic jurists in periodicals and academic journals.

Books of Council and Session. The common name for the Register of Deeds and Probative Writs kept at Register House, Edinburgh, in which may be recorded for preservation or for preservation and execution deeds and writings containing warrants consenting to such registration. Once registered a deed is kept for ever, but extracts, i.e. photocopies, may be obtained.

Books of Entries. Books of the late mediaeval period containing collections of precedents of pleading. Until English was made the language of the law they are all in Latin, and are nearly all arranged in alphabetical order of the writs which begin the actions to which the pleadings are incidental. They are all, to modern eyes, bulky, verbose, and technical, and allegations are confused with evidence.

The best known of this species of book are *Intrationes* or *Liber Intrationum*, printed by Pynson in 1510 and not arranged alphabetically; *Intrationum Liber* (1545 or 1546) a quite different book; William Rastell's *Collection of Entries* (1566 and later editions), sometimes called *Antient Entries, Old Book of Entries*, or *New Book of Entries*; Sir Edward Coke's *Book of Entries* (1614 and 1671), sometimes called *New Entries* and *New Book of Entries*; R. Aston's *Placita Latiné rediviva: A Book of Entries* (1661, 1673); William Browne's *Formulae bene placitandi, A Book of Entries* (1671, 1675); Sir Humphrey Winch's *Le beau-pledeur: A Book of Entries* (1680); Andrew Vivian's *The Exact Pleader: A Book of Entries* (1684); Richard Brownlow's *Latine redivivus; A Book of Entries* (1693); Henry Clift's *A New Book of Declarations, Pleadings, Verdicts, Judgments and Judicial Writs; with the Entries thereupon* (1703, 1719); and John Lilly's *A Collection of Modern Entries* (1723 and later editions).

Books of Sederunt. A minute book kept by the clerk of the First Division of the Court of Session in Scotland recording the installation of new judges, the swearing in of law officers, the admission of advocates, Acts of Sederunt (q.v.), and certain other matters. Previous volumes of this book are kept in H.M. General Register House. Each Division of the Inner House of the Court of Session keeps a Sederunt Book recording the names of the judges composing the bench and the cases disposed of.

Books, Prohibition of. The practice of the Catholic church declaring that certain books cannot be read as being dangerous to faith and morals, a practice followed in the exercise of that church's belief that she is divinely appointed guardian and teacher of the revealed word of God. Control is exercised by prior censorship of writings that pertain to faith and morals, and by forbidding the reading of works published but judged to be offensive. Formerly certain books were prohibited by name and placed in the Index of Forbidden Books, but now only general categories of books are listed which cannot be read without ecclesiastical permission. The Code of Canon Law lists twelve broad categories of books forbidden by the law itself. For good reason a Catholic may obtain permission from his bishop to read forbidden books.

Boon-days. Days of work which tenants on a feudal manor were obliged to give their lord if and when called upon to do so.

Booth, Sir Robert (1626–81). Called to the English Bar in 1650 and the Irish Bar in 1657. He became a judge of the Common Pleas in Ireland in 1660, Chief Justice in 1670, and Chief Justice of the King's Bench in Ireland in 1679.

Bootless crimes. In early Germanic society those crimes which could not be atoned by payment of money (*bot*) to the relatives of the victim but which were punishable by the King.

Booty. In international law, property captured in war on land. Formerly private property might be taken as booty of war, but now only military arms, equipment, material, and stores may be so taken. It belongs to the Crown or may be granted by the Crown to the forces capturing it. Private property at sea may be captured but must be adjudicated on by the Prize Court (q.v.) to determine whether or not it is lawful prize.

Border warrant. A warrant granted by a magistrate on either side of the border between England and Scotland for arresting the person or property of a person living on the other side of the border. By statute a constable for any one of the border counties may execute within any other of those counties a warrant for apprehension or recovery of stolen goods.

Borough, St. John (?–1643). Keeper of the Records in the Tower of London and later Garter King-of-Arms, author of *The Sovereignty of the British Seas* (1651) and a *Commentary on the Formulary for Combats before the Constable and Marshal* and of minutes and notes of treaties and negotiations of state.

Borough. From Anglo-Saxon times places, frequently fortified or where a market was held, were recognized as boroughs and, though subject to the sheriff of the county, had special customs and privileges and a special court. After the Norman Conquest charters from kings or other lords tended to separate boroughs from other areas; they developed distinct privileges, and distinct bodies of customary law often codified in their custumals. Many new boroughs were founded after the Norman Conquest and they developed rapidly in the twelfth and thirteenth centuries. From the time of Henry III borough representatives were summoned to the King's parliaments. A court, communal or franchise, was always a feature of a borough. If a borough had the right to exercise criminal jurisdiction, it had right to hold a court leet, which in some cases developed into the governing body of the borough. In most boroughs the court leet disappeared in the seventeenth and eighteenth centuries. From the mid-sixteenth century charters commonly made the mayor and some aldermen J.P.s, sometimes giving the right to hold only petty sessions, sometimes also the right to hold quarter sessions. In 1835 the more important boroughs were allowed to get a court with a paid lawyer to preside and a separate court of quarter sessions. Other boroughs lost their criminal jurisdiction. All boroughs had courts of civil jurisdiction, and different charters conferred various titles and great variety of jurisdiction. From the thirteenth century the judges were generally a group of aldermen, or the mayor, aldermen, and leading officials. From the end of the eighteenth century these courts were decadent. The majority are now disused. By the nineteenth century many boroughs were corrupt and the Municipal Reform Act, 1835, swept away many obsolete and unrepresentative institutions, establishing councils elected by the ratepayers. In 1974 boroughs ceased to have local government functions, being absorbed into new districts, but they remain the main centres of population and business.

In the U.S. boroughs existed in colonial Virginia and in 1722 borough charters were given to Williamsburg, Norfolk, and Richmond. The name disappeared at the Revolution though it is used to designate the five sub-divisions of New York City and applied to some incorporated towns in some eastern states.

For Scotland see BURGH.

J. Tait, *Medieval English Borough*; F. W. Maitland, *Domesday Book and Beyond* and *Township and Borough*; A. Ballard, *British Borough Charters* (1913–23); M. Bateson, *Borough Customs* (1904–06).

Borough courts. From very early times English boroughs had special privileges including courts and sometimes customary law. When granted charters boroughs not infrequently copied the charters and customs of existing boroughs, but all were subject to the royal judges and the common law. There developed a great variety of borough courts.

All boroughs had a court with civil jurisdiction, resembling a manorial court baron. According to the different charters such courts had very varied jurisdiction, and very varied names. From the thirteenth century the judges were generally a group of aldermen, or under a charter, the mayor, aldermen, and chief officials. By the late eighteenth century in most boroughs these courts were inactive or even in disuse. In 1835 statute extended the jurisdiction of surviving courts provided they had a barrister as judge, but of the courts which survived that Act most are now disused. Only a few reformed by local Acts, were still active into the twentieth century, notably the Mayor's and City of London Court, the Liverpool Court of Passage, Salford Hundred Court, and Bristol Tolzey Court (qq.v.). Elsewhere the modern county courts superseded them. All, except the Mayor's and City of London Court were abolished in 1974.

Many boroughs early acquired criminal jurisdiction, obtaining the right to hold a court leet, which in some cases developed into the governing body of the borough. The borough court leet generally disappeared in the seventeenth and eighteenth centuries, giving way to the justices of the peace. From the sixteenth century borough charters frequently made the mayor, some officials, and aldermen, justices of the peace; some gave the right to hold quarter sessions, others only petty sessions. Where a borough had a right to hold quarter sessions the court leet merged with or was superseded by quarter sessions. In some boroughs the right to hold quarter sessions was limited and did not exclude county quarter sessions.

In 1835 the Municipal Corporations Act enabled specified boroughs to obtain a separate court of quarter sessions presided over by a qualified judge, the Recorder, while other boroughs lost their quarter sessions. Quarter sessions were abolished in 1971.

In large towns, first in London and later elsewhere, the practice developed of appointing stipendiary magistrates instead of justices in petty sessions.

Borough-English. In English common law, a customary tenure of land under which land descended on death to the youngest son or, failing sons, to the youngest daughter to the exclusion of all other children. This custom of ultimogeniture was very general in English boroughs and gained the name of Borough-English after a case in 1327 had drawn attention to the rule of ultimogeniture in the English borough of Nottingham, whereas in the nearby French borough the rule of primogeniture applied,

The custom continued in many places until the twentieth century, but was abolished in 1925.

Borstal institution. A custodial institution, named from Borstal in Kent where the first such institution was established in 1908, intended to segregate young offenders from adult criminals and to provide a regime of discipline and training. Borstals are commonly organized in houses, like public schools; some are 'open', some are camps. There is strong emphasis on work and vocational training. Youths or girls within the requisite age limits may be sent for Borstal training, the maximum period being fixed by law, and in practice they are released earlier when deemed to have made sufficient progress to justify release.

Boswell, James (1740–95). Son of a Scottish judge (Lord Auchinleck) and destined for the law, he betook himself to London in 1760 but was prevailed upon to return to Scotland where he became acquainted with Kames, Hume, and other notable men of the time. In 1763 he set out for Utrecht, where he studied Roman law, and then made a grand tour of Europe, meeting Voltaire and Rousseau. Before leaving for Utrecht he had, in 1763, made the acquaintance of Dr. Johnson and thereafter they met regularly and Boswell noted much of Johnson's conversation and dicta, much of which was later incorporated in his *Life of Johnson*. In 1765 he visited Corsica and became friendly with the Corsican patriot, Paoli, a visit which was the basis of his successful *Account of Corsica* (1768).

In 1766 Boswell had been called to the Scottish Bar and practised there steadily and with reasonable success till 1785 when his father's death enabled him to gratify his desire to live in London. In 1773 Johnson came to Scotland and the two made a tour of the north of Scotland and the Western Isles. Johnson published his account of the trip in 1775, while Boswell published his own account in 1785.

In 1786 he was called to the English Bar and for two years, 1788–90, was recorder of Carlisle but made no mark in the law. He was a man of letters, not a lawyer or jurist.

Boswell wrote a great deal but his reputation stands on the *Account of Corsica* (1768) and his *Journal of a Tour of the Hebrides* (1785), but, above all, on his *Life of Dr. Johnson* (1791). Recently his extensive journals have been found and are being published, casting much light on his character and life, and incidentally on the legal and literary life of the time.

F. Pottle, *James Boswell: The Earlier Years; Journals* (volumes under various titles).

Bot, wer, and wite. In early law bot was compensation for harm done, at first an alternative to and later in substitution for the exaction of harm or blood in return, by way of blood-feud. Some offences, such as treason, were botless, i.e. non-compensable. Wer or wergeld was the money value set on a man, according to his rank and status, for the purposes of compensating various kinds of wrongs to him. Wite, later called amercement, was a penalty exigible by the King, in addition to bot payable to the injured party. If a wrongdoer failed to pay bot and wite he became an outlaw.

Botany Bay. A place 8 km. south of Sydney, N.S.W., the scene of Captain Cook's first landing on Australian soil in 1770 and named for the flora collected there. Not long after it was suggested as a suitable place for a convict settlement and the name has become used for the destination of transported convicts.

Bottomry bond. A contract whereby the owner or master of a ship borrows money in circumstances of unforeseen necessity or in case of distress to enable him to pay for repairs and despatch of the vessel for the completion of her voyage, and pledges the keel or bottom of the vessel, as a symbol of the whole vessel, for repayment. The master must, if possible, communicate with the owner before resorting to bottomry. If the ship is lost, the lender loses his money, but if she survives, he recovers the loan with interest, and may bring an action *in rem* to enforce his claim. It is now almost obsolete. Such loans are known from the fourth century B.C. and were elaborately regulated by both Greek and Roman law as they represented an important form of capital investment. See also *Respondentia*.

Bouchard, Alexis-Daniel (1680–1758). A French theologian and humanist jurist, who edited the *Institutes of Gaius*, and Paul's *Sententiae* and wrote *Juris Caesarei seu civilis Institutiones breves* (1713) and *Summula conciliorium generalium romanae catholicae Ecclesiae* (1717).

Bouhier, Jean (1673–1746). French jurist, author of *Observations sur la Coutume de Bourgoyne* and many other works and originator of a project for uniform legislation for the whole of France.

Boule. In ancient Greece, the council of a city or other political organization, as distinct from the popular assembly. At Athens Solon is thought to have created a new boule in 574 B.C. to guide the work of the assembly. The new boule was later connected by Cleisthenes with his 10 new tribes and increased to 500. Thereafter the boule was elected by lot every year, 50 from each tribe. Its main task was to draft matters for discussion and approval in the assembly.

The boule could also impose fines and initiate a trial in the courts, mainly for crimes against the State.

It never wholly replaced the Areopagus in political importance and in Roman times the Areopagus and the boule shared the government of the city.

Boundaries. The limits of an area, particularly between political communities. Delimitation of boundaries between States is partly a matter of long-continued possession and frequently a diplomatic procedure. In 1919 the Paris Peace Conference fixed the boundaries of European States. Various factors have influenced the delimitation of boundaries, notably strategic, geographic, historical, economic, and ethnic factors, the importance of each varying in different circumstances. Once a boundary line has been delimited it must be demarcated, sometimes by setting up monuments, buoys, or other markers.

Particular problems arise on the sea and in the air. Round its coast a state's boundaries are fixed by base-lines (q.v.) beyond which is a belt of territorial sea (q.v.) over which it also claims jurisdiction. Boundaries on the continental shelf are fixed by agreements. In the air jurisdiction extends upwards to include air space at least as high as aircraft can fly, but beyond that is outer space which is open to all. The boundary between the two is not precisely defined.

Within states boundaries have to be delimited for such purposes as parliamentary constituencies and local government areas and statutes have established Commissions to effect such delimitation.

As between pieces of land in separate ownership the boundaries are defined by the titles by reference to natural features, measurements, and plans. Fences, ditches, or walls along and marking a boundary frequently belong to the owners of both properties in common. Disputes as to boundaries may have to be settled by action for trespass, or an action to ascertain the boundaries. Any going on the surface of the land of, or excavating under the land of, another, beyond the boundary between them, is a trespass or encroachment and actionable.

Bounty, Queen Anne's. A fund for augmenting the maintenance of poor clergy in England, formed by the transfer in 1703 of all the revenue of first-fruits and tenths belonging to the Crown to trustees. In 1925 there was added all ecclesiastical rent charges. In 1947 Queen Anne's Bounty was united with the Ecclesiastical Commissioners under the name of the Church Commissioners.

Bourgeois, Léon Victor Auguste (1851–1925). French statesman, prime minister in 1895. He distinguished himself at the Hague Peace Conference of 1899, became a member of the Permanent Court of Arbitration in 1903 and was a delegate to the Hague Conference of 1907. He was one of the first to propose a League of Nations, represent France in negotiations for the League in 1919, and was French representative to the League Council and Assembly till 1924. In 1920 he received the Nobel Peace Prize.

Bourjon, François (?–1751). French jurist, author of *Droit commun de la France et la Coûtume de Paris reduite en principes* (1747), an attempt to formulate a common law for France on the basis of the customs, influential on the drafting of the Civil Code.

Boutillier, Jehan (or Bottelgier) (*c.* 1335–*c.* 1395). French jurist, bailli of Vermandois, author of a learned treatise entitled the *Somme Rural*, stating the whole of French law at the end of the fourteenth century, containing customary, Roman, and canon law and founded on the customs of the north of France and the decisions of the courts. It was very useful in practice, was translated into Flemish, had considerable influence on the development of the law of the Netherlands and northern France, and remains of interest as one of the most important sources of law of the Middle Ages.

Bouvier, John (1787–1851). American jurist, was in business as a printer before being called to the Pennsylvania bar in 1822 and later held minor judicial office in Philadelphia. He published *A Law Dictionary* in 1839, which has since then been a standard reference book, an edition of *Bacon's Abridgment of the Law* (1841–45) and an *Institutes of American Law*, based on Pothier, in 1851.

Bovate. In mediaeval law, a measurement of land, being one-eighth of a carucate (q.v.), and nominally as much land as one ox could keep under the plough, and hence also known as an oxgang or oxgate.

Bovill, Sir William (1814–73). Though originally articled to a solicitor, he transferred his interests and was called to the Bar in 1841. As an advocate he was outstanding. He became M.P. in 1857 and Solicitor-General in 1866, and shortly thereafter became Chief Justice of the Common Pleas. Though he had a good practical acquaintance with the law he was not a great judge, but rather excelled in jury practice.

Bow Street Magistrate. The chief metropolitan magistrate and other magistrates sitting at Bow Street, London, have special powers in relation to the extradition from the United Kingdom of fugitive criminals from certain foreign states and from Commonwealth countries, and certain other powers beyond those exercised by other metropolitan stipendiary magistrates.

Bow Street Runners. When Henry Fielding (q.v.) was appointed police magistrate at Bow Street, London, he established in 1750 a small force of paid detectives, Mr. Fielding's People, who became known by 1785 as the Bow Street Runners, and fore-runners of the modern C.I.D. and Special Branch officers. They were a police aristocracy, achieved a great reputation, and were usually selected for confidential duties. They attended banks and guarded royal palaces, and travelled abroad in the course of their duties. The Bow Street Office ceased to exist as a police unit only in 1839, 10 years after the New Metropolitan Police was established.

G. Armitage, *History of the Bow Street Runners*; P. Fitzgerald, *Chronicles of Bow Street Police Office.*

Bowen, Charles Synge Christopher, Lord Bowen of Colwood (1835–94). An outstanding scholar at Oxford, was called to the Bar in 1861. His practice was mainly commercial, but in 1871–74 he was junior counsel in the Tichborne cases. In 1879 he became a judge of the Q.B. division, in 1888 a Lord Justice of Appeal and in 1893 a Lord of Appeal in Ordinary. His judgments are notable for precision and refinement of language, and lucidity and accuracy of expression.

Cunningham, *Lord Bowen.*

Bowes, John, Lord (1690–1767). Called to the English Bar in 1718 and began to practise at the Irish Bar in 1725. He became Chief Baron of Exchequer in Ireland in 1741, Chancellor of Ireland 1757, and a peer in the following year. He was important in the administration of the penal laws.

***Bowles* v. *Bank of England*,** [1913] 1 Ch. 57. The Bank of England deducted income tax from dividends payable to a stockholder, the imposition of income tax at that rate having been authorized by a resolution of the Committee of Ways and Means of the House of Commons. It was held that tax was not deductible until or unless imposed by statute. The Provisional Collection of Taxes Act, 1913 and 1968, however, now give statutory authority to the practice for a limited period.

Boxhornius, Marcus Zuerius (1602–53). A classical scholar, historian and writer on politics, professor at Leyden, and author of *Institutionum politicorum libri duo* and other works.

Boycott. The refusal to have social or business dealings with one on whom persons wish to bring pressure. The word is derived from Captain Boycott, agent for the Earl of Erne's estates in County Mayo. He was threatened, his servants had to leave him, his letters were intercepted, and his food supplies interfered with because he declined to accept rents at figures fixed by the Earl's tenants. Boycotting became an essential of the Irish Nationalists' plans and was struck at by the Crimes Act, 1887. Boycotting may be lawful but is frequently an illegal conspiracy to injure the person, property, or business of the victim and actionable.

Boyle, David, of Shewalton (1772–1853). Passed advocate in 1793, became solicitor-general for Scotland in 1807, and a judge of the Courts of Session and Justiciary in 1811. Later the same year he was promoted Lord Justice-Clerk and in 1841 Lord President. He resigned in 1852. He was a man of distinguished personal appearance and was well regarded as a judge.

Boyle, Michael (1610–1702). A clergyman, became Bishop of Dublin in 1663, and Lord Chancellor of Ireland 1665 but was superseded in 1686. He became Archbishop of Armagh in 1679 and acted several times as a Lord Justice.

Brachylogus iuris civilis (or *Summa Novellarum*, or *Corpus Legum*). A learned manual for the teaching of Roman law, probably originating in France in the early twelfth century. It is based on Justinian's *Institutes*, but draws largely on the *Digest*, *Code*, and *Novels* and some passages show acquaintance with *Breviarium Alaricianum.* The name appears first in a Lyons edition of 1553. Savigny attributed it to Irnerius himself. Its value is mainly historical, showing that knowledge of Justinian's legislation was always maintained in northern Italy.

Brackley, Viscount. See *Egerton, Thomas.*

Bracton or Bratton, Henry de (d. 1268). He was probably born at Bratton Fleming in Devonshire, became an ecclesiastic and probably learned law as a royal clerk, possibly to William Raleigh, later a royal judge. He is recorded as an itinerant justice in the Midlands in 1245 and in the West of England from 1248 to 1260. He was a judge of the King's Bench from 1248 to about 1257. In 1264 he was appointed archdeacon of Barnstaple and, later that year, chancellor of Exeter cathedral.

His fame rests on the treatise *De Legibus et Consuetudinibus Angliae*, which gives evidence of acquaintance with the Roman law, particularly the *Institutes and Digest* of Justinian, and the canon law, and the *Summa* of Azo (q.v.), and dates from about 1250–59. It was undertaken, he tells us, from fear that perversion of the law by foolish and ignorant judges then on the bench would corrupt the knowledge of younger men, and was accordingly based on study of the judgments of better men, particularly Pateshull and Raleigh, and their reduction to order and brevity. It was published by Tottell in 1569. Vinogradoff, in 1884, found in the British

Museum a manuscript containing notes made by him of cases decided by the royal judges, and other materials. This was published by Maitland in 1887 as *Bracton's Note Book*.

The work, though unfinished, is the first systematic statement of English law. The major part of the law is the practice and procedure of the courts, and this is dependent largely on decided cases. The introductory part shows clear influence of Roman law, the remainder is based mainly on the practice of the courts. It has had not only great immediate but permanent influence, as well as being an historical record of the highest importance, and was abridged by Thornton (q.v.) and summarized in Britton and Fleta (qq.v.). It was first printed by Tottell in 1569, and has subsequently been edited, notably by Prof. Woodbine (1915–42) and by Prof. Thorne (1968–).

F. W. Maitland, *Bracton's Note Book* (1887), *Bracton and Azo* (1895).

***Bradford, Mayor of* v. *Pickles*,** [1895] A.C. 587. The defendant, with a view to forcing the plaintiffs to buy his land at a high price, abstracted water flowing through the subsoil beneath his land and thereby affected the supply to the corporation's reservoir. It was held that his act, lawful in itself, did not become unlawful because of his wrongful motive.

Bradlaugh, Charles (1833–91). A radical, atheist, free-thinker, and politician who frequently clashed with the law over the publication of allegedly blasphemous or indecent works. From 1874 to about 1885 he was associated with Mrs. Annie Besant in advocating unpopular causes including birth-control. He is remembered for his long struggle with the House of Commons (1880–86) over the right to affirm instead of taking the oath on the Bible. His claim was resisted because of his heterodox opinions, and he was four times excluded from the House and unseated but five times re-elected. In 1888 he secured the passage of the Oaths Act, which authorized an affirmation in lieu of oath in all cases where either was necessary.

Bradlaugh* v. *Clarke (1883), 8 App. Cas. 354. A common informer sought to recover a penalty of £500 for Bradlaugh's having sat and voted in the House of Commons without having taken and subscribed the oath in accordance with the Parliamentary Oaths Act, 1866. It was held that the penalty was exigible only by the Crown.

Bradlaugh* v. *Gossett (1884), 12 Q.B.D. 271. Bradlaugh was elected M.P. for Northampton but the House of Commons resolved that he should not be permitted to take the oath and should be excluded from the House. In this action Bradlaugh sought to have the resolution declared void and the Serjeant-at-Arms restrained by injunction from enforcing it. The Queen's Bench Division refused to grant the declaration sought, holding that the court would not take cognizance of matters arising within the House and would accept the Commons' interpretation of a statute regulating their internal proceedings.

Bradlaugh* v. *Newdegate (1883), 11 Q.B.D. 1. Bradlaugh sat and voted as an M.P. without having taken the oath and the defendants procured Clarke to sue to recover the statutory penalty of £500, indemnifying Clarke against his costs and expenses. Bradlaugh recovered damages for maintenance as the defendant and Clarke had no common interest in the result of the action to recover the penalty. (This tort has subsequently been abolished.)

Brady, Sir Maziere (1796–1871). Called to the Irish Bar in 1819, he became Solicitor-General in 1837, Attorney-General 1838, Chief Baron of Exchequer in Ireland 1840, and Lord Chancellor of Ireland in 1846-52, 1852–58, and 1859–66. He has been described as a good chief baron spoiled to make a bad chancellor.

Brampton, Lord. See HAWKINS, HENRY.

Bramwell, George William Wilshere, Lord Bramwell (1808–92). Worked in a bank and practised as a special pleader before being called to the Bar in 1838. He was a member of the commissions on common law procedure, companies, and the Judicature Act. In 1856 he became a Baron of Exchequer, in 1876 a Lord Justice of Appeal and, on his retiral in 1881, a peer. He frequently sat thereafter in the House of Lords. He was a complete master of the common law and highly regarded by the Bar, and made substantial contributions to the development of the law. Many of his judgments are splendidly concise and forceful.

Fairfield, *Some Account of Baron Bramwell*.

Brand, Sir David (1837–1908). Called to the Scottish Bar in 1864 he became Sheriff of Ayr (1885) and Chairman of the Crofter's Commission, 1886–1908.

Brandeis, Louis Dembitz (1856–1941). A successful Boston lawyer, admitted in 1878, was deeply concerned with problems of social justice, and normally appeared for individuals and small stockholders against railroads, public utilities, and large corporations. In 1908 he made history by introducing evidence of social and economic facts in his arguments in *Muller* v. *Oregon* before the Supreme Court (the Brandeis Brief). He influenced President Wilson's adoption of the New Freedom, stressing the prevention of monopoly rather than trust-busting.

In 1916 he became an Associate Justice of the Supreme Court, the first Jew to sit, and served till 1939, earning a high reputation for scholarly judgment and liberal thinking.

A. Mason, *Brandeis: A Free Man's Life.*

Brandeis brief. When Louis D. Brandeis (q.v.) was to argue the case of *Muller* v. *Oregon* (1907), 208 U.S. 412 before the U.S. Supreme Court, in which a state statute was attacked as unconstitutional, as being an unreasonable interference with life, liberty, or property, he successfully contended that evidence should be admitted from social study investigations which cast light on the dangers which the statute was intended to counter. This enabled the Court to appreciate the social background as well as the legal logic involved. It has since been used in other similar cases such as *Bunting* v. *Oregon* (1916), 243 U.S. 426 and *Adkins* v. *Children's Hospital* (1923), 261 U.S. 525.

Branding. Burning a mark into the flesh of a person or animal with a hot iron. Among the Greeks and Romans runaway slaves were branded. The canon law allowed branding as a punishment and it was used in England. After 1487 persons who claimed benefit of clergy (q.v.) were to be branded on the thumb with M in the case of murder or T in the case of theft. Under the Statute of Vagabonds, 1547, men and women who could but would not work were to be branded on the breast with a V and to be adjudged a slave for two years. If such person escaped he was to be branded on the forehead or cheek with an S. Under an act of 1551 a person brawling in church or churchyard was in certain cases to be excommunicated and branded with the letter F. Under an Act of 1694 persons dealing in filings from current coin were to be branded with R. All these provisions have been repealed and were long obsolete before repeal.

Branding was abolished in 1829 except for deserters from the army who were tattooed rather than branded with D below the left armpit and soldiers of bad character were branded BC. These were abolished in 1879. In France galley-slaves could be branded until 1832.

Branks (or scolding bridle). A contrivance formerly used in Britain, apparently without express legal authority, to punish scolding women. It was used in Scotland in the sixteenth century but not till later in England. It consisted of an iron headpiece with a flat piece of iron projecting inwards to enter the mouth and press down the tongue. Later it became more like a cage. The offending woman was marched through the streets with this on her head or chained to the market cross, there to be gibed at by passers-by. It is now long obsolete.

Braun, Conrad (Brunus) (1491–1563). Professor at Tubingen, and statesman, author of *De Legationibus libri quinque* (1548) and many other works, mainly on religion.

Brawling. Creating a disturbance in consecrated ground or a consecrated building. Prior to 1551 it was dealt with under canon law. In that year statute provided for penalties, including excommunication, and loss of the ear or branding on the cheek. Such conduct is now a summary offence.

Braxfield, Lord. See McQueen, Robert.

Breach. A breaking, or, more generally, the violation of a legal obligation, by failure to implement a legal duty.

Breach of arrestment. In Scots law, a contempt of court committed by an arrestee who disregards an arrestment (q.v.) served on him and, in defiance thereof, pays or delivers property to the debtor.

Breach of close. A legally unjustifiable entry on another's land involving an actual or notional breaking through a boundary.

Breach of condition. The non-implement of a term of a contract so material that the breach is treated as a repudiation of the contract and entitles the other party to treat the contract as at an end and to claim damages on that basis. Contrast Breach of Warranty.

Breach of confidence. Failure to observe a duty of secrecy undertaken expressly or imposed by law in particular relationships.

Breach of contract. An unjustifiable refusal or failure by one party to a lawful and enforceable contract to implement any of the duties incumbent on him under the contract, normally by refusing to perform, failing to perform, performing late, or performing badly. Refusal may be made before (anticipatory breach) or at, the due date for performance. Breach gives the other party a claim of damages for the loss sustained in consequence of the breach and may also justify him in rescinding or terminating the contract. In some cases the aggrieved party is entitled to have the contract implemented by obtaining a decree of specific performance.

Breach of duty. An unjustifiable refusal or failure to implement, a duty incumbent by agreement or by law, but particularly the latter. The commonest cases are breaches of the duty generally incumbent so to act as not to cause loss or injury to other persons foreseeably likely to be injured, such as other road-users, employees, visitors to premises, consumers of one's manufactured goods, and the like. Cases which come under the general principle but are considered separately include breach of contract and breach of trust.

Breach of privilege. See PRIVILEGE.

Breach of promise. The unjustifiable refusal or failure of one party to mutual promises between a couple to marry each other to implement that promise. At common law it was actionable for damages, but since 1970 it has not been actionable in England.

Breach of the peace. Any disturbance of public order and the Queen's peace, including bawling, shouting, and swearing in public, quarrelling and fighting, causing a public disturbance, and any conduct likely to provoke retaliation or a disturbance. It includes the particular offences of affray, assault and battery, and riot (qq.v.). If a breach of the peace is reasonably feared, as where one person threatens another with violence, he may be bound over to keep the peace.

Breach of trust. Any deliberate or negligent failure to implement the duties incumbent by law on a trustee or person in a fiduciary position. It covers not only, e.g. converting the trust funds to the trustee's own purposes but neglecting to invest them, investing them in an improper way, paying the wrong beneficiary, doing anything unauthorized by the trust deed or failing to do anything required by the trust deed or by law in the circumstances. Prima facie a trustee is personally liable to make good any loss to the trust resulting from breach on his part. The courts have traditionally demanded a high standard of care and diligence of a trustee and visited severely any breach. Modern statutes empower courts to excuse a trustee where he has acted honestly and reasonably and ought fairly to be excused.

Breach of warranty. A failure by a party to a contract to satisfy the standard required by a warranty (q.v.), common law or statutory, express or implied, applicable to that contract. A warranty is in English law, in general, an undertaking of a non-fundamental nature, for the breach of which an action lies but for which the contract cannot be declared at an end. Contrast Breach of Condition.

Breaking bulk. Separating some goods from the bulk of them, at common law a requisite for the conviction for larceny of goods originally lawfully obtained by the accused but subsequently wrongfully converted to his own use. Since 1861 it has been immaterial whether bulk was broken or not.

Breathalyser (or breath analyser). A device used by police to test for alcohol the breath of motor vehicle drivers suspected of having taken alcohol in excess of a prescribed proportion of milligrammes of alcohol per 100 millilitres of blood. The device works on the principle that an amount of air (the suspect's breath) passing through a chemical solution causes a colour-change in the solution proportional to the amount of alcohol in the air sample.

Breath test. A police constable in uniform may, if he has reasonable cause to suspect that the driver of a motor vehicle has taken alcohol, or has committed a traffic offence while the vehicle was in motion, require the driver to provide a specimen of breath by causing him to blow into a breathalyser (q.v.). Failure without reasonable cause to provide a specimen of breath is an offence.

Breda, Declaration of (1660). A manifesto drawn up and issued on behalf of Charles II while still in exile, based on advice from General Monck, who was effectively in control in London. Charles promised an amnesty to all save those expressly excluded by Parliament, a degree of religious freedom, payment of arrears of pay due to Monck's soldiers and their absorption into the forces of the Crown, and reference of the land problem to Parliament. In the result the declaration was largely but not quite completely implemented. The Clarendon Code (q.v.) did not allow religious freedom and the King's own attitude to dissenters was not entirely permissive.

Brehon laws (from *Breiteamh*, a judge), or properly *Feinechus*. The ancient laws of Ireland. These laws applied throughout Ireland prior to the Norman invasion and in Ireland outside the Pale until the early seventeenth century. A Brehon was an arbitrator and expositor of the law rather than a judge in the modern sense. Every king or chief of a substantial area had an official Brehon whose studies occupied 20 years and who had free land for his maintenance. The biggest and most important of the extant materials of Brehon law is the *Senchus Mor* or Great Old Law Book of *c.* A.D. 440. The *Book of Aicill* deals with, in substance, criminal law. They are in very archaic Gaelic and had evolved over a long period. By reason of their antiquity they are of great archaeological, anthropological, and philological interest.

The extant laws show a system in which assemblies, national, provincial, and local discharged legal, legislative, and administrative functions. Clan and kinship were important and kinship determined status and consequential rights and obligations. Fostering was common. Land seems to have been held by the *fine* or joint family group of four generations. Land was not sold or let, but the right to graze cattle was let, frequently with the cattle.

The law of contract was undeveloped. In the criminal law capital punishment did not exist, save in the last resort. Ordinary crimes were redressed by civil claim, compensation being calculated by reference to the value of the person killed or injured. Revenge and retaliation were discouraged. Compensation was determined by the status of the person injured. Distress or seizure of property was the mode of satisfaction for breach of contract, crime, or any other cause, and hence is an extensive and important branch of law.

Brehon Law Commission: *Ancient Laws of Ireland* (1865–1901); P. Joyce, *Social History of Ancient Ireland* (1903); Arbois de Jubainville and P. Collinet, *Études sur le droit celtique* (1895); S. Bryant, *Liberty, Order and Law under Native Irish Rule* (1923).

Brett, William Baliol, Viscount Esher (1815–99). Called to the Bar in 1846 and soon developed a practice, particularly in mercantile cases. He became Q.C. in 1861, M.P. in 1866, and was appointed Solicitor-General in 1868. He became a judge of the Common Pleas in 1868, in 1876 a Lord Justice of Appeal, and in 1883 succeeded as Master of the Rolls. He became a peer in 1885 and Viscount when he retired in 1897. He was a strong judge but hardly a profound, nor a scholarly, lawyer, who relied much on common sense as opposed to technicalities.

Bretton Woods (New Hampshire, U.S.A.). Scene of an international conference in 1944 which set up the International Monetary Fund with a capital of $8,800 million subscribed by the members on a quota basis, the members agreeing to stabilize exchange rates and to confer with the fund before making any major change in parity, and the International Bank for Reconstruction and Development with an initial capital of $10,000 million to assist long-term reconstruction of war-devastated economies. While the conference established major new international economic agencies, many of the hopes for co-operation envisaged there failed to materialize.

Bretts and Scots, Laws of. The name given in the thirteenth century to a body of laws in use among the Celtic tribes in Scotland north of the Forth (the Scots) and the British inhabitants of Cumbria (Bretts). It has a similar origin to ancient Anglo-Saxon, Irish, and Welsh laws and similarly deals with the price payable (*cro*) for crimes of violence.

Breve. Originally applied to any sort of letter in England after the Conquest, but shortly confined to writs as distinct from charters, and to writs directed to a person ordering something to be done. By Glanvill's time (*c.* 1190) the royal writ (q.v.) had become a fundamental element of the developing system of justice and collections of forms of writs began to be made in the form of many versions of the *Registrum Brevium* (q.v.).

Brevia magistralia. A class of writs recognized by Bracton which could be freely issued to meet new cases, in which it was thought expedient that an action should be granted. It was the power to issue these writs which caused the rapid development of the law in the early thirteenth century.

Brevia placitata. A book, in 75 sections, in law French, dating from about 1260, mostly consisting of forms of pleadings, notes appended thereto, but including also some exposition of law and some reports of cases. It states the division of pleas in the King's court, then pleas pertaining to the great court of the King, writs, declarations, and defences appropriate to particular actions. It was edited by Turner and Plucknett for the Selden Society in 1947.

G. Turner and T. Plucknett, *Brevia Placitata* (Selden Society).

Breviarium Alaricianum (or *Lex Romana Visigothorum* or *Breviarium Aniani*). A compilation of law for the Roman population of the kingdom of the Visigoths, compiled by a commission under Anianus appointed by King Alaric II (485–507) and approved by the popular assembly in A.D. 506. It was intended to epitomize the whole mass of Roman law and was founded on the *Codex Theodosianus*, some later Novels, the *Epitome Gaii*, an abbreviated version of Gaius' *Institutes*, the *Liber Gai*, substantial extracts from Paul's *Sententiae*, several constitutions from the *Codex Gregorianus* and the *Codex Hermogenianus*, and a *responsum* of Papinian. Historical and jurisprudential matter was omitted and an *interpretatio* provided a commentary. The Visigothic kings later sought to incorporate the Roman law element with that of their own customs, and the *Breviary* was repealed in the mid-seventh century, when Recesswinth promulgated in 654 a uniform law for Goths and Romans.

Though the *Breviarium* was later replaced by the *Lex Visigothorum* (see Visigoths, laws of the) in the Visigothic kingdom, it continued in use in southern France and Lombardy, which had meantime passed under the dominion of the Franks. Its qualities made the *Breviarium* a book of high authority throughout the whole of western Europe during the Middle Ages and it was one of the main channels through which Roman law entered western European law prior to the Reception. Abridgements were made of it between the sixth and ninth centuries. A defective abstract was the basis of the *Lex Romana Curiensis* and a ninth-century version made for use in Lombardy was preserved in the *Codex Universalis*.

E. Haenel, *Lex Romana Visigothorum*; Conrat, *Breviarium Alaricianum*.

Breviarium extravagantium. A twelfth century canon law text compiled by Bernard of Pavia.

Brewster, Abraham (1796–1874). Called to the Irish Bar in 1819, and became Solicitor-General, 1846–47, Attorney-General, 1853–55, and Lord Chancellor of Ireland, 1867–68.

Brewster sessions. The special sessions held by justices for the renewal or grant of licences for the retail sale of intoxicating liquors.

Bribery. The crime of giving money or other valuable things or rights to another as an inducement to him to act in a certain way, or to reward him for having so done. It is penalized by statute in various contexts, such as bribing judges, voters, members of local authorities or public officers.

There have been a number of instances in English legal history of judicial corruption, notably in 1289 and 1350. Other notable cases were Lord Chancellor de la Pole in 1387, Lord Chancellor Bacon in 1621, Cranfield, Earl of Middlesex in 1624, and Parker, Lord Macclesfield, in 1725. In Queen Anne's reign a speaker of the House of Commons was expelled for bribery.

Brideprice (or bridewealth). In various primitive societies, the payment made by the bridegroom or his kin to the wife's kin in order to ratify marriage. It is usually as much a social and symbolic exchange as well as economic reciprocity, a pledge that the wife will be well treated as much as compensation for her loss. Similarly a marriage is not deemed terminated until the return of the brideprice has been made, indicating acceptance of the divorce.

Bridewell. A district in London between Fleet Street and the Thames, so-called from the well of St. Bride or St. Bridget. It contained a disused royal residence, established as a hospital in 1557, which provided employment for applicants for relief, to which idle vagrants could be sent and where they could be compelled to work. The palace was demolished in 1863. The name survives in common usage for a house of correction.

Bridgeman, Sir Orlando (?1606–74). Called to the bar in 1632, he sat in the Long Parliament and was a supporter of monarchical government. During the Commonwealth he withdrew from practice but made his name as a conveyancer; he invented the strict settlement of land. At the Restoration he won immediate favour. He was made successively Chief Baron of the Exchequer (1660) in which capacity he presided at the trials of the regicides, and then C.J. of the Common Pleas (1660–67) when he became Lord Keeper (1667–72). As Chief Justice he was highly regarded for learning and integrity, and he assisted in developing the modern rule against perpetuities. As Lord Keeper he was less successful and the Court of Chancery fell into endless delay. In 1672 he was dismissed for refusal to seal grants for the King's mistresses. He also compiled a volume of Common Pleas reports not published till 1823.

Brief. In England, a document of instructions to counsel to appear in court, including also the papers relevant to the case, copies of pleadings, proofs (q.v.) of the evidence of witnesses, etc. prepared and sent by the instructing solicitor. It is counsel's authority to appear and represent the client. From this comes the verb 'to brief counsel', i.e. to instruct him. A 'watching brief' is an instruction to counsel to attend a hearing on behalf of one not a party to the case in case the latter's interests are affected.

In the U.S. it is a written legal argument presented to a court to assist it in reaching a decision on the issues involved in the case, particularly in appellate courts where the issues are purely legal and there is little or no oral argument. Courts have rules governing the format and content of briefs. Occasionally, and usually in constitutional cases, such a brief may include extensive social, economic and sociological data. See BRANDEIS BRIEF.

Brief, papal. A letter from a pope, usually less formal and less significant than a papal bull, signed with an impression of the fisherman's ring.

Brief-bag. A sack in which counsel's papers are carried to and from court. In the early thirteenth century such bags were possessed only by counsel who had been given one by a Queen's counsel, who prior to 1830 had pens, paper, and purple bags as perquisites of their rank. Brief bags in England are now blue or red. A blue bag is bought by a barrister on being called as part of his outfit and, by custom, is not allowed to be visible in court. When a junior has rendered assistance of special value to a Q.C. in a major case the latter may present the junior with a red bag. In the U.S. brief-bags are green.

Brierley, James Leslie (1881–1955). Had a distinguished academic career and was called to the Bar in 1907. In 1922 he became Chichele Professor of International Law at Oxford and after his retiral in 1947 professor of international relations at Edinburgh, 1948–51. He served on League of

Nations committees and national and public bodies, and became a member of the Institute of International Law.

He wrote *The Law of Nations*, a brilliant introductory book, and many papers on international law, many reprinted in *The Basis of Obligation in International Law*, and edited Zouche's *Juris et Judicii Fecialis Explicatio* and several editions of Anson on *Contract*. He also edited the *British Yearbook of International Law*, 1929–36.

Brieves. In mediaeval Scots law, writs issued from Chancery directed to justiciars, sheriffs, barons, and others, as the means of initiating proceedings before such judges. They contained an issue or question, which was submitted to an inquest, i.e. a civil jury. They were simple, flexible, and free from excessive technicality, but became stereotyped by the end of the fifteenth century, though about that time they began to be replaced by a summons in substantially modern form. Some, 'pleadable' brieves, were not returnable to Chancery; in non-pleadable brieves, the finding had to be 'retoured', or returned, to Chancery, where the decision was confirmed by recording in the King's presence, or by royal charter. A few forms survived to modern times.

Brieves, Registers of. Collections of Scottish mediaeval procedural styles, conveyancing writs and administrative precepts probably made for King's clerks or legal practitioners. They declined from the fifteenth century and disappeared by the sixteenth.

Brinz, Alois von (1820–87). German jurist, professor at Munich, author or an important *Lehrbuch der Pandekten* (1857).

Brisson, Barnabé (Brissonius) (*c.* 1530–1591). A distinguished French humanist jurist, Advocate-General of the Parlement of Paris, and author of a legal dictionary, *De Verborum Significatione* (1557), a systematic collection of the principal provisions in the Ordonnances in force under Henry III, published under the title of *Code du roy Henry III*, (1584) and other works.

Bristol, Red Book of. An anonymous treatise of about the fourteenth century on the law merchant and containing a copy of the laws of Oléron (q.v.).

Bristol Tolzey Court. This ancient court had jurisdiction in all personal and mixed actions to any amount. It was held by a barrister of five years standing, who might be the Recorder or his deputy. There was also a Court of Pie Poudre held in September, during the sittings of which the Court of Tolzey was suspended, and there were formerly also a Mayor's Court and a Court of the Staple. The Tolzey and Pie Poudre courts were abolished in 1971.

British Commonwealth of Nations. The term given from the 1920s to about 1950 for the association of countries, originally Great Britain and its colonies and dependencies, which came to be called the British Empire and then, as the major elements became recognized as independent and equal, the British Commonwealth of Nations. In the late 1940s, and particularly when India became a republic in 1950, the term 'British' was dropped and the association became simply the Commonwealth (q.v.).

British Commonwealth. See COMMONWEALTH.

British Isles. The group of islands off the north-western shores of the Eurasian continent, comprising Great Britain (itself comprising the countries of England, Wales, and Scotland) and adjacent lesser islands and groups of islands, Northern Ireland (comprising the part of the island of Ireland, which includes the six counties constituting Northern Ireland), the Isle of Man, and the Channel Islands (comprising Jersey, Guernsey, Alderney, and Sark). The British Isles is a geographical term and they are not a state but contain the United Kingdom of Great Britain and Northern Ireland, the Isle of Man, and the Channel Islands, of which the latter two are parts of Her Majesty's dominions, but neither parts of the United Kingdom nor colonies or dependencies thereof. The Republic of Ireland (the island of Ireland other than Northern Ireland) is independent and a separate state and would object to being included in the British Isles.

See also CHANNEL ISLANDS; MAN; ENGLISH LAW; IRISH LAW; NORTHERN IRISH LAW; SCOTS LAW; WELSH LAW.

British law. By reason of history there is no such thing as British law. England, Wales, Scotland, Ireland, Man, and the Channel Islands all have distinct legal and general histories. England and Wales became a single state in 1295; they united with Scotland in 1707 and this unity united with Ireland in 1801 to form the United Kingdom, but in 1921 Ireland had to be divided into Northern Ireland, which remained part of the United Kingdom, and the Irish Free State, later Eire and later Republic of Ireland, which became independent.

The term British law is sometimes used of those branches and principles of law which are substantially common to the legal systems of England and Wales, Scotland, and Northern Ireland, or sometimes to the first two of them only, but it has no strict or accurate meaning.

British nationality. Prior to 1948 British nationality was a category of status enjoyed by all persons born in countries owing allegiance to the British Crown. Since 1948 this common allegiance has disappeared, being replaced by the principles that the people of each of the countries of the Commonwealth have the status of citizens of their respective countries and also by reason thereof the common status of British subjects or Commonwealth citizens (these terms being synonymous). In consequence a citizen of any Commonwealth country is not an alien in the U.K. In the United Kingdom the category of citizenship is called 'citizen of the United Kingdom and Colonies'. Citizens of the Republic of Ireland are not automatically British subjects or Commonwealth citizens, but in certain circumstances they may be so, and Ireland is not deemed a foreign country and its citizens are not aliens.

British North America Acts. A series of Acts of the British Parliament regulating the government of the British colonies of North America remaining after American independence. The Act of 1840, following the Durham Report, united Upper Canada (Ontario) and Lower Canada (Quebec) into the province of Canada under one legislature, but the two divisions continued at loggerheads. Under the Act of 1867 Canada was divided into Quebec and Ontario and formed into a federation along with Nova Scotia and New Brunswick under the name of the Dominion of Canada. Other provinces have subsequently acceded to the Dominion. Numerous proposals have been made, all unsuccessfully, for the transfer to Canada of full powers to amend the Dominion constitution, but modern British North America Acts are only passed at the request and with the assent of the Dominion governments.

British protected person. One of a class of persons declared by Order in Council made under statute in relation to any protectorate, protected state, mandated territory, or trust territory to be, for the purpose of British nationality law, British protected persons by virtue of their connection with that territory. They are not aliens and may become naturalized or registered as British subjects or citizens of the U.K. and Colonies.

British subject. Formerly this meant a natural born British subject, or a person naturalized, or a person who became a subject of the Crown by annexation of territory. By the British Nationality Acts, 1948–65, the term is synonymous with Commonwealth citizen and comprises every person who is a citizen of the United Kingdom and colonies or a citizen of one of the countries of the Commonwealth as defined by the relevant Commonwealth country's own law.

A person becomes a citizen of the U.K. and colonies by birth, descent from a father who was such a citizen, registration, naturalization and, in certain cases, marriage. The status may be lost where a dependent territory becomes an independent country or ceases to be part of the Commonwealth, by renunciation of citizenship or deprivation.

Britton. A book on English law written about 1290, based on borrowings from Bracton (q.v.) and Fleta (q.v.), but with the subject-headings and the matter thereunder somewhat differently arranged. It may have been compiled by John le Breton, Bishop of Hereford, or by another of the same name, a justice of trailbaston. It was in law-French, seems to have been popular, and purports to be a direct codification and enactment of the law by Edward I. The subject-matter is common law and it was a practical book for lawyers practising in the royal courts. First printed about 1530 the standard edition is that of F. M. Nichols (1865).

Broad arrow. A mark showing the barb of an arrow with a short intermediate shaft, for long used as a mark on government property and at one time stamped on clothing worn by persons in prison. This latter use is now long discontinued. The arrow mark may have a Celtic origin, indicative of superiority or authority.

Broadmoor. A special hospital in Berkshire opened in 1863 at the instigation of the Lunacy Commissioners as the first State Lunatic Asylum, for patients requiring secure treatment by reason of dangerous, violent, or criminal propensities. It took patients from Bethlem and other asylums and prisons and from its opening a majority of the country's criminal lunatics were in Broadmoor, though it also held many dangerous, though not criminal, lunatics. In 1948 the term Broadmoor patient superseded that of 'criminal lunatic', but that in turn was abolished in 1959. Similar establishments are at Rampton, Moss-side and, in Scotland, Carstairs.

R. Partridge, *Broadmoor: A History of Criminal Lunacy and its Problems.*

Brocard. A term, said to be derived from Burchard (q.v.) who digested the canon law, for legal maxims borrowed from Roman law or based on ancient custom.

Broderick, Alan, Viscount Midleton (1656–1728). Called to the English Bar in 1678, became Solicitor-General (1695–1704), Attorney-General for Ireland (1707), and Chief Justice of the

Irish Queen's Bench in 1709 but was superseded in 1711. He was also speaker of the Irish Parliament. In 1714 he became Chancellor of Ireland and later sat in the British Parliament. He resigned in 1725.

Broker. A mercantile agent who in the ordinary course of his business is employed to make contracts for the purchase or sale of property or goods, but who is not entrusted with the possession thereof or of the documents of title thereto. Specialized kinds of brokers deal in particular kinds of property, e.g. stockbrokers, or in services, e.g. insurance brokers and shipbrokers.

Bromley, Sir Thomas (1530–87). Called to the Bar, becoming recorder of London in 1566, Solicitor-General in 1569, and Lord Chancellor in 1579. He proved a good equity judge and decided *Shelley's case* ((1581) 1 Co. Rep. 93b). He was concerned in the prosecution of Babington for his plot against Queen Elizabeth, and presided at the trial of Mary, Queen of Scots, at Fotheringhay Castle in 1586. The worry over the execution of that Queen hastened his death.

Bronchorst, Everard (1554–1627). Famous Dutch jurist, author of *Methodus Feudorum* (1613), *Enantiophanon centuriae sex* (1621), and *In titulorum Digestorum de regulis iuris antiqui enarrationes* (1624).

Brooke (or Broke) Robert (?–1558). Became Recorder of London in 1545, sat in Parliament and became Speaker in 1554, and in 1554 Chief Justice of the Common Pleas. He was renowned for his probity and learning. He compiled an *Abridgement* (1568 and later editions), digesting nearly 21,000 cases, notes, and points under 404 alphabetical titles, drawn from the Year Books, Fitzherbert's *Natura Brevium*, and other works.

A selection from Brooke's *Abridgement* and known as *Brooke's New Cases, Little Brooke* or *Bellewe's Case temp. Henry VIII, Ed. VI et la Roygne Mary*, was published in 1578. It was compiled by Richard Bellewe, and was in chronological order but a later edition contained the matter under alphabetical heads.

Brothel (or bawdy house). One variety of disorderly house, a place resorted to by persons of one or both sexes for lewd heterosexual or homosexual practices, particularly for reward. It is not a brothel where one person alone receives persons for sexual practices. In modern law, though prostitution is not an offence by itself (though soliciting is), it is an offence, at common law and under statute, to keep a brothel, or manage one, or let premises to be so used.

Brougham, Henry Peter, 1st Lord Brougham and Vaux (1778–1868). Became an advocate of the Scottish Bar in 1800 but, having little success there, was called to the English Bar in 1808. He had already begun to make a name as a writer in the *Edinburgh Review*, and in politics, and he entered Parliament in 1810. There he rapidly came to the front, but lost his seat in 1813, returning to the House in 1816. All his life he supported free trade, law reform, Parliamentary reform, and liberal causes generally. He acted for Queen Caroline in the proceedings against her in 1820 and was mainly responsible for their abandonment.

In 1828 he made a great speech in the House on the reform of the law, criticizing the existing structure of the courts and urging the need for local courts for quick and cheap redress in lesser cases, condemning obsolete modes of procedure and rules of pleading, and exposing defects in the rules of evidence. This resulted in the appointment of two commissions to enquire into the common law courts and the law of real property, whose reports resulted in very extensive changes. He also supported parliamentary reform, the abolition of slavery, public education, and helped to found the University of London.

In 1830 he became Lord Chancellor, in which capacity he treated the peers like dirt and aroused bitter feelings. As chancellor he worked hard to expedite cases in chancery, but as a judge he was not a success, paying inadequate attention to arguments, not studying the papers, and dealing with cases in a careless fashion. The legal profession had a poor opinion of his knowledge of equity though he was not ignorant of the branch. His contribution to the law lay in promotion of legislative reforms rather than in judicial work. He was responsible for the abolition of various obsolete courts and for the creation of the Judicial Committee of the Privy Council and the Central Criminal Court, and helped to maintain and forward the movement for reform of the law by statute which gained impetus later in the nineteenth century.

After his dismissal from office in 1834 he sat frequently in the House of Lords and the Judicial Committee and gave important judgments in some cases. After 1850, assisted by two lay lords, he alone undertook the appellate work of the House of Lords, but the results provoked unfavourable criticism.

Though a very able lawyer Brougham's chief interest was in politics, but his actions were too often motivated by ambition and he lacked stable political beliefs. As a law reformer he achieved much of lasting importance.

W. Bagehot, *Biographical Studies*; A. Aspinall, *Lord Brougham and the Whig Party*; Hawes, *Henry Brougham*; H. Brougham, *Life and times of Henry, Lord Brougham*, 3 vols. (1871).

Brouwer, Hendrik (1625–83). Dutch jurist, author of a *De iure connubiorum* (1665).

Brown v. Board of Education of Topeka (1954), 347 U.S. 483. In this case there was challenged, in the field of elementary education, the doctrine that provision of 'separate but equal' facilities (see *Plessy* v. *Ferguson*) was consistent with the Fourteenth Amendment. The Supreme Court held unanimously that segregated public schools were not and could not be made equal; in the field of public education 'separate but equal' had no place.

Brunnemann, Johann (1608–72). German jurist, author of *Commentarius in libros Codicis* (1663), *Commentarius in libros Pandectarum* (1670), *Tractatus Juridicus de processu fori* (1659), and other works.

Brunner, Heinrich (1840–1915). German legal historian, whose work laid the foundations of the systematic study of early German law and institutions. His major works are *Deutsche Rechtsgeschichte* (1887) and *Grundzuge der deutschen Rechtsgeschichte* (1901).

Bruns, Karl Eduard Georg (1816–80). German jurist, author of *Recht des Besitzes im Mittelalter* (1848), *Fontes Juris Romani Antiqui* (1860) and other works.

Brussels Treaty, 1965 See MERGER TREATY: TREATY OF BRUSSELS, 1965.

Brussels Treaty, 1972. See ACCESSION TREATY: TREATY OF BRUSSELS, 1972.

Brutus, Marcus Junius (2nd cent. B.C.). One of the earliest Roman jurists, author of a *de iure civili* in dialogue form.

Brutus, Stephanus Junius. Pseudonym of, probably, Hubert Languet (1518–81) author of *Vindiciae contra tyrannos* (1579).

Bryce, James, Viscount (1838–1922). Author, jurist, and statesman. Had a distinguished career in classics at Oxford, becoming a Fellow of Oriel in 1862, was called to the Bar in 1867, and obtained a respectable practice. In 1863 he won the Arnold historical prize at Oxford with an essay which became *The Holy Roman Empire* (1864). In 1870 he became Regius Professor of Civil Law at Oxford and began the revival of the study of Roman law there; he held that post till 1893 and did much to revive the study of civil law. He sat in Parliament from 1885 to 1906 and held various offices, including Under-Secretary for Foreign Affairs, Chancellor of the Duchy of Lancaster, President of the Board of Trade, and Chief Secretary for Ireland. He was ambassador to Washington, 1907–13, where he was highly respected. He received the O.M. in 1907, was made a peer in 1914 and G.C.V.O. in 1918. During the First World War he strongly supported the promotion of a League of Nations and, at its conclusion, was chairman of a joint conference on the reform of the House of Lords. He travelled widely and the fruits of his observations were published in *Transcaucasia and Ararat* (1877), *The American Commonwealth* (1888), *Impressions of South Africa* (1897), *South America: Observations and Impressions* (1912), and *Modern Democracies* (1921).

He was associated with the foundation of the *English Historical Review*. The best of his legal writings are contained in his *Studies in History and Jurisprudence* (2 vols., 1901) but he published work on a great variety of topics of public affairs. He diffused his abilities and energies widely, and attained considerable distinction in several fields.

H. A. L. Fisher, *James Bryce* (1927).

Bubble Act (1720). The Bubble Act, passed in consequence of the South Sea panic, made it difficult for joint stock companies to assume a corporate form, and contained no rules at all for the conduct of such societies if and when they assumed corporate form. Thenceforth there was to be no confusion between a corporate and a non-corporate commercial society, and the privilege of possessing transferable stock was to belong to a corporate society only. It inhibited the development of company law and was repealed in 1825.

Buckland, William Warwick (1859–1946). Called to the Bar and became a Fellow of Gonville and Caius College, Cambridge, in 1889. In 1914 he became regius professor of civil law there and held the office till 1945. As a tutor he taught all branches of English law, but his field of interest and achievement was Roman law on which he wrote extensively: *The Roman Law of Slavery* (1908), *Equity in Roman Law* (1911), *Elementary Principles of Roman Private Law* (1912), *A Manual of Roman Private Law* (1925), *The Main Institutions of Roman Private Law* (1931), *Roman Law and Common Law* (with A. D. McNair, 1936), *Studies in the Glossators of the Roman Law* (with H. F. Kantorowicz, 1938), and *Some Reflections on Jurisprudence* (1945), and his masterpiece, *A Textbook of Roman Law from Augustus to Justinian* (1921), the most important and valuable treatise on the subject yet published in England, and a standard text ever since. His work has been of the first importance for the study of Roman law in Britain.

Buckley, Henry Burton, Lord Wrenbury (1845–1935). Called to the Bar in 1869 he occupied his early years in writing *Buckley on the Law and Practice under the Companies Acts* (1873) which rapidly became and has since remained a standard authority, but had a sound grasp of all equity matters. He became a chancery judge in 1900, in 1906 a judge of the Court of Appeal, and on his retirement in 1915, a peer and a Privy Councillor. As such he sat regularly for years in the House of Lords and the Judicial Committee; his judgments were highly respected.

Buckmaster, Stanley Owen, Viscount Buckmaster (1861–1934). Called to the Bar in 1884 he slowly built up a chancery practice and in 1906 entered Parliament. In 1913 he was made Solicitor-General and on the outbreak of war in 1914 head of the Press Bureau. In 1915 he was made Lord Chancellor but went out of office in 1916. Thereafter he spoke frequently in favour of social and legal reform and sat regularly in the House of Lords and the Privy Council. In 1933 he became a Viscount. Contemporaries deemed him an acute, learned, and very fair judge, and a number of his judgments are valuable.

Buddeus, Joannes Franciscus (1667–1729). German jurist, author of *Selecta iuris naturae et gentium* (1704), and works on history and theology.

Budé, Guillaume (Budaeus) (1467–1540). A distinguished French polymath and humanist jurist, secretary to King Louis XII of France, and later royal librarian and founder of the College de France. He was one of the group which developed a more rational and scientific method for the study and understanding of Roman law. His main legal work, the *Annotationes in XXIV libros Pandectarum* (1508) had great influence on the study of Roman law, but he also contributed greatly to the study of Greek literature in France, notably by his *Commentarii linguae graecae* (1529).

Buggery. The crime of a man having intercourse *per anum* with a man or a woman (sodomy), or of a person of either sex having intercourse *per anum* or *per vaginam* with an animal (bestiality). Consent is no defence, and both participants are guilty. Since 1967 it has not been an offence in England for a man to commit buggery or gross indecency with another male if it is done in private, both parties are aged 21, and both consent, but in other circumstances it continues to be a crime.

Building lease. A lease of land for a substantial period on the terms that the lessee will build thereon one or more buildings of agreed size and value and that on the expiry of the lease it or they will fall to the landlord with the land.

Building society. A society of persons formed to gather money from the deposits of members, on which they are paid interest, which can then be lent to other members to enable them to purchase houses therewith. Borrowing members pay interest on money lent and mortgage the property bought to the society until they have paid off the money borrowed with interest. The formation and management of building societies has been regulated by statute since 1874 and is now regulated by Act of 1962. Today building societies have assets of many millions of pounds and have been the means whereby thousands of persons have been enabled to buy their own homes, while deposits of money with building societies are a very important form of saving.

Bulgarus (?–1166). Italian jurist, one of the 'four doctors', immediate successors of Irnerius at Bologna, and an early glossator (q.v.). He was the leader of a 'school' of jurists, opposed by another school headed by Martinus Gosia (q.v.). The former adhered more to strict law, the latter to equity. Among his pupils were Rogerius and Bassianus who in turn taught Hugolinus and Azo. He took a leading part at the Diet of Roncaglia in 1158 and was a trusted adviser of the Emperor Frederick I. His chief works were a commentary, *De Regulis Juris*, *De judiciis*, and *Quaestiones*.

Bull. Bulla was a leaden seal used for authenticating documents used by the papal chancery from the sixth century and by the royal chancelleries of Europe, though sometimes using gold or silver in place of lead. The lead was impressed with devices on both sides. From the thirteenth century the documents with such seals were themselves called bulls. One class, *bullae maiores* or *privilegia*, dealt with the conferment or corroboration of rights without limit of time. This type was discontinued in the fourteenth century. Another class, *bullae minores* or *litterae*, dealt with lesser matters and were later classified as rescripts or executive documents.

Bulls were formerly dated by locality and date of issue according to Roman calculation, but since 1908 are dated by the civil calendar. They are in Latin. Since 1878 the leaden seals have been reserved for the most solemn bulls; for all others a red ink stamp is used. Bulls are commonly known by the first words of the text, e.g. *Rerum Novarum*.

Bundesvertrag, Der. The Swiss federal constitution ratified in 1815.

Bundesgerichtshof. The Supreme Federal court of the Federal Republic of Germany, sitting at Karlsruhe, having jurisdiction in civil cases to review judgments passed by regional courts, and in criminal cases to review judgments of large criminal divisions and intermediate appellate courts. In treason cases the court is the sole court competent.

Bundesrat. The Federal Council of the Federal Republic of Germany, an assembly of representatives of the Länder republic. It has in general a power to delay, but not wholly to block legislation.

Bundestag. The federal Diet of the Federal Republic of Germany under the Basic Law of 1949. The deputies are elected by popular direct vote for a four-year term. Its principal functions are election of the President of the Federal Republic and of the Federal Chancellor, parliamentary control of the federal government and legislation.

Bundesverfassungsgericht. The Federal Constitutional Court of the Federal Republic of Germany which sits at Karlsruhe. Its main jurisdiction is as to the interpretation of the Basic Law of the Republic in disputes concerning the rights and obligations of any of the highest federal organs; the constitutionality of laws; the comparability of Land legislation with federal legislation; disputes between Federation and Länder; and constitutional complaints.

Bundesverwaltungsgericht. The Federal Administrative Court of the Federal Republic of Germany which deals with disputes under public law of a non-constitutional nature, unless expressly assigned to another court.

Burchard (?965–1025). Bishop of Worms, author of a compilation of decretals of the church, *Decretorum libri XX* or *Brocardus* (*c.* 1020), commonly referred to before Gratian's *Decretum* and a material source of that work. It was based on the Collection of Regino of Prum, the *Anselmo dedicata*, the *Dionysio Hadriana*, the False Decretals, the councils of the ninth century, the episcopal *Capitula*, the *Collectio Hibernensis*, some penitentials, and extracts from other works. There is little matter from Roman law, but some from Carolingian capitularies. Burchard used his materials freely and modified texts to suit the times and promulgate his own ideas.

Burden. That which is borne, hence the duty to discharge some obligation. The burden of a contract is the liability to perform; the burden of an easement the liability to allow it to be exercised.

Burden of Proof. See ONUS OF PROOF.

Burdens. In Scots law any restriction, limitation, or encumbrance affecting person or property, but particularly one requiring payment of money. Such a burden may be personal, where a person is taken bound to pay money to the granter of a deed or to a third party; or real, where such an obligation is made to affect and be payable out of land. The term is also used in the phrase public burdens, of sums payable to public authority.

Burdick, Francis Marion (1845–1920). Admitted to the Bar in 1872, became one of the first professors of the Cornell Law School in 1887 and then taught at Columbia, 1891–1916. He wrote a number of books, including the *Law of Sale of Personal Property* (1897), *The Law of Partnership* (1890), and *The Law of Torts* (1905). He was also one of the commissioners on uniform laws.

Bureaucracy. Government by a professional corps of officials organized in an hierarchical structure and functioning under impersonal and uniform rules and procedure. The word usually has a critical or pejorative overtone. Bureaucratic government has been encouraged by the development in Western countries of widespreading governmental control of and intervention in many spheres; health, industry, commerce, building and planning, education, and state concern with many matters left to private initiative prior to the twentieth century. Socialist policies encourage the growth and power of bureaucracy as more and more areas of activity are brought under control. In democracies the corps of officials is theoretically the servant of the people, but with the growing complexity of administration and the development of a highly skilled body of career officials the power of the permanent official over the general public is great and even against the Cabinet minister is considerable. Thus the resistance of the Treasury to reform in tax laws is powerful. Though the officials are subject to the rule of law there are enormous difficulties in obtaining any remedy against neglect, unfairness, or other maladministration. A significant development in many countries is the appointment of an official (see PARLIAMENTARY COMMISSIONER) to investigate alleged injustice and maladministration.

Burgage. The mode of feudal tenure of tenements of land in any ancient borough, or burgh, held of the King or a mesne lord by the burgesses for fixed rents or services. This tenure ranked as a form of socage, and was subject to various local customs, notably that of borough-English, whereby the burgage tenement descended not to the eldest son but to the youngest. In Scotland the tenure was proper only to royal burghs, the lord was always the King and the return was the watching and warding of the burgh against enemies and sometimes a burgh-mail or payment to the King. It remained a distinctive tenure until modern times. Booking (q.v.) was a variant of burgage tenure. A few royal burghs did not have burgage tenure.

Burgage boroughs. Borough constituencies in the unreformed British parliament, where the vote was held by the owner or the tenant of lands held by burgage tenure.

Burgage tenure. In Scots law, the manner under the feudal system of holding lands and buildings within royal burghs, though not all royal burghs had burgage tenure. In these royal burghs the only feudal superior is the Crown, and the community hold their common property and the individual burghal owners their private property directly from the Crown in each case. In burgage tenure there is no feu-duty or casualties, the burgh holding its lands in return for watching and warding them, and greater freedom of alienating the existing estate, though no liberty to sub-feu.

Burger, Warren Earl (1907–). Admitted to the Minnesota Bar in 1931 and later served as an assistant Attorney-General of the U.S. 1953–55 and judge of the U.S. Court of Appeals for D.C. Washington, 1955–69. In 1969 he was appointed fifteenth Chief Justice of the U.S. Supreme Court. This was something of a break with tradition in that previous Chief Justices had been men of considerable public reputation and it was generally believed that he was appointed to lead the court away from intrusion into politics and away from excessive concern for the rights of accused or suspected criminals.

Burgerliches Gesetzbuch (BGB). The German civil code. See GERMAN LAW.

Burgersdijck, Franco Petri (1590–1639). Dutch philosopher, author of *Idea economicae et politicae doctrinae* (1644) and other works.

Burgess. Originally one of the inhabitants of a borough and later one of the inhabitants who were entitled by payment of rates to be enrolled to elect the council. The term was sometimes restricted to the magistrates. A burgess member was formerly an M.P. who sat for a city or borough constituency as distinct from a country member.

Burgess, John William (1844–1931). Admitted to the Massachussetts Bar in 1869 and then studied in Germany. He was successively professor of history, political science, and political economy at Amherst, and of constitutional law and political science at Columbia, and became its first dean. He founded the *Political Science Quarterly* and wrote many books, including *Political Science and Comparative constitutional Law* (2 vols., 1890–91). He had a high reputation, not only in the U.S.A. but in Europe.

Burgesses, House of. The representative assembly in colonial Virginia and the first elective governing body in a British overseas possession, composed of two burgesses from each settlement in Virginia. It was the lower house of the legislature established at Jamestown in 1619. The other house comprising the governor and a council all appointed by the proprietors of the colony, the London Company.

Burgh, Hubert de (?–1243). Chamberlain of King John, and later sheriff of several counties, he became chief justiciar in 1215. He headed the party supporting the young Henry III, becoming justiciar for life in 1228. He was deprived of office in 1232. He was the last of the great chief justiciars and at the height of his power was enormously powerful and rich. His administrative policies were beneficial but roused much opposition and this, coupled with the hatred and envy of others, brought about his fall.

Burgh, Walter Hussey (1742–83). The name Burgh was added in 1762. Called to the Irish Bar in 1769 and attained fame as an orator in the Irish Parliament. He was Chief Baron of Exchequer in Ireland 1782–83.

Burgh. In Scotland burghs are towns whose inhabitants have become incorporated by royal charter or under statute for civic government. They developed in the twelfth and thirteenth centuries as administrative, economic, and military centres, and in time came to have duties of parliamentary attendance and contribution to taxation. From the fifteenth century distinctions developed between royal burghs, burghs of barony, and burghs of regality, chartered respectively by the King, a baron, or a lord of regality, usually situated near the King's or a baron's castle, or a cathedral or monastery, and all having burgh courts (q.v.). All burghs eventually obtained financial autonomy, paying their superior a lump sum for protection and privileges. Burghs were self-governing, the burgesses choosing their own councillors and magistrates, though from 1469 to 1833 councils were self-perpetuating oligarchies, but in 1833 statute introduced election by residents, while the exclusive rights of burgesses were abrogated.

From the early nineteenth century there developed also parliamentary burghs, participating in the election of M.P.s, and police burghs, not chartered but having rights of local government. Since 1929 the distinction was between the four burghs (Edinburgh, Glasgow, Aberdeen, Dundee) dignified as cities and having county powers (counties of cities), large burghs, and small burghs, all having different local government functions and burgh police courts (q.v.). The distinction on the basis of the original chartering has only historical but not legal significance. Every burgh had an elected town council comprising a provost (Lord Provost in the cities), bailies, and councillors, a town clerk, and other officials. From 1975 burghs ceased to have distinct local government functions, being absorbed into local authority districts.

Burgh court. When burghs were constituted in Scotland the magistrates had jurisdiction as well as administrative functions. All burgesses were supposed to attend the three head courts in the year. The civil jurisdiction was mainly concerned with ownership and possession, trading and markets; criminal jurisdiction extended to slaughter, the hanging of thieves and adulterers, but became more modest as time went on. From the seventeenth century the competence of these courts was eroded by the development of the central courts, the increasing importance of the sheriffs, and the establishment of justices of the peace. In the eighteenth and nineteenth centuries jurisdiction was limited by statute, and abandoned to other courts, especially the sheriff court. After the reform of the Scottish burghs in 1833, they were empowered to appoint 'police magistrates' to sit in burgh police courts (q.v.) with a limited criminal jurisdiction. Burghs also had Licensing Courts (for licensing premises for the sale of excisable liquor) Licensing Appeal Court, and a Dean of Guild Court (q.v.).

Burgh police court. A minor court in Scottish burghs, having a criminal jurisdiction similar to that of the J.P. Court in counties. It was held by the provost or a bailie of the burgh or by an ex-magistrate as Judge of Police, all of whom were persons elected by the town councillors from among themselves, or in a few cases by a stipendiary magistrate. Prosecution was by the burgh prosecutor. The jurisdiction was confined to minor offences, particularly breaches of the peace. In 1975 they were abolished and the jurisdiction transferred to district courts (q.v.).

Burglary. At common law and under statute in England, the crime of breaking and entering the dwelling-house of another person in the night with intent to commit some felony therein, whether the intent is executed or not; or to break out of the dwelling-house of another in the night, having entered it with intent to commit a felony therein, or having committed any felony therein. The common law and statutory crimes were abolished in 1968 and replaced by the new statutory crime of burglary, which is constituted if a person enters any building or part of a building or inhabited vehicle or vessel, as a trespasser and with intent to commit any of the offences of stealing anything, or inflicting on any person any grievous bodily harm, or raping any woman, or doing unlawful damage to the building or anything therein; of having entered any building or part of a building as a trespasser, he steals or attempts to steal anything in the building or that part of it or inflicts or attempts to inflict on any person therein any grievous bodily harm. Aggravated burglary is constituted by burglary wherein the burglar has with him any firearm or imitation firearm, any weapon of offence or any explosive. It is also a crime to have with one any article for use in the course of or in connection with any burglary. The Scottish counterpart is housebreaking.

Burgos, laws of. Thirty-two laws promulgated in December 1512, the first Spanish attempt to establish a recognized legal code for the treatment of American Indians, produced in response to friars' complaints of ill-treatment of Indians. Royal officials in Spanish America were instructed to act as protectors of the Indians and Spaniards were ordered to treat their Indian servants with humanity. But enforcement was ineffective and continuing complaints resulted in a second code, the New Laws of 1542.

Burgundian code. See BURGUNDIANS, LAWS OF THE.

Burgundians, laws of the. In 501 Gundobad, king of the Burgundians (A.D. 467–516), issued a code known as the *Lex Gondebada*, or *Lex Gombata*, or *Lex Burgundionum*, or *Liber Constitutionum*, based on his and his predecessors' decrees. It was supplemented by novellae. It applied to Burgundians and between them and Romans, and was largely Teutonic with strong traces of Roman law influence. Roman law remained in force between Romans. When in 534 the Burgundian kingdom fell to the Franks the Code was opposed by the Frankish clergy, but it long preserved its validity for Burgundians as their personal law. The Burgundian law is mentioned as late as the tenth and eleventh centuries, but probably the references are to customs founded on, rather than provisions of, the code.

About A.D. 510 Gundobad also issued for Burgundian subjects of Roman race, a law known as the *Liber Responsorum*, or *Responsum Papiani*, or *Papianus*. It covered the civil and criminal laws and procedure and is partly based on the Burgundians' Germanic laws and partly on Roman sources, the *Codex Theodosianus*, Novels, Gaius, and Paul, the *Codex Gregorianus* and the *Codex Hermogenianus*, possibly copied from the *Breviarium Alaricianum*. When the Franks overthrew the Burgundian kingdom it was superseded by the Breviary.

Burgundio (?–1194). Italian jurist and Greek scholar, said to have translated the Greek materials in the Pandects into Latin. These were accepted as part of the Vulgate text of the Pandects.

Burke, Edmund (1729–97). Statesman, orator, and political thinker and writer. He studied law but disliked it and devoted himself to literature. He founded the *Annual Register* (1759) and became private secretary to Rockingham, the prime minister (1765–66). In 1766 he became M.P. and till 1790 was one of the inspirers of a revived Whig party. His speech to the electors of Bristol in 1774 stands as the classic statement of the relation of a Member of Parliament to his constituents before the development of rigid party discipline. He led the inquiry into the affairs of the East India Company and opened the case for the impeachment of Warren Hastings (1788). His political writings include *Thoughts on the Cause of the Present Discontents* (1770) and *Reflections on the Revolution in France* (1790). His major interests were the emancipation of the House of Commons from the influence of George III; freeing India from the misgovernment of the East India Company; and opposition to the French Revolution.

P. Magnus, *Edmund Burke*; Cone, *Burke and the Nature of Politics.*

Burke, William (1792–1829) and Hare, William (*fl.* 1828). Burke was an Irish navvy who lodged in Hare's lodging-house in Edinburgh in 1827. Burke and Hare sold the body of a fellow lodger who died to a surgeon for dissection. They then inveigled unknown wayfarers into their lodging-house and murdered at least fifteen, selling their bodies to a school of anatomy. In 1828 the body of a victim was found in the surgeon's premises. Burke was tried for murder and Hare turned King's evidence. Burke was hanged on 28 January 1829, the mob screaming 'Burke him. Burke him'. The verb 'to Burke' accordingly means to kill, or shut up. The incidents gave rise to the Anatomy Act, 1832, which regulated the supply of bodies to medical schools.

Burke's Peerage. A book of reference giving a complete genealogical and heraldic history of the peerage, baronetage, and knightage of the U.K., alphabetically arranged. Founded in 1826 by Sir John Burke, it was extended by his son Sir John Bernard Burke, a voluminous writer on heraldry and genealogy and it has appeared regularly ever since 1847; the 100th edition was in 1953.

Burlamaqui, Jean Jacques (1694–1748). Swiss jurist, was professor of civil and later of natural law at Geneva and a member of the council of state. His major works were *Principes du droit naturel* (1747) and *Principes du droit politique* (1751), both very influential. His main achievement was to put Pufendorf's theories into symmetrical and intelligible form, and had considerable influence on Blackstone and Woodeson's lectures on jurisprudence and on legal thought in England and America.

Burn, Richard (1709–85). A clergyman who became a J.P. for Westmorland and Cumberland and chancellor of the diocese of Carlisle, but was also learned as lawyer and antiquarian. He edited three editions of *Blackstone* and in 1755 published a book on justices of the peace, *The Justice of the Peace and Parish Officer*, which eventually, in the hands of editors, reached a thirtieth edition, enormously enlarged. He dealt with the topics under alphabetical heads and founded his remarks solidly upon authority and the careful study of legal books. He also wrote a large book on *Ecclesiastical Law* (1760) which went through nine editions. It also is arranged alphabetically, the titles containing the history of the topic, the statutes, decisions, and other relevant materials. It is based on diligent and accurate research and judicious selection of materials, and is the clearest and most successful of all the great treatises upon English ecclesiastical law. His other works, of much less importance, included *A New Law Dictionary* (1792).

Burnaby's Code (1765). A codification by Admiral Sir William Burnaby of the usages of the settlers on the Bay of Honduras, which later served as the basis of the first constitution of British Honduras (now Belize).

Burnell, Robert (?–1292). A clerk to King Edward I of England, later a regent of the kingdom, and Chancellor, 1274–92, and Bishop of Bath and Wells, 1275. He was much employed in state and diplomatic business, was probably the first great Chancellor of the kingdom and probably had a hand in most of Edward's great legal reforms, such as the Statute of Westminster I and II, the Statute of Wales and others. He is said to have lived a covetous and ambitious life and to have amassed great wealth.

Burnett, James, Lord Monboddo (1714–99). Having studied at Aberdeen, Edinburgh, and in Holland, was called to the Scottish Bar in 1737, and was leading counsel in the Douglas cause (q.v.). In 1767 he became a Lord of Session taking the title Lord Monboddo. A good scholar, with an enquiring mind, he wrote *The Origin and Progress of Language* (1771–76), which is learned but somewhat eccentric, and *Ancient Metaphysics* (1779–99).

E. Cloyd, *James Burnett, Lord Monboddo* (1972).

Burnett, John (?1764–1810). He passed advocate in 1785, and became sheriff of Haddingtonshire in 1803. In 1810 he became Judge-Admiral of

Scotland and died later in the same year. His book on the *Criminal Law of Scotland* was published in 1811. It is still of some authority and value and is one of the earliest attempts to compile a satisfactory collection of Scottish criminal cases.

Burning to death. An ancient mode of capital punishment used by the Romans as the penalty for arson and among the Germanic peoples for adultery. In England it was used in cases of petty treason (q.v.) or high treason (q.v.) committed by a woman. The punishment was last inflicted for petty treason in 1726 and abolished in 1790, but it had in practice rarely been actively inflicted.

The burning of heretics was known in Europe from the thirteenth century and there were occasional instances in England. At least in 1400 the Crown issued a writ for the burning of one Sawtry and statutory authority was given by the Act *de haeretico comburendo* passed in the same year. This was repealed, revived, and finally repealed in 1558. There were later, and probably illegal, instances of burning of heretics, the last being in 1610.

Burning was also used in cases of witches in Scotland and in Europe, the last being in 1708, though in England they were generally hanged.

Buron* v. *Denman (1848), 2 Exch. 167. A naval officer, instructed to suppress the slave trade, attacked a Spanish slave-trader in foreign territory and liberated his slaves. The British Government adopted and ratified his acts. It was held in an action by the Spaniard that the action would fail as the officer's act were an act of state (q.v.) in respect of which no action would lie against him or the Crown.

Burton, John Hill (1809–81). Scottish lawyer and historian, practised at the Scottish Bar and for a time edited the *Scotsman*. He wrote a *Manual of the Law of Scotland* (1839), *Narrative from Criminal Trials in Scotland* (1852), and a *Treatise on the Law of Bankruptcy in Scotland* (1853). He also published a *Life of David Hume* (1846), biographies of Lord Lovat and Duncan Forbes (1847) and a *History of Scotland* (1853–70).

Bury, Thomas (1655–1722). He was called to the Bar in 1676, became a Baron of Exchequer in 1701 and Chief Baron of Exchequer, 1716–22. He appears to have been competent but undistinguished.

Bushe, Charles Kendal (1767–1843). Called to the Irish Bar in 1790 he entered Parliament and opposed the Union of 1800. He was Solicitor-General, 1805–22, and the Chief Justice of the King's Bench in Ireland, 1822–41.

Bushell's case (1670), 1 St. Tr. 869; Vaughan 135. Bushell was foreman of a jury which acquitted Penn and Mead, two Quakers, on a charge of having preached to a large assembly of people in London, contrary to the Conventicle Act. The trial judge fined the jurors 40 marks each; Bushell refused to pay and, having been committed to prison, brought a writ of *habeas corpus*, and Vaughan, C.J. held that the return to the writ, that Bushell had found a verdict against the evidence and against the direction of the court, was inadequate. This decision established the freedom of jurors to find a verdict independently of the views of the presiding judge.

Bushrangers. Outlaws of the Australian bush or outback who in the late eighteenth and nineteenth centuries harassed settlers, miners, and aborigines and whose exploits provide some of the folklore of Australia, being celebrated in folk songs. They acted individually or in small groups and specialized in robbing coaches, banks, and small settlements. The earlier bushrangers, from 1789 to about 1850, were mostly escaped convicts. Thereafter, until they disappeared about 1880, they were mostly free settlers who had been in trouble. The first was John Caesar about 1789 and the last Ned Kelly (1855–80). Some were ruthless killers, but others, such as Matthew Brady and Edward Davis (Teddy the Jew Boy), treated their victims humanely and even gave some of their loot to the poor.

Business. An indefinite term covering most kinds of commercial relations, particularly on a substantial scale. It is generally considered wider than trade, which is confined to shopkeeping and the exercise of such skilled trades as joinery and plumbing. Business may be distinct from or may in certain contexts include the practice of certain professions.

Business associations. Groupings of individuals associated for business purposes with a formal system of management and assets distinct from those of the members. The main forms of business association are the partnership and the company or corporation.

Business law. A term of indeterminate connotation, signifying not a distinct division or branch of the law of a community but rather a convenient compendious term, mostly used as the title of academic courses or textbooks, for those topics of law relevant to persons training for or engaged in business, including such topics as contract, agency, sale, hiring, bills and cheques, bankruptcy, etc. Anglo-American business law originated in the law merchant as incorporated by Lord Manfield in the common law of the eighteenth century. Then, and in the nineteenth century, many seminal works on branches of business or commercial law were published such as Pothier on *Obligations* and *Sale*, Blackburn and Benjamin on *Sale*. It is generally synonymous with commercial law and mercantile law (qq.v.).

Business name. The name, if other than the true name or names, of persons carrying on a business. In the U.K. it is necessary to register business names under pain of inability to enforce contracts made in relation to the business, and to disclose in correspondence the true names of the person, persons, or corporations trading under the business name.

Bustamante y Rivero, Josè (1894–). Peruvian jurist, judge, and statesman, President of Peru 1945–48, judge of the International Court 1961–70, and President 1967–70, and author of many books on law and politics.

Bustamante y Sirven, Antonio Sanchez de (1865–1951). Cuban jurist, author of *El Tribunal Permanente de Justicia International* (1925), an important *Derecho international publico* (1933–38), and other works. In 1908 he became a member of the Permanent court of Arbitration of The Hague and in 1921–39 a judge of the Permanent Court of International Justice. He drew up the Bustamante Code of Private International Law which was adopted by 15 Latin American countries in 1928.

Butrigarius, Jacobus (*c.* 1274–1348). A Bolognese commentator, author of *Lectura super Codice* (1516), *Lectura in Digestum Vetus* (1606), *Quaestiones et Disputationes* (1557), and lesser works.

Butrio, Antonio di. See ANTONIO DI BUTRIO.

Butt, Sir Charles Parker (1830–92). Called to the Bar in 1854 and practised chiefly in mercantile matters. In 1883 he became a judge of the P.D.A. Division, becoming President in 1891. Latterly his health was poor and he failed to achieve the reputation his abilities might have justified.

Buxton, Sir Thomas Fowell (1786–1845). Philanthropist, belonged to a firm of brewers, but studied literature and economics and interested himself in philanthropic and charitable works in London. He was a brother-in-law of the prison reformer Elizabeth Fry. By reason of visiting prisons he wrote *An Inquiry whether Crime and Misery are produced or prevented by our present system of Prison Discipline* (1818). This led to the formation of the Prison Discipline Society. He was an M.P. 1818–37 and concerned himself with criminal law and penology, and sat on committees concerned with penal matters. He also played a leading part in the movement for the abolition of slavery, succeeding Wilberforce as leader of the campaign which resulted in the Abolition Act of 1833.

Byles, Sir John Barnard (1801–84). Practised as a special pleader (q.v.) and was called to the bar in 1831. In 1857 he became a Queen's serjeant, and he was the last survivor of that order. In 1858 he was made a judge of the Common Pleas, where he proved himself a strong and learned judge, especially in commercial cases. He resigned in 1873, but was made a member of the Privy Council. About 1830 he published a book on *Bills of Exchange* which has repeatedly been re-edited and remains a standard text, and various other works.

Bye-law (or By-law). A form of subordinate legislation made by a corporation or other authority less than Parliament on a matter entrusted to that authority, having the force of law but applicable only to the particular area or scope of the responsibility of the authority. A common-law corporation has an implied power to make bye-laws incidental to and within the purposes of its constitution, and statutory corporations are normally given powers to make bye-laws within stated limits and for stated purposes. Local authorities have wide powers to make bye-laws, sometimes subject to the approval of a minister or other external authority. Unlike statutes, bye-laws must be published before they can be operative, and they may be challenged as being *ultra vires*, or contravening the general law of the land, or uncertain, or unreasonable, or not having been made in accordance with prescribed procedure.

Bynkershoek, Cornelis van (1673–1743). Dutch jurist and international lawyer. He was successively an advocate, judge of the Supreme Court of Holland at The Hague (1703), and president of that court (1723). He collected the decisions of his court, which were published as *Observationes tumultuariae* in 1926–62, and were continued by Pauw's (q.v.) *Observationes*. He wrote extensively, his most notable works being *De Dominio Maris* (1702); *De Foro Legatorum*, on the rights and duties of ambassadors (1720); *Quaestiones juris publici* (1737) part of which deals with international law and the customs of wars; and *Observationum juris Romani* (1710–33) on Roman law. An unfinished *Quaestiones juris privati* appeared posthumously (1744). His work on Dutch municipal law, *Corpus Juris Hollandici et Zelandici*, was never published. His works on public and international law had begun to exercise influence on legal and political thinking in his own time, and have subsequently been regarded as entitling him to a high place among international jurists. Though he dealt with a smaller range of subjects than many others, he treated those discussed fully and thoroughly and supports his reasoning with reference to history and actual practice. He was the originator of the positive school of international jurist; the actual usage and practice of nations was more important than deductions from natural law.

Numan, *Cornelis van Bynkershoek* (1869).

Bysse, John (?–1680). Called to the Irish Bar in 1632, he sat in the Irish Parliament, was a judge of assize, and was chief Baron of the Irish Court of Exchequer, 1660–80.

Byzantine law. The law of the Byzantine or Eastern Roman Empire, centred on Byzantium or Constantinople. The origin of the Byzantine Empire is usually dated from A.D. 330, when Constantine I, the first Christian Roman emperor, moved the capital from Rome to Byzantium on the Bosphorus, which he renamed Constantinople. From 395 the empire was ruled by two or more colleagues, one in the east and one in the west, with equal powers, though in theory it remained a single unit. There were two main senates, at Rome and Constantinople. In the west the European provinces were subjected to raids and invasions by barbarian Germanic peoples from the east. The Visigoths crossed the Danube in 376 and settled in Moesia but later moved westwards; in 410 they captured Rome; by the mid-fifth century most of southern Gaul was occupied by Visigothic and Burgundian federated tribes. In 406 the Vandals, Suevi and Alans crossed the Rhine, ravaged Gaul, settled in Spain and in 439 created an independent state in North Africa. After 455 Ostrogoths settled in Pannonia and in 476 Odoacer deposed the last western emperor and became in effect king of Italy though admitting the overlordship of the western emperor. By the end of the fifth century the east had successfully warded off the Germans while the western provinces had been largely lost to them. Nevertheless, Roman institutions, administration, and laws continued to operate throughout the west though under new masters.

In the sixth century Justinian I made a bold attempt to recover the lost western provinces which was not, however, permanently meaningful. After his death the Lombards invaded Italy and Slavs penetrated into the Balkans while the Persians pressed on the eastern frontier. The Heraclians (610–711) reorganized their territories but in the seventh century they were challenged by the rising power of Islam. Constantinople itself was besieged in 674–78 and North Africa, Egypt, Palestine, and Syria were lost to the Muslims. A smaller but more manageable empire survived. The Byzantines still, however, held south Italy until the southern provinces fell to the Normans in the eleventh century.

The Isaurians (717–813) continued constructive statesmanship and reorganization, but were occupied mainly by seeking to resist the Bulgars and Muslims. The diminished prestige of the Roman Empire was reflected in Pope Leo III's coronation of Charlemagne as Holy Roman Emperor in 800, which symbolized the separation of the two parts of the empire.

Under the Armorians and Macedonians (813–1025) the empire was on the defensive, but there were notable developments internally with spiritual and cultural vitality. From the eleventh century there was internal dissension and decline. The Normans took south Italy. The empire was weakened by civil wars and vigorous invasions from outside. Finally in 1453 the Ottoman Turks took Constantinople and what was left of the imperial territories.

Though in theory the Roman Empire remained a unity, from the death of Theodosius in A.D. 395 the Eastern and Western halves of the empire were in fact separate until the fall of the Western Empire in 476, after which the Eastern empire alone remained until 1453 when Constantinople fell to the Turks. But from A.D. 330 onwards the Roman law nominally extended over the whole empire. By this time legislation by the emperor was the sole effective mode of law-making and two unofficial collections of imperial constitutions, the *Codex Gregorianus* of about 291 and *Codex Hermogenianus* of about 295, were made, almost certainly in the East, possibly at Beyrout.

The need for an official collection of imperial constitutions was first met by the *Codex Theodosianus*, compiled by a commission appointed by Theodosius II (east) and Valentinian III (west), and given the force of law from A.D. 439. Constitutions from the time of Constantine I (306–77) onwards were generally invalid unless included in the *Codex*. Theodosius is thought also to have planned a compilation of juristic writing as well as of legislation, but this was not attempted until Justinian. The *Codex Theodosianus* was superseded in the east by Justinian's legislation, but it continued to be used in the west thereafter.

A collection of later constitutions was made under Majorian, and some other constitutions have also survived.

Justinian (A.D. 527–65) found the law in great confusion and uncertainty, and immediately after his accession appointed a commission to deal with the imperial legislation. This commission produced the *Codex constitutionum* which greatly reduced the bulk of the body of constitutions, removed many obscurities and inconsistencies and repealed obsolete matter. It was promulgated in 529, repealing all imperial ordinances not included in it.

He then issued a series of constitutions, the Fifty Decisions, *Quinquaginta decisiones*, which settled many of the points on which older jurists had been divided in opinion, and appointed a fresh commission of 16 under the chairmanship of the jurist

Tribonian (q.v.) to examine all the writings of the authorized jurists, to extract from them whatever was of permanent and substantial value, to avoid repetitions and contradictions, and to modify the expression of the author where necessary to adapt his views to the state of the law in Justinian's time. The commissioners did their work by making a series of 9,123 excerpts from 39 authors, distributing these in 50 books, subdivided into titles, following the order of the Perpetual Edict. This collection, the *Digest* or *Pandects*, was published as an imperial statute in A.D. 533. In enacting it he repealed all the other law contained in the writings of the jurists, but went too far in forbidding commentaries on the *Digest*.

The *Digest* is of immense value as in many cases the only authority for statements of law made by jurists of the classical period of Roman law, but suffers from unscientific arrangement. The order of the Edict was known and convenient, but was accidental and historical rather than logical. Within each title the order of the extracts depended on the way in which the sub-committees of the commission found their material.

It was thought that a new elementary textbook was needed and there was then published the *Institutes of Justinian*, a work which owes much in style, arrangement, and substance to the much earlier *Institutes* of Gaius (q.v.).

In view of the many changes made in the law since the publication of the *Codex*, a further commission was appointed, comprising Tribonian and four others, to revise the *Codex* and incorporate new constitutions therein. In 534 the revised *Codex (Codex repetitae praelectionis)* was promulgated and declared to be alone authoritative. It consisted of 12 books containing 4,652 constitutions from the time of Hadrian to the time of Justinian, the majority being of one or other of the great eras of imperial legislation; the times of Diocletian (285–305), Constantine (313–37), Theodosius II (408–50), and Justinian. This is of great historical as well as legal value, though the extracts have less scientific value than the excerpts collected in the Digest.

Thereafter Justinian continued to legislate, issuing new constitutions, the Novels (*Novellae constitutiones post Codicem*) which in many cases materially altered the law. There are three extant collections of Novels, all apparently privately compiled and it is uncertain how many Novels were promulgated. Most of them were issued in Greek, translations being made available for the western provinces. The three collections comprise one containing 168 constitutions, some by successors of Justinian, another, called the *Epitome Juliani*, containing 125 Novels in Latin, and the third, the *Liber Authenticarum* or *Vulgata versio* containing 134 Novels, also in Latin.

These four works, the *Digest*, *Institutiones*, the second *Codex* and *Novels*, were later known as the *Corpus Juris Civilis* and this body of law dominated the Greek East right down to the fall of Constantinople.

Justinian's work is not a codification in the modern sense, but two collections, a textbook and further legislation. Nevertheless they were of immense value to the empire by reducing the bulk of the legislation and juristic law, consolidating it and making it more readily available. The work continued to be the chief law-book of the Roman empire and came into widespread use in Italy, which had been made into a new prefecture, with its capital at Ravenna, after its conquest by Belisarius in 535–53. Despite the later Lombard conquest it remained law for the bulk of the Italian people.

Apart from its contemporary importance the subsequent importance and value of the *Corpus Juris* has been immense. It preserved for mediaeval Europe and modern times a vast amount of information about earlier Roman law which would otherwise have been lost. After its rediscovery and the commencement of close study of the Roman law in the legal renaissance of the eleventh century principles and rules contained in one or other of the books of the *Corpus Juris* provided the basis for legislation, decision, and juristic thinking in very many spheres of law over most of western Europe.

After Justinian's legislation had been promulgated the first task needed to make it available for practical use was to translate it into Greek, which was done at the two law schools of Beyrout and Constantinople. In the middle years of the sixth century the legislation was extensively studied in those schools and there was an extensive literature of notes, commentaries, and summaries of the legislation. Of these there have survived Theophilus' *Paraphrase of the Institutes* and Julian's *Epitome Novellarum* and fragments from various other commentaries. Under Justinian's immediate successors more collections of excerpts from Justinian novels were made, and this whole body of literature was the basis of practically all later Byzantine law and largely replaced the actual text of Justinian'a codification.

It is material to remember that in the Byzantine Empire of the seventh and eighth centuries ecclesiastical canons were also being issued and there were occasional conflicts between legislation of Church and of State. *Nomocanones* began to be compiled from Justinian's time, at first merely setting down ecclesiastical and secular law side by side, but later works sought to make them a unity.

It is difficult to know how far legal science was cultivated after Justinian, but it seems certain that practice was modified in accordance with the opinions of Christian society and ecclesiastical canons. In the Quinisext or Trullan council of 692 various rulings based on ecclesiastical doctrine and

Mosaic principles were made and sanctioned by the emperor as laws.

In 726 Leo III the Isaurian issued the *Ecloga*, a legal handbook in Greek to replace the *Corpus Juris*, incorporating changes in customary law and legislation since Justinian's day. It was heavily influenced by Christianity and both ecclesiastical and oriental influence are noticeable in the criminal law, in which mutilation was largely substituted for the death penalty. The Church also exercised a great influence on the operation of the criminal law. The *Ecloga* had an important influence in the later development of Byzantine as well as Slav law. To the seventh or eighth century may be dated certain other surviving codes, such as the *Nomos georgicos* (Farmer's law) and the *Nomos Rhodion nauticos* (Rhodian sea law). The latter seems to have been a private collection of legal rulings bearing on shipping and maritime commerce.

The last great period of legislative activity was under Basil I and Leo VI the Wise (867–912) and to some extent represented a reaction against the *Ecloga* and a return to Justinian. Basil abrogated the *Ecloga*; he aimed at a revival of legal studies treating the text of Justinian or Greek versions thereof in the same way as Justinian's commission had treated the classical jurists. A handbook of extracts from the *Institutes*, *Digest*, and *Code* was issued in 879, the *Procheiros Nomos* in 40 titles, to fulfil a function comparable to the *Institutes*. The *Epanagoge*, also in 40 titles, was really only an improved edition of the *Procheiros*. Thereafter two commissions prepared a collection of all the laws of the empire and this was issued by Leo II as the *Basilica* (q.v.), in which the *Digest*, *Code* and *Novels* were amalgamated into one work in 60 books. In many matters of civil law, but not criminal law, the principles of the *Ecloga* were replaced by principles of older law.

The *Basilica* soon ousted the sixth and seventh century commentaries on the law and Leo VI issued a collection of 113 novels making further corrections and reforms, but these were the last important reform of civil law.

After the *Basilica* there was no large-scale legislation, but there was a notable revival of legal study under Constantine IX who in 1045 founded a faculty of law in the reorganized university of Constantinople and important commentaries were written by such as John Xiphilin (eleventh century), Theodore Balsamon (twelfth century), and Harmenopoulos (fourteenth century) (qq.v.). A valuable anonymous commentary is the *Ecloga librorum I–X Basilicorum* published in 1142. These works were influential on the developing Slav states and modern Greece professed to base its civil law on the edicts of the emperors as contained in Harmenopoulos' *Hexabiblos* (*c.* 1345), though its formal model was the Napoleonic civil code. Many scholars also wrote scolia on the *Basilica*. After the early eleventh century private law did not show any important change and only public law showed any further development. It developed substantially with the changes in the structure of the Byzantine state between 1000 and 1453.

While over the period 330–1453 there were many changes in the administration of justice in the Byzantine empire, the eparch (prefect) of the city generally controlled the police organization and administration of justice in the capital and, in the emperor's absence, presided in the imperial court of justice. Appeals reached the emperor via the bureau of petitions; he might deal with the case immediately or refer it to the imperial court of the divine judges, a court instituted by Justinian. In the twelfth century there were four imperial courts of justice in Constantinople. Later there was a single imperial court, but thereafter the power of the Church increased and in the fourth century two of the four chief justices were clerics. There were also ecclesiastical courts to which were secured certain kinds of cases. The association of church and state continued to the end and the distinction between civil and ecclesiastical jurisdiction became increasingly unclear. In the provinces ordinary justice was administered by judges distinct from the governors of the themes.

In the lands of Eastern Europe beyond the frontiers of the Empire there emerged in the early Middle Ages rulers anxious to receive the laws of Byzantium to supplement their own inadequate customary laws. This occurred in the Balkans where a tradition of Romano-Byzantine law, antedating the Slav invasions, survived in the towns and among the clergy. Moreover to the rulers of Eastern Europe, who were seeking to increase their power, Romano-Byzantine law tended to support their ascendancy over their subjects. So too adoption of Christianity made it necessary to obtain collections of canon law. As ecclesiastical legislation was mixed with imperial decrees on church matters the adoption of canon law involved a reception of Roman law. Not least in the eyes of neighbouring rulers Byzantine law enjoyed immense prestige, greater than any barbarian code could have.

The lawyers of the East European countries outside the empire never attained any great level of juristic skill, but they were able to select the books of Byzantine law which were most useful, to secure their translation into Slavonic, and sometimes to omit passages or to insert into them Slavonic versions of Byzantine law items derived from Latin sources. Sometimes there was resistance by Slavonic customary law to Byzantine legislation.

The *Nomocanones*, or Byzantine collections of canon law, seem to have been accepted unquestioningly and without alteration in mediaeval Eastern Europe. The oldest one attributed to John Scholasticus, patriarch of Constantinople, 565–77, was translated into Slavonic in the ninth century and used in Bulgaria. Later a second Slavonic version was made and was used in Bulgaria and Russia until the thirteenth century. About 1219 St. Sava, archbishop of Serbia, transcribed a *Nomocanon* with commentaries, added a translation of the *Procheiron* and several later edicts, and created St. Sava's *Nomocanon*, a work known to the Slavs as the *Book of the Pilot*, which became the basic constitution of the Serbian, Bulgarian, and Russian Churches.

Byzantine secular legislation also had an influence, but one which was difficult to measure. The earliest Byzantine law book adopted in Eastern Europe was the *Ecloga*, and though it was later superseded in Byzantium by the *Basilica*, it remained in Slavonic versions the most popular Byzantine law book in the East European lands of the commonwealth until the fourteenth century. There were two Slavonic versions of the *Ecloga*, one a fairly faithful translation probably made in tenth century Bulgaria, but in some respects deviating from the original; the other, entitled the *Judicial Law for Laymen*, deviating to a greater extent and largely concerned with criminal law. It is thought to have originated in ninth-century Moravia. It was current in mediaeval Bulgaria and was probably accepted in the pro-Mongol period in Russia. By 1280 at latest it formed part of the *Nomocanon* used by the Russian Church and remained so to modern times.

In the fourteenth century East European peoples acquired another collection of Byzantine law, the *Syntagma*, compiled in Thessalonica in 1335 by Matthew Blastares. It included material from the *Nomocanon* and secular law books such as the *Procheiron* and the *Basilica*, was translated into Slavonic in Serbia at the instigation of Stephen Dusan and had great authority in Serbia, Bulgaria, Rumania, and Russia.

Byzantine law does not seem to have materially influenced the earliest indigenous Russian code, the *Pravda Russkaya* of the eleventh century, nor the late mediaeval *Statute* of the St. Vladimir, which also adopted the concept of tithes from Western law. It also deeply influenced late mediaeval Serbian law. In 1349 Stephen Dusan published a famous code, the *Zakonik*, and issued an enlarged version in 1354. The earlier version is preceded in all the manuscripts by a shortened form of Blastares' *Syntagma* and a collection named *Justinian's Law*, thought to be an abridgement of the *Farmer's Law*, a code issued, probably by Justinian II, for provincial use in the seventh or eighth century. The *Zakonik* supplements these codes in many ways. Moreover many of the populace in Dusan's Serbo-Greek empire had long lived under Byzantine law and it is clear that he based his administration on Byzantine law. The *Zakonik* certainly has some basis in Serbian customary law but is predominantly derived from Byzantine sources.

It is apparent accordingly that Byzantine law was to a considerable extent adopted by nations of Eastern Europe and carried over also the principles of the emperor as the source of all legislation and of the universal validity of his ordinances throughout the Christian world. Byzantine secular legislation was most readily accepted by the Balkan States and Rumanians, its canon law most influential in Russia, but the degree of authority varied from country to country.

K. E. Zachariae von Lingenthal, *Historiae iuris graeco-romani delineatio* (1839); *Geschichte des griechische-römischen Rechts* (1892); J. A. Montreuil, *Histoire du droit byzantin*, 3 vols. (1843–46); *Jus Graeco Romanum*, ed. K. E. Zachariae von Lingenthal, 7 vols. (1856–84); *Jus Graeco Romanum*, ed. J. & P. Zepos, 8 vols. (1931).

C

C. (Lord) Chancellor.

C. B. Chief Baron of the former Court of Exchequer (q.v.); also Commander of the Order of the Bath.

C. i. f. contract. A contract of sale of goods for export, whereby the seller quotes a price for the goods including cost, insurance, and freight to the destination. He must effect an insurance available for the buyer, make a contract of affreightment with a carrier to the contractual destination, deliver the goods to the carrier and forward the invoice, insurance policy, and bill of lading to the buyer. The property passes by delivery of the documents and the buyer's liability to pay the price then arises. It may be provided that payment shall be made by the buyer accepting a bill of exchange forwarded with the other documents; if he does not accept the bill the property in the goods does not pass to him; if the bill is dishonoured the seller's rights as an unpaid seller revive.

D. Sassoon, *C. i. f. and f. o. b. Contracts.*

C. J. Chief Justice.

C. J. C. P. Chief Justice of the (former) Court of Common Pleas (q.v.).

C. J. K. B. (or Q.B.). Chief Justice of the (former) Court of King's (or Queen's) Bench (q.v.).

Cabassut, Jean (1604–85). French canonist and theologian, author of a *Iuris canonici theoria et praxis* (1675), which went through many editions, and *Notitia Conciliorum* (1685).

Cabinet (U.K.). The Cabinet evolved from the Privy Council, from the practice after the Restoration of the King's most trusted members of the Privy Council meeting as a committee and taking decisions before or without consulting the full Privy Council. In the time of Queen Anne it became the main machinery of executive government and the Privy Council became formal. The name probably developed from its meeting in the King's cabinet or private room. From about 1717 George I ceased to attend and thereafter the Cabinet met independently. By the late nineteenth century its collective responsibility had replaced the authority of monarch and ministers for carrying on the government of the country.

In modern practice the Cabinet is a committee of persons from the majority party in the House of Commons, chosen and led by the Prime Minister, and mostly entrusted by him with the headship of major departments of State. It has the function of deciding, in association with the Prime Minister, on government policy, of controlling the executive, and co-ordinating the activities of the executive departments. The members of the Cabinet are collectively responsible for all government policy and action and no member may publicly disagree or dissociate himself; if he disagrees he should resign. This convention was broken sometimes in the 1970s in relation to entry to the European Communities. They depend on the support of the House of Commons but being chosen from the majority group therein can generally count on such support.

The size of the Cabinets has varied; during both World Wars very small Cabinets had supreme policy-making functions; prior to 1939 the Cabinet included most of the Ministers, whereas since 1945 the number of ministries has increased so that not all heads of departments are included in the Cabinet, others being summoned to attend as required. Since 1945 the size has usually been about 25. No Minister has an absolute right to be included in the Cabinet but the holders of such major offices as Chancellor of the Exchequer, Foreign Secretary and Home Secretary would always be included and the holders of offices ranked as Secretaries of State will probably always be included. It frequently also includes two or three members without departmental responsibilities such as the Lord President of the Council, Lord Privy Seal, and Chancellor of the Duchy of Lancaster, who are available for particular tasks, co-ordination of work of several departments and chairmanship of Cabinet committees. The Lord Chancellor is always a member. All Cabinet Ministers are members of the Privy Council and bound by the Privy Councillor's oath of secrecy.

The business is highly confidential, and strict secrecy is maintained as to subjects discussed and decisions; it is understood that no formal vote is taken.

Prior to 1916 Cabinet business was conducted in a very unsystematic way, without secretariat, agenda, or minutes. Lloyd George established a Cabinet Secretariat, which serves both the Cabinet and its committees, preparing agenda, relative papers, and minutes.

Much Cabinet work is delegated to committees for detailed study and the preparation of proposals. Some committees are permanent, such as the Defence Committee, the Legislation Committee concerned with the legislative programme, the Future Legislation Committee, while others are temporary for special purposes. The number, names and membership of Cabinet Committees are normally not revealed during the life of the government. The more important are normally composed of Cabinet and other Ministers, the less important may include Parliamentary Secretaries and civil servants, normally with a Cabinet Minister as Chairman.

W. I. Jennings, *Cabinet Government*; J. P. Mackintosh, *The British Cabinet*; H. Morrison, *Government and Parliament.*

Cabinet (U.S.). The U.S. Cabinet has no foundation in constitutional law or statute save that under the Constitution (Art. II, sec. 2) the President may require the advice of such heads of departments as Congress may create. Washington utilized a cabinet of four, the Secretaries of State, Treasury and War, and the Attorney-General, and the custom has continued. The number has increased as new Departments have been created, and it now includes the Secretaries of State, Treasury, Defense, the Attorney-General, Postmaster-General, Secretaries of the Interior, Agriculture, Commerce, Labor, Health, Education and Welfare, and Housing and Urban Development. In recent years the Vice-President has frequently been invited to attend meetings, and Under-Secretaries or Assistant Secretaries of departments may be invited to attend. The U.S. Ambassador to the U.N. has recently been accorded cabinet rank at the pleasure of the President, though not as a right.

The President's selection of members is a personal prerogative, though the persons chosen as heads of the Departments are subject to approval by the Senate. They serve at the pleasure of the President but by custom resign after a four-year term in office or on the death of the President. The President's Cabinet merely advise him, and are responsible only to him, not to the Congress, nor are they members of either House of Congress. They are usually chosen with a view to maintaining party support, from various areas of the country and various wings of the party.

Presidents have sometimes been advised by persons other than the official Cabinet. F. D. Roosevelt was advised on the New Deal by a group of economists and lawyers known as the 'Brain Trust', and J. F. Kennedy had a number of academic advisers.

Cachet, lettres de. Letters issued and signed by a king of France under the *ancien régime*, and countersigned by a secretary of state, authorizing the imprisonment of a person named, usually in the Bastille. They were abolished in 1789.

Cadit assisa. A thirteenth-century tract abridging what Bracton wrote on the assize of *mort d'ancestor*.

Woodbine, *Four Thirteenth-Century Law Tracts.*

Caduciary. From Roman law *caduca*, relating to the lapse, or escheat, of an estate to the Crown, as distinct from transmission by succession.

Caecilius Africanus, Sextus (*fl.* second century A.D.). Roman jurist, author of *Quaestiones* (nine books) containing records of discussions on legal problems with his master, Salvius Julianus (q.v.) and valuable as reporting the latter's views.

Caepolla, Bartholomeo (*c.* 1420–77). An Italian jurist and commentator, author of *Bartholomei Caepollae de Servitutibus* (1660).

Caesar, Sir Julius (1557–1634). Was called to the Bar in 1580. He became Judge of the Admiralty Court (1584), Master in Chancery, Master of Requests, Chancellor of the Exchequer (1606), Privy Councillor, member of the Court of Star Chamber, and Master of the Rolls (1614). He also wrote a large number of papers on legal matters, particularly on international law. His son Charles also became Master of the Rolls in 1639.

Caesaropapism. The concept of government in which supreme lay and sacerdotal powers are combined in one lay ruler. It is particularly applied to the kind of government exercised by the Eastern Roman emperors at Constantinople, based on the transfer of the ancient function of the emperor as *Pontifex maximus* to the Christian Roman emperor. The Emperor Zeno in his *Henoticon* of 482, in disregard of the Council of Chalcedon, ordained the faith for his subjects and the theory reached its highest point under Justinian who to all intents and purposes acted both as king and high priest. It remained the governmental principle of Byzantium throughout its whole existence and contributed to the breach between eastern and western churches, and continued thereafter in Russia where Peter the Great transformed the church into a department of State. In the west it reappeared after the Reformation in Germany in the principle *cuius regio eius religio* and was manifested also in Gallicanism and in Henry VIII's assumption of the office of Supreme Head of the Church.

Cahiers de doleances. Addresses to the King of France embodying complaints and demands for reform. The most notable were those presented in 1789. Particular cahiers were combined in a general one presented at Versailles and evidence the grievances felt at the time.

Cairns, Hugh McCalmont Cairns, 1st Earl (1819–85). Was called to the Bar in 1844 and rose quickly at the Chancery Bar. In 1852 he entered Parliament and quickly won success in the House. In 1858 he became Solicitor-General. When his party regained power in 1865 he became Attorney-General but shortly after (1866) became a Lord Justice of Appeal and a peer in 1867. In 1868 he held office briefly as Chancellor. He held that office again from 1874, and having been chairman of the committee on judicature reform, which reported in 1869, was responsible for the Appellate Jurisdiction Act, 1876, which established the appellate jurisdiction of the House of Lords. He lost office in 1880. He became an earl in 1878, was a most devout churchman, strict, ungenial, and not popular. He was a great debater and orator, more successful in the Commons than in the Lords. He sponsored or assisted much valuable legislation, including the Conveyancing Acts of 1881 and 1882 and the Settled Land Act, 1882, and as a judge made great contributions to the development of the law, having a great grasp of principles and capacity to apply them, and many of his decisions are leading cases, lucid and clearly reasoned. Many regarded him as the first lawyer of his time.

Brief Memoirs of the First Earl Cairns; J. Bryce, *Studies in Contemporary Biography.*

Calas, Case of Jean. Jean Calas (1698–1762). Was a French Protestant merchant condemned to be tortured, broken on the wheel, and burned for having, it was alleged, strangled a man who had sought to turn Catholic. Voltaire took up the case and made it famous all over Europe. Finally the King and council annulled the decision of the *parlement*, declared Calas innocent and made payments to the family.

F. H. Maugham, *The Case of Jean Calas* (1928).

Caldecote, Viscount. See INSKIP, T.

Calderinus, Joannes (*d.* 1365). Italian canonist, author of *Consilia sive Responsa* (1472), *Repertorium Utriusque Juris* (1475), and numerous other treatises.

Calendar. The means of grouping days within the solar year for regulating civil life, religious observances, and similar purposes. Many different calendar systems were known in ancient times. The Julian Calendar, based on the Roman Republican Calendar, was adopted in the first century B.C. but was inaccurate in that the year of 365·25 days, every fourth year having 366 days, was longer than the solar year and a correction was made by Pope Gregory XIII who in 1582 promulgated the Gregorian Calendar. But this new style calendar was adopted in different countries at different dates from 1582 onwards, Protestant states generally declining to adopt the Gregorian Calendar.

By 1751 the Julian Calendar differed from the Gregorian by 11 days and the Calendar (New Style) Act, 1751 brought the British Calendar into line with the Gregorian by omitting the days between 2 and 14 September 1752. The change gave rise at the time to considerable public outcry.

In England under the Julian Calendar the year began, at first on 25 December, but from the fourteenth century, on 23 March. The Gregorian Calendar established January 1 as the beginning of the year.

Distinct calendars are maintained by the Jewish and Muslim faiths.

Calendars. Books published by the Public Record Office summarizing documents of various categories held there. Thus there are Calendars of the Patent Rolls and the Charter Rolls.

Call. In company law, a demand by the company, or its liquidator, on shareholders to pay part or all of the price of the shares not already paid. In limited companies the calls are limited by the total amount unpaid on each share; in unlimited companies a liquidator may make calls on solvent shareholders, to the extent of the company's debts.

Call of the House. A summons to every Member of the House of Commons to attend on an occasion when the view of the whole House is considered necessary. Members not attending are reported as defaulters and may be committed to the custody of the Serjeant-at-Arms. No call is known to have been enforced since 1836 and the practice is now superseded by the power of the Whips to secure the attendance of Members.

Call to the Bar. The ceremony, held in each Inn of Court (q.v.) several times a year, at which students of the Inn who have passed the examinations and satisfied the other requirements are called or admitted by the Benchers (q.v.) to the degree of the Outer Bar, i.e. are made barristers. They may thereafter plead at the bar of any court at which barristers have audience. When barristers become Queen's Counsel (q.v.) they are sometimes said to be 'called within the Bar' because in an English court only Q.C.s plead within the Bar.

Callières, François de (1645–1717). French diplomat, author of an important *De la manière de negocier avec les souverains*, 1716.

Calling list. In Scotland the list of cases exhibited on the walls of the Court, as having been served and now brought into court as actions to be heard in due course. Originally new cases were called out by a clerk of court on certain days, but the printed calling list was substituted in 1820.

Callistratus (*fl.* third century A.D.). A late classical Roman jurist, interested in the law of the Greek provinces, and author of *Institutiones*, *Quaestiones*, *De Jure Fisci*, *De Cognitionibus*, and a book on edictal law *Ad Edictum Monitorium*.

Calumny, oath of. Originally, in Scots law, an oath which could be required of both parties that the facts set out by them were true. It was till 1977 required of pursuers in all actions of divorce, to the effect that there has been no collusion or suppression of facts to enable a false case to be substantiated or a just defence withheld.

Calvin's case (1608), 7 Co. Rep. 1; 2 St. Tr. 559. After the personal union of the Crowns of England and Scotland in the person of James VI and I the King wished to have *Post-nati*, i.e. persons born after his accession to the English throne, mutually naturalized, and the *ante-nati* naturalized by enabling Act. The English Parliament being unwilling, a collusive action was brought by Calvin, a child born in Edinburgh in 1605, against Smith, a person who had allegedly deprived him of his land. A majority of the judges in the Exchequer Chamber held that the Scottish *post-natus* was a natural born subject of the King of England for all purposes of descent and inheritance and not an alien, and that allegiance to the sovereign and protection by the sovereign were reciprocal rights.

Calvinus, Johannes (Kahl) (*c.* 1550–*c.* 1610). German jurist, author of books on politics, Jewish and Roman law, particularly *Themis Hebraeoromana*, *Jurisprudentiae Romanae synopsis methodica*, *Jurisprudentia feudalis*, and particularly *Lexicon Juridicum juris Caesarei simul et Canonici* (1600).

Calvo, Carlos (1824–1906). Argentinian diplomat and jurist, one of the founders of the Institute of International Law, author of and important *Le droit international théorique et pratique* (1887–96), *Manuel de droit international public et privé* (1881) and *Dictionnaire du droit international public et privé*

(1885), as well as of a great collection of treaties of the South American republics.

Calvo clause. A clause, named after the Argentinian jurist who devised it, frequently used by Latin-American governments when making concession contracts with aliens, that the alien agrees not to seek the diplomatic protection of his own state and submits matters arising from the concession to the local jurisdiction. The validity of such a clause is not certain because a clause in a private agreement cannot deprive a state of the right under international law to give diplomatic protection to a national.

Calvo doctrine. A doctrine, named after Calvo (q.v.) declaring improper the use of diplomatic pressure or military force by one state against another to obtain the payment of claims made by its nationals for losses suffered as a result of civil war or internal disturbance. Calvo based his opposition to such action on abuse of power, inequality of treatment of nations and foreigners, and infringement of the territorial jurisdiction of independent states. The Drago doctrine (q.v.) is a special and limited form of this doctrine.

Cambacérès, Jean Jacques Regis de, Duke of Parma (1753–1824). French jurist and statesman. As chairman of the committee on legislation of the French Convention of 1792–95, he presented in 1793 a plan for a Civil Code in 695 articles. Though inspired by the revolutionary spirit of the time, it was not revolutionary enough for the convention. Later he submitted a shorter, second draft in 297 articles which was also rejected while a third draft in 1798 was not debated. He was later Minister of Justice (1799) and Second Consul of France (1800) and high in Napoleon's counsels; later he was Arch-Chancellor of the Empire (1804), President of the Senate and a Prince of the Empire. As Napoleon's chief legal adviser he took an important part in co-ordinating and drafting the French civil and other codes between 1801 and 1810.

Cambridge University law school. The Regius Chair of Civil Law was founded in 1540 and the Downing Chair of English Law in 1800 but the Law School did not really take shape until the Law Tripos was established in 1858. From 1870–74 law and history were linked but again separated in 1874 and since then there has been substantial development. The chairs have been held by some outstanding scholars, the Chair of Civil Law (1540) by Maine, Clark, and Buckland, the Downing Chair (1800) by Maitland, Kenny, Hazeltine, E. C. S. Wade, and Ivor Jennings, the Whewell Chair of International Law (1867) by Maine, Westlake, Oppenheim, Pearce Higgins, McNair, and Lauterpacht, the Rouse Ball Chair of English Law (1927) by Winfield and Glanville Williams, the Chair of Comparative Law (1934) by Gutteridge and Hamson, and the Chair of Criminology (1959) by Radzinowicz (qq.v.).

Camden, Lord. See PRATT, CHARLES.

Camden, William (1551–1623). English antiquary, author of *Britannia* (1586) a great compendium of information about the country of that time, a selection of which appeared also in *Remains concerning Britain* (1605). He also wrote *Annals of the reign of Elizabeth* (1615) and was responsible in 1622 for founding the chair of ancient history at Oxford. The Camden Society was founded in 1883 in his memory.

Camera. The judge's chambers in Serjeants' Inn as contrasted with the bench in Westminster Hall, or, now, his private room. A trial takes place *in camera* when it is held in private, the public being excluded. No criminal trial may be held *in camera*, save where specially authorized by statute. In civil cases interlocutory proceedings may be taken in private but the Divisions of the High Court have only limited powers to hear particular kinds of business in private.

Camera Stellata. The Star Chamber (q.v.).

Campbell, Sir Alexander of Cessnock (1676–1740). Became an advocate in 1696, was Lord Justice-Clerk 1689–92, and became a Lord of Session in 1704 but resigned in 1714. He later became Lord Clerk-Register, 1716–33, and second Earl of Marchmont in 1724.

Campbell, Archibald, Earl of Ilay, 3rd Duke of Argyll (1682–1761). Studied civil law at Utrecht, became Treasurer of Scotland in 1705, was one of the commissioners for the Union of 1707 and sat as a representative peer at Westminster. He was an Extraordinary Lord of Session, 1708–61. In 1710 he became Lord Justice-General of Scotland, holding the office till his death. He was also Lord Clerk-Register from 1714, Keeper of the privy seal from 1721 and manager of Scottish affairs and in 1733 Keeper of the Great Seal, and associated with President Forbes in forming the Highland regiments. When the heritable jurisdictions were abolished in 1747 he was awarded £21,000 for the offices of Justiciar of Argyll and the Isles, of Sheriff of Argyll and of Regality Lord of Campbell. He was a competent lawyer, collected a fine library and built Inveraray Castle.

Campbell, Sir Ilay, of Succoth (1734–1823). Was called to the Scottish Bar in 1757, became

Solicitor-General in 1783 and, after an interval out of office, Lord Advocate in 1784. In Parliament he was unimportant. In 1789 he became Lord President of the Court of Session, holding that office till 1808. In 1794 he presided over a commission of oyer and terminer for trials of treason cases at Edinburgh. On retirement he was created a baronet. After his retirement he presided over two Royal Commissions on the reform of legal procedure in Scotland and published *Decisions of the Court of Session*, 1756–1760 (1765) and a collection of the old Acts of Sederunt, 1532–53 (1811). He was respected as a great lawyer and a good judge, though no orator.

Campbell, James Henry Mussen, Lord Glenavy (1851–1931). Was called to the Irish Bar in 1878 and to the English Bar in 1898. He was in Parliament 1898–1900 and 1903–16, being Attorney-General for Ireland 1905 and 1916, Chief Justice of Ireland, 1916–18, and Lord Chancellor of Ireland, 1918–21. He became Baron Glenavy in 1921 and sat in the Irish Free State Senate 1922–28 and was chairman thereof.

Campbell, John, Lord (1779–1861). Destined originally for the ministry in Scotland, got work first as a tutor and then as a reporter in London, and read law in Tidd's chambers, being called to the Bar in 1806. In his early years he edited a book on partnership and started a series of *nisi prius* reports (1807–10) which have a high reputation. His practice developed rapidly and he took silk in 1827, became chairman of the Real Property Commission, and in 1830 entered Parliament. In 1832 he became Solicitor-General and in 1834 Attorney-General, and succeeded in carrying several measures important in the law of real property. He was now the leader of the common law bar, and participated in many famous cases. In 1834 and 1836 he was disappointed of the office of Master of the Rolls, but was placated by his wife being ennobled as Lady Stratheden. Eventually, after some discreditable transactions, he was made Lord Chancellor of Ireland with a peerage in 1841; he held office for six weeks only. While in opposition, he wrote his *Lives of the Chancellors* (1845–47) and *Lives of the Chief Justices* (1849–57) which brought him a considerable reputation. Though eminently readable, and containing valuable material, they are in parts biased and inaccurate. In 1846 he became Chancellor of the Duchy of Lancaster, and while leading the House of Lords, secured the passage of many important bills. In 1850 he succeeded Denman as Chief Justice of the King's Bench, and held that office till 1859. He showed himself a very able lawyer and many of his decisions have been taken as settling the principles ever since.

In 1859 he was promoted Chancellor and held that office till his death. As an equity judge he was competent but not distinguished. Throughout his career he was always seeking his own advancement, and tactless in dealing with colleagues in the courts and Parliament, while views expressed in his *Lives* provoked quarrels with many other judges and lawyers. Yet he was always advocating reforms and was largely responsible for the passage of many important reforming measures.

Mrs Hardcastle, *Life of Lord Campbell.*

Campegius, Johannes (1448–1511). Italian jurist, reputed one of the most learned of his time, author of *Consilia*, *Tractatus de Statutis*, *de Dote* and other works.

Canadian law. The first settlements in what is now Canada were made by the French, first Champlain's settlement at Port Royal, now Annapolis, Nova Scotia, and at Quebec (1608). New France, in the St. Lawrence Valley, initially settled by a chartered company, became a royal province in 1663. By the Treaty of Utrecht (1713) France ceded to Britain her claims to Hudson Bay, Newfoundland, and Nova Scotia, but still held the shores of the St. Lawrence. In 1759 in the course of the Seven Years War the British captured Quebec and by the Treaty of Paris, 1763, Canada was finally ceded to Great Britain.

In 1774 the Quebec Act of the British Parliament made all the territory France had claimed, including as far as the Mississippi and the Ohio, the province of Quebec. The French civil law was established, along with English criminal law. At the American Revolution large numbers of loyalists left the United States and settled in Ontario and Nova Scotia.

In 1791 the Constitutional Act separated Lower Canada (Quebec) from Upper Canada (Ontario) giving each its own government, Lower Canada, chiefly French, retaining its old laws, Upper Canada being purely British.

In 1840, following Lord Durham's Report, the two provinces were united, but this did not give political stability and in 1867 the British North America Act created a federal union, first of four provinces, Quebec, Ontario, Nova Scotia and New Brunswick, to which others later acceded (British Columbia, 1871; Prince Edward Island, 1873; Manitoba, 1870; Alberta, 1905; Saskatchewan, 1905; Newfoundland, 1949) with a Parliament of Canada.

The Statute of Westminster, 1931, freed the Canadian Parliament from the limitations on legislative power to which colonial legislatures are subject, apart from constitutional amendments.

The British North America Act, 1867 and amending Acts do not provide a comprehensive constitution, and it is supplemented by Canadian legislation, common law principles and conventions and case law. The British North America (No. 2)

CANADIAN LAW

Act, 1949, conferred on the Dominion Parliament power by ordinary legislation to amend the Constitution of Canada with certain exceptions, but constitutional amendments affecting provincial rights and matters still require legislation by the United Kingdom Parliament, which is never refused if agreed on in Canada and requested. A provincial legislature has exclusive power to amend the constitution of the province.

The constitution as contained in the British North America Acts, 1867 to date, assigns particular legislative powers to the provinces and vests the residue of legislative powers in the federation. In particular the Dominion Parliament has exclusive legislative authority in relation to 29 enumerated classes of subjects, which include taxation, the armed forces and defence, regulation of trade and commerce, the criminal law, marriage and divorce, and bankruptcy. Each provincial legislature has exclusive power to make laws in relation to matters within 16 enumerated classes including direct taxation within the province, education, property, and civil rights in the province, and the administration of justice in the province, including provincial courts and civil procedure. The two groups of enumerated powers are mutually exclusive. The provisions have been frequently interpreted by the Judicial Committee of the Privy Council in appeals from Canada.

Canadian constitutional law also includes substantial sections of English common law, constitutional usages and conventions and general principles transplanted from Britain, notably the general principles of Cabinet government.

There is no Bill of Rights in the constitution but the major rights were confirmed by the Canadian Bill of Rights in 1960. The right to use either English or French is protected constitutionally and by statute.

Federal government

The head of the executive is the Governor-General, appointed by the Crown on the advice of the Canadian Government. He is assisted by the Privy Council of Canada. The Cabinet is selected by the Prime Minister from the Privy Council but must be representative of various provincial, religious, and sectional interests.

The federal legislature consists of the Queen, represented by the Governor-General, the Senate, persons nominated formerly for life, now till age 75, by the Governor-General in numbers roughly proportionate to provincial populations, and the House of Commons, in which each Province is represented in proportion to its population. The Houses have equal powers, but most legislation is initiated in the House of Commons and it has exclusive power in financial matters. Both English and French have official status as languages at federal level. The procedure is generally similar to that of the United Kingdom Parliament.

Provincial governments

Each province has a Lieutenant-Governor, appointed by the Governor-General in Council on the advice of the Canadian government, and an Executive Council. Each has an elected Legislative Assembly, and Quebec has also a Legislative Council. The Lieutenant-Governor may reserve bills for the Royal pleasure to be signified by the Governor-General, and the latter may, on the advice of the Privy Council of Canada, disallow Provincial Acts, and has been resorted to where provincial acts were wholly or partly unconstitutional, contrary to Commonwealth or Dominion policy, or unjust and improper. Subject thereto the provinces enjoy self-government and autonomy.

In 1918 there was created the Conference of Commissioners on Uniformity of Legislation in Canada which has drafted a number of model statutes for adoption by several or all provinces.

The Yukon territory and the North-West Territory are administered directly by the federal government.

Judicial system

In the eighteenth century the governor and his Council frequently originally acted as courts. In Nova Scotia in 1749 the Governor and Council assumed full judicial authority, civil and criminal, legal and equitable, but it was superseded by a Supreme Court in 1754.

In the other maritime provinces there was initially separate administration of common law and equity, the Governor not infrequently acting as Chancellor. In Prince Edward Island separate administration still survives. In Newfoundland a Court of Common Pleas was established in 1790 and a Supreme Court in 1793.

In Quebec an Ordinance of 1764 provided for a Court of Common Pleas for civil causes, and a Court of King's Bench for civil and criminal causes. An ordinance of 1777 established Courts of Common Pleas for Montreal and Quebec and a Court of King's Bench with criminal jurisdiction only and after various changes there was created in 1849 a Court of Queen's Bench as civil court of appeal and criminal court of trial, and a Superior Court with mainly civil jurisdiction. In Ontario a Court of King's Bench was established in 1794 with civil and criminal jurisdiction and appeal to the Governor or Chief Justice and two of the Council. In 1837 a Court of Chancery was established, and in 1849 a Court of Common Pleas, and a Court of Error and Appeal. An Exchequer Division was created in 1903, and a separate Court of Appeal. It now consists of a High Court of Justice, not divided into divisions, and a Court of Appeal. The court organization in the western provinces is generally similar. There also exists, save in Quebec, systems of county and district courts. Courts have been established for the Yukon and Northwest Territories.

The Supreme Court of Canada was created in 1875; it hears cases of disputes between federal and provincial legislatures, and appeals in all matters, civil and criminal, from provincial courts. It has also an advisory jurisdiction. It does not, however, have the constitutional status of the U.S. Supreme Court or the Australian High Court.

There is also a Federal Court of Canada (1970), replacing the former Exchequer Court of Canada, comprising an Appeal Division and a Trial Division; it is a court of law, equity and admiralty, and a superior court of record having civil and criminal jurisdiction. Appeal lies to the Supreme Court of Canada. There are also various courts of special jurisdiction such as a Court Martial Appeal Court.

Federal courts, other than the Supreme Court, can deal only with actions involving federal matters, but provincial superior courts may administer federal law.

Until 1949 ultimate appeal lay to the Judicial Committee of the Privy Council.

The judiciary was initially recruited mainly from England and from the American colonies after the Revolution there, some having no legal qualification at all. Appointment was initially in British hands as part of the royal prerogative, authority to appoint being commonly delegated to local Governors, but latterly the power of appointment was vested in the Ministry of the province, on whose advice the Governor acted.

After confederation judicial appointments to the superior courts were made federally from the bars of the respective provinces. Since then also tenure during good behaviour has been constitutionally guaranteed to superior court judges. Though the judges of the superior, county and district courts of the provinces are federally appointed the courts themselves are constituted and organized by the provinces.

Criminal law and procedure.

The Quebec Act, 1774, provided for the application of the criminal law of England into Quebec. Elsewhere English criminal law was adopted without express statute.

On confederation in 1867 exclusive jurisdiction in respect of criminal law and procedure was conferred upon the federal Parliament, and since 1893 there has been a federal Criminal Code. It was based on the English draft code of 1878, Stephen's *Digest of Criminal Law*, Burbige's *Digest of the Canadian Criminal Law* and Canadian statutes. It was not exhaustive. A new Criminal Code was adopted in 1955 and has subsequently been amended.

Jury trial was introduced on English lines, but is now mandatory only in a few serious indictable cases; in other cases the accused may elect jury trial, but the great majority of cases are tried summarily. In some provinces the grand jury still exists.

Civil law and procedure.

Prior to the cession of Canada to Britain in 1763 French law applied in what is now Quebec. A Royal Proclamation of 1763 guaranteed the inhabitants enjoyment of their property and freedom of worship 'so far as the laws of Great Britain permit'. Pending elections and the summoning of an assembly, they were to have 'the enjoyment of the benefit of the laws of our realm of England'. An Ordinance of 1764 established a judicial system on English lines, the judges to determine agreeably to equity, having regard nevertheless to the laws of England so far as the circumstances would permit. This resulted in the continued application of French civil law in disputes among French-speaking Canadians, and the Quebec Act, 1774, affirmed that in all matters of controversy relative to civil rights and property resort was to be had to the French law that had been in force in the country, but that the criminal law of England should continue to be observed as law in Quebec.

Under the Constitutional Act, 1791, French-speaking Lower Canada (Quebec) was separated from English-speaking Upper Canada (Ontario) and the Act of the Upper Canada legislature introduced English civil law instead of French there as at 15 October 1792, and in 1800 Upper Canada adopted English criminal law as at 17 September 1792.

English civil law was adopted in Nova Scotia (including till 1784, New Brunswick) as at 1758, Prince Edward Island as at 1763, Newfoundland as at 1833, British Columbia as at 1858, Manitoba, Alberta, and Saskatchewan (all formed from Hudson's Bay Company lands) as at 1870 (though in the last three cases the adoption was retrospective from 1874, 1905, and 1905 respectively). In consequence whether some parts of English law are parts of the law of a province depend on its date of adoption of English law. After adoption the English law could be and was modified by local legislation, subject, down to the Statute of Westminster of 1931, to the overriding power of the British Parliament, and by local judicial decisions. But much legislation, both federal and provincial down to mid-twentieth century has copied from United Kingdom legislation.

In 1866 Quebec adopted a Civil Code based on the Code Napoleon, but there too English law has been influential.

British Privy Council decisions were binding, at least until appeal to the Privy Council was abolished in 1949. Other English decisions have tended to be followed as expositions of the common law adopted in most of the provinces and still the basis of the law of most of the Dominion. The Supreme Court no longer considers itself bound by its own decisions. The principle of *stare decisis* has never been rigidly followed within the provinces, but provincial courts are expected to follow decisions of the Supreme Court of Canada. Since confederation there has been an increasing tendency to look to U.S. decisions.

Jury trial was early adopted from English practice, but the general rule has come to be that there is no jury unless one is asked for, though jury trial is common in certain provinces, particularly Ontario. The number on the jury varies in the several provinces, and the unanimity rule does not apply.

In Quebec there is a Code of Civil Procedure, but even there English influence has been considerable.

There is a federal Law Reform Commission of Canada with the function of keeping the law under review and recommending improvements.

Legal profession.

The legal profession is organized on a provincial basis; it has never been differentiated into barristers and solicitors, though in Quebec the separate branches of advocate and notary exist. The practice of appointing Q.C.s was adopted from England.

Law societies have been incorporated in all the Provinces, commencing with Law Society of Upper Canada in 1797. They control admission to practice and administer legal education; there is no separate federal bar.

The earliest law schools were the École de Droit (1851), and those of McGill University (1853), Laval University (1854) and Dalhousie University (1883). By the 1920s legal education in Canada had swung towards the American pattern of professional education in university law schools.

There long was in Canada, and to a considerable extent has continued to be, heavy reliance on English books and reports, though Blackstone does not seem to have been very influential. Very little Canadian legal writing appeared before the mid-nineteeth century.

R. Boult, *Bibliography of Canadian Law*; W. P. M. Kennedy, *Constitution of Canada*; R. M. Dawson, *The Government of Canada*; A. G. Doughty and others, *Canadian Constitutional Documents*; W. R. Lederman, *The Courts and the Canadian Constitution*; B. Laskin, *Constitution of Canada*; B. Laskin, *The British Tradition in Canadian Law*; E. McWhinney, *Canadian Jurisprudence*; J. Clarence-Smith, *Private law in Canada*; Canada Year Book.

Candlemas. The feast of the Purification of the Blessed Virgin Mary, held on February 2, so called from processions with lighted candles and the consecration on that day of candles for the service of the following year. In Scotland it is one of the quarter-days of the year.

Canisius, Henricus (1548-1610). Dutch canonist and historian, author of *Summa iuris canonici* (1594), *Commentarius in regulas iuris libri VI Decretorum* (1600), and *De Differentiis iuris canonici et civilis* (1620) as well as several volumes of historical documents.

Canisius, Petrus (1521–97). Saint and Doctor of the Church, papal nuncio in Germany, author of *Summa doctrina christianae* (1585) and a catechism of the Christian church.

Canon. (1) A rule of conduct, particularly in ecclesiastical law denoting a rule made by Pope or Council included in one of the authoritative collections of canon law (q.v.).

(2) A rule of conduct promulgated by the Convocation of the Church of England, whether having legal force or not.

(3) A clergyman holding an appointment as one of the chapter of a cathedral.

(4) In the phrase 'canon of construction' a guiding principle in the construction or interpretation of statutes.

(5) In the phrase 'canons of inheritance', the rules regulating inheritance on death.

Canon law. The body of law constituted by ecclesiastical authority for the organization and government of the Christian Church. A canon is a rule, norm or measure and the Church, as much as a civil society, has found it necessary to prescribe rules setting standards and norms for conduct in many situations, and the body of these is the canon law. But by reason of the nature and purpose of the Church its rules deal with different subjects from the rules of a civil society.

The history of canon law is a history of borrowing and adaptation form various legal systems, particularly Roman law. Its sources are the New Testament and tradition, legislation by popes and general councils of the Church, and decisions in particular cases.

From at least the early third century the Christian community, wherever it existed, had some machinery for its government. The bases of its organization are in many passages of the New Testament, supplemented by writings of apostolic fathers. In the third century Tertullian and Cyprian began to mould the framework and vocabulary of canon law and there were many pseudo-Apostolic constitutions and juridico-liturgical documents which be-

CANON LAW

came widely diffused. They included the *Didaché* or *Doctrina XII Apostolorum* (*c.* A.D. 100) from Syria, the *Traditio Apostolica* of Hippolytus (*c.* 218) which was the basis for subsequent constitutions, the *Didascalia Apostolorum* (*c.* 250–300), the first attempt at a corpus of canons, the *Constitutiones Apostolorum* in eight books (*c.* 400) and many more. Many of these were in Greek but were soon translated into Arabic, Coptic, Latin, and other languages, and re-edited.

Development in the East

The first Greek collection preserved is the *Synagoge Canonum* in 50 titles by John the Scholastic III (*c.* 570). There were numerous decrees of various Church Councils, many collected as the *Corpus Canonum* of Antioch (*c.* 380), supplemented by canons of later councils, notably that of Chalcedon, the whole collection being known as the *Syntagma Canonum Antiochenum.* In the sixth century the canon of later councils and the 85 *Canones Apostolorum* were added. The Council in Trullo or Quinisext Council limited the sources of law to the general and local councils, the patristic canons and the Canon of Cyprian and this list constitutes the foundation of all Eastern canon law.

After Constantine recognized Christianity the emperors often legislated on ecclesiastical matters and Justinian exercised decisive influence on the development of canon law by his religious constitutions. These imperial laws were added to the collections of canons, making such collections as the *Collectio LX titulorum* (*c.* 535) and John the Scholastic's *Collectio L titulorum* (*c.* 570) which made mixed collections, preparing the way for the later *Nomocanon.*

Development in the West

In the West canons of Nicaea and Sardica were collected in translation in the *Vetus Romana* (*c.* 410) while the *Isidoriana* or *Hispana Collectio-Versio* was probably compiled in Rome in the early fifth century. The *Prisca* or *Itala Collectio-Versio* differs from the *Isidoriana* in the ordering of the canons. In the sixth century much work was done seeking to unify and co-ordinate the councils and decretals so as to produce a body of legislation subject to papal authority and to make it universally obligatory. A notable work is the *Collectio-versio* of Dionysius Exiguus or *Dionysiana*, of which three editions are known, the *Prima* (*c.* 500), *Secunda* (early sixth century) and *Tertia* (before 523). Dionysius also made a *Collectio Decretalium* comprising decretals from 385 to about 500. The two Dionysian works, known also as the *Liber canonum* and the *Liber decretorum*, are known as the *Dionysiana Collectio.* About the same time were made the *Quesnelliana Collectio* (*c.* 500), the *Frisingensis* collection (*c.* 500), the *Vaticana*, the *Sanblasiana* and the *Teatina* or *Collectio Ingilrami*, all seeking to collect, organize and unify the materials.

Each country tended to produce its own collections. There were several in Italy, the *Thessaloniensis* (*c.* 531), the *Avellana* (*c.* 555) and the *Mutinensis* (*c.* 601). In Spain there was the *Collectio Hispana Chronologica* or *Isidoriana*, falsely attributed to Isidore of Seville, based on the *Dionysiana*, formulated at the Council of Toledo in 633 and later enlarged and developed into the *Hispana Systematica* (*c.* 800). In Gaul many collections were made, including the *Statuta Ecclesiae Antiqua* (*c.* 485) the *Andegavensis I* (after 450) and possibly the *Quesnelliana*, and many more. In Africa the *corpus canonum orientale* was translated and became the *corpus canonum Africanum.* There were collections of canons such as the *Breviarium* of Hippo (393) and the *collectio concilii carthaginensis* (419). From the British Isles numerous penitentials appeared from the sixth century onwards containing scales of penances for sins.

The Eastern Church to 1054

The Quinisext Synod (in Trullo) met at Constantinople in 691–2 and revised and completed the legislation of the earlier ecumenical councils. It also defined the sources of Eastern canon law to the exclusion of all other materials, and consequently collections produced on this basis are the common origins of Eastern canon law. After the council of Constantinople (869) there were no general councils and ecclesiastical questions were decided by the patriarch at Constantinople. Systematic collections began about 535 with the *Collection in 60 titles* and various *Nomocanones.* These were continued by the *Nomocanon in 14 titles* composed by Enantiophanes, a revised edition of which by Photius in 883 was in 920 recognized as the official collection of the Church of Constantinople. There were also notable systematic commentaries, giving paraphrases of texts, scholia and other comments. The first commentator seems to have been Theodore Prodromus (eighth century) and the most famous John Zonaras, Alexis Aristenes and Theodore Balsamon.

Under the Carolingian Empire

In the eighth century Pepin initiated ecclesiastical reforms and sought to achieve unity by reforming councils and legislation. Many ecclesiastical capitularies were issued. The *Admonitio generalis* of 798 was important for church discipline, but there were no official collections of the numerous capitularies. Important private collections included those of Ansegisus of Fontanelle (*d.* 833) and the pseudo-Isidorian collection of Benedict Levita which claimed to continue it. General collections became of more importance. Pope Adrian I in 774 sent to Charlemagne a model code containing the councils and the collection of decretals of Dionysius Exiguus, the *Dionysio-Hadriana*, and this became the standard source of materials. About 800 the *Dacheriana* was compiled from the *Hadriana* and the *Hispana* and can be considered the real achievement of Frankish reform. There was however in Carolingian times a lack of systematic and theoretical investigation of the canon law. The most prolific scholar seems to have been Archbishop Hincmar of Rheims, who was, however, more concerned to strengthen the power of metropolitans and synods.

About 850 the pseudo-Isidore or False Decretals were put into circulation, a collection of genuine texts with apocryphal papal letters and false decretals interpolated. They supported episcopal power, made general the appeal to Rome in important cases, regularized judicial procedure, emphasized the sacred character of ecclesiastical property and had considerable influence on canonical literature.

Canon law was in decline in the tenth and early eleventh centuries, by reason of the weakening of papal authority and the near-anarchy of Europe. The most notable literature were the private local collections, selecting and adapting materials to the needs of the occasion. The most noteworthy of these were the *Collectio Anselmo dedicata* (*c.* 890), and *Collectio libri quinque* (*c.* 1015) in Italy, in Germany the *Libri duo de synodalibus causis et disciplinis ecclesiasticis* of Regino of Prum (*c.* 900), in France the *Collectio* of Abbo of Fleury (*c.* 980) and most importantly, the *Decretum* of Burchard, Bishop of Worms, in 20 books (1020-25) setting out the principles which should govern imperial reform.

In the early eleventh century the Gregorian reform of the church was founded on its independence, the primacy of the papacy, and fidelity to tradition. The Pope was recognized as the primary source of ecclesiastical law, exercising authority through councils and decretals. Collections of materials of this period include the *Dictatus Papae*, attributed to Gregory VII, the *Breviarium* of Atto of Vercelli, the *Collectio Libri duo*, the *Collection of 74 titles* (*c.* 1175), the collections of Anselm II of Lucea (*c.* 1082) and the *Collectio Deusdedit* in four books (*c.* 1085) of Cardinal Deusdedit, the *Liber de vita Christiana* of Bonizo of Sutri (*c.* 1090), the *Collectio Britannica* (*c.* 1090), the *Polycorpus* of Cardinal Gregory (*c.* 1105), the *Liber Tarraconensis* (*c.* 1090). The most important was the *Collectio* of Ivo of Chartres, comprising the *Tripartita*, *Decretum* and *Panormia*. In Spain there was the *Collectio Caesaraugustana* (*c.* 1115). There were many other lesser compilations and an enormous polemical literature.

Ivo's work sought to harmonize the traditional texts with the newer authorities, utilizing the principles of interpretation and concordance of texts of Bernold of Constance and this marked a step forward in canonistic science.

The number of these collections and the variety of rules contained in them produced a need for harmonization, and this gave rise to concordance treatises and collections such as by Bernold of Constance (late eleventh century), the *Liber de misericordia et iustitia* (*c.* 1105) of Alger of Liège and the *Sententiae Sidonenses* (*c.* 1135), all of which paved the way for the work of Gratian.

The classical period

The classical period of canon law was the period 1140–1350, which witnessed the beginning of the scientific study of the canons by Gratian and the first official collections of universally binding laws of the church. The reformers were intent on unifying and reforming the mass of written traditional materials. But there were numerous contradictions between texts and varying types of collections. The first need was for a system of interpretation of texts which would make for unity in the Church and reconcile the Roman and Franco-Germanic traditions. Important were the works of Bernold of Constance (*De excommunicatis vitandis* (*c.* 1091), Ivo of Chartres (prologue to the *Panormia*) (*c.* 1096) and Alger of Liège (*Liber de misericordia*) (*c.* 1105) who all sought to distinguish general principles from particular decisions. At the same time theologians were working on similar lines, notably Peter Abelard's *Sic et non* (*c.* 1115).

About 1140 Gratian, applying Abelard's dialectical methods to the texts, set side by side texts for and against particular propositions and produced the *Concordia discordantium canonum* or *Decretum* (*c.* 1140) the foundation of the classical law. This gave canonists a great mass of material bringing together patristic, conciliar and papal teaching on the organization of the church, the structure of Christianity, the Sacraments, worship, and liturgy. It was at once adopted by the Curia and the Schools of canon law, but was never adopted by the Church as an authentic collection. Commentaries and glosses at once began to appear, composed by decretists (q.v.) notably by such as Paucapalea, Rolando Bandinelli (later Pope Alexander III), Stephen of Tournai, Rufinus, Sicardus of Cremona, Huguccio and many others.

The Decretum gave rise also to many questions and the papacy was asked for authoritative answers to many of them. It also gave rise to a great deal of canonistic activity, in the form of collections of material unknown to Gratian, and of later decretals. The first systematic collection was that of Bernard of Pavia (*Collectio Parisiensis II*) (*c.* 1179) comprising decretals from 1127 to 1170. Others were the *Appendix Concilii Lateranensis III*, the *Collectio Bambergensis* and the *Collectio Lipsiensis*. In 1191–92 Bernard of Pavia published the *Breviarium Extravagantium*, a collection of 900 decretals from 1140 to 1191, arranged in five books labelled *iudex*, *iudicium*, *clerus*, *connubia*, *crimen*. This became known as *Compilatio Prima* of the Quinque *Compilationes Antiquae* and itself became the subject of a literature of apparatuses by Peter of Spain, Richard de Mores, Bernard himself, Alanus Anglicus, Lawrence and Vincent of Spain, Tancred, John of Wales, and others. The *Quinque Compilationes Antiquae* were the five most important decretal collections in the period 1140–1234, though there were several others. *Compilatio prima* (1187–91) was composed mainly of post-Gratian decretals and included the canons of the Lateran Council of 1177. *Compilatio tertia* was compiled by Peter of Benevento for Innocent III, comprised decretals of the first twelve years of Innocent's pontificate and was promulgated in 1210, becoming the first official collection. *Compilatio Secunda* was compiled by John of Wales (1210–15) and was deemed second because it included decretals from the period between *prima* and *tertia*. It drew on the works of Gilbert and Alan. *Compilatio quarta* may be attributed to Joannes Teutonicus or to Alan and contains decretals of the later years of Innocent III and the canons of the Lateran Council of 1215. It is not certain that it was promulgated. *Compilatio Quinta* was made, possibly by Tancred, at the instigation of Honorius III from decretals of his pontificate from 1216 and promulgated in 1226.

All five were the subject of important commentaries and glosses, particularly (on *prima*) by Richard de Mores, Bernard himself, and Tancred (*glossa ordinaria*), (*on secunda*) by Alanus Anglicus, and Tancred (*glossa ordinaria*), (*on tertia*), John of Wales, Joannes Teutonicus and Tancred (*ordinaria*), (*on quarto*), Joannes Teutonicus (*ordinaria*), Peter of Spain and William Vasco, and (on *quinta*) James of Albenga (chief glossator) Tancred, Lawrence of Spain and Joannes Teutonicus.

These collections were of importance in that when the *Decretals* of Gregory IX were being compiled the plan of *Compilatio prima* was adopted and 1,771 of the 1,971 chapters of the five collections were incorporated.

In the early thirteenth century there was much activity on study of the *Decretum* culminating in 1216 in the publication of Joannes Teutonicus' gloss on the *Decretum* which summed up learning on the subject and became the *glossa ordinaria*.

By 1230 there was great profusion of decretals and of collections and Gregory IX saw the need for a definitive collection of decretals not included in, or subsequent to, Gratian's *Decretum*. In 1230 he commissioned Raymond of Penaforte to collect in one volume the constitutions and decretals of Gregory's predecessors scattered through various earlier collections as well as Gregory's own constitutions and decretals outside the usual collections. Raymond based his work on the *Quinque Compilationes Antiquae*, taking most of his materials from them and dividing and dispersing materials under the various headings. It was promulgated in 1234 and became the first authentic general collection of decretals and constitutions. It became known as the *Decretales Gregorii IX* or *Liber Extravagantium* and superseded all previous collections, private or authentic. This gave rise to a great flood of glosses, commentaries and other writings on the decretals, notably by Sinibaldo dei Fieschi (later Pope Innocent IV), Godfrey of Trani, Bernard of Parma, author of the glossa ordinaria, (*c.* 1240), Hostienis (*Summa*, 1253; *Lectura*, 1270), Duranti (*Speculum judiciale*, 1272) and others. Bartholomew of Brescia brought Joannes Teutonicus' *glossa ordinaria* into line with the new law of the Church in 1240–45. From this time onwards the spread of a uniform law was possible and activity concentrated on the development of a centralized and strictly regulated society.

After 1234 decretals continued to be issued and the Councils of Lyons issued constitutions. Collections of this new material were made and glosses composed on them, as by Bernard of Compostella on the *Novellae* of Innocent IV (*c.* 1254) and by Duranti the Elder on the *Novellae* of Gregory X (*c.* 1275).

In 1298 Boniface VIII promulgated the Sext (*Liber Sextus*) which consisted largely of legislation passed to meet new needs or to modify legislation since 1234. It was intended to supplement the *Decretals* of Gregory IX. It included decretals from Gregory IX to Nicholas III, canons of the Council of Lyons and chapters of Boniface VIII. The *Sext* was the last great collection of the classical period and gave rise to a great flood of apparatus, glosses, and other commentaries from such canonists as Guido de Baysio, Albericus of Rosciate, and many others. The *glossa ordinaria* on the Sext was composed by Joannes Andreae in *c.* 1301, but other commentators included Guido de Baysio, and Albericus of Rosciate.

In 1317 John XXII promulgated the *Constitutiones Clementinae*, the last collection, which completes the official *Corpus Iuris Canonici*. The glossa ordinaria on the Clementines was produced by Joannes Andreae in 1322. He also, in 1338, in his *Novella Commentaria* on the *Decretals* of Gregory IX surveyed the whole decretalist literature from the *Quinque Compilationes Antiquae* and reduced the mass of glosses to a coherent body. His death in 1348 brought the classical period to an end.

The *Extravagantes* of John XXII in 14 titles (1325) were not officially issued but were also the subject of apparatus and commentary. The *Extravagantes Communes* is a collection of 70 decretals from 1260 to 1484, and a few others. They were never received in the schools. The *Liber Septimus decretalium* of Petrus Matthaeus (1590) had little influence and the *Decretales Clementis VIII* (1598) were not authenticated.

The *Corpus Iuris Canonici* (a name which dates from its first use by Chappuis and de Thebes, making a private edition in 1499–1502, and by Pope Gregory XIII in 1580 when approving as textually authentic the edition made by the *Correctores Romani*), comprises the *Decretum*, the *Decretals of Gregory IX*, the *Liber Sextus*, the *Clementinae*, the *Extravagantes Ioannis XXII*, and the *Extravagantes Communes*. The Roman edition of 1580 remained the only authenticated *Corpus* until 1917.

From the time of Gratian the study of the canon law was taken up in the nascent universities of Europe. Distinct courses, and degrees, in civil and in canon law developed: Bologna was as great a centre of canonist studies as of civilian studies, and it was being taught at Paris and Oxford before the twelfth century ended. A great literature of commentaries, summaries, questions, and the like developed.

The work of the classical period gave an impetus to canonist studies all over Europe and from the early thirteenth century a Faculty of canon law was to be found in most of the universities of Europe. Bologna remained the outstanding centre and an enormous volume of literature was produced there, but Paris, Oxford, and other centres were notable. The canon law took its place along with the older civil (Roman) law and theology as one of the major studies and there was much overlapping of study.

In the twelfth century the outstanding canonists were Rufinus and Huguccius, in the following century Tancred, at Bologna, author of the *Ordo-Judiciarius*, Hostiensis (Henry of Susa, Cardinal of Ostia) who wrote a *Summa Decretalium*, commonly called the *Summa Aurea*, Durandus, writer of a vast compilation on practice, the *Speculum Judiciale* (1271), and in the fourteenth century, Johannes Andreae, professor at Bologna, John of Torquemada, in Spain, and in Italy Panormitanus. In the next century the greatest names were Antonius Augustinus, Archbishop of Tarragona, Pierre Pithou in France, who codified the *Libertés de l'Eglise gallicane* (1594), work continued by P. Dupuy, and in the next century by Fevret and Van Espen. Other notable scholars were Hauteserre and Reiffenstuel, Thomassin who wrote an *Ancienne et Nouvelle Discipline de L'Eglise* (1678), Héricourt who wrote a *Lois Ecclesiastiques* and Durand de Maillane who compiled a *Dictionnaire de droit canon*.

The canon law had widespread influence on other bodies of law. The community of ideas and rules created by the church contributed to the formation of international law, the idea of systematic organization and of sovereignty to the development of public law, it exercised a humanizing influence on systems of criminal law and contributed much to the acceptance of the idea of public prosecution by its development of inquisitorial procedure. In the private law of all Christian countries, it had immense influence, on the law of matrimonial relations, on the protection of possession, the importance given to good faith in obligations and the desire to repress fraud.

The Reformation and after

The Reformation had wide-ranging effect in removing whole countries from allegiance to Rome and from the jurisdiction of canon law. But even in countries which adopted varieties of reformed Christian faith and repudiated papal authority the influence of the pre-Reformation canon law continued, particularly in relation to marriage, and the terminology, concepts and principles of the older law persisted though modified.

In the fifteenth century there was an enormous literature of commentary on various parts of the canon law and on the current problems of the Church, on schism, on church and state relations, on war, and other problems.

The Council of Trent (1545–63) laid the foundations for reform of the church. It sought to restore, renew and supplement the older canon law and to find a basis for modern development. The work of codifying canon law was begun under Gregory XIII (1572–85). The text of the *Corpus Iuris Canonici* was reviewed by a commission of cardinals and experts (*Correctores Romani*) and an official Roman edition was published in 1580–82. Clement VIII had all the scattered decretals of the older canon law collected in a *Liber Septimus* but it was not officially approved. Benedict XIV (1740–58) however, had his decrees published as authentic sources of canon law.

In 1904 Pope Pius X announced a plan for codification of canon law and the new work, known as the *Codex Iuris Canonici* was promulgated in 1917 and became effective in 1918.

Its interpretation was entrusted to a permanent commission known as the Commission for the Interpretation of the Code of Canon Law, which has the functions of watching over the structure of the code and making modifications, and of declaring the authentic meaning of canons presenting difficulties. In 1959 Pope John XXIII announced his decision to constitute, and in 1963 constituted, a Commission for the revision of the Code of Canon Law.

J. F. Schulte, *Geschichte der Quellen und Literatur des canonischen Rechts*, 3 vols. (1875–83); A. F. Tardif, *Histoire des sources du droit canonique* (1887); J. Gaudemet, *L'Eglise dans l'Empire romain* (1958); B. Kurtscheid and F. Wilches, *Historia Iuris Canonici*, 2 vols. (1941–43); I. Zeiger, *Historia iuris Canonici*, 2 vols. (1940–47); G. Le Bras, *Histoire du droit et des institutions de l'Eglise en Occident* (1955); W. M. Plöchl, *Geschichte des Kirchenrechts*, 3 vols. (1953–59); C. A. Bachofen, *A Commentary on the New Code of Canon Law*, 8 vols (1918–81); A. G. Cicognani,

Canon Law (1934); Naz, *Dictionnaire du droit canonique.*

The canon law in England

In England down to the Norman Conquest there were decisions of Anglo-Saxon Councils and penitentials, based largely on local custom but also containing biblical texts, extracts from the Fathers and canons of other councils. William the Conqueror had Lanfranc create a code for the English church, which consisted of large extracts from the False Decretals, and another work, the *Capitula Angilrami*, from the Pseudo-Isidorian collection. The full canon law of the Western Church was regarded as binding in England but there were some matters in which local custom warranted divergence from the general canon law. The major examples of such customs, which acquired the force of law in England, were the right of spiritual courts to a more extensive jurisdiction in the probate of wills, the transfer of jurisdiction in cases of ecclesiastical patronage from spiritual courts to temporal, and the refusal to accept the general principle of legitimation *per subsequens matrimonium.* But these were mere local variations and pre-Reformation England was otherwise subject to the general canon law of the West. Church authorities and church law ruled in all matters of faith and morals as well as in purely ecclesiastical matters.

After Gratian's time some canons were made in England, but only to emphasize the general rules and apply them to local needs or to devise new modes of enforcement. Thus canons were made in 1195 by the Archbishop of Canterbury, as papal legate, at York, and there were many other instances, the most important being the canons made by the papal legates Otho and Othobon in 1237 and 1268 respectively.

In the fifteenth century Lyndwood, chief legal official of Archbishop Chichele, compiled, in his *Provinciale*, a collection, with legal commentary, of the provincial constitutions of successive Archbishops of Canterbury. It deals only with a small part of the law and assumes the binding nature of the Decretals.

When the English Parliament at the behest of Henry VIII repudiated the papal supremacy, Henry asserted that the canon law had not been applicable in England by virtue of the papal authority but because it had been accepted by the English kings and peoples, and further, that the King in future was to be the source and fount of canon law. He thus converted the canon law into national law, and made it the ecclesiastical law of England; accordingly, more than ever it was binding on all the King's subjects, and the Tudor and Stuart monarchs sought to use it, as much as the common law, as a means of ruling laity as well as clergy. Parliament and Convocation both requested a revision of the canon law but stated that until this was done the old canon law should remain in force.

The revision of the canon law was entrusted to a commission of 32 appointed by the King; a fresh commission was authorized under Edward VI; the matter dropped under Mary Tudor but was revived under Elizabeth, but that Queen in 1571 stopped Parliament's discussion of the revision, known as *Reformatio Legum Ecclesiasticarum*, and Convocation never ratified it, so that it has never had any authority. The canon law long stood to the extent that 'canons not contrary or repugnant to the laws, statutes and customs of the realm, nor to the damage or hurt of the King's prerogative royal, shall still now be used and executed as before': Submission of the Clergy Act, 1533, s. 7, repealed 1969. Pre-Reformation canons accordingly still stand, only so far as consistent with common law and statute.

After 1571 canons were formulated by Convocation on several occasions and new canons added in 1604, and in a limited range are a major attempt at canonical legislation by the Church of England. Further canons of 1640 never became operative, but some of those of 1604 were amended in the nineteenth and twentieth centuries. Parliament has, however, made extensive changes in the law and on the jurisdiction of the ecclesiastical courts, restricted the operation of the canon law to a limited range and put parts of the canon law into desuetude. Thus the exclusive jurisdiction of ecclesiastical courts over clerics has been abolished, supervision and administration of the probate of wills was in 1857 transferred from the ecclesiastical courts to the Court of Probate, marriage, legitimacy and separation were, also in 1857, transferred to the Court for Matrimonial Causes, civil marriage was established alongside church marriage, and more exclusively ecclesiastical issues have been affected, such as the abolition of tithes, restoration and repair of churches, rights of patronage modified, excommunication shorn of its civil consequences, and so on. The discipline of the clergy is regulated by post-Reformation canon law.

Z. Brooke, *The English Church and the Papacy;* F. W. Maitland, *Roman Canon Law in the Church of England*; E. Gibson, *Codex Juris Ecclesiastici Anglicani*; R. Burn, *Ecclesiastical Law*; R. J. Phillimore, *The Ecclesiastical Law of the Church of England.*

The canon law in Scotland

The early Church in Scotland looked to Iona and Ireland rather than to Rome or elsewhere but came to adopt Latin rites and in 1188 was made directly subject to the Apostolic See, whose special daughter she was declared to be. Between 1177 and 1221 at least six Church Councils were held in Scotland by legatine authority. In 1225 the Pope authorized the Scottish bishops to hold provincial councils without a legate or even a metropolitan. A Provincial

Council was instituted and probably met annually for many years. So far as it legislated it merely sought to enforce the rules of the Roman Church by applying them to the special needs of Scotland; decisions of the Fourth Lateran Council and other General Councils were codified and decrees of English councils were sometimes adopted. Its work was mainly administrative.

In 1472 St. Andrews was made an archbishopric and in 1487 primatial, the archbishop being declared to be *Legatus Natus* of the Apostolic See. In 1549 a Provincial Council met to extirpate heresies and passed many canons regarding the character and duties of the clergy, some being repetitions of reforming statutes of the Council of Trent. The Provincial Council ceased to exist at the Reformation in 1560 and its authority lapsed. But down to the Reformation the institutions, organization, liturgy, doctrine and laws of the Church in Scotland were those of the Western Church.

In 1560 the Scottish Parliament abolished the jurisdiction and authority of the Pope, and legalized the creed of the Reformers. In 1567 it abrogated all the laws, acts and constitutions, canon, civil, or municipal, contrary to the reformed religion, and passed legislation, particularly affecting marriage at variance with the former practice. But the new Commissary Court, which replaced those of the Officials of the dioceses, continued to rely very largely on the pre-Reformation canon law, though it had no force in face of Act of Parliament and, like the Roman law, was regarded as a body of principles valuable as guidance but not binding. In the course of time reliance on it diminished with the growth of native law, partly legislative, partly created by decisions.

The canon law in the United States

The Protestant Episcopal Church in the United States succeeded to the Anglican Communion in the colonies before the Revolution, which was subject to the laws of the Church of England. But much of the Reformation and post-Reformation English ecclesiastical legislation was inapplicable to the U.S. Some of the principles developed in the pre-Reformation canon law accordingly passed into U.S. church practice but, no church being established in the U.S., the ecclesiastical law is not part of the law of the land. The Roman Catholic Church in the U.S. has continued to be subject to the canon law.

Significance of canon law.

Throughout its history the canon law has played an important role in the organization and discipline of the Christian church, its liturgy, preaching, and other activities, and in influencing the attitude of Christian communities and members of the church to law, morals, and behaviour generally. The canon law also played an essential role in the transmission of ancient Greek and Roman law to modern times and in the reception of Roman law throughout Western Europe in the Middle Ages. The Middle Ages were dominated by ecclesiastical issues which cannot be fully understood without an understanding of canon law. The canon law had enormous influence on secular law, particularly of marriage, property, succession, crime and punishment, proof, and evidence. It powerfully influenced the law of the Protestant churches which arose after the Reformation. It influenced the development of international law while the thinking and writing of canonists and theologians has contributed to the modern idea of the State, to the perennial problem of church and state relations, and to many legal ideas and concepts, possession, good faith, legal persons and others. It continues in Roman Catholic countries to be a major form of law and elsewhere to be the governing body of rules for clergy and adherents of that faith.

Eastern canon law,

The canon law of the Eastern and Western Christian churches was much the same in form until they separated in 1054. Important early sources of rules and customs were the *Doctrina duodecim Apostolorum* (possibly second century), the *Didascalia duodecim Apostolorum* (third century) and the *Traditio Apostolica* (*c.* 218) attributed to Hippolytus, Bishop of Rome, but extensively used in the East. From these sources there was developed about A.D. 400 the *Constitutiones Apostolicae*, including 85 *Canones Apostolicae*. Many councils issued legislation, notably the Council of Chalcedon (451), which had a *syntagma canonum* or *corpus canonum orientale* comprising the collected canons of various fourth century councils. This was later supplemented by legislation of later fourth and fifth century councils, the *Constitutiones Apostolicae*, *Canones Apostolicae*, and letters of Cyprian, third century Bishop of Carthage, and of various Greek fathers. The Council of Trullo (692) accepted the whole of this material, except the *Constitutiones Apostolicae*, together with the canons it enacted as the official code of law of the Eastern churches. This in turn was supplemented by the canons of the ecumenical council of Nicaea II (787) and certain later councils.

There were many collections of materials, canons and laws of the Eastern Empire relating to ecclesiastical matters. The first known one which is extant is the *Collectio 50 titulorum* of the patriarch Johannes Scholasticus of *c.* 550. He also collected from Justinian's *Novels* the *Collectio 87 capitulorum* and the *Collectio Tripartita* which comprised the whole of Justinian's ecclesiastical legislation. The *Nomocanon 50 titulorum* of about 580, composed of the works of Johannes Scholasticus, remained in use until the twelfth century. The *Nomocanon 14 titulorum* was completed in 883 and in 920 was taken as law for the whole Eastern Church.

In the law schools particularly those of Constantinople and Beirut canon law was studied as much as civil law. Notable commentators on the canon law in the twelfth century were Johannes Zonaras and Theodore Balsamon, from whose works Matthew Blastares compiled in 1335 a *Syntagma Alphabeticum*, or alphabetical digest of both church and imperial law.

Various churches within Eastern Christianity separated from allegiance to the patriarchate of Constantinople from the fifth century onwards. These include the Syrian, Armenian, Coptic, and Ethiopian churches. They each developed bodies of canon law, but knowledge of these systems is very imperfect.

Canon law of Church of England. The canon law of the pre-Reformation Roman church as applied in England has continued into the post-Reformation period, but further developments in it have not affected England, and parts of it have been varied or repealed by or since the Reformation. It is now rather English ecclesiastical law, of which canon law is an historical source, rather than a branch of the general canon law of the Christian Church.

From the Reformation to 1919 the sources of new legislation were canons of the Convocations of Canterbury and York, and Acts of Parliament. In 1919 Parliament empowered the Church Assembly to pass measures which might, with the consent of Parliament be presented for royal assent and thereafter have the full force of statute. In 1970 the Church Assembly was replaced by the General Synod of the Church of England which has the same legislative power.

Canonist. A mediaeval scholar devoting attention to the study of the canon law of the Roman Church. The earliest canonists were concerned chiefly to collect the canons of church councils, ecclesiastical capitularies, and other materials. But from the early sixth century the work extended to ordering the materials into a single corpus to unify and co-ordinate the legislation and make it unilaterally obligatory, and the most famous work is the *Dionysiana* compiled by Dionysius Exiguus in the early sixth century. Similar collections were made in France, Spain and Africa. Repeatedly efforts were made to unify and reform the confused mass of written traditions, the major attempt being the *Decretum* of Burchard of Worms (*c.* 1023) though this did not satisfy the reforming party in the Church at that time. Most of Burchard's work reappeared in the *Decretum* of Ivo of Chartres (*c.* 1096). In the early twelfth century Gratian, stimulated by contemporary theological works using dialectical methods, produced the *Concordia discordantium canonum* (*c.* 1140) in which he presented texts for and against selected propositions, sought to define terms and to apply rules of interpretation. His work provided canonists with a mass of source-materials and the basis for study of the law.

The classical period of canonist study was the period from Gratian's *Decretum* (*c.* 1140) to the death of the canonist Joannes Andreae in 1348, in which period there were promulgated the official collections of universally building laws, the *Decretals* (1234), *Liber Sextus* (1298), and *Clementinae* (1317). The *Decretum* provided canonists with the collected materials; from 1150 it was in use in the schools and glosses, commentaries and summaries of it began to be written, notably by Paucapalea, Roland Bandinelli (later Alexander III), Stephen of Rufinus, Sicardus of Cremona, Huguccio, and others. From about 1160 canonists began to make collections of other patristic, papal, and conciliar materials, notably decretals (q.v.), a tendency which culminated in the *Breviarium Extravagantium* or *Compilatio prima* of Bernard of Pavia (1191–92).

The flow of decretals and the collections of them in turn gave rise to commentaries on the decretals and in 1216 Joannes Teutonicus published the standard *Glossa Ordinaria* on the *Decretum*, which summed up many years of canonist learning.

In the years 1250 to 1300 there was a flood of glosses, commentaries, *summae*, *lecturae* and other works, notably by such as Sinibaldus Fieschi (later Innocent IV), Godfrey of Trani, Bernard of Parma, Bernard of Montmirat, Hostiensis, and William Durant.

The final phase of the classical era of canon law was that following the promulgation of the *Liber Sextus* in 1298 and it led to a flow of glosses, apparatus and other works from such as Guido de Baysio, John le Moine, William of Mont Lauzun, Zenzelinus de Cassianis, Petrus Bertrandus, and Albericus of Rosciate. The classical age came to an end with the two *Glossae Ordinariae* of Joannes Andreae on the Sext (1301) and on the Clementines (1322) respectively. He also, in a *Novella Commentaria* on the Decretals (1338), surveyed the whole of the decretalist literature and arranged a century of glosses into a coherent apparatus.

By the end of the twelfth century there were schools of canon law all over Europe, notably in the new *studia generalia*, and canonists were as numerous as civilians and as productive of glosses, summae and other literature. Subsequently the publication of the *Extravagantes Joannis XXII* and the *Extravagantes Communes* opened up new fields of study for canonists but there was a lack of originality in their volumes of *commentaria*, *summae*, *quaestiones*, and others.

Many canonists wrote on the problems of schism and other contemporary problems such as the relations of church and state, but there were numerous commentators on one or more of parts of the *Corpus Juris Canonici*. Some discussed both civil and canon law.

Though the Reformation destroyed the formal authority of canon law in much of Europe and canon law declined as a university study, it has continued to the present day in a much more limited sphere as a branch of legal study and there have continued to be distinguished canonist scholars.

Canterbury, Archbishop of. The Primate of All England, metropolitan of the ecclesiastical province of Canterbury and chief dignitary of the Church of England. He has extensive ecclesiastical patronage and many powers and duties and was given by statute in 1833 the power of granting dispensations in cases, not contrary to the Holy Scriptures and the law of God, where the Pope had formerly granted them. He is a spiritual Lord of Parliament and superior ecclesiastical judge within his province, and may grant 'Lambeth degrees' (usually only in divinity) equivalent to those of a university.

Canute, Laws of. A code of law promulgated by King Canute (or Cnut) at Winchester, with the consent of the Witan, between 1016 and 1035. It comprises rules dealing with ecclesiastical matters, secular affairs and the forest laws. The authenticity of the third group has been doubted. They are printed in the *Ancient Laws and Institutes of England* (ed. B. Thorpe).

Capacity. One of the attributes of a person or entity having legal personality, denoting legal ability to bear and exercise rights or to be affected by legal duties or liabilities. According to a person's status (q.v.) he may have full legal capacity or capacity qualified in certain respects. Thus an infant or minor or person of unsound mind has restricted legal capacity as compared with the person of full age and sound mind.

Capacity or incapacity is frequently fixed by different rules in different parts of the legal system; a person may be capable of marriage at a different age from that at which he acquires capacity to vote and so on.

Capacity has to be considered both actively and passively. Active capacity is concerned with capacity to have and exercise rights and to enter into and undertake legal transactions, e.g. contracts. Passive capacity is concerned with capacity to be subjected to legal liability or to be held responsible for torts or crimes. Thus very young children are not capable of being held responsible criminally.

Active capacity is also distinct from power; a person of full age has legal capacity to make contracts but, if acting in particular roles such as agent or trustee, his power to make contracts of particular kinds may be restricted by the terms of his agency or of the trust deed or the general law.

Capias (Latin, 'that you take'). The general name of several varieties of writs formerly used in English law, addressed to the sheriff directing him to arrest the person named therein. Such writs were used for various purposes, to commence an action, in execution and otherwise.

Capita, per (by heads). The principle in the law of succession that each entitled person takes in his own right, and equally with other claimants, as distinct from succession *per stirpes* where, when claimants of different generations take, they take as representing their branches of the family and division is equal as between branches.

Capital. A corpus of property or assets, as distinct from periodical produce, return or income. The distinction may be important in relation to accounting and taxation.

In company law nominal, authorized, or registered capital is the total nominal value of shares which may be issued. Issued capital is the nominal value of shares which have been actually allotted to subscribers for cash or assets or services. Unissued capital is the nominal value of shares not yet allotted. Uncalled capital is the portion of the nominal value of shares actually issued which has not yet been required to be paid. Paid-up capital is the amount of money actually paid or deemed to have been paid on shares allotted. The powers of companies to alter their share capital is regulated by the Companies Acts.

Capital crimes or offences. Those crimes or offences, conviction of which incurred liability to the death penalty. In England, until the early nineteenth century, all felonies except mayhem and petty larceny were in theory capital, though the harshness of the law was tempered by benefit of clergy (q.v.), the substitution in some cases of transportation for the capital penalty, and the reluctance of juries to convict. In the early nineteenth century the number of capital offences was greatly reduced. In Britain now the only capital crime is treason, the death penalty for murder having been abolished in 1965.

Capital Gains Tax. A tax charged on gains realized or assets disposed of after 1965, regardless of how long the assets have been owned and of when they were acquired or brought into existence. It is distinct from income tax and corporation tax and in principle is a tax on receipts which do not come under income profits or gains. It is charged at a specified rate on the total amount of chargeable gains accruing to a person in a year of assessment after deducting allowable losses. The requisites for liability are the disposal of an asset, the accrual therefrom of a chargeable gain, and accrual to a person chargeable to capital gains tax. There are exemptions for small disposals, small gains, certain kinds of assets, and disposals by charities and bodies set up for the public benefit. On a person's death assets of which he was competent to dispose are deemed to be acquired by his personal representatives but are not deemed to be disposed of by him on his death.

Capital punishment. The infliction of death by an authorized public authority as a punishment. It was recognized by ancient legal systems, Babylonian, Assyrian and Hittite. Ancient Hebrew law prescribed death for homicide and for some religious and sexual offences, including bearing false witness, kidnapping, sexual immorality, witchcraft, idolatry, blasphemy, and sacrilege. Greek law generally regarded homicide, treason, and sacrilege as capital. Roman law recognized the death penalty but regarded hard labour and banishment (*interdictio aquae et ignis et tecti*) as lesser capital punishments, as banishment involved serious loss of civil status (*capitis diminutio*). In republican times death was mainly imposed for military crimes. Under the dominate and the empire it became more common as the penalty for a wider range of offences.

In Anglo-Saxon England murder was punishable by a fine of which part went to the King and part to the relatives. By the thirteenth century death was the common law punishment for all felonies except mayhem and petty larceny but the law was mitigated by the benefit of clergy (q.v.).

In early modern English law, the law imposed the death penalty for a wide range of offences, some quite petty, and a major change effected in the nineteenth century was to restrict it to treason and murder. In 1957 the United Kingdom restricted it to treason and capital murder, as defined by statute, and in 1965 abolished it for murder entirely.

Various methods of inflicting the death penalty have been utilized. The Babylonians used drowning, the Hebrews stoning. The Greeks might allow a free man to take poison but a slave would be beaten to death. Roman usages included precipitation from the Tarpeian rock, strangulation, exposure to wild beasts, crucifixion and for patricide, the *culeus* (drowning a condemned man in a sack with a cock, a viper and a dog). In mediaeval and early modern Europe hanging and beheading were the usual ways; burning at the stake was used for religious heretics; breaking on the wheel and slow strangulation were also used. In modern times in the U.K. only hanging was used, though the punishment for treason is hanging, or, if the Crown thinks fit, beheading. In some states of the U.S. the electric chair or the gas chamber have been used.

In the nineteenth and twentieth centuries attention has increasingly been devoted to the religious and moral justifications for imposing death as a penalty. Beccaria (q.v.) gave impetus to a rejection of the death penalty as unjustifiable and also as cruel and ineffective. In England Bentham and Romilly influenced a great reduction in the list of capital crimes. The death penalty was abolished in Tuscany in 1786 and Austria in 1787 and since then many Western countries have done so. Communist countries in general have not done so. Moreover even where the depth penalty is retained there have in many countries have steep drops in the numbers of actual executions. The favourite argument of abolitionists that the death penalty is not a deterrent may be true, but nothing is a complete deterrent.

A problem consequential on abolition is of the alternative. Life imprisonment may be even more cruel than death and long imprisonment, short of life, may be as bad. Neither have any more deterrent value; modern experience is that the deterrent value of life imprisonment is inadequate.

Capital Transfer Tax. A tax introduced in 1974 on gifts and other transfers of value made during lifetime and on death on a cumulative basis, subject to an exempt amount and small annual exempt amounts. It replaced Estate Duty. Primary liability for the tax falls on the transferor or on a deceased's estate. Transfers between husband and wife, and transfers to charities made on or within a year before death are exempted.

Capitant, Henri (1865–1937). French jurist, author with Ambrose Colin of a famous *Introduction à l'étude de du droit civil* and a *Cours du droit civil.*

Capite, tenure in. Tenure in chief or immediately. It means generally the relation of tenure (q.v.) between a tenant and his immediate lord, but particularly the relation between a tenant and the Crown, holding directly of the Crown. Such tenure in chief might be *ut de corona*, where the tenant held of the king as lord paramount; or *ut de honore*, where the tenant held lands of an honour which came into the hands of the Crown, as by escheat; or *ut de persona*, where the tenant held lands of a lord whose seignory escheated to the Crown. These distinctions have now disappeared.

Capitis diminutio. In Roman law, a reduction in status. There were latterly three degrees, *maxima*, loss of liberty and enslavement, involving loss of citizenship and family rights; *minor* or *media*, loss of citizenship, as by *deportatio*; and *minima*, change of family position, involving break of agnatic ties but leaving liberty and citizenship unaffected, as by *adoptio*, *adrogatio*, emancipation and entry into or manumission from civil bondage. The development of the concept is fairly late and also obscure.

Capito, Gaius Ateius (*temp.* Augusti). A notable Roman jurist, consul in A.D. 5. He was a contemporary and rival of Labeo (q.v.) and their political and legal rivalry gave rise to the rival schools of jurists, the Proculians and Sabinians (q.v.). He wrote a comprehensive *De Jure Pontificio*, a monograph *de Jure Sacrificiorum*, a book *de officio senatorio*, and a collection entitled *Coniectanea* but is rarely cited by later jurists though more frequently by non-legal authors.

Capitula legis regum Langobardorum. A compilation of the legislation of the Lombard kings of Italy from Rother's Edict (643) to that of Astolf (750), arranged under six heads and chronologically within each, made at Pavia about A.D. 830.

Capitulare Italicum. A compilation made at Pavia about 1090 of the Lombard statutes and capitularies subsequent to those in the *Capitula legis* (q.v.).

Capitularies. Enactments or ordinances of the Frankish kings in Western Europe in the eighth to tenth centuries. The name was derived from *capituli*, the canons of a church council and ecclesiastical ordinances. They were of very wide scope and had unlimited authority, being applicable to all the subjects of the Frankish empire. They might be ecclesiastical, frequently merely giving the sanction of the state to the decrees of popes or councils, or secular, applicable to a part only of the Empire or to the whole of it, and independent—*capitula per se scribenda*—and applicable to all subjects with full territorial validity, or supplementary—*capitula legibus addenda*—issued to amend or supplement existing laws, such as the *Lex Salica*, the *Lex Ripuaria*, the *Leges Langobardorum*, or the existing general body of such laws. If supplementary to a specific law, they were applicable only to those to whom the principal law was applicable, but if not, they were applicable to all inhabitants of the country named therein. Among the independent capitularies the *capituli missorum* had a distinct character; these were the king's instructions to royal magistrates, *missi dominici*, sent periodically into different regions to inform themselves on the conduct of his vassals or other public matters, and containing decisions or replies to inquiries by *missi* on the execution of their duties, or warrants of authority to do particular acts outwith their ordinary authority.

Capitularies were proclaimed at a national assembly, by public reading in churches or at other popular gatherings, by the despatch of copies to high officials who communicated them to their subordinates, and by circulation by the *missi*, when on their tours of inspection. The originals were recorded by deposit in the royal archives, but many escaped this recording.

Failing official collections of capitularies, various private collections were made, notably that of Ansegisus of Fontanelle, completed in 827, containing in four books the ecclesiastical capitularies of Charlemagne and of Louis the Pious and the secular capitularies of those kings. Within each book they were arranged chronologically. Three appendices contained *capituli missorum* and others omitted from the four books. This collection had great vogue and is reliable. The other well-known collection bears the name of Benedict Levita, of about 850, *Capitularia regum Francorum*. It professed to continue the work of Ansegisus by collecting in three books and four appendices capitularies unknown to Ansegisus. In fact the compilation is spurious, originating in Western France; and the supposed capitularies are a farrago of well-known legal sources, church canons, and patristic writings.

In Italy, now an autonomous kingdom within the Frankish empire, there were not only the general capitularies, valid throughout the whole Empire, but local Italian capitularies. In 832 Lothar I submitted a revision of the whole body of capitularies, beginning with those of Charlemagne, to the Assembly at Pavia, and a selection, known as the *Constitutiones Papienses*, was enacted. It included many of the general capitularies, and this became a local law for Italy. It comprised both *capitula per se scribenda*, which applied throughout the territory, independently of personality, and *capitula legibus addenda* modifying the *leges Langobardorum*, having the effect of protecting the interests of the church, reforming the administration of justice, and containing improvements in both public and private law. Collections of capitularies peculiar to Italy were made, notably the *Capitulare Italicum*, compiled by jurists of Pavia about the latter half of the tenth century.

The capitularies were in their time the most important legal sources and had longstanding influence in Europe.

Monumenta Germaniae historica, Leges; Seeliger, *Die Kapitularien der Karolinger* (1893).

Capitulations. (1) Grants by treaty by a State to the subjects of another State resident in the granting State of commercial privileges, and in particular of exemption from the jurisdiction of the local courts and of the privilege of having courts of the grantee State to administer justice to them. Such concessions are known from the second millennium B.C. and were made from the early Middle Ages. Notable examples were grants by Ottoman sultans to French in Turkey from 1536 onwards allowing them to have French consuls judge the civil and criminal affairs of Frenchmen in Turkey according to French law and to invoke the aid of the Sultan's officers in the execution of their sentences. Later nearly every European power obtained capitulations in Turkey, and they survived until well into the twentieth century in China, Egypt, Turkey, and Morocco. They came to an end as eastern countries improved their legal systems and became conscious of their independence.

(2) In international law, conventions between armed forces of belligerents stipulating the terms of surrender of defended places, ships, or bodies of troops. The terms are a matter of negotiation, and may allow the surrendering force to march out with full honours. Only the commanders of the opposing forces may conclude capitulations. If they contain arrangements other than of a local and military character concerning surrender these are not valid unless ratified by political authorities of both belligerents. Once a capitulation has been signed it is unlawful to destroy weapons or munitions, as these belong to the victors.

Caption. The formal heading of an indictment, information, affidavit, or similar document; also, formerly, arrest under civil process; also, in older Scots law, a warrant for the apprehension of the person of a debtor or obligant for the non-performance of an obligation, founded on the fiction that the debtor in the obligation had refused obedience to the sovereign's letters charging him to pay or perform and was accordingly a rebel and liable to imprisonment.

Capture. In international law, the right of a belligerent to seize an enemy ship, or a neutral vessel if she or her cargo is liable to confiscation by reason of breach of blockade, carriage of contraband, unneutral service, or grave suspicion demanding further inquiry. But the effect of capture differs in that an enemy ship may be appropriated as enemy property whereas a neutral ship, or cargo thereof, is captured to enable a Prize Court to determine whether ship or cargo or both should be confiscated. As a rule, captured neutral vessels may not be sunk, burned, or destroyed, and should be released.

The term is also used of property, other than private property, seized on land by a belligerent. It is 'booty' and belongs to the Crown.

In civil law it is the mode of acquiring property in fish, fowl, or wild animals, not already the object of ownership by another.

Caput. A head, hence in many contexts the chief place of a county, barony, or other area.

Caracalla, Marcus Aurelius Antoninus. Roman emperor A.D. 211–17, issued the *Constitutio Antoniniana de civitate* which conferred Roman citizenship on almost all free inhabitants of the Empire. In other respects he is generally regarded as a bloodthirsty tyrant.

Cardozo, Benjamin Nathan (1870–1938). U.S. judge and jurist, graduated from Columbia and from 1891 practised law in New York until elected to the New York Supreme Court in 1913. He became a member of the New York Court of Appeals in 1917 and Chief Justice in 1926. He was an Associate Justice of the U.S. Supreme Court from 1932 to his death and supported liberal views on social and economic matters. His judgments were learned, eloquent, and liberal, and many have been often cited on both sides of the Atlantic. He had great influence on his colleagues. His published lectures, particularly *The Nature of the Judicial Process* (1921), *The Growth of the Law* (1924), *Paradoxes of Legal Science* (1928) and *Law and Literature* (1931) have been highly regarded and reveal him as a perceptive and learned jurist and one of the very greatest American judges, probably second only to Holmes in making the judicial process creative yet evolutionary.

Care, Duty of. A fundamental concept of the common law of negligence. One person who has harmed another is legally liable to him on the ground of negligence only if the court holds that the defendant had in the circumstances owed the plaintiff a legal duty of care, not to have done or allowed to happen the acts which caused the harm, and had in the circumstances failed to implement the duty by not taking the kind and standard of care and precautions required by law. If in the circumstances there was no duty of care then any harm suffered is *damnum sine injuria* (q.v.) and the defendant is not liable. Whether there was or was not a duty of care in particular circumstances is a question of mixed fact and law, frequently determined by precedent failing which by consideration of whether in reason and justice the defendant should have taken precautions, or better than he did.

Care, Standard of. Where one person owes another a duty of care (see CARE, DUTY OF) law prescribes what the standard of that care should be. It is normally reasonable care, such care as the hypothetical reasonable man would take in the circumstances (and the reasonable man takes greater care with dangerous things than with safe ones) but in cases of strict liability (q.v.) the standard required is to ensure safety, subject to being able to rely on one or another of a few limited defences, and in cases of absolute liability (q.v.) created by statute the standard of care required is to ensure that accidents do not happen. Whether the defendant's conduct comes up to or falls short of the required standard of care is a question of fact, determined by evidence.

Carey Street. The street in London, off Chancery Lane, parallel to the Strand and behind the Law Courts, in which were situated the offices of the Chief Bankruptcy Registrar of the High Court. The name has accordingly become a synonym for bankruptcy.

Carleton, Hugh, Viscount (1739–1826). Was called to the Irish Bar in 1764. Later he was Solicitor-General (1779) and Chief Justice of the Common Pleas in Ireland (1787). In 1800 he retired and became a representative peer for Ireland.

Carlsbad decrees. Resolutions of a conference of ministers of German States held at Carlsbad in 1819 and called to take action against revolutionaries. The decrees were enforced with varying severity in the German States but failed to stifle German nationalism or to curtail liberal developments.

Carmichael, Sir James, of Hyndford, Lord (?1578–1672). Was introduced to the court of James VI of Scotland, became Sheriff of Lanarkshire in 1632 and was Lord Justice-Clerk from 1634 to 1636 when he resigned on being made Treasurer-Depute. In 1639 he became in addition an ordinary Lord of Session. He was made a peer in 1647. He was deprived of his offices under the Act of Classes in 1649. He was again Lord Justice-Clerk 1649–54 but was dismissed by Cromwell.

Carnal Knowledge. Coitus, sexual intercourse, or the penetration of the human female vagina by the human male organ.

Caro, Joseph ben Ephraim (1488–1575). An authority on the Talmud and codifier of Jewish law, author of the *Beth Yoseph* (1550–59), a thorough analysis of Talmudic and post-Talmudic sources, the *Shulhan Arukh* (1564–65), a code of Jewish law, an abridgment of the earlier work, and other works.

Carolina (or *Constitutio Carolina Criminalis*). A criminal code enacted for the Holy Roman Empire by Charles V in 1522. The project was resolved on by the Diet of Worms in 1521 and the first draft was based on the Bamberg code (q.v.) of 1507. The fourth draft was adopted by the Diet of Regensburg in 1532. It was primarily a code of criminal procedure with substantive law interpolated, and the first true code in which German and foreign law were reconciled. It obtained general force, despite a clause saving local law, and dominated German law for two centuries.

Caroline, Queen, case of. Caroline of Brunswick (1768–1821) married the Prince of Wales (later George IV) in 1795 but they separated in 1796. She went abroad in 1814 and consorted, probably adulterously, with an Italian, Pergami or Bergami. When her husband became King in 1820 she returned to England to claim her place as queen-consort whereupon a bill was introduced to dissolve the marriage and deprive her of the title of Queen; it was abandoned. She was refused admission to the coronation in 1821 and died shortly thereafter. She had great popular support at the time, which was probably unjustified, but it contributed to the discredit of the monarchy. See *Queen Caroline's Case* (1821), 1 St. Tr. (N.S.) 949.

Carpzov (Carpzovius). A Saxon family which included several distinguished statesmen and jurists. Benedikt (1565–1624) was professor at Wittenberg and Chancellor of Saxony. He published *Disputationes Juridicae*. His second son, Benedikt (1595–1666) was professor at Leipzig and privy councillor at Dresden. He is deemed the founder of German legal science, did much to systematize German law, and his work had great influence in securing recognition of the legal force of German law and custom. He wrote: *Commentarius in Legem Regiam Germanorum sive capitulationem Imperatoriam* (1635), *Practica nova Imperialis Saxonica rerum criminalium* (1635), *Definitiones forenses* (1638), *Opus decisionum illustrium Saxonicae* (1646) *Processus juris Saxonici* (1657), *Jurisprudentia ecclesiae seu consistorialis* (1649) and *Jurisprudentia forensis Romano-Saxonica* (1703). The fifth son, August (1612–83) was a diplomat and also wrote treatises on jurisprudence. Other members of the family were distinguished in theology and philosophy.

Carré de Malberg, Raymond (1861–1935). French jurist, who introduced Kelsen's thinking into French jurisprudence. He is the author of *Contribution à la Theorie générale de l'état* (1920–22) and *Confrontation de la Théorie de la formation du droit* (1933).

Carriage of goods by air. International carriage of goods by air is almost everywhere regulated by international conventions, notably the Warsaw Convention of 1929 amended by the Hague Protocol of 1955, and incorporated in British law by the Carriage by Air Act, 1961. This provides for standard documentation and limitation of liability for loss of or damage to goods. Domestic carriage is subject to the international rules, applied to domestic conditions by regulations.

Carriage of goods by land. In relation to carriage by land the basic distinction is between common carriers, who hold themselves out as ready to carry for anyone goods of the kinds they profess to carry to any place to which they profess to carry and who are liable for loss of or damage to the goods while in their custody unless they can establish that the damage or loss was attributable to act of God, act of the Queen's enemies, default of the consignor or inherent vice of the goods, and private carriers who carry under special contract only and are liable only for loss or damage through negligence. Limitation of liability by contract was limited by the Carriers Act, 1830. Since the Transport Act, 1962, British Railways have not been a common carrier and the contract of carriage is regulated by contract, normally in one of the forms prescribed by British Railways Book of Rules. Most carriers by road are private carriers and use contracts incorporating the Road Haulage Association's standard terms of contract.

International carriage by rail is regulated by international conventions of Berne of 1890, now of 1961, and international carriage by road by a Geneva Convention of 1965, both incorporated in British law by statute.

Carriage of goods by sea. A carrier by sea may hire a complete ship for a time or for a voyage by a time or voyage charter-party (q.v.). In other cases he contracts for the carriage of his goods in another's ship under a bill of lading (q.v.). It serves as an acknowledgment that goods described therein have been received for shipment, as a contract or evidence of the contract of carriage, and as a symbol of the goods.

In 1924 an international convention for the Unification of Certain Rules of Law Relating to Bills of Lading sought to unify rules by subjecting all bills of lading to standard clauses. This has been incorporated in British law by the Carriage of Goods by Sea Acts, 1924 and 1971. The Act confers extensive exemptions from liability for loss of or damage to goods carried.

Carriage of passengers by air. International carriage of passengers is regulated by international convention, the Warsaw Convention of 1929, amended at The Hague, 1955, incorporated in British law by the Carriage by Air Act, 1961, and domestic carriage by similar rules applied by municipal law. The convention makes provision for documentation and limitation of liability in the event of death or injury.

Carriage of passengers by rail. British Railways are common carriers of passengers but their liability in the event of injury or death is for negligence only, though in such cases as a collision the principle of *res ipsa loquitur* will raise a presumption of negligence. The Board may not exclude or limit their liability for death of or injury to a passenger save in exceptional cases.

Carriage of passengers by road. Common carriers of passengers must carry any passenger who offers himself and is willing to pay the fare, if the carrier has accommodation and the passenger is not objectionable. Private carriers carry under special contract only. In both cases liability for injury depends on failure to take reasonable care for passengers' safety.

Carriage of passengers by sea. The carriage of passengers by sea depends on individual contract. Liability for injury or death may be excluded or limited by contract.

Carrier. A person who, gratuitously or for hire, carries goods or passengers from place to place by land, sea or air. At common law carriers are distinguished into common carriers, who hold themselves out as willing to carry any person or goods to the places to which they ply, must accept persons or goods offered for carriage if they have space, and are, in the case of goods, strictly liable for loss of or damage to the goods in transit unless they can bring themselves within limited exceptions; and private carriers, who carry only under contract in each particular case and who may accept or reject any request to carry, e.g. furniture removers. The liability of common carriers of goods by land was at common law strict but has been limited by the Carriers Act, 1830. Common carriers of passengers are liable only for failure to take reasonable care. Under modern legislation British Railways are not common carriers, nor are publicly-owned carriers by road, while carriers by sea and carriers by air are not normally common carriers.

Carson, Edward Henry, Lord, of Duncairn (1854–1935). He was called to the Irish Bar in 1877 and rapidly made progress, took silk in 1889, and became Solicitor-General for Ireland in 1892, in which year he entered Parliament. In 1893 he was called to the English Bar and took silk in 1894. In 1900 he became Solicitor-General for England, an office he held till 1905. From 1910 he was leader of the Irish Unionists in Parliament and of the movement to keep Ulster in the United Kingdom if Irish Home Rule were established. In 1915 he became Attorney-General but resigned in 1916, being, however, appointed first Lord of the Admiralty in the Lloyd George Cabinet, but he resigned in 1918 on the proposal to introduce a home rule bill for the whole of Ireland. In 1921 he was made a Lord of Appeal in Ordinary and sat as such till 1929. He was a great advocate, orator, and fighter for Ulster but ordinary as a judge.

E. Marjoribanks and I. Colvin: *Life of Lord Carson.*

Carta de Foresta. A charter granted in 1217 and confirmed by the Statute of Marlborough, 1207, disafforesting many forests (q.v.) which had been made unlawfully and mitigating the harshness of the forest laws. The name is also applied to certain other early statutes.

Carta Mercatoria. A charter granted in 1303 to certain foreign merchants, giving them extensive trading rights, free from various municipal dues, and rights to jury trial in non-capital cases, in return for certain customs duties. It is also known as the *Statutum de Nova Custuma.*

Cartels. In international law, conventions between belligerents for permitting certain kinds of non-hostile intercourse, e.g. to arrange postal communication, or exchange of wounded. Cartel ships are vessels commissioned for the carriage by sea of exchanged prisoners to their own country. Customary rules secure their protection and employment solely for the purpose of their mission.

The term is also used in industrial contexts of arrangements between companies to divide markets, share export orders or otherwise interfere with competition.

Carter, James Coolidge (1827–1905). Was admitted to the New York Bar in 1853 and practised with great success. He became President of the American Bar Association and appeared for the United States in the Behring Sea Arbitration. When David Dudley Field's proposed Civil Code for New York was being debated Carter led the fight against codification, notably in his pamphlet *The Proposed Codification of Our Common Law* (1883), and followed the same line of thought in his lectures, posthumously published as *Law: Its Origin, Growth and Function* (1907).

Carter, Thomas (1690–*c.* 1760). Studied in Dublin and London, and became depute Master of the Rolls in Ireland in 1725 and Master of the Rolls in 1731 but was removed in 1754. He was thereafter a joint secretary of state.

Carting a jury. An alleged ancient practice at assizes in England, that in cases of life and limb, if the jury could not agree before the judges departed for another town, they were to be carried in carts after them and give their verdict out of the county. Whether the practice was ever used is doubtful. It is said to have been followed in Ireland in the early nineteenth century.

Carting London whores. An alleged custom of London that a whore might be carted, i.e. carried in a cart, out of town and dumped there.

Cartulary. Originally a place where books and records are kept, and particularly a register or record-book of charters and deeds. Cartularies of abbeys and monasteries accordingly contain much evidence of landownership and conditions of tenure and of the actual working of the law and are important sources of legal information for the thirteenth and following centuries. Many have been printed by record societies.

Carucage. A tax at one time levied on every plough (*caruca*). In the time of King John it was three shillings on each carucate.

Carucate (from *caruca*, a plough). In mediaeval law, a measurement of land, a ploughland, nominally as much land as could, in one year, be kept under the plough by a plough-team of eight oxen. It amounted in different parts of the country to between 60 and 120 acres.

Cas de demandes. See CASUS PLACITORUM.

Cas de jugement. See CASUS PLACITORUM.

Cas royaux. Serious cases, e.g. *lèse-majesté*, sedition, forgery of the royal seal, which under the legal system of the *ancien régime* in France had to be tried by royal judges in the bailiwicks and on appeal by the *parlement* of Paris.

Casaregis, Guiseppe Maria Lorenzo (1670–1737). A humanist jurist specializing in commercial law, author of *Tractatus de commercio* and *Discursus legales de commercio* (1719) and other works.

Case. An individual legal dispute, particularly one involving litigation. The word is also used of the body of pleadings, arguments, and submissions for one of the parties. In the House of Lords and Privy Council a case is a printed statement prepared by a party to an appeal setting out the facts on which he relies and the grounds of his appeal; it contains in an appendix any evidence relied on and the judgments in the courts below.

A case stated is a written statement by an inferior court or judge of the circumstances and decision of a case in that inferior court, submitted to a superior court and formulating a question of law for the superior court's opinion. It is accordingly a mode of appeal on a point of law from the inferior to the superior court. A case to counsel is a statement of facts submitted to counsel for his opinion.

The word is also used of an action, or a trial, or the body of pleadings, arguments, and evidence submitted on behalf of a party to an action. In legal literature a case is the report of an action, including the opinions of the judge or judges who decided the case, considered as an illustration of the law as a topic and as a possible precedent (q.v.) for use in later cases. See also CASEBOOK; CASE-LAW.

Case, action on the. As the action for trespass (q.v.) developed in mediaeval English law writs for assault, asportation of chattels and unlawful entry on land became matters of common form, but in other cases the writ had to be specially drafted to suit the circumstances. Such writs were said to be 'upon the case', and frequently narrated a custom of the realm or a statute, breach of which had caused harm to the plaintiff. The older view, that the action on the case developed by reason of the clause *In consimili casu* in the Statute of Westminster II (1285) is now regarded as unfounded. It was important to distinguish instances of trespass from instances of case, because the defendant could be arrested or outlawed in trespass, but until 1504 not in case. At least latterly the important distinction was that trespass lay only for harm which was direct and immediate, whereas indirect and consequential harms were redressable by case.

A. K. Kiralfy, *Action on the Case.*

Casebook. A kind of book much used in legal studies in Anglo-American countries, where the principles of law are frequently contained in judgments delivered in decided cases. The first casebook, Langdell's (q.v.) *Cases on the Law of Contract* taken into use at Harvard in 1871 was a selection of judgments in decided cases printed at length but later casebooks have included also material from statutes, government reports, articles, questions and points for discussion, and are accordingly now frequently called *Cases and Materials.* The casebook method, as a paedagogic device, is based on the premise that students should not be told by textbook or lecture what the principles of the law are but should be led by reading the casebook and being questioned on the cases to discover the principles for themselves. The method swept American law schools from 1871 and was even used in courses on legal history, jurisprudence and other unlikely subjects but in the twentieth century is used only selectively. Casebooks and the casebook method have been used in other Anglo-American jurisdictions but have never been so popular as the systematic textbook.

Case law. The general term for principles and rules of law laid down in judicial decisions, for generalizations based on past decisions of courts and tribunals in particular cases. In Anglo-American law case-law has been a fundamental, and still is an important factor in the development of the law, and a major source of principles. What is fundamental about case law is not that previous decisions are reported, nor that judges and other adjudicators in later cases look to previous decisions for help or guidance, but that the previous decisions are treated as normative and looked to for principles or rules which by convention should, and in some circumstances must, be followed and applied. Furthermore the higher courts in reaching decisions and issuing opinions do so in the knowledge that their decisions are laying down rules which will be, and sometimes must be, followed in future by later courts, and sometimes issue opinions with the clear intention of laying down a rule for the future, overruling a previous decision deemed wrong, or otherwise settling the law.

The major differences between case law and legislation are that case law results from the accidents of litigation, whereas legislation may be passed in anticipation and deal with general cases, not only the particular case; case law is not expressed in exactly formulated verbal statements; what is authoritative is the principle, not the verbal formulation, though some decisions have been formulated as carefully as many statutes, and have sometimes even been interpreted with the same deference to the words used as statutes have been; legislation on the other hand is verbally authoritative.

Case law is a major source of legal rules in the U.K., U.S.A. and other 'common law' jurisdictions, but in civil law countries it is only a subsidiary source of law.

See also PRECEDENTS; REPORTS; STARE DECISIS; YEAR BOOKS.

Case stated or stated case. A statement prepared by an inferior court for a superior, such as in England a magistrates' court for the High Court, containing a statement of the facts admitted or held to be proved and one or more questions of law concerning the rightness or wrongness of the inferior court's decision, as the basis for the superior court's decision on the rightness or wrongness in law of the inferior court's disposal of the case. It is accordingly commonly the material for decision of an appeal. The final decision is that of the inferior court, but in accordance with the superior court's statement of the relevant law.

Cassation. A mode of review of judicial decisions found in civil law countries under which a decision may be brought up to a superior court and its rightness in law challenged. If the court of cassation upholds the challenge it does not, as in an appeal, substitute its own ruling on law and alter the decision of the court below, but strikes down (*casser*) the decision as incorrect and remits the case to another inferior court of the same grade to decide the case afresh, the latter court not being bound to apply the ruling of the court of cassation.

See APPEAL; COUR DE CASSATION.

Cassin, Rene Samuel (1887–1976). French jurist, national commissioner for justice in the French government in exile (1941–43) and after the liberation of France Vice-President of the Conseil d'Etat (1944–60), president of the École Nationale d'Administration and the Cour Supreme D'Arbitrage and, in 1960, member of the Conseil Constitutionnel. As president of the U.N. Commission of Human Rights he was one of the initiators of the Universal Declaration of Human Rights and later president of the European Court of Human Rights. In 1968 he received the Nobel Peace Prize.

Cassius Longinus, Gaius (first century A.D.). Roman jurist, proconsul of Asia 40–41, legatus of Syria 45–49, and exiled to Sardinia by Nero in 65 (but recalled by Vespasian). He wrote a *Libri iuris civilis* in 10 books. He was a disciple of Sabinus (q.v.) and succeeded him as leader of the Sabinian school, so that it was sometimes called the *Schola Cassiana*, and had a high reputation as a teacher and jurist.

Castberg Frede (1893–). Norwegian jurist, professor at Oslo, member of the Institute of International Law, author of *Problems of Legal Philosophy* (1939), *La Philosophie du Droit* (1970), and many other books mostly on jurisprudence and international law.

Castle Chamber, Court of, in Ireland. A court established by Queen Elizabeth I for the suppression of disorders in Ireland. It was to consist of at least one of the Lord Deputy, Lieutenant Justice or Justices, Lord Chancellor or Keeper or Lord Treasurer of Ireland, and at least two of any of the lords spiritual and temporal and any of the Privy Council or Justices of any of the courts in Ireland, and had power to deal with disorders in the same way as the Court of Star Chamber in England.

Castle guard or castle-ward. Under the feudal system an arrangement whereby the duty of finding knights to guard royal castles was imposed on certain baronies and divided among their knights' fees. Greater barons made similar arrangements with their knights. Both kind of obligations were early commuted into fixed payments, 'castle-guard rent' which lasted for many centuries.

Castro, Paolo di (Paulus Castrensis) (?–1441). An Italian jurist and commentator, professor at Vienna, Avignon, Padua, and elsewhere, and a very famous teacher of canon law, author of *Commentaria in Digestum* (1548), *Consilia* (1582), *Commentaria in Codicem* (1594), and *Singularia* (1596).

Casual ejector. In the former action of ejectment (q.v.) the nominal defendant, Richard Roe, was called the casual ejector because he was supposed casually or by accident to have come on the plaintiff's land or premises and turned out the lawful possessor.

Casualty. In Scots Law, a payment due from a feudal vassal to his superior not regularly but on the occurrence of particular events. In the tenure of ward-holding (q.v.) the casualties were ward, the superior's entitlement to possession of the vassal's estate if and so long as the vassal was under age, to enter on the lands and draw the rents thereof, though the right might be commuted to a fixed sum, in which case it was known as taxed ward, marriage, the superior's right to find a suitable wife or husband for the heir and to claim the marriage portion which the heir might have received; the sum payable was called the avail, and became fixed as three years' rent, or later two years', but if the heir married another against the superior's wishes he became liable to double avail; and recognition which was the total forfeiture of the feu to the superior if the vassal alienated all the lands without the latter's consent.

In all modes of tenure other than burgage (q.v.) the casualties of non-entry, relief, and liferent-escheat were recognized. When feus became hereditary an heir became entitled to claim the lands held by his ancestor on applying to the superior for renewal tendering the customary dues. The superior could enforce his right to have the successor enter by bringing a declarator of non-entry which entitled him to the fruits of the land so long as they were not held by an entered vassal. Relief was the sum payable by the heir of the last vassal who died entered in the feu for the superior's acceptance of his as a vassal, a sum later fixed at one year's feu-duty. Liferent-escheat (as distinct from single escheat, or forfeiture to the Crown of a man's moveable estate, which was part of the penalty of a capital sentence) involved forfeiture to the superior of the annual profits of the vassal's lands during his life or the duration of the sentence when sentence of outlawry had been pronounced against him.

Composition was the payment due to the superior when a vassal sold his rights in lands to a stranger, and came to be fixed at a year's rent. It was a statutory fine rather than a true feudal casualty.

No casualties were due from land held burgage, or udal lands; lands held blench were liable to the same casualties as were feus.

Ward, marriage and recognition were abolished in 1747 and non-entry in 1874; all casualties then existing were, by statute of 1914, to be redeemed by 1935 or no longer exigible.

Casualty insurance. Provision by insurance against such happenings as sickness, accidents, or the incurring of legal liability for negligence, professional negligence, or other chance event.

Casus belli. A situation alleged by a State to justify it in making war. According to the U.N. Charter warlike measures are permissible, apart from any authorized by the Security Council or the General Assembly, only if necessary for individual or collective self-defence against armed attack.

Casus omissus. A case or situation overlooked, omitted and not provided for in a deed or statute. Where it appears that there is a *casus omissus* a court is not entitled to supply the lack or fill the gap.

Casus placitorum (or *Cas de demandes*, or *Cas de jugement*). A tract said to be made up of judgments and decisions of prominent judges, of leading principles, and of rules of procedure, and compiled to show the rulings of judges as to the law of a period before the early legislation of Edward I, and including some passages of *Fet asavar*.

See Woodbine, *Four Thirteenth-century tracts*.

Casus regis. The succession of King John in 1199 on the death of his elder brother Richard I in preference to Arthur son of Geoffrey, a brother intermediate between Richard and John who had predeceased Richard. The incident was regarded as settling the principle that a junior member of an elder generation succeeded in preference to the representative in the next generation of a more senior member, and interrupting the development of the principle of representation, which was, however, re-established by the Great Cause (q.v.) in which Edward I preferred Balliol to Bruce in the contest for the Scottish Throne.

Catching bargain. A bargain for the payment or loan of money made on extortionate or otherwise unconscionable terms with a person having property in expectancy or in reversion. Equity has long granted relief to the borrower in such a case in so far as the bargain is unfair or unreasonable, and the principle has been extended by legislation.

Catholic emancipation. A movement in Great Britain and Ireland in the late eighteenth and early nineteenth centuries to release Catholics from civil and political disabilities. Ultimately Relief Acts for England (1778), Ireland (1791), and Scotland (1793) removed many liabilities and full emancipation was achieved in 1829.

Cathrein, Victor (1845–1931). German philosopher and jurist who derived natural law from Christian revelation and held all positive law contrary to natural law to be void. His principal work is *Recht, naturrecht und positives Recht* (1909).

Cato, Marcus Porcius (?–152 B.C.). Roman jurist, son of Cato the Censor, author of a *Commentarii iuris civilis* based on his *responsa*.

Cato Street Conspiracy. A plot devised by a radical, Thistlewood, and others in 1820 to assassinate the Cabinet, seize London and promote a revolution. The plan was betrayed and Thistlewood and others hanged. The failure diverted radicals to less violent courses of action and helped to reconcile the country to repressive government.

Cattle trespass. At common law the owner of cattle must take care that they do not trespass or stray on the lands of another. An action for cattle-trespass lay at the instance of the party damaged by failure in this duty. This was replaced in 1971 by a right to detain and sell trespassing livestock.

Cauchy, Eugène-François (1802–77). French civil servant, author of *Le droit maritime international* (1862) and other works.

Caulfield, St. George (?–1778). Called to the Irish Bar in 1723, he was successively Solicitor-General (1739), Attorney-General (1741), and Chief Justice of the King's Bench in Ireland, 1751–60, when he resigned.

Causation. The concept of relationship between two events based on the belief that they have not only happened fairly close in time and space but are related as cause and consequence in that the one has brought about or necessitated the happening of the other.

Causation was first investigated by Aristotle who distinguished material, formal, efficient, and final causes, and this distinction was later much elaborated. Hume contended that causal necessity was merely a projection of the habit of expecting certain consequents to follow certain antecedents, on the basis of previous experience. Kant concluded that causality was one of the ultimate *a priori* forms in which the understanding ordered its experience and required to make orderly experience itself possible.

In legal discourse causation is particularly important in relation to questions of assigning liability for injuries, harms, losses, and other happenings. The courts concern themselves not with physical causes but with causation from the standpoint of attributing liability; thus a medical inquiry might say that the cause of a death was multiple fractures and a legal inquiry that the cause of the same death was the defendant driver's negligence, or, more fully, an impact brought about by the defendant driver's negligence or actings in breach of legal duty.

In investigating the potency of certain facts, particularly a defendant's conduct, as causal factors in bringing about a result, courts have frequently distinguished between a fact which is *causa sine qua non*, a mere prerequisite, and *causa causans*, the efficient cause, while a factor supervening on an initial causal factor and possibly replacing it as the efficient cause of the matter under investigation is frequently called a *nova causa interveniens* or *novus actus interveniens*. Thus an injured person's presence at the locus of an accident is mere *causa sine qua non*, in that he could not have been injured if he had not been there, but the *causa causans* is the driver's negligence (breach of duty of care) or possibly the injured person's own carelessness, or possibly both: and if having been injured death is caused in hospital by injection of a wrong anaesthetic, that factor is *nova causa interveniens*.

One event cannot be said to be the cause of another if the latter would equally have occurred without the first. To be legally significant as a cause the first event must have at least been a *sine qua non* or prerequisite of the happening of the second. But not every prerequisite is a significant cause; If X had not been in the car he would not have been injured, but his presence, though prerequisite, is not the cause of his injuries. To be a cause the first event must in some way have brought about the second.

In civil law the problem arises most commonly in the context of liability for tort or delict; was the defendant's conduct, in breach of legal duty, the 'cause' of the injury or harm suffered by the plaintiff?

In practice several causal factors may contribute to a single result; in particular a harmful event may be caused partly by the defendant's breach of duty and partly by the plaintiff's own lack of care (the problem of contributory negligence), or partly by the breaches of duty of two or more defendants (the problem of joint fault). Again, where one act has started a chain-reaction of consequences, another act, of the plaintiff or of the defendant or of a third party, or an event of nature, may supervene (*novus actus interveniens*) and divert or accelerate the chain-reaction so that the ultimate consequence is other than the person who started it foresaw, or could have foreseen. The presumption is that the party who started the reaction continues responsible despite the intervention of a new factor, but it may be so unexpected and so substantial as to be held to have superseded the initial cause and become the main course of the ultimate harm.

In criminal law similar problems arise, in relation to, e.g. the cause of a death.

Causation is a question of both fact and law; whether an act which is a *sine qua non* of some event, which may be a tort or a crime, is also a cause of it is a question of law. Whether the act was a *sine qua non* of the event is a question of fact.

H. L. A. Hart and A. M. Honoré, *Causation in the Law*.

Cause. (1) A legal suit or action. Before 1875 cause was the generic term for civil proceedings both at law and in equity, but excluding statutory proceedings in equity commenced by petition, motion or otherwise, which were known as 'matters'. Since 1875 'cause' has generally been superseded by 'action'. The Daily Cause List at the Royal Courts of Justice lists the cases to be heard on the following day.

(2) That which brings about some consequence or result. See CAUSATION.

(3) In the theory of contract developed by the canon law it was a requisite of validity that there should have been a reasonable and lawful cause, or moral justification, for making the promise. This theory was widely adopted in Europe but in England at least from the sixteenth century it became established that, in general, a contract was unenforceable if there was no consideration (q.v.) for the promise.

Causes célèbres. A work containing reports of decisions of the French courts of the seventeenth and eighteenth centuries of great interest and importance. It comprises 22 volumes by Gayot de Peteval and 15 volumes (*Nouvelles Causes Célèbres*) by Des Essarts. Hence *Cause célèbre*, a cause arousing great public interest and notoriety, such as the Tichborne case (q.v.).

Cause of action. The fact or set of facts which give a person a right to bring an action. A cause of action is sometimes a wrongful act by itself, e.g. trespass, but in other cases, e.g. negligence, a cause of action is not complete until there is both wrongful act and resultant harm. Questions frequently arise as to where, or when, a cause of action has arisen.

Causidicus. A term for the professional pleaders at Rome, who were not scholarly lawyers and indeed rather despised legal learning but orators, knowing only enough law to understand the advice they got from the jurisconsults.

Caution. (1) A warning to be careful. When a police officer suspects a person of having committed an offence he must administer a caution, i.e. warn him that anything he says may be taken down and given in evidence at his trial. Failure to do so may render any statement made inadmissible.

(2) An intimation which may be lodged with the Land Registry to the effect that no dealing with any registered land, or charge may be made until notice has been given to the person lodging the caution.

(3) In ecclesiastical law, a security for the performance of a duty.

Caution (pron. káyshun). In Scots law, security. It has many forms, the commonest now being the requirement that a judicial factor or curator or executor-dative (qq.v.) should find caution for his due administration of the estate in his charge. Caution in such cases is found by a single-premium policy granted by a guarantee company.

Cautionary obligation (pron. káyshunary). In Scots law, an accessory obligation undertaking to guarantee or answer for the default of another, corresponding to suretyship (q.v.) in English law. It takes two main forms, proper cautionary where the cautioner, or cautioners, are bound expressly as such, and improper cautionary, where the principal debtor and his cautioners are all taken bound, jointly and severally, and are, *ex facie* of the obligation, all joint and several debtors, though without prejudice to their rights of relief *inter se*. At common law proper cautioners had the benefit of discussion (*beneficium ordinis*) whereby the creditor had to seek to exact payment from the principal debtor first, but this right has been removed by statute, and the benefit of division, each cautioner being liable for his *pro rata* share only. All cautioners have a right to total relief against the principal debtor and a right of contribution or mutual relief *inter se*. A cautioner's liability is extinguished if the principal obligation is extinguished, by lapse of time, and if the creditor does anything which prejudices their position as cautioners.

Cavalier Parliament. Also called the Pensionary Parliament and the Long Parliament of the Restoration, it was the second Parliament of Charles II. It sat from May 1661 to January 1679. It was initially strongly Royalist in sympathy and reversed much of the legislation of the Long Parliament, but its attitude changed over the years and an organized opposition developed, and it latterly asserted the principle of ministerial responsibility to Parliament, parliamentary control of finance, and the establishment of Protestantism.

Cave, Edward (1691–1754). A printer and publisher who was the first to issue systematic reports of parliamentary debates for the public. In 1731 he started the *Gentleman's Magazine* which from 1732 included reports of the debates of both Houses based on information supplied by Members and friends. After a debate in 1738 in which the House of Commons condemned the practice of reporting debates, Cave resorted to concealing the identity of speakers and the place of debate, as being those of 'The Senate of Great Lilliput'. After further trouble, he discontinued the reports in 1747, beginning again in 1752.

Cave, George, 1st Viscount (1856–1928). Was called to the Bar in 1880, practised on the Chancery side, took silk in 1904 and was made Recorder of Guildford. In 1905 he entered Parliament and in the early part of World War I worked on the Contraband Committee. In 1915 he became Solicitor-General, in 1916 Home Secretary, and in 1918 a Lord of Appeal and a Viscount. He was rated a very good judge. In 1922 he became Chancellor but lost office in 1924 being, however, reappointed while the government again changed later that year and holding the office till 1928. During this tenure of office the great mass of real property legislation of 1925 was passed. As Chancellor, Cave showed the same sober learning and accuracy which he had displayed as a Law Lord.

Mallett, *Lord Cave*.

Caveat. An entry caused to be made in the books of a registry to the effect that no steps of certain kinds, e.g. a grant of probate, may be taken without previous notice to the caveator.

Caveat emptor ('let the buyer beware'). A basic principle in the law of sale of goods to the effect that, in the absence of fraud or breach of an express or implied warranty, the buyer buys at his own risk, relying on his own estimate of the quality and suitability of the goods. Modern legislation has, however, increasingly introduced exceptions to the principle so as to give ordinary consumers protection against sellers of defective goods.

Cay, John (1700–57). Was called to the Bar in 1724 and later became one of the judges of the Marshalsea. He published *An Abridgement of the Publick Statutes in force from Magna Carta to 1738* in 1739, later continued by his son, and *The Statutes at Large from Magna Carta to 1757* (6 vols., 1758) a learned and accurate edition, continued by Ruffhead to 1773.

Cecil of Chelwood, Edgar Algernon Robert Gascoigne-Cecil, Viscount (1864–1958). British statesman, played a large part in drafting the Covenant of the League of Nations in 1919 and was several times British representative to the League. In 1937 he was awarded the Nobel Peace Prize.

Celsus, Publius Juventius (*c*. A.D. 75–*c*. 140). Distinguished Roman jurist, succeeded his father as head of the school of Proculians (q.v.) and was a member of Hadrian's *consilium*. He had a profound knowledge of earlier legal literature and wrote *Epistulae*, *Commentarii*, and *Quaestiones*, and a *Digesta* in 39 books, and was an independent thinker and a severe critic of the views of others.

Celtic law. The Celts were a numerous people who emerged from central Europe in the first millennium B.C. and spread westwards into Gaul and Britain, Spain and south-eastern Europe. They were an ethnic group rather than a political entity.

Knowledge of their social institutions is largely derived from classical writers and ancient Irish literature. The Irish laws are probably the oldest surviving in Europe and are unaffected by Roman and other Mediterranean law. They are the laws of a pastoral people, superimposed on even earlier tradition. Early Irish society was hierarchical and the country was divided into many kingdoms. Kingship was common, most kings being elected, and in Gaul in Caesar's time the power of the aristocracy was predominant. Later this increased and the freedom of ordinary people was diminished by the system of clientism, whereby lesser men put themselves under obligation to a powerful lord in return for protection. In Ireland customary law recognized kings, warrior aristocracy, and craftsmen and freemen farmers. The family was patriarchal and kinship was determined by agnatic descent. Land was vested in the family. The obligations of rights of freemen were defined by customary law. Parties to legal disputes seem to have agreed in advance to accept the rulings of professional jurists. The kinship group was ultimately responsible for the actions of individual members.

In ancient times the Welsh laws had a similar basis to the Irish though they were more complex and have been overlaid by the later Teutonic system. The Welsh laws are all prefaced by a statement that they were enacted by King Hywel Dda, as an organized statement of the ancient customs of the race and henceforth only alterable by a representative assembly. Three main versions of the Code of Hywel have survived, thought to have been compiled from the original version for particular districts. They are commonly called the Venedotian Code, the Goventian Code and the Dimetian Code. They reveal that justice was administered by tribes and that in some areas there were professional judges.

The few traces of Celtic law which have survived in Scotland suggest that society and law was similar to that of Ireland.

The interest of Celtic law so far as known is its revelation of a society and legal systems unaffected by either Roman or Teutonic influences.

E. MacNeill, *Early Irish Laws and Institutions*; J. Cameron, *Celtic Law*; A. Owen (ed.), *Ancient Laws and Institutes of Wales*; W. M. Hennessy and others (ed.), *Ancient Laws of Ireland.*

Censor. A Roman magistracy, first held in 443 B.C., with the function of reviewing the rolls of citizens, equites, and senators, regulating morals and ritually purifying the Roman people. There were two censors, possibly originally elected for a five-year term but from 433 their term of office was fixed at 18 months though held at four-yearly or, from 209, five-yearly intervals. The office enjoyed high prestige and authority and it was the summit of a political career. After 22 B.C. the office became a prerogative of the Emperor.

Censorship. A practice of scrutinizing and, when thought fit, prohibiting and suppressing acts or the communication of ideas. The justifications commonly invoked for the practice have been the security of the state, the prevention of revolutionary or dangerous ideas, prevention of corruption of the young, and the imposition of conformity of views.

It has been practised in different forms by many societies, in ancient China and India, in Israel and Greece. In Greece the death of Socrates resulted from condemnation of free enquiry which led to unpalatable views. Greek philosophers developed the theoretical arguments which have subsequently been used to support censorship of freedom and Greek lawyers set out the practical grounds for censorship. The Hellenistic kingdoms exercised a censorial function in the editing and interpreting of texts of authors. At Rome the office of censor was established in 443 B.C., their functions including regulation of morals. In Augustus' time and later imperial censorship was directed to rooting out treason and checking immorality.

The Renaissance, the Reformation, and the spread of printing had effects on censorship. The theological factions condemned each other's writings and proscribed their study and lists of prohibited books were issued. The Index of Prohibited Books was issued by Pope Paul IV in 1559 and the Index of the Council of Trent appeared in 1564, including general rules for the control of literature settled at that council.

In England a requirement that books should be licensed for printing by the Privy Council was introduced in 1538. Mary Tudor forbade the printing of any book without licence. After 1557, when the Stationers Company was incorporated, only its members or persons holding a special licence might print any work for sale. In 1559 Elizabeth required that no book be published unless licensed by her, the Privy Council, or certain Church dignitaries. The Star Chamber tried to limit the number of printing presses and again forbade the printing of unlicensed works. The licensing system continued under the Commonwealth. Newsletters were frequently suppressed. Under the Restoration control and licensing of the press were continued. The licensing system came to an end in 1694. Thereafter prosecution and punishment of the publishers of seditious libels were commonly resorted to. The famous incident of Wilkes and the North Briton arose from such proceedings.

Though executive interference with the press diminished in the eighteenth century it continued until 1771, in the view of Parliament, to be a breach of privilege to print without permission accounts of the debates and proceedings.

The licensing of dramatic performances began in 1549 and from about 1574 no play might be performed in England without the licence of the Master of the Revels. Towards the end of the seventeenth century he was forbidden to license anything contrary to religion or good manners. In the early eighteenth century control of drama seems to have been assumed by the Lord Chamberlain and in 1737 the Licensing Act made the Lord Chamberlain licenser of theatres and censor of plays.

The criminal law, particularly prosecution for seditious libel, was extensively used between 1790 and 1832 to try to check the spread of dangerous ideas from France but the Reform Act of 1832 removed many of the causes of complaint and the practice of prosecution diminished. In the nineteenth century checks on publication concentrated more on preventing obscene matter. A society for the Suppression of Vice was established in 1802 and an Obscene Publications Act passed in 1857. There were relatively few prosecutions under the latter Act. In 1959 a more liberal Obscene Publications Act was passed. The Lord Chamberlain's control of drama disappeared in 1968 and since about 1950 society has become much more permissive, while only the most extreme seditious material or conduct is now prosecuted but prosecution is frequently made against dealers in pornographic books.

Movements for toleration and freedom developed in the seventeenth century, argued notably in Spinoza's *Tractatus Theologico-Politicus*, Milton's *Areopagitica*, and Locke's first *Letter concerning Toleration*. The toleration movement gained ground with the lapse of the Licensing Act in 1694 and in the eighteenth century with the success of Wilkes, and increasing defiance of the law, by Wilkes and others. In the nineteenth century the question of freedom and control was again considered, notably by Herbert Spencer who argued against government restraint in *Social Statics* and J. S. Mill, who contended for total freedom of expression in *On Liberty*. Moreover wherever there has been censorship there has developed an 'underground' press and organizations for the dissemination of the banned ideas. The Roman Church has abandoned The Index.

In the United States in the colonial period censorship of various kinds was enforced in many colonies. The First Amendment has however been interpreted as prohibiting pre-publication censorship and control of books has been by administrative action or by pressure from individuals and groups. There have been organized movements for decency in literature and the performing arts. Sedition and obscenity have been prosecuted.

In other countries conditions vary, but in Russia strict censorship has been applied in the interest of protecting Communism.

Legally censorship is one aspect of the continuing problem of regulating the respective spheres of freedom and control, in the knowledge that total freedom includes freedom to propagate every kind of filth and corruption, and a perennial problem of public law and human rights.

Controls and restrictions on what may be communicated and published are also imposed by the possibilities of actions for libel and prosecutions for seditious libel (qq.v.).

Centlivres, Albert van de Sandt (1887–1966). South African judge 1922, Judge of Appeal, 1939–50 and Chief Justice of South Africa, 1950–57. He was Chancellor of the University of Cape Town and highly regarded as a judge.

Central Arbitration Committee. A body established in 1975, replacing the Industrial Arbitration Board (originally the Industrial Court), consisting of a legal chairman and one representative of each of employers and employees, not being connected with the industry in the case before the committee. Though the members are nominated by the Advisory, Conciliation and Arbitration Service (q.v.) it is independent of that Service, the Department of Employment, and the Government, and acts in judicial form and with judicial independence. It has the principal function of arbitrating with the consent of both parties in industrial disputes, but in certain circumstances it can act compulsorily or unilaterally.

Central Criminal Court. The Central Criminal Court was constituted in 1834, superseding the sessions long held at the Old Bailey (q.v.) under special commissions of gaol delivery for Newgate and of oyer and terminer for London and Middlesex. In practice it superseded quarter sessions for London and Southwark. It is now the sitting of the Crown Court for London and sits under the general powers applicable to that court (q.v.).

The judges of the court are the Recorder of London, the Common Serjeant, additional judges of the Central Criminal Court, any judge of the High Court, and any circuit judge or recorder. The Lord Mayor or any of the Aldermen of the City of London may sit with the trial judge. Sittings are held regularly and the court may sit in more than one division simultaneously. In practice the Recorder, Common Serjeant and additional judges sit regularly, and judges of the Queen's Bench Division attend to try the most serious cases.

It has jurisdiction to try all indictable crimes committed within Greater London, and any offences committed on the high seas or elsewhere within the jurisdiction of the Admiral of England, and also homicides of and by persons subject to military law where the Q.B.D. so orders. Trial is on indictment only. It has no appellate jurisdiction.

Central Office. The Central Office of the Supreme Court of Judicature established in 1879 to combine the offices of the Common Law Divisions, the Crown Office of the Queen's Bench Division, and the offices of the Chancery Division.

Centumviri. A civil court in ancient Rome, constituted possibly about 150 B.C. Originally there were 105 judges (three from each of the 35 tribes) but under Augustus this was increased to 180. In the first century B.C., they were elected by the *comitia tributa* and later selected by lot. The court was presided over by ex-quaestors but later by *decemviri stlitibus iudicandis*. The praetor who heard the proceedings *in iure* decided whether trial was to be before a single judge or before the centumviral court. The court dealt with a wide range of civil claims, particularly as to inheritance. It disappeared in the third century A.D.

Ceorl. The free peasant at the base of Anglo-Saxon society. Originally he paid dues to the King, owed suit of court and in wartime had to serve in the fyrd. In time the class became depressed and had to give labour services to lords and after the Conquest many were reduced to the status of unfree villeins. The wergeld for slaying a ceorl was one-sixth of that of a thegn. The variant form of the word, churl, was associated with a subject and depressed peasant and early acquired the connotation of a rude, coarse fellow.

Certainty of law. Involves two issues: how far the law should be, and how far the law is, certain. Certainty means the possibility of accurate prediction of the legal results and implications of a particular course of action, and the more certain the law the easier it is to plan conduct and advise on conduct. But certainty does not automatically mean or imply justice; indeed certain rules may be unjust by taking too little account of the circumstances in which, or the mental states with which, something is done. A rule that to cause an abortion is always criminal is certain, but not necessary always just.

Certainty is not the same thing as fixity or rigidity of rules. A rule may be certain but not fixed or rigid in that it may leave room for regard to circumstances. There is need for some flexibility in rules and for some scope for the exercise of discretion. But reasonable certainty is far preferable to total uncertainty, or decisions depending on the whim of the judge or on whatever he chose to think, right or wrong.

The need for certainty varies from one branch of law to another. For reasons of business convenience or security of rights in property rigid rules are frequently necessary. In matrimonial and personal relations, decisions must have regard to the particular circumstances of different persons in different situations. In substantive criminal law there is need for considerable certainty so that people can know what is punishable and what is not. But it is very plain that no system of law is completely certain, nor can one ever be wholly certain, for many reasons. Frequently the principle or rule laid down is uncertain, unclear or ambiguous by reason of drafting or verbal uncertainty. This is particularly so where the principle has to be derived from the *rationes* of a number of decided cases. Sometimes it may be very doubtful which of several possible rules should be applied to a set of facts. The principle or rule may be uncertain because of unforeseen or unprecedented circumstances, different from any contemplated when the rule was formulated, or because of changes in social philosophy, views or attitudes.

The result of a litigation may be uncertain because it is always doubtful how evidence will come out and whether it will convince or influence a court or jury. Some evidence may be lost or inadmissible.

Belief in absolute certainty is accordingly an illusion. Legal advice must consequently frequently be a statement or prediction of probabilities only, not of certainties.

Certification. In Scots law, a warning to a party to judicial process of the course which will follow if he disobeys the will of the summons or other writ, or the order of the court. The certification in a summons is now merely that if the defender ignores it the pursuer may proceed to decree in absence against him. In an action of reduction the certification is that the deed called for will be held void until produced.

Certiorari. Formerly a prerogative writ, now an order, which may be granted by the High Court in England, directed to an inferior tribunal, requiring the record of the proceedings in some cause to be transmitted to the High Court, to be dealt with there, so that the applicant may have speedy justice. It originally issued from Chancery only, but from the sixteenth century the Queen's Bench had established an equal right. Like the orders of mandamus and prohibition (qq.v.), the main use of *certiorari* is a means of controlling inferior courts and persons and bodies having authority to determine issues affecting the rights of individuals and having a duty to act judicially in so doing, such as administrative tribunals, but it may be used to remove an action into the High Court for trial there, and in certain other cases.

It lies only to persons and bodies having judicial or quasi-judicial functions and duties, including, in certain respects, magistrates' courts, not to bodies exercising administrative functions, and where there appears on the face of the proceedings that the inferior body has acted without jurisdiction or determined an issue wrongly in law, but not on the ground that it had misconceived a point of law if it had jurisdiction and the proceedings are *ex facie* regular, nor on the ground that its decision is wrong in fact. It is accordingly not an appeal. It may be used to quash a decision if the person or body deciding failed to observe the rules of natural justice, or acted without or in excess of jurisdiction.

In the U.S. *certiorari* is used to review questions of law or to correct errors or excesses by lower courts, and in exceptional cases where an immediate review is required.

Cess. A shortened form of 'assess', a term formerly applied in Scotland to land-tax and elsewhere to local taxes.

Cessio bonorum. In Roman law and old Scots law the surrender by an insolvent of his property to his creditors as the only relief from imprisonment for debt. In Scots law it took the form of an action in court in which all creditors were called. With the general abolition of imprisonment for debt in 1880 it became largely unnecessary and was wholly superseded from 1913 by procedure for sequestration.

Cessio in jure. In Roman law, a means of transfer of property by fictitious action, in which the person who was to acquire the thing claimed it. The other party acknowledged the justice of the claim, and the magistrate adjudged it to be the property of the claimant.

Cession. In international law, the transfer of sovereignty over a tract of territory by one state to another. It is effected by treaty and the cause of ceding may be exchange, gift, sale, lease, pledge, merger of the ceding state with the other, or surrender after defeat in war. The treaty of cession must be completed by actual transfer to and the taking of physical possession by the other state. Third-party states have no title to object to cession, save where the cession has been exacted by force or illegal means, though they may object for political reasons. Persons domiciled in the ceded territory automatically become subjects of the acquiring state, unless the treaty of cession gives them the option of retaining their former citizenship, in which case they may be expelled from the ceded territory.

Cestui que trust (pron. setty-key-trust). The person who has the equitable right to property, the legal title to which is vested in a trustee, and who is the beneficiary of the trust.

The plural is *cestuis que trust*, not *cestuis que trustent*.

Chafee, Zechariah (1885–1957). After a short period in practice, was a Professor of Law at Harvard, 1916–56, best known for his advocacy of civil liberties and author of *Freedom of Speech* (1920), *Free Speech in the United States* (1941), *Government and Mass Communications* (1947), *Documents on Fundamental Human Rights* (1951), *Freedom of Speech and the Press* (1955), and many other works.

Chaffwax. An officer of the old Court of Chancery whose function was to chafe or heat and affix the wax used to seal writs, commissions, and other deeds. The office was abolished in 1852.

Chain of causation. A series of events each caused by the preceding one, such as accident–injury–depression–maltreatment–suicide. The question frequently arises of how far the person responsible for the original event should be held liable and of whether at any point some independent factor has intervened to make later events the result of it rather than of earlier events.

See CAUSATION; NOVUS ACTUS INTERVENIENS.

Chain-gang. A former American method of controlling convicts, particularly on road work, quarrying, and similar work in the Southern states, so-called because the men in the gang had an iron cuff riveted round the right ankle to which was attached a chain. When going to and from work and confined at night the other end of the chain was attached to a master chain. Sometimes chains linking both ankles were used.

Chairman of Committees, House of Lords. A peer who, at the start of each session, or on the occurrence of a vacancy, is appointed Chairman and first of the Deputy Speakers in the House of Lords. He takes the chair in all committees of the Whole House, and is also Chairman of all other committees of the House, unless the House directs otherwise. He has an important function in the supervision of Private Bills.

Chairman of Ways and Means. The Member of the House of Commons who presides when the House goes into Committee of the Whole House. He is also the Deputy Speaker and has the same functions and powers as the Speaker but cannot inflict any serious punishment. Formerly he was Chairman of the Committee of Ways and Means only, but since 1841 he has acted as Chairman of all other committees of the whole House also. He is appointed at the beginning of each Parliament, normally from the government party. He also has responsibilities in connection with Private Bills.

Since 1902 there has been also a Deputy Chairman who may exercise all the Chairman's powers in his absence.

Chaldron, chaldern or chalder. A customary measure formerly used for coal, the weight varying in different parts of the country. Legally a chaldron was a measure of capacity, being 36 bushels each of eight gallons. In Scotland a chalder was a measure of quantity, amounting to six bolls, used in measuring ministers' stipends.

Chalking the door. A former Scottish mode of giving notice of eviction. A burgh officer in the presence of witnesses chalked the door 40 days before the term at which the tenancy was to end and made out an execution or certificate of chalking. It has been made obsolete since at least 1907.

Challenge. An objection taken to jurors. It may be a challenge to the array, or objection to all the jurors returned collectively, for some fault in the sheriff who arranged the panel. Such a challenge may be a principal challenge, on such ground as that the sheriff is related to a party, or be a challenge for favour, showing ground for believing that there may be favour or partiality to one party. Alternatively it may be a challenge to the polls, i.e. to individual jurors, which in turn may be a peremptory challenge, without cause shown, which is only allowed in criminal trials and limited in numbers, normally to not more than seven jurors, or challenge for cause, which may be made to any number.

Chalmers, Sir Mackenzie Dalzell (1847–1927). Was called to the Bar in 1869 and after service as an Indian civil servant entered practice. In 1878 he published his *Digest of the Law of Bills of Exchange* which was the basis of the Bills of Exchange Act, 1882. He was a county court judge from 1884 to 1896 and while thus acting turned to the task of codifying the law of sale of goods, work which resulted in the Sale of Goods Act, 1893. In 1896–99 he was legal member of the Viceroy's Council in India, where he was concerned with revision of the code of criminal procedure. In 1899 he became assistant and in 1902 First Parliamentary Counsel, and in 1903 Permanent Under-Secretary of State for the Home Department. In 1906 his work on the subject resulted in the Marine Insurance Act, 1906. He retired in 1908, but thereafter sat on various commissions and was a delegate to conferences. He published various editions of his *Digest of the Law of Bills of Exchange*, *The Sale of Goods Act, 1893*, *A Digest of the Law of Marine Insurance*, all of which are still standard books on their subjects, and various other works.

Chamavian Code. This code, known as the *Lex Francorum Chamavorum*, in 48 articles, is a memorandum made by royal judges sent among the Chamavi, a Frankish people, by Charlemagne.

Chamber, King's (or *camera regis*). In England, part of the royal household and the earlier royal financial office. In the Anglo-Norman period a distinction developed between the *camera* or bedchamber, the *camera curiae*, which was the financial office of the household, and the exchequer or great office of account. Later the *camera* developed also administrative and secretarial functions and acquired a seal of its own, the small or privy seal. After a period of diminished importance the chamber was revived in the fifteenth century and from about 1487 the treasurer of the chamber was the chief financial officer of the Crown. The chamber's importance was reduced in the 1530s and from about 1550 was merely a household department.

Chamberlain. Originally an officer charged with domestic affairs. The *camerarius* of the Carolingian emperors had charge of the royal treasury. In England after the Conquest the Chamberlain had charge of the administration of the royal household and as such had considerable financial responsibility. In time the office became the hereditary and sinecure offices of *magister camerarius* or Lord Great Chamberlain (q.v.), the domestic office of *camerarius regis*, or Lord Chamberlain (q.v.) and the Chamberlains of the Exchequer.

A similar office survived in France down to the nineteenth century. The Apostolic Chamberlain in the papal curia is head of the treasury and during a vacancy in the papacy is head of the administration of the Roman Church.

In British local authorities the name chamberlain has frequently been given to municipal treasurers.

Chamberlain ayre. The Chamberlain, as chief financial officer of the Scottish kings from the twelfth to the fifteenth centuries, held an ayre or circuit court, to ascertain whether burgh magistrates were doing their duty, trading regulations being observed, and rents and duties being paid. The court seems also to have dealt with some cases affecting burghal law. The last ayre was in 1512 and the Chamberlain's powers vanished shortly thereafter.

Chamberlain, Great, of Scotland. A high officer of state, Keeper of the King's Treasury until the office of treasurer was introduced. He had a universal jurisdiction over burghs in matters of police, customs and law, and weights and measures, and held circuits, or chamberlain-ayres, in the burghs to examine their accounts and see that their revenues were properly expended. Appeal lay to him from the burgh courts and from the chamberlain-ayre to the Chamberlain Court or Court of the Four Burghs, of which he was convener and president. There was a revision of the Chamberlain laws as late as 1633 but within the next 50 years the office had ceased to exist.

G. Crawfurd, *Officers of State in Scotland*; *Handbook of British Chronology*, 177.

Chamberlain, Lord. An officer of the royal household responsible for the care and maintenance of the royal premises. Under statute he formerly had control of theatres in London and of the licensing of plays.

Chamberlain, Lord Great. An hereditary officer having certain ceremonial duties at coronations, duties of care for the Houses of Parliament and of introducing peers and bishops when they first take their seats in the House.

Chamberlain's court. The Chamberlain of the City of London may hold a court, the judges being the Chamberlain and the Controller of the Chamber (the Vice-Chamberlain) to determine disputes between masters and apprentices. Appeal lies to the Recorder of London and a jury in the Mayor's and City of London court.

Chambers. The private rooms of a lawyer. At the English Bar groups of barristers rent a set of chambers, usually in one of the Inns of Court, which they use collectively as their office, studies and work-rooms and library. Provincial chambers are chambers in major towns other than London, e.g. Liverpool and Manchester, where there is a local Bar. Each chambers has a clerk who acts as intermediary between the barristers and solicitors wishing to brief one of them and arranges their work. Some chambers specialize in particular branches of legal work.

The term is also applied to a judge's private room adjacent to the court room. Certain kinds of business may be, or are directed to be, dealt with in chambers and accordingly in private and less formally than in court. The term is given also to the rooms in which the Masters of the Supreme Court sit to dispose of business which does not need to be dealt with in open court.

Chambers of the King. Certain bays on the coast of Britain falling within lines drawn from one promontory to another and traditionally held to be within the territorial jurisdiction of the Crown.

Chambers, Sir Robert (1737–1803). Succeeded Blackstone in the Vinerian Chair at Oxford in 1766. In 1773 he was made second judge of the Supreme Court in Bengal but did not resign the Chair till 1776. In 1789 he became Chief Justice and in 1797 president of the Asiatic Society. He wrote a work on *Estates and Tenures* published in 1824 which is readable but not a classic.

H. G. Hanbury, *Vinerian Chair*.

Chambers* v. *Florida (1940), 309 U.S. 227. A group of negroes had been convicted of murder and robbery on the basis of confessions extorted by pressure of police questioning. The Supreme Court unanimously asserted the right of the defendants to the protection of the Fourteenth Amendment against conviction on the basis of extorted confessions.

Chambre ardente. An extraordinary court of justice in France mainly held for the trial of heretics discovered by inquisitorial tribunal. The first such court was held in 1535; it was abolished in 1682.

Chambre des Comptes. In France under the *ancien régime* a sovereign court which separated from the King's court in 1320 charged with many aspects of the financial administration of the country but also having administrative and legislative functions. In 1807 it was replaced by the *Cour des Comptes*.

Chambre des Enquêtes. In France before the Revolution a chamber of the *Parlement* of Paris, organized in the fourteenth century, responsible for conducting investigations ordered by the Grand Chambre of the *Parlement*. It was abolished at the Revolution.

Chambre des Requêtes. In France under the *ancien régime* a chamber of the *Parlement* of Paris responsible for examining the petitions of parties seeking to bring a case before the *Parlement* and for acting in certain cases as a court of first instance. It disappeared at the Revolution.

Champerty (Lat. *campi partitio*). A kind of maintenance (q.v.), namely maintenance of an action in consideration of a promise to give the maintainer a share in the subject-matter sued for or the proceeds thereof. It is illegal both by common law and statute.

Champion. One who fights on behalf of another. In trial by battle (q.v.) clergymen, women, children, and persons disabled by age or infirmity might nominate champions to represent them. In the case of bishops and abbots this function was part of the duties of the *advocatus*.

The King's Champion is an hereditary office dating from the fourteenth century connected with royal coronations. Originally the King's Champion rode, clad in armour, into Westminster Hall during the coronation banquet and thrice threw down the gauntlet challenging anyone who disputed the King's title to rule to engage with him in mortal combat. This was last done in 1821. There is no record of the challenge ever being taken up. Since 1902 the King's Champion has carried the standard of England.

Chance. That which is unintended and unplanned, an accident. Causing harm by chance is not criminal, nor even tortious, unless the chance was so substantial that precautions should have been taken against its occurrence.

Chancellor (*cancellarius*). A title given in many European countries to various officers, usually of an administrative, secretarial, or legal character. Roman *cancellarii* were minor legal officials later employed in the imperial writing offices. Hence later the chancellor was commonly head of a chancery or office of a secular or ecclesiastical dignitary; he was usually an ecclesiastic. In mediaeval European kingdoms he developed into an important official, having custody of the great seal used to authenticate royal documents. In England he became head of the Court of Chancery and ultimately of the whole judiciary and also chairman of the House of Lords. The office was abolished in Austria (1806), France (1848), and Spain (1873). In Germany and Austria the title has attached to the office of Prime Minister. The title is also attached in the U.K. to the titular head of a University, the Vice-Chancellor being the working head. In Scotland the chancellor of a jury is the spokesman or foreman who announces the verdict.

Chancellor, Lord High. See LORD CHANCELLOR.

Chancellor of the diocese. The title of the person who, in every diocese of the Church of England, holds the court of the archbishop or bishop, as his official principal, for the trial of ecclesiastical causes in the diocese. In the Diocese of Canterbury the official principal is called the commissary. He must be a barrister or a person who has held high judicial office.

Chancellor of the County Palatine of Durham. The sole judge of the former Chancery Court of the County Palatine of Durham which exercised the old jurisdiction of the High Court of Chancery within the county palatine. The court was merged with the High Court in 1971.

Chancellor of the Duchy of Lancaster. The estates and jurisdiction of the Duchy and County Palatine of Lancaster have been vested in the Crown since 1399. The Chancellor nominally controls the estates, though in practice the Vice-Chancellor and his staff do the work. The Chancellor, however, has certain powers which in other counties fall to the Lord Chancellor or the Home Secretary. The office is now one of ministerial, frequently of Cabinet, rank but with only slight departmental duties which leaves the Chancellor free to perform chairmanship of government committees or other duties which may be given him. Nominally he was judge of the Chancery Court of Lancaster, the judicial duties of which were performed by the Vice-Chancellor.

Chancellor of the Exchequer. The Minister in charge of finance and taxation in the United Kingdom. The Chancellor was originally clerk to the Lord Chancellor and was appointed in the twelfth century to have charge of the Seal of the Exchequer, and also had judicial functions in the Exchequer account. The meetings of the Treasury Board, which consisted of the King, the Lords of the Treasury, and the Chancellor of the Exchequer, became formal in the nineteenth century, and ceased in 1856, and the Chancellor became steadily more important. The Chancellor is now always one of the Commission of the Treasury and also holds the office of Under Treasurer.

His office has become of greatly increased importance as taxation has increased and the national financial position became of greater importance. He is now responsible for national revenue and expenditure, taxation, economic planning, public investment, and international co-operation in financial and economic matters. Each year, in April, he presents to Parliament his 'Budget', reviewing the country's financial position and making his proposals for changes in taxation to meet the requirements of government for the following year.

He is now assisted by the Chief Secretary to the Treasury, and the Parliamentary and Financial Secretaries to the Treasury, the former of whom is the Government Chief Whip. There have also been at various times other Ministers with related duties such as the Minister for Economic Affairs.

Chancellor's court. Cambridge University had a Chancellor's court with civil jurisdiction in personal actions to any amount and criminal jurisdiction, both latterly excluded in cases where a person not a member of the university was a party. It appears to have fallen into abeyance since the mid-nineteenth century.

Oxford University has a chancellor's court with jurisdiction in contracts and trespasses in cases where a member of the university is a party. It is to meet from time to time as determined by the Vice-Chancellor before the Vice-Chancellor or his deputy assisted by the proctors if they wish to attend. The vice-chancellor's deputy, the 'assessor', must be a barrister, and there is a registrar. This court only rarely sits.

Chancellor's foot. Selden (q.v.) in the seventeenth century, wrote that 'Equity is a roguish thing, for law we have a measure to know what to trust too. Equity is according to the conscience of him who is

Chancellor as it is larger or narrower so is equity. Tis all one as if they should make the standard for the measure we call a foot to be the Chancellor's foot.' The phrase accordingly signifies the highly personal character of equity in the seventeenth century. By 1818 Lord Chancellor Eldon could say 'Nothing would inflict upon me greater pain, in quitting this place, than the recollection that I had done anything to justify the reproach that the equity of this court varies like the Chancellor's foot'. The change indicates how by 1818 equity had become a regular body of principles almost as settled as those of law.

Chance-medley (French *chaude melée*). A casual affray in which a man is assaulted in a sudden brawl and, having declined further fighting, kills his assailant in self-defence.

Chancery. The common name in Western countries for the office of the Chancellor, which was also frequently the repository of public records and public archives. The apostolic Chancery is the bureau of the Roman Catholic Curia responsible for the preparation and preservation of official documents.

In England the Chancery was originally the office for the issue of writs, including original writs by which all common law actions were originally commenced. The Chancellor had the duty of sealing charters and other instruments passing the Great Seal and when questions arose proceedings were taken before the Chancellor. This was the basis of the ordinary or common law or 'Latin side' jurisdiction. The chief part of the common law side was its offices, notably the Petty Bag Office, the Hanaper Office and the Enrolment Office.

There also developed the extraordinary or equitable jurisdiction, or 'English side', of the Chancery which became the Court of Chancery (q.v.) or of equity.

In Scotland the Chancery was the office in which were recorded all charters, patents of dignities, gifts of offices, brieves, returns, and other writs passing the great or quarter seals. The Sheriff of Chancery had jurisdiction in all questions relating to the service of heirs.

Chancery, Court of. Under the Norman and Angevin kings the Chancellor was the secretary of state for all departments and the Chancery was his departmental office. He acted in close touch with the *Curia Regis*, from which the Chancery separated from 1238. The Chancellor was Keeper of the Great Seal and royal justice was initiated by original writs sealed by the Chancellor. Similarly applications for justice and petitions to King, Council, or Parliament had at some stage to apply to the Chancery for a writ.

The power to issue original writs gave the Chancellors wide powers over the remedies granted by the royal courts, but the growth of Parliament practically excluded the power to create new writs, though the Statute of Westminster II (*In consimili casu*) gave the Chancellor power to vary the forms of writs so that justice might be done in cases similar to those previously recognized. This facet of the Chancellor's work became more ministerial with the growing power of Parliament and the courts of common law, and his jurisdiction became chiefly connected with writs affecting the King's interest. This came to be known as his common law jurisdiction.

The Chancellor and his Chancery had always been closely associated with the King's Council, which retained some extraordinary judicial powers even after the common law courts had branched off from it. In the fourteenth and fifteenth centuries it became apparent that an exercise of extraordinary jurisdiction was necessary to correct the injustices and mitigate the rigidities of the common law. It was common for petitions for relief to be addressed to the King or the Council, particularly when a remedy was sought on grounds of justice. The Chancellor was close to the King, and in charge of the Chancery which had the machinery to give a remedy if thought fit. From the latter fifteenth century there was a tendency to delegate such matters to the Chancellor and in the next century petitions were frequently addressed to the Chancellor and Council or even to the Chancellor alone. The Chancery naturally tended to assume some of the characteristics of a court. In 1340 it is mentioned in statute as a court. The tendency was for the Chancellor to act as a judge, first as delegate of Council or Parliament, acting sometimes with the advice of councillors or common law judges, but in 1474 the Chancellor is recorded as making a decree by his own authority.

In the fifteenth century the Chancellor and Council had jurisdiction over cases outside the common law, such as concerning alien merchants, maritime or ecclesiastical law, cases where the common law gave a remedy but the courts could not act because the country was disturbed or the defendant too powerful, and cases which the ordinary courts could not dispose of by reason of deficiency in the law.

Under the Tudors the judicial powers of the Council came to be exercised in the Court of Star Chamber and from the early fifteenth century the Chancery had become a distinct court, with both a common law and an equitable jurisdiction, the latter being the more important. In the next couple of centuries its work increased greatly while its equitable jurisdiction was becoming settled. Parallel with this development came a change in the kind of persons acting as chancellors. In the Middle Ages, they were mainly statesmen and ecclesiastics, such as Wolsey, but Sir Thomas More and his two immediate successors were lawyers. The last ecclesiastical Chancellor was Bishop Williams (1621–25) and from the accession of Elizabeth most were lawyers, and frequently distinguished lawyers such as Nicholas Bacon, Bromley, Egerton, and Francis Bacon. Egerton and Bacon settled the relations of the court and the common law courts, and its procedure.

During the Commonwealth the abuses of the Chancery were heavily attacked and a bill was introduced for its reform. An Ordinance of 1654 was put into force but was unworkable, and the Restoration brought back the old system, which continued with little change until the impeachment of Lord Macclesfield. A commission of enquiry into the officials of the courts and their fees in 1740 had no results.

A longstanding source of complaint was the abuses among the official staff, who were paid by fees. This gave rise to sinecure offices, multiplication of useless papers and fees, delay, and corruption. The office of Master in Chancery was bought and sold at a steadily rising price. In the nineteenth century great reforms were introduced in the official staff, many useless offices being abolished.

Another longstanding defect of the Court was the inadequate judicial staff which remained at one from the time the court first emerged, despite a steady increase in the volume of business. In 1688 the idea of putting the seal permanently into commission was seriously considered. Cases might be heard at any stage before the Master of the Rolls and reheard before the Lord Chancellor and be reheard by him, and before the House of Lords, and at any stage before the House of Lords fresh evidence could be introduced.

Under Lord Chancellor Eldon the delays became intolerable, partly because of his dilatory habit of mind, partly because of his desire to consider every case carefully so as to be sure that he did justice.

In 1813 a Vice-Chancellor was appointed, but could decide only cases delegated to him by the Chancellor, and appeal lay to the Chancellor. In 1831 the bankruptcy jurisdiction was transferred to a Chief Judge in Bankruptcy assisted by three judges and six commissioners, with appeals to a court of Review and then to the Chancellor. In 1833 it was provided that the Master of the Rolls should sit continuously and his jurisdiction was extended. In 1842 the equity jurisdiction of the Court of Exchequer was transferred to the Court of Chancery, and two additional vice-chancellors were appointed. In 1851 a Court of Appeal in Chancery was created to hear appeals from the Vice-Chancellors and the Master of the Rolls, with a further appeal to the House of Lords. It consisted of two Lord Justices in Chancery and the Lord Chancellor, if he chose to sit, and might be assisted by the Master of the Rolls, the Vice-Chancellors or any of the judges. In 1869, the bankruptcy jurisdiction was further transferred to the London Court of Bankruptcy, the judges being the Chief Judge in Bankruptcy, assisted by such other judges of the superior courts as the Chancellor might appoint, with appeal to the Court of Appeal in Chancery.

By the Judicature Act of 1873 the Court of Chancery, the London Court of Bankruptcy and the common law courts were consolidated into the Supreme Court of Judicature, comprising the High Court and the Court of Appeal. To the High Court was assigned the jurisdiction of the Court of Chancery including the jurisdiction of the Master of the Rolls, and of the London Court of Bankruptcy. For convenience the High Court was divided into five divisions, to one of which, the Chancery Division, were assigned the kinds of matters formerly dealt with in the Court of Chancery. To the Court of Appeal was assigned, *inter alia*, jurisdiction of the Lord Chancellor and the Court of Appeal in Chancery, both as a court of equity and as a court of appeal in bankruptcy.

The jurisdiction of the Chancellor and the Court of Chancery was of three kinds, common law, equitable, and miscellaneous. The common law jurisdiction was never important. It covered jurisdiction connected with the issue of and procedure on various writs, cases in which the King or a grantee from the King was concerned, and personal actions by or against officers of the court.

The equitable jurisdiction was much more important. The distinction between strict law and modification on grounds of equity, morality, or justice is common but only in England did the two become distinct bodies of rules administered by separate courts. In the thirteenth and fourteenth centuries the discretionary remedies for the defects of common law developed into a body of principles though at that stage equitable principles were still administered by the judges of all the common law courts, but by the end of the fourteenth century this had ceased. The degree of fixity early acquired by common law principles confined the administration of equity to the Court of Chancery, and the narrowness and technicality of the common law made the Chancellor's intervention to give justice necessary.

In the sixteenth and seventeenth centuries the Chancery jurisdiction was concerned mainly with the recognition and protection of uses and trusts, with the grant of specific performance of contracts, with remedies for fraud, forgery, duress, and undue influence, with remedies in cases where the common law could not act at all, or not with effect, and with special remedies such as injunctions, and the taking of accounts.

A sharp conflict between the common law and

equitable jurisdictions developed in the late sixteenth and early seventeenth centuries, principally out of the practice of the Chancery granting injunctions against proceeding under a common law judgment. In 1616 James I made an order in favour of the Chancery view, largely settling the controversy, but it revived after the Restoration.

In the late seventeenth, eighteenth and early nineteenth centuries the principles of equitable jurisdiction were settled; the principles on which the Chancellors acted became transformed from the Chancellor's personal idea of fairness and justice to increasingly settled principles. Cases were reported, and increasingly cited as precedents; discretion, it was held, was to be exercised in accordance with precedent. A number of great Chancellors, notably Nottingham, Hardwicke, and Eldon, did much to systematize and settle the principles on which the court acted, and define the circumstances in which remedies would be granted, so that by Eldon's time it could be said that the doctrine of the court ought to be as well settled and as uniform almost as those of the common law. Consequently in the nineteenth century, many of the doctrines of equity could be formulated in codifying Acts.

The miscellaneous jurisdiction of the court arose from the common practice of conferring special statutory jurisdictions on the Chancellor; the most important were the jurisdiction in bankruptcy conferred finally by statute in 1732, and greatly enlarged by the Chancellors thereafter, and transferred to the London Court of Bankruptcy in 1869, and the jurisdiction in lunacy, which developed from delegation by the Crown to him of the Crown's powers and duties over persons of unsound mind; this was altered in the nineteenth century and is now vested in the Lords Justices of the Court of Appeal.

Under the Jurisdiction Acts 1873–75 the Court was abolished but its jurisdiction continued to be exercised by the Chancery Division of the new High Court of Justice.

W. S. Holdsworth, I, 395; D. M. Kerly, *Historical Sketch of the Equitable Jurisdiction of the Court of Chancery*; G. Spence, *Equitable Jurisdiction of the Court of Chancery*.

Chancery, Court of, in Ireland. A chancery was established in Ireland in 1232 and seems to have developed on lines similar to those of the chancery in England. The Chancellor and Master of the Rolls developed an equity jurisdiction.

In 1867 a vice-chancellor was appointed but the office disappeared under the Irish Judicature Act of 1877, being replaced by that of Judge of the Chancery Division.

Chancery Court of the County Palatine of Durham and Sadberge. A court having within the County Palatine the jurisdiction of the Chancery Division. It was merged in the High Court in 1971. The judge was the Chancellor of the County Palatine.

Chancery Court of the County Palatine of Lancaster. A court having in the County Palatine the jurisdiction of the Chancery Division. It was merged in the High Court in 1971. The judge was the Vice-Chancellor of the County Palatine.

Chancery Court of York. The court of ecclesiastical appeal from diocesan courts within the ecclesiastical province of York. The judges are the Auditor, or Official Principal, who is also Dean of the Arches (q.v.) and four other persons.

Chancery (Scotland). This was the department of the Lord Chancellor of Scotland, but unlike the Chancellor in England he developed no court for redress of grievances which could not be redressed by common law process. For a time before 1532 he presided in the Court of Session. Originally issues of property and possession were tried on brieves issued from Chancery and directed to an inferior judge and this continued in respect of service of heirs after brieves had been superseded by summonses as modes of initiating actions.

The office of Sheriff of Chancery was created in 1847 to deal with petitions for service as heir of a deceased owner of lands, a jurisdiction co-ordinate in most cases with that of the sheriff of the county where the lands lay. With the abolition of the special position of heir in heritage in 1964 petitions for service are disappearing. The Director of Chancery issues certain crown writs and commissions.

Chandler, Richard (?–1744). Compiler of a collection of House of Commons debates from the Restoration published in 1742–44 in 14 volumes, and an important source of Cobbett and Wright's *Parliamentary History*.

Channel Islands. The Channel Islands are part of the British Islands (q.v.) but neither parts of the United Kingdom nor colonies thereof. They are the only portion of the Duchy of Normandy still possessed by the Crown as successors to the Duke of Normandy who acquired the Crown of England in 1066 and subsequent years. When continental Normandy was lost by the English Crown they were retained, and still retain the ancient custom of the Duchy as their law. They consist of two independent bailiwicks, Jersey and Guernsey, to the latter of which Alderney and Sark are dependencies.

Legislation of the United Kingdom Parliament can extend to the Channel Islands but must be made applicable to one or all by express words or by necessary implication. In such a case the practice is to transmit it to the Royal Court of the Island for registration. Legislation may also be effected by Order in Council.

Jersey

The Lieutenant-Governor, appointed by the Crown, may refuse assent to minor legislation, subject to the consent of the Home Secretary. The Crown also appoints the Bailiff, who is chief magistrate and judge, presides over the States, and is the channel of communication between the Lieutenant-Governor and the administrative departments. The States consist of the Bailiff, the Attorney-General, Solicitor-General, the Dean of Jersey, 12 senators, 28 deputies, and the 12 constables of the twelve parishes. The Bailiff has a casting vote and the Dean no vote. The States may enact, amend, and repeal provisional laws having force for three years, so long as not repugnant to superior legislation, nor affecting the royal prerogative, without the necessity of the royal assent, and may renew them if relating to purely municipal and administrative topics. These laws, and laws more fundamental, if assented to by the Queen in Council, have permanent validity.

The Royal Court consists as a court of first instance of the Bailiff and two or three jurats appointed by the Electoral College, and as a court of appeal of the Bailiff and seven jurats. When sitting as Criminal Assizes the Bailiff is assisted by a *grande enquête* or jury of twenty-four.

The law is not English common law but is based on the customs of Normandy as at 1205, contained in the *Ancienne Coutume de Normandie* (ed. De Gruchy, 1881) and the *Grand Coutumier du Pays et Duché de Normandie* (with commentary by Le Rouille, 1539). The best evidence of the old custom of Normandy is Terrien's *Commentaires du Droit Civil, tant public que privé, observé au Pays et Duché de Normandie* (1574), and the *Grand Coutumier*.

Reference is also made to the *Coutume Reformée de Normandie* compiled shortly after 1585 and to commentaries thereon.

But the ancient customary law has been largely altered by local legislation and local customs and the strongest authorities are the decisions of the Royal Court and, on appeal, of the Privy Council. The latter has described Le Geyt's *Constitutions, Lois et Usages de Jersey* (*c.* 1710, pub. 1846) as the most authoritative work on Jersey law; he also wrote a *Privileges, Loix et Coustumes de L'Isle de Jersey* (pub. 1953), known as '*Le Code le Geyt*'.

Guernsey

Guernsey has a Lieutenant-Governor, appointed by the Crown, and a Bailiff who is chief magistrate and judge. The States of Deliberation consist of the Bailiff, the two Law Officers, 12 Conseillers, 33 People's Deputies, and 10 douzaine representatives. The Bailiff has a casting vote and the Law Officers no vote. There is a States Legislation Committee, consisting of the Bailiff and seven members of the States of Deliberation, which has the function of reviewing for submission to the States *Projets de Loi* presented by the Law officers in implement of resolutions of the States and draft ordinances, subject to annulment by resolution of the States. Ordinances do not require the approval of the Crown in Council.

The Royal Court consists of the Bailiff and 12 jurats, the Bailiff and two jurats sitting as a civil court of first instance, and to hear civil appeals from Alderney and Sark, the Bailiff and three jurats sitting as the *Cour des Plaids d'Héritage*, the Bailiff and seven jurats sitting as matrimonial court, civil court of appeal, as licensing court, as criminal court, as court of criminal appeal from the Court of Alderney, or from the Court of the Seneschal of Sark.

The law is based on the customs of Normandy so far as adopted in Guernsey, usually as contained in Terrien's *Commentaires du Droit Civil* (1574) as interpreted in the *Approbation des Loix, Coustume et Usages de l'Isle de Guernezey differentes du Coustumier de Normandie d'ancienneté observés en la dite Isle*, approved by the Queen in Council in 1583.

Alderney

Alderney is a dependency of Guernsey. The States comprise a President and nine members. Most responsibilities are now vested in Guernsey, the legislation of which now extends to Alderney.

The Court of Alderney consists of jurats appointed by the Home Secretary.

Sark

Sark is a dependency of Guernsey. It has a Chief Pleas consisting of the Lord of Sark or Seigneur, the Seneschal (appointed by the Seigneur with the approval of the Lieutenant-Governor for three years), 12 deputies and the holders of the 40 tenements. It has legislative power in local affairs in matters of public order. Taxation, with exceptions, requires the consent of the Crown in Council. The Seigneur may veto ordinances, but this may be overruled by the Chief Pleas. Ordinances may be annulled by the Royal Court of Guernsey if unreasonable or *ultra vires*, but the Chief Pleas may appeal against this to the Crown in Council.

Herm and Jethou

These islands are dependencies of Guernsey and have no independent institutions.

Channell, Sir Arthur Moseley (1838–1928). Was called to the Bar in 1863 and specialized in local government work, gradually developing a good practice. In 1897 he became a judge of the Queen's Bench Division and proved an admirable judge. He retired in 1914 but was sworn of the Privy Council and sat regularly there, especially in prize appeals, as late as 1927.

Chapter. (1) The group of ecclesiastics, canons, and prebendaries, who form the ecclesiastical staff of a cathedral church. They are headed by the Dean. All are subordinate to the Bishop.

(2) In the system of citation of statutes long followed in Britain each statute was numbered as a 'chapter' in the legislation of that session of Parliament, the session itself being designated by the regnal year or years of the monarch over which it extended. Since 1963 statutes are numbered as 'chapters' in the legislation of a calendar year. In very long statutes the term chapter is sometimes used of a subdivision of a Part of the Act dealing with a distinct group of topics.

Chapters of the eyre. The list of matters put by justices in eyre (q.v.) holding the General Eyre before presenting juries listing the matters on which they were to enquire. The list extended to over a hundred matters.

Character. The general reputation of a person. It is, in general, a fact irrelevant to a civil claim or liability, or to his guilt of a criminal charge. But in actions for libel or slander a plaintiff's general bad reputation may be proved in mitigation of damages, and this may be rebutted by general evidence of good character. In criminal trials the accused's bad reputation may be proved only if he has sought to establish his good character.

Charge. (1) A general name for a burden attaching to a certain property making it security for the payment of a debt or the performance of an obligation. The term includes mortgages, liens and many security rights not having special names. Charges are of many kinds; a fixed charge attaches to specific property whereas a floating charge attaches to whichever property happens to be in the possession of the party at the time when the charge crystallizes. Some charges are capable of subsisting at law, including a rent charge in possession charged on land, a charge by way of legal mortgage, and land tax or similar charge on land not created by an instrument, while others are equitable charges only, such as a mortgage by deposit of title deeds and equitable assignments by way of security.

(2) The making of an accusation or the taking of proceedings against a prisoner.

(3) A judge's explanation to a jury, at the conclusion of the hearing of evidence, of the relevant law, the issues for their consideration, the onus of proof, the effect of finding certain facts to have been proved and other matters which they must consider when reaching their verdict on the evidence.

(4) In Scots law, a written judicial demand or requisition at the instance of a creditor calling on the debtor to pay a debt or perform an obligation within a stated time, on pain of defined legal consequences. It is the first step of diligence (q.v.) to enforce a decree, and must proceed on an extract decree of court or certain equivalents thereto. It is delivered to the debtor personally, or left with a person at his dwelling-house, or affixed to the door thereof, by a messenger-at-arms or a sheriff-officer. Various other forms of charge were formerly known, including a charge to enter as heir to one's predecessor and a charge by an heir or adjudger or purchaser on a superior to enter the successor as vassal of the lands.

Chargé d'affaires. A grade of diplomatic representative ranking below ambassadors and ministers plenipotentiary or resident. He is normally accredited to the Foreign Minister of the country in which he operates, not to the Head of State, and acts in the absence of the ambassador or other head of mission. He may be head of mission in a case where diplomatic relations between his country and the receiving state are limited.

Charging order. An order which may be made by a court on the application of a judgment creditor ordering that certain funds are to be charged or burdened with the payment of an amount due or to become due under a judgment or order. There are various types of charging orders available in different circumstances against stock or shares, land, an interest in a partnership. After the lapse of prescribed time the unpaid judgment creditor may take proceedings to secure a sale of the charged property so that he may be paid out of the proceeds.

Charitable uses and trusts. Uses and trusts the objects of which are legally charities. They are known in England from the time of Henry VI and became more important after the suppression of the monasteries. The court's equitable jurisdiction in such cases is probably derived from the Crown's prerogative to act as trustee of funds devoted to charity, where no trustees or objects have been selected. The Court of Chancery always regarded with peculiar favour trusts deemed charitable. The modern law originates from the Statute of Charitable Uses of 1601 and was affected by two statutory doctrines, those of superstitious uses and of mortmain. Superstitious uses had as their purpose the maintenance or propagation of religious rites or usages not tolerated by the law; this doctrine was mitigated by statutes removing disqualifications of various categories of dissenters, so that many trusts originally for superstitious uses became valid as charitable trusts. The doctrine of mortmain provided that no gift or conveyance whereby lands were alienated to a corporation, lay or ecclesiastical, was valid, but the land thereupon became forfeit to the overlord. It was frequently evaded by the use of trusts and in time the Crown came to have the power to dispense with the doctrine by granting licences, while charters and statutes also frequently contained exemptions from the Mortmain Acts. Most of the earlier legislation was replaced by the Mortmain and Charitable Uses Acts, 1888–91.

The differences between the law as to charitable and private trusts are that if the trust shows a general intention to benefit charity only it will not be allowed to fail, in the last resort the court being always ready to give effect to the settler's intention by ordering the preparation of a scheme for its administration; that if a testator specifies an object which becomes impossible or impracticable, the court will devote it to a similar object which is nearest to the testator's original intention, provided always that there has been a clear indication of a general intention to benefit charity; that charitable gifts are not within the rules on perpetuities; that restrictions on the conveyance of property to charities are imposed by various statutes and that there are special rules relating to the alienation of property by a charity; and that where the income of property is applicable for charitable purposes only and is so applied, no income tax is payable thereon.

Charity. In English law, a term of technical meaning, covering generally an object of benevolence, but in some ways wider and in some narrower than the everyday meaning of the word.

A use or trust is charitable only if its purposes fall within the spirit of the preamble to the Charitable Uses Act, 1601, grouped in modern authorities into four divisions, viz. the relief of poverty, the advancement of education, the advancement of religion, and other purposes beneficial to the community or analogous in spirit and purpose to one of these divisions, and it is also directed to the public benefit. The persons to be benefited must be a substantial body of the public. Not every object of public utility is necessarily a good charitable object. Whether particular and specific trust purposes are charitable is a question of law for a court, determined largely by precedent and analogies to precedents.

In Irish law, though the Act of 1601 did not apply, the legal meaning of charity was almost exactly the same as in England.

In Scots law charity has no technical meaning but has the basic meaning of something generally beneficial to the public, and is narrower than the meaning in English law, though wider than simply the relief of poverty. There is no distinction between charitable gifts or trusts and gifts or trusts for the benefit of the public or sections thereof. The Charitable Uses Act, 1601, has no application, but for income tax purposes the wider meaning borne by 'charity' in English law is accepted.

Charitable trusts and gifts have always been given benign construction and minor errors and ambiguities have never been allowed to defeat the donor's wishes. If the general intention is apparent the court will find means of carrying it into effect. If there is a general charitable intention the court will not allow a gift to fail but will invoke the *cy-près* principle (q.v.) and apply it as nearly as possible to the furtherance of the truster's objects.

J. Jordan, *Philanthropy in England*, 1480–1600; G. Jones, *History of the Law of Charity*, 1532–1827; G. W. Keeton, *Modern Law of Charities*; O. Tudor on *Charities*.

Charity Commissioners. A Board of Commissioners established under the Charitable Trusts Act, 1853, reconstituted under the Charities Act, 1960, with the responsibility of regulating the management of charitable trusts in England and Wales. They also have the function of acting so as best to promote and make effective the work of charities to meet the needs for which they were established. The Commissioners are appointed by the Home Secretary and report annually to him, and he answers in Parliament for their work. All charities with certain exceptions are required to be registered with the Charity Commissioners.

Charlemagne (A.D. 742–814). King of the Franks from 768 and Emperor of the West from 800, united most of Western Europe into a superstate and effected a cultural renaissance therein. He established a well-organized administrative system and sought to raise the level of the administration of justice. He issued capitularies, partly as complement to tribal laws, partly as regulations applicable to many aspects of public and private affairs, and partly as specific instructions to royal officials. His successors were unable to maintain the unity of his empire or the ideals and standards he had set and his reign is a notable highlight in the Dark Ages.

Charles River Bridge* v. *Warren Bridge (1837), 11 Peters 420. In 1785 the Massachusetts legislature authorized the Charles River Bridge Company to bridge the Charles River and levy tolls from users. People grew tired of paying tolls and the legislature chartered the Warren Bridge Company to build another bridge, levying tolls only till the bridge was paid for, it then becoming a free bridge. When this happened the public ceased to use the Charles River Bridge. The old bridge owners first sought an injunction against the construction of the new bridge, which was denied, and then contended that their charter was intended to protect their exclusive rights as bridge-operators, and that the legislature, in chartering the new bridge, had unconstitutionally impaired the obligation of its grant to them. The Supreme Court held in favour of the new bridge, laying down that legislative grants were to be strictly construed in favour of the public and that property rights must be limited in the public interest.

Charondas (sixth century B.C.). Celebrated lawgiver of Catana in Sicily. His laws were adopted by other Chalcidic colonies in Sicily and Italy and were noted for their precision.

Charondas le Caron (1536–1617). French jurist, whose real name was Loys le Caron but adopted the name Charondas, author of *Le Grand Coutoumier de France* (1598) and *Coutoumier de Paris* (1598).

Charter. A deed, in modern practice granted only by the Crown, in the form of letters patent under the Great Seal, of special powers, rights, privileges, and immunities. It is usually made to public institutions, universities, and similar bodies, but was formerly more widely granted, particularly to boroughs. In Scots law it is also the deed whereby in feudal land law the Crown, or a subject-superior, creates a new feudal fee, conveying lands to be held by the grantee of and under the granter as feudal vassal. An original charter is one by which the first grant of the subjects is made. A charter by progress was one renewing the grant in favour of the heir or singular successor of the original or a succeeding vassal; the latter have been abolished.

Charter of Liberties. A charter issued by Henry I on his accession, abolishing the illegal exactions of Rufus' time, and granting to the people generally the laws of Edward the Confessor with the amendments made by William the Conqueror. The charter is important as a recognition of national rights and of limitations on royal power. It seems to have been re-issued at various times, and also by Stephen and Henry II, and in 1215 was the basis of Magna Carta, but Henry himself frequently disregarded it.

Charter of Liberties and Privileges. The first enactment of the first popular assembly of New York Colony, passed in 1683. Under it the assembly was established according to English custom, composed of freemen representing towns and populace. It had the power to summon and adjourn itself and to determine its own membership. All taxes were to be levied through the Assembly. Freemen were granted civil rights and ownership of property. Judicial procedure was established, including indictment by grand jury, trial by jury, and release on bail. When the Duke of York, proprietor of the colony, became King as James II, the Lords of Trade disallowed it as in conflict with the superiority of Parliament. New York was thereafter brought under New England and general distrust of James II was fostered.

Charter of the United Nations. The constitution of the United Nations Organization, drafted at the U.N. Conference on International Organization, held at San Francisco in 1945. It comprises a preamble and 70 articles in 19 chapters, dealing respectively with (1) purposes and principles, (2) membership, (3) organs, (4) The General Assembly, (5) The Security Council, (6) pacific settlement of disputes, (7) action with respect to threats to the peace, breaches of the peace, and acts of agression, (8) regional arrangements, (9) international economic and social co-operation, (10) the Economic and Social Council, (11) declaration regarding non-self-governing territories, (12) international trusteeship system, (13) the trusteeship council, (14) the International Court of Justice, (15) the Secretariat, (16) miscellaneous provisions, (17) transitional security arrangements, (18) amendments, and (19) ratification and signature.

Chartered company. A company granted a charter by the Crown under the royal prerogative or special statutory powers. Such a charter normally confers corporate personality. The earlier chartered companies were trading associations chartered to develop trade in particular directions. They included the Merchant Adventurers, the Eastland Company (1579), the Russia Company (1553), the Turkey Company (later the Levant Company) (1592), the East India Company, the Hudson's Bay Company, and the South Sea Company (1711). The Crown was empowered by the Trading Companies Act, 1834 and the Chartered Companies Act, 1837 to confer by letters patent all or any of the privileges of incorporation, except limited liability, without actually granting a charter. The latter Act provided that the letters patent might expressly limit the personal liability of members to a specified amount per share. Towards the end of the nineteenth century there was a revival of chartered companies such as the British South Africa Company. Only a small number of trading companies now exist as chartered companies.

Charterparty (Lat. *carta partita*, divided parchment). So-called because it was originally written on one sheet and divided, a copy being given to each party, a document evidencing a contract whereby a ship or some principal part thereof is let to a charterer until the expiry of a fixed period (time charter) or for one or more voyages (voyage charter) or where the ship, with or without crew, passes to the charterer for his use (charterparty by demise). Various standard forms, varying according to the type of trade concerned, are commonly used. Each kind of charterparty contains a large number of usual and customary stipulations, including various exceptions from liability for loss or damage to cargo or other non-performance of the contract of carriage. In voyage charters the shipowner impliedly undertakes that his ship shall be seaworthy, that she shall proceed with reasonable despatch, and shall proceed without unjustifiable deviation. The charterer impliedly undertakes that he will not ship dangerous goods.

Charter Rolls. Charters were the means of evidencing grants from the Crown and other solemn acts. They were executed in the presence of various witnesses and delivered open, with the wax impressed with the Great Seal hanging at the foot. Before being issued from the Chancery transcripts were made for future reference and rolls were made by the separate charters, being sewn end to end into one continuous strip, which was then rolled up. Portions of the Charter Roll from 1200 to 1515 have been edited, indexed or calendared.

Chartists. A working-class organization founded in 1838 with the object of bringing about parliamentary reform. It arose from dissatisfaction with the Reform Act of 1832 which benefited the middle classes only. It took its name from the People's Charter setting out the organization's six claims, namely, adult male suffrage based on residence only, abolition of property qualifications for the franchise, voting by ballot, division of the country into equal electoral districts, payment of Members of Parliament, and annual Parliaments. All the points were conceded in the course of time, except the last which is unlikely ever to be conceded (though in practice Parliament meets every year). The movement sought in 1839 to petition the House of Commons to accept its Charter; the petition was rejected but the organization was unable to call the threatened general strike, and it got into the hands of the more revolutionary among the leaders. A second petition in 1842 was rejected and some strikes and rioting took place. The movement declined and disintegrated as the condition of the working classes improved with successive reforms after the repeal of the Corn Laws in 1846.

T. F. Tout, *The Chartist Movement.*

Chartulary. A mediaeval manuscript register containing copies of charters and other deeds relating to the foundation, privileges, property, and rights of the owner.

Chase (from Norm. Fr. *Chace*, from *chacier*, Lat. *captare*, to hunt). A district of land set apart for wild animals, objects of the chase, with the exclusive privilege of hunting therein. It is a kind of franchise (q.v.) and is normally less than a forest (q.v.) and not so provided with officers, courts and laws, but is more than a park, and differs from a park in not being enclosed but yet must have bounds. A chase might be in the hands of the Crown (chase royal) or of a subject (frank or free chase), but the latter has now been abolished.

Chase, Salmon Portland (1808–73). Was admitted to the Bar in Washington, D.C., in 1829 and Ohio in 1830, and published an edition of the *Statutes of Ohio* (3 vols., 1833–35). He became a Senator from Ohio, 1849–55 and 1860–61, Governor of Ohio, 1856–60, and Secretary of the Treasury, 1861–64, and finally sixth Chief Justice of the Supreme Court, 1864–73. He several times sought nomination for the presidency. He was a zealous abolitionist and as Chief Justice was much concerned with problems of the aftermath of the Civil War and with the reconstruction legislation. He presided at the early stages of the trial of Jefferson Davis, and later at the impeachment trial of President Johnson, when he asserted his prerogative as presiding judge to ensure that the proceedings were properly conducted.

He was a man of the utmost intellectual integrity and the highest ability, energy and efficiency, but opinionated and difficult to work with. His judicial opinions show an emphasis on principles rather than intellectual brilliance or deep legal learning.

A. B. Hart, *Salmon Portland Chase*; Schuckers, *Life and Public Services of S. P. Chase.*

Chase, Samuel (1741–1811). The only justice of the U.S. Supreme Court to be impeached. A signatory of the Declaration of Independence, he became Chief Judge of the Maryland General Court in 1791 and an associate justice of the Supreme Court in 1796. He was impeached in 1804 because of his activities on behalf of the Federalist party on eight articles. The trial took place between 2 January and 1 March 1805. On three charges there was a majority of votes against him but on none was he convicted. He resumed his place on the bench and sat till his death. The trial clarified the constitutional provision that judges should hold office during good behaviour.

Smith and Lloyd, *The Trial of Samuel Chase* (1805).

Chasseneux, Barthelemy de (*c.* 1480–1541). French jurist and counsellor of the *Parlement* of Paris, author of *Commentaria in consuetudines ducatus Burgundiae* (1517) and *Repertorium consilorium* (1531).

Chastity. Continence or abstention from sexual intercourse. An imputation against a woman's chastity is defamatory. Provision is sometimes made in a will, settlement, or separation order for a woman *dum casta*, i.e. so long as she remains chaste.

Châtelet or Grand Châtelet. The ancient castle of Paris, the seat of the local court and latterly a prison, demolished 1802–10. In the twelfth century it was the seat of the royal *prévôt* who had jurisdiction, civil and criminal, over the Paris district, and heard appeals from all royal and seigniorial courts within his area of jurisdiction. It was accordingly the principal site of common law jurisdiction from the Middle Ages till the Revolution. The procedure is recorded in Jacques d'Ableiges' *Grand Coutumier de France* (fourteenth century). The *prévôt* also had financial and administrative functions. There was a staff of *commissaires-examinateurs* who were responsible for public order, security and the supervision of the Bastille and other prisons. In 1667 many of the powers of the *prévôt* were transferred to the lieutenant general of police. The jurisdiction was abolished in 1790.

Chattel mortgage. The use of chattels as security for a loan, possession being retained by the debtor. Because of the danger that the debtor will mislead other creditors it is commonly provided in the U.S. that the mortgage instrument be filed in a public office or that the transaction is of limited validity until the creditor takes possession. In the U.S. chattel mortgage is widely used to finance the raising of crops or stock. In England the institution is less commonly used, the requisite form being the bill of sale (q.v.).

Chattels personal. Strictly speaking, a term for movable things or goods, but commonly used in a wide sense to denote any kind of property other than real property (q.v.) and chattels real (q.v.), and accordingly comprising both corporeal and incorporeal chattels, or choses in possession and choses in action (qq.v.). Corporeal chattels or choses in possession include, e.g. animals, vehicles, books, and other things which can be physically possessed; incorporeal chattels include, e.g. debts, shares in companies, copyrights, and other things which exist only as claims recoverable by action at law.

Chattels real. A term for rights derived out of real estate (land and buildings) but classified as chattels because at common law they devolved on the personal representatives of the deceased owner, not on his heir. The main instance of chattels real were terms of years (q.v.). The main differences between real estate and chattels real were in mode of devolution on death, rights of succession on intestacy and legal remedies, but these disappeared in 1925 and the distinction is now only of historical interest.

Chatterton, Hedges Eyre (1820–1910). Was called to the Irish Bar in 1843, became Solicitor-General in 1866 and Attorney-General in 1867 and then became Vice-Chancellor of Ireland. He retired in 1904.

Chaud melle. A term of old Scots criminal law for an act of homicide committed suddenly and without premeditation, in the heat of blood, as distinct from deliberate murder. Before the Reformation the person guilty of this had the benefit of sanctuary. Later it came to be merged in the crime of culpable homicide. Cf. CHANCE-MEDLEY.

Cheating. At common law the crime of obtaining the property of another by fraud if effected by a deceitful and illegal practice, not amounting to felony, which directly affected or might affect the public. Since 1968 the common law offence has been abolished, except in relation to the public revenue, and been replaced by the statutory crimes of, by deception dishonestly obtaining property belonging to another with the intention of permanently depriving the other of it, and of, by deception dishonestly obtaining for himself or another any pecuniary advantage.

Checks and balances. The principle of government which distinguishes legislative, executive, and judicial powers and seeks to ensure that any misuse or excessive use of power by any one can and will be checked and balanced by one or both of the other branches of government.

Chelmsford, Lord. See THESIGER, FREDERIC.

Cheque. A bill of exchange (q.v.) drawn on a banker and payable on demand, and a common way of paying a debt. Save as otherwise provided a cheque is regulated by the law of bills of exchange. It is negotiable by endorsement by the payee and delivery to a third party. Cheques may be open, or crossed. A crossed cheque can only be paid into a bank account and a bank will not normally cash or accept a cheque marked 'account payee' except for that payee. Bankers are given special statutory protections when collecting cheques for customers. A cheque bearing to have been paid by the bank on which it is drawn has the effect of a receipt.

Chequers or Chequers Court. A country house in Buckinghamshire given by Lord Lee of Fareham to the nation as a weekend country retreat for U.K. Prime Ministers. The house is basically Elizabethan but redecorated in neo-Gothic. It contains an important collection of portraits and relics of Cromwell.

Cherry, Richard Robert (1859–1923). Was called to the Irish Bar in 1881, became Professor of Constitutional and Criminal Law in Dublin 1888 and Attorney-General 1905. He was later a Lord Justice of Appeal in Ireland, 1909 and Chief Justice of Ireland, 1913–16. He wrote several works on Irish land law.

Cheshire, Geoffrey Chevalier (1886–1978). Was successively All Souls lecturer in Private International Law, All Souls Reader in English Law, and Vinerian Professor of Law (1944–49) at Oxford. He wrote three outstanding textbooks, *Modern Real Property* (1925), *Private International Law* (1935) and, with C. H. S. Fifoot, *The Law of Contract* (1945), all of which have gone through many editions.

Chester. From the time of the Conquest Chester was a county palatine and down to 1830 it had its own palatine courts, a Court of Session and a Court of Exchequer. These were abolished in 1830 and the jurisdiction transferred to the Courts of Chancery and Exchequer in London.

Chief Baron. The chief judge of the Court of Exchequer down to 1875 and of the Exchequer Division thereafter. The office was abolished in 1880 and the Exchequer Division merged with the Queen's Bench Division. For holders of the office see Appendix.

Chief Justice of the Common Pleas. The chief judge of the Court of Common Pleas until 1875 and of the Common Pleas Division thereafter until 1880. The office disappeared when Lord Chief Justice Coleridge was appointed Lord Chief Justice of England in 1881 and the Common Pleas Division was merged with the Queen's Bench Division. For holders of the office see Appendix.

Chief Justice of the King's (or Queen's) Bench. The chief judge of the Court of King's (or Queen's) Bench until 1875 and the Queen's Bench Division until 1880, when Lord Chief Justice Coleridge was appointed Lord Chief Justice of England. For holders of the office see Appendix.

Chief Pleas. See CHANNEL ISLANDS.

Chief, tenants in. Persons who under the feudal system held lands immediately of and under the King in right of the Crown.

Child. A person under the age of majority, particularly in relation to a particular other person as being the parent of the child. The term is commonly defined in different ways by particular statutes for their particular purposes. Children are distinguished for some purposes into legitimate children, illegitimate children, legitimated children, and adopted children.

A child under 10 cannot be guilty of any offence; between 10 and 14 a child is presumed incapable of criminal intent but evidence may overcome the presumption. There is a large volume of law concerned with the education, employment and protection of children. See also INFANT; MINOR.

Child-bearing. The process of conceiving, carrying to full term and giving birth to, a child. A question frequently arises whether a particular woman is or is not still capable of child-bearing. There are no legal presumptions but the courts may, on the basis of common knowledge, regard the possibility as being in some cases so remote as to be negligible.

Child Destruction. In English law, the statutory crime of killing any child capable of being born alive (i.e. after 28 weeks gestation) before it has an existence independent of its mother, provided that the act causing the death was not done in good faith for the purpose only of preserving the life of the mother. It overlaps with the offence of procuring the miscarriage of a viable child. The offence was created because at common law it was not murder to cause the death of a child in the womb or while in the process of being born. See also INFANTICIDE.

Child stealing. The offence of unlawfully, by force or fraud, leading or taking away or enticing away or detaining any child under 14 with intent to deprive any parent, guardian, or other person having lawful charge, of the possession of the child.

Children and young persons. Large bodies of legislation exist in both England and Scotland to protect children from cruelty, corruption, exploitation, indulgence in smoking or alcohol, and to regulate their employment in various capacities. Children in need of care and protection may be committed to the care of a fit person, which may be the local authority.

Children deprived of normal home life may be taken into the care of a local authority, or committed to such an authority by the court, and the authority may assume all parental rights in certain circumstances. It must provide accommodation and maintenance and seek to further the best interests of the child.

Children and young persons who commit crimes are dealt with separately from adult offenders. See JUVENILE COURTS.

Chiltern Hundreds. A Member of Parliament cannot directly resign his seat but by the Succession to the Crown Act, 1707, no Member of Parliament may occupy an office of profit under the Crown, and if he does, he automatically vacates his seat. The offices of steward or bailiff of Her Majesty's three Chiltern Hundreds of Stoke, Desborough, and Burnham, and of steward of the Manor of Northstead are nominally existing offices of profit in the gift of the Crown, though in fact they do not now exist, and since about 1750 a Member desiring to resign applies for one of these posts. Appointment therefore implies vacating his seat. Each office is held until the Chancellor of the Exchequer receives an application from another member wishing to resign. Each is a total sinecure and unpaid. Appointments are made to each post alternately. The principle is now laid down by the House of Commons Disqualification Act, 1975. The disqualification formerly extended to the stewardship of the Manors of East Hendred and Hempholme, which were last used for these purposes in 1840 and 1865.

Chirograph (Gk. *cheir*, hand, and *grapho*, I write). A deed or other instrument in writing attested by signatures and the marks or signatures of witnesses, and particularly the kind of deed written in two parts, each the counterpart of the other, written head to head. Between the heads of the two the word CHIROGRAPHUM was written, or sometimes the capital letters of the alphabet, and the deed divided into two by cutting in an irregular and angular or wavy way through the word *chirographum* so that the authenticity of each part might be tested by fitting it to the other. This technique was also used in final concords, in which the deed was made out and executed in triplicate like a letter T, later divided along indented lines each cutting through the word *chirographum*, into three parts, two, the indentures (the arms of the T) being given one to each party and the third (the shaft of the T), the foot of the fine, being retained by the *custos brevium*. The chirographer was the officer appointed in the Court of Common Pleas to engross fines. The office disappeared in 1833.

See FINES; FEET OF FINES.

Chisholm* v. *Georgia (1793). Two citizens of South Carolina sued Georgia for recovery of confiscated property. Georgia denied the jurisdiction of the Supreme Court. The majority held that the Supreme Court had jurisdiction, but the Eleventh Amendment adopted the opposite view, protecting States from suit by citizens of another State or a foreign State.

Chitty, Joseph (1776–1841). Practised as a special pleader (q.v.) before being called to the Bar in 1816. He had an enormous junior practice till illness forced him to turn to writing. His works included treatises on bills of exchange, the law of nations, criminal law, commercial law, reports, works on practice, an edition of Blackstone, and *A Collection of the Statutes of Practical Utility* (1829) (later continued). His son Joseph (*c.* 1800–38) wrote on the prerogatives of the Crown, bills of exchange, and *Chitty on Contracts* (1841). His second son Thomas (1802–78) practised as a special pleader and trained many lawyers later eminent, edited Archbold's *Practice* and Burn's *Justice of the Peace* and wrote *Forms of Practical Proceedings* (Chitty's Forms) (1834). His third son, Edward (1804–63) was a legal reporter and compiled Chitty's *Equity Index* (1831), an *Index to Common Law Reports* (1841) and the *Commercial and General Lawyer* (1834).

Chivalry (Lat. *caballarius* and connected with cavalier and cavalry). Knights equipped for battle, mediaeval men-at-arms and more generally, knights and gentlemen, and hence the knightly system of feudal times with its social, moral, and religious code of conduct. Chivalry and the feudal system gave rise to the concept of knight's fees and to the modern Orders of Chivalry or of Knighthood.

See KNIGHTHOOD, ORDERS OF.

Chivalry, High Court of. A court which sits very occasionally to take cognisance of questions of rights to arms, precedence, descent and kindred matters of honour which are outside the jurisdiction of the ordinary courts. In the Middle Ages it was held jointly by the Lord High Constable and the Earl Marshal but since the sixteenth century by the latter alone. It sat in 1954 for the first time for over 200 years.

See also CONSTABLE AND MARSHAL, COURT OF.

Choppin, Rene (1537–1606). French jurist and pamphleteer, author of *De domanio Franciae* (1574), *De privilegiis rusticorum* (1575), *De sacra politia forensi* (1577), *De legibus Andium municipalibus libri III* (1611) and *Monasticon, seu de jure coenobitorum* (1601).

Choses in action or things in action. Personal rights of property claimable or enforceable by legal action, as distinct from choses in possession, things capable of physical possession. This class of rights includes a great variety of rights of an intangible character. They are divided into legal and equitable choses in action, according as they may be recovered or enforced by action at law or, formerly, only by suit in equity. The former includes such as debts, claims under insurance policies, and shares in companies, the latter includes equitable rights in property such as interests under trusts, arising from those kinds of rights as to which the Court of Chancery formerly had exclusive jurisdiction. Choses in action arise from contracts, will and trusts, by virtue of statutes and from other sources. They are in general assignable by the person entitled to another by legal or equitable assignment, intimated to the debtor or fundholder, or in accordance with statute or, in the case of negotiable instruments, by delivery of the document of title, and pass on death or bankruptcy. Certain kinds of choses in action, such as certain salaries and pensions, are not assignable. The rights may be discharged in various ways, notably by payment.

Choses in possession. Tangible objects of property rights which can be reduced to actual possession, such as animals, books, or furniture.

Christian, Edward (*d.* 1823). A brother of Fletcher Christian of the Mutiny of the Bounty, was called to the Bar in 1786 and became Downing Professor of the Laws of England in 1788. Under the new Charter of incorporation of Downing College he was named first Professor of Laws. In 1790 he became Professor of General Polity in the East India College. He published an edition of Blackstone and numerous other legal works now forgotten.

Christian marriage. Defined as the voluntary union for life of one man with one woman to the exclusion of all others. This stresses the permanent (though not necessarily indissoluble) and monogamous nature of marriage, but does not limit it to marriages between Christians or to marriages by the rites of a Christian denomination.

Christian name. The baptismal name or forename of a baptized Christian, or, more generally, a person's first name. In England the Christian name may, with the consent of the bishop, be changed at confirmation, and may doubtless be changed by statute or assumption and use of a different name. See also NAME.

Christianity. A world religion founded in Judaea (modern Israel) 2,000 years ago by Jesus of Nazareth. After three centuries of persecution it became the official religion of the Roman Empire and spread rapidly throughout the world. It is divided into three major branches, Roman Catholic, Protestant, and Eastern Orthodox, and many subdivisions. The influence of Christianity on Western law has been enormous, principally because until modern times one or another branch of it has been the official state religion, and also the predominant one, in all Western countries, and its basic beliefs have generally been accepted and held by a majority of legislators, judges, and jurists. The Roman Catholic church built up a large and complex legal system of its own (see CANON LAW) which prior to the Reformation, and even to a substantial extent thereafter even in countries which rejected the primacy of Rome, regulated much of the lives not only of the clergy but of the laity, notably in relation to marriage and succession. General concepts of the Christian faith, such as the worth of individual men, respect for personality, have played material parts in influencing law. Thus Christianity has frequently exercised strong influence against slavery. The Christian principle of loving one's neighbour is a basis of the law of tort—you must not injure your neighbour. Churches and church leaders have frequently pressed for legal changes of a humanitarian character. Conversely, the promotion of religion, particularly Christianity, has generally been a favoured object of the law.

It was said in the past that Christianity was part and parcel of the law of the land but this is hardly so today, as the law recognizes and does not discriminate against non-Christians, agnostics, or atheists. Nevertheless, the sovereign must be a Christian and a Protestant and the (Anglican) Church of England is represented in Parliament, in the House of Lords.

The influence of Christianity on State, government and law has declined since 1800 with states becoming increasingly secular, the growth of agnosticism and atheism among governors and decreasing regard for the values represented by Christianity. Marriage is now regarded as a dissoluble civil relationship which can be made and unmade without the intervention of the Church.

Christinaeus, Paul (Christynen) (1533–1631). Belgian jurist, author of *Practicarum quaestionum rerumque in supremis Belgarum curiis actarum et observationum decisiones* (1626) an important collection of decisions and *In leges municipales civium Mechlinensium notae seu commentationes* (1624).

Chronological Table and Index to the Statutes. Two works, proposed by Lord Cairns (then Lord Justice) in 1867, and first published in 1870. Editions of both works have been published, at first periodically and now annually under the supervision of the Statute Law Committee, and the form of both works has been greatly improved in time.

The *Chronological Table* records by regnal year and chapter number every Act printed in the *Statutes of the Realm* (of England) from 1235 and, from 1797, every Public General Statute passed by the Parliaments of Great Britain and the United Kingdom, down to the end of the year preceding publication, showing in respect of each, if repealed, the short title and the repealing enactment and, if wholly or partly in force, the short title and which sections have been repealed or amended by which subsequent legislation. Changes made by the Parliament of Northern Ireland and Dail Eireann on pre-1923 United Kingdom Acts extending to Ireland are not shown.

The *Index to the Statutes* digests the living statute law, other than pre-1707 Scottish and pre-1801 Irish Acts, under recognized legal headings and sub-headings in accordance with instructions laid down by Lord Thring in 1876. The scheme groups, under titles descriptive of recognized branches of the law, index entries setting out in general terms the subject-matter of the enactments which constitute those branches, and against each entry appear regnal year citations of the Acts and sections thereof supporting each proposition indexed. It contains also a chronological list of all the Acts and Church Assembly Measures, showing against each the titles under which the Act appears, in whole or in part, in the Index. Some titles are divided into, e.g. River, and River, Scotland, the latter containing additional matter applicable to Scotland, or into Public Health, England and Public Health, Scotland, where the bodies of statute law are distinct. Some titles have prefixed a list of the statutes relevant to the title and indexed in the following headings and sub-headings.

Church and State, relations of. In many societies problems have arisen from, and because of, the relations between church and state, or superior spiritual and temporal powers in the community. Is either superior to the other, or are they equal? To which, in case of conflict, does the individual owe prior duty? This has been described as the most serious practical problem in politics down to the Industrial Revolution. At Rome the church was a branch of the state and *ius sacrum* a part of *ius publicum*. The problem arose sharply with the growth of Christianity, and particularly after Christianity was first tolerated and then established as the official religion of the Roman state by the Emperor Constantine in A.D. 313. This was a concession made by the state by secular law but made to a large group united by a common faith and having an extensive organization with disciplinary powers.

The establishment of Christianity involved recognition by the State of the organization, the Church, and its spiritual authority, but also the submission of Christians to the Emperor's sovereignty.

In the late fifth century Pope Gelasius I made the most sweeping and comprehensive statement of the view that the world was ruled by two authorities, bishops and royal power, the former being heavier since they must account to God even for the kings of men. The Church stood beside and was independent of the State but closely related; spiritual and temporal powers were entrusted to different orders, independent, each supreme in its own sphere but subordinate in the other sphere. Later a unitary view of society prevailed, the higher power being the spiritual.

Augustine believed that justice was incomplete if not based on Christian law as well as the law of nature and from this view it was a short stage to Gregory VII's view that the duty of securing divine justice must ultimately rest on the Pope and to Innocent IV's claim that the Pope was the Judge Ordinary of all men.

Accordingly, by the early Middle Ages some of the problems had arisen, the immunity of clergy from secular justice, the limits of spiritual and temporal jurisdiction, the obligations of church property, and religious uniformity enforced by secular authority.

In the period between the eighth and tenth centuries law became territorial; many of the mediaeval law books were written and feudalism began to develop.

By the eleventh century the competing claims of Empire and Papacy had produced the first of the great contests, the Investiture question, whether the King was to have any say in the choice of a bishop, whether bishops were to be invested with their lands and civil authority, and to do homage, before or after consecration, whether a king could choose and control his vassals without clerical interference, and whether the Church could exclude those whom it considered unworthy. In the Empire the struggle took place notably between Pope Gregory VII and the Emperor Henry IV and was not ended until the Concordat of Worms in 1122. In France lay investiture was abandoned by the King in the early twelfth century.

In England the struggle was chiefly between Henry I and Anselm, and was compromised by 1107. The King abandoned investiture by pastoral staff and ring but retained the right of feudal homage from the bishops. The murder of Becket caused a setback to royal claims and freedom of election of bishops was confirmed in Magna Carta. The violence and disorders provoked by the controversy led some prelates, notably Ivo of Chartres, to advocate conciliation, holding that harmony of Church and State was necessary for social order.

Point was lent to the controversy by the revived study of Roman law and of Aristotle's ethical and political works from the eleventh century. The earlier formulations of the extreme papal claims were made by canonists, notably Innocent IV and Hostiensis, and the earlier denials were by civil lawyers. In a sense the contest was between canon law and civil law. One of the most distinct assertions of papal authority came from John of Salisbury who claimed for the Church supreme authority in temporal matters as well as spiritual, so that kings and princes were nothing but deputies of the bishop.

In the thirteenth century, the period of the Papacy's greatest power, such views were widely held by the clergy. Innocent III asserted an almost unlimited authority over the Empire, especially in matters of election and coronation and his exercise of it was unparalleled in extent and effectiveness. In England John surrendered his kingdom to the Pope as feudal overlord and the Pope annulled Magna Carta on the ground of ecclesiastical right, and claimed to dictate terms of peace to England and France. In his Decretals Innocent III gave voice to the extreme view of papal authority in temporal matters and this was followed by writers who sought to prove that imperial authority was derived from the Pope. St. Thomas Aquinas advocated a breadth of power as extensive as the Popes claimed, but the canonists were the chief authors of the theory of the direct power of the Pope in temporal matters, and they developed this theory rapidly in the thirteenth century. The complete and final exposition was in Hostiensis' *Summa Aurea* and Durandus' *Speculum Juris*; the spiritual power was prior to, and superior to, the temporal.

Thereafter decline set in, notably in the struggle between Boniface VIII and Philip the Fair of France. Boniface VIII asserted that clerical property must be exempt fron taxes levied by the King with common consent of the realm, but had to moderate his claims. In 1302 the Pope issued the Bull *Unam Sanctam* asserting that for every human creature it was absolutely necessary for salvation to be subject to the Pope. He was supported by Egidius Colonna in his *De Ecclesiastica Potestate* (1301) and James of Viterbo in his *De Potestate Ecclesiastica* and answered by various writers, notably John of Paris's *Tractatus de Potestate Regia et Papali.*

Far reaching claims for papal supremacy were contained in the Donation of Constantine, a document purporting to be one by which Constantine, in view of the removal of his capital to Constantinople, relinquished to Pope Silvester and his successors power and jurisdiction over Rome, Italy or the West. Though usually accepted as valid at this time, the document was shown by Valla in 1439 to be a fabrication, probably of the eighth century. The extent of the grant and the area over which it applied were stretched during the Middle Ages.

In England in the Middle Ages the supremacy of the Pope and the binding force of canon law were accepted but as all major litigation went on at Rome, the great canonists and literature on canon law were foreign. Lyndwood's *Provinciale* deals only with the provincial constitutions which were no more than an appendix to the common law of the Church. In the canon law the Pope was the supreme legislator and supreme judge, with both original and appellate jurisdiction.

Disputes, however, naturally arose when there were co-existing two systems of courts and law. It gave rise to the investiture controversy in England, the dispute over criminous clerks.

The *Articuli Cleri* of 1315 was an attempt to delimit the spheres of temporal and spiritual jurisdiction, and were the basis of all subsequent legislation down to the Reformation. They made clear the King's intention to be the dominant partner in the state, and the growth of national feeling gave rise to various statutes seeking to regulate abuses of rights by the Church and to secure the supremacy of the English State and its law against papal pretensions. Thus the Statute of Carlisle (1306–7) forbade religious houses to send money out of the kingdom to pay taxes imposed by foreign superiors; the Statute of Provisors sought to protect spiritual patrons against the Pope; the first Statute of Praemunire punished those who drew any cause out of the realm in plea whereof the cognizance pertaineth to the King's court, and the second was aimed at those who pursued in the court of Rome or elsewhere sentences of excommunication, bulls or other things which touched the king, whereby his courts were hindered in their jurisdiction.

In the next stage of the controversy between Empire and Papacy in the early fourteenth century, the protagonists were John XXII and the Emperor Lewis of Bavaria, and the principal supporting writers on the papal side Agostino Trionfo (*Summa de Potestate Ecclesiastica*, *c.* 1325) and Alvaro Pelayo (*De Planctu Ecclesiae*, 1332), and on the imperial side Lupold von Bebenburg (*Tractatus de Juribus Regni et Imperii Romani*, 1340), William of Occam, principally in his *Dialogus* and *De Imperatorum et Pontificum Potestate* and Marsiglio di Padua's *Defensor Pacis*, in the last of which the secular state claimed superiority and treated the church as a department of the state.

The mediaeval phase of the conflict between Church and State came to an end by the mid-fourteenth century, the claims of the Church being materially weakened by the Papacy at Avignon and the Great Schism, which made the principle of unlimited papal power untenable. The unity of the church was restored by the conciliar movement but the Papacy triumphed over the councils and negotiated concordats with the states of Western Europe. These in fact acknowledged the sovereignty of national states and virtually ended the mediaeval theory of the co-existence of Empire and Papacy. France alone declined to make a concordat but by the Pragmatic Sanction of Bourges in 1438 asserted a measure of independence.

The Reformation finally ended the former controversy of Church and State by destroying the identity of Christendom and the Catholic Church, but that and the rise of nation-states raised the problem in a new form, of the relations of and struggle for supremacy between Church and State in particular kingdoms. Thus by the English Reformation the Pope was ejected from his position in relation to the Church of England and the king became the Supreme Head on earth of that Church. But conformity was enforced as much as if it had been under Catholicism and this continued, after the reversal of power under Mary Tudor, until the end of the seventeenth century.

In Geneva under Calvin a theocratic polity was established from 1541, based on the principle that every member of the community was also under the discipline of the Church, which was exercised by the clergy and elders. He recodified the Genevan laws and constitution, but his ecclesiastical discipline provoked much opposition. Nevertheless he was, if not director, at least consultant on every matter of government of Geneva in these years. Calvinism spread to parts of France and to Scotland where in the seventeenth century the clergy had excessive power in government even in the direction of military operations. In Puritan England and the New England States the church had similar influence on the state and on economic life generally.

In the British colonies in America Church and State were originally not separated in nine of the 13 colonies. After the Declaration of Independence separation came about generally though in New England dis-establishment was delayed until the early nineteenth century. The First Amendment to the Constitution prohibits Congress from making a law respecting an establishment of religion or prohibiting the free exercise of religion. The trend of Supreme Court decisions on this matter has been towards a strict separation of Church and State.

In Britain in the nineteenth century there were several conflicts between Church and State authorities. In Scotland controversy over lay patronage of ecclesiastical livings gave rise to conflicts of jurisdiction between Parliament and General Assembly of the Church of Scotland and between civil and ecclesiastical courts, and resulted in the Disruption (1843) which divided the Church of Scotland for nearly a century. In England the introduction of judicial divorce in 1857 marked a major transfer of authority in social relations from Church to State.

In modern Britain there is no dispute or conflict between Church and State. In England the Church of England is established by law but has the adherence of only a minority of the population. The sovereign is by law the Governor, final court of appeal in ecclesiastical causes, and on the Prime Minister's advice, nominates the archbishops and bishops and has extensive church patronage. Legislation affecting the church is effected not by Parliament but by the General Synod of the Church. The Church in Wales was dis-established in 1920.

In Scotland the Church of Scotland is presbyterian and established but independent of the state; the State has no control over legislation, ecclesiastical courts, or patronage. It has the adherence of a majority of the population. In all parts of the United Kingdom numerous dissenting churches are tolerated and no disabilities now attach to members of such churches.

The Roman Catholic hierarchy was re-established in England in 1850 but exists by toleration not by law.

In the U.S.A. in the colonial period the Church of England was established in Virginia, the Carolinas, and Georgia; New Haven–Connecticut was a theocracy, but in Rhode Island the earliest fundamental code denied civil magistrates authority over spiritual matters, in Pennsylvania there was no established religion, and in Delaware there was toleration. By the time of the Revolution persecution of dissenters had ended and a measure of toleration existed but in 10 of the 13 colonies there was preference and support for one religion over others. The Bill of Rights struck at any establishment of religion or prohibition of the free exercise thereof. But remnants of the state establishments existed well into the nineteenth century. In the period 1834–1900 there remained in some States religious tests for holding public office. Some of the original States were slow to remove such requirements but nearly all later States guarantee religious freedom. Since 1940 the Supreme Court has in various contexts enforced religious freedom and the neutrality of the state on religious issues.

Church Assembly. The National Assembly of the Church of England was created by the Church of England Assembly (Powers) Act, 1919, to facilitate the passage of Church of England legislation through Parliament. The Assembly consisted of the Convocations of Canterbury and of York (each comprising a House of Bishops and a House of Clergy), and the two elected Houses of Laity for these two provinces. The Church Assembly met normally three times a year for four days at a time. It was concerned with administration and not with faith or doctrine.

It debated proposed legislation (known as 'Measures') in much the same way as does Parliament. Once a Measure had passed through all its stages and been approved by the majority of the Church Assembly it was referred to the Legislative Committee, which determined when and how to present it to the Ecclesiastical Committee, which was a committee of 15 members of the House of Lords nominated by the Lord Chancellor and 15 members of the House of Commons nominated by the Speaker, with a quorum of 12. It examined the Measure and reported thereon to the Assembly's Legislative Committee which might recommend withdrawal or amendment of the Measure. If the

Measure were proceeded with, both Measure and Report were laid before Parliament, which might debate or reject but might not amend the Measure. If both Houses of Parliament approved the Measure might be presented for the Royal Assent; on receiving the Royal Assent a Measure had the same force and effect as an Act of Parliament.

Devolution of legislative powers in church affairs has been very successful in enabling the initiative to come from the church and the details of legislation to be worked out in the Church Assembly, while retaining ultimate Parliamentary control. Parliament has only rarely rejected Measures though two of those rejected concerned proposed revision of the prayer book.

In 1970 the Church Assembly was renamed and reconstituted as the General Synod of the Church of England.

Church Commissioners. The Ecclesiastical Commissioners Acts, 1836 and 1850, established a permanent commission, augmented in 1850 by three Church Estates Commissioners of whom two were appointed by the Crown and one by the Archbishop of Canterbury. At least one of the former two was an M.P. appointed by the Government. The income was chiefly used to augment the income of the clergy. In 1948 the Church Commissioners were established by the amalgamation of the Ecclesiastical Commissioners with Queen Anne's Bounty, a corporation established by Queen Anne in 1703, the income of which was used mainly to augment the maintenance of clergy of the Church of England. There are also a large number of commissioners, some *ex officiis* and others appointed and nominated in various ways. The Second Church Estates Commissioner represents them in the House of Commons. The Church Commissioners' principal functions are to hold and administer various endowments, make additional provision for the care of souls in parishes, approve dealings with glebe and parsonage houses and prepare schemes as to pastoral reorganization.

Church in Wales. The Church of England so far as it existed in Wales and Monmouth was disestablished in 1920 and all ecclesiastical corporations were dissolved, ecclesiastical courts ceased to exercise jurisdiction, and the ecclesiastical law of the Church in Wales ceased to exist as law. The existing ecclesiastical law continued to bind members of the Church in Wales on a consensual basis. No bishop of the Church in Wales sits in the House of Lords and priests thereof may sit in the House of Commons.

The Province of Wales comprises six dioceses and the bishops, clergy, and laity may hold synods and frame constitutions and regulations for the government of the Church in Wales. There is a representative body of the Church in Wales in which is vested all ecclesiastical property. There are various courts but they are not courts of the land.

Church of England. The branch of the Christian Church which was separated from the Roman Catholic Church under Henry VIII who, by the Act of Supremacy of 1558, constituted himself the Supreme Head of the Church. The Church is established in the sense that it, its hierarchy and institutions, its courts and law are recognized by Parliament as part of the constitution and general law of the land, though limited as to the persons to which these courts and law apply. The bishops are nominated by the Crown. The sovereign must be in communion with the Church of England and a number of the bishops are Spiritual Lords of Parliament and sit in the House of Lords. The archbishops take high precedence in the State and play a major part in coronations. The General Synod of the Church of England has been granted delegated legislative power in relation to matters affecting the Church. The standard of doctrine and practice, the Thirty-Nine Articles, was agreed on by convocation in 1562 and confirmed by Statute in 1571 and the Prayer Book was ratified by the Act of Uniformity, 1662. Tithes (q.v.) were originally payable from all land for the support of the Church but tithe was in England commuted in 1836 for a rent charge depending on the price of grain and in 1936 tithe rent-charge was abolished, provision being made for compensation to persons interested in tithe rent-charge. The Church of England is accordingly treated differently from other churches, which are voluntary associations.

Church of Ireland. An Anglican Church was established in Ireland and declared by the Act of Union in 1801 to be perpetually united with the Church of England, but it never had the allegiance of or ministered to more than a small fraction of the population and its existence was a constant irritation in Anglo-Irish relations. It was dis-established and largely dis-endowed in 1869, separated from the Church of England and made into a voluntary society.

Church of Scotland. The branch of the Christian Church which, the authority of the Pope having been abolished in 1560 after a religious Reformation movement, was finally established with a presbyterian form of government in 1690. It is established in the sense that the civil power accepts the duty to give support and assistance to the church, but unestablished in that it is free from State control as regards government, discipline, and doctrine. It is governed by a General Assembly presided over by an elected Moderator and at which the sovereign is represented by a Lord High

Commissioner, by synods, presbyteries and kirk sessions. These are courts of the Church and are part of the legal system of Scotland but have jurisdiction only over members and only in spiritual matters. Other branches of the Protestant Church in Scotland, the Roman Catholic and other churches, are voluntary associations only. The Church is wholly independent of the state in government, worship, discipline and doctrine and enacts its own legislation by Acts of Assembly.

Ecclesiastical property and endowments are now held by the General Trustees of the Church.

Churchill, John (?–1685). A collateral of the Marlboroughs, was called to the bar in 1647 and practised in Chancery. In 1685 he was promoted Master of the Rolls but died shortly thereafter.

Churchwardens. In Anglican churches, lay members who act as the guardians or keepers of the parish church and representatives of the general body of parishioners. In ancient parishes any householder even if not conforming with the doctrine of the Church of England might be appointed and, if appointed, was bound to serve, but in modern practice a person is only chosen if willing to serve. Two churchwardens are chosen annually by a parochial church meeting. If the minister of the parish and the parishioners cannot agree, each chooses one. They are a quasi-corporation for the purpose of holding the goods of the church. As guardians of the parish church they own the movables in the church and maintain order in the church, arrange the seating of parishioners and provide the necessary requisites for divine service such as books and linen. As officers of the Ordinary they must present at his visitation whatever is irregular or amiss in the parish. Most of the other functions formerly falling on churchwardens, notably finance, and care of the fabric of the church and the churchyard, have been transferred to the parochial church council.

Churl or ceorl. In Anglo-Saxon law the ordinary freeman. After the Norman Conquest churls became included within the great class of villeins (see VILLEINAGE) who were personally unfree. Nevertheless in Anglo-Norman law a churl retained the wergild of a free man and long remained subject to the duty to attend the hundred court. Later however such persons became a depressed class and came to be regarded as legally dependent on their lords.

Cicero, Marcus Tullius (106–43 B.C.). Roman politician and orator, lawyer and writer, studied philosophy at Athens, and engaged in political and forensic life. He was consul in 63 and in 52 was sent as proconsul to Cilicia. He was murdered as a political opponent of Octavian in 43. He was one of the most learned men of his time and left a large collection of works, speeches, both political and forensic, rhetorical, notably his *De Oratore* and *Orator*, epistolary, comprising various series of letters, and political and philosophic treatises. The last are generally in the form of dialogues; they include a *de Republica* and a *de Legibus*, both much influenced by Plato, *de Finibus, Tusculan Disputations*, and others. Though an adequate lawyer he was not a great jurist, but his political and philosophic works had great importance, not because of their originality but because he invented a philosophical terminology for the Romans and by way of free translations and adaptations did much to introduce Greek philosophic ideas to Rome and, not least by reason of their literary qualities, transmitted much of this learning to later times.

As a legal philosopher his view of law was dominated by the Stoic conception of true law; there is a true law, right reason, consistent with nature, external and unchanging and existing among all men. Cicero sought to give concrete meaning to his concept of true law, and later Roman jurists worked out the idea in practice using the doctrine of the law of nature. He distinguished from this true law both the law of nature and positive law. Justice he believed to be the supreme virtue; it was the sentiment which assigned to each his own and maintained human solidarity; it had its source in nature. If positive law were to be true law it must seek to realize the rules of justice and reason, not merely express the desires of the dominant class. He contributed substantially in his rhetorical works to the understanding of interpretation of law.

J. L. Strachan-Davidson, *Cicero*; M. Gelzer, *Cicero: Ein biographischer Versuch*; A. H. J. Greenidge, *The Legal Procedure of Cicero's Time*.

Cino da Pistoia (Cino Sinibaldi) (*c.* 1270–1336). An Italian poet and jurist, friend of Dante, one of the earliest commentators, professor at Siena, Perugia, and Florence, and master of Bartolus and Joannes Andreae. He introduced the French method of legal dialectic into Italy. His lectures on the Code (*Lectura in codicem*) and *Lectura in Digestum vetus* were highly regarded by Savigny.

Cinque Ports, Court of the Lord Warden of the. A franchise court, known from the fourteenth century, having a mainly Admiralty jurisdiction but, in the sixteenth and seventeenth centuries, also an equitable jurisdiction. It was held by the Lord Warden of the Cinque Ports (Dover, Sandwich, Romney, Hastings, and Hythe, to which Winchelsea and Rye, the 'Ancient Towns', were later added) or by a person sitting as his official and commissary. The office of Lord Warden has in recent times been held by distinguished elder statesmen such as Sir Winston Churchill and Sir

Robert Menzies. In 1978 it was conferred on Queen Elizabeth, the Queen Mother. The admiralty jurisdiction is concurrent with the Admiralty Court within geographical limits, from Shore Beacon in Essex to Redcliff in Sussex. In its civil capacity it was known as the Court of Shepway. In 1855 the general civil jurisdiction was abolished, but the Admiralty jurisdiction was preserved and in 1869 it was provided that appeals in Admiralty cases from the county courts within the Lord Warden's jurisdiction should lie to him, and this jurisdiction still exists.

Distinguished holders of the office of judge of the court have included Sir R. J. Phillimore and Sir Frederick Pollock (qq.v.). Down to 1949 special provisions related to the appointment of justices and the powers of justices and quarter sessions of those towns which were boroughs. The Courts of Brotherhood and Guestling were courts, originally separate federal institutions of the ports, later combined and rarely held, which had cognizance of matters affecting the Cinque Ports and Ancient Towns and the Ports and Ancient Towns and their members respectively.

Cinque Ports Salvage Commissioners. A body appointed by the Lord Warden of the Cinque Ports or the Deputy Warden of the Cinque Ports and Lieutenant of Dover Castle, having the function of determining questions as to the salvage of anchors and chain cables found at sea or supplies to ships and as to salvage rendered to ships within the jurisdiction, and to goods wrecked or stranded therein, provided that the master or owner of the salved ship or the owner of the goods salved, or their representatives, are present. Appeal lies to the Admiralty Court of the High Court or the Court of Admiralty of the Cinque Ports (q.v.).

Circuit judges. Full-time judicial officers appointed in England since 1971. They must be barristers of 10 years standing or have been Recorders (q.v.) for three years. They sit as county court judges in civil cases and as presiding judges in the Crown Court (q.v.) trying criminal cases of medium importance. One or more circuit judges are assigned to each county court district, but they may sit and act for any district. They may at the request of the Lord Chancellor sit as judges of the High Court and in some cases as deputy High Court judges. The holders of various offices of comparable standing were made circuit judges when the office was created in 1971.

Circuits. Groups of towns visited in succession by one or more superior court judges. From the twelfth century judges of the English superior courts were sent regularly into every county to try actions and criminal cases. The Judicature Act, 1873, fixed seven circuits. Until 1971 the judges acted under commissions of the peace, of oyer and terminer, of gaol delivery, and of assize, and the persons named in the commission were usually judges of the Queen's Bench Division, but judges of the other Divisions, Queen's Counsel, county court judges and retired judges might be included. When exercising the jurisdiction a commissioner had the powers of a High Court judge. Both civil and criminal business was dealt with but the latter had priority.

In 1971 assizes were abolished, the courts were reorganized, and England and Wales were divided into six circuits, each having a circuit administrator with court administrators in charge of groups of courts. The circuit administrator is responsible for the conduct of court business in the High Court, Crown Court and county courts in the circuits. These circuits are Midland and Oxford (Birmingham), North Eastern (Leeds), Northern (Manchester), South-Eastern (London), Wales and Chester (Cardiff), and Western (Bristol). For each circuit two judges of the High Court are appointed as presiding judges to exercise a general responsibility for the way business is conducted on each circuit.

There are resident counsel in the chief towns of the different circuits and many London counsel attach themselves to one or other of the circuits and belong to the circuit bar mess. They may then appear at the High Court and Crown Court held at towns on the circuit.

County court circuits were created in 1846. Each is a district including several places at which the circuit judge of the district holds the county court.

Circuit courts. Criminal courts held by the judges of the High Court of Justiciary in Scotland, when on circuit from Edinburgh to towns in the north, west, and south of Scotland.

Circuits, judicial (U.S.). When the Federal judiciary system was created in 1789 two Supreme Court justices were assigned to each of three circuits, Eastern, Middle, and Southern, and required to hold courts there twice a year, sitting with district judges. Though onerous by reason of bad roads and poor accommodation, these circuits provided employment so long as there was not much work for the Supreme Court. In 1793 only one judge was required to ride each circuit and in 1800 circuit duties were abolished and new circuit courts established. Other circuits were added as population spread westwards.

In 1869 the system was changed by the appointment of Federal circuit judges. Since 1891 the judicial districts in states and territories are served by U.S. District Courts, having federal (and in certain districts also local) jurisdiction, and grouped into 11 circuits, each served by a District Court of Appeals of three or more judges.

Circumspecte agatis (that you act cautiously). The name given to a document of 1285, later included among the statutes, concerned with prohibitions on judges interfering with the bishop's courts in matters of spiritual discipline and defining the respective jurisdictions of the lay and ecclesiastical courts in certain spheres.

Circumstantial evidence. Evidence of circumstances which themselves provide indirect or presumptive evidence of some fact of which direct evidence, e.g. by observation, is not available. Thus a man's clothing left at the locus of a crime is circumstantial evidence of his involvement in or at least presence at the crime. For such evidence to be cogent, the circumstances proved must be ones which are necessarily or at least normally connected with the fact in question. Circumstances equally capable of other explanation are little or no evidence of the fact in issue. Much scientific evidence, e.g. of fingerprints, is circumstantial, showing the connection between the circumstance and the fact in issue.

Circumstantial letter. The letter of complaint, addressed to the convening authority, stating in detail the facts on which a charge is made, which initiates proceedings in a naval court-martial.

Circumvention. In Scots law, conduct which if used to a person who is facile or susceptible to influence, persuades him to act to his detriment and renders the transaction voidable.

Citation. (1) The process of calling a person not a party to an action, to appear before the court in that action. It applies particularly to procedure in matrimonial, probate, and ecclesiastical cases.

(2) A reference in argument or judgment to an authority (statute or case) in support of an argument or judgment. The word is also sometimes used of the notation by which a case or statute referred to may be found, e.g. [1921] 3 K.B. 560, or 1973, *c.* 99.

Citation (Scotland). In Scots law, (1) the procedure whereby the commencement or existence of an action in court is intimated to parties who have, or may have, an interest to oppose the pursuer's claim. (2) The process for requiring a witness to attend and give evidence in an action.

Citation of Statutes. See ACT OF PARLIAMENT.

Citations, Statute of. An Act of 1531 under which no man may be cited, save for a spiritual offence or for heresy, to appear before any ecclesiastical court outside the diocese in which he resides.

Citations, Theodosian Law of. In fifth century Rome there was uncertainty as to the authority of the works of various earlier jurists. In A.D. 426 Theodosius II enacted that the works of Papinian, Paul, Gaius, Ulpian, and Modestinus (qq.v.) should be primary authorities, it being expressly provided that Gaius' works were to have the same authority as the others, and that the works of those jurists cited by any of the five might also be quoted, though such quotations had to be confirmed by comparison of manuscripts. If different views were expressed, the majority view was to be adopted; if the divergent views were equal in number, the view adopted by Papinian was to prevail; if there was equal division of opinion and Papinian's view was not expressed, the decision was left to the discretion of the judge. These rules evidence a serious decline in legal thinking, for jurists' views should be weighed by quality, not number. These rules remained, at least in theory, in force till the time of Justinian.

Citations, Valentinian Law of. An alternative name to the Theodosian Law of Citations, the constitution having been issued by Theodosius II and Valentinian III.

Citizenship. The legal link between an individual and a particular state or political community under which the individual receives certain rights, privileges, and protections in return for allegiance and duties. Whether an individual has citizenship of a particular state depends on its own legal system and by reason of differences between legal systems some individuals may be stateless and others have citizenship of more than one state.

In ancient Athens only some of the population were citizens; resident aliens, women, and slaves were excluded. The Romans similarly initially had a restricted concept of citizenship, but gradually extended it until in A.D. 212. Caracalla's *Constitutio Antoniniana* gave citizenship to most of the freemen of the Empire. The concept was in abeyance in the Middle Ages until city dwellers became a third force in politics, with the nobles and the clergy. Citizenship was the relationship to a city implying certain liberties. The American and French Revolutions gave a new meaning to citizenship, contrasting it with 'subject', while in the twentieth century the movement for women's rights has further extended the concept.

In modern practice what rights and duties attach to citizenship depends on the municipal law of each state. Citizenship of the United Kingdom and Colonies, which carries the consequent status of British subject, is acquired by descent, by birth within the United Kingdom and Colonies, by registration, by naturalization, or by the incorporation of territory, but not merely by marriage to a British subject. The citizenship of Commonwealth countries is governed by the law of the country concerned. A person may be a citizen of more than one unit of the Commonwealth.

Citizenship of the U.K. may be renounced, but is not now lost automatically by naturalization in a foreign country, but may in such circumstances be renounced. Citizens by registration or naturalization may be deprived of British citizenship in certain circumstances. Citizenship of the U.K. is not lost by marriage to an alien.

Citizenship (U.S.). By the U.S. Constitution residents in the states became citizens of the U.S., but for long the general view was that citizenship of a state was prior to citizenship of the U.S. The Fourteenth Amendment established that a member of any race born in a place subject to the jurisdiction of the U.S. is an American citizen. Citizenship was extended to negroes by the Civil Rights Act of 1866, and to American Indians by Congress in 1924.

Citizenship may be acquired by naturalization, subject to laws made by Congress and may be acquired collectively, as when the inhabitants of the Louisiana Territory became U.S. citizens by the transfer of the territory from France to the U.S., or individually. No person can be denied naturalization on account of race.

City. A term applied in common usage to many large urban centres, but properly applied only to an urban area incorporated by charter or statute or, in the U.S., by act of a state legislature. The status is commonly conferred on places which are sites of episcopal sees or universities. Some towns, the seats of ancient bishoprics, are deemed to have prescriptive right to be called a city, but being the seat of a bishop or the site of a university does not automatically confer the rank of city on a place. Apart from the City of London a city in the U.K. does not have any special standing in local government, though it is frequently a major part of a local government area.

City of London Court. A court known prior to 1855 as the Sheriffs' Court of the City of London and in 1920 amalgamated with the Mayor's Court of London to form the Mayor's and City of London Court (q.v.).

City-manager plan. A form of municipal government first adopted in 1908 and now employed by many cities in the U.S., whereby a small elected council appoints a city manager or chief executive officer, who appoints the city officials and manages the municipal affairs. The council acts collectively only and its members have no administrative functions. It has been generally very successful so long as the office is kept out of politics and the manager is both able and politically neutral.

City-state. A form of political organization found most notably in ancient Greece in which a city and its immediately surrounding district formed an autonomous state.

Civil and military relationship. In every society there is a potential conflict between civil authority and military power and repeatedly in history military power has been used to bolster or overthrow civil governments, to promote or limit ideologies, or to defend or annex territory. Even when fundamentally united against a common adversary there have repeatedly been disputes between civil and military leaders over methods of defence or of attack.

In modern western civilization there have been three distinguishable phases. From the Middle Ages till the late eighteenth century, the nation-state subdued feudal powers and gained control of military power in its territory. From the late eighteenth century till 1918 with the rise of popular democracies the military power became predominant in some states and could use the whole power of the state for aggressive purposes. Since 1918, particularly in totalitarian states, every activity of the State has reference to its miltary effect; the exercise of military power is political.

In the U.K. formal control of military power is in the hands of a civilian minister and the service chiefs are his advisers only. Parliament exercises control through political control of senior appointments, legislation, and financial control. In the U.S. there is also civilian control, but Congress has direct contacts with senior military officers.

Civil bill court. A former court in Northern Ireland with a jurisdiction similar to that of an English county court.

Civil Constitution of the Clergy (1790). An attempt during the French Revolution to reorganize the Catholic Church in France on a national basis after the National Assembly had confiscated church lands and abolished tithes. The main features of the proposed Civil Constitution of the Clergy were to make dioceses correspond to départements, to have bishops and priests elected, and the clergy paid by the State. The Civil Constitution provoked much opposition and about half of the clergy refused to take an oath supporting the new state constitution and the reorganization of the Church. The Pope condemned the Civil Constitution in 1791 and the divisions were not healed until Napoleon's Concordat with the Pope in 1801.

Civil Courts. Courts exercising civil jurisdiction, e.g. declaring parties' rights, awarding damages, or other remedies, as distinct from administrative tribunals which determine disputes arising incidentally to the performance of public services; and from criminal courts exercising criminal jurisdiction, trying persons for crimes and in case of conviction imposing penalties. In the context of military law (q.v.), however, a civil court means any court other than a service court or tribunal and includes criminal courts.

Civil death. A man was formerly said to be civilly dead (*civiliter mortuus*) when he abjured the realm, went into a monastery, or had been attainted of treason or felony. The concept is now obsolete.

Civil disobedience. A symbolic or ritual, non-violent, violation of law, done deliberately as a protest against some injustice or social situation. It has been extensively employed by nationalist movements, civil rights movements, anti-war movements, and similar groups. By violating the law and submitting to punishment the members of the group hope to set a moral example which will bring about the change they desire. The philosophical basis of civil disobedience can be found in Cicero, Thomas Aquinas, Locke, and Jefferson, all of whom sought to justify conduct by its harmony with an antecedent and superhuman moral principle. The concept was clearly formulated by Mahatma Gandhi and used by his followers in quest of equal rights and freedom, and it has also been used by American negroes seeking civil rights.

Civil interests. The interests as civilians of persons called up for certain compulsory or in some cases voluntary service in the armed forces, e.g. security of tenure of places of residence. Statutory provision is made for the protection of such interests during service in the forces.

Civil law. A term used in various different senses:

(1) In that the Roman law (q.v.) civil law (*ius civile*) was that applied to Roman citizens (*cives Romani*) only, as contrasted with *ius gentium*, law applicable to *peregrini*, immigrants and foreigners at Rome.

(2) Civil law (*ius civile*) was also distinguished from the rules of law developed by the praetors and other Roman magistrates (*ius praetorium*, *ius honorarium*).

(3) As the privileges of the *ius civile* were extended and more and more persons in the ancient world came to be granted Roman citizenship, *ius civile* came to mean Roman law, the law of the Roman world and of Roman citizens everywhere.

(4) From being the law of the Roman State (*civitas*) the term civil law was extended to the law which any state had constituted for itself.

(5) With the development of the canon law (q.v.) of the Roman Church, civil law came to mean the whole body of state law as opposed to the body of church law, and the two great bodies of legal materials were named the *corpus iuris civilis* and the *corpus iuris canonici*.

(6) In countries whose legal systems have been founded closely on the Roman civil law, civil law means the law applicable to ordinary citizens, as distinct from commercial law, applicable to merchants or persons in trade, which comprises rules derived from the law merchant, the law maritime, and the customs of traders; thus France has a Civil Code and a Commercial Code.

(7) In the context of comparative law (q.v.) a civil law system is one mainly based on the Roman law, as contrasted with a common law system, one mainly based on the common law of England.

(8) Civil law is commonly contrasted with criminal law (q.v.) and administrative law (q.v.), the distinction depending on the sanctions for contravention of the rules, and the courts in which and procedure by which the bodies of rules are enforced; it is the body of law which regulates the conduct of persons in ordinary relations with one another.

(9) Civil law is also contrasted with military law (q.v.), as including all rules, civil and criminal, not part of the exclusively military code, and with ecclesiastical law as including all lay or temporal law.

(10) Civil law is sometimes used to designate the national or municipal law of a state as contrasted with public international law.

See also COMMON LAW; ENGLISH LAW; ROMAN LAW.

Civil law systems. The general term for the group of legal systems which developed mainly in Western Europe in the countries which rose from the ruins of the Western Roman Empire and in which concepts and principles drawn from Roman law have had powerful influence. After the collapse of the Western Roman Empire, western and central Europe were dominated by Germanic peoples and their customs tended in most areas to prevail over the law established by the Roman Empire. But the canon law of the western Roman Church kept alive the legal ideas and concepts of Rome, and in the eleventh century the Roman civil law was rediscovered and became the subject-matter of intense study, particularly in northern Italy. From there the teaching and study spread, encouraged by the demand for judges and administrators, so that Roman law spread north of the Alps and powerfully influenced the administration of justice particularly in Germany and the Low countries.

This reception of Roman law was facilitated in

the Holy Roman Empire because the emperors regarded themselves as successors of the Roman Caesars and emphasized the continuity of their succession, because the Romano-canonical procedure was superior to customary and irrational modes of procedure and proof such as ordeal and battle, and because the training of jurists in Roman law was intellectually superior to the undeveloped qualities of local and untrained judges.

Nowhere did Roman law wholly supersede local law, but in many places it produced various mixtures. In many places it strongly influenced the law of contracts and delicts, while canon law strongly influenced matrimonial law. Both, along with other elements, such as feudalism and Germanic traditions, influenced the law of property and succession. Roman law strongly influenced legal concepts, thinking, and argument and everywhere it was the basis of study, training, and discussion.

After the development of the distinct nation-states the movement for codification tended to differentiate legal systems into distinctive national varieties, albeit with an underlying unity. The earlier codifications were in Denmark (1683), Norway (1687), Sweden-Finland (1784), and Prussia (1791), but by far the most influential were the groups of French codes enacted under Napoleon, notably the Code Civil of 1804. The French codes were enforced on Belgium and Luxembourg, copied in the Netherlands, Italy, and Spain and in Latin America. In Germany codification came later, the civil code (BGB) not becoming effective till 1900. But throughout the nineteenth century German legal scholars, working on the modernized Roman law in force in Germany, had immense influence on legal science and scholarship everywhere. The German civil code was similarly taken as a model by Switzerland and much of eastern Europe. Accordingly modern civil law systems are normally codified systems on either the French or German model.

Most of the legal systems of continental Europe are civil law systems. From Europe the pattern spread with colonization to Latin America and elsewhere and to countries which remodelled their law on Western lines, such as Thailand, Turkey, Ethiopia, and Japan. Scandinavian countries have a distinctive kind of law, more like civil than common law. In a few countries, Scotland, Quebec, Louisiana, the Philippines, South Africa, Rhodesia, Sri Lanka, and Mauritius, originally civil law systems have been overlaid by common law principles and are now of a mixed character.

The characteristics of civil law systems are, normally, the existence of codes covering large areas of the law and setting down the rights and duties of persons in fairly general terms, the use of terminology and concepts and frequently of principles which can be traced back to the Roman law, a less strict regard for judicial precedents, and a greater reliance on the influence of academic lawyers to systematize, criticize, and develop the law in their books and writings.

Civil liberties. The compendious term in the U.S. for the main liberties, privileges, and immunities of citizens.

See Civil Rights and Liberties.

Civil list. The financial provision made in the U.K. by Parliament for the salaries and expenses of Her Majesty's Household and for the personal incomes of the sovereign and other members of the Royal Family. It is made in lieu of the revenues from royal patrimonies, which are assigned to the Treasury for public use. The first Civil List Act was passed in 1697; certain revenues were assigned to the King to pay the expenses of the royal household, the royal personal expenses, and the civil services generally, but not defence. Under George III the hereditary revenues of the Crown were assigned to the Exchequer, but a fixed allowance was paid to the King. In the time of William IV the Civil List was freed from all charges for government service other than those for the court and the royal family and this practice has continued. Various alterations were made to the Civil List provision on the commencement of each of the succeeding reigns. The sums appropriated to the Civil List are charged on the Consolidated Fund.

Royal trustees are appointed to manage the provisions and to report to the Treasury thereon.

In addition to the Civil List the Queen receives the revenues of the Duchy of Lancaster. The Prince of Wales receives the revenues of the Duchy of Cornwall.

For long the practice has been to settle a provision by statute at the commencement of a reign and not thereafter to vary it, but in 1972 it was found necessary by reason of inflation to increase the provision made for Queen Elizabeth's Household in 1952 and to give power to the Treasury to increase certain financial provisions by order.

Civil power. The justices of the peace, police, and others charged with the preservation of the public peace. The armed forces may, very exceptionally, be used in aid of the civil power to preserve the peace.

Civil procedure. The body of rules laid down regulating the mode of presentation to, and adjudication by, civil courts, of various kinds of claims, controversies, and disputes by persons against one another. The essential elements of civil procedure are provision for: (a) a statement of a claim, or matter of doubt or dispute, by one person against another or others; (b) communication of the claim to those whose interest is to contradict the claim made and a statement of a defence, answer, or other contrary view, by some or all of them; (c) the ascertainment of any facts in doubt or dispute; (d) the decision of any doubtful or disputed matter of law raised by the controversy; (e) the decision of the claim or dispute, on the basis of the facts found to exist and of the rules of law held relevant, and the issue of a judgment decisive of the issue; and (f) provision for an appeal, rehearing, or reconsideration of the case. There are many incidental elements of procedure, but all ancillary to one or other of the main elements mentioned. Procedure is a matter peculiar to each legal system and there are great variations in detail between different systems.

Civil proceedings. An application or claim made in a court with the object of declaring or enforcing a right for the advantages of the person claiming, or of recovering money or property, as contrasted with administrative or criminal proceedings which have the objects of securing a benefit or the punishment of a public offence respectively.

Civil remedies. The rights which an aggrieved individual may exercise, and the orders which civil courts may make, for the redress of grievances and the securing of an aggrieved individual's rights. They are distinct from political remedies such as appeal to the Parliamentary Commissioner (q.v.) or a local authority; from administrative remedies such as the award of a social security benefit, and from criminal prosecutions. Civil remedies include awards of damages, decrees of divorce, orders for specific performance, injunctions, and many other kinds of orders (qq.v.).

Civil rights or civil liberties (or the liberties of the subject, personal and group liberties recognized and protected by law). They overlap with individual rights or liberties, but belong more to the area of social and public interests than do individual rights, which belong mainly to the area of individual interests. They are concerned essentially with what individuals and groups may do within the law, e.g. stand for election to a public authority, rather than with what they may exact, e.g. social security. Civil rights may be regarded as attempts to give meaning to the ideal of equality under laws, and civil liberties as flowing from the ideal of freedom.

Civil rights protect certain general human needs and interests, but sometimes conflict with other human needs and interests, conflicts which have to be adjusted and resolved in the courts or by legislation. A civil right or liberty exists only in so far as it is legally recognized and protected, not merely if it is asserted, or even proclaimed by a government or political party.

Civil liberties are distinguishable from moral liberty or freedom of the will; from political liberties, such as the right to elect and to stand for election; from 'human rights' or 'natural rights'; from economic liberties, such as freedom of contract, trade, competition, of organizing, and of striking; from religious liberties, such as freedom of belief and of worship; and from academic freedom; though there is considerable overlapping, and as law in general protects each of these groups of liberties, some would class all of these within the general group of civil liberties. Civil liberties were justified by seventeenth- and eighteenth-century philosophers as inherent or inalienable rights. Historically most of them arose by way of successful resistance to kings, harsh employers, unrepresentative parliaments, and the like.

The question what civil rights or liberties citizens enjoy under a particular system of government and law, depends partly on what rights or liberties are conferred by constitution, code, statute, and case, but also, and frequently more particularly, on how these rights or liberties are actually interpreted, and how far they can be and are secured and enforced, particularly by minority groups.

The civil liberties now recognized in Anglo-American civilizations are based on philosophic doctrines, historical events, enactments, and cases, and the common law. Their development is one aspect of the intellectual and legal history of Britain and U.S.A. Among philosophic doctrines may be mentioned the Stoic doctrine of the moral autonomy of the individual, the doctrine of natural rights, Christian doctrines of the freedom of conscience, and the equality of men. Among classic political texts there are Milton's *Areopagitica* (1644), Paine's *Common Sense* (1776) and *Rights of Man* (1791), and Mill's *On Liberty* (1859). Historical events include the Reformation, the constitutional struggles of the seventeenth century, and the Puritan Revolution. Major events, enactments, and cases include Magna Carta (1215), the Statutes of Westminster (1275), the Petition of Right (1628), *Bushell's case* (1670), the Habeas Corpus Act (1679), the Bill of Rights (1688), the Toleration Act (1689), the Acts of Settlement (1700–1), Fox's Libel Act (1792), the Reform or Representation of the People Acts (1832, 1867, etc.).

In America the earlier colonists brought with them the traditions of England and the turmoil of ideas which underlay the seventeenth-century political struggles. In New England social contract ideas were prevalent and later natural rights were invoked and ideas imported from French revolutionary doctrines. Early assertions of liberties included Maryland's Toleration Act (1649), the Charter of Rhode Island (1662), Otis' *The Rights of the British Colonies Asserted and Proved* (1764), and Samuel Adams' *The Declaration of the Rights of Man*

CIVIL RIGHTS

(1772). The First Continental Congress issued a Declaration of Rights and Liberties in 1774, and the Declaration of Independence of 1776 is based on the ideas of natural rights and civil liberties. Several states adopted bills of rights, notably the Virginia Declaration of Rights (1776) and those of Maryland (1776), New York (1777), and Massachusetts (1780). More important have been the Ordinance of 1787 which organized the Northwest Territory and inaugurated the principle of development of areas into territories and states, the Bill of Rights of 1791 (the first ten amendments to the Constitution), the post-Civil War amendments (1865–70), and various decisions of the Supreme Court.

Since the French Revolution declarations of the rights of man have frequently been prefixed to or included in many constitutions of newly created or independent or developing countries, but the crucial question is whether such rights can be enforced and vindicated in case of need. Periodically there are demands in Britain for the formulation of a declaration of rights, but this seems likely to be productive more of dispute as to drafting than anything else.

Civil liberties are usually advocated by or for the underdogs in the community because they are, in the particular respect in question, the minority or hold the unpopular ideas, fear repression and seek to propagate their views or at least to be allowed to do what they wish. The most striking instances of denial of civil rights and struggle to have them made effective has been the resistance of the Southern States of the U.S. to the principle of negroes having equal rights with whites. The Fifteenth Amendment of 1870 gave negroes the vote, but it was not effective for nearly a century. Other aspects were the notable school segregation cases, e.g. *Brown* v. *Board of Education* (1954), 347 U.S. 483. A Civil Rights Act of 1964 contained provisions against discrimination and was followed by a Voting Rights Act of 1965.

Precisely what civil rights or liberties are recognized in a given state at a particular time may be difficult to determine as it depends partly on the constitution and other public law, and very largely on how the law is secured and enforced. Traditionally British courts were zealous in protecting liberties, but many twentieth-century cases show reluctance to do so in face of widely drawn legislation. In Anglo-American countries the major civil rights are freedom from subjection, serfdom, or slavery; freedom of movement within and from the state; freedom of speech, writing, the Press, and broadcasting; freedom to co-operate, organize, and combine in groups for all manner of lawful purposes; freedom of assembly and meeting; freedom of belief, worship, and religious observance; freedom of access to the courts. Some liberties are imperfectly recognized and protected, notably privacy or freedom from interference.

Liberties cannot be absolute without leading to anarchy and excessive interference with the liberties of others. The crucial questions are of the points at which certain liberties must be limited by regard to the liberties of others. The balance between reasonable and excessive liberty in particular respects must be struck partly by morality, custom and public opinion, and reasonableness, and partly by legislation and judicial decision. Thus freedom of speech and of the Press is limited by the rules of libel, obscenity, sedition, and the like; freedom of assembly by the needs of public safety and order, of passage in the streets. Systems of liberties and freedoms are functions of the balance of power between political and economic groups and depend on current social and politico-economic thinking. Thus in the nineteenth century the prevailing philosophy favoured the maximum freedom of contract. But today freedom of contract is increasingly being restricted in the interest of the consumer or party in the weaker bargaining position.

In wartime and other national emergency states normally consider it right and necessary to restrict various liberties, notably freedom of the Press, of movement, and of criticism or of opposition to government.

The abuse of civil liberties always creates a demand for the limitation or abolition of the liberty abused. The right to organize in trade unions has been used to monopolize, exploit, oppress, intimidate, and ruin as much as to represent or bargain collectively; the right to strike is often used as a political weapon against society as a whole. The liberty to propagate certain kinds of social, or political, or other opinions, e.g. sexual licence, or communism, or drug-taking, may be denied by many and there may be strong movements to deny or restrict a liberty in so far as used in a pernicious way. Must a democracy allow Communists to use liberty of opinion, speech, and writing to advocate and encourage the overthrow of that democracy and substitute a denial of liberty of opinion, speech, and writing? Again a dominant political party is apt to deny to a minority group liberty not to conform, e.g. to run an independent school, in the interest of equality of opportunity.

There are numerous arguments both for and against civil liberties and it cannot be assumed that continued extension of the sphere of such liberties is either automatic or necessarily beneficial to anyone. Increasingly the protection of the individual, not only against the power of the State, but against the power of large organizations and groups within the State, requires to be safeguarded.

A. V. Dicey, *Introduction to the Study of the Law of the Constitution* (1912); Whipple, *The Story of Civil Liberty in the U.S.* (1927); H. Laski, *Liberty and the*

Modern State (1930); Z. Chafee, *Free Speech in the United States* (1940).

Civil Rights Acts. Legislation passed at various times in the U.S. since the Civil War, usually as efforts to secure the position of negroes. The 1866 Act conferred citizenship on negroes but was held unconstitutional by the Supreme Court in 1866, but similar Acts of 1957, 1960, and 1964 have gone far to prevent racial discrimination in voting, employment, and facilities.

Civil Rights Cases (1883), 109 U.S. 3. The Fourteenth Amendment to the U.S. Constitution sought to confer on negroes full rights of citizenship and Congress enacted broad protections of civil rights, as in the Civil Rights Act of 1875. When cases reached the Supreme Court it held that the Fourteenth Amendment has not given Congress power to protect civil rights but only power to correct abuses by the States. The decision accordingly relieved Congress of its basic obligation for the protection of the civil rights of negroes. The Court's attitude was not reversed till cases in the 1950s following on *Brown* v. *Board of Education* (q.v.).

Civil service. The body of full-time officers employed by a state in the administration of civil affairs as distinct from members of the armed forces. It is distinct from politicians who may for the time being be appointed heads of executive departments. It may or may not include provincial and municipal officers. It comprises both general administrators and professional, scientific, and technical specialists. In most countries a distinction exists between the home civil service and the diplomatic or foreign service. Everywhere civil services are organized in hierarchical structures of offices, powers, and duties.

Civil services have existed from the earliest large-scale organized governments, being indeed necessities of such organizations. The Chinese had a civil service from the seventh century and the Roman empire had one.

The foundations of modern European civil services were laid in Prussia in the late seventeenth and eighteenth centuries and by Napoleon's development of a highly organized hierarchy, which was copied by many countries in the nineteenth century. In the U.K. civil servants were originally the monarch's personal servants and members of the King's household. Considerable stimulus was received from Clive's creation from 1765 of a civil service to govern India and the genesis of the modern system is to be found in Macaulay's report on recruitment to the Indian Civil Service which inspired the report of 1854 on the organization of the permanent civil service in Britain which recommended recruitment by open competitive examination, the selection of higher civil servants on the basis of general intellectual attainment rather than technical capacity, and the establishment of a Civil Service Commission to ensure proper recruitment.

In the U.K. the civil service provides the staff of all the Ministries and Departments, but does not include the judiciary, staffs of local authorities, public corporations or universities, nor the police. The Civil Service Department, the Minister for which is the Prime Minister, manages the civil service. The everyday running of the Department is handled by the Lord President of the Council. The Permanent Secretary of the Department is Head of the Civil Service.

Senior civil servants have important functions in guiding and advising Ministerial heads of departments, in shaping policy and proposing developments, as well as in applying and executing policy. The middle and lower grades actually administer and apply the policies fixed by Parliament. The civil service as a whole is politically neutral and generally honest and fair. Persons in senior grades are debarred from political activity.

In the U.S.A. patronage was long the basis of recruitment. Entry by open competitive examination was not accepted until after the Civil War and, in 1883, the U.S. Civil Service Commission was established. At first its work was restricted to the lower grades, but its sphere has gradually expanded though many of the principal policy-making offices continue to be filled by presidential nomination. Each state, county, and city has its own civil service, and recruitment by merit has increased since 1900. The judiciary is not part of the civil service.

The French civil service was founded by absolute monarchs before 1789 and reorganized by Napoleon. Notable modern features include competitive entry, the high standard of the various professional schools such as L'École Nationale d'Administration and L'École Polytechnique, and the powerful control exercised by the Conseil d'État checking administrative abuses and injustices and giving the aggrieved individual rapid and effective relief against maladministration.

The German civil service developed out of feudalism under a series of powerful kings culminating in Frederick the Great. It developed traditions of integrity, industry, and obedience. Until 1933 each German state had a separate civil service, but all were strongly influenced by the Prussian system. Since 1945 there has developed a modernized system with a hierarchy of administrative courts controlling the exercise of administrative powers.

Everywhere civil servants in fact possess enormous power, and questions have arisen of the adequacy of ministerial control and judicial power to control abuses of power or maladministration.

Their powers have given rise to the term 'bureaucracy', indicating fear that governing power belongs to the men in the office, not to the M.P.s or ministers. In the U.K. the office of Parliamentary Commissioner for Administration (q.v.) has been created to inquire into alleged maladministration. Many European countries have modelled an administrative jurisdiction on that of the French Conseil d'État (q.v.) which has wide powers to check administrative mistakes and injustices.

In the twentieth century the growth of international organizations has given rise to the creation of international civil services drawn from a wide range of countries, owing allegiance to the organization and its aims and not being representatives or agents of their own country. Thus substantial bodies of civil servants are employed by the U.N.O. and its specialized agencies, by the European Communities, and by specialized international bodies.

E. Barker, *The Development of Public Services in Western Europe*, 1660–1930; W. J. Mackenzie and J. Grove, *Central Administration in Britain*; B. Chapman, *The Profession of Government*; M. H. Bernstein, *The Role of the Federal Executive*.

Civil Service Commission. A body established in the U.K. in 1855 to be an autonomous, semi-judicial body charged with the recruitment of persons for the public service. It seeks to select on the basis of character and ability, and recruits both persons of general ability for general administration and persons with professional or technical qualifications for specialized branches of the service.

Civil Service Commission (U.S.). This was the first Federal Regulatory Commission, created in 1883, to make appointments to many Federal governmental posts, initially mostly in the lower grades, on the basis of merit. The merit system has subsequently been greatly extended and applied to many more categories of appointments. The principal policy-making offices are still outside its ambit and filled by presidential nomination.

Civil war. Large-scale and sustained hostilities (as distinct from mere revolt or rising) between the organized armed forces of two or more factions within one state, or between the government and a rebel or insurgent group. Notable examples have been the English Civil War or Great Rebellion (1642–51), the U.S. Civil War (1861–65) and the Spanish Civil War (1936–39). In international law a state of civil war exists only if two sides are recognized by other states as belligerents.

Civil wrong. A breach of legal duty causing harm considered as giving rise to a civil claim for a civil remedy, normally damages, as contrasted with a breach of legal duty considered as inferring liability to punishment or penalty as being criminal. In undeveloped law the distinction between civil and criminal wrongs is unclear. In developed law it is clearer, but there are some kinds of conduct which infer civil liability only, e.g. negligent personal injuries; others which may infer both civil and criminal liability, e.g. assault; and others which infer criminal liability only, e.g. many road traffic offences. To be a civil wrong the conduct need not be morally wrong, or in any sense illegal; it is wrong only as being in breach of duty not to have done or allowed that kind of conduct.

Civilian. (1) A person learned in the civil law of Rome. Originally the term was opposed to canonists, or decretists, or decretalists (qq.v.), who were learned in the canon law, and also to common lawyers, learned in the common law of England.

In England the term was particularly applied to advocates of Doctors' Commons (q.v.) who practised in the ecclesiastical and admiralty courts, the law of which courts had been much influenced by Roman law. In the later sixteenth century the civilians in England had extensive practice in these courts, in diplomacy, in arbitrations ordered by the Council in which points of foreign commercial or maritime law were involved, in some cases before the Star Chamber, Chancery or Court of Requests, in the courts of the Constable and Marshal and of the two old universities, and in various matters arising in connection with state business. They made important contributions to the development of modern law in developing commercial law and the prize jurisdiction in the admiralty courts and in ecclesiastical (including marriage) law which remained in their hands until 1857, and also in early writing on jurisprudence and comparative law.

(2) A person other than a member of one of the armed forces.

Claim. A general term for the assertion of a right to money, property, or to a remedy. It is also used of certain kinds of pleadings, notices in legal proceedings, and other statements.

Claim of liberty. A suit in the former Court of Exchequer to have a liberty or franchise confirmed by the Attorney-General.

Claim of Right. The Scottish counterpart of the English Declaration of Right of 1689, which declared that James VII had forfeited his right to the throne and offered it to William and Mary.

Claims, Court of. See COURT OF CLAIMS.

Clandestine marriage. Originally a marriage contracted secretly, which originally was possible by exchange of oral consent, without any religious solemnities. The fourth Lateran Council (1215) required the proclamation of banns and marriage in church as requisites of every regular marriage, but marriages though clandestine were valid if blessed by a priest. The Council of Trent, however, put an end to clandestine marriages by declaring invalid marriages not celebrated by the priest of the parties' parish.

The term was also applied to marriages celebrated by a priest but without prior proclamation of banns, or not celebrated in church. Later still it was applied also to marriages celebrated by priests or ministers not of the denomination established for the time being, marriages without consent of parents, not celebrated on Sunday, not celebrated in church, or otherwise irregularly.

Clandestine Outlawries Bill. A bill which is given a first reading in the House of Commons at the commencement of every session of Parliament before the debate on the Address, but which is never debated nor proceeded with further. The sole purpose is the continuing assertion of the right of the House of Commons to have redress of grievances before granting supply, and to debate whatever business they think necessary in priority to any measures proposed in the sovereign's address. In the House of Lords the corresponding bill is the Select Vestries Bill (q.v.).

Clapham omnibus, man on the. A phrase used by Lord Bowen and frequently used as a colourful variant of the 'reasonable man', the hypothetical person commonly taken as the standard for judging whether the conduct of the defendant in a negligence action comes up to or falls short of the standard of care which the law requires in the circumstances. He has not the courage of Achilles, the wisdom of Ulysses, or the strength of Hercules, though Lord Bramwell occasionally attributed to him the agility of an acrobat and the foresight of a Hebrew prophet.

Clapmar, Arnold (1574–1604). Professor of history and politics at Altdorf, author of *De Arcanis rerum publicarum libri sex* (1641) and works on education.

Clare, Earl of. See FITZGIBBON, JOHN.

Clare constat, precept of. In Scots law, a deed formerly granted by a feudal superior to the heir of a deceased vassal setting out that it clearly appears that the claimant is the lawful heir on which the heir might take sasine. A writ of clare constat was a later statutory version which also confirmed all deeds necessary for a good title.

Clarendon, Earl of. See HYDE, EDWARD.

Clarus, Julius (1525–75). An Italian humanist jurist, specializing in criminal law, author of *Practica criminalis sive Sententiarum Receptarum Libri Quinque.*

Clarendon, Assize of. A code of 22 articles promulgated by Henry II in 1166 and stating the principles on which the administration of justice was to be effected. It contained instructions for the use of the judges going on circuit for dealing with criminals. In every county and every hundred the 12 most lawful men of each hundred and the four most lawful men of each vill should be sworn to present any man who was suspected of serious crime to the King's justice or to the sheriff. If he had been captured, he was to be handed over with two men to state the circumstances of the capture.

Clarendon, Constitutions of (1164). A record in 16 articles issued by Henry II of the customs ascertained to have regulated the relations of church and state in the time of Henry I, acknowledged by the bishops in presence of the barons of the kingdom and containing their promise to observe them. They were intended to curb the power of the ecclesiastical courts and restrict ecclesiastical privileges. The major points were that all clerics accused of crime should be tried first by the King's justices, to determine whether the case should be tried in the lay or spiritual courts, and if in the latter, the accused, if found guilty, was not to be protected by the Church; tenants-in-chief and officers of the King's household were not to be excommunicated without consent of the King or justiciar; the King was to have custody of vacant sees and abbeys and the revenues paid to him; new bishops and abbots were to be elected by the clergy with the King's consent, and before consecration the person elected must do homage to the King, saving the rights of his order; appeals should proceed from archdeacon to bishop and then to archbishop, and in the last resort to the King; and the sons of villeins were not to be admitted to holy orders without the assent of the lord on whose land they were born. The constitutions provoked the quarrel between Henry II and Becket. The bishops led by Becket at first reluctantly swore support of the constitutions, but Becket soon repudiated his oath and was forced into exile. His martyrdom in 1170 forced Henry to moderate his attack on the clerical powers though he did not withdraw any of the constitutions.

Clarendon Code. The common name of four Acts passed in England when Edward Hyde, Lord Clarendon, was King Charles II's chief adviser, all intended to curb the powers and liberties of Dissenters and Nonconformists. The Acts were the Corporation Act (1661), excluding them from municipal offices; the Act of Uniformity (1662), excluding them from church office; the Conventicle Act (1664) and the Five Mile Act (1665) restricting their liberty of worship.

Clarke, Thomas (1703–64). He was probably called to the Bar in 1729 and soon developed a large practice. He sat in Parliament from 1747 to 1761. Through the favour of Hardwicke he was made Master of the Rolls in 1754, and held that office till 1764. He was a sound and learned lawyer, is said to have refused the chancellorship in 1756, and published an edition of *Fleta* in 1735.

Classification. (1) The establishment of an hierarchical system of categories and classes of principles and rules of law, and the allocation of a principle or rule of law to the appropriate branch of the legal system. Classification may be made on various bases, factually, by bringing together all the rules about animals, or books, or houses, and so on; but this would do very little to systematize a mass of legal rules. Or it could be done functionally, by bringing together all the rules relating to banking, or building, or medical treatment, and so on. In fact the basis of classification normally adopted is by reference to the general legal concepts and relations involved, to such concepts as contract, ownership, succession; and such relationships as employer and employee, landlord and tenant, husband and wife. This is exemplified in such works as the *Index to the Statutes*, the main *Digests* of Case-law and the major encyclopaedias such as *Halsbury's Laws of England.* There has developed a fairly well-settled scheme of headings and sub-headings. Classification is very important for the discovery of the principles of law applicable to a particular case and for the statement of principles and rules of law in encyclopaedias and textbooks.

The first purpose of classification is academic, to arrange the rules of law and the problems they cover in a rational structure, bringing together those which are related and connected. This is a task traditionally performed by the academic jurist and textwriter.

The second purpose, which follows from the first, is to facilitate study, understanding, and discovery of the rules relevant to a particular problem; thus cases and statutes are classified by editors of reports and reference books under recognized legal headings, such as contract, trusts, negligence, and so on. Under these heads the trained lawyer will look, rather than under 'kick by a horse' or other factual headings.

The third purpose is practical; it may be relevant for such purposes as jurisdiction to enquire whether a claim is founded on contract, or tort, or trust. In particular, in cases involving issues of international private law a court must frequently classify the cause of action, or determine the juridical nature of the question requiring decision, before it can apply the appropriate rule for the choice of law. In this context classification is frequently called 'characterization'. Thus a court must decide whether a foreign widow's claim to a share of her late husband's land is a question of matrimonial rights (determined by the *lex domicilii*) or a question of succession to land (determined by the *lex situs*).

How legal principles and rules are classified is a matter for each particular system, and particularly for the academic jurists working in that system. Much confusion and difficulty has been caused by attempts to classify English law by categories of Roman law as if those categories were of universal application, and there is probably no scheme of classification universally applicable. For practical purposes, as in an action raising a question of international private law, a judge must classify the problem facing him in accordance with the scheme of classification, and the heads and sub-heads, accepted in his own legal system, not, however, interpreting any of the categories too narrowly, lest there be no room for a rule or institution of foreign law which has no counterpart in the judge's own system. Thus community of property between husband and wife, unknown in English law, must be allocated to the sphere of contract or of succession by an English court.

The classification used in practice in English law has been determined mainly by legal history, by the development of the forms of action, the importance of land, the growth of equity, and so on. In Scotland the classification used owes much more to jurists who adopted and built on the categories of the Roman law.

(2) In prison administration, the sorting of prisoners into various classes for which different prisons or different parts of prison are reserved. The purposes of classification are to enable untried prisoners to have privileges denied to convicted prisoners, to segregate the sexes, to separate unsophisticated prisoners from the influence of experienced criminals, to enable prisoners sentenced to special régimes to be differently treated in certain ways, to ensure stricter security, to keep those subverting discipline apart.

Claudius, Appius (fifth century B.C.). The leading member of the commission which published the *Law of the Twelve Tables*, the first authoritative statement of Roman law (q.v.).

Clause. The numbered divisions of a bill presented to Parliament which, if the bill is passed, become sections of the Act. At the committee stage each clause is debated separately, and new clauses and amendments may be proposed and carried.

Clauses Acts. The general term for a number of statutes, mostly passed in 1845–47, the purpose of which was to settle the form of, and bring together in one Act, many detailed provisions which commonly appeared in special Acts authorizing various kinds of public undertakings, such as the construction of railways. Thereafter the provisions of the appropriate Clauses Act could be incorporated by reference in subsequent special Acts, thereby promoting uniformity and consistency and shortening and simplifying the special Acts. Some Clauses Acts are deemed incorporated in later special Acts unless varied or excluded thereby; others apply only in so far as expressly incorporated.

Clauson, Albert Charles, Baron (1870–1946). A nephew of Buckley, Lord Wrenbury (q.v.), was called to the Bar in 1891 and developed a large junior practice in Chancery. He took silk in 1910. He became a judge of the Chancery Division in 1926, a Lord Justice of Appeal in 1938, and on his retirement in 1942, a baron; he thereafter sat occasionally in the House of Lords and Privy Council. He had a clear and logical mind and his judgments were lucid.

Clausula rebus sic stantibus. In international law, the doctrine that a treaty is intended to bind the parties only so long as there is no fundamental change in the circumstance which existed at the time of the negotiation of the treaty and were taken as the basis thereof.

Clayton, Richard (1702–70). Called to the English Bar in 1729 and went on to become Chief Justice of the Common Pleas in Ireland in 1765 but resigned in 1770.

Clayton-Bulwer Treaty. A treaty between the U.S. and Great Britain in 1850 binding both parties not to obtain or maintain any exclusive control of the area of the proposed Panama Canal, guaranteeing the neutralization of the canal, and binding both parties not to occupy, colonize, or fortify that part of Central America. The Hay-Pauncefote Treaty of 1902 replaced the Clayton-Bulwer Treaty but affirmed the neutralization of the Panama Canal.

Clayton's case (or properly *Devaynes* v. *Noble, Clayton's Case* (1816), 1 Mer. 572). The leading case on the appropriation of payments in a current account, as with a banker, laying down that, in the absence of express appropriation by the creditor to particular debts, credit payments are presumed appropriated to discharge those debts in the order in which they had been incurred. The rule does not apply where a person in a fiduciary position pays money acquired in that character into his own account, in which case he is presumed to withdraw his own money, not the trust money.

Clean hands. A traditional maxim of equity (q.v.) is that a man must come to a court of equity with clean hands, that is, with a claim unsullied by fraud or other unfair conduct.

Cleisthenes (sixth century B.C.). Athenian statesman and the founder of Athenian democracy. He secured the change of basis of political organization from the family and class to the locality, to membership of a deme or township. Then new tribes were founded, townships from each of the urban, inland, and coastal areas being included in each tribe. Solon's Council of 400 was increased to 500, 50 from each tribe. His reforms led to wider participation in public life.

Clementines (or *Constitutiones Clementis V*). The authentic collection of papal legislation of Clement V and of the Council of Vienne (1311–12) promulgated by John XXII in 1317. They did not abrogate all other legislation after the promulgation of the *Sext* (q.v.) in 1298. The collection is, like the *Decretals* and the *Sext*, divided into 5 books, 52 titles and 106 chapters. As early as 1319 commentaries began to appear, notably those of William of Mont Lauzun and Gesseliu de Cassanges (1323), and a gloss by Joannes Andreae (1326).

Clement's Inn. One of the Inns of Chancery (q.v.) named from the Church of St. Clement Danes, to which the Inn originally belonged.

Clergy. The general name for the class of persons distinguished from the laity as having a special status to administer spiritual counsel and religious instruction and to perform religious ceremonies. Prior to the Reformation clergy were divided into regular clergy, subject to the rules of a religious order, particularly abbots, priors, and monks; and secular clergy, not subject to such rules, including bishops, deans, and parsons. The term now comprehends persons of all branches of the Christian Church who have been ordained to the ministry of God's word, and of all ranks in hierarchical denominations. Numerous statutes confer particular powers or privileges, or impose duties or disabilities on clergy, while a substantial part of ecclesiastical law is concerned with discipline of the clergy.

Clergy, Benefit of. The principle of benefit of clergy developed from the contest between Henry II and Becket as to jurisdiction over clerics accused of crime. An accused who pleaded his clergy could not be punished by a lay court but had to be handed over to church authorities who were supposed to try but usually released him. Later benefit of clergy was available only in cases of felonies and statute in the fourteenth century extended it to secular clerks and probably to everyone who could read. To satisfy themselves of the validity of a claim to clergy, judges introduced the practice of requiring accused to read (or recite) the 'neck verse'—Psalm 51, verse 1. It was later decided that only persons actually in orders could claim clergy a second time and under Henry VII it was provided that anyone convicted of a clergyable offence should be branded on the thumb, M for murder or T for theft. Elizabeth provided that laymen who relied on it should be liable to imprisonment for not more than a year and in the eighteenth century, despite a plea of clergy, a prisoner could be transported for 7 years instead of being branded. Originally nuns could take advantage of it, but from the late seventeenth century all women could do so. From the sixteenth century various felonies were made punishable 'without benefit of clergy'. The 'neck verse' was abolished in 1706, the branding in 1779, and the principle altogether in 1827.

Clerk, Sir John, of Penicuik (1676–1755). Called to the Scottish Bar in 1700, and achieved success and sat in the Scottish Parliament and in the British Parliament after 1707. In 1708 he became a baron of the Court of Exchequer in Scotland. He was the author of various antiquarian works and, with Baron Scrope, of the *Historical View of the Form and Powers of the Court of Exchequer in Scotland* (1726, published 1820).

Clerk of Arraigns. Formerly an officer of the Central Criminal Court and courts of assize who, in the absence of the clerk of assize, opened the assize, arraigned the prisoner and put to the jury the formal question to ascertain their verdict.

Clerk of Assize. Formerly the principal officer of assizes (q.v.) who read the commission, kept the formal records, arraigned the prisoner, and asked the jury the questions requisite to ascertain their verdict.

Clerk of the Council in Ordinary. The official who summons meetings of the Privy Council or committees of it.

Clerk of Justiciary. The offical responsible for arranging the sittings of the High Court of Justiciary in Scotland, and for keeping the books and records of the Court.

Clerk of Session. The clerk to the Court of Session in Scotland. There have at different times been different numbers of principal clerks, and depute-clerks. The office was formerly held by advocates or writers to the signet and notable holders of the office included Sir Walter Scott and George Joseph Bell (qq.v.).

Clerk of the Crown in Chancery. Head of the permanent staff of the Crown Office, which is headed by the Lord Chancellor. The office dates back to 1349 and the Clerk participates in many acts of state, such as reading the title of Bills in the House of Lords when they receive the Royal Assent, and sending out the writs of summons to peers and the writs instructing returning officers to hold elections and return the names of persons elected to the Crown Office.

Clerk of the Hanaper or Hamper. An officer on the common law side of the old Court of Chancery who saw that the writs were sealed in bags, later to be issued. The writs and the returns thereto were originally left in a hamper (*in hanaperio*). He also had functions in connection with patents and grants which passed the great seal.

Clerk of the House. The chief permanent officer of the House of Commons. He assists and advises the Speaker and members on matters of procedure, has the custody of all records and is head of one of the three administrative departments of the House of Commons, that concerned with the conduct of business, which is itself divided into five offices, Public Bill Office, Private Bill Office, Committee Office, Journals Office, and Table Office. He sits at the table below the Speaker's Chair, wearing wig and gown, and signs addresses, Bills sent to the Lords, and the like.

The Clerk is assisted by the Clerk Assistant and the Second Clerk Assistant, who keep minutes, receive notices and motions, questions and amendments and prepare the Order Book.

The office has been held by many who have contributed notably to the literature of Parliament and its procedure, notably Sir T. Erskine May (q.v.).

Clerk of the Markets, Court of. In mediaeval times the Clerk of the Markets of the King's Household had jurisdiction to inquire whether weights and measures were according to the King's standards or not. The charters of many boroughs included grants of a court of the clerk of the markets, which grant excluded the jurisdiction of the Clerk of the Markets of the King's Household and gave the borough corporation jurisdiction in its court to inquire as to weights and measures.

Clerk of the Parliaments. The chief permanent officer of the House of Lords and its adviser and authority on procedure. He is responsible for the minutes of the House, and for the custody of documents, and controls four administrative departments, Public Bill Office, Private Bill and

Committee Office, Judicial Office, and Accountant's Department. He participates in the ceremony of signifying the Royal Assent to Bills. He is assisted by a Clerk Assistant.

Clerk of the Peace. An officer originally appointed to keep the county records and assist the justices of the peace in drawing indictments, issuing process, entering judgments, and in managing administrative business. The office came to an end when quarter sessions (q.v.) were abolished in 1971.

Client. In Roman law a person dependent on a rich and powerful man to whom he looked for protection, advice, and assistance. In modern law the term is used of a person who consults a solicitor or instructs him to carry through any legal business. The relationship is highly confidential and is also one of a fiduciary character so that any transaction between solicitor and client, while the relationship exists, is viewed with suspicion as possibly affected by undue influence and is challengeable. A barrister's client is the solicitor who instructs him, not the layman who had consulted the solicitor.

Clifford's Inn. Originally the London home of the Barons Clifford and demised in 1345 to a body of students of law. It was one, and the most important, of the Inns of Chancery (q.v.). The society was dissolved in 1902 and the property sold.

Clog on the equity of redemption. Any provision in a mortgage with the effect of preventing redemption by the mortgagor on payment or performance of the obligation for which the mortgage was granted. Any such clog or fetter is void.

Close. A field or piece of land separated from others or from common land and enclosed by a bank or hedge. Legally unjustified entry on another person's close is a trespass, known as trespass *quare clausum fregit*, for breaking a man's close.

Close of pleadings. The stage of written pleadings when written averments have been concluded and issue is joined.

Close company. A company which is under the control of five or fewer participators or of participators who are directors, but not a company if shares carrying at least 35 per cent of the voting power are beneficially held by the public and have, within the past 12 months, been the subject of dealings on a stock exchange. The concept is relevant for taxation of the company's profits.

Close rolls. Chancery records extending from 1205 to the present, containing commands addressed by kings to one or more specified individuals, closed and sealed, as contrasted with open letters recorded in the patent rolls. They are preserved in the Public Record Office.

Close seasons. The periods of the year, defined in various statutes dealing with various kinds of fish and game, during which it is unlawful to take or kill the kind of fish or game in question.

Closed shop. A place of work in which trade union members working there refuse to allow non-union members to be employed, and consequently the system of that principle throughout an industry or industry generally. The germ of the closed shop can be found in mediaeval gild regulations preventing the working of non-members, and the principle was developed extensively in the nineteenth century. In the twentieth century as unions have become larger and more powerful the securing and maintaining of the closed shop has become a vital principle of trade union ideology and many disputes and strikes have been provoked by this demand. In 1975 in the U.K. the closed shop was statutorily authorized. An employer may on pain of dismissal compel an employee to belong to a trade union; the only excuse is genuine objection on grounds of religious belief to belong to any trade union whatsoever.

Several principles conflict in relation to the closed shop, the interest of the union or unions concerned to maximize membership, and to secure employment for their members exclusively, their solidarity, and the convenience to the employer of negotiating with representatives of all the workers; on the other hand, the exclusion from employment of individuals who for any reason disagree with, or cannot, or will not, join the union. Recognition of the closed shop makes the union and its officials complete masters of whether particular men can work or not and therefore infringes a fundamental civil liberty as much as did compulsory membership of a church, as in the Middle Ages. Any restrictions on access to jobs should be imposed by society and not by organizations not publicly responsible.

Closure. The procedure by which a debate in the House of Commons may be ended, even though some Members still wish to speak. It was introduced in 1882. Any Member may move the closure by moving 'That the question be now put'. If the motion is accepted, and it is in the Speaker's discretion to accept or reject it, it must be at once put to the vote; if there are at least 100 votes for the motion and it is carried, the debate must be ended and the question before the House put to the vote.

Club. A society of persons associated together for any purpose other than profit, such as for social intercourse, the promotion of arts, politics, sport, or other lawful purpose. The main legal division is into members' clubs, which are associations of persons, each of whom contributes to the funds of the club and jointly with the others owns the club property and assets, the rights and liabilities of members being settled by the rules; and proprietary clubs, where a proprietor owns the property, the members being entitled in return for their subscriptions to such rights as their contract with the proprietor provides. Either kind of club may be incorporated as a company, or be unincorporated. The relations between clubs and members are regulated by their constitutions and rules, and by general principles of contract and agency. Working men's clubs and shop clubs are registered Friendly Societies.

Clyde, James Avon, Lord (1863–1944). Having passed advocate of the Scottish Bar in 1887, he steadily developed a practice, took silk in 1901, became Solicitor-General for Scotland in 1905, M.P. 1905–20, and Lord Advocate 1916–20. He was also Dean of the Faculty of Advocates 1915–19. He became Lord President of the Court of Session in 1920, till his retiral in 1935. As such he showed himself a fine lawyer as well as a good head of the court; his judgments are among the best of the period. He published a translation of Craig's *Jus Feudale* (2 vols., 1934), an edition of Hope's *Major Practicks* (2 vols., 1937–38), and an edition of the *Acta Dominorum Concilii*, 1501–3 (1943). His elder son, J. L. M. Clyde (q.v.) also became Lord President.

Clyde, James Latham McDiarmid (1898–1976). Son of Lord President J. A. Clyde (q.v.) was called to the Scottish Bar in 1924, became Q.C. in 1936, was an M.P. 1950–54, Lord Advocate 1951–54, and Lord President of the Court of Session and Lord Justice-General, 1954–72.

Coastguard. A body concerned with policing the coasts, preventing smuggling, and reporting wrecks and other incidents at sea. In 1660 the defence of the coast was transferred from the maritime counties to the militia. In 1822 there were consolidated as the Preventive Service three bodies concerned with preventing smuggling, the revenue cutters, riding officers, and the preventive waterguard. This became the Revenue Coastguard in 1829, the riding officers becoming the Mounted Guard. The latter disappeared in 1831 and the Coastguard Service was transferred to the Admiralty and, in 1925, to the Board (now Department) of Trade.

Coat and conduct money. An illegal exaction imposed by Charles I in 1640, for clothing and paying the travelling expenses of the recruits whom he had pressed into service against the Scots.

Cobbett, William (1763–1835). A journalist, compiler from various older sources of the *Parliamentary History of England*, which constitutes a legislative history of England, Britain, and the United Kingdom from the Norman Conquest to 1803 in 36 volumes. The work was completed by J. Wright. Cobbett began the *Parliamentary Debates* from 1803 and incorporated them in his weekly *Political Register* (1802–35), and in 1812 they were taken over by T. C. Hansard (q.v.). He was a zealous advocate of Parliamentary reform and entered the House in 1832 after several earlier defeats. He also originated the series of *State Trials*, known as Howell's, from the editor employed by Cobbett. Among his many other writings, the best known are his *Rural Rides*, a picture of the English countryside in the years 1820–30.

Co-belligerency. A relationship between states distinct from a military alliance, where the troops of one country co-operate with those of another but the former country is not fully independent and not a full ally.

Cocceji, Heinrich von (1644–1719). German jurist, professor at Heidelberg and Frankfurt, and influential on the study of natural law and public law, author of *Deductiones, consilia et responsa in causis illustrium, Juris publici prudentia* (1695); *Grotius Illustratus seu commentarii ad Hugonis Grotii De Jure Belli ac Pacis* (1744); *Prodromus juris gentium*; *Hypomnemata juris feudalis*; and *Hypomnemata juris ad seriem Institutionum Justiniani.*

Cocceji, Samuel von (1679–1755). Son of H. von Cocceji, professor at Frankfurt and later engaged in judicial activities and reform in Prussia, author of *Disputatio de principio juris naturalis unico, vero et adaequato* (1699); *Tractatum juris gentium*; *Elementa* (or *Novum Systema*) *jurisprudentiae naturalis et Romanae* (1740); and editor of his father's *Grotius Illustratus* (1744). As High Chancellor of Prussia he drafted the Prussian Code for Frederick II (1749–51) (*Codex Fredericianus*) which was not adopted till 1794.

Cockburn, Adam, of Ormistoun (1656–1735). Sat in the Scottish Parliament from 1678 and, though not an advocate, was appointed Lord Justice-Clerk in 1692, holding office till 1699 when he became Treasurer Depute. In 1705 he was again appointed Lord Justice-Clerk and a Lord of Session, but was superseded in the former office in 1710 though he held the latter till his death. In 1714 he regained the office of Lord Justice-Clerk and this time held it till his death. He was deemed very zealous in suppressing the rebellion of 1715 and almost universally hated.

Cockburn, Sir Alexander James Edmund, Bart (1802–80). Was called to the Bar in 1829 but was slow in getting a practice. In 1847 he became an M.P. and was Solicitor-General in 1850 and Attorney-General in 1851–52 and again 1852–56. In the courts he was a brilliant advocate and was equally successful in the House of Commons. In 1856 he became Chief Justice of the Common Pleas and in 1859 Chief Justice of the Queen's Bench, which post he held till his death. From 1874 he was officially styled Lord Chief Justice of England, though that title had been used informally by chief justices of the Queen's Bench from the time of Coke (1613–16). He was both a great lawyer and a great judge and gave decisions over a wide range of legal issues. He presided at the Tichborne Trial and was particularly able in his conduct of criminal trials. As an arbitrator in the *Alabama* arbitration in 1872 he dissented and his judgment reveals profound knowledge of international law. He wrote two small but learned books on nationality and on the Judicature bill.

Cockburn, Henry Thomas (1779–1854). Son of a Baron of Exchequer in Scotland, was called to the Scottish Bar in 1800 and won a place among a distinguished bar. He became Solicitor-General for Scotland in 1830, played an important part in drafting the Scottish Reform Bill, and became a Lord of Session in 1834, and a Lord of Justiciary as well in 1837. He contributed to the *Edinburgh Review* and wrote a *Life of Lord Jeffrey* (1852), but is best remembered for his *Memorials of his Time* (1856), *Journals* (1874), *Circuit Journeys* (1888), and *Trials for Sedition in Scotland* (1888), of which the first three are a superb evocation of Edinburgh and the legal and political life of his time. He had an intense love of Edinburgh and fought to safeguard its charm.

Cockpit. A building erected on the site of the old cock-pit (used for cock-fighting) of the Palace of Whitehall. In the reign of William III it became the Privy Council offices and is frequently referred to in literature of and about the eighteenth century.

Code. The term code is commonly used of various ancient bodies of legal rules, e.g. Code of Hammurabi, though these bodies of rules are manifestly not a complete statement of the rules of that king's lands on any branch of the law. The term is also used of the bodies of regulations found in the Old Testament, particularly the Book of the Covenant or Covenant Code (Exodus 20 and 23), the Deuteronomic Code (Deut. 12–26), and the Priestly Code in Exodus, Leviticus, and Numbers, of which a major section is the Code of Holiness (Lev. 17–26), where also the bodies of rules are not complete statements.

In Justinian's legislation there were two Codes, the Code of A.D. 529, which was a new collection of Roman imperial constitutions compiled from the *Codices Gregorianus*, *Hermogenianus*, and *Theodosianus* and subsequent enactments, revised and arranged in titles according to subject matter. It was in force till A.D. 534 only and only fragments are extant. In 534 there was published a second Code (*Codex repetitae praelectionis*) in 12 books, divided into titles. It includes imperial constitutions from the three older *Codices* and more recent legislation, amended and rearranged.

The term code in also frequently applied to the bodies of law known as the Barbarian or Germanic laws and to the collections of maritime customs and usages widely accepted throughout Europe.

From the fifteenth century onwards the term came to be applied to a more or less comprehensive systematic statement in written form of major bodies of law, such as the civil law or the criminal law of a particular country, superseding the mixture of customs, decisions, and bits of legislation which had previously applied. This movement was stimulated by the growth of nation-states, the growth in power of rulers and the spirit of nationalism. It culminated in the five Napoleonic Codes of 1804–10 (civil code, civil procedure, commercial code, criminal procedure, and penal code). See CODE NAPOLÉON. This was introduced in countries under French control and conquered by the French.

During the nineteenth century it was adopted or copied in numerous European and Latin American countries. The Louisiana Civil Code of 1825 was closely connected with it. At the beginning of the twentieth century the German Civil Code (1900) and the Swiss Civil Code (1912) appeared and were also extensively copied.

Codification has been urged in the U.S., notably by Livingston and David Dudley Field, but apart from Louisiana (1825, revised 1870) has not made much headway save in the field of procedure.

Codification has repeatedly been urged in Britain, by Bacon and under the Commonwealth. Bentham advocated it and Lord Cranworth contemplated a Code Victoria, though all that resulted were various statute law revision and consolidating Acts. Codification was, however, used to a substantial extent in India, in the various Anglo-Indian Codes, such as Macaulay's Penal Code. It is a function of the Law Commissions (q.v.) established in 1965, but little progress has been made with this task.

The possibility of codification of even parts of English or Scottish law is small, because of the preoccupation of Parliament and draftsmen with current problems, the difficulty of getting principles agreed, the British habit of legislating at great length and in enormous detail rather than by stating principles baldly and simply, and the endless futile debates which would ensue in Parliament.

Code Louis. The name sometimes given to the collection of Ordinances enacted in France at the instance of Louis XIV and Colbert. These comprised the Civil Ordinance (1667) (to which alone the term Code Louis is sometimes given) dealing with civil procedure; the Criminal Ordinance (1670) dealing with criminal procedure; and lesser Ordinances dealing with commerce, marine affairs, and colonial affairs. All had great influence on the French codifications of the Napoleonic period.

Code Michau. A French royal ordinance of 1629 in 461 articles, dealing with justice, *parlements*, universities, military organization, and hospitals. It was the work of Michel de Marillac, keeper of the seals under Louis XIII.

Code Napoléon. The name given in 1807 to the French Civil Code enacted in 1804 and, as amended, still in force. Since 1870 it has correctly been called the Code Civil. Prior to the French Revolution, in Paris and the northern provinces of France, there had developed bodies of customary law based on Frankish and Germanic institutions, though feudalism had been very influential, and marriage and family life were regulated by canon law and by the Church. Each area had its own body of customs and despite attempts in the sixteenth and seventeenth centuries to codify the bodies of customary law little headway had been made. In the south, law based on Roman law still obtained. From the sixteenth century an increasing number of matters were governed by royal decrees, ordinances, or by rules developed by the *parlements*.

Codification had been used elsewhere and had often been urged in France. Colbert's ordinances went some way towards codification, and Lamoignon's and D'Aguesseau's works comprised large drafts, but the Revolution gave the opportunity and the impetus towards it and rendered it desirable, if not indeed necessary. Cambacères (q.v.) produced two drafts of a civil code for the Convention and a third for the Directory. Napoleon in 1800 appointed a commission consisting of Tronchet, Bigot de Préameneu, Portalis, and Malleville to prepare a draft, which they did in four months. It was enacted in the form of 36 laws between 1803 and 1804, which were then united into a single civil code.

The code was based on the premises of creating a purely rational body of law, of the equality of all citizens, the separation of civil from ecclesiastical regulation, and on the principles of freedom of person, freedom of contract, and inviolability of private property. In general the draftsmen succeeded in fusing ancient French law and the modern revolutionary approach.

The code is in six introductory articles and three books; Book I (Arts. 7–515) deals with the law of persons, and the law of domestic relations and the family; Book II (Arts. 516–710) deals with the law of things and property; and Book III (Arts. 711–2281) with the acquisition of rights, succession, and obligations. The order was based on that of the Roman law and the commissioners founded heavily on Pothier. As compared with Anglo-American legislation the articles are extremely brief, frequently consisting of a single sentence, and very general, leaving great scope and need for judicial interpretation in the application of the provisions to specific cases. Hence while interpretative case-law is not binding in the same way as it is in Anglo-American countries, study of the judicial interpretations of the brief sentences of the Code is essential to appreciate what the various Articles have been held to mean.

The code was introduced in 1804 not only into France but into areas then under French control, Belgium, Luxembourg, parts of western Germany, and northwestern Italy and was later introduced into other countries conquered by France, such as the Netherlands in 1809, and before 1815 it had been accepted voluntarily by various smaller European states.

In the nineteenth century it was adopted or copied in various European and Latin American countries. In Louisiana the Civil Code of 1825 (revised 1870) was closely connected with the Code Napoléon. The Albertine Code of Piedmont (1848) and the Italian Civil Code of 1865 similarly owed much to it. After 1900 the German Civil Code (BGB) (1900) and Swiss Civil Code (1912) were adopted or copied by many countries and diminished the influence of the Code Napoléon.

Napoleon was responsible also for the enactment of the four other codes, Civil Procedure (1806), Commercial Code (1808), Criminal Procedure (1808), and Penal Code (1810), but these are all much inferior to the Code Civil.

In the course of time many articles have been amended and developed by judicial interpretation and many matters are regulated by laws outside the Civil Code so that it is not a complete statement of French civil law.

In 1945 a Commission of Reform of the Code Civil was established, but its labours have not resulted in a new or revised code.

Code Noir. A code of 1685 issued by Louis XIV ordering the expulsion of the Jews from French colonies, the banning of non-Catholic practices there, and providing a framework for the government of French slave societies. It was superseded by the Code Napoléon.

Code Savary. One of the major French ordinances formulated by Colbert, promulgated in 1673 and laying down a code of commercial procedure.

Code of Canon Law. The modern revised version of the canon law of the Roman Catholic Church, undertaken in 1904, promulgated in 1917, and effective in 1918. It comprises five books, viz.; I, General Norms; II, Persons; III, Things; IV, Procedure; and V, Crimes and Penalties. These are subdivided into parts and sometimes into sections, titles, and chapters. There are 2,414 canons in all. The Code applies only to Latin Catholics, the Oriental Catholic Church having its own Code. Pre-existing laws contrary to the Code are revoked. A commission of cardinals, the Commission for the Authentic Interpretation of the Code, was established in 1917 to ensure the stability of the Code; its interpretations have the force of law. A commission of cardinals was established in 1963 to undertake a revision of the Code.

Code (of Justinian). See CODEX JUSTINIANUS.

Codes, biblical. Collections of ancient Hebrew law found in the Old Testament. They may have been influenced by the earlier codes of Lipit-Ishtar of Eshnunna and of Hammurabi (see BABYLONIAN LAW) and are conceived as laws given by the God of the Israelites to his chosen people. There are two main forms: regulations in the form of commands, such as the Ten Commandments (apodictic law); and casuistic or case law, containing conditional statements, with specifications of relative punishments. The major codes in the Old Testament are the Book of the Covenant or Covenant Code, the Deuteronomic Code, and the Priestly Code, which contains a section known as the Code of Holiness (qq.v.).

Codex. A form of book in the later Roman Empire contained not in a papyrus roll but in leaves gathered into quires, chiefly used, and possibly invented, by the Christian community in Egypt. The name was later applied particularly to books containing collections of legal materials. The earliest collections of Imperial constitutions were the private and anonymous compilations known as the *Codex Gregorianus* (*c.* A.D. 291) and the *Codex Hermogenianus*, a supplementary collection (A.D. 295). The *Codex Theodosianus* was an official collection (A.D. 438) (qq.v.).

When Justinian became emperor in A.D. 528 he ordered the compilation and publication of a *Codex* containing a new collection of constitutions drawn from the three older codices and later enactments, omitting obsolete materials, revised and rearranged. This was published and effective in 529 and thereafter previous enactments with some exceptions might be cited only as appearing only in the Codex. It remained in force only till 534 when a revised edition was enacted (*Codex repetitae praelectionis*). It superseded both the earlier edition and the original form of the subsequent constitutions and the editorial work seems to have been extensive. It is divided into 12 books treating of ecclesiastical law, sources of law, and duties of higher officials (Book I), private law (Books II–VIII), criminal law (Book IX), and administrative law (Books X–XII), and subdivided into titles within which constitutions are arranged chronologically. The arrangement is based on the older *codices* and hence ultimately on the Edict.

Text in Mommsen, Krueger, and Scholl, *Corpus Juris Civilis.*

Codex Eurici (or *Euricianus*). The formulation of the tribal laws of the Visigoths, formulated by King Euric (A.D. 466–84). It was later revised by Leovegild (568–86) and by Recceswinth (649–72) when the materials were rearranged in a systematic law-book modelled on Justinian's *Codex* and entitled the *Lex Visigothorum* or *Liber Judiciorum.*

Codex Gregorianus. A collection, made unofficially and anonymously, possibly by one Gregorius, probably in the East at Beyrout, and probably published *c.* A.D. 291, of imperial constitutions down to that date. They were reproduced in full, and the work was divided into books and titles, arranged in the way traditional for digests compiled by classical jurists. It has survived only in quotations.

Codex Hermogenianus. A collection, made unofficially and anonymously, possibly by one Hermogenianus, probably in the East at Beyrout, and probably published *c.* A.D. 295, intended to supplement the *Codex Gregorianus* by giving a fairly full collection of imperial constitutions from Diocletian's time. It was divided into titles only and has survived only in quotations in later works.

Codex Justinianus. The Emperor Justinian in A.D. 528 appointed a commission of 10 to compile a new collection of imperial constitutions from the *Codices Gregorianus, Hermogenianus*, and *Theodosianus* (qq.v.), and subsequent imperial enactments, with powers to omit obsolete and unnecessary material, to remove contradictions and repetitions, and where necessary to make additions and amendments, and to amalgamate several enactments. The constitutions were to be grouped under titles and, within each title, arranged chronologically. The Code was published in 529 but remained in force only until 534. Save for fragments the text has not survived.

After the compilation of the Digest and Institutes (qq.v.) the first Code was no longer reliable and a commission of five was appointed to prepare a revised version, again with wide powers of alteration and omission, powers which appear to have been extensively used. It is impossible to know how far the earlier Code was modified, but the changes seem to have been substantial. It was published under the title *Codex repetitae praelectionis* in 534 and superseded both the earlier Code and constitutions subsequent to it. It is divided into 12 books, each subdivided into titles and the constitutions therein in chronological order. The arrangement is founded on the older *Codices* and hence indirectly on the praetorian Edict. The constitutions have inscriptions stating the emperor who promulgated them and the addressee, and subscriptions giving the dates of issue and publication and sometimes the place. Some are undated.

Book 1 deals with ecclesiastical law, the sources of law, and the duties of higher officials; 2–8 with private law; 9 with criminal law; and 10–12 with administrative law. In later times the last three books became separated from the first nine and in some manuscripts were combined with the *Institutes* and the *Authenticum* in a volume called *Volumen* or *Volumen Parvum*. The Code was extensively commented on by mediaeval and early modern jurists.

Codex Maximilianus. The Bavarian Civil Code, drafted by Kreitmayer and published in 1756.

Codex Theodosianus. An official compilation of all *leges generales* enacted since Constantine, whether obsolete or in force, made at Constantinople by a commission appointed by Theodosius II in 429, and completed by a second commission of 435 which had power to alter but not omit constitutions. The work was published in 438 and a copy was sent to and approved by the Roman Senate. It was provided that no reference might thereafter be made to any constitution, with a few exceptions, save as enacted in the *Codex*. It was divided into 16 books, these into titles, and, within titles, chronologically. The titles were generally arranged as in the jurists' *Digesta*. It has not survived completely but in parts in later works. In the East it was superseded by Justinian's legislation, but continued to be used in the West. It was edited by Gothofredus, and by Mommsen and Meyer.

Codicil. A writing supplementary to a will (q.v.) intended to supplement or modify some or all of the provisions of the will. It requires to be executed in the same way as a will and is construed as part of the will, revoking the will only so far as expressly or impliedly inconsistent therewith.

Codification. The verbal formulation of the whole law of a territory having a distinct legal system, or of some distinct branch of that law, such as the criminal law or the law of succession, in a statement, replacing and superseding all existing rules of law, customary, settled by cases, or statutory, in the field covered by the code. The term was invented, or at least introduced into English, by Bentham, who saw codes as complete and self-sufficient and not to be supplemented or modified save by legislation. In Germany Thibaut urged the need for a General Civil Code for Germany, and Savigny opposed it on the ground that the law should be developed by the internal force of national consciousness.

Two major factors have operated in favour of codification, the desire to rationalize a more or less chaotic volume of pre-existing law and to provide a legal system with a fresh basis for development, and the desire to provide a unifying element for a newly-created or developing state by creating a unified legal system.

In France in the late seventeenth and eighteenth centuries a number of *grandes ordonnances* were enacted, each of which is a kind of codification of a branch of the law. The chief of these were the *Ordonnance civile touchant la reformation de la justice* of 1667, in substance a code of civil procedure, the *Ordonnance Criminelle* of 1670, the *Ordonnance du Commerce* of 1673, the *Ordonnance de la Marine* of 1681, the *Ordonnance sur les Donations* of 1731, the *Ordonnance sur les Testaments* of 1735, and the *Ordonnances sur les Substitutions* of 1747.

After the outbreak of the Revolution there was widespread desire to get rid of the local *coûtumes* and variations and to foster national unity by having a national legal system. After several attempts at codification had failed Napoleon appointed a commission in 1800 and its product, the *Code Civil* or *Code Napoleon*, was published in 1804, and followed by the Code of Civil Procedure (1806), the Commercial Code (1808), the Penal Code (1810), and the Code of Criminal Procedure (1808). All have been subsequently amended and supplemented by legislation. The Civil Code was not in fact conspicuously novel but firmly based on the Ordonnances, the former *coûtumes*, especially the Coutûme de Paris, and the writings of the great jurists, notably Pothier.

Ever since the *Code Civil* was promulgated, it has been copied elsewhere, and most of the countries which achieved statehood in the nineteenth century modelled their revision and codification of their laws on the French Civil Code.

In the States forming the Holy Roman Empire the common law, which was a modified form of the Roman law embodied in Justinian's *Digest*, finally received in Germany in 1495, was subject to many varieties of local customs and in many states displaced or supplemented by local codes. The major of these were the Bavarian Codex Maximilianus of 1756, the Prussian Landrecht of 1794, the Code Napoleon in several states and a slightly modified German version in Baden, the Badische Landrecht of 1810, the Austrian Civil Code of 1811, and the Saxon Civil Code of 1863. Purely

Germanic ideas remained influential in criminal law and procedure, and in these fields there was a famous codification in the *Carolina* of 1532. In addition states had uniformly adopted the German Bills of Exchange Code in 1848–50 and the German Commercial Code in 1861–66.

In Germany there was no machinery, by reason of the existence of a multitude of independent states, for the enactment of general law prior to the creation of the North German Confederation in 1866. A Criminal Code passed for the North German Confederation in 1870 was re-enacted for the German Empire in 1871, while Codes of Civil and Criminal Procedure and certain others were enacted in 1879. A commission was appointed in 1874 to draft a Civil Code and completed their work in 1887. In 1890 a second commission was appointed to revise the first draft and it was enacted in 1896 and came into force in 1900 along with revised editions of the Commercial Code and Code of Civil Procedure. Much law remained outwith the scope of the codes, and many important statutes supplement it. The Civil Code was based on the great systematic textbooks on modern Roman law (*Pandektenrecht*) which the Code superseded.

The German Code was taken as a model by other Central European countries. It was extensively supplemented and altered during the Nazi regime.

Other European countries in general enacted codes in the second half of the nineteenth century.

In the United States the Civil Code of Louisiana, based on French law, was passed in 1808 and an amended code in 1824. In 1821 Edward Livingston (q.v.) was appointed to draft a Code of Criminal Law and Procedure for Louisiana, which was partially adopted in 1824 and was very influential in English-speaking countries subsequently. David Dudley Field (q.v.) and others were appointed commissioners in New York to codify the law of that State. Their Code of Civil Procedure was enacted in 1848, their Code of Criminal Procedure in 1881 and their Penal Code in 1882, but the Civil Code never became effective. The two procedure codes have been adopted with modifications in several states. California enacted four codes in 1872 and Dakota seven codes in 1877.

Codification was successfully pursued in British India. The Charter Act, 1833, provided for the appointment by the Governor-General of India in Council of a Law Commission to inquire into the courts and laws in British India, to report thereon and to suggest alterations. The first Indian Law Commission was appointed in 1834 and comprised T. B. Macaulay and three civil servants. It published Macaulay's draft Indian Penal Code in 1836 but lost vitality after Macaulay left India and, though it published reports, it achieved nothing. The Penal Code, as revised by Sir Barnes Peacock and others, was ultimately enacted in 1860.

An Act of 1853 provided for the appointment of a new Commission which sat in London till 1856 and presented four reports. A Code of Civil Procedure was passed in 1859, Macaulay's Penal Code revised in 1860, and a Code of Criminal Procedure in 1861.

A third Indian Law Commission was constituted in 1861 to prepare a body of substantive law for India, and it drafted a law of succession and inheritance for others than Hindus and Muslims which was put through as the Indian Succession Act, 1865, by Maine (q.v.) then Law Member of the Governor-General's Council. The Commissioners then prepared drafts on contracts, sale, bailments, agency, partnership, bills and cheques, evidence and transmission of property *inter vivos*, but controversy arose over the contract code and most of the Commissioners resigned though the Indian Contract Act was passed in 1872. While Sir James Fitzjames Stephen (q.v.) was Law Member there were passed the Limitation of Actions Act, 1871, the Evidence Act, 1872, the Code of Criminal Procedure was revised and the Indian Contract Act passed.

In the time of the next Law Member, Mr. (later Lord) Hobhouse (q.v.) the emphasis was on consolidation of legislation and the publication of English Statutes relating to India (2 vols.), General Indian Acts (13 vols.), and Provincial Codes (10 vols.), the main work for which was done by Whitley Stokes, who became Law Member in 1877, but the Specific Relief Act, 1877, was also passed.

In 1879 Stokes and others were appointed Commissioners and their labours resulted in the passage of the Negotiable Instruments Act, 1881, the Indian Trusts Act, 1882, the Transfer of Property Act, 1882, and the Indian Easements Act, 1882, while the Codes of Civil and Criminal Procedure were revised in 1882. Under the next Law Member, Courtenay Ilbert, there were passed Tenancy Acts for several provinces, and a general system of law and procedure for Upper Burma, annexed in 1886, while an Indian Civil Wrongs Bill was drafted by Sir Frederick Pollock. Accordingly all the major topics of substantive law and procedure were codified in half a century, several having been superseded and revised.

In British colonies Quebec adopted a Civil Code in 1866, based on French law, and a Civil Procedure Code in 1867, and Canada a Criminal Code in 1882. New South Wales adopted a Criminal Code in 1883, Victoria a Crimes Act in 1890, New Zealand a Criminal Code in 1900. Ceylon adopted some of the Indian Codes.

In Britain itself codification has made little headway. Lord Chancellor Westbury contemplated a Digest of English law and a Royal Commission was appointed in 1866 to examine the expediency of a digest of law. It reported in 1867, and the project lapsed. Subsequently Sir James Fitzjames Stephen drafted a Code of the Law of Evidence, which received only a first reading in Parliament, and a Code of Criminal Law and Procedure in 1879. The part relating to Procedure was introduced into Parliament, but dropped. Codification is stated as one of the objects of the Law Commissions appointed in 1965 but no material progress has been made.

The term has also been used of the Bills of Exchange Act, 1882, the Partnership Act, 1890, the Sale of Goods Act, 1893, and the Marine Insurance Act, 1906, but these do not deserve to be called codifications on account of their small scale.

Coemptio. In Roman law a form of marriage by a fictitious sale of the bride to the husband utilizing the forms of *mancipatio* (q.v.). It had disappeared before Justinian's time.

Coercion. Pressure or influence amounting to compulsion. An act done while subject to overpowering physical coercion is legally ineffective, but this is not so in the case of an act done while subject to moral coercion. The term coercion is however generally restricted to a special defence available at common law to a wife who has committed certain crimes in the presence of her husband. It was presumed that she acted under such coercion as to entitle her to be excused. At first confined to felonies it appears later to have been extended to misdemeanours. The practical reason for its application to felonies was that it saved a woman from the death penalty if she could not plead benefit of clergy (q.v.) but the rule lasted after that presumption was abolished in 1925; it still remained a defence to prove that a crime, other than treason or murder, was committed in the presence of and under the coercion of the woman's husband.

Cognate. A person related to another through the mother. See also AGNATE.

Cognisance. Knowledge which a judge is presumed to have without the matters having to be established by evidence, such as the structure of government, the powers and privileges of Parliament, the jurisdiction of the courts, the existence of war and peace with foreign states and the like.

Cognisance or conusance. A privilege granted to a place to hold pleas of all matters within the area of the franchise, and entitling the mayor of the place to demand that the plea be heard before him.

Cognitio extraordinaria. A system of Roman civil procedure introduced during the principate. It was an extension of the powers of command which the praetors formerly had, and a kind of administrative procedure in which an official used his powers of investigation and compulsion to decide an issue between private parties in the same way as a dispute arising in the course of his administrative duties. But the process remained judicial and many of the rules of ordinary procedure still applied. Though in some cases a *formula* (q.v.) was used to define the issue, the important difference was that the judge was appointed by an official, not by agreement between the parties. The case might be heard by the official before whom it came, or by another person to whom he delegated the hearing of the whole case, or a two-stage procedure as in the older Roman civil procedure but in which the *iudex* derived his powers from the official and not from the parties.

Cognition. In Scots law a judicial process, now obsolete, whereby a man might be found insane and a curator appointed to manage his property.

Cohabitation with habit and repute. A mode of irregular marriage in Scotland, constituted by parties cohabiting sufficiently long and openly as to give rise to the reputation of being married on the basis that an exchange of matrimonial consent must be presumed. It is still competent.

Cohen, Felix S. (1907–53). U. S. jurist and government lawyer, son of Morris Raphael Cohen (1880–1947) a distinguished philosopher and author of *Law and the Social Order* (1933). He wrote *Ethical Systems and Legal Ideals* (1933) and with his father edited *Readings in Jurisprudence and Legal Philosophy* (1957). A collection of his papers was published as *The Legal Conscience* (1960).

Cohen, Lionel Leonard, Lord Cohen of Walmer (1888–1973). Called to the Bar in 1913, he took silk in 1929 and was a judge of the Chancery Division 1943–46, a Lord Justice of Appeal 1946–51, and a Lord of Appeal in Ordinary 1951–60. He was chairman of the Company Law Committee 1943–45, the Royal Commission on Taxation of Profits and Income, and the Royal Commission on Awards for Inventors, and a member of the Council on Prices, Productivity, and Incomes 1957–59.

Cohens* v. *Virginia (1821), 6 Wheaton 264. This decision of the Supreme Court reasserted the principle of *Martin* v. *Hunter's Lessee* (q.v.) and interpreted judicial power to mean that appellate courts might accept suits against a State if the State had originally instituted the suit.

Coif. A close-fitting white lawn or silk cap fastened under the chin, represented latterly by a patch on the wig, worn by serjeants-at-law (q.v.) as the badge of their order. On top of it a smaller black skull-cap was worn. When wigs came to be worn they had to be made so as not wholly to conceal the coif and at first it was visible at the back of the wig. In time this was replaced by a space on top of the wig on which were attached a white crimped border, less than an inch wide, surrounding a small round smooth black silk patch about two inches in diameter. This represented the coif and black skull-cap.

Coif, Order of the. The order of serjeants-at-law (q.v.) at the English Bar.

A. Pulling. *Order of the Coif.*

Coke, Sir Edward (1552–1634). Called to the Bar in 1578, he soon built up a large practice. By the favour of Burghley he rose rapidly in public office, becoming Recorder of London in 1591, Solicitor-General in 1592, Speaker of the House of Commons in 1592, and Attorney-General in 1594, which office he held till 1606. As such he championed the crown and its prerogative powers. He had the conduct of many of the great trials of the time, prosecuting Essex and Southampton (1600–01), Raleigh (1603) and the Gunpowder Plot conspirators (1605). As a prosecutor he was frequently savage and tyrannical. He had a lasting hatred of Catholics and prosecuted them with vehemence.

In 1606 he became Chief Justice of the Common Pleas and came into conflict with the King over his views on the extent of jurisdiction of the common law courts, first quarrelling with the ecclesiastical courts and asserting the power of the common law courts to determine limits of ecclesiastical jurisdiction. In 1610 Coke gave an opinion that a royal proclamation cannot change the law, and in 1611 came in conflict with the Court of High Commission and had to evade being put on the Commission himself. He had now become the embodiment of the common law and an opponent of royal power.

In 1613 he was transferred to the Chief Justiceship of the King's Bench in the hope that he would protect the rights of the Crown, and made a member of the Privy Council, but trouble continued. He maintained that, if the judges were consulted on a pending case, they should be consulted in a body. Then he quarrelled with Lord Ellesmere as to the jurisdiction of the court of chancery, a dispute finally settled by James I in favour of the chancery, and finally he quarrelled over the King's right to stop or delay proceedings in the common law courts. In a case of 1616 the court refused to delay hearing a case and in a letter to the King Coke evidenced an inclination to deny the King's prerogative to control the conduct of pending cases where royal interests were involved. Coke was summoned before the Council and dismissed (1616). The King, James I, was unable to trust Coke in office again because of fundamental differences between them in political views but by 1617 he was back in the Privy Council.

In 1620 Coke appeared as leader of the Parliamentary opposition, completely changed front, and forged the alliance between Parliament and the common law against the royal prerogative. He quickly assumed the lead in all the developing resistance to the King and his policy. In 1623 he was removed from the Council and for a time was a prisoner in the Tower, but was shortly restored to favour. In the Parliament of 1628 he took the lead in framing the Petition of Right, but thereafter took no part in public affairs.

While a law officer he began to accumulate the materials for his published works. He was a master of mediaeval law and the old books and records, though on non-legal literature he was uncritical and fathered many inaccuracies. He was also a lucid expositor and a good writer. His earlier works were his *Readings*, a small treatise on Bail and Mainprize, and the *Complete Copyholder*. More important is his *Book of Entries* (1614), a collection of pleadings, and his *Reports*, in 11 parts (1600–15) with two posthumous parts (1655 and 1658). The reports are in some cases bases for a summary of the law on the subject, or for a collection of cases on a topic, or use the case as a peg for a disquisition on the law. The cases form a corpus of the common law, civil and criminal, as it stood in the sixteenth and early seventeenth centuries; the old learning is restated, the newer developments justified and explained. Above all, in his *Institutes* he wrote the first textbook on the modern common law. Of the four books, the first known as *Coke on Littleton* contains Littleton's (q.v.) *Tenures* with an elaborate commentary (1628), the second a commentary on Magna Carta and other mediaeval statutes and certain later statutes (1641), the third deals with the criminal law (1641), and the fourth gives an account of the various courts which had jurisdiction in England (1641). The first is the most elaborate and is the fruit of a lifetime's study, and utilizes all Coke's learning and experience. It is virtually a legal encyclopaedia, the entries hung on pegs suggested by the sentences and words of Littleton. The last three parts were accepted when they came out in 1641 as justifying the claim that Parliament and the common law were supreme in the State, and they became the basis of much of modern British constitutional law. Later criticism has pointed to historical inaccuracies and defects of arrangement and classification, but nobody since Bracton had tried to give a complete exposition of English law. His writings have great merits, in that they covered the whole field and restated English law from the point of view of the sixteenth century, deduced positive rules of law from the dicta in the Year Books, brought Year Books and mediaeval literature into line with reports and modern literature, familiarized the common lawyers with the developments in other fields, such as Admiralty and Chancery, and ensured the continuity of the development of the common law in the period of change.

His influence has been enormous in fixing the form and development of modern English law; the work has always been recognized as of the highest authority. It doubtless tended to preserve ancient forms in the law, but largely underlies the constitutional idea of the rule of law. He was more learned and practical than the philosophic Bacon, and his work has had far greater influence in the law.

Johnson, *Life of Sir Edward Coke* (1837); C. D. Bowen, *The Lion and the Throne* (1957); S. E. Thorne, *Sir Edward Coke, 1552–1952*.

Colbert, Jean Baptiste (1619–83). French statesman. Destined initially for a commercial career he rose under Cardinal Mazarin, was entrusted with diplomatic missions, and on Mazarin's death (1661) became intendant of finances, Comptroller-General of Finance in 1665 and in 1669 Secretary of State. He reformed the finances, encouraged industry, improved communications, established the French navy, patronized art and literature, stimulated public building, and controlled every aspect of activity in France. Among very many governmental functions he planned the amendment and codification of the law of France, then partly written, partly customary, partly Roman, partly legislation, partly decisions of local Parlements. He established a Conseil de Justice to discover defects and devise remedies. The results of their labours were *L'Ordonnance Civile*, a code of procedure (1667), *L'Ordonnance des Eaux et Forêts* (1669) dealing with water and forest rights, *L'Ordonnance Criminelle* (1670) which became the French criminal procedure code for 120 years, *L'Ordonnance pour le Commerce* (1673) regulating business affairs, *L'Ordonnance pour la Marine* (1681), dealing with maritime and admiralty law, and *L'Ordonnance Coloniale* (1685) regulating the French West Indian colonies and the treatment of negroes. After Colbert's death several further ordinances were issued. This great work of codification in various fields not only greatly simplified and rationalized the law but paved the way and indicated the methods for later reforms, culminating in the *Code Napoleon*. It is an immense achievement and of enormous value to France.

E. Clément, *Vie de Colbert*; G. Mongredien, *Colbert 1619–1683*.

Coldbath Fields. The Middlesex House of Correction or local prison in Clerkenwell, under the control of the Middlesex magistrates, catering for local prisoners. It dated from 1794.

Long unreformed and a scene of inefficiency, corruption, and squalor, it was reorganized on the 'silent system' in 1834. It was generally regarded as a severe prison and hard labour was regular. Charles Dickens knew and greatly admired Chesterton, the governor from 1830, and was influenced by what he saw and heard of the system there.

Colepeper, John, Lord (?–1660). His origins are uncertain, but he sat in the Long Parliament, became a close adviser of Charles I and, in 1643, Master of the Rolls, and in 1644, a peer. He was in exile with Charles II and in 1660 resumed his judicial office but died within a month. He made no mark on the law.

Coleridge, John Duke, Lord (1820–94). Son of a judge who published an edition of Blackstone, he was a good scholar and of literary taste, and was called to the Bar in 1846. He took silk in 1861 and became an M.P. in 1865. In 1868 he became Solicitor-General and Attorney-General in 1871. In 1873 he became Chief Justice of the Common Pleas and in the following year a peer. In 1880 when Lord Chief Justice Cockburn and Chief Baron Kelly died and the Common Pleas and Exchequer divisions were amalgamated into the Queen's Bench Division, he became Lord Chief Justice of England, an office he held till 1894. As an advocate he was very successful; as a judge he was, though not learned, a lucid expositor of law. His son became a judge of the King's Bench Division 1907–23.

H. Coleridge, *Life and Correspondence of Lord Coleridge* (1904).

Collar of SS. A collar of gold links, 28 being the letter S, 27 garter-knots, one the Beaufort badge of the portcullis and rose, and a single portcullis. It is worn round the shoulders and chest by the Lord Chief Justice of England. The significance of the letters S is not certainly known and has been variously explained as standing for Sanctus, or Seneschallus, or Souvenez-vous de moi, but best explained as signifying 'Souvenez'. The portcullis and rose is the badge of Henry VII and the Beauforts.

Various such livery collars were worn in the fourteenth century. This is the Lancaster collar; there was also a York one with falcons and fetter locks. It is said to have been first used by John of Gaunt, was later worn by various persons of high degree and gradually became confined to the chiefs of the three common law courts, and the officers of the Heralds' College, and the Lord Mayor of London. From the time of Elizabeth it seems to have been invariably worn by the three heads of the courts. The royal collar was discontinued during the Commonwealth though Cromwell's Chief Justice wore a collar of a similar description. The collar was resumed at the Restoration. The collar of the Chief Justice of the Common Pleas, said to be the one worn by Coke, certainly passed from Chief

Justice to Chief Justice to Lord Truro in 1846. The collar of the Chief Justice of the King's Bench transmitted from Sir Matthew Hale to Lord Ellenborough, who kept it. Lord Tenterden provided a new one which passed to Lord Denman who gave it to the Mayor of Derby. Lord Campbell provided a new one and kept it, and Sir Alexander Cockburn also had to provide a new one. The Exchequer one passed from Chief Baron to Chief Baron till 1824 when Chief Baron Richards' widow kept it. Chief Baron Alexander got a new one but Chief Baron Abinger's son kept it and Chief Baron Pollock had to purchase a new one in 1844.

In Scotland the Lord Lyon King of Arms wears a similar chain composed of S's only with the royal arms badge worn at the centre front.

Collateral. In kinship, a person of the same ancestry but not necessarily of the same direct line of descent, e.g. brothers, sisters, cousins, and their children. In other legal contexts collateral indicates something distinct from but existing alongside or contemporaneous with another principal subject. Thus a collateral security is given in addition to the principal security, a collateral agreement or warranty is distinct from but subordinate to a principal agreement, and a collateral issue is a question not immediately or directly in issue. In U.S. usage collateral is the security which a borrower gives for repayment of the loan to him.

Collatio bonorum. The bringing into hotchpot (q.v.) of a portion or money advanced by a parent to a child, in order to achieve equality of distribution of personal estate among the children.

Collatio legum Mosaicarum et Romanorum. A work possibly dating from *c.* A.D. 400, apparently intended to prove that Mosaic law contained all the essentials of Roman law, consisting of 16 titles dealing with crime, delict, and succession and each containing an extract from the Pentateuch and extracts from the Roman law on the same subject, taken from the jurists mentioned in the Law of Citations (q.v.), the *Codex Gregorianus* and the *Codex Hermogenianus.* The introduction and conclusion are missing and the text seems to have been worked on and built on in various ways.

Collation. The ceremony by which a bishop institutes and admits a clerk to a church or benefice which is in the bishop's gift, corresponding to the acts of presentation and institution, where the patron and bishop are separate. Also the comparison of a copy with the original to determine the accuracy of the copy.

Collation *inter haeredes.* The principle in Scots law, when the heir in heritage was determined by primogeniture and preference for males over females, and was entitled to succeed to the whole heritable estate, that if he were also one of the next-of-kin and claiming a share of the moveable estate, he must lump the heritage and moveables into one mass and share the combined estate. See HOTCHPOT.

Collation *inter liberos.* The principle in Scots law that, in the division among the children of a deceased of the *legitim* (q.v.) fund, or portion of the estate which must be left to children, that advances made in the parent's lifetime must be brought into account to equalize the shares of the children.

Collecting society. A kind of friendly society (q.v.) which carries on the business of industrial assurance, collecting premiums from assured persons.

Collectio Anselmo Dedicata. A ninth century collection of canons dedicated to Anselm, bishop of Milan, divided into the books in which the materials were distributed according to subjects. It was widely used and had great influence on later collections down to Burchard's *Decretum.*

Collective bargaining. The process of negotiating terms and conditions of employment between representatives of one or more trade unions or other organizations of employees and representatives of one or more employers or associations of employers. In industrialized countries the practice has developed greatly in the twentieth century with the growth of trade unions. It requires the employees to have become organized and their organization to be accepted as a bargaining agent representing all or at least most employees in a particular category. The process of negotiation is frequently bedevilled by express or implied threats of strikes to enforce demands not conceded. Agreements reached by collective bargaining often need to include a grievance procedure to resolve disputes on interpretation and decide on alleged violations. The terms of a collective bargain are not terms of the contract of employment, between individual employer and individual employee, except in so far as incorporated therein by statute, agreement, reference, or by custom.

Collective responsibility. The principle, accepted in respect of the British Cabinet, that each and every member of the Cabinet accepts responsibility for all that is agreed and decided on in Cabinet. Each member must support and advocate every branch of Cabinet policy; if he does not, he must resign. In 1932 for a short time there was an 'agreement to differ', Ministers being allowed to differ from their colleagues on the subject of import duties without resigning, but this lasted only a short time. In 1975–77 the convention was again departed from on certain topics.

In primitive legal systems, groups such as tribes and families, were commonly held collectively responsible for the wrongs of members of the tribe or family. In the Babylonian law and the Old Testament punishment of the children for their father's fault is frequently mentioned.

Collective security. Arrangements by which states have sought to prevent or stop wars. By a collective security agreement between a number of states, aggression by a third party against any one is deemed aggression against all, and all unite to repel the aggression. The League of Nations and the U.N. were both founded on the principle of collective security, but neither has been able to apply the principle successfully because of the conflicts of interest among the powers; under the U.N. Charter any permanent member of the Security Council may veto collective action.

Collective title. A title frequently authorized by statutes whereby a whole series of statutes on one general theme may be cited as a group, e.g. the Merchant Shipping Acts, 1894 to 1978. The Short Titles Acts, 1892 and 1890, applied collective titles to many groups of Acts not previously so titled. Each succeeding Act bearing on the general theme will provide that it may be cited with the earlier ones under the revised collective title. Sometimes one part of an Act may be cited with certain others under one collective title and another part with different Acts under another collective title.

Collectivism. The name for any of various types of social organization in which the individual is viewed as subordinate to a group such as a class, race, nation, or state. Collectivist theories emphasize the rights and interests of the group whereas individualist theories emphasize those of the individual. The first modern expression of collectivist ideas is Rousseau's *Social Contract* (1762), which contended that the individual found his true being and freedom in submission to the general will of the community. Hegel expounded the view that the individual realizes his true being and freedom in unqualified submission to the institutions of the nation-state. Marx held that social being determined the consciousness of men. Collectivism in various forms and different degrees underlies socialism, communism, and fascism.

Collector-General. One of the Officers of State of Scotland, with some responsibility for the raising of royal revenue. The office seems later to have been joined to that of Comptroller; it fell into disuse with the Union of 1707.

College. A company, society, or group of persons, incorporated with certain privileges, particularly one created for the pursuit of learning. The learning may be 'pure', as in the cases of the Oxford and Cambridge colleges, or have a vocational aspect, as in the cases of the Royal Colleges of Physicians and of Surgeons. Some schools and other educational institutions, though called colleges, may not be incorporated, particularly institutions which do not have the rank of a university and do not have independent degree-granting powers. In Oxford and Cambridge a college is a society within the university and one becomes a member of the university by becoming a member of a college. In some other universities a college is a set of residential buildings within the university but not academically distinct therefrom.

In the U.S. the term is generally applied to degree-conferring institutions, frequently having a curriculum confined to liberal arts and sciences. They may be independent of, or be the undergraduate division of, a university.

College of Advocates. The society of practitioners in civil and ecclesiastical law which later purchased the premises known as Doctors' Commons (q.v.). Members had a monopoly of practice in the Admiralty and ecclesiastical courts. It was incorporated in 1768 and abolished in 1857.

G. D. Squibb, *Doctors' Commons.*

College of Arms. The Kings of Arms (Garter, Clarenceux, and Norroy and Ulster), heralds (Windsor, Chester, Lancaster, Somerset, York, and Richmond), and pursuivants (Rouge Croix, Rouge Dragon, Bluemantle, and Portcullis) incorporated by royal charter of 1556. Heralds Extraordinary and Pursuivants Extraordinary are occasionally appointed but are not members of the College of Arms and have only ceremonial duties. The College building in Queen Victoria Street is also known as the College of Arms or Heralds' College. It was formerly the function of the Kings of Arms to visit the different counties in England and Wales and record the arms and pedigrees of all who could establish a right to the title of esquire or gentlemen. The last visitation was in 1687. The College is now concerned, apart from ceremonial functions, with the grant of coats of arms to applicants, genealogical investigations, peerage claims, and similar matters.

College of Justice. A collective name for those who participate in the administration of justice in the Court of Session in Scotland. They comprise the Lords of Council and Session, i.e. the judges, who are Senators of the College, advocates, writers to the signet, clerks of session and the bills, clerks of the Exchequer, directors of the Chancery, their deputy and two clerks thereof, the Auditor of the

Court of Session, and certain other officers, who are members of the College. Solicitors, not being writers to the signet, are not members. Members of the college formerly enjoyed certain privileges including a general immunity from local taxation, but this has long been disused.

Colligenda bona, letters ad. Letters granted by the court of probate to a person empowering him, in default of relatives or creditors, to collect the goods of the deceased, and giving him while so acting the full powers of an administrator.

Colli, Hippolyt von (*Hippolytus à Collibus*) (1561–1612). Swiss jurist, who wrote works entitled *Nobilis* (1588), *Princeps* (1593), *Consiliarius* (1596), and *Palatinus sive Aulicus* (1600).

Collier, Sir Robert Porrit, Lord Monkswell (1817–86). Called to the Bar in 1843, he entered Parliament in 1852 and became Solicitor-General in 1863 and Attorney-General in 1868. In 1871 he was made a judge of the Common Pleas for a few days to give him the statutory qualification for appointment as a paid member of the Judicial Committee of the Privy Council, a device for which the government narrowly escaped the censure of both Houses of Parliament. As a judge Collier was very competent. In 1885 he became a peer as Lord Monkswell.

Collinet, Jean-Joseph Paul (1869–1938). French jurist, professor at Lille and Paris, author of *Études historique sur le droit de Justinian* (5 vols., 1912–47) and of many other works on legal history.

Collins, Richard Henn, Lord (1842–1911). Called to the Bar in 1867, and developed a reputation as a sound lawyer; he took silk in 1883 and became a judge of the High Court in 1891, a Lord Justice of Appeal in 1897, Master of the Rolls in 1901, and a Lord of Appeal in Ordinary in 1907. From 1894 he was Chairman of the Railway and Canal Commission, and Chairman of the Historical MSS. Commission, 1901–07. He was one of the arbitrators in the Venezuela boundary dispute in 1897.

Collisions at sea. Liability for damage done to one ship by another coming in contact with her is determined by ordinary principles of negligence, but particular importance attaches to the observance or otherwise of the International Regulations for Preventing Collisions at Sea, 1948, though nothing in the regulations exonerates any vessel from neglecting to take any seamanlike precaution. From 1873 to 1911 there was a statutory presumption of fault on the part of a ship disobeying the collision regulations, but this was abolished in 1911. If both vessels have been in fault and partly caused the damage, each is liable to make good the other's damage in proportion to the degree in which his vessel was at fault; if it is not possible to establish different degrees of fault, liability must be apportioned equally.

R. G. Marsden, *Collisions at Sea.*

Collision on land. A person who, or whose vehicle or other property, has been injured or harmed by a collision on land with an animal, vehicle, or object must in general prove negligence, or breach of the duty of reasonable care, in order to recover damages. The person in charge of a vehicle must maintain proper look out and manage his vehicle at a speed and in a manner which, having regard to the light, surface, traffic, and other factors is reasonable in the circumstances. Failure to comply with any of the prescriptions of the *Highway Code* is not automatically fault, but may be adduced as evidence of fault.

Collusion. An agreement between two or more persons to act to the prejudice of a third party or for an improper purpose. Collusion in judicial proceedings is a secret agreement that one will sue the other to obtain a judicial decision, where either the facts put forward as the foundation for the decision do not exist, or exist but were arranged for the purpose of obtaining the decision; the decision in such a case is a nullity. It was, till 1969, a bar to proceedings in England for divorce or judicial separation that a suit for divorce was procured by agreement, express or implied, between the parties. It remains a bar in nullity proceedings. In Scotland it is a bar that there has been any agreement to allow a false case to be substantiated or a just defence withheld.

Colonial law. By English law a colony acquired by discovery, settlement, and occupation is governed by the law of England; and a colony acquired by conquest or by cession from another power by its laws existing at the date of conquest or cession until the U.K. Parliament thought fit to change them. But it was frequently held that rules of English law depending on circumstances peculiar to English conditions did not apply to territories conquered or ceded. After the grant of legislative power to an assembly in a colony the legislative power of the British Parliament is excluded, save as regards Acts expressly or by clear implication made applicable to the colony or to colonies generally.

K. Roberts-Wray. *Commonwealth and Colonial Law.*

Colonial Laws Validity Act, 1865. An Act providing that any colonial law in any way repugnant to the provisions of any Act of the United Kingdom Parliament extending to that colony, or to any

Order or Regulation made under its authority, is to be absolutely void and inoperative to the extent of the repugnancy. Repugnancy to a rule of English common law does not avoid the colonial law. 'Colonial law' includes laws made for a colony by the Queen in Council, or by the colonial legislature.

By virtue of the Statute of Westminster, 1931 (q.v.), the 1865 Act no longer applies to post-1931 legislation of the fully self-governing members of the Commonwealth nor of the provinces of Canada. The 1865 Act is excluded by modern Acts granting independence to parts of the Commonwealth.

Colonies. Tracts of land, either (1) settled by British subjects and recognized as British territory, or (2) conquered by force and made part of H.M. dominions, or (3) ceded by another state, tribal chiefs, or inhabitants. The constitutional position of a colony depends on its mode of acquisition. Colonies are administered by the government of the United Kingdom. In settled colonies the prerogatives of the Crown and the rights and immunities of British subjects are similar to those enjoyed in the U.K. In conquered or ceded colonies the Crown has full power to establish such legislative, executive, and judicial authorities as it thinks fit. As colonies have developed many have acquired representative and then responsible governments, and have finally been granted independence. Others remain dependent territories.

Colquhoun, Archibald Campbell (?–1820). Passed advocate in 1768 and became Lord Advocate in 1807–16. He sat in Parliament from 1807 till his death. As Lord Advocate he took part in the reform of the Court of Session and sat on the commission which enquired into the administration of justice in Scotland. In 1816 he became Lord Clerk-Register.

Colquhoun, Patrick (1745–1820). Lived for a time in Virginia, but returned to Glasgow in 1766 and became provost there in 1782 and founded the Glasgow Chamber of Commerce. In 1789 he settled in London and began to agitate for a more efficient police system. In 1792 he was made a city magistrate. In 1796 he published a *Treatise on the Police of the Metropolis* and in 1800 a work on *The Police of the River Thames*, which so roused public attention that an effective river police was at once instituted. He initiated charitable schemes and in 1814 published *A Treatise on the Population, Wealth, Power and Resources of the British Empire*. He was an indefatigable reformer and citizen of great public spirit.

Colvile, Sir James William (1810–80). Called to the Bar in 1835 and served as a judge in India, 1848–59. By reason of his knowledge of Indian law he was appointed a Privy Councillor, and sat on the Judicial Committee, and in 1871 was appointed one of the paid judges of that body, a capacity in which he served till his death.

Colour. Formerly an important concept in pleading, now obsolete. It signified an apparent or prima facie right and was found in novel disseisin and later in trespass. It might be express or implied. Express colour was a feigned matter pleaded by the defendant in an action of trespass, making it appear that the plaintiff had a good cause of action whereas he had in reality only an appearance or colour of a cause. It was abolished in 1852. Implied colour was an allegation apparent only by implication from the terms actually pleaded. It disappeared in 1873.

Combination Acts. Acts of 1799 and 1800 which made trade unionism illegal, imposing penalties on any working man who combined with another to gain an increase in wages, a decrease in hours, who solicited anyone to leave work, or objected to working with another workman. Other clauses forbade employers' combinations but these were never enforced. Jurisdiction was entrusted to magistrates and appeal was made difficult. At common law trade unions were illegal in so far as their purposes were in restraint of trade. The Acts were repealed in 1824 and this was followed by an outburst of strikes, and an attempt was made in 1825 to revive the Acts. In 1871 the Trade Union Act recognized the legality of trade unions.

See also TRADE UNION.

Comitatus. In Germanic tribal society, a band of chosen warriors forming a fraternity of military adventure under a chief and having a close personal bond with him. It has been regarded as one of the root institutions from which feudalism developed.

Comitia. An assembly of the Roman people summoned by a magistrate. It voted only on business presented by the magistrate, the majority in each group determining the vote of the group, and could not amend the proposal. As the three main divisions of the Roman people were the 30 curiae, the 193 centuries, and the 35 tribes, there were three forms of *comitia* and the *concilium* (or *comitia*) *plebis*.

(a) *comitia curiata*, the most ancient, which confirmed the appointment of magistrates, witnessed the installation of priests, adoptions and the making of wills, but was politically unimportant by the time of the republic.

(b) *Comitia centuriata*, instituted in 450 B.C., which enacted laws, elected magistrates, declared war and peace, and inflicted the death penalty. It was organized in a way which reflected military organization and gave most weight to the wealthier classes.

(c) *Comitia plebis tributa*, which developed after *plebiscita*, i.e. decisions of the *concilium plebis*, were given full validity as laws in 287 B.C. In it trials for non-capital offences were held and many forms of business transacted. In the later republic this was increasingly the usual legislative assembly.

(d) *Comitia populi tributa*, founded at an uncertain date in imitation of the *comitia plebis tributa*, which elected quaestors, tribunes, enacted laws, and held minor trials.

In republican times the latter three were all equally capable of passing legislation. Which of the latter three was summoned for particular business depended on which magistrate wished to put a proposal.

The legislative and judicial functions lapsed under the empire though *comitia* were nominally in existence until at least the third century A.D.

Comity. A principle of community spirit, friendliness, and good manners between states. It is invoked by some jurists as justifying the recognition and enforcement in one state of rights acquired in another state and under its legal system, and hence as underlying international private law. Sometimes it is regarded as a stimulus to the pattern of behaviour which develops into international custom.

Command papers. Papers presented to Parliament by a minister of the Crown, legally by command of Her Majesty. They include reports on the work of government departments, evidence to and reports of Royal Commissions and Departmental Committees, and materials relative to matters of policy which may be the basis for debate or legislation. Originally printed as appendices to the House of Commons Journal, they have since 1833 been printed separately and numbered in five series as follows:

No. 1	to No. 4222,	1833 to 1868–69;
C. 1	to C. 9550,	1870 to 1899;
Cd. 1	to Cd. 9239,	1900 to 1918;
Cmd. 1	to Cmd. 9889,	1919 to 1956;
Cmnd. 1	to Cmnd.	1956 to date

See also PARLIAMENTARY PAPERS.

Commander-in-Chief. A military office created by patent in 1793 to remove the army from political influence. From 1854 the army was under the dual control of the Commander-in-Chief and the newly created Secretary of State for War and in 1870 the latter became the responsible Minister, the Commander-in-Chief being his subordinate. The office was abolished in 1904, the functions being divided between the Army Council and the Inspector-General of the Forces. The Army Council is now merged in the Ministry of Defence.

Commandité, société en. In French law, a partnership in which some of the partners are lenders of capital to the firm only, not responsible for losses beyond the amount of their capital, and not permitted to participate in the management of the firm's business. The principle was introduced in the U.K. by the Limited Partnership Act, 1907.

Commendam. A benefice or ecclesiastical living committed to the charge of a clerk to be supplied until it can be provided with a regular pastor. Formerly, if a parson were made a bishop he held his former benefice *in commendam* if the Pope or the King gave him power to retain his benefice. Commendams have been void since 1836. The person who received the fruits of a benefice during a vacancy was a commendator. Properly he was a trustee, but the office was sometimes conferred for life.

Commendams, case of (or *Colt and Glover* v. *Bishop of Coventry* (1616), Hobart, 140). A commendam was a papal prerogative which passed to the Crown at the Reformation of the grant of a living to be held along with another ecclesiastical office. A bishop who held the See of Lichfield was granted a living to be held along with his bishopric. Colt and Glover brought an action against the bishop claiming that the presentation was theirs and that the King's grant was invalid. The case was heard before all the judges in the Exchequer Chamber. The King commanded the judges not to deal with the case till they should hear his pleasure; he harangued the judges on the prerogative and intimidated all the judges save Coke into promising to act according to the royal wishes. Coke was then dismissed from the office of Chief Justice. As a result the common law judges held office at the whim of the monarch.

Commendation. In feudal law, where an owner of land placed himself and his land under the protection of a lord and constituted himself a feudal tenant or vassal. It was one of the primary modes whereby feudalism was developed.

Commentarii. Memoranda or notes to assist the memory. At Rome the *commentarii principis* were the register of official acts of the emperor. The *commentarii senatus* were probably the same as the *Acta Senatus.*

Commentary. A term frequently applied to a kind of legal literature. In classical Roman law a commentary frequently took the form of a separate roll from the text containing observations on passages and phrases in the text. There were also commentaries which developed into separate books, on *leges* and *senatusconsulta*, such as Gaius' commentary on the *Lex Julia et Papia Poppaea*, and his commentary on the *SC Orfitianum*, and commentaries on the praetorian and other edicts, such as Labeo's on the urban praetor's edict and Gaius' commentary *Ad edictum provinciale libri xxxii*. The biggest commentary on the edicts was that of Pomponius in at least 83 books.

In the mediaeval period the work of the commentators (q.v.) were known as commentaries.

In modern times the term is sometimes used for a systematic textbook, as in the case of Story's (q.v.) Commentaries on various topics of American law.

Commentators (or Post-glossators). The name given to the group of legal scholars and teachers who flourished chiefly in Italy from *c.* 1250 to *c.* 1400, following and reacting against the Glossators (q.v.). They worked mainly by dialectic, logical reasoning, and discussion, expounding by division and subdivision, syllogisms and dilemmas. The major defect of their work was deference to authority, degenerating into mere counting of authorities. Nevertheless their work was of great importance in relating and adapting Roman law to city statutes, feudal and Germanic customs, and the principles of the canon law; Roman law was transformed into Italian law. Their work took the form of large and systematic commentaries and independent treatises and tracts. The school originated in France with Jacques de Revigny (1215–96), followed by Pierre de Belleperche (d. 1308).

Among the more notable of the school were Cino da Pistoia (1270–1336) who introduced the method into Italy, Albericus de Rosciate (d. 1354), Bartolus of Sassoferrato (1314–57), the greatest of them all, who wrote commentaries on all titles of the Digest, opinions, and essays on almost every topic of the law, Baldus de Ubaldis (1327–1400), Lucus di Penna (d. 1390), Bartolomeus Salicetus (d. 1412), Raphael Fulgosius (d. 1427), Paulus di Castro (d. 1441), Marianus and Bartolomaeus Socinus, Lancelottus (d. 1503), Philip Decius (d. 1536), Johannes Faber, Matthaeus de Afflictis, and Jason Mainus. Many of their writings were published in the sixteenth century. Their influence extended all over Europe, and their writings influenced the law in many countries. They prepared the way for a natural law school of the seventeenth century and laid the theoretical foundation of the modern state. Much later, such jurists as du Moulin, Bodin, and Gentilis belonged to the Bartolist School of thought, though they worked in other fields.

Savigny, *Geschichte des römischen Rechts im Mittelalter* (1850); Weiacker, *Privatrechtsgeschichte der Neuzeit* (1952); Calasso, *Medio evo di diritto*; W. Ullman, *Mediaeval Idea of Law* (1946); Camb. Med. Hist., V, 697.

Commerce. The exchange of commodities and all the arrangements involved in effecting such exchanges. It accordingly involves advertising agency and making contracts, sale, carriage, insurance, guarantees, banking and finance (including bills, cheques, and credits), and bankruptcy.

Commerce clause. Art. I, Sec. 8 of the U.S. Constitution, authorizing Congress 'to regulate commerce with foreign nations, among the several states and with Indian tribes'. The Supreme Court has traditionally interpreted the power broadly and made this the principal basis for the extension of the power of the federal government over the economy, and it has come to be interpreted as allowing Congress to regulate business, labour, and agriculture when involved, even indirectly, in interstate commerce.

Commercial arbitration. See ARBITRATION.

Commercial Court. In 1895 commercial actions, arising out of the ordinary transactions of merchants and traders, in the High Court in London were assigned to a 'commercial list' and tried under simplified procedure in expeditious trials by a specialist judge. In 1970 a Commercial Court was constituted as part of the Queen's Bench Division of the High Court to take such causes as might be entered in the commercial list. The judges are such of the puisne judges of the High Court as the Lord Chancellor may nominate to be Commercial judges. A judge of the Commercial Court may accept appointment as sole arbitrator or as umpire by or by virtue of an arbitration agreement, where the dispute appears to him to be of a commercial character, if he can be made available to do so.

Commercial law. A general and rather indeterminate term for those topics of law relevant in commerce, such as contract, agency, sale, negotiable instruments, documents of title, bankruptcy. In the United Kingdom and common law countries generally it is not a distinct body or branch of the law but simply a compendious term for topics, mostly drawn from the law of contract and of property, relevant to business and commercial practice, the law on the topics included being the same whether the transaction is a commercial one or one between friends. There is no distinct legal category of commercial men or of acts of commerce.

In some European countries on the other hand the civil law and commercial law are distinct bodies of law, frequently set out in distinct codes, applicable to distinct sets of persons, or distinct and defined kinds of transactions. Commercial transactions are in many respects subject to particular rules, which differ from those applicable to ordinary consumer transactions. Accordingly there must be definition of a 'merchant' as in German law, or of a 'commercial transaction', as in France. This distinction owes something to the Roman law distinction between the *jus civile* and the *jus gentium*, the latter of which applied to foreign, normally trade, relations, but more to the mediaeval law merchant.

Commercial paper. A term covering all kinds of short-term negotiable instruments which call for payment of money and which may be used for borrowing, and which arise from and in the course of commercial dealings.

Commercial treaties. Treaties between states relating to trade. There is no clearly defined distinction between commercial treaties and political treaties and many treaties contain clauses on both subjects. Treaties dealing with commerce are mentioned by Aristotle and are known to have been concluded between Rome and Carthage in 509 and 348 B.C. The earliest English commercial treaty is one with Norway in A.D. 1217.

Hertslet's *Commercial Treaties*, 33 vols. (1840–1925).

Commissaires du gouvernement. Officers appointed from among members of the *Conseil d'Etat* (q.v.) whose function is to present objective, impartial, and independent arguments to that body's judicial section in administrative matters, not representing the department concerned nor any other party, but acting as the conscience of the court and seeking to assist the tribunal to arrive at a right decision. He does not participate in the giving of judgment.

Compare ADVOCATE-GENERAL.

Commissary courts. At the Reformation in Scotland the courts of the diocesan officials (q.v.) were replaced in 1563 by a Commissary Court of Edinburgh with exclusive jurisdiction in marriage, divorce, and bastardy and a general jurisdiction in the same field as the old officials' courts. There was a limited right of appeal to the Court of Session. There were four Commissaries. There were inferior commissary courts with a jurisdiction mainly in testamentary matters. These courts largely adopted the canon law except in so far as contrary to the reformed religion. In 1823 the inferior commissary courts' powers were transferred to the sheriffs, though the Edinburgh local court survived till 1836. All were abolished in 1876. The Commissary Court of Edinburgh came to an end in 1830, its powers being transferred to the Court of Session.

The name also attaches to the ecclesiastical court of the diocese of Canterbury corresponding to the consistory court of other dioceses, as distinct from the Court of Arches in which the archbishop exercises jurisdiction as metropolitan.

Commission. (1) An order or authority, particularly to a factor or agent, to do some act. The word was also formerly given to the authority given to Commissioners of Assize, by commissions of assize, of oyer and terminer, and of gaol delivery, to hold courts at assizes.

(2) The remuneration paid to an agent, particularly one related to the volume or value or importance of the business negotiated.

(3) A person or body of persons authorized by the Crown, or a court, or other authority to do some act, particularly to enquire into certain facts or to manage certain businesses. Thus Royal Commissions are appointed to enquire into some matter and report; the office of Lord Treasurer has long been entrusted to a commission or body of persons. The Civil Service Commission is the body charged with recruitment of persons into the civil service.

(4) In the U.S. the name is given to various kinds of major administrative bodies, such as the Interstate Commerce Commission (q.v.).

Commission, Assent to bills by. See ROYAL ASSENT.

Commission and diligence. In Scots law, the means whereby a civil court delegates part of its duties to a commissioner, to take the deposition of witnesses on oath and report them to the court, or to examine documents relevant to an action but in the hands of a third party and report them to the court.

Commission for examination of witnesses. A commission which may be issued to a person to take the evidence of a witness whose evidence is required but who is outside the jurisdiction or whose personal attendance should for any sufficient cause be dispensed with. The evidence is taken on oath *viva voce* or by interrogatories, or in both modes and the depositions by the witness are read at the trial.

Commission of anticipation. A commission under the Great Seal issued in the Tudor and early Stuart period to collect a tax or subsidy before it actually became due. The practice was revived or copied in the nineteenth century when it became the practice for the Committee of Ways and Means to pass a resolution assenting to certain new taxes and for the revenue authorities to proceed to collect them before they had been authorized by Parliament. The practice was successfully challenged in *Bowles* v. *Bank of England*, [1913] 1 Ch. 57, but then legalized by the Provisional Collection of Taxes Acts, 1913 and 1968.

Commission of appraisement and sale. An authority granted to a marshal, where property has been arrested in an Admiralty action *in rem* and ordered to be sold, to inventory the property, have it valued and sold by public auction, and the proceeds paid into the Admiralty registry.

Commission of array. A commission formerly issued to send into all counties officers to muster the inhabitants and put them in military order.

Commission of assize. A commission formerly issued to judges, Queen's counsel, and others, authorizing them to deal with civil business at assizes (q.v.).

Commission of eyre. The authority granted to itinerant justices to hold the eyre (q.v.).

Commission of gaol delivery. One of the commissions formerly granted to judges or other commissioners of assize, authorizing them to try and, if they acquit, deliver from gaol delivery every prisoner held in custody or on bail for an alleged crime, whatever the locus of the alleged crime.

Commission of oyer and terminer. One of the commissions formerly granted to judges or other commissioners of assize, authorizing them to try persons charged with treasons, felonies, and misdemeanours within the area named in the commission.

Commission of the European Communities. The main executive organ of the Communities which, since 1967, has replaced the former High Authority of E.C.S.C. and the Commissions of Euratom and E.E.C. Since 1973 it comprises 13 members, two from each of the four larger states and one from each of the five smaller states, chosen for their independence and general competence. They are appointed for four-year renewable terms. They elect their own President and five Vice-Presidents for two-year renewable terms. The Commission as a whole may be dismissed by a vote of censure by the European Parliament. It is located in Brussels. Its functions are to provide the general staff of the Communities and, subject to the general directions of the Council of Ministers, to initiate policy, co-ordinate action, and organize and supervise the working of the Treaties. Each member of the Commission is responsible for one or more of the main branches of Community activities. The administration is divided into departments each under a Director-General.

The Commission makes an annual report on the activities of the Communities which is debated in the European Parliament.

The Commission meets regularly and is collectively responsible for all decisions. It has the duty of ensuring that the treaties are applied, which in substance imposes the initiative on the Commission and the Commission has great influence on the way in which Community issues are considered by the way it formulates proposals. In general the Commission, after consulting experts and national officials, formulates proposals for the Council of Ministers which may, or sometimes must, consult the European Parliament and the Economic and Social Committee. The Commission itself also directly makes much legislation of a routine and technical character.

Each member of the Commission is assisted by a *cabinet*, or private, or departmental office staff. The large staff of the Commission is representative of all the member states, many being seconded from national civil services and universities.

Commission of the peace. A commission issued by the Crown appointing a number of persons to act as justices of the peace within a stated area. Prior to 1835 commissions were issued only to counties, justices in boroughs having jurisdiction by borough charter or prescription. From 1835 separate commissions were issued for each county and each county borough and for some non-county boroughs. Since 1974 there are separate commissions of the peace for each county.

Commission plan. A system of municipal government adopted in many cities of the U.S. since 1900, under which all legislative and executive power is vested in a small elective board or commission, usually of five. In many places it has been superseded by the city manager plan.

Commissioner for oaths. A solicitor, or in exceptional cases another fit and proper person, appointed by the Lord Chancellor to administer oaths, affirmations, make declarations, and take affidavits. When doing so the commissioner must satisfy himself of the deponent's identity and that he is in a fit state to understand what he is doing.

Commissioners in bankruptcy. Persons formerly appointed by the Lord Chancellor to exercise jurisdiction over bankrupts' persons and their estates, under his supervision. In 1831 a court of bankruptcy was established with four judges, exercising the Lord Chancellor's supervisory function, and six permanent commissioners. The Court of review was abolished in 1847. Commissioners were abolished in 1869.

Commissioners in Lunacy. Originally two persons, appointed under statute of 1842, to whom commissions were addressed to inquire into the state of mind of an alleged lunatic and to certify the result by inquisition. Under statute of 1845 they became Masters in Lunacy and 11 new officials were appointed Commissioners in Lunacy. The number was later increased, the whole body constituting the Board of Control. It was dissolved in 1959, the judicial functions passing to the Court of Protection (q.v.).

Commissioners in sequestration. In Scots law, persons appointed in a sequestration to advise the trustees.

Commissioners of Crown Lands. A body established in 1927 to manage Crown lands, replaced in 1956 by Crown Estate Commissioners.

Commissioners of Customs and Excise. The department of state which collects customs and excise duties and value added tax.

Commissioners of Inland Revenue. A body appointed by letters patent under the Great Seal, charged with collecting internal taxes, particularly income tax, capital gains tax, capital transfer tax, development land tax, and stamp duties.

Commissioners of Justiciary, Lords. The judges of the High Court of Justiciary, the supreme criminal court in Scotland. All the judges of the Court of Session have, since 1887, been Lords Commissioners of Justiciary also.

Commissioners of Sewers. Persons appointed by Act of 1571 and later Acts to see that ditches and drains were well kept in the lower parts of England and water channelled to the sea. Their functions were transferred in 1930 to catchment boards and river boards.

Commissioners of Supply. Persons possessing stated qualifications as landowners and established by statute of 1667 on a county basis throughout Scotland to levy and collect the land tax for the 'supply' of the Sovereign. Prior to 1798 Commissioners were expressly named for each county in the annual Acts of Supply. In time further duties were added by statutes, including the preparation of the valuation roll, provision of a police force, sheriff court houses and prisons, lunacy, militia, reformatories, and registration of voters. Most of their powers and duties were transferred to the new county councils in 1889, but their powers in relation to Land Tax survived.

Commissioners of Teinds. The judges of the Court of Session in Scotland in the exercise of their judicial and ministerial functions in relation to teinds and the stipends of the parochial ministry. See TEIND COURT.

Commissioners of Woods, Forests and Land Revenues. A board established to supervise and manage these kinds of public property. They later became the Commissioners of Crown Lands (q.v.).

Commissioners of Works and Public Buildings. A board appointed in 1874 to manage these kinds of public property. Their functions were transferred in 1942 to the Minister of Works and have now passed to the Department of the Environment.

Commissions for Local Administration. Two bodies, for England and Wales respectively, created in 1974 to deal with complaints of maladministration by local authorities, police, and water authorities. The Parliamentary Commissioner (q.v.) is a member of each.

Commit, committal and commitment. The verb 'commit' and its derivatives are used of a person being sent to prison, normally temporarily or for a short period, by a court or judge, or being sent to a higher court for sentence. Committal for trial is the determination of examining justices who have heard the evidence that there is sufficient evidence to justify putting the defendant on trial by jury for an indictable offence. Committal will be in custody or on bail. Where all the evidence before the court consists of written statements, examining justices may in certain circumstances commit without consideration of the evidence.

Committee. A person or group of persons to whom some matter is committed for investigation, discussion, recommendation, or otherwise as may be determined when it is established. In Parliamentary practice there are committees of the whole House, select committees comprising certain members appointed to report on a matter for the information of the House, sessional committees appointed each session to deal with particular business, and standing committees which consider in detail bills referred to them. There are also joint committees of equal numbers of members of each House.

Committee, Judicial. See JUDICIAL COMMITTEE OF THE PRIVY COUNCIL.

Committee for Privileges. A committee of the House of Lords, elected each session, to consider claims for peerages and precedence and matters relating to the orders, customs, and privileges of the House. It consists of any Lords who attend, but must include at least three Lords of Appeal. It may call for papers, hear counsel, and examine witnesses on oath.

Committee of Both Kingdoms. A Puritan committee established in 1644 charged with the supreme strategic direction on the Parliamentary side in the English Civil War.

Committee of General Security. An organ of the French revolutionary government. It was established by the National Convention in 1792, directed the secret police and revolutionary justice and administered the terror of 1793–94.

Committee of Inspection. A committee of creditors appointed by the creditors to superintend the administration by the trustee of the bankrupt's property, or appointed by the creditors and contributories of a company which is being wound up.

Committee of Lunatic. A person to whom the custody of the person or property of a lunatic so found by inquisition was granted by the Lord Chancellor. These duties are now performed by the Court of Protection (q.v.).

Committee of Privileges. A committee of the House of Commons, usually of 12, constituted each session, to consider complaints of alleged breach of privilege. When a complaint of breach of privilege is made the Speaker must consider whether or not there is a prima facie case; if he rules that there is, the Leader of the House or the member making the complaint must make a motion to refer the case to the Committee of Privileges or otherwise disposing of the matter. Once a case has been referred to the committee the House takes no action until the committee has reported. It may call for papers and records, summon persons and examine witnesses, but does not hear counsel.

Committee of Public Accounts. The Public Accounts Committee, originally established in 1861, is an important committee of the House of Commons, composed of senior members from all parties in proportion to their representation in the House. The chairman is a senior member of the opposition and it is assisted and advised by the Comptroller and Auditor General (q.v.), and two Treasury Officers of Accounts. The committee's function is to ensure that all public money is spent as Parliament intended and not otherwise. It considers the Appropriation Accounts which detail the expenditure of the armed forces and civil departments, and the Accounting Officers of the departments have to appear and explain difficulties. Overspending, wastage, and extravagance are investigated and criticized, frequently in very strong terms. It reports to the House and its reports are debated. Its recommendations are treated with great respect by the Treasury and its existence is none the less effective though it is dealing with transactions which are past.

Committee of the Whole House. A committee of all the members of the House of Commons presided over by the Chairman of Ways and Means, or the Deputy Chairman, not by the Speaker, and under rather less formal procedural rules. The House may refer any matter to a Committee of the Whole House, but the committee stages of bills involving finance are always so taken. Distinct types of Committee of the Whole House were the former Committee of Supply, dealing with estimates, the Committee of Ways and Means, and Committee of the Whole House on a financial resolution.

A Committee of the Whole House may also be appointed in the House of Lords, when the chair is taken by the Lord Chairman of Committees.

Committee on Administrative Tribunals and Enquiries. A committee, commonly called the Franks Committee from its chairman, Sir Oliver (later the Rt. Hon. Lord) Franks. Its report (Cmnd. 218, 1957) was largely given effect to by the Tribunals and Inquiries Act, 1958, later replaced by a similar Act of 1971. This provided for the establishment of the Council on Tribunals (q.v.) as an advisory body and with general oversight over tribunals and inquiries, not as a court of appeal, which may make general recommendations on the membership of many kinds of tribunals, and which must be consulted before procedural rules are made for them. A right of appeal to the High Court on a point of law was given in the case of various kinds of tribunals and a right was given to a reasoned decision from any of certain listed tribunals.

The Act fell short of the Committee's recommendations in certain respects, but has effected considerable rationalization and improvement of the law and procedure in relation to both tribunals and inquiries. Most of the Committee's recommendations for improvements in the practice as to public inquiries were also accepted and put into force.

Commodatum. In Roman law, proper loan, the gratuitous loan of a thing to be returned after the period specified, or after the purpose of the loan has been served, or in any event after the lapse of a reasonable time or on request, in the same form and in as good a state as when lent, save for necessary wear and tear. Possession only, and not property, is transferred to the borrower. Contrast *mutuum*.

Common. (1) A right of the nature of a profit which a man has in the land of another, a species of *profit à prendre*, so called from the community of interest between the claimant and the owner of the land or between the claimant and others having the same right. It is the right to take any part of the natural produce of the land or water belonging to another in common with the claimant. Most rights of common may be claimed by prescription or created by grant.

There are four principal kinds of common; common of pasture, or the right to pasture one's animals on the land of another; common of estovers, or the right of taking a reasonable portion of timber for use in the commoner's tenement; common of turbary, or the right of taking peat or turf from another's waste land for fuel in the commoner's house; and common of piscary, or the right of fishing in another man's water. These rights may be appendant, being the right of every freehold tenant of a manor to use the lord's waste land, or appurtenant, being a right annexed to certain lands, or in gross, being unconnected in any way with the tenure or occupation of land, by reason of vicinage, being rights of commoners of adjoining land to let cattle stray across the boundary. Common of estovers or of turbary cannot be appendant. Certain other kinds of common are recognized, common in the soil, the right to take sand, gravel, stones, or clay from another's land, and common of taking wild animals or wildfowl.

(2) Is a piece of land subject to rights of common. The inclosure of many commons has been effected by legislation, but provision was made for safeguarding the rights of commoners. Since 1965 all land which is common land and all rights of common over such land have to be registered, failing which the rights cease to be exercisable.

Common Agricultural Policy. A major policy of the E.E.C. intended to increase productivity, to ensure a fair standard of living for the agricultural community, to stabilize markets, ensure the availability of supplies, and reasonable prices for customers. In pursuance of this policy there have been set up various Common Market organizations, e.g. for cereals, poultry-meat, fruit, and vegetables, with structures varying according to the product concerned and the strength of the production industry. The organizations operate mainly by controlling and directing the internal market and removing obstacles to the free circulation of goods, and providing protection against imports by imposing levies on the commodity.

Common assault. An assault (q.v.) not aggravated in any particular way.

Common Bench. The old name for the Court of Common Pleas.

Common calamity. An incident in which two or more related persons die. Such an incident gives rise to problems of succession. In English law there was no presumption of survivorship until the Law of Property Act, 1925, enacted that the younger should be deemed to have survived the elder. In Scots law the Succession (Scot.) Act, 1964, provided that where the persons were husband and wife, it should be presumed that neither survived the other and in other cases that the younger survived the elder unless the elder left a will in favour of the younger, whom failing, in favour of a third person, and the younger died intestate, in which case the elder is presumed to have survived.

Common calling. At common law one of the occupations which carried a special status with consequential special legal liabilities. The main examples were carriers, innkeepers, and farriers. They were liable without having expressly accepted liability in the event of harm resulting, and it was later established that their liability was strict, and did not require proof of negligence.

Common carrier. At common law a person who made a public profession of willingness to carry between any specified places for anyone who offered, goods or passengers. By custom there attached to the common carrier an absolute liability in the event of loss of goods carried, which could only be rebutted by proving that the loss was attributable to the act or default of the consignor, act of the Queen's enemies, or inherent vice of the goods themselves. No such liability attached to the common carrier of passengers who was liable for negligence only.

Common Council. The council of the City of London.

Common counts. The counts in an action for *indebitatus assumpsit* (q.v.) prior to the Judicature Act, 1873, in a declaration for goods bargained and sold, or sold and delivered, for money lent, for work done, for money received to the use of the plaintiff, for money due on an account stated, or for interest.

Common employment. The doctrine that an employer was not liable to an injured employee if the injuries were caused by the fault of a fellow employee engaged in common employment with the injured man for and under the same employer. Invented by Abinger, C. B., in *Priestley* v. *Fowler* (1837), 3 M. & W. 1 and approved by the House of Lords in *Bartonshill Coal Co.* v. *Reid* (1858), 3 Macq. 266, it was modified by the Employers' Liability Act, 1880, fell into disfavour and was limited by the House of Lords in *Radcliffe* v. *Ribble Motor Services*, [1939] A.C. 215, and was abolished by the Law Reform (Personal Injuries) Act, 1948.

Common good. In Scots law, the entire property of a burgh held by the corporation for behoof of the community, including lands held and taxes, tolls, and dues exigible. Some of the common good, such as lands and houses, is alienable, but some, such as public lands and buildings are held in trust for the community and are not alienable.

Common informer. A person who took proceedings for infringements of certain statutes solely for the penalty or share thereof which by the Act fell to one who gave information of a breach. Such proceedings were abolished generally in 1951 and in relation to parliamentary disqualification in 1957.

Common injunction. An injunction granted by the Court of Chancery to prevent the institution or continuance at common law of proceedings which were contrary to equity. With the fusion of law and equity by the Judicature Acts, 1873–75, it became obsolete.

Common jury. A jury composed of ordinary jurymen, as contrasted with a special jury. Save in the case of City of London special juries in commercial cases, special juries were abolished in 1949.

Common law. A term used in various distinct senses:

(1) It was originally used by the canonists (q.v.) *jus commune*, as denoting the general law of the Church, as distinct from divergent local customs which in particular areas modified the common law of Christendom;

(2) As the powerful centralized system of justice of the English kings developed in the twelfth and later centuries, the royal justices increasingly developed and administered general rules common to the whole of England, the common law of England, as distinct from local customs, peculiarities, and variations, such as gavelkind (q.v.);

(3) The common law accordingly came to mean the whole law of England, including ecclesiastical law and maritime and mercantile law, as administered in England, as distinct from that of other countries, particularly those based on the Roman law;

(4) Hence, in the context of comparative law, a common law system is one based fundamentally on English common law, as distinct from a civil law system, based on the civil law of Rome;

(5) With the development of equity (q.v.) and equitable rights and remedies, common law and equitable courts, procedure, rights, remedies, etc., are frequently contrasted, and in this sense common law is distinguished from equity; thus at common law a person aggrieved by a breach of contract could claim damages only, but in equity he could claim specific performance (q.v.);

(6) Common law was similarly distinguished from ecclesiatical law;

(7) Common law rights, powers, remedies, crimes, etc., are frequently distinguished from statutory rights, powers, remedies, crimes, etc., according to the formal source of the particular right, etc., in principles of common law or in the prescriptions of statute.

In French and German law common law (*droit commun, Gemeinrecht*) mean law common to the whole area of the State as distinct from local or regional customs or peculiarities.

See also CIVIL LAW.

Common law marriage. In Western Europe prior to the Council of Trent in 1563 no religious ceremony was necessary; the only requisites were declaration by the parties that they took each other as husband and wife, *per verba de praesenti* in which case the marriage was binding immediately, *per verba de futuro*, in which case it became binding as soon as consummated. But it early became customary for preliminary publication of banns or notice of intention to marry and for the marriage to be contracted at the church in the presence of the priest and to be followed by nuptial mass. In time the emphasis shifted to the presence of the priest or clerk so that in *R.* v. *Millis* (1844), 10 Cl. & F. 534, the House of Lords laid down that a valid marriage at common law could be contracted only by exchange of consent *per verba de praesenti* in presence of an episcopally ordained clergyman. Legislation later regulated the formalities of marriage. In England Lord Hardwicke's Act of 1753 required marriages (except for Quakers and Jews) to be solemnized according to the rites of the Church of England in the parish church of one of the parties in the presence of the clergyman and two other witnesses. 'Gretna Green' marriages were struck at by the Marriage Act, 1823 (replacing the 1753 Act), itself superseded in time. In some U.S. states common-law marriages were still valid in the 1960s.

The term common law marriage is now applied in English law to a marriage celebrated according to a common law form in a place where the local forms of marriage cannot by utilized, e.g. desert island, or are morally unacceptable to the parties, e.g. Mohammedan country, or no clergyman is available. In Scotland it applies also to persons married by cohabitation with habit and repute (q.v.). In both countries the term 'common law wife' is sometimes applied to a concubine or mistress where the relationship is of some duration and stability.

Common law systems. The group of legal systems founded mainly on the common law of England, as contrasted with civil law systems (q.v.). It comprises the laws of Wales and Ireland, of most of the English-speaking parts of the world originally colonized from England, including notably the U.S., Canada, Australia, and New Zealand. Various other countries, notably Scotland, South Africa, and

Rhodesia, have been influenced by English common law but are not common law systems in the full sense. The characteristics of common law systems, following the English model, are: the historical basis is the customary law of England as unified and developed by itinerant royal justices from the twelfth century; while legislation has played a part, and is today playing an increasing part and in some areas a predominant part, in developing and stating the principles and rules, it is frequently regarded as an interloper, an interference; the major form in which law is stated is the judicial opinion, which is commonly regarded as binding or persuasive on a later court dealing with the same or a closely related problem (see CASE-LAW; JUDICIAL PRECEDENT; *Stare decisis*); a distinction was formerly and in certain respects is still drawn between principles and rules of law developed by courts of law and principles and rules of equity developed by courts of chancery or equity; the importance of judicial opinions and even dicta is high, more so than that of academic lawyers seeking to rationalize, criticize, and develop the law; the classification and terminology of the law owe little to the categories and concepts of Roman law; procedure is fundamentally adversary or accusatorial rather than inquisitorial, and extensive use is made of the jury (q.v.) as an instrument for finding the facts.

See CIVIL LAW SYSTEMS; ENGLISH LAW; U.S. LAW.

Common Market. The common term for the European Economic Community (q.v.), though properly speaking the common market is an economic consequence of the activities of the three European communities (q.v.).

Common Market Law. The common term for the body of law of the European communities (q.v.).

Common pleader in the City of London. Four barristers who, under a system operative from early times till mid-Victorian times, were entitled to the monopoly of appearance in the Mayor's, Sheriff's, and Hustings Courts of the City of London. Their places were obtained by nomination and purchase. Similarly only the four attorneys who were known as clerks of the Mayor's Court could practise in any of these three courts. In the nineteenth century practice in these courts was opened generally.

Common Pleas, Court of. By the early thirteenth century there was a clear division in England between the court which moved with the King and had jurisdiction over pleas of the Crown and common pleas, and that fixed at a certain place, henceforth usually Westminster, to hear common pleas, and their proceedings were from 1223 recorded in separate *Coram Rege* and *De Banco* rolls. Magna Carta, Art 17 provided that the court of common pleas should not travel with the King but should be held in a certain place. In 1272 a separate Chief Justice of the Common Pleas was appointed. But the court was inferior to that of the King, the King's Bench, since error lay from it to that court. Its jurisdiction was over common pleas between subject and subject, and in it alone real actions and the old personal actions of debt, detinue, account, and covenant (qq.v.) could be commenced. This gave it a monopoly of important work in the mediaeval period and at this time it was the most active though not the highest of the common-law courts. It exercised supervision of local and manorial courts, but its judgments were subject to review by the Court of King's Bench. Though its exclusive jurisdiction over real actions lasted until the nineteenth century, the old forms had become largely obsolete long before, and the King's Bench and the Exchequer had encroached on the jurisdiction of the Common Pleas by legal fictions, and by the development of the actions of ejectment and of trespass on the case.

It also had jurisdiction to supervise or correct the errors of the older local courts, by writ of Pone, of False Judgment, and possibly by writ of Error, and in the seventeenth century it acquired a general jurisdiction to issue the writs of Prohibition and of Habeas Corpus.

The serjeants-at-law (q.v.) early established a sole right of audience in the Common Pleas which made the Court unpopular owing to its expense.

It was abolished in 1875, being merged in the new High Court, but the name survived till 1880 in the Common Pleas Division and all actions formerly within the exclusive jurisdiction of the former Court were assigned to that Division. The Common Pleas Division was amalgamated with the Exchequer Division, into the Queen's Bench Division in 1880.

Common Pleas at Lancaster, Court of. A court formerly existing in the County Palatine having a local common law jurisdiction. Its jurisdiction was transferred to the High Court by the Judicature Act, 1873.

Common Pleas in Ireland, Court of. In the later thirteenth century the Bench was a court sitting at a fixed place, as distinct from the court which travelled with the justiciar. It developed from a differentiation of function in the itinerant justices, and emerged with a Chief and three puisne justices in 1276. Its justices later served also as justices in eyre.

In 1655 it was renamed the Lower Bench, but the old name returned at the Restoration. In 1877 under the Irish Judicature Act it became the Common Pleas Division of the High Court in

Ireland, and in 1887 that division was merged with the Queen's Bench Division.

Common seal. A seal used by an incorporated body in the execution of deeds; the right of possession of such a deed is conferred by incorporation.

Common Serjeant. The Common Serjeant is a barrister appointed by the Crown, who acts, like the Recorder of London, as an adviser and advocate of the City of London Corporation. He is a circuit judge (q.v.) and acts as judge of the Mayor's and City of London Court and of the Central Criminal Court.

Common, tenancy in. An interest in land existing where several persons have concurrently distinct interests, which may be of the same or of different quantity, and of equal or of unequal shares, in an object of property by separate distinct titles, as contrasted with the joint title held by joint tenants. Tenants in common hold the property undivided but have separate and distinct interests therein, which may be devised or transferred independently. On the death of one tenant in common his interest does not accresce to any other. Tenancy in common may be terminated by voluntary partition or by union of all the titles and interests in one person.

Commons. Open and uncultivated land which certain occupiers of adjacent enclosed land do not own, but over which they have certain rights. Rights of common go back to Anglo-Saxon times. They comprise mainly common of pasture, the right to pasture animals on land; of piscary, to fish in another's water; of turbary, to take turves or peat from another's land; and of estovers, to take wood from another's land to maintain the commoner's house, but there is also recognized a right of common in the soil, to take sand, stone, or gravel from another's soil and certain other miscellaneous rights. Common lands usually comprise manorial waste, lands which are or were royal forests, and there are also commonable lands held severally during a portion of the year but in common after the crop has been gathered. Common land and the ownership of rights over that land have now under statute to be registered. Rights of common are created by statute, by express grant, prescription, or custom. Members of the public have access for exercise to most commons. The lord of a manor has ownership of the whole lands of the manor subject to the commoners' rights.

Common stock. The U.S. term for what are called in the U.K. ordinary shares.

Commons, House of. See HOUSE OF COMMONS.

Commons Supplication against the Ordinaries (1532). An attack by the Reformation Parliament on the ordinaries, i.e. bishops or their deputies acting as judges in ecclesiastical courts, on clerical legislation as being independent of royal control and on clerical jurisdiction as being expensive and uncertain, and on various misdemeanours by ecclesiastics. It evidenced current anti-clerical feeling and, in consequence of the Supplication, the clergy abandoned their legislative independence.

Commonwealth. A term generally signifying a form of government which has regard to the common weal or good and in which the general public have a say. But the term is particularly applied in the U.K. to the form of government in force in England, Scotland, and Ireland between 1649 and 1653. (See COMMONWEALTH AND PROTECTORATE.) The term is the official designation of the federation of Australian colonies created in 1900 and in the U.S. of the states of Massachusetts, Pennsylvania, Virginia, and Kentucky. The association of the self-governing communities of the former British Empire came to be known as the British Commonwealth of Nations and subsequently as the Commonwealth.

The name 'Commonwealth' has wholly replaced the older term British Empire as British colonies and possessions have become independent, and the words 'British' and 'of Nations' have tended to be dropped; the customary modern usage is simply to speak of 'the Commonwealth'. The Commonwealth is largely the product of convention and of the resolutions of the former periodical Imperial Conferences and subsequent meetings of Commonwealth Prime Ministers and later meetings of heads of Commonwealth Governments. The Report of the Inter-Imperial Relations Committee of the Imperial Conference of 1926 stated that members 'are autonomous communities within the British Empire, equal in status, in no way subordinate one to another in any respect of their domestic or internal affairs, though united by a common allegiance to the Crown, and freely associated as members of the British Commonwealth of Nations'. In 1971 the heads of Commonwealth Governments issued the Commonwealth Declaration, which stated that 'the Commonwealth of Nations is a voluntary association of independent sovereign states, each responsible for its own policies, consulting, and co-operating in the common interests of their peoples and in the promotion of international understanding and world peace'. There is no formal constitution and no written rules of membership.

The original members were the United Kingdom, Canada, Australia, New Zealand, South Africa, the

Irish Free State, and Newfoundland. The membership is not fixed; Newfoundland became a province of Canada; Ireland and South Africa have left the Commonwealth; and there have been added India (remaining a member despite having in 1949 become a republic), Pakistan (now Bangladesh only), Ceylon (now Sri Lanka), Ghana, Malaysia, Cyprus, Nigeria, Sierra Leone, Tanganyika, Jamaica, Trinidad and Tobago, Uganda, Zanzibar, Kenya, Malawi, Zambia, Malta, the Gambia, Guyana, Singapore, Botswana, Lesotho, Barbados, Swaziland, and certain other states. A number of these are republics. The admission of a colony to full membership on attaining independence must be preceded by consultation with existing members. It does not follow automatically from independence. Colonies such as Hong Kong are not members in their own right.

Generally the binding factor is that Commonwealth countries recognize the Queen as the Head of the Commonwealth, but India and some others have remained members despite becoming republics and their peoples no longer owing allegiance to the Crown. The Queen as Head of State acts through different legislative, executive, and judicial agencies in each non-republican country of the Commonwealth. The Governor-General in the countries which continue to owe allegiance is appointed by the Crown on the advice of the government of the Dominion, and represents the Crown, not the United Kingdom government. Thus in World War II the United Kingdom and some of the Commonwealth countries declared war separately and on different dates, and Eire (then still a member of the Commonwealth) remained neutral. An important consequence of the common allegiance to the Queen, where it continues, is the concept in U.K. law of British nationality or Commonwealth citizenship, which is not, however, inconsistent with citizenship of particular Commonwealth countries, which makes such persons members of the community of that country only.

Formerly a binding factor was the existence of appeals to the Privy Council, but this has been discontinued by most of the Commonwealth countries.

There has since 1887 been a practice of holding periodical Commonwealth (till 1937 Imperial) Conferences at Prime Minister, or now Heads of Government, level to discuss matters of common interest. A Commonwealth Secretariat was established in 1965. It has no executive functions but acts as a centre for information and oversees many kinds of Commonwealth co-operation.

Among themselves Commonwealth countries treat each other as not fully foreign in that their diplomatic representatives to each other are High Commissioners rather than ambassadors, and there was, at least formerly, a practice not to refer disputes between members to international tribunals or organizations.

Halsbury's Laws of England, s.v. COMMONWEALTH.

Commonwealth and Protectorate. After the execution of Charles I in 1649 the remnant of the House of Commons abolished the monarchy and the House of Lords and elected a Council of State.

In 1653 Cromwell expelled the Members of the House of Commons and informed the Council of State that its powers had ceased. Cromwell then summoned a body of 140 nominees, variously known as the Little Parliament, the Assembly of Nominees, and Barebones Parliament. It adopted the title of Parliament of the Commonwealth of England, but in December 1653 surrendered all its powers to Cromwell. The military leaders persuaded Cromwell to accept the written constitution known as the Instrument of Government (q.v.) whereby executive power was vested in a Lord Protector and legislative power in a Parliament. Cromwell was installed as Lord Protector and issued various reforming Ordinances. The first Parliament of the Protectorate challenged the Instrument of Government and Cromwell dissolved it and reverted to military government.

The second Parliament of the Commonwealth was summoned in 1655 and in 1657 it formulated proposals known as the Humble Petition and Advice (q.v.) offering the Crown to Cromwell. He rejected the Crown but accepted the rest of the proposed scheme and the new constitution was inaugurated in June 1657, but dispute soon arose over the recreation of a second chamber, and in 1658 he dissolved this Parliament also.

Cromwell died in September 1658; he appointed his third son, Richard, to be his successor, and Richard summoned a Parliament in 1659 but dissolved it under pressure from the army leaders, who then recalled the Long Parliament, on the basis that it had never been legally dissolved. Richard Cromwell resigned the Protectorship and a new Council of State was elected. Further disputes between army and Parliament resulted in the army again expelling the Long Parliament, but in December 1659 it was again recalled and a new Council of State elected. General Monk marched on London and forced Parliament to issue writs for a new Parliament. It did so in March 1660. The tide was now running in favour of a restoration of the monarchy. Monk negotiated with Charles II and the Commonwealth vanished, as the Restoration took place.

S. R. Gardiner, *History of the Commonwealth and Protectorate*; S. R. Gardiner, *The Last Years of the Protectorate*; G. Davies, *The Early Stuarts, 1603–60.*

Commonwealth citizen. See BRITISH SUBJECT.

Commonwealth Law. There is no body of law which can be called the law of the (British) Commonwealth, but the name is a convenient one for those common elements found in the legal systems of most of the countries which are or were members of the (British) Commonwealth, and for those legal links between them and between any one and Britain itself.

Much of the law of countries of the Commonwealth is explicable only historically. The basic principle was that the British Parliament had jurisdiction over colonies settled by Englishmen, and the King had the prerogative power to control colonies. His prerogative was more extensive in colonies acquired by conquest, though he could not take away constitutional rights which he had granted.

The Crown appointed the Governor of a colony, and could disallow legislation passed by the colonial legislature which affected the prerogative, or conflicted with fundamental constitutional principles, or with established principles of English law or with British statutes. The Governor had power to refuse assent to bills submitted by the legislature of the colony, to reserve bills until the Crown's pleasure on the matter was known, and the Crown had power to disallow or annul a colonial Act.

In settled colonies the main principle was that when they settled abroad, British settlers were held to take with them the common law of England and the statute law of England as existing at the time of settlement, so far as applicable to the circumstances and suitable to the condition of a new colony. Subsequent statutes did not apply, unless expressed to apply to the colony in question or accepted by long and uninterrupted usage. The date at which English law was deemed to have been introduced was sometimes fixed by statute. In some cases of settled colonies native law has been retained for aboriginal residents, e.g. Maoris in New Zealand and Islamic law for Mohammedans in Ceylon.

In colonies already settled and acquired by the British Commonwealth by conquest or by cession, the principle is that the existing legal system was retained unless or until abrogated or altered by the Crown; thus Roman-Dutch law continued in the South African colonies. But existing law has been held abrogated if contrary to fundamental British constitutional principles, or the fundamental religious or moral principles of Europeans, or to British statute law extending to that colony. In some cases English common and statute law as existing at the time or at a stated date have been applied expressly by statute to colonies, e.g. to Cyprus in 1882. Subsequent British statutes do not apply, unless expressly applied. Full effect has frequently been given to existing land law. By virtue of the principles of continuing existing legal systems Roman-Dutch law was preserved in the South African colonies (now the Republic of South Africa) and in Ceylon, the Code Napoleon in Mauritius, the Coûtume de Paris in Quebec (now the Quebec Civil Code), and Spanish law in Trinidad until superseded by local Ordinances.

Legislation for colonies might be effected by the Crown under the prerogative or under statutory powers, by the United Kingdom Parliament, or by the legislature of the colony. The royal prerogative of legislation was, however, lost when a colony acquired a representative legislature, unless, in the case of a conquered or ceded colony, it was expressly reserved in the grant of representative government. Parliament had legally unrestricted power to legislate for any colony, but from the mid-nineteenth century there developed a convention not to legislate for the colonies which had become self-governing. Colonial legislatures were subordinate to Parliament, their powers depending on the statute or other instrument granting the powers.

The common law rule was that a colonial act was invalid if repugnant to English law, but the Colonial Laws Validity Act, 1865, provided that a colonial act was to be void only to the extent of any repugnancy extending to that colony, or of any subordinate legislation made thereunder.

In many colonies representative government developed into responsible government, where the executive was not appointed by the Governor but by, and responsible to, the legislature. They became accordingly self-governing and, after 1907, some were called Dominions. The power to reserve legislation for royal approval has practically vanished and, if exercised at all, the power would be exercised in accordance with the practice of the dominion concerned; disallowance has disappeared in relation to independent territories.

The Statute of Westminster, 1931, provided that the Colonial Laws Validity Act, 1865, should not apply to any law made by the Parliament of one of the self-governing dominions, and that no Act passed by the United Kingdom Parliament should extend to a Dominion as part of the law thereof unless it was expressly declared in that Act that that Dominion had requested and consented to the enactment thereof.

Since then, except in relation to Canada, the term 'dominions' has been discarded as misleading, possibly implying subordination to the United Kingdom.

Courts of justice have been created in Commonwealth countries by the Crown in the exercise of the royal prerogative, and by the legislatures of the countries concerned. There is a general tendency to follow the nomenclature of the English legal system.

Among the prerogative rights of the Crown is the right to entertain appeals from courts in Her Majesty's dominions outside the U.K. and accordingly an appeal lies in all cases, civil and criminal, from the highest court in each dominion, colony, or posssession of the Crown, except in so far as the prerogative right has been surrendered or abrogated. Since 1931 the former Dominions and, since 1947, all independent members may abolish all appeals to the Privy Council and most have done so. Colonies cannot abolish such appeals. The Judicial Committee of the Privy Council advises Her Majesty on the disposal of the appeal. The Judicial Committee is a Commonwealth tribunal, the common apex of the judicial hierarchies of the parts of the Commonwealth and colonies, not an English court, and judges from Commonwealth countries sit. They apply the law of the country where the case originated.

Appeals to the Judicial Committee are of two main classes, appeals as of right and appeals at the discretion of the local court, and appeals by special leave of Her Majesty in Council, where the lower court has no power to grant or has refused leave to appeal. The former class may be, and have in many cases been, abolished by local legislation; the latter rests on the royal prerogative and cannot be abolished save by authority of statutes of the Parliaments of the Members of the Commonwealth. The conditions under which appeals of the former kind may be brought are regulated by Orders in Council. In granting leave in the latter class of case the Judicial Committee considers whether in criminal cases there would be serious injustice in refusing, and in civil cases whether the case involves important constitutional issues or other substantial questions, such as of status or property of great value. Leave will not be granted where the power to grant leave has been surrendered, or validly abolished by the Parliament of the Commonwealth country.

Cambridge History of the British Empire; J. Marriott, *Evolution of the British Empire and Commonwealth*; K. C. Wheare, *The Constitutional Structure of the Commonwealth*; K. Roberts-Wray, *Commonwealth and Colonial Law*; Annual Survey of Commonwealth Law.

Communal courts. A generic term for the courts of the different communities into which mediaeval England was divided and which represent a national system of government and jurisdiction. They were chiefly the county court and the hundred court, while related to the vills or townships was the frankpledge system (qq.v.).

Communio bonorum. In older Scots law, moveable goods held for husband and wife in common. The wife's moveable property, with certain exceptions, was assigned by law to the husband by virtue of his *ius mariti* and under his entire power during the marriage. Lands and buildings did not fall under the *communio*. Despite the name it was not a system of community property but a transfer of all the wife's moveable property to the husband. On the husband's death the widow was entitled to half of the moveables as her legal rights, or to one-third if there were children, the children also getting one-third.

Communis error facit jus (common error makes law). The maxim expressing the fact that an erroneous view of what the law is, if long persisted in and accepted as the basis of practice and ruling, may be held to have established the law in the erroneous sense even when the error has been established. The maxim is particularly applicable where a wrong decision or misunderstanding has been for long made the basis on which rights have been established and arrangements as to property made.

Communis opinio doctorum. A term introduced by the thirteenth century canonists and used by later civilians to supplement the gloss (q.v.) on the texts of the Roman law. The fifteenth century civilians interpreted *communis* as meaning a majority view, particularly if the majority were the more weighty jurists so that greater weight tended to be given to the views of such as Bartolus or Baldus and the authority of such a jurist's reputation tended to outweigh the reason of his views.

Communities. In Wales, since 1972, a subdivision of a local government district. There are community councils and community meetings. By passing a resolution a community may become a town.

Community homes. Homes provided by local authorities for children, superseding former children's homes, approved schools, approved homes, detention centres, and other similar places.

Community land. Land acquired by local authorities under the Community Land Act, 1975, as being needed for development of stated kinds.

Community law. The law of the European Communities.

Community property. A system of treating the property of married persons as belonging to both of them as distinct from separate property in which each spouse owns his or her own property individually. Generally community-property systems distinguish separate property, owned by one spouse before marriage, which does not generally become community property, or used exclusively by one spouse, and community property, which comprises most property acquired after marriage. Most community-property systems provide what property acquired after marriage is to be considered separate, all other acquired property being deemed community property. The problem also arises of right to administer community property. Sometimes this has been vested in the husband, but some community-property statutes modify this. The history of matrimonial property law indicates a general spread of the community-property system.

Community service order. An order which a court may make, instead of dealing with an offender in any other way, requiring him to perform a specified number of hours of unpaid work. It requires the consent of the offender and provision in the area for him to do work. An offender made subject to such an order must perform the number of hours' work as may be instructed by the relevant officer. If the offender fails to comply with the requirements or satisfactorily to perform the work he has been instructed to do, he may be brought before a magistrate's court and fined or dealt with as he might have been dealt with originally. Orders may be amended or revoked and the offender dealt with otherwise.

Commutation. Alteration of a legal right or liability, as where a death sentence was commuted to one of life imprisonment. It is distinct from reprieve which is merely a delay or suspension of sentence, and from pardon which wholly extinguishes both conviction and sentence. Also conversion of a right to a periodical or variable payment into a fixed payment. Thus under the Tithe Acts a rentcharge was substituted for the right to take tithes in kind.

Company. An association of a number of persons for a common purpose, frequently for the carrying on of a business with a view to profit, and a mode of organization suitable for associations too large to operate as partnerships. The term is, however, often used of associations which are legally partnerships, or even of sole traders. Ancestors of modern companies include chartered trading enterprises, joint stock companies, unincorporated associations formed under a deed of settlement, and companies incorporated by letters patent under the Trading Companies Act, 1834. The Joint Stock Companies Act, 1844, provided for incorporation by mere registration and in 1855 the Limited Liability Act provided for the limited liability of the members of companies, subject to certain conditions. Unincorporated companies of individuals carrying on business under a corporate name, and having some of the qualities of incorporated companies may still exist. Incorporated companies may be formed with unlimited liability of members. The basis of modern company law is The Companies Act, 1862. A major characteristic of the modern company is that it is incorporated, i.e. is a legal entity quite distinct from all the persons who comprise its membership and it continues notwithstanding changes in membership.

Companies may today be created by royal charter, by special Act of Parliament (which normally incorporates the detailed provisions set out in the Companies Clauses Consolidation Act, 1845), by registration under the general provisions of the Companies Acts, 1948 to 1976 (or their predecessors from the Joint Stock Companies Act, 1856, onwards), and cost-book companies (confined to tin-mines in Devon and Cornwall).

The Companies Acts, 1948 to 1976, provide a legal framework within which individual companies have considerable freedom to determine their own regulations. They are concerned also to provide various safeguards for members of the company, persons lending to the company, and persons doing business with the company. The Acts do not contain all the law relative to companies, much of which is provided by general principles of law, particularly of contract and property. In certain respects there are differences between English and Scottish law.

These Acts provide for registration of various forms of companies, with the liability of the members unlimited, or limited by the nominal value of their shares, or limited by their guarantee, as a public company the shares of which are dealt in on Stock Exchanges, or as a private company, the membership of which is limited and the shares not so dealt in. The great majority of companies are private as this is a form of organization whereby two or more persons can carry on business with the great advantage of limited personal liability for trading losses. Certain professions cannot, however, be carried on by way of a company and the persons must be in partnership only.

Registration is effected by submitting to the Registrar of Companies a Memorandum of Association, setting out the name of the company, the country of the United Kingdom where the registered office is to be situated, the objects of the company, that (if it be the case) the liability of the members is limited, and the amount of share capital and the division thereof into shares of fixed amount, Articles of Association, prescribing regulations for the management of the company, and certain other documents. Incorporation is effected by the Registrar registering the company and issuing a certificate to that effect. The Memorandum and Articles bind the company and its members and constitute a contract between the company and each member. Both may subsequently be altered by the company.

Companies normally obtain working capital by issuing shares to members of the public who subscribe for them, and may borrow, particularly by issuing debentures. There may be different classes of shares, with different rights attaching to them. Capital may be increased or, with the sanction of the court, reduced. Stringent statutory provisions regulate the issue of a prospectus to induce inventors to take shares or debentures, and the publication of audited accounts, specifying what information must be disclosed.

Membership of a company is constituted by subscribing the Memorandum and Articles, or by subscribing for shares, or by purchasing an existing member's shares. Members are entitled to be summoned to an annual general meeting, to see accounts, to elect directors, and to receive dividends on their shares. In general the views of the majority of the members must rule, but statutory provision is made for protecting the interest of the minority against oppression. Extraordinary general meetings may be called and, for certain purposes, meetings of classes of shareholders must be called.

Companies are normally managed by a Board of Directors elected by the members in general meeting. They are trustees of the company's interest and property, but not trustees for shareholders or creditors. They are also the company's agents for transacting business with other parties. Many statutory duties are imposed on the directors. Detailed administration is done by the secretary, who must also maintain statutory records and make certain statutory returns to the Registrar of Companies, such as the Annual Return. There must by an independent auditor to examine the books and accounts and to report thereon to the members.

The Department of Trade has wide powers to inspect a company's books and investigate its affairs.

Debenture holders have power, if the interest on or capital of their loans are not duly paid, to appoint a receiver to take possession of the company's assets, manage it in the interest of the creditors and realize assets so as to pay the debenture-holders.

A company may be reconstructed, as by selling its undertaking to a new company, or entering into a scheme of arrangement between it and its creditors or members or any class of either.

Stringent statutory requirements are imposed as to the keeping of accounts and the disclosure of accounts and information relative thereto to the members of the company.

A company is brought to an end by winding-up, which may be done by the court, or voluntarily, or subject to the supervision of the court. Under all three modes a person is appointed liquidator, whose functions are to collect the company's assets, ascertain what creditors' claims are outstanding, ascertain who, as present members or persons who ceased to be members within a year of winding-up, are liable to contribute to the company's assets, to pay secured claims, preferential claims, and other claims and, if there be any surplus, to divide it among the members. A committee of inspection, of creditors and contributories, may be appointed, to assist the liquidator. When the company's affairs have been wound up it will be removed from the Register, though its dissolution may be avoided in certain circumstances.

R. Formoy, *Historical Foundations of Modern Company Law* (1923); W. R. Scott, *Joint Stock Companies to 1720* (1910–12); A. B. Dubois, *The English Business Company after the Bubble Act* (1938); B. C. Hunt, *Development of the Business Corporation in England* (1936); H. B. Buckley on *Companies*; F. B. Palmer, *Company Law*; L. C. B. Gower, *Principles of Modern Company Law* (1969).

Companies, Chartered. The Crown has frequently in the past granted under the prerogative, and may still grant, a charter incorporating a number of persons into a company. Thus charters were granted to the African Company, the East India Company, the Hudson's Bay Company, the Merchant Adventurers, and certain other companies, to trade in certain parts of the world. The Virginia Company and the Massachusetts Company were incorporated by charters to found the colonies of these names. Some other companies such as the Bank of England and the London Assurance Company were later incorporated by royal charters which could not have been granted under the prerogative but which were authorized by statutes. The Chartered Companies Acts, 1837 and 1884, empower the Crown to charter a company for a definite period.

G. Cawston, *Early Chartered Companies, 1296–1858*.

Companies, City of London. The city companies originated from the craft guilds which monopolized and regulated the practice of crafts. In time practically every trade and craft had its guild. From about 1300 admission to the freedom of the City of London was confined to members of recognized trade or craft guilds and their apprentices. From 1376 the guilds were given the right to select the freemen who were to elect the Lord Mayor, sheriffs, and officials and the M.P.s for the City. In 1384 it was established that the Court of Common Council should consist of freemen, so that in time the guilds had established a monopoly control of all elective offices in the city. Over the period 1300–1725 many of the guilds had obtained royal charters of incorporation as companies. There are now 12 'great' companies and a large number of lesser ones. Though maintaining names indicative of trade, e.g. Mercers, Grocers, Drapers, Merchant Taylors, these companies are now mainly charitable bodies and most of the members have no connection with these trades at all, though qualifications for admission continue to include relationship or apprenticeship to a member. The companies have now no privileges in relation to elections to Parliament, but still have influence in the election of the Lord Mayor and officers of the City. They make ordinances for their own government, the admission of members, and the control of finances. Some particular companies have special rights; thus the Goldsmiths are empowered to assay and mark gold and silverware, the Society of Apothecaries to grant qualifications as medical practitioner.

A company became a livery company by the adoption of a livery by some or all of its members, but after 1560 this required the consent of the Court of Aldermen. Livery companies are usually incorporated, normally by royal charter, and this may have preceded or followed recognition as a City company. A grant of livery is now made by letters patent under the mayoralty seal.

Comparative law. Despite the phrase, comparative law is not a principle or body of rules of law, but an approach to, or method or technique of studying, law or any branch or topic of it, arising from the diversity of legal systems in the world and the different approaches adopted to common problems, by examining side by side and in relation to one another the attitudes, institutions or rules on any matter of any two or more legal systems. Law may be studied comparatively for various purposes, from disinterested curiosity, with a view to discovering general trends or majority views on any matter, with a view to discovering generalizations applicable to law generally, or a group of systems generally, with a view to law reform, or with a view to the unification of law and the achievement of a universal law shared by all civilized humanity. Not least, comparative study is valuable for putting the study of any one legal system into perspective and deepening the understanding of it. It is also invaluable for jurisprudential studies, of the nature, function, and general ideas of and about law. The study is distinct from the study of foreign law, which is simply the study of any system of law other than the one with which the student is primarily concerned, and does not necessarily involve any element of comparison.

It was only in the nineteenth century that substantial regard began to be paid in European countries to foreign legal systems. The first International Congress of Comparative Law was held in 1900. The growth of international trade and international relations stimulated the need to know something of foreign legal systems, and also the idealistic concept of a common body of law for the world.

Comparative studies have grouped the main legal systems of the world into families, the members of which show marked family resemblances in institutions, forms of statement, content, and methods, usually resulting from common historical background or from borrowing. The main family groups are:

(a) The civil law, or Romano-Germanic, group, comprising legal systems based ultimately in the main on the civil law of Rome; these include French, German, and most Western European and Central and South American legal systems;

(b) The common law group, comprising legal systems based chiefly on the common law of England; these include English, Irish, Canadian (except Quebec), United States (except Louisiana), Australia, New Zealand.

In addition there are:

(c) Hybrid systems, such as Scots, Israeli, Louisiana, Quebec, and South African law, basically civil law systems, but heavily influenced by common law ideas and rules;

(d) The socialist group, countries which formerly had legal systems of the Romano-Germanic group, and have preserved some traces thereof, but under political and economic influence, have created new legal systems based on Marxist-Leninist philosophy. This group includes Russian law, since the 1917 Revolution, and the laws of the people's republics of Europe and Asia.

(e) The group of religious systems, which regulate the personal relations of members of particular faiths, rather than the lives of the citizens of particular countries; these are Islamic or Muslim, Hindu and Jewish law.

Within each of the first four groups there are fundamental similarities of institutions, concepts, and rules, which transcend language barriers.

There are certain other groups which have affinities with, but do not readily fall into, one or other of the main family groups. These include Scandinavian, Far Eastern, and African Legal Systems.

H. C. Gutteridge, *Comparative Law*; J. Wigmore, *Panorama of the World's Legal Systems*; R. David and E. Brierley, *Major Legal Systems in the World Today; International Encyclopaedia of Comparative Law.*

Compassing. Imagining, contriving, or planning, a term used in the Treason Act, 1351, which makes it criminal to compass or imagine the death of the King or Queen.

Compellability. The name given to the question whether a person can be compelled to attend a court to give evidence. The general rule is that a witness who is competent to give evidence is compellable but not a professional adviser in respect of confidential communications, a husband or wife in respect of communications by the other spouse during the marriage, nor, in criminal cases, any witness in respect of any question which might tend to incriminate him by exposing him to criminal proceedings. A defendant is not compellable, nor is his or her spouse.

Compensation. (1) A sum of money paid to persons injuriously affected, e.g. by having their land compulsorily acquired, or by having to surrender a tenancy after making improvements to the land. Statutes make provision for compensation in many circumstances, for disturbance of a tenant, for improvements made by him, for compulsory purchase.

(2) A reward, particularly for apprehending a criminal.

(3) In Scots law, the process introduced in 1592 whereby, where parties are mutually debtor and creditor, one may set-off the debt due by the other to the effect of extinguishing or diminishing *pro tanto* the debt due to him.

Compensation for victims of violence. See CRIMINAL INJURIES COMPENSATION.

Compensation order. An order which an English criminal court may make, in addition to dealing with an offender in any way, requiring him to pay compensation for any personal injury, loss (particularly by theft), or damage resulting from that offence or any other offence taken into consideration by the court in determining sentence. Regard must be had to the offender's means. No compensation may be ordered for loss suffered by a person's dependants by reason of his death, nor, with exceptions, in respect of loss, injury, or damage arising out of the presence of a motor vehicle on a road. An order may be subsequently reduced or discharged if the loss is held to be less than thought, or property is recovered. If damages are subsequently awarded in a civil action, they must be assessed without regard to the order but must be reduced by any amount paid under the order. While the principle of making compensation orders is good, the practical difficulties of enforcement are great, with offenders disappearing, delaying to pay, or paying by trivial instalments over long periods. See also CRIMINAL INJURIES COMPENSATION.

Competence. The authority of a court to deal with specific kinds of matters. It may be distinguished from or be synonymous with jurisdiction (q.v.).

Competency (or admissibility). In the law of evidence, the question whether evidence on a particular matter can be allowed to be received. It is also used of a witness, denoting whether he is legally permitted to give evidence at a trial or not. The general rule in modern law is that all persons who can give evidence are competent. See also COMPELLABILITY.

Competition. An economic situation in which there are a number of potential suppliers of particular goods or services and a number of potential customers thereof, so that no one on either side can dictate the supply, price, or other factors of the market, as contrasted with monopoly, where there is only one, or a very few, buyers, in each of which cases the natural operation of a market is modified.

The attitude of the law to competition has varied at different times and in different contexts. In the U.K. nationalization of various industries and public utilities has involved that in many contexts there is a monopoly situation, alleged to be justifiable in the public interest and supposed to be controlled by Parliament. In certain other contexts competition is assumed to be beneficial and there are elaborate systems of investigation by the Monopolies and Mergers Commission, of control by the Restrictive Practices Court and otherwise, to try to maintain a competitive situation and to restrict monopoly.

In the law of the European Communities the promotion of fair competition is a major purpose of the E.E.C. and there is an extensive body of law based on three sets of rules, rules applying to undertakings prohibiting agreements which may seek to or effect the prevention, restriction, or distortion of competition, rules against dumping or exporting goods for sale at low prices in order to capture a market, and rules against state aid which distorts competition by favouring certain undertakings or the production of certain goods.

Compilacion de Canellas (or *de Huesca*). A compilation of laws issued by King James of Aragon in 1245.

Compilationes Antiquae. See CANON LAW.

Complaint. The writ by which proceedings before magistrates' courts to obtain an order for payment of money is commenced.

In Scottish summary criminal procedure it is the writ charging a person with having committed a stated offence.

Completion. In a contract for the sale of land, completion is the act of the vendor in conveying the estate contracted for with a good title and delivering up actual possession, and the act of the purchaser in accepting the title, taking possession, tendering a conveyance for execution by the vendor, and paying the price.

Complicity in crime. A person may be involved in the commission of crime as a principal or perpetrator, who by his own act or omission brings about the criminal act, or as one, formerly called an accessory, who aids, abets, counsels, or procures the commission of the crime. Both are equally liable to be tried and punished as a principal offender, but a person who renders assistance after the commission of a crime, formerly called an accessory after the fact, does not thereby become a party to it. A secondary party must have intended that the crime be committed or have known that a crime was contemplated and a crime of the same kind is committed.

Composition. In Germanic law, money given to a person injured by the person responsible, in substitution for the blood-feud and personal revenge. The amount was determined by the man's *wer* or worth as determined by his status in society.

In modern law it is an offer by a debtor who is unable to pay all his creditors in full to pay them all a stated proportion, e.g. 50p in the £1, in full satisfaction of their claims and thereby to be discharged of his debts. It is effected by a composition agreement in writing by the debtor to pay the agreed composition by instalments on specified dates, usually with sureties to pay if the debtor fails to pay them.

Composition contract. In Scots law, an agreement whereby a debtor offers to pay a proportion of all his debts and his creditors accept this as full discharge of his obligations. It is frequently offered as an alternative to sequestration. Alternatively, a person who has been sequestrated or his friends may offer the creditors a composition on the whole debts. If accepted and approved by the Court, the bankrupt is reinvested in his property and discharged except as regards payment of the composition.

Composition deed. A deed complying with the Deeds of Arrangement Act, 1914, providing for an arrangement between a debtor unable to pay his debts and his creditors for discharge of his liabilities to them by partial payments.

Compounding a felony or misdemeanour. Agreeing, in return for the restoration of goods stolen or other benefit, to abstain from prosecuting the person guilty of the theft or other crime in question. It is itself criminal. It was no offence to compound a misdemeanour unless the offence was in substance an offence against the public. When felonies were abolished in 1967 the offence of compounding an arrestable offence was created; it is committed by accepting or agreeing to accept a consideration for not disclosing information relating to the arrestable offence, but not if the consideration is no more than making good the loss or injury caused by the offence or the making of reasonable compensation for it.

Compromise. A settlement of a disputed claim by agreement, or an agreement by parties to refer a disputed matter to arbitration.

Compter (or Counter). The debtors' prison attached to the court of the mayor or sheriffs of a borough. There were two in London, the Poultry Compter, demolished in 1817, and the Giltspur Street Compter, which replace the Wood Street Compter, which itself had in 1555 replaced the Bread Street Compter, and which was closed in 1854.

Comptroller or Controller. An officer charged with accounting and financial control. There are several comptrollers in the Royal Household, the chief being the Comptroller of the Queen's Household, who is a political appointee changing with the government and having only nominal duties.

In the old Court of Chancery there were Controllers of the Hanaper, who entered in books all documents sealed by the Clerk of the Hanaper, Controllers of the Pipe, and Controllers of the Pell.

The Comptroller in Bankruptcy was responsible for receiving from a trustee in bankruptcy his accounts and for calling the trustee to account for misfeasances, neglects of duty, or omissions in the discharge of his duties.

Comptroller. One of the Officers of State of Scotland with duties mainly in the financial management of the royal household, but also in ingathering royal rents. The office seems to have been combined with those of Collector and of Treasurer in 1644. It did not survive the Union.

Comptroller and Auditor General. An officer of Parliament and head of the Exchequer and Audit department. The office was created in 1866. He is appointed by letters patent and his salary is charged on the Consolidated Fund and not provided for in the annual Estimates. His main duty is to examine the accounts of the departments of the armed forces and civil departments and other public accounts, and to satisfy himself that moneys appropriated have been spent as intended and authorized by Parliament. He submits an annual report to Parliament which is considered by the Committee on Public Accounts, and the Comptroller sits with that committee to assist and advise it.

Comptroller of the Pipe. An official anciently known as *Clericus Cancellarii*, who in mediaeval times annually made up the Chancellor's Rolls, which were counter-rolls intended to check the Pipe Rolls (q.v.) and were duplicates of them. A series of such rolls in the Public Record Office runs, with gaps, from 1163 to 1831.

Comptroller-General of Patents, Designs, and Trade Marks. The official in charge of the Patents, Designs, and Trade Marks Office and responsible for the grant or refusal of patents and of registration of designs and trade marks. He has numerous statutory powers and duties under the relevant statutes, including quasi-judicial functions, from the exercise of which appeal lies to the courts.

Compulsive behaviour. Behaviour influenced by recurrent impulses or longings to perform irrational acts, sometimes associated with a diagnosable psychosis. It may be harmless, or be anti-social, a common form being compulsive stealing, possibly of quite useless things by persons well able to pay for them. This used to be called kleptomania. Deviant sexual desires may become compulsive. Giving way to the compulsive desire gives some relief, but frequently results in remorse, self-disgust, or fear of the consequences.

Compulsory acquisition (commonly called 'compulsory purchase'). The taking of land or an interest in land from the owner without his agreement. The right to do so is one enjoyed by the Crown and, though prerogative powers still exist, nearly always exercised under statute. Originally it was conferred on statutory undertakings, such as railway companies, in individual Acts, but later Acts incorporated large parts of the general codes enacted in the Lands Clauses Consolidation Act, 1845, and other 'Clauses Acts'. Many modern statutes confer powers on road authorities, housing authorities, education authorities, and similar bodies to acquire or affect land for purposes within their general powers and for their functions, which power is exercised by making a compulsory purchase order, subsequently submitted to and confirmed by a Minister of the Crown or other confirming authority. The authorizing statute may prescribe the procedure or may apply the standard procedure contained in the Acquisition of Land (Authorization Procedure) Act, 1946.

The right to compensation for land taken or injuriously affected depends on the statute or order in question, both as to existence and basis of compensation.

Compurgation or wager of law. The mediaeval procedure whereby an accused might clear himself of a claim or charge by his own oath, fortified by the oaths of compurgators, or oath-helpers, usually 12 in number and commonly his relatives or neighbours, who testified not to the facts but to the character of the person for whom they took the oath, that the oath which he had taken was clean. If the defendant could not get the required number of oath-helpers, or they did not swear in proper form, he lost. The oaths of different men varied in legal value, that of a thegn being equivalent to that of six ceorls. The institution was not of Roman law origin, but common among many of the barbarian tribes who invaded the Roman empire. By reason of its commonness the Church adopted it and used it considerably under the title of 'Canonical Purgation'. In the twelfth century it came to be thought that compurgators should be required to swear only to their belief in the truth of the defendant's assertions, and it began to fall into disfavour about this time, though in the twelfth to fourteenth centuries the rules were laid down more precisely. Only in 1342 was it settled that the number of compurgators must be 12. Compurgation was used frequently in manorial courts, in ecclesiastical courts, where it survived long after the Reformation, and in borough courts, as an alternative to trial by battle. Some towns, notably London, stubbornly retained it as a defence, even in cases of felony. In debt and detinue it lasted as a defence until the nineteenth century. It was not allowed where the Crown was a party, and in real actions the assizes (q.v.) superseded it. The last instance of its use was in 1824 and it was abolished in 1833.

Computers and law. The applications of computer science and technology having and likely to have relevance to law seem to lie mainly, if not entirely, in the area of information storage and retrieval. The computer has obvious uses for, e.g., storing large quantities of information about known criminals and retrieving it quickly on request, and it is already being used extensively for such purposes. Computers also have value for holding and indexing the kinds of available information at present available in such books as the *Index to the Statutes*, *Guide to Government Orders*, the various Digests of case-law, bibliographies, indexes to encyclopaedias and major text books and so on, but hitherto not much work has been done on their applications in these fields. In particular determining what is to be stored for retrieval, under what headings or keywords, and so on, is not a simple task. It would not be helpful if a computer were to print out a list of all the places in legal books where the word 'negligence' was used. Existing indexes and digests could not simply be put on tape, because all are more or less defective.

It is questionable if there is much scope for the use of computers in solving legal problems, because so much depends on a court's interpretation of facts found, on interpretation of the relevant principles, and rules of law in relation to the facts found, and in many cases on the exercise of judicial discretion. Doing what is right and just in the circumstances cannot, and probably should not, be determined wholly mathematically. Research on the use of computers in law is continuing, but a very great deal remains to be done.

C. Tapper, *Computers and the Law*.

Comte, Auguste (1798–1857). French philosopher, disciple of Saint-Simon, and author of the *Course of Positive Philosophy* (1830–42) and *System of Positive Polity* (1851–54). He was one of the first writers to apply sociological method to the study of society and to relate politics to its total environment.

In his view each branch of knowledge passed through three phases and there were three ways in which the human mind explained phenomena. The phases were the theological, metaphysical, and positive. The great aim of the positivist philosophy was to move the study of society into the third stage and to remove social phenomena from theological and metaphysical speculation and to introduce to them the same scientific observation of their laws as has produced the natural sciences. Positive philosophy also had the object of giving science the generality of philosophy and giving philosophy the rigour of science. The arrangement of all the sciences and the law of the three states explained the course of human thought and knowledge. Comte also emphasized authoritarian government and a new religion, the cult of humanity, but political superstructure is unimportant being limited to security and well-being, and government should be the concern of specialists, bankers being at the top of the structure.

H. Hutton, *Comte's Life and Work*; E. Caird, *The Social Philosophy of Comte*; R. Fletcher, *Comte and the Making of Sociology*.

Comyns, Sir John (1667–1740). Called to the Bar in 1690 and sat in Parliament for many years. He early attained a reputation for industry and learning, and was considered by contemporaries the best lawyer at Westminster.

He became successively a Baron of Exchequer (1726–36), Judge of the Common Pleas (1736–38), and Chief Baron of Exchequer (1738–40), but is remembered today for his reports, covering the period 1695–1741, and particularly for his *Digest of the Laws of England* (1762–67), both written in law French and later translated, and the latter several times re-edited. It is an exposition of the law rather than merely a digest of it, the division and subdivision is logical, and it has been described as a book of very excellent authority.

Concealment. Action taken to prevent another party from ascertaining a fact; it is more active and positive than mere non-disclosure. While in contracts there is, in general and subject to exceptions, no duty to disclose defects, so that non-disclosure is permissible, concealment is equivalent to positive misrepresentation, and accordingly may be a species of fraud. Concealment may also in some circumstances be a criminal offence.

Concealment of birth. The statutory crime of, by secret disposition of the dead body of a child, endeavouring to conceal the birth thereof. The object of the statute was to catch women who might have escaped on a charge of murder because of the difficulty of proving live-birth and subsequent killing of the child. There must be concealment of the fact of birth, carried out by secret disposition of the body. In Scotland there is a corresponding statutory crime of concealment of pregnancy.

Concepts and conception. Terms applied to general ideas, abstractions formed by the mind, derived from and distinct from particular observed instances. The idea is the concept and the mental process of arriving at it by abstraction of the general features from a large number of instances is conception, but the latter word is also used as an equivalent for concept, as in Hohfeld's *Fundamental Legal Conceptions*, though strictly speaking it is a mental process contrasted with perception. Thus 'contract' is the legal concept abstracted from various instances of legal relationships which are called contracts. Concepts are authoritative categories to which types or classes of transactions, cases, or situations are referred, in consequence of which a series of principles, rules, and standards become applicable. Thus Austin enumerated the notions of duty, right, liberty, injury, punishment, and redress, their various relations to one another and to law, sovereignty, and independent political society as among the necessary notions and conceptions of jurisprudence. Hohfeld (q.v.) in his *Fundamental Legal Conceptions*, examined what he called the lowest common denominators of the law, rights and duties, privileges and no-rights, powers and liabilities, immunities and disabilities. Hart (q.v.), in *The Concept of Law*, sees the idea of obligation as at the core of a rule; rules of obligation are distinct from other rules operating in society and rules are distinguishable into primary rules which regulate behaviour and secondary rules which identify primary rules. The union of primary and secondary rules creates a legal system. Analytical jurists (q.v.) devote much attention to the analysis of legal concepts, such as rights, duties, possession, ownership, property, liability, responsibility, and the like. To a substantial extent the study of legal concepts is linguistic, the examination of what particular words convey in legal discourse in a particular system, and it may be much influenced by interpretation provisions in statutes, providing that for the purpose of the statute in question the word X is to mean certain things. Thus the definition of the word 'goods' in the Sale of Goods Act, 1893, makes that a concept which may differ from the ordinary understanding of the word.

Conceptualism (or the jurisprudence of concepts). The practice of treating legal concepts and terms, e.g. negligence, as having fixed and invariable meanings in all contexts, so that the law applicable to a particular case can be discovered by simple syllogistic reasoning. It is founded on failure to appreciate that legal concepts and terms are English words, not having fixed and settled definitions and meanings like, e.g. ampere, and conceptualism can lead to wrong decisions or inhibit the development of legal rules in different contexts.

Concert of Europe. An informal organization of the major European powers, Austria, Britain, France, and Russia, to maintain peace and order in Europe on the basis of the settlement reached after the Napoleonic Wars. It was assumed to be the right and duty of the major powers to impose their collective will on the lesser powers, and the members took the view that a disturbance anywhere in Europe was the legitimate concern of the major powers. The powers met in congress at Vienna in 1814–15, Aix-la-Chapelle in 1818, Troppau and Laibach in 1820, and Verona in 1822. The Concert was shaken by the revolution in France in 1830 but allowed the independence of Belgium in that year, and by the revolutions of 1848 and the Crimean War, but even after that the Ottoman Empire was admitted to the Concert of Europe. It perished when Italy and Germany achieved their unification and Germany defeated France in 1870 and created the German Empire.

Concession. A grant by a public authority to a person of authority to so something, such as to work land, extract minerals, operate an industry, or the like.

Concessions, extra statutory. Relaxations of the strict law, or interpretations of the law favourable to the taxpayer which the Inland Revenue from time to time intimates that it proposes to make.

Conciergerie. A mediaeval jail in Paris, now part of the Palais de Justice, named from its chief officer, the concierge, an official nominated by the *parlement* of Paris.

Conciliation. (1) The process of seeking to settle disputes between employers and workmen. The Conciliation Act, 1896, provided for the establishment of bodies or persons appointed for these purposes, but they could only bring the parties together and seek to influence them to make a settlement. There was established in 1976 an Advisory, Conciliation and Arbitration Service intended to promote the improvement of industrial relations.

(2) In international law, the process of settling a dispute by referring it to a commission of persons whose task it is to elucidate the facts and issue a report containing proposals for a settlement, but which does not have the binding character of an award or judgment. The practice developed out of international commissions of enquiry into facts. After 1920 many treaties provided for conciliation but few references were made to the machinery established thereby. The U.N. may set up Commissions of Conciliation in fulfilment of its function of settling disputes and ensuring peace.

Concilium plebis. At Rome the popular assembly, usually called a *comitia* (*comitia plebis tributa*) after 287 B.C. when *plebiscita* were given equal validity with laws enacted by the *comitia centuriata.*

Conclusion. (1) An irrebuttable presumption or rule of law.

(2) In old common law practice the part of a pleading by which a party offered to have the issue raised tried by a jury.

(3) In old Admiralty practice a pleading denying the previous pleading and praying that the pleadings might be brought to conclusion and issue joined.

(4) In Scottish practice the part of a summons which states what remedy the pursuer requests the court to grant him.

Concord. An agreement for the compromise of a right of action arising out of a trespass, or an agreement between the parties to a fine (q.v.) of land, whereby the deforciant or defendant acknowledged that the lands belonged to the complainant or plaintiff.

Concordats. Public treaties or agreements having force in international law, between the Holy See and one or more secular states, regulating matters of common interest. The first true concordat was probably the Concordat of Worms (1122) between Calixtus II and the Emperor Henry V which settled the Investiture Contest. Other notable ones have been that of 1801 between Pius VII and Napoleon regulating Church-State relations in France and that of 1929 with Italy settling the Roman question. On the papal side the contracting party is the Catholic Church personified by the Holy See. The making of concordats is reserved almost exclusively to the Pope as head of the Catholic Church. In the past general councils of the Church have exercised the power to negotiate and sign concordats.

Concordia de singulis causis (or *Capitula legis regum Langobardorum*). A compilation of the general rules of Lombard law in systematic form dating from about A.D. 830.

Concordia discordantium canonum. See CANON LAW; GRATIAN.

Concubinage. A continuing state of cohabitation and sexual relations between two persons of opposite sexes not linked by legitimate marriage. The legal status of a concubine has varied in different societies. In ancient Hebrew society a concubine was a true wife but of secondary rank, concubinage being an offshoot of polygamy. In Roman law concubinage was a recognized legal status so long as the parties were not married and had no other concubines. In the Western Empire it continued to be recognized even by Christian emperors and the early Christian Church allowed a wife or concubine. The institution was recognized by many early European codes. The Salic law allowed 'morganatic marriages' between nobles and women of lower rank. When the Church was trying to enforce clerical celibacy the Papacy generally spoke of priests' wives as concubines and concubinage tended to be tolerated. The Council of Trent tightened the requirements of marriage and imposed heavy penalties on concubinage.

See also HANDFASTING.

Concurrent jurisdiction. The topics of law which, prior to the Judicature Acts, 1873–75, were within the jurisdiction of both common law and equity courts. Under the Judicature Acts the whole jurisdiction became nominally concurrent though in practice distinctions continued. See also AUXILIARY JURISDICTION; EXCLUSIVE JURISDICTION.

Concurrent sentence. A sentence to a criminal penalty which is to operate at the same time as another sentence, as distinct from a consecutive sentence. Where concurrent sentences of imprisonment are imposed the prisoner, in effect, serves the longest one of the sentences.

Condemnation. The act of a prize court (q.v.) in determining that a vessel captured was justifiably captured. The effect is to vest the property in the vessel or cargo condemned in the captor, unless the court orders the property to be sold or delivered up on bail, in which case the money realized or the appraised value is substituted for the vessel or cargo.

Condescendence. In Scottish civil pleading, the statement of facts, set out in consecutive numbered paragraphs, which the pursuer relies on as justifying the claim he has made in the conclusion (q.v.) for the remedy or relief sought.

Condictio. In Roman law, originally a formal notice to appear on a stated day to choose a *iudex*, but usually the general term for a personal action claiming repayment of money. Thus *condictio indebiti* was the claim for the recovery of money or other property handed over by mistake. Other varieties were *condictio ob rem dati*, also called *condictio causa data causa non secuta*, for the recovery of property where the consideration had failed, *condictio ob turpem causam*, where money had been paid for an illegal or immoral purpose, *condictio furtiva*, for the recovery of stolen property, *condictio sine causa*, for causeless enrichment, and various others.

Conditio si institutus sine liberis decesserit. In Scots law, a condition implied in a will where the testator is making a provision of the nature of a family settlement to a class of beneficiaries as the parent of, or *in loco parentis* to the beneficiaries, that if a designated beneficiary should predecease, his children should take the provision in preference to any substitute. The *conditio* may be excluded by indications that the bequest was personal to the predeceasing beneficiary.

Conditio si testator sine liberis decesserit. In Scots law, a condition implied in a will that, if the testator left a general settlement making no provision for children subsequently born to him, the omission to provide for such children must be deemed unintentional and the settlement will be held revoked unless there is evidence rebutting the presumption.

Condition. A clause, common in contracts and wills, prescribing an act or event on which the existence or quality or effectiveness of some obligation or payment is dependent. Conditions are distinguished as conditions precedent or suspensive conditions, which hold up or postpone the operation of the obligation till the condition is satisfied, and conditions subsequent or resolutive conditions, which extinguish the obligation on the condition being satisfied. Conditions are also sometimes distinguished as protestative, within the power of the party, casual, dependent on chance happening, and mixed, dependent on elements of both choice and chance. Conditions which are illegal, immoral, impossible or contrary to public policy are generally void.

The term condition is also used, particularly in the context of sale of goods in English law, as meaning a stipulation as to some important matter, the breach of which gives rise to a right to treat the contract as repudiated, as contrasted with a warranty (q.v.) which is a stipulation, the breach of which gives rise to a claim for damages but not to a right to treat the contract as repudiated.

In the sale of goods in Scots law the distinction is between fundamental or material stipulations, usually called warranties, breach of which justifies rescission of the contract and non-fundamental stipulations, sometimes called conditions, breach of which justifies damages only. Conditions in this sense may be express, set out in the parties' contract, or implied by law or custom.

Conditional discharge. An order which may be made by a criminal court if it does not think that it is expedient to impose a punishment and a probation order is not appropriate. If the offender commits a further offence during the period fixed, not exceeding three years, he may be sentenced for the original offence.

Conditional fee. At common law a limitation of land to a man and a special class of heirs was a conditional fee simple. The condition was held satisfied by the birth of issue of the specified class and thereafter the grantee had a fee simple absolute which he could alienate, but if he died without having alienated the land it descended according to the prescribed course. The Statute of Westminster II (*De Donis Conditionalibus*) enacted that thereafter the grantor's intention should be observed so that the grantee should not be able to alienate but that it should remain to his issue after the grantee's death, or in the event of failure of the prescribed issue, revert to the grantor. The statute accordingly prevented the land ever becoming a fee simple absolute and converted it into what came to be known as a fee tail.

In modern law an estate in fee simple may be granted on condition and if the event provided in the condition happens the granter can re-enter and determine the estate. To defeat an estate on condition there must be actual re-entry.

Conditional limitation. A qualification attached to the grant of an estate or interest in land, providing for its determination on the happening of a particular contingency, e.g. so long as the grantee continues unmarried.

Also a future use or interest in land limited to take effect on a stated contingency.

Conditional sale. A contract of sale subject to a condition, frequently relating to payment or to the passing of the property. Such contracts are specially defined for the purposes of consumer credit (q.v.) legislation. It differs from a hire-purchase contract in being an outright sale, albeit subject to conditions, whereas hire-purchase is a hiring coupled with an option ultimately to buy.

Conditions of Sale. The terms on which offers to buy may be made and accepted. They are commonly stated in writing, e.g. in an auctioneer's catalogue or along with the particulars of an estate to be sold. In sales of land in England the conditions commonly used are the Law Society's Conditions of Sale and the National Conditions of Sale. By statute certain conditions are void and statutory conditions made applicable to contracts of sale of land made by correspondence.

Condominium. A governmental regime in which two powers jointly exercise governmental functions, as in the Anglo-Egyptian government of the Sudan, 1899–1955. Also applied to the individual ownership of one dwelling-house in a multi-dwelling building combined with common ownership of the parts of the building used in common.

Condonation. Prior to 1969 it was a bar to proceedings in England for divorce or judicial separation that the innocent spouse had condoned or forgiven all known matrimonial offences, conditional on no further matrimonial offence occurring. Condonation might be express or implied, and was cancelled if a subsequent offence should arise. In Scots law condonation or forgiveness of known matrimonial offences is a bar; it cannot be conditional.

Conduct. A general term for the acts, abstentions, and omissions of a person which are of legal significance. It is frequently used in the context of infamous conduct, disgraceful conduct, conduct unbefitting a member of a profession, and similar phrases.

Conduct money. Money paid to a witness to defray expenses incurred in coming to, staying at, and returning from the place of trial.

Confarreatio. In Roman law the oldest and most solemn form of marriage, confined to patricians and obligatory for the three *flamines maiores* and the *rex sacrorum.* So far as known the ceremony required bride and bridegroom to sit with veiled heads on joined seats covered with a sheepskin. A cake of spelt (*far*) was used, possibly eaten, and the ceremony was in honour of Jupiter Farreus. Stated witnesses had to be present. The marriage could be dissolved only by an elaborate ceremony, *diffareatio.* It had fallen into disuse by the end of the Republic.

Confederacy. In international law, a combination of two or more states for mutual aid, benefit and protection.

Confederate States of America. On 8 February 1861 delegates from the States of South Carolina, Georgia, Florida, Alabama, Mississippi and Louisiana, which had all seceded from the United States, met at Montgomery, Alabama, and adopted a Constitution for the Provisional Government of the Confederate States of America. On

11 March this Congress adopted a constitution modelled on that of the U.S.A. but having special States' rights provisions. The seceding states, together with Texas which had by then also seceded, ratified it. The Confederacy was later joined by Virginia, North Carolina, Tennessee, and Arkansas. Jefferson Davis (Miss.) was elected President and Alexander Stephens (Ga.) Vice-President. After the surrender of the Confederate forces the Confederacy came to an end. Military rule replaced civil, and under this regime newly enrolled electorates chose constitutional conventions which drafted new state constitutions. Reconstructed governments had been set up in eight states by 1868 and in the other three in 1870. After their legislatures had ratified Amendments XIV and XV Congress readmitted them to the Union and withdrew the military garrisons.

Confederation. An association or league of sovereign states with certain common organs for common purposes, e.g. the German *Bund*, established by the Congress of Vienna in 1815, and Switzerland. It differs little from a federation, though a confederation stresses the independent sovereignty of the states (cf. German *Staatenbund*) and a federation lays more stress on the supremacy and power of the central government (cf. German *Bundesstaat*). Down to 1789 the U.S. was a confederation; then the term federation or federal state was adopted, as signifying closer union. The southern states, which seceded in 1861, formed a confederation.

Conference. (1) A coming together for discussion of representatives of bodies of a particular kind, or of delegates of sovereign states. International conferences have been used in diplomacy since the sixteenth century, becoming important with the Congress of Westphalia (1645–48) and regularly used since the nineteenth century. Conferences may be called for various purposes, political or non-political. Preliminary business includes the election of president, and chairmen of committees, examination of the credentials of delegates, settlement of rules of procedure and voting rules. The original rule of international law required unanimity, but practice since 1945 has favoured majority decision. From the seventeenth century to 1918 French was the diplomatic language. Thereafter French and English predominated. After 1945 French, English, Spanish, Chinese, and Russian were accepted as official languages. Simultaneous translation is now provided at major conferences. Conference decisions are incorporated in treaties, conventions, declarations, recommendations, and protocols. A final act may be adopted which summarizes the proceedings and results.

(2) In parliamentary practice, a mode by which either House communicates important matters to the other, by each sending deputations of their own members. The common instances are the inability of either House to agree to the other's amendments to a Bill.

(3) A meeting between counsel and solicitor, with or without the client, to discuss and consider some matter.

Confession. In criminal law, a judicial or extra-judicial statement admitting guilt of the offence charged. An extra-judicial confession, e.g. to the police shortly after the offence, is admissible in evidence so long as it was not obtained by coercion, threats, or otherwise unfairly. It is not, however, normally sufficient evidence without corroboration by other evidence. A confession admits the whole charge, whereas an admission covers only certain facts. In civil procedure a confession is a formal admission.

Confession and avoidance. A pleading which admits the truth of a fact alleged in the preceding pleadings but seeks to avoid its effect by alleging some new matter.

Confession of Faith (1647). An exposition of Presbyterian doctrine, prepared by the Westminster Assembly of 1643–52 and ratified by the Long Parliament in 1648. It affirmed government by presbytery and the dogma of predestined election. It has been adopted by English-speaking Calvinist churches and remains a fundamental statement in the Church of Scotland.

Confession of villeinage. In early mediaeval law if a man once confessed himself the serf of another in a court of record he was subsequently held bound thereby and was deemed a villein as much as if he had been born such.

Confidence, breach of. A failure to maintain the confidentiality which the person should have maintained as to some matter.

Confidential communication. Communications of information between one person and another which the courts will not require to be divulged. They include communications between husband and wife, party and solicitor, or solicitor and counsel, made during and with reference to judicial proceedings, and possibly between parishioner and clergyman but not between doctor and patient.

See also CROWN PRIVILEGE.

Confirmatio Cartarum. A charter confirmed by Edward I later reckoned as an Act of 1297, 25 Edw. I, and now mostly repealed, which enacted a series of new provisions intended to deprive the Crown for the future of the right of arbitrary taxation which it had assumed. It was not merely a re-issue of Magna Carta and the Charter of the Forest, with special provisions for their distribution throughout the realm and annual publication but the enactment of new provisions.

Confirmation. (1) An act by a person ratifying what was voidable, as by a wife ratifying a settlement made during minority, or by a superior authority ratifying the proposals of an inferior as where a Minister approves a compulsory purchase order.

(2) In the Christian Church the rite whereby the vows made on a young person's behalf at baptism are renewed by him personally and he is admitted an adult member of the church.

(3) In ecclesiastical law, the ratification by the archbishop of the election of a bishop by the dean and chapter prior to consecration of the bishop. Objection has frequently been made to confirmation but it was held in *R.* v. *Archbishop of Canterbury*, [1902] 2 K.B. 503 that confirmation could not be opposed on the ground of alleged heterodoxy.

(4) In Scots law, the ratification by the appropriate sheriff court of an appointment of executors by a deceased in his will (testament testamentar) or, failing that, of an appointment by the court (testament-dative), in either case conferring on the person confirmed title to uplift, realize, administer, and dispose of the whole estate of the deceased as contained in an inventory given up by the executor, on which confirmation proceeds. The executor takes an oath or makes affirmation to an inventory of the deceased's estate. Confirmation in respect of estate omitted from the first inventory or under-valued is obtained in the form of an Eik to confirmation, or by another person in the form of confirmation *ad omissa vel male appretiata.* If an original executor does not complete administration of the estate another may obtain confirmation *ad non executa.* A creditor of the deceased may obtain confirmation as executor-creditor to part only of the estate, to enable him to exact payment of his debt therefrom.

Confirmed credit. If it is a condition of a contract of sale for export that the buyer shall pay by means of a confirmed credit, the buyer must get his bank to issue an irrevocable credit in favour of the seller, undertaking directly or through the seller's bank, to accept drafts drawn on it for the price of the goods against tender by the seller of the shipping documents, and adding the bank's undertaking to pay to that of the buyer. The banker normally retains the shipping documents until reimbursed by the buyer. The arrangement creates a new contract between the bank and the seller and offers the latter greater certainty of payment.

Confiscation. The act of appropriating private property for State or sovereign use. As an exercise of state power it has been known since the Roman Empire and has usually been the result of the doing by the owner of some prohibited act. It is distinct from expropriation, which was originally the voluntary surrender of property rights by an individual to the State, but is now normally the compulsory surrender in return for nominally fair compensation. It is also the penalty for trying to carry contraband or supplies to a place blockaded or besieged.

In criminal law confiscation of smuggled property or property used or intended for use for purposes of crime, such as housebreaking instruments, is part of the penalty for certain offences.

Conflict of Laws. See INTERNATIONAL PRIVATE LAW.

Conformity, bill of. A bill formerly filed in Chancery by an executor or administrator, who found that he could not safely administer the deceased's affairs save under the direction of the court by reason of their being involved, against the creditors to have their claims adjusted and a decree made to have the estate administered under the direction of the court.

Confusion or intermixture. In Roman law, the mode of acquiring ownership of goods which have become mixed and undistinguishable. If mixed by agreement the proprietors own the mixture in common in proportion to the shares they contributed. If mixed by accident or by a third party the proprietors own the mixture in common in proportion to the amounts contributed. If one without the knowledge of the other mixes his goods with those of the latter the whole belongs to the latter. If the things mixed be capable of separation, e.g. sand and salt, the property in them is unchanged. See also SPECIFICATIO.

Confusion is also a mode of extinguishing a debt, as when either party succeeds to the title of the other by inheritance or otherwise.

Congé d'élire (permission to elect). In England since 1534 the right of nomination of bishops has been vested in the Crown. On the occurrence of a vacancy the King sends a *congé d'élire*, issued under the great seal, to the dean and chapter of the cathedral of the bishopric with a letter missive containing the name of the person to be elected. If there is delay to elect the King may nominate by letters patent. After the election, the chapter inform the Crown of the result and the Crown signifies royal assent by letters patent addressed to the archbishop of the province, who then proceeds to confirmation and consecration.

Congress. An assembly of representatives from different countries met to consider matters of common concern. The word came into use in the seventeenth century. It is usually applied only to meetings of the first importance attended by sovereigns or their foreign ministers, e.g. the Congress of Vienna, 1815, the Congress of Berlin, 1878, less important meetings being called conferences, but the dividing line is unclear. In the twentieth century the word has fallen into disuse being superseded by conference.

Congress of the United States. The Constitution of the United States, by Article I, vests all legislative power thereby granted in a congress of the United States, consisting of the Senate, composed of two senators from each State elected for six-year terms, one third of the House retiring every two years, and the House of Representatives, composed of 435 representatives elected every two years by the people of the several States, based on the relative populations of the States, each State having at least one representative. Each House is judge of the elections, returns and qualifications of its own members. The Vice-President is President of the Senate, and the Speaker chairman of the House. Its legislative powers are to make all laws necessary and proper for carrying into execution the powers listed in the Constitution, Article I, section 8 (which include taxation, borrowing, commerce, war and peace, and the armed forces). But Congress has no power to legislate on the matters specified in Article I, section 9, nor to make legislation conflicting with the Bill of Rights. Amendment X to the Constitution reserves powers not delegated to the U.S. to the States.

The First Congress under the Constitution met in New York on 4 March 1789. Since then annual meetings have convened; since 1933 Congress convenes annually on 3 January.

The management of Congress is largely centred in each House on the chairmen of the Standing Committees, who are chosen by seniority in office, each committee having a majority from the party having a majority in the House. Congressional committee hearings have frequently made valuable investigations and disclosed the need for legislative change or other reform.

The power of Congress has varied and varies with the times and the power and prestige of the Presidents, being probably at its highest during the Reconstruction period after the Civil War.

Congressional Globe. A privately issued record of speeches and statements made in Congress, 1834–73, replaced thereafter by the *Congressional Record* (q.v.).

Congressional Record. A verbatim but unofficial report of all speeches and statements made on the floor of Congress, or prepared but not delivered, or any further material a Congressman may desire to have inserted. It has been published daily during sessions of Congress since 1873. The official records of proceedings are the *Journals* of the Senate and the House, published annually since 1789.

Conjugal rights. The rights which spouses have in each other, notably to society and marital intercourse. They cannot be enforced by act of either party. A suit for restitution of conjugal rights might, prior to 1970, be brought if either spouse lived separate from the other without sufficient reason, but decree would not be enforced by attachment but only by an order for periodical payments of money. See also CONSORTIUM.

Connan, François de (Connanus) (1508–51). A distinguished French jurist, who began the general classification of law later carried on by Domat and other systematizers, author of *Commentarii juris civilis* (1538).

Connivance. Knowledge that a wrong is being done, and either assisting it or taking no steps to prevent it. It was till 1969 a bar to proceedings in England for divorce or judicial separation that the petitioner had been accessory to or connived at the respondent's adultery, by promoting or encouraging the initiation or continuance of the adultery. In Scotland it continues to be a bar.

Connotation. In logic, the term describing the sum of the qualities of a given thing or class of things and indicated by the name by which such things are known, and practically equivalent to full definition of the thing. Controversy about such general terms as 'liberty', 'property', and the like frequently flows from dispute as to the connotation of the terms.

Conquest. In international law, the acquisition of territory by a victorious State from a defeated State in warfare. The State acquiring by conquest is regarded as the successor to the rights and duties previously applicable to the territory. The modern assertion of the principle that aggressive war is contrary to international law logically results in the denial of legal recognition to the territorial changes made by such war, though this consequence cannot be said yet to be fully recognized. Some jurists distinguish conquest from subjugation, the former referring to military occupation, the latter to transfer of title where the defeated state is totally annexed.

Conquest. In feudal law, a term for purchase; in mediaeval English law and Scots law down to 1874, heritable property of kinds, title to which had to be completed by infeftment, acquired by gift, purchase, or other singular title from a person to whom the acquirer would not have succeeded by law, as distinct from heritage, heritable property to which a person succeeded as heir. There were, prior to 1874, distinctions as to the succession to heritage and to conquest but these were then abolished.

Conring, Herman (1606–81). One of the great polymaths of his time, professor at Helmsted of medicine and later of law, and writer also on theology and politics. His numerous works include *Dissertationes de Republica* and *De Origine Juris Germanici* (1643) which gave a detailed history of German legal sources and showed that the existing common law was based on the Reception; hence he is sometimes called the father of German legal history.

Consanguinity. The blood relationship existing between persons descended one from another or from a common stock. The legally important issues are how consanguinity is determined, the different degrees thereof, and the consequences. It is determined by the existence of a common stock or root, from which descend direct lines (e.g. son, grandson, etc.) and collateral lines (brother, cousin, etc.). The degree is the measure of distance between any two persons descended from the common stock. The Civil (Roman) law aggregated the steps from each person to the common ancestor; thus cousins were related in the fourth degree, and a man to his niece in the third degree. The canon law counted as many degrees as there were generations in the longer of the lines of descent from the common ancestor; thus cousins were related in the second degree, as was a man to his niece or to his aunt. In Western societies marriage had generally been prohibited between persons within certain degrees of consanguinity. The basis of this may be folk knowledge of the undesirability of inbreeding. In the early Christian Church the degrees within which it was prohibited varied but the Fourth Lateran Council (1215) prohibited relationship to the fourth degree and this remained unchanged in the Catholic Church till 1918. In each legal system the matter is now one of positive law. Similarly intercourse between persons within certain degrees of consanguinity is punishable as incest.

See also AFFINITY; INCEST; MARRIAGE.

Conscience. The mental faculty or capacity which judges the moral quality of one's own or another person's motives or conduct, approving or condemning them as good or bad. A conscientious objector is one whose objection is that what is in question is morally wrong.

From an early time the Chancellor was deemed Keeper of the King's Conscience, a fact which explained and justified his exercise of an equitable jurisdiction in the court of chancery, and his reliance in its exercise on what was consistent with good conscience. Thus Selden (q.v.) said that equity was according to the conscience of him who was chancellor and varied as did the length of the chancellor's foot, but since his time the principles of equity have become settled and there are fairly defined standards for what is or is not according to conscience.

See also EQUITY.

Conscience, Court of. Courts, constituted by statute in the City of London and other towns, for the recovery by summary procedure of small debts and deciding substantially by reference to what was consistent with equity and good conscience. They were superseded from 1846 by the establishment of the modern county courts.

The Court of Chancery was sometimes referred to as a court of conscience because its jurisdiction was originally founded on relief granted by the Chancellor, as Keeper of the King's Conscience, in circumstances where equity and justice demanded it.

Conscription. The compulsory enrolment of persons for military or other national service. It was practised by the Egyptians and Romans and in modern times by most European States. In the U.K. a kind of conscription was long in force under the Militia Acts, but in both World Wars conscription was imposed by statute. After World War II it continued for a time under the name National Service.

Conseil d'Etat. The supreme French administrative body, partly advisory to the executive, partly judicial. It is composed of the cream of the French civil service and belongs to the executive branch of government. It was created by Napoleon in 1799 on the model of the older Conseil du Roi, but has always been distinct from the Council of Ministers. It has important functions in planning and advising the government on the problems of proposed administrative action, but also judicial functions in checking and controlling abuses of power by administrative officials on such grounds as *vice de forme, incompetence, détournement de pouvoir* or *violation de la loi*. It operates both in general assembly and in sections, one of which is concerned with contentious legal business, and individuals aggrieved by official decisions may require them to be justified before this section on pain of being annulled. In certain matters it acts as final tribunal, in others as court of appeal, in others again as tribunal of first instance. The law applied is largely judge-made by the practice of the Conseil d'Etat itself. Within the Conseil d'Etat there are a number of Commissaires du Government who in a disputed case have to consider the issues impartially beforehand and reach an individual conclusion as to what should in law and justice be done. After the parties have made their submissions he states the facts, the relevant law, and proposes what conclusion should be reached and on what principle. His conclusions do not bind the tribunal but are influential. A single judgment is delivered after private discussion. Since 1875 the *Conseil d'Etat statuant au contentieux* has greatly extended the ambit of its control of administrative actings and has been increasingly recognized as a valuable protection of the citizen against the executive and defects in administration.

M. Waline, *Manuel Elémentaire de Droit Administratif*; A. de Laubadère, *Traité Elémentaire de Droit Administratif.*

Consensus in idem. The state of agreement between negotiating parties, particularly in making a contract. The validity and enforceability of the contract depends on, *inter alia*, the parties having reached *consensus in idem* on all the essential terms of their bargain.

Consent. An act of the human will acquiescing in a mental judgment or deciding to implement it. Consent always implies freedom of judgment, deliberation and freely given acquiescence in what is considered desirable. There is free consent only if the person is not blinded by anger or intoxicated, or ignorant or deceived, subject to duress or overreached.

Consent is a material element of many relationships, notably marriage and contracts which, in many other cases, a person may validly take some legal step only with the consent of another designated person. Consent is in general no defence to a criminal charge, but may be in some cases, e.g. rape.

Consequences. The results which flow from and follow on a particular act or event, which is deemed the cause of them. The consequences of a particular act or event may be material in law as determining the legal quality or nature of that act or event; thus if A stabs B, whether B dies or not in consequence will determine whether A may be charged with murder (or manslaughter) or merely with assault. A problem of difficulty in many cases is whether a particular state, e.g. death, is a consequence of a particular act, e.g. stabbing, or of something else, e.g. negligent medical care or independent natural cause.

In many cases a distinction must be drawn between the immediate consequences of an act or event, e.g. sinking of a ship in consequence of collision, and remote or indirect consequences, e.g. loss of profit to the owners of goods carried, bankruptcy of the shipowner, unemployment of his employees, loss of opportunity in life for their children, and so on. Thus both in cases of breach of contract and of tort the question of remoteness of damage arises, i.e. if a party is in breach of duty and has caused harm, for what consequences thereof is he to be held liable? For immediate consequences, or foreseeable consequences, or all consequences traceable to the breach? On what principle is the line limiting his liability to be drawn? Different answers are given in different systems to these questions.

See also REMOTENESS OF DAMAGE.

Conservation laws. Laws adopted to conserve and manage the natural resources of a country so as to maintain and enhance the quality of life by rational utilization of the environment. They include rules for the management of natural resources, water, soil, and minerals, for the management of living resources, wildlife, fish, plants, for the reduction of pollution and waste.

Conservator of Scots Privileges. In the fourteenth and fifteenth centuries Scottish trade with the Low Countries demanded a base and after competition from Bruges Campvere was recognized as the staple port, which it remained till 1799. The Conservator of Scots Privileges was the resident official of the Convention of Royal Burghs and guardian of mercantile interests. He had jurisdiction in cases between merchants, and also in criminal cases.

Conservator of the Truce and Safe Conducts. An officer appointed during the reigns of Henry V and Henry VI to inquire into all offences committed on the high seas against the truce with France or safe conducts for sailing to and from the Cinque Ports. He was assisted by two men learned in the law and the judges were authorized, if infringement took place, to cause restitution and amends to be made to the party injured.

Conservators of the peace. Officers who, before justices of the peace were appointed, had at common law powers for the maintenance of the public peace. *Custodes pacis* were named in the mid-thirteenth century and from 1285 were associated with enforcement of the Statute of Winchester. They were later chosen by landholders of the county and finally appointed by royal writ. They had to take indictments of felonies and misdemeanours and hold the accused until trial by royal judges. Some had this power as an element in offices held by them, including the Lord Chancellor, judges of the Queen's Bench Division, and coroners and sheriffs, and these are still conservators of the peace. Others had it by itself and these were superseded from 1327 by the appointment of justices of the peace.

Consideration. That which is given, done or forborne in return for the promise or act of another party. In the English law of contract a promise, unless made by deed, is not enforceable by action unless made for valuable consideration, a rule not definitely established until the late eighteenth century. Consideration may be a promise to do or forbear from doing something, or some loss or detriment suffered at the request of the promisor. It is normally a payment, transfer of goods, doing of services or surrender of another legal claim. It cannot be anything which the promisee is already under an obligation to do. It must be of some value, but not necessarily value equivalent or adequate to the promise. Past consideration, i.e. something done before the promise, is insufficient. The doctrine of consideration is peculiar to Anglo-American common law but its functions in evidencing the contract, ensuring that it is not lightly undertaken and discouraging transactions of doubtful validity are performed in other ways in other systems.

See also CAUSE.

Consignation. In civil law and Scots law, the deposit of a thing owed or disputed with a third party under the authority of the court.

Consiliatio Cnuti. An almost complete Latin translation of Cnut's laws, made by an unknown, probably a Norman, who was not, however, a practical lawyer, probably between 1100 and 1130, and based on an Anglo-Saxon work since lost.

F. Liebermann, *Gesetze der Angelsachsen*, I, 618.

Consilium principis. Any Roman magistrate might summon advisers to assist him. Augustus established a *consilium* which prepared business for the Senate. This ended under Tiberius. Under the Empire the body of the Emperor's advisers acquired the character of a privy council. Initially it was an unofficial group of those summoned to act as assessors in a judicial enquiry or as advisers on a matter of administration. Alexander Severus established a regular consilium of 70, 20 being jurists, to assist in framing constitutiones. Under Diocletian the body was reorganized, renamed the *sacrum consistorium*, and membership made permanent.

Consimili casu. A clause in the Statute of Westminster II, c. 24 (1285) empowering the clerks in the Chancery to frame a writ for a new case involving the same law and requiring the same remedy as a former case. The primary object seems to have been to extend the remedies already available between parties to be available to their heirs or successors. It did not give the clerks powers to create new forms of action. The only major extension was the creation of the writ of entry *in consimili casu*. It was formerly thought that this provision gave rise to the development of the action of case out of the action of trespass but this view is now discredited.

See also ACTION ON THE CASE.

Consistorial action. In Scots law, actions which were formerly competent only in the consistorial courts (which were originally the courts of the bishops of dioceses and after the Reformation the Commissary courts (q.v.)), and are now mostly competent only in the Court of Session; they are actions of declarator of marriage, of nullity of marriage, of legitimacy and bastardy, actions of separation *a mensa et thoro*, of divorce, adherence, putting to silence and actions of aliment between husband and wife.

Consistory. From the time of Diocletian the *consistorium* was the name of the Roman imperial consilium since the members stood in the Emperor's presence. It functioned both as a general council of state and as a supreme court of law. It normally comprised the principal civil and military officers and legal experts. It was still an active council of state in the fourth century but by the fifth century its proceedings seem to have become purely formal. The word was also applied to the chamber in which the emperor received petitions and gave judgment, then to the persons who took part and those who formed the tribunal. Later still it was used of a gathering of ecclesiastical persons for administering justice or transacting other business. The term is still used in the Roman Church for various kinds of meetings of superior clergy.

Consistory court. Prior to the Reformation the bishop of each diocese in England held a consistory court for the diocese. From the mid-twelfth century it was usually presided over by the Chancellor or Official of the bishop, who might exercise all jurisdiction competent to the bishop, except in such cases as the bishop might reserve for himself. In time he became the permanent judge of the court, though sometimes the bishop delegated jurisdiction over parts of the diocese to his commissary. Appeal lay from the consistory court to the archbishop's provincial court. There continues a consistory court in each diocese (in the diocese of Canterbury called the commissary court) held by the Chancellor as official principal (in the diocese of Canterbury called the Commissary-General). The Chancellor acts as an ordinary or independent judge according to ecclesiastical law, uncontrolled by the bishop. The court has original jurisdiction to hear proceedings on articles charging an offence by a priest or deacon holding preferment in the diocese, but not in matters of doctrine, ritual, or ceremonial, and to determine a cause of faculty for authorising any act relating to land in the diocese, the sale of books in a parochial library, and certain other proceedings. Appeal lies to the Court of Arches (q.v.) or to the Chancery Court of York (q.v.).

Consolato del Mare. A collection of maritime customs and ordinances, published in Catalan at Barcelona, in the fifteenth century. It embodied the rules of law and practice commonly followed by the commercial judges known as consuls in the chief maritime towns around the Mediterranean, and comprised a code of procedure issued by the Kings of Aragon for the guidance of the courts of the consuls of the sea, a collection of ancient customs of the sea and a body of ordinances for the regulation of cruisers of war. It became a maritime common law of the Mediterranean and was translated into many languages.

L. Pardessus, *Collection des lois maritimes* (1834); T. Twiss (ed.) *Black Book of the Admiralty* (1874).

Consolidated Fund. Prior to 1786 each public service was paid from the revenue from taxes assigned to that purpose. In 1786 Pitt established the Consolidated Fund; at first only the Civil List (q.v.) and charges in respect of the National Debt were paid out of the Consolidated Fund, but after 1854 all public revenues were paid into it and all public payments made therefrom. Since 1968 there is also a National Loans Fund. The account of the Consolidated Fund is kept at the Bank of England and, on Parliamentary authority, money is transferred from it to the account of the Paymaster-General who in turn pays to departments the sums approved by Parliament in the Estimates.

Certain expenses, Consolidated Fund Services, are charged on the Consolidated Fund under permanent legislation, and are consequently not liable to annual scrutiny in the Estimates. These include the Civil List, payments of interest on the National Debt, and the salaries of high officers, including the Speaker, the Comptroller and Auditor-General, the Parliamentary Commissioner for Administration and the judges.

At least two Consolidated Fund Acts are passed each Session authorizing the Treasury to make certain sums of money available for the public service, and to borrow money by the issue of Treasury bills, one to give legislative authority to the Supplementary Estimates, the other near the end of the financial year to authorize the issue of sums required to cover any outstanding Supplementary Estimates, and Excess Votes for the previous financial year, and sums for the Votes on Account for the new financial year.

Consolidation Act. An Act intended to bring together into one Act statutory provisions contained in a number of existing Acts, without altering the substance of the law, and to supersede and repeal the statutory provisions consolidated. Consolidation has been done since the reign of Elizabeth I and it was proposed repeatedly thereafter, by Lord Keeper Bacon, James I, Sir Francis Bacon, and by committees under the Commonwealth. Parliament effected some consolidation in 1816 and Peel consolidated portions of the criminal law in 1826–32. Lord Cranworth appointed a Board for the revision of the Statute Law in 1853 which was superseded by a Statute Law Commission in 1854, and in 1861 there were passed seven Criminal Law Consolidation Acts dealing with accessories and abettors, criminal statutes repeal, larceny, malicious damage, forgery, coinage offences, and offences against the person. After the establishment of the Statute Law Committee in 1868 and the office of the Parliamentary Counsel to the Treasury in 1869 many consolidation Acts were passed, including such major Acts as the Merchant Shipping Act, 1894.

Prior to 1949 consolidation had normally to be preceded by an Act making any desired corrections and improvements in the legislation to be consolidated.

In 1949 the Consolidation of Enactments (Procedure) Act provided that consolidation may include making corrections and minor improvements in the enactments consolidated. The Lord Chancellor lays before Parliament a Memorandum proposing such corrections and minor improvements as he thinks expedient and the Joint Committee from both Houses appointed to consider Consolidation and Statute Law Revision Bills informs him, before reporting the Bill, which they approve. If the Lord Chancellor and the Speaker concur, the corrections and improvements become law for the purpose of further proceedings in Parliament on the Bill.

Consolidation has proceeded regularly in modern times and contributed greatly to the improvement in form and utility of the statute book.

Consolidation of actions. The procedural device of, when several actions raise the same issue, allowing one to proceed and staying the others or of having all the actions carried on as one.

Consortium. The right of each spouse to the companionship, affection, and assistance of the other. To a limited extent the law recognizes interference with or impairment of consortium by a third party, as by injuring one spouse, as wrong actionable by the other spouse.

Conspiracy. An agreement between two or more persons to effect some unlawful purpose, whether a crime or a tort, or a public mischief, though probably not a mere breach of contract, or to do a lawful act by unlawful means. A person cannot conspire by himself. In criminal law a person is by statute guilty of conspiracy if he agrees with any other person or persons that a course of conduct will be pursued which will necessarily amount to or involve the commission of any offence or offences by one or more of them if the agreement is carried out in accordance with their intentions. The common law offence of conspiracy still exists in certain cases. The crime is complete when the parties have agreed and it is immaterial that they do not put their plan into effect, but the fact of agreement is usually evidenced by actings in pursuance of the common plan or attempt to carry it into effect.

In tort conspiracy is tortious if the agreement is to injure another and if damage is in fact caused him, unless the major and predominant purpose of the conspirators was to advance their own lawful interests in a matter in which they believed their interests would otherwise suffer, rather than the pursuit of spite or malevolence, so long as unlawful means are not used in the furtherance of the predominant purpose. Hence conspiracy is actionable if the predominant purpose is disinterested harm, or if unlawful means are used to secure it, but not if the predominant purpose is legitimate, such as to secure a larger share of a market.

Constable. The title of a high officer of state found from mediaeval times in many western countries, derived from *comes stabuli*, head of the stables at the court of the Byzantine emperor. Later under the Franks the *comes stabuli* was in charge of the royal stud, his subordinate being the marshal (*marescallus*). In France he became one of the great officers of state with some jurisdiction and with command of the cavalry. In England the office of constable was known in Henry I's reign, and was forfeit to the Crown in 1521. Since then appointments have been made for coronations only.

The title was also given to officers with military commands and in charge of castles and garrisons, e.g. the Constable of the Tower of London. Sometimes the appointment was linked with that of conservator or justice of the peace.

In every local area a chief or high constable had responsibility for suppressing riots and violences and under him in each village or tithing had petty or parish constables. These officers remained the executive legal officers until the public Acts of the nineteenth century authorized the establishment of paid bodies of police constables. There are still bodies of high constables in Scotland, notably those of Edinburgh and Holyroodhouse, with functions on state occasions only.

In the U.S. constables had in rural districts functions similar to English parish constables but were largely superseded by uniformed state police in relation to criminal matters.

See also POLICE.

Constable and Marshal, Court of, or Court of Chivalry. In the Middle Ages the rules for the discipline of the army were administered by the court of the Lord High Constable and Earl Marshal. Its jurisdiction extended to allied matters, contracts relating to deeds of arms, hiring of soldiers, and prisoners or prize. Statutes in the fourteenth century limited the jurisdiction to matters, civil or criminal, relating to war and done outside the realm. The office of Lord High Constable was forfeited to the Crown in 1521 and thereafter the right of the Earl Marshal to hold the court alone was confirmed by letters patent in the seventeenth century. The Tudors extended the court's jurisdiction and in the later seventeenth century codes of military law were issued but were enforced by army officers, the jurisdiction of the Constable and Marshal's court being obsolete. Military law and courts martial were legalized and developed by the Mutiny Act, 1689 and later similar Acts.

Another branch of the jurisdiction of this court was over heraldic cases and slander on men of noble blood, and this jurisdiction was regularly exercised till the early eighteenth century when it fell into disuse. A sitting again took place in 1955.

Constable, Lord High. In feudal times one of the great officers of state who attended the King at his coronation and on military expeditions. Along with the Earl Marshal he exercised the jurisdiction of the Court of Chivalry (q.v.) and presided at trials by battle (q.v.). The office was held hereditarily but on the attainder of Stafford, Duke of Buckingham, in 1521 the office was forfeited to the Crown. The duties are now limited to coronations, for which a peer is appointed.

Constable of Scotland, Lord High. A Scottish office of State, originally responsible for keeping the peace outside the King's hall and hence having jurisdiction in all cases of slaughter and disorder within four, later three, miles of the king's person. He sat with the King and the Marshal in a court of chivalry, and claimed his right to hold court in the nineteenth century which gave rise to the practice of appointing the magistrates of Edinburgh and certain other persons as constables-depute when the sovereign is at Holyrood. The office has been hereditary in the family of Hay, Earls of Errol, since 1315.

Constantine the Great (*c.* 280–337). Roman emperor A.D. 306–337, became converted to Christianity in 312, being the first Christian emperor, and thereafter supported that religion. In 324 he rebuilt Byzantium, changed its name to Constantinople and decided to make it the new Rome and his permanent capital. It was dedicated in 330, given a Senate and designated the Second Rome. His legal reforms, though not radical or extensive, were influenced by Christian and Stoic views in the direction of humanity. He issued numerous laws relating to Christian practice, abolished crucifixion and the practice of branding certain criminals in the face, and extended privileges to the clergy. His main achievement was the establishment of a Christianized imperial governing class.

Constituency. In democratic government the unit represented by one or more elected persons. Basically a constituency is normally a defined geographical area. All constituencies should be roughly equal in number of electors, but cannot be equal in size or character, and to attain this regular revisions of boundaries must be undertaken. For certain kinds of elections, e.g. to the governing body of a profession, constituencies may be other than geographical, e.g. by categories of members.

The term constituent was formerly used of one who appointed another to act for him by power of attorney. It is now chiefly used of a voter in a constituency.

Constitutio Antoniniana. An edict of the Roman emperor Caracalla of A.D. 212 which effected a wide extension of Roman citizenship, previously limited to inhabitants of Italy and Cisalpine Gaul, and to communities and individuals granted the privilege. The extent of the extension is uncertain. Some authorities suggest that it made all inhabitants of the Empire citizens, but others suggest that it was less complete than that. The citizenship had, however, lost political importance and was valued mainly as a symbol of imperial unity.

Constitutio de Feudis. A law promulgated by the Holy Roman Emperor Conrad II in 1037 protecting the land rights of lesser vassals (vavasours) in Lombardy.

Constitutio Deo auctore. The order of 15 December 530 by the Emperor Justinian for the compilation of the *Digest* (q.v.).

Constitutio Imperatoriam majestatem. The order of Justinian of 16 December 533, giving the *Institutiones* the validity of an imperial statute with effect from 30 December of that year.

Constitutio Tanta circa. The order of Justinian of 16 December 533, which gave the *Digest* (q.v.) the force of law from 30 December of that year.

Constitution. (1) A name sometimes given to particular kinds of laws or ordinances, such as the constitutions issued by Roman emperors, the *Novellae Constitutiones* of Justinian, the Constitutions of Clarendon etc. In this sense the word never took root in English law, such words as Assize, charter, and later, statute and ordinance being used.

(2) The body of rules prescribing the major elements of the structure and organization of any group of persons, including clubs, associations, trade unions, political parties and the citizens of a state. Sometimes a constitution has a special form and name such as the charter of a society or university, or the memorandum and articles of association of a company.

(3) The fundamental political and legal structure of government of a distinct political community, settling such matters as the head of state, the legislature, executive and judiciary, their composition, powers and relations. Every state has a constitution, since every state functions on the basis of certain fundamental principles and rules. In old states the constitution is to some extent a natural or customary development, and it is only since the Renaissance and Reformation that it has been accepted that the people of a country can deliberately settle and adopt a constitution for themselves as the citizens of a state.

Constitutions may be classified in many ways into unwritten constitutions, such as that of the U.K., where the principles and rules as to the structure are scattered through various statutes, judicial decisions and unwritten usages, practices or conventions, and written constitutions, where the major principles and rules are embodied in a written and enacted document, albeit interpreted by judicial decisions and supplemented by usages and conventions as well as by rules on particular subjects; into autocratic, authoritarian, or despotic constitutions, where the main power is in the hands of a person, or small group, who can declare and enforce their will, and democratic constitutions, where the main power is in the hands of delegates or representatives of the community in general; into monarchical or republican according as the head of state is a hereditary monarch or an elective President; into fixed or rigid and flexible constitutions, according to the difficulty or ease of amending them and the need for special machinery or legislative procedure; into unitary constitutions, which provide a single structure of legislature, executive and judiciary for the whole community, and federal constitutions, which provide for a central structure of government, and also, co-existing therewith, for separate structures of government in particular states, provinces or other divisions of the community, and for the distribution of powers and functions between federal and state or provincial governments.

It is not always easy to classify the constitution of

a particular country, and there are many hybrids and specimens which defy classification. Moreover, in every case one must consider the realities behind the words found in documents, whether, for example, the rights of the subject declared in the constitution can really be effectively secured.

The term Constitution is also given to the authoritative text or verbal formulation of the constitution for the time being of a particular political community, e.g. the Constitution of France. The Constitution of the U.S. is the oldest written constitution in this sense, being adopted by the Federal Constitutional Convention on 17 September 1787, ratified by the 13 original States between December 1787 and May 1790, and becoming effective when nine had ratified it. Washington was inaugurated as President, Congress first met, and the Constitution was declared in effect on 30 April 1789. Some countries, e.g. France, have had a series of constitutions, new ones being adopted after political upheavals.

There cannot be said to be any essentials of a constitutional document, but it commonly makes provision for a Head of State, and his appointment and powers, a supreme legislative body, its composition, meetings and powers, a chief executive, his appointment and powers, the major departments of administration, the supreme judicial body and its powers, and so on. It should make provision for its own amendment. It frequently lists certain rights of the individual and guarantees them against being abridged or altered save under special conditions.

J. Bryce: *Studies in History and Jurisprudence*; C. J. Friedrich: *Constitutional Government and Democracy*; A. J. Peaslee: *Constitutions of Nations*; J. Hawgood: *Modern Constitutions since 1787*; K. C. Wheare, *Modern Constitutions*; E. S. Corwin, *The Constitution and What it Means Today*; C. H. Pritchett, *The American Constitution.*

Constitutional convention (U.S.). The assembly of delegates which met at Philadelphia in 1787 and drafted the constitution of the U.S., which replaced the Articles of Confederation. By 1787 it had become doubtful whether the Articles of Confederation could be modified to create an effective central power to handle foreign relations and regulate inter-state commerce and the convention had been called as a result of the demand for a stronger central government. Seventy-four delegates were chosen, but only 55 took part and only 39 ultimately signed the constitution. The convention early decided to frame a new constitution rather than try to amend the Articles of Confederation and after long debate agreed on the document which still stands. It was referred to the states and by July 1788 had been approved by the state conventions of 11 of them. It came into force in 1789.

M. Farrand, *The Framing of the Constitution of the U.S.*; C. Warren, *The Making of the Constitution.*

Constitutional government. The government of a state or politically organized community in accordance with defined principles and rules, as distinct from autocratic, dictatorial or arbitrary government. They cannot be wholly fixed and unchangeable or unchanging rules but cannot on the other hand be changeable at the mere whim of a person or group. The principles and rules underlying and influencing the structure of the constitution of a particular state are in every case partly a product of that state's history and partly the product of the social and political philosophy of the dominant group at the times when the constitution was adopted or revised and when particular rules or conventions were accepted.

The theoretical foundations of modern constitutional government where laid down in the writings of Hobbes, Locke, and Rousseau and their thinking powerfully influenced the great period of constitution-making exemplified by the American Declaration of Independence and Bill of Rights and the French Declaration of the Rights of Man.

Among the major features of constitutional government are a division of power among several organs, between legislature, executive, and judiciary, between central and provincial governments, between central and local authorities; representation, whereby those in office act as representatives of their constituents, whether elected by them or not, accountability, that those who govern are accountable to at least part of the general body of those governed, as by having to offer themselves periodically for re-election; openness, that decisions on public issues should, in general, be reached openly and subject to public comment; and procedural stability, that fundamental rules should not be subject to arbitrary or frequent change. Most importantly, in constitutional government there are standards set by which the legality and legitimacy of the actings of the organs of government are judged and means whereby it can be checked. The U.S. Constitution by interpretation and some other constitutions, e.g. Germany and Italy, by express provisions, provide for judicial review of the constitutionality of legislation. Even the legislature and the executive are subject to the law as laid down by the judiciary. So far as statutes are concerned this is not so in the U.K., and judicial control of the executive has been weak and uncertain. Judicial review has been adopted in some modern European constitutions, notably Germany and Italy.

The survival of constitutional government in a community depends ultimately on the extent to which the fundamental rules and practices are agreeable to the behaviour, customs and ways of thinking of the majority of the people. This depends in part on the extent to which the constitution is capable of development, modification and change by amendment, interpretation and practice. The most successful constitutions have been those which have permitted development, frequently without changing the letter of the constitution.

C. J. Friedrich, *Constitutional Government and Democracy*; H. J. Spiro, *Government by Constitution*; W. S. Livingston, *Federation and Constitutional Change*; S. H. Beer and A. B. Ulam, *Patterns of Government; The Major Political Systems of Europe.*

Constitutional history. The study of the origins, evolution and historical development of the constitution of a community, considering how various parts of it have been developed and influenced, and by what factors.

See, e.g. W. Stubbs, *Constitutional History of England*; D. L. Keir, *Constitutional History of Modern Britain*; C. B. Swisher, *American Constitutional Development*; A. H. Kelly and W. A. Harrison, *The American Constitution; Its Origins and Development.*

Constitutional law. The body of legal rules determining the essential and fundamental elements of the legal structure of government of a particular political community, their relations, powers and operation. In many countries the scope of constitutional law is emphasized and clarified by the existence of a written constitution, which is frequently embodied largely or entirely in a document called the Constitution, as in the U.S.A. In such countries constitutional law is concerned with the content and interpretation and application of the rules stated in or derived from the basic document. In the United Kingdom, on the other hand, there is no single document called the Constitution, and the principles and rules of constitutional law are contained in a large number of statutes of varying dates, in the *rationes* of a large number of cases, and in an amorphous body of customs, usages, and practices commonly called the conventions of the constitution.

Constitutional law can be said to deal with the structure and the major principles of the functioning of the sovereign power in the State. While many matters, such as the title to the Crown, and the structure and powers of Parliament, would universally, or at least commonly, be accepted as topics of British constitutional law, there can be much more doubt and dispute about other topics, and whether they are included in or excluded from a discussion of constitutional law is largely a matter of choice. Large parts of the civil liberties of individuals in the United Kingdom are not secured or declared by any constitutional document, but have been upheld by the courts at various times in the context of particular controversies and thereafter been generally accepted.

Particularly in a country not having a written constitution, constitutional law shades off into details of the structure of the state, the organization of subordinate bodies and the particular and detailed rules applicable to each, much of which falls under the heads of local government law and of administrative law while many cases establishing points of constitutional law were not great state trials but ordinary actions or prosecutions.

E. C. S. Wade and G. G. Phillips, *Constitutional and Administrative Law*; A. E. Sutherland, *Constitutionalism in America.*

Constitutiones. The generic name for legislative enactments of Roman emperors. These took various forms. The emperors, like all higher magistrates, had the *ius edicendi* and could therefore issue *edicta*. Imperial *edicta* were normally used for legislation of a general character, e.g. Caracalla's edict on citizenship. They might also issue *decreta*, decisions in civil or criminal cases and *rescripta*, imperial decisions on specific points in reply either to petitions (*libelli*) by litigants or to references (*relationes*) by judicial officers. *Rescripta* might be called *epistulae* when in the form of a letter or *subscriptiones* when appended to a *libellus*. Also among *constitutiones* were *mandata*, instructions to officials, such as provincial governors, usually dealing with administrative matters, but sometimes with private law. All forms of *constitutiones* could be legislative, in that even answers to specific cases frequently took the form of general statements and laid down new rules which were accepted as authoritative and binding in all similar cases. By the middle of the second century *constitutiones* were recognized as having the force of law with ratification by the Senate. From the Antonines onwards imperial legislation was an increasing source of law, and from the time of Diocletian the imperial legislative power was unlimited.

The volume of *constitutiones* became such that collections were essential; for these see CODEX GREGORIANUS, C. HERMOGENIANUS and C. THEODOSIANUS.

Constitutiones Clementinae. The collected legal decisions of Pope Clement V promulgated by Pope John XXII in 1317 and a part of the *Corpus Juris Canonici* (q.v.).

Constitutiones Cnuti regis de foresta. A compilation by a high official of about A.D. 1185, designed to make the Norman forest laws less unpopular by falsely attributing them to King Cnut, but though a forgery nevertheless giving some idea of what law was like in the time of Henry II.

F. Liebermann, *Gesetze der Angelsachsen*, I, 620.

Constitutiones dominii Mediolanensis. A collection of the constitutions of the rulers of Lombardy made in 1541 which was the basis of the law of that region until the end of the eighteenth century.

Constitutiones Marchiae Anconitae (or *Collectio Aegidiana*). A revision of the law of the Marches in Italy, made by a commission of jurists appointed by Cardinal Albornoz (q.v.) published in 1357 and authoritative for some two centuries.

Constitutiones principum. Legislation issued by the Roman emperors. They might take the form of *Edicta* or proclamations, which the emperor, like any magistrate, might issue; *mandata*, or instructions to subordinates, particularly provincial governors; *rescripta*, or written answers to officials who had consulted the emperor, particularly on a point of law; and *decreta*, or decisions issued by the emperor when sitting as judge. The volume of *constitutiones* were several times collected particularly in the *Codex Gregorianus*, *Codex Hermogenianus*, *Codex Theodosianus* and *Codex Justinianius* (qq.v.).

Constitutiones Sirmondi. A collection of 16 constitutions almost all concerning ecclesiastical matters, so called from having been published by J. Sirmondus in 1631. The last is dated A.D. 425 and they must have been collected before A.D. 438.

Construction. The process of interpreting a deed, statute, will, or other writing and determining its meaning in the circumstances which have happened and given rise to controversy. The object is to ascertain the true intention of the maker of the writing and to give effect to it. The function is controlled by various guiding principles, some statutory, some common law, such as that the writing is to be construed as a whole, that words are to be given their plain, ordinary meaning unless this leads to absurdity, but may be given a technical meaning if used in a technical context; technical legal terms must have their technical legal import. It is assisted by various presumptions such as that express stipulations oust implied stipulations, that equivocal phrases are to be construed against the interest of the party who used them, and that expression of one thing implies the exclusion of others of the same kind not mentioned. In general oral evidence is not admissible to add to, contradict or vary the terms of a deed or writing. Some additional principles apply to the interpretation of statutes. See ACT OF PARLIAMENT.

E. Beale, *Cardinal Rules of legal Interpretation*; R. F. Norton on *Deeds*; F. V. Hawkins and E. C. Ryder on *Construction of Wills*.

Construction of law. A fixed rule that a particular legal consequence follows from certain acts or words without reference to the intention of the parties. Sometimes it may mean simply an implication.

Constructive. Denotes that an act, statement, or other fact has an effect in law though it may not have had that effect in fact. See CONSTRUCTIVE FRAUD, MALICE, NOTICE, TOTAL LOSS, TREASON and TRUST.

Constructive fraud. Conduct not involving actual intent to defraud but such that a court will not enforce a contract or transaction, as it would be unconscientious to allow a person to take advantage of the situation. This includes transactions between persons between whom there is a fiduciary relationship, and cases where one person has taken unfair advantage of another's inexperience.

Constructive knowledge. Knowledge which a person could have had, and must be taken to have had, whether in fact he had it or not.

Constructive malice. The former rule of English law that if a man caused death in the course of committing another crime such as rape or robbery, he was guilty of murder, the malice or intent requisite for murder being supplied by the intent to commit the prior crime of which the death was the, possibly unintended, consequence. It was abolished by statute in 1957.

Constructive notice. Knowledge which is imputed by law to a party, whether he had it or not, as where he fails to make ordinary and proper inquiries or to investigate some fact which should have put him on his inquiry, or deliberately refrains from some inquiry.

Constructive total loss. In marine insurance the term applied where the subject insured has not been totally lost or damaged but has been so affected by one of the perils insured against that its survival or chance of being recovered or repaired is highly doubtful. In such a case the insured, by giving notice to the insurers, may relinquish all his right as for a total loss. This arises where a ship is so damaged that the cost of repairs would exceed her value if repaired.

Constructive treason. The doctrine that a conspiracy to do some act which might endanger the King's life was an overt act of compassing the King's death and accordingly treason.

Constructive trust. A trust arising by legal construction or interpretation of certain actings of parties. It arises in many cases, notably when a person purchases property subject to a trust or equitable interest with either actual or constructive knowledge thereof, in which case he acquires the property still subject to a constructive trust for the persons beneficially interested therein.

Consuetudines Cantiae (the Customs of Kent). An alleged enactment attributed to the time of Edward I, included under these names in the Statutes of Uncertain Date in the *Statutes of the Realm*. The chief custom of Kent was gavelkind (q.v.).

Consuetudines et Assise de Foresta (the Customs and Assize of the Forest). An alleged enactment attributed to 1278, included under the Statutes of Uncertain Date in the *Statutes of the Realm*.

Consul. In the ancient Roman republic the consuls were the two chief magistrates, elected annually for the year. In the later Middle Ages merchants in the commercial towns of Italy, France, and Spain used to elect one or more of their number to act as arbitrators in commercial disputes, known as *Juges Consuls* or *Consuls Marchands*. When merchants from these countries settled in Eastern countries they carried the institution with them. The powers of consuls became extended by treaties between their home countries and the countries where they traded and came to comprise the whole civil and criminal jurisdiction over and protection of the lives, privileges, and property of their countrymen. The institution spread westwards also but in the west the jurisdictional powers of consuls decayed from the seventeenth century as foreigners were brought under the jurisdiction of the local courts. Their functions came to be limited to protecting the commercial interests of their countrymen. They are now agents of states for purposes mainly concerned with commercial and maritime interests of the appointing state. They may be local residents acting for another state. There are usually consuls in all the major cities. Consuls are not diplomatic representatives and do not enjoy diplomatic privileges, unless also accredited as *chargés d'affaires*. There are generally recognized four classes of consuls, consuls-general, in charge of several consular districts, consuls, vice-consuls, and agents-consular who may be appointed to exercise consular functions in certain towns.

A state is not obliged to admit any consuls, but if it admits consuls of one state it is normally obliged to admit from all states. Consuls are appointed by a *Lettre de provision* of the sending state, which is communicated to the Head of the host state and an *exequatur* granted thereon. Vice-consuls and agents-consular are appointed by the consul.

Their functions are to promote commercial, economic and scientific relations, by helping merchants and reporting to their home state, to supervise navigation, and assist the shipping and seamen of the home state, to assist subjects of the home state, to issue passports, and to exercise numerous registral and notarial functions for nationals of the home state resident in the host state. Consular premises may not be used for purposes incompatible with consular functions.

Though they do not enjoy the privileges of diplomatic agents, consular premises are inviolable and exempt from taxation, and consular officers are personally inviolable, immune from local jurisdiction and from taxation, from customs duties and personal services and contributions.

Consular office terminates on the death, recall or dismissal of the consul, withdrawal of the *exequatur* or war between the appointing and the host state.

Consulado de Burgos. A tribunal of merchants founded at Burgos in Spain in 1494 to promote trade. It was directed by a prior, two consuls and a treasurer, and had jurisdiction, independent of both royal and civil courts, over mercantile cases. It served as model for other *consulados* established at Madrid, Bilbao, and Seville. After overseas trade declined it was in 1679 converted into the Council of Coinage and Commerce.

Consultatio (or *veteris cuiusdam iurisconsulti consultatio*). A work of the fifth or early sixth century A.D., edited by Cujas in 1577, consisting partly of answers by a jurist to an advocate, and partly of theoretical disquisitions. Both quote as authorities Paul's *Sententiae* and constitutions from the *Codices Gregoriani*, *Hermogeniani*, and *Theodosiani*.

Consultation. A meeting of two or more counsel and the solicitor instructing them for discussion and advice. See also CONFERENCE.

Consumer credit. The provision of shorter and medium-term loans to enable individuals to purchase goods and services which they could not, or do not wish to, pay for at once. It includes moneylending, pawnbroking, credits, small loans, hire-purchase, mail order and instalment payment. In the U.K. legal control of such transactions developed in rather haphazard fashion from the later nineteenth century, but was largely rationalized by the Consumer Credit Act, 1974.

Consumer hire. The letting on hire of goods of moderate value for more than very short terms. In the U.K. it is controlled by the Consumer Credit Act, 1974.

Consumers and consumer protection law. Economically consumers are those who buy, obtain, and use all kinds of goods and services (including housing). As social standards rise it is increasingly recognized that they require better and more extensive protection against defective or unsatisfactory goods and services, frequently provided by large and economically much more powerful suppliers. Some legal protection has long been given by, e.g. rules making fraud criminal and tortious, by implied terms in contracts providing for goods of merchantable quality. But the trend has been strongly for greater protection. This has given rise in the U.K. to a distinction in many kinds of contracts, e.g. sale of goods, between consumer contracts, where the buyer is a private person, and commercial contracts, where he is not, the buyer in the former case getting much more extensive protection. In the cases of hiring, loan of money, hire-purchase and related transactions extensive protection is given by the Consumer Credit Act, 1974, and regulations thereunder, all controlled by the Office of Fair Trading (q.v.). Protection is also given by some principles of tort, e.g. imposing on manufacturers a duty of care owed to the ultimate consumer of the product not to cause him injury by defect in the manufacture, and by criminal provisions, e.g. imposing safety standards, or penalizing false and misleading trade descriptions. There has also been a development of local authority consumer protection departments and consumer advice centres, while the Consumer Council and the Consumers' Association draw attention to defects in products and encourage complaints. The whole of this area of law, comprising elements of contract, tort and criminal law is sometimes described as consumer protection law or consumer law.

Consummation of marriage. Full normal sexual intercourse between married persons after their marriage, irrespective of whether conception is effected or is precluded, or is impossible for any reason. Inability to consummate by reason of impotence is a ground of nullity of marriage, and wilful refusal to consummate is also a ground of nullity in England.

Contemporanea expositio. A principle in the construction of documents, particularly old ones, that they should be construed as they would have been when made and that the interpretation placed on them at or about that time should be adopted unless there are strong reasons to the contrary.

Contempt of court. Criminal contempt consists of words or acts obstructing or tending to obstruct the administration of justice, civil or criminal, and is punishable by summary process of committal or by fine or order to give security for good behaviour. It may take place in the face of the court, as by insulting the court, or outside the court, as by speech or writing attacking the court, commenting on pending proceedings, publishing material which may prejudice fair trial, obstructing officers of court the parties or witnesses, or otherwise.

Civil contempt or contempt in procedure is deliberate disobedience to an order of the court, or breach of an undertaking given to the court, and is punishable by committal to prison or writ of sequestration or fine. It is essentially a wrong to the person entitled to the benefit of the order or undertaking.

Each of the Houses of Parliament has power by law and custom to protect its freedom, dignity and authority against insult or violence. Superior courts have inherent power to commit for contempt, but inferior courts have power only to punish contempts committed in face of the court or by disobedience to its lawful orders.

Contempt of Parliament. Both Houses of Parliament have power to punish for contempt similar to that possessed by the superior courts of law. Contempt is not limited to breach of one of the recognized privileges, nor to conduct for which there is precedent, and any act or omission which obstructs or impedes any member or officer of the House in the discharge of his duty, or tends to do so may be treated as contempt. The dividing line between contempt and breach of privilege is narrow but the latter covers properly only infringement of the collective or individual rights or immunities of one of the Houses.

Contentious business. Legal business which is challenged and disputed by another party, in particular the obtaining of probate (q.v.).

Context. The surrounding words, phrases and clauses of a word or phrase in a deed, will or statute, the meaning of which is disputed. The general principle is that a word or phrase must be interpreted in its context and, where possible, given a meaning consistent with the surrounding words, phrases and clauses.

Contiguous zone. A belt of water outside the territorial sea of a state within which the coastal state may take action to protect itself in revenue, police, health and similar matters. It is not accepted by all states, and the modern tendency is to extend the territorial sea to include the contiguous zone. It is usually accepted as extending to 12 miles from the baselines on the coast, but certain states have asserted a wider zone. The U.K. in 1964 established a 12-mile belt for exclusive fisheries.

Continental Congress. The First Continental Congress, called simply 'The Congress', met at Philadelphia 5 September to 26 October 1774, primarily to consider joint action for the restoration of rights lost under the coercive legislation of the British Parliament. It adopted a declaration of rights and a series of resolutions calling for commercial non-intercourse with Great Britain.

The Second Continental Congress met at Philadelphia on 10 May 1775, after hostilities had taken place at Lexington and Concord. It was not a legislative or superior authority but a gathering of delegates from the several colonies giving expression to common feelings and to public opinion. In June it created the Continental Army and for six years acted as provisional government and exercised general direction of the War of Independence.

On 2 July 1776 it resolved to discuss the resolution 'that these United Colonies are, and of right ought to be, free and independent States' and on 4 July it adopted the Declaration of Independence. Thereafter, to give the States a governmental structure, in November 1777, it adopted the Articles of Confederation (q.v.) which in 1781 became the first constitution of the U.S. and it continued to function under that constitution until 1789 when it was replaced by the congress under the Constitution adopted in 1788.

Continental shelf. The shallow part of the seabed contiguous to land areas sloping gently under the sea, where the depth of water is not so great as to prevent exploitation of the natural resources thereof, a depth sometimes stated as 200 metres. The breadth varies from place to place. The whole of the North Sea is part of the European continental shelf. It extends beyond the seabed underlying the territorial sea of a particular state but constitutes a natural extension thereof. At its outer edge the earth falls away more sharply to the deeper floor of the oceans. Since 1945 interests of defence and in fisheries, and oil or minerals thereunder, have prompted countries to assert claims, frequently claims akin to sovereignty, over the natural resources of the seabed and subsoil of the shelf, even beyond their territorial sea, though some countries make claims to the waters and airspace above as well. The seabed is not *res nullius* and incapable of appropriation; it is appurtenant to the adjacent territory but must be claimed by it. The boundary of the continental shelves of different states is the median line between the baselines of their respective territorial seas.

Contingency. An uncertain event on which the existence or enforceability of some claim, right interest or liability depends. Thus money may be payable under an insurance policy on the occurrence of various contingencies, e.g. accident, death.

Contingent remainder. In the English law of real property a remainder is an estate in land limited to take effect in possession on the determination of all the preceding estates, and is contingent if limited so as to depend on an event or condition which may never happen or be performed, or may not happen or be performed until after the determination of the preceding estate.

There was a great deal of obscure and difficult learning about contingent remainders which formed the subject of a celebrated book, *Fearne on Contingent Remainders.*

Continuity of statehood. In international law the problem of whether a state continues to exist notwithstanding change, absorption into another state, conquest or other change. The answer is important in relation to liability for state debts, to treaty obligations and other purposes. It is fairly certain that states continue notwithstanding even drastic change in the internal form of government. During World War II it was accepted that states continued despite German conquest and occupation, and the liberation of many of them restored independence but continuity was considered uninterrupted. But the continuity of Germany and the relations of the German Federal Republic and the German Democratic Republic to pre-1945 Germany pose many difficulties.

Continuous voyage. The principle, in relation to the carriage of contraband (q.v.), that a neutral vessel carrying contraband to a neutral port but intending to carry it from there to an enemy port, must be deemed to be carrying contraband during both stages of her voyage, both being treated as a single continuous voyage.

Contius, Antonius (Le conte) (*c.* 1525–86). French jurist and humanist, an opponent of Duarenus and Hotman, an editor of the *Corpus Juris* and author of *Notae in libros Institutionum* and *Disputationes juris civilis.*

Contra proferentem. In interpretation of a document, the principle that an ambiguity should be interpreted in the sense unfavourable to the party who drafted and put forward the document.

Contraband of war. The origins of contraband (*contra bandum*, in defiance of injunction) may be found in the repeated promulgation by mediaeval Church councils of canons forbidding Christians to export arms to Saracens. From this developed the attitude of belligerents claiming that even if they did not interfere with trade between neutral and enemy, they would do so in cases of warlike material and stores. Sometimes they conceded that they would not seize but buy such goods. It accordingly now means such goods as each belligerent forbids to be carried to the enemy, as enabling him to wage war more vigorously. Grotius sought to distinguish between goods with warlike, doubtful and peaceful uses. Commercial treaties of the eighteenth century frequently made provisions as to contraband. The practice developed considerably during the Napoleonic wars and Lord Stowell developed the concept of conditional contraband, i.e. goods liable to condemnation only if actually destined for military use. As industry became more important to waging war the category of goods deemed contraband grew larger.

The London Naval Conference of 1908 produced three agreed lists of goods, military goods which were absolute contraband, liable to confiscation, conditional contraband, confiscable only if bound directly for enemy territory and destined for purposes of war, and non-contraband articles, but the declaration was not ratified. In the First World War the British added very largely to the lists of absolute and conditional contraband and in 1916 abolished the distinction. In modern total warfare the distinction is almost useless.

Whatever their nature goods are contraband only if destined for the use of a belligerent in warfare, not if destined to a neutral. A neutral vessel may herself be contraband if fitted for use in war.

Belligerents may prohibit and punish the carriage of contraband by neutral ships, though neutral states are not bound to prevent the manufacture, sale or carriage of such goods. This applies both where the ship is carrying to an enemy port, or is ostensibly bound for a neutral port but intending that the goods be carried on from there to an enemy port. By the doctrine of continuous voyage, the whole in the latter case must be treated as an indivisible voyage. Similarly, by the doctrine of continuous transportation goods are deemed bound for the enemy if being carried to a neutral port and to be forwarded by land or sea to the enemy. A belligerent may seize a vessel carrying contraband and confiscate the contraband and the vessel herself if she sailed under false papers or the goods were a substantial part of the whole cargo.

Contraception. The use of any means of preventing sexual intercourse from resulting in conception. The practice raises difficult theological, moral, social, and psychological problems. In law the practice of contraception by any means does not render a marriage voidable, nor is it a ground of divorce unless it gives rise to marital disharmony causing breakdown of the marriage.

Contract, law of. A contract is an agreement between two or more persons intended to create a legal obligation between them and to be legally enforceable. The law of contract is accordingly the body of principles relative to those kinds of agreements which are intended to, and do, create binding and enforceable bilateral obligations, conferring rights *in personam* on each party and which are therefore legally contracts. It exists accordingly to regulate all kinds of business and economic relationships, sale, hiring, employment, construction and many more. The law of contract comprises both principles, mainly derived by induction from many particular cases, which are applicable generally to all or most kinds of contracts, such as about capacity to contract, consideration, the effect of fraud or mistake, and the like, and other principles and rules which are peculiar to particular kinds of contracts, such as to sale or hire-purchase.

Law of contract is a product of business civilizations and a market economy. Roman law recognized various types of contracts and agreements, not all being legally enforceable. Contract law developed as a result of the political, economic, and commercial revival after the Dark Ages. A major factor was the development of the practices of merchants and traders, and of merchants' courts whereby their agreements could be promptly enforced. In the twelfth and thirteenth centuries the development of contract law began to proceed on different lines in England and on the Continent. In England two forms of action developed which gave effect to contractual liabilities, debt, which lay for a liability to pay a fixed sum of money, and covenant, for breach of a promise under seal. Later the action of assumpsit evolved for the enforcement of undertakings. On the Continent the revived study of Roman law encouraged the enforcement of formal bargains while the canon law encouraged the view that informal bargains should be binding. Later, by the eighteenth century, it was generally accepted that obligations freely and voluntarily undertaken were binding and enforceable, unless special considerations, such as public policy, prevented this.

In modern times there has been increased recognition of inequalities of knowledge, understanding, and bargaining power and consequently of the need to protect persons in a weaker position. Similarly, it is recognized that in many cases parties have little option but to take the terms offered them, and in such cases also there is need for society's legal system to intervene to ensure at least a reasonable fairness between the parties by regulating the standard terms for which suppliers of goods and services are willing to do business.

The detailed rules of different systems vary substantially on many matters, determined partly by history and partly by different approaches to questions of policy. Thus Anglo-American law in general requires that there be consideration, i.e. that one party's promise to perform is being paid for by the other, whereas civil law systems in general hold a contract enforceable even without consideration if an appropriate formality has been complied with.

In modern times business and trade are increasingly international and attempts have been made to limit the effect of divergencies in the contract law of different legal systems. In the U.S. the Uniform Commercial Code applies in nearly every State. Internationally conventions have been negotiated on many branches of transport law regulating international carriage.

The essentials of a valid contract are that both or all parties must have at the time of contracting had contractual capacity and legal power or authority to make the kind of contract contemplated, must have reached agreement on all the material terms of their bargain, normally by communication of offer and acceptance thereof, must have intended a legally enforceable, as distinct from a merely social or extra-legal agreement, and the agreement must not be objectionable as intended to effect an impossibility or anything illegal or contrary to public policy or otherwise unenforceable. In certain cases contracts require certain formalities of constitution or proof, without which they are unenforceable or even wholly void. Thus English law distinguishes simple contracts and contracts under seal, simple contracts requiring consideration (q.v.) for their validity. Contracts may be vitiated by mistake, misrepresentation, fraud, duress, undue influence, or similar factors, factors which in some cases render an apparent contract wholly void, and in other cases leave it voidable, or challengeable, by the party imposed upon.

A contract defective in some respect may be wholly void, in which case it has no legal effect at all, or be voidable, in which case it is valid and effective but contains a defect entitling one party to challenge it and have it set aside.

The mutual undertakings of the parties are contained in the terms, conditions, and other stipulations of the contract, relative to such elements as quantity, quality, delivery, price, and the like. In many cases statute or custom imply certain stipulations into contracts of particular kinds, e.g. of sale of goods or carriage by sea.

In general a contract is enforceable by and against only the parties thereto and not third parties, but exceptions exist, as where one party has contracted as agent, or where rights or liabilities have been assigned by a party or by operation of law. Some contractual rights and duties are not assignable.

Breach of contract is the failure by either party to implement any of the duties incumbent on him by the contract and gives rise to a claim of damages in compensation for the breach, or if in a fundamental respect, to a right to treat the contract as at an end and to recover damages for its total failure.

A contract is discharged and terminated by agreement, by due performance or payment, by waiver of performance, by replacement by a new contract, by supervening impossibility or illegality of performance, by frustration through the happening of some event which renders performance, if not impossible, at least fundamentally different from what was contemplated and by breach in a fundamental respect, if the innocent party treats the breach as a repudiation of the contract.

Many specific kinds of contracts, such as sale or employment, are regulated in certain respects by particular statutes.

W. R. Anson on *Contract*; F. Pollock on *Contract*; G. C. Cheshire and C. H. S. Fifoot on *Contract*; J. Chitty on *Contract* (English law); W. M. Gloag on *Contract*; D. M. Walker on *Contracts* (Scots law); A. L. Corbin on *Contracts*; S. Williston, *Treatise on the Law of Contracts* (U.S.).

Contract, inducement of breach of. The wrong of unjustifiably inducing a breach of a lawful contract between two other persons. Exceptionally inducing such a breach has been held justifiable. Trade unions and their officers and members are by statute protected against liability for inducing such a breach if acting in contemplation or furtherance of a trade dispute.

Contract labour. A system, distinct from slavery, in which an individual surrendered his freedom to leave his master or his work, usually for a period. It was sometimes involuntary, as in the cases of debtors or criminals, or voluntary, as in the case of indentured servants who gained some advantage in return, such as transport to a new land. North American indentured labour began with the foundation of the colonies and continued along with free labour and slavery until the American Revolution. The persons bound were mainly from Britain or Western Europe, and the terms might be harsh or light, the harshest terms being imposed on criminals whose sentences were commuted on condition of entry into indenture. The system operated also in other parts of the world, particularly from poverty-stricken parts of Asia, and has survived in some parts of the world to the present time.

Contract of record. A transaction which has been entered in the records of a court of record and is accordingly conclusive proof of the facts appearing thereon and which could formerly be enforced by legal action in the same way as if contained in a contract. These transactions included judgments, recognizances, and statutes staple.

Contracting out. The possibility by contract of renouncing certain statutory benefits. Statutes increasingly prohibit the practice of contracting out from the provisions of the statute, as such nullifies the benefit the statute is intended to confer. The phrase is also used of the decision of a member of a trade union not to contribute to the union's political fund.

Contractor. In a general sense any party who engages by contract, but particularly and usually a person, frequently called an independent contractor, who contracts to do work and perform services for another but in so doing remains independent, does not become a servant or employee of the person for whom he does work, and is not subject to detailed control and direction in how he does the work required. Contractors accordingly include persons offering professional services, e.g. architects, lawyers, and persons or firms undertaking to bring about desired results, e.g. building, repairing, installing. A contractor is paid by a fee, an employee by salary or wage. The consequence of the distinction is that, in general, an employee is not liable to a third party injured by the fault or negligence of an independent contractor but he is liable for the fault or negligence of his own employee.

Contravention. An act done in violation of a condition or rule, particularly in older Scots law, an act by an heir of entail in breach of one of the provisions of the entail.

In French law a *contravention* is a minor offence.

Contribution. Payment by one of several persons liable of his share of what is due. Thus in general average (q.v.) where loss has been sustained by one, each of the other interests involved must pay his share according to his interest in the property preserved. Among wrongdoers, each of several wrongdoers must contribute to the damages awarded to the injured person.

The term is also used of the compulsory payments to social security funds and, in international law, to a money payment exacted by an occupying army from the inhabitants of the place occupied.

Contributory. A person who is bound to contribute to the assets of a company when it is being wound up in order to pay its debts. There are two categories, the A list, comprising members at the commencement of the winding-up, and the B list, comprising persons who had been members within one year prior to that date. The B list is required to contribute only if the existing members cannot satisfy the contributions required of them.

Contributory negligence. Contributory negligence is carelessness by a plaintiff which has contributed to and is in whole or in part the cause of the injury or harm he complains of, as having been caused him by the defendant's fault. While negligence (q.v.) implies the existence of a legal duty to take precautions against harm contributory negligence does not do so; it means carelessness or lack of reasonable care for himself which is a contributory or partial cause of the injury or harm suffered by the plaintiff.

At common law proof that the plaintiff had himself to any extent at all contributed to the injury complained of by some negligence on his own part completely exculpated the defendant. Since the Law Reform (Contributory Negligence) Act, 1945, if the plaintiff is partly in fault his claim is not defeated but the damages recoverable are to be reduced to such extent as the court or jury thinks just and equitable having regard to the claimant's share in the responsibility for the damage. Responsibility is determined not only by the causative potency of his acts but also by their blameworthiness.

Conduct which would constitute contributory negligence by an adult may not do so in the case of a child, and not every careless or inadvertent act by an employee will amount to contributory negligence. Nor is an instinctive reaction or a voluntary action to save another necessarily contributory negligence.

Conventicle Acts (1664 and 1670). Parts of the Clarendon Code (q.v.) aimed at suppressing meetings of dissenters. The first imposed penalties for attendance at non-Anglican services. The second imposed penalties for allowing conventicles to be held in persons' homes and was followed by an intensive campaign against dissenters.

Convention. (1) An international meeting to discuss some matter of common interest.

(2) An agreement, particularly one between states, such as the Hague Conventions and Geneva Conventions. Many conventions are given legal force by statute.

(3) In constitutional law, one of the practices which are recognized as obligatory though not laid down by statute or case law, such as the rules that the Queen's Ministers are drawn from the party having the largest number of seats in the House of Commons, that a Prime Minister defeated on a motion of confidence or one which he indicates that he regards as such will resign, and that the Queen must accept the advice of the Ministry for the time being.

(4) The name given to the French assembly elected in 1792 after the collapse of the monarchy and the 1791 constitution, which proclaimed a republic, executed the King and established committees to rule France. It remained the final authority until it established the Directory and dissolved itself in 1795.

Convention Parliament. Parliaments which met initially without the summons of the sovereign and were later converted into Parliaments. The two Convention Parliaments have been those which restored Charles II to the Throne in 1660 and which offered the Crown to Prince William and Princess Mary of Orange in 1689 after the abdication of James II. In Scotland a convention Parliament in 1689 declared that James VII (and II) had forfeited the Scottish Crown and offered it to William and Mary.

Convention of Royal Burghs of Scotland. A very ancient Scottish institution which has developed from the twelfth century Court of the Four Burghs (of Edinburgh, Stirling, Berwick, and Roxburgh). It later consisted of delegates appointed by all the burghs in Scotland who met annually to discuss matters of common concern. In 1975 it transformed itself into the Convention of Scottish Local Authorities, a body comprising representatives of the regional, district and islands authorities, meeting to consider matters of common concern, but with no executive powers.

T. Pagan, *The Convention of Royal Burghs of Scotland.*

Convention of Scottish local authorities. See CONVENTION OF ROYAL BURGHS OF SCOTLAND.

Conventional law. Rules, having legal effect and recognized by the courts, which have their origin in the agreement of those subjects thereto to be bound thereby. Thus persons joining a club, trade union or partnership, or becoming members of a company are bound by the rules made by the body, or the governing body thereof. The general law of the land permits the making of such rules, and will recognize and enforce them if validly made and not contrary to any rule of the general law of the land.

Conversion. (1) A wrongful act of dealing with goods in a manner inconsistent with the rights of the person entitled thereto, with the intention of denying his title or asserting a right inconsistent with it. It may be constituted by wrongful sale and delivery of goods, wrongful refusal to return, wrongful taking, misdelivery or wrongful receipt of them. The plaintiff must have had a right of property or possession in the goods and not have voluntarily relinquished or abandoned his rights. Conversion was originally remedied by the action of detinue (q.v.) and later by trover, now called the action for conversion.

(2) The changing of land into money or money into land. This was important in connection with the devolution of property on the owner's death intestate before 1926, when realty and personalty devolved on different persons. In equity money directed by a will to be employed in the purchase of land, and land directed to be sold, were deemed already to be the species of property into which they were directed to be converted. Conversion might also be operated by contract or by statute. By reconversion a previous notional conversion is cancelled and the property restored in the contemplation of equity to its original actual quality.

Conveyance. A mode by which property is transferred from one person to another by written instruments and related formalities, and also such an instrument itself, such as a lease, mortgage, or vesting instrument.

Conveyancer. A barrister or solicitor mainly concerned with the practice of conveyancing. It was formerly competent in England for a member of an Inn of Court to practise as a conveyancer 'below the Bar', i.e. without having been called to the Bar.

Conveyancing. The science and art of validly creating, transferring, and extinguishing rights in property, particularly in or over land, by written deeds of various kinds. It is accordingly a major branch of legal work and lawyers' business. It is based on the knowledge of what rights can exist in or over particular kinds of property, of what ends can be secured within the existing rules of law, and of what machinery, such as vesting in trustees, can appropriately be employed to achieve particular ends. It includes investigation of title and the preparation of agreements, wills, private Acts and other instruments which operate as conveyances. Conveyancing depends to a large extent on practice, on the professional practices, customs, and usages which experience and prudence have dictated as necessary or appropriate in particular circumstances, and on precedents or styles, forms of words which have been found appropriate in the past, or held by the courts to be capable of securing particular desired ends.

Conveyancing counsel. Three to six barristers appointed by the Lord Chancellor to assist the Court of Chancery. The opinion of any of them may be received and acted on by the court or a judge in chambers in the investigation of the title to an estate or the settlement of a draft conveyance.

Convicium. In Scots law, a branch of the wrong of verbal injury (q.v.), which consists in communicating hurtful ideas to or about a person and thereby bringing him into hatred, contempt or ridicule. The idea communicated need not be untrue (*veritas convicii non excusat*), nor properly defamatory (q.v.), and may be merely derogatory, e.g. imputation of meanness or impotence, but is nevertheless actionable.

Convict. A person who has been convicted of any criminal offence by a criminal court, but particularly one convicted of an indictable offence and sent to prison. At common law a person convicted of treason or felony became civilly dead, suffered corruption of blood so that he could neither inherit nor transmit property and his estate was forfeited to the Crown. These consequences have been abolished.

Conviction. A finding by a criminal court, on an accused person's confession and plea of guilty, or after trial, that he is guilty of the crime charged. An ordinary conviction takes place in a prosecution on indictment, a summary conviction in a magistrates' court.

Convocation. Representative assemblies of the clergy of the provinces of Canterbury and of York of the Church of England, summoned by the respective archbishops, pursuant to royal writs issued formerly at the same time as writs for Parliament. In each Convocation there are Upper and Lower Houses, the Upper consisting of the archbishop and diocesan bishops, the Lower consisting of the deans of cathedral churches, certain archdeacons, proctors for each diocese elected by the clergy thereof, and certain university representatives. There may be a House of Laity. Convocations are presided over by the archbishops, the president of each House being the prolocutor. Convocations are dissolved by royal writ to the archbishop, or after the expiry of five years, or formerly by the dissolution of Parliament. In 1970 the legislative functions, authority, rights, and privileges of Convocations were transferred to the new General Synod of the Church of England (q.v.) and this greatly diminished the character and importance of the convocations. The Upper and Lower Houses of the two convocations form the Houses of Bishops and of Clergy of the General Synod. Each convocation may, however, still meet separately to consider matters concerning the church in its province.

Convoy. A group of vessels sailing together with an armed escort. The practice developed as a protection against pirates. Since the seventeenth century neutral states have claimed immunity from search for their vessels sailing in convoy under escort. In the Armed Neutralities of 1780 and 1800 Russia sought, recognized, and formalized the right of neutral convoy. During World War II the right was not exercised, neutral ships obtaining navicerts issued by a belligerent; these were of the nature of passes for approved cargo and exempted from visit and search.

Cook, Walter Wheeler (1873–1943). Studied at Columbia and in Europe, and then returned to Columbia to teach mathematics. He then turned to law and taught it at various law schools, particularly Columbia and Yale. At Johns Hopkins he established an Institute of Law as a research school devoted to the objective study of law as a social institution. He wrote extensively, notably *The Logical and Legal Bases of the Conflict of Laws.*

Cooke, John (1666–1710). Became an advocate of Doctors' Commons in 1694, was knighted in 1701, was a commissioner to negotiate the Union with Scotland in 1701 and became Dean of the Arches in 1703 and Advocate-General to William III.

Cooley, Thomas McIntyre (1824–98). Early established a reputation for scholarship and compiled the statutes of Michigan and became reporter of the decisions of the State Supreme Court. Later he was professor of American Law (1859–84) and of American History (1884–98) at Michigan, a Justice of the State Supreme Court (1864–85) and first Chairman of the Interstate Commerce Commission (1887–91). His writings, notably *A Treatise on Constitutional Limitations* (1868) and *General Principles of Constitutional Law in the United States* (1880) are classics, and he ranks with Story among the foremost commentators on the Constitution. He also edited an edition of Blackstone with notes indicating differences between American and English law, an edition of Story's *Commentaries on the Constitution* (1876) and wrote a *Treatise on the Law of Torts* (1879), long deemed authoritative, and other books.

Cooley* v. *Wardens of Port of Philadelphia (1852), 12 Howard 299. A Pennsylvania Statute provided for the regulation of pilotage in the port of Philadelphia, and raised the question whether Congress's power to regulate commerce was exclusive or could be shared with the States or regulated by the States until such time as Congress legislated.

The Supreme Court distinguished between commerce which, though interstate or foreign, was essentially local and subject to local regulation, and commerce which was national and required regulation by a national rule, laid down by Congress, and that in the circumstances the State law was valid.

Cooper, Anthony Ashley Cooper, 1st Earl of Shaftesbury (1621–83). Though a member of Lincoln's Inn, he never practised but devoted himself to politics, at first on the King's, and then on the Parliament's, side. He actively promoted the Restoration and became Chancellor of the Exchequer in 1660. He opposed Clarendon, and became Lord Chancellor in 1672 but discovery of Charles II's intentions made him the leader of the opposition and the first great demagogue and party leader in

English history. The Seal was taken from him in 1673. The rest of his life was spent in opposition.

As an equity judge he was unimportant but brushed aside technicalities to do speedy justice and might have effected useful reforms in Chancery procedure.

Cooper, Charles Purton (1793–1873). Called to the Bar in 1816, became a leading Chancery practitioner, and a leader in the movement for the reform of the Court of Chancery. In 1828 he published a book containing his criticisms and proposals for reform.

He was also secretary of the second Record Commission and widely known as a scholar in matters of English legal history. He was also responsible for some volumes of reports and many pamphlets and lesser works.

Cooper, Thomas Mackay, Baron, of Culross of Dunnet (1892–1955). Was called to the Scottish Bar in 1915, took silk in 1927 and entered Parliament in 1935. He became Solicitor-General for Scotland in 1935, Lord Advocate in the same year, and in 1941 Lord Justice-Clerk of Scotland. In 1947 he was promoted Lord President.

As a judge he was very quick and acute, and his judgments clear and well-founded on principle; he was certainly the outstanding judge of his time. A man of wide learning and interests, he had a deep love of Scottish law and history, and was Chairman of Council of both the Scottish History Society and the Stair Society, and did much to encourage these studies. He received a peerage in 1954. He wrote *The Scottish Legal Tradition* and edited Skene's *Regiam Majestatem* for the Stair Society. A posthumous volume of *Selected Papers* was published in 1957.

Co-operative. An organization owned by and run for the benefit of those using its services. Co-operatives operate in many fields notably marketing of farm products, purchasing of equipment and raw materials, and retail supply of goods and services. Consumer or retail co-operatives originated with the Rochdale Society of Equitable Pioneers in 1844, on the principle of open membership, sales at market prices and the return of some earnings to members by way of dividend. The movement developed widely in the nineteenth century, particularly in industrial towns and among the working class and spread to most northern European countries and to the U.S. A modern co-operative society is an industrial and provident society (q.v.) and regulated by the law applicable thereto.

Co-ownership. Two forms of co-ownership of property exist, joint property or joint tenancy, where two or more persons jointly hold the same title for the same duration and the seisin or possession is vested in all; and common property or tenancy in common, where each co-owner has his own title and interest to a share in the property though it is held undivided. Another form, ownership in coparcenery, was formerly possible, where two or more persons took hereditaments by the same title by descent.

Coparcener (or parcener or tenants in coparcenery). One who, prior to 1926, was one of several inheriting land on intestacy as co-heirs, e.g. several daughters. Since 1925 the legal estate in such a case vests in trustees for sale, coparceners retaining their rights as equitable interests.

Copilacion de Canellas (or *de Huesca*). A collection of Spanish laws prepared by Bishop Vidal de Canellas and promulgated by James of Aragon at Huesca in 1245.

Copley, John Singleton, 1st Lord Lyndhurst (1772–1863). Son of a portrait painter and born in Boston, U.S.A., Copley studied in Tidd's chambers, became a special pleader, and was called to the Bar in 1804. He entered Parliament in 1818 and became Solicitor-General the next year. As such he prosecuted the Cato Street conspirators. In 1824 he became Attorney-General and in 1826 Master of the Rolls only to become Lord Chancellor in the next year. He remained Chancellor till 1830 when, on the change of government, he became Chief Baron of the Exchequer. In 1835 he again became Chancellor and held the office again from 1841 to 1846. In the House of Lords he opposed the successive Reform Bills but was not a diehard opponent of reform and supported many law-reforming measures. As a judge he won high regard, his judicial appointments were good and his use of patronage discriminating and fair. After 1846 he sat occasionally to hear appeals and retained his abilities to the end. He had a great memory and a very clear and logical intellect but his influence on the judicial development of the law was less than his influence as legislator.

T. Martin, *Life of Lord Lyndhurst.*

Copyhold. A tenure of land in England, abolished after 1925. It originated in the system of manors, whereby tracts of land were assigned to lords and occupied and worked as economic units. When the feudal system of landholding was introduced after the Conquest the lord of a manor was treated as a tenant-in-chief and the freemen who held land under him became his free tenants. Occupiers of land who were not free ranked as villeins but were frequently allotted some land to occupy and cultivate on certain terms, originally held entirely at the will and pleasure of the lord, but subsequently on the terms customary in that manor as recorded in the records of the manorial customary court, tenants having copies thereof as evidence of their holding and its terms. Copyhold tenants owed suit of court to the customary court of the manor. Copyhold tenure accordingly developed as the tenure by which unfreemen, liable to render services uncertain in nature to their lord, held land. Copyholds passed from one tenant to another by surrender and admittance enrolled in the rolls of the memorial court, and devolved according to the custom of the manor. After 1925 all copyhold land was enfranchised and became freehold land, all customary modes of descent and customs being abolished.

P. Vinogradoff, *Villeinage in England; Growth of the Manor.*

Copyright. The group of rights relative to original literary, dramatic, musical, or artistic work, whether published or unpublished, vested normally in the creator thereof, giving him monopoly rights in the reproduction, multiplication, performance, and other exploitation of the verbal or other expression of his ideas, though not in the ideas themselves. Copyright was a by-product of the Renaissance and the development of printing, as rulers made monopoly grants of the right to publish. In 1556 the Stationers Company was chartered and given a monopoly right in the books published. All books had to be submitted for approval and enrolled in their register. In 1710 a statute recognized the right of the author to protection but gave him protection for 28 years only. Litigation later established that this statutory right superseded any common law right in published works, so that once copyright expired published work fell into the public domain. In 1790 the U.S. Congress enacted a federal copyright law closely modelled on the British Act of 1710. While it lasts copyright is a proprietary right.

In the United Kingdom it now exists by virtue of statute only. The various rights comprised in copyright are acquired without registration, by mere creation of the work, are choses in action, capable of assignment, and transmit on death. The owner of copyright may licence another to do what would otherwise be an infringement. Infringement of copyright, by unauthorized publication or performance, is actionable for damages or an account of profits, and in certain cases is also an offence. In general copyright subsists for the author's life plus 50 years, but indefinitely if the work was unpublished at his death. There are special provisions for works deposited in libraries, copyright in broadcast and televised materials and films.

Certain libraries have the privilege of receiving gratis from the publisher a copy of every book published in the United Kingdom. The Crown has copyright in statutes, Orders in Council and other state documents, and in the Bible and Book of Common Prayer, but in the last cases customarily licenses commercial publishers to publish editions thereof.

Copyright, originally confined to literary work, has gradually been extended to dramatic, musical and artistic works (including works of architecture) and, as technology has developed, to sound recordings, cinematograph films, and sound and television broadcasts. The acts restricted by the existence of copyright vary somewhat in each case, and an owner of copyright in fact possesses a variety of rights, e.g. to publish, to translate, to change into a play, which can be dealt with separately.

In the U.S. statutory copyright protects an author's right of publishing, adapting, performing, and recording. Unpublished work is protected by the common law of the various states. To secure statutory copyright there must be publication with the copyright notice and registration in the Copyright Office. The protection lasts for 28 years and is renewable for a second 28-year term.

Since the (Berne) International Convention of 1886 any work first published in a convention country enjoys copyright protection in all states parties to the convention. The Convention has subsequently been revised several times. A number of Pan-American copyright conventions were also developed. In 1952 a Universal Copyright Convention was formulated at Geneva under the auspices of UNESCO.

Under it published works by authors who are citizens of member countries are protected in the countries of all parties to the Convention, wherever first publication has taken place, provided all copies bear the symbol © the name of the copyright proprietor and the year of first publication.

W. Copinger and F. E. Skone James on *Copyright* (U.K.); T. Howell on *Copyright Law* (U.S.).

Coquille, Guy (Conchylius) (1523–1603). French jurist and administrator, author of *Les Coutoumes du pays et duché de Nivernais* (1605), *Institution au droit français* (1607), *Questions et responses sur les Coustumes de France* (1611), and other works.

Coram rege (in presence of the King). Under Henry II the *Curia regis* (q.v.) developed offshoots, namely the officials who sat as the Exchequer and from 1178 five persons were appointed to hear all the complaints of the kingdom. From 1234 there were two distinct courts with their separate rolls, one held before the justices of the Bench at Westminster, the *de banco* rolls, the other following the King, its records being the *coram rege* rolls. The court *coram rege* was from the start superior to the bench for it could correct errors made there, and at least one aspect of it came to be styled the King's Bench. The early plea rolls were accordingly classified as *Curia regis* or *coram rege* rolls and assize rolls, while *coram rege* and *de banco* rolls separated about 1234.

Coras, Jean de (Corasius) (1515–72). French humanist, killed in the St. Bartholomew massacre, author of *Variae interpretationes* (1546), *Enarrationes in Codicem et Digesta, Commentarii*, and *Miscellaneorum juris civilis libris sex* (1549).

Corbin, Arthur Linton (1874–1957). A distinguished American law teacher, professor at Yale. He was the author of a massive *Treatise on the Rules of Contract Law*, reporter on the *Restatement of Contracts*, and a most stimulating teacher.

Coren (or Cooren), Jacob (?1570–1631). Dutch jurist, author of *Observationes XLI rerum in suprema senatu Hollandiae, Zelandiae et Frisiae judicaturum* (1633), an important source of Roman-Dutch law, and *Observatien van den Hoogen Raad.*

Co-respondent. Prior to 1969 when a husband sought to divorce his wife on the ground of her adultery he normally had to cite the other man as a co-respondent. The co-respondent might be held liable in damages to the husband.

Corn-laws. A body of laws governing the import and export of grain, important in English history particularly in the early nineteenth century. The French war raised the price of grain after 1791, there was crisis in 1799–1801 and bad harvests in 1805–13. An attempt was made to control prices after 1815 and prices fluctuated, protection being unpopular in consequence. From 1839 the Anti-Corn Law League led by Cobden sought to unite the opposition of the middle classes and Peel reached the conclusion, particularly after the failure of the Irish potato crop in 1845, that protection must go, and in 1846 the Corn Laws were repealed. They forbade the export of corn when the home price was high but encouraged it by a bounty when the price was low.

Cornage. A tenure by which land might be held, the duty being to blow a horn to given warning that Scots or other enemies were coming to invade England. It was later paid in money. It was a variety of grand serjeanty or knight service. It has also been explained as being originally a payment of so much per horn or head of the beasts which the tenant put out on the common pasture. It was abolished in 1660.

Cornwall, Duchy of. The Duchy of Cornwall is a royal duchy created in 1337 vested by inheritance in the eldest son of the sovereign of the U.K. or on his death not survived by a son, in the next surviving son; this happened in the cases of Henry VIII, Edward VI, and Charles I. But if the eldest son dies survived by a son, the Duchy reverts to the Crown. If the sovereign has only daughters, the eldest daughter, though heiress presumptive to the throne, does not become Duchess of Cornwall. The Duke usually appoints a Lord Warden of the Stannaries in Cornwall and Devon, the Keeper of the Privy Seal, Attorney General, Receiver-General, and certain other offices.

See also STANNARIES.

Cornwallis Code (1793). The body of enactments by which Lord Cornwallis, Governor-General of India, shaped the administrative framework of British India. Originating in Bengal the system spread over all northern India and was the basis of the government of British India until 1833. The system divided the East India Company's service into three branches, revenue, commercial, and judicial. In the third branch there were appointed district judges responsible in civil cases to provincial courts and in criminal cases to circuit judges. The law administered was Hindu and Muslim personal law and a modified Muslim criminal law. The system gave social and political stability to Bengal but neglected the rights of lesser landholders and agricultural tenants and excluded Indians from any place in the administration.

Corody or corrody. An allowance of money, accommodation, food, clothing or other necessaries allowed by a house of religion to a servant of the Crown at the King's request, which the King was articled to demand only from houses of which he was founder by rights of the Crown. It differed from a pension in that a pension was given to a royal chaplain for his maintenance until he received a benefice.

Coronation. The public ceremony at which a king or queen regnant of the United Kingdom is crowned and, *inter alia*, takes the coronation oath to observe the laws, customs and privileges of the kingdom and to maintain the Protestant reformed religion as prescribed by the Bill of Rights 1688, modified in 1910 and 1937. It is normally held about a year after the accession. In the case of Edward VIII it was never held. Coronation is not a requisite of performing the royal functions. In connection with a coronation various claims to perform certain services have grown up and before each coronation a Court of Claims (q.v.) is constituted which investigates and adjudicates on the claims which have been made. The forms and ceremonies observed have varied at different coronations but in 1953 they comprised presentation of the sovereign to the people, taking of the coronation oath, presentation to the sovereign of the Holy Bible, anointing by the archbishop, investiture with the *colobium sindonis*, the Supertunica or close pell of cloth of gold with girdle of cloth of gold, presentation to the sovereign of the spurs and award and an oblation of the sword by the sovereign, investiture with the armills, the stole royal and robe royal, and delivery to the sovereign of the orb, investiture with the ring, glove, sceptre with the cross and rod with the dove, putting on of the Crown, benediction, enthroning, fealty by the clergy and homage by the peers, oblation by the sovereign of a pall or altar cloth and an ingot of gold, and the celebration of Holy Communion.

W. Wickham Legg, *English Coronation Records*.

Coronation cases. Edward VII's coronation was fixed for 26 June 1902 but immediately before that date he contracted appendicitis and had to be operated on and the coronation had to be postponed. A number of cases resulted from the postponement, dealing with the hire of rooms to view the procession, to see the review of the Royal Navy and similar issues, elucidating several points of contract law.

Coroner. The holder of a very ancient English office. The office was first referred to as *custos placitorum* in 1194, but may have existed earlier. The coroner was originally elected by the freeholders of the county and responsible for safeguarding the King's property; he also served as a check on the sheriff. The office is held *ex officio* by all the judges of the High Court, who are coroners for the whole of England, but most coroners are appointed by local authorities. There were formerly franchise coroners, and there are still a Queen's Coroner and Attorney, an office long combined with that of Master of the Crown Office, and the Coroner of the Queen's Household. The person appointed must be a barrister, solicitor, or legally qualified medical practitioner of at least five years' standing. The coroner's duties are to inquire into deaths, particularly by holding an inquest or having a post-mortem examination made, to hold inquests on treasure trove (q.v.) and, on occasion, to act in place of the sheriff. Inquests are necessary where the deceased died violently, unnaturally or unexpectedly, in prison, or the cause of death is unknown, or where an inquest is required by statute. They are held in the coroner's court and in cases prescribed by statute a jury may be summoned, and the coroner or jury may return a verdict and certify it by an inquisition in writing setting out the findings. If the verdict is that the death was due to murder, manslaughter, or infanticide the inquisition formerly stated the offence and named the persons found to be guilty of the crime but this is not now competent. The coroner of the Queen's household is appointed by the Lord Steward and has jurisdiction in relation to deaths within the limits of any palace or house in which the Queen actually is. The Queen's Coroner and Attorney is appointed by the Lord Chancellor, is *ex officio* a master (q.v.) of the Supreme Court and for long the office has been held by a master. He is head of the Crown Office and Associates' Department of the Central Office of the Supreme Court.

The office also exists in the U.S., where it is frequently an elective one. The functions are generally similar to those of the English coroner.

Coroner's court. In England a coroner's court is a common law court having a special jurisdiction. Its function is to inquire into the cause of death of any person whose body is lying within the coroner's jurisdiction, if there is reasonable cause to think that the person died a violent or unnatural death or from unknown causes, in prison or in such circumstances as require an inquest pursuant to any statute. The coroner's inquisition might formally charge a person with murder, manslaughter, or infanticide. The court may sit with a jury. Formerly the coroner's jury functioned like a grand jury (q.v.) in that it did not try but named a suspect.

Corporal punishment. Punishment of persons, particularly criminals and wrongdoers by inflicting bodily pain, as by flogging, branding, or mutilation. It was formerly frequently used in cases not thought to justify the death penalty. Since the late eighteenth century its use has declined greatly, surviving longest as a disciplinary measure for those in prison for breach of prison discipline. Considerable controversy has attached to its use by parents or school teachers.

Corporate personality. A corporate body or corporation is treated by the law as an entity capable of bearing rights and of being subject to duties, just as if it were a natural person, and its personality is quite distinct from those of all the persons who are at any time members of the incorporation. Such is accordingly frequently called a juristic person. But while it is accepted that personality attaches to an incorporation controversy has long raged as to the nature of the legal entity and the justification for its having personality attributed to it. One view is that 'juristic person' is merely a symbol to help effectuate the group's purposes and in effect brackets the members in order to treat them collectively as a unit. Another, the 'fiction' theory, is that corporations are not legal persons but are only treated as if they were human beings. Another, the 'concession' theory, proceeds on the basis that the status of being a juristic person is conceded by the State and by its system of law. The 'realist' theory, however, asserts that juristic persons enjoy a real existence as a group.

Akin to this is the 'organization' theory which treats corporations as organisms similar biologically to natural persons; they have real lives and group wills. Other theories are the purpose theory; corporate personality is a device to achieve a collective purpose; and the theory of the 'enterprise entity', which bases the corporate entity on the reality of the underlying enterprise. Many of the theories have been developed to explain particular kinds of corporations and fail adequately to explain others; no single theory deals with all the cases and British law at least is not committed to any theory.

Corporate state. A society composed of and governed by economic and functional groups rather than by individuals. Members of the central governing body represent interests rather than localities. Corporativism was an important feature of Italian Fascism and played a lesser part in German and Austrian ideology. In practice, corporate states have usually reflected the will of a dictator rather than a balance of interests of groups.

Corporation. A group of individuals, or a series of the holders of an office, who are deemed in law to be a single legal entity. A corporation is accordingly sometimes called an artificial or juristic or corporate person, as distinct from a natural person. The corporate entity is legally distinct from all the individuals who compose it, has legal personality in itself and can accordingly sue and be sued, hold property and transact, incur liability, and generally act as if it were a natural person. Thus the factory of the AB Company Ltd. belongs to it and not to its shareholders. It has perpetual succession and continues indefinitely notwithstanding changes in the persons who compose it.

Corporations fall to be distinguished from groups of individuals who form a club or society or trade union, persons joined in partnership, persons acting as trustees, and other unincorporated groups. Some groups, e.g. trade unions, are not full corporations but so similar thereto as sometimes to be called quasi-corporations.

The earliest recognized corporations were boroughs, universities, and ecclesiastical orders and by the fifteenth century the distinctness of the corporation from its members was recognized. Business enterprises were first created as corporations in the late sixteenth and early seventeenth centuries. Thus the East India Company, the Royal African Company and the Hudson's Bay Company were incorporated by charter.

English law recognizes two main kinds of corporations, corporations aggregate, composed of incorporated groups of natural persons, such as the members of a company, and corporations sole, which are bodies politic constituted in a single person who, by virtue of some office, has power to hold property taken to him and his successors in office for ever, the succession being perpetual though not necessarily continuous. There were also a few instances of corporations aggregate of which the head was a corporation sole. Scots law does not recognize corporations sole, save in cases created by statute applicable also to Scotland.

Corporations aggregate are of many kinds and created for many purposes; common kinds are corporations having statutory powers and functions in relation to local government, public corporations (q.v.) created by statute to provide public services or manage enterprises in the public interest, such as the National Coal Board, trading corporations, including chartered companies, companies incorporated by special Acts of Parliament, companies registered under the Companies Acts, and societies registered under the Industrial and Provident Societies Acts, or the Building Societies Acts. A corporation aggregate can, in general, express its will only by deeds under its common seal.

Corporations sole were originally mainly ecclesiastical, and an archbishop, bishop, dean, parson and vicar are all corporations sole. The Sovereign is also a corporation sole at common law and some Ministers of the Crown have been made corporations sole by statute. The consequence of being a corporation sole is that official property passes on the holder's death to his successor in office as if he and his successor were the same person.

A corporation is created by royal charter, letters patent, or statute, but may also exist by prescription, a charter from the Crown, now lost, being presumed, or by custom. Statutory corporations have such powers and rights as are authorized directly or indirectly by the creative statute; other corporations have all such powers and rights as private persons have, unless restricted by statute.

In a corporation aggregate there are certain offices which may be filled in the manner prescribed by the constituting instrument. Every corporation has power to make byelaws relative to the purposes for which it is constituted which, in general bind members only.

Its corporate acts must be done at a corporate meeting, unless the constitution authorises otherwise, and must be authenticated by the seal of the corporation.

A chartered corporation is dissolved by surrendering its charter to the Crown, or its charter may be forfeited for abuse of its powers and privileges, or revoked by a later charter or by statute, or it may be dissolved on proceedings on a *scire facias* on the Crown side of the Queen's Bench Division. Corporations incorporated by or under statute may be dissolved by subsequent statute or by the statutory procedure of winding-up.

J. P. Davis, *Corporations; A Study of the Origin and Development of Great Business Combinations*; W. R.

Scott, *English, Scottish and Irish Joint Stock Companies to 1720.*

Corporation Act (1661). A statute, part of the Clarendon Code (q.v.), whereby no person might be elected to office in any corporate town who had not within one year previously taken the Lord's Supper according to the rites of the Church of England. In 1828 an obligation to subscribe a declaration was substituted as the requirement and the Act was repealed in 1871.

Corporation boroughs. Borough constituencies in the unreformed British Parliament, where the mayor and corporation chose the members to represent the borough in Parliament.

Corporation tax. A tax introduced in 1966 and levied at a rate fixed annually on the profits, income, and capital gains of bodies corporate and unincorporated associations, other than partnerships, resident in the United Kingdom or carrying on trade in the U.K. through a branch or agency. The tax is administered by the Commissioners of Inland Revenue.

Corporeal property. Objects of rights of property which have physical existence and form, such as land and goods, as distinct from incorporeal property, which comprises legal rights only. Corporeal hereditaments are land itself and such physical subjects as are annexed to and form part of it or are treated as forming part of it. Corporeal chattels are goods and physical objects not annexed to land.

Corpus delicti. The facts which constitute a crime, not, as is often thought, the body of a murdered person or another thing which is the subject of a crime.

Corpus Juris. A large American legal encyclopaedia edited by William Mack and W. B. Hale, published in 1914–24, in effect a new edition of the *Cyclopaedia of Law and Procedure* edited by Mack and H. P. Nash in 1901–12. It has been superseded by *Corpus Juris Secundum.*

Corpus Iuris Canonici. A term used unofficially from the twelfth century, and officially adopted by the Council of Basle in 1441, for the body of canon law consisting of Gratian's *Decretum* (1141–1150), Raymond of Pennaforte's *Liber Extra* (or *Decretales Gregorii IX* or *Liber Extra Decretum*), (1234), the *Liber Sextus* of Boniface VIII (1298), the *Clementines* of Clement V (1313) and the *Extravagantes Joannis XXII* (1325) and *Extravagantes communes.* After the reforms of the Council of Trent the whole of the foregoing collections were in 1582 re-edited by a commission of cardinals and doctors and the title *Corpus Iuris Canonici* given to the whole. In 1563 the Institutes of Lancelot, prepared by one Lancelot by papal order, were added to the C.I.C. with official sanction in 1580 as *Appendix Pauli Lanceloti.* There is also a *Liber Septimus* of Petrus Matthaeus of 1590 but it is outside the C.I.C. Modern canonists refer to the law prior to Gratian as *ius antiquum*, to the C.I.C. as *ius novum*, and to the decretals of the Council of Trent and later legislation as *ius novissimum.* It remained the only authenticated collection of material for the Western Church until the promulgation of the Code of Canon Law in 1917. The standard edition is that of E. Friedberg (Leipzig, 1879 and later editions).

Corpus Iuris Civilis. A collective name for Justinian's works on the Roman law. This name is not ancient but probably first applied by Gothofredus in the 1583 edition of the Accursian gloss, to distinguish the Roman law texts from the *Corpus Iuris Canonici* (q.v.). It normally comprised five volumes, containing respectively the *Digestum Vetus* (Books 1–24.2), the *Infortiatum* (Books 24.3 to 38), the *Digestum Novum* (Books 39 to 50), the *Codex*, Books 1–9, and in the fifth volume the *Institutes, Novels* (in the form of the *Authenticum*) the remaining three books of the Code (*Tres Libri*), the *Libri Feudorum* and some imperial statutes. Volume 5, being smaller, was often referred to as *volumen parvum.* The standard modern edition is the Berlin one, comprising the *Institutions* edited by P. Krueger, the *Digest* by T. Mommsen, the *Codex* by P. Krueger and the *Novels* by R. Schoell and W. Kroll.

Correction, House of. One of the public workhouses first established by the Poor Law Act, 1576. They were originally intended for the confinement and setting to work of rogues and vagabonds and idle paupers. Later they were used for persons who had deserted their families and the mothers of bastard children. After 1706 they were used for the confinement of persons convicted of felonies which had the benefit of clergy and thereafter it was common to prescribe that minor offenders be imprisoned in houses of correction. Later, in the case of any indictable offence, committal to a house of correction to await trial was authorized. In 1865 the distinction between houses of correction and prisons was abolished.

Corrective training. A special regime instituted in 1948 for prisoners thrice convicted of offences punishable with imprisonment for two years or more with a view to their reformation and the prevention of crime. The duration was from two to four years. It was abolished in 1967.

Correlatives, opposites and contradictories. Terms of logical relationships of concepts.

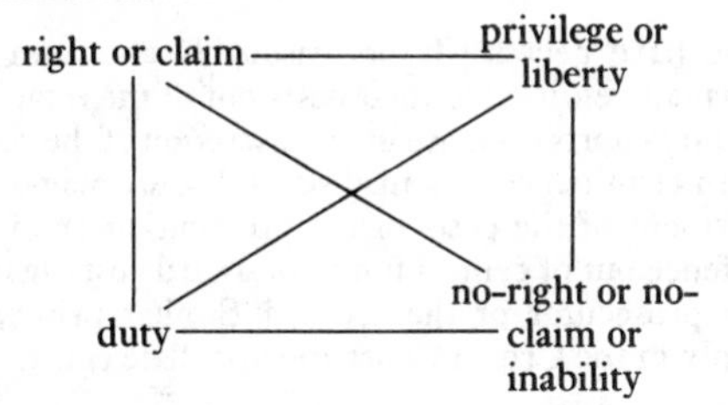

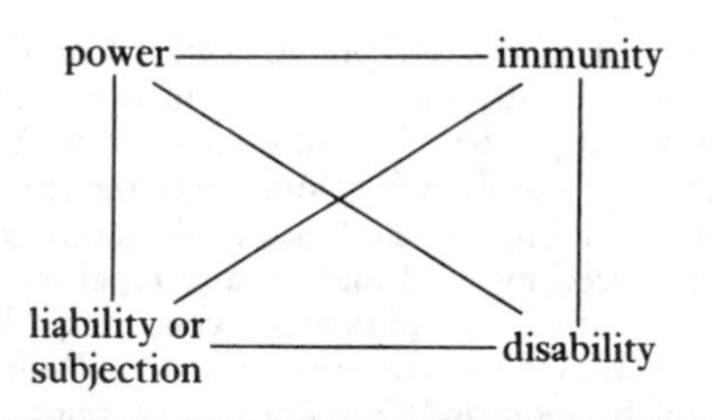

The analysis of the terms 'right' and 'duty' by such jurists as Windscheid, Salmond, and Hohfeld (qq.v.) has revealed in the first place that each of these terms is sometimes used for one or other of four distinguishable ideas, as represented in the diagram. (There are many variants in the terminology employed.)

It also revealed various logical relations between the terms. In each square the vertical lines indicate ideas which are correlatives of each other, in that in any legal relationship the presence of one, e.g. power, in one party to a relationship, implies the presence of the other, i.e. liability, in the other party, and vice versa.

The diagonal lines indicate ideas which are opposites of one another, in that in any legal relationship the presence of one, e.g. right, in one party implies the absence of its opposite, i.e. inability, in himself, and vice versa.

The horizontal lines indicate ideas which are contradictories of one another, in that in any legal relationship the presence of one, e.g. right, implies the absence of its contradictory, i.e. privilege, in the other party, and vice versa.

See also DUTY; RIGHT.

Corroboration. Evidence from an independent source confirming and strengthening other evidence. In English law, generally, evidence of a single witness, if accepted, suffices to establish a fact, in both civil and criminal cases, but in certain cases, e.g. perjury, corroboration is required and in other cases, e.g. sexual offences, as a matter of practice the court will require corroboration. In Scots law evidence on any material point must, in general, be corroborated but in certain cases statute provides that evidence from one witness, if accepted, is to be sufficient.

Corrupt and illegal practices. Statutory offences involving interference with the liberty of the individual freely to exercise his right to vote at an election, and the commission of which may render the election void. Corrupt practices include bribery, treating, undue influence, personation, making a false declaration as to election expenses and incurring certain expenses without the authority of the election agent. Illegal practices include making false statements about candidates, failing to make, or making inaccurate, returns as to election expenses, or making unauthorized payments. Election expenses have been checked since the Corrupt Practices Prevention Act, 1854, and limited in amount since the Corrupt and Illegal Practices Prevention Act, 1883.

Corruption. The perversion of anything from its original pure state, used particularly of accepting money or other benefit in consideration of showing favour to or benefiting the donor, and of the degrading influence of obscene publications.

Corruption of blood. One of the consequences of attainder for treason or felony, the effect being that the person attainted could neither inherit lands or hereditaments, nor continue to possess those in his possession, nor transmit them to any heir; they escheated to the lord, subject to the King's superior right of forfeiture.

Cortes. The Spanish term for courts, and particularly the parliaments of the mediaeval Spanish kingdoms and later of Spain. Cortes were in existence in Castile and Leon from the early thirteenth century and appear a little later in the other kingdoms. They developed as elected representatives began to participate in the work of the King's council. The Cortes were composed of nobles, clergy and *procuradores* or town clerks of the fortified municipalities with written instructions from their electors. During the fourteenth century the last group dominated the Cortes since only they could authorize the special taxation required by the Crown. Though these assemblies continued to exist until the eighteenth century they had ceased to exercise any real power. Since the nineteenth century the term has been applied to the national assembly of Spain.

Corvée or statute labour. Corvée signified originally the regular work which both serfs and freemen owed their lord under feudalism, but later became synonymous with statute labour, the system under which peasants had by law to give a specified amount of labour on public works. Corvée was revived in France in 1726 for road work but was abolished at the Revolution. It survived in Austria till 1848 and in French colonies into the twentieth century. In England and Scotland statute labour was enforced for road making and repairing from the sixteenth to the nineteenth centuries, when it was superseded by the turnpike system. Both were general, periodic and short-term obligations.

Corwin, Edward Samuel (1878–1963). Professor of Politics (1911–18) and of Jurisprudence (1918–46) at Princeton, was a great authority on American constitutional law and history. His writings include *The Doctrine of Judicial Review* (1914), *The Constitution and What it Means Today* (12 edn., 1958), *The President, Office and Powers* (4 edn., 1958) and he was responsible for the official *The Constitution Annotated* (1953).

Cos, Johan (eighteenth century). A Dutch jurist, author of works on matrimonial law, community of property, and succession.

Cosenage or cosinage. Cousinship or consanguinity. It was also the name of a writit which lay where great-grandfather of the heir was seised of land in fee at his death and a stranger had entered and disseised the heir. It was abolished in 1833.

See also AIEL; BESAIEL; TRESAIL.

Cost book mining company. A partnership formed to work a mine under the local customs of the district. They were found mainly in the High Peak and Wirkworth districts of Derbyshire and in Devon and Cornwall.

Costs. Subject to any relevant statute and to the Rules of the Supreme Court the costs of and incident to all proceedings in the Supreme Court in England are in the discretion of the court or judge, with full power to determine by whom and to what extent such costs are to be paid. But this discretion is exercised judicially in accordance with practice. In the absence of special circumstances a successful litigant is awarded his costs against the unsuccessful party. Costs are taxed by the Taxing Master of the Supreme Court on one or other of three bases, in general as between party and party, which covers costs reasonably incurred in prosecuting or defending the action, on either of two scales, higher and lower, but sometimes on the common fund basis, formerly called as between solicitor and client, a rather more generous allowance, or more liberally still, as between solicitor and own client. Costs incurred or increased by over-caution, mistake, negligence or unusual expenses will be disallowed by the taxing master. In administrative proceedings, e.g. to ascertain the construction of a will, all parties who have necessarily or reasonably appeared are normally entitled to their costs out of the estate. In county courts costs are in the discretion of the court.

In criminal cases a magistrates' court may order payment of the costs of the prosecution or of the defence out of central funds, or award costs against the prosecutor or the accused. Similar principles apply in the Crown Court and appellate courts.

Cottenham, Lord. See PEPYS, C. C.

Couch, Sir Richard (1817–1905). Called to the Bar in 1841, he became a puisne judge, 1862–66, and Chief Justice of the High Court of Bombay, 1866–75. In 1875 he became a P.C. and in 1881 was appointed to the Judicial Committee where he sat for many years, proving himself a very sound, though hardly brilliant judge.

Council. An assembly of persons taking counsel together, particularly on affairs of government. In the Christian Church it is a meeting of bishops and other leaders to debate and rule on matters of doctrine, discipline, and administration. The term is applied particularly to provincial councils, the decision of which, canons, early began to form the series of rulings which ultimately took shape as the canon law (q.v.), and to general or ecumenical councils. According to the generally accepted list there have been 21 ecumenical councils, from that of Nicaea (325) to the second Vatican Council (1962–65). Their legislation has contributed largely to the canon law.

In England the body of advisers attending the King which with and under him governed the realm was the Council. After reorganization by Thomas Cromwell about 1536 it became the Privy Council. See PRIVY COUNCIL.

The elected governing body of local government areas is also called the council.

Council of Europe. A body formed by a number of European states in 1949 to achieve a greater unity between its members for the purposes of safeguarding and realizing the ideals and principles which are their common heritage and facilitating their economic and social progress by discussion of questions of common concern, and by agreements and common action in economic, social, cultural, scientific, legal, and administrative matters and in the maintenance and further realization of human rights and fundamental freedoms. Every member state must accept the principles of individual freedom, political liberty and the rule of law as the basis of its internal policy. It is governed by a Committee of Ministers, one from each member State, being the Foreign Minister or his alternate, which has the duty of concluding conventions or agreements and concerting a common policy.

Member states are permanently represented at Strasbourg by Ministers' Deputies. The Consultative Assembly is a deliberative body, consisting of representatives in proportion to member States' population and meets for several weeks each year. It may discuss any matter within the scope of the Council and may agree on an opinion, resolution, or recommendation. The official languages are English and French. It appoints various committees.

A European Convention for the Protection of Human Rights and Fundamental Freedoms came into force in 1953 and a Commission on Human Rights was established in 1954 to hear complaints between and against contracting parties. In 1959 there was established the European Court of Human Rights (q.v.), sitting at Strasbourg.

Other bodies established include a Council for Cultural Co-operation and a Committee on Legal Co-operation.

Council of Law Reporting, The Incorporated. Prior to 1865 law reporting in England was in the hands of commercial publishers, some of whose series of reports were 'authorized' and enjoyed the privilege of exclusive citation in the courts, while others were unauthorized. In 1863 a meeting of the Bar was held, at which a committee was appointed to prepare a plan for the alteration of the system. After overcoming the opposition of the publishers and reporters of the authorized reports, the Council of Law Reporting, later incorporated by royal charter, was established, reporters were appointed and the first numbers of the Law Reports were published in 1865. The Law Reports absorbed and superseded all the older series of so-called authorized reports and established a planned set of semi-official reports over which the profession through its representatives on the Council had control. Since 1865 the Council has published reports regularly in three series, namely the First Series (1865–75), the Second Series (1876–90) and the Third Series (1891 to date), each comprising at least one annual volume for the House of Lords and for each Division of the Court. The reports do not have any monopoly of being cited but are usually preferred to other reports. The judgments are revised by the judges which is not always the case with other series.

See also LAW REPORTS.

W. Daniel, *History and Origin of the Law Reports* (1884).

Council of Legal Education. A body established in 1852 in pursuance of resolutions of the four Inns of Court, reconstituted in 1967 as a Committee of the Senate of the Four Inns of Court and in 1974 as a Committee of the Senate of the Inns of Court and the Bar (q.v.). It is responsible for the training and examination of student members of the Inns of Court and manages the Inns of Court School of Law.

Council of Ministers of the European Communities. A body of ministers, one from each member state, representing the governments thereof. It may vary in composition according to the matter under consideration. It meets regularly, normally in Brussels. The chairmanship passes in rotation round the representatives of each government for six-month periods. It is assisted by the Committee of Permanent Representatives (or Ambassadors) which prepares and co-ordinates the work of the Council, and has its own secretariat and permanent staff. The Commission of the European Communities is represented at all stages of the Council's work.

The Council takes the final decision on all major Community problems. In general it can act only on a proposal from the Commission. Certain major decisions require unanimous decision. Most decisions are taken by a qualified majority, each state having a number of votes proportioned to its size, and 40 out of the possible 58 votes representing a qualified majority. Proposals of the Commission can be adopted or rejected by a qualified majority but amended only unanimously. Difficulty is sometimes found in reaching decisions where vital interests of one or two states are involved. Commission proposals when received are referred to a committee or working party, the work of which is co-ordinated by the Committee of Permanent Representatives, and members of the Commission take part in Council meetings when their proposals are under consideration.

Council of State. The chief organ of government in the system established by the Instrument of Government (1653). It was intended as a check on the Lord Protector and, along with him, controlled the executive power. Councillors were appointed for life. In most matters the Protector had to consult the council and act on its advice, but it could meet in his absence and he was not always involved in its discussions. It was renamed the Privy Council by the Humble Petition and Advice (1657) and disappeared at the Restoration in 1660.

Council of the North. A body probably originating in the private council of Richard, Duke of Gloucester, who administered the north of England for his brother, Edward IV. It was re-established by Henry VIII as a body mainly of lawyers and administrators and administered the Tudor lands in the north and also exercised wide civil and criminal jurisdiction. Later it was restricted to Yorkshire only. After the Pilgrimage of Grace (1536) it was reorganized on a permanent basis, as a body of administrators and judges under a lord president, responsible for the whole of the north. It was an instrument of government, based on royal commission and not a delegate of the Privy Council. It had common law jurisdiction, civil and criminal, and an equity jurisdiction developed; it had a Star Chamber jurisdiction in riots and the supervision of administration. It was generally subordinated to the Privy Council. In the years 1570–1600 it was an ordinary part of the administration of England and generally effective in enforcing the Crown policy and providing justice. It came to an end about 1640.

R. Reid, *The King's Council in the North.*

Council of the West. A body set up in 1537 by Henry VIII modelled on the reformed Council of the North (q.v.). It administered the four western counties, enforced the law and gave civil remedies. It disappeared before 1547.

Council of Wales (or, properly, Council in the Marches of Wales). A body which grew out of the council which administered Edward IV's marcher lands and was revived under Henry VII as a council for Arthur, Prince of Wales, with competence over Wales and the Tudor lands in the marches. It was reconstituted in 1525 and in the 1530's a formal council was created for the maintenance of order in Wales and the marches. It was based at Ludlow and controlled Wales and the border counties. Its authority rested on the King's commission and it was not an offshoot of the Privy Council but came under that Council's general supervision. It exercised conciliar jurisdiction, being both a local Star Chamber and Chancery, and had jurisdiction in common law matters also, but this provoked disputes with the courts at Westminster. It had also many administrative tasks. In abeyance during the civil war and Protectorate, it revived at the Restoration and was not abolished until 1689.

R. Flenley, *Register of the Council in the Marches*, 1569–91; H. Owen, *Administration of English Law in Wales and the Marches*; C. Skeel, *The Council in the Marches of Wales.*

Council on Tribunals. A standing body established in 1958 to keep under review the constitution and working of specified tribunals, to receive and investigate complaints from the public about tribunals, to be consulted by the responsible minister before procedural rules are made for tribunals and inquiries and to make general recommendations to the appropriate Minister on the appointment of members of tribunals. It consists of up to 15 members appointed by the Lord Chancellor and the Secretary of State for Scotland. The Parliamentary Commissioner for Administration (q.v.) is an *ex officio* member. It is not an appellate body from tribunals but supervisory.

Counsel. In the U.K. the general term for members of the Bar collectively, both senior, i.e. Queen's Counsel (q.v.), and junior, i.e. barristers and advocates (qq.v.), as distinguished from solicitors, and also for an individual member, as in such phrases as 'the plaintiff's counsel' and 'to instruct counsel'.

Counsel and procure. Words charging a person with advising and securing the commission of a crime though not present at or participating in its commission.

Counsellor or counsellor-at-law. A term obsolete in England but still sometimes used in Ireland and the U.S., where it signifies an attorney admitted to practice in all the courts.

Counsellor of State. A member of the Privy Council, and particularly a member of the Royal Family appointed to act in place of the sovereign during his illness or absence from the country.

Count (Lat. *comes*, Fr. *comte*, Sp. and Port. *conde*, Germ and Russ. *graf*). The Roman *comes* was originally a household companion of the Emperor; under the Franks he was a city commander and a local judge. Counts were later incorporated into the feudal structure and gradually became hereditary, becoming a title of nobility instead of an official. The rank was below that of duke though some countships were as great as duchies. The English equivalent was the earl, the *vice-comes* being the viscount.

In older civil procedure the count originally meant the declaration in a real action, or, in a personal action, a section of the declaration, alleging a distinct cause of action. The common, or money, counts were a number of claims for money set out at the end of the declaration, all analogies to that arising from the real cause of action.

In criminal procedure a count is a distinct paragraph of an indictment charging a distinct offence. One indictment may contain any number of counts charging separate crimes.

Count of the House. In the House of Commons if notice be taken that less than 40 members (the quorum of the House) are present the Speaker directs strangers to withdraw, and members are summoned by ringing the division bells. The Speaker counts Members and if a quorum is not established within four minutes he adjourns the House. Special rules apply before 4 p.m. (1 p.m. on Fridays). The House cannot be counted out between 7.30 and 8.30 p.m. or after 10 p.m.

Counterclaim. A cross-claim which can conveniently be examined and disposed of in an action initiated by another party. It is not necessarily a defence, but a substantive claim against the plaintiff which could have grounded an independent action. If the counterclaim is one which can conveniently be disposed of in the pending action the judge gives judgment on both original claim and counterclaim. If it cannot it may be struck out, in which case the defendant must bring a cross-action.

Counterfeiting. The manufacture of false money with intent to deceive. The making of counterfeit money has long been criminal and the law has been largely uniform by reason of a convention resulting from a Geneva conference of 1929. Interpol (q.v.) was established initially to organize international measures against counterfeiting.

Counting-house of the King's Household. The formal name of the body commonly called the Board of Green Cloth (q.v.) in which the Lord Steward and Treasurer of the King's House, the Comptroller, Master of the Household, Cofferer and two clerks of the Green Cloth sat for taking the accounts of all expenses of the household and authorizing payment.

Countors or contors. The name in medieval times given to serjeants-at-law (q.v.) who argued and settled the pleadings in a case, so called because they dealt with counts (q.v.) and declarations.

Country. In the old phrases 'trial by the country' and 'putting himself on the country' signifies a jury.

Countryside Commissions. Two bodies for England and Scotland respectively with powers relating to the preservation and conservation of the countryside, reasonable public access thereto, and facilities for the enjoyment and use of the countryside by the public.

County or shire. In the Roman Empire a count (*comes*) was a companion, counsellor, and minister of the Emperor. In the Frankish Kingdom counts were frequently appointed to govern administrative areas of the kingdom, having executive and judicial functions and by the eighth century a *comitatus* began to signify a geographical area. As feudalism developed official counts developed into feudal vassals holding as fiefs the counties entrusted to them.

In Anglo-Saxon England, by the early eleventh century, Mercia was divided into administrative districts known as shires, each named from a defensible town. In Wessex a group of shires had come into being before the Danish wars. In eastern Mercia the shires represent the districts occupied by the Danish armies between which this country had been divided. The western midlands had been divided by a strong king into shires which disregarded the boundaries of ancient peoples. By the Norman Conquest in all England south of the Humber and East of Wales the shires existed with substantially their modern boundaries. Apart from the Danish shires of north-east Mercia every county south of the Mersey and Humber was divided into hundreds, but Kent was divided into lathes and Sussex into rapes. In the Danish shires the subdivisions were wapentakes.

In each shire the King's representative was the earl (*comes*) who, along with the diocesan bishop, presided in the shire court, but tended to become remote from the life of the district. The gap thus created was met by the appointment in each shire of a reeve, *scir gerefa* or sheriff (*vice-comes*) chosen by the King and responsible to him alone for local finance and the execution of justice. By 1066 the sheriff had become the King's chief executive agent in every branch of local government.

After the Conquest William took over the administration of Edward the Confessor and tried to govern through the men who had held high office in Edward the Confessor's time but latterly he had to rely more on Normans of the baronial class as keepers of royal castles, and sheriffs. After the Conquest shires accordingly came to be called counties. County and shire are accordingly the same thing, those terms deriving from the Latin and Norman, and from the Anglo-Saxon, words respectively.

From 1327 justices of the peace were appointed in every county, with administrative and judicial functions. As time went on more and more functions were entrusted to them and they came to exercise the whole of the preventive powers and police jurisdiction formerly wielded by the feudal courts, but subject to the control of the Privy Council.

When the machinery of the poor law was established by the Poor Relief Act, 1601, the poor law overseers were made subordinate to the justices, and this was extended to roads and bridges. In the seventeenth century Privy Council control was established and the council of local government transferred to Quarter Sessions.

The Tudor system of local government was very little altered until 1888. Popularly elected county councils were established in 1888 and took over many of the J.P.'s administrative powers and duties. Some counties were subdivided into two or more administrative counties, each with its own council and administrative structure.

From the fourteenth century certain cities and boroughs were granted the privilege of being counties by themselves, and from 1888 these became county boroughs.

The modern administrative county was divided until 1974 into boroughs, urban districts and rural districts, and since then into districts. Both counties and districts have elected councils. Certain populous counties are metropolitan counties divided into metropolitan districts.

In Scotland sheriffs were introduced on the English model from the early twelfth century and the Scottish sheriffdom appears to have been an artificial unit, not based on earlier divisions of the land nor on natural boundaries. They were mostly based on the King's castles, and denoted the surrounding and dependent district. Between the thirteenth and sixteenth centuries many small sheriffdoms were united into larger ones. Long before the Heritable Jurisdiction Act, 1747, abolished hereditary sheriffships, the number and headtowns of the sheriffdoms had been settled on substantially the modern basis, though detached portions of shires survived in some cases into the twentieth century. A shire in Scotland was accordingly basically a sheriffdom.

From the seventeenth century shires or counties were areas for local government. Commissioners of Supply were instituted in 1667 to levy and collect the land tax, but their duties were later extended to include police, prisons, and valuation of lands. Justices of the Peace were also appointed on a county basis, and had administrative as well as judicial functions. In 1867 parochial boards were established for public health purposes. In 1878 county road trustees were established to maintain and manage highways.

Popularly-elected county councils were established in 1889, and superseded commissioners of supply, justices of the peace, and county road trustees, while a boundary commission rectified many county, parish, and borough boundaries. As subsequently developed in 1929 county councils became the most important agency of local government, exercising all powers entrusted to local authorities, save that within counties large and small burghs had respectively substantial and small powers within their own areas. The cities of Edinburgh, Glasgow, Aberdeen, and Dundee were made 'counties of cities' and given county powers. After 1929 a number of responsibilities were transferred to national boards.

In 1975 counties disappeared, local government being reorganized into regions, each comprising a number of districts, many of which generally correspond to former counties or shires.

County boroughs. Those boroughs which from 1888 to 1974 were administrative counties in themselves for local government purposes and practically exempt from the jurisdiction of the council of the county in which they were situated.

County court. At the time of the Norman Conquest all England was divided into shires or counties. Thereafter the chief official of the shire was the sheriff, who as royal reeve enforced the King's fiscal rights, as deputy of the ealdorman fulfilled his police and judicial duties, and was agent and representative of the central government. His court, the county court or shire moot, dealt with administrative, financial, and military business, and had general jurisdiction in civil and criminal cases. In theory it was attended by many persons, but as the duty of suit of court came to be attached to pieces of land the suitors became a defined class; they were the judges, under the presidency of the sheriff. It generally sat once a month or once every six weeks.

As a representative body its importance diminished with the rise of Parliament; as a judicial body it sank as the court of quarter sessions grew in importance. In the twelfth and thirteenth centuries the county court was very important. However, as royal justice grew in elaboration the courts held by itinerant justices came to be related more to the central courts and less to the county courts. By the end of the thirteenth century most of the civil jurisdiction was passing to the itinerant justices and the common law courts, and the Assizes of Clarendon and Northampton reserved to the royal justices jurisdiction over major crimes, so that by the early thirteenth century its criminal jurisdiction had disappeared. By 1307 the county court was relatively unimportant and, despite later legislative attempts to revive it, it passed completely into disuse.

In the seventeenth century attempts were made to establish local courts for minor actions, and in 1750 courts were established in every hundred in Middlesex held by the county clerk and in the eighteenth century Courts of Conscience or Courts of Request were established in many large towns. In 1846 statute established an entirely new set of courts, having only the name of the old county courts. England and Wales are divided into county court districts each having one or more judges. Courts are held in each district at least once a month.

The county court judges are the persons appointed by the Lord Chancellor as circuit judges (q.v.), who must be barristers of at least 10 years' standing or recorders who have held that office for at least three years. Deputy judges may be appointed *ad interim*. Trial may in certain cases take place with a jury of eight, who must return a unanimous verdict.

The jurisdiction of the county courts, originally limited to claims for debt or damages not exceeding £20 has frequently been extended, and now extends to any action founded on contract or tort where the sum claimed is not more than £2,000, certain actions for the recovery of land and certain kinds of equity proceedings with limit of value. They have a limited probate jurisdiction and jurisdiction to determine any undefended matrimonial cause. Certain county courts have admiralty jurisdiction in respect of particular claims to a limited extent. County courts have also been assigned jurisdiction under many statutes such as the Rent Acts. An action commenced in the High Court may be remitted to the county court. Appeal lies to the Court of Appeal (q.v.) on a question of law or the admission or rejection of evidence, but not on a question of fact.

W. A. Morris, *The Early English County Court.*

County of a city or of a town. Down to 1974 certain cities e.g. Chester, and towns e.g. Nottingham, were, by grant from the Crown or by statute, counties by themselves governed by their own sheriffs, magistrates and council, and exempt from the jurisdiction of the officers of the county.

County palatine. A county the lord of which formerly had the same royal powers and jurisdictions as those held by the Crown elsewhere in the kingdom. In England there were three such counties, Chester, Durham, and Lancaster, the first two existing by prescription from time immemorial, the last created by Edward III.

The Palatinate of Chester reverted to the Crown in 1237 and the court of the county palatine was abolished in 1830.

The jurisdiction long vested in the Bishop of Durham was in 1836 vested in the Crown as a separate franchise and royalty and the jurisdiction of the Duke of Lancaster was vested in the Crown by Henry IV. The Chancellor of the Duchy of Lancaster is a member of the government with nominal duties in relation to the Duchy.

The distinct Court of Common Pleas at Lancaster and the Court of Pleas at Durham were merged in the High Court in 1873 but the separate Chancery Courts of each existed till 1971.

The Isle of Ely was not a county palatine but a royal franchise, the bishop of which had been granted *jura regalia* by Henry I.

J. B. Yates, *Rights and Jurisdiction of the County Palatine of Chester*; G. T. Lapsley, *The County Palatine of Durham*; H. Fishwick, *Pleadings and Depositions in the Duchy Court of Lancaster*, 1489–1558.

Cour de Cassation. The highest court of civil and criminal appeal in France, with the power to quash (*casser*) the decisions of inferior courts. Its major function is to secure uniformity in interpretation of the law among French courts and it considers its decisions only from the point of view of correct application of the law. It does not deal with facts nor with appeals on fact, nor has it power of judicial review of legislation.

Its origins lie in the Middle Ages and from the sixteenth to the eighteenth centuries appeals for cassation were dealt with by the privy council section of the King's council. By the late seventeenth century that council's powers were limited to setting aside decisions which were contrary to law. The Cour de Cassation was established on this basis at the Revolution.

The court now comprises a criminal and five civil chambers. It has a premier president and chief prosecutor (*procureur général*) and each chamber has a president and counsellors, 17 in the criminal and 15 in each civil chamber. The chamber which hears the case does not retry it nor entertain an appeal but merely hears argument on the point of law in question. It may uphold the decision, or quash it, in which case the case is remitted to another court of the same rank as that from which it came, which court is not bound to follow the view of the Cour de Cassation. If it differs in view the case comes back to the Cour de Cassation and is heard by a special chamber composed of the presidents and certain senior counsellors. If the Cour de Cassation adheres to its view it again remits the case to a lower court which in this case is bound to apply the ruling on law of the Cour de Cassation.

Court. A court was originally the King's or a great lord's palace or mansion, the place where he stayed with his retinue of friends and advisers, e.g. the Court of St James's, and the name became transferred to the group of confidants, advisers, and chief administrators, to those who, singly or in groups, in the sovereign's or lord's name exercised judicial functions and also to the places where justice is administered. In the last senses a court is accordingly a person or group of persons having authority to hear and determine disputes in accordance with rules of law. Tribunals or adjudicators who exercise adjudicative functions by virtue of contract or of the voluntary submission of persons to their decisions, such as arbitrators, disciplinary bodies or committees of clubs, trade unions, and the like are not, strictly speaking, courts because they do not exercise jurisdiction by force or authority of law. Nor is a person or body a court merely because he or it has to hear evidence, to act fairly and impartially, to reach decisions, or because it is subject to appeal. Bodies, frequently called tribunals, may have many of the characteristics of courts, while bodies called courts may in fact have mainly or entirely administrative functions.

The function of courts is to decide disputes, authorize actings for which application is made, e.g. to adopt a child, to impose punishments, penalties and sanctions, to award remedies for infringements of rights, and to exercise various administrative functions. In the course of exercising these functions courts apply rules of law and also give interpretations of rules, and in so doing may modify the general understanding and operation of the law for the future. This is particularly so in Anglo-American legal systems under the principle of *stare decisis* (q.v.) but is largely true in codified European systems also where judicial interpretation gives content to the bare words of a section of a code.

Modern states have many different kinds of courts passing under different names, but certain classifications are common and useful.

One fundamental distinction is between civil courts and criminal courts. The former deal with controversies between two or more parties at issue over such matters as a breach of contract, a commercial dispute, liability for harm done, rights in property, a familial dispute, in which in general one party is claiming some remedy, relief or redress against the other or others. The latter deal with accusations, frequently brought by a person representing the state or public interest, against another or others, that he or they have infringed the rules for social behaviour set by the community, e.g. by committing theft, in which the accused, if adjudged to have committed the behaviour charged, is subjected to correction or punishment, usually by way of fine or imprisonment.

A second basic distinction is between courts of general and courts of special jurisdiction. Courts of general jurisdiction, whether civil or criminal, may deal with practically any kind of case, civil or criminal as the case may be, brought before them. Courts of special jurisdiction may deal only with stated and limited kinds of issues. This latter class includes ecclesiastical courts, courts-martial, commercial courts, industrial or labour courts, probate courts, juvenile courts, traffic courts, and many others.

A third distinction is between superior courts and inferior courts. A superior court deals with more important and difficult cases. An inferior court deals with less important and more ordinary cases; it frequently has jurisdiction limited in area and sometimes in value; inferior criminal courts frequently have limited powers of punishment. Judges of superior courts normally have higher status and qualifications. Judges of inferior courts may be laymen, part-time judges, or not legally qualified.

A further distinction is between trial courts or courts of first instance and appellate courts. A court of first instance has to hear evidence, decide the facts, ascertain and apply the relevant law and thus reach a decision. An appellate court considers appeals against the decision of a court of first instance and may amend or alter the court of first instance's decision. Sometimes there is more than one level of appellate court so that a further appeal can be taken, as in England, from the High Court or county courts to the Court of Appeal and thence to the House of Lords. Appeal may be limited to appeal on a question of law only. Appellate courts are commonly composed of several judges. An appellate court may be empowered or required to retry the case, or may decide on the basis of the record made in the court below of its findings of fact and conclusions in law, or may proceed solely on written records of the proceedings in the court below.

In a federal state there is the further complication of the distinct but overlapping bodies of federal and state law. In the U.S. there is a set of courts for each state and also a federal set, headed by the Supreme Court. In Canada and Australia each state or province's courts handle both state and federal law with a supreme court to decide issues of the relationship of the central authority to the state or provincial authorities and between the state or provincial authorities themselves.

In England the distinction is also recognized between courts of record and courts not of record. A court of record is one the acts and proceedings of which are enrolled as a permanent record and which has power to fine and imprison for contempt of its authority. The superior courts are courts of record, but some courts of special jurisdiction are made courts of record by statute.

Thus in England and Wales the major courts may be classified as follows:

Civil	*Criminal*	*Special* (examples only)	*Administrative* (examples only)
	Superior		
House of Lords	House of Lords	—	—
Court of Appeal (Civil Division)	Court of Appeal (Criminal Division)	Courts Martial Appeal Court	—
High Court	Divisional Court of Q.B.D. of High Court	Restrictive Practices Court	Employment Appeal Tribunal
	Crown Court		National Insurance Commissioner
	Inferior		
County Court	Magistrates' Court	—	Industrial Tribunals

In Scotland the major courts may be classified as follows:

Civil	*Criminal*	*Special* (examples only	*Administrative* (examples only)
	Superior		
House of Lords	—	—	—
Court of Session (Inner House)	High Court of Justiciary (appeals)	Courts Martial Appeal Court Lands Valuation Appeal Court	—

Court of Session (Outer House)	High Court of Justiciary (trials)	Restrictive Practices Court	Employment Appeal Tribunal National Insurance Commissioner
		Inferior	
Sheriff Court	Sheriff Court District Court	—	Industrial Tribunals

Court Baron. A court which was an incident of a manor and inseparable therefrom, established for the maintenance of the services and duties stipulated for by lords of manors as due from their tenants and for determining disputes where the debt or damages did not exceed 40 shillings. It was usually held once a year, before the freeholders who owed suit of court (q.v.) to the manor, the lord's steward acting as president rather than as judge.

In time the manorial court was regarded as having two natures, the court baron for freeholders and the court customary, in which the steward was judge and the copyholders were the parties. The function of the court customary was to deal with matters concerning the rights of copyholders among themselves, and particularly to register transfers of copyhold land by entry on the court rolls.

Courts baron do not appear to have been abolished but have long been in disuse.

W. Greenwood, *Authority, Jurisdiction and Method of Keeping County-Courts, Courts-Leet and Courts-Baron*; F. W. Maitland and W. P. Baildon, *The Court Baron.*

Court, borough. See BOROUGH COURT.

Court, burgh. See BURGH COURT.

Court, Central Criminal. See CENTRAL CRIMINAL COURT.

Court Christian. An ecclesiastical, as distinct from a lay, court. In medieval times the ecclesiastical courts in England claimed wide jurisdiction, over all matters of ecclesiastical status and spiritual functions, church property, matrimonial causes, testamentary matters, promises made by oath or pledge of faith, and over clerics and all personal causes, civil or criminal, in which a cleric was accused or defendant.

Court, Crown. See CROWN COURT.

Court, customary. See COURT BARON.

Court, European. See COURT OF JUSTICE OF THE EUROPEAN COMMUNITIES.

Court for Crown Cases Reserved. A court, created in 1848, for the decision of questions of law arising on the trial of a person convicted of treason, felony, or misdemeanour and reserved by the judge for the consideration of this court. The court consisted of five or more judges of the High Court, whose decision was final, and the question was raised by a case stated and signed by the trial judge setting out the facts and posing the question. The jurisdiction was transferred to the Court of Criminal Appeal in 1907.

Court for Divorce and Matrimonial Causes. A court established in 1857 and given the jurisdiction previously exercisable by any ecclesiastical court in England in matrimonial matters, and also power to grant divorce, declaration of legitimacy, and certain other powers. It consisted of the Lord Chancellor, the Chief Justices of the Queen's Bench and Common Pleas, and the Chief Baron of Exchequer, the senior puisne judges of these three courts, and the Judge Ordinary. The last-named in fact heard almost all cases at first instance. By the Judicature Act, 1873, the court was abolished and the jurisdiction transferred to the new High Court, in which it was exercised in the Probate, Divorce, and Admiralty Division, now the Family Division.

Court hand. A variety of script derived from Caroline minuscule script of the ninth and tenth centuries, and so-called because the records of the superior courts of law were one of the chief kinds of documents in which it was used. It was at its best in the later thirteenth century, but later degenerated.

C. T. Martin, *The Record Interpreter* (1910); C. Johnson and H. Jenkinson, *English Court Hand, 1066–1500* (1915); H. Jenkinson, *Palaeography and Court Hand* (1915).

Court, International. See INTERNATIONAL COURT OF JUSTICE; PERMANENT COURT OF INTERNATIONAL JUSTICE.

Court leet. Leet was an East Anglian term for a territorial and jurisdictional area, and court leet was a court held by right of franchise in which a private lord exercised the jurisdiction of the sheriff's tourn. In the fourteenth century it spread over England. Leet jurisdiction was frequently exercised by lords of manors and by boroughs, and was both a governmental and a judicial body. It was frequently difficult to distinguish it from the jurisdiction of feudal or manorial courts.

The court met twice a year and was presided over by the lord's steward who from an early date was a lawyer and acted as judge. Its main functions were to hold the view of frankpledge (q.v.) and to receive notices of accusation of crimes made by juries. Increasingly, serious cases were reserved to itinerant justices and courts leet were restricted to petty misdemeanours.

From the sixteenth century courts leet declined in face of the growing importance of the J.P.s. They were largely impotent, sat only once a year, could only fine or amerce but not imprison. Decline was, however, slow and their existence was recognized until well into the nineteenth century. In theory some still exist.

W. G. Greenwood, *Authority, Jurisdiction and Method of Keeping County-Courts, Courts-Leet and Courts-Baron*; F. Hearnshaw, *Leet jurisdiction in England*; W. A. Morris, *The Frankpledge System.*

Court of Admiralty. See ADMIRALTY, HIGH COURT OF.

Court of ancient demesne. A court in which the freeholders of land in ancient demesne (q.v.) were the suitors and judges.

Court of Appeal. The Court of Appeal is a branch of the Supreme Court of Judicature; it was created by the Judicature Act, 1873, and exercises the jurisdiction formerly exercised by the Court of Exchequer Chamber, the Lord Chancellor and the Court of Appeal in Chancery, and by the Privy Council in Admiralty and lunacy cases. It consists of the Lord Chancellor, ex-Lords Chancellor, the Lord Chief Justice, the Master of the Rolls, the President of the Family (formerly Probate, Divorce, and Admiralty) Division, and the Lords Justices of Appeal. Lords of Appeal and puisne judges may also be asked to sit. The Court normally consists of the Master of the Rolls and the Lords Justices sitting in courts of three but may sit as a full court of five judges.

In 1966 the Court of Criminal Appeal (q.v.) was abolished and since then the Court of Appeal sits as Civil Division or as Criminal Division. A court of the criminal division consists of an uneven number of judges, at least three in number, and one is usually a judge of the Queen's Bench Division. The Lord Chief Justice frequently presides.

The Civil Division has jurisdiction to hear appeals from the High Court and county courts, the Criminal Division to hear appeals from persons convicted on indictment by the Crown Court (q.v.) and to deal with cases referred to the court by the Home Secretary.

Any number of courts of each Division may sit simultaneously.

Court of Appeals in Cases of Capture. The chief U.S. federal court, 1780–87, prior to the establishment of the Supreme Court, established to deal with questions of naval prize arising during the War of Independence.

Court of Appeal in Chancery. A court set up in 1851 consisting of two Lords Justices of Appeal and the Lord Chancellor if he chose to sit, assisted if necessary by the Master of the Rolls, the Vice-Chancellors (q.v.) or any of the judges, to hear appeals from the Master of the Rolls and the Vice-Chancellors sitting individually as Chancery judges of first instance, and from the London Court of Bankruptcy. Further appeal lay, in most cases, to the House of Lords. The court was merged in the new Court of Appeal (q.v.) in 1875.

Court of Arches. See ARCHES, COURT OF THE.

Court of Assize. See ASSIZE.

Court of audience. An ecclesiastical court in which the two archbishops formerly personally exercised jurisdiction. It is probably now obsolete, save possibly in the case of the trial of a bishop.

Court of Augmentations. A court established in 1535 to see that Henry VIII was fully paid his share of the property of the religious houses forfeited to the Crown earlier that year. The management of church lands retained by the Crown was transferred to the Exchequer in 1554 and the court dissolved.

Cout of Chancery. See CHANCERY, COURT OF.

Court of chivalry. A judicial body dealing with military disputes and questions of the law of arms. In most countries courts-martial regulating the conduct of persons involved in war came first. They were presided over by monarchs or great officers of state such as earls marshal. In time heralds became the principal officers of and practitioners before these courts and the law of arms became more and more concerned with rights to armorial bearings. In England the Court of Chivalry or Court Military, also known as the Court of the Constable and Marshal, was held before the Lord High Constable and the Earl Marshal of England. It had jurisdiction, civil and criminal, in deeds of arms and war, armorial bearings, matters of precedence, and, held before the Earl Marshal alone, as a court of honour. Long in abeyance it was revived in 1955. In Scotland the court of chivalry is the Court of the Lord Lyon King of Arms which has both civil and criminal jurisdiction. In many other countries, such as Switzerland and Spain, there are heraldic officers and a heraldic judicial function and even in republican France the civil courts protect arms legally recorded under the monarchy.

Court of Claims. A court established after the accession of a new sovereign to judge the validity of the claims of persons to perform certain honorary services at the coronation of the new monarch. The first recorded such court was held in 1377 before the Lord High Steward. Since the time of Henry VII commissioners have been appointed in lieu, and the court now consists of a royal commission appointed under the Great Seal and termed the Committee of Claims. The court may refer any claim to the sovereign pleasure, and the sovereign may withdraw a claim from the commission and transfer it to another tribunal.

Court of Claims, U.S. A special court which deals with claims brought against the government of the U.S.

Court of Common Council. This court is also called the Court of the Lord Mayor, Aldermen and Commoners of the City of London in Common Council assembled and dates from at least 1275. It consists of the Lord Mayor, aldermen, and a large number of common councillors elected by the wardmotes of the wards of the City. It shares the administration of the City with the Court of Aldermen.

Court of Common Hall (or Meeting or Assembly of the Mayor, Aldermen and Liverymen of the several Companies of the City of London in Common Hall assembled). It consists of the Lord Mayor or one of the sheriffs, at least four aldermen, the sheriffs and all the freemen of the City who are Liverymen of the City Companies. It nominates two aldermen for the office of Lord Mayor and elects the Sheriffs and certain other city officers.

Court of Common Pleas. See COMMON PLEAS, COURT OF.

Court of Conscience. Also known as the Court of Requests, a court established in London in 1517 consisting of two aldermen and four, later 12, common councilmen, as commissioners to decide matters of debt between citizens and freemen of the City. It was abolished in 1847 and merged in the Sheriff's Court of the City of London (q.v.) which was in turn absorbed in the county court system. There were also some other courts, passing under this name, held by members of various corporations for the recovery of debts. The Court of Chancery (q.v.) was also sometimes called a court of conscience.

Court of Criminal Appeal. See CRIMINAL APPEAL, COURT OF.

Court of Customs and Patent Appeals, U.S. A court which hears appeals from decisions of the Customs Court over duties on imported goods, judgments of the Tariff Commission and of the U.S. Patent Office.

Court of Delegates. Commissioners appointed by the Crown by special commission of delegacy, first issued in 1533, to hear appeals from the Court of the Admiral or to hear an ecclesiastical cause in place of the archbishop of the province, or by way of appeal from him. The court comprised six delegates, three common law puisne judges and three doctors of civil law. The court was abolished in 1832 and its jurisdiction transferred to the Judicial Committee of the Privy Council. A Commission of Adjuncts was issued if for any reason it was necessary to increase the number of delegates and a Commission of Review if a case heard by the Court of Delegates was to be reconsidered.

Court of Ecclesiastical Causes Reserved. A court of five, two who hold or have held high judicial office and three who are or have been diocesan bishops, established in 1963, acting for both the provinces of Canterbury and York, having original jurisdiction in ecclesiastical offences involving doctrine, ritual, or ceremonial committed by a priest or deacon or by an archbishop or bishop and appellate jurisdiction from consistory courts in causes of faculty involving doctrine, ritual or ceremonial.

Court of equity. A court exercising jurisdiction founded on the principles of equity, justice, and fairness, as distinct from the application of principles of strict law. In England the major court of equity was for long the Court of Chancery (q.v.) but there were minor courts of equity, such as the Court of Requests (q.v.). In other jurisdictions a court of equity is not necessarily a distinct court from a court of law, but rather a court which takes cognizance of equitable claims and pleas as well as strictly legal ones.

Court of error. A court empowered to rectify on appeal errors in law or fact on the part of another court. Originally the King's Bench was predominantly the court which corrected the errors of other courts including, till 1783, the King's Bench in Ireland and, till 1870, the Common Pleas. Until 1875 error from one of the superior courts lay to the Court of Exchequer Chamber and then to the House of Lords; error from inferior courts of record lay to the Queen's Bench. The Judicature Act, 1875, replaced error from the Superior courts by appeal to the Court of Appeal.

In criminal cases the court of error was the Court for Crown Cases Reserved. Writ of error in criminal cases was replaced by appeal to the Court of Criminal Appeal.

Court of Exchequer. See EXCHEQUER, COURT OF.

Court of Exchequer Chamber. See EXCHEQUER CHAMBER, COURT OF.

Court of Faculties. The office of the Archbishop of Canterbury from which are granted faculties, including licenses to marry. The judge of the provincial courts of Canterbury and York is *ex officio* Master of the Faculties.

Court of First Fruits. A court established in 1540 for the collection of first fruits or annates and tenths. It was abolished in 1554. Thereafter the Governors of Queen Anne's Bounty exercised the functions formerly vested in this court until the functions of that body were in turn transferred to the Church Commissioners in 1947.

Court of Great Sessions in Wales. A court established in 1543 which performed in all Welsh counties the functions then performed by judges of assize in England. It was abolished in 1830.

C. Abbot, *Jurisdiction and Practice of the Court of Great Sessions of Wales.*

Court of High Commission. A body created under statutory authority in 1559 to exercise some of the powers assumed by the Crown when it assumed the position of supreme head of the Church in England. The commissioners had wide powers to enforce the Acts of Supremacy and Uniformity, to deal with ecclesiastical offences generally and to suppress movements dangerous to the Church. Their powers were exercised according to the instructions and under the supervision of the Council, and their procedure followed the civil rather than the common law. The council referred to this commission petitions on ecclesiastical matters so that it began to act as a court, developing forms and rules of procedure. The powers were later specified in detail. It stood towards the Church much as the Star Chamber stood to the State and the ordinary courts, and its relations with Council and Star Chamber were close. It not only supervised the ordinary ecclesiastical courts but had an original jurisdiction, almost as wide as they had. In the seventeenth century it attracted the animosity of the common lawyers and it was abolished in 1641.

R. G. Usher, *Rise and Fall of the High Commission*, 1535–1641.

Court of Human Rights, European. A court established by the European Convention on Human Rights, sponsored by the Council of Europe, consisting of one judge from each country which is a member of the Council, elected by majority vote of the Consultative Assembly from jurisconsults of recognized competence. They are elected for nine years, one third retiring after each three years and serve part-time. The Court sits in Strasbourg and meets at least once a year. It has to decide cases on the interpretation of the Convention. The Commission on Human Rights must assist the court in having the case for an aggrieved individual put before the court. Judgment is by a majority and is final and binding on the parties. It is for the State concerned to take action to implement the judgment.

Court of Hustings. The highest and most ancient court of the City of London. About the thirteenth century much of its jurisdiction passed to the Sheriff's Court and to the Mayor's Court, only real and mixed actions being retained. It later became divided into the Court of Hustings for Pleas of Land and the Court of Hustings of Common Pleas. The Lord Mayor or an ex-Lord Mayor and the Sheriffs are nominally the judges but the Recorder of London is the actual judge. Courts of Hustings of a similar kind existed also in a number of ancient boroughs, such as Winchester and York.

Court of inquiry. A tribunal appointed *ad hoc* by authorized persons in the armed services to ascertain the facts in some matter. Disciplinary action may follow on its findings.

Court of Justice of the European Communities. Originating as the Court of Justice of the European Coal and Steel Community, established by the Treaty of Paris of 1951, it was extended to serve also the EEC and Euratom when they were created by the Treaties of Rome in 1958. It sits at Luxembourg. Its function is to ensure that the law is observed in the interpretation and implementation of the Treaties and its jurisdiction is entirely derived from the Treaties. The contentious jurisdiction is concerned with actions against member states and actions against Community institutions. It is composed of nine judges unanimously elected by the governments of the member states and holding office for renewable six-year terms. The judges elect their own President for a renewable three-year term. The Court may for certain purposes sit in two chambers and there are two Vice-Presidents each presiding over a chamber, elected for one-year renewable terms. The court is assisted by four advocates-general who have to consider cases impartially and make reasoned submissions, in open court, proposing to the Court a solution to the problem. This is not binding on the court but is frequently adopted.

The procedure is regulated by the Court's own Rules of Procedure and comprises a written stage, in which pleadings are exchanged, an enquiry stage in which a *juge-rapporteur* considers whether enquiry or the taking of proof is necessary, and an oral stage in which the *juge-rapporteur* reports on the case, oral argument is heard, and the advocate-general makes his submissions.

An important function of the Court is, on a reference made to it by a court in one of the member states, to give a preliminary ruling on the interpretation of a provision in one of the Community Treaties to be adopted by the court of the member state in deciding the case before it.

The Court issues a single judgment. Its decisions are final and not subject to appeal but in special circumstances may be revised by the court itself.

The law applied is being developed by the Court itself by interpretation of the Treaties but relies substantially on analogies with similar concepts found in existing European systems, particularly French law.

Court of King's (or Queen's) Bench. See QUEEN'S BENCH, COURT OF.

Court of last resort. A colloquial expression for any court from which no appeal lies.

Court of Passage. An inferior court in Liverpool with a very ancient jurisdiction and down to 1835 also called the Borough Court of Liverpool. It was abolished in 1971.

Court of Peculiars. A branch or function of the Court of Arches (q.v.) having jurisdiction over the parishes in the province of Canterbury which were exempt from the jurisdiction of the diocesan bishop and responsible to the metropolitan only. It was probably abolished in the nineteenth century.

Court of pie poudre or piepowder (*pedes pulverosi*). In early times the right to hold a court was incidental to the right to hold a fair or market, and the court was so called because it was frequented by chapmen with dusty feet who travelled from fair to fair. These courts of the fairs were held by a lord's steward, or the mayor or bailiffs of a borough, but the judges were the merchants who attended the fair. It dealt summarily with civil disputes and minor offences committed at the fair. These courts declined in importance from the fifteenth century.

Court of Pleas at Durham. A court of the County Palatine of Durham with a common law jurisdiction in the county. It was abolished in 1873 and its jurisdiction transferred to the High Court.

Court of Probate. A court established in 1857, taking over from the ecclesiastical courts the jurisdiction to make grants of probate and administration. It had a single judge, who was the same person as the judge of the Court of Admiralty. Appeal lay to the House of Lords. The court was abolished in 1875 and its jurisdiction transferred to the High Court and exercised by the Probate, Divorce, and Admiralty Division. The contentious jurisdiction is now exercised by the Chancery Division and the non-contentious by the Family Division.

Court of Protection. The department of the High Court which deals with the property of persons of unsound mind. The judges of the Chancery Division are the judicial authorities and the Master and Deputy Master of the Court exercise power in ordinary cases. Appeal lies to the Court of Appeal and the House of Lords.

Court of Quarter Sessions. See QUARTER SESSIONS.

Court of record. A court whose acts and proceedings are enrolled for permanent record and which has power to punish for contempt of its authority. This category includes not only the superior courts but the county courts and certain special courts established by statute.

Court of Referees. The Court of Referees consists of the Chairman and Deputy Chairman of Ways and Means in the House of Commons, Counsel to the Speaker, and not less than seven members of the House appointed by the Speaker. They are concerned with determining the right of petitioners to be heard in opposition to Private Bills (q.v.), if their *locus standi* (q.v.) is challenged by the promoters of the Bill. Their decisions are legal decisions and no appeal lies against them. The committee dealing with a particular bill also has power to decide such questions arising incidentally in the course of their proceedings. The name was also given to a former body which decided questions relating to unemployment benefit.

Court of Requests. A minor court of equity which flourished during the Tudor and early Stuart periods, in origin closely related to the Court of Star Chamber. Originally it exercised jurisdiction at the suit of poor men or the King's servants. The Lord Privy Seal was its president, but control was assumed by the legal assessors, the Masters of Requests. It supplemented both Star Chamber and Chancery (qq.v.) and was a popular court. It was attacked by the courts of common law which denied that it was a legal court. It was practically abolished in 1641.

The name was also applied to inferior small claims courts established by statute in various places, and all superseded in 1864 by the modern county courts.

I. S. Leadam, *Select Cases in the Court of Requests*, 1497–1569.

Court of Review. A part of the Court of Bankruptcy under statutes of 1831 and 1842 which had a supervisory jurisdiction over the commissioners in bankruptcy. It was abolished in 1847.

Court of Session. The supreme Scottish civil court. Civil justice was originally administered by justiciars but by the end of the twelfth century a royal council had emerged and heard cases and complaints. After the War of Independence the council both in and out of Parliament seems to have heard cases, and in the following century repeated delegation was made to auditors to hear causes.

In 1425 provision was made for the Chancellor and certain members of the three estates sitting to determine complaints and causes and provision for judicial sessions was repeatedly made in subsequent years. When sessions were not held the Council seems to have tried to cope with the business. In 1491 it was enacted that the Chancellor and certain lords of council or else the lords of session were to administer justice at three fixed terms in the year. In the early sixteenth century the distinction between privy council and session widened and in the 1530s there was a tendency for members of the sessions to be a permanent group from among the council.

In 1532 King James IV obtained from the Pope a subsidy to establish a College of Justice, consisting of a clerical President and fourteen lords, half lay, half clerical, but this new court was but the old session in permanent form, and for a time lords of council continued to act in civil causes.

From 1532 to 1808 the Court was substantially unchanged. It was a unitary court in which the 15 judges sat together, though at any time three might be absent in the Outer House supervising the preparatory stages of actions and a fourth in the Bill Chamber (q.v.), authorizing exceptional forms of actions. But the decisions were reached by most, if not all, of the whole court. In 1808 the Court was divided into a First Division, of the Lord President and seven Lords Ordinary, and a Second Division, of the Lord Justice-Clerk and six Lords Ordinary. In 1810 the five junior judges became permanent Lords Ordinary sitting only in the Outer House and in 1825 the court was settled on the basis which has remained, of an Inner House of two Divisions each of four, the remaining judges (seven till 1830, five till 1948, and increasing since then to 14 in 1978) sitting only in the Outer House. Transfer from Outer House to Inner House is made, as vacancies arise, by seniority.

The court inherited the jurisdiction of the Scottish Court of Admiralty (*c.* 1450–1830), the Commissary Court of Edinburgh (1563–1836), the Scottish Court of Exchequer (1708–1856) and the Jury Court (1815–30). Judges of the Court of Session also sit as Commission for the Plantation of Kirks and Valuation of Teinds, as Lands Valuation Appeal Court and Election Petition Court. The Court has no criminal jurisdiction but the judges also sit as Lords Commissioners of Justiciary in the supreme criminal court, the High Court of Justiciary (q.v.).

In modern practice the court is both a court of first instance and a court of appeal. Lords Ordinary, sitting alone, or sometimes with a jury of 12, have jurisdiction in all kinds of civil cases, save those expressly reserved to other courts or tribunals, and correspond generally to justices of the High Court in England.

The Inner House corresponds generally to the Court of Appeal in England. Each Division of the Inner House has co-ordinate jurisdiction. The Divisions hear certain kinds of cases at first instance, but their main jurisdiction is to hear appeals from the Lords Ordinary, from sheriffs-principal and sheriffs in the sheriff courts (q.v.), from certain courts of special jurisdiction and tribunals, and under numerous statutes.

The Inner House may direct a rehearing of a case before five, seven or a larger number of judges, or even the whole court, and the decision of such a sitting may overrule difficult precedents (q.v.) or give very authoritative rulings on difficult points.

Further appeal lies, under certain conditions, to the House of Lords (q.v.).

Court of sewers. The Commissioners of Sewers (q.v.) sitting for the discharge of their duties.

Court of summary jurisdiction. In England and Wales a magistrates' court exercising criminal jurisdiction summarily, i.e. without a jury, but not including justices holding a preliminary inquiry into an indictable offence. In Scotland it is a sheriff court exercising criminal jurisdiction summarily, or a district court.

Court of survey. A court consisting of a judge and two assessors empowered to hear appeals from decisions of certain officials under the Merchant Shipping Acts.

Court of Teinds. See TEIND COURT.

Court of the Clerk of the Markets. The Clerk of the Markets of the King's Household formerly had jurisdiction to enquire whether weights and measures accorded with the King's standard or not. The charters of certain boroughs contained a grant of a court of the clerk of the markets, which excluded the jurisdiction of the Clerk of the Markets of the King's Household.

Court of the Constable and Marshal. See COURT OF CHIVALRY.

Court of the Council of the Marches of Wales. Henry VII established a council at Ludlow to control the lords marchers and it was ratified by statute in 1543. It exercised functions like those of the Star Chamber (q.v.) but also acted as a court of appeal from the Court of Great Sessions (q.v.). Cheshire and Flint were excluded from its jurisdiction in 1569. Its extrajudicial functions were abolished in 1640 and its judicial functions, in abeyance from 1640 to 1660, were abolished in 1688. See MARCHERS, LORDS; WALES.

Court of the Duchy Chamber of Lancaster. A court held by the Chancellor of the Duchy of Lancaster or his deputy with jurisdiction in equity as to lands held by the Crown in right of the Duchy of Lancaster, whether within the county palatine or not. The court is believed not to have sat since 1835.

Court of the Lord High Steward. A court constituted by commission to a peer as Lord High Steward to try a peer indicted for treason or felony while the House of Lords was not sitting. The Lord High Steward was sole judge of law, but all the peers were summoned as judges of fact. The right of peers to trial before the other peers was abolished in 1948 so that this court is impliedly abolished. A peer was also appointed Lord High Steward to act as speaker when the House of Lords was trying a person impeached by the Commons.

Court of the Lord Steward of the Queen's Household. A court which existed to punish offences committed by the King's servants against any member of the King's council. The jurisdiction was abolished in 1828.

Court of the Lords Justices. A name for the Court of Appeal in Chancery (q.v.).

Court of the Marshalsea. Despite the name it was unconnected with the Marshalsea prison and was a court held before the Seneschal or Steward and the Marshal of the King's Household, with jurisdiction where either party was a member of the royal household and within a radius of 12 miles of the King's residence for the time being. In the seventeenth century it began to sit permanently at Westminster. It was abolished in 1849.

B. Morice, *Essay towards an History of the Ancient Jurisdiction of the Marshalsea of the King's House*; W. Buckley, *Jurisdiction and Practice of the Marshalsea and Palace Courts.*

Court of the Star Chamber. See STAR CHAMBER.

Court of Wards and Liveries. A court established in 1540 to give the King an effectual means of asserting his rights to the incidents of feudal tenure by knight service, including wardships and liveries, relief and primer seisin, etc. It was very unpopular and was abolished finally in 1660.

H. E. Bell, *History and Records of the Court of Wards and Liveries*; *Decrees and Cases in the Court of Wards and Liveries*, 1553–81.

Courtesy. In Scots law, the right of a surviving husband to a liferent of the heritage in Scotland in which his wife died infeft, exigible only if a child of the marriage had been born and heard to cry and been, for however short a time, the mother's heir. It was abolished in 1964.

See also CURTESY.

Courts, ecclesiastical. See ARCHES, COURT OF.

Courts, feudal. See COURT OF CHIVALRY; COURT OF CLAIMS; FOREST COURTS; HUNDRED COURTS; MANORIAL COURTS.

Courts-martial. Courts-martial are courts held within the armed forces under the disciplinary codes applicable to the several services. Naval courts-martial consist of from five to nine officers of the naval forces, assisted by a judge-advocate and a clerk of the court. The prosecution is conducted by the commanding officer of the accused's ship or by a civilian lawyer. The accused may be assisted by an officer, rating, or civilian lawyer. Finding and sentence must be reviewed by the Admiralty, unless appeal is lodged with the Courts-Martial Appeals Court. Army and Air Force courts-martial may be either general, district or field-general, the composition of the courts varying, and their jurisdiction and powers of punishment varying. A judge-advocate may be appointed by the Judge Advocate-General or the convening officer. He advises the court on law and procedure, sums up the evidence to the court and advises the court on the law applicable. Finding and sentence must be confirmed by a confirming officer who may direct a revision of any finding of guilt, and the confirmed finding promulgated to the accused. The accused may petition for review, and appeal to the Courts-Martial Appeal Court.

C. Clode, *The Administration of Justice under Military and Martial Law* (1872); *Queen's Regulations.*

Courts-Martial Appeal Court. The Courts-Martial Appeal Court, established in 1951, consists of the *ex officio* and ordinary judges of the Court of Appeal and such of the judges of the Q.B.D. as the Lord Chief Justice may nominate, such of the Lords Commissioners of Justiciary in Scotland as the Lord Justice-General may nominate, such judges of the Supreme Court in Northern Ireland as the Lord Chief Justice of Northern Ireland may nominate, and such other persons of legal experience as the Lord Chancellor may appoint. It is a United Kingdom court and may sit in two or more divisions and at any place within or without the United Kingdom. The court consists of an odd number of members, at least three, and at least one of whom is a judicial member.

A person convicted by court-martial may appeal to the Court, with its leave, against conviction, but not against sentence, though the Court may in certain cases substitute another sentence. Certain authorities may also refer the finding of a court-martial to the Court. The decision of the court is that of the majority of those sitting. The court has power to call for documents, evidence, and reports, refer a question for inquiry, and authorize a new trial where a conviction is quashed in the light of fresh evidence.

Appeal lies with leave to the House of Lords at the instance of accused or the Defence Council on a point of law of general public importance.

Courts, Naval. See NAVAL COURTS.

Courts, U.S. Being a federal state, the U.S. courts form two systems, federal and state, each having exclusive jurisdiction in certain matters, though where there is overlap the Federal courts have superior powers. The Federal Courts comprise the Supreme Court of the U.S. (q.v.), the 11 Circuit Courts of Appeal and the Federal District Courts, together with courts of special jurisdiction such as the Court of Claims and the Court of Tax Appeals.

The judicial system of each State is determined by the constitution and laws of that State and normally comprises a State Supreme Court or Court of Appeals, sometimes an intermediate level of appellate courts, a level of district or county courts, and finally a level of magistrates', municipal, and justice of the peace courts for minor cases. There are also frequently special courts for domestic relations, juvenile courts and probate courts.

Cousin. A collateral relation other than the brothers and sisters of any ancestor, and one's brothers or sisters and their descendants. A first cousin is the child of one's uncle or aunt, and the child, grandchild, etc. of such a cousin is a first cousin once, twice, etc. removed. A second cousin is the grand-child of one's great-uncle or great-aunt, and the child, grandchild, etc. of such a cousin is a second cousin once, twice, etc. removed. The word is sometimes used of any collateral relative and used officially when the sovereign addresses an earl as cousin though not related to him.

Coutumes. Customary law and general custom, particularly the bodies of law in force in parts of northern France prior to the Revolution. Mediaeval local customary law was based on Frankish law, Roman law, canon law, feudal law, and royal legislation. Initially each manor had its own body of custom. In the south the rediscovery of Roman law had such influence that a body of written law, modelled on Roman law but influenced by customs, became accepted as the general customary law of the area. In the north after 1300 customs became more uniform within larger areas such as Normandy, but the customs differed very substantially from one another. At the time of the Revolution there were still about 60 provincial *coutumes* and about 300 local *coutumes.*

A movement early developed to record the various bodies of customs and in the thirteenth and fourteenth centuries various scholars recorded the customs of their provinces in *coutumiers* (q.v.) and in the fifteenth century some provinces sponsored editions of their own customs. In 1453 Charles VII ordered compilations of all the *coutumes.* This proceeded slowly and in 1495 Charles VIII established a commission to oversee the enterprise. Each *coutume* was to be collected, examined by commissioners and presented to representatives of the three estates of the area for consideration. The final version was recorded by the local *parlement.* Thereafter the *coutumes* had the force of statute and could not be changed save by a new edition. Most of the *coutumes* were compiled in the early sixteenth century, and many were severely criticized by legal scholars. Accordingly in the second half of that century a project was initiated of subjecting them to a fresh editing and such revision continued until the Revolution. By that time there were 60 *grandes coutumes* covering the provinces and 300 local *coutumes.*

The local *parlements* used the local *coutumes* to prevent the development of a common body of law for France. The *coutume* of Paris gradually, however, developed as the *coutume* to be consulted in default of local custom or if it was inconclusive. There was also a gradual development from about 1650 of the idea that the common law of France was contained in the consensus of opinion of the *coutumes* and books adopted on that standpoint. Such books as well as the major *coutumes* themselves, particularly that of Paris, were important sources for the work of codification which culminated in the adoption of the Code Napoléon.

A. Gouron and O. Terrin, *Bibliographie des Coutumes de France.*

Coutumes de Beauvaisis (*c.* 1280). A work compiled by Phillipe de Remy, lord of Beaumanoir, judge and deputy-governor of Clermont. It is a compilation of the customs and also a treatise on Customary law, based on sound knowledge of Roman law. It had great influence on the legal usages of northern France.

Coutumiers. The generic name for compilations of the customs of particular regions in northern France made in the thirteenth and fourteenth centuries. The more important include the *Très Ancien Coutumier de Normandie*, the *Grand Coutumier de Normandie*, the *Coutumiers de Picardie et de Vermand*, the *Anciens Usages d'Artois*, the *Livre de Justice et de Plet*, the *Etablissements de Saint Louis*, Beaumanoir's *Coutumes de Beauvaisis*, the *Somme Rural*, the *Grand Coutumier de France* (qq.v.) and custumals of Anjou, Berry, Brittany, Maine, Orléans, and Poitou.

Covarruvias y Leyva, Diego (1512–77). Known as the Spanish Bartolus, professor of Canon Law at Oviedo, Bishop of Ciudad Rodrigo and Segovia, a humanist, and one of the chief jurists of his time, author of *Variarum ex jure Pontificio, Regio et Caesareo resolutionum libri IV* and *Practicae Quaestiones*. He concerned himself with the justification of the Spanish conquest of America.

Covenant. One of the early forms of personal action in English law originally used in cases of leases of land for terms of years. About the late thirteenth century the rules became established that it could not be used where an action of debt would lie, and that covenant would lie only on a deed under seal, and thereafter it ceased to be of great importance. It gave rise to the action of covenant and the writ of covenant.

The word is also applied to a promise or agreement under seal, and to a particular undertaking contained in a deed, or implied by law in a deed of the particular kind, as in, e.g. 'covenants running with the lands' or 'covenants in restraint of trade'.

F. W. Maitland, *The Forms of Action at Common Law*.

Covenant, Book of the (or Covenant Code). An ancient collection of law contained in Exodus 20, 22, 23, 19 and 23, 20–33. It shows considerable similarity to the Code of Hammurabi and is divided into prologue, laws on the worship of Yahweh, laws as to persons, laws as to property, laws concerned with the continuation of the Covenant between God and his people, and an epilogue containing warnings and promises.

Covenant of the League of Nations. The charter, forming the first part of the Treaty of Versailles (1919), establishing the League of Nations (q.v.).

Covert. In old law a married woman, during the subsistence of the marriage, was in Norman French *feme covert*, as being under the protection of her husband.

Coverture. The state of being a married woman and under the protection of her husband.

Covin. A secret agreement of two or more persons to defraud or prejudice another. It seems originally to have avoided the transaction affected but did not give rise to liability as did fraud or conspiracy.

Cowell, John (1554–1611). Became a member of Doctors' Commons in 1584, Regius Professor of Civil Law at Cambridge, 1594, and Master of Trinity Hall, 1598, and reputed the most learned civilian of his time. In 1607 he published *The Interpreter*, a dictionary of legal terms, which was suppressed by the Commons because of the monarchical character of some of its definitions. He resigned his chair in 1611. Several later editions of the book were published subsequently. He also wrote an *Institutiones Juris Anglicani ad methodum institutionum Justiniani* (1605).

Cowper, William, Lord (1664–1723). He was called to the bar in 1688 and at once allied himself with William of Orange. He soon acquired a large practice and also became leader of the Whig party. He became Lord Keeper in 1705, a peer in 1706 and in 1707, having taken a leading part in the negotiations for the Union with Scotland, the first Lord Chancellor of Great Britain. He resigned with his party in 1708 but was reappointed when George I succeeded and held the Seal again 1714–18, being the new King's chief adviser and taking a leading part in securing the passing of the Riot Act and the Septennial Act. In 1718 he became an earl. He was an eloquent speaker, an honourable and fair judge, and a sound statesman. As a chancery judge he won high approval for learning and ability, though he tended to be conservative, and he made considerable contributions to the formation of the modern system of equity.

Cox, Sir Richard (1650–1733). Was called to the English Bar in 1673 and the Irish Bar in 1674. He became a judge of the Common Pleas in Ireland in 1690, Chief Justice in 1701, Lord Chancellor of Ireland, 1703–07, and Chief Justice of the King's Bench in Ireland 1711–14. He published a number of works including a History of Ireland from the English Conquest.

Cozens-Hardy, Herbert Hardy, Baron Cozens-Hardy (1838–1920). He was called to the Bar in 1862 and developed a good chancery practice. He sat in Parliament from 1885 to 1899 when he became a chancery judge. In 1907 he was promoted to the Court of Appeal and in 1907 he became Master of the Rolls. He became a peer in 1914 and retired in 1918. He was a sound and industrious judge.

Craft union. A trade union combining workers who though employed at different places and by different employers are engaged in a particular craft or trade. They were first formed on a permanent basis in Britain in the nineteenth century and their power is a result of their skills and their control of the supply of craftsmen, through regulation of apprenticeship and entry.

Craig, Sir Thomas (1538–1608). Studied in France and was admitted to the Scottish Bar in 1563 and in the next year became Justice-Depute under the hereditary Justice-General of Scotland, and in 1573 Sheriff of Edinburgh. He practised at the Bar for many years but devoted much time to his learned *Jus Feudale* (*c.* 1603) (published 1655, translated as *Craig's Jus Feudale* by Lord President Clyde in 1934), which is a milestone in the development of Scots law. It is the earliest reasoned examination of Scots law, relating the customary law to the Roman, canon, and feudal laws, the common legal heritage of Western Europe. In 1603 he wrote a *Treatise on the Succession* to further the claim of James VI of Scotland to succeed to the Throne of England. He went to England with James VI and I in 1603. In the following year he was one of the Scottish commissioners to negotiate on the King's project of a closer union between Scotland and England, and in 1605 wrote a work urging this course, *De Unione Regnorum Britanniae Tractatus* (Scottish History Society, 1909). His last work related to the controversy regarding the homage which England asserted was due from Scotland. This was published by George Ridpath in 1695 as *Scotland's Sovereignty Asserted, or a Dispute concerning Homage.* He also wrote much Latin verse. He is still respected as one of the major Scottish jurists and the *Jus Feudale* is one of the great classics of Scottish legal literature. Many of his descendants were notable in the law.

P. F. Tytler, *Life of Sir Thomas Craig* (1823); Intro. to Lord Clyde's translation of *Jus Feudale.*

Craig* v. *Missouri (1830), 4 Peters 410 (U.S. Sup. Ct.). The question was whether a Missouri law authorizing the issue of loan certificates was a justifiable use of the State's power to borrow or contravened the clause in the Constitution (Art. I, Sec. 10) prohibiting states from issuing bills of credit. The Supreme Court by a majority, took the latter view, one which aroused some resentment at the time.

Craigie, Robert, of Glendoick (? 1685–1760). Called to the Scottish Bar in 1710, he developed a busy practice and was appointed Lord Advocate in 1742. He was very busy during the Jacobite Rebellion of 1745–46 but lost office in the latter year, but in 1754 became Lord President of the Court of Session and held that office till his death. A man of only moderate ability but of industry and perseverance, and with a profound knowledge of law, he seems to have been a competent head of the court, with a considerable reputation in the field of land law.

Cranworth, Lord. See ROLFE, R. M.

Crassus, Lucius Licinius (140–91 B.C.). Roman lawyer and politician, along with Marcus Antonius the greatest forensic orators before Cicero. He sponsored the *Lex Licinia Mucia*, which provided for prosecution of any who falsely claimed Roman citizenship but later supported attempts to enfranchise Rome's Italian allies and to reform the courts.

Craven. A term of disgrace applied to a combatant in trial by battle (q.v.) who admitted defeat, by crying 'cravent'. The penalty for such a person was to become infamous and be deemed unfit to be a witness or serve on a jury. Hence 'to cry craven' is to admit defeat.

Credibility. In a witness, the quality of deserving to have his evidence believed. Factors affecting credibility are the witness's knowledge of the facts, disinterestedness, fairness, and apparent veracity. A witness may be cross-examined for the sake of impugning his credibility.

Credit (Trust or reliance). The practice of obtaining or supplying goods or services in return, not for immediate payment, but for a promise of future payment. There may, or may not, be a charge for the deferment of payment in the form of interest. Credit is of great importance in modern society and business. Whether one party will deal with the other on credit depends on his estimate of the debtor's character, capacity to pay, and any security offered. Much modern credit is extended by specialized financial institutions, banks, finance houses, and others.

Credit agency, mercantile. An organization engaged in recording and supplying information on the financial standing and credit-worthiness of business firms. Such agencies may be general, providing information on all kinds of businesses, or specialized, concerned only with firms in a particular business or particular district or with supplying information only to members of a trade protection society. Such an agency must be licensed by the Office of Fair Trading.

Credit brokerage. The business of putting person seeking loans of money or credit in touch with sources of finance. It includes, e.g. estate agents or solicitors putting prospective buyers in touch with building societies, and shops which introduce prospective purchasers of goods on hire-purchase to finance companies. Such businesses must be licensed by the Office of Fair Trading.

Credit bureaux. Organizations providing information to traders on the credit-worthiness of customers. They may be private enterprises or co-operative enterprises organized by traders in a district. Sources of information include reports from traders who have dealt with the customer, his employers, landlords, court records, newspapers, and other investigations.

Credit card. A small card issued to a person, containing a means of identification such as a photograph or specimen signature, enabling him to obtain goods and services on credit and charge those obtained to his account which he must settle periodically. The first universal credit card, which could be used at a wide variety of premises, was introduced by Diners Club Inc. in 1950. Co-operating traders pay a service charge and the company charges cardholders an annual fee and submits a consolidated bill monthly or annually. A further development is the bank credit-card which operates in the same way but the cardholder may obtain credit from the bank for part of the balance due, paying interest on the unpaid balance. By reason of this interest element bank credit-cards can be issued without annual fee to the holder.

Credit reference agencies. Businesses which maintain records of information, and, on request, supply traders with information, about the credit-worthiness of individuals. They obtain information from court records of unsatisfied judgments, records of bankruptcy proceedings, and information from traders about bad debts and defaulting customers. A check by a trader on a prospective customer's credit-worthiness may prevent dishonest or unreliable customers from getting goods on credit and cut down bad debts. Serious problems arise as to the accuracy and privacy of the information and consumers can insist on having a copy of the information held about them and request correction.

Credit sale. A contract of sale in which credit is given for the whole or a material part of the price. The property in the goods sold passes at once, the seller being a personal creditor only for the unpaid price. Such a transaction is specially defined in consumer credit legislation.

Credit unions. In Canada and U.S.A. non-profit-making co-operative credit associations organized to afford members an opportunity of saving and of obtaining loans at moderate interest, usually for short terms for the purchase of consumer goods. They originated in societies founded in the nineteenth century in Germany and Italy and started in Canada in 1900, since when development has been rapid.

Creditor. A person to whom another, the debtor, owes money, or sometimes more generally, the person in right of any obligation or debt due by another. According to the mode in which the debt is constituted a creditor may be a contract creditor, a bond creditor, a judgment creditor or an execution creditor. According as he does or does not hold security he may be designated a secured or an unsecured creditor. In bankruptcy and liquidation there are rules regulating the ranking of different classes of creditors.

Cremation. The disposal of the body of a deceased by burning. It has been recognized as legal in England since 1884 but there are special rules about death certificates to minimize the risk of cremation being used to conceal death caused criminally.

Cresswell, Sir Cresswell (1794–1863). Was called to the Bar in 1819, soon attained a practice, and was an M.P. from 1837 to 1842, when he was made a judge of the Common Pleas. In 1858 he became the first judge of the newly established Court of Probate and Matrimonial Causes. In this office he discharged successfully the task of adapting the older ecclesiastical law and procedure to the new court's jurisdiction. The work proved heavy but he dealt with cases expeditiously and very acceptably. He was learned, diligent, and had great powers of accurate thought and exposition. He was part-editor of a series of King's Bench reports.

Crime. An act or omission which, whether or not it is morally wrongful or is deemed a wrong to an individual and civilly actionable by him for compensation for the harm done to him, is legally deemed an offence against the State or the community or the public and is punishable, as a deterrent to the offender and others, for the sake of public order, peace and well-being and in the interest of society. Any kind of conduct may be declared criminal by law, and it may be made criminal at one time and cease to be criminal at another without change in its characteristics. But, in general, conduct is made criminal if it threatens the security or well-being or good order of society. It frequently is of a kind generally regarded as morally wrong, but not always. Thus infringements of traffic regulations or of administrative rules, are not inherently criminal or immoral but are merely prohibited under penalty. The only essential difference between criminal conduct and conduct deemed to be merely a tort, breach of contract or other

CRIME

breach of legal duty, is that a crime is dealt with in criminal courts, under criminal procedure, and results in the imposition of a penalty or punishment or other treatment, whereas other conduct is the basis of civil proceedings for the recovery of property, damages, or other civil remedy. Many kinds of conduct are both crimes and breaches of civil duty, while other are crimes only or torts or breaches of contract only.

Classification of crime

Crimes are distinguished into common law crimes and statutory crimes, according to whether the legal source of, and authority for, the statement that particular conduct is criminal is found in common law or statute. Statutory crimes are usually called statutory offences, but beyond the general understanding that an offence is of lesser gravity than a crime, the distinction between crimes and offences has no legal significance. The distinction between *mala in se*, conduct deemed contrary to the rules of God and nature, and inherently criminal, and *mala prohibita*, conduct criminal only because it has been prohibited, is also of no legal significance, although most common law crimes would commonly be deemed *mala in se*.

In English law crimes were formerly distinguished into treasons, felonies, and misdemeanours. Treasons (q.v.) were certain crimes particularly directed against the sovereign and the safety of the State. Felonies (q.v.) were those crimes made such by common law or statute, and misdemeanours covered all the rest.

Elements of crime

The commission of crime generally requires, firstly, that the accused person committed an overt act of a kind deemed unlawful and criminal, or allowed to happen what he could and should have prevented, or omitted to do something which by law he should have done and the failure to do which is deemed unlawful and criminal. The act or omission must be voluntary, and in general there is no criminal liability for conduct which is involuntary by reason of sleep, physical or mental defect or illness, or physical coercion.

The commission of crime generally requires, secondly, that the act or omission was done with a legally blameable condition of mind, usually called *mens rea.* In common law crimes the *mens rea* required varies; it may be a specific intention, or either specific intention or utter recklessness as to the consequences of conduct, or, sometimes, serious negligence or disregard of foreseeable consequences. In statutory crimes the requisite mental element may be defined by the statute, or the statute may be silent, in which case the court may infer that common law *mens rea* is required, or that liability exists for the doing, irrespective of *mens rea* (strict liability). Accordingly, where a particular intent or state of mind is requisite for the crime, the accused may excuse himself by disproving that he had the requisite state of mind, but did the act justifiably, or accidentally, or in ignorance, or had an honest belief in the existence of facts which, if they had existed, would have made this act innocent.

A person is not deemed legally responsible for his conduct if at the time he was subject to such defect of reason, from disease of the mind, as not to know the nature and quality of his act, or as not to know that what he was doing was legally wrong. The onus is on the accused pleading insanity to establish that fact. Voluntary drunkenness is not an excuse, unless it has produced temporary insanity. In the case of murder a mental state bordering on insanity may be deemed diminished responsibility and reduce the crime to manslaughter. In general a person is responsible for his own criminal conduct only, but in some statutory cases a person has been held vicariously liable, i.e. for the crime of one for whom he is responsible.

The nature of the motive for, or absence of motive for, the conduct is irrelevant to criminal guilt, but may be relevant to punishment.

An attempt to commit a crime is itself a crime. It is frequently difficult to distinguish the stages of planning, preparation, attempt, and complete perpetration of a crime; to amount to attempt conduct must have gone beyond preparation and be immediately connected with and directly tending to the commission of the complete crime. To incite or solicit another to commit or attempt to commit crime is also criminal, as is conspiracy, or the agreement of two or more persons to do something contrary to law, or wrongful or harmful towards another person, or to use unlawful means in the carrying out of something otherwise lawful.

The most convenient grouping of specific crimes is into crimes against the State, the government and the community, the administration of justice, and public order generally, such as treason, treason felony, unlawful assembly, rout and riot, perjury, bigamy (qq.v.), and a vast number of statutory offences, many connected with road traffic and the use of vehicles; crimes against the person, such as murder, manslaughter, assault, and sexual crimes against the person such as rape, indecent assault, incest, abduction, and cruelty (qq.v.); and crimes in respect of property, such as theft, embezzlement, fraudulent misappropriation, robbery, extortion, burglary, housebreaking, receiving stolen goods, obtaining property by fraud, forgery, malicious damage to property, arson (qq.v.).

The causes of crime

The search for the cause of crime has been generally abandoned as fruitless, and many theories have been propounded of causal factors. Some concentrated on finding biological, constitutional, or other peculiarities in offenders, other on social, economic, or other external factors. All such, and other, factors may operate in individual cases but cannot validly be generalized. Among factors which influence a person towards crime or at least militate against law-abiding conduct are lack of family control and discipline, family disorganization, low-intelligence parents, large families, low income, irregular employment, and alcohol. Parental and school influence can do much to prevent resort to crime. Suitable instincts, including greed, lust, jealousy, hatred, often play a part.

It is recognized in Western societies that some behaviour infringing criminal codes requires psychiatric treatment, but the view that all criminals are sick and not bad cannot be sustained.

The incidence of crime

The measurement and analysis of the incidence of crime in a society or locality depends on notoriously unreliable statistics, dependent on discovery of the commission of crime, covering-up, reporting, non-reporting, and misreporting, classification and other factors. Even making allowances for defects and deficiencies in the statistical picture and ignoring merely technical contraventions, it appears that crime has increased substantially since 1945 in most Western countries.

Analysis of statistics indicates that, in general, crimes are much more prevalent in urban than in rural areas, that much more crime of nearly every kind is committed by men than by women, that young persons under 25 are responsible for a very disproportionate amount of crime, and that particular districts and even streets are much more associated with crime than others. Particular days and hours are also favoured.

Classification of criminals

A useful classification of criminals by reference to their career patterns is the casual offender, the habitual petty offender, who offends frequently but commits only minor crimes such as small thefts, the conventional or career offender or recidivist, who has a long record of convictions, has progressed from juvenile offences to adult criminality and has been in every kind of custodial institution, and professional criminals, who often undertake major crimes involving skill, planning, and sophisticated equipment, such as bank robberies. Distinct from these are white-collar criminals who engage in fraudulent practices, dishonesty in business, embezzlement, and the like. Particularly in the U.S., there are large-scale rackets, organized, long-term and protected by corruption of politicians and the police; they operate particularly in such areas as drug traffic, prostitution, gambling, and protection.

Control and repression of crime

The control and repression of crime requires everywhere an extensive criminal investigation organization, supported by technical and scientific staff skilled in finger-printing, photography, and other scientific skills, and a prosecuting organization. In England most prosecutions are done by the police though some require the consent of the Attorney-General or are conducted by the Director of Public Prosecutions, but in Scotland the procurators-fiscal (q.v.), subject to the direction of the Crown Office, conduct prosecutions and in most European countries there are agencies for public prosecution.

Crime, délit and contravention. A common classification of crimes in civil law countries, denoting respectively a serious crime, major offences, and petty offences. The classification is frequently correlated with types of tribunals and procedure. Many countries have, however, reduced the categories of crimes to two, *délits* which involved criminal intent and directly infringed the rights of individuals and groups, and *contraventions*, which were acts forbidden by law even without criminal intent.

Criminal Appeal, Court of. A superior court of record in England, created in 1907 and superseding the former Court for Crown Cases Reserved (q.v.) and proceedings in the Queen's Bench Division on writs of error from courts of assize and the Central Criminal Court. It had jurisdiction to hear appeals by persons convicted on indictment, criminal informations, and coroners' inquisitions, on grounds involving a question of law alone, or, with the leave of the court on a question of fact or of mixed fact and law, or, with such leave, against sentence. The Home Secretary might refer a case, or any point, to the court. It had power to quash a conviction, substitute a verdict for that returned, vary the sentence, but not to order a new trial, unless the trial had been a legal nullity. Further appeal lay, on conditions, to the House of Lords. The judges were any three or more of the Lord Chief Justice and all the judges of the Queen's Bench Division, and the court might sit in two or more divisions. In 1966 the court was abolished and the jurisdiction transferred to the Court of Appeal, Criminal Division (q.v.).

Criminal bankruptcy. A form of bankruptcy introduced in England in 1972 in relation to persons convicted of crime to make available to those who have, at the hands of persons convicted of an offence, suffered loss or damage, using the machinery of the Bankruptcy Act, 1914 to facilitate the recovery of assets to make good that loss or damage. A 'criminal bankruptcy order' made by the Crown Court is equivalent to an 'act of bankruptcy' on which a criminal bankruptcy petition is based. The Director of Public Prosecutions (q.v.) acts as official petitioner and the Official Receiver (q.v.) becomes trustee in bankruptcy and all the bankrupt's property vests in him.

Criminal compensation. See COMPENSATION ORDER; CRIMINAL INJURIES COMPENSATION.

Criminal conversation (adultery). A common law action for criminal conversation formerly lay at the instance of a husband against one who had committed adultery with his wife, for damages. After the institution of judicial divorce in 1857 this was replaced by a claim for damages against a co-respondent.

Criminal court. A court having jurisdiction in criminal cases. In England such courts are the House of Lords, the Court of Appeal (Criminal Division), the Divisional Court of the Queen's Bench, Crown Court, and Magistrates' courts. In Scotland they are the High Court of Justiciary, the Sheriff Court and the District Court.

Criminal information. A proceeding in the Queen's Bench Division filed by the Attorney-General or the Master of the Crown Office at the suit of a relator, and without a presentment by a grand jury or indictment. Most criminal informations were abolished in 1938 and the remainder in 1967.

Criminal Injuries Compensation. A scheme established on a non-statutory basis in 1964 whereby application may be made to the Criminal Injuries Compensation Board for an *ex gratia* payment of compensation where the applicant sustained personal injury directly attributable to a crime of violence, arrest, or attempted arrest of an offender or suspected offender, prevention or attempted prevention of an offence, or the giving of help to a constable engaged in arresting or attempting to arrest an offender or preventing or attempting to prevent an offence. Traffic offences are in general excluded, as are offences between persons living together, trivial offences, and circumstances not reported to the police. Compensation is assessed on the basis of common law damages but loss of earnings is limited to twice the average of industrial earnings and there is no element of exemplary or punitive damages. Compensation is reduced by the value of any entitlement to social security benefits and may be reduced by any pension accruing as a result of the injury. It is reduced by any sum received under a compensation order (q.v.) made by a criminal court and must be repaid from any damages or compensation recovered. Compensation is normally a lump sum payment.

Claims may also be made by a spouse or dependants of a person killed by a crime of violence, but no compensation is payable for the benefit of his estate. It is reduced by the value of any entitlement to social security benefits.

The Board is appointed by the Home Secretary and the Secretary of State for Scotland. All members are legally qualified. It sits in London, Edinburgh, and elsewhere as necessary. Hearings are in private. Applications are normally decided by one member but a dissatisfied applicant is entitled to a hearing before three other members. No appeal lies to the courts.

Criminal intent. The state of mind which is a necessary ingredient of a criminal offence. It varies from one crime to another and the specification of the requisite criminal intent is part of the definition of each crime. At common law the requisite criminal intent was generally the intentional or at least reckless doing of the forbidden kind of conduct. Under statute the requisite criminal intent is indicated by many different words in different contexts; these include 'with intent to . . .', 'recklessly', 'maliciously', 'wilfully', 'knowingly', 'fraudulently', 'dishonestly', 'corruptly', 'causing', 'allowing', 'permitting', and others. Many of these words require, and have received, judicial interpretation in particular cases. Increasingly, in modern statutory offences, the offence is one of strict liability, the doing inferring guilt, so that the only intent involved is the allowing or permitting of the prohibited conduct to happen.

Criminal law. The body of law dealing with the definition of what kinds of conduct, in what circumstances and with what concomitant mental or other factors, amount to crimes, i.e. to conduct declared to infer liability to prosecution and punishment, and with the prescription of kinds of punishment. More than many branches of law, what is criminal, and consequently within the ambit of the criminal law, depends on the general views of the community as interpreted by the legislature. Conduct may readily be made, or cease to be, criminal. The major topics of criminal law are the constituents of criminal conduct in general, and of each particular kind of crime, the mental states necessary for particular kinds of conduct to be criminal, the effect of abnormal mental states, such as insanity, and the validity of defences, such as ignorance, provocation, or self-defence. Lastly, there is the question of what penalties a person found guilty is liable to suffer.

Liability to criminal prosecution and to criminal sanctions may overlap with civil liability, i.e. liability to be sued for wrongful conduct or harm done and to incur liability in damages. Some kinds of conduct may be both criminal and tortious; others may be criminal or tortious depending on circumstances; thus an intentional or reckless injury to another will be criminal, while a merely negligent or purely accidental injury will normally be, at most, tortious.

The criminal law is a major form of social control. The older view was that it had to repress kinds of conduct which were major disturbances of peace, order, and good government, and wicked, such that the offender must suffer retribution for harm done and expiate his moral guilt. Under the influence of scholars of the Enlightenment it came to be thought that the main function of the criminal law was to prevent crime, protect the public and reform the offender, though these are unattainable ideals.

The form and content of the criminal law differs markedly from one jurisdiction to another.

In England the criminal law developed over centuries through judicial decisions, though a number of important statutes were passed in 1861 and in the twentieth century substantial modifications have been made by legislation. In Scotland the criminal law is still mainly common law, in many respects different from English common law. U.S. criminal law was derived from English common law, adapted in some respects to American conditions. In a number of States penal or criminal codes have been enacted but, though these supersede the common law, they are frequently codifications of common law or consolidations of criminal statutes enacted at various times. Since 1945 both in the U.K. and the U.S. there have been extensive examinations of the criminal law and moves towards codification.

In Western European countries the major modern influence has been the Napoleonic *Code d'Instruction Criminelle* (1808) and *Code Pénal* (1810) and later the German penal code (1871) and code of criminal procedure (1879).

In most countries certain fundamental principles are accepted, namely (1) the principles of *nullum crimen sine lege* or *nulla poena sine lege*, that there is no crime unless law specifies the conduct done as such; hence immoral, antisocial, or evil conduct is not criminal unless it falls within a category of crime, as defined by the courts in common law countries or by the code, and even in common law countries the function of declaring criminal what has not previously been found so is now normally left to the legislature; (2) criminal provisions must be clear and unambiguous to give fair warning of what conduct is liable to penalty; (3) criminal statutes must be interpreted strictly and not extended by analogy; and (4) unless the statute plainly so directs, a criminal provision must not be retrospective, penalizing what was innocent when it was done.

L. Radzinowicz, *History of English Criminal Law* (4 vols.); G. L. Williams, *Textbook of Criminal Law*; J. C. Smith and B. Hogan, *Criminal Law*; G. H. Gordon, *Criminal Law of Scotland*; J. Hall, *General Principles of Criminal Law* (U.S.).

Criminal liability. Liability to be made subject to penalty, punishment, or treatment for having contravened a provision of the criminal law, as distinct from civil liability (q.v.). Criminal liability depends on confession, or being found guilty by a court or jury of having committed an act falling within a known category of crime with the mental element required for guilt thereof. In general, a person does not incur criminal liability unless he intended to bring about, or recklessly brought about, those elements which constitute the crime but occasionally a person incurs criminal liability if he has negligently brought about the constituent elements and in some cases, principally statutory cases, a person may incur criminal liability by reason of having allowed or permitted a happening, even though he acted without intention, recklessness or negligence. Such are usually called crimes of strict or absolute liability. There is a presumption in favour of the requirement of intention or recklessness and this presumption must be overcome before strict liability is held imposed. The definition of every crime accordingly, expressly or by implication, includes specified elements which make up its *actus reus* or criminal act, such as, in the case of murder, unlawfully to cause the death of a human creature in being and within the Queen's peace, and a specified mental element or *mens rea*, which in the case of murder is malice aforethought (q.v.) either express or implied by law.

If criminal liability is incurred the extent of that liability depends on various factors, particularly the competence of the court dealing with the case, the statutes regulating the powers of courts in relation to penalties for cases of that crime, the age and mental capacity of the defendant, the gravity and circumstances of the particular offence, the defendant's age, criminal record, and suitability for particular kinds of correctional measures, possibly the recommendations of social workers, psychiatrists, and others, and the judge's view of what is most appropriate in the circumstances.

Criminal libel. Libel (q.v.) made the subject of indictment as a public offence, as tending to provoke the person defamed to commit a breach of the peace. The defendant in a prosecution for libel was never allowed at common law to allege truth by way of justification unless it were also for the public benefit. The prosecution is not bound to prove that the libel is unusually likely to provoke a breach of the peace.

Criminal lunatic. The former term for a Broadmoor (q.v.) patient.

Criminal offence. See CRIME.

Criminal procedure. The body of rules prescribing the steps to be gone through for the determination of the liability of a person to the sanctions of the criminal law for alleged conduct on his part. The rules must seek to reconcile the need to maintain public order and repress crime, with the need to give fair protection and a fair trial to persons accused. The essential elements of criminal procedure are (a) provision for securing the presence of the person accused; (b) provision for acquainting him with what conduct on his part is alleged by the prosecuting agency to be a contravention of the criminal law; (c) an opportunity to the accused to prepare his defence; (d) ascertainment of facts bearing on the accused person's guilt of the crime charged, including opportunity to the accused to challenge the facts alleged and to state his defence; (e) decision of any issue of law doubtful or challenged; (f) decision of the accused person's guilt of the crime charged, on the basis of the facts held to have happened and of the rules of law held relevant; (g) determination of the appropriate sanction; and (h) provision for appeal on fact, or law, or disposal of the accused. Other elements of procedure are ancillary to one or more of these main elements.

Criminal procedure is peculiarly a matter for each system of law and there are substantial variations of detail between the procedure of one system and that of another. In Anglo-American procedure the police are responsible for the execution of warrants, apprehension, search and seizure. In England most charges are preferred and most prosecutions conducted by the police, nominally acting as private citizens. In Scotland all prosecutions are initiated by public officials, the procurators-fiscal (q.v.), proceeding on information given by members of the public, police and others, or by other public officials such as factory inspectors. In European countries the police generally act as agents of the prosecuting authorities or examining judges.

Anglo-American law generally provides for pre-trial hearings, formerly by hearing before a grand jury and now by a hearing before an examining magistrate, to determine whether or not there is a *prima facie* case, to justify a trial. In Scotland there is no hearing and the decision is taken by the procurator-fiscal in consultation with the Crown Office. In European countries the pre-trial inquiry is conducted by the prosecuting authority, in France by a *juge d'instruction*.

A principle generally accepted in Anglo-American and European criminal practice is that in criminal cases the burden of proof is on the prosecution and that guilt must be established beyond reasonable doubt.

In Anglo-American law trial by jury was traditionally regarded as the standard and proper mode of trial, as the great protection of the accused against unjustified conviction, and the U.S. Constitution guarantees jury trial in criminal cases. In England however some offences may be tried either before a jury or summarily and many petty offences are triable summarily only, i.e. by magistrates.

In the U.S. the defendant may waive trial by jury. In European countries the jury has been almost entirely abandoned. In England and Scotland a jury may return a majority verdict; in the U.S. a jury is usually required to be unanimous.

Legal systems generally make provision against the danger that a person be prosecuted more than once for the same conduct. In Anglo-American law the major problem has been whether a second prosecution is for the 'same' or a 'similar' offence; it has generally been held that an acquittal on one charge precludes subsequent trial for any lesser crime comprehended in that charge. In European systems the question is usually whether the subsequent charge arises out of the same factual situation as the former one or not.

G. Mueller and Le Poole Griffiths, *Comparative Criminal Procedure*; J. Archbold, *Pleading and Procedure in Criminal Cases*; M. Arguile, *Criminal Procedure*; L. Orfield, *Criminal Procedure and the Federal Rules*.

Criminalistics. The name, particularly in European countries, for the study of crime detection, including photography, toxicology, fingerprinting, analysis of blood, and other techniques. It is sometimes included in criminology, sometimes in forensic science and medicine.

Criminology. The study of the phenomena of criminal conduct in society, including the causes of criminal conduct, the kinds of crimes committed, their prevalence in particular districts or age-groups, and the general pattern of crime in society, with the object of developing a body of general and verified principles and other knowledge regarding the processes of law, crime, and treatment.

In a wide sense of the word, criminology includes penology (q.v.) but the two, properly speaking, must be distinguished. In some countries the term includes criminalistics, or the science of criminal investigation. It may include study of the prevention of crime.

Criminology received little serious study until the eighteenth century. The classical school of criminology originated about 1775, when Beccaria (q.v.) applied the hedonistic psychology, that man regulates his behaviour by balancing pleasures and pains, to penology, and urged that all persons who violated a particular law should receive identical punishments. Beccaria and his English followers

Romilly, Howard, and Bentham, were all mainly interested in penology and in humanizing penal methods. The cartographic school emerged about 1830; the leaders, such as Quetelet and Guerry in France, were concerned primarily with the geographical and social distribution of crimes. The socialist school of criminology, based on the writing of Marx and Engels, emerged about 1850 and emphasized economic determinism, but proceeded by factual and statistical studies, and elicited much material as to variations in crime rates associated with variations in economic conditions. Thereafter three typological schools developed, all based on the view that criminals differ in personality traits from non-criminals. The Lombrosian or Italian school originated in the work of Cesare Lombroso (q.v.), who contended that criminals were by birth a distinct type with physical anomalies and recognizable by those characteristics. This view has subsequently been disproved. The mental-testing school held the view that inherited feeble-mindedness caused crime, but subsequent work has largely disproved this thesis also. The psychiatric school emphasized psychoses, mental instability, and emotional disturbances. In the twentieth century crime has been analysed in a sociological manner and is frequently interpreted as a function of social environment and as related to social organization. Most recently criminologists have insisted that crime in a product of a large number and great variety of factors, each crime being caused by a particular combination of causal factors.

Modern criminological study depends largely on statistics of many kinds, studies of particular kinds of crimes, and individual case studies, and utilizes elements from the work of most of the earlier schools of criminology but is increasingly cautious about formulating general hypotheses of causation. Criminology is inter-disciplinary and draws background material, methods and techniques from many sciences.

H. Mannheim, *Comparative Criminology*; H. Mannheim, *Pioneers in Criminology*; E. Sutherland and D. Cressy, *Criminology*.

Cripps, Charles Alfred, Baron Parmoor (1852–1941). Had a brilliant academic career, was called to the Bar in 1877 and established a good practice. He entered Parliament in 1895 and sat, with intervals, till 1914. In 1914 he became a peer and sat in the Judicial Committee of the Privy Council, especially in prize appeals during World War I.

In 1924 and 1929–31 he was Lord President of the Council; he was a leading layman in the Church of England, a pacifist, and advocate of international amity. He wrote a *Treatise on the Principles of the Law of Compensation*.

Crofter. The tenant of a small-holding or croft in the north-west Highlands and Islands of Scotland. The status of crofters was greatly improved by statute in 1886 while in 1955 there was re-established a Crofters' Commission to develop and improve the crofting way of life.

Crofton, Sir Walter Frederick (*fl. c.* 1860). Chairman of the Board of Directors of the Irish prisons, a notable prison reformer of the mid-nineteenth century.

Croke, Sir George (1560–1642). Called to the Bar in 1584 but does not seem to have acquired much practice for some years although he had begun reporting law cases as early as 1581. He was a judge of the Common Pleas, 1625–28, and of the King's Bench, 1628–41. In the great constitutional cases of the time he always resisted royal interference with judicial procedures. His reports, covering 1580–1640, are in law French but were later translated by his son-in-law, Sir Harbottle Grimston (q.v.).

Cromarty, Earl of. See MACKENZIE, GEORGE.

Crosbie, John, Earl of Glandore (1752–1815). Sat in the Irish Parliament, became a Privy Councillor and was joint Master of the Rolls in Ireland 1789–1801. He appears to have been a scholarly man.

Cross, Geoffrey, Lord Cross of Chelsea (1904–). Called to the Bar in 1930, he took silk in 1949, became a judge of the High Court in 1960, a Lord Justice of Appeal in 1969 and was a Lord of Appeal 1971–76.

Cross-examination. The questioning of a witness by a party other than the party which adduced the witness and normally after he has been examined in chief by the party adducing him. Cross-examination is intended to cause the witness to alter, qualify, amend, or retract evidence given, to discredit his evidence, and to elicit from him evidence favourable to the party cross-examining. Leading questions (q.v.) are allowed in cross-examination and greater latitude is usually allowed to the questioner than in examination in chief. Failure to cross-examine a witness on any matter generally implies acceptance of his evidence on that point. If evidence has already been given or is to be given to a different effect from that stated by the witness the effect of that evidence must be put to him in cross-examination to enable him to admit, deny, or explain it.

Crossed cheque. A cheque (q.v.) on which the drawer has marked two parallel lines diagonally across the instruction to pay, having between them the words 'and company' or any abbreviation thereof, with or without the words 'Not negotiable', or the two parallel lines with or without the words 'not negotiable' (general crossing) or with the name of a banker written across the instruction to pay, with or without the words 'Not negotiable' (special crossing to that banker). Where a person takes a crossed cheque marked 'Not negotiable' he does not have, and cannot give, a better title to the cheque than that which the person from whom he took it had, but the cheque remains negotiable, i.e. transferable. The effect of crossing is that the bank on whom the cheque is drawn can pay it only to or through another bank and this gives some security against the cheque being cashed if it is lost.

The words 'Account payee' are not a crossing by themselves or in conjunction with one of the authorized crossings. They are a mere direction to the receiving banker and should put a collecting banker on his enquiry if other than the named payee seeks to be credited.

Crown. The term 'crown' has been extended in British usage from denoting the crown which the monarch wears on his or her head on state occasions as the badge of royal office, rank and power to the wearer himself or herself, to the monarch for the time being in his or her official capacity, and further to many contexts where there was originally and still is in theory a connection between the institution and the sovereign personally. The King or Queen for the time being who wears the crown is the personification of the dignity, powers and duties attaching to the Crown or monarchy as the institution of government. The term 'the Crown' accordingly frequently means the Queen acting as head of State on the advice of her Ministers, as distinct from the Queen personally. In the last, most abstract sense, the term 'the Crown' is roughly synonymous with the term 'the State' in some other countries. Thus one finds mention of Crown Colonies, Crown courts, Ministers of the Crown, Crown proceedings, Crown lands, Crown copyright, allegiance to the Crown, and so on. A representation of a royal crown is used in very many contexts as indicating that the thing or person so marked belongs to the Crown, the royal establishment, the government, the royal forces, etc.

Title and Succession to the Crown

The title to the Crown of the United Kingdom was regulated at common law by the feudal principles of inheritance of land, with preference of males over females, and of elder over younger, with the qualification that in the event of succession of females the eldest succeeded alone. The succession has, however, been settled by statute by and since the Act of Settlement, 1700, the Union with Scotland Act, 1706, and the Union with Ireland Act, 1800, on the heirs of the body of Princess Sophia, Electress of Hanover, grand-daughter of James VI of Scotland and I of England, being Protestants. The present order of succession follows the principles of the descent of land under the feudal system, the succession opening successively to descendants, collaterals and ascendants with at every stage priority for males over females and for age over youth. It is accordingly (1) Prince Charles, (2) Prince Andrew, (3) Prince Edward, (4) Princess Anne, (5) her children in order, (6) Princess Margaret, (7) Viscount Linley, (8) Lady Sarah Armstrong-Jones, (9) the Duke of Gloucester, (10) his children in order, (11) the Duke of Kent, (12) his children in order, (15) Prince Michael of Kent, (16) Princess Alexandra, (17) her children in order, (18) the Earl of Harewood, (19) his children in order.

When a female succeeds as Queen Regnant she has the same powers as a King. By convention a King's wife takes the rank and style of her husband and becomes Queen Consort, but the converse does not apply and no constitutional position attaches to the husband of a Queen Regnant.

Royal style and titles: family name

The present royal style and titles, under statute, are: Elizabeth the Second, by the Grace of God of the United Kingdom of Great Britain and Northern Ireland and of Her other Realms and Territories Queen, Head of the Commonwealth, Defender of the Faith. The form of the royal title varies for the other member-countries of the Commonwealth which owe allegiance to the Crown to suit the particular circumstances of each. Since 1917 the Royal House and Family is styled the House and Family of Windsor, and all descendants in the male line of Queen Victoria, being British subjects, other than females married or on marriage, bear the name of Windsor. A female descendant, on marriage, takes her husband's surname in the ordinary way, e.g. the Princess Alexandra, the Hon. Mrs. Angus Ogilvy.

The sovereign's family

The sovereign's consort is a subject and may sue and be sued as a private person. A Queen-consort is, however, protected as to life and chastity by the law of treason. The sovereign's children are subjects, but the eldest son and eldest daughter are protected by the law of treason. The eldest son succeeds at birth to the rank of Duke of Cornwall and the Principality of Scotland, carrying the titles of Prince of Scotland, Duke of Rothesay, Earl of Carrick, Baron of Renfrew, and Lord of the Isles. He is customarily created Prince of Wales; the younger sons are customarily created Dukes. The eldest daughter is sometimes created Princess Royal.

Demise of the Crown

On the death of a sovereign the office is at once transmitted to the person next in succession. Demise of the Crown formerly caused a dissolution of Parliament but this rule has been altered, nor does it now have any effect on the holding of any office under the Crown.

Accession

There is no interregnum between the death of one sovereign and the accession of the next. The new sovereign is proclaimed at once at an Accession Council to which all members of the Privy Council are summoned and at which the Lords Spiritual and Temporal, the Lord Mayor and aldermen of London, and certain other persons are present.

Regency

If the person succeeding to the throne is under age or thereafter becomes totally incapacitated statutory provision exists for the royal powers being exercised by a Regent.

Coronation

The coronation follows the accession after about a year. As shown in the case of Edward VIII it is not legally necessary for the performance of any royal function. It takes place in Westminster Abbey in the presence of representatives of the Lords, the Commons, the great public interests of the United Kingdom, and representatives of Commonwealth and foreign countries.

Under the coronation oaths the duties of the sovereign are (1) to govern the people of the United Kingdom of Great Britain and Northern Ireland and the dominions thereto belonging according to the statutes in Parliament agreed on and the laws and customs of the same, (2) to cause law and justice in mercy to be executed in all judgments, to the utmost of the sovereign's power, (3) to maintain the laws of God, the true profession of the Gospel, and the protestant reformed religion established by law; (4) to maintain and preserve inviolably the settlement of the Church of England, and the doctrine, worship, discipline, and government thereof, as by law established in England, (5) to preserve unto the bishops and clergy of England, and to the Church therein committed to their charge, all such rights and privileges as by law do or shall appertain unto them or any of them, and (6) to preserve the Presbyterian Church in Scotland.

Powers and functions

The whole government of the United Kingdom is carried on in name of the Crown. The Queen is an integral part of the legislature, head of the judiciary in each of the parts of the United Kingdom, commander-in-chief of the armed forces and temporal governor of the established Church of England but in all these, and most other respects, the Queen acts on the advice of the Ministers appointed from the party having for the time being a majority in the House of Commons, and in particular on the advice of the Prime Minister.

Among the more important powers and functions of the Queen are to summon, prorogue and dissolve Parliament, to open its sessions with a Speech from the Throne and to give Royal Assent to bills passed by Parliament, to confer honours and dignities, to make appointment to important offices of State and leading positions in the Church of England, to commute criminal sentences or pardon offenders, to conclude treaties, accept or cede territory, and declare war or make peace.

Of the royal powers some exist by custom but others have been conferred by statute, being usually allocated to a Minister of the Crown who is responsible for their exercise. The customary powers constitute the prerogative (q.v.), 'the residue of discretionary or arbitrary authority, which at any given time is legally left in the hands of the Crown'.

Crown Agent. Solicitor to the Lord Advocate's (q.v.) department and in charge of the Crown Office, the headquarters of the administration of criminal justice in Scotland. He acts as solicitor in all legal proceedings in which the Lord Advocate appears as representing his own department. He issues all general instructions from the Lord Advocate for the guidance of Crown counsel, procurators-fiscal, sheriff clerks and other public officials, transmits instructions from Crown counsel to procurators-fiscal about prosecutions, and, in consultation with the Clerk of Justiciary, arranges sittings of the High Court of Justiciary. At trials in the High Court in Edinburgh he attends as instructing solicitor. He is appointed by and holds office during the pleasure of the Lord Advocate and accordingly goes out of office on a change of Lord Advocate.

Crown Agents. The commercial and financial agents in the U.K. of colonies which do not have representative government and of protectorates. They are appointed by the Foreign Secretary but responsible to the governments of the colonies for which they act.

Crown Appointment Commission. A body established in 1977 to choose bishops of the Church of England. It comprises the two Archbishops and representatives of the General Synod of the Church and of the diocese, with the Prime Minister's Patronage Secretary and the Archbishop of Canterbury's Appointment Secretary. On a vacancy occurring the Commission decides on two names, possibly in order of preference, for submission to the Prime Minister for nomination by the Queen, but the Prime Minister may disapprove the

Commission's suggestion and put forward his own nomination to the Queen. When an Archbishop of Canterbury is to be chosen the Commission will be presided over by a layman nominated by the Prime Minister and the Secretary-General of the Anglican Consultative Council will also be a member.

Crown Cases Reserved, Court for. For long there was no form of appeal from decisions of the King's Bench in criminal cases, but the practice developed of judges reserving points of law for informal discussion with their brethren. In 1848 the Court for the Consideration of Crown Cases Reserved was established, having as members all the judges, with a quorum of five, and jurisdiction to hear questions of law arising during trial and reserved by the trial judge. It remained a matter for the discretion of the trial judge whether or not to reserve a point. The Court considered many leading cases, such as *Bailey* (1800), *Prince* (1875), and *Ashwell* (1885). If the judges were equally divided in opinion, as in *Ashwell*, the conviction was affirmed. The Court was replaced in 1907 by the Court of Criminal Appeal, when appeal by the person convicted was also permitted, and the jurisdiction is now vested in the Court of Appeal, Criminal Division.

Crown colony system. A pattern of colonial government developed by Britain after 1815 under which power was in the hands of a governor and executive council. Later, legislative assemblies were added, though the governor could still veto the legislation passed. It was a suitable system when the populace was not sufficiently advanced to participate in fully democratic self-government, or where the populace was in general opposed to the British attitude to government. The system declined from the 1880s as more and more territories acquired elected assemblies and moved towards self-government.

Crown Court. The superior English criminal court, created in 1971 and replacing assizes and quarter sessions (qq.v.). It is a part of the Supreme Court of Judicature and may sit at any place in England and Wales. The judges are all the judges of the High Court, circuit judges (both categories being full-time judges and the latter acting also as county court judges) and recorders, who are part-time judges appointed on a temporary basis. Two, three or four Justices of the Peace must sit with a professional judge to hear appeals from magistrates' courts and proceedings on committal for sentence, and may sit in certain other cases. The Crown Court has exclusive jurisdiction in trials on indictment; offences are classified as triable only by a High Court judge, by a High Court judge unless released, by a High Court judge, circuit judge or Recorder, or normally by a circuit judge or recorder, in each case with a jury. The Court also has jurisdiction to hear appeals by persons convicted summarily in magistrates' courts, to sentence persons committed for sentence after summary conviction, and a limited civil appellate jurisdiction. Procedure is regulated by Crown Court rules. The Court incorporates the Crown Courts established in Liverpool and Manchester in 1956 and presided over by the Recorders of these cities as full-time judges, with both original and appellate jurisdiction. It also incorporates the Central Criminal Court (q.v.), though that has been retained in name in that when the Crown Court sits in London it is known as the Central Criminal Court. Appeal from conviction on indictment lies to the Court of Appeal, Criminal Division, and to the House of Lords.

Crown Courts. In 1956 Crown Courts were established at Liverpool and Manchester by commissions of oyer and terminer and gaol delivery. They exercised the jurisdiction usually conferred on courts of assize. The commissions were addressed to the judges of the High Court and the Recorders of Liverpool and Manchester and all had the powers of a High Court judge. Each court was a superior court of record and was treated as a court of assize; it might sit in more than one division simultaneously. When the commissions were in force they superseded general criminal assizes for the areas covered, and also acted as quarter sessions for the cities. In 1971 these Crown Courts were absorbed in the new Crown Court (q.v.) then created.

Crown debts. Debts due to the Crown. By English law Crown debts have preference over all other debts affecting personal estate.

Crown, demise of the. See CROWN.

Crown Estate Commissioners. A body of eight commissioners established by statute in 1956, superseding the Commissioners of Crown Lands, to manage the landed estates of the Crown and to maintain and enhance its value, having due regard to the requirements of good estate management.

Crown lands. In the United Kingdom, land belonging to the Crown. In Anglo-Saxon times there were private estates of the King, crown demesne, comprising palaces and the like, and rights in the folkland of the kingdom. After the Norman Conquest all three became merged in the estate of the Crown, held by the King as king. The Crown lands were increased from time to time by escheats, confiscations, and similar means and from time to time Crown lands were granted to favourites. At the beginning of his reign George III surrendered the hereditary revenues of the Crown including income from the Crown lands, to Parliament in return for a fixed income, the Civil List (q.v.), Parliament assuming liability for the armed forces and the expense of government. This practice has been followed in every subsequent reign. The ownership of the Crown lands remains vested in the sovereign and they are managed, under statute, formerly by the Commissioners of Woods, Forests, and Land Revenues, and later by the Commissioners of Crown Lands and now by the Crown Estates Commissioners.

Crown law. The criminal law.

Crown Office. Originally a department belonging to the Court of King's Bench, it was made part of the Central Office of the Supreme Court in 1879. It deals with the business of the Crown, i.e. criminal, side of the court and much ministerial business dealt with by single judges of the Queen's Bench Division.

Crown Office in Chancery. One of the offices of the old Court of Chancery, now part of the Central Office.

Crown paper. Prior to 1875 a list of the cases on the Crown side of the Queen's Bench.

Crown private estates. Estates vested in the sovereign in her natural capacity rather than in her in right of the Crown. Law however attributes to the body natural of the sovereign all the qualities of her body politic so that the private estates can at common law be dealt with only subject to the same incidents and formalities as estates vested in her in right of the Crown. But in some cases estates vested in the sovereign are not affected with prerogative privileges and land acquired before succession or descending from collaterals vests in the sovereign in her natural capacity but thereafter descends with the Crown. Private estates as statutorily defined consist of land purchased out of money issued for the use of the privy purse.

Crown privilege. The right of the Crown and Ministers acting on its behalf to object to producing a document in court on the ground that it would be contrary to the public interest to do so. The plea is not confined to proceedings in which the Crown is a party. The rule has attracted much criticism because privilege was claimed for whole classes of communications and in cases where it was impossible to see any possible public interest. In 1968 the House of Lords held, following for England the previous Scottish practice, that the courts could question a Minister's claim of Crown privilege and, having inspected the documents, could order their production.

Crown proceedings. The general term in British law for proceedings by and against the Crown, or the Queen, not in a personal capacity, but as the embodiment of the State. At common law, as the King personally and his personal servants originally administered the government, privileges attaching in feudal law to the monarch came to attach to the Crown and government departments. The main privileges were that an action would not lie in contract against the Crown (though a petition of right (q.v.) might be brought in such a case) and in accordance with the maxim, The King can do no wrong, no action lay in tort. At common law the Crown also had special methods of enforcing its own claims in the courts and Crown proceedings followed special forms of procedure.

The Crown Proceedings Act, 1947, subject to exceptions, enabled civil proceedings to be taken against the Crown in the same circumstances as they could be taken against a private individual. It does not apply to the sovereign in her private capacity, in which case the former procedure remains applicable.

Thus the Crown was made subject to liability in tort in respect of H.M. government in the U.K. generally to the same extent as a private defendant. The Act, however, preserved the common law immunity of H.M. ships or aircraft from proceedings *in rem* and there are savings for acts done under prerogative and statutory powers.

The Act also, subject to exceptions, abolished the special forms of procedure which previously affected civil proceedings by and against the Crown but left unaffected certain prerogative privileges, such as that the Crown is not bound by statute unless expressly or by necessary implication bound thereby, and excluded remedies such as injunction which could particularly interfere with governmental processes, the courts being empowered in lieu to make declaratory orders. Civil proceedings may be instituted by or against an authorized government department in its own name, or by or against the Attorney-General.

The 1947 Act does not, in general, affect proceedings on the Crown side of the Queen's Bench Division, concerned with the High Court's supervisory jurisdiction, exercised through the prerogative writ of habeas corpus and the orders of *certiorari*, mandamus and prohibition (qq.v.).

In Scotland there had never been any objection to actions against the Crown founded on breach of contract, though actions were directed against the Officers of State and latterly against the Lord Advocate, and the rule that the King could do no wrong was not originally the rule of Scots law but crept in in the late nineteenth and early twentieth centuries. Nor were there any special procedural rules. The Lord Advocate represented the Crown and all Departments of State. The Crown might, however, rely on certain special pleas. The 1947 Act applies in part to Scotland also and effected changes similar to, but less drastic than, those unable in England.

G. Robertson, *Civil Proceedings by and against the Crown*; G. L. Williams, *Crown Proceedings Act*.

Crown servants. Persons employed by the Crown. The special features of the relationship are that, save by special statutory provision, contracts of service under the Crown are terminable without notice on the part of the Crown, even if the contract expressly provides to the contrary because the Crown cannot by contract deprive itself of the liberty of dismissing a servant at will. Modern legislation is however limiting this liberty. At common law Crown servants cannot be made personally liable on contracts entered into by them in their official capacity, as they are presumed to contract as agents for the Crown. Superior Crown servants cannot be made liable for the wrongful acts of their subordinates unless they authorized or ratified these acts. Crown servants may however be made personally liable for wrongs or crimes committed by them in their official capacities.

Crown side. The prerogative and criminal jurisdiction of the Queen's Bench Division, as contrasted with the plea side which handles the civil business.

Crucifixion. A form of punishment, common in the ancient world, used by the Romans, probably borrowed from Carthage. It was normally confined to slaves and not used for soldiers and was considered particularly suitable for political or religious agitators, pirates, slaves, and persons with no civil rights. The general practice was to flog the condemned man, and require him to carry a cross-beam to the place of execution where a stake was fixed in the ground. He was stripped and fastened to the beam by nails or cords and it was pulled up till his feet were clear of the ground and fastened to the upright. The weight of the body may have been taken in part by a ledge projecting from the upright. There is little evidence for a foot-rest and the feet were sometimes tied or nailed. Over the criminal's head was placed a notice stating his name and crime. Death probably resulted from exhaustion or heart failure but was sometimes hastened by breaking the leg bones. The penalty was abolished by Constantine about A.D. 315.

Cruelty. The unjustifiable causing of pain or distress to an animal, child, or other person by deliberate conduct or by neglect. In most western societies cruelty to animals or to children is punishable. Cruelty to the other spouse has frequently been a ground for separation or divorce.

Cruelty to animals was reprobated by some from early times but no positive legal steps were taken until in 1822 when one Richard Martin secured the passing through Parliament of an anti-cruelty Act. Other European countries followed suit and the scope of legislation has steadily been extended. The English Society for the Prevention of Cruelty to Animals was founded in 1824 and other countries also founded animal welfare societies. Both legislative protection and anti-cruelty societies exist in nearly every country.

Cruelty to children was similarly outlawed in Britain from 1884 when also the National Society for the Prevention of Cruelty to Children was founded, and similar developments have followed in many countries.

Cruise, William (?–1824). A learned conveyancer, author of *An Essay on the Nature and Operation of Finds and Recoveries* (1783) and *A Digest of the Laws of England respecting Real Property* (7 vols., 1804, and later editions) a book setting out real property law, not under alphabetical headings, but on the basis of a scientific classification largely based on Blackstone. An adaptation to American conditions was made by Greenleaf.

Cucchi, Marco Antonio (?–1565). Italian canonist, who introduced a logical and concise system in the treatment of canonical questions in contrast to the exegetical method of the time. He published *Praelectiones iuris canonici.*

Cujas, Jacques (Jacobus Cujacius) (1522–90). French jurist, studied law and other subjects at Toulouse and became professor at Cahors, Bourges, Valence, and Turin. His wanderings were partly prompted by civil and ecclesiastical disturbances. He diligently sought and examined manuscripts and studied his materials historically, comparatively, and critically. His works, notes on the *Institutes* and Paul's *Sententiae*, commentaries on various titles of the Digest, and *Observationes*, won for him the reputation of the leading jurist of the day. Other writings seek to restore the text and form of, and establish the meaning of, the classical jurists whose works are excerpted in the *Digest.* He also published amended texts of works previously published, and treatises on feudal and ecclesiastical law. He recovered and published the *Codex Theodosianus* and procured the manuscript of the *Basilica.* The *Observationum et emendationum libri XXVIII* is a great collection of suggestive restorations of texts and amendments. His *Commentaries on Papinian* were published posthumously and he composed a commentary on the *Consuetudines Feudorum* and on parts of the *Decretals.* His work is comparative, critical and historical, and secured the triumph of the historical humanist school of jurists over glossators and commentators.

Masson, *Jacobi Cuiacii* (1590); Spangenberg, *Jacob Cujas und seine Zeitgenossen* (1822).

Culpa. In Roman law, neglect resulting in harm. Jurists distinguished *culpa lata*, gross neglect verging on deliberate or reckless harm, *culpa levis*, lesser fault, and *culpa levissima* or trivial, venial fault. In Scots law, it is a term roughly synonymous with negligence, in the sense of failure to exercise the degree of diligence and care which the reasonably careful and prudent person would have exercised in the circumstances, but sometimes used more widely for the mental state with which any harm is done or allowed to happen.

Culpability. Blameworthiness.

Culpable homicide. In Scots law, homicide not amounting to murder. It includes voluntary culpable homicide, which is homicide in extenuating circumstances, such as under serious provocation, subject to intoxication, or by a person of diminished responsibility, and involuntary culpable homicide where death is caused unintentionally but with a very serious degree of negligence, or in the course of unlawful conduct, such as assault or rape.

In certain circumstances, such as exceeding lawful duty, necessity, coercion, or infanticide, homicide is in practice charged as culpable homicide rather than murder.

Culprit. A prisoner at the bar, having pled not guilty and awaiting trial; the word was used in the phrase: 'Culprit, how will you be tried?'. The word is probably a combination of *culpable* and *prist* or *prit*; when a prisoner pled not guilty, the clerk answered 'culpable' and said he was ready (*prist*) to join issue. This was entered on the roll as *cul. prist*, which became corrupted into culprit. But there are other explanations. It is colloquially used for a prisoner, an offender or a guilty person.

Cumberland, Richard (1632–1718). English bishop and antiquarian, author of *De Legibus naturae Disquisitio philosophica* (1672), provoked by his opposition to Hobbes. In it he sought a more certain philosophical foundation for the law of nature than the general agreement of civilized nations, on which Grotius had based it, and this he found in consciousness of the law of nature from our own natural reason, observation, and experience. He set out to show that there are firmly established laws of nature which make it desirable for men to pursue the common good rather than their own particular advantage. He is important in the history of English ethics, influenced Shaftsbury and Hutcheson, and is sometimes regarded as the father of Utilitarianism.

Cum sit necessarium (or *Modus componendi brevia*). A tract in Latin dating from about 1290, possibly by Ralph de Hengham (q.v.) or another judge, John de Metingham. It deals with the writ of right, the conditions of demanding a tenement in demesne, the writ of utrum, the four modes of claiming dower, the writ of right close, writs relating to advowsons, courts and judges, false judgment, and the remedies for it, common of pasture and exceptions. It is printed in Woodbine's *Four Thirteenth-century Tracts* (1910).

Cumulative. Constituted by addition; thus if legacies are given twice in the same will to the same legatee, they may be held cumulative, or substitutional, one instead of the other, according to the apparent or presumed intention of the testator.

Cuneiform law. The kinds of law developed by various people of ancient Mesopotamia and adjacent areas, written in cuneiform script.

See Assyrian Law; Babylonian Law; Hittite Law.

Cuneo, Gulielmus de (?–1335). A French jurist and commentator on the civil law, who later became a bishop, author of a *Tractatus de materia securitatis*.

Cunningham, William, 9th Earl of Glencairn (1610–64). He early became a Privy Councillor in Scotland (1641) and in 1647 was appointed Lord Justice General but was deprived of office in 1649. He lived in retirement till the Restoration when, in 1661, he was appointed Lord Chancellor of Scotland for life but enjoyed this office only till 1664.

Curator. A person charged at Rome with a particular duty, e.g. the *cura annonae*. Augustus instituted boards of *curatores* to take charge of administrative tasks. The term was later applied to imperial officials appointed to supervise the finances of towns in Italy and the provinces. In Roman law a *curator* was a person, appointed in many different circumstances, to care for and protect minors, persons of unsound mind and others.

In Scots law a *curator* is a parent or other person whose function is to advise and consent to the actings of a minor (q.v.). A *curator bonis* is a person appointed by the court to manage the affairs of a person incapacitated from doing so in person, a *curator ad litem* one appointed to look after the interests of a party to a litigation, if unable to do so for himself.

Curia. A name given at Rome to the assembly places of each of the *curiae* (which were the most ancient divisions of the Roman people) and of many other corporations, and particularly to the Senate House. The word accordingly became transferred to the council. Under the Empire *curia* is the usual name for the municipal senates.

In mediaeval Latin *curia* came to be used in the general sense of court, and of bodies such as the *Curia Regis* (q.v.). In ecclesiastical contexts there are known diocesan curiae, metropolitan curiae, patriarchical curiae and the Roman curia (q.v.).

Curia advisari vult. Abbreviated to *cur. adv. vult* or C.A.V. (the court wishes to consider the matter), a phrase which, prefixed to the judgment in the report of a case, signifies that judgment was reserved and given later, after consideration.

Curia Regis. In a wider sense the *curia regis* was the feudal, Norman, version of the national or great council of England, superseding the older, Anglo-Saxon, Witenagemot, which met thrice a year and was attended by the bishops, earls, and tenants-in-chief.

In a narrower sense it denotes the smaller select group of members of the larger body which met much more frequently to assist the King and carry on the government. The *Curia* in both senses was itinerant, following the King on his travels, and both had the same undefined powers of government. The distinction between the two kinds of *Curia* developed and became more important, the larger *Curiae* being feudal councils and the smaller rather administrative or legislative boards and courts of law.

It had the political function of giving counsel and consent to legislation and other important acts, an administrative and supervisory authority over the feudal society, but also, as a feudal court, had an instance jurisdiction in all disputes between the King's tenants-in-chief and other great men, and had the old appellate jurisdiction of the Witenagemot. Suitors were allowed, on payment of a fine, to remove their plaints to it from the older courts of the shire and hundred. It also superintended the assessment and collection of the royal revenue.

The financial and judicial work tended to become specialized. From the time of Henry I, a branch or committee of the *Curia Regis* was engaged in fiscal matters and when thus engaged sat in the Chamber called the Exchequer and itself became called that (*Curia Regis ad Scaccarium*). There were two divisions of the Exchequer, the lower, in which money was received, and the upper, in which accounts and disputes were settled. Twice a year each sheriff had to appear at the Exchequer and account for the money due from his shire, and legal disputes as to payment were decided. In 1198 the Chancellor withdrew from the Exchequer and his clerk became an important official, later entitled Chancellor of the Exchequer. The higher officials were the Barones Scaccarii and were often called on to decide difficult matters, and also employed on judicial business even though not arising from revenue matters.

Down to the time of Henry II, the *Curia Regis* continued as a single undifferentiated court, though some of the judges, selected from time to time and varying in number and personnel, were in practically continuous sessions at the Exchequer for financial business. Under Henry II the business increased enormously and the King's court acquired wide civil and criminal jurisdiction, not confining its attention to great men and great causes. The number of judges rose to 18, but in 1178 the King reduced the number of judges to five, but reserved for hearing by himself in council cases in which this court should not do justice. This court became the Court of Common Pleas; it continued to move about with the King, and long remained virtually indistinguishable from the *Curia*, but Magna Carta, c.17, provided that common pleas be held in some certain place, which became settled as Westminster. This made it both a stationary and a purely judicial body, and c.18 provided that the possessory assizes should be tried in the counties where the cases arose, four times a year by two judges assisted by four knights of the county.

Differentiation of function gradually developed but by the end of Henry III's reign (1272) the *Curia Regis* as a superior court had become permanently divided into three courts, each with a defined sphere of competence, the Exchequer, dealing with fiscal matters, the Common Pleas, dealing with civil disputes between private individuals, and the King's Bench, at first actually and later in theory a court held *coram rege ipso*, charged with all remaining business including criminal business, and with jurisdiction to correct the errors of the Common Pleas. A trace of the original unity remained in the practice of all the judges meeting to discuss difficulties in the Exchequer Chamber (q.v.).

Curia, Roman. The complex of bureaux which assists the Pope to implement his legislative, executive, and judicial work as head of the Roman Church. The body of bureaux grew up gradually and was regulated by bull in 1588. Its specific powers are contained in the Code of Canon Law and certain later documents. It consists of congregations with administrative powers, tribunals having judicial powers and offices possessing ministerial power. The tribunals are the Sacred Penitentiary for matters pertaining to the form of conscience, the Roman Rota, a collegiate court of appeals for hearing cases from diocesan or metropolitan tribunals, and the Apostolic Signatura, the highest court in the Roman Church.

Curran, John Philpot (1750–1817). Was called to the Irish Bar in 1775 and gained a great reputation as an orator, both at the Bar and from 1753 in the Irish Parliament. He was a great champion of Irish liberties, supported Roman Catholic emancipation and championed many victims of religious and political oppression. He became Master of the Rolls in Ireland, 1806–14.

Cursitor baron. One of the members of the old Court of Exchequer having ministerial but not judicial functions and latterly a sinecure post. The barons of the Exchequer were originally officials of the fiscal side of the Exchequer and only from 1579 were they lawyers ranking equally with the judges of the King's Bench and the Common Pleas. But it was then found necessary to associate with them one with knowledge of fiscal matters and a cursitor baron was appointed from 1610. Latterly the office was a sinecure. The office was abolished in 1856 on the death of the last holder.

Curtesy (or tenancy by the courtesy). By the law of England, a tenancy of land which formerly devolved on a man from his wife on her death, provided that she had been seised of her land and had a child of the marriage who was born alive capable of inheriting the wife's lands. It extended to an estate in the whole of the wife's lands for his life. If there were no child of the marriage the wife's real estate descended to her heirs. It was abolished in 1925. A similar right, known as courtesy, subject to the same conditions, was recognized in Scots law down to 1964.

Curtilage. A courtyard, yard, garden or piece of ground near and belonging to a house.

Curtis, George Ticknor (1812–94). Practised law in Boston and later in New York and before the Supreme Court. He wrote biographies of Daniel Webster and James Buchanan and *The Constitutional History of the U.S. to the Close of the Civil War* (1889–96) which is the classic Federalist interpretation of the theme, as well as books on copyright, patents, and *Commentaries on The Jurisdiction, Practice, and Peculiar Jurisprudence of the Courts of the United States* (1854–58). His brother Benjamin Robbins Curtis (1809–74) prepared *Decisions of the Supreme Court and a Digest of its Decisions to 1854*, and served as a justice of the Supreme Court 1851–57.

Cushing, Caleb (1800–79). Served in the Massachusetts legislature and in Congress, and negotiated the first treaty between the U.S. and China (1844). Later he was Attorney-General (1853–57) and was nominated as Chief Justice of the Supreme Court but his name was withdrawn on account of opposition in the Senate. Later he was U.S. Minister to Spain.

Cushing, Luther Stearns (1803–56). Held minor judicial office in Boston, but is remembered for his legal writings, especially *Elements of the Law and Practice of Legislative Assemblies in the United States* (1856) and his *Manual of Parliamentary Practice* (1844). He also translated Savigny on *Possession*, Pothier on *Sale*, and Mittermaier on *Criminal Responsibility*.

Custodes Judaeorum. Judges of the Exchequer appointed first in 1198 to decide cases affecting Jews. They were not barons of the Exchequer and constituted the Exchequer of the Jews.

Custodes libertatis Angliae auctoritate Parliamenti (guardians of the liberties of England by the authority of Parliament). The title under which writs passed in England between the execution of Charles I (January, 1649) and the proclamation of Oliver Cromwell as Lord Protector (December 1653).

Custody (guardianship). In divorce, nullity or domestic proceedings courts have wide powers to determine which parent is to have custody of the children of the marriage. In relation to chattels custody is frequently distinguished from possession (q.v.) as a lesser right, particularly that of a person such as an employee who holds only for and on behalf of another, who is deemed to have legal possession, and not for himself.

Custom. In most developed communities, general or widespread custom has frequently been an important historical source of law in that general customary practices have in time been so recognized that they have become customary law by judicial acceptance, adoption, and subsequent application to other cases, or have been stated by text-writers to have the force of law, or have been adopted and recognized in legislation. Thus in English law rights similar to easements (q.v.) may by custom exist in the lands of another. In Scots law the shares of a deceased's property falling to his spouse and children have been settled by ancient custom, accepted by text-writers, courts, and legislation. To drive on the left of the road is an ancient custom now settled by decisions and statutes. Much common law is based on the acceptance and generalization of national or widespread customs; common and general custom of the realm becomes common law.

In modern law a custom is a particular rule which has long been recognized in a particular locality and has attained the force of law, which is distinct from but not contrary to the general common law of the realm and which, where it applies, replaces the general common law. To be recognized as a custom, a practice must have existed from time immemorial, a time long settled as 1189, the accession of Richard I, or at least it cannot be recognized if it is shown that it could not have existed at any time since 1189, must have continued without interruption since then, be confined to a limited and defined locality and be certain and reasonable.

Customary law includes also conventional customs, that is, customs which are legally binding because deemed to have been expressly or impliedly incorporated in agreement between the parties. A person who contracts in a recognized trade or a recognized market, such as the stock exchange, is presumed to have agreed to contract in accordance with the established customs and usages of that trade or market and is accordingly bound by those usages. Thus a large part of the law of bills of exchange and negotiable instruments, of bills of lading and marine insurance, originated in the customs of persons engaged in those activities. Similarly customs of particular professions have been recognized in many cases. Customary law of this kind has at first to be proved, as matter of fact, in each case as an established custom or usage of the trade or market in question, but subsequently judicial notice may be taken of the custom, sufficiently proven in previous cases. In some cases rules which originated in proven customs, and were subsequently recognized judicially as customary rules have been embodied in statute. Thus the principles applicable to bills of exchange have passed into statute.

Particular customs or usages, such as of trades or professions, may always be proved in a particular case to have been taken by the parties as providing a rule of law in a particular case. For any such custom to be upheld by a court, the party relying on the custom must prove what the alleged custom is, for if it is not reasonably certain it cannot be held a custom, that it is reasonable, that, while it may be an exception to or qualification of an established general rule of common law, it is not inconsistent with or in contradiction of such a general rule, that it is not contrary to any statute, that it is notorious, such that everybody in the area or trade or profession enters into it with that usage as an implied term, that it has been observed for a long period and that it has been observed as a matter of legal right and not by force, contract or otherwise, *nec vi nec clam nec precario.* Usage of many particular trades and professions, of particular seaports and other maritime dealings have in many cases been considered by the courts.

Custom in international law. Custom in international law is usual or habitual practice and it continues to play a significant role in the development of international law. International custom as evidence of a general practice accepted as law is one of the sources of law to be applied by the International Court. There must accordingly be evidence of general practice, and acceptance of this as legally obligatory. The evidence of custom is found in state legislation, international and national decisions, treaties, textbooks, and many other sources. To amount to custom there must be a practice of substantial duration, uniform and consistent and general, if not universal.

A difficult issue is whether customary law, *ex hypothesi* developed by other states, can be held binding on a newly independent state. The older view was that states attained independence on terms of accepting the existing international legal order, but recently liberated states have shown reluctance to accept this proposition.

Customary court baron. A court required by custom to be held in the manor for which it was held. It was distinct from the court baron of the manor but might be held at the same time.

Customary freehold. Privileged copyholds (q.v.) or copyholds of frank tenure, estates held by customs and not in this respect differing from copyholds, held at the lord's will. The incidents and services attaching to them were similar to those relating to copyholds properly so called. A distinction was drawn between a kind called free copyhold, of which the freehold was in the lord, and that called customary freehold, of which the freehold was in the tenant.

Customary law. Customs, usages, and practices deserve the name of customary law when they are sufficiently fixed and settled over a substantial area, known and recognized and deemed obligatory, as much as are systems of law based on written formulations of rules. Customary law has been widespread in the world. Where and when it applies legislation and case-law are of secondary importance. In mediaeval Europe the Germanic tribes were mostly regulated by customary practices and laws, and the existing 'barbarian laws' regulated only parts, frequently small parts, of their social relationships. Customs were largely superseded by the progress of Roman law, which was halted only where confronted by major statements of customary law, such as the *Coutumier* of Beaumanoir in France, the *Sachsenspiegel* in Germany and, later, by new legislative compilations such as the *Siete Partidas* in Spain, or the redaction of customs by the French *Ordonnance de Montil-lez-Tours* of 1454. In France Roman law was influential in the south (*pays du droit écrit*) but less so in the north (*pays du droit coutumier*) where customary law prevailed.

In Scandinavia in the thirteenth and fourteenth

centuries there were compilations of law based mainly on customary law, which made Roman influence less strong. The Russian customs of the Kiev region were collected over the eleventh to fourteenth centuries (*Russkaia Pravda*) and reflect in greater detail a rather more evolved society than that of the Germanic or Scandinavian tribes at the time of the barbarian laws; the law is territorial rather than tribal and in some respects anticipates feudalism. The peasantry continued to live according to custom down to the late nineteenth century, recognizing, for example, only communal or family ownership rather than the individual ownership provided by legislation.

In England until the growth of the power of the central monarchy and the work of the itinerant justices in building up a common law had made progress, what law there was existed only as customary law, much of it local and peculiar to the tribes which inhabited particular areas; vestiges of this survived for centuries. Glanvill and Bracton both wrote of the Laws and Customs of England. The old county or hundred courts applied only local customs, and even the feudal courts which replaced them applied predominantly local customary law. When the royal justices were over the years building up, decision by decision, the common law it was on the basis of the immemorial custom of the realm, but in fact the only customs which existed were local ones, and the general custom was the creation of the judges, generalizing on the basis of particular local customs.

Customary law is still widespread in the world. Much of Africa is subject to customary law, with in some places a veneer of law of a western type. In Indonesia customary law (*adat* law) has been a basis for the influence of Muslim law and civilian ideas imported by Dutch colonization. In China it is probably still the main effective law.

H. Maine: *Early Law and Custom.*

Customs and excise. Customs duties are duties or toll payable on goods exported from or imported into the U.K. They are of two main classes, protective duties and revenue duties. Since 1932 the majority of goods imported into the U.K. are subject to customs duties, but reduction of rates has been effected pursuant to various international, commonwealth, and supranational agreements, noteably those concerned with membership of the E.E.C. Goods are classified according to the Customs Tariff. Thus customs duties are levied on foreign wines, cameras, and tobacco.

Excise duties are duties or toll payable on goods produced in the U.K. and used or consumed there, and excise licence duties, on mechanically propelled vehicles, on the exercise of certain trades, businesses or activities. Thus excise duty is charged on beer, spirits, tobacco, excise licence duty on vehicles, and on such trades as bookmaking, retailing liquor, pawnbroking, dealing in game, and on operating a television set.

The Commissioners of Customs and Excise are persons appointed by letters patent under the Great Seal and holding their appointments at the will of the Crown, charged with the duty of collecting, accounting for and managing the revenues of customs and excise. They have extensive powers to search persons, premises, vehicles, ships and aircraft, and to seize goods liable to forfeiture.

Custos rotulorum. From about 1400 every commission of the peace named a keeper of the rolls. He was first justice of the peace and first civil servant of the county. By statute of 1545 he was appointed clerk of the peace, and he was nominally keeper of the rolls or records under the commission of the peace. The office was normally held by the Lord Lieutenant of the county or another person of rank.

Custumals. Records of ancient manorial customs.

Cynus de Pistoia. See CINO.

Cy-Près (approximation). From *ici-près* or, probably, *aussi-près*, a doctrine evolved in English equity in relation to charitable trusts whereby, if a gift is clearly for charitable purposes only, it will not be allowed to fail because the precise object to be benefited, or the mode of application of the fund is uncertain. It must be evident that the truster had a general charitable intent, but the precise purpose is impossible, or has never existed, or has ceased to exist before the testator's death, or the purpose or institution has ceased to exist after the gift has taken effect, or in certain other cases where the question of general charitable intent is less material. If the conditions are satisfied the court will settle a scheme for the application of the funds to another purpose as near as possible to that prescribed by the truster.

L. Sheridan and V. Delany, *The Cy-Près Doctrine.*

Cyprus, law of. Cyprus has been occupied at various times by all of the peoples who have exercised power in the Eastern Mediterranean. The present inhabitants are mainly Greeks, descendants of the earliest inhabitants and of colonists from Greece, and Turks, descendants of the Ottoman Turks, who occupied the island from 1571 to 1878. From 1878 to 1960 the island was held by Britain. Since 1960 the island has been an independent republic but both major racial groups aim at union with their respective home countries, though the constitution forbids this. It has a Greek Cypriot president and a Turkish Cypriot vice-president, both elected by universal suffrage by their respective communities, a Council of Ministers (seven Greek and three Turkish) selected by them, and an elected House of Representatives, elected every five years by universal suffrage. The major units of local government are districts under district officers appointed by the Cabinet. The Constitution incorporates a Treaty of Guarantee naming Greece, Turkey and the U.K. as guarantor powers.

There is according to the constitution a Supreme Constitutional Court with a neutral foreign judge as President, and a High Court having a neutral foreign chief justice. These were amalgamated in 1964. Other judges are appointed by the president. There are now a Supreme Court (appeals court), district Assize Courts having criminal jurisdiction and District Courts exercising summary criminal jurisdiction. There are also a Tribunal for Labour Disputes and boards exercising professional disciplinary jurisdiction. There are also communal courts dealing with disputes as to the personal status of members of the Turkish community. The Attorney-General is in charge of all criminal prosecutions. The legal system is based on Roman law.

G. Hill, *History of Cyprus*, 4 vols. (1940–52).

Cyrillus (fifth century A.D.). A Greek jurist, professor in the law school at Berytus, one of the founders of the oecumenical school which paved the way for Justinian's legislation.

D

D.P.P. DIRECTOR OF PUBLIC PROSECUTIONS (q.v.).

Dabin, Jean (1889–). Belgian jurist, professor at Louvain, author of *La Philosophie de L'ordre juridique positif* (1929), *La Technique de l'elaboration du droit positif* (1935), *Doctrine Générale de l'Etat* (1939), *Theorie Générale du Droit* (1944), and other works. As a Catholic he sought to harmonize the scholastic belief in the superiority of immutable laws of nature, as interpreted by neo-Thomists, with the recognition that civil societies have a monopoly of power to establish rules of conduct sanctioned by force.

Dagger-money. Money in the form of a Jacobus, or gold sovereign, and a Carolus, formerly presented to the senior and junior judges respectively when they left Newcastle having held the Northumberland assizes, to enable them to buy daggers to defend themselves against Scottish raiders.

Dail Eireann. The Assembly of Ireland, parliament of the Republic of Ireland, first established in 1919 when the Irish republicans elected to Westminster in the 1918 election proclaimed an Irish state. It was not recognized and in 1920 the British held elections in Ireland for the separate parliaments for Northern and Southern Ireland which had been established under the Government of Ireland Act. The Irish members refused to participate in this Parliament and formed the second Dail. An Anglo-Irish Treaty was negotiated and presented to the Dail in 1921. In 1922 a republican insurrection was raised by the I.R.A. which recognized only the republican minority in the Dail. The Dail then proclaimed the Irish Free State. Under the modern constitution the Dail is the lower house of the Oireachtas. The Dail consists of 148 members elected in three, four and five-member constituencies. The maximum life of a Dail is five years. It is the more important house and the government is responsible to it, and it in effect passes money bills.

Dallas, George, of St. Martin's, Ross-shire (1630–?1702). Became a writer to the Signet and in 1660 Deputy Keeper of the Signet. He is known as author of *A System of Stiles* (pub. 1697 and 1774) or conveyancing forms, long indispensable in Scottish legal offices.

Dallas, Sir Robert (1756–1824). Called to the Bar in 1782, he was one of the counsel for Warren Hastings in 1787, and entered Parliament in 1802. In 1813 he became Solicitor-General and, later, a judge of the Common Pleas. In 1818 he succeeded as Chief Justice and as such presided at the trial of the Cato Street conspirators. He resigned in 1823. He was a very good lawyer and many of his judgments show a great grasp of legal principle.

Dalrymple, Sir David, 3rd baronet, Lord Hailes (1726–92). A great-grandson of James Dalrymple, Viscount Stair (q.v.), studied in Holland, and was called to the Scottish Bar in 1748 and gained a reputation as a learned and accurate lawyer. In 1766 he became a judge of the Court of Session as Lord Hailes and in 1776 became also a judge of the High Court of Justiciary. He was unsurpassed in legal knowledge, and was noted for his humanity in criminal trials. He was a friend of Dr. Johnson and from Hailes Boswell acquired the desire to meet Johnson. He published much, notably *Memorials and Letters relating to the History of Britain in the Reign of Charles I* (1766), *The Canons of the Church of Scotland* (contributed to Wilkin's *Concilia Magnae Britanniae* and published separately in 1769), a *Catalogue of the Lords of Session, with Historical Notes* (1767), an *Examination of some of the Arguments for* *the High Antiquity of Regiam Majestatem* (1769), *Decisions of the Court of Session*, 1766–91 (1826), and, above all, the *Annals of Scotland* (1776, later continued) a very valuable study of mediaeval Scottish history, which won Dr. Johnson's approbation and is still worthy of note.

Dalrymple, Sir Hew, of North Berwick (1653–1737). Third son of James Dalrymple, Viscount Stair (q.v.), was admitted advocate in 1677, became one of the commissaries of Edinburgh and in 1695 Dean of the Faculty of Advocates. His elder brother, Sir John (1648–1707) (q.v.) was Lord Advocate 1687–88 and 1690–91 and Hew acted as substitute for his brother in 1690–91. In 1698 he was made Lord President of the Court of Session, an office he held till his death. He also sat in the last Scottish Parliament, and was a commissioner for the Union of 1707. He was regarded as a judge of complete integrity and a very sound lawyer. He published a collection of the *Decisions of the Court of Session* 1698–1720 (1758). His younger brother David, first baronet of Hailes (?–1721), was Lord Advocate 1709–11 and 1714–20, when he became auditor of the Scottish Exchequer, 1720.

Dalrymple, James, 1st Viscount Stair (1619–95). Educated at Glasgow University and subsequently (1641–47) regent in philosophy there, was called to the Scottish Bar in 1648. In 1657–60 he was one of the commissioners for the administration of justice in Scotland under the Protectorate, and was continued in office under Charles II. In 1663 he retired to avoid taking the oath abjuring the right to take up arms against the King, but resumed office in 1664. In 1671 he became Lord President of the Court of Session but again retired in 1681 rather than take the oath under the Test Act. In 1682 he had to take refuge in Holland, but returned with William of Orange in 1688 being reappointed President and made a Viscount (1690). He remained in office till his death.

In 1681 he published his *Institutions of the Law of Scotland* and *Modus Litigandi, or Forms of Process delivered before the Lords of Council and Session*, and in 1684 and 1687, while in exile, two volumes of *Decisions of the Court of Session*, 1661–81. In 1693 an enlarged edition of his *Institutions* appeared. Stair's *Institutions* has always been regarded as the supreme text on the law of Scotland, in which rules are derived from their underlying principles and their sources in Roman, canon, and feudal law. It remains of the highest authority and in default of contrary later authority is deemed to settle the law.

His eldest son, John, became Dean of the Faculty of Advocates, Lord Justice-Clerk (1688–89), Secretary of State for Scotland and an earl in 1703. His second son, James, was author of historical works. His third son, Hew (q.v.) became Lord President. His fifth and youngest son, David, Lord Advocate (1709–11 and 1714–20) and more remote descendants were also distinguished in law and historical pursuits.

A. J. G. Mackay, *Memoir of Stair*.

Dalrymple, Sir John, 2nd Viscount and 1st Earl of Stair (1648–1707). Eldest son of the 1st Viscount Stair (q.v.), became an advocate in 1672 and distinguished himself, particularly in the defence of the Earl of Argyll in 1681. For some time he was subjected to considerable persecution and was imprisoned.

In 1686 he became Lord Advocate, but by reason of his toleration of conventicles was deprived of the office in 1688 and appointed Lord Justice-Clerk. He acted strenuously in the cause of William of Orange and from the first had the chief management of Scottish affairs and the duty of representing the government in the Scottish Parliament. He was again Lord Advocate 1690–91. In 1691 he was appointed joint Secretary of State. As such he knew and approved of the arrangements for the massacre of Glencoe. After the report of the royal commission which inquired into the massacre was issued he resigned, but later received a remission for his part in the affair. After a period out of public life he was made a privy councillor in 1702 and Earl of Stair in 1703. Thereafter he continued chief adviser of the government in Scottish affairs, and zealously promoted the union of 1707.

His younger brother Hew (q.v.) was Lord President of the Court of Session, and a still younger brother David was Lord Advocate, 1709–11 and 1714–20.

His son, John (1673–1747), became a distinguished general and diplomat.

Dalton, Thomas (1682–1730). Called to the English Bar in 1711 and was Chief Baron of Exchequer in Ireland, 1725–30.

Damage. Loss or harm suffered by a person in circumstances recognized as legally actionable. It may take the form of damage to person, to reputation, to economic interests, to property, or otherwise. It may be caused by breach of any kind of legal duty. Not all damage is actionable; it must be *damnum injuria datum* (q.v.), i.e. caused wrongfully, and not *damnum sine injuria* (q.v.), without legal wrong, for which no action lies.

Damage may be general, or damage in law, where a person's right has been infringed though no actual damage done to any interest, which justifies nominal damages only, or actual or special damage where the infringement has caused loss assessable in money and which justifies compensatory damages. The term special damage is also used of loss peculiar to the plaintiff, differing in kind from that suffered by the community generally, as where a person suffers damage from a public nuisance.

Damage may be direct, i.e. caused immediately by the conduct done in breach of duty, e.g. injury to a person in an accident, or indirect, or consequential, when it results from the conduct only after the intervention of other factors, e.g. pecuniary loss resulting from a vehicle being damaged. Whether and, if so, to what extent a person in breach of duty is legally liable to another to whom his breach has caused indirect or consequential damage raises the problem of remoteness of damage (q.v.).

See also DAMAGES; REMOTENESS.

Damage, criminal. The intentional or reckless destroying or damaging of any property belonging to another without lawful excuse. If done by burning it is chargeable as arson.

Damage feasant (or faisant). At common law damage done on a man's land by the beasts or fowls of another which had got there without leave, or licence, or fault, as by eating crops or damaging trees, which justified him in distraining or impounding them until the owner compensated him for the harm done. Animals impounded had to be cared for and might not be sold. The remedy was abolished in 1971.

See DISTRESS.

Damage, malicious. Deliberate or wilful damage.

Damages. Pecuniary compensation payable by one person to another for injury, loss, or damage caused by the one to the other by breach of legal duty, normally by breach of contract or commission of tort. They are distinguished into general damages, compensation for the loss presumed to flow from a breach of contract or a tort, and special damages, compensation for particular losses not presumed but which in fact have followed in a particular case; the latter must be specially claimed and strictly proved. Damages may also be unliquidated, i.e. unascertained till fixed by the judgment of the court, or liquidated, a sum agreed on by the parties as payable in the event of breach of duty by the party in breach. The underlying principle of damages is restitution, to restore, so far as money can do so, the plaintiff to the position he would have been in if the contract with him had been duly implemented, or if the tort in question had not been done to him, but in a few cases English law sanctions the award of punitive or exemplary damages, intended not merely to compensate the plaintiff but to punish the defendant and mark the outrageous nature of his conduct. It is always necessary to prove that the loss was caused by the defendant's breach of duty, not by another factor.

The principle of restitution is limited by the rule of remoteness, that compensation will not be given for loss deemed too remote from and too remotely connected with the defendant's breach of duty. In breach of contract cases consequences are too remote to be taken into account in awarding damages if they were not a sufficiently likely result of the breach of contract to be held to have followed naturally from that breach, regard being had to the defendant's knowledge at the time the contract was made. In tort cases consequences are too remote if they could not reasonably have been foreseen by the defendant at the time of his wrongful act.

In all cases the plaintiff is under a duty to take reasonable steps to mitigate or minimize his claim of damages.

Much dispute arises in tort cases whether particular factors should or should not be taken into account, and how much, in terms of money, should be given for particular kinds and degrees of losses. Such valuation must have regard to the current value of money and to imponderables such as a person's expectation of life and his future prospects. Awards are determined in many cases not by any absolute standard but by the level of awards being made by other courts in generally comparable cases.

Damages may be reduced where the plaintiff was himself partly to blame for the loss or injury (contributory negligence) (q.v.), and may be apportioned when the loss has been caused by two or more joint wrong-doers.

In British law damages must be assessed once and for all and cannot be subsequently modified on change of circumstances or because the original award has proved inaccurate, and payment takes the form of a lump sum, not of an annuity or periodical payment.

Damages, measure of. The computation in money of what is fair and proper compensation in the circumstances for the legally admissible heads of damages.

Damasking. When a new Great Seal of England is required (as always happens on the accession of a new monarch, and sometimes at other times) it is usual for the old silver discs which form the Great Seal and the Counter-seal (and which make the impressions on the obverse and reverse of the lump of wax which is appended to documents to which are affixed the Great Seal) to be 'Damasked' or formally rendered useless by being tapped with a hammer by the Lord President of the Council. They are then the perquisite of the Lord Chancellor, though he has sometimes handed one of the discs to a predecessor as a memento. Lord Halsbury acquired two old Great Seals in this way, when a new Seal was made in 1897 and when another new one was required in 1904 after the accession of Edward VII.

Damhouder, Jodocus (Joost) (1507–81). A Flemish jurist and magistrate, author of *Praxis rerum criminalium* (1555) which appeared in Latin, Dutch, and French and won wide acceptance over Europe and was for long deemed a high authority (but was in fact plagiarized from Philip Wielant's (1439–1519) *Practycke Crimineele*), *Praxis rerum civilium* (1567), and many other works.

Damnosa haereditas. In Roman law an inheritance which brought loss rather than profit, and accordingly a term applied to anything acquired by or devolving on a person which is disadvantageous.

Damnum absque injuria. See *Damnum sine injuria.*

Damnum fatale. Loss due to an event which, if not inevitable, was at least wholly exceptional and not preventible by any normal or reasonable care of precautions.

See also ACT OF GOD.

Damnum infectum. Loss not yet suffered but only apprehended.

Damnum injuria datum. In Roman law one of the recognized classes of *delict* (q.v.), denoting loss (*damnum*) caused (*datum*) by *injuria* (wrongful or culpable conduct whether intentional or merely negligent). The basis of this delict was the *Lex Aquilia* (q.v.) of *c.* 287 B.C., dealing with the killing of slaves and cattle, and with damage to property generally. The elements of the wrong were the subject of interpretation and extension by the jurists and the praetors, and on the basis of this concept the later civil law built a general theory of civil wrongs, of liability for loss caused wrongfully.

Damnum sine (or *absque*) ***injuria.*** Loss (*damnum*) befalling a person without *injuria*, without legal wrong or breach of duty having been done to him. Not all harm or damage done is necessarily legally wrongful or actionable. Where this principle applies the person damnified has no right of action against the person responsible for causing the loss because the latter was not, in causing or allowing the harm to befall, been in breach of legal duty to him. This applies, e.g. where the injured person was not one of those to whom the legal duty was owed, or where there was no legal duty to refrain from causing that sort of harm. Thus in competitive trade one may deliberately capture another's trade and ruin him. A landowner may exercise numerous rights on his own land which, though harmful to, are not deemed infringements of the rights of, his neighbours. This principle is the major exception to the principle of liability for *damnum injuria datum* (q.v.), i.e. loss caused by legally wrongful conduct, the principle which underlies nearly all tortious liability.

Dana, Richard Henry (1815–82). As a young man sailed round Cape Horn and later wrote his account in *Two Years Before the Mast* (1840). He was admitted to the Bar, specialized in admiralty cases, and wrote a manual on maritime law, *The Seaman's Friend* (1841). He failed to attain higher public office, as he desired, and devoted himself to international law, producing an edition of Wheaton's *Elements of International Law* with valuable notes in 1866.

Dane, Nathan (1752–1835). Graduated from Harvard and was admitted to the Bar in 1782. He was later a delegate to the Continental Congress and in 1795 a commissioner to revise the laws of Massachussetts. Deafness caused his withdrawal from practice, but he wrote *A General Abridgment and Digest of American Law with Occasional Notes and Comments* (8 vols., 1823, and Supp., 1829), the first comprehensive compendium of law compiled and produced in America. A bequest to the Harvard Law School enabled the establishment of the Dane Professorship and the founding of Dane Hall there.

Danegeld. A tax of two shillings on every hide of land, originally imposed under Ethelred II of England in 991 to raise money for tribute exacted by the Danes, repeated several times later, and turned into a permanent levy for national defence by the Norman kings. It is usually applied to payments made between 991 and 1016. After 1162 it dropped out of use, but was revived as *donum* or hidage. The exaction was revived by Richard I under the name of carucage, a tax on all holders of land of whatever tenure.

Danelaw. The Danes first attacked the east coast of England towards the end of the eighth century, returned repeatedly and began to make permanent settlements in the ninth century. Despite Alfred the Great's efforts he was unable to dislodge them and by a treaty of 885 the limits of Danish occupation were defined as the Thames, the Lea, to Bedford, and along the Ouse to Watling Street. In the north the limits were the Tyne and the hills of Cumberland and Westmorland. In this area the Danish law was in force and the Danish occupiers governed the Anglian peoples. Under Alfred's successors the Danes were reduced to submission and in 959 Edgar became King of all England. Edgar granted autonomy to the inhabitants of the Danelaw, permitting them to enjoy their own customs and laws. Thus England was roughly equally divided into counties under the laws of the West Saxons, the

Mercians, and the Danes, the former two being similar and the last quite distinct from them.

By the eleventh century the customary law of the shire courts within the Danelaw had distinct individuality from the Danish influences. The Danelaw fell into four main regions, each having variations in detail, in law, and custom, and several statements of Danish customary law have survived, notably a list of wergilds prevailing within the Scandinavian Kingdom of York, the Northumbrian Priests' Law, and the custom of the Five Boroughs, a code issued at Wantage in Berkshire by Aethelred II, intended to give the authority of the King of all England to the existing provincial customs of local courts within the Danelaw.

In the area of the Danelaw the hundred (q.v.) was generally called a wapentake, and a hide (q.v.) a plowland. The law had many procedural differences, including the use of an aristocratic jury of presentment to lay charges against persons suspected of crimes.

F. Stenton, *Danelaw Charters* (1920); *The Danes in England* (1928).

Danger. A risk of harm of some kind. It is in general an actionable wrong if a person causes or permits something to exist which creates a foreseeable danger or risk to others and if harm of a foreseeable kind results. Many kinds of conduct such as keeping wild animals, operating vehicles, installing and operating machinery create obvious dangers and consequently are held to impose duties of care to take reasonable precautions against the danger resulting in harm.

Danger, alternative. The principle that if a person is, by the fault of another, placed in the dilemma of having to chose between the risk of one danger and of another, e.g. staying on or jumping off a train about to crash, he is not himself in fault if he chooses the danger which, as things turn out, brought about the greater harm.
See also AGONY RULE.

Dangerous driving. The driving of a motor vehicle on a road in a manner liable to cause accidents or injury to the driver, to other road-users, or to property. Whether driving is dangerous or not is a question of fact in the circumstances.

Dangerous goods. Goods which are likely to cause injury to persons who take or use them. A high standard of duty is imposed in many circumstances on persons who put into circulation goods likely to cause injury or damage.

See also DANGEROUS THINGS, ESCAPE OF.

Dangerous Things, escape of. A person who brings on land and collects and keeps there, save in the course of natural user of the land, anything which is liable to do harm if it escapes, is bound to keep it in at his peril and is strictly, but not absolutely, liable for damage, at least to property, which is the natural consequence of its escape. He is not liable if the escape were due to the plaintiff's consent or default, the consequence of *vis major* or Act of God, or the act of a stranger, or there is statutory indemnity. The thing need not be dangerous in itself, but must be dangerous if it escapes, e.g. water in a reservoir.

See *Rylands* v *Fletcher*; STRICT LIABILITY.

Danish law. In Denmark the earliest laws were customary. It is uncertain whether, as in other Scandinavian countries, there were Law-men who recited the laws at the assemblies and advised on the disposal of cases. The early assemblies seem to have exercised legislative powers and there developed by the thirteenth century a General Assembly of the kingdom, Danehof, which in time acquired legislative functions. The Land-Thing became the supreme tribunal about 1300.

The earliest written laws appeared in the twelfth century, all probably private and unofficial; there were also ecclesiastical codes. In 1241 the civil code, the Lov-bog, was enacted by a king with the consent of a national assembly, and other statutes and ordinances were regularly passed thereafter. This, the Jutland Code or Jydske law, was ratified in 1326 and a later revision was introduced in 1590. It is the oldest Danish code and continued in effect until 1683.

Denmark was united with Norway from 1380 to 1814 and Denmark was the dominant partner, Norway being governed from Copenhagen. In 1604 a redrafted version of the National Law codification of Norway of 1274 was translated into Danish as King Christian IV's Norwegian Law.

A Supreme Court was established in Copenhagen in 1661 which acted for Norway also.

After 1661 several commissions were appointed to reform the laws and prepare codes. A draft prepared by Rasmus Vinding based on the previous legislation of Denmark was revised by him and others and proclaimed in 1683 as King Christian V's Danske Lov. It comprises six books and includes criminal law. There were subsequent additional statutes. A revised version of this code was enacted for Norway (Norske Lov, 1687) and was supplemented by certain provisions particular to Norway. These codes were lacking in theory and general principles and rules. Accordingly textbooks had great influence in developing the law, German legal literature being an important source. In this way in the eighteenth and nineteenth centuries hitherto unknown principles, sometimes of Roman origin, were admitted. But reaction at the end of the nineteenth century diminished their importance.

No great changes took place between 1683 and 1814 and statutes and legal literature dealt with Danish–Norwegian law as one system. In 1814 Norway broke away and united with Sweden till 1905. About 1840 proposals were made for a revised Danish code, but the influence of Savigny's school caused it to fail and most of the codes of 1683 survived till about 1900.

From about 1870 Denmark, Norway, and Sweden collaborated extensively in uniform legislation particularly on commercial topics. In 1891–93 a unified Scandinavian maritime law was adopted.

The present constitution is of 1953 and provides for a constitutional monarchy, legislative authority being vested in the Crown and Parliament (Folketing).

In modern law there is no general code, though criminal law is dominated by the criminal code of 1930 and procedure by comprehensive statutes. In public and criminal law legislation is detailed and supplemented by administrative regulations. Decisions of the superior courts are normally treated as settling the law, but do not have the importance they do in Anglo-American law. Legal texts are of importance for filling gaps in legislation. Law may contain express provision that it must give way to custom, and custom may be a valuable supplement where enacted law is silent, but the courts may refuse effect to unreasonable custom.

The Supreme Court sits in Copenhagen for hearing appeals only, usually in two chambers. There are two High Courts for East and West, and a large number of district courts. High Courts and district courts have both civil and criminal jurisdiction. There is a Director of Public Prosecution. There is also a Maritime and Commercial court.

S. Juul, A. Malmström and J. Søndergaard, *Scandinavian Legal Bibliography*; J. Søndergaard, *Danish Legal Publications* in *Scand. Studies in Law*, 1963; *Dansk juridisk bibliografi*; L. Orfield, *Growth of Scandinavian Law*; A. Boeg, *Danish and Norwegian Law*; H. Herlitz, *Elements of Nordic Public Law*; H. Munch-Petersen, *Den borgerlige ret*; K. Waaben, *The Danish Criminal Code.*

Dans Causam (or locum) ***contractui*** (giving rise to the contract). A requisite of fraud or misrepresentation justifying rescission of a contract. The contract must stand if it cannot be established that the misrepresentation induced the plaintiff to make the contract.

Danube Commission. An organization established by riparian states in 1948 to regulate the use of the Danube and make improvements to it.

D'Anvers, Knightley (*fl.* 1700). Author of *A General Abridgement of the Common Law* (3 vols., 1705), a partial translation of Rolle's *Abridgement* into English with references to later reports.

D'Argentré, Pierre (1488–1548). French jurist, seneschal of Rennes, one of the royal commissioners for the reform of the customs of Brittany and author of a work on the practice of the court of Rennes, *Le Stille et usage gardé et observée en la court de Rennes* (1536).

Darling, Charles John, Lord (1849–1936). Called in 1873 he took silk in 1885 and was an M.P. 1888–97. In the latter year he became a judge and overcame early doubts of his fitness for judicial office. His witticisms attracted much publicity. He retired in 1923 and was made a baron in 1924. He published *Scintillae Juris* (1877), *On the Oxford Circuit* (1909), and various lighter works.

Darnel's, or the Five Knights', case (1627), 3 St. Tr. 1. In 1626 letters were issued under the privy seal assessing certain individuals for contributions to a forced loan. Darnel and four other knights refused to pay and were committed to prison. Darnel obtained a writ of habeas corpus to the Warden of the prison to show the cause of his imprisonment. The return showed that he was detained under a warrant from the Privy Council certifying that he had been committed by the special command of the King. Hyde, C.J., decided for the King, accepting that the King had absolute power to commit, subject to no appeal, that the King could do no wrong and that the committal was no violation of Magna Carta. The imprisonment was an abuse of power, but the court could not go behind the return. After the release of the knights in 1628 the issue continued to be discussed and gave rise to the Petition of Right (1628) (q.v.).

Darrein presentment, assize of. A procedure instituted by Henry II to determine disputes about the right to present a cleric to a church, under which a jury of 12 would enquire about the last presentment (*de ultima presentatione*) and the person who last presented should do so again, without prejudice to the question of the right to present, that is, to the proprietary right to the advowson, an action for which had to be commenced in the King's court by royal writ and in which the claimant had to offer battle, the defendant being entitled to choose between battle and the grand assize. The procedure survived till 1833.

Darrow, Clarence Seward (1857–1938). A leading American criminal lawyer who specialized in criminal and labour organization cases. He was a well-known speaker and debater and wrote many books on social questions, including *Crime, Its Cause and Treatment* (1922).

Dartmoor. A famous English prison on an isolated site on high moors in Dartmoor, Devonshire. Originally built in 1806 to hold French prisoners of the Napoleonic Wars and then used as a convict prison, it was converted in 1850–51 into a central prison for long-term prisoners, and has since then been the principal English prison for serious offenders. Though the buildings are old, the site provides good opportunities for training and work on the land and escape is difficult.

Dartmouth College* v. *Woodward (1819), 4 Wheaton 518 (U.S. Sup. Ct.). Dartmouth College had been chartered in 1769 to promote education in New Hampshire. In 1816 after political and ecclesiastical disputes the State legislature proceeded to change the name of the college and to transfer control to other hands. The college contended that the legislature had unconstitutionally impaired vested rights and violated contract obligations. The Supreme Court classified the college with private rather than public corporations, held that the charters of private corporations were contracts, and that the State could not impair its charter, and that the State laws were unconstitutional. The decision protected private endowments from political interference and encouraged charitable gifts, gave added stability to the rights protected by private charters and facilitated investment of private resources in private corporate enterprises, but gave corporations an immunity from judicial interference of state police power.

Date. The day, month, and year on which a deed or writing is granted, or a transaction effected, or an event happens. A deed is good though not dated if the date of delivery can be proved. The date on which something is done or happens is frequently material in determining, e.g. priority, or whether limitation has run.

Davey, Horace, Lord Davey of Fernhurst (1833–1907). Called to the Bar in 1861 and practised with great success in Chancery, taking silk in 1875. He had a very subtle mind and was an effective pleader. He was Solicitor-General in 1886 and 1892, but was not a success in the House of Commons. He became Lord Justice of Appeal in 1893 and a Lord of Appeal in 1894 as Lord Davey of Fernhurst. His judgments are lucid, clearly arranged but concise, and many regarded him as the most accomplished lawyer of his time.

Davis, George Breckenridge (1847–1914). Served in the U.S. Army from 1867 and as an instructor at West Point, wrote *Outlines of International Law* (1887). Later he served in the judge Advocate-General's department and rose to be Major-General. He published in 1897 *The Elements of Law: An Introduction to the Study of the Constitutional and Military Law of the U.S.* (1897) and a *Treatise on the Military Law of the U.S.* (1898) which became the standard work on the subject.

Davys, Sir William (?–1687). Called to the English Bar in 1657 and the Irish Bar in 1661, acted as a justice of assize, and appointed Chief Justice of the Irish King's Bench in 1681.

Dawkins* v. *Paulet (1869), L.R. 5 Q.B. 94. An officer complained to the Adjutant-General that the plaintiff, a Lieutenant-Colonel, was incompetent and unfit to command a battalion in barracks. The plaintiff's action for libel failed as the defendant had acted in the course of his duty and enjoyed qualified privilege (q.v.).

Day. Generally the period of twenty-four hours between midnight of one night and midnight of the next (natural day), but sometimes the period between the rising of the sun and the setting of the sun (civil day). A statute which comes into force on a stated day comes into force on the expiration of the previous day. Rules frequently require certain things to be done by a given day or within a stated number of days.

A 'business day' is a day other than a Sunday, Good Friday, Christmas Day, a bank holiday, or a day appointed as a public feast or thanksgiving, or a day declared by order to be a non-business day. 'Clear Days' are days which must completely elapse in the computation; thus seven days' notice means that notice expires on the eighth day; seven clear days' notice means that it expires on the ninth day.

See also HOLIDAY; LAY DAYS; RUNNING DAYS.

Day, year, and waste. A concise statement of the rights which the King formerly had in the lands of a person attainted of petty treason or felony, to have the profits of his land for a year and a day, and the liberty to waste (q.v.) them.

Days of Grace. Days allowed for making a payment, e.g. of an insurance premium, or doing an act though the legal time for doing so has elapsed.

***De bonis non (administratis)*.** A grant of administration made where an executor or administrator has died without having completed his administration, to enable the assets not administered to be completed.

***De Diversis Regulis Juris Antiqui*.** The title of Digest 50, 17, the last title of Justinian's Digest, containing a large number of brief observations of the kind which would today be called legal maxims, and some of which have survived into modern collections of maxims, such as *Nemo plus juris ad alium transferre potest quam ipse habet*, or *Nuptias non concubitus sed consensus facit*, or *Imperitia culpae* *adnumeratur*. In mediaeval and early modern times many books were devoted to commenting on this title.

De Donis Conditionalibus. A statute of 1285, intended to secure the interests of the donor and his issue where land had been given to a man and the heirs of his body, which provided that conditional donees should have no power to alienate so as to prevent the land falling to their issue, failing whom reverting to the donor, and thereby created the interest in land known as the interest in fee tail. The statute provided remedies, writs of formedon, whereby the lord could recover the land on the failure of issue, or the issue could recover the land on the death of the ancestor.

De facto. As a matter of fact, contrasted with *de jure* (q.v.). Recognition of a foreign government may be granted as the government *de facto*, i.e. having control in fact and exercising power. This does not acknowledge the government as being entitled to possess the powers of sovereignty, but only as actually possessing the powers, though the possession may be wrongful or precarious. The U.K. has frequently granted recognition in two stages. Thus it recognized the Soviet régime in Russia *de facto* in 1921 and *de jure* in 1924, the Italian conquest of Ethiopia *de facto* in 1936 and *de jure* in 1938. International law is indifferent to the form of recognition so that the difference between *de facto* and *de jure* recognition exists only in municipal law. The distinction is frequently dictated by politics, by the approval or disapproval of the new regime.

de Grey, Sir William (1719–81). Called to the Bar in 1742, he entered Parliament in 1763, became Solicitor-General in 1763, and Attorney-General in 1766. In 1771 he succeeded as Chief Justice of the Common Pleas but resigned through ill-health in 1780. Just before his death he was ennobled as Lord Walsingham. He had a reputation of being a most accomplished lawyer with great powers of memory.

De Heretico Comburendo. An English statute of 1401 passed for the suppression of the Lollards and condemning the teaching or writing of doctrines contrary to Catholicism. It empowered the bishops to arrest and imprison persons reported or strongly suspected of holding heretical doctrines, until they should make canonical purgation, and if convicted, to punish them with fine and imprisonment. Any who refused to abjure these doctrines or abjured and relapsed to be taken into custody and burned as an example to others. The initiative lay with the bishops and the Act provoked resentment. An Act of 1414 entrusted the arrest of heretics to secular officers, but preserved the Church's jurisdiction in the resulting trial. The Act was repealed by Henry VIII, restored by Mary and again repealed by Elizabeth.

De homine replegiando. An ancient writ available in certain cases to secure the release of a freeman from unlawful detention.

***De Jager* v. *Attorney-General of Natal*,** [1907] A.C. 326. During the Boer War the appellant, an alien, was resident in Natal. The Boer forces occupied Natal and he aided them. The Privy Council held that as he had been resident in British territory he owed allegiance to the Crown and by his aiding the enemy he had been guilty of high treason.

De jure. As a matter of legal right, contrasted with *de facto* (q.v.). Thus Charles II is deemed to have been King *de iure* from 1649 though *de facto* only from 1660.

De Lolme, Jean Louis (1740–1806). Swiss constitutionalist, author of a notable *Constitution de l'Angleterre* (1771, Eng. trans. 1772) which gave many on the continent their ideas of the British constitution. He also wrote an *Essay on the Union of Scotland with England* (1787) and lesser works.

De minimis non curat lex (or *praetor*). The law (or the court) does not take account of trifles, the general principle that some breaches of duty, or mistakes in procedure or in deeds, are too trivial to be made the ground for giving a legal remedy, or dismissing a claim, or as the case may be. Where the liberty of the subject is concerned, quite small mistakes may not be capable of being ignored under the *de minimis* rule.

De odio et atia. A writ addressed to a sheriff commanding him to enquire whether a prisoner charged with murder had been committed on general cause of suspicion or *propter odium et atiam*, on account of hatred and ill-will, so that in the latter case another writ might allow him bail.

De praerogativa regis. A statute attributed to 1323 providing reinforcement of the common law, that the King should have ward of the lands of natural fools, taking profits without waste or destruction, but finding them necessaries, and on their deaths conveying the estate to their heirs.

de Smith, Stanley Alexander (1922–74). Downing professor of the Laws of England at Cambridge, 1972–74, author of *Judicial Review of Administrative Action* (1959), *Constitutional and Administrative Law* (1971), and other works on public law.

De tallagio non concedendo. A statute of 1297, enacting that no tallage or aid is to be levied without the consent of clergy, peers, burgesses, and other freemen of the land, and traditionally relied on as a basis of the constitutional principle of no taxation without Parliamentary assent.

De Verborum Significatione. The title of Digest, 50, 16, containing notes on the meanings and uses of many words and phrases commonly used. Also a short dictionary by Sir John Skene (q.v.) appended to his edition of the Scots Statutes containing an exposition of the terms and difficult words contained in *Regiam Maiestatem* (q.v.) and used in practice in Scotland.

de Villiers, John Henry, Baron (1842–1914). Called to the English Bar and began a practice in Cape Town in 1866. He entered the Cape Parliament, became Attorney-General in 1872 and Chief Justice of the Cape in 1874. In 1910 he became first Chief Justice of the Union of South Africa and a peer. He sat also as a member of the Privy Council. He was one of the greatest South African judges of his time, with a masterly grasp of legal principles and great knowledge of the Roman, Dutch, and English authorities. His great work was to adapt the Roman–Dutch law to modern conditions.

Dead freight. The amount payable by the charterer of a ship for the part of the cargo-carrying capacity of the ship which he does not fill. It is in substance damages for loss of freight on the empty space. Save by contract, a shipowner does not have a lien for dead freight over cargo actually shipped.

Dead man's part. The part of an intestate's moveable estate beyond that which belonged by right to his widow and children. By the custom of London, part of the estate of a deceased intestate freeman of the city was spent in saying masses, and later seems to have been claimed as a perquisite by the executor, though from 1685 all assets were required to be distributed in accordance with the Statute of Distribution of 1670. In Scotland 'dead's part' or 'dead's part of gear' is the proportion of a man's moveable estate not required for the payment of legal rights (q.v.) and which he can dispose of by will or which will be distributed under the rules of intestate succession.

Dean of Arches. The judge of the Court of Arches. See ARCHES, COURT OF.

Dean of Faculty. The elected or sometimes appointed head of a group. In Scotland the term is applied to the elected leaders of the Faculty of Advocates and of various local legal societies. In academic usage the Dean of a Faculty is the head of the group of scholars and teachers concerned with a particular field of learning.

Dean of Guild court. Merchant guilds were recognized in Scottish burghs from the twelfth century and from the fifteenth century the Dean of Guild and his court came to be recognized as parts of the normal constitution of royal burghs and of some burghs of barony. As head of the merchants the Dean tried cases between merchants, and between merchants and mariners. From the sixteenth century the court began to exercise an aedilic jurisdiction over questions of neighbourhood and from this developed a jurisdiction over new buildings, demolitions, additions, and repairs, which became the sole, but important jurisdiction. The Dean, latterly usually elected by the town council from its own members, and assisted by technical staff, continued through his court to act as statutory buildings authority for the burgh and regulate all building developments. These courts were abolished in 1975.

Dean of St. Asaph's case (or *R.* v. *Shipley* (1783), 21 St. Tr. 847). Shipley, Dean of St. Asaph, was indicted for seditious libel for having published his brother-in-law's pamphlet in favour of parliamentary reform. On the direction of the trial judge the jury found him guilty of publishing only, but whether a libel or not the jury did not find. On application for a new trial Lord Mansfield C.J., held that this limitation on the jury's function was in accordance with law. Fox's Libel Act (Libel Act, 1792) changed the law to entitle the jury to return a general verdict of guilty or not guilty on the whole matter put in issue on the indictment.

Deane, Joseph (1674–1715). Called to the Irish Bar, he became Chief Baron of Exchequer in Ireland in 1714 but survived only six months.

Dearle* v. *Hall ((1823), 3 Russ. 1). ***Rule in.*** The rule that a mortgagee or assignee of an equitable interest in personalty, who gives notice of the transaction to the legal owner of the personalty, gains priority over an earlier mortgage or assignment of which he had no knowledge and which had not been intimated to the legal owner.

Death. An event of legal importance in many contexts, though it may be difficult to say when it takes place. For most legal purposes it can be said to take place when heartbeat, pulse, and breathing stop. Absolute signs of death are algor, rigor, and livor mortis. Determination of whether death has occurred and precisely when it occurred is important for such legal purposes as removal of an organ for transplanting to a living person, and for termination of marriage, for succession and inheritance, insurance and social security benefits, claims by surviving relatives, and other purposes.

Determination of the cause of death, whether by natural causes, accident, suicide, or homicide, is very important for many principles of liability, whether civil or criminal, e.g. for claims of damages for tort, or for prosecution for murder, manslaughter, or other crime. If the cause of death is natural, the law is rarely concerned with whether it be attributable to one clinical factor rather than another.

The fact that a person's death has taken place has consequences for his claims and liabilities under contract or tort. Some claims and liabilities may lapse on death, others transmit to his personal representative (executor or administrator) for the benefit of or against his estate.

A person who has disappeared may in many legal systems be presumed dead after the lapse of a fixed period, frequently seven years. Such a provision is necessary to enable property and other rights, e.g. marriage, to be determined.

See also CIVIL DEATH.

Deathbed. A rule in Scots law that a person on his deathbed might not competently make a disposition of his heritable property to the prejudice of his heir, and any such deed was challengeable. To be on deathbed a person had to be suffering from the illness from which he died, and that fact was rebutted by evidence that he had survived the deed for 60 days, or gone to kirk, or market unsupported. The ground of challenge was abolished in 1871.

Deathbed (or dying) declarations. Statements formally made by dying persons which are admitted as exceptions to the rule excluding hearsay evidence, and admissible in evidence though no cross-examination was or is possible. They may preserve important evidence and it is presumed that the prospect of death is a powerful compulsitor to tell the truth.

Death duties. The general term in the United Kingdom for confiscatory duties levied on a person's property on his death. These included legacy duty, payable between 1796 and 1949 on personal property devolving under a will or intestacy; succession duty, payable 1853–1949, on all other benefits arising on death; probate duty or, in Scotland, inventory duty, payable 1801–94, in respect of personal estate; account duty, payable 1881–94, in respect of certain personal property devolving under gratuitous titles; temporary estate duty, 1889–96; settlement estate duty, 1894–1914; and estate duty 1914–75. Estate duty was levied on the principal value of all property, real and personal, settled or not, which passed on death after 1 August 1894, or was legally deemed to pass if the deceased was, at the time of his death, competent to dispose of it, and including property in which the deceased or any other person had an interest ceasing on the deceased's death to the extent to which a benefit accrues or arises by the cesser of the interest, and donations *mortis causa*. Small estates were exempted, as were gifts to reduce the National Debt, gifts of land to the National Trust or the National Trust for Scotland, gifts of land to a government department, local authority, or non-profit-making body, if the Treasury so directed, and objects of national interest provided they are kept in the United Kingdom.

Duty was levied on the aggregate value of all property passing on death in respect of which duty was leviable, but property in which the deceased never had an interest was not aggregated and funeral expenses, debts, and encumbrances were deducted. The rate of duty was according to a graduated scale which was revised from time to time.

In 1975 Capital Transfer Tax (q.v.) replaced Estate Duty; under it duty is levied, subject to qualifications, not only on property passing on death but on all property transferred during life.

In the U.S. the federal death tax was introduced in 1916 but many states had already introduced such taxes.

Death—qualified jury. In U.S. law a trial jury held fit to decide a case involving the death penalty. The practice long was to excuse from serving any juror who stated opposition in principle to the death penalty, but more recently it has been held that opposition in the abstract to capital punishment did not disqualify a juror, but only those who indicated an absolute refusal to bring in a verdict implying liability to capital punishment and those whose impartiality would be affected by their views were excluded.

Debate on the Address. The Queen's Speech at the opening of each session of Parliament is now used by the Cabinet to set out in general terms its policy and programme of legislation for the session. Though read by the Queen it is prepared by the Cabinet. In both Houses an address in reply is moved thanking Her Majesty for Her Gracious Speech, but the debate on this provides an opportunity for general discussion of the government's policy, and amendments are frequently proposed regretting this or that proposal or the absence of proposals for this or that. It is regarded as an honour to be selected to move and second the address.

Debenture. A document which creates or acknowledges a debt, normally one owed by a company. It normally expresses, and always implies, an undertaking to repay; it may be a single document or one of a series all ranking equally; it may be unsecured, but is frequently secured by a charge over certain property, or by a floating charge over the whole property and undertaking of the company. A trust deed may be granted by which property is specifically mortgaged to trustees for the debenture holders, as security for payment of the interest and capital. Debentures may be redeemable or irredeemable. Debenture stock is generally constituted and secured by a trust deed containing a charge on the company's property, stock certificates being issued to each holder of stock stating the conditions on which the stock is issued and held. A debenture to bearer is negotiable and transferable by delivery, a registered debenture is transferable by instrument of transfer, and debenture stock is transferable in the same way as shares in or stock of the company. Various statutes authorize the issue of what they call debentures, e.g. by local authorities under the Local Loans Act, 1875.

Debita fundi. In Scots law, debts attaching to the soil, such as feu-duties, or debts and money burdens heritably secured, such as ground annuals. The right to these creates a lien over the lands and the superior or creditor is entitled to recover them by an action of poinding of the ground. This category of debts does not include teind, land-tax, or public burdens or assessments.

Debrett, John (?–1822). A publisher who compiled a record of parliamentary debates, 1743–74, and a *Register of Proceedings of both Houses from 1780 to 1803*, both utilized by Cobbett and Wright in their *Parliamentary History*. From 1781 till his death he edited *The New Peerage*, re-named in 1802 *Debrett's Correct Peerage*, and its companion *The Baronetage of England*, now combined as Debrett's *Peerage, Baronetage, Knightage and Companionage*, long recognized as a standard reference book.

Debt. (1) That which is owed by one person to another, and particularly money payable arising from and by reason of a prior promise or contract, but also from and by reason of any other ground of obligation, e.g., statute or order of court. The moral and legal obligation is on the debtor to pay his creditor, but in many cases the existence or extent of the obligation to pay must be determined judicially.

English law distinguishes simple contract debts and specialty debts, the latter being those created by deed or instrument under seal. A debt of record is a debt proved to exist by the official records of a court, particularly judgment debts, fixed as being recoverable by a judgment of a court.

Debts may be secured, i.e. for which the creditor holds some security, e.g. by mortgage, in addition to the debtor's personal liability, or unsecured.

(2) One of the earliest forms of personal actions in English law. It could be brought on debts acknowledged by deed under seal, or to recover the price on a sale, loans or rent, though there was no deed under seal, and even statutory penalties. The action was fairly comprehensive but its defects were that trial was by compurgation (wager of law) and particularly complicated rules of pleading attached to it. Accordingly from the early fifteenth century there was a tendency to avoid using the writ of debt and to utilize other forms of action.

See also COVENANT.

Debt, imprisonment for. Originally a creditor could have his debtor arrested for owing any, even trivial, sum. The debtor would be kept in the Marshalsea or Fleet prison until friends found the money to have him released. The Imprisonment for Debt Act, 1827, provided that no person could be arrested on mesne process for debt of less than £20 and only on an affidavit that the sum was owed. The Judgments Act, 1838, abolished arrest in mesne process in civil actions, apart from certain exceptions, and the Debtors Act, 1869 almost completely abolished imprisonment for debt.

Deceit. Deception, fraud, or cheating, the actionable wrong of making a false statement, knowing it to be false, or without honest belief in its truth, or recklessly, not caring whether it be true or false, with the intention that another should rely thereon and act to his detriment and with the result that the other does so act. An action on the case for loss resulting from deceit was recognized in 1789 (*Pasley* v. *Freeman*, 3 T.R. 56). If the detriment is entering into an unprofitable contract, the contract is voidable on account of the deceit, as well as damages being recoverable. Statute has made company directors liable for even innocent misrepresentation resulting in the plaintiff taking shares.

See also *Derry* v. *Peek* (1889).

Decemviri. The Roman term for a commission of 10 members, particularly the body of patricians (*decemviri legibus scribundis*), who prepared 10 tables of the law in 451 B.C. and who, or a second board, prepared two further tables in 450 B.C. completing the laws of the Twelve Tables (q.v.). The *decemviri stlitibus judicandis* judged cases to determine whether a man was free or slave, and were probably made presidents of the centumviral tribunal by Augustus.

Deception. In English law the offence of dishonestly obtaining property belonging to another, or a pecuniary advantage by words or conduct as to fact or law, or as to his or anyone else's present intentions, done deliberately or recklessly.

Deciners (*decenniers* or *doziners*). Persons so-called because they had supervision of 10 households, who had the oversight of the view of frankpledge (q.v.) for the maintenance of the King's peace.

Decio, Filippo (1454–1535). An Italian jurist and commentator, auditor of the Rota, author of commentaries on the *Digest* and *Code*, the *Usus Feudorum* (1483), *Decretals* (1540), *De Actionibus* (1483), and other works.

Decio, Lancelloto (?–1503). An Italian jurist and commentator on the *Digest*.

Decisions. (1) A general term for the results of judicial consideration of cases brought to courts or tribunals, submitted to arbitration or Ministers. What is sought by contending parties, who are in dispute about the interpretation of the facts and/or about the principle or rule of law which is relevant thereto, or its application to the facts or its interpretation is a judicial decision. The actual judgment or decree of the court decides the matters in dispute, but particularly in Anglo-American countries more interest attaches to the judicial opinion or opinions, reviewing the relevant law, and bringing out why the judge reached the decision he did, because this opinion may contain statements valuable for use in future cases.

See JUDICIAL PRECEDENTS; *Stare Decisis*.

(2) A kind of administrative act of the Council or Commission of the European Communities, ranking below regulations and directives (qq.v.). A decision may be addressed to member states or to individuals and take effect on notification to the addressees. It is binding in its entirety on addressees, and leaves no discretion as to the manner in which it is to be carried out. It is an act originating from the competent organization intended to produce judicial effects, constituting the ultimate end of the internal procedure of the organization and according to which such organization makes its final ruling in a form allowing its nature to be identified. Whether an administrative act is a regulation, directive, or decision is determined by content and object rather than by form. General decisions are quasi-legislative made by a public authority and having a rule-making effect as regards everyone; individual decisions are decisions in which the competent authority had to determine concrete cases submitted to it.

Declaration. In general a formal statement intended to create, assert, or preserve a right. In common law and probate practice before the Judicature Acts, the declaration was the first pleading delivered by the plaintiff, stating his claim. It was divided into counts and framed in highly technical language.

In modern practice, particularly in Chancery cases, the decision of the courts is frequently a declaration of the rights of the parties.

In Scottish criminal procedure it is a statement which a prisoner apprehended on suspicion of crime gives to a magistrate, usually a sheriff; it is taken down in writing and signed by declarant and magistrate. It is not obligatory to emit a declaration and in modern practice it is rarely done. If one is made, it is admissible in evidence at the trial.

See also DECLARATORY JUDGMENT.

Declaration of Independence. At the Second Continental Congress of delegates from the British colonies in North America, which met in Philadelphia on 10 May 1775, after open hostilities between colonists and units of the British Army had taken place, some delegates had already been empowered to press for independence from Britain. On 7 June Lee called for a resolution 'that these United Colonies are, and of right ought to be, free and independent States'. On 11 June the Congress appointed a committee to prepare a declaration, which was drafted by Jefferson, and reported to the Congress on 28 June. On 2 July the congress adopted Lee's resolution, and on 4 July the amended declaration was adopted without dissent, the New York delegation, under instructions, abstaining. The Declaration was engrossed on parchment and signed by the 56 members of the Congress.

Jefferson based his draft on the philosophy of natural rights, and asserted the right of the colonists to sever their link of allegiance to the British monarchy on the ground that no person or group have moral right to govern any other group of persons without the latter's consent.

The Declaration has subsequently been regarded as a classic statement of the claim to self-government and freedom from oppression.

Declaration of Indulgence. A declaration was issued by Charles II in 1672 suspending the penal legislation directed against Roman Catholics and dissenters, leaving as the only restrictions on religious freedom that Protestant ministers must be licensed and that Catholics must not worship in public. When Parliament next met, the Commons asserted that statutes could be suspended only by subsequent statutes, not by royal edicts, and the King had to withdraw the Declaration. Parliament then passed the Test Act (q.v.).

A further declaration was issued by James II in 1687, in less general terms. It suspended the penal laws and ordered that the oaths and declaration required by the Test Act, 1673, of all office-holders should not be tendered, and further offered individual dispensations. The King reissued this Declaration in 1688 and ordered the Anglican clergy to read it from their pulpits. The Archbishop of

Canterbury and six bishops petitioned against this order on the ground that Parliament had declared the suspending power illegal; they were indicted for seditious libel (see SEVEN BISHOPS, TRIAL OF THE) and acquitted. The Bill of Rights, 1689, abolished the suspending power and condemned the dispensing power, as it had been assumed and exercised of late.

Declaration of London (1909). A proposed international agreement intended to settle doubts about the application of the doctrines of contraband, continuous voyage, and neutral destination. Eleven countries signed the convention, but the House of Lords refused to ratify it. It was in practice followed by Britain until 1916.

Declaration of Paris. A State Paper agreed upon at the end of the Crimean War, 1856, by the representatives of the powers engaged to the effect that privateering was abolished, the neutral flag covered enemy's goods except contraband of war, neutral goods under the enemy's flag were not subject to seizure unless they were contraband of war, and that blockades, to be binding, must be effective. It was signed by most of the then great powers, but ceased to be binding during World War I.

Declaration of right. The only means of redress open to a subject against the Crown prior to the Crown Proceedings Act, 1947, for breach of contract, was by petition of right, on which the court might make a declaration, whereupon the Crown might grant redress. The modern statutory claim is still subject to the conditions attaching to a petition and declaration of right.

Declaration of Rights. (1) Issued by the Convention Parliament of England (1688) to resolve the constitutional crisis resulting from the flight of King James II, laid down the terms on which the Crown was to be offered to William of Orange and Mary which were later incorporated in the Bill of Rights (q.v.).

(2) A resolution which embodied the views of the delegates to the First Continental Congress (of the American colonies) held at Philadelphia in October, 1774, on the violation of their rights by the legislation of the British Parliament, and which was adopted in an effort to secure the repeal of that body's coercive legislation. The failure of this policy to secure any concessions led to the Declaration of Independence in 1776, which was in part modelled on this Declaration of Rights.

Declaration of the Army. A statement issued by the Council of the Army in 1647 by which the Army intervened in the negotiations between Charles I and the Long Parliament, contending that it was more representative of the people than the parliament, demanding that parliament expel members hostile to the Army, provide for its own dissolution, and for succeeding parliaments. Some members in fact withdrew but returned when parliament did not comply further with the declaration.

Declaration of trust. The normal mode whereby a person vested with property creates a trust and constitutes himself holder of that property in trust. It must be manifested and proved by writing, signed by the person able to declare the trust.

Declaration of uses. From about the thirteenth century it was competent in England to create a use by declaring that one held property to the use of another, or conveying property to another, declaring the uses or purposes in writing or verbally. After the Statute of Uses of 1536 such a declaration vested the land in the persons in whose favour it was made. Declarations of uses were frequently employed to achieve the purposes of modern settlements.

Declaration of war. A communication by one state to another that a condition of war now exists between them and of its intent to commence hostilities. In earlier times declarations were attended with considerable solemnities, but today a simple message suffices. According to Hague Convention III of 1907 a declaration must be unequivocal and must give the reason for resort to war. But a declaration is in practice frequently not made, or only made after an attack, and this does not affect the legality of the war. Sometimes a severance of diplomatic relations takes the place of a declaration or an ultimatum is delivered with a conditional declaration of war, as by Britain in 1914 and 1939. No delay between declaration and actual commencement of hostilities need take place. The existence of a state of war following a declaration must be notified to neutrals.

Declaration, statutory. A form of declaration introduced to avoid the need to resort to declarations on oath in extrajudicial proceedings, and required in statute in many cases. The form is provided by the Statutory Declarations Act, 1835.

Declarator. In Scots law, a form of action by which some status or right is declared to attach to the pursuer, but in which nothing is demanded to be paid or done. Decree confers no new right, but declares authoritatively that the pursuer has the status or right in question. Instances of declarator include declarators of nullity of marriage, or bastardy, of servitude, of trust, and various others. Declarator is not competent to decide theoretical questions but only issues which are in doubt or dispute and which the pursuer has a practical interest to have determined. It corresponds to the English action for a declaration.

Declaratory Act, 1719. This Act of the British Parliament, affirmed its right to legislate for Ireland, the Irish Parliament being treated as a mere dependency.

Declaratory Act, 1766. Act of the British Parliament, passed after the passing of the Sugar Act, 1764, and the Stamp Act, 1765, imposing taxation on the American colonies for revenue and the repeal of the Stamp Act as a concession to protests, asserted the right of Parliament to make laws binding on the American colonies in all cases whatsoever. This attitude confirmed the resistance which ultimately led to the Declaration of Independence.

Declaratory Act or Statute. One which for the avoidance of doubt or dispute declares or states formally the law on a matter but does not purport to change it.

Declaratory judgment. A judicial declaration of the rights of a party without reference to their enforcement, or a decision on a state of facts which has not yet arisen. The declaration must relate to some legal right and confer some benefit on the plaintiff; the courts will not answer merely academic doubts. The power to make a declaratory order is discretionary. In certain cases the courts may make declaratory judgments against the Crown where some other judgment would have been granted against a subject.

Declinature. The privilege and duty of a judge to decline to exercise jurisdiction in a case in which, as by reason of relationship to a party or possible pecuniary or other interest in the result of the case, he might be thought to be biased.

Decree (originally *decreet*). From Roman law *decretum* (a decision of the emperor as judge), an order of a court made after consideration of a case. The term was formerly used in equity cases and still is in matrimonial cases, but the term judgment has since the Judicature Acts, been the standard one and includes a decree.

In matrimonial suits decrees for dissolution of marriage are in the first place decrees *nisi*, which are made absolute after the lapse of a period if there is no contrary reason.

In the Court of Arches a decree is a process by which a suit is commenced.

In Scottish practice a decree (in older practice 'decreet') is the term for a final order of the court granting or refusing the remedy sought. A decree arbitral is the order made by an arbiter, and a decree conform is a decree granted to enable the order of a foreign court or other tribunal to be enforced.

Décret. In French law an act of the government corresponding roughly to a British statutory instrument or an American executive order.

Décret-loi. In French law an executive decree based on a statute which confers full powers on the executive to legislate for a limited period with respect to matters normally requiring an act of Parliament. The Conseil d'Etat has recognized the validity of this mode of legislation even when used to amend the Civil Code, but has stated general limitations on its use.

Decreta. In Roman law decisions of magistrates given after investigation of a case by *cognitio* (q.v.) and in particular, decisions of the emperor as judge of first instance after trial by *cognitio*, or as a judge of appeal. As the highest authority in the State the emperor could interpret the law freely and even introduce new principles. Consequently imperial decisions were authoritative interpretations of the law or even innovatory and regarded as statements binding for the future, and as such quoted by the jurists. They were as much creative of law as edicts or rescripts. They were not only communicated to the parties but recorded in the records of the imperial court and private persons might obtain copies of them. The term is sometimes used of imperial enactments generally.

Decretalist. A scholar of canon law whose main study was papal decretals, especially in the formative stages of canonist science *c.* 1200–34. They are distinguished from decretists, whose main object of study was the tradition of the Church contained in Gratian's *Decretum* (q.v.). Their main concerns were the discovery and collection of decretals and the systematic and scientific exposition of the canon law contained in the decretals. In the first period of decretalist activity (*c.* 1160–1200) the main concern was the collection of decretals and numerous collections were made of which the most notable was the *Breviarium extravagantium* (*Compilatio prima*) of Bernard of Pavia (1191–92) which was made a subject of teaching at Bologna in 1196. But, particularly in England, there was activity in the writing of glosses and *summae*.

In the second period (*c.* 1200–34) many collections of decretals were made at Bologna to supplement the *Compilatio prima* and four were accepted as basic texts; the *Tertia Antiqua* of Petrus Collivaccinus (1210), the *Secunda Antiqua* of John of Wales (1210), the *Quarta Antiqua* of Joannes Teutonicus (1215), and the *Quinta Antiqua* of

Tancred (1226). Collectively these were known as the *Quinque Compilationes Antiquae*. These and other lesser collections stimulated commentaries, such as *Glossae* and *Summae*, and particularly the *apparatus glossarum* or systematic continuous gloss on the text. Among the leading commentators were Tancred, Joannes Teutonicus, and Raymond of Penaforte. There were also monographs such as Tancred's *Ordo judiciarius*.

In the third period (1234–1348) the age of decretal research came to an end with the definitive collection of decretals made by Raymond of Penaforte for Gregory IX in 1234, and canonists were free to concentrate on scholarship and interpretation. This period gave rise to many volumes of *apparatus, glossae, summae, lecturae, quaestiones*, and other works. There was also a great volume of commentary on post-Gregorian collections of decretals, constitutions, and conciliar legislation.

Decretals (*epistolae vel litterae decretales*). Letters containing a papal ruling, particularly one relating to matters of canonical discipline, and most precisely a papal rescript in response to an appeal. They might be restricted to an immediate problem, or have limited or universal applications. Some mandates relating to other matters are found in decretal collections. They were widely used from the fourth century in matters of canonical discipline and greatly increased in number as papal power increased, as appeals from all parts of the Western Church to the Curia multiplied and as there was increased need to resolve doubtful questions. *Gratian's Decretum* (*c.* 1140) was accepted as an adequate summary of the older laws of the Church and thereafter decretals formed the principal element in collections and became the subject of more precise definition and classification in commentaries. Also, from an early date the binding authority of decretals was recognized, Gratian putting them on the same level as *canones* and *decreta* and above the writings of the Fathers.

From an early date papal decretals were included in collections of canons, and collections of decretals were made from the fifth century. Outstandingly important were the translations and collections of the Scythian Dionysius Exiguus (*Dionysiana Collectio*) (*c.* 514) and there were regional collections in Africa, Italy, and the Frankish Kingdom. In Italy there were many, notably the *Hadriana Collectio* sent by Adrian I to Charlemagne in 774, which became the *liber canonum* for the Frankish Kingdom. The *Hispana Collectio* (sixth century), associated with Isidore of Seville, was the most important Spanish collection. The False Decretals, of Pseudo-Isidore assembled in France about 850, included a large number of spurious texts designed to enhance clerical status and privilege and consolidate papal authority and jurisdiction, and found their way into later collections. Among later collections important specimens are the Italian *Anselmo Dedicata Collectio* (882–896), the Frankish *Collectio Abbonis Floriacensis* (988–996), the German collection of Regino of Prum (906), and the *Decretum* of Burchard of Worms (*c.* 1012).

The *Decretum* of Gratian (*c.* 1140) summarized the best of the existing tradition and became the standard text. It marked the end of the old style of collection and laid the foundation for the classical period of decretal codification, which by the late twelfth century had become the central interest of collectors of decretals. This was followed by numerous collections, the most important of which was the celebrated *Quinque Compilationes Antiquae* (q.v.) which in practice superseded all other decretal collections. They were themselves superseded by the *Decretals of Gregory IX* (1234) in five books divided into 185 titles and 2154 chapters, to which in time further collections were added, namely the *Liber Sextus* of Boniface VIII (1298), and the *Constitutiones Clementinae* of Clement V (1317) which completes the official *Corpus Juris Canonici*. The *Extravagantes* of John XXII and the *Extravagantes communes* of various popes from 1260 to 1480, though included in the *Corpus Juris Canonici*, were private compilations prepared by Jean Chappuis in 1501 and never issued officially, and other supplementary collections were not of lasting importance.

Decretals, False. See FALSE DECRETALS.

Decretals of Gregory IX. The first authentic general collection of papal decretals and constitutions, compiled by Raymond of Penaforte at the request of Pope Gregory IX in 1230–34 and promulgated in 1234. It is a systematic collection based on the *Quinque Compilationes Antiquae*, but reproducing the errors, mutilations, and methods of the earlier compilers, and shortening some passages and eliminating others. Following the earlier collections the division into five books deals with *Iudex, iudicium, clerus, connubia*, and *crimen*. It was really accordingly a fresh edition of decretal law, and was exclusive of all other decretals and collections apart from the *Decretum*. It gave rise to a vast amount of commentaries and literature.

Decretists. Mediaeval jurists whose main object of study was laws of the Roman Church (*decreta*), particularly those of the period between Gratian's *Decretum* (*c.* 1140), and the *Decretals of Gregory IX* (1234), whose main interest was the study of the *Decretum*. The centre of the study was Bologna but other centres were Cologne, Paris, and Oxford.

In the first period (*c.* 1140–1200) the principal methods were evolved, the writing of *glossae*, marginal or interlinear annotations; *summae*, concise, ordered rendering of, and commentary on the principal contents of the *Decretum*; and quaestiones, questions designed to interpret, propound or clarify the content of the *Decretum* by dialectical techniques; and numerous scholars occupied themselves with all three modes of elucidation. Outstanding decretists included Paucapalea, Rufinus, Huguccio, Honorius and Peter of Blois.

In the second period (*c.* 1200–20) an outburst of activity resulted in four great *apparatus*, the *Jus Naturale* of Alanus Anglicus (*c.* 1202), the *Apparatus* of Lawrence of Spain (1210–15), the *Glossa Palatina* (1210–15), and the *Apparatus* of Joannes Teutonicus (1216), a revised form of which was received at Bologna as the *glossa ordinaria*.

Simultaneously there appeared a vast number of works by many decretists. The *glossa ordinaria* of Joannes Teutonicus marks the end of the decretist era and though after 1234 the study of decretals took primary place the *Decretum* continued to stimulate writing for generations.

Decretum. The name given to several important works in canon law. As the mass of materials comprising the canon law multiplied, the need was increasingly found for collections putting the materials into order. The *Decretum* of Burchard of Worms (*c.* 1033) was intended to promote reorganization of the Church and to maintain the discipline of the clergy and the Christian people. Much of his work reappeared in the influential *Decretum* of Ivo of Chartres (*c.* 1096). Peter Abelard's *Sic et Non* sought in the context of theology to arrive at a balanced tradition by setting off contradictory views one against the other, and this stimulated Gratian who, by setting out texts for and against various propositions and applying principles of interpretation, compiled the *Concordia Discordantium Canonum* or *Decretum*, the foundation of the classical law of the Roman Church (*c.* 1140). Study of the *Decretum*, and summaries of and commentaries and glosses on it, began the great period of canonist scholarship.

See also DECRETISTS.

Dedication. The setting apart and devoting of a thing to a particular purpose, as where a person expressly or impliedly opens to public use a road on his land and the public avail themselves of it. A dedication may be limited in time or qualified.

Deed. An instrument written or otherwise marked on paper or similar substance expressing the intention or consent of some person to make, confirm, or concur in some assurance of some interest in property, or some obligation, duty or agreement, or otherwise to affect the legal relations of some other person, signed, sealed, and delivered as that party's act and deed. A deed is in principle necessary for every transaction which common law requires to be evidenced by writing, and in many cases under statute. Deeds are either deeds poll or indentures. A deed poll, so called because the parchment has been polled or shaven even at the top, is a deed made by and expressing the intention of one party only. An indenture so called because the parchment was indented or cut with a wavy line at the top, is a deed to which two or more persons are parties, evidencing some act or agreement between them. In modern practice a deed is described by a name, e.g. settlement or mortgage, indicating the nature of the transaction. A deed must be signed, but, statute apart, need not be attested by any witness, though it has long been customary to have a witness or witnesses and append a statement that the deed was signed, sealed, and delivered and for the attesting witness to sign his name, adding his address and description. In principle it must also be sealed, the party to be bound doing some act expressly or impliedly acknowledging the seal to be his, and delivered, at least by uttering the words 'I deliver this as my act and deed', but today the seal, if any, is usually a small circle of gummed paper stuck on the deed, and the only necessary formality is signature in the presence of a witness.

In Scots law the term deed is used generally of any writing authenticated according to the statutory solemnities of execution. In Scots law sealing is not and has for centuries not been a requisite, save in the case of corporations, signature before two witnesses being requisite, and in some cases delivery also.

Deed of arrangement. A class of instruments comprising an assignment of a debtor's property, a deed of or agreement for a composition, a deed of inspectorship for the carrying on or winding up of a business, a letter of licence authorizing the debtor to manage or dispose of a business with a view to the payment of debts, or an arrangement for the purpose of carrying on or winding up a debtor's business. Any of these may be entered into by a debtor as an alternative to bankruptcy proceedings. To be valid it must be registered with the registrar of the Department of Trade.

Deed of settlement. A deed whereby a joint-stock company was usually formed before 1844. It made certain persons trustees of the partnership property and laid down regulations for the management of the business. There might also be a private Act or letters patent. Between 1844 and 1862 companies registered under the Joint Stock Companies Act, 1844, had to file a deed of settlement containing prescribed particulars and other documents before they were fully incorporated.

Deed-poll. A deed so called because it is cut evenly or polled at the top (whereas an indenture or deed between two parties was formerly indented or cut at acute angles), and containing a single person's declaration to all of his act and intention, e.g. to grant a power of attorney or to change his name.

See also DEED; INDENTURE.

Deeming. A common modern kind of legal fiction. Particularly in statutes it may be provided that one thing shall be 'deemed to be' another, e.g. that a dog shall be deemed to be a natural person, in which case the 'deemed' thing must be treated for the purposes of the statute as if it were the thing it is statutorily deemed to be.

Deemster. In the Isle of Man there are two Deemsters who are High Court judges.

See also DEMPSTER.

Defamation. The general legal term for the wrong of damaging another's reputation by falsely communicating to a third party matters bringing the complainer into unjustified disrepute. In Roman law abusive chanting was punishable. In early Germanic law insults were punishable by cutting out the tongue. In English law defamation has developed in the two forms of libel (q.v.), which is defamation in permanent form, such as writing, and slander (q.v.) which is spoken. Libel is both a civil wrong or tort and a crime (criminal libel).

In Scots law defamation is the branch of the wrong of verbal injury (q.v.) which deals with damage to a person's self-esteem, or to his public reputation, or honour, corresponding to the English libel and slander combined. It may be effected by any means by which an idea may be communicated, and is actionable even if communicated to the person defamed only, there being no need for wider publication. An idea is defamatory if injurious to the person's self-esteem or reputation, judged by the standard of right-thinking members of society generally; it includes imputations of crime, bad character, unfitness for office, and the like. Malice is presumed from the utterance of a defamatory imputation and need not be proved. Relevant defences are that the statement complained of was true in fact, spoken in heat, a fair retort in self-defence, fair comment on true facts, a fair report of public proceedings, or was protected by absolute privilege or qualified privilege.

Default. A failure to do something which the person should have done. Judgment by default is competent against a defendant who, intentionally or by mistake or neglect, does not properly take the steps necessary to bring his defence before the court.

Defeasance. That which defeats, particularly a condition which, if performed, avoids the deed containing it.

Defectum sanguinis. Failure of issue.

Defence. In pleading, a denial by the defendant of the allegations made against him. In civil matters a defence may be a traverse, or denial of the plaintiff's claim or an allegation of counter-claim, or a confession and avoidance, which admits the plaintiff's allegations but alleges other facts which justify or excuse the defendant, or a statement of defence raising an objection in point of law to the effect that the facts alleged, even if established, do not disclose a good cause of action. In criminal matters there is no written defence; a defence is raised by pleading not guilty.

Defendant (or in Scotland, defender). The person against whom civil proceedings are brought. It is also used of the person charged in English summary criminal proceedings.

Defender of the Faith. A title of the sovereign of England, and now, of the United Kingdom. It was conferred by Pope Leo X on King Henry VIII in 1521 in appreciation of the King's having written a tract against Martin Luther. By statute of 1543 the title was given by Parliament and has continued to be used by all succeeding sovereigns, notwithstanding the repeal of the Act of 1543 in 1554. It is represented by the letters F.D. or Fid. Def. (*Fidei Defensor*) on coins.

Definitions, legal. A legal system formulates its principles and rules in verbal terms, and it needs to define many of its terms and concepts to enable courts and lawyers to know with some approach to certainty whether a particular set of facts amount to a particular category for the purposes of legal consequences, e.g. whether an association is a 'partnership', or a building a 'factory', or a death in particular circumstances 'manslaughter'. Statutes define many terms and concepts, frequently assigning meanings narrower or more extensive than the dictionary or the layman would give to the particular word or phrase. Sometimes statutory definitions are very unhelpful merely saying that X includes or does not include Y, but giving no assistance on the meaning of X. In default of statute decided cases may give, if not verbal definitions, at least guidance on the meaning to be assigned to a term or concept, but decided cases always discuss terms and concepts in the light of the facts and circumstances then before the court, and these interpretations may be unhelpful if the term or concept arises again in a very different context. The more general terms and concepts, e.g. posssession, are discussed, though seldom defined, in textbooks and juristic writings.

F. Stroud, *Judicial Dictionary*; J. Saunders (ed.), *Words and Phrases Legally Defined.*

Deforcement. In English feudal property law, is the wrongful taking and possession by one man of land belonging to another. It included disseisin and ouster or the act of a stranger forcing a lawful heir from his inherited land, but deforcement did not require that the person against whom the land was wrongfully withheld had once had possession of the land. Deforcement included intrusion and abatement, or the wrongful entry into and occupation by a stranger of vacant land belonging to another. In Scots law it is the crime of resisting an officer of law in order to prevent him carrying out his duty.

Degradation. A dismissal from or deprivation of a status or dignity, and particularly an ecclesiastical penalty by which a clergyman is dismissed from holy orders. It might be summary, by words; or solemn, by stripping the person of the insignia of his rank. It has been practically obsolete since the Reformation, deposition being employed instead. A lord or knight might be degraded at common law or by statute and an hereditary peeress in her own right, is said to incur degradation by marrying a commoner.

Degree (from *gradus*, a step or grade). A step in a line of descent, the number of degrees indicating proximity to or remoteness from another. In the old law of real actions the nature of the action which might be brought by a person disseised depended on the degrees between the original disseisor and the person in possession. It is also an academic rank conferred by a university or equivalent place of learning as a mark of having attained a certain standard of knowledge, degrees ranking *inter se* in a definite order. Thus a person is 'called to the degree of the Outer Bar', i.e. promoted from being a student. In criminal law a distinction was drawn between a principal in the first degree, who committed a felony himself, and one in the second degree, who was an accessory.

Del credere. An agreement by which a factor (q.v.), called a *del credere* agent, undertakes for an extra commission, called a *del credere* commission, when he sells goods on credit, to guarantee to his principal the purchaser's solvency and that he will perform the contract. In effect he is a surety to his principal for each party with whom he deals.

Delectus personae (choice of person). The principle that where the personal qualities of a party are important, as in partnership, contractual relations involving that party cannot be altered without the consent of all parties to the relationship.

Delegated legislation. Legislation made not by Parliament but by persons or bodies on whom Parliament has conferred power to legislate on specified subjects. The practice has been quite common since the fourteenth century, being common under the Tudors, notably under Henry VIII's Statute of Proclamations which gave the King a limited power to issue proclamations having the force of Acts of Parliament, while his Statute of Sewers appointed Commissioners with power to make laws, ordinances, and decrees with statutory effect. Thereafter delegated legislation was frequent, either designed to meet emergencies, or conferring power on local government authorities. There was much less of this in the eighteenth century, but the practice continued.

The practice has developed greatly since 1832, particularly as Parliament has come to interfere more and more in social and economic affairs, and it received great impetus during the two World Wars. It is now recognized as essential that Parliament should define the principles and the details be settled and amended by the departments of state concerned. But delegated powers are frequently very wide and may include powers to amend statutes, though usually only consequential amendments.

The practice has evoked criticism and was considered by the (Scott-Donoughmore) Committee on Ministers' Powers in 1932 which recommended certain safeguards.

Power is commonly conferred on Ministers of State, public corporations and boards, local authorities, courts, universities, and other bodies.

It is exercised by Orders in Council, Departmental Rules, Regulations and Orders, public corporations or local authority bye-laws, Rules of the Supreme Court, University ordinances, and under other names.

From 1890 onwards annual volumes of Statutory Rules and Orders were published, renamed Statutory Instruments in 1948, and there was published an edition of S.R. & O's Revised to 1948.

The Rules Publication Act, 1893, made some provision for the publication of statutory rules and orders, but was replaced by the Statutory Instruments Act, 1946, which gives the name statutory instrument to most forms of delegated legislation which is of a legislative rather than executive character, particularly orders made by departments of state.

Parliamentary control is inadequate, many ministerial orders not being examined or discussed at all. Some require affirmative resolution of Parliament before coming into force, others come into force unless annulled by negative resolution, while others again must be laid in draft before one or both Houses. Many kinds of orders require confirmation by a Minister or other official before becoming effective. A Select Committee on Statutory Instruments, first by the Commons and now by the Houses jointly, has been appointed annually since 1943 with the duty of drawing Parliament's attention to unusual or excessive uses of delegated power. A Select Committee on Delegated Legislation in 1953 could recommend little change. Since 1973 there has been a Joint Committee on Statutory Instruments and Standing Committees may consider statutory instruments referred to them. There are also committees of the two Houses to consider E.E.C. secondary legislation.

Challenge of delegated legislation in the courts is frequently hindered by provisions making the regulations as valid as the empowering Act, or that confirmation of a scheme by a minister is to be conclusive evidence that the requirements of the Act have been complied with.

Particularly during the Second World War the practice developed of sub-delegating legislative powers, which sometimes resulted in several tiers of legislation.

C. K. Allen, *Law & Orders*; C. T. Carr, *Delegated Legislation*; J. Kersell, *Parliamentary supervision of delegated legislative power*; M. Sieghart, *Government by Decree*.

Delegates, High Court of. Prior to the Reformation practically all cases within the jurisdiction of the ecclesiastical courts might be appealed to Rome. In 1532 the Ecclesiastical Appeals Act prohibited all appeals to Rome and provided that certain appeals should go from the archdeacon to the bishop, and from him to the Courts of the Arches or of Audience (q.v.) and from them to the Archbishop, whose decision was to be final save in cases affecting the King. A year later statute created the High Court of Delegates as a new court of appeal for all ecclesiastical causes. As it consisted of persons commissioned to hear an appeal, it had no permanent membership. The statute had declared it to be final, but it was nevertheless held by lawyers in Elizabeth's time that the King could, like the pope, issue a commission of review to have the whole case reheard. Lawyers from Doctors' Commons were usually included in the commission, but bishops, sometimes included at earlier times, were entirely excluded by the late eighteenth century. It is said to have consisted usually of three puisne judges and three civilians, sometimes with temporal peers and judges added. It was an unsatisfactory court and in 1832 the jurisdiction was transferred to the Privy Council, and since 1833 has been exercised by the Judicial Committee of the Privy Council.

J. Duncan, *The High Court of Delegates*.

Delegatus non potest delegare (a person who is a delegate cannot himself delegate his powers). The general principle that certainly in matters involving judicial powers, trust, the exercise of discretion, or where the qualities of the delegate are material, a delegate must act himself and may not, unless expressly authorized, delegate his functions to another. The principle does not extend to merely executive functions. It applies also in administrative law.

Delict. In Roman law an obligation to pay a penalty by reason of wrong committed. The Roman law was essentially penal though fines were paid to the injured person. Four major types of delict were recognized: *furtum* (theft), *rapina* (robbery), *injuria* (injury), and *damnum injuria datum* (damage to property). In the Twelve Tables, *injuria* was in a transitional state from private vengeance to compensation. Later *injuria* included insults and defamation. In the third century B.C. the *Lex Aquilia* restated the law on *damnum injuria datum.* If a slave or grazing animal were killed unlawfully, damages equal to the highest value it had had in the past three years were given; if other kinds of property were burned, broken, or smashed, damages at its highest value in the last 30 days were given. In each case the damage must have been caused by a wilful or negligent act.

The term delict and variants thereof are used in civil law systems for civil wrongs, corresponding to Anglo-American torts (q.v.).

In Scots and South African law delict is the branch of law, corresponding generally to the law of torts in the common law, which deals with civil liability for causing unjustifiable harm or loss. It differs sharply from the law of torts in having a broad general principle of liability for harm or loss caused, based on developments of the Roman *actio injuriarum* and *actio legis Aquiliae* and this principle covers the majority of cases. The general principle is of liability to compensate for loss or harm caused by deliberate or negligent breach of duty to avoid or prevent conduct foreseeably likely to cause harm to a person in the position of the injured person.

The individual delicts are not, as in the common law, particular kinds of wrongs for which law has given a remedy, but particular classes of cases falling within the general principle of liability, to which particular names have for convenience been attached. Roman law distinguishes between delicts and quasi-delicts but this distinction has come usually to be equated with that between intentional harms and negligent harms, but is unimportant in these modern systems.

As in the law of torts (q.v.), failure to implement the requirements of statutory provisions may give rise to statutory obligations, co-existing with and overlapping into the field of delictual obligations, so that a given harm or loss may be a ground of action both for common law delict and for breach of statutory duty.

D. M. Walker, *Law of Delict in Scotland*; R. G. McKerron, *Law of Delict in South Africa*.

Delinquency The commission of crimes, or sometimes a minor crime. The term 'juvenile delinquency' denotes the whole subject of the commission of crimes by children and young persons.

Délit. In French law a crime of medium gravity.

Deliverance. A verdict, as in the old expression of a clerk of court when a prisoner elected trial, 'May God send you a good deliverance.' This disappeared in 1827.

In Scotland it is used of an order of the court in sequestration proceedings.

Delivery. Transfer of the possession of a moveable thing from one person to another. It may be actual, by handing over the thing; or constructive, by operation of law; which in turn may be symbolic, where a thing, such as a bill of lading is transferred as a symbol of the goods to which it relates; or by attornment, where a person alters the capacity in which he has actual possession, as where a seller of goods continues to hold the goods but as baillee for the buyer.

Delivery of a deed. Normally a requisite of its validity. It may be effected by acts, or by words, or both. It may be absolute where it it intended that the deed take effect at once and unconditionally, or be conditional on the performance of stated conditions. Deeds take effect only from delivery and have precedence according to the times of delivery, though their effect may vary accordingly.

Delivery order. An order addressed by the owner of the goods to a person holding them on his behalf, such as a storekeeper, to deliver them to a person named in the order.

Delos, Joseph Thomas (1891–). French jurist, professor of sociology and law and later professor of social theology and a diplomat, author of *Société internationale et les principes du droit public* (1929) and other works.

Delusion. A belief in facts which no rational person could believe, e.g. the deluded person thinks that he is Napoleon or John the Baptist. It is a manifestation of mental illness. In civil law a person suffering from delusions is probably not in any exceptional position, e.g. as regards capacity to contract or liability in tort, unless the delusion is one which affects the legal relationship in question. A delusion does not affect a person's capacity to make a will unless the delusion is such as was calculated to influence the person when making it. In criminal law being subject to a delusion is no defence unless it prevented the person conceiving the intention or having the other mental element necessary to commit the crime in question.

del Vecchio, Giorgio (1878–1970). Italian jurist, professor at Rome, was author of *Il concetto di diritto* (1906), *Il concetto di natura e il principio del diritto* (1908), *La Giustizia* (1923), *Lezioni di filosofia del diritto* and other works on legal philosophy, which have been widely influential, and founder and editor of the *Rivista internazionale di filosofia del diritto* from 1921.

Demembration. In old Scots law, the offence of maliciously cutting off any limb or member from the body of another.

Demesne lands (or lands in demesne (*in dominio*)). In mediaeval law, those which were retained by the owner of the manor and tilled at the cost and by the labour of the owner, and those held from him by villeins, but not including lands which belonged to the lord but which had been let by him as fiefs to vassals in return for services. When villein tenure developed into copyhold, the lord's demesne came to be restricted to the lands immediately surrounding his house. The royal demesne was all land in the realm which had not been granted to private persons and from which the Crown derived rents and other revenues.

Ancient demesne was land vested in the King in 1066, the tenants of which enjoyed certain privileges such as freedom from tolls and duties and also the tenure by which such lands were held. Land held in ancient demesne is sometimes called customary freehold.

Deminutio capitis. In Roman law, loss of civil capacity. Classical Roman law distinguished three degrees, *maxima* when the individual lost his freedom by enslavement; *media* or *minor*, when he lost citizenship with consequent dissolution of family ties, as when a Roman became *peregrinus*; and *minima*, when a man lost the membership of his family, as by *adoptio* or *emancipatio*. The consequences of the last kind diminished in the classical period with the weakening of the institution of the Roman family and in late law it for the most part lost its actuality.

Demise. A grant, particularly of lands for a term of years, but the term is also applied to the grant of an estate in fee or for life. In the old common law action of ejectment (q.v.) the action bore to be by 'Doe d. (or dem.) Able v. Baker' where Able was the real plaintiff and Doe a fictitious person to whom Able had made an imaginary lease; d. or dem. meant 'on the demise of'.

Demise of the Crown. The death of the sovereign, whereby the kingdom is transferred to his successor. Until 1867 Parliament was automatically dissolved, on the theory that the sovereign's invitation to the House to consult with and advise him lapsed on his death. Statutes subsequently provided that Parliament, unless earlier dissolved, should last for six months after the sovereign's death, and later that Parliament should not be affected by the demise of the Crown. On the demise, however, Parliament must meet at once. The Demise of the Crown Act, 1867, provided for the continuance of military commissions and the similar Act of 1901 for the continuance of the holding of any office under the Crown.

Democracy. Rule by the people, the form of government in which the general body of the people ultimately exercise the power of government. It is distinguishable into direct democracy, in which political decisions are made by the whole body of citizens, the majority view being accepted, and representative democracy, in which political decisions are made by persons chosen to represent, and be responsible to, the whole body of citizens. The term is extended to social and economic contexts where the emphasis is on minimizing inequalities of power, rights, privileges, and property.

Democracy originated in the small city-states of ancient Greece where all adult male citizens were eligible for most executive and judicial offices, appointment sometimes being by election, sometimes by lot.

Modern democratic ideas were influenced by the concept of law as a restraint on autocratic powers of monarchs and by the need to consult representatives of the community particularly to secure assent to the raising of money. They were furthered by the growth of representative assemblies, and by the development of the concepts of natural rights and of the equality of men. The American and French Revolutions furthered the idea of democracy and in the nineteenth and twentieth centuries the cardinal features of democracy came to be representative assemblies freely elected on a very broad franchise, to which the executive is nominally responsible and from which it must obtain its powers.

The major defects of democracy are the incapacity of the majority of citizens to understand the extremely difficult and complicated issues of social and economic policy involved in modern government, the constant danger of their being deluded by popular leaders to support courses which are attractive and easy, the low intelligence of the great mass of voters and their liability to be influenced by motives of greed, jealousy, and selfishness.

A democracy can be monarchical or presidential, according as the head of state is a monarch or a president, but in either case the powers of the head of state are limited and subject ultimately to the representative assembly or to the body of citizens; he is neither an autocrat nor a dictator.

A democratic political system can be capitalist, collectivist, or mixed in any proportions, according as it accepts or rejects private ownership of the major organizations of production, exchange, and distribution, but in Western Europe the tendency in the twentieth century has been to move steadily towards collectivism.

Other features normally associated with democracy are regular and free elections, an independent judiciary, large measures of freedom of the Press, of speech, of assembly and of association, and the idea of the rule of law, namely that the executive and the administrators are subject to rules of law and can be called to account for infringements thereof.

Since 1945 there have been established, notably in Eastern Europe, so-called 'people's democracies' which are in fact Communist autocracies or oligarchies maintaining a façade of democratic institutions.

Demonstrative legacy. A legacy which indicates from what fund payment is to be made. If the fund fails, the legatee is entitled to his legacy out of the general assets of the estate. If there is a deficiency of assets, a demonstrative legacy does not abate unless there is nothing from which to pay general legacies.

Dempster. In old Scots law an official who originally declared the law, but in time came to be merely the declarer of the law or doom (judgment) determined by the judge. The office further declined as time went on and by the seventeenth century he was the common hangman. The office disappeared in the seventeenth and eighteenth centuries.

See also DEEMSTER.

Demurrage (from Fr. *demourer*). A sum payable by the party freighting a ship for detaining the ship in port longer than the lay, or running, or working days, stipulated as allowed for loading or unloading. Demurrage is of the nature of liquidate damages for delay, and if any specified number of demurrage days are exceeded, damages for delay are payable, usually measured by the rate of demurrage.

Demurrer. In English common law pleading, the plea by one party that the other party's pleadings, even if proved, did not entitle him to succeed and that he was himself entitled in law to succeed on the facts alleged and admitted by the other, and was content to rest (*demourer*) on that plea. In substance the parties were at issue on a point of law only, which the court had then to decide after argument. Several kinds of demurrer were distinguished, particularly special demurrers which alleged defects in the pleadings, and general demurrers in which the court's judgment was sought on the record as a whole. Demurrers were abolished in 1883 and replaced by provision for hearing objections on points of law.

Demurrer was also competent in criminal prosecutions; it is now probably obsolete.

Denizen. Originally a natural-born subject of a country, and later a person born an alien to whom the sovereign has by letters patent under the prerogative granted the status of a British subject. The grant is in the discretion of the sovereign and is now almost obsolete, being replaced by naturalization. A person granted letters of denization must take the oath of allegiance and probably enjoys all the rights of native-born subjects of the Crown save that he may not, under the Act of Settlement, hold public office or obtain a grant of lands from the Crown.

Denman, Sir Thomas, Lord (1779–1854). Practised as a special pleader before being called to the Bar in 1806. He appeared in many of the important trials about 1815, being regarded as second only to Brougham and Scarlett as an advocate, and entered Parliament in 1819. He defended Queen Caroline and in consequence was out of royal favour for years, but became Attorney-General in 1830. He drafted the first and second reform bills and pursued an active policy of promoting law reform. In 1832 he became Chief Justice of the King's Bench and in 1834 a peer; his health broke down in 1849 and he resigned in 1850. He was a man of high moral character, great intellectual power, and outstanding eloquence, but an able rather than distinguished judge. His fourth son, George (1819–96) was a judge of the Queen's Bench Division, 1872–96.

Denning, Alfred Thompson, Baron Denning of Whitchurch (1899–). Called to the Bar in 1923 he became a judge of the High Court in 1944, a Lord Justice of Appeal in 1948, a Lord of Appeal in Ordinary in 1957, and Master of the Rolls in 1962. He made many notable and some controversial decisions, being more concerned to achieve a just situation than to apply the accepted law. He identified himself strongly with legal education and was a popular lecturer. He edited, jointly, Smith's *Leading Cases*, and delivered the lectures printed as *Freedom Under The Law*, *The Changing Law*, and *The Road to Justice*.

Dennis, James, Lord Tracton (1721–82). Called to the Irish Bar in 1746 he was Chief Baron of Exchequer in Ireland, 1777–82.

Denunciation. A mode by which an international treaty may be terminated. A treaty may provide by a denunciation clause for its unilateral termination but, if there is no such clause, denunciation must, to be effective, be agreed on by all parties. The term is frequently applied to what is truly repudiation, a unilateral declaration by one party that it considers itself no longer bound by the treaty.

Deodand (*Deo dandum*, to be given to God). A personal chattel which had been the immediate cause of the death of a living being and which at common law was forfeit to the Crown to be applied to pious uses. The principle of attributing responsibility to animals or things is very ancient, and in the middle ages sometimes led to judicial trials of animals or things for manslaughter. The rights to deodands were commonly granted by the Crown to manorial lords and others, and in indictments the thing causing the death and its value were presented and found by the grand jury so that the Crown or grantee of the right of deodand might claim it. In the nineteenth century a railway engine was claimed as deodand. Latterly such forfeitures were unpopular and juries found deodands of trivial value so as to defeat inequitable claims. The right was abolished in 1862.

Deontology. The title of a book by Jeremy Bentham (q.v.) who introduced the term to denote a utilitarian system of ethics. It has subsequently been applied to a system which emphasizes duty rather than goodness or rights.

Département. The principal French unit of a local government, created in 1790. In each *département* there is a prefect representing the central government and acting as executive, and a council with responsibility for local services.

Departmental Committee. A committee established by the Minister in charge of one of the departments of state to enquire into some matter and to report. It is a less important kind of investigation than a Royal Commission (q.v.). The Committee's report to the Minister is published as a Command Paper and frequently comes to be commonly known by the name of the chairman. Where more than one department is concerned with the subject matter under investigation the inquiry may be by an Inter-Departmental Committee appointed by both or all the Ministers concerned jointly.

Department of State. The branch of the federal government of the U.S. concerned with foreign relations and headed by the Secretary of State.

Departments of State. The various branches of the central government of the U.K. staffed by civil servants and headed by political heads called Secretaries of State, Ministers and, in some cases, for historical reasons, ministers holding ancient offices with distinctive titles such as the Chancellor of the Exchequer. The older departments are known by traditional names such as the Home Office, the Foreign Office, the Treasury, but the more modern ones tend to be called Ministries such as the Ministry of Defence, or departments, such as the Department of the Environment, the Department of Education and Science, and the Department of Health and Social Security.

Dependency. A region not annexed by the Crown but subject to British jurisdiction. It is a wider term than colony. The term 'dependent territory' is now frequently used of the colonies and territories under the jurisdiction of the Crown as distinct from independent countries of the Commonwealth.

Deportation. The expulsion of an alien whose presence in a country is deemed undesirable or, more widely, including banishment, exile, and transportation of criminals (q.v.). In modern practice an undesirable alien is normally returned to his own country. It differs from extradition in which a person accused of crime in another country is returned to that country to be tried there. It has been extended to 'non-patrials' (see PATRIAL).

Deposit. In Roman and Scots law, the contract whereby goods are transferred to another to be kept by the latter gratuitously and returned on demand or after the lapse of a specified time.

The term is also used of a sum paid as security for the performance of a contract and which may be forfeited if the contract is not completed, of the lump sum payment made under hire-purchase contracts as distinct from the part of the price paid by later instalments, and of sums held to a person's credit by a bank or building society.

Deposition. (1) A declaration or statement of facts made by a witness under oath and reduced to writing for subsequent use in court proceedings.

(2) In ecclesiastical law, the penalty by which a clergyman is deprived of holy orders.

Deprivation. The ecclesiastical penalty by which a clergyman's patronage, vicarage, or other dignity is taken away from him.

Deputy-judges. The Permanent Court of International Justice (q.v.) as originally constituted consisted of 11 judges and four deputy-judges. The function of the latter group was to replace judges when for any reason a sufficient number of those were unable to be present. The Protocol for the Revision of the Statute of the P.C.I.J. of 1929 provided for the abolition of deputy-judges and the appointment of a bench of 15, but deputy-judges were elected at the second elections of candidates for the Court, in 1931. The office vanished when the Court was replaced by the International Court of Justice. Since 1971 persons may be appointed as deputy-judges of the High Court or deputy circuit judges.

Derelict. A thing abandoned, particularly a ship abandoned on the high seas; if salved, it belongs to the owner, unless he has abandoned it to the underwriters, but salvage reward is payable.

Dereliction. The abandonment of a chattel; also the leaving behind of land by a river changing its channel.

Derivative acquisition. Acquisition from another, by gift, sale, or bequest, as distinct from original acquisition on birth or by creation.

Dernburg, Heinrich (1829–1907). German jurist, professor at Zurich, Halle, and Berlin, a leading Pandectist and author of *Die Institutionen des Gaius* (1869), *Lehrbuch des preussischen Privatrechts* (1871–96), *Pandekten* (1884), and *Das bürgerliche Recht des Deutschen Reichs und Preussens* (1898).

Derogate. Derogation from a grant or right is to prejudice or destroy it. It is an established rule that a grantor cannot be permitted to derogate from his grant.

Derry* v. *Peek [1889], 14 App. Cas. 337. Promoters of a tramway company announced that they had powers to use steam for their trams, though in fact they only confidently expected to be granted such powers and the powers were refused. A shareholder sued the company for fraudulent misrepresentation. It was held that misrepresentation was fraudulent only if made knowingly or without belief in its truth or recklessly, careless whether it be true or false, and that in this case the misrepresentation was not fraudulent. It is the leading case on what constitutes a fraudulent misrepresentation.

Descent. The system of kinship relationship, generally based on common ancestry, which enables a person to assert rights, privileges, or status in relation to another and which is important in relation to inheritance and succession.

In English real property law prior to 1926 descent was one of the two main methods of acquiring an estate in land as distinct from acquisition by gift or limitation, and consisted in relationship by consanguinity. Descent had to be traced from the 'purchaser', i.e. the last person who acquired otherwise than by descent, and was determined by rules based on preference for males over females and among males, on primogeniture. There was recognized also customary descent, regulated by the custom of the place where the land was situated, of which the notable examples were gavelkind and borough-English (qq.v.). The Duchy of Cornwall also had a special mode of descent.

Desegregation. The general term for the process, begun in the U.S. in the 1950s, for securing effective and not merely nominal freedom and equality for coloured persons in the U.S., by outlawing the provision of segregated facilities for them in transport, education, and other spheres. In *Plessy* v. *Ferguson* (1896), 163 U.S. 537 the Supreme Court had held that if separate but equal facilities were provided for coloured persons they were being given equal protection of the laws. In *Brown* v. *Board of Education* (1954), 347 U.S. 483 the Supreme Court held that the concept of 'separate but equal' had no place in public education. This was the most notable victory of the campaign for desegregation and for integration of white and coloured peoples.

Desertion. (1) In matrimonial law, the separation of one spouse from the other with the intention of bringing cohabitation to an end, without legal justification or the consent of the other spouse. It may be done by leaving or physically departing from the other spouse, or constructively, by one spouse so treating the other that he or she is driven to, and held justifiably to, leave the deserting spouse. Desertion justifies application to a court for separation and/or for maintenance. It was formerly a ground of divorce and may now evidence breakdown of the marriage.

(2) In the armed forces, the offence of absenting oneself from one's unit or one's post without orders of permission.

Desertion of the diet. In Scottish criminal procedure, the abandonment of proceedings under a particular libel. It may be *pro loco et tempore*, in which case the prosecutor may bring a fresh libel, or *simpliciter*, in which case the abandonment is absolute and the prosecutor may not bring a fresh libel.

Design. Those features of shape, configuration, pattern, or ornament applied to an article by an industrial process which appeal to and are judged solely by the eye. Such a design is registrable under the Registered Designs Act, 1949, and once registered is protected against copying or unauthorized exploitation by another.

Despagnet, Frantz Clément René (1857–1906). French jurist, author of *Cours de droit international public* (1894), *La Diplomatique de la troisième république et le droit des gens* (1904), *Essai sur les protectorats* (1896), and other works on international law.

Despatch money. A payment promised for more prompt execution of a contract than the contract provides for.

Destination. In Scots law the nomination in a deed or will of the persons to whom the rights given are to pass.

Desuetude (disuse). In Roman law it was recognized that disuse could abrogate a statute. Acts of the old Scots Parliament may be held to have fallen into desuetude if long disregarded in practice. Similarly in international law an old treaty may fall into disuse if long disregarded. The principle has never been accepted in relation to legislation of the English, British, or United Kingdom Parliaments which remain of full effect however long disused in practice.

Detainer. Unlawful detainer is the unjustified keeping of a person's goods, even if the original taking of them was lawful.

Detainer was also the continued holding of a person in arrest in a civil proceeding under a second writ for his arrest though he was entitled to be discharged under the first writ. This disappeared when arrest on mesne process was abolished in 1838.

Detention centres. Places in which young persons may be kept for short periods under discipline suitable to their age and description. Separate premises are provided for the sexes, and there are junior and senior centres dealing with different age-groups. Inmates of compulsory school age are given full-time, and others may be given part-time, education within the working week. The régime emphasizes strict discipline and a vigorous training.

Determinable interest. An interest in property liable to be ended by satisfaction of a condition, lapse of time, defeasance, remarriage, or other specified event.

Deterrence. Commonly regarded as one of the main objects of criminal law and punishment. Deterrence has two aspects, deterring the individual from offending again, and deterring others from offending in the same or any similar way. The most powerful deterrent would be the certainty that an offender would be caught and punished and get no benefit from his offence but only detriment, but this is unattainable in practice, and even certainty of conviction would probably not deter all offenders. Much argument has taken place on whether particular penalties, particularly capital punishment for murderers, are effective deterrents or not. It is impossible to reach a certain conclusion on such an argument since the interpretation of much of the evidence can only be subjective. Probably no regime of punishment or treatment is a certain deterrent as some criminals always think they will evade conviction and some crimes are committed in excitement, passion, or other circumstances which negative rational consideration of chances of capture and conviction. But deterrence remains a purpose underlying many penalties, most obviously so in the case of exemplary sentences.

Detinue. One of the earliest common law forms of action in English law, which lay for the recovery of chattels wrongfully detained by the defendant, as by bailee holding against bailor, or for their value and for damages for the detention. The basis of the action is the unlawful failure to deliver up the goods when demanded. Originally connected with Debt, it became differentiated from debt in the thirteenth century. The action was abolished by the Judicature Acts, 1873–75, but an action for the return of a specific chattel is still called an action of detinue.

Détournement de pouvoir. A concept adopted from French administrative law into European Community law, the concept of misuse of power, or disregard of legal objectives, the use by a public authority of its powers for an objective other than that for which the power was conferred on it. It is a principal ground of appeal to the European Court for annulment of an act of an organ of the Community and the only ground on which a private party may appeal against a general act or regulation. The Court views a *détournement* primarily with reference to the general fundamental objectives of the act complained of, and whether it is committed in good or bad faith. The illegal objectives need not be actually attained.

Deusdedit, Collection of (or *Collectio Canonum*). A major collection of canon law materials designed to buttress the status and rights of Rome in the Church, compiled about 1085. It comprises 1,173 texts in four books drawn both from earlier collections and from papal archives. Though of limited influence the collection is valuable evidence of the aims and methods of eleventh century papal reform.

Deuteronomic Code. A body of ancient Hebrew Law found in Deuteronomy 12–26. It is a reinterpretation or revision of Hebrew law based on historical conditions as interpreted by historians of the seventh century B.C. It sought to purify the worship of Yahweh from Canaanite or other influences. Apostasy was the greatest sin and carried the death penalty. It is divided into ordinances and regulations, relating particularly to dealings with the Canaanites and to worship in the temple in Jerusalem alone to the exclusion of the high places, sabbatical laws concerned with the year of release from obligations, regulations for leaders, various civil, ethical, and cultic laws, and an epilogue of blessings and curses.

Devastavit (he has wasted). The wasting of the property of a deceased by his executor or administrator, by misapplying the assets. It renders him personally liable to creditors and legatees having claims against the estate by proceedings to make him responsible.

Deviation. In carriage by sea, a deliberate and unnecessary departure from the due course of a voyage for even the shortest time, or delay in sailing or prosecuting the voyage for unjustified purposes. Deviation is justifiable to save life or property and in certain other cases. If unjustifiable, it discharges underwriters from liability and disentitles the shipowner from relying on exception clauses in the charter-party and from claiming the contractual freight.

The ordinary course of a voyage is by the most direct and safe route, or the usual and customary route. If it is to a series of ports they must be taken in the order named in the contract, or in natural geographical order. A contract may reserve liberty to deviate or to call at ports in any order.

Devilling. The well-established custom at the English Bar of handing over a brief and the responsibility for it to another counsel, with the instructing solicitor's permission, or of obtaining assistance from another barrister in drafting and researching the relevant tract of law, in which case the barrister instructed remains responsible for the work done. A barrister who employs a devil must pay adequate and reasonable remuneration for the work actually done.

Devil's Island. A small island in the Atlantic off the coast of French Guiana and long used by the French as a penal settlement. Captain Dreyfus was among those confined there.

Devise. A disposition or gift of property made by will. Properly speaking the word is applicable only to real property (q.v.), 'bequeath' being appropriate to personal property, but the two terms are sometimes used indifferently. Devises may be specific or residuary.

Devlin, Patrick Arthur, Baron (1905–). Called to the Bar in 1929, became a judge in 1948, a Lord Justice of Appeal 1960, and a Lord of Appeal in Ordinary 1961–64 when he retired, thereafter serving as Chairman of the Press Council and a judge of the administrative tribunal of the I.L.O. He was also Chairman of the Restrictive Practices Court, 1956–60. He published *Samples of Lawmaking, The Criminal Prosecution in England, Trial by Jury*, and *The Enforcement of Morals*.

Devolution. The political device of delegating or transferring certain powers to another body, while retaining at least in theory full power over the same matters. Thus in the U.K. Parliament has devolved certain powers on the Scottish and Welsh Assemblies (qq.v.). Devolution differs from federalism in that under federalism both central and regional authorities exercise certain powers, each being supreme in certain matters.

The word is also used of the transmission of an interest in property from one person to another by operation of law, e.g. on death or bankruptcy.

D'Ewes, Sir Simonds (1602–50). A member of Parliament 1640–48, and author of a diary giving an account of the proceedings of the Long Parliament and of a *Journals of all the Parliaments during the Reign of Queen Elizabeth* (1682) used by Cobbett and Wright in compiling their *Parliamentary History*.

Dharmasastra. The body of ancient Indian jurisprudence, still in force, subject to statutory changes in India and as family law of Hindus living outside India. It is as ancient as Jewish law and older and more continuous than Roman law. It is fundamentally concerned with the right course of action rather than a body of legal precepts and some principles of dharmasastra are known to most Hindus. The literature is in Sanskrit and very voluminous. It comprises sutras, concise maxims, smrtis, treatises in verse, nibandhas, digests of smrtis verses, and vrttis, commentaries, the last two being legal works appropriate for legal advisers. The approach of these texts are normally to state the maxim, stanza, or text and explain it, reconciling divergent principles by interpretation. The oldest sutra, Gautama, may date from the eighth century B.C. and the most famous smrti, the Manu-smrti, may be dated 100 B.C.–A.D. 200. Vrttis began to be written about the sixth century and nibandhas from the eleventh to the nineteenth centuries. Among famous treatises are the Mitaksara, a twelfth-century commentary on legal verses, and the Dayabhaga, an eleventh-century work dealing particularly with succession. Western jurists became acquainted with the Dharmasastra chiefly through the work of Sir William Jones and by later scholars, while modern Indian scholars have studied it thoroughly. The principles were affected by the British administration of India and were applied in rigid fashion. Since Indian independence many of the principles have been statutorily modified.

P. Kane, *History of Dharmasastra*; N. Sen-Gupta, *Evolution of Ancient Indian Law*.

Dialogus de Scaccario. A book in Latin compiled before 1179 by Richard Fitzneal or Fitznigel, Bishop of London 1189–98, and Treasurer of England 1158–98, from personal knowledge, and giving a complete and detailed survey, in the form of an imaginary dialogue of the business of the Exchequer in the twelfth century. It gives much legal information, such as how sheriffs had to account for the fines and forfeitures to which the administration of justice gave rise. It was possibly intended for the use of clerks in the Exchequer and is of immense interest and historical value. It was printed in Madox's *History of the Exchequer*, but the best printed edition is by Hughes, Crump, and Johnson (1902).

Dicastery. A judicial body in ancient Athens. From the time of the reforms of Cleisthenes (*c.* 508 B.C.) when the Heliaea was made a court of original jurisdiction, dicasteries were divisions of the Heliaea. Six thousand citizens were chosen each year, probably representing a cross-section of the free population, and groups of 500 constituted a court for the year from which panels were chosen for the trial of cases. The dicasts were judges of law and fact and decided by majority. Verdicts could be overturned by the Ecclesia. In important cases several dicasts might be combined.

Dicey, Albert Venn (1835–1922). Read classics at Oxford, became a fellow of Trinity and won the Arnold Prize with an essay on the Privy Council. In 1863 he was called to the Bar and practised with considerable success. In 1870 he published a *Treatise on the Rules for the Selection of the Parties to an Action* and in 1879 *The Law of Domicile*. These works made his reputation and led to his election in 1882 as Vinerian Professor of English law at Oxford, a post he held till 1909; from 1910 to 1913 he also held a lectureship in private international law. As

Vinerian Professor, Dicey revived the importance of the chair and published three works of enduring value, the *Introduction to the Study of the Law of the Constitution* (1885 and later editions), his *Digest of the Law of England with reference to the Conflict of Laws* (1896, developed from his work on *Domicile*), and his *Lectures on the relation between Law and Public Opinion in England during the Nineteenth Century* (1905, based on Lowell Lectures delivered at Harvard in 1898). He took a prominent part in political controversy, and wrote extensively against Irish Home Rule. In his later years he collaborated with R. S. Rait in *Thoughts on the Union between England and Scotland*, an event which he believed to be a great achievement of statesmanship.

R. S. Rait, *Memorials of A. V. Dicey*

Dicta (or *obiter dicta*). See *Dictum*.

Dictator. In the Roman Republic, a magistrate with extraordinary powers appointed in times of military or, later, civil crises. He was nominated by a consul on the recommendation of the Senate and confirmed by the Comitia Curiata. The office lasted for six months, though the dictator usually laid down office when the crisis was past. The other magistrates remained in office but were subject to the dictator. No dictator was chosen after 202 B.C. At the end of the Republic Sulla and Caesar were dictators with practically unlimited powers. Caesar acquired dictatorial powers for 10 years in 46 B.C. and shortly before his death was given the power for life.

In modern times a dictator is a person having complete governmental powers unchecked by constitutional limitations, though the office he holds may be variously named. His power essentially depends on power exercised through his control of military, police, and other forces. It may be an authoritarian dictatorship which has the consent of many of the governed as being preferable to civil war, or a totalitarian dictatorship, such as those of Hitler and Stalin, seeking to create a new society and denying the liberty of anyone to be outside its control.

Dictatus Papae (Dictates of the Pope). A legal collection compiled by Pope Gregory VII about 1075. It consists of 27 titles affirming the spiritual headship of the bishop of Rome over all Christians and his official superiority to all clergy and temporal princes.

Dictionaries, legal. Legal dictionaries define and illustrate the meanings and usages of words and phrases which have a technical legal meaning or are used in legal contexts with a distinct meaning. Such usages are of course, also treated in the major English and American general dictionaries, such as the *Oxford English Dictionary*. Akin to legal dictionaries are books dealing with words and phrases which have been judicially discussed and defined in particular contexts.

Among dictionaries of the languages formerly used in English law valuable works are Toller's edition of Joseph Bosworth's *Anglo-Saxon Dictionary* and his *Supplements* and Godefroy's *Dictionnaire de l'ancienne langue francaise*.

The first major law dictionary proper was John Rastell's *Expositiones Terminorum Legum Anglorum* of 1527, translated by his son William in 1567 as *Termes de la Ley*, which, orginally written in law-French, went through many editions. In 1605 John Cowell published *Nomothetes: The Interpreter, or Booke containing the Signification of Words*, which cited authority for its definitions. It, too, appeared in numerous editions, some later ones enlarged by Thomas Manley. Thomas Blount's *Nomo-Lexikon: a Law Dictionary and Glossary* (1670) gave more space to etymology and the description of ancient customs. A revised edition by Nelson came out in 1717. In 1729 was published Giles Jacob's *New Law Dictionary*, later (from 11th edn., 1797) edited and enlarged by Thomas Tomlins; it contained writs, precedents, forms, and the like and was more of an encyclopaedia than a dictionary. Timothy Cunningham's *New and Complete Law Dictionary* (1764–65) aimed at the inclusion of all legal terms.

Among the modern dictionaries the principal are John J. S. Wharton's *Law Lexicon* (1848 and many later editions), Charles Sweet's *Dictionary of English Law* (1882), and W. G. Byrne's *Dictionary of English Law* (1923), and now Jowitt's *Dictionary of English Law* (1959) based on Wharton and Byrne.

Among books discussing legal maxims the principal have been Francis Bacon's *Maxims of the Law* (1636) and Herbert Broom's *A Selection of Legal Maxims*.

Among books dealing with words and phrases the chief have been Frederick Stroud's *Judicial Dictionary* (1890, 3rd edn. 1952) and Roland Burrow's *Words and Phrases Judicially Defined* (1943–45 and supplements), revised by Saunders as *Words and Phrases Legally Defined*.

In Scotland Sir John Skene (q.v.) published the *De Verborum Significatione* in 1597, which is invaluable for old and obsolete terms. W. Bell's *Dictionary and Digest of the Law of Scotland* (1838) founded on R. Bell's *Dictionary of the Law of Scotland* (1807–8) is rather an encyclopaedia than a dictionary, as it does much more than define legal terms. Peter Halkerston produced a *Collection of Latin Maxims* in 1823.

In the U.S. John Bouvier published in 1839 *A Law Dictionary Adapted to the Laws of the United States, and of the Several States of the American Union; with Reference to the Civil and other Systems of Foreign Law*, the first such book having special reference to American Law and conditions. There is also a *Words and Phrases* published by the West Publishing

Co., while maxims and words and phrases are included in *American Jurisprudence* and *Corpus Juris Secundum* (on which see ENCYCLOPAEDIAS, LEGAL).

In Europe a famous dictionary was Calvinus' *Lexicon Juridicum Juris Romani Simul et Canonici* (1600 and later editions) while Charles Du Fresne Du Cange's *Glossarium ad Scriptores Mediae et Infimae Latinitatis* (1638 and later editions) and *Glossarium ad Scriptores Mediae et Infimae Graecitatis* (1688) are great mines of information.

Dictum (or *obiter dictum*). A statement made by a judge in the course of his judgment which is not necessary for the decision or goes beyond the particular occasion and deals with a point or lays down a rule which is unnecessary for the purpose in hand. It is to be distinguished from statements which are held to be, or to be part of, the *ratio decidendi* (q.v.) of the case, and it has no binding authority for another court, though it may have persuasive value. Whether a particular statement is, or is part of, the *ratio decidendi* or merely a dictum is a matter to be decided by a subsequent court considering the judgment as a precedent, possibly giving guidance for the decision of the later case.

Dictum de Kenilworth. A declaration of the terms agreed between Prince Edward (late Edward I) and the barons who had been in arms against him. It is, however, included in some editions of the statutes as a statute of 1266.

***Dies cedit* and *dies venit*.** Expressions taken from Roman law used in relation to performance of a conditional obligation. *Dies cedit* is used when the obligation is due; *dies venit* whenever implement may be demanded.

Dies fasti vel nefasti. In Roman law, days on which the praetor was entitled to administer justice in courts and days on which neither could he do so, nor could the *comitia* meet; in effect business days and holidays.

Diet. (1) In Scottish criminal procedure, the dates on which a party must appear in court. In procedure on indictment there are two diets, the pleading diet, when the accused is called to plead, and the trial diet when, if he has pled not guilty, he is tried.

(2) The deliberative assembly of the Holy Roman Empire. Originally composed of lay and ecclesiastical princes it had evolved by the fourteenth century into three colleges, the seven electors, the princes and nobles, and the representatives of the towns, and had legislative, taxing, and other powers. From the treaty of Westphalia (1648) the diet ceased to be an assembly and became a congress of envoys from the states constituting the loose confederation of Germany. From 1663 it became totally ineffective and ended with the Empire in 1806.

P. Schroder, *Lehrbuch der deutschen Rechtgeschichte* (1902).

Differential association. A theory formulated by E. H. Sutherland of the origin of criminal behaviour. It states that a person learns criminal behaviour through contact with other people, chiefly in intimate personal groups, and becomes criminal if the personal associations he makes are predominantly favourable to law-breaking. The crimogenic significance of these differential associations depends on their frequency, duration, and intensity, and on how early in the individual's life they occur. The theory is consistent with much of the evidence on the incidence of criminal behaviour and is supported by experimental work on the psychology of small groups.

Digamy. Remarriage after the dissolution of a previous marriage. In some societies it is sometimes obligatory, as under the Hebrew levirate law. Under Christianity it is permitted, but certain Bible texts discourage it.

Digest. (1) Classical Roman jurisprudence applied the name *Digesta* to systematic comprehensive treatises on the law as a whole, and works under that title were composed by various jurists.

The name is particularly applied to the collection made by order of Justinian. By constitution of December 530 Justinian instructed Tribonian (q.v.) to compile a synthetic collection of passages from the work of classical jurists. The work was done by an editorial committee of 16, comprising Tribonian, five professors and 11 advocates, in three years. Justinian states in the constitution Tanta, which authorized publication and contains information on the origin and conduct of the work, that the compilers read 2,000 books containing 3 million lines, but that only one-twentieth were selected for inclusion.

The excerpts chosen are from jurists covering more than three centuries, from C. Mucius Scaevola and Labeo (contemporaries of Augustus) to Modestinus, Hermogenianus, and Charisius of the third century A.D. The Florentine manuscript gives a (not absolutely accurate) list of the authors and works excerpted. There are 2,465 excerpts from Ulpian, 2,081 from Paul, 601 from Papinian, and 535 from Gaius.

In view of the haste with which the work had to be done it is imperfect and repetitions and contradictions, though forbidden by Justinian, are found.

The excerpts are divided into 50 books and these into titles with headings taken mainly from the praetorian edict. The order of books and titles generally follows that of commentaries on the edict.

The compilation was intended to be a codification of law, not a selection of readings on legal history, and the compilers were instructed to alter and reform anything imperfect or inconvenient. Accordingly in many excerpts there are omissions, alterations, and additions, as by excising mention of institutions, rules, and procedure obsolete or repealed by Justinian's time, and replacing them by current expressions, to adapt classical texts to newer conditions. In the modern study of Roman law much attention has been paid to detecting these interpolations, so as to ascertain the original state of the classical text and to discern what alteration was made.

By the constitution 'Tanta' of 30 December 533 the Digest was brought into force.

This part of Justinian's legislative work is also known as the Pandects, *Pandectae*, from the Greek counterpart of the Latin name.

In the Middle Ages the Digest was commonly divided into the *Digestum Vetus* (Books I–XXIV, tit. 2), the *Digestum Infortiatum* (Books XXIV, 2–XXXVIII), and the *Digestum Novum* (Books XXXIX–L) because it became known to the Bolognese jurists in three bits, but these divisions probably resulted from Justinian's instructions that in the first three years of legal study teachers were to expound Books I–XXIII, to which XXIV, 1 and 2, came to be attached, and the rest up to Book XXXVIII in the fifth year.

The Digest is usually now cited by book, title, *lex* or fragment and paragraph e.g. D. 42, 2, 1, 1. Many titles have an introductory paragraph (*principium*, abbreviated as 'pr', e.g. D.42.2.,.pr.) but mediaeval practice was to cite *leges* and fragments by their opening words and some older books sometimes cite the *lex* first followed by an abbreviated reference to the title in which it appears, e.g. l.1.pr. *de adquir. vel amitt. poss.* (i.e. *de adquirenda vel amittenda possessione*).

(2) In modern law, a collection of summary statements of the points raised and decided in cases adjudicated on by the courts, particularly, and frequently only, the superior courts, arranged under alphabetical headings and sub-headings descriptive of the main branches and topics of law, and giving the names of the parties and the references to the published series of reports where full reports of the cases, and the text of the judgments, can be found. In legal systems where decided cases have authority in subsequent decisions (see PRECEDENTS) Digests are essential tools for ascertaining the law.

The Digest of case-law is one of the modern successors of the Abridgments (q.v.) of English law and of the Practicks (q.v.) of Scots law. Its necessity, importance and value all depend on the practice of courts following precedents and the rule of *stare decisis*.

The modern Digests give brief synopses of all cases reported from the superior courts classified under recognized legal headings alphabetically arranged, and sub-heads analytically arranged, with numerous cross-references, and construed, previous cases discussed, followed, overruled, or otherwise commented on, and of words and phrases judicially defined or discussed. There are also normally alphabetical indexes of cases by the names of the parties. The principal Digests of English case-law are the *English and Empire Digest* (2nd ed. 56 vols. with supplements) which deals briefly with Scottish, Irish, and Commonwealth cases in footnotes; and *Mews' Digest of English Case Law* (2nd edn.) with supplementary volumes which is confined to English law and is more selective.

There are also the *Law Reports Digest of Cases and Statutes* (1865–1950, 12 vols.), followed by *Law Reports Index* (1951–60 and 1961–70), and the *All England Law Reports Indexes* (1558–1935 and 1936–76), and supplements thereto, both of which are Digests of and Indexes to cases in these series of reports only.

In Scotland there are Morison's *Dictionary of Decisions* (1540–1808), with Brown's *Supplement*, Brown's *Synopsis*, and Tait's *Index*. For the more modern cases there are Shaw's *Digest* (1800–68), followed by the *Faculty Digest* (1868–1922) with decennial supplements, and the *Scots Digest* (1800–1947).

In the U.S. the *American Digest System* covers all reported state, supreme, and federal courts in the U.S. as well as certain other lower courts. It comprises the *Century Digest* (1658–1896, in 50 vols.), and *Decennial Digests* (1897 to date). There are also separate Digests for the various reporters.

(3) In modern law, the term Digest is sometimes given to a textbook setting down principles and rules of law in something the same form as might be enacted in a code, e.g. Stephen's *Digest of the Criminal Law*, Jenks' *Digest of English Civil Law*. Such books have no higher authority than other textbooks.

Dignity. The right to bear a title of honour or of nobility. Dignities may be hereditary, such as peerages (q.v.) and baronetcies, or for life, such as life peerages and knighthoods. The dignities of peerages and baronetcies are created by writ or letters patent, that of knighthood by dubbing as knight. A dignity of inheritance may also exist by prescription. Dignities of inheritance are incorporeal heritaments having been originally annexed to the possession of certain lands or created by a grant of those lands and are generally limited to the grantee and his heirs or his heirs of the body. If heirs are not mentioned, the grantee holds for life only. The heirs are determined by the rules which governed the descent of land prior to 1926.

Dilhorne, Viscount. See MANNINGHAM-BULLER, R.

Diligence. (1) Care, attention, and application. In Roman law two levels of diligence were recognized, *exacta diligentia*, such as a careful man would take, and *diligentia quam in suis rebus*, which a man displays in his own affairs.

(2) In Scots law, the generic term for the forms of law by which a creditor endeavours to exact payment of a debt, proceeding normally on the basis of a decree of court for payment. It comprises diligence against the person by civil imprisonment (now almost obsolete), diligence against heritable property, comprising adjudication, inhibition, poinding of the ground, action of maills and duties, and the process of ranking and sale, and diligence against moveable property, comprising arrestment and furthcoming, personal poinding and sale, sequestration and sale of a tenant's effects, and mercantile sequestration of an insolvent.

(3) The word is also used of the process to compel the attendance of witnesses before a commissioner or to produce documents.

Dillard, Hardy Cross (1902–). U.S. lawyer, taught at and became Dean of University of Virginia Law School, and was a Judge of the International Court 1970– .

Dillon, John Forrest (1831–1914). Practised medicine before studying law; he was admitted to the Bar in 1852 and in 1862 was elected to the Iowa Supreme Court. From 1869–79 he was U.S. circuit judge for the eighth circuit. From 1879 to 1882 he was a professor of law at Columbia College. His reputation rests on *Municipal Corporations* (1872) and *The Laws and Jurisprudence of England and America* (1894). He also edited the Marshall Centenary addresses and papers prepared for the Marshall Centenary in 1901.

Diluvion. The process of suddenly leaving land bare in the narrow seas or in rivers as far up as the tide flows and reflows. Such land prima facie belongs to the Crown, but to a subject if the place on which the land is situated belongs by grant or prescription to the subject. If the diluvion takes place gradually the land left bare belongs to the adjacent owner and the Crown is apparently not entitled where the owner of land left bare, which was formerly covered, can recognize any portion which originally belonged to him, or where an island is created by a sudden flood of water in cutting off land or by a sudden recession of water.

Diminished responsibility. A partial defence in criminal cases, possibly only in murder, introduced judicially in Scotland in the nineteenth century, to the effect that if the accused were suffering from infirmity or aberration of mind or impairment of intellect to such an extent as not to be fully accountable for his actions, a state bordering on but not amounting to insanity, the crime would be reduced from murder to culpable homicide. It was introduced by statute in England in 1957 in murder cases and, if established by the accused, entitles him to be found guilty of manslaughter only. In both cases the object was to avoid persons not fully entitled to be acquitted on the ground of insanity suffering the capital penalty for murder. With the abolition of capital punishment for murder the doctrine has lost much of its importance.

Dino da Mugello (Dynus) (*c.* 1240–99). One of the foremost Italian jurists of his time, author of commentaries on various parts of the Digest and Code, *Commentaria in Regulas Juris Pontificii* and an *Ordo judiciarius* (*c.* 1298).

Diocese. A geographical area subject to the ecclesiastical supervision of a bishop. In each diocese in England there is a court of the bishop known as the consistory court (in the diocese of Canterbury the commissary court) of the diocese. The judge is the chancellor (in Canterbury the commissary general) of the diocese. The chancellor must be a person who has held high judicial office or a barrister, and a communicant; most chancellors are laymen. The chancellor though appointed by the bishop derives his authority from the law and acts as an independent judge, uncontrolled by the bishop, to whom no appeal lies. He acts as official principal of the bishop and is also vicar-general of the diocese.

Dionysiana Collectio. A collection of canons of church councils compiled by Dionysius Exiguus in the sixth century, intended to pave the way for a reconciliation of the churches of the West and East. It sought to stress the supremacy of Rome. It was probably prepared at Rome about 520 and soon became a favoured collection. Later many additions were made to it and it was the basis for the *Hadriana Collectio* (q.v.) and for other collections such as the *Dionysiana* of Bobbio.

Dionysiana-Hadriana (or Hadriana). A collection of legal texts sent by Pope Adrian I to Charlemagne in 774 and accepted at a synod held at Aachen early in the ninth century.

Dionysius Exiguus (early sixth cent.). A Scythian monk who about A.D. 550 translated into Latin the canons of the major ancient Councils of the Church down to the Council of Carthage (419) and made a compilation of the Decretals of the popes from Siricius (385) to Anastatius (498) known as the *Collectio Dionysiana*. These compilations were published by Charlemagne as a law of the Empire in 802 and had great authority in the West. He had a high reputation as a theologian and astronomer and is said to have introduced the method of reckoning the Christian era.

Diplock, William John Kenneth, Baron (1907–). Called to the Bar in 1932, he was a judge of the Queen's Bench Division, 1956–61, a Lord Justice of Appeal, 1961–68, and became a Lord of Appeal in Ordinary in 1968. He was a member, and later president, of the Restrictive Practices Court, 1960–61, and Chairman of the Institute of Advanced Legal Studies from 1973 to 1977.

Diplomacy. The art of conducting international negotiations. Modern diplomacy dates from the development of permanent missions from one country to another, which began in the fifteenth century.

Diplomatic representation and privileges. Diplomatic representation of one state at the capital of another is known from a very early date and was highly developed among the Greeks and at Rome, though only in the form of envoys, not permanent embassies. The Emperors at Byzantium, the Popes, and Charlemagne and his successors all sent embassies to many rulers. Early on the Popes sent personal representatives to the various national churches, sometimes for long periods, and princes began to have their interests represented at Rome. Venice was the first state to create a diplomatic service, with ambassadors in all the main capitals in the sixteenth century. Thereafter France and Spain established continuous representation at the main courts and Henry VIII sent many envoys on special missions; Elizabeth had almost continuous representation in Paris. In these times and until much later ambassadors were not only representatives, but spies and fomenters of plots and even rebellions. In the seventeenth century permanent embassies became commoner and it later became common in treaties with oriental states to provide for the establishment of resident representatives.

Diplomatic representatives also came to have important functions in promoting alliances for war, such as the League of Cambrai in 1508, and congresses for peace, such as that of Cambrai in 1529 and Cateau-Cambresis in 1559, though the most notable of the latter was the Peace of Westphalia in 1648.

In the seventeenth century diplomatic representatives became divided into recognized grades. The Congress of Vienna distinguished (a) ambassadors, legates, and nuncios; (b) ministers plenipotentiary and envoys extraordinary; and (c) chargés d'affaires accredited to foreign ministers. Later the category of ministers resident was provided for, ranking between (b) and (c). Although their privileges are materially the same, they differ in rank and honours. Ambassadors are considered the personal representatives of the heads of their states.

The earlier type of special ambassador was entrusted with a definite mission, whereas the later type of permanent ambassador is a permanent source of information about the country in which he is stationed and a regular channel of communication with its government.

In modern law the existence of a community of states implies a right of legation, of sending envoys, and a duty to receive envoys from other states, though not permanent envoys. Envoys may be distinguished into special envoys for particular occasions and political envoys; the latter may be sent to represent their state at a conference or congress, or be permanently accredited to a state for the purpose of maintaining communications and negotiations with it.

The appointment of an individual as a diplomatic envoy is conveyed by papers to be handed by the envoy to the receiving state, a Letter of Credence (*lettre de créance*) if a permanent ambassador or Full Powers (*pleins pouvoirs*) if sent to negotiate a special treaty or convention. These are handed over at an audience with the head of state. A state may decline to receive a certain individual as envoy.

A permanent ambassador represents his own state in all international relations and receives communications to be reported to his home state, observes all matters which may affect the interests of his state and reports them, and protects the persons, property, and interests of his state within the territory of the state to which he is accredited. He may not interfere in the internal politics of the host state.

Diplomatic privileges traditionally comprehended *droit de chapelle*, the entitlement of the diplomatic corps to practise their own religion; *droit de quartier*, immunity from being subject to the local police; and *droit de l'hotel*, extra-territoriality, immunity of the embassy from local jurisdiction and taxation. In addition they enjoy the same legal protection by the local law as the heads of states whom they represent, immunity from local criminal jurisdiction, immunity from civil jurisdiction unless they waive it, exemption from police orders and regulations, and from personal and local taxation. Members of diplomatic staff enjoy the same privileges as the ambassador himself. A Vienna Convention on Diplomatic Relations, 1961, settled the modern practice on many matters. It was given effect to in the U.K. by the Diplomatic and Other Privileges Act, 1971.

Diplomatic privileges are extended also to officials of the U.N., N.A.T.O., E.E.C., judges of the I.C.J. and certain major international organizations.

All the diplomatic envoys accredited to a particular state form a body styled the Diplomatic Corps, headed by the Doyen, who is the Papal

Nuncio, or the oldest ambassador or oldest minister plenipotentiary. The corps has no legal functions, but guards the privileges and honours due to all diplomatic envoys.

Among diplomatic usages, an important one is language. The language of diplomatic intercourse was formerly Latin but during the political ascendancy of France it came to be French. In the twentieth century English and French have been treated as equal. In all organs of the U.N. other than the I.C.J. the working languages are English and French and the official languages Chinese, English, French, Russian, and Spanish.

A diplomatic mission terminates if the object of the mission has been accomplished, or the envoy is recalled on the expiry of his tour of duty, or on a cessation of diplomatic relations between his and the host state, or at the request of the host state. On the death of a monarch who is head of either state all missions sent and received by him terminate and all envoys must receive new letters of credence. Missions are also terminated by a revolutionary change of government in either sending or host state, or the extinction of the existence of either state.

Diplovatatius, Thomas (1468–1541). Born in Corfu, he studied in Italy and became a magistrate at Pesaro. He was author of a treatise *De Praestantia Doctorum* of which the most valuable part is book IX, *Liber de Claris Jurisconsultis*, giving biographical information about jurists from Roman times to his own day, and other works, many biographical.

Direct applicability. In European Community law, a concept developed mainly by the Court of Justice, to the effect that an article of the Treaties which is directly applicable has the force of law in member states without the need for municipal legislation or other action to give the force of law to the article in question or to give individuals rights which they may enforce by action before municipal courts. Direct applicability, like the supremacy of Community law over national law in matters where both apply, is a manifestation of the new legal order in the international field constituted by Community law. In the U.K. this is recognized by the European Communities Act, 1972. To be directly applicable Treaty provisions must be clear and precise, unconditional, and leaving no discretion to the Community or to the member states. By the Treaties regulations (q.v.) of the Council and the Commission are directly applicable in all member states. It is questionable whether directives or decisions are directly applicable.

Direction. Instructions on a matter of law given by the presiding judge to a jury, normally contained in his summing-up or charge to the jury. If the direction is erroneous in point of law, this is a misdirection which may be a ground of appeal and a ground for setting aside the verdict.

Directions, summons for. A procedural step in actions commenced by writ of summons in the High Court. On hearing the summons the court or Master must determine as many interlocutory matters as possible and decide the appropriate way for having the action heard and disposed of.

Directives. The second rank of administrative acts (inferior to regulations, superior to decisions) made by the Council or Commission of the European Communities in order to carry out their tasks in accordance with the Treaties. They must be addressed to states, not individuals, but many create rights for individuals or allow the directive to be pleaded before municipal courts. Directives are binding as to the result to be achieved on each member state to which they are addressed, the choice of the method of achievement being left to the state concerned. Unlike regulations, directives are not meant to be an instrument of uniformity even if a directive is addressed simultaneously to several states with the same objective. In practice they are used to effect approximation of national laws. Like regulations, directives have to be reasoned or motivated and be based on the Treaties. They take effect on notification to the addressees. As sources of community law they are less prominent than regulations but have considerable potential as sources of domestic law.

Director. A person elected by the shareholders of a company to determine the policy and supervise the management of the company's business. Directors may or may not also have executive or managerial functions. The directors are agents of the company and bind it by all acts done by them within the scope of their authority. They are not trustees of the company's property or assets in a strict sense, because the company owns its assets and the shareholders jointly own the company, but have a responsibility akin to that of trustees to act honestly and in the interests of the company and of the whole body of shareholders. The Companies Acts, 1948 to 1976, impose numerous duties and responsibilities on directors.

Director-General of Fair Trading. The official who heads the Office of Fair Trading and is concerned with consumer protection, consumer credit, fair trading generally, monopolies and mergers, and trade descriptions. He has power to make numerous determinations and decisions, appeal lying to the Secretary of State and to the High Court.

Director of Public Prosecutions, England and Wales. This office was created in 1879 with the duty, under the general supervision of the Attorney-General, of instituting, undertaking or carrying on criminal proceedings, and of advising the police and others conducting prosecutions. In 1884 it was united with the office of Solicitor to the Treasury, but the offices were again separated in 1908. He instructs counsel, subject to the directions of the Attorney-General.

The police must report numerous kinds of crimes to the Director, though the prosecution is frequently left in the hands of the police, but in other cases the staff of the department prepare the case and present it themselves or instruct solicitor and counsel to present it on behalf of the Director. In certain cases, by statute, prosecution has to be undertaken by, or requires the consent of, the Director. In general the investigation of crime and the discovery of evidence is done by the police, not by the Director.

The existence of the office trenches only to a small degree on the general principle of English criminal law that private persons, including the police, or local authorities, or other public bodies, normally prosecute, and only in a small proportion of serious cases does the Director institute or take over a prosecution, though he may take over the conduct of any prosecution at any stage. He has no functions in Scotland or Northern ireland.

J. Edwards, *The Law Officers of the Crown.*

Diriment impediments. In canon law *impedimenta dirimentia,* those bars to marriage which, if contravened, render the marriage null and void *ab initio.* They include existence of a prior marriage, relationship within the prohibited degrees, lack of sufficient age, lack of mental or physical capacity, absence of true consent or impetration of consent by force, fraud, and mistake.

Disability. The legal concept of lack of legal power to do something. Thus, persons are normally under a disability from transferring property which they do not own. The correlative (q.v.) concept is immunity. Disability may be general when the person is disabled from acting legally altogether, as in the case of persons of unsound mind, or special, when the person is disabled only temporarily or in relation to a particular matter, as where a person lacks authority to make a particular kind of contract. Disability may also be personal, attaching to the person himself, or absolute, affecting his descendants and successors also, as formerly happened in cases of attaint for treason or felony, which resulted in corruption of blood. The word is also used in the special sense of a disability to enter into a marriage.

Disaffection. Sedition, disloyalty, discontent with established authority.

Disbarring. The action of the benchers of an Inn of Court in expelling a barrister from the Inn. A barrister may be disbarred on his own petition, and must be so disbarred if he wishes to become a solicitor or to practise any trade or profession deemed inconsistent with membership of the Bar. Or disbarring may be imposed by the benchers as the ultimate punishment on a barrister guilty of conduct unbecoming the profession. If the barrister is a bencher, he must be disbenched as well as disbarred and, if a Q.C., he must be dispatented. (See KENEALY, E.) Reasons must be given by the benchers for disbarring or disbenching. The term is also used for the corresponding act in other countries.

Discharge. As noun—the termination of liability by and under contract, a receipt for a payment.

As verb—to release a person from a debt or obligation, to release a prisoner, to free a person adjudicated bankrupt from his debts and the disabilities of bankruptcy, to dispense with the further services of a juror or to let a whole jury go because they cannot agree, or because of some procedural difficulty, to set aside an order of a court.

Discharged prisoners' aid societies. Such societies were established in England in the early nineteenth century on a voluntary basis to assist persons discharged from prison to find employment and to enable them to live by honest work. Statute of 1862 authorized the local justices to give them money for the benefit of prisoners. The Prison Commissioners from their formation in 1878 promoted the foundation of such societies and they were given increased grants. A Central Discharged Prisoners' Aid Society was instituted in 1918 to pool ideas and co-ordinate efforts, and reconstituted as a National Association in 1936. Over the years the Association and its constituent societies have done much with limited resources to settle discharged prisoners. At local prisons the Association appoints one or more prison welfare officers, while local societies' officers concentrate on home visits.

Disclaim. To renounce a claim, to refuse to accept, to decline any participation in or connection with something.

Disclaimer of peerage. Succession to a peerage disqualifies for membership of the House of Commons and consequently, in practice, for the major offices in a government, since it is now accepted that such ministers as the Prime Minister, Chancellor of the Exchequer, Home Secretary, and Foreign Secretary must be in the Commons. This may be avoided by disclaimer of the peerage under the Peerage Act, 1963, which must be done, if an M.P., within one month and in other cases within

12 months of succeeding to the peerage or attaining majority.

Disclamation. In feudal law, a vassal's disavowal or repudiation of a person as being his feudal superior. It resulted in forfeiture of the fee.

Disclosure. Making known that which is not apparent. A person contracting is not bound to make any disclosure as to the subject matter of contract unless asked, in which case he must make honest disclosure, but in the special cases of contracts *uberrimae fidei* (q.v.) a party must make disclosure of all facts which might influence the other party, whether asked or not.

Discovery. The process by which parties to a civil cause in England may, within limits, obtain information of the existence and contents of all documents relevant to the matters in dispute between them. The object is to elicit documents before the trial so as to eliminate surprise at the trial and promote fair disposal of the case. It originated in the Court of Chancery and the practice and procedure are now governed by the Rules of the Supreme Court, failing which by the former practice of the Court of Chancery. Save in actions begun by writ, discovery is granted in the exercise of judicial discretion, but in most actions commenced by writ, every party is prima facie entitled to discovery.

Discredit. To discredit a witness is to cast doubts on the truth or accuracy of his evidence by showing that he has previously made statements inconsistent with his evidence, by giving evidence of his general bad reputation for truth, or by proving conduct, such as having been bribed, which casts doubts on his veracity.

Discretion. The faculty of deciding or determining in accordance with circumstances and what seems just, fair, right, equitable, and reasonable in those circumstances. Rules of law frequently vest in a judge the power or duty to exercise his discretion in certain circumstances, sometimes if he finds certain requisites satisfied, and sometimes a discretion within stated limits only.

A question of judicial discretion is accordingly a question not determined, like a question of fact, by evidence, nor one determined, like a question of law, by authorities and argument, but one determined by an exercise of moral judgment. In cases where the discretion has long been vested in judges there is, however, a strong tendency for the ways in which judges have exercised their discretion to be reported and for subsequent judges to exercise their discretion consistently with the ways in which it has been exercised in the past, so that the discretion comes to become, not unfettered, but limited by precedents.

Vesting discretionary power in judges is one of the commonest ways of individualizing the application of law and making it flexible and adaptable to circumstances; without it law would much more often be criticized as harsh, unfeeling, and unjust.

In administrative law a discretion or discretionary power is frequently conferred on a Minister. Such must be exercised reasonably.

Discretionary trust. A trust in which the trustees have discretion as to when, to whom, and how much to disburse.

Discussion. In Scots law, the right of a cautioner (q.v.) unless renounced, that diligence (q.v.) be done first against the principal debtor before being done against him. The privilege was taken away in 1865 but may be expressly conferred.

Disentail. The process by which a tenant in tail bars the entail (q.v.) so as to convert it into a fee simple or a base fee and defeat the rights of all claiming after and under him.

Disestablishment. The severance of the connection between a particular branch of the Church and the State, removing its privileged legal status and reducing it to the level of a voluntary association. The Anglican Church of Ireland was disestablished in 1870, surplus church funds being used to endow higher education, relieve poverty, and encourage agriculture and fisheries. The Welsh Church was disestablished in 1920.

Dishonour. To neglect or refuse to accept or to pay a bill of exchange when presented for acceptance or for payment.

Disorderly conduct. A general term for conduct frequently of a kind amounting to a breach of the peace (q.v.).

Disorderly house. A brothel, common gaming house, or a disorderly place of entertainment. The term is most commonly used of premises used as a brothel.

Disparagement. In feudal law, the act of a lord entitled to give an heir in marriage, of pairing the heir in marriage with one of inferior degree.

Dispensation. In ecclesiastical law, the action of an authorized person in allowing conduct notwithstanding a general prohibition of that kind of conduct. In the Roman Church dispensations were originally granted only where it would be for the good of the Church as a whole that the rule be departed from, but later they came to be granted for the benefit of individuals. A famous instance was the Pope's refusal to grant Henry VIII dispensation from his marriage to Catherine of Aragon, which prompted the English Reformation and the breach with Rome. In the Roman Church the principle remains that the authority which made a rule may dispense from it, as may that authority's superior, the ultimate authority being vested in the Pope.

Dispensing Power. A royal power, frequently used in the Middle Ages, exempting particular persons from the operation of particular penal laws. It was copied from papal practice, which issued bulls, '*non obstante* any contrary law', and this was copied in grants and writs by Henry III and his successors. Parliament occasionally assented to the device, and sometimes expressly excluded it, but the Crown continued to exercise the right by virtue of the royal prerogative and did so repeatedly in the fourteenth and fifteenth centuries. In Henry VII's time it was held that the King could dispense with penalties if the act were *malum prohibitum*, but not if *malum in se*. In 1685 James II issued dispensations to Catholics to hold certain offices, notwithstanding the Test Act, and in *Godden* v. *Hales* (q.v.) subservient judges held the dispensation valid, and the power was thereafter vigorously exercised. The Bill of Rights, 1689, declared that the pretended power of dispensing with laws or the execution of laws by royal authority, as it had been assumed and exercised of late, was illegal, and section 12 thereof enacted that thenceforth no dispensation by *non obstante* of or to any statute or any part thereof should be allowed, but be void and of no effect except a dispensation be allowed if in such statute or in such cases as be specially provided for by bills to be passed during that session of Parliament. No such bill was passed.

Disqualification. Any fact which has the legal effect of making a person previously qualified or entitled to do something no longer qualified or entitled to do so. Thus, as part of the penalty for certain driving offences a court may impose disqualification for holding or obtaining a driving licence for a stated period. Persons may be disqualified from holding various offices by stated facts. Thus, a person is disqualified for membership of the House of Lords by being an alien, under full age, bankruptcy, conviction by treason, or expulsion from the House; a person is disqualified for membership of the House of Commons by being an alien, under age, a peer, a clergyman (except non-conformist churches), by unsoundness of mind, bankruptcy, conviction of treason, or corrupt practices at elections. The House of Commons Disqualification Act, 1975, disqualifies the holders of any of a long list of offices, including particularly judges, civil servants, and members of the armed forces.

Disseisin (or dissasine). The act of wrongfully ejecting a person actually having seisin or lawful possession of a freehold, or of wrongfully depriving a person of the seisin of lands, rents, or other hereditaments. The person disseised had both a right of entry and a right of action. Disseisin differed from abatement or intrusion which was wrongful entry where the possession was vacant. Disseisin by election took place when a person alleged or admitted that he was disseised when this had not in truth happened, so that he could invoke the assize of novel disseisin. Disseisin of incorporeal hereditaments can take place only by disturbing the owner in his enjoyment of them, and only took place by election. Equitable disseisin took place where a person was wrongfully deprived of the equitable seisin of land, i.e. of the rents and profits. Disseisin technically became obsolete with the abolition of the forms of action.

Dissenter. A person not adhering to the established Church, particularly an adherent of other Protestant denominations. The Toleration Act, 1688, first allowed dissenters to worship according to their own forms in certified meeting places. The Universities Tests Act, 1871, enabled dissenters to take any degree other than in divinity in any of Oxford, Cambridge, or Durham Universities.

In Scotland the established Church is Presbyterian and all other denominations are dissenters.

Dissenter, The Great. Mr. Justice O. W. Holmes (q.v.) of the U.S. Supreme Court. He in fact dissented less frequently than many of his colleagues on the Supreme Court, but his dissents were powerful and reflected fundamental differences of view from those of the majority.

Dissenting opinion. The opinion delivered by any judge or judges of a bench who disagree, in whole or in part, with the judgment of the majority. An opinion agreeing with the majority but for different legal reasons is not a dissenting opinion, but a separate concurring opinion. It is a matter of the practice of each individual court whether the expression of dissenting opinions is permitted or not. Not till 1966 were dissenting opinions allowed to be expressed in Privy Council cases. Reports of decisions may indicate such opinions by noting, e.g. Lord Blank dissenting, or *dissentiente*, or diss. Lord Blank. A dissenting opinion cannot be taken into account in determining the *ratio decidendi* (q.v.) of the decision, but a dissenting opinion, particularly a powerful one, can greatly weaken the value of the majority decision, and it may be influential in a later case. Dissenting opinions are sometimes approved and upheld in later cases which overrule the majority view, so that from being heretical they are transformed into orthodoxy.

Dissolution of Marriage. A marriage is dissolved by the natural death of either spouse, by decree of divorce (q.v.), or, if a person satisfies the court that reasonable grounds exist for believing that his or her spouse is dead, in which case the court may make a decree of presumption of death and dissolution of the marriage. The continuous absence of the respondent for seven years or upwards and absence of reason to believe that the other party has been living within that time is evidence that he or she is dead unless the contrary is proved.

Dissolution of Parliament. Parliament is dissolved by Royal Proclamation in the exercise of the royal prerogative. Since 1701 the sovereign has not dissolved Parliament on his own responsibility, and the dissolution is now effected on the advice of the Prime Minister. By convention a Prime Minister's request for a dissolution is not refused; it has not been done since 1858, but circumstances may exist where it could be proper to refuse. The choice of the time for making the request is usually determined by his judgment of whether he thinks it electorally suitable or not to his party. Failing a request for a dissolution, Parliament would, under the Parliament Act, 1911, come to an end after five years.

Dissolution of partnership. The bringing of a partnership to an end, which may be effected by the partners, by the court, or by operation of law in such cases as death or bankruptcy of a partner.

Distinguishing. The technique of pointing to differences in the facts or the point of law raised between those of a prior case referred to as a precedent and those of the case presently under consideration. The question always is whether the differences are so material or significant as to justify a different decision. If the precedent is held to be distinguishable, the principle of the decision therein, from whatever height in the hierarchy of courts it may come, is not binding on the court dealing with the present case, but it may still be persuasive. Distinguishing is accordingly resorted to where a judge does not think that he is, or does not wish to be, bound by a particular precedent, and some cases are so frequently distinguished in later cases as being decisions on their own facts or otherwise as to be almost completely deprived of authority. If a precedent is indistinguishable, the later court will be bound or persuaded thereby according to the rules of *stare decisis* (q.v.).

Distraint. The seizure of goods by way of distress (q.v.).

Distress. Under English law, a summary remedy by which a person may, without legal process, take possession of the personal chattels of another, and hold them to compel the performance of a duty, the satisfaction of a debt or demand, or the payment of damages for cattle-trespass. Statutory rights to distrain have been introduced in certain other cases. The most important case is distress for rent, whereby, at common law a landlord may seize the goods and chattels found on the premises rented to secure the payment of rent or the performance of obligations due to him. Payment of local rates is also enforceable by distress, and income tax is similarly recoverable.

Distress damage feasant. A common law remedy for the recovery of compensation for trespass by animals taken while doing the damage (damage feasant). The animals had to be seized while on the land and in course of the trespass; they could not be followed if they left the land. To justify the seizure there had to be actual damage done. Once seized the animals had to be impounded, i.e. confined in a manor or common pound or in the distrainor's premises and they can then be detained till satisfaction is made for the damage and charges are paid. The owner could estimate and tender the value of the damage done. This remedy was abolished in 1971.

Distribution. In English law the transfer of the property of a deceased intestate to those entitled thereto, formerly under the Statute of Distribution, 1670, or according to local customs of distribution prevailing in certain places, and now under the Administration of Estates Act, 1925, and the Intestates' Estates Act, 1952.

Distributive justice. Justice as expressed in the distribution of rights, privileges, duties, and liabilities among classes and members of the community. In modern law this is done by Parliament and by legislation, as by taxation and social security provisions.

District council. In England and Wales since 1974 the lowest and smallest areas of local government administration, sub-divisions of counties. Each has an elected council, which has numerous statutory powers and duties. In Scotland districts are sub-divisions of regions and have elected councils with statutory powers and duties.

District court. The lowest grade of Scottish criminal court, exercising summary jurisdiction only in respect of minor offences. It consists of a stipendiary magistrate or one or more justices of the peace, and prosecution is by a procurator-fiscal.

In the U.S. district courts have federal jurisdiction in each of the 50 states, and in the District of Columbia, Canal Zone and certain other places have federal and local jurisdiction. They are courts of trial in cases raising federal issues, appeal lying to circuit courts of appeal.

District registries. In England and Wales local offices situated in the principal county towns, normally in the county court office, established to issue writs of summons in High Court actions and to deal with proceedings in an action generally down to the stage of entry for trial.

Distringas (that you distrain). A writ so-called because it commanded the sheriff to distrain on a person for a certain purpose. Formerly *distringas* might be issued to enforce appearance to an indictment, information, or inquisition in the Queen's Bench Division, or to command the sheriff to distrain jurors named in the panel to enforce their appearance to make a jury (*distringas juratores*). These and other varieties of the writ were abolished in 1852. A *distringas* in detinue was a special writ of execution to compel a defendant to deliver goods wrongfully detained by repeated distresses against him. In equity a *distringas* was issued against corporations to compel appearance or to restrain a transfer of stock by the Bank of England or a public company.

***Distringas*, notice in lieu of.** A notice given to a bank or public company to stop the transfer of stock or payment of dividends, particularly to stop unauthorized dealing by the holder. The recipient of the notice must, if they receive a request to deal with the stock, notify the party lodging the notice and may only deal with the stock if no order or injunction is obtained to restrain them from doing so.

Dittay. In old Scottish criminal law, the matter of a charge or ground of indictment against a person accused.

Diversité de courts et lour jurisdictions, et alia necessaria et utilia. A short tract, probably of Henry VIII's reign, dealing with the superior courts, the inferior courts, trial in writs of right, and other matters. There are several sixteenth-century editions and it is frequently found bound up with other smaller tracts.

Dividend. The portion of the profits of a company which is distributed to and among shareholders according to the class and amount of their holdings. It is not permissible to pay dividend out of capital. The sum available for dividend is appropriated first to payment of dividend on preference stock at the rate fixed for such stock, and thereafter to payment on ordinary stock at such rates as the directors, subject to any government control in force, recommend as possible and provident. The term is also applied to periodical payments to creditors from the realization of a bankrupt's assets.

Divine Right of Kings. The political doctrine that a king derived his authority direct from God and accordingly had absolute power which overrode all other authorities. It originated in the mediaeval concept of God's award of temporal power to civil rulers and spiritual power to the Church. It was claimed by the earlier Stuart kings in England, and explains many of their attitudes in the struggle which developed between them and Parliament for political sovereignty. If the King had authority from God the people could have no right to disobey or dissent or resist. A clash was inevitable as Parliament became conscious of its growing power, and asserted the view that the power of kings and Parliaments rested on the consent of the governed, and that the king was subject both to God and to the law of the land. The principle of divine right was submerged during the Commonwealth but re-emerged under James II, but disappeared with his flight and abdication. Both Hobbes and Filmer, however, sought in their writings to justify this approach to government, and Locke's *First Treatise on Civil Government* (1689) was written to refute Filmer's arguments. In France Bossuet contended that the King's person and authority were sacred and that his power was like that of a father, but absolute, being derived from God and controlled only by reason. The powers implied by the doctrine were claimed by Louis XIV, but the theory was obsolete before the Revolution.

J. N. Figgis, *The Divine Right of Kings.*

Divine service, tenure by. A tenure (q.v.) of land, one of the spiritual tenures, under which the tenant was bound to perform certain divine services, such as to sing a stated number of masses for the lord's ancestors. Laymen could hold by this tenure and the lord could distrain for non-performance of the agreed services.

Divisible contract. A contract is divisible if it provides for a number of distinct acts of performance with separate consideration for each. In this case breach by either party in respect of any one or more parts cannot entitle the other to treat the whole contract as discharged nor entitle him to refuse performance of his part in relation to the other parts of the contract. If the consideration is unitary, a contract is not divisible merely because performance is to be made in instalments or by a number of distinct acts.

Division. In both Houses of Parliament, the physical separation into two lobbies of those voting for and against a motion. This mode of voting has been used since 1554. Members wishing to vote leave the chamber and go into one or other of the division lobbies. Two tellers from each side take post at the exits from the division lobbies and members, having given their names to clerks, file out of the lobbies and are counted by the tellers. Members who abstain do not leave the Chamber. The senior teller for the majority reads out the figures to the Speaker or Chairman who announces the result of the division. The division list printed in *Hansard* (q.v.) records the names of the members who voted Aye and No.

Divisional court. A court consisting of two or three judges of the Queen's Bench, or Chancery, or Family, Divisions of the High Court in England and Wales. Divisional courts have both original and appellate jurisdiction. Divisional courts are quite distinct from the Divisions of the High Court (q.v.).

A divisional court of the Queen's Bench Division hears applications for the issue of the prerogative writs of habeas corpus and the orders of mandamus, prohibition, and certiorari, and hears certain civil appeals from magistrates' courts, and appeals, by way of case stated, from magistrates' courts exercising criminal jurisdiction, and appeals from various tribunals and under various statutes.

A divisional court of the Chancery Division deals with appeals from county courts in bankruptcy and land registration matters.

A divisional court of the Family Division hears appeals from magistrates' courts in domestic and matrimonial proceedings, from county courts in relation to guardianship and adoption, and from the Crown Court in relation to affiliation proceedings.

Appeal lies from a divisional court exercising civil jurisdiction to the Court of Appeal.

Divisional system. Between 1898 and 1948 a court in England sentencing an offender to imprisonment directed whether the prisoner should serve his sentence in the first, second, or third division. The first division was a privileged form of detention for persons guilty of seditious libel or other political offence, the second was for those not of depraved or criminal habits and the third was for the general body of prisoners and might be aggravated by hard labour. The system of judicial classification had before its formal abolition been superseded by an administrative classification.

Divisions of the High Court. When the High Court of Justice was created by the Judicature Act, 1873, replacing the former separate courts of: Queen's Bench; Common Pleas; Exchequer; Chancery; Probate, Divorce, and Matrimonial Causes; Admiralty (qq.v.), and certain other courts; it was for the more convenient despatch of business divided into five divisions: the Queen's Bench; Common Pleas; Exchequer; Chancery; and Probate, Divorce, and Admiralty Divisions. Though it was provided that all business competent to the High Court might be transacted by any division, the divisions in practice assumed the kind of business which the courts which they had replaced had transacted. The first three divisions were amalgamated into the Queen's Bench Division in 1880 and the Probate, Divorce, and Admiralty Division was renamed the Family Division in 1971, some of its business being reassigned to Queen's Bench or Chancery Divisions. All jurisdiction belongs to all the divisions alike and all the judges have equal power, authority, and jurisdiction. The divisions exist for the convenient distribution of business and the Lord Chancellor may redistribute business among the divisions. The Queen's Bench Division deals with most common law business, criminal work, and includes an Admiralty Court and a Commercial Court. The Chancery Division deals with equity business, including partnership, company, bankruptcy, mortgages, wills, trusts, settlements, and property business. The Family Division deals with matrimonial causes, adoption, guardianship, and related business. Judges are assigned to a division and may be transferred from one division to another.

Divisions of the law. In every legal system which has evolved to any reasonable degree of maturity and developed any substantial quantity of principles and rules jurists have, for facility and convenience of exposition and study, divided the corpus of rules into a number of divisions and subdivisions and have sought to classify and group the rules in some appropriate way.

Much attention was devoted by analytical jurists, particularly in England in the nineteenth century, to classifying the main principles and rules of a legal system, to bringing them under main headings and to establishing the major divisions and subdivisions of a legal system. It is questionable whether any classification is applicable to all legal systems, and within any one legal system the divisions and classifications adopted may be determined partly by legal history, partly by purely practical considerations. Different classifications may be appropriate for different purposes. But if a system of law is to be a system and not a mere aggregate of rules, prescriptions, and prohibitions, some classification is essential for the statement of the rules of a particular legal system, in textbooks and encyclopaedias, for the study and understanding of the rules, and to facilitate finding the rule applicable to a particular problem.

Even the simplest classification, alphabetical, presumes the existence of recognized divisions or headings, such as crime, contract, or matrimonial relations, under which particular rules may be placed; and of words, such as treason or adultery, which have distinct legal connotations and which will serve as sub-divisions. Functional classification, by reference to the kinds of human relations involved, is impracticable without abstraction of the rules common to the various relations considered. The most useful basis of classification is based on the nature of the legal relations involved rather than on the economic or functional content of these relations. Many such juristic classifications of principle have been made, by Hale, Blackstone, Salmond, and Pound among others.

The first clear attempt to divide and classify the principles of a legal system was made by Gaius and other Roman jurists, who distinguished between public and private law, and divided the latter into law relating to persons, to things, and to actions, though the bases of both divisions are to some extent matters of dispute.

This distinction between public and private law, and the classification of private law has been of continuing influence in all legal systems which have been materially influenced by the Roman law, such as Scots, South African, and most European and South American systems of law.

Another Roman classification was into *jus civile,* law applicable to Roman citizens, and *jus gentium*, law applicable to non-Romans; this came to be transformed from the sixteenth century into another basic division, that between the law of a particular political community and the law of nations or public international law, applicable between independent political communities.

Stemming from the Roman distinction between law of persons and law of things on the one hand, and law of actions on the other, there developed another basic division, between substantive law, which states and defines the legal rights and duties of persons in various situations, in relation to other persons and to things, and adjective law or procedure, dealing with the means whereby rights and duties may be declared and vindicated and enforced.

Another division is by reference to the limited or unlimited extent of the persons affected by or against whom a legal right is available, whether the right in question avails only against a determinate person or persons, or is available against anyone, against persons generally. The former is traditionally called a right *in personam*, the latter a right *in rem*, based on the Roman law distinction between different classes of stipulations, pacts, actions, exceptions, and edicts, and analogous to the description of judgments as *in rem* or *in personam* and to the mediaeval distinction between *statuta personalia* and *statuta realia*.

Lastly rights have been divided into those which exist by virtue of the rules of the legal system, such as the right to have delivery of goods bought, variously called primary or antecedent rights, and those which exist only if there is default in a primary right, such as the right to enforce delivery of goods bought, or to recover damages for non-delivery, variously called secondary, or remedial, or sanctioning rights, or rights of redress. The latter are part of the machinery for enforcing and securing the former kind of rights.

To a considerable extent these divisions cut across one another. Thus private law includes both rights *in personam*, such as for delivery of goods under a contract, and rights *in rem*, such as not to be defamed or injured, and both primary or antecedent rights, such as to delivery or not to be injured, and secondary or remedial rights, such as to recover payment, or damages, or obtain an injunction.

Yet another division is by reference to the source of the particular rules in statute or case-law, yielding the distinction between statutory rights, claims and duties and common-law rights, claims and duties.

Legal systems of the civil law or Romano-Germanic family are commonly divided into branches on Roman law lines, primarily into public and private law, and jurists and lawyers specialize in, and their books are about, one or other of these branches or their major divisions, notably civil, commercial, criminal, and administrative law.

In English law historical development played a considerable part and, certainly down to 1875, the major distinctions recognized were common law and statutory rights and into common law and equity. The distinction between public law and private law is practically unknown. Another factor which was very influential in English law was the early development of distinct forms of action, or sets of circumstances which gave rise to recognized remedies. Hence, in English common law a recognized division was into actions for debt, for trespass, on assumpsit, and so on, which later developed into the distinction between actions based on contract and actions founded on torts.

Hence English law has always been resistant to logical division and the classification of principles and rules, and history, the practice of the courts, and practical convenience have played more important parts than the logical scheme of jurists. The older books dealt with topics such as Pleas of the Crown or Tenures (of land). Blackstone's division was into Rights of Persons, Rights of Things, Private Wrongs, and Public Wrongs. The absence of attempts since Blackstone's time to write general treatises on English law has largely absolved jurists from dealing with problems of division and classification, and the major modern works, such as *Halsbury's Laws of England*, treat the law under topic headings, alphabetically arranged, some of the headings containing rather miscellaneous contents. Similarly the major treatises deal with headings such as contract, tort, trusts, real property, rather than with logical divisions of law such as obligations.

In commonwealth countries and the U.S.A. generally the same divisions are adopted, with some variations arising from history or from different circumstances, and the structure of these legal systems is similar to that of the law of England.

Scots law, on the other hand, under heavy Roman influence has from the first been divided and its principles classified under categories known to the Roman jurists and later civilian commentators. The distinction of public and private law is recognized. The latest general textbook divides private law into introductory part, international private law, persons, obligations (including mercantile and industrial relations), property, trusts, succession, bankruptcy, remedies.

The codes adopted in many European countries since 1800 and the great text-writers on the major European legal system divide private law substantially on Roman lines, though there is much variation in details. The German Civil Code has been the model of division and classification since 1900 and has been repeatedly copied; its major divisions are General Principles, Obligations, Things, Family, and Succession.

Divorce. The dissolution of a marriage, otherwise than by death, normally permitting each party to remarry. The religious authority of a country or the convictions of the parties may forbid divorce, but its possibility is increasingly being recognized in the Western world. The degree, and kind, of formalities required also varies, down to the case of divorce at will or by consent.

In Roman law divorce was permitted from the time of the Twelve Tables, but Justinian under Christian influence limited the circumstances in which it was competent.

In the canon law marriage was indissoluble, but ecclesiastical courts granted decree of divorce *a mensa et thoro* which did not completely sever the marriage bond nor permit remarriage. After the Reformation in England and particularly after 1700 divorce *a vinculo matrimonii*, permitting remarriage, began to be granted by private Act of Parliament, but this procedure was slow, cumbrous, and expensive. In 1857 the Matrimonial Causes Act introduced divorce *a vinculo matrimonii* by judicial process, and continued the former ecclesiastical practice in relation to nullity of marriage and judicial separation (the former divorce *a mensa et thoro*). The grounds of divorce were adultery and cruelty (adopted from ecclesiastical law) and desertion (added by the 1857 Act). A husband could obtain divorce for adultery, but a wife only for adultery with cruelty or desertion. Not till 1923 could a wife divorce her husband for adultery alone.

The Matrimonial Causes Act, 1937, introduced further grounds, divorce becoming competent to the petition of either spouse if the other spouse had committed adultery, deserted the petitioner for three years, treated the petitioner with cruelty, was and had been for five years continuously under care and treatment as an insane person, or if a husband-defendant had been guilty of rape, sodomy, or bestiality.

A husband might also till 1970 claim damages from a person guilty of adultery with his wife.

By the Divorce Reform Act, 1969, the sole ground for divorce was made that the marriage had broken down irretrievably, which may be held established only if the court is satisfied that the respondent has committed adultery and the petitioner finds it intolerable to live with the respondent, or has behaved in such a way that the petitioner cannot reasonably be expected to live with the respondent, or that the parties have lived apart for a continuous period of at least two years and the respondent consents to decree, or have lived apart for a continuous period of at least five years.

Either party may be ordered to pay maintenance pending suit to the other party and, on decree being granted, either may be ordered to make periodical payments to the other, or to secure periodical payments, or to pay a lump sum. Provisions for children may also be ordered. Orders for financial provision may be varied or discharged.

A decree for divorce is in the first instance a decree *nisi* and not made absolute until six weeks have elapsed, or in special circumstances a lesser period. Remarriage is competent only after decree has been made absolute and the time for appealing has elapsed or an appeal has been dismissed.

A petitioner may also have a marriage dissolved if there are reasonable grounds for supposing that the other spouse is dead and if the court presumes his death.

In Scotland power to grant divorce for adultery was assumed by the courts after the Reformation in 1560 and confirmed by a statute in 1573 which also introduced divorce for four years' malicious desertion. In 1939 the period required for desertion was reduced to three years and divorce was made competent also on the grounds of cruelty, sodomy or bestiality, or incurable insanity, while it also became possible to decree dissolution of marriage on the ground of the other spouse's presumed death. Condonation, collusion, and connivance (or *lenocinium*) are bars to divorce. No distinction between decree *nisi* and decree absolute has ever been recognized. Until 1964 divorce had the same effect on property rights as the death of the guilty spouse, and the innocent spouse could exact from the estate of the guilty spouse what he would have been entitled to on that spouse's death. Since 1964 the court may award a capital payment or a periodical allowance or both. In 1977 the sole ground of divorce was made the breakdown of the marriage, established by proof of adultery, desertion, intolerable conduct, or separation.

Divorce Court. Or properly Court for Divorce and Matrimonial Causes, established in 1857, taking over from ecclesiastical court jurisdiction in nullity and separation cases and given the new jurisdiction to grant divorce established by statute in that year. It consisted of the Lord Chancellor, the chief justice, and senior puisne justice of each of the common law courts, and the judge of the Court of Probate, the last being appointed the Judge Ordinary of the Court. Appeal lay from the Judge Ordinary to the full court and then, in divorce cases, to the House of Lords. In 1875 it became part of the Probate, Divorce, and Admiralty (now Family) Division of the High Court.

Divorce Division. A common name for the Probate, Divorce, and Admiralty Division of the High Court, 1875–1971. It is now the Family Division.

Dixon, Sir Owen (1886–1972). Called to the Victorian Bar in 1910, he took silk in 1922, became an Acting Judge of the Supreme Court of Victoria in 1926, a Judge of the High Court of Australia in 1929, and was Chief Justice of Australia, 1952–64. He became a P.C. in 1951 and received the O.M. in 1963. In 1942–44 he was Australian Envoy in Washington and in 1950 U.N. mediator between India and Pakistan. He was one of the most distinguished jurists yet produced by Australia, and his interpretations of the Australian constitution are famous.

Dock (Flemish *docke*, or bird-cage). The enclosed space in a court-room in which a prisoner is placed during a criminal trial.

Dock brief or dock defence. In England the instruction of a barrister directly from the dock by a person arraigned for trial on indictment. With the introduction of legal aid (q.v.) the practice is now virtually obsolete.

Dock warrant. A document of title to goods granted by a dock owner to the owner of goods stored in a dock warehouse to the effect that goods listed therein are held by him deliverable to the person named therein or to the latter's assigns by indorsement. Indorsement and delivery probably operates as constructive delivery until the dock owner has agreed to hold the goods for the indorsee. The holder can at any time lodge the dock warrant and take actual possession of the goods.

Doctor. The title given to a person who has attained an academic grade in a particular field of learning. In the middle ages the title of doctor and master were not clearly distinguished. In the mediaeval universities recognized teachers were usually called *magistri, magistri artium, sacrae theologiae, legum* (i.e. civil law) *decretorum* (i.e. canon law), but in the twelfth century the lawyers of Bologna preferred to be called *doctor*, and not until the sixteenth century were master theologians, lawyers, and physicians commonly addressed as doctor.

In the fifteenth century, at least in England, the title *doctor* became appropriated to the senior faculties of theology, law, and medicine; *magister* being confined to the lower faculties, of arts and grammar.

In English speaking countries the hierarchy of academic grades is now generally bachelor, master, and doctor though not every branch of learning or university utilizes all the grades and the titles vary within some branches of learning. Thus the degree above Bachelor and Master of (the Liberal) Arts is usually Doctor of Letters or Literature rather than Doctor of Arts. Universities organized on French models usually confer the titles of *licence* and *docteur*.

The practice of conferring a doctor's degree as the first degree in a subject merely because the candidate has been required to have a bachelor's degree in Arts or Science as a pre-requisite of entry to studies in that subject is academically unwarranted.

In the mediaeval Roman Church the title doctor was frequently given as an honorific title to outstanding scholars. Thus Aristotle was the Philosopher, and Thomas Aquinas was variously called Doctor Angelicus, Doctor Beatus, Doctor Cherubinus, and by other titles. The title Doctor Decretalium was given to Bongratia of Bergamo, Doctor Decretistarum to Peter Quaesnet.

The title Doctor of the Church has been given by the R.C. Church to certain canonized saints who were ecclesiastical writers on account of their sanctity, great learning, and the benefit to the Church from their teaching. The group includes Ambrose, Augustine, Jerome, Gregory the Great, Aquinas, Bonaventure, Bede, and others.

The use of the term doctor in English-speaking countries denoting a physician sprang from the fact that a physician was originally one who had become a doctor of medicine, and this usage became ingrained even when in later times physicians were ordinarily bachelors of medicine and surgery only.

Doctor and student. See SAINT GERMAIN, CHRISTOPHER.

Doctors' Commons. About 1495 Richard Blodwell, Dean of the Arches, formed an Association of doctors of law and of the advocates of the Church of Christ at Canterbury, and it obtained premises in Paternoster Row, later known as Doctors' Commons. The Association was incorporated in 1768 as the College of Doctors of Law exercent in the Ecclesiastical and Admiralty Courts, and in 1782 purchased the premises. The conditions for admission were to have obtained the degree of doctor of civil law at Oxford or Cambridge, have obtained a rescript or fiat from the Archbishop of Canterbury and have been admitted as an advocate by the Dean of the Arches, and have attended court for a year. Such an advocate might then be elected a fellow. Doctors' Commons comprised all those licensed to practise as advocates before the courts, the principles and practice of which were founded on Roman law, which comprised the Court of Admiralty and the ecclesiastical courts with jurisdiction in matrimonial, testamentary, and probate matters. It regulated its branch of the profession, but was not an educational body. It was dissolved in 1858 after the society's exclusive rights to practise in the Admiralty and ecclesiastical courts had been abrogated.

G. D. Squibb, *Doctors' Commons*; N. Nys, *Le droit romain, le droit des gens, et la collège des docteurs en droit civil*; W. Senior, *Doctors' Commons and the Old Court of Admiralty.*

Doctrina placitandi. A book in law French, ascribed to Samuel Ever, serjeant-at-law, published in 1677, containing not a collection of forms, but a series of head-notes under alphabetical headings, illustrated by cases from the Year Books and the early reports. It was judicially described as containing more learning than any other book.

Doctrines of law. Systematic formulations of legal principles, rules, conceptions, and standards with respect to particular situations, or types of cases, or fields of the legal order, in logically interdependent schemes, whereby reasoning may proceed on the basis of the scheme and its logical implications. Examples are the doctrine of consideration in contract, the doctrine of personal bar, and the doctrine of *respondeat superior* (qq.v.). The development and formulation of doctrines are the work of judges and jurists, not of legislation, which treats of particular rules only.

Document. Anything on which signs have been marked to record or transmit any information, a category including books, letters, deeds, title-deeds, maps, plans, drawings, photographs, and the like. Public documents record matters noted for the public benefit and include statutes and statutory instruments, judgments, public records and registers, and are frequently proved by production of official or authenticated copies. Private documents are privately and unofficially produced, and include letters, contracts, wills, and must generally be proved before being admissible in evidence.

Document of title. A document used in the course of commerce as proof of the possession and control of goods and the transfer of which, by mercantile custom long recognized as law, gives the transferee, or enables the transferee to complete, a right of property in the goods described in and represented by the document of title. The main recognized classes of document of title are bills of lading, which pass the property in the goods described therein by indorsement and delivery, or by delivery alone, delivery orders issued by the owner of goods and addressed to the keeper of a warehouse wherein they are stored, and store warrants and dock warrants, by which a custodier acknowledges that he holds goods and undertakes to deliver them to a person named in the order. Documents of title are negotiable instruments (q.v.) in that a transferee does not acquire a better title to the goods than the transferor had.

See also TITLE DEEDS.

Documentary evidence. Evidence provided by a document such as official records, archives, statutes and other public documents, certificates, surveys, assessments and reports made by public authority, registers of births, deaths and marriages. judicial records, histories, scientific books and records, maps and private papers. Save where statute regulates particular cases it must generally be established by whom the document was made, so as to establish its authenticity, before its contents can be considered for the evidence they give as to facts in issue.

Dodd, Sir Samuel (1652–1716). Called to the Bar in 1679, was counsel for Sacheverell in 1710, and held office as Chief Baron of Exchequer, 1714–16. He left little trace in the law.

Doderidge or Dodderidge, John (1555–1628). A judge of the King's Bench, 1615–26, author of various legal works. The work known as Sheppard's *Touchstone of Common Assurances* is attributed to him and said to have been published by Sheppard as his own.

Dod's Parliamentary Companion. A work of reference started in 1832 by Charles Dod and now published annually in pocket size, containing information about the government, public offices, biographical information about members of both Houses of Parliament, list of constituencies with details of elections, and other useful matters.

Dodson, Sir John (1780–1858). Became an advocate of Doctors Commons in 1808 and a barrister in 1834. He was appointed admiralty advocate in 1829, King's Advocate in 1834, and judge of the Prerogative Court and Dean of the Arches from 1852 till 1857. He proved himself a sound and learned ecclesiastical lawyer. He also compiled a useful volume of decisions in the admiralty and ecclesiastical courts, 1811–22 (1815–28).

Doe, Charles (1830–96), Judge, 1859–74, and Chief Justice 1876–96 of the Supreme Court of New Hampshire, and a great name in American law, particularly for his enlightened views on mental illness and legal responsibility.

Doe, John. A name sometimes given to a fictitious legal character in pleadings, and particularly to the fictitious plaintiff in the old action of ejectment (q.v.) who was supposed to have demised land to the real plaintiff.

See also ROE, RICHARD.

Doherty, John (1785–1850). Called to the Irish Bar in 1805 and became Solicitor-General, 1827, and Chief Justice of the Common Pleas in Ireland in 1830–50.

Dole. A portion or lot; a boundary mark; a share in waste or common land; a colloquial term for unemployment benefit. In older Scottish criminal law dole (*dolus*) was the wrongful intention generally required to render conduct criminal.

***Doli incapax*.** Mentally incapable of committing crime. A minor between 10 and 14 is presumed *doli incapax*, but this presumption may be overcome by evidence of mischievous disposition or of knowledge that he was doing wrong.

Dolus. In Roman law, intent to injure, defraud, or harm. In the phrase *dolus dans causam contractui* (fraud giving rise to the contract) it emphasizes that if a contract is to be voidable by reason of fraud or misrepresentation, the misrepresentation must not only have been made but have induced the contract on the terms on which it was made.

Domain. A concept derived from Roman law *dominium* and importing the fullest and most complete right of property in land. The term is also used of the law itself. Eminent domain denotes the sovereign power of the King or the State to expropriate privately owned land for public purposes. In the U.S. the public domain is land occupied by federal government or buildings, or not granted to private owners.

Domat, Jean (1625–96). A great French jurist and intimate of Pascal, a systematizer, author of a *Traité des Loix* and particularly *Les lois civiles dans leur ordre natural* (1689), a great work combining in one system the materials of Roman law and French legislation and decisions, which was several times translated. He sought to found all laws on ethical principles. He also published under the title *Legum delectus* a selection of the most common laws in the collections of Justinian.

Domesday Book. A record in two volumes, containing respectively 382 folios covering 30 counties (sometimes called Great Domesday) and 450 smaller folios covering East Anglia (called Little Domesday), the latter volume being possibly the earlier and giving more but less good information than the former. Not all counties in England are covered, some being included in other counties and the northern counties not being included at all. It was made in 1086 by seven or eight panels of royal commissioners sent round the country to collect information which they obtained from sworn inquests and making returns which were checked by a second set of commissioners. The results were summarized and put in order by royal clerks at Winchester.

The main purpose of the inquiry was to ascertain the extent and value of the lands of the King and his tenants in chief, and fiscal, to afford a basis for assessing Danegeld, but the book is an important authority for English law of the time. Facts recorded therein would often be, and later often were, decisive of matters litigated.

The book was printed by the Record Commission in four volumes in 1783–1816.

J. Round, *Feudal England* (1895); F. W. Maitland, *Domesday Book and Beyond* (1897).

Domesday of Ipswich. A local record containing *inter alia* the Ipswich code of sea-laws, printed in the *Black Book of the Admiralty*, vol. II.

Domestic jurisdiction. In international law those matters which fall entirely within the limits of the jurisdiction of courts of a particular state as contrasted with that of an international court. The term is also sometimes applied to jurisdiction in matters of domestic relations.

Domestic proceedings. In English law proceedings in magistrates' courts under Matrimonial Proceedings (Magistrates' Courts) Act, the Guardianship Acts, Summary Jurisdiction (Separation and Maintenance) Acts, Maintenance Orders Acts, and certain related statutes.

Domestic relations, law of. A branch of law largely coincident with family law, dealing with the constitution, consequences, and dissolution of the domestic relationships of husband and wife, parent and child, and guardian and ward.

Domestic tribunal. A quasi-judicial body established within the organization of a society or association for the determination of disputes between members and the enforcement of the discipline and rules of the society, such as the disciplinary committees of professions.

Domicil (or domicile). The term in international private law (q.v.) for the territory having a distinct legal system in which a person has, or is deemed to have, his permanent home, a connection which determines what legal system regulates many of the legal questions affecting him personally. It may be relevant to jurisdiction or to the choice of which system of law governs the individual in certain contexts. Domicile is distinct from nationality, and from country of actual residence, and is related to legal systems rather than states, so that a British subject may be domiciled in France, and a person is domiciled in England or in Scotland, not in the United Kingdom.

A person's domicile is a matter of law determined partly by inference from facts and partly by evidence of his expressed intention. Every person must have one domicile, and no person can have at any time more than one.

The law attributes to every person a domicile of origin, which persists until and unless he acquires a domicile of choice elsewhere, by residence elsewhere with the intention to reside with some permanence, and which revives if he abandons the domicile of choice. Dependent persons, such as children, have their domicile determined by their father.

Anglo-American legal systems generally prescribe that an individual's personal law is that of his domicile, but European systems commonly prescribe the law of his nationality.

Dominant tenement. The term for the tenement of land in favour of which and for the benefit of which an easement or servitude exists over another, the servient, tenement.

Dominate. In the history of Rome and the Roman law a term for the period from the re-establishment of the emperor's authority by Diocletian after A.D. 284 to the division of the Empire from A.D. 395 or to the end of the Western Empire in A.D. 476. The characteristics of this period were the transformation of the imperial power into a monarchy on Oriental lines, the reorganization of the administrative machinery, and the division of the empire into areas under co-rulers. In the Eastern Empire autocracy continued to the time of Justinian and long after.

Dominion status. The dominion of Canada attained, in effect, self-government in and after 1867, by virtue of the British North America Acts, and by 1914 Newfoundland, Australia, New Zealand, and South Africa had attained a similar state. In 1919 they were separately represented at the Peace Conference, and came to be known as the British Dominions. The Inter-Imperial Relations Committee of the Imperial Conference, 1926, described the Dominions as 'autonomous communities within the British Empire, equal in status, in no way subordinate to one another in any aspect of their domestic or external affairs, though united by a common allegiance to the Crown, and freely associated as members of the British Commonwealth of Nations'. Dominion status was accordingly the status of a territory which had attained the position of one of these five countries. The Irish Free State was accorded Dominion Status in 1922 but asserted independence in 1949 and is the Republic of Ireland. Newfoundland became a province of Canada in 1949.

The term 'Dominion' was never given to any of the dependent territories given independence from 1947 onwards. The concept is probably now obsolete, as the Commonwealth now includes countries which are republics and do not owe allegiance to the Queen, though accepting her as Head of the Commonwealth, and South Africa has left the Commonwealth. The Secretary of State for Dominion Affairs became in 1947 the Secretary of State for Commonwealth Relations (since 1966 for Commonwealth Affairs and since 1968 for Foreign and Commonwealth Affairs).

Dominium. In Roman law and feudal land law absolute ownership of land or goods, with full rights of possession and use. As the feudal theory of land law, under which lord and tenant or superior and vassal simultaneously enjoyed certain rights of ownership in the land, was elaborated, a distinction was drawn between *dominium directum*, the rights reserved to the lord or superior, and *dominium utile*, the rights transferred to the tenant or vassal, such as to possess, occupy, and use.

Dominium eminens. Eminent domain, the right vested in the supreme power in a state to acquire compulsorily on making compensation the property of subjects when required by necessity or at least substantial need for public purposes.

Dominium ex jure Quiritium. Ownership under the old Roman *jus civile*, open to Roman citizens only. A quiritary owner could transfer the ownership of *res mancipi* only by *mancipatio* (q.v.) or *in jure cessio* (q.v.), but the praetors later protested ownership not acquired in these ways; a praetorian owner became a civil one by *usucapio*, uninterrupted possession of a movable object for one year or of immovable for two years. The concept had disappeared by Justinian's time.

Dominus legum. A mediaeval title sometimes given to Azzo dei Porri, a thirteenth-century jurist.

Dominus litis (master of the litigation). In Roman law the party to a case as distinct from his procurator or legal representative; in English law, a person who controls an action and can dispose of it as he wishes, including the party bringing a test case, the decision of which will determine a number of claims; in Scots law, a person who has truly though not nominally initiated a case, and who may be held liable in the expenses thereof.

Domus conversorum (house of the converted). A house in what is now Chancery Lane set aside by Henry III for the use of Jews converted to Christianity. After the expulsion of the Jews it was used as depository for Chancery records and the site is now covered by the Public Record Office.

Domus Procerum. Abbreviated as Dom. Proc. or D.P. an old name for the House of Nobles, or House of Lords.

Donatio inter vivos. The gift by a person, not under any legal obligation to make it, of any form of property capable of transfer by him to another. It requires proof of, or circumstances implying, *animus donandi*, and transfer of the subject from donor to donee by whatever means is required for the kind of property in question.

Donatio mortis causa. A transfer by a person, not under any legal obligation to make it, of any form of property capable of transfer by him, to another, in contemplation of the donor's death, so that the property is immediately transferred to the donee, but subject to revocation if the donor recovers, and subject to lapse if the donee should predecease the donor. It requires proof of or circumstances implying *animus donandi* and also transference of the subject in appropriate manner. It differs from a legacy in not being contained in a will and being preferable to legacies in the event of deficiency of assets.

Donation of Constantine. A document setting out the supposed grant about A.D. 730 by Constantine the Great to Pope Sylvester I and his successors of supremacy over all the other patriarchates, over matters of faith and worship, and of temporal dominion over Rome and the Western Empire. It was regularly appealed to in the Middle Ages in support of papal claims to power, but has now long been accepted as a forgery put together from various sources in the eighth century. It was, however, included in the False Decretals in the ninth century and in Gratian's *Decretum*.

Doneau, Hugues (Hugo Donellus) (1527–1591). A French humanist jurist, professor at Orleans, Bourges, and Heidelberg and a great systematizer, an opponent of Cujas and almost equally distinguished, who sought the actual significance of the Roman law texts, and was author of *Commentarii juris civilis*, a methodical exposition of the Roman rules which was important in Germany till the fifteenth century, and works of religious controversy.

Donnellan, Sir James (?–1665). Called to the English Bar in 1623 and the Irish Bar not long after. He acted as a judge of assize and became a judge of the Irish Common Pleas in 1637 and Chief Justice in 1660.

Donnellan, Nehemiah (1649–1705). Son of Sir James Donnellan (q.v.), was called to the English Bar in 1689 but had practised in Ireland before then. He became a baron of Exchequer in Ireland in 1695 and Chief Baron in 1703.

***Donoghue* v. *Stevenson*,** [1932] A.C. 562. A person drank part of the contents of an opaque bottle of ginger-beer and then, according to her allegation, when pouring the remainder into her glass, a decomposed snail which had been in the bottle floated out. In consequence she became ill. She sued the manufacturers. It was held that a manufacturer of food or drink who put out his products in the form in which they were to be used by the ultimate consumer, with no reasonable possibility of their intermediate inspection, owed a duty of care, independent of contract, to the ultimate consumer to take reasonable care for the ultimate consumer's health in supplying the product.

Donoughmore Committee. See MINISTERS' POWERS, COMMITTEE ON.

Donovan, Terence Norbert, Baron (1898–1971). Served in the Civil Service 1920–32 though called to the Bar in 1924, and sat in parliament 1945–50. He became a judge of the King's Bench Division in 1950, Lord Justice of Appeal in 1960, and a Lord of Appeal in Ordinary in 1963. He was chairman of the Royal Commission on Trade Unions and Employers' Associations, 1965–68.

Doom (Anglo-Saxon dom, judgment). A judicial sentence or judgment. The Anglo-Saxon laws were generally called dooms. In the ancient law of Scotland the term was used in both civil and criminal cases, and falsing of dooms was a name for appealing to a higher court. Hence domesmen, deemster, dempster.

Doomsmen. Among mediaeval Germanic peoples, legal disputes were determined by a court of all the free men of the district acting in accordance with custom. These or a representative group of them were the doomsmen. Charlemagne introduced groups of permanent doomsmen (scabini) in each county, appointed for life by peripatetic royal envoys. In Norman England the suitors of the local communal courts were its doomsmen and, under the presidency of the sheriff or bailiff of the hundred, found its judgments. Similarly in the feudal courts the lord's steward held the courts with doomsmen.

Dormant funds. Funds in court which have not been dealt with for 15 years.

Dormant partners. Persons who are not known and do not appear as partners of a firm but who participate in the profits and are deemed partners at least in relation to third parties.

Dorotheus (sixth century A.D.). Byzantine jurist, law teacher at Berytus, and one of the commission which compiled the *Digest* and the second *Code* and, along with Tribonian and Theophilus, writer of the *Institutes*. Fragments of his *Index*, a commentary on the *Digest*, probably based on his lectures on the *Digest* given by him at Berytus, are preserved in the *Basilica*.

Douaren, Francois (Duarenus) (1509–59). French humanist jurist, professor at Bourges and rival of Cujas, author of *Pro libertate ecclesiae gallicae* (1551), *Commentarius in consuetudines feudorum* (1558), and of a Commentary on the *Digest*.

Double jeopardy. To be prosecuted twice for substantially the same offence. The general principle is that it is not permitted to try a man twice for the same offence, nor can a person be convicted of different crimes arising out of the same conduct unless the crimes are by definition of significantly different kinds, e.g., murder and robbery. The fifth amendment to the U.S. Constitution provides that no one shall be 'twice put in jeopardy'. But a member of a profession or of the armed forces may by committing crime also infringe the code of discipline of the profession or force and be liable to be dealt with under its disciplinary code as well as by law.

Douglas, Charles, 3rd Duke of Queensberry (1698–1778). Took part in public life under George III, being made a Privy Councillor and Keeper of the Great Seal of Scotland, 1761, and Lord Justice General of Scotland, 1763–78.

Douglas, William, 3rd Earl and 1st Duke of Queensberry (1637–95). Took part in public affairs and was appointed Lord Justice-General of Scotland, 1680–82. He was also Lord High Treasurer of Scotland, 1682–86, during which time he held the chief power in Scotland, and an extraordinary Lord of Session, 1681–86. He supported repression of the Covenanters and was King's Commissioner to James VII's first Scottish Parliament, but was later deprived of his offices. He was again made an Extraordinary Lord of Session in 1693–95. He built Drumlanrig Castle.

Douglas, William Orville (1898–). Admitted to the New York Bar in 1926, taught law at Columbia, 1927–28, and then became a professor of law at Yale, 1928–34, and then a member of the Securities and Exchange Commission. There he initiated a vigorous policy of reform, and continued it as chairman (1936–39). In 1939 he was made an Associate Justice of the U.S. Supreme Court, where he showed consistently liberal attitudes. He retired in 1976, after the longest tenure in the Court's history. He also wrote extensively on travel and legal matters.

Douglas cause. A litigation which aroused enormous interest in Scotland in the eighteenth century. Archibald, 3rd Marquis and 1st Duke of Douglas, owner of vast estates, died without issue in 1761. His sister, Lady Jane Douglas, born in 1698, in 1746 married Colonel John Steuart, but kept the marriage a secret from her brother, went abroad, and, allegedly, in Paris in 1748 gave birth to twin sons. She returned to Scotland in 1752 and in 1753 the younger twin died; Lady Jane died in poverty in 1753. In 1754 the Duke of Douglas settled his estates on the Duke of Hamilton as his heir male, but revoked this in 1760, and in 1761, 10 days before his death, named Archibald Douglas Steuart, Lady Jane's elder twin son, as his heir. Archibald was served as heir and the Duke of Hamilton raised actions against him founding on certain deeds of entail, but they failed. In 1762 the Duke of Hamilton and others sought to have Archibald's service as heir set aside on the ground that he was not Lady Jane's son, but a supposititious child. In 1764 Steuart (now Sir John Steuart of Grandtully) died, having made a solemn declaration that Archibald was his son by Lady Jane. Much evidence was taken in France by both sides. In 1766, amid intense popular excitement in Scotland, argument was heard and long Memorials prepared and considered.

In 1767 the full Court of Session, by eight to seven, decided in favour of the Duke of Hamilton;

this provoked riots in Edinburgh, the judges on the majority side having their house windows broken; Andrew Stewart, the Duke of Hamilton's tutor, fought a duel with Thurlow. An appeal to the Lords was at once marked.

In 1768 the Douglas Cause reached the House of Lords which, without a division, reversed the Court of Session, though five peers protested against the decision. The decisive speeches were by the only law lords present, Lord Chancellor Camden and Lord Mansfield. This result provoked wild joy in Scotland; in Edinburgh the judges' windows were smashed, the Hamilton apartments in Holyroodhouse plundered, and troops had to restore order. The Cause delighted society in London; everyone enjoyed the revelations of mad and bad life in high places. Dr. Johnson and Boswell argued about it, and Boswell published a pamphlet, *Dorando*, and poetry about it.

Archibald Douglas, the successful defender, was twice married, was created Lord Douglas of Douglas in 1790 and died in 1827. Though he had eight sons all were childless, and through his eldest daughter the Douglas estates descended to the present Earls of Home.

Steuart, *The Douglas Cause*.

Doujat, Jean (1609–88). French canonist, author of many works including *Specimen juris ecclesiastici apud Gallos recepti* (1671), *Histoire du droit canonique* (1677), *Historia juris civilis Romanorum* (1678), and other works.

Dowager. A widow endowed or in enjoyment of her dower (q.v.). The term is used mainly in connection with the widows of kings and nobles.

Dower. In mediaeval English law, the right of a wife on her husband's death, to a third of the land of which he was seised for her life, of which she could not be deprived by any alienation made by him but only in certain defined and limited ways. It probably replaced older forms of dower, by gift of the husband, or made at the church door, and dower secured to the widow by common law or by custom. The rules later developed that a jointure (q.v.) would bar dower, and dower was not allowed out of a trust. After 1833 a husband could deprive his wife of dower, and it arose only where he died intestate. Dower disappeared in 1925.

A similar right was recognized in Scots law under the name of terce (q.v.) until 1964.

See also Dowry.

Downes, William (1751–1826). Called to the Irish Bar in 1776 and became a judge of the King's Bench in Ireland in 1792 and Chief Justice in 1803. He resigned in 1822 and was given a peerage as Baron Downes.

Downing Chair of Law. The senior law chair in the University of Cambridge, founded by Sir George Downing (?1684–1749) who founded Downing College and provided by its charter of 1800 for Professors of Law (Downing Professorship of the Laws of England) and of Medicine. The chair has been held by Edward Christian (1800–23, acting from 1788), Thomas Starkie (1823–49) Andrew Amos (1849–60), William Lloyd Birkbeck (1860–88), F. W. Maitland (1888–1906), Courtney Stanhope Kenny (1907–19), Harold Dexter Hazeltine (1919–42), Emlyn Capel Stewart Wade (1945–62), Sir William Ivor Jennings (1962–65), Richard Meredith Jackson (1960–70), Stanley Alexander de Smith (1970–74) and Gareth Hywel Jones (1975–).

Dowry (or *maritagium*). Property given to a woman on marriage by her family and which, until recent times, accordingly accrued to her husband. In earlier societies it has served not only to consolidate friendship between the two families, but to compensate the husband and his kin for the expense incurred in payment of the bride-wealth (q.v.). In Europe dowry has frequently been an important factor in marriages, enhancing the wealth and power of rising families and even affecting state frontiers. Contrast Dower.

Doyne, Robert (1651–1733). Studied law in England and Ireland and became Chief Baron of Exchequer in Ireland in 1695 and Chief Justice of the Common Pleas in Ireland in 1703 but was superseded in 1714.

Draco (seventh century B.C.). An Athenian lawgiver who in 621 B.C. prepared a code of law with fixed penalties and stated rules of procedure. Draco seems to have reduced to writing the customary law of his time, particularly procedure, and may have incorporated decisions of magistrates. It may have been the first comprehensive code of Athenian law. The purpose was to allay plebeian discontent and limit the ability of aristocratic magistrates to alter or apply the law in a way favourable to the interest of their class, and this was largely achieved. It was also necessary to provide courts to deal with persons committing homicide, to limit blood feuds, and because persons dealing with a murderer were regarded as polluted. The penalties provided in his code were severe and Draconian became a by-word for severity. On the ground of their severity Solon (q.v.) modified or repealed all Draco's laws except those relating to homicide. A constitution attributed to Draco in Aristotle's *Constitution of Athens* is not genuine.

Draft. An order drawn by one person on another for the payment of money, and including bills of exchange and cheques. Also a first version of a deed, document, etc. prepared as a basis for discussion and consideration, and for amendment and revision from which a fair copy will later be engrossed or copied for execution.

Draftsman. A person who drafts or prepares deeds, bills, and other legal documents. A barrister on the Chancery side who drafted Chancery pleadings was formerly called an equity draftsman. The draftsmen of Government Bills are the Parliamentary counsel to the Treasury (q.v.) or, if affecting Scotland, the Lord Advocate's Legal Secretaries and Parliamentary Draftsmen.

Drago, Luis Maria (1859–1921). Argentinian statesman. He enunciated the Drago doctrine (q.v.), represented Argentina at the Hague Peace Conference of 1907, was a member of the North Atlantic Coast Fisheries arbitration tribunal, and was invited to draft the statute of the Permanent Court of International Justice but died before he had done this.

Drago doctrine. A restatement of the Calvo doctrine (q.v.) made by an Argentine foreign minister called Drago (q.v.) in 1902 asserting that indebtedness by an American state could not authorize armed intervention by European nations to exact payment. The U.S. government assented to the Drago version at the Second Hague Peace Conference (1907) but, though opposed to European intervention, reserved for itself the right to intervene with armed force in any Latin American state where conditions appeared to threaten U.S. interests.

Draw. To draw is to draft or frame a document or bill. To draw a bill of exchange is to write and sign it; the person who does so is the drawer and the person to whom it is directed the drawee and, if and after he accepts, the acceptor.

Drawing and quartering. Part of the penalty formerly prescribed in England for treason. The full penalty included that the condemned man be dragged to the place of execution, hanged by the neck but not to the point of death, drawn or disembowelled while still alive and his entrails burned before him, decapitated, and his body divided into quarters. This penalty was inflicted on the Welsh prince David (1283) and on the Scottish Sir William Wallace (1305), and on Despard and others in 1803 for conspiring to assassinate George III. Latterly the whole process was not in fact carried through. The sentence was last passed in 1867 but not carried out.

Dred Scott case. See *Scott* v. *Sanford.*

Drenches, Drenges or Drengs. Said to be tenants in Northumbria holding by a tenure, partly military and partly servile, which ante-dated the Norman Conquest.

Dress, legal. In Western European countries a distinctive dress for judges and lawyers in court has long been known.

Throughout Western Europe an organized legal profession with ranks and professional discipline came into being in the fourteenth and fifteenth centuries. Clergy disappeared from the ordinary courts about that time and the influence of ecclesiastical dress was slight.

In England judicial robes seem to have been green in the fourteenth century, but other colours were used in the next century. There was associated especially with judges an *armelausa*, a loose mantle fastened on the right shoulder, originally part of the dress of nobles in France and originating in the *lorum* worn in the later Roman and Byzantine Empire. In the fifteenth century judges' robes assumed the general shape they have possessed ever since, and scarlet displaced green as the universal colour for their full dress. They also wore a skull-cap and a coif. Chief justices wore a scarlet tabard or sleeved *supertunica*, with an *armelausa* on important occasions and a hood and shoulder-piece or cape, all lined and trimmed with miniver. Inferior judges seem to have worn the same kind of dress of a mustard colour, though the *supertunica* was always worn and the sleeveless tabard never.

In the sixteenth century the hood developed and black square caps were worn. From the late sixteenth century the chief justices and chief baron wore the Collar of SS. (q.v.). By the seventeenth century the ordinary uniform of judges consisted of the *supertunica* and shoulder piece, both usually black or violet, with a white coif and a black skull-cap and soft square cap.

In 1635 a decree was made by all the judges, which is the foundation of all later dress. During term judges were to wear black or violet gowns with a hood of the same colour and a mantle such that the end of the hood hung down from under the mantle behind. The head-dress was a lawn coif, a velvet skull-cap, and above these a soft square cap. In summer the robes were faced with taffeta but in winter faced with miniver. Scarlet dress was worn on holy days and special occasions. On circuit judges were to wear scarlet. The gown was a closed *supertunica* with close sleeves.

Wigs came in after the Restoration as an article of ordinary dress and the coif and skull-cap became smaller until they were given up in the eighteenth century. At first wigs were of natural colour, but white and grey powdered wigs were introduced in the early eighteenth century. Between 1720 and 1760 the judge's wig became a badge of office as judges maintained a fashion which had passed out of general use. About 1770 many took to wearing smaller wigs for ordinary occasions. White bands at the collar became invariable for judges, barristers, and clergymen.

King's or Queen's counsel from their first creation in the early seventeenth century wore an open black silk gown of the kind affected by men of importance generally in Elizabethan and Jacobean times. They adhered to large wigs in the mid-eighteenth century when barristers took to the short wig. Barristers from 1500 wore the ordinary Tudor lay gown without sleeves. By the eighteenth century they wore what is still the style, an open gown of black stuff with bell sleeves drawn up to the elbow with buttons and vertical cords.

In Scotland it was laid down in 1609 that Lords of Session were to wear purple cloth robes faced with crimson velvet or satin. Judges of the High Court of Justiciary were to wear red robes faced with white, the Lord Justice-General's to be lined with false ermine. Clusters of ribbons were attached to the shoulder piece, being originally bows for fastening up the robes. Later they became rosettes. Advocates wore black gowns. Coifs were never worn in Scotland. Full-bottomed wigs were adopted in the eighteenth century but by 1750 had been given up by judges save for ceremonial occasions. Bands were worn down to the mid-eighteenth century but thereafter judges and senior counsel wore long linen falls.

In Ireland legal dress was the same as in England save that Irish serjeants did not have a distinctive dress until 1639.

W. N. Hargreaves-Mawdsley, *History of Legal Dress in Europe.*

Drink. For legal purposes normally means alcoholic beverages or intoxicating liquor. The sale of drink is controlled by the requirements of premises for the sale and consumption of drink being licensed. To be under the influence of drink is an offence in many circumstances, particularly if while driving, attempting to drive, or in charge of a vehicle on the road.

See also DRUNKENNESS.

Droit. In French signifies both law in general and its major divisions, e.g. *droit civil*, *droit maritime*, and a legal right, e.g. *droit de passage*, right of way.

In English law the word is used in the phrase *droits of admiralty*, which are certain rights assigned by the Crown to the Lord High Admiral. The chief of these are ships and goods taken in port in the time of war, flotsam, jetsam, lagan, treasure, deodand, and derelict (qq.v.) within the admiral's jurisdiction, fines, forfeitures and ransoms, sturgeons, whales, and other large fishes. In prize law droits of admiralty are distinguished from droits of the Crown formerly granted to the captors of ships and cargoes captured at sea.

Droit d'aubaine. In mediaeval feudal law a stranger who entered a seigneur's domains and lived there became the latter's serf, and a serf had very limited power to own property against the lord. In France by the fourteenth century it was accepted that a stranger might acquire and possess but not inherit or transmit by will or on intestacy. In 1386 the French king assumed the seigneurial *droit d'aubaine* or right to inherit. In treaties in the seventeenth and eighteenth centuries the right was frequently renounced. Louis XVI in 1787 abolished the right as against subjects of Great Britain without reciprocity. The Constituent Assembly abolished the right in 1790 and it was commonly abolished elsewhere in the early nineteenth century. A variant was the *droit de détraction*, the levy of a tax on a stranger's succession; this also disappeared about the same time.

Droit de seigneur (or *jus primae noctis*—right of the first night). An alleged right of feudal lords in mediaeval Europe to sleep the first night with the bride of any one of his vassals. There is some evidence of such a right in some primitive societies. The only evidence of its existence in Europe is of payments by a vassal in lieu of enforcement of the right, and it is probable that it was merely a kind of tax like the avail or redemption payment in lieu of the lord's right to select a bride for his vassal.

Drugs. Chemical substances which affect the functions of living things. They are extensively used in medical treatment, but some are also used by individuals to stimulate, to relieve pain, to change their moods, to produce oblivion or a feeling of well-being, and for other purposes. The danger of abuse of drugs has produced difficult social, moral, and legal problems. It is established that many drugs can be physically and mentally dangerous, particularly when used improperly or excessively; driving while under the influence of a drug is dangerous and an offence; acts of aggression or suicide may be provoked; and the user may do himself and his descendants permanent damage. It appears that legal control of drug use is necessary. Many kinds of drugs and preparations including them are obtainable subject to restrictions or on medical prescription only. The mere possession of various kinds of drugs is an offence, while trafficking in prohibited drugs, supplying them or using them are all distinct offences.

Drummond, James, 4th Earl and 1st Duke of Perth (1648–1716). Became a member of the Scottish Privy Council in 1678, Lord Justice-General in 1682–84, an extraordinary Lord of

Session in 1682, and Lord Chancellor of Scotland in 1684–88. He turned Catholic to follow James VII and II, but fled at the Revolution, was captured and imprisoned and not allowed to leave Scotland till 1693. He spent the rest of his life in France.

Drummond, James Eric, 16th Earl of Perth (1876–1951). Entered the Foreign Office in 1900, assisted A. J. Balfour at the Peace Conference of Paris in 1919 and was first Secretary-General of the League of Nations 1919–33. Thereafter he was British Ambassador in Rome till 1939. He established the basic principle, later adopted by nearly all international organizations, that the members of the League Secretariat must form an international civil service, owing allegiance to and working for the organization, and should not be representatives of national interests. He became a representative peer from Scotland in the House of Lords in 1941 and was deputy leader of the Liberals in the Lords from 1946.

Drunkenness. To have taken intoxicating drink to an extent which affects conduct or judgment, has various legal consequences. It may be criminal by itself particularly if it renders the taker disorderly, incapable, or a nuisance in public, or if he was driving or attempting to drive a motor vehicle and if the proportion of alcohol in his blood exceeds a fixed limit.

Drunkenness, total or partial, is frequently given as an explanation or excuse for having committed a crime. If a person commits a crime, it is in general no defence that he was intoxicated at the time unless the degree of intoxication is so great as wholly to deprive him of reason and control and to be equivalent to insanity, though it may amount to diminished responsibility (q.v.). In such cases drink has excluded the capacity to form the special intent to commit the crime. It is certainly no defence if he formed the intention and took drink to fortify his resolution, or if drink made him quarrelsome or led him to do something wrong or stupid. Under many statutes it is an offence to be drunk in stated circumstances, e.g. while on duty, even if not totally incapable.

In civil law drunkenness at the time of making a contract may be sufficient to render the contract void if it indicates total absence of genuine contractual intent.

Dualism. A theory of the relation between public international law and the municipal law of any state, which holds, in opposition to monism, that the two are distinct legal orders and that neither has the power to create or alter rules of the other, though the municipal law of a state may provide that international law will be applicable in part within its jurisdiction. This is not deferring to international law but adopting it or applying municipal rules which are identical with those of international law. Dualism tends to be the theory of positivists who deny the validity of law not laid down by a state or who deny the validity of sources of international law apart from the practice of states.

Duarenus. See DOUAREN, F.

Du Bellay, Pierre (sixteenth century). French lawyer who first expounded thoroughly the theory of the divine right of kings. In *De L'Authorite du Roi* (1587) he maintained that authority was created and conferred by God and that the King, as God's lieutenant, was responsible to him alone.

Dubitante (abbreviated dub., doubting). The term used in a law report of a judge who doubts a proposition of law but does not go so far as to repudiate it as bad or wrong, e.g. Held, *dubitante* Blank, J., that . . .

du Cane, Sir Edmund Frederick (1830–1903). Served in the Royal Engineers, 1848–87, attaining the rank of Major-General. In 1851–56 he was employed in organizing convict labour on public works in Western Australia, and in 1863 was appointed director of convict prisons and inspector of military prisons. In 1869 he became chairman of the board of directors of convict prisons. In 1873 he put forward a scheme for the transfer to the government of all local prisons and this was effected in 1877 when he was made chairman of the prison commissioners. In consequence the number of prisons was reduced, the staff co-ordinated into a single disciplined service and useful employment of prisoners developed. He also inaugurated the registration of criminals, and encouraged the use of finger-printing in identification. He retired in 1895.

Duces tecum, subpoena (that you bring with you under penalty). An order served on a person having in his possession some document which it is desired to have put in evidence, to attend and bring the document with him. This the witness must do unless he has an excuse deemed satisfactory by the presiding judge.

Duchy. The estates of a Duke.

Duchy Court of Lancaster (or Court of the Duchy Chamber of Lancaster). A court formerly held by the Chancellor of the Duchy or his Deputy and dealing with equity matters relating to lands held by the Crown in right of the Duchy, even though not in Lancashire, and with matters of the revenue of the Duchy. It was distinct from the Chancery Court of the County Palatine and the Court of Chancery had concurrent jurisdiction. It does not appear to have sat since 1835, but has not been abolished.

Duchy of Cornwall. A royal duchy created in 1337 vested during the life of a sovereign in his eldest son; if that eldest son dies leaving a son the Duchy reverts to the Crown, but if the eldest son dies without leaving a son, the next surviving son of the sovereign succeeds. The duchy property consists of 11 manors and of all mines and minerals under the whole of the 17 manors comprised in the original grant. The officers of the Duchy include the Lord Warden of the Stannaries in Cornwall and Devon, the Keeper of the Privy Seal and others, and the duchy estates are managed under statutory powers.

Duchy of Lancaster. A royal duchy vested in the Crown, comprising the former county palatine of Lancashire and certain lands discontiguous therefrom, including parts of London, is the estates originally the patrimony of the earls and dukes of Lancaster. It became vested in Henry VII and was settled on him and his heirs separate from the Crown. The Duchy is managed by the Chancellor of the Duchy. There is a Court of the Duchy Chamber of Lancaster (q.v.).

Ducke, Arthur (1580–1648). Became a Fellow of All Souls in 1604 and an advocate of Doctors' Commons in 1612. He is said to have held the office of master of requests and was later chancellor of the diocese of London. In 1633 he was made a member of the ecclesiastical commission and in 1645 a master of chancery. He is remembered mainly by his *De usu et auctoritate Juris Civilis Romanorum* (1653, translated 1724).

Ducking Stool. A mode of punishment of scolds, witches, and prostitutes from the early seventeenth to early nineteenth centuries. It consisted of a wooden chair in which the culprit was seated, held by an iron band, which was fixed to one end of a long beam at the edge of a pond or river. Magistrates when sentencing stated the number of duckings the woman was to have. Sometimes the chair had wheels so that it could be wheeled through the streets. Sometimes too the chair was on wheels with long shafts fixed to the axles; it was pushed into the pond and the shafts released so that the chair tipped up backwards.

Dudley, Edmund (*c.* 1462–1510). Minister of Henry VII and president of the King's council, a body which helped to re-establish the payment of feudal dues. He was executed for treason in 1510 and while in the Tower wrote *The Tree of Commonwealth* (1509) urging moderation in the use of royal powers, attacking administrative abuses allowed by law and urging strict performance of their duties by all members of society.

Duel. A prearranged fight between persons using weapons, according to settled rules and as an alternative to judicial decision. The earliest form was the judicial duel or trial by battle (q.v.), which is first reported among the Germanic peoples and became established in Europe in the early Middle Ages. Judicial duel was introduced into England by William the Conqueror in the eleventh century, but was early superseded by the grand assize in civil cases and by indictment at the instance of the Crown in criminal cases. If a person asserted before a judge that another was guilty of crime and the other denied it, the judge ordered them to meet in a duel and fixed the time, place, and weapons. The throwing down of a gauntlet was a symbolic challenge which was accepted by picking it up. This form of trial was open to all free men and only women, ecclesiastics, the sick, and persons under 20 or over 60 could claim exemption. In certain cases persons could appoint professional champions to represent them. Both parties had to find sureties that they would appear. As it was believed that God would not allow the right cause to be defeated, the defeated party was dealt with by law. Trial by battle was abolished in 1819 after the attempt in *Ashford* v. *Thornton* (1818), 1 B. & Ald. 405 to revive a virtually obsolete procedure. In many countries duels were also used to decide other issues. In such cases the procedure was laid down in detail and the fight took place in the presence of the court and judicial and ecclesiastical dignitaries. Thus in Spain in 1085 a duel was fought to decide whether the Latin or the Mozarabic rite should be used in the liturgy at Toledo.

From about the fifteenth century duels of honour were fought arising from real or imagined insults or affronts. They sprang from the customary wearing of swords and were at first no more than attacks or assaults. But it became usual to issue a challenge and for the parties to be accompanied by friends or seconds who later themselves also fought to show their belief in their principals' causes. Many famous duels were fought in the reign of Louis XIV. Duelling did not begin in England till the later sixteenth century, but was common in the eighteenth century. Duelling survived into the nineteenth century; Canning and Castlereagh fought a duel in 1809; the Duke of Wellington and Lord Winchelsea fought in 1829.

Political duels were frequent in France in the nineteenth century and even, rather nominally, in the twentieth century. Student duels were long recognized in German universities, and evidence of wounds has been regarded as badges of courage.

The Nazi and Fascist regimes encouraged duelling. As a matter of law duelling has probably always been criminal and in the nineteenth century duellists were frequently tried and even sentenced to death.

In modern British law duelling is not specifically criminal but a challenge to fight is criminal, and active fighting with weapons would be an affray (q.v.), certainly if in a public place, while killing or wounding in a duel would be treated as murder, or attempted murder, or unlawful wounding.

Due process. The conduct of legal proceedings according to established principles and rules which safeguard the position of the individual charged. The concept of due process is rooted in English common law and expressed in Magna Carta, Art. 39 (1215), whereby the King promised that 'No freemen shall be taken or imprisoned or disseised or exiled or in any way destroyed ... except by the legal judgment of his peers or by the law of the land'. This was later interpreted to require trial by jury. In later statutes and books the phrase was used with or in lieu of the phrase, the law of the land. The concept was adopted by the U.S. Constitution in the fifth amendment (1791) which was extended to state action by the fourteenth amendment (1867) and has been said by the Supreme Court to mean the same as the law of the land.

Due process clauses. The fifth (1791) and fourteenth (1867) amendments to the U.S. Constitution contain provisions denying the federal and state governments respectively any right to deprive persons of 'life, liberty or property, without due process of law'. The original purpose and scope of these provisions was to ensure fair legal procedure before life, liberty, or property was taken away or interfered with. These provisions have given rise to much litigation and have been invoked in circumstances not envisaged when they were adopted, a situation worsened by the development of industrialization, the growth of big business, and the increase in interstate commerce. For a time the provisions were interpreted as restrictive of state regulation of individual enterprises, but latterly interference with individual rights and regulation for social and economic reasons have been held justifiable. By invoking the fourteenth amendment the Supreme Court has exercised substantial control over the administration of criminal justice in state courts and rather lesser control over civil and administrative justice. Due process has been held to require the provision of counsel, freedom from being questioned without an attorney present, non-admissibility of illegally obtained evidence. It has been given wide interpretation in connection with civil liberties, in particular making applicable to the states the guarantees of liberty in the Bill of Rights.

Duff, Sir Lyman Poore (1865–1955). Called to the Ontario Bar in 1893 he became a judge of the British Columbia Supreme Court in 1904 and of the Supreme Court of Canada in 1906. He was chief justice 1933–44, and Canada's most distinguished judge, having great intellectual power and capacity for legal analysis. His contributions were most notable in constitutional law. He did much extra-judicial work and acted as administrator of Canada in the absence of Governors-General. He was also a member of the Judicial Committee of the Privy Council, 1919–46.

***Duff Development Company* v. *Government of Kelantan*,** [1924] A.C. 797. The Sultan of Kelantan was sovereign ruler of an independent but British protected state, the government of which granted certain mining rights to the appellants. The Sultan applied to have set aside an award by an arbitrator under a clause in the contract, and was held not to have waived his immunity from being made subject to the jurisdiction of the High Court. It was also held that a certificate by the Secretary of State for the Colonies that Kelantan was an independent state and the Sultan its ruler was conclusive.

Duffy, Sir Frank Gavan (1852–1936). Called to the Victorian Bar in 1874, lectured on law and edited law reports. In 1913 he became a justice of the High Court of Australia and in 1931–35 was Chief Justice. He showed himself opposed to an extended interpretation of the powers of the Commonwealth. He became a member of the Privy Council in 1932.

Dugdale, Sir William (1605–86). Legal antiquary. He early attracted the notice of Spelman and Hatton and they secured for him a post in the College of Heralds where he eventually (1677) became Garter King-of-Arms. His chief works are the *Antiquities of Warwickshire* (1656) and *Monasticon Anglicanum* (1655–61–73) a collection of records of mediaeval English religious houses, but of greater legal importance are the *Origines Juridiciales* (1666, 1671, and 1680) which gives much information about the origins of English law and legal institutions and about holders of legal offices and the legal profession. Abridgments of it, continued to date, appeared in 1685 and 1737. In 1675 he published *The Baronage of England* and the *Chronica Series*, a chronological table of the chancellors, judges, law officers, and King's serjeants from the Conquest to his own time.

Duguit, Léon (1859–1928). French jurist, professor of constitutional law at Bordeaux, author of *Transformations Générales du Droit privé depuis Le Code Napoléon* (1912) *Transformations du Droit Public*

(1913), *Traité du Droit constitutionnel* (1921–25), and other works. He rejected intuitive and metaphysical theories of justice, and such concepts as sovereignty and right as well as the division between public and private law. His idea of law was of something based on social solidarity through division of labour, a spontaneous product of circumstances. The interdependence of groups and classes created a spontaneous self-regulation of human behaviour. For individual rights he substituted the protection of social functions; the role of the law was part of the observed fact of social solidarity, which determined the content of the rule of law. Law existed to ensure fulfilment of the needs of this social solidarity. But social solidarity is a vague and uncertain guiding principle and Duguit did not enumerate the specific dictates of his principle, and in many areas of law different rules would seem to be equally consistent with it.

Duke, Henry Edward, Baron Merrivale (1855–1939). worked as a journalist and was called to the Bar in 1885. He was successful, took silk in 1899 and became one of the best first instance advocates of his time. He sat in Parliament for two periods and acquired a reputation there too; in 1916 he became Chief Secretary for Ireland, but his task was impossible and he resigned in 1918. Thereafter he was appointed a Lord Justice of Appeal and in 1919 President of the P.D.A. Division, in which office he was dignified and efficient. He was made a peer in 1925 and retired in 1933.

Duke. The title attaching to a member of one of the higher orders of nobility in European countries. In the later Roman empire a *dux* had military functions only but later acquired civil functions also, and as great civil and military officers *duces* were known in the Frankish empire. By about the tenth century the number of dukes became fairly settled and title and office became hereditary.

Important duchies were Bavaria, Franconia, Lotharingia, Saxony, and Swabia, but other duchies were later recognized. There were also Teutonic and Lombard dukedoms, the dukes of which were chosen for military prowess and leadership, but this kind of dukedom also became hereditary. The Norman Kings of England were Dukes of Normandy, and the title was introduced into England by Edward III in 1337 when he created his son, the Black Prince, Duke of Cornwall (whence the Duchy of Cornwall has always been held by the sovereign's eldest son, from his birth or his parent's accession). In Scotland Robert III in 1398 created his sons Duke of Rothesay (a title now attaching to the sovereign's eldest son) and Duke of Albany. Since then it has been the highest rank in the British Peerage, and Dukes rank next to the blood royal, the archbishops, and the Lord Chancellor. A duke normally bears also titles of lower ranks in the peerage (thus the Duke of Marlborough is also Marquis of Blandford and Earl of Sunderland) and during his life his eldest son and eldest grandson bear two of the lower titles. His younger sons have the courtesy title of Lord and his daughters that of Lady. Courtesy titles do not prevent the bearer sitting in the Commons. In France the title exists but only as a mark of distinguished family, descended either from old feudal aristocracy or from ducal creations of the nineteenth century, many being creations of Napoleon I or III.

Duke of Exeter's daughter. A fifteenth century instrument of torture resembling the rack, said to be so-called from the Duke of Exeter, a minister of Henry VI, who introduced it into England.

Dum bene se gesserit (so long as he shall have conducted himself well). A term applicable to those offices from which the holder cannot be removed at the will of the Crown, such as judicial offices, as contrasted with offices held *durante bene placito* (q.v.). See also *Quamdiu se bene gesserit.*

Dum casta vixerit (so long as she shall have lived chaste). A clause, frequently called the *dum casta* clause, often incorporated in separation agreements providing that an allowance shall be payable only so long as the wife lives chaste. Similar clauses were *dum sola*, so long as single, *dum sola et casta*, so long as single and chaste, and *dum vidua*, so long as a widow.

Dumont, Jean, Baron de Carlscroon (1667–1720). French diplomat and scholar, author of *Corps universel diplomatique du droit des gens* (1726–31) later continued by Rousset, and works of military history including a life of Marlborough.

Dumont, Pierre Etienne Louis (1759–1829). Swiss cleric, secretary to Bentham, whom he assisted to prepare his works for publication and whose works he translated into French. During the Revolution he was in France and worked for Mirabeau. In 1809 Alexander I made him a member of the commission charged with codifying Russian law. In 1814 he returned to Geneva and was elected to the Representative Council for which he drafted rules of procedure. He also drafted a proposed penal code on utilitarian lines.

Dumoulin, Charles (Molinaeus) (1500–66). A famous French humanist jurist and writer on law, sometimes called the French Papinian, or the prince of jurisconsults, author of *Commentarii ad Codicem*, works on the feudal law (*De Feudis*, 1539), canon law (*Commentarius in regulas juris pontificii* (1548)), the French monarchy, and other subjects, particularly on the customary law of Paris. He did much to bring together a large body of French customs and usages and to systematize and extract the principles of it, notably in his *Le Grand Contumier du royaume de France et des Gaules* (1567), and thereby he largely prepared the way for Pothier and the French Civil Code.

Dumoulin, Jean (Joannes Molinaeus) (1525–75). Belgian canonist, editor of Ivo of Chartres' *Decretum* (1562).

***Duncan* v. *Jones*,** [1936] 1 K.B. 218. Duncan, in face of a police officer's warning, began to address a public meeting in the street. Held that she was rightly convicted of obstructing the respondent in the execution of his duty, because he had reasonably apprehended a breach of the peace and had been therefore under a duty to prevent it, as he had done by warning her to desist.

Dundas, Henry, Viscount Melville (1742–1811). Son of the first Lord President Robert Dundas (q.v.), passed advocate of the Scottish Bar in 1763 and at once distinguished himself in the courts and in the General Assembly of the Church of Scotland. In 1766 he became Solicitor-General for Scotland and in 1774 an M.P. Thereafter he lived mainly in London. In 1775 he became Lord Advocate and began to attain equal importance in Parliament. He was also Dean of Faculty, 1775–85. In 1782 he became, in addition to Lord Advocate, Treasurer of the Navy, Keeper of the Scottish Signet for life, and was given the patronage of all places in Scotland; this made him the most powerful man in Scotland, but he lost the Treasurership in the following year and was replaced as Lord Advocate by Erskine. After the short interval of the Fox-North coalition he became Treasurer of the Navy again (1783–1800). For years thereafter Dundas was Pitt's chief subordinate and held the offices of Home Secretary, Secretary for War (1794–1801), and First Lord of the Admiralty (1804–5). He took a prominent part in the proceedings against Warren Hastings. In 1801 Pitt's ministry fell and in 1802 he became Viscount Melville and Baron Dunira. When Pitt resumed power in 1804 he became First Lord of the Admiralty. In 1805, however, a Report cast doubts on financial transactions while he had been Treasurer of the Navy, and he was impeached in 1806 (the last use of impeachment (q.v.)) but, by a majority, acquitted. For 30 years he was the effective ruler of Scotland, but he did not subordinate the public interest to private solicitations.

Dundas, Robert (1685–1753). Son of Sir Robert Dundas, a Lord of Session (Lord Arniston, 1689–1720) and grandson of Sir James Dundas, a Lord of Session (Lord Arniston, 1662–71), passed advocate in 1708. Talent and interest secured his advance and in 1717 he became Solicitor-General for Scotland, Lord Advocate in 1720, Dean of the Faculty of Advocates in 1721, and an M.P. in 1722. He lost office in 1725 but in 1737 became a Lord of Session and in 1748 Lord President, an office he held till his death. He was alike distinguished for legal knowledge and eloquence and for integrity and honour. Of his sons one, Robert (1713–87) (q.v.), also became Lord President, while another, Henry (q.v.), became Viscount Melville.

Dundas, Robert (1713–87). Son of Robert Dundas, Lord President Arniston (q.v.), studied at Utrecht, and was called to the Scottish Bar in 1738, and soon showed that he had inherited the family legal abilities. In 1742 by political favour he became Solicitor-General for Scotland but lost office in 1746, though in the same year he was elected Dean of the Faculty of Advocates. In 1754 he was elected to Parliament and was appointed Lord Advocate, and as such was largely responsible for settling the highlands after the uprising of 1745; he held that office until in 1760 he was appointed Lord President of the Court of Session. He was dignified and diligent as a judge and head of the court, anxious to improve the administration of justice, and, though not a scholar, a very sound lawyer. As such he presided at the trial of the Douglas cause (q.v.) and became very unpopular in consequence. One of his sons, Robert (1758–1819) (q.v.) became Chief Baron of Exchequer in Scotland, 1801–19.

Dundas, Robert (1758–1819). Son of the second Lord President Dundas, became an advocate of the Scottish Bar in 1779, and successively Solicitor-General (1784) and Lord Advocate (1789–1801). As such he tried to reform the Scottish burghs and prosecuted the Friends of the People. He was also Dean of Faculty, 1796–1801. In 1801 he became Chief Baron of Exchequer in Scotland. Though he lacked the great abilities of his father and grandfather, he was well-liked and highly esteemed as a person and as a judge. His son, William Pitt Dundas, became Deputy Clerk Register, 1856–80.

Dunedin, Viscount. See GRAHAM MURRAY, A.

Dunfermline, Baron. See ABERCROMBY, JAMES.

Dunfermline, Earl of. See SETON, ALEXANDER.

***Dunlop Pneumatic Tyre Co. Ltd.* v. *New Garage Motor Co. Ltd.*,** [1915] A.C. 79. Manufacturers supplied tyres to dealers under an agreement whereby, if the dealers contravened certain conditions, they were to pay £5 per tyre 'by way of liquidated damages and not as penalty'. It was held that this was enforceable as a genuine pre-estimate of loss sustained and not unenforceable as a penalty *in terrorem* of the party in breach.

Dunning, John, Lord Ashburton (1731–1783). English lawyer who defended Wilkes (q.v.) on charges of seditious libel. He became Solicitor-General in 1768 but resigned in 1770. He moved the famous motion 'that the influence of the Crown has increased, is increasing and ought to be diminished' in 1780. He became a peer in 1782.

Duoviri. A Roman magistracy of two. Such pairs existed for many purposes and included *duoviri perduellionis*, two judges selected by the chief magistrate who tried cases of treason, while the chief magistrates in colonies and *municipia* were often called *duoviri juri dicundo*.

du Parcq, Sir Herbert, Baron (1880–1949). Called to the Bar in 1906, he took silk in 1926, and became a judge of the K.B. Division in 1932. In 1938 he was advanced to the Court of Appeal and in 1946 to the House of Lords. He was a very sound and respected judge.

Duplicity. The name for a defect in an older pleading in that it, or a portion of it, contained more than one claim, charge, or defence. The rule against duplicity has now been much relaxed.

Durand (or Duranti or Durantis), Guillaume (Durandus). Known as Speculator (1237–96), governor of Romagna and Ancona and later Bishop of Mende, a canonist, author of *Speculum Juris* (or *Judiciale*) (1271) a vast compilation concerned with practice and procedure in civil, criminal, and canon law, which went through many editions and had lasting influence in the courts and schools. It synthesized Roman and Canon law. He also wrote a *Rationale divinorum officiorum*, *Commentaria in Gratiani Decretum*, on Christian ritual, a *Repertorium sive Breviarium juris canonici*, and a *Commentarius in sacrosanctum Lugdunense concilium*, inserted in the *Sextus* of the canon law. His nephew of the same name (?–1330) was also a bishop and a canonist.

Durand, Guillaume (Durandus of St. Pourçain) (?–1332). Known as Doctor Resolutissimus, French scholastic theologian and bishop, author of *inter alia* a *De Jurisdictione Ecclesiastica et de Legibus* (1506).

Durante bene placito (during the good pleasure, of the Crown). The term applied to offices such as Lord Chancellor the holders of which hold at the pleasure of the Crown only, as contrasted with those who hold *dum bene se gesserit* (q.v.).

Duress. Force applied or threatened to an individual to coerce, and actually coercing, him to act in a particular way. It may be actual or threatened physical harm, but commonly takes the form of threat of unlawful or at least unwarranted pressure, such as of dismissal from employment. It includes threats to wife, parent, or child. An act done under duress is in general legally ineffective. Any contract, gift, or renunciation of rights obtained by duress is voidable. In criminal law duress is a defence, because threats of death or personal violence such as to overbear the ordinary powers of human resistance should be a justification for acts which would otherwise be criminal. In Scots law the same principle is called force and fear and harks back to the Roman law *actio quod metus causa*.

See also COERCION; THREATS.

Durham, Chancery Court of County Palatine of. The claim of the bishops of Durham to exercise palatine powers and jurisdiction was confirmed by the Norman kings and survived largely unimpaired till 1836 when the *jura regalia* were transferred to the Crown, save for the Chancery Court of the County Palatine. The court exercised within the county palatine the jurisdiction of the old High Court of Chancery and an extensive statutory jurisdiction, concurrent with the corresponding jurisdiction of the High Court. Appeal lay to the Court of Appeal and House of Lords. The judge was the Chancellor of the County Palatine of Durham, who was appointed by the Crown. There were an Attorney-General, a Solicitor-General, and other officers of the court. The court was merged with the High Court in 1971.

Durham, County Palatine of. The Bishop of Durham formerly exercised palatine powers in the county, but these were transferred to the Crown by statute in 1836 and vested as a separate franchise and royalty in the Crown.

The County Palatine formerly had a separate Court of Pleas which was abolished in 1873 and its jurisdiction transferred to the High Court. The Durham Court of Chancery had jurisdiction co-extensive with the Chancery Division of the High Court, but was merged with the High Court in 1971.

Durie, Lord. See GIBSON, ALEXANDER.

Durkheim, Emile (1858–1917). French sociologist, author of *The Division of Social Work* (1897), *The Rules of Sociological Method* (1895), *Suicide* (1897), *Socialism, its Definition and Beginnings* (1897), *The Elementary Forms of the Religious Life* (1912), and other works. He examined the relations of the individual and the State, of society to government, and the conflicting theories of positive and customary law, but abandoned attempts at total explanation of society. The institutions and mentality of a society is embodied in law, language, public opinion, and other ways; human institutions are organic and alive.

Dutch auction. A kind of auction in which the property is offered at a stated price and offered at successively lower prices until a bid is made.

Dutch law. See NETHERLANDS LAW.

Duty. (1) A duty, from *debitum*, *devoir*, is a legal disadvantage, that which is owed or due to another, and should be satisfied. Legal duties are accordingly prescriptive formulations of conduct which by law should be done or observed, normally for the future but, as in Acts of Attainder or retrospective legislation, duties may be formulated with reference to past conduct, i.e. what should have been done. A statement of legal duty is a statement of conduct to which persons ought to conform, but to which they may, or may not, in fact conform. The statement of duties and the imposition of sanctions, civil or criminal, for failure to implement them, is in fact the principal mode by which law regulates conduct. Frequently duties are not formulated expressly but only implied, e.g. by a statement that X is punishable or failure to do Y will be visited with an award of damages.

Duties may be non-legal (moral, natural, religious, or other) or legal, and some kinds of injunctions may state both non-legal and legal duties, e.g. not to cheat, but not all moral or natural duties are also legal, e.g. not to be ungrateful or uncharitable. Narrow questions may arise in relation to, e.g. spitefulness or infringement of privacy. Conversely many legal duties arise in circumstances where there is no moral duty or go beyond moral duty. But the existence of moral duty has been a powerful factor in influencing systems of law to recognize a legal duty, e.g. to avoid injury to one who obviously might be injured by failure to take care, or to seek to rescue persons in trouble.

Duties may be stated positively, e.g. the duty to perform a contract, or negatively, e.g. the duty to avoid injuring another, positive formulation implying approval and negative implying disapproval. Breach of duty is always disapproved by law and normally visited by civil or criminal penalties.

Again legal duties may be so formulated as to require positive action for their implement, e.g. to perform lawful contracts, or so formulated as to require abstention from conduct, e.g. not to kill, injure, or steal, or so as to require efforts to avoid certain results, e.g. to take reasonable care for the safety of one's employees.

The content of legal duties varies greatly, depending on time, place, and circumstances and the particular legal system, and the trend of legal development has been to extend both by legislation and by the course of decisions the catalogue of legal duties. Thus, the duty of care towards persons voluntarily intervening to try to effect a rescue has been recognized. It is impossible in any developed legal system to catalogue all the legal duties recognized, not least because whether a duty exists in a particular case may depend entirely on the precise circumstances of the case.

As has been mentioned, an express or implied prescription by law of a duty is normally supported by prescription of a sanction, civil or criminal, for breach of the duty. But sanction is not an essential of a duty. A duty is a duty even if there be no sanction, or it be not enforced. In an ideal society persons would implement duties without need for sanctions. There are many instances of sanctionless duty. Sanction is accordingly not the test of the existence or validity of legal duty. The test is whether a court would recognize the duty.

The existence of a duty in a particular case is also important in that it may transform action or inaction from innocent to culpable. Inaction becomes culpable if, in the circumstances, there was a legal duty to act.

The existence or recognition of a duty frequently implies the recognition of a right in some other person to have the duty performed in relation to him or to recover damages for non-performance, i.e. it creates a beneficiary of the duty. This person may not be identifiable until or unless the duty is breached, e.g. the pedestrian injured by breach of the duty to drive carefully. In other cases the duty is owed to the State or to the whole community, and only the State, normally acting through the criminal law, can complain of non-implement. Accordingly rights (q.v.) and duties are commonly said to be correlative, in that the presence of one in one party to an obligation implies the presence of the other in the other party. Austin, however, maintained that there were also some absolute duties to which no rights corresponded. In this category he included duties not to persons but, e.g. to God or to animals, duties owed to persons indefinitely or to the community, self-regarding duties, and duties owed to the sovereign. The first are not legal duties, the second are in truth a mass of duties to particular individuals, the third are probably impossible, there probably being no legal duties to oneself, and the last depend on Austin's view that a right-duty relation cannot exist unless enforceable by a superior, and there is no superior to the sovereign; though these are of a special kind it is not always impossible to regard the sovereign or state as conceding rights to the subject.

It has long been recognized that the term duty comprised several distinct ideas, which would frequently be better expressed by different words, such as liability and disability, particularly when contrasted with various synonyms for the word 'right'. This has been worked out in relation to the analysis of the concept of right. See RIGHTS.

(2) A tax or import levied on commodities, transactions, or estates rather than on persons, e.g. estate duty, stamp duty, excise duty.

Duval, Claude (1643–1670). A famous English highwayman, remembered for the daring of his robberies and his gallantry to women. He was captured in London and hanged at Tyburn.

Dwight, Theodore William (1822–92). Admitted to the New York Bar in 1845 and taught law at Hamilton College till 1858, then becoming professor of municipal law at Columbia Law School, and head of the Law School till 1891. His method was expository and he resigned when the case-method of instruction was introduced in 1891. He was interested in prison reform and worked for years in the N.Y. State Prison Association. He also served as a member of the Commission of Appeals set up to assist the N.Y. Court of Appeals in 1873–75, and was an active member of the movement to expel the Tweed Ring from power in New York.

Dyarchy. A term applied by Mommsen to the Roman principate (q.v.), a period in which he held that sovereignty was shared between the princes and the senate. The term has also been given to a system of government, promoted as a constitutional reform in India by Montagu and Chelmsford and introduced by the Government of India Act, 1919. It marked the introduction of democracy into the executive of the British administration of India by dividing the provincial executives into authoritarian and popularly responsible sections composed respectively of councillors appointed by the Crown and ministers appointed by the governor and responsible to the provincial legislative councils. Subjects of administration were classified as reserved subjects, appropriated to the councillors, and transferred subjects. The system ended when full provincial autonomy was granted in 1935.

Dyer, Sir James (1512–82). Chief justice of the Court of Common Pleas from 1559, initiated law reports of the modern pattern in place of the Yearbooks by publishing three volumes of cases decided in the King's (or Queen's) Bench and Common Pleas covering 1513–82. They were in law French and did not appear in English till 1794.

Dying declaration. A verbal or written statement made by a dying person, which although not made on oath or in the presence of the accused, is admissible in evidence on an indictment for murder or manslaughter of that person, provided the person making it had a belief, without hope of recovery, that he was about to die shortly.

Dymoke. The English family holding the office of king's champion, the functions of which were to ride into Westminster Hall at the coronation banquet and challenge all comers who sought to impugn the King's title to the throne. The ceremony is recorded at the coronation of Richard II, but having been performed by Dymokes many times was allowed to lapse after the coronation of George IV in 1820. At the coronation of Edward VII a Dymoke bore the standard of England.

E

Earl. The third grade of the British peerage (after duke and marquess), but the oldest, a rank going back to Saxon times. The title is Scandinavian in origin, the title *jarl* being modified to *earl*. The earliest creation by charter was that of Essex in 1140. Under Norman and Angevin kings earls were rulers of counties and presided in the county courts, the dignity becoming hereditary in time. Palatine earldoms, notably Chester and Durham, had certain royal privileges and the earls were petty sovereigns. In the fourteenth century some earldoms were created without reference to a county, but in many cases grants of lands were made for the support of a new earl. The earldom of Chester is granted to the Prince of Wales on his creation. In Scotland the seven ancient provinces were each under a *mormaer*, later called *jarl* by the Norsemen and *comes* or earl under Norman influence. The earldom of Sutherland probably dates from the thirteenth century. The earldom of Carrick is held by the sovereign's eldest son.

In modern times the title has frequently been conferred for outstanding public services, e.g. Earl Cairns, Earl of Selborne, Earl of Halsbury, without necessary connection with the holding of land. The title carries the right to sit in the House of Lords and passes to the grantee's heirs under the limitation in the grant until, on their failure, it becomes extinct. The earl's wife is a Countess, and his eldest son by courtesy bears the next highest title attaching to the family, usually a viscounty or barony. Other sons are by courtesy designated 'the Honourable' and daughters 'Lady'.

Earl Marshal. One of the great officers of state in England, head of the College of Arms; he appoints the kings-of-arms, heralds and pursuivants, arranges state ceremonials, and attends the sovereign at the opening and closing of Parliament. The office dates from the twelfth century and since 1672 has been held by the Howards, Dukes of Norfolk. Under the Normans the Earl Marshal and the Lord High Constable were the chief officers of the feudal army and jointly held the Court of Chivalry (q.v.).

See also MARSHAL.

Earldorman. In Anglo-Saxon England a high official who administered justice, commanded the military forces, and executed law in a shire, or latterly group of shires. Under the Danish kings, the title was superseded by that of earl. In each shire his deputy was the shire reeve or sheriff who gradually acquired all the civil administrative functions, leaving to the earl the command of the local forces. The office of ealdorman tended to become hereditary and disappeared about the eleventh century.

Earmark. A mark of identity or ownership, derived from the practice of marking cattle on the ear. Property is earmarked when distinguished from other property of the same kind. Money, in general, cannot be earmarked, unless it is paid to a person for a special purpose and kept in a separate fund or separately invested by him, in which case it can be recovered by the person entitled thereto.

Earnest (or earnest money or arles, (Roman *arrha*)). A coin or other nominal sum of money sometimes given by buyer to seller as a token that the parties are bound or in earnest about the bargain. It is legally unnecessary.

Easements and profits. An easement in English law is a right attaching to one piece of land (the dominant tenement) entitling the owner thereof to exercise some right over adjacent land in other ownership (the servient tenement) though not to take any part thereof or take any of the natural produce thereof, or to prevent the owner of the other land from utilizing his land in some particular manner. There must accordingly be two tenements of land owned by different persons and there must be a right capable of forming the subject matter of a grant affecting the servient tenement and benefiting the dominant tenement.

Easements may be positive or negative, according as the right is to do something or to prevent another from doing something. They may be created by statute or by grant, express or implied, or by prescription, based on long use or presumed grant. They may be extinguished by statute, release, express or implied, or unity of ownership of dominant and servient tenements.

Among the common recognized easements are right of way, right to light, watercourse, support of buildings but there is no confined list of easements and many miscellaneous cases have been recognized, including easement to do what would otherwise be a nuisance, to erect signboards or name-plates on another's premises, and the like. Interference with an easement is a private nuisance and actionable.

Profits à prendre are rights to enter the land of another person and to take therefrom some profit of the soil or a portion of the soil itself. By analogy with easements the lands are called the dominant and servient tenements. That which may be taken may include animals, fish and fowl, turf and peat, heather and litter, stone, sand and shingle and other things capable of separate ownership.

The right may be created by statute or grant or prescription, and extinguished by statute, release, unity of ownership, or exhaustion of the subject-matter. Profits à prendre are rights of a possessory nature and the owner may bring an action of trespass for their infringement.

Easements and proftis à prendre together constitute servitudes (q.v.).

East, Sir Edward Hyde (1764–1847). Was called to the Bar in 1786, entered Parliament in 1792 and in 1813 became Chief Justice at Calcutta, retiring in 1822. He then re-entered Parliament and sat in the judicial committee of the Privy Council. He is remembered for his King's Bench Reports (1785–1809, and 1800–12), the first to be published at the end of each term, and for his *Pleas of the Crown* (2 vols., 1803).

East India Company. A company incorporated by royal charter in 1600 for trading purposes, with a monopoly for 15 years of the East India trade. It established factories, i.e. trading depots, in India and gradually assumed control of substantial areas of India. In 1772 financial difficulties led to the Regulating Act of 1773 and increased control under Pitt's India Act of 1784. A dual authority of government and company was established. The company gradually lost its political power and became more involved in the tea and opium trade with China. After the Indian Mutiny its governing powers were transferred to the Crown and the company expired when its charter lapsed in 1873.

Easter. A Church feast commemorating the resurrection of Christ. Easter Day is the first Sunday after the full moon occurring on or next after the vernal equinox (March 21) and if that fall on a Sunday, the next Sunday. It can accordingly fall on any day between March 22 and April 25. The Easter Act, 1928, to make a fixed date for Easter, has never been brought into force.

It gives rise to the Easter vacation and Easter sittings of the courts (which runs from Easter to Whit Sunday).

Easter Rising. An insurrection in Ireland commencing on Easter Monday, 24 April 1916, against the British government there. The British had arrested Sir Roger Casement in County Kerry when attempting to smuggle arms in from Germany. The leader of the Irish Volunteers cancelled the rising but some of the rebels, the Citizen Army, some of the Irish Volunteers, the Irish Republican Brotherhood, and the Sinn Fein party went ahead. Dublin G.P.O. and certain other places were seized; fighting continued for about a week until the rebels surrendered. Fourteen of the leaders were hanged and others imprisoned and this did much to antagonize Irish feeling, so that thereafter Irish Home Rule, passed into law in 1914 but suspended by reason of the World War, no longer would satisfy Irish feelings.

Eaton, Theophilus (?1590–1658). Co-founder and governor 1639–58 of New Haven colony in America. He helped to establish a Puritan church and government and to prepare a new law code in 1655.

Eavesdrop (or stillicide). An easement (q.v.) or servitude imposing on the servient tenement the burden of receiving rainwater from the eaves of a building.

Ecclesia. An assembly of the citizens in ancient Greece, particularly at Athens. Its powers were first defined by Solon in 594 B.C. and enlarged by Cleisthenes in 508 B.C. From about 450 B.C. the ecclesia comprised all male Athenians of citizen birth on both sides and, along with the Boule, which initiated business, exercised complete sovereignty. It heard appeals, took a part in the election of magistrates and financial matters. It met about 40 times a year and decided all State business by voting by show of hands. Later it proved unstable and attempts were made to limit its membership and to introduce rules of procedure. In the fourth century there was payment for attendance. Under the Roman Empire the powers of all ecclesiai vanished. Among the early Christians the word was adopted for a church.

Ecclesiastical Causes Reserved, Court of. A court with original and appellate jurisdiction, for both the provinces of Canterbury and York of the Church of England, established in 1963, consisting of two persons who hold or have held high judicial office and three persons who are or have been diocesan bishops. It has original jurisdiction to hear cases charging ecclesiastical offences involving doctrine, ritual or ceremonial committed by a priest or deacon who held preferment or by an archbishop or bishop who was a diocesan or suffragan.

It has appellate jurisdiction from consistory courts in cases of faculty involving doctrine, ritual or ceremonial. On petition to the Queen a commission may be appointed of five persons, three Lords of Appeal and two Lords Spiritual of Parliament to review a finding of this court.

Ecclesiastical Commissioners. A body established in England in 1835, comprising the archbishops and bishops, Lord Chancellor, Chancellor of the Exchequer and a secretary of state, to manage the property and finances of the Church of England. It did much to adapt that church to the growing industrial society, building and restoring many churches and reallocating revenue between richer and poorer parishes. In 1948 the body was, along with Queen Anne's Bounty, remodelled as the Church Commissioners (q.v.) to manage the estates, trusts and revenues of the Church.

Ecclesiastical corporations. Corporate bodies created for the promotion of religion and the securing of the rights of the Church, the members being entirely spiritual persons. They include corporations sole, viz. bishops, parsons, and vicars, and corporations aggregate, viz. deans and chapters.

Ecclesiastical courts. Courts established by churches to deal with disputes between clerical authorities and with disputes involving spiritual matters affecting either clerics or laity. Such courts have been found in Christian denominations, among Jews, Muslims, and other religious faiths. In the Middle Ages in Europe the courts of the Roman Catholic Church had wide jurisdiction, extending to many temporal matters, and often rivalled the secular courts.

As time has gone on and societies have generally become more secular, the trend has been for the jurisdiction of ecclesiastical courts to be limited to church property and matters of ecclesiastical discipline.

Ecclesiastical courts (England). In accordance with canon law, as the Christian Church was organized in England a hierarchy of ecclesiastical courts was developed. By the twelfth century they claimed a wide jurisdiction, which was at no time completely conceded by the State, and in time most of the jurisdiction has been transferred to the State. It comprised criminal jurisdiction in all cases where a cleric was accused, cases of offence against religion, and a corrective jurisdiction over both clergy and laity for the good of their souls; civil jurisdiction over all questions of marriage, divorce, and legitimacy, over grants of probate, and to supervise executors and administrators; and administrative jurisdiction over all matters ecclesiastical in nature such as ordination, consecration, ecclesiastical persons and property, land held in frankalmoign and the like.

In respect of the criminal jurisdiction the Church claimed in the twelfth century that 'criminous clerks' should be exempt from all secular jurisdiction and subject only to ecclesiastical jurisdiction. Henry II proposed that the cleric should be charged before a temporal court, plead his clergy, be tried by an ecclesiastical court and if found guilty and degraded from holy orders, be returned to the temporal court for punishment. Becket objected to a cleric being accused before a temporal court, and to any royal officer being present at his trial to take him back if degraded and these objections were supported by the canon law, and also to a cleric suffering both degradation and punishment by a temporal court. After Becket's murder the temporal courts maintained their claim to bring the accused before them but abandoned the claim to punish the degraded clerk. This gave rise to the doctrine of Benefit of Clergy (q.v.).

The ecclesiastical courts also had a wide control over clergy and laity in respect of religious belief and morals, entitling them to punish heresy (by death by burning) and immoral conduct, including adultery, defamation, witchcraft, blasphemy, drunkenness, and usury. The criminal jurisdiction of the ecclesiastical courts was abolished in 1641 but restored in 1661, but since then many of these kinds of conduct have been dealt with by the temporal courts, while others are now longer deemed immoral.

From the twelfth century the ecclesiastical courts had exclusive jurisdiction in matrimonial causes. In each diocese the bishop, acting through his chancellor, official principal, or surrogate sitting in the consistory court, could decide issues of the validity of a marriage and grant separation, but not divorce which required an Act of Parliament. Appeal lay to the Archbishop's court, the Court of Arches, or the Chancery Court of York. In 1857 divorce by judicial process was introduced and the jurisdiction was transferred to the newly-created Divorce Court (q.v.). The ecclesiastical courts also had jurisdiction over grants of probate or of administration, in respect of personal estate, and over the conduct of executors and administrators. In 1857 this jurisdiction was transferred to a new Court of Probate (q.v.). In 1875 these courts were amalgamated with the Court of Admiralty and formed into the Probate, Divorce and Admiralty (now Family) Division of the High Court.

The administrative jurisdiction of the ecclesiastical courts over doctrine and ritual, ordination, divine service, ecclesiastical property and the like has largely remained in their hands.

As reorganized in 1963 the ecclesiastical judicial system is that in each diocese there is a court of the bishop called the consistory court (or at Canterbury the commissary court) presided over by the chancellor (q.v.) of the diocese, having original jurisdiction only in respect of offences by priests or deacons, the grant of faculties, and certain other matters.

In each of the provinces of Canterbury and York there is a court of the archbishop, namely the Arches Court of Canterbury and the Chancery Court of York, presided over in the case of Canterbury by the Dean of the Arches, two clerics and two laymen and in the case of York by the Auditor (who is the same person as the Dean of the Arches) two clerics and two laymen, in each case with appellate jurisdiction from consistory courts.

In each province a commission may be appointed by convocation with original jurisdiction for the trial of bishops.

In the provinces together the Court of Ecclesiastical Causes Reserved has original jurisdiction in proceedings for offences involving matters of doctrine, ritual or ceremonial, and appellate jurisdiction from consistory courts in faculty cases involving doctrine, ritual or ceremonial and a commission may be appointed with original jurisdiction in the trial of archbishops, while commissions may be appointed by the Queen to review the findings of the Court of Ecclesiastical Causes Reserved, or of a commission trying an archbishop. and a final appellate jurisdiction is reserved to the Queen in Council.

Ecclesiastical courts (Scotland). Before the Reformation ecclesiastical jurisdiction was vested in the bishop of each diocese and exercised by 'officials' and 'commissaries', but judicial functions were also exercised by archdeacons and rural deans, or by commissioners appointed *ad hoc* to hear particular cases. Appeal lay from subordinate commissaries to the officials, and after 1472 to the archbishop. Appeal lay also to Rome, which sometimes entrusted the case to judges delegate, an *ad hoc* tribunal of three. During the Great Schism the Conservator of the General Council or Provincial Synod had the right to hear appeals which would otherwise have gone to the Pope. The jurisdiction extended to matrimonial cases, executry disputes, and the enforcement of obligations fortified by oaths. From the thirteenth century testaments had to be confirmed by the bishop.

At the Reformation, in 1560, statute abrogated

the papal authority and thereafter some bishops exercised consistorial jurisdiction while Protestant superintendents and kirk sessions both assumed the former episcopal jurisdiction. The Court of Session also began to exercise jurisdiction in matters formerly ecclesiastical. In 1564 the Privy Council created the Commissary Court of Edinburgh with jurisdiction over actions relating to defamation, wills, teinds, and matrimony. Judicial divorce *a vinculo* had been admitted since 1560. Local jurisdiction was exercised by local commissaries with appeal to the Commissary Court of Edinburgh and then to the Court of Session.

The Commissary Court of Edinburgh fell into disuse in the eighteenth century. Its appellate jurisdiction and divorce jurisdiction were transferred to the Court of Session in 1830, though it survived as a local court till 1836 and was not abolished till 1876. In 1823 the inferior local commissaries were ended, each sheriffdom becoming a local commissariot with the sheriff as judge. But the official dealing with the testaments of persons dying outside Scotland is still called the Commissary Clerk of Edinburgh, and executry business is still called commissary business.

Even after commissary courts were established presbyteries and kirk sessions continued to grant decrees of adherence as essential preliminaries to actions of divorce for desertion, but in the course of time their jurisdiction came to be limited to internal discipline of the church and moral correction of members, and even the latter is now in disuse.

The Reformed Church developed four grades of court, the kirk session, the presbytery, the synod, and the General Assembly, in all of which both clergymen and laymen sit. Since 1592 the courts of the Church of Scotland have had a statutory jurisdiction in spiritual and ecclesiastical matters, separate from the civil and criminal courts, and are equally supreme in their own sphere. Other churches, including the Episcopal Church in Scotland and the Roman Catholic Church have no legal jurisdiction but only a jurisdiction founded on voluntary submission by their members.

Ecclesiastical law. In a general sense, covers all laws relating to a church, whether derived from state law, divine law, the law of nature and reason, or the rules of independent societies. In the context of English law it means the law of the Church of England as administered by the Ecclesiastical courts, the constitution of the church, its property, clergy, benefices, and services. In England, prior to the Reformation, ecclesiastical courts administered the general canon law of the Western Christian Church. At the Reformation this became English ecclesiastical law. In England the Church of England is established by law and its law is a part of the law of the land. The Church in Wales was disestablished in 1920; the Church of Ireland was separated from the Church of England and disestablished in 1871. Other Churches, though recognized by law, are fundamentally voluntary associations and their codes of law are binding on their members only by virtue of voluntary submission. The ecclesiastical law of the Church of England consists of Acts of Parliament affecting the Church and of Measures passed by the Church Assembly, and later by the General Synod, and of principles adopted from Roman law and pre-Reformation canon law and incorporated into the statute or common law of England. The accepted legal doctrine is that the Church of England is a continuous body from its establishment in Saxon times, and before the Reformation its constitution was determined by the general canon law. Before the Reformation the Scottish Church had some measure of independence of papal authority and formulated canons of its own. The Reformation in 1560 abolished papal authority and in 1567 Presbyterianism was established. The Crown was not Head of the Church. Since 1592 the Church Courts have had a statutory jurisdiction in matters ecclesiastical and spiritual independent of the civil and criminal courts, and in their sphere as supreme as the civil and criminal courts. Other denominations are tolerated, but their ecclesiastical law is binding on their members only by submission thereto. In Scotland the Church of Scotland, which is presbyterian in government, was finally established in 1690, is free from interference by civil authority and has power to legislate, and adjudicate finally, on all matters of doctrine, worship, government and discipline in the Church.

In some other countries, e.g. U.S.A., all churches are tolerated but none has any of the privileges of establishment.

Ecclesiastical law (English). Prior to the Reformation the Church in England was a branch of the universal Catholic Church of the West which had its supreme legislature, executive and judiciary in Rome. The canon law of the Roman Church was binding on all members of the Church (which at that time comprised everyone) and provincial constitutions were only supplementary to the general canon law. Ecclesiastical jurisdiction covered discipline of the clergy, offences against religion such as heresy, offences against morals, and matrimonial and testamentary matters.

The Reformation effected great changes. By statute in 1534 Henry VIII became Supreme Governor on Earth of the Church of England. Provision was made for revision of the existing canons though in fact this was never done, existing canons continuing in force so long as not conflicting with God's law or the King's, but later statutes subordinated Church to State. The study of the canon law declined.

In the time of Elizabeth more definitely dogmatic changes were made. Under the Act of Uniformity the ecclesiastical law became part of the law of England, and it has steadily shrunken in application with the growth of dissent, agnosticism, humanism and the like.

In 1857 the jurisdiction of the ecclesiastical courts in probate and matrimonial matters was transferred to civil courts which, however, for long carried forward the law developed in the pre-1857 courts, though that law has gradually moved away from its origins.

Today ecclesiastical law is concerned only with the constitution and property of the Church of England, the provincial, diocesan, and parochial system, the clergy, services, doctrine, ritual and practice. It is made by the General Synod of the Church and adjudicated on in the ecclesiastical courts (q.v.). It does not affect members of other religious denominations. Such churches, and their internal relationships, are regulated by general principles of contract and trust.

R. Phillimore, *Ecclesiastical Law*; H. W. Cripps, *Law of Church and Clergy*.

Ecclesiastical law (Scottish). Prior to the Reformation the church in Scotland was a branch of the universal Catholic Church of the West, and the canon law was binding on all members of the Church but supplementary canons were made by the Scottish clergy in provincial councils and remained in force till after the Reformation. Church courts and officials, as elsewhere in Christendom, exercised wide jurisdiction, particularly in matrimonial and testamentary affairs.

At the Reformation in 1560 the authority and jurisdiction of the Pope were abolished. Presbyterian church government was not finally established till 1690. Secular courts dealt with matrimonial causes, wills and executors from shortly after the Reformation but to a large extent utilized concepts derived from the earlier law and relied on the pre-Reformation canon law. In 1609 bishops were given jurisdiction in wills and divorces, but this did not long survive. The established church, the Church of Scotland, and its courts had great moral influence, but splits in the Church in the eighteenth and nineteenth centuries took many persons outwith the establishment.

The numerous separate presbyterian churches, the Episcopal Church of Scotland, and other churches regulate their affairs on the bases of contract and trust. The Roman Catholic Church remains subject to the general ecclesiastical law of that church.

Ecclesiastical Titles Act, 1851. Was passed to counter Pope Pius IX's plan to reestablish a Catholic diocesan hierarchy in England and prohibited the assumption of ecclesiastical titles in use by the Church of England. It was, however, circumvented by the use of other titles, e.g. Archbishop of Westminster, and the Act was repealed in 1871.

Echt, Bachoff. See BACHOFF.

Eck, Cornelius van (1662–1732). Dutch jurist, author of *De Septem Legibus Pandectarum* (1682), *Principia juris civilis secundum ordinem Digestorum* (1689), and many other works on civil and canon law.

Eck, Jacobus (1691–1757). German jurist, author of *De genuinis fontibus jurisprudentiae forensis* (1715) and other works.

Ecloga. A piece of legislation published in Greek by Leo III the Isaurian, Emperor of the East, *c.* A.D. 740, described as a selection from the works of Justinian altered in the direction of humanity, but in fact containing much new law and intended to make the law more agreeable to Christianity. It is the most important Byzantine work after that of Justinian and had strong influence on later Byzantine legislation and also on the development of law in Slav countries. Notable features are the enhancement of the rights of women and children at the expense of the father's rights, the restriction of capital punishment and the substitution of mutilation, equality of punishment for all classes and the attempt to eliminate bribery and corruption.

École Nationale d'Administration. A French school set up in 1945 for training professional administrators from which most of the higher civil servants are recruited. It was established partly as a counter-weight to the older École Libre des Sciences Politiques (1871) which was after 1945 taken over by the State and renamed the Institut d'Études Politiques, and has acquired a high reputation.

Economic and Social Committee of European Communities. A body of 144 persons representing various categories of trades and professions, with membership roughly proportioned to the population of the member-states, with advisory functions only. It may be consulted by the Council of Ministers or the Commission when thought appropriate; in some cases consultation is obligatory. Its views, though not binding, are influential as being the views of groups vitally affected by Community policy.

Economic and Social Council of U.N. One of the principal organs of the U.N. This body is composed of 27 members, nine elected annually by the General Assembly for three years. There are no permanent members but permanent members of the Security Council are in fact normally elected to the Economic and Social Council. Its functions are to be responsible for the economic and social activities of the U.N., to make reports, recommendation, etc. on international economic, social, cultural, educational, health, and related matters, to promote respect for and observance of human rights and fundamental freedoms, to call international conferences on matters within its competence, to negotiate agreements with the specialized agencies defining the terms of their relationship with the U.N., to co-ordinate the activities of the specialized agencies, to perform services for members of the U.N. and the specialized agencies on request, and to consult with non-governmental agencies concerned with matters with which it deals. The Council functions through comissions, sub-commissions and committees. It supervises the U.N. Children's Fund (UNICEF).

Economic Commission for Europe. A commission established by the U.N. in 1947 for exchanging information and statistical data on coal, electricity, and transport in Western and Eastern Europe. As the division between the Western and Eastern blocs deepened it failed as an instrument for co-operation.

Economic theories of law. Economic theories of law are mainly associated with the socio-economic doctrines of Marx and Engels. They generally reflect the views that economic facts are independent of and antecedent to law and law is a superstructure on the economic system. It is moreover an instrument used by the economic rulers to keep the masses in subjection and after the proletarian dictatorship is established will be the instrument used to crush and eliminate the capitalist minority. Law along with the State forms an apparatus of compulsion in capitalist societies but when the communist or classless society arrives there will be no domination or inequality and its instruments, law and the state, will wither away. It is certainly true that many rules of law have been shaped by economic factors and reflect the interests of economically dominant groups, but the experience of no country has justified dispensing with law, but rather the contrary. Everywhere the more socialist a society becomes the more law is an instrument of domination and compulsion, and of oppression, and the less it does to protect the liberties of individuals. The State is everything and the individual only a serf.

Edge, Sir John (1841–1926). Was called to the Irish Bar in 1864 and the English Bar in 1866. He became Chief Justice of the Western Provinces, India, 1886–98, a member of the Council of India, 1898–1908, and a member of the Judicial Committee of the Privy Council in 1909.

Edict. Higher Roman magistrates (praetors, aediles, quastors and censors, the government in provinces, and latterly the emperors) had the right to issue proclamations notifying their orders in their several spheres. As most of these had jurisdiction, edicts were also issued to state legal rules which they intended to apply in the course of their administration. Originally these were oral but were later posted up in the Forum. Particularly important was the edict of the *praetor urbanus* who, by granting actions and remedies in cases not permitted or not provided for by the strict municipal law, created in time a great body of new law (*ius praetorium* or *honorarium*) supplementing and modifying the *ius civile*.

The edict of the *praetor peregrinus* was probably more free since he dealt with foreigners to whom the *ius civile* did not apply, and it may have adopted features from the *ius gentium* and foreign law, some of which may have been adopted by the urban praetor later. From 67 B.C. magistrates were strictly bound by their edicts. The edict of the *curule aediles* dealt with fairs and markets, buildings, and police regulation generally.

Originally edicts were limited to the issuing magistrate's year of office and ceased to be binding thereafter, but the practice developed of successors repeating and confirming their predecessor's rules, making only necessary amendments, so that the rules were for the most part of continuing validity (*edictum tralaticium*). Accordingly in time the various edicts became a fairly complete and settled body of magistral law, virtually a codification, capable moreover of being revised annually. They were of an essentially procedural character; the praetor gave rights of action, remedies and defences.

Under the Principate the innovatory vigour of praetorian developments slackened and in *c.* A.D. 130 Salvius Julianus (q.v.), on the instructions of the Emperor Hadrian, consolidated and revised the praetorian edicts and this, the *Edictum Perpetuum*, confirmed by *senatus consultum*, became permanent and the right of the praetors to modify it was abolished. The Edict was the subject of commentaries by many later jurists and its arrangement and headings influenced the compilers of the Digest. (See also ROMAN LAW.)

O. Lenel, *Edictum Perpetuum*.

Edict nautae, caupones, stabularii. The provision of the praetorian edict (Inst. IV,5,3; Dig. 44,5,6; 47,5), whereby shipmasters, innkeepers, and livery-stable keepers were made strictly answerable for the safety of property travellers brought into the ship, inn, or stable, and liable if they were lost or damaged, even without proof of fault on the part of the shipmaster, innkeeper, or stable-keeper himself. This principle has been adopted into Scots law.

Edict of Atalaric. An edict issued by Atalaric, King of the Ostrogoths (*c.* A.D. 535) aimed at wrongful disseisin, concubinage, and other wrongs.

Edicta. Under the Roman principate the Emperor was the supreme magistrate and accordingly had the power all magistrates had to issue edicts, and his powers were wider-ranging than any inferior magistrate. Accordingly *edicta* issued by the emperor were under the principate a major source of law. Imperial edicts could deal with very varied matters and some were of major importance; the *constitutio Antoniniana* (q.v.) of A.D. 212 was an edict. Edicts might be promulgated orally but were normally issued in writing. It seems, though this is doubtful for the earlier principate, that, unlike the *edicta* of ordinary magistrates, imperial *edicta* continued effective after the death of and notwithstanding the execration of the memory of the Emperor.

Edicta Justiniani. Thirteen novels of Justinian appended to the Greek novel-collection, and printed as an Appendix in Scholl and Kroll's edition of the Novels.

Edictal citation. In Scots law, citation in an action in the Scottish courts of a person out of Scotland or whose whereabouts are unknown but who is subject to the jurisdiction. It is effected by delivering a copy of the summons to the Keeper of Edictal Citations.

Edictum. See EDICT.

Edictum perpetuum. The part of the body of edicts (q.v.) issued by praetors or other Roman magistrates which were to be in force for the whole period of the magistrate's tenure of office. This part included the jurisdictional edicts stating the principles which the magistrates proposed to follow in granting remedies and recognizing defences.

Edictum Theodorici. A code in 154 sections, promulgated possibly by Theodoric the Great, King of the Ostrogoths in Italy about A.D. 500, and intended to apply both to barbarians and Roman subjects, or possibly in name of Theoderic II, a western Gothic ruler who reigned A.D. 453–66 by Magnus of Narbo, prefect of the Gauls, about A.D. 458–59, and applicable to both barbarian and Roman subjects. The sections contain statements of legal rules drawn from the *Codices Gregorianus Hermogenianus* and *Theodosianus*, the later Novels, Paul's *Sententiae* and possibly some other works. It was not intended to supersede other sources, but to make enforcement of the existing law more certain.

Edictum tralaticium. The part of the edict (q.v.) issued by a praetor or other Roman magistrate stating the principles which he intended to observe in the exercise of his jurisdiction, which was carried forward from the edicts of his predecessors, possibly modified and amended, as distinct from any parts of his edict which were wholly new. If a new part of the edict, such as the grant of a new remedy or the allowance of a new defence, proved useful and valuable, it could be repeated and be absorbed into the *edictum tralaticium* for the future; if not, it would be dropped.

Edictus (or *Liber Edictus*). The book recording the laws and edicts of the Lombard kings from the *Edictum Rothari* (A.D. 641) onwards. It had considerable influence on later Italian law and with later work, the *Liber Papiensis*, the *Walcausina* and the *Expositio* (all eleventh century), became the *Liber Lombardae*.

Edmund-Davies, Herbert Edmund Davies, Lord (1906–). Called in 1929, he became Q.C. in 1943, judge of the High Court, 1958, Lord Justice of Appeal, 1966, and Lord of Appeal, 1974.

Education, legal. See LEGAL EDUCATION.

Edward I (1239–1307). King of England 1272–1307, sometimes called the English Justinian, was early invested with lands in Wales, Gascony, and the earldom of Chester. In 1264–65 he took a prominent part in the fighting between the King (Henry III) and the barons and played a leading role in the defeat of the barons at Evesham. In 1268 he went on a crusade and was in Sicily when in 1272 he learned of his father's death. In his first years as King he strove to achieve administrative reforms and enacted much important legislation, such as the statute of Gloucester (1278) which gave better remedies for the termors, the statute *de Viris Religiosis* (1279) which introduced the law of mortmain, the statutes *De Donis* in 1285 and *Quia Emptores* (qq.v.) in 1290. The statute of mortmain of 1279 forbade the further grant of lands to ecclesiastical corporations without royal consent and *Circumspecte Agatis* of 1285, limiting the church courts to clerical business only, produced clerical opposition.

Over the years 1267–83 he conquered most of Wales and reduced the Prince of Wales to a petty chieftain; he set up an English system of government in the areas ceded and in 1284 issued the statute of Wales laying down a scheme for the government of the principality. In 1286–89 he was in Gascony and affairs in England got into a state of confusion so that when he returned in 1289 he had to dismiss many ministers and judges for corruption. In 1290 he expelled the Jews from England.

In 1291–92 he was asked to, and undertook to, resolve the disputed succession to the throne of Scotland but his attitude provoked Scottish resistance which continued until, after his death, the English conceded Scottish independence by the Treaty of Northampton (1328). At the same time he became involved in war with France. In 1295 Edward realized that he could not continue without general support from his subjects and he convoked a parliament comprising representatives from the three estates, later called the Model Parliament, as the model for all later assemblies. He had also to face clerical and baronial opposition; the barons forced on him a confirmation of the charters of liberties, the *Confirmatio Cartorum*, with additional clauses safeguarding people from arbitrary taxation.

The last years of his reign were taken up with ceaseless war against the Scots, and his ambition to conquer them was frustrated.

Edward was a vigorous ruler and achieved much, in administration and legislation, by conquering Wales and defeating his barons and ecclesiastics and the French king. His Model Parliament is a very significant constitutional development. In his reign many institutions acquired the forms they maintained for centuries, parliament, convocations of the clergy, exchequer, chancery, learned judges and skilled lawyers, king's bench and common bench. From the early years of his reign come the earliest Year Books (q.v.), and his reign is one of the periods in which statutory developments have been most important. But his restless ambition led him into profitless wars. He would have been greater if he had concentrated on internal reforms.

W. Stubbs, *Constitutional History*; T. Tout, *Edward I*; T. Plucknett, *Legislation of Edward I.*

Edward the Confessor, laws of. A collection of laws, in Latin, dating from about 1130–35, which became popular and was sometimes regarded as authoritative but is now regarded as thoroughly unreliable. Its account of William the Conqueror getting evidence of the law of Edward the Confessor, with a view to confirming it, from 12 Anglo-Saxon jurors, is a fabrication.

Edward the Elder, Laws of. Two series of laws or of ordinances made by Edward the Elder, King of England 901–25.

Effective cause (or *causa causans*). Is distinguished from a merely prerequisite causal factor or *causa sine qua non*. Thus a person's presence at the time and place where an accident befalls him, is a prerequisite or *causa sine qua non* of his being injured but the effective cause is his own, or the other party's, negligence.

Effective occupation. In international law, a concept corresponding generally to possession in municipal law. It is important in relation to extension of state sovereignty to new land, territory abandoned or not possessed by a political community recognizable as a state. It involves the intention and will of a state to act as sovereign and some effective and continuous display of state authority.

Effectiveness. An important principle in relation to the jurisdiction of a court. In general, a court can only claim jurisdiction over a person or subject-matter of dispute if it can make any order it pronounces effective, by coercion of the individual, seizure of the subject-matter, or otherwise. Also more generally a principle followed by courts generally in that they seek to make law actually regulate the relations of parties and their rights and not be merely statements of pious aspirations. Thus the International Court has shown determination to secure a full degree of effectiveness of international law in general and in particular of the obligations undertaken by parties to treaties, declining to have obligations negatived by strained interpretation and holding that the maximum of effectiveness should be given to an instrument creating an obligation consistently with the intention of the parties.

Egerton, Sir Thomas, Baron Ellesmere and Viscount Brackley (?1540–1617). He was called to the Bar in 1572, acquired a large chancery practice, and became Solicitor-General in 1581 and Attorney-General in 1592. In 1594 he became Master of the Rolls (till 1603) and in 1596 Lord Keeper. He enjoyed the Queen's confidence and was a member of all her commissions. In 1603 he became Baron Ellesmere and Lord Chancellor, and gave judgment in *Calvin's case* (1607). In the struggle with Coke over the position of the courts of common law and of equity Ellesmere maintained the supremacy of his own court. He became Viscount Brackley in 1616 and retired in the next year shortly before his death. He left various manuscript works of which one, *The Privileges of Prerogative of the High Court of Chancery*, was published in 1641. Two others published in 1651 and 1710 are doubtful.

Ehrlich, Eugen (1862–1922). Austrian jurist, Professor at Czernowitz. His major interest was in expounding the social basis of law: law was derived largely from social facts and depends not on state authority but on social compulsion, its real source being in the activities of society itself. Emphasis had to be put on exploring the real foundations of legal rules, their scope and meaning, rather than on analytical jurisprudence. His major works were *Grundlegung der Soziologie des Rechts*, trans. as *Fundamental Principles of the Sociology of Law*, and *Juristic Logic*. He was a leader among sociological jurists and is generally regarded as the founder of the sociology of law.

Eichhorn, Karl Friedrich (1786–1854). German jurist, professor at Frankfurt, Berlin, and Gottingen, a great authority on German constitutional law and author of *Deutsche Staats-und-Rechtsgeschichte* (1808–23), and other works. With Savigny he published the *Zeitschrift für geschichtliche Rechtswissenschaft* (1815–43).

Eire. See IRELAND, REPUBLIC OF; IRISH LAWS.

Eisenhart, Johann Frederic (1720–83). German jurist, author of *Institutiones Juris Germanici privati* (1752) and many other works.

Eisenhower Doctrine (1957). A pronouncement by U.S. President Dwight D. Eisenhower promising military or economic aid to any Middle East country needing help to resist Communist aggression, as part of the post-war U.S. policy of resisting and limiting extension of Soviet influence.

Ejection. In Scots law, an action to remove persons occupying land or buildings without title, and also the executive warrant following on decree in an action of removing against a tenant whose lease has expired. It is also used of the unlawful removal of a possessor from heritage, entitling him to bring an action of ejection to have the dispossessor ejected and himself restored.

Ejectment. In older English law a special form of trespass, *de ejectione firmae*, at the instance of a lessee against anyone who had ejected him from his term of years. Initially it gave him damages only, but from the fifteenth century it enabled him to recover his term also and later this enabled ejectment to be used instead of most of the old forms of real action to try a question of title to the freehold of land. Thus when two persons had to try the title to a piece of land, one of them fictionally leased it to the imaginary John Doe and the other to the imaginary Richard Roe or William Styles. One lessee was said to have ejected the other and the court, in the guise of trying the rights of the lessees, determined the rights of the lessors. The real plaintiff appeared as lessor, and the action was entitled *Doe d. Bloggs* v. *Snooks*, i.e. Doe on the demise by Bloggs. In the seventeenth century the lessee replaced the fictitious Richard Roe or William Styles as defendant, having been fictitiously asked by Roe or Styles to defend his interest. In 1852 John Doe was abolished by statute and in 1875 the whole form of action was abolished. In the U.S. ejectment was a part of colonial law but was early reformed by statute in most states to make it an action for determining title which could be used by any landowner.

Ejudem generis. Of the same kind, a canon of interpretation to the effect that where general words follow an enumeration of particulars the general words are understood as limited to general categories of the same general kind as the particulars. The principle is inapplicable if the particulars do not belong to any general category, or if the general words are followed by a word such as 'whatsoever', emphasizing their complete generality.

Elastic clause. A term applied to Article I, Section 8, of the U.S. Constitution, giving Congress authority 'to make all laws which shall be necessary and proper for carrying into execution' the powers vested in the federal government, because it gives the Congress implied powers which can be stretched to extend to circumstances for which legislative power is not expressly conferred.

Elchies, Lord. See GRANT, PATRICK.

Eldon, Lord. See SCOTT, JOHN.

Election. (1) In public law, the machinery whereby in democratic states persons entitled to vote choose one or more representatives of the electoral district or constituency to represent them in national or local government. The term is also used for the occasion on which the election of representatives is held. The elements of an election, as to each of which there are bodies of rules, are the title to stand as a candidate, the qualifications and disqualifications for voting, the returning officer, the mode of voting, the prevention of corrupt and illegal practices, canvassing of voters, and limitation on expenditure.

In Western democratic states there is generally freedom of candidature, subject to an age qualification, but, in general, only the candidates put forward by the major political parties stand any chance of election. The right to vote is today commonly open to all residents, of either sex, of full age, and sound mind.

The mode of voting is today normally by ballot, i.e. by marking a voting paper and depositing it in a sealed box, later to be opened and the votes counted, but the mode of marking depends on whether the voter may vote for only one candidate, the candidate securing the most votes being elected, (plurality system) or whether the voter may indicate preferences for several candidates in order (alternative or preferential system) or preference for several candidates in order, the returning officer distributing the surplus preferences of candidates who have obtained a sufficient quota of votes to be elected, to other candidates so as to secure that there is representation in proportion to the votes cast for each party within the constituency (proportional representation). Corrupt and illegal practices, such as bribery, treating, undue influence, and personation of a voter are commonly punishable criminally, while limits are placed on expenditure permitted for advertising, printing, and other expenses.

Cognizance of disputed elections was originally vested in the King and council or, if it were alleged that the sheriff was in fault, in the Exchequer. In 1384 and 1404 Parliament itself intervened to investigate disputed returns. In 1604 the Commons vindicated their right to determine contested elections against the attempt of James I to transfer such issues to his Court of Chancery. This was recognized by the courts: *Barnardiston* v. *Soame* (1674), 6 St. Tr. 1092 and (1689), 6 St. Tr. 1119, and by statute. Since 1868 petitions against the validity of an election have been decided by an election court of two judges; their judgment is a report to the Speaker which the House resolves to record in the Journals of the House.

The term is also applied to an occasion when an election takes place, as in the phrase 'the election of 1945'.

(2) An equitable principle in the interpretation of wills, based on the principle that a man may not accept and reject the same instrument, to the effect that where a testator purports to give to A property which belongs to B and to confer benefits on B, B many not accept the benefit unless he is prepared to give effect to the testator's intentions, and must elect whether to accept in whole or reject in whole the will. In Scots law the same principle applies, under the name of 'approbate and reprobate'.

Election agent. A person who must be appointed by each candidate in a parliamentary or local government election, who is responsible for controlling, and making a return of, all expenses incurred on the candidate's behalf.

Election commissioners. Persons being barristers but not being members of Parliament nor holders of an office of profit under the Crown who might be appointed to inquire into allegations that corrupt practices had extensively prevailed at an election. They reported to Parliament and the Attorney-General decided whether or not to prosecute anyone in consequence. The office was abolished in 1969.

Election petition. A petition calling for inquiry into the validity of an election of a member to the House of Commons, based on his ineligibility for election, corrupt or illegal practices or on other grounds inferring invalidity. The petition is heard by an election petition court.

Election petition court. A court consisting of two judges of the Queen's Bench Division, or in Scotland of the Court of Session, or in Northern Ireland of the Supreme Court, which since 1868 has sat as required to consider a petition lodged by a voter or a defeated candidate challenging the validity of an election on such grounds as the candidate's eligibility. Its judgment is a report to the Speaker which the House of Commons is by statute bound to accept. But it preserves the form of dealing with petitions itself by resolving that the report be recorded in the Journal of the House.

Elector. (1) A prince of the Holy Roman Empire who had the right to participate in the election of the Emperor. From about 1273 and as confirmed by the Golden Bull of 1356 there were seven electors, the archbishops of Trier, Cologne, and Mainz, the duke of Saxony, the count palatine of the Rhine, the margrave of Brandenburg and the king of Bohemia. Later other electorates were created, namely Bavaria (1623–1778), Hanover (1708) and Hesse-Kassel (1803). The office disappeared when the Empire was abolished in 1806.

(2) A person having a vote at any election.

Electoral College. Article II of the Constitution of the U.S. established an electoral college composed of electors from each state equal in number to the total representatives of that state in the two Houses of Congress. Since 1961 the District of Columbia also has three electors. The electors meet at their state capitals after Election Day in presidential election years, to vote for persons for the offices of President and Vice-President of the U.S. In theory the electors elect the most suitable persons. In practice the candidates for President and Vice-President who win a majority of the popular vote in a state obtain all the electoral votes from that state, so that it is possible, by winning a narrow majority in the most populous states (which have the largest number of electoral votes) to win the election for President and Vice-President, despite losing by a heavy majority in the less populous states. Accordingly a President may be elected on a minority of the popular vote and in the cases of Hayes (1876) and Harrison (1888) the President had received less of the popular vote than his chief opponent. This possibility is even more likely where there are more than two serious contenders for the Presidency, and there have been fourteen such instances, including Lincoln (1860), Wilson (twice, 1912 and 1916), Truman (1948) and Kennedy (1960). Whether or not this happens the electoral vote usually seriously exaggerates the President's real support among the electorate.

Electoral Commission. In the U.S. Presidential election of 1876 the Democrats ran Samuel J. Tilden on a platform calling for reform and the Republicans ran Rutherford B. Hayes. Tilden carried states giving him 184 electoral votes, Hayes rather less, but South Carolina, Florida, Louisiana, and Oregon all submitted two sets of electoral votes.

Congress set up an Electoral Commission of eight Republicans and five Democrats chosen from Senate, House, and Supreme Court. There is little doubt that the Republicans made a deal with Democrat leaders in the three southern states to secure their support. The Commission on a strictly party vote rejected the Democratic returns from the four states and this gave Hayes 185 electoral votes, though Tilden had a plurality in the popular vote. There was considerable public indignation, and the commission had undoubtedly acted dishonestly and in extreme partisan fashion.

Electrocution. A method of judicial execution used in the U.S. in which the condemned person is made to sit in a wired chair and electrodes are fastened to his head and one leg and a heavy charge of electric current passed through his body. It was first used at Auburn State Prison, N.Y., in 1890 and was later adopted in many states. It is believed to cause painless and instantaneous death.

Elegit. A writ of execution (q.v.) under which a judgment creditor might obtain possession of the debtor's land and hold it until the debt was satisfied, whether from the rents or otherwise. It was a clumsy remedy and was abolished in 1956. The court may instead impose a charge (q.v.) on the judgment debtor's land.

Ellenborough, Lord. See LAW, EDWARD.

Ellesmere, Lord. See EGERTON, THOMAS.

Elliot, Sir Gilbert, of Minto (1693–1766). Son of the first Lord Minto (judge, 1705–18), became an advocate in 1715, a Lord of Session in 1726, a Lord of Justiciary in 1733, and Lord Justice-Clerk in 1763. He was not specially eminent as a judge but a man of letters, a musician, and an agricultural improver. His son, Gilbert (1722–77) became Treasurer of the Navy in 1766, and his grandson, Gilbert (1751–1814) was Governor-General of India, 1806–14.

Elliott, Walter Archibald, Lord (1922–). A member of the Scottish and English Bars he became a Scottish Q.C. in 1963, President of the Lands Tribunal for Scotland in 1971, and Chairman of the Scottish Land Court in 1978.

Ellsworth, Oliver (1745–1807). Studied both theology and law before being admitted to the Bar in 1771, serving intermittently as a member of the Continental Congress, particularly as a member of the Committee of Appeals, and becoming a judge of the Superior Court of Connecticut in 1784 and a U.S. Senator and Federalist leader of the Senate in 1789. He was chief author of the Judiciary Act of 1789 which established the federal court system. He was a principal draftsman of the conference report on the first 12 proposed amendments to the Constitution, including the Bill of Rights. At this time his public reputation developed greatly. In 1796 he became third Chief Justice of the U.S. Supreme Court, holding that office till 1800 and serving as special envoy to France in 1799–1800. His record as Chief Justice was satisfactory but he left no mark on the law and was not a learned lawyer. His decisions show commonsense rather than learning.

Brown, *Life of Oliver Ellsworth.*

Elmira system. An American penal system, named after Elmira Reformatory, N.Y., opened in 1876. The system utilized the mark system but added moral, physical, and vocational training, classified prisoners, giving them individualized treatment with an emphasis on vocational training and industrial employment. It used also indeterminate sentences, rewards for good behaviour, and release on parole under supervision. The practices utilized at Elmira brought to general notice the importance of special treatment for young offenders and of recognition that at least for a time there was potential for good in every criminal, and subsequently had substantial influence on penal thinking.

Elsynge, Henry (1598–1654). A great scholar, renowned in his own time as an authority on the history and law of Parliament, and Clerk of the House of Commons, 1632–48. His father, of the same name, had earlier been Clerk of the House of Lords, and was author of *The Ancient Method and Manner of holding Parliaments in England.*

Ely. The Isle of Ely was sometimes described as a county palatine and had a distinct judicial organization assimilated to that of the rest of England in 1837. It was now part of the county of Cambridge.

Elyot, Sir Thomas (*c.* 1490–1546). Became Clerk to the Privy Council under Wolsey and was later a friend of Sir Thomas More. He wrote a *Boke named the Governour*, a treatise on moral philosophy for the use of princes, which owes something to Erasmus' *Institutio Principis Christiani* and was modelled on Patrizzi's *De regno et regis institutione*, a Latin Dictionary, and many translations. There is an edition of the *Governour* (1880) by H. H. S. Croft.

Emancipation. The act of freeing a person, such as a serf, slave, or person subject to the power of another, such as a child, from the control of the master or parent and making him a free person. In

Roman law the important cases were the release of a child from *patria potestas*, which was effected by a triple mancipation (q.v.) or fictitious sale to a third party and manumission or release by him, and manumission of slaves, which was effected by collusive action, by will or by enrolment among the citizens or, later, by informal methods.

Politically important instances of emancipation have been the emancipation of the serfs in Russia, effected over the years 1861–81, and the emancipation of slaves in the seceded Southern States of the U.S. effected by Lincoln's Emancipation Proclamation of 1863.

Emancipation, Catholic. The process whereby Roman Catholics in the U.K. were freed from legal disabilities. Under various penal laws passed in the sixteenth century Catholics were subjected to disabilities and penalties. Toleration began by allowing Catholics to avoid penalties for religious observance. An Act of 1778 enabled Catholics to own land and hold long leases, provided they took an oath of allegiance to the Hanoverian succession and denied the temporal power of the Pope. The Scottish Bill provoked resistance and was not passed until 1798. In that year Catholics were given the Parliamentary franchise in Ireland, though not the right to stand for election.

The Roman Catholic Relief Act, 1829, abolished the oaths formerly required as part of the qualification for most public offices, but continued to require an oath of allegiance similar to that of the Irish Act of 1774. Most of the remaining anti-Catholic laws were repealed in 1844 and 1926 but only in 1974 was it declared that the office of Lord Chancellor was open to a Catholic. There remains the rule that the King and Queen be in communion with the Church of England.

Emancipation Proclamation. In the early months of the Civil War President Lincoln refused to decree emancipation of slaves, as he wished to prevent the border states from joining the South.

In 1862 Lincoln, who alone had prepared it, announced his intention to his Cabinet and after the victory of Antietam put the Confederate armies on the defensive and blocked foreign recognition of the Confederacy, he issued his preliminary declaration on 22 September 1862, affirming that the restoration of the Union, and not the issue of the abolition of slavery, was his aim in the war, and that he supported compensated emancipation, i.e. a subsidy to states to enable them to buy, free and settle slaves. The declaration was merely to the effect that persons held as slaves in areas in rebellion against the United States would be free as of 1 January 1863. On that date the definitive proclamation was issued. The proclamation exempted certain border territories. It could and did become effective only with a Union victory, but the proclamation was speedily recognized as a wise and humane act, which roused support for Lincoln and the Union cause, save in the South where it was regarded as an attempt to raise a slave rebellion. Full emancipation was effected nationally by the Thirteenth Amendment to the U.S. Constitution, and the Fourteenth denied any right to compensation for property in slaves.

Emancipists. Convicts in Australia in the late eighteenth and early nineteenth centuries who had been pardoned and were struggling to attain full civil rights. The term was also used of convicts who had served their full sentences ('expirees'). Emancipists were given land but were excluded from social and political life. Governor Macquarie sought to give them full rights but free settlers generally opposed this trend. By 1842, however, they were recognized as fully entitled to participation in government.

Embargo (civil). An order from the Crown to British ships not to leave British ports, sometimes ordered for reasons of national safety, as where the use of the ships will require to be requisitioned.

Embargo (international). A kind of reprisal (q.v.) meaning the detention of ships of a delinquent state in a port of the injured state for the purpose of compelling the delinquent state to make reparation for the wrong done. See also ARRÊT DE PRINCE; RIGHT OF ANGARY.

Embargoes have also been imposed by belligerent states on neutral ones and, conversely, by international organizations. Thus, in 1951 the U.N. prohibited states from shipping arms and strategic materials to China and North Korea.

Embargo Act (1807). An Act passed by Congress at the request of President Jefferson as a reply to British and French molestation of U.S. merchant ships carrying or suspected of carrying war material to countries engaged in the Napoleonic War. It closed all U.S. ports to exports in U.S. or foreign ships and placed restrictions on imports from the U.K. It did not have the full effects hoped for and caused hardship to U.S. agriculture and trade. In 1809 it was modified by the Non-Intercourse Act permitting U.S. trade with countries other than France and the U.K.

Embassy. The mission on which an ambassador of one state is sent to another, or the body of persons attached to such a mission, or the official residence and office of the ambassador.

Embezzlement. The crime committed by a clerk or servant who stole any money, chattels, or valuable security belonging to or in the possession of his employer, or who intercepted them before they came into his employer's possession. It was abolished as a distinct crime in 1968 in England and now comes under theft. In Scotland it is the misappropriation by a person of the property of another which had been entrusted to him with the owner's authority. In general it differs from thefts in that in embezzlement the criminal frequently obtains possession of the property lawfully and misappropriates it rather than obtains possession unlawfully.

Emblements. The growing crops of those plants which are produced annually by cultivation. They are deemed personal property even before being cut or gathered notwithstanding that they are attached to the soil.

Embracery. The offence of attempting to corrupt, influence or instruct a jury to reach a particular conclusion by any means other than evidence and argument in court, or of giving jurors money after their verdict or of their taking it.

Emergency powers. Powers which may be invoked in cases of emergency, national danger, and other exceptional circumstances. Under the Tudors the royal prerogative was freely used, as in Queen Elizabeth's measure for meeting the threat posed by the Spanish Armada. Under the Stuarts the Crown seized on the need for such powers to build up a general doctrine of the prerogative and sought to apply this both to normal and to emergency times. In the eighteenth century it was contended that the normal jurisdiction of a secretary of state included some emergency powers. In the nineteenth century there was increasing resort to legislation. Thus during the Napoleonic Wars there were numerous statutes passed for the defence of the realm, including provisions suspending the Habeas Corpus Act in some cases. The years after 1815 resulted in legislation of the same kind. The Defence of the Realm Acts, 1914–15, conferred very wide powers on the government to legislate by orders, regulations, and other delegated legislation. The Emergency Powers Acts, 1920 and 1964, established a similar permanent reserve of power for use in emergency in peacetime, e.g. during major strikes, and the Emergency Powers (Defence) Acts, 1939 and 1940, again conferred almost unlimited powers by delegated legislation to control everything during the Second World War. Most constitutions contain provisions for the exercise by government of emergency powers.

Emerigon, Balthazar-Marie (1716–84). French jurist, author of *Nouveau commentaire sur l'ordonnance de la marine de 1681* (1780) and *Traité des assurances* (1783), both works of great importance and value.

Eminent domain. The power, usually deemed inherent in sovereign states, to take private property for public use, subject to making reasonable compensation, as distinct from mere seizure. It comprehends compulsory purchase.

The power was recognized by the natural-law jurists such as Grotius and Pufendorf. In seventeenth-century England Parliament frequently authorized the taking of property, prescribing compensation, or providing judicial machinery for assessing compensation, but not always allowing the owner to be heard on the question. The supremacy of Parliament implies that it is possible in any case for Parliament to authorize the taking of property without compensation, or with only such trivial compensation as it cares to give, and this has happened under nationalization legislation.

In the U.S. the practice was known in colonial times, judicial machinery being established to let the expropriated owner be heard on compensation. The Fifth and Fourteenth Amendments recognized the right, but subject to there being due process of law. The question has arisen, but remains undecided by the Supreme Court, whether the federal government may exercise the right of eminent domain in a state without that state's consent.

Emperor (Lat. *imperator*). A title applied to certain monarchs. There are no fixed criteria for its use but it generally implies superiority over ordinary kings. In Republican Rome (500–27 B.C.) *imperator* was the term applied to a victorious general. Under the Roman Empire the term was regularly applied to the rulers and came to be the title of the office. The title was used by Charlemagne and his successors as Holy Roman Emperors from A.D. 800 to 1806 and has been assumed by others, e.g. by Napoleon as Emperor of the French (1804), Kaiser Wilhelm I of Prussia as German Emperor (1871), and Queen Victoria as Empress of India (1877) (all of these cases now disused). The title has also been used of certain Oriental monarchs.

Emphyteusis (or *ius emphyteuticarium*). In Roman law, a lease for agricultural purposes granted for a very long period or in perpetuity, carrying most of the rights of ownership but requiring the payment of a regular annual rent (*canon*) to the owner. It originated in the early imperial period and at first could be granted by the State only but such grants were soon made by private landowners also. The grant was heritable, transferable, and recognized and protected by law. This kind of grant has certain analogies in some civil law countries. *Superficies* was a similar grant for the erection of buildings.

Empiricist theories of justice. Theories which seek to impose obligation by virtue of their derivation from social facts and identify the criterion of justice with the observed facts of social life. Such enthroning of the facts may be seen in Bentham's utilitarianism and very clearly in Duguit's insistence that his criterion, the objective law, emerged spontaneously from the social solidarity manifested in the facts of social life. It is assumed by these theories that when the facts are ascertained and analysed, the just solution will emerge. But none of the jurists adopting an empiricist standpoint show how the just solution does emerge nor how a just solution emerging from the facts can be reconciled with a clear rule of positive law to a contrary effect.

Employers' association. Any permanent or temporary organization which consists of employers or individuals owning an undertaking, or constituent or affiliated organizations or representatives thereof whose principal purposes include the regulation of relations between workers and employers or associations of either. It may be a body corporate or an unincorporated association. A list of such associations is maintained by the Registrar of Friendly Societies.

Employer's liability. A particular, but common and important case of liability for negligence, namely that of an employer to his employees. It is now settled law that an employer is liable at common law to an employee injured by reason of his failure to take reasonable care for the safety of his employees, and liable under numerous statutes for injuries resulting from his failure to satisfy the standard of care set in many circumstances by various statutes and regulations made thereunder. The employer is liable both personally, for his own faults, and vicariously, for the faults of his employees, including fellow-employees of the employee injured.

Employment. The contractual relationship known in the older books and cases as that of master and servant, the salient characteristics of which are that one person, the employee, engages for a period to give his service to and do work in a stated capacity for another, the employer, and in so doing to act subject to the direction and control of the employer, in return for remuneration normally by way of salary or wages calculated by the hour, day, week, month, or year.

In Roman law the contract was *locatio operarum*, letting on hire of service, as distinct from *locatio operis faciendi*, letting of work to be done or contract for the rendering of services. The distinction between a contract of employment and a contract for the rendering of services or the securing of a result for a fee is frequently difficult to draw and does not necessarily depend on whether the person employed has to use practical or professional skill, knowledge, and discretion in the work or not. A solicitor or an architect may be an employee of, e.g. a local authority, or be an independent contractor instructed by clients to bring about a result e.g. to buy a house.

In most modern states the employment relationship is today heavily overlaid with terms negotiated by collective bargaining and by statutory regulation. In the U.K. modern legislation tends to regard an employee having something akin to a right of property in his job.

In many cases an employer must give an employee a written statement of particulars of the terms of his employment. Terms settled by collective bargaining between employers and trade unions may be incorporated in individual contracts of employment.

During employment an employee must serve faithfully and obey lawful instructions, not wilfully absent himself from work, and perform his duties with reasonable skill and care. The employer must pay wages and in some cases provide work, ensure, so far as reasonably practicable, the health, safety, and welfare at work of all employees, provide competent fellow-workers, suitable plant and appliances, provide and maintain a reasonably safe system of working and effective supervision, and maintain a reasonably safe place of work. A large number of statutes impose stringent duties of care in respect of employments in particular trades. Employers must insure against liability in damages to injured employees.

Statutory minimum periods are prescribed for notice to terminate employment and in any event a reasonable period is required. A written statement giving particulars of the reason for dismissal must in some cases be given. An employee may claim damages at common law if dismissed in breach of contract or under statute if dismissed unfairly. In the event of dismissing staff on account of redundancy an employer must first consult the relevant trade union. Employees dismissed are entitled to payments from a redundancy fund.

Employees are entitled to time off work for carrying out trade union duties or activities, for public duties, to look for other work, or make arrangements for training.

Wages are settled by negotiation, by collective bargaining, by rates fixed by statutory wages councils, but frequently subject to overriding statutory control.

R. Rideout, *Principles of Labour Law*.

Employment Appeal Tribunal. A United Kingdom court, composed of judges seconded from the High Court and Court of Session and lay members representing both employers and employees, established in 1976 and replacing the National

Industrial Relations Court (q.v.), which hears appeals on questions of law from tribunals handling disputes arising under a wide range of industrial and labour legislation, concerned with, e.g. redundancy payments, equal pay, contracts of employment, sex discrimination, employment protection, and trade unions. An appeal lies to the Court of Appeal or the Court of Session with the leave of the Tribunal or of either Court.

Emptio spei (the purchase of a chance). In a contract of sale of goods the buyer obtains a chance of getting the goods. The contract is conditional on the seller's and absolute on the buyer's part; the seller must deliver the goods if he gets them, but the buyer must pay the price in any event. A common instance is the purchase of a lottery ticket which confers the chance of a prize.

Emptio-venditio. In Roman law the contract of sale and purchase, classified as a consensual contract *bonae fidei.* It was a contract only and not a conveyance, and passing of the property required in the case of *res mancipi* a *mancipatio* or *in iure cessio* or in the case of *res nec mancipi* a *traditio.* With the formation of the contract the risk passed to the buyer. The seller was liable for certain defects whether or not he knew or could have known of them and in such a case the buyer might return the thing and recover the price, or, certainly in later law, demand a proportional reduction of the price.

F. de Zulueta, *Roman Law of Sale* (1945).

Emslie, George Carlyle, Lord (1919–). Was called to the Scottish Bar in 1948, took silk in 1957, and was Dean of the Faculty of Advocates 1965–70 before becoming a Lord of Session, 1970–72, and Lord President of the Court of Session in 1972.

En ventre sa mère (in his mother's womb). The state of an unborn child, In general, where such a decision is for a child's benefit, an unborn child is treated for the purposes of injury to it, or of right in property, as if born. To cause the death of an unborn child may be the crime of procuring abortion (q.v.) or, if the child is capable of being born alive, of child destruction (q.v.).

Enabling Act. A statute conferring powers, the exercise of which is *prima facie* discretionary but which may be held mandatory.

Enact. To pass an Act and establish some rule by law.

Enactment. A general term for a statute or Act of Parliament, statutory instrument, by-law or other statement of law made by a person or body with legislative powers by the appropriate means.

Enclosure Acts. Enclosure was the process of amalgamating mediaeval open fields and common lands into compact areas enclosed by hedges and fences, and this was a feature of agrarian development in Britain, chiefly in England between the sixteenth and nineteenth centuries. It could be done by agreement but private legislation was usually resorted to after 1760. In 1774 it was provided that notice of an enclosure scheme must be posted at the parish church for three successive Sundays. Enclosure Acts were frequent during the Napoleonic Wars when high prices encouraged large-scale farming. In 1801 the Enclosure Consolidation Act (the General Enclosure Act) provided a common form for Enclosure Bills. Enclosure Acts have often been criticized for overriding claims of small proprietors and causing loss of common rights but on the whole they dealt fairly with vested rights.

Encroachment. A trespass of substantial duration on the land of another as when one tenant takes in land adjoining his own land so as to make it appear to be part of his own. In England an encroachment acquiesced in for 12 years is considered annexed to the original tenement. The term is also used of an alteration of a dominant tenement of land, having an easement over other land, so as to impose an additional burden on the servient tenement.

Encumbrance. A burden or liability, particularly a charge or mortgage on land.

Encyclical, papal. A letter communicated by the Pope to the whole Catholic Church dealing with some matter of morals, doctrine, or discipline. They are normally addressed to the bishops and take their titles from the first words of the official text, which is normally in Latin. Though issued from an early period the first encyclical normally so-called dates only from 1740 and it is only from the mid-nineteenth century that they have frequently been issued. They are not necessarily deemed infallible.

Encyclopaedias, legal. A legal encyclopaedia is a statement in narrative form under headings, usually alphabetical, of the whole law of a particular country or territory having a distinct legal system. That legal system is divided analytically into a large number of well-known titles or headings, such as Criminal Law, Landlord and Tenant, Patents, which are arranged alphabetically and each subdivided into subheads and with numerous cross-references. Each heading is, in effect, a brief treatise on the law of the topic, with narrative, statutes, cases and other authoritative sources for the principles and rules stated.

The modern encyclopaedia is one of the lineal successors of the later alphabetical Abridgments, such as Comyn's *Digest*. In England there are the *Encyclopaedia of the Laws of England* (12 vols. 1897–98; 2nd edn. 15 vols. 1906–9; 3rd edn. 1938–40) but much better is *Halsbury's Laws of England* (31 vols. 1907–17; 2nd edn. 37 vols. 1931–42; 3rd edn. 43 vols. 1952–64; 4th edn. 1973–) with Annual Supplements and monthly service.

In Scotland there are R. Bell's *Dictionary of the Law of Scotland* (1807–8) expanded by W. Bell into *Dictionary and Digest of the Law of Scotland* (7th edn. 1890) and Green's *Encyclopaedia of the Laws of Scotland* (14 vols. 1896–1904; 3rd edn. 16 vols. 1926–35 with supplementary volumes to 1952).

In the U.S. the first encyclopaedia was Thornton's *Universal Cyclopaedia of Law* (1883) which was followed by others, particularly the *American and English Encyclopaedia of Law* (31 vols., 1887–96) and the *Cyclopedia of Law and Procedure* (40 vols., 1901–12). There are now two general legal encyclopaedias, *American Jurisprudence* (72 vols., 1936–52) and *American Jurisprudence, 2nd series* (probably 70 volumes) and *Corpus Juris* (72 volumes, 1914–37, with Permanent and Annual Annotations volumes), the text (but not the notes) of which is superseded by *Corpus Juris Secundum* (116 volumes, 1936–64, with supplements). It is also an annotated legal dictionary.

Endowment. Primarily the assigning or giving of dower (q.v.) to a woman. Also the setting apart of a portion to a vicar for his perpetual maintenance, and more generally property of any kind set apart for particular purposes, particularly charitable purposes.

Endowment policy. A policy of assurance providing for the payment of money to the assured on his having attained a stated age or survived a stated number of years, frequently also providing for payment on prior death.

Enemy character. A quality attaching to certain persons and property in time of war. Determination whether it attaches in particular circumstances is important but difficult. In general, it attaches to subjects of belligerent states on the opposite side in a war and to their property, and even to British subjects resident or trading in enemy territory, but not subjects of neutral states and their property. But subjects of neutral states inhabiting enemy territory must be treated as enemies, though they do not lose the protection of their home state against arbitrary treatment unjustified by the laws of war, and trading with them is trading with the enemy. The character of ships is generally determined by their flag, whatever the nationality of the owner; a vessel under a neutral flag bears neutral character unless the vessel had no right to fly that flag, or takes a direct part in hostilities, or forcibly resists the legitimate exercise of the right of visit and capture, or possibly if the owners can be shown to have enemy character. Goods found on board an enemy ship are presumed to be enemy goods unless the contrary is proved by neutral owners.

Enfranchise. To make free or to confer a liberty, as by conferring the franchise or liberty to vote at elections. Copyhold land was said to be enfranchised when freed from the customs attaching to that tenure and made subject to the same law as freehold land. This might be done at common law, by conveyance of the freehold to the copyholder, or under statutory provisions. Similarly leaseholds may be enfranchised and made into freeholds.

Engagement. (1) An agreement between Charles I and Scottish commissioners in 1647 by which the King agreed to establish Presbyterianism in England for three years and to suppress Independency in all its branches in return for Scottish support against the Army. This gave rise to the second civil war which ended when Cromwell defeated the Scottish army at Preston in 1648.

(2) A term for certain kinds of contracts, particularly the mutual undertakings of persons to marry at a later date, i.e. *sponsalia de futuro*, the contract of service in the armed forces, and the contract of service of the crew of a fishing boat.

England. The part of the island of Britain south of the frontier with Scotland and east of that with Wales. For most legal purposes England, by the Wales and Berwick Act, 1746, includes Wales and the town of Berwick on Tweed. In statutes after 1967 references to England do not include Wales. For governmental purposes Monmouth is in Wales. It does not include either Scotland or Northern Ireland (still less the Republic of Ireland), nor the Isle of Man nor the Channel Islands, nor any part of the Commonwealth, all of which are independent jurisdictions with their own courts and systems of law. England is not synonymous with Britain or the United Kingdom. See also English Law.

English information. A process by which a suit in equity was initiated by the Crown on the Revenue side of the Queen's Bench Division, so called to distinguish it from Latin informations used to recover debts or damages for tort or chattels. Four classes of English informations were known, money claims, claims to foreshore, claims to other lands or minerals, and proceedings under the Marriage Acts. An English information was inquisitorial in nature in that it could be put in interrogatories which the defendant had to answer, subject to penalty, and the Crown could after the answers amend its case and deliver fresh interrogatories and so on. Though intended as a means for the Crown to obtain an accounting from its servants or others, it could be a most oppressive proceeding. English informations were abolished by the Crown Proceedings Act, 1947, but the Crown retains the right to deliver interrogatories, but not more than two without leave of the court.

English Law. English law is the system of law which has grown up and is applied in England (i.e. the part of the United Kingdom and of the island of Britain which comprises the former kingdom of England) and in Wales but not in Scotland. From the fairly early development of a law common to all England English law is frequently called 'the common law' (see also COMMON LAW).

By reasons of its history and development (for which see *infra*) English law is an unsystematic body of law and has never been systematized or its rules thoroughly classified. Not only is it uncodified, so that the discovery of the provisions on any matter is frequently a long, complicated, and difficult task, involving digging into old books and decisions, but the branches and divisions of the law are indefinite.

A feature of great importance, both in historical development and in daily application of the law, is the standing and prestige, and the deference paid to the decisions, of even single judges of the superior courts. In default of statutory provision, and it is only since 1830 that statute has played a major role in laying down the law, a single decision of even a single judge is taken as establishing a rule of law; and *a fortiori* in the case of decisions by benches of appellate judges. Another basic principle underlying the common law until recently, and still influential, is that a common law right existed only if there was a recognized means for remedying the complaint: *ubi remedium ibi jus*. Substantive law was bound up with and developed in the interstices of procedure.

Another feature of great importance has been the development of parallel bodies of principles, of common law and of equity (q.v.) formerly sometimes in conflict, now usually collaborating, formerly administered in separate courts and by distinct procedure, now both administered in courts empowered to apply principles derived from either body of principles or from both. Yet there is still not complete union of the bodies of principles. In case of conflict the rules of equity prevail.

Conversely, differing in this respect from many continental systems of law, no distinction exists, nor has for long existed, between civil law applicable to ordinary citizens and commercial law applicable to merchants or to mercantile transactions. The same rules, generally speaking, apply whether the parties are ordinary people or commercial agents doing business.

I. *Historical development*

For three and a half centuries (A.D. 43–407) England was under Roman rule but that occupation has left no legal marks. Later, Jutes, Angles, and Saxons from the European mainland occupied large parts of England over the years 400 to 600 and during the Anglo-Saxon period, down to about 1100, there are numerous bodies of Anglo-Saxon law (see ANGLO-SAXON LAWS). Danes and Norsemen later occupied parts of the country and King Cnut ruled England, Norway, and Denmark and provided a body of legislation long popular in England. Other Norsemen settled in Normandy in the tenth century and made it a well-governed state.

In 1066 William, Duke of Normandy, invaded and conquered England and began to introduce orderly methods into government and law. The Norman Conquest superimposed ideas drawn from Frankish and continental law on the substratum of Anglo-Saxon law, which belongs to the Teutonic family of folk-laws. William established the Exchequer as a strong, central financial department with a firm control over the sheriffs and local government generally. His last years were taken up with the great survey of the Kingdom known as *Domesday Book*. He systematized landholding and made England the most completely organized feudal state in Europe, in which every piece of land was held of a feudal superior, who held of a higher superior, and so up to the King. This development of a clear doctrine of landholding, ultimately from the Crown, is of great importance for the development of the common law. It stimulated the development of primogeniture in succession. The earliest civil business which the royal courts dealt with were cases relating to title to land. In criminal cases the Norman Conquest introduced trial by battle.

In reign of Henry I there was elaborated an efficient governmental organization at Westminster; he sent justiciars on circuit to deal with pleas of the Crown, a jurisdiction which they elaborated by asserting that any matter concerning the King's peace could be treated as a plea of the Crown. Legislation became more common and in 1100 the King issued a Charter of Liberties. But the law was still substantially Anglo-Saxon and administered locally by the King's sheriff according to ancient custom, which differed in various parts of the country.

Even at this stage legal literature developed, in the form of a group of treatises, notably the so-called *Leges Henrici Primi*, mostly founded on Canute's laws and the older English dooms, attempting to state the old Anglo-Saxon law in a form suitable for Norman times.

By the eleventh century there had been recognized a range of royal writs directed to the sheriffs of counties and initiating litigation in royal courts without regard to the rights of feudal lords. Litigation was begun by purchasing a royal writ suitable to the circumstances, and the chancery developed expertise in framing writs for different circumstances, and the writs became formalized.

Henry II (1154–89) was ruler not only of England but of Normandy, Anjou, and Aquitaine. Early in his reign he clashed with Archbishop Becket who asserted vigorously the rights of the church which since 1066 had developed a large body of canon law and acquired wide jurisdiction. The Constitutions of Clarendon (1164) effected a compromise. But also in the reign there was a growing definition of the courts of law; an extension of the system of itinerant justices; the establishment of the petty assizes, particularly novel disseisin (1166) (q.v.) as a rapid method of trying cases of recent dispossession of land, and of the Grand Assize which enabled the defendant in a suit concerning law held by freehold tenure to claim trial by jury in the royal courts as an alternative to battle; and a great outburst of legislation in the form of charters and assizes, some of which established new forms of trial by inquisition or jury, or established new forms of action in real property; among these was the remodelling of criminal procedure by the Assize of Clarendon (1166) which established the jury of inquest to discover alleged criminals.

Not least in Henry II's reign Ranulf de Glanvill (q.v.), the justiciar, or possibly his nephew, Hubert Walter, wrote the first treatise on the common law, *De Legibus et Consuetudinibus Angliae*, in the form of a commentary on the different writs, thereby setting the pattern for English textbooks for centuries. The main features of English law are apparent in Glanvill; the law is royal, laid down by the King's court, common to the whole country, and strongly procedural, being much determined by the form of the procedural writs which initiated claims. In about 1179 came that other masterpiece of the age the *Dialogue of the Exchequer* of Richard Fitz Nigel or Fitz Neal, expounding the fiscal system.

It was well that the administrative machinery had been firmly established because between 1189 and 1272 great strains were placed on it. Under John (1199–1216) maladroit rule and oppression provoked rebellion, and in 1215 John sealed Magna Carta (q.v.) conceding the demands of the rebellious barons. The Great Charter was subsequently modified several times and reissued but many of its principles still stand as important grants of fundamental freedoms. The reign of Henry III (1216–72) was also a period of baronial revolt against the Crown but the period is significant for the summoning by Simon de Montfort in 1265 of a Parliament including two knights from each shire, two citizens from each city and two burgesses from each borough, an event commonly taken as marking the genesis of the English Parliament as a genuinely representative assembly.

Also, during the reign Henry of Bracton (q.v.) wrote his great treatise on the *Law and Customs of England*, founded, significantly, largely on notes made by him (published long afterwards as *Bracton's Note Book*) of decisions by royal judges in cases coming before them but, like Glanvill, written in the form of a series of commentaries on the various forms of action and the writs initiating each. Bracton uses the cases as evidence of custom, not as precedents, but may have originated or stimulated the recording of interesting cases. Moreover there had been rapid development of the law; Bracton knew many writs unknown to Glanvill.

About this time, too, the forms of action had become so numerous that copies of the formulary, the Register of Writs, began to be made.

During the period down to 1272 the old communal courts, notably the county court and the hundred court, were being steadily overshadowed by the feudal courts, such as the court leet, which had developed since the Conquest, and then by the royal courts at Westminster. Between the Conquest and 1272 the royal courts were created and equipped with bodies of law and procedure, and the idea had developed that all justice was exercised centrally or by delegation from the Crown.

At this stage, too, the royal household began to throw out branches which were to develop into independent offices of State. The Exchequer early achieved independence, so early that a treatise on Exchequer procedure, the *Dialogus de Scaccario*, could appear by 1179, and the Chancery became an independent office for the management of the Great Seal. The control of financial interests of the Exchequer naturally led to its decision of disputes arising incidentally. In 1236 there appeared the first roll of the Exchequer of Pleas, and the Exchequer had definitely become a court. Originally having a jurisdiction in revenue disputes, it later acquired jurisdiction over disputes between individuals.

There developed also, as well as the royal court moving about the country, groups of officials touring the country, as Justices in Eyre. They were concerned to safeguard royal interests of all kinds and conducted investigations into the whole local administration of justice.

Probably in 1178 Henry II created a new court, the Court of Common Pleas, which initially perambulated the country with the King, and was not fixed permanently at Westminster till Magna Carta. It was subordinate to the great legal advisers constantly attendant on the King, who were concerned with matters touching the King, trial of the pleas of the Crown, and the correction of error in the Common Pleas. The King's Bench was the last court to develop out of the Council; it became a separate institution early in the reign of Edward I. Above the King's Bench, but working in close contact with it, was the King's Council. From its close association with the Crown the King's Bench acquired the jurisdiction to issue the prerogative writs of mandamus, *certiorari*, and prohibition. Its principal jurisdiction was in civil matters.

Soon after the Norman Conquest the inquisition (q.v.) was introduced as a means of obtaining information, the most notable use being the inquiries recorded in Domesday Book. Henry II extended the use of the inquisition or assize to ascertain whether land was held by ecclesiastical or lay tenure, whether the claimant had recently been disseized of his land, and finally in 1179 it was allowed in lieu of battle for trying the writ of right. In 1166 the inquisition was established as a means of inquiring into suspected criminals, the basis of the grand jury, and about half a century later a jury began to be utilized to decide the guilt of a prisoner. Later, in the thirteenth century, the practice had developed of taking the verdict of a small (petty) jury and of imposing jury trial in all cases, not giving accused persons the option of submitting to the ordeal.

Furthermore, by the thirteenth century there was recognizable the origins of the modern judiciary in the shape of royal servants regularly employed in judicial work, such as Pateshull and Raleigh, both of whom were revered by Bracton for their decisions.

Thus within the first three centuries after the Conquest the King's courts had assumed control of the administration of justice throughout England, through the centralized system of courts and judges. As they established their ascendancy, the old local courts such as those of the sheriff declined. The Statute of Gloucester of 1278 limited the sheriffs' civil jurisdiction to 40 shillings.

The period 1272–1399. The reign of Edward I, the 'English Justinian', was marked by a great outburst of reforming legislation. By this time the courts were tending to leave deliberate law-making to Parliament and to confine themselves to application of the law, but on the other hand the judges were becoming more independent of the court.

In 1284 after the conquest of Wales, there was passed the *Statutum Walliae* restating the fundamentals of the common law for the information of sheriffs seeking to apply it in Wales, and in 1285 the Statute of Westminster II. Of this chapter 24 laid down a procedure for the development of the law by extension of the available writs in defined circumstances, and chapter 30 established the *nisi prius* (q.v.) system, whereby the facts in most contested cases could be found by juries in the country towns, rather than juries and parties being required to come to Westminster. In 1290 the statute *Quia Emptores* forbade continued subinfeudation and thus perpetuated the feudal tenures and the incidents thereof.

In 1357 the Exchequer Chamber was created as a court to hear errors in the Exchequer, which gradually developed into a body of all the judges and serjeants sitting to determine points of difficulty. Decisions of this body naturally enjoyed exceptional prestige and by the seventeenth century it was settled that a decision of the Exchequer Chamber was a binding precedent.

In 1362 French was replaced as the language of the courts by English and from about the same time there was increasing rigidity in the rules of pleading, a practice which developed till it became an impediment to doing justice.

From this time onwards the function of the jury was becoming better settled and defined.

From the end of the twelfth century, also, commissions began to be issued to local knights and gentry to co-operate with the sheriffs in enforcing the law, and from 1368 these justices of the peace are found exercising judicial powers by themselves. The justices met four times a year in quarter sessions, while two or more justices sat informally to deal with minor cases in a summary manner. The establishment of the justices of the peace marked the end of the old communal jurisdictions.

The Year Books, records of points of pleading in cases, collected by the terms and years, originated early in the reign of Edward I and continued, being a fairly complete and continuous record from about 1350, to the mid-sixteenth century, forming an invaluable record of the growth of the law. Even in the fourteenth century it seems to have been recognized that a decision might lay down a general rule and evidence or even settle the custom of the court on a point.

In the time of Edward I many institutions took on the form which they kept for many centuries, Parliament comprising the three estates, King's council, convocations of the clergy, the Chancery, the Exchequer, King's Bench, Common Bench, commissioners of assize, professionally skilled judges and skilled lawyers.

1399–1485. A notable feature of the rule of the Lancastrian Kings was the growth in importance of the Council. It sometimes initiated semi-judicial proceedings against magistrates too powerful to be reached by the ordinary courts.

The practice of litigants who were unable to get a remedy because of the formalism and inflexibility of the common law, petitioning the King, Council, or Parliament became more common; it was normally the Council which took action, and it tended to delegate its functions to the Chancellor, who had moreover an organized office and staff and had for long exercised the power of issuing writs, judicial and administrative, to all the King's officials.

In 1474 there is the first recorded instance of the Chancellor issuing a decree in his own name. Thereafter there was a steady development of what came to be known as equity. The major equitable remedies were recognized by the sixteenth century.

In the fifteenth century there was some conflict between common law and chancery but considerable tolerance and co-operation. Not much later, in St. Germain's *Doctor and Student* a philosophical justification was found for equity in the canon law, in conscience, and discretionary granting of remedies based upon conscience.

By the end of the fifteenth century the conveyance of land by one to the use of another had become common and recognized, and was being used increasingly as a device to dispose of lands after death, to hold land for charities, and to defraud feudal lords of the feudal incidents of tenure. Actions based on debt, covenant, and account were well recognized.

The most notable legal treatise in the century is Littleton's *Tenures* (*c.* 1481), which seeks to reduce the complicated law of real property to system, which for three centuries was the standard introduction to real property law.

By this time, too, there had developed a good working scheme of legal training in the Inns of Court (q.v.), in effect legal colleges in which law was studied in a practical way. Moreover, these societies had suceeded in making admission to their degrees—the call to the Bar—the only gateway to practice in the King's courts.

1485–1603. The sixteenth century witnessed all over Europe the waves of new ideas sometimes called the Renaissance, the Reformation, and the Reception. The first brought in new ideas and new knowledge of ancient theories; the second brought about in England a quarrel between Henry VIII and the Papacy, the royal assumption of supremacy over the Church of England and the acceptance of Protestantism in England; the third the great expansion of interest in and knowledge of the Roman law, though this had much less effect in England than in France, Germany, or Scotland.

The Tudor monarchs also created a range of courts of special jurisdiction, such as the Courts of Augmentations, First Fruits, Wards, Liveries (later combined), and Surveyors. All these were primarily administrative departments. The Court of Star Chamber originated in the Council's criminal powers, and its gradual restriction of its business to the handling of petitions, especially those raising questions of public order, and the administration of criminal law. By the reign of Henry VIII the old Council had become the Court of Star Chamber and he had to create a new institution, the Privy Council, and to revive the executive and deliberative functions of the old Council. The Court of Requests first appears in 1483 and under Henry VII it was in effect a committee of the Council for the hearing of poor men's causes and matters, relating to the King's servants, with a jurisdiction mainly civil. Concurrent jurisdiction with the Council and the Star Chamber were exercised by the Council of Wales and the Council of the North.

In consequence of the Reformation there were created new ecclesiastical courts, the High Court of Delegates, to hear appeals which had formerly gone to the Pope, and the Court of High Commission, concerned mainly with the criminal side of the former papal jurisdiction.

In the sixteenth century there developed a distinct separation of the judiciary from the executive, with even Chief Justices frequently not being members of the Privy Council.

By the mid-sixteenth century the Court of Chancery had acquired a large amount of business and developed an establishment. The Master of the Rolls was coming into prominence as an assistant, and latterly as deputy to the Chancellor, while masters in Chancery dealt with minor matters.

In 1536 the famous Statutes of Uses made the nominal owner of property held by one person to the use, i.e. for the benefit of another, the legal owner, but within a century the Chancery had begun to develop the modern law of trusts.

There were considerable developments in contract and tort in the sixteenth century. The liability of one who, being liable to an action of debt, subsequently expressly undertook to pay, was extended by presuming the subsequent undertaking, and in *Slade's case* (1602) it was settled that the older action of debt and the newer *indebitatus assumpsit* were alternatives, the undertaking to pay being presumed. Soon a large variety of implied contracts were remedied by *indebitatus assumpsit.*

There was also a great reception of Italian mercantile law, though for a further two centuries the common law applied mercantile custom only when it had been proved in a particular case. Thus Italian practice as to bills of exchange was recognized by the mid-sixteenth century.

In 1585 another court was created in the Exchequer Chamber, consisting of all the judges of the Exchequer and the Common Pleas, to hear errors from the King's Bench, but subject to further proceedings in error in Parliament.

During this period the King's Bench adapted the old prerogative writs of mandamus, *certiorari*, and prohibition to legal purposes, and assumed the task of directing them to and thus using them as a means of control of various branches of central and local government.

The Year Books came to an end in 1535 but thereafter there began to appear reports of decided cases. Early reports of outstanding importance are those of Plowden (1550–80) and Coke (1572–1616) who appended to each case a summary of all authority upon the point down to his time. Alphabetically arranged 'Abridgments' of mediaeval cases, such as those of Fitzherbert and Broke were among other works of permanent value.

Under Henry VIII the academic study of the canon law, which has been pursued in the ancient universities was prohibited; regius professorships of the civil law were founded and civilians were to sit as judges in the ecclesiastical courts. But no full reception of the civil law took place. The common law was too firmly rooted.

1603–1688. In 1603 England and Wales, and Scotland became linked in a personal union, the two kingdoms having the same king. The keynote of the seventeenth century is the clash between the royal power and the doctrine of the supremacy of law over King and people equally. Clumsy monarchs, an accumulation of grievances, and increasingly assertive Parliaments resulted in civil war, the execution of King Charles I, and the rule of Cromwell before, in 1660, King Charles II and Parliament were restored. The disputes gave rise to numerous cases and developments of fundamental constitutional importance, and the steady growth in importance of Parliament. Parliament abolished most of the prerogative courts established by the Tudors. A sharp struggle ensued between common law and chancery, but chancery survived because of the value of its remedies. Many of the changes made under the Commonwealth and Protectorate, though reversed at the Restoration, foreshadowed modern reforms.

At the Restoration there was considerable legal reform, and much of the mediaeval feudal law disappeared, though this was balanced by technical developments in conveyancing. The most notable development in public law was the passing of the Habeas Corpus (q.v.) Act in 1679.

From the Restoration down to the early eighteenth century equity became settled as a distinct and consistent body of rules; it was accepted that there was little place for mere exercise of discretion. Henceforth the relation of law and equity were amicable and the two systems were increasingly complementary.

The short reign of James II (1685–88) witnessed clashes between Crown and people based on religious differences and a conflict between claims of the royal prerogative and of Parliament. William of Orange and his wife Mary were invited to become King and Queen, the terms of the settlement being embodied in the Bill of Rights (1689) (q.v.) and the Act of Settlement (1701) (q.v.).

Coke not only played a great part in the struggle between Parliament and Crown, but in his *Institutes* (1628–44) set out most of the law of England in commentaries, on Littleton's *Tenures*, on the principal statutes notably Magna Carta, on the pleas of the Crown, and on the jurisdiction of the courts. The same period included the lives of Selden, the great antiquary, Prynne, a historian with a vast knowledge of the public records, and Hale, who wrote a *History of the Common Law* and *History of the Pleas of the Crown.* In Chancery Ellesmere and Bacon secured the independence of the chancery and developed its practice, while Nottingham made great strides in fixing and settling the principles of equity.

From Coke's time the citing of precedents is common, though for another century the function of citations is mainly that of proving a settled practice or policy.

1689–1790. The eighteenth century witnessed the industrial revolution and the rise of commerce in importance and as a source of wealth, and a consequent change in law, as the emphasis moved away from land to partnerships, joint-stock companies, contract, banking, and employment. In 1707 the Parliaments of England and of Scotland, which had remained separate despite the personal Union of the Crowns since 1603, amalgamated to form the Parliament of Great Britain, but the two legal systems remained separate, particularly in private law.

One outstanding feature of the eighteenth century was the tenure of the office of Lord Chief Justice of the King's Bench 1756–88 by Lord Mansfield, who greatly developed commercial law and incorporated it into the English law, and tried unsuccessfully to supersede the requirement of consideration in contract and make moral obligation the factor which made promises actionable. Lord Chancellor Hardwicke (Chancellor, 1737–57) restated most of the principles of equity and largely completed it as a system. An equity bar began to form and equity reports began to be compiled. Another notable event was the publication of Blackstone's *Commentaries* (1765), a masterly exposition of the whole of the law, which powerfully influenced English and American law.

1790–1832. The French Revolution, the wars against Napoleon, and the aftermath thereof had the main effect of delaying constitutional and legal reform for a generation.

Private law was hindered by the conservative and reactionary attitudes of Lord Chief Justice Ellenborough and Lord Chancellor Eldon, though the conditions of the Napoleonic War gave Lord Stowell the opportunity to develop Admiralty, and particularly, prize law and the work of Eldon substantially completed the statement and systematization of the principles of equity.

1832–1914. The nineteenth century witnessed great changes in the judicial system, stimulated chiefly by such reformers as Bentham, Romilly, and Brougham.

In 1846 the modern county courts were created, the ancient county courts having completely declined.

The Court of Chancery was still a one-judge court, though formal or detailed business was delegated to subordinates. Enormous arrears of business accumulated and in 1813 a Vice-Chancellor was appointed and in 1833 the Master of the Rolls was authorized to sit as a court. In 1842 two more Vice-Chancellors were appointed, but appeal lay from all to the Chancellor. In 1851 the Master of the Rolls and two Lords Justices were constituted the Court of Appeal in Chancery with a further appeal to the House of Lords.

In the common law courts King's Bench, Common Pleas, and Exchequer had co-ordinate jurisdiction in many classes of case. In 1830 a new court of Exchequer Chamber was constituted consisting of the judges of all three common law courts, appeals from any one court to be heard by the judges of the other two.

On the criminal side the practice had developed of trial judges reserving difficult points for discussion with their colleagues, and in 1848 this was sanctioned by the creation of the Court for Crown Cases Reserved. Not till 1907 was the Court of Criminal Appeal created.

In 1857 the jurisdiction of the ecclesiastical courts in probate was transferred to the Court of Probate and judicial dissolution of marriage was introduced by the Matrimonial Causes Act, 1857.

A series of statutes introduced procedural reforms; the Uniformity of Process Act introduced one form of writ for commencing actions in all three common law courts; the Real Property Limitation Act 1833 abolished almost all the real actions; the Civil Procedure Act 1833 created a new system of pleading and the Common Law Procedure Acts, 1852, 1854 and 1860, abolished the forms of action and allowed equitable pleas to be considered in common law actions. Reforms in chancery procedure enabled that court to award damages, and even try cases by jury.

The Judicature Act, 1873, abolished in 1875 all the old central courts and created the Supreme Court of Judicature, consisting of the Court of Appeal and the High Court, which was organized in five divisions (Queen's Bench, Common Pleas, Exchequer, Chancery and Probate, Divorce and Admiralty). In 1881 the first three were merged in the Queen's Bench Division. Appeal to the House of Lords, abolished by the Act, was reinstated before the Act became effective, and in 1876 the Appellate Jurisdiction Act authorized the appointment of Lords of Appeal in Ordinary to sit both in the House of Lords and in the Judicial Committee of the Privy Council. A formal fusion of law and equity was effected by the provision that all judges might give effect to both legal and equitable rights and remedies, the equitable to prevail in cases of conflict, but this has affected the administration of both systems, rather than their principles, which remain largely distinct.

The system of precedents and the rule of *stare decisis* became fully established in the nineteenth century with the settling of the hierarchy of courts and by the establishment of the House of Lords as a court with a professional staff.

The nineteenth century also saw the writing of a large number of treatises and texts, many of which, in revised editions and rewritten versions, have served generations of lawyers, and some of which, such as Pollock's works on *Contract* and *Torts* have considerably influenced the development and rationalization of the law.

Numerous attempts were made in the nineteenth century to restate parts of the law in more systematic form, the most successful being the Larceny Act, 1861, and the Offences against the Person Act, 1861. In 1878 a draft Criminal Code was discussed in Parliament, but was never enacted.

Later, partial codification achieved its few successes with the Bills of Exchange Act, 1882, the Partnership Act, 1890, the Sale of Goods Act, 1893, and the Marine Insurance Act, 1906.

The extension of commerce and industrialization had their influence on many branches of law, resulting in great developments in many kinds of contracts and the growth of liability for negligent harms caused in industrial, mining, and railway accidents. Vicarious liability, common employment, strict liability under the rule in *Rylands* v. *Fletcher* all developed greatly. Company law developed, largely out of agency and partnership but from mid-century regulated by statute. The principle of limited liability was accepted in 1855.

In the latter half of the nineteenth century a great number of important principles were settled and stated by distinguished judges, in many cases in terms ever since accepted as authoritative.

Since 1914. Among the most significant developments since 1914 have been the growth of administrative law, of delegated legislation, administrative tribunals, and the quasi-judicial powers of administrators. Vast bodies of social welfare legislation have been created. Taxation has steadily become more onerous and more important and influential. Increasing crime, not least by reason of the growth of motor transport, has brought difficulties of detection and enforcement, but the treatment of criminals has become steadily more humane, varied, and adapted to varying kinds of criminals. The amount of legislation and the amount and complexity of control of the individual have steadily multiplied.

In judicial organization the Court of Appeal and the Court of Criminal Appeal were merged in 1966 into the Court of Appeal with Civil and Criminal Divisions, and a Restrictive Practices Court established in 1956 and the Employment Appeal Tribunal in 1975. The Crown Court replaced Assizes and Quarter Sessions in 1971; many ancient local courts disappeared, while a great number of statutory tribunals appeared.

The establishment, first in 1934 of the Law Revision Committee, of the Criminal Law Revision Committee in 1959, and eventually, in 1965, of the Law Commission has marked a determined effort to subject the law to systematic reconsideration and revision. In consequence while common law remains important, English law is more and more a system of statute law, but not a codified system.

F. W. Maitland, *Constitutional History of England* (1908); J. Jolliffe, *Constitutional History of Mediaeval England* (1954); D. L. Keir, *Constitutional History of Modern Britain* (1975); T. F. T. Plucknett, *Concise History of the Common Law* (1956); F. Pollock and F. W. Maitland, *History of English Law Before the time of Edward I* (1898); W. Holdsworth, *History of English Law* (1936–66); L. Radzinowicz, *History of English Criminal Law* (1948–68); E. Foss, *Judges of England* (1870).

II. *Sources and literature*

The major historical sources of English law are common law, equity, and legislation. The common law, based on custom, has been developed by royal justices and courts since the Norman Conquest as a body of rules common in nearly all respects to all England. Equity developed because, by the fourteenth century, the common law was becoming formal, rigid, and inflexible, and there was need for some exercise of discretion in the application of the law; petitions to the King in Council for such discretionary remedies were delegated to the Chancellor and gradually Chancellors built up a system of principles in accordance with which they granted remedies outwith and supplementary to the common law. Legislation, or deliberate law-making and law-changing, developed as first the royal power and subsequently the power of Parliament increased; from the fifteenth century statutes have been recognized as of absolute and literal authority. Other historical sources of lesser influence have been proven trade or local customs, the general mercantile and maritime law, the canon law of the Western Church, and Roman law. As time has gone on legislation has become increasingly important and has bulked ever more largely in English law, and has come to be the major means of legal development.

The formal sources of modern English law, which yield authoritative statements of rules of law, are legislation, judicial precedents, and the principles laid down in books of authority.

Legislation. Legislation comprises Statutes or Acts passed by the Parliaments of England down to 1707, of the Parliaments of Great Britain from 1707 to 1800, and of the Parliaments of the United Kingdom from 1801 to date, all so far as unamended and unrepealed, and except in so far as the legislation is applicable only to Scotland or to Ireland (since 1921 Northern Ireland). It also comprises delegated legislation, namely Rules, Regulations and Orders (now generally known as Statutory Instruments (q.v.)) made by Ministers of the Crown and others to whom authority is delegated by Parliament in particular Acts, to make rules having statutory authority and force on matters of detail for the working out in practice of the particulars laid down in the Acts.

Before legislation can supply a rule to determine a dispute or difference of view, a legal adviser, or ultimately in case of doubt, a court, must interpret the relevant legislation, i.e. determine what it means in the context of the dispute.

Judicial precedents. Judicial precedents are the recorded decisions of judges of the superior courts in cases brought before them, stating their decisions and their reasons for arriving at them, which are referred to in later cases as influencing, and sometimes compelling, the judge in the later case to decide in accordance with the principle or rule applied in the precedent case. A precedent is not merely a previous decided case, nor an example of the application of a rule, but a decision containing a principle or rule which may, or sometimes must, be used subsequently for the decision of a later case raising the same or a closely related point. The system of following precedents has developed because of the judicial habit of not only deciding a case but giving a reasoned exposition of the relevant law applied, and because of the desire of later judges to utilize the wisdom of their predecessors and to decide a case consistently with previous decisions in similar cases.

These recorded decisions are mainly contained in the Year Books (q.v.) and then in the various series of law reports (q.v.) which began in the sixteenth century and have continued ever since, but decisions not reported in any series of reports, but only in a book or newspaper or transcript, if vouched by a member of the Bar who was present when judgment was delivered, are equally accepted as precedents. By no means all decisions are reported. The law reports, which are published for the information of lawyers, are distinct from the official court record, which treats the case as a dispute disposed of and not as an example of the development or application of some principle and a possible source of a rule on which to decide a future case.

Precedents are of value in subsequent cases only in so far as they disclose the principle or ground of law on which the decision was reached; the bare decision is of little value and binds only the parties, but the ground of decision may be fundamental and rule many subsequent cases raising the same or closely related questions.

Precedents are distinguished as binding precedents and persuasive precedents. A binding precedent is one which a subsequent judge considers himself bound to apply to a case before him, a persuasive precedent one which exercises persuasive influence on him in a case before him. Whether a particular precedent is binding or persuasive in a particular case depends on the standing in the judicial hierarchy of the court which decided the precedent case in relation to the standing of the court now deciding the later case. There has developed in England a very rigid doctrine of precedent, as follows:

The House of Lords, until 1966, considered itself bound by its own previous decisions in English cases; since then it has declared its power in exceptional circumstances to reconsider its own decisions; it is not bound by the decisions of any other court, but such decisions may be persuasive. Its decisions bind all lower English courts, whether or not appeal lies to the House.

The Court of Appeal is bound by decisions of the House of Lords in English cases and, probably still, at least in general, by its own previous decisions. The civil and criminal divisions probably regard themselves as each bound by the other's decisions though each will not often have to deal with other's principles of law. Its decisions are binding on all lower English courts.

Divisional courts (q.v.) of the High Court are bound by decisions of the House of Lords or of the Court of Appeal, and their decisions in civil matters bind judges of the same Division of the High Court and in criminal matters bind lower criminal courts.

Single High Court judges' decisions, though persuasive, are not binding on other single judges, but bind lower courts. Decisions of the Crown Court on circuit are in the same position.

The decisions of lower courts are only occasionally reported and not binding on other lower courts.

Persuasive precedents include all decisions on similar facts which are not binding, including decisions of the superior courts of Commonwealth countries, the U.S., Scotland, and Ireland, on matters on which the law is common to, or very similar to English law. The precise extent to which a precedent is persuasive depends on such factors as the branch of law in question, the importance of the court and the eminence of the judges who decided the precedent, the age of the precedent and the intrinsic quality of the judicial reasoning in it. The advice of the Judicial Committee of the Privy Council is not binding on any English court or judge, but the decisions may be of high persuasive value, as the Judicial Committee includes persons who also sit as Lords of Appeal in the House of Lords.

Precedents are binding or persuasive only in respect of the *ratio decidendi* or the principle or principles of law which the judge relied on as justifying his actual decision on the facts of the precedent case. This principle is not always explicitly stated and it is sometimes not clear what the *ratio* was, on what principles the judge reached his decision. Decisions on facts, such as that certain conduct was negligent, are not precedents. The actual decision binds the parties to the case, but it is the principle on which the case was decided that may be valuable as a precedent for the future. There may be several points decided in one case, in which case there must be a *ratio* in respect of each point, and several judges' opinions, each of which may proceed on a different *ratio*.

There is no fixed or invariable way of discovering the *ratio* of a decision, but it is the principle of law which justifies the actual decision of the precedent on the facts which the court held in that case to be admitted or proved and to be necessary for or material to the decision. The *ratio* is no wider than the principle necessary for the decision of the precedent. Judges frequently make statements which are much more general than are necessary for the decision of the case, as where in a case as to liability for a harmful bottled drink judges made statements about liability for food or drink, or for manufactured products.

Distinct from the *ratio decidendi* are *obiter dicta*, which are judicial statements of, or observations on principles of law which are, however, not the basis of nor necessary for, or are wider than necessary for the decision of the precedent, or statements of the law based on facts not found to exist or not material to the case or analogies, or merely gratuitous observations of matters of law not relevant to the decision. Statements in dissenting opinions are entirely *obiter*. Such *dicta* vary in weight and importance from carefully formulated statements not necessary to the decision to mere observations made in passing. Though never binding *obiter dicta* may be persuasive, even highly persuasive, and may be adopted on their merits in a later case.

It is for a judge in a later case, dealing with a precedent, to determine whether in the circumstances it is binding on him or merely persuasive, and to find what the *ratio* of the decision is and what statements are to be disregarded as mere *obiter dicta*. It may be difficult to determine the *ratio* of a precedent, particularly where it was decided by an appellate court and different members of the court reached their conclusions on different grounds. Courts have occasionally expressed their inability to discover what the *ratio* was of a precedent or to say on what principles it was decided.

Precedents may be handled in various ways in later cases. A precedent, or rather the *ratio* thereof, may be 'applied' or 'followed', where the same principle is used, to decide the later case; or it may be 'distinguished' where the judges in the later case point to differences in the facts between the precedent and the current case which they say are sufficiently material to justify them in not following the precedent but applying some other principle. Sometimes later courts draw very fine distinctions so as to be able to distinguish a precedent by which they do not wish to be bound; or it may be criticized, or 'doubted', or views reserved on its soundness; or it may be 'overruled', which is only possible where the later case is before a court higher in the judicial hierarchy than the court which decided the precedent, where the later court declares the precedent to have been wrongly decided or to contain an erroneous statement of law. When a precedent is overruled the effect is retrospective; what it stated is now declared never to have been good law. A precedent may also, of course, be in effect overruled by subsequent statute, but the effect in this case is prospective only. Apart from distinguishing a precedent a later court may decline to follow a precedent if satisfied that it was decided *per incuriam*, such as without having considered some relevant statute or precedent or otherwise by mistake, but not merely that it was not fully argued or appears to take a wrong view of a statute or precedent.

Books of authority. Books of authority are books, all of some antiquity, which, though setting out the views of the writer on the law rather than stating principles and rules laid down by the authority of Parliament or of the courts when deciding actual disputes, have been traditionally accepted as containing authoritative statements of legal principles and rules, in many cases because they contain the earliest statement of a rule consistently applied subsequently and not subsequently changed by statute or precedent. In civil law the recognized books of authority are Glanvill's *De Legibus et Consuetudinibus Angliae* (1187), Bracton's *De Legibus et Consuetudinibus Angliae* (*c.* 1250), Littleton's *Tenures* (*c.* 1481), Coke's four *Institutes of the Laws of England* (1628–44), and Blackstone's *Commentaries on the Laws of England* (1765–69) (qq.v.).

In criminal law the books of authority are Coke's third part of the *Institutes* (1644), Hale's *Pleas of the Crown* (1678), Hawkins' *Pleas of the Crown* (1716–21), Foster's *Crown Law* (1762), the fourth book of Blackstone's *Commentaries* (1769), and East's *Pleas of the Crown* (1803) (qq.v.).

Legal literature. Legal literature includes all textbooks other than the recognized books of authority; they are not sources of the law but only repositories of information on the law which may be adopted in argument. Nevertheless many of them have survived a long time, been repeatedly edited and used, attained a high reputation and have considerable persuasive authority, though a statement in a textbook, however much respected, is never authoritative by itself and must always give way to statute or judicial decision, and courts have sometimes criticized or even stated to be erroneous statements in standard textbooks. Similarly articles in legal periodicals have sometimes had persuasive value.

Other sources. In default of guidance from statute, books of authority, or textbook, a judge may derive guidance on the decision of a case from examination of the approach of other legal systems within the U.K., or other systems within the Commonwealth or the U.S. founded on the same basis of the common law, or Roman law, or in the last resort from his moral beliefs and his concept of justice.

W. S. Holdsworth, *Sources and Literature of English Law* (1925); C. K. Allen, *Law in the Making* (1964); R. J. Walker and M. G. Walker, *The English Legal System* (1976); R. Cross, *Precedent in English Law* (1968); W. F. Craies, *Statute Law* (1975).

III. *Judicial organization*

Superior courts. The highest court is the House of Lords, which exercises the judicial function of Parliament. In theory appeal to the House of Lords is an appeal to the whole House but in practice, particularly since the Appellate Jurisdiction Act, 1876, created a group of salaried life peers, the Lords of Appeal in Ordinary, or 'law lords', there is an established convention dating from 1844 that lay peers do not participate in judicial sittings of the House. Appeals are referred to an Appellate Committee of the House. By that Act an appeal must be heard by at least three of the Lord Chancellor, the Lords of Appeal in Ordinary, and such peers as hold or have held high judicial office. The House has almost entirely appellate jurisdiction only, in civil and criminal cases from the Courts of Appeal in England and in Northern Ireland and in civil cases only from the Court of Session in Scotland.

The Court of Appeal sits in both civil and criminal divisions. The Civil division hears appeals from the High Court, county courts, the Restrictive Practices court, certain special courts, and certain tribunals, such as the Lands Tribunal. The Criminal division hears appeals by persons convicted on indictment in the Crown Courts.

The High Court in its civil jurisdiction is divided into three Divisions (Queen's Bench, Chancery, and

Family (formerly Probate, Divorce and Admiralty)) to each of which certain kinds of cases are assigned. Divisional courts (q.v.) of each of the divisions, consisting of two or more judges, have limited appellate jurisdiction in certain cases. The main civil jurisdiction is exercised by single judges hearing cases of the kind appropriated to the divisions to which the judges belong.

The criminal jurisdiction of the High Court is exercised exclusively by the Queen's Bench Division. A divisional court of two or three judges of that Division deals with appeals from a Crown Court and magistrates' courts, and also exercises the supervisory jurisdiction of the court, issuing the prerogative writ of habeas corpus and to ensure that magistrates' courts and inferior tribunals exercise their power properly, by granting orders of mandamus, prohibition and *certiorari.*

The Crown Court, created in 1972, replaces the former assizes and quarter sessions. It exercises criminal jurisdiction and sittings are held regularly at major towns throughout England and Wales. It comprises judges of the Queen's Bench Division of the High Court, circuit judges and Recorders (part-time judges). They sit singly with juries trying persons charged on indictment with crimes. A judge of the Crown Court sits with two to four justices of the peace to hear appeals from magistrates' courts and proceedings on committal by magistrates to the Crown Court for sentence.

The Central Criminal Court, known as the Old Bailey, is a sitting of the Crown Court, having criminal jurisdiction only, over indictable offences committed in Greater London or on the high seas. The court consists of *ex officio* judges and in practice consists of judges of the Queen's Bench Division, the Recorder of London, the Common Serjeant, and certain additional judges of the Central Criminal Court.

Inferior courts. County courts have exclusively civil jurisdiction, which is limited in extent and in area, and which is entirely statutory. The judges are persons who also hold office as Circuit judges of the Crown Court.

Magistrates' courts consist of a stipendiary magistrate, or of from two to seven (usually two or three) lay justices of the peace; a single lay justice has a very limited jurisdiction. Magistrates' courts have civil jurisdiction in relation to certain debts, licences, and domestic proceedings. In the exercise of criminal jurisdiction one or more justices may sit as examining magistrates to conduct a preliminary investigation into an indictable offence. A magistrates' court may try summarily many minor statutory offences, and also certain offences if the prosecutor applies for the case to be heard summarily, the court agrees it is a suitable mode of trial and the defendant does not elect jury trial. Cases may be appealed to the Crown Court or defendants remitted for sentence to the Crown Court.

Coroners' courts are principally concerned with inquests into the deaths of persons in prison, from unknown causes, or which appear to have been violent or unnatural. A jury of seven to 11 may be summoned and the procedure is inquisitorial.

Courts of Special Jurisdiction. Courts of special jurisdiction, which have jurisdiction in special fields only outside the range of the ordinary civil and criminal courts, include the Restrictive Practices Court, which considers whether agreements are restrictive of the conditions of supply of goods and in the public interest, and applications for exemption from the statutory ban on the maintenance of resale prices; Courts-martial, which exercise jurisdiction over members of the armed forces, with appeal to the Courts-Martial Appeal Court; the Court of Protection, which consists of a Chancery judge dealing with the property of persons of unsound mind; Ecclesiastical courts, dealing with discipline of clergy within the Church of England; the Court of Chivalry, concerned with disputes over the right to use armorial bearings and ensigns; the Employment Appeal Tribunal which handles appeals from industrial tribunals; the Patents Appeal Tribunal, and certain others.

Administrative tribunals. There is a vast number and variety of tribunals created by various statutes adjudicating on questions and disputes, principally those arising from or in connection with the working of public administration, nationalized industries and services and social services. Important ones include industrial tribunals, rent tribunals, rent assessment committees, National Insurance Commissioners and local tribunals. Some have a legally qualified chairman and from some appeal lies on a question of law to the ordinary courts. A Council on Tribunals was established in 1958 to act, not as a court of appeal from tribunals, but to keep under review and report on the working of tribunals.

Domestic tribunals. In the case of certain professions tribunals have been established under statute to exercise disciplinary jurisdiction, and in other cases similar non-statutory tribunals have been established to exercise discipline over members or resolve disputes between members. In some cases appeal lies to the ordinary courts.

Halsbury's *Laws of England*, s.v. Courts.

IV. *Public law*

There is no written constitution and the major principles of constitutional law are found in various statutes and the principles underlying various decided cases, in the law and custom of Parliament supplemented by conventions, or customs and usages. The monarchy, Parliament, and the Cabinet are common to the English and Scottish legal systems, though in certain details the law applicable differs in the two systems, and the rules prescribing the rights and duties of Ministers, and public and local authorities, in certain respects differ materially.

Since 1947 the Crown and Departments of State no longer enjoy their former near-immunity from civil action, but are generally liable in contract, tort, and as occupier of property in the same way as a private individual (see CROWN PROCEEDINGS). The Crown, however, still retains certain procedural advantages.

A notable feature is the use by the courts of the prerogative writs (now orders) of mandamus, prohibition and *certiorari* (qq.v.) to control the acts of inferior officials and courts, but ordinary civil remedies, such as action for declaration, injunction or damages, are also extensively employed against public authorities.

The fundamental liberties of the subject are secured not by constitutional guarantee but by ordinary courts and ordinary legal actions; liberties are not guaranteed but interferences with liberties are restrained. These liberties can be abridged or abolished by ordinary legislation and the courts may be powerless in face of oppressive legislation.

The application of public law is partly in the hands of Departments of State, partly of public corporations, and partly of local authorities. Departments of State are responsible for, *inter alia*, finance, foreign affairs, national defence, the promotion and general direction of industry and trade, the national health service, and the system of pensions and social security benefits. Some of these departments, such as Energy, Education and Science, operate largely through public corporations and local authorities rather than providing and controlling services directly.

Since 1945 numerous public corporations have been established by statutes to run nationalized industries, such as the steel industry, operate public utilities, such as gas and electricity, manage public services, such as the postal system and groups of hospitals, and perform similar functions. These are sometimes financed by Parliament, sometimes by charges for their services, but are always accountable to Parliament, a particular Minister having considerable ultimate power. The members of the boards of public corporations are nominated by Ministers, not elected.

Local government is effected by county councils, (certain of which are metropolitan counties) and district councils, all composed of members popularly elected but served by large qualified staffs, and having in different degrees in their areas numerous and important statutory rights and duties, particularly in such fields as housing, health and sanitation, town planning, roads, and education. Local government services are financed partly by grants from the national Exchequer, partly from rates levied on land and buildings, and partly from charges for services rendered and other sources. Much local authority activity is controlled and checked by departments of state. A distinct system applies to London which is divided into London boroughs with major functions reserved to the Greater London Council.

The police is not a national service, but a number of forces organized and maintained by groups of local authorities, the tendency being to amalgamate smaller forces into larger, while there is much mutual assistance and co-operation and regional crime squads. The Home Office has important supervisory functions.

The system of national taxation is very complicated comprising, as main elements, customs and excise duties, income tax, capital gains tax, corporation tax, capital transfer tax, and value-added tax (formerly purchase tax and selective employment tax). The annual taxes require statutory authority each year. The rates of taxes are normally fixed each April when the Chancellor of the Exchequer makes his Budget speech, which is followed by the annual Finance Act.

V. *Criminal law*

English criminal law has developed over a very long period. Though it is uncodified, many major crimes have, particularly since 1860, been defined by statute, and in the twentieth-century statutes and regulations thereunder have created many offences while road traffic offences under statute have become the most numerous crimes. There remain, however, many crimes defined only by common law.

The distinction of crimes into treasons, felonies and misdemeanours (qq.v.) existed at common law and was not abolished till 1967. From the procedural point of view the major distinction is between indictable offences (i.e. those which must be tried by jury) and summary offences, (i.e. those which must be tried by a stipendiary magistrate or lay justices without a jury), though many offences may be tried either on indictment or summarily, as the court thinks appropriate, and in certain cases an accused may elect jury trial.

There is no State or public system of prosecution. It is the duty of the Attorney-General to institute prosecutions for crimes which tend to disturb the peace of the State or endanger the government. The Director of Public Prosecutions has the duty of instituting criminal proceedings in cases referred to him by government departments or in cases of difficulty or importance, or as may be directed by the Attorney-General. Some statutes require that criminal proceedings be taken only by order of a judge, or by the direction or with the consent of the

Attorney-General, the D.P.P., or some other person or body. But apart from these, and of statutory provisions, any person may on his own initiative, and without any consents, institute criminal proceedings with a view to an indictment by laying an information before a justice of the peace. In practice the police, theoretically as private individuals, initiate most criminal proceedings.

The basic principle of the criminal law is that a person is guilty of crime only if he has, by act or omission, caused or permitted conduct, or brought about a result, which falls within the definition of a recognized crime, with the mental state requisite for the particular crime (usually intention or recklessness). In some cases of statutory offences, however, the accused may be convicted if he caused, or allowed, the prohibited thing to happen, irrespective of his state of mind, and in some other cases the statute indicates the state of mind requisite for the commission of the crime.

Persons under 10 cannot be convicted of crime and a person of unsound mind is excused if he can establish that he did not know the nature and quality of the act, or if he did, that he did not know that it was legally wrong.

The major groups of crimes are offences against the safety of the State (e.g. treason, sedition) or public order (e.g. riot, breach of the peace), offences against the person (e.g. murder, manslaughter, assault, sexual offences), offences against property and possession (e.g. robbery, theft, receiving, malicious damage to property) and miscellaneous offences (e.g. forgery, road traffic offences, and contraventions of many statutes and regulations). Treason is still based on the Treason Act, 1351.

Down to the early nineteenth century punishments were cruel, the death penalty being common. Capital punishment for murder was abolished in 1969, being replaced by a mandatory sentence of life imprisonment. Other normal penalties are imprisonment, suspended sentence of imprisonment, probation, and fine.

Special measures are competent in the case of young offenders, including Borstal training, detention in a detention centre, committal to community homes, attendance centres and lesser penalties. The emphasis is on reformation of the offender though frequently the hope is unjustified.

J. F. Archbold, *Criminal Pleading, Evidence and Practice* (1976); J. C. Smith and B. Hogan, *Criminal Law* (1978); W. O. Russell on *Crime* (1964); G. L. Williams, *Textbook of Criminal Law* (1978).

VI. *Civil (private) law*

There are no accepted divisions of civil (or private) law and the divisions used are made primarily for purposes of study and exposition. Until 1875 the major distinction was into common law and equity; technically this no longer exists though it underlies and explains many rules of law and practice; the distinction between civil and commercial law has never been important in English law and has not existed since principles of mercantile law were absorbed into common law in the seventeenth and eighteenth centuries.

There are no civil codes. A large part of the civil law is still founded on principles stated in judicial precedents but statutes have, in various particulars, altered or restated, or innovated upon, the principles worked out by the courts, sometimes to a major extent, and in many spheres most of the law consists of the judicial interpretation and application of statutory principles.

Law of persons or family law. Family law was mainly in the hands of the ecclesiastical courts till 1857, when jurisdiction in matrimonial causes was transferred to the new Court for Divorce and Matrimonial Causes, which in 1875 was amalgamated into the Probate, Divorce and Admiralty (now Family) division of the High Court. The court may grant divorce, dissolution of marriage on the ground of presumed death, nullity of marriage, judicial separation, and restitution of conjugal rights. Lesser matrimonial remedies can be granted by magistrates' courts.

Marriages may be contracted civilly, in a registrar's office, or in church, accompanied by ecclesiastical ceremony. The age for marriage is 16 for both sexes but young people under 18 require parental consent, or the consent of a magistrates' court, to marry. A marriage may be declared a nullity if vitiated by lack of true consent, unsoundness of mind, the existence of venereal disease, pregnancy *per alium*, wilful refusal to consummate, incapacity to consummate or on certain other grounds.

At common law, in general, the property of both spouses was vested during marriage in the husband but a series of statutes in the nineteenth century changed the rules to one of each spouse owning his or her property quite separately. Married women have full independent contractual and tortious powers and liabilities.

Divorce was competent only by private Act of Parliament till 1857 when the court was empowered to grant it for adultery. Other grounds were added later. In 1970 the former grounds of divorce (adultery, three years' desertion, cruelty, insanity, rape, sodomy or bestiality) were replaced by the sole ground of the breakdown of the marriage, which may be evidenced by statutorily specified indicia.

In both nullity and divorce cases the court has wide powers to order the payment of maintenance and to order the variation of ante-nuptial or post-nuptial settlements. The court has power to deal with the custody of children of parties to nullity or divorce proceedings.

The position of illegitimate children has been much improved over the last century and for nearly all purposes they are in the same position as those born legitimate. Legitimation of children *per subsequens matrimonium* was introduced in 1926; such children rank for all purposes as if born legitimate. Adoption of children was introduced in 1926 and requires various consents and the authority of the court. An adopted child ranks as the child of its adoptive, and not of its natural, parents.

Companies. Most industrial and commercial enterprises are carried on by incorporated companies of which the commonest kind are companies with the liability of members, i.e. shareholders, limited to that of the nominal value of the shares taken. A major distinction is into large public companies and small private companies with not more than 50 members. Successive Companies Acts have included increasingly stringent provisions against fraud by promoters or directors, requirements of disclosure of accounts, of auditors' reports, of the identity of large shareholders, and the like.

In theory companies are democratically controlled, the members electing the directors, and the wishes of the majority of members, but in practice companies are controlled by the directors, who are only rarely ousted by shareholders.

Law of contract. The law of contract is still mainly judge-made. The essence of contract is an enforceable bargain produced by the agreement of parties, by their arrival after negotiation at *consensus in idem* (q.v.).

An important principle is that of consideration, namely that no simple contract, i.e. one not under seal, is enforceable unless each party can show that, by entering into the contract, he either confers a benefit on the other party or incurs some detriment himself. Consideration must be real but need not be adequate, must have been furnished by the promisee, and must have been given in respect of the other party's promise and not be 'past'.

Mistake, fraud and undue influence as grounds for challenging a contract have all been developed entirely by judicial decisions, in which equitable ideas have played a considerable part.

In the sphere of discharge of contract judicial empiricism has developed the doctrine of frustration, that a contract is terminated by unforeseen events, or fundamental change of circumstances, even though performance has not been rendered absolutely impossible.

Among specific contracts, several, notably sale and hire-purchase of goods, are now mainly regulated by statute, but others are regulated by common law. To a large extent contracts are in practice in standard form, or use terms designed to obviate the consequences of inconvenient judicial decisions, the result being to minimize the area for application of the idea of agreement. Sale of goods, hire-purchase, marine insurance, and the contracts implied by bills of exchange are all chiefly regulated by statutes.

Nationalization has transferred most inland carriage into the hands of public corporations who carry on standard terms and common law principles are largely excluded, but carriage by sea is still largely common law only modified by statute.

The contract of employment, basically one of common law, has been greatly developed by statute and affected by statutes, such as the Employment Protection Acts and Factories Acts, passed to secure proper standards in employment. Wages are partly regulated by statute, but largely fixed by collective bargains between trade unions and employers. Strikes and union pressures are affected by a complicated mixture of common law and *ad hoc* statutory provisions. Arbitration to settle industrial disputes is not compulsory.

Law of torts or civil wrongs. Torts (q.v.) have been heavily influenced by historical development, and to a large extent individual kinds of wrongs have been recognized and harms not falling within a recognized category were not regarded as actionable. Most have developed from trespasses, which were direct and forcible injuries, and actions on the case, which were harms resulting otherwise from the defendant's act or omission. The whole branch of law is predominantly judge-made with some statutory corrections.

Today there are recognized trespasses against the person, to goods, and to land, conversion and detinue, fraud, nuisance, libel and slander, and negligence, which consists in the defendant's failure to observe the standard of care required by law not to harm the plaintiff in person or property.

There are also recognized certain cases of strict liability where the defendant is liable, independently of intention or negligence, if harm results, and in modern law statute has in many circumstances been held to impose liability for harm, which liability may be more strict than at common law, or even absolute in that no defences are admissible. The major theoretical difference of opinion is between the cases of liability based on fault, and strict liability for having created the risk of harm.

Vicarious liability, notably that of employer for harm done by his employee, is extensively developed, and liability of employer for even his independent contractor has tended to extend.

In certain fields, notably the liability of occupiers of premises for injury to persons thereon, the common law had become so complicated that it has been replaced by reforming statute. In others, such as defamation, statute has materially amended the common law.

There remain several questionable torts, such as infringement of privacy, and torts the boundaries of which are uncertain and vague; their development still leaves scope for judicial ingenuity and boldness.

A point of particular controversy is of the extent of liability, of how far a defendant is liable for consequences resulting more or less remotely from his initial wrongful act or omission.

At common law the contributory negligence of the plaintiff was a complete defence, but since 1945 that fact justifies reduction of damages having regard to the plaintiff's own share in responsibility for the harm.

In cases of industrial injuries an injured person may claim industrial injury benefit under the social security scheme as well as claiming damages, though if he recovers damages account is taken of the benefits recovered. In cases of injuries caused by criminal conduct an injured person may claim criminal injuries compensation (q.v.).

Law of property. Property is divided into real property (land and buildings) and personal property (leaseholds and moveables, and rights such as shares in companies).

Down to 1925 the land law was basically feudal, parcels of land being held from superior tenants who held from the King, each tenant owing duties to his superior for his land. No man, save the King, owned his land outright; he owned an estate in the land, certain defined rights, never amounting to full ownership. Freehold (q.v.) estates in land might be fee simple (q.v.) estates, estates in fee tail (q.v.), or life estates. There also developed other interests in land, notably the copyhold, whereby tenancies were held by copy of the record in the court roll of the manor in which the land was situated and the term of years, which developed into the leasehold.

Both common law and equity contributed many principles and rules to the elaboration of a very complicated body of law, and latterly rights in land might be either legal or equitable. In particular equity developed the idea of the use, later the trust, under which much land was held by one person for the benefit of others. Settlements of land tied up land in a family but from 1864 the powers of a life tenant of settled land were greatly enlarged.

In 1925 sweeping changes were made in real property law especially for conveyancing purposes and in consequence much of the modern law is statutory. All land is now held by freehold tenure, and the feudal incidents of tenure have disappeared, so that the tenant in fee simple is for all practical purposes the absolute owner of the land. Copyhold has gone. The only legal estates in land now possible are the estates in fee simple absolute in possession (q.v.) and terms of years absolute (q.v.). Some other interests, such as easements and profits (q.v.), may still be legal, but all other interests are equitable only.

Registration of title to land was begun in 1862 and is spreading over England and Wales as areas are made subject to compulsory registration, but is still far from being complete. This has greatly simplified conveyancing. Under the Land Charges Act, 1925, third party rights in land are registrable. Despite the 1925 reforms land law is complicated, and an increasing variety of charges exist, not all of which are registrable.

The use of land is severely restricted by legislation on Town and Country Planning, Access to the Countryside, and similar Acts, while compulsory purchase and expropriation are common under legislation on New Town, Housing, and for many specific projects such as new motorways.

The relations of landlord and tenant have been increasingly controlled by statute in modern times, in the case of urban tenancies by Rent Acts controlling rents and protecting occupancy and by Landlord and Tenant Acts, and in the case of rural tenancies by Agricultural Holdings and Agriculture Acts. The landlord can exercise very few contractual rights and the tenant has numerous statutory protections and immunities.

Prior to 1925 mortgages of land normally were effected by way of transfer of the freehold title to the lender, with provision for reconveyance after six or 12 months. In 1925 this form was abandoned in favour of two others, a long lease, usually for 3,000 years, without possession, extinguishable on repayment of the principal, or a legal charge conferring on the person in whose favour it was granted most of the rights enjoyed by a mortgage by way of lease. A notable equitable development was the 'equity of redemption' which permitted a borrower to redeem notwithstanding the occurrence of events which at law operated forfeiture and carried the mortgaged lands absolutely to the lender. Equity consistently declined enforcement to any provision which operated as a clog on the equity of redemption.

Personal property is divided into choses in possession, i.e. leaseholds and moveables which can be possessed, and choses in action, which are intangible and rights to which can only be asserted by action, including negotiable instruments, shares in companies, copyrights, etc. Choses in possession are largely regulated by common law, modified by statutes on such topics as sale, hire-purchase, and bills of sale, whereas choses in action are mainly regulated by the particular statutes dealing with bills of exchange, companies, copyrights, etc. developed by decisions. This branch is far from being a unified body of law.

Law of trusts. From the fifteenth century the Chancery courts recognized uses, now called trusts, whereby property was conveyed to a person on the understanding that he was to hold it on behalf of the grantor, or of a third party, and enforced the mutual rights and duties of parties to the relationship. The trust has many applications, as a means whereby property is held for persons under age, for persons in succession, for unincorporated associations, for charitable purposes, and so on.

Trusts are either private, enforceable at the interest of the beneficiaries, or charitable, enforceable at the suit of the Attorney-General on behalf of the Crown. Private trusts may be express, expressly created by the settlor by deed, writing, will, or even orally, constructive, when imposed by law in particular circumstances independently of anyone's intention, or resulting, when persons must be presumed to have intended that a trust be created.

Charitable trusts are trusts for the relief of poverty, advancement of education, the advancement of religion, or other purposes beneficial to the community. Charitable trusts are favourably regarded by the law, being largely exempt from taxation, and, provided it is clear that the settlor had a general charitable intention, charitable gifts which fail because they cannot be applied as directed will be applied *cy-près*, i.e. to another charitable purpose as near as possible to the original purpose.

Trustees are usually individuals who act gratuitously and must allow no conflict between personal interests and those of the trust. The courts view strictly the conduct of trustees and hold them personally liable for breaches of trust and unauthorized acts. Their rights, powers, and duties are largely defined by statutes, but leaving considerable room for the exercise of discretion; in particular their powers of investment are statutorily prescribed in some detail. A Public Trustee was established in 1906, and corporate bodies, such as banks and insurance companies, may also act as trustees.

Law of succession. Prior to 1926 there were separate sets of rules regulating the devolution of property on intestacy (q.v.), freeholds descending to the heir-at-law under feudal rules, and leaseholds and other personal property descending to the next-of-kin. Since 1926 all property descends uniformly, and a simple table lays down what persons have rights in the estate.

Regulation of succession by will has long been known but until the Wills Act 1837 wills of land were a matter for the royal courts and wills of personalty (q.v.) for the ecclesiastical courts. That Act established a form of will applicable to both types of property. So long as in writing, signed and attested by two witnesses, there are no formalities attaching to wills, but many problems attach to interpretation. A will is revoked expressly, by a later inconsistent will, by destruction with that intent or by marriage.

Until 1938 there was no provision to prevent testators excluding their relatives from all benefit in succession. Since that year dependents excluded may apply to the court for reasonable provision from the estate.

On death all property vests in personal representatives, who, if appointed in a will, are called executors, and if appointed by the court, are administrators. Executors cannot act until they have obtained a grant of probate of the will from the court, nor can administrators until they have obtained letters of administration. Their duties are to collect debts due to the estate, pay debts, and distribute the estate according to the will or the rules of intestacy.

Bankruptcy. Bankruptcy (q.v.) is regulated by statute. The statute prescribes circumstances constituting acts of bankruptcy, following on which a petition may be filed. When the debtor is adjudicated bankrupt the creditors appoint a trustee in bankruptcy and a committee of inspection. The debtor's whole property is vested in the trustee, who must get in and administer the bankrupt's estate, and, after meeting certain costs and charges, distribute the assets among the creditors. Certain debts have priority and a secured creditor may realize his security and prove for the balance or surrender it and prove for the whole debt.

Evidence. The rules of evidence have all developed since the seventeenth century, and developed largely in the context of the jury system; they are mainly common law, with certain statutory amendments. At common law many classes of witnesses were treated as incompetent but most of these restrictions have been removed.

The rules seek to confine evidence to the facts in issue and to exclude other matters as irrelevant, but relevant facts may be excluded at the courts' discretion if the Crown claims Crown privilege from disclosure for certain official documents, or professional privilege is claimed. The hearsay rule excludes evidence of matters not seen or heard by the witness himself, but there are many exceptions, notably admissions and confessions, and declarations by deceased persons.

There is no general requirement of corroboration of oral evidence but there are exceptions and in some cases practice requires corroboration.

Written documents (q.v.) are generally treated as containing the exclusive and conclusive evidence of the matter recorded therein, but there are many exceptions to this rule also.

Civil procedure. Though, since the Judicature Act, 1873, all Divisions of the High Court have power to grant any remedy certain matters have been exclusively assigned to each Division, though in some other cases the parties have a choice. Procedure is largely regulated by the Rules of the Supreme Court. In the Queen's Bench Division an action is normally initiated by a writ of summons, served on or accepted by the defendant, accompanied by a statement of claim.

Pleadings should contain a brief statement of the party's case alleging only the facts relied on, and neither the evidence by which they will be established nor the principles of law relied on. Each party must prepare his own evidence and submissions in law independently of the other.

Interlocutory (q.v.) matters are dealt with mainly by Masters of the Supreme Court, subject to appeal to a judge, who may give directions as to future procedure.

Apart from defamation actions, juries have almost completely disappeared in civil actions, and hearing on both facts and law is normally before a judge alone in open court. The points at issue are decided after evidence, if any, and argument on law. It is in the discretion of the court to determine by whom and to what extent the costs (q.v.) of proceedings are to be paid, but the practice is to award a successful litigant his costs.

In the Chancery Division much less emphasis is placed on oral evidence, much being dealt with on affidavit (q.v.) evidence. There are almost never juries. Most Chancery business is largely based on documentary evidence and turns on the interpretation of documents.

In the Family Division there is more reliance on oral evidence in divorce cases, and less in maintenance and custody cases.

An appeal to the Court of Appeal is by notice of motion, giving the grounds of appeal. At the hearing further evidence may be received, but the Court of Appeal will not normally allow a party to raise a point not taken in the court below. The Court considers the materials which were before the trial judge and then makes up its own mind.

P. Bromley, on *Family Law* (1974); W. Anson on *Contract* (1975); G. C. Cheshire and C. Fifoot on *Contract* (1976); J. Salmond on *Torts* (1973); P. H. Winfield on *Tort* (1975); R. E. Megarry and H. W. R. Wade on *Real Property* (1975); C. Vaines on *Personal Property* (1973); T. Lewin on *Trusts* (1974); D. Hughes Parry on *Succession* (1966); H. Theobald on *Wills* (1963); Halsbury's *Laws of England, passim.*

VII. *Professional organization*

In England and Wales, the legal profession is sharply divided into barristers and solicitors (qq.v.) a division known under various names since the fourteenth century when professional representation developed. Those with right of audience in the common law courts were the serjeants and barristers, while the preparatory stages of an action were the responsibility of officers of court known as attorneys. Solicitors first appeared in the fifteenth century and practised in the Court of Chancery. By the eighteenth century the same persons frequently held the positions of attorney and solicitor: the distinction was abolished in 1873, as also was the rank of serjeant. Barristers had to be members of one of the Inns of Court which after the sixteenth century refused to admit attorneys or solicitors to membership. By 1739 the latter had formed the Society of Gentleman Practisers in the Courts of Law and Equity. From the seventeenth century senior barristers were promoted to the rank of Queen's Counsel.

The main distinctions are that barristers act mainly as advocates in court and advisers on problems of difficulty, and have exclusive right of audience in the superior courts, whereas solicitors mainly practice in chambers (q.v.) in non-contentious business and giving general legal advice, and deal also with the preparatory stages of litigation, having a right of audience limited to county courts, magistrates' courts and other inferior courts. Solicitors do conveyancing, administration of estates and much commercial work.

Barristers must have been called to the Bar by one of the four Inns of Court (q.v.) and are subject to the disciplinary jurisdiction of the Senate of the Four Inns of Court and the Bar. They must normally be instructed by a solicitor, not by the lay client directly, and the relation between barrister and solicitor is mandate, not employment. Barristers' fees are honoraria, not contractual fees, and cannot be recovered by action. But by ancient custom and for reasons of public policy barristers are not liable for negligence or misconduct, whether in litigious or advisory matters.

Solicitors have been admitted by the Law Society and are officers of the Supreme Court. Their remuneration is governed by orders made under statute. Solicitors are subject to the Solicitors Disciplinary Tribunal, which is a statutory body independent of the Law Society, and are liable to actions of damages for professional negligence.

VIII. *Legal education and training*

It is not a requisite of admission to either branch of the legal profession that a person has graduated at a university nor, if he has, that he has studied law there, but graduation, particularly in law, is becoming increasingly common. As the two branches of the profession are quite distinct with different requirements for qualification the provision of training and examination has developed independently.

Training for the Bar is controlled by the Council of Legal Education (q.v.) on behalf of the Inns of Court. Intending barristers must join one of the four Inns of Court (q.v.) as students, 'keep terms' (a requirement which is a relic of the former collegiate character of the Inns and of residence there, now held satisfied by dining in the Inn Hall a requisite number of times for the requisite number of terms), and pass the Bar examinations, from which limited exemptions may be obtained on the basis of university studies in law. These stages may be and often are completed while the person is a student at a university. This period of study requires normally three years. Thereafter they must, if going to practise in England, undergo a course of practical training or work in the chambers of a practising barrister as a pupil for a time.

The Council of Legal Education provides courses of instruction but graduates in law normally make little use of these facilities in respect of basic legal subjects and the courses in these are mainly geared to the needs of non-graduates. Attendance at the instruction in the more practical subjects is compulsory for those who intend actually to practise in England and Wales.

Training of intending solicitors is controlled by the Law Society. Intending solicitors must enrol with the Law society, serve as an articled clerk with a practising solicitor for five years, attend courses of instruction and pass the Solicitors' Qualifying Examinations. The requirement of attending a course of instruction may be satisfied by taking a university degree or by attending a school recognized by the Law Society, and graduates in law may be exempted from examination in certain subjects. Graduates also need serve only a shorter period under articles. Courses of instruction are provided by the College of Law, which is sponsored by the Law Society.

Halsbury, *Laws of England*, tit. Barristers; Solicitors; H. Cohen, *History of the English Bar*; *Report of the (Ormrod) Committee on Legal Education*, 1971.

English Pale. A term used from the fourteenth century for the area in Ireland which, from Henry II's expedition of 1171–72, was directly subject to the English authority in Dublin. It comprised the area round Dublin and existed until the subjugation of Ireland under Elizabeth.

There was also an English Pale round Calais during the period of English occupation, 1347–1558.

Englishry. A plea that a man found murdered was an Englishman. It is said that Canute passed a law, for the protection of his Danish followers, that the town or hundred in which a man was murdered was to be liable to a fine, it being presumed that the dead man was a Francigena or foreigner, unless it was established that he was an Englishman. Presentment of Englishry under which the fine was exacted was abolished in 1340.

Engross. Formerly to write the fair copy of a deed in a special script founded on ancient court-hand, now to write or type a deed or agreement with all formal clauses, ready for execution by the parties.

To engross was also formerly the crime of buying up large quantities of corn or other commodities, for resale at a profit. Engrossing, like the related practices of forestalling and regrating, was long regarded as a serious offence in restraint of trade and punishable both by common law and statute. All the statutes against these offences were repealed in 1844.

Enjoyment. The exercise of a legal right, corresponding to possession of corporeal property. Enjoyment if open, peaceable, continuous and of right is a mode of acquiring by prescription the right thus enjoyed. Adverse enjoyment is exercise of a legal right as if the party were entitled to do so but without the permission of the true owner thereof.

Enlargement. In real property law the development of a person's rights into a greater right. It is a course of action competent, subject to conditions, to a mortgagee who has obtained a title to land free from the mortgage by remaining in possession for 12 years, by deed to convert his title into a fee simple.

A fee tail or base fee in possession may also be enlarged into a fee simple, formerly by suffering a recovery (q.v.) and now under statute.

A long term of years may equally be enlarged into freehold.

Enlightenment. A philosophical movement, originating in the rational approach to society, religion and science of Locke and Newton, and spreading in the late seventeenth and eighteenth centuries over most of Europe. It influenced many thinkers, notable Voltaire, Montesquieu, Helvetius, Beccaria, and Rousseau, who all had the common belief that man was a rational being and could work out his salvation without the intercession of the Church. Accordingly they were opposed to tradition, feudalism, absolutism, and the dead hand of the Catholic Church. By creating this anti-clerical and rationalist attitude these and other thinkers greatly influenced thought throughout Europe in the direction of reforms and had some influence on the ideas of the French Revolution, religious toleration and a more rational approach to justice and law. Man-made law had to be examined in the light of reason and the law of nature; law was not made by edicts but was discoverable by right reason, and the means of reforming the law and restating it in accordance with reason was the code. This was the impetus to the Prussian, Austrian, and French codes.

Enquête par turbe. A special procedure of the nature of a group inquest, rather like an early English inquiry by jury, by which in mediaeval France doubtful or disputed customs could formally be proved. It was used by the agents of the French crown for various kinds of enquiries, including proof of local custom, in the twelfth and early thirteenth centuries. In 1270 it was adopted by ordinance as the normal mode of establishing local custom in litigation before the Parlement of Paris. The verdict had to be unanimous. Other courts then adopted the procedure and it became a standard procedure in northern France for proof of customs which were doubtful or disputed, and came to have important effects in preserving and defining customary rules which had previously survived by oral tradition only.

Enrol. To enter or copy a document into an official record, so called because such records were originally in the form of rolls of parchment. The Enrolment Office was a department of the Court of Chancery and later of the Chancery Division. Various statutes have at different times required particular kinds of documents to be enrolled in Chancery.

Formerly a party to a suit in equity who wished to prevent a rehearing before the judge who had pronounced the decree or order he held or to prevent an appeal to the Lord Chancellor or the Court of Appeal in Chancery could do so by enrolling the decree.

For practically all purposes enrolment was abolished by the Judicature Acts 1873–75.

Ensign. A flag which is the insigne of connection with the United Kingdom. The ensigns normally flown are the white ensign (the cross of St. George with a Union Jack in the upper canton near the staff) flown by H.M. ships, the blue ensign (a blue flag with the Union Jack as before) flown by ships in the public service other than the Royal Navy and merchant ships commanded by officers of the Royal Naval Reserve, the red ensign (a red flag with the Union Jack as before) flown by all other ships belonging to Her Majesty's subjects, the civil air ensign (light blue, a double blue cross edged with white, the Union Jack in the first quarter) flown on British aircraft and at certain airports and buildings, and the Royal Air Force ensign (light blue, in the centre three roundels red on white on blue and the Union Jack in the first quarter) flown at R.A.F. stations.

Entail or fee tail (*foedum talliatum*). In English law, an interest in land existing where land was given or devised to a grantee and to the class of his heirs specially mentioned in the gift e.g. 'to A and the heirs of his body', not to his heirs-general. The land could not be disposed of by the tenant in tail and was not forfeitable to the Crown. It could not be removed from the family so long as lineal heirs of the class specified were in existence. It served accordingly to maintain estates and pass them down undivided through a family. The entail was probably based on the *maritagium* (q.v.). Methods were devised of barring estates tail and converting them into fee simple, notably by suffering a common recovery or levying a fine. Since 1925 an interest in tail may be created in any form of property, real or personal, by way of trust, and may be barred by disentailing assurance or with the consent of the protector of the settlement. An entailed interest may be devised by will.

Entailed estates were also known in the American colonies but they were abolished in Virginia in 1776 and in most other states subsequently.

Entail (Scots law). A settlement of land to a prescribed series of successive heirs who are not the heirs-at-law of the entailer, to keep the land in the family. The entailing of lands was recognized by statute of 1685, which provided that lands might be entailed and that the deed of entail must contain prohibitions, including the three cardinal prohibitions (prohibiting alienation of the estate from the specified heirs, contracting of debt affecting the estate, and alteration of the order of succession), irritant clauses declaring null all deeds granted in contravention of the prohibitions, and resolutive clauses forfeiting the right to the estate of any heir of entail who contravened any of the prohibitions. Strict deeds of entail had also to be registered in the Register of Entails or Tailzies. Under this system many estates were entailed in the eighteenth century. But under the Act an heir's powers were so limited as to give rise to hardship, and to hinder beneficial estate management. Consequently various statutes from 1747 relaxed the fetters of entails and empowered heirs in possession to improve their estates and exercise various powers. A statute of 1848 first gave the heir in possession power to disentail. Finally a statute of 1914 prohibited the making of new entails. But large tracts of land continue to be held under existing entails.

Enticement. The wrong, now abolished, of a person attracting another away from his or her matrimonial obligation to the plaintiff, or of attracting a child away from its parent or guardian.

Entireties, tenancy by. The former tenancy of land existing where land was conveyed or devised to a husband and wife during marriage, each being deemed seised of the whole estate. After the Married Women's Property Act, 1882, husband and wife took as joint tenants and tenancy by entireties was abolished in 1925.

Entick* v. *Carrington (1765), 19 St. Tr. 1030. King's Messengers, under a warrant of the Secretary of State to search, seize and carry away papers, broke into and entered the plaintiff's house and seized his papers. Entick brought action of trespass and recovered damages. On appeal Lord Camden, C.J., held that the Secretary of State had no power to grant such a general warrant. Every invasion of private property was a trespass unless some positive law had empowered or excused it. A warrant had to be specific as to person and property. Nor was there recognized any doctrine of 'state necessity'.

Entrenched clause. A clause in a constitution or statute which can be amended or repealed only by special procedure or by a stated majority and is thereby to some extent protected from hasty or casual amendment. It normally deals with fundamental institutions or rules. Thus the (British) South Africa Act, 1910, which established the Union of South Africa, provided that no person who, under the existing laws of the Cape of Good Hope, was capable of being registered as a voter should be disqualified only on the ground of race or colour, and that the English and Dutch languages should have equal status, and that any constitutional amendment altering either of these sanctions should be valid only if passed by a two-thirds majority of both Houses of Parliament sitting together. The Statute of Westminster, 1931, conferred on the Union Parliament the power to legislate repugnantly to British imperial legislation. In 1948 the South African government proposed to put the Cape coloured voters on a separate roll and a Bill to this effect was passed by a simple majority of each House sitting separately. The Appellate Division of the Supreme Court of South Africa in 1952 held this Act invalid. The government ultimately had its way only after by statute enlarging the Appellate Division of the Supreme Court and reconstituting the Senate so as to ensure that it had a two-thirds majority, acts of obvious political manipulation.

G. Marshall, *Parliamentary Sovereignty and the Commonwealth* (1957).

Entrapment. The instigation or inducement of a person to commit a crime by a law-enforcement officer. Persons thus induced are generally treated as not guilty of the offence. It is not constituted by tricks or deception to catch criminals nor by merely providing an opportunity for committing a crime, as by leaving property unguarded but watched.

Entries, Books of. The name in mediaeval and early modern English law for collections of precedents of pleadings. All are in Latin, until English, by statute of 1731, became the official legal language. All are generally similar, usually being arranged in the alphabetical order of the writs which began the actions. The main ones are:

- *Intrationes* or *Liber intrationum*, printed by Pynson in 1510;
- *Intrationum Liber* (The 'Old Book of Entries') printed by Smythe in 1545 or 1546, or by Middilton in 1546;
- William Rastell: *Collection of Entries* (sometimes called the *New Book of Entries* or *Antient Entries* or *Old Book of Entries*), arranging alphabetically matter gathered from four older books, printed in 1566 and later editions. This superseded the older books, being larger, better arranged, up to date and provided with notes and references;
- Sir Edward Coke: *Book of Entries* (sometimes referred to as *New Entries* and *New Book of Entries* to distinguish from Rastell's book), 1614 and 1671;
- Sir Edward Coke: *Declarations and Pleadings contained in his Eleven Books of Reports*, 1650;
- R. Aston: *Placite latine rediviva: A Book of Entries*, 1661, 1673;
- William Browne: *Formulae bene placitandi: A Book of Entries*, 1671, 1675;
- Sir Humphrey Winch: *Le beau-pledeur: A Book of Entries*, 1680;
- Andrew Vivian: *The Exact Pleader: A Book of Entries* (pleadings in the King's Bench in Charles II's time), 1684;
- Richard Brownlow, (Chief Prothonotary of the Common Pleas): *Latine redivivus: A Book of Entries*, 1693;
- Henry Clift: *A New Book of Declarations, Pleadings, Verdicts, Judgments and Judicial Writs; with the Entries thereupon*, 1703, 1719;
- John Lilly: *A Collection of Modern Entries*, 1723.

Entry. The act of going on to land with the intention of asserting right in the land. It may be made by the claimant or by someone representing him. It may take the form of actual entry, or of constructive or fictitious entry, which is entry in law. Where a person has taken possession of lands without right, the person entitled to the land may make a formal and peaceable entry on the lands, declaring that he thereby takes possession of them. With the substantial disappearance of seisin, the rules about entry are of diminished importance, but peaceable entry remains a competent means of regaining possession of land from a person wrongfully in possession.

It is an offence without lawful authority to use or threaten violence to secure entry into any premises if there is someone present who is opposed to the entry, or to enter as a trespasser and fail to leave on being required to do so by or on behalf of a displaced occupier or an intending occupier.

Many modern statutes confer on public authorities right of entry on land or into buildings. In these cases entry signifies physical entering.

In relation to customs duties entry is the declaration to the customs officers of the nature of a ship's cargo.

Entry, writs of. In mediaeval English law an important group of royal writs addressed to the King's court, which alleged that a tenant of land had obtained entry by a particular stated means, thereby acquiring only a defective title. By the Statute of Marlborough, 1267, demandants were allowed to allege that the tenant had no entry save after a particular defective title, so that dealings in the land thereafter did not need to be specified in the writ. Sometimes these writs supplemented the assize of novel disseisin, and they all gave the claimant to land a better opportunity to establish his claim to the land. They were all abolished in 1833.

Epanagoge. A legal code compiled at Byzantium about A.D. 879 under the Emperor Basil I. Based on Justinian's works and on the *Ecloga* (q.v.) it was intended to be the introduction to a comprehensive collection of laws in Greek. It is original insofar as it deals with the rights and duties of the emperor, the patriarch and the other lay and ecclesiastical dignitaries. It regards Church and State as a unity presided over by the Emperor and the patriarch who collaborate for the general benefit, the former seeking to promote material prosperity, the latter caring for spiritual welfare. This theory of separatism of lay and religious powers is thought to be attributable to the patriarch Photius. The whole work seems never to have received legislative sanction. It served as a basis for the *Basilica* (q.v.) and extracts from it can be found in Slavic codes and in the later Russian Book of Rules.

Eparch. The chief government official at Byzantium, between the sixth and eleventh centuries, ranking second only to the Emperor and responsible for maintaining good order and public safety in the capital, supplies for the city, supervision of the law courts and of the conduct of industry and trade. The *Book of the Eparch*, a work of the ninth–tenth centuries, explains his jurisdiction over the corporations and tradesmen's guilds of the city which had monopolies of the supply of various commodities to it. In the twelfth century the Eparch's functions passed to other officials and by the following century the title survived as a title of honour at the court only.

An eparchy was a civil administrative district in the Byzantine Empire and the term survives for a provisional division in Greece and for an ecclesiastical administrative division in the Eastern churches.

Ephor. The title of the chief magistrates in ancient Sparta, five in number, who along with the King formed the executive. They were elected annually from adult male citizens, presided over meetings of the *gerousia* or council of elders and *apella* or assembly, and were responsible for executing their decisions. They had police powers and might in emergency arrest, imprison and take part in the trial of a king.

Epistulae. In Roman law of the classical period, imperial letters, normally sent to governors, officials, provincial assemblies or communities, containing advice or rulings on points of law. These rulings were frequently framed by one or more of the Emperor's legal advisers, but the growth of this practice of issuing imperial rulings brought about the decline of the independent advisory activity of the jurists, so that by about the early third century A.D. jurists could participate in developing the law only as officials of the Emperor. These imperial rulings from the latter second century began to be collected and cited by the jurists as authoritative; many, collected about the end of the third century, survive in Justinian's *Codex*. A very famous *epistula* is that in which Trajan replied to Pliny's requests for instructions about the treatment of Christians.

Epitome Gaii (or *Liber Gaii*). A part of the *Lex Romana Visigothorum* (or *Breviarium Alaricianum*), comprising an abridgment of Gaius' *Institutes*, derived probably not directly therefrom but from an epitome thereof, and probably a product of a law school in the West.

Epitome Hispanica. A Spanish abridgment of Roman canon law known in Italy from the sixth century.

Epitome Juliani. An abridged Latin version of 124 laws of A.D. 535–55 compiled about the same time by one Julian, a teacher of law in Constantinople, probably intended for use in Italy reconquered by Justinian's armies. It continued to be known there throughout the Middle Ages.

Equal Opportunities Commission. A body established in the U.K. under the Sex Discrimination Act, 1975, to work towards the elimination of sex discrimination, to promote equality of opportunity between men and women generally, to keep under review the working of the Equal Pay Act, 1970 and the 1975 Act and other statutory provisions which may require men and women to be treated differently. It may conduct formal investigations and make reports, undertake or assist research and educational activities. It may serve non-discrimination notices on persons committing acts contravening the legislation.

Equal Pay. The principle, long urged by feminists, that men and women should be given equal pay for equal work. The principle was incorporated in the original constitution of the International Labour Organization and reiterated in the Declaration of Philadelphia, 1944, which renewed the mandate of the I.L.O. The Council of Europe's European Social Charter restated the principle and in the U.K. the Equal Pay Act, 1970 enacted the principle, more generally by requiring equal treatment as regards terms and conditions of employment, the cardinal difficulty being to decide whether different pieces of work can be rated as equivalent.

Equality. The quality of persons or things being all of the same standard or level and all treated alike. In legal contexts equality is commonly deemed among 'natural rights', or one of the ideals and one of the attributes of justice. The basis of this is that all men are equally free-willing beings. Thus Equality before the Law is proclaimed as an ideal to be sought. The U.S. Declaration of Independence asserts as a self-evident truth that 'all men are created equal'. But such assertions ignore the obvious facts that all men are very unequal in many respects, e.g. in physical and mental powers, moral judgment, self-restraint, conduct and many other ways.

Equality has many facets of application in legal contexts. The most general application is the principle that rules of law should apply equally to all members of the community engaging in the relevant activity, and that nobody should be exempt or treated differently save for good and obvious reasons. Most, if not all, systems of law have found some privileges and immunities to be justifiable in particular circumstances.

The idea of equality has increasingly been used as a basis for legal reforms, such as extension of the franchise, the rights of women and equality of the sexes, equal opportunities in education and training, equal rights and protection for members of ethnic and other minorities, and in other respects. But there are many respects in which legal means cannot secure equality, e.g. equality of hereditary endowments, intellectual and physical capacities.

Nor does equality of rights require or justify that all should have the same quantity of rights: all men have equal rights, but not rights to equal things; all men have equal rights to own property, but no right to an equal amount of property.

Equality is not moreover a standard to be rigidly applied, but deviations from equality require rational justification. Equality in all respects would, in many cases, effect what many would consider injustice, rather than justice. Thus minors and persons of unsound mind have different legal capacity from that of adults, and young children have restricted legal liabilities. The punishment for different crimes varies according to their presumed gravity and the punishments inflicted on different individuals for the same, or the same kind of, crime, may vary according to age, degree and mode of participation, previous record, the impact of the punishment, and other circumstances. Equality of impact, not equality of kind, of punishment is the ideal. Different amounts of damages must be awarded to different plaintiffs, having regard to their different losses. Taxation is unequally imposed on different people.

Equality before the law accordingly means that all persons, save for obvious and special reasons, must be treated as having an equal capacity for acquiring and enjoying rights, and equal liability for becoming subject to legal duties and liabilities, but the actual quality and extent of the rights depend on the individual, his capacities in other respects and his conduct. This principle was formerly mainly a procedural one, and only more recently has it become, in relation to substantive law and rights, a principle of legislation.

Equality of States. In international law, the doctrine that communities recognized as states are equal in law, though not necessarily in political or economic power. The consequences are that international law and the rights and duties of states all apply as much to one state as to any other. All have sovereignty, personality in international law and the duty to respect the personality, political independence and territorial integrity of all other states. But in international organizations equality may be qualified; certain states have permanent membership of the Security Council, others only a chance of election thereto; an organization may weight voting rights according to state population, or another, unequal, basis.

Equitable. What is fair, reasonable and right. Also, what was recognized, regulated, and enforced by the courts of equity or chancery in accordance with the principles of equity (q.v.) as developed and understood by these courts, differing in many respects from what was recognized, regulated and enforced by courts of common law. Thus an equitable mortgage was and is different from a legal mortgage.

Equitable apportionment. The apportionment of a rent or other payment by persons entitled to it without the consent of the party liable to pay, binding only the persons making the apportionment. It applies also to apportionment as between tenant for life and remainderman in accordance with the rules of equity as opposed to statutory apportionment.

Equitable assets. Assets which could be made available to a creditor only in a court of equity; the distinction between them and legal assets is now unimportant.

Equitable assignment of debt. An assignment constituted by the debtor merely being informed that the debt has been transferred to a third party, without need for assignment at law.

Equitable charge. A security where the lender does not get a legal estate in the property charged, but a right recognized in equity only, as by deposit of title deeds.

Equitable claims and defences. Originally claims and defences which could be advanced only in a suit in equity. From 1854, it was competent for a defendant to plead in a common law action facts which would have entitled him, if judgment passed against him, to relief in equity therefrom on equitable grounds, and for a plaintiff to avoid such a defence by replication on equitable grounds. After the Judicature Act, 1873, both legal and equitable claims and defences can be pled in any court and both legal and equitable remedies granted.

Equitable estates and interests. Rights in property existing where the legal ownership is vested in another person, or in the holder in another legal capacity. Among the major instances of such estates and interests are those existing under a trust, where the legal estate is vested in the trustee and the beneficial owner has only a personal right in equity to enforce the obligations of the trust. Many statutes have assimilated many of the incidents of ownership in equity to legal ownership. Conversely under the property legislation of 1925 the concept of being equitable estates and interests was extended to certain estates and interests which had previously been recognized at law.

Equitable interests. Interests in property created and enforced by courts of equity, such as the interest of the beneficiary under trust property, the legal title to which is vested in the trustee alone. All estates, interests, and charges in or over land, other than those specified in the Law of Property Act, 1925, s. 1, take effect as equitable interests. These include entailed interests, life interests, rentcharges, and many other interests.

Equitable jurisdiction. The jurisdiction of the Court of Chancery, and now of the High Court and particularly of the Chancery Division, as successor thereto, and to a limited degree of the county court. It comprised an exclusive jurisdiction, whereby it enforced rights not, in general, recognized at law, as in trusts, equities of redemption and married women's settled property, a concurrent jurisdiction, whereby it enforced rights also recognized at law and gave an alternative, and frequently a more efficient remedy, as in contract or partnership, such as specific performance of a contract for the sale of land, or supplied a remedy to replace a legal remedy which the plaintiff has lost, and an auxiliary jurisdiction whereby the court gave the parties the benefit of its special procedure, as by discovery of facts or documents, or protected property by the appointment of a receiver. This triple distinction, though historically important, has no significance today, but is still sometimes referred to. In modern practice the High Court exercises both a legal and an equitable jurisdiction, and may grant all remedies, legal or equitable, to which the parties may appear properly entitled.

Equitable mortgage. Mortgage under which the mortgagee does not acquire a legal estate in the subject of mortgage. The category includes a written agreement to grant a mortgage, a deposit of title deeds of land, and the mortgage of an interest under a trust or of an equity of redemption.

Equitable waste. Those kinds of waste (q.v.) which courts of equity would always prevent, even in the case of a tenant of land unimpeachable for waste.

Equites. In ancient Rome, originally the cavalry of the army, but later also signifying a social and political class, the equestrian order. By the first century B.C. the equestrian order tended to be restricted to posts of officers or members of the staff. Under the principate the rank could be conferred on any man of free birth, good character, and the necessary wealth. They had become a class distinct from the senatorial order and could and did engage in finance, commerce and government contracts, in tax farming and exploitation of public lands, and were employed as procurators and civil servants. Later the equestrians held such offices as prefect of the praetorian guard and prefect of Egypt.

Equity. The basic meaning of equity is evenness, fairness, justice, and the word is used as a synonym for natural justice. In a secondary meaning the term is used as contrasted with strict rules of law, *aequitas* as against *strictum jus* or *rigor juris*; in this sense equity is the application to particular circumstances of the standard of what seems naturally just and right, as contrasted with the application to those circumstances of a rule of law, which may not provide for such circumstances or provide what seems unreasonable or unfair. A court or tribunal is a court of equity as well as of law in so far as it may do what is right in accordance with reason and justice. The opposition between equity and law is frequently minimized by rules of law laying down flexible standards and conferring discretionary powers, but in some cases the conflict between what is fair and just and what is lawful may arise. This distinction, sometimes opposition, between law and equity was recognized in Roman law where the action of the praetors in granting remedies in situations for which the *jus civile* provided no remedy was well recognized.

In a third sense equity is the body of principles and rules developed since mediaeval times and applied by the Chancellors of England and the Courts of Chancery and, since 1875, chiefly by the Chancery Division of the High Court, as contrasted with the body of principles and rules of the common law developed and applied by the common law courts, of King's (or Queen's) Bench, Common Pleas, and Exchequer, and, since 1875, chiefly by the Queen's (or King's) Bench Division of the High Court. Thus there developed in England, and to a substantial extent still exist, equitable and legal ownership, equitable and legal rights, equitable and legal remedies, and so on. This distinction also exists in states outside England which adopted from England the distinction in principles and in administration of justice between law and equity, and which in some cases maintain the distinction. In this third sense equity is unknown in legal systems based mainly on Roman law, including Scots law, though the rights and remedies existing in equity in England are normally available in those other legal systems through ordinary legal forms.

Origins of equity in England

The common law developed quickly and early in England but developed within the framework of certain recognized forms of action; outside the prescribed limits of these forms of action no action lay and no remedy could be given at law. The common law judges early abandoned discretionary powers. Persons who wanted extraordinary relief petitioned the King and his Council, which in time tended to delegate the function of dealing with these petitions to the Chancellor who already had an office familiar with judicial work and with issuing writs to the King's officials. The machinery utilized, by bill and *sub poena*, was that of the administrative machinery of the Council. Sometimes Parliament itself granted relief of a kind essentially equitable, and in the fourteenth century there were many petitions to Parliament against Council and Chancery, but from about 1400 statutes accepted and even enlarged the jurisdiction of the Chancellor and Council. The common law judges seem originally to have been willing to cooperate with Chancellor and Council, and many rules later distinctive of equity emerged in the common law courts. The new tribunal did not wholly originate equity, but carried on the work of the older courts by developing through different means the equity inherent in royal justice. Equity grew up to supply the defects and correct the injustices of the common law.

In the early sixteenth century an underlying theory for equity was propounded by St. Germain (q.v.), in his *Doctor and Student*, in the idea of conscience. The mediaeval Chancellors were ecclesiastics and naturally acted in a way consistent with morality if not with strict law. They would grant a remedy if it were contrary to good conscience to let the defendant escape. Equity acts on the defendant's conscience. This became the characteristic idea of classical equity, and the *Doctor and Student* provided a definite theoretical basis for the development of the jurisdiction.

In the fifteenth and sixteenth centuries the development of the rules of equity was largely determined by the defects of common law, its slowness, technicality, inefficiency, and antiquated methods of proof, defects in its substantive content, and its failure to take account of uses and trusts or to grant specific performance. In the sixteenth century the common law courts sometimes adopted principles developed in Chancery.

Development of English equity

By the sixteenth century the Chancellor's jurisdiction was recognized as the Court of Chancery with a great deal of business. But it had but one judge, the Chancellor, who had many other duties, and the Master of the Rolls was becoming recognized as the assistant and sometimes deputy of the Chancellor, while the Masters in Chancery handled minor matters. By the end of that century the court was transacting a large volume of judicial business on fairly settled lines and the original arbitrary character of equity had been largely eliminated. Moreover many of the chancellors were not only laymen but common lawyers.

Discord between common law and Chancery emerged in the seventeenth century, prompted by the jealousy of Coke and the common lawyers of any other system and by alleged association between Chancery and the royal prerogative. The common lawyers attacked the Chancery on the unfairness of granting specific performance or of preventing, by injunction, the enforcement of a judgment obtained by fraud. In 1616 James I personally adjudicated between the two jurisdictions and, on the advice of Ellesmere and Bacon, upheld the power of the Chancery. The necessity for a court granting equitable remedies was obvious when the Chancery was not abolished under the Commonwealth.

The most notable Chancellor of the time was Egerton (q.v.), later Baron Ellesmere and Viscount Brackley (Keeper 1596–1603, Chancellor 1603–17). He maintained that equity was law and not discretion. His successor, Bacon, was distinguished in many spheres, helped to restore harmony between Chancery and common law, cleared off arrears of cases, and created a code of Chancery procedure which lasted for two centuries. Finch, later Earl of Nottingham (Keeper 1673–75, Chancellor 1675–82) began the work of systematizing the principles on which the Court of Chancery acted and thus equity began to assume its final form. He also made many fundamental contributions to principles by his decisions.

By the late seventeenth century the chancellors had practically ceased to administer a vague and indefinite equity; there was a distinct tendency to formulate the principles on which relief would be granted, to define its limits, and to make equity a system. By this time equity had been enforcing trusts, intervening in mortgages, exercising jurisdiction over infants, supervising matters of accounts and the administration of estates and granting equitable relief against fraud, accident, mistake, and undue influence, and developing family settlements. Equity was becoming a distinct and consistent body of rules.

Reports of equity decisions commence in the late sixteenth century but they did not become continuous till the eighteenth century, but the development shows that equity was being administered on principle and there was a growth of the practice of following precedents.

In the eighteenth century relations between law and equity were harmonious, each system becoming more involved with the other and to some extent helping each other. Mansfield, as Chief Justice, tried to infuse some equitable principles into the common law, but without success.

During the eighteenth century the greatest chancellor was Yorke, later Earl of Hardwicke (Chancellor 1737–56), who restated the basic principles of equity, harmonizing the principles with the ideas of the time and practically completed the system; many of his judgments were and have remained classics. He firmly established the general rule that an equity judge should follow existing principles established through a trend of precedents rather than by individual ones.

Final settlement of equity and fusion with law

Equity received its final settlement through the work of Scott, Earl of Eldon (Chancellor, 1801–06, 1807–27) (q.v.), who completed the process of making equity a system almost, but not as fixed and settled as, the common law. It became a defined and limited system, in which precedent was followed. The defect of his work was his care and slowness which caused arrears of work to build up in the court.

As finally developed, equitable jurisdiction was frequently distinguished into the exclusive jurisdiction, as over trusts and married women's separate property, the concurrent jurisdiction (concurrent with common law) as over specific enforcement of contracts, fraud, mistake and accident, and the auxiliary jurisdiction, notably to grant injunctions and appoint receivers.

The Judicature Acts, 1873–75, abolished the Court of Chancery and effected a fusion of the common law and equity jurisdictions with the proviso that in all matters not specifically provided for or in the event of conflict the equitable rules should prevail.

Though equity as an independent body of principles has disappeared from English law, its influence remains not only in principles developed by it, such as the enforcement of trusts, but in the dualism of many concepts, such as legal and equitable mortgages. The subjects which are matters of equity jurisdiction are still for the most part dealt with in the Chancery Division, and equity is still frequently regarded as a distinct branch of English law for study purposes.

The modern applications of equity

Though the Judicature Act, 1873, united the legal and equitable jurisdictions by amalgamating the common law courts and the Court of Chancery into the new Supreme Court of Judicature and thereby enabled any Division of the High Court to do justice by giving whatever remedy, legal or equitable, was appropriate, it did not provide for a merger of the recognized kinds of legal and of equitable rights, estates and interests and, subject to changes made subsequently, as by the Law of Property Act, 1925, they still exist.

The main applications of equitable principles in modern law are the recognition and enforcement of (1) equitable interests in property, notably those arising under a trust, the mortgagor's equity of redemption (q.v.), equitable mortgages and charges, equitable interests in personal property and under contracts, restrictive covenants creating equitable interests in land, and equitable assignments; (2) equitable doctrines, doctrines affecting property, such as conversion (q.v.), election (q.v.), satisfaction and ademption (qq.v.), performance (q.v.), marshalling (q.v.), merger of estates and charges (q.v.), and subrogation (q.v.); (3) equitable relief against penalties and forfeitures and in cases of fiduciary relationship; (4) equitable defences, such as equitable set-off, release and waiver, acquiescence and laches (q.v.).

W. Buckland, *Equity in Roman Law* (1911); G. Spence, *Equitable Jurisdiction of the Court of Chancery* (1846); D. Kerly, *History of Equity* (1890); W. S. Holdsworth, *History of English Law* (1956); H. G. Hanbury, *Modern Equity* (1976); E. Snell, *Principles of Equity* (1973).

Equity, maxims of. A number of maxims which conveniently epitomize various general principles on which courts of equity have founded their decisions. They do not cover the whole of equitable jurisdiction and overlap. The list of maxims is not completely settled and there are variant formulations of some of the maxims. They include: Equity will not suffer a wrong to be without a remedy; Equity follows the law; Where there is equal equity the law shall prevail; Where the equities are equal, the first in time prevails; He who seeks equity must do equity; He who comes to equity must come with clean hands; Delay defeats equities; Equity looks to the intent rather than to the form; Equity looks on that as done which ought to be done; Equity imputes an intention to fulfil an obligation; Equity acts *in personam.*

Equity of a statute. The general intent and spirit of a statute rather than the strict letter thereof. It is founded on a theory of the sixteenth and early seventeenth centuries found in St. Germain's *Doctor and Student* and in Coke, of a special 'equity' which controlled statutes by extending them in some cases and restricting them in others. A certain amount of learning grew up around this and some of the Abridgements treat it as an aspect of statute law. In the late seventeenth and eighteenth centuries the theory shrank and has now long since disappeared.

Equity of redemption. In English law the right of one who had mortgaged his property to redeem it on payment of principal, interest and costs was, until 1926, not merely a right, which usually subsisted for 6 months, but an equitable interest or estate in the property mortgaged, entitling the mortgagor to deal with his property in any way consistent with the rights of the mortgagee and, in particular, to redeem the mortgaged property at any time despite agreement to the contrary and though the time for repayment stated in the mortgage had passed. A clog or fetter on the equity of redemption is void, and no agreement between mortgagor and mortgagee can make a mortgage irredeemable. Since 1925 the right to redeem is no longer strictly an equitable estate or interest, though in the nature of an equitable interest. So too even though a mortgagee of chattels had seized them the mortgagor can still redeem so long as they are in the mortgagee's hands.

The right to redeem is lost on a sale or release of the equity of redemption by mortgagor to mortgagee, or by sale by the mortgagee under his power of sale, or by sale by process of the court.

Equity to a settlement. Before the Married Women's Property Act, 1882, if a husband became entitled in possession in right of his wife to property not recoverable at law, and if the intervention of a court of equity was in any way invoked, that court would allow the husband to take the property only subject to the wife's right or equity to a settlement, that is to have part of the property settled on the wife. Since the Act of 1882 the concept is obsolete as a wife can take and hold property on her own account.

Erastianism. The doctrine not held by but derived from views of Thomas Erastus (1524–83), Swiss theologian, that the State should direct and control the affairs of the Church in a specific area and should punish all offences, ecclesiastical and civil, where all the citizens adhered to a single religion. It was founded on views of some mediaeval monarchs and received considerable impetus from the Reformation. The word acquired its modern meaning from debates during the Westminster Assembly of 1643.

Erenbergk, Waremundus de (*Eberhard von Weyhe*) (*c.* 1553–1633). German jurist, who under the name Durus de Pascolo published an *Aulicus Politicus* (1596) and, under the name Waremundus de Erenbergk a treatise *de regni subsidiis* (1606).

Erie Railroad Co.* v. *Tompkins (1938), 304 U.S. 64. In this case the Supreme Court, overruling the contrary view laid down in *Swift* v. *Tyson* (1842), 16 Peters 1, held that it was not open to it to take a different view on a point of the law of one of the states from that accepted by the supreme court of that state itself. The decision denied the existence in the U.S. of any federal general common law; there are as many common laws as there are states, though there is a fundamental underlying unity of common law principles in the U.S.

Erle, Sir William (1793–1880). Was called to the Bar in 1819 and gradually developed a practice. He was in Parliament 1837–41 but never spoke. In 1845 he became a judge of the Common Pleas, in 1846 transferred to the King's Bench, and in 1859 returned to the Common Pleas as Chief Justice, where he sat till his resignation in 1866. He was an impartial and strong judge and decided many important cases, but he was not a great legal scholar and he lacked subtlety and flexibility of mind. He was a member of the Royal Commission on Trade Unions in 1867 and published a book on the law of trade unions in 1868.

Error. In Scots law, an erroneous belief as to some matter relevant to liability, corresponding to mistake (q.v.) in English law. Error may be as to some matter of fact, e.g. the identity of a person, or as to some matter of law, e.g. whether conduct is criminal, or as to the legal consequences of some act or deed, though the classification of error as to fact or as to law is sometimes difficult.

In criminal law error as to fact, if shown to be honestly and justifiably made, will generally result in the accused being treated as if the circumstances had been as he believed them to be. Error as to law is generally irrelevant; as *ignorantia juris neminem excusat* so too *error juris*.

In civil law the chief importance of error is that it may vitiate a contract, where one or both parties have contracted under error. Error as to law is only exceptionally relevant. Among cases of error in fact the general principles appear to be that merely clerical errors may be corrected or ignored by the court, that if both parties to a contract have made the same error when contracting and it is substantial the contract is void, if each is mistaken as to the other's intention, the contract may be void, or may be enforceable by one party if that party's understanding of the contract is held to be correct, if only one party is mistaken his error is ineffective, but if one party has contracted under error induced by the misrepresentation of the other the contract is at least voidable, or may be wholly void, depending on the materiality of the error.

Error of law. *Certiorari* (q.v.) can be invoked to ensure that inferior tribunals and statutory bodies apply the law correctly only if an error of law is disclosed on the record of the decision. Error of law generally means error in applying principles of statute or common law. The record may disclose an error without expressly stating it, if error can be inferred from the decision reached. The error need not be confined to jurisdiction. The power of the court to review for error of law may be excluded by statute, but a mere provision that the tribunal is to be 'final' does not exclude review.

Error, proceedings in. The former mode of having erroneous decisions reviewed. In criminal cases it was long held that a jury verdict could not be reviewed though a writ of error could be brought if the Crown admitted that a trial was unsatisfactory. But in 1705 it was held that a writ of error must be issued in cases of misdemeanour. Later the judges resorted to the practice of reserving difficult cases for informal discussion with their brethren, which gave rise in 1848 to the creation of the Court for Crown Cases Reserved. In 1907 this Court and the writ of error was abolished and appeal allowed to the new Court of Criminal Appeal and thence, exceptionally, to the House of Lords. In civil cases in mediaeval times a writ of error lay to remove the record of a case in the Common Pleas into the King's Bench to be reviewed on the law. Error lay from the Exchequer to the Exchequer Chamber. Error lay originally from the King's Bench to Parliament but later to a Court of Exchequer Chamber, and in 1830 a new Court of Exchequer Chamber was established, error from any one common law court (King's Bench, Common Pleas, Exchequer) being heard by the judges of the other two. In the Middle Ages error lay from the Irish courts to the Westminster Parliament and statute in 1719 enacted that error lay to the English and not the Irish Parliament. This continued till 1800. After 1707 error lay also from the Court of Session in Scotland to Parliament. In 1876 appeal from the new Court of Appeal or the Court of Session was permitted to the House of Lords.

Error, writ of. The original writ formerly directed to a court requiring it to send the record and proceedings to a superior court for review. It lay for substantial error in law disclosed by the record and not for error in fact, so that inferences from evidence could not be challenged. It lasted in criminal cases till 1907 but had disappeared earlier in civil cases.

Erskine, Charles, of Tinwald (1680–1763). Became a regent, and latterly Professor of Public Law, in the University of Edinburgh, before he was called to the Scottish Bar in 1711. In 1722 he entered Parliament and became Solicitor-General for Scotland in 1725, being the first holder of that office to be given the privilege of pleading within the Bar. He became Lord Advocate in 1737 but resigned in 1741. In 1744 he became a Lord of Session and in 1748 Lord Justice-Clerk, an office which he held till his death. He was esteemed an able civilian. His son, James (1722–66) was successively a baron of exchequer in Scotland (1754), Knight-Marshal of Scotland (1758) and a judge of the Court of Session as Lord Barjarg (1761) or, later, Lord Alva.

Erskine, Henry (1746–1817). Brother of Thomas, Lord Erskine (q.v.), and David, 11th Earl of Buchan, and grandson of Sir James Stewart of Goodtrees (q.v.), and great-great-great-grandson of Sir Thomas Hope (q.v.), was called to the Scottish Bar in 1768 and early showed great powers of oratory in the Forum, a debating society, and in the General Assembly of the Church of Scotland. Being also a very sound, though possibly not outstanding, lawyer, he rose rapidly at the Bar and became Lord Advocate in 1783 but his party went out of office in the same year and in 1795 he was even voted out of the office of Dean of the Faculty of Advocates for his liberal sentiments. He was again Lord Advocate, and an M.P. in 1806–07, but had to retire from professional life in 1812 owing to ill-health. In his day he had an outstanding reputation for eloquence, humour, and charm of manner, and at the cost of his own advancement he never hesitated to expose injustice, protect the oppressed, and uphold liberty. It was several times generally thought that he would be appointed to the bench.

A. Fergusson, *Henry Erskine* (1882).

Erskine, James, of Grange (1679–1754). Was called to the Scottish Bar in 1705 and became a Lord of Session and of Justiciary in 1707 and Lord Justice-Clerk in 1710. The last office he lost in 1714. He was in high favour with the strict presbyterians and asserted the utmost freedom of presbyteries and ministers from control by government or lay patrons. Determined to participate in politics, he resigned his judicial offices in 1734 but, though elected to Parliament, made little impact there and he returned to the Scottish Bar, but passed his last years in London. He is remembered for having had his wife abducted to and confined on St. Kilda from 1732 till her death in 1745.

W. Roughead, The Husband of Lady Grange, in *The Riddle of the Ruthvens.*

Erskine, John, of Carnock (1695–1768). Son of Colonel John Erskine of Carnock, grandson of Lord Cardross and great-grandson of Sir Thomas Hope (q.v.) and of a collateral line to the Earls of Buchan and the family of David, Henry, and Thomas Erskine (q.v.), was called to the Scottish Bar in 1719 and practised without great distinction till 1737 when he was elected Professor of Scots Law in Edinburgh university. In 1754 he published his *Principles of the Law of Scotland in the order of Sir George Mackenzie's Institutions of that Law* intended chiefly for the use of his students. The book was an immediate success and, regularly revised, it continued to be a standard students' text in the Scottish Universities till about 1925. He resigned his chair in 1765 and occupied his last years in preparing his major work *The Institutes of the Law of Scotland*, in four books, which was published posthumously in 1773 and reached an eighth edition in 1870. Its merits were immediately recognized and it has been regarded as a Scottish legal classic ever since. Though it lacks the philosophical depth of Stair and deals briefly with mercantile topics, it is a clear and accurate exposition of the law in its classical stage of common law, before legislation spoiled the symmetry and the principles. His eldest son John Erskine, D.D., was a distinguished divine and a prolific writer of sermons and discourses.

Erskine, Thomas Erskine, 1st Lord (1750–1823). Probably the greatest advocate who ever practised at the English Bar, served in both the navy and the army before being called to the Bar in 1778, where he at once made his way by his eloquence. He took silk in 1783 and entered Parliament but lost his seat the next year and did not return till 1790. He defended the Dean of St. Asaph and Stockdale on charges of seditious libel, and Paine, Hardy, Horne Tooke, Thelwell and those others indicted for high treason in 1794. His liberal opinions stood in the way of his advancement. In 1806 he was appointed Chancellor, though he had never appeared in Chancery, but lost office in 1807 and, apart from slight participation in the House of Lords, took little part in legal or public life thereafter. As Chancellor he presided at the trial on impeachment of Lord Melville. His powers lay in oratory, and ability in examination of witnesses, though he would probably have made his mark as a Chief Justice and might, as Chancellor, if he had had a longer tenure of that office.

His elder brother David, Earl of Buchan and Lord Cardross (1742–1829) deprived himself of much to further Thomas Erskine's education. For a time he was secretary to the British Embassy in Spain but was devoted to the history, antiquities, and literature of Scotland and was the founder of the Society of Antiquaries of Scotland (1780). He contributed to various journals. Another elder brother was Henry Erskine (q.v.). A son of Lord Erskine was a judge of the Common Pleas, 1839–44.

H. Brougham, *Historical Sketches*, 236; W. C. Townsend, *Twelve Eminent Judges*, I, 398; II, 1; W. S. Holdsworth, XIII, 580.

Erskine May, Sir Thomas, Lord Farnborough (1815–86). Spent his whole life in the service of the House of Commons, from being assistant librarian in 1831, clerk assistant 1856–71 and clerk 1871–86; on his resignation he was created Lord Farnborough. He was called to the Bar in 1838. In 1844 he published his *A Practical Treatise on the Law, Privileges, Proceedings and Usage of Parliament* which has ever since, in many later editions, been recognized as the authoritative work on the rules, conventions, and customs of Parliament, and The Bible of parliamentary procedure. 'Erskine May' or 'May's *Parliamentary Practice*' has also been widely used in Commonwealth countries with legislatures on the Westminster model. In 1854 he reduced to writing for the first time the *Rules, Orders and Forms of Procedure in the House of Commons* and also published a *Constitutional History of England since the Accession of George III, 1760–1860* (1861–63), continuing the work of Hallam, *Democracy in Europe: a History* (1877), and other works. He also served on the Digest of Law Commission and was President of the Statute Law Revision Committee from 1866 to 1884. He was sworn of the Privy Council in 1885.

Escape. Getting away from restraint or custody. It is criminal to escape from lawful custody, to aid a prisoner to escape, or to permit a prisoner to escape.

Escape of dangerous things. A doctrine of the English law of torts, enunciated in *Rylands* v. *Fletcher* (1868), L.R. 3 H.L. 330, that a person who for his own purposes brings on his lands and keeps there anything not naturally there and likely to do mischief if it escapes, must keep it in at his peril, and if he does not do so, is *prima facie* answerable for all the damage which is the natural consequence of its escape. Escape infers strict liability and negligence need not be proved. This liability applies in such cases as accumulating water, starting fires, causing earth-shaking vibrations or explosions. Defences are Act of God, fault or consent of the plaintiff and act of a third party.

Escheat and forfeiture. In feudal land law land reverted to the lord if the tenant died without heirs (escheat *propter defectum sanguinis*) or committed any gross breach of the feudal bond (escheat *propter delictum tenentis*). In early times a conviction for felony entailed escheat of the felon's land to the lord. In all cases of escheat the lord's right was subject to the Crown right to 'year, day, and waste' (q.v.). The former kind of escheat came to be of little value when all free tenants acquired the right to devise their lands and was abolished in 1925 being replaced by a right of the Crown or the Duchy of Lancaster or the Duchy of Cornwall to *bona vacantia* (q.v.) on a person's death intestate and without heirs. The latter was abolished in 1870.

Forfeiture did not depend on feudal tenure. It was a prerogative right of the King to all the lands of a person convicted of high treason, of whatever lord they were held, and the lord lost his claims to escheat. It was abolished in 1870.

In the U.S. it is generally provided that land will escheat to the state or county if an owner dies intestate and no heirs are discoverable.

Escrow. A written instrument evidencing obligations between two or more parties, given to a third party on condition that he deliver it to the other party, only on the happening of a stated condition, such as payment of a price or the death of a person, which being done it takes effect as a deed. If the condition is not performed it never becomes a deed.

Esher, Viscount. See BRETT, WILLIAM B.

Eshunna, laws of. Legal rules found on two tablets found near Baghdad, stating early rules of Babylonian law (q.v.).

Esmein, Adhémar (1848–1913). French jurist, an outstanding legal historian who excelled in reconstruction of institutions. He wrote *Cours élémentaire d'histoire du droit français* (1892), *Précis élémentaire d'histoire du droit français de 1789 à 1814* (1908) and *Éléments de droit constitutionnel français et comparé* (1896), *Le mariage en droit canonique* (1891), *Études sur les contrats dans le très ancien droit français* (1883), and *Histoire de la procédure criminelle en France* (1882).

Espen, Zeger-Bernard van (1646–1729). Famous Belgian canonist, author of *Jus ecclesiasticum universum, De officiis Canonicorum*, and other works on canon law.

Espionage or spying. In international law the act of a person who, during war, clandestinely or under false pretences seeks to obtain information concerning one belligerent in order to communicate it to the other. Such a person, if caught, is punishable as a war criminal. A spy should be tried by court-martial. If he is a member of the forces of one side and rejoins those forces he may not, if later captured, be punished for his previous spying. Spying is distinct from scouting, reconnaissance, or observation made by a uniformed member of the forces of one belligerent in the theatre of the enemy's operations or even behind his lines. It is also distinct from carrying of messages, escape from besieged place, or raiding, even though the person concerned also observes and intends to communicate information. Such persons, if captured, must be treated as prisoners of war.

In British laws the Official Secrets Act, 1911, creates offences of engaging in espionage and related activities which might be useful to a possible enemy and therefore injurious to the security of the State, as by passing information about defence secrets to a foreign power.

Essay on Crimes and Punishment. See BECCARIA, CESARE.

Essence of the contract. A provision in a contract, such as for performance by a stated date, may be stated therein, or be held judicially, to be of the essence of the contract when it is an essential or fundamental term, non-performance giving rise to a right not merely to damages but to rescind the contract. In the sale of goods unless a different intention appears from the terms of the contract, stipulations as to time of payment are not deemed to be of the essence of the contract; whether any other stipulation as to time is of the essence of the contract or not depends on the terms of the contract. In other contracts the position formerly was that a time stated for performance was deemed at law to be of the essence, but in equity performance within a reasonable time sufficed unless it was apparent that time was intended to be of the essence. The equitable rule now applies generally.

Essex affair (1804–5). An Anglo-American dispute arising from the British capture of the U.S. vessel *Essex*. Britain, with the intention of damaging neutral trade between France and the French West Indies, applied the Rule of the War of 1756 to the effect that a belligerent could not open to neutrals trade with its colonies which was closed to them in time of peace. U.S. ships sought to evade this rule by carrying cargoes from the West Indies to a U.S. port and then to France. In the *Essex* case the High Court of Admiralty declined to treat such as a broken voyage but treated it as a continuous voyage, and this resulted in a considerable increase in British seizure of U.S. ships.

Essoin (or *essonzie*) (from Lat. *exoneratio* and *essonium*, Fr. *essoine*). In mediaeval law, an excuse offered by or on behalf of one summoned to appear in court to perform suit or to answer to an action, by reason of sickness, infirmity or other admissible cause of absence. Many causes of excuse were recognized in English courts, particularly *de infirmitate veniendi*, illness while coming to court, *de infirmitate resiantiae*, illness involving confinement to bed, and *de ultra mare*, absence overseas. By utilizing essoins it was possible to delay a case for a considerable time. The first general return day of a law term was for long called the essoin day because the court sat then to receive essoins but this was abolished in 1830.

Established church. A church legally recognized as the official Church of the State or nation and having a special position in law. That church normally has special legal privileges, but frequently also certain liabilities arising from its connection with the state. Thus the (Anglican) Church of England is established in England and its law is part of the law of England but its archbishops and bishops are nominated by the Queen and it is only in the twentieth century that it has acquired power to regulate its doctrine, order, and worship, independent of Parliament. In Scotland the (Presbyterian) Church of Scotland is established but free of Parliamentary control. Establishment formerly carried the right to tithes or teinds payable by landowners for the support of the church but in the U.K. these have been redeemed or commuted into redeemable annuities. There is no established church in the U.S. but the R.C. Church is established in Italy and Spain, Judaism in Israel and the concept is found in other countries also. In countries where the concept is recognized other churches are merely voluntary associations.

Establishment, right of. Under the law of the European Economic Community, the right for a person or body, corporate or unincorporate, having the nationality of one member state to establish himself or itself in another member state either as a business or by establishing an agency, branch, or subsidiary. Hence restrictions on the freedom of establishment of nationals of a member state in the territory of another member state have been progressively abolished, and nationals of member states can pursue activities under the same conditions as are prescribed by domestic law for nationals. Accordingly, English or Scottish lawyers can practice at the French Bar and conversely, but subject to acquiring the necessary qualifications and competence. In the sphere of legal practice having sufficient linguistic competence and acquiring the qualifications is for practical purposes a greater obstacle to establishment than any former barriers of nationality. There are certain exceptions to the principle of freedom of establishment, particularly where the work involves the exercise of official authority and allowing for the special treatment of aliens on grounds of public policy. The EEC Treaty provides for the mutual recognition of qualifications obtained in member states in other states.

Estate (Lat. *status*, Fr. *état*). The condition or circumstance of standing of a person or a class of persons. Accordingly the estates of the realm (q.v.) or 'the three estates' are the major groups in the same condition, namely the barons or peers, or Lords Temporal, the clergy or Lords Spiritual, and the commons. Thus in France the Estates-General or States-General was the assembly of the three estates of the realm. The term was transferred to the standing of a person in relation to land, designating the qualities of rights which the person had in the particular land (see ESTATES IN LAND) and colloquially it is used of the tract of land itself. The term is also used as meaning property, as in the phrases partnership estate, trust estate, estate of a deceased.

Estate agents. Persons who, as agents for sellers or lessors of land and houses, endeavour by advertisements and other means to find persons willing to buy or rent the parcels of land or houses. They are normally remunerated by a percentage commission on the price obtained for the sales or leases procured by their efforts.

Estate contract. A contract by the owner of a legal estate, including a tenant from year to year, or by a person entitled at the date of the contract to have a legal estate conveyed to him, to convey or create a legal estate, including a contract conferring a valid option to purchase, a right of pre-emption, or other similar right. It is a kind of land charge (q.v.).

Estate duty. A tax imposed by the Finance Act, 1894, on the principal value of all property, real or personal, settled or not, passing or deemed to pass on death. Duties imposed on the estate left by a person on his death, were foreshadowed by casualties incidental to feudal tenure, but first appeared as such on grants of probate and administration in 1694. In 1780 stamps were required for receipts for legacies and in 1796 this was converted into a duty on the legacies. In 1805 this was extended to testamentary gifts of the proceeds of sales of real property. In 1853 succession duty was imposed on real property and settled property. In 1888 succession duty replaced legacy duty. In 1881 account duty was imposed on kinds of property which escaped other duties. In 1889 temporary estate duty was introduced in addition to probate and legacy or succession duties. In 1894 estate duty replaced probate or inventory duty, account duty, and temporary estate duty but legacy and succession duty remained. Estate duty was chargeable at percentage rates rising with the value of the estate on all property passing on death, including property of which the deceased was competent to dispose and gifts made within limited periods before death. The detailed rules were frequently modified and the duty became steadily more confiscatory. In 1975 estate duty was replaced by Capital Transfer Tax.

Estates in land. In feudal land law each tenant in the chain of tenure (q.v.) between the Crown and a particular piece of land owned in contemplation of law an abstraction, called an estate in the land, which comprised all the rights and interests which he had in the land. By reason of the principle of tenure several persons might simultaneously hold different estates in one piece of land, such as the King, mesne lord, tenant for life, and remainderman; no one was absolute owner of the land.

In English land law before 1926 the different estates were classified according to the time for which they might endure into (a) estates of freehold, of indeterminate duration, and (b) estates less than freehold, of duration fixed or capable of being fixed. The former class comprised (i) freeholds of inheritance which might devolve on heirs indefinitely, namely estates in fee simple, estates in fee tail and estates in frankmarriage, and (ii) freeholds not of inheritance, namely estate for life, estate *pur autre vie,* estate of tenant by the curtsey, and of tenant in dower. Class (b) comprised leases for a fixed term of years or from year to year. Since 1925 the only estates in land capable of subsisting at law are an estate in fee simple absolute in possession, and a term of years absolute. Certain interests or charges in or over land may subsist at law, but in all other estates, interests and charges in or over land take effect as equitable interests only.

In Scots law the recognized estates are those of the Crown (*dominium eminens*) a subject superior (*dominium directum*), the vassal who occupied the land (*dominium utile*), heir of entail in possession, a liferenter and a tenant, but the doctrine of estates has never been so developed or elaborated as in English law.

Estates of the Realm. The status-groups of a mediaeval Parliament. Three such groups tended to be identified, the clergy, the barons, and the commons, now represented by the Lords Spiritual and Temporal (in the House of Lords) and the commons (in the House of Commons). The phrase seems to have been adopted in the fifteenth century on a mistaken analogy with the French Estates or States-General.

Estimates. Statements put before the House of Commons each year stating the grants of money which it is estimated will be required for the civil and defence services. They are drawn up on the basis of departments of state, but all the services administered by a single department are not necessarily included in the same 'vote'. The 'votes' number about 170, are grouped into a number of classes and subdivided into subheads according to the description of the service rendered. The main classes are (1) Government and Finance, (2) Commonwealth and Foreign, (3) Home and Justice, (4) Communications, Trade, and Industry, (5) Agriculture, (6) Local Government, Housing, and Social Services, (7) Education and Science, (8) Museums, Galleries, and the Arts, (9) Public Buildings and Common Governmental Services, (10) Other Public Departments, (11) Miscellaneous, and (12) Defence. Each vote specifies the services and the net amount of the grant requested, and the items of expenditure and receipts making up the amount requested. Each vote is votable separately. Parliament debates the estimates and the result of its debates is embodied in the annual Appropriation Act which authorizes the expenditure of the total sum of the estimates, appropriating stated sums to each vote. The 'supply business' of the House includes debates on the main estimates, supplementary estimates, excess votes, and reports from the select committees of the House dealing with expenditure.

Estimates Committee. Formerly one of the major select Committees of the House of Commons, with the function of examining such of the estimates as it thought fit, reporting how the policy implied by these estimates may be carried out more economically and, if thought fit, considering variations between the last year's and the current year's estimates. It received written and oral evidence and could require attendance of witnesses and production of paper. Much of its investigation was done by sub-committees. It was replaced by the Select Committee on Expenditure.

Estoppel. The principle which precludes a party from alleging or proving in legal proceedings that a fact is otherwise than it has appeared to be from the circumstances. There are three kinds of estoppel, estoppel of record, which arises where a fact has been judicially determined by a tribunal and it comes again in issue in another matter between the same parties; estoppel by deed, where there is a statement of fact in a deed made between parties and verified by their seals in which case a party is not permitted to deny any fact which he has thus asserted; and estoppel *in pais*, where a party has, expressly, or impliedly by conduct or negligence, made a statement of fact, or so conducted himself, that another would reasonably understand that he might act in reliance thereon, and has so acted, when the party who made the representation is not allowed to allege that the fact is otherwise than he has represented it to be. Examples of the last category include holding another out as, or allowing him to appear to be, one's agent, holding out as a partner, and negligence which facilitates the fraud of another.

Estovers (Lat. *fovere*, to keep warm, cherish, or sustain; Fr. *estevoir*, what is necessary; *estoffer*, to furnish). Any kind of sustenance, particularly wood, a reasonable amount of which any tenant for years or for life of freehold may by common right take from the land for fuel or repairs. This is also sometimes called bote. Estovers or botes are distinguished into house-bote, for repairs, fire-bote, for fuel, plough-bote, for making and repairing agricultural implements, and hay-bote, for repairing fences.

Estrays. Valuable animals found wandering and ownerless; if kept for a year and a day and no owner has appeared they fall to the Crown or, by grant from the Crown, to the lord of the manor. The category does not extend to wild animals unless ownership had been acquired at common law. Of birds only swans and cygnets are included.

Estreat (Lat. *extractum*). A copy of a record of a court. If a condition of a recognizance (q.v.), e.g. on a grant of bail, is broken it was formerly estreated, whereupon the cognizors become debtors of the Crown. In modern practice in the superior courts the court has a discretion whether whole or part of a recognizance will be estreated or not. Magistrates courts may similarly order forfeiture of the recognizance with which they are concerned.

Établissements de Saint Louis. A fourteenth century French compilation, the first part drawn from the ordinances of King Louis of France, the rest dealing with the custom of Touraine-Anjou, and the second book dealing with the customs of Orleans. It contains also extracts from the *Corpus Juris*, and had considerable influence in its time.

Eternal law (or *lex aeterna)*. A concept developed first by Stoic philosophy and distinguished from *ius naturale* and *ius humanum*. It is the law of reason of the cosmos which rules the universe. Human reason, an emanation of cosmic reason rules the lives of natural men, so that natural law partakes of eternal law. The concept was adopted and elaborated by Cicero and later by St. Augustine, who identified eternal law with the reason or will of God. St. Thomas Aquinas saw divine law as the part of eternal law which God made known through divine revelation, not to be grasped by human reason but given to man as an Eternal Truth. Natural law was the part of eternal law which man could apprehend with his unaided reason but could neither create nor change by reason or will and within the limits which natural law prescribed human or positive law was created by human reason for the common good. Later still Suarez maintained that law had rational foundations, *lex aeterna* providing the outer matrix for all law and *lex naturalis* being still common to all men and accessible to their reason. Since the decline of the intimate connection between religious and legal thought jurists have been less inclined to discuss the concept of eternal law, as comprehending or distinct from natural law.

Ethics or moral philosophy. The systematic study of the ultimate problems of human conduct, a study vigorously pursued by philosophers from the times of the Greeks to the present. Those ultimate problems turn chiefly on the issues of the ultimate ideal aim or ultimate standard of right conduct, the origin and source of knowledge of the highest good or of right and wrong, the motives which prompt right conduct, and the sanctions of moral conduct. On the overlap and interaction of legal thought and moral philosophy see JURISPRUDENCE AND ETHICS; on the overlap and interaction of rules of law and practical morality see LAW AND MORALS.

Ethics, professional. The standards of right and honourable conduct which should be observed by members of learned professions in their dealings one with another and in protecting the interests and handling the affairs of their clients. The legal profession has developed a substantial body of principles of professional ethics, much of it not formulated in writing but absorbed by younger lawyers from older ones, but some formulated in rulings by such bodies as the Bar Council or the Solicitors' Discipline Committee. The American Bar Association has tried to formulate a code of professional ethics.

Among the main principles of ethical conduct expected of a lawyer are that, in relation to his client, he will maintain confidentiality about everything revealed to him during the professional relationship, will always use his skills and knowledge to the best of his ability to secure the client's interests, but will not knowingly be a party to any fraud, dishonesty or underhand dealings.

In relation to other members of the profession, a lawyer must deal honestly with them and implement undertakings to them.

In relation to the courts a lawyer should not participate in any fraud on the court as by putting forward false evidence or withholding evidence, should not put forward any misleading arguments or misrepresent or conceal any authorities, relevant to his case, whether in favour of or against his client, and must honour any undertakings he gives on behalf of his client.

H. Drinker, *Legal Ethics.*

Etiquette, legal. The code of honour and customary rules of behaviour within the legal profession. To a large extent this is unwritten, customary, and assimilated by entrants to the profession and inculcated by the example of their seniors as much as by precept. In England the Bar Council and the Council of the Law Society, and their counterparts in other jurisdictions, make and modify etiquette to a small extent by publishing rulings on particular points and by dealing with complaints reported to them.

W. Boulton, *Conduct and Etiquette at the Bar.*

Euric. King of the Visigoths, A.D. 466–83, said to have been responsible for the first compilation of written law of the Visigoths, a body of rules which were later revised and added to by later Visigothic rulers and which influenced later Germanic codes.

European Atomic Energy Community (Euratom). An organization created by a Rome Treaty between Belgium, France, Germany, Italy, Luxembourg, and the Netherlands in 1957 as a specialist market for atomic energy, to develop nuclear energy, distribute it within the Community and sell any surplus.

The Treaty established an Assembly, a Council of Ministers, a Commission, and Court of Justice modelled on those of the E.C.S.C. but in 1969 these were merged with the corresponding authorities of the European Economic Community. In 1973 the U.K., Denmark, and Ireland joined Euratom.

European Coal and Steel Community (E.C.S.C.). An organization constituted by a Paris Treaty between Belgium, France, Germany, Italy, Luxembourg, and the Netherlands in 1951. The political aim is to establish a united Europe on the basis of economic unity. The economic aim is to stimulate an expanding economy by creating and maintaining a common market for coal, iron ore, scrap, and steel throughout the territories of the parties, removing customs barriers, quotas, restrictions and all discrimination, by rendering restrictive practices within the area illegal and limiting excessive concentrations of economic power. It does not direct the production of coal or steel. The Treaty established a common Assembly, a Council of Ministers, representing the states, a High Authority as permanent executive, and a Court of Justice, but in 1957 the Assembly and the Court and in 1969 the Council and the Commission were merged with the corresponding authorities of the European Economic Community (E.E.C.) (q.v.) and of Euratom. In 1954 the U.K. became associated with E.C.S.C. and in 1973 the U.K., Denmark, and Ireland joined the E.C.S.C.

European Commission of Human Rights. See EUROPEAN CONVENTION ON HUMAN RIGHTS.

European Communities. The collective term for the European Atomic Energy Commission (Euratom), the European Coal and Steel Community (E.C.S.C) and the European Economic Community (E.E.C.) (qq.v.). Each community is a distinct legal entity but all are closely allied. From their foundation in 1957 Euratom and E.E.C. shared a common Parliamentary Assembly and a common Court of Justice with E.C.S.C., these institutions replacing those originally created for E.C.S.C. only. The Brussels Treaty of 1965 (the Merger Treaty) established, from 1969, a single Commission and a single Council for all three Communities. The U.K., Denmark, and Ireland joined the Communities in 1973. The E.E.C. is the most important of the three and has the broadest social and economic spheres of activity, the others being specialized.

The three Communities are supra-national entities in that they have legal existence separate from each other and from the states which are members of them, yet are neither international institutions like the U.N., of which any state in the world may become a member, nor confederations of the member states, because in many and important respects the members remain wholly independent. The founding treaties confer independent legal personality on the Communities.

By the constituent treaties in each of the member states each Community is to enjoy the most extensive legal capacity accorded to legal persons under their laws and, in particular, may acquire or dispose of movable and immovable property and may be party to legal proceedings. Moreover in continental legal systems the Community personality is governed by public law; they are public corporations. But the institutions of the Communities do not have legal personality in the laws of the member states, though the Communities may be represented by the Commission in legal matters. The Communities also enjoy in the territories of the member states such privileges and immunities as are necessary for the performance of their tasks, such as are usually accorded to the premises and officers of international organizations, including inviolability of premises, buildings and archives, exemption of assets and revenues from taxation, and from custom duties and restrictions, immunity of official communications from censorship.

The Communities have contractual liability in the law of member states in accordance with the law applicable to the contract in question, and are liable in damages for damage caused by their institutions or servants in the performance of their duties.

The Council has authority, with the assistance of the Commission, to conclude agreements between the Communities and international organizations and non-member states, and the Communities have made association agreements with certain states intended to create a customs union between the Community members and the associated states, and external trade agreements with other states, usually providing for preferential reductions in tariffs in relation to these states. Multilateral association agreements have been entered into with those non-European states which have special relationships with member states and relations are maintained with international organizations, particularly the U.N. and its specialized agencies concerned with economic affairs.

In each of the Communities the Council, which consists of representatives of the governments of all the member states, is the final expression of the political will of the members. It has a large degree of influence in drawing up the Community budget and over expenditure, the final word in international negotiations and no new departure or important decision can be taken without the Council's agreement. The Commission is the executive of the Communities, having wider powers under the E.C.S.C. treaty than under the Euratom and E.E.C. treaties; it has powers of initiative, preparation and decision; its power of decision is limited to executive matters, though in Euratom and the E.E.C. nearly all important decisions by the Council must be taken on the basis of proposals made by the Commission. The European Assembly or Parliament (q.v.) has no direct legislative powers but must be consulted on some matters and, in practice, the Commission refers most of its proposals of any importance to it. The Commission is responsible to it alone; it consists of representatives of the peoples of the member states, weighted according to their populations. The Court of Justice has jurisdiction to ensure respect for the law in the interpretation and application of the Treaties, and it is the custodian of the Treaties, the watchdog of legality and executor of the supremacy of Community law over the national laws of member states.

The acceptance of the powers of Council, Commission, Assembly and Court implies some surrender of sovereignty by the member states, but this is little different from the partial surrender of sovereignty made by joining the U.N. or other international organizations. No state has for long enjoyed complete sovereignty.

The Communities are required by the constituent treaties to establish co-operation with the U.N., its specialized agencies, G.A.T.T., the Council of Europe and O.E.C.D. and have established such relations, chiefly in commercial fields.

European Community Law. The body of law created by and consequent on the association of nine European countries (Belgium, France, Germany, Italy, Luxembourg, Netherlands since 1951, and also the U.K., Denmark, and Ireland since 1973) in the three European Communities. The treaties creating the Communities created supranational entities, distinct from the member states, with subordinate institutions now common to all three, the Council of Ministers, the Commission, the Assembly, and the European Court of Justice, and through these institutions the communities have developed a very extensive body of law. Community law is a legal order quite distinct from the municipal legal systems of each of the member states and has some resemblance to federal law in a federal state, where there are institutions and a body of law distinct from those of constituent provinces or states yet directly applicable thereto and to persons therein and enforceable within the constituent states. By creating the Communities and endowing them with the powers they have done, the member states have, to that extent within particular fields, restricted their national sovereignty.

By creating the Communities the member states also created in the law of each of them legal entities having the most extensive legal capacity accorded to legal persons under their laws, entities akin to public corporations and governed by public law, and having extensive powers, privileges, and immunities. Also, community law is part of the internal law of each member state, and as the process of approximation of laws proceeds a growing field of law, mainly in commercial matters, will be both Community law and municipal law of each member state.

Legal sources

The authoritative sources of Community law are not specified, but in fact are (a) the treaties establishing the three Communities (E.C.S.C.—Paris Treaty, 1951; E.E.C.—Rome Treaty, 1957; Euratom—Rome Treaty, 1957) with their annexes and protocols, the treaty establishing a single Council and a single Commission (the Merger Treaty—Brussels, 1965), a treaty amending the four foregoing treaties, and the treaty concerning the accession of Denmark, Ireland, and the U.K.

(the Association Treaty—Brussels, 1972), and also some inter-state conventions supplementing the treaties; (b) the Acts of the institutions of the Communities, namely regulations, directives, and decisions; (c) judgments of the Court of Justice of the Communities, which may be persuasive but are not binding, and (d) legislation and judgments of municipal courts of member states on matters of Community law. Recommendations and opinions of the Council and the Commission are not legislative and probably not sources of law.

In some cases the Court has applied principles of general public international law, and general principles of law held on the basis of comparative analysis of the legal systems to be common to the laws of the member states or most of them. Occasional reference is made to writings of jurists of high reputation.

Among the acts of Community institutions, regulations are published in the Official Journal, have general application and are directly applicable in all member states; directives are binding in substance on the member states to which they are addressed but allow national authorities to choose the method and form of implementation nationally; decisions are binding in their entirety on those to whom they are directed, who may be governments, or companies, and are enforceable in national courts; directives and decisions are usually published in the Official Journal; recommendations and opinions have no binding force.

Application of community law

Community law not only applies within the communities as organizations, but it binds member states and also applies within them and creates rights and duties for legal persons therein. By the treaties community laws have the force of law in the member states, and overrule domestic laws inconsistent therewith, whether earlier or later than the principle of community law in question. If and insofar as there is conflict, accordingly, community law is supreme. In the U.K. these principles were accepted by the European Communities Act, 1972.

The treaties require member states to take all appropriate measures to ensure fulfilment of the obligations arising out of the treaties, or resulting from action taken by Community institutions, to facilitate achievement of the Community's tasks, and to abstain from any measures which could jeopardize the attainment of the objectives of the treaties.

The treaties, moreover, create laws directly applicable in all member states, without need for adoption or ratification by municipal legislation, and creating rights which nationals of member states could enforce and rights which municipal courts would have to protect. This is only where the treaty provisions are clear and precise and the obligation is unconditional, and it leaves no discretion to the Community or to the member state. At least some of the Acts of Community institutions, particularly regulations, are also directly applicable, but doubt attaches to decisions and directives.

Community law in member states

Community law binds not only the member states but their citizens, immediately and directly. By joining the communities each member state has delegated a portion of its law-making authority to the Community and must accept and apply the law made within its sphere of jurisdiction by the Community. Some provisions are directly enforceable in the courts of member states. In other matters where Community law applies the European Court has jurisdiction though it cannot directly apply Community law within member states, but only declare that it applies, leaving the manner of application to member states and their courts and national courts must take cognizance of the rules of Community law. In case of doubt national courts must apply to the European Court for a ruling on the applicability or interpretation of Community law arising in a case in a member state's courts. In matters where Community law does not apply each member state's own courts alone have jurisdiction. Theoretically a monist doctrine explains the relationship, Community law and the laws of the member states forming for each state a single body of law and there being neither dualism nor conflict at any point. If there is conflict the Community rule must receive effect as being supreme.

The enforcement of the treaties in member states is secured by various political, economic and legal means. The Commission on discovering an infringement of a Treaty may direct the state to comply and if it fails to do so may refer the matter to the European Court. Its ultimate sanction is the pressure of public opinion within the Community and of persuasion by other member states.

Harmonization of laws

The Council is under a duty to issue directives for the harmonization of such provisions in member states as directly affect the establishment or functioning of the Common Market, and provisions concerning taxes on trade, and also for overcoming the differences between the laws of member states which are distorting the conditions of competition. The range of matters affected by harmonization provisions is wide, notably weights and measures, food standards, motor vehicle safety standards, and tax laws.

Language of community law

The E.C.S.C. treaty was drawn up in French only but the other treaties were in German, French, Italian, and Dutch, all texts being equally authentic.

The Accession Treaty was drawn up in eight languages, all equally authentic. Discrepancies in the meaning of words used for legal concepts in a variety of equally authentic texts may give rise to difficulties of interpretation. In principle, it is necessary to find a meaning compatible with all the texts, or in accordance with the spirit of the treaty. The treaties all have an unmistakable stamp of French law and the technical language of that system is frequently the best guide to meaning.

The European Court has as official languages English, Danish, French, German, Italian, and Dutch and any one of these may be chosen as procedural language; if the defendant is a member state or a subject thereof that state's official language will be the procedural language, or if the parties desire, the court on joint application, may designate another official language as the procedural language, or exceptionally may authorize use of another official language. The procedural language is used in pleas and written evidence; witnesses and experts should use the procedural or an official language.

European company. An idea proposed by the Commission of the European Communities in 1966 for the establishment by convention of a system of company law under which European companies might be incorporated. After investigation the Commission in 1970 proposed that European companies should come into being by the merger of companies incorporated under the laws of two or more member states. No clear pattern has won acceptance nor been adopted.

European Convention on Human Rights. A convention drawn up by the Council of Europe (q.v.) in 1950 and inspired by the U.N. Universal Declaration of Human Rights (q.v.) of 1948. It is on the classic Western liberal model and declares the essential human rights which the parties considered would be generally accepted in liberal democracies and should be protected by law in each state. It does not cover the whole field, and omits social and economic rights. The rights declared include freedom from slavery or forced labour, the right to liberty and security of person, to a fair trial, freedom of thought, conscience, and religion, of peaceable assembly and association, and the right to marry and found a family. Many rights are subject to exceptions and qualifications. Some states have but others, including the U.K., have not incorporated the Convention into domestic law, so that it is not enforceable as British law.

To enforce the rights declared the Convention created the European Commission of Human Rights, comprising a member from each state elected by the Committee of Ministers of the Council of Europe, and the European Court of Human Rights, comprising a similar number of judges elected by the Consultative Assembly of the Council of Europe. The function of the Commission is to enquire into alleged breaches at the instance of a state or, if a state has recognized the competence of the Court to receive petitions, on petition from an individual or non-governmental organization alleging violation of rights by a state. If it considers the petition admissible the Commission must investigate the facts and seek to secure a friendly settlement, failing which it reports to the state and to the Committee of Ministers. A case may be brought before the Court only if the states concerned have accepted the compulsory jurisdiction of the Court or consented to a case being heard, and only a state or the Commission can refer a case to the Court. The Court sits at Strasbourg and its decision is final. If it finds that an infringement has taken place and the domestic law of the state does not give full reparation it may afford just satisfaction to the injured party.

Since the Convention came into force in 1953 some thousands of individual petitions have been lodged and a small number of inter-state complaints have also been lodged. The Convention assumes that the main protection of human rights is by the domestic law of the member states but cases have shown defects in various domestic legal systems. British cases have sometimes resulted in changes in British domestic law, such as the creation of immigration appeal tribunals.

I. Brownlie, *Basic Documents on Human Rights*; J. E. S. Fawcett, *The Application of the European Convention on Human Rights*; F. G. Jacobs, *The European Convention on Human Rights*; *Yearbook of the European Convention on Human Rights.*

European Court of Justice. The common name for the Court of Justice of the European Communities (q.v.). It was established as the Court of Justice of the European Coal and Steel Community (q.v.) by the Paris Treaty of 1951 and consisted of seven members with the function of securing the observance of the Treaty, examining the decisions of the E.C.S.C. High Authority in the light of the Treaty and adjudicating on alleged breaches of the Treaty. In 1957 it became the Court of Justice of the European Communities, its jurisdiction being extended to include Euratom and E.E.C. as well.

It is not an international court like the International Court of Justice but an internal court of the Communities and in its relation to national legal systems in some ways resembles the position of the chief federal court in a federal state. It is comparable to a constitutional or administrative court in a municipal legal system rather than to a civil court and its practice and powers are modelled on those of the French Conseil d'Etat (q.v.).

With the accession of the U.K., Denmark, and Ireland to the Communities in 1973 the number of judges was increased to 9. These must be persons of proven independence and qualified to hold the highest judicial offices in their respective countries. They are nominated to the court by the member states for six-year renewable terms, five and four judges retiring alternately every three years. The judges have the usual immunities and privileges and may not hold any administrative or political office. Judges may resign or be moved if in the unanimous opinion of the other judges he no longer fulfils the conditions of tenure or meets the obligations resulting from his office. The Court elects its own President for three-year terms and two Vice-Presidents. Save where the Treaties require a plenary session of at least seven judges, the court sits in two chambers, each presided over by a Vice-President.

The Court's jurisdiction is created by and is derived entirely from the Community treaties. It has compulsory and exclusive jurisdiction over disputes between member states concerning the application of the terms of the treaties and a voluntary but exclusive jurisdiction over disputes concerning the object and purpose of the Communities in general. It may determine whether, if the Commission alleges so, a state is in breach of any treaty obligation. It has jurisdiction at the instance of member states, community institutions or private parties to control the acts of the Community institutions, by way of actions for annulment of any of the legally binding acts of the Commission or activities of the Council on any of the grounds of lack of competence, infringement of an essential procedural requirement, infringement of the Treaties or of any rules of law relating to that application, or misuse of powers, and to take action at the insistence of states, other community institutions or, in some cases, individuals, against the Council or Commission for violation of Treaty through inactivity. The Court also has plenary jurisdiction to examine parties' claims and substitute its own judgments for those of the Communities' institutions in cases of non-contractual liability, but only for *faute de service* and not for *faute personnelle*.

The Court is assisted by, originally two, now four, advocates-general, who must possess the same qualifications as the judges and are also appointed for six-year renewable terms by unanimous decision of the Council of the Communities. They represent the law, not any institution of the Community, nor any state or party. In every case one of the Advocates-General must make an independent and impartial assessment of the issues involved and of the relevant law and present a reasoned submission thereon to the Court. This is read in public and later published but does not bind the Court. The Court's decisions are binding on parties and not subject to appeal.

The procedure of the Court resembles that of a superior court in a civil law rather than in a common law country. It is based on Rules of Procedure drawn up by the Court and has three main stages.

The written stage consists of the lodging of a complaint, to which the defendant may make a defence, the complainer a reply, and the defendant a rejoinder. As soon as a complaint is lodged the President appoints one judge to be *juge-rapporteur*; he prepares a preliminary report for the Court's consideration and on the basis of the pleadings decides whether the case requires *instruction*, i.e. enquiry or hearing of evidence. If so, it may be held before the court, a chamber thereof or the *juge-rapporteur* alone, and the inquiry is carried out by the Court, chamber or *juge-rapporteur*, and the Advocate-General may participate. The Court and the Advocate-General may summon additional witnesses. After the pleadings and *instruction*, if any, there is an oral stage. The *juge-rapporteur* reports, outlining the case, summarizing the arguments, and stating the facts, and this is followed by oral argument. At this stage parties must be represented by a legal adviser or a member of the bar of a member state. The Court and the Advocate-General may question the parties' agents and counsel, and may order further *instruction*. Parties' representatives then address the Court and the Advocate-General makes his submissions. The judges deliberate in private and return a single collegiate judgment; dissents are not disclosed.

The procedural language must be one of the official languages of the Community, chosen by the complainer, but if a member state is defendant, its language is used. The judgment is published in each of the official languages but the copy in the procedural language is the authentic one. The judgment is binding from the date of delivery.

Though the principle of *stare decisis* does not apply the Court has tended to develop a consistent body of case-law and has regard to its own prior decisions in developing the law.

The Court's decisions are final, but a party may request revision of the judgment on the ground of discovery of a material fact previously unknown both to the party and to the court, if made within three months of discovery of the fact and ten years of the judgment. Any party may also ask the Court to interpret the essential part of its judgment if obscure or ambiguous.

G. Bebr, *Judicial Control of the European Communities*; D. Valentine, *The Court of Justice of the European Communities*; E. Wall, *The Court of Justice of the European Communities*; F. G. Jacobs and A. Durand, *References to the European Court.*

European Court of Human Rights. A court established by the European Convention for the Protection of Human Rights and Fundamental Freedoms (q.v.) and set up in Strasbourg in 1959.

Cases may be brought before the Court by the Commission of Human Rights and by states parties to the Convention. Its jurisdiction is compulsory only for states which have made express acceptance of jurisdiction. The U.K. accepted compulsory jurisdiction in 1965. The aggrieved individual does not have *locus standi*.

See also EUROPEAN CONVENTION ON HUMAN RIGHTS.

European Defence Community. An organization sought by Western European states with U.S. support to be established to control a supranational integrated European army as a counter to the Russian forces. The plan was put forward at the Council of Europe in 1951 and a treaty was concluded in 1952 but was not ratified by France and instead Western European Union was established in 1955, a plan for interdependent national contingents in a European force. In fact military co-operation continued mainly within the North Atlantic Treaty Organization (q.v.).

European Economic Community (E.E.C.). Commonly called the Common Market, it is an organization constituted by a Treaty of Rome between Belgium, France, Germany, Italy, Luxembourg, and the Netherlands, signed in 1957 and operative in 1958. The U.K., Denmark, and Ireland joined the E.E.C. in 1973.

The basis of the community is a customs union, or common market, having a prohibition on customs duties as between member states and a common external tariff as regards third countries. The community seeks to give effect to the principles of the free movement of goods, persons, services, and capital, and to ensure that conditions are equal as between member states and wide divergences in national policy eliminated. Provision is made to restrain competition between undertakings, on the harmonization or approximation of differing national laws, on economic policies, particularly external trade policies, social policies and transport.

The E.E.C. is controlled by an Assembly, a Council of Ministers, a Commission, and a Court of Justice. From 1957 it shared a single Assembly and Court of Justice with E.C.S.C. and Euratom and under the Merger Treaty of 1965 from 1967 a single Council and a single Commission for all three communities superseded the E.E.C. Council and Commission. There are also a number of special committees to handle particular matters of policy, a European Investment Bank, a European Social Fund, and a European Development Fund.

A common agricultural policy was established in 1962 providing for common guaranteed prices for producers. A common system of value-added tax was also adopted by 1972.

European Establishment Convention. A multilateral treaty of 1955 between most members of the Council of Europe dealing with the entry, stay, and removal of nationals of member states to, in, or from the territory of another, civil rights, legal and administrative protection, and economic activities.

European Free Trade Association (E.F.T.A.). An association of states formed in 1960 after the failure of the European Economic Community to link with other European states which were in the Organization for European Economic Co-operation. It was in effect a group outside the E.E.C., but not aiming at any kind of political unity. Its aims are economic expansion, fair competition, and equality of conditions of raw material supply for countries of the Association, and the removal of barriers to world trade. It is controlled by a Council consisting of one member from each state.

European Investment Bank. A body created by the E.E.C. Treaty, but an independent institution with separate legal personality. Its members are the member states and its task is to contribute from its own resources and access to other sources of capital to the development of the E.E.C. Thus it may make loans and give guarantees to finance projects for developing the less developed regions of the E.E.C., for modernizing or converting undertakings or developing fresh activities required by the E.E.C., and projects of common interest to several member states. Its capital is subscribed by the member states. Its seat is in Luxembourg and it is directed and managed by a board of governors, consisting of the Finance Ministers of the member states, a board of directors appointed for five-year renewable terms by the Board of Governors to manage the Bank in accordance with the E.E.C. Treaty, the Statute of the Bank and the general directives of the Board of Governors, and a management committee consisting of a president and two vice-presidents appointed for six-year renewable terms. The Bank obtains funds, primarily from loans floated on capital markets. The European Court has jurisdiction over certain kinds of disputes concerning the Bank but disputes between the Bank and its creditors or debtors are decided by national courts unless jurisdiction has been conferred on the European Court.

European Law. As a cultural, as distinct from a geographical, area, Europe can only be said to have an existence from the break-up of the Roman Empire in the fourth century, and though there developed, increasingly as time went on, a largely common civilization and a largely common social, religious, intellectual, and cultural life, this coexisted with great variety of constitutional and legal systems, of legal institutions, concepts and rules connected with the great number and variety of political entities. The development of this is considered in the articles on ANCIENT LAW; CANON LAW; MEDIAEVAL LAW; ENGLISH LAW; FEUDAL LAW; ROMAN LAW.

Despite the great variety of municipal and local laws which have developed, there have always been a few unifying factors which have made the European systems recognizably similar. Among these factors have been the common civilization imposed by the Roman Empire and the imposition of Roman law which, even after the break-up of the Empire in the west left a substratum of Roman law, the revival and spread over most of Europe of the study of Roman law from the tenth century, the creation and spread over most of Europe of a legal science founded on the study of the rediscovered Roman law, the general acceptance of Christianity and of the authority of the Church of Rome which developed its own courts and law, the canon law, the widespread acceptance of feudal practices, the growth of trade which gave rise to mercantile practices and maritime law, and the long-continued use of Latin as the *lingua franca* of scholars.

Similarly later influences, such as the Renaissance, the Reformation, the Reception of Roman law, the Enlightenment, philosophic movements such as the rise of natural law thinking, Idealism, the move for codification, the ideas of the American and French Revolutions, Marxism, and others had European and not merely localized or even national influence.

Accordingly, while from about the fifteenth century Europe became increasingly sharply divided into distinct nation-states, each developing its own system of government, law and courts, it was a process of differentiation from a largely common base and substantially parallel development of fundamentally similar systems. But as the nation-states increased in power different trends became more evident. France was unified early and the great ordinances of the seventeenth century paved the way for the Napoleonic codifications which were imposed on or copied by many of the other states of Europe. Germany was not unified, politically or legally, till the nineteeenth century and codification at the very end of that century was both a consequence of and a force for that unification. Italy similarly was late in being united and in codifying.

In the development of European law, England (with Wales and Ireland, but not Scotland), stood apart, having early developed a distinctive legal system, little influenced by Roman or canon law, and based on decisions of the royal courts and legislation, aided by juristic writing, notably of Bracton. The Scandinavian countries were never part of the Roman Empire and further away from the intense mediaeval and early modern ferment of legal science and the influences flowing from southern Europe and also to some extent have stood apart.

In modern times forces for the harmonization and ultimate unification of the legal systems of Europe have been the increased study of comparative law and of foreign law, stimulated by increasing international intercourse and co-operation, the activities of the Hague Convention on Private International Law and of the International Commission for the Unification of Private Law, many particular conventions promoting uniformity of law and, not least, the development of the European Communities which are both creating a body of supranational law enforceable in the member states and promoting uniformity of law. Indeed the term European law is sometimes given to the law of the European Communities.

European Monetary Agreement. An agreement providing for the establishment within the Organization for European Economic Co-operation of a European Fund and multilateral system of settlements, superseding the European Payments Union Agreement. The Fund's purposes are to enable discretionary credits to be given to parties to the agreement for up to two years and to facilitate the operation of the settlements system.

European Parliament. The European Coal and Steel Community, established by the Treaty of Paris of 1951, established as one of its governing institutions a Common Assembly. The European Atomic Energy Community and the European Economic Community, set up by separate Treaties of Rome in 1957, created similar institutions, the Assembly being common to both. The European Parliamentary Assembly was thus one of the institutions common to the three communities and first met in Strasbourg in 1958. In 1962 it changed its name to the European Parliament.

In consists of representatives nominated by the national parliaments of the member states, though in 1960 the Assembly adopted a draft convention providing for direct election of its members by the populace of the member states. Sessions are normally held at Strasbourg. It appoints its own President, eight vice-presidents, and officers, and has its own rules of procedure; it takes decisions by an absolute majority of votes cast. It sits, in practice, for 11 sessions a year, the annual session commencing in March. Members sit in political, not national, groups. The Parliament verifies the credentials of members appointed from national parliaments. Members enjoy the customary immunities and privileges. The President and vice-presidents constitute the Bureau which is the principal organizing committee of the Parliament.

Voting is by show of hands, or by sitting and standing, or by roll-call.

Under the treaties its function is to exercise advisory and supervisory powers. It may debate any matter within the aims and scope of the Communities as defined by the treaties.

It is directed to discuss in public the annual general report submitted to it by the Commission; if a vote of censure is supported by a majority of members and two-thirds of the votes cast, the Commission must resign as a body. It may not remove or censure individual commissioners. It has some control over the Ministers who form the Council but not absolute control. Members of the Commission may attend meetings of the Parliament to answer questions or be heard at their own request on behalf of the Commission. Members of the Council may be heard under conditions laid down in the Council's rule of procedure. The Parliament may also comment on the framing of Community policies when the President of the Commission presents his annual programme of the Communities' future activity.

It must be consulted by the Council of Ministers before the Council acts on proposals from the Commission to make regulations or issue directives which have the force of law in member states. Any member may table a motion for a resolution on matters within the sphere of activity of the Communities. A committee may take the initiative in studying any matter within its field of responsibility and make a report to the Parliament.

Members may ask parliamentary questions on the progress of the Commission's work calling for written answer, oral answer with or without debate, or table questions to the Council. A large number of committees exist, whose reports form the basis of most debates. The Parliament must be consulted on many kinds of proposed legislation and return an opinion, based on a committee examination and report, to the Council and the Commission. Its views have frequently been influential on the Council and the Commission. From 1975 the Parliament has qualified control over the administrative expenditure of the Communities.

The Commission of the European Communities exchange reports with the Committee of Ministers of the Council of Europe and joint meetings are held of members of the European Parliament and the Consultative Assembly of the Council of Europe. There is little formal contact between the European Parliament and W.E.U. and it has no regular contacts with the North Atlantic Assembly.

The Parliament publishes a wide range of documents in the languages of the member states, including verbatim reports of debates.

In 1976 agreement was reached on a directly elected Parliament of 410 members, France, Germany, Italy, and the U.K. to have 81 each, the Netherlands 25, Belgium 24, Denmark 16, Ireland 15, and Luxembourg 6 seats, and arrangements were made to have direct elections in 1979.

Eustace, Maurice, Baron (?1590–1665). Called to the English Bar in 1625, became Speaker of the Irish Parliament and, in 1660, Lord Chancellor of Ireland.

Euthanasia (or mercy killing). The causing or hastening of death, particularly of incurable or terminally ill patients, and at their own request. It raises difficult religious and moral problems. Plato regarded it as justifiable but Christianity has rejected it. Generally it is treated as illegal and not distinguishable from murder, largely because of the difficulty of distinguishing in legal rule and in fact between criminal and justifiable causing of death. A narrowly distinguishable case is of refraining from seeking to prolong life in cases of great pain or inevitable death, which is generally considered morally and legally permissible.

Evans, Sir Samuel Thomas (1859–1918). Practised first as a solicitor being called to the Bar only in 1891; he entered Parliament in 1890 and sat till 1910. He was Solicitor-General 1908–10 and President of the P.D.A. Division 1910–18, in which capacity he decided a long series of important prize cases during World War I. Many of these judgments are of outstanding importance, as Evans not merely adopted but logically developed doctrines of that branch of law and skilfully applied them to new conditions. In other cases he showed great shrewdness and ability in unravelling complicated facts.

Events. Happenings which take place without the intervention of the will of any human being, as contrasted with human acts. Events may be brought about by natural forces, such as the fall of a tree or a flood. A happening of such kind does not necessarily infer civil or criminal liability of anyone; but it may, if the event was foreseeable, and should in the circumstances have been foreseen and prevented, e.g. by felling a dangerous tree or cutting a culvert for flood water. But some events, such as deaths, explosions, and accidents, may result from a prior exercise of human will and human act or omission, and cannot be regarded as pure events. They infer liability for the consequences if such consequences should have been foreseen and prevented.

Everardus, Nicolaus (1462–1532). President of the court of Holland (1509–28) and the Great Council of Mechlin (1528–32) and the leading Dutch jurist of his time, author of *Topica* (1516) and *Responsa sive consilia* (1554) and one of the earliest authorities on Roman-Dutch law.

Evershed, Francis Raymond, Lord (1899–1966). Called to the Bar in 1923, he took silk in 1937 and was a judge of the High Court, 1944–47, a Lord Justice of Appeal, 1947–49, Master of the Rolls, 1949–62, and a Lord of Appeal, 1962–65, and also a member of the Permanent Court of Arbitration at The Hague from 1950. He presided over a Committee on Supreme Court Practice and Procedure (1953).

Eviction. Dispossession of an occupier from land occupied by him. It may be lawful, particularly under court order, as when a squatter is evicted from buildings he has occupied or a tenant whose term has expired from premises let to him, or unlawful as when a trespasser ejects a lawful possessor or a landlord evicts a tenant in circumstances legally unjustified.

Eviction by title paramount. Dispossession of land by reason of the fact that the lessor had no title to grant the lease to the occupier.

Evidence. Facts, inferences from facts, and statements which tend to convince a court or other inquiring body that certain facts, the state of which is unknown but being inquired into, are to a certain effect. While the task of the court, to ascertain if at all possible the true state of the facts being inquired into, would suggest that everything which casts light on these facts might be brought to the notice of the court, the rules of evidence, based partly on experience and partly, in some cases, on rulings applicable to circumstances which no longer apply, frequently restrict the kinds of evidence which may be adduced. The development of the law of evidence is however on the whole a movement from reliance on non-rational grounds for decision to rational grounds.

Non-rational grounds for decision are found in the appeal to God which underlay trial by ordeal and trial by battle. The belief was that God would ensure that right prevailed. Rather more rational was compurgation whereby the accused could clear himself by his oath, if supported by the oaths of a sufficient number of compurgators or oath-helpers swearing not to the truth of the facts but to their belief in the truth of the oath sworn. In effect they guaranteed the oath sworn. Ordeals disappeared in the thirteenth century and trial by battle by the fifteenth century. By this time the law of evidence, like other branches of law, was influenced by the Roman and canon law worked out in the universities. Under this influence evidence was evaluated hierarchically; not all persons were competent or admissible witnesses and to prove a fact there must be two or more witnesses. The judge was in substance required to accept evidence if the required number of witnesses testified to the same effect. The interrogation of witnesses was done privately. Moreover, it was believed that confession was the best evidence and that reliable confessions could be extracted by torture. Nevertheless, despite its defects, Romano-canonical law went some way towards the exclusion of non-rational evidence from legal procedure. Rational grounds for decision of facts came in when it came to be recognized that witnesses, documents, and the real evidence of physical objects were more likely than other means to establish facts.

In Anglo-American law the rules as to evidence were materially influenced by the jury system and the system of examination and cross-examination. Thus the general exclusion of hearsay, which is frequently relied on in everyday life, is attributable to the desire to have before the jury the actual author of a statement and to have him cross-examined thereon. It is in fact less strictly excluded in civil cases today.

The first question is of the standard to be attained by evidence. If the courts required that facts under investigation had to be established with certainty, few cases would ever be held proved. Accordingly common law countries have generally accepted that in civil cases a fact can be held established if the evidence renders it more probable than not that the fact was to a particular effect and that in criminal cases a fact can be held established if the evidence satisfies beyond reasonable doubt. In both cases accordingly evidence must establish probability, in different degrees, but need not establish certainty. In civil law countries such a high degree of probability is demanded as to exclude reasonable doubt.

The second question is of the burden of proof: on whom is incumbent the duty of establishing certain facts. This burden depends largely on the substantive law applying to the matter in issue; and whether the facts to be proved concerned a substantive claim or the establishment of an exception, qualification, or exemption. In general, in civil cases the party asserting certain facts must prove them, and in criminal cases the prosecution must prove the facts necessary to bring home the crime charged to the person accused.

The third question is of the competency or admissibility of particular kinds of evidence, such as evidence of similar facts, evidence of character, hearsay, admissions and confessions, statements in public documents, books, maps and many other kinds of statements. For long parties to a cause were deemed inadmissible.

Fourthly there is the question of the relevancy of evidence, that is of its pertinency to the matter in dispute. If wholly irrelevant it should be excluded, but if relevant and accepted the question remains of what weight should be given to it.

The final question is of the weighing or evaluation of the evidence, the determination of which version or explanation is to be preferred and which discarded as inaccurate. In civil cases a fact may be held established on balance of probabilities; in criminal cases a charge must be proved beyond reasonable doubt before conviction is justified.

The kinds of evidence usually produced are physical objects (real evidence), written statements and documents (documentary evidence), testimony of the parties (personal evidence), and testimony of expert witnesses on the conclusions or inferences to be drawn from facts found by measurement, testing or other examination (expert or opinion evidence). Facts must be established by the best evidence which the nature of the fact will allow.

The ways in which evidence is elicited and brought to the notice of the judge vary. Two major systems are recognized in legal history. The inquisitorial system originated in Romano-canonical procedure. In it the judge himself investigates the facts and may call for evidence additional to that put before him by counsel. This system underlies criminal procedure in most Continental legal systems and systems derived from them. The other is the accusatorial, or adversary, system in which the parties and their counsel put before the court the evidence which they think necessary and sufficient. The judge does not himself make investigation of the facts save exceptionally, though he may ask questions of witnesses in supplement of those put by counsel.

In practice, knowledge of the rules of evidence is highly important because problems may arise in the course of a trial calling for immediate objection or other action.

History: H. Brunner, *Entstehung der Schwurgerichte* (1871); M. Bigelow, *History of Procedure in England* (1880); F. Pollock and F. W. Maitland, *History of English Law* (2 ed. 1898); J. Thayer, *Preliminary Treatise on Evidence at the Common Law* (1898); W. Holdsworth, *History of English Law*.

Textbooks: W. Best, *Principles of the Law of Evidence* (1922); S. Phipson, *Law of Evidence* (1952); J. Archbold, *Pleading evidence and Practice in criminal cases* (1976); J. Wigmore, *Treatise on the Anglo-American System of Evidence*.

Evolution of law. So little is known of law, legal systems, and legal relations in preliterate societies and prehistoric times that it is very difficult and dangerous to formulate general statements about the evolution of law in human societies. Not all societies have followed the same lines of development. Even for those societies where the development of law in comparatively modern times is substantially documented, much remains uncertain. Clearly there is some connection and correlation between the law and the other social and cultural institutions of a particular society but there tends to be a considerable time-lag between developments in material culture such as the progress of the Industrial Revolution, and developments in legal institutions and principles. In some cases evolution seems to have been slow and steady. In other cases there have been enormous and sudden jumps in legal development, as in the cases of Eastern countries adopting developed European codes almost complete.

A point of some importance is that different societies pass through particular evolutionary stages at different times, so that in the world at any particular times some are at relatively primitive and some at relatively advanced stages. Evolution accordingly cuts across chronology.

Again, nineteenth century liberals saw legal evolution as inevitably developing in the direction of greater freedom for the individual whereas twentieth century experience in Western societies has been precisely the opposite, evolution towards increasing state control and extinction of individual rights and liberties. Evolution is not accordingly always in the same direction.

Subject to these qualifications a broad distinction can be drawn between primitive law, middle-stage, and mature or developed law. Primitive law is characterized by reliance on custom, social pressures and undeveloped court systems. The blood-feud is a common institution and as courts develop they tend to make the acceptance of compensation customary and ultimately compulsory. Conscious rule-making is found but is not a predominant form of law.

Middle-stage law tends to be more strict, rigid and formal; courts and legal institutions are found and the rules gradually become more precise; procedure becomes technical.

Mature or developed law is characterized by a complex of courts and other institutions, a developed body of practitioners, jurists and legal scholars, a body of literature, a variety of modes of law-making and law-changing, and a sophisticated and technical body of knowledge, far beyond what the ordinary man can know or understand.

In some cases societies reach a stage of decadence when law becomes confused, judges and jurists fail and the whole system collapses. This happened to Roman Law in the Dark Ages. It is happening to British law as it becomes increasingly a matter of mere irrational regulation and incomprehensible complexity.

It is important to bear in mind that different societies move from one of these stages to another at very different times. The Babylonians had reached at least the middle-stage before the Romans had ever been heard of, and the Romans were in the stage of decadence before English law had reached the middle stage. An historically early code, that of Hammurabi, is much more mature and developed than historically later bodies of law such as the Pentateuch or the Twelve Tables, which themselves are earlier, but more mature than the Salic Law and other *Leges Barbarorum*. Evolution, that is, is concerned with stages of development and maturity, not with chronology, but it is more difficult to chart evolution than to establish chronology.

B. Malinowski, *Crime and Custom in Savage Society* (1926); L. Hobhouse L. Wheeler and M. Ginsberg, *The Material Culture and Social Institutions of the Simpler Peoples* (1930); A. S. Diamond, *Evolution of Law and Order* (1951); *Primitive Law, Past and Present* (1971); E. A. Hoebel, *The Law of Primitive Man* (1954).

See also ANCIENT LAW; MEDIAEVAL LAW; PRIMITIVE LAW.

Ewart, William (1798–1869). British M.P. and penal reformer who induced Parliament to abolish the penalty of hanging criminals in chains (1834) and capital punishment for minor offences (1837). He long urged the abolition of capital punishment and secured the appointment in 1864 of a select committee on the topic.

Ex aequo et bono (from fair and right). The Permanent Court of International Justice was, and the International Court of Justice is, entitled, with the consent of the parties, to decide an issue *ex aequo et bono*, i.e. on the basis of what is fair and right. This is distinct from the basis of equity, as understood in Anglo-American law, because equity is now a fairly settled body of principles and rules. Conferment of such a power is tantamount to endowing the tribunal with a legislative power, and it endows the court with wide legislative discretion.

Ex contractu (out of contract). According to the Roman law one of the major divisions of obligations, those arising from agreement, as contrasted with obligations arising from other sources such as *ex delicto*.

Ex debito justitiae (from debt to justice). That which an applicant is entitled to as of right, as distinct from what may be granted in the exercise of discretion.

Ex delicto (out of delict). According to the Roman law, one of the major divisions of obligations, those arising from the commission of delict or civil wrong, as contrasted with those arising from other sources, such as *ex contractu*.

Ex dolo malo non oritur actio (action does not arise from fraud). A maxim expressing the principle that an action cannot be brought on the basis of a fraud.

Ex gratia (as a matter of grace). A payment made though not, or without admission that it is, legally due, such as a payment made to settle a claim rather than go to the expense of contesting it.

Ex nudo pacto non oritur actio (action does not arise from a bare agreement). A maxim expressing the principle that a right of action does not arise on a mere agreement without consideration.

***Ex officio* information.** A criminal information filed by the Attorney-General on behalf of the Crown in respect of crimes affecting the government or peace and good order of the country. It was utilized in cases of seditious writings or speeches, seditious riots, libels on foreign ambassadors, and the obstruction of public officers in the course of their duties. It was abolished in 1967.

***Ex officio* oath.** An oath under which a person could be compelled to answer questions the answers to which would expose him to censure or punishment. Such an oath was administered by the Court of High Commission (q.v.) but the administration of such an oath by any ecclesiastical court was forbidden in 1661.

Ex parte (on behalf of). In relation to applications to court, indicates a person not a party to the proceedings but having an interest therein. More commonly it means an application made by one party without notice to and in the absence of the other, as where an interim injunction is sought *ex parte*.

***Ex post facto* law.** A retrospective law such as one which makes criminal what was not criminal when it was done, increases the penalty, or alters the procedure disadvantageously to the accused in such circumstances. The principle of the objection to such a law is expressed in the rule *nulla poena sine lege*; it is the recognition of the unfairness of such a law. In the U.K. Parliament can pass such legislation but the courts will not interpret legislation as having retrospective effect unless that is clearly the intention of the relevant provision. The U.S. Constitution forbids Congress and the states to pass any *ex post facto* law but since 1798 it has been settled that this applies to criminal laws only. The prosecution of the Nazi leaders after World War II for crimes, some defined specifically only by the charter creating the International Military Tribunal for war criminals, gave rise to considerable discussion of the justice of *ex post facto* criminal laws.

Ex relatione (from information or communication). A phrase used of a report of a decided case when the reporter derives his knowledge not from personal presence but from information from another. Criminal informations and *quo warranto* informations were formerly filed *ex relatione* of an individual known as the relator.

Examination. The interrogation of a person, orally or by written interrogatories (q.v.). Normally this takes place in court by way of examination in chief, cross examination, and re-examination (qq.v.). If a receiving order has been made against a debtor he must attend for a public examination, when he is questioned as to his property, dealings, and conduct generally; the official receiver, the trustee if any, and any creditor may question the debtor. A preliminary examination is one where a person is accused of an indictable offence and evidence is heard by one or more magistrates to enable him to decide whether there is a *prima facie* case sufficient to justify committing the accused for trial on indictment.

Examination in chief or direct examination. The questioning of a witness in court by the counsel for the party who called the witness to elicit his evidence on the matters in dispute of which the witness has knowledge. It is followed by cross-examination (q.v.) and that in turn by re-examination (q.v.) of the witness.

In European legal procedure a witness is commonly examined first by the judge and only thereafter by counsel for the parties.

Examiners of Private Bills. Two examiners are appointed, one from the staff of each House of Parliament, to ascertain whether all the preliminaries to consideration of a private bill, required by Standing Orders, have been complied with. Notice of the examination is given to the promoters, who must appear, failing which the petition will be struck out.

Exarch. In the Byzantine Empire, a title given at first to high civil and ecclesiastical dignitaries and later to high officers commanding forces on land or sea. Later it was conferred on the officials governing the Exarchate of Ravenna (584) and the Exarchate of Africa (591), who exercised full civil, military, and judicial powers over the areas committed to them. The title long continued in the Orthodox Church to be used for ecclesiastical dignitaries.

Excambion. In Scots law, the contractual exchange of land or buildings for other land or buildings, possibly with a monetary element to equalize the exchange.

Excepted risk. In insurance, a risk which is excepted from the undertaking to indemnify, and accordingly not covered.

Exceptio. In the Roman law, a plea taken by way of defence. Thus the *exceptio doli* was that the contract had been induced by the other's fraud. An *exceptio* could be included in the *formula* (q.v.) worked out for a particular case, to the effect that the judge should find for the plaintiff unless he held that the point raised by the *exceptio* was established.

Exception d'illégalité. A form of challenge known in French law and the law of the European Communities whereby a defence of illegality can be raised notwithstanding the expiry of the time within which an act of authority can be challenged by way of action for annulment. If the defence is successful it declares the act illegal in relation to the complainer only and not in terms of its general application. Before the European Court individuals, corporations, and even member states may invoke the defence of illegality.

Exceptiones ad cassandum brevia (or *Exceptiones contra brevia)*. A Latin tract, possibly a second part of the tract *cum sit necessarium* (q.v.), but possibly not by the same author. It is printed in Woodbine's *Four Thirteenth-century Tracts* (1910).

Exceptiones Legum Romanorum (or *Exceptiones Petri)*. A tract on Roman law, dated from the later eleventh century and originating in Dauphiné, apparently designed to help a magistrate exercise his office and forming a short manual for practical use. It is drawn from the *Corpus Juris Civilis.*

Exceptions, Bill of. In civil jury trial prior to 1875 if counsel thought that the judge had misstated the law in his instructions to the jury, he might require the judge to sign a bill of exceptions stating the point, which was in the nature of an appeal to the appellate court. It survives in Scottish civil jury practice.

Exceptions clause. A clause in a contract excluding liability for certain breaches of the contract or certain kinds of loss. The courts do not favour such clauses and interpret them narrowly and statutes limit their effect.

Excess profits tax. An additional tax on the profits of trades and businesses levied from 1939 to 1946, charged on the excess of the actual profits order a standard figure calculated on one of several bases.

Exchange, bill of. See BILL OF EXCHANGE.

Exchequer, Barons of. The Exchequer originated in the twelfth century as an aspect of the *Curia Regis,* in one division of which legal disputes as to payments to the Crown were decided. All the great Officers of State were present at the Exchequer. There was no separation between the purely financial and judicial sides of the department for most of the thirteenth century, but by the end of that century the judicial side was beginning to be definitely separated from the administrative side. The higher officials of the Exchequer were the Barons of the Exchequer and as it became more a department of State and relevant knowledge became more technical, the functions of the Barons became differentiated; from 1234 the term was used of a special set of officials with duties of mixed judicial and administrative kind. In the fourteenth century they were recruited from among Exchequer and eminent practitioners in the common law courts. A Chief Baron is known from the late thirteenth century and he was usually a lawyer; the other Barons were not always lawyers. Judges of the Exchequer were long deemed inferior to those of the other common law courts, they did not appear among the judges of assize, were not members of Serjeants' Inn, but took rank after the serjeants, and were not summoned to Parliament with the other judges. In 1579 on the appointment of a Baron it was declared that he should have the same rank as judges of the other courts. From this time Barons were appointed from serjeants and went on circuit. The cursitor Baron was one appointed to acquaint the rest with the course of Exchequer business, which lawyer-Barons did not have. The office dates from 1323 and was abolished in 1857.

Exchequer, Court of. In the twelfth century the Exchequer became distinct from the King's Council and the earliest department of State. Initially there was no separation between the financial and judicial branches of the department. By the end of the thirteenth century the two branches were separating and the judicial branch was becoming similar to the two common law courts, Common Pleas and King's Bench. The term Baron of Exchequer ceased to denote any of the high officials of the Exchequer and came to be used for officials with judicial and administrative duties only. At the end of the thirteenth century a Chief Baron is mentioned. The close connection of Exchequer with the *Curia Regis* was maintained all through the thirteenth century, and down to the mid-sixteenth century the jurisdiction was almost entirely limited to revenue cases, but thereafter it developed a supplementary equitable and common law jurisdiction. From the fourteenth century the Chief Baron was usually a lawyer and some of the Barons lawyers, though not till the sixteenth century were the Barons regarded as of the same rank as judges of the other courts. The Exchequer was firstly a revenue court held before the Treasurer and Barons deciding questions between the Crown and accountants to the Crown, and between Crown and taxpayer, and in the *cursus scaccarii* equitable and legal pleas could both be considered; secondly, it was a court of common law held before the Treasurer and the Barons, with exclusive jurisdiction over the officials of the court, a privilege later extended to other persons; moreover by the use of fictions such as *quominus* (q.v.) it acquired jurisdiction between individuals in all actions except real actions; and thirdly, it was a court of equity held before the Treasurer, the Chancellor of the Exchequer and the Barons, having certain equitable powers under the *cursus scaccarii*, and in the end the Exchequer obtained a general equitable jurisdiction by simple assumption of powers. In 1842 the equitable jurisdiction was transferred to the court of Chancery.

In 1875 the court was abolished and merged in the new High Court.

G. Gilbert, *Historical View of the Court of Exchequer*; *Treatise on the Court of Exchequer*; H. Hall, *Antiquities and Curiosities of the Exchequer.*

Exchequer, Court of, in Ireland. The Exchequer was the oldest department of administration which developed in Ireland but its development is unclear. The Irish Exchequer developed a legal side, held before the Treasurer and Barons of the Exchequer. Barons are known of from 1251 and a chief Baron from 1309. In 1877 it became the Exchequer division of the High Court in Ireland, and in 1897 that division was merged with the Queen's Bench Division.

Exchequer, Court of, in Scotland. From the thirteenth century the King's Council both exercised judicial powers and sat in the Exchequer. It used legal forms to compel royal officers to account for the revenues they recovered. The audit was a sitting of the Council in Exchequer, and in the fifteenth century Exchequer business and judicial business were hardly differentiated. By the end of that century there was more specialization but auditors of Exchequer frequently dealt with non-financial matters and were simply a body sharing the judicial functions of the council with a special responsibility for the royal revenues.

A statute of 1708 established a new Court of Exchequer on the English model, with a Chief Baron and ordinary Barons, with the same authority in revenue cases as the English Court of Exchequer and saddled with English forms and English procedure. From 1832 statutes restricted its functions and in 1856 it was abolished, the jurisdiction being transferred to the Court of Session, which, however, still treats revenue cases as Exchequer causes, and till 1947 English forms of process were maintained in revenue cases.

Exchequer Chamber, Court of. The practice early developed of judges and Exchequer officials meeting in a chamber at the Exchequer to discuss legal and other business difficulties. This gave rise to the idea that such meetings were a court and Parliament sometimes conferred powers on meetings of judges in the Exchequer Chamber. Prior to 1873 four distinct courts were called Court of Exchequer Chamber.

When the Court of Exchequer developed as a Court, questions arose as to the correction of errors therein. In 1357 statute established a court consisting of the Chancellor and Treasurer, with the judges as assistants, to hear errors in the Exchequer. The composition of this court was modified by several later statutes. In the *Bankers Case* (1696) it was held that the Lord Keeper might reverse a decision notwithstanding the contrary view of the judges.

In 1585, to surmount the difficulty that Parliament, which alone could amend errors of the King's Bench, was rarely summoned, a new Court of Exchequer Chamber was established, consisting of any six or more of the justices of the Common Pleas and the Barons of the Exchequer, to hear error from the King's Bench in certain kinds of cases, but without prejudice to a further appeal to Parliament. Error still lay to the House of Lords in proceedings not mentioned in the Act.

In 1830 the two courts created by these statutes were merged in a court of Exchequer Chamber, in which judges of any two common law courts (King's Bench, Common Pleas, and Exchequer) heard appeals against a decision by the third one, with a further appeal to the House of Lords. This court was replaced by the new Court of Appeal in 1875.

The name was also given to a court of debate, composed of an assembly of judges of all three common law courts, sometimes with the Lord Chancellor as well, to which cases of great importance might be adjourned for discussion, before judgment was given in the court in which the case had arisen. This meeting became important only in the fifteenth century. It never became a court of law in the technical sense, and was never formally constituted, but great weight attached to the general view of all the judges. At an earlier time more cases seem to have been referred by Parliament, Council or Chancery, but from the end of the fifteenth century most references came from the common law courts. Among important cases discussed in Exchequer Chamber were *Hampden's Case,* on ship money, and *Godden* v. *Hales.* This court became of less importance in the eighteenth century but may have given rise to the practice of the judges in criminal cases reserving difficult cases for the Court for Crown Cases Reserved.

Exchequer Division. One of the Divisions of the new High Court of Justice created by the Judicature Act, 1873, to which was assigned the business of the Court of Exchequer, abolished by that Act. In 1881 it was merged in the Queen's Bench Division.

Exchequer and Audit department. The department of government headed by the Comptroller and Auditor-General (q.v.) which controls the issues of public money from the Consolidated Fund and also conducts a continuous audit within departments of State and carries out special test audits. Its staff work closely with the departmental accounting officers who are, in their departments, personally responsible to Parliament for the regularity of any expenditure from each vote and for rendering the final account of each year's receipts and expenditure which, after audit, is laid before Parliament. Departmental accounting officers are usually the permanent heads of the departments.

Exchequer of the Jews. A separate branch of the Exchequer, existing from about 1195 till about 1290, concerned with the Jews, a class who were tolerated in England at the time but disliked. The Exchequer of the Jews was established to protect them better. Registries for Jewish bonds were established in London and other towns where there were important settlements of Jews, and there copies of loan bonds, assignments, and discharges had to be enrolled. In 1198 *Custodes Judaeorum* were appointed by the King, and associated with the Barons of Exchequer. They had jurisdiction in the affairs of Judaism, subject to the control of the Barons of Exchequer with whom they conferred in cases of difficulty. Popular feeling against Jews increased in the thirteenth century and restrictive ordinances were made against them, and in 1290 Edward I banished them. They did not begin to return to England till after the Restoration.

Exchequer records. A vast collection of documents evidencing the financial activities of the English Exchequer and including Domesday Book, the *Dialogus de Scaccario*, the Chancellor's Rolls, the Pipe Rolls, the Black Book of the Exchequer and the Red Book of the Exchequer (qq.v.).

See generally T. Madox, *The History and Antiquities of the Exchequer of England* (1711 and 1769); F. S. Thomas, *The Ancient Exchequer of England* (1848).

Exchequer, Stop of the. The act of Charles II in 1672 of deferring by proclamation payment out of the Exchequer of any money due on certain kinds of existing security but promising to add the interest outstanding to the capital and to allow 6 per cent interest on the whole sum. The King thereby acquired the use of about £1.3 million. Interest was paid till 1683 and in 1699 an Act was passed, to come into force in 1705, charging the excise with interest at 3 per cent on the sum of £1.32 million. No compensation was ever paid for total loss of interest 1683–1705.

Excise. See CUSTOMS AND EXCISE.

Exclusive jurisdiction. The part of the jurisdiction of the former Court of Chancery (q.v.) which it alone exercised, as distinct from that in which common law courts also had jurisdiction. It comprised particularly jurisdiction over trusts and administration of assets.

Excommunication. The ultimate sanction in the Catholic Church, excluding a member of the Church from the Christian communion and all its benefits. It is distinguishable from suspension, which applies only to clergy and denies them some of their rights and from interdict which does not exclude from the communion but forbids certain sacraments and sacred offices. The Catholic Church distinguishes two kinds of excommunication, one which renders the person *toleratus* and the other which renders him *vitandus*. The latter is reserved for the most serious offences, but both bar the excommunicated person from the sacraments and from Christian burial. If an excommunicated person confesses his sin and undergoes penance he may be absolved, but sometimes only by a bishop or even by the Pope. In mediaeval times it was frequently employed very effectively, e.g. against King John of England in 1209–13 but was later abused and lost much of its effect.

Excommunication is known in the Anglican Church, but very rarely used. In other reformed churches similar penalties are known usually under another name.

Execute. To complete, perform or carry through. To execute a deed is to sign, seal, and deliver it. To execute a judgment is to enforce it or put it into effect.

Executed consideration. An act done in return for a promise of counterpart performance.

Executed and executory contracts. A contract is executed when the transaction is completed immediately after the contract is made, as in the case of a cash sale. An executory contract is one where the agreement has been made but material part of performance is yet to be made, such as to build a house or deliver goods at a future time.

Execution. In English law, the general term for the enforcement by a public officer of the judgments or orders of courts of justice. The present system is based on the methods in use before the Judicature Act, 1873. Writs of execution are issued by the office of the court to the judgment creditor. They are directed from the Queen to the sheriff, commanding the latter what he is to do and, usually, to make a return of his execution of the writ. Writs are executed by bailiffs under warrant from the sheriff or under-sheriff, normally by bailiffs who are regular officers and obliged with sureties for the due execution of their office, but occasionally by any other person, called a special bailiff.

The common modes of execution are writ of *fieri facias* (*fi. fa*) where money or costs are payable and under the writ the debtor's goods and chattels may be seized and sold to satisfy the judgment, writ of *elegit,* which was the usual method of execution against land and is now replaced by power to impose a charge on land, writ of *capias ad satisfaciendum* (*ca. sa.*) under which a debtor might be imprisoned (today almost obsolete), writ of delivery, for the recovery of property other than land or money, writ of possession, whereby a party may recover possession of land, writ of sequestration, where a party is in contempt for disobedience of the court, under which commissioners called sequestrators must take possession of all the contemnor's real and personal property, and writs of *fieri facias de bonis ecclesiasticis*, directed to a bishop for execution against an ecclesiastical person.

Analogous proceedings have been developed to obtain satisfaction out of property which the creditor could not reach by these writs, such as attachment in the hands of a third party, charging orders, stop orders, the appointment of a receiver, and other means.

Execution of criminals. Putting to death under judicial authority. The method accepted in the U.K. was by hanging, causing dislocation of the neck and instant death. Until 1868 this took place in public but thereafter within prison walls. In other countries other methods have been used, such as gassing, electrocution, and beheading.

Execution of deeds. The rules defining in what way a deed or will must be authenticated by a party to be legally valid and effective. In English law a deed must be signed by the individuals executing it, and sealed, and delivered as the act and deed of the person expressed to be bound thereby though the sealing and delivery are now unimportant. Apart from statute, attestation by witnesses is not necessary though this is customary. A will must be signed at the end by the testator or by some other person in his presence or by his direction in the presence of two or more witnesses who attest and subscribe the will in the presence of the testator. In Scots law deeds must be signed at the end in the presence of two witnesses who also sign, an attestation or testing clause giving their names, addresses and designations. Wills must be signed on each page and at the end and similarly attested, unless they are holograph when the testator's signature without witnesses suffices.

Executive. The branch of the government of a state concerned with securing the carrying out of policy and applying principles and rules of law to cases, as contrasted with the legislature, whose function is to settle policy and make rules of law, and the judiciary, whose function is to decide disputes as to the meaning or application of rules of law. The executive branch in the U.K. accordingly comprises the Crown, the Ministers in charge of the various departments of State, and the civil servants and other permanent officials of different branches and grades down to those who deal with particular cases and individuals. The chief members of the executive are the Prime Minister and members of the Cabinet, who are also members of the legislature. To some extent local authorities belong to the executive branch of government in that while the authorities have some discretion as to how they act, and how they deal with individual cases, they have no discretion as to whether or not to provide, e.g. education, housing, and other services, and they must work within guide-lines set by Parliament.

Executive agencies. Organizations established in the U.S. to execute legislation passed by Congress.

They include the National Security Council, Central Intelligence Agency, Bureau of the Budget, Council of Economic Advisors, National Aeronautics and Space Agency, Office of Economic Opportunity, Office of Emergency Planning, Office of Science and Technology, Civil Service Commission, Federal Regulatory Commissions, Veterans Administration, Federal Communications Commission, The General Services Administration, Small Business Administration, U.S. Information Agency, the Civil Rights Commission, and Federal Aviation Agency.

Executive agreements. Agreements between the President of the U.S. and another head of state. They may be made by the sole authority of the President as commander-in-chief, or by virtue of his constitutional executive power, or some have been made under special powers conferred by statute or joint resolution of Congress. They have the validity and effect of a treaty but do not require Senate ratification, and largely for that reason have frequently been resorted to. They frequently deal with matters of lesser importance than do treaties, but include such agreements as those reached at Yalta and Potsdam in 1945.

Executive certificate. In international law a certificate issued by the executive branch of government to a court certifying some fact which is in issue before the court. Thus in British practice whether a body claiming to be the government of a territory is recognized as such by the British government, or whether a person is entitled to diplomatic status or not, is determined by a certificate from the Foreign Office. Such a certificate is normally conclusive and the fact thereafter not open to dispute. The justifications for the practice are that the relevant branch of the executive is likely to know best and that the courts and the exeutive should not adopt different views on a matter which may be politically delicate.

Executive orders of the President of the U.S. Rules issued to executive and administrative agencies to put into effect the provisions of legislative policy. As the President is required by the Constitution to execute the laws of the land, such orders are legally enforceable.

Executor. In England a person appointed by a will or codicil to administer the testator's property and execute the provisions of the will, as distinct from an administrator who is the person appointed by a court to administer the estate of a deceased, usually in cases of intestacy. Both are comprehended under the title of personal representative.

An executor is normally nominated by a testator in his will, or may be nominated by another, authorized by the testator to do so, or may be the person whom the general tenor of the will indicates is to act. A person nominated may renounce the office. A person may be appointed subject to limits or for limited purposes only. His authority to act is a grant of probate of the will by the Probate Division of the High Court, which gives him a right to make title to the deceased's property and administer it, but he may act as executor before probate.

An administrator acts under a grant of administration by the Probate Division of the High Court. Relatives have title to obtain a grant in a settled order of priority. The person appointed must give a bond for the due collecting and administering the estate. After the grant of administration he has generally the same rights and liabilities as an executor.

All property, real and personal, belonging to the deceased devolves on his representative, and is held by the latter as such representative and not in his own right. The representative's duties are to get in the deceased's assets, pay debts and estate duty, pay legacies and residue or distribute the estate according to the rules of intestacy. In general a personal representative is not bound to distribute the estate before one year after the deceased's death.

A personal representative must act gratuitously and may not make any profit from the office. He has for administrative purposes the powers and duties conferred by law on a trustee for sale, must keep accounts, and is liable for unauthorized payments or loss of assets.

A person who without authority intermeddles with the estate of a deceased is an executor *de son tort* and liable for the whole estate coming into his hands.

In Scotland an executor nominated in a will is an executor nominate. In the case of intestacy certain relatives, according to an order of priority, have title to petition the sheriff to be appointed executor-dative. An executor-dative must find security for his intromissions with the estate. Either kind of executor must then obtain from the sheriff confirmation of his appointment, which vests in him all the deceased's property and gives him title to ingather assets, pay debts, deal with the estate, and pay legacies and residue or distribute the estate in accordance with the rules of intestate succession. Failing either kind of executor a creditor of the deceased may obtain appointment as executor-creditor to so much of the estate as is needed to satisfy his debt; this is really a form of diligence (q.v.) rather than administration of the estate. A person who without title deals with a deceased's property is a vitious intromitter and liable for all the deceased's debts.

Executor *de son tort*. A person who without legal authority takes it on himself to act as executor or administrator as by any acting or dealing with any of the deceased's property other than actings necessitated by humanity or necessity. Such a person renders himself liable to all the burdens of executorship without any of the advantages. He may discharge his liability by obtaining a grant of probate or accounting to the personal representative or the court.

Executory interests. Future interests in land or personal estate, other than reversions and remainders. They comprise executory devises, created by will in land, and executory interests in personalty, created by conveyance *inter vivos* or by will (executory bequests).

Executory trust. A trust in which the limitations of the equitable interest are not created precisely but only generally and which can become operative only when some step is taken by the truster or the trustee, such as a further document executed, as contrasted with an executed trust where the truster's intentions are completely set out in the deed establishing the trust.

Exemplary damages. Very substantial damages or damages additional to compensatory damages, sometimes awarded to mark the court's or jury's disapproval of the defendant's conduct as a deliberate, aggravated, or outrageous tort. They are competent in cases of oppressive or arbitrary conduct by officials, where the defendant's conduct is calculated to profit himself to an extent greater than compensation payable to the plaintiff and, in certain cases, under statute, as for defamation.

Exemplification. An official copy of a document made under the seal of a public official or of a court. Thus an exemplification of a private Act is made under the Great Seal.

Exemption. A privilege or provision taking a particular case out of the general principle otherwise applicable to it as to other cases.

Exequatur. A permission granted in the discretion of the government of a country to a consul or commercial agent of another country to enter the former country and perform there his functions on behalf of the latter country. It is revocable at any time.

Exhibit. A document or thing shown to a witness and referred to by him in evidence or in an affidavit.

Exhibitionism. The exposure, usually by a male, of the genital organs to a person of the other sex as a means of obtaining sexual pleasure and not as a threat of or invitation to intercourse. It is usually treated as a breach of the peace or statutory nuisance.

Exile. Prolonged absence from one's own country. It may be voluntary, for reasons of business or health, or as refuge from danger, oppression or uncongenial government. Or it may be imposed as a kind of punishment. The Greeks imposed exile in cases of homicide and ostracism was a kind of exile imposed for political reasons. At Rome a citizen might originally escape the death penalty by voluntary exile, but later exile might be imposed on offenders of higher rank. This was ratified when a decree of the people imposing *aquae et ignis interdictio* was introduced; it was deportation for life to an island, with loss of citizenship, all civil rights and confiscation of property, and was imposed for serious offences. *Relegatio* was a milder form of *deportatio*. In later texts *exsilium* is used as an equivalent of *interdictio, deportatio* and *relegatio*.

Expatriation. Renunciation of allegiance to a person's native country on becoming a subject of another state. In the U.K. this was impossible until the Naturalisation Act, 1870. It is now competent to make a declaration of renunciation of citizenship of the United Kingdom and Colonies on the registration of which by the Home Office the person ceases to be such a citizen. This does not necessarily make the person cease to be a British subject as he may have become a citizen of another Commonwealth country. Naturalization in a foreign country does not cause automatic loss of citizenship of the U.K. and colonies.

Expectant heir. A person who has a hope of succession or reversionary right to any property, particularly one who by reason of lack of present assets is in danger of selling or mortgaging such interests for an inadequate sum and is protected by equity from such danger. The onus is on the buyer of an expectancy to show that the transaction was entered into in good faith and the expectancy fairly and independently valued and generally fair.

Expectation of life. The number of years which, in the ordinary course, a particular person might have expected to survive. Damages awarded for personal injuries or death may include a modest sum for loss of expectation of life where the evidence is that the injured person in consequence of his injuries will not live, or has not lived, as long as he would normally have done.

Expenditure, Select Committee on. A committee of back-benchers of the House of Commons, replacing in 1971, the former Estimates Committee, with the function of scrutinizing the expenditure incurred on various heads of the public service. It is an instrument of general administrative review and a valuable source of information on how departments operate. It operates through sub-committees dealing with broad areas of government, such as defence and external affairs, and within these each sub-committee inquires into a particular branch of government activity. The committee may appoint experts to assist it and each sub-committee has power to summon witnesses and usually hears evidence in public from senior civil servants and others. Its aim is to inform the Government about particular areas of the Government's work and it considers the implications for the national economy of the level of public expenditure adopted. The reports usually contain recommendations based on the evidence, and once or twice each session selected reports from the Committee are debated by the House.

Expert evidence. Evidence given to a court by a person skilled and experienced in some professional or technical sphere of the conclusions he has reached on the basis of his knowledge, from facts reported to him or discovered by him by tests, measurements or similar means. This kind of evidence is commonly given by, e.g. doctors, psychiatrists, chemists, architects, finger-print experts, and the like.

Expiring Laws Continuance Act. An Act passed each session by the U.K. Parliament to continue for a further period Acts which have been passed for a limited period only, either to deal with some emergency, or expected to be needed for a limited time only. If not scheduled to the Expiring Laws Continuance Act such an Act lapses automatically at the end of the period for which it was passed.

Express. The communication of an idea is express when it is made by words or actings which directly convey the idea communicated, as distinct from an implied communication where the idea is communicated by implication, or by inference from what is said or done.

Expressio unius est exclusio alterius (Express mention of one thing is the exclusion of the other). A maxim sometimes invoked in the interpretation of deeds and statutes, where some persons or things are mentioned individually but others are not, that the others were intended to be excluded.

Expressum facit cessare tacitum (Express mention of one thing excludes anything not mentioned). A maxim to the effect that a thing not mentioned in an enumeration is intended to be excluded.

Expropriation. Deprivation of a person of his property, particularly by public authority, as by compulsory purchase.

Expulsion from Parliament. The House of Lords has the right, when a peer has been found guilty on impeachment, to disqualify him from sitting in the House, temporarily or permanently. This is now obsolete.

The House of Commons has the right to expel a Member for offences against the House or for serious crime. Expulsion does not disqualify the Member from being re-elected to the same or a subsequent Parliament but the House cannot be compelled to allow him to take his seat.

Extent, Old and New. In Scotland, two valuations of land made in the reign of Alexander III, *c.* 1280, and under a statute of 1474.

Extent, writ of. A remedy competent prior to 1947 to recover debts of record due to the Crown. They were of two kinds, extent in chief and extent as aid. Extent in chief was a proceeding by the Crown for recovery of a debt due to it, issued from the Exchequer directing the sheriff by an inquisition of lawful men to ascertain the lands and goods of the debtor and seize them for the Crown. There was also an extent in chief for the second degree, which lay only against the debtor of a Crown debtor against whom an extent in chief had been issued.

An extent in aid was sued out at the instance and for the benefit of a debtor to the Crown for recovery of a debt to himself, the Crown being nominal debtor. The Crown debtor had to make oath that the debt would otherwise be lost. There was also a special writ of extent issued in the event of the death of a Crown debtor, by which the sheriff had to enquire by a jury concerning the lands and goods of the deceased debtor and seize them in the Crown's hands.

Extenuating circumstances. Circumstances which lessen the seriousness or culpability of criminal conduct and can accordingly be taken into account in mitigation of punishment, e.g. provocation of the accused by the victim, or the accused's diminished responsibility (q.v.). They are distinct from circumstances which provide a complete defence to the charge, such as self-defence or insanity at the time. Some, such as diminished responsibility, operate rather illogically by reducing the offence of murder to that of manslaughter; this is because the penalty for murder is fixed. In civil law countries the minimum penalties for various crimes are more commonly fixed so that more formal rules of extenuating circumstances have been developed.

Exterritoriality. The principle that certain persons and premises, though in the territory of one state, are legally deemed outside it and therefore not subject to the local law. It applies to foreign sovereigns and heads of states, diplomatic envoys and other persons enjoying diplomatic privileges embodied in the privileges of immunity of domicile, exemption from civil and criminal jurisdiction, exemption from subpoena as witnesses, from police orders and regulations, from local taxation, and liberty to practise their own religion, and applies in lesser degree to visiting forces on the territory of another state and to warships and public vessels in foreign territorial waters.

Extinction of rights. Rights may be extinguished in many different ways, which vary according to the particular kinds of rights, in what they consist, and accordingly to the particular legal system. They may be extinguished (1) by acts of one or other or both parties concerned, such as by performance of a duty, release by the other party of a party bound, agreement of both parties, or circumstances satisfying a condition. Rights may also be extinguished (2) by operation of law, by such rules as prescription, which extinguishes many rights after the lapse of various periods of time, merger of one right in another, discharge of a bankrupt, frustration of a contract or the occurrence of facts which make a contract impossible or illegal to perform.

Exton, Sir Thomas (1631–88). Was son of John Exton (?1600–68), judge of the Court of Admiralty (1649–65) and author of *The Maritime Dicaeologie, or the Sea Jurisdiction of England*, 1664, contending for an extended jurisdiction for the Admiralty court but containing a good history of the court and of its jurisdiction, and of some of the principles of maritime law.

He was knighted in 1675 and appointed one of the judges of the Admiralty. Later he is described as Advocate-General.

Extortion. The crime of dishonestly obtaining from a person some payments or benefit not lawfully due, particularly by threats. The term is also used of stringent terms in a civil contract, particularly as to interest, which are extremely heavy and harsh.

Extract. In Scots law, a certificate obtainable from the office of a court signed by the proper officer, the Extractor, that a decree in terms stated is contained in the records of the court. It contains warrant to do the ordinary forms of diligence (q.v.) and is a necessary prelude to enforcement of the decree. The term is also applied to certified copies, authenticated by the proper official, of private deeds recorded in public registers. This extract may, according to the purpose of the registration, be a mere copy of the deed, or be a warrant for diligence to enforce payment or performance in terms of the obligation contained in the deed.

Extradition. The delivery by one state to another of a person whom the latter state desires to proceed against for crimes triable in the latter state. A person found in the United Kingdom may be surrendered only under statute, provided a reciprocal arrangement has been made with the other state concerned, and extradition treaties have been entered into with many countries for this purpose. The statutory provisions apply to fugitive criminals who have been convicted or accused of the commission within the jurisdiction of a crime which, if committed in England, would be a crime listed in the Extradition Acts, 1870–1935, and is within the treaty with the other state concerned. Extradition is not competent for political offences.

If another state desires extradition of a person from the U.K., a diplomatic representative of that state makes a requisition to a Secretary of State, who may require a police magistrate to issue a warrant for the apprehension of the criminal. After arrest the criminal is brought before the police magistrate, who hears the case against him as if he were charged with an indictable offence. If there is prima facie proof of guilt the magistrate must commit him to prison; the prisoner may apply for habeas corpus. Not less than 15 days after committal the fugitive criminal may be surrendered, under warrant of a Secretary of State. Modified procedure applies in British possessions and under the Fugitive Offenders Act, 1967, between different parts of the British Commonwealth.

Conversely persons desiring the extradition of criminals from foreign or Commonwealth countries apply to a Secretary of State to have the person extradited to this country for trial there.

The conditions for extradition vary considerably from one country to another. Many states decline to extradite their own citizens and it is common to refuse to surrender for political offences, though the definition of a political offence is a matter of great dispute and uncertainty.

Within the U.S. interstate extradition is governed by the interstate comity clause (Art. IV, sec. 2) which entitles the governor of one state to return a prisoner to the state whence he fled on the application of the state having jurisdiction over the crime. This provision has been held by the Supreme Court to be discretionary and to permit refusal to extradite, without need to state a reason.

Extrajudicial. Outwith the ordinary course of judicial proceedings. Thus distress (q.v.) is an extrajudicial remedy.

Extravagantes (*extra*, outside; *vagantes*, wandering). In canon law, a term applied to decretals outside the Decretum (q.v.), particularly to the Decretals of Gregory IX collected by Raymond of Penaforte and published in 1234 and most commonly to decretals published after the Clementines (1317) (q.v.) and not included in any of the prior collections. They continued to be called by this name even when collected and published as the *Extravagantes* of John XXII and the *Extravagantes Communes*. The term is also sometimes given to *decretales extravagantes* of Innocent III of the thirteenth century only some of which were included in later compilations.

Extraordinary lords. When the Court of Session (q.v.) in Scotland was founded in 1532 it comprised a President, 14 Ordinary Lords and three or four Extraordinary Lords. They were nominees of the King, not necessarily qualified, not salaried, and free to sit or not as they pleased. This may have been a device to conciliate the barons, but it facilitated royal interference, and these Lords tended to sit only in cases where they had a personal interest. The number rose to eight in 1553 but after protest was reduced to four and continued at or about that level till 1723 when it was provided that no future vacancies should be filled. Archbishop Burnet was the last cleric to hold judicial office, being an Extraordinary Lord 1664–68, and John Hay, Marquis of Tweeddale, the last Extraordinary Lord holding office 1721–62.

Extrinsic. A term applied to evidence in aid of interpretation of a statute or document drawn from a source outside that statute or document, as opposed to intrinsic.

Eye for an eye. The phrase drawn from Exodus 21:24 and elsewhere, usually used as signifying legally permitted retaliation in kind for harm done, that is allowing the victim to inflict a similar injury on the wrongdoer. It was permitted in ancient Babylonian, Jewish, Roman (*lex talionis*), and Islamic law but, in practice, was often satisfied by the payment of money or other satisfaction. The acceptance of composition later became compulsory.

Eye-witness. A witness giving evidence of facts seen by himself, a kind of testimony usually regarded as peculiarly trustworthy. Modern psychological investigations, however, cast doubt on the trustworthiness of observation and of eye-witness's evidence.

Eyre, Sir James (1733–98). Was called to the Bar in 1755, became Recorder of London in 1763, a baron of Exchequer in 1772, and Chief Baron in 1787. In 1794 he became Chief Justice of the Common Pleas where he presided till his death. He was a good lawyer in common law, equity and criminal cases, and presided at the trials of Hardy, Horne Tooke, and Thelwell for high treason with conspicuous patience, impartiality and fairness.

Eyre, Sir Robert (1666–1735). Called to the Bar in 1689 and sat in Parliament from 1698, became Solicitor-General in 1707, a judge of the King's Bench in 1710, Chief Baron of Exchequer in 1723 and Chief Justice of the Common Pleas in 1725. He had the reputation of being an able lawyer though haughty.

Eyre, General. The most important and extensive of the commissions granted to itinerant justices empowering them, as justices in eyre, to hear all pleas. In the thirteenth and fourteenth centuries the general eyre was important in the government of the county, comparable to the sheriff's tourn in the hundreds. The eyre's business fell into three main divisions, pleas of the Crown, pleas of juries and assizes, and gaol delivery, of which the first in particular enabled justices to inquire into the way all courts and officials had been fulfilling their duties since the last eyre. For this enquiry a writ was directed to the sheriff to summon the leading civil and ecclesiastical persons of the county, the reeve and four men from each borough and all others bound to attend the justices in eyre, a very large gathering. During the eyre all the local courts were suspended and cases before the Common Pleas were sent to be heard before the justice in eyre, but cases in King's Bench or Exchequer were not adjourned thereto. The main business was to obtain answers from juries to the Articles of the Eyre, which were intended to obtain information on every subject which might require a fine or an amercement from somebody, but some civil and criminal cases were also heard. The eyre was very unpopular and it came to be a rule that an eyre should be held not more often than once in seven years. With the rise of Parliament the general eyre became unnecessary and the regular courts and itinerant justices acting under other commissions gave remedies for the misconduct of officials. General eyres ceased to be used in the early fourteenth century.

W. C. Bolland, *The General Eyre.*

F

F.A.S. contract A contract for the sale of goods, frequently for export, whereby the seller undertakes to deliver the goods free alongside the ship, i.e. at his expense, the buyer bearing the cost of freight, insurance, and export and import duties. Prima facie the property and risk pass and the price becomes payable when the goods are delivered alongside the ship.

F.C. & S. clause A clause commonly found in a policy of marine insurance, to the effect that the goods are warranted free of capture, seizure, and detention, and the consequences thereof.

F.o.b. contract. A contract for the sale of goods for export whereby the seller quotes a price for the goods delivered free on board, i.e. put on board ship at his expense, for carriage to the buyer. The contract of carriage may be made by the buyer, or by the seller on the buyer's behalf, the buyer being liable for freight. The seller must give notice to the buyer to enable him to insure the goods. The property and risk pass on shipment. Payment is due on delivery of the goods to the ship, but may be made by acceptance by the buyer's agent of a bill of exchange.

F.o.r. contract. A contract for the sale of goods, whereby the seller undertakes to deliver the goods free on rail, i.e. into railway wagons or at the railway loading yard at his expense, the buyer bearing the cost of carriage and insurance. Prima facie the property and risk pass and the price becomes payable when the goods are put on rail.

Faber, Antonius. See FAVRE, ANTOINE

Faber, Johannes (?–1340). French jurist, professor and seignorial judge, collector of materials on the *coutumes* and an authoritative writer, author of *Breviarium in Justiniani Codicem* (1480) and *Commentarii in Institutiones Justinianeas* (1488).

Faber, Petrus (1540–1600). A distinguished French humanist jurist, author of a *De diversis regulis iuris antiqui* (1566).

Fabric lands. Land given for repair or maintenance of the fabric or building of a church.

Fabrot, Charles Annibal (1580–1659). French jurist who continued the work of Cujas and Doneau, and edited Cujas' works, Theophilus and the *Basilica* (7 vols., 1647).

Facility and circumvention. The principle of Scots law that a contract, gift, or will is voidable if it is established that granter was facile, i.e. weak and easily imposed on, and was circumvented, i.e. influenced and persuaded to act as he did, rather than permitted to exercise his own will and judgement.

Fact (Lat. *factum*, a thing done). As distinguished from act, an occurrence, or happening not immediately brought about by human activity; more generally, anything cognizable by any of the senses, which may have legal implication, or which may be a matter of enquiry as a basis for a conclusion of law; in the context of the phrases accessory before or after the fact, a crime; as contrasted with theory or speculation, that which exists, can be seen or measured. In pleading and litigation the facts are the circumstances, deeds, sayings, and inferences from them as distinct from the legal consequences, rules applicable thereto, and legal conclusions.

Matters of fact are accordingly matters, circumstances, acts, and events, which in legal controversy are determined by admission or by evidence, as distinct from matters of law which are determined by authority and argument. Thus what A did is a matter of fact; whether it constituted a particular crime or not is a matter of law. A matter of fact is also distinct from a matter of belief or opinion, which may not be susceptible of proof or disproof.

A question of fact may be any question which has to be determined by admission or evidence rather than by authority and argument, and by the jury, if any, rather than by the judge, or any question which is not determined by a rule of law but depends on the circumstances, such as the appropriate measure of damages to award for loss or injury.

Such a question is distinct both from a question of law and from a question of judicial discretion, which is concerned with a decision of what is right and reasonable or just and equitable in the circumstances.

See also *Fact and law*.

Fact and law. In legal inquiries this distinction is frequently involved. A matter of fact or a question of fact concerns the existence, or some state, at some past time relevant for the inquiry, of some person, or thing, or state of affairs, ascertainable by the senses or by inference from conduct or happenings. Matters of fact thus include e.g. time, place, weather, light, speed, colour, identification of persons, what was said, done, heard, and so on, and such inferred facts as a person's intention, sanity, state of mind, knowledge, and the like. Matters of fact have to be ascertained, failing admission, by competent and relevant evidence given by witnesses, experts, or provided by deeds, records, reports, etc.

Matters of law or questions of law on the other hand include what are the rules of law applicable to some issue, what their proper formulations are, and what they require or permit. Matters of law have to be ascertained, failing admission, by interpretation of statutes, cases, and other authoritative sources of the law, aided by the arguments of the parties' counsel. The interpretation of documents is always a question of law.

In particular cases either the matters of fact or those of law may be admitted or undisputed, but in many cases both the matters of fact and those of law involved may be uncertain and be contested.

The distinction is frequently important: an appeal on matters of fact allows investigation at the hearing of the appeal of the evidence and the proper inferences from it, whereas an appeal on a point of law limits consideration at the appeal to such questions as whether facts admitted or held proved justify or permit, by the rules of law, a particular decision or disposal of the case before the court.

The distinction also arises in the interpretation of decided cases; thus a case may be thought to have decided that particular conduct in question fell within a category such as of lack of reasonable care for an employee's safety, a decision of fact, or that that particular kind of conduct was itself legally negligent, a decision of law.

In a secondary sense any matter to be decided on evidence and inference therefrom is called a matter of fact, and other matters are matters of law.

In a further sense any matter to be decided by the jury, if there is one, or by the judge performing the jury's function, is called a matter of fact, and any matter reserved to the presiding judge is called a matter of law. This is so notwithstanding that a matter reserved may be factual in nature.

In many circumstances questions of mixed fact and law arise; thus whether X is guilty of manslaughter depends on what he is held to have done (fact) and whether such conduct in such circumstances amounts to manslaughter, as that concept has been defined and explained by authoritative cases (law). Whether Y is liable for slander depends on whether he published matter of and concerning the plaintiff (fact) and whether it fell within the category of slander as defined in law (law). Accordingly a jury verdict is normally a mixed finding of fact and law and the decision of a judge of first instance is, normally, also a mixed finding.

Fact-finding. The process which a court must undertake in every case where all the facts which have to be held established or found before the law applicable thereto can be determined and applied so as to result in a judicial decision have not been admitted. The process of fact-finding is essentially one of reconstruction of what was said and done. It is sought to be done by each party adducing evidence of admissible kinds and each other party cross-examining the witnesses adduced, each then contending that the effect of the evidence adduced is to establish certain facts or facts to a particular effect, and by the trial judge expressly or impliedly making findings of fact, holding that certain facts have been established and others not established. This process involves considering the veracity and accuracy of witnesses' evidence, the weight to be attached to some as against others, the likelihood of this or that, and so on. The process is rendered difficult by such factors as missing or deceased witnesses, faulty recollection, prejudice, unconscious inaccuracy, dishonesty, missing letters, and similar difficulties, and even by the prejudices of the judge. The facts 'found' may accordingly not be a correct reconstruction of what actually happened. The difficulties may be even greater in front of juries, particularly as they seldom make explicit findings of fact, but return general verdicts for or against the plaintiff. Jerome Frank's *Courts on Trial* (1950) is a classic analysis of the divergences between the actual facts and the facts as 'found'.

Factor. A mercantile agent who in the ordinary course of business is entrusted with possession of goods or of the documents of title thereto with a view to their sale. By statute a factor may in certain cases pass to another a good title in property not his own but entrusted to him. A factor differs from a broker in that he may sell in his own name.

The term factor is also used in Scotland for a

property manager, who acts on behalf of the owner of the property, receiving rents and arranging for repairs. An estate factor is the general manager of the estate.

Factoring. The commercial practice whereby a factor accepts responsibility for credit control, credit risk, and debt collection on behalf of a customer. It may take the form of collecting debts for the customer and assuming the credit risk, the factor receiving the customer's invoices and paying money to the customer at intervals, or of providing finance as well, in which case the factor pays the customer a large part of the price at once, recovering in due course from the debtor. The factor charges a commission for his services, having regard to the risks undertaken.

See also INVOICE DISCOUNTING.

Factory and Commission. The counterpart in Scots law of a power of attorney, a deed whereby one empowers another to act in his stead.

Factory legislation. It commenced in Britain with Peel's Health and Morals of Apprentices Act (1802) which limited daily hours of work to 12. An Act of 1819 banned children under 9 from cotton mills and limited daily hours for all under 16 to 12, but both of them were inadequately enforced, there being no provision for inspection until an Act of 1833 which banned the employment of children under 9 in textile factories, limited the hours of children from 9 to 13 to 9, plus 2 hours' schooling, and to 48 hours in the week, and those of children from 13 to 18 to 12 per day and 69 in the week. Adult male workers were not protected and campaigned for a Ten Hours Bill. An Act of 1844 restricted female labour in textile factories to 12 hours per day and children to 6½ hours, but allowed them to be employed at 8. An Act of 1847 imposed a 10-hour limit for women and young persons but was frustrated by the introduction of shift working and was modified in 1850 to a 10½ hour day with 1½ for meals, mills to remain open for only 12 hours. These provisions were gradually extended to other kinds of premises and in 1867 an act extended the 10-hour limit to all factories and workships of stated sizes. Since then factory legislation has devoted more attention to provisions for safety, health, and welfare, hours of work being determined largely by trade union negotiation.

B. Hutchins and A. Harrison, *History of the Factory Act* (1926); A. Redgrave, *The Factories Acts.*

Factum praestandum, ad. In civil and Scots law, an obligation to perform an act, such as to deliver goods sold, which can in general be specifically enforced.

Faculties, Court of. A court or office of the Archbishop of Canterbury with jurisdiction over the appointment and removal of notaries public and the granting of such faculties as the Archbishop may grant in both provinces. It is presided over by the Master of the Faculties, an office now held by the Dean of the Arches (q.v.) *ex officio.*

Faculty (Lat. *facultas*, a power). A licence, authority, or permission. In English ecclesiastical law it is episcopal authority for any alteration to consecrated land or buildings, including the fabric, furnishings, ornaments, and decorations, obtained by petitioning the consistory court of the diocese. In deciding whether to grant or refuse a faculty the court has a discretion and will consider the view of the incumbent, churchwardens, and parochial church council and of persons interested or affected, but is also influenced by legal, doctrinal, liturgical, and artistic considerations. Deviation from a faculty is a contempt of court and disobedience will be censured by a monition.

In Scots law a faculty is another name for a power (q.v.).

In universities a faculty is the group of scholars concerned with studying and teaching a particular major branch of learning, such as arts, science, law, or medicine, probably so-called because they have licences to teach. Hence sometimes it is used of a department of knowledge. It is also used sometimes of a professional body, e.g. the Faculty of Advocates (q.v.).

Faculty of Advocates. The society which constitutes the Scottish Bar, corresponding to the Inns of Court.

Faggot voter. A person who formerly was entitled to vote in respect of land just sufficient to confer a vote and of which he had acquired actual, or even nominal, ownership merely to give him the vote. The category came into being about 1700 but was greatly affected by the Reform Act of 1832 and finally disappeared in 1918.

Fair comment. One of the defences to a claim of damages for libel (q.v.) to the effect that the words complained of as defamatory, so far as they were comment on facts, were fair comment on a matter of public interest. Fair comment is expression of view honestly held, not necessarily restrained or moderate, still less favourable, comment. The defence may be rebutted by proving that the comment was made maliciously, i.e. spitefully. The questions of fairness and of malice are for the jury, that of public interest or not for the judge.

Fair Labor Standards Act (1938). The first permanent legislation in the U.S. establishing maximum working hours and minimum wages for industries engaged in inter-state commerce. Later legislation has modified the provisions.

Fair Trading. A concept introduced in British Law in 1973 with the general intention of protecting consumers. An Office of Fair Trading is established headed by Director General of Fair Trading, to keep under review the carrying on of commercial activities relating to goods supplied to consumers, and to collect information about these activities. He may refer matters to the Consumer Protection Advisory Committee, publish information and advice to consumers, make monopoly references to the Monopolies and Mergers Commission, and acts as Registrar of Restrictive Trading Agreements, and directs the working of the Consumer Credit Act. He may obtain undertakings or initiate action in the courts against persons carrying on business, who persist in conduct which is detrimental to the interests of consumers, and take action on reports of the Commission on a monopoly or merger reference.

In the U.S. 'fair trade' is associated with the Federal Trade Commission Act by which it seeks to prevent unfair competition and deceptive practices in inter-state trade.

Fair Wages clause. In 1891, 1909, and 1946 the House of Commons passed a Fair Wages Resolution agreed by the government, the British Employers Confederation, and the T.U.C., which is incorporated as a standard condition in contracts between government departments and contractors. It requires the contractor, by machinery of negotiation of employers and trade unions representative of substantial proportions of the employers and workers in the trade or industry in the district, to recognize the freedom of his workpeople to be members of trade unions and to pay wages not less favourable than those established for the trade or industry in the district where the work is carried out. These resolutions were originally directed against the use of 'sweated labour'. Only contractors undertaking to observe these conditions will be placed on the list of firms invited by a department of government to tender for a contract.

Fairs. In mediaeval Europe fairs were centres of periodic meetings, between merchants from different regions. Some fairs were specialized, e.g. wool and cloth at Boston, wine at Winchester, but others dealt with a wide range of goods, and the great international fairs developed into opportunities for the exchange of a great range of commodities. Thus the Fairs of Champagne developed into two-monthly meetings for trade. Fairs did not develop at some of the major centres of trade, e.g. Venice, because they had a high and steady level of business all the year.

In England a fair at common law and under statute is a franchise conferring a right to hold a concourse of buyers and sellers. There seems to be at common law no distinction between a fair and a market, but a fair is usually understood to be a great kind of market held once or twice a year. In the past the occasions at which farm servants offered themselves for employment for the succeeding year or half-year were hiring fairs, and amusements and side-shows are not infrequently held at fairs. The Crown had by royal prerogative the power of granting to a subject the right to hold a fair or market, in which case the right was a franchise, but the right might also exist by statute or by long-continued usage. A franchise granted by the Crown is liable to forfeiture for non-use or neglect of use, outrageous tolls, or on certain other grounds, and may be extinguished by statute or a new charter.

Fairs and Boroughs, courts of. In the Middle Ages the right to hold a fair and levy tolls was a franchise granted by the King and an incident of this was the right to hold a court, which came to be known as a court of piepowder or pie poudre because it was frequented by chapmen with dusty feet (*pedes pulverosi, pieds poudrés*) who wandered from fair to fair. These were of the same type as the courts of similar fairs held all over Europe. Where the lord of the fair was an individual it was sometimes held by the lord, or his steward, and where the fair was held by a borough, the court was held by the mayor or bailiffs, but in either case in the thirteenth and fourteenth centuries, the judges were the merchants who attended the fair.

The jurisdiction of such courts extended to all cases except those concerning land and the pleas of the Crown, and included cases of contract, trespass, and minor criminal offences done at the time of fair or market and within the precincts thereof. It was in continuous session during the fair and the procedure was summary. It administered the customary law merchant.

In boroughs the piepowder court was often regarded as merely a special aspect of the ordinary borough court, though sometimes the two were distinct.

In the later fifteenth century, piepowder courts declined in importance, and statute and decisions diminished the usefulness of such courts, restricting the jurisdiction to matters arising within the limits of the fair and while the fair was being held. In consequence in the sixteenth and seventeenth century these courts decayed and disappeared.

Falconer, Sir David, of Newton (1640–86). Became an advocate in 1661 and a judge of the Court of Session in 1676, a Lord of Justiciary in

1678, and, in 1682, Lord President of the Court of Session. He sat in the Parliament of 1685 and was a member of various committees thereof. He collected the decisions of the Court from 1681 to 1685 and these, along with those of Lord President Gilmour, were published as Gilmour and Falconer's *Decisions* (1701). He was a grandfather of David Hume.

Fallacy. Strictly any violation of the conditions of logically valid inference, but more generally any unsound argument or inference. Material fallacies involve misstatement of facts. They were classified by Aristotle into: (1) fallacy of accident, confusing what is accidental with what is essential; (2) *secundum quid*, arguing from a general rule to a particular case without regard to specialties affecting the particular case, or conversely arguing from a particular to a general case; (3) *ignoratio elenchi* or irrelevant conclusion, where attention is diverted to an irrelevant fact; this includes appeals to sentiment, fear, hatred, personal antipathies, and the like; (4) *petitio principii* or begging the question, where a conclusion is sought to be demonstrated from premises which presuppose that conclusion. A common mode is by using such terms as 'undemocratic' and 'unconstitutional'; (5) fallacy of the consequent, or arguing from a consequent to its condition, e.g. if A is bankrupt he defaults on his obligations; hence, if he defaults on his obligations he is bankrupt; (6) *non sequitur*, or fallacy of false cause, where a conclusion is drawn from an insufficient reason; (7) *post hoc, ergo propter hoc*, where because X follows Y it is inferred that Y caused X; (8) fallacy of many questions, where two or more points are embodied in one question, e.g. when did you stop beating your wife.

Verbal fallacies involve the drawing of a false conclusion because of ambiguous or improperly used words. Forms of this are: (1) equivocation, where a word is used in two senses; (2) amphiboly, where ambiguity results from grammatical usage or word order; (3) composition, resulting from confusing use of collective terms; (4) division, the converse of composition; (5) accent, which may alter the meaning according to the word in the sentence to which it is applied; (6) figure of speech, misinterpreting a form of expression or inflection.

Formal or logical fallacies comprise violations of the formal rules of syllogistic reasoning.

Falsa demonstratio non nocet (False description does not harm). The principle that if the subject-matter of a conveyance, gift, bequest, or devise is identifiable, it does not matter if it has been misdescribed in the deed.

False Decretals. (or Pseudo-Isidore Decretals). So-called because they passed as a Collection of St. Isidore of Seville, a Spanish historian and encyclopaedist, or sometimes as the Collection of Isidorus Mercator, a ninth century collection of ecclesiastical legislation including many forgeries, purporting to be a collection of decrees of councils and papal decretals from the first seven centuries of the Christian era. It includes the forged Donation of Constantine (q.v.) which purports to give the Pope spiritual supremacy over all matters of faith and morals and superiority over all temporal rulers, and the whole compilation was intended to ensure the freedom of the Church from states, and to maintain the independence of bishops against the growing power of archbishops.

They were generally accepted as authentic down to about the fifteenth century, but the falsity of many of the items was exposed by the seventeenth century. While accepted they did have considerable influence, particularly by reviving the powers of the clergy and upholding that of the bishops.

False imprisonment. Both a common law crime and a tort, consists in unlawfully and either intentionally or recklessly restraining another's freedom of movement from a particular place, which may be a house, or hotel, or vehicle and need not be a prison. It must be a total restraint for a time, however short, not merely a partial restriction. There is no false imprisonment if a person arrests in circumstances where arrest is justifiable or if there is otherwise reasonable and probably cause for the restraints, or where a constable arrests under warrant or under statutory powers. An aggravated species of the crime is kidnapping and probably every kidnapping is a false imprisonment.

False pretence. A false representation by words or conduct that a particular fact or state of facts exists or has existed. The statutory offence of obtaining property by false pretences has now been replaced by that of dishonestly obtaining property by deception.

False trade description. An indication, direct or indirect, given by any means, of the quantity, method of manufacture, composition, fitness for purpose, or other similar fact with respect to any goods or parts thereof, which is false to a material degree or misleading. The application of any such description is an offence under the Trade Descriptions Act, 1968, but without prejudice to the validity of a contract affected thereby.

Falsing of dooms. The mediaeval means in Scotland of challenging the judgment of a court in a higher court, though the falsed dooms of the court of the justiciar could be judged only in Parliament. In the fourteenth century a group was several times named in Parliament to deal with falsed dooms and complaints. This mode of appeal seems to have been competent until about 1550.

Familiae emptor. In classical Roman law the common form by which a will was made was for the testator to write out his will and then, in the presence of five witnesses and a balance-holder, go through the form of conveying his estate by *mancipatio* (q.v.) to a *familiae emptor* and confirm generally the detailed bequests contained in the will writted on wax tablets. Originally the *familiae emptor* seems to have acted like a modern executor and disposed of the testator's property in accordance with his instructions. Later, however, the heir nominated in the will disposed of the estate and the *familiae emptor* took no part in the matter, being merely a part of the formalities of making the will. Many unsettled questions attach to the institution.

Families of laws. While each political society in the world has at least one legal system and body of laws operating within it, many of these have family resemblances and comparative jurists have grouped the world's legal systems into families, mainly on the basis of general similarities of sources, forms, methods, and techniques, based on historical origins and borrowing. On this basis the major families of law are: (a) The civil law or Romano-Germanic family, countries whose legal science has developed on the basis of Roman Law; this family originated in Western Europe and is mainly represented by French, German, Italian and Spanish law, but has spread to Central and South America and influenced parts of Africa and the Middle and Far East which were affected by European, particularly French civilization.

(b) The common law family, comprising the law of England, Ireland, most of the states of the U.S.A., and of many Commonwealth countries which have been largely developed by borrowing from English law.

Certain hybrid systems embody substantial elements from both civil and common law families; these include the laws of Scotland, Israel, Quebec, Louisiana, and South Africa. Scandinavian law is indigenous but has been influenced by Roman legal science.

(c) The socialist family, comprising states which formerly had Romano-Germanic law and have preserved some ideas from that, but since the Russian Revolution of 1917 have developed new principles based on Marxism-Leninism. This group includes Soviet Russian law and the law of the people's republics of Eastern Europe and Asia.

(d) The group of religious systems, which regulate the relations of members of particular faiths, rather than the lives of the citizens of particular countries; these are Islamic or Muslim law, Hindu law, and Jewish law.

(e) There are certain other groups which have affinities with, yet do not readily fall into, one or other of the main family groups. These include Far Eastern and African legal systems, and the Adat law of Indonesia.

R. David and J. Brierley, *Major Legal Systems in the World Today.*

Family A fundamental social unit consisting basically of two adults, a male and a female, living in one household co-operating in many activities and frequently producing and caring for children. This gives rise to various legal relationships of great social importance, particularly those of husband and wife (including questions of the creation and dissolution of that relationship), parent and child, legitimation, adoption, and problems of individual, joint and community property, and of inheritance. The term 'family' and legal relationships attaching thereto is variously defined in different bodies of legislation and has no settled legal connotation. Variants on the basic pattern of family are the extended family, which includes married children, their spouses and offspring, and the 'one-parent' family, where one spouse is missing, dead or divorced. In some cultures polygamy is permitted, either polyandry (more than one husband) or polygny (more than one wife). These variants raise different legal problems.

Legal systems commonly treat the family as a social institution to be preserved and supported. In many systems at different times rules have been sought to encourage family life, favouring marriage, legitimate children, and succession within the family. This object has justified provisions in such fields as property, succession, evidence, and criminal law as well as in family law itself.

Family allowances. Since 1946 weekly allowances were paid in the United Kingdom to families including two or more children, in respect of each child other than the eldest. They were paid out of funds provided by the Treasury and not related to contributions. The allowances were for the benefit of the family as a whole. The payments continued so long as there were at least two children under school leaving-age or under 18 and undergoing full-time instruction. In 1975 they were superseded by Child Benefit.

Family arrangement. Any arrangement arrived at between members of a family for the benefit of the family as a whole, such as to preserve the family property, or to avoid disputes. It may be contained in a deed or be implied from conduct. The courts treat such arrangements favourably and uphold them if at all possible, if generally for the benefit of the family estate or of all parties, taking account of considerations which would not be taken into account when dealing with strangers, such as the avoidance of disputes, safeguarding the honour of the family, and continuing the family property in the family.

Family Division. A Division of the High Court in England constituted in 1971, mainly by renaming the former Probate, Divorce, and Admiralty Division but transferring contentious probate business to the Chancery Division and Admiralty business to the Admiralty Court within the Queen's Bench Division. Its jurisdiction comprises matrimonial causes, matrimonial property, children, and non-contentious probate work.

Family law. The body of laws relating to the organization of the family and the legal relations of its members. In English law it has been recognized as a distinct branch of law only in the twentieth century, by bringing together topics naturally related to one another, but previously separated largely by the practice of the courts assigning them to different divisions of the courts. It comprises the law of marriage, its contracting and dissolution, the relations of parents and children, guardianship and rights in property as affected by dissolution or marriage, death, and rights of children.

R. H. Graveson and F. R. Crane, *A Century of Family Law* (1957); P. Bromley, *Family Law* (1976).

Family provision. Provision for the maintenance of the dependants of a person dying domiciled in England which the court may by statute award out of his estate if it is of the opinion that the provision made by the deceased or by the law of intestate succession does not make reasonable financial provision for the maintenance of that dependant. Any provision ordered may be in the form of periodical payments, a lump sum payment, a transfer of property, a settlement of property, or a variation of settlements.

Farinacci, Prospero (Farinacius) (1544–1618). An Italian humanist jurist, specializing in criminal law, procurator general to Pope Paul V, author of a vast *Praxis et Theorica criminalis* (1616) which is the definitive work on torture, and numerous other works, notably a *Decisiones Rotae Romanae.*

Farm, or ferm (Anglo-Saxon feorme, food-rent). A fixed sum or rent, usually due annually. In mediaeval English law the sheriffs' farm was the fixed sum payable annually by way of composition for all regular royal revenues derived from the shire. The sheriff farmed the shire, i.e. contracted in advance to pay a fixed sum and profited in so far as he could collect more than this sum.

Farmer's law. A Byzantine legal code of the eighth century, probably prepared under Leo III the Isaurian. It was intended for the growing class of free peasants, a class augmented by Slavonic peoples who migrated and settled within the empire. It was concerned mainly with farmer's property, property damage, misdeeds of village people, and taxation. The village was treated as a fiscal unit and all its members were responsible for paying the tax levied on the community; anyone who paid was entitled to seize the lands and crops of those who defaulted. The later influence of the code was considerable, particularly in Serbia and among the southern Slav peoples.

Farnborough, Lord
See ERSKINE MAY, THOMAS.

Farwell, Sir George (1845–1915). Called to the Bar in 1871, he became Q.C. in 1891 and a judge in 1899. He was a Lord Justice of Appeal 1906–13 and is remembered for his *Concise Treatise on the Law of Powers* (1874).

Faryndon's Inn. The name, prior to about 1484, of the Inn later known as Serjeant's Inn, on the east side of Chancery Lane and next to Clifford's Inn. The name is said to be derived from the person who leased the inn on behalf of the judges and serjeants and may have become attached to Farington Street.

Fasti (from *fas*, divine law). In ancient Rome sacred books or lists of the *dies fasti* and *nefasti* or days of the month on which it was competent or forbidden to do legal business. The *fasti* were exhibited in the Forum from 304 B.C. and contained information about festivals, lists of magistrates, military victories, and other matters. They are known to have been carved in stone as well as recorded in writing. The term is also applied to registers giving lists of consuls (*fasti consulares*) of triumphs (*fasti triumphales*).

Fatal accident. The everyday expression for an unintended and undesired occurrence in consequence of which a person or persons die. It may give rise to various kinds of legal claims and may be an inevitable accident (q.v.), e.g. death from lightning, or a happening resulting from the legal negligence, or breach of legal duty of some person, e.g. careless driving. The latter kind of accident may give rise to criminal prosecution and also to claims of damages by or on behalf of surviving relatives against the person who caused the accident on the ground of breach of duty at common law and/or breach of statutory duty. A fatal accident of either kind may also give rise to claims under policies of insurance and for certain social security benefits.

Fatal Accidents Acts. At common law in England surviving relatives had no civil claim if a person was killed by the fault or negligence of another. The development of railways and industrialization, with the rise in the number of deaths caused by negligence, provoked change, effected by Lord Campbell's Act (Fatal Accidents Act 1846) later extended, whereby if a person is killed in circumstances in which, if he had merely been injured, he would have had a cause of action against the wrongdoer, a claim may be brought against the wrongdoer for the benefit of specified groups of dependant relatives. The claim is for pecuniary loss, actual or prospective, resulting from the death, not for solatium for grief or injured feelings.

Fatal Accident Inquiry. An inquiry held under statute in Scotland before a sheriff, formerly with a jury of seven, into deaths in industrial employment, due to accident, and other cases of sudden and suspicious death, including road accidents.

Evidence is adduced by the procurator-fiscal and witnesses may be examined and cross-examined. The finding cannot be founded on in subsequent civil or criminal proceedings.

Father of the House. The title given in the House of Commons to the Member with the longest unbroken period of membership (even though for different constituencies).

Fauchille, Paul (1858–1926). French jurist, founder of the *Revue Générale de droit international public* and author of *Traité de droit international public* based on the work of Bonfils (q.v.).

Fault. A mental element frequently required as a condition of tortious or delictual liability, corresponding to Roman law *culpa* (q.v.). It is essentially a moral concept, importing that the defendant is liable because he could have and should have avoided or prevented the conduct causing harm, but as the law has developed is increasingly imputed by legal rules, as in cases of vicarious liability, or cases of statutory liability, where the fault is technical only. Liability without fault is urged by reformers who point to the difficulty faced by a plaintiff in many cases in establishing fault.

Faute de service. In European Community law, default of the administration, a ground for judicial redress. Common cases have been of grievances of Community employees such as failure in an irregular way to renew a contract of employment.

Favre, Antoine (Antonius Faber), (1557–1624). Savoyard humanist jurist author of *Conjecturarum juris civilis libri XX* (1580), *De erroribus pragmaticorum et interpretum juris* (1658), *Rationalia in Pandectas* (1604) and *Codex Fabrianus definitionum forensium* (1610), a famous work based on decisions of the court of Savoy.

Fealty (Lat. *fidelitas*). An oath of fidelity, taken by a feudal tenant to his lord, distinct from homage, though both were performed together when a vassal received a fief from a lord. Fealty to the Crown overrode all other obligations, even homage to a mesne lord. Fealty comprised the obligations to do no bodily harm to the lord, to do no secret damage to him in his home, not to harm his reputation, not to injure him in his property, and to facilitate the lord's doing good. Fealty was done by the tenant putting his right hand on a book and saying to the lord: 'I shall be to you faithful and true and shall bear to you faith for the lands and tenements which I claim to hold of you, and truly shall do the customs and services that I ought to do to you at the terms assigned: so help me God.' He then kissed the book. Fealty was also an incident of tenure due from copyhold tenants to the lord of the manor and consisted in swearing to be faithful in performing the services of the tenancy.

Fearne, Charles (1742–94). A man of classical and mathematical attainments and of inventive powers, developed a substantial chamber practice and in 1772 secured his place in legal history by the publication of his book *Essay on the Learning of Contingent Remainders and Executory Devises*, which at once became the classic text on its subject. It went through many editions and was later greatly enlarged. It was notable for its systematic arrangement, analytical power, and consistency. Other interests and a love of ease later distracted him and his practice diminished.

C. Butler's 'Preface' to 7th edn. of *Essay.*

Featherstone Riots Inquiry. On 7 September 1893, after disturbances at a colliery at Featherstone, Yorkshire, a small party of soldiers was sent to maintain order, police being occupied elsewhere. A magistrate read the proclamation in the Riot Act, 1714, and later gave the officer in command written orders to fire on the crowd. This was done and two persons were killed. A committee consisting of Lord Justice (later Lord) Bowen, R. B. (later Lord) Haldane, and Sir Albert Rollit, M. P., investigated the circumstances, found the shooting justifiable, and the report contains a leading statement of the circumstances in which troops may use arms to suppress civil disorder. Military regulations required an order from a magistrate but this was of no legal effect.

Federal Bureau of Investigation. A division of the U.S. Department of Justice, created in 1908 to conduct investigations for that Department, but reorganised in 1933 to investigate violations of

Federal legislation. It is responsible for internal security in the U.S. in respect of treason, conspiracy, espionage, and sabotage and handles alleged violations of a large number of Federal Statutes. Its investigations are reported to the Attorney-General and district attorneys.

Federal Communications Commission. A U.S. federal agency established in 1934 to regulate inter-state and overseas cable, telegraph, telephone, radio, and television communications. The members are appointed by the President and confirmed by the Senate.

Federal Constitutional Convention. In 1786 disputes between the North American states which had declared their independence from Britain led Virginia to invite all states to send delegates to a convention at Annapolis 'to take into consideration the trade of the United States'. Five states sent delegates to this Annapolis Convention (Sept. 1786). Hamilton and Madison persuaded the other delegates to propose that all 13 states send delegates to a convention to devise further provisions to render the constitution of the federal government adequate to the exigencies of the Union. On 21 February 1787 Congress invited the states to send delegates to Philadelphia in May for that purpose and to revise the Articles of Confederation.

The Federal Convention which met at Philadelphia on 25 May 1787, consisted of 55 delegates from 12 states (Rhode Island having declined the invitation). Debate extended to 17 September, and resulted not in revisions of the Articles of Confederation, but in a practicable and workable Constitution. Only 39 delegates signed it, and there was heated argument and bitter debate in the State legislatures before it was eventually ratified by all 13 states.

Farrand, *The Framing of the Constitution of the U.S.*; Mitchell, *A Biography of the Constitution of the U.S.*

Federal Constitutional Court (Bundesverfassungsgericht). The court of the Federal German Republic having jurisdiction to review legislation and administrative and judicial decisions to determine whether they contravene the Basic Law or constitution of the republic, to ensure that the rights of the individual against the government guaranteed by the Basic Law are protected, to resolve disputes between the federal and a state government, to determine whether a political party is pursuing aims conflicting with the spirit of the Basic Law, and which also serves as a court for the impeachment of the President or federal judges. It has two chambers each of eight members, some being career judges, some academic jurists, and some practising lawyers. They are elected by the Bundestag by means designed to prevent any party having predominant influence on the court. The federal or a state government, or one-third of the members of the Bundestag may petition the court on the constitutionality of a law and cases may be brought by a lower court or by an individual.

Its decisions have full legal force and are binding on all other courts.

Federal government See Federalism.

Federal Register. The official journal of the U.S. Government, authorized by Congress in 1935. It is a daily record of executive and administrative orders and includes such documents as the President may direct or Congress may require to be published. It is produced and distributed by the division of the Federal Register, National Archives.

Federal regulatory commissions. Some 60 agencies established at various times by the U.S. Government mainly to regulate various enterprises. These bodies commonly have delegated legislative and judicial powers. The choice of members is intended to be as free as possible from political bias.

The first such body was the Civil Service Commission (1883) established to recruit staff for the federal government on the basis of merit. More recent creations have been the Interstate Commerce Commission (1887), the Federal Trade Commission (1914), the Federal Power Commission (1920), the Tennessee Valley Authority (1933), the Federal Communications Commission (1934), and the Atomic Energy Commission (1946).

Federal Trade Commission. A body created by the U.S. Congress in 1914 intended to prevent unfair trade in interstate commerce. It is a non-political body, of five members, appointed by the President for seven-year terms, with power to investigate and control monopoly practices, the maintenance of resale prices, the including of false claims for patents, and power to call for reports from corporations, and to investigate business practices. It is authorized to issue 'cease and desist' orders against corporations guilty of infringements, and its operations have had a considerable effect on regulating business.

Federalism. A system of government of a country under which there exist simultaneously a federal or central government (legislature and executive) and several state or provincial legislatures and governments as contrasted with a unitary state. Both federal and state governments derive their powers from the federal constitution, both are supreme in particular spheres and both operate directly on the people; the state governments accordingly are not exercising powers delegated by the federal government, nor are they subordinate to it (though they may deal with less important matters). Federalism is appropriate to large countries where government from one centre would be complicated and difficult and could readily be out of touch with the needs and desires of widely separated areas, and to countries where particular parts have racial, linguistic, legal, or other peculiarities which they desire to have safeguarded.

A federal state requires a written constitution or basic constitutional act apportioning governmental responsibilities between the central government and the state government, and conferring the balance on the states, or conversely. Thus in Australia the Commonwealth Parliament has only such jurisdiction as is expressly given it or not expressly withdrawn from the parliaments of the states, whereas in Canada the Dominion Parliament has jurisdiction over all matters not specially assigned to provincial parliaments. Usually some powers are concurrent. Provision must also be made for the possibility of conflict between federal and state government, and between state and state. Federalism also necessarily implies judicial review (q.v.), as the possibility of *ultra vires* legislation must be considered. The advantages of federalism are that it frees the federal government for consideration of national and international issues, and gives state or provincial governments a measure of independence and the ability to devise their own solutions for local problems. The disadvantages are that it creates a fairly complicated and legalistic structure, with much opportunity for disputes between federal and state governments on powers and their exercise. Finance and taxation, in particular, give rise to difficulties. The prime example of a federal state in the modern world is the U.S.A., which comprises a federal government and 50 state governments, all 51 having legislative, executive, and judicial organs.

Within the Commonwealth federalism exists in Canada, Australia, and India, but in all three certain states have shown tendencies to secede, and in other parts of the Commonwealth federal constitutions have come to grief for various reasons. It is also found in West Germany, Switzerland, and elsewhere, and some believe that the most satisfactory solution to the problems of English, Scottish, Welsh, and Northern Irish differences within the United Kingsom would be the establishment of a federal structure there.

Federal states are variously called Confederations, Federations, Unions, and sometimes by other names but little turns on the precise name used.

Federalist, The. A collection of 85 essays, 77 of which originally appeared in New York newspapers in 1787–88 over the signature Publius, the remainder being added later, and which were published in collected form as *The Federalist* (1788, and later editions). About 50 were written by Alexander Hamilton, about 30 by James Madison and about 5 by John Jay. They were intended to stress the inadequacy of the Articles of Confederation and to promote the adoption of the Federal Constitution, then before the State legislatures for ratification. They were widely reprinted and were influential in securing support for the Constitution. But not only have they been a tract for the times, they have provided a basic treatise on the underlying theory of the Constitution, and repeatedly furnished the Supreme Court with principles as a basis for its exposition of the application of the Constitution to particular circumstances and for its interpretation. The papers have also been widely studied as profound essays in political science.

Fee (Late Latin, *feudum*) (cf. in Scotland *feu*). (1) The quality of an estate in land that it descended through the heirs of the owner for the time being indefinitely unless he, having the power to do so, in fact alienated it. Fees were in origin feudal grants made to a man and his heirs in return for stated services. In English law a fee simple is the fullest estate or interest recognized in land, and for practical purposes the owner of an estate in fee simple is an absolute owner. A fee simple may be absolute, or determinable by condition or by limitation. A fee tail, now an equitable interest only, is an estate of fee limited to a series of heirs other than those who would inherit by operation of law.

The term was also formerly applied to land itself. A knight's fee was the extent of land which sufficed for the maintenance of a knight and such as might be granted in fee for the service of providing one knight.

In Scotland a fee or feu is similar to a fee simple, save that the feudal relationship with the superior has continued to the present. Fee is commonly distinguished from entail (q.v.) and from liferent (q.v.).

(2) The payment made to a lawyer, in particular to a barrister, for his professional services, or to another professional man such as an architect or surgeon.

Fee-farm rent. A payment due to the lord in perpetuity from the holder of lands held in fee simple, reserved when the original grant was made.

Fee simple. In the English law of real property an estate (q.v.) in land, held heritably (fee), and descending to heirs generally, without restraint to any particular class of heirs (simple), as contrasted with a fee tail (q.v.), or estate tail, or entailed interest. An estate in fee simple comes as close to absolute ownership as the system of tenure (q.v.) will allow; when absolute (not liable to divestiture or determination) and in possession it is the only estate of freehold which may now exist at law. Other estates in land exist as equitable interests only.

Fee tail or estate tail. In the English law of real property an estate in land held heritably (fee) but descending to heirs according to a defined line, different from those who would inherit by law, e.g. the heirs of the body of the grantee. Thus estates in tail may be in tail general, in tail male general, in tail female general, or in tail special.

Feet of fines. The term fine is sometimes used for a final concord or agreement ending actual or fictitious litigation and acknowledging lands to belong to one of the parties. In form the compromise of an action, it is in substance a conveyance. It was recorded in (a) a writ of covenant, alleging a non-fulfilment of agreement by one of two parties; (b) the concord or agreement signed by the parties and acknowledged in open court; (c) the note of the fine, made out from the concord; and (d) the foot of the fine, being a document executed in triplicate and divided into three by cuts forming an inverted letter T, down the middle and along a line parallel to the bottom, of which the two main parts were given one to each party, and the foot retained by the keeper of the writs. Collections of the feet of fines exist, chiefly by counties, from as early as 1175 and in profusion from 1200, until 1834 when fines and recoveries ceased as a means of conveying lands. Some early Feet of Fines have been published by the Pipe Roll Society.

Fehmgerichte (hooded courts). Also called free courts, of the mediaeval period, presided over by masked judges. They were common in parts of Germany and Switzerland from the eleventh century onwards and were originally instruments of local self-government. They operated in the emperor's name but outside his control and maintained order locally in rather rough and ready fashion, sometimes perpetrating outrages and injustices. They generally disappeared in the sixteenth century.

Fellow-servant, doctrine of (or doctrine of common employment (q.v.). The principle that an employer, though liable for injuries done to his employee by reason of the employer's fault, was not liable for injuries done to one employee by the fault of another employee engaged on common work with the employee injured. The principle, invented in *Priestly* v. *Fowler* (1837), 3 M. & W. 1, was restricted by legislation and by decisions and abolished in 1948.

Felo de se (a felon as to himself). A person who committed suicide. To commit or attempt to commit suicide was in England a misdemeanour at common law, but ceased to be criminal in 1961. But it is an offence to aid, abet, counsel, or procure the suicide or attempt suicide of another. It is manslaughter to kill another in pursuance of a suicide pact with that other. Originally persons who committed suicide had to be buried in a place where four roads met with a stake driven through their bodies. This was abolished in 1823 and later legislation allowed burial and, with the bishop's permission, a religious service.

Felony (Lat. *felonia*, from fallere, to deceive). At common law in England a felony was a serious crime which involved forfeiture of the felon's land and goods to the Crown, and any offence involving forfeiture, such as murder, wounding, arson, rape, and robbery, was accordingly felony. In 1870 forfeiture was abolished, but crimes classed as felonies continued to be called such and thereafter some statutory crimes were declared felonies, which continued to have procedural differences from misdemeanours. The distinction did not coincide with one between grave and minor crimes. In 1967 all distinctions between felonies and misdemeanours were abolished, the procedural rules applicable to misdemeanours being made applicable to both categories, so that the term is now obsolete. The distinction was replaced by that between arrestable and non-arrestable offences. The distinction never existed in Scotland.

In the U.S. most jurisdictions distinguish between felonies and misdemeanours, the distinction usually depending on the penalties attaching to the crime, a felony frequently being defined as a crime punishable with imprisonment for at least one year.

Feme covert. A married woman.

Feme sole. An unmarried woman, including spinster, widow, and divorced.

Fenwick, Charles Ghequiere (1880–1971). U.S. international lawyer, Director of the Department of International Law of the Pan American Union, 1947–62, author of works on *American Neutrality* (1940), *The Organization of American States* (1963), President of the American Society of International Law and an Associate member of the Institut de Droit International.

Feoffee to uses. The person to whom, prior to the Statute of Uses of 1535, land was granted to the use of another, and in whom the legal seisin of land was vested, the beneficial ownership being in the *cestui que use*. The feoffee was regarded as owner by the common law, the *cestui que use* as the true owner by equity. That Act 'executed the use' and transferred the legal possession of the land to the *cestui que use*, that is, turned the *cestui que use*'s former equitable estate into a legal estate carrying common law seisin and taking away the feoffee's common law seisin entirely.

See also TRUST.

Feoffment (Fr. *feoffer*, to give a fee). In mediaeval English law the normal and regular mode of creating or transferring a freehold interest in land of free tenure. The essential part of it was the livery of seisin (q.v.) or delivery of corporal possession by giving a clod or twig as symbol of the land, made with the intention of transferring part of the granter's interest. At first writing was unnecessary, but after the Normal Conquest became more frequent, and there developed the elaborate charter or conveyance as a record of the transaction. After the Statute of Frauds of 1677 writing was necessary in every case and transfer by livery of seisin became obsolete.

Fergusson, George, Lord Hermand (1743–1827). A son of Sir James Fergusson, Lord Kilkerran, became a Scottish advocate in 1765, became one of the four commissaries of Edinburgh in 1775 and held that office till raised to the bench. In 1799 he was made a Lord of Session and in 1808 also a Lord of Justiciary. He resigned both offices in 1826. He was impatient and sarcastic, but a good lawyer and a popular judge. He is best remembered for his capacity for liquor, which has given rise to many stories. He also had a great love of Scots law and a veneration of its traditions and dislike of change. His collection of decisions of the consistorial court *Consistorial Decisions in the Order of a Dictionary*, 1684–1777, were published as *Lord Hermand's Consistorial Decisions* by the Stair Society in 1940.

Fergusson, Sir James, Lord Kilkerran (1688–1759). Passed advocate in 1711 and became a judge of the Court of Session in 1735, as Lord Kilkerran, and also a Lord of Justiciary in 1749. He was regarded as one of the ablest lawyers of his time. He collected, and digested in the form of a dictionary, the *Decisions of the Court of Session*, 1738–52 (1775).

Fergusson, James (1769–1842). Called to the Scottish Bar in 1791 and became successively one of the judges of the consistorial court, a principal clerk of session, and keeper of the register of entails. He wrote on the reform of the courts, on entails, but chiefly *Reports of Some Recent Decisions by the Consistorial Court of Scotland in Actions of Divorce*, 1817, and *A Treatise on the present state of the Consistorial Law in Scotland with reports of decided cases*, 1829.

Ferrers, George, Case of. In 1543 George Ferrers, M.P., was arrested for debt while on his way to the House. The House sent the Serjeant-at-Arms with the mace to demand the release of Ferrers, but the Sheriffs of London refused the request. The House then summoned the Sheriffs and those who had initiated the claims against Ferrers to appear at the Bar of the House and they were committed to prison. The action of the House was confirmed by Henry VIII and the case is held as establishing the privilege of Members of the House of Commons of immunity from civil arrest during session and that the Commons through its officers could release a Member so arrested. The principle is of less importance since civil arrest for debt has been almost completely abolished.

Ferretti, Emilius (originally *Domenico*) (1489–1552). An Italian jurist and humanist, papal secretary, displomat and professor, wrote many works on law.

Ferretti, Giulio (1480–1547). Italian jurist, author of *De Re et Disciplina militari* (1575), *De Jure et Re navali* (1579), and *Consilia et Tractatus varii* (1562).

Ferri, Enrico (1856–1929). Italian criminologist, lawyer, and professor, leader of the positive school of criminal science, investigated the causal factors of crime, sought to group and classify criminals, emphasized social defence as the purpose of criminal justice, and campaigned for the scientific training of judges and prison officers. He wrote extensively, notably *Sociologia Criminale* (1892).

Ferrini, Contardo (1859–1902), Italian jurist and scholar in Roman and Byzantine law, author of an edition of the Greek *Paraphrasis*, a *Manuale di Pandette* (1900), and *Esposizione del diritto penale romano* (1899).

Ferry. A right of ferry is a franchise right, unconnected with the ownership or occupation of land, created by royal grant, statute, or by prescription at common law, conferring the exclusive right to carry by boat passengers, animals, or goods over a river or arm of the sea from town to town, or roads leading thereto. The right to charge tolls is usually incident to a ferry, but the sovereign, crown vehicles and servants, and certain other classes of persons are exempt. The person having the franchise must provide a reasonable service, take only reasonable tolls, and carry all persons willing to pay the tolls and provide a safe landing place. He is liable in respect of goods carried in the same way as a common carrier (q.v.).

A ferry owner may by injunction prevent infringement of his rights by another person establishing a ferry, but not if the other erects a bridge.

Fet asaver. A tract in law French dating from about 1260, so-called from the words used at the start of the tract and of the paragraphs. It may have been written by Ralph de Hengham (q.v.) as it is akin to his *Summa Magna*, while parts resemble *Brevia placitata*. It deals with procedure in the chief real actions, and was popular, being concise and useful, and frequently copied. It was printed in Woodbine, *Four Thirteenth-century Tracts* (1910).

Fetiales. Roman priestly officials who conducted international relations such as the declaring of war and the making of treaties. There was a college of 20 *fetiales* and they seem to have operated in pairs, certainly when making a treaty with another state. War was declared by a *fetialis* crossing the border and asking for satisfaction from the other state concerned. If this were not given the Senate was consulted and if it voted for war, the *fetialis* went to the boundary, formally declared war in the presence of three adults and cast a spear or stake sharpened and hardened in a fire into the enemy territory. The *ius fetiale* was accordingly what would now be called the law of diplomacy or international relations.

Feu. In Scotland a piece of land granted by one person to another to be held of and under him by feudal tenure, paying in return an annual payment known as feu-duty. To feu land is to grant land on such a basis. The annual payments of feu-duty may since 1970, and in some cases must be, redeemed by payment of a capital sum and feu-duties cannot now be stipulated for.

Feud. A continuing animosity or enmity between two persons, groups, or families frequently productive of repeated violence, commonly in revenge for harm done by the other group on a previous occasion. In particular in early law a significant development was the replacement of the blood-feud or execution of retribution for causing death by compensation to the family of the victim and sometimes also a fine to the King.

Feudalism (Late Lat. *feodum* or *feudum*, a fief or fee). Feudalism developed from the weakness of government in late Roman and early mediaeval Europe and the need of small landowners for protection against enemies, invaders and other, more powerful, landowners. This need invoked the aid of two institutions already recognized in law, the *patrocinium* or patron and client relationship, whereby a small landowner or landless freeman offered his services to a powerful lord in return for shelter and support, which developed into a transaction called commendation, in Germany commonly created by written contract with counterpart prestations of protection and support, and of service, and the *precarium*, a form of letting land as a reward or to secure a debt. The *precarium* gave rise to the *precarium fundorum*, whereby the poor landowner surrendered his lands to the lord and received them back as a *precarium*, to be held under the protection of the lord.

When the Franks conquered Roman Gaul they were already familiar with the institution of the *comitatus*, a personal relationship between a lord and a group of men, with mutual duties of faith and service, and seem to have adopted the idea of the *precarium*, which developed rapidly and took root as standard institutions of the Frankish kingdom. Moreover, the great men of the land were already of the *comitatus* of the King and members of his household and bodyguard and began to adopt the *precarium* tenure, becoming clients of the King or his greatest lords. The Church also seems to have employed the *precarium* tenure. Gradually the tenure came to be more strictly defined and to be the subject of written contract. The duration of the *precarium* came to be for life, and then for a series of lives, and the services due in return were more closely defined. Under the Carolingian kings there was developed the duty of military service with mounted soldiers when called upon as the price of a grant of land. Under Charlemagne the army developed into the feudal host as each lord came under the duty to produce his band of soldiers and himself lead them. The lords also came to be regarded as the overlords of their districts and the duties and payments which the landowners had originally owed the State became obligations to their lords and conditions of their tenure of lands from those lords. In this process the right of doing justice among the landowners passed to the lords, they secured immunity from royal agents, and their courts acquired civil and criminal jurisdiction. Feudal counties developed, corresponding roughly to the old administrative divisions of the country, the feudal lords exercising administrative and judicial functions.

It came to be accepted as convenient for the King to be lord of the lords, holding them responsible for the public duties of his vassals. The rule became established that a man who received a fief from a lord owed a vassal's duties and one who owed vassal's duties was given a fief. A fief was usually land but might be, instead or in addition, the right to collect a toll, to operate a mill, to do some honourable service to the King or lord. The practice of subinfeudation was recognized, whereby the great lord granted much of his land to dependants, to be held from him for services.

Feudalism thus came to be recognized over much of western Europe as a system of administration, jurisdiction, military service, and land tenure. But it was not equally or systematically developed everywhere. It was unknown, at least in any developed degree, in England until the Norman

Conquest when William introduced it, appropriating the extensive royal domains and giving many of the lands of the defeated Saxons to his chief supporters. On each insurrection by native English William confiscated the estates of the rebels and granted them to his Norman nobles to be held by feudal tenure. By the end of the eleventh century all tenure seems to have become feudal. William moreover was determined to reign as king and broke the practice of the exclusive dependence of a vassal on his lord by exacting in 1086 the Salisbury Oath whereby not only tenants-in-chief but mesne lords and all landowners were obliged to swear fealty to him.

The introduction of feudalism was of immense significance for the legal and political structure of England for centuries. Once it was established all lands were held subject to obligations due to the King from his tenants-in-chief as original grantees of lands, or to them from mesne lords, and to them in turn from vassals who actually occupied the lands. There might be any number of steps in the chain of tenure. The commonest obligation was that of knight's service, the duty to provide the King with a stated number of knights armed and equipped to serve in his forces for 40 days. Tenure by knight-service was subject also to certain incidents or incidental burdens, namely aids, relief and primer seisin, wardship, marriage, and escheat (qq.v.). Sometimes religious bodies held lands in frankalmoign, or free alms, the duty being spiritual, to put up prayers for the soul of the granter. Other tenures were serjeanty and free socage (qq.v.). When a fief was granted the tenant was invested in his lands by livery of seisin, actual or symbolical delivery to him, and he did homage to his lord therefor and took the oath of fealty to him.

Feudalism was introduced into Scotland by David I largely on the English model, though it developed variations in nomenclature and in substance.

The most enduring element of feudalism was the land law which, down to 1925 in England, was purely feudal, developed and elaborated and modified in many respects by equity. In Scotland feudal land law continues to survive, albeit much modified, though not by equity, and heritable jurisdictional rights lasted till 1746.

Feudalism was also the basis of much of the older nobility of Europe, giving them their lands, their titles, their prestige, and powers. It is at the basis of such institutions as knighthood, chivalry, and the manor, and made a lasting impression by its castles and feudal warfare.

M. Bloch, *Feudal Society* (1961); F. Ganshof, *Feudalism* (1952); W. Stubbs, *Constitutional History of England* (1897); J. Round, *Feudal England* (1895); F. W. Maitland, *Domesday Book and Beyond* (1897); F. Pollock and F. W. Maitland, *History of English Law* (1895); P. Vinogradoff, *Villeinage in England* (1892); F. M. Stenton, *First Century of English Feudalism* (1932).

Feudal courts. One aspect of the institution of feudalism (q.v.) was that a feudal lord had the right to hold a court for his tenants. In England these courts declined in importance in the thirteenth century, largely because of the growth of the jurisdiction of the royal courts. Manors (q.v.), however, came to imply jurisdiction, borrowing this idea from feudalism, and developed courts baron and courts customary, serving freeholders and copyholders respectively. In Scotland the heritable jurisdictions (q.v.) of great feudal landlords continued until abolished in 1747.

Feuerbach, Paul Johann Anselm von (1775–1833). An important writer on German criminal law, author of a *Revision of the Criminal Law* (1799), and famous *Textbook of general criminal law in Germany* (1801), Minister of Justice for Bavaria (1805–14) and draftsman of the Bavarian Criminal Code of 1813 which became the model for criminal legislation in many of the German states thereafter. His works were a strong protest against vindictive punishment and greatly furthered the reformation of German criminal law. Later he was second president of the court of appeal at Bamberg (1814–17) and first president of the court at Ansbach (1817–33).

Feus, or Fiefs, Books of the. The development of feudalism gave rise to a literature relative thereto. The Books of the Fiefs or Feus, *Libri Feudorum*, *Usus Feudorum* or *Consuetudines Feudales* were originally composed privately by Milanese jurists. The original collectors are unknown, but they collected statutes, decisions, and customs of various epochs and their collections suffered many revisions before attaining their final form in the hands of jurists of Bologna. Authorship at least of the introductory essay to Book I was formerly attributed to Gerard Capagisti consul of Milan, and of the essay at the commencement of Book II to Oberto dall'Orto, written to his son Anselm, a Bolognese lawyer.

As now known the Books of the Feus comprise two books based mainly on feudal enactments of Conrad II, Lothair III, and Frederic I and each divided into title and rubrics. A rearrangement is attributed to Ardizo who used it as the basis for his *Summa Feudorum*, and a further one to Accursius. A much later rearrangement in five books was made by Cujas, but this has not superseded the older arrangement. It embodies the later feudal law which emphasized property and family rights rather than political and military rights and duties.

Though unofficial the Books of the Feus were widely used in the law schools and the courts and acquired considerable authority in Italy, France, and Germany. Ugolini, in the thirteenth century, included the Books in the *Corpus Juris Civilis,* after Justinian's Novels and forming part X of the *Collationes Novellarum* along with the enactments of the Swabian emperors. A translation of the Books of the Feus is appended to Lord Clyde's translation of Craig's (q.v.) *Ius Feudale.*

Another compilation of feudal law is the Assizes of Jerusalem (q.v.).

Fiar. In Scots law the person in whom a right of property in land is heritably vested as contrasted with a life-renter who has rights of use for his lifetime only.

Fiars prices. The values of each of the sorts of grain grown in the various counties of Scotland ascertained annually by sheriffs with juries at fiars courts in order to determine the stipends of the parochial ministry, which were fixed in terms of grain, and formerly the value of grain due to feudal superiors, titulars of teinds, and others to whom dues quantified in grain were due. They are now obsolete. Formerly the price was struck by the boll, but later by the imperial quarter.

Fiat (Lat. let it be done). A decree or warrant of a judge or public officer. It is commonly used to designate the leave which the Attorney-General must give before proceedings can be taken under certain statutes.

Fiat justitia (Lat. let justice be done). At one time a warrant to bring a writ of error in Parliament was granted only after the King had endorsed *fiat justitia* on a petition for such a warrant. Similarly a petition of right (q.v.) could be heard by the courts only after the King or the Home Secretary on his behalf had endorsed the petition with these words.

Fichte, Johann Gottlieb (1762–1814). German philosopher, worked as a tutor and, under the influence of Kant, wrote a *Critique of Revelation* (1792) before becoming professor of philosophy at Jena. He was later rector of the new Berlin University. His major views are contained in the works *On the Notion of the Theory of Science,* and *Outline of what is Notable in the Theory of Science* (both 1794), and the *System of the Theory of Morals* (1798). He sought to complete Kant's work and to expound the complete system of reason. His principal legal work is *The Science of Rights* (1889) and his theory of law is an approximation to Kant's system but restating it in natural law terms. The basis of law was the idea of legal relations, and from this basis the rules of positive law were to be deduced. His theory of law is highly abstract, but in the notion of legal relations and in his conception of the necessary requirements of an international order he enunciated ideas of great value.

R. Adamson, *Fichte*; R. Kroeger, *Fichte's Science of Rights.*

Fiction, legal. Any assumption which conceals or affects to conceal the fact that a rule of law has undergone alteration, its letter remaining unchanged, its operation being modified. In short, case A is pretended to be and treated legally as if it were an instance of case B. Fictions have often been used to circumvent obstacles in code or statute. Fictions were extensively used in Roman and English law as means of legal development, extending rules to cases not originally covered by them and Maine (q.v.) elevated observation of these systems into a generalization that law was developed by fictions, equity, and legislation; this is a great overstatement. English courts resorted extensively to fictions to extend their jurisdiction, e.g. by the Bill of Middlesex (q.v.) Many rules based on fictions have, however, survived to modern times though many have been swept away by reforms, such as the rule that a father might sue for the seduction of his daughter on the fiction of the loss of her services in consequence of the seduction.

Fiction is extensively used in legislation by the verbal device of stating that X is to be treated 'as if' it were Y.

See also EJECTMENT; *Latitat*; *Quominus.*

Fiction theory. One of the theories of corporate personality, developed in modern times chiefly by Savigny. The theory is briefly that artificial or juristic persons, such as corporations, are not real legal persons but are treated as legal persons only by fiction of law; they are treated in many ways 'as if' they were legal persons. The consequences of the theory are that the corporation has no will, or mind, or ability of its own to act, but only such as the law imputes to it as if it was a person. The theory, though developed in great detail by Savigny, will not bear some of the deductions drawn from it.

Fideicommissum. In Roman law a *fideicommissum* was a simple, informal request by a testator to any person who benefited by his will or on his intestacy to give certain objects to a third party, the *fideicommissarius.* Neither writing nor formal words was necessary, if the intention was clear. From the time of Augustus *fideicommissa* became actionable in special courts where the official had wide discretion in interpreting the testator's gift. *Fideicommissa* might be of the whole estate in which case the *fideicommissarius* was in effect the heir. In Justinian's law the provisions about legacies and *fideicommissa* were amalgamated and the distinction disappeared. The institution was used to tie up property for generations and a Novel of Justinian's time limited the period to four generations.

The concept was adopted into German common law, where *fideicommissa* resembled entails. Permanent and profit-yielding estates or collections of assets are by declaration of the donor's will given *inter vivos* or *mortis causa* entrusted to a party to possess but restricted by the real rights of parties entitled in expectancy. Modern German law voids any appointment of successive heirs which has not taken effect within 30 years. It has also been adopted in the Roman-Dutch law. Fideicommissary substitutions were recognized in French law, limited from 1560 to two generations and forbidden altogether, with limited exceptions, in the Civil Code of 1804.

The concept has obvious similarities to the trust recognized by Equity, though the general opinion is that the English trust was not derived from the Roman *fideicommissum*, and the concept did not develop the dualism of ownership, legal and equitable, characteristic of the English trust. Also *fideicommissa* could not be created *inter vivos*.

Fideiussio. In Roman law a form of suretyship, developed later than *sponsio* and *fidepromissio* and free of nearly all the restrictions applicable to these forms. Though entered into by *stipulatio*, like the others, this form could be used to guarantee any kind of obligation whether created by *stipulatio* or in any other way.

Fidepromissio. In Roman law a form of suretyship. *Sponsio* was open to Roman citizens only but *fidepromissio* to peregrines also. Both enabled the creditor to sue either the principal debtor or the surety, where the principal debt had been created by *stipulatio*. Both were subjected from about 200 B.C. to legislation apparently designed to limit the burdens on sureties, but none of these restrictions applied to *fideiussio* (q.v.).

Fiducia. In Roman law an agreement subsidiary to a conveyance by *mancipatio* or *in iure cessio*, which imposed on the transferee a trust with respect to the thing conveyed, particularly for its reconveyance in certain circumstances. An important form was *fiducia cum creditore*, a form of pledge by conveyance to the creditor with a *fiducia* for reconveyance if and when the debt was paid. Originally unenforceable, the fiduciary obligation became enforceable by *actio fiduciae* some time in the first century B.C.

Fiduciary. A person in a position of trust, or occupying a position of power and confidence with respect to another, such that he is obliged by various rules of law to act solely in the interest of the other, whose rights he has to protect. He may not make any profit or advantage from the relationship without full disclosure. The category includes trustees, company promoters and directors, guardians, solicitors and clients, and others similarly placed.

Fief. A fee or feu, or piece of land held as a feudal estate by a tenant of and under a lord.

Fiefs, Books of the (or *libri feudorum*). See FEUS, BOOKS OF THE.

Field, David Dudley (1805–94). Admitted to the New York bar in 1828 and achieved renown in arguments before the Supreme Court. His greatest achievements were in the sphere of law reform. He was a consistent supporter of codification, being made a member of a New York codifying commission in 1846. Between 1860 and 1865 complete political, civil, and penal codes were prepared, but only the Penal Code was adopted in 1881. After a great struggle the Civil Code was finally rejected in New York. But his code of civil procedure has been adopted by many states, and the criminal procedure code by almost as many. California adopted all five Field codes. He urged the preparation of a code of international law, and published a *Draft Outline of an International Code* (1872, 1876).

His brother, Stephen Johnson Field (1816–99) removed to California where he drafted codes of civil and criminal procedure, served as Chief Justice of the Supreme Court of California (1859–63) and an Associate Justice of the U.S. Supreme Court (1863–97).

H. M. Field, *Life of D. D. Field*.

Field, Sir William Ventris, Baron Field of Bakeham (1813–1907). Practised as a solicitor and as a special pleader before being called to the Bar in 1850. He obtained a large practice and in 1875 became a judge of the Court of Queen's Bench, the last to be appointed before that court was merged with the High Court. As a judge he displayed learning, ability, and sometimes bad temper. He retired in 1890, was sworn of the Privy Council and ennobled, and for a time sat often in the House of Lords.

Fielding, Sir John (1721–80). Half-brother of the novelist Henry Fielding, he became a London magistrate in succession to his half-brother, co-founder of the Bow Street Runners, and was very influential in reforming the administration of criminal justice in London and in suppressing crime. He was blind but was reputed to be able to recognize 3,000 thieves by their voices. He published *An Account of the Origin and Effects of a Police* (1758), *Extracts from such of the Penal Laws as particularly relate to the Peace and Good Order of the Metropolis* (1768), and *A Plan for Preventing Robberies within 20 miles of London* (1775).

Leslie-Melville, *Life and Work of Sir John Fielding* (1934).

Fieri facias (usually abbreviated as fi. fa.) (that you cause to be done). A writ directed to a sheriff commanding him to cause a levy to be made of the debtor's goods and chattels real, subject to certain exemptions, for a sum contained in a judgment with interest. To execute the writ the officer enters the debtor's premises (but may not break the outer door of a house to do so), and makes a formal seizure of them. If the judgment is not satisfied, the officer gets an auctioneer to inventory the goods and to sell them or remove them and sell them, to an amount reasonably sufficient to satisfy the debt and expenses, any balance being paid to the judgment debtor. When the sheriff has performed his duty he makes the return of the writ *fieri feci*, or, if applicable, *nulla bona*, or goods seized but unsold. Other forms of return are sometimes applicable. Variants of the ordinary writ are *fieri facias de bonis ecclesiasticis*, where the defendant is a beneficed clerk having no lay fee, and *fieri facias de bonis testatoris*, where the defendant is an executor sued for a debt due by the testator.

Fifoot, Cecil Herbert Stuart (1899–1975). Oxford legal scholar, author of *English Law and its Background* (1932), *Lord Mansfield* (1936), *Law of Contract* (with G. C. Cheshire, 1945), *History and Sources of the Common Law* (1949), *Judge and Jurist in the Reign of Queen Victoria* (1959), *F. W. Maitland, a Life* (1971), and other works.

Fifty decisions (*Quinquaginta decisiones*). A series of constitutions issued by Justinian (q.v.) in A.D. 530 to settle controversies and disputes and formally to abolish certain obsolete rules. They have not survived, though the substance of many must be contained in his revised Codex (q.v.) of A.D. 534.

Filacers (Lat. *filum*, a thread). Officers of the old common law courts. Their duties were to file original writs and to issue process thereon. The offices were abolished in 1837.

Filangieri, Gaetano (1752–88) An Italian jurist who continued the work of Vico and wrote *La Scienza della Legislazione* (1780) urging reforms and denouncing the abuses of his time. The third book (1783) deals with the principles of criminal jurisprudence and the fourth (1785) with education and morals. The work was widely translated and studied and had great success and influence throughout Europe.

Filibustering (from the Dutch for 'freebooter'). The practice of obstructing legislative business by dilatory tactics, particularly by speaking at inordinate length so as to waste time. In this sense the word originated in the U.S. where long filibusters have taken place in the U.S. Senate. Since 1900 it has been prevented in the House of Representatives by rules of procedure and since 1917 in the Senate by a closure rule. In the House of Commons the Speaker has power to direct a member to discontinue a speech which is irrelevant, repetitive, or merely time-wasting.

Filius nullius (son of nobody). A bastard.

Filiusfamilias. In Roman law a person subject to the *patria potestas* of another, as contrasted with one *sui iuris*.

Filmer, Sir Robert (1588–1653). English political writer who defended the royal authority in his *Patriarcha, or the Natural Power of Kings* (*c.* 1650, published 1680) which treated government as a development from the family, exercising authority originally derived from God. The theory was revived in the 1680s and attacked by Locke (q.v.).

Financial year. In statutes with reference to finance or moneys provided by Parliament means the 12 calendar months ending on 31 March, being the date fixed by statute of 1854 as that up to which the annual public accounts should be made. For income tax purposes since 1832 the year runs to 5 April. From early times the financial year ended at Michaelmas (29 September). When the calendar was reformed in 1752 the calendar year lost 11 days but the financial year was not shortened and accordingly ran to 10 October. At the end of the eighteenth century it was decided that the financial year should run from 6 January to 5 January (being the reformed calendar date corresponding to the old Christmas Quarter day to bring it into line with the annual accounts of the Customs and those of commerce and navigation generally). In 1832 supply was taken for five quarters and since then 1 April has been the date for annual grants of money. The financial year of public corporations, companies and other bodies is fixed by statute or internal resolution.

Finch, Sir Heneage Finch, Earl of Nottingham (1621–82). One of a family prominent in public life he was called to the Bar in 1645 and built up a practice and a reputation under the Commonwealth. He became Solicitor-General in 1660 and Attorney-General in 1670, Lord Keeper in 1673, Lord Chancellor in 1675, and Earl of Nottingham in 1681. He believed in an extensive royal prerogative and maintained the validity of the royal suspending power, but at times asserted the privileges of the House of Commons. He took a leading part in passing the Statute of Frauds, but his great work was as an equity judge. He developed many principles and enunciated them clearly, and displayed the qualities of a broad intellect and a fine grasp of the facts of cases and the relevant rules. Thus he classified trusts, originated the modern rule against perpetuities, and contributed greatly to developing equity into a regular system, services for which he has traditionally been called the Father of Modern Equity. He left various writings, including reports of some of the cases he heard.

Finch, Sir Henry (1558–1625). An ancestor of Sir Heneage Finch, Lord Chancellor Nottingham (q.v.), was called to the Bar in 1585 and was knighted and made a serjeant in 1616. He was engaged with Bacon, Noy, and others on an abortive attempt at codifying the statute law. His main work was *Nomotechnia* (1613) in four books, treating of jurisprudence, common law, procedure, and special jurisdictions. It was in law French; an English version came out in 1627 under the title, *Law, or a Discourse Thereof in Foure Bookes* and there were later editions. Another and closer translation appeared in 1759 as a *Description of the Common Laws of England according to the Rules of Art compared with the Prerogatives of the King*. An abridgement called *A Summary of the Common Law of England* appeared in 1673. As an exposition of the common law it was not superseded, on common law, till Blackstone, and, on jurisprudence, till Austin.

Finch, John, Baron Finch of Fordwich (1584–1660). Called to the Bar in 1611, he sat in Parliament and became Speaker in 1628 and then Chief Justice of the Common Pleas, 1634–40, and as such a strong upholder of Charles I's personal rule. He gave the leading judgment in Hampden's case (the case of ship-money) (q.v.) and became very unpopular. He was Lord Keeper 1640–41, when he was impeached but escaped to Holland, returning only at the Restoration.

Finding. (1) The discovery of a thing, the owner of which is not known. The finder of a thing lost obtains a good title of possession to it as against everyone except the true owner, but no title of ownership. If having found a thing, the finder knows to whom it belongs, or knows that the owner can be found, but keeps it for himself, he commits theft. There are many decisions, hard to reconcile, on whether property found in public places, or in private premises, belong to the finder or to the occupier of the premises.

Where the thing found is money, gold, silver, plate, or bullion which has been hidden or buried by a person unknown, it is treasure-trove and belongs to the Crown, or to a person, such as the lord of the manor, to whom the Crown has made a grant of treasure trove. The coroner must hold an inquest to determine whether things found are treasure or not, as defined, but cannot decide who is entitled to it.

(2). A conclusion on an inquiry of fact, as by the verdict of a jury, or after the trial of an issue of fact by a judge alone. Findings of fact include simple findings such as, e.g. that the defendant drove at a particular speed, and more complex findings such as the conclusion, e.g. that the defendant could not in the circumstances have foreseen that fire would result from his action. The findings in fact made in a trial court may have great influence when the case is appealed because, in general, an appellate court can decide a case only on the basis of the facts 'found' by the trial court. Findings in law are findings as to the effect in law of certain circumstances, e.g. that an alleged contract was void and ineffective.

Fine (1) In criminal law, a sum of money ordered to be paid to the Crown by an offender by way of punishment. At common law a fine was one of the penalties for a misdemeanour. In 1861 statute permitted a fine for certain felonies but only in 1948 were courts empowered generally to fine persons convicted of felonies. Magistrates' courts have limits set on the amount of fine imposed for various kinds of offences and statutory offences normally provide for fines within stated limits as the, or a, penalty. Payment of a fine may be enforced by the Crown suing the offender in the civil courts, or, after judicial inquiry into his means, by imprisoning for default in payment.

No general limit has been set for fines though Magna Carta and the Bill of Rights prohibit excessive and unreasonable fines. It is now the commonest kind of penalty particularly in non-indictable offences. The fine imposed must be related to the offender's ability to pay.

(2) In feudal tenure, a fine was a money payment due by a feudal tenant to his lord. In tenure in chief by knight service a fine was due to the King on every conveyance by the tenant to a third party. This kind of fine was abolished in 1660. In copyhold land a fine was normally due to the lord on a change in the tenancy.

Fine (or *finalis concordia*). In mediaeval English law an action compromised in court and by leave thereof on terms approved, utilized as a means of conveying land. A *praecipe* of writ usually of covenant was read out; parties appeared and the action was compromised, with the leave of the court, on the basis that the lands were adjudged to belong to one of the parties. This agreement was put in writing and enrolled in the records of the court. A fine consisted of the writ, the licence to agree given by the court on payment of a fine to the King, the concord or agreement, a note of the proceedings drawn up by the chirographer and the 'foot' or 'chirograph' of the fine which narrated the whole proceedings. The feet of fines, the bottom part of the deed preserved in the Treasury, were the main record of the transaction. It contained a record of the whole transaction. It was a simple and speedy form of conveyance and an easy way of effecting a family settlement. Fines were abolished in 1833.

Fine rolls (or Oblate rolls). Rolls, covering 1199–1641, recording payments of money to the Crown for various purposes. Many payments are merely payments for writs or brieves purchased. Later there are more prominent payments for licenses and pardons, for the alienation and acquisition of lands, for charters and confirmations thereof, and so on, but the most important and numerous are documents, whether letters patent or letters close, issued under the Great Seal relating to matters in which the Crown had a direct financial interest, including writs ordering the delivery of goods to executors for administration, writs to sheriffs and others as to lands in the King's hands by forfeiture, grants of wardship, marriage, licenses to marry, appointments of escheators, sheriffs, and others who would be responsible to the Exchequer. By the end of the fourteenth century the Fine Rolls tend to specialize in escheats and the disposal of wardships. Some of the Fine Rolls have been edited or calendared.

Fingerprints. Marks made on any firm surface by the ridges on the skin at the tips of the thumb and fingers. The marks can be made more visible and photographed and compared with the marks made by the fingers of known persons, forming accordingly a valuable means of identifying the maker of the marks. This identification proceeds on the basis of the discovery that no two persons in the world have exactly the same pattern of ridges and of marks, and that the patterns can be classified and filed. In 1907 Scotland Yard adopted the Galton-Henry system of fingerprint classification and this was followed by similar law-enforcement agencies in many parts of the world. Such agencies maintain extensive collections of the fingerprints of criminals. The taking of the fingerprints of suspected criminals is accordingly a recognized part of criminal investigation, though regulated by statute. Criminals frequently, however, resort to devices, such as wearing gloves, to avoid leaving fingerprints on objects handled in the course of committing crime. One of the first methods of scientific identification was devised by Alphonse Bertillon and the Indian Police early adopted fingerprinting for the identification of criminals.

Finland, law of. Finland was under the suzerainty of Sweden from the thirteenth century till 1809, then a Russian grand duchy until it attained independence in 1917. It became a republic in 1919 when it adopted its constitution.

The legislature is a unicameral Parliament elected for four-year terms. Executive power is exercised by the President, who is elected for a six-year term, and the Council of State. Ministers are responsible to Parliament. The President can dissolve Parliament and has decree-making powers; he is responsible for foreign policy and is head of the armed forces.

The country is divided into 12 provinces, each under a provincial governor, divided into communes each having an elected council and executive board.

The judiciary is independent of the executive and the legislature. It comprises a Supreme Court, four Courts of Appeal in the principal towns, which also act as courts of first instance for cases of treason and offences by high officials, and district courts, held by a qualified judge with five to seven lay assessors, or town courts, held by a burgomaster (qualified) and assessors or a judge and jury of 12. The chancellor of justice has both the supervision of the administration of justice and also acts as public prosecutor. In district courts prosecution is by the local sheriff and in town courts by the city attorney.

There is also a supreme administrative court, a land partition court, and an insurance court.

The substantive law has the same origin as that of Sweden and did not develop independently until after the separation of 1809. A Swedish civil code was adopted in 1734. A penal code was adopted in 1891. Both have been extensively altered since 1919.

J. Uotila, *The Finnish Legal System*; V. Merikoski, *Précis du droit public de la Finlande.*

Finlay, Sir Robert Bannatyne, 1st Viscount Finlay (1842–1929). Graduated in medicine, then turned to law and was called to the Bar in 1867. He made steady but unspectacular progress and took silk in 1882 and in the following year entered Parliament. He was out of the House from 1892–95, but was building up his position at the Bar. In 1895 he became Solicitor-General and in 1900 Attorney-General. In those years he made a considerable reputation in international law, in problems arising out of the Jameson Raid, the Boer War, the Dogger Bank incident, and other similar incidents, and his opinions show a sound and scholarly knowledge of the subject. For this work he was honoured with the G.C.M.G. In 1905 he went out of office though he had become a Privy Councillor. He was again out of Parliament from 1906 to 1910 but in 1916 he became Lord Chancellor. As a judge he was in the first rank, and his judgments are sound, careful, and cautious. In 1918 he was dismissed, but promoted viscount.

In 1921 he was elected a judge of the Permanent

Court of International Justice at the Hague and won high regard among colleagues trained in very different legal systems, as much by reason of his classical and literary culture and knowledge of modern languages as by his profound knowledge of international law. His son became a judge of the King's Bench division and Court of Appeal.

Fiore, Pasquale (1837–1914). Italian jurist, an authority on international law, author of *Il diritto civile italiano*, *Trattato di diritto internazionale publico* (1865), and other works chiefly on international law.

Fire-raising. In Scots law the setting on fire of the property of another, or of one's own property with intent to defraud insurers. It is distinguished into wilful fire-raising, which is the setting fire to buildings, stacked crops, growing wood and coal-heughs, and culpable and reckless fire-raising, which applied to other subjects, even if set on fire intentionally.

Firm. A common term for a partnership, considered as a group. In England the firm is not, but in Scotland it is, a body distinct from the individual partners, but in neither country is it a corporate body or distinct entity with full legal personality. This is important in bankruptcy. For some procedural purposes the firm is treated as a body; thus an action may be brought by or against a firm by its firm name.

Firma. **Rent.** *Firma burgi* was the right to receive the rents, tolls and other income of a borough, which might be granted by the King or lord of the borough to the burgesses on payment of a fixed sum.

First Fruits and Tenths, Court of. Set up in 1540 to enforce payment by the clergy of dues originally paid to the Pope and annexed to the Crown in 1534, consisting of the first year's income from a benefice and of a tenth of the income thereafter. The court was amalgamated with the Exchequer in 1554 and in 1703 the income was assigned to Queen Anne's Bounty (q.v.) (now the Church Commissioners) to supplement the poorer ecclesiastical livings.

First impression, case of. A case which presents for the court hearing it a fresh question of law, for the decision of which there is no precedent.

First instance, court of. A court before which an action is heard for the first time, as distinct from an appellate court.

First Lord of the Treasury. The nominal political head of the U.K. Treasury. The office is now always held by the Prime Minister, the Chancellor of the Exchequer being the working political head of the Treasury. As early as the time of Queen Anne the Lord Treasurer had become the pre-eminent minister of the Crown and, after 1714, this place was emphasized because the office carried much larger amounts of Crown patronage than fell to any other minister, and the use of patronage was the means of ensuring support in the Commons. Since 1714 the office of Lord High Treasurer has been exercised by commissioners, the First Lord, the Second Lord or Chancellor of the Exchequer, and five Junior Lords who are M.P.s assisting the Chief Whip in marshalling the government majority in the Commons. Two of the junior Lords are a quorum and sign certain formal documents on behalf of the Board. Accordingly all the lengthy ministries of the eighteenth century were led by First Lords of the Treasury in the Commons. The offices were occasionally separated in the late nineteenth century. The development was strengthened by the absence of any distinct position of Prime Minister, so that he has to hold, at least nominally, a departmental office, and by the steadily increasing importance of finance in government.

First offender. A person who has not previously been convicted by a criminal court. There are now statutory restrictions on imprisonment of a first offender, and a strong tendency to deal with such offenders by methods other than custodial sentence.

First President. The senior legal officer appointed by the French kings, who was the real head of the *parlement* of Paris and presided there at plenary sessions and, in the absence of the chancellor, at ceremonial occasions. He was a person of great importance, a permanent member of the King's council, and having important judicial functions.

Fiscal (from Latin *fiscus* (q.v.)). Pertaining to the national revenue. Thus the fiscal year is the financial year. For income tax this runs from 6 April to 5 April in the next year. The word is also used in Scotland as a contraction for procurator fiscal (q.v.), who is the public prosecutor in the inferior Scottish criminal courts.

Fischer Williams, Sir John (1870–1947). Called to the Bar in 1894 he took silk in 1921 and became Legal Adviser to the Home Office and British Legal Adviser to the Reparation Commission (1920–30). He wrote *Chapters on Current International Law and the League of Nations* (1929), *International Change and International Peace* (1932), *Some Aspects of the Covenant of the League of Nations* (1934), and *Aspects of Modern International Law*

(1939) and contributed to journals on international law.

Fiscus (Lat. basket or chest). The private treasury of the Roman emperors as distinct from the *aerarium*, or public treasury. It was probably created by Augustus and received income from imperial provinces, confiscated property, and unclaimed lands. Under Vespasian the *fiscus Alexandrinus* and *fiscus Asiaticus* received money from Egyptian and Asiatic revenues which had formerly gone to the *aerarium*. It was under the *a rationibus*, usually a freedman down to Hadrian's time, and thereafter an *eques*. Nerva appointed a special praetor to try disputes between the *fiscus* and the public and Hadrian instituted a department of *advocati fisci*, or counsel to the treasury, to appear in the courts. As time went on it became more important and the *aerarium* less so. Money from the *fiscus* went to pay the army and the fleet and the salaries and expenses of the imperial administration.

Fish, royal. Whales, porpoises, and sturgeon, if caught near the coast or cast ashore, belong to the Crown or, in the Duchy of Cornwall, to the Prince of Wales. Ancient writers say that in the case of a whale the head belongs to the King and the tail to the Queen.

Fisheries. Save where the Crown or a subject has acquired proprietary right exclusive of public right, the public has at common law the right to fish in territorial tidal waters up to mean high-water mark of ordinary tides. But the public has no right to fish in non-tidal waters even though they are navigable. Private fisheries in tidal waters might be created by the Crown prior to Magna Carta, or now by statute. In non-tidal waters the right to fish, if not retained by the Crown, is attached to riparian lands. A private right of fishery in either tidal or non-tidal waters may be either exclusive or a liberty shared with other persons. In ponds and lakes the right to fish belong to the owner of the land or lands in which the waters are situated and the general public have no right to fish. There is no right of property in fish, unless confined in a net, tank, or fish-pond, but the taking of fish by anyone other than the person having the right of fishery is a statutory offence.

Fisheries on the high seas outside the territorial waters of any state are open to vessels of all nations.

Fisheries in the open sea are increasingly the subject-matter of extensive statutory regulation in the interests of reasonable conservation of fish.

Fitting, Heinrich Hermann (1831–1918). German jurist, notable chiefly for his studies in early mediaeval legal history. He edited the *Summa Codicis des Irnerius* (1894) and *Quaestiones de iuris subtilitatibus des Irnerius* (1894) and wrote *Juristische Schriften des früheren Mittelalters* (1876) and *Die Anfange der Rechtsschule zu Bologna* (1888).

Fitton, Alexander, Baron Gawsworth (?–1698). Called to the English Bar in 1662. Having turned catholic he became Chancellor of Ireland in 1687 and followed James II to France in 1690.

Fitzgerald, John David, Lord (1816–89). Called to the Irish Bar in 1838 and soon won a reputation as the best pleader at the Bar. In 1847 he took silk and in 1852 became an M.P., in 1855 Solicitor-General for Ireland and Attorney-General for Ireland in the next year. After a year (1858–59) out of office he became in 1860 a judge of the Court of Queen's Bench in Ireland. In 1882 he became a Lord of Appeal and a peer, the first Irish judge to be appointed to that office. He sat regularly thereafter and won a high reputation. In 1885 he declined the office of Lord Chancellor of Ireland.

Fitzgerald, William Robert, Duke of Leinster (1749–1804). Sat in the Irish Parliament and was briefly Master of the Rolls in Ireland 1788–89.

Fitzgibbon, John, Earl of Clare (1747–1802). Called to the Irish Bar in 1772. In 1789 he became Chancellor of Ireland and a baron, in 1793 a Viscount, and in 1795 an earl. He played a large part in promoting the Act of Union of 1800.

Fitzherbert, Sir Anthony (1470–1538). Became a serjeant in 1510 and a judge of the Common Pleas in 1521, one of the commissioners to hear cases in chancery in place of Wolsey in 1529 and one of the tribunal which tried Fisher and More in 1535. He had a profound knowledge of English law and great powers of lucid exposition. In 1514 or 1516 he published in three volumes *La Grande Abridgement*, a digest of the Year Books under appropriate titles in alphabetical order and including some cases not in the Year Books. It abridges over 14,000 cases under 260 titles, though the arrangement of cases within titles is disorderly. It was reprinted repeatedly. It was the first serious attempt to set the law down in systematic shape, and served as a model to later writers, such as Brooke and Rolle (qq.v.). In 1517 there appeared a *Tabula*, or digest index to the *Abridgement*. He also compiled *La Novelle Natura Brevium*, a manual of procedure (1534 and many later editions), long regarded as of the highest authority and utility, *L'Office et Auctorité de Justices de Peace* (1538), *L'Office de Viconts, Bailiffes, Escheators, Constables, Coroners* (1538), and *A Treatise on the Diversity of Courts* (1646). There is also attributed to him *A Boke of Husbandrie* (1523) and a *Boke of Surveyinge and Improvements* (1523).

Fitzmaurice, Sir Gerald Gray (1901–). British jurist and international judge. Called to the Bar in 1925 he was legal adviser to the Foreign Office (1953–60) and to U.K. delegations to many post-war conferences, becoming a judge of the International Court (1960–73) and a judge of the European Court of Human Rights (1974–). He was also a member of the International Law Commission (1955–60) and President of the Institute of International Law (1967–69).

Fitznigel or Fitzneal, Richard (?–1198). Bishop of London, administrator and treasurer of England, 1159–96. He also served as an itinerant justice. He was author of the *Dialogus de Scaccario* (q.v.), an invaluable account of the working of the Exchequer in this interesting earlier period in the form of a dialogue between the author and an anonymous young man. He seems to have relied on oral tradition and his own experience because some details are inaccurate.

Five Knights' Case (or Darnel's case) (1627). Charles I, having failed to obtain money from his Parliament of 1626, decided to levy a forced loan. Five knights who were M.P.s were imprisoned for refusing to contribute and applied for a writ of habeas corpus, which was granted, but the return to it stated that they were imprisoned by special command of the King. They contended that they should be released on bail, but the court held that in doubt the verdict must always be for the King in the king's courts. The knights were not released until 1628 and clauses of the Petition of Right of that year condemned taxation without consent of Parliament and imprisonment without cause shown.

Five members, arrest of the. In 1642 Charles I determined to impeach a lord and five members of the House of Commons for attempting to subvert fundamental laws, to deprive the monarch of his rightful power, and to alienate the affections of his people. The lords refused to order the arrest of the accused and the King then sent the serjeant-at-arms to the Commons to arrest the five members and, the next day, came to the House himself to arrest them, but found that they had taken refuge in the City of London. This provoked a declaration of the House touching the breach of their privileges, and heightened feeling against the King.

Five Mile Act (1665). The last element of the Clarendon Code (q.v.) seeking to suppress Puritanism, forbidding Puritan ministers to teach in a school or come within five miles of a town in which they had previously held office, unless they took an oath not to seek to overturn the Church or the constitution. The name is also given to an Act of 1592 which forbade popish recusants convicted of failing to attend Anglican service moving more than five miles from their usual residence.

Fixed charge. A charge (q.v.) attaching to particular property, such as a mortgage, as contrasted with a floating charge (q.v.).

Fixture. Any thing in itself movable but so attached to or connected with land or a building as to become part of it and to pass with it. Mere placing in, or on, or in juxtaposition with realty does not make a thing a fixture but regard must be had to the degree or extent of attachment, and whether the thing can readily, or at all, be severed or removed, and to the intention of attachment, whether for the better enjoyment and use of the realty or rather for convenience in use of the movable object itself. Apart from contract the right to remove movable is greatest in the case of a tenant as against his landlord, less in the case of a life tenant against the remainderman, and least in the case of executor against heir.

Flag. The Royal Standard is the personal flag of the Sovereign and can properly be flown only with Her Majesty's permission, which is granted only when the Sovereign is present. In Scotland the Lion flag (lion rampant) is similarly a royal standard. The Union flag (Union Jack) may be flown on government buildings and by every citizen in the Commonwealth. The national flags of England, Scotland, Wales and Northern Ireland are the St. George's cross, St. Andrew's cross (saltire), red dragon, and hand of Ulster respectively. The white ensign is worn by Her Majesty's ships in commission and the Royal Yacht Squadron, the blue ensign by merchant ships commanded by officers of the Royal Naval Reserve, and the red ensign by all other ships belonging to H.M. subjects. The R.A.F. ensign is flown at R.A.F. stations and formation headquarters. The civil air ensign may be flown at certain aerodromes and on civil aircraft.

Each state has a national flag worn by its ships. In wartime whether a ship has enemy or neutral character is usually determined by the flag which the ship is entitled to fly, though Anglo-American practice considers also the commercial domicile of the owners in the case of a neutral flag, and that fact may even cancel the effect of entitlement to fly the flag of the captor state or of a co-belligerent. In private international law the law of the country whose flag a ship wears may be taken to be the law governing carriage in that ship. The law of the ship's nationality, which is usually the law of its flag, governs internal matters of a merchant ship on the high seas, and may do so even in the territorial waters of a foreign state.

Flags flown merely as decoration may fall within town and country planning controls.

Flag of truce. A flag, traditionally white, used as a symbol by a person who during hostilities approaches the enemy to negotiate with them. It is a recognized duty not to fire on the bearer of a flag of truce nor to make him a prisoner, but there is no duty to receive or negotiate with the bearer. If received, the bearer must not be attacked or imprisoned, and allowed to withdraw when his mission is complete. A bearer loses his inviolability if he abuses his mission, e.g. by spying, and it is an abuse of the flag of truce to send a feigned mission as a ruse, or cover for an attack.

Flagrante delicto (when the wrong is flagrant). A person may be said to be caught *in flagrante delicto* if found or apprehended in the act of committing a crime or wrong.

Flambard, Ranulf (?1050–1128). Son of a priest from Bayeux, came to England with William the Conqueror, became chaplain to William Rufus and was made justiciar. As such he controlled the whole fiscal and judicial business of England and elaborated the feudal rights of the King, turning the incidents of tenure into systematic exactions, and extended them to ecclesiastical fiefs also. On the death of Rufus he was committed to the Tower but escaped, and lived out his days as Bishop of Durham.

Flash-houses. In early nineteenth-century London a term covering low public-houses, lodging-houses, and coffee-shops, which were headquarters of gangs and places of general resort for criminals, where they could get assistance, rest, information, and get rid of stolen property. They were well-known to the police and often visited but not suppressed because of the convenience to the police of knowing where criminals or information could be picked up.

Flavius, Gnaeus (fourth century B.C.). Son of a freedman, secretary to Appius Claudius Caecus, notable for having first, about 304 B.C. published the technical rules of legal procedure, previously a secret of the pontifices and patricians, whereby they maintained their supremacy over the plebeians. This became known as the *Jus Flavianum*. In this way the Roman people first learned the *legis actiones*, the *dies fasti* and *nefasti*, and other intricacies of procedure. Flavius later became *curule aedile* and a senator, despite his humble origins, and was author of a *De Usurpationibus*, on interruption of prescription.

Fleet prison. A famous London prison, situated in the neighbourhood of Farringdon Street, near the Fleet river, dating from the twelfth century and later used to house victims of religious persecution, persons committed by the Star Chamber, and subsequently debtors and persons convicted of contempt of court. It was closed in 1842 and demolished in 1848. The liberties or rules of the Fleet were the limits within which certain prisoners were allowed to reside outside the prison walls, subject to certain conditions. The head of the prison, the warden, frequently farmed out the prison to the highest bidder, which gave rise to cruelty and ill-treatment. It was well known in the seventeenth and eighteenth centuries for clandestine marriages, celebrated in the prison by persons who were, or at least pretended to be, in holy orders. The first was recorded in 1613 and the practice continued down to 1753 when it was stopped by statute. The 'Fleet Books' record the marriages and baptisms celebrated in and about the prison between 1674 and 1754.

T. Ashton, *The Fleet, Its River, Prison and Marriages* (1888).

Fleischer, Johann Lorenz (1689–1749). German jurist, author of *Institutiones juris naturae et gentium* (1722); *Institutiones juris feudalis* (1724); *Dissertatio de juribus ac judice competente legatorum* (1724); and other works.

Fleta seu Commentarius luris Anglicani. A Latin treatise in six books of about 1290 and possibly so-called from the unknown author having been committed to and written his work in the Fleet prison, which, though treated by Coke as an authority, has been stigmatized as little better than an ill-arranged epitome of Bracton. It was printed in 1647 and edited by Richardson and Sayles (Selden Socy.) in 1953. Selden (q.v.) wrote an *Ad Fletam Dissertatio*, appended to the 1647 and 1685 editions containing much information about Bracton, Fleta, and other ancient writers on English law.

Fletcher* v. *Peck (1810), 6 Cranch 87 (U.S. Sup. Ct.). A corrupted Georgia legislature had sold a huge tract of land taken from the Indians to a group of speculators for a small sum. It had then been broken up and resold at enormous profit. A subsequent legislature repealed the grant, but its power to disturb the recently acquired rights of innocent third parties was challenged before the Supreme Court. The Court held unanimously that the original grant was a contract within the meaning of the Constitution and that the original granters had right to make the initial sale, and that the repealing act was unconstitutional as a violation of the contracts clause and also of the natural rights principle. The case is the main precedent for the Court invalidating a state law as contrary to the Constitution, and did much to stabilize contract and property rights, though it did allow the speculators to reap the profits of their dishonesty.

Fletcher, Andrew, Lord Milton (1692–1766). Passed advocate in 1717 and became a judge of the Court of Session in 1724, a Lord of Justiciary in 1726, and Lord Justice-Clerk in 1735. As such he presided at the trial of Captain Porteous in 1736. He resigned the last office in 1748, retaining his other appointments till his death. During the 1745 Rebellion he abstained from severity, and also took a prominent part in promoting trade and manufactures and in the suppression of heritable jurisdictions. Under his advice most of the government patronage was dispensed. He was a man of acute judgment, with an extensive knowledge of Scottish laws and customs.

Floating charge. A charge or security over the assets for the time being held by a business. The business may be continued and the assets dealt with in the ordinary way until the charge 'crystallizes' and becomes fixed and enforceable, which happens when the creditor intervenes or the undertaking ceases to do business or goes into liquidation.

Flogging. The major method of corporal punishment, utilized as a mode of enforcing discipline in the armed forces and among slaves, and under many legal systems as a mode of punishment. In the U.K. it survived for crimes of violence till 1948 and for some time thereafter for mutiny, incitement to mutiny, and gross personal violence to prison officers. Elsewhere it is still competent in some countries.

Of the means used the lash or cat-o'-nine-tails has been the commonest; it is made of a handle and nine knotted cords or thongs of rawhide. Variants of this are the Russian knout, which has thongs interwoven with wire, sharpened, and hooked, and the oriental bastinado, which consists of blows on the soles of the feet with a rod or knotted lash.

Flood, Warden (1694–1764). Called to the Irish Bar in 1720. He became Solicitor-General (1741), Attorney-General (1751), and Chief Justice of the King's Bench in Ireland, 1760–64. His son was Henry Flood, the prominent Irish politician.

Florentinus (second century A.D.). A Roman jurist, author of an *Institutiones* in 12 books which was founded on in Justinian's Institutes.

Flotsam. Goods lost at sea and still floating on the sea. If not claimed by the owners within a year and a day, it belongs to the Crown.

See also JETSAM; LAGAN; WRECK.

Foedera, Conventiones, Literae et cujuscunque Generis Acta Publica. A collection of treaties and other public documents made by Thomas Rymer (1641–1713) and (vols. 16–20) by Robert Sanderson and first published 1704–35. The documents comprising treaties, agreements, letters, and other public acts between the kings of England and other sovereigns and states from 1066 to 1654, are an invaluable collection of source-material for history and international relations.

Foelix, Jean Jacques Gaspard (1791–1853). French jurist, founder of the *Revue Étrangère de legislation*, and author of a *Traité de droit international privé* (1843).

Folkland. one of the modes of land ownership known in Anglo-Saxon law, importing ownership by private persons according to folk or customary law, without written title, though cultivated in common by members of a village community. It could be given to individuals, though not in perpetuity, and on the expiry of a term reverted to the community. It was subject to various financial burdens.

See also BOOKLAND; LAENLAND.

Folkmoot. An assembly of the people in mediaeval times to deal with matters of common concern. Such gatherings included the shire moots, or shire court, the hundred court, and the sheriff's tourn (qq.v.).

Following assets. An important doctrine of equity to the effect that, where a fiduciary relationship exists between parties, property which the one party has, in breach of his fiduciary obligation, allowed into the hands of another may be recovered not only from anyone who has acquired a legal title to that property but also from volunteers, though not from a bona fide purchaser for valuable consideration without notice of the fiduciary obligation. The doctrine has many applications, including following trust property alienated by a trustee in breach of trust, following trust funds which a trustee has mixed with his own funds, and recovering trust assets from persons who have been wrongly paid or overpaid.

Fonblanque, John de Grenier (1762–1837). Called to the Bar in 1783 and developed a chancery practice. He edited the *Treatise on Equity* ascribed to Henry Ballow and made it almost a new work, and it went through several editions. His son, John Samuel (1787–1865), was a commissioner in bankruptcy and joint author of a book on *Medical Jurisprudence* (1823).

Food and Agricultural Organization. One of the specialized agencies of the U.N. established in 1945 in succession to the International Institute of Agriculture (1905) to disseminate information about food and nutrition and to improve the provision of food and agricultural products.

Forbearance. The abstention from doing or intentional not doing of an act as distinct from the unintentional or negligent omission to do it. A forbearance may be a ground of civil or criminal liability if in the circumstances the person should have done the act in question but not if he was legally entitled to refrain from doing it.

Forbes, Duncan, of Culloden (1685–1747). Called to the Scottish Bar in 1709 and entered Parliament in 1722 and became Lord Advocate in 1725. As such he was mainly responsible for the management of Scottish affairs and did much to encourage trade and manufactures between 1725 and 1737. While in office he had to repress the Malt tax riots in Glasgow and deal with the consequences of the Porteous riots in Edinburgh. In 1737 he became Lord President of the Court of Session and held that office till his death. He greatly expedited the business of the court and raised the level of its work. During the Jacobite Rebellion of 1745–46 he played a considerable part in maintaining support for the government, and after its failure sought to mitigate the brutality of the English occupation of the north of Scotland. He was also author of some works on theology.

G. Menary, *Life and Letters of Duncan Forbes of Culloden* (1936); J. Hill Burton, *Lives of Simon Fraser and Duncan Forbes* (1847).

Forbes, William (?–1745). Son of a professor in Padua, he was called to the Scottish Bar in 1696, and served as Clerk to the Faculty of Advocates. He was the first of the official Reporters of Decisions, appointed by the Faculty of Advocates and his reports, covering the period 1705 to 1713, were published in 1714 under the name of *A Journal of the Session*, the Preface of which gives an historical account of the Session and proceedings therein. In 1714 he was appointed Regius Professor of Law in the University of Glasgow, a post which he held till 1745. Forbes wrote extensively and was one of the first in Scotland to write on specific branches of the law. His works include *On Elections* (1700); *The Duty and Powers of Justices of the Peace in Scotland* (1703); *Bills of Exchange according to the analogy of the Scots Law* (1703); *Treatise on Church Lands and Tithes* (1705); and a small *Institutes of the Law of Scotland: I Private Law* (1722) and *II Criminal Law* (1730). He also devoted himself to writing a large work, a *Great Body of the Law of Scotland* (*c.* 1720) which though valuable evidence of the state of the law at the time, still remains unpublished.

Force. The strength or power of coercion or compulsion. It is a necessary element of any system of law in that principles and rules must be capable of being in the last resort enforced and individuals subject to them compelled by organs of society to conform by physical compulsitors, e.g. imprisonment or proprietary sanctions, e.g. fines, penalties, awards of damages, expropriation. No system of law, however just, could secure the voluntary acceptance of all subject to it and voluntary compliance with every liability.

As between individuals subject to a particular system of law an essential role of the legal order is to control the force which one person may legitimately apply to another, both physical force and moral or economic pressure, and as between states an essential role of the international legal order is to control the force which one state may apply against another, in the shape of warfare or economic sanctions. Similarly in both national and international law whether and in what circumstances an individual or a state may justifiably use force by way of self-defence or retaliation are frequently difficult issues of law.

Force and arms, with (a translation of *vi et armis*). A phrase originally, together with the phrase 'against the peace' (*contra pacem*), a necessary though often fictitious part of every count for misdemeanour or pleading for civil trespass, until rendered superfluous by statutes of 1851 and 1852 respectively.

Force majeure. Irresistible compulsion or coercion, a phrase used in commercial documents and having a wider meaning than Act of God (q.v.). A *force majeure* clause must always be construed with close attention to the words preceding and following and with due regard to the nature and terms of the contract.

Forced labour. Work done by persons under compulsion, as by slaves in the ancient world. The term is applied in modern times particularly to the extensive use made by Germany during World War II of foreign labour drafted into Germany and civilian personnel in occupied territories. It was utilized in concentration camps, labour camps, ghetto workshops, in war industry, and elsewhere, and was frequently indistinguishable from slavery, with the concomitants of inadequate food, insanitary housing, brutal guards, and total disregard of standards of safety, health, and welfare. The death rate was high, frequently deliberately so, and the use of forced labour was condemned by the International Military Tribunal in 1945–46.

Forced loans. A method sometimes resorted to of raising money without Parliamentary assent.

They were extensively used under Richard II, but rarely under the Lancastrian kings. Henry VIII had recourse to forced loans in 1522, promising to repay from the next subsidy granted by Parliament, and in 1544, but in 1529 and 1544 Parliament released the King from his debts, thereby converting the loans into gifts. Mary extorted loans in 1557 in face of outcry and resistance. Elizabeth occasionally had recourse to forced loans from the wealthy, but these were punctually repaid. James I resorted to forced loans in 1614–20 as did Charles I in 1627. *Darnel's* (or the *Five Knights'*) *case* (q.v.) is a notable one arising from resistance to the loans, and the matter was a serious grievance when Parliament met again in 1628. Notwithstanding his assent to the Petition, Charles again resorted to forced loans after dissolving the Short Parliament in 1640. No later cases are recorded, the Bill of Rights declared them illegal, and by the end of the seventeenth century control of taxation had passed into the hands of Parliament.

Forcible detainer. The refusal to restore another's goods, originally lawfully taken, after amends have been tendered. If the original taking was unlawful, forcible detainer was criminal and against the public peace. It is also the offence committed by a person who has entered peaceably on any lands but detains them by violence or threat, such as would make entry in such fashion a forcible entry (q.v.).

Forcible entry. The former crime of entering on any land in a violent manner to take possession of it. The violence may be by threats, by breaking in, by gathering a large number of people to make the entry, or by applying actual force to any person. It has been prohibited and made criminal by statute since 1381 and the concept is now established. Forcible entry by a person entitled to possession does not give rise to a civil action and there is no wrong when the rightful owner enters to take possession from an occupier or from a trespasser.

Foreclosure. The process whereby a mortgagee may make effectual his ownership of the lands mortgaged, which arises as soon as the day fixed for redemption is past. On bringing an action for foreclosure a further day is appointed for payment, and, if the money is not then paid, the property is adjudged to belong to the mortgagee absolutely for the whole interest of the mortgagor, and any subsequent legal mortgage is extinguished. The effect is to bar the mortgagor's equity of redemption (q.v.).

Foreign enlistment. By statute of 1870 it is an offence for any British subject, without the licence of the Crown, to accept a commission or employment in the military or naval service of any foreign state at war with another foreign state which is at peace with this country, or to build or equip any ship for the service of any such state, or to assist in increasing her armament or warlike force.

Foreign judgment. In certain circumstances a judgment obtained in a foreign country will, in a British court, be held decisive of the issue decided, and be recognized and enforced by a British court. In general judgments as to the status of a person, e.g. decreeing divorce, pronounced by the courts of the country of his domicile, or by the courts of a country recognized by the courts of the country of his domicile as having jurisdiction so to decree, are recognized. Within the United Kingdom judgments by the courts of one country for payment of money may under statute be registered in another U.K. country and enforced in the same way as a judgment of the courts of that country, and similar provisions apply to certain Commonwealth and foreign countries. But judgments of other foreign courts are merely causes of action on which an English, Scottish, or Irish judgment may be obtained, and the foreign judgment must have been granted by a court of competent jurisdiction, been final and conclusive, and for a definite sum of money. They are not good causes of action if obtained by fraud, in disregard of British ideas of natural justice, or contrary to British ideas of public policy. Foreign judgments enforcing fiscal liabilities or imposing penalties will not be enforced by British courts.

Foreign jurisdiction. The jurisdiction which a state has over its own subjects when they are, or in relation to acts done, beyond the boundaries of its internal jurisdiction. It comprised jurisdiction exercised in the near East and in China by virtue of capitulations (q.v.) and the jurisdiction exercised by consuls (q.v.).

Foreign law. The law of any jurisdiction having a different system of law from that considering the issue. Thus Scots law, French law, and the law of Inner Mongolia are all equally foreign law in an English Court, and conversely. Foreign law may arise in an English court where some foreign element is involved in the case and the English rules of conflict of laws or international private law (qq.v.) direct that some issue in the case must be determined by foreign law. In an English court foreign law is a matter of fact, to be proved by evidence of practitioners competent in the legal system in question, though in certain circumstances statutory provisions exist to enable a court to obtain a statement from a court in a foreign country of its law on the matter. European Communities law is not foreign law in the United Kingdom but part of U.K. law. As an academic subject foreign law is the study of any legal system other than that of the jurisdiction in which the study is being done. It differs from Comparative law (q.v.) in not being deliberately comparative though study in any jurisdiction of the law of another jurisdiction inevitably promotes comparative observations.

Foreign state. Any state other than the state exercising sovereignty in the place where the question arises. The general principle is that a state applies its own law throughout its own territory but not over the territory of other states, nor over the heads of foreign states or their representatives, the public ships or aircraft of foreign states, while within the jurisdiction. Such persons and property are held entitled to immunity from the local jurisdiction.

See DIPLOMATIC IMMUNITY.

Foreman. The member of a jury who acts as chairman when the jury is deliberating on its verdict and speaks for the jury when asked for its verdict. Any member of the jury may be chosen, but the juror who is chosen first in the ballot often acts. In Scotland the foreman is sometimes called the chancellor.

Forensic Science, medicine, and psychiatry. The branches of science, medicine, and psychiatry concerned with the applications of those bodies of knowledge to legal purposes, particularly to eliciting and interpreting facts which may be of significance in legal inquiries. Forensic science deals with such matters as weapons and ballistics, explosives, and examination of altered documents.

Forensic medicine (or legal medicine or medical jurisprudence) deals with such topics as the cause of death or wounding, blood-tests, questions of nullity of marriage, pregnancy, poisoning, sexual offences, the influences of drugs or alcohol, fingerprints, and identification.

Forensic psychiatry is concerned with mental capacity to transact or commit crime, states of mind, such as amnesia, fitness to stand trial or to testify, and the treatment of persons suffering from mental abnormality or disease.

Foreshore. The shore and bed of the sea and of tidal inlets to seaward of the line reached by high tides, or according to another formulation, the line reached by the medium high tide between spring tides and neap tides and to landward of low water mark. Property in the foreshore is vested in the Crown, unless vested in a subject proprietor by prescription or by grant from the Crown, and is held by the Crown for the benefit of the public generally, but cannot be used so as to interfere with the public liberties of navigation and fishing. There is no general right of loading or unloading boats on the foreshore. A right to use the foreshore for bathing may be gained by the inhabitants of a place by custom or prescription. The right to shoot and take birds on a foreshore is a profit *à prendre* which can be acquired in the same way as other profits. The control and management of the foreshore is vested in the Crown Estate Commissioners. The foreshore forms part of the parish and county on the landward side thereof.

Foresight. A capacity which is frequently notionally postulated of a hypothetical 'reasonable man' when a court is seeking to determine whether a particular defendant should or should not be held liable for a particular happening. 'Reasonably foreseeable' is frequently used as a standard, e.g. where a court states that a defendant will be liable if he should reasonably have foreseen that consequences of a certain kind were likely to result, or states that a defendant is liable for the reasonably foreseeable consequences of his conduct. How far this foresight of a reasonable man would extend in particular circumstances is a policy decision not determined by any logical criterion, and it has been said that Lord Bramwell used to attribute to his reasonable man the foresight of a Hebrew prophet. Generally, however, it is restricted to the kind of things which a sensible person should have realized might well happen.

Forest. In early times forests were areas, not entirely or necessarily wooded, set apart to provide exercise and hunting for the King and his friends. A right of forest was an incorporeal hereditament consisting of a right to keep for hunting all animals pursued in field sports in a certain area set apart for the purpose; the right may extend over the holder's own lands or the lands of another. A royal forest was one belonging to the Crown; a forest in the hands of a subject was a franchise (q.v.). The forests were regulated by special rules of law, enforced by special officials (see FOREST LAWS), but these decayed and in 1829 some surviving powers were vested in the First Commissioner of His Majesty's Woods, Forests, and Land Revenues, latterly the Commissioners of Crown Lands and now the Crown Estates Commissioners. In 1923 woods and forests generally, now managed chiefly for the growth of timber, were transferred from the Commissioners to the Forestry Commissioners, a body established in 1919, though some forests, particularly old royal forests, such as the New Forest, are regulated by special legislation. The Commission now has extensive power of control over forests generally, not only over those it itself manages.

Forest laws. The forest laws in England originated under King Cnut, though he expressly recognized the right of every man to hunt on his own ground. Prior to the conquest hunting was a means of getting food more than a sport. William I had a passion for field sports and introduced harsh forest laws to protect open ground and wild animals for sport, which was the exclusive privilege of the King and his guests. The killing of wild boars and hares was forbidden, and the penalty for killing a hart or hind was loss of sight. Under William II the capture of a stag was a capital offence.

Under Henry I the forests were extended and the forest laws became even more harsh and were probably always enforced by special justices of the forest.

Henry II by the Assize of Woodstock (1184) established courts of the forest, which had supreme jurisdiction over all forests, whether within the royal demesne or not, and appointed justices to visit the forests in the same way as the justices itinerant. About 1217 Henry III issued the Charter of the Forest, embodying the forest clauses of Magna Carta, which were the first mitigation of the harshness of the forest laws. The Charter removed the heaviest burdens of the forest laws, reduced the area subject thereto to what had existed before 1154, abolished death and mutilation as punishment for breakers of the forest law, and abolished the separate forest jurisdiction. In 1238 the forests were divided into two north and south of Trent, each having a justice responsible for the whole administration of the forests. Each forest or group had a Warden and all had officers called Verderers appointed in the county court. There were also Foresters and Agisters. The officials of the forest exercised their powers in the court of Swanimote, which originally met three times a year, and the court of Attachment held every 40 days dealing with minor trespasses, in Special and General Inquisitions held whenever an offence was discovered, in the Regard, an enquiry held every three years by 12 knights chosen for the purpose, and in the Eyre held by the Justices in Eyre of the forest, sometimes called the court of Justice Seat. These courts all administered the forest law which, in the thirteenth century, was regarded as a body of law distinct from the common law. Complaints that the Charter of the Forest was not observed were included in a list of grievances presented to Edward I in 1297 and concessions were made in the *Confirmatio Cartarum* of that year. The forest organization was in decay by the end of the sixteenth century, largely by reason of the desuetude of the eyre, and the common law courts strove to limit the jurisdiction of the forest courts.

Under Charles I the boundaries of the royal forests were greatly extended and adjacent landowners heavily fined for alleged encroachments thereon, but the Long Parliament reduced the boundaries to their limits in 1623, and provided that no place at which a forest court had not been held for 60 years before 1625 was to be accounted forest.

In the sixteenth to eighteenth centuries the royal monopoly in sporting rights came to be the privilege of the landed gentry, and the forest laws gave rise to the game laws, which for long restricted the right to take and sell game, even on a man's own lands, unless he had a stated qualification in real property. By the seventeenth century also the forests were being valued, not for sport but for timber, and the machinery of control of the forests was transformed in the nineteenth century and vested in the Commission of Woods and Forests.

The forest law has now been almost completely abrogated and franchises of forest, free chase, park or free warren abolished.

Forestall. To get in first and obstruct a person's progress by force and arms, or to buy goods on their way to market and thereby to raise the price, or to dissuade sellers from bringing their goods to market. It was formerly criminal as an interference with freedom of trade.

Forfeiture. The loss of some right, property, or privilege, by reason of some specified conduct. It has been prescribed by many rules of law as a penalty. Among such rules notable specimens are:

Forfeiture of goods on conviction of treason, misprision of treason, felony, murder, suicide, *praemunire* (q.v.), and striking or threatening a judge. The Forfeiture Act, 1870, removed this penalty in cases of treason, felony, and suicide. Sentence of death or imprisonment for more than 12 months for treason causes forfeiture of a public office, save in certain cases.

In feudal land law forfeiture was a common penalty for many kinds of acts which implied disregard of the duties owed by feudal tenants to their lords.

Alienation of land to a corporation in mortmain (q.v.) without licence by the Crown results in forfeiture to the Crown or the mesne lord.

Companies normally provide for forfeiture of shares for non-payment of calls thereon.

Provision is frequently made in trust instruments, leases, and other deeds for the incurring of a forfeiture of some party's rights in stated circumstances; in statutes penalizing smuggling, poaching, and the like for the forfeiture of the goods involved; and in other statutes for forfeiture of weapons or instruments with which a crime was committed.

Forgery. The statutory crime of the making of a false document in order that it may be used as genuine with intent to defraud or deceive, or the counterfeiting of certain seals or dies with the same intent. A document is false if the whole or any material part purports to be made by or on behalf of or on account of a person who did not make it or authorize its making; or if any material alteration has been made in it, or part of it purports to be made by or on behalf of a fictitious or deceased person, or if it is made to pass as having been made by a person other than the maker. Documents cover wills, deeds, banknotes, documents of title, and the like. It is forgery even if the document is incomplete or ineffective in law. False paintings are not documents and not forgeries. Uttering a forged document is in England a distinct offence. In Scotland (where the law is common law) uttering is an essential of the crime, forgery by itself not being criminal; in Scotland a forged document is one which pretends to be genuine in the sense of pretending to be authenticated by a particular person.

Forgetfulness. The quality of having forgotten something. It may influence a person's conduct in many circumstances with legal consequences. If it be shown that a mistaken representation of fact resulted from forgetfulness, the representation will be treated as an innocent misrepresentation. Payment by mistake resulting from forgetfulness justifies a claim for repayment. Forgetfulness of events leading up to an alleged crime, if proved by the accused on balance of probabilities to be genuine and not feigned, can be treated as equivalent to insanity barring trial.

Fori regni Valentiae. A law code promulgated by James I of Valencia in 1240 and replacing codes derived from Aragon and Catalonia previously in use. Based on Roman law and originally in Latin it was translated into Valencian dialect in 1261 and several times revised. It gives a complete picture of mediaeval Valencian law, legal institutions, and customs.

Forisfamiliation. The liberation of a child from family tutelage. In Roman law it was achieved by emancipation or the death of his *paterfamilias*. In modern law it is effected by his attainment of full age, now 18 in the U.K., but for many practical purposes may be effected earlier, by leaving home or, in the case of a daughter, by marrying, even before 18.

Formalities. Particular actions, words, kinds of writings, or steps in procedure, which have in various circumstances to be done, said, or made as a condition of the validity of some legal transaction. Formalities are more common and more strictly insisted on in immature legal systems though some formalities continue to be required even in mature systems. They are frequently required to ensure by their formality and solemnity that the transaction effected is public and likely to be remembered. Thus in Roman law the formalities of the fictitious sale (*mancipatio*) by copper and scales (*per aes et libram*) were utilized for many purposes, such as a form of marriage, adoption, emancipation of a child or a slave and a form of will as well as for its primary purpose of transfer of ownership. So too certain kinds of contracts had to be made by the formal question and answer of the *stipulatio* (q.v.). In feudal law the transfer of land had to be made by the formal livery of seisin, the actual handing over of parts of the land or symbols thereof in the presence of witnesses.

In modern law, the formalities most commonly required as evidence of serious intent to be bound and of its own authenticity is the deed, signed, sealed, and witnessed, or a variant thereof such as a document in prescribed form signed by the party to be bound, but in various particular cases such as marriage other formalities are required. Accordingly a distinction must be drawn in such transactions between the formal validity or invalidity of the transaction and its essential or substantial validity or invalidity. Formalities have place also in the legislative process, and in court procedure, where various steps have to be taken in prescribed order and at prescribed times.

Formedon (*forma donationis*). A group of common law writs brought by persons who claimed land under a gift in tail when the land was in the possession of a person not entitled to it. They included formedon in the descender, brought by the issue in tail, formedon in the reverter, brought by the reversioner, and formedon in the remainder, brought by the person entitled in remainder after the determination of the estate tail. All were abolished in 1833.

Forms of action. See ACTION.

Forms and precedents, books of. In the drafting of legal deeds the value of using a form of words settled by an experienced practitioner, or approved by a court, or interpreted as having a particular effect, is great and does much to allay doubts and promote certainty. The provision in book form of forms, precedents, and styles for all kinds of deeds and writs is accordingly long-established and such books are indispensible parts of legal practitioners' working libraries. Historically such collections go back to mediaeval formularies, the Registers of Writs, the *formulae* of the Roman law, and even earlier specimens. The published collections are normally restricted to particular spheres, such as forms for use in court practice, conveyancing forms, forms for wills, company deeds, etc. Some treatise on substantive law also include some forms appropriate to the topic.

In England among the principal have been Bythewood and Jarman's *Selection of Precedents in Conveyancing* (1821, 4th edn. 1884–90), Prideaux's

Forms and Precedents in Conveyancing (1853, 24th edn. 1948–52), Key and Elphinstone's *Compendium of Precedents in Conveyancing* (1878, 15th edn. 1953–54).

On a larger scale are the *Encyclopaedia of Forms and Precedents other than Court Forms* (3rd edn. 1939–50, 20 vols. and supplements) and the *Encyclopaedia of Court Forms and Precedents in Civil Proceedings* (2nd edn. 1963– , vols. and supplements).

There are also volumes of precedents on particular branches of law such as Company Law.

In Scotland there have been successively the *Juridical Styles* (3 vols. 1787–94) and later edns.), *Scots Style Book* (8 vols. 1902–11) and the *Encyclopaedia of Scottish Legal Styles* (9 vols. 1935–48).

In the U.S. there are, among many, *Modern Legal Forms* (1938–50) and *American Jurisprudence Legal Forms Annotated* (1953–).

Forms of Law. The different forms in which precepts for the regulation of social conduct and the conduct of states in international society are found. The relative importance and bulk of different forms differ at different times and in different societies. The first form historically, and consequently most important in immature societies, is custom, and indeed it is frequently a question whether in a particular context certain usual practices are observed merely as traditional and customary or because they are accepted as obligatory or are enforceable by some sanction. The second form is judicial decision, where it is accepted that the decision of a dispute has to be taken as laying down the law for that dispute, and particularly where there is a tendency for the decision to become generalized and treated as a rule customary for disputes of that kind. Judicial decisions will in many cases in earlier law be declaratory of custom, and also creative of custom. The third form is legislation, rules consciously made by a lawgiver, autocrat, or assembly to supplement or alter customary rules, or to supply wholly new rules; and the fourth is conventional law, legal norms laid down by agreement and binding the parties thereto. The fifth is textbook law, formulated by jurists on the basis of historical or comparative study, analysis of accepted rules and inductive generalization from them, suggestion on theoretical bases, e.g. the basis of natural law.

These forms can be seen exemplified in most legal systems of any maturity, though the importance of the different forms varies. Thus in Roman law there was legislation (XII Tables, *leges*, *plebiscita*, *senatus consulta*, magistrates' edicts, and imperial constitutions) and *responsa* and treatises of the jurisconsults. In European systems there is legislation (codes, statutes, orders), interpretative decisions, and textbooks. In Anglo-American law there is legislation (constitutions, statutes, orders, and regulations), judicial decisions, or case-law (frequently declaratory of customs, historically prior to and in many fields of law still more important than legislation), conventional law adopted by churches, companies and trade unions, and authoritative books (now of limited importance). In international law custom is still material, but law is increasingly being made by international legislation and agreement (treaties, conventions), and by judicial decisions, while textbooks, formerly important, are now subsidiary.

The study of what forms of law are recognized in a given legal system is closely connected with the study of what are the recognized formal sources (q.v.) of the law in that system, or from what kinds of materials can one draw statements of law which are authoritative and binding.

Formulae. In Roman law the name for the issues which the praetors (q.v.) allowed the parties to settle in writing, focussing the matters in dispute between the parties and entitling the judges of fact (see *iudex*) to condemn or acquit the defendant according as they found the contentions of the plaintiff established or not. The typical formula was an instruction to the judge of fact, if he found certain facts established, to condemn the defendant to pay money, e.g. 'Let X be *iudex*. If it appears that the defendant ought to pay 100 sesterces to the plaintiff, let the *iudex* condemn the defendant to pay 100 sesterces to the plaintiff. If it does not so appear, let the *iudex* absolve him.' If the defence was that there had been an agreement that the defendant need not pay, the formula would include, after the If clause, 'unless there was an agreement between the parties that the defendant need not pay him, let the *iudex* . . .'. When the formula had been settled in terms which allowed for all the contentions of the parties, it was put into writing and finally approved by the praetor. The *iudex* was then chosen from the official list and he heard the case and decided on the basis of the formula. The formulary system (q.v.) of procedure superseded the older *legis actiones* (q.v.).

Formularies. Books of forms of words used in Roman times for the composition of laws, imperial constitutions, rescripts and private charters and in the Middle Ages for the composition of charters, letters, and other grants. Only fragments of Roman formula books survive, but there are many specimens of mediaeval formularies. They are important sources of knowledge of mediaeval law and its application. Most of them came from Frankish territory. The most important are the *Formulae Marculfi* (eighth century), replaced about 830 by the *Formulae Imperiales*, but there are many other unofficial collections, mainly used for drafting private charters. In Italy Cassiodorus, jurist of the Ostrogothic period, left a collection of chancery forms. Rather later there are Lombard forms in treatises which give instructions for lawsuits; these include the *Liber Papiensis* and the *Chartularium Langobardicum* which collects notarial forms. In France and Germany many formularies were compiled from 1100 onwards, notable specimens being the *Summa de Arte Prosandi* of Konrad von Mure of Zurich (*c.* 1275) and the Baumbartenberg *Formularius de Modo Prosandi.* The Papal Chancery produced the *Liber Diurnus Romanorum Pontificum* (ninth century), containing formulae going back to about A.D. 600. Formularies were in common use in the papal chancery in the fourteenth and fifteenth centuries and similar works were found in royal chanceries. Similarly, from the twelfth century practical books on actions also contained formulae or clauses for use in litigation; this kind of literature reached its climax in Durandus' *Speculum judiciale*, probably the most comprehensive mediaeval legal handbook.

Formulary system. In Roman law the system of civil actions which was the procedural framework in which Roman law developed and operated throughout its classical period. The characteristics of the system were that for each cause of action there was an appropriate form of action expressed in a formula or set of words which constituted the pleadings in the action and formulated the issue for decision. It is uncertain how and when the formulary system was introduced, but the decisive step was the *Lex Aebutia* of *c.* 125 B.C. In each case the plaintiff and his advisers asked the praetor (q.v.) as magistrate in charge of litigation to grant the formula or form of action appropriate to the kind of claim the plaintiff had, which was settled, put into writing, and signed by the praetor. Subsequently the parties appeared before the *iudex* (q.v.), adduced evidence, and he gave judgment according to the evidence on issues put before him by the formula.

The principle that each cause of action had its own formula gave the praetors great opportunities for development of the law. By recognizing new formulae they could in substance create new causes of action. This development was mainly effected by the praetorian Edict (q.v.) which contained specimen formulae both for those already existing to enforce the *ius civile* and for those remedies promised by the Edict, but praetors could at any time, either on general grounds or to suit the circumstances of an unprecedented case, allow a new formula. Such would frequently be made permanent in the following year's Edict. Thus praetorian modifications of the formula, by allowing particular *exceptiones* or *replicationes*, enabled a *iudex* to take account of such circumstances as that the parties had agreed that money should not be repaid, or that an agreement had been secured by fraud or fear. Another praetorian development was the recognition of the contractual principle of good faith by allowing a formula which directed the judge to determine the plaintiff's claim not according to strict statutory law but according to the principle of good faith. In this way there arose a group of good faith actions which materially modified the law of contract. Another was the extension of the civil law to circumstances to which it was not literally applicable, by allowing a formula which directed the judge to assume that certain conditions, normally necessary for the civil claim concerned, were actually satisfied (*formulae ficticiae*). The praetors thus obtained the key position in every action, being able, doubtless with the advice of jurist friends, to settle the issue which the judge hearing the evidence had to have in mind and to answer.

In the post-classical law the formulary system was replaced by the *cognitio extraordinaria* (q.v.) procedure.

Fornication (Lat. *fornix*, a brothel). Sexual intercourse between two unmarried persons (as distinct from adultery), who are not closely related (when it is incest) and consensual (as distinct from rape). Originally the court leet imposed fines and under the Commonwealth it was punishable by imprisonment. The ecclesiastical courts can impose penance but this power is in desuetude, save in the case of clergymen. Fornication is not now criminal unless the female is under 16, or it is committed in a public place and involves indecency or brothel-keeping.

See also PROSTITUTION.

Forster, John (1667–1720). Called to the Irish Bar, was Solicitor-General (1709) and Attorney-General in 1709–11, and in 1714 became Chief Justice of the Common Pleas in Ireland.

Fortescue, Sir John (?1394–?1476). Became a serjeant in 1429 or 1430. In 1442 he was made Chief Justice of the King's Bench and knighted. In 1461 he was attainted for adherence to the house of Lancaster and had to leave England probably being Lord Chancellor of the government in exile. His attainder was reversed in 1471, after he had retracted arguments published by him when in exile. He wrote a *De Natura Legis Naturae*, in Latin, in support of the claim of the house of Lancaster, 1461–63, and, more importantly, *De Laudibus Legum Angliae*, written in Latin for the instruction of Edward, Prince of Wales, while in exile about 1470. It takes the form of a dialogue between Fortescue and the prince, in which the latter is urged to become acquainted with the laws of England, and the English legal system is sketched and eulogized. It was popular and has been repeatedly edited. Other works include a treatise on *The Governance of England*, first published in 1714.

It is a small book, developing the comparative merits of despotic and constitutional government by contrasting contemporary France and England. It was edited by Plummer in 1885.

C. Plummer, in edition of *Governance of England* (1885); S. B. Chrimes, in edition of *De Legibus* (1942).

Fortescue, William (1687–1749). Called to the Bar in 1715 he was a friend of Gay, Pope, and Swift and Secretary to Walpole. He became a Baron of Exchequer in 1736, a judge of the Common Pleas in 1738, and was Master of the Rolls from 1741 till his death. He was a sound but ordinary judge.

Forty days. A period frequently mentioned in the Bible. Moses was 40 days in the mountain; the rains of the flood fell for 40 days; Noah opened the window of the ark after 40 days; Elijah was fed for 40 days by the ravens; Nineveh had 40 days to repent; Jesus fasted 40 days; He was seen 40 days after the resurrection. Similarly alchemists looked on 40 days as the charmed period when the philosopher's stone and the elixir of life were to appear. Superstition attached to 40 days; St. Swithin's day foreshadowed 40 days' rain or fine weather. Hence 40 days came to be treated as a standard, almost sacred, period and came to be incorporated in many legal prescriptions. In the Mosaic law the Jews were forbidden to inflict more than 40 lashes on an offender. Thus, at common law, 40 days was the limit for the payment of a fine for manslaughter; sanctuary lasted for 40 days; a widow might remain 40 days in her husband's house after his death; the King required 40 days' service of his feudal tenants and they 40 days from their tenants; after 40 days a stranger had to be enrolled in a tithing; Members of Parliament were protected from civil arrest for 40 days before the House sat and 40 days after it was prorogued; a new-made burgess had to forfeit 40 pence unless he built a house within 40 days. In Scotland the sheriff court has jurisdiction if the defender has resided 40 days and has ceased for less than 40 days to reside there.

Forty shilling freeholder. A category of small landholders in England who became qualified for the Parliamentary franchise in the English counties from about 1430 onwards. Their importance disappeared with the nineteenth-century extensions of the franchise.

Forum. The name of several open places in ancient Rome, of which the most important, the Forum Romanum, was the location of, *inter alia*, the law courts. Hence the term is used of the court or area of jurisdiction appropriate for particular purposes, as in the phrases *forum non conveniens* (an inappropriate court or jurisdiction), *lex fori* (law of the forum or court in which the proceedings are taken), and *actor sequitur forum rei* (the plaintiff follows or goes to the forum of the defendant). A *forum domesticum* is one which decides matters internal to the body which established it, such as matters of professional discipline.

Foss, Edward (1787–1870). Solicitor, a founder and later President of the Incorporated Law Society, 1842–43, Fellow of the Society of Antiquaries, and member of Council of the Camden Society. He collected materials for Lord Campbell's *Lives of the Chancellors*, and wrote *The Grandeur of the Law or the Legal Peers of England*. His major work is the *Judges of England*, in nine volumes, 1848–64, not an entertaining or lively work, but one solidly based on authorities, accurate and dispassionate, giving, under each reign, the biographies of chancellors, masters of the rolls, and judges of the courts and much other biographical information. It is a work of reference of the utmost value for the legal historian. He also produced *Tabulae Curiales*, Tables of the Superior courts showing the judges who sat in each from 1066 to 1864 (1865) and *Biographia Juridica*, a biographical dictionary of the judges from 1066 to 1870 (1870), based on his *Judges* and various papers on antiquarian topics.

Memoir, prefixed to *Biographical Dictionary*.

***Foss* v. *Harbottle* (1843) 2 Hare 461, rule in.** A member of a company took proceedings against the directors to compel them to make good losses sustained by the company by reason of their fraud. Held that the action must fail as the company was the proper plaintiff. The rule accordingly established is that the proper plaintiff in an action in respect of a wrong done to the company is the company and that no individual can bring an action where the alleged wrong is a transaction which might be made binding on the company by a majority of members.

Foster, Anthony (1705–1779). Called to the Irish Bar in 1732, he became Chief Baron of Exchequer in Ireland in 1766, retiring in 1777.

Foster, Sir Michael (1689–1763). Called to the Bar in 1713, he became recorder of Bristol in 1735 and a judge of the King's Bench in 1745. He was recognized as a learned, impartial, and independent judge, especially in criminal law. He is best remembered for his collection of reports published in 1762, to which he annexed learned notes and four discourses on particular topics of criminal law, *A Report of some Proceedings on the Commission of Oyer and Terminer and Gaol Delivery for the Trial of the Rebels in the year 1746 in the County of Surrey, and of other Crown Cases: To which are added Discourses upon a few Branches of the Crown Law*. It is known as

Foster's *Crown Cases* and Foster's *Crown Law*. The clarity of his discussion of the principles involved has made the book a classic on its subject, though it is limited to high treason, homicide, and other capital offences.

J. Dodson, *Life of Sir Michael Foster*, 1811; Wallace, *Reporters*, 440.

Foster, Robert (?1589–1663). Called to the Bar in 1610, he became a judge of the Common Pleas in 1640, an office to which he was restored at the Restoration. Shortly thereafter he was promoted Chief Justice of the King's Bench, 1660–63, in which capacity he was much engaged in the trial of Fifth Monarchy men and other conspirators.

Fostering. Caring for a child by a person who is neither a parent by blood nor by adoption. In modern times the commonest instances of fostering are where local authorities under statutory powers assume parental rights over children orphaned, abandoned, or neglected by parents, and delegate the actual care of the child to persons whom they pay to look after the child along with their own or in the same way as their own. Fostering may be short-term, covering a period of home difficulties, or long-term.

Foulis, Sir James, of Colinton (?–1688). Sat in the Scottish Parliament in 1645–48 and in 1651; he was appointed a Lord of Session in 1661, a Lord of Justiciary in 1671, and a member of the Scottish Privy Council in 1674. In 1684 he became Lord Justice-Clerk and held that office till his death. He had a son who became a Scottish judge as Lord Reidford in 1674.

Foundation. The establishing of an institution, particularly one for educational or charitable purposes, by endowing it with funds for its continued existence, and consequently an institution thus established with an endowment fund.

Founders' shares. Shares issued, normally fully paid up, and used to recompense the founders or promoters of a company. They were normally deferred to preference and ordinary shares, not carrying any dividend until ordinary shares had received a stated rate of dividend. They are now almost obsolete.

Four Burghs, Court of the. The four burghs were originally Berwick, Roxburgh, Edinburgh, and Stirling, the first two being later replaced by Lanark and Linlithgow. By the fourteenth century this court, consisting of the Chamberlain of Scotland and four burgesses from each of the four burghs, met annually at Haddington and heard appeals from the Chamberlain ayre (q.v.). Its judicial powers remained though it tended to develop into a consultative body for all the burghs. As a court it disappeared shortly after 1500, but as a consultative body it continued, and in 1552 it became the Convention of Royal Burghs, a mainly legislative and executive body, which as a body for joint consideration of matters of common concern to Scottish burghs still exists.

Four Doctors. The name for the four twelfth-century Roman law scholars, Bulgarus, Martinus, Jacobus, and Hugo, who succeeded Irnerius as leaders of the school of glossators.

Four freedoms. Four aspects of freedom mentioned in a declaration of objectives made by President F. D. Roosevelt in a message to Congress urging support for the Lend Lease Bill during World War II, namely freedom from want, freedom of worship, freedom of speech and expression, and freedom from fear.

Fourteen Points. A statement by President Wilson in an address to Congress on 8 January 1918 of the essential bases for a peace settlement after World War I.

The points were: (1) open covenants openly arrived at; (2) freedom of the seas; (3) removal of trade barriers; (4) reduction of armaments; (5) adjustment of colonial claims with the interest of the populations weighing equally with that of the claiments; (6) evacuation of Russian territory and independent determination by Russia of her own national policy; (7) evacuation and restoration of Belgium; (8) evacuation and restoration of France, and the return of Alsace–Lorraine; (9, 10 and 11) readjustment of Austro–Hungarian, Italian, and Balkan frontiers along historically established lines of nationality; (12) freedom of the Dardanelles and self-determination for the peoples under Turkish rule; (13) an independent Poland, with free access to the sea; and (14) establishment of a general association of nations under specific covenants.

Their formulation aroused great enthusiasm in the world and gave President Wilson a leading position among the leaders on the allied side, but the Paris Peace Conference and the Treaty of Versailles did not achieve all of them, while Congress refused to ratify the Treaty and U.S. membership of the League of Nations, formed in pursuance of the fourteenth point.

Fourteenth Amendment. This amendment to the U.S. Constitution, ratified in 1868, was intended to provide means for securing the natural rights and liberties of citizens from infringement by states. Its origins were in the emancipation of the slaves which developed into a movement to secure civil rights generally. It is largely declaratory of rights, in the form of prohibitions against depriving any person of life, liberty, or property without due process of law, or denying anyone the equal protection of the laws, and specifically limits the powers of the States in the same way as the federal government is limited by the first eight amendments. It has been repeatedly construed by the Supreme Court and its modern significance flows from judicial interpretation. In the *Slaughter House* cases (83 U.S. 36) the Court gave it a narrow interpretation, limiting it to the protection of negroes. In 1896 in *Plessey* v. *Ferguson* (163 U.S. 537) the Court sanctioned the provision of equal but separate provision for education, and this permitted state segregation statutes for 50 years. But in the *School Segregation Cases* (347 U.S. 483) the Court overruled *Plessey* v. *Ferguson* and required not legal equality but actual equality.

Fourth estate. The daily press (the other three being the Lords Spiritual, the Lords Temporal and the Commons). Edmund Burke is credited with first referring to the reporters' gallery in the House of Commons as the fourth estate.

Fox's Libel Act. In the eighteenth century criticism and political discussion were limited by the law of criminal libel, not least by the rule that in a prosecution for libel the jury had to pronounce on the fact of publication or not and the truth or falsity of the innuendoes (q.v.), but not on the question whether a paper published with the meaning alleged by the prosecution was in law a libel. The view that the jury had a right to pronounce on libel or not was urged from 1731 in such cases as *R.* v. *Almon* (1770), 20 S.T. 870; *R.* v. *Miller and Woodfall* (1770), 20 S.T. 895; *R.* v. *Horne Tooke* (1777), 20 S.T. 651; and, above all, in the *Dean of St. Asaph's Case* (*R.* v. *Shipley* (1783), 21 S.T. 847) where the jury found the accused guilty of publishing only. Lord Mansfield upheld the view that the jury was not entitled to pronounce on the issue of libel or no libel, this being reserved to the judge; this view was probably historically correct but out of accord with the political ideas of the day. But other lawyers held the opposite view. After various attempts to legislate, Fox's Act of 1792 empowered the jury to give a general verdict of guilty or not guilty on the whole matter in issue, and provided that if the jury found the defendant guilty, he might still move the court in arrest of judgment on the ground that the writing is not a libel. The Act did not change the law of libel, but proved an important safeguard for free political discussion in entitling the jury to decide not merely whether the accused had the intention to publish what the law regards as seditious but whether he published with seditious intention.

Franchise. (1) A right held by royal grant. In the early Middle Ages jurisdiction had in many cases been assumed by private persons. In 1274 Edward I enquired into franchise jurisdictions, and sent commissioners to enquire by what warrant different landowners were exercising such rights. Franchises could be established if granted by charter, by prescription, or by proof that the grantee had paid rent to the King for it. The rule that no franchise could exist save by royal grant became law, and the general eyre exercised strict control over franchises. The most important franchise jurisdictions were those of the King over the royal forests, those of landowners, boroughs, guilds of merchants and craftsmen, of bodies of commissioners to superintend matters related to commerce or industry, and of members of particular professions and groups.

Franchises enjoyed by landowners varied from the wide-ranging powers of Lords Marchers and lords of counties palatine. Lesser franchises included the rights to hold fairs, to appoint coroners, to hold the hundred court, to hold the court leet, and to hold the view of frankpledge. The lesser franchises became of less importance and they became unpopular, as the common law courts, the judges of assize and justices of the peace became better known and more efficient.

Franchise courts is accordingly a generic term for courts held by individuals by virtue of royal grant, such as courts of manors or boroughs, as distinct from courts held by virtue of feudal jurisdictional powers, where the jurisdiction depended on tenure.

(2) The right to vote at elections. In the United Kingdom the franchise now extends to all persons of either sex who are British subjects or citizens of the Republic of Ireland, are resident in a Parliamentary constituency, have attained 18 or will attain 18 within a year of the publication of the register of electors, and are registered as electors for the district in which they may vote.

The franchise does not extend to aliens (not including Irish citizens), minors, peers (except peers who have disclaimed their hereditary peerages), persons of unsound mind, persons convicted and detained in a penal institution and persons convicted of corrupt or illegal practices in connection with elections. Nor does it extend to legal persons such as companies notwithstanding that they are very much affected by the law made or changed by the representatives elected and are heavily taxed, more so than very many of the natural persons who do have the right of franchise.

Franchise courts. Courts held by private persons by virtue of a franchise or grant from the King, as distinct from courts held by the King's representative. The most permanently important was that of the borough, but others were those of maritime and commercial courts, and of commissioners superintending various aspects of industry and commerce. Franchise jurisdictions attained very large dimensions in the Palatine counties. Edward I found that the assumption of private jurisdiction had gone so far that in 1274 he sent commissioners to enquire into usurpations of royal rights, on the basis of which he founded his *Quo Warranto* enquiries. Franchise-holders could show title by charter, by proof of having paid rent to the Exchequer for the franchise, or by prescription under the statutes of Gloucester of 1290, which provided that possession without interruption since Richard I's reign should confer a good title.

Frank, Jerome New (1889–1957). U.S. jurist, held high office in administrative agencies, was Chairman of the Securities Exchange Commission and a judge of the U.S. Court of Appeals, (1941–57). He wrote *Law and the Modern Mind* (1930), *If Men were Angels* (1942), *Courts on Trial* (1949), and *Not Guilty* (1957). Basically a legal realist he developed the concept of fact-scepticism or continual questioning of the factual assumptions so as to expose the uncertainties of the judicial process, which led him to doubt the reality of justice.

Frankalmoin (Norman fr. *fraunche aumoyne*, free alms). The feudal tenure in England by which an ecclesiastical body held land, in return for saying prayers and masses for the soul of the granter. Not only was secular service frequently not due but in the twelfth and thirteenth centuries jurisdiction over land so held belonged to the ecclesiastical courts. It fell into disuse because on any alienation of the land the tenure was converted into socage, and no fresh grants in frankalmoin, save by the Crown, were possible after *Quia Emptores* (1290). This tenure was so uncommon by 1660 that it was not formally abolished then. In 1925 the tenure was converted into common socage.

Frankfurter, Felix (1882–1965). Born in Vienna, he came to the U.S. in 1894, was admitted to the Bar and served as an assistant district attorney, 1906–09, and law officer of the War Department's Bureau of Insular Affairs, 1910–14, before becoming a professor at Harvard, 1914–39. He also acted as assistant to the Secretary of War and was Chairman of the War Labor Policies Board in 1918 and was long a friend and adviser of President F. D. Roosevelt. He was then an Associate Justice of the U.S. Supreme Court, 1939–62. He was a strong advocate of civil liberties but favoured judicial self-restraint and non-intervention by the Court in political issues.

His publications include *The Case of Sacco and Vanzetti* (1929); *The Business of the Supreme Court* (1928); *The Public and its Government* (1930); *Mr. Justice Holmes and the Supreme Court* (1938); and *Of Law and Men* (1956).

H. S. Thomas, *Felix Frankfurter: Scholar on the Bench*; L. Baker, *Felix Frankfurter.*

Franking of letters. The privilege, claimed by both Houses of Parliament in 1660, of sending letters by post free of charge on merely signing them on the outside. The privilege was recognized by statute in 1763 but restricted, but members frequently gave friends envelopes signed by them for use at their convenience. The privilege was further restricted in 1837 and abolished in 1840, but members of either House can send unstamped (official prepaid) letters from the Houses of Parliament. In the U.S. the privilege of franking was originally granted to soldiers fighting in the War of Independence and later extended, but it was subsequently restricted to congressmen and government officials.

Frankmarriage. Land given to a man and his wife by the father or a blood relative of the wife 'in frankmarriage' gave the land to them and their issue, despite the absence of reference to heirs or issue in the gift. They and their issue to the fourth degree had the land free of services to the donor but had to bring it into hotchpot before taking any other land by descent in fee simple from the donor.

See also MARITAGIUM.

Frankpledge. A system in Norman England, replacing the local responsibility of the tithing, or group of 10 men, for producing an offender, whereby every freeman over 12 years of age had to be enrolled in a friborh, or frankpledge, or tenmannetale, an association of 10 men who formed collective bail for the appearance of any one of them before a court, when necessary. If an accused man was condemned, he had to make reparation from his own property, but if he fled, the other members of the tithing were liable, unless they could exculpate themselves from all part in his crime or escape. Twice a year the hundred court had the duty of seeing that every man was duly enrolled in his tithing and this continued, as the sheriffs tourn, or the two occasions in the year when the sheriff held courts to take a view of frankpledge. In many cases, too, the owner of the hundred had the right of view of frankpledge to the exclusion of the sheriff, and for this purpose held a court leet.

W. Morris, *The Frankpledge System.*

Franks Committee. See COMMITTEE ON ADMINISTRATIVE TRIBUNALS AND ENQUIRIES.

Frantzke, Georg (1594–1659). A German jurist and administrator who wrote commentaries on the

Digest (1644), *Institutes* (1658), and other works on law.

Fraser, Patrick, Lord Fraser (1819–89). Called to the Scottish Bar in 1843 and rose rapidly in the profession. He became Dean of the Faculty of Advocates in 1878. He was a judge of the Court of Session as Lord Fraser, from 1880 till his death. He was a legal scholar of considerable repute. His first book, *The Law of Personal and Domestic Relations* (1846) was later split into three works, *The Law of Parent and Child* (1866), *The Law of Husband and Wife* (1876), and the *Law of Master and Servant*, of which the first two remained authoritative for about a century.

Fraser, Walter Ian Reid, Lord Fraser of Tullybelton (1911–). Called to the Scottish Bar in 1936, became Q.C., 1951, Dean of the Faculty of Advocates, 1959, a judge of the Court of Session, 1964–75, and a Lord of Appeal in Ordinary in 1975.

Fraternity or co-operation. The third (along with liberty and equality) of the qualities or principles commonly deemed an element of justice; unlike the others it is a psychological fact rather than a principle, and connotes rather solidarity, community spirit or brotherhood, mutual aid, and communal provision for common needs. The legal implications of the quality include recognition of the function of the State or society to provide and maintain systems of external defence, and internal law and order and, increasingly, a wide range of social services.

Fraud. In civil law fraud is a misrepresentation or contrivance to deceive, commonly by way of a statement knowingly made falsely, or without honest belief in its truth, or recklessly, careless whether it be true or false, and intended to be, and in fact, relied on, by the person deceived, but it may equally be made by concealment, or deliberate omission to make a statement where one should have been made, or by actings. If a contract is induced by misrepresentation made fraudulently, the contract may be held totally void, but is at least voidable at the instance of the party deceived, he also having a claim of damages for loss sustained. If the fraud had resulted in detriment other than by entering into a contract, the party deceived may treat it as the tort of deceit (in Scotland the delict of fraud) and claim damages for loss resulting from the fraud. Equity (q.v.) gave an extended meaning to fraud, extending it to any conduct less than fair and honourable. Constructive fraud (q.v.) covered many cases of failure to make full and frank disclosure.

In criminal law fraud is deliberate dishonesty and an element in many crimes, such as cheating and obtaining by false pretences.

In Scots law fraud was formerly known as 'falsehood, fraud and wilful imposition' and is a very broad crime extending to any bringing about of a practical result by false pretences.

Apart from common law fraud there are many kinds of statutory frauds. Dispositions of property in fraud of creditors are challengeable in England under statute of 1571, 1584, and 1579, now superseded by the Law of Property Act, 1925, and under bankruptcy legislation, and in Scotland under statutes of 1621 and 1696 and under bankruptcy legislation.

Fraud, Constructive or equitable. At common law in England fraud was limited to actual deceit. Equity, however, gave a more extended meaning to the concept of fraud, by developing the doctrine that if one man enters into a transaction with another in circumstances in which the latter is not a free agent and not fully able to look after his own interests, the latter is entitled to the protection of the court and may rescind the transaction. This principle applies where a confidential or fiduciary relationship exists between the parties, where a party has placed himself in such a position that it has become his duty to act fairly towards and have regard to the interests of the other party, where the bargain appears clearly unconscionable and inequitable, and where the principle of undue influence applies. This is to the effect that if one party obtains benefit from another, by contract or gift, by exerting an influence over him which, in the view of the court, prevented him from exercising an independent judgment, the latter may set aside the contract or recover the gift. Undue influence is presumed in many cases where a confidential relationship exists between the parties, and in any other case may be proved.

Frauds, Statute of. An Act of 1677 intended to prevent frauds and perjuries by making written or other adequate evidence essential to certain transactions, notably conveyances of interests in land, wills of real estate, declarations or assignments of trust, and certain classes of contracts, which were declared unenforceable by action unless evidenced by a note or memorandum in writing signed by the party to be charged therewith. The Act was necessary partly because jury trial was in a transition stage and the rules of evidence were rudimentary and excluded the evidence of parties. It has given rise to much litigation and interpretation. It was copied in Ireland and many British colonies but never applied in Scotland. Since the early nineteenth century much of the Act has been repealed, but some of it has been replaced by modern legislation.

Fraudulent misrepresentation. A misstatement by act or concealment, made with intent to deceive. It gives rise to an action of damages for deceit provided it was in the circumstances material, induced the plaintiff to alter his position to his detriment, and caused him actual or temporal damage or loss, most commonly loss of money or loss arising out of a contract entered into on the faith of the representation. A contract induced by fraudulent misrepresentation is voidable and may be annulled by a court on proof of the misrepresentation.

Fraudulent preference. A payment or transfer by a person about to go bankrupt to a creditor rather than surrender of the money or goods to his trustee for behoof of the general body of creditors and which is accordingly a fraud on the other creditors. The concept is now defined by statute and affects transactions where the debtor was unable to pay from his own money his debts as they fell due and is adjudged bankrupt on a bankruptcy petition presented within six months after the transaction sought to be impeached. It must have been done voluntarily and not under pressure.

Free Church case. In 1843 a substantial body of the ministers and members of the (Presbyterian) Church of Scotland left that Church and formed the Free Church of Scotland. In 1900 the majority of the Free Church decided to unite with the United Presbyterian Church to form the United Free Church. The minority, however, elected to carry on the Free Church and claimed to be the true Free Church, adhering to the fundamental tenets of that Church and, as such, entitled to all the property held by and for the Free Church. The House of Lords, reversing the Court of Session, by a majority, upheld their claim (*Free Church* v. *Lord Overtoun*, [1904] A.C. 515) on the basis of their adherence to the trust purposes. The result, legally right or wrong, was absurd in awarding a large amount of property to a minority of the Church, and Parliament had to intervene to resolve the situation by appointing a statutory commission to allocate the disputed properties according to rules laid down in the Act. The minority in the result was treated fairly and even generously. The decision of the Lords was probably legally right in that the majority had abandoned what had been a fundamental principle on which the Free Church had been based when founded in 1843.

Free movement of goods. One of the foundations of the E.E.C. is the principle that to ensure that goods may move freely economic barriers between countries of the Community shall not be erected and customs duties on imports and exports shall be prohibited. Quantitative restrictions on imports and exports are also prohibited.

Free movement of persons, services, and capital. One of the basic freedoms essential for the E.E.C. is that member states are bound not to permit obstacles to the free movement of any of these categories. Manual workers are free to move and self-employed persons are entitled to establish themselves in any member state without discrimination.

Free scientific research (*libre recherche scientifique*). The term for an approach to the interpretation of legislation advocated by Gény (q.v.) in his *Méthodes d'interprétation et sources en droit privé positif* (1919), whereby he sought to have French lawyers appreciate that lacunae in statutory provisions must be filled by use of reasoning other than the interpretation of texts. It encouraged judges to take into account a wider range of factors than they had previously done, but did not make them acknowledge this openly.

Free will. The philosophic concept of the power or capacity in man to choose between alternatives and to act in particular contexts on the basis of his reason and not wholly determined by restraints. The concept figures extensively in philosophic and theological argument, but is important also in some theories of law. Thus Kant's basic tenet in the field of practical reason, which includes morals and law, is the freedom of human volition; he stated his categorical imperative in terms of the harmonization of the conduct of each free-willing individual with that of all others, and this has been developed by later jurists such as Stammler.

From a practical standpoint free will is assumed by many basic concepts and rules of law. Responsibility for crimes or torts or other actions, and liability to punishment or for damages, imply the assumption that the individual could exercise his will, and could have avoided doing what he did; if he is not a free-willing individual, it is not just to hold a man responsible or legally liable. Hence one of unsound mind, or in a fit, cannot be held responsible. The concept underlies the distinction between voluntary and involuntary conduct. But it is also obvious that in the context of a system of law a person is rarely wholly free, in that he knows that in many circumstances the choice of one of the possible courses of conduct will result in some unpleasant legal sanction. And in other cases, e.g. social or economic reasons, a many may not be free to do what he wishes.

Freebench. A widow's interest due from copyholds (q.v.) to which she was entitled by the custom of some manors, corresponding to dower (q.v.) from freeholds. The quantity, quality, and duration of the estate which the widow was entitled to take depended on the custom of the particular manor; by analogy with common law dower it was usually a third for her life, but might be more or less. In most manors the right was subject to any alienation of the land made by her husband, *inter vivos*, or by will, but was superior to her debts. It was enforceable by an action similar to an action of dower. It has now been abolished. In some manors the widower of a woman copyholder had a customary estate in his wife's lands called 'man's freebench' or customary curtesy, analogous to an estate by the curtesy (q.v.) in freeholds. Similarly there was a custom in some boroughs called frankbank, whereby the wife had as dower all the tenements which belonged to her husband. Freebench disappeared with copyhold in 1925.

Freedmen. Emancipated slaves. In Greece freed slaves ranked with resident foreigners, not citizens, and most remained in humble positions, only a few rising to positions of wealth or influence.

In Roman law a slave by manumission became a citizen, but did not have full civil rights, being unable to serve in the army or to hold public office; his son was, however, a full citizen. Many freedmen remained in humble positions, but under Augustus freedmen became state officials and high civil servants.

In mediaeval Europe the position of freedmen varied. Under Charlemagne a freedman's descendants could claim the rights of a freeborn subject only after three generations, and elsewhere a freedman was usually not on an equality with freeborn subjects. In the U.S. racial and social differences still distinguish the descendants of freed slaves from other citizens, though nominally and legally there is no distinction.

Freedom, academic. The liberty of scholars and teachers in every kind of research, scholarly, or teaching institution, to seek truth in their field of investigation and to communicate it to others, however unpalatable that may be to government or church or to the authorities governing the institution in question, and not to have to modify findings or views to accord with governmental, religious, or other orthodoxies. In the Middle Ages, academic and doctrinal censorship was not uncommon. Theodoric of Chartres was summoned before a council to answer for the theological consequences of his cosmology. The Council of Paris in 1210 prohibited the teaching of Aristotle's works on natural philosophy. The University of Paris in 1215 prohibited certain of Aristotle's works together with the works of three heretics and these prohibitions were repeated periodically. The mediaeval European universities laid the foundations of academic and intellectual freedom, but faculties regularly condemned or expurgated the writings of their colleagues. Teaching was also controlled by regulation of university booksellers. From the fifteenth century the major control was the paper condemnation of various works and the establishment of the Index of Forbidden Books. Academic freedom is today generally recognized in Western liberal democratic countries, though some teachers utilize it to indoctrinate students by presenting socially and politically biased versions of facts and events. In these countries it is not a freedom specifically secured or protected by law, but is merely an aspect of the general freedom of thought and expression.

Freedom, intellectual. The liberty of an individual to inquire, think, and reach conclusions without interference or dictation, and to publish and communicate his thoughts or findings for the information or pleasure of other people. A particular, but important, branch of intellectual freedom is academic freedom (q.v.). At many times and in many societies intellectual freedom has been sought to be controlled, repressed, or limited by political or religious authorities. In Western Europe in modern times the most far-reaching religious system of restriction was that operated by the Catholic Church through the Index of Forbidden Books and the Congregation of the Index, while most states operated systems of censorship and licensing. In England pre-publication censorship of books ended in 1695, but controls by the law of sedition, blasphemy, obscenity, and defamation continued. In the U.S. though the First Amendment (1791) prohibited Congress from making any law abridging freedom of speech or the Press, the law of sedition, obscenity, and defamation had controlling force, supported in many cases by organized movements of public opinion. In modern Western liberal-democracies the restrictions on intellectual freedom are accordingly represented by limited liability to prosecution or to civil action for expressing ideas which are seditious, obscene, or defamatory; non-conformity or blasphemy are practically obsolete as offences. It is far otherwise, however, in totalitarian regimes where there is no freedom for thought or expression contradictory or even critical of the regime.

See also CENSORSHIP.

Freedom, personal. One of the major civil liberties, connoting freedom to come and go, to participate or to abstain, and generally to do as one pleases so long as what one does does not infringe some rule of positive law. A material aspect of personal freedom is the right of privacy, the right to be left alone.

Another very important aspect is the right not to be arrested or imprisoned save on defined and limited grounds, a liberty which underlies much of Magna Carta, habeas corpus, and many rules of criminal law and procedure. It is the belief in this liberty which underlies opposition to excessive police and governmental powers, to concentration camps and secret police, to vaguely defined offences and retrospective provisions.

Some restrictions on personal freedom are necessary and these increase as society becomes more developed and has more regard for the protection of others; there can be no liberty to drive a vehicle while under the influence of drink or drugs to the danger of others. One cannot maltreat animals, or use one's property in an anti-social way.

Freedom, political. The rights and liberties of the individual within the politically organized community to participate in running its affairs. It comprises two distinguishable notions, the right to participate in government at all levels, whether as candidate or member, or merely as elector, and the right of both electors and candidates to be free from intimidation and compulsion and threats of detrimental consequences in urging policies and seeking votes, and on the other hand in weighing-up competing claims and casting votes. It accordingly normally implies voting by secret ballot, and elements of the liberties of association, assembly, free speech, and freedom of the Press.

Freedom from arrest or molestation. One of the privileges claimed for members of the House of Commons by the Speaker at the commencement of each Parliament. It extends to freedom from arrest in all civil actions during the time of Parliament and during the times of travelling to or from Parliament. This is practically valueless as imprisonment in civil process has been practically abolished since 1838 and for debt since 1869. It does not extend to exemption from criminal justice or emergency legislation, but the Speaker must be informed of the arrest and the cause, and of any sentence for a criminal offence. The privilege extends to illegal molestation as well as the legal process of arrest. A similar freedom is claimed for officers of either House, witnesses summoned to attend before either House and others in personal attendance on the business of Parliament. Akin to this freedom are the privileges of M.P.s from being summoned as a witness, from having to serve on a jury.

Freedom from discrimination. A kind of civil liberty only fully recognized in the twentieth century under the influence of the U.N. Universal Declaration of Human Rights and the European Convention on Human Rights (qq.v.). Discrimination by employers and providers of goods or services may exist on many grounds, notably age, sex, race, or colour. Discrimination in employment on grounds of sex has been in general outlawed by the Sex Disqualification (Removal) Act, 1919, the Equal Pay Act, 1970, and the Sex Discrimination Act 1975. More generally it is illegal under the Race Relations Act, 1976, to discriminate against a person on the ground of colour, race, ethnic, or national origin in many respects. Discrimination is in some cases still justifiable on specific and limited grounds.

Freedom of access. At the commencement of every Parliament it is customary for the Speaker to claim for members, *inter alia*, freedom of access to Her Majesty whenever occasion shall require. This privilege has been claimed since at least 1536 and was regularly claimed from the end of the sixteenth century. It is exercised by the House as a body through the Speaker. In Parliament it is done by attending the Queen on summons to the House of Lords for purposes prescribed by the Queen. Out of Parliament the right of access is exercised for the purpose of presenting addresses, which may deal with any matter of public policy. The privilege does not attach to individual members. Ministers, particularly the Prime Minister, and formerly Privy Councillors, have a right of individual access.

Freedom of assembly. A major civil liberty. The basic proposition is that a number of people may legitimately assemble together for any lawful purpose, but to do so on private premises requires the invitation or consent of the owner or occupier of the premises. To do so on public open spaces is in many cases lawful only with the consent of a local authority, the police or some other person or body empowered to give or withhold permission. The fact that many members of the public may be at or in a place, such as Trafalgar Square, does not entitle any one to use it for a public meeting. To do so on a public highway is a trespass against the owner (usually the local authority) unless it has been authorized and obstruction of the highway is also a public nuisance. The police may stop or disperse an assembly if it seems liable to cause a breach of the peace. Similarly, a procession so substantial as to be an obstruction of the ordinary use of a highway or to be a public nuisance requires prior authority and may be prohibited if liable to give rise to serious public disorder. The law on the matter has to reconcile possible conflicting interests, in expression of views and in public order, and it is not always clear or certain.

Freedom of association. One of the major civil liberties. It involves the liberty to join with others for the pursuit or furtherance of any of a wide variety of ends, social, artistic, literary, scientific, cultural, political, religious, or other. The crucial question in every society is whether or how far the governing body of the society will permit associations for particular purposes to exist and to seek to further their purposes. Prior to the Reformation associations promoting religious views contrary to those of the established Church were generally repressed. In modern times the problem is whether a democratic society can allow to exist Communist or Fascist or other organizations which have as objects the overthrow of the democratic society. Trade unions were unlawful associations at common law as being in restraint of trade and of men's liberties to sell their labour at whatever price they would, but they were legalized by statute in 1871. Since then freedom to associate in trade unions for the purpose of collective bargaining has been fully recognized.

It remains criminal under the Public Order Act, 1936, to take part in the management of a body organized to usurp the function of the police or armed forces or to display force in promoting any political object.

The forms which associations may take are very varied, including unincorporated associations, friendly societies, trade unions, chartered societies, and bodies incorporated as companies, industrial and provident societies, or building societies. Subject in many cases to complying with the guide-lines laid down by statute, associations have wide freedom of promoting their objects and managing their own affairs, free from governmental interference or control. They are subject to the general restrictions on liberty imposed by the law of sedition, libel, malicious damage, and criminal conspiracy, but not to special restrictions.

An aspect of freedom of association is freedom of meeting, to meet in private, or in public, for purposes connected with the objects of the association, to propagate its aims and win support.

See Freedom of Assembly.

Freedom of belief and opinion. A liberty which may be thought to be secured by any legal system at least so far as concerns the holding of the beliefs and opinions, since the mere holding can never be interfered with by legal process. But beliefs and opinions are almost invariably expressed by conduct or words, food, ceremonies, dress, membership of groups, and in other ways, in which case they may be checked on such grounds as being seditious or treasonable, or blasphemous, or immoral, or defamatory. In the past the expression and even the holding of beliefs and opinions, particularly on religious matters, was frequently denied and sought to be repressed by massacres, persecution, and every form of tyranny, while the removal of the legal disabilities of non-conformists, Catholics, Jews, atheists, has been more recent. In modern Britain beliefs and opinions are unchecked, but conduct expressive of beliefs and opinions may be restrained if treasonable or seditious, defamatory, a public nuisance, or conducive to breach of public order.

Freedom of communication. One of the fundamental liberties of individuals is to have freedom to communicate information and comments orally, in newspapers and books (see Freedom of the Press), on radio and television, and to receive information and comments from others by the same means. This implies permitting the dissemination of seditious, subversive, immoral, and other undesirable matter.

The extent to which this freedom exists in different countries has varied, and varies. Nowhere has it been nor can it be completely uncontrolled. The laws of sedition, libel and contempt everywhere impose some limitation. In Britain the main limitation is on the advocacy of, or incitement to, violent or unconstitutional changes in the social and political structure of society.

Freedom of conscience. One of the major civil liberties, the right to hold what principles of conduct one pleases and to live accordingly. It is closely connected with, but not identical with, freedom of religion, because conscience, belief in the rightness or wrongness of conduct, is distinct from religious belief. It is also connected with, but distinct from, freedom of assembly and association, and freedom of speech and expression.

The only instances in which conscience and beliefs based thereon come sharply in contact with the law are where individuals assert a conscientious objection to serve in the military forces of the State or to join a trade union. In Britain in the former of these cases, if the individual could convince a tribunal set up to investigate such cases of the genuineness of his belief, he might be exempted from service. In Britain in the latter case the forced dismissal of an employee who will not join a trade union, where a closed shop applies, is fair if his refusal is based on religious belief, but not if merely based on conscientious grounds. But many countries do not permit the closed shop and compulsory union membership.

Freedom of contract. The general principle that individuals of full age and sound mind should have the utmost liberty to contract, or not, and as to the terms on which they contract, and that performance of their bargains will be enforced by the courts, save only in limited and exceptional cases. The principle has been long and widely accepted in the law of contracts generally as implying that a person could make whatever kind of contract he pleased and on whatever terms, and that the law would not interfere to protect him save on limited grounds, such as that he had been defrauded or otherwise taken advantage of. But since the mid-nineteenth century it has been recognized that in many cases individuals have no real freedom of contract, in that they have to choose between contracting, or not contracting, with a monopoly or a dominant supplier, or are presented with a standard set of contractual terms which they have not negotiated, and cannot reject or change, or are so weak in bargaining power compared with the other party that in effect they must take whatever terms are offered. These evils are being countered by compulsorily implied terms in contracts, statutory minimum standards, the restriction of exemption clauses, and the growth of trade union bargaining power on behalf of employees. On the other hand individuals have in some cases no choice but to contract with state monopolies, such as the Post Office, gas and electricity boards, and others on whatever terms the statutory monopoly chooses to impose.

Freedom of expression. The general principle of English law, that an individual may express any views at all without legal hindrance, is subject to a large number of restrictions and qualifications. There are various prior restraints such as the need to obtain a licence before being able to broadcast any material. There are liabilities to civil action or criminal prosecution if one expresses ideas which are defamatory, blasphemous or seditious, or publishes or possesses obscene literature. There is no general press censorship and prior censorship of theatrical productions by the Lord Chamberlain ended in 1968, but a non-statutory voluntary body, the British Board of Film Censors, grants certificates of various kinds to films, and local authorities may refuse to permit exhibition of a film even though given a certificate by the Board. The Official Secrets Act and civil service regulations restrict the communication of particular kinds of material.

There are also many practical, though not legal, obstacles to the expression of views of various kinds. Editors and printers of newspapers and journals may decline to publish a person's views and the police or public authorities may decline to permit meetings in public places. Freedom of expression may also be restricted in particular cases by express or implied terms of an individual's contract of employment.

Freedom of religion. One of the major civil liberties. This liberty denotes freedom to believe in and to practice the outward observances of any religious faith, or none. A question may, of course, arise whether certain beliefs are a faith or not. In mediaeval Europe freedom of religious belief was not in general tolerated, conformity to the Catholic Church being generally enforced both by church and state. The position was not materially changed by the Reformation as in general each ruler and state sought to impose conformity on the residents of that state, but everywhere there were groups of dissenters, frequently persecuted or at least subjected to disabilities.

In the U.S. freedom of religion was secured from independence and recognized by the Constitution. In the U.K. freedom was recognized from the eighteenth century, but disabilities attached to various groups of dissenters and full civil equality of believers in any faith, or none, was not achieved until the nineteenth century.

An important aspect of freedom of religion is freedom of meeting, in private or in public, to worship, propagate the aims of the religious group, teach children, win adherents, or otherwise, and religious freedom is incomplete if freedom of meeting for religious purposes is limited.

Freedom of speech and expression. One of the major civil liberties. It denotes the freedom to communicate orally, in writing, print, by broadcasting or otherwise, statements or views on any matter. It is qualified by the need to regard the interests of others, which are to some extent secured by the law of libel, contempt and other means, and the interests of the public which are sought to be secured by the prohibition of obscene publications.

Freedom of speech in debate is one of the privileges claimed for members of the House of Commons by the Speaker, at the commencement of every Parliament. This has been claimed since at least 1541 and was finally confirmed by Art. 9 of the Bill of Rights 1689, though the extent of the privilege is not wholly certain. From the principle several conclusions follow: One is that a Member must not enter into any contractual agreement with an outside body controlling or limiting his own or other Members' freedom. Another is that Members enjoy absolute privilege from liability for defamation for anything said in debate. On the other hand Members have frequently been punished by the House for offensive words. Yet another is the right to exclude strangers and debate in secret session, and another the right to control publication of their debates and proceedings, though in practice publication is tolerated.

Freedom of the borough. See FREEMEN.

Freedom of the Press. One of the major civil liberties. From the development of printing the authorities in England took steps to check the spread of sedition and heresy. A requirement that books be licensed for printing by the Privy Council or other royal officials was introduced in 1538. Mary Tudor forbade the printing of any book without her special licence and endeavoured to prevent the importation of the works of reformers.

In 1557 the Stationers' Company was incorporated and thereafter only members of that company or other persons holding a special patent might print any work for sale. The wardens of the company seem to have exercised some discretion in allowing printing. In 1559 Elizabeth issued Injunctions requiring that no book should be printed unless licensed by her, some of her Privy Council or certain church dignitaries; this seems to have been largely ineffective. In 1586 an ordinance of the Star Chamber directed that, apart from one in each of Oxford and Cambridge, no printing press should be set up in any place other than London and forbade the printing of unlicensed works. In 1558 Whitgift appointed 12 persons to licence books for printing, hoping thereby to check the flow of Puritan tracts.

A stringent decree of the Star Chamber of 1637 imposed heavy penalties on offending printers, and in 1643 Parliament issued an ordinance requiring that no book be printed unless licensed by an officer appointed by them; this provoked Milton's *Areopagitica*, but the licensing system continued. Though in 1622 authority was given to certain stationers to issue pamphlets about foreign wars the Star Chamber prohibited these in 1632, substituting in 1638 a monopoly grant to two persons to print foreign news. There were many unauthorized newsbooks at this time but there were many suppressions and imprisonments and for a time policy fluctuated between suppression and licensing.

Under the Restoration control and licensing continued. A Licensing Act was passed in 1662 and extended to 1679 forbidding the printing of works contrary to the Church or government, requiring all books and pamphlets to be licensed, authorizing search for and seizure of unlicensed books, and regulating the conduct of the book trade. It was revived in 1685 and continued to 1694 when the licensing system finally ended by lapse of the statute.

Thereafter the detection of libels was entrusted to the Solicitor to the Treasury, aided by informers and examiners of the Press. Punishments were severe and seizure of papers and type were frequent. In connection with such searches general warrants were often issued; their use for the seizure of unspecified papers was declared illegal in Wilkes's case in 1763, and in 1792 Fox's Libel Act conferred on the jury the right to determine whether an accused was guilty of libel or not.

Governmental interference with the Press diminished in the eighteenth century though there was an outburst of prosecutions in the early nineteenth century, and there was a vigorous pamphlet war, the government sometimes subsidising pro-government papers. In this period also began the reporting of debates in Parliament, initially in journals, such as the *Gentleman's Magazine* and the *London Magazine*. The House of Commons reaffirmed in 1738 that it was a breach of privilege to print without permission the proceedings of either House, but this did not stop unauthorized reports and a final and unsuccessful attempt to assert the privilege was made in 1771. The last special restriction on newspapers, the Stamp Duty, was abolished in 1855.

The liberty of the press has always been restricted by the general law of sedition and libel, the law of contempt of Parliament and of the courts, by the Official Secrets Act and, in wartime, by special temporary censorship. But these apart, it has been competent to publish whatever the editor wills. Trade union authority now constitutes a different kind of threat to the freedom of the Press, by the demand that editors belong to a union which might permit them to publish only what the journalists' or printers' union might approve, and by the refusal to print matter of which the union disapproves. Provision was made in 1976 for a charter on the freedom of the Press to deal with complaints on this score.

In other countries experience has been that the most stringent control of the Press cannot prevent some unauthorized and subversive materials being published.

Freedom of the Press has another side, in that it may be abused by intrusive investigations, sensational or inaccurate reporting, refusal or failure to publish explanations, and publication of material better ignored. In 1953 the newspaper industry established a Press Council; it was reconstituted in 1963 with an independent chairman and members from outside the industry. It seeks to maintain good standards and is ready to consider complaints against the Press. It will consider a complaint only provided the complainer has first tried to obtain redress from the editor of the paper and provided no legal proceedings are being brought. It has no disciplinary powers and many of its decisions seem merely to whitewash newspapers.

Freedom of the seas. The principle that the high seas, outside territorial waters, are free and not subject to the sovereignty of any state has not always been accepted. When Spanish and Portuguese navigators explored the Atlantic, Spain obtained Papal bulls in 1493 conferring on it all territories south and west of a line 100 leagues west of the Azores; this was revised by the Treaty of Tordesdillas to a line 370 leagues west of the Cape Verde Islands; thereafter the Spaniards claimed the Caribbean and the Pacific, the Portuguese the South Atlantic and Indian Ocean. Venice claimed the Adriatic and this was asserted by Julius Pacius in *De Dominio Maris Hadriatici* (1669); Genoa claimed dominion over the Ligurian Sea. Grotius in *Mare Liberum* (1608) asserted the freedom of the seas, but Selden wrote *Mare Clausum* (1635) at the instance of James I to justify British control of the adjacent seas from Spain to the Arctic and without limit to the West. Bynkershoek (*De Dominio Maris*, 1702) conceded that there was nothing in natural, Roman, or international law in the way of sovereignty over the sea, and down to the end of the eighteenth century Britain demanded a salute to the British flag in the British seas. Denmark long claimed dominion of the Baltic and exacted Sound dues until the mid-nineteenth century. By the late eighteenth century claims to sovereignty over large areas of sea declined and in the following century the major powers supported the principle of the freedom of the seas outside the limits of the territorial sea.

In modern law it is accepted that no part of the high seas may be appropriated by state or person and that persons of all nations have liberty of navigation, of fishing outside countries' exclusive areas, of laying submarine cables and pipelines, and of flying over the high seas. Each state is responsible for the maintenance of good order and security on the high seas by ships which it has registered and to which it has granted the right to fly its flag and, save by international agreement, no state may exercise jurisdiction over ships flying foreign flags.

The exceptions to the principle of the freedom of the high seas are the right of ships of any nation to suppress piracy *jure gentium* and other illegal acts committed by ships on the high seas, as by ships under the control of insurgents or illegal attacks by ships under the authority of a lawful government, the right of warships to approach to verify the nationality and identity of ships, the right of warships to board and search a foreign merchant ship if there are reasonable grounds for suspecting that she is engaged in piracy, slave-trading, or concealing that she is of the same nationality as the warship, the right of hot pursuit of ships which have infringed the law within the territorial sea, and, in time of war, blockade of the enemy's coast.

Exceptions exist also by treaties conferring powers of visit and capture in such cases as suspected slave-trading, smuggling, or trading in arms.

Ships on the high seas are subject to the jurisdiction only of the State whose flag they fly, save in cases of piracy, hijacking, hot pursuit, slave trading, and the right of approach by warships. Penal or disciplinary action may be taken against master or crew only in the country of the flag or the country of which the person is a national.

Freedom to hold private property. One of the major civil liberties. It denotes the recognition of the right of an individual to acquire and hold by and for himself alone land and goods, and not to be restricted in his use and enjoyment of them save for obvious reasons of public utility, nor to be deprived of them unjustly, and to leave property as one pleases. All socialist and communist polities deny this liberty and thereby reduce the individual to serfdom. Liberal-democratic regimes profess to recognize this freedom, but impose so many restrictions, by way of taxation, restrictions on use, liability to compulsory expropriation, and the like, as largely to empty the freedom of content.

Freedom to work. One of the major civil liberties. It denotes the freedom of the individual to work or not, and at what trade or profession he chooses, as contrasted with being bound to follow the trade of one's father, or being liable to be directed by some authority. Social, economic, and intellectual constraints exist which make it difficult to enter certain careers and many persons find that they have not the ability or qualities to enter certain careers. Sex or marriage is now a disqualification for entering certain careers only very exceptionally, e.g. the priesthood in the Roman Church, and in general no legal discrimination can exist preventing a person from qualifying to practise any trade or profession.

Restrictions on freedom of work exist in various forms. There are restrictions imposed by society and by professional organizations on the freedom of persons to practice various skilled professions; they must have undergone specified training and attained prescribed standards of competence. Trade unions commonly impose restrictions on the employment of persons who have not served apprenticeships in the trade in question.

More serious from the point of view of legal liberties is trade union insistence on the 'closed shop' and the refusal of members to work with, or permit the employment of, a non-union man, a policy frequently enforced by strikes or threats of strikes. The closed shop is an unjustifiable denial of the legal freedom to work.

The counterpart of the right to work is freedom not to work or to withhold or withdraw one's labour. This right takes the form of the right to leave employment or, in association with others, to strike. In wartime and other serious emergency this right is frequently abridged by statutory powers requiring service in the forces or in industry and withdrawing the right to strike. The right to strike is so grossly abused, by being invoked in any and every case of actual or imagined dissatisfaction or dispute, as to give rise to a serious question whether it can be tolerated, particularly where, as in strikes in the so-called public services or public utilities, the strike is directed against the whole community or where strikes in large-scale national industries, such as the railways, are damaging to the whole national economy and thereby to both parties involved in the strike.

Freedom to work does not imply any legal entitlement to have a job or any particular kind of job. The alleged 'right to work' is a liberty or freedom to work, not a claim to have a job provided; this is something which is unattainable by law, and probably unattainable in fact.

Freehold. A term originally designating the holding of an estate in land by free tenure, by a man having the rights of a free man as contrasted with a villein. Such tenure included knight service, serjeanty, free socage, and frankalmoign (qq.v.). Contrasted with these were the unfree tenures of villeinage (later copyhold), and customary modes of tenure, such as gavelkind, borough English, and ancient demesne (qq.v.). A person holding a freehold estate owed services which were free from servile incidents. In land held by freehold tenure there were three recognized estates of freehold, where the length of duration was uncertain, namely, estate in fee simple, estate in fee tail, and estate for life or *pur autre vie.*

Distinct from these are estates less than freehold, where the duration is certain, comprising leases for terms of years and tenancies from year to year.

In modern English law freehold has absorbed copyhold and is contrasted with leasehold, or land held for a term of years, and with the disappearance of all the incidents and dues payable to a lord is for all practical purposes complete ownership of the land.

Freeman, Richard (1646–1710). Called to the English Bar in 1674, became Chief Baron of Exchequer in Ireland in 1706 and Chancellor of Ireland in 1707.

Freeman. A term originally denoting a man who was not a serf or slave, not tied to a master nor to a piece of land. In the U.K., however, the term came to mean a person having the full privileges of a city or borough, or a company, especially one of the Livery Companies, which he has obtained by birth, apprenticeship, or purchase. Prior to 1835 the freemen of boroughs had trading privileges, and frequently had tax exemptions and influence in or even monopoly of the Parliamentary and local government franchise. The Municipal Corporations Act, 1835 (England), and the corresponding legislation for Scotland abrogated these privileges, and the status of freeman is now honorific only. It is the practice to confer the freedom of the city or borough on distinguished individuals for services to the place or district.

Freemen boroughs. Borough constituencies in the unreformed British parliament, where the vote was held by freemen by apprenticeship, birth, or purchase. In some freemen boroughs freeholders also had the vote, and these were much more politically open than freemen boroughs, where the number of freemen was frequently carefully restricted to keep it manageable.

Freight. The money paid by a charterer or shipper to a shipowner for his letting the ship or space therein for the carriage of goods. Advance freight is freight paid in advance; lump sum freight is freight payable irrespective of the amount of goods loaded; back freight is freight payable when delivery is not taken at the port of discharge and the master deals with the goods at the owner's expense; dead freight is freight payable for the space unused where the charterer or shipper fails to load a complete cargo or the complete amount of space taken; and *pro rata* freight is reduced freight payable where goods are delivered, by agreement, short of the previously agreed port of discharge.

Freirechtslehre (free law school). A movement in German law which gained strength after the promulgation of the German Civil Code in 1896. It was supported by such scholars as Ehrlich and Kantorowicz (qq.v.) and has some parallels to Gény's *libre recherche scientifique* in France. It represented a reaction against the excessive reliance on logic in German juristic and judicial activity down to that time. The major tenet of the movement was that in the decision of cases, where statute was not clearly applicable, there should be a search for a just rule for decision without the compulsion of conceptions formulated in advance and deductions rigorously made from them. It was in substance a movement in favour of flexibility in the growth of principles.

French Community. An association of states created by the French Constitution of 1958 and adopted by referendum by most of the then French colonial territories. The fields of the Community's competence included foreign policy, defence, economics and financial policy, higher education, and certain other matters. Some of the territories have since become independent within the Community, then called the 'renewed Community', while the others have become totally independent. France has, however, co-operation agreements with all these states. The Community organs (Executive Council, Senate, and Court of Arbitration) have fallen into abeyance and relations with the members of the Community are conducted by the Ministries of Foreign Affairs and Co-operation.

French law. The system of law developed in France is an important member of the Civil law or Romano-Germanic family of legal systems and has had widespread influence outside France, partly because of the Napoleonic codifications and partly because of the early establishment in France of a democratic political system, and to a lesser extent because of the wide diffusion of French as a language in the world.

I. History

Under the Roman empire Roman law applied in Gaul and at the fall of the Western Empire the Gallo-Romans had a body of written law comprising the writings of Gaius, Papinian, Paul, Ulpian, and Modestinus, given the force of law by Valentinian III in 426, and imperial constitutions.

After the collapse of Roman power in the West, Gaul was overrun by barbarians; the Visigoths established a kingdom from the Loire south, the Burgundians another in the Rhone valley, the Ripuarian Franks round Cologne and the Moselle, and the Salian Franks round Cambrai. The kings of the Burgundians and Visigoths were initially representatives of the Western empire and the barbarian kings sought to use rather than to destroy the Roman legal and administrative system. For 500 years the *Breviarium Alaricianum* was the main source of law for the people of Gaul while the Germanic tribes continued to live under their individual customary laws.

Under the Merovingians (486–700) the administration of justice seems to have been organized partly on a Roman, partly on a Germanic basis. Justice was administered by the count of the district on the verdict of notables (*boni homines* or *rachimburgii*) assembly. The King also administered justice and issued general commands or prohibitions. Judicial and fiscal rights frequently passed to landowners, while the Church gained many privileges and bishops administered civil and criminal jurisdiction over clergy. Under the mayors of the palace there appeared some of the precursors of feudalism, grants of lands as *precaria* or *beneficia*, initially for years but subsequently for life. Not only lands but offices such as of count or *dux* were granted and these became property, initially for life and later hereditary.

Gradually the Frankish Kingdom became supreme and in the late eighth century Charlemagne eventually asserted his power over all of what is now France. He reorganized the administration of justice, fixed the jurisdictions of the count and his deputy, the *centenarius*, and established permanent *scabini* in place of the *rachimburgii*, chosen by the counts, to assist them; he established *missi dominici*, itinerant judges, and founded two general assemblies a year, attended by the chief officials, lay and ecclesiastical, at which capitularies (q.v.) were drawn up and promulgated. He maintained firm authority over the Church and nominated the bishops and abbots. In each district count and bishop owed each other mutual support and the *missi* on a circuit were frequently a count and a bishop.

In the Frankish monarchy personality of laws applied, each man being subject to the law of his own race. The barbarian customs existed alongside the Roman law and as those customs came to be written down in Latin they were very much Romanized. The Gallo-Romans and even the Church adopted the barbarian methods of trial, such as compurgation and ordeal, but the capitularies were legislation for all and tended to produce some unity and men of barbarian race assimilated a certain amount of Roman law. Slavery, under the influence of the Church, tended to evolve into serfdom, but there were many grades intermediate between freeman and serf, notably Roman *coloni* and Germanic *liti*, both attached perpetually to a piece of land which they cultivated in return for rent. Free ownership of land continued under the name of *alod*, but the occupation of much land tended to acquire the character of tenure, land held of and under a lord for an annual payment.

The rise of feudalism. In the tenth and eleventh centuries there was anarchy, legislation ceased, old capitularies fell into disuse but territorial customs, initially both local and ill-defined, arose and feudalism took more definite shape, as men associated in groups for mutual defence. Men attached themselves to a chief offering him fidelity and armed assistance in return for his protection. To some he gave lands, initially for life, later to be held heritably. Fiefs did not become universally hereditary till the later eleventh century. Though the monarchy continued to exist it exercised power only where dukes, counts, or other lords did not hold sway. The power of the Church continued to grow, as did its jurisdiction. Each lord consulted his feudatories and they formed a court of justice, each claiming to be subject to the judgment of his peers only. Under the presidency of the lord they sat in judgment on the villeins and serfs. Thus there tended to develop a nobility, holding land and devoting themselves to arms, and able to recruit by conferring knighthood and giving fiefs.

The twelfth and thirteenth centuries were the period of the feudal monarchy. The Crown recovered strength and authority and acted through hierarchically organized feudalism. It began again to legislate and to govern. The study of Roman law spread rapidly, except in Paris, which was closed to it by the papacy, and this assisted the reconstitution of the royal power on the Roman imperial model. Feudalism became finally organized as a social structure, a method of landholding and in the administration of justice. Legislation was again made, and royal *baillis* and seneschals were appointed as representatives of the King in the provinces. Feudal courts were influenced increasingly by Roman and canon law and the practice developed of appeal from seignorial courts to royal courts. The *curia regis* was gradually transformed into a regular court of appeal and came to be called the *parlement.*

It became settled in Paris and originally had jurisdiction over the whole kingdom. Criminal law was also influenced by Roman law, arbitrary penalties replaced the fixed, frequently cruel, ones of the customs and the inquisitorial procedure of the canon law became normal, including secrecy, absence of defence, and torture.

Under feudalism territoriality of law became established. This led to the division of France into two regions, the south, *le pays du droit écrit* where the Roman law as codified by Justinian was the law of the majority and was applied to everyone as the custom of the region, and the north (as far south as an east–west line from about Bordeaux to near Geneva), *le pays du droit coutumier* where the Capitularies, the canon law and local Germanic usages made up the customary law and Roman law was subsidiary only. The line of demarcation corresponds roughly to that between the Langue d'oc (south) and the Langue d'oïl (north). The customs were very numerous, some 300 in all, but of those about 60 extended to provinces or larger territories and the rest to only single cities or small territories. In the south every village had municipal statutes modifying the Roman law. In the north the Roman law still had great authority and was taught in the universities with the canon law, but it was not 'received' nor applied as law; it had a juristic or theoretical authority. It was extensively studied in the Universities, but from 1219 was officially banned at Paris. The relations of seigneurs and vassals were regulated by the *Libri Feudorum* incorporated in the *coutumes*, though in the south, though the *Libri Feudorum* were consulted, the law consisted chiefly of the customs of the province as settled in the decisions of the Parlements. The *parlements* were judicial sessions of the *Curia Regis* of the Capetian kings. Royal ordinances fixed the *parlement* in Paris though the royal court was itinerant. Originally the jurisdiction of the *parlement* of Paris covered the whole royal domain, but later provincial *parlements* were created though that of Paris remained the most important. They were essentially courts of appeal from feudal courts and minor royal courts, not legislative bodies, and royal legislation in France was impeded by the existence of customs and feudal rights.

Private editions of the customs of particular places were made by jurists. Among these are the *Très Ancien Coutumier de Normandie*, comprising two treatises, the first of about 1200 concerned with private law, procedure, criminal law, and the seignorial courts, the second, of about 1220, dealing with a suit involving title to property; the *Grand Coutumier de Normandie*, of about 1255; *The Advice of Pierre de Fontaines* (*c.* 1255) a compilation of the customs of Vermandois, chiefly procedural; the *Anciens Usages d'Artois* or *Coutumier d'Artois* of about 1290 which contains citations from Roman and canon law; *The Livre de Justice et de Plet*, of about 1260, largely a translation from Roman texts, but concerned with the usages of Vermandois; the *Établissements de Saint Louis*, i.e. a compilations of the legislation of that king; and Philippe de Beaumanoir's *Customs of Beauvaisis*, at once a compilation of the customs and a treatise on customary law.

Rather later there appeared compilations of customs from other places. For the very north of France there was a kind of encyclopaedia in which Roman, canon, and customary law mingle, the *Somme Rural* of Jehan Boutillier of Tournay (*c.* 1380), for the Paris region the *Stylus Curiae Parliamenti* of William du Breuil (*c.* 1330) dealing with procedure, and the *Grand Coutumier de France* or *Coutumier de Charles VI* (*c.* 1385) by Jacques d'Ableiges, dealing with the feudal law of the Ile de France and with procedure. There was, however, no complete record of the *coutumes* until in 1453 the *Ordonnance de Montil-les-Tours* made provision for a revised publication by royal commissioners of all the *coutumes*. The work was still in progress a century later and some of the less important *coutumes* disappeared because they were not preserved. The first Custom of Paris was published by decree in 1510.

In this same period clerks began to record notable decisions, such as the old decisions of the exchequer of Normandy and the *Olim* registers of the *parlement* of Paris.

Though there was such a variety of civil customs, there was great uniformity of commercial and maritime customs. The *Roles d'Oleron* is a collection of decisions allegedly rendered by the sea-judges of Oleron dating from the eleventh century. They became the common law of the western seas and were reproduced in part in many of the other western collections of maritime laws.

The limited monarchy. The limited monarchy extended over the fourteenth and fifteenth centuries. The *curia regis* disappeared, its deliberative functions passing to the new *conseil du roi* and its judicial to the *parlement*. A group of royal ministers and secretaries began to be important. About the same time emerged the States General. Provincial estates and *parlements* arose in imitation of that of Paris. The *parlements* which had become a judicial committee nominated annually became, during the fourteenth century, a body of magistrates permanent but removeable. Provincial *parlements* developed in place of superior jurisdictions in formerly independent districts or districts which had been one of the great fiefs. They acquired the right of opposing the promulgation of laws, of revising them and making representations to the King when they refused registration.

Various other royal jurisdictions arose, notably

the provosts of the marshals of France, who exercised criminal jurisdiction while the *procureurs du roi*, originally the King's representative in court, came to take over public prosecutions. The King was recognized as having a reserve and overriding power of justice whereby he could reverse decisions of even courts of final appeal. He could reprieve or pardon; the extreme royal intervention was by *lettres de cachet* which ordered imprisonment without trial or exile.

The absolute monarchy. From the sixteenth century to the Revolution absolute monarchy ruled. The States General were not called after 1614; the *parlements* lost their power to refuse to register laws in the seventeenth century, but later recovered them. The *conseil du roi* remained an important organ of the central government, but the only great officer of state to survive was the chancellor; the King's real ministers were the Secretaries of State. In the seventeenth century the council tended to divide into sections, one of which was concerned with legal affairs. But in this time the major principles of public law were defined, notably the fundamental laws of the realm, which were constitutional principles binding even the King, and the theory of the rights and liberties of the Gallican Church. In 1771 the *parlement* of Paris was purged for having resisted the King and a new *parlement* constituted.

The general customs of the *pays coutumiers* were nearly all recorded officially in the sixteenth century. Drafts made by royal officials were promulgated in the chief towns of the districts by royal commissioners to the lords of the district. Each article was discussed and voted on. Those approved were decreed by the commissioners. The *coutumes* accordingly in substance became statute law, and could be modified only by formal revision.

Royal legislation, by *ordonnances*, *édits*, *déclarations*, or *lettres patentes* became more common. *Grandes ordonnances* had previously been chiefly reformatory of previous law, frequently issued after discussion in the States General, but in the seventeenth and eighteenth centuries became essentially codes, giving a detailed and systematic statement of a major branch of law. There were two groups of such codifying *ordonnances*, those inspired by Colbert (q.v.), including those of 1667 on civil procedure, of 1670 on criminal investigation, of 1673 on mercantile commerce, and of 1681 on shipping, and those promoted by d'Aguesseau (q.v.) of 1731, 1735, and 1747, on deeds of gift (1731), wills (1735), and family settlements (1747).

After 1500 the influence of the Humanist scholars of the Roman law was felt in France, particularly Alciatus from Milan, Udaric Zasius, a German, and Guillaume Budé, a Frenchman. But these were followed by a great school of French Humanists, notably Cujas (1522–90?) (q.v.) who wrote Commentaries on Papinian and *Observationum et Emendationum libri XXVIII*, Doneau (1527–91), Douaren (1509–59), Baudouin (1520–73), and Hotman (1524–90) who wrote *Antitribonianus* and *Franco-Gallia*, a manifesto of Protestantism.

They were followed in the seventeenth century by Denys Godefroy (1549–1622) who edited the *Corpus Juris* and his son Jacques (1587–1652) who reconstructed the text of and wrote a commentary on the *Theodosian Code*, Antoine Favre (1557–1624) author of *Rationalia ad Pandectas* and *Codex Fabrianus*, Fabrot (1580–1659) editor of the *Basilica*, and others.

After the official compilation of the customs which took effect in the sixteenth century each province had thenceforth its local code, and many bodies of local customs were absorbed in General Customs of larger areas so that there was a reduction in number, but it is said that there were still 60 General Customs and 300 Special Customs. This official compilation gave rise to the deduction of general principles, a common customary law of France, and treatises such as Guy Coquille's *Institution au Droit Français* and Loysel's *Instituts Coutumiers*, followed by Argou's *Institution au droit français*. From the sixteenth century the *Coutume de Paris* began to take a leading place, partly on its merits, partly because Paris was the capital, and partly because of the eminence of some of the commentators thereon, notably Dumoulin. When French law was carried abroad it was the *Coutume de Paris*.

Many notable jurists worked in the sixteenth century. Among those who devoted themselves to the customs the most notable were Michel de L'Hopital (d. 1573), Charles Dumoulin (1500–66), Guy Coquille (1523–1603), Antoine Loysel (1536–1617), while those who devoted themselves to the Roman law included Jacques Cujas (1522–90), François Baudouin (1520–73), Hugues Doneau (1527–91), and François Hotman (1524–90).

The seventeenth century is notable for Guillaume de Lamoignon (1617–77) who laid down the principles of the customary law in the form of a Code, Jean Domat (1625–96) who composed *Les Lois civiles dans leur ordre naturel*, the Chancellor D'Aguesseau (1668–1751), and Robert Joseph Pothier (1690–1772) who wrote *Pandectae in novum ordinem digestae*, a systematic treatise on each title of the Digest, the texts being rearranged to bring together those related by subject, and a series of treatises, notably the *Traité des Obligations*. Pothier exercised great influence on Napoleon's codifying commission.

The revolution and codification. The French Revolution effected great changes in the law, radically changing the constitution, making a new division of the country into *départments*, and reorganized the judiciary. In private law the major changes were enfranchisement of the land, reorganization of property-rights, and introduction of the principle of equality of inheritance. Unification of the private law had been contemplated repeatedly and steps towards it had been taken by Colbert's and d'Aguesseau's Ordinances. In 1790 the Assembly planned a codification of all the law of France. In 1793 Cambacères presented the Convention with a plan for a Civil Code. In 1800 Napoleon as First Consul appointed a commission of four to prepare a draft, Tronchet, Portalis, Bigot du Preameneux, and Malleville. It was finally approved in 1804 and superseded all the older law. It was based on the customs, especially the Custom of Paris, the Roman law, especially as embodied in Pothier's works, the canon law, royal Ordinances, and revolutionary laws enacted since 1789. To the commissioners written reason was represented by traditional law and they did not innovate greatly. Napoleon participated in the debates and personally influenced some parts of the Code. It comprises a preliminary title and three books, dealing with Persons (Arts. 7–515), Property (Arts. 516–710), and Various Methods of Acquiring Ownership (Arts. 711–2281). It has the merits of unity, system, precision, and clarity. Subsequently other Codes were enacted, viz.: Civil Procedure (1806) (based on Colbert's ordinance of 1667), commercial Code (1808) reproducing Colbert's ordinances, Criminal Procedure (1808), Penal Code (1811), and all have subsequently been extensively amended. The Code of Criminal Procedure was revised in 1832 and replaced in 1959. The Penal Code was revised in 1832 and 1863 and the Code of Commerce modified in 1832.

The Code was extended to the territories which were united to the French Republic before the Peace of Amiens, such as Belgium, Savoy, and Piedmont, to countries later conquered by Napoleon, such as Holland, while other countries such as Westphalia and Hanover, voluntarily adopted it. It was imitated and copied in Belgium and Italy, Louisiana and Haiti, Greece, Egypt, Quebec, and generally in South America. It was influential on the codes of the Netherlands, Spain, and Portugal.

The judicial system was completely reorganized by the Constituent Assembly on the basis of a *juge de paix* in each canton and a civil court of five judges in each district, while for criminal cases there was in each department a court with its own president and public prosecutor but with judges borrowed from the district courts which tried cases with a jury. *Juges de paix* as correctional tribunals dealt with petty offences. Commercial courts were continued. The *Tribunal de Cassation* had the function of preserving uniformity of interpretation of the law; it dealt with appeals on law only, and, if violation of the law were established, sent the case back for retrial by another court of the same rank.

Extensive changes were however made by Napoleon. The civil courts comprised a court of first instance in every *arondissement*, and a number of courts of appeal, each covering several departments. The separate criminal courts were abolished, but there was introduced the *juge d'instruction*. The *Tribunal*, renamed the *Cour*, *de Cassation* was retained.

Since 1815 a great deal of legislation has been passed in many respects amending the Civil Code and in others supplementing it, and the code has been the basis of numerous commentaries and general treatises.

Revision of the Civil Code was contemplated in 1904 and again after 1945 and the Code of Criminal Procedure was replaced in 1959. In 1910 a *Code du Travail et de la Prévoyance Sociale* was promulgated, codifying a great mass of previous legislation on industrial or labour law. The court structure has also been much modified since 1815.

Just as the common law of England has provided a model on which other countries have built their legal systems and has provided many concepts and principles, so French law has been a model for many countries, both those which were once part of the French colonial system and others which were not parts thereof, and its concepts and principles have been widely influential.

II. Sources

The structure of the modern law depends fundamentally on the basic distinction between public and private law, with distinct supreme courts, the *Conseil d'État* and the *Cour de Cassation*. The former may review administrative acts and has developed to a high degree judicial control of the exercise of administrative powers. The latter is not a court of appeal, but reviews findings of law and may quash a decision and remit the case to a court of the same rank; it has a major task in keeping the law uniform throughout the country. Private law contains distinct bodies of rules on *droit civil* and *droit commercial*, the test for applying one or the other being the concept of *acte de commerce* as defined in the Commercial Code. Criminal law is akin to private rather than public law. No distinction between law and equity has ever existed.

The formal sources of modern law are *loi*, legislation, including the Codes, statutes, decrees and administrative ordinances, sometimes interpreted in the light of custom and usage, *jurisprudence*, or decisions of the courts, important as evidence of the way provisions are understood and applied, though these are never more than persuasive, and *doctrine*, the body of opinion expressed in books and periodicals, which is much more influential than in Anglo-American systems though not a binding guide to decision. Its persuasive authority flows from regard for legal scholarship. Enacted law is of prime importance and legislation is the primary source of law. Decisions must all be based on some enacted law, but the enactment is frequently so terse as to require considerable elucidation and interpretation by cases and criticism. The general principles of administrative law depend, however, entirely on the jurisprudence of the *Conseil d'Etat* and private international law consists primarily of judicial decisions.

In view of the primacy of enacted law, particularly in private and criminal law, interpretation is of great importance, and four distinct approaches are recognized: grammatical, stressing the meaning of the terms used; logical, considering the text in the context of other rules in the same field; historical, ascertaining the intention and purpose by research into the legislative history of the statute; and teleological, seeking to fit the language to conditions and necessities, to find the meaning which best corresponds to the contemporary view of social welfare and justice.

An important distinction in private law is between imperative and suppletive rules, the former of which must be applied without regard to the intentions of parties, the latter of which apply only if parties have not shown their intention to exclude the rule or regulate their relationship otherwise. This distinction is not drawn in public law contexts where all laws are regarded as imperative.

Custom may be *consuetudo secundum legem* (custom supporting law) where custom gives meaning to words and ideas of everyday life used in the law, e.g. as to what is meant by a 'marriage', *consuetudo praeter legem* (custom preceding law) which establishes new rules independent of but not inconsistent with legislation, e.g. in kinds of contracts unknown at the time of the Civil Code, and *constuetudo adversus legem* (custom contrary to law), whereby the courts have held certain transactions not governed by the Code but by custom.

Apart from some collections of notable decisions, reports of decided cases did not appear in France until 1799, but there are now regular series of reports of significant decisions. But, unlike Anglo-American practice, courts cannot and do not create new rules of law by their decisions; decisions are examples of the interpretation put by the court on a particular provision. Also no precedent is ever absolutely binding, though there is a strong tendency to reach decisions consistent with prior decisions. The *Cour de Cassation* can freely decline to follow its own precedents though it is reluctant to do so. Lower courts normally follow the line adopted by the *Cour de Cassation* but are not bound to, and attach less weight to decisions of other courts.

The role of *doctrine* is also different from that of the Anglo-American textbook. To the French jurist the law stands above judicial decisions and jurists must guide judges and try to bring order and coherence to their decisions. An essential function is criticism of judicial decisions and opinion, principally done by case notes appended to reports but also in treatises and manuals.

In the event of there being uncovered a gap in the law, or exceptionally to help correct existing legislation, reference may be made to one or other of a few supereminent or overriding principles, notably equity, abuse of rights, public order, and good morals. Similarly in administrative law reference may be made to general legal principles to ensure the supremacy of law and justice.

III. Administration of justice

The supreme court is the *Cour de Cassation*, to which appeal lies on law only. It sits in criminal and civil chambers, the latter comprising civil, commercial, and social sections. It is composed of a First President, four presidents, 60 judges, a procureur-général and advocates-general, and may annul any judgment as being contrary to law and return the case to the court below for retrial. The *Conseil d'Etat* sits as supreme court in administrative cases from decisions of administrative courts. It also has administrative and advisory functions. There is also a Court of Audit.

There are 28 Courts of Appeal hearing appeals from Courts of Grand Instance and Correctional Courts. The decisions of Courts of Appeal and Courts of Assize are, subject to reference on law to the *Cour de Cassation*, final.

Tribunaux de grande instance, usually comprising several judges, handle ordinary civil business, and *tribunaux d'instance* minor civil business. Courts of Assize, which have a jury of nine, deal with very serious criminal cases and Correctional Courts with serious criminal cases. In such cases of crime a preliminary enquiry is held in secret by an examining magistrate (*juge d'instruction*) and a public prosecutor (*procureur*) presents the case in court. Petty criminal cases (*contraventions*) are dealt with in *tribunaux de Police* by a judge of the *tribunal d'instance*.

Commercial courts (*tribunaux de Commerce*) composed of tradesmen and manufacturers elected for a period deal with commercial litigation. Social courts (*conseils de prud'homme*) deal with employment disputes. Juvenile offenders go before special courts and judges, and Conciliation Boards deal with small trade and industrial disputes.

The *Conseil Supérieur de la Magistrature* nominates magistrates to the President for appointment and ensures their independence, discipline, and directs the administration of their tribunals.

In public law there are various *jurisdictions administratives* in each *département*, a *tribunal admin-*

istratif in the region and the *Conseil d'Etat* in Paris. The *tribunal des conflits* decides conflicts of jurisdiction between the courts and the administration.

Public law. The present constitution was adopted in 1958. There is a Constitutional Council to ensure that the constitution is respected. A Council of State advises the government on drafting bills, decrees and ordinances, and also acts as the supreme administrative court. There is an Economic and Social Council which is an advisory body. The constitution embodies the principle of separation of powers (q.v.).

The head of state is the President, elected by direct vote for a seven-year term. He appoints the Prime Minister and, on the latter's recommendation, the other ministers, negotiates treaties, presides over the Council of Ministers, signs ordinances, promulgates law, can institute constitutional reform, and can call for a referendum on important bills. In emergency he assumes exceptional powers.

The legislature comprises a Senate, elected by an electoral college for nine-year terms, one-third retiring every three years. The National Assembly is elected by direct suffrage for five-year terms. The legislature passes legislation, may delegate some or all of its powers to the government, and checks the government by questions and censure motions. It can force the government to resign.

The executive consists of the President and a cabinet, responsible to the National Assembly. The Prime Minister directs the government's actions, makes appointments, ensures the execution of law.

For administrative purposes France is divided into 95 *départements*, in each of which a prefect acts as chief official of government and supervises the offices of the ministries in his area. The unit of local government is the commune, each of which has an elected Municipal Council; these are grouped into cantons and these into *arondissements*.

Public law includes constitutional law, though many of the topics within that head in Anglo-American law belong to the other spheres of law; administrative law, which deals with the powers of the administration and litigation involving state agencies and private parties (*contentieux administratif*); and tax law. Constitutional law in France corresponds more to political science of Anglo-American law schools and many of the issues which arise under this head in Anglo-American law do not arise in France. There is no question of judicial review or of power to refuse to apply a rule as being in conflict with the Constitution. Public law also covers relationships where the State or one of its agencies is a party, e.g. employment of a teacher, supply of equipment to the navy. Since a law of 1790 the courts have been prohibited from interfering with the executive or hearing litigation to which the government was a party and since then the *Conseil d'Etat* has been the principal agent in the development of public law.

Public law is of more recent formation than private law and lacks its basis in tradition. It is uncodified. It is more liable to be influenced by political factors and more changeable. But it also appears that the distinction between public and private law is becoming less sharp and certain.

Criminal law. Criminal law does not have the basis in traditional development possessed by the civil law, but has long been treated as akin to civil rather than to public law. Hence, in the administration of criminal law, the liberal interpretation customary in civil law has tended to encroach on the established principle of restrictive interpretation of penal statutes. There is a tendency for the public interest to have precedence over the protection of the accused. Increasingly today it is however truly a part of public law though still the concern of privatists rather than publicists. There are special disciplinary codes for the armed forces and merchant marine. The existence of codes in the sphere of criminal law is regarded as essential to safeguard the principle of *nullum crimen sine lege*.

Private law. Private law for historical reasons is sharply distinguished into civil law and commercial law, based historically on Roman law and on mercantile customs respectively and dealt with in separate codes. The distinction is whether or not a transaction is an 'act of commerce' as defined by the commercial code, and it is a matter of importance and difficulty.

The civil law is concerned with physical and moral persons, family law, property law, obligations (contract, torts, and unjust enrichment), matrimonial community property, and succession on death. Specialized topics outwith the Civil Code but falling under civil law include literary and artistic property, rent control and similar topics, and private international law.

The civil law is the core of French law; as formulated it is characterized by abstraction and generality as compared with the more specific and detailed formulation found in Anglo-American legislation. In classification and terminology the influence of Roman law is very obvious. It leaves much to be filled in by interpretation. Case-law to the French lawyer is more illustrative examples than sources of principles to be followed.

Labour law is increasingly becoming a body of law distinct from the civil law and the volume of legislation which has developed in the later nineteenth century has been largely consolidated in the *Code du travail*.

Commercial law. The Commercial Code of 1867 provides the basis, but a mass of later legislation is the chief source of modern commercial law. The fundamental concept is the *acte de commerce* as the kinds of which 14 kinds of activities are enumerated. It comprehends commercial legal institutions, business organizations and relationships, banking, exchange, and bankruptcy. It includes maritime law and marine insurance. The same contract may be commercial as regards one party and civil as regards the other. The requirements of written evidence are relaxed in commercial contracts.

Increasingly commercial law has become the legislative regulation of business and business enterprises and has come to assume some of the characteristics of administrative law.

Profession and training. The judiciary are members of the judicial civil service, a career structure which is entered on qualification from the Bar and not recruited therefrom, and judges are less known as individualists and less creative than Anglo-American judges. Moreover, judges normally sit in groups and when they do always issue opinions from the court, dissenting opinions being forbidden. Judges' affinities are with academic lawyers rather than with advocates.

All courts include also agents of the Public Ministry, procurators, procurators general, advocates general and their assistants, whose function is to represent the interests of society and the law. This also is a career structure.

The legal profession was formerly divided into *avocats au Conseil d'Etat et à la Cour de Cassation, avocats, avoués, notaires, agréés*, and *conseils juridiques*. Since 1972 *avocats*, most *avoués* and *agréés* have been amalgamated as *avocats*, doing all litigious and non-litigious legal work other than pleading before the highest courts and the conveyancing and estate business done by notaries. *Conseils juridiques*, business legal advisers, were at the same time regulated and restricted to advising and the drafting of documents. *Notaires* act as family counsellors and often as informal arbiters of disputes, and are authorized to draw up and record ante-nuptial agreements, notarial wills, mortgages and gifts *inter vivos*, and are usually retained to frame contracts of sale of land, and act in winding up estates and the formation of companies.

Legal education is primarily undertaken in universities, professional training coming later.

In French universities law faculties cover a wide range of studies and law is studied by many not intending to take up a career in any branch of law. Law is considered not only or even primarily as lawyers' techniques, but as an aspect of a just social order and indivisible from politics and economics. Professors belong to a career structure and have high standing, fully comparable to judges. Their opinions, though not binding on the courts, have great weight in influencing judicial decisions.

Generally: C. Sladitz, *Guide to Foreign Legal Materials—France, Germany, Switzerland*; M. Amos and F. Walton, *Introduction to French Law*; R. David, *Le Droit français*; R. David and de Vries, *The French Legal System* (1958). *History*: F. Olivier-Martin, *Histoire du droit français* (1948); J. Declareuil, *Histoire générale du droit français* (1925); A. Esmein, *Cours élémentaire d'histoire du droit français* (1925); J. Brissaud, *Manuel d'histoire du droit français* (1898–1904). *Public law*: G. Burdeau, *Traité de science politique*; G. Burdeau, *Manuel de droit constitutionnel* (1947); M. Waline, *Traité élémentaire de droit administratif* (1957). *Criminal law*: H. Donnedieu de Vabres: *Traité de droit criminel* (1947); P. Bouzat, *Traité Théorique et pratique de droit pénal* (1957); G. Vidal et J. Magnol, *Cours de droit criminel* (1949). *Private law*: G. Marty et P. Reynaud, *Droit civil* (1966); Carbonnier, *Droit civil*; H. L. & J. Mazeaud, *Leçons de Droit civil* (1955); A. Colin, H. Capitant et L. de la Morandière, *Traité de droit civil français* (1950–53); J. Escarra, *Cours de droit commercial* (1952); G. Ripert, *Traité élémentaire de droit commercial* (1954); J. Hamel, G. Lagarde et M. Jauffret, *Traité de droit commercial* (1954).

Freund, Ernst (1864–1932). Became Professor of jurisprudence and public law at Chicago in 1902 and was a leading authority on American law. His major writings were *Police Power* (1904), *Standards of American Legislation* (1917), and *Administrative Powers over Persons and Property* (1928).

Friedberg, Emil Albert (1837–1910). German jurist and canon law scholar, professor at Leipzig, editor of the standard modern editions of the *Corpus Juris Canonici* (1878–81) and *Quinque Compilationes Antiquae* (1882).

Friedmann, Wolfgang Gaston (1907–72). German-American jurist, taught at London, Melbourne, Toronto, and Columbia and was a leader in introducing the teaching of sociological jurisprudence. He wrote *Law and Social Change in Contemporary Britain* (1951), *Legal Theory* (5th edn., 1967), *Law in a Changing Society* (1931), and *The Changing Structure of International Law* (1964).

Friendly Societies. In British law voluntary societies existing to raise by subscriptions or voluntary contributions funds for the mutual relief of the members in old age, sickness, and infirmity, or for the relief of the widows and children of members, and one of several classes of voluntary benefit and provident associations. There are various types of friendly society operating in slightly different ways. Similar functions were performed by some mediaeval guilds, and some friendly societies were founded in the seventeenth century and in the following century they became widely established, though many early societies were financially unsound. The 'orders' such as the Order of Oddfellows (1745) and the Independent United Order of Mechanics (1756 or 1757) were probably founded in imitations of freemasonry.

Since 1793 legislation has sought to ensure the good management of societies and the protection of their funds. Under early Acts the rules of societies were allowed by the justices of the peace and enrolled at quarter sessions. Later a barrister was appointed to examine and certify rules and later he was constituted Registrar of Friendly Societies. In 1875 the Registry of Friendly Societies was established. Societies may be registered or unregistered, the former having certain privileges. Societies are not corporate bodies. Branches may be separately registered. Any seven or more persons may apply for registration and submit their rules for scrutiny. All property is vested in trustees for the society. The rules must provide for all matters relating to membership. Societies operate by receiving small sums of money regularly from members, accumulating and investing these and paying members' claims. Their assets and liabilities must be valued quinquennially. Some societies engage in industrial assurance (q.v.) business. A society may convert itself into a limited company, amalgamate with another society, and dissolve itself. The Chief Registrar of Friendly Societies has important supervisory functions over societies. In the U.S.A. fraternal insurance organizations cover the activities of British friendly societies. Trade unions are not friendly societies and are regulated by distinct legislation though they have some functions akin to those of friendly societies.

F. Eden, *Observations on Friendly Societies* (1801); *Reports of Commissioners Appointed to Enquire into Friendly and Benefit Building Societies* (Brit. Parl. Pap. 1871–74); Frome Wilkinson, *The Friendly Society Movement* (1886); E. Brabrook, *Provident Societies and Industrial Welfare* (1898); W. Beveridge, *Voluntary Action* (1948); F. Fuller, *Law Relating to Friendly Societies.*

Friends of the People. A society formed in 1792 by Gray, Erskine, and others to support the movement for Parliamentary reform. It was the more aristocratic and restrained wing of the reform movement as contrasted with the radical wing, represented by Tom Paine. Their report on the State of the Representation (1793) brought out in detail the defects of the existing system and they sponsored the unsuccessful motion on parliamentary reform of 1793.

Fright (or emotional shock). Fright caused to a person is as much a ground of action for damages by him as is physical injury. To be actionable the fright or shock must be not merely a momentary upset but a distinct mental lesion. The conduct bringing about the fright may be a deliberate act, a negligent act, or a failure to keep a danger under control. Initially it was thought that fright was actionable only if it arose from reasonable fear of immediate personal danger to oneself, but later this was extended to fright arising from fear for a spouse or fiancée, or for a child, but probably not for an animal, nor does it extend to shock at an accidental disturbance of the corpse of a relative. The fright may be conveyed by seeing or hearing and possibly even by being told, e.g. of a relative's death. The boundaries of this tort are still being determined. To cause death by fright is manslaughter.

Frisian code. This code comprises a *Lex Frisionum*, in 22 titles, containing general rules applicable to the people of the area now the Netherlands, with mention of variants for the northern or southern districts of this area. It dates from about 734 with the variants about half a century later. Twelve further titles, the *Additiones Sapientum*, are later, and date from about 800. The code contains a mixture of laws made by king or assembly, private writings, and some general maxims, and was probably compiled unofficially. It also contains many inconsistencies such as pagan customs and rules for enforcing Christianity.

Frivolous and vexatious proceedings. A judge has power to stay or dismiss any action which is frivolous or vexatious. This power will be exercised if the pleadings disclose no reasonable cause of action or answer, or are so plainly frivolous that to put it forward would be an abuse of the process of the court. See also VEXATIOUS ACTIONS.

Front Bench. In the House of Commons, the front bench of seats on the Speaker's right, sometimes called the Treasury bench, is customarily occupied by ministers and senior members of the government, and the bench opposite by the leading members of the main opposition party. The distance between the two front benches, or rather between the two white lines behind which members sit, is by tradition said to be two swords' length to ensure that members on the opposite sides could not get within sword distance of their opponents.

Frontager. A person owning or occupying lands which abut on or have a frontage to a highway, river, shore, or other feature.

Fructus industriales. Crops which do not spring up naturally but are the product of cultivation and labour. Growing crops if *fructus industriales* are deemed chattels, but if *fructus naturales* are deemed part of the soil until severed.

Frustration. The premature termination of a contract by circumstances which render performance impossible or at least so different from that contemplated that it would not be reasonable to hold the parties bound by the contract. The principle has developed from the principle that supervening factual or legal impossibility of performance, not attributable to the default of either party, discharges the contract. The court may hold a contract frustrated if what has happened has truly taken away the basis or substratum of the contract even if performance is still possible.

Fry, Elizabeth (née Gurney) (1780–1845). Philanthropist and a leading promoter of prison reform. In 1817 she formd an association for the improvement of the female prisoners in Newgate, and did much to alleviate conditions there. Her methods were extended elsewhere. In 1818–19 she visited prisoners in the north and in Scotland, and the publication of her notes of the tour and recognition of her work by Parliament led to correspondence with prison authorities abroad. Later she visited many European countries and made recommendations to the authorities, many of which were adopted. Many of her ideas have since her time become standard practice, such as classification of prisoners, provision of instruction and employment, female supervision for women prisoners and the like.

E. Fry, *Memoirs, with Extracts from her Journals and Letters* (1847).

Fuero (from Latin *forum*). In Spanish law the term for charters of townships, giving special privileges, special statutes granted to social groups and institutions such as the Castilian nobility, and latterly for sets of laws of state-wide application (*fueros generales*) which differed from kingdom to kingdom and of which the *Fuero Juzgo* (q.v.) was a prototype. Municipal *fueros* became as important a source of law as customary law, embodied many rules derived from ancient custom and became in some cases voluminous compilations. The main grants of municipal *fueros* were in the eleventh to thirteenth centuries during and after the reconquest from the Moors, to promote resettlement of the country. The *fueros* fall into related groups and in many cases were borrowed by one kingdom or town from another. They also became longer and more elaborate as time went on. Latterly the word came to be used in a general sense, as a general code of law. As the power of the Spanish crown grew local *fueros* slowly declined, but the Basque *fueros* were not extinguished until 1876 after the Carlist rising.

Fuero de las Leyes. A code of law, based on Roman law and intended to apply to the whole of Castile, prepared on the order of Alfonso X the Wise about 1255. It was the basis of the later *Siete Partidas* (q.v.) but the attempt to impose it provoked opposition from nobles and towns and they forced the King to abandon the attempt to create a single body of law and to confirm their traditional customs and privileges.

Fuero de Leon. The oldest body of laws in Spain, comprising laws applicable to the city of Leon particularly but also to the kingdom in general, granted about A.D. 1020.

Fuero de los Espanoles. A charter granted by General Franco in 1945 granted stated personal freedoms to citizens of Spain provided that no attack was made on the government. It was a move towards a more democratic regime.

Fuero Juzgo (or *Liber Judiciorum*, or *Lex Barbara Visigothorum*, or *Lex Visigothorum Reccesvindiana*). A body of Visigothic law enacted by Visigothic kings of Spain in the seventh and early eighth century, intended to serve as a common body of law for both Hispano-Romans and Visigoths. There were editions in A.D. 654, 681, and 693. It contains a mixture of rules on various subjects and is less advanced than the legislation of Justinian, but it evidences a well-developed sense of justice. It continued despite the Moorish conquest of much of Spain and was never wholly abrogated. As Spanish colonies were established its influences extended more widely and it has had some influence in the law of much of central and south America, in Puerto Rico, and in some states of the U.S.A.

Fuero Real (or *Fuero de las Leyes*). A code of law issued by King Alfonso X (the Wise) of Castile in 1255, covering a wide range of subjects but neither complete nor systematic. It was based on the *Fuero Juzgo*, local *fueros*, customary law, and Roman law, and was given to various cities as their municipal law.

Fuero viejo de Castilla (or *Ordenamiento Real de Alcala*). A famous Spanish code of laws dealing with the rewards and privileges of the nobility, originally having the force of law in Old Castile but later in the whole kingdom. It was abrogated in 1254, but the nobles forced restoration of their privileges. In 1348 in the *Ordenamiento de Alcala* it was provided that nobles who had special *fueros* should have the benefit of them so far as they were being observed, and in 1356 a new compilation of regulations constituting the *Fuero Viejo* was issued.

Fueros de Aragon. The collection of mediaeval legal codes of the kingdom of Aragon. It includes the Code of Huesca, dealing with customary law, the *Privilegio General* and various *fueros* granted to particular municipalities.

Fugitive offenders. Legislation, principally the Fugitive Offenders Act, 1967, provides for the apprehension of persons who have committed a 'relevant' offence in the U.K. or other parts of Her Majesty's dominions in the place where they are found, and their return to the jurisdiction in which the offence was committed for trial there.

The 1967 Act largely assimilated the law applicable to the Commonwealth to that contained in the Extradition Acts applicable to the U.K. and foreign countries. As in extradition, offences of a political nature are excluded. There are substantial safeguards for the fugitive in that he may not be returned if his extradition would on stated grounds be unjust or oppressive.

Fugitive Slave Acts (1793 and 1850). U.S. legislation providing for the return of runaway slaves to their masters. The Act of 1793 embodied provisions usual in earlier colonial statutes, but was largely ignored by people in the northern states who frequently assisted runaway slaves. In 1850 a new Act provided for more rigorous enforcement and heavier penalties. Both Acts gave rise to much opposition and bitterness, and some northern states even passed legislation to counter the federal Acts by protecting personal liberty. These Personal Liberty laws were among the grievances which led to the secession of the South.

Fulbeck, William (1560–?1603). Wrote various legal and other works of which the best known are a *Direction or Preparative to the Study of the Law* (1600), *A Parallele or Conference of the Civil Law, the Canon Law, and the Common Law of England* (1601), and *The Pandectes of the Law of Nations* (1602).

Fulgosio, Raphael (1367–1427). A famous Italian jurist and commentator, author of *Commentaria in Codicem* (1547) and *Commentaria in Digestum Vetus et Novum* (1499).

Full age (or majority). At common law the age of 21, reduced by statute as from 1970 to 18. At common law in England majority was attained at the commencement of the day preceding the twenty-first birthday but by statute it is now attained at the commencement of the eighteenth birthday. There remain some instances, such as membership of Parliament, for which a person is disqualified until he attains 21.

Full Bench. In Scotland, in the High Court of Justiciary, a court composed of more than the quorum (three) required for the hearing of criminal appeals. A Full Bench is frequently five or seven judges or even more, but is not necessarily literally the full body of judges of the High Court.

Full Court of Divorce. A court composed of the Judge Ordinary with two other members of the former Court for Divorce and Matrimonial Causes, having partly an original, partly an appellate, jurisdiction, which was transferred to the Court of Appeal in 1881.

Full faith and credit clause. Article IV, section 1 of the U.S. Constitution, requiring each state to give full faith and credit to the public Acts, Records, and Judicial Proceedings of every other state, i.e. making them effective in other states also. Congress is also given power to prescribe the manner in which such Acts, Records, and Proceedings are to be proved, and the effect thereof.

Full powers. The documents customarily required to be produced by diplomatic agents concerned in the negotiation of a treaty as evidence of their authority to negotiate and proceed to signature. The form of the Powers and the extent of authorization may vitally affect the nature and extent of the obligations undertaken. Formerly the consideration of the Full Powers occupied a substantial portion of conference time, it being important that agents' authority be precisely defined beforehand, as engaging in excess of Full Powers was the basis for a justified refusal of ratification. In modern practice, ratification is discretionary and the extent of the agent's power to negotiate and sign is unimportant, since his signature creates no obligation to ratify. In consequence scrutiny of Full Powers at conferences is now cursory. Modern Full Powers comprise preamble, designation of the negotiators, authority to negotiate, and final guarantee, reserving ratification. A distinction has sometimes been drawn between General and Special Full Powers. In the U.K. the Foreign Secretary on assuming office obtains a General Full Power enabling him to sign any number of comprehensive treaties.

J. Mervyn Jones, *Full Powers and Ratification.*

Fuller, Melville Weston (1833–1910). Practised law in Augusta, Maine, and Chicago, and was briefly a member of the Illinois legislature. He developed a reputation for quick but thorough work, and attained a high professional standing. He was eighth Chief Justice of the U.S. Supreme Court from 1888 to 1910. Fuller was a cultured gentleman, who pursued a successful legal practice and for the most part kept out of politics. He was the first Chief Justice to have had any academic legal training, having attended Harvard Law School in 1854–55, but had taken no degree in law. But he was not a great jurist, his learning being extensive but not deep. He successfully controlled a court including several justices who overshadowed him in intellect and forcefulness. In constitutional law he was uncertain and his ideas were stationary rather than reactionary, and he failed to exercise great influence as a judge, though as head of the court he was dignified, tactful, and enjoyed the respect of the Bar. He was a member of the Venezuelan Boundary Commission, 1897–99, and a member of the Permanent Court of Arbitration at The Hague, 1900–10.

W. L. King, *Melville Weston Fuller.*

Fumage (fouage or fuage). A tax mentioned in Domesday Book, sometimes called smoke farthings, levied on every house having a chimney. A statute of 1662 imposed a similar tax called hearth-money, which was abolished in 1688 as oppressive. Fumage also meant manure and manuring of land.

Functional jurisprudence (or functional school of jurists). The group of jurists, sometimes called sociological jurists, whose concern is with what law does and how it operates in society, rather than with law as a system of axioms and logical deductions, or an overriding theoretical principle. They regard law as a social institution which may be improved, developed, and remodelled. The school can be traced from Montesquieu but has later drawn on social philosophy, sociology, and psychology, and its members look on law as part of, and in relation to, the whole process of social control of the conduct of individuals living in society. In the nineteenth century the jurist most closely approaching this approach was Jhering, who analysed law on the basis of the interests protected, and in the twentieth the leading proponents have been Roscoe Pound and Julius Stone (qq.v.). More extreme attitudes are sometimes adopted by realist (q.v.) jurists such as Holmes and Frank, with their emphasis on the uncertainty of law, the uncertainty of judicial fact-finding and the influence of social and other non-legal factors on decisions. There is undoubtedly a great deal of substance in the contentions of jurists of this school, who present a picture of the legal system which the everyday working lawyer understands and appreciates and whose general standpoint substantially accords with that of intelligent laymen and legislators.

Functus officio (having performed his function). Used of an agent who has performed his task and exhausted his authority and of an arbitrator or judge to whom further resort is incompetent, his function being exhausted.

Fund in court. A fund paid into or deposited in court to the credit of any cause, matter or account in any division of the High Court and vested in the Accountant-General on behalf of the Supreme Court. The kind of funds thus lodged include funds belonging to persons of unsound mind, funds the entitlement to which is in dispute, and other funds which should be kept in neutral custody.

Fundamental Constitution of Carolina. A scheme of government for Carolina framed at the instance of Sir Anthony Ashley Cooper (later Lord Shaftesbury) by John Locke. It mixed romanticized feudalism with modern concepts. There was to be a Grand Council with executive and judicial authority, to work in collaboration with a Governor, hereditary nobles, and deputies, who collectively comprised a provincial assembly. Cooper and the other Carolina proprietors were to form the Palatine Court. The ranks of society were to comprise landgraves (counts with baronies of land), caciques or chiefs, lords of manors, freeholders, serfs, and slaves. Ownership of land and social rank were equated. The scheme never became effective.

Fundamental law. A term frequently used in the seventeenth century in England, usually denoting rules of law which could not be infringed by Parliament or courts. Sometimes the common law was described as fundamental law, sometimes Magna Carta was, and frequently natural law or reason was so described, but there is no clear case of statute or rule of common law being nullified or disregarded by a court as contrary to natural law or reason, and this idea disappeared in the eighteenth century.

Appeal to fundamental law was common during the Great Rebellion, and seems generally to have meant the traditional rights of liberty and property; thus the arrest of the five members was condemned as being 'against the fundamental liberties of the subject' and Lilburne published in 1649 *The Legal Fundamental Liberties of the People of England* which equated fundamental liberties to Parliament and the common law. Prynne advanced a concept of fundamental law which amounted to the traditional rights of personal liberty and property, and parliamentary government. About the end of the seventeenth century the idea of the sovereignty of Parliament superseded invocation of the idea of fundamental law, and the idea disappeared in the eighteenth century. What fundamental law essentially stood for was the principle that politics were subordinate to ethics, and the preservation of the rights of the individual, notably his liberty and property.

In modern political science fundamental law designates some framework of government such as the U.S. Constitution, or some provisions in a system of law which are 'entrenched', and which cannot be repealed or amended by ordinary legislative procedure. It is generally associated with judicial review, by which a supreme court has the last word where the validity of legislative enactments is challenged and has to decide whether they are or are not in conformity with the constitution. In this sense there is no fundamental law in the U.K.

J. Gough, *Fundamental Law in English Constitutional History*.

Fundamental Orders of Connecticut and of New Haven. The two groups of emigrants from Massachusetts who in 1636 and 1638 respectively founded towns at Hartford, Windsor, and Wethersfield, which became the nucleus of Connecticut, and at New Haven, each in 1639 drew up written constitutions with these titles, providing representative governments which served until they were combined as the Colony of Connecticut under Charles II's charter of 1662, which in turn lasted till 1818. The Connecticut Fundamental Orders were framed by Thomas Hooker, John Haynes, and Roger Ludlow, emphasized the welfare of the community rather than of the individual and conferred a more extensive franchise than did Massachusetts, though still considerably short of full democracy. They were a clear and compact statement of government and were utilized by later framers of state constitutions.

Fundamental rights. An imprecise term generally used to mean the basic liberties of the subjects, or the natural rights asserted by political theorists, notably of the revolutionary periods in America and France. They were formulated in the American Declarations of Independence (1776) and the Bill of Rights (1791), in the French Declaration of the Rights of Man (1789) prefaced to the Constitution of 1791, and have in various versions been included in many constitutions since then. In these cases they are fundamental rules of law, entrenched and unalterable. Many are also enunciated in the U.N. Universal Declaration of Human Rights. In British law the term has no legal meaning, and it can only be taken as a general expression for the rights or liberties of the subjects generally regarded as essential and fundamental, including personal freedom, freedom of speech, of conscience, of the Press, of religion, of association, of family rights, of property, and of equal treatment by the law. In every legal system the crucial question is not whether the constitution or other law provides for or guarantees a particular right or liberty, but whether it is in fact recognized and protected, whether an individual can effectively assert it.

Fundamental rights of states. The essential and most important rights which states enjoy under international law by virtue of being thereby recognized as states. The philosophical theory was that states participated in the society of nations on the basis that their fundamental rights were recognized and accepted. The development of the idea of state sovereignty reinforced this theory. Recently the ideas of fundamental rights have had to be modified by reason of the development of international and supranational institutions. The fundamental rights usually asserted are legal independence in the sense of freedom from the domination of other states and from liability to intervention by any other state, with a consequential liberty of self-defence in case of need, and legal equality with all other states. Both fundamental rights are legal and do not exclude political, economic, and other dependence of some states on others and gross inequalities in many respects.

Fungibles. Movable goods which may be measured or valued by number, weight, or other measure and which are consumed by use, such as food and fuel.

Furca et fossa (gallows and pit). A clause frequently contained in ancient grants of jurisdiction, denoting the rights to hang male felons and to drown female felons.

Furnival's Inn. One of the former Inns of Chancery (q.v.).

Furtum. Theft in Roman law. In the XII Tables the penalties combined private justice with composition and were graduated according to the age of the thief and the gravity of the act. A distinction was also drawn between *furtum manifestum* (thief caught in the act) and *furtum nec manifestum*, and this long survived. Praetorian reforms treated *furtum* as a delict rather than a crime and subjected the thief to an action for a penalty, either double or quadruple the value of the stolen thing, the *actio furti*, and also to a claim for recovery of the stolen thing, either by a special claim, the *condictio furtiva*, or by the ordinary procedure for the protection of property. In developed Roman law theft was broadly defined as any dishonest handling of a thing for gain, including in handling the use of another's property lawfully held by the dishonest handler. This is *furtum usus*, as where a lawful borrower appropriates to his own purposes the use of the thing borrowed.

Fustel de Coulanges, Numa Denis (1830–1889). French historian. His masterpiece *La Cité Antique* (1864) is of great importance for legal history, though it over-emphasizes the place of religion in the evolution of Greece and Rome. Of his other works the most important is *Histoire des institutions politiques de l'ancienne France* (1875–92), later expanded and recast.

Future estates or interests (expectancies). At common law there were two kinds, remainders and reversions. Remainders were the remnant of an estate, limited to continue for a remainderman after a prior estate for life had determined; they might be vested, or contingent on the ascertainment of the person entitled or on the occurrence of some event. Reversions were residual interests which would revert to the granter on the determination of a particular estate granted to another, e.g. for a term of years. Since 1925 all future interests in land are equitable only, and may be vested or contingent.

Future goods. Goods to be manufactured or acquired by the seller after the making of the contract of sale, as distinct from goods then existing. A contract to sell future goods operates as an agreement to sell the goods, and the property in the goods passes to the buyer when the goods come into existence or are acquired, and the seller delivers them to the buyer or otherwise appropriates them to him, or the buyer takes possession of them by the authority of the seller.

Future property (or after-acquired property). Property which the party does not at the time actually possess. An assignment of future property is in equity an agreement to assign it when he does actually acquire it. Marriage-settlements frequently provide that future property is to be brought into settlement.

Fyfe, Sir David Patrick Maxwell, 1st Viscount Kilmuir (1900–1967). Called to the Bar in 1922 and practised at Liverpool before moving to London and taking silk in 1934. He entered Parliament in 1935, served as Solicitor-General 1942–45 and as Attorney-General in 1945. He was deputy chief prosecutor at the Nuremburg War Crime Trials, 1945–46. After being out of office till 1951 he was Home Secretary 1951–54 and Lord Chancellor 1954–62.

Fyrd. The English militia during the Anglo-Saxon period. The ealdorman of the shire was probably charged with calling out and leading the fyrd. It was probably composed of all freeholders. It was gradually superseded by the gathering of the thegns and their retainers, but seems to have been called occasionally even after the Norman Conquest.

G

Gabelle. A kind of tax in France before the Revolution, originally any tax on the sale of consumer goods, but particularly, from the fifteenth century, the salt tax. It was unfairly allocated, being based in some areas on real or estimated consumption, in others on production, and elsewhere areas were exempt. The nobility, clergy, and certain privileged persons were exempt. It gave rise to widespread smuggling, and was very unpopular, figuring high on the list of grievances presented to the States-General in 1789. It was accordingly abolished in 1790.

Gag. A colloquial term for the closure in the House of Commons.

Gag laws. Laws passed by the U.S. Congress in 1836 in violation of the First Amendment providing that petitions relating to slavery should be put forward without reference to committee or being printed. They were repealed in 1844.

Gage of land. The handing over of land in security for debt. It might originally take several forms, but the gagee was always given possession. If the profits of the land were applied in reduction of the debt it was lawful, but if not it was usurious and sinful and called *mortuum vadium* or mortgage. Later attempts were made to fit gages into the pattern of estates in land as by selling a term of years for a down payment or a transfer with a condition that if the debt were not paid the lessee should hold absolutely. By the fifteenth century the settled form was a feoffment (q.v.) with a provision for re-entry on payment, or later with a covenant for reconveyance on payment. Later still the mortgagor was allowed to remain in possession. From the sixteenth century the Chancery intervened in mortgages and began to develop the doctrine of the equity of redemption (q.v.).

Gail(l), Andreas von (1526–87). A German judge and jurist, author of *Practicarum Observationum libri duo* (1578) which influenced the Dutch law of procedure and was itself edited and annotated many times.

Gaius (*c.* A.D. 115–*c.* 180). A famous Roman jurist. Very little is known about him personally. He probably lived at Rome as a teacher and writer on law, but probably never had the *ius respondendi* (q.v.). Though an adherent of the Sabinian School, he sometimes adopts opinions of the other school. He seems to have been little appreciated in his own time and is not cited by Paul or Ulpian. He wrote a commentary on the *Edictum Provinciale* (30 books), one on the *Edictum Urbicum*, six books on the Twelve Tables, 15 books on the *Leges Juliae et*

Papia, *Res cottidianae vel Aurea*; *Liber singularis regularum*; a book *de casibus*, and several monographs, and, above all, his *Institutionum iuris civilis commentarii quatuor*, an elementary textbook on Roman private law, dealing partly with legal doctrine and partly with legal history, written about A.D. 161, a book which became and was for 300 years the standard textbook for students of Roman law. The work is well-arranged, clearly expressed, and the exposition plain and perspicuous. As his reputation grew, his other works came to enjoy similar authority. Under the Valentinian Law of Citations (A.D. 426) he was included as one of the five jurists all of whose writings were to have binding authority. The compilers of Justinian's *Institutes*, officially promulgated in A.D. 533, founded largely on Gaius' work, in order and substance. In the *Digest* there are 535 excerpts from Gaius. The compilers of the *Breviarium Alaricianum* (or *Lex Romana Visigothorum*) of A.D. 506 also used Gaius, condensing the first three books into two and omitting the fourth completely. A palimpsest containing the *Institutes* was discovered and identified at Verona in 1816, and new fragments were found in Egypt in 1933. The so-called Autun Gaius is a commentary on Gaius found at Autun in 1898, intended for teaching purposes, and there is an epitome of Gaius in the *Lex Romana Visigothorum* (q.v.).

A. M. Honoré, *Gaius* (1962); Editions: ed. Kubler in Huschke's *Jurisprudentia antejustiniana* (1908); ed. Baviera in *Fontes Juris romani antejustiniani*; ed. de Zulueta, *Institutes of Gaius* (1946–53).

Galiani, Ferdinando (1728–86). Italian economist, author of many works on political and economic subjects and of *Dei Doveri dei principi neutrali* (1782) an application of utilitarian principles to the problems of international law.

Gallicanism. The body of doctrines and practices designed to limit the papal power in France while yet maintaining the country within the Catholic Church.

Gallows. A beam projecting from a post or joining two posts, from which persons were hanged. Hence colloquially a term for hanging or capital punishment.

Gambling. The hazarding of money or other valuable thing on the outcome of a game, contest, or uncertain event, the result of which may be determined by chance or skill, with consciousness of risk of loss and hope of gain. The activity on the outcome of which the result depends may be of the nature of a game or sport, or be a serious activity attended with some risk or uncertainty, such as buying and selling stock or commodities, or an ocean voyage or other business venture. The outcome of the activity may be determined mainly by chance, e.g. tossing a coin, or by skill, e.g. a football match. Gambling has been from early times a widespread human activity, a great social evil and has been condemned by Church and State for centuries. It is also connected with and gives rise to various forms of criminality. Legal aspects relate mainly to efforts to suppress it as harmful and to regulate it and try to ensure that persons gambling are not cheated, but questions also arise of the civil validity and enforceability of gambling transactions. From Edward III English kings issued prohibitory laws designed to repress gambling and promote manly sports and military training.

Within the general category of gambling, English law distinguishes betting, wagering, gaming, and lotteries. Betting is hazarding a sum of money or other bet on the outcome of an uncertain event. It may and frequently does, involve more than two parties, e.g. a football pool, and this differentiates it from wagering.

Wagering is a kind of betting, the contract by which two persons only, professing to hold opposite views as to the outcome of a future uncertain event, agree that according to the outcome one shall win from the other and the other pay or transfer to him, money or other stake. There cannot be more than two parties to a wager. A contract by way of wager is void by statute.

Gaming is playing a game of chance or skill for stakes hazarded by the players and a gaming contract is one of mutual promises to transfer the sum or thing staked on the result of the game. Some games, e.g. horse-racing, are lawful, others, e.g. baiting animals, unlawful. Gaming may be done by a machine. Gaming is not an offence, but gaming at unlawful games is unlawful gaming. Contracts by way of gaming are void.

A lottery is a scheme for distributing prizes by lot or chance. A prize competition is not a lottery if the result depends to a substantial degree on skill.

All these kinds of gambling are specially defined for the purposes of certain statutes.

Statute has intervened to regulate betting, wagering, gaming, and lotteries and to penalize transactions outwith the permitted premises. Betting shops have been legalized, subject to conditions. Betting may also be conducted, subject to conditions at licensed race courses. Gaming is permitted, under conditions, only at licensed clubs. Special statutory provisions regulate gaming machines. Lotteries are lawful only in statutory defined circumstances.

In Scotland betting, wagering, and gaming transactions are not void, but are deemed *sponsiones ludicrae*, not seriously intended, and legally unenforceable.

Game. An indefinite term covering many categories of wild land animals frequently killed for sport or for food, such as rabbits, hares, pheasants, partridges, grouse, and the like; many particular statutes contain their own definitions. At common law the Crown had right to game in forests, chases, and manors and these rights either survive in the Crown or have been granted as privileges to certain subjects; elsewhere the right to take and kill game is incidental to the ownership or occupation of land. An owner of land may kill and take game on his own land as he pleases, or let the shooting rights with or without the land. An agricultural tenant may kill and take ground game on the land occupied, whether any other person may do so or not, and has right to compensation from his landlord in certain cases for damage done to his crops by certain kinds of game. There is no right of property in the game until it is killed or reduced to possession.

The Norman kings were great lovers of field sports and greatly extended the forest laws (q.v.) making hunting and killing animals a sport reserved to the King and those whom he allowed to share it. Though mitigated by later kings this gave rise to the game laws, a system which down to the time of William IV made it illegal for a man to take or sell game, even on his own land unless possessed of a real property qualification of at least £100 a year.

There are extensive bodies of legislation in England and Scotland directed to the protection of animal and bird life by provision of close seasons and days, to prohibiting certain modes of killing of various kinds of game and imposing restrictions on purchase, sale or possession of game at particular times. Trespass in pursuit of game is actionable and also punishable criminally.

Game licence. An authority to kill game of stated kinds during the open or permitted seasons and on lands where the holder of the licence has permission to go to shoot, issued under statute.

Gamekeeper. The name given to two classes of persons, persons appointed by a lord of a manor, lordship or royalty or steward of the Crown of any manor, lordship or royalty appertaining to the Crown, appointed by writing and authorized to seize dogs, nets, and other instruments used by persons taking or killing game without a game licence, and also persons employed by ordinary landowners to look after game.

Gaming. Playing a game, whether of skill or of chance or of both combined for stakes hazarded by the players. It may be a bipartite, but it is commonly a multipartite, arrangement. If bipartite, it may also be described as a wager (q.v.). At common law games for stakes were lawful.

Commercial gaming is now restricted to premises licensed and registered under statute, and gaming elsewhere is restricted by statute. Outwith these limits gaming is unlawful. Contracts by way of gaming are void.

Gaming-houses. Premises to which a substantial number of people resort habitually for the purpose of gaming. To keep a common gaming-house is an offence at common law.

Gandino, Albertus de (*c.* 1245–*c.* 1310). An early writer on criminal law, author of a *Tractatus de Maleficiis* (1491), long deemed valuable in Germany.

Gang. A group of individuals who in collaboration participate in some activity, particularly in many kinds of crimes such as extortion of protection money, assaults, thefts, burglaries, and malicious damage. They are commonly associated with large housing areas, public houses and shiftless existence and regularly claim particular areas as 'theirs' and fight with any outsiders who come in. Carrying weapons is accordingly common. The term has also been used of, in the past, groups of young persons who collaborated in crime under the leadership or instructions of an adult or older boy, and of groups of convicts assigned to particular places or work.

Gangster. A member, particularly in the U.S., of an organization devoted to profit from such activities as extortion, gambling, drugs, and prostitution. He may be an organizer who plans the gang's operations or a participant. The worst recorded period of gangsterism was in the U.S., particularly New York and Chicago during the Prohibition era, when gangsters engaged openly and violently in struggles for the control of the making and distribution of illicit liquor. Murders were commonplace. Since then gangsterism has been less prominent but still underlies much crime, and legal process has proved comparatively ineffective to repress their activities.

Gans, Edward (1798–1839). German Jewish jurist who sought to raise acceptance of Jewish culture and initiated some important studies in that field. He became a Christian in 1825 and professor of law at Berlin in 1826. His major work is *Das Erbrecht in Weltgeschichtlicher Entwicklung* (Property law in historical development), and he also wrote on Roman law and history.

Gaol (Lat. Gaola, Fr. Geole, a birdcage). A prison or place of detention; hence gaol-bird or jail-bird, a prisoner. In England gaols were originally common or county gaols as distinct from private prisons kept by nobles or bishops. They were not purpose-built but any kind of building served and the authorities farmed the gaols out to private persons on a profit-making basis. The gaolers exacted fees for everything, admission, hammering on prisoners' irons, meals, better rooms, provision of alcohol or women, and so on. No health precautions were taken so that gaol fever was common, and sometimes spread to the courts killing judge, jurors, and witnesses. Conditions were ameliorated by the work of John Howard, Fowell Buxton, Elizabeth Fry (qq.v.), and others and in Parliament by Romilly, Fowell Buxton, Mackintosh, and Peel whose Gaol Act of 1823 began the modern prison system.

Gaol Delivery, commission of general. A commission addressed to certain persons directing them to deliver the gaol of a stated place and try the prisoners therein. It authorized the trial of every prisoner charged with any treason, felony, or misdemeanour no matter when committed. Latterly the distinction between this commission and that of oyer and terminer (q.v.) became slight as both were granted to and executed by the same person. It was one of the commissions given to judges and commissioners of assize.

Gaol sessions. Under the Gaol Sessions Act, 1824, a court consisting of all the justices of the county or any two of them and being the prison authority managing the gaols of the county. The functions were transferred to the Home Secretary in 1877.

Gardiner, Gerald Austin, 1st Baron (1900–). Called to the Bar in 1925 he took silk in 1948 and was made a life peer in 1963 and Lord Chancellor in 1964–70. His main interest was law reform and his period of office was noted for the establishment of the Law Commission and the Scottish Law Commission as standing bodies to promote law reform.

Garnish. (1) Money exacted from prisoners on their admission to gaols and bridewells, until such exactions were prohibited in 1815.

(2) To attach a debt. A garnishee is a debtor whose debt to his creditor has been arrested in his hands by a garnishee order which warns him not to pay his creditor but a third party who holds a judgment against his creditor.

Garofalo, Raffaele (1852–1934). Italian criminologist, a leader of the positive school, saw criminals as person whose altruistic sensibilities were lacking or inadequately developed and examined the theories and measures of crime prevention and repression. He wrote *Criminology* (1885) and other works.

Garrati, Martino (fifteenth century). Italian jurist, author of many works including a *Tractatus de Bello* and *Tractatus de principibus eorumque legatis et consiliariis.*

Garrotting. The crime of robbery from a person by partially throttling him, but not normally doing any permanent injury. There were outbreaks of this crime in Manchester in 1850 and in London in 1850 and 1862, and the Garrotters Act, 1863, imposed the penalty of flogging in addition to penal servitude and this almost entirely stopped the crime.

Garrotting was also used in some countries, notably Spain, as a method of executing criminals. It utilized a garrotte which took the form either of an iron collar attached to a post, which was put around the victim's neck and tightened by a screw until asphyxiation occurred or of a length of wire with wooden handles at the ends, held by the executioner and pulled to effect throttling.

Garter, The Most Noble Order of the. An English order of knighthood founded about 1350 by Edward III. It is the highest such order in Britain and probably the most ancient and distinguished of all such orders. It comprises the sovereign, members of the royal family, and 25 knights companion. The insignia are a blue sash, a collar of gold, a badge bearing the cross of St. George and the motto *Honi soit qui mal y pense* ('Shame to him who thinks ill of it'), a silver star, and a blue garter edged with gold worn below the left knee. The chapel of the order is St. George's Chapel, Windsor Castle. The motto is alleged to commemorate the founding of the order, when a lady of the court dropped her garter and Edward III returned it, rebuking bystanders with the phrase.

Gas chamber. A method of capital punishment used in some U.S. States. The prisoner is strapped in a chair in a sealed chamber into which highly poisonous fumes are released. A deep breath causes almost instantaneous death but attempts by prisoners to hold their breath or take shallow breaths may cause slow and painful death.

During the Second World War large gas chambers were extensively used by the Nazis to kill large numbers of Jews and other unwanted people.

Gavelkind. A special kind of tenure of land, practically confined to Kent (in which county land was presumed to be so held in default of contrary proof), whereby the lands were held by fealty and suit of court. A tenant in gavelkind had rights lost by the common law of England, such as to devise land by will, while on intestacy lands descended to all sons equally, whom failing to all daughters equally, and were not liable to escheat for felony, and a youth could alienate his land once he had attained 15. It was abolished in 1925.

T. Robinson, *Common Law of Kent or Custom of Gavelkind.*

Gavelkind Act (1704). One of the Irish penal laws, by which on the death of a Catholic landowner his estates were to be divided among all his sons, unless the eldest conformed to the Church of Ireland within one year or on coming of age, in which case he inherited the whole estate. It was intended to, and did, reduce the number of Catholic landowners.

Gazette. The official government news sheet. It is said to have been published first at Oxford in 1665 and later renamed the *London Gazette.* It contains intimation of State Business, proclamations, appointments to offices under the Crown, Orders in Council and other orders required to be published therein, notices of bankruptcy and other similar matters required by statute to be published therein. There are similarly an *Edinburgh Gazette* and a *Belfast Gazette.* By various statutes production of a copy of the Gazette is made *prima facie* evidence of facts printed therein.

Gazumping. The withdrawal by a vendor from a contract of the sale of land to which he has agreed 'subject to contract' so as to accept a higher offer from another party.

Geilhoven, Arnold (fifteenth century). Dutch jurist, author of *Remissorium utriusque juris*, an encyclopaedia of Roman and canon law, and other works.

Gemot. In Anglo-Saxon times a mote or moot or meeting or assembly. There were various kinds. The folk-gemot or general assembly of the people of a town or shire usually held annually; the shire-gemot, held thrice a year; the hundred-gemot or hundred court which originally met 12 times a year, but in Norman times was to be held twice a year as the Sheriff's tourn or view of frankpledge; the halle-gemot or court baron; and the wardemote.

Gendarmerie Nationale. In France a military force controlled by the Ministry for Armed Forces but made available to the Ministry of the Interior for police work. It polices rural areas and small towns and acts also as military police.

General Act for the Peaceful Settlement of International Disputes (1928). An attempt by the League of Nations to create a comprehensive system of international arbitration. It was based on a series of model treaties devised by the Committee on Arbitration and Security set up to prepare for the 1927 Disarmament Conference, and proposed as alternative methods of settling disputes the Permanent Court of International Justice for legal disputes and Conciliation Commissions and Arbitral Tribunals for different stages of non-justiciable disputes. It could be signed as a whole or in respect of particular chapters and came into force between states accepting parts of it. The U.K. adhered to it in 1931, but reserving intra-Commonwealth disputes and the principle that disputes might still be submitted to the arbitration of the League Council.

General Agreement on Tariffs and Trade (G.A.T.T.). An international agreement signed in 1947 comprising a set of bilateral agreements concerned with customs and commercial policy. Its main aim is the liberation of world trade from quotas, restrictions, and controls but it also tries to encourage economic development and to reduce and stabilize tariffs. It also sets out general rules providing for most-favoured-nation clauses for all G.A.T.T. countries and for the obligation of each member to negotiate for tariff cuts at the request of another. It covers about 80 per cent of world trade and is an important forum for exchange of information and views on trade. It is one of the specialized agencies of the U.N., which co-operates with the Economic and Social Council and maintains close liaison with the European Communities. The G.A.T.T. International Trade Centre was established in 1964 to assist the less developed countries to promote their export trade.

General Assembly (of U.N.O.). The General Assembly comprises all member states of the U.N., each having one vote and not more than five representatives. It meets once a year and in special sessions convened at the request of the Security Council or of a majority of members. Its functions are to consider and make recommendations on the principles of general co-operation in the maintenance of international peace and security, to discuss any problem affecting peace and security, or any question within the scope of the Charter affecting the powers and functions of any organ of the U.N., to initiate studies and make recommendations to promote international political co-operation, the development of international fundamental freedoms for all, and co-operation in economic, social, cultural, education and health fields, to receive and consider reports from the Security Council and other organs of U.N., to make recommendations for the peaceful settlement of any situation which might impair friendly relations, to supervise the execution of the trusteeship agreement, to elect members of the other organs of the U.N. and appoint the Secretary-General, and to consider and approve the budget of the U.N.

It operates through seven main committees, Political and Security, Special Political, Economic and Financial, Social Humanitarian and Cultural, Trusteeship, Administrative and Budgetary, and Legal committees. There are also two procedural committees and power to appoint joint and *ad hoc* committees. It does not and cannot legislate for member states though its recommendations may have considerable moral and political persuasive power.

General Assembly of the Church of Scotland. The supreme court and legislature of the Church of Scotland. It meets annually in Edinburgh for 10 days in May. The Queen is represented by the Lord High Commissioner, a layman appointed to represent the Queen for the occasion. In 1969 the Queen attended personally and there was no Lord High Commissioner. It is presided over by the Moderator, a clergyman elected for a year. He is assisted by Principal and Depute Clerks of Assembly and the Procurator of the Church, who is a practising Q.C.

General average. Where a loss arises to one or more of the three interests involved in a maritime adventure, the ship, the cargo, and the freight, in consequence of extraordinary sacrifices or expenses incurred for the preservation of the several interests involved, it must be borne in due proportion by all and is called a general average loss. The danger must have been real and the property in respect of which the claim is made must have been voluntarily and reasonably sacrificed, as by jettison of cargo to lighten a ship, for the benefit of the adventure and the property must have been saved by reason of the sacrifice. The liability to contribute may be enforced by lien or by action. A general average loss differs from a particular average loss which is a loss fortuitously caused by a maritime peril and which has to be borne by the party on whom the loss originally fell.

R. Lowndes and G. Rudolf on *General Average and the York–Antwerp Rules.*

General Council of the Bar (or Bar Council). A body established in 1895 to represent the English Bar as a whole, dealing particularly with the honour and independence of the Bar, professional standards, conduct and etiquette, the furtherance of legal education, the system of legal aid, and related matters. It now operates within the framework of the Senate of the Inns of Court and the Bar, but is autonomous in relation to its separate powers and functions. It consists of a Chairman, Vice-Chairman, the Attorney-General, the Solicitor-General, elected members, and certain additional and co-opted members. It operates through standing and special committees. Its annual reports frequently give guidance on points of professional conduct. It has no disciplinary powers.

See also SENATE OF THE INNS OF COURT AND THE BAR.

General or Oecumenical Council. A meeting of the cardinals and bishops of the Roman Catholic Church, the decrees of which are valid for the Church everywhere and are important sources of Roman canon law. Twenty-three Councils have been held, as follows: Nicaea I (325), Constantinople I (381), Ephesus (431), Chalcedon (451), Constantinople II (553), Constantinople III (680), Nicaea (787), Constantinople IV (869), Lateran I (1123), Lateran II (1139), Lateran III (1179), Lateran IV (1215), Lyons I (1245), Lyons II (1274), Vienna (1311–12), Constance (1414–18), Basel–Florence (1431–45), Lateran V (1512–17), Trent (1545–63), Vatican I (1869–70), Vatican II (1962–64).

General Court. The term commonly applied in the New England colonies to their bicameral legislatures, consisting of the Governor and his Assistants (Upper House) and the deputies (Lower House). It remains the official name of the legislatures of Massachusetts and New Hampshire.

General damages. Monetary damages which may be awarded in compensation for general damage or the kind of harm and loss which naturally and normally follows from the wrong and which does not need to be specifically pleaded and proved, as in cases of personal injuries for pain and suffering, loss of limbs and damage to health. It is distinct from special damages (q.v.).

General Dental Council. The body in the United Kingdom charged by statute with the general government of the dental profession, the promotion of a high standard of professional education and conduct, and maintenance of the register of dental practitioners. It consists of members nominated by the Queen on the advice of the Privy Council or by the Governor of Northern Ireland, or by universities and royal colleges, and members elected by registered dentists from among themselves. Through its disciplinary committee it regulates the professional conduct of registered practitioners and may erase from the register the name of a person guilty of infamous or disgraceful conduct in a professional respect. Appeal lies to the Judicial Committee of the Privy Council.

General issue. In English common law pleading, the plea by the defendant throwing the whole onus of proof of the facts required to establish the plaintiff's claim on him, as distinct from a special plea such as that a contract was voidable. The terms of the general issue varied from one form of action to another; in debt on a simple contract, it was *nil debet*, which denied the existence of the debt, or *non assumpsit*, which denied the existence of the contract; in debt on a speciality, it was *non est factum*, in trespass, it was not guilty. Pleading the general issue was abolished in 1873.

General jurisprudence. In the legal philosophy of John Austin (q.v.) general (or comparative) jurisprudence or the philosophy (or general principles) of positive law is the science concerned with the various principles, notions and distinctions common to maturer systems and forming analogies or likenesses by which such systems are allied. Among the necessary principles, notions, and distinctions, the examination of which occupies a large part of Austin's work, were the notions of duty, right, liberty, injury, punishment, redress, their relations one to another and to Law, Sovereignty and Independent Political Society, and the distinction of rights, obligations, and injuries into their sub-groups. Valuable as Austin's analytical examination of these notions has been he was too ready to infer from the few legal systems he considered, almost entirely Roman and English law, that such distinctions as obligations *ex delicto* and *quasi ex delicto* were necessary and generally found in mature legal systems. Some distinctions, e.g. into law and equity, civil law and commercial law, are peculiar to particular systems and neither general nor necessary and the number of ideas which are general, even in mature systems of law, is probably very small. It is possible to have a system of law in which no person has any rights, but only duties.

General Medical Council. The body in the United Kingdom charged by statute with the general government of the medical profession, supervision of medical education and maintenance of the register of registered practitioners. It consists of members nominated by the Queen, members appointed by universities and royal colleges, and members elected by registered practitioners. Through its disciplinary committee the G.M.C. exercises strict discipline over the conduct of registered practitioners and it may direct that a practitioner's name be erased from the register or impose lesser sanctions. An appeal lies to the Judicial Committee of the Privy Council.

General Nursing Council. Two councils, for England and Wales and for Scotland, maintain, and prescribe qualifications for admission to the registers of nurses. They consist partly of appointed and partly of elected members. The councils regulate professional discipline through disciplinary committees and may remove from the registers the names of persons convicted of crimes. Appeal lies to the Divisional Court of the Queen's Bench Division or to the Court of Session.

General Optical Council. The body in the United Kingdom charged with promoting high standards of professional education and conduct, approving training institutions and qualifications, regulating optical practice, and maintaining registers of ophthalmic opticians. It consists partly of nominated, partly of elected members. Through its disciplinary committee it exercises control over registered opticians and may direct the removal of the name of any person or body convicted of certain offences or contravention of the rules made by the council such as to render the person unfit to have his name on the register. An appeal lies to the Judicial Committee of the Privy Council.

General Part. The introductory part of the German Civil Code, copied by some other codifications, dealing with the concepts and principles which are common to all parts of the Code, such as legal persons and the effects of lapse of time.

General principles of law as recognized by civilized states. One of the sources from which the International Court may, under its statute, draw rules for application to particular cases. The kinds of propositions which have been held to be general principles have been such as that any breach of an engagement involves an obligation to make reparation, and that individual members of a corporate body have no right of independent action in matters affecting the organization. It has been suggested that the intention was to allow the Court to apply the general principles of municipal law, particularly of private law, so far as applicable to the relations of states. It will frequently be a matter of doubt whether a particular proposition can be called a general principle of law, and the Court has, in practice, not often invoked general principles, and when it has done so, has refrained from resorting to them by that name. Arbitral tribunals have however, frequently resorted to municipal law analogies.

B. Cheng, *General principles of law as applied by international courts and tribunals.*

General Privilege. The Spanish counterpart of Magna Carta, a legal charter compiled in Aragon and Valencia in 1283 which laid down limitations of the royal power in both kingdoms. It was accepted by King Pedro III of Aragon in 1283. It received the name of General Privilege because it ratified the privileges (*fueros*) of the nobles and municipalities, made justice a function of the Cortes or Parliament, and made annual sessions of the Cortes mandatory. After the acceptance of the charter there was constant struggle between king and nobles and in 1348 Pedro IV, having defeated the nobles, abolished the Privilege.

General sessions. A court of record formerly held by two or more justices of the peace. When held at stated times in the four quarters of the year, they were known as quarter sessions (q.v.) and general sessions held at other times were restricted to London and Middlesex. They were abolished in 1971.

General ship. A ship in which the owners and master contract with various consignors to ship their goods, normally under bills of lading (q.v.), as contrasted with a ship chartered to one or more persons under a charter-party (q.v.) for the carriage of their goods only.

General strike. A strike by a large number of workers covering a number of industries as an organized attempt to secure political or economic ends, as distinct from single-factory or single-industry strikes. In the 1820s and 1830s radicals and union leaders advocated general strikes as means of achieving the reform of Parliament and improvements in the conditions of working people. In 1842 the Chartists tried to extend strikes in Lancashire into a general strike but the attempt failed in the face of governmental reistance.

General strikes occurred in Belgium in 1893 and 1902 on the question of manhood suffrage and a similar strike occurred in Sweden in 1902. In 1909 there was a general strike in Sweden over wages, which lasted a month. In 1926 in the U.K. a strike in the coal industry spread to railways, steel, building, and some other industries and was supported by the T.U.C. It lasted nine days but after a settlement had been negotiated the miners remained on strike for six months. The legality of this general strike was disputed.

There were wide-ranging strikes in France and Italy in 1967, starting with demand by students for reforms in education and taken up by industrial workers.

In Asian and African countries general strikes have frequently been used as weapons in campaigns for independence, though these have frequently been only partially effective because of the small proportion of the population organized in unions.

The general strike is frequently more a political than an economic weapon and may be hardly distinguishable from a revolt or even revolutionary movement.

General Surveyors of the King's Lands, Court of. Between 1512 and 1515 Henry VII appointed two general surveyors of Crown lands responsible for the revenue of the Treasurer of the Chamber. The appointment was later regulated by statute. In 1542 the general surveyors were made a court, administered the Crown lands acquired by Henry VII and Wolsey, and adjudicated on disputes arising out of the revenue under its charge. In 1547 it was combined with the Court of Augmentations and in 1554 absorbed in the Exchequer.

General Synod of the Church of England. A body created in 1969 reconstituting and replacing the Church Assembly. It consists of the House of Bishops, the House of Clergy, and the House of Laity, of which the first two are the Upper and Lower Houses of the convocations of Canterbury and of York; the third consists of elected members from each diocese with certain *ex-officio* and co-opted members. It appoints a legislative committee to which are referred all Measures (q.v.) passed by the Synod which it is desired should be given the force of statute in accordance with the procedure prescribed by the Church of England Assembly (Powers), Act, 1919. It considers matters concerning the Church and provides for them by Measures, canons, or orders and may consider and express its opinion on any matter of religious or public interest.

General verdict. A jury verdict finding generally for the plaintiff (or, in such cases as seditious libel, for the prosecution) or for the defendant, as distinct from a special verdict, finding certain facts but leaving their effect in law to be determined by the court.

General warrant. A warrant formerly issued by a Secretary of State authorizing the arrest of persons not named but merely designated as those responsible for a particular act, such as the publication of a libel. This left it to those charged with executing the warrant to determine whom to arrest. Such a warrant could also authorize the search for and seizure of unspecified papers. They were common in the seventeenth century and authorized by statute in the Licensing Act, 1662. In the famous cases involving John Wilkes (q.v.) (*Entick* v. *Carrington* (1768), 19 St. Tr. 1030; *Wilkes* v. *Wood* (1763), 19 St. Tr. 1153; and *Leach* v. *Money* (1765), 19 St. Tr. 1001) they were held illegal at common law. In 1766 the House of Commons resolved that general warrants were illegal.

General will. A concept of Rousseau, to the effect that a community has a collective good which is not the same thing as the private interests of its members. In accordance with the analogy of an organism, it may be said to have a will of its own, the general will (*volonté générale*).

In a democratic society the State represents and gives effect to the general will of the citizens, that which is accepted by at least most of them, and consequently in accepting the State's laws and rules, each citizen is really pursuing his own interest. Rousseau said repeatedly that the general will is always right, but was vague as to the criterion of right.

Généralité. An administrative unit in pre-Revolution France, established in the fourteenth century to manage the collection of the royal revenues. Properly speaking, the name applied only to the districts where the amount of tax was determined and apportioned by royal officials, but was extended to areas where taxes were granted and apportioned by provincial assemblies, and there were also *généralités* established in lands annexed in the eighteenth century. The number, originally four, increased in time. In the seventeenth century the *généralités* became the areas of authority of the *intendants* who were royal agents having financial, judicial, and police powers. By 1789 there were 33 *généralités*, the boundaries of which had no historical or rational basis. After the Revolution they disappeared being replaced by the new *départements*.

Genet, Raoul (twentieth century). French jurist, author of *Traité de diplomatie et de droit diplomatique* (1931–32), *Précis du droit maritime pour le temps de guerre* (1937), and other works on international law.

Geneva Arbitration. The arbitration held to determine the extent of British liability in the *Alabama case* (q.v.).

Geneva Conventions. A series of international treaties concluded at Geneva between 1846 and 1949 with the object of mitigating the harm done by war to both service personnel and civilians. Dunant, founder of the Red Cross, started international discussions resulting, in 1864, in a Convention for the Amelioration of the Wounded in Time of War. It was widely ratified, amended and extended in 1906 and extended to maritime warfare by the Hague Conventions of 1899 and 1907. A Convention Relating to the Treatment of Prisoners of War was agreed in 1929.

The earlier Conventions were flouted during the World Wars and in 1949 four Conventions were agreed extending and developing the principles earlier settled. These dealt with (1) Amelioration of the Condition of the Wounded and Sick in Armed Forces in the Field; (2) Amelioration of the Condition of the Wounded, Sick and Shipwrecked Members of Armed Forces at Sea; (3) Treatment of Prisoners of War; (4) Protection of Civilian Persons in Time of War. These restated and elaborated the earlier conventions and have been ratified by the majority of states.

Geneva Protocol for the Pacific Settlement of International Disputes (1924). An agreement to submit all justiciable disputes to the Permanent Court of International Justice, and non-justiciable disputes to the League Council or to an arbitral committee formed by the Council or by agreement between the contending parties. Any state refusing to submit to arbitration, or to carry out an arbitral award, was to be deemed an aggressor and the victim of aggression was to receive assistance from the other signatories. The protocol was unanimously approved by the Assembly and by some countries, but rejected by the U.K. and the project collapsed.

Genocide. The crime under international law of seeking systematically and deliberately to extinguish a race of people or ethnic group by mass murder, as was practised against the Jews by Germany during the Second World War. In 1948 the U.N. General Assembly adopted a Convention on the Prevention and Punishment of the Crime of Genocide which confirmed that it was a crime under international law in peacetime as much as in war, inferring state responsibility and personal responsibility of individuals concerned in the policy. This came into effect in 1951. Parties to the convention undertook to take the necessary measures under their own laws to ensure the effective application of the Convention. The Convention has established that genocide is not a private matter for a state but a matter of international law and general concern. Any state can call on the U.N. to intervene and take action considered appropriate for the prevention and suppression of genocide.

Genossenschaft. In German law, a corporate association formed by a combination of persons. There are various kinds regulated by special laws. The minimum number of members is seven and their liability depends on the kind of association. They have the rights of a corporation, if formed in compliance with the laws and are commonly used for various lawful trade purposes.

Gentili, Alberico (1552–1608). Italian jurist, studied at Perugia, but as a Protestant had to leave Italy and, in 1581, became D.C.L. of and Reader at Oxford and in 1586 Regius Professor of Civil Law where his teaching was famous. He was employed on embassies abroad and, in practice, in London and his fame rapidly grew. His publications include *De Legationibus* (1585), *De jure belli commentatio prima* (1588), revised as *De jure belli libri tres* (1598), *Advocationis Hispanicae libri duo* (1613), and treatises on Roman law such as *Dialogi sex de veteribus iuris interpretibus* (1582), and on ancient war practices as well as a *De Unione Angliae et Scotiae* (1605). He has increasingly been recognized as the most notable forerunner of Grotius in establishing public international law and the first true scholar and writer on modern international law, as the law of the community of states. He first grasped as a whole the relations of states with one another and suggested solutions of problems in reliance on law other than Roman, particularly on the law of nature, and placed the subject on a non-theological basis, being one of the first to separate secular law from theology and canon law.

Gentili, Scipio (1563–1616). Younger brother of Alberico Gentili (q.v.), an Italian jurist and humanist, editor of Donellus and author of works on Roman law.

Gentleman (Lat. *gentilis homo*, a man of ancestry). Originally a man of good birth, above the rank of yeoman (q.v.). The word has no technical or exact legal meaning, but is sometimes used to designate a person having private means and no occupation, yet not entitled to be designated esquire as a member of a learned profession. It is also used more colloquially to designate a person of very good manners and of a high standard of conduct.

Gentleman's agreement. An agreement between two or more parties which, by express provision or technical legal defect, is not legally enforceable as a contract. It is said to have been judicially defined as 'an agreement which is not an agreement, made between two persons, neither of whom is a gentleman, whereby each expects the other to be strictly bound without himself being bound at all'. Another judge has observed that 'a gentleman's agreement usually ends in each party calling the other no gentleman'.

Gény, François (1861–1959). French jurist, author of *Méthode d'interprétation et sources en droit privé positif* (1899) and *Science et technique en droit privé positif* (1914–1924), the first a plea for revision of the traditional French methods of interpreting the law, emphasizing the element of judicial discretion, and *libre recherche scientifique*, the latter dealing with the problem of providing a new method of free scientific research to replace the traditional lines of legal development and interpretation. Both are now classical. He was one of the first and most important of the jurists contending for a new sociological conception of the law. To him the social matter with which the lawyer operates is as much a part of legal science as the legal form which is part but by no means whole of the juristic technique. His work has been very influential on legal philosophy.

Gerard de Rayneval, Joseph Matthias (1746–1812). French diplomat and author of *Institutions au droit de la nature et des gens* (1803) and *De la liberté des mers* (1811).

Gerber, Karl Friedrich Wilhelm von (1823–91). German jurist, author of *System des deutschen Privatrechts* (1848–49) and *Grundzüge eines System des deutschen Staatsrechts* (1865), which became the foundation of the system of public law for the German Empire.

German Confederation. A federation created in 1815 to replace the Holy Roman Empire (q.v.). There was a federal constitution, providing for a Bundestag or Federal Diet; Austria dominated the confederation until 1858, but it never attained any substantial unity. It was superseded in 1867 by the North German Confederation of 35 states and four free cities.

German Empire. Created at the end of the Franco-Prussian War (1871) to succeed the North German Confederation. Its constitution was based on that of the Confederation, though enlarged to include more states. It provided for a Bundesrat in which each state was represented and a Reichstag elected by universal male suffrage. The only federal minister was the Chancellor who was appointed by the King of Prussia who was Emperor. Each of the 26 states retained its own constitution and administered both federal and state law and justice, and some responsibilities, such as posts, remained with the states. It came to an end in 1918.

German (Federal Republic of Germany) law

History

In Germany the laws of the barbarian tribes and the Capitularies of Charlemagne were early forgotten and superseded by unwritten tribal law, feudal law administered in feudal courts, and city law. The emperors enacted some legislation, *constitutiones pacis* and constitutional statutes such as the Golden Bull of 1356 regulating the imperial election. Towns had privileges and frequently the right to make private rules, and there came into being unofficial treatises on city law. There were manorial customs which were occasionally put in enacted form.

Written law appeared from about 1200 alongside customary law, in the form of textbooks of territorial and feudal law. Of these the most famous is the *Sachsenspiegel*, a record of Saxon law composed about 1230 by Eike von Repkow. It acquired great authority, was translated and was furnished about 1330 with a gloss by Johann von Buch, later enlarged. Based on it was the *Spiegel der deutschen Leute* or *Deutschenspiegel* of about 1250, which tried to present Germanic common law. It was soon displaced by the *Kaiserliches Land und Lehnrecht*, miscalled the *Schwabenspiegel*, of about 1260. Both books were re-edited to meet the needs of various districts. Later about 1320, an unknown jurist wrote a *Kleines Kaiserrecht*, which aimed to present common law as fixed by Charlemagne for all Christendom, and there were various books on feudal procedure and practice.

There were also statutes enacted in the Imperial diets notably 'public peaces' and constitutional statutes, such as Charles IV's Golden Bull of 1356, and some local ordinances. Written town law stemmed from privileges conferred by the King or town lord, and was frequently adopted by one town from another. There were also treatises on city law and compilations of the judgments of lay-judges in town courts.

The development of a single legal system in Germany was delayed for centuries by the absence of a strong central monarchy. Each district had its local law, which prevailed over the regional, and regional over general law.

From about 1400 there took place the Reception of the Roman, canon, and Lombard feudal laws. From about 1100 there had been a conviction that the Roman law had a claim to validity in Germany, that the Roman Empire of the German nation continued the old Roman Empire. Later Roman law began to be received, in practice, by the appearance of judges and jurists schooled in Roman law. When the Imperial Chamber of Justice (Reichskammergericht) was established in 1495, half of its members had to be learned in the law, and all were to judge according to the common laws of the Empire. This led to a reception of Roman law in the lower courts. The reception proceeded because of the underdeveloped state of the native law. It was encouraged by the universities and the appearance of popular legal literature, such as Raymund von Wiener-Neustadt's *Summa Legum* (*c.* 1345), the book of pleas published in 1516 under the name *Klagspiegel* and Ulrich Tengler's *Laienspiegel* (1509) dealing with private, criminal, and procedural law. But the general lack of system and principle in the native law led scholars to the study of the Roman law. The Reception was the outstanding event in German legal history and effected a breach with the older tradition.

The clergy brought canon law into Germany for use in the administration of the Church but the efficiency of the church courts was such that by 1200 they had acquired an importance equal to that of the civil tribunals. It, and its study, presupposed Roman law. Various books of canon law and procedure appeared in Germany, such as Johannes de Stynna's *Speculum Abbreviatum* based on the *Speculum Juris* of Durantis, and numerous *Summae confessorum*.

The canon law was a principal subject of study in German universities from their foundation, even more so than the Roman law. Only from about 1500 was the latter studied equally.

The Roman law, however, triumphed because of its systematic character and basis in principle over the multitudinous German rules, written and unwritten, mostly applicable in limited territories only. The influence of Roman law is seen in the anonymous *Der richterliche Klagspiegel* (Mirror of Judicial Plaints) (*c.* 1400), which deals with the civil law and criminal law and procedure, and thereafter the lay-judges increasingly looked to Roman law to fill gaps, and then not the *Corpus Juris* but the results of the work of the Commentators and since that time academic teaching and research have had much greater influence on the development of German law than they have in Anglo-American countries.

The Roman law was most influential in the fields of obligations and property and canon law in domestic relations, succession and civil procedure but native Germanic institutions were never completely ousted, notably in criminal law and procedure.

In the sixteenth century the Imperial Diet passed much legislation of a constitutional nature and concerning the Imperial Chamber of Justice, but little concerning private law. The most important statute was the criminal code of Charles V of 1532, the so-called Carolina, i.e. *Constitutio Carolina Criminalis*, adopted by the Diet of Regensburg. It was based on Roman law as developed by Italian legal science, adopted in the Revised Ordinances of Worms of 1499 and the two criminal codes provided by Maximilian I for Tyrol (1499) and Radolfzell (1506), and in the criminal code prepared for Bamberg in 1507. The *Bambergensis* was the basis of the Brandenburg-Franconian criminal code of 1516 and of the Carolina. The Carolina was primarily a code of criminal procedure with substantive law interpolated, and dominated German law for 200 years.

In the states there was also much legislation and territorial codes are found in many states, some mainly Germanic, others largely Romanistic. Such included the *Constitutio Joachimica* in Brandenburg–Prussia in 1527.

Among the jurists of the Reception in Germany notable figures were Johannes Apel (1486–1536), Gregor Haloander (1500–31), and Bernhard Walther (1516–84), Joachim Mynsinger and Andreas Gail.

The influence of Roman law was broken in the seventeenth century by the rise of the two schools of natural law jurists and of German national jurists. The former grew from Grotius' *De Jure Belli ac Pacis* (1625) and came to exercise a dominant influence in legal thinking for the better part of 200 years. It had been foreshadowed by the work of such jurists as Johann Oldendorp (1480–1567) and Johannes Althusius (1557–1638) and was taken up and developed by Samuel Puffendorff, whose most original work was in German public law, Leibnitz, and others.

The latter group comprised Herman Conring (1606–81) whose historical investigations (*De Origine juris Germanici*, 1643) gave legal science in Germany a national basis, Benedict Carpzov (1595–1666), David Mevius (1609–70), and Christian

Thomasius (1655–1728), who all in different ways had great influence in promoting German law, and Georg Beyer (1665–1714) who gave the impulse to the systematic study of German law, separate from the Roman. The term *Usus modernus Pandectarum* was coined by Samuel Stryk (1611–1701).

In the seventeenth and eighteenth centuries there was little general legislation but abundance of particular legislation. In the eighteenth century the German empire broke into a group of independent states. The idea of codification of the civil law was, in many minds, influenced by the school of natural law, and this was often a means of confirming a state's reality. A draft of a *Corpus Juris Fredericianum* (1749–51) prepared by Cocceii was ineffective. In Bavaria the *Codex Juris Bavarici Criminalis* (1751), the *Codex Juris Bavarici Iudiciarii* (1753), and the *Codex Maximilianeus Bavaricus Civilis* (1756) were all prepared by von Kreittmeyr. The last retained the principle that the Pandects were in force as subsidiary to native law. In Prussia Frederick the Great established a commission to codify the whole substantive law, and the first book of the *Corpus Juris Fredericianum*, the code of civil procedure, was enacted in 1781 and, after various revisions, the code was published in 1794 as the *Allgemeines Landrecht fur die Preussischen Staaten*. The other states all codified separately, but all collections embodied much from the general law of Germany, as well as special rules.

In Austria a criminal code (*Constitutio Criminalis Theresiana*) was enacted in 1781, and later part of a code of private law, the *Josephanisches Gesetzbuch*. In 1811 the *Allgemeines burglerliches Gesetzbuch* completed the Code.

In the eighteenth century Brandenburg assumed the primacy among the German states and developed into the kingdom of Prussia. The French Revolution brought about the final collapse of the Empire. It also gave rise to the Code Napoleon, a specimen of a successful national codification and German versions of the French Code were introduced in some territories east of the Elbe. But from the early nineteenth century voices were raised in calling for uniform German legislation, particularly in private law. In 1814 Thibaut published a pamphlet calling for a General Civil Code for Germany, which was answered by Savigny who denied the desirability of such as course. The controversy gave rise to the Historical and Philosophical Schools; the Romanists sought to cultivate the Roman law in its utmost purity, the Germanists to study the law between the Dark Ages and the Reception, and too little attention was paid to current law. But as time went on the difference between the groups of jurists diminished and works of scholarship of the greatest importance were published such as Savigny's *System des heutigen römischen Rechts* (1840–49) and Windscheid's *Lehbuch des Pandektenrechts* (1862–65). Other leading jurists of this time were Bruns, Vangerow, Dernburg, Brins, Beseler, and Brunner.

In the nineteenth century there was codification in many German states. Prussia made a revision of the Code of Judicature (*Allgemeine Gerichtsordnung*) of 1793 in 1833 and 1846, and reformed her criminal procedure in 1849–52. There were codes of criminal law enacted in Bavaria in 1813 and 1861, in Prussia in 1851, and in Austria in 1852–53. A Civil Code was issued for Saxony in 1863, and there were codes of civil procedure for Hanover (1850) and Bavaria (1869).

The North German Confederation (1866) had power to legislate on certain matters for the whole federal territory and on its formation the general bills of exchange law (1848–50) and the commercial code of 1861–66, adopted earlier as uniform states' laws, were re-enacted as federal law. A federal supreme commercial court helped to promote uniformity.

National unity was not fully achieved until the proclamation of the German Empire in 1871. All the laws of the North German Confederation were re-enacted as Imperial laws and much legislation followed. An Act of 1877 established uniform procedure in the courts and created a Supreme Imperial Court at Leipzig. But till 1900 the French Civil Code applied west of the Rhine and in Baden; most of eastern Germany and Westphalia was governed by the Prussian Code of 1794; Saxony has its own Code of 1863; but the rest of Germany was governed by the *Gemeines Recht*, by received Roman and canon law varied by regional and local customs. The preparation of a civil code was begun in 1874. The first draft was submitted in 1887 and provoked a mass of comment and criticism. The draft was entrusted in 1890 to a fresh commission for recasting, and the second draft was published in 1896, promulgated as the *Bürgerliches Gesetzbuch* (B.G.B.) and came into force in 1900.

The arrangement of the B.G.B. is based on that of the textbooks on Pandektenrecht: it comprises five books, general principles (arts. 1–240), obligations (arts. 241–853), things (arts. 854–1296), family law (arts. 1297–1921), and inheritance (arts. 1922–2385). It has been the model for the codes of various other legal systems, such as those of Austria, Switzerland, Poland, and Japan.

Germanic elements are more prominent in the books on property and family law, Roman elements in the book on obligations. It has proved a systematic piece of legislation of the utmost value and has been copied elsewhere. Revised editions of the Commercial Code and Code of Civil Procedure were enacted at the same time.

Criminal law had been much affected by the reforming movement of the eighteenth century and individual German states had codified their criminal law, Bavaria being the first in 1751. The federal criminal code was issued in 1871 and has subsequently been extensively altered.

Modernization and rationalization of the law continued under the Weimar Republic but there was substantial judicial sabotage of democracy and liberalism. Under the Nazi regime legislation was voluminous and far-reaching, and Nazi ideology affected every sphere of law; the power of the central government was enormously increased; arbitrary power was entrusted to many officials; and rules were interpreted and applied in the light of party views. Since 1945 German law had been largely purged of these elements.

Sources and literature

The major division of the modern law is into public law, comprising, *inter alia*, constitutional, administrative and criminal law, and civil and criminal procedure, and private law, the latter comprising the topics dealt with in the civil and commercial codes and legislation. Commercial law deals with professional merchants and their acts. Some branches such as industrial law do not fit easily into either field. The distinction of law and equity has never existed.

The hierarchy of authorities comprises the codes and other legislation, as interpreted linguistically, analogically, or historically, and custom. The writings of textbook writers have some persuasive authority and the rules established by the practice of the Courts similarly. Decisions are not, save in certain statutory cases, absolutely binding, but frequently treated with the greatest deference. Decisions are anonymous, and any dissents are not recorded.

German law, more than French, and much more than an Anglo-American system, accepted a mechanical, doctrinal, conception of law and the judicial process, but this approach has altered in recent years.

In interpretation considerable attention is given to preliminary work done by legislative assemblies and committees and the official 'motives' of any law may be referred to in aid of interpretation.

Public law

The Federal Republic established since 1945 is based on the *Grundgesetz* or Basic Law of 1949, which established a federal government headed by a president, who nominates the federal chancellor and cabinet ministers, and appoints federal judges, the Bundesrat or upper chamber consisting of members nominated by the *Länder*, which has wider powers. Certain functions are assigned to the 10 *Länder* and West Berlin; the governments of the *Länder* are generally parallel to those of the federation. In the Basic Law many features were adopted from U.S. and other federal constitutions but the *Länder* have greater autonomy than state governments in the U.S. The Basic Law also contains a formal declaration of human rights and of bases for the government of men, and clearly established the courts as independent.

The judiciary is, in general, a career structure with promotion determined largely by seniority.

Constitutional cases go to the Federal Constitutional Court (*Bundesverfassungsgericht*) at Karlsruhe, and there is a federal Administrative Court (*Bundesverwaltungsgericht*) at Berlin. Below it are Higher Administrative Courts and Administrative Courts, deciding on all disputes under public law.

The highest regular civil court is the Federal Court (*Bundesgerichtshof*), located at Karlsruhe, sitting in several senates, which reviews decisions of the regional supreme courts (*Oberlandesgerichte*), of which there is at least one in each *Land*, sitting in several senates, and which reconsider cases heard in the regional courts (*Landgerichte*). These are the normal superior courts and have both first instance jurisdiction and appellate jurisdiction from *Amtsgerichte*, inferior or local courts. Commercial chambers of *Landgerichte*, composed of a judge and two elected local merchants, handle commercial business. *Landgerichte* and *Amtsgerichte* also exercise criminal jurisdiction. There are distinct hierarchies of courts for fiscal, labour, and social legislation disputes culminating in the Federal Finance Court (*Bundesfinanzhof*) at Munich, the Federal Labour Court (*Bundesarbeitsgericht*) at Kassel, and the Federal Social Court (*Bundessozialgericht*) at Kassel. Admiralty courts sit at *Land* level and there is a Federal Admiralty Court (*Bundesoberseeamt*) at Hamburg.

Criminal law

The individual German states codified their criminal law, the first being Bavaria in 1751. The federal criminal code was issued in 1871.

In the 1960s extensive reform in criminal law and treatment of offenders was undertaken.

Private law

Like French, and unlike English law, German law distinguishes between civil law (*Bürgerliches Recht*) and commercial law (*Handelsrecht*), the latter being the part of private law applicable only to commercial affairs. It is governed by the *Handelsgesetzbuch* (H.G.B.) which also came into force in 1900, but is not identical with the old commercial code of 1861. Commercial law is based on the concept of 'merchant' and is applicable only to professional merchants and their dealings; commercial acts are those acts of a merchant which pertain to his enterprise. Since 1900, various supplementary statutes have been passed, dealing with particular topics. The H.G.B. is in four books, dealing with legal definitions, commercial associations, commercial transactions, and admiralty law.

Procedure

Codes of civil and criminal procedure have been in operation since 1879 and 1877 respectively.

Legal profession

The legal profession is not divided, as in the U.K., the functions of both barristers and solicitors being performed by the one kind of lawyer. A person seeking to qualify will take a degree in law, and undergo a substantial period of practical training in the courts and various kinds of offices as a *Referendar* and finally pass a state examination to be admitted to the profession as a *Rechtsanwalt*. He may then enter the judicial service, public prosecution, the civil service, or private practice. A *Rechtsanwalt* is admitted to a specific court, so that the profession is divided territorially. The notary is a member of a special branch of the profession.

General: C. Sladitz, *Guide to Foreign Legal Materials—France, Germany, Switzerland*; *History:* F. C. von Savigny, *Geschichte des römischen Rechts im Mittelalter*; H. Planitz, *Deutsche Rechtsgeschichte* (1961); H. Mitteis and H. Lieberich, *Deutsches Privatrecht* (1963) and *Deutsche Rechtgeschichte* (1966); E. Wolf, *Grosse Rechtsdenker der deutschen Geistesgeschichte. Public law:* T. Maunz, *Deutches Staatsrecht. Criminal Law:* A. Schönke and H. Schröder, *Stafgesetzbuch. Private Law;* H.M.S.O. *Manual of German Law*; L. Enneccerus, T. Kipp and M. Wolff, *Lehrbuch des Burglerlichen Rechts*; O. Palandt, *Kurzcommentar zum BGB*; A. Baumbach and K. Duden, *Handelsgesetzbuch* (1970).

Germanic law. The laws of peoples of Germanic race from the time that their tribes came into contact with the Romans until they developed into national territorial laws. The first observation of the legal practices of these peoples was recorded in Caesar's *Gallic War* and Tacitus' *Germania*. Among the Germanic tribes law was customary and personal, not territorial, applicable to the members of a nation only. Accordingly the collections of written law, known originally as the *Leges Barbarorum*, are not legislation, still less comprehensive codes, but restatements or modifications of law on particular topics, and assume the existence of large tracts of unwritten and unrecorded customary law. Also they were influenced by contact with Roman civilization and law and written down in Latin, unlike the Anglo-Saxon laws which are in the vernacular. When Germanic peoples settled in parts of the former Roman Empire after the collapse of the imperial authority, in the fifth and later centuries, their customs and recorded laws long remained personal, being inapplicable to Roman citizens in the former provinces, who were still governed by Roman law. Thus the Germanic code of Euric, King of the Visigoths, applied to Visigoths only and the *Lex Romana Visigothorum* (*Breviarium Alaricianum*) applied to Roman subjects of the Visigothic kingdom. Not till the seventh century was Visigothic law equally applied to both races, who by that time had become largely merged.

The reduction of parts of their customary law to writing seems, in some cases, to have been initiated by the kings but, however originating, seems always to have required the approval of the popular assembly of the people. Hence it was sometimes called a *Pactus* or agreement. The Visigothic laws seem, from the start, to have been monarchical legislation, and this practice was followed by the Lombard kings, who issued edicts from the seventh century onwards, and the Frankish kings who issued *edicta* or *Praecepta* and, under the Carolingians, *capitularia* or edicts divided into articles (*capitula*). Charlemagne and his successors asserted the power of the Roman emperors to legislate without the consent of nobles or popular assembly.

The surviving collections, known as Germanic or barbarian laws, comprise four groups, usually known as the Gothic, Frankish, Saxon, and Suabian groups. The Gothic group comprises two bodies of law, the Visigothic law of Euric (466–83), which were later revised and enlarged by Leovigild (539–86) or his son Reccared (586–601) and which came to be merged with the *Lex Romana Visigothorum* into the *Leges Visigothorum* of 634, and the *Lex Burgundionum* (or *Liber Constitutionum* or *Lex Gundobada* or *Lex Gombata*) of King Gundobad (474–516), supplemented by his successors, originally inapplicable as between Roman subjects. The Frankish group includes the *Lex Salica*, the law of the Salian Franks, which had great authority and wide influence in Western Europe, and was revised and amended into the time of Charlemagne, the *Lex Thuringorum*, the *Lex Francorum Chamavorum*, and the *Lex Ribuaria*. The Saxon group comprises the *Lex Saxonum* of *c.* 802, of which the later part was influenced by Charlemagne and Frankish law, and the *Lex Frisionum* probably of the eighth century. The Suabian group from south Germany includes the fragmentary *Pactus Alamannorum* of the early seventh century and the later *Lex Alamannorum*, and the *Lex Baiuvariorum*, which embodies elements from several earlier codes, probably merged into one code in the mid eighth century.

The society governed by custom, supplemented and possibly modified by one or other of those bodies of written law, comprised freemen, nobles and ordinary freemen, and unfree men. The tribes were ruled by kings, councils of nobles and elders, and tribal assemblies of the free members of the tribe. Land seems originally to have belonged to the family collectively, which developed into ownership by the head of the family who for long could not alienate without the consent of the nearest heirs. On death, property passed to descendants with a preference for males; the prohibition in the Salic law of inheritance by females was later invoked by French lawyers to justify the exclusion of women from inheriting the crown. Succession by will does not seem to have been known.

Wrongs against the group, such as treason and theft were punished by outlawry imposed by the assembly. Wrongs against individuals such as murder or robbery became the basis of a blood feud if the parties belonged to different family groups, but peace could be secured by the payment of compensation (known as *wergild* in cases of homicide and *bot* in other cases). At first this was voluntary, but later became compulsory. As royal power developed the blood feud became much restricted, being replaced by the idea of punishment for breach of the King's peace. Proof might be by oath or compurgation.

The Germanic laws gradually declined in importance with the growth of the Church and the canon law, and of the feudal relationship, under which each lord held a court for his feudal tenants and applied the same law to them all without regard to their race. Hence the principle of personal law gave way to the application of the customary law of the region, which was usually mainly Germanic but influenced by Roman law, frequently the *Breviarium Alaricianum.*

E. Canciani, *Barbarorum leges antiquae* (1781–89); T. Walter, *Corpus iuris germanici antiqui* (1824); *Monumenta Germaniae Historica, Leges*; F. Stobbe, *Geschichte des deutschen Rechtsquellen* (1860–64); H. Brunner, *Deutsche Rechtsgeschichte* (1906); H. Conrad, *Deutsche Rechtsgeschichte*; (1962); K. von Amira and K. Eckhardt, *Germanisches Recht* (1960); E. Jenks, *Law and Politics in the Middle Ages* (1913); P. King, *Law and Society in the Visigothic Kingdom* (1972).

Germonius, Anastasius (1551–1627). Italian canonist, author of *Paratitla in libros quinque Decretalium* (1586), *De legatis principum et populorum libri tres* (1617), and other works.

Gerontocracy. The dominance of the elderly, a factor universally found in past human societies in consequence of the greater knowledge, experience and balance in judgment. It gives rise in legal contexts to such phenomena as 'councils of elders' and 'seniority' and while nowhere a formal principle of government or law, it has been in fact a powerful influence in many contexts. In modern times the value of experience is too frequently underrated.

Gerousia. The Council of Elders at ancient Sparta, comprising the two kings and 28 elders aged at least 60. They were elected by acclamation of the citizens and held office for life, but later the office became annual and under the Romans re-election became common. The Council had both deliberative and judicial functions and prepared business for the Apella, the decisions of which it could reverse. It heard cases involving death, exile or disgrace, and could try even the kings, and collaborated with the ephors in sentencing those charged by the ephors. The name was also given to similar city councils in Greek towns of Asia Minor in Hellenistic and Roman times.

Gerrymandering. The division of a county into electoral districts so as to ensure political advantages to the party in power, derived from Governor Gerry of Massachusetts who resorted to this device in 1812.

Gerson, John (Jean Charlier de) (1367–1429). French Church reformer and political thinker, sought to establish peace in Church and State and to emphasize the superiority of the Church over the papacy, notably in *De Potestate ecclesiae et origine iuris* (1417).

Gesellschaft mit beschränkter Haftung (G.m.b.H.). One of the main forms of corporations known in German law. (For the other see AKTIENGESELLSCHAFT.) It is suitable where the capital is small or the number of shareholders limited. The corporation comes into being by registration on application of the managers. There need be only two members. After registration all the shares may be held by one person. The annual balance sheet and accounts need not be published. There may be special restrictions on the transfer of shares. A board of directors is optional.

Gestation. The process of carrying young in the womb. In humans the period of gestation is that between conception and the birth of a child, normally about nine lunar months or 280 days, but considerable variations from the norm are known. Its duration may be material in questions of adultery or legitimacy and the courts have frequently, on the evidence in particular cases, declined to hold that exceptionally short or exceptionally long periods have been impossible.

Get (or *gett*). A Jewish document of divorce, written in Aramaic in accordance with a prescribed formula, granted by a Rabbinic court. Orthodox Jews regard it as the only means of dissolving a marriage; Reform Jews accept the decision of a civil court as sufficient; and outside Israel Rabbinic courts require a civil divorce before a *get* is issued. It becomes effective when the husband delivers it to his wife in the presence of three members of the court and two witnesses. Court officials then record the divorce and issue documents to both parties.

Gibbons* v. *Ogden (1824), 9 Wheaton 1 (U.S. Sup. Ct.) A New York statute gave Fulton and others a monopoly of the right to operate boats powered by steam on the waters of the State. The development of steamboats enriched the holders of the monopoly, who sought to extend their rights to waters connecting New York with other states. Ogden, who had a licence from the New York monopoly holders, challenged the right of Gibbons, who had a United States licence to operate in the coasting trade, to engage in navigation between New York City and New Jersey. The Supreme Court gave a broad interpretation to the power given to Congress to regulate commerce with foreign nations and among the several states, which was the basis for later very wide interference, as by the Interstate Commerce Act of 1887. Regulation, the court held, included fostering, protecting, developing, and inhibiting.

Gibbs, Sir Vicary (1751–1820). Became an accomplished special pleader (q.v.) before being called to the Bar in 1783. He took silk in 1794 and in 1805 became Solicitor-General, and in 1807 Attorney-General. He was a failure in the House of Commons and was unpopular for his campaign against the press for seditious libels. In 1812 he became a judge of the Court of Common Pleas, Chief Baron of Exchequer in 1814, and later the same year Chief Justice of the Common Pleas, where he presided till 1818. Though able, he was a severe and harsh judge but his decisions are clear and show a good grasp of the principles involved.

H. Brougham, *Historical Sketches*, I, 124; W. Townsend, *Twelve Eminent Judges*, I, 239.

Gibraltar, law of. Gibraltar is an almost completely self-governing colony of the U.K. captured in 1704 and ceded by Spain in 1713. It has a Governor appointed by the U.K. government, a chief minister and council of ministers, and an elected House of Assembly. Executive authority is exercised by the Governor, normally on the advice of the Council. The Governor and House of Assembly constitute the legislature. The judicial system comprises a Court of Appeal, Supreme Court, Court of First Instance, and Magistrates' Court. The law was originally Spanish but the law of England as at 31 December 1883 was introduced in 1884 in so far as it did not conflict with other overriding provisions and has subsequently been developed.

Gibson, Sir Alexander, Lord Durie (?–1640). Became a Lord of Session as Lord Durie in 1621. In 1633 he was appointed a commissioner for reviewing the laws and collecting the customs of the country. When the office of president of the Court of Session was elective Durie was elected head of the court, in 1642 and 1643, and was frequently vice-president. He was a great student of the law and had the reputation of penetrating wit and clear judgment. He made a collection of *Decisions of the Court of Session* for the period 1621–42, the earliest collected decisions, but not published till 1690. His son, also Alexander (*d.* 1656), became a judge as Lord Durie in 1646, and his grandson, Alexander (*d.* 1693), was a Clerk of Session and edited the *Decisions.*

Gibson, Edmund (1669–1748). Bishop successively of Lincoln and of London, and antiquary, edited the *Anglo-Saxon Chronicle*, translated Camden's *Britannia* and edited Spelman's works. He published *Synodus Anglicana* in 1702, the most authoritative work on the constitution and procedure of convocation, and in 1713 appeared his greatest work, *Codex Juris Ecclesiastici Anglicani*, a collection of all the statutes and other materials relating to English ecclesiastical affairs from the earliest times, grouped under 52 titles, with an abridgment and commentary, which has always been of the highest authority, though some of his claims in the Introductory Discourse were challenged by common lawyers. It is a great monument of research and still of high authority. He also published sermons and pastoral tracts and was a great patron of scholars.

Gibson, Edward, Lord Ashbourne (1837–1913). Was called to the Irish Bar in 1860 and was Attorney-General for Ireland, 1877–80, and Lord Chancellor of Ireland and a peer, 1885–86, and again in 1886–92 and 1895–1905. He was much associated with land purchase reform.

Gibson, Robert, Lord (1886–1965). Called to the Scottish Bar in 1918, he sat in Parliament, 1936–41, and was Chairman of the Scottish Land Court, 1941–65.

Gidel, Gilbert Charles (1880–). French jurist, author of *Le droit international public de la mer* (1932–34) and other works on international law.

Gierke, Otto Friedrich von (1841–1921). German jurist and legal historian, professor at Berlin, a leading figure in the Germanist Group of the historical law school, author of an important *Das deutsche Genossenschaftrecht* (1868–1913), giving a critical account of the general theory of human society, of politics, constitutional law and law of associations, developed by the school of natural law and concerned with the nature of groups and their life and authority, of a *Deutsches Privatrecht* (1895–1917), and of a book on Johannes Althusius (q.v.). He played an important part in the preparation of the German Civil Code in defending the German tradition of the law against the Romanists and strongly criticized the first (1888) draft of the Civil Code for adding Roman elements to indigenous Germanic principles, and this attitude was reflected in his *Deutsches Privatrecht*. He was a founder of modern German constitutional law and stimulated the sociological approach to law.

Giffard, Sir George Markham (1813–70). Was called to the Bar in 1840 and rapidly developed a busy equity practice. In 1868 he became Vice-Chancellor and later in the same year a Lord Justice of Appeal in Chancery. His reputation is of one quick and learned, unsurpassed knowledge of legal principles, and of sound judgment.

Giffard, Hardinge Stanley, 1st Earl Halsbury (1823–1921). After Oxford Giffard helped his father to edit a paper and was called to the Bar in 1850. He practised initially on the Chancery side and only later made his name at the Old Bailey. He appeared for the Tichborne (q.v.) claimant and took silk in 1865. In 1875 he became Solicitor-General and got into Parliament in 1877. From 1880 to 1885 he was out of office but consolidated his position as one of the foremost jury advocates of the day. In 1885 he became Lord Chancellor and Lord Halsbury, but went out of office in the next year, but later in 1886 resumed the office, holding it till 1892, and again from 1895 to 1905. He sat judicially in the House of Lords as late as 1916. Conservative both in politics and by temperament he sponsored only two important legal reforms but supported reform introduced by others. As a judge he was plain and direct but cannot be ranked among the greatest common law judges. He was not a jurist or legal scholar nor profoundly interested in the law as a system, but some of his judgments are of undoubted authority. His name is also remembered for having acted as General Editor of the first edition of Halsbury's *Laws of England*, 31 volumes, 1907–17.

Giffen, Hubert van (Giphanius) (1534–1616). A Dutch humanist, jurist and philologist, author of *In quattuor libros Institutionum commentarius* (1611), *Lecturae in Digesta et Codicem* (1605), *Oeconomia juris* (1612), and other works on law and classical literature.

Gifford, Robert, Lord Gifford of St. Leonards (1779–1826). After spending some years in an attorney's office and as a special pleader (q.v.), he was called to the Bar in 1808. In 1817 he became Solicitor-General and entered Parliament, and Attorney-General in 1819, in which capacity he prosecuted the Cato Street conspirators. He held the office of Chief Justice of the Common Pleas for three months in 1824, and was made a peer, before being appointed Master of the Rolls. He seems to have been an able lawyer, with a considerable knowledge of equity, and his decisions are clearly expressed.

Gift (or donation). The voluntary and gratuitous transfer of any property from one person to another. It may be conditional but, condition apart, is not revocable nor terminable. Acceptance is presumed unless dissent is signified, but a gift may be rejected when the donee becomes aware of it. The title to the subject of gift must be transferred in whatever way is necessary for the kind of property concerned. Gifts may be made *inter vivos*, or on death, by will (q.v.) or donation *mortis causa* (q.v.).

Gift over. A gift to one in succession to a prior gift to another on a certain contingency, e.g. to A if he attains 21, whom failing to B.

Gilbert, Sir Jeffrey (1674–1726). Called to the Bar in 1698, he was a judge of the Irish King's Bench (1715), Chief Baron of Exchequer in Ireland (1715–22), a baron of Exchequer in England (1722–25), and Chief Baron (1725–26). He made no particular reputation as a judge, but he published two sets of reports (1734 and 1760) and wrote many books on law, all published posthumously and some altered by editors. These include *The History and Practice of the High Court of Chancery*, the two parts being entitled *Forum Romanum* and *Lex Praetoria* (1758), a book on *Uses and Trusts* (1734), on *Devises, Revocations and Last Wills* (1756), books on the history and procedure of the *King's Bench* (1763), the *Common Pleas* (1737) and the *Exchequer* (1758), on the *Law and Practice of Ejectments* (1734), the *Law and Practice of Distress and Replevin* (1757), and the *Law of Executions* (1763), on the *Law of Evidence* (1754), a *Treatise of Tenures* (1730) and a *Treatise on Rents* (1758). All his books exhibit precision and lucidity, and mastery of the subject. They went through various editions, and he made a material contribution in his time to professional understanding of the law. The *Treatise on Evidence* was the standard authority throughout the eighteenth century, and has been called a classic. He was probably the ablest legal writer of the century, apart from Blackstone.

Gild. A kind of association formed for various purposes in England from Anglo-Saxon times. Soon after the Norman Conquest appeared the Gild Merchant (later there arose various trade guilds) governed by a governor and associates, a form of organization which much influenced the trading companies of the sixteenth and seventeenth centuries. Membership was usually by birth or apprenticeship; members had to take an oath of fidelity and participated in feasts.

Gilmour, Sir John, of Craigmillar (1605–71). He passed advocate in 1628 and was nominated Lord President when the Court of Session was restored in 1661. He became also a Privy Councillor and one of the Lords of the Articles and, in 1664, a member of the Court of High Commission. He resigned in 1670. He made a collection of decisions of the Court of Session, 1661–66, published, with those of Lord President Falconer, 1681–86, as Gilmour and Falconer's *Decisions* in 1701.

Gin Act (1751). An Act, stimulated by public outcry at the prevalence of gin drinking and consequent drunkenness, which gave magistrates powers to enter gaols, workhouses, and houses of correction and to search them for illegally introduced spirits. It proved much more effective than earlier legislation which imposed excise duties on gin and heavy charges for retailers' licences.

Girard, Paul Frédéric (1851–1962). French jurist working particularly in the field of Roman law, translated Mommsen (q.v.) and was author of *Manuel de droit romain* (1896), *Mélanges de droit romain* (1912–23), and *Histoire de l'organisation judiciaire* (1913).

Gladstone Committee Report. The report issued in 1895 of a departmental committee under H.J. (later Lord) Gladstone on the English penal system. It condemned uniformity of treatment and emphasized the need to consider prisoners individually and to seek to reclaim, train, and improve prisoners. It considered the primary objects of imprisonment to be deterrence and reformation, condemned unproductive penal labour and recommended the employment of prisoners on useful industrial work. The Prison Act, 1898, gave effect to many of its recommendations and its approach underlay prison administration for at least 50 years.

Glanvill(e), Ranulph de (*d.* 1190). Was successively sheriff of Yorkshire (1160–70) and of Lancashire (1173–74), and was reappointed to the former post after winning a decisive victory over the Scots at Alnwick. He was a justice in eyre in 1176, Ambassador to the Court of Flanders in 1177 and, in 1180, became chief justiciar of England. Thereafter he was a principal assistant of Henry II, leading armies, negotiating, and judging. He may have resigned, or been deposed, in the early years of Richard I, and died at Acre when on crusade with Richard.

He must have had a hand in carrying through the great legal changes of Henry II, and tradition ascribes to him the assize of novel disseisin (q.v.). To him has also been ascribed a *Tractatus de Legibus et Consuetudinibus Angliae* (*Treatise on the Laws and Customs of England*), the oldest English legal classic, but the author is probably his nephew and secretary, Hubert Walter.

The Latin treatise known as Glanvill, the earliest treatise on the common law, which was written in the latter part of Henry II's reign, seems to have been completed between 1187 and 1189, and its purpose is to describe the procedure of the King's courts. There is much information about litigation, and some 80 writs are included in the text. It is in 14 books. The writer is keenly interested in legal problems and practical difficulties and not afraid to admit ignorance of some of the answers. The book must have done much to define the common law and settle the procedure of the royal courts. It became a venerated authority among English lawyers, and later writers acknowledged their debt to it. A Scottish adaptation was made about a century later, known from its opening words as *Regiam Majestatem* (q.v.) and an attempt was made to make a revised edition about 1265. First printed by Tottel in about 1554, there were later editions, the best modern one being that by Prof. Woodbine (1932).

Glaser, Julius (1831–85). Austrian jurist, head of the Austrian Ministry of Justice and author of the Austrian Code of Criminal Procedure of 1873, which was influenced by British procedure. He also wrote extensively on criminal law and procedure.

Glebe. (1) All land, houses, or other buildings forming part of the endowment of an ecclesiastical benefice. An incumbent may cultivate the glebe and, subject to conditions, mortgage it, let it on lease, sell or exchange it.

(2) In Scotland, the portion of land adjacent to the church to which a parish minister is entitled, apart from stipend in grain or money. The glebe cannot be alienated but may be feued or leased.

Gloss. Textual interpretation by marginal or interlinear note, the basic method of the jurists who revived the study of Roman law at Bologna, and then of the canonists after Gratian's *Decretum* and the beginning of papal decretal legislation. It was a long established traditional method in the study of classical literature, and also of Biblical texts. The Bolognese jurists were, however, the first to apply the method to legal texts.

Initially, the gloss was a brief interpretation of a difficult word by a more familiar equivalent, but was soon expanded to become an explanation of a whole passage or legal principle embodied in the text. The single-word gloss was usually interlined, the longer comment usually inserted in the margin. Glosses were initialled or otherwise marked with an identification mark and the term Glossator came to be used of a scholar who composed glosses. Glosses gradually became more elaborate and developed into a full commentary on the text. Irnerius was the founder of the glossatorial school, but Accursius was the most influential and his elaborate gloss became the *glossa ordinaria* which, in the schools and courts, replaced all others: *quicquid non agnoscit glossa nec agnoscit curia.*

When canon law began to flourish after the publication of the *Decretum*, canonists borrowed the method of glossing. Initially glosses were simple citations of concordant or discordant headings, later parallel passages were added and summaries of the contents of the chapter. Glossing reached a more advanced stage when it proceeded to the interpretation of the law itself, and to an analytic and systematic commentary.

About the end of the twelfth century there began collection of the now extensive glosses, revision, and systematization, so that there came into being a continuous comprehensive commentary, the *apparatus glossarum*, which itself was subject to constant revision and expansion.

The schools and courts recognized the *glossa ordinaria* as the one deemed to have made the best selection and to have offered sound opinions, and also be complete yet succinct. The first canonical *glossa ordinaria* is that of Joannes Teutonicus, in the form of a marginal gloss on Gratian's *Decretum*, and later included in the printed texts of the *Decretum* and adapted to the later decretals by Bartholomew of Brescia in about 1245. The work which came to be considered the *Glossa ordinaria* on the *Quinque Compilationes Antiquae* is that of Tancred of Bologna, *c.* 1220. Joannes Teutonicus wrote the standard gloss on the *Compilatio quarta* and James of Albegna that on the *Compilatio quinta*. The *glossa ordinaria* on the Decretals was produced by Bernard of Parma, later added to by Joannes Andreae who also produced the *glossa* on the *Liber Sextus* (*c.* 1301) and the Clementines.

Glossators. A category of mediaeval legal scholars. When the study of Roman law was revived at Bologna in the eleventh century, the main method of study was by the gloss or textual interpretation. This was partly because Justinian had prohibited altering his enactments by liberal interpretation and the tradition persisted, partly because legal study had developed from the schools of dialectic and grammar, partly because the ascertainment of accurate texts of Justinian was still a matter of importance. The gloss was originally an explanation of a difficult word or phrase, interlined or written in the margin, but it developed into an exposition of the whole passage or principle concerned. Different glossators attached their initials or other mark to their glosses. As the gloss developed it became a compendium of knowledge to the point, including critical notes on variant readings of the manuscript, parallel passages, conflicting passages, and finally it became a full commentary comprising a summary, illustrative cases, deduction of general principle, and discussion of actual problems.

The period of the Glossators extends from *c.* A.D. 1050 to *c.* 1250. The first glossator was Pepo but the founder of the school was Irnerius (q.v.), who was succeeded by the four doctors, Bulgarus, Martinus, Jacobus, and Hugo, and then by Bulgarus' pupils, Johannes Bassianus and Rogerus, who in turn taught Azo and Hugolinus. Azo composed the *Summa Azonis*, on the Institutes and Code, a work which surpassed all previous works and was used by Bracton. Hugolinus left glosses on the Digest, Code and other parts of the *Corpus Juris*. Placentinus established the school of law at Montpellier and Vacarius the school at Oxford. Others included Burgundio, Anselmus of Orta, Carolus of Tocco, and Roffredus of Benevento. The greatest name is, however, Accursius (q.v.) (1182–*c.* 1260), who made a fairly comprehensive collection and synthesis of most of the preceding glosses, summaries, and other works and thereby almost completely superseded them. After him the method of study declined, largely because study of the gloss was substituted for study of the text. Among the last were Odofredus, Rolandinus Passegerius who wrote a *Summa artis notariae*, Albertus of Gandino, and Gulielmus Durantis who wrote a vast treatise on procedure, *Speculum Judiciale*. The works of many of the glossators were published in the sixteenth century.

The glossators laid the foundation of the study of Roman law on which the commentators and humanists built, and they established the study of Roman law as the basis of legal training, which attracted countless students to Bologna and had widespread and lasting influence on European legal systems.

Similarly there arose a school of glossators on the canon law, notably Joannes Teutonicus who wrote a gloss in the form of a marginal gloss on Gratian's *Decretum*, Bartholomew of Brescia, who adapted it to the later Decretals, Tancred of Bologna who glossed the *Quinque Compilationes Antiquae*, Bernard of Parma who glossed the *Decretals*, and Joannes Andreae who added to these and also glossed the *Sext.*

P. Viora (ed.), *Corpus glossatorum juris civilis*; F. C. Savigny, *Geschichte des Römischen Rechts in Mittelalter*; H. Fitting, *Die Anfange der Rechtsschule zu Bologna*; H. Kantorowicz and W. Buckland, *Studies in the Glossators of Roman Law*; F. Koschaker, *Europa und das römisches Recht*; P. Vinogradoff, *Roman Law in Mediaeval Europe*, Ch. 2; H. D. Hazeltine, in *Encyc. Soc. Sc.*, VI, 679; *Camb. Med. Hist.* V, 697.

Gloucester, Statute of (1278). A statute by which for the first time a plaintiff recovering damages was given a right to costs.

Gloves. Gloves figure in a variety of miscellaneous legal customs. Thus it was customary for a man convicted of murder or manslaughter but pardoned by the King, to present gloves to the judges and court officers as fees. It is said that in the seventeenth century an outlawry could be reversed only if the outlaw appeared in the King's Bench in person, made a present of gloves to the judges, and begged them to reverse it. At a maiden assizes or sessions, when there were no prisoners to try, it was long customary for the sheriff to present white gloves to the judge who sat to deal with criminal business.

Gloves were sometimes the gift to which persons performing some service were entitled; sometimes the gloves were lined with money or glove-money was given in lieu of gloves.

Glück, Christian Friedrich von (1755–1831). A German Pandectist jurist, author of works on Roman Law in Germany, particularly *Ausführliche Erläuterung der Pandekten* (1790–1830) and *Handbuch zum systematischen Studium des neusten römischen Privatrechts* (1812).

Glueck, Sheldon (1896–) and **Eleanor** (1898–1972). Together they worked in criminology, Sheldon Glueck being latterly (1950–63) Roscoe Pound Professor of Law at Harvard. They developed prognostic tables to predict post-offence behaviour of criminals. Their work has yielded much information on criminal careers and been of value in predicting social maladjustment and in sentencing. Their publications include *One Thousand Juvenile Delinquents* (1934), *Crime and Justice* (1945), *Unravelling Juvenile Delinquency* (1950), *Predicting Delinquency and Crime* (1959).

Gneist, Heinrich Rudolph Hermann Friedrich von (1816–95). German jurist, pupil of Savigny, professor at Berlin, and also judge of superior courts until 1850, when he threw himself into politics. He was a member of the North German parliament and the Reichstag, and influential on legislation, particularly on judicial and penal matters. He was also, for a time, a member of the German supreme administrative court. He greatly admired English constitutional law and wrote extensively on it, notably his *Das heutige englische Verfassungs- und Verwaltungsrecht* (1857–63) and *Englische Verfassungsgeschichte* (1882).

Goddard, Rayner, Baron (1877–1971). He was called to the Bar in 1899, took silk in 1923, became successively a judge of the High Court, 1932–38, a Lord Justice of Appeal, 1938–44, a Lord of Appeal 1944–46, and Lord Chief Justice of England, 1946–58. He was a strong, stern judge with little faith in lenient treatment of criminals and frequently increased sentences in frivolous appeals. He opposed the abolition of capital punishment. He was probably the greatest Chief Justice since Russell.

Godden* v. *Hales (1686), 11 St. Tr. 1165. Hales, as colonel of a regiment, was obliged by the Test Act to take the sacrament as directed by the Act and to take the oaths of allegiance and supremacy. He did not do so and Godden, a common informer (q.v.), sued to recover the sum forfeited by Hales for his breach of the Act. Hales pleaded that King James II had, by letters patent, dispensed him from taking the oaths and other obligations imposed by the Test Act, and the question was whether the dispensation barred the claim. The majority of the judges of the Court of King's Bench held that it was a prerogative of the King to dispense with penal laws for particular necessary reasons, of which he was the sole judge. The Bill of Rights (q.v.) abolished the King's alleged power of suspending laws and dispensing with the operation of statutes, save where this was authorized by Parliament.

Godefroy, Denis (Dionysius Gothofredus) (1549–1622). A French humanist jurist, professor at Geneva and Heidelberg, was author of a great *Corpus Juris Civilis cum notis* (1583) which was long a standard work, and other works on Roman law and classical antiquity.

Godefroy, Jacques (Jacobus Gothofredus) (1587–1652). A French humanist jurist, second son of the foregoing, editor of a volume of ante-Justinian texts of Roman law, *Quatuor fontes juris civilis*, in which he reconstructed the Twelve Tables, and author of a great edition and commentary on the *Codex Theodosianus* (1665), a masterpiece of scholarship, and long the definitive edition. He also wrote a *Manuale iuris* (1654) and *Opera iuridica*.

Godefroy, Theodore (1580–1649). Eldest son of Denis, became historiographer of France and wrote a classic work on royal ceremonial in France. His son, Denis II, succeeded him as historiographer.

Goffredo da Trani (?–1245). Italian canonist, author of *Glossae* on the *Decretals* and a *Summa super rubricis Decretalium* (1241).

Going through the Bar. The name of an obsolete custom. Before 1873 the Chief Justice or Chief Baron of each of the common law courts, at the sitting of the courts on days other than Special Paper Days and days on which motions were not taken, was accustomed to ask each barrister in court, according to seniority, whether he had any motion to make.

Going to the country. In common law pleading, a defendant was said to 'go to the country' when he concluded his written pleading with the phrase 'and of this he puts himself upon the country' meaning that he asked for trial by jury. A plaintiff similarly would state 'and this the plaintiff prays may be enquired of by the country'.

Golden Bull. The general title of any charter decorated with a golden seal or bulla, but applied particularly to a few charters of notable importance politically. The earliest was a charter granted in 1222 by King Andrew II of Hungary under pressure from nobles enraged by his excesses, setting out the limits of the King's powers and defining the basic rights and privileges of the Hungarian nobility and clergy. It comprised 71 articles, partly affirming ancient rights, partly granting new ones. It required him to summon the diet regularly, forbade him to imprison a noble without proper trial before the palatine, who was an official exercising the chief administrative functions during the King's absence, and denied his right to tax nobles' and Church estates, released nobles from liability for unpaid service in the army abroad and prohibited foreigners from owning land. It extended the nobles' authority in their counties, ensured that the King's officers in the counties should not hold hereditarily and should be dismissible for misconduct. The nobles and clergy were given a right to resist if the King or his successors should violate the Golden Bull, without being guilty of treason. After 1222 all Hungarian kings had to swear to uphold the Golden Bull. The right of resistance was abolished in 1687.

Golden Bull of the Empire (1356). A constitutional statute promulgated by the Emperor Charles IV in 1356, dealing with the constitutional status of the electoral princes, fixing their number at seven, and the election of the Emperor but containing also provisions regarding the national peace, grants of municipal citizenship to rural residents and other matters.

Golden Bull of Rimini (1226). An imperial order of the Emperor Frederick II which granted sovereignty over Prussia to the Teutonic Knights.

Golden Bull of Sicily (1212). An order of the Emperor Frederick II which confirmed the autonomy of Bohemia.

Goldschmidt, Levin (1829–97). German jurist and the leading authority on commercial law, judge of the German federal supreme court of commerce, 1870–75, and thereafter professor at Berlin. His major work is *Handbuch der Handelsrecht* (1864–68) but he also wrote on Roman law.

Gomez, Antonio (?–1568). Spanish jurist, one of the most famous of his time, author of *Commentariorum Variarumque Resolutionum Juris Civilis, Communis et Regii Tomi Tres* (1552).

Gomez, Luiz (?–1543). Spanish jurist, bishop and auditor of the Rota, author of *In Regulas Cancellariae Apostolicae Commentaria* (1526) and *Decisionum Rotae Libri II* (1546).

Good behaviour, security for. Becoming bound with one or more sureties in a recognizance or obligation to the Crown, taken in court by a judicial officer, under which the parties acknowledge themselves to be indebted to the Crown in the sum stated, subject to its being void if the party bound is of good behaviour, either generally or for a set time.

Good faith. An act is done in good faith if done honestly, even though negligently.

Good offices and mediation. In international law the intervention in a dispute of a third state to procure a settlement. This intervention may have been requested by one or both parties, or been offered spontaneously. Several states may jointly act as mediators. Good offices is, in theory, distinguishable from mediation. It consists in various kinds of action tending to bring about negotiations between disputing states whereas mediation is direct conduct of negotiations between parties at issue on the basis of proposals by the mediator. But treaties and practice do not always distinguish them. There is no general duty on parties to ask for or accept a third state's good offices or mediation, not on a third state to offer, nor, if requested, to give its services. The Charter of UN gives the Secretary General and every state a right to invoke the collective mediation of the UN by drawing the attention of the Security Council of General Assembly to any dispute.

Good Parliament. The Parliament of Edward III which assembled in 1376, so called by reason of its measures against the corruption of the court and the government. It is the first Parliament in which there was certainly a speaker of the House of Commons, the first in which impeachment (q.v.) was used, and the first in which there was any account of the debates. In it the Commons openly opposed John of Gaunt and displayed independence and initiative.

Goodhart, Arthur Lehman (1891–1978). Born in U.S.A., taught law at Cambridge from 1919, and was thereafter Professor of Jurisprudence at Oxford, 1931–51, and Master of University College there, 1951–63. He became an honorary K.B.E., Q.C. and F.B.A. and was a generous benefactor to

Oxford. He served on many committees connected with legal matters. He edited the *Cambridge Law Journal*, 1912–25, and the *Law Quarterly Review*, 1926–70, being thereafter Editor-in-Chief till 1975, and published *Essays in Jurisprudence and the Common Law* (1931), *English Law and the Moral Law* (1953), and numerous papers on legal subjects. His greatest power was in the critical analysis of judicial decisions, but he also had much influence in improving the mutual understanding by English and American lawyers of their respective systems.

Goodnow, Frank Johnson (1859–1939). Taught administrative law at Columbia from 1883 till 1914 when he became president of Johns Hopkins University. He was a leader in the study and teaching of public administration. His writings include *Comparative Administrative Law* (2 vols., 1893), *Principles of the Administrative Law of the United States* (1905), *Municipal Problems* (1897), *Municipal Government* (1909), *Politics and Administration* (1900), and *Principles of Constitutional Government* (1916), and he also served on many commissions.

Goods. In an old broad sense goods includes all chattels real and personal (q.v.). In a modern and narrower sense it is limited to chattels personal or personal property, excluding chattels real. In many contexts the double phrase 'goods and chattels' is used. It is specifically defined, in different terms, in various statutes.

Goods in communion. In older Scots law, the part of the husband's moveable property which could be actually enjoyed by the spouses while the marriage subsisted and of which part devolved on the widow on his death, as *ius relictae*. It did not signify a system of community of property. The concept became obsolete in the nineteenth century when each spouse became capable of owning moveable property independently.

Goodwill. The advantage attaching to a business from its reputation and connection with customers or clients and the circumstances which tend to make that connection continuing. It cannot be severed from the business to which it attaches and may attach particularly by reason of the personality or personal qualities of the proprietors or managers of the business, or by reason of its site, or partly for both reasons. It is a kind of personal property which may be bought and sold, given by will or charged and the owner may vindicate his right thereto by an action for passing-off. On the discontinuance of a business the goodwill is part of the assets and passes on bankruptcy to the trustee. In general, it includes unregistered trade marks and trade names.

Goos, Carl (1835–1917). Danish jurist and Minister of Justice, author of *Forelaesninger over den almindelige retslaere* (1885–92), a general theory of law, and *Den danske strafferet* (1875–96), an analysis of Danish criminal law.

Gordon, Edward Strathearn, Lord Gordon of Drumearn (1814–79). He was called to the Scottish Bar in 1835 and developed a fair practice. He became Solicitor-General for Scotland in 1866–67, an M.P. in 1867, and Lord Advocate 1867–68. In 1869 he became Dean of the Faculty of Advocates and M.P. for the Scottish Universities and in 1874–76 Lord Advocate again. While in Parliament he secured the passage of a Court of Session Act, a Titles to Land Act and an Act abolishing patronage of ecclesiastical livings. In 1876 he was appointed the first Scottish Lord of Appeal under the Appellate Jurisdiction Act but did not live long enough to make much impression as a judge, though the judgments he delivered were sound and carefully considered.

Gordon, Sir George, of Haddo, 1st Earl of Aberdeen (1637–1720). Studied law abroad, was called to the Scottish Bar in 1668, and attained a great reputation as a lawyer. He sat in the Scottish Parliament from 1669. He became a judge in 1680 (Lord Haddo), Lord President of the Court of Session in 1681, and Lord Chancellor of Scotland, 1682–84, being created Earl of Aberdeen in 1682. His administration was firm but not severe. He was dismissed in 1684 in consequence of intrigue, having enforced the religious conformity laws severely, but sat in Parliament again in Queen Anne's time and supported the Treaty of Union of 1707.

Gore, John, Lord Annaly (1718–84). Was called to the Irish Bar in 1742, and became Solicitor-General (1760), Chief Justice of the King's Bench in Ireland (1764), and Baron Annaly (1766).

Gorell Barnes, John, Baron Gorell (1848–1913). Was articled to a solicitor but was called to the Bar in 1876 and developed a practice in admiralty and mercantile cases. In 1892 he was made a judge of the P.D.A. Division, and in 1905 succeeded as President. In 1909 he was made a peer and thereafter sat in the House of Lords. He was chairman of the Royal Commission on Divorce of 1909–12 and chairman or member of various other committees. The foundation of the commercial court was due to him and Mathew, J., whose pupil he had been. He was a very sound judge, of good sense and industry.

Goring, Charles Buckman (1870–1919). English prison doctor and criminologist, carried out extensive examinations and statistical analyses of data about convicts and largely refuted the views of the positive school of criminology. His major work was *The English Convict* (1913).

Gortyn, Laws of. Among inscriptions found at Gortyn, one of the most important towns of Dorian Crete, is the Code of Gortyn (*c.* 450 B.C.) including many older laws or references to them and large parts of a codification supplementary thereto. It appears to be a mixture of primitive and more developed regulations. It constitutes an important source of early Greek law, and reveals a relatively developed state of legal conceptions. It includes rules of civil law only, in no systematic order, dealing with the family and family-property, slaves, surety, mortgage, donations, procedure, and various other topics. Private and family property were clearly distinguished, particularly in matters of inheritance. The rules relative to adoption and to the rights of slaves were distinctly liberal. Criminal law was closely connected with family law, though monetary penalties had superseded feuds and harsher punishment, and the fines were frequently payable to the State. Proof was by witnesses and compurgation, and by the oath of the party.

J. Kohler and E. Ziebarth, *Das Stadrecht von Gortyn.*

Goudelin, Pierre (Petrus Gudelinus) (1550–1619). Professor at Louvain and author of *De iure novissimo* (1620), the first attempt at a systematic exposition of the law of the Netherlands. His work is thought to have influenced Grotius. He also wrote a *De jure feudorum* (1624) and *Syntagma regularum juris* (1640).

Goudy, Henry (1848–1921). Was called to the Scottish Bar in 1872, and wrote a standard book on *Bankruptcy* (1886). He became Professor of Civil Law at Edinburgh in 1889 and at Oxford in 1893, and published an edition of Muirhead's *Private Law of Rome.*

Gould, James (1770–1838). Studied law at Litchfield, Conn., under Tapping Reeve (q.v.), 1795–98, joined him in managing the school, 1800–20, and managed the school alone, 1820–33. He also engaged in practice and was a judge of the superior court of Connecticut 1816–18. He had a profound knowledge of the common law and great ability to impart his knowledge. He wrote *Principles of Pleading in Civil Actions* (1832).

Govea, or Gouvea, Antoine de (Gouveanus) (1505–1566). An eminent Portuguese humanist jurist who spent all his life in France and became a distinguished Romanist. He is regarded as ranking next to Cujas in his application of the humanist method, seeking to reconstruct the Roman law from the sources and tracing its evolution in relation to the historical development of the Roman Empire. His many works included *De jure accrescendi* (1549), *De jurisdictione* (1550), and works on classical authors.

Government. The process and practical machinery of ruling and directing the affairs of a state or part thereof. Secondly, the body of persons having the function of ruling and directing, particularly the Cabinet or ministry, other Ministers, the civil service, and the executive officials generally. Thirdly, in Parliament, the party from which the Cabinet and other ministers are chosen, as contrasted with the Opposition.

The process of government involves three distinguishable functions, legislation or law-making, execution or carrying into effect of policies and laws made, and decision of disputes and controversies arising incidentally out of the other functions. The way in which these functions are to be exercised and by whom is determined by the constitution of the particular state. Political scientists from Plato onwards have distinguished and classified governments. Aristotle distinguished good and bad forms of rule by one, a few, and many, namely monarchy and tyranny, aristocracy and oligarchy, polity and democracy (or mob-rule). But this is not exhaustive; other forms are gerontocracy or rule of the elders, theocracy or priest-rule, bureaucracy or officials' rule, rule by generals, or by trade-union bosses. There are many varieties. In modern times the major classification of governments is into liberal-democratic and totalitarian. Whether a government is monarchical or republican matters little as monarchical or republican are each consistent with both liberal-democratic and totalitarian government. The determining factor is the predominant social and political philosophy of the ruling group, whether liberal, subject to checks and balances, as by Bills of Rights and recognizing rights in individuals, or socialist-communist, subordinating individuals to the interests of the dominant group. All claim to be democracies. The prevailing social, political, and economic philosophy has important consequences for the legal system and the institutions and rules of law, as can be seen in Britain where a liberal philosophy prevailed until 1914, being gradually replaced by a socialist one in which the power of the Prime Minister and Cabinet is enormous, that of Parliament negligible and the electorate almost irrelevant, being limited to deciding periodically which group of leaders are supposed to be in power. It is the underlying philosophy which is enacted and applied and which sets the limits within which disputes are to be determined.

The connection between governments and law is very close. The form of government and the powers and functions of its organs are defined by law and every organ operates by making, applying, and enforcing rules of law on individuals in particular cases. Moreover, in the twentieth century governments have increasingly intervened to regulate and control countless relationships, such as employment, landlord and tenant, trading, and others, which were formerly regulated mainly by principles founded on custom and practice.

S. E. Finer, *Comparative Government* (1970); H. Finer, *Theory and Practice of Modern Government* (1949); C. J. Friedrich, *Constitutional Government and Democracy* (1950).

Government departments. The departments of state or ministries which under the direction of their political heads (Secretaries of State, or Ministers) execute the policy determined by the Cabinet and Parliament. The names and precise areas of responsibility of departments are changed from time to time and departments are periodically created, merged, or abolished. The principal U.K. departments are the Civil Service Department, the Treasury, the Lord Chancellor's Office, the Ministry of Defence, the Home Office, the Foreign and Commonwealth Office, the Department of Education and Science, the Department of Health and Social Security, and others. Government departments are quite distinct from public corporations (q.v.) and local authorities, not least in that they are in theory Crown servants, act as agents of the Crown and enjoy the privileges and immunities of the Crown.

Government printer. Her Majesty's Stationery Office and any printer authorized to print statutes and other public Acts. It is the Queen's Printer who classifies statutes as public, general, local, or personal, and numbers statutory instruments, and prints copies of each.

Governmental functions and organs. The major functions of government are legislative, the making of rules of law, executive or administrative, the application of rules of law in particular cases, including sometimes the exercise of a measure of discretion, and judicial, the decision of disputes and controversies, including the ascertainment of the facts and the rule of law relevant thereto. Where the application of rules of law in particular cases permits little or no exercise of discretion it is sometimes classified as ministerial. Important consequences may follow from classification of a particular function as legislative, executive or administrative, ministerial or judicial (q.v.) but the functions overlap, shade into one another and are not always clearly distinguishable. The acts of a person or body may be within one class for some purposes and within another for other purposes. A particular person or body may act in an administrative capacity at one time and in a judicial capacity at another. Under some constitutions a fairly strict separation of governmental powers and functions exists. In Britain there is only a general separation and differentiation. Legislative functions are exercised by Parliament but also, by delegation, by Ministers, departments, local authorities, Rules Committees, and other bodies. Executive or administrative functions are exercised by Ministers of the Crown, civil servants, local authority officers, and many others. Judicial functions are exercised by courts and tribunals but also by Ministers, club committees, and other bodies.

Gown. A loose outer garment, particularly the open-fronted robe worn by lawyers and university teachers. Gowns have been worn by lawyers in all Western European countries from the early Middle Ages. In England, by the eighteenth century, barristers' gowns had settled in what is still the accepted shape, open and of black cloth with bell sleeves. King's (or Queen's) Counsel seem, from the start, to have worn open gowns with flap collars and winged sleeves of the pattern worn by gentlemen in Jacobean times, and made of silk. Attorneys seem to have worn a variety of shift gown. Scottish advocates seem, from the eighteenth century, to have worn a long black gown with bell sleeves gathered at the elbows.

In modern practice Q.C.s wear gowns of black silky material and barristers and advocates stuff gowns. As a rule, counsel wear wigs and gowns at all sittings of courts and of tribunals which are robed, but not at sittings in chambers, before tribunals which do not robe, before magistrates or arbitrators and a court may dispense with the wearing of gowns where there is good reason to do so. Where solicitors have a right of audience, in circumstances where counsel would wear their gowns, they wear black gowns, but not wigs.

See also ROBES; WIGS.

Grace. A faculty, dispensation or licence. The designation Act of Grace was given to statutes giving a general or free pardon.

Graham, James, 3rd Duke of Montrose (1755–1836). Sat in Parliament from 1780 and secured the repeal of the statutory prohibitions on wearing the highland dress. He held various political offices under Pitt. He succeeded to the peerage in 1790 and became a Privy Councillor, and was Lord Justice-General of Scotland, 1795–1836. He was the last lay holder of the office, statute having provided, in 1830, that on the death of the holder the office was to be combined with that of Lord President of the Court of Session, and accordingly

Lord President Hope (q.v.) became Lord Justice-General also in 1837.

Graham Murray, Andrew, Viscount Dunedin (1849–1942). Was called to the Scottish Bar in 1874 and rapidly established himself. He was successively Advocate-Depute (1888), Sheriff of Perthshire, M.P. (1891–1905), Solicitor-General (1891–92, 1895–96), Lord Advocate (1896–1903), and Secretary for Scotland (1903–05), before being appointed Lord Justice-General of Scotland and Lord President of the Court of Session in 1905, becoming Baron Dunedin. In 1913 he was made a Lord of Appeal. He became a Viscount in 1926, and retired in 1932. Dunedin has been credited with being the greatest Scottish judge of the century, with a fine grasp of principle and a great knowledge of case law, particularly of feudal land law and conveyancing. His reputation is of the very highest. He also was active in expediting the business of the Court of Session and acted as Consulting Editor of the *Encyclopaedia of the Laws of Scotland*. He received many honours, including the G.C.V.O. in 1923.

Grand Assize. A species of jury trial introduced in the reign of Henry II whereby the tenant or defendant, in a writ of right, could have either trial by battle or trial by his peers. It was abolished in 1833.

Grand Conseil. A French royal court, an outgrowth from the royal council. It was established in 1497 with rather vague jurisdiction; it heard cases involving bishoprics and ecclesiastical benefices, cases removed by the King from the *parlements*, disputes between the *parlements* and *présidiaux* and cases transmitted by the *conseil des parties*, the royal court of appeal. Its judgments were enforceable throughout France and there was no further appeal. It sat alternately at the Louvre and at Versailles, and was abolished at the Revolution.

Grand Coutumier de France (or Coutumier de Charles VI). A compilation by Jacques d'Ableiges of about 1385, dealing with ordinances, the feudal law of the Île de France, rules of practice, procedure and forms.

Grand Coutumier de Pays et Duché de Normandie. A collection, probably compiled in the thirteenth century, of the common or customary law of the ancient Duchy of Normandy as at 1205, which still remains the law of Jersey, except in so far as modified by later local usage, legislation, or judicial decision.

Grand days. Formerly one day in each law term was designated Grand Day, and these days were observed with considerable ceremony in the Inns of Court and of Chancery. The courts did not sit on Grand Days. Now Grand Day is a day in each of the terms when the Benchers of each Inn entertain guests to dinner and a more sumptuous dinner than usual is served in Hall.

Grand jury. A body which had its origins in the Assize of Clarendon in 1166 and at first both accused and tried suspects, but later accused them only, the trial being heard by a petty jury. In England grand juries were abolished, with some exceptions, in 1933, the surviving instances disappearing in 1948. They never seem to have existed in Scotland. In jurisdictions where the institution still exists it comprises a group of laymen summoned from a district to enquire into alleged crimes. The members, usually between 12 and 23 in number, are chosen by ballot. They are instructed by a judge as to their inquiry. Public officials give information and summon witnesses, who normally must testify or be held in contempt of court. The proceedings are usually secret and conducted in private but minutes of the proceedings are made available to the court, the prosecutor and, sometimes, to the accused. The proceedings result in a presentment or indictment, i.e. a formal criminal charge against a named person. This is 'finding a true bill'. Or the grand jury may find 'not a true bill' in which case the person accused is discharged, though a fresh bill may later be found against him by another grand jury. Originally a standard part of the common law system of criminal procedure, the grand jury has declined in importance in modern times. In many jurisdictions its use has been abandoned.

Grand larceny. Originally the theft of goods exceeding 12 pence in value, a crime punishable by death. Any distinction between grand and petty larceny was abolished in 1827.

Grand Remonstrance. A document of protest prepared and passed by the House of Commons in 1641 and purporting to put before Charles I a 'faithful and lively representation of the state of the kingdom'. It was provoked by Charles' duplicity in dealing with Parliament, and contained 206 clauses setting out the state of the nation, the King's defaults, the grievances of the Commons, and the reforms which Parliament had effected, and demanding further changes, including improvements in the administration of justice, the appointment of Ministers whom Parliament could trust, and the enforcement of the laws against Roman Catholics. The debate on it was the first major split between the Royalist and Parliamentary parties, and resulted in only a small majority for the latter. It was published and in turn provoked the King to attempt to arrest the five members responsible for the document. See FIVE MEMBERS, ARREST OF THE.

Grandfather clause. A name given to provisions in the post-Civil War suffrage laws of several southern U.S. states exempting from educational tests and property qualifications persons who had voted, or whose ancestors had voted, before 1867. The Supreme Court in 1915 declared such provisions unconstitutional as violating the 15th Amendment.

Grand jours. French provincial assizes, composed of magistrates of the *parlement* of Paris, sitting during the summer vacation to dispense justice in a time of rebellion or lawlessness.

Grand serjeanty. In feudal law, a mode of tenure of land. A tenant holding by grand serjeanty held by such honourable services as he should perform personally for the sovereign, such as by carrying his banner or his lance, acting as his marshal or leader of his forces, his butler or his carver, or any one of numerous offices about the court. Grand serjeanty was distinguished from petty serjeanty because the services were nobler and more honourable, and were not at all menial. Tenure by grand serjeanty has survived being now a form of free and common socage but with the honourable services preserved.

Grant. At common law a conveyance, originally confined to incorporeal hereditaments, of which livery of seisin could not be given. After 1845 it was provided that corporeal hereditaments should be deemed to lie in grant as well as in livery, and this was extended in 1925 to all lands and interests therein. The term is also often applied to rights created by the Crown, e.g. charters, and to sums of public money given to local authorities or other bodies for designated purposes.

Grant, Patrick, Lord Elchies (1690–1754). Passed advocate in 1712 and became a judge of the Court of Session in 1732 and also a Lord of Justiciary in 1737. He wrote Notes to Stair's *Institutions* (1824) and made a collection of the *Decisions of the Court of Session*, 1733–57, published in 1813.

Grant, William (1909–72). Called to the Scottish Bar in 1934, he sat in Parliament, 1955–62, became Solicitor-General for Scotland in 1955, Lord Advocate in 1960, and Lord Justice-Clerk in 1962.

Grant, Sir William (1752–1832). Educated at Aberdeen and Leyden, was called to the Bar in 1774. In 1775 he went to Canada and became Attorney-General of Canada, but returned to England in 1787 and became an M.P. in 1790; his success in the Commons brought success at the Bar and in 1799 he became Solicitor-General and in 1801 Master of the Rolls. He resigned in 1817. Both as a judge of equity and in the Privy Council he proved himself to be a judge of exceptional merit and an outstanding lawyer. In prize law he stands second only to Lord Stowell and he was deemed a perfect model of judicial eloquence.

Grant of probate or of letters of administration. The giving of authority by an appropriate court to an executor (q.v.) or administrator (q.v.) to act as such in relation to the estate and affairs of a deceased. A grant may be general, if unrestricted, or limited when confined to part only of the deceased's property, or to a period of time, such as during the executor's widowhood, or to a particular object, such as *pendente lite* or *ad colligenda bona*. If an executor or administrator dies or becomes incapacitated before he has administered the estate the court appoints a new personal representative by granting probate or letters of administration *de bonis non* (*administratis*) to enable him to complete the administration.

Grassum. In Scots law, a capital sum exacted by a superior on the grant of a feu-right or by a landlord on the grant or renewal of a lease, in each case in addition to the periodical feu-duty or rent.

Graswinskel, Dirk (or Theodore) (1600–66). Dutch jurist, associated with Grotius, and author of *De Jure majestatis* (1642), *Maris Liberi vindiciae* (1652), and other works opposing Selden's *Mare clausum*.

Gratian (Graziano, Francisco) (*c.* 1090–1159). A Camaldolian monk of Ravenna and later of Bologna, and said to have been later Bishop of Chiusi, compiler, about 1140, of the *Concordia discordantium canonum* or *Decretum Gratiani* (1151), a collection of about 4,000 texts of canon law drawn from previous collections down to the Lateran Council of 1139. It was not the first collection of materials bearing on canon law, but at once superseded earlier collections because of its systematic arrangement, based on the work of jurists on Roman civil law. It sought to resolve the discrepancies and contradictions in the texts of Church law drawn from different sources and of very different dates. It became the major text for the study and teaching of mediaeval canon law and the first part of the accepted corpus of canon law texts. It became the subject for glosses and commentaries by all the most distinguished canonist scholars and laid the foundation for the study of canon law as distinct from theology and ecclesiastical history. See also CANON LAW.

Gravina, Gian Vincenzo (1664–1718). An Italian humanist jurist, author of a famous *Origines juris civilis libri III* (1701) tracing the derivation of Italian from Roman law and long a leading authority on legal history, and other works on Roman law and classical antiquity.

Gray, John Chipman (1839–1915). Was admitted to the Massachusetts Bar in 1862 and served in the Civil War. In 1866 he assisted in founding the *American Law Review*. In 1869 he joined the Faculty of the Harvard Law School, becoming Story Professor (1875) and Royall Professor (1883–1913). His main field of interest was property and his books on that branch of law are still important, notably *Restraints on the Alienation of Property* (1883) and *The Rule against Perpetuities* (1886). He also compiled *Select Cases and Other Authorities on the Law of Property* (6 vols., 1888–92) and gave the lectures published as *The Nature and Sources of the Law* (1909).

R. Gray, *John Chipman Gray.*

Gray's Inn. See INNS OF COURT.

Great Britain. The political entity formed in 1707 by the union of England and Wales with Scotland (which had been united in a personal union from 1603 when James VI of Scotland succeeded Elizabeth I of England, and became James I of England) and which became the United Kingdom in 1801 by the union with Ireland. The name 'Great Britain' is said to have been suggested by Bacon. See ENGLISH LAW; IRISH LAW; SCOTS LAW; UNITED KINGDOM; WELSH LAW.

Great Charter. Magna Carta (q.v.).

Great Council of the Realm. The assembly of tenants in chief of the King and greater clergy summoned by the Norman kings and replacing the Anglo-Saxon Witan. It later developed into the House of Lords, but even after that House was well developed, the name was sometimes applied to an assembly of the peers meeting outside Parliament. Such a Great Council was held by Charles I at York in 1640.

Great Contract. A financial reform proposed in England in 1610 by which Parliament was to vote James I and his successors £200,000 per annum as compensation for his renouncing the royal right to levy certain feudal exactions. Negotiations were lengthy but eventually broke down and feudal dues were not got rid of until abolished by parliamentary ordinance of 1643, confirmed by statute in 1660.

Great Officers of the Realm. According to the Act for the placing of the Lords in Parliament 1539, there are 12 Great Officers of the Realm. They are, in order of precedence, the Lord Chancellor, Lord President of the Council, Lord Privy Seal, Lord Great Chamberlain, Lord High Constable, Earl Marshal, Lord Steward of the Household, and Lord Chamberlain, together with the offices of Ecclesiastical Vice-regent, Lord Treasurer, Lord Admiral, and King's Secretary which no longer exist, though Queen Elizabeth II resumed the office of Lord High Admiral. The first three are political offices, but the others which survive are hereditary.

In Scotland there are, in some cases, corresponding Officers known as Officers of State (q.v.).

Great Rolls of the Exchequer. See PIPE ROLLS.

Great Seal. The Great Seal is affixed to proclamations, writs, letters patent, and documents giving power to sign and ratify treaties. Authority for affixing the Great Seal is given by Royal Warrant under the Queen's sign manual and countersigned by the Minister on whose advice the document is being issued. The seal attached to documents is a piece of wax six inches in diameter hanging from the document by a piece of ribbon and has impressions made by the Great Seal on both obverse and reverse. By tradition the obverse depicts the sovereign on horseback and the reverse the sovereign robed and enthroned. Accordingly the Great Seal consists of two silver discs, carrying the impressions for the obverse and reverse of the wax.

The Lord Chancellor originally had the custody of the Great Seal but in the intervals between Chancellors it was entrusted to a temporary Lord Keeper and later it was frequently held by a Lord Keeper for substantial periods, there being no Chancellor.

From the time of Sir Nicholas Bacon the offices of Lord Keeper of the Great Seal and Lord Chancellor have always been held by the same person, but it was for long common not to confer on a Lord Keeper the title of Lord Chancellor at once, e.g. Sir Robert Henley, Lord Keeper 1757, Lord Chancellor 1761–66 (Earl of Northington, 1764), and on a number of occasions for periods, sometimes of several years, the seal was in commission, i.e. several commissioners were appointed for executing the office of Lord Keeper. The last occasion was in 1835–36. Under the Commonwealth and Protectorate the parliamentary and protectorate Great Seal was in the hands of various commissioners who were neither Keepers nor Chancellors. In modern practice the person is created Lord Chancellor by delivery to him of the Great Seal.

A new Great Seal is adopted when the old one is worn out, or at the commencement of a new reign, or if the royal arms are changed. When this happens the old seal is 'damasked', or rendered useless by being tapped with a hammer by the Lord President of the Council. The old seal becomes the property of the Lord Chancellor, and this has given rise to disputes, between Lords Lyndhurst and Brougham in 1830 and between Lords Cairns and Hatherley in 1878. In each case the Seal was divided. On two occasions the Great Seal has been lost, when James II dropped it in the Thames in 1688 and when burglars stole it from the house of Lord Chancellor Thurlow in 1784.

There was a separate Great Seal in Scotland kept by the Lord Chancellor of Scotland (q.v.). The Treaty of Union 1707 appointed a seal to be used in Scotland in place of the Great Seal of Scotland. The Secretary of State for Scotland is Keeper and the Keeper of the Registers of Scotland Deputy Keeper of this seal.

H. C. Maxwell-Lyte, *Historical Notes on the Great Seal* (1926).

Great Sessions, Courts of. By statute of 1543 superior courts, called Courts of Great Sessions, were established for Wales, with jurisdiction independent of the English courts. The 12 Welsh counties were divided into groups of three, and a court of Great Sessions was to be held twice a year for six days in each group. The judges were to be the Justice of Chester, the Justice of North Wales, and two persons nominated by the Crown. Great Sessions had power to exercise all the jurisdiction exercisable by the King's Bench and Common Pleas, and it acquired or assumed an equitable jurisdiction by the seventeenth century. Error lay to the Council of Wales in personal actions, to the King's Bench in real and mixed actions.

By the late eighteenth century the courts at Westminster had, in almost all actions, established rights to exercise jurisdiction concurrent with that of Great Sessions and in 1830 statute abolished the court of Great Sessions, extended the English circuit system to Wales and Chester and transferred the equitable jurisdiction to the Chancery or the Exchequer.

Greek law (ancient). Ancient Greek civilization was too extensive in space and time and too diverse to produce any single body of law. The Mycenaean civilization of Greece (*c.* 2500–1100 B.C.) and the Minoan civilization of Crete (*c.* 1600–1000 B.C.) have left no known legal records. The Homeric poems (*c.* 750 B.C.) depict scenes of legal relevance but give no indication of a legal system beyond indications that kings were judges in peace as well as generals in war and that justice was an aspect of tribal government.

In the eighth century B.C. cities grew up in Greece, replacing villages as the ordinary settlements. Government of Greek city-states was originally monarchical but between 700 and 500 B.C. aristocracies nearly everywhere replaced kings. Law seems first to have been written down about the seventh century B.C. According to tradition in many cities law-givers framed codes of laws. The chief of these were Lycurgus at Sparta (*c.* 875 B.C.) (possibly mythical); Pheidon at Corinth (*c.* 850 B.C.), Zaleucus at Locri (*c.* 675 B.C.) whose laws were accepted by many cities in Italy and Sicily, Charondas in the Chalcidian cities (*c.* 660 B.C.): Draco at Athens (*c.* 625 B.C.); Philolaus at Thebes (*c.* 600 B.C.); Solon at Athens (*c.* 595 B.C.) and Pittacus at Mitylene (*c.* 580 B.C.). In all cases it is uncertain how much of the bodies of law were truly attributable to such law-givers. Since each city-state had its own law there was no such thing as Greek law, but there were elements common to many cities.

Public law

For the first time in legal history we have substantial knowledge of the public law and constitution of some of the city-states. In the eighth century tribes had chieftains who were assisted by councils of elders. As city-states developed leadership was transferred to an aristocracy; power became centred in a class which alone had knowledge of law and religion, and frequently in large part owned the land. This was replaced in some states by oligarchies, usually based on wealth, in others by tyrannies. In the cases of Athens and Sparta we have more detailed knowledge of the constitutions. Athens in the fifth and fourth centuries was a democracy, headed by a Council of Five Hundred and nine archons. These were numerous subordinate boards of magistrates. The 10 generals were the most important officers of the State. Sparta was governed originally by two kings and a council of elders. The principal magistrates were five ephors.

The Code of Gortyn (q.v.) in Dorian Crete (*c.* 450 B.C.) is an important source of knowledge of Greek law and reveals considerable maturity of legal conceptions. In criminal law money penalties had replaced earlier punishments; they varied with personal status and were frequently payable to the State. In civil law there were rules relative to family, property, slaves, mortgages, donations, and there was a settled procedure. The Code is probably a restatement rather than new legislation. It illustrates the aristocratic type of legal system, as contrasted with the democratic type illustrated by the law of Athens, and the power and influence of Athens helped to spread through Greece the principles of law, which its own system exemplified.

Courts and procedure

Only the Athenian system of courts is adequately known. Justice was administered by magistrates, popular courts, and the Areopagus. Some of the magistrates had special jurisdiction, the King archon over the religious cases and murder, the archon over family and succession cases, the general over the metics (resident aliens) and the *thesmothetai* ('customs deciders') over other cases. The popular courts or dicasteries, sections of the popular assembly of all the free males numbering 201 in civil cases and 501, 1001 or 1501 in criminal cases, chosen by lot, were the judges. In some cases of great political importance the whole popular assembly organized as a court of 6001 was convened. These courts were expressions of the democratic principle that all the citizens should have a say in judging the affairs of individual citizens. The Areopagus, a court composed of former archons, heard murder cases.

Civil and criminal procedure were little differentiated. The main distinction seems to have been into special actions, notably impeachment, private suits including claims for homicide, and public suits to indicate a public interest. The initial step was the seizure of the person or property of the defendant in pursuance of a private claim, or a public assertion that the defendant was guilty of crime, and in either case the trial was to determine whether this had been justifiable or not. In either case the initial step was to summon the defendant before the appropriate magistrate and to lodge a written complaint, which the magistrate examined. If the claim were a civil one for money the magistrate referred the case to a public arbitrator; if his award was not accepted or if the case was not one for arbitration, it was referred to a popular court presided over by a magistrate which, having heard evidence and speeches, voted by secret ballot whether to accept the one party's proposal or the other. This decision was final; a successful plaintiff had to enforce the judgment himself by seizing property of the defendant.

Private law

Citizenship was limited to adult males descended on both sides from citizens and naturalization was permitted only occasionally. Resident aliens had rights to freedom and property but did not have political rights. In general, they had no right to sue in the court but in Athens non-resident foreign merchants might sue and be sued in special courts at the Piraeus.

The status of women seems to have varied from city to city. In Gortyn they might own property; in Athens they were subject to the power of the head of the house. Slavery was recognized everywhere, the precise condition varying from one state to another. At Athens the slave was the property of his master; elsewhere he seems to have been nearer the status of the villein. Everywhere manumission seems to have been competent.

In Athens the family seems to have been hierarchic, property and authority being vested in the father, but in other states family property seems to have been vested in the whole family, including women.

Marriage was monogamous, restricted to citizens and involved solemn ceremony. The husband became master of the wife and her property, including her dowry. Divorce was open to both spouses and was effected by mere separation without legal process or need for justification.

Contract law was relatively well developed, as many states engaged extensively in commerce. Contracts depended not on consent but on consideration, on the incurring of loss if the other party did not implement his part of the contract. The fixing of what is now called liquidate damages was competent. The rules for constitution and proof of contract were not complicated; for most purposes an oral contract, proved by witnesses, was adequate, but from the fourth century written agreements were favoured. Maritime loans were always constituted in writing. A large range of specific contracts was recognized. In sale the payment of a deposit and the forfeiture of a multiple thereof, if the seller failed to perform his bargain, were recognized. Loan of money at interest was common, as was the maritime loan, secured by ship or cargo and repayable when the voyage was safely accomplished.

Both personal and real security were recognized. In personal security the surety vouched for the fulfilment of the debtor's undertaking or for the appearance of a defendant in court; if the principal failed the surety was liable to the same penalty as he was. In real security the standard form was the hypothec, without transfer of possession, but if the debtor defaulted the creditor could seize the thing given in security.

At Athens individual ownership of property was the rule whereas at Sparta, control of the helots was vested in communal ownership of land. In many states transfer of title to land required publicity by way of announcement by heralds, statement of transfer in the presence of witnesses, or recording in a public register.

The chief private wrong was the causing of damage to another's property; if it were done deliberately the penalty was double the value of the damage done. Homicide was regarded more as a tort than a crime in that the onus of pursuing the killer was on the victim's relatives and indeed restricted to them; this attitude is a relic of a much older custom of the blood feud. Hybris or outrageous conduct was both criminal and tortious, giving rise to claims for physical injury and for some kinds of libel. Other recognized torts included forcible entry into another's property and prevention of lawful seizure of property.

Criminal law

Legislation specified a large range of offences as criminal, including conduct prejudicial to the State and the community, such as treason, cowardice, impiety, falsely pretending to be a citizen and acting prejudicially to the existing democratic order, conduct contrary to the public peace, such as theft, and various kinds of misconduct, such as outrageous behaviour or improper treatment of parents, orphans, and the like. Any citizen was entitled to bring a criminal prosecution.

General

Greek law singularly failed to produce any jurists, or scholars who devoted themselves to systematizing, analysing, and expounding the legal system or any branch thereof. Or if there were any, their work has perished. There were no rules of statutory interpretation and no developed technical terminology. There were orators but not lawyers, apart from those who did the notarial work of preparing legal documents for commercial purposes. Even the philosophers were much more interested in abstractions such as justice than in analysing positive law. It was only with the Roman conquest that law became a matter of professional study and training.

On the other hand, Greece produced philosophers who examined many of the concepts underlying law and order in society, though they seem to have regarded the details of law as unworthy of their attention. Occasionally, as in the work of Theophrastus, pupil and successor of Aristotle, the work of philosophers had some influence on legislation. He made a review and comparison of the rules and institutions in force in the different Greek cities. His pupil Demetrius was ruler of Athens 317–307 B.C. and wrote on law and legislation. It is the work of the philosophers not the lawyers, jurists or orators that has influenced later law and thinking about law.

Greek legal practice survived in Egypt to the fourth century A.D. and penetrated to Greek colonies, but had very little influence on Roman law, though Greek thinking had an influence.

Public law: G. Busolt and H. Swoboda, *Griechische Staatskunde*; Hermann, *Lehrbuch der Griechischen Staatsaltertümer*; G. Glotz, *The Greek City and its Institutions*; R. Bonner and G. Smith, *Administration of Justice from Homer to Aristotle* (1930–38); C. Hignett, *History of the Athenian Constitution* (1952); G. Calhoun, *Growth of criminal law in ancient Greece* (1927). *Private law*: P. Vinogradoff, *Outlines of Historical Jurisprudence*, vol. II (1922), J. H. Lipsius, *Das attische Recht und Rechtsverfahren* (1905–15); L. Beauchet, *Histoire du Droit privé de la République Athénienne* (1897); F. Pringsheim, *The Greek Law of Sale* (1950); J. Kohler and E. Ziebarth, *Das Stadrecht von Gortyn*; A. R. Harrison, *Law of Athens, 2 vols.* (1968–71); J. W. Jones, *Law and Legal Theory of the Greeks* (1956); G. Calhoun, *Introduction to Greek Legal Science*; D. MacDowell, *The Law in Classical Athens.*

Greek law (modern). Greece gained independence from Turkey in the years 1821–29 and became a kingdom in 1830. Until 1973 it was a constitutional monarchy but in that year became a republic.

The constitution of 1911 was amended in 1951 and a new constitution was adopted in 1968. There is a State Council having jurisdiction in administrative disputes and contraventions.

Justice is administered by a Supreme Court sitting in four sections (three civil and one criminal) with jurisdiction in appeals and the most important cases, 11 Courts of Appeal and some 58 courts of first instance. There are also minor courts for the trial of petty offences and juvenile courts.

A new civil code was adopted in 1940; it reflects the influence of the German civil code.

Green, Thomas Hill (1836–82). English philosopher who exercised great influence on philosophy and social thought in the later nineteenth century. He based his ethics on the spiritual nature of man; freedom is the power to identify oneself with the true good. His political philosophy was based on his ethical ideas. Ideally political institutions embodied the community's moral ideas and helped to develop the character of individual citizens. His social attitude was influenced by seventeenth-century Puritan ideals, and he attempted to reformulate political liberalism to emphasize the need for positive action by the State and diminish the negative rights of the individual. His *Liberal Legislation and Freedom of Contract* (1881) advocated state intervention in education, land ownership, factory legislation, and licensing laws. His major works were *Prolegomena to Ethics* (1883) and *Lectures on the Principles of Political Obligation* (1883).

Greene, Wilfrid Arthur, Baron (1883–1952). Was called to the Bar in 1908, and became a Lord Justice of Appeal, 1935–37, Master of the Rolls, 1937–49, and a Lord of Appeal in Ordinary, 1949–50.

Greenleaf, Simon (1783–1853). Was admitted to the Bar in 1806, but initially had much time for the further study of law, and was reporter of the supreme judicial court of Maine, 1820–32, and also took a leading part in framing the initial legislation of the newly admitted state. In 1833 he became Royall Professor at Harvard and later succeeded Story as Dane Professor there, being responsible along with Story for the rise of Harvard to a leading position. He wrote a *Treatise on the Law of Evidence* (1842–53) and *Cruise's Digest of the Law of Real Property, Revised and Abridged for the Use of American Students* (7 vols. in 5, 1849–50), which superseded the English original. He was deliberate and thorough, and highly esteemed as a teacher.

Greenshields' case (1711). Established the right of the House of Lords to hear appeals from the Court of Session in Scotland in civil cases after the Union of the Parliaments of 1707. Greenshields sought to conduct an Episcopalian meeting-house in Edinburgh; the Court of Session held he was not entitled to do so. He appealed to the House of Lords which entertained the appeal, and reversed the Court of Session. Since then the right of appeal has continued, though it is questionable if it was intended by the Treaty of Union 1707.

Gregorius Tolosanus (Pierre Grégoire) (*c.* 1540–97). French jurist, a humanist, author of *Syntagma juris universi* (1582), *Universi Iuris Methodus parva* (1582), and other works.

Gregory IX (1145–1241). Ugolino of Segni, later cardinal-bishop of Ostia and Pope, 1227–41. A distinguished theologian and canon lawyer, and part author of the rule of the Franciscan order, he began the restoration of Aristotle as the basis of scholastic philosophy and was responsible for the publication, in 1234, of the five books of *Decretals* which form the second major part of the canon law. He was a vigorous head of the Church, defender of papal privileges and founder of the Papal Inquisition.

Gretna Green marriages. Prior to 1940 marriage could be contracted in Scotland by mutual declarations of consent to take each other as husband and wife in the presence of two witnesses, or proved by writing. No clergyman needed to be present or to participate in any way. Couples in England, determined to escape parental refusal to permit their marriage, frequently eloped to Scotland and were married by declaration of consent at the first convenient place in Scotland, frequently the village smithy at Gretna Green (north of Carlisle) (though such a mode of marriage could equally have been effected anywhere in Scotland). Such runaway marriages were struck at by Lord Brougham's Act of 1856 which required that at least one party to such a marriage must have had his or her usual residence in Scotland for at least 21 days before the marriage. It did not otherwise affect marriage by declaration. Since 1940, save for cases of marriage by cohabitation with habit and repute (q.v.) all marriages in Scotland must be conducted by a clergyman or an authorized civil registrar. Couples sometimes continue, however, to resort to Gretna or elsewhere in Scotland to avoid the requirements of their own legal system requiring parental consent to the marriage of persons under particular ages, there being no requirement of parental consent under Scots law, and this being treated as a formality regulated by the *lex loci contractus* and not as an element of capacity to contract marriage, but they must reside 21 days and they must be married by religious or civil ceremony, not by bare declaration.

Grey, Anchitell (?–1702). A member of Parliament, who compiled for his own use a record of the Debates of the House of Commons, 1667–94, published in 1769, which was the first continuous record of contemporary debates to be published and used as a source by Cobbett and Wright in their Parliamentary history.

Griffith, Sir Samuel Walker (1845–1920). Australian politician and judge. Twice Prime Minister of Queensland, he was Vice-President of the Federal Constitutional Convention of 1891 and played a major part in drafting the Commonwealth Constitution Bill. He was Chief Justice of the Supreme Court of Queensland, 1897–1903 and influenced the Queensland Criminal Code, and in 1903 became first Chief Justice of the High Court of Australia. As such he dominated the court. He was influenced by Marshall and the attitude of the U.S. Supreme Court and interpreted the constitution in a way opposed to the expansion of the federal power but the High Court later abandoned this attitude.

Grimston, Sir Harbottle (1603–85). He was called to the Bar about 1630 and sat in Parliament from 1628, being an opponent of the court party. He was a strong Protestant and a supporter of constitutional monarchy and became speaker of the Convention Parliament. In 1660 he became Master of the Rolls and held office till 1684, when the King dismissed him. During these years he continued to sit in Parliament. He translated and edited Croke's Reports (1657–61 and later editions) (Croke was his father-in-law) and had the reputation of being an impartial judge.

Griswold, Erwin Nathaniel (1904–). Worked in government service, and was later professor at and then Dean of Harvard Law School (1946–67) and thereafter Solicitor-General of the U.S. He was an authority on tax law and author of several books.

Groenewegen van der Made, Simon à (1613–52). Dutch jurist, annotated Grotius' *Introduction to Dutch Law* (1644) and wrote the *Tractatus de legibus abrogatis et inusitatis in Hollandia* (1649) an important work, discussing what parts of the *Corpus Juris* had been abrogated in Holland.

Groening, Johann (1669–1730). German jurist, author of works on history, politics, and other subjects, and including *Navigatio libera* (1693) and *Bibliotheca juris gentium* (1701).

Gronovius, Laurentius Theodore (seventeenth–eighteenth century). Dutch jurist, author of an *Emendationes Pandectarum* (1688) and works on archaeology.

Groot, Huigh de (Hugo Grotius) (1583–1645). Famous Dutch jurist, studied at Leyden and became Advocate-General of Holland, Zealand and West Friesland in 1607. He was ambassador to London in 1613 but in 1618 he was arrested, tried for treason and sentenced to perpetual imprisonment in the Castle of Louvestein, whence he escaped in a chest ostensibly containing books after 20 months. While there he wrote his *Introduction to Dutch Legal Science* (1631) giving a synopsis of a legal system based on a combination of Roman law and Germanic customary and local laws. He went into exile in France and there wrote his *De Jure Belli ac Pacis* (1625). In 1631 he returned to Holland but religious bigotry forced him to leave again. In 1634 he entered the Swedish diplomatic service and served for 10 years as Swedish Ambassador at the French court. During this time he completed a *History of the Netherlands*, translated Procopius' *History of the Goths and Vandals*, and wrote a work on *The Origin of the American Nations*. In 1645 he applied for recall but resolved to quit Sweden and, *en route* from Stockholm to Lübeck, he was shipwrecked and died from the exertions and exposure.

Among his other works the most notable is *Mare Liberum* (1609) in which he maintained that the high seas are free and cannot be appropriated by any country; it was vigorously attacked by Selden in *Mare Clausum*. He also wrote poetry, edited classical authors, and collected select passages from classical literature.

His reputation rests chiefly on the *De Jure Belli ac Pacis* (based on an earlier *De Jure Praedae*) subsequently repeatedly issued and translated, which established modern public international law. He insisted that actions were bound by natural law, which was based on man's own nature and independent of God, and that on the basis of the law of nature it was possible to formulate a coherent code suitable for all times and places. Natural law was distinct from positive law and could be used to test its rightness. Grotius' work was enormously influential on the later development of international law and of natural law thinking generally.

W. Knight, *Life and Works of Hugo Grotius*; A. Lysen, *Hugo Grotius: Essays on his Life and Works.*

Groot Placaet-Boeck. An authoritative compilation of charters, ordinances, and laws of the Netherlands from 1097 to 1795, published between 1658 and 1796, an important source of Roman-Dutch law.

Gros, Karl Heinrich von (1761–1840). German jurist, author of *Lehrbuch der philosophischen Rechtswissenschaft oder des Naturrechts* (1802).

Gross, Hans (1847–1915). An examining justice and later public prosecutor in Austria, founded criminalistics, the scientific study of criminal investigation, utilizing photography, microscopy, and other aids. Later he taught criminal law at Prague and Graz. He wrote extensively on the investigation of crimes, notably his *Encyclopädie der Kriminalistik* (1901) and *Kriminalistische Tätigkeit und Stellung des Arztes* (1908) and founded the *Archiv für Kriminologie* in 1898.

Grotius. See GROOT, HUIGH DE.

Ground annual. In Scots law, an annual payment, like a perpetual rent charge, secured over and forming a real burden on certain lands. It might be created by grant, or, more usually, by reservation on a grant of the lands to another.

Ground-rent. A rent reserved on a lease, usually for a substantial term of years. The lessee may dispose of the lease during the term but subject to the ground rent. It is usually stipulated for where land is let on a building lease where the tenant erects buildings on the land which at the termination of the lease fall to the landlord along with the reversion of the land.

Group counselling. A psychotherapeutic technique based on the discovery that the exaggerated or abnormal reactions of neurotic patients to their personal difficulties can often be effectively altered by discussions under the guidance of a psychotherapist, more so than by individual interviews. The technique is now extensively used in prisons and takes the form of regular discussions between small groups of inmates with, and under the leadership of, a member of staff on any matter of the institution, their own offences, their difficulties and problems. Participants have to rely on each other to treat the discussion as confidential and are believed to come to do so.

Groups. In considering the working of law in a society regard must be had to the kinds of groups which the individuals of the society form. Groups exist for many purposes from the group who attend a theatre or football match, united only by the wish to see the spectacle and coming together largely by chance and never re-assembling, to close groups which meet and interact regularly. Among important kinds of groups membership of which is more or less permanent are families, kinship groups, young people, old people, the poor, men, women, vocational groups (workmates, unions, professional associations), voluntary groups (religious, cultural, philanthropic) pressure groups and political groups, minorities, ethnic groups, criminal and delinquent groups, and sexual deviant groups. For many of these kinds of groups special legal provision is made in different ways. These include the rules about marriage and the family, guardianship and education of the young, protection and welfare of the old and the poor, special rules against racial discrimination (or in some societies enforcing discrimination) and, above all, the large body of law concerning the conviction and treatment of members of criminal and delinquent groups. Sociological jurists have emphasized the importance of groups in law, the way in which they have influenced it and the way it has influenced them.

Grundnorm. In Kelsen's (q.v.) philosophy of law, the basic, fundamental postulate, which justifies all principles and rules of the legal system and from which all inferior rules of the system may be deduced. It is such a principle as: The will of the Queen in Parliament must be obeyed, or the Constitution and rules dependent on it must be obeyed.

Grupen, Christian Ulric (1692–1767). German historian and jurist, author of works on matrimonial law, and editor of works on early European law, notably *Deutsche Altertüme zur Erlauterung des Sächsischen und Schwäbischen Land- und Lohnrechts* (1746).

Guarantee and indemnity. A guarantee is an accessory contract whereby one party undertakes to be answerable for the debt, default, or miscarriage of another, who is primarily liable to a third party. The surety's liability does not arise until the principal debtor has defaulted and the duration and extent of that liability depend on the terms of the contract. Before recourse can be had to him, any conditions precedent to his liability must be fulfilled.

When a surety has paid he is entitled to be subrogated to all the creditors' rights in respect of the debt or default to which the guarantee relates, including a right to the benefit of all securities which the creditor obtained from the principal debtor, to be indemnified by the principal debtor, to compel contribution from any co-sureties.

The surety is discharged by payment by the principal debtor, by material variation in the terms of the contract between creditor and principal debtor or transaction between them in a manner at variance with the contract guaranteed, or departure by the creditor from the terms of his contract with the surety, or an agreement by the creditor to give time to the principal debtor, or any loss by the creditor of securities held or interference by him with them, or discharge or death of the principal debtor, or discharge of one of several joint or joint and several sureties.

An indemnity is a contract whereby one party undertakes to keep the other free against loss. Unlike guarantee, it is an independent and substantive, not accessory obligation. In a general sense it includes most contracts of insurance but specifically it connotes a distinct undertaking to indemnify. This may arise by express contract, or by implication of law; thus an agent or servant is entitled to be indemnified against liabilities incurred in the performance of the agency or employment. The extent of the right to indemnity depends on the terms of the contract. An indemnity may be unenforceable for illegality. The right becomes enforceable in general only when the claimant has had to pay the third party's claim.

Guardianship, or guardian and ward. A concept or relationship arising from the natural incapacities of infants and persons of unsound mind, and sometimes other categories of persons, to manage their own affairs. In ancient Greece and Rome guardianship began as a right to preserve and protect the ward's property in the interest of the whole kin-group, who would inherit if the ward died, but developed into a duty to preserve the property for the benefit of the ward himself. In Greek law guardians existed for both boys and girls till majority or registration of boys in the citizen-list, or till marriage or for life in the case of girls. Guardians were appointed by will, whom failing the nearest male relatives, whom failing a person appointed by an official (in Athens, the Chief Archon). The guardian's administration was subject to the control of the magistrates.

In Rome children not subject to *patria potestas* or *manus* were subject to guardianship, *tutela* of boys under 14, girls under 12, and women, *cura* of *minores* (persons under 25) lunatics and spendthrifts. Failing a parent the Twelve Tables allowed a father to appoint a tutor by will, whom failing the nearest agnate as *tutor legitimus*, whom failing a guardian appointed by a magistrate, there being for long a special *praetor tutelaris*. *Tutela mulierum* and *cura furiosi vel prodigi* also passed to the nearest agnate. Later, guardianship became a public duty.

Initially the tutor managed the ward's property but gradually the power of young wards to act with the authority of their tutors was admitted. Gradually also, the supervision of tutors by the magistrates was developed and they could be removed for misconduct. The tutor of a married woman did not administer her property but had to give authority for important transactions only. *Tutela mulierum* fell into decay in the last century of the Republic. From about 200 B.C. there developed a curatory entirely in the interest of minors. *Cura minorum* was introduced to supply his lack of experience and judgment. Curators were appointed by magistrates, originally for one transaction and later for general purposes, and the rights of permanent curators were extended to the detriment of the minor's independence. In post-classical and Byzantine law *tutela* and *cura* were largely assimilated, though not completely.

In English law a person may be guardian of a minor by nature, custom, for nurture, by parental right, by parental appointment, by court appointment, or by appointment by the minor himself. Guardianship may be of the minor's person or his estate, or both. A father is guardian by nature over his minor child's person, and a father, whom failing, the mother, is guardian for nurture of a minor child to 14, and by parental right guardian of the person of a minor under full age. By statute the surviving parent, on the other's death, is guardian alone or jointly with a guardian appointed by the other or by the court. Both parents have power by deed or will to appoint persons to act as guardians of a minor child after their respective deaths, and a guardian appointed by a deceased parent acts as guardian jointly with the surviving parent unless the latter objects. The court may, failing other persons, appoint a person to be guardian of the minor, having regard to the minor's welfare as the first and paramount consideration. In the last resort a minor may elect a guardian himself.

A guardian of the person is entitled to the custody and control of the person of the infant, and to regulate his education. A guardian has the right to receive the profits of lands and to manage the personal estate of the ward for the latter's benefit until the infant attains full age, and must apply the income for the ward's maintenance.

In Scots law pupil (q.v.) children have tutors, who govern the person and manage the estate of the ward, but minors (q.v.) have curators who advise and consent to the minor's actings. The classes of tutors at common law, failing the father, are tutors nominate or testamentary, the nearest male agnate as tutor of law, and tutors-dative, appointed by the court, or the court may appoint a judicial factor (q.v.) *loco tutoris*. Under statute the surviving parent is tutor, alone or jointly with a tutor appointed by the predeceased parent. A tutor has control of the pupil's person and complete management of his estate. The classes of curators, failing a parent, are curators nominate, and curators chosen by the minor for himself. A curator has no control of the minor's person and only advises and consents to the minor's administration of his property. A minor need not have a curator at all. The mother is the natural custodier and tutor of a bastard child but cannot nominate a guardian. A bastard minor has no curator unless he chooses one for himself.

Gudelinus (Pierre Goudelin). See GOUDELIN.

Guerilla warfare. Hostilities conducted in territory occupied by troops of one state by armed bodies not forming part of the organized forces of another state. Provided such bodies comply with the terms of the Hague Regulations and observe the laws of warfare, they are not war criminals. Such activities are not illegal merely because their activities are hopeless defiance. Members of guerilla bands are entitled to be treated as prisoners of war if they bear arms openly, conduct operations in accordance with the laws of war, bear a distinctive sign, and are commanded by a responsible person.

The term is also applied, inaccurately, to methods of warfare used by the regular forces of one state, characterized by small-group raids, penetration behind the enemy's lines and disruptive activities as distinct from conventional military operations.

Guernsey. See CHANNEL ISLANDS.

Guerrero, José Gustavo (1876–) San Salvador lawyer and statesman, judge of the Permanent Court of International Justice 1930 (President 1937–39) and of the International Court (1946–55) (President 1946–49) and author of many writings on international law.

Guest, Christopher William Graham, Baron Guest of Graden (1901–). Was called to the Scottish Bar in 1929. He became Dean of the Faculty of Advocates in 1955, a Lord of Session in 1957, and was a Lord of Appeal in Ordinary from 1961 to 1971.

Guggenheim, Paul (1899–). Swiss jurist, member of the Permanent Court of Arbitration and notable in international law, author of *Lehrbuch des Völkerrechts* (1948–51), *Traité de droit international public* (1953–54), and other works on international law.

Guidon de la Mer. A French work by an unknown author, written in the latter sixteenth century and published in 1607, dealing with maritime law, and believed to have been compiled for the use of merchants at Rouen. It is one of the earliest works on insurance, bottomry, and average. It may be regarded as a supplement to the Laws of Oléron though it never had such great authority in France. Much set out in it was subsequently absorbed into French and other systems of law, notably by the French *Ordonnance de la Marine* of 1681.

Guido di Baysio (?–1313). Italian canonist, author of *Apparatus ad Decretum* or *Rosarium* (1300), a classic of canon law and indispensable for knowledge of older canonical writings, including material not used in or later than the gloss on the *Decretum*. He wrote also an *Apparatus ad Sextum*

(1306–11), one of the classical commentaries on the *Sext*, and other works.

Guilds or Gilds. Permanent voluntary associations for the promotion of common interests. Roman guilds or *collegia* were mainly professional associations of merchants or workmen, and social and religious associations. Under the Empire the Roman professional colleges were transformed into obligatory and hereditary closed corporations. At Byzantium the prefect of Constantinople supervised a large number of guilds, membership of which was a privilege not a duty, and which were largely religious and humanitarian in purpose. Guilds based on Roman ones survived in the barbarian West and social and religious associations were common in Germanic society. The oldest guild statutes in Europe are found in England from the eleventh century.

Merchant and artisan guilds appeared in northern Europe from about 1000; they were concerned to promote the social, religious, and professional interests of members, and helped members struck by disaster and assisted their dependants. From the thirteenth century at least, guilds regulated standards of workmanship, arranged training and, to preserve peace among members, held courts and developed their own bodies of customary law; they fixed hours of work, methods of selling, prices and output. Guilds developed corporate organization, which usually took the form of a patron, frequently the lord or bishop of the city, elected officials, variously called aldermen, wardens, consuls, *jurati*, and the assembly of masters.

As time went on guilds grew in power and importance, and were frequently recognized and employed by the State as agents for regulation of economic life. Colbert sought to bring all workers into guilds and to standardize guild regulations. The Statute of Artificers in England, 1562, shows the kind of regulation favoured at that time.

Guilds of the older kind declined with the Enlightenment, the Industrial Revolution and the growth of *laissez-faire* social and economic philosophy. In France guilds were abolished in 1791. In England obligatory apprenticeship was abolished in 1814 and guild monopolies were abolished by the Municipal Corporations Act, 1835. Other European countries generally suppressed guilds in the nineteenth century.

Guilds gave rise to co-operative mercantile enterprises, such as the Merchant Staples and Merchant Adventurers and to the sixteenth century joint-stock company. They contributed something to the ideas of friendly societies and modern professional associations and gave rise to the still surviving London livery companies and corresponding bodies elsewhere. They did not contribute much directly to modern trade unions. They had great influence on the development of education and trade training; the mediaeval universities were themselves guilds; and they gave rise to courts and bodies of law which had extensive jurisdiction in commercial and maritime relationships.

C. Gross, *The Gild Merchant*; Clune, *The Medieval Gild System*.

Guildford, Lord. See NORTH, FRANCIS.

Guillotine. (1) A means of inflicting capital punishment by decapitation. It consisted of two upright posts held parallel by a cross-beam at the top and acting as guides for a heavy oblique-edged knife, which was raised by a rope and allowed to fall on the victim's neck. An instrument of this kind had been used in England, in Scotland (under the name of 'the maiden') and elsewhere in the later Middle Ages, but it came into common use after a French doctor, Joseph-Ignace Guillotine had the French National Assembly pass a law in 1789 requiring death sentences to be carried out by a machine, to ensure that decapitation would no longer, as in the past, be confined to victims of noble birth, and be painless. He was himself later guillotined. It was first used in 1792 and was the standard means of carrying out executions during the French Revolution. At first it was called Louisette or Louison but later came to be called *la guillotine*.

(2) A procedure, intended to expedite the passage of a Bill through the House of Commons, by laying down a time-table allotting a certain number of days to each stage and, in respect of committee and report stages, stipulating the number of clauses which must be considered on each day or portion of a day. At the end of the time allotted the question must be put and clauses not discussed are passed without consideration. A motion to apply the guillotine procedure must be discussed and passed prior to the debate to which the procedure is to be applied. The guillotine procedure is justifiable only if deliberate delaying tactics are being employed, but it is sometimes invoked merely because of the fear that prolonged debate, particularly on a long and complicated Bill, will delay the government's legislative programme for the session. Its use in such circumstances is a denial of the rights of Members and of democracy, as it results in large portions of the Bill being passed undiscussed. The procedure may also be applied to proceedings in Standing Committees, though the motion to apply the guillotine must be passed by the House itself.

Guilt. The concept of having committed some failure of duty, usually a crime or offence, and consequently being liable to some penalty. A person accused may admit guilt, or be found guilty on the evidence. In the case of common law crimes guilt normally coincides with having committed moral fault, but in the case of some statutory offences guilt may arise by merely having done something or allowed it to happen without any moral fault at all.

Guilty but insane. The form of verdict introduced, it is said at Queen Victoria's request, by the Trial of Lunatics Act, 1883, where a person tried for crime was found by the jury to have done the act charged but at the time to have been insane so as not to be legally responsible for his actions. From 1800 to 1883 the form of verdict was, and since 1964 has again been, Not guilty by reason of insanity. The practical effect of both verdicts was the same, committal to a mental hospital at the pleasure of the Crown, i.e. until the Home Secretary otherwise directs.

Guilty mind. The *mens rea* requisite for guilt of a particular crime.

Gumplowicz, Ludwig (1838–1909). Austrian jurist and influential sociologist, author of *Rasse und Staat* (1875), *Grundrisse der Sociologie* (1885), and *Allgemeines Staatsrecht* (1907). He saw the history of civilization as an unending struggle between groups, then between states and finally between classes within states.

Gunpowder Plot. A conspiracy of a group of Catholics to blow up the Houses of Parliament in 1605. The true history of the plot is obscure, but it is generally thought that it sprang from the repressive measures used against Catholics in England since 1558. The plot was probably originated by Robert Catesby and later joined by others. In March 1605, one of the conspirators leased a vault underneath the Lords' Chamber of the Houses of Parliament and barrels of gunpowder were concealed there. In October a Catholic peer was warned by anonymous letter not to attend the opening of Parliament if he valued his life; he communicated the letter to the King's Ministers; on 4th November, the day before King James I was to open Parliament, the vaults were searched, the gunpowder found and a minor conspirator, Guido or Guy Fawkes, arrested. Eight conspirators were caught, tried and executed, and others were killed when attempting to escape. It is probable that the government had knowledge of the plot and had awaited their chance to catch the conspirators.

The plot is commemorated by the traditional searching of the vaults by the Yeomen of the Guard before the opening of each session of Parliament, and by bonfires and the setting off of fireworks on 5 November.

Günther, Carl Gottlob (1752–1832). German jurist, author of *Grundriss der europäischen Völkerrechts* (1777) and *Europäisches völkerrecht* (1787).

Gutteridge, Harold Cooke (1876–1953). Was called to the Bar in 1900 and was Professor of Industrial and Commercial Law at London 1919–30 and then Reader and Professor of Comparative Law at Cambridge till 1941. He had an international reputation and was highly regarded abroad. In *Comparative Law* (1946) he made the first systematic statement of the case for a new branch of legal science. He also wrote on *Bankers' Commercial Credits* and published many papers.

Gwyer, Sir Maurice Linford (1878–1952). Became a Fellow of All Souls and was called to the Bar in 1903, joined the legal staff of the National Health Insurance Commission and became legal adviser to the Ministry of Health. He also edited Anson's *Contract* and *Law and Custom of the Constitution*. He then became Treasury solicitor (1926) and in 1930 first Parliamentary counsel to the Treasury, and in 1937 Chief Justice of India (the first Chief Justice under the Government of India Act, 1935) where he remained till 1943. He was also vice-chancellor of the University of Delhi 1938–50 and did much to develop it and to foster good race relations in India. He also played an advisory role in the drafting of the constitution of India of 1950.

H

Habeas corpus. In England, a prerogative writ (q.v.) for securing the liberty of the subject, affording an effective means of securing release from unjustifiable detention in prison, hospital, or private custody. It originated in a command to a person detaining another to have that person brought before a court, or sometimes to bring an accused person before a court.

At common law in England any freeman was entitled, if detained other than after criminal conviction or for civil debt, to demand from the King's Bench a writ of *habeas corpus ad subjiciendum* directed to the keeper of the prison, commanding him to bring up the body of the prisoner with the cause of his detention so that the court could judge of its sufficiency and remand the prisoner to prison, admit him to bail, or release him. The writ is older than Magna Carta and was established by the seventeenth century as the appropriate process for checking illegal imprisonment. It issued as a matter of right but had various defects. The prisoner might be moved to another prison; it was doubtful whether the Common Pleas could issue the writ and the Exchequer never did; it was doubtful whether a single judge of the King's Bench could issue the writ in vacation. In *Darnel's case* (1627) (q.v.) the sufficiency of the return 'by special command of the King' was challenged, and upheld, but the Petition of Right, 1628, declared against its validity. The Habeas Corpus Act, 1640, gave a statutory right to the writ where detention was ordered by the King or his Council, but was not wholly effectual. Other cases after the Restoration showed other defects in the procedure and the Habeas Corpus Amendment Act, 1679, sought to remedy them. It declares no principles and gives no new rights but greatly improved the machinery by which the right to liberty might be enforced. It provided that on complaint by or on behalf of any prisoner on a criminal charge, the Lord Chancellor or any judge of the superior courts should issue a habeas corpus and release the prisoner on bail; the prisoner must be brought up within at most 20 days; no person delivered is to be recommitted for the same offence; habeas corpus may be sought from Chancery or Exchequer as well as from King's Bench or Common Pleas; and no person is to be sent prisoner to Scotland, Ireland, or places beyond the seas. The defect of not limiting bail was remedied by the Bill of Rights (1689) which declared that excessive bail should not be required, while in 1816 the remedy was extended by the Habeas Corpus Act, 1816, to non-criminal charges, and the judges were also empowered to determine the truth of the facts set out in the return. The operation of the 1679 Act has on several occasions, been temporarily suspended by Act of Parliament in times of public danger on grounds of political necessity, e.g. in 1794–1801 and in 1817.

The writ is today available as a remedy in most cases of unlawful deprivation of personal liberty, civil or criminal, enabling the court to inquire into the justification for the detention. The view that an application might be made to one judge after another has been held incorrect; a fresh application can only be brought on fresh evidence. Appeal lies against both an order to release and a refusal to release, in both civil and criminal cases. It lies against the Crown, its ministers and officials, the judge being bound to safeguard the liberty of all subjects and aliens, against anyone, and the writ has frequently been invoked to test the validity of acts of the executive, including committal for extradition. It is a writ of right, grantable *ex debito justitiae*, but may be refused if there is an alternative remedy available. It does not lie where a person has been sentenced to imprisonment by a court of competent jurisdiction. The writ formerly lay to any part of the dominions of the Crown but since the Habeas Corpus Act, 1862, it does not lie to colonies where the local courts may issue it.

The 1679 Act applies to Wales, Berwick-on-Tweed, Jersey and Guernsey and the 1816 Act applies also to the Isle of Man. The writ was sometimes issued from the King's Bench into Ireland but does not now do so. The 1679 Act does not apply to Ireland but corresponding provision was made by an Act of 1781–82 of the Irish Parliament. The writ does not issue from an English court into Scotland and it does not exist in Scots law but statutory provision was made in 1701 for preventing wrongful imprisonment and to avoid undue delays in trials and there are provisions to the same effect in the modern Criminal Procedure Act, 1975. The Court of Session has power to order the release of a person unlawfully detained in other circumstances.

It will not be granted where the effect would be to review the judgment of a superior court which can otherwise be reviewed. It does not lie to secure the release of a member of the House of Commons committed by that House for breach of privilege, nor of a non-member committed by either House for breach of privilege until after prorogation or dissolution of Parliament.

A prisoner discharged from illegal custody under habeas corpus cannot be again imprisoned or committed for the same offence.

Obedience to the writ is enforceable by fine or imprisonment for contempt, which may be granted in cases of disregard of the writ, or making a false or insufficient return to it.

There are also other writs of habeas corpus, *ad testificandum*, to enable a person in custody to be brought before a court to give evidence, *ad respondendum*, to bring up prisoners detained in custody before magistrates or courts for trial or examination on any other charge, *ad deliberandum*, and *recipias*, now obsolete, to enable the removal of prisoners from one custody to another for trial.

R. J. Sharpe, *Law of Habeas Corpus*.

Habeas corpus (in U.S.). The writ of habeas corpus (q.v.) was transplanted to the American colonies as common law and was frequently invoked before the War of Independence. It is available to contest detention by public or private authority. The Supreme Court can issue the writ as an original proceeding only in a case within its jurisdiction, but under its appellate jurisdiction it may issue the writ to enquire into the validity of a detention by order of an inferior court to determine whether such court has acted without jurisdiction or in excess of authority. The U.S. Constitution provided (Art. 1, sec. 9) that the writ shall not be suspended unless, when in case of rebellion or invasion, the public safety may require it. It does not provide who may suspend the issue of the writ. In 1861 Lincoln suspended it in parts of Maryland, in view of the Civil War emergency, but in the *Merryman* case,

Chief Justice Taney, sitting as circuit judge, laid down that only Congress had the right, but Lincoln ignored the decision and acted on the view that the President could act in the absence of legislation. In 1833 Congress authorized federal courts to issue the writs to state officials, but in *Ableman* v. *Booth* (1859) (q.v.), the Supreme Court decided that a state judiciary had right to interfere in federal cases, to inquire into any commitment for an act done in pursuance of a law of or authority under the U.S. Various states have shown divergent views as to the accused's right of appeal from an order denying the issue of *habeas corpus*.

In modern practice the principal use of the writ is as a post-conviction remedy whereby prisoners convicted in state courts can have the proceedings reviewed in the federal courts on the ground that a right guaranteed by the Constitution has been violated.

Habendum. The name, derived from the Latin term *habendum* (to be had) in older charters, for the part of a conveyance determining the quantity of interest conveyed thereby. Though now expressed by the phrase 'To hold', the Latin term is maintained for it.

Habit and repute. A phrase in Scots law signifying 'held and reputed', found most often in the context of cohabitation with habit and repute, a surviving mode of irregular marriage whereby a couple may be held to be married if there is evidence of their cohabitation for a substantial period with the acquisition of the reputation of being married. It was also formerly an aggravation of certain crimes that the accused was habit and repute a criminal, i.e. was known as such.

Habitatio. In Roman law a personal right to live in a house, a lesser right than a life interest therein.

Habitual offender. A person who has frequently been convicted and sentenced for criminal offences and has shown that crime is a normal part of his career. One of the major problems of penology is how to deal with such an offender. Provision has frequently been made either for more prolonged detention of such offenders or special regimes of treatment. It is frequently considered that reformation of such offenders is unlikely and that their removal from society for prolonged periods for the protection of society is the best course. In the U.K. two regimes were devised in 1948 for persistent adult offenders, corrective training for at least two and not more than four years, and preventive detention for at least five and not more than 14 years, intended to keep the offender out of society for a considerable period. These have since 1973 been replaced by a power to impose an extended term of imprisonment. This is intended as a last resort.

Hackworth, Green Haywood (1883–1973). American lawyer, legal adviser of the Department of State (1925–46) and later judge of the International Court (1946–61) (President 1955–58), and author of a great *Digest of International Law* (1940–43).

Hadley* v. *Baxendale (1854), 9 Ex. 341. A mill was stopped by the breakage of the driving shaft, and it was sent to the makers for repair but the carriers were not informed that by reason of the lack of the shaft the mill was at a standstill and profits were being lost. Delay took place on the carriers' part in returning the shaft and it was held that only such damages were recoverable as arose naturally and in the ordinary course from the breach of contract and not for any loss occurring in the special circumstances not brought to the notice of the carriers and therefore, not within their knowledge. Since its decision this has been the leading case on the remoteness of damage for breach of contract.

Hadriana Collectio. An official collection of canons of the Roman Church sent by Pope Adrian I to Charlemagne in 774 and received officially at the Assembly of Aix-la-Chapelle in 802 as the Code of the Frankish Church. It is based on the *Dionysia Collectio* (q.v.), with additions. In the ninth century, it was adapted to practical requirements in the *Breviarium ad inquirendas sententias* and the *Epitome Hadriani*, and linked to the *Hispana Collectio* (q.v.) to make a composite collection known as the *Hadriano-hispanica*. It is found in many later collections and many later scholars, including Burchard, Ivo, and Gratian, included in their compilations the versions of texts found in the *Hadriana*.

Haeretico comburendo, De. A common law writ lying against a person who was convicted of heresy by a bishop, abjured it, but later fell into the same or another heresy, and was then handed over to the secular authorities to be burned. The writ was abolished in 1677 but saving the jurisdiction of the archbishops, bishops, or other judges of ecclesiastical courts to punish atheism, blasphemy, heresy, schism or other opinions by excommunication, deprivation, depradation, or other ecclesiastical censure not amounting to death. There is now no ecclesiastical jurisdiction over laymen for heresy.

Hagerström, Axel (1868–1939). Swedish philosopher and jurist, author of *Inquiries into the Nature of Law and Morals* and works on Roman law, and the leading figure of the group known as the

Scandinavian realists who have regard to the actual uses of legal terms and concepts and analysis of the mental attitudes involved. Hagerström sought to destroy the notion that legal obligations and right-duty relations have any objective existence; they are merely words which express feelings.

See also LUNDSTEDT; OLIVECRONA; ROSS.

Hague Academy of International Law. Founded in 1923 with the support of the Carnegie Foundation for International Peace as a centre of higher studies in international law, public and private, and cognate sciences to facilitate the examination of questions bearing on international juridical relations. It arranges annual courses of instruction in international law given by distinguished invited scholars, and publishes the Hague *Recueil des Cours.* Since 1949 it has conferred a Diplôme de l'Académie. Participants have organized themselves into the Association des auditeurs et anciens auditeurs de l'Académie (A.A.A.).

Hague Conference on Private International Law. Initiated in 1893 to work for the unification of the rules of private international law. It publishes *Actes* and *Documents* relating to each session. These conferences have resulted in a considerable number of conventions and substantial movement towards harmonization of various countries' rules on private international law. Only since 1951 have British delegates attended as full participants. In 1951 the Conference was put on a permanent footing by the establishment of a Bureau under a Secretary-General. It works under the general direction of the Standing Government Commission of The Netherlands, established in 1897, with the object of promoting the codification of private international law, and its chief functions are to prepare proposals for the unification of private international law and to keep in touch with other organizations such as the International Law Association.

Hague Conventions. See HAGUE PEACE CONFERENCES.

Hague Peace Conferences. International conferences held at The Hague in 1899 and 1907. The first was convened by Russia and 26 states participated. It failed to achieve its first objective, a limitation on armaments, but agreement was reached on several conventions, defining the conditions of a state of belligerency and other customs relating to war on land or sea, prohibiting the use of asphyxiating gases, prohibiting the use of expanding bullets and prohibiting the discharge of explosives from balloons. Most importantly the conference adopted a Convention for the Pacific Settlement of International Disputes and created the Permanent Court of Arbitration (q.v.). The second conference was proposed by the U.S. but convened by Russia and was attended by representatives of 44 states. It adopted a number of conventions such as the rights and duties of neutrals during war, the laying of automatic submarine contact mines, the status of enemy merchant ships, bombardment by naval forces in wartime, and establishment of an international prize court. Some conventions of the earlier conference were renewed. The conferences were important in establishing the principle that international problems could and should be resolved by periodic conferences, and this influenced the post-war creation of the League of Nations.

Hague Rules. Rules adopted by a conference at Brussels in 1923, amended in 1968, and embodied in the U.K. Carriage of Goods by Sea Act, 1924, now 1971. They apply to outward shipments from U.K. ports, and make important changes in the common law rights and liabilities of shipowners. They contain certain exceptions and stipulations in favour of the shipowner, moderate the shipowner's undertaking to provide a seaworthy ship to an obligation to exercise due diligence to make the ship seaworthy, extend permitted deviation, relieve the shipowner from liability for loss or damage due to certain excepted perils and contain provisions limiting the amount of the carrier's liability in certain cases. The rules are frequently embodied in charter-parties and bills of lading which would not otherwise be regulated by them.

Hailsham of St Marylebone, Lord. See HOGG, QUINTIN.

Hailsham, Viscount. See HOGG, DOUGLAS.

Haines, Charles Grove (1879–1948). Professor of Political Science at California (Los Angeles), was an authority on American constitutional law. His writings include *The Conflict over Judicial Powers in the U.S.* (1909), *The Revival of Natural Law Concepts* (1930), *The American Doctrine of Judicial Supremacy* (1932) and, most important of all, *The Role of the Supreme Court in American Government and Politics* (1944).

Hakewill, William (1574–1655). Was called to the Bar and sat in several Parliaments between 1601 and 1629. In 1614 he was one of six lawyers appointed to revise the existing laws. In 1647 he became a Master in Chancery and sat with the Commissioners of the Great Seal to hear causes. He was author of *The Liberty of the Subject against the Pretended Power of Imposition* (1641) and of *The Manner how Statutes are enacted in Parliament by passing of Bills* (1641). The latter was enlarged in *Modus Tenendi Parliamentum ... together with The*

Privileges of Parliament and the Manner how Lawes are there enacted by passing of Bills (1659). He also wrote various papers on points of legal antiquities.

Haldane, Richard Burdon, 1st Viscount (1856–1928). After attaining distinction as a student of philosophy was called to the Bar in 1879 and started slowly but was sufficiently established to take silk in 1890. He practised mainly in Chancery, but had a large practice in the House of Lords and Privy Council. In 1885 he won a seat in Parliament and became closely associated with Asquith (later Prime Minister) and Grey (later Foreign Secretary). In 1905 when the Liberals secured office he became Minister for War; as such he effected a reorganization of the Army of supreme importance, establishing the General Staff at the War Office and organizing the Army into an Expeditionary Force, a Special Reserve, and the Territorial Army, organized on a county basis, designed to reinforce the Regular Army. He accordingly shaped the Army with which Britain entered World War I. In 1911 he was made a Viscount to help the government in the House of Lords and in 1912 succeeded Loreburn (q.v.) as Lord Chancellor. As a judge he ranks high, and little short of the highest. He excelled in equity and Dominion constitutional law, and showed an awareness of the social, political, and economic implications of judicial decisions. As a law reformer he started the process culminating in the reform of English real property in 1925.

Haldane had frequently visited Germany and in 1915 he was driven to resign by unfounded aspersions of friendship for Germany, but was simultaneously appointed to the Order of Merit. In 1917–19 he was Chairman of a Committee on the Machinery of Government whose report has had less influence than it should. After the War he moved to the Labour Party and served again as Lord Chancellor in the Labour Government of 1924, sat on the Committee of Imperial Defence and instituted proposals for a Ministry of Justice; thereafter he led the Labour Opposition in the Lords.

He was a philosopher of some standing and gave a series of Gifford Lectures, published as *The Pathway to Reality*, and also published *The Reign of Relativity*, on Einstein's Theory, and *Human Experience*, and numerous other works, mainly on philosophy or education. He was also much interested in higher education and took a large part in the refoundation of London University in 1898.

R. B. Haldane, *Autobiography*; F. Maurice, *Haldane*; Sommer, *Haldane of Cloan.*

Hale, Bernard (1677-1729). Called to the English Bar in 1705, he served as Chief Baron of Exchequer in Ireland, 1722–25, and then as a baron of Exchequer in England.

Hale, Sir Matthew (1609–76). As a young man Hale studied widely and became acquainted with Selden, who encouraged the breadth of his studies. He was called to the Bar in 1636 and soon made his name; he probably advised Strafford and certainly advised Laud and others. He became a judge of the Common Pleas in 1654 but resigned in 1658 and then sat in Parliament. He assisted the Restoration and became successively Chief Baron of Exchequer (1660–71) and Chief Justice of the King's Bench (1671–76). He was a member of the special tribunal which adjudicated on questions arising out of the Great Fire of London. He was a man of sincerity, honesty, and religious conviction and was notable for his high professional honour and humanity. Apart from Bacon, he was the most scientific jurist yet known in England, and a master of all branches of English law. In his studies he was a disciple of Selden and gathered a fine collection of manuscripts on matters of legal and constitutional history.

His writings are numerous and important; they include a *History of the Common Law* (1713), the first history ever written of the common law of England, *A Treatise on the Original Institution, Power and Jurisdiction of Parliaments* (1707), edited by Hargrave as *The Jurisdiction of the Lords' House* (1796) dealing with many of the political issues of the time, various treatises published in *The Hargrave Law Tracts*, *Pleas of the Crown, or a Methodical Summary of the Principal Matters relating to that Subject*, merely a series of head-notes and possibly intended as a plan for his *History of the Pleas of the Crown.* It dates from about 1650 and was printed in 1678 and several times thereafter. He also wrote an unfinished *History of the Pleas of the Crown* (1736), the first attempt at a history of English criminal law, an anonymous edition of *Rolle's Abridgment* (1668), annotations on Fitzherbert's *New Natura Brevium* (1730) and on Coke's *First Institute* (printed 1787), and some writings still unpublished. He was the greatest common lawyer since Coke, and his work was based on first-hand knowledge of the sources, allied to great ability to develop a scientific arrangement of legal principles. In this Blackstone acknowledged his indebtedness to Hale.

G. Burnet, *Life and Death of Sir Matthew Hale* (1682); J. Williams, *Memoirs of Sir Matthew Hale* (1835).

Half-blood. Relationship of one person to another through one only of his parents or other ancestors and not through both. Half-blood is accordingly either consanguinean or uterine. Until 1833 a relative of the half-blood could not inherit real estate. Until 1925 a relative of the half-blood could inherit personal estate but since then they succeed only in the complete absence of relatives of the whole blood in equal degree.

Half-quarter days. In England the days halfway between one quarter-day (q.v.) and the next. They are 2 February, 9 May, 11 August, and 11 November.

Halifax Gibbet Law. A right, probably a survival of the right of infangtheft, or possibly derived from a forgotten royal grant, held by the Burgesses of Halifax, to execute anyone taken within the liberty of Halifax found guilty of the theft of goods to the value of more than 13 pence. The execution was carried out on a hill outside the town by decapitation with an instrument resembling a guillotine. The last instance is believed to have been in 1650.

Halimoot (or Hallmote). A court kept by a lord, such as a great abbey, on each of its manors, in addition to a central court, *libera curia*, kept for its greater freehold tenants.

Halkerston, Peter (*d.* ?1833) became a Solicitor in the Supreme Court of Scotland and Bailie of the Abbey of Holyrood. He wrote a *Treatise on the History, Law and Privileges of the Palace and Sanctuary of Holyrood House* (1831), *A Compendium or General Abridgment of the Faculty Collection of Decisions of the Lords of Council and Session, 1754–1817* (1819–20), *A Collection of Latin Maxims* (1823), and some lesser works.

Hall, Arthur, case of. Hall was expelled by the House of Commons in 1581, having published an offensive book questioning the authority of the House and slandering various members. It is the leading case on the right of the House to expel a Member.

Hall, Sir Charles (1814–83). Was articled to a solicitor but called to the Bar in 1838 and worked for a time as assistant to Duval the conveyancer. He became a leader of the Chancery Bar and the chief authority of his time on real property law. In 1873 he was knighted and made a vice-chancellor, but resigned owing to ill-health in 1882.

Hall, William Edward (1835–94). Was called to the Bar in 1861, but was more interested in travel and study. Almost by accident, he came to concentrate attention on international law and in 1880 the publication of his *International Law* marked an epoch in the literature of the subject. It passed through eight editions in the next 50 years. He also wrote a *Treatise on the Foreign Powers and Jurisdiction of the British Crown* (1894).

Hallam, Henry (1777–1859). Was called to the Bar and practised for a time, but his tastes were historical and he is remembered for *The View of the State of Europe during the Middle Ages* (1818), *Introduction to the Literature of Europe in the 15th, 16th and 17th Centuries* (1838–39) and particularly the *Constitutional History of England* (1827) which deals with the period from 1485 to 1760 and was long a standard authority.

Halleck, Henry Wager (1815–72). Served in the U.S. army 1839–54, when he resigned and became head of a law firm in California. During this time he wrote books on mining law and *International Law, or Rules Regulating the Intercourse of States in Peace and War* (1861), a work which was highly regarded. In the Civil War he rejoined the army and commanded the Department of Missouri and then the Department of the Mississippi, and became, in 1862, military adviser to the President and general-in-chief, an office altered in 1864 to Chief of Staff. He served in various other high offices till his death, displaying great administrative ability but having little success as an army commander.

Hallmote (or hallimote). An Anglo-Saxon court corresponding to the later court baron. The name has survived for the courts of manors forming part of the endowments of the see of Durham, and for meetings of the London City Companies in their halls.

Hallmarking. The marking on articles of gold or silver of certain symbols denoting that the article has been tested at an authorized assay office and contains at least specified proportions of pure gold or silver. In the U.K. hallmarking dates from 1300. The Worshipful Company of Goldsmiths has been responsible for the assaying and marking of plate in London since then and there are assay offices in certain other towns also. The fineness of gold is expressed in carats, 24 carats being pure gold. Sterling silver contains at least 92.5 per cent fine silver. The marks actually impressed include a number of symbols indicating the maker, the standard, the date and the symbol of the office which adhibited the mark. In the U.S. there is no hallmarking but local regulations existed in certain towns in the eighteenth and nineteenth centuries, though there is no consistent system of symbols.

Haloander, Gregor (1501–31). German jurist, edited the *Corpus Juris* from original manuscripts and published *Novellae graecae cum Haloandri interpretatione latina* (1530); *Digestorum seu Pandectarum Libri L* (1529), and other works on Roman and Canon law highly esteemed by later editors.

Halsbury, Lord. See GIFFARD, HARDINGE S.

Hamel, Gerard Anton van (1842–1917). Dutch criminologist, a leading scholar in criminal law and criminology. He wrote an *Inleiding tot de Studie van het nederlandsche strafrecht* (1889), was a founder of the *Tijdschrift voor strafrecht* (1886), and was prominent in work for criminal law reform, particularly in the treatment of juveniles.

Hamesucken (hamesoken or hamsocn). The old crime of seeking a man out in his house and assaulting him there. It was known in Anglo-Saxon law and survived in Scots law to modern times. At one time capital, it is now treated as an aggravated assault.

Hamilton, John Andrew, Viscount Sumner (1859–1934). Called to the Bar in 1883, he slowly built up a large practice, principally in commercial cases, and supplementing his earnings by journalism. He wrote numerous entries for the *D.N.B.* He took silk in 1901, became a judge of the King's Bench Division in 1909, a Lord Justice of Appeal in 1912 and a Lord of Appeal in 1913. He became a viscount in 1927. He was undoubtedly one of the greatest judges of his time, a master of legal principles with outstanding ability in expressing his judgments.

Hamilton, Sir William, of Whitelaw (?–1705). Passed advocate in 1664 and became a Lord of Session, as Lord Whitelaw, in 1693, and later was Lord Justice-Clerk, 1704–05.

Hammarskjold, Dag Hjalmar Agne Carl (1905–61). Swedish economist and statesman, Secretary-General of the United Nations, 1953–61, acted boldly in the interests of peace particularly in the Middle East and in the Congo. His leadership enhanced the prestige and effectiveness of the U.N. He was posthumously awarded the Nobel Peace Prize in 1961.

Hammer* v. *Dagenhart (1918), 247 U.S. 251. Congress, in the exercise of the commerce power, had enacted federal regulation of interstate aspects of food and drug shipments, narcotics, and other matters, but in this case the Supreme Court, by a majority, evidencing its *laissez-faire* philosophy, held unconstitutional Congressional prohibition of the shipment in interstate commerce of goods produced by child labour. Ultimately, in *U.S.* v. *Darby* (1941), 312 U.S. 100, the Court overruled this case.

Hammond, William Gardiner (1829–94). Was admitted to the Brooklyn Bar in 1851, and later was partly responsible for the Iowa Code of 1873. He opened a private law school which became the law department of the University of Iowa in 1869. In 1881 he became head of the law school at Washington University, St. Louis. He did much to raise the standard of legal education, was chairman of the legal-education committee of the American Bar Association, and edited Sanders' *Justinian* (1876), Lieber's *Legal and Political Hermeneutics* (1880), and published an edition of Blackstone's *Commentaries* (1890).

Hammurabi. Ruler of Babylon from 1792 to 1750 B.C., who unified Mesopotamia and made its northern region of great importance. He is notable in legal history as the ruler who caused the promulgation of the body of law usually called the Code of Hammurabi.

See BABYLONIAN LAW.

Hammurabi, Code of. See BABYLONIAN LAW.

Hampden's case (or the Case of Ship Money) (1637), 3 St. Tr. 825. In 1634 and 1635 Charles I, requiring money for the Navy, issued writs commanding seaport towns to furnish ships and instructing towns to raise money for the purpose. There having been some resistance, Charles obtained a judicial opinion favourable to his exaction. In 1636 John Hampden, a gentleman of Buckinghamshire, not a seaport town, refused to pay the sum assessed on him. After argument in the Exchequer Chamber seven judges found for the King, holding the exaction justifiable, and five for Hampden. The majority judges in varying terms upheld the absolutist view that the King cannot be fettered, particularly in emergency, by the need to obtain Parliamentary consent for taxation. The decision was reversed by the Long Parliament and the Bill of Rights declared that it was illegal for the Crown to raise money without Parliamentary approval.

Hamsocn. In Anglo-Norman law and, in the form hamesucken in old Scots law, the crime of violently breaking into a man's house and assaulting him there. It was one of the early pleas of the Crown.

Hanaper office. An office concerned with the common law jurisdiction of the old Court of Chancery, so-called because all writs relative to such business and the returns thereto, were kept in a hamper (*in hanaperio*) in contrast with those relating to the Crown which were kept in a small bag in the Petty Bag office. The office was concerned with the sealing of charters, patents, and writs which passed the Great Seal.

Hand, (Billings) Learned (1872–1961). Practised law in Albany and became a judge of the U.S. Southern District of New York (1904–24) and of the U.S. Court of Appeals, Second Circuit (1924–56). His judgments were noteworthy for eloquence and lucidity, and he acquired a reputation as an outstanding liberal. He wrote papers collected under the name *The Spirit of Liberty* (1952) which defends freedom against extremism. Though never a justice of the Supreme Court he is reckoned as great a judge as the best of them.

Handcuffs. Instruments for holding together the wrists of persons under arrest or for attaching such a person to a constable to prevent him escaping. They take three forms, two rings which fit round the wrists and are connected by a short chain, a ring twisted like a figure 8, and nylon or plastic cord. All are self-locking. They are sometimes called shackles, manacles, or by other names.

Handfasting. A custom, known in the English and Scottish borders, of betrothal by joining of hands, or sometimes of marriage though subject to a canonical impediment not yet purged.

The name was given in Scotland to a kind of temporary marriage which entitled the couple to live together for a year and a day and then make their marriage permanent or separate.

In the Danish code of Valdemar II (in force from 1280 to 1683) it was provided that a concubine kept only for three years should thereby become a lawful wife and Scots law still recognizes marriage by cohabitation with the habit and repute (q.v.) of being a wife.

Handling stolen goods. In English law the crime of, otherwise than in the course of the stealing, dishonestly receiving goods knowing or believing them to be stolen, or dishonestly undertaking or assisting in their retention, removal, disposal, or realization by or for the benefit of another person, or arranging to do so. 'Stolen' includes obtained by blackmail or deception.

Handwriting. The act of marking letters on paper by a pen or similar instrument, and also the distinctive features of size, shape of letters and other characteristics which distinguish the handwriting of one person from that of another. Each person's handwriting tends to be distinctive and difficult, if not impossible, for another to copy perfectly. Identification of handwriting as being that of a particular person, and the detection of copying or alteration of the writing of another are sometimes material elements in evidence, when the authenticity of a deed or signature is in doubt.

Hanging. A common mode of effecting capital punishment. Originally it seems to have been an indignity practised on criminals' corpses and was only later utilized as a means of causing death. The Anglo-Saxons introduced it from Teutonic customs and by the time of Henry II it was established as the punishment for homicide. Grants of jurisdiction to lords of manors included the 'right of pit and gallows' and in the Middle Ages towns, abbeys, and most lords of manors had the right to hang. From the end of the twelfth century the jurisdiction of the royal courts became exclusive. In 1770 hanging was substituted for burning as the punishment for petty treason and in 1814 for beheading as the penalty for male traitors.

Hanging is effected by attaching the noose of a rope, fastened to a beam above, round the victim's neck. In an older method the victim was pulled off the ground by the rope, which tightened and he was asphyxiated. In the later method, as used judicially in Britain and the U.S., the condemned man stands on a trapdoor, the noose is adjusted round his neck, and the trap is released; he then falls several feet and the jerk breaks the cervical vertebrae and causes immediate death. It is still competent in Britain for treason, but not for any other crime. Hanging is occasionally resorted to as a means of suicide. In the U.S. it was largely replaced by the electric chair or the gas chamber.

Hanging Cabinets. The term for Cabinet meetings in the late eighteenth and early nineteenth centuries at which decisions were taken whether or not to exercise the prerogative of mercy in the case of prisoners sentenced to capital punishment. At that time responsibility for deciding had not devolved on the Home Secretary and the final decision was taken by the King personally at a Cabinet meeting, aided by the Recorder of London in Old Bailey cases and a report from the presiding judge in cases tried at assizes. The Judgment of Death Act, 1823, allowed judges, in cases deemed fit to be recommended for the royal mercy, merely to record sentence of death and not pronounce it, though in murder cases the sentence had to be pronounced. (The 1823 Act was extended to murder cases in 1837.) In practice 'hanging Cabinets' seem usually to have had to decide only in London cases. It was only from about 1825, and particularly from Victoria's accession in 1837, that the decision came to be the responsibility of the Home Secretary alone.

Hanging, drawing, and quartering. The ancient sentence pronounced on persons found guilty of high treason, involving that the criminal be drawn on a hurdle to the place of execution, there hanged but not until dead, his bowels taken out while he was still alive, and thereafter his head to be cut off, and his body divided into four quarters. The practice developed of hanging until death before carrying out the further parts of the sentence, and the Treason Act, 1814, made this mandatory. The Forfeiture Act, 1870, made the penalty hanging alone.

Hanging in chains. The court sometimes, in very bad cases of murder, as by pirates, ordered the murderer's body to be hanged in chains, in which case the body, covered with pitch, was hung by chains after death from a gibbet in a public place. The practice was recognized by statute in 1752, when the judges were authorized to order that a murderer's body be hanged in chains or given to surgeons to be dissected. The alternative of dissection was abolished in 1832 and that of hanging in chains in 1834.

Hannen, James, Lord (1821–94). Was called to the Bar in 1848, did well as a junior, and became a judge of the Queen's Bench in 1868 and, in 1872, judge of the Probate and Divorce Court. In 1875 he became President of the new Probate, Divorce and Admiralty Division of the High Court. In 1888 he presided over the Parnell Commission, and in 1891 became a Lord of Appeal in Ordinary. In 1892 he was an arbitrator in the Bering Sea fishery dispute between Great Britain and the United States. He was grave, dignified, and courteous and a man of much ability and learning. He did much to settle the altered jurisdiction of his court.

Hansard. The official verbatim report of debates in the United Kingdom Parliament. It takes its name from the Hansard family. In 1811 T. C. Hansard bought Cobbett's interest in the reports of Parliamentary debates contained in his *Political Register*; initially the reports were compiled from newspaper reports and appeared monthly. In 1890 the interest was sold by the Hansard family, and in 1908 the reporting was taken over by the Government. Since 1909 there has appeared the Official Report, a complete, accurate, and verbatim record of the debates. In 1943 the name Hansard was restored to the title page, having lapsed since 1892.

Hansard is published by H.M. Stationery Office in daily and weekly parts and bound volumes are published periodically. Weekly and sessional indices are also published. There is a separate report for each House and separate parts and volumes contain the debates in Standing Committees of the House of Commons.

In addition to a complete record of every word spoken in the House, Hansard gives the answers to questions, both written and oral, put at Question Time, and prints the division lists showing how Members voted.

The debates are recorded verbatim by skilled shorthand writers who have reserved places in the Press Galleries of the Houses; revision is limited to corrections of mistakes, and members may not revise their speeches.

The accepted modern rule is that Hansard may not be cited in argument when a court is seeking guidance on the interpretation of a statute.

Hansard, Luke (1752–1828). Entered the printing house of John Hughs, printer to the House of Commons. He became a partner and in 1800 owner of the business. He printed the Journals of the House of Commons, 1774–1828. He was highly regarded by the officials of Parliament and by men of letters, and introduced technical improvements in printing. His two younger sons, James and Luke, succeeded him as printers to the House of Commons and were followed by their sons.

Hansard, Thomas Curson (1776–1833). Eldest son of Luke Hansard (q.v.); he began to print Parliamentary debates for Cobbett's *Political Register* in 1803, and in 1812 bought Cobbett's interest in publishing the debates. In 1855 the Hansard business was saved only by a sales guarantee by the Treasury and in 1878 the business received a Treasury subsidy, but in 1890 it was sold to the Hansard Publishing Union (which soon went bankrupt).

Hanse (or Hanseatic League). An association founded by north German towns and commercial groups to defend their trading interests. It was an important force in northern Europe from the thirteenth to the fifteenth century, particularly in trade with the Low Countries, England, and the Baltic, and grew into a power capable of successfully waging war on its neighbours. The members acted in co-operation to make trade safer and to establish bases abroad and secure monopolies. Among their bases were the Steel Yard in London. The League declined from the fifteenth century, but its assembly did not cease to meet until 1669.

Hanseatic laws of the sea. A collection of maritime law of the Hanse towns made and published by the Hanseatic League in 1591 and widely accepted as authoritative in northern Europe.

Hanworth, Lord. See POLLOCK, ERNEST.

Harbour. A place suited or adapted for ships to ride at anchor. The Crown has the monopoly power to constitute harbours or ports and to settle their limits but, for long, this has been granted to landowners or local authorities. In modern practice, harbours are always created by statute. The Crown is also conservator of all ports and harbours, its powers now being exercised by the Ministry of Transport.

Harbouring. The offence of concealing various kinds of persons, e.g. thieves, spies, to prevent them being caught and punished. It is also tortious to take an employee into and keep him in employment knowing that he was in breach of contract with a former employer.

Harcourt, Simon Harcourt, Viscount (*c.* 1661–1727). He was called to the Bar in 1683, became an M.P. in 1690 and Solicitor-General in 1702. He was one of the commissioners on the Union with Scotland in 1707, and a draftsman of the Treaty of Union of 1707, and Attorney-General in 1707–08, and again in 1710. In 1710 he became Lord Keeper and a Privy Councillor, in 1711 a peer, and Lord Chancellor in 1713. Though dismissed in 1714, he became a viscount in 1721, was readmitted to the Privy Council, and acted as a Lord Justice during the King's absences from England. Though a powerful speaker, an able statesman, and an honest lawyer, he did not attain greatness in either sphere. His decisions make a contribution, though not a notable one, to the development of equity.

Hard labour, imprisonment with. An English regime of imprisonment first defined in 1865, comprising at least three months (reduced in 1877 to one month) of first-class hard labour, including the heavier forms of exercise (treadwheel, crank, shot drill) and thereafter second-class hard labour, later called useful labour, which included such other kinds of bodily labour as might be appointed by the justices. First-class hard labour was abolished in 1898 and all hard labour in 1948.

Hardwicke, Lord. See YORKE, PHILIP.

Hardwicke's (Lord) Act. An Act of 1753 which stopped clandestine marriages by providing that marriages celebrated without publication of banns or licence should be void, and that a marriage celebrated in other than a parish church or public chapel in which banns were usually published should be void unless the Archbishop of Canterbury had granted a special licence for its celebration elsewhere. Though the Act stopped such scandals as Fleet Prison marriages, it invalidated marriages of dissenters and Roman Catholics under their respective rites, which had been valid at common law, by making only Church of England marriages valid, and this lasted until 1837.

Hardy, Sir Thomas Duffus (1804–78). Became a clerk in the Records Office in the Tower of London and edited several publications of the old Record Commission. In 1861 he became Deputy Keeper of the Public Records. As such he secured the appointment of the Historical Manuscripts Commission in 1869 and edited various volumes for the Rolls Series. In 1843 he prepared a *Catalogue of the Lords Chancellors, etc.*, a useful list of holders of legal offices.

Hare, John Innes Clark (1816–1905). Was admitted to the Bar in 1841, sat in the district court of Philadelphia in 1851–75 and was president judge of the city court of common pleas, 1875–96; he was also a professor of law in the University of Pennysylvania, 1868–88. A man of wide knowledge and ability, he was an outstanding judge, and contributed greatly to the introduction of equity into Pennsylvania. He edited a number of collections of cases, and *Select Decisions of American Courts in the Several Departments of the Law* (1847–48, later editions being known as *American Leading Cases*), and wrote *The Law of Contracts* (1887), *American Constitutional Law* (1889) and other works.

Hargrave, Francis (?1741–1821). Recorder of Liverpool (1797), a great historical lawyer, and collector of legal historical materials, but also a competent modern lawyer. He published a scholarly, annotated edition of Coke upon Littleton, completed by Charles Butler (1787), the fourth edition of the *State Trials* (11 volumes, 1775–81), but, more important, *Hale's* (q.v.) *Jurisdiction of the Lords House of Parliament* (1796), *Law Tracts* (1787), which included some of Hale's writings, and *Collectanea Juridica*, consisting of Cases, Tracts, etc. (1791–92), both of which made available some hitherto unpublished tracts which he had collected. His library was purchased and deposited in the British Museum.

Harlan, John Marshall (1833–1911). Was admitted to the Kentucky Bar in 1853. He took an active part in politics and, in 1877, became an associate justice of the Supreme Court. He had enormous respect for the Constitution and was a stern defender of civil liberties; he dissented frequently, and vigorously, but retained the respect of his colleagues and of the bar, and in some respects, his view has been adopted in later Court decisions. In 1893 he was a member of the Bering Sea arbitration tribunal.

Harmenopoulos, Constantinos (1320–83). Greek jurist, legal adviser to the Byzantine Emperor John VI, Cantacuzenus and holder of high judicial offices. He was responsible for the *Hexabiblos* or *Procheiron ton nomon*, a compendium, based on the *Basilica*, and subsequent Greek laws, made to suit the needs of fourteenth-century Byzantium, which was taken by the Greek assembly in 1828 as the basis of the civil law of Greece and survived till 1941, and an *Epitome der Kanones*.

Haro, Clameur de. A cry of alarm or distress, in full *Haro, Haro, Haro, à l'aide mon prince, on me fait tort*, or more recently of *Haro* repeated three times, known in Channel Islands law from the earliest times and still in force. In Jersey and Guernsey it is the ancient form of protesting against trespass to land. The fact that the cry has been made is registered at the local Register Office. The cry must be respected and the trespass discontinued; the effect is accordingly that of an interim injunction. It was recognized in the Grand Coutumier de Normandie. Though mentioned in England, 'to cry harrow' seems there merely to mean to express alarm or distress and to have no special significance. A similar custom was recognized by Saxon law, under the name of Clamor violentiae, but only in criminal charges. This may have been in observance at the time of the Norman Conquest and been extended to civil wrongs, later disappearing completely. The call for aid is connected with the idea of the hue and cry (q.v.).

Harmonization of laws. In European Community law, an alternative term to approximation of laws (q.v.), both being used as translations of *rapprochement* and *Angleichung*. In English approximation goes further than harmonization or co-ordination but falls short of unification.

Harpprecht, Johann (1560–1639). German jurist, author of *Commentarius de publicis judiciis* (1599), *Tractatus de processu judiciario* (1602), *Tractatus criminalis* (1603), and other works.

Harrington, James (1611–77). English political philosopher. A convinced republican, he sought to have a republic established on the fall of the Protectorate in 1659. His major work, *Oceana* (1656), described a Utopian State with a written constitution, providing for a bicameral legislature, rotation in office, the indirect election of a president, secret ballot, and other ideas much in advance of their time, some of which influenced the framers of the U.S. Constitution.

Harrison, Frederic (1831–1923). Called to the Bar in 1858, he worked under Lord Westbury on codification, was an adviser to the Royal Commission on Trade Unions, 1867–69, secretary of the Commission for the Digest of the Law, 1869–70, and Professor of Jurisprudence and International Law to the Council of Legal Education, 1877–89. He strove to introduce Comte's philosophy into England and wrote extensively on history, biography, and criticism, and a book on *Jurisprudence and the Conflict of Laws* (1919).

Harsh and unconscionable. A term for a bargain which is hard and such that a court of equity would regard it as contrary to good conscience to enforce.

Hart, Anthony (*c.* 1754–1831). Was called to the Bar in 1781 and practised in Chancery. He took silk in 1807 and became Vice-Chancellor of England in 1827 but, five months later, was made Lord Chancellor of Ireland (1827–30). In his short term of office he proved himself an able judge, his judgments being sound and impartial.

Hart, Herbert Lionel Adolphus (1907–). He practised at the Chancery Bar and then became Professor of Jurisprudence at Oxford 1952–68, and later Principal of Brasenose College. He was author of *Causation in the Law* (with Honoré) (1959), *The Concept of Law* (1961), *Law, Liberty and Morality* (1960) and other works on legal theory, and editor of some of the works of Bentham. In *The Concept of Law* he contends that a defining characteristic of a legal system is that it includes a fundamental rule for the identification of the other rules of the system. Rules of law are primary, which are prescriptive of behaviour, and secondary, which relate to the identification, creation, change and application of primary rules.

Hartmann, Adolph (nineteenth century). German jurist, author of *Institutionen der praktischen Völkerrechts in Friedenszeiten* (1874).

Hatherley, Lord. See PAGE WOOD, WILLIAM.

Hatsell, John (1743–1820). Clerk of the House of Commons (1768–1820) and a great authority on parliamentary practice. His major works are *A Collection of Cases of Privilege of Parliament, from the earliest records to 1628* (1776) and *Precedents of proceedings in the House of Commons* (4 volumes, 1781).

Hatton, Sir Christopher (1540–91). A favourite of Queen Elizabeth and member of her privy council, he participated in the trials of Babington and Mary, Queen of Scots, and was made Lord Chancellor in 1587. As such he appears to have been an adequate judge.

Hauriou, Maurice (1854–1929). French jurist and teacher of administrative law. He wrote a *Précis du droit administratif et de droit public général* (1892), *Principes du droit public, Précis du droit constitutionnel*, and *La science sociale traditionelle* (1896). He developed a theory of institutions seeking to reconcile the State as a concert of free wills and a determinist conception of social science, which was further explored by Renard, Delos, and others. It has been influential on thinking about public law and general jurisprudence.

Hautefeuille, Laurent Basile (1805–75). French jurist and author of *Les droits et les devoirs des nations neutres en temps de guerre maritime* (1848–49) and other works.

Hauteserre (or Alteserre), Antoine Dadin de (1602–82). French canonist and historian, author of a major four-volume *Dissertationum juris canonici libri quattuor* (1651). Two volumes were added in 1654. He also wrote commentaries on the decretals of Innocent III and other works.

Hawkers and pedlars. A hawker is a person who travels from place to place, or to other persons' houses, or to hired premises, with a horse or other beast of burden, carrying for sale, or exposing for sale, any goods or merchandise or exposing samples or patterns of goods to be delivered later. A hawker must take out an annual excise licence. A pedlar is a person who goes from place to place, or to other persons' houses, carrying to sell, or exposing for sale, any goods, wares, or merchandise, or procuring orders for goods, wares, or merchandise immediately to be delivered or selling, or offering for sale, his skill in handicraft. A pedlar requires a pedlar's certificate.

Hawkins, Sir Henry, Lord Brampton (1817–1907). Worked as a special pleader and was called to the Bar in 1843. He was very successful with juries and was a master of cross-examination. He became a judge of the Queen's Bench Division in 1876, being almost immediately transferred to the Exchequer Division. He excelled in criminal cases, and had an unwarranted reputation for severity, but was less successful in civil cases, his judgments being verbose and unclear. He resigned in 1898, was sworn of the Privy Council and, in 1899, made a peer as Lord Brampton. Thereafter he sometimes sat in the House of Lords or Privy Council.

Harris, *Reminiscences of Sir Henry Hawkins, Baron Brampton.*

Hawkins, William (1673–1746). Became a barrister and, in 1723, a serjeant-at-law and achieved fame as a writer on criminal law. His great book, *A Treatise of the Pleas of the Crown*, appeared in 1716 and in many later editions. It deals both with substantive criminal law and with the courts exercising criminal jurisdiction and criminal procedure. It is clearly arranged, well written, and firmly based on the authorities, and has always enjoyed a high reputation. His first book, an abridgement of Coke upon Littleton (1711) also went through many editions. He also published an abridgement of his *Treatise* under the title *A Summary of the Crown Law* (1728) and a collection of statutes (1735).

Hay, Sir George (1715–78). Became an advocate of Doctors' Commons in 1742 and an M.P. in 1754. He was King's advocate from 1755 to 1764. In 1764 he became Dean of the Arches and, in 1773, judge of the Court of Admiralty. Despite his irregular private life he was held to be a capable lawyer and an excellent judge.

Hay, John Hay, 2nd Earl and 1st Marquis of Tweeddale (1626–97). Fought both for and against Charles I. After the Restoration he became restored to favour, became a member of the High Commission in 1663, an Extraordinary Lord of Session in 1664 and, in 1692, became Lord Chancellor of Scotland, but was superseded in 1696 for having supported the Darien scheme. He was characterized as an able and worthy man and sought to mitigate the severity of the government towards the covenanters.

Hay, John Hay, 2nd Marquis of Tweeddale (1645–1713). Son of John Hay, 1st Marquis (q.v.), adhered to William of Orange at the Revolution and became a privy counsellor. He became Lord Chancellor of Scotland in 1704–05 and supported the Union of 1707. Thereafter he was a Scottish representative peer at Westminster. He had a good reputation as a man of honour who sought to promote the trade and welfare of the country.

Hay, John, 4th Marquis of Tweeddale (?1695–1762). Studied law, became an extraordinary lord of session in 1721, and was a Scottish representative peer for many years. From 1742 to 1746 he was principal Secretary of State for Scotland and in the Cabinet. In 1761–62 he held the office of Lord Justice-General. On his death the office of extraordinary Lord of Session came to an end.

Headborough. The chief of the 10 men who comprised a frankpledge (q.v.), elected by the court leet with responsibility for the keeping of order in the area for which he was elected. They were gradually replaced by petty and parish constables.

Head-courts. Formerly, in Scotland, the three courts at which all freeholders who owed suit and presence at the lord's court, required to be present under pain of fine. They were later reduced to one, at Michaelmas, and by statute of 1747 fines for non-attendance at head-courts were abolished.

Head-note. The part of a modern law report, composed by the reporter and the editor of the series of reports, which attempts to state concisely the point or points of law discussed and decided or illustrated by the decision and the judicial opinions delivered justifying it. It commonly comprises a series of catch-words indicating the main headings and sub-headings under which the case falls in a logical classification of the law, a statement of the facts found relevant to the decision, the substance of the decision, frequently prefaced by the word 'Held ...', a note of which counsel appeared for each party, and a summary of the arguments presented with a mention of the cases, statutes, books, and other sources referred to. It is not authoritative and may be criticized as inaccurate.

Heads of state. As a state is an abstract idea, a legal and political concept, it must have a living person as head of state to represent and for some purposes embody it. How the head of state is chosen depends on the constitution of each individual state. There are three main categories of heads of state: monarchs, who hold office by hereditary right, presidents, who are elected for stated periods, and dictators, who obtain and hold power by force, though they may nominally be entitled presidents and nominally be elected or otherwise chosen. Sometimes a dictator may be the real controlling force but have a monarch, president or other nominal head of state above him. Whether a head of state is a figurehead, or a person also having executive powers, depends on the constitution and practice of each state. Thus in the U.S.A. the President is the chief executive and heads the Cabinet; in France there is a separate Prime Minister who heads the Cabinet. By international law heads of state represent their states and their acts, so long as legally competent by their own municipal law, are acts of their states. By custom each state accords various ceremonial honours to the head of each other state when visiting, grants him special protection, exterritoriality (q.v.) and immunity from the jurisdiction of the local courts. Presidents may be entitled to lesser honours than monarchs.

Heads of the Proposals. Conditions formulated by the council of the Army in 1647, and offered to Charles I. The proposals were that Parliament should set a date for its own termination, that there should be biennial Parliaments lasting for not less than 120 days and having control over army and navy for 10 years; that no royalist should be permitted to hold any office in the state for five years, or to be elected until after the second biennial Parliament; that there should be a council of state of persons to serve for seven years to conduct all foreign negotiations but requiring the consent of Parliament to make war or peace; that Parliament should nominate great officers of state, for 10 years directly and indirectly thereafter; that all coercive powers should be taken away from bishops and ecclesiastical officers; that all acts enjoining the use of the Book of Common Prayer should be repealed; that no one should be compelled to take the covenant; and that royalists be allowed to compound on easy terms for their delinquency. The Proposals angered the King. They were eventually implemented, with some changes, in the Instrument of Government of 1653.

Heard, Franklin Fiske (1825–89). Was admitted to the Bar at Concord in 1850 and later practised in Boston. He wrote numerous legal works, including the first American book on libel and slander (1860), a book on *Equity Pleading*, and a work on *Shakespeare as a Lawyer* (1883).

Hearing. A non-technical and general term for an occasion on which a person may present argument or adduce evidence or both, or otherwise be heard by a judge or tribunal on some matter under legal investigation. Various kinds of hearings usually have technical names, such as preliminary investigation, trial, appeal, but the term hearing may be used for cases to which no technical term has been applied. A point of importance in relation to the justice of proceedings is whether an individual is entitled to a hearing, i.e. to present submissions or arguments orally to the person making the decision, or not. There is no universal rule as to the right to a hearing and natural justice does not invariably require that the parties be entitled to an oral hearing.

Hearn, William Edward (1826–88). Was successively professor of Greek in Galway and professor of modern history in Melbourne, and then (1873) Dean of the Faculty of Law there. He was a member of the Irish (1853) and Victorian (1860) Bars, and a Q.C. of Victoria (1886). In the legislative council for Victoria he devoted his energies to codification of the law but his draft bills failed to pass. Of his writings the most important are *The Aryan Household, its Structure and Development; an Introduction to Comparative Jurisprudence* (1879), *The Theory of Legal Rights and Duties: an Introduction to Analytical Jurisprudence* (1885), and the *Government of England* (2 edn. 1886).

Hearsay. In the law of evidence, testimony not of what the witness himself saw, heard, or otherwise observed, but of what he heard others tell him or say about the matter under investigation. The general rule of Anglo-American law is that hearsay is not admissible evidence, because it is not the best evidence, in that the actual observer is not giving the evidence nor subject to cross-examination, and the witness tendering hearsay evidence is rendering second-hand evidence; even if he reports it accurately, the actual observer may have been mistaken and his credibility is not tested. There are, however, many exceptions to the general rule against hearsay, as where the actual observer is dead and it has been largely abolished in civil cases in England. In European countries, on the other hand, judges may generally consider any evidence tendered, the question whether it is of immediate observation or second hand being a matter to be weighed when evaluating the evidence.

Heck, Philipp (1858–1943). German jurist, concerned primarily with the jurisprudence of interests, which regards it as the task of legal science to facilitate a judge's task by preparing a proper decision through an investigation of the law and the relevant situation of life. Law is not merely a logical construction but a balancing of interests. He was author of *Schuldrecht* (1929), *Sachenrecht* (1930), *Begriffsbildung und Interessenjurisprudenz* (1932), *Interessenjurisprudenz* (1933), and other books covering a wide range of subjects.

Hedges, Charles (?1650–1714). Became an advocate of Doctors' Commons in 1675 and was a judge of the Court of Admiralty from 1689 to 1714. He sat in Parliament, was also Secretary of State, 1700–06, and, in 1711, became also judge of the Prerogative Court. He was a statesman of character and a notable judge of the Court of Admiralty. Many of his decisions in constitutional and international law, and maritime and prize law are noteworthy.

Hedley Byrne & Co. Ltd. **v.** ***Heller & Partners, Ltd.***, [1964] A.C. 465. H.B., through their bankers, made enquiries from H. as to the financial stability of E. The reply was negligently prepared, was inaccurate, and H.B. acted on it and lost money. H.B. sued H. for damages for having negligently misrepresented the facts enquired into. The House of Lords laid down that where, in the course of business, a person gives advice or information in response to enquiry in circumstances in which a reasonable man would know that his answer would be relied on by a third party, a duty arises to take such care as is reasonable in the circumstances in answering, and if it is not taken an action lies for any loss which has resulted from mistaken reliance on the information.

Hedonist utilitarianism. A term descriptive of the theory, which goes back to Helvetius, Montesquieu, and Beccaria, underlying the work of Jeremy Bentham and his followers, that the rightness or goodness of the law on any matter was to be determined by whether or how far it promoted the greatest happiness of the greatest number. Happiness was to be determined by computing the pleasures and pains which result or would result from a particular state of the law on any matter. Bentham meant the aggregate of individual surpluses of pleasure over pain. He enumerated pleasures and pains and sought to determine the method of ascertaining the quantity of pleasure or pain resulting from a given act. In his view the legislator's power to control conduct lay in his use of pleasures and pains as sanctions, though he admitted that there were also sanctions imposed by nature, public opinion and other forces.

The theory is open to many objections and criticisms. Nevertheless, in the nineteenth century Bentham and his followers achieved a vast amount of practical law reform, in criminal law, prison reform, reform of the lunancy laws, liberalization of commerce and investment, personal incapacities and inequalities, and in evidence and court procedure.

Heffter, August Wilhelm (1796–1880). German jurist, author of *System des römischen und deutschen Zivilprocessrechts* (1843), *Das Europäische Völkerrecht der Gegenwart* (1844), and other works.

Hegel, Georg Wilhelm Friedrich (1770–1831). German philosopher. Initially he studied theology and later economics and government. He taught at Jena and Heidelberg and, from 1818, he was Professor of Philosophy at Berlin in succession to Fichte. his writings include the *Phenomenology of the Spirit* (1807), *Science of Logic* (1812–16), *Outlines of the Philosophy of Right* (1821), and lectures on aesthetics, philosophy of religion, philosophy of history, and history of philosophy. His philosophy was later very influential in Britain and from 1850 to 1914 numerous thinkers expounded and developed it. His philosophy of law is concerned with the Idea of right, the concept of right and the actualization of that concept. Right, in general, has its foundation in the will, which is free, and the system of right or law is the province of actualized freedom. Personality is the abstract will consciously self-contained. From the ideas of will and personality Hegel develops the categories of the sphere of abstract right, namely possession or property, contract, and wrongdoing or crime. The individual will become objective in property. Contract he conceives in terms of property; it depends on a unity of differing wills. Wrong and crime spring from conflict of wills; the will of the criminal must be penalized in order to annul the crime. The transition from abstract right to positive law is accomplished by the idea of relationship; positive law is the objective sphere of relationships which exist in civil society. The State is the synthesis of family and civil society. Hegel's thought has been widely influential, notably on Marxists and modern analytic philosophy but less so in law than in other spheres. It underlies doctrines of the supremacy of the State over the individual. But in the sphere of law his philosophy has been less important.

E. Caird, *Hegel*; W. Stace, *The Philosophy of Hegel* (1924); T. Knox, *Hegel's Philosophy of Right* (1942) J. H. Stirling, *Lectures on the Philosophy of Law*; E. Lasson, *Einleitung zu Hegels Rechtsphilosophie*.

Heilborn, Paul (1861–1932). German jurist, author of *Recht und Pflichten der neutralen Staaten* (1888), *Der System des Völkerrechts entwickelt aus den völkerrechtlichen Begriffen* (1896), and *Grundbegriffe des völkerrechts* (1912).

Heimbach, Gustav Ernst (1810–51). German jurist, author of *Observationes Juris Graeco-Romani* (1830) and other works on Byzantine law.

Heimbach, Karl Wilhelm Ernst (1803–65). Brother of the foregoing, also worked on Byzantine law and is remembered for his great edition of the *Basilica* (1833–50).

Heineccius, Johann Gottlieb (1681–1741). A renowned German jurist, professor at Halle, a great expositor who tried to treat law as a rational discipline and not merely an empirical art, author of *Elementa juris civilis secundum ordinem Institutionum* (1725), *Elementa juris civilis secundum ordinem Pandectarum* (1727), *Historia juris civilis Romani ac Germanici* (1733), *Elementa juris naturae et gentium* (1737), and other works.

Heir. In a general sense one who succeeds to, or takes by succession from, an ancestor, but in a stricter and more accurate sense, before 1926 in England and 1965 in Scotland, the person designated by the feudal rules of succession as entitled to succeed to a deceased person's real or heritable property, as contrasted with the executors and next-of-kin entitled to succeed to the personal or moveable property. The feudal rules of succession applied to find the heir were based on the succession of males in preference to females, on primogeniture among males, and on representation, whereby a descendant of a pre-deceased heir, e.g. a grandson, took in preference to a collateral of the deceased, e.g. a brother. An heir general or heir at law is such an heir; an heir apparent is one whose right to succeed is indefeasible provided only that he outlive the ancestor, e.g. an eldest son; an heir presumptive is one whose right is defeasible by the birth of a nearer heir, e.g. a younger brother, defeasible by the birth of a son to the ancestor.

Heirlooms. In English law, such personal chattels as, by force of special custom, cannot be disposed of by will but pass on the owner's death to his heir rather than to his executor. They included the best bed, table, cart, and the like. Other chattels which pass to the heir and are said to be of the nature of heirlooms are things incidents of the tenure of land, or naturally falling to an owner of an estate, or necessary for maintaining his title or dignity such as a nobleman's coat of armour or other ensigns of honour, and certain other articles. In common parlance heirloom is used of many articles, such as pictures, plate, jewellery, and the like which are assigned or bequeathed to trustees to be held subject to being used by the persons for the time being in possession of a house or estate, or even to something which is bequeathed subject to a condition that it is to be kept in the family.

Heirship moveables. In Scots law, certain articles of a moveable character which contrary to the general rule passed on death to the heir of the deceased rather than to his executors to ensure that he did not inherit a house and land without the means of working them. Old statute gave the heir the best one of each item of furniture and plenishing in the house, and custom gave him certain indivisible articles such as the family seal of arms. The claim to them was abolished in 1868, but in 1964 the concept of 'heirloom' was introduced, defined as any article which has associations with the intestate's family of such nature and extent that it ought to pass to some member of that family other than the intestate's surviving spouse.

See HEIRLOOM.

Heirs-portioners. In the pre-1964 Scottish law of succession to heritage, when the succession opened to the female line heritage devolved not on the eldest female but to all females in the same degree of relationship, who inherited equally and *foro indiviso* as heirs-portioners.

Heliaea. At ancient Athens, the judicial assembly of Athenian citizens over 30 years old, and also the name of its meeting-place. Appeal to this body from the archons' decisions was introduced by Solon, and it may have been defined by Cleisthenes as 6,000 citizens chosen by lot from the demes. Only after 462 B.C. were the archons confined to preliminary enquiries and the *heliaea* became the supreme court of Athens. In practice the *heliaea* was divided into several dicasteries, each representative of the whole.

Hellenistic law. The body of law, Greek in origin, which applied to Greek and Hellenized inhabitants of Asia Minor, Syria, Palestine, Egypt, and other parts of the ancient Near East between the time of Alexander the Great (323 B.C.) and the absorption of the kingdoms formed on the break-up of his Empire in the Roman Empire, down to 30 B.C. Thereafter Roman law gradually superseded it, though Hellenistic law lasted and continued to develop until A.D. 212 when Roman citizenship was conferred on all the inhabitants of the Roman Empire. Hellenistic law is not a single uniform body of law because it was drawn from the laws of various Greek city states, which were not uniform, and developed in widely separated areas. Knowledge of it is uncertain and far from complete because no codes or treatises are known and most of our knowledge is derived from materials found in Egypt.

When Alexander the Great's armies moved eastwards some Greek citizens settled in conquered territories and there seems to have gradually developed a common body of private law, based on common features of the legal systems of Greek city states, but with local variations and subject to influences from local law. Public law, however, varied more because Hellenistic rules, to a large extent, adopted the existing governmental and administrative systems of the territories they ruled. Hellenistic law, however, lacked clear and defined terminology, clear concepts and well-defined rules. A notable feature, common to all the Hellenistic legal systems, was the extensive reliance on writing for proof or extinguishing an obligation.

Everywhere, Hellenistic law was superimposed on and did not supersede pre-existing local law, and this continued right down to the Roman period. Systems of personality of laws existed so that persons of each race were governed by their own laws and there were concurrently different courts administering different bodies of law for peoples of different races. Later the principle of personality disappeared so that each person became free to choose by what system of law he wished to be governed. This gave rise to provisions in contracts for choice of law, and it enabled persons of one race to enter into transactions under another system and forbidden in their racial system. Thus Jews utilized both Jewish and Hellenistic law and Egyptians both Egyptian and Hellenistic law. In the third century, administration and judiciary became mixed, minor officials taking on many adjudicative functions.

A great deal is known about the highly organized governmental system of Ptolemaic Egypt in the third century B.C., the salient feature of which was the extent of nationalization. The King was the State and claimed to own the entire land of Egypt, including the lands of the old nobility and the temple lands. All the major products, corn, oil, and textiles, were royal monopolies, production for private sale and use being only additional to the quota for the State. Many other businesses and trades could be carried on only by royal licence. Taxation was heavy, of many kinds, and there were elaborate registers of population, land, and animals. There were armies of officials and as time went on the bureaucratic system began to break down as the greed and brutality of officials grew ever greater. Euergetes II about 118 B.C. strove to reform the system but with little success.

Among the fellahin, the basis of the system was that each man had his own place, which he could not leave except by official order or permission. There is evidence of numerous withdrawals of labour, strikers frequently taking refuge in a temple with the right of asylum, and it is apparent that there were no liberties of the subject.

Subject to the qualification that there can be no certainty that the same rules were accepted and applied everywhere from the Bosporus to the Nile nor that there were no developments within the period a good deal is known of the general character of Hellenistic law. In relation to persons both sexes seem to have had almost equal legal capacity, women being less restricted than under Athenian law, but still frequently requiring the intervention of a guardian at least as a formality. Marriage seems to have been created by cohabitation and divorce, competent to either spouse, by ending cohabitation. Polygamy and marriage between close relatives were known. Paternal and maternal authority over children were limited and did not extend to anything like the powers of the Roman *paterfamilias*. Adoption was known. Slaves existed, though they were not numerous, and could own property and participate in legal transactions, thus being better off than slaves in earlier Greek or contemporary Roman society.

Knowledge of the law of contract is mostly of sale and loan. Sale of goods seems to have been normally a ready-money transaction. Sale of land was recognized.

Loans and the use of credit were common, and interest was regulated in Egypt about 250 B.C. It could be paid in money, goods, or services, and the device of stating as repayable the total of capital and interest thereon was known.

Tort law was undifferentiated from criminal law, so that actions and prosecutions were not distinguished. Distinctions were, however, drawn between premeditated and negligent harm and aggravated offences, such as those done at night or by armed men, were treated as more serious. Proceedings for wrongs were frequently for compensatory damages, but also, particularly in the case of offences against the community, for penalties such as confiscation of the offender's property or sometimes death. Corporal punishment seems generally to have been restricted to slaves.

Private property in land seems to have been recognized, together with rights of leasing and liabilities to taxation by the King, though the King was said to own all the land; his right seems in fact to have been truly a right to tax. There were several classes of land, notably royal land, temple land, and land held or originally held, by soldiers in return for military and other services. But the different concepts and terminology of later law were undeveloped, so that the different rights of persons in property are not clear.

As regards succession the tendency was for property to be divided among the heirs, who were primarily the natural or adopted children, both male and female, of the deceased. Capacity to make a will was recognized though at least in Alexandria the Greek incapacity of women to test remained.

The differences between original Egyptian and Greek law tended to disappear in the Hellenistic period.

In Egypt jurisdiction was exercised not only by judges but sometimes by important individuals. There were Egyptian, Greek, and Jewish courts and, in at least the first two of these, a royal official could supervise the courts and had the functions of preparing the cases and executing sentences. Execution of civil decrees was the function of an executive office of the court and might proceed against the debtor's person or his property.

Hellenistic law survived, though increasingly overlaid by Roman law, at least until Roman citizenship was conferred on all the inhabitants of the Roman Empire in A.D. 212 and thereafter had some continuing influence in matters not covered by Roman law.

J. Modrzejewski, 'Monde Hellénistique' in *Introduction bibliographique à l'histoire du droit et à l'ethnologie juridique* (ed. Glissen), vol. A/8; R. Taubenschlag, *Law of Greco-Roman Egypt in the Light of the Papyri, 332 B.C.–640 A.D.*; E. Seidl, *Ptolemaische Rechtsgeschichte.*

Hellfeld, Johann August (1717–82). German jurist, author of *Historia juris romani* (1740), *Historia juris germanici et canonici-pontificii* (1741), and *Jurisprudentia forensis secundum Pandectarum ordinem proposita* (1764).

Hemmingsen, Nicolas (Hemmingius) (1513–1600). Danish theologian, notable for his *De Lege naturae apodictica methodus* (1562) a forerunner of the work of Grotius. He also wrote extensively on theology and was a leader of the Reformed Church in Denmark.

Hengham, Ralph de (?–1309). Became Chancellor of Exeter Cathedral in 1275 and seems to have been a judge of the King's Bench from 1270, of the Common Pleas in 1272, and Chief Justice of the King's Bench in 1274. He was removed from office and fined in 1290, but was restored to the bench 10 years later and again became Chief Justice in 1301, holding office till his death. He may have been the author of the tract *Cum sit necessarium* (q.v.) and is normally regarded as the author of the small books called *Summa Magna* and *Summa Parva*, both in Latin. *Magna* deals, in 13 chapters, with the writ of right. It dates from 1270–75, is founded on *Registrum brevium*, Glanvill, and largely on Bracton, and is akin to *Fet asaver*. *Parva* deals, in eight chapters, with procedure by assize. It dates from 1285–90 and may have been founded on Bracton, but much abbreviated.

Both *Summae* were printed in Fortescue's *De Laudibus Legum Angliae* (1616 and 1660 editions), with notes by Selden.

Henley, Sir Robert, 1st Lord Henley and 1st Earl of Northington (*c.* 1708–72). Henley became a fellow of All Souls before his call to the Bar in 1732. He developed a practice, became an M.P. in 1747, and attached himself to Frederick, Prince of Wales. He became Attorney-General in 1756 and Lord Keeper in 1757. He was the last holder of that office. He became Lord Henley in 1760, Chancellor in 1761, and Earl of Northington in 1764, but resigned for health reasons in 1766 and took the office of Lord President of the Council. As a judge he was able, with a good grasp of principle and powers of rapid and lucid decision, though not outstanding.

Henley's *Memoir of L. Ch. Northington* (1831).

Henn, Henry (?–1708). Called to the English (1653) and Irish (1669) Bars, he became a baron of Exchequer in Ireland (1673) and Chief Baron (1680–87).

Henogamy. The custom by which only one member of a family is permitted to marry, springing from the need to keep property intact and limit the number of heirs. It is found in some tribes in south India.

Henry II (1133–89). King of England 1154–1189, a very able king whose major achievements were to strengthen financial administration, to rationalize the tenure of land and better ensure protection of possession, and to establish better and regular provision for the administration of justice. He succeeded to realms ill-governed, in disorder and neglected and his first task was to restore firm government. This he quickly did, and he began to improve the administration of justice, assisted by a group of able barons in his council and some able civil servants at the Exchequer. By the Assize of Clarendon (1166), an improved procedure of criminal justice was established; 12 lawful men of every hundred and four of every village, acting as a jury of presentment, had to declare on oath whether any local man was a robber or murderer. Trial of persons thus accused was reserved to the King's justices. Regular visits to all parts of the country by the King's justices became a normal part of the administration of justice and there developed distinct groups of judges, one on the bench at Westminster, and one itinerant with the King when out of London, apart from justices visiting counties 'on eyre'. Furthermore, the group of officials sitting at the Exchequer acted, not only as accountants and auditors but judicially, deciding disputed revenue matters. The genesis of the three common law courts, Exchequer, King's Bench, and Common Pleas lay in these arrangements.

Wrongful dispossession of land was a common complaint and Henry established the possessory writs, in the form of an order from the Exchequer to the local sheriff to convene a local jury to determine whether dispossession had been effected and, if so, to reinstate the plaintiff pending the trial of the issue of right at the grand assize. The first such writ was that of novel disseisin; similar, later, writs of *mort d'ancestor* decided whether the plaintiff's ancestor had previously possessed the land, and of darrein presentment decided who had last presented a cleric to a particular benefice. The possessory assizes not only did much to right wrongs but enhanced the reputation of royal justice.

In seeking to impose justice, Henry came into conflict with Archbishop Becket and the bishops over the issue of criminous clerks, the question whether clerics accused of crime should be tried by a royal or an ecclesiastical court. The Constitutions of Clarendon (1164) sought to define royal and Church rights in these and related matters but Becket, after initial compliance, refused to accept them and this contributed to discord, severance of relations and Becket's withdrawal into exile. This in turn had bad effects on public life and caused many difficulties for Henry.

His reign is also significant for the earliest writings on English law. Richard FitzNigel or Fitzneal wrote the *Dialogus de Scaccario*, a description of the work of the Exchequer (1179) and Ranulf Glanville (or more probably Hubert Walter) in *De Legibus et Consuetudinibus Regni Angliae* produced the first reasoned account of legal proceedings (1180). The plea-rolls are known to have been kept from 1183.

W. Warren, *Henry II*; F. Pollock and F. W. Maitland, *History of English Law*; F. Stenton, *English Justice between the Norman Conquest and the Great Charter, 1066–1215*.

Henry, Sir Denis Stanislaus, Bart. (1864–1925). Was called to the Irish Bar in 1885, took silk in 1896, became Solicitor-General for Ireland, 1918–19, Attorney-General for Ireland, 1919–21, and first Lord Chief Justice of Northern Ireland, 1921–25.

Henry VIII clause. A clause, so named in disrespectful commemoration of Henry VIII's tendency to absolutism, occasionally found in legislation conferring delegated legislative power, giving the delegate power to amend the delegating Act or, usually, any other Act, in order to bring the enabling Act into full operation, or otherwise by Order to remove any difficulty. It seems to date from the Local Government Act, 1888, but was included in some later Acts. The Committee on Ministers' Powers recommended the abandonment of the practice save for the purpose of bringing the Act into operation but instances of its use continue.

Henry system. A classification by letters and numbers of fingerprint patterns, treating the 10 fingers as a unit, which is the basis of most systems of fingerprint classification used in Anglo-American countries.

Heptarchy. The seven Anglo-Saxon kingdoms established in England towards the end of the fifth century and mostly destroyed by the Danes in the ninth century, namely Northumbria, Mercia, East Anglia, Essex, Kent, Sussex, and Wessex.

Her Majesty's dominions. Territories under the sovereignty of the British Crown. Apart from special statutory definition it does not include those independent countries within the Commonwealth which do not regard the Queen as their Head of State, nor protectorates, protected states, or trust territories. Thus, the term includes Australia, Canada, and New Zealand but not India, Nigeria, and Tanzania. It is quite distinct from the term 'Dominion' as used in the now-obsolete phrase 'Dominion status'.

Her Majesty's Forces. The armed forces of the Crown authorized by royal prerogative and statute to be raised and maintained. They are the Royal Navy, the Army, and the Royal Air Force, together with the reserve and volunteer reserve units of these services, the Women's Royal Army Corps and the Women's Royal Air Force. It is uncertain if the phrase extends to forces raised in Commonwealth countries which are deemed within Her Majesty's dominions (q.v.).

Her Majesty's pleasure, detention during. The sentence which is imposed on persons under 18 in circumstances where an adult must be sentenced to life imprisonment. In practice, such offenders are detained in borstal, young offenders' institution or prison as the Home Secretary or the Secretary of State for Scotland may determine for such period as may seem right in the circumstances. On discharge, conditions may be imposed.

The Criminal Lunatics Act, 1800, provided that persons insane at the time of a crime were to be acquitted on the ground of insanity and on such a verdict the court had to order that the accused be kept in custody until His Majesty's pleasure be known. In 1964 the Criminal Procedure (Insanity) Act substituted detention in a hospital specified by the Secretary of State.

Herald. Originally a member of a royal or noble household, later one employed to convey messages and act as ambassador between different courts. Corporations or colleges of heralds came to be established, first by the King of France in 1407 and by the King of England in 1484, as the College of

Arms. In Scotland the Lord Lyon King of Arms is the successor of the sennachie or official bard of the old Celtic kings. In Ireland an Ireland King of Arms (1382) was replaced by Ulster King in 1553.

The English College is under the Earl Marshal, an office long now hereditary in the Dukes of Norfolk. It comprises three Kings of Arms (Garter, Norroy and Ulster, and Clarenceux), six heralds (Windsor, Richmond, York, Lancaster, Chester, and Somerset) and four pursuivants (Rouge Croix, Bluemantle, Rouge Dragon, and Portcullis). The Scottish college comprises the Lord Lyon King of Arms, three heralds (Albany, Marchmont, and Rothesay) and four pursuivants (Carrick, Kintyre, Unicorn, and Ormond). In both countries there may be heralds extraordinary. All are members of the royal household. The office of Ulster King of Arms was united with that of Norroy and a Chief Herald of Ireland was appointed for the Republic of Ireland. There were heralds also in Sweden, the Low Countries, and Portugal.

In modern times, the English and Scottish colleges participate in royal ceremonials, grant arms, record arms and genealogies and deal with matters of precedence. The Lord Lyon has more extensive powers than the English college in that he is a great officer of state, holds a court and is controller of the messengers-at-arms and accordingly head of the executive department of Scots law.

Heraldry. The knowledge and work belonging to a herald, but from his work with armorial bearings, particularly the science and art relating to armorial bearings, or identification devices originally used on shields and seals and now used as badges by individuals, corporations, and institutions. Heraldic designs came into general use in western Europe from about 1150, used particularly on shields in warfare and on seals as marks of authenticity. Since the use of arms was associated with the higher grades of feudal society, heraldry became associated with the concept of gentility, and every gentleman sought to have armorial bearings. Their use was extended early to superior ecclesiastics and guilds such as the great livery companies. It early became recognized that armorial bearings were a kind of incorporeal real property, transmissible hereditarily. In modern times arms are borne not only by individuals but by ecclesiastics and many kinds of institutions and corporations.

The science of heraldry is in general principles common to the Western European countries but has developed variations in each country.

Heralds are known from the thirteenth century; at first all great nobles had their heralds, the royal being distinguished only by the greater importance of kings than nobles. The King of France formed his heralds into a college in 1407 and the King of England established the College of Arms in 1484. An Ireland King of Arms existed in 1382, and Ulster King was created in 1553. From about 1400 the English heralds have made grants of arms on behalf of the sovereign and from 1530 to 1686 royal commissions were granted to the heralds to visit countries to inspect the arms in use there.

Treatises on heraldry began to be written early. Bartolus (q.v.) published a short *Tractatus de insigniis et armis* about 1356. English works include John of Guildford or Joannes de Bado Aureo (*Tractatus de Armis*, 1394), a Welsh treatise *Llyfr Arfan*, and Nicholas Upton (*De Studio Militari*, c. 1440). Later works of importance include Ferne's *The Blazon of Gentrie* (1586), Mackenzie's *Science of Heraldry treated as part of the Civil Law and Law of Nations* (1680) and Dugdale's *Ancient Usage in bearing of ... Arms* (1682).

A. Wagner, *Heralds and Heraldry in the Middle Ages*; *Historic Heraldry of Britain*; A. C. Fox-Davies, *Complete Guide to Heraldry*; *Armorial Families: A Directory of Gentlemen of Coat Armour*; J. Woodward and G. Burnett, *A Treatise on Heraldry, British and Foreign*; J. Burke, *General Armory*; T. Innes, *Scots Heraldry*; R. Mathieu, *Le Système héraldique français*; J. Asensio y Torres, *Tratado de heraldica y blason*.

Herauld, Didier (Heraldus) (1579–1649). French jurist, author of *De rerum judicatarum* (1640), *Observationum et emendationum liber* (1644), and other works.

Herbert, Edward (?1648–98). Son of Lord Keeper Sir Edward Herbert. After being called to the Bar he became Attorney-General of Ireland and successively Chief Justice of the King's Bench (1685) and of the Common Pleas (1687). He fled with James II, under whom, in exile, he was titular Lord Chancellor and Earl of Portland. He is said to have been a virtuous man but an indifferent lawyer and in *Godden* v. *Hales* expressed the view that there were no limits to the King's dispensing power.

Herbert, William (1771–1851). Became librarian of the Guildhall Library in 1828. Best known for his *History of the Twelve Great Livery Companies of London* (1836–37) he also published an *Antiquities of the Inns of Court and Chancery* (1804).

Hereditament. A term, now only used in connection with rating, for every kind of property which could be inherited, both corporeal (land) and incorporeal (the rights and profits issuing out of land).

Hereditary revenues of the Crown. Income derived from such land as is or may become vested in the sovereign in her body politic in right of the Crown, and from various prerogative rights and privileges relating to property enjoyed by the

Crown, such as to wreck and treasure trove. Since 1760 it has been customary for each sovereign to surrender the hereditary revenues for his or her lifetime to the Exchequer in return for an annual income known as the Civil List (q.v.) and certain allowances paid to certain members of the Royal Family. The revenues of the Duchies of Lancaster and Cornwall were excluded from the original surrender and have been retained; the latter fall to the heir apparent when there is one, as do the revenues of the Principality of Scotland, which were also reserved. Certain revenues from taxation, formerly settled by statute on the Crown, have been surrendered. The sovereign also enjoys the revenues from such estates as the sovereign enjoys in her natural capacity, i.e. as an individual and not as Queen, such as the Sandringham and Balmoral estates, and which she may freely alienate, although the manner in which these estates may be dealt with is largely regulated by statute.

Hereditas iacens. In Roman law the inheritance or estate of a deceased after his death but before the entry of the heir. It raised the question of who could authorize acts done on behalf of the estate. Roman views were that the inheritance represented the person of the deceased, or that the heir's entry had retrospective effect, or that the inheritance was a quasi-person in itself. Modern continental law generally adopts the third solution. In British law the problem does not arise because there is an executor or administrator responsible for the estate; till one is appointed the estate vests in England in the President of the Family Division of the High Court.

Heresy. The holding of an opinion or doctrine at variance with the orthodox doctrine of the Christian Church. In 1382, following the preaching of Wycliffe and the Peasants' Revolt, statute ordered sheriffs to arrest all persons certified by the bishops as preachers of heresy, but the statute was repealed in the next year. In 1388 the Crown issued commissions to search for and seize heretical books and the King and council took over jurisdiction in heresy cases from the Church courts. In 1401 the statute *De Haeretico Comburendo* was passed giving the bishops power to arrest persons guilty of propagating heretical views and, if they refused to abjure or relapsed after abjuration, the sheriff was to have them burned. By canon law this statute was regarded as unnecessary and though the statute was repealed, burning for heresy was still competent and is said to have happened in the time of Edward VI.

Heresy is still an offence by a clergyman cognizable by the ecclesiastical courts but there are many points of doctrine not authoritatively decided and on which clergymen may legitimately hold opinions differing from those more generally held. It is not an offence on the part of a layman.

Heriot. In feudal law, the right of a feudal lord to take a tenant's best beast or other chattel on the tenant's death. It is thought to have originated from the custom under which the lord lent horses and armour to those of his tenants who followed him in war, the horses and equipment being returned when the tenant died. Later when tenants provided their own equipment, the lord nevertheless claimed heriot. Heriot service was an incident of freehold tenure, eventually becoming a kind of rent, and heriot custom was a payment due under copyhold tenure on alienation of the holding as well as on death.

Heritable and moveable. The distinction in Scots law corresponding closely but not exactly to that between real and personal (qq.v.) in English law, heritable being those kinds of things, particularly land and all rights in and connected with land, which prior to 1964 devolved on intestacy on the heir at law, and moveable being those kinds of things which were physically moveable and devolved on the executor for division among the next of kin.

Heritable bond. In Scots law, a bond undertaking payment or repayment of money coupled with a conveyance of heritage (land) to the creditor to be held by him in security of the money due.

Heritable jurisdictions. Under feudalism one element in the feudal relation of superior and vassal was that the superior had jurisdiction over and was entitled to hold a court for the vassals who held lands of and under him, and they were subject to his judicial powers. In Scotland the heritable jurisdictions of lords and other feudal superiors survived until 1747 when they were abolished by statute.

Heritable property (or heritage). In Scots law, those kinds of property which formerly passed undivided on death to the owner's heir-at-law as contrasted with moveable property which formerly passed to the executor for division among the next of kin. (Since 1964 both kinds, with exceptions, pass to the executor.) It is further distinguished into corporeal heritable property and incorporeal heritable property; the former of these groups corresponds closely to what would be classified as physically immoveable (including land and buildings) and the latter includes those rights which are necessarily associated with land or buildings, such as feuduties and ground annuals, leasehold rights, servitudes, and rights to salmon fishings, or in their nature must pass to one heir, such as titles of honour and coats of arms. But arrears of feuduties, rents and interest on heritable securities are moveable.

So-called heritable securities, i.e. real securities secured over heritage, were originally deemed heritable but since 1868 have been deemed moveable in the creditor's succession, with certain exceptions. Whether certain bequests are heritable or moveable in succession may be affected by the doctrines of conversion and reconversion, and whether certain things are heritable or moveable may depend on the rules as to fixtures.

Heritable securities. In Scots law, real securities secured over heritable property, and constituted by bond and disposition in security, bonds of annual rent, bonds of annuity, real burdens, ground annuals, dispositions *ex facie* absolute but truly in security, and standard securities. Since 1868, by statute, certain heritable securities are in fact deemed moveable in the creditor's succession in certain circumstances.

Heritor. In Scots law, the proprietor of heritable subjects, particularly such as are liable for public burdens.

Hermaphrodite. A person or other biological organism having both male and female reproductive organs. This is extremely rare in humans. Individuals having the chromosomal constitution and reproductive organs of one sex but the secondary sexual characteristics and external appearance of the other sex are called pseudo-hermaphrodites, whereas intersexuals have physical characteristics intermediate between those of the true male and the true female, or sometimes a mixture of sexual parts or male on one side and female on the other, all stemming from chromosomal or hormone abnormality, and transsexuals have the chromosomal constitution and physical characteristics of one sex combined with the belief that they truly are, the psychological attitudes of, and the desire to be and to pass as, one of the other sex. Persons suffering from one or other of these anomalous constitutional defects may require surgical and psychological treatment, and their conditions may give rise to problems of sex-determination, naming and education, and of marriage.

Hermogenian Code. See CODEX HERMOGENIANUS.

Herschell, Farrer Herschell, 1st Lord (1837–99). He was called to the Bar in 1860 and had little practice initially, but progressed rapidly and was able to take silk in 1872. In 1874 he entered Parliament and achieved prominence there. In 1880 he became Solicitor-General and worked hard in the House and in the courts. He lost office in 1885 but, in 1886, his party was again in the majority and he became Lord Chancellor for a few months. While in opposition, in 1886–92, he took a prominent part in debates in the House of Lords and sat in the House to hear appeals. He resumed the great seal in 1892 and held office till 1895. Thereafter he served as chairman of several Royal Commissions and represented the United Kingdom in the international arbitrations relating to the Venezuela boundary and to the disputes between Canada and the U.S. on fishing rights and boundaries. His judgments are of a high order showing mastery of the relevant law and clear exposition.

Herslet, Lewis (1787–1870). Became librarian of the Foreign Office. He was responsible for *A Complete Collection of the Treaties and Conventions between Great Britain and Foreign Powers relating to Commerce and Navigation* (1820) and, with his son Edward, for *A Complete Collection of the Treaties and Conventions between Great Britain and Foreign Powers* (16 vols. 1827–85).

Hert, Johann Nikolaus (Hertius) (1652–1710). A leading German jurist of the seventeenth century, Professor of Law at Giessen, author of *Elementa prudentiae civilis* (1689), *Commentationum atque opusculorum volumina II* (1700), and many other works.

Heusler, Andreas (1834–1921). Swiss jurist and historian, professor at Basel and President of the Basel Court of Appeal. His major work was an *Institutionen des deutschen Privatrechts* (1881–86), which emphasized the importance of mediaeval German law on the modern law of Teutonic countries.

Hewart, Gordon Hewart, Viscount, of Bury (1870–1943). Worked as a journalist before being called to the Bar in 1902. He took silk in 1912 and became an M.P. in 1913. He was Solicitor-General, 1916–19 and Attorney-General, 1919–22. In 1921, for political reasons, he could not be spared from his post to take the office of Lord Chief Justice which, as Attorney-General, he claimed as of right, and the office was given to Mr. Justice A. T. Lawrence, to keep the seat till Hewart could be spared. In 1922 Lawrence was retired (it is said that he read in The Times of his own resignation) and Hewart succeeded as Lord Chief Justice. As a judge he was not a great success. In 1929 he published a book *Bureaucracy Triumphant*, inveighing against the delegated legislative and judicial powers of Ministers and departments of state, which provoked the appointment of the Committee on Ministers' Powers (q.v.). He has been characterized as perhaps the worst Chief Justice since the seventeenth century, not as being dishonest but as lacking dignity, fairness, and sense of justice. He was no jurist. Latterly his health declined, and the administration of the courts got out of hand, and he resigned in 1940.

Hewitt, James, Viscount Lifford (1709–89). Called to the English Bar in 1742, he became a judge of the English King's Bench in 1766, and went to Ireland as Chancellor early in 1768, becoming a baron and (1781) a Viscount in the Irish peerage. He remained Chancellor till his death and seems to have been a competent judge.

Hexabiblos (or *Procheiron ton nomon*). A compendium of law in six books adapted to the requirements of fourteenth century Byzantium compiled by Harmenopulos (q.v.) on the basis of the *Basilica*, and later collections and laws affecting Byzantium. It was adopted by the Greek Assembly in 1828 as the basis of Greek civil law and survived until 1941.

Hibernensis Collectio. The comprehensive collection of canons of the Irish Church, dating from about A.D. 700. It draws on the Bible, the Church Fathers, the *Statuta Ecclesiae Antiqua*, but not extensively on papal decretals, and on acts of Irish synods and Irish ecclesiastical writers. It is in 65 chapters arranged according to subject matter. It had considerable influence in Europe after the eighth century and was frequently utilized by compilers of canonical collections before Gratian.

Hide. An Anglo-Saxon term which survived into mediaeval England, commonly used for a measurement of land, roughly equivalent to a carucate (q.v.), but more properly a unit of measurement, e.g. for taxation. It was also the basis for mustering the fyrd or militia.

Hierarchy. Sacred rule, primarily the power given by Christ to His apostles and their successors to govern and rule the Church; secondarily, a term for the grades or ranks of ecclesiastical persons according to the degrees of spiritual power which they possess, which are conferred on persons by ordination; thirdly, and most generally, any system of grades or ranks, such as of courts, judges, or sources of law, in which some are superior to others who in turn are superior to others.

Higgins, Alexander Pearce (1865–1935). English international lawyer, adviser on international law to the Procurator General and Treasury Solicitor during World War I, and, from 1920–35, Whewell Professor at Cambridge. He was a member of the Permanent Court of Arbitration at The Hague and President of the Institute of International Law, 1929–31. He wrote extensively on international and maritime law and history, and was one of the founders of the *British Yearbook of International Law*.

High bailiffs. Officers appointed under statute of 1888 by the judges of county courts to attend the sittings of the courts and personally or by their assistants to serve summonses and execute warrants. The office has fallen into abeyance and has, in substance, been replaced by the registrar.

High Commission, Court of. After Henry VIII made himself Supreme Head of the Church of England in 1534 he delegated various ecclesiastical powers to commissioners. In 1535 Thomas Cromwell was appointed vicegerent with the plenitude of royal authority in ecclesiastical affairs and power to delegate. The first general commission was issued in 1549 and was followed by others. The commissioners had wide powers, to enforce the Acts of Supremacy and Uniformity, to deal generally with ecclesiastical offences and to suppress movements dangerous to the Church. They exercised their powers under the supervision, and on the instruction of the Council and procedure tended to be on civil rather than on common law lines.

As general commissions were regularly issued to the same persons they became a regular institution of the Church, and the Council began to refer to them petitions praying for relief not competent in the ordinary ecclesiastical courts, and the commission took on judicial functions. By 1570 it was regarded as a court, the title appearing about 1580. As it developed into a court, the common law judges ceased to attend.

The jurisdiction of the commissioners was practically confined to the province of Canterbury, but during most of the period 1559–1640 a northern commission sat at York, and similar commissions were issued to Ireland and, in 1579, to Wales.

Under the Stuarts the existing constitution and procedure of the court was recognized and other powers specified in detail. On occasions, provision was made for another commission to review the main commission's decisions. It also greatly increased in size, reaching 108 in 1625. Members of Commissions sat anywhere in the country where it was needed and this secured a uniform discipline and administration of ecclesiastical law.

It had close relations with the Star Chamber and the Council and like the Star Chamber the High Commission's power and efficiency made it a very popular court, which attracted business away from the ordinary ecclesiastical courts. It had an original jurisdiction in separation, alimony, immorality, heresy, and non-conformity and, from 1613, it enforced the Star Chamber rules as to the censorship of the press. Procedure had been left to its own discretion and a notable feature was the oath *ex officio*, imposed by the judge by virtue of his office. The penalties inflicted were normally fine and imprisonment. There is no warrant for the allegations that it utilized torture, mutilation, or the death penalty.

Towards the end of the sixteenth century it came under attack from the Puritans, both as to its law and its procedure. The reliance of the Stuart kings on the prerogative united against them the House of Commons, the common lawyers and the Puritans. Coke attacked it strongly and, in 1641, the relevant portion of the Act of Supremacy was repealed, the court was abolished, and it was provided that no similar court should be again set up. It was however re-established by James II in 1686 but finally condemned by the Bill of Rights of 1689.

R. G. Usher, *The Rise and Fall of the High Commission* (1913).

High Commission, Court of, in Scotland. In 1610 James VI (I of England), in furtherance of his policy of imposing episcopacy on Scotland, created two Courts of High Commission for the punishment of ecclesiastical offences, each under an archbishop with clerical and lay membership and a wide and indefinite jurisdiction. They were united in 1615 and became a byword for tyrannous oppression and coercion. In 1634 Charles I established a new court of the same name to enforce episcopacy on Scotland and gave it powers even beyond those of the older court. In 1638 he had to concede the abolition of the court, and the Glasgow Assembly of 1639 decreed its abolition.

In 1664 the Court was revived to deal with ecclesiastical offences but its powers soon lapsed and the Privy Council exercised ecclesiastical discipline.

High Commissioners. The titles of the representatives in the United Kingdom of Commonwealth countries and conversely. They rank with and enjoy the immunities of ambassadors.

High Court of Admiralty. Originally the court of the Lord High Admiral, having jurisdiction in both civil and criminal cases of maritime character, and as a court of prize. The court was abolished by the Judicature Acts, 1873–75, and the jurisdiction transferred to the Probate, Divorce, and Admiralty Division of the High Court and in 1971 to the Admiralty Court of the Queen's Bench Division.

High Court of Chancery. See CHANCERY.

High Court of Delegates. Before the Reformation an appeal lay in all cases from the English ecclesiastical courts to the Pope. In 1533, all appeals to the Pope were prohibited and it was provided that certain appeals should go to the bishop and then to Courts of Arches or Audience, thence to the archbishop. In the next year statute created the High Court of Delegates to hear appeals from the archibishop's courts, and from peculiar jurisdictions exempt from archiepiscopal control. The appeal was heard by commissioners, usually including lawyers from Doctors' Commons. In 1833 the jurisdiction was transferred to the Judicial Committee of the Privy Council.

High Court of Justice. The main constituent, along with the Court of Appeal and the Crown Court (qq.v.) of the Supreme Court of Judicature (q.v.), created in 1875 and replacing the former separate courts of the High Court of Chancery, Court of Queen's Bench, Court of Common Pleas, Court of Exchequer, High Court of Admiralty, Court of Probate, Court for Divorce and Matrimonial Causes, and the London Court of Bankruptcy. It was originally divided into five Divisions, namely, Chancery, Queen's Bench, Common Pleas, Exchequer, and Probate, Divorce and Admiralty Divisions. In 1880, the Common Pleas and Exchequer Divisions were combined with the Queen's Bench Division. In 1971, the Probate, Divorce and Admiralty Division was renamed the Family Division, contentious probate business being transferred to the Chancery Division and Admiralty business being transferred to the Queen's Bench Divisions. All jurisdiction vested in the High Court belongs to all the divisions alike and all the judges have equal authority; all divisions and courts constitute one High Court.

The judges of the High Court are the Lord Chancellor, the Lord Chief Justice of England, the President of the Family Division and a staff of puisne judges attached to the several Divisions. A judge of the Court of Appeal may sit as a judge of the High Court and a circuit judge or recorder must sit as a judge of the High Court if requested by the Lord Chancellor. The Lord Chancellor is president of the Chancery Division but may nominate one of the puisne judges to be Vice-Chancellor to manage the business of the Division. The Lord Chief Justice is president of the Queen's Bench Division and the President is president of the Family Division. Judges of the High Court may sit in other courts or tribunals as required, including the Court of Appeal, Restrictive Practices Court, and other courts of special jurisdiction.

The High Court inherited, in 1875, the jurisdictions possessed by all the courts it replaced, together with certain other courts. Its jurisdiction covers England and Wales only and is both original and appellate. Its original jurisdiction extends to all causes of action and is unlimited in amount, but in certain matters is concurrent with other courts. The exercise of the jurisdiction in particular matters is assigned to particular Divisions by rules of court. The Queen's Bench Division includes an Admiralty

Court and a Commercial Court. Its appellate and supervisory jurisdiction is exercised, in some cases, by a single judge but frequently by Divisional Courts of two or three judges in the Queen's Bench Division, Chancery Division, or Family Division. Important instances of this jurisdiction are appeals to the Divisional Court of the Q.B. Division from inferior courts or tribunals and applications for the issue of the writ of *habeas corpus* (q.v.) or an order of mandamus, prohibition or *certiorari* (qq.v.).

Apart from particular instances, such as appeals to a Divisional Court of the Queen's Bench Division from magistrates' courts, the jurisdiction is a civil one only, extending to actions founded on contract or tort, commercial or property rights or matrimonial or family rights. The remedies granted include damages, injunctions, declarations of rights, dissolution of marriage, and others.

See also APPEAL, COURT OF; CROWN COURT; HOUSE OF LORDS; SUPREME COURT OF JUDICATURE.

High Court of Justice (Northern Ireland). A branch of the Supreme Court of Judicature of Northern Ireland (q.v.), established in 1920 and restructured in 1978. It is the superior civil court of first instance and exercises, in relation to Northern Ireland, the jurisdiction of the former High Court of Justice in Ireland.

In consists of the Lord Chief Justice of Northern Ireland and five judges of the High Court.

It is divided into Queen's Bench, Chancery, and Family Divisions. The Queen's Bench Division has jurisdiction on the Crown side in respect of the special remedies of *habeas corpus*, mandamus and *certiorari*, and on the civil side in all common law, bankruptcy, probate, matrimonial and Admiralty proceedings. The Chancery Division deals with trusts, estates, wills, partnerships, companies and related matters.

High Court of Justiciary. The superior criminal court of Scotland, instituted in its present form in 1672 and replacing the Court of the Justiciar. It consists of the Lord Justice-General of Scotland as president, the Lord Justice-Clerk as vice-president, and the Lords Commissioners of Justiciary, who are the same persons as hold the offices of Lord President of the Court of Session, Lord Justice-Clerk, and Lords of Council and Session in the supreme civil court, the Court of Session (q.v.).

It exercises both original and appellate jurisdiction and its jurisdiction extends over all Scotland.

The original jurisdiction extends to all crimes and offences and is exclusive in respect of treason, murder, rape, official secrets, breach of duty by magistrates and a few other crimes, and concurrent with lower courts in respect of all other crimes, unless statute specially prescribes. This jurisdiction is exercised at sittings on circuits to the major towns in Scotland, normally by a single judge and always on indictment before a jury. The court has unlimited power of sentence, save as restricted by statute.

The appellate jurisdiction is exercised in sittings at Edinburgh only, normally of benches of three judges presided over by the Lord Justice-General or Lord Justice-Clerk, though a larger bench may be convened in cases of difficulty or importance. It takes two forms. A bench of the High Court may sit as Court of Criminal Appeal to hear appeals against conviction or sentence from persons convicted after trial on indictment in the High Court or Sheriff Court. A bench of the High Court also sits to hear appeals by the Crown or person convicted against decisions of the Sheriff Courts, District Courts or stipendiary magistrates in the exercise of their summary criminal jurisdiction. No further appeal lies to the House of Lords. Akin to appellate jurisdiction is the *nobile officium* or extraordinary equitable power of the High Court to provide a remedy in exceptional circumstances for which there is no express power.

High Court of Parliament. See PARLIAMENT.

High seas or open sea. All parts of the seas covering the surface of the world except those parts included in the internal waters or territorial sea of any state, but including contiguous zones (q.v.) and the waters above the continental shelf (q.v.). By custom or acquiescence seas virtually land-locked are sometimes treated as parts of the high seas, e.g. the Baltic and Black seas. Waters land-locked, however large, are part of the territory of a state or states. In the Middle Ages and down to the seventeenth century, claims to sovereignty over parts of the high seas were made. Since about 1800 it has been accepted that the high seas are not open to acquisition or claim to sovereignty over them by any state or states, nor may any state prevent passage by the ships of any other. In modern practice, the freedom of the high seas comprises freedom of navigation on, of fishing in, of laying submarine cables and pipelines under, and of flying over, the high seas, but most of these matters are regulated in detail by multilateral conventions. A ship cannot be arrested on the high seas even for violation of municipal law within a territorial jurisdiction but can be arrested for piracy or other illegal acts, or on suspicion of slave-trading.

Each state must have a maritime flag and lay down the conditions under which vessels may fly that flag, and require ships sailing under that flag to carry certain ship's papers. Public ships at sea are deemed to be floating parts of their home states and private ships to be floating parts similarly only while under the exclusive jurisdiction and protection of the flag state. Warships have a general right to verify the nationality of any merchant ship it may meet. International agreements have provided for rules for signals and the avoidance of collisions and for the safety of life at sea, and for other similar purposes.

In time of war, any state may blockade enemy ports and territorial sea and visit and search neutral vessels for contraband or blockade-running. Warships may, in case of suspicion, require any ship on the high seas to show her flag, and may stop her to verify that flag, or pursue from the territorial sea into the open sea and bring back any ship which has violated the law of the littoral state while in its territorial waters.

No state may appropriate or take possession of vessels abandoned or goods abandoned at sea, though a salvor need not restore them without payment of salvage reward.

International conventions have been entered into designed to prevent pollution of the sea by oil, or by radiation, as by dumping of radio-active waste, and to promote conservation of stocks of fish and protect them from over-exploitation.

High Sheriff. A person appointed annually for every county in England and Wales. The office is a compulsory one of great antiquity and importance, and a person appointed is bound to serve. (In practice persons are asked beforehand if they are willing to serve.) Three fit persons are nominated for each county at a meeting at the Royal Courts of Justice by the Lord Chancellor, the Chancellor of the Exchequer, the Lord President and others of the Privy Council, and the Lord Chief Justice, or any two or more of them with any two or more of the judges of the High Court. The names are engrossed on parchment and presented to the Queen in Council and the Queen appoints the persons to serve by pricking with a bodkin one name for each county. The sheriffs of the counties of Greater Manchester, Merseyside, and Lancashire are appointed by the Queen in right of the Duchy of Lancaster from a list submitted by the Chancellor of the Duchy. The sheriff of the county of Cornwall is made by the Prince of Wales as Duke of Cornwall. The person appointed must have sufficient land within the county to answer the Queen and her people and, in practice, the sheriff must be a person of considerable means. Certain persons are exempt from liability to serve. Persons appointed receive warrants of appointment from the Privy Council. During the term of his office the sheriff, as keeper of the Queen's Peace, is the first man in the county and superior to any nobleman therein. The Corporation of London annually elects two sheriffs for the city.

Most of the sheriffs' ancient civil and criminal jurisdiction has been transferred to the regular courts but they still have some judicial and important ministerial powers and duties.

Every sheriff appoints an under-sheriff, who may be reappointed by the next high sheriff and who performs most of the functions of the high sheriff, and bailiffs or sheriff's officers for the purpose of collecting fines and executing writs and processes.

As conservator of the Queen's Peace it is the sheriff's duty to suppress unlawful assemblies and riots, apprehend offenders, and defend his county against invasion by the Queen's enemies. It was formerly his duty to pursue and arrest felons in the county and for that purpose to raise the hue and cry (q.v.). He was formerly bound to receive, make arrangements for, and attend on the judges of assize when they visited the county.

Writs of execution on judgments and orders of the Supreme Court are directed to the Sheriff and it is his duty to execute them.

High treason. Formerly distinguishable from petit or petty treason, but since the latter was abolished in 1828, known simply as treason (q.v.).

Higher law. The concept of a rule more authoritative than mere prescriptions of positive law. It comprehends the eternal law and the law of nature as recognized by Aquinas and, down to the seventeenth century, was vaguely recognized in England. The omnipotence of Parliament and the supreme authority of its legislation was not the accepted doctrine in the time of Bracton, or Coke, or Holt. It was adopted, and even then with lip-service to natural law, in Blackstone. Even in the seventeenth and eighteenth centuries, when the doctrine of the supremacy of Parliament was asserted, apparently without reservation, it was often accompanied by the qualification of the overriding authority of some higher law. In *Calvin's case* (1608), and *Bonham's case* (1610) Coke qualified his statement of the supremacy of Parliament by making the interpretation of statutes subject to an overriding law of nature. There are later instances, too, before the principle of the unqualified omnipotence of Parliament came to be accepted by the end of the seventeenth century. The nature of this higher law is nowhere very clearly set out, but it is pretty clearly the major principles of the law of nature, those which are today more clearly formulated in statements of human rights.

J. W. Gough, *Fundamental Law in English Constitutional History* (1955); E. S. Corwin, *The Higher Law Background of American Constitutional Law* (1928–9).

Highness. The title from the twelfth century of kings of England. Henry VIII adopted the title Majesty and after both had been in use for about a century, James I adopted the latter term for himself,

Highness and Royal Highness being used of the remoter and nearer members of the Royal Family. Since 1917, the title Royal Highness is confined to the sovereign's children, the children of any of his sons and the eldest living son of the eldest living son of the Prince of Wales, other descendants of any sovereign not being entitled to be Royal Highness, Highness, Prince or Princess. The title of Royal Highness is borne by the Duke of Edinburgh.

Highway. A highway is a path over which all members of the public have liberty to pass and repass for business or pleasure, and includes footpaths, bridle paths, driftways or drove-roads, and carriage roads. The public right is, at common law, one of passage only and does not imply ownership of the soil but in new roads the soil is frequently acquired by the road authority. The existence of a public highway is established by proving dedication of the highway to the public and acceptance of the highway by it, or under statute. Statute introduced presumptions of dedication by use for periods of 20 or 40 years. A highway cannot be extinguished or diverted by the public releasing their rights, but many statutes, in particular the Highway Act, 1835, conferred powers to extinguish the right.

At common law in England, the liability to repair highways was on the inhabitants at large or on the parish. The first general legislation, the Highway Act, 1835, imposed this duty on the surveyor of highways, but it was later transferred to borough and urban district councils, and in rural areas, to highway districts under highway boards and, subsequently, to rural district councils and later to county councils. The main roads were from the eighteenth century mainly controlled by various turnpike trusts, which charged tolls from users under statutory authority, but after 1878 certain roads were transferred to county quarter sessions and, from 1888, to county and county borough councils. In the twentieth century main roads were placed under, first the Road Board, then the Ministry of Transport, and then the Department of the Environment, and trunk roads and special roads were established. At common law, the inhabitants at large were not liable for injury sustained in consequence of non-repair of a highway, but were for misfeasance or negligence in repairs done, but this rule was abrogated in 1961. Highway authorities have now extensive statutory powers to widen and improve highways. Unlawful obstruction of highways is a public nuisance.

In Scotland, a highway is similarly a public right of passage over the lands of a person. Early legislation established a statute-labour system imposing on local residents the liability to give an annual quota of labour to repair of the roads. From 1713 turnpike roads developed, built with borrowed money and maintained by tolls taken at specified places along the roads, and managed by a trust established by statute. A comprehensive national system was introduced in 1878, vesting the highways in county road trustees, subsequently replaced by county councils and, in the case of trunk roads, the Secretary of State for Scotland.

Highway code. A code issued from time to time by the Department of the Environment containing rules for good conduct and safe driving. Failure to observe the rules is not an offence, nor tortious negligence, but is a factor which may be taken into account to establish or negative liability, civil or criminal.

Highwayman's case, *Everet* v. *Williams* (1725). Reported in (1893) 9 L.Q.R. 197. Everet brought a Bill in the Equity side of the Exchequer against Williams, narrating that he was a dealer in certain commodities, that they entered into an oral partnership and had dealings with various parties but that the defendant would not come to a fair accounting with the plaintiff touching the partnership. In truth, both parties were highwaymen and the dealings with various parties were by way of highway robbery. The court held the bill to be scandalous and impertinent, the plaintiff's solicitors were attached for contempt and fined, the defendant was executed at Maidstone in 1727, the plaintiff at Tyburn in 1730, and one of the plaintiff's solicitors was later convicted of robbery and transported. Knowledge of the case was maintained by tradition and it was sometimes thought to be fictitious but in 1893 it was established that it was true.

Highwaymen. Persons who committed robbery by stopping coaches at isolated places and robbing the travellers. Their heyday was the first half of the eighteenth century and, at the time, considerable glamour attached to their exploits; among the most notable were Dick Turpin and Claude Duval, who were both caught and hanged. Their activities were stopped by the introduction of mail-coaches, the increase of traffic, the extension of the turnpike-road system, and the increased speed of travel.

Hijacking. The offence of, while aboard an aircraft, unlawfully, by force or threats of any kind, seizing it or exercising control of it. It has been frequently done, usually by political extremists seeking to attain some political end. An international Convention for the Suppression of Unlawful Seizure of Aircraft was agreed at The Hague in 1970 and given effect to in Britain by the Hijacking Act, 1971.

Hilary rules. A new set of rules of pleading drawn up by the judges under the Civil Procedure Act, 1833, and coming into force in Hilary Term,

1834. The rules sought to steer a middle course between the extreme precision of the older special pleading (q.v.) and the vagueness of the general issue, merely denying the plaintiff's allegations. The result of the rules was, unfortunately, to extend the necessity of conforming to the very strict rules of special pleading to many cases previously not affected by it, and this, in turn, emphasized changes which had developed in substantive law. Thus, the distinction between different forms of action was emphasized anew. In consequence, the development of pleading had to be restricted by subsequent Common Law Procedure Acts until the Judicature Acts, 1873–75, and the Rules of the Supreme Court, made thereunder, substituted a new and more informal system.

Hill, Matthew Davenport (1792–1872). Brother of Sir Rowland Hill of penny postage fame, reported in the House of Commons, worked as a journalist, and was called to the Bar in 1819. He appeared for the defence in many political trials, became friendly with Bentham, and entered Parliament after the Reform Act of 1832. He took silk in 1834 and became Recorder of Birmingham (1839–65) and bankruptcy commissioner for Bristol district (1851–69). As such, he helped greatly to reform the criminal law and his views, expressed as charges to grand juries, were published as *Suggestions for the Repression of Crime* (1857). He devoted much attention to the problem of juvenile delinquents. A brother, Frederic Hill (1803–96) was first inspector of prisons in Scotland, remodelled the gaol system there and wrote *Crime: Its Amount, Causes and Remedies* (1853).

Hilliard, Francis (1806–78). Practised law for some years but devoted most of his life to writing legal textbooks, of importance in showing how far English common law had been followed or modified by American courts, and, by citing cases from all states, encouraging the development of general rather than local doctrines. His books included works on real property, personal property, mortgages, vendors and purchasers, torts, contracts, bankruptcy, injunctions and *American Law: A Comprehensive Summary of the Law in its Various Departments* (2 vols., 1877–78). His book on *Torts* (1859) was the first treatise in English on the subject. His work, though useful, was, however, inferior to that of Story in power of analysis and criticism.

Hindu law. A specimen of the class of religious laws followed not by the citizens of a particular state, but by the adherents of a particular religion. It is the personal law of members of the Hindu community in India, East Africa, and South-east Asian countries, for whom religious precepts play some of the part which, in Western societies, is undertaken by law.

The sources of Hindu law are in very ancient religious texts, the *Smrtis*, which include the *Rig-Veda*, the *Vedangas* and the *Upanishads*. These are conceived as the source of all knowledge and, in addition to purely philosophical and religious matter, describe principles of organization, morality, and behaviour. A full understanding of the teachings of these books requires reference to the *Sastras*, which teach men how to conduct themselves in a way pleasing to God, which is the science of *dharma*. The purpose of *dharma* is to set out what a man must do if he is to be righteous, to avoid losing caste, or has thought for his after-life. There are many treatises on the *dharma*, the *dharmasastras*, which include the 'laws' of Manu, Yajnavalkya, Narada, Brihaspati, and Katyayama, believed to have been written between 100 B.C., or even earlier, and about A.D. 400, and themselves the product of attempts to collate and systematize earlier legal propositions and maxims, and commentaries thereon, and also the *nibandhas*, by various writers, dating from the twelfth to the seventeenth centuries, the purpose of which is to elucidate the *dharmasastras* and reconcile contradictions between them. The *dharmasastras* do not constitute revealed truth; they are private compilations and their authority is traditional.

Life in the world is regulated by *dharma*, and also by expediency (*artha*) and pleasure (*kama*) and these three together produce custom. Positive Hindu law is a customary law dominated by Hindu religious doctrine.

There are two main schools of Hindu law, that of Mitakshara (which is subdivided into four) and that of Dayabhaga, both attached to particular areas, which have preferences for particular *dharmasastras*. Furthermore, different principles may apply to persons according to their caste, of which there are two thousand, each having its own customary rules.

Despite first the Muslim and later the British supremacy in India, Hindu law survived, first as religion and custom, and later as law applicable particularly to personal and family relations, though even in this respect this represented a restriction of the ambit of Hindu law, in that law on Western lines came to apply to other relationships.

British judges in India were hampered by inability to discover the principles of Hindu law and long invoked the aid of the pandits, religious teachers, to interpret the relevant rules. Latterly popular customs were collected and applied, but English judges and jurists, in their ignorance, frequently seriously distorted Hindu law, and sometimes imported ideas from English law with similar effect. The effect of British rule was, generally, to limit the sphere of Hindu law to family organization, land tenure, and succession. The resulting amalgam of

Anglo-Hindu law was expounded in various treatises, as by J. D. Mayne and D. F. Mulla.

Since Indian independence in 1947 the caste system has been rejected and parts of a Hindu code enacted, which in many respects has made fundamental changes, and the tendency in India is to replace law related to religion by secular law on western lines.

P. Kane: *History of Dharmasastra* (1930–62); N. Sen-Gupta, *Evolution of Ancient Indian Law* (1962); J. Derrett: *Hindu law past and present* (1959); *Introduction to Modern Hindu Law* (1963); N. Raghavachariar, *Hindu Law, Principles and Precedents* (1965).

Hinojosa y Naveros, Eduardo de (1852–1919). Spanish legal historian. His major works were *Historia general del derecho español* (1887), *El elemento germanico en el derecho español* (1915), and *El regimen señoriel y la cuestion agraria en Cataluña durante la edad media* (1905).

Hinschius, Paul (1835–98). German jurist and member of the Reichstag. His major works are the *Decretales Pseudo-Isidorianae et Capitula Angilrami* (1863), the first critical edition of the False Decretals, now challenged, and *Das Kirchenrecht der Katholiken und Protestanten in Deutschland* (1869–77), a major study of ecclesiastical government.

Hippolytus de Marsiliis (1450–1529). Italian jurist, specializing in criminal law, author of a *Practica causarum criminalium* (1528).

Hire-purchase. A hybrid contract developed in the late nineteenth and early twentieth centuries to enable persons to have at once the use of goods which they wished to buy but could not at once pay for completely. The essence of the contract is that the owner of goods lets them on hire to another, at a rental which provides for payment of the full price by an initial payment followed by instalments over a period with allowance for interest on the unpaid balances, and also confers on him an option at the end of the hiring to purchase the goods outright for a final payment. It differs from hiring (q.v.) in which the right of property in the goods never passes, from credit sale, in which there is a sale and the right of property passes at once but no payment is then made, and from sale with payment by instalments, in which there is a sale and the right of property passes at once but payment is made by instalments. The increasing popularity of hire-purchase resulted, for a time, in government control of minimum initial payments and maximum periods for payment of the balance of the price. The frequency of hire-purchase dealers taking advantage of individuals has resulted in complicated legislation controlling advertising and dealing with hire-purchase, credit-sales, and conditional (instalment) sales within certain financial limits, designed to protect purchasers against dealers. In modern practice, the transaction is for commercial reasons frequently tri-partite, the dealer in the goods selling the goods outright to a finance company which then hires them, under a hire-purchase agreement, to the customer. The transaction may be regarded as, in essence, one of moneylending, the price being advanced to the customer to pay the dealer and recovered with interest by instalments, the finance company retaining a security right over the goods though not having possession thereof. The finance company may require the hirer's undertakings to be guarantee by a surety, either by a separate guarantee, or incorporated in the hire-purchase agreement.

Agreements within the financial limits of the legislation contain statutory stipulations as to title, the quality and fitness of goods and protections for the hirer who has to discontinue the agreement before completing the purchase.

The hirer must take reasonable care of the goods, and notify the owner of the place where he keeps them. He may not, in general, assign his interest in the agreement, and cannot sell or pledge the goods.

In the event of breach of contract by the hirer the owner may, at common law, retake possession of the goods, but this right is restricted by statute and possession is recoverable by action only. If the hirer terminates the hiring prematurely provision is usually made for payment by him to ensure that a minimum proportion of the total price has been paid. Hire-purchase contracts are now elaborately regulated by the Consumer Credit Act, 1974, and relative subordinate legislation.

R. Goode, *Hire-Purchase Law and Practice*; A. Guest, *Law of Hire-Purchase.*

Hiring. The contract whereby one person lets to another, for consideration in money or money's worth, the possession and use of some goods for a period or for a particular purpose. In English law it is considered a variety of bailment, in Scots law a variety of *locatio-conductio*, and it corresponds to the *locatio-conductio* of Roman law. The right of property remains vested in the owner (lessor) and the hirer does not acquire, and can never exercise, rights of ownership. The lessor must put the hirer in possession, and maintain him in the use and enjoyment of it, it must be reasonably fit for the purpose for which it is hired, and he must keep it in repair. The hirer must take reasonable care of the thing hired, not put it to any forbidden or exceptional use, pay the hiring and restore it at the termination of the contract in the same condition as received, fair wear and tear excepted.

In the broader sense, hiring or letting to hire comprises letting a thing on hire (*locatio rei*); letting of service, or employment (*locatio operarum*); letting of work to be done, or the doing of services (*locatio operis faciendi*); letting to hire of care and custody (*locatio custodiae*); and letting to hire of the carriage of goods (*locatio operis mercium vehendarum*).

See also BAILMENT.

Hispana Collectio. The most extensive and important collection of canon law down to its time. It appeared in Gaul early in the eighth century and spread rapidly. It comprises all the traditional legislation of the Church, conciliar and decretal. The conciliar part has suffered three recensions, the Isidorian, to 633, the Ervigian, to 685, and the Vulgate, to 694. It was, in substance, the official collection of the Spanish Church, superseding all previous ones. It was copied and combined with other collections, such as the Hispana–Adriana and the Dacheriana, and utilized to support Carolingian attempts at reform of the Church. The *Hispana Chronologica* recension was the basis of the *Excerpta*, a collection of summaries of the canons of the Spanish councils, made about 656 and the *Hispana Systematica*, in which summaries were substituted for the complete texts of the canons.

Hispana Versio. A canonical collection of the fifth century, known in Africa as the *Corpus canonum Africanum*, thought to have been the work of Dionysius Exiguus, and found in three versions, the *Antiqua-originalis*, the *Vulgata*, and the *Isidoriana-gallica*.

Historic bays and waters. In international law, by general acquiescence, bays may become a part of the internal waters of a state even though the closing line of the bay extends the limits permitted by the general law. Thus Canada claims Hudson Bay though the mouth is 50 miles wide. The concept of historic title may apply also to the territorial sea.

Historic rights. A state's title to territory created in derogation of international law through historical means whereby a state has asserted a jurisdiction originally illegal but this has come to be acquiesced in by other states generally. The state's claim is validated only by custom expressing universal or at least general consent and by absence of protest or counter-claim. Thus in the *Minquiers and Ecrehos* case (1953) title was held established after examination of historical facts and without regard to the time for which the U.K. had exercised sovereignty over the islands.

Historical interpretation. Interpretation of a treaty or statute by reference to the circumstances existing when it was entered into or passed and to the circumstances leading up to it.

Historical jurisprudence. The general term for the attitude of those jurists who see law primarily an historical development. Vico, Montesquieu, Hegel, Kohler, and others all developed in different ways a legal philosophy from a philosophy of history. But the term is usually associated with the movement which invoked history in the name of tradition, custom, and national spirit as the major factor in legal development in opposition to conscious law-making, a movement always associated with the name of Hugo, Puchta and, particularly, Savigny. Their school of thought was a reaction against eighteenth century rationalism with its belief in natural law, reason, and the establishment of a legal theory by deduction without regard to historical fact and national peculiarities, and against the spirit of the French Revolution and its revolt against tradition and authority.

Montesquieu was the forerunner of the school in collecting comparative materials on the laws of different peoples, and deducing the dependence of law on natural and social factors, including the 'spirit of the nation'. Edmund Burke originated the main beliefs of the school by asserting that law could only be the produce of a gradual and organic development. Hugo insisted on the non-rational elements in a legal system which militated against deliberately-made rules. But the precipitating factor was the question of codification of German law after the Napoleonic wars. Thibaut advocated the unification and rationalization of the variety of laws then applicable in different parts of Germany and the codification of German law. Savigny opposed this: his basic doctrines were that law was found, not made, and legislation was less important than custom, that as law became complicated popular consciousness was not manifested directly but through lawyers who put into technical shape the feelings of the people, and that each people developed its own legal system and practices as it developed its own speech, manners and habits. Savigny stressed the need for the scientific study of the development of a system, whereby each generation adapted the law to its needs, and he saw in the development of Roman law a model for juristic influence in developing a body to meet changing times and needs. To him legal science was better than law reform or legislation.

In Germany the movement influenced a revival of Germanistic influence in the law, as opposed to Romanistic and excessive reliance on the Roman law received in Germany, led by Beseler, Eichhorn, and Gierke. National socialism adopted some ideas from the historical school, but perverted them. In Britain, the movement influenced Maine's application of comparative studies to a legal theory based on the principle of historical evolution. In the U.S., there was a parallel to the Thibaut-Savigny controversy in the Carter-Field controversy on codification. The historical school's attitude is hardly a theory of law, but it has been very influential, in promoting the study of legal history, in emphasizing the influence of custom, practice and feelings of right in the development of legal systems, and in leading on to modern sociological theories.

J. C. Carter, *Law, Its Origin, Growth and Function*; H. Maine, *Ancient Law*; R. Pound, *Interpretations of Legal History*; P. Vinogradoff, *Historical Jurisprudence.*

History. See HISTORICAL JURISPRUDENCE; LAW AND HISTORY; LEGAL HISTORY.

Hittite law. The Hittites were a great power in central and southern Asia Minor about the middle of the second millennium B.C. There exist, on clay tablets, treaties, administrative decrees, state documents and a body of Hittite laws. The main tablets, containing laws, deal with a wide range of civil and criminal topics, and different versions indicate stages of development and amendments. Some deal with special cases and may be based on decided cases. The tablets are really a collection of judicial formulae and decisions. Some bear to be revisions of older rules. There was probably in addition a body of customary law. We have little knowledge of courts and judges though there is evidence of very detailed inquiries into disputed facts. Restitution is more important that retribution, and there were few capital offences. Collective responsibility attached to a family for wrongs. Though there is some attempt to put the rules in order there is no evidence of legal science or thinking.

J. Garstang, *The Hittite Empire* (1929); J. Gurney, *The Hittites* (1952); E. Neufeld, *The Hittite Laws* (1951); F. Hrozny, *Code hittite* (1922); E. Cuq, *Etudes sur le droit babylonien, les lois assyriennes et les lois hittites* (1929).

Hlothere (?–865). King of Kent. He seems to have shared power with his nephew Eadric and a code in their joint names is one of the oldest and most important sources of knowledge of Anglo-Saxon law. It is an enlarged and revised version of the code of Aethelberht and reflects a primitive Germanic social organization, laying heavy emphasis on crime and penalties.

Hobart, Sir Henry (?–1625). Called to the Bar in 1584, he became Attorney-General in 1606 and succeeded Coke as Chief Justice of the Common Pleas in 1613. He was regarded a learned judge and a volume of his reports covering 1603–25 was later published.

Hobbes, Thomas (1588–1679). English philosopher. After leaving Oxford he became friend and tutor to William Cavendish, later second Earl of Devonshire, and latterly to his eldest son. He travelled extensively and met many of the great thinkers of the age. He was, for a time, intimate with Francis Bacon and his legal ideas were probably influenced by Bacon and by Bodin. He was also intimate with Selden. Though not a lawyer, he displays extensive knowledge of English legal literature and law. In 1640 he left England and about this time, stimulated by the constitutional struggle, completed a treatise on *The Elements of Law, Natural and Politique* (pub. 1650), in which he urges a social contract view of government; sovereignty is derived from the people. His work *De Cive* was completed in 1641 but not published till later (1642, 1647, 1651). While in exile (1640–1660) he wrote *Leviathan or The Matter, Form and Power of a Commonwealth, Ecclesiastical and Civil* (1651), a comprehensive statement of his view of sovereignty. He was influenced by the sectarian animosities of his time to think of the subordination of ecclesiastical to secular authority. States, he held, had been formed as the only alternative to the state of nature which, in his view, led to anarchy. The supremacy and unity of the sovereign power was therefore the essential condition of civilized life and to this even moral law must be subordinated. Law is dependent on sanction. Though he retained the ideas of natural law and natural right he defined natural law with reference to self-preservation as the only guiding principle. Later in life, he devoted more attention to mathematics and published extensively on this topic. Later he wrote *Behemoth: the History of the Causes of the Civil Wars of England* (1679) and the unfinished *Dialogue between a Philosopher and a Student of the Common Laws of England* (1681), a criticism of the constitutional theory of English government upheld by Coke. He was a founder of modern psychology and of sociology and aimed to give a complete account of matter, man and the State. He distinguished law and right, which he considered mutually opposed, and natural law from natural rights, and was the predecessor of later analytical and positivist legal philosophers, such as Austin, who made a modified version of Hobbes's views the basis of his jurisprudence. The legal order derived authority from politically organized society and law was a rule for which the state had commended obedience.

Leslie Stephen, *Hobbes*; H. Warrender, *Political Philosophy of Hobbes.*

Hobhouse, Arthur, Baron Hobhouse of Hadspen (1819–1904). Called to the Bar in 1845, he practised in Chancery and in 1866 became a charity commissioner, and was legal member of the council for India in 1872–77. In that office he made less progress than his predecessors with codification and consolidation of legislation. In 1881 he became a member of the judicial committee of the Privy Council, in which capacity he delivered many judicial opinions, and, in 1885, he was made a peer to assist in the judicial work of the House of Lords but only by statute in 1887 was he, as a P.C. who had not been a judge of a U.K. court, entitled to sit in the Lords. All his judicial work was done gratuitously. He retired in 1901. He devoted much energy to local government in London. As a judge he was careful and painstaking.

L. T. Hobhouse and J. Hammond, *Lord Hobhouse, A Memoir.*

Hodson, Francis Lord Chorlton Hodson, Lord (1895–). Was called to the Bar in 1921, became a judge of the P.D.A. Division in 1937, a Lord Justice of Appeal in 1951, and a Lord of Appeal in Ordinary in 1960.

Hoen, Philipp Heinrich von (Hoenonius) (1576–1640). German jurist, teacher at Herborn, and author of *Disputationum politicarum liber*, a work on public law, and numerous works on private law.

Hofacker, Carl Christoph (1743–93). German jurist, author of *Institutiones juris romani methodo systematica adornatae* (1773), *Principia Juris Civilis Romani-Germanici* (1788–94), and other works.

Hog, Sir Roger, Lord Harcarse (?1635–1700). Passed advocate in 1661 and became a judge of the Court of Session in 1677, and of the Court of Justiciary in 1678. In 1688 he was removed from the bench by James II. He compiled a *Dictionary of Decisions, 1681–1692* (1757).

Hogg, Douglas McGarel, 1st Viscount Hailsham (1872–1950). Initially spent years in British Guiana managing a sugar estate, then served in the Boer War, was called to the Bar in 1902 and made rapid progress. In 1922 he was appointed Attorney-General and entered Parliament, being at once recognized as a powerful debater. After his party was out of office for 10 months in 1924 he resumed the office and entered the Cabinet. As such, he had much to do during the General Strike of 1926. In 1928 he succeeded as Lord Chancellor and became Lord Hailsham (Viscount Hailsham, 1929) but was out of office between 1929 and 1931, when he became Secretary of State for War in the National government. In 1935 he again became Chancellor. As such, he presided as Lord High Steward at the trial before the House of Lords of Lord de Clifford, the last occasion of the trial of a Lord by his peers. In 1938 he resigned on account of health but was made Lord President of the Council for a few months. As a judge, he was sound, careful and satisfactory but he did not have a deep interest in legal principles and cannot be deemed outstanding.

Hogg, Quintin McGarel, Baron Hailsham of St. Marylebone (1907–). Son of Viscount Hailsham, he had a brilliant career at Oxford. He was called to the Bar in 1932, and sat in Parliament, 1938–50. He was First Lord of the Admiralty, 1956–57, Lord President of the Council, 1957–59 and 1960–64, and Lord Privy Seal, 1959–60. He succeeded, unwillingly, to the Viscounty of Hailsham in 1950 but disclaimed that peerage in 1963 under the Peerage Act and was re-elected to the Commons. In 1970 he was made Lord Chancellor and received a life peerage, as Baron Hailsham of St. Marylebone. He published *The Door Wherein I Went* (1975) and *The Dilemma of Democracy* (1978).

Hohfeld, Wesley Newcomb (1879–1918). After a short period in practice he taught law at Stanford and Yale. He is remembered for the posthumously collected papers entitled *Fundamental Legal Conceptions as Applied in Judicial Reasoning* (1919) pointing out the confusion arising from the use of some legal terms, such as right and duty, having multiple or indefinite connotations without proper differentiation, and stressing the need for precise and accurate use of terms in legal discourse and the analysis of the different senses of similar terms.

He worked out jural relationships as follows:

Jural correlatives	Right	Privilege	Power	Immunity
	Duty	No-right	Liability	Disability
Jural opposites	Right	Privilege	Power	Immunity
	No-right	Duty	Disability	Liability

and this has been developed by later jurists.

The significance and utility of his analysis has been increasingly recognized since his death and, despite some criticisms of the terminology, it is now a standard part of legal thinking.

Hold. To have certain lands as tenant. Also used of a court, meaning to have arrived at a particular view on a point of law. Thus, a report of a case frequently states: '*Held* that...', and this view is the court's 'holding'. It is said that, strictly, a court 'holds' but a single judge 'rules'.

Holder. A payee or indorsee of a bill of exchange or promissory note who possesses the bill or note.

Holder in due course. A holder who has taken a bill of exchange, cheque, or promissory note, complete and regular on the face of it, who became the holder of it before it was overdue, and without notice that, if such be the case, it had been dishonoured, and took it in good faith and for value, having, at the time it was negotiated to him, no notice of any defect in the title of the person who negotiated it to him.

Holding. The land let to a tenant; in Scots law also the tenure or kind of right given by the superior to the vassal, as feu-holding or blench-holding; in relation to a judgment of a court, the view adopted on a point of law disputed in the case.

Holding company. A company which is a member of and controls the composition of the board of directors of another company, its subsidiary, or holds more than half in nominal value of the subsidiary's equity share capital and can accordingly control it by votes at general meetings.

Holding out. Conduct whereby a person represents, or allows it to be thought, that he has a particular status, position, or authority, such as of agent for another, partner of another, or otherwise. Where a person is deemed to have held himself out to be a person of a particular kind, or to have a particular authority, he is held to be estopped or barred from later denying those facts and incurs the liabilities which he would have done if his representation had been true. Thus, a person who is not a partner may, by holding himself out to be a partner, incur liability to third parties as though he actually were a partner.

Holding over. A lessee's keeping possession of land after the expiry of his term, whereby he either becomes a trespasser and liable to be ejected, or, if the landlord consents, becomes a tenant at will or a licensee, tolerated on sufferance.

Holiday. Originally holy day, a feast day on which work cannot be exacted. Traditionally, and by old statutes, all Sundays, Good Friday, Christmas Day, and certain other days have been holidays. Holidays now comprise Sundays, days appointed by the Banking and Financial Dealings Act, 1971, as Bank holidays, days appointed as public holidays, weekly half-holidays in certain trades in lieu of Saturday afternoons free, and periods in each year when by custom and statute employees are entitled to be absent from work.

Holdsworth, Sir William Searle (1871–1944). Read both history and law, and was also called to the Bar. He was Professor of Constitutional Law at London, concurrently with holding an Oxford Fellowship, 1903–8, and Vinerian Professor of Law at Oxford, 1922–44. He sat on the Indian States Committee and the Committee on Ministers' Powers. Like Blackstone, he had immense learning and great industry. He received many honours including the Order of Merit. He wrote *The Sources and Literature of English Law* (1925), *An Historical Introduction to the Land Law* (1927), *Some Makers of English Law* (1938), *The Historians of Anglo-American Law* (1927), *Some Lessons from our Legal History* (1928), *Charles Dickens as a Legal Historian* (1928), and a large number of papers in legal periodicals, some reprinted in *Essays in Law and History* (1946). He edited three volumes for the Selden Society. His *magnum opus* is the *History of English Law*, originally in three volumes, but as rewritten, and finally completed by his literary executors, in 16, giving a complete picture of the development of English law from the Conquest to 1875. It deals with sources and literature, courts and institutions, judges and writers, substantive and adjective law, and is a mine of information, wholly superseding all previous work, save that of Pollock and Maitland in the period prior to 1300.

H. G. Hanbury, *Vinerian Chair*, 186.

Holiness, Code of. The collection of secular, ritual and moral regulations in Leviticus 17–26, so named because of the concern of these rules for sanctity. It includes regulations as to sacrifices, ritual cleanliness, priestly regulations, blasphemy, sexual regulations, sacrifices to Moloch, days to be held holy, and laws concerning the sabbatical or jubilee year in which Israelite slaves were to be freed and interest was to be prohibited.

Holland, law of. See NETHERLANDS, LAW OF.

Holland, Sir Thomas Erskine (1835–1926). Was a descendant of Lord Chancellor Erskine. At first he taught philosophy in Oxford but was called to the Bar in 1863 and practised for some years. In 1874 he was elected Vinerian Reader in English Law and later in the same year Chichele Professor of International Law and Diplomacy. He held this chair till 1910. His inaugural lecture on Alberico Gentili (q.v.) revived interest in that jurist, and in 1877 he published an edition of Gentili's *De Jure Belli*. Later he published editions of Zouche's *Jus et Judicium Feciale* (1911) and Legnano's *De Bello* (1917). For the Admiralty he rewrote the *Manual of Naval Prize Law* (1888) and for the War Office a handbook on the *Laws and Customs of War on Land* (1904). A posthumous volume of *Lectures on International Law* appeared in 1933. His *Elements of Jurisprudence* (1880 and later editions) was long a standard work and contributed to the continued vitality of the Austinian analytical jurisprudence in England though he substituted enforcement by a determinate authority for Austin's command of the sovereign as the criterion of a law. He was one of the founders of the *Law Quarterly Review* and contributed to it regularly, as well as to other journals, served on royal commissions, and was a very active member of the Institut de Droit International. He was knighted in 1917.

Holmes, Oliver Wendell (1841–1935). U.S. judge and jurist, son of the writer of the same name, graduated from Harvard in 1861 and served in the Civil War. He was admitted to the Bar in 1867 and worked hard to master his craft. He edited the *American Law Review* (1870–78), edited an edition of Kent's *Commentaries* (1873), and began to systematise his ideas in the lectures published as *The Common Law* (1881), a book which made his reputation. From teaching law at Harvard (1882) he was promoted to Associate Justice (1882–99) and Chief Justice (1899–1902) of the Massachusetts Supreme Judicial Court, and then to Associate Justice of the U.S. Supreme Court (1902–32). As a judge he was labelled The Great Dissenter, for in his earlier days he repeatedly disagreed with the conservative attitude of the Supreme Court but had the satisfaction later of seeing his views frequently recognized. His opinions are the product of deep knowledge and profound thought, and he has been characterized as the greatest intellect in the history of the English-speaking judiciary. His opinions are supreme for their penetrating character and originality of composition.

From his first visit to England in 1866 he enjoyed the friendship of Bryce, Dicey, Pollock, and other great intellectual figures of the time and, in particular with Pollock, corresponded with them for many years and came to be well known and admired in England. His fame as a jurist rests on his judicial opinions, on *The Common Law* (1881), which is a classic, and on some very perceptive and stimulating papers and addresses published as *Collected Legal Papers* (1920).

M. Howe, *Justice Oliver Wendell Holmes—The Shaping Years*, 1841–70, and *The Proving Years*, 1870–82; C. D. Bowen, *Yankee from Olympus*; F. Biddle, *Mr Justice Holmes*; M. Howe (ed.) *Holmes–Pollock Letters*, 1874–1932; M. Howe (ed.) *Holmes–Laski Letters*, 1916–35; H. Shriver, *The Judicial Opinions of O. W. Holmes.*

Holograph. That which is written wholly, or at least, in all essential words, in the granter's own hand. In Scotland deeds and wills thus written are deemed valid though the signature is not attested by witnesses and, for most purposes, it is equal to a formally attested writing.

Holt, Sir John (1642–1710). Called to the Bar in 1663, he appeared, usually for the defence, in many of the State trials of the next few years. In 1686 he became recorder of London, but was dismissed for giving opinions contrary to the King's views. In 1689, having played a leading part in the Revolution, he became Chief Justice of the King's Bench, which office he held till his death. He declined the Chancellorship in 1700. Holt was a learned common lawyer and was able to develop legal rules to the needs of changing conditions. His judgment in *Coggs* v. *Bernard* (1703) has ever since been regarded as a classic restatement of the Law of Bailment, founding on principles of Roman law. He revolutionized criminal proceedings by his fairness to the accused, developed the doctrine of employer's liability, recognized the principle of negotiability of bills of exchange and the assignability of bills of lading, and gave judgments in many important cases involving the liberties of the subject, and the rights of the citizen in relation to Parliamentary privilege. He established a new era of judicial purity and freedom, when learning, independence, and fairness became normal judicial attributes.

Holtzendorff, Johann Wilhelm Franz Philipp von (1829–89), German jurist, editor of a major *Encyclopädie der rechtswissenschaft* (1880–82), part-author and editor of *Handbuch des Völkerrechts* (1885–89), and author of many other works on law.

Holy days. See HOLIDAYS.

Holy orders. A grade or rank in the Christian ministry; hence 'in holy orders' is to have the rank or status of an ordained clergyman. In episcopal churches, the orders are those of bishops (and archbishops), priests and deacons; in presbyterian and other non-episcopal churches the only order is that of minister. The principle was long accepted that ordination was indelible and that even if deposed or degraded an ordained person remained in orders. But it is now possible to relinquish orders and resume the status of laity.

Holy Roman Empire. The Roman imperial title lapsed in western Europe in the fifth century but was revived in 800 and conferred by Pope Leo III on Charlemagne, King of the Franks. It impliedly asserted continuity with the older Roman Empire founded by Augustus. It was borne by German kings from the mid-tenth century until abolished in 1806. The Latin title dates only from 1254. The territory of the empire included what are now the Low Countries, eastern France, Germany, Austria, western Czechoslovakia, Switzerland, and parts of northern and central Italy. The German lands were always the chief part and the emperor was usually the German King. From the fifteenth century the imperial title and the German kingship were practically hereditary in the House of Habsburg, though elections continued to be held. The empire became involved in a long struggle with the papacy for the leadership of Christian Europe between the eleventh and thirteenth centuries and was weakened by the Reformation in the sixteenth century, which gave rise to a division between the Emperor and the Catholic princes and on the other hand the Protestant princes. The Thirty Years' War devastated Germany in 1618–48 and thereafter the Empire was a loose collection of semi-independent states.

In 1804 Napoleon declared himself Emperor of the French and set himself to assert the primacy among European monarchs, and Francis II adopted the title Emperor of Austria and in 1806 resigned the office of Holy Roman Emperor.

The German Empire of 1871–1918 was sometimes called the Second Reich, asserting its continuity with the older Empire, and Nazi Germany 1933–1945 was called the Third Reich.

J. Bryce, *The Holy Roman Empire.*

Homage. Under feudalism, a grantee of lands was publicly invested with his lands by actual or symbolic delivery (*livery of seisin*) and then did homage, so called from the words used. The grantee knelt before his lord, placed his hands between those of his lord, and said (in Norman French) 'I become your man (homme), from this day forward, of life and limb, and of earthly worship; and unto you shall be true and faithful, and bear to you faith for the tenements I claim to hold of you'. Homage was accordingly the acknowledgment of the bond of tenure between lord and tenant. It made the military tenant under the duty of aiding and protecting his lord and of serving him faithfully against all men. The lord was under the duty of protecting the tenant and of defending him against attacks, by force or by legal process. The breach of the obligations involved in homage was the most heinous offence known, and the original essence of felony. As time went on homage declined in importance. The duty of allegiance to the King took precedence over each vassal's duty to his lord, and as the property element in land law superseded the personal bond of duty the importance of homage declined. The lord then kissed the vassal and received the oath of fealty (q.v.). Homage was abolished in 1660 but the oath of fealty survived.

A bishop still does homage to the Queen for his temporalities or barony in similar fashion, by taking an oath and acknowledging that he holds his temporalities of her.

Home, Henry, Lord Kames (1696–1782). At first apprenticed to a Writer to the Signet, Home was called to the Scottish Bar in 1724. Though not very successful at first, he became a judge, as Lord Kames, in 1755, and in 1763 also a Lord of Justiciary. As a judge he had a strong feeling for justice and a good understanding of law. He acquired a considerable reputation as an agricultural improver and his *Gentleman Farmer* (1776) was a useful book which went through several editions. He achieved fame not only as a writer on law, but on history, criticism and morals. He published *Remarkable Decisions of the Court of Session, 1716–28* (1728); *Remarkable Decisions of the Court of Session 1730–52* (1766); *The Decisions of the Court of Session, abridged and digested under proper heads in the form of a Dictionary* (2 vols., 1741) (Vols. 3 and 4 by Fraser Tytler, 1797, and Supplementary volume by McGrugor, 1804); *Selected Decisions of the Court of Session, 1752–68* (1780); *Essays upon Several Subjects in Law* (1732); *Essays upon Several Subjects Concerning British Antiquities* (1747); *Essays on the Principles of Morality and Natural Religion* (1751); *Principles of the Law of Scotland* (1754); *Statute Law of Scotland Abridged* (1757); *Historical Law Tracts* (1758); *Principles of Equity* (1760); *Sketches of the History of Man* (1774); *Elucidations respecting the Common and Statute Law of Scotland* (1777). Many of these works went through several editions.

A. Tytler, *Memoirs of the Life and Writings of Henry Home of Kames* (1807); W. Lehmann, *Henry Home, Lord Kames, and the Scottish Enlightenment* (1971); J. Ross, *Lord Kames and the Scotland of his Day* (1972).

Home, Sir John, of Renton (?–1671). Became a Lord of Session in 1663 and also Lord Justice-Clerk for life. He is said to have been a great zealot for the prelates in Scotland.

Home, Patrick, Lord Polwarth. See HUME.

Home Building and Loan Association* v. *Blaisdell (1934), 290 U.S. 398. In 1933, during the great depression, Minnesota enacted a statute postponing foreclosures on mortgaged property. In *Bronson* v. *Kinzie* (1843), 1 Howard 311, the Supreme Court held unconstitutional an Illinois statute postponing foreclosure and forbidding sale for less than two-thirds of the appraised value of the property as impairing the obligation of contracts. In this case the Supreme Court, by a majority, upheld the Minnesota statute in the light of the emergency economic situation.

Home Office. The Department of State, headed by the Secretary of State for the Home Department, which is concerned almost exclusively with England and Wales, but also with relations between the U.K. Government and the Isle of Man and the Channel Islands. It is concerned with the general government of England and Wales, in so far as not falling within the responsibility of other departments, including particularly civil defence, public order, aliens, naturalization and immigration, the supervision of the police and the prison system, extradition, the fire service, gaming, cruelty to animals, broadcasting, and many other subjects. The Home Secretary is the proper means of communication between Crown and subject and notifies many matters of State intelligence to local officials. Addresses and petitions to the Crown, such as for the exercise of the prerogative of mercy, are addressed to him. He is also the means of communication with the Church of England.

Home Rule. The general title for the movements in Ireland from 1870 to 1921 and in Scotland from 1850 for national government largely or entirely independent from Westminister, reversing the centralizing policy of the Acts of Union (with Scotland) 1707, and (with Ireland) 1800. In Ireland, the Home Rule party was founded in 1870 and down to 1916 acted mainly constitutionally; from 1916 to 1921 the growing physical force party took over. The Home Rule Association was formed by Isaac Butt in 1870 but in the years 1879–86 the driving force was Michael Davitt's Land League. Gladstone unsuccessfully introduced Home Rule Bills in 1886 and 1893 and Asquith a Home Rule Bill in 1912. It finally passed in 1914 but was never put into operation by reason of the World War. The Government of Ireland Act ('Home Rule Act') 1920 divided Ireland into Northern Ireland with an independent Parliament and remaining areas but a majority of the persons elected for Irish constituencies styled themselves Dail Eirann, the Assembly of Ireland, and formed a Provisional Government. The Anglo-Irish Treaty of 1921 gave Ireland, under the name of the Irish Free State, the status of the dominion of Canada, with liberty to Northern Ireland to opt out, which it did. Home Rule was accordingly achieved in most of Ireland. The Irish Free State became an independent republic in 1937. Northern Ireland had its own Parliament from 1921 to 1972, when it was prorogued indefinitely and rule from London resumed.

In Scotland, dissatisfaction with the administration of Scottish affairs by United Kingdom, i.e. English, departments resulted in the establishment of the office of Secretary for Scotland in 1885. Thereafter powers were increasingly conferred on him and distinct Scottish Boards were created to deal with education, health, and agriculture. In 1926 the Secretary became a Secretary of State and, by 1939, there was an Edinburgh headquarters of the Scottish Departments (Home, Health, Education, and Agriculture, now Home and Health, Education, Agriculture and Fisheries, Development, and Economic Planning). Increasingly distinct organizations have been established for the management of affairs in Scotland, though some matters are managed by U.K. Departments. Agitation after 1945 culminated in the appointment of the Royal Commission on the Constitution (1970–73) and in the Acts of 1978 to create separate Scottish and Welsh Assemblies.

E. Curtis, *History of Ireland* (1936); W. Gladstone, *The Irish Question* (1886); N. Mansergh, *The Irish Question* (1965); R. Coupland, *Welsh and Scottish Nationalism* (1954); H. Hanham, *Scottish Nationalism* (1969).

Homeyer, Karl Gustav (1795–1874). German jurist and member of the Prussian diet. His major works are an edition of the *Sachsenspiegel* (1827), *Genealogie der Handschriften des Sachsenspiegel* (1859), and *Die Stadtbücher des Mittelalters* (1860).

Homicide. The generic term for the causing, or accelerating, the death of a human being by another human being. It raises difficult problems of responsibility and of liability to civil and criminal sanctions. In undeveloped law homicide is commonly primarily a wrong to the kindred of the person slain, giving rise to the blood-feud. Only later is it primarily an offence against the community and public order. In England, under the Normans, homicide became a plea of the Crown and the right of the kindred to exact private vengeance was superseded by the right of the King to exact forfeiture where the homicide was a felony. It was, however, long before it was clearly settled which forms of homicide were criminal and which merely tortious or even legally justifiable.

Homicide is distinguished into murder, manslaughter or, in Scotland, culpable homicide (qq.v.), justifiable homicide, and accidental, negligent, or excusable homicide. The first two categories are criminal, the latter two not. Special cases of manslaughter are causing death by dangerous driving, infanticide (q.v.), and killing in pursuance of a suicide pact. To kill an unborn child *in utero* is neither murder nor manslaughter but may be abortion or child destruction (q.v.). In some U.S. jurisdictions the distinction is drawn between first and second degree murder.

Justifiable homicide is causing death in circumstances authorized by law, including executing a person condemned to death, killing in the course of dispersing a riot, killing a prisoner who is attempting to escape, and killing in self-defence and in prevention of a violent attack.

Accidental, negligent, or excusable homicide is causing death unintentionally, by misadventure or without gross or culpable negligence, as in a street accident or an unsuccessful attempt at surgery. It may be tortious but is not criminal.

European codes generally group all non-justifiable killings as homicide but provide different penalties for crimes arising in different circumstances. In all systems the important distinction in sentencing is between socially dangerous homicide and merely reckless conduct or acts of passion.

Homine replegiando, de. A mediaeval writ obtained from Chancery directing the sheriff holding a prisoner to deliver him up, unless he had been taken at the special command of the King or his chief justiciar, or for homicide or a forest offence, or some other cause which made him not repleviable. In 1275 statute laid down fairly strict rules which became the settled law of later times.

Homologation. In Scots law, an act approbatory of a deed, the effect of which is to render the deed, though itself defective, binding on the person by whom it is homologated.

Homosexuality (or sexual inversion). Sexual attraction of one person to another of the same sex, leading to physical contact and sexual pleasure. Male homosexuality, or sodomy, is a widespread phenomenon and is not infrequent where groups of men are isolated for long periods from all women, e.g. in prison, on shipboard, but not confined to such circumstances. The causes are obscure, whether biological or psychological. The social dangers involved are considerable, including corruption of young persons, blackmail, and psychological disturbance. For long male homosexuality was strongly socially reprobated and criminally punishable as a form of buggery (q.v.). In England the Sexual Offences Act of 1956 made it permissible in private between consenting adults aged at least 21.

Female homosexuality is often called lesbianism (from the poetess Sappho who was leader of a group of women living on Lesbos). It is thought to be less common and does not appear ever to have been criminal in England.

Honeste vivere, alterum non laedere, suum cuique tribuere (to live honourably, to injure no man, to render to each his own). According to Ulpian and adopted by Justinian's *Institutions*, the fundamental precepts of the law. The third phrase repeats the earlier sentence defining justice as the constant and lasting wish to render to each his own.

Honestiores. Persons belonging to the upper classes of Roman society as distinguished from *plebei* or *humiliores*. In Imperial times, particularly from the third century, they were privileged in criminal law, being subject to lesser penalties for some crimes. Thus capital punishment was only exceptionally imposed on them, and even then not by crucifixion or being cast to wild beasts, and they were less liable to corporal punishment, forced labour in the mines, torture, and other penalties. They also had certain privileges in procedure on appeal. Some of these privileges applied to all classes of *honestiores*, but some only to particular categories of them.

Honor (or *Honour*). In mediaeval law the group of estates, normally scattered about England, from which the greater tenants-in-chief of the Crown derived their prestige and status, and on which inferior lordships were dependent, e.g. the Honour of Huntingdon. An honour was governed by a court consisting of all the barons who held land of it, and its jurisdiction and procedure resembled that of the King's Court, and sometimes important issues of property were litigated and determined in honorial courts.

Honour. Good standing, reputation, and esteem in the eyes of others, a quasi-proprietary attribute which most individuals value and cherish and for unfounded aspersions on which, by way of libel or slander, a claim of damages lies. The term is also used of that which normally confers honour on the recipient, the conferment of a degree of nobility, knighthood, membership of an order of chivalry or similar distinction (see HONOURS). Thus membership of the Privy Council confers the title 'The Right Honourable . . .'. The designation of a circuit judge is 'His Honour Judge X'. In the U.S. members of Congress, judges of superior courts and persons of similar standing are usually designated 'The Honorable . . .'. In the U.K. this designation is given by courtesy to justices of the High Court, judges of the Court of Session, and the children of hereditary peers above the rank of baron.

The verb 'honour' is used of the drawee of a bill of exchange when he accepts it and of the acceptor when he pays it when due; either dishonours the bill by non-acceptance or non-payment.

Honour, Court of. See COURT OF CHIVALRY.

Honour and vital interests. Matters formerly commonly excepted in treaties conferring jurisdiction on international arbitral or judicial tribunals.

Honours. All degrees of the peerage, baronetcies, knighthoods, membership of the various orders of knighthood, and other similar distinctions. Conferment of an honour is an act of the Queen. Appointments to the Order of Merit and the Orders of the Garter and the Thistle are made in her discretion, and membership of the Royal Victorian Order is granted for personal services to the monarchy, but other honours and dignities are conferred on the advice of the Prime Minister of the country concerned, though many of the lesser awards have no political significance at all. Peerages may be created for purely political reasons, such as to force a Bill though the House of Lords and this has been threatened several times, notably in 1830 and 1910. Since the Life Peerages Act 1958 only life peerages have been conferred though the power to confer hereditary peerages continues to exist. It is doubtful if a peerage of Scotland can now be created and peerages of Ireland cannot. In law a subject cannot refuse to accept a dignity or honour conferred by the Crown but in practice an individual may decline. A Prime Minister usually appoints a Political Honours Scrutiny Committee to consider the character of persons on whom it is proposed that honours be conferred and to report if any be unsuitable.

Hooker, Richard (1553–1600). English divine, Master of the Temple, 1585–91, and author of the *Eight Books of the Laws of Ecclesiastical Polity* (1593–97; books VI–VIII were posthumously published and probably altered by editors). The work is an answer to Presbyterian attacks on the Episcopalian polity and practices. He bases his reasoning on the unity and all-embracing character of law, as operative in nature and as regulating man's character and actions, and as a manifestation and development of the divine order. It is a reinterpretation of the theories of Aquinas to fit them to Anglicanism, and explores the theological foundations of legal philosophy. Natural law is eternal and immutable whereas positive law varies according to external necessity and expediency and its application must be determined by reason aided by every variety of knowledge and experience. His theory of government was the basis of Locke's *Treatise on Civil Government* though Locke develops the idea in his own way. Taken as a whole his theory is the first philosophical statement of the principles which have subsequently regulated political progress in England. He asserted the royal supremacy in religion and identified Church and commonwealth as different aspects of the same government.

Horton, Earnest Albert (1887–1954). American physical anthropologist who examined the relationship between physical type and personality, particularly in relation to criminal conduct. He wrote *The American Criminal* (1939) and *Crime and the Man* (1939), as well as several works on anthropology.

Hope, Charles, of Granton (1763–1851). A descendant of Sir Thomas Hope (q.v.), he was called to the Scottish Bar in 1784 and became Lord Advocate in 1801 and an M.P. in 1802. He became Lord Justice-Clerk in 1804 and, in 1811, Lord President of the Court of Session. He was sworn of the Privy Council in 1822. He was a fine speaker and a man of imposing appearance. In 1836 he also became Lord Justice-General on the death of the last hereditary holder of that office. He resigned in 1841. For many years he was an active officer of volunteers. A son, John, became Lord Justice-Clerk.

Hope, John (1794–1858). Eldest son of Lord President Charles Hope (q.v.), he became an advocate in 1816, Solicitor-General for Scotland in 1822, and Dean of Faculty in 1830. In 1841 he became Lord Justice-Clerk, an office he held till his death, and, in 1844, a Privy Councillor. He was a diligent and able judge and presided at many important cases during his tenure of office.

Hope, Sir Thomas, of Craighall (?1580–1646). Was called to the Scottish Bar in 1605 and soon rose to prominence. In 1625 he prepared the deed of revocation whereby Charles I recalled all erections of church lands and all deeds executed by the King during his minority. He became joint Lord Advocate in 1626, sole Lord Advocate and a baronet of Nova Scotia in 1628, and was in high favour with the King. He held office till his death. He compiled a *Minor Practicks,* a concise manual on the law of Scotland, published in 1726 and 1734, and the *Major Practicks,* published by the Stair Society in 1937, an extensive collection of notes of statutes and cases and practical observations on the whole range of Scots law as it stood about 1633, though not a connected treatise, both of which are important records of the law at the time. Among descendants were Lord President Hope (q.v.) and Lord Justice-Clerk Hope (q.v.).

Höpfner, Ludwig Julius Friedrich (1743–97). German jurist, annotator of Heineccius, and author of works on Roman law.

Horn, Andrew (d. 1328). Was Chamberlain of London, 1320–28. He made a compilation of City of London laws and customs, entitled *Liber Horn.* He was also author or editor of the *Mirror of Justices* (q.v.).

Horning, letters of. In Scots law, an old form of diligence, superseded but still competent, in the form of letters in the sovereign's name, proceeding on a decree of court, directed to messengers-at-arms ordering them to charge the debtor to pay or perform within a certain time, under pain of being held in rebellion and being put to the horn as outlaws. (A person was declared rebel by the messenger-at-arms giving three blasts on a horn and publicly proclaiming the fact.) The letters were registered in the Register of Hornings.

Hospital. A charitable corporation for the education of impecunious scholars, e.g. Christ's Hospital, or the relief of persons, particularly by affording them food or shelter or medical treatment. In modern times the word is more often used of a place providing medical care and treatment, either private or more commonly, managed under the National Health Service.

Hospitallers (or Order of the Hospital of Saint John of Jerusalem, or Sovereign and Military Order of the Knights Hospitaller of Saint John of Jerusalem). A religious and military order founded in the twelfth century by the Pope. It built a hospital for sick pilgrims in Jerusalem and hostels in various cities in Provence and Italy on the route to the Holy Land. When the Knights Templar were suppressed in 1314, most of their property was transferred to the Hospitallers. Later, in the fourteenth century, the Hospitallers demised the Temple, or London monastery of the Templars, to lawyers who later formed the societies later known as the Inner Temple and the Middle Temple. (See INNS OF COURT.) The Hospitallers' other estates in England continued to be held until the order was dissolved and its property transferred to the Crown under the Grantees of Reversions Act, 1540. In the Middle East the Hospitallers continued their work in Cyprus and Rhodes and ruled that island as an independent state. Later the order owned Malta until 1798. The order continues in several countries as a humanitarian society.

Hospitia Cancellariae. The Inns of Chancery (q.v.).

Hospitia Curiae. The Inns of Court (q.v.).

Hostage. A person, usually of importance, taken from, or surrendered by agreement by one belligerent to another, to be held as security. The practice was formerly common as a means of securing legitimate warfare. In the twentieth century it has been resorted to so as to secure the safety of forces in occupied territory against hostile attacks by civilians, and in the Second World War hostages were frequently taken by German forces in occupied territory and shot to prevent or punish civil disorder. A Geneva Convention of 1949 has prohibited the taking of civilian hostages. The seizure of persons and holding of them as hostages has more recently been resorted to, wholly illegally, by hijackers of aircraft or trains as a compulsitor on governments to give in to the terrorists' demands.

Hostiensis (Enrico Bartolemei, Henricus de Segusio) (*c.* 1200–1271). Cardinal, canonist and diplomat. He taught possibly at Bologna and certainly at Paris and became a distinguished clerical diplomat. He wrote a famous *Summa*, sometimes called *Summa Copiosa* (1253), inspired by Godfrey of Trani and Azo, which was a synthesis of Roman and canon law and in constant use until the seventeenth century, *Lectura in Novellas Innocentii IV* (1253), and *Lectura in Quinque Libros Decretalium* (*c.* 1239). His work had great subsequent influence.

Hostile witness. A witness who, when giving evidence, conducts himself in a manner hostile to the party calling him, such that the party calling him may, by leave of the presiding judge, cross-examine him as if he were a witness for the other side. A witness is not hostile merely because he gives evidence unfavourable to the party calling him.

Hot pursuit, right of. In international law, the right of a vessel, when another vessel or someone on board her, while within the territorial sea of the first vessel, has committed an infringement of the law of the littoral state, to pursue that other vessel into the open seas and there arrest her. The pursuit is deemed a legitimate extension of an exercise of jurisdiction which has been commenced and justifiable to enable the territorial jurisdiction to be efficiently exercised. It may be maintained only so long as uninterrupted and ceases when the vessel pursued enters the territorial sea of its own or a third country. It is questionable whether an infringement in the contiguous zone confers a right of pursuit, but if there is infringement in the territorial sea pursuit may probably be commenced in the contiguous zone. Pursuit may be made by warships, military aircraft, or other authorized vessels on government service.

Hotchpot (Fr. *Hochepot*, a dish shaken up). The throwing together of land and chattels into a single fund for distribution; it is comparable to the Roman *collatio bonorum* and the Scottish *collatio inter haeredes* (q.v.) and *collatio inter liberos* (q.v.). At common law, lands given in frank-marriage had to be divided equally with lands descending on death to others in the family. At common law children were entitled to equal shares of their intestate ancestor's goods and chattels and advances made were accounted part of a child's portion. In the case of a person dying intestate after 1925 the deceased's property, not falling to any surviving spouse, is held on statutory trusts for the surviving issue, but children must bring into account any money or property they received from the deceased in his lifetime by way of advancement or on marriage if they wish to share in the estate.

A hotchpot clause in a will may contain a direction that advances made by a testator to his children in his lifetime are to be brought into account in determining their shares.

Hotel. At common law an inn is premises where a person holds himself out as willing to receive any travellers who come and to provide food and lodging for them, their attendants and animals. At common law special liabilities attached to an innkeeper. See INNKEEPER. Under the Hotel Proprietors Act, 1956, a hotel is an establishment held out by the proprietor as offering food, drink and, if so required, sleeping accommodation to any traveller presenting himself who appears able and willing to pay a reasonable sum for the services and facilities offered and who is in a fit state to be received. Such an establishment is an inn and no other establishment is an inn, and all the duties, liabilities and rights which are attached to an innkeeper as such, now attach to a hotel proprietor and to no other person, subject to modification of the common law liability for guests' property. A hotel may, accordingly, be an inn, but is not if it offers residential accommodation and food to lodgers, but where nobody other than a lodger is supplied with food. A public house, tavern, or coffee-house is not a hotel, nor an inn, nor is a restaurant. A private hotel or boarding-house or lodging-house is probably not a hotel because the owner does not hold himself out as willing to receive anyone who comes. Whether premises are a hotel, and therefore an inn, within the meaning of the Act, is a question of fact.

Hotoman or Hotman, François (Hotomannus) (1524–90). A distinguished French jurist who succeeded Cujas at Bourges and took a leading part in legal, political and religious controversies. He studied law in the light of archaeology and philosophy and was chiefly concerned with interpretation. He wrote on Roman law, feudal law, and on French public law, in such works as *De jure regni Franciae* and *Franco-Gallia* (1573), in which he showed that there was no historical foundation for the growth of royal absolution in France, and argued for representative government and an elective monarchy. In *Antitribonian* (1567) he argued for unity of legislation and a union of practice, with historical and synthetic treatment of law, and for a French code not borrowing excessively from Roman law.

House of Commons. The lower, but in fact the more important and powerful of the two Houses of the United Kingdom Parliament. It originated in the thirteenth century in the practice of the knights of the shire and the city and borough representatives sitting and debating separately from the barons and greater clergy. New boroughs were created and the number of members rose steadily, reaching 513 in 1707, to which were then added 45 members for Scotland and, in 1800, 100 members for Ireland.

The representative system was, from the start, very imperfect and, as time went on, became more so; as ancient towns decayed, new ones arose and populations grew and shifted. In English counties, the franchise was limited to forty-shilling freeholders, and these were subject to the influence of the nobles and local landowners. In English boroughs the franchise was generally restricted to corporations or to freemen, and many were under the influence of the Crown (or landowners) which could buy up sufficient of the properties in the borough having the right to vote and so control the election. In Scotland, the franchise in counties was restricted to the owners of feudal superiorities, and in boroughs, it was vested in self-perpetuating corporations. In Ireland, the electorate in counties were more numerous but less independent than in England or Scotland, and the boroughs were under the control of landowners.

Agitation for the reform of the Commons to make it more genuinely representative of the people was urged, from the mid-eighteenth century, by Lord Chatham, John Wilkes, the Younger Pitt, and increasing popular agitation. This culminated in the passing of the Reform Act of 1832, which disfranchised many decayed boroughs, created many new ones, and extended the franchise. Similar Acts were passed for Scotland and Ireland. Thereafter, many reforms followed. The property qualification for members was modified in 1838 and abolished in 1858. A major extension of the franchise was made for England in 1867 and for Scotland and Ireland in 1868. Voting by secret ballot was introduced in 1872. The Representation of the People Act, 1884, gave the vote to rural labourers and many industrial workers, and that of 1918 extended it to all male residents or occupiers of business premises and to women over 30. In 1928 the voting age for women was reduced to 21 and in 1969 the age for both sexes was reduced to 18. Redistribution of seats has been effected several times and stringent controls have been introduced to prevent corrupt and illegal practices, and to limit the permitted amounts of election expenses. After considerable variations, in practice, the whole country is now divided into constituencies of approximately equal population, each returning a single member.

The House now consists of 630 members elected to represent the voters in the Parliamentary constituencies of England (511) and Wales (36), Scotland (71) and Northern Ireland (12). The electorate consists of British subjects and Irish citizens of either sex, aged 18 or over, who are subject to no legal incapacity and are resident (on a particular date) in, and registered in the register of parliamentary electors for a particular constituency. Many classes of persons are disqualified from standing for election, including peers, judges, members of the civil service, and holders of many kinds of offices and places of profit under the Crown.

Candidates nearly all stand for election as representatives of one or other of the political parties, and Members are nearly always adherents of a party. Each party has its Whips in each House, to ensure that its members are present to vote when necessary.

Disputed elections, formerly decided by the King and Council and later by committees of the House, are now decided by two judges of the High Court or Court of Session, who report to the Speaker. Payment of members was known in the Middle Ages, but lapsed. In 1911, provision was made for a salary of £400 to members and this has been raised periodically. Secretarial expenses are also paid and there are allowances for travel, telephones, and postages.

The main functions of the House may be summarized as:

(1) The representation of popular opinion.
(2) The control of finance, through the Budget resolutions and the annual Finance Bill.
(3) The formulation and control of policy through:
    (a) the address in reply to the Queen's Speech;
    (b) consideration of the estimates and Consolidated Fund Bills;
    (c) substantive and adjournment motions;
    (d) questions to Ministers.
(4) Legislation by way of debates on:
    (a) Public Bills;
    (b) Private Bills;
    (c) Delegated legislation.
(5) The scrutiny of administration by means of select committees.

In so doing it normally sustains the Cabinet in office, but on occasion may, by failing to give it a vote of confidence or rejecting its proposals, turn it out.

The Officers of the House are the Speaker, who is elected at the commencement of each Parliament and is the channel of communication between the Commons and the Lords, and the Commons and the Queen. (Whoever occupies the chair, also maintains order, rules on procedure, and guides the House on practice and privileges.) The Chairman of Ways and Means acts when necessary as Deputy-Speaker and takes the chair when the House is in committee; there is also a Deputy-Chairman of Ways and Means. The Clerk of the House prepares the journals of the House. The Serjeant-at-Arms attends the Speaker, maintains order, executes warrants for contempt, and brings persons in custody before the Bar of the House.

House of Commons Papers. A series of papers bearing to be 'ordered by the House of Commons to be printed' and including reports and returns required to be presented to the House by a provision in statute, reports from Government officials prepared in compliance with an order from the House, such as returns from the Treasury and revenue departments and the financial statistics which accompany the annual Estimates, reports of Select Committees and Minutes of Standing Committees. They are numbered serially within each session.

House of Keys. The elected branch of the Court of Tynwald, the legislature of the Isle of Man, comprising 24 members elected for five years by adult suffrage.

House of Laity. See GENERAL SYNOD.

House of Lords (as court). Down to the early thirteenth century the *Curia Regis* had legislative, executive, and judicial powers. By the end of that century, the Council, sitting in Parliament, was regarded as the highest court of royal justice and, even in the fourteenth century, the dispensing of justice was the main function of Parliaments. It had an original and appellate jurisdiction, hearing important cases, and answering petitions.

In the fourteenth and fifteenth centuries Council, the royal officials and advisers, became distinguished from Parliament and the latter divided into two Houses, with the growth of the idea that the magnates, specially summoned to Parliament, were peers and a distinct class. In the early fourteenth century, a great business of Parliaments was to receive and answer petitions, and these were dealt with by committees at first but, from late in the fourteenth century, by the House itself. The volume of petitions was so great that, in the fifteenth century, the Lords abandoned them to the Council and in the sixteenth century this work was divided between Council, Chancery, and Star Chamber.

Lords and Commons, however, allied against the Council to maintain control over the law, and this united them with the judges and common lawyers, who resisted the view that errors by the common law courts were open to review by the Council. After 1366 the jurisdiction to amend errors by the common law courts was left to Parliament. During the fourteenth century it was generally recognized that this jurisdiction in error was vested in the Lords only, and this has never since been questioned. Only in the seventeenth century, however, were the judicial powers of the House of Lords finally ascertained and settled.

In the thirteenth century, Parliament often exercised original civil jurisdiction but, in the next two centuries, there was a growing tendency not to exercise it, though on a few occasions in the seventeenth century it was exercised. *Skinner* v. *East India Company* (1666), finally decided that the House had no original civil jurisdiction. In the fourteenth century Parliament sometimes interfered in a case to settle it or direct the court below how to dispose of it. More important was the jurisdiction in error (q.v.), recognized in the fourteenth and fifteenth centuries as belonging to the Lords, and repeatedly invoked since the early seventeenth century. Error lay from the King's Bench, Exchequer Chamber, and common law side of the Chancery in England, till 1783 from the Irish Parliament, and after 1707 from the Court of Session in Scotland. In the seventeenth century, the House established the right to hear appeals from the equity side of Chancery, and in the next century, except for 1783–1800, appeals from the Court of Chancery in Ireland.

The House had original criminal jurisdiction in that the Crown could at one time accuse great offenders in Parliament; after Charles I resorted to similar procedure in 1641 in relation to the Five Members, it was declared illegal. One subject could also appeal against, or accuse another of, crime before Parliament. It was resolved in 1663 that this was incompetent. The most important criminal jurisdiction was in cases of impeachment, where a person, if a peer impeached for treason or felony, was charged by the Commons and tried before the Lord High Steward and the whole House of Lords, or in other cases, before the Lord Chancellor or Lord Keeper and the whole House. Though still legally possible it has not been used since 1805 and it is probably obsolete. The other criminal jurisdiction, over peers charged with treason or felony was last used in 1935 and abolished in 1948.

After 1707, the House assumed appellate jurisdiction from the Scottish Court of Session, though this had not been conferred by the Treaty of Union. The Act of Union with Ireland, 1800, conferred on the House appellate jurisdiction in civil cases from the Irish courts (now Northern Ireland courts). Appeal in criminal cases was provided when the Court of Criminal Appeal was established in 1907 and from Northern Ireland in 1921. There is no appeal from the High Court of Justiciary in Scottish criminal cases.

In theory, an appeal was, and is, to the whole House and sittings of the House for judicial business were not distinct from those for legislative business. In the seventeenth and eighteenth centuries peers, and even bishops, sometimes used their powers to carry a decision, even contrary to the opinion of the judges. The Chancellor, or an ex-Lord Chancellor, was normally the only legally skilled peer present and, in 1834, the House decided a case without any lawyer present, while regularly down to 1867 Chancellors and other peers with knowledge only of English law decided Scottish appeals. In *O'Connell* v. *The Queen* (1844), the Lord Chancellor ignored the votes of the lay peers and since then there has been a convention that only legally qualified peers take part when the House is sitting to hear appeals. But a third person must have been present with Lord Cairns and Lord Cranworth at the hearing of *Rylands* v. *Fletcher* (1868), to make the quorum of three (see 86 L.Q.R. 160, 311). In 1883 in *Bradlaugh* v. *Clarke* a lay peer sought to vote but his vote was ignored.

In 1856 an attempt was made, by the exercise of the prerogative, to confer a life peerage on Parke, B., with the right to sit and vote in the House, to assist with judicial business. This having been held impossible by the Committee of Privileges (*Wensleydale Peerage Case*, 1856) he was given an ordinary peerage, and some other judges were subsequently ennobled to enable them to assist with the work.

In 1873 the Judicature Act abolished the appellate jurisdiction of the House, but before the Act became effective it was repealed and provision made by the Appellate Jurisdiction Act, 1876, for two (now 11) Lords of Appeal in Ordinary, who must be practising barristers or advocates, of at least 15 years' standing or have held high judicial office for at least two years. They are given life peerages and may sit and vote in the House, in any kind of business, for life. In practice, they participate in legislative business only on legal matters. The Act of 1876 provided that an appeal must be heard by at least three of the Lord Chancellor, the Lords of Appeal in Ordinary, and such peers of Parliament as hold or have held high judicial office. By custom, the Lords of Appeal normally include two from Scotland and appointments have been made from Northern Ireland. Appeals are usually heard by five Lords but larger numbers have occasionally sat in very important cases.

It was formerly the practice in case of difficulty, to summon the judges of the Queen's Bench Division to assist their Lordships with advice, but this has rarely been done since the 1876 Act, the last cases being *Allen* v. *Flood* (1898) and *Free Church* v. *Lord Overtoun* (1904).

The House has jurisdiction to hear appeals in civil cases from the Court of Appeal in England, by leave of that court or of the House, subject to statutory restrictions, direct from the High Court in England, from the Court of Session in Scotland, with, in certain cases, the leave of that court, from the Court of Appeal in Northern Ireland, by leave of that court or of the House and, subject to statutory restrictions, direct from the High Court in Northern Ireland.

Under the Criminal Appeal Act, 1907 (now 1966), the House of Lords can hear appeals, formerly from the Court of Criminal Appeal and now from the Court of Appeal, Criminal Division, and from the Court of Criminal Appeal or High Court in Northern Ireland, but not from the High Court of Justiciary in Scotland. It may also hear appeals from a Divisional Court of the Queen's Bench Division in a criminal cause and from the Courts-Martial Appeal Court.

The House appoints two appeal committees (distinct from the appellate committees mentioned hereafter) each consisting of three Lords to consider and report to the House on petitions for leave to appeal and petitions incidental to causes depending or formerly depending.

The House formerly sat for judicial business in the House of Lords chamber when the House was not sitting for legislative business but now sits as an appellate committee in one of the committee rooms. It may sit when Parliament is prorogued or dissolved, and may sit in two divisions. The appellate committee hears appeals and reports its conclusions to the House.

The Lords, other than the Lord Chancellor, do not wear wigs and gowns. The Lord Chancellor, if present, presides; in his absence the senior Lord of Appeal in Ordinary who is present presides.

The origins of the jurisdiction are preserved in that the appellant petitions for review for the judgment and for 'such relief as to Her Majesty the Queen in her High Court of Parliament may seem just'. Judgment is given by the House at a judicial sitting following the House's agreement to the committee's report. The House usually sits in the forenoon to give judgment before the afternoon sitting of the House for public business. The Lords give their opinions in the form of speeches and vote on the motion that the appeal be allowed. The case then returns to the courts below that it may give judgment in accordance with the motion carried in the House.

The whole House has jurisdiction in case of impeachment (q.v.), the lords being judges of both fact and law.

The House also had jurisdiction to try peers (including peeresses, and Scottish and Irish peers who were not Lords of Parliament, but not bishops) for treason or felony. If Parliament were sitting the accused was tried before the Lords as judges of fact and law, a person (usually the Lord Chancellor) being appointed as Lord High Steward to preside. If Parliament were not sitting the accused was tried before the Lord High Steward (normally the Lord Chancellor) and a jury of peers. The last case was that of Lord de Clifford (1935). The privilege of trial by one's peers was abolished by statute in 1948 at the instance of the Lords.

The House has jurisdiction in privilege cases, including contempt of the House, and peerage cases, including cases referred to it by the Crown concerning disputed claims to old peerages and as to the validity and effect of new peerages.

In 1898 the House accepted judicially the view that its decisions were binding on itself as well as on all lower courts, but in 1966 it was announced in an extra-judicial Practice Statement that the House asserted the power in exceptional circumstances to depart from its own previous decisions.

L. Blom-Cooper and G. Drewry, *Final Appeal* (1972).

House of Lords (as legislative body). The House of Lords, as upper House of the United Kingdom Parliament, originated in the practice of the barons and greater clergy, summoned to the mediaeval *Curia Regis*, sitting together, separate from the lesser clergy (who early abandoned attendance altogether) and from the knights of the shire and burgesses. The House was abolished in 1649 and restored in 1660. In earlier times the House was small; only 28 peers were summoned to the first Parliament of Henry VII in 1485. By 1707 the hereditary peerage had increased and was further increased by 16 representatives, elected by themselves from their own number for each Parliament, of the peers of Scotland. There were numerous creations of peerages in the eighteenth and nineteenth centuries. It was held, in 1782, that a Scottish peer might sit by virtue of a British peerage; thus the Duke of Hamilton (Scottish peerage, 1643) sat by virtue of being also Duke of Brandon (British Peerage, 1711) and only under the Peerage Act, 1963, were all Scottish peers enabled to sit. From 1800 the Irish peerage was represented by 28 peers, elected from among themselves for life, the rest being entitled to stand for and sit in the House of Commons for non-Irish seats; the total number of Irish peers was to be restricted to 100. No election of Irish representative peers took place after the Irish Free State was created in 1922, the last one died in 1961, and the relative legislation was repealed in 1971. All holders of hereditary peerages of Great Britain or of the United Kingdom have always been entitled to sit. To sit a peer must be 21 and not disqualified as being an alien, bankrupt, lunatic, or serving a sentence for treason or felony. In 1876 there was created the category of judicial life peers, i.e. the Lords of Appeal in Ordinary, and, from 1958, life peerages may be conferred generally, on women as well as men. By the Peerage Act, 1963, hereditary peeresses in their own right may sit in the House of Lords. A peer who disclaims his peerage under the Peerage Act, 1963, cannot sit unless given a life peerage (see, e.g. LORD HAILSHAM). The foregoing constitute the Lords Temporal.

The Lords Spiritual comprise the Archbishops of Canterbury and of York, the bishops of London, Durham and Winchester and the next 21 bishops of the Church of England in order of seniority of appointment. Other churches are not represented. The total membership of the House is now nearly 1,000, but only about 200–300 attend regularly. There is no limit to its size as peerages can be created at the wish of the sovereign, exercised in modern practice on the Prime Minister's advice.

Several times the question has arisen of creating a number of peers to carry a measure threatened by defeat in the House. This was done in 1712 and 1832 and, in 1910, it was known that the Prime

Minister had obtained the royal undertaking, if necessary, to create sufficient peers to carry, in the one case, the Reform Bill, and in the other, the Parliament Bill.

The officers of the House are the Speaker, now almost always the Lord Chancellor, the Chairman of Committees, who takes the chair when the House is in committee, the Gentleman Usher of the Black Rod, who executes warrants of commitment, and is now also the Serjeant-at-arms, who carries the mace, and the Clerk of the Parliaments, who keeps the journal of the House.

As a legislative House it has coordinate power with the House of Commons save that it cannot introduce or amend but may discuss financial bills, subject to the Parliament Acts, 1911 and 1949 (q.v.) and, particularly since 1832, the standing and influence of the Commons have been greater because the life of the Cabinet depends on keeping the confidence of the Commons. The Parliament Acts, 1911 and 1949, limited the time for which the Lords can delay the passing of legislation.

The functions of the House are the provision of a forum for debate on matters of public interest, the revision of Bills sent up by the Commons, the initiation and first consideration of less controversial public bills, consideration of subordinate legislation, scrutiny of the activities of the executive and scrutiny of private legislation. While legislation must be considered and passed by the House of Lords as well as by the House of Commons, the House of Lords has no power in financial matters and its unwillingness to pass a bill sent up by the Commons is a delaying power only (under the Parliament Acts, 1911 and 1949 (q.v.)) and not a power of veto. Nevertheless, it is a valuable legislative chamber and the intellectual quality of its members is high. The Speaker of the House is the Lord Chancellor who may, however, leave the Woolsack and speak for the government. In committee the chair is taken by the Lord Chairman of Committees.

Disagreement may arise between the House of Lords and the House of Commons as where either House refuses to accept the other's proposed legislation or makes amendments deemed unacceptable by the other. In practice, amendments are usually accepted, or not insisted on by the Lords, or a compromise reached after discussion by party leaders in the two Houses. A direct confrontation could be resolved, as has several times been threatened, in favour of the House of Commons by the creation by the Crown at the government's request of sufficient new peers to vote through the disputed Bill.

The House is not representative in any systematic way and proposals have been made repeatedly for reform of the composition and powers of the house but the House of Commons is fundamentally afraid of creating a strong House of Lords or a truly independent body.

House of Representatives. The lower house of the Congress of the U.S. (q.v.), comprising representatives elected from single-member constituencies in the several states in numbers related to their respective populations. The numbers are adjusted following each decennial census. Members are elected every two years. They must be at least 25, citizens of the U.S. for at least seven years and resident in the state for which they are elected. The House was intended to represent the popular will and shares with the Senate equal reponsibility for lawmaking. It also shares the power to override the presidential veto by a two-thirds vote. The Constitution vests certain powers in the house exclusively, notably the right to initiate money bills and the right to initiate impeachment proceedings.

The organization of the House has been materially influenced by the development of a two-party system. Much importance attaches to the positions of Speaker, and majority and minority leaders. It operates largely through committees organized mainly round major policy areas, and having subcommittees. Each committee is controlled by the majority party. They play an important role in the control which Congress exercises over government agencies and departmental heads, and senior officials are frequently summoned before them to explain policy.

The procedure of the House is based on the Constitution, Jefferson's manual, rules adopted by the House itself, and rulings by Speakers and chairmen of committees. Each party has its caucus or conference, floor leader, party whip, and several committees.

Housebreaking. In England, formerly the statutory crime of breaking and entering any dwelling-house or other building otherwise than between 9 p.m. and 6 a.m. and committing any felony therein, or, having committed any felony in such a building, of breaking out therefrom. In 1968 it was replaced by statutory burglary (q.v.). In Scotland it is an aggravation of theft and not criminal unless done to steal, or with intent to steal, or to do malicious damage.

Household, Royal. In all the mediaeval monarchies of western Europe the royal household developed from the band of companions chosen as the personal following of the Teutonic chieftain. In time these companions developed into offices of state and the chief administrators of the kingdom. After the Norman Conquest the ducal household of Normandy became the royal household of England. The principal offices in the Norman household appear to have been the treasurer, chamberlain, steward, butler, constable and marshal. Under feudal influence these offices became hereditary, being held as grand sergeanties of the Crown. Hence, they tended to become separated from practical functions which came to be exercised by other officials. Thus, the steward and the chamberlain came to be superseded in their domestic functions by the steward and chamberlain of the household and in their political functions by the justiciar and the treasurer. In time the holders of the original great offices of the household ceased to attend court, save on special occasions. The chief instance of a separation of offices for state and domestic purposes is the offices of Lord Chamberlain and Lord Great Chamberlain.

The royal household of modern times is a development from the mediaeval household but the duties and functions are for the most part performed by persons appointed for the purpose and selected in part from members of the political party in power. Until the early twentieth century it was settled constitutional practice that the principal officers of the Household changed on a change of government, but under an arrangement of 1924 the only officers to change on a change of government are the Treasurer, Comptroller and Vice-Chamberlain of the Household, who are government whips, and Lords-in-Waiting, the remaining offices being in the Queen's personal gift.

Houses of Correction. In 1552, King Edward VI gave Bridewell, an old royal palace in London, as an institution for the reception of vagabonds, idle and dissolute. It was alternatively known as a House of Correction. An Act of 1576 made provision for the assignment of houses of correction in every county for those who refused work or were punishable as rogues. Initially they seem to have had some success in dealing with idle rogues. The distinction between gaols and houses of correction disappeared by the end of the eighteenth century. Houses of correction were also erected on the continent, a famous one being the House of Correction erected at Amsterdam in 1575, which dealt with vagrants and thieves. Disorderly women and neglected children, who at first were dealt with there, were very shortly transferred to other institutions. There was compulsory work at low wages, part being saved against discharge. The Hanse towns in Germany followed the Dutch example and similar premises were erected in Switzerland. By the end of the eighteenth century there were 60 houses of correction in Germany. A similar development was the correctional quarters for boys in the Hospice of St. Michael in Rome, started in 1703, dealing with juvenile delinquents, while, in 1775, a House of Correction erected at Ghent in 1627 was reorganized as the famous de Force with an emphasis on industrial labour.

T. Sellin, *Pioneering in Penology: The Amsterdam Houses of Correction* (1944).

Houses of Parliament (or Palace of Westminster). The buildings in Westminster, London, occupied by the United Kingdom Parliament. A royal palace is thought to have existed on the site since the time of Canute (*c.* 1030). Buildings were erected by Edward the Confessor and enlarged by William the Conqueror. William II (Rufus) built Westminster Hall (q.v.) in which the courts of justice sat from the twelfth to the nineteenth century. In 1512 the palace was badly damaged by fire and ceased to be used as a royal residence. Henry VIII was the last king to live in the Palace of Westminster. From 1547 the House of Commons used St. Stephen's chapel as a meeting place, the lords meeting in another part of the palace, first the White Chamber and later the Painted Chamber. In 1834 the whole palace, excepting Westminster Hall and St. Stephen's chapel, was destroyed by fire. The present buildings were erected in 1840–60 to the designs of Sir Charles Barry and A. W. N. Pugin, in late Perpendicular style with Gothic details. The Central Lobby gives access to the House of Commons (at the Westminster Bridge or Big Ben end) and the House of Lords (at the Victoria Tower end) and there are also libraries for each House, a residence for the Speaker, and numerous committee and other rooms and offices. The buildings were badly damaged by bombs during World War II and the Commons chamber was destroyed and, as redesigned by Sir Giles Gilbert Scott, not reoccupied until 1950. The Lords Chamber is 97 feet long, the Commons Chamber 70 feet long. The Clock Tower, 329 feet high, contains the famous clock with its 13 ton bell, known as Big Ben. A flag flies and, after dark, the light above Big Ben burns till the Commons adjourns, when it is extinguished. The Victoria Tower is a repository for parliamentary records. Since 1950 alterations have been made, but the Palace is basically unchanged since the rebuilding in the nineteenth century.

See also HOUSE OF COMMONS; HOUSE OF LORDS; PARLIAMENT.

Howard, John (?1726–90). Philanthropist and prison reformer. In 1773 he became high sheriff of Bedfordshire and, in that capacity, the defective arrangements of the prisons came to his notice. From that time he devoted himself entirely to philanthropy. He travelled all over the country unearthing abuses and scandals in the prison system, and extended his tours to Ireland and the Continent. His great work, on the *State of the Prisons in England and Wales* was published in 1777. He made further tours abroad and published, in 1780, an Appendix to his work. The abuses exposed in his writings gave the initial impetus to the movement for improvement in the building and management of prisons. In 1774 he secured the passage of legislation stopping remuneration of gaolers by prisoners' fees rather than by salaries and partly inspired another Act, of 1779, never implemented, authorizing the establishment of prisons intended to reform prisoners by religious instruction, supervised labour and solitary confinement at night. His name was taken for the Howard League for Penal Reform.

Howe* v. *Lord Dartmouth (1802), 7 Ves. 137. This case lays down the principle that, it being assumed, in the absence of evidence of contrary intention, that the testator leaving residuary personal estate settled, had intended that his legatees should enjoy the same thing in succession, such parts of the estate as are of a wasting or reversionary character should, as between tenant for life and remainderman, be converted and invested in permanent investments of a recognized character. Certain exceptions are recognized to the rule.

Howel Dda (or Hywel the Good) (?–950). A famous king of most of Wales, who maintained peace and good relations with England and often attended the English court. A collection or codification of Welsh law is attributed to him and he appears to have achieved some coordination of previous Welsh customary law. They cover a wide range of subjects and have affinity with Anglo-Saxon laws.

A. Owen (ed.), *Ancient Laws and Institutes of Wales*; A. Owen Jones *et al.* (eds.) *Myvyrian Archaeology of Wales*; M. Richards, *Laws of Hywel Dda.*

Howell, Thomas Bayly (1768–1815). Was called to the Bar in 1790 and became editor of the *State Trials*, vols. 1–21 (1809–15). His son Thomas Jones Howell (?–1858) edited the *State Trials*, vols. 22–23 (1815–26).

Huber, Eugen (1849–1923). Swiss jurist, Professor of Law at Basel from 1880, and at Berne from 1892, author of *System und Geschichte des schweizerische Privatsrechts* (1886–93) a systematic history and exposition of the principles of Swiss law, and other works on law, and draftsman of the Swiss Civil Code (*Schweizerisches Zivilgesetzbuch*) which came into force in 1912.

Huber, Max (1874–1960). Swiss lawyer, Professor of Law at Zurich, and writer on international law. He was a representative to the League of Nations and various governments, and was later a judge of the Permanent Court of Arbitration, and of the Permanent Court of International Justice, 1921–30, and its President 1925–28.

Huber, Ulricus (1636–94). A famous Dutch jurist, professor at Franeker, Utrecht, and Leyden and a judge in Friesland, author of *De jure civitatis* (1682), *Heedendaagse Rechtsgeleertheyt* (1686) later edited by his son and now translated as *Jurisprudence of my Time* (1939), *Praelectiones juris civilis* (1687), containing an important chapter on conflict of laws, and *Digressiones Justinianiae* (1670). His works have been important in South Africa.

Huber, Zacharias (1669–1732). Son of the foregoing, edited his father's work and compiled an important collection of *Decisiones Frisicae* (1723).

Hubert Walter (?–1205). Archbishop of Canterbury (1193–1203) and Chancellor of England. He was brought up in the household of his uncle Ranulf de Glanville (q.v.). He was justiciar of England 1194–98 and, as such, he governed England, devising new taxation, new record-keeping, and improved local government. He may have been the author of the treatise *De Legibus et Consuetudinibus Angliae* ascribed to Glanville.

Hudson, Manley Ottmer (1886–1960). Served as a commissioner on Uniform State Laws and was a member of the American delegation to the Paris peace talks in 1919, and later a legal adviser to International Labour Conferences. He became Bemis Professor of International Law at Harvard in 1923, devoted himself to the advancement of the rule of law among nations, and published several collections of materials on that subject. He directed the Harvard Research in International Law between 1927 and 1939. He became a member of the Permanent Court of Arbitration at The Hague, 1933–45, and judge of the Permanent Court of International Justice in 1936–46, campaigning to bring the U.S. into the Permanent Court. In 1948 he became Chairman of the U.N. International Law Commission. He edited the *World Court Reports*, 1922–46 (4 vols. 1932–42), the volumes of *International Legislation*, 9 vols. 1931–50, and wrote *Progress in International Organisation* (1932), *Cases in International Law* and *International Tribunals* (1944), and *The Permanent Court of International Justice*, a standard authority (2 edn. 1943). He also edited the *American Journal of International Law*, 1924–60.

Hue and cry. A principle of early English criminal law was that if a robbery or felony was committed the inhabitants of the hundred where it was committed were liable for the damage sustained by the person injured, unless they raised the hue and cry, i.e. made an outcry calling for pursuit and captured the criminal within 40 days. If the hue and cry did not effect a capture the inhabitants would be amerced. But neighbouring hundreds were uncooperative and the liability fell on the more substantial inhabitants. A statute of 1584 imposed partial liability on neighbouring hundreds and provided for its recovery. The relevant statutes were repealed in 1827 but later legislation (now repealed) provided that every person in a county must be ready at the command of the sheriff and the cry of the county to arrest a felon on pain of fine.

Huesca, Code of. The most important legal code of mediaeval Aragon, prepared by Bishop Vidal de Canellas and promulgated by King James I of Aragon in 1247. It is named from the town of Huesca. Its original function was to define the territorial boundaries of Aragon. In the original Latin version it extended to eight books, but in the Aragonese translation it expanded to 12 volumes and not only served its original purpose but served as a source of civil and criminal law until the fifteenth century.

Hughes, Charles Evans (1862–1948). Attended Colgate and Brown Universities and Columbia Law School, and was admitted to the New York Bar in 1884. He was, thereafter, a Professor of Law at Cornell 1891–93 and a Special Lecturer there, 1893–95, and at New York, 1893–1900. He was Governor of New York, 1906–10, and then an Associate Justice of the Supreme Court, 1910–16, from which he resigned to run for President and nearly won the election against Woodrow Wilson. He was Secretary of State, 1921–25, under President Harding, in which capacity he secured approval of the peace treaties and of the Washington disarmament treaties. He was chairman of the Washington Disarmament Conference, 1921–22, a member of the Permanent Court of Arbitration at The Hague, 1926–30, a Judge of the Permanent Court of International Justice, 1928–30, and eleventh Chief Justice of the U.S. Supreme Court, 1930–41.

When he became Chief Justice, Hughes was the acknowledged leader of the American Bar and a lawyer of international reputation. As a judge his judgments made a thorough analysis of the facts and the authorities. He excelled in the many difficult constitutional cases arising from the increase of social welfare legislation, generally favouring the exercise of federal power. He also greatly increased the efficiency of the Federal Court system and showed a high degree of statesmanship in managing the Court during the debate on Roosevelt's reorganization plan of 1937. He wrote several books including one on *The Supreme Court of the United States* (1928).

M. J. Pusey, *C. E. Hughes*; D. Perkins, *C. E. Hughes*; Ranson, *C. E. Hughes*; Hendel, *Charles Evans Hughes and the Supreme Court.*

Hugo (?– *c.* 1170). One of the 'four doctors' who succeeded Irnerius in the Law school at Bologna and developed the revised study of the Roman law by glossing the text.

Hugo, Huguccio (or *Ugo*) **(Hugh of Pisa)** (*c.* 1145–1210). Famous canonist and decretist, Bishop of Ferrara, author of grammatical and theological works, including a great dictionary of legal and ecclesiastical terms, *Liber Derivationum*, and of a *Summa super decreta* (1188–90), the most complete of all commentaries on Gratian's *Decretum*, founded on extensive knowledge of civilians, canonists, and theologians. It is not merely a compilation of glosses but a personal exposition, discussion and elaboration of the texts utilizing Hugo's own opinions, which played an important part in the development of the legal doctrines of the Church.

Hugo, Gustav von (1764–1844). German jurist, professor at Gottingen, founded on Leibnitz and became founder of the historical school of jurisprudence continued by Savigny. He wrote on Roman law and *Lehrbuch eines zivilistischen Kursus* (1792–1821), *Zivilistisches Magazin* (1790–1837), and extensively on Roman law history.

Hugolinus (?–*c.* 1233). A famous Bolognese jurist, one of the leaders of his time, reputed to be the author of the tenth collection of Novels and of glosses on the Digest, Code, and of various legal opinions.

Hulks. Old warships moored at various places off the coast which, from the late eighteenth century, when the American colonies would no longer take criminals, received prisoners sentenced to transportation. Many of these prisoners served their sentences in the hulks and were released without ever being away from this country. Their use ended in 1853, when penal servitude replaced transportation.

W. Branch-Johnson, *The English Prison Hulks.*

***Hulton* v. *Jones*,** [1910] A.C. 20. An article in a newspaper made defamatory remarks about a character Artemus Jones, believed to be fictitious. A respectable person named Artemus Jones (later a county court judge) recovered damages for libel, it being held no defence that the defamation was quite unintentional, if the matter could be, and in fact was, understood by some other persons as referring to the plaintiff. The Defamation Act, 1952, now allows a defendant in such circumstances to tender amends.

Human act. An act done by a human being as distinct from a happening by force of nature. For legal purposes a distinction must be drawn between voluntary or willed human acts and involuntary acts, such as sneezing, or where the will does not direct the act but external physical force compels the act or where the will is affected by disease, drink, or drugs. A willed human act involves perceiving an end, making a judgment and a choice of means, and deciding to attain the desired end by carrying out the chosen course of action. It involves interplay between the intellect (apprehending, judging, choosing) and the will (intending, deciding). Among willed acts the distinction must be taken between acts willed with good intention and those willed with evil intention, though law is concerned not only with the intention with which an act is done but with its nature and its consequences. Willed acts normally give rise to legal liabilities and other consequences of many kinds, and on many legal grounds.

Unwilled acts do not, in general, subject to liability, civil or criminal, even for harmful consequences resulting, though the actor must generally show that his act was not willed by him but was one in which his will was overcome, and serious difficulties arise with cases of 'losing one's temper' or 'irresistible impulse'.

Human rights (or rights of man or fundamental freedoms). Claims asserted as those which should be, or sometimes stated to be those which are, legally recognized and protected to secure for each individual the fullest and freest development of personality and spiritual, moral and other independence. They are conceived of as rights inherent in individuals as rational, free-willing creatures, not conferred by mere positive law, nor capable of being abridged or abrogated by positive law.

The origins of assertions of human rights are to be found in the ideas of natural law and natural rights. These ideas were developed by the Greek and Roman Stoics, by the Roman lawyers and the Christian fathers, and by Aquinas and some of the mediaeval English jurists, as the basis of the beliefs in the freedom and equality of all men; the crucial case was slavery, which positive law commonly recognized but natural law condemned. The ideas were based partly on speculation and partly on observation of the laws and customs followed by the generality of mankind. In the Middle Ages, natural law underlay much legal and political thinking and emphasized the subordination of rulers to the rule of law, and also came to be identified with the law of God.

Ancient bodies of law did not seek to define any kinds of individual freedom protected from state interference and the first declarations of rights of at least the nobles of the land, are found in the mediaeval accords between kings and barons or feudal assemblies. Thus, in 1188 the Cortes of Leon obtained from King Alfonso IX confirmation of a series of rights, including the right to the inviolability of life, honour, home, and property, and the right of an accused to a regular trial. In 1222 the Golden Bull of King Andrew II of Hungary guaranteed that no noble would be arrested or ruined without being first convicted in accordance with judicial procedure. In 1215 King John of England assented to the demands of his barons contained in Magna Carta. This was a set of baronial demands and not an assertion of the rights of all individuals but some of its clauses, notably clause 39 to the effect that no freeman shall be taken or imprisoned or exiled or in any way destroyed except by the lawful judgment of his peers and the law of the land, expressed an idea capable of extension and repeatedly invoked subsequently as applicable to all men.

From the seventeenth century, the law of nature was utilized as a basis of modern international law, a body of principles developing from and necessitated by the growth of the near sovereign nation-states arising from the ruins of the mediaeval spiritual and temporal unity of Christendom. The founders of modern international law stressed the value and importance of the natural rights of man. Vitoria asserted that the primitive peoples of America were entitled to the protection of law and justice. Grotius originated the idea of international humanitarian intervention for the protection of fundamental rights. Vattel, Wolff, and Pufendorf also exercised powerful influence on the notion of the inalienable rights of man.

Similarly, from the mid-seventeenth century the practice began of safeguarding, by international treaties, the rights of religious freedom and the rights of aliens.

In early modern times natural law was invoked by Puritans, Levellers, and Parliament men against the King, by Milton and Hale, by Grotius and Locke, and later by Blackstone. Human rights, under the name of 'The immemorial rights of Englishmen' were in issue in the English Civil War and were asserted in such documents as The Petition of Right (1628) and the Bill of Rights (1689).

Human rights were clearly asserted in the Virginia Declaration of Rights of 1776, which commenced: 'That all men are by nature equally free and independent, and have certain inherent rights of which, when they enter into a state of society, they cannot ... deprive or divest their posterity; namely, the enjoyment of life and liberty, with the means of acquiring and possessing property, and pursuing and obtaining happiness and safety.' This was followed in various terms, by the Bills of Rights of other states, notably Pennsylvania and Massachusetts, and in the American Declaration of Independence of 1776, which asserted it to be 'self-evident, that all men are created equal, that they are endowed by their Creator with certain inalienable rights, that among these are Life, Liberty and the Pursuit of Happiness', and the Bill of Rights, namely the first 10 amendments to the U.S. Constitution (1791).

The French Declaration of the Rights of Man and of the Citizen of 1789 was inspired by the writings of Tom Paine (q.v.) and the American example as much as by European philosophers of the Enlightenment.

In the nineteenth and twentieth centuries, assertions of certain fundamental rights of man became common in the constitutions of developed states. These include the constitutions of Sweden, 1809, Spain, 1812, Norway, 1814, Belgium, 1831, Sardinia, 1848, Denmark, 1849, Prussia, 1850, most of the German states, 1871, and Switzerland, 1874. After 1918, Germany and most of the new European states incorporated similar statements in their constitutions and the Basic Law of the Federal Republic of Germany (1949) firmly restated the rights of individuals. Latin-American and Asiatic states have done the same. But experience has shown, in many cases, that the real question is not whether fundamental rights are asserted on paper but whether they are recognized in practice and can be secured.

The law of nature was, however, sometimes invoked to justify opposition to universal suffrage and to the equality of the sexes, to state interference in economic affairs, freedom of contract and rights of private property. Particularly in the U.S., the ideas of the law of nature and natural rights were invoked to curb state interference with rights of private property and freedom of contract.

Great Britain has remained an exception to the trend to codify the basic rights of the individual. The constitution as a whole, including the rights of the citizen, have to be gathered from a long series of enactments and cases, any of which could be abolished or altered by an ordinary statute, or even by a decision of a superior court. These include Magna Carta, the Petition of Right, the Habeas Corpus Act, the Bill of Rights, the Act of Settlement and other statutes, and many leading cases. Many of these statutes and cases powerfully influenced later generalized formulations such as the American and French Declarations.

Moreover, down to the eighteenth century, the doctrine of the absolute supremacy of Parliament, unrestrained by any higher law, was not accepted: it originated in Blackstone's explanation of the position since 1688. But Coke and Holt in *obiter dicta* recognized the overriding authority of a higher law, and even Blackstone asserted the inalienable and absolute right of man, which it was the primary end of human laws to maintain. Locke asserted the rights of man against those of the State and Paine did little more than elaborate on Locke.

In developed democratic countries, recognition of some of the rights asserted as human rights has been steadily proceeding for many years. Thus, in the U.K. the Married Women's Property Acts, the Representation of the People Acts, the Equal Pay Act, the Race Relations Act, and similar measures, have all sought to secure various human rights.

Another aspect of regard for human rights has been the intervention of states by international action seeking to secure human rights for inhabitants of other states. Thus, in 1827 Britain, France, and Russia took action against the Ottoman Empire to end the grievances of the people of Greece suffering under Turkish government, which led to the independence of Greece in 1830. Similarly governments have, on various occasions, sought to end massacres of Christians in Syria (1860), to end persecution in Crete (1866–68) and to secure fair treatment for Jews in Russia and elsewhere.

International treaties have also been utilized to protect religious and other minorities from the seventeenth century; thus the Treaty of Westphalia (1648) sought to establish equality of rights for Catholics and Protestants in Germany. By the Treaty of Kutchuk Kainardji (1774), the Turks undertook to protect the Christian religion in its territories. The Congress of Vienna of 1815 provided for the free exercise of religion and for equality of religions in Switzerland and Germany. When various Balkan states achieved independence from Turkey they had to guarantee religious freedom and equality of rights to all inhabitants.

Another aspect of international activity was the movement for abolition of slavery and of the slave trade. The movement gradually acquired momentum and was largely achieved during the nineteenth century, though as late as 1926 the League of Nations approved an International Slavery Convention binding the signatories to prevent and suppress the slave trade and achieve its total abolition.

Similarly, later in the nineteenth century, there were moves to establish international regulation of social and labour issues. In 1905–6, two multilateral labour conventions were concluded at Berne, which were also the first international conventions for protecting personal safety. While not far-reaching they were noteworthy stages and foreshadowed the valuable work done by the International Labour Organization since 1919. It has made material contributions to the securing of human rights in relation to industrial health, safety, and welfare, hours of work and holidays with pay, the abolition of forced labour, the elimination of discrimination in employment, freedom of association and equal pay for equal work.

The nineteenth century also witnessed the major moves to protect victims of warfare. The Convention for the Amelioration of the Condition of the Wounded in Time of War was signed at Geneva in

1864, and the Declaration of St. Petersburg followed in 1868. These were the bases of other conventions at the Hague Conferences (1899 and 1907), and of the Geneva Conventions of 1949.

The Covenant of the League of Nations, though not containing any statement of human rights, contained various provisions making for the international recognition of the rights of man. Members of the League accepted an obligation to secure and maintain fair and humane conditions of work for all persons, the just treatment of natives in colonies, entrusted the League with supervision of agreements regarding traffic in women and children, and pledged themselves to take steps in matters of international concern for the prevention and control of disease. In some of the post-World War I peace treaties certain states had to accept obligations towards racial, religious, and linguistic minorities.

International protection of human rights was probably first found in the treaties after World War I, which provided for protection of life and liberty of members of minority groups and free exercise of their religion without discrimination on grounds of language, race, or religion. Similarly, territories subject to League of Nations mandates or U.N. trusteeship have special protection for human rights and fundamental freedoms. The work of the International Labour Organization has also been of practical importance in ensuring certain human rights.

The Charter of the United Nations proclaims, in its preamble, a reaffirmation of faith in fundamental human rights, and such rights figure in the statement of the purposes of the U.N., the functions of the General Assembly, and in the objects of the organization in the field of economic and social co-operation and the Trusteeship system. Members of the U.N. are under an obligation to act in accordance with its purposes. The charter does not, however, define the human rights and freedoms which members are bound to observe, and there is no express provision by which members agree to respect human rights and fundamental freedoms. Moreover, the 'domestic jurisdiction' clause of the Charter prevents the organization from intervening in matters essentially within the domestic jurisdiction of a member state and from requiring members to submit such matters to settlement under the Charter.

By Article 68 of the Charter, the Economic and Social Council is required to set up, *inter alia*, a Commission on Human Rights and this was done in 1946, though the Commission itself denied that it could take effective action on complaints of violations of human rights.

An International Bill of the Rights of Man was proposed in the course of drafting the U.N. Charter and the Charter of the U.N. gave further impetus to the protection of human rights. The Economic and Social Council set up a Commission on Human Rights in 1946. In 1948 the General Assembly adopted a Universal Declaration of Human Rights, but it is not legally enforceable, though some of its provisions constitute general principles of international law. The Commission on Human Rights prepared Covenants on Civil and Political Rights and on Economic, Social, and Cultural Rights, which were opened to signature and ratification in 1966. Other international conventions with objects of the same kind include the Genocide Convention of 1948, the Supplementary Convention on the Slave Trade of 1956, and Conventions against Forced Labour (1957), Discrimination in Employment and Occupation (1958), Discrimination in Education (1960), Racial Discrimination (1956), and others.

The Statute of the Council of Europe requires each member to accept the principles of the rule of law and of the enjoyment by all persons of human rights and fundamental freedoms, and in 1954 the members adopted a Convention on Human Rights providing for a European Commission of Human Rights and a European Court of Human Rights, the former to act as a commission of inquiry with limited advisory and mediatory functions, the latter to adjudicate on cases concerning the interpretation and application of the Convention brought before it by a High Contracting Party or the Commission. Any person, organization, or group may petition the Commission, alleging violation of the rights set out in the Convention. The Organization of American States established, in 1959, an Inter-American Commission on Human Rights on similar lines with an Inter-American Court of Human Rights.

U.N. Yearbook on Human Rights (1946–) *Yearbook of the European Convention on Human Rights* (1955–); E. Lauterpacht, *International Law and Human Rights*; A. Robertson, *Human Rights in Europe*; Weil, *European Convention on Human Rights*; Morrison, *The Developing European Law of Human Rights*; J. E. S. Fawcett, *Application of the European Convention on Human Rights.*

Humanists. Legal scholars, particularly of the years 1400–1600, influenced by the revival of ancient, particularly classical, learning which affected letters, manners, and thought in every sphere. Their approach was generally one of renewed respect for the Roman law, desire to arrive at the correct text and to restore the classical Roman law itself, and they succeeded the glossators (q.v.) and commentators (q.v.) who had concentrated respectively on the meaning of the texts and their application to contemporary conditions. Scholarly editions of texts were produced with commentaries which gave new interpretation of many of the concepts and principles.

The new approach was heralded in Italy by Andreas Alciatus (1492–1550) who taught at Avignon, Milan, Pavia, and Bologna and who wrote a famous commentary on the Digest, followed by Scipio Gentilis (1563–1616), Lelius Torelli, Emilius Ferretti, Hippolytus di Colle, Turamini di Siena, Julius Pacius, and others. In France, Guillaume Budé (1467–1540) published his notable *Annotationes in Pandectas* in 1508, and in Germany, Ulrich Zasius (1461–1535) published his *Lucubrationes* in 1518.

In this period, moreover, the nation-states of modern Europe were developing and law was growing apart on national lines. In France, where humanist jurisprudence flowered most, the influence of Alciatus at Bourges led to the work of Jacques Cujas (1522–90), who had an enormous knowledge of the sources and whose guiding purpose was the reconstruction of the texts of the jurists utilized in Justinian's compilations, followed by Molinaeus, Brissonius (1531–91), Duarenus (1509–59), Denis (1549–1622) and Jacques (1587–1652) Godefroy. The breadth of their learning and the acumen displayed led to the humanist jurists in France being called the 'Elegant School of Law' and their period the 'Golden Age of Jurisprudence'.

The Bartolist faction was, however, led in opposition to Cujas by Donellus (1527–91) and Hotman (1525–90) who maintained that law must be studied in its relations to everyday needs, and whose work led on to that of Domat (1625–95) and Pothier (1699–1772).

Though the group includes many scholars with differing emphases and approaches the Humanist jurists as a group are important for their application of philological and historical methods to the study of Roman law and as having provided the foundation for the modern study of legal history.

In the seventeenth and eighteenth centuries the Humanists were replaced by the natural law school but the value of their work remains.

Koschaker, *Europa und das römische Recht* (1947); Gilmore, *Humanists and jurists* (1963).

Humble Petition and Advice. In 1657, after England had been ruled by military force for several years, the second Protectorate Parliament presented the Humble Petition and Advice which was, in substance, a revision of the constitution established by the Instrument of Government (q.v.). In its first form it offered the Crown to Cromwell who, after hesitating, refused it. In its final form, amended by the Additional Petition and Advice, it left him as Lord Protector. The Lord Protector was to have power to nominate his successor and to call Parliaments of two Houses at least once in three years. The Council of State was re-designated the Privy Council and legislation against bishops and other ecclesiastical officers was to be unaffected. The upper House was to consist of 40 to 70 members nominated by the Lord Protector. The government was granted a fixed revenue with a specific military allocation.

The changes show a distinct swing of opinion back towards monarchy and a bicameral legislature.

Hume, David (1711–76). British empiricist philosopher, historian and economist, and a grandson of Lord President Falconer, who studied at Edinburgh and in France and published his *Treatise of Human Nature* in 1739–40. He returned to Scotland and, as secretary to General St. Clair, participated in an embassy to Vienna and Turin, publishing, in 1748, his *Philosophical Essays, concerning Human Understanding*, later renamed *An Enquiry concerning Human Understanding.* From 1751 to 1763 he lived mostly in Edinburgh as Librarian of the Advocates' Library, publishing *Political Discourses*, a revision of Book III of the *Treatise* called an *Inquiry concerning the Principles of Morals* (1751), *Essays and Treatises on Several Subjects* (1753), the *History of England* (1754–62), the first attempt at a comprehensive treatment of historical facts, and *Four Dissertations—The Natural History of Religion, Of the Passions, Of Tragedy*, and *Of the Standard of Taste.* In 1763–69 he was at the Embassy in Paris and at the Foreign Office in London but thereafter lived in Edinburgh.

Hume contended that only knowledge obtained by mathematical reasoning was certain; knowledge obtained from other sciences was only probable. His theory of justice is that it serves both an ethical and a sociological function. Public utility is the sole origin of legal justice and the sole foundation of its merit. It is approved because of its utility. A legal system to be useful must adhere to its rules even though it cause injustice in particular cases. He did not make a formal analysis of law but distinguished equity or the general system of morality, the legal order, and law, as a body of precepts. The authority of civil law modifies the rules of natural justice according to the particular convenience of each community and the laws must always make reference to the circumstances of the particular society. He was a pioneer in attempting to explain law in psychological terms. His work was influential on Bentham and Mill, and on Comte, and remains significant, though not always accepted.

E. C. Mossner, *Life of Hume*; Leslie Stephen, *English Thought in the Eighteenth Century*; C. Hendel, *Studies in the Philosophy of David Hume*; N. Kemp Smith, *The Philosophy of David Hume.*

Hume, David (1757–1838). Nephew of philospher David Hume (q.v.), passed advocate in 1779, and was Professor of Scots Law in the University of Edinburgh, 1786–1822. His *Lectures* delivered in that capacity have been published by the Stair Society and give a valuable picture of the law at that time. In 1811, he became a Principal Clerk of Session and, in 1822, became a Baron of the Scottish Court of Exchequer. He published a valuable volume of *Reports of Decisions*, 1781–1822, and *Commentaries on the Law of Scotland respecting Crimes* (1797–1800 and later editions). This is founded on the most thorough investigation of the records of the High Court of Justiciary and has ever since been the classic Scottish text on criminal law, one which is still of the highest authority and value.

Life in Hume's *Lectures* (Stair Socy.) vol. 6.

Hume, Patrick, 1st Lord Polwarth and 1st Earl of Marchmont (1641–1724). Studied law in France and entered the Scottish Parliament in 1665. His resistance to government policy against the Covenanters led to his imprisonment in 1675–79. He had later to take refuge in England and abroad. He returned with William of Orange in 1688, sat in the Convention Parliament and in 1693 became an Extraordinary Lord of Session. In 1696 he became Lord Chancellor of Scotland but was superseded in 1702. Thereafter he became a leader of the squadrone party but came into favour again after the accession of George I.

Hundred (or wapentake). A group of townships in pre-norman England, usually called a hundred or, in Danish districts, a wapentake, in the north a ward, in Kent a lathe, in Sussex a rape, and forming a division of a shire. It is uncertain on what principle a hundred was established and whether it represented the area settled by 100 households of the Anglo-Saxon invaders or that comprising 100 hides (q.v.) of land. It had a hundred moot or assembly, held every four weeks and attended by the thegns of the hundred and representatives from each township. The hundred court was held monthly, or latterly three-weekly, presided over by the Sheriff or his deputy. In some cases the owner of the hundred held a court leet and enjoyed a right of view of frankpledge to the exclusion of the sheriff. Until 1886 the most important of the surviving duties of the hundred was the liability to make good damage caused by rioters.

Hundred court. The court of a hundred (q.v.). In the twelfth century it was held monthly but from the thirteenth to the fifteenth centuries it usually sat 17 times a year, sometimes at a fixed place, sometimes at various places in the hundred. Twice a year, a specially full court was held, *hundredum magnum* or *hundredum legale,* this being at the sheriffs' tourn and held after Easter and Michaelmas. The chief purpose of these special sessions was to ensure that all who should be were in the frankpledge (q.v.) and that the articles of the view of frankpledge had been duly observed. Each township of the hundred was represented by a number of suitors who were bound to appear at these sessions without individual summons; if they did not appear then the township was amerced.

All the courts had jurisdiction in debt, covenant, and trespass not exceeding 40 shillings; in these cases the freeholders of the county who were present were judges. At the full courts, criminal business was transacted and the sheriff or lord of the hundred was the sole judge. By reason of the fall in the value of money hundred courts decayed rapidly under the Tudors and most of them were extinguished in 1867. The last survivor, the Salford Hundred Court, was abolished in 1971.

Hundred rolls. Rolls recording the results of an inquiry made in 1274–75 by royal commissioners from juries of hundreds, boroughs, and liberties to ascertain what franchises were exercised. They record also specification of the services due from various categories of tenants to their lords. The returns were nicknamed Ragman Rolls, from the strips of parchment with seals attached which hung from them. Many returns were printed in *Rotuli Hundredorum* (1812–18).

Hung jury. A jury which has been unable to reach a verdict.

Hurdle. A sledge on which, prior to 1870, traitors were dragged to their execution.

Hurst, Sir Cecil James Barrington (1870–1963). Was called to the Bar in 1893 but, in 1902, became first Assistant Legal Adviser (1902–18) and then Legal Adviser to the Foreign Office (1918–29) and played a large part in the Paris Peace Conference of 1919 and the creation of the League of Nations. He also frequently represented Britain before the Permanent Court of International Justice.

In 1929 he became a judge of the Permanent Court of International Justice, was President, 1934–36, and Vice-President, 1936–46. He was not a judicial innovator but a very respected and able judge. He was also British member of the Permanent Court of Arbitration at The Hague, 1929–50, and first President of the United Nations War Crimes Commission, 1943–44. He was a leader in establishing the *British Year Book of International Law* in 1922, and contributed some important papers to that journal, was President of the Grotius Society for some years, and took a large part in the meetings of the Institute of International Law.

Husband and wife. The legal relationship between a man and a woman created by a legally valid marriage, from which flow numerous legal consequences, including the duties to adhere to each other and cohabit at bed and board, to maintain each other, and to be sexually faithful to each other. The husband has the prior right to determine the place of residence and the style in which they will live. The spouses may agree to separate and live apart, or either may petition the court for separation, or, if deserted, for maintenance. In certain circumstances the marriage may be declared legally null and void, or be dissolved by decree of dissolution.

See DIVORCE; NULLITY OF MARRIAGE.

At common law, both in England and Scotland, a married woman's legal personality was submerged in that of her husband and she was, in general, incapable of acquiring, holding, or alienating any property, independently of her husband. This position was alleviated in the wife's favour by equity and since the latter part of the nineteenth century has been completely altered by statute, so that a married woman has now full contractual and proprietary powers independently of her husband. Each is alone liable for his or her ante-nuptial obligations. Each owns his or her own property and may administer it independently, but property may he held jointly or in common and some property frequently is so held. Either party may contract with the other, employ the other, be tenant of the other, be a partner of the other, act as agent for the other, or sue the other in contract or tort. Savings made by the wife from an allowance made to her for housekeeping purposes belonged, at common law, to the husband but by statute belong to both in equal shares. In England a deserted spouse has a right to remain in the matrimonial home, though it is owned by the other, for so long as the court thinks just, but this right terminates when the marriage is dissolved. But certain traces of the former legal unity of the spouses survives, in that their incomes are aggregated for tax purposes and the wife's domicile was, until 1973, determined by and changed with that of her husband.

The children of a married couple are deemed legitimate.

Interference with the relationship of husband and wife gives rise, in England, to a claim by the husband for damages for loss of consortium, and formerly, for damages for enticement and for damages for adultery, and in Scotland to claims for enticement and, possibly, for loss of consortium. If the death of either spouse is caused by the fault of a third party, the surviving spouse has a right of action against the third party for the pecuniary loss caused by the death, under the Fatal Accidents Acts (q.v.) in England and at common law in Scotland.

Neither spouse is criminally liable for the act of the other unless it is committed at the instigation or command of the other.

On the death of either spouse, the survivor has rights in the estate of the deceased under the rules of intestate succession.

The relationship is terminated only by death of either spouse or by divorce. See DIVORCE.

Hustings. Prior to the Ballot Act, 1872, candidates for election to Parliament were nominated on a platform known as 'the hustings', and they addressed the electors from there. Subsequently the word has come to be used, colloquially, to designate the whole proceedings of an election.

Husting, Court of. The most ancient and highest court in the City of London, originally having jurisdiction over all actions to any amount, but later restricted to real and mixed actions, personal actions and ejectments being dealt with in the Mayor's Court and the Sheriffs' courts. It was subsequently divided into the Court of Husting for Pleas of Land and the Court of Husting for Common Pleas, held in alternate weeks. The Lord Mayor and sheriffs are nominally the judges, but in fact the Recorder of London is the judge. Wills and deeds may be enrolled in the court but this is rarely now done.

Hybrid Bill. A hybrid bill is a Bill before Parliament which is general in application and, accordingly, a public Bill, but which in its application particularly affects private or local interests and which is accordingly referred to the Examiners of Private Bills, and is treated in its intermediate stages generally in the same way as a Private Bill but ultimately becomes a public Act.

Hyde, Charles Cheney (1873–1952). Was admitted to the Illinois Bar and began to specialize in international law. He became a Professor of Law at Northwestern University in 1907 and Solicitor to the Department of State, 1923–25. In 1925 he became Professor of International Law and Diplomacy at Columbia, retiring in 1945. He was a member of the Permanent Court of Arbitration at The Hague, 1951–52. He wrote a major book, *International Law, Chiefly as Interpreted and Applied by the United States* (1922).

Hyde, Edward, 1st Earl of Clarendon (1609–74). Nephew of Sir Nicholas Hyde, chief justice of the King's Bench, 1627–31, was called to the Bar in 1633 but devoted himself, at first, to history and letters rather than law. From 1640 he sat in Parliament and sided with the popular party, but later abandoned it and so obstructed the Root and Branch bill of 1641 that it was dropped. He recommended to Charles I that he should adhere to strictly legal and constitutional methods and increasingly became a servant of the King. He was knighted and admitted to the Privy Council in 1643. He left England with the Price of Wales in 1646 and did not return to England till 1660, being

Charles II's most trusted adviser during the 1650s. At the Restoration he became Chancellor and Baron Hyde and, in 1661, Earl of Clarendon. He was removed from office in 1667 and the House of Commons sought to impeach him; he fled to France and was banished by Parliament, and spent the rest of his life in exile. He was firstly a statesman, secondly an historian, and only subsidiarily a judge. As such, however, he endeavoured to reform the abuses of the court and acted very impartially, never, it is said, making a chancery decree without the assistance of two judges. He was a great lover of books and collected a fine library. His *History of the Rebellion* (1702–04) is the most valuable contemporary account of the civil wars and its printing gave a name to the Clarendon Press, Oxford. He also wrote his own *Life* (1759) and various minor works, and left a huge and valuable body of correspondence.

Hyde, Robert (1595–1665). Cousin of Clarendon (q.v.) and nephew of Sir Nicholas Hyde (q.v.), he was called to the Bar in 1617. He sat in the Long Parliament and adhered to the King's party. At the Restoration he became a judge of the Common Pleas (1660) and then Chief Justice of the King's Bench, 1663–65, but attained no great reputation as judge or jurist, though he is said to have been an authority on the pleas of the Crown.

Hypothec. In Roman law, a kind of security for debt in which the creditor had neither ownership nor possession, found in such a case as a tenant pledging his tools or equipment needed for working the land in security of the rent due. Possession could be taken only if the debt were not paid. Hypothecs are found in modern Scots, French, and other civil law countries. In Scots law there are conventional hypothecs, namely bonds of bottomry and of respondentia, and legal hypothecs, as of a landlord over his tenant's moveable goods in the premises let in security of the rent. In France there are three kinds of hypothecs, contractual, judicial, and legal; the first are created by notarial act, the second by a court over all the property, present and future of the debtor, and the third include rights given to married women over the husbands' property and to children over their guardians' property. Some kinds of liens, notably maritime liens, are truly hypothecs, in that they exist without possession of the subject of security.

Hypothetical facts. Facts which did not and do not exist but are supposed or postulated. Courts normally decline to decide questions on hypothetical facts, e.g. if A had happened, and anything in a judgment in relation to hypothetical facts can be obiter dictum only.

Hywel Dda, laws of. See HOWEL DDA, LAWS OF.

I

I.O.U. A written admission of indebtedness in a stated amount, addressed to one person by, and signed by, another, so called because the letters I.O.U. have by custom been substituted for the words, 'I owe you'. It is not a bill of exchange, promissory note, or other negotiable instrument, but may be assigned by the creditor.

Iavolenus Priscus (*c.* A.D. 60–125). Notable Roman jurist. He held the appointments of *legatus consularis* of Britannia, Germania Superior, and Syria, and proconsul of Africa. He was famous as a law teacher, head of the Sabinian school, and highly regarded by his contemporaries though less so by later jurists. His works are mainly epitomes of the works of older jurists (*libri ex Cassio, libri ex Plautio, libri ex Posterioribus Labeonis*) and his chief original work is his *Epistulae* (15 books) which evidence original thought and were excerpted by the compilers of the Digest.

Ibn-Hanbal, Ahmad Ibn-Muhammad (780–855). Mohammedan theologian and jurist, the fundamental authority of the Hanbalite school of Islamic law (q.v.). His collection of traditions, the *Musnad*, comprised about 28,000 items and was edited and added to by his son. He left his followers an attitude of rigorousness in rituals and social relations, and fanatical intolerance. His doctrine is today that least accepted in Mohammedan studies.

Iceland, law of. Iceland was a dependency of Norway from 1263 to 1381, and with Norway a dependency of Denmark from 1381 to 1814. It was granted a constitution in 1874, then till 1918 was an independent state with the same king as Denmark, and has since 1944 been an independent republic. It is a republic with a parliamentary democratic regime. The Constitution of Ulfliot (930) appointed a central moot (Althing) for the whole island. Later the island was divided into four quarters, each with a head-court and the Innovations of Skapti the Law-Speaker (*c.* 1030) set up a fifth court as the ultimate tribunal in criminal matters. Law and politics in early times is illustrated in the sagas, two collections of law-scrolls (*Codex Regius*, *c.* 1235 and Stadarhol's Book, *c.* 1271) embodying earlier redactions of customs, and other sources. Law was very largely customary.

The head of state is the President who can summon and dissolve Parliament and may refuse to ratify legislation, in which case it is submitted to a referendum.

The Parliament (Althing) established in 930 is divided into upper and lower houses with equal powers; the members serve for four years or until dissolved. The executive consists of a Prime Minister and cabinet appointed by the President, which must have majority support in the Althing. For administration the island is divided into 17 provinces each with a chief executive, and 34 judicial districts. Towns, counties, and districts administer local affairs through elected councils. The judicial system comprises a Supreme Court of Appeal and a system of lower courts (provincial magistrates and town judges), most of which have both civil and criminal jurisdiction; all judges are appointed by the Minister of Justice and the jury system is not used. There are also some special courts, in urban districts, such as Maritime and Commercial Courts and Court of Labour Disputes. There are no constitutional or administrative courts. Judges of lower courts also act as tax collectors, control police, and direct local government.

S. Juul, A. Malmström and J. Søndergaard, *Scandinavian Legal Bibliography*.

Icelandic lawspeaker. The sole official of the mediaeval Icelandic republic, elected by the priest-chieftains for three years during which time he recited the whole law at meetings of the Althingi, the Icelandic court, one-third at each meeting. He also repeated the whole procedural law each year and supervised the actings of the Althingi. Unless it was challenged at the time, everything pronounced as law by the lawspeaker had full force. The office came to an end in 1271 when Iceland came under Norway.

Ictus. On the title pages of old books, an abbreviation for *iurisconsultus* (q.v.).

Ideals. Ends conceived of as most desirable of attainment, but very possibly unattainable. The ideals held by legislators, law reformers, and judges and others undoubtedly influence their making, interpreting, and applying of law. These ideals in turn are influenced by their social, moral, political, and other philosophies, by their religion, experience, and other factors. To say that the ideal of a legal system and of each principle is justice is unhelpful because it merely converts the question into one of what justice is and what it requires in different circumstances.

Different schools of jurists have, consciously or unconsciously, set up different ideals for legal systems. Austin and the analytical jurists seem to have entertained as the ideal of a legal system a logically structured and self-consistent body of propositions. Others have seen as ideals such factors as certainty, flexibility, stability, uniformity, order, and reasonableness. Lawyers as a class tend to be conservative and to favour such ideals as certainty, predictability, stability, and uniformity, not least because they render predictions of the legal results of a course of action easier. Flexibility and reasonableness can easily render law as uncertain as the Chancellor's foot (q.v.).

Identification. The process of showing that a particular person or thing is the person or thing concerned in certain events under investigation. Thus, in a criminal trial identification of the person accused as being the person responsible for the crime is of importance. This may be done by the evidence of eye-witnesses, by fingerprint evidence, by evidence of mud, blood, etc. from the locus being on the accused's person or clothing, and in other ways. In investigating title to land the purchaser is entitled to proof that the land described in the title is truly the land which he has contracted to buy.

Identification parade. A means of criminal investigation in which a suspect is included in a group of persons of generally similar size and appearance and these are inspected by a complainer or witness to see if he can identify anyone in the group as the assailant or person concerned in the crime. The dangers of mistaken identification are great and the Home Office has issued guidance to the police on the conduct of such parades. All suspects are given a leaflet describing the purpose and nature of an identification parade, their rights in respect of it, and the consequences if they refuse to take part. The guidance warns against such practices as showing the complainer or witness photographs beforehand or other doubtful practices.

Idle, John (*c.* 1690–1755). Called to the English Bar (Middle Temple, 1715, Lincoln's Inn, 1728) was Chief Baron of the Scottish Court of Exchequer, 1741–55.

Idle and disorderly person. Under the Vagrancy Acts, a general term comprehending unlicensed pedlars, common prostitutes, and persons wandering about and begging.

Ignorance. The state of not knowing. Ignorance is closely akin to error or mistake (q.v.), which is inaccurate knowledge though not absence of knowledge. In legal contexts a distinction is usually drawn between ignorance of fact and ignorance of law. Ignorance of fact is frequently an excuse or a defence, particularly where it negatives intention or other state of mind requisite for commission of a crime or wrong. But in other cases a man is not excused by the plea of ignorance of fact; if he buys a defective thing, his ignorance is no defence.

Ignorance of law on the other hand is in general no excuse or defence; a person is presumed to know the law, or at least cannot be permitted to excuse himself on the ground of that ignorance, because otherwise many would plead this defence and escape, benefiting from their own ignorance or carelessness of the legality of their conduct. But ignorance of law may mitigate punishment.

Ignorance is distinguishable from nescience, which is absence of knowledge in one by nature not capable of having it, such as an infant or a person severely mentally incapacitated.

Ignorantia iuris neminem excusat. Ignorance of the law excuses no man. See IGNORANCE.

Ilbert, Sir Courtenay Peregrine (1841–1924). Had a brilliant classical career before being called to the Bar in 1869. Thereafter he assisted Sir Henry Thring in the office of parliamentary counsel. From 1882 to 1886 he was law member of the viceroy's council in India and did important legislative work. On his return to England he became assistant parliamentary counsel and in 1899 head of that office, but in 1902 he became clerk of the House of Commons, an office he held till 1921. He was a consummate draftsman with a great grasp of law. He wrote *Legislative Methods and Forms* (1901), *Parliament, its history, constitution and practice* (1911), and the *Mechanics of Lawmaking* (1914); he received the K.C.S.I. in 1895 and the G.C.B. in 1911.

Illegal. A very general term for what is contrary to law, with no precise meaning or consequence. It may mean what is positively contrary to law, or prohibited, punishable and criminal. Or it may mean merely what is in breach of legal duty, or contrary to public policy and unenforceable.

Illegal contract. An agreement to do anything prohibited by statute, or contrary to common law, but sometimes extended to cover contracts which are objectionable and void as being contrary to public policy.

Illegitimacy. A major social and moral problem, the causation and incidence of which give rise to much argument. Briefly, it is the status of a person born outside wedlock, to parents not married to each other, or begotten of a married woman by one other than her husband. The child of a void marriage was at common law illegitimate, but is now legitimate if at least one parent believed the marriage was valid. The child of a voidable marriage which has been annulled is legitimate. Generally legitimacy is presumed unless the facts clearly indicate the contrary. The primary onus of maintaining an illegitimate child is on the mother, but by an affiliation order or by contract she may obtain contribution from the father. At common law such a person was *filius nullius* save for certain purposes of the relationship with his parents, and could in general not inherit property but could acquire it by gift or will. Statutes have materially altered the common law position and gone a considerable way towards assimilating the position of the illegitimate to the legitimate. He may acquire property by gift or will, succeed on intestacy, and recover damages for the natural parent's death.

In England at common law legitimation was not recognized, except by statute. Since 1926 legitimation by the subsequent marriage of his parents has been recognized in most cases. At common law in Scotland, following the canon law principle of legitimation *per subsequens matrimonium*, the subsequent marriage of the parents of an illegitimate child legitimated it from the date of the marriage, provided the parents could have married at the time of conception, or now, from the date of the marriage. The effect of legitimation is generally to place the child in the same position as one born legitimate.

Adoption (q.v.) is in practice a common solution to a case of illegitimacy. An illegitimate child may be adopted by either natural parent or both, or by third parties.

Immigration. The entrance into a country of persons not born there but intending to make it their more or less permanent residence. To what extent it is to be allowed raises difficult moral, social, political, and economic problems; how it is to be controlled raises legal problems. In the U.S. about 35 million persons entered from Europe between 1820 and 1970, and large numbers have entered Australia and New Zealand. In the U.K. acute problems have been caused since 1945 by the immigration of persons from Commonwealth countries, particularly the West Indies, India, and Pakistan. The U.S. introduced a quota system in 1921 and the U.K. introduced in 1962 restrictions, now contained in the Immigration Act, 1971, on the entrance of citizens from the independent Commonwealth countries and British colonies. Immigration of aliens could at common law be controlled under the royal prerogative and has been controlled by statute and regulations since at least

1914. Within the European Community a national of any member state may establish himself in any other, subject to limitations on grounds of health, public security, or public policy.

See also RACE RELATIONS.

Immorality. Arises in many legal contexts, such as the object of a contract, the consideration for a contract, copyright, trusts, and others. In general it means sexual misconduct, by way of fornication, adultery, and similar practices, and not any or every other kind of conduct which a puritan might criticize.

Immovable. That which cannot physically be moved, particularly land and things attached to land, such as buildings. In foreign legal systems and in international private law the basic distinction of objects of proprietary rights is into immovable and movable, corresponding generally but not exactly to the distinction into real and personal of the common law. Thus leaseholds are a right in immovable property though, by English law, classified as personal property.

Immunity. (1) A state of freedom from certain legal consequences or the operation of certain legal rules. In Roman law immunity meant exemption from *manus*, i.e. from any obligation arising from law, custom, or authority. The correlative concept is disability. In municipal law for particular reasons particular categories of persons are immune from civil or criminal liability in particular cases. Thus young children are immune from criminal liability and persons protected by absolute privilege immune from liability for defamation.

Foreign sovereigns and the governments of foreign states have immunity from the jurisdiction of municipal courts in other states. This is variously explained by the comity of nations, the independence and equality of sovereigns and states, the dignity of sovereigns and states, the principle of extraterritoriality, and on other grounds. British law has generally accepted the view that immunity is absolute, but some states take the view that foreign sovereigns may object to legal proceedings only in respect of governmental activities or property, though what is covered by these categories may be disputed. Exceptions are recognized in the case of equity actions where the Sovereign claims a trust fund or where the Sovereign submits to the jurisdiction. The immunity attaches in British law also to public ships, both warships, government ships and state trading ships, to government agencies, and public corporations, and to visiting armed forces (though the last matter is usually regulated by treaty).

Diplomatic representatives have immunities in the country to which they are accredited, but the details vary in each legal system. Under a Vienna Convention on Diplomatic Relations of 1961, diplomatic immunity is enjoyed by heads of missions, diplomatic staff of an embassy, servants, and the family of diplomats so long as part of their households, and administrative and technical staff only in respect of official acts. The immunity under the Convention is complete except in respect of real actions concerning private immovable property in the receiving state, actions relating to any professional or commercial activity exercised by the agent outside his official functions and in certain other cases. It is an immunity from liability to legal process, not liability under the law.

Immunity extends to taxation and excise, subpoena to give evidence, arrest and criminal proceedings, bankruptcy. It covers personal property and incorporeal property. Diplomatic premises and their contents are inviolable and exempt from local taxation. Immunity may be waived by the diplomat or by his superiors.

Consular immunity depends chiefly on consular conventions but is supplemented by a Vienna Convention on Consular Relations of 1963. Immunity is usually conceded to a consul in respect of his official activities and inviolability of consular premises recognized.

International organizations and their officials enjoy similar immunities; members of the I.C.J. enjoy diplomatic immunities while engaged in the business of the court.

(2) In the analysis of legal terms developed by Hohfeld (q.v.), a freedom from the power of another, who is under a disability in relation to the person immune. Thus a diplomat enjoys immunity from police regulations in the country to which he is accredited and the police are under a disability in respect of charging him.

Imola, Alexander da. See TARTAGNUS, ALEXANDER.

Imola, Giovanni da (Johannes de) (?1375–1436). An Italian commentator on civil and canon law, professor at Ferrara, Padua, and Bologna, author of *Super primo et secundo Decretalium* (1475), *In tertium Decretalium commentaria* (1475), and *Commentaria in Digestum Infortiatum et Novum* (1475).

Imola, Paulus de (twelfth century). An Italian jurist and commentator.

Impalement. A method of execution mentioned in the Babylonian and Assyrian laws whereby the living body was pierced in the abdomen by being thrust on a spike fixed in the ground. It was prescribed for adultery and self-induced abortion. It may have been practised by the ancient Israelites also.

Impartiality. An aspect of equality, a quality generally accepted as desirable in judges and administrators of law, connoting determination to deal equally with both or all parties to a dispute, and not to favour any, but to apply the law equally and fairly to all. Partiality of an arbitrator would justify setting his award aside.

Impeachment. (1) The mode whereby a person was accused by the Commons and tried before the Lords for 'high crimes and misdemeanours'. On a motion by the House of Commons to impeach a person, articles of impeachment were framed by a committee of the House and exhibited to the House of Lords. Trial was by the House of Lords, all peers being judges both of fact and of law, the verdict being by a majority. It was partly a means of attacking ministers before the doctrine of ministerial responsibility had developed, and partly simply a weapon of political warfare. Its use declined after 1689 because the criminal law extended to cover more ordinary crimes and the development of ministerial responsibility gave Parliament adequate means of attacking ministers for misconduct in office.

The earliest case seems to have been that of Lord Latimer in 1376. It was well established by the time of Richard II. Other notable impeachments were of the Duke of Buckingham in 1626 for having monopolized several of the high offices of state; Bacon in 1621 for having taken bribes; the Earl of Strafford (1640) for high treason, in reality for having executed Charles I's policy; Archbishop Laud (1641) for high treason; Hyde, Earl of Clarendon, (1667) for high treason; Osborne, Earl of Danby (1679) for high treason; the Earls of Portland, Oxford, Halifax, and Lord Somers (1701) for high treason for their share in promoting the Spanish partition treaties in 1700; Dr. Sacheverell (1710) for two sermons on the doctrine of unlimited passive disobedience; Oxford, Bolingbroke, and Ormond (1715) ministers impeached for their share in negotiating the Peace of Utrecht, the last instance of purely political impeachment; Warren Hastings (1788) for misgovernment in India; and Viscount Melville (1804) for alleged malversation while acting as Treasurer of the Navy, the last case. In Danby's case it was decided that the King's power to pardon could not stop the impeachment of a minister. In 1681 the House of Commons passed a resolution declaring it to be their undoubted right 'to impeach any peer or commoner for treason or any other crime or misdemeanour'. But a royal pardon may be given after the Lords have given judgment.

In the U.S. the Federal Constitution (Art. I, Sec. 5) provides that the House of Representatives may by a majority adopt a resolution to impeach, and the Senate, which tries the impeachment, may convict by a two-thirds majority. The House appoints managers to conduct the prosecution. The Senate is presided over by its president, unless the President of the U.S. is on trial when the Chief Justice presides. Conviction does not preclude criminal proceedings in the ordinary courts, and the President's power of pardon does not extend to conviction on impeachment.

Impeachment has been used frequently, usually against inferior court judges, who cannot otherwise be removed, but usually unsuccessfully. President Grant's corrupt Secretary for War, Belknap, was impeached but resigned before the Senate convened for trial. President R. M. Nixon would have been impeached if he had not resigned. It has also been utilized as a political weapon, as in the cases of President Andrew Johnson (1867–68) and Justice Samuel Chase (1804), both of which cases demonstrated the drawbacks of the system. Many of the States provide for removal of executive or judicial officers by impeachment.

(2) The defence in Scottish criminal procedure, that the crime was not committed by the accused but by another named person.

Impeachment of waste. A life tenancy given 'without impeachment of waste' allows the life tenant to cut trees and open mines, but not to demolish a mansion house, fell ornamental timber, or do any of the kinds of acts known as equitable waste.

Impediments to marriage. Canon law developed a system of impediments to marriage, directed to protecting the sanctity of marriage and promoting the Christian religion. They were classified as diriment (*impedimentum dirimens*), which prevented a marriage from coming into existence at all, or prohibitive (*impedimentum impediens*), which made the marriage valid though sinful and unlawful. The diriment impediments were non-age, lack of baptism, impotence, existing marriage, vow of virginity, being in holy orders, absence of consent, force or fear, error excluding consent, consanguinity, affinity, public decency, qualified adultery, and abduction; the prohibitive impediments were incest, abduction of the wife or bethrothed of another, and certain others. Notwithstanding the Reformation many of these impediments have survived in modern English and Scots law.

Impeding apprehension. In English law the offence of doing any act with intent to impede the apprehension or prosecution of a person who has committed an arrestable offence. This replaces the former offence of becoming an accessory after the fact. Common instances are concealing the offender, providing him with means of escape, and destroying incriminating evidence. It is immaterial that the accused is unaware of the offender's identity.

Imperative theory. The general term for the theory of law associated with the names of Bodin, Hobbes and, particularly, Austin and his followers which views law and legal propositions as imperatives, that is as normative statements laying down rules to guide human conduct, as distinct from statements of fact. Frequently statements of such a kind are referred to as commands, though this word is probably too narrow in connotation. Law is a command which obliges a person to a course of conduct. In its crude form the imperative theory is open to many criticisms, for example that in modern law many of the propositions do not order anybody to do anything but confer claims or powers. But interpreted by reference to their logical status it is undeniable that legal rules are imperative or prescriptive, rather than descriptive. This approach to law postulates in each legal system a superior power who or which can command, and some sanction which it can impose in the event of disobedience. That the crude command view is inadequate is seen by considering the instance: The law prescribes that you may regulate the devolution of your property by making a will; if you do not, the sanction is that your property will be distributed under the rules of intestacy. The legal prescription is a power, not a command, and the sanction is merely that if the power is not exercised the law will arrange disposal. Other criticisms of the Austinian approach are: Should law be defined by reference to the State, which commands? Is a sanction an essential of law? Should not some reference be made to the purpose of law?

Imperfect obligation. Obligations of the nature of moral duties, not binding in law but only in honour or morals.

Imperial Chamber (Reichskammergericht). The supreme judicial court of the Holy Roman Empire, 1495–1806. It replaced the Hofgericht, in which the emperor presided with assessors, which ceased to sit in 1450. In the Reichskammergericht the official court of the Empire, the emperor appointed the president but the members were nominated from the empire, and as they had to be skilled in Roman law, the court contributed greatly to the Reception of Roman law in Germany. The court sat at Spires until 1693 and then till 1806 at Wetzlar. Its jurisdiction, as gradually defined by statute and usage, extended to branches of public order, arbitrary imprisonment, violation of the emperor's decrees, and other matters. It acted also as a court of appeal from territorial courts, mainly in civil matters, but had no authority in territories having *privilegium de non appellando.* It suffered from competition from the Aulic Council, the emperor's personal advisers, a body which lost executive functions and took on judicial ones, a competition which was regulated by the Treaty of Westphalia, and it was not an efficient court.

O. Stobbs, *Reichshofgericht und Reichskammergericht* (1878).

Imperial standards. The weights and measures statutorily defined and authorized to be used in place of all previously recognized customary or local weights and measures. Metric standards are now equally lawful.

Imperium. In ancient Rome, the supreme military and judicial power which belonged to the kings and later to the consuls, military tribunes with consular powers, praetors, dictators, and *magistri equitum.* Under the Republic *imperium* was limited by the existence of several magistrates sharing power, by laws guaranteeing prisoners an appeal to the people against a capital sentence and by a proconsul's or propraetor's *imperium* being limited to his province only. Exceptionally *imperium* was conferred on private individuals, and in the last century of the Republic extraordinary grants for periods of years were made, e.g. to Pompey against the pirates. In 23 B.C. Augustus received a grant of *imperium* superior to other magistrates and which he was entitled to exercise from within the city of Rome. The Senate granted it to subsequent emperors on their accession, though it was formally ratified by a *lex curiata. Maius imperium* was sometimes conferred on others than the emperor, for the creation of a single military command.

Impey, John (d. 1829). An attorney who wrote books which were the first systematic accounts of the practice of the two major common law courts, namely *The New Instructor Clericalis, stating the Authority, Jurisdiction and Practice of the Court of King's Bench* (1782) and *The New Instructor Clericalis, stating the Authority, Jurisdiction and Practice of the Court of Common Pleas* (1784), both of which went through many editions.

Implication. That which is not expressly declared in but is logically inherent in what is said or written or done. Also the logical process of importing a logical consequence into some act or statement. Thus, if a person buys goods the contract implies, or attaches by legal implication, the term that the goods are of merchantable quality.

Implied authority. Authority not expressly conferred but legally deemed to have been intended to be conferred.

Implied contract. A contract not made expressly but created by force of law in circumstances which make the creation of a contract just and reasonable.

Implied powers of Congress. The Constitution of the U.S. endows Congress with various specified powers (Art. I, sec. 8) and finally with the power to make 'all laws which shall be necessary and proper for carrying into execution the foregoing powers, and all other powers vested by this Constitution in the Government of the United States, or in any Department or Office thereof.' Initially this provoked differences of view between those favouring narrow construction of this power, limiting the powers of the central government, and those favouring a liberal interpretation. The Federalists, led by Alexander Hamilton, and supported by decisions of the Supreme Court under Chief Justice Marshall, made wide use of the power to strengthen the power of the central government. Since the Civil War decisions of the Supreme Court have permitted great expansion of federal power and of governmental intervention and regulation.

Implied term. A term in a contract which is an element of the contract, not because it has been expressed or included in the contract by the parties but because it is added to the contract by force of law. A term or terms to any specified effect may be implied in various ways, by common law settled by judicial decisions, by statute, by the settled course of dealing between the parties, and by the court as being a term which, if the parties had thought of the possibility of what has happened, they would pretty certainly have expressed in their contract. Except in cases of statutorily implied terms there may be dispute in particular cases whether any term should be implied, and, if so, what the term should be.

Implied trust. A trust not expressly created but arising from an equitable interpretation of the facts and conduct of parties. Such a trust is implied in two classes of cases, where it seems just to hold that such was the presumed intention of parties, and where there has been fraud or notice of an adverse equitable claim.

Impositions. In 1606 James I, being in need of money, by his own authority, directed the collectors of customs to demand an additional duty on imported currants. John Bate refused to pay and this gave rise to *Bate's Case* or the *Case of Impositions.* The Court of Exchequer unanimously found for the King on the basis that regulation of trade came within the royal prerogative as to foreign affairs, but merchants appealed to the House of Commons which requested that the impositions should cease to be levied. In 1610 the House argued the legality of impositions and, being opposed to them, presented a remonstrance to the King and passed a bill enacting that no imposition beyond those already in existence should be laid without consent of Parliament, but the bill was rejected by the Lords and *Bate's Case* accordingly justified these arbitrary exactions.

Impossibility. A situation in which something cannot be brought about or accomplished. There are at least three kinds of impossibility of legal relevance. An act may be physically or factually impossible, where it cannot be done; legally impossible, when legal machinery does not exist for it being done effectively; and logically impossible, when there is a repugnancy or internal contradiction in the act, such as a gift to A for his own benefit on condition he transfers the gift to B. A distinction is also drawn, particularly in relation to contracts, between initial impossibility, existing at the time of contracting, and subsequent or supervening impossibility, arising subsequently. If a contract was impossible *ab initio*, it is a nullity, but if the impossibility arises only subsequently, the parties are excused from further performance. Legal impossibility is distinct from illegality; a course of action may be legal and unobjectionable, but legally impossible in that legal machinery may not exist to enable it to be done.

Impotency. The inability to have complete normal sexual intercourse. It may arise from physical defect in either partner, or from psychological barrier amounting to invincible repugnance on the part of one to sexual relations with that partner. Sterility is irrelevant and does not imply impotency. At common law, following canon law, impotency was a ground for decree of nullity of the marriage, non-consummation of the marriage leaving it voidable and subject to annulment, and statute has affirmed this view.

Imprescriptible rights. Rights which, however long disused, cannot be lost by reason of prescription (q.v.) or lapse of time.

Impressment. A power vested in the Crown to take persons or property for the defence of the realm, with or without the consent of the persons concerned. Though used to find men for the army, the commonest use was to take seamen sailing on merchant ships, or even landsmen, for service in the Royal Navy. Pressing was authorized by Orders in Council and executed under warrant by press gangs. This practice was permitted at common law under the plea of necessity and was common down to 1815; but was thereafter abandoned in favour of voluntary enlistment. The Admiralty granted protections against impressment in certain cases, sometimes under statutes protecting certain classes of persons. It is probably still legal, being impliedly recognized by certain later statutes. During the two World Wars compulsory service in the armed forces was authorized by statute. In early times impressing of merchant ships was also practised.

Imprimatur. A permission required by Roman canon law to be granted by a bishop for the publication of a work on Scripture, dogmatic or moral theology, or any writing important for religion or morality. The permission is signified by the word *Imprimatur* (Lat: Let it be printed) on the verso of the title page. This permission must be preceded by a certificate from a censor, indicated by the words *Nihil obstat* (Lat: nothing hinders it), and accordingly the term *imprimatur* has come to mean ecclesiastical approval of the publication, though it does not guarantee the integrity of the contents, nor endorses them in any way.

Imprisonment. Restraint of a person's liberty of movement. It extends to confinement in a house or even in a street or public place, because in such cases the person is not free to go as and when he pleases. The term is, however, most commonly used of confinement or detention in a gaol or prison as a punishment. Originally imprisonment was used more to hold persons for trial than as a mode of punishment, but punitive imprisonment was known in ancient civilizations, though not in Greece or Rome, and was used by the early Christian Church, first against delinquent clergy but later against laity also. It appeared in the laws of Alfred, but was rarely used as a primary punishment until the thirteenth and fourteenth centuries. Thereafter it became common. There was and is a constant conflict between imprisonment as a punishment and as a means to reformation. Formerly imprisonment might be imposed with or without hard labour, and in the latter case it might be in the first, second, or third division. Prisoners in the first division were treated in very much the same way as persons in custody awaiting trial. Prisoners in the second division were allowed visits and letters. Prisoners in the third division were allowed fewer visits and letters. Hard labour and the divisions were abolished in 1948.

The powers of imprisonment of various grades of courts are specified by statute, and there are restrictions on the imprisonment of first offenders and of young persons. Imprisonment may be imposed on failure to pay a fine.

Imprisonment was formerly a common penalty for failure to pay debts, but this was abolished in 1869 save for limited exceptions. It is still competent as a means of enforcing obedience to an order of court or to ensure a person's presence at a trial, as an accused or as a witness.

Improvement commissioners. *Ad hoc* bodies established by local Acts to supplement the limited powers of local authorities. The first body was established in 1662 to supervise cleaning, lighting, and sewerage in London and Westminster. By 1800 there were about 300 such bodies, many in the London area. The membership varied, but usually included the mayor and corporation, the justices and some leading citizens, and sometimes some elected representatives. They had limited rating and borrowing powers. From 1835 no further bodies were created and gradually their powers were transferred to the new elected local government authorities.

In banc. Prior to the Judicature Acts, 1873–75, a sitting of one of the old common law courts (Queen's Bench, Common Pleas, Exchequer) to decide questions of law as distinct from sittings of single judges at *Nisi Prius* (q.v.) or on circuit (q.v.).

In camera. Proceedings are held 'in camera' if heard in a judge's private room, or in a court the doors of which have been closed and persons, other than those concerned in the case, excluded. This course is adopted where publicity would be very undesirable, and in certain matrimonial cases. Only very exceptionally, as where young persons are involved, can a criminal case be heard 'in camera'.

In chief. Examination in chief is the examination of a witness by counsel for the party who has called him and constitutes his basic statement of his evidence. It may be followed by cross-examination (q.v.) by counsel for the opposing party and, if necessary, re-examination by counsel for the party who called him.

A tenant in chief is one holding land feudally immediately of and under the Crown.

In consimili casu. The Statute of Westminster II (1285), c. 24 enacted that if it happened in the Chancery that there was found a writ in one case, but not in another case though involving the same law and requiring the same remedy, the clerks of the Chancery should frame a writ or adjourn the matter to the next Parliament. The only substantial extension of the law which resulted was the creation of the writ of entry *in consimili casu* for use by reversioners and remaindermen against the alienees of tenants for life and by the courtesy immediately after such an alienation. It was formerly thought that the development of trespass on the case (q.v.) out of trespass (q.v.) was connected with this statutory provision, but this view is now considered unfounded.

In fee simple. A phrase used of an estate in land, denoting that it descended to the heir of the owner for the time being and was capable of existing as long as there were heirs of such an owner, that is, in theory, infinitely. An estate in fee simple is now created by a conveyance of freehold land to any person without words of limitation, unless any contrary intention appears.

In forma pauperis. In the character of a pauper. By statutes of 1495 and 1532, if a poor person wishing to bring an action could swear that, apart from his clothing and the subject matter of his claims, he had not property worth £5, he was allowed to have counsel and solicitor assigned to him who acted without fee, and he was exempted from payment of court fees. These Acts were repealed in 1883 and provision for such cases is now made by the Legal Aid Scheme (q.v.).

In integrum restitutio. In Roman law a remedy which might be granted by a praetor, to restore a party injured to his original position, in order to annul a legal result which he considered inequitable. An important case was to allow it to a minor if his inexperience had led him into a disadvantageous transaction.

In iudicio (or *apud iudicem*). The second stage of proceedings in a Roman lawsuit in Republican times, at which the parties appeared before the *iudex* appointed by the magistrate and sought to convince him by evidence that he should condemn or absolve the defendant in accordance with the *formula* or issue on which he was required to try the action. The *iudex*, though neither a magistrate nor a professional lawyer, was chosen from a list (*album iudicum*) of suitable persons kept by the magistrate. Later, proceedings *in iudicio* might take place before the *centumviri*, or rather a panel selected from them, or the *decemviri stlitibus iudicandis.*

See also *In iure.*

In iure. The first stage of proceedings in a Roman lawsuit in Republican times, at which the parties appeared before a magistrate who heard the formal claim (under the *legis actio* procedure) or who settled the *formula* (q.v.) (under the formulary system) or issue to be tried and assigned a judge to try the case. The judges appointed for the trial were private persons but differed from arbitrators in that they were appointed by the magistrate, not chosen by the parties.

See also *In iudicio.*

In iure cessio. In Roman law a form of conveyance taking place before a magistrate, whereby the transferee in the magistrate's presence asserted title to the thing and the transferor did not deny it. It seems to have been a kind of collusive action but was effective against third parties. It was probably not so ancient as mancipation (q.v.) but probably existed at the time of the XII Tables. Later it was a competent alternative to *mancipatio* for conveyance of *res mancipi* and to *traditio* for conveyance of *res nec mancipi*, but its main application seems to have been to incorporeal things to which mancipation was inapplicable.

In iure, non remota causa sed proxima causa spectatur (in law the proximate and not the remote cause is looked at). A maxim recorded in Bacon's *Maxims of the Law*, frequently invoked in problems of causation of harm, signifying that the nearest and immediate case is treated as the factor responsible, notwithstanding that an earlier and more remote causal factor was a prerequisite of the nearer cause and of the ultimate harm. It is invoked in cases of liability for negligence, marine insurance, and other cases.

In ius vocatio. In Roman legal procedure the summons of a defendant which initiated a litigation. It seems to have been done orally.

In personam. A term used of a right or claim vested in a person availing against a specific person only, and imposing on him a personal liability such as a claim for payment of money, as contrasted with a claim *in rem* (q.v.). A person may have several distinct but similar rights *in personam* against several other persons arising from some entitling fact, such as a claim against several joint tort feasors.

In relation to judicial proceedings an action *in personam* is one directed against another legal person. A judgment *in personam* is made against a legal person. Violation of a right *in personam* probably always gives rise to an action *in personam*, e.g. for debt or damages.

The equitable maxim 'Equity acts *in personam*' meant historically that the ultimate sanction for enforcement of equity decrees was the imprisonment of the defendant for contempt.

In rem. A term used of a right of claim vested in a person availing against other persons generally and imposing on everyone a legal liability to respect the claimant's right, such as a right of ownership or the right not to be injured, as contrasted with a right *in personam* (q.v.).

In relation to judicial proceedings an action *in rem* is one directed against an object or property, commonly a ship, such as to have it realized to satisfy the plaintiff's claim, or directed to the recovery of property. A judgment *in rem* determines rights in relation to the object which is in contention. Violation of a right *in rem* accordingly does not necessarily give rise to an action *in rem*; it frequently gives rise to an action *in personam*, e.g. for debt or damages.

In tail. Words which, prior to 1926, in a limitation of real property by will, had the effect of creating an estate tail or entailed interest.

Inability or no-right. In Hohfeld's (q.v.) analysis of legal terms, the legal concept of having no right or legal ability to object to or prevent the exercise by another of a privilege or liberty. Thus, if a party is in right of an easement over land, such as a right of way, the other party has an inability or no-right to prevent the exercise of that right of way. The correlative concept is privilege or liberty.

Inaction, appeal against. In European Community law, by analogy with practice in the original member states, an appeal may be taken to the European Court against illegal inaction by a public authority, to prompt the authority to take the action desired. There must have been an express or implied refusal to act, in violation of one of the Treaties.

Inalienable rights. See HUMAN RIGHTS; NATURAL RIGHTS.

Incapacity. Lack of ability in some legal respect. It may be physical incapacity, as by non-age; or mental incapacity, as by unsoundness of mind; or legal incapacity, as by bankruptcy; and is relevant in relation to contracting, suing, and in other respects.

Incest. Prohibited heterosexual relations between persons within a culturally or legally defined kinship group. Such prohibitions are almost universally found in human societies and have generated extensive sociological and psychological literature. It is usually thought that genetic considerations underlie this taboo. The general principle is that sexual relations are prohibited between persons so related that marriage between them is not legally permissible. In England it was, until 1908, not a crime but dealt with by the ecclesiastical courts. It is now the crime of a male having sexual relations with his own granddaughter, daughter, sister, or mother, or of a female having sexual relations with her own grandfather, father, brother, or son. The relationship may be of full or half-blood, and legitimate or illegitimate. In Scotland it is the crime of sexual intercourse between persons related within the forbidden degrees defined by statute in 1567 by reference to Leviticus, Ch. 18, 6–18. The offence is committed whether or not there is consent.

Inchoate. Something incomplete and ineffective.

Incident. That which depends on or pertains or attaches to another, principal, thing. In relation to feudal tenure incidents of tenure included fealty and homage, aids, relief, wardship, marriage, escheat, and forfeiture (qq.v.).

Incitement. The common law crime of persuading and encouraging another to commit a crime. The crime may be of any kind and it is irrelevant that it is impossible of commission. Mere knowledge of the other's intent to commit a crime is not enough. One may even incite another to attempt to commit crime. If the crime incited is actually committed, the person who incited is guilty as a participant. Incitement to violence against the laws and institutions of the State or against a section of the community is the main element of the crime of sedition. Incitement to disaffection is the offence of maliciously endeavouring to seduce a member of the forces from his duty or allegiance to the Crown. Incitement to racial hatred is the offence of publicly or by publication communicating ideas likely to cause breach of the peace.

Income tax. Income tax was introduced in Britain in 1799 and again in 1803. It continued till 1815 when it lapsed but was reintroduced in 1842 and has continued ever since. Since 1860 it has been imposed annually for a single financial year only, though the legislation is permanent (Income and Corporation Taxes Act, 1970), being amended as required in each annual Finance Act. It has several times been consolidated. In general the Acts impose the tax on a person residing in the United Kingdom in repect of his income from every source, wherever situated, and on a person not so resident in respect of his income from any source in the United Kingdom. Corporate bodies and unincorporated associations are subjected to corporation tax.

The tax is charged at a specified basic rate per pound of taxable income, and where total income exceeds a stated amount the excess bears tax also at higher rates but there are various personal allowances and reliefs allowed by deduction of specified amounts from total income. Surtax, an extra tax on higher incomes, was combined with income tax from 1973–74. Since 1842 the lowest rate of tax was 2d. per £1 in 1874–75 but it may now reach 98 per cent on the highest 'slices' of income.

The tax is imposed on income, which gives rise to many difficulties as to whether a particular receipt is to be deemed income or capital. Investment income, exceeding a certain amount is charged at additional rates.

The Acts impose charges under six Schedules, viz: Sched. A: property in land (discontinued in 1963); Sched. B: occupation of land (restricted to woodlands); Sched. C: profits from interest, dividends, and annuities payable out of public revenue; Sched. D: profits from trades and professions, interest and annuities, and capital gains made on the disposal of an asset; Sched. E: emoluments from any office or employment, and pensions; Sched. F: company dividends not charged under Sched. D or E. The tax is imposed on the income of the year of assessment, though the basis of the assessment may be another year or period. Thus under Sched. D. the measure of income is generally the income received in the year preceding the year of assessment, and in the case of trades, professions, or vocations the income of the year for which accounts are prepared, and under Sched. E the general basis is the income of the year of assessment. Tax under Sched. E is mostly paid under a P.A.Y.E. (pay as you earn) scheme under which amounts are deducted by employers from weekly or monthly payments of wages or salary in accordance with tax tables supplied by the Inland Revenue and the sums deducted are accounted for by the employer to the Inland Revenue.

Certain bodies, notably university colleges and halls, friendly societies, trade unions, savings banks, and bodies established for charitable purposes only are exempt from income tax, and specific types of income, e.g. undergraduate's scholarship, are exempt from income tax.

Income tax is managed by the Commissioners of Inland Revenue through a decentralized system of inspectors of taxes. Persons and bodies must make returns of income when required to do so, normally annually. Tax is then assessed, and a taxpayer may appeal against an assessment to the General Commissioners or Special Commissioners of Income Tax. A further appeal lies on a point of law, in England to a judge of the High Court and to the Court of Appeal and House of Lords, and in Scotland to the Court of Session and House of Lords.

Incorporated Council of Law Reporting for England and Wales. A body set up in 1865 in accordance with the general view of the Bar, to publish regular and accurate reports of the more important and legally significant decisions of the superior courts of England and Wales. Its reports superseded those of the private reporters who attached themselves to particular courts. Since 1865 it has published series of reports recognized by the courts and the profession as of high authority but not official nor entitled to any exclusive privilege of citation. The Council consists of certain *ex officio* members, two representatives from each Inn of Court, and two solicitors, and it appoints an editor and reporters. The text of the judgments is revised by the judges. The reports are issued in monthly parts for each court and the volumes for each division of the court subsequently bound up, containing the reports of that court for the year. Thus there are each year one or more volumes for cases in the House of Lords and Privy Council (Appeal Cases), cases in the Queen's Bench Division and, on appeal therefrom, in the Court of Appeal, cases in the Chancery and Family Divisions similarly. There are also occasional volumes for the Restrictive Practices Court and Employment Appeal Tribunal. There are separate Councils of Law Reporting for Scotland and for Ireland.

W. Daniel, *The History and Origin of the Law Reports.*

Incorporated Law Society. See LAW SOCIETY.

Incorporation. (1) The creation of a succession of the holders of a stated office (corporation sole), or of a group of persons associated for a common purpose (corporation aggregate), into a distinct legal entity or corporation (q.v.). It may be effected by royal charter, by royal letters patent, by special Act of Parliament, or by registration in accordance with the requirements of an Act of Parliament such as the Companies Act.

(2) The addition of one thing to another so that they are treated as one. Thus a codicil (q.v.) can be regarded as incorporating, or as being incorporated in, a pre-existing will. A statute may by reference incorporate the whole or parts of a prior statute; the former is then treated as if the latter were written into it, and made part of it.

(3) The doctrine that international law is part of the law of a state only if and so far as made part of it by statute, decision, or equivalent act.

Incorporeal chattels or property. Legal rights relative to objects of property which have no corpus or physical existence, e.g. patent rights, copyright.

Incorporeal hereditament. Interests of a heritable character in property not corporeal. They include franchises, profits *à prendre* and heritable rights not necessarily connected with land, such as offices.

Indebitatus assumpsit (being indebted, he undertook). A variety of the common law action of *assumpsit* in which the plaintiff alleged that the defendant had contracted a debt to him, and in consideration thereof had undertaken to pay. At first the undertaking (*assumpsit*) had to be alleged and, if traversed, proved. Later it was held that where a debt existed, a subsequent *assumpsit* would be presumed in law and need not be proved in fact, and finally, in *Slade's case* (1602) it was held that every contract executory imported in itself an *assumpsit.* In effect this made *indebitatus assumpsit* equivalent to the common law action for debt. It has been obsolete since the Judicature Acts, 1873–75.

Indecency. Any act between two persons involving the handling or use of the genitalia of either. An exception exists in the case of medical examination or treatment.

Indecent assault. The crime of making an assault (q.v.) or an actual physical contact with a person involving an element of sexual affront or harm, e.g. by exposing the male organ to a woman, or by a person attempting to touch, or touching, the sexual organs of another without that other's consent and against the other's will.

Indecent exposure. The common law offence of exposing the body in public in such a way that more than one person can see the act of exposure and so as to outrage public decency. It may be committed by a person of either sex. Sexual motive is unnecessary. Under statute it is an offence for a male to expose his genitals with intent to insult any female.

Indemnification. The making good of a loss which a person has suffered in consequence of the act or default of another.

Indemnity. An undertaking to compensate for loss, damage, or expense; marine and fire insurance contracts are contracts of indemnity; statutory provisions for contribution by one tortfeasor to damages for which he and another tortfeasor have been held liable may amount to a complete indemnity; and provision used commonly to be made in trust settlements for trustees to be indemnified against losses caused by their agent.

Indemnity, military. The recovery by a military conqueror of part or all of the costs involved in his military operations. Since 1800 it has usually been imposed at the end of a war and regulated by the peace treaty. It is distinguishable from pillaging, and looting and from requisitions made on occupied territory and seems to have emerged as an alternative to the idea that the victors were compensated by annexation of territory and stimulated by mercantilist ideas of securing treasure. Napoleon imposed indemnities on defeated kingdoms with little regard to capacity to pay or the justice of the demand. The most famous indemnity was that imposed on France by Germany in 1871, largely punitive but in fact paid quickly. In 1919 the peace of Versailles imposed reparations, which were never paid to any material extent.

Indemnity Acts. Statutes passed to relieve persons of penal consequences incurred by particular conduct. From 1727 to 1828 such an Act was passed annually to enable nonconformist Dissenters who achieved office by taking the Anglican communion after, instead of before, election, to avoid the penal consequences of the Corporation Act of 1661. The need for such Acts ended when the Test and Corporation Acts were repealed. Indemnity Acts have also been passed to protect persons engaged in suppressing revolts or riots from liability for harm done. Indemnity Acts are still occasionally necessary when a person has sat in the House of Commons while disqualified or failed to lay regulations before the House or otherwise acted irregularly though in good faith.

Indenture. A deed made between two or more parties, e.g. a lease or conveyance. It was formerly the practice in such a case to write two copies on one sheet and then divide them by an indented or jagged line so that each could fit the other and so be identified as genuine, but such is not now necessary. The deed is however still called an indenture.

Independence Hall. The former Pennsylvania State House in Philadelphia, in which the Second Continental Congress met in 1775–77 and the debates on the Declaration of Independence and on the Articles of Confederation took place. It also housed the Philadelphia Convention (1787) and was the site of the Federal Government from 1790 to 1800. It is now preserved as a memorial to the American Revolution.

Independence. An essential attribute in international law of a state as a distinct legal person. It implies freedom to provide for its own well-being and to develop without being subjected to the domination of other states, provided that in so doing it does not impair or violate the legitimate rights of other states. Independence implies that no other state has the right to intervene, directly or indirectly, in the state's affairs, as by armed intervention or threats or fomenting civil strife, though a state may take belligerent or non-belligerent action against another in pursuit of legal redress. The U.N. by its Charter may not intervene in matters essentially within the domestic jurisdiction of any state except in the application of enforcement measures in the maintenance of peace. A corollary of the right to independence is the right of self-defence.

Independent contractor. A person who undertakes to bring about a desired result for an employer, but in so doing is not an employee of or subject to the detailed direction and control of the employer, but is free to use his own judgment and discretion in bringing about the desired result, e.g. professional men, builders, haulage contractors, etc. The contract with the employer is one for rendering services (*locatio operis faciendi*) and not one of service or employment (*locatio operarum*). In consequence the employer is not in general vicariously liable for wrongs of the independent contractor or his men.

Indeterminate reference, categories of. A term for classes of criteria of judgment which are uncertain, indefinite, or do not allow of being applied to a set of facts to produce a certain decision. The commonest criterion of this kind is the legal standard, requiring a court to decide by use by such standards as fairness, reasonableness, due care, sufficient cause, just cause or excuse, unconscionableness, exorbitancy, or similar quality. These standards are indeterminate in that the decision depends on what the judge thinks is fair, reasonable, or otherwise in the circumstances. The converse is the fixed standard, determined by such measurable factors as time, speed, value in money, or the like. The category of indeterminate reference allows a wide range for variable judgment, approaching compulsion only in extreme cases, which are obviously fair or unfair.

Indeterminate sentence. A sentence of imprisonment prescribing no definite duration, but often with a prescribed maximum. Such sentences originated as a means of securing social protection against habitual offenders; they were occasionally used in the Middle Ages and appeared for this purpose in the *Carolina* (1532), the *Theresiana* (1768), the colonial laws of Connecticut (1769), and the Prussian *Landrecht* (1774) (qq.v.). Such sentences were favoured by prison reformers in the nineteenth century on the principle that release was to be determined by progress in rehabilitation. In modern practice eligibility for parole is determined by the parole authority by reference to the offender's record and his apparent suitability for parole.

Index Florentinus. An index of authors and books from which excerpts had been taken in the compilation of Justinian's *Digest* (q.v.). It is found only in the Florentine manuscript and hence is called the *Index Florentinus* and is not wholly accurate in that it omits some works used, mentions others not used, and confuses some distinct works. It may have been compiled not from the inscriptions of headnotes of the fragments excerpted but from the lists of books prepared for the use of the compilers. Despite the inaccuracies it adds something to knowledge of the compilation.

Index Interpolationum. Works surveying the allegations made in many books and articles of interpolations (q.v.) in the text of Justinian's legislation. There are the *Index Interpolationum quae in Justiniani Digestis inesse dicuntur* by Mitteis, Levy, and Rabel (1929–), the *Index* for the *Code* by Broggini (1969), and for the juristic writings by Volterra in *Rivista di Storia del diritto italiano.*

Index Librorum Prohibitorum. The catalogue of books deemed dangerous to the faith or morals of Roman Catholics or heretical, formerly by the Holy Office and from 1587 by the Congregation of the Index, and which accordingly they might not read. The first Papal Index was commissioned by Pope Paul IV in 1559 but was too severe, and in accordance with a recommendation of the Council of Trent Pope Pius IV commissioned a new Index in 1564. The *Index* continued in revised editions till 1966 when it was discontinued.

Index to the Statutes. See CHRONOLOGICAL TABLE AND INDEX TO THE STATUTES.

Indian law. The British occupation of India between the eighteenth century and 1947 brought about the political unification of India and the creation of a governmental and legal system heavily influenced by English ideas. The Indian Independence Act, 1947, created the separate states of India and Pakistan (now Pakistan and Bangladesh) out of what had been British India.

The law existing in 1947 comprised English law adopted in India, law made expressly for India by the British Parliament, law made for India by bodies having legislative authority there, and three bodies of religious personal law, Hindu law (q.v.), Islamic law (q.v.), and Burmese Buddhist law. Courts were established in India in the eighteenth century as the East India Company's authority came to extend to substantial territories in Bengal and south India. They were directed to apply English law so far as applicable to Indian conditions in the presidency towns (Calcutta, Madras, and Bombay), and in the country Hindu and Mohammedan law to persons of those faiths. But gaps in the law were filled and adaptation to new needs effected by recourse to principles of justice, equity, and good conscience, frequently as understood in English law. Hence, much English common law and equity came into the practice of the Company's courts. Much law, mostly of a constitutional character, was enacted by the British Parliament, particularly from 1773. When legislative authorities were established in India, their legislation was substantially influenced by English law. From 1833 to 1870 various commissions drafted legislation which was enacted by the Governor-General's legislative council and in the period 1859–82 great legislative activity produced many codes and important statutes such as the Code of Civil Procedure (1859, now 1908), Criminal Code (1860), Code of Criminal Procedure (1861), Contract Act, 1872, and many others.

British government of India did not mean a general imposition of English law, but it brought Hindu law to light by officially recognizing its authority and value, but also limited its application to certain relationships whereas other relationships were regulated by new and generally applicable territorial law.

Accordingly there had been before 1947 a very substantial reception of English law, not least after 1861 when a new structure of courts was set up staffed by judges trained in the common law.

Following independence the movement towards unification and codification of Hindu law continued, and major parts of a Hindu Code were adopted, and there was promised a general Civil Code, a national secular body of law. But the former law remained until modified or replaced. Moreover in language, concepts, terminology, and techniques, such as regard for precedent, Indian law had become by 1947 and still is one of the common law family.

Constitution

The Constitution of India of 1950 created India a federal republic, called a Union, comprising states some of which had been provinces of British India, some Indian states or unions of such states, and some formerly Chief Commissioners' provinces and Indian States administered as such. There is provision for the creation of new states. Secession from the Union is not possible.

The constitution is superior and paramount law changeable only by special means, and legislation conflicting therewith is void. Amendment is in general possible by the vote of two-thirds of those present and a majority of total membership in each House of Parliament. Alteration of parts of the Constitution requires in addition ratification by the legislatures of at least half of the major states, and of other parts merely a simple majority in both Houses of Parliament. The Supreme Court is guardian and interpreter of the Constitution.

Legislative power is distributed partly to the Union, partly to the states, and some subjects are competent to both.

The Head of State is the President, elected by the elected members of Parliament and of the lower houses in the State legislatures, the latter having votes related to the population of their states, by secret ballot on a system of proportional representation. He holds office for five years and may be re-elected; he is liable to impeachment. There is also a Vice-President.

Parliament consists of the President, the Council of States, and the House of the People. Legislation is normally passed by both Houses sitting separately and on receiving Presidential assent. The House of the People is elected for five years by adult suffrage.

The President may dissolve the House of the People. At the commencement of each session he addresses both Houses together, and may send messages to Parliament which it must consider. He must lay before Parliament the Budget and certain reports. He may veto bills passed by Parliament and return a bill, other than a money bill, for reconsideration, but if it is returned he may not again veto it.

The President appoints various high officers including State Governors, judges of the Supreme Court and High Court, the Attorney-General and the Comptroller and Auditor-General, may grant pardons, suspend and commute sentences, and has various emergency powers, including power to legislate by ordinance in certain circumstances.

The President is Commander-in-Chief of the Indian Defence Forces.

The Indian Constitution deals with the Constitutions of the main states of the Union, and their own powers to change their constitutions are limited. Each state has a Governor or a Rajpramulch, and a legislature; in some states this is bicameral, Legislative Councils and Legislative Assemblies, but other states have the latter House only. In the states, formerly Chief Commissioner's provinces, the executive is a Lieutenant-Governor or Chief Commissioner appointed by the President and the Indian Parliament has full legislative powers.

The Constitution lays down Directive Principles of State Policy, statements of aspirations and exhortation to government, not legally enforceable, and Fundamental Rights, many being similar to those laid down in the American Bill of Rights, some guaranteed to all persons and others applicable to Indian citizens only. The Constitution also forbids discrimination as between citizens on the ground only of any of religion, caste, sex, or place of birth. Seven freedoms are defined, freedom of speech, assembly, association, movement, place of residence, property, and to follow an avocation, but all are subject to qualifications and restrictions, and may be suspended by a Proclamation of Emergency.

The High Courts may issue writs of *habeas corpus*, *mandamus*, prohibition, *certiorari*, and *quo warranto*.

Executive

The executive is headed by the President, who acts on the advice of Ministers, who are members of, and must have the support of a majority in, Parliament. The Prime Minister is chosen by the President and advises on the appointment of other Ministers. The President has a choice of Prime Minister only if no party has an absolute majority.

Courts

When the East India Company's factories were established, charters gave the Governor and Council in each settlement power to judge persons subject to them by the laws of England. A court was established in Bombay in 1672, in Madras in 1654, and in Bengal in 1694. In 1726 the E.I. Company secured the establishment of Mayor's Courts and Courts of Oyer and Terminer in the Presidency towns and this involved the introduction of English law, common and statute, as at that date. In Bengal from 1772 the Company, as delegate of the Mogul emperor, set up Moffassal courts, and these were established in Bombay in 1799 and Madras in 1802. Meanwhile in 1773 a Supreme Court was established in Calcutta, in Madras in 1801, and in Bombay in 1823. There was thus a dual system of courts and law, the Supreme Courts' jurisdiction being both civil and criminal within the Presidency towns, the Moffassal courts' jurisdiction covering the whole Presidency. There followed much change and reaction, before, in 1858, the Crown assumed responsibility for justice in India and in 1861 began to create High Courts in various parts of the country. In 1833 a Law Member was added to the Governor-General's Council and a Law Commission recommended in 1840 that, saving the personal law of Hindus and Muslims, the general law should be based on English law. A later Law Commission decided in 1855 that there should be a body of substantive civil law for India as a whole. There resulted the Indian Penal Code (1860), Codes of Civil Procedure (1859), Criminal Procedure (1861), Indian Succession Act (1865), Evidence Act (1872), Contract Act (1872). In 1875 the fourth Law Commission proposed further codification.

Under the Constitution of 1950, there is a Supreme Court of India, High Courts, and various inferior courts.

The Supreme Court has functions comparable both to the U.S. Supreme Court and the British House of Lords and Privy Council. It is interpreter of the Constitution, has jurisdiction over disputes between state and state, and between the Union and states, and has appellate jurisdiction from judgments of a High Court, in civil and criminal cases, if a substantial question of interpretation of the Constitution is involved. It is also a general court of appeal from High Courts, subject to certain conditions. The President may refer any question of fact or law to the Supreme Court, but the Court is not obliged to give its opinion thereon.

There are High Courts for all the major states, with jurisdiction defined by Parliament, frequently based on the jurisdiction of the courts the High Courts replaced. A High Court has superintendence over all tribunals within its territory.

There are many subordinate courts, including district and sessions courts, city civil courts, small cause courts.

Criminal law and procedure

The main crimes are prescribed in the Indian Penal Code of 1860, which is founded on English law adapted to Indian conditions. There is no distinction between treasons, felonies, and misdemeanours. Persons may participate in crimes as principal or abettor. The main division of crimes is into those done with a specified intent, those done with interest that, or knowledge that, certain consequences will follow, and rash or negligent acts.

The Penal Code is supplemented by both Central and Provincial legislation, such as those regulating the distribution and sale of alcohol.

The first Code of Criminal Procedure was enacted in 1861 and replaced in 1872, 1882 and by the present Code in 1898 which applies to all Indian courts. A magistrate may try minor offences summarily, or examine witnesses and, if satisfied that there is a prima facie case, commit the accused for trial before a Court of Session or the High Court.

Private law

Most legal principles and rules were introduced during the British occupation, and as all laws in force when the Republic was inaugurated remained in force until repealed or amended, the main principles and general body of law built up prior to 1950 continues. These were based on English law though the activities of successive Law Commissions since 1833 have given India a large number of Codes and major statutes. The 1950 Constitution aims to secure for all Indians a uniform Civil Code.

Persons

There is no general law relating to marriage. Marriage between Hindus or between Muslims are regulated by their personal law. Legislation deals with Parsi and Christian marriages.

Hindu law and Islamic law are applicable as personal law to members of those faiths in matters of marriage, minority, family relations, religious usages, and certain other matters, partly as laid down in certain statutes and partly according to the courts' decision of justice and equity.

Contract

The Indian Contract Act, 1872, was a simplified statement of English law with modifications for Indian conditions, some taken from the draft New York code.

Consideration is not a requisite, but a contract without consideration must be in writing and registered, or satisfy certain other conditions. A contract is voidable if consent was obtained by fraud, misrepresentation, coercion, or undue influence.

In place of the rules as to penalties and liquidated damages, there is a rule that in every case only reasonable compensation, not exceeding the sum specified, can be recovered.

Specific performance is regulated by the Specific Relief Act, 1877.

The principles of quasi-contract are wider than the corresponding principles of English law.

The Indian Sale of Goods Act, 1930, follows closely the British Act on the subject, but there are important differences.

Property

The transfer of immoveable property must, in general, be effected by an instrument, registered with the registrar for the area where the land is situated. There are six types of mortgages. The law of easements is based on English law. The law of agricultural tenures varies from one part of India to another, the modern policy being to make farmers hold directly from the State.

Trusts

The concept of trust was well known in Hindu and Islamic law though the English law of trusts is inapplicable in India. The Indian Trusts Act, 1882, does not apply to religious or charitable endowments. The rights and duties of trustees are generally similar to those of English law, and investment powers are prescribed.

Succession

The Indian Succession Act, originally of 1865, now of 1925, saves the personal law of Hindus and Muslims, but applies to all immoveable property in India and to all persons domiciled in India. The provisions are basically a simplified version of English rules. All property descends according to the one scheme.

A will must be executed by signature or mark, and attested by two witnesses.

A grant of probate or letters of administration is necessary before a deceased's estate can be administered.

Evidence

The Indian Evidence Act, 1872, was not closely based on English law, but introduced a simpler and clearer version of rules founded on English law.

Civil procedure

Various Codes of Procedure were prepared by successive Indian Law Commissions, notably those of 1859, 1877, 1882, and the present Civil Procedure Code of 1908 based on the English Judicature Act and Rules made thereunder. A suit is commenced by a plaint, on which a summons to appear and answer is issued to the defendant.

Legal profession

The movement has been towards a single class of practitioners at the Bar of all courts in all states, namely advocates, superseding all the former categories of vakils, pleaders, and others. In Bombay and Calcutta the profession has continued divided into advocates and attorneys, advocates being instructed by attorneys as barristers are instructed by solicitors, but elsewhere the profession is undivided. The Bar Council of India exercises supervision of the State Bar Councils which admit advocates to the Bar in a particular state.

C. H. Alexandrowicz, *Bibliography of Indian Law* (1958); A. Gledhill, *India, the Development of its Laws and Constitution* (1964); A. B. Keith, *Constitutional History of India, 1600–1935*; H. Cowell, *History and Constitution of the Courts and Legislative Authorities in India*; M. C. Setalvad, *The Common Law in India* (1960); J. Fawcett, *First Century of British Justice in India* (1934); C. Rankin, *Background to Indian Law* (1946); Whitley Stokes, *Anglo-Indian Codes*; D. D. Basu, *The Constitution of India*; H. M. Seervai, *Constitutional Law of India* (1967).

Indian Law Commissions. A series of commissions which framed major pieces of legislation for British India. The first was appointed in 1833, Lord Macaulay being a member, and was responsible for the Indian Penal Code finally enacted in 1860. It also planned three codes, of Hindu law, Muslim law, and general territorial law. The second commission of 1853–56 sat in England and was responsible for the Code of Civil Procedure (1859) and the Code of Criminal Procedure (1861). The third commission (1861–70) prepared draft codes on succession, contracts, bills and cheques, evidence and other topics, some of which were in substance enacted after the commissioners' resignation. These commissions included a number of very distinguished lawyers and their work in framing codes was of enormous benefit to the development of law in India.

Indictment. A written accusation, running in England in the name of the Queen or, in Scotland, in the name of the Lord Advocate, against a person, charging him with serious crime triable by jury. In England a bill of indictment may be preferred only if the person charged has been committed for trial by an examining magistrate, or by the direction or with the consent of a High Court Judge. It is signed by an officer of the Crown Court and preferred by being delivered to the proper officer of that Court. Indictment was the common law remedy for treasons, felonies, misprisions of either, and misdemeanours of a public nature, and now for any serious crime, other than an offence over which courts of summary jurisdiction have exclusive jurisdiction. In Scots law an indictment is brought if the Lord Advocate decides to prosecute on indictment which may be done in any case except where the offence is one which may be prosecuted summarily only.

In English law an indictment runs in the following form:

THE QUEEN v. A.B.
Crown Court held at Blanktown
Statement of Offence
Common Assault
Particulars of Offence

A.B. on the day of , in the county of assaulted C.D.

In Scots law an indictment is in more narrative form, thus:

A.B. Prisoner in the prison of Edinburgh, you are indicted at the instance of the Right Honourable X.Y., Her Majesty's Advocate, and the charge against you is that on [date] in [place] you did attack C.D. with a razor or other sharp instrument and inflict wounds on him from which he subsequently died and you did murder him.

By Authority of Her Majesty's Advocate,
P.Q.,
Advocate Depute.

This is followed by a List of Productions, i.e. of articles to be produced in evidence and spoken to, and a List of Witnesses who will or may be adduced as witnesses.

In the United States an indictment is a formal written accusation of crime affirmed by a grand jury and presented by it to the court for trial of the accused. There is doubt as to when an accused must or may be tried on indictment. The Fifth Amendment provides for trial of capital or other infamous crimes on indictment, but some states permit waiver of the indictment by the defendant, and as grand juries have declined there is a tendency to allow the prosecution of all crimes on information. In the U.S. under the Federal Rules there are no formal requirements of an indictment provided it sets out the essential facts constituting the offence.

Indies, Laws of the. The body of law promulgated by Spain in the sixteenth to eighteenth centuries for the government of its colonies, particularly in America, and particularly the series of collections of decrees compiled by royal authority and culminating in the *Recopilacion de las Leyes de los Reinos de Indias* (1680). From the earliest colonizations in America the Spanish kings legislated specially for the Indies, i.e. America, in matters of public law, but in private law the main principles were taken from the law of Castile. A substantial part of the legislation was accordingly the adaptation of Castilian administrative and judicial institutions to new conditions. The Laws of Burgos of 1512 regulated relations between Spaniards and Indians, in particular to secure the spiritual welfare of the latter. The New Laws of the Indies of 1542 which were intended to cure deficiencies in the former code, were opposed, and reissued, amended, in 1552. The powers and procedure of colonial courts were defined in 1563.

After abortive attempts at codification the *Recopilacion* was commenced in 1524 under the supervision of the jurists Rodrigo de Aguiar y Acuma and Juan Solorzano Pereira and it was promulgated in 1681; it comprises nine books containing 218 chapters and over 6,000 laws. Legislation of the eighteenth century, particularly concerning commerce and administration, soon made the *Recopilacion* obsolete. Recodification was begun in 1805, but never completed, and the last editions contained only supplementary matter. In this form the code applied to Cuba, Puerto Rico, and the Philippines until 1898. Despite its faults the *Recopilacion* was by far the most comprehensive law code ever prepared for a colonial empire and laid down humane principles for the treatment of Indians, though these were not always implemented.

Individualism. The philosophical standpoint which emphasizes the liberty of each person to exercise his will freely. It was foreshadowed in later natural-law thinking, but became prominent in two lines of nineteenth-century legal thinking, Kantian ethico-juristic thinking, and Benthamite utilitarianism. Kant developed a metaphysical individualism based on his categorical imperative, from which he derived the one innate right belonging to everyone by nature, independence of the arbitrary will of another. Bentham developed individualist utilitarianism, that men should act so as to maximize pleasure and minimize pain.

The ideal of the free-willing individual had numerous manifestations in the nineteenth century. The most notable were belief in freedom of contract, in vested property rights and freedom of testamentary disposition, and immunity from liability for harm unless fault were established. These ideas have survived though all have been very much eroded by legislation and are less accepted in principle than they were. The ideal of the free willing individual, moreover, contained many ambiguities as a concept. There is no agreement on the content of the freedom to perfect or express oneself, nor on how to make the liberty of one consistent with the equal liberty of others.

Indorsement (or endorsement). Anything written on the back of a deed or writing. Bills of exchange and cheques are negotiated by the payee indorsing them, by writing his name on the back, and transferring them. Indorsement may be general, or conditional, or without recourse, or restrictive, or qualified.

Indorsement of claim. Writs of summons in the High Court must be indorsed with a statement of the claim made, or remedy or relief sought.

Inducement. That which motivates or brings about something; the introductory part of a pleading. Inducement by a third party of a breach of a lawful contract between two others by one of them is tortious.

Indulgence, Declaration of. Royal announcements to relieve the harshness of the laws against dissenters so as to secure their support. Charles II issued one in 1662, intimating his intention to seek parliamentary approval for the use of the royal dispensing power, but it elicited a very angry reaction from the Commons. He issued a second one in 1672, in accordance with his undertakings in the secret Treaty of Dover, suspending the penal laws against both Catholics and nonconformists, but Parliament bribed the King by a grant of money to withdraw the declaration and passed the Test Act (1673) which strengthened the law against the Catholics.

James II issued a declaration of indulgence in 1687, seeking by use of the royal prerogative to suspend the Test Act for all dissenters and to destroy the privileged position of the Anglican Church. He issued a second one in 1688 and it was ordered to be read from all parish church pulpits. This led to the protest of the seven bishops, their trial and acquittal, and the invitation to William of Orange to come over to deliver England from unconstitutional government.

Industrial and provident societies. A name originally given to societies established to trade profitably and to distribute the profits by way of provident provisions for their members' future. An Act of 1834 enabled friendly societies to be established for any purpose not illegal, and an Act of 1846 permitted such societies to be established for the frugal investment of the savings of members. Till 1852 industrial and provident societies were established under the legislation relating to friendly societies, but in 1852 the first Industrial and Provident Societies Act was passed but was restricted to co-operative societies. One third of the profits made from the carrying on of the members' business could be divided among the members, but the rest had to be retained or applied for the provident purposes authorized in respect of friendly societies. Since 1862 registered societies are incorporated and the members have limited liability.

Under the modern legislation societies for carrying on any industries, businesses, or trades authorized by their rules, and consisting of at least seven members or of two registered societies may register with the Chief Registrar of Friendly Societies, provided the Registrar is satisfied that the society is a bona fide co-operative society or that in view of the purposes of the society there are special reasons why it should be registered under the Acts rather than under the Companies Acts. The name must contain the word Limited, and once a society is registered it becomes a body corporate, and all property is vested in the society. The rules of the society must make provision for certain matters; they bind the society and all members and all persons claiming through them. A society is managed by a committee of management, which has statutory powers and functions. Members may be individuals or other registered societies and have statutory rights; general meetings of members must be held periodically. The rules must provide for the appropriation of profits, and there are requirements of annual returns and audits. Societies may amalgamate or transfer their engagements to another society, convert themselves into companies, wind-up voluntarily, or at the instance of creditors or under the order of the court.

Industrial arbitration. The reference of disputes arising out of industrial employment to an independent person or persons to determine. In the U.K. arbitration can be adopted only with the consent of both parties to the dispute and there is no general system of arbitration. The Industrial Court (q.v.), established in 1919, provided a standing arbitration tribunal but references are only by consent and awards unenforceable. *Ad hoc* arbitral tribunals have been set up by consent to adjudicate on many industrial disputes. In many Commonwealth and European countries compulsory arbitration is in force.

Industrial Arbitration Board. An institution established as the Industrial Court in 1919, which may sit in two or more divisions, composed of panels of independent members, persons representing employers, and persons representing employees, all nominated by the Secretary of State for Employment. A matter referred to the Court may be heard and determined by a single member. If the members of the Court cannot agree, the matter is decided by the chairman.

It has jurisdiction under various statutes, principally to deal with a trade dispute referred, with the consent of both parties, to it for settlement. The Secretary of State may refer to the Court for advice any matter relating to or arising out of a trade dispute, or trade disputes in general, or trade disputes of any class, or any other matter which in his opinion ought to be referred. In 1974 it was renamed the Industrial Arbitration Board.

Industrial assurance. The business of effecting assurances on persons' lives, the premiums being collected by collectors at weekly or monthly intervals. Such business may be carried on only by assurance companies registered as companies or industrial and provident societies, or by registered friendly societies, and is regulated by special legislation. The Chief Registrar of Friendly Societies, in the capacity of Industrial Assurance Commissioner, is the authority controlling industrial assurance business, and has administrative and judicial functions under the regulating statutes; he reports annually to Parliament. The classes of policy which industrial assurance companies and collecting societies may issue, and the amount thereof are limited. It is in general no longer lawful to insure for funeral expenses. There is extensive statutory control of the mode of conduct of business, such as protecting against forfeiture of a policy.

Industrial Court. See INDUSTRIAL ARBITRATION BOARD.

Industrial diseases. Diseases commonly contracted by workers engaged in certain industrial processes. Disability resulting therefrom is compensable as being an industrial injury.

Industrial injuries. Injuries sustained by persons at work, arising out of and in the course of their employment. Compensation is payable by way of industrial injury benefit and if tortious liability can be established, damages may also be recoverable at common law or for breach of statutory duty.

Industrial law (or Labour law). Not a separate branch of law but a compendious term for the topics of law dealing with employment and the relationship of employer and employee (known down to the twentieth century as master and servant), employers' associations and trade unions, strikes, disputes, and the legal aspects of industrial relations. Despite the name this branch of law includes not only industrial employment but employment in agriculture, trade, and otherwise. Analytically it is not a distinct branch of law, but rather a convenient grouping of topics of law drawn from several branches, mainly from the law of contract and of tort.

Industrial property. A term sometimes applied to those kinds of rights valuable in industry, notably patents, trade-marks, designs, and 'know-how'.

Industrial relations and disputes. Industrial relations includes all relations between persons and groups or organizations of persons engaged in any kind of industry, in particular between employers and managers on the one hand and employees, their representatives and trade unions on the other, and including all negotiations about such matters as hours of work, wages, and conditions of employment generally. Industrial disputes are disputes arising in the course of such relations and frequently resulting in strikes. Industrial relations have in Britain traditionally depended on bargaining at plant, district, or national level largely outwith the legal system, but the Industrial Relations Act, 1971 (repealed 1974), sought to subject such relations to a greater extent to law, by making collective bargains generally legally enforceable as contracts. Indeed a striking feature of industrial relations in the U.K. has been the refusal of employees and trade unions to have their bargains or disputes regarded as justiciable at all, and their refusal to be bound by law. Trade unions have very large immunities from legal liability for actings done in the course of or in contemplation of an industrial dispute.

Industrial schools. Places established by statute in 1857 which received children who had committed less serious offences or no offence at all, but who lived in such circumstances that they seemed to be in danger of becoming criminals.

Industrial tribunals. Bodies established in Britain from 1964 onwards, consisting of three members of whom the chairman alone need be legally qualified, and exercising jurisdiction under various statutes over various kinds of disputes arising out of employment, particularly complaints of unfair dismissal and claims for redundancy payments. Appeal lies to the Employment Appeal Tribunal (q.v.).

Industry. A general term for all persons and enterprises engaged in productive work or manufacture, but extending also to the provision of services, e.g. hotels and catering, entertainment, as contrasted with trade and commerce.

Ine (*c.* 660–*c.* 726). Anglo-Saxon king of Wessex, 688–726, and the first West Saxon king to issue a code of laws. They deal mainly with judicial procedures, listing punishments for various offences. They are an important source for the structure of early English society.

Inevitable accident. An unintended and undesired calamity not avoidable by the taking of any care or precautions reasonable and practicable in the circumstances, such as a death or a fire caused by lightning.

Infamia. The discredit or diminution of personal dignity which in Roman law resulted to persons condemned for some private delicts (e.g. theft, fraud) or in some civil actions, where the defendant was guilty of breach of trust or confidence. It equated him to persons practising some shameful professions. Similarly dishonourable or shameful actions or bankruptcy involved *infamia.* The praetorian edict listed categories of persons thus tainted and Justinian enlarged the list. It entailed different disabilities for different kinds of persons, exclusion from office, incapacity to act for another, to bequeath by will, and others.

Infamous conduct. Any conduct in the course of the practice of a profession which competent practitioners of good repute would regard as disgraceful or dishonourable. It is a ground under various statutes for removing a practitioner's name from the professional register.

Infamy. Public disgrace or loss of reputation, especially in consequence of criminal conviction. In early common law whether a crime was infamous or not depended on whether or not it stamped the offender as untrustworthy, and it was accordingly originally limited to *crimen falsi*, originally perjury, but extended to any crime involving fraud or corruption. In time all felonies came to be treated as infamous. Conviction for an infamous crime resulted in disqualification for giving evidence. This result has been abolished in England and generally in the U.S. The Fifth Amendment to the U.S. Constitution requires indictment by a grand jury for any capital or otherwise infamous crime, and any crime punishable by imprisonment in a state penitentiary is deemed infamous, as is any federal crime punishable by imprisonment for more than a year. In canon law infamy is a penalty provided as a punishment for certain offences enumerated in the penal code, and brands those guilty as unworthy of the good esteem and honour of their fellows. It is automatically incurred by apostasy, heresy, schism, bigamy, and sexual offences, and may be imposed by judicial sentence. The effects of incurring infamy are to deny the guilty person all honour and all the benefits of ecclesiastical rights. The penalty may be remitted only by the Pope.

Infangthef and outfangthef. Words used in grants of jurisdiction in early mediaeval law. The former is the right to hang a thief under one's own jurisdiction caught redhanded on one's land, the latter the right to do so whether or not the capture is made on one's own land.

Infant. In English law an infant is a person of either sex under the age of 21, or since 1970, of 18. The term minor is now used alternatively. Certain statutes also use the terms 'child' and 'young person' for those under 14 and aged 14–17 respectively.

At common law an infant's contracts are generally voidable at his instance though binding on the other party, but by the Infants Relief Act, 1874, accounts stated, and contracts for repayment of money lent or to be lent or goods supplied, apart from necessaries, are wholly void, though not entirely ineffective. But contracts of apprenticeship and service are binding. An infant over seven is generally liable for tort. There is a conclusive presumption that a child under eight cannot be guilty of crime and a rebuttable presumption that a child between eight and 14 does not have capacity to form guilty intent to commit crime. An infant cannot hold an estate in land, or be a trustee, or exercise the office of executor.

He is generally incapable of exercising rights of citizenship or holding public or private offices. He may be a peer (but cannot sit in the Lords until attaining 21), or be made bankrupt.

Special statutes deal with particular functions such as entitlement to drive a motor vehicle, and certain statutes confer particular powers on infants over 16.

Infanticide. In English law the crime under the Infanticide Act, 1938, committed by a mother who causes the death of her child, under the age of 12 months, but the balance of whose mind was disturbed by not having fully recovered from birth or by the effect of lactation. It is punishable as manslaughter. This crime was created in 1922 to avoid the public distaste for treating such killing as murder.

Inferior court. A general term for a court of lower rank than another, or other courts, the jurisdiction of which is frequently limited, the decisions of which are normally subject to appeal to a superior court, and which is subject to the control of a superior court. Thus county courts and magistrates' courts in England and sheriff courts and district courts in Scotland are inferior courts.

Infeudation. The process under the feudal system of landholding of putting a person into legal possession of lands.

Information. An initial writ by which certain kinds of proceedings were or are initiated.

At common law the prerogative process of information was a species of civil claim at the instance of the Crown instituted to recover damages for injury to Crown property or recover goods or money due to the Crown. It was initiated by the Attorney-General. Other varieties of this were the information of intrusion, in the nature of the action for trespass *quare clausum fregit*, brought for any trespass on Crown lands; the information of debt, brought on a contract for moneys due to the Crown; the information *in rem* or real information, where goods were supposed to become Crown property as derelict or were forfeited for non-payment of excise duties; and the information of *devenerunt*, in the nature of an action of *detinue* to recover goods belonging to the Crown but in the hands of another.

There were also proceedings called English informations in the Court of Exchequer, in the exercise of its equitable jurisdiction, which was generally similar to a bill of complaint in equity.

Proceedings by information in the Court of Chancery were brought by the Attorney-General on behalf of the Crown or persons under its particular protection (information *ex officio*), frequently on the relation of some person, whose name was inserted in the information, and who frequently had a personal interest in the matter in issue (Att.-Gen., *ex rel* A.B. . . .).

Informations were used also in criminal procedure, some brought partly at the instance of an informer, to exact a penalty under a penal statute. In the Queen's Bench Division informations might be filed by the Attorney-General *ex officio* in cases of crimes which specially affect the sovereign or disturb or endanger the government, or by the Master of the Crown Office on the complaint of a private individual for the punishment of riots, misconduct by magistrates, and other serious offences. Only informations by the Attorney-General *ex officio* are still competent.

Criminal proceedings in magistrates' courts are initiated by informations, which state briefly the alleged criminal conduct.

Informer. A person who takes legal proceedings for the recovery of a penalty, the whole or part of which falls to him, or a person who prosecutes for any offence which affects the public generally, the right to prosecute for which has not been confined to persons of any particular class.

A common informer was one who sued for a penalty payable by an offender for some contravention of the law, and recoverable in whole or in part by the informer himself for having given the information. Such proceedings cannot be brought since 1951.

Infortunium, homicide per (homicide by misadventure). At common law a person found guilty of homicide *per infortunium* was liable to forfeiture of goods, but as a matter of course he received a pardon and a writ of restitution. Accidental causing of death became accepted as not involving guilt of homicide and in 1861 the forfeiture was abolished.

Infringement. A violation of a statute or of a right vested in another. The word is mostly used in the contexts of infringement of copyright, patent, trade-mark, or design.

Inglis, John, Lord Glencorse (1810–91). Called to the Scottish Bar in 1835 soon made a reputation; he became Solicitor-General for Scotland in 1852 and Lord Advocate later in the same year. He was also elected Dean of the Faculty of Advocates 1852–58. He again became Lord Advocate in 1858 and entered Parliament but in the same year became Lord Justice-Clerk and in 1867 was promoted Lord President, an office he held till his death.

As Lord Advocate he had promoted a bill for the reform of the Scottish Universities, and while a judge he was chairman of the Commission appointed to give effect to the Act. He was Rector of the Universities of Aberdeen (1857) and Glasgow (1865) and, from 1868, Chancellor of Edinburgh University. On the bench he was dignified, patient, and courteous. As a judge Inglis was unquestionably the greatest of the century and laid down law in classic fashion in what have, ever since, been deemed leading cases. He had great reverence for Scots law as a national institution. His contributions to the development and modern formulation of the law have been immense and his judgments are of the highest authority.

J. Crabb Watt, *John Inglis.*

Inhabited house duty. A tax assessed on the rent or letting value of houses, but not business premises, levied between 1808 and 1924.

Inheritance. The devolution of property on the death of its owner to other persons. The existence of the concept and of legal rules about it depend on the recognition of individual or group property rights in objects, such as land and goods. Psychologically, the recognition of inheritance and the possibility of leaving property to one's family is a powerful incentive to conserve and save. Inheritance rules are accordingly attacked by persons holding particular social and political views based on jealousy, and these views are to some extent implicit in the existence of inheritance taxes. Where, however, inheritance of private rights in property is recognized, two further issues of principle arise; to what extent shall owners have the power to determine who are to inherit, and, if not, or failing expression of wish on the matter, who are to inherit by operation of law.

The former of these has given rise to the very widespread recognition of the will (q.v.), or expressed wish of a property-owner, and to the principles of testate succession (q.v.), the latter to the rules of intestate succession (q.v.). In both cases the question arises whether any dependants are to have any guaranteed entitlement to any part of the property. There must also be rules for the machinery of transfer, the administration of the estate in orderly fashion, settlement of the deceased's affairs, and transfer of the free balance to those entitled to it.

The word is also frequently used in a concrete sense, in the sense of hereditament (q.v.), or for the estate which at common law descended to the heir on the owner's death intestate.

See also SUCCESSION.

Inhibition. (1) In the English law of land registration, an order or entry on the register prohibiting dealing with the land registered for a given time or until further order. In ecclesiastical law it is an order forbidding an incumbent to perform any service until he obeys a certain order, or a writ forbidding a judge from proceeding further with a case before him, as in case of appeal. In Privy Council practice prior to the Judicature Acts, an inhibition might be issued in admiralty appeals prohibiting the inferior court and the appellant from proceeding further in the cause.

(2) In Scots law, a personal prohibition at the instance of a creditor restraining the person inhibited from contracting any debt or granting any deed whereby any part of his lands may be alienated to the creditor's prejudice. It may be used in security, or in execution preparatory to enforcement of payment, to the creditor. It must be served on the debtor and notified to the public by being recorded in the Register of Inhibitions and Adjudications.

Initiative. The political device whereby a group of persons of a defined size may themselves introduce legislation, without waiting for the State legislature to do so. It was developed in Switzerland in 1860–1920. Introduced by South Dakota in 1898, it is recognized also in a number of other American states.

Injunction. A remedy of an equitable nature, whereby a person is ordered to refrain from doing (restrictive injunction), or ordered to do (mandatory injunction) a particular act or thing. Injunctions may also be interlocutory, to preserve the *status quo* until the relevant facts can be ascertained, or perpetual, based on final determination of the rights of parties and intended permanently to prevent infringement thereof. In granting injunction the court acts *in personam*, even though the act is done outwith the jurisdiction. Ancillary relief may also be granted, such as an account of profits, and damages may be awarded instead of, or in addition to, injunction, and the jurisdiction to grant injunction is not excluded by the existence of a special statutory remedy unless the statute so provides. Interlocutory injunctions will be granted only where a strong prima facie case can be shown, to restrain a threatened or apprehended injury, or very substantial harm is likely to be done. In doubtful cases the court considers the nature of the injury and the balance of convenience to the parties, the conduct of the parties, whether there has been acquiescence or delay, and generally what is equitable in the circumstances. The court will not normally grant perpetual injunction to restrain an actionable wrong where damages is the proper remedy, but only where injury is irreparable and could not be adequately remedied by damages, and is continuous. Injunction may be granted even where no damage had been caused and where the injury complained of had ceased when action was brought, but where inconvenience does not entitle a party to an injunction, nor bare interference with a legal right. A plaintiff is entitled to an injunction to restrain a threatened infringement of a legal right. Mandatory injunctions are appropriate where the injury done cannot be adequately compensated by damages, or is in breach of an express agreement, or so serious and material that restoration of the *status quo ante* is the only way whereby justice can be adequately done.

An injunction must be implicitly obeyed and observed to the letter. The remedy for breach of an injunction is committal to prison for flouting the order of the court.

The name injunction was also given to a new type of royal instrument introduced under the Act of Supremacy, 1534, which replaced the old canons of convocation in giving minute directions to the clergy as to their duties. In 1536 and 1538 Thomas Cromwell issued injunctions attacking traditional practices concerned with images, relics, and pilgrimages.

In the U.S. the injunction has retained its equitable character, and is applicable to a wide variety of cases of harmful or potentially harmful conduct.

An application of the injunction which has caused particular problems has been its use in labour disputes, but legislation has limited its use in these contexts. Considerable extension of the use of injunctions has been in the field of governmental regulation, and in the enforcement of federal and state statutes, injunctions are frequently sought as alternatives to criminal convictions. Injunction has also been extensively used to ensure protection of rights guaranteed by the Constitution, particularly to prevent violations of the rights of free assembly, free speech, and denial of equal rights and opportunities on racial grounds.

Injuria. In Roman law, that which was contrary to law (*in ius*) and in breach of legal duty to refrain from such conduct. In the classical law it was also a nominate delict, involving any contumelious disregard of another's rights or personality, including assaults, insults, abuse and affronts to dignity or reputation, always so long as done wilfully and with contumelious intent.

Injuria sine damno. Legal wrong or breach of duty (*injuria*) without *damnum*, resultant harm or loss. In general wrong or breach of duty not resulting in ascertainable harm or loss justifies an award of nominal damages and no more. But in certain cases, notably the tort of negligence, some ascertainable loss is an essential of the actionability of the wrong, and if there is no loss no claim is competent at all.

See also *Damnum sine injuria.*

Injurious falsehood. A category of wrongs constituted by the communication of a false statement of a prejudicial nature, affecting a person not, or at least not directly and immediately, in his honour or reputation, but in his property. It comprehends the individual wrongs of slander of business, slander of goods, and slander of title.

Injurious reliance. A principle which is sometimes invoked as justifying the claim to have a contract enforced, the principle that where one person has, to his detriment, altered his position in reliance on another's undertaking that he shall make good the undertaking or compensate the first person for his detriment.

Injury. (1) An infringement of a legal right vested in the complainer, done deliberately or by taking care inadequate to prevent it, though unintentional. It is distinct from damage, and there may be injury in this sense without damage or harm.

Scots law distinguishes real injuries, inflicted by actings, e.g. negligent physical harm; and verbal injuries, inflicted by words, e.g. defamation. A distinction is sometimes drawn between public injuries or infringement of the rights of the State or community and private injuries, infringements of the rights of individuals. This distinction corresponds generally to that between crimes and torts.

(2) Especially in the plural, actual damage or harm, particularly to a person.

Injustice. Since Aristotle the sense of injustice felt by individuals at decisions affecting them, or at cases attracting public attention, has been recognized as an index to the idea of justice. It is easier to recognize what one regards as injustice than the converse. But an injustice is not a negation of justice in general but only an instance in which at least one person feels that justice has not been realized in practice. Attempts to elucidate instances of injustice have not, however, been productive of general ideas, nor have these contributed materially to the understanding of justice.

Inland Revenue. The portion of the public revenue derived from the taxation of income and capital in the U.K. as contrasted with excise duties and customs duties on imported goods. It is supervised by the Commissioners of Inland Revenue.

Inner barrister. A student member of one of the Inns of Court. See also OUTER BAR.

Inner House. The part of the Court of Session (q.v.) comprising the First and Second Divisions, which are mainly concerned with exercising appellate jurisdiction. It is thought to have been so-called as having been further from the entrance to the courts than the Outer House (q.v.), which deals with business at first instance.

Inner Temple. One of the Inns of Court (q.v.).

Innes, Cosmo (1798–1874). Became an advocate of the Scottish Bar in 1822, was sheriff of Moray and from 1852 principal clerk of Session. He was an acute and learned antiquary and student of ancient Scottish records, and edited various cartularies and other works. He completed the work of Thomas Thomson (q.v.) in editing the folio Record edition of the *Acts of the Parliaments of Scotland* (1124–1707) and compiled the General Index thereto. From 1846 till his death he was professor of constitutional law and history at Edinburgh. He published also *Origines Parochiales Scotiae* (1850), *Scotland in the Middle Ages* (1860); *Sketches of Early Scottish History and Social Progress* (1861), *Ancient Laws and Customs of the Burghs of Scotland* (1868), *Lectures on Scotch Legal Antiquities* (1872), and other works.

Innocent III (Lothar of Segni) (1160–1216) (Pope 1198–1216). One of the greatest mediaeval popes, a great theologian and canonist. He issued a large number of decretals, which led to the compilation of a number of collections. In 1209–10 he issued his own collection of relevant decretals for the period 1198–1208, the *Compilatio Tertia*, prepared by Petrus Collivaccinus, which was the first officially promulgated body of canon law, and sent it to the University of Bologna, where it greatly stimulated canonist studies. He sat regularly in consistories dealing with legal matters. He was author of numerous reforms and much reorganization, and supervised the Church vigorously and was largely responsible for the attainment by the papacy of its enormously important position in mediaeval Europe.

Innocent IV (Sinibaldo Fieschi) (*c.* 1200–54) (Pope 1243–54), a teacher of canon law and an auditor of the papal Curia before being elected Pope, he wrote a major *Commentaria super libros quinque Decretalium* (pub. 1590) a brilliant and important commentary on canon law.

Innocent misrepresentation. A misrepresentation (q.v.) which is made in the honest but mistaken belief that it was correct and with no intention to deceive or defraud.

Innocent passage. The right of a ship of one state to sail in and through the territorial sea of another state, so long as the passage is not prejudicial to the peace, good order, or security of the coastal state. The right of innocent passage for merchant ships has been recognized, subject to local regulations as to navigation, pollution, health, and to criminal jurisdiction. A ship which infringes the local law while exercising the right of innocent passage may be arrested by the local state. Warships possibly have a similar right, but may not be arrested even if they violate local regulations.

Innominate contracts. In Roman law contracts other than those in the categories of real, verbal, literal, and consensual contracts. The more important of these had names such as *permutatio* (barter) and *aestimatum* (sale or return) and the importance of the innominate contracts is that by Justinian's time there had developed a general action most often called *actio praescriptis verbis* which lay wherever a bargain fell outwith any of the recognized categories and one party had performed his side of the bargain but the other party had failed to do so. This was important in that it admitted the general principle that an agreement performed on one part is a legal contract and enforceable, even though the transaction was nameless and outwith the quadripartite classification. In Scots law the term survives for contracts which cannot be pigeonholed under one of the recognized headings of named kinds of contracts.

Inns and innkeepers. At common law an inn was premises offering food, drink, and accommodation to any traveller requesting such facilities, without previous contract, so long as he was in a fit state to be received and appeared able and willing to pay a reasonable sum for the services provided. Distinct from inns were and are lodging-houses and boarding-houses, where persons may stay, usually for an agreed time, by previous contract; and public houses and ale-houses, which do not offer sleeping accommodation. The innkeeper was bound at common law to receive all such persons if he had room for them and might not pick and choose between them. The obligation to receive included receiving the traveller's horse, carriage or motor-car, and luggage. An innkeeper was not bound to receive other than travellers, and might refuse to accept lodgers, or turn out travellers who had by length of stay in effect become lodgers. He was, and is, liable to an action if he refused without lawful excuse to receive and lodge a traveller. He has a lien over any property, except vehicles, animals, and their equipment brought by the guest, in security of an unpaid bill.

Under statute only an hotel offering food, drink and, if required, accommodation, without prior contract, to any traveller fit to be received and apparently capable of paying for the services is an inn, and residential or private hotels, boarding-houses, restaurants, and public-houses, whatever called, are not inns. The modern Hotel Proprietors Act, 1956, confers certain safeguards against liability on proprietors of inns which fall within the definition of hotel.

At common law and under statute an innkeeper or, today, a hotel proprietor, must take reasonable care for his guests in respect of safety of the premises, and wholesomeness of food and drink supplied.

At common law and by the custom of the realm in England, and at common law, founded on the praetorian edict *nautae, caupones stabularii* (q.v.), in Scotland, an innkeeper was strictly liable to guests for property lost or stolen within the inn or its precincts, without proof of fault or negligence. An innkeeper was not liable for loss caused by the misconduct or negligence of the guest, nor where the guest assumes exclusive care of the goods, nor for loss arising from Act of God or of the Queen's enemies. Now under statute an hotel proprietor is liable to make good to any guest loss not only by property lost or stolen but by damage to property, but only where sleeping accommodation had been arranged for the traveller and the loss occurred between the midnight before and the midnight after the guest's stay. Where the hotel proprietor is liable as innkeeper, his liability to one guest is not to exceed £50 for one article or £100 in aggregate except where the property was stolen, lost, or damaged through the default, neglect, or wilful act of the proprietor or a servant, or the property was deposited by the guest for safe custody, or it was offered for deposit and not accepted. This limitation applies only if a statutory notice is exhibited at or near the reception desk or entrance. The hotel proprietor's liability does not extend to loss of or damage to any vehicle or any property, or any horse or animal, or its harness or equipment.

This strict liability does not attach to premises which are not statutory inns, where guests can recover for property lost, stolen, or damaged only on proof of fault on the proprietor's part.

Inns of Chancery. A number of Inns similar but subordinate to the Inns of Court (q.v.). The Inns were Thavie's Inn and Furnival's Inn, attached to Lincoln's Inn; Staple Inn and Barnard's Inn, attached to Gray's Inn; Clifford's Inn, Clement's Inn, and Lyon's Inn, attached to the Inner Temple; and New Inn, attached to the Middle Temple. Certain others were sometimes mentioned. All were of great antiquity. Originally they were collegiate houses, in which students resided when learning the rudiments of law and preparing for admission to the Inns of Court.

In the seventeenth century the Inns of Court sought to exclude attorneys and solicitors from membership and this drove them to the Inns of Chancery, which by the eighteenth century were exclusively in the hands of attorneys. But these Inns were little more than dining clubs and never exercised professional discipline or control over their members as did the Inns of Court, nor did they give the educational training which the Inns of Court did and which, after faltering, revived in the nineteenth century.

The Inns of Chancery were all dissolved in the latter nineteenth century, the last to go being New Inn and Clifford's Inn. Only fragments remain of the buildings.

Inns of Court. Voluntary unincorporated societies which have existed in London from very ancient times (at least the thirteenth century) in which originally apprentices-at-law lived a semi-collegiate life, subject to common discipline and government, and pursuing a common system of education and, after call to the Bar, of professional life. They probably developed out of the practice of aspirants to the legal profession seeking to live with or near distinguished lawyers and learning from them, and have similarities to the Oxford and Cambridge colleges and, like them, evolved from the medieval gilds.

Originally there were a number of Inns of Court, the chief of which are the four Inns now remaining, namely, The Honourable Societies of Lincoln's Inn, the Inner Temple, the Middle Temple, and Gray's Inn, and subordinate to them were a number of Inns of Chancery (q.v.), all of which have ceased to exist. The serjeants-at-law (q.v.) had two Inns of their own, Serjeants' Inns, and when a barrister became a serjeant he left his former Inn and joined Serjeant's Inn. As till 1875 the degree of serjeant was a necessary qualification for being a judge of one of the common law courts, all the judges belonged to Serjeants' Inns.

From their foundation a major object of the Inns was to provide practical legal education, by readings, moots or mock trials, and exercises. This system declined in the seventeenth century and became a formality. But in Tudor and Stuart times, particularly, the Inns rivalled the older Universities for their full social and fashionable life; they were a kind of finishing school for young men, many of whom were not destined for the law. They are sometimes referred to as the university of the law. In 1852 the Inns jointly established the Council of Legal Education which provides instruction for students and conducts the Bar examinations.

The legal education in the Inns of Court was subject to the control of the judges who granted audience in court only to members of the Inns who had satisfied the conditions. Hence, today the Inns of Court have the sole power of calling persons to the Bar, and of censuring or even disbarring a barrister.

As now constituted the Inns comprise three grades of members, Masters of the Bench or benchers, who are the governing body, co-opt to their own number, exercise a disciplinary jurisdiction over the members and call students who have satisfied the requirements to the Bar; barristers; and students. Each Inn annually elects a Bencher to be Treasurer and titular head of the Inn. While the Inns are independent, they have common regulations for admission and qualification, and jointly manage the Council of Legal Education, which arrange the instruction and examination of students before certifying them as fit to be called to the Bar.

Many groups of counsel have their professional chambers in the buildings within the Inns, and the Inns are the focus of professional lives. All the Inns maintain fine libraries and their Halls are of outstanding architectural beauty and interest.

In Ireland King's Inns, Dublin, performs functions similar to the Inns of Court in London, and in 1925 an Inn of Court of Northern Ireland was established. In Scotland the Faculty of Advocates (q.v.) has a corresponding position and exercises similar functions.

D. S. Bland, *Bibliography of the Inns of Court and Chancery* (1965); T. Cunningham, *History and Antiquities of the Four Inns of Court and of the Nine Inns of Chancery* (1780); W. Herbert, *Antiquities of the Inns of Court and Chancery* (1804); G. Goldsmith, *The English Bar* (1843); R. R. Pearce, *History of the Inns of Court and Chancery* (1848); H. H. Bellot, *The Inner and Middle Temple* (1902); J. Williamson, *History of the Temple* (1924); H. H. Bellot, *Gray's Inn and Lincoln's Inn* (1925); B. T. Duhigg, *History of the King's Inns in Ireland* (1806).

Innuendo. In libel (both as a crime and a tort) and slander the words or acts connecting a statement with the plaintiff, or the secondary meaning alleged to have been conveyed by an *ex facie* innocent or ambiguous statement which renders it defamatory. If a statement is not *ex facie* defamatory of the plaintiff, an innuendo must be pled and evidence led to the effect that the statement was in fact understood in this secondary, defamatory, sense. If there are no facts and circumstances alleged capable of supporting the innuendo, the case must be withdrawn from the jury.

Inofficious testament. A will which entirely passes over relatives having substantial moral claims on the testator without any sufficient reason being given. In Roman Law such a will raised a presumption that the testator was suffering from loss of reason or loss of memory and it could be challenged, though even a trivial legacy was enough to rebut the presumption. In English law the doctrine never applied, though an unfounded belief that it applied led to the practice of giving a merely nominal legacy, 'cutting off with a shilling', where the testator did not wish to give any material benefit.

In Scots law the doctrine never applied because relatives were provided for by the doctrine of legal rights (q.v.).

Inquest. An inquiry held by a coroner as to the cause of a death. Deaths in various premises and in various circumstances require to be reported to the coroner, and any death may be reported. The coroner must decide whether to hold an inquest and must do so if there is reasonable cause to suspect that the deceased died a violent, or unnatural, or sudden death of which the cause is uncertain. The inquest may be held at any place within the coroner's jurisdiction, and must be in public unless interests of national security require that it be held in private. A jury of between seven and 11 is summoned in cases specified by statute. The coroner must, and the jury may view the body. The coroner examines as witnesses all persons who wish to give evidence and all whom he thinks it expedient to examine. Any person who has an interest may examine witnesses. The evidence and proceedings are directed to ascertaining who the deceased was and how, when, and where he died. At the conclusion the coroner must sum up to the jury and take its verdict, and embody it in an inquisition under prescribed headings, which is signed by coroner and jurors. Amendment to an inquisition may be made by the Queen's Bench Division, or it may be quashed and a fresh inquest ordered. Inquests are also held on treasure trove (q.v.).

Inquest of office. A prerogative remedy for the benefit of the Crown in the form of an inquiry by a sheriff, coroner or the officer of the Crown, and a jury, as to any matter which entitled the Crown to the possession of lands or chattels by reason of escheat, forfeiture, or idiocy. These inquests were either offices of entitling, used where the Crown became entitled only after an office had been held, or offices of instruction used where the office was necessary only for the better instruction of the officer before seizure. The abolition of escheat and forfeiture has made such inquests almost obsolete.

Inquest of sheriffs. An inquiry instituted by Henry II in 1170 into the conduct of the sheriffs whom he had removed from their offices, into their receipts and wrongful exactions. They had to give pledges to answer, and make compensation for, all wrongful exactions.

Inquiry. In actions in the High Court an investigation of facts subsidiary to the main issue by an official, frequently a Master, to whom matters are remitted for inquiry and report. Such matters may include, in Queen's Bench business, the amount of damages, or, in Chancery business, the extent of an estate, the next-of-kin, and the like.

Under the Tribunals of Inquiry (Evidence) Act, 1921, a tribunal may, on a resolution by both Houses of Parliament, be set up, having all the powers of the High Court or Court of Session, to inquire into a definite matter of urgent public importance.

Departmental inquiries are directed to be held under many statutes. Among the commonest are those held to inquire into and report on objections to compulsory purchase orders and on appeals against refusals of planning permission by a local planning authority. In many other cases a Minister is empowered by statute to hold an inquiry, as into a railway or aircraft accident.

Inquisitio post mortem. An inquest of office (q.v.) held on the death of a Crown tenant to ascertain in what lands he died seised and who is his heir and the heir's age, so as to know the Crown's entitlements to marriage, wardship, and other aids. After feudal tenure disappeared such inquisitions were limited to cases of escheat and forfeiture.

Inquisition. (1) A formal inquiry to ascertain whether or not a person was of unsound mind and incapable of managing himself or his affairs, in which case the management of his affairs was entrusted to a committee. The procedure is now obsolete.

The term was also used of the investigation held before the sheriff or coroner to assess the value of land taken under compulsory powers.

The term is also used of the document recording the result of an inquiry, or coroner's inquest. In the latter case it states who the deceased was, how, when, and where he died, and if his death was caused by murder or manslaughter, the persons held by the jury to have been guilty of such murder or manslaughter.

(2) A special permanent tribunal established by Pope Gregory IX in 1231 to combat heresy, taking its name from the new procedure created by Pope Innocent III permitting the searching out of persons accused of heresy. At first limited mainly to Germany and extended to Aragon in 1231, it became general in 1233. The judges or inquisitors were recruited almost entirely from Dominicans and Franciscans, each tribunal consisting of two judges. Initially they went on circuit, but soon became stationary tribunals, summoning suspects before them. Local authorities and even bishops were often reluctant to co-operate with the Inquisition and common people were often hostile.

Inquisitors were provided with authentic guides, such as the *Processus inquisitionis* (1248), the *Practica officii inquisitionis heretice pravitatis* (*c.* 1320), and the *Directorium inquisitionis* (*c.* 1375).

According to the new procedure the judges could

ex officio bring suit against anyone the victim of even rumour. When questioned the accused had to take an oath to tell the truth. The accused did not know the witnesses for the prosecution, and evidence was accepted from persons normally disqualified as witnesses, criminals, infamous, excommunicates, and even heretics. Lawyers were not allowed to assist the accused, for that would make them accomplices.

The suspect was questioned and refusal to answer or equivocation was treated as guilt. Though torture was alien to canon law, the Italian inquisitors seem to have adopted it from about 1252, and in the fourteenth century complaints against excessive use of torture were so widespread as to result in papal intervention.

The right to appeal to the Pope was excluded, but in fact appeals were frequent.

Before pronouncing a definitive sentence inquisitors had to obtain the approval of the bishops and of a number of assessors, ecclesiastical and lay. Later, only major penalties required the assent of the bishop.

The inquisitors could at their own hand impose minor penalties such as scourging, pilgrimages, or wearing the yellow cross of infamy on the clothing. Imprisonment for life was the normal punishment for the converted heretic, though this was sometimes commuted and many prisoners escaped. When it was impossible to extract an abjuration the tribunal surrendered the prisoner to the secular court for death by burning, inflicted in a public place. In some cases the remains of a convicted heretic were exhumed and burned. Death and life imprisonment involved also total confiscation of the condemned person's property, and a special bureau existed for the administration of confiscated property.

The Inquisition reached its highest point in the latter thirteenth century and at this time the tribunals were almost entirely free from control, even papal, and abuses of power went unchecked. In the fourteenth century measures were taken to check abuses and the powers of the Inquisition were attacked and restricted, both by the popes and, in France, by the kings and the Parlements.

At the end of the fifteenth century a new tribunal was created in Spain when the Pope granted the Catholic Kings the right to appoint inquisitors, and this became an independent institution. Ferdinand V and Isabella organized a Supreme Council of the Inquisition under Torquemada as Inquisitor General. Under him it acted energetically both against heretics and against converted Moslems. The Spanish Inquisition spread to America and was not finally suppressed until 1834.

Especially as developed in Spain the Inquisition was a disgrace to the Christian Church, and in modern terms a total denial of the human right of freedom of conscience and religious belief. It acted chiefly in areas influenced by Roman law and was never effective in Britain or Scandinavia.

Inquisitorial system. The system of criminal procedure in which the detection and prosecution of the culprit are not left to private initiative. It originated in the later Roman Empire and was adopted by the Roman Church, and the influence of these sources made the system common in Europe by the sixteenth century. Some forms of this system have involved secret inquiries and the use of torture. In all forms the judge's investigation is not limited to the evidence put before him, but he proceeds with an inquiry on his own initiative. The alternative is the accusatory system (q.v.).

Inrollments of Acts of Parliament. The original authority for the English and British statutes from 1483 onwards, apart from the period 1642–60. They are records containing the Acts of Parliament, certified and delivered into Chancery. Down to 1849 a transcript of the whole Acts of the session was engrossed on parchment, signed and certified by the Clerk of the Parliaments and deposited in the Rolls Chapel and arranged with the other records. Since 1849 two prints on vellum of every Act are made by the Queen's Printer and authenticated by the Officers of each House and then preserved, one in the House of Lords and one in the Public Record Office.

Insanity. Not a strictly defined legal term, but commonly used in criminal law to denote a condition of mental defect or disorder so severe as to justify holding that the person is not responsible for crime committed. The justification for this is the belief that responsibility for crime is based on and assumes mental capacity to make moral judgments and to distinguish right from wrong, and on understanding of what is legal or illegal, permissible or forbidden, by the legal code of the community, and also the capacity to control one's conduct and to refrain from doing wrongful or prohibited conduct. If a man genuinely cannot do these things, he should not be held guilty and punished. The most difficult question is how to settle a verbal test which can be applied by judges and juries. The test for insanity for this purpose is not one of whether the accused suffers from any mental disorder or not, but rather of whether his mental state is such that he should fairly be held irresponsible. Anglo-American systems generally base the test on the answers returned by the judges to questions by the House of Lords in *R* v. *McNaghten* (1843), 10 Cl. & F. 200, to the effect that 'to establish a defence of insanity it must be clearly proved that, at the time of the committing of the act, the party accused was labouring under such a defect of reason, from disease of the mind, as not to know the nature and quality of the act he was doing; or, if he did know it, that he did not know he was doing what was wrong'. This formulation has been much criticized and many alternatives have been suggested. The American Law Institute's Model Penal Code suggested that it should be a defence if the accused, at the time of the act, lacked substantial capacity to appreciate the criminality of his conduct or to conform his conduct to the requirements of the law.

A particular problem is that of the 'irresistible impulse', whether prolonged, or merely temporary as by losing one's temper. In the U.K. this has not been accepted as a defence, but in some U.S. courts it has been. U.K. law also accepts the concept of 'diminished responsibility' (q.v.) resulting in liability to diminished or modified punishment.

If a prisoner is held irresponsible, the verdict is one of Not Guilty by reason of insanity, and he is not punished but committed to a custodial mental institution.

A question of insanity may arise at any one of three other stages, whether or not it is alleged that the accused was insane at the time of the crime. The accused may have been committed for trial, but medical evidence may be that it is not practicable to try him or would be injurious to his mental state to do so; in this case he may be committed to mental hospital. Or when the accused is brought up for trial he may be found unfit to plead on arraignment, that is, unable to understand the charge and to instruct his defence; in this case the issue of fitness to plead is determined by a jury, and if the accused is found unfit to plead, he is ordered to be detained in hospital until discharged with the Home Secretary's consent. Or when the accused is sentenced he may be unable to comprehend the sentence being passed on him. In this case too he will be ordered to be detained in hospital. When capital punishment was competent a convicted person was never executed if there was doubt as to his sanity.

In Scots law the McNaghten Rules are not strictly or literally applied, but the question is whether the accused was of sound or unsound mind. The other stages at which insanity may be relevant are called respectively insanity in bar of trial, insanity at the time of trial, and insanity at the time of sentence.

Nigel Walker, *Crime and Insanity in England.*

Inskip, Sir Thomas Walker Hobart, 1st Viscount Caldecote (1876–1947). Called to the Bar in 1899 he took silk in 1914 and entered Parliament in 1918. He served as Solicitor-General from 1922 to early 1924 and again from late 1924 to 1928 and as Attorney-General 1928–29, and also developed a reputation as a Parliamentarian. He was out of the House between 1929 and 1931 and was again Solicitor-General 1931–32 and again Attorney-General 1932–36 when he became Minister for the Co-ordination of Defence, a difficult if not impossible duty, which he discharged satisfactorily. He supported the policy of appeasement in 1938. In 1939 he became Secretary of State for Dominion Affairs and on the outbreak of war Lord Chancellor, becoming Viscount Caldecote. He held that office only eight months and made no mark therein. In 1940 he again became Secretary of State for the Dominions and later that year Lord Chief Justice of England. His second tenure of the Dominions office was more successful and he made a good impression. As Lord Chief Justice he was dignified, impartial, and fair, but his health failed and he resigned in 1946. He also took an active interest in the work of the Council of Legal Education. He was a devout Christian and a strong supporter of many evangelical causes, and a man of the highest integrity and principles. He had little opportunity to make a reputation as a judge, and his reputation is sound rather than brilliant.

Insolvency. The state of inability to pay debts in full. By itself insolvency has no legal consequences. In England the filing of a declaration of inability to pay debts is an act of bankruptcy on which a receiving order may be made. In Scotland it is one of the *indicia* of notour bankruptcy (q.v.), which is a prerequisite of sequestration (q.v.). Prior to the Bankruptcy Act, 1861, a distinction was drawn between insolvents and bankrupts, in that only a trader could be made bankrupt and obtain an absolute discharge from his debts and Insolvent Acts were passed from time to time discharging insolvent non-traders from their debts on surrendering their property to their creditors. To administer this system there was a Court for the Relief of Insolvent Debtors.

Inspection. In English civil procedure the obligation on a party, if ordered by the court, to produce for the inspection of the other party every document in his possession relating to any matter in question in the cause and material and relevant to that cause, unless the document is one of the classes of protected documents.

Inspector. The title given to many categories of persons whose functions are to examine and inquire, and to report their findings. Thus there are inspectors of factories, mines and quarries, ancient monuments, schools, and numerous other classes of subjects, all acting under various statutory powers.

The Department of Trade may appoint inspectors to investigate the affairs of a company which is thought not to have been conducted in accordance with the Companies Acts.

The name inspector is also given to the officials appointed to conduct a statutory inquiry. These may be officers of the Department concerned, which does not give the impression of impartiality, or independent persons appointed *ad hoc*. It was long settled that a party to proceedings before a statutory inquiry had no right to see a copy of the inspector's report, but in certain cases this rule has been reversed.

Instalment purchase and selling. The general term for transactions whereby property is sold, the whole or part of the price being paid subsequently in stages, e.g. weekly or monthly, in small amounts. It is an ancient practice, being known at Rome, but developing rapidly in the nineteenth century. It takes a number of distinct legal forms, including conditional sale, credit sale, and hire-purchase (qq.v.), all applicable to goods, and purchase through a building society (q.v.) mortgage, applicable only to buildings. An important distinction is between the pure form of instalment sale, in which the seller stands out of his money, takes the risk of the purchaser's default, and is paid directly in instalments by the purchaser, and the developed form in which, when the agreement is made, a finance house takes over ownership of the goods and the risk of default, pays the seller in full less an agreed discount, and is then paid by the purchaser in instalments. The seller may, as agent for the finance house, receive the instalments. There are important technical differences between the legal forms, notably in that in credit sale the property in the goods passes at once to the purchaser, but in conditional sale it does not pass until the condition, i.e. payment of all instalments, has been satisfied, and in hire-purchase it does not pass until all instalments have been paid and the option exercised to convert the hire into a purchase. Similarly, in relation to buildings, it is competent to pay the seller by instalments, but more convenient to have a building society pay him in full and then to repay the building society by instalments.

All forms of instalment purchase involve an element of interest on the money unpaid, or advanced to the buyer by the finance-house or building society. Despite the cost, the convenience of paying by instalments while having the use of the thing has made instalment purchase very common and of great economic importance in that restrictions on the size of initial payments and on the period for repayment produce substantial changes in the volume of trade. Most cases of instalment purchase of goods in the U.K. are now regulated under the Consumer Credit Act, 1974.

Instance. The Admiralty Court, in the exercise of its jurisdiction in other than prize cases, was formerly known as an instance court. The term 'court of first instance' means a court in which proceedings are heard for the first time, as contrasted with a court of appeal.

Instituta Cnuti aliorumque regum Anglorum. A compilation, in Latin, apparently by a cleric, made about A.D. 1100, of Anglo-Saxon laws, the first two parts containing mainly passages from Cnut's code and the third excerpts from the laws of Alfred, Ine, and other Anglo-Saxon sources.

Institute. A term derived from Justinian's *Institutiones* (q.v.) frequently given to a general treatise on the law or major parts of it. Coke's (q.v.) *Institutes* comprise four volumes, the first an extended commentary on Littleton's *Tenures* and commonly known as Coke on Littleton or Co. Litt., and including a great wealth of common law learning culled from Year Books and older cases but put together in disorderly fashion, the second a commentary on Magna Carta, the third a treatise on the pleas of the Crown, and the fourth an account of the different courts, including the High Court of Parliament, the House of Lords, and the House of Commons. The volumes, particularly the later ones are often cited as 1, 2, 3, or 4 Co. Inst. J. Cowell's *Institutiones Juris Anglicani* (1605) was modelled on Justinian's work and T. Bohan's *Institutio Legalis* (1708) was an introduction to legal study which had considerable success. T. Wood published *An Institute of the Laws of England* in 1720, probably the most systematic book before Blackstone's (q.v.) *Commentaries.*

In Scotland Forbes, MacDouall (Lord Bankton) and Erskine (qq.v.) all wrote major texts entitled *Institutes* while Stair and Mackenzie (qq.v.) wrote *Institutions.*

See also INSTITUTIONAL WRITINGS; *Institutiones.*

Institute of International Law. Formed in 1873 to promote the development of international law by endeavouring to formulate general principles in accordance with civilized ethical standards, and by giving assistance to genuine attempts at the gradual and progressive codification of international law. Membership is limited to 60 members and 72 associates. It publishes the *Annuaire de l'Institut de Droit International.*

Institution, Legal. This term has many usages which overlap with the meanings of social, political, and economic, etc. institutions. It is sometimes used of an established set of customs, usages, or practices round which legal rules cluster, e.g. marriage, the family, contract, and property. Or it may be used of an establishment or organization having legal functions and powers, e.g. for associations, courts, judges, companies, and trade unions.

In ecclesiastical law institution is the process whereby a bishop commits the cure of souls to an incumbent of a benefice.

Institutional theories of law. Theories having reference to institutions as legally personified sets of social relations. An institution in this context is a complex of external circumstances calling for human action together with the attitudes of men towards those circumstances, having continuity though not necessarily identity over a period of time. Such institutions are the social reality underlying law; new kinds of institutions, e.g. trade unions, are constantly arising as adjustments of human relations in the face of new situations and are expressed more or less adequately in definite legal rules and principles. The social reality is, however, anterior to and more permanent than the rules. The common convictions of the group form the institutional structure underlying the law of the group.

Institutionalism was developed in France by Hauriou and Renard (qq.v.). Law and justice according to Hauriou sprang from the organization of forms and processes which tend towards permanence. It was adapted in Italy by Santi Romano but has made little headway elsewhere.

W. I. Jennings, *The Institutional Theory*, in *Modern Theories of Law*.

Institutional writings. The name given in Scotland to those writings, modelled on Justinian's *Institutions*, and themselves called Institutions, Institutes, or an Institute, which deal systematically and usually at great length with the whole law, or at least with the whole civil or criminal law. Thus it is applied to Stair's (q.v.) *Institutions*, Mackenzie's (q.v.) *Institutions*, Forbes' (q.v.) *Institutes*, Bankton's (q.v.) *Institute*, and Erskine's (q.v.) *Institute*. The term is extended also to include books of similar general, or at least extensive, coverage though not entitled Institutions. Thus it covers Craig's (q.v.) *Jus Feudale* and Bell's (q.v.) *Commentaries* and *Principles*, and probably Kames's (q.v.) *Equity*, and, in the sphere of criminal law, Mackenzie's (q.v.) *Laws and Customs of Scotland in Matters Criminal*, Hume's (q.v.) *Commentaries*, and Alison's (q.v.) *Principles and Practice of the Criminal Law*. All these are deemed of considerable authority, and some, notably Stair, Erskine, and Bell and, in criminal law, Hume, of the highest authority, and have had great influence in shaping and developing the law.

From the authoritative character of these books, and the deference shown to them by later judges and jurists, the term has come to correspond to the English 'books of authority'. Accordingly a statement in an institutional work is as good evidence of the law and as authoritative as an appellate decision of the Court of Session. There are some works, e.g. Mackenzie's *Institutions*, Forbes' *Institutes*, and Kames's (q.v.) *Equity*, which are sometimes deemed institutional in this sense but not always.

The privileged status of institutional writing does not extend automatically to all of a particular jurist's writings. Thus Erskine's *Principles*, written as a textbook for students, is not deemed institutional, and Hume's *Lectures on Scots Law*, which were not finally revised nor published by him, are not entitled to the same weight as his *Commentaries on Crimes*.

Textbooks on particular subjects can never, technically, be institutional, though they may have persuasive influence on courts not far short of that of an institutional work.

Institutiones (from *instituere*, to instruct). Elementary textbooks of Roman law, the most famous of which are those of Gaius (q.v.) and of Justinian (q.v.) (compiled by Tribonian, Theophilus and Dorotheus), founded on Gaius' *Institutiones* and *Res cottidianae* and on the *Institutiones* of such other jurists as Marcian, Florentinus, and Paulus, but referring to Justinian's reforms. It was published in A.D 533 as an introduction to the *Digest*.

Justinian's *Institutiones* are divided into four books dealing respectively with introductory matters and the law of persons; with things and wills; with succession and obligations; and with wrongs, security, and actions.

The name was copied in Scotland by Stair and Mackenzie who wrote *Institutions* and gave rise to the collective title of institutional writers (q.v.).

Instruction. In continental legal procedure and in that of the Court of Justice of the European Communities, an inquiry or proof-taking stage. In the European Court the *juge-rapporteur* examines the pleadings and considers whether the case requires an instruction and the Court also hears the Advocate-General on this point. If *instruction* is necessary, it may be held by the full court, a chamber, or the *juge-rapporteur*. It takes the form of an appearance by the parties and witnesses and the production of documentary evidence, the examination being conducted by the court, chamber, or *juge-rapporteur*. After the *instruction* the court may allow the parties to submit observations on matters which have arisen in the course thereof. Thereafter parties proceed to oral argument before the court.

(2) The term for giving a barrister the requisite information to enable him to prepare and present a case, and the request that he does so. A barrister must in general be instructed by a solicitor.

Instrument. A formal legal writing or document under which some legal right or liability exists such as a charter, deed of transfer, or agreement. Some statutes define the term as including an Act of Parliament. A statutory instrument (q.v.) is a piece of subordinate legislation, most frequently a set of regulations or orders made by a Minister under statutory authority, but including also other kinds of subordinate legislation. In Scots law many varieties of instruments are known, notably the Instrument of Sasine and the Notarial Instrument.

Instrument of Government. A constitution for England, Scotland, and Ireland drawn up by some officers of the Army in 1653. It provided that legislative authority should be in the hands of Cromwell as Lord Protector and Parliament, and that the executive should be controlled by the Lord Protector and a council not exceeding 21. Parliament was to meet by September 1654 and thereafter once in every third year, and when in session to share in the control of the military and naval forces. It was to have power to make laws and to prevent any tax without its consent. A redistribution of seats in Parliament was made. All who had fought against Parliament in the civil war were to be disabled from voting or being elected to the first four Parliaments and Roman Catholics to be disqualified permanently. The Protector was to hold office for life, but the office was to be elective; he was to have only a suspensive veto over legislation. The Instrument was the first attempt in English history at a written constitution and was a complete breach with the immediate past. It was defective in making no provision for the co-ordination of the authorities, Protector, council, and Parliament, nor for amending the Instrument itself.

Instrument under hand. A document in writing or similarly represented which creates or affects legal rights or liabilities and is authenticated by the granter's signature, but not sealed by him. The term is not confined to formal documents, and documents of this kind are used in a wide variety of transactions. Such documents are required only by statutes, or by direction or agreement of parties. In general attestation is not required, and in general a signature by an agent suffices.

Insurance. A contract whereby one party, the insurer, undertakes, in return for a consideration, the premium, to pay the other, the insured or assured, a sum of money in the event of the happening of a, or one of various, specified uncertain events. Insurance was known in ancient Greece and among maritime peoples trading with the Greeks. It developed first as a means of spreading the huge risks attendant on early maritime enterprises, and dates as a distinct contract from the fourteenth century, when it evolved in the commercial cities of Italy. It is found in the Admiralty Court in England in the sixteenth century. Life and fire insurance developed later. Fire insurance was stimulated by the Great Fire of London. Lloyds of London began in the seventeenth century in a coffee-house patronized by merchants, bankers and underwriters; it was reorganized in 1769 as a group of underwriters accepting marine risks. The first U.S. company was organized by Benjamin Franklin in 1752. Since the mid-nineteenth century insurance against other kinds of risks has developed greatly. For this reason reference is frequently made in all branches of insurance to basic principles formulated in marine insurance cases and codified in the Marine Insurance Act, 1906. Insurance is carried on as a business by insurance companies, which are subject to special statutory regulation, and, in relation to marine risks, by underwriters. Companies and underwriters have developed many varieties of policies to suit the various kinds of insurance commonly sought by the public.

The main classes of insurance are life and other personal insurance, in which a sum becomes payable on death, injury, or illness; marine insurance when a sum becomes payable on the happening of a marine peril; accident or property insurance, in which the sum becomes payable on the happening of an accident, such as fire, theft, or flood; and liability insurance, when the sum becomes payable when legal liability is incurred, as for personal injuries or professional negligence to another. A particular policy may insure against risks within several or all classes. Industrial assurance (q.v.) is a specialized form of life insurance.

Some contracts of insurance are contracts of indemnity, in which the amount recoverable is determined by the pecuniary loss sustained by the assured; this includes marine, accident, and liability insurance. Others are not contracts of indemnity in that the agreed amount is payable independently of the extent, if any, of loss sustained; this includes life, personal accident, and sickness insurance. Life insurance is in fact mainly a method of investment and saving, and of providing for dependants.

The assured must always have an insurable interest, in that he will be prejudiced by the happening of the peril insured against, which must have a pecuniary value. This is particularly so in marine, fire, and life insurance. Thus an assured has an insurable interest in his ship, if it be lost, and in his property, if it be burnt. But he has no insurable interest in another's house. Insurance without interest is void. But interest need not, in general, be specified in the contract.

There must be uncertainty as to whether and, if so, when, the event insured against will happen, and the event must be one, e.g. damage, injury, death, legal liability, detrimental to the assured.

It is a fundamental principle of insurance law that each party must observe the utmost good faith and disclose honestly all facts relevant to the risk and consequently to the premium.

The contract, known as the policy, sets out the terms and conditions on which money is payable and the validity of a claim may raise issues of the interpretation of the policy. In marine insurance there are statutory requirements as to the policy; in other cases there are not. The contract is commonly for a period, such as a year, renewable thereafter for further periods. In the event of a loss occurring, whether a claim is competent under the policy or not depends on the interpretation of the policy.

In all non-marine contracts of insurance which are contracts of indemnity the principle of subrogation applies, so that the insurers are entitled to all rights and remedies of the assured in respect of the subject matter and the assured must do nothing which might prejudice the right of subrogation.

Insurgency. A state of revolt or rebellion falling short of the status of belligerency allowed in the case of revolting subjects carrying on civil war against the government of a state.

Insurrection. A term generally used as meaning an uprising against the constituted authority of a state lesser in scope and purpose than what is designated a revolution or a rebellion.

Intendant. A royal official in France under the *ancien régime*, chiefly concerned with matters of justice, finance, and police in the provinces on behalf of the central government. The office appeared first as an extraordinary commission in the sixteenth century, chiefly concerned with the collection of the *taille*, but became permanent about the end of that century. Later the *intendants* came to have very extensive powers and duties, including representation of the Crown at provincial estates, advice to the Crown on various local matters, administrative, fiscal, and other, and enforcement of laws which were being flouted. Richelieu used the *intendants* to counterbalance the authority of the nobility in the provinces and increasingly they appeared to be the local agents of royal authority. In 1648 in consequence of the Fronde their commissions were revoked, but they were re-established by Colbert and supervised a great range of matters, including justice and police. They were usually professional judges drawn from the *maitres des requêtes* and as time went on they rose in the social scale. The office disappeared at the Revolution.

Intention. A mental formulation involving foresight of some possible end, and the desire to seek to attain it. By itself an intention is normally of no legal significance, but when a person performs certain acts the question of with what intention he did them may be highly relevant to his legal liability, civil or criminal, for those acts or their consequences. Thus A's intention to kill B by shooting is irrelevant so long as it is only his intention. But if A steals a weapon and ammunition, or shoots at B, or shoots at, hits, and causes B's death, what his intention was in each case may be crucial. Intention is rarely expressed or admitted and usually has to be inferred from circumstances and conduct. It may be inferred if the actor admittedly actually foresaw, or must as a reasonable man be taken to have foreseen, the consequences which resulted as inevitable or very probable or even, in some cases, possible, and the actor desired, or must be taken to have desired, those consequences; a person must be taken to have desired the certain or inevitable consequences of his actings, and may be taken to have desired consequences which are probable or even possible. A person is normally presumed to intend the natural and probable consequences of his acts, so that his intention will be inferred from what he does, but he may establish that he did not foresee or did not desire those consequences.

In relation to crimes it is an element in the definition of each crime whether any intention in the doer is necessary to infer guilt thereof or whether there is absolute liability for the mere doing, and in the former case what intention is necessary, intention to bring about a particular result, e.g. permanently to deprive the owner of his property, or a result of a defined general class.

See also MOTIVE; RECKLESSNESS.

Inter-American Conferences. A series of meetings of the states in the Americas to show solidarity and resolve difficulties. The first attended by all the Latin-American countries was held in 1889. The major achievements have been the establishment of the Pan-American Union, which passed under the control of the Organization of American States when it was founded in 1948, and which acts as the administrative agency of the O.A.S. promoting cultural and informational exchanges.

Interdict. (1) In Roman law, orders or prohibitions issued by a magistrate without long investigation and normally addressed to a particular person, intended to give immediate protection to a threatened interest of a complainer. The interests concerned were those required for the maintenance of public order, such as possession, rights of succession, servitudes, mortgages, and some personal rights. If the defendant did not obey the interdict, the case would be tried in a way unfavourable to him. Interdict could also be sought for the protection of public interests, such as in relation to roads or rivers. Particularly important were the interdicts by which the praetors protected possession.

In mediaeval canon law it was the penalty of suspension of the sacraments and a denial of the public services of the Church. Rome was placed under an interdict by Adrian IV in 1155, Scotland by Alexander III in 1180, and England by Innocent III in 1208. Excommunication, on the other hand, involved exclusion of the offender from the sacraments and from Christian burial, and, in the case of the greater form, major excommunication, from intercourse with fellow Christians. It was a ceremony carried out with high solemnity, but as time went on mediaeval princes became accustomed to and ceased to be frightened by such papal denunciations.

(2) In Scots law a remedy roughly comparable to the English injunction. Its main use is to stop apprehended or threatened, or the continuance or repetition of, infringement of some right vested in the pursuer, such as by nuisance, pollution, infringement of patent, or copyright. It may be sought as interim interdict, the grant or refusal of which is very much in the discretion of the court having regard to the balance of convenience or hardship in the circumstances; or as perpetual interdict, to regulate matters between the parties permanently. It is also used to stay execution of diligence, and as a means of regulating possession of heritable property. Disobedience to interdict is treated seriously and punishable by imprisonment or fine.

Interdiction. In Scots law a process, now obsolete, whereby a person of weak and facile disposition might be disabled from signing any deed to his prejudice without the consent of his curators, called interdictors. A person might place himself under this restraint voluntarily, by granting a bond of interdiction, or it might be imposed judicially, by the Court of Session. Recall might be effected with the consent of the interdictors, or by the Court of Session.

Interesse termini. In real property law, an executory interest, being the right of entry which the grant of a lease conferred on the tenant and which was converted into an estate by actual entry by the tenant on the lands. The term vested at once if the lease had been created by bargain and sale. The doctrine was abolished in 1925.

Interessenjurisprudenz (or the jurisprudence of interests). The name for a group of jurists whose attitude developed from Jhering's attack on conceptual thinking and adopted his idea of law as the resultant and solvent of conflicting human interests. Premises and analogies in the codified law had to be sought by reference to the interests involved and not merely by regard to verbal concepts and logical deduction. The leader was Philipp Heck (q.v.), but other significant writers were Max Rumelin, Paul Oertmann, Heinrich Stoll, Julius Binder, and Hermann Isay.

M. Schoch (ed.). *The Jurisprudence of Interests* (1948).

Interest. (1) The profit which can be made from the use of money, and consequently the further sum of money recoverable when a payment of money has been delayed or wrongfully withheld. Interest may be due in any case by express agreement and is due in many cases by legal implication and in other cases by statute.

(2) The connection a person has with a thing entitling him to make a claim in respect thereof. Thus in insurance a person has an interest in property or in the life of another if damage to that property or the death of that person would cause him pecuniary loss. A person may, independently of property, have an interest in another; thus a husband has an interest in his wife.

(3) In relation to property, rights in respect of property which entitled or might subsequently entitle the holder to make some claim or exercise some power, e.g. charges, easements, and profits *à prendre*. The words sometimes included, sometimes excluded, estates (q.v.). Since 1925 interests in land are all such claims or rights as a person may have in respect of land, other than the recognized estates of fee simple absolute in possession and term of years absolute. Interests accordingly include easements, rent charges in possession, charges by way of legal mortgage, and other interests capable of subsisting at law, and all equitable interests in or over land such as equitable mortgages.

(4) In evidence a declaration made by a deceased person as to facts within his personal knowledge and consciously against his pecuniary or proprietary interest is admissible, being an exception to the rule against hearsay.

Interests. Those claims, wants, desires, or demands which persons individually or in groups seek to satisfy and protect, and of which the ordering of human relations in society must take account. The legal system of a country does not create interests; these are created or extinguished by the social, moral, religious, political, economic, and other views of individual, groups, and whole communities. The legal system recognizes or declines to recognize particular interests as worthy of legal protection, and jurists classify those recognized and argue about whether certain others should be recognized. If the legal system does recognize an interest, it does so by saying or providing that persons or groups have certain legal rights, enabling them to secure protection for their interests, or redress, or compensation for infringement of them, or by providing that other persons or groups or the community as a whole is under certain duties. Rights and duties are accordingly the means whereby certain interests are secured and protected.

The interests commonly recognized in modern developed communities may be classified as: (1) individual interests, including interests in the physical person, freedom of movement, honour and reputation, privacy and sensibilities, freedom of belief and opinion, in the domestic relations, in business relations, in liberty to own and use property, in power to gift and bequeath, freedom of industry and contract, and the like; (2) public interests, desires looked at from the standpoint of life in an organized community, including the interests of the State, as an entity, in its security, dignity, and reputation, in its territory and natural resources, in peace and order, the security of institutions, the conservation of resources, and the general welfare and improvement of life of its citizens; and (3) social interests, those wider demands or desires involved in or looked at from the standpoint of social life in civilized society, including the maintenance of law and order, safety and health, liberty of association, freedom of trade and business, security of acquisitions and transactions, the maintenance of general morality, social economic and cultural progress, the care of the aged, sick, dependents and delinquents, the care of the environment, the improving of general conditions of life, and others.

As societies have become more civilized, the tendency has been for the legal system to recognize and protect more interests.

It may be a matter of doubt and argument how far certain interests, e.g. in privacy, can be, or are, or should be protected. At different times the weight attached to certain interests varies: thus in wartime the public interest in the general security justifies limitations on the interests of the individual, on personal liberty, freedom of use of property, and the like. Similarly the replacement of democratic government by fascist or communist government would effect changes in the list of interests recognized and the values attached to various ones.

Particular interests are not correlated to particular rights and duties; various rights and duties exist for the protection of a particular interest. Thus the interest of an individual in his personal integrity (freedom from hurt or injury) is protected by separate rights not to be assaulted or injured, and to damages if he is so assaulted or injured, rights to social security if off work by reason of injury, to criminal injuries compensation if injured by criminal act, to invoke police protection and to have an assailant prosecuted. Civil, criminal, and administrative processes may all be involved in protecting a particular interest.

The consideration of law as the instrument of recognition and protection of certain interests and as the means of resolving disputes arising from conflicts of interests is mainly the work of Jhering and Pound (qq.v.), developed by Stone, Paton, and others in the twentieth century. Stone distinguishes only social and individual interests, concerned respectively with the conditions of the social life of individuals and the conditions of individual life in society. The enumeration and classification of interests recognized is a substantial matter; possibly more important is the consideration of interests not yet or imperfectly recognized, and of whether they should be recognized. Is there, or should there be, a recognized individual interest in privacy, in being left alone?

The value of the theory of interests is that it cuts across traditional analytical classifications of principles and rules of law into public and private, civil and criminal, relating to persons, obligations, or property, and the like, and brings out that rules from different areas of the law have similar functions and protect the same interest.

Intergovernmental Maritime Consultative Organization. A specialized agency of the U.N. finally established in 1957 to provide machinery for co-operation among governments in relation to shipping in international trade, and to encourage maritime safety and efficiency.

Interlocutor. In Scottish practice a judicial finding, decision, or order disposing of any part, or the whole, of a case. It is the operative part of a judicial decision, as distinct from the opinion which contains the judge's reasons for his decision.

Interlocutory. Incidental, partial, or interim. Thus an interlocutory application, e.g. for discovery or for an interim injunction, is a stage only in the prosecution of a claim.

Interlocutory judgment. A decision which is not final or which deals with only part and not the whole of the matter in controversy. Thus in England a decree *nisi* of divorce provisionally dissolves the marriage, becoming final and complete only after the lapse of a period.

Intermixture. The mixing of the goods of one person with those of another. Unless done deliberately the resulting mixture belongs to both in common, in the proportions contributed by each to the combination, unless the property of each is distinguishable and the mixture separable.

Internal or national waters. Stretches of water which are not part of the high seas nor of the territorial waters of a state but are part of the surface of that state, being surrounded by its land or so substantially enclosed as to be equally within its exclusive jurisdiction, e.g. the mouth of the Thames or of the Hudson. Internal waters lie to the landward side of base-lines following low-water mark and crossing the mouths of rivers and bays so long as not exceeding 10 miles in length.

International administrative law. The branch of international law concerned with the relations between international persons, such as the U.N., and their officers and staff. The U.N. Administrative Tribunal deals with many such controversies, relying largely on the rules of administrative law accepted in Western Europe.

International agreements. Agreements made between states and other subjects of international law to regulate matters of common concern. They are frequently embodied in treaties and are consequently regulated by international law. Other kinds of instances are memoranda of agreement, conventions, protocols, charters, covenants, exchange of notes, and so on. Little turns on the name if the parties treat the agreement as binding. Treaties may be bilateral or multilateral, and contractual or law-making. Contractual treaties or agreements deal with some dispute or matter of business, whereas law-making treaties lay down principles of agreed rules for future relations with regard to, e.g. trade, tariffs, transport, or other matters of common concern. The major functions of international agreements are the resolution of actual or potential disputes or conflicts, the establishment of new fields of co-operation and integration between states, and the development and codification of international law. Some international agreements set up organizations, regional or world-wide, for particular purposes. Many modern international agreements are of immense complexity; among such are the Bretton Woods Agreements of 1944 establishing the International Monetary Fund and the International Bank for Reconstruction and Development, and the General Agreement on Tarriffs and Trade.

International Association for Penal Law. Founded in 1924 to promote co-operation between persons who are engaged in the study or practice of criminal law, to study crime, its causes and cure, and to further the theoretical and practical development of international penal law. It publishes the *Revue Internationale de Droit Pénal.*

International Association for Philosophy of Law and Social Philosophy. Founded in 1909 for scientific research in philosophy of law and social philosophy at an international level. It publishes the *Archiv für Rechts- und Sozialphilosophie.*

International Association of Comparative Law. A body founded in 1960 to promote and improve comparative legal studies.

International Association of Democratic Lawyers. Founded in 1946 to facilitate contacts and exchanges of views between lawyers and lawyers' associations, and to foster understanding and goodwill; to work together to achieve the aims of the Charter of the U.N. It publishes the *Revue de Droit Contemporain.*

International Association of Lawyers. Founded in 1927 to promote the independence and freedom of lawyers, and defend their ethical material interests on an international level; to contribute to the development of international order based on law.

International Association of Legal Science. Founded in 1950 and sponsored by U.N.E.S.C.O. to promote mutual knowledge and understanding of nations by encouraging the study of foreign legal systems. It operates through national committees of comparative law in many countries.

International Atomic Energy Authority. An agency of the U.N.O. established in 1957 with responsibility for research into and the development of atomic energy for peaceful purposes. Its headquarters are in Vienna.

International Bank for Reconstruction and Development (or World Bank). A bank established in 1945 after the Bretton Woods Conference and affiliated to the U.N.O. It exists to assist economic development by providing loans. Its capital is provided by member countries' subscriptions and by loans raised on the open market. It supports only projects believed to be economically viable.

International Bar Association. A body founded in 1947 to advance the science of jurisprudence particularly in regard to international and comparative law, to promote uniformity of law, and the administration of justice, and to promote in their legal aspects the principles and aims of the United Nations.

International Bill of Rights. A United Nations' agreement of 1966 on governments' obligations in questions of civil rights.

International Civil Aviation Organization. A body created by the Chicago Convention, 1944, to develop the principles and techniques of international air navigation, to foster the planning of international air transport, to decide disputes as to the interpretation of the Convention and adopt international standards and practices on matters relating to civil aviation, such as Rules of the Air,

Nationality and Registration, Airworthiness, Telecommunications and Accident Investigation. It consists of a permanent Council to which states are elected triennially, an Assembly meeting annually, and certain subordinate bodies.

International claims. Claims by one state or other entity having international personality against another. They may arise from international wrongs, such as breach of a treaty, insult to the complaining state, damage to its property, or from wrongs done by a national of one state to another state or its nationals. In such circumstances the aggrieved state must, save in cases of direct injury to the State itself, first exhaust all remedies available under the local law. Thereafter, or in the first place in the case of injuries to the State, it may have resort to an international tribunal, claiming damages or a declaration of the rights of parties.

International Commission of Jurists. Founded in 1952 to promote and strengthen the Rule of Law in all its manifestations and to defend it through the mobilization of world legal opinion in cases of violation of or threats to such principles of justice. It publishes a *Review*.

International conventions. Whether general or particular, establishing rules expressly recognized by the contesting states, are, under its statute, the prime source of the law to be applied by the International Court of Justice in a dispute between two states. Such international conventions or agreements are generally contained in treaties, protocols, or other written form.

International conventions have in the twentieth century become an increasingly important source of municipal law in circumstances where it is desirable to have uniform rules in most or all countries. Hence carriage by air and shipping are largely regulated by legislation implementing international conventions. Similarly conventions promoted by the International Labour Organization ensure uniform minimum standards in certain employment contexts.

International Court of Justice. The major judicial organ of the U.N., established in 1946 at The Hague in succession to the Permanent Court of International Justice (q.v.). It is constituted by a Statute annexed to the Charter of the U.N. and is composed of 15 judges nominated by the national groups in the Permanent Court of Arbitration and by groups representing members of U.N. not represented at that court, and chosen by majority votes of the General Assembly and Security Council, voting separately, but simultaneously. All judges must be from different countries and have judicial independence, not being representatives of their countries, and diplomatic privileges, and immunities. They serve for nine years, five judges being elected every three years. Casual vacancies are filled in the same way, the judge elected completing his predecessor's term. The judges elect their own President and Vice-President for three-year terms. The quorum is nine. For list of judges see APPENDIX I. If the country of a judge is a party to a case he is not debarred from sitting, but a country party to a case and not having a judge on the court may nominate a judge *ad hoc* for the case.

All member-states of U.N. are automatically parties to the Statute of the I.C.J. and other states may adhere thereto, and some have done so.

The court's jurisdiction is limited to civil cases brought by and against sovereign states; it can adjudicate with the consent of parties only. Its jurisdiction covers all cases which the parties refer to it by special *ad hoc* agreement and cases covered by special clauses in treaties, by treaties of pacific settlement providing for judicial settlement or by the 'Optional Clause' (q.v.) of the Statute, whereby members might declare that they recognized as compulsory in relation to another state recognizing the same obligation the jurisdiction of the court in all or any of the classes of legal disputes concerning: (a) the interpretation of a treaty; (b) any question of international law; (c) the existence of a fact which, if established, would constitute a breach of an international obligation; and (d) the nature and extent of the reparation to be made for the breach of an international obligation. It can also consider cases laid before it for advisory opinions by the General Assembly, Security Council, and certain specialized agencies. It can consider individuals' claims only if they are presented by a state.

It administers international law derived from international treaties binding on the parties; international customary law; general principles of law recognized by civilized nations; case law embodying decisions on international law and the works of distinguished jurists. It is not bound by its own previous decisions nor by any other decisions. It may decide a case *ex aequo et bono* if the parties agree thereto.

The official languages of the Court are French and English, but the court must at the request of the parties authorize another language. Procedure is by exchange of written memorials, counter-memorials, and replies and by oral hearing, normally in public. The Court may invite any individual body, bureau, commission, or other organization, to carry out an enquiry or give an expert opinion.

All questions are decided by a majority of the judges who hear the case and the President or other presiding judge has a casting vote. The judgment is that of the majority, but any judge may write an individual opinion and dissenting judges may write one or more dissenting opinions. The judgment is final and not appealable, but may be reconsidered by the Court for interpretation, or for revision on the discovery of a new decisive factor, unknown at the time of the decision.

The Court may set up one or more special chambers of three or more judges to deal with particular categories of cases such as labour cases or transit and communications cases, or to decide a particular case, or to hear and determine cases by summary procedure.

Judgments are binding on the parties. If a state fails to perform the obligations incumbent on it under a judgment the other party may have recourse to the Security Council which may make recommendations or decide on measures to be taken to give effect to the judgment. But such recommendations are not binding.

The Court publishes its Judgments and Opinions, Acts and Documents concerning the Organization of the Court, and Annual Reports.

S. Rosenne, *The Law and Practice of the International Court*; E. Hambro, *The Case Law of the International Court*; H. Lauterpacht, *The Development of International Law by the International Court.*

International Criminal Police Organization (Interpol). Founded in 1923 with the object of rendering mutual assistance between police authorities in different countries in the suppression of crimes. The General Secretariat in Paris coordinates the activities of police authorities of member-countries in international affairs and centralizes documentation relative to the movement of international criminals. Interpol deals particularly with international criminals whose operations take place in more than one country, or are committed in one but affect another country, or who commit crime in one country and take refuge in another. Each affiliated country has a domestic clearing-house through which its individual police forces can communicate with the general secretariat or with the police of another country.

International delinquency. Any injury to another state committed, in violation of duty imposed by international law, by the head of state or government of a state, or by officials or individuals commanded or authorized by the head or government. The category includes acts ranging from criminal acts in violation of international law to failure to implement a treaty obligation. It is distinct from the category of crimes against the law of nations, such as piracy or slave-trading, which are acts by individuals which are criminal by international law and normally also by the law of their own states, and from the category of merely discourteous or unfriendly acts by states. The category of international delinquency includes war crimes, responsibility of individuals acting in pursuance of such orders, and abuses of rights enjoyed by virtue of international law, such as arbitrary expulsions of aliens, and internationally injurious acts by legislatures or judicial functionaries, such as delay or denial of justice.

In general an international delinquency requires to have been done wilfully and maliciously or with culpable negligence, so that an act done by right or for self-preservation in necessary self-defence is outwith the category. A state cannot evade international obligations by pleading its own municipal legislation.

In the event of international delinquency there arise liabilities to make reparation for the moral and material wrong done. Reparation may, according to the circumstances, include pecuniary damages, interest on money not paid, and apologies. States may also incur criminal responsibility.

International Development Association. An organization distinct from the International Bank for Reconstruction and Development, but affiliated therewith, instituted in 1960, to make loans to further the development of member nations. Its resources come mostly from the subscriptions and contributions of member nations. Many of its loans have been used to finance such projects as transportation, agriculture, and fisheries.

International disputes. Disagreements between two or more of the states in the world, or groups of such states. They are sometimes distinguished into legal, in which parties base their claims on grounds recognized by international law, and political disputes. This distinction is also called the distinction between justiciable and non-justiciable disputes. Formerly such disputes were frequently resolved by threats of, or actual use of, force.

Since 1815 attempts have been made to avoid the use of force by one state or group against another by various means. Under the U.N. Charter the use of force by individual states as a means of settling disputes is not permissible and peaceful settlement is the only permissible means.

Settlement may be effected by agreement between the parties, or by political means, as by organs of international organizations, such as the Security Council or the General Assembly of the U.N., or by legal means, such as arbitration (q.v.), or judicial settlement. Judicial settlement includes many *ad hoc* arbitral tribunals, some of them semi-permanent such as the Arbitral Commission on Property, Rights, and Interests in Germany and the International Centre for Settlement of Investment Disputes between states and nationals of other states. Provision for settlement of disputes by arbitration is made in the constitutive instruments of many international organizations. The most strictly judicial modes of having disputes determined are by resort to, formerly, the Permanent Court of International Justice and now to the International Court of Justice (qq.v.).

Arbitration and judicial settlement are usually deemed suitable for legal controversies or disputes as to rights, and conciliation for political disputes, and some treaties have provided for disposal of disputes of these two kinds by distinct kinds of machinery, but in some treaties conciliation has been a preliminary only to judicial settlement.

Under the Charter of the U.N. members are obliged to settle their disputes by peaceful means and refrain from threats or use of force, but retain the right of individual or collective self-defence against armed attack.

International Federation for European Law. Founded in 1961 to advance studies on European Law among members of the European Community by co-ordinating activities of member societies and organizing colloquia.

International Finance Corporation. An organization founded in 1956, separate from but affiliated to the International Bank for Reconstruction and Development, founded to stimulate the economic development of member states by providing capital for private enterprises when sufficient private capital was not available. It finances only predominantly private enterprises and makes loans without government guarantee of repayment. The share capital is subscribed by member countries and loans have been made to promote agriculture, publishing, tourism, and other enterprises.

International Grotius Foundation for the Propagation of the Law of Nations. A body founded in 1945 for the study and development of international law.

International Institute for the Unification of Private Law (UNIDROIT). Founded in 1926 to undertake studies in comparative law, to prepare for the establishment of uniform legislation and prepare drafts of international agreements on private law. It publishes a *Yearbook*, *Digest of Legal Activities of International Organisations*, and other literature.

International Institute of Space Law. Founded in 1960, to make studies on juridical and sociological aspects of astronautics. It publishes an *Annual Bibliography of Space Law* and other reports.

International Juridical Institute. Founded in 1918 to supply information in connection with any matter of international interest respecting international, municipal, and foreign law.

International Labour Organization (I.L.O.). I.L.O. was established in 1919 under the Treaty of Versailles and was associated with the League of Nations. In 1946 it became the first specialized agency of the U.N. It was founded to advance the cause of social justice and a primary function has always been to raise standards by creating an international code of labour law and practice. These are set by the International Labour Conference in the form of Conventions and Recommendations. Conventions are legal instruments conceived as models for national legislations; recommendations are designed to guide governments.

The supreme body is the International Labour Conference which meets annually in Geneva and which elects the Governing Body as executive council.

The main fields of activity are human resources development; employment promotion and planning; social institutions development; and promotion of fair working conditions, including remuneration and occupational safety and health. It provides technical co-operation and operates an International Institute for Labour Studies at Geneva.

International law, private. An alternative name for conflict of laws (q.v.) and international private law (q.v.). This name, though common, is unfortunate in suggesting that the body of principles and rules is a body of international law, or is in some way a part of or a body parallel to (public) international law (q.v.), whereas it is a branch of the private law of each distinct legal system.

International law (public) (law of nations, *droit des gens, Völkerrecht*), is the law of the international community, or the body of customary and treaty rules accepted as legally binding by states in their relations with each other, or certain rules of conduct which modern civilized states regard as being binding on them in their relations with one another, or the body of rules of conduct, enforceable by external sanction, which confers rights and imposes obligations, though not exclusively, upon sovereign states and which owe their validity both to the consent of states as expressed in custom and treaties and to the fact of the evidence of an international community of states and individuals.

Every definition of international law is open to challenge. Austin and some other positivist jurists, by reason of their definition of law as a command, have denied that international law is properly called law because it is not made by a determinate sovereign person or body, and accordingly that it can impose rights or obligations. But if law is otherwise defined, e.g. as rules of conduct accepted and applied in the regulation of disputes, international law can properly be called law. In practice most states accept much of international law as legally binding, and as law, not merely honourable practice.

Undoubtedly international law differs from municipal law in the absence of a sovereign lawmaker, uncertainty of many rules, and in the uncertainty of enforcement, but it is increasingly accepted, made or modified by agreement, and acted upon, in much the same way as municipal law. Increasingly in the twentieth century international law is being declared by judicial decision. But the legal quality of a rule does not depend on whether it is infringed, or enforced, and much international law is in fact habitually obeyed.

The existence of international law depends on there being a community of states interacting with and transacting with one another on many different planes: cultural, scientific, commercial, industrial, humanitarian, and other. These inter-state relations give rise to questions and disputes and, for the avoidance of anarchy, require understandings, customs, and rules for regulating the conduct of states *inter se*.

Public international law and private international law

The former is a body of principles and rules applying between states and is, in theory, a single body of rules of world-wide acceptance, though many rules are differently understood and interpreted by different states.

Private international law, international private law, or conflict of laws is on the other hand a branch of each distinct system of municipal law in the world concerned with the principles and rules which that system applies to regulate transactions involving also another legal system, and when determining whether or not to give effect to rights acquired under another system of law. It is not a single system, but parts of a multitude of systems, nor has it world-wide acceptance because the rules vary from one municipal system to another.

There is, however, some overlap, in that increasingly rules of international private law are being unified by multilateral or international conventions.

History of international law

Though the name 'international law' was coined by Jeremy Bentham only in 1780, in his *Principles of Morals and Legislation*, law governing relations between states has been recognized from very early times.

In the ancient Jewish states the 12 tribes distinguished between their circle of kindred tribes and outsiders, and among the latter between the seven nations who had originally occupied the Promised Land and other foreign communities. With the latter sub-group only the Jews established friendly relations. With other nations the Jews had no relations and waged cruel war with them. The Jews, however, had the hope and ideal of an international society of peoples united in peace.

Among the city-states of ancient Greece, there was recognition of a law of the Hellenes and a different, less humane, law of all mankind. Among the Hellenes common language, religious feeling, and propinquity encouraged friendly relations, but disputes and war were common. International arbitration was frequently resorted to and there were widely accepted practices as to declarations of war, prisoners, and the like.

Rather ill-defined general rules of conduct were recognized in relations between Greek and barbarians.

In Italy, Rome early became a dominant city-state exercising sovereignty over dependent towns and districts. The Senate entered into treaties with other peoples and Rome early became a great commercial centre to which foreigners were drawn. At an early stage provision was made by the appointment of a *praetor peregrinus* for the adjudication of disputes between Roman and foreigner, and between foreigners at Rome. More importantly the jurists began to distinguish the *jus civile* applicable to Romans only from the *jus gentium* or those more general and less formal rules applicable where a foreigner was involved. *Jus gentium* included rules of an international character. They also recognized a *jus fetiale*, a body of rules relative to the making of war and peace and the framing of treaties, and some rules later called *jus belli*, settling limits to the savagery of war.

Rome later became a continental power negotiating and warring with other major powers such as Carthage, Macedon, Persia, and Egypt as well as with many lesser states. Among the more civilized of these states there were recognized laws of war and common laws of mankind, the latter protecting ambassadors, condemning breach of faith, and protecting the lands of allies.

In the time of the Empire the extent and power of the Roman state almost extinguished the concept of international law; there was only one people in the civilized western world. Throughout the Dark Ages there is little evidence of law restraining force and brutality in Europe. International law has little place in Justinian's compilations.

In the Dark Ages following the collapse of the Western Roman Empire there is little trace of law in inter-state relations, and there again seemed little need for it with the rise of the Frankish Empire and the Roman Church. The rising power of the Church had humanizing influence and from the thirteenth century the popes sometimes adjudicated in international disputes, but failed to establish any general practice. In the Middle Ages warfare was savage, even in the Crusades, though the practices of the Arabs, influenced by the law of war in the Koran, compared favourably, and chivalry introduced elements of honour and even courtesy into knightly warfare.

Modern international law is, however, a product of Christian civilization and of the disintegration of the mediaeval world and the development of distinct nation-states and the subsequent expansion of the system as new states came into being. The old Christian states of Western Europe were the founder-members of the community of states, and in time there were added the states which developed from colonies sent out by the European States. Turkey was expressly admitted to the family of nations only in 1856, and since then other non-Christian countries have been accepted. Other factors which prepared the way for the development of international law after 1500 included the revival of legal studies by the civilians and canonists, the growth of international trade and of bodies of international maritime law, the growing practice of states sending embassies and establishing permanent legations, the growth of standing armies, and the Renaissance and Reformation.

The period of the Reformation was one of warfare, and religious fervour led to atrocities; there survived, however, from the age of chivalry the employment of heralds and the taking and ransoming of prisoners. The rules were well settled that if a place were carried by storm the defenders could be massacred and no distinction was made between combatants and non-combatants. The Thirty Years' War was noted for its inhumanity and brutality. Neutrality was hardly recognized, but more precise notions of neutrality began to appear at the end of the sixteenth century. James I of Britain proclaimed neutrality in the contest between the United Provinces and Spain in 1604.

The sixteenth and early seventeenth centuries were times of growing national power and of almost constant warfare. The Reformation and the religious wars brought to an end the old dualism of Papacy and Empire, and Europe became a collection of more or less independent monarchies, principalities and dukedoms, the major ones being coincident with national groupings. The inter-state relations of Italian city states gave rise to the beginnings of writing about international law in some of the writings of Bartolus and Baldus. There was increasing growth of national consciousness in the sixteenth century. Jurists such as Vitoria, Vasquez, and Solorzano denied the temporal power of the Papacy and the breadth of imperial claims to sovereignty. Sully in 1607 proposed the division of Europe into 15 states linked federally under a General Council.

The concepts of *jus naturale*, *jus gentium*, and *jus civile* had been inherited by the philosophers and jurists, both civil and canon, from the Roman law and Vasquez postulated a group of independent states with rights *inter se* regulated by *jus naturale et gentium*. Suarez in his *Tractatus de Legibus et Deo Legislatore* (1612) presented *jus gentium* as not only law common to all nations but law between nations; there was a society of states, independent but interdependent.

Increased international trade brought to the forefront the question of territorial waters; in mediaeval times many states had claimed dominion in respect of various parts of the seas. Some mediaeval civilians had claimed as much as a hundred miles of sea as within the area controlled by a maritime state.

Grotius later in *Mare Liberum* (1609) denied the existence of private property in the sea, but James I of Britain issued a proclamation forbidding persons, unless British subjects, to fish on any of the coasts or seas of Britain or Ireland without a licence, and later had Selden write his *Mare Clausum* seeking to uphold his pretensions.

Similarly closer international relations brought an increase in permanent diplomatic representation at foreign courts, and state etiquette and protocol became a matter of growing importance; such questions as the exemption of ambassadors from the law of the country to which they were accredited became important in the sixteenth century. In general the sanctity of his embassy was both theoretically recognized and respected in practice.

The literature of international law began to emerge at the time of the Reformation. In mediaeval times legal writers had continued the discussion of the relations of *jus naturale*, *jus gentium* and *jus civile* and the writings of Irnerius, Bartolus, Baldus, Alciatus, Hotoman, and Cujas also commented incidentally on Roman texts bearing on war and peace among nations. Canonists had also long discussed such issues in the light of Christianity. Tertullian, Augustine, and Isidore of Seville all dealt with such matters and their views found their way into Gratian's *Decretum*. Succeeding commentators amplified the discussion; Hostiensis in particular classified truces and distinguished seven species of war; Aquinas recognized that war might be just under certain conditions.

In the fourteenth and fifteenth centuries many small works were written on war, reprisals, and related subjects, such as the *De Bello et represaliis et de duello* of Joannes de Lignano of 1360 (published 1477). More's *Utopia* (1516) denounced unnecessary war though Machiavelli's *Prince* (1532) commended war as an instrument of policy. The Spanish jurist-theologians began to make solid contributions to a science of international law. Franciscus a Vitoria's *Relectiones Theologicae* (1557) denied the judicial power of the Pope in temporal matters, and admitted as just war if waged for the good and utility of the commonwealth; he questioned the right of the Spaniards to dominate the Indies and treated of many problems of the law of war. Bodin's *Six Livres de la Republique* (1577) Book I contains much of importance about international law.

Among the early writers on international law was Conrad Brunn or Braun who wrote a *De Legationibus* (1548), Ferdinand Vasquez, an *Illustrium Controversiarum aliorumque usu frequentium Libri tres* (1564) which recognized a composite *jus naturale et gentium* as governing the relations of princes and free peoples *inter se*; he also challenged the idea of dominion over the seas by maritime nations. Belli published in 1563 his *De Re Militari et de Bello.* Balthazar Ayala treated of the laws of war in his *De Jure et officiis bellicis et disciplina militari* (1581) and Albericus Gentilis found his way to Oxford and there wrote *De Legationibus* (1585) and *De Jure Belli* (1588–89); in 1605 in his *Hispanicae Advocationis Libri Duo* he dealt with neutrality. Gentilis marked an advance on all earlier work in his attention to current problems and practical approach and is the major forerunner of Grotius. John of Carthagena's *Propugnaculum Catholicum de jure belli Romani Pontificio adversus Ecclesiae Jura violantes* (1609) treats a special problem. Francisco Suarez' *Tractatus de Legibus et Deo Legislatore* (1612) insisted on the existence of a society of states and a law for it supplied by natural reason and general custom by, in the last resort, dependence on God.

Grotius in 1609 in *Mare Liberum*, following the approach of Vasquez, asserted the general right of nations to navigate and trade and denied the Portuguese monopoly of trade with the Indies. In 1625 he published the famous *De Jure Belli ac Pacis*; though not conspicuously original in matter or arrangement, the principles were more elaborately worked out than ever before; the work won instant recognition and modern international law is always dated from it. The leading principles of his work were recognized in the Peace of Westphalia (1648) and became the foundation of the new public order of Europe which began to be established thereafter. Zouche's *Juris et Judicii Fetialis sive Juris inter Gentes* has been called the first manual of positive international law, resting it mainly on customary international practices.

In the seventeenth and eighteenth centuries three schools of writers developed, the naturalists, the positivists, and the Grotians.

Pufendorf is the leader of the naturalists; he was appointed to a newly-created chair of Natural Law at Heidelberg in 1661 and expounded and developed the natural law view in *Elementa Jurisprudentiae universalis* (1666), *De Jure Naturae et Gentium* (1672), and *De Officio Hominis et Cives juxta legem naturalem* (1673). The law of nations was that part of the law of nature which dealt with the relation of states to one another, and the law of nature was identified with the law of God, discoverable by reason from the tendency of actions to promote the happiness of society. His most celebrated follower was Thomasius (*Institutiones Jurisprudentiae* (1648) and *Fundamenta Juris Naturae et Gentium* (1705)), but the same approach was adopted by Rutherforth, Francis Hutcheson, Barbeyrac, and Burlamaqui (*Principes du droit de la nature et des gens*).

The positivists, who were more important in the eighteenth century, stressed the positive aspects of the law of nations as the product of custom and international treaties. Of these the major writers were Rachel, Textor, and particularly Bynkershoek (*De Dominio Maris* (1702), *De Foro Legatorum* (1721), and *Quaestionum juris publici libri II* (1737)), Moser, whose books contain enormous stores of facts, and G. F. von Martens, who began the valuable collection of treaties and wrote a *Précis du droit des gens moderne de l'Europe* (1789).

The Grotians are in an intermediate position, attaching roughly equal importance to natural law and to positive rules of custom and treaty. The most noteworthy scholars were Wolff (*Jus Gentium Methodo scientifica pertractatum* (1749) and *Institutiones Juris Naturae et Gentium* (1750)) and Vattel (*Le Droit des gens* (1758)), the latter's work being very influential.

In the nineteenth and twentieth centuries, from the Congress of Vienna onwards, the major powers have tried repeatedly to submit their relations to regulation by law, efforts which have been repeatedly rebuffed by Germany under Bismarck and Kaiser Wilhelm I, by Germany under Hitler, and by Russia under the successors of Stalin.

In the writing of publicists, moderate positivism has attained supremacy over other schools, though it is recognized that gaps in positive law may be filled by recourse to general ideas of justice and general principles of law.

From the Declaration of Paris of 1856, part of the Treaty of Paris which concluded the Crimean War, there have been a series of pure law-making treaties. The Geneva Convention of 1864 for the amelioration of the lot of the sick and wounded in warfare followed. The Declaration of St. Petersburg 1868, the Convention of 1888, which provided for the neutralization of the Suez Canal, are others. The First Hague Peace Conference of 1899 resulted in three Conventions and three declarations and the Second Hague Conference of 1907 produced 17 conventions.

The Second Hague Peace Conference, 1907, adopted 13 conventions clarifying important topics of international law. The London naval conference of 1908–9 produced a code, the Declaration of London, dealing with prize law, but it was not ratified.

A conference for the codification of international law at The Hague in 1930 made little headway.

General instruments such as the Covenant of the League of Nations, the Statute of the Permanent Court of International Justice, the General Act for the Pacific Settlement of International Disputes of 1928, the General Treaty for the Renunciation of War, and a great number of conventions, many concluded under the aegis of the International Labour Organization, contributed to the development of positive law in the inter-war period. The judgments of the P.C.I.J. operated in the same way.

Again, after the Second World War, the major powers by an important treaty provided for the punishment of major war criminals and adopted the Charter of the United Nations, while the work of the International Court of Justice and numerous international conventions have continued the conscious development of rules of international law.

C. Phillipson, *International Law and Custom of Ancient Greece and Rome*; E. Nys, *Les Origines de Droit International*; C. Kaltenborn, *Die Vörlaufer des Hugo Grotius*; T. Holland, *Studies in International Law*; F. Laurent, *Histoire du droit des gens*; T. A. Walker, *History of the Law of Nations*; R. Ward, *Enquiry into the Foundation and History of the Law of Nations*; G. Butler and S. Maccoby, *Development of International Law*; A. Nussbaum, *Concise History of the Law of Nations*; W. Simons, *Evolution of International Public law in Europe since Grotius.*

Basis of international law

The real basis of international law is common consent, the recognition of states in the world that it is generally better to regulate their relations by a body of principles which have developed by custom and usage, treaty and convention, arbitral and judicial decision, than to live in anarchy, and that habitual flouting and disregard of commonly accepted principles will lead to a state being distrusted, disliked, and even outlawed by the international community.

Binding force of international law

Hobbes and Pufendorff denied that the rules of international law were legally binding and Austin and his followers, having defined law as rules set by a sovereign political authority, had to deny that international law was law; it was only positive morality. But in modern practice international law is constantly appealed to and recognized as law, though less enforceable than is municipal law; states which violate it more frequently try to interpret it as permitting their conduct than try to deny its existence.

Relations of international and state or municipal law

The theoretical question of this relationship is whether international law is superior and national systems inferior or subordinate branches of law, or whether national or state law is superior and international law exists by recognition by municipal law, or whether international and the different systems of municipal law co-exist in different spheres without any question of relative supremacy. The practical questions are whether international law binds state courts and individuals within the jurisdiction of that state, without its rules being incorporated expressly or impliedly into state law, whether, if so, a state legislature can modify a rule of international law and whether a state may invoke or apply its municipal law as an excuse for refusal to comply with international law.

The theoretical question is the subject of two rival doctrines, the monist and the dualist or pluralist doctrines. According to the former in both the international and municipal spheres law is a power binding the subjects of the law, and both international and municipal law are manifestations of a single conception of law. According to the latter the two kinds of law differ so radically in sources, subjects, and substance that international law cannot be part of the law of any country except in so far as incorporated; the will of the State is the exclusive and decisive source of legal obligation for it and its subjects; this will may be exercised by participation in a collective act of will by several states, or by a manifestation directed to its subjects only; hence international law and municipal law are different spheres of legal action. A compromise view, the harmonization theory, sees international law and municipal law as concordant bodies of doctrine, each autonomous in its own sphere but harmonious in that the totality of the rules aims at human good.

The practical question has been treated by international courts on the basis that they must apply international law and that municipal laws are merely facts which express the will and constitute activities of states, which have legal effect in the sphere of their own competence but have no binding effect on any international tribunal.

The practical question depends in each jurisdiction on the constitution and the attitudes of the executive and courts. Blackstone contended that the law of nations was fully adopted into English law and, subject to the supremacy of statute, this is modern English law in so far as it concerns customary international law. This principle of adoption has resulted in the rules that statutes are not to be presumed to derogate from international law, but must be given effect to even if contrary to international law, though they are not to be so interpreted unless the intention is clear, and that the substance of international law is a matter of judicial notice, not, like foreign municipal law, a matter of fact to be proved by experts. But it is otherwise with international law established by treaty or conventions. Treaties affecting private rights of persons and internationally binding by ratification are not enforceable by British courts unless an Act of Parliament has enabled or required effect to be given to them in municipal law. This is because in Britain the conclusion and ratification of treaties is within the prerogative of the executive. In practice enabling legislation is frequently enacted prior to ratification of or accession to a treaty.

In the U.S. customary international law is, subject to contrary Act of Congress, part of the law of the U.S. and as Congress participates in treaty-making no enabling Act of Congress is necessary to make treaties enforceable save where treaties expressly or by implication require further legislative action. Treaties equally with congressional statutes are part of the law of the U.S. but treaties contrary to constitutional provisions are not valid as law of the U.S.

In France the constitution of 1958 incorporates customary international law into French law; presidential ratification generally suffices for the internal operation of treaties but ratification by legislation is necessary for certain classes of treaties, and in only some cases is treaty recognized as superior to legislation. In Germany the Basic Law of the Federal Republic makes the general rules of public international law part of the federal law and gives them precedence over municipal law. It is a question of construction whether a particular treaty is productive of internal law rules or not.

In Italy the constitution of 1948 states that the Italian legal system conforms to the generally recognized principles of international law; customary international law forms part of Italian law. Other civil law systems appear generally to adopt the harmonization principle.

Sources of international law

The philosophical source or basis of international law is the existence of a community of states, the need for them to regulate their mutual relations and their general willingness to accept rules which aid them to do so. The formal sources, or places from which authoritative statements of law can be drawn, were formerly custom and usage of nations, writings, accepted as authoritative, of jurists, recording the usage of nations or general opinions respecting their conduct, treaties and other international agreements, declaring, defining, or modifying law, adjudications of international tribunals, statutes or ordinances of particular states, and the opinions and advice of jurists given to their own governments.

As international law has developed the weight attached to the different sources has changed. The Statute of the International Court of Justice provides that the sources from which that court may draw its law are international conventions, general or particular, establishing rules expressly recognized by the contracting states, which constitute the bulk of international law as a growing body of rules; international custom as evidence of a general practice accepted as law, which embodies a substantial, and the most ancient and fundamental, portion of international law; general principles of law recognized by civilized nations, that is those principles of law, public and private, which examination of the legal systems of developed countries leads one to regard as maxims of a general and fundamental character; judicial decisions and the teachings of the most highly qualified publicists of the various nations, as subsidiary means for the determination of rules of law, that is decisions of international courts and tribunals and of municipal courts on matters of international law, and equity in the general sense of fairness, good faith, and moral justice. Decisions of the I.C.J. have no binding force as precedents. These sources are probably stated in order of importance. International custom itself is evidenced by state papers, treaties, historical records, opinions of law officers and jurisconsults, and other materials.

Subjects of international law

The subjects of international law, the entities on which the law confers rights, and imposes duties, are normally, but not exclusively, states; they included also organizations of states, collective bodies other than states, political bodies not having full statehood, such as the Holy See, and international public bodies such as the specialized agencies of the U.N., e.g. the I.L.O. Individual persons have in general no position in international law, but are subject to duties under the law of war, and may have rights thereunder, e.g. to human rights and fundamental freedoms.

Content of international law

The content or subject-matter of international law can be indicated here only briefly, the topics being dealt with under their respective headings; they include personality, statehood, recognition of states and governments, state territory and territorial sovereignty, nationality and alienage, extradition, asylum, human rights, the sea, comprising territorial waters, high seas, and seabed, air and space, rights in foreign territory, state jurisdiction, state responsibility, state succession, injuries to aliens and their property, diplomatic and consular relations, international transactions, treaties and agreements, international disputes, war and neutrality, international organizations and tribunals, and the judicial settlement of international disputes.

Codification of international law

Conviction that the authority of international law and its certainty of statement required codification has prompted many efforts to that end. Bentham proposed a detailed scheme; Grégoire, a member of the French Convention, proposed a Declaration of the Rights of Nations in 1795 but it was abandoned; Lieber in 1863 drafted the Laws of War as a body of rules; Bluntschli published *Das moderne Völkerrecht als Rechtsbuch dargestellt* in 1868. Other attempts were made by Mancini, David Dudley Field, Levi, Fiore, and Guerrero. The Institute of International Law and the International Law Association have from time to time adopted various drafts. The major private attempt was made by Manley Hudson at Harvard. The work of the First (1899) and Second (1907) Hague Peace Conferences produced a substantial number of codifying conventions. After the First World War there were important pieces of partial codification in the Covenant of the League of Nations, the General Act for the Pacific Settlement of International Disputes, the General Treaty for the Renunciation of War and numerous specific conventions, while the League of Nations called the Hague Codification Conference of 1930, which had only limited success. After the Second World War the U.N. Conference on the Law of the Sea adopted four conventions and in 1947 the U.N. established the International Law Commission charged with the task of codifying and developing international law.

J. Delupis, *Bibliography of International Law*; J. Robinson, *International Law and Organization: General Sources of Information*; I. Brownlie, *Principles of Public International Law*; L. Oppenheim, *International Law*; D. P. O'Connell, *International Law*; G. Schwarzenberger, *International Law*.

International Law Association. Founded in Brussels in 1873 as the Association for Reform and Codification of the Law of Nations, it adopted the present name in 1895. Its objects are the study, elucidation and advancement of international law, public and private, the study of comparative law, the making of proposals for the solution of conflicts of law, and for the unification of law and the furthering of international understanding and goodwill. At its conferences the Association has considered many topics and made many proposals for improvements in law.

International Law Commission. By the U.N. Charter, Art. 13, the General Assembly of U.N. was given power to initiate studies and make recommendations for the purpose of encouraging the development of international law and its codification and in 1949 set up the International Law Commission. The Statute of the I.L.C. states its objects as the promotion of the progressive development of international law and its codification, and provides that while primarily concerned with public international law it is not precluded from entering the field of private international law. It consists of 25 members, persons of recognized competence in international law, no two from the same state, elected by the General Assembly for three year renewable terms from candidates nominated by governments of member-states. The Commission as a whole is intended to represent the main forms of civilization and the principal legal systems of the world. It has made various recommendations to the General Assembly and its work has been the basis of various conferences and resulting multilateral conventions. Its headquarters is in Geneva and it publishes an annual *Yearbook* and makes an annual report to the General Assembly.

International Law, Institute of. Founded at Brussels in 1873 it is an Academy limited to a small number of distinguished scholars in international law with the object of influencing the development of international law. At its foundation it resolved to establish rules to resolve conflicts of laws.

International legal persons. Those entities which are endowed with legal personality by international law and have legal rights and duties thereunder. International legal personality attaches to every independent and fully sovereign state recognized by the community of nations, including states linked by personal union, having the same head of state, e.g. Great Britain and Hanover, 1714–1837, but not to states not fully sovereign but subject to the sovereignty or under the protection of another, e.g. Andorra, nor to states linked in a real union, e.g. Austria-Hungary, 1723–1918, nor states confederated, e.g. the German Confederation, 1815–1866, nor states which are member-states of a federal state e.g. the states of the U.S.A. (though member-states are sometimes recognized for some purposes as international persons).

It is a matter of argument whether a state becomes an international person by coming into existence, or only on being accorded recognition by other states.

A state remains the same international person despite changes of name, territory, form of government, or of dynasty or of headship of state. But if two states enter into a real union they disappear and there appears a single new international person. A state ceases to be an international person if it ceases to exist, as by merger with another state, annexation after being conquered by war or breaking up into several states.

International personality also attaches to the Vatican City, the United Nations Organization (q.v.), its principal organs, and the major international organizations such as the International Civil Aviation Organization, though they do not necessarily all have the same legal powers and capacities, nor the same as those of a sovereign state. These depend on the purposes and functions of the organization in question.

It is questionable whether international personality attaches to the Commonwealth, which is a loose association of sovereign states, or the Organization of American States. It probably does to the

European Economic Community, which is akin to a federal state in having a distinct legal structure.

Difficult questions of whether a body is justifiably treated as a distinct legal person or not arise in cases of insurgent governments; if a body of insurgents is deemed entitled to be recognized as a *de facto* government of a territory, that government acquires legal personality, though the legitimate government also has legal personality.

The consequences of being recognized as an international person are recognition of what are sometimes called the fundamental rights of states, namely the rights of existence, self-preservation, equality, independence, territorial supremacy, holding and acquiring territory, intercourse *inter se*, and good name and reputation. From these fundamental rights follow many subsidiary rules such as the immunity of states from the jurisdiction of other states' courts and the acceptance of the validity and legality of the official acts of other states.

Individuals have legally protected interests and can enjoy rights and be the subjects of duties under municipal law derived from international law and accordingly, though their capacities are different from those of states, they are to that extent persons in international law.

International Maritime Committee. Founded in 1897 to contribute to the unification of maritime law by means of conferences and publications and to encourage the creation of national associations for these ends.

International military tribunals. Judicial bodies established by international agreements of 1945 for the trial as war criminals of the leaders of the German and Japanese states. The two major tribunals were that which sat at Nuremberg in 1945–46 and that which sat at Tokyo.

International Monetary Fund. An international organization founded at the Bretton Woods Conference in 1944 and affiliated to the U.N.O., intended to secure international monetary co-operation, stabilize exchange rates, and expand international liquidity. Operating funds are subscribed by member governments according to national income, volume of international trade, and international reserve holdings. It acts as a means of consultation and a centre for information and research on monetary matters.

International organizations. Bodies of various kinds set up by multilateral agreements between states for various co-operative purposes. Since the Congress of Vienna there have been numerous international conferences and multilateral treaties setting up bodies of various kinds, many concerned with technical, social, and economic co-operation between states. Many are composed of states, albeit associated for particular purposes, such as peace-keeping or defence, or health, but some, e.g. the Universal Postal Union, are organizations of postal administrations.

Such organizations are normally created by multilateral treaty, but may be created in other ways, such as by resolution of the U.N. General Assembly.

There is no legal machinery in the international community corresponding to the incorporation of an entity in municipal law, and whether an organization has legal personality or not depends not so much on the mode of creation, still less on any formalities, but rather on whether it is in fact treated as such. The major relevant factors appear to be whether or not there is a permanent organization with lawful objects and organs to effect its purposes, whether the purposes and powers are distinct from those of the member states, and whether the powers are exercisable on the international plane or solely within the national systems of law of one or more states. Applying these tests, the U.N. and its specialized agencies, the I.L.O., the E.E.C., E.C.S.C. and Euratom certainly have legal personality, but the British Commonwealth does not. Many subsidiary organs and agencies of organizations, though they have some powers and functional independence, probably do not have full legal personality in international law.

Organizations which have legal personality have various rights, privileges and immunities, powers and duties, and liabilities. Much depends in particular cases on the constitutive treaty or other instrument, and the position in fact conceded to the organization by individual states. In the case of the U.N. a General Convention on the Privileges and Immunities of the U.N. was concluded in 1946. The U.K. conceded various privileges to the U.N. by the International Organizations Act, 1968, and the U.S. and the U.N. concluded a Headquarters Agreement in 1947.

Organizations may have treaty-making powers, by express or implied authority, or established by interpretation of the constitutive document, may bring claims against either member or non-member states for injuries to the organization, and may incur responsibility. The distribution of power between an organization and its members may raise difficult questions of interpretation.

In relation to member-states an organization may take decisions within the sphere of its functions binding on the members and even issue instructions to members.

In relation to non-member states, an organization has no powers or responsibilities and cannot bind them, though the U.N. is required to ensure that non-member states act in accordance with its Principles so far as may be necessary for the maintenance of international peace and security. Sometimes an organization makes an agreement with a non-member state.

Organizations as legal persons in international law, may enter into legal relations with one another and make agreements with each other and one organization may even become a member of another, though this may give rise to problems of co-ordination, a problem for which some constitutive instruments make special provisions. Organizations have responsibility under international law for damage done to the interests of other organizations and of states.

The political control of an organization depends on the powers of the executive and deliberative organs of the organization, the limitations thereon, and the powers conferred on it by the constituent states or reserved to themselves. There is no general judicial power of control by the International Court or other body. In the case of the European Communities a Court of Justice has power to control the acts of organs of the Communities if one of them has contravened a provision of one of the constitutive treaties.

An organization inevitably comes in contact with various systems of municipal law. The first problem is of the extent to which its legal personality is recognized in a particular system, and what rights and immunities, duties and responsibilities each particular system concedes. The constitutive instrument may provide for these matters, at least in relation to member states, but may not, in which case the organization may be treated in much the same way as a foreign corporation. In some cases, such as the E.E.C., the organization has a delegated power to make law which is effective as part of the legal system of member states, and in other cases regulations made by an organization will be effective unless a member enters a reservation or declines to apply the regulation.

Incidentally to their functioning, international organizations have importance as creative of international law. They may sponsor multilateral treaties, elucidate and develop customary international law by declarations, resolutions and practice, sponsor drafts or projects for development of international law; their judicial organs, such as the International Court or the European Community Court, may issue important decisions; their controlling organs may give rise to important decisions or statements on international law or practice by government representatives; and the actual practice of an organization and its organs, by acting, recommending, interpreting, and in other ways, may have influence in shaping, or evidencing international custom and practice.

International personality. A status recognized in international law consisting of recognition by the international community of a state or organization as being a subject of public international law. International personality is accorded to every independent sovereign state, and to such special entities as the Holy See, but not to provinces or states within a state, to insurgent governments having *de facto* control of territory but not to any small group of rebels, and also to various other entities, notably the United Nations and its specialized agencies, and to other major international organizations. Natural persons do not in general have international personality though in certain respects, human rights, war crimes and genocide, piracy, slavery, and the protection of minorities, they are treated as subjects of international law, having rights and duties thereunder and being subject to its protection.

See INTERNATIONAL LEGAL PERSONS.

International Private Law. International private law, or private international law, or conflict of laws, exists because there exist in the world a large number of separate and independent legal systems having in many respects different rules to regulate similar kinds of activities and legal relations, and because individuals may act in ways having legal consequences and effect legal transactions in countries other than their own, or having transfrontier consequences. Thus a Briton may negligently injure a German in Italy; what system regulates the German's claim against the Briton, and why? The rules of the international private law branch of any legal system come into play when an issue arises in one country which contains a foreign element and when, accordingly, reference to that country's legal system alone might be unjust or unreasonable. In this context 'foreign' means 'pertaining to any other legal system', so that England and Scotland are mutually foreign, as much as either is to France or Argentina.

It is noteworthy that the rules of international private law are not international in the sense of being universally or even generally recognized in all legal systems, but are a branch of and peculiar to each system of private law, being that part of the private law of any one legal system which must be invoked when an issue involves a trans-frontier or foreign element, and requires consideration of the rules of a foreign system as well as of the system in which the issue has arisen.

This branch of law was called Private International Law by Story, Westlake, and many other scholars, and this is probably the commonest name, but this name suggests some parallel with (public) international law, and may lead to confusion between the two, though the one is a part of the private law of each legal system, each system having its own rules, and the other, at least in theory, a single world-wide common body of principles not peculiar to any country. This branch has been called Conflict of Laws by some jurists, but this is misleading if it implies that the systems involved are in conflict with one another. Other names have been suggested but none has found favour, and international private law is on the whole the best as bringing out that it is a branch of the private law of a legal system, that branch dealing with problems having an international or foreign aspect.

When the rules of the branch of any legal system which deals with international private law are invoked, they do not directly or immediately decide the issue. They decide only three preliminary points: (a) jurisdiction, whether the courts administering that system of law can justifiably and validly take cognizance of the issue, or must decline jurisdiction in favour of another system's courts; (b) if those courts have jurisdiction, whether they must apply their own internal rules of law or the rules of another system involved, the question of choice of law; and (c) whether they must recognize as decisive any foreign judgment already pronounced on the issue, or enforce any right vested in a party under a foreign judgment, or on the other hand treat such judgment or right as nullities. It follows that the rules of any one legal system's international private law branch may require that system to decline jurisdiction, or apply rules of foreign law, or accept as valid and apply a foreign judgment.

A question involving a matter of international private law may arise in relation to any kind of issue, domestic relations, contract, bankruptcy, succession, and, strictly speaking, in every issue which arises for legal advice or court decision consideration should be given first to the question whether there is any foreign element requiring the application of principles of international private law before it is assumed that only the internal law of the country is involved.

While the rules of international private law are peculiar to each distinct legal system, on some topics there is little disagreement and the areas of disagreement are slowly being diminished by the adoption of international conventions, as on carriage of goods by sea.

G. C. Cheshire, *Private International Law*; J. H. C. Morris, *The Conflict of Laws*; A. V. Dicey and J. H. C. Morris, *The Conflict of Laws*; A. E. Anton, *Private International Law*.

International regimes. Restrictions on a state's sovereignty over part of its territory by another state, or others, to secure rights benefiting the international community generally or one or more particular states. Thus the regimes established over the Suez and Panama canals are for the benefit of other nations generally. In many cases states have rights in water from rivers, mining seams, or the like within the territory of another state for their own benefit.

International rivers. Rivers which geographically and economically affect the territory and interests of two or more states. Basically rivers separating the territories of states are subject to the territorial jurisdiction of each *ad medium filum aquae*, but in many cases the rules as to rivers are regulated by treaty. In the case of navigable rivers customary law is generally deemed not to recognize a right of free navigation, though this may be secured by treaty, including navigation rights for non-riparian states.

The first organised system of international river control was that set up for the Rhine by the Treaty of Munster (1648) later modified by the Congress of Vienna. The European Commission of the Danube was created by the Treaty of Paris of 1856.

International Social Science Council. Founded in 1952 and reshaped in 1973 as a federation of organizations, including the International Association of Legal Sciences and the International Law Association (qq.v.) to further the advancement of the social sciences and their application, and the spread of co-operation therein. It co-operates with national Social Science Research Councils.

International Telecommunications Union. Established in 1961, replacing earlier unions dating from 1865, to maintain and extend international co-operation in the use of telecommunications and promote the development and efficiency of services. It allocates the radio frequency spectrum to avoid interference.

International tribunals. Tribunals having jurisdiction under international law to adjudicate on disputes between states. They are of several kinds, arbitral tribunals established by treaties, such as the *Alabama* (q.v.) *Claims* arbitral tribunal, the Permanent Court of Arbitration at The Hague, *ad hoc* arbitral tribunals, and mixed commissions such as the U.N. Tribunal for Libya, the Court of Justice of the European Communities, and the International Court of Justice. Some other international bodies such as commissions of inquiry set up by the I.L.O. to investigate breaches of I.L.O. conventions have a judicial aspect.

International waterways. The great inter-sea canals are by agreement open to the ships of all nations, the Suez Canal by Convention of 1888 and the Treaty of Lausanne of 1923, the Panama Canal by various treaties between the U.S. and Panama and the Kiel Canal by the Treaty of Versailles. A

Convention of 1923 attempted to internationalize the Dardanelles and the Bosphorus by establishing an international commission.

Internment. The detention of alien enemies in restricted areas in time of war to minimize the risk of their acting in the interests of the enemy.

Interpleader. The English legal process whereby a person in possession of, or liable for, property in which he claims no interest but to which other parties make competing claims, may have the other parties seek to establish their claims to the satisfaction of the court at their own expense. Originally mainly an equitable remedy, it is now more readily available at common law. The relief is discretionary and cannot be claimed as of right. The subject matter of interpleader is generally a debt, money, goods, or a chose in action which was then legally assigned.

Interpol. See International Criminal Police Organization.

Interpolations. Alterations made by the commission which prepared Justinian's *Digest* (q.v.) in the texts of the classical materials which they were collecting. The commissioners were instructed to make change to adapt the texts to contemporary requirements and facilitate codification. The humanist jurists, notably Cujas and Faber, sought to discover these changes so as to recover the original text of the classical law. The quest was not of great importance in early modern Germany since Justinian's text was of interest as the basis of modern law, but when the Pandect law lost its practical utility the search for interpolations and the recovery of the classical texts was taken up. It began with the work of Alibrandi in Italy and of Gradenwitz in Germany and reached its zenith in the 1920s in the work of von Beseler and Albertario. Since then it has declined. Some of the allegations of interpolation have been found unjustified or questionable, but the quest has been valuable in helping to distinguish the ideas and views of classical, post-classical, and Justinianian writers, and the modern tendency is to discover the courses and stages of development and the differences of opinion, even within the classical period, apart from subsequent misunderstandings and changes.

See also *Index Interpolationum*.

Interpretation or construction. The process of determining the meaning of a text, such as that of a statute, or deed or will, in the circumstances which are under consideration. Hardly any form of words can be thought of which is not, in some circumstances, ambiguous and requiring interpretation. There are two broad approaches to interpretation, the literal, which stresses the literal meaning of the words used, and the liberal, which stresses the general intention and purpose of the statute or deed, but the task of interpretation may be assisted by interpretation sections (q.v.), by various presumptions, and by reference to previous cases in which the phrases in question have been considered and interpreted.

The first principle of interpretation is that words should be given the meaning they ordinarily bear in English, and technical terms their ordinary meanings in the branch of science, or profession, or trade in question. The second is that words must be read in their context and given a meaning which makes sense in that context, and makes sense of the sentence, paragraph, or section.

In interpreting statutes there are various presumptions such as that the Crown is not bound unless that is expressed, that taxing and penal statutes are interpreted strictly, that injustice, inconvenience, or absurdity is not intended, that retrospective operation is not intended, and that a violation of international law is not intended.

P. B. Maxwell, *Interpretation of Statutes*; C. E. Odgers, *Construction of Deeds and Statutes.*

Interpretation Act. An Act, in successive revised editions of 1837, 1889, and 1978, laying down certain general principles for the interpretation of statutes and assigning to a large number of terms commonly found in statutes, such as 'Secretary of State', 'Court of Appeal', 'person', or 'committed for trial', standard meanings to be attached to these words in interpreting all statutes, unless the particular statute or the context or other circumstances indicate otherwise.

Interpretation clause. A clause in a deed stating the meanings to be given to particular words and phrases in the context of the particular deed and for the purposes of that deed only.

Interpretation section. A section included in all substantial modern Acts of Parliament stating the meanings to be given to particular words used in the Act, for the purpose of interpreting that particular Act only, unless the context requires otherwise. Such sections may attach unusual meanings to words for the purposes of the particular Act and not generally.

Interpreter, The. A law dictionary published in 1607 by Dr. Cowell, regius professor of civil law at Cambridge, which aroused the wrath of the House of Commons by reason of the extravagant assertions in support of the King's absolute power contained in the entries King, Parliament, and Prerogative. Before any action was taken the King expressed displeasure at the language used and a proclamation was issued prohibiting the purchase, uttering, or reading of the book and calling in all copies issued. But the proclamation put the royal dignity on a level with God which somewhat blunted the effect of the condemnation. The book was reprinted in 1708, with the proclamation in the Preface.

Interregnum. The interval between the reigns of two kings, and particularly in English history the period 1649–60 between the execution of Charles I and the Restoration of Charles II, covering also the period of the Commonwealth and Protectorate. The legislation of the period has been printed as *Acts and Ordinances of the Interregnum* (q.v.). There was also an interregnum between the flight of James II and the acceptance of the Crown by William III and Mary. In mediaeval times there had been several interregna and not till the accession of Edward II (1307) did the new King's reign begin the day after the late King's reign ended.

Interrogation. The process of questioning of suspected criminals to determine whether they should be charged with crime or can be released. In England interrogation is largely regulated by the Judges' Rules, which prescribe that if a police officer decides to charge a person, he should caution that person before asking questions or further questions, i.e. warn him that he need not say anything but that if he does it will be taken down in writing and may be given in evidence at a trial. A prisoner may make a voluntary statement, but no threats or promises may be made to try to induce a statement, and a statement not truly voluntary will not be submitted in evidence. In the U.S. elaborate safeguards have been developed by the Supreme Court to regulate the interrogatory practices of the police.

Interrogatories. Questions in writing framed by one party to a civil action in England, which the court, i.e. a Master in chambers, may allow him to require his opponent to answer on oath, relating to matters in question in the cause between them, and necessary for disposing of the cause fairly. The power to administer interrogatories is based on old Chancery practice and is always subject to the discretion of the court. Questions may not be put on matters which the questions have to prove, or to secure admissions. They must be relevant to matters in issue.

Interruption. A breach in the continuity of enjoyment of a right, a matter of importance in relation to easements, profits, franchises, and similar rights. If it evidences interruption of the right itself, this may prevent the acquisition of the right by prescription, or cause the acquisition of a merely qualified right, or destroy a right which has been acquired.

Interstate commerce. In U.S. constitutional law a term denoting commerce which crosses state boundaries or involves more than one state. The Constitution gives Congress power to regulate commerce with foreign nations and among the several states. Commerce has been interpreted as including both trade and transportation, and as extending to all kinds of businesses and means of transportation not in existence when the Constitution was adopted.

Interstate Commerce Commission. A Federal Regulatory Commission created by Congress in 1887 to regulate commerce between states, particularly railroads engaged in interstate or foreign commerce. The practical results of its establishment were at first small since its orders did not have the force of decrees of court, but from the start of the twentieth century its authority has been greatly increased. Since 1948 it has had the regulation of all monopoly business practices as well as over all carriers of interstate commerce.

Intervener. A person who voluntarily but with the leave of the court interposes in a proceeding in the Family Division, usually to challenge accusations of adultery or unnatural practices made against him in the petition.

Intervening cause (or *nova causa interveniens*). Any causal factor which, a series of consequences having been initiated by the act of one party, supervenes thereon and modifies or aggravates the consequences. The new intervening causal factor may be an act of the victim, or of a third party, or a natural force. It may be held to be ineffective to release the party initially in fault from liability, or be held to supersede his fault as being the major cause of the ultimate harm. Thus if A by careless driving injures B and B while being conveyed to hospital is further injured or killed by his own act, or C's fault, or by a tree blown down, these factors are intervening causes and may or may not be held to absolve A.

Intervention. In international law the dictatorial interference of a third state in a dispute between two other states for the purpose of settling the dispute in the way demanded by the intervening state. It is much stronger than proffering good offices or mediation (q.v.) or offering advice. It may be made by one or several states and can take place at any time after a dispute arises and before hostilities have broken out between the disputing states or after. In some cases several powers have intervened in the interest of the balance of power and of humanity. Or intervention can take place in the internal affairs of another state. Intervention violates the external independence or the territorial supremacy of those subjected to it.

One state may have a right of intervention against another state, as where a state is restricted by international treaty in its independence or territorial supremacy, or a state has violated the rules of international law universally recognized by custom or laid down by law-making treaties, e.g. asserting sovereignty over part of the high seas.

Intestacy. The condition of a person dying without leaving a valid will to regulate the disposal of the whole of his estate after his death. It may accordingly be total or partial. In such circumstances the court appoints a personal representative (administrator in England, executor-dative in Scotland) to administer the deceased's estate, and statutes lay down which surviving relatives, in what order and in what proportions, are entitled to the intestate estate.

Intimidation. The tort and crime of applying legally unjustifiable threats or force to a person to compel him to do something, or to refrain from doing anything which he is legally entitled to do.

Intolerable Acts (or Coercive Acts, 1774). Name given in U.S. history to measures enacted by the British Parliament in retaliation for acts of defiance by the colonists. They comprised the Boston Port Act, closing the harbour of Boston until restitution was made for the tea lost in the Boston Tea Party; the Massachusetts Government Act, abrogating that colony's charter of 1691 and reducing it to the level of a crown colony; the Administration of Justice Act protecting British officials charged with capital offences while enforcing the law; the Coercive Act, making new arrangements for housing British troops in America; and the Quebec Act, which removed the territory and trade between the Ohio and the Mississippi from colonial jurisdiction and transferred it to Quebec. These attempts to reimpose British control came too late and provoked the summoning of the First Continental Congress later in 1774.

Intolerance. The unwillingness or refusal to recognize or permit the existence of views other than those held by a dominant group. It may spring from dedication to a fixed goal, such as a social or political purpose, but infringes the liberty of the dissenting individual to hold the views he does. It can take many forms, of which religious and political intolerance have been most important. Thus the Roman Catholic Church has shown intolerance of other views by reason of its claim to be custodian of the eternal and unalterable truth on which the destiny of man in the next life depends. Socialist and communist states and parties practise political intolerance, and trade unions by insistence on the 'closed shop' practise intolerance against those who do not wish to belong to their, or a, union. The practice of toleration and the suppression of intolerance should be favoured by the law but frequently it fails to do so. In the U.K. religious toleration was achieved only in the nineteenth century by such legislation as the Catholic Emancipation Act, 1829, the Jews Relief Act, 1858, the Promissory Oaths Act, 1868, and the Oaths Act, 1888, but political and economic intolerance have been growing.

Intra vires. Within the powers, as contrasted with *ultra vires* (q.v.). An act is *intra vires* if within the powers or authority of the person doing it.

Intrationum. See ENTRIES, BOOKS OF.

Intrusion. In Scots law the wrong of entering into possession of land without title.

Invecta et illata. In Scots law the ordinary equipment of premises, furniture, stock, etc., brought by a tenant on to land let to him, and over which the landlord has a hypothec (q.v.) for the rent.

Investitive fact. A fact which invests a person with a particular legal right. Thus the breach of a contract or the doing of a tort invest the party injured with a claim for damages, the occurrence of a fire invests the insured with a claim under his fire insurance policy. Investitive facts may be either original titles, creative of rights, or derivative titles, transferring rights already existing from one person to another, as on sale. The opposite are divestitive facts.

Investiture. Generally the induction of a person to an office or dignity; in feudal law, the public delivery of seisin or legal possession by a lord to his tenant; in ecclesiastical law, a formality in the confirmation by the archbishop of the election of a bishop. Investiture is distinct from homage (q.v.). The former was concerned with the proprietary aspects of the relationship, the conferment of title, the latter with the personal aspect, the creation of a personal bond between lord and vassal, and homage normally came first. In mediaeval times several disputes arose as to the investiture of the clergy. Under Henry I the compromise was reached that the ring and crozier, symbolizing spiritual jurisdiction, were to be conferred by the Pope; fealty and homage, as civil dues, were to be rendered to the King for the temporalities of the See.

Investment. The expenditure of money on the acquisition of things or rights which may be expected to retain their value, or appreciate, or to yield income. Commercial and industrial investment is mainly by way of acquiring new buildings, machinery, or equipment. Financial investment is by way of acquiring shares, debentures, or other claims. Persons holding the money of others, such as trustees, are expected to invest the money of which they have charge. Trustees are subject to detailed statutory regulation of the ways in which they may invest, subject in certain cases to the authority of the court.

Investment company or trust. A company which pools the funds contributed by its shareholders and invests them in a variety of other enterprises, thus securing for each shareholder an interest in a wide variety of companies, minimizing the risk of loss by decline or failure of any particular company and giving the shareholder the benefit of professional management of his investment. Investment companies originated in England and Scotland about 1860. The managers may have limited or unlimited discretion as to the ways in which they may invest their funds, and some concentrate on capital growth, some on maximizing income.

Invitee. A person expressly or impliedly invited to enter on certain premises.

Invoice discounting. The commercial practice of assigning book debts (i.e. monies receivable) to a finance company, which advances a proportion of the nominal value of the sums due on the security of, e.g. goods, in the case of goods let on hire-purchase, or of bills of exchange accepted by the borrower. The borrower collects the monies due and repays the finance company. The finance company charges a commission, depending on the risks involved and the borrower is responsible for bad debts.

See also FACTORING.

Irish law. Ireland, originally inhabited by Picts, was colonized by Celts in the Bronze Age and became a land of small states and petty kings, which tended to coalesce into five major groups of states. From a very early date regular courts and judges existed and Brehon laws (q.v.) existed in oral tradition and are said to have been put in writing in the fifth century. In their existing form they are not, however, so early. One Cenwfaelad is said to have written down the mnemonic formulae of Irish law about 650.

Ireland was untouched by Roman occupation, but fell a victim to Scandinavian raiders who established an ascendancy until overthrown at Clontarf in 1014. Norman invaders came from England in the twelfth century, but complete subjection of the country to English rule was not achieved till much later. English legal institutions began, however, to be imported from late in the twelfth century, and eventually it became settled that English common law applied throughout the country. Under John the lordship of Ireland was united with the English Crown and he organized a permanent government under a justiciar. Some English statutes were applied to Ireland by being sent over there under seal, and from the late thirteenth century there was an Irish Parliament on the English model. From 1310 there was a Parliament with representatives of knights, clergy, shires, and towns, but the native race had no voice. It was, however, subject to some control from London and did little more than enact statutes already in force in England and which could be applied to Ireland, and other measures to further the subjugation of the country. A serious attempt to create a real central authority was made under Edward I, and a series of justiciars extended the common law and sought to reduce the feudal liberties of the chiefs. In 1495 Poyning's Law enacted that all statutes of the realm of England concerning or belonging to the common and public weal should apply to Ireland, and that no parliament should be held in Ireland until the King and council were informed of the necessity of it, of the Acts proposed to be passed, and the royal approval was previously obtained. Much later, Yelverton's Act, 1781, was to a similar effect.

Under the Tudors the Reformation was pushed through and a determined effort made to introduce sheriffs, garrisons, and English law in face of the bitter opposition of the native chiefs and the English settled in Ireland.

In the time of James I an attempt to make the English land system applicable provoked revolt in the north and English and Scottish settlers were settled in confiscated lands. Under Cromwell, a policy of extermination of the native Irish was followed and their land parcelled out among adventurers, and a new English and Puritan landlordry imposed on Ireland.

After the defeat of James II at the battle of the Boyne in 1690 a Test Act of 1691 excluded Catholics from the Dublin Parliament. A body of penal laws passed between 1702 and 1715 sought to repress the Irish and Catholic majority. The Dublin legislature became a mere dependency of Westminster, hardly representative even of the Anglican minority. There was, however, some moderation of the repressive policy in the latter part of the eighteenth century.

In 1782 the Parliament of Great Britain renounced legislative power over Ireland, but in 1800, following dubious manipulation, the Irish Parliament united with that of Great Britain to form the Parliament of the United Kingdom, and thereafter Ireland elected members to the latter Parliament and was administered as a part of the United Kingdom. But within 30 years there was an organized Home Rule movement based on opposition to absentee landlords, and on famine and misery. Latterly there were roughly parallel constitutional Home Rule movements and revolutionaries, and the middle years of the nineteenth century are a confused tale of conciliation, repression, and reforms.

Gladstone introduced a Home Rule Bill in 1886 which would have transferred Irish administration to an executive appointed by an Irish Parliament. It failed, however, as did later measures until 1914 when a Home Rule Bill was passed but suspended by reason of the War. In 1918, though elected to the U.K. Parliament, 73 Sinn Fein supporters refused to take their seats there, organized themselves as the Dail Eireann, passed a declaration of independence, and appointed ministers. In 1920, after a rising in Dublin in 1916 and during a period of guerilla warfare and terrorism, 1919–21, the Government of Ireland Act, 1920, was passed which divided Ireland for political purposes into two areas, the 26 southern and western counties and six north-eastern counties, the major part of Ulster, each with a Parliament. For the legal structure of the latter see *Northern Irish Law*. There was also a Council of Ireland to co-ordinate the work of the two Parliaments.

An Anglo-Irish treaty of 1921 between Parliament and the Dail and resultant Act of 1922 gave Ireland the political status of the dominion of Canada, under the name of the Irish Free State (Saorstat Eireann), with liberty to the six counties of the north-east to opt out, which they did. The Irish Free State was subsequently called Eire and in 1949 Eire became the Republic of Ireland and ceased to be part of Her Majesty's dominions, but it is not in British law a foreign country.

Constitution

The Irish Free State came into being as a British dominion in 1922. In 1937 it became Eire, a sovereign, independent, and democratic state, and in 1949 the Republic of Ireland. The original constitution was framed in 1922 and comprised a Governor-General representing the King, advised by an executive council, and a bicameral legislature (Oireachtas), consisting of the Senate (Seanad Eireann) and the chamber of deputies (Dail Eireann). The Senate was abolished in 1936.

A new written constitution was adopted in 1937 and ratified by plebiscite; it has been influenced by Irish nationalism, Catholic philosophy, and English and Irish parliamentary traditions. It declares Eire to be a sovereign, independent, and democratic state. For three years the constitution might be amended by ordinary legislation, but thereafter a popular referendum has been necessary to amend the constitution.

The constitution recognizes the family as the unit of society with rights superior to all positive law. Private property is recognized as a natural right of man, though this and other personal rights may be limited by regard to the common good. The constitution enumerates directive principles of social policy, intended for the guidance of the Oireachtas, including the right of every citizen to a decent livelihood, the establishment of families in economic security on the land, and the distribution of property in the interests of the common good.

The Head of State is the President, elected by direct vote for seven years and eligible for re-election for a second term. He appoints the Prime Minister and, with the latter's advice and the approval of the Dail, appoints the other members of the government. He is advised by a council of state consisting of ex-presidents, present and former prime ministers and chief justices, and certain others. He may refer certain bills to the Supreme Court for a ruling on their constitutionality.

The legislature consists of the President and two houses, Seanad Eirann, which consist of 11 members nominated by the President and 49 elected by the Dail and county councils from panels representative of various interests, and the Dail Eireann, consisting of 144 members elected by adult suffrage under proportional representation, for a maximum term of seven years. The Dail alone can initiate money bills and the Seanad may make recommendations only on them. The Seanad cannot veto legislative proposals and has a limited time for consideration or amendment of bills passed by the Dail.

The executive comprises from seven to 15 ministers of whom the Prime Minister, his deputy, and the minister of finance must be members of the Dail; the others must be members of one or other house, but not more than two of the Senate.

Local government is organized through county councils, county borough and borough corporations, urban district councils, and town commissions.

The Civic Guard is a nationwide police force under a commissioner responsible to the minister for justice.

Judicial organization

Courts were early established on the English model, of King's Bench, Common Pleas, Exchequer, and Chancery. Under the Irish Judicature Act of 1877 these were amalgamated into the High Court and Court of Appeal.

Under the constitution of 1922 the office of Lord Chancellor of Ireland was abolished and new courts, the Supreme Court and the High Court were constituted. The structure under the Constitution of 1937 now comprises the Supreme Court, which has appellate jurisdiction from all decisions of the

High Court and, in certain cases from decisions of the Court of Criminal Appeal, a Court of Criminal Appeal consisting of a Supreme Court judge and two High Court judges; the High Court which has original jurisdiction in all matters, civil and criminal, and which in the exercise of criminal jurisdiction is known as the Central Criminal Court, consisting of a high court judge and jury and having unlimited criminal jurisdiction; eight Circuit Courts which deal with civil business of medium importance and most criminal offences other than treason or murder; and 24 District Courts dealing with lesser civil business and lesser crimes. The Circuit Court acts also as an appeal court from the District Court. The district judges have replaced the former justices of the peace. Juries are in use in both civil and criminal cases, except minor ones.

The Supreme Court may also be requested to determine whether a Bill is repugnant to the Constitution.

There are also a number of administrative tribunals and a Land Commission dealing with the purchase and subsequent settlement of land.

Criminal law

The criminal law is basically the English common law system, with divergences and statutes special to Ireland.

Private law

Prior to 1922 Irish law was basically a common law system, basically similar to but differing in certain particulars, particularly affecting real property, from English law. Since then there has been increased divergence from English law in detail, since Westminster no longer has power to legislate for Ireland whereas the Oireachtas has. Legislation made by the Parliament of Ireland (to 1800), and the Parliament of the United Kingdom (1801–1922) is supplemented and modified by legislation by the Oireachtas.

Legal profession

A separate legal profession in Ireland developed early. King's Inns, Dublin, originated about 1300, taking its name from a gift by Henry VIII of the monastery of Friar's Preachers for the use of the professors of the law in Ireland. That site having been occupied by the Four Courts, the present hall was built in 1800.

Down to 1866 both intending barristers and intending attorneys and solicitors studied at King's Inns together.

The Irish Bar maintains a distinction between barristers and solicitors, but has changed the dignity of King's (or Queen's) Counsel to that of Senior Counsel.

As a small jurisdiction Ireland is not well provided with legal literature but has to rely heavily on old books or on English books.

E. Curtis, *History of Mediaeval Ireland, 1086–1513*; E. Curtis, *History of Ireland*; R. Bagwell, *Ireland under the Tudors*; R. Bagwell, *Ireland under the Stewarts*; J. O'Connor, *History of Ireland, 1798–1924*; W. Lynch, *View of the Legal Institutions established in Ireland during the reign of Henry II*; H. Richardson and G. O. Sayles, *The Irish Parliament in the Middle Ages*; J. MacNeill, *Constitutional History of Ireland*; J. MacNeill, *Early Irish Laws and Institutions*; F. E. Ball, *The Judges in Ireland, 1221–1921*; B. T. Duhigg, *History of the King's Inns in Ireland*; J. McNeill, *Studies in the Constitution of the Irish Free State*; L. Kohn, *The Constitution of the Irish Free State*; N. Mansergh, *The Irish Free State*; P. O'Siochain, *The Criminal Law of Ireland*; R. Sandes, *Criminal Law in Ireland*; Sweet and Maxwell, *Legal Bibliography*, vol. 4: *Irish Law*.

Irish system. A penal regime developed in the 1850s by Sir Walter Crofton. It emphasized training and performance as means of reform, and consisted of three phases, solitary confinement, work in association with advancement for industry and good behaviour, and finally a period with minimal supervision during which the prisoner showed his reliability and capacity for employment in the outside world. Release depended on continued good conduct and the offender could be recalled to prison if this seemed necessary.

Irnerius (1055–*c.*1125). Italian jurist, sometimes called "lucerna iuris", the most famous of the early teachers at the law school of Bologna, whose teaching greatly stimulated the revival of the study of Roman law, and founder of the school of Glossators (q.v.). He seems to have been the first systematically to collate and compare texts and draw conclusions therefrom. Under him Bologna became the leading law school in Europe. He is believed to have taught Bulgarus, the greatest of the next generation of legal scholars. He wrote numerous glosses, summaries of Lombard law and Justinian's Code (*Summa Codicis*), and treatises on legal formulae and interpretation have been attributed to him.

Irregular marriage. In Scots law a marriage constituted without formalities or ceremony, now competent only by cohabitation with the habit and repute of being married, but formerly also by exchange of consent to take each other as spouse, or by promise of marriage followed by intercourse on the faith thereof (promise *subsequente copula*).

Irresistible impulse. An impulse to do an act which the actor cannot control, even though he may know the nature and quality of his act and possibly also that it is legally wrong. If the act is criminal, e.g. killing, the person so acting is not under British law exempted from liability on the ground of insanity, nor is it even a symptom of insanity. It is impossible to distinguish an impulse irresistible by reason of mental illness and one irresistible because of jealousy, hatred, or revenge, or from one not adequately resisted.

Irritancy. In Scots law the determination or forfeiture of a right in consequence of a failure to comply with, or acts in contravention of, the express or implied conditions attaching to the right, as a compulsitor for observance of these conditions. Legal irritancies exist in certain cases, such as the forfeiture of a feu right for five years' non-payment of feu-duty, and in other cases a conventional irritancy may be incorporated in a contract to provide against such contingencies as the parties may agree, such as forfeiture of a lease if the tenant becomes bankrupt. Enforcement of irritancy requires a decree of court and legal, but not conventional, irritancies may be purged up till final decree by payment of the arrears due.

Isaacs, Sir Isaac Alfred (1855–1948). Called to the Victorian Bar in 1880 he entered the state legislature in 1892 and was Solicitor-General (1893) and Attorney-General (1894–99 and 1900–1). He became a member of the Australian Federal Convention in 1897 which framed the Australian Constitution. He entered federal politics, was Attorney-General of Australia 1905–6, when he became a judge of the High Court of Australia and in 1930 Chief Justice. As a judge he established a great reputation, particularly in constitutional cases. In 1931 he became the first Australian Governor-General of the Commonwealth and served till 1936.

Isaacs, Sir Rufus Daniel, Earl and Marquess of Reading (1860–1935). Left school at 14 and worked as a ship's boy and a stock-exchange jobber before being called to the Bar in 1887. He rapidly became busy and took silk in 1898, and was thereafter in the first rank of counsel. In 1904 he entered Parliament, and in 1910 became Solicitor-General and, six months later, Attorney-General. He was also heavily engaged in government business, such as the Health Insurance Act, 1911. In 1913 he was appointed Lord Chief Justice and in 1914 Baron Reading. In 1915 he led a mission to the United States seeking American credits for war supplies. For much of the rest of the war he was employed as a negotiator with the United States and in 1918 he became ambassador to that State. In the intervals of this public service he continued as Chief Justice. In 1919 he resumed work as Chief Justice, but in 1921 he was appointed Viceroy of India, with the main function of putting into effect the early stages of introducing Indians to positions of responsibility. He returned in 1926 having discharged his duties with distinction in difficult times; he received a marquessate for his services in India. Thereafter he undertook business posts but in 1931 held office briefly as foreign secretary. A man of outstanding ability, his manifold public services in other spheres prevented him from showing his potentiality as a judge to the full.

Lord Reading, *Rufus Isaacs, first Marquess of Reading.*

Isidorus Ispalensis (Saint Isidore of Seville) (570–636). Bishop of Seville, restorer of the Church in Spain, theologian and a great scholar, author of *Etymologiae*, an encyclopaedia in 20 books of all secular and religious knowledge then available, histories, commentaries on scripture, and other works. He may have been responsible for the original edition of the *Hispana collectio*, sometimes known as the *Collectio canonum Isidoriana*, the collection of the canons of the Spanish Church. In the ninth century an enlarged edition of the *Hispana* was falsely attributed to Isidore, and is now known as the *Pseudo-Isidorian Decretals.* He was canonized in 1598 and declared a doctor of the Church in 1722.

Islamic law (sometimes called Mohammedan law). Islamic law (Shariʿah) is a leading example of the group of religious laws, which apply to persons not by virtue of citizenship of or domicile or residence in a country, but by virtue of professing the religion. Islamic law is a branch of Muslim Theology, giving practical expression to the faith, which lays down how a Muslim should conduct himself in accordance with his religion, both towards God and towards other men. It is accordingly applicable between Muslims only. It is based on man's duties or obligations rather than on his rights, and the sanction is that he will fall into sin. It has a wider application than any secular system of law since it claims to regulate all aspects of a man's life, his duties to God, to his neighbour and to himself. It originated in the decisions of Mohammed himself sitting as a judge at Medina.

Muslim theologians and jurists have developed a very detailed body of law on the basis of divine revelation and this has been a powerful influence in developing the community and social life of Muslim peoples. By the tenth century there was a corpus of doctrine which had been formulated in legal manuals and was later systematized and elaborated in numerous commentaries.

The sources of Islamic law are: (a) the Koran, the sacred book of Islam, (b) the Sunna, or model behaviour of the Prophet, (c) the Ijma, or consensus of scholars of the Muslim community, and (d) the Kiyas, juristic reasoning by analogy. The Koran is the collection of the revelations of Allah to Mohammed, the last of his prophets. It is the main source of Islamic law and contains many precepts of a moral character, but is not a code of Islamic law. Authoritative interpretation of the Koran has been made by the doctors, and the Muslim judge must refer to their works for interpretation. The Sunna contains the way of life of the Prophet; it consists of the collected tradition of the acts and statements of Mohammed, particularly as collected by two great doctors in the ninth century. The Ijma is the unanimous agreement of the legal scholars on solutions to problems not provided for by the Koran or the Sunna. By merging traditional custom and practice to give a principle, or rule, or an institution, the consensus of Islamic doctors and legal scholars gives a legal solution the force of truth. It was not required of the earlier scholars that they be wholly unanimous, and different schools, called 'rites' have been admitted, each constituting a particular interpretation of Islamic law. There are four orthodox rites, the Hanafi, the Maliki, Shafi'i, and the Hanbali, which generally prevail in different countries. Among the heretical rites are the Shi'ite, the Wahhabi, and the Abadi in other countries. The rites resemble each other on most matters of principle, but differ on many points of detail. Of these sources the Koran and Sunna are fundamental, but are today only historical sources, and in modern practice only Ijma need be sought; it is the dogmatic basis of Islamic law, and it is the duty of Muslims to obey the doctors. In consequence the functions of legal science consists of expounding the classics and the solutions of the great doctors of the past. Finally, analogical argument and reasoning are accepted as means of interpreting and applying the law, which can usually find a solution for a particular case. There is no question of developing or modifying the rules, but only of reasoning from them. Though immutable, Islamic law has room for custom and contractual rights of parties and outmoded solutions may be avoided by legal stratagems and fictions.

Diversity in legal practice developed as Islam spread and local cadis or judges placed different interpretations on particular texts, and variations in local customary practices were recognized.

Some states, chiefly in North Africa, affirm in their constitutions adherence to Islamic law, but no country is governed exclusively by Islamic law. Even predominantly Muslim countries have their bodies of secular law as well and these vary very much, some being basically of the civil law family, others of the socialist family. The law of ritual and religious behaviour, of persons and of the family, and of succession, have always been considered closely connected with religious belief, but in other respects secular law has been accepted, and both secular and religious courts exist in such countries.

In some Islamic countries codification, even of family and personal law, has been effected though such codes have generally been based on fundamental concepts of Islamic law. In some cases the dualism of jurisdiction has been abolished. In many countries the law has been substantially influenced by western forms and concepts but, except where the influence has been socialist law, it is unlikely that the Muslim tradition or element in the law will be eliminated.

N. J. Coulson, *History of Islamic Law* (1964); J. Schacht, *Introduction to Islamic Law* (1964) and *Origins of Muhammadan Jurisprudence* (1950); J. Anderson, *Islamic Law in the Modern World* (1959) and *Islamic Law in Africa*; A. Fyzee, *Outline of Muhammadan Law* (1964); Abdur Rahim, *Muhammadan Jurisprudence* (1911); F. Ziadeh, *The Philosophy of Jurisprudence is Islam* (1961).

Islands. Areas of land entirely surrounded by water. Each has its territorial sea and its continental shelf. In international law, problems arise of whether pieces of land are sufficiently free from tidal inundation to rank as distinct islands, and of the effect the existence of islands has on defining the base-line dividing the internal waters of a state from its territorial waters. Islands may also be independent states, e.g. Iceland, Australia, or mere detached parts of the territory of a state such as the Hebrides. They may also be divided between two or more states, one or more of which may have other territory, e.g. Ireland.

Isle of Man. See MANX LAW.

Issue. An outcome or consequence; the legitimate children and other descendants of a person.

In pleading, parties are said to be 'at issue' when each has answered the other's pleadings and they have arrived at certain points on which they are not in agreement. Such points are called the issues in the action. Under the old common law system of pleadings, where the defendant denied all the material allegations in the declaration or the main fact on which it was founded, he was said to plead 'the general issue'. The general issue in most of the common kinds of actions has a special name, such as *Non assumpsit.* In criminal procedure a prisoner is said to plead the general issue when he pleads Not Guilty.

In Scottish civil practice an issue is a question formulated in writing to be answered by a jury, the answer to which is decisive of the action.

Italian law. Down to the collapse of the Western Roman Empire in A.D. 476, the history of early Italian law is the history of Roman law (q.v.). Thereafter it is the history of the Ostrogoths, and then of the Lombards. As masters of Italy the Ostrogoth Theodoric and his successors issued the

Edictum Theoderici, and the subsequent minor Ostrogothic edicts, which sought to merge Goths and Romans. That kingdom ended with the reconquest of Italy by Byzantium and the extension of Justinian's legislation to Italy. But Italy soon again fell to invaders, the Lombards (A.D. 567), who in turn soon enacted legislation, though it was influenced by the superior culture of the conquered people. The *Edict* of Rothar (643) is a work of some maturity, and was supplemented by his successors, the whole becoming known as the *Edictum regum Langobardorum*. The principle of personality of law applied, and Roman law survived in the areas of Italy remaining under Byzantine rule. In Lombard Italy, the Church and much of the subject population continued to live under Roman law as being their personal law.

Later when, after 774, Italy was an autonomous kingdom within Charlemagne's empire, both the general capitularies of the Frankish empire and special capitularies for Italy were enacted, many of which were collected in the *Capitulare Italicum*. It came to be joined to the Lombard *Edict* in the eleventh century. Despite these changes much Roman law persisted, as native custom.

From an early date an important school of Lombard law developed at Pavia. The Lombard *Edict* was edited there, about 830, as the *Capitula legis regum Langobardorum* and again, about 1090, as the *Capitulare Italicum*. The last collection, the *Lombarda*, appeared about the same time. About 1050 a systematic treatise, the *Liber legis Langobardorum* or *Liber Papiensis* was written containing the *Capitulare Italicum* and the Lombard *Edict*, not merely a collection but a commentary and a treatise for practice. It rapidly attained vogue all over Italy. Lombard law became the dominant law of Italy and even prevailed against Roman law. Generations of legal study at Pavia produced the *Expositio ad Librum papiensem*, a large collection of glosses consolidated in the eleventh century. A treatise on the differences between Roman and Lombard law was written as late as 1541.

The building of the canon law (q.v.) is also part of the history of Italian law. The home of the papacy in Italy for practically all its history has made the development of that body of law particularly connected with Italy. Moreover, Roman law sources were increasingly used in the compilations of canon law from about the ninth century; a notable example was the *Collectio Anselmo dedicata*.

The rediscovery of the Roman law and the rise of the law school of Bologna is the genesis of both Italian law and the *jus commune* of Europe. Based on Roman law, as developed and interpreted by the Glossators and Commentators (qq.v.), and on canon law it was later received by much of Europe as the civil law of the Holy Roman Empire.

Italian law, as a distinct system, developed with and from the revival of the study of the Roman law from about the twelfth century, in the labours of the Glossators and Commentators. As the feudal period and the Germanic empire came to an end in Italy the principalities and dukedoms in Italy gained political independence. Customs varied from place to place. City-state legislation was characteristically by means of *Statuti*, based on local custom, frequently framed by one or more commissioners and approved by the council or popular assembly. Thus Genoa recorded its customs in a code of statutes in the tenth century, Pisa in its *Constitutum usus Pisanae civitatis* (1160), and Milan in its *Consuetudines Mediolanenses* (1216). Later statutes innovated on customs. Cities frequently modelled their statutes on those of another place. In the northern cities statutes were largely occupied with public law, consuetudines with private law, but in the south city codes dealt more with private law. Elements of Roman, canon and Germanic law were merged in city statutes. From about 1200 they were in Italian rather than Latin.

In particular regions there were city statutes, statutes of the city administration, a statute of the common people, and statutes of the arts, trades and crafts, notably of the merchants.

The maritime ports of Amalfi, Pisa, Genoa, Trani, Ancona, and Venice, developed codes on maritime law of great importance for later maritime law. The *Capitula et ordinationes curiae maritimae nobilis civitatis Amalphae*, or *Tabula Amalfitana*, was compiled between 1100 and 1350. Trani had an *Ordinamento et consuetudo maris* of between 1063 and 1200. Venice's *Capitulare nauticum* of 1255 was revised periodically. Pisa and Genoa had maritime codes reduced to writing in the twelfth and later centuries.

Though the *Libro del consolato del mare* was published at Barcelona in 1494 it was based largely on statutes of Italian cities.

There was also legislation from the states to which some cities were subordinate and from the Empire, but the imperial authority waned rapidly after the thirteenth century and thereafter only exceptionally did imperial legislation affect Italy, though the *Constitutio Carolina* of Charles V had force in northern Italy as a code of criminal procedure.

The legislation of the Italian states differed greatly in different regions, and was largely devoted to eliminating ancient contradictions and substituting more general rules for local custom and special privileges. Thus, in Naples and Sicily, the Lombards were succeeded by Normans who enacted the *Assisae regum regni Siciliae* or *Assisae di Ariano Irpino* in the twelfth century, and in turn by Swabian emperors who enacted the *Constitutiones regni Siciliae*, frequently called the *Liber Augustalis*, in 1231. This survived in force until the eighteenth century. Later the Angevins at Naples introduced *Capitula* for French subjects there, and the Aragonese their *Pragmatica* and *Capituli*. In the Papal states, Roman and canon law were the chief sources, but localities and provinces had their own legislation. Thus, in the Marches, there was a body of *Constitutiones Marchiae Anconitae* (1357) promulgated by the papal legate Cardinal Albornoz, which applied for two centuries. The principalities later united in Savoy were given Decreti by Amadeus VIII in 1430, a forerunner of the comprehensive code, the *Constitutiones*, of Victor Amadeus II of 1723, while in Lombardy *Constitutiones dominii Mediolanensis* of 1541 continued in force till the end of the eighteenth century. Venetian law owes much to the Doge Iacopo Tiepolo who revised the criminal code, compiled the civil statutes, and the code of procedure in 1232–44. These, revised frequently, lasted till the end of the eighteenth century.

In the fifteenth and sixteenth centuries, humanism gave rise to a reaction against the methods of the Commentators, and Italian legal science lost its pre-eminence. The centres of humanist study were in France. Law in Italy generally remained mediaeval, supplemented by a disorderly, unsystematic, mass of legislation.

In the period since 1700 there are three distinguishable strands: private compilations of law to alleviate the evils of the confused masses of customs and legislation of different periods; official compilations of common law; and finally codification.

Beccaria (q.v.) sought to reform the criminal law and remove its cruelties and other defects; Filangieri to build a complete and rationalized science of legislation; Spedalieri tried to reconcile the philosophical beliefs of his times with the doctrines of the Catholic church; and Romagnosi sought to coordinate the social sciences. The latter part of the eighteenth century saw a widespread tendency of governments to reform government on the principles of such social and legal thinkers; in consequence there were numerous tentative measures of codification. Thus, in Tuscany, a penal code was enacted in 1786; in Modena, a code in 1771; in Lombardy, a code of practice in 1786; in Venice a criminal code (1751), a feudal code (1770), and a maritime code (1786).

The French Code Napoléon of 1804 was extended to all the Italian states as they came under French domination after 1796, and when Napoleon was overthrown his legislation, in some measure, survived. In the various duchies, codes were produced at various times in the next 50 years, all deriving something from the Napoleonic legislation. The Austrian Code of 1811 was put into force in Venice and Lombardy in 1816. In Piedmont, codes were enacted, some being extended to all Italy after unification, though Tuscany retained its own laws for a number of years. After the political unification of Italy in 1861 there were promulgated, in 1865 and subsequent years, a Civil Code (1865), Code of Commerce (1882), Penal Code (1889), and Codes of Civil (1865) and Criminal Procedure (1865). A new criminal code was enacted in 1890. A new Civil Code, comprehending both civil and commercial law, was adopted in 1942. Strictly speaking, it is only since 1861 that it is possible to speak of Italian law.

Modern constitution

Italy became a republic in 1946 and, in 1948, adopted a new constitution watched over by a Constitutional Court. The head of state is the President, elected for a seven-year term. Legislative power is vested in the government and the two chambers of the legislature: Senate (elected regionally), and Chamber of Deputies (elected by universal adult suffrage for five years). The principle of cabinet responsibility applies.

Local government is exercised through regions, provinces, and communes. The regions have certain legislative and administrative rights in local matters exercised through regional councils, provincial councils, and communal councils.

Judicial system

The judicial system forms a unified network, headed by the Court of Cassation (*Corte Suprema di Cassazione*), having appellate jurisdiction and dealing with points of law only; below it are 23 appeal courts districts (plus three detached appeal court districts), 35 Assize Courts of Appeal districts, 94 Assize Court districts, 159 tribunal districts, and nearly 900 magistrates' courts areas. Civil cases are initiated in magistrates' or district courts (*tribunali*) according to value and may be appealed to the appropriate court of appeal and, on a point of law, to the Court of Cassation. Criminal cases are brought in magistrates' or district courts or Assize Courts, and may be appealed to the Court of Appeal or the Assize Court of Appeal and, on a point of law, to the Court of Cassation.

There are also district administrative courts concerned to protect individual interests connected with public interests, and to supervise and control public funds. The superior administrative courts are the judicial sections of the Council of State (*Consiglio di Stato*). Courts-martial have jurisdiction over members of the forces, and there are special tax courts and a Court of Audit.

A Constitutional Court of 15 judges pronounces on the constitutionality of laws and decrees.

The judiciary are career judges who are civil servants, and there is interchangeability in the lowest courts between the bench and the position of prosecuting counsel. There is a Public Prosecutor's Department with functions in criminal and some civil cases. In the superior courts it is represented by Procurators General and Procurators of the Republic.

The law

The sources of the modern law are the Constitution, statutes, regulations, corporative norms such as collective agreements, and usage. But these are supplemented by doctrinal interpretation by scholars, authentic interpretation, and judicial decisions, which are, formally, never binding. In fact, the work of legal scholars dominates Italian legal process, and the reception of German legal science has profoundly influenced Italian legal doctrine.

Italian law is divided into public law and private law: the former comprising constitutional, administrative, ecclesiastical, financial, criminal, and procedural law; the latter, civil and commercial law. Much of the law is codified—the conceptual basis being Roman law. The Civil Code of 1942 consists of some general provisions (Arts. 1–31) and six books dealing respectively with Persons and the family (Arts. 1–455), Succession (Arts. 456–809), Property (Arts. 810–1172), Obligations (contract and particular contracts, unjust enrichment and torts) (Arts. 1173–2059), Labour, companies and partnership, industrial property and competition (Arts. 2060–2642), and Protection of rights (evidence, security rights, prescription) (Arts. 2643–2969). There are also a Code of Civil Procedure (1940), Penal Code (1930), Code of Criminal Procedure (1930), and Navigation Code (regulating maritime and air transport) (1942). There is no longer a separate commercial code. Much public law is uncodified and the codes cover only part of the law. There are several Unified Texts of Law containing consolidations of statutes on specified topics. Other matters are governed by individual statutes.

Legal profession

The profession is divided into judges, lawyers, and notaries. Lawyers qualify initially as attorney (*procuratore legale*) and after a period may become an *avvocato* entitled to practice before all courts except the Court of Cassation, for which a longer period of experience is required. Judges are career civil servants, not promoted from the bar. The governing body of the judiciary is the Superior Council of the Magistracy. Notaries draft and authenticate many kinds of legal instruments and are frequently appointed by court to look after property. There is almost no lateral movement between the branches of the profession. Law is very widely studied in Italy since a legal training is commonly the mode of approach to careers in the civil service, administration, and industry.

General: M. Cappelletti, J. Merryman and J. Perillo, *The Italian Legal System*; *Enciclopedia del Diritto*; *Enciclopedia Forense*; A. Grisoli, *Guide to Foreign Legal Materials—Italian*. History: C. Calisse, *Storia di diritto italiano* (1903–30); A. Pertile, *Storia del diritto italiano* (1890–1903); G. Salvioli, *Storia del diritto italiano* (1930). Public law: C. Mortati, *Istituzioni di diritto pubblico* (1962); A. Tesauro, *Istituzioni di diritto pubblico* (1960). Criminal law: F. Antolisei, *Manuale di diritto penale* (1959–60); V. Manzini, *Trattato di diritto penale* (1950–52). Criminal procedure: V. Manzini, *Trattato di diritto processuale penale italiano* (1956). Private and commercial law: F. Messineo, *Manuale di diritto civile e commerciale* (1959); A. Trabucchi, *Istituzioni di diritto civile* (1962); R. de Ruggiero and F. Mario, *Istituzioni di diritto civile* (1947); A. Torrente, *Manuale di diritto privato* (1960); C. Vivante, *Trattato di diritto commerciale* (1929); L. Mossa, *Trattato del nuovo diritto commerciale* (1942–51). Civil procedure: F. Carnelutti, *Istituzioni del processo civile italiano* (1956).

Iter Camerarii. An old Scottish law-text dealing with the Chamberlain's administrative and economic surveys of the activities of the burgesses.

Itinerant justices. Shortly after the Norman Conquest William I began to send commissioners round the country as delegates of the *Curia Regis* to deal with various kinds of business: fiscal, legal, and other. From the time of Henry II justices regularly travelled the country, and by Glanvill's time the distinction between the *Curia Regis* and itinerant justices was well marked. They acted under various commissions. They heard civil and criminal pleas of the Crown under the assizes of Clarendon and Northampton, and had other administrative and fiscal duties. A set of *Capitula placitorum Coronae Regis* of 1194 lists a very wide and varied jurisdiction, foreshadowing that of the later justices in eyre and of the future judges of assize.

In the thirteenth and fourteenth centuries the various commissions issued to itinerant justices became settled, and they were issued at regular intervals. Those which early became obsolete were the General Eyre and the Commission of Trailbaston; those of continuing importance were the commissions of Oyer and Terminer, of Gaol Delivery, of the Peace, and of Assize (qq.v.).

The courts held by itinerant justices became integral parts of the English system for various reasons. One was that the practice developed of issuing the commissions of assize, oyer and terminer, and gaol delivery, at regular intervals. Thus, in 1272, England was divided into groups of counties, to each of which two justices were assigned. In 1285, it was enacted that assizes should be held three times a year, and in 1330 that they should be held three times a year or oftener if need be. Commissions of oyer and terminer came to be issued when the justices of assize were visiting the counties.

It was also partly because of the development of the *nisi prius* system (q.v.) which gave justices of assize a general jurisdiction over civil cases begun in the King's Bench or Common Pleas, and because the exercise of the various commissions directed to itinerant justices require the competence of the judges on the commission, who in time entirely superseded the laymen.

The justices of assize were, in practice, the judges of the common law courts, and since the commissions were issued regularly the courts held by them became as regular as the common law courts themselves. Moreover, under the *nisi prius* system judges represented the courts in which the actions had begun, and the same procedural rules came to apply to trials at assizes and to trials *in banc.*

The judges of assize could always adjourn a case into the appropriate court. In civil cases, a writ of error or a bill of exceptions lay to the court of error for that court. In criminal cases, a point might be reserved for the Court of Crown Cases reserved, or a writ of error lay for errors on the record.

The system of itinerant justices was and is the only way in which the highly centralized judicial system of England was satisfactorily workable, and in which uniform administration of justice throughout the country was practicable.

Iudex. In Roman civil procedure, actions proceeded in two stages: *in iure*, and *apud iudicem*. The first, before a magistrate who conducted the proceedings to the stage of *litis contestatio*, which was a formal agreement between the litigants under the magistrate's supervision and with his co-operation, and containing a formulation of the principal points of the matter in issue; and the second, before a *iudex*. The *iudex* was a private person from the higher social classes, the senators or, later, the *equites*, and not having legal qualifications. For a particular trial the *iudex* was designated by both parties or, failing agreement, chosen from a panel of persons entitled to sit. The parties' choice, if approved by the magistrate *in iure*, bound the *iudex* to act, unless he could produce acceptable reasons for exemption. The stage *apud iudicem* was conducted without any prescribed formalities; it comprised the pleading of the parties and their advocates, and the hearing of the evidence. The functions of the *iudex*, assisted during the trial by legal advisers, were to hear and evaluate the arguments and evidence and reach a decision, which he announced orally in the presence of the parties. He was liable to a party wronged by an unjust decision arrived at by negligence or unfairness; the praetorian edict gave a special remedy against the careless or corrupt judge (*judex qui litem suam fecit*).

In the last kind of Roman civil procedure, *cognitio extra ordinem* (q.v.), the judge was appointed by an imperial law officer alone or, later, by the emperor; he was now called *iudex datus, pedaneus* or *specialis*, and his jurisdiction depended on the terms of his commission.

Under the later Empire the term *iudex* was applied to any official having jurisdiction or administrative power.

Iudicium. In Roman law, the second stage of a litigation (that held before the *iudex* (q.v.)), though the term is often used of the whole trial, or of the written *formula*, or of the final act of the trial—the sentence.

Iulianus Aemilianus, Lucius Octavius Cornelius Publius Salvius (*c.* A.D. 100–*c.* 169). A Roman jurist and last recorded leader of the Sabinian school. At the age of 30 he was commissioned by the emperor Hadrian to reduce the praetorian edict to its definitive form. This was a work of consolidation rather than originality. He had a distinguished official career and was a member of the imperial consilium. He published commentaries on some lesser jurists, a monograph *De Ambiguitatibus*, and principally *Digesta*, a systematic treatise on civil and praetorian law in 90 books, frequently quoted in later literature and in Justinian's *Digest*; later jurists made new editions of it, with their own notes. He was a lucid and authoritative expositor and with Julian Roman legal science reached the height of its development. His restatement of the Edict won high praise.

Buhl, *Salvius Julianus*; Guarino, *Salvius Julianus.*

Iuridicus. In Roman law, a judicial functionary of praetorian rank, introduced by Marcus Aurelius in Italy (except Rome) in A.D. 163. Their jurisdiction extended to civil cases only, and was confined to one or more districts. Procedure was by *cognitio extra ordinem* (q.v.). They disappeared under Diocletian. The term is also used later, and in the *Digest*, for the *iuridicus Alexandreae*, an officer of extensive jurisdiction in Egypt. In other provinces, there were *iuridici* or *legati iuridici* with limited jurisdiction, also appointed by the emperor.

Iuris praecepta sunt haec: honeste vivere, alterum non laedere, suum cuique tribuere (The precepts of the law are these: to live honourably, to injure no man, to render to each his own). A quotation from Ulpian, in Justinian's *Institutions*, I, 1, 3, setting out some basic precepts or principle of law.

Iurisprudentes, iurisperiti, iurisconsulti. In Roman law these were jurists who advised individuals, magistrates and judges, assisted parties to litigation with pleadings and procedure, drafted forms of wills and deeds, and frequently also instructed younger men by teaching and discussion of problems. They were a class distinct both from the magistrates and judges, and from the advocates who represented parties before tribunals. The Republican jurists were of senatorial families and frequently held high offices in the state. Under the principate they often held high administrative offices. They exercised considerable influence on the development of the law, as members of the *consilia* or groups of advisers of *praetors* and other magistrates, advising them on their edicts and on the allowance of new *formulae*, and later as legal advisers of the emperors. As teachers and writers they created and developed the science and literature of the law.

Iurisprudentia. In Roman law, defined by Ulpian as the knowledge of divine and human things, the science of the just or unjust: the former indicating the knowledge of law when it was a monopoly of the priestly class; the latter defining the later jurisprudence, or the science practised by the lay jurists from the publication of the XII Tables to the decay of legal science in the fourth century A.D.

Jurisprudence was long a vocation of the nobility at Rome, and involved a number of different aspects: firstly the practical work of jurists in assisting litigants, giving opinions and drafting legal forms; secondly, in acting as advisers of magistrates, particularly *praetors*, and advising them on their edicts and the *formulae* they might approve, and later as advisers of emperors; thirdly, as advisers of *iudices* in deciding actual cases; and, fourthly, by teaching and writing, systematizing and developing legal science. The last is the function of most permanent importance and value, and reached its height in the first two centuries A.D. From about A.D. 250 it began to decline, though there was some revival particularly in the East in the fifth century.

See also JURISPRUDENCE.

Ius. In Roman law the term for law in general, or any major branch thereof, e.g. *ius civile* (a particular statute is a *lex* (q.v.)) or law as a subject of study, or a legal right, or a collection of rights, or the place where the praetor sat to administer justice, or judicial proceedings themselves. It fell to be distinguished from *fas*, divine law.

Ius was distinguished in many ways: into *ius publicum* and *ius privatum* (qq.v.), into *ius naturale, ius gentium* and *ius civile* (qq.v.), and into *ius scriptum* and *ius non scriptum.*

In Roman civil procedure, proceedings *in iure* were proceedings in the court of the praetor, as distinct from proceedings *in iudicio*, before the *iudex* whom he appointed to hear and decide the case.

Ius ad rem. A right to claim something from the debtor in an obligation, but not to claim it against persons generally.

Ius civile. A term in Roman law with various meanings. Originally the *ius civile* or *ius Quiritium* was the law for Roman citizens only and not applied to foreigners. It sprang from *leges* enacted by the *comitia, plebiscita* enacted by the *concilium plebis*, and juristic interpretation and custom. It was distinguished from the *ius honorarium* or *praetorium* developed by the magistrates, principally the praetors (*see* EDICT), and including law based on other sources such as *senatus consulta*, foreign law and principles deemed common to all free men (*ius gentium*).

Ius civile, comprehending both *ius Quiritium* and *ius praetorium*, was distinguished from *ius naturale* (q.v.), as positive law from natural justice, and from *ius gentium* (q.v.) as law applicable exclusively to Romans from law common to all peoples. It is also occasionally used as distinguished from criminal law.

In early modern times, as in the phrase *Corpus Iuris Civilis*, it means the whole law of the Roman state as contrasted with the *Corpus Iuris Canonici*, the law of the Roman church.

Ius cogens. A term derived from Roman law designating rules which may not be altered by contracting parties, as contrasted with *ius dispositivum*, those which may be so altered. In modern usage the term is most met with in international law, where there are no such peremptory rules imposed by customary international law, though there is nothing to prevent sovereign states from creating peremptory international law by treaty, and to a small extent this has been done, e.g. by the United Nations Charter. Limitations on the freedom of state action imposed by self-interest, common sense, or the desire to be well-regarded, are wholly distinguishable from *ius cogens.*

Ius disponendi. The right of disposition or of alienation. The phrase is used in relation to sale of goods, in the question whether the seller has reserved the right of property to himself notwithstanding that he has parted with possession of the goods.

Ius edicendi. In Roman law, the right belonging to certain higher magistrates of issuing public and imperative notices on matters within their jurisdiction, or forming part of their official business. Thus the censors issued edicts as to a forthcoming census; the consuls issued edicts to summon the senate and the comitia, and the aediles issued edicts regulating the public markets. The most important exercise of these powers was by the praetors, who had the supreme civil jurisdiction, and it became customary for them each, at the beginning of his year of office, to announce in this way the principles which, apart from the established rules from the *ius civile* (q.v.), he intended to apply in the administration of justice. It became the obligation of every praetor to state the rules which he would apply during the year, and the practice was to continue the practices from year to year; the part of the edict derived from that of his predecessor was *edictum tralaticium*. This became the *edictum perpetuum*, as distinct from *edicta repentina* (isolated orders issued from time to time during the year), and the edict became a permanent body of law. The praetor had no legislative power, but his control of the courts and of the forms of civil procedure enabled him to effect major developments in substantive law.

Ius gentium. In Roman law, was originally the principles deemed common to all peoples as distinct from those peculiar to Roman citizens. At first such principles were matters of morality only and not legally obligatory but, as commercial relations with non-Romans developed, *praetores peregrini* admitted non-Romans to the courts and adopted principles from foreign systems and generally accepted practices and customs. There thus grew up a great body of law free of the technicalities of the *ius civile*, a general system of rules regulating relations between free men, independently of nationality. About the same time Rome acquired provinces, and governors had to exercise jurisdiction over non-Romans. Thus there developed, through the edicts of the peregrine praetors and the provincial governors, a general body of rules governing relations between free men without reference to nationality—rules basically Roman but influenced by foreign ideas and free from formal elements. This body of law had influence on the law of contract more than on any other branch. In time the two bodies tended to coalesce, and after all peregrines were admitted to Roman citizenship (*Constitutio Antoniniana de civitate*, A.D. 212) it ceased to have any independent importance.

There was also a tendency to equate the *ius gentium* with the *ius naturale*, law established by natural reason among all men, independently of the specialties of any particular system.

The term *ius gentium* was also used later and, until modern times, of the principles applicable to relations between independent states, that is, what is now public international law.

Ius honorarium. *See* IUS PRAETORIUM

Ius in personam. A right availing against a determinate person only, such as a claim under a contract.

Ius in re aliena. A right which one person has in an object of another's property, such as an easement over his land, a right in security over his property, or a charge over it. Such a right derogates from or limits the proprietor's rights in his own property.

Ius in re propria. A right in an object of one's own property, such as to use, sell, consume or bequeath it. The totality of such rights as can be exercised in any object of property amount to the owner's right of ownership.

Ius in rem. A right availing against persons generally, such as a right of ownership, that nobody shall interfere with his ownership.

Ius mariti. The right of a husband to his wife's chattels or moveables which, at common law, he acquired on marriage. But a gift to or settlement on the wife might be made to her separate use and exclusive of her husband's *ius mariti*. The right was ended by the Married Women's Property Act, 1882.

Ius naturale. A conception originating in Greek philosophy and taken up by the Roman jurists, covering those principles of right and justice which natural reason instals in the minds of men, consistent with natural justice, *aequum et bonum*, as distinct from man-made rules. Natural law was supposed to be the same everywhere, common to all men, and consistent with life according to nature. The instances given in Greek and Roman philosophers as precepts of the law of nature are only vague moral maxims of conduct. The concept tended to be confused with *ius gentium*, the concepts common to men of all nations, and the two were frequently identified. But slavery, though reckoned as *ius gentium*, was usually deemed contrary to *ius naturale*.

See also NATURAL LAW.

Ius Papirianum. According to tradition, some of the earliest Roman laws attributed to the kings and later published by a pontifex named Sextus Papirius.

Ius praetorium vel honorarium. In Roman law the practice of praetors issuing an edict stating how they would exercise their jurisdiction, and what remedies they would give in particular circumstances, gave rise to a body of magisterial law, *ius praetorium vel honorarium*, existing along with the original *ius civile* and having the function of aiding, supplementing and correcting the *ius civile*. The praetors' real power to introduce new legal principles dates from the *Lex Aebutia* of possibly 241 B.C. which practically superseded the old strict forms of *legis actiones* by new forms, in which the praetor was enabled to grant *actiones honorariae* not based on the civil law.

Ius primae noctis. An alleged mediaeval custom, said by some to have existed in the Pyrenees, said by others to be a more widespread custom, whereby the lord of the district was entitled to enjoy a tenant's bride on her wedding night. There is no evidence in any authentic record for the existence of any such custom. It has been suggested that it was truly a dispensation from the obligation of continence for the first three nights of marriage imposed by the early Church on newly married couples.

Ius privatum. In Roman law, private law, which pertains to the convenience of individuals, as distinct from public law. In the Roman view, private law comprised the law of persons, things, obligations and succession, and also civil procedure. It was said to be derived from precepts of the *ius naturale*, or of the *ius gentium*, or the *ius civile* (qq.v.).

Ius publicum. In Roman law, public law, which regards the standing of the Roman state, as distinct from private law. In the Roman view, public law comprised constitutional and administrative law, dealing with the location and exercise of sovereign power, the *ius sacrum*, or law concerning religion, and criminal law.

Ius quaesitum tertio. In Scots law, the right possessed by a third party to a contract, if the terms of the contract indicate that it was an object of the contract to benefit the third party, and he is named or referred to therein, giving him a title to enforce the contract as against an original contracting party.

Ius relictae. In Scots law, one of the legal rights (q.v.) in succession, recognised at common law, being the right of a widow to one third of her late husband's moveable estate, if he left widow and children, or to one half if he left widow alone. Certain statutory rights now have priority to the widow's claim of *ius relictae*.

Ius relicti. In Scots law, one of the legal rights (q.v.) in succession, introduced by the Married Women's Property (Scotland) Act, 1881, giving a widower the same rights in his wife's moveable estate as she would have had in his (see *ius relictae*). Certain statutory rights now have priority over his claim of *ius relicti*.

Ius respondendi. A major activity of learned lawyers at Rome under the Republic was the giving of *responsa*, advice to private individuals on the law applicable to some matter, or to lay judges in doubt as to the rule to apply. A litigant might have taken a jurist's opinion and give evidence of it to the judge trying his case. Such opinions were naturally influential in developing the law. *Responsa* were normally given in relation to real cases but might be in relation to a hypothetical case. From the early years of the Empire *responsa* became a more important source of law, and Augustus distinguished some of the more eminent of the jurists by giving them the *ius (publice) respondendi*, the right to give responses with the emperor's authority. Most of the great jurists of the principate had this right, but it is known that Gaius did not have it. Probably all those who published works called *Responsa* had it, and most of the jurists quoted in the *Digest* did. It probably did not make their *responsa* binding on a judge but inevitably made them more highly persuasive. Moreover, *responsa* were now in writing and given sealed so that they could be transmitted unopened to the judge. Indeed, Gaius defines the right as pertaining to those who have been given permission 'to lay down the law', and it may be that Hadrian made *responsa* legally binding on judges, or that he decreed that the common opinion of authorised jurists in their writings was to be taken as law. In this way the jurists were the principal agents of legal development during the principate. The last jurist known to have the *ius respondendi* was Modestinus (q.v.), and as the Empire became more autocratic there was no place for lawmaking by authorised opinion.

Ius scriptum. In Roman law, all the parts of the law which were reduced to written form, including *leges*, *plebiscita*, *senatus consulta*, and imperial constitutions, themselves comprising *edicta*, *mandata*, *decreta*, and *rescripta*.

Ius non scriptum. In Roman law, the unwritten part of the law comprising the customary law of Rome and originally the whole of the *ius gentium*, and subsequently such parts of it as did not become *scriptum* by being included in the edict or other legislation.

Ius tertii. The claim by a person prima facie liable to one, in defence to a claim by him, that the money or property claimed does not belong to him but belongs by superior right to another party. In general, a wrongdoer cannot set up *ius tertii*; thus an agent cannot refuse to account to his principal by setting up *ius tertii*, unless he does so by authority of a third party.

Iusiurandum. An oath. In some trials in Roman law, a litigant's oath could end litigation, which was at the stage of *in iure*, if the plaintiff offered an oath to the defendant and the latter swore that he did not owe anything to the plaintiff. The defendant might, however, refer the oath back to the plaintiff, who then had to swear to his claim, losing the case if he refused. In special cases a *iusiurandum calumniae* was required, whereby on the demand of one party the other was compelled to swear to the good faith of his claim or defence. Justinian made this obligatory for the parties' advocates. At the stage of proceedings *apud iudicem*, the oath was useful only as evidence or for the valuation of the object in issue. An extra-judicial oath by a party affirming the truth of his claim gave rise to a special action based on this fact alone.

Iustitium. In Roman law a temporary suspension, which might be proclaimed by the Senate, of all exercise of jurisdiction and all actings by magistrates, judges, and courts in both civil and criminal causes, on account of any event disturbing all public life, such as a national calamity, riots, or the like.

Ivo of Chartres, Saint (*c.* 1040–1116). Bishop of Chartres and deemed the most learned canonist of his age. There are attributed to him three works. The *Tripartita* (*c.* 1093–94) comprises a group of 655 fragments of decretals, 789 conciliar canons or patristic texts, and 861 fragments of his own *Decretum*. The *Decretum* (*c.* 1094) is a compilation of 3760 chapters in 17 parts based on Burchard's *Decretum*, his own *Tripartita*, and other collections. It was very successful and valuable by reason of its order, brevity, and the precision of the summaries, many of which were copied by Gratian. It was, in fact, a compendious encyclopaedia of canon law at the end of the eleventh century.

J

(For Roman law terms and Latin words frequently spelled with J, e.g. *Jus*, see under I.)

J. Justice. The title of one of the ordinary judges of the old Courts of Queen's Bench or Common Pleas, and of the modern High Court.

J.C. (or *Jctus*). *Jurisconsultus*, a jurist or learned lawyer.

J.P. Justice of the peace.

Jack the Ripper. Murderer of seven prostitutes in Whitechapel, London, between 7 and 10 August, 1888. In each case the throat was cut and the body was usually mutilated in a way indicating extensive knowledge of anatomy. The criminal authorities received taunting notes from a person calling himself Jack the Ripper and claiming to be the murderer. The case remains unsolved and gave rise, at the time, to great public outcry, and has subsequently provided themes for various books and films.

Jacob ben Asher (*c.* 1280–*c.* 1340). Son of Asher ben Jehiel, chief rabbi of Toledo, and author of the standard codification of Jewish law, the *Arbaʿah Turim*. He sought to restore the legal casuistry destroyed by Maimonides (q.v.). His work remained the standard one until the publication of that of Joseph Karo in 1555, and became the leading work after the Bible and the usual basis for rabbinic decisions. It based most of its statements on post-Talmudic rabbinical authorities, rather than on the Talmud itself.

Jacobazzi, Domenico (1444–1527). Italian cardinal and canonist, auditor, and later, Dean of the Rota. He wrote a comprehensive *De concilio* during the Fifth Lateran Council, defending the papacy against conciliar power.

Jacobinus de San Georgio. (Thirteenth century.) A commentator, author of *De Feudis* (1584); possibly the same as Jacopo di Colombo, who glossed the *Libri feudorum*.

Jacobs, Giles (1686–1744). A diligent compiler of legal works, was author of *Lex Mercatoria: or the Merchant's Companion* (1718 and 1729) dealing with the law of shipping, factors, prize, piracy, and related topics. It was intended as much for merchants as for lawyers and included precedents, a collection of leading cases, a summary of the laws of Oleron and other information. Though small, it was a useful book. He also wrote *A Treatise of Law* (1721) giving a general introduction to the Common, Civil and Canon laws, *The Modern Justice* (1716), *Lex Constitutionis* (1719) surveying the main features of the constitution, *The Student's Companion* (1725), a valuable *Law Dictionary* (1729), and a *Law Grammar* (1744). Of these the Law Dictionary has been of the most permanent value. His fondness for alphabetical arrangement lessened the value of some of his works. Other works included the *Compleat Court Keeper* (1713), a practical book for the guidance of stewards of manors, *The Accomplished Conveyancer* (1714), and *The Compleat Chancery-Practiser* (1730).

Jacobus (?–1178). One of the 'four doctors' who followed Irnerius in teaching Roman law at Bologna, establishing the glossatorial method. He wrote glosses on parts of the *Corpus Iuris*.

Jacobus de Ardizzone (Thirteenth cent.). Italian commentator and author of a *Summa Feudorum* (1518) and a *Summa* on part of the Code.

Jacobus de Arena (thirteenth cent.). A commentator and author of *Super jure civili commentarii* (1541) and various other treatises.

Jacobus de Balduino. See BALDUINUS.

Jacobus de Belvisio (1270–1335). Italian jurist, teacher at several French and Italian universities and author of commentaries on the *Authenticum* (1511), the *Libri Feudorum* (1511), an *Aurea Practica Criminalis* (1515), and other works.

Jacobus de Colombo (thirteenth century). Author of glosses to the *Libri feudorum*.

Jacobus de Revigny (1230–96). French jurist, first to apply the scholastic method to law. He wrote extensively, commentaries on the Institutions, Digest, and Code, a *Summa de feudis*, disputations, and a kind of legal dictionary, *Lumen ad revelationem gentium*.

Jacovacci, Domenico (?–1527). Italian canonist, auditor of the Rota and later a cardinal. Author of a *De computatione dotis in legitimam* and a *De concilio*.

Jactitation of marriage. A suit against a person who falsely boasts that he or she is married to the petitioner, a claim which might be prejudicial to the latter's reputation, and decree in which pronounces a person free from the alleged matrimonial bond. The petitioner also sues to have the respondent enjoined to maintain perpetual silence on the matter thereafter.

James, Henry, Lord James of Hereford (1828–1911). Was called to the Bar in 1852 but rose only slowly. In 1869 he became a Q.C. and entered Parliament. There he came to the fore more quickly. In 1873 he became Solicitor-General and then Attorney-General, and resumed the latter office in 1880–85. He declined the Lord Chancellorship in 1886. In 1895 he became a peer and Chancellor of the Duchy of Lancaster, and a member of the judicial committee of the Privy Council in 1896. Though a good advocate he was not a profound lawyer, and made no great mark as an appellate judge.

James, Sir William Milbourne (1807–81). Was called to the Bar in 1831 and practised in chancery. He advanced steadily and appeared in many important cases, though he was not an outstanding speaker. He became a Vice-Chancellor in 1869 and a Lord Justice and a P.C. in 1870. He was a master of the law, a shrewd judge, and enunciated principles tersely and clearly. He was a member of various commissions on equity procedure, and of the judicature commission.

James of Albenga (*c.* 1190–*c.* 1250). Teacher of canon law and decretalist, and author of glosses on the *Decretum*, supplementary glosses to Tancred's disquisition, to *Compilatio Prima* and a famous commentary to *Compilatio Quinta*.

James of Viterbo (*c.* 1255–1308). Variously known as *Doctor gratiosus, doctor inventivus*, and *doctor speculativus*, he was an Italian archbishop, theologian, and political theorist, author of *De regimine christiano* (*c.* 1302), the earliest scientific treatment of the Catholic doctrine of the Church and the papacy; the Church is the true and perfect state and the social community of mankind.

Jason Maino (1435–1519). Famous Italian commentator on the Roman law, author of *Commentarii ad Digestum et Codicem* and *Consilia sive Responsa*.

Jay, John (1745–1829). Son of a prosperous New York merchant, studied at King's College (now Columbia University), and was admitted to the New York Bar in 1768. He was a member of the first Continental Congress, in which he showed a capacity for drafting state papers; a delegate to the Second Continental Congress, and to the New York Provincial Congress, at which he was virtually the draftsman of the New York State Constitution; Chief Justice of the Supreme Court of New York, 1777–79; President of the Continental Congress in 1778, before becoming Minister to Spain, 1779–80; one of the Commissioners, who at Paris negotiated peace with Great Britain, 1782–84; and Secretary for Foreign Affairs and interim Secretary of State, 1784–90. Before becoming Chief Justice, he had accordingly a national reputation for ability and integrity. In 1789 he became the first Chief Justice of the U.S. Supreme Court, holding office till he became Governor of New York State 1795–1801. In 1794–95 he was a special envoy to Britain. He resigned in 1795, finding judicial duties irksome and being disappointed with the position of the Court. He was again appointed Chief Justice in 1800 but declined the appointment. In its first years the Supreme Court had little business and Jay wrote only one important opinion. Most of the time of the Justices was taken up riding the circuits twice a year. He wrote some of the papers in *The Federalist*, dealing chiefly with the Constitution in foreign affairs. He was a very able man, of great intellectual vigour, and of the utmost moral rectitude.

G. Pellew, *John Jay* (1890); W. Jay, *Life of John Jay* (1833); H. P. Johnston, *Correspondence and Public Papers of John Jay* (1890–93); F. Monaghan, *John Jay* (1935).

Jebb, Sir Joshua (1793–1863). An officer in the Royal Engineers, first Surveyor-General of Prisons in England (1837), architect of Pentonville Prison, and first chairman of the Directors of Convict Prisons (1850), the body which controlled, from the Home Office, the main part of the English prison system and indirectly exercised considerable influence over the rest of the system. He worked vigorously to develop large convict prisons and got rid of the hulks (q.v.). The major ideas of his administration were separate confinement and hard labour, but the ideas of ticket-of-leave, use of prison labour on public works, and progressive stages were also adopted. He was succeeded by Sir Edmund du Cane (q.v.).

Jeddart (or Jedburgh justice). Named from Jedburgh in Roxburghshire near the English border, where disputes frequently arose between border peoples, and denoting the procedure whereby a person is sentenced first and tried later. Summary justice was doubtless done in border troubles but whether this form of procedure was ever used is doubtful. Some accounts say it is founded on an incident in the late sixteenth century when Sir George Home summarily hanged a gang of robbers. It differs from 'lynch law' in that it was done by a kind of summary court, not by persons wholly unauthorized.

Jefferson, Thomas (1743–1826). Principal author of the American Declaration of Independence and third president of the U.S., was admitted to the Virginia Bar in 1767 and, in 1769, entered the Virginia legislature. He became a delegate to the Second Continental Congress and was principal author of the Declaration of Independence of 1776. Thereafter he secured abolition of the laws of primogeniture and of entail in Virginia, so as to discourage concentration of property in the hands of a few great landowners, and secured the passing of the statute of Virginia for religious freedom (1786). He also urged a comprehensive plan of public education, of which the only part he secured was the University of Virginia. In 1784 he succeeded Franklin as resident minister to the French government. He was critical of the Constitution adopted in 1789 for failing to include a bill of rights. In 1789–94 he was Secretary of State, in 1797 Vice-President and in 1801 President. In 1803 he negotiated the Louisiana Purchase which nearly doubled the size of the U.S. and during his presidency the power and prestige of the Supreme Court grew under Chief Justice Marshall (q.v.). Jefferson opposed the power of the Court as final interpreter of the Constitution. His last major public service was the foundation of the University of Virginia.

H. Randall, *Life of Thomas Jefferson* (1853); D. Malone, *Jefferson and His Time* (1948–).

Jeffrey, Francis Jeffrey, Lord (1773–1850). Passed advocate of the Scottish Bar in 1794 and, in 1802, still unsuccessful, he joined Sydney Smith and others in establishing *The Edinburgh Review*, which he edited from 1803 to 1829, continuing thereafter to contribute essays to it. It supported the Whigs, who came into power in 1830, and Jeffrey became Lord Advocate and introduced the Scottish Reform Bill into Parliament. He became a judge of the Court of Session in 1834. In his time *The Edinburgh Review* was widely read and influential throughout the whole English-speaking world.

Jeffreys, George, Lord Jeffreys of Wem (1648–89). He was called to the Bar in 1668. He was a lawyer of considerable ability but, particularly latterly, he was notorious for bullying, invective, coarse and overbearing conduct. He cared only for his own advancement and had no scruples as to means for promoting himself. Though Common Sergeant (1671) and Recorder (1678) of London he realized that the royal favour was the way to the heights, and he made himself useful to the Crown. After the Popish Plot he acted as prosecutor or judge in many of the trials of suspected Catholic conspirators and achieved notoriety by his ridiculing and bullying of defendants. In 1685 he sentenced Titus Oates to flogging and imprisonment. His management of the trials consequent on the Ryehouse Plot and Monmouth's Rebellion (the 'Bloody Assizes' of 1685) are exemplars of injustice for which his rewards were respectively the Chief Justiceship of the King's Bench in 1683 and the Chancellorship in 1685. He was drunken, coarse, and brutal as a judge, unscrupulous in appointments to the bench, and never has the bench been so degraded. As Chancellor he forced the Church of England to accept James II's pro-Catholic policies. He has been characterized as the worst judge that ever disgraced Westminster Hall though some authorities rate him an able lawyer. When William of Orange came to London Jeffreys tried to flee the country but was arrested and died in the Tower of London.

G. Keeton, *Jeffreys and the Stuart Cause.*

Jekyll, Sir Joseph (?1663–1738). He was called to the Bar in 1687 and was a Whig M.P. for 40 years. He participated in the impeachment of Sacheverell and Oxford, in the suppression of the rebellion of 1715, and, in 1717, became Master of the Rolls, which office he held till his death concurrently with membership of the House of Commons. He was a very competent lawyer and many of his judgments show great knowledge of the authorities, mastery of the principles, and ability in exposition.

Jellinek, Georg (1851–1911). German jurist, professor at Vienna and Heidelberg, particularly concerned with government and public law, author of *Die Erklärung der Menschen-und Bürgerrechte* (1895), *System der subjektiven öffentlichen Rechte* (1892), and *Allgemeine Staatslehre* (1900). He emphasized the overlap of the State and society; the State forms social groups and all social groups are influenced by the State. The State can be justified by the ends it accomplishes, and its purposes may be reduced to protection, regulation, and support. The final purpose of all government activity is to further the progressive development of the individual members of society, and of the race. He approached law by its study as a subjective phenomenon of human expression.

Jenkins, Sir David Llewelyn, Baron (1899–). Called to the Bar in 1923, he was a judge of the Chancery Division, 1947–49, a Lord Justice of Appeal, 1949–59, and a Lord of Appeal in Ordinary, 1959–63.

Jenkins, Sir Lawrence Hugh (1858–1928). Was called to the Bar in 1883 and, in 1896, became a judge of the High Court at Calcutta. In 1899 he became Chief Justice of the High Court at Bombay and in 1909 he became Chief Justice of Bengal. He retired in 1915 and was appointed to the Judicial Committee of the Privy Council but played little part in the work of that tribunal. He had a fine grasp of Hindu law.

Jenkins, Sir Leoline (1623–85). Principal of Jesus College, Oxford, deputy to the Regius Professor of Civil Law, assisted Sheldon to found his theatre and the university printing house, and was admitted to Doctors' Commons in 1604. With others he compiled a body of rules on the prize jurisdiction of the Admiralty Court and, after being assistant to John Exton (q.v.), judge of the Court of Admiralty, 1665–68, was himself judge of the Court of Admiralty, 1668–85, and judge of the prerogative court of Canterbury from 1669. He served on diplomatic missions and was a commissioner for the projected union with Scotland in 1669, was an M.P. and Secretary of State 1680–84. He was a master of Roman law, international law, maritime and prize law, and ecclesiastical law. As judge of the prize court, he laid down some fundamental principles on the exercise of that jurisdiction and assisted in some of the major legal developments of his time, being responsible for the Statute of Distributions and, in part, for the Statute of Frauds.

W. Wynne, *Life of Sir Leoline Jenkins* (1724).

Jenkins, Roy Harris (1920–). Was an M.P. 1948–77, Home Secretary 1965–67 and 1974–76, and Chancellor of the Exchequer 1967–70 and a strong supporter of British participation in Europe. In 1977 he became President of the Commission of the European Communities. His writings include biographies of Attlee and Asquith.

Jenks, Clarence Wilfred (1909–73). Legal Adviser and later Director-General of the International Labour Organization and author of many books on international law, including *International Protection of Trade Union Freedom* (1957), *Human Rights and International Labour Standards* (1960), *International Immunities* (1961), *The Proper Law of International Organisations* (1962), *The Prospects of International Adjudication* (1964), *Law in the World Community* (1967), *The World beyond the Charter* (1968), and *A New World of Law* (1969).

Jenks, Edward (1861–1939). Qualified as a solicitor, then read law and history at Cambridge and, in 1887, was called to the Bar. In 1889 he went to Melbourne as Professor and Dean of the Faculty of Law. In 1891 he went to the Queen Victoria chair of law at Liverpool and, in 1895, to the Readership in English law at Oxford. In 1903 he moved to London as principal and director of legal studies of the Law Society, a post he held till 1924 when he became Professor of Law in the University of London at the London School of Economics. As a teacher he was concerned, not merely to make students technically competent, but to develop their abilities and critical faculties. In 1909 he took a prominent part in founding the Society of Public Teachers of Law. He wrote *Law and Politics in the Middle Ages* (1898) and was the editor of *A Digest of English Civil Law*, arranged in the form of a code on the lines of the German civil code (1905–17).

Jenner (later (1842) Jenner-Fust), Sir Herbert (1778–1852). Was called to the Bar in 1800 and became a member of Doctors' Commons in 1803. He was King's Advocate from 1828 to 1834 and judge of the Prerogative Court and Dean of the Arches from 1834 till his death. He was an authority on international law and also a sound ecclesiastical lawyer.

Jeofails. In the period of oral pleading down to the fifteenth century, a pleader who realized any mistake in his oral allegation acknowledged it by saying *j'ay faillé*, and sought leave to amend. Hence

Statutes passed from 1413 to prevent technical objections being taken later were called Statutes of Amendment and Jeofails.

Jersey, law of. See CHANNEL ISLANDS LAW.

Jervis, Sir John (1802–56). A second cousin of Lord St. Vincent, he served in the army before being called to the Bar in 1824. He had an extensive knowledge of practice, acted as a reporter in the Court of Exchequer, and produced an edition of Archbold's Criminal Practice, a book on Coroners, and a work on the rules of practice. He founded *The Jurist* and contributed frequently to it. He became an M.P. in 1832, became Solicitor-General briefly in 1846 and was then Attorney-General till 1850 when he became Chief Justice of the Common Pleas. His judicial powers were rated highly, particularly in criminal cases, but in civil cases also he pronounced many important judgments notable for terse and lucid language. He was president of the commission for inquiry into the practice of the common law courts.

E. Manson, *The Builders of our Law*, 28.

Jessel, Sir George (1824–83). Was called to the Bar in 1847 and soon gained a large practice in the rolls court. He became Q.C. in 1865, M.P. in 1868 and Solicitor-General in 1871, and, in 1873, Master of the Rolls, an office which he held till his death. As such, he sat in the Rolls Court till 1881 and thereafter presided in the Court of Appeal, the Judicature Act of 1881 having made the M.R. the president of that court. He was a notable jurist, and had a profound knowledge of equity coupled with capacity to explain, rationalize, and expound its principles, and soon acquired a reputation for quick and clear judgment. He was the most distinguished holder of his office since Grant, one of the most distinguished equity lawyers of all time, and many of his decisions have settled important principles of equity. He was impatient of technicalities and sometimes brusque and impatient of argument and disposed of business very rapidly. He had been educated at University College, London, and took a life-long interest and part in the government of the University of London. He was the first Jew to be a Law Officer, a P.C. or a judge.

J. Bryce, *Studies in Contemporary Biography*, 170.

Jessup, Philip Carlyl (1897–). U.S. jurist, judge of the International Court, 1961–70, and author of *A Modern Law of nations* (1948), *The use of international law* (1954), and other works.

Jetsam. Goods jettisoned or cast into the sea from a ship, which sink and remain under water. If recovered they belong to the Crown, unless claimed by the original owner. See also FLOTSAM; LIGAN.

Jettison. The throwing overboard from a ship of cargo in time of danger in an attempt to save the ship and remaining cargo. It is doubtful if it extends to throwing overboard in any other circumstances, e.g. of cargo which is rotting. Jettison, if justifiable in the circumstances, gives rise to a claim for general average (q.v.) contribution.

Jeune, Francis Henry, Baron St. Helier (1843–1905). He was called to the Bar in 1868, and was junior counsel in the *Tichborne case* (q.v.) but practised largely in ecclesiastical cases. He took silk in 1888 and, in 1891, became judge of the Probate, Divorce and Admiralty Division, and, in 1892, President of that division. From 1892 to 1904 he held, in addition, the office of Judge-Advocate-General, an office then transferred to the War Office. He resigned in 1905 and was created a peer, but died shortly thereafter.

Jewish law. The Hebrews were a group of tribes who created, in what is now Palestine, a kingdom under David (*c.* 1011 to 972 B.C.) and Solomon (971–932 B.C.), but later broke into a northern kingdom of Israel and a southern kingdom of Judah, both of which fell to greater powers. Israel fell to the Assyrian Empire in 721 B.C. and Judah to the Babylonian Empire in 587 B.C. There survives a large body of Hebrew literature, much of it preserved in the Old Testament, which includes legal materials.

Torah, the Hebrew word for law, includes all the directions, religious, moral, ceremonial, judicial, and civil, issued by priests and scribes, and the *Torah* became the designation of the combination of narrative, poetry, and law found in the first five books of the Hebrew Bible, the Pentateuch.

There are five stages in Israelite legal history. In the nomadic period prior to the settlement in Canaan, the people were probably governed by the customary Semitic law of the desert and there was no need of written law. Their customs were probably derived from a common stock from which Babylonian, Assyrian and Hittite law also came. Then from the settlement in Canaan (*c.* 1100 B.C.) to the revolt of Jehu in 842 B.C., the Hebrews largely absorbed the civilization and laws of the Canaanites. At this time, some rules for the guidance of judges and people may have been put in writing. The tendency was to attribute the origins of legal institutions to the beginnings of Israel's history, and to the giving of the law to Moses. The original ten commandments, said to have been given by God to Moses on two tablets of stone and contained in Exodus 20 and repeated in Exodus 34, the second being an older version, may date from *c.* 1000 B.C. In Exodus 21 to 23 there is a characteristic code of fairly developed secular law suitable for an agricultural community, sometimes known as the Covenant Code. It is probable that some of the provisions are later amendments or interpolations from the religious writings of the Pentateuch; these include the statement of talion, 'an eye for an eye and a tooth for a tooth'. These amendments may have been made as late as 400 B.C., and in this form the code was promulgated in the Pentateuch. But the original version is much older, dating possibly from 800 B.C. or earlier—old enough to be ascribed to Moses—and it may, through the Canaanites, have absorbed something from the Babylonian laws of Hammurabi. This probably represents the legal customs current in Palestine before the advent of Israel and adopted along with other elements of Canaanite culture.

In the third period, from 842 to 586 B.C., the Prophets were active, impressing their principles on the Hebrews, and during this period the great moral and humane laws probably developed. To this period belong the mass of laws in Deuteronomy, possibly to about 630 B.C., which formed the basis of the reforms of Josiah in 621 B.C., and is known as the Book of the Covenant. The mass comprises seven loosely defined groups of laws without logical order or arrangement, and does not consist of one code. It is rather a collection of minor codes later brought together but it represents the development of Hebrew law as influenced by the great prophets and teachers of the period 750–600 B.C.

The fourth period extends from 586 to about 300 B.C., and by the end of this period, the canon of the law was probably closed though, in practice, the law continued to be expanded in the schools of the Scribes, though the results of their labours were preserved in oral tradition until after the fall of Jerusalem in A.D. 70. The Holiness Code of Leviticus 17–26 is of about this time possibly 597–586 B.C. It consists of 10 or 11 groups of law, either disarranged or interpolated by later editors and seems to have been intended as a book for the people. The code found in Ezekiel 40–48 dates from about 572 B.C. and is largely concerned with religious ritual. Though never adopted this Code was influential with later lawgivers and in modified and more practical form.

After the fall of Jerusalem in 586 B.C. there developed a large body of priestly regulations, mostly relating to ceremonial observances, now scattered through Exodus, Leviticus, and Numbers and known as the Priestly Code. They come from several hands and various dates, spread over a century or more, and there is much internal evidence of revision and additions. After 586 B.C., Palestine was under Babylonian, Persian, Greek, Roman, and other domination, and the Jewish people were scattered over the world.

The final stage of Hebrew law consisted in the collection of decisions on new problems. These were originally preserved only in oral form, but by A.D. 200 they were committed to writing. The Mishna records the majority of them and, in turn, there grew up around it the body of comments and traditional decisions ultimately collected in the Babylonian Talmud about A.D. 600.

Through the Bible and Christianity Hebrew law has influenced Western law. It was taken to represent the revealed will of God. There were various conscious attempts to copy the laws of Israel in later times, in Alfred's code, Calvin's theocratic state, Puritan England, and colonial America, where judges were commanded to inflict penalties according to the law of God. Jewish law also had influence on natural law thinking, in controversies on the boundaries between Church and State, in canon law, ecclesiastical law, and in commercial law where Jewish practices frequently underlie modern rules.

Today, Jewish law is the body of religious duties regulating worship and ritual, private and social behaviour, and indeed all aspects of the life of Jews. It is the personal law of all Jews irrespective of citizenship or domicile, a Jew being a person born of a Jewish mother or converted to Judaism. Orthodox Jews everywhere consider themselves bound by Jewish law as part of their observance of the Jewish faith, and doubts and disputes are resolved by a rabbi, and it is deemed binding on even apostate Jews.

The sources of Jewish law are, firstly, divine revelation and, secondly, divinely inspired interpretation, but subordinate legislative power was vested, at first in the assembly of elders and later in the central academies of Palestine and Babylonia.

Jewish law originated in the Pentateuch, and other legal prescriptions in the Prophets and the Writings, which all have to be considered in the light of other ancient legal systems, such as the Babylonian, Assyrian and Hittite laws. Scripture was later interpreted in the light of the Oral Law, which also was ascribed to divine revelation, and transmitted at first orally, but between A.D. 200 and the Middle Ages committed to writing. The *Midrash* consists of commentaries on the legal parts of the Pentateuch, and other ancient sources. From about 200 B.C. the Oral Law had increased greatly in quantity and there appeared the *Mishnah*, compiled about A.D. 200 by the patriarch Judah from earlier similar books, a collection of the traditional materials of legal relevance systematically arranged. Materials left out of the *Mishnah*, called *Baraita*, is collected under the title *Tosefta*, (third or fourth century A.D.) which is partly supplement to, partly commentary on, the *Mishnah*. The Palestinian *Talmud* and the Babylonian *Talmud* both contain discussions of the meaning of the *Mishnah* text, with relevant citations, controversies, and decisions, and were compiled by the Palestinian and Babylonian academies in the fifth and sixth centuries respectively. There are also numerous commentaries on this literature and many collections of rabbinical opinions (*Responsa*) on particular problems used as guides to the decision of similar problems. Many innovations on the law were introduced by interpretation of texts particularly by the central academies of Palestine and Babylonia down to the eleventh century, and thereafter by rabbinical leaders of each community or group. Custom is also recognized as a source of law. Among systematic treatises and commentaries the major code is Moses Maimonides' (1135–1205) *Mishnah Torah* and the most authoritative one Joseph Qaro's (1488–1575) *Shulchan 'Arukh*, while in modern Israel collections and digests of the material continue to be made.

A feature of Jewish law is that the study of the law has never been the monopoly of the rabbis, but has been the duty of every individual, and is done in synagogues and privately all over the world.

As Jewry became more scattered over the world customs and practices came to differ, each being preserved by the different synagogues. The Jewish community in Israel has tended to create a custom of its own, in some cases sanctioned by state legislation.

Marriages and divorces effected under Jewish law, if valid according to the *lex domicilii* of the parties, are accepted in U.K. law.

Jewish law is part of the law of modern Israel, being the personal law of Jews in matters of personal status, and referred to in relation to the interpretation of religious concepts, services and charities. It has also influenced some of the concepts utilized in, and the interpretation of terms in, modern Israeli law. In Israel only oriental and Eastern European orthodoxy is recognized and the leaders thereof aim at making Israel a theocratic state, though a majority of the population do not support their view, and have opposed various reforms of private law.

C. Kent, *Israel's Laws and Legal Precedents*; C. North, *Laws in the Pentateuch*; A. Jirku, *Das weltliche Recht im alten Testament*; Z. Falk, *Hebrew Law in Biblical Times*; G. Horowitz, *The Spirit of Jewish Law*; I. Herzog, *The Main Institutions of Jewish Law*; J. Rabinowitz, *Jewish Law*; B. Cohen, *Jewish and Roman Law*.

Jews. In post-Biblical times the Jewish people were largely scattered after the revolts against Rome were suppressed and there came to be Jewish communities in most of the important centres of the Roman Empire.

Some Jews are thought to have come to England in the wake of William the Conqueror and, by the mid-twelfth century, Jewish communities existed in all the major towns, as far north as York, but particularly in London. They were tolerated and, under Henry I, a charter of liberties may have been granted to the Jews. In the twelfth century anti-Jewish feelings became evident and heavy taxes were exacted from them. In the period of crusading enthusiasm there were riots and murders and, in 1194, Richard I ordered the establishment of a chirograph chest in the principal cities to preserve records of debts contracted with Jews. As co-ordinating authority, there were established the Exchequer of the Jews, with judicial and financial functions, and the office of archpresbyter, as officially appointed Crown representative on Jewish matters. Under Henry III, the Jews initially recovered some degree of prosperity but later they were mercilessly taxed and ecclesiastical enactments were enforced against them and many of their communities were sacked. Finally in 1290 Edward I expelled the Jews from England.

A few Jews continued to live in England and, in 1655, Cromwell's Council of State assented to their return. Thereafter, the community developed and Jews increasingly entered various aspects of the country's life. The Toleration Act, 1688, was extended to Jews in 1846 so that the Jewish religion has since then been a recognized and tolerated dissenting denomination and places of worship, property, and trusts are recognized and protected in the same way as any other.

By the nineteenth century Jews were well established, particularly in finance and commerce. They were admitted to the offices of sheriff (1835) and municipal offices (1845) and to Parliament (1858). Minor disabilities were removed by the Religious Opinions Relief Bill (1846) and the highest offices were open to them by Acts of 1866, 1871, and 1878. Only in 1890 were religious restrictions on practically every political position and dignity removed. With industrial development, Jewish communities not only increased in size but developed in many cities. They entered the professions: Jessel (q.v.) became Solicitor-General in 1871 and, subsequently, Master of the Rolls. Isaacs (q.v.) became Attorney-General, Lord Chief Justice, and Viceroy of India. Nathaniel de Rothschild received a peerage in 1885.

In Europe, Jews were generally oppressed until the Enlightenment and the French Revolution; they were given civil rights in France and French-occupied countries under Napoleon, in Germany in 1870–71, and in Austro-Hungarian countries in 1867. But anti-Semitism erupted periodically in practically every country.

Jews began to settle in Central and South America from an early date but were frequently subjected to persecution by the Inquisition and some appeared among the English settlements in New England and Canada. By 1776 there were said to have been 2,000 in the American colonies and they benefited from the Virginia Bill for Establishing Religious Freedom (1785) and the exclusion of religious tests in the U.S. Constitution. In the nineteenth century German and Eastern European Jewish immigrants came in large numbers and in the mid-twentieth century New York was the greatest urban Jewish centre in the world. After 1918 Jewish immigration was checked to some extent but it continued in Canada and Latin America.

The study of law had been a tradition among Jews but they were generally prevented from practising law in most of the countries in which they settled until the nineteenth century. In that century, however, many won distinction as judges, jurists, or practising lawyers. Among notable Jewish lawyers may be mentioned: in England Jessel, Isaacs (Lord Reading), Lord Cohen, and Lord Salmon among judges, Oppenheim and Lauterpacht among international lawyers, Jolowicz, Kahn-Freund, and Goodhart among jurists; in the U.S. J. P. Benjamin and, in the Supreme Court, Brandeis, Cardozo, Frankfurter, Goldberg, and Fortas and among jurists Frank, Cohen, and Radin; in France Cassin, Levi-Ullman and others; in Germany Dernberg, Laband, Jellinek, Arndt, Kantorowicz, and Kelsen; in Italy Volterra, Ascarelli, Sereni, and many more; in the Netherlands Carel Asser, Tobias Asser, Jitta, and others. In Australia Sir Isaac Isaacs became Chief Justice and Governor-General; in Canada Bora Laskin became Chief Justice; and in South Africa several Jews have attained the bench.

Jhering, Rudolph von (1818–92). German jurist, became a law teacher at Berlin in 1843 but, having taught in various universities, settled at Gottingen in 1872, teaching initially principally Roman law but with a very wide range of interests and activity in legal science. His extensive writings include *Geist des römischen Rechts* (*The Spirit of Roman Law*) (1852–58), *Die Jurisprudenz im täglichen Leben* (*Jurisprudence of Everyday Life*) (1870), *Der Kampf ums Recht* (*The Struggle for Law*) (1872) and *Der Zweck im Recht* (*Purpose in Law*) (1877–83), his most ambitious work. He developed a philosophy of social utilitarianism which differed from that of Bentham in putting more stress on the needs of society and is sometimes called the father of sociological jurisprudence. The individual is the starting point of his philosophy but the service of the individual must be enlisted for the benefit of society.

Jim Crow laws. Jim Crow was the name of a minstrel routine called 'Jump Jim Crow', performed from 1828 by its author, Thomas D. Rice, and many others, and the name was adopted as a derogatory name for negroes. From the 1870s legislatures in the southern United States passed laws enforcing racial segregation in public transport, later extended to schools, parks, theatres, restaurants, and such racial segregation laws came to be known as Jim Crow laws. These laws were directed against 'persons of colour', a term which was extended to anyone with known or even suspected black ancestry in any degree. The era of such legislation declined with the growth of the civil rights movement in the 1950s and was ended by decisions of the Supreme Court declaring segregation in schools and other public places unconstitutional.

Jitta, Daniel Josephus (1854–1925). Dutch jurist and latterly member of the Netherlands Council of State, author of *La Méthode du Droit International Privé* (1890), *The Substance of Obligations in Private International Law* (1906–7), and *International Privaat Recht* (1916), which are significant contributions to the theory of that branch of law.

Joannes Andreae (*c.* 1270–1348). The most distinguished lay canonist of the Middle Ages, taught canon law at Bologna, and was author of the *Apparatus* on the *Sext* (1301) and the *Clementines* (1308), both of which were accepted as the *glossa ordinaria*, and more importantly of the *Novella* on the *Decretals* of Gregory IX (1338), the *Quaestiones mercuriales* (1338–40), and the *Novella* on the *Liber Sextus* (1342–46). The second of these comprises a *Novella Commentaria* on the *De Regulis Juris* of the *Liber Sextus* and a collection of over 100 *quaestiones disputatae*, arranged according to the rules of law concerned. His last work was *Additiones* to Duranti's *Speculum* (1346–47).

Joannes Calderini (*c.* 1300–65). Italian canonist, author of *Breviarium Decretorum*, and writings on the Clementines.

Joannes de Deo (*c.* 1200–67). Portuguese canonist, author of *Principium Decretalium*, *Apparatus decretorum* (*c.* 1240), *Summa super quattuor causis decretorum* (1243), and many other works.

Joannes Faventinus (John of Faenza) (*c.* 1120–87). Italian canonist and author of an important *Summa* on the *Decretum*.

Joannes Garsias (thirteenth century). Spanish canonist and commentator on the *Decretum*.

Joannes de Lignano (*c.* 1320–83). Italian jurist and canonist, author of *Commentaria in Decretum*, *Commentaria in Decretales*, *Commentaria in Clementinas* and, most importantly, *De Bello* and *De Pace*, systematic expositions of problems of war and peace which entitled him to be considered a precursor of public international law.

Joannes Scholasticus (or John of Antioch) (*c.* 503–577). Patriarch of Constantinople, theologian, and ecclesiastical lawyer. He compiled a Collection of Canons, the earliest surviving collection of

Byzantine church legislation, which collated imperial ecclesiastical legislation with that of the theologian-legislator Basil of Cappadocia, and a Collection of 87 Chapters, a synthesis of Justinian's later legislation on ecclesiastical subjects. He also wrote theological works.

Joannes Teutonicus (*c.* 1175–1245). One of the outstanding scholars of Bologna and one of the most important glossators of his time. He wrote an *Apparatus* for the constitutions of the Fourth Lateran Council, the *Compilatio IV antiqua* (1215), and the *Apparatus* thereto which was accepted as the *Glossa ordinaria* thereon, *Apparatus* for *Compilatio III antiqua* (1217), and glosses on *Compilatio V antiqua*. His chief work is the *Glossa ordinaria* on Gratian's *Decretum* (*c.* 1217) which became the standard *apparatus* on the *Decretum* in the schools of canon law and is still a basic source for research on the history of canon law.

Jocelyn, Robert, Viscount (1688–1756). Was called to the Irish Bar in 1719, became Solicitor-General (1727) and Attorney-General in 1730, and finally Chancellor of Ireland (1739) and a peer (1743).

John of Acton (*c.* 1280–1349). English canonist, remembered chiefly for his gloss on the Legatine Constitutions of the legates Otho and Othobon (qq.v.) (*c.* 1335), which is printed as an appendix to Lyndwood's *Provinciale*.

John of Wales (Johannes Galensis) (thirteenth century). A Welshman who became a canonist at Bologna, author of an *apparatus* to *Compilatio tertia antiqua* (1215) and compiler and glossator of *Compilatio secunda antiqua* (1210–15), a collection based on those of Gilbertus Anglicus and Alanus Anglicus.

Joint adventure. In Scots law, a partnership entered into for a particular voyage, adventure, or enterprise only and not constituting any firm.

Joint and Common property. Two modes in which a number of persons may concurrently have rights of ownership of one object of property, whether real or personal. In joint ownership or joint tenancy of land there is a single title, interest, time of commencement and unity of possession. The right of survivorship exists so that, on the death of any joint owner, his rights are extinguished and vested in the survivors until the last survivor becomes sole owner of the property. Joint ownership of real property may be severed and the right converted into tenancy in common.

In ownership in common, each owner has a separate title to and interest in his share and the shares need not be equal, though the several owners have undivided possession of the object of property, There is no right of survivorship and on the death of any one title to his share passes under his will or on intestacy to his heirs.

Joint and several liability. Liability which is imposed on each of several parties jointly, and also severally, so that a creditor can claim against all, or against any one, leaving it to that one to work out any relief he can from his co-debtors.

Joint liability. Liability which is imposed on several persons, e.g. partners, together, so that if one is sued he can insist on the others being joined, and on the death of one the liability devolves on the survivors or survivor.

Joint resolutions of Congress. A joint resolution of both Houses of the Congress of the U.S. is a part of the law. Apart from joint resolutions proposing constitutional amendments, the resolution must pass both Houses and receive the President's signature. The procedure is the same as for bills. Joint resolution is the normal vehicle for stating policy.

Joint stock companies. Business associations in which the common stock or capital was contributed by a large group of persons. The largest and oldest was the East India Company, but later foundations included the Royal African Company and the Hudson's Bay Company. In 1662, statute protected participants by limiting their legal liability to the nominal value of the capital contributed and these companies were able to attract capital from a wide public. Their monopoly privileges were often criticized. In 1694, the Bank of England was formed and by statute was given a monopoly of joint stock banking in England, a situation which lasted until 1826. By 1717 the capital of the biggest joint stock companies ran to millions, that of the South Sea Company being £10 million. Its crash, in 1720, cast doubts on the whole principle of joint-stock enterprise and, after 1720, the foundation of such companies was checked by the 'Bubble Act', which prohibited the formation of new joint stock companies unless authorized by Crown charter or private Act, but did not affect the existing public companies. This gave rise to the growth of large unincorporated joint-stock associations. In the nineteenth century there developed the principles and law of incorporated companies, which from 1862 could have the liability of members limited.

W. R. Scott, *Constitution and Finance of Joint-Stock Companies to 1720*.

Joint tortfeasors. Persons whose respective acts have combined to contribute to one wrongful act causing harm and who are accordingly held jointly and severally liable to the injured person. Any one joint tortfeasor who has to pay damages has a right against each other joint tortfeasor for contribution.

Jointure. Properly a joint interest given to a husband and wife, but commonly used to mean a sole interest given to a wife only, normally by a provision made by a husband for the support of his wife after his death. Such a jointure could formerly be either legal, or equitable, in the latter case generally consisting of a rentcharge or annuity payable by the trustees of a marriage settlement to the wife if she survived her husband. Acceptance of an equitable jointure barred a wife's claim to dower.

Jones, Leonard Augustus (1832–1909). Practised law in Boston, and published many legal treatises, several on securities, and others on real property. He also published an *Index to Legal Periodical Literature* (2 vols, 1888–99) covering Anglo-American journals down to 1899, a work later continued by others, and edited the *American Law Review* (1884–1904). His *Forms in Conveyancing* was commonly used for many years. Latterly he was judge of the Massachusetts land court.

Jones, Sir Thomas (1614–92). Was called to the Bar in 1634. He became a judge of the King's Bench (1676–83) and Chief Justice of the Common Pleas (1683–86) until dismissed for denying the dispensing power. He left a volume of reports of decisions. He was a learned judge and a student of records, but showed severity and harshness when presiding at political trials.

Jones, Sir William (1746–94). Became a good classical scholar and a great orientalist. He published translations and a Persian grammar, and became intimate with Dr. Johnson, Burke, and Gibbon. Being in quest of a career, he was called to the Bar in 1774 and became a jurist rather than a profound English lawyer. In 1776, he became a commissioner in bankruptcy.

His *Essay on the Law of Bailments* (1781) is remarkable for the wealth of other legal knowledge, ancient and modern, on which he draws for examples. The book examines the subject analytically, historically, synthetically, and critically, and has ever since been regarded, as by Story, as profoundly illuminating.

In 1783, he became a judge of the High Court at Calcutta and performed his duties with great ability, though his main interests were literary. He studied many aspects of Indian culture, mastered Sanskrit and foresaw the importance of comparative philology. He decided to prepare a complete digest of Hindu and Mohammedan law as observed in India of which only the first parts, the *Institutes of Hindu Law, or the Ordinances of Manu* (1794) and *Mohammedan Law of Succession* (1792) appeared. He was an outstanding scholar with a deep understanding of and sympathy with Indians and did much to make Indian learning known to Europeans.

Joufroy, Henri (1786–1859). Author of *Le droit des gens maritime universal* (1806) and *Catechisme de droit naturel* (1841).

Jougs (joggs or juggs). An instrument of punishment known in Scotland, Holland, and elsewhere, in the form of an iron collar, placed round the offender's neck and fastened with a padlock, and the whole attached by a short chain to a wall, often of the church, or a tree. It was used for ecclesiastical as well as ordinary offences.

Journals of the Houses of Parliament. The House of Lords Journals begin in 1 Henry VIII (1509) and the Commons' Journals in 1 Edward VI (1547). They replace the Parliament Rolls.

Journal of the House of Commons. The official and permanent record of the proceedings of the House, as distinguished from the verbatim record of the debates in *Hansard* (q.v.). The Journal was begun in 1547 and, save for 1584–1601, has been continuous since then. It is uncertain when it became the official record of the proceedings of the House. It is prepared in the Journal Office of the House from the Votes and Proceedings, and the minute books of the Clerks-at-the-Table, is published annually with an index, and usually records the proceedings of a single session. It was first ordered to be printed as a result of a Select Committee Report in 1742. It takes the form of a continuous narrative save that proceedings in Committee of the Whole House are recorded in the words of the Votes and Proceedings. Since 1580 it has been treated as an authoritative source of precedent and is treated as legal evidence. An *Index to the Journals 1547–1714* was compiled in the nineteenth century by Erskine May.

Journal of the House of Lords. The official permanent record of the proceedings of the House of Lords, distinct from the verbatim record of debates contained in *Hansard* (q.v.). The Journal, apart from fragments of a record from the fifteenth century, runs from 1509 and it has been recognized as official from 1620. It is compiled in the Journal Office from the Minutes and Proceedings made by the clerks at the table during sittings of the House, and includes the letters patent and writ of summons of a newly created peer and, if he desires, the letters patent by which the title of a peer succeeding by descent was granted. It includes proceedings in Committee of the Whole House and Select Committee Reports in full. It is published annually.

Jowitt, William Allen, 1st Viscount (1947) and 1st Earl (1951) (1885–1957). Called to the Bar in 1909, he took silk in 1922 and also entered Parliament. He was Attorney-General 1929–32 and Solicitor-General 1940–42, and thereafter Paymaster-General, Minister without portfolio and Minister of National Insurance. He became Lord Chancellor in 1945, holding office till 1951 and was responsible for the Crown Proceedings Act, 1947.

Joy, Henry (1763–1838). Called to the Irish Bar in 1788, became Solicitor-General in 1822, Attorney-General in 1827 and Chief Baron of Exchequer in Ireland, 1831–38.

Joyce* v. *Director of Public Prosecutions, [1946] A.C. 347. Joyce was born in America and was an American citizen, but lived in England from 1921 to 1939. When applying for a British passport he described himself as a British subject. At the outbreak of war in 1939 he went to Germany and broadcast enemy propaganda in English. It was held by the House of Lords that until his passport expired he owed allegiance to the King and, having been adherent to the King's enemies beyond the realm, was guilty of treason. He was accordingly executed.

Joyeuse Entrée. In the Middle Ages, the visit of a prince to the territory of a new possession, usually the occasion for a grant or confirmation of liberties. The name is given particularly to the Joyeuse Entrée of Brabant (1356), a charter granted by Duchess Johanna and Duke Wenceslas on their accession to the Duchy of Brabant confirming the rights and privileges of its assemblies and towns. It guaranteed the indivisibility of the Duchy, maintenance of free trade, due process of law, and the right to give advice and consent on matters of war and peace. It also recognized the right of citizens of the Duchy to abjure their allegiance to any prince who failed on his part to adhere to it. This charter was appealed to repeatedly by the provinces of the Netherlands, when defending their freedom against the Spaniards in the sixteenth century and against Austria in the eighteenth century.

Judge. The general name for a person whose function is to adjudicate on disputes and other matters brought before courts for decision. Judges may be, and in higher courts always are, persons qualified and experienced in law and legal practice, but in some lower courts may be unqualified laymen. A distinction must also be drawn between judges sitting as a court, or bench, or college of judges, as is common in appellate courts, where two or more judges sit together; judges sitting with lay colleagues, as a court, usually one with a special jurisdiction; judges sitting with lay assessors; judges sitting with a jury; and judges sitting alone. Where judges sit together, particularly in an appellate court, the court may give a single judgment, or each judge may give his own judgment. Where judges and laymen compose a court, the judge commonly presides but has no more weight in the decision than his colleagues. Where judges sit with assessors, they may all equally be judges of fact but the qualified judge alone may decide matters of law, or all may have equal powers on both fact and law. Where judges sit with a jury, the jury's function is to decide questions of fact and the judge's function is to preside, to decide questions of law and, generally, to instruct the jury on the relevant points of law, and on what they must consider when considering their verdict. Where judges sit alone they must decide both issues of fact and questions of law. The composition of courts varies in different jurisdictions and at different levels.

Judges attain their posts in different ways. In Britain, the judiciary is not a career structure. In England, Lords of Appeal and Lord Justices of Appeal are appointed by the Queen, usually from judges of the High Court, but occasionally direct from the Bar, on the advice of the Prime Minister after consultation with the Lord Chancellor, and High Court judges and many inferior judges are appointed, usually from practising barristers, occasionally from solicitors, by or on the advice of the Lord Chancellor. A Lord of Appeal must have had 15 years' experience at the Bar or have held high judicial office for two years, a Lord Justice of Appeal have had 15 years' experience or one year's service as a judge of the High Court, and a judge of the High Court have had 10 years' experience at the Bar. These minima are usually greatly exceeded. Prior to the Judicature Acts, 1873–75, judges of the superior courts had to be serjeants-at-law. In the past political affiliation sometimes played a part in appointment. It was commonly thought, until 1945, that the Attorney-General was entitled to first refusal if there were a vacancy as Lord Chief Justice but since 1945, this tradition has been departed from. The office of Lord Chancellor is, however, a political one and rarely held by a politically neutral figure. In Scotland, judgeships are filled on the advice of the Lord Advocate, frequently by Law Officers or Deans of the Faculty of Advocates. In Britain, there is little promotion from county court to High Court and promotion from the High Court to the Court of Appeal carries little material rewards. A superior court judge may resign, but in modern times a judge has rarely resigned and returned to practice. Judges are removable only by a resolution of both Houses of Parliament assented to by the Queen. England relies heavily on part-time lay judges (magistrates or justices of the peace) who deal with all except the most serious crimes and some matrimonial business. Scotland relies much less on lay judges. Justices of the peace are appointed for life, in England by the Lord Chancellor and in Scotland by the Secretary of State for Scotland, in each case on the advice of local committees.

The constitutional position of judges in Britain is somewhat anomalous. Except for the Lord Chancellor, they hold office during good behaviour and not during the royal pleasure. They are appointed by the Queen and paid out of the Consolidated Fund, but are not Crown servants in that they cannot be controlled or given directions by the Queen or her Ministers, nor by Parliament, nor any government department. They are completely independent and enjoy total immunity from action based on anything done or said in the exercise of the judicial function, even though done in bad faith. Even a judge of an inferior court is immune from civil action if he exceeds his jurisdiction. In the U.S., the position is similar, the independence of the judiciary from political interference or pressure being recognized as fundamental.

The titles attaching to judges are: Lord of Appeal, the title Lord; Lord Justice of Appeal, the title Lord Justice, abbreviated as, e.g. Smith, L.J., and judges of the High Court the title Mr Justice or Mrs Justice, abbreviated as, e.g. Brown, J. Circuit judges are known as His Honour Judge Jones and addressed in court as 'Your Honour'. Deputy circuit judges and Recorders are addressed in courts as 'sir' or 'madam'.

In Scotland, judges of the Court of Session are called, e.g. Lord Cameron, though they are not peers, and sheriffs-principal and sheriffs are called, e.g. Sheriff Stewart. All are addressed in court as 'Your Lordship'.

In the U.S., Justices of the Supreme Court are appointed for life by the President, subject to confirmation by the Senate. They may resign and return to practice, or be removed by impeachment. In other federal courts and some state courts judges are appointed, usually by the chief executive, and politics plays an important part in selection. In some states judges, particularly of lower courts, are popularly elected, sometimes on a political ticket, sometimes on a non-political basis. Favour is growing for the Missouri Plan, a means of selecting judges independently of politics; a nominating commission considers candidates for judicial office and submits to the appointing authority a list of qualified persons from which one must be chosen; the appointee serves a probationary term and then submits himself to election, but on his record, not as an opposed candidate.

In Western European countries of the civil law tradition, the judiciary is a career structure; young men, having qualified by study and examination, enter the judicial service direct and serve in lower courts, being promoted by age, ability, and experience to higher courts, as may be recommended by a council of senior judges or the Minister of Justice. Judges in the common-law world enjoy high prestige and independence of government and their affinities are with the Bar from which they have been promoted. In the civil-law world, judges, save of the highest courts, enjoy lower prestige and their affinities are with the civil service of which they are a part; the highest prestige attaches to the leading professors.

The judicial function is to apply the law to disputes and controversies but, particularly in superior courts in Anglo-American jurisdictions, by reason of the traditional practice of not only deciding the issue but giving an opinion setting out the reasons for the decision and explaining the application of the relevant rules of law, and by reason of the principle of *stare decisis*, judges have considerable influence in making, developing, and restating the law, and large tracts of common law were entirely or largely made, and are still being developed and restated, by judicial decisions.

Similarly, in interpreting legislation, judges have a creative function; the approach they adopt to the legislation can very substantially effect its application and effect.

Judge Advocate, Judge Advocate-General. The Judge Advocate General was, prior to 1793, the secretary and legal adviser of the board of general officers at the head of the army. Thereafter he became principal legal adviser to the Commander-in-chief and, until 1892, was an M.P. and a junior member of the government. Thereafter till 1905, the office was conferred on the President of the Probate, Divorce and Admiralty Division, though exercised by deputies, but in 1905, was created the Department of the Judge Advocate-General at the War Office. The Judge Advocate-General is now an officer appointed by the Lord Chancellor and is adviser to the Army Department in relation to courts-martial and military law. All general courts-martial are assisted by the Judge Advocate who advises the court on law, evidence, and procedure and ensures that the prisoner has full opportunity to state his defence. The proceedings of courts-martial are transmitted by the Judge Advocate attending at the trial to the Judge Advocate-General.

Judge Advocate of the Fleet. An officer of the former Court of the Lord High Admiral, whose office was united with that of Counsel to the Admiralty in 1837. He now performs functions for the Navy similar to that of the Judge Advocate-General for the Army. A Deputy Judge Advocate may, when required, attend naval courts-martial to advise on law and procedure.

Judge in Lunacy. The judges of the Chancery Division, acting under the former Lunacy Act, 1890. The jurisdiction now belongs to the Court of Protection (q.v.).

Judge Ordinary. The name of the judge of the Court for Divorce and Matrimonial Causes established in 1857. The judge of the Court of Probate was made also Judge Ordinary and was judge of first instance in all matters within the court's jurisdiction. In 1875 he was made President of the Probate, Divorce and Admiralty Division of the High Court (now the Family Division).

Judge-made law. Recognition in common-law systems of the rule that judicial precedents or case-law are not merely decisions of past cases but provide rules which, in certain cases, may be, or should be, or even must be applied in subsequent cases, implies recognition of the fact that the judges make law just as Parliament or other legislature does, though not in the same way, not to the same extent and, in case of conflict, not with the same authority. Thus, no principle or rule, even laid down by the House of Lords, is of any avail if in conflict with a rule of statute; judicial decisions only make law when they state, extend, narrow, redefine or otherwise apply a principle, not when it is applied again to a familiar situation; decisions cannot make law save where a litigation has raised the point for decision; and they do not do so in verbally authoritative terms. Nevertheless, it is accepted today in common-law jurisdictions that the judicial function is not only to apply the law but by interpretation, analogy, and reconsideration to develop it, restate it and, frequently, to create it. Even in civil law jurisdictions where, nominally, there is no judge-made law it is increasingly recognized that judges can materially affect the law on a subject and, indeed, the whole system of law by the way in which they interpret provisions of a code.

Judge's notes. In Anglo-American courts it is customary for a judge, even when evidence is being recorded in shorthand, to make notes of evidence and arguments for his own use. In English criminal appeals notes made by the judge must be furnished to the Court of Appeal (Criminal Division).

Judges' Rules. The Judges' Rules probably originated in a letter of 1906 from L. C. J. Alverstone to the Chief Constable of Birmingham in answer to a request from the Chief Constable for advice on the proper conduct of police inquiries and the mode of questioning of suspects. Four rules were formulated and approved by the King's Bench judges in 1912 and five more in 1918. A new set of rules was approved by a meeting of all the Queen's Bench judges in 1964. A person must be cautioned (q.v.) when a police officer wishes to question him and again formally cautioned when he is charged or informed that he may be prosecuted. The rules have no statutory or other force, but state, generally, the principles relative to the cautioning of suspects and their interrogation, and police who follow the rules are unlikely to be criticized or to have information they elicit held inadmissible. If the rules are not observed the court may decline to admit an accused's statements in evidence. See [1964] 1 W.L.R. 152; [1964] 1 All E.R. 237.

Judge's trials. The Court of Session was, by Act of 1579, given power to test the suitability of a royal nominee to the Bench. In 1674, the Act of Sederunt prescribed the practice followed till 1933. The new judge, as Lord Probationer, heard several cases and reported his decision on them to the Court. A dispute arose in 1721 relating to one Patrick Haldane whose nomination was later withdrawn. A statute of 1724 provided that if the Court were dissatisfied it should merely certify its dissatisfaction to the Crown, but should be obliged to accept the nominee if the Crown insisted. The practice became a formality long before it was abolished.

Judgment. The intellectual process of reaching a conclusion on a dispute, and also an actual decision of a court determining the rights of parties in an action brought before it. A judgment has legal effect only if the court had jurisdiction over the parties and in relation to the kind of dispute litigated before it, matters regulated by rules of the legal system of the particular court in question.

Judgments are distinguishable into interlocutory and final judgments, the former deciding only particular issues or stages of the case, and the latter finally adjudicating on the matters in issue. Depending on the procedural code of the particular system, appeals to a higher court may be competent against interlocutory judgments or may be competent only against final judgments.

Judgments are also distinguishable into judgments *in personam* and judgments *in rem*, the former imposing a personal liability on a defendant to pay money, implement a contract or undertaking, desist from a wrong, or otherwise do something, the latter determining an issue of right, status or property in a way binding persons generally.

In the Anglo-American system, placing much reliance on law developed by courts, while the judgment is the operative part of the court's decision, as much or greater interest attaches to the opinion or opinions given by the judge or judges giving their reasons for giving judgment in the sense in which they have done, because these opinions, insofar as they disclose the *ratio decidendi* or principle on which the decision was founded, have influence in determining the law applicable to similar cases in the future.

In some cases, such as of the International Court and the constitutional courts in some countries, courts may be asked to give, and may give, not only judgments decisive of disputes, but advisory judgments on the application or legal effect of some rule of law in certain circumstances. An appeal to the Judicial Committee of the Privy Council is similar to this in that, technically, the judgment of the Committee is only advice to the Queen to dispose of an appeal in a particular way, but it is, in effect, decisive of the matter appealed.

In courts comprising several judges the judgment is generally that of the majority of the judges and no account is taken of the views of dissenters.

See also JUDICIAL OPINION.

Judgment creditor. The person named in a judgment or order as entitled to the benefit thereof and who may issue execution (q.v.) against the judgment debtor.

Judgment debtor. One against whom a judgment has been given ordering him to pay money and who has not yet paid it. He might formerly be committed to prison under a judgment summons and may now be adjudicated bankrupt.

Judgment *in personam*. A judgment determining the rights of persons *inter se* in or to any money or property in dispute, but not affecting the status of persons or things or determining any interest in property except between the parties. They include all judgments for money.

Judgment *in rem*. A judgment of a court of competent jurisdiction determining the status of a person or thing, or the disposition of a thing by way of declaration of status or title, condemnation, forfeiture, or order for sale or transfer. Examples are judgments establishing or dissolving a marriage, revoking a patent, condemning a ship as prize, and determining that a street is a public highway.

Judgments and orders. Decisions reached and issued by a court on a question at issue between the parties to a matter before that court. The terms overlap and cannot always be clearly distinguished. Judgments may be *in rem*, determining the status of a person or the disposition of a thing, or *in personam*, determining the right of a party in relation to some claim. They may again be interlocutory, dealing with a point or stage in a case, or final, disposing of and determining the principal matter in issue.

Judgments Extension Acts. Statutes of 1868 and 1882, enabling persons holding a judgment or decree for debt, damages, or costs against another, obtained in the courts of one country in the United Kingdom, to register it in the corresponding court of another country thereof and enforce it as if it were a judgment of the latter court. Later statutes make similar provision for the enforcement in the U.K. of judgments of superior courts in certain Commonwealth countries and certain foreign countries.

Judicature Acts. Statutes of 1873–75, which recast the judicial machinery of England, by abolishing nearly all of the then existing superior courts and establishing in lieu, the Supreme Court of Judicature, comprising the High Court of Justice (divided into Queen's Bench, Common Pleas, Exchequer, Chancery, and Probate, Divorce and Admiralty Divisions) and the Court of Appeal, provided a uniform code of civil procedure and pleading for the court thus established, and merged the administration of law and equity by providing that each body of principles might be administered in all courts and that, in case of conflict, the rules of equity should prevail.

Judicial. Pertaining to a judge or judges, distinguished in many contexts from legislative and administrative, and in other contexts from extra-judicial, i.e. what is done outwith courts and without the intervention of a judge.

Judicial combat. See BATTLE, TRIAL BY.

Judicial Committee of the Privy Council. The tribunal created by the Judicial Committee Act, 1833, for hearing appeals which could then be brought before the King or the King in Council. Various other kinds of appeals have been directed by various statutes to be heard by the Judicial Committee.

The Committee consists of the Lord President and ex-Lords President of the Privy Council, the Lord Chancellor, the Lords of Appeal in Ordinary, and such other members of the Privy Council as are or have been Lords of Appeal in Ordinary, judges of the Supreme Courts of England or Northern Ireland or of the Court of Session in Scotland, and judges or ex-judges of the superior courts of the Dominions or of any other British possession fixed by Order in Council, who are Privy Councillors. In practice, the composition of the Committee for hearing an appeal is three or five judges, usually mostly Lords of Appeal, so that the composition of the Committee is very similar to that of the House of Lords constituted for judicial purposes.

The Committee sits in the Council Chamber in Whitehall, the members not wearing wig or gown, and argument is heard as in other appellate courts. Two or more committees may sit simultaneously.

Not more than two counsel will be heard on any side. The committee is reluctant to entertain any point not raised in the court below, and will not give opinions on hypothetical questions. It is very hesitant to disturb findings of fact.

The jurisdiction is to hear appeals from courts in colonies and dependencies (most Commonwealth countries, on attaining independence, having abolished appeals; countries leaving the Commonwealth have also done so), from the Admiralty court in prize, certain English ecclesiastical appeals, and under various statutes. The Committee is not an appellate court from United Kingdom courts. The Queen may, through the Attorney-General, refer to the Committee for hearing or considering any other matter, not necessarily arising from a judicial decision, and seek an advisory opinion from the Board.

The result of the appeal is announced in open court on a day appointed. The reasons are not read out in court but printed copies of the opinion are distributed. The decision of the Committee is technically not a judgment. As the Privy Council are advisers of the Queen, the decision is couched in the form that the Committee will humbly advise her Majesty that the appeal be allowed, or dismissed, for reasons given. At the next Council the report is submitted to Her Majesty, embodied in an Order in Council, and this becomes the final decree. For the same reason, and on the theory that advice to Her Majesty must be single, dissenting opinions were, for long, not allowed to be disclosed, but since 1966 such opinions are permitted.

Decisions and opinions of the Committee are reported in the law reports. They are not, strictly speaking, binding precedents for any court in the U.K., but if the subject matter of an appeal and the law concerned are similar to that of the U.K. decisions of the Privy Council are of high persuasive value, particularly in view of the similarity of composition of the Committee to that of the House of Lords in its judicial capacity and of the personal eminence of members of the Committee.

Judicial discretion. The faculty of deciding various issues not according to a strict rule of law, but by reference to what is fair, just, reasonable, convenient, or otherwise appropriate in the circumstances. Many matters are referred to a judge's discretion, rules of law and prior decisions at most indicating what factors he should take into account in determining how to exercise his discretion.

Judicial factor. In Scots law, a person appointed by the Court of Session or sheriff court to preserve and administer property which is in dispute, the subject of litigation, or from any cause without proper legal control and administration. Common instances are *curators bonis* to persons of unsound mind, and judicial factors on trust estates. A person appointed becomes an officer of the court, must find caution (security) for duly accounting for his care of the estate, lodge an inventory with the Accountant of Court, and submit annual accounts to him. He has the powers of a trustee and may obtain from the court other or special powers, and is subject to stringent penalties for default in duty.

Judicial functions. One of the major groups of functions of government involving the ascertainment of facts, the ascertainment of law relative thereto, and the application of the relevant law to the facts to the effect of determining claims, controversies, and disputes. This may involve the exercise of some, or substantial, discretionary power, e.g. in regard to the penalty imposed, or the remedy granted. The judicial function is essentially decisive, i.e. to decide controversies, not to advise on the law applicable or as to a course of action permissible or proper in a particular case. Judicial functions are commonly entrusted to courts of one or more legally qualified persons, but may be to laymen, or to Ministers.

Judicial Law for Laymen. The title of a Slavonic version of excerpts and adaptations from the *Ecloga*, chiefly concerned with criminal law. It may have originated in ninth-century Moravia or Bulgaria. It was current in mediaeval Bulgaria and was probably accepted in the pre-Mongol period in Russia. By 1280, it formed part of the Nomocanon used by the Russian Church as it still does.

Judicial legislation. Strictly speaking, judges have no legislative function or power to make or alter law, save in the limited sphere of making rules of court under authority delegated by Parliament, but the term judicial legislation is sometimes applied to judicial decisions which, by reason of the principles of *stare decisis* and binding precedents, in substance materially restate or modify hitherto understood principles of law and have the same practical effect as Parliamentary or delegated legislation. To apply an old rule to new facts, to declare that a rule does not apply to particular kinds of cases, to extend a rule to cover also other sets of facts, all have the effect of making or altering the law. But courts can do so, even if they wish to, only in limited circumstances, within the limits of the case which has arisen, and if put before them for decision and not, as a legislature may do, before a problem arises for decision. Courts sometimes decline to reach a particular conclusion on the ground that to do so would amount to judicial legislation.

Judicial method. Judicial method or the method or techniques followed by a judge or bench of judges in hearing and deciding a litigated controversy and announcing his or their decision varies from judge to judge, court to court and jurisdiction to jurisdiction. It is affected by many factors, the practice and procedure of the court, the different sources of law which are authoritative in that jurisdiction, and the personal attitude and temperament of the particular judge. Wide differences exist between the long discursive individual opinions of common-law judges and the terse anonymous judgments of Continental courts. Much importance attaches to whether, and how far, and how a judge may adapt old rules to new situations, or how far he may only deduce from actual rules in force.

In the first place, there is a question of whether a judge truly finds certain facts and then seeks the appropriate rule of law decisive of those facts, or whether he intuitively feels that the plaintiff or defendant is in the right and then 'finds' facts and law which justify a decision in that sense. Does a decision proceed on the basis of fact-finding and legal reasoning or of a 'hunch'?

In trial courts, there is normally, secondly, the problem of finding the facts, which depends on assessment of the veracity and accuracy of evidence tendered, and of trying to reconstruct from the various bits of evidence tendered, what was actually said and done by the parties and what the state of things was, e.g. the position and speed of vehicles. This is largely a matter of experience and of knowledge of human nature, and what is normal and likely in human conduct.

In the third place, there is the question of the mode of discovery from the authoritative sources of the legal materials for a decision, statutory provisions, precedents, statements in text-books, etc., their evaluation and the discovery from them of the rule decisive of the case before the the judge. This may be by deduction, or by analogy, or by inductive discovery from several decisions of a more general principle and the deduction therefrom of a rule applicable to the particular case. In superior courts judges can, generally, rely on counsel bringing to their notice in argument most, if not all, of the relevant authorities, but how they shape the principles and rules derived therefrom into an opinion is a highly individual matter.

Judicial notice. The knowledge, which is attributed by law to judicial persons, and which does not require to be given them in a particular case by evidence. It includes all matters of the law of his own system (though a court or judge may legitimately be reminded of the existence or terms of particular rules of law or hear argument as to the proper interpretation in the circumstances of a particular rule of law) and those matters of fact which are undisputed and of everyday knowledge, including all public matters of government, the ordinary facts of geography and nature generally, the use and meanings of common words of the language, and the like. In particular cases, it may be a difficult question whether a court is bound, or entitled, to take notice of a fact without evidence. This sometimes explains such judicial questions as: 'Who is X (a film star or sporting hero or current idol of the masses)?' The circumstances that a fact has been proved in one case does not enable the courts to take judicial notice of that fact in later cases.

Judicial opinion. In deciding disputes, judges commonly do two things, give judgment in favour of a particular party or in a particular sense, e.g. order that the defendant pay the plaintiff £X, and also issue an opinion reviewing the relevant law and giving legal reasons for giving judgment in that way. The opinion is of great importance for deciding whether or not it should be appealed. Particularly in the Anglo-American system the value of a decision as a precedent (q.v.) for future use depends more on the opinion than on the bare judgment, because it is from the opinion that one can discover the *ratio decidendi* (q.v.) or principle underlying the decision.

In appellate courts consisting of more than one judge, there may be a leading opinion and bare statements of concurrence by one or more other judges, or several full concurring opinions, or opinions concurring in the judgment or decision, but for partially or completely different reasons, or one or more dissenting opinions (which normally imply dissent also from the actual judgment or decision). It was, until 1966, the rule that only one opinion was delivered in the Judicial Committee of the Privy Council and no dissents were uttered because, technically, the Privy Council was not deciding the appeal but merely advising Her Majesty that the appeal be allowed or dismissed. The practice was changed in 1966 and now there is a single majority opinion and may be a dissenting opinion.

Judicial proceedings. Causes in being before any court, tribunal, or person having by law power to bear, receive, and examine evidence given on oath.

Judicial process. As distinct from legislative and administrative processes, the function of determining a controversy by ascertaining the facts and applying the law to the facts so found. The distinction of the three processes is not sharp or clear, nor always easy to draw in particular cases. Some jurists have defined law by reference to the judicial process: Holmes, J. said that the prophecies of what the courts will do in fact were what he meant by law. Cardozo, J. said that a principle so established as to justify a prediction that it will be enforced by the courts if its authority is challenged is a principle of law.

Judicial review. The doctrine that the validity of acts of a legislature may be challenged before, and adjudicated on, by a judicial body. It is not fully recognized in the United Kingdom, where Acts of Parliament are unchallengeable, though the validity of delegated legislation may be examined. It is best exemplified by the power of the Supreme Court of the U.S. to review the constitutionality, or otherwise, of Acts of Congress and of the legislatures of the states. The Constitution does not expressly confer any such power. The argument that the Supreme Court could exercise this power seems to have been based on the power of the British Privy Council, in colonial times, to deem acts of a colonial legislature void, as being contrary to the laws of England, and on analogy with the power, which had, in fact, sometimes been exercised by state courts in the U.S., to declare acts of state legislatures void as being contrary to natural justice, or to the state constitution.

But the problem is inherent in any written Constitution which confers certain powers on, and withholds others from, its legislature, or distributes power between federal and state legislatures. If a particular legislature is not to be allowed, by inadvertence or deliberately, to contravene the limits of its powers, some person or body must have the power to decide whether or not particular legislation contravenes the Constitution, and it is natural to commit this function to the judiciary.

The Supreme Court first held an Act of Congress void in *Marbury* v. *Madison* (1803), while in *Fletcher* v. *Peck* (1800), a state law was held contrary to the Constitution. The power of review of state legislation was extended to state courts by the decisions in *Martin* v. *Hunter's Lessee* (1816), and *Cohens* v. *Virginia* (1821). The danger inherent in the power is that of opposing the social and political views of the Court to that of the Congress. In 1935–37, the Court, in 12 decisions, invalidated much of Roosevelt's New Deal legislation and prompted his proposal to alter the composition of the court.

E. S. Corwin, *The Doctrine of Judicial Review*.

Judicial review of administrative action. The power of the superior courts to review the legality and validity of the actings and decisions of persons and bodies exercising administrative powers, whether of a legislative, executive or judicial or adjudicatory character. This is quite distinct from political and administrative challenge, e.g. through an M.P. or by complaint to the Parliamentary Commissioner (q.v.), and from any right of appeal which may exist in a particular case. In U.K. law, this is a very complicated subject and a very difficult area of law to chart.

The leading principle is, that if a person or body to whom authority has been entrusted exceeds that authority, or exercises a power without authority, the exercise may be pronounced invalid by a court. This includes compliance with the conditions attaching to exercise of the power, acting in good faith and excluding irrelevant considerations from the decision. If the repository of a statutory power has a discretion whether to act, it must apply its mind to whether the discretion should be exercised, and how, free from dictation by outside influence and bias, and must not disable itself from exercising the discretion. Accordingly, a major ground for control is that a body entrusted with a power has exceeded its jurisdiction in the way it exercised the power. Another, which is difficult to establish, is that the power, or discretion, has been exercised fraudulently, or in bad faith, or for an improper purpose.

The principles of natural justice, notably that a person shall not be judge in his own cause and that no man shall be condemned unheard, must be observed by all persons and bodies which have a duty to act judicially but not necessarily if the duty is merely executive.

The remedies which may be sought against administrative action are, in England, one or more of habeas corpus, *certiorari*, prohibition and mandamus, injunctions, declaratory orders, and damages (qq.v.). Since 1978, the main remedy is by way of application for judicial review. In Scotland, they are one or more of the common law remedies of declarator, reduction, interdict, damages (qq.v.) and, in certain cases, statutory orders to perform statutory duties.

S. A. de Smith, *Judicial Review of Administrative Action*.

Judicial separation. Before the Reformation, marriage was indissoluble but, from an early period, ecclesiastical courts in England granted decrees of divorce *a mensa et thoro*, which permitted spouses to live apart but did not break the marriage bond nor permit remarriage. When judicial divorce *a vinculo matrimonii* was introduced in England in 1857 this became judicial separation. The grounds for obtaining judicial separation are the same as those which ground a petition for divorce. If decree is granted it is no longer obligatory for parties to cohabit. The decree may be reversed or rescinded, and does not prevent a subsequent petition for divorce on the same facts.

In Scotland, though judicial divorce *a vinculo matrimonii* was introduced about 1560 and has been granted ever since, divorce *a mensa et thoro*, or judicial separation, continued to exist side by side with divorce. Since before the Reformation the grounds have been adultery or cruelty; by statute habitual drunkenness is cruelty. Since 1977, the grounds are the same as for divorce. A claim for separation is usually conjoined with a claim for aliment, so long as the separation lasts. Decree does not prejudice a subsequent action for divorce on the same adultery or cruelty.

Judicial trustee. A court may, in its discretion, appoint any fit person a judicial trustee to be a trustee of a trust jointly with another trustee or as sole trustee, or in place of any or all existing trustees. Such a person must give security for due application of the trust property, furnish the court with a statement of the trust property, and have his accounts audited annually, may seek directions of the court as to the trust or its administration, and may receive remuneration for his work and outlays from the trust estate. He may be suspended or removed by the court, or be discontinued in office.

Judiciary. The whole body of professional judges in any legal system.

Judiciary Acts. Legislation in the U.S. governing the structure of the Federal courts. The 1789 Act created the federal court structure and defined the jurisdiction of the courts, providing, in certain cases, for appeal from state courts to federal courts. The 1801 Act was an attempt by the Federalist party to retain control of the federal judiciary, creating new district and circuit courts and ending the riding of the circuits by Supreme Court Justices, but the Jeffersonians, in 1802, passed a new Circuit Court Act, re-establishing six circuit courts and renewing participation by Supreme Court Justices. The 1914 Act increased the opportunity to appeal state court decisions to the Supreme Court and that of 1925, increased the authority of the Supreme Court to dismiss certain cases and required that petitions for writs of *certiorari* to affirm lower courts decisions be granted if they obtained the affirmative votes of at least four justices.

Judicium. See IUDICIUM.

Judicium essoniorum. A tract written, probably about 1275, possibly by Ralph de Hengham (q.v.), and despite its title, dealing not only with essoins, but with the form of oath in the grand assize, defaults, jurors, entry, exceptions, and other procedural points. It seems to have been popular; it is printed in Woodbine's *Four Thirteenth-century Tracts* (1910).

Juge d'instruction. In France, a magistrate whose function is to conduct the investigation prior to a criminal trial at which the main evidence is collected and presented, witnesses heard, and their depositions taken. Though preliminary investigation was known from the seventeenth century, the office of *juge d'instruction* was created only in the mid-nineteenth century. The *juge* may institute proceedings only when criminals are caught in the act of committing crime or immediately afterwards, or if ordered to do so by the *procureur* (public prosecutor), or if requested to do so by a private citizen. When the investigation has been started, the accused must have the assistance of counsel, and the latter is entitled to access to all evidence and documents. The accused may be heard or may keep silent. In the conduct of the investigation the *juge* has wide powers to have the accused's residence searched and evidence seized, to have witnesses or experts appear to give evidence. The evidence is collected in a dossier on the basis of which the *juge* decides whether or not to commit for trial. If, after the investigation, the *juge* is not convinced that there is fully adequate evidence of guilt to warrant trial, the trial will not take place. The dossier is also made available to the presiding judge at the trial and to counsel for the defence, though the verdict at the trial must be based on the evidence then adduced.

The office has certain analogies to the function of the examining justice in English procedure, though the latter will only discharge the accused if satisfied that there is no such evidence against him as could result in a conviction.

Junior (or junior counsel). Any barrister, irrespective of age or length of standing at the Bar, who has not been appointed a Queen's Counsel (q.v.) or senior counsel. A standing junior counsel to a department or body is one appointed to act as junior counsel in all matters requiring counsel's attention arising in the department to which he is appointed.

Jura regalia. See REGALIA.

Jural correlatives and opposites. See JURAL RELATIONS.

Jural postulates. In the legal thinking of Roscoe Pound (q.v.), a set of basic hypotheses concerning a legal order which are presupposed by the claims actually made by persons for the legal protection of demands or interests. Pound, following Kohler, (q.v.), worked out what he saw as the jural postulates of the civilization of the place and time, namely mid-twentieth century U.S.A. They are, stated briefly, that men must be able to assume that others will commit no intentional aggressions upon them, that they may control what they have discovered or created or acquired, that others will carry out their undertakings, that others will not cast an unreasonable risk of injury on them, and that others will keep dangerous things within bounds. In short, they are:

freedom from aggression, private property, enforceability of bargains, reparation for negligence or harm of strict liability. The list is incomplete; there is no reference to freedom of inheritance. Clearly, also in differently-structured societies, different jural postulates must be formulated. A socialist society would reject at least the second postulate and probably add several others, and even a capitalist society, at a different place and time, might modify some of the postulates. Nevertheless, the formulation of the jural postulates is a valuable attempt to state the basic assumptions of the particular legal order. It cannot have universal validity.

Jural relations. A term for the legal relations of legal persons viewed in the most general way. Starting from the recognition that, in general, where any two legal persons have any legal relationship, one is said to have a right against the other, e.g. to payment of the price of goods sold, and the other is said to be under a correlative duty to the first, e.g. to hand over the goods sold, analytical jurists, notably Austin, Windscheid, Thon, Bierling, Salmond and Hohfeld, have distinguished other jural relationships. The main jural relationships recognized are those of right and duty, e.g. of payment, which can be described as I claim: you must; of privilege or liberty and no-right or inability, e.g. wear a scarlet gown, i.e. I may: you cannot; of power and liability, e.g. arrest, i.e. I can: you must; and of immunity and disability, e.g. utter possibly defamatory matter in my judicial capacity, i.e. I can: you cannot.

The sets of terms signifying these jural relationships, if put down in order indicate other logical relationships. (See the diagram below.)

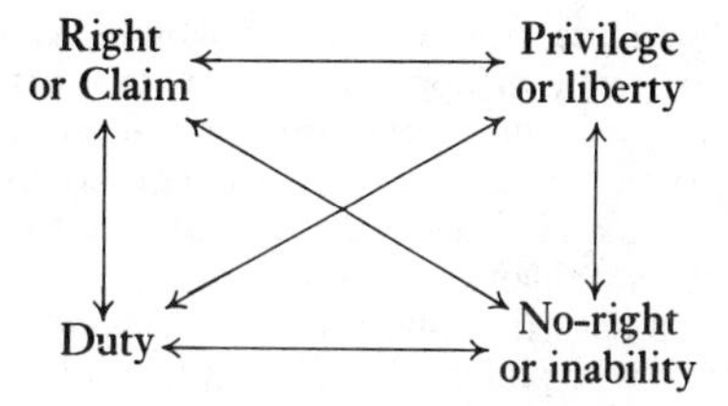

The vertical arrows indicate jural correlatives, in which the presence of either term in one party implies the presence of the other term in the other party to the relationship; the diagonal arrows indicate jural opposites, in which the presence of either term in one party implies the absence of its contradictory in the other party.

This analysis is of considerable value in clarifying the true meanings and usages of words, and in exploring the relationships implicit in a legal principle or rule. It makes clear that not every legal right is of the same kind, nor involves the same subjection in the other party. My right (claim) to payment of a debt has as its correlative the duty to pay me. My right (privilege) to sit in Parliament has as its correlative a non-member's no-right or inability to sit. My right (power) to arrest has as its correlative the wrongdoer's liability to be arrested. My judicial right not to be sued (immunity from being sued) for a wrong decision has as its correlative your disability to sue me. Much legal thinking would be clarified if every time the speaker was disposed to use the term 'right' or 'duty' he substituted whichever was appropriate of the other terms.

Jurat. The statement, at the end of an affidavit, of the names of the parties swearing it, the officer before whom it is sworn, the date and any other necessary particulars.

The assistants of the bailiffs in the legal systems of Jersey and Guernsey are called jurats and the name was probably adopted from there when used in the Cinque Ports and some other places for a municipal official rather like an alderman.

Jurimetrics. The scientific investigation of legal problems, including application of computers and other machines to legal purposes, such as for retrieval of source-materials for advice or decisions, e.g. passages in legislations, decisions and books, discovery of data from non-legal bodies of knowledge, and possibly, also to prediction of decisions on statistical or other mathematical bases, and solving problems of logical interrelations of meanings of words and phrases.

Jurisdiction. In international law jurisdiction, or the power of a sovereign to affect the rights of persons by legislation, by executive decree or by the judgment of a court, is intimately connected with the concepts of independence and of territory. Internally, jurisdiction is almost exclusive and, extra-territorially, it may be exercised only so far as other states allow.

An important distinction exists between jurisdiction to prescribe laws for persons and jurisdiction to enforce those laws. The former is unlimited but the latter is limited in that a state cannot, in general, enforce its laws against persons not physically present on its territory.

The jurisdiction of a state extends over the whole of the land, including internal waters, recognized as falling within its boundaries.

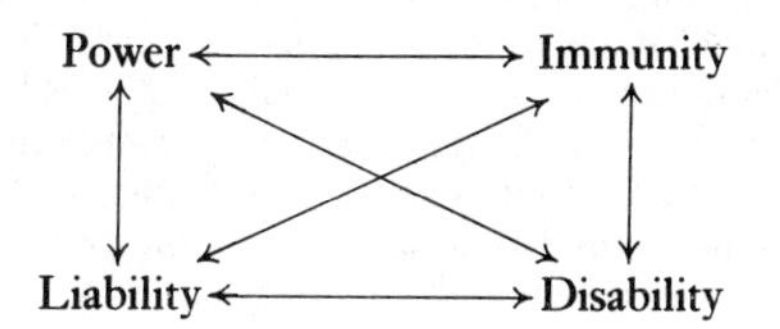

In respect of ships, the flag state has plenary jurisdiction over all the world, not because the ship is treated as a floating part of national territory, but as a matter of practical convenience. Municipal laws vary as to the extent of jurisdiction asserted over foreign merchant ships in internal waters of particular states, but it is generally accepted that a vessel does not wholly subject herself to local jurisdiction by taking refuge in a port by reason of stress of weather.

State jurisdiction extends over the territorial sea, and includes criminal and administrative jurisdictions over merchant ships passing innocently, as they are entitled to do, through the territorial sea. Hence regulations as to navigation, pilotage, pollution, and public health apply.

Personal jurisdiction is based primarily on nationality, which is granted or withheld according to the municipal law of various states, and a state may exercise jurisdiction over all its own nationals, both within and outwith its territory. A state cannot exercise powers extra-territorially in respect of aliens, but may do so over aliens entering its territory, and may treat them differently from its own nationals.

Personal jurisdiction is qualified by the rights of asylum (q.v.) and the claim for extradition (q.v.).

In municipal law, jurisdiction is the power or authority of a court to take cognizance of matters put before it and to decide such matters. In the United Kingdom, the leading principle is that all jurisdiction is derived from royal authority and judges exercise jurisdiction in name and by authority of the sovereign.

Limits may be set by the statute, charter or commission under which the court is held, by reference to area, subject-matter or value of cases, powers of punishment which may be exercised, or otherwise. An exercise of jurisdiction beyond that possessed by a court is a nullity and, in England, the Queen's Bench Division can, by order of prohibition, restrain an inferior court from exceeding its jurisdiction. If judgment has been given by the inferior court, the Queen's Bench Division may, by order of *certiorari*, have the record brought up that the judgment may be quashed.

Jurisdiction may also be divided into civil, criminal, or special (including, e.g. court-martial, ecclesiastical courts, and the like), and in any of these cases may be unlimited or limited in various ways, and into original (or first instance) and appellate jurisdiction. Some courts have both civil and criminal jurisdiction, and some both original and appellate jurisdiction. Thus, the High Court in England has unlimited original civil jurisdiction, and limited appellate jurisdiction, exercised normally by Divisional courts. The Court of Appeal has very limited original jurisdiction and mainly appellate jurisdiction. The county court has original civil jurisdiction, limited in point of area, subject matter and value. Civil jurisdiction must, generally, be founded on the power of the court to make its judgments effective against the defendant and criminal jurisdiction on the commission of a crime within the area for which the court exercises jurisdiction.

Jurisdictional facts. If a court is appealed to to review the decision of an administrative agency, a major ground for doing so is that the agency has acted *ultra vires*, or beyond the jurisdiction conferred on it in certain defined factual situations. To enable it to effect the review, the court must, accordingly, be satisfied that the essential elements of the defined factual situation are all present (these are the jurisdictional facts); if one is not satisfied, the agency can be said to have acted without jurisdiction. Thus, if a Ministry is empowered to make orders to demolish houses for reasons of public health, the court may review its decision to do so in a particular case if it can be alleged that the premises in question were not 'houses' as defined in the empowering Act. Their being 'houses' is a jurisdictional fact, and if the premises are not 'houses', the Ministry had no jurisdiction to make a demolition order. This issue of *vires* or jurisdictional facts is applicable to legislative, executive, and judicial functions of administrative agencies.

Jurisprudence. The term is used in various senses, firstly as *juris prudentia* or *juris scientia*, covering the study and knowledge of law in the widest sense, in which sense the word is synonymous with the broadest sense of the phrase legal science (q.v.). Secondly, as the term for the division of legal science concerned with thinking about law generally, as distinct from making, explaining, expounding, criticizing or applying the law of a particular system, and with examining the most general, abstract and fundamental ideas and questions about law in general. This sense is sometimes equated with theory of law or legal theory, science of law or legal science (in a more restricted sense of that term), or with philosophy of law or legal philosophy. Thirdly, the word is sometimes used as a more grandiloquent synonym for law, e.g. medical jurisprudence, but particularly in contexts where the word 'law' might seem inappropriate, e.g. equity jurisprudence. In French law, *la jurisprudence* means the course of decisions, the body of case-law on a topic, and this usage is sometimes also found in English law.

Jurisprudence in the second sense, thinking about law and its problems, has been a subject of study from early times, pursued at least as much by philosophers in the context of their speculations about man, society and the state, as by lawyers Philosophers have tended to look at the more abstract and general ideas, and lawyers at the general issues raised by particular legal systems, principles, and problems. Among the questions commonly discussed in works on jurisprudence are such as: What are the sources of law? What is the connection between law and morality? Why should law be obeyed? How are the principles of a legal system grouped and classified? How does law control conduct and protect persons? What is meant by saying that X has a 'legal right'? What is law? But the content of a book on jurisprudence commonly reflects the writer's standpoint as much as anything else.

In the English language, the term jurisprudence has been used in this second sense only since the writings of Bentham and Austin and, because of their (particularly the latter's) emphasis on the analysis of the structure, ideas, and terms of English law, jurisprudence was thought of in England until the mid-twentieth century mainly as analytical jurisprudence (q.v.).

The literature of jurisprudence is voluminous, and comes in many languages and styles. Some writers on the subject have soared so high into abstract speculation that their theories and views appear to have no relation whatever to any actual system of law, or to any concrete legal problem, while others have kept so down to earth that their views are no more than a rather generalized account of the rules of a particular system, frequently of Roman or English law, and show no awareness of the existence of other systems of law, or other approaches to the same problem. In some respects, the latter approach is justifiable, for an analysis of such ideas as malice, or possession, or trust can possibly only usefully be made by close examination of one or two legal systems, the problem being largely linguistic, of word usage in that system.

A distinction used to be drawn between general, particular and comparative jurisprudence, the first purporting to ascertain and examine ideas found, generally, in legal systems, the second examining the materials provided by one system, and the third utilizing materials drawn from several systems examined in relation to one another. It is, however, questionable whether any ideas are found generally in legal systems and, if so, whether they operate in the same way in all cases; the same is so in lesser degree of comparative jurisprudence, but particular jurisprudence is in constant danger of identifying a characteristic of one legal system, e.g. the distinction between law and equity, with a natural, necessary, and normal factor in legal systems instead of an historical accident of one.

Developed legal systems may be looked at from at least five main points of view:

(*a*) *philosophical* or ethical, examining the philosophical bases of the institutions and doctrines of law, seeking to understand its fundamental principles, to understand and organize the ideal element in the formal sources, and to evaluate and criticize law in terms of the ideals and ends it sets itself; this approach is sometimes called theory of justice;

(*b*) *historical*, investigating the origin and development of the system, of the growth and change of its institutions and doctrines, to discover its spirit and general principles, and organizing the source-materials on the basis of the principles as developed historically;

(*c*) *comparative*, examining systems, institutions, structure, concepts, and rules by reference to their growth, scope, application, and use in different legal systems at comparable stages of development;

(*d*) *analytical*, examining the sources, structure, subject matter, concepts, and rules to ascertain the theories, principles, and ideas the system logically presupposes, and to organize the authoritative materials for judicial and administrative determination on this logical basis;

(*e*) *sociological* or functional, studying systems of law functionally as systems of social control of conduct, and studying their institutions, doctrines, and methods with reference to the social ends sought to be attained.

Definitions of jurisprudence tend accordingly to reflect the attitude or approach of the particular thinker, e.g. the science of the human will, in the distinction of the particular from the universal, and in the relation of the particular to the universal (Herkless); scientific knowledge of the history and system of right (Puchta); the formal science of positive law (Holland); the science of the first principles of the civil law (Salmond); the lawyer's examination of the precepts, ideals, and techniques of the law in the light derived from present knowledge in disciplines other than the law (Stone).

It is, however, mistaken to approach the study of law solely from any one standpoint; full understanding requires that all be utilized.

Jurisprudence originated in Greek philosophy, in speculation about justice and the social order, notably by Plato and Aristotle, and by the Stoics who developed the concept of natural law, an emanation of the law of reason of the cosmos. It continued in a more practical way among some of the Roman thinkers, notably Cicero, who was much influenced by Stoic ideas and began to analyse some of the general ideas found in Roman law and, in particular, developed the ideas of the *jus civile*, the rules applicable to Roman citizens, the *jus gentium*, rules generally applicable among civilized men, and hence applicable to foreigners as well as Romans, and the *jus naturale*, a body of principles derived from reason and worked out philosophically, potentially applicable to all men everywhere, and serving, sometimes, as a basis for legislation, for decision, and for criticism of existing laws. The Roman jurists were, however, generally more practical legal scholars than speculative theorists about law.

In the Dark Ages, St. Augustine reintroduced Stoic philosophy beside Christian thought, putting divine reason beside divine will as the highest source of the divine law binding both man and all other creatures. Below divine law he placed natural law and the positive law of the State. This helped to preserve the idea of government under law until the Dark Ages had passed.

The modern science of law began with the revival of the study of Roman law in the twelfth century and proceeded down to the seventeenth century on two parallel lines—legal and philosophical. The legal line began with Irnerius, Accursius, and the other glossators, continued with Bartolus and the other commentators, and culminated in the great humanists, Alciatus, Cujacius, and Donellus.

The philosophical line began with St. Thomas Aquinas, and the schoolmen with whom jurisprudence was subordinate to theology, and temporal law subordinate to natural and eternal law. A late example of this line of thinking is St. Germain's *Doctor and Student*. But jurisprudence broke away from theology in the work of Hemmingius, Belli, Gentilis, Soto, and Suarez.

In the seventeenth century, thinking about the fundamentals of law became freed, both from connection with theology and from almost exclusive consideration of the text of the *Corpus Juris Civilis*, by the work of such as Hotman and Conring. The law-of-nature school, which sought to deduce a complete system of principles of universal validity from the nature of man in the abstract, arose in the works of Grotius and Pufendorf.

The positivists, insisting on the supremacy of human lawgivers, represented a different divergence from theology; they are represented by Machiavelli, Bodin, and Hobbes.

The law-of-nature school continued influentially throughout the eighteenth and into the nineteenth century, as exemplified in the work of such as Burlamaqui, Wolff, Rutherforth, and Vattel. It influenced Blackstone and some decisions; in America it provided a background for some of the earlier decisions on judicial review. As late as the work of Lorimer, it was seen by some as the basis of law.

Since the early nineteenth century, a variety of schools of thought about legal problems have flourished. There was a reaction against both natural law thinking and Kantian Idealism. One major development was analytical positivism, concerned with positive law, as laid down in particular communities and with the analysis of concepts, logical examination of terms and principles, classification and exegesis, not with the ethical evaluation of rules of law or their effects. This approach has been extremely influential in England and the U.S.A. Starting with Bentham and Austin, it was the basic standpoint of Markby, Holland, Salmond, and Gray.

Another development was historical jurisprudence, in which the leading figure was Savigny, who saw law as a development of the popular consciousness of a people. In England, Maine's broader interests and the influence of ideas of biological evolution, and of anthropological studies led him to a comparative historical jurisprudence and to the recognition of legal evolution.

The study of society, social evolution, and the development of the social sciences generally, gave rise to sociological jurisprudence which is more concerned with the social effects of legal institutions and rules, and effectiveness and effects of law, and with law in society. It was influenced in its early stages by Comte and Jhering, by biology and psychology, and developed by Kantorowicz, Pound, and Stone. The approaches and methods of the social sciences, and their findings are increasingly invoked in the making and application of law.

The Marxist economic interpretation of history emphasizing economic factors has had most influence in Eastern Europe but, combining with English utilitarianism, has had some influence in England. Some jurists, such as Renner and Weber, without accepting economic determinism, have shown how influential economic factors have been in law-making.

In America, a modern development has been realism, concentrating attention on what courts and legal officers do about disputes. Holmes, Frank, and Llewellyn have been the most notable figures. A distinct kind of realism has developed in Scandinavia, in the work of Hagerstrom, Lundstedt, and Olivecrona.

In the twentieth century, there has been a revival of natural law thinking, springing from disillusionment with positive law; it was initiated by Gény and Stammler, and followed by Radbruch. Linked to this has been the French institutionalist school, notably Hauriou and Renard, who regarded social institutions as expressing the social reality underlying law.

Another noteworthy development has been Kelsen's pure theory of law, which seeks to segregate law from ethical, social, and other factors. The quest is for the Grundnorm or fundamental norm from which all the other norms in the hierarchy of norms in a particular legal system are derived. The sole object of study for jurisprudence is the nature of the norms or standards which are set up by law.

In modern jurisprudence, there is increasing recognition that all approaches to legal thinking have something to contribute to understanding law, and there has been increasing acceptance that, particularly in English-language jurisdictions, excessive attention was devoted in the nineteenth century to law as a logical structure of norms and too little attention to the factors which gave rise to and influenced the development of law and to the actual effects and consequences of particular rules and categories of rules. Much more than in the past law is looked on in relation to society and the other social sciences, as an instrument of social policy. Law is being consciously modified on the basis of research into the operation of rules and their consequences and deficiencies. The major problems of jurisprudence include clarification of the relations between law and the State, authority, liberty, conscience, and justice. There is need for consistent and continuing reappraisal of these issues in the light of contemporary problems.

H. Cairns, *Legal Philosophy from Plato to Hegel*; C. J. Friedrich, *Philosophy of Law in Historical Perspective*; R. Pound, *Introduction to the Philosophy of Law*.

Jurisprudence and anthropology. Anthropology studies the biological characteristics and distribution of races and tribes of men throughout the world; the branch thereof designated social or cultural anthropology is concerned with the social organization, institutions, culture, customs, and modes of life of such groups, particularly those living in primitive or undeveloped conditions today. The study of developed societies belongs rather to sociology. Studies of such primitive groups and peoples frequently reveal that non-literate peoples have a highly complicated network of customs and practices recognized as binding and as coercive as a system of law. Taboo, ridicule, and other group pressures are as coercive as judges and gaolers in developed societies. In studying the legal aspects of savage societies, Durkheim's sociological theory of early culture and Kohler and the German school of historical jurists' doctrine of primitive legal cohesion have been important.

From such studies, it is reasonable to infer the nature of and mode of operation of customs and practices in developed societies at a much earlier stage of their development, before the existence of either written law, legislation, courts and judges, or of written records of law or disputes, and to see the probable origins of and understand the justifications for rules and practices of early law. Particularly interesting and instructive are the institutions of marriage and the family, while study of primitive peoples gives insight into the psychology of such peoples and of other peoples when they were at a similarly primitive stage.

E. Westermarck, *History of Human Marriage*; L. T. Hobhouse, *Morals in Evolution*; R. Lowie, *Primitive Society*; B. Malinowski, *Crime and Custom in Savage Society*; H. Hogbin, *Law and Order in Polynesia.*

Jurisprudence and criminology. Criminology, as the study of criminal behaviour, is initially dependent on law in that law defines what is criminal as distinct from merely morally wrongful or socially unpleasant and unacceptable, and investigation and legal disposal of cases provides the materials for criminological investigation. But that done, criminology, as the study of why persons commit crime, its incidence, and pattern, depends on psychology, demography, statistics, and other sciences much more than on law. But better knowledge of the factors in man and society which lead to the commission of crimes provides important information for legislators, judges, and prison officials which may lead to changes in the law, in its administration, and in more understanding application of the law.

The other aspect of criminology, penology, or the study of punishment and treatment, draws on philosophy and psychology and provides knowledge and arguments for modifying the law relative to punishment, and utilizes the experience of particular punitive methods, enforced according to the law in force at a particular time and place, as providing experimental data for further proposals.

Jurisprudence and economics. Economics is the study of the production, exchange, and distribution of wealth in and between communities. On the philosophical plane, legal and economic philosophy overlap and interact and the organization of society aimed at, *laissez-faire* or dirigiste, public or private enterprise, determines the legal structure and system as much as the economic. To a substantial extent, the sciences of law and economics deal with the same topics, but viewing institutions and relationships from different standpoints. Law is largely concerned with economic relationships, such as buying and selling, lending and borrowing, with means of credit, security, payment and interest, defining the incidents of the relationships and the consequence of default. As societies become more developed, law becomes increasingly concerned with controlling economic activities, such as monopolies, and economic organizations, such as corporations and trade unions, even with regulating prices and wages, rent and interest. The implications of taxation for many activities and possible courses of action must be considered.

Economics, in turn, influences law in that the needs of business, trade, and employment constantly give rise to new needs and legal institutions, and rules have to be adapted or developed to take account of them. Thus the development of negotiability, of commercial credits, of insurance, of new forms of business organization, of limited liability trading, have all been stimulated, and even forced, by economic practices and needs. The increasing control of economic activities by modern governments and the increased importance, range and variety of taxation, increase the utilization of legal means for achieving economic ends.

Jurisprudence and ethics. Ethics, or moral philosophy, is concerned with what is morally good or bad, right or wrong, and clearly overlaps into the sphere of the study of what is legally right or wrong, but while there is an area of overlap, there are also points of distinction. Law is concerned with social relations, ethics with the conduct of individuals. Law is concerned mainly with external conduct as measured by external standards, ethics with intentions and the motive for action. But ethics must consider the consequences of action and the obligations of a man to his fellow-men.

Law in a society accordingly tends to lay down only those standards which are deemed essential, and cannot seek to enforce the highest ethical standards. Also, it has to deal with many situations, such as traffic regulations, where ethical principles give little or no guidance for the solution of the problem.

Ethics helps to provide standards for criticism of positive law by formulating ends which the legal system should seek to promote among individuals. Thus, in England, Benthamite Utilitarianism was the basis of criticism and reform of whole tracts of law.

The major area of common concern to jurisprudence and ethics is the problem of justice, a subject which has been studied more, and more effectively, by moral philosophers than by jurists.

Jurisprudence and history. These studies are closely interwoven. In the first place, law in general, legal theory, particular systems of law, particular doctrines, principles, and rules can all be studied historically and their courses of development charted. This is, moreover, not only interesting but frequently an essential step to understanding when, how, and why evolution proceeded and particular changes took place, and appreciation of the historical development of a principle or rule may be essential to its true understanding. Secondly, the development of law in society generally, and in a particular country or community, is merely one facet of the general history of that community, and is inextricably woven in with the other historical threads, political, constitutional, economic, ecclesiastical, and other, which go to make up the total pattern of events; legal history must be looked at as part of general history. Thirdly, the development of law, both its theory and its content, is frequently powerfully influenced, or even totally determined, by events which belong primarily to another thread of history; thus the French Revolution brought about enormous changes in the law of France, the Industrial Revolution in the law of the U.K.

Jurisprudence and philosophy. Philosophy, a term originally comprehending practically all higher learning and knowledge, or culture in general, is now limited to the study of the most general ideas and problems raised by more specific sciences, such as psychology, physics, politics, or law. It is a questioning attitude of mind, an approach to problems, rather than a body of knowledge, and philosophers differ radically among themselves on the questions they discuss and the answers they offer.

It is distinguishable into several branches, particularly, metaphysics, concerned with the ultimate nature of reality, epistemology, with the theory of knowledge, logic, concerned with truth and reasoning, ethics, with goodness and rightness, and aesthetics, with beauty, but the methods of philosophy are applied to the most general issues raised by the particular sciences, giving rise to the subjects of legal philosophy, political philosophy, social philosophy, philosophy of religion, of science, of education, and others.

The interaction between law and several branches of philosophy is close and extensive because many of the most famous philosophers have devoted attention to philosophic problems of law and because, the more a lawyer thinks about legal issues, the more he is led to an essentially philosophic approach, and to ask philosophic questions about law, such as: What is the authority of law? What is a right? Why should people obey law? and the like.

Legal philosophy is particularly related to ethics and to social and political philosophy and overlaps with each of these various respects. In Plato and Aristotle, ethics and politics are closely related and the State is an organ of morality and itself an expression of ethical ideas or qualities. The problem of justice inevitably involves consideration of social and ethical reflections on that subject, and on law.

Legal argument and reasoning is connected with, and to a substantial extent, relies on logic, on deduction of a rule for a case from a more general principle, on inductive discovery of a general principle from several particular decisions, on reasoning by analogy and other applications of logic. Analytical jurisprudence is essentially a critique of law in terms of logic.

In modern jurisprudence, much attention is devoted to classification and better understanding of language and terminology and this interest has been stimulated by and draws extensively on the work of modern linguistic analysis philosophers.

Major contributions to legal thought have been made by philosophers incidentally to their more general analyses of society and human conduct, of knowledge, reasoning, and thinking.

Jurisprudence and political science. The studies of legal and political science have a largely common origin and development, and a substantial area of overlap, both on the theoretical (legal and political philosophy) and practical (legal and political systems and organizations) sides. The ideas of law and of the State are almost inextricably enmeshed and constructive thinking about the ideals, ends and aims, concepts and forms of either inevitably involves the other and involves thought about the other. Every legal system, theoretical or which has existed or exists, which has matured beyond mere custom and usage, assumes the existence of some kind of State organization which, if it does not create, at least amplifies, modifies, supports and enforces the legal ordering of society. Every political system, imagined or actual, has operated and largely depended on its legal system. The structure and attitudes of the legal and the political order in a society, repressive or liberal, react on and influence one another. Major concepts, such as the constitution, the separation of powers and functions, the relations of Church and State, the exercise and control of power are central to both bodies of knowledge.

Most of the great thinkers, who have devoted attention to the problems of how communities should organize themselves for regulating the lives of their citizens, from Plato and Aristotle to Marx and Lenin, Sartre, Durkheim, Wallas, and others, have devoted attention to the legal structure which makes a society what it is.

In modern times, as the power of the State, in both communist dictatorships and nominally liberal-democratic states, extends and develops, questions have increasingly arisen: How can individuals control the power of the State? How can they assert rights against the State? Can men have any liberty in a State?

Jurisprudence and psychology. Jurisprudence is much concerned with the regulation of human conduct, while psychology studies the behaviour of man and may, therefore, cast light on what makes men act in ways legally significant, on motivation, emotion, intelligence, and cognitive processes. Particular interest attaches from the jurisprudential angle to social psychology, the behaviour of individuals in their social and cultural setting, their attitudes and beliefs, and their interactions in groups, and to the psychology of abnormal behaviour, deviations and delinquency. In cases where behaviour is materially disturbed it is more a concern of psychiatry than psychology. Such disturbances frequently lead to conduct which contravenes social and legal norms.

Jurisprudence and sociology. In origin, sociology was intended by Comte as a study of the impact of the new scientific and historical knowledge which had developed between Hume's *Treatise of Human Nature* of 1739 and his own day on the conceptions of Human Nature and Society. It has later come to be used for the study of living societies, their evolution and change, structure, stratification, social institutions, organizations and phenomena.

The connection with law is close in that a system of law and legal institutions, such as courts, lawyers, marriage, contract and inheritance are also social institutions, and because law and legal institutions grow out of the needs of and exist to serve the needs of the society in which the system exists. Those who make, amend, and apply the legal system must understand the institutions and organizations of their society, and, particularly, such factors as social conformity and deviation. Sociology studies many topics of legal relevance, such as divorce, alcoholism, crime, attitudes to colour, and religion.

The connection has given rise to two developments, the growth of sociological jurisprudence, which views law functionally, as a mode of social control rather than as a hierarchy of norms, and the sociology of law, which examines law and legal institutions from the sociological standpoint and concerns itself with how law operates rather than with what it is.

Jurisprudence and theology. Theology is thinking about God, or what is conceived to be the Supreme Being or controlling force in the universe, about the world, man, faith, and related themes. It is the attempt of the adherents of a faith to formulate their beliefs consistently. Similarly, jurisprudence is the attempt of the thinkers of a legal system to make a consistent scheme of the aims, dogmas, and methods of that system. In primitive societies, religious beliefs and beliefs about what conduct is obligatory or forbidden is hardly differentiated, and one of the earliest conceptions of law is as rules ordained by the gods, and interpreted to men by priests; law is indeed frequently stated to have been delivered to men by a god, as in the story of the delivery of the tablets to Moses. Sin and legal wrong may be hardly distinguishable. In less primitive but early societies, many aspects of human relations were regulated by religious beliefs and cultic forms and practices.

To the Christian scholastic philosophers, particularly Aquinas (q.v.), God was the source of religious belief, morals, and law alike and rules of law were human formulations of rules to promote conformity of action to God's will. This view influenced natural law thinking and modern scholastic views and maintains a connection in the view of some jurists between theology and jurisprudence.

Later, the Protestant jurist theologians, notably Hemmingsen and Oldendorp, sought to emancipate jurisprudence from theology though, in the sixteenth and early seventeenth centuries, the Spanish jurist theologians, Soto and Suarez, continued to find in theology the bases of law. From the seventeenth century, however, jurisprudence has steadily moved away from theology.

The institutions of, beliefs in and study of theology and jurisprudence have many parallels; to the priests correspond the judges; to vestments correspond robes; to the theologians correspond the jurists; to the Fathers of the Church correspond the lawgivers and writers, Henry II, Bracton, Coke and others; to the ritual corresponds procedure; sin requires penance and absolution, crime or wrong requires restitution or compensation, and ultimate absolution, and so on. Not least, both theology and jurisprudence are great bodies of learning which have both fascinated, and been illuminated by the thought of, some of the most powerful intellects which have ever lived.

Jurisprudence and the social sciences. The social sciences is the collective name given to the group of sciences, or branches of learning, which deal predominantly with aspects of the conduct, relations, and interactions of men living together in societies. The divisions between the humanities, or literary and philosophic studies, social sciences, and natural sciences, the three main categories of sciences or branches of knowledge, are neither clear nor rigid and the classification of a body of learning or knowledge in one of the three divisions can only be by regard to its predominant characteristics and concerns. The main social sciences are generally taken to be anthropology, criminology, economics, history, law, politics, and sociology, but there are other sciences, such as geography, psychology, and statistics, which have substantial social aspects and many other sciences, such as architecture, medicine, public health, town planning, with substantial social implications. Indeed, no branch of intellectual investigation is without social implications as soon as it proceeds beyond the mind of its discoverer, and his library or laboratory. Philosophy has both social aspects, in social and political philosophy, and penetrates all the social sciences, in that their more ultimate questions are essentially philosophical. History is both a study in itself and an approach to both the theoretical and practical studies of each of the social sciences.

Legal science or jurisprudence (in its broad sense) is justifiably grouped among the social sciences. Moreover, it is one of the oldest and among the first of them to be cultivated as a science, a body of knowledge systematically organized and cultivated as a subject of study, analysis, and exposition. Written laws have been found dating from many centuries B.C. Serious thinking about law, its aims, ideals, nature, and structure began with Plato and Aristotle. To look at law as a social science emphasizes its character as a body of principles for the regulation of the conflicting interests of states and individuals, and as a corpus of knowledge capable of being analysed and expounded as a rational body of doctrines, principles, and rules, rather than its character as a body of practical knowledge utilized by several professions in their everyday work for the practical purposes of deciding disputes, making agreements, transferring property and attaining other practical ends. These two aspects of law are, broadly speaking, the pure or theoretical and applied or practical aspects; they run into one another and cannot be sharply distinguished, and the more difficult a practical problem the more the lawyer is thrown back on theory, principles and abstract ideas. The state of the law of a country on any matter is a matter of general social concern, though persons generally may be more interested in what is criminal, or the grounds of divorce, than in bills of exchange, or settlements of land. Even the most technical and practical issues of law, such as the modes of conveying land, court procedure, representation and legal aid, costs, execution, and distress are of general social concern, indeed frequently more so than a controversy over some matter of legal theory.

H. Cairns, *Law and the Social Sciences.*

Juristenrecht (jurists' law). A term applied by some scholars to systems of law, such as English and French law, strongly influenced in their development by the principles worked out by jurists, including especially judges and practitioners, in actual cases, as contrasted with *Professorenrecht* (professors' law) systems, such as German law worked out largely by academic theorists and accordingly highly doctrinal and theoretical.

Juristic act (or act in the law). An expression of will or declaration of intention by a person having legal power to bring about a particular legal result, or produce a legally possible and permissible result. It is by juristic acts that legal persons create, modify, or extinguish rights and duties, and affect legal relations between legal persons. Major examples of juristic acts are making promises or wills (unilateral juristic acts) and making agreements (bilateral juristic acts). They have to be distinguished from acts of the law where the law itself affects the rights and duties of a person, or persons, independently of their volitions.

Juristic conceptions, Heaven of (Begriffshimmel). A place which lies far beyond the solar system in complete darkness, the site of von Jhering's (q.v.) experience described in his satirical *Scherz und Ernst in Jurisprudenz* (1885), where juristic theorists are rewarded for their service on earth to the examination of pure legal conceptions by learning the answer to some of the endless civilian problems, such as the nature of possession. The abode of the theorists includes the greased pole of difficult juristic problems, the museum of pure conceptions which has no doors and is entered by running one's head against the brick wall, the apparatus for 'restoring' difficult texts, and other useful equipment. At the end of the satire Jhering awakes, to find that it has all been a dream. The satire delightfully exposes the excesses of theoretical conceptualizing by German civilian jurists.

Juristic persons. Those groups of natural persons or things which a particular legal system endows with legal personality and treats as being, in law, kinds of persons and, accordingly, being able to sustain legal rights and duties. In U.K. law the major kinds of groups, which are treated as juristic persons, are corporations (including chartered and statutory bodies and companies), building societies and industrial and provident societies, but not trustees, partnerships, trade unions, friendly societies, clubs, and associations. In other legal systems, however, foundations, temple idols, and other things may be deemed juristic persons.

Juristic writing. See BOOKS OF AUTHORITY; INSTITUTIONAL WRITINGS; SOURCES OF LAW.

Jurists. A term without a precisely defined meaning but generally used as designating persons very learned in the law. A jurist may also be a judge or practising lawyer, but the term is normally applied principally to recognized legal scholars or writers or consultants.

In Roman law, the terms *iurisperiti*, *iurisprudentes*, and *iurisconsulti* were used of persons who advised the magistrates on the composition of their edicts, drafted legal forms, assisted in the formulation of new or modified *formulae* for the trial of actions, and gave opinions (*responsa prudentium*) to parties, which were commonly influential with judges, to individuals and to judges themselves. They also gave instruction to young men aspiring to legal or administrative careers.

In republican time jurists belonged to senatorial families and often held high offices as consul and praetor. In the classical age the jurists were a recognized class of persons of high rank, mostly of senatorial rank and many of whom held magistracies and offices in the imperial service. Those who had the *ius respondendi* (q.v.) had an official position with imperial authority. All were concerned with advising on law, giving opinions, and guiding magistrates and litigants on the formulation and decision of practical problems. Many were also extensive writers on law, almost entirely on practical issues and matters closely related to everyday problems, and only incidentally on theoretical issues. As teachers and writers they promoted the development of legal science.

Their educational function became more definite and academic and there is evidence of two law schools at Rome in the first and second centuries—the Proculians and Sabinians. Roman legal science reached the heights of its development under Trajan and Hadrian with the later members of the two schools.

Their extensive output of legal literature comprised *institutiones* or *enchiridia*, elementary manuals, *regulae*, *definitiones* or *sententiae*, handbooks for practitioners and students, general works on the *ius civile*, commentaries on the edict, *digesta*, treatises on the law as a whole, *responsa*, opinions on issues presented to them, *quaestiones* and *disputationes*, discussions of problems raised in practice or in argument, and commentaries on individual *leges* or *senatus consulta*. The careers of many individual jurists are known.

Under Justinian, jurists belonged to professional groups, the bureaucratic jurists, a group including the jurists who belonged to the imperial council and framed the imperial rescripts and statutes and the *Codices Gregorianus*, *Hermogenianus*, and *Theodosianus*, the academic jurists, and the practical advocates. But the dead hand of autocracy restrained any original exposition of law.

From the revival of the study of the Roman law, the category of jurists is mainly associated with the academic lawyers, who have taught and written in the universities, though some who have been primarily judges or statesmen or philosophers can justifiably be called jurists. The term is inappropriate for the general run of academic lawyers and appropriate rather for those who, by their teaching and particularly by their writings, have made a name and a reputation, have put order into large tracts of legal rules, have explored the history and philosophy of law, have influenced judges and legislators and made real contributions to scholarship. In Europe, the term is appropriate for Pufendorf, Pothier and Savigny, to name only a few, in England for Hale and Blackstone, in Scotland for Stair and Erskine, in America for Kent and Story. Among the moderns the title is apt for such as Maine, Dicey, Anson, Pollock, Maitland, Salmond, Holdsworth, and Cheshire. Judges who merit also the title of jurist include such as Atkin, Lindley, Scrutton and Holmes.

In modern times, the functions of jurists are various. They include the critical examination of legislation and decisions of the courts and the publication of comments, bringing out the meaning and effect, and importance, of the legislation or decisions, pointing to the confusions, mistakes, and misunderstandings, and to the effects of adopting one interpretation or another of the ruling; the systematization of the principles and rules of particular branches of the law, or on particular topics, as by extracting from the statutes and cases on contracts or negligence, the general principles, the special rules, and the exceptions; the writing of books and articles commenting on or criticizing the law, pointing to its deficiencies and suggesting remedies; the exploring of the philosophic and theoretical bases of principles, rules, tracts of law, institutions, and the whole idea of law itself; the elucidation of the historical origins and development of law in general, particular legal systems and particular bodies or principles and rules; and the comparative examination of legal systems, doctrines, and rules. In modern times, jurists are not infrequently appointed to Law Commissions and committees charged with examining particular branches of law and recommending changes.

F. Schulz, *History of Roman Legal Science*; H. Rashdall, *Universities of Europe in the Middle Ages*; J. Macdonnell (ed.), *Great Jurists of the World*; C. H. S. Fifoot, *Judge and Jurist in the Reign of Queen Victoria*; J. P. Dawson, *The Oracles of the Law*.

Jury. A group of laymen summoned to assist a court by deciding a disputed issue of fact on evidence heard.

An ordinance of King Ethelred II (*c.* A.D. 1000) provides that in the gemot of every wapentake the reeve and the 12 senior thegns should go out and present, on oath, all whom they believed to have committed any crime. The guilt or innocence of the accused had to be determined by ordeal or compurgation. The institution probably continued in use until reconstituted by Henry II and is probably the original of the former criminal grand jury.

The origin of the jury is probably to be found in the importation, from Normandy, of a system of inquisitions in local courts by sworn witnesses. This was found in England shortly after the Norman conquest, and from the first, combined with the existing procedure of the shire-moot. The earliest jury was a body of neighbours summoned by a public officer to give an oath as answer to some question.

The sworn inquest seems, first, to have been used in non-judicial matters, such as the inquiry which is reported in Domesday Book, but was early applied to strictly legal matters to ascertain by the recognition on oath of a number of *probi nomines*, selected to represent the neighbourhood, and to testify to facts of which they had personal knowledge. Henry II applied recognition by jury to every kind of fiscal and legal business and down to the time of Edward I it was the most usual machinery for the assessment of taxation. By the Constitutions of Clarendon (1164), sheriffs were directed, if no one would accuse a powerful layman, to swear 12 lawful men of the neighbourhood to tell the truth, and the chief justice was required to decide all disputes as to the lay or clerical tenure of law by the recognition of 12 lawful men.

Henry II, in the institution of the Grand Assize (1179), applied the principle of recognition by jury to the decision of actions to try the right to land. By the Grand Assize, the defendant had a choice between wager of battle and the recognition of a jury of 12 sworn knights of the district chosen by four other knights summoned for that purpose by the sheriff.

Actions dealing, not with absolute right to land, but with recent change of possession, were determined by 12 knights or freeholders (*legales homines*) of the district, chosen by the sheriff and sworn to try the question. In both cases the recognitors had to found their verdict on their own knowledge, gained from their fathers or from eye-witnesses, or credible information. The remedy of assize was improved by later statutes and this contributed to the decline of the old local courts, as all actions on the assize were tried by justices itinerant or in the King's court.

The principle of recognition by jury was extended to criminal cases by the Assize of Clarendon (1166), possibly reviving or reconstituting Ethelred II's accusation by 12 thegns, which provides that, in every county, 12 lawful men of each hundred with four lawful men from each township, should be sworn to present all reputed criminals in each county court. Richard I, in 1194, issued Chapters of the Eyre as instructions to the itinerant justices. This further regulated the jury of presentment and assimilated it to the system already in use for nominating recognitors of the Grand Assize. This is the origin of the grand jury. Presentment was only the first stage; the trial of the accused had to be by ordeal or by combat. But, in 1215, the fourth Lateran Council in effect abolished trial by ordeal, and combat was applicable only if an injured person came forward as appellant to claim it, and the practice increased of summoning a second or petty-jury to affirm or reject the view of the first set of jurymen. This became the general usage in the latter thirteenth century.

As originally established, the function of the jury was not to weigh evidence but to decide on the basis of their own knowledge, or the general belief of the district, and for this reason they were always selected from the hundred or district where the question for decision arose. If they did not have the knowledge they could readily ascertain it. They were accordingly witnesses to rather than judges of facts.

Originally 12 was the usual number of jurors, and if 12 did not concur in the verdict, at least in civil cases, the jury was afforced, i.e. other recognitors were added until 12 were found who could give a verdict. For a time, a majority verdict was accepted but, by the mid-fourteenth century, the rule was established that the unanimous verdict of 12 was needed.

The institution of the jury developed several varieties.

In criminal cases there developed, from the early type of jury called to discover and present facts, the jury of presentment or grand jury, summoned to discover and present to the royal officials, persons suspected of serious crime. This was regularly used in the royal courts from the Assize of Clarendon. Such juries could present, from their own knowledge or from the information of others. After the general eyre ceased, the selection of the grand jury was by the sheriff summoning 24 persons from the body of the county, of whom 23 were chosen, a majority of whom decided whether to find a true bill or ignore the examination. It was gradually recognized that the function of the grand jury was merely to decide whether there was a probable ground of suspicion.

At the end of the twelfth century, a person accused of crime by another could, on payment, obtain the right to be tried by jury. The presenting jurors decided as to the mode of proof. Compurgation and the ordeal were discredited in the early thirteenth century. The need to find a new means of determining guilt gave rise to the petty jury. Gradually, the practice developed of selecting a trial jury of 12. The accused had to consent to jury trial. This consent had to be obtained, if need be, by *peine forte et dure* (q.v.).

In civil cases, the Grand Assize was a special kind of jury introduced to try which of two parties to a writ of right had the better right to lands held by free tenure, as an alternative to trial by battle. The plaintiff had to obtain a writ whereby four knights of the county chose 12 knights to say, on oath, which of the parties had the better right. The tenant could still be decided by battle or compurgation. In the possessory assizes (q.v.) jury trial was obligatory, and remained so till the abolition of the real actions in 1833.

In civil cases generally, the assise, or body of persons summoned to answer a specific question, became in time altered to a *jurata* or jury, to which, by agreement of parties, other relevant questions were also submitted. The jury was the necessary mode of trial in all actions of trespass and other new forms of action developed in the thirteenth century.

From the mid-fourteenth century, witnesses were occasionally added to the jury to give them the benefit of their testimony, and later the jury came to decide on evidence formally produced. In the following century, all evidence was required to be given at the bar of the court, and the judges were enabled to exclude improper evidence. This gave rise to the distinction between law and fact, and the distinct functions of the court and the jury, to the law of evidence, and to the functions of counsel in presenting and marshalling evidence and winning verdicts. For a long time, however, jurors were still entitled to rely on their own knowledge in addition to the evidence, but in the early eighteenth century, decisions put an end to this ancient power of jurors.

So long as jurors were supposed to make findings from their own knowledge, they would be guilty of perjury if they gave a wrong verdict. They were liable to the writ of attaint whereby the cause was tried again by a jury of 24. If the verdict of the second jury differed from that of the first, the original jurors were arrested and imprisoned, their lands and goods forfeited, and they became infamous. As the jurors ceased to be recognitors, attaint fell into disuse, though it was not abolished till 1825. It did not lie on the proceedings in a criminal appeal.

The Tudor and Stuart kings sometimes employed, illegally, fine and imprisonment by the Star Chamber as a means of punishing a jury for an illegal false verdict, as where it had failed to convict. After the abolition of the Star Chamber, the Crown made use of judges to intimidate juries. Finally in *Bushell's case* (q.v.) the immunity of juries who give an honest verdict was established.

In the nineteenth and twentieth centuries, the use of juries in criminal cases has been restricted to the more important crimes tried on indictment. The great majority of crimes are tried summarily, or are cases in which jury trial may be waived. In civil cases, the use of juries has steadily declined in the twentieth century and they survive only exceptionally, usually in libel cases.

In Scotland, jury trial in civil cases was known in early times but vanished completely in the sixteenth century, although revived by statute, under English influence, in 1815. A special tribunal, the Jury Court, was established, but in 1830, it was merged with the Court of Session where jury trial survives in limited spheres.

For long, in England, there continued to be property qualifications for jury service and only recently have these disappeared. Various classes of persons, such as police, lawyers and doctors, are exempted from service. Since 1969, in U.S. federal courts, the idea has been of genuine random selection from the population.

It has long been permissible to challenge some of the jurors chosen by ballot peremptorily, and others on cause shown. In the U.S., a great deal of trial strategy, and much time, is often devoted to selection of jurors and their examination by trial counsel to ascertain their prejudices and suitability or unsuitability.

Traditionally, the Anglo-American jury was required to be unanimous, though in Scotland a simple majority seems always to have been sufficient. In 1968, England adopted the acceptability of verdicts by 10 out of 11 or 12, or 9 out of 10. Some states of the U.S. allow majority verdicts.

Where jury trial is utilized, the jury acts under

the direction of the judge presiding at the trial. The division of functions between judge and jury varies in different jurisdictions. The judge, generally, has to preside and decide what evidence is admissible and what is inadmissible. Generally, he has the power, in appropriate cases, to decide that no sufficient evidence has been adduced on which a jury could convict or, in a civil case, find for the plaintiff and, accordingly, to withdraw the case from the jury. Frequently, he may, but sometimes he may not, on the conclusion of the evidence, summarize it to the jury, and normally he must explain to the jury the points of law involved in reaching their verdict, e.g. what facts have to be held established before the jury is entitled to convict of a particular crime.

A jury normally returns a general verdict, that is, one of guilt or innocence of a particular crime, or that a defendant is or is not liable. Sometimes, however, it may be instructed to, and may, return a 'special verdict', a series of answers to specific questions, which answers are then utilized by the presiding judge to construct a finding. Reasons are not given by the jury.

In criminal cases the jury is not normally concerned with sentence, but in civil cases claiming damages, the jury normally has to determine the amount of the award as well as the issue of liability, and sometimes, also the degree of contributory negligence on the plaintiff's part.

The modes of challenging a jury verdict as unsatisfactory differ from those applicable in trial by a judge alone. In criminal cases it is exceptional for it to be possible to order a new trial; in civil causes this is normally the only course open if a verdict is set aside as unjustifiable.

Apart from ordinary civil and criminal cases, older statutes frequently invoked juries for exceptional kinds of proceedings, such as coroner's juries, juries to determine compensation for land acquired compulsorily, and to determine causes of death in fatal accidents and sudden deaths cases.

The use of the jury system spread, with the expansion of the English common law, to territories colonized from England, but in England its use had declined considerably, whereas in the U.S. it has flourished. In 1968, in *Duncan* v. *Louisiana*, the Supreme Court declared that jury trial was a constitutional right in all criminal cases in which the penalty may exceed six months' imprisonment. Whether there is a constitutional right in civil cases is uncertain but it is widely available. The actual extent of use of juries depends on its availability in different actions under state laws and the extent to which parties in fact utilize it.

In Europe, the French Revolution and its consequences brought it in as a symbol of popular government, particularly in cases of political and major crimes, and introduced it to all countries conquered by Napoleon. But, from the mid-nineteenth century, it has been increasingly dispensed with in Europe, so that it survives only in a few countries. France, Germany, and Italy have wholly abandoned it. The jury is accordingly, in origin and modern use, exclusively an institution of Anglo-American law.

The merits or demerits of jury trial have long been debated. The principal arguments in favour of the system have been the value of the participation of ordinary citizens in the administration of justice; the advantage of 12 heads over one; the near-certainty that if all, or at least a majority, of a jury are convinced, that the case has been fully established; the protection of the liberties of the subject afforded by entrusting decisions to a group chosen at random; the impossibility of bribing or intimidating a group which does not exist until the trial commences; and the ability of a jury to temper legalism by common-sense.

The principal arguments against are the inexperience, sometimes ignorance and even stupidity, of jurors; the uncertainty whether they have properly appreciated the issues or the evidence put before them, or properly understood or applied the judge's directions; the greater expense and delay involved in jury trial; and the inadequacy of appeal. The jury does not give reasons, and it is sometimes thought that if they had to, the system would collapse when the inadequacy of the reasons became apparent.

In America, in the 1960s, a major investigation of jury trials was undertaken, which indicated that, in a substantial number of cases, juries acquitted where judges would have convicted but, that in three-quarters of both civil and criminal cases, judges thought that the jury had acted rightly.

Jury trial must be considered as one aspect of the larger issue of lay judging or lay participation in justice.

See LAY JUDGES.

W. Forsyth, *History of Trial by Jury*; W. S. Holdsworth, *History of English law*; G. Williams, *The Proof of Guilt*; P. Devlin, *Trial by Jury*; W. Cornish, *The Jury*; H. Kalven and H. Zeisel, *The American Jury*; J. P. Dawson, *History of Lay Judges.*

Jury of matrons. A jury composed of matrons empanelled on a writ from Chancery to determine whether a woman was pregnant or not, where a succession was involved, or where a woman was condemned to be executed and pleaded pregnancy as a ground for delaying the execution. From 1931, a pregnant woman would not be sentenced to death but only to imprisonment; she had, however, to satisfy a jury that she was pregnant.

Jus. For title *Jus* and all phrases commencing with the word *Jus* see under *Ius.*

Just and equitable. A phrase used in several statutes as describing the circumstances in which the court may exercise a particular power. They are nowhere defined, save inferentially in the cases where the power has been considered, but, generally, require the court to be satisfied of the fairness and rightness in the circumstances of what it is going to do.

Just war (or *bellum iustum*). A concept discussed by older writers on international law, raising the question whether in any, and if so in what, circumstances making war on another state is morally and legally justifiable. Canonists and theologian-jurists discussed the problem extensively. Aquinas held that to be just a war must be waged by a party with sufficient authority, there must be a just cause of offence and there must be a just intention to make war solely for the establishment of peace, the assistance of the good or the repression of the wicked. Some of the later canonists considered as just wars against infidels, rebels, in self-defence, and in virtue of a right cause. Much discussion took place on the notion of a 'just cause'. In practice, there was total inability to agree on whether one party was wholly just and another wholly unjust in making war. The canonists allowed the wager of a 'just' war extensive rights over the persons and property of his 'unjust' enemy. From the seventeenth century, most wars were the product of ambition, jealousy, and other inadequate motives, but the Vienna Settlement of 1815 restored the concept of just war, and nationality and self-determination became just causes of war in the nineteenth century.

In both World Wars, the Allied Powers claimed that theirs was a 'just war', being fought to free various national groups from subordination to an alien power, to secure their self-determination, and to resist being themselves overrun, subordinated and exploited by another power. In modern practice, hostilities conducted with the sanction of the U.N. would be rated just, but the issue has no practical effect. Any state conducting hostilities, for whatever real reasons, will publish justifications designed to prove the justice of its actings which will be accepted by those who agree with its policies anyway.

F. H. Russell, *The Just War in the Middle Ages* (1975).

Justice. A moral value commonly considered to be the end which law ought to try to attain, which it should realize for the men whose conduct is governed by law, and which is the standard or measure or criterion of goodness in law and conduct, by which it can be criticized or evaluated. The close connection between justice and law is seen in such titles, names, and phrases as Justiciar, Mr. Justice, Lord Justice, Courts of Justice, administration of justice, requirements of justice, and so on. Theories of justice are concerned to determine what justice is, settling its status as an ethical standard and to settle practically what the requirements of this standard are. Discussion of issues of justice have been the concern of ethical, social, and political philosophers, as well as jurists from the earliest times.

Justice is distinguishable into social, political, economic, moral, legal and probably other kinds of justice. The only standard for judging whether an act or course of conduct is just from the social, political, or economic standpoint is its consistency or otherwise, with the social, political, or economic philosophy of the person judging. Justice is what is pleasing or approved. Thus, the justice or otherwise of expropriation without adequate, or even without any, compensation is judged differently by persons holding a conservative or liberal philosophy and by persons holding a socialist philosophy. But a statute passed to effect such a purpose, if implemented and applied strictly according to its terms, will be legally just and condemnation of the actings of the courts in giving effect to the purpose is condemnation by an extra-legal standard of justice. Similarly, the justice or injustice of abortion or capital punishment is judged largely by a moral and religious standard. In modern democratic and quasi-democratic countries, it is impossible to resist the conclusion that justice or injustice depends on consistency with the views of the stronger, the party or group in society who can, in fact, have their views enacted and enforced. Justice is, accordingly, not solely a legal concept or value and concentration on legal aspects, and exemplifications of it give an incomplete picture.

Among the Greeks, justice first appeared as a kind of metaphysical cosmological principle regulating the operation of the forces of nature on the elements of the universe, securing balance and harmony. The Pythagoreans evolved the idea of justice as equality. In Plato, justice became an ethical principle for human conduct, a human virtue, and he formulated the principle of justice that each should do what pertains to him, perform his function, and do what is required of him; justice was the correct placing of all of the virtues in relation to one another. Aristotle distinguished justice as comprehending virtue in general, and justice in a narrower sense, as equality, the right proportion; within this he distinguished distributive justice from corrective justice, and natural from conventional justice. Concurrently with, and tangled with notions of justice, there appeared in Greek philosophy the concept of natural law (q.v.).

In Hebrew thinking the classical teachings were more practical than speculative; justice and righteousness were to emerge from practical issues of life and the Mosaic law more clearly and completely than the Greek evolved the idea of justice according to law; law and justice were the same because both proceeded from God. The Hebrew tradition emphasized the importance of law for the doing of justice.

At Rome, law originated as a rather primitive law of a kin-organized society, applicable to citizens only. The idea of natural law was adopted from the Greeks and, particularly through the work of the *praetor peregrinus* and the concept of *jus gentium*, adapted the old *jus civile* to a broader role. The *praetors*, advised by jurisconsults, set out to discover and, in fact, created a more general and simpler body of principles applicable to Mediterranean peoples generally, and adapted the law of a small agricultural society to the needs of a great empire. The Greek concept of a law of nature offered an intellectual basis for this development. It was also connected with divine law and provides a connection between law as of divine origin and law of human origin.

Stoic philosophy arose about 300 B.C. and evolved the distinction between *lex aeterna*, *ius naturale* and *ius humanum*, which was to be come a central theme of natural law thinking, the distinction between the law of reason of the cosmos, natural law which partook of, but was more limited in ambit than the eternal law, and the positive law of men. Stoic ideas entered Roman culture and influenced Cicero and later thinkers and jurists.

Augustine distinguished *lex aeterna*, *lex naturalis*, and *lex temporalis*, but even temporal law, though changeable according to place and time, must be warranted by the unchangeable eternal law. His natural law is a transcript of eternal law, inborn in man's soul, through which God speaks to us in our conscience.

Aquinas saw eternal law, divine law, and natural law all as unalterable, all based on divine reason, and as setting the standards for human law and justice while Duns Scotus, who exalted free will above reason, considered that justice was not to be willed as a means to an end but solely an end for itself, and that law proceeded from the divine will itself, not from the divine reason.

A secularization of natural law set in in the sixteenth century. Suarez, however, maintained that law had rational foundations: *lex aeterna* provided the pattern for all law and there was still a *lex naturalis* common to all men and discoverable by their reason; the State, being committed to justice, must function within the framework of the *lex aeterna* and the *lex naturalis*. Grotius carried into the new age of sovereign states, a version of universally valid natural law transcending any human law and yet providing practical norms for new situations.

After Grotius, the connection of justice with natural law and the Graeco-Roman-Christian course of development became weaker as law became increasingly secular, positivist, and estranged from broader philosophical movements. Jurists had to find a rational basis for law which transcended the fact of state power but yet recognized it and its power to make law as it willed. Grotius offered an essentially secularized natural law. Similarly, natural law, in the eighteenth century underlay the nascent law of nations, developing out of and away from natural law, the detailed working out of a natural law system by Rutherforth and the natural law codes of Prussia and Austria. At this time, moreover, a tendency became noticeable to rest natural law partly on the human pursuit of utility or happiness, and this became predominant as utilitarianism became more widely accepted.

At the end of the eighteenth century, Kant formulated his philosophical system. In the field of practical reason, a sphere which included moral and legal philosophy, his basic tenet was the freedom of human volition; but human action was right only if it conformed to the supreme principle of the categorical imperative. Though he based natural law on an *a priori* basis, rather than an intuitive basis, his teaching was rapidly interpreted in natural law terms, notably by Fichte and Ahrens. His views led to the attitude that freedom should be maximized and legal restraints minimized in the name of justice. The dominant juristic influence from Kant was individualist and *laissez-faire*.

About the same time, Bentham developed the theory of utilitarianism; he reacted vigorously against natural law thinking and held that the ultimate test of justice was whether the law, on balance, promoted human happiness or human misery. Utility was the criterion of good law. The legislator's power to control conduct lay in his use of pleasures and pains as sanctions. His reforming zeal had enormous influence on the reform of English law throughout the nineteenth century and influenced both Fabian socialists and liberals.

In the mid-nineteenth century, Marx propounded the thesis that men's ideas of justice, law, and ethical duty were reflexes of the economic and particularly the class structure of society; Marx denied the independent reality of ideals of justice.

More recently, Jhering and Duguit have both rejected intuitive conceptions of justice and found justice, in the securing and protection of individual social and public interests, a social-utilitarian theory of justice, and in social solidarity respectively. Kohler developed a theory based on the Hegelian philosophy of history; the general function of law is to preserve further and transmit civilization. It must delimit, for the time being, the sphere of individual rights as against civilizational values. The jural postulates of a civilization are the criterion of its just law. Stammler redirected Kantian philosophy on to the problems of justice and law, rejecting both natural law and positivist approaches; the norms of justice must be found in an understanding of what is presupposed when men think about justice, the social ideal.

More recently, there has been a revival of natural law, the relativist theory of justice of Radbruch, to the effect that law is a reality, the meaning of which is to serve justice, at the core of which is equality, though justice must have regard to individual cases, and the pragmatic theory of Roscoe Pound, that justice is such an adjustment of relations and ordering of conduct as will make the goods of existence go round, as far as possible, with the least friction and waste. A legal system attains the ends of the legal order by recognizing certain interests and defining the limits within which those interests shall be recognized.

Theorizing about justice has, accordingly, been continuous since the earliest speculative thinking and the most widely divergent theories have been propounded. Among the most commonly held theories of justice have been natural law theories, raising the issue whether justice must be thought of in terms of norms transcending human experience, metaphysical theories, claiming to impose obligation by virtue of deduction from a pure idea as distinct from theories derived from social facts, theories which seek absolute standards as compared with those which insist of the relativity of the criterion to time, place and social context, and theories which seek to define justice by a general or universal formula rather than see the uniqueness of justice in each particular case.

It is sometimes said that the attributes of justice are liberty, equality, and fraternity or co-operation. These principles do help to give a content to the concept of justice, but the former two attributes, at least, cannot be understood as absolute or unqualified, and only liberty consistently with the liberties of others and equality so far as inequality is not justifiable can be accepted. These attributes are certainly commonly found in mature and developed legal systems. Or, it is said, that justice links liberty, equality, and reward for services contributing to the common advantage in the principles, that all participating in an institution have equal right to the most extensive liberty compatible with a like liberty of all, and that inequalities are arbitrary, unless they can be expected to work to everyone's advantage and the positions to which inequalities pertain are open to all. Or, it is said that justice is the adjustment, reconciliation and synthesis of a number of political values, which are necessary to an organized system of human relations, and these are principally liberty, equality, and fraternity or co-operation; they are present in different degrees at different times and places and there is constant adjustment and readjustment of their relative claims. Justice is the final principle which controls the distribution of rights and the principle of their distribution. It is a general right ordering of human relations in and by the State. The law of the State can be said to be right, and to partake of the quality of justice, if it secures and guarantees, for the greatest possible number of its citizens, the external conditions necessary for the greatest possible development of the capacities of their personality, and in so doing it acts in accordance with the general principles of liberty, equality and fraternity or co-operation.

The realization of justice in rules of law passes through various stages, from the appreciation by men of the ideal of a system of justice for the right ordering of the relations of persons living in society, through the development of associations for the realization of the ideal, to the stage of formulation in verbal form of concrete rules intended to make the ideal a reality. Law emerges from social thought about justice.

In practice, accordingly, men's inherent appreciation of what is just leads to the growth in custom and practice, then in decisions by leaders and kings, and finally, in express formulation, of principles and rules regulating action in particular circumstances. It becomes accepted that it is right that a wrongdoer be punished by the injured man, or his kin, or the community, and this becomes a rule, an embryo law.

The practical importance of justice is regard for it as a governing principle in human action, both in general and in detail, as against such other principles of wide application as convenience, expediency, benefit or advantage to some object which one approves. It would undoubtedly be more convenient to have fixed fines for particular offences, or a tariff of damages for particular kinds of losses, but would such rules be just? It may be expedient that certain land be acquired compulsorily, but is it just?

Theoretically, law is not essential to the administration of justice, and justice in disputes could be achieved by judges and courts deciding individual issues on their merits and in accordance with the dictates of justice. Plato's Republic was a commonwealth without law or judges, the philosopher-kings' knowledge and judgment ensuring justice. In practice, however, this is probably unattainable and the merits of deciding most issues in accordance with generally applicable, known and published, pre-determined rules are greater, notwithstanding that general rules may be unsuitable or unfair in particular cases. Interpretation, equity and judicial discretion do much to mitigate injustice.

Legal justice, accordingly, requires that there be a system of law and that justice be done in accordance with pre-ordained and publicly-known principles and rules. This does not make the administration of justice purely mechanical because, whether the facts do or do not require the application of a particular rule may have to be decided, the precise formulation and effect of the relevant rule may be doubtful and, in many cases, there is room for, and need for, the exercise of judicial discretion, consideration of, and regard for the precise circumstances and needs of the particular case.

Furthermore, to achieve justice, the system of law must, in theory, be universal and capable of furnishing a rule for the determination of any issue at all which may require decision, even if the rule be that the judge does what he thinks is right. The judge cannot decline jurisdiction on the ground that there is no law appropriate. Hence, every system must make provision for the *casus improvisus* and leave scope for constructive judicial enterprise in an appropriate case.

Particular rules of law may often be criticized as unjust. In the first place, any rule is made for the general and great mass of cases and may be inappropriate and productive of injustice in a particular, exceptional, case. Secondly, rules formulated at one time or place may effect something less than, or other than, justice at a later time and another place, or when circumstances or social views have changed. Views of the justice of the rules relating to married women changed very much between 1850 and 1950, so that rules generally accepted and approved at the earlier date were generally deemed unjust at the later date (and had in fact almost completely been changed).

Legal justice can be administered and realized in various ways and all of these ways have been utilized at various times and places. It may be achieved legislatively, as has been illustrated by Greek trials before popular assemblies, Roman capital trials before the people, Germanic administration of justice by assemblies of free men, judicial exercises of power by the English and U.K. Parliaments, such as by granting relief against duress and fraud, hearing petitions of error and appeal, impeachments, parliamentary declarations of treason, bills of attainder, bills of pains and penalties and divorce bills, and judicial exercises of power by American colonial legislatures and state legislatures immediately after the Revolution, legislative appellate jurisdiction, equitable relief, jurisdiction in divorce or bankruptcy, and impeachment, even of Presidents. But, in practice, legislative justice has proved expensive, unequal, uncertain, and influenced by prejudice, political considerations and even corruption.

Justice may again be achieved by the executive or administration, and this is largely done in modern states, where decisions on many matters are made by officials, inspectors, commissioners, boards, or Ministers. The advantages claimed are speed, secrecy, expert knowledge, detailed knowledge of the relevant rules, regard for policy, freedom from rules of evidence or procedure, and the need to avoid overwhelming the courts. But cases taken to courts by appeal show that administrative justice sometimes fails to observe the requirements of natural justice, tends to confuse policy with fairness, to overstep jurisdiction, or to act arbitarily for administrative convenience. In some cases appeal lies from administrative justice to judicial justice, to some court, and in other cases administrative tribunals have acted very like courts, issuing reasoned judgments which are reported and followed.

Justice may again be achieved judicially, by entrusting the function to persons chosen for knowledge, experience, impartiality, and permanently devoted to deciding disputed issues. The advantages are that, by experience, training and habit, judges seek to discover and apply general rules fairly; their decisions are public and subject to appeal, public comment and professional scrutiny and criticism; on the whole judicial justice combines reasonable certainty and predictability of rules with discretionary moderation better than any other form of administering justice. But it is criticized on such grounds as that judges are out of touch with common men and do not understand their attitudes, problems, and difficulties, that the structure of legal concepts and rules is too rigid and does not give sufficient scope to non-legal factors, that the premises of legal reasoning are too pedantic, antiquated and fixed and take too little account of social changes and are irresponsive to developments, and that there is a tendency to reduce everything to rules and to make cases fit into the categories of the rules. While there may be defects in the judiciary, and in the system and the rules, there can be little doubt that, at least in the developed democracies, the system of judicial justice is a serious and honest, and on the whole, reasonably successful attempt to provide a system which secures justice to individuals.

R. Pound, *Jurisprudence*; J. Stone, *Human Law and Human Justice*; J. Rawls, *A Theory of Justice.*

Justice, College of. College of Justice.

Justice, Department of. One of the Departments of State of the U.S. The office of Attorney-General of the U.S. was established in 1789, but only after the civil war did his duties become so manifold and burdensome as to require the creation of a department. The office of Solicitor-General of the U.S. was created when the department was organized and has the chief responsibility for representing the Federal government in cases before the Supreme Court. The Department has expanded and now comprises such divisions as Anti-Trust, Civil Rights, Federal Bureau of Investigation, Immigration and Naturalization, Internal Security, Lands, Prisons, and Tax, headed by assistant attorneys-general.

Justices. The name given, since Norman times, to judges of the superior English courts. In the old records it appears as *justitia*, and it is not a derivative of *justiciarius*. It was originally the general term for the official class who staffed the *Curia Regis* and the Exchequer. As the work of the *Curia* developed they acted in the *Curia*, or the Exchequer, or as itinerant commissioners. The title barones tended to be given when they were doing fiscal business, justices when doing judicial business. As judicial tribunals branched out of the *Curia Regis* the justices were employed in conducting the judicial business thereof, and the term came to be associated exclusively with administration of judicial business. Accordingly the judges of the Courts of Queen's Bench and Common Pleas were called justices, the heads of the two courts being the Chief Justices. Since the Judicature Acts, 1873–75, the ordinary judges of all the divisions of the High Court have been called justices.

The term has been carried all over the world with the expansion of the common law, and judges of superior courts all over the English-speaking world are commonly called justices. Though, in England, all judges of the High Court are knighted on appointment, the official title is Mr (or Mrs) Justice Doe, and the customary abbreviation in books and reports is Doe, J., or in the plural Doe and Roe, JJ.

Justices of the Peace (q.v.) are frequently referred to as 'the Justices'.

Justices in eyre. In the time of Henry I, some of the justices of the *Curia Regis* were sometimes required to visit counties, to collect revenue, determine disputes as to amounts, punish frauds by sheriffs, and hear pleas, civil and criminal. Henry II re-established the practice after disuse. In 1176, Henry divided England into six circuits for fiscal and judicial purposes, assigning to each three itinerant judges. The justices in eyre were unpopular as they acted as the royal agents to exact taxation. The eyre system was abandoned at the end of the fourteenth century. The justices in eyre sat in the full county court when they visited a county and this formed a link between the royal courts and the shire courts.

The system of justices in eyre was superseded by the justices of assize and *nisi prius*, appointed by the Statute of Westminster II (1285), to visit counties to try matters of fact, which has survived to the present as the assize system.

Justices of Appeal. The title, under the Judicature Act, 1873, of the members of the new Court of Appeal in England, shortly thereafter altered to Lords Justices of Appeal, abbreviated as L.J. or, in the plural, L.JJ.

Justices of the Peace. *Custodes pacis* were nominated in England several times from 1252 and later appear to have been chosen by the landholders of the county and, later, by royal writ. Justices of the Peace were first appointed by a statute of 1327, under which the King assumed the right of appointing all conservators of the peace. In 1344, it was enacted that a few in each county should be assigned keepers of the peace by the King's Commission. Initially their functions were to take indictments and hold the accused for trial by royal judges, but, in the fourteenth century, they were given powers to try prisoners. Subsequent statutes regulated the numbers and authority of the justices. In the eighteenth century, numerous functions of essentially local governmental character were imposed on the justices, but most of these functions were transferred to elected local authorities when they developed in the later nineteenth century.

Justices act under a commission of the peace issued by the Crown for any county, any of the five commission areas for Greater London, and the City of London, and addressed generally to all persons holding office as justices for the area. The justices for any commission area are appointed by the Lord Chancellor or in the appropriate areas by the Chancellor of the Duchy of Lancaster. Every commission names the Lord Chancellor, the justices of the High Court, and certain other officers. Mayors and aldermen are no longer justices *ex officio* save in the City of London and there is now no distinction between borough and county justices. Justices aged over 70 are placed on a supplemental list and these do not normally exercise judicial functions.

Each commission area is divided into a number of petty sessions areas. The justices for each petty sessions area elect their own chairman.

Unless specially authorized by statute, at least two lay justices must be present to constitute a magistrates' court, but not more than seven. (A stipendiary magistrate sitting alone constitutes a magistrates' court.)

The judicial functions of justices are imposed by numerous statutes; they are exercised in magistrates' courts and are both civil and criminal. Their civil jurisdiction is varied, including certain civil debts, e.g. rates, adoption orders, affiliation proceedings and matrimonial proceedings, i.e. applications for matrimonial orders on such grounds as adultery, desertion, cruelty and wilful neglect to maintain.

Their criminal jurisdiction is (a) to enquire into alleged indictable offences committed by persons brought before them and to commit alleged offenders for trial, if it appears that there is sufficient evidence to put them on trial by juries for indictable offences; this may be exercised by one justice as examining magistrate; and (b) to try and adjudge matters which may be dealt with summarily, including certain offences which may be tried summarily if that mode of trial appears more suitable. They also receive information or complaints, issue warrants or summonses, and deal with matters preliminary to hearing.

Their administrative powers are chiefly the grant or renewal of licences for the sale of liquor and for music, dancing and entertainments.

Justices of the peace also sit in the Crown Courts (q.v.); two, three, or four sit along with a legally qualified judge to hear appeals from magistrates' courts or proceedings on committal to the Crown Court for sentence. Any jurisdiction or power of the Crown Court may be exercised by a regular judge of that Court with not more than four justices. When they do so sit, the decision of the court is by a majority though, in case of equal division, the regular judge has a casting vote.

The office of justice of the peace was instituted by statute in Scotland in 1587 and their duties extended in 1609. The office has never, however, really taken root in Scotland and the justices have never enjoyed the prestige, respect, or jurisdiction which they have done in England. Appointments to the commission of the peace are made by the Secretary of State for Scotland on the advice of advisory committees. Until 1975, the justices exercised minor civil jurisdiction in the J.P. Small Debt Court and criminal jurisdiction in courts of summary jurisdiction in counties.

The justices now have no civil jurisdiction. Their criminal jurisdiction is exercised by one or more justices sitting in a District Court (q.v.) and extends to lesser summary offences only. They have also some administrative functions, particularly in relation to licensing.

In the U.S.A., justices are usually elected but sometimes appointed. They form the lowest state courts, with limited civil and criminal jurisdiction, and commonly have such functions as issuing warrants for arrest, holding inquests and performing marriages.

In France and Italy, there are persons with names corresponding to justices of the peace but they are the lowest level of the career judiciary.

J. H. Gleason, *The Justices of the Peace in England*, 1558–1640; C. A. Beard, *Office of Justice of the Peace in England*; R. Burn, *Justice of the Peace and Parish Officer* (1755); S. Stone, *Justices' Manual* (annual).

Justicia. An institution of mediaeval Aragon to protect subjects against royal or private injustice. The *justicia*, or chief judge, was at the head of a court of 21, five appointed by the Crown and 16 nominated by the *cortes*. The office of *justicia* became hereditary and a strongpoint of opposition to royal authority but, after 1591, Philip II made the office held at the pleasure of the Crown and all the judges royal nominees.

Justiciability. The quality of being capable of being considered legally and determined by the application of legal principles and techniques. This concept is of importance in international law, where it is sometimes argued that certain disputes are not justiciable, being political and not legal in their nature. 'Justiciable' is frequently equated with 'legal'. Many disputes between states are political in that they involve issues of independence, national honour, vital interests, or the like, but they are not necessarily, on that account, non-justiciable and some other disputes are certainly justiciable. In truth, a state will contend that a dispute is not justiciable if it does not, for reasons of policy, want to submit to judicial or arbitral decision of the issue. The only clear cases for treating an issue as non-justiciable are where there are no legal criteria by which the issue can be judged, e.g. whether a state applying for membership of the U.N. is or is not 'peace-loving'.

Justiciar (England). The office of justiciar originated in the frequent absences of William I from England. This new officer represented the King in all matters and acted as regent in his absence, was chief minister, and administered the legal and financial business of the country. The holders can be traced from about 1116 to 1234. Sometimes two persons held the office concurrently. The office had different functions at different times and the emergence of the exchequer board for the auditing of accounts gave the justiciar one of his chief functions, while the growing importance of the central courts under Henry II gave greater prominence to judicial duties. Among notable holders of the office were Ranulf Flambard, Glanvill (q.v.), Hubert Walter, and Hubert de Burgh.

The dignity of the office was high until the death of King John when the justiciar Hubert de Burgh was besieged in Dover Castle and the barons made the Earl of Pembroke *Rector regis et regni*, de Burgh retaining his office. After his fall in 1232, the office lost its importance and became temporarily extinct, but was of sufficient importance for the barons, in 1258, to demand that the justiciar be chosen annually with their approbation. It continued to be filled until 1265. Edward I dispensed with the office altogether, the Chancellor succeeding to many of the rights and dignities of the justiciar.

West, *The Justiciarship in England*, 1066–1232; List of office-holders in *British Chronology*, 67.

Justiciar (Ireland). From 1172, English kings appointed justiciars in Ireland. In 1226, the justiciar was ordered to keep the laws and customs of England and, in 1227, a register of writs was despatched for use in Ireland. The early justiciars had original jurisdiction, both civil and criminal, and might also hear cases brought from other courts.

The justiciar was the King's *alter ego* in Ireland and the channel of communication between King and Irish administration.

About the mid-thirteenth century, a common bench emerged, settled in Dublin and, by the end of that century, it consisted of a Chief Justice and several puisne judges.

Under Edwards I and II, the justiciar was the chief governor of Ireland, superior to the Chancellor and Treasurer and the justiciar's court moved with him. From the later thirteenth century, there was an assistant, the justice of the justiciar's pleas, and in 1324, a second such justice was added. The justiciar's court dealt with pleas of the Crown, assizes, and common pleas. It also exercised a review jurisdiction over judgments in lower courts, the Dublin bench, the eyre, sessions of justices of assize and the steward's courts of the great liberties.

From 1250 cases were appealed to the king's court in England, both direct and via the justiciar's court.

After 1324, the justiciar's court was substantially a replica of the King's Bench at Westminster.

From 1308, the chief governor was sometimes designated King's Lieutenant, the latter term becoming commoner and being, in turn, superseded by Lord Justice, Lord Deputy, and Lord Lieutenant. The term justiciar disappeared after 1478.

Justiciar (Scotland). An Officer of State in Scotland who, from a very early date, held the King's Court, which had supreme cognizance of cases of every kind. The King was head of the court and the Justiciar, Justice Deputes, and Justiciar's Clerk acted with him. At various times, there were several justiciars for different districts. Later the Justiciar, if his commission empowered him, could create deputes to act with or for him. Apart from an interregnum during the Commonwealth the appointment was hereditary in the Argyle family from the late sixteenth century till 1628 when the office was resigned into the King's hands, with the exception of Argyll and the Western Isles which remained in the family till 1747. Encroachment on the Justiciar's jurisdiction was made by the institution of the College of Justice (1532) which relieved him of civil business, and by the growth of regalities. From the fifteenth century the name Justice-General is used instead of Justiciar.

In 1672 statute suppressed the office of justice-depute and made the Justice-General, Justice-Clerk and five of the judges of the Court of Session Lords Commissioners of Justiciary with supreme jurisdiction in criminal causes.

See HIGH COURT OF JUSTICIARY.

Justiciary, High Court of. The supreme criminal court of Scotland. From the twelfth century, Scottish kings appointed justiciars who held ayres, visiting sheriffdoms in turn, and holding courts for both civil and criminal business, but in the fifteenth and early sixteenth centuries the council and committees thereof became more active in handling civil cases, and this culminated in the establishment of the Court of Session (q.v.) in 1532, and thereafter the justiciar's jurisdiction became solely criminal. It was always confined to pleas of the Crown, too important to be heard by the sheriff. In 1514, the Earl of Argyll was given a commission as Lord Justice-General but in 1628 the Argylls resigned the office to the King, retaining only the office of Justiciar of Argyll and the Western Isles, which office they continued to hold heritably until 1747. From the early sixteenth till the late seventeenth century, criminal justice was administered by local deputes appointed by the Lord Justice-General or under special commissions of justiciary. The Justice-Clerk, originally clerk to the court, was regularly made an assessor to assist the lay Lord Justice-General or his depute and, in 1683, he was elevated to the bench and began to preside in the Court of Justiciary.

In 1672, the office of Justice-Depute was abolished and the High Court of Justiciary established, consisting of the Lord Justice-General, Lord Justice Clerk, and five Lords of Session as Lords Commissioners of Justiciary. In 1830, it was provided that the Lord President of the Court of Session was to become Lord Justice-General also on the death of the then holder of the commission, which happened in 1836. In 1887 it was provided that all the Lords of Session were to be also Lords Commissioners of Justiciary.

The court's jurisdiction extends to any crime committed in Scotland, or within territorial waters, or piracy on the high seas. It is the only court competent to try cases of treason, murder, rape, breach of duty by magistrates, corrupt disclosure of official secrets, deforcement of messengers, and a few other cases.

The Court sits in three capacities, (a) as a court of trial, when, normally, one judge sits with a jury of 12 to try cases of persons charged on indictment. (Sittings are held regularly in Edinburgh and on circuit; there are three circuits West, North and South, in each of which sittings are held, if there are cases, in several towns. There are monthly sittings in Glasgow but elsewhere courts are held as required.) (b) It sits as a court of appeal from summary trials in sheriff courts and other inferior courts, at which three or more judges sit. (The presiding judge has no vote unless the other judges differ in opinion. A single judge may deal with incidental applications.) (c) It also sits as a court of appeal from trials on indictment in the High Court or Sheriff Court, when three or more judges sit, the view of the majority determining the issue. (A single judge may deal with incidental applications.)

Judgments of the High Court are of great importance in the formulation of Scottish criminal law and regulation of its procedure. The judges of the High Court have power, by Act of Adjournal, to make rules for the regulation of criminal procedure.

Justifiable homicide. Causing the death of a human being in circumstances legally justifiable and hence not inferring any criminal guilt. It particularly occurs in the case of the execution of a convicted criminal, and in cases of death caused when arresting or attempting to arrest a person who resists, when dispersing a riotous mob, and similar cases.

Justification. The defence in the law of torts and criminal law of showing that the act complained of was lawful and justifiable. Accordingly in the tort of defamation to show the truth of the alleged statement is justification; in the crime of assault or manslaughter to establish that it was done in self-defence is justification.

Justinianus, Flavius Petrus Sabbatius (A.D. 483–565). Emperor of the East Roman or Byzantine Empire, was born in Illyricum, adopted by his uncle Justin I (Emperor 518–527), and educated at Constantinople. He succeeded his uncle in 527, determined to revive the glories of the older Rome in law and empire. The law consisted of some *leges* of the Roman republic and empire, *senatus consulta*, the edicts of magistrates, and a great body of juristic writing, particularly of the first three centuries of the Christian era, supplemented by imperial constitutions from Augustus onwards. On his accession, he appointed commissioners to collect and revise the surviving constitutions; this work resulted in the *Codex* (q.v.) promulgated in 529. He then promulgated constitutions—the Fifty Decisions—resolving the main issues on which the earlier jurists had been divided in opinion, and other amending ordinances and, in 530, appointed Tribonian and 15 others to select, from the writings of the authorized jurists, matter of permanent value, editing it to eliminate obsolete matter, repetitions and contradictions, and arrange it in 50 books. This was completed in 533 and published as the *Digest* (q.v.) or Pandects, and it superseded all the other matter in the jurists' writings. He then authorized the publication of the *Institutes* of Justinian, a new elementary manual based on, but superseding, the *Institutes* of Gaius (q.v.). The *Codex* was then revised to take account of changes since the first edition, and the revised edition (*Codex repetitae praelectionis*) was promulgated in 534. Thereafter, Justinian issued many new ordinances, known as Novels, *Novellae constitutiones post Codicem*, and though a promise to prepare an official collection of these remained unfulfilled, there were three unofficial collections: see *Novellae*. Much of this reforming activity was inspired and supervised by Tribonian (q.v.). The four works, *Institutes*, *Digest*, *Code* and *Novels* came later to be known collectively as the *Corpus Juris Civilis* and, though undertaken as a measure of law reform and restatement, they have preserved much earlier Roman law which would otherwise have been lost, and powerfully influenced legal thinking, study, and the actual law of many European countries from the tenth century onwards. Concurrently with, and continuing after, his legal reforms, Justinian sought to recover the old Roman empire in the West from the barbarians. Belisarius defeated the Vandals and recovered much of North Africa, Sicily, and southern Spain, and Belisarius and Narses waged a long war in Italy, ultimately recovering Italy from the Ostrogoths in 562. He hoped to restore the social and economic well-being of Italy by various measures embodied in the Pragmatic Sanction of 554. But resources had been excessively spent in recovering two countries of little value and which could not be held. He also waged a long war with the Persians, which only weakened the Roman Empire.

Throughout his reign, he pursued an active ecclesiastical policy and, at times, persecuted particular sects. He summoned a general council of the Church, which produced schism rather than agreement. He also spent lavishly on churches, notably St. Sophia in Constantinople.

In administration generally, he moved towards further centralization, and his policy generally was very expensive, resulting in heavy taxation, which provoked insurrections. He was genuinely concerned to promote the well-being of his subjects by providing access to justice and eliminating corruption. He was assisted by some outstanding ministers, notably John of Cappadocia and Peter of Barsymes. Though not a great statesman, he was, nevertheless, a ruler of ability, foresight, and industry.

J. B. Bury, *History of the Later Roman Empire*; A. H. M. Jones, *The Later Roman Empire*; P. Ure, *Justinian and His Age*; R. Browning, *Justinian and Theodora*.

Juvenile court. A special sitting of a magistrates' court consisting of the J.P.'s who are members of the juvenile panel, held for the hearing of a charge against a child or young person, or to determine the best treatment for a neglected, abandoned, or ill-treated child or young person. Juvenile courts must sit in different rooms or buildings from adults courts, or on different days, and the general public are not permitted to be present. The general principle underlying the establishment of such courts is that young persons need to be specially treated and, wherever possible, rehabilitated rather than punished and that care proceedings should replace criminal proceedings. These courts have powers to order a parent or guardian to exercise proper control or to make supervision, care, hospital, and guardianship orders.

Juvenile delinquency. The term for crimes, offences, and other infringements of law perpetrated by children and young persons below the age of adulthood. In all Western countries the amount of such law-breaking is large and includes a material proportion of serious crime as well as truancy, petty theft, and malicious mischief. It has sharply increased in modern times. Attempts to control and reduce juvenile delinquency by special courts and means of dealing with reported cases, special penalties, and places for dealing with juvenile delinquents, have been conspicuously unsuccessful.

Special measures have long been used to deal with juvenile offenders, with strong emphasis on reclamation and reform. A School for the reform of young criminals was established in London in 1788. In 1838, Parkhurst was established for the detention and correction of juvenile offenders; it combined industrial training with religious and educational instruction. In 1854, reformatory schools were established though, until 1899, a short period of imprisonment preceded the period in the reformatory. Industrial schools were established from 1857 as training schools for children needing care and protection. In 1908, the Borstal system was formally established. In 1933, the distinction between reformatory and industrial schools was abolished and they were renamed approved schools. From 1948, a wider variety of measures became available for young offenders, remand homes for observation, remand centres for more unruly and depraved young persons, attendance centres, probation, detention centres, approved schools, and Borstals. Not all of these measures are available for particular age-groups. Detention centres are intended to provide a short, sharp shock to the offender, community homes and Borstals opportunities for constructive training over a substantial period.

K

K.B. King's Bench.

K.C. King's Counsel (see QUEEN'S COUNSEL).

Kachenowski, Dimitri Ivanovich (1827–72). Russian jurist and author of *Prize Law* (1867).

Kahrel, Hermann Friedrich (1719–87). German jurist and philosopher, author of *Das Recht der Natur* (1746), *Völkerrecht* (1750), *Jas Publicum universi et Germanici* (1765) and other works.

Kain. A word used in ancient Scottish grants of land signifying fowls or animals deliverable by the vassal to his superior as part of the return for the land, and latterly signifying poultry and eggs deliverable by a tenant to his landlord under his lease.

Kaltenborn von Stachau, Karl von (1819–66). German jurist and author of *Kritik des Völkerrechts* (1847) and other works on natural and maritime law.

Kames, Lord. See HOME, HENRY.

Kamptz, Karl Christoph Albert Heinrich von (1769–1849). German statesman and author of *Neue Litteratur des Völkerrechts* (1817) and other works.

Kangaroo. The name of a means for abbreviating debate on amendments in the House of Commons sitting as Committee of the Whole House. The Chairman, if authorized by resolution of the House, selects the amendments to be discussed and leaps over the rest which are left undebated. Similarly, at Report Stage, the Speaker may select amendments to be debated.

Kant, Immanuel (1724–1804). German philosopher, tutor, librarian, professor of philosophy, and author of the *Critique of Pure Reason* (1781), *Prolegomena* (1783), *Fundamental Principles of the Metaphysics of Ethics* (1785), *Critique of Practical Reason* (1788), *Critique of Judgment* (1790), *Religion within the bounds of Reason alone* (1793), and *Metaphysical Principles of the Science of Right* (1797). His philosophy was immensely influential in Britain and Germany in the nineteenth century, and his legal philosophy hardly less so than his general philosophy. It has influenced Stammler, del Vecchio, and Kelsen. He claimed to have put forward a rational system of jurisprudence, both complete and comprehensive. His theory of law was an application of his general system of ethics to the sphere of jurisprudence. Right comprehended the whole of the conditions under which the voluntary actions of any one person could be harmonized with the voluntary actions of every other person, according to a universal law of freedom. This formula eliminated the idea of eternal natural rights. From his conception of right, Kant developed his universal law of right: Act externally in such a manner that the free exercise of your will may be able to co-exist with the freedom of all others according to a universal law. His principle of right is, in fact, the rule in terms of which legal principles are to be measured, the formal principle by which the first principles of jurisprudence may be tested. He discussed natural or private right and civil or public right and many of his views have, subsequently, been influential, notably on the theory of contract. His conception of the criminal law turned on the idea of retributive justice. Possibly the greatest merit in the Kantian theory was the doctrine that knowledge must be based on principle, on the ability to recognize the particular in the universal, in the struggle to reduce to order the mass of legal phenomena.

E. Adamson, *Philosophy of Kant*; E. Caird, *Critical Philosophy of Kant*; W. Hastie, *Kant's Philosophy of Law* (1887); H. Cairns, *Legal Philosophy from Plato to Hegel* (1949).

Kantorowicz, Hermann (1877–1940). German jurist who taught law in Germany till 1933 and then in England. He developed a doctrine of free law (*Freirechtslehre*), which contributed to the development of the sociology of law. Judicial decision-making is a kind of legislative activity and judges should develop new rules from custom and social usage to fill gaps left by statutes. His writings include *Der Kampf um die Rechtswissenschaft* (1906), *Studies in the Glossators of the Roman Law* (with W. W. Buckland) (1938), *The Definition of Law* (1958), and other works.

Karlowa, Otto (1836–1904). German jurist and author of *Beitrage zur Geschichte des römischen Zivilprozess* (1865), *Der römisches Zivilprozess zur Zeit der Legisaktionen* (1872), *Die Rezeption des römischen Rechts in Deutschland mit besonderer Rüchsicht auf Kurpfalz* (1878), and *Römische Rechtsgeschichte* (1885–1901).

Keatinge, John (?1630–91). Called to the Irish Bar in 1673, he became Chief Justice of the Common Pleas in Ireland in 1679, but was superseded in 1691 and shortly thereafter committed suicide.

Keelhaul. To drag a person under the keel of a ship by means of ropes rigged from the yardarms, a form of punishment at one time used in the Navy.

Keener, William Albert (1856–1913). Admitted to the Bar in 1879, he taught law at Harvard, 1883–90, and at Columbia, 1890–1903 (Dean 1891–1901), thereafter returning to practice. Though not a great scholar nor a very learned lawyer, he was a good teacher. He was a pioneer of the case method of teaching. His writings were *A Selection of Cases on the Law of Quasi-Contracts* (2 vols., 1888–89) and a *Treatise on the Law of Quasi-Contracts* (1893).

Keeper of the Great Seal. See LORD KEEPER OF THE GREAT SEAL.

Keeper of the Great Seal of Scotland. The Great Seal was, from the earliest times, held by the Chancellor of Scotland.

At the Union of 1707 it was provided that there should be one Great Seal for Great Britain but that a seal be used in Scotland in all things relating to private rights which had usually passed the Great Seal of Scotland.

The Secretary of State for Scotland is now Keeper of the Great Seal of Scotland. The Register of the Great Seal is extant from the beginning of the fourteenth century.

Keeper of the King's conscience. A name formerly sometimes given to the Lord Chancellor because of his function in exercising, in the court of Chancery, the royal power of deciding on petitions to the King to do justice, a power which, prior to the development of fairly settled rules for doing equity, was exercised according to conscience.

Keeper of the Peace. See CONSERVATORS OF THE PEACE.

Keeper of the Privy Seal. See LORD PRIVY SEAL.

Keeper of the Privy Seal. An office of State in Scotland, created in the fifteenth century, to relieve the Chancellor. The affixing of the Privy Seal became a preliminary to the fixing of the Great Seal and in some cases sufficient by itself.

Keeper of the Signet. The Signet was the King of Scotland's personal seal, and, at least from the foundation of the Court of Session in 1532, summonses and other writs in litigation, had to pass the Signet, and these writs had to be drawn and signed by the clerks in the Office of the King's Secretary of State, who became the Society of Writers to the Signet. Their monopoly of initiating litigation has disappeared, though summonses must still pass the signet.

The Keeper is appointed by the Crown and the office is now combined with that of Lord Clerk-Register. He is nominal head of the Society of Writers to Her Majesty's Signet, and appoints a Deputy Keeper who is the actual head of that Society.

Keeper of the Wardrobe. In the thirteenth century, the English Keeper of the Wardrobe was chief clerical officer of the royal household and one of the most influential royal officers. Under the Yorkist and Tudor kings, the wardrobe was eclipsed in importance by the royal chamber. The records kept by the wardrobe are important sources of knowledge of the working of the monarchy in the thirteenth to fifteenth centuries.

Keeping terms. The acts of a student member of one of the Inns of Court in taking dinner in the Hall of his Inn a sufficient number of times to make the term count for being called to the Bar.

Keessel, Dionysius Godefridus van der (1738–1816). Professor of Law at Groningen and Leiden, author of *Theses selectae iuris Hollandici et Zelandici* (1800), a work to supplement Grotius' *Introduction* to the Roman–Dutch law, frequently cited in South Africa, and of *Praelectiones juris hodierni ad Hugonis Grotii Introductionem ad jurisprudentiam Hollandicam* (1961–7), *Dictata ad Justiniani Instutionum libros quattor* (1965–7) and *Dictata ad Jus Criminale* (1967–73), based on lectures, but valuable expositions of the finally developed Roman–Dutch law.

Keith, Arthur Berriedale (1879–1944). Barrister and advocate, Professor of Sanskrit at Edinburgh, and author or editor of a large number of books on Indian myths and religion, law, international law, and, particularly, on constitutional and Commonwealth law, including *Responsible Government in the Dominions* (1912), *Constitutional History of the First British Empire* (1930), *The Governments of the British Empire* (1935), and *Constitutional History of India, 1600–1935* (1939).

Keith, Henry Shanks, Baron Keith of Kinkel. (1922–). Son of Lord Keith of Avonholm (q.v.), he was called to the Scottish Bar in 1950 and the English Bar in 1958. He took silk in Scotland in 1962, became a judge of the Court of Session in 1971, and a Lord of Appeal in 1977.

Keith, James, Baron Keith of Avonholm (1886–1964). Passed advocate of the Scottish Bar in 1911, took silk in 1926, and became Dean of the Faculty of Advocates in 1936. He was a judge of the Court of Session, 1937–53, and a Lord of Appeal in Ordinary, 1953–61.

Kellogg, Frank Billings (1856–1937). Was admitted to the Bar in 1877 and later served as special counsel to the U.S. Attorney General engaged in the prosecution of Trusts. He entered Congress in 1917 and was later Ambassador to Great Britain, 1924–25, and Secretary of State, 1925–29. He was largely responsible for the Kellogg–Briand pact outlawing war (1928) and, in 1929, received the Nobel Peace Prize. Later (1930–35) he served as a judge of the Permanent Court of International Justice.

Kellogg–Briand pact (or Pact of Paris) (1928). Based on a proposal by Briand, the French premier, to the U.S. Secretary of State, Kellogg (q.v.), for a treaty outlawing war between those countries, Kellogg proposed a multilateral treaty and, in 1928, 15 countries signed an agreement condemning recourse to war for the solution of international controversies. Eventually, 62 states adhered to the pact, but it rested on moral force and the support of public opinion only and had no sanction or real effect. Many signatories made qualifications and reservations, excluding wars of self-defence, certain military obligations arising from the League Covenant, the Monroe doctrine, and other exceptions.

Kelly, Sir Fitzroy (1796–1880). Called to the Bar in 1824, he rapidly developed an extensive practice and acquired a great reputation as an advocate. He was an M.P., 1837–41, 1843–47, and 1852–66, Solicitor-General, 1845–46, Attorney-General, 1858–59, and was appointed Chief Baron of Exchequer in 1866. Despite his age, he proved himself an able judge, though latterly cases proceeded very slowly before him and he made no great contribution to the law.

Kelsen, Hans (1881–1973). Austrian jurist, Professor of Law at Vienna, author of the Austrian constitution (1920), and judge of the Supreme Constitutional Court of Austria (1920–30). He came to England and later became professor in American universities. He was author of *General Theory of Law and State* (1945), *The Pure Theory of Law* (1934, revised 1960), *Principles of International Law* (1952), *What is Justice?* (1957), and many other works. His fame is owed, chiefly, to the Pure Theory of Law or Doctrine of Pure Law, first set out in his *Hauptprobleme der Staatsrechtslehre* (1911) which rigidly separates law from metaphysics, politics, and sociology and challenges both philosophical and natural law theories of law. It is a theory of positive law in general, and a general theory, not an interpretation of specific national or international legal norms. It was 'pure' in that it must be self-supporting and not depend on extra-legal values. Each system of positive law consists of coercive norms created by the State; the validity of these norms rests on the *Grundnorm*, or basic norm generally accepted, such as that the Queen in Parliament is supreme, or the Constitution is supreme. The basic norm is the one the validity of which cannot be derived from a higher one. This leads to a glorification of the state and the view that the value of peace is greater than that of justice. He was an important constitutional and international lawyer and, possibly, the most influential jurisprudent of the twentieth century.

Kelyng, John (?–1671). He was called to the Bar in 1632 and, as an ardent royalist, seems to have been imprisoned during much of the Commonwealth and Protectorate. He was essentially a criminal lawyer. He was made a judge of the King's Bench in 1663 and Chief Justice of the King's Bench, 1665–71, and got into trouble for fining and imprisoning jurors. He compiled a volume of reports of King's Bench and criminal cases, which were later edited by Holt, C.J., but attained no reputation as a lawyer.

Kenealy, Edward Vaughan Hyde (1819–80). Was called to the Irish Bar in 1840 and the English Bar in 1847. In 1873, he acted as leading counsel for Orton, the Tichborne claimant, and conducted the case violently, aggressively, and rudely. Following the verdict, in which the jury censured his language, he started a scurrilous journal to plead Orton's case, in which he attacked the characters of L. C. J. Cockburn and Sir John Holker, S.G. In consequence, he was dispatented, and disbenched and disbarred by Gray's Inn in 1874. He then entered Parliament but made no figure there. He was a man of considerable learning and a voluminous writer, and it must be supposed that his mind had been affected by the Tichborne case.

Kennedy, Hugh (1879–1936). Was called to the Irish Bar in 1902, became Law Officer of the Irish Provisional Government, 1922, first Attorney-General of the Irish Free State, 1922–24, and first Chief Justice of the Irish Free State, 1924–36.

Kennedy, Neil John Downie, Lord (1854–1918). Was called to the Scottish Bar in 1877 and held the posts of Professor of Scots Law at Aberdeen 1901–7, Chairman of the Crofters Commission, 1908–12, and Chairman of the Scottish Land Court, 1912–18.

Kenning to a terce. The procedure, now obsolete, by which the sheriff in Scotland fixed the lands falling to a widow under her right of terce (q.v.).

Kenny, Courtenay Stanhope (1847–1930). Practised as a solicitor before going to Cambridge, where he had an outstanding career, winning the Yorke Prize three times. He was called to the Bar in 1881 and, in 1888, succeeded Maitland as Reader in English Law. In 1907, he again succeeded Maitland as Downing Professor of the Laws of England, holding the chair till 1918. He was unquestionably the most successful Cambridge law teacher of his time. His chief work is the *Outlines of Criminal Law* (1902 and many later editions) which has become a classic, and he was one of the first to prepare collections of cases as an adjunct to study: *Cases on Criminal Law* (1901) and *Cases on the Law of Torts* (1904). He sat in Parliament from 1885 to 1888 and his Yorke Prize essay on charitable uses influenced the Mortmain and Charitable Uses Act, 1891.

Kent, James (1763–1847). Graduated from Yale in 1781, was inspired to become a lawyer by reading Blackstone's *Commentaries*, was admitted to the Bar in 1785 and practised in Poughkeepsie and New York. He acquired a reputation for legal knowledge and became the first Professor of Law at Columbia in 1794–98, and again in 1824–26. He was, successively, Recorder of New York (1797) and a judge of the New York Supreme Court (1798). He was Chief Justice of the New York Supreme Court (1804–14) and Chancellor of New York (1814–23), in which capacity he modified English Chancery rules and practice and, by his decisions, laid the foundations of American equity. His reputation rests mainly on his *Commentaries on the American Law* (4 vols., 1826–30), modelled on Blackstone's *Commentaries* and based on his Columbia lectures, which was America's first legal classic and exercised immense influence thereafter. Its six parts deal with the law of nations, the government of the United States, the sources of the municipal law, the rights of persons, personal property, and real property. He did not deal with crimes. It was immediately accepted as an authoritative exposition of English common law as adopted and modified in America and also a standard Federalist interpretation of the Constitution. It went through many editions. He also prepared a revision of New York Laws, an annotated edition of the Charter of New York City, and a *Commentary on International Law*.

Duer, *Discourse on the Life of James Kent* (1848); W. Kent, *Memoirs of James Kent* (1898); Curtis, *James Kent, The Father of American Jurisprudence*; Horton, *James Kent: A Study in Conservatism* (1763–1847); J. Goebel, *History of the School of Law, Columbia University* (1955).

Kenyon, Lloyd, Lord (1732–1802). Called to the Bar in 1756 after serving articles to an attorney, Kenyon was an industrious and able lawyer with a practice in both common law and chancery cases. He was never a good speaker in court, nor in Parliament, but became Attorney-General in 1782. He was Master of the Rolls, 1784 to 1788, and then became Chief Justice of the King's Bench, which office he held till his death. He was a sound and learned lawyer, not a statesman, and his qualities were industry, long study of the law, and great capacity for quickly grasping the correct solution of a problem. Many of his decisions have become leading cases. He enforced a high standard of honesty and morality, but his early life left him mean and boorish, and wholly lacking in refinement.

Kenyon, *Life of Lord Kenyon.*

Ker, Charles Henry Bellenden (?1785–1871). Was called to the Bar in 1814, became a member of the public records commission and, in 1833, of the commission on consolidating the statute law and on the criminal law. In 1854, he became the chief working member of the royal commission on consolidation of the statute law, from whose labours resulted the first edition of the Statutes Revised, the chronological tables of the statute law and the criminal law acts of 1861.

Ker, Robert, 4th Earl and 1st Marquess of Lothian (1636–1703). Educated abroad, was sworn a Privy Councillor in 1686, but removed by James VII (II of England) in 1687, and restored in 1690. He was Lord Justice-General of Scotland, 1689–1703, and a Commissioner to treat of Union with England.

Ketch, John. Public executioner in the reign of Charles II and James II and hence a name sometimes applied to another filling that office.

Keys, House of. The legislative assembly of the Isle of Man. See MANX LAW.

Kidd, William (Captain Kidd) (?1645–1701). A famous pirate who lived at Boston, Mass., was captain of a privateer, imprisoned for piracy in 1699, and then sent to England, tried and hanged.

Kidnapping. The common name for the common law offence of carrying away, or secreting, of any person against his will, or against the will of his lawful guardians. It may be constituted by false imprisonment, which is total restraint of a person and his confinement without lawful authority or justification, or by carrying him away to another place against his will. The carrying away of females is generally termed abduction and is punishable by statute. The motives may be to subject the victim to slavery or other involuntary servitude, to expose him or her to danger of the commission of criminal acts, or to extort ransom in return for the victim's release.

In modern times, kidnapping has frequently been employed by political extremists as a means of extorting concessions of some kind from a government, such as the release of political prisoners.

Kilbrandon, Lord. See SHAW, C. J. D.

Kilkenny, Statutes of (1366). Statutes in Norman-French passed by an Anglo-Irish Parliament, summoned at Kilkenny by Lionel, Duke of Clarence, King Edward III's Lieutenant, forbidding intermarriage with the Irish, the use of Brehon Law, selling arms or armour to the Irish, excluding the Irish from cathedrals, abbeys, and benefices, requiring the use of English surnames, English speech and customs, and other divisive enactments. The purpose of the statutes was to preserve as large an area as possible of Ireland for England at the cost of abandoning the rest. The independent Irish were treated as enemies and rightless. These statutes were supported by excommunication of all who contravened them. They were reissued many times and remained in force for over two centuries. They marked the failure of the Conquest of Ireland, and divided the conquered part of Ireland in law, custom, speech, and other respects, from native Ireland.

Kilmuir, Lord. See MAXWELL FYFE, D. P.

Kindersley, Richard Torin (1792–1879). Was called to the Bar in 1818 and practised in Chancery, being latterly a leader in the Rolls Court. He became a Master in Chancery in 1848, a Vice-Chancellor in 1851, and retired in 1866. He won respect as a sound equity judge.

Kindly tenancy. A form of landholding, widespread in pre-Reformation Scotland, whereby land was let, on favourable terms, to the successors of the ancient possessors of the land or to those whom the lord wished to favour. It preserved ancient customary rights to land and may be likened to copyhold, as the right was constituted by writ, frequently by entry in the King's Steward's or the landlord's rental book, but not by charter.

In a few districts, notably at Lochmaben in Dumfriesshire, the tenure survives, but in most cases the tenure has been altered to feudal tenure.

King, Sir Peter, 1st Lord (1669–1734). After early historical and theological studies, King was called to the Bar in 1698, soon acquired a practice, and entered Parliament in 1701, and by 1712, had become leader of the Whig party and recorder of London. In 1714, he became Chief Justice of the Common Pleas and proved himself a great lawyer, learned and impartial. He tried many persons implicated in the rising of 1715. In 1725, he was made Speaker of the House of Lords and presided at the trial of Lord Macclesfield. In the same year, he became a peer and Lord Chancellor but, though an admirable common lawyer and competent in equity, he never fully mastered either the principles or the practice of that branch. He was diffident and dilatory. Nevertheless, he gave some important decisions and carried through reforms which deprived the Masters in Chancery of control of suitors' money. He resigned in 1733.

King. The name given to the male sovereign over a nation or territory, and usually reserved for important rulers, the only higher title being that of emperor (q.v.), to whom one or more kings may be subordinate. A king is deemed of higher rank and more important than a count, duke, or prince. In origin, kingship has sometimes been connected with religion, and in Egypt, the Hellenistic period, and among the Roman emperors the king was deemed divine. The Christian Roman emperors claimed to have authority from God and, in medieval thought, the kingship was regarded as similar to the priesthood. This approach was also seen in the theory of the Divine Right of Kings (q.v.). Kingship may be elective, as in mediaeval Germany, but is more usually hereditary; it is usually monarchical, but dyarchies have been known, as in ancient Sparta, where two kings ruled jointly; the office may also be absolute and unfettered, or constitutional, where the king's powers are defined and limited by constitutional rules and practices, and he reigns but rules only within limits in accordance with the constitution.

The office is now found only in some Western European countries, in all of which royal power is now regarded as flowing from the people rather than from God and in which the king is more a non-political head of state than an active ruler, and at least as much fettered by constitutional rules as the heads of the executive branch of government.

Where the office may be held and the powers exercised by a woman she is a Queen Regnant. The wife of a king is a Queen Consort.

King can do no wrong, The. A principle of feudal law justified by the fact that the King, as the peak of the legal structure of a state, could not be impleaded in the King's, i.e. his own, courts, and be judged by his own servants. Later, it was sought to be justified on the basis of the unaccountability of the monarch to any human agency. In England before the Crown Proceedings Act, 1947, the King was immune from legal liability for the wrongful acts of his servants, high or low, though the Crown (in the sense of the government carried on in name of the king or queen) would, as a matter of grace, pay damages awarded against a Crown servant for tort committed in the course of his employment. The 1947 Act made the Crown liable in the same manner as a private person in contract and for specified classes of torts committed by its servants or agents; certain public bodies may be and others may not be servants or agents of the Crown. The Crown is also liable for some breaches of statutory duty but not for acts of State (q.v.) or acts done by virtue of the royal prerogative. The Act does not authorize actions against the Queen in her private capacity. The principle of Crown immunity was not accepted in older Scots law but the English rule was later followed; the 1947 Act applies, in part, to Scotland also.

King-of-Arms. See HERALD.

King's (or Queen's) Bench, Court of. The Court of King's Bench did not become a distinct body with its own judges and records till about a century after the Common Pleas (q.v.) had become a distinct court. Being held *coram Rege* it was not located in a definite place but moved about the kingdom with the King. It developed as the body of justices which sat in the King's hall and heard cases concerning the King, or cases affecting great persons who had the privilege of being judged only by the King himself, and which could correct the errors of all the other justices. The court's jurisdiction in error, and in criminal cases, originated from the close connection with the King.

In the fourteenth century, the court became distinct and lost its close connection with the King and the King's Council. It thus became simply a common law court, but retained quasi-political powers from the time when the court *coram Rege* was both Council and King's Bench. From the fifteenth century it had a criminal jurisdiction over ordinary cases, over indictments and other proceedings from inferior courts removed into the King's Bench by writ of *certiorari* on account of some difficult point of law, and in cases where the correctness of the decision of an inferior court was questioned by writ of error, or subsequently by motion for a new trial or reservation of a point for discussion with the other judges and ultimately by the creation of the Court of Criminal Appeal.

Its civil jurisdiction was originally concurrent with the Common Plea in cases of trespass, and by the fictitious Bill of Middlesex, it acquired concurrent jurisdiction in all personal and mixed actions. In the thirteenth and fourteenth centuries, it acquired jurisdiction by writ of error on the record over, till 1783, the Court of King's Bench in Ireland and, till 1830, the Court of Common Pleas. It might also correct errors by writ of *Audita Querela*, or by motion for a new trial. It also developed an important supervisory jurisdiction, exercised mainly by means of the prerogative writs of habeas corpus, *certiorari*, prohibition, mandamus, *quo warranto*, and *ne exeat regno.*

In 1875, the court was abolished and merged in the newly-established High Court in which the name is perpetuated in the King's (or Queen's) Bench Division.

King's (or Queen's) Bench in Ireland, Court of. The Irish King's Bench developed from the court of the justiciar in Ireland, being an itinerant court hearing pleas held before the King's justiciar. It had jurisdiction over pleas of the Crown, possessory assizes and some common pleas. In 1290, a justice was appointed to hold the pleas following the justiciar and, in 1324, a chief justice and, in 1391, the court became the King's Bench. In 1655, it became the Upper Bench but the old name returned at the Restoration. In 1877, it became the Queen's Bench Division of the High Court in Ireland. It came to an end when the courts of Ireland and of Northern Ireland separated.

King's briefs or King's letters. Letters patent issued by the King from Chancery, usually addressed to the clergy and church-wardens and to be read during service. They were first issued after the Reformation and were frequent from the Restoration down to about 1835. The main object was to obtain contributions for alleviating distress caused by widespread disasters or other purposes likely to evoke general sympathy.

King's Council. The group of persons in the royal household, usually including the Officers of State, some of the judges, bishops, and barons, which gave the English kings advice and, notwithstanding the breaking away of the Exchequer, Common Pleas and King's Bench (qq.v.) and their development into regular courts, advised him on the exercise of his personal jurisdiction, both appellate and, at first instance. It gradually developed into the modern Privy Council (q.v.) but, to some extent, also into the modern House of Lords.

J. F. Baldwin, *The King's Council in the Middle Ages*; T. P. Taswell-Langmead, *Constitutional History*.

King's Counsel. See QUEEN'S COUNSEL.

King's Inns. An Inn of Court (q.v.), established in Dublin, by Henry VIII, as a means of making Ireland an English dependency and the centre of the Irish legal profession, without which English law could not spread throughout Ireland. Despite changes of government, it remains the Inn of Court of the Republic of Ireland.

B. T. Duhigg, *History of the King's Inn* (1806); G. E. Hamilton, *The Society of King's Inn, Dublin* (1915).

King's peace. In early mediaeval English law, particular individuals or institutions, such as the Church, had a peace, and wrongs were looked upon as a breach of his or its peace. For these a fine was due. In the eleventh century the King's peace was local and personal, and confined to certain places and seasons of the year, and died with him. Between the death of one king and the coronation of his successor, disorder and crime was not a breach of the King's peace and entailed no liability, though they might be breaches of an earl's or lord's peace. When Edward I succeeded to the throne, his peace was at once proclaimed and this has been the practice ever since. The King's Peace extended until it covered all England at all times and when this happened, a man could not commit crime at any time without being liable to a fine at the instance of the King. The King sought to extend the concept of breaches of the King's peace because they were a source of revenue to him, and the list of these came to be the catalogue of pleas of the Crown.

C. K. Allen, *The Queen's Peace*.

King's printer. By letters patent from 1530, the Crown granted to various persons the office of King's printer, with sole rights to print the statutes and, later, the abridgements thereof, though it sometimes granted concurrent rights to the universities of Oxford and Cambridge. Oxford never used this right but Cambridge published Pickering's edition of the statutes. The King's Printer accordingly had a monopoly. He drew the distinction between public and private statutes, and introduced numbering by sections, marginal notes, and punctuation. Today the Controller of H.M. Stationery Office is King's (or Queen's) Printer of Acts of Parliament and holds the copyright in all government publications, but in respect of the Bible and Prayer Book Eyre & Spottiswoode have royal monopoly shared with Oxford and Cambridge University Presses. See *Oxford & Cambridge* v. *Eyre & Spottiswoode* [1964], Ch. 736. For holders of the office see *Basket* v. *Cambridge University* (1758), 1 W. Bl. 105.

King's widow. The widow of a tenant *in capite* of the King. Under the statute *De Praerogativa Regis* she required the King's licence to remarry and, if she did so without licence, she was liable to a fine, generally of a year's value of her dower. When the Court of Wards and Liveries was established in 1540 the Master of the Wards was empowered to compound with King's widows for having remarried without licence.

King-at-Arms. The principal officers of arms and heralds of the monarch. In England, there are three king-at-arms, Garter, Clarenceux and Norroy; in Scotland, Lyon; and in Northern Ireland, Ulster. See also HERALD.

Kings, divine right of. A political theory that legitimate monarchs derive their authority and power from, and are responsible to, God alone, and that, by divine ordinance, this authority is hereditary in a certain line of succession. The theory is probably due to oriental influences and to the Old Testament and is connected with the anointing of kings at coronation. In England, the doctrine of divine right was developed to its limit in the seventeenth century, notably by Hobbes and in most extreme degree by Filmer, who was answered by Locke. It was held by James I and Charles I and was at the basis of much of the dispute between Crown and Parliament, which led to the Civil War. It was a principle of the Anglican Church after the Restoration but was hit when James II made it impossible for the clergy to obey their King and their consciences. It ceased to be a tenable theory at the Revolution of 1688, but continued in France, notably as exemplified by Louis XIV, until the theory was shattered there by the Revolution of 1789.

J. N. Figgis, *The Divine Right of Kings.*

Kingsdown, Lord. See LEIGH, PEMBERTON.

Kinnear, Alexander Smith, Baron (1833–1917). Passed advocate of the Scottish Bar in 1856 and specialized in property and trust law. In 1881, he was elected Dean of the Faculty of Advocates and, in the next year, became a judge of the Court of Session. He received a peerage in 1897 and retired in 1913 and occasionally sat in the House of Lords after his retiral. As a judge he was very distinguished and his judgments were respected. He was also chairman of the Scottish Universities Commission and a member of the royal commission arbitrating between the Church of Scotland and the Free Church in 1905.

Kinship. The system of human relationships derived from marriage and descent. In tribal and simpler societies, it is a major factor in social and economic life. In modern industrial societies, it is important in social and political relations. Legally, it underlies much of the law concerned with the family, marriage, children, family, property, and succession though, in Western European countries, the forms of family and marriage which have been recognized, have been strongly influenced by Christianity. Anthropological studies show the wide varieties of kinship relations found at various places in the world and bring out that those recognized in Western society, monogamy, criteria of legitimacy, patriarchal succession, and so on, are not universal nor necessarily right, but only those accepted in particular parts of the world. In law, kinship is determined by descent from a common ancestor. For many purposes degrees of kinship are important. The Roman law counted these by counting the generations from the one person up to the common ancestor and down to the other person; the canon law counted the number of generations from the person further removed from the common ancestor. Thus, first cousins are related in the fourth degree by Roman law computation and in the second degree by canon law computation. Agnates were, according to Roman law, persons related to one through males but, in Scots law, related through the father; cognates were according to Roman law persons related through one or more females but, in Scots law, those connected with a person through his mother.

Kipp, Theodor (1862–1931). German jurist, author of *Geschichte der Quellen des römischen Rechts* (1896), *Lehrbuch des bürgerlichen Rechts* (1912), and other works.

Kirk-session. In Presbyterian churches in Scotland, the church court of the parish, comprising the minister and elders, and having some administrative and judicial functions, the latter now practically in abeyance.

Kissing the book. The practice, peculiar to English courts and discontinued in 1909 after being in decline for at least a century, of kissing the New Testament after taking the oath in court. The original practice had been for the witness to put his hand on the Testament when taking the oath. The practice was never followed in Scotland.

Kitchin, John (?1520–?90). A double reader and Treasurer of Gray's Inn, author of *Le Court Leet et Court Baron* (1580), a book useful for those holding such courts and designed to induce lords of manors to appoint, as bailiffs, men learned in the law. It is elaborately annotated and contains forms and notes for pleaders.

Klaestad, Helge (1885–1965). Norwegian judge, judge of the Supreme Court of Norway, 1931–46, and of the International Court, 1946–61 (President 1958–61).

Klagspiegel, Der richterliche (The Mirror of Judicial Plaints). A compilation made by an unknown author in the fifteenth century from the writings of a small number of Italian jurists, which was important in introducing Roman law and Italian legal learning into Germany. It comprises two treatises dealing respectively with civil law and with criminal law and procedure and includes forms of pleadings.

Klang, Heinrich (1875–1954). Austrian judge and jurist, head of a division of the supreme court from 1945, and editor of the *Kommentar zum Allgemeinen Buergerlichen Gesetzbuch* (1931–35).

Kleptomania. A neurotic disorder consisting in the obsessive impulse to steal, frequently, objects of little value. The act of stealing is satisfying and releases tension; the motive is not gain. The things stolen may be all of one kind, or of related kinds; they are sometimes hoarded, sometimes disposed of. They may symbolize some unconsciously desired sexual object. This compulsive theft is more common among women than men. The person concerned is aware of what he is doing and of the legal quality and consequences of his acts, but nevertheless, is under this compelling urge to steal. It is accordingly distinguishable psychologically though not legally from simple theft for gain. The word is sometimes, though inaccurately, used as a synonym for shoplifting.

Kluber, Johann Ludwig (1762–1837). German public lawyer and author of many works, including *Le droit des gens moderne de l'Europe* (1819) (also issued as *Europäisches Völkerrecht*).

Knight, Sir James Lewis (later (1827) Knight Bruce) (1791–1866). Called to the Bar in 1817, he soon confined himself to equity practice and, in 1841, was made one of the two additional Vice-Chancellors authorized in that year. When the Court of Appeal in Chancery was created (1851), he was one of the first Lord Justices, and sat in that court till his retiral in 1866. He had a profound knowledge of law and was always anxious to shorten procedure and avoid technicalities. His judgments were terse and lucid. He frequently sat in the Privy Council, where he utilized his extensive knowledge of foreign law. A man of learning, literary gifts, industry, and ability, he was a most capable judge with, moreover, a reputation for humour and epigram.

Knight. Originally, in feudal Europe, a mail-clad and mounted warrior. He was one granted an estate in land to be held of the King, or of an overlord, on condition of performing military service with horse and armour for a period, normally 40 days a year but, from about the thirteenth century, actual knight service was commonly commuted to a money payment, scutage. The rank of knighthood was conferred by the King at a ceremony at which he was girded with a sword and dubbed with a tap by a sword on the shoulder. Latterly, kings had to put pressure on persons of the appropriate wealth and social standing to accept knighthood and pay the fees then exigible. Compulsory knighthood was ended in 1640.

In modern Britain, knighthood survives as a honour conferred by the Crown, normally for distinguished services in some sphere. It is normally conferred on judges of the High Court on their appointment. It is a personal dignity, conferred for life and not inheritable, and entitles the knight to the prefix 'Sir' to his name. Such knights are knights bachelor, as distinct from knights of the orders of knighthood (q.v.). A knight may, very exceptionally, be degraded, as for treason, as in the case of Sir Roger Casement in 1916.

Knighthood. A personal dignity conferred for life by the Queen, normally on the advice of the Prime Minister. The sole privileges are the title 'Sir' and the right to take precedence of untitled persons. Knights bachelors are ordinary knights who belong to none of the particular orders of knighthood.

The origin of knighthood is obscure; it springs partly from feudalism, designating the military tenants of a king or noble, partly from cavalry warfare, the word being rendered as caballarius, eques, ritter, chevalier and the like, and from both sources it drew the requisite of formal initiation to the rank and implied superiority over ordinary people. Later, the concept of knighthood was recognized as separate from feudalism, and analogies with monasticism recognized, resulting in the rise of the religious orders such as the Knights Templars and Knights Hospitallers, the Teutonic Order, the Order of St. Lazarus and others.

The mediaeval age of chivalry produced societies, membership of which was confined to knights, some of which have survived, while others have been founded to reward public services. The modern Orders are not confined to knights, and admission to the lower classes does not confer knighthood. The earliest in England was the Order of the Garter, founded about 1348, consisting of the sovereign, 25 knights, and additional royal knights *extra numerum*. Membership is a high honour and the recipients are usually peers, not ordinary knights. Similar orders in European countries included the Burgundian Order of the Golden Fleece (1429) and the French Order of St. Michael (1469). The Order of the Thistle was founded or revived by James VII in Scotland in 1687 and re-established in 1703; there are the sovereign and 16 knights. The Order of St. Patrick was founded in 1783 and revived in 1883; no appointment has been made since 1934. The Order of the Bath (1725) is now divided into two branches, military and civil, and three classes, Knights Grand Cross, Knights Commander, and Companions. The Order of the Star of India (1861), and the Order of the Indian Empire (1878), are now declining, no appointment to either having been made since 1947. The Order of St. Michael and St. George (1818), used to reward colonial service, consists of three classes, Knights Grand Cross, Knights Commander, and Companions. The Royal Victorian Order (1896) used to reward personal service to the sovereign, comprises five classes, Knights and Dames Grand Cross, Knights and Dames Commander, Commanders, and Members of the fourth and fifth classes. The Order of the British Empire (1917) is divided into two sections, civil and military, and five classes, Knights and Dames Grand Cross, Knights and Dames Commander, Commanders, Officers and Members. The Distinguished

Service Order (1886) and the Imperial Service Order (1902) have each one class only, Companions. Membership of any class of any of the foregoing orders entitles the holder to certain precedence over ordinary persons. Membership of certain other orders, such as the Order of Merit (1902) and the Order of Companions of Honour (1917), though highly esteemed, confers no precedence.

H. Nicolas, *History of the Orders of Knighthood of the British Empire.*

Knight-service. The main mode of tenure of land under feudalism whereby, after 1066, the King granted tracts of land to his followers to be held by them on terms of providing a stated number of knights, armed and accoutred, to serve in the army at their own expense for 40 days in the year. The number of knights was commonly a multiple of five or 10, being fixed by relation to the unit of the feudal host, the *constabularia* of 10 knights. About 1166, money payments called *scutage* began to be exacted instead of the production of a quota of knights.

Tenure by knight-service was also subject to incidental burdens, namely (1) *aids*, whereby the tenant was expected, and later obliged, to make contributions to assist his lord in emergency. Latterly, they were limited to three occasions, to ransom the lord from captivity, to knight his eldest son, and towards the expense of the marriage of his eldest daughter. (2) An heir on taking up his ancestor's fief had to pay a *relief*, the amount of which was not fixed till Magna Carta defined it as £100 for a barony and 100s. for a knight's fee. Tenants-in-chief also had to pay *primer seisin*, the right of the King on the tenant's death, leaving an heir of full age, to one year's profits of the land inherited. (3) The lord was entitled, if the heir were under age, to *wardship*, the custody of the person and lands of the heir, without liability to account for the profits. When the heir attained full age he was entitled to ousterlemain, or require release of the lands from the guardian's hands, on payment of half-a-year's profits in lieu of relief and primer seisin. (4) The lord has also the right of disposing of the *marriage* of a female ward, a right intended as security against the lord having to receive an undesirable tenant. If the female ward rejected a suitable match, she forfeited the value of the match, such sum as the suitor was willing to pay as the price of the match. If she married without the lord's consent, she forfeited double the market value of the marriage. The right extended to female wards, and heiress-presumptive of living vassals, and was later extended unjustifiably to male wards.

Alienation of land held by knight-service was permitted only after the statute of *Quia Emptores* (1290) forbade subinfeudation and permitted freemen to alienate their land to new vassals to hold it of and under the lord from whom the alienator had held it. Tenants-in-chief required a licence from the King and had to pay a fine but, from 1327, they could alienate at will on payment of a reasonable fine to the King.

On failure of legal heirs to the vassal (*propter defectum sanguinis*) or on his conviction for felony or treason (*propter delictum tenentis*) the fief escheated and reverted to the lord who had granted it. The lands of a convicted felon were liable to forfeiture to the Crown for ever, if convicted of treason, for a year and a day if convicted of felony.

The system began to disintegrate when the obligation of military service began to be commuted for money. Even in Bracton's time liability for scutage (q.v.) made the tenure of land military. But the knight fee remained and the incidents long continued to be a source of revenue to the Crown.

Tenure by knight-service was abolished in 1660 and lands thereafter were held by socage (q.v.).

Knights of the shire. In the older English and British Parliaments, those members elected to represent shires or county districts as contrasted with burgesses elected to represent boroughs. In early parliaments, the writs calling the parliament directed four (1217) or two (1254) or three (1261) or four (1264 and sometimes thereafter) knights to be chosen from each county. At that time these representatives had to be knights chosen in, and by, the county court but, later, this does not seem to have been insisted on and commoners could also be elected.

Knock-for-knock agreement. In motor insurance, an agreement that, in the event of an accident involving more than one insured vehicle, each insurer carries the risk so far as concerns damage to the car he had insured, whoever may be legally responsible for causing the damage. Such an agreement cannot prevent an assured person making a claim on his own part for damage to his car, even though he is being compensated by his own insurer.

Know-how. The information, practical knowledge, techniques, and skill required for the achievement of some practical end, particularly in industry or technology. It has come to be recognized as virtually a kind of incorporeal property, saleable like a patent right.

Knowledge. Awareness of certain facts, the converse of ignorance. Knowledge of certain facts is frequently relevant to various kinds of legal liability. Thus knowledge of certain dangers is relevant to liability for harm caused by the thing containing the danger. In criminal law, guilty knowledge or knowledge of facts which render conduct criminal, is frequently required for conviction. Statutes sometimes create offences of 'knowingly' doing something, in which case, proof that that thing was done intentionally is usually required. Knowledge may include not only actual knowledge, i.e. actual awareness of the facts relevant, but constructive knowledge, i.e. knowledge attributed by law to the party in the circumstances, whether he actually had the knowledge or not, and knowledge may be attributed to a person who has sought to avoid finding out, or has shut his eyes to obvious means of knowledge, e.g. the man who is offered valuables cheaply in circumstances which suggest that they may well have been stolen, but who refrains from enquiry.

Kocourek, Albert (1875–19). U.S. jurist, professor at Northwestern, Chicago, and editor, with Wigmore, of the *Evolution of Law* series (1915–18). His major work in analytical jurisprudence, *Jural Relations* (1927) utilizes a technical vocabulary developed for the purpose but produces valuable insights into legal reasoning. He also wrote *An Introduction to the Science of Law* (1930).

Kohler, Josef (1849–1919). German jurist, professor at Berlin, author of ethnological studies, and a leader among comparative legal historians. In his *Lehrbuch der Rechtsphilosophie* (1908), a study of the theory of justice based on Hegel's philosophy of history, he originated the idea of jural postulates of civilization, later taken up by Pound, and made an effort to solve the problem of the relation between social facts and the norms of justice. Other major works were *Der Ursprung des Rechts* (1876), *Die Anfänge des Staats- und Rechtslebens* (1878), *Die Grundlagen des Rechts* (1884), and *Grundriss der ethnologischen Jurisprudenz* (1894–5).

Korematsu* v. *U.S. (1944), 323 U.S. 214. By Act of Congress and executive order of the President, there was imposed, after the Japanese attack on Pearl Harbour, a curfew programme for Japanese-Americans and, subsequently, their removal from critical areas and, in some cases, incarceration in relocation centres. In 1943, the Supreme Court held that the curfew order had been a proper exercise of the war power and, in 1944, that detention in relocation centres had been illegal, not being authorized by the statute and order in question. In the present case the Court upheld the removal of persons of Japanese extraction from critical areas.

Kotzé, Sir John Gilbert (1849–1940). Called to the English Bar in 1874, he practised at the Cape Bar and, in 1877, became judge of the High Court in the Transvaal, then Chief Justice of the South African Republic in 1881 till dismissed in 1898. He returned to the bench in 1904 and, from 1922 to 1927, was a judge of the Appellate Division of the Supreme Court of South Africa. He had a profound knowledge of the sources of South African Law, translated Van Leeuwen's *Commentaries on Roman Dutch Law* with his own commentary, and also wrote a series of articles on the history of the Roman-Dutch law in the *South African Law Journal*. He was undoubtedly one of the most distinguished jurists that South Africa has produced, and his judgments continue to rank as of the highest importance.

Krabbe, Hugo (1857–1936). Dutch jurist, author of an important *The Modern Idea of the State* (1922), who saw the law as both the force which restrains the absolute power of the state over its citizens and the force which leads mankind from the stage of state sovereignty to the reign of law in international relations.

Krause, Karl Christian Friedrich (1781–1832). German philosopher whose main work *Grundlage des Naturrechts* (1803) marks the transition from the natural law standpoint to the modern philosophical view of law and government. Law is a postulate of reason and refers to the establishment of the external conditions of reason which exist and which should be maintained independently of the freedom of the will and of natural forces. He strongly influenced Ahrens (q.v.). Among his other works were *Das Urbild der Menschheit* (1811) and *Abriss der Systemes der Philosophie des Rechtes oder des Naturrechtes* (1828).

Kreise. Circles, notably the imperial circles or administrative districts of the Holy Roman Empire from the early sixteenth century till its dissolution in 1806. They were established by the Emperor Maximilian I and the Diet of Augsburg of 1555 gave them powers of law-enforcement, including the right to execute the decisions of the imperial chamber or *Reichskammergericht*. The circles, particularly in southern and western Germany, gave a measure of regional solidarity during the political and religious upheavals of the Reformation. The circles were Burgundian, Lower Rhine-Westphalia, Lower Saxon, Upper Saxon, Electoral Rhenish, Upper Rhenish, Franconian, Swabian, Bavarian, and Austrian.

Ku Klux Klan. The first organization of this name was a secret association of whites in the southern states of the U.S., formed in 1865 for self-protection against the newly emancipated negroes and in opposition to the reconstruction measures (1865–76) of Congress. In some cases, the societies became bands of outlaws. In many, they had the effect of giving protection to the whites, minimizing the coloured vote, expelling the worst of the carpet-baggers, and nullifying federal legislation by keeping the formerly dominant groups in control of the states. The organization declined after 1870. The second organization was formed in 1915 and was directed to white supremacy with some antagonism to Catholics and Jews; it fostered racial and religious prejudices and had some influence in local politics but declined about 1928.

Kunz, Josef Laurenz (1890–1970). U.S. international lawyer, author of *Die Völkerrechtliche Option* (1925–28), *Die Anerkennung von Staaten und Regierungen im Völkerrecht* (1928), *Die Staatenverbindungen* (1929), *Kriegsrecht und Neutralitaetsrecht* (1935), *The Changing Law of Nations* (1968), and other works.

L

L.C. Lord Chancellor.

L.C.J. Lord Chief Justice.

L.J. Lord Justice of Appeal.

L.S. (Printed within a circle on a form.) (*locus sigilli*, the place of the seal.)

Label. In Scottish criminal procedure a synonym for a piece of real evidence, e.g. a weapon, produced at a trial, because a label is attached to it and signed by each witness who speaks to it in evidence, as identifying the particular thing.

Labeo, Marcus Antistius (*c.* 50 B.C.–A.D. 18). A notable Roman jurist, very learned and of great independence of mind. He was a contemporary of Capito and their personal and political antagonism gave rise to the schools of jurists known as Proculians and Sabinians (q.v.), though they were not the founders of these schools. He was a voluminous writer, spending six months a year in writing, and is said to have written about 400 volumes. His works, known only from quotations by later jurists, included a treatise *De Iure Pontificio* (15 books), *Libri ad Edictum* (a commentary on the praetorian edict), *Responsa, Epistulae*, *Pithana* (collected decisions), and *Posteriora* (posthumous works). He was highly regarded by contemporaries and by classical jurists.

A. Pernice, *Labeo* (1873–92).

Labes realis. In Scots law an inherent defect or taint in a title to property attached, e.g. by its having been stolen, which affects it notwithstanding its transmission to an innocent third party.

Laborans (?–*c.* 1190). Canonist and cardinal, author of several works including the *Codex Compilationis* (1182), an attempt to rearrange the materials of Gratian's *Decretum* (q.v.) in more logical order with new materials incorporated.

Labour law. A term, roughly synonymous with industrial law, for the bodies of legal principles and rules connected with the employment of labour. It deals accordingly with the contract of employment, including wages, holidays, conditions of work, duties of care for the employee's health and safety, employers' associations and trade unions, collective bargaining, strikes and trade disputes, and the legal aspects of labour or industrial relations. Though nowadays regularly accepted as a branch of law for academic study, it comprises topics selected from a number of the divisions of law recognized by analytical classification.

Modern labour law is essentially the product of industrial development since the eighteenth century. In Britain the first important stage in modern labour law was the Health and Morals of Apprentices Act, 1802. The stages in the legal recognition of trade unions are parallel but distinct.

In the twentieth century many Western governments have established ministries of labour or employment with substantial responsibilities in the fields of labour law, and the establishment of the International Labour Organization in 1919 was a substantial step towards international co-operation and generally accepted minimum standards. The detailed provisions, however, vary substantially from one state to another.

Labourers, Statute of. The statute passed in 1349 after a large part of the population had died of the Black Death, providing that everyone under 60, except traders, craftsmen, those with private means, and landowners, should work for anyone willing to employ them at the wages paid between 1340 and 1346. Justices of labourers were constituted to administer the Act. It was repealed in 1863.

Laches. Unreasonable delay or negligence in pursuing a right or claim, particularly an equitable one, which may be held to disentitle the claimant to relief. It proceeds on the basis of the maxim *vigilantibus non dormientibus jura subveniunt.* The defence is allowed only when there is no statutory time-bar applicable; if there is, a claimant is entitled to the full statutory period before his claim becomes unenforceable. In considering whether delay amounts to laches and defeats a claim, a court must consider whether there has been any acquiescence on the defendant's part as a result of the plaintiff's delay.

Lachs, Manfred (1914–). Polish lawyer, statesman, and diplomat, a judge of the International Court (1967–85) and President, 1973–76.

Lacuna. A gap in a document, particularly a case not provided for in a statute.

Ladd, William (1778–1841). An American who became a sea-captain and then a farmer, but in 1819 became interested in the cause of international peace and turned to writing peace propaganda and founding peace societies. In 1828 he formed the American Peace Society and devoted the rest of his life to urging its principles. He published *A Brief Illustration of the Principles of War* (1831) and *An Essay on the Congress of Nations* (1840) the latter of which, founded to some extent on Bentham's work, urged a congress of nations for formulating the principles of international law and for providing for the general welfare of nations, and a court for settling disputes by judicial decision. This was a distinct contribution to international thought and has had importance in subsequent international thought and organization, and in essentials has been realized in the twentieth century.

Lading, Bill of. See BILL OF LADING.

Lady. The courtesy title of the wife of a baron, baronet, or knight, and of anyone entitled to the courtesy title of Lord. By courtesy it is given to the daughters of dukes, marquises, and earls. Except formally a marchioness, viscountess, or countess is usually designated Lady.

Lady Day. In the older calendar 6 April, and now 25 March, the Annunciation of the Blessed Virgin; it was, in mediaeval times, the first day of the legal year and still is, in England, one of the usual quarter days for the payment of rent.

Laenland. A mode of landownership in Anglo-Saxon law, distinguished from bocland and folkland (q.v.), by which land was granted by temporary loan or gift for one or more, frequently three, lives. The grantee was sometimes bound to perform services or to pay rent or a lump sum for the land. Such grants were commonly made by the Church.

Laesae fidei. In canon law and ecclesiastical law, breach of faith, the ground justifying the claim of those systems of law to have jurisdiction in matters of breach of contract.

Laesae maiestatis, crimen. The crime of injured majesty or treason. The French *Lèse-majesté* is much wider and includes offences less serious than treason.

Lagan. Goods sunk in the sea but attached to a buoy. See also FLOTSAM; JETSAM; WRECK.

Laferrière, Édouard Louis Julien (1841–1901). French jurist, president of the judicial section of the *Conseil d'Etat*, vice-president of the Conseil, and procurator-general at the Court of Cassation. He greatly influenced the development of French administrative law by his *Traité de la jurisdiction administrative et des recours contentieux* (1887–88).

Lainé, Armand (1841–1908). French jurist, one of the creators of modern private international law and author of an historical study, *Introduction au droit international privé* (1888–92), *La Théorie du renvoi* (1909), and *Étude critique d'un projet de convention concernant la solution des conflits de lois* (1902).

Laissez-faire. An expression which may have originated in a reply by an industrialist to a question by Colbert, the French Minister of Finance in the late seventeenth century, enquiring what the government could do for industry. The answer was *Laissez nous faire*, leave us alone, let us get on with it. The term has passed into a label for non-intervention by government in economic affairs and in related legal controls. The theory of *laissez-faire* or non-intervention was accepted and taught with different emphases by such economic thinkers as Ricardo and the Mills in England, Say and Bastiat in France, and by such political thinkers as Bentham, de Tocqueville, Tarde, and Spencer. The excesses of the application of the theory, which led to largely unrestrained competition, provoked, from the mid-nineteenth century, a reaction supported by socialists, emphasizing the need for protection of employees and consumers, state control, and regulation. In the legal context *laissez-faire* emphasizes freedom of contract and of use of and dealing with property as contrasted with legal restriction.

Lakes and land-locked seas in international law. Lakes may be national, if entirely within the territory of one state, or belong in part to the riparian states, or, if navigable from the open sea, international, though, save by treaty, this may not imply free right of navigation for vessels of all states. The Black Sea and the Sea of Marmora are parts of the open sea, but such seas as the Caspian are national territory.

Lambarde, William (1536–1601). Became a barrister but is remembered as an antiquary. He became a master in chancery in 1592 and keeper of the records at the Rolls Chapel in 1597, and in 1601 keeper of the records in the Tower. In 1568 he published *Archaeonomia*, a collection and translation or paraphrase of the Anglo-Saxon laws. In 1581 there appeared *Eirenarcha, or the Office of the Justices of Peace*, which was for long the standard authority on that subject and was often reprinted, and in 1635, *Archeion, or a Commentary upon the High Courts of Justice in England*. He also compiled the earliest county history, the *Perambulation of Kent* (1576).

Lammas. In the old calendar 12 August and in the reformed calendar 1 August; it is the festival of the wheat harvest and one of the quarter-days of the year.

Lammas lands. Arable and meadow lands held by owners individually during part of the year and in common after the crop has been removed, both by the owners and by other commoners. The name is derived from the former practice of keeping them open from Lammas Day (12 August) to the succeeding Lady Day.

Lammasch, Heinrich (1853–1920). Austrian jurist and supporter of international arbitration, member and President of The Hague Court of Arbitration, and international arbitrator. His major work is *Das Völkerrecht nach dem Kriege* (1917) and he also wrote a *Grundriss der Strafrechts* (1899).

Lamoignon, Guillaume de (1617–1677). A great French jurist, Chief President of the Parlement of Paris, prominent in the drafting of the Civil Procedure Ordinance of 1667 and the Criminal Ordinance of 1670; author of *Arrêtes, ou Lois projetées* in which he laid down the principles of French customary law in the form of a code.

Lampredi, Giovanni Maria (1732–1793). Italian jurist, author of *Juris gentium universalis* (1776) and *Del commercio dei popoli neutrali in tempo di guerra* (1788).

Lancaster, Chancery Court of County Palatine of. In 1351 Edward III granted to the Duke of Lancaster a Chancellor, Court of Chancery, and such other *jura regalia* as pertained to a county palatine. Since 1399 the County Palatine has been held by the sovereign but independently of the Crown. The County Palatine retained its Court of Chancery, and there was an Attorney-General of the County Palatine and Duchy.

Within the County Palatine the court had the same powers and jurisdiction as the Chancery Division of the High Court, concurrent with the High Court. Proceedings might be transferred to or from the court by other courts. It had the statutory jurisdiction of the Chancery Division under many statutes. Appeal lay to the Court of Appeal and House of Lords. It was merged in the High Court in 1971.

The Chancellor of the Duchy is now a political office, and the Vice-Chancellor, appointed by the Chancellor, performs all the judicial functions.

Till 1873 the County Palatine had also its own Court of Appeal in Chancery, Common Pleas and commissions of assize.

Lancaster, Court of Duchy Chamber of. This court might be held before the Chancellor of the Duchy of Lancaster or his deputy and had jurisdiction in respect of equities relating to lands held by the sovereign as holder of the Duchy of Lancaster, whether within the County Palatine or not. Proceedings were conducted as on the equity side of the old courts of Exchequer and of Chancery. The court last sat in 1835.

Lancelotti, Giovanni Paolo (1522–90). Italian canonist, author of the famous *Institutiones juris canonici* (1563), based on Justinian, which was very influential in the study of canon law, *Institutionum juris canonici commentarium* (1560), and *De comparatione juris pontificii et caesarei* (1574).

Land. The solid and dry part of the surface of the earth, including, however, streams traversing tracts of land and enclosed, or semi-enclosed, areas of water. Land is a major factor of economic production, an economic asset as a source of wealth and a store of wealth, and possession of it has in many cases been a source of social, political, and economic power and a great status symbol. The law of each country applicable to land tends, by reason of the permanence of land, to be old at least in origin and to be complicated, to reflect social and economic patterns which may have changed very substantially. In consequence, at many times in history and in many countries, major reforms of land-tenure have been made, frequently forcibly, and often without benefit to the utilization and management of the land for productive purposes.

From the legal point of view the material questions are whether land is owned by the State, or by communities, or by groups jointly, or by individuals, and what rights and interests can be held in it, whether it is heritable or not, whether freely alienable or not.

Important in all Western European countries was the feudal theory that all the land in a country belonged to the King, who had granted tracts of it on certain conditions and in return for certain services or payments to lords, who in turn sub-granted smaller tracts to subordinate tenants, and so on. Under this system several superiors, king, lord, possibly several intermediate superiors, and the tenant who actually possessed the land, all simultaneously owned certain rights and interests ('estates') in the land, and each was connected to the person above him and the person below him in the chain by links of tenure, defining the terms and conditions on which each held the land from the person immediately above him. Landholding was accordingly pyramidal in structure. In Western European countries this feudal pattern has everywhere been abolished or modified, but in some systems remains important for the theoretical understanding of institutions, principles, and rules of law. In France feudalism was abolished at the Revolution. In England the system, greatly modified, survived until 1926.

In general, ownership of land includes the airspace above and the subsoil below down to the centre of the earth, as expressed in the maxim *Cujus est solum ejus est usque ad caelum et ad inferos*, but the mineral strata in the subsoil may be sold or leased or otherwise belong to another than the owner of the surface, and modern legislation renders the flight of aircraft through an owner's airspace not a trespass.

For legal purposes land normally includes trees and crops growing on the soil, and also buildings built on and permanently attached to the soil.

Land Acts, Irish. A series of Acts of the U.K. Parliament intended to give Irish tenant farmers control of their land and to restrict, and ultimately to end, landlords' ownership of land altogether. Gladstone's Act of 1870 gave compensation to tenants for improvements; his Act of 1881 guaranteed fixity of tenure, fair rents, and freedom to sell the tenant's interest. Between 1885 and 1903 a series of land purchase Acts allocated money to enable tenants to acquire their own land; as a result much of Ireland was transferred from tenant farming to peasant proprietorship.

Land charges. In English law certain rights or claims affecting land, frequently of kinds not discoverable either by inspection of the property or by the normal investigation of title, such as rents, annuities, or monies payable by an owner and exigible from, and a burden on, his land.

The Land Charges Act, 1925, established a system of protecting certain rights or claims affecting land by registration, consolidating enactments dating from 1776. There is no statutory duty to effect registration, but it is the effective means whereby persons entitled to the benefit of such burdens as may be registered may make their claims known to the world, and whereby purchasers are protected from such burdens as ought to be but are not registered, since interests are not enforceable against a purchaser for value of the land unless registered. There are six classes of land charges, viz.: Class A—rents or sums of money charged upon land pursuant to the application of some person, under the provisions of any statute to secure money spent on the land under the Act; Class B—charges on land of any of these kinds not created pursuant to the application of any person; Class C—puisne mortgages, limited or statutory owner's charges, general equitable charges, and estate contracts; Class D—charges for death duties, restrictive covenants, and equitable easements; Class E—certain annuities; and Class F—a charge on land as a matrimonial home. Local registers are also maintained by local authorities in which charges over land acquired by these bodies may be registered.

Where title to land has been registered the system is superseded thereby, except in respect of local land charges.

Land registers. Until the late nineteenth century conveyances of land in England and Wales could not be and were not registered in any way, save that certain instruments affecting land in Yorkshire have been capable of registration since 1706. The Land Registry Act, 1862, was the first attempt to introduce registration of title. It operated on a voluntary basis and subject to conditions. The system now operative is based on that of the Land Transfer Act, 1875. The Land Registration Acts, 1925 and 1936, consolidating earlier enactments, provided for compulsory registration of title to land in specified areas on a sale taking place and voluntary registration elsewhere, but the compulsory areas are being extended and will in time cover the whole country. The essential features of the system are that once a purchaser's title to a plot of land has been investigated and registered, the proprietor has a title thereto guaranteed by the State and backed by a state indemnity to compensate for any loss occasioned by mistake, rectification of title, or otherwise. A purchaser may by an official search ascertain the state of the register at the time of completion of a disposition by registration.

In Scotland the General Register of Sasines (q.v.) has, since 1617, provided a public register of deeds affecting land anywhere in Scotland in which transfers, most kinds of burdens, and restrictions are recorded. It is a register of deeds affecting land, not of land itself. Official photostat extracts of recorded deeds are equivalent to the originals. Persons concerned with land may obtain searches showing the state of the register at the time of effecting a transaction affecting the land.

Land tax. The first land tax in England was an assessment of 4s. in the pound on real estate, on personal estate and certain offices and pensions, imposed in 1692 for one year. In 1698 it was imposed as an annual grant of a fixed sum, the proportion to be contributed by various local areas being specified. Thereafter a similar act was passed annually until 1797. The Land Tax Perpetuation Act, 1798, made the land tax perpetual and made permanent the quota payable by each parish as at that date. Provision was made for redemption. In 1949 provision was made for the stabilization of the charge to tax, exoneration of certain properties, and the compulsory redemption of land tax still unredeemed.

In Scotland taxes were levied on land in the Middle Ages and land tax or cess had before the union practically been accepted as an annual tax. In the Act of Union the proportion payable in Scotland was fixed at about one-fortieth of the English contribution. It was to be levied from the land rent of Scotland for ever, subject to a power of redemption. It was payable partly from burghs and partly from shires, burghs being assessed according to their rents and income by stentmasters, and the inhabitants of counties according to the yearly revenue of their land, by commissioners of supply. It was a tax which attached to the land, the owner for the time being being liable as the tax fell due. When redemption was introduced in 1802 the Scottish contribution was finally fixed. After 1896 it was payable only in counties.

Land tenure—England and Wales. After the Norman Conquest the system of land tenure which developed in England and Wales was feudal. All land in theory belonged to the Crown, but it made grants of tracts of land to tenants in chief in return for certain services, not necessarily military. Tenants in chief in turn made grants of smaller tracts of land, and in theory there was no limit to the amount of subinfeudation which was possible. But subinfeudation was stopped by the statute *Quia Emptores* in 1290 and thereafter no new sub-tenancies could be created.

Feudalism led to complications and the development of different kinds of tenure, namely the free tenures (frankalmoin or spiritual tenure, tenure by knight service, by serjeanty and socage tenure) and non-free tenure or villeinage, later called copyhold. In addition there were a few customary modes of tenure in particular districts, notably gavelkind, borough English, and ancient demesne. In 1660 tenure by knight service was abolished and tenure by serjeanty converted into socage save that certain honorary and dignified services were in certain cases preserved. Frankalmoin fell into desuetude and from 1660 tenures were, for practical purposes, reduced to socage and copyhold. The money payments due to the lord under socage tenure diminished in value in time and were not collected, and in most cases mesne lords ceased to be able to prove their lordship, in which case it was presumed that the land was held direct of the Crown.

Legislation of 1925 converted copyhold into socage and abolished the other modes of tenure. All the incidents of tenure, such as duties and payments to a lord, were abolished and, while in theory tenure still exists, for practical purposes a tenant in fee simple can be regarded as the outright owner of the land.

Land tenure—Scotland. The principles of feudal land tenure were adopted in Scotland in the eleventh and twelfth centuries and it came to be accepted that all land belonged ultimately to the Crown, and was held of the Crown, mediately or immediately, by tenants in chief, mid-superiors, or vassals. No statutory check on subinfeudation was ever imposed and it remains competent. The tenures recognized were the spiritual one of mortmain or mortification, and the lay ones of wardholding, socage, feu or feu-farm, blench and burgage. There were one or two other exceptional forms. Of these feu is by far the most important, though blench and burgage still exist. Practically all the incidents of these tenures have been commuted or abolished, but feu-holding in return for annual payments of feuduty still survives, though statutes have prohibited the creation of new feuduties and made provision for the redemption of existing ones, but without abolishing the tenure.

Land tenure—U.S.A. When the first settlements were made, land was held commonly and only thereafter was private property in land accepted. The institution of private property contributes to the success of Virginia in the seventeenth century. In colonial times, land was commonly held freehold. In New England the colonial governments granted land to towns, which in turn granted them to settlers. New York favoured large grants to proprietors and discouraged settlement, whereas New Jersey and Pennsylvania granted land in small lots to settlers. In the South in the seventeenth century, under the system of headrights, each person who transported an immigrant at his own expense received about 50 or 100 acres per head. In Maryland, Lord Baltimore used the headright system and granted manors to those who brought out sufficient people to rate 2,000 acres.

Large grants of back country land were made by colonies to such companies as the Ohio Company (1747) and the Transylvania Company (1775).

Land was inheritable, usually by children equally, with a double portion for the eldest son, but this died out before 1800, as did the system of primogeniture, also commonly accepted in the colonies. In New York and in the South accumulation of land and its retention in the family was aided by the device of entailing lands, but Virginia abolished entails in 1776, and most other states followed suit soon after.

As the frontier moved westwards land was acquired by annexation, purchase, and cession for the public domain, but from 1785 onwards there had been extensive alienation of the public domain.

From 1796 to 1820 the law for the sale of public land required alternate townships to be sold in stocks of eight sections, the intervening townships in single sections of 640 acres, all at auction for an upset price of $2 per acre, payable by instalments within three years. In 1800 the unit of sale was reduced to a quarter-section and in 1820 credit was stopped, but the upset price was reduced and the minimum unit to 80 acres. Between 1796 and 1820 the U.S. sold nearly 20 million acres for nearly $48 m.

When Ohio, the first state to be carved from the public domain, was admitted a state in 1803 the principles were established that the federal government retained all ungranted land within the new state's borders and that it donated one section of each township still unsold to a state education fund.

In 1862 the Morrill Land Grant Act gave each state in the Union 30,000 acres per congressman to endow at least one mechanical and agricultural college in each state; states with insufficient land were allowed to obtain title to undeveloped frontier regions. About 17 million acres of public domain were in this way transferred to the states, which sold the holding for about $8m. This permitted the founding of some 70 land grant institutions.

Land values duties. Duties imposed by the Finance (1909–10) Act, 1910, consisting of increment value duty, reversion duty, undeveloped land duty, and mineral rights duty. All, except the last, cost more money and trouble to collect than they were worth, and were abolished in 1920.

Landed men. Formerly in Scots law, if the accused in a criminal trial was a landed proprietor, he was entitled to have a jury the majority of whom were composed of landed men. No title in land lesser than that of property sufficed. An accused wishing to claim this privilege had to prove his entitlement by production of his title of infeftment.

Landlord and tenant. The relationship whereby an owner of land or buildings, minerals, fishing or sporting rights, or other interests in land, grants another the exclusive possession and use of particular interests for a determinate time, or at the will of the parties, normally in return for a payment known as rent. At common law the general principles applied equally to all kinds of leases, but increasingly legislation has introduced specialties into particular kinds of leases, so that leases of agricultural land, or unfurnished houses, of furnished houses, of shops and business premises, of mines, and of sporting rights have statutory peculiarities. Land may be also leased for a long period, the tenant being entitled to build thereon and to sell the unexpired portion of the lease with the buildings. The relationship is created by agreement for a lease, completed by a formal lease containing the terms and conditions of the contract.

The tenancy created may be at will, from year to year, for a week, month, or other period, for a term of years, or for a term determinable on notice after death or marriage. The rent is settled by agreement, and must be certain.

A lease may be determined by surrender, disclaimer, merger, the occurrence of an event subject to which the term was created, forfeiture, lapse of time, or notice to quit.

Numerous statutes have introduced provisions regulating rents, protecting tenants from eviction, giving relatives rights to continue tenancies, providing for compensation on departure, and otherwise controlling the free operation of contract.

W. Woodfall, *Landlord and Tenant*; H. Hill and J. Redman, *Landlord and Tenant*; (Scotland) J. Rankine on *Leases*; C. Paton and J. Cameron on *Landlord and Tenant.*

Landrecht. In a general sense the common law of a *land* or country. In the Middle Ages the term was applied to a variety of written sources of law, compilations of the customary law of a district, the liberties and privileges which a prince granted to his subjects, and the statutes and regulations passed by local or provincial authorities with or without the approval of their prince. From about the fourteenth century *landrechten* tended gradually to become territorial legislation, promulgated by princes with the consent of their estates. In the Low Countries *landrechten* were known in Friesland from the eleventh century and by the fifteenth century these had shaded into bodies of provincial legislation.

Lands Clauses Acts. Acts of 1845, later amended, which consolidated into two Acts for England and Scotland respectively, provisions previously usually inserted in special Acts authorizing the taking of lands for public purposes, which provisions could thereafter be included by reference in future Acts authorizing particular public works, with consequent uniformity and shortening of the legislation.

Lands Tribunal. A tribunal set up in 1949 to determine questions relating to compensation for the compulsory acquisition of land, the modification or extinction of restrictive covenants, and certain other matters.

Lands Tribunal for Scotland. A tribunal set up in 1949, but not made effective till 1970, to determine questions relating to compensation for the compulsory acquisition of land, alteration or extinction of restrictions on the use of land, and certain other matters.

Lanfranc (*c.* 1005–1089). Born at Pavia, studied law and was associated with Irnerius as a pioneer in the revival of Roman law. He moved to France, and opened a school at Bec which became famous, first opposed and then obtained the Pope's approval of the uncanonical marriage of William of Normandy, and secured the Pope's support for the Norman invasion of England. In 1070 he was made Archbishop of Canterbury in which capacity he supported the King but upheld the independence of the Church, introduced Norman practices and reforms, and influenced the separation of ecclesiastical from secular courts and the introduction of the pre-Gregorian collections of canon law into England. He was also noteworthy as a dialectician and theologian.

Langdale, Lord. See BICKERSTETH, HENRY.

Langdell, Christopher Columbus (1826–1906). Graduated from Harvard in 1854 and practised law in New York (1854–70). He then returned to Harvard Law School as Dane Professor and Dean and taught for 30 years. He introduced examinations and the case method of study of law, publishing his *Cases on Contracts* in 1871. This was based on the belief that the law must be discovered inductively by the student from examination of decided cases. It was widely taken up and revolutionized legal study, and started the proliferation of case-books, as contrasted with British and European practice, where the basis of study is the expository textbook. He published also *Cases on Sales* (1872), *Cases on Equity Pleading* (1875), *A Summary of Equity Pleading* (1877), *A Summary of the Law of Contracts* (1879), and *A Brief Summary of Equity Jurisdiction* (1905).

J. B. Ames, *C. C. Langdell* in *Lectures on Legal History*; C. Warren, *History of Harvard Law School*, II, 359.

Language, Legal. Prior to the Norman Conquest both English and Latin were used in England. Royal dooms or law were published in English and court proceedings were in English. Charters and landbooks were usually in Latin but sometimes in English. After the Conquest Latin continued to be used in formal documents, in the Plea Rolls and Charter Rolls, and in legal treatises, such as Glanvill and Bracton. But in the royal courts French was spoken and by the end of the thirteenth century, with the rise of the royal courts and the decline of the local courts, French had superseded English. The newer governmental records, such as the Parliament Roll and the Statute Roll, were in French. French, however, though predominant in the fourteenth century was decadent by the sixteenth and obsolete by the seventeenth century, and the later years include such mixtures as the famous prisoner who 'ject un brickbat que narrowly mist' (1631). Thus Coke's *Institutes* appeared in English and some of the seventeenth-century reports were also in English. From the fourteenth century English was winning; Fortescue's *Governance of England* (*c.* 1475) was in English and there are numerous legal books in English in the sixteenth century. An Act of 1650 required reports, resolutions of the judges, and law books to be translated into English and new ones to be published in English. Though this lapsed at the Restoration it had its effect, and law French by the end of the seventeenth century was practically dead. In 1731 Parliament made English the official legal language.

In Scotland Latin was used for charters, but from the earliest records, statutes, pleadings, and books were in Scots which gradually became closer to standard English. Many words are, however, derived from Latin or French, but the French of France rather than the law French of England.

In modern usage legal language contains a substantial proportion of words and phrases derived from Latin and Law-French. It is composed partly of words having specifically legal meanings and little, if at all, used in everyday speech, e.g. bailment and tortfeasor, partly of words of everyday speech which have special legal connotations, e.g. malice, negligence, trespass, and partly of everyday words. Sometimes everyday words, e.g. factory, may have special connotations attached in particular contexts, as by the interpretation section of a statute. Much legal work consists in seeking to draw meaning from statutes, contracts, correspondence, and other documents, and in seeking accurately to embody one's or one's client's meaning in words. Words are not symbols having fixed and certain denotations like some scientific technical terms, and ambiguities and uncertainties of connotation are inevitable. Law would be much better, not worse, if it had a much larger technical vocabulary. In many cases also legal phrases, e.g. reasonable care, deliberately allow scope for flexibility of application in circumstances, but consequently for uncertainty.

Latin maxims are still frequently quoted as embodying principles concisely and conveniently, such as *actus non facit reum nisi mens sit rea* and many rules are conveniently though not always completely or accurately embodied in Latin phrases, such as *volenti non fit injuria* (q.v.).

Further problems arise in international contexts. It is extremely difficult to render a legal text in one language into another. In the European Court the languages of all members are recognized and any one may be used as the procedural language, with consequent need in many cases for translation.

F. Philbrick, *Language and the Law*; D. Mellinkoff, *The Language of the Law*; H. Broom, *Selection of Legal Maxims.*

See also DICTIONARIES; MAXIMS.

Lansing, Robert (1864–1928). Admitted to the Bar in 1889 and became an authority on international law. He frequently acted as counsel for the U.S. before international tribunals. He was founder of the *American Journal of International Law* (1907) and was an editor of it till his death. He served as Secretary of State under President Wilson, 1915–20, and attended the Paris Peace Conference in 1919. He did not regard the Covenant of the League of Nations as part of the peace treaty, and differences with the President led to his resignation.

Lant, Matthew (?–1741). Chief Baron of the Scottish Court of Exchequer, 1726–41.

Lapide, Hippolytus à (pseudonym of Bogestas-Philippe Chemnitz) (1645–78). German historian and jurist, author of *De ratione status in imperio nostro Romano-Germanico* (1647).

Larceny (from *latrocinium*). At common law in England the crime of stealing, constituted by, without the consent of the owner, fraudulently and without a claim of right made in good faith, taking and carrying away anything capable of being stolen, intending, at the time of the taking, permanently to deprive the owner thereof. Taking included obtaining by a trick, by intimidation, under a mistake on the owner's part, and by finding. It was replaced in 1968 by the crime of theft (q.v.). At one time larceny was distinguished into grand and petty larceny, according as the value of the property taken did or did not exceed 12 pence, but this was abolished in 1861.

Laskin, Bora (1912–). Canadian jurist and judge, taught law at Toronto and Osgoode Hall before becoming a judge of the Ontario Court of Appeal in 1965, a judge of the Supreme Court of Canada in 1970, and Chief Justice of Canada in 1973. He wrote a standard *Canadian Constitutional Law*.

Lask, Emil (1875–1915). German philosopher, author of a *Philosophy of Law* (1905) and works on general philosophy.

Last court. An administrative assembly in the Cinque Ports, now obsolete. It consisted of the bailiffs and jurats and could make orders to impose and levy taxes and impose penalties for the preservation of the marshes of Kent.

Last heir. The person, being the King or the lord, to whom land formerly devolved in default of heirs.

Last resort, court of. A court from which no further appeal lies.

Latent. That which is hidden and not patent or obvious on examination. A latent ambiguity exists where an instrument appears to be clear but is found, when sought to be applied to the persons or things covered, to be equally applicable to two or more persons or things. Evidence must then be sought from other sources of the author's intention. A latent defect is one not discoverable by reasonable examination and only emerging in use. In the case of sale of land there is no implied warranty of quality or other matters affecting the value of the property. In the case of sale of goods, if the buyer has made known his purpose so as to show reliance on the seller's skill and judgment, there is an implied condition of reasonable fitness for the purpose which covers latent defects.

Lateran Councils. The name of five major and various lesser councils of the Roman Church. The five major ones all produced legislation important for canon and municipal law. The first (1123) was concerned largely with discipline in the Church; the second (1139) with reform of the Church, the third (1175) marked an important stage in the development of papal authority; the fourth (1215) legislated extensively on penal and procedural matters, prohibiting clerical participation in judicial ordeals; while the fifth (1512–17) was concerned with abuses in the Church.

Lateran Pacts. The treaties and agreements between the Holy See and the Kingdom of Italy, signed in 1929, creating the Vatican City as an independent state under the Holy See and regulating the position of the Roman Church in Italy.

Lathe. A division of land peculiar to Kent, intermediate between the shire and the hundred and including three or four hundreds.

Latin information. An information and action in the nature of civil action brought by the Crown, the Duchy of Lancaster or the Duchy of Cornwall, to recover a debt or damages for a tort or otherwise. It was abolished in 1947.

See also ENGLISH INFORMATION.

Latin side of Chancery. The common law or ordinary side or branch of the jurisdiction of the old Court of Chancery (q.v.) (which ante-dates the equitable or extraordinary jurisdiction), so-called because the proceedings were enrolled in Latin, whereas the proceedings on the equity side were in English.

Latini Iuniani. In Roman law a class of inhabitants of the plain of Latium who, under the *Lex Iunia* (prob. 17 B.C.), if informally manumitted, statutorily acquired freedom and defined rights. They had *ius* *commercii* but could not make a will nor receive a legacy, and on death their property reverted to their patron. Their children were freeborn Latins and enjoyed *Latinitas* which could lead to Roman citizenship. Manumissions of slaves, which did not comply with certain conditions of the *Lex Aelia Sentia* (A.D. 4), conferred only *Latinitas Iuniana*, unless it were repeated, in which case it gave citizenship. The category was abolished by Justinian.

Latitat. The Court of King's Bench about the fifteenth century invented a process which gave it concurrent jurisdiction with the Common Pleas. It was based on the fact or fiction that the defendant was in the custody of the Marshal of the Marshalsea prison, and therefore within the court's jurisdiction. This was done by the plaintiff filing a Bill of Middlesex against the defendant, stating that he was guilty of trespass *vi et armis*, and stating fictitiously that he had given bail to appear. If the sheriff of Middlesex could not find the defendant, a writ of *latitat* (he is hiding) was issued to the sheriff of an adjoining county, which ordered the sheriff to catch him. When the defendant was known not to reside in Middlesex the issue of a *latitat* was the first step in the action. It was abolished in 1832.

Lauder, Sir John, Lord Fountainhall (1646–1722). Studied abroad and passed advocate in 1668. He became a member of the Scottish Parliament in 1685 and supported the revolution of 1688. In the next year he became a judge and in 1690 also a Lord of Justiciary. Though respected he was not an outstanding judge. He is more noteworthy as a chronicler. His works include *The Decisions of the Lord of Council and Session 1678–1712* (1759–61), which contains also many historical points and anecdotes, and notes on contemporary affairs, particularly *Historical Observes of Memorable Occurrents*, on which Sir Walter Scott founded his *Chronological Notes of Scottish Affairs, 1680–1701* (1822) and which was published in full in 1840. *Historical Notices of Scottish Affairs, 1661–88*, based on his works, was published in 1848.

Launay, François de (1612–93). French jurist, professor of law at the Collège de France and first teacher of French law, author of *Institution du droit romain et du droit français* (1686), Commentaries on Loisel's Institutes, and other works.

Laurent, François (1810–87). Belgian jurist and historian, professor at Ghent, author of a monumental *Études sur l'histoire de l'humanité* (1861–70), *Principles of French Civil Law* (1867–79), *International Civil Law* (1880–82), *Communes of the Middle Ages* (1885), and *Historical Study of the Church and State in Belgium* (1860). He also worked on reform of the Belgian civil code.

Lauterbach, Wolgang Adam (1618–78). An eminent German jurist, author of *Collegium Theorica-Practicum ad L. Pandectarum libros* (1690), *Compendium juris* (1679), *Consilia civilia et criminalia* (1731), and many other works.

Lauterpacht, Sir Hersch (1897–1960). International lawyer, came to England from Poland in 1927 and became Whewell Professor of International Law at Cambridge, 1938–55, and a judge of the International Court of Justice, 1955–60. He was also a Bencher of Gray's Inn, a Fellow of the British Academy, and a Member of the Institute of International Law.

He lectured at many centres and was a member of many committees and governmental bodies. His publications include *Private Law Sources and Analogies of International Law* (1927), *The Function of Law in the International Community* (1933), *The Development of International Law by the International Court* (1934, revised 1958), *International Law and Human Rights* (1950); and he edited Oppenheim's *International Law*, the *International Law Reports* and the *British Yearbook of International Law, 1944–55*. His *Collected Papers* were posthumously published.

Law, Edward, Lord Ellenborough (1750–1818). Commenced his legal career as a special pleader, but was called to the Bar in 1780 and led for the defence at the trial of Warren Hastings. He acquired the reputation of being a good advocate and a fine lawyer. He took silk in 1787 and became Attorney-General and entered Parliament in 1801. In 1802 he succeeded as Chief Justice of the King's Bench, becoming also a peer. He took a large part in debate in the House of Lords and in 1806, though he refused the Chancellorship, he sat in the Cabinet, the last occasion on which this happened. He was an opponent of legal reforms, particularly of Romilly's attempts to mitigate the severity of the criminal code, and of procedural changes. He was a man of strong and upright character, impartial but sarcastic, ill-tempered and arrogant, and severe in criminal cases. His judgments made a considerable contribution to the law particularly in commercial cases. His eldest son became Governor-General of India.

Law, Hugh (1818–). Called to the Irish Bar in 1840, and became Solicitor-General, 1872, Attorney-General, 1873 and 1880, and Lord Chancellor of Ireland, 1881–83.

Law. Examination of the nature of the concept of law, of law in the most general sense of that word, is a central problem of legal philosophy or theory. Numerous attempts have been made at verbal definition but probably no definition is satisfactory or would secure universal acceptance.

To confuse the issue the words 'law' and 'laws' are also used in English not only in juristic but also in scientific contexts as designating generalizations explaining the instances of observed invariable phenomena, stating what in certain circumstances always happens, e.g. the law of gravity, the laws of thermodynamics, and the like. Despite this established usage the significant difference is that these statements are descriptive of natural happenings, and not, like laws of political societies, prescriptive of human conduct.

In the first place a distinction can be drawn between 'a law' and its plural 'laws', which words are used of one or more particular and concrete instances of legal precepts, and 'the law' which phrase signifies something more general and abstract. Thus particular precepts, such as those of the Sale of Goods Act, the Theft Act, the Sherman Anti-trust Act, and the like can each be called 'a law'. On the other hand 'the law' is used for something much wider and more general, sometimes in conjunction with words descriptive of a recognized branch of legal science, e.g. the law of torts, or with words descriptive of a particular system of law, e.g. the law of England. Sometimes too 'the law' means or includes the institutions and persons who represent and administer the law, the complex of courts and prisons, judges, lawyers, clerks, and police.

This distinction between particular-concrete and general-abstract senses of the term is more marked in those other languages which have one word for the particular-concrete sense (*νόμος*, lex, loi, Gesetz, legge, ley, etc.) and another for the general-abstract sense (*το δίκαιον*, ius, droit, Recht, diritto, derecho). It is noteworthy, moreover, and may be significant, that in these languages the term for law in the general-abstract sense also denotes the ideas of right and just, though this in turn introduces further possibilities of confusion of meaning.

But, at an even more abstract and general level, what is the nature of what is called law, of which the laws of England and of Rome, the laws of tort and property, and the most specific and particular rules of any branch of any legal system, all partake? What is the nature of 'law', without any kind of article or adjective? What is the nature of the idea of law, of the general concept, distinct from all the manifestations of it in this material world?

The working lawyer's main concern is to have a standard by which to distinguish statements or precepts of law from statements of non-law. Thus, are the statements: 'A person who commits a breach of contract must pay damages', and 'You must give up your seat on a bus to a lady', statements of law? The test for distinguishing statements of law from statements of non-law is whether the statements emanate from what is recognized as an authoritative source, for the legal system in question, of legal precepts which are enforceable. If the statement cannot be found to be laid down, directly or inferentially, by an authoritative source, it is non-law in that system. See also SOURCES OF LAW.

The legal philosopher's concern is, however, to ascertain the true nature of all the precepts and norms which are deemed to be law, and how they interact with norms of non-law, e.g. of morality.

Law, in general, is a regime of adjusting relations and ordering human behaviour through the force of a socially organized group.

The particular specimens of the general concept which are found in any particular system of body of laws are not all of the same kind. There are (a) rules defining the authoritative sources, or rules for the recognition of rules of law as distinct from maxims of non-law. Thus, in the U.K., it is a rule that what is prescribed by Act of Parliament is law; what is prescribed by a social group, or morality, or fashion, is non-law. (b) There are rules of substantive law, prescribing the rights and duties, powers and disabilities, of persons in particular circumstances. Thus, an employer must take reasonable care for the safety of his employees; if he does not and an employee is injured in consequence, the employee may claim damages from the employer. (c) There are rules of adjective law (q.v.) prescribing how legal rights and claims can be claimed, declared, enforced, and worked out.

The detailed content of a particular system or body of laws will vary from one society to another, being influenced by the history and the religious, social, moral, political, economic, and other philosophy, predominant in the society at the time of the development or reformation of particular rules. The existence of this diversity gives rise to the study of comparative law and to the problems of international private law and law reform.

The first characteristic of law is that it is an attribute of human beings and is found only when groups of such beings have associated and organized themselves into a political society, that is, one for governing themselves. Robinson Crusoe had no need for law, but as soon as Man Friday appeared the possibility of conflicts of claims, e.g. to scarce food, emerged, and in larger and more complex societies there is even greater likelihood of such conflicts.

Secondly, law is not descriptive of happenings like the so-called laws of the natural sciences, but prescriptive of human conduct or normative; it is concerned not with the Is but with the Ought, with what a man should or must do in particular circumstances, and with what is to happen if he deviates beyond the permitted limits of conduct. It is a principle of order, a rule, and measure of human acts and relations.

Thirdly, it is concerned with regulating human relations, defining men's permissible and required conduct to each other; but on what basis? The basis laid down by the strong? or by the most respected elder? or by the holy man professing to interpret the view of the god of the tribe?

Greek thinking generally saw law as a guideline for conduct, and Roman jurists as the art of what is right and equitable. In the Middle Ages Aquinas described a law as an ordinance of reason for the general good, emanating from him who has the care of the community, and promulgated. Grotius and those others who severed law from its religious basis, distinguished natural law, the dictates of reason applicable to human society, and positive law, the application of reason to the civil relations of men. Hobbes saw chiefly the imperative element: 'civil law is to every subject those rules which the commonwealth hath commanded him ... to make use of for the distinction of right and wrong, that is to say of what is contrary and not contrary to the rule', and Blackstone described a law as a rule of civil conduct prescribed by the supreme power in a state, commanding what is right and prohibiting what is wrong.

The metaphysicians of the late eighteenth and early nineteenth centuries defined law in very abstract terms, such as the sum of the circumstances according to which the will of one may be reconciled with the will of another according to a common rule of freedom (Kant), or the organic whole of the external conditions of life measured by reason (Krause). The English analytical school on the other hand were concerned with positive law. 'Of the laws or rules set by men to men, some are established by political superiors, sovereign and subject; by persons exercising supreme and subordinate goverment in independent nations or independent political societies.... To the aggregate of the rules thus established, or to some aggregate forming a portion of that aggregate, the term law as used simply and strictly is exclusively applied' (Austin), or: 'the general body of rules which are addressed by the rulers of the political society to the members of that society, and which are generally obeyed' (Markby), or: 'a general rule of external human action enforced by a sovereign political authority' (Holland), or: 'the sum of the rules of justice administered in a state and by its authority' (Pollock), or: 'the rules recognized and acted on in courts of justice' (Salmond).

Sociological jurists prefer to look at realities: 'the prophecies of what courts will do in fact and nothing more pretentious, are what I mean by the law' (Holmes); 'a principle or rule of conduct so established as to justify a prediction with reasonable certainty that it will be enforced by the courts if its authority is challenged' (Cardozo); 'what these officials do about disputes is, to my mind, the law itself' (Llewellyn).

There have been said to be a dozen main conceptions of what law is: the divinely ordained rules for human action; the tradition of the old customs acceptable to the gods and pointing how man may walk with safety; the recorded wisdom of the wise men of old who have learned the safe or divinely approved course; a philosophically discovered system of principles expressing the nature of things to which man should accommodate his conduct; a body of agreements of men in politically organized society as to their relations with each other; a reflection of the divine reason governing the universe; a body of commands of the sovereign authority in a politically organized society as to how men should conduct themselves therein; a system of precepts discovered by human experience whereby the individual human will may realize the most complete freedom possible consistently with the like freedom of will of others; a system of principles discovered philosophically and developed in detail by juristic writing and judicial decision whereby the life of man is measured by reason or harmonized with those of his fellow men; a body or system of rules imposed on men by the dominant class for the time being in furtherance of its own interest; and the product of the dictates of social or economic laws with respect to the conduct of men in society, discovered by observations, and expressed in precepts worked out through experience.

Many attempts have been made at formulating verbal definitions of law; many are unsatisfactory in not making clear whether the definition is of 'law' or of 'a law' or 'laws'. Kantorowicz arrived at the definition: a body of social rules prescribing external conduct and considered justiciable.

According to the definition adopted, some bodies of rules which may be commonly called branches of law may or may not fall within the ambit of law as defined. Thus Austin's definition of law as a command from a political superior drove him to exclude much of constitutional law and all of international law from 'law'. Rules of morality may or may not be distinguished from rules of law according to the definition adopted. Mere social rules are commonly thought of as distinct from legal rules, but also may or may not be excluded according to the definition adopted: emphasis on 'external conduct' and 'enforcement by external coercion' are usually invoked to exclude these kinds of rules from law.

The most general characteristics of law in developed systems are that the precepts of which it is comprised are general rather than particular in statement, universal in application, and introduce a factor of predictability into the outcome of doubtful and disputed cases.

It may be more valuable than seeking to formulate a verbal definition to look for characteristics of what we understand by law. These include the existence of a legal order in society, a system for ordering conduct and adjusting relations, the existence of legislative, administrative, and judicial processes, and the existence of a body of authoritative grounds of or guides to administrative action and judicial decision, and of techniques of finding therefrom grounds of decision and of applying them and adapting them; these authoritative grounds are normally general rather than particular, universal or at least general in application rather than partial and predictable, so that many issues can be resolved without formal adjudication.

It is also useful in order to clarify the concept of law to look at some particular problems of law. Among these are the relations of law and morality, or jurisprudence and ethics, and of law and the State (qq.v.).

A fundamental question about law is its end or purpose; what do we believe law should seek to do in society? Can or should law be defined by reference to its end or purpose? That end is commonly accepted as being to assist men to achieve justice in international, inter-group, and interpersonal relations, though what is interpreted as just in particular circumstances depends very much on the individual's social, moral, and political standpoint and other personal factors. It is reasonable to include in the concept of law reference to its purpose in the working of justice, while recognizing that a particular rule of law may be questionably just. Moreover, some contend that the end of law is security, or the greatest happiness of the greatest number, or the general good, or the reconciliation of the will of one with the liberty of another, and that the realization of this is justice, or, it may be, is the proper end of law.

Again there is the question of the ultimate source of the authority of a system of law and its individual rules. Is law authoritative because it expresses the will of the Supreme Being, or what jurists believe to express that will, or because it is the command of a political superior who can enforce compliance, or because it accords with a conception of the forces which rule the universe and the living creatures therein, or has some other end?

Among other questions is the question of validity. What is it that makes a statement of law, e.g. damages are not recoverable for hurt feelings when a contract is breached, a *valid* statement of law? A doctrine, principle, rule, or other formulation is valid in a particular legal system only if it has been made, or adopted, or incorporated into that particular structure of law, e.g. English law, by one or other of the methods which that legal system itself recognizes as conferring validity on the verbal formulation, or has been stated by source recognized as authoritative. Thus, in English law validity is conferred by enactment by Parliament or authorized subordinate legislative body or person, by statement as part of the *ratio* of certain classes of decided cases, by statement in a book of authority, and in certain other ways, but not by, e.g. proclamation, or statement in a textbook. See SOURCES OF LAW. A statement may of course also be invalid if it is so inaccurately stated that a person competent in that system of law would challenge it as wongly formulated.

Similarly, there is the question of enforcement. Does a rule of law cease to be such a rule, or part of the law, if it is in fact not enforced, regularly or invariably, or even if it is unenforceable in fact, or in law, no machinery existing for its enforcement? The general attitude of developed systems is that so long as it stands unrevoked it is part of the law. There may indeed be difficulties in securing the repeal of, e.g. Sunday observance laws, though they are rarely, if ever, enforced. Prohibition in the U.S.A. was largely unenforceable by reason of widespread public opposition to it, but it was undoubtedly law until altered.

Then there is the question of obedience. Does a law, or the law of a country, or particular rules thereof, cease to be valid if it is commonly or even regularly disobeyed? Clearly it does not, because if invalidity by reason of common regard were the rule, common regard or disregard, obedience or disobedience would be the test, and the sole test, of the validity or authoritative quality of every rule of law, which is an intolerable situation. Common disregard may be a good reason for altering a rule of law, but cannot by itself abrogate it. An exception to this is recognized in the doctrine of desuetude, which is to the effect that a rule of law may be deemed abrogated if not applied or enforced for a long time and deemed out of keeping with modern views on the subject; in Scots law this doctrine is recognized in relation to pre-1707 Scots Acts; in English law it is not so recognized.

In relation to obedience three other questions arise: why do people in general obey law? why should people obey law? and in what, if any, circumstances is a person entitled to disobey law? Why people in general do, or sometimes do not, obey law, is a question of psychology. Various factors operate in favour of obedience, acceptance of the rule as just and necessary, the habit of conformity, fear of the sanction, the wish to be thought law-abiding, and to be well regarded and similar factors; against obedience are such factors as greed, hatred and lust, disbelief in the social, political, and economic order which has created and is maintained by the present rules of law, and occasionally religious or moral dissent from the present rules.

More important is the moral question why people should obey law. The fundamental reason is that in a modern developed society a detailed set of guidelines for conduct of practically every kind is essential for minimizing friction and conflict between individuals, and that generally speaking the law, the set of guide-lines in force at the time, has been created by representatives of the community and must be deemed to have the assent of the community. In short, people should obey law because this is the way to avoid chaos and civil strife, rioting, and anarchy, even civil war. Obedience to a bad law and agitation to change it by constitutional means is better than disobedience and its consequences. If, on the other hand, the society is dictatorial or tyrannical, disobedience to law may be necessary to force a change, if need be by bringing down the government.

The question in what circumstances a person is entitled to disobey law is difficult. In the first place there can be said to be a moral, as well as a legal, duty to obey the law, even if one disagrees with it, and persons cannot be allowed to disobey the law merely on the plea that they disagree with it because that puts obedience to law, any and every law, on an optional basis, but it is possible to imagine occasions, and occasionally to find actual instances, where a rule of law is so flagrantly antithetic to a common moral belief that an individual may feel justified in disobeying the law in obedience to the dictates of conscience. The problem has arisen in relation to members of the armed forces ordered, e.g., to kill prisoners or refugees. It has arisen in relation to anti-strike legislation and from objections to military service. It could arise if abortion or euthanasia were not merely permitted but ordered in certain circumstances, or in relation to capital or corporal punishment. Disobedience to law may be morally justifiable, but can never be legal.

There is also the question of the relation of law to justice. Ignoring for the present the difficult question of the nature and definition of justice, it seems that the law of a community as a whole and each individual rule of the law is doubtless intended to achieve or to promote justice, as the dominant class in the community at the time when the rule was adopted saw justice to lie, but by reason of the passage of time, changed circumstances and unforeseen contingencies, a particular rule may not, or, in the view of some or many, not, promote justice. The rule is not thereby affected, but these views may lead to a movement to change the rule. There is little doubt that many legal norms in a particular legal system may be valid and enforceable, but generally or even universally deemed unjust.

More fruitful than trying to define law verbally are the efforts of jurists who have sought to clarify the concept in its most general and abstract sense, by considering such problems as the origins and development of law, the interaction of law with morality and religion, the relations of law with the organized political community, the state and the supranational and international orders, the purpose and objects of law, and the modes whereby it can and does attain those objects. Many of these topics are discussed in works on jurisprudence.

The notion which emerges is that law is a controlling factor in all human societies which have progressed much beyond barbarism, like custom, morality, and religion. As it develops it is increasingly consciously formulated, developed, and modified by the dominant persons in, or section of, each community; it increasingly takes the form of precise and specific rules formulated in codes, statutes, rules, and in books; it increasingly becomes coercive, enforceable, and enforced; it becomes increasingly voluminous and complicated; it extends to an ever-widening range of the wants, desires, and claims of persons and groups. It is an essential condition of life in society being peaceful and orderly rather than nasty, brutish, and short. It takes different forms in different societies and everywhere develops and changes its detailed prescriptions in response to social, economic, and political changes.

R. Pound, *Introduction to the Philosophy of Law* (1954); H. Kantorowicz, *The Definition of Law* (1958); H. L. A. Hart, *The Concept of Law* (1961); and generally books on Jurisprudence.

Law agents. The name formerly applied to legal practitioners in Scotland not being advocates, now called solicitors. In the latter seventeenth century three classes of law agents developed in the Court of Session, advocates' first clerks, writers to the signet, and solicitors. The first group included young men studying for the Bar, who acted as law agents negotiating with clients, but this group disappeared in the mid-nineteenth century. The second group, writers to the signet (q.v.) were long prohibited from acting as law agents in the Court of Session, but their right to do so was recognized in 1754. In the meantime the third group, solicitors, first mentioned about 1600, had gradually become established, and their right to practise was recognized in 1754. They became incorporated as the Society of Solicitors in the Supreme Courts of Scotland (q.v.). In the local inferior courts the practitioners formed themselves into societies of writers or procurators, with the monopoly privilege of practising in the particular court and of testing and admitting others to their membership. In 1825 the Court of Session, without prejudice to the rights of chartered bodies, made regulations as to training and admission. The Procurators Act, 1865, permitted county law societies to obtain incorporation and control training and examinations.

The Law Agents Act, 1873, established a general examination for law agents throughout Scotland and abolished exclusive privileges. Persons wishing to be admitted had to serve an apprenticeship of five years with a practising law agent (three years for a graduate) and pass the prescribed examinations. In 1883 the Incorporated Society of Law Agents was formed, to form a society comprising all law agents. In 1925 it was renamed the Scottish Law Agents Society. In 1933 the term solicitor became the correct term for all such practitioners in Scotland, but the older term is still sometimes used, and in 1949 the Law Society of Scotland (q.v.) was established.

Law and custom. Custom or habitual practice and usages may be in the first place an important historical source of law in that custom, practices habitually followed, what is normally done, commonly develops into what is expected, what is always done, and what must be done. Custom is anterior to kings and courts. At a rather later stage of development customs may provide the materials for the rules laid down by a law-giver or codifier, or at least be alleged to be the basis for his rules. In Celtic, Frankish, and Anglo-Saxon law, tribal custom was the main form of law.

Secondly, custom has been, and in some countries still is, a material form of law in that the courts and the law books accept as obligatory and binding, and make legally enforceable, particular rules of conduct which have for long been so accepted. English common law was largely general custom accepted by judges and stated as obligatory rules and subsequently enforced as such. In northern and central France down to 1789 the law was largely bodies of coutumes. At first each seignory had its own customs based on a mixture of Roman law, canon law, Frankish law, feudal law, and legislation, but after about 1300 customs became more uniform in larger areas such as Normandy and Britanny. In the south codified law based on the *Corpus Juris Civilis* served as the general coutume for the area. After 1650 the idea developed that the common law of France resided in the bodies of provincial coutumes, and books on the coutumes gave assistance to those who framed the *Code Civil.* In Germany custom is still, along with legislation, the major source of legal rules. In Scandinavia custom is still a major source.

Thirdly, even in systems where rules are made largely by legislation or by courts, custom may still be a valid source of legal rules. Thus in English law a special custom, an exception to the general law of the land, may be accepted by a court as providing a rule for the decision of a case if it be shown to be certain, reasonable, not in conflict with any other custom, or with a fundamental principle of common law, have been continuously observed and peaceably enjoyed, and have existed from time immemorial, a time conventionally set at 1189, or at least be such as would not have been impossible in 1189. Similarly, an accepted or proven custom of trade may be an implied term of a contract.

But, these cases apart, custom is distinct from law. It states what is normally, even invariably, done but not what should be done, and must be done subject to some sanction. Custom may be flouted and may change, even quite rapidly, or remain unchanged for long periods.

A rather special case is the custom of Parliament, which is the long-standing habitual practice of the Houses.

Law and equity. From about the fifteenth century there developed in England two bodies of legal principles, rules of law developed by the courts of common law and to some extent by legislation, and principles of equity (q.v.) developed by Chancellors in the Court of Chancery. By the eighteenth and nineteenth centuries these had developed into two distinct and sometimes conflicting jurisdictions, applying distinct and sometimes inconsistent bodies of principle and rules. Equity, which had arisen as a flexible body of principles to supplement the deficiencies of and correct the injustices of common law, had become a complicated body of technical rules very early, as fixed and settled as those of common law. In 1873–75 the Judicature Acts merged the administration of common law and equity by enacting that the new courts, established in that year, could all equally give both common law and equitable remedies, and that in case of conflict between the two systems of rules the rules of equity should prevail. But many consequences of the former dualism continue and in many areas of the law there are still recognized legal rights and equitable interests, legal and equitable mortgages, and other parallels. The trust is entirely a creation of equity. The dual system of principles was carried overseas with English settlers, took root in English colonies and long survived in places but has now almost everywhere disappeared, though in many places leaving substantial surviving traces.

Law and fact. A distinction constantly recurring in legal problems and analysis of legal cases, but difficult to formulate generally. Thus, courts are sometimes required to make findings of fact and of law separately; appeals frequently lie on a matter of law, but not on fact; distinctions are drawn between mistake of fact and mistake of law, and so on. In general it may be said that a question of law concerns the existence, correct formulation or application of some rule of the legal system, a question of fact concerns the existence, nature, or state of some matter cognizable by the senses, by evidence or by inference therefrom. Thus whether adultery is a ground of divorce is a question of law; whether adultery took place in a particular case is a question of fact. Whether the accused did certain things is fact; whether they amount to manslaughter is a question of law.

The principal practical distinctions are: questions of law are determined by the court or judge on consideration of norms drawn from authoritative sources of law and after hearing argument; questions of fact are determined by the jury, if there is one, and, if not, by the trial judge on the basis of admissions and after taking evidence into account. Following from this, matters reserved for determination by the court or judge are deemed questions of law, those entrusted to the jury, if any, or to the court acting as such, are questions of fact. These distinctions are important in relation to pleading and proof, division of function in trials, appeals, precedent, and *res judicata*.

The distinction cannot always be sharply drawn because the relevance of a fact may depend partly on the rule of law in question, and the rules of law relevant depend in part on the facts in issue.

In many circumstances questions of mixed fact and law arise; thus the determination of A's guilt of the crime charged depends on both the legal constituents and requisites of the crime, and the facts held to have been established, and a jury verdict that A is guilty of that crime is a finding of mixed fact and law.

Law and morals. Law and morality, in the sense of the actual standards of conduct accepted and approved in the life of a particular community, have a common origin when communities at an early stage of development are regulated by an undifferentiated mixture of custom, religious beliefs, taboos, beliefs of a moral character and practices deemed obligatory. As development proceeds, customs and beliefs become crystallized and differentiated into beliefs as to what is deemed good or right and bad or wrong, and rules deemed obligatory, whether thought right or wrong. Mosaic and Islamic law comprise largely undifferentiated codes or moral and legal prescriptions for conduct.

The distinguishing factors are readily apparent: law is formulated and imposed by the State; morals by conscience, beliefs, and the attitudes of society. Law has fixed and powerful sanctions, morality is sanctioned, if at all, by public opinion, disapproval, possibly ridicule and exclusion from a particular society.

While in modern developed societies a clear line can be frequently drawn between moral beliefs, operating on the minds of individuals, and legal rules, consciously enacted or developed by courts, there remains a large measure of overlap. Morality sets limits to the operation of law. Even positive law cannot ignore morals, in that positive rules deemed by a substantial section of the community contrary to morality are in danger of being disregarded. The recurring importance of natural law as a touchstone for good law is essentially a reference to moral beliefs. In the first place, moral beliefs and standards, e.g. as to sexual relations, regard for women, children, and animals, as to saving or preserving life, as to not avoidably doing harm to another, are frequently the basis for legislation, or a basic premiss for legal reasoning. Secondly, moral claims influence interpretation; a moral claim may not found a legal claim but may be held to bar a counter-claim for repayment. Thirdly, moral standards influence judges in setting legal standards; thus the scrupulous trustee, the *bonus paterfamilias*, even the reasonable man of the law of tort, are all beings with moral standards; the concepts of good faith, constructive fraud and unjust enrichment, all bear witness to moral standards. A large part of equity is founded on morality, and statutes as much as common law refer to 'just and reasonable', 'equitable', and other formulations of moral standards.

Not least, every case in which some matter is left to the discretion of the judge is essentially a remit to his moral standards.

The distinction between law and morals in respect of object-matter and application lies in such facts as that law has regard to acts, morals to thought and feeling, legal rules are of general and absolute application, whereas moral principles must be applied with reference to individuals and circumstances.

Law and politics. Just as the theoretical studies of jurisprudence and political theory are intimately connected the more practical rules of law and practical politics are inextricably enmeshed. Rules of law are made and unmade by politicians and political assemblies in pursuance of political theories, beliefs, and aims; much of the legal system is the machinery of giving affect to political, i.e. governmental, policies; practical politics in turn is to some extent influenced and controlled by what is legally possible or permissible, and by legal procedures, machinery, and stages. The conventions (*q.v.*) of the constitution are not, strictly speaking, rules of law but political understanding and practices, but they are frequently very little different in effect from rules of law, and quite as binding and compelling. The whole area of public administration is a mixture of legal and political considerations, machinery, and interaction. While in public law the interaction and influence of politics is most apparent, political influence increasingly invades private law and such relationships as employer and employee and landlord and tenant are more and more determined by political considerations.

Law and procedure. Procedure is at once a branch of law and of a legal system, and a body of rules distinguished from, and sometimes even treated as opposed to, rules of law or of substantive law. In primitive and underdeveloped legal systems substantive law and procedure are intermixed and inseparable; substantive law is secreted in the interstices of procedure. Relics of this are found in mediaeval English law where rights of action existed only if an appropriate form of action was recognized, and each form of action might have, and commonly had, procedural peculiarities. It is only at a mature stage of legal development, such as in Roman law at the end of its development and in English law since 1875 that it is possible to analyse the relations of parties as a matter of rights and duties, mainly independently of a largely generalized machinery for giving redress where a right has been infringed or a duty breached. Blackstone's *Commentaries* were the first attempt to state, so far as then possible, English law as a body of rights and duties. Under Roman law influence Stair's *Institutions* had done this for Scots law a century earlier. In modern law the distinction is fairly clear, though sometimes influenced by conceptions drawn from earlier law. It remains difficult, however, to formulate the distinction verbally. It can only be expressed in such a form as that substantive law states what rights and duties attach to a particular individual in particular circumstances, whereas procedure states the steps which have to be taken and the machinery which has to be utilized to determine or declare and enforce the claimant's rights and the respondent's duties and, where appropriate, to give the appropriate redress. In many systems, particularly those largely codified, the distinction is sharpened by the existence of civil and criminal codes and distinct codes of civil and criminal procedure. In England and Scotland the rules of procedure are generally contained in rules made by judges and practitioners in consultation under statutory authority, and the sources (legislation, subordinate rules, case-law) are the same as of substantive law. It is quite wrong to equate the distinction between law and procedure with that between rights and remedies, because remedies are kinds of rights and there are rights in the sphere of procedure. There is in fact no certain way of drawing the line between substantive law and procedure, and the line probably cannot be drawn in the same place for all purposes.

Law and religion. Religion involves belief in, and conciliation of, powers deemed superior to man which are believed to regulate and control the course of nature and of human life. It involves elements of belief, a body of dogma, acts of conciliation or worship, and ritual. Religious beliefs and practices are very widespread among humans and are found from very primitive stages of development upwards. As the social order becomes more developed and complex, ritual observances, beliefs, and doctrines, tend to grow and to take on the specific character of religious observance.

Maine stated that every system of law when it first emerged was entangled with religious ritual and observance, but later investigations have disproved this and many ancient bodies of law are entirely secular. Nevertheless there are many instances where the functions of priests, lawgivers, and judges have overlapped. Law springs from custom, and many of the customs regulate relations depending on religion. Promises, contracts, and treaties are frequently made by invoking religion. When legal rules take the form of written records embodying what is believed to have been delivered to a people directly by the deity, or through sages inspired by a divine power, the sanctity of law attains its maximum and it is virtually a part of religion, and those who know it have both a legal and a religious function. The Roman *pontifices* early had legal powers. Offences against the community may be deemed offences against the deity, and punishment may take the form of devoting the culprit to the wrath of the powers of the underworld. Trial by ordeal, the appeal to the judgment of God, has been in various forms a widespread institution in underdeveloped communities. In some cases religious revelations and sacred writings have included rules for secular conduct.

The codes of law contained in the Pentateuch bear in part to have been given by God to Moses and in their present form are an amalgam of religious and secular rules.

To the Mohammedan the Koran provides both a body of beliefs and ritual and a comprehensive code of behaviour. It prescribes both religious and legal observances. The Koran is the primary expression of Islamic law, though not a code of law, as well as the basic formulation of religious ethics. Jewish law comprises rules of worship and ritual, private and social behaviour and, as the personal law of Jewish citizens in questions of status, is part of the law of modern Israel.

In Christian Western Europe in the Middle Ages the powers claimed for the spiritual and lay authorities came into sharp conflict on many occasions, notably in the investiture contest. The quarrel between Henry II and Becket and that between the Emperor Henry IV and Pope Gregory VII. The long conflict between Church and State (q.v.) is essentially a conflict between the authorities in the religious and legal spheres of national life.

Religious ideas and beliefs, particularly the belief that there is a God who in various ways reveals what is the right conduct, powerfully influenced the mediaeval idea of natural law. God's law, law revealed by reason, the law determined by nature are hardly distinguishable. Aquinas and the scholastics posited the law of nature as being the law of God; the sharing in the eternal law by a rational creature is natural law. Later the Spanish jurist-theologians, such as Soto and Suarez, found a religious basis for systems of law, national and international.

In some cases again as religions have developed and become institutionalized, they have developed their own bodies of law regulating not only the organization of the Church ritual, practice, and beliefs, but large parts of the secular lives of their adherents and, when a religion has secured a controlling hold on the State authority, law dictated by the State religion shares with law dictated by the secular State the total area of social control by law. In Western Europe the great example has been the canon law (q.v.) which became a major part of the legal system of every Christian country and even after its formal authority was abolished at the Reformation, as it was in many countries, it long continued to exercise powerful influence on the law of, e.g. marriage, nullity and divorce, and wills, and on the attitude of the courts.

There have also been instances where the state has been theocratic and religious persons, beliefs, attitudes, and rules have been the major controlling force in the community, acting through state law; examples have been Calvin's Geneva and, at times, Puritan England and Presbyterian Scotland.

The Reformation in England produced profound changes in the law, not least in that supreme exercise of parliamentary power which overthrew the papal supremacy and made Henry VIII supreme head under God of the Church in England. As a result of these changes English ecclesiastical law, even in purely spiritual matters, is part of the law of the State, and it is only in the twentieth century, by delegation by Parliament to the Church Assembly and latterly to the General Synod, that the Church of England has had practical legislative independence.

In America separation of church and state did not come about till after the Declaration of Independence. In colonial times citizenship depended on church membership and Massachusetts, though not a theocracy, was a Bible Commonwealth in which the magistrates sought counsel from the ministers. The Massachusetts General Laws and Libertyes (1648) contributed to a decline in Puritanism.

Not least, it is necessary to consider the influence of Christianity and the Bible on Western legal systems. These factors have influenced law in at least five distinct ways; firstly, by influencing the law of nature theory; secondly, by directly supplying rules which were enacted or followed; thus the Scots statutes regulating the degree of relationship for marriage and incest expressly incorporate the relevant verses of Leviticus 18; thirdly, by reinforcing ethical principles and providing an underlying justification for rules of statute or common law, e.g. the punishment of murder, theft, and adultery; fourthly, by influencing law in a humanitarian direction, emphasizing the worth of the individual, the protection of dependants and children, and the sanctity of life; and fifthly, justifying and emphasizing the maintenance of moral standards, notions of honesty, good faith, fairness, and others. The commonest single aspect of Christian influence on law is the oath, which many persons must take on various occasions, the sovereign to uphold Church and State, the judge to administer law justly, and, most commonly, the witness to tell the truth. In the last case the underlying belief in the oath is that the force of religious belief will compel the witness to tell the truth, and that, if he does not, God will punish him.

The nineteenth century was in the United Kingdom the high-water mark of the force of religious belief and practice, as a general factor of social control; it was undoubtedly profoundly influential in very many ways.

In modern times atheism, agnosticism, and widespread decline in religious beliefs have weakened the influence of religion. The oath to tell the truth is repeatedly broken. Persons may decline to take an oath and affirm instead. Yet religion remains for many people in most parts of the world an important form of social control, along with law, a powerful regulating factor.

Law and the state. A state is the whole community of persons living in a determinate part of the earth's surface viewed as a legal entity, a body of persons legally organized for their own government. It is a permanent political organization. A requisite of a full state as distinct from a dependent territory, is that it is supreme within its territory and independent of legal control from without. It has full sovereignty (q.v.). It comprises the Head of State, the legislative body, the executive and administrative bodies, the judicial bodies, and the mass of other citizens who, as well as being members of the community, form many subordinate groups for different purposes, including parties, churches, trade unions, employees of X or Y, tenants, consumers, taxpayers, and so on.

A legal system or systems is an essential of a state, defining its own organization and upholding its own existence, maintaining order, and providing a body of norms for the regulation of the conflicting interests and claims of the citizens, the provision of a system of law and order being one of the fundamental functions of a state.

The legal system or systems obtaining within the territory of a state are normally largely the creation of the State; and constantly being altered and modified by the legislature and to a lesser extent by the executive, by delegated rule-making power, and by the judicial agencies, by interpretation and case-law. The State is accordingly the immediate practical source of the principles and rules of law.

The political structure of the particular state powerfully influences the form and content of the legal system, as can be seen by considering the French and Russian Revolutions. An alternative view is that law precedes the State and binds it from the time it comes into being. This is the approach of natural law thinkers, who hold that the State, though it may make positive law, is bound by natural law; while some modern thinkers deny that the State creates law, and assert that it has its source elsewhere, for example in the subjective sense of right of the community, or in the fact of social solidarity.

Yet another view is that law and the State are the same thing; the State is only the legal order looked at from another point of view, the centralized legal order regulating the conduct of the inhabitants of the territory, the set of legal norms and the acts performed by legislators, judges, administrators, and others acting as organs of the State.

Law Commissions. Two permanent bodies for England and Scotland respectively, established by statute in 1965, each consisting of a judge, seconded from the Bench, as chairman and four commissioners. They have the functions of keeping the whole law of their respective jurisdictions under review, preparing programmes for, and carrying through inquiries and research directed to, the systematic revision, reform, and restatement of the two legal systems. Their recommendations have been the basis of much legislative reform of these two systems since 1965.

Law Courts. The common name for the building, correctly called the Royal Courts of Justice, on the north side of the Strand, London, between St. Clement Dane's Church and Chancery Lane which houses the court-rooms and offices of the Supreme Court of Judicature in England and Wales. The building was opened in 1882. There are physically separate extensions behind, in Carey Street. Prior to erection of this building the Chancery Division occupied courts at Lincoln's Inn, and the Queen's Bench and Probate, Divorce, and Admiralty Divisions occupied courts adjoining Westminster Hall. Not all courts sit in that building. The House of Lords sits in the Houses of Parliament at Westminster and the Judicial Committee of the Privy Council in Downing Street. The Central Criminal Court sits at the Old Bailey. Sittings of the High Court and the Crown Court may be held at any place in England and Wales and regularly are held at various towns on the six circuits (q.v.). Sittings of county courts and Crown Courts presided over by circuit judges or recorders are held at numerous other places in London and many towns up and down England and Wales. Scottish courts and Northern Irish courts sit in these distinct jurisdictions.

Law French. Until about 1400 the language of English lawyers, like that of the court, Parliament and learned work generally, was Anglo-Norman. But after about 1400 Law French continued to be used after French had ceased to be generally used by laymen. The fifteenth and sixteenth centuries are the period of true Law French and it is in use throughout Year-Book (q.v.) period (1292–1535). It was rare by 1600. It was abolished by Cromwell in 1650 but restored at the restoration and was finally abolished by statute from 1733. Law French was always a highly technical language and preserved many old Anglo-Normanisms, but with the introduction of English forms, inflections, word-order, and construction, as in the notorious 'fuit assault per prisoner la condemne pur felony que puis son condemnation ject un brickbat a le dit justice que narrowly mist'.

Law Libraries. The nature of legal studies, the multiplicity of legal systems, and the variety of sources (codes, treaties, statutes, cases, commentaries, textbooks, and others) which may have to be searched make library collections essential for legal studies and work. Not much is known about legal libraries in the ancient world. Aristotle had a collection in the Lyceum. There were a number of libraries, public and private, at Rome and later at Constantinople. The library of the law school at Berytus survived till the destruction of that city in A.D. 551 and Justinian's Imperial Library is said to have had 600,000 volumes.

In the mediaeval period the libraries of Western Europe were mainly ecclesiastical, but these could include canon and civil law. From about the twelfth century the monasteries as centres of learning began to be replaced by the early universities and as soon as the Inns of Court began to develop they and their members must have had collections of books. They are known to have had libraries from the sixteenth century. The Advocates' Library in Edinburgh was established in 1683.

In the twentieth century legal libraries have expanded in numbers and size with the great proliferation of legal materials, writing about law, and need to hold law-related materials, as on criminology, economics, and government. In England noteworthy collections are those in the British Library, the House of Lords, House of Commons, and certain governmental departments, the libraries of the four Inns of Court, the Bodleian Law Library in Oxford, the Squire Law Library at Cambridge, and of the Institute of Advanced Legal Studies in London. In Scotland the most valuable are the Advocates' Library and the Signet Library in Edinburgh. In Northern Ireland the Queen's University of Belfast collection is predominant and in the Republic of Ireland the Library of King's Inns, Dublin.

In the U.S. the largest holdings are those of the Library of Congress, U.S. Supreme Court Library,

U.S. Department of Justice Library; and among law schools, those of Harvard, Yale, and Columbia are outstanding, those of Michigan, North-western, and Minnesota of great importance. There are also numerous bar association and county libraries.

In Canada the major collection is the Supreme Court of Canada Library, but also noteworthy are those of the Law Society of Upper Canada and York University. In Australia there are the National Library of Australia and the High Court of Australia Library.

In France the largest law libraries are those of the Cour de Cassation and of the Ordre des Advocats à la Cour d'Appel de Paris.

The Deutsche Bibliothek, Frankfurt, the Federal Court Libraries at Berlin and Karlsruhe and the libraries of the major universities such as Heidelberg are the chief ones in Germany. In the Netherlands a great collection is that at the Peace Palace in The Hague, but other notable collections are in the Universities of Leiden, Utrecht, and Amsterdam.

An International Association of Law Libraries was founded in 1960, which has done much to advance standards and communicate information in this field, and there are Associations of Law Librarians.

Law List. A list of barristers, solicitors, and other legal practitioners, with addresses and dates of qualification, published annually from 1801 to 1976. By statute the list is prima facie evidence that any solicitor whose name appears therein holds a practising certificate for the current year. From 1977 the annual *Law List* was superseded by the *Bar List*.

Law Lords. A colloquial term signifying sometimes the Lords of Appeal in Ordinary (q.v.) and sometimes the rather larger body of persons who may sit in the House of Lords (q.v.) in its judicial capacity to hear appeals.

Law-men. In mediaeval Scandinavia men who were learned in the ancient customary rules and who acted as instructors, private counsellors, and judges in disputed cases. About A.D. 930 the law-saga man in Iceland was entrusted with custody of the law and had the duty of reciting the existing rules before the Althing or General Assembly, completing the entire body of the law in three years. This functionary to some extent superseded the law-men but in some cases was bound to consult a number of them. Similarly, the Norwegian law-men appeared at local and general Things down to the end of the eleventh century to declare the law applicable to particular cases and interpret it. In Sweden the law-man early became an elected public official with the duties of enforcing the administration of justice, presiding at the Thing and giving advice on provisions of the law applicable to cases which arose. They not only had supreme judicial authority but also political and administrative functions. The chief law-men of various districts gave the regular public addresses stating the law of the land to the assembled people. It is uncertain whether the institution was known in Denmark.

Law Merchant. The bodies of customs and law which grew up in the Middle Ages in Western Europe among merchants to regulate their relations with each other. These bodies were of an international or cosmopolitan character. There was not a single Law Merchant all over Europe but variations between different states and towns, though there was a general similarity and various towns modelled their charters and laws on other more famous places.

In Roman law one body of principles applied to merchants and to other persons, and rules of a mercantile character are scattered throughout the *Corpus Juris*. In the Middle Ages Italy was the centre of commercial and legal life in Europe and in Italy a unique system of settlement of commercial disputes developed. Many towns had gained their independence and merchants had considerable liberty to organize the government. Commerce and industry developed and *consules mercatorum* arose with extensive powers and jurisdiction in commercial matters, and latterly with courts recognized as parts of the judicial organization; their enactments became parts of the law of the city-states.

From the thirteenth century onwards Italian influences were directly affecting every country in Europe and this was facilitated by the general reception of Roman Law, although the principles which became influential represented mercantile custom more than rules of the civil law, but customs developed into legal principles with the help of concepts derived from the civil and canon law. This body of mercantile law accordingly became acclimatized readily in countries which had received the technical rules of Roman law.

The organized bodies of merchants commonly issued detailed orders governing the conduct of commerce and industry. They had wide powers, subject to the general conditions that rules they made must not contravene a law of the civil authorities, must relate to mercantile matters, and must be just and reasonable, conditions which, through the writings of jurists, passed into the general law of Western Europe, and regulation covered such topics as banking, dealings in bills, pawnbroking, registry and loading of ships, and underwriting. Commercial relations between towns gave rise to the posting of consuls to advise and assist foreign citizens visiting places on business. Those relations also enabled the towns to develop a substantially uniform body of mercantile and maritime rules. The Italian Law Merchant became comprehensive, detailed, and uniform. The merchant gilds acquired wide jurisdiction and applied the rules which had developed as commerce developed. Roman law was valuable in supplying a technical language and concepts of obligations; and the canon law supplied moral ideas, the idea of good faith, of fair dealing, of keeping bargains; but the commercial principles were largely free of technicalities.

The commercial law of the states and cities of Northern Europe originated in the franchises and privileges granted to persons holding or resorting to markets. Permanent trade grew up at markets and gilds of merchants and traders arose. As towns with markets developed, the court and law of the market became part of the regular organization of the town. The law depended very much on the municipal constitution of the particular town, though similarity of problems and direct borrowing led to similarities between the laws of different market centres.

In some market towns, especially on the main trade routes through Europe, there developed great international fairs in the twelfth to sixteenth centuries, and at those gatherings law developed to meet the needs of international business transactions, and this was influenced by the more advanced concepts of Italian commercial law, while companies of foreign merchants obtained the privilege of governing themselves by their own laws. Accordingly, at the great international fairs a uniform commercial law was largely made and administered by merchants for themselves. To attract foreign merchants to the fairs, states were frequently ready to abandon claims and rights which hindered trade.

While in Italy a factor in the development of commercial law was the growth of special commercial courts presided over by the merchants, in Northern Europe the privileges of merchants depended on a grant by king or lord of a franchise of fair or market, but the lord kept the administration of the law in his own hands, though frequently assisted by merchants as assessors. At the great fairs the courts sprang from the powers which the associations of merchants were allowed to exercise over their menbers. By about 1500 the principle had been generally accepted that commercial courts should be presided over by merchants and that they should have as great weight as lawyers in them. The constitution of commercial courts differed in detail in different countries. Through these courts of the great international fairs the principles of law adapted to commercial transactions, which had evolved in the Italian and other southern European trading cities, were spread over the rest of Europe. The fairs were the meeting places of Germanic and Roman worlds of law and commerce and became financial clearing houses.

In Roman law commercial and maritime law were developed as part of the ordinary law, but on the Continent generally they were developed by a separate set of courts, and are generally now dealt with in separate codes. In Central and Northern Europe generally towns, following the example of Italian cities, began to acquire privileges and franchises to modify their commercial organization of gilds and their commercial customs on the Italian models. In Italy and in some towns of southern France and Spain the merchants actually obtained control of the government. Both factors operated to give merchants peculiar status, privileges and powers and to emphasize the distinction between ordinary and commercial law.

This separation did not arise in England because England early developed a strong and centralized common law and an active legislature, because its cities never attained the independence they did in many European countries, and because substantial commercial development, and the need for commercial law, did not develop till after the Reformation. When the need arose doctrines were adopted from continental commercial law but not so completely as in Europe, and the pattern of separate commercial courts was not adopted. In England, alone among mediaeval and early modern states, commercial law became simply a branch of the ordinary law, founded on the principles and rules of the cosmopolitan law merchant, but developed by the machinery of the courts of common law and equity, and by lawyers who knew too little of the law and legal theory which had been and was being developed on these topics on the continent.

In England, as on the Continent, commerce tended to be limited to permanent centres, specially protected by law, where markets arose, and the customary law in the boroughs adapted itself to the needs of merchants. Mercantile law was administered down to the fourteenth century in local courts. The Gild Merchant sometimes adjudicated on mercantile disputes. A common incident of the holding of a fair was the holding of a court of piepowder (q.v.) of which the major characteristics were that the law must be administered summarily and speedily, and ordinary procedure was relaxed. Trial might be by jury or by compurgation. The questions most commonly arising related to sale, agency, and partnership. The Statute of the Staple, 1353, established Staple Towns, each having a special court for the merchants who resorted to it, which applied the Law Merchant, not the common law.

In the sixteenth century the common law courts assumed jurisdiction over most cases concerned with internal commerce, and the expansion of their remedies rendered common law proceedings adequate for most cases, and by the end of the mediaeval period the future development of mercantile law was obviously in the central courts rather than in independent courts.

Most of modern English mercantile law is attributable to a reception of continental principles in the sixteenth and seventeenth centuries, partly by Council and Parliament, partly by the courts of Admiralty, Star Chamber, Chancery, and of common law. The main source of the principles was the usage and practice of merchants themselves, as is seen from the writings of Malynes, Marius, and Molloy. When the common law courts began to take cognizance of customary commercial law they had to admit the Law Merchant as part of the laws of England, though in the early seventeenth century they insisted on the custom of merchants being pleaded specially, and even thereafter the recognition of particular customs as parts of the common law depended on evidence of the custom alleged to be applicable to the case before the court.

Finally, Lord Mansfield in a series of cases formed the rules deducible from the various commercial customs which had come before the courts into a coherent system and completely incorporated it in the common law. Since his time the Law Merchant has been only the historical source of some of the principles of the common law.

G. Malynes, *Consuetudo vel Lex Mercatoria* (1622); W. Beawes, *Lex Mercatoria Rediviva* (1752); W. Mitchell, *Early History of the Law Merchant* (1904); W. A. Bewes, *Romance of the Law Merchant* (1923); C. Gross, *The Gild Merchant*; C. Gross and C. Hall, *Select Cases on the Law Merchant* (Selden Soc.).

Law of citations. An enactment of the Roman emperors Theodosius II and Valentinian III of A.D. 426 providing that reference might be made to the works of Papinian, Paul, Ulpian, Modestinus, and Gaius as primary authorities, and also to the works of those cited by the five primary authorities if their texts were found to be reliable by comparison of manuscripts. If there were difference of opinion, the majority view was to be accepted; if numbers were equal, the side on which Papinian was numbered was to prevail; if Papinian had expressed no view, the judge had a discretion. While of interest as showing what jurists were most highly regarded, the edict shows the depths to which legal science had sunk that the views of commentators were merely counted and not evaluated.

Law of Nations (or *ius gentium*). The older name for public international law.

Law of nature. See NATURAL LAW.

Law Officers of the Crown. In feudal theory the King could not appear in his own courts to plead his own cause, so from an early date he utilized an attorney or agent to appear for him. From the time of Edward I Attorneys-General are known in England and from the time of Edward IV Solicitors-General (qq.v.). The Solicitor-General is the assistant and deputy of the Attorney-General. In modern practice both are M.P.s and appointed from supporters of the party in power and are expected to defend the legality of the actings of government, ministers, and departments and assist in piloting legally technical bills through Parliament.

The law officers are the legal advisers and representatives of the sovereign, and, in modern practice, of the government, and are frequently called on for joint opinions on matters of international and constitutional law, and on other matters at the requests of government departments. They may not engage in private practice.

The Attorney-General is head of the Bar; he also directs the office of the Director of Public Prosecutions and himself prosecutes in cases of constitutional importance.

The Law Officers have long, like the judges, received writs of attendance to the House of Lords at the commencement of each Parliament, but both are today normally M.P.s and appointed from among the supporters of the government in the House of Commons. In modern practice the A.G. is and the S.G. may be a Privy Councillor.

Formerly there was also numbered among the Law Officers the Queen's Advocate-General (q.v.), but his office lapsed in 1872, and there are an Attorney-General of the Duchy of Lancaster and an Attorney-General and Solicitor-General of the County Palatine of Durham who represents the interests of the Crown in respect of the Duchy and County Palatine respectively, but are not usually called Law Officers of the Crown.

In Scotland a King's Advocate is known from 1483 and a Solicitor-General from 1729. The Lord Advocate (q.v.), who is appointed by the party in power from its supporters (or, failing a suitable person, on a non-party basis), is not only the principal legal adviser of the Government but head of the Scottish system of public prosecution of crime and the person who acts on behalf of all government departments in Scotland. Though he takes precedence in court he is not the leader of the Scottish Bar; that distinction belongs to the Dean of the Faculty of Advocates. The Solicitor-General is his deputy; in public prosecution he is assisted by several Advocates-depute, who represent the Crown in prosecutions, and each government department has a standing junior counsel appointed to it by the Lord Advocate. The Solicitor-General is also similarly normally appointed on a party basis but the Advocates-depute are not necessarily so.

The Lord Advocate is frequently but not always an M.P., and the Solicitor-General for Scotland somewhat less frequently so. On a vacancy in the office of Lord Advocate the Solicitor-General is normally promoted. The Lord Advocate almost invariably and the Solicitor-General very frequently is promoted to the bench.

There was also until 1972 an Attorney-General for Northern Ireland but no Solicitor-General. He was a member of the Privy Council and of the Parliament of Northern Ireland, but not of the Northern Ireland Cabinet. The Attorney-General is now *ex officio* A.G. for Northern Ireland and the S.G. acts for him when necessary as in England.

See J. L. Edwards, *Law Officers of the Crown* (1964); G. Omond, *The Lord Advocates of Scotland* (1883–1914); Lists of holders of the offices in Appendix.

Law Quarterly Review. A journal founded by Holland, Anson, Bryce, Dicey, Markby, and Pollock (qq.v.) in 1884 for the promotion of legal science. It appeared first in 1885 and has ever since held a leading place for critical comment on recent decisions, articles of high quality on a wide range of topics, and book reviews. The editors have been Sir Frederick Pollock, 1885–1919; A. E. Randall, 1919–26; A. L. Goodhart, 1926–70 (thereafter, till 1975, Editor-in-chief); and P. V. Baker, 1971–

Law reform. The alteration of the law in some respect with a view to its improvement. Until the mid-nineteenth century little serious effort was made to effect reform of the law and reforms depended mainly on occasional efforts by individual statesmen.

The earliest substantial efforts were the appointment of a Royal Commission in 1833 to digest the common law and statute law affecting crimes each into one statute. A second commission was appointed in 1845. On the basis of their reports seven English criminal law consolidation Acts were passed in 1861. The passing of Statute Law Revision Acts began in 1861 and in 1868 the Statute Law Committee was established. It effected the publication of successive editions of *Statutes Revised* (first edition 1878–85, second edition 1888–1901, third edition 1950) and the *Statutes in Force*, in loose-leaf form, replacing the *Statutes Revised*, from 1974. In 1853 Lord Chancellor Cranworth contemplated a Code Victoria and appointed a Board for the Revision of the Statute Law which was later superseded by a Statute Law Commission. At the end of the nineteenth century there were passed the only codification Acts enacted in Britain, the Bills of Exchange Act, 1882; the Partnership Act, 1890; the Sale of Goods Act, 1893; and the Marine Insurance Act, 1906.

In 1934 a Law Revision Committee was established in England to consider doctrines referred to it by the Lord Chancellor. After 1945 it was reconstituted as the Law Reform Committee. Separate Criminal Law Revision Committees and Private International Law Committees were established in 1959 and 1952. The Law Reform Committee for Scotland was established in 1954.

In 1965 there were established by statute the Law Commission and the Scottish Law Commission, each consisting of a judge as chairman and four other commissioners and each having the duty of keeping under review the whole of their respective systems of law with a view to their systematic development and reform. Both have produced numerous proposals for reform and draft Bills and many of their proposals have been enacted. The practical difficulty is to secure Parliamentary time for enacting reforms which frequently have little popular appeal. Moreover, all the time the law is being complicated, confused, and thrown into disorder by legal changes pushed through for political ends by mere politicians.

Law reports. Reports of decisions of the superior courts on disputed points of law, published for the information of the profession, The existence of series of published reports is an essential of the practice, accepted in all English-speaking countries, of referring to previous decisions as precedents and utilizing them in argument to persuade later courts to reach decisions consistently with the relevant precedents. See PRECEDENTS; *Stare decisis.* Modern reports consist of various parts, a series of catchwords indicating the main headings under which the decision falls, e.g. Contract—sale—implied warranty; a rubric, or short narrative stating the salient facts and the legal issue, and the court's decision thereof, the last frequently introduced by the words '*Held* that . . .'; a summary of the facts; a summary of the main arguments advanced; the names of the counsel who argued the case; the opinions announced by the judge or judges; and the disposal of the case.

In European countries reports vary somewhat. In particular the judgment of a court is normally anonymous, and is precisely and concisely framed, not stated in the discursive manner of an Anglo-American judgment (which may have been taken down as spoken *extempore*), and the published report is frequently furnished with long scholarly notes by academic lawyers.

Law reports—Australian. High Court of Australia decisions are reported in the *Commonwealth law reports* (1903–) and states reports cover the decisions of the Supreme Court in each state.

Law reports—Canadian. The *Reports of the Supreme Court of Canada* (1874–1922), *Canada Law Reports, Supreme Court of Canada* (1923–), *Reports of the Exchequer Court of Canada* (1891–1922), and *Canada Law Reports, Exchequer Court of Canada* (1923–) cover the major federal courts. *Canadian*

Criminal Reports Annotated (1898–1946) and *Criminal Reports (Canada)* (1946–) cover criminal cases. The *Dominion Law Reports* (1912–) contains reports from all courts and there are numerous sets of special-subject reports. Provincial courts are mostly covered by the *Maritime Provinces Reports* (1929–) and the *Western Weekly Reports* (1911–).

Law reports—English. The earliest English law reports were the Year Books (q.v.). They came to an end in 1535 and were replaced by the early reports, which were of very varied nature and quality. The earliest were Dyer (1513-82), Plowden (1550–80) and Coke, whose reports cover 1572 to 1616 and contain a great deal of antiquarian learning on the point under consideration by the court in the case reported. Thereafter there were a great many volumes of reports by individual judges or lawyers. A new standard was set by Burrow's Reports (1756-72) which first distinguished facts, arguments, and decision. By the end of the eighteenth century, moreover, judges had adopted the practice of looking over draft reports and agreeing the text of their judgments, which led to certain reporters in each court being regarded as 'authorized'.

In the nineteenth century reporting continued to be unsystematic, reports frequently being compiled by young briefless barristers, some of whom later achieved fame. The practice developed of one reporter or pair of reporters being 'authorized' in each court, though this did not prevent the compilation of other reports and some periodicals such as the *Law Journal* (1831–1949) and the *Law Times* (1859–1947) long published reports. In 1865 the Incorporated Council of Law Reporting for England and Wales was established and began to publish the monthly reports entitled *The Law Reports*. There are separate series of volumes for each court and since 1875 for each Division of the High Court. These reports have no monopoly and there are other series of reports, both general, such as the *All England Reports*, from 1936 (and *All England Reports Reprint*, selected reports reprinted, 1558–1935), and special, dealing with particular categories of cases, e.g. criminal cases, patent, design and trade mark cases, tax cases, industrial cases, and so on; and reports and notes continue to be published in legal journals and newspapers. The rule is that any report made by a member of the Bar may be cited in argument.

Only a selection of decided cases are reported, the proportion being high in the House of Lords, fairly high in the Court of Appeal, and quite low in the High Court and Crown Court. The selection is made by the reporters and the editors of the series on the basis of the importance of the judgments and of the principles stated in the judgments. Judgments proceeding purely on the facts of cases are rarely reported.

J. W. Wallace, *The Reporters* (1882); W. S. Holdsworth, *History of English Law*; Sweet and Maxwell, *Guide to Law Reports and Statutes.*

Law reports—France. The decisions of the Cour de Cassation are published in official reports. These, and other courts' decisions, are contained in the *Recueil général des lois et des arrêts fondé par J. B. Sirey*, the *Recueil Dalloz*, the *Gazette de Palais*, and *La Semaine Juridique* or *Juris Classeur Périodique*.

Decisions of the Conseil d'État are published in the *Recueil des Arrêts du Conseil d'État Statuant au contentieux* known as the *Collection Lebon.*

Law reports—Germany. There are semi-official reports of the Federal Supreme Court, *Entscheidungen des Bundesgerichtshof* in two series, *Zivilsachen* and *Strafsachen*, of the Federal Constitutional Court, *Entscheidungen des Bundesverfassungsgericht* (1952–) and of the Federal Administrative Court, *Entscheidungen des Bundesverwaltungsgericht* (1954–), and many private collections of reports.

Law reports—Italy. The major, general series of reports, *Foro Italiano* and *Giurisprudenza Italiana*, contain reports of cases in all the major branches of the law. There are also several special series of reports dealing with cases in one or another of the branches, constitutional, criminal, etc., and collected volumes entitled *massimari*, *repertori*, and *rassegni.*

Law reports—Scottish. The earliest Scottish reports were brief notes made by judges for their own and their colleagues' use, and known as Practicks (q.v.). The earliest published volume containing substantial notes is Stair's *Decisions* (2 vols., 1683–87, covering the period 1661–81); Durie's *Decisions* covering 1621–42 were not printed till 1690. The practice of judges collecting, and later publishing, decisions, was continued by Falconer, Dalrymple, Kames, and others down to Lord Hailes (1826, covering 1766–91).

In 1705 the Faculty of Advocates appointed William Forbes (q.v.) official reporter of decisions and his reports (*A Journal of the Session*, 1705–13) appeared with the approbation of the Court. He was succeeded by Bruce, Edgar, and David Falconer, to 1752, but the work was carelessly done between 1714 and 1752. From 1752 to 1825 the Faculty published reports collected by teams of reporters and known as the Faculty Collection or Faculty Decisions.

Lord Kames (q.v.) when at the Bar collected in his Folio Dictionary all the reports for the period 1540–1728 and reprinted them in two volumes, and this work was continued to 1796 by Alex. Fraser Tytler (Lord Woodhouselee) (2 vols., 1797) with a supplementary volume by McGrugar. W. M. Morison compiled a Dictionary of all the decisions of the Court of Session from 1540 down to 1808, in 19 (or 38 volumes) together with a Synopsis or Digest in two volumes of the cases contained in the Dictionary and a Supplementary volume. An Appendix containing the latest cases is usually bound in with each title. He also published in two volumes (1814–16), also entitled Synopsis, a continuation of the Dictionary for 1808–12 and 1812–16. M. P. Brown published (1826) a five-volume Supplement containing cases (1628–1794) omitted by Morison from his Dictionary and a four-volume Synopsis (1829) of all the cases in Morison's Dictionary, Brown's Supplement, and other reports down to 1827. Tait published a valuable Index to these series by the names of the parties. The older cases are now more frequently cited from these collections.

Reports published annually by Patrick Shaw from 1821 began the series of Session Cases. Six series have been published, the first five commonly known by the editor's name (Shaw, 1821–38; Dunlop, 1838–62; Macpherson, 1862–73; Rettie, 1873–98; and Fraser, 1898–1906). In 1906 Session Cases were taken over by the Faculty of Advocates and in 1957 by the Scottish Council of Law Reporting.

Collateral reports include the *Scottish Jurist* (45 vols., 1829–73) the *Scottish Law Reporter* (61 vols., 1866–1924) and the *Scots Law Times* (1893 to date).

Justiciary (i.e. criminal) cases were reported by private reporters from 1819 and have been included in Session Cases from 1874 and also in collateral reports.

Cases appealed to the House of Lords have been reported since 1707 and have been included also in collateral series and in English reports.

Law reports—U.S. The first U.S. reports were unofficial, Kirby's *Connecticut Reports 1785–88.* Unofficial reporters later published reports covering earlier periods. Official reporting began when the Supreme Court appointed a reporter in 1790, which resulted in 1 U.S. Reports (1 Dallas) which in fact contained only Pennsylvania reports. Since then official reporting has spread to all states. Lower federal court cases have mostly been reported only unofficially. Till the mid-nineteenth century reports were known by the reporters' names, but since then series have been named by the jurisdiction, though the *Supreme Court Reports*, volumes 1–90, are still known by the reporters' names.

Decisions of the Supreme Court are contained in the *U.S. Supreme Court Reports* and in various unofficial series. Decisions of the U.S. circuit courts of appeals are published in the *Federal Reporter.* Decisions of the State supreme courts are published in official series, but sometimes only in unofficial series, the Reporters system, various series covering groups of adjacent states. There are also sets containing selected cases from all jurisdictions, *American Law Reports.*

Law schools. The earliest systematic study of law and the earliest resorting to places for guidance in understanding law must be sought in the Greek schools of philosophy. At Rome in the third century B.C. Tiberius Coruncanius, *pontifex maximus*, gave public legal instruction. Under the Republic some jurists are known to have been active as teachers; they admitted young men to consultations and had discussions with them.

In the second century A.D. there were at Rome numerous *stationes ius publice docentium aut respondentium*, while the leading jurists had groups of pupils, though the famous Proculian and Sabinian 'schools' were not organized educational establishments. From the third century schools, which might claim to be legal universities, were recognized at which the commentaries, such as of Paul and Ulpian, were studied and interpreted. There were schools at Rome, Carthage, and probably in Gaul in the West, and, in the East, at Beyrout, the most famous, Constantinople, Alexandria, Caesarea in Palestine, Athens, and Antioch. They were particularly flourishing in the fifth century. Officially appointed professors of law are known from A.D. 425 when Theodosius II laid down that there were to be at Constantinople, *inter alios*, two professors of law. In Justinian's time there were eight at Constantinople and Beyrout, probably four in each centre. At Beyrout in the fifth century the law school had a fixed curriculum arranged in annual courses, the subjects of study being classical legal literature and the imperial constitutions. Justinian suppressed all the schools with the exception of those at Beyrout and Constantinople, and both sent members to the commission which compiled the *Corpus Juris.* The Beyrout school was transferred to Sidon in 551 after an earthquake destroyed the city. The School at Rome continued to exist even after the fall of the Western Empire, and after the reconquest of Italy Justinian ordered the payment of the professors to continue.

Latin was used as the language of instruction till the early fifth century when Greek was substituted. The method of instruction seems to have been for the teacher to take a text, and comment thereon, giving references to parallel passages, imperial constitutions, general principles, difficulties, and practical examples. The course normally lasted four years. In the later fifth century a certificate of attendance and of adequate knowledge, furnished on oath by the professors, was necessary for admission to practise in the courts of the *praefecti praetorio.*

According to Justinian, prior to his time, students in the first year learned the Institutes of Gaius and four *libri singulares*; in the second, *prima pars legum* (probably parts of the commentaries on the Edict) and selected titles from the *partes de rebus* and *de judiciis*; in the third they did the remainder of these parts and eight of the 19 books of Papinian's *responsa*; in their fourth they read the *responsa* of Paulus. Sometimes a fifth year was given to the study of imperial constitutions.

Justinian introduced a new syllabus, under which first-year students had lectures on the Institutes and the preliminary parts of the Digest; in the second year they dealt with the *pars de judiciis* or the *pars de rebus* and the four books of the Digest (23, 26, 28, and 30) corresponding to the topics of the former *libri singulares*; in the third year they took the *pars* not taken in the second year; in the fourth year they had to study by themselves what was left of the fourth and fifth *partes* of the Digest; and in the fifth year they studied the Code privately. Digest 37–50 was not part of the course but was studied privately.

From an early date embryo lawyers and civil servants had to follow a special course of study. In fourth-century Egypt an official went to elementary school, Latin school, and Law school and obtained a certificate of qualification to take up his profession. Law students were early set apart from others.

The University of Constantinople may have been founded by Constantine. Certainly in 425 state professors were appointed and, together with Rome and Beyrout, it was a centre of legal studies. In 1045 Constantine IX Monomachus refounded the University and created a school of Law with John Xipilinus as nomophylax. From about 1150 the post of director of the Law school was generally held by one of the clergy attached to St. Sophia. The last important nomophylax was Harmenopoulos in the fourteenth century.

Universities began to develop with the intellectual revival as Europe emerged from the Dark Ages. At Pavia there was a famous school of Lombard law in the tenth century, and at Ravenna a school of Roman law. From about A.D. 1000 the fame of Bologna developed with the teaching of Roman law (though this had never wholly lapsed during the Dark Ages) by Pepo and Irnerius. The epoch of Irnerius marks the beginning of the systematic study of the whole *Corpus Juris Civilis* as the basis of all legal education. This was reinforced by the appearance of Gratian's *decretum* in 1151, which made it a centre of canon law studies also. When Azo lectured at Bologna about 1200 there were said to be 10,000 students, the majority foreigners. Reggio Emilia and Modena also had flourishing schools of civil law before 1200. Pavia was long famous for Roman law; Perugia specialized in law; Siena had a faculty of jurisprudence in the thirteenth century; and Roma had schools of civil and canon law for foreign students in the fourteenth and fifteenth centuries.

Paris was strong in philosophy and theology and had a faculty of canon law but not of civil law which was forbidden by Honorius III in 1219. Montpellier had separate universities of medicine and of law in the thirteenth century, while in the fourteenth century Orleans had great fame as a law school; in the Middle Ages it was the greatest law university in France.

Salamanca was founded in 1243 with the main emphasis on civil and canon law and the other early Spanish foundations were the same.

The study of civil law was taken up in Heidelberg from its foundation in 1385, as it was at Cologne, while Erfurt had a great reputation as a school of jurisprudence in the fifteenth century. In the sixteenth century the fame of Louvain was very high. Probably every university in mediaeval Europe had a faculty of canon law.

After the Renaissance and the Reformation the canon law was greatly reduced in importance but the civil law flourished in northern Europe, particularly in German and Dutch universities. Increasingly, however, it had to compete with the study of the municipal law of the country in which each university was situated. The study of Swedish law at Uppsala dates from the early seventeenth century.

In Oxford and Cambridge law was studied early; Vacarius taught civil, i.e. Roman, law at Oxford in the twelfth century; Trinity Hall, Cambridge, was founded in 1350 as a school of civil and canon law; Henry VIII founded chairs of Civil Law at both universities. But in England the study of law remained academic and the major, and practical, study of law was done in the Inns of Court. Only after Blackstone (q.v.) was elected Vinerian Professor at Oxford in 1758 did law have a major place in the two older English universities.

In Scotland civil and canon law were studied in the universities from their foundation, but these studies languished and the study of law became serious only from the beginning of the eighteenth century, when Queen Anne founded Regius Chairs in Edinburgh (1707) and Glasgow (1712) and Edinburgh established chairs of Civil Law (1710) and Scots law (1722). Even thereafter anything like an adequate range of legal studies did not develop until the mid-nineteenth century.

In the U.S. systematic legal studies began with the private law schools, notably that of Tapping Reeve at Litchfield, Conn., from 1784 and that of Peter Van Schaack from 1786. George Wythe lectured on law at William and Mary from about 1780 and James Kent (q.v.) at Columbia from 1793. The Dane (q.v.) Chair at Harvard was established in 1819.

H. Rashdall, *Universities of Europe in the Middle Ages* (1936); A. J. Harno, *Legal Education in the United States* (1953); J. Goebel, *History of the School of Law, Columbia University* (1955); A. Sutherland, *The Law at Harvard*, 1817–1967 (1968); F. H. Lawson, *The Oxford Law School*, 1850–1965 (1968). See also LEGAL EDUCATION; LEGAL SCIENCE.

Law Society, The. Originally the Society of Attorneys, Solicitors, Proctors, and others, not being Barristers, practising in the Courts of Law and Equity of the United Kingdom, established in 1825 and chartered in 1831, later known as the Incorporated Law Society (1860) and since 1903 as the Law Society, it is the organization representative of, and regulating admission to practise as, solicitors in England and Wales. Its functions include maintaining the roll of solicitors of the Supreme Court, prescribing the requirements for admission, conducting professional examinations, issuing practising certificates, maintenance of a compensation fund from which loss caused by a solicitor's dishonesty may be relieved, and the making of regulations as to accounts, professional practice and discipline. It is governed by a President, Vice-President, and Council.

Independent of, yet working in close association with, the Law Society are a large number of London and provincial law societies. The Worshipful Company of Solicitors of the City of London (1908) became a livery company in 1944.

Law Society of Scotland. A body created by statute in 1949 to comprises all practising solicitors in Scotland. It acts as Registrar of Solicitors, controls examinations and admission, practising certificates, and the operation of the legal aid scheme. It makes and enforces rules as to solicitors' accounts and administers a guarantee fund from which to compensate persons who suffer loss by reason of a solicitor's incompetence or dishonesty.

Lawburrows (from law, and borgh or borrow, a pledge or security). In Scots law an ancient remedy, still competent and sometimes invoked, whereby a person who has, or thinks he has, reason to apprehend danger to his person or property from another, can have that other required by the court to find caution (security) not to trouble or molest him.

Lawgivers. Ancient and primitive legal systems not infrequently ascribe the initial statement, or the first codification, of their law to a particular lawgiver. This person is frequently said to have been divinely inspired. This is said of Hammurabi, Urukagina, Moses, and Zaleucus of Locri, but divine inspiration is not attributed to the decemviri who drew up the Twelve Tables at Rome, still less to later legislative or codifying commissions. The role of lawgiver frequently overlapped with that of judge; Hammurabi and Draco were judges as well as lawgivers.

The kinds of law given have varied greatly, some religious, e.g. the Pentateuch, the Koran; some largely penal, e.g. Draco; or even largely moral, e.g. Manu.

In truth such lawgivers, where they have existed at all and are not legendary, have been mainly formulators, codifiers, reformers, or expositors of pre-existing customary law, and where there is genuine historical knowledge of the person concerned his role can be seen to be truly one or more of these.

The term is not usually applied, though it might be, to rulers who promoted major restatements and codifications such as Justinian, Napoleon, Mohammed, and others, nor to jurists who have in their writings been creators as much as expositors.

Lawrence, Alfred Tristram, Baron Trevethin (1843–1936). He was called to the Bar in 1869 and soon acquired a practice; he took silk in 1897 and became a puisne judge in 1904. In this capacity he was respected and his judgments were sound.

In 1921 he was appointed Lord Chief Justice to keep the seat warm for the Attorney-General, Hewart (q.v.) who could not then be spared from Parliament. In 1922 Trevethin learned from *The Times* of his resignation. His tenure of office enhanced his sound reputation, but gave him no chance to make an independent reputation as head of the court. He also acted as president of the Railway and Canal Commission and certain other boards. A son became Lord Oaksey.

Lawrence, Geoffrey, Lord Trevethin and Oaksey (1880–1971). Son of Sir A. T. Lawrence, Lord Trevethin (q.v.), was called to the Bar in 1906, attained distinction during the First World War, became K.C. in 1925, a judge of the High Court, 1932–44, Lord Justice of Appeal, 1944–47, and a Lord of Appeal, 1947–57, taking the title Lord Oaksey; he succeeded his brother in the Trevethin barony in 1959. He was President of the War Crimes Tribunal at Nuremberg, 1945–46.

Lawyer. A general term for one professionally qualified to practise law in some capacity, and including judges (other than lay judges), legal practitioners, and law teachers. It does not include administrative officers or clerks, unless they are qualified. The term is sometimes restricted to those enrolled as practitioners, excluding judges and law teachers. The category of legal practitioners may be undifferentiated, as in the U.S. and Canada, or differentiated into barristers (or advocates) and solicitors as in the U.K. and some other jurisdictions. Other subdivisions of practitioners are met with in continental countries.

Lay days. See CHARTER-PARTY.

Lay judges. Persons not legally qualified or trained who exercise judicial functions, or participate with trained judges in exercising judicial functions. Lay judges participate in various ways in different legal systems. Two major kinds of lay judges exist, unqualified individuals who sit alone or in small groups exercising jurisdiction, and unqualified individuals who sit as assessors with, or co-judges with, one or more legally qualified judges. Thus in English magistrates' courts a bench of laymen have wide judicial powers, and in sittings of the Crown Court to hear appeals from magistrates' courts lay judges sit with a legally-qualified judge.

In France, West Germany, and some Scandinavian countries qualified and lay judges sit in mixed tribunals in both civil and criminal cases, and also in appeals.

A distinct variety of lay participation in the administration of justice is the jury (q.v.), but jurors are fact-finders rather than judges and function differently from lay judges proper.

J. P. Dawson, *A History of Lay Judges.*

Lay observers. Persons, not being lawyers, appointed in England and Scotland since 1974 to consider complaints made by members of the public about the disposal, by the Law Societies of the two countries, of complaints about solicitors or their employees. The Law Societies must consider reports or recommendations made to them by lay observers and notify them of any action taken. Lay observers make annual reports on their discharge of their statutory functions.

Laymann, Paulus (1574–1635). One of the leading canonists of his time and a prolific author, notably of a *Jus Canonicum seu Commentaria in libros decretales* (1666–98).

Le Roy* (or *La Reine*) *le veult. The King (or Queen) wills it, the traditional formula expressing the Royal Assent (q.v.) to Bills which have passed through all their legislative stages in Parliament

Le Roy* (or *La Reine*) *remercie ses bons sujets, accepte leur benevolence et ainsi le veult. The King (or Queen) thanks his (her) good subjects, accepts their benevolence and wills it accordingly, the formula expressing the Royal Assent (q.v.) to a Supply Bill, granting money.

Le Roy* (or *La Reine*) *S'avisera. The King (or Queen) will consider it, the formula used refusing the Royal Assent (q.v.) to a Bill.

Leach, Sir John (1760–1834). Called to the Bar in 1790 and became an M.P. in 1806 and P.C. in 1817. In 1818 he was appointed Vice-Chancellor of England and succeeded as Master of the Rolls in 1827. Though he had been a good advocate, he was a failure as a judge. He was irritable, hasty, and dictatorial, had an inadequate knowledge of the law, was lacking in understanding and judgment, and his decisions were repeatedly reversed.

Leach* v. *Money (1765), 19 St. Tr. 1002. A Secretary of State issued a warrant to search for the authors, printers, and publishers of *The North Briton* alleged to contain seditious libels. Leach, along with John Wilkes (q.v.), was arrested by Money, a King's Messenger. As he was neither author, printer, nor publisher he recovered damages in the Common Pleas and, on the case coming before Lord Mansfield, he condemned the use of general warrants, which were discontinued thereafter.

Lead. To lead evidence is to call or adduce evidence.

Leader. The senior counsel for a party in a case.

Leader of the Opposition. The leader of the party in the British House of Commons, other than the government, which has the largest representation in the House. In case of doubt the Speaker decides who is Leader of the Opposition. He or she has a recognized position and is paid a salary as Leader of the Opposition apart from a salary as an M.P.

Leading case. A judicial decision always regarded as the chief precedent (q.v.) or judicial statement of principle on a particular point. Such cases are always referred to in relevant textbooks and later cases, and various collections of leading cases have been published, notably J. W. Smith's *Leading Cases in various branches of the Law* (1837 and many later editions) and F. T. White and O. D. Tudor's *Leading Cases in Equity* (1849 and later editions), which collect the leading cases on particular points and show by annotations how the principles stated in the leading cases have been developed and applied in later cases.

Leading question. A question put to a witness which tends to lead him to a particular answer, e.g. 'When you went in did you see X?', or 'Do you not agree that what Y did was dangerous?' In general, leading questions may be put only as to formal and preliminary matters of a witness's evidence which are not disputed, such as his name, address, and qualifications, and in cross-examination (q.v.). If, however, a witness shows himself to be hostile to the party calling him, the court may allow him to be examined as if he were being cross-examined.

League of Nations. A draft scheme for a league for the avoidance of war was worked out by Viscount Bryce and others in 1915, and a similar movement called the League to enforce Peace was founded under ex-President Taft in the U.S.A. in the same year. At the Peace Conference in 1919 the Covenant of the League was adopted and formed part of the Peace Treaties. Its members were the states who were signatories of the peace treaty, states which acceded to the Covenant, and states later admitted. The principal political organs of the League were the Assembly and the Council, assisted by the Secretariat. The seat of the League and the office of the Secretariat were at Geneva. Two institutions largely independent but organically connected with the League were the International Labour Organization and the Permanent Court of International Justice (qq.v.). There were three technical organizations, the Economic and Financial Organization, the Organization for Communications and Transit, and the Health Organization, and various international bureaux came under the direction of the League. Among institutes set up for purposes within the province of the League, by agreement between the League and particular governments, was the International Institute for the Unification of Private Law.

The Assembly was the conference of members and met at least once a year, each member-state having one vote. The Council was the executive and was composed of the permanent members (the Great Powers who were members of the League) and 11 elected non-permanent members.

The main purposes of the League were the maintenance of international peace and security and the promotion of international co-operation. Members undertook to respect and preserve, as against external aggression, the territorial integrity and existing political independence of all members. This was a collective guarantee.

The Covenant also established the mandate system for the administration of overseas territories.

From the start the League was weakened by the refusal of the U.S.A. to join, and its authority undermined by the failure of members to implement their obligations in such cases as the Japanese attack on China (1932) and the Italian attack on Abyssinia (1935).

The League was dissolved by resolution of its last Assembly in April 1946, the United Nations Organization (q.v.) having been founded and having declared its willingness to assume various functions previously entrusted to the League.

F. Pollock, *The League of Nations* (1922); A Zimmern, *The League of Nations and the Rule of Law, 1918–1935* (1939).

Lease. A letting or grant of property, and particularly a deed giving effect to the contract between a proprietor of lands, buildings, fishings or other real property, the landlord, and a tenant, to whom the proprietor grants the possession and use of stated subjects for a term of years in return for a stipulated rent or annual payment. Leasehold property may be sub-leased. Distinct sets of conditions have developed according as the subjects of lease are, agricultural land, houses for habitation, fishings, and other kinds of property; in modern times extensive bodies of legislation affect different kinds of leases.

Lease and release. A former mode of conveying land in England secretly, without enrolling the conveyance in a court of record as required by the Statute of Enrolments, 1535, by concluding a bargain and sale of the land for a term of years, followed by a release of the reversion by the lessor to the termor. It was recognized as valid in 1621 and was in general use thereafter. It was the normal method of conveyance until 1841, when a release alone was made effectual, and in 1845 a deed of grant was submitted for the release. It is now impracticable and unnecessary.

Leasehold. The mode of holding land consisting in having a lease of it for a term of years, frequently long, as contrasted with freehold, when the land may be held in perpetuity. Leaseholds are in English law deemed chattels real and accordingly personal property, not real property, though certain statutes treat leaseholds as included in real property; they are lesser interests in land than estates of freehold, or even than life interests.

Leasing-making (or verbal sedition). In Scots law, the old and probably obsolete crime of uttering calumny of the sovereign out of malice and evil disposition, by inventing rumours to her dishonour, or by impeaching her title to the throne, or by speaking contemptuously of her.

Le Conte, A. See Contius.

Lee, Robert Warden (1868–1958). Served in the Ceylon Civil Service before being called to the Bar in 1896. He was Dean of the Faculty of Law at McGill University, 1914–21, when he became first (and only ever) Professor of Roman-Dutch law at Oxford. He taught also at the Inns of Court. He wrote an *Introduction to Roman Dutch Law* (1915), edited and translated Grotius' *Jurisprudence of Holland* (1926–36), wrote an *Elements of Roman Law* (1944), and edited the *Digest of South African Law*. He was a good linguist and insistent on the importance of Roman and comparative law.

Lee, Sir William (1686–1754). Called to the Bar in 1711, he became in 1730 a judge, and in 1737 Chief Justice of the King's Bench, holding that office till his death. He was competent, impartial, and learned, but not a great lawyer or judge. He presided at the trials of various Jacobite rebels in 1746, some of which are still of importance in the law of treason, and had to deal with a large number of new issues of mercantile law.

Lee* v. *Bude and Torrington Junction Railway Co. (1871), L.R. 6 C.P. 576. Solicitors sought payment of the costs of securing the passage of private Acts establishing the railway company from the company and its shareholders. The latter contended that the company was a sham and that Parliament had been induced fraudulently to pass the Act. In rejecting the defence the Common Pleas held that the courts could not question the validity of an Act of Parliament or the process by which it had been obtained; so long as it stood unrepealed the courts were bound to obey it, even though it might be contrary to natural justice.

Leeuwen, Simon van (1626–82). A famous Dutch lawyer, inventor of the name 'Roman-Dutch law', author of *Paratitula iuris novissimi* (1652), *Censura Forensis* (1664) dealing with Roman civil law, and, above all, *Het Roomsch Hollandsche recht* (1664) dealing fully with the law of Holland and briefly with Roman law. Though never highly regarded in Holland, his works have always been extensively cited in South Africa, and were given statutory place as subsidiary authority in the South African Republic in 1859.

Lefroy, Augustus Henry Frazer (1852–1919). Canadian jurist, whose main work lay in the systematization of the law of distribution of legislative power in the Canadian federation, and in the interpretation of the Canadian constitution. He wrote *The Law of Legislative Power in Canada* (1898), *Canada's Federal System* (1913), *A Short Treatise on Canadian Constitutional Law* (1913), and other works.

Lefroy, Thomas Langlois (1776–1869). Called to the Irish Bar in 1797, he several times declined judicial preferment before accepting the post of baron of Exchequer in Ireland in 1841 and that of Chief Justice of the Queen's Bench in Ireland in 1852; he resigned in 1866.

Legacy. A gift made by a person in and by his will and effective on his death. Strictly speaking a legacy is a gift of personal or moveable property only, a gift of land being a devise, but the term legacy is commonly used for both.

Legacies are distinguished into general legacies, such as of a sum of money, and special or specific legacies, such as of a specific thing or investment. Demonstrative legacies are special legacies expressed to be payable out of or secured on a particular fund or security. Residue, or the balance of the testator's estate after satisfaction of debts and legacies, may be spoken of as a legacy but is properly distinguished from legacies.

If the legacies left exceed the available estate, there is nothing to form a residue and general legacies must abate proportionally or even completely, and then special legacies. A special legacy is adeemed or revoked if the subject of the legacy had ceased to be part of the testator's estate before his death.

Problems may arise of identification of the proper legatee, and of the subject of the legacy, and many questions have arisen as to the meaning of particular words, such as 'money', 'my wordly goods', and the like.

A legacy may be conditional, e.g. being postponed, or subject to forfeiture if the legatee has not satisfied some condition. If subject to a condition, the condition is deemed satisfied if the legatee does what is in his power to satisfy the condition, even if he does not in fact satisfy it. Conditions illegal, impossible, or immoral, or contrary to public policy are disregarded.

If the same legacy is given twice by a will or codicil, it is presumed mere repetition, but there may be the intention to give twice; if the legacies are different or unequal, both are due. Legacies to the same person given in different wills are prima facie cumulative, but the later legacy may be shown to be intended to be in substitution for the earlier one.

Legal. (1) That which is lawful, or according to or consistent with law, and not contrary to law.

(2) Legal or legal period of redemption, in Scots law the period of 10 years within which a debtor whose heritage has been adjudged for debt, i.e. transferred to the creditor in satisfaction, is entitled to pay the debt and recover the land, disencumbering it, failing which the creditor may obtain a declarator of the expiry of the legal and convert his judicial security into a right of property.

Legal Aid and Advice. In England the right of poor persons to sue *in forma pauperis* was recognized from 1495 and legal aid to persons unable to pay the costs of civil litigation was granted in the High Court and Court of Appeal under the Rules of the Supreme Court. In Scotland a statute of 1424 established the poor's roll whereby persons qualified by poverty, but having *probabilis causa litigandi* were entitled to have their cases conducted gratuitously by counsel and agents for the poor. In England provision was made in 1903 for counsel in criminal cases on indictment.

The modern system was introduced under statute in 1949. Schemes are operated by the Law Society and the Law Society of Scotland, and the categories of courts in which legal aid is available has been considerably extended. To be entitled to legal aid a person must have disposable income and capital less than stated amounts and satisfy a local committee that he has a prima facie case. If he obtains a certificate entitled him to legal aid, he may select a solicitor and counsel from the panels of those participating in the scheme. They carry through the case in the usual way and are paid a proportion of the fees allowed on taxation out of the legal aid fund. A person receiving legal aid may be required to make a contribution to the legal aid fund towards the cost of his case. A successful unassisted opponent is entitled to costs out of the legal aid fund. Legal advice is given to persons of small means by solicitors in ordinary practice who have intimated their willingness to do so, and who subsequently recover fees from the legal aid fund for so doing.

In criminal cases examining justices might grant a person a legal aid certificate and a person committed for trial was entitled to free legal aid in the preparation and conduct of his defence, and to have solicitor and counsel assigned to him if a defence certificate were granted. In Scotland the legal bodies appointed persons from their own number to undertake the defence of persons accused. In both countries the Legal Aid Scheme was extended to criminal cases and persons charged may be granted legal aid, on their resources and commitments being assessed and sometimes on condition of making a contribution to the costs, in which case they are entitled to representation by solicitor and counsel.

Legal bibliography. The literature of and about law, and every branch of it is enormous. This article can only mention the main guides to the literature in different fields of law which will themselves point to individual books. Moreover, printed materials of law comprise both primary materials (codes, statutes, regulations, cases, etc.) and secondary materials (treatises, commentaries, textbooks, encyclopaedias, periodicals, etc.) as well as a vast literature of books and articles examining, discussing, and elucidating matters great and small of and about law.

A guide to bibliographies and catalogues is in Besterman's *World Bibliography of Bibliographies*, s.v. Law. It lists many older bibliographies and catalogues of collections. The catalogues of major law libraries always deserve examination. The most general bibliography is U.N.E.S.C.O. *Register of Legal Commentaries in the World* (2nd edn. 1957) which lists, for each country, the sources of law, reports of decisions, periodicals and bibliographies. The older *Bibliographie Générale des Sciences Juridiques* by Grandin (1926–48) and the *Bibliografia Guiridica Internazionale* (1932–41) are still useful. For Anglo-American law Hicks, *Materials and Methods of Legal Research* (1942), Beardsley and Orman, *Legal Bibliography and the Use of Law Books* (1947), and Price and Bitner, *Effective Legal Research* (1953) are useful. Friend, *Anglo-American Legal Bibliographies* (1966) is a guide to the same ground. Stollreither, *Internationale Bibliographie der Juristischen Nachschlagewerke* (1955) is wider-ranging.

The *Annual Legal Bibliography* (Harvard Law School) gives a selected list of works acquired by that library, and is supplemented by *Current Legal Bibliography*. The *Index to Legal Periodicals* (1907–) and the *Index to Foreign Legal Periodicals* (1960–) deal with articles in periodicals.

Legal history

A useful bibliography is the *Bibliographie en langue française d'histoire du droit 987–1875*, also useful is Gilissen, *Introduction bibliographique à l'historie du droit et à l'ethnologie juridique* (1965–).

On Roman law there is Caes and Henrion's *Collectio bibliographica operum ad ius Romanum pertinentium* (1949–).

On canon law Naz, *Dictionnaire de droit canonique* (1935–) is helpful.

On older English law there are Holdsworth, *Sources and Literature of English law* (1925), Winfield, *Chief Sources of English Legal History* (1925), Beale, *Bibliography of early English law books* (1926), and Cowley, *Bibliography of abridgments, digests, dictionaries and indexes of English law to 1800* (1932).

On older Scots law see the Stair Society, *Sources and Literature of Scots law* (1936).

Jurisprudence

Dias, *Bibliography of Jurisprudence* (3rd edn. 1978) is selective but useful. Champliss and Seidman produced *Sociology of the Law: A research bibliography* (1970).

Comparative law

U.N.E.S.C.O. has published a *Register of Legal Documentation in the World* (2nd edn. 1957), listing legal bibliographies, collections of laws and reports, legal journals, and other materials. The International Association of Legal Science has produced a series of bibliographies of various countries for the use of foreign jurists. It has also published *Cahiers de Bibliographie juridique* which lists legal works published in Western European countries and is publishing an *Encyclopaedia of Comparative Law*. Szladits' *Bibliography on Foreign and Comparative Law* (2 vols., to 1960, and supplements) lists books and articles in English.

International law

Harvard Law School Library *Catalog of International Law and Relations* is a very exhaustive compilation. Schlochauer's *Wörterbuch de Völkerrechts* (1960–62) documents every topic discussed. There are also Olivarts' *Bibliographie du Droit Internationale* (1905) and Delupis' *Bibliography of International Law* (1975).

On international organizations the most useful works are Speeckhaert's *Bibliographie sélective sur l'organisation internationale, 1885–1964* (1965) and Colliard, *Institutions internationales* (1966). On the U.N. there is Brimmer's *Guide to the Use of U.N. Documents* (1962).

The library of the International court publishes an annual *Bibliography of the International Court of Justice* (1947–).

Supranational law

Apart from catalogues of publications of the European Communities and the Council of Europe there are Paklow's *Bibliographie européenne* (1964) and Pehrsson and Wulf's *The European bibliography* (1965) and the Conference on the Atlantic Community, *The Atlantic Community: An Introductory Bibliography* (1962).

Systems of municipal or national law

Useful bibliographies include;

Australia: Sweet & Maxwell Ltd., *Legal Bibliography of the British Commonwealth of Nations*, vol. 6 (1955).

Belgium: Graulich and others; *Guide to Foreign Legal Materials: Belgium, Luxembourg, The Netherlands* (1968); Dekkers, *Bibliotheca Belgica Juridica* (1951).

Canada: Sweet & Maxwell Ltd., *Legal Bibliography of the British Commonwealth of Nations*, vol. 3; Boult, *Bibliography of Canadian Law* (1966).

Denmark: Iuul, *Scandinavian Legal Bibliography* (1961).

Finland: see Denmark.

France: Szladits, *Guide to Foreign Legal Materials: France, Germany, Switzerland* (1959); David and de Vries, *The French Legal System* (1958); David, *Bibliographie du droit français 1945–60* (1964).

Germany: see France. *Bibliographie des deutschen Rechts* (1964); Cohn and others, *Manual of German Law*.

Italy: Grisoli, *Guide to Foreign Legal Materials: Italy* (1965).

Luxembourg: see Belgium.

Netherlands: see Belgium.

Norway: see Denmark.

South Africa: Roberts, *A South African Legal Bibliography* (1942).

Spain: de Villavicences and de Sola Canizares, *Bibliografia juridica española* (1954).

Sweden: see Denmark.

Switzerland: see France.

United Kingdom: Bibliographical Guide to U.K. Law (2nd edn. 1973); Sweet & Maxwell Ltd., *Legal Bibliography of the British Commonwealth of Nations* (2nd edn.), vols. 1, 2, and 5 (1955).

United States: Andrews, *Law of the U.S.A.* (1965).

Legal charge. A deed in English law whereby a beneficial owner of land passes a legal interest in the land to the chargee as security for money lent him, giving the same protection, powers, and remedies as if he had been granted a mortgage by way of lease for 3,000 years. It may be used for freeholds and for leaseholds. It is distinguishable from an equitable charge which may be created by any agreement in writing, however informal, by which any property is to be security for a sum of money.

Legal concepts. General and abstract notions arrived at inductively by thinkers about law from examination of particular rules and cases. Analytical jurists, like analytical and linguistic philosophers, devote much attention to clarifying our ideas of the uses of general terms and their connotations. Concepts serve as the bricks of legal thought and the bases of reorganization of the masses of particular instances into categories. Examples are duties, contract, tort, possession, ownership, trust, bill of exchange, and many more. Some, e.g. ship, are developed from perceptions of fact, sometimes aided by definitions in statutes or cases, some, e.g. corporate personality, developed from long evolution and speculation. The more abstract and general the concept, the more useful it is in legal thinking and developing legal principles. The abstract concept of the 'estate in land' has been extremely important in the development of real property law. Different people do not own the land or bits of it, but they own different estates in the one piece of land.

Many concepts are extremely difficult to define or to understand and their analysis occupies a substantial part of books on particular topics in which they are relevant and, in the case of the most general concepts, e.g. rights and duties, a substantial part of books on jurisprudence or legal theory. Moreover, they do not have settled or unchangeable meanings, but these are subject to change as the whole body of law develops. Frequently they are carelessly used or misused. The word 'possession' may mean different things in different branches of the law.

It is probable that the more technical legal concepts become and the more remote from everyday language, e.g. contingent remainder, the more useful they are and the more scientific the legal system will be. But too many concepts, e.g. possession, property, and negligence, suffer from having the same verbal form as ideas of everyday usage and rather indefinite meaning, frequently materially different from the connotation of the legal conception. In British law the formation of concepts is hindered by the practice of statutes formulating their own definitions of terms, valid for that statute only.

Important analyses of the concepts 'right' and 'duty' were made by Salmond and, in particular, Hohfeld, and later by Kocourek (qq.v.). Savigny's analysis of possession has been very influential but possibly detrimental to understanding. For the fundamental question in all analysis of legal concepts is: How is the word used in the particular legal system and what are its connotations?

Legal doctrines, principles, and rules.

These three terms have no fixed or settled meanings or usages. All refer to kinds of legal precepts, but are generally used as implying different degrees of scope and generality in the expression of the precepts. A 'doctrine' is the most general; it is usually used of a substantial body of related precepts with a common theme, a systematic fitting together of principles, rules, and standards with respect to particular situations or types of cases or fields of the legal order, in logically interdependent schemes, whereby reasoning may proceed on the basis of the scheme and its logical implications; examples are the doctrine of consideration in the Anglo-American law of contract and the doctrine of public policy. A 'principle' is normally used of a general precept which justifies, unifies, and explains numerous more specific formulations and applications and serves as an authoritative starting point for further legal reasoning; examples are the principle that an employer must take reasonable care for the safety of his employees, and the principle that a party in breach of contract is liable for those consequences which should have been within his contemplation at the time he made the contract. A 'rule' is normally used of a more particular and specific precept, attaching a particular legal consequence to certain facts; examples are the rule that, to be valid, a will must be attested by two witnesses, and the rule that claim must be made for all losses resulting from one cause of action in one action.

Legal education and training (Canada).

In Canada, there was established in 1797 the requirement of five years' articles or clerkship before admission to legal practice, latterly reduced for holders of college degrees. In each province admission was, and is, regulated by the Law Society of that province.

An École de Droit opened as a private enterprise in 1851 (and lasted till 1967) and law schools at McGill in 1853 and Laval in 1854, a branch of the latter becoming the University of Montreal in 1920. The Law Society of Upper Canada maintained a law school at Osgoode Hall, 1873–78, and 1881–1968 when it became the faculty of law of York University, Toronto, but the senior university law school in the common law provinces is Dalhousie (Halifax, N.S.).

By the latter part of the twentieth century the pattern had become established of attendance on university law school followed by a short period of articles. Canadian law schools have been heavily influenced by the best U.S. law schools and are more similar to them than to British law schools.

The contribution to legal scholarship by Canada has been disappointing. The earliest large work was Beamish Murdoch's *Epitome of the Laws of Nova Scotia* (1832–33); there were several adaptations of Blackstone, or parts thereof, to Canadian conditions, and some standard English texts have been published with Canadian notes. In the twentieth century, apart from the work of J. D. Falconbridge, some monographs on matters peculiar to Canada have supplemented reliance on English texts.

Legal education and training (England).

The aims of legal education are varied and have differed at various times and places. It may be aimed broadly at understanding the functioning of law in society, as a training for administrators and civil servants as much as for legal practitioners; or it may be more narrowly aimed at training persons for legal practice.

At Rome and latterly at Constantinople, Beyrout, and elsewhere there were regular schools of law and in the mediaeval universities Roman and canon law were extensively studied. The study of national legal systems in European universities developed much later, mostly in the eighteenth century. Accordingly by that time it had been well established that law was a proper, and important, subject for university study.

The aims, methods, and content of legal education have differed in different countries, affected particularly by whether there were or were not professional schools concentrating on practical branches of law, the relative importance of legal treatises and decisions of courts, the relative standing of professors and judges and other factors.

In England before the Reformation the universities studied only Roman and canon law, and education and training for the practice of English law was provided in and by the Inns of Court (q.v.). But by the later seventeenth century there was virtually no organized education for English lawyers until an apprenticeship for solicitors was introduced in 1729.

Wood's *Institute of the Laws of England* of 1720 was the first modern attempt to produce a systematic textbook, and had some success as is evidenced by eight editions in 34 years, but the work is a dull statement of the law in categorical propositions, not going far towards making the law a science. In 1708 Wood suggested that the English universities offer courses on the common law, in his pamphlet *Some Thoughts concerning the Study of the Laws of England in the Two Universities.*

In 1753 Blackstone (q.v.) was a candidate for the Regius Chair of Civil Law but was not appointed; instead, at the instigation of William Murray, the Solicitor-General and later Lord Mansfield, he began a course of lectures on English law. He put rational order into the legal system and his lectures became very popular, and sets of students' notes began to find a ready market.

In 1756 Charles Viner (q.v.), compiler of the *Abridgment*, died leaving his copyright, unsold copies, and net personalty to Oxford to endow a professorship of the common law of England and fellowships and scholarships in the same subject. In 1758 Blackstone was appointed to the Vinerian Chair and in 1765–68 published his four volumes of *Commentaries* founded on his lectures. His achievement was, firstly, to set down the main rules of the common law in a systematic way, so as to appear a reasoned system; secondly, to set it down in readable English prose; thirdly, to make the common law an intellectually respectable subject of study, as proper for a university as the civil law of Rome. After Blackstone, however, legal studies faded and did not revive until the creation of the school of Law and Modern History at Oxford in 1850. They were divided in 1872. See OXFORD UNIVERSITY LAW SCHOOL.

At Cambridge the Downing Chair was founded in 1800 and the Law Tripos started in 1858; from 1870 to 1874 it was joined with History. See CAMBRIDGE UNIVERSITY LAW SCHOOL.

In 1852 the four Inns of Court jointly established the Council of Legal Education and began to provide courses, and in 1872 instituted a compulsory examination for call to the Bar.

A fundamental difficulty for a long time was the absence of good analytical textbooks. There were many dull and heavy practitioners' books, and Stephen's *Commentaries*, founded on Blackstone, appeared from 1841 onwards and Broom's *Commentaries on the Common Law* from 1856. But from the mid-nineteenth century this defect began to be remedied with the writings of Anson (*Contract*, 1879; *Constitutional Law*, 1886); Pollock (*Contract*, 1876; *Torts*, 1887; *Partnership*, 1877); Dicey (*Law of the Constitution*, 1885; *Conflict of Laws*, 1896, founded on his *Law of Domicil*, 1879); Holland (*Jurisprudence*, 1880); Moyle (Justinian's *Institutions*, 1883); Markby (*Elements of Law*, 1871), and others. Williams on *Real Property* appeared in 1845 and *Personal Property* in 1848, Challis on *Real Property* in 1885, and Digby's *History of the Law of Real Property* in 1875.

In the later nineteenth and twentieth centuries legal studies have been undertaken in many of the newer universities and latterly in some of the polytechnics, and there is now an overwhelming number of textbooks suitable for students on every subject.

The methods of legal education have been lectures and tutorials, essays and written work, much recommended reading, and moot courts at which students argue a case as if before an appellate court.

There has been a strong tendency to broaden the curriculum to include jurisprudence, legal history, comparative law, international law, foreign law, European Communities law, and also non-traditional topics such as taxation, labour law, and social security law. But no attempt is made to cover all branches of law, nor to equip the student to cope with all the problems he may meet in practice. It is now accepted that provided it included the essential or fundamental subjects, any law degree will exempt a student from the basic or academic stage of professional examinations.

The older universities did not, and do not, seek to teach professional or practical subjects, that being left to the Inns of Court or the Law Society's School and to reading in chambers or service under articles. The younger universities have more regard for such subjects. But there has been a strong movement to introduce organized practical teaching of professional subjects, such as conveyancing, evidence, and procedure, but this has given rise to disputes whether universities or the profession should do the teaching.

H. G. Hanbury, *The Vinerian Chair and Legal Education* (1958); F. H. Lawson, *The Oxford Law School 1850–1965* (1968).

Legal education and training (Scotland). Before the foundation of the older Scottish universities there is no record of legal studies or education. Many of the judges and lawyers were clerics, and knowledge of canon law was probably acquired in monasteries and cathedrals, and in some cases clerics are known to have studied on the Continent. Because of the recurring warfare between Scotland and England from 1295, Scots could not and did not go to Oxford or Cambridge. Scots resorted extensively to Paris and Orleans and later to Cologne in the fifteenth century. It seems that the need for local training schools for lawyers as well as for clerks underlay the foundation of the pre-Reformation Scottish universities.

In St. Andrews (1411) canon law was studied from the outset, and there is thought to have been some study of civil law.

In Glasgow (1451) founded on the model of Bologna, both civil and canon law had been taught before 1500, and general legal principles were long included in the study of moral philosophy. The Regius Chair of Law was founded by the Crown in 1713 and William Forbes (q.v.) was appointed to teach the civil, feudal, canon, and Scots law. The holders of the chair largely confined themselves to Scots law though John Millar (1761–1801) taught a wide range of legal subjects. Not till 1861 were there any other teachers of law.

In King's College, Aberdeen (1494) the founder, Bishop Elphinstone, aimed at making his university a school of law.

At the Reformation the Reformers, in the *Book of Discipline*, provided for the study of law in each of the then-existing three Universities, a four-years course under the Readers in Roman law and Municipal law; this scheme never became effective.

In Edinburgh, Robert Reid, Bishop of Orkney and a judge, and later President of the Court of Session, left money in 1558 to found a college for, *inter alia*, teaching the civil and canon laws. The money was eventually paid to the Town Council which applied it to the uses of the Town's College, i.e. later to become the University. In 1590 the College of Justice endowed a Professor of the Laws, but the first incumbents taught only Latin and it became diverted to that purpose. In 1707 the Crown endowed a Professorship of the Public Law and the Law of Nature and Nations.

The lack of adequate university facilities led some, such as John Spottiswoode, to conduct small private schools of law, and in 1710 Edinburgh Town Council made one such private teacher, James Craig, Professor of Civil Law, though initially without stipend. In 1722 provision was made also for a Professor of Scots Law and Alexander Bayne, who had been teaching privately outside, was appointed.

During the seventeenth and early eighteenth centuries Scots lawyers, if desiring academic study of law, resorted to Utrecht, Leyden, Groningen, Halle, or elsewhere in northern Europe, and there studied civil law. Hence there is heavy reliance, down to the end of the eighteenth century, on such authorities as Grotius, Voet, and Vinnius.

Solicitors learned their craft by apprenticeship, but in 1793 the Society of Writers to the Signet in Edinburgh created a private lectureship in conveyancing which, in 1825, became the Chair of Conveyancing. In Glasgow the Faculty of Procurators similarly controlled apprenticeship and from 1817 provided extra-mural teaching in Conveyancing, and throughout Scotland local societies did the same. Apprentices frequently attended university lectures on law, but could not and did not graduate, nor did they have the opportunity of studying more than professional topics.

Scots lawyers had, however, even at this time, a good supply of books giving systematic expositions of the law, including the major *Institutions* of Stair (1681), *Institute* of Bankton (1751), Erskine's *Institutes* (1774) and Bell's *Commentaries* (1800) and *Principles* (1829), and, among smaller books, Mackenzie's *Institutions* (1684) and *Criminal Law* (1678), Forbes' *Institutes* (1722), Erskine's *Principles* (1754), as well as numerous lesser works and books on practice. Bell, as Professor of Scots Law at Edinburgh, not only prepared his *Principles* originally as a students' textbook but compiled a companion, *Illustrations from Adjudged Cases of the Principles of the Law of Scotland* (3 vols., 1836–38).

In all the Scottish Universities progress was slow until the executive Commissioners under the Universities (Scot.) Act, 1858, framed their Ordinance (applicable to all four universities), creating the degree of LL.B. to be considered as a mark of academical and not professional distinction, requiring passes in six legal subjects, and obtainable only by graduates in Arts. In fact only at Edinburgh did professors of all the requisite branches of law exist. The great majority of students, however, in all the universities attended only a few courses for professional purposes, and many did not have the prerequisite Arts degree. Accordingly in 1874 the Universities created the degree of B.L. for which an Arts degree was not a prerequisite; it required preliminary passes in Arts subjects and passes in four legal subjects.

By the 1860s prospective advocates had to pass examinations in a substantial range of legal subjects and from 1874 a law degree exempted in whole or in part from professional examinations.

In 1873 the Law Agents Act regulated the entry of persons as solicitors, requiring five years' apprenticeship, and the passing of certain examinations in law; attendance on university classes became a requirement, and from 1878 graduates in law were exempted from examination, save in procedure.

Neither branch of the profession created any professional law school, but relied on the universities.

A continuing feature of Scottish legal studies, reflecting the civil law basis of Scots law, its affinity with continental practice, and the individual court system has been the practice of teaching Scots law as a large unit, rather than dividing it into courses on contract, property, trusts, etc. Similarly from the great institutional writings of Stair, Erskine, and Bell and the lesser writings of Mackenzie and Bankton, down to the present, the major students' and practitioners' textbooks have been on the whole law, though large practitioners' works on particular topics have appeared.

The first major book on a distinct branch of law was David Hume's *Commentaries on Crimes* (1797–1800), followed by Alison's *Principles* (1832) and *Practice* (1833) *of the Criminal Law of Scotland.*

Later in the nineteenth century there appeared Lord Fraser's books on *Domestic Relations* (1846), later divided into *Parent and Child* (1866), *Master and Servant* (1872, 1882), and *Husband and Wife* (1876–78) and Lord Justice-Clerk Macdonald's *Criminal Law* (1867, 5th edn. 1948), Lord McLaren's *Trusts* (1862), later revised as *Wills and Succession* (1868, 1894), and the *Lectures on Conveyancing* of Menzies (1856, 4th edn. 1900), Montgomery Bell (1867, 3rd edn. 1882) and Wood (1903). Other notable works were Professor Sir John Rankine's *Landownership* (1879, 4th edn. 1909), *Leases* (1887, 3rd edn. 1916), and *Personal Bar* (1921).

The twentieth century was rather a lean time for scholarly writing, save for the work of Professor W. M. Gloag, whose *Contract* (1914, 1929), *Law of Rights in Security* (with Irvine, 1897), and *Introduction to the Law of Scotland* (with R. C. Henderson, 1927, 7th edn. 1968), were valuable, but since 1950 there has been an efflorescence of sound and valuable textbooks. On many subjects, e.g. companies, shipping, frequent reference is made, often misunderstood and to the detriment of true principles of Scots law, to English books.

Legal education and training (U.S.). In colonial America lawyers were few and unimportant, and it was not until the latter part of the eighteenth century that they became influential in public affairs.

Before Blackstone's *Commentaries* were published, young men learned their law by working as apprentices or clerks to established lawyers, learning in a rule of thumb way, by example and by experience. Some read more widely, books on political science and philosophy, the older books on English law, of which the best, for instructional purposes, was probably Thomas Wood's *Institute of the Laws of England* (1720), even some Roman law. A few went to the Inns of Court in England. Story and Daniel Webster started by being given *Coke on Littleton* to read. But young Americans had no textbooks on their municipal law, still less a systematic curriculum or organized instruction. Shortly after Blackstone's *Commentaries* were published, sets were being imported into America; an American edition appeared at Philadelphia in 1771–72, and from that time *Blackstone* was a staple of American legal studies.

Professorships of law were proposed in American colleges before independence was fully won. One was planned for Yale in 1777, where President Stiles proposed four courses on Roman law, English law, the codes of the thirteen states, and comparative government, political theory, and international law. In all of these the student was to concentrate on theory, learning the practical application in practice. The first American Professorship of law was created at William and Mary College, Virginia, in 1779. It was here that Jefferson set up a chair of law and police and added to the duties of the chair of moral philosophy instruction in the law of nature and nations. George Wythe, later chancellor of Virginia, was appointed, and in 1779–80 John Marshall, in an interval of duty in the War of Independence, studied under him. Wythe was succeeded in 1789 by St. George Tucker, who produced an edition of *Blackstone* in 1803.

In 1784 Judge Tapping Reeve founded his private law school at Litchfield, Conn., and before it closed in 1833 it had taught about 1,000 students from all over the Union, including many who later became distinguished. Reeve, and his later associates, James Gould and Jabez Huntingdon, seem to have given systematic instruction by well-prepared lectures, with references to *Blackstone* and decided cases, weekly examinations, probably oral, and moot courts. At the time it was the best instruction available. The peak enrolment was 55, in 1813. Other private law-schools sprang up, but none lasted or achieved the reputation of the Litchfield Law School.

In 1790 Philadelphia College appointed James Wilson, an Associate Justice of the Supreme Court, as professor of law, and he planned a course on constitutional law, international law, and civil law, the origin and rules of the common law, law merchant, and maritime law, over a three-year period. The course did not arouse interest.

Between 1793 and 1798 James Kent lectured on law at Columbia College; his audience fell away, but in 1824, having been a justice and Chief Justice of the New York Supreme Court, chancellor, and reviser of the state's constitution, he lectured again. The fruits of his lectures were his *Commentaries on American Law*, 4 vols., 1826–30, a work which went through nearly as many editions as had *Blackstone*.

In 1816 David Hoffman, professor of law in the University of Maryland, published *A Course of Legal Study* and re-stated the view that law was a science, and must be founded on principle, and that a teacher of law must discover and systematize the principles.

There were proposals for a Harvard chair of law in 1778 and in 1780s, but they came to nothing. In 1781 Isaac Royall, a loyalist refugee during the War of Independence, by his will endowed a chair of law there but, by reason of the war and its consequences, the chair was not established till 1815 when, after John Lowell had refused it, Isaac Parker, Chief Justice of Massachusetts, was appointed. In 1817 Harvard adopted a proposal by Parker to create a true law school, and appointed Asahel Stearns as University Professor of Law: candidates could obtain the LL.B. after five years' study, or three years' study if already a college graduate, or eighteen months' study if followed by training in the office of a practitioner.

Initially, and for a long time, many students did not graduate. Instruction was by lecture but moot-courts were frequently held; there were examinations and dissertations were required. Parker resigned in 1827 and Stearns in 1829 and the Harvard Law School reached a very low ebb.

In 1829 Nathan Dane, who had compiled a *General Abridgment and Digest of American Law* in eight volumes, with a final one in 1829, proposed the establishment of a new professorship, the Dane Chair, for lectures on natural law, commercial and maritime law, and constitutional law, to be held by Joseph Story (q.v.) who had published editions of several standard English lawbooks and had, since 1811, been a justice of the Supreme Court. Dane prescribed a series of publications as part of the professor's functions. In the next dozen years, Story, as well as judging and lecturing, wrote the wonderful Series of *Commentaries*, which did more than any other man of his century to make the law luminous and easy to understand. He died in 1845. In Story's time Harvard began notably to teach American law, not Massachusetts law, and to attract men from all over the Union. In 1833 Greenleaf joined him, as Royall Professor.

At Yale the law school came into being when Judge David Daggett was appointed professor, the school adopting a private law school which had operated in New Haven since 1800, and in 1826 a chair of Law and Politics was created in the new University of Virginia.

Parallel to these developments were the growth of bar admission requirements. Between 1767 and 1829 at least 17 jurisdictions adopted regulations prescribing requirements for admission to the Bar, usually requiring three to five years of preparation; some distinguished between attorneys and counsellors, and some shortened the period required of college graduates. But those developments were hindered by Jacksonian democracy with its distrust of formal education and of prerequisites for the practice of law.

By 1850, however, there were only 15 law schools in the U.S., including Yale (1843), Tulane (1847), Pennsylvania (1850), Albany (1851), and Columbia (1858), but the next 20 years saw the foundation of 16 more. Law school training was far from being, it was steadily becoming, the accepted means of preparation for the Bar; but the law school faculties were still mainly staffed by judges and practising lawyers and the curriculum was normally of only one or two years; there were no entrance qualifications; there were no serious examinations; and instruction was by lecture, assisted by reading.

A new era dawned with the appointment of Christopher Columbus Langdell as Dean of the Harvard Law School in 1870. Langdell introduced the 'case-method' of study and instruction, and in his time entrance requirements were stiffened and examinations at the end of each year were obligatory. In 1873 he was joined by Ames, the first appointment of an outstanding young man who had not practised law, and shortly after by Thayer and Gray and then by Beale, Williston, and Wambaugh. The case-method based on the premise that the law of any topic consists of a comparatively small number of principles, contained in and discoverable from cases, swept the United States, and still largely holds the field, particularly for first-year courses, though modern case-books contain more straight exposition and reference to non-case and even non-legal materials. Also the growth of legislation has gone a considerable way to disturb the premise that the law can be discovered from the cases, and increasingly the need is recognized to see legal problems in their social, political, and economic context, not merely as logical exercises.

About the same time the foundation of local bar associations and of the American Bar Association in 1878 gave an impetus to raise standards, a trend accentuated by the foundation of the Association of American Law Schools in 1905. Also in the twentieth century law study and teaching came to be a distinct branch of the legal profession, whereas previously law teachers had normally been judges or practitioners as well, or even primarily. In 1921, the A.B.A. adopted resolutions that, while admission to the Bar should be in the hands of states, every candidate for admission should have graduated from an approved law school, and in 1923 published its list of approved law schools. In 1931 a National Conference of Bar Examiners was organized.

A further significant development was the Law Reviews, journals edited and partly written by students, containing articles, comments, criticism of cases and book reviews. The *Harvard Law Review* appeared in 1887. Others followed, and university law journals have become a nation-wide academic institution, copied all over the English-speaking world.

Subsequent changes have been the general adoption of a college degree as a prerequisite for law school, the multiplication of courses and the extension of the curriculum to at least three years, and the general adoption of the title of doctor for the first degree in law, as distinct from doctorate for advanced work. Great widening of the scope of legal studies has taken place and legal research, writing, and scholarship have passed almost entirely into the hands of academic lawyers. It is, however, significant that university law schools in the U.S. seek to prepare students for the practice of the law to a much greater degree than do law faculties of universities in the U.K.

A. J. Harno, *Legal Education in the U.S.* (1953); C. Warren, *History of the American Bar* (1911);

C. Warren, *History of the Harvard Law School* (1908); J. Goebel, *The School of Law, Columbia University* (1955); A. Sutherland, *The Law at Harvard* (1967).

Legal estate. In English real property law, the estates, interests, and charges which may subsist or can be created or conveyed at law, as distinct from equitable interests. The only estates in land which are capable of subsisting at law are an estate in fee simple absolute in possession and a term of years absolute; the only interests or charges are an assessment, right, or privilege in or over land; a rent-charge in possession perpetual or for a term of years absolute; a charge by way of legal mortgage; land tax, or similar charge on land not created by an instrument; and certain rights of entry.

Legal ethics. The ethical principles which members of the legal profession are expected to observe in the practice of the law. In some jurisdictions legal professional associations have sought to put the principles into written form, but they may equally exist by general understanding. Opinions given by committees are useful indications of what is or is not deemed ethical. In England the rulings of the General Council of the Bar, in the U.S. the opinions of the committee of the American Bar Association, have been influential. Among major principles are that a lawyer must seek to promote his client's interest, but in doing so must not act dishonestly or fraudulently. He must assist and not mislead a court. He must not seek to represent two conflicting interests. He must maintain strict confidence about all matters communicated to him. In criminal cases, if defending, he is entitled to, and should, insist on the prosecution proving its case beyond reasonable doubt even though he may privately think his client is a worthless scoundrel.

Advertising is in Anglo-American countries considered unethical, as it emphasizes gain rather than service to the client. Fees are frequently regulated by statute or by custom, and it is part of the ethos of the profession to keep them moderate and related to the time, trouble, and difficulty of the work. It was long considered a professional obligation to serve poor clients without reward, but this has been much modified by the introduction of legal aid schemes.

H. Drinker, *Legal Ethics* (1953); M. Orkin, *Legal Ethics* (1957); T. Lund, *Professional Ethics* (1970); W. Boulton, *Guide to Conduct and Etiquette at the Bar of England and Wales* (1970).

Legal executive. In England a person not fully qualified as a solicitor but who has satisfied the requirements of the Institute of Legal Executives and is employed by a solicitor in private practice or as an assistant to a solicitor employed by a public body. The term has replaced the older 'managing clerk'.

Legal history. One of the major divisions of legal science, concerned in its general aspect with the origins and development of legal institutions, systems, principles, and thought about law from the most ancient times. Early on law, like other elements of civilization, was found in the Middle East, in Mesopotamia, and Asia Minor (see ANCIENT LAW, ASSYRIAN LAW, BABYLONIAN LAW, HITTITE LAW).

Rather later there is considerable evidence of the existence of legal ideas and rules in Asia Minor and Palestine (see JEWISH LAW).

Later still, law was well known in Greece and in Greece also there is the first philosophic speculation about the ends and purposes of law, about law and morality, and other theoretical issues (see GREEK LAW).

In Italy the city-state of Rome rose to prominence and power, and long before the commencement of the Christian era had developed a legal system of remarkable maturity, range, and scope (see ROMAN LAW). Roman law reached its climax about A.D. 200. After the move of the capital of the Roman Empire to Constantinople, Roman Law declined for a time until given its final form in the codification of Justinian (q.v.). The decline of the Roman Empire and the irruption of the barbarians into Europe and Africa gave rise to mixed systems of customary and Roman law, from which the legal systems of most Western European countries today are derived (see MEDIAEVAL LAW).

After the Dark Ages Roman law was rediscovered and subjected to commentary and analysis by numerous scholars all over Europe (see GLOSSATORS, COMMENTATORS, HUMANISTS).

In England law evolved largely independently of this development based on Roman law (see ENGLISH LAW).

As the Christian Church grew in power and importance, throughout Europe it developed a body of law dealing both with its own constitution, organization, and spiritual matters and regulating the lives of individuals in many spheres, notably matrimonial relations (see CANON LAW). After the Reformation, in countries adopting reformed doctrines, papal authority was repudiated, but canon law principles remained influential in their systems of ecclesiastical law and parts of their systems of civil law.

The growth of feudalism (q.v.) gave rise to the feudal law, affecting the structure of society and particularly the tenure of land.

Developing commercial intercourse from the mediaeval period also gave rise to customary practices among merchants, frequently of different nationalities, and traders by sea, resulting in the Law Merchant and the Law Maritime (q.v.).

From the early seventeenth century, as the great explorations indicated hitherto unknown and largely uninhabited areas of the world, trickles of colonists left European countries for America, Canada, and the Cape of Good Hope, taking with them the salient principles of English, French, and Dutch law. Later colonists went to Australia and New Zealand and other parts of Africa, while British and French chartered companies contended for India, and in the process acquired sovereignty and had to introduce systems of administering justice.

The eighteenth century was notable for early essays in codification, the break-away from Great Britain of the American colonies and, consequently, the genesis of distinct American law.

The nineteenth century witnessed at its outset the Code Napoleon in France, a development copied repeatedly all over the world, and at its end the German codification, the BGB, also widely copied. In between there were enormous developments in English and Scots law, a rapid growth to maturity of American law, and the development of independent legal systems in many Commonwealth and European countries.

Legal history also includes the study of the historical development of particular legal systems, of particular institutions therein, such as the jury, of particular branches of law, such as of equity or matrimonial law. Such particular legal history is one facet of the general history of the country or community in question, and overlaps on the political, constitutional, social, economic, and other facets of that country's history, because facts primarily political, social, economic, or other contantly react upon and influence legal development and in turn are influenced by it. Legal history is particularly closely connected with constitutional history; in a broad sense legal history may include constitutional history, but more usually constitutional history is concerned with developments in public law and legal history with developments in private law and criminal law.

The study of particular legal history is not only intrinsically interesting but frequently of practical utility, as it is only by examining the origins, course of development, and reasons for particular developments, that the true scope and rationale of a particular rule can be found or understood. (See under particular legal systems.)

The history of jurisprudence and of thinking about law goes back to the Greek philosophers and since then, particularly since the rise of scholastic philosophy in the Middle Ages, there has been a continuous tradition of philosophizing about legal issues, partly by philosophers and partly by jurists (see JURISPRUDENCE).

International law was known in rudimentary form in the Greek states, but emerged in the seventeenth century with the development of the nation-states of Western Europe (see INTERNATIONAL LAW).

Legal institutions. A term of rather uncertain connotation. It sometimes means established and significant elements in a system of law, e.g. marriage, property, inheritance, and courts, as distinct from individual specimens of each, to each of which a large number of specific principles and rules attach, defining how and when an individual instance comes into being or terminates, its characteristics, functions, and attributes, and the legal consequences in various circumstances of its existence. Thus, in relation to property, there are principles and rules defining the kinds of rights of property, what they involve, how they are created, transferred, given in security and redeemed, and extinguished. See e.g. Noyes, *The Institution of Property*.

The term is also used of organizations of people which exist despite changes in the actual personnel, and many of which are important in legal contexts, such as universities, Inns of Court, parliaments, police forces, trade unions, and the like, though these are possibly more correctly social institutions having legal functions, powers, and importance. Legal institutions overlap with and run into social, political, and other institutions. Thus marriage is a social and religious institution as well as a legal one.

Legal literature. Legal literature comprises all books and journals on or about law. Most legal literature is unauthoritative as compared with the authoritative statements of law in codes, statutes, and case-law. But some legal literature is of authority. In international law writings of highly qualified publicists are recognized as potential sources of authoritative rulings. In common law countries a special class of legal literature consists of those older books which are called in England books of authority (q.v.), or in Scotland institutional writings (q.v.), and which are themselves regarded as of authority and sources of principles and rules of law as valid and authoritative as principles and rules laid down by a superior court. But, apart from this special class, legal literature includes encyclopaedias, treatises and textbooks, books of reference, books of forms and precedents, students' manuals, legal journals and periodicals, and other writings on or about law. Such writings are not authoritative sources of law, but only literary sources, or sources of information on law, and statements therein are not authoritative and have no standing in face of legislation or rules laid down by a superior court. But in many cases the views of eminent jurists, such as Anson, Cheshire, or Pollock, have had an influence on courts not far, if at all, short of judicial pronouncements, because they express the considered views of jurists of the highest ability, and seek to synthesize and reconcile the rules laid down in various statutes and cases and to extract the ruling principle from them. Even articles in periodicals have in some cases had persuasive value.

Following the unsystematic character of English law, English legal literature has, since *Blackstone*, taken the form of books on specific branches of law, e.g. contract, tort, trusts, and even of particular topics, e.g. libel and slander, rather than general treatises on public or private law or on obligations or property. In Scots law, on the other hand, reflecting its civil law connections and more systematized character, the major textbooks have dealt with the whole law, or the private law as a whole, though there have also been texts on specific branches.

Works of legal scholarship (*la doctrine*) on the other hand were for long the fundamental source of law in civil law systems; the surviving body of Roman law consists largely of excerpts from a huge body of literature, textbooks, and commentaries, of which Gains' *Institutes* is the best preserved example. In the universities the principles of law emerged between the thirteenth and nineteenth centuries, largely worked out by many scholars in very many books, but subsequently the primacy of legal writing has given way to enacted law. *Doctrine* remains a very important source of law, by establishing the methods by which law is understood and statutes interpreted, by influencing legislatures, by indicating the evolution and tendencies of the law, by criticizing or justifying the law and by giving the courts a guide in the application of enacted law. French jurists have rather preferred to write systematic treatises, German jurists rather to produce annotated codes, and in neither country is there anything like the wealth there is in common law countries of textbooks on specific branches of law.

Legal medicine. See FORENSIC MEDICINE.

Legal memory. In English law the commencement of the reign of Richard I (1189), a date adopted because a statute of 1275 fixed 1189 as the period of limitation for the recovery of land by a writ of right. At common law prescription (q.v.) is based on a presumed grant, which the law assumes to have been made, and an easement (q.v.) to be held established by prescription has to be proved to have been enjoyed from a time whereof the memory of man runneth not to the contrary, i.e. during legal memory. But in practice the courts are willing to presume that enjoyment has lasted from 1189 if there is proof of enjoyment for as far back as witnesses can say and there is nothing making enjoyment since 1189 impossible.

Legal mortgage. In English real property law a mortgage made by way of a demise for a term of years absolute, or a charge by deed expressed to be by way of legal mortgage. A legal mortgage of personal property is a conditional assignment to the mortgagee of the mortgagor's legal interest therein.

Legal order. A term for the complex of persons, institutions, relationships, principles, and rules existing in a particular society viewed from the legal standpoint and considered in their legal functions. The legal order co-exists with social, political, economic, religious, and other orders; it is society viewed as legally organized. The legal order accordingly consists of persons considered as spouses, parents, tax-payers, employers, tenants, bankrupts, and criminals, of institutions considered as corporations, unincorporated associations, trusts, or other, as having powers and immunities, duties and liabilities, or relationships as being husband and wife, parent and child, employer and employee, criminal and victim, and so on, and of principles and rules as norms for conduct. The phase is used by some jurists in different senses, as synonymous with institution, or with legal system or even with law.

Legal philosophy (or philosophy of law or philosophy of legal science). Phrases sometimes used as synonyms for jurisprudence (q.v.) (in the narrower sense of that word) and sometimes for a branch of jurisprudence (in the narrower sense), namely the topics considered in an examination of law from a philosophic standpoint or by the application of philosophic methods to legal problems, such as the nature and definition of law, the relations of law and morality, of law, society and the State, the ends to be attained by law, obedience to law, the analysis of legal concepts and words, the nature and validity of legal reasoning, and the like. Legal philosophy is necessarily intimately connected with and overlaps on the spheres of social, moral, and political philosophy, and some of the questions dealt with are the same though possibly looked at from a slightly different standpoint. A philosophy of law necessarily reflects the particular scholar's philosophic standpoint, idealist, utilitarian, or other, and is part of a general philosophy. Philosophies of law have sometimes attempted to give a rational account of the law of a particular time and place, or to formulate a general theory of law to meet the needs of a particular period of legal development.

From the earliest speculations of a philosophic character in ancient Greece, philosophers have expressed views on legal issues and problems, on the nature and characteristics of justice and law. Among the Greeks Plato and Aristotle made notable contributions. Aristotle distinguished the dual character of man as part of nature and also master of it through his capacity to will freely; he distinguished also between distributive and corrective justice, between legal and natural justice, or positive and natural law, and between law and equity. Later the Stoics saw the highest good in living according to nature, and the greatest virtues in justice, magnanimity, prudence and temperance.

Roman law gave rise to many great practical lawyers but to no legal philosopher. The concept of natural law adopted from the Greeks was applied to widen the applicability of Roman legal principles. Cicero was the leading figure in Roman legal philosophy as the intermediary who made accessible, and adapted to Roman law, the traditional philosophies of Greece, particularly that of the Stoics. As the Roman Empire declined Christian ethics became influential.

As Europe emerged from the Dark Ages the dominance of the Roman Church determined the philosophy of law. St. Augustine saw in 'Pax' the regulating principle which brought sacred order to every organism, including the State. Tertullian, Origen, and Cyprian also made contributions to legal thinking. The legal philosophy of the Middle Ages culminated in St. Thomas Aquinas who postulated a supreme reason derived from divine reason, and expressed in the principle of temporal rule; natural law developed from the absolute rule of God, and positive law carried out in detail the principles of natural law. Positive law comprised Roman law, canon law, local law, and customary law. Law was an ordering of reason for the common good promulgated by him who had care of the community. Justice was one of the cardinal virtues and gave rise to particular obligations. Different trends were, however, seen in Dante's De Monarchia, William of Occam, and particularly Marsilius of Padua who advocated the sovereignty of the people.

The Renaissance and Reformation turned legal philosophy in new directions. Bodin defined law as the command of the sovereign and analysed the concept of sovereignty, and Althusius the State and society, seeing in sovereignty the foundation of public authority and of the coercive nature of law. Grotius separated natural law from its religious basis and founded an independent and purely rationalistic system of natural law as a basis for government and law; the State is a human institution and law a human creation arising from the rational nature of man. Natural law again assumed a constructive and practical function in giving rise to the principles of international law. Pufendorf, Thomasius, Wolff, and Selden all reasserted the supremacy of the law of nature as the basis of their general theory, and in particular as the basis of the law of nations. Hobbes saw in the need for protection and security the basis for the State and law; the State must provide security by keeping faith. Pufendorf saw concern for the future rather than fear as the justification for the State; the State was the result of a series of contracts. Hobbes denied any authority to natural law; law was dependent on sanction, and all real law was that commanded and enforced by the sovereign. Hobbes is accordingly positivist and utilitarian.

In England Hooker reinterpreted the theories of Aquinas to fit them to Anglicanism and made the foundation on which Sir Thomas Smith based his interpretation of the Elizabethan political and governmental systems.

Thomasius completed the secularization of legal philosophy; natural law was founded on admonitions proceeding from God as father and Counsellor.

Spinoza contributed to philosophy the concept of pantheism which led to a determinist position, but one nearer to Locke than to Hobbes.

Locke based his philosophy of law on an empirical and not on a rationalistic basis. God gave man the world as a common right; men created states by agreement to secure by law the common welfare and protect their freedom and property. The authority of government is derived from the people; they make the laws and are the supreme power; the monarch is only the executive. The idea of the social contract goes back to Marsilius of Padua and Occam. Locke's philosophy was influential in developing the English constitutional state. Later Bentham and Mill supplemented and modified Locke's views, seeing in the maximum summation of happiness the supreme object of policy and law. Austin developed the legal basis of utilitarianism: a strictly imperative theory of law, law as the commands of the sovereign, and law and morals clearly distinguished.

Rousseau's social contract theory required that the law serve common interests and promote the common will; by a social contract men unite to secure their rights of freedom and equality through the State. His views have been much more important in politics than in law.

Montesquieu's major contribution was insistence on division of authority of the State into legislative, executive, and judicial functions, a doctrine based on the view of the State as a protection against tyranny and one of great subsequent practical importance. Though he adhered generally to the law of nature doctrine, law must be influenced in his view by environment and the conditions of the people to whom it applied.

In Germany Leibnitz founded his legal philosophy on Roman law and associated ethics intimately with law. His influence owes much to the form given it by Wolff in his work on natural law; he saw natural law as arising on an ethical basis and particularly from the sense of duty, and the major moral principle as self-perfection.

Hume's demolition of the basis of natural law opened the way to an era of anti-metaphysical, utilitarian, and scientific law; though natural law was revived and restated by Ahrens, Krause, Lorimer, and others, and more recently by Geny and Del Vecchio.

The beginning of the nineteenth century saw in Germany the rise of metaphysical legal philosophy. Kant's philosophy of law was influenced by Thomasius, Leibnitz, and Wolff. He sought to define morals so widely as to include the concept of law; good will or intent, which alone might be considered unreservedly good, was expressed in the concept of duty; loyalty to the law was the only proper motive for the will to act upon, and law was the aggregate of conditions under which the wills of individuals might be reconciled under a general inclusive law of freedom.

Fichte reasserted the Kantian concept of the constitutional state as the embodiment of law, but later moved to the concept of the State as having a mission of culture. Under the influence of Rousseau he made human freedom central to law; the law should be so formulated that each member of society restrict his individual freedom by the exercise of an inner freedom so that all may be free; a law-abiding community depended on mutual restraint of freedom. Schopenhauer also took Kant as his starting point. Hegel found the justification of government and law in that they were the absolute demands of the practical reason. The source of the law is the will; a legal system was a realm of realized freedom. The State was the supreme expression of morality. Hegel's work was followed by many thinkers in Germany, Krause, Gans, Stahl, Trendelenburg, and Lasson.

Schelling was important as supplying the foundation for the historical school and completing the break with natural law thinking. The historical school was founded by Hugo, who rejected natural law and foreshadowed the organic conception of the genesis of law and government, a view developed by Savigny, Puchta, Niebuhr, and Eichhorn. Jhering may be regarded as the last of the historical school; he sought to show how a concrete national spirit was realized in the legal institutions of a people.

The development of socialism and communism in the nineteenth century also included views on the nature of law. The forerunners were Saint-Simon, Fourier, Proudhon, Louis Blanc, and the founder Karl Marx, supported by Engels and followed by many lesser thinkers. Law is not oriented to the idea of justice but is a means of dominance, and a tool of the exploiters who use it in the interests of their class. The consequences of their views are a great increase in the power of the State and in public law, modifications of private law in the interest of the socially and economically depressed. In Britain socialism combined with Benthamite utilitarianism to effect enormous practical changes and reforms in the legal system.

Sociological thinking originated with Ferguson and Comte; on Comte's view the concept of society absorbed both law and government. In Herbert Spencer's sociology more was owed to Darwinian evolution, but it also absorbed law and government. More recently this line of thinking has been adopted by Ehrlich, Max Weber, and others.

Broader historical and comparative knowledge of law gave rise to ethnological study of law in the work of Bachofen, Kohler, Post, and others, including Maine and Morgan. The approaches of Kant and Hegel were revived by such as Stammler and Kohler respectively. Stammler saw law as one factor in, and the necessary condition of, social life and sought to develop the idea of a right law in contrast to merely positive law. The right law is just law, but not natural law nor law derived from ethics or morality.

In England the positivist approach, concerned with the analysis of positive law, without regard to ethical content, rather than with the nature of law, began with John Austin, who was followed by Markby, Holland, Salmond, Pollock, and others, and in America by Gray, and in English speaking countries remains the predominant attitude to law. Positivist in quite a different way, but equally opposed to all natural law, is the pure theory of law developed by Kelsen, under which the validity of every legal norm is tested until one gets back to the basic norm of the particular legal system.

In the twentieth century the most significant developments have been the sociological jurisprudence of Pound, Stone, and others, who view law as social engineering, the reconciliation of conflicting claims and interests, and Kelsen's pure theory of law. But there has been a revival of natural law thinking represented by such jurists as del Vecchio, Geny, and Krabbe.

In the U.S.A. realism represented by Holmes, Frank, Arnold, Llewellyn, and others takes a realistic, almost cynical, view of law; the law is how judges, jailers, and others deal with disputes.

Legal philosophy based on Marxism-Leninism has flourished in Russia and Eastern Europe.

Throughout these centuries the philosophic problems of law have been constant though the answers given have been very varied; the nature of law, the end to be sought by law, its relations with custom, morals, religion and political authority, its place and function in society, the problems of its authority and validity, of obedience to it, its form and modes of expression, the analysis and classification of its precepts, and similar issues have all been examined from many standpoints. No finality, no answers satisfying everyone, have been reached nor ever will be.

M. Hamburger, *Awakening of Western Legal Thought* (1942); F. Berolzheimer, *The World's Legal Philosophies* (1912); H. Cairns, *Legal Philosophy from*

Plato to Hegel (1949); C. Friedrich, *Philosophy of Law in Historical Perspective* (1958); J. W. Jones, *Historical Introduction to the Theory of Law* (1940); R. Pound, *Introduction to the Philosophy of Law* (1954); W. Friedmann, *Legal Theory* (1953); P. Sayre (ed.), *Interpretations of Modern Legal Philosophies* (1947); J. Stone, *Human Law and Human Justice* (1965); W. I. Jennings (ed.), *Modern Theories of Law* (1933).

Legal profession. In all countries of the United Kingdom the legal profession has always been divided into 'the Bar' or counsel or barristers (called advocates in Scotland), and solicitors, with distinct governing bodies, methods of qualification and training. Traditionally members of the Bar enjoy higher prestige and, formerly, social standing. Counsel are divided into Queen's Counsel or senior or leading counsel, and junior counsel, who have not attained that rank. The functions of counsel are advice on law, the preparation of pleadings, and the representation of litigants in court. Members of the Bar have a monopoly of appearing in the superior courts and, partly by law and partly by custom, appointments to the bench of the superior courts are made from the Bar. They are concentrated in London and the major cities and may not practise in partnership, though they may share sets of chambers. The functions of solicitors (formerly called attorneys, proctors, or, in Scotland, writers or law agents) are advice and guidance on business, property dealings and family affairs, and possibly some appearance in inferior courts. Solicitors live and practise in every town up and down the country, frequently in partnership. Within each branch of the profession there is frequently some specialization, in common law, or chancery, or criminal, or commercial work at the Bar, in conveyancing or probate and trust work or company and commercial work in solicitors' chambers. Many members of both branches of the profession are employed by government departments, local authorities, companies, and the like as legal advisers and administrators.

In the United States the legal profession is organized by states, each state having its own prescription for admission to practice. The profession is undivided, though in the larger towns many lawyers in fact do only court work ('trial lawyers') or chamber work.

There are two systems of courts, federal and state, each supreme in their own jurisdiction. Judicial appointments are made sometimes by election, sometimes by appointment by a state governor. The Chief Justice and Justices of the Supreme Court are appointed by the President, subject to confirmation by the Senate. All appointments are made from practising lawyers.

In France there were avocats, whose function is oral argument in court but whose services are not required save before the *Cour de Cassation*, the *Conseil d'Etat*, the criminal courts, and avoués, limited in numbers, appointed by the State and having a monopoly in certain matters, who act as agents of litigants and are concerned with procedural details and represent litigants in civil courts and lesser appellate courts but not in oral argument, though counselling on legal matters is often done by persons not members of any regulated legal profession. These classes have largely been merged. Notaires are trained lawyers authorized to draw up and record antenuptial agreements, notarial wills, and authenticate writings which must be formally executed, and are commonly employed to draft contracts of sale of real property, and in administration of property in deceased persons' estates.

The judiciary is a civil service recruited from young law graduates, appointed to subordinate courts and possibly subsequently promoted to superior judgeships. The judiciary accordingly have affinity with the civil service and academic lawyers rather than with the practising bar. Conversely, whereas in the United Kingdom and the United States appointment to the bench is the crown of successful barrister's career, the French lawyer does not and cannot look for such appointment.

In Germany the legal profession is undivided, the organization, professional outlook, and character of the legal profession being akin to that of the British solicitor rather than to members of the Bar. The civil service is largely staffed by persons trained in law, and the higher-ranking civil servants and professors in the universities enjoy the prestige and position which in Britain attach to judges and senior members of the Bar.

Judges in Germany are civil servants and are rarely appointed from the Bar. Interlocked with the judicial service is the public prosecution service, and members of the one not infrequently transfer to the other. Both come under the Minister of Justice of the Land.

R. Bonner and T. Burgess, *Lawyers and Litigants in Ancient Athens* (1927); F. Schulz, *History of Roman Legal Science* (1946); H. Cohen, *History of the English Bar and Attornatus to 1450* (1929); M. Birks, *Gentlemen of the Law* (1960); R. E. Megarry, *Lawyer and Litigant in England* (1962); B. Abel-Smith and R. Stevens, *Lawyers and the Courts* (1967); Q. Johnstone and D. Hopson, *Lawyers and their Work* (1967); A. Weissler, *Geschichte der Rechtsanwaltschaft* (1967).

Legal research. Systematic investigation of problems of and connected with law. Research may be pursued to obtain better knowledge and understanding of any problem of legal philosophy, legal history, comparative study of law, or any system of positive law, international or municipal. It is an essential of text-writing and teaching, to ascertain the correct formulation of, the qualifications and limitations on, the principles or rules on any topic, e.g. the rules of vicarious liability in tort, and in modern times it is extensively pursued as a preliminary to law reform. Such research frequently involves an essentially sociological investigation into the actual operation of particular institutions, e.g. courts or juries, or particular bodies or principles of law, e.g. laws controlling the sale of liquor, and is necessarily inter-disciplinary.

Research is pursued in every major law school and also in special institutes, frequently connected with law schools, such as Harvard's Civil Liberties Research Bureau, New York University's Institute of Judicial Administration, Columbia's Project for Effective Justice, Birmingham's Institute of Judicial Administration, or Cambridge's Institute of Criminal Sciences. The Institute of Advanced Legal Studies in London assists and promotes much individual research. In recent years Royal Commission investigations into the operation of legal institutions and rules, and law-reform bodies such as the Law Commissions, have conducted and commissioned research.

Research in limited areas, though sometimes deep and exhaustive, is also frequently necessary in the preparation of advice or opinions, of arguments for submission to courts, and by judges in preparing their judgments.

Legal rights. In Scots law claims which the surviving spouse and/or children of a deceased have to shares of his or her estate, whether the deceased left a will or not. Prior to 1964 a widower had a claim to courtesy (q.v.) from heritable estate and *jus relicti* (q.v.) from moveable estate, a widow to terce (q.v.) from heritable estate and *jus relictae* (q.v.) from moveable estate. Children, and since 1968 more remote descendants, have a claim to *legitim* (q.v.) from each parent's moveable estate. Legal rights cannot be excluded by disinheriting, nor by leaving all the estate to others, but can be by discharge by the person entitled, or by satisfaction, i.e. making a testamentary provision in lieu of legal rights. A will ignoring legal rights is valid but challengeable to the extent of the claims. Prior to 1964 legal rights were also exigible from the estate of a divorced spouse as if he or she were dead.

Legal scholarship. Systematic research into and thinking and writing about any division or subdivision of legal science. It is mainly the function of the legal scholar or jurist, frequently in a university, but valuable contributions to legal scholarship have been made by philosophers, historians, and scholars in other disciplines, and by judges and practising lawyers. The work may be pursued from any one or more of at least four standpoints; philosophical, historical, comparative, and analytical or expository, and may be directed to any body or system of law. Its purpose may be highly theoretical or severely practical, to elucidate some abstract matter or to reduce to order and make understandable and usable the prescriptions of a particular statute.

This activity is sometimes called 'legal science', though that phrase seems more appropriate for the total body of knowledge, understanding of which is advanced by legal scholarship.

History of legal scholarship

Among ancient and primitive communities there may be, and frequently undoubtedly are, bodies of law, but nothing that can be recognized as legal science. The Greeks are the earliest people who made any contribution to legal scholarship, and that was mainly by philosophers. We know of law givers, such as Draco and Solon, orators, courts and juries, and lawyers. The nature of justice, and natural law were extensively discussed. Aristotle made a beginning of comparative law with his collection of constitutions. Plato in his *Republic* and *Laws* considered what legal regulations were necessary to a well-ordered state. But there were no great legal scholars or jurists systematically examining and expounding law; legislation was empirical and judicial work that of amateurs. The only known jurists were Theophrastus, successor of Aristotle, whose Nomoi appears to have been a systematic review and comparison of the institutions and rules in the different Greek cities, and Demetrius of Phalerum, who was supervisor of Athens 317–307 B.C. and who wrote books on law and legislation inspired by Plato's *Laws*.

Roman legal scholarship

The earliest private law jurists were the pontiffs, who gave legal opinions and supplied formulae for use in legal actions. In the last two centuries B.C. there developed lay jurists practising as consultants in private law, giving responses, drafting wills, contracts and formulae for actions, and advising lay magistrates. Roman jurisprudence was influenced by Greece and distinctions and classifications adopted from Greek dialectical philosophers were applied to law. But Roman legal science was predominantly practical and applied. The evolution of the praetorian and aedilician edicts were products of jurists' work. In this period law, especially private law, was first expounded in books. Among the most notable jurists were Granius Flaccus, M. Manilius, and, particularly, Q. Mucius Scaevola.

In the classical age the law reached full maturity; the jurisconsults not only drafted contracts and wills but also discussed more theoretical problems in teaching and writing. Legal development by the growth of new formulae ceased. Responsa were still given and some jurists were authorized to give responsa by the emperor's permission. Some still served as advisers to magistrates. Legislation and imperial edicts were common. There were numerous notable jurists, Labeo, Capito, Cocceius Nerva, Massurius Sabinus, Proculus, Gaius, Javolenus, Salvius Julianus, Papinianus, Paulus, Ulpianus, Modestinus, and others. The classical jurists continued to apply the dialectical method, to draw distinctions, and formulate generalizations in their commentaries. Two schools of jurists, Proculians and Sabinians, are known. But there remained large blanks in legal philosophy and history, and a lack of monographs on specific branches of law.

In the post-classical period, down to the end of the Empire in the West, legal scholarship declined as bureaucratic government became predominant. The chief jurists were the members of the imperial council and chancery who framed the statutes and rescripts, and the *Codices Gregorianus*, *Hermogenianus*, and *Theodosianus*. Academic jurists, law teachers at Rome and Berytus, now became important.

There was a strong tendency to rely on the classical sources, particularly in the law schools, and to make adaptations, epitomes, and versions of earlier works. A closed canon of authoritative, standard, juristic works was arrived at, as in the Theodosian Law of Citations of 426, and more clearly by the publication of the *Institutes* and the *Digest*. An interest in Greek dialectic revived, and new juristic categories, rules, and definitions were evolved, but there is an almost complete absence of commentaries from these centuries.

Mediaeval period to 1500

While legal scholarship declined at Rome after A.D. 476, it remained at a higher level at Byzantium. Justinian's great restatement fixed the final form of the Roman law, and was the most notable example of reliance on antiquated but classical authorities. Everything else was uncanonical, apocryphal. The bureaucratic jurists remained the most important class, but academic jurists were also notable. The *Institutes* were an official educational manual and law was a recognized study. Greek influences were again influential on the law, and Christian influence was felt also.

In the west the barbarian laws exhibit a very low level of legal scholarship, the *Edictum Theodorici* a rather higher level, because of its derivation from Roman law. The *Breviarium Alaricianum* (*Lex Romana Visigothorum*) (A.D. 506) was compiled by a commission of jurists.

The Lombard laws, from Rothar's Edict onwards, had form and system, attempt to formulate definitions and develop principles, and this foreshadows the later importance of the school of law at Pavia in the years 850–1050. There were treatises, commentaries, formularies, and other works. Similarly the legislation of Charlemagne evidences a moderately developed standard of legal science. The *Book of the Fiefs* is significant as having been composed by practical lawyers, not bureaucratic legislators. In all these spheres a modest standard of legal learning developed parallel to the law itself with practical treatises, a beginning of legal learning and schools of law, notably that of Pavia in the tenth century.

Legal science took a big step forward from about the year 1000 with the revival of the study of Roman law, particularly at Bologna. The first group of scholars, the Glossators (q.v.), concentrated on textual interpretation and exposition of particular passages and principles. Among them Irnerius and Accursius are the greatest names.

They were succeeded by the school of Commentators (q.v.) who reintroduced dialectal methods, reasoning and discussion. With them are always associated the names of Bartolus and Baldus.

In the fifteenth century the Humanists (q.v.), led by Alciatus, Budaeus, Zasius, and others, sought to set the Roman law in historical perspective, to reconstruct texts, to write general treatises. From Italy humanist methods and the study of Roman law spread all over Europe and everywhere was influential on the native legal system.

The influence of the Christian Church and the importance of the Papacy had grown steady from about the fourth century and as the Church developed it developed its own canon law, which gave rise to parallel series of scholars, compilers, and writers, the outstanding work being Gratian's *Decretum*, the basis of the *Corpus Juris Canonici*, itself the basis of vast erudition and commentaries.

In France customary law prevailed throughout the period and there were collections of the customs such as Beaumanoir, and treatises on the law. Later royal enactment became common, and there was an official compilation of the customs in the fifteenth century. In Spain Roman law added a new layer to earlier laws and resulted in the great *Siete Partidas* in 1245. In Germany law was territorial, manorial, city, feudal, but there were many law books trying to state coherently the law of their town or district. From the fourteenth century a great Reception of Roman law was under way and the study of Roman law developed strongly. In Scandinavia written codes appeared from the twelfth century and kings made and improved laws for their peoples.

In England an indigenous legal system had been developing and good practical lawyers had helped able kings to create a system based on royal writs. From the twelfth century civil and canon law were studied in England, and Glanvill wrote the first treatise on the common law administered in the royal courts. In the following century Bracton compiled his collection of cases and wrote his great treatise on the *Laws of England*. From the fourteenth century a native development of English law was ensured, and a steady flow of practical texts followed. At the end of the mediaeval period appeared

Fortescue's *De Laudibus Legum Angliae* and Littleton's *Tenures*, which summed up the mediaeval development of the land law.

The canon law developed into a complex, widespread, and practically important system, a factor linking all Western countries and relating all of them to the Roman Church. Among the most famous canonists were Rufinus and Huguccius, Tancred, Hostiensis, Durandus, and Panormitanus.

Maritime and commercial law developed from the statutes of the merchants in various ports and trading towns, Amalfi, Trani, Ancona, Venice, and others. The *Consolato del Mare* was first reduced to writing in Barcelona about 1300. The *Rules of Oleron* date from about the eleventh century.

Legal philosophy was much influenced by the scholasticism of St. Thomas Aquinas and by early political philosophers such as Marsilius of Padua.

Early modern period—1500–1800

The modern period was ushered in by the Renaissance, the religious Reformation and the Reception of Roman law in Germany and northern Europe. The nation-states were developing increasingly on independent lines and law also was growing on national lines. In France a great compilation of the customs was made, and a movement towards the reduction of a common customary law of France.

The school of jurists founded by Alciatus, Bude, and Zasius included Cujas, Baudoin, Doneau, Hotoman, and Dumoulin in the sixteenth century, Domat in the seventeenth and Pothier in the eighteenth. Their work and that of the legislators, Colbert and D'Aguesseau, led naturally to codification for which the Revolution provided the occasion and Napoleon the impetus.

In Italy both academic and practical jurists were at work in his period. Faber, Clarus, Farinacius, Straccha, and Gravina were only a few of many who worked in every field of law.

In the Netherlands humanist jurisprudence gave rise to the work of Grotius and Vinnius, and their successors Voet and Noodt.

The same spirit permeated legal thinking and writing in Germany, notably in the works of Carpzov and Conring, Mevius, Thomasius, and Beyer. Large-scale legislation began with the Bambergensis and the Carolina, both criminal codes, and numerous lesser territorial codes, and in the eighteenth century several substantial and important codes were undertaken, such as the Prussian General Code. Notable theoretical jurists were Oldendorp, Althusius, and Wincler.

In Spain headway was made in unifying the law of the peninsula, and a great series of jurists wrote on every branch of law, notably on international law. Such were Suarez and Vitoria.

In England legal science developed increasingly independently. The rise of Parliament and the constitutional struggle of the seventeenth century gave rise to much legislation, and to notable judicial law-making by Lords Nottingham, Mansfield, and others. Among the outstanding treatises were those of Coke, Hale, and Hawkins and of Blackstone, the most influential of all English legal books.

Simultaneously in Scotland another independent system was developing, much influenced by European developments, of which the highlights are the great treatises of Stair and Erskine in the seventeenth and eighteenth centuries.

All over Europe in the eighteenth century there was a movement for codification.

International law dates from the seventeenth century, and this period saw many of the great classical texts written, those of Grotius, Puffendorf, and others.

Legal philosophy was mainly developed by political theorists, reacting against scholasticism, Machiavelli and Bodin and later under the influence of natural law thinking, by such as Gentilis, Vico, Grotius, and Pufendorf, while Leibnitz stood apart from this line with independent thinking about law and its study.

Later modern period—1800 to date

In this recent period legal scholarship has been increasingly national. Roman law has had its practical influence, and is now almost entirely a matter of academic study only.

Over much of continental Europe codification, which had made headway in the eighteenth century, and had been given an impetus and a model by the French Code Civil of 1804, was accepted. Most countries, as they achieved independence and unity, made codes for themselves. At the end of the nineteenth century the other great codification, the German Civil Code, appeared. Both have been extensively copied, and though modified in their native countries have not been replaced.

In France the work of the jurists has been chiefly the study, explanation, and exposition of the Codes. In Germany the nineteenth century witnessed the struggle between those urging a code and those resisting it, and also the final development of Roman law, adapted to modern German conditions and expounded in detail. At the end of the nineteenth century codification was finally accomplished and Romanism became a historical factor only. In Spain there were various partial codifications culminating in the Codigo Civil of 1889. In Scandinavia also codes were adopted, though customary law remains important and uniform legislation on various matters has been adopted by the countries.

In England the law remained uncodified though more and more of it came to be incorporated in statutory form. Judicial law making remained very important, particularly by such giants as Lords

Eldon and Blackburn, Willes, J., Jessel M.R., Lords Sumner, Atkin, and Wright, Scrutton L.J., and many others. Increasing influence was exerted by treatises on individual branches of the law, notably the works of Anson, Dicey, Pollock, Chitty, Russell, and, in the twentieth century, Cheshire.

In Scotland codification would have been a natural development but was impossible by reason of its attachment to the English (nominally United Kingdom) legislature. Judicial law making was done notably by Lords Watson, Dunedin, and Macmillan, Lords Presidents Inglis and Cooper. Text-writing was the work of Bell and Hume, and a number of hardly lesser writers, such as Gloag.

Among the Commonwealth countries notable contributions were made by Salmond of New Zealand, and by judges of these countries, both in the Privy Council and in their own countries.

In the U.S. the law developed from English common law, but became increasingly independent as time went on. Judicial fame attaches internationally to Holmes, Brandeis, Cardozo, and others. Kent and Story wrote books of permanent value and importance, and were followed by a host of others, Corbin, Wigmore, and many more. In legal teaching Langdell's case-book method initiated a revolution which has, however, had lesser influence outside America.

International private law has grown rapidly in importance, and been developed almost entirely by judges and text-writers, until the latter part of the twentieth century when international conventions have begun to be of importance in its development also.

Public international law developed enormously and increasingly came to be created and developed by treaties and conventions, and by decisions of international tribunals and courts. The writings of text-writers, though more numerous as time has gone on, have become of diminishing importance as bases for decision and sources of law, as distinct from expositions.

The formation of original groupings of states for defence or economic purposes gave rise in the second half of the twentieth century to supranational law, taking priority in certain spheres over the national law of the member states. The most developed is that of the European Community.

Legal philosophy has always been international and transcended national frontiers. Since 1800 the predominant school in English-speaking countries has been the analytical, concerned with examination of positive law; its classification, concepts, and methods, primarily Anglo-American law. Started by Bentham and Austin, it has been developed by Markby, Holland, Salmond, and others. Variants have been the American sociological school, led by Pound, and the American realist school.

In Europe legal philosophy has always been more abstract, theoretical, and philosophical, and writers of every philosophic school have made contributions to legal thought. The concept of natural law had great vogue from the eighteenth century and has never been wholly extinguished. It provided a basis for public international law from the time of Grotius onwards. In the nineteenth century a strong nationalist historical school developed in Germany and was influential more widely, but later gave way to anthropological and sociological schools of thought.

Methods of legal scholarship

The methods of legal scholarship have varied and still vary very much according to the branch of legal science under consideration or the structure of the legal system being considered and its sources and forms. At least six methods may be distinguished: philosophical, studying the philosophic bases of the institutions and doctrines; historical, investigating the origins and development of institutions, principles, and rules, and trying to discover the reasons for, line and direction of evolution; analytical, examining structure, subject-matter, and doctrines on linguistic and logical bases, seeking to ascertain the most general principles of a legal system; sociological or functional, examining a legal system as a mode of social control and with regard to the ends it seeks to achieve and how it effects its aims; expository or dogmatic, seeking to state the principles and rules actually in force in systematic order, to arrive as generalizations and to determine consistency of rules and decisions one with another; and critical, subjecting the principles and rules actually in force to criticism from many points of view. A broad distinction exists between the Anglo-American systems and the European civil law systems, the former being in general more pragmatic and concerned with details, practical cases and results, and the latter generally more theoretical, logical and concerned with abstract rights and principles. There is no method equally applicable to all countries and subjects of law.

Achievements of legal scholarship

The achievement of legal scholarship has been, within the last 3,000 years, to work out in most parts of the world, and to an increasing extent in the relations between states also, systems of rules for the regulation of human conduct which to a large extent protect individuals in their lives, liberty, and property, and enable them to live harmoniously and to resolve their disputes and conflicts of interests without resort to violence. Legal scholarship has thus been a powerful civilizing factor and its work a prerequisite of peaceable and orderly living in society. At no time and nowhere have these achievements been total or complete, and in too many instances even in the last century, there have been failures. In particular, rules of law are not yet adequately influential in the international society of nation-states.

In countries, mostly of the Anglo-American common law, having uncodified legal systems, the contribution of legal scholarship to the form and development of the law has been immense, particularly by judges developing and restating doctrines, principles, and rules by judicial decisions, which become guides, if not rules, for the future, and by jurists in putting in rational and coherent order the rules of statutes and the generalizations extracted from numerous decisions so as to create more or less systematic statements of the principles of different branches of the law. This has been the contribution, on the judicial side, of such as Mansfield, Nottingham, Eldon, and on the literary side, such as Blackstone, Anson, Pollock, Kent, Story, Bell, and many more.

In countries mostly of the continental civil law family, having today codified legal systems, the contribution of legal science has been mainly in the preparation for and formulation of the codes, and in the writing of the commentaries on the codes and other legislation. Judicial work, though not unimportant, has played a lesser influence.

P. Vinogradoff, *Outlines of Historical Jurisprudence* (1922); G. Calhoun, *Introduction to Greek Legal Science* (1944); M. Hamburger, *Awakening of Western Legal Thought* (1942); H. F. Jolowiz and J. Nicholas, *Historical Introduction to the Study of Roman Law* (1974); F. Schulz, *History of Roman Legal Science* (1946); P. Vinogradoff, *Roman Law in Mediaeval Europe* (1929); *General Survey of Continental Legal History* (1912); *Great Jurists of the World* (1912); F. C. Savigny, *Geschichte des römischen Rechts im Mittelalter* (1815–31); J. Brissaud, *Manuel d'histoire du droit français* (1898–1904); F. Olivier-Martin, *Histoire du droit français* (1948); P. Hubner, *Institutionen des deutschen Privatrechts*; W. S. Holdsworth, *History of English Law*; *Sources and Literature of English Law* (1925); Stair Society, *Introduction to Scottish Legal History* (1958); F. Berolzheimer, *World's Legal Philosophies* (1912); C. J. Friedrich, *Philosophy of Law in Historical Perspective* (1958); H. Cairns, *Legal Philosophy from Plato to Legel* (1949); T. A. Walker, *History of the Law of Nations* (1899); G. Butler and S. Maccoby, *Development of International Law* (1928).

Legal science (or science of law, or jurisprudence (q.v.) in the broader sense of that word). Systematized and organized knowledge, from every standpoint, philosophical, historical, comparative, expository, critical, or other, of and about law in the widest sense of that word, its development, transmission, making, exposition, and application. Like every other science it has two main branches, pure or theoretical legal science, devoted to the investigation of relevant materials and the attainment and refinement of knowledge and understanding of the subjects investigated; and applied legal science, concerned with the ascertainment of the principles and rules relevant to given problems, their applications, and the results. The former branch is the sphere of the legal scholar or jurist, the latter the sphere of the legislator, judge, and legal practitioner. By reason of the distinct character of legal institutions, concepts and rules in different legal systems, work in both branches tends to be particular and related to individual legal systems rather than be general or universal. No man could know adequately more than a few of the legal systems which have existed or now exist in the world, not least because all living ones are constantly evolving and being changed both by external forces and consciously by legal reform. The actual concepts, principles, and rules of any individual legal system, and the conditions for which it has been developed and in which it operates, are the main subject-matter of legal science; the data to be studied, systematized, criticized, expounded or used, applied or otherwise handled.

Pure or theoretical legal science has close connections with philosophy, history, and other social sciences, applied legal science with government and politics, administration, business, finance, industry and commerce, social work and welfare, and every kind of human and inter-personal activity and relationship.

The subject matter of legal science may be divided into seven divisions, namely: theory and philosophy of law; history of law and of individual systems; comparative studies; international law; supra-national law; the State, national or municipal law of every locality having a recognizably distinct system (comprising legislation, courts and profession, and bodies of substantive and adjective rules); and subjects ancillary to law, such as forensic science, medicine, and psychiatry. Each of these may be sub-divided under many heads and sub-heads, public law, criminal law, private law, and other, so that the subject of a particular study or book or articles may be a minute element of the whole, greater, corpus of knowledge of and about law. Its value may be permanent or ephemeral. Of these divisions the first three are theoretical branches, the next three are applied branches, and the last comprises learning not itself legal but relevant to problems raising legal issues.

The term 'legal science' may also be limited to systematic thinking and writing about law, as distinct from law making and application of law to practical problems, what might be better described as legal scholarship (q.v.).

Legal standards. The verbal formulation of principles and rules of law makes much use of standards by reference to which contravention of the principles or rules is to be judged. The standard may be fixed, e.g. not to drive at more than a stated speed; or may be flexible, e.g. to show the care of a reasonable man in the circumstances, or a prudent trustee, and the use of the flexible standard has been one of the major ways in which legislation and case-law have stated principles and rules in ways not rigid and unbending, but adaptable and elastic, capable of taking account of new circumstances and allowing courts to make allowance for exceptional cases.

Legal system. In a theoretical sense the phrase is applied to the set of all the laws enacted directly or indirectly by one sovereign, or by the exercise of powers conferred, directly or indirectly by one basic norm, for one society. It is the totality of the laws of a state or community.

The phrase is more commonly used qualified by an adjective, e.g. English Legal System, in which uses it designates the complex of institutions (legislative, executive and judicial, courts, judges, administrators, etc.), bodies of principles and rules (constitutional law, criminal law, law of property, etc.), ideas, theories, methods, procedures, techniques, traditions, and practices which collectively go to make up a distinct organization, existing to apply law to the problems of a distinct society or state. In this sense it is wider than "law" in that it includes institutions, traditions, persons, techniques and practices, and other elements which are questionably part of the "law", though not among the precepts for conduct. In this sense every state has and, within some states, provinces, or distinct divisions of the State, have, a distinct legal system. Every legal system probably must include a fundamental rule for identification of the other rules of the system and distinguishing rules of law from non-legal norms.

Legal tender. The mode of offering payment of a debt which a creditor is entitled to demand and in which alone a debtor is entitled to make payment. Payment offered in other than legal tender need not be accepted. In the U.K. money of legal tender comprises Bank of England notes, for payment to any amount (but only Bank of England notes for less than £5 are legal tender in Scotland), gold coins, to any amount; cupro-nickel or silver coins of more than 10p, to any amount not exceeding £10; cupro-nickel or silver coins of not more than 10p, to any amount not exceeding £5; bronze coins to any amount not exceeding 20p; other coins, if made current by proclamation, to the amount specified therein. Scottish bank notes are not legal tender in the U.K. but are only promissory notes.

The Crown has power at common law to legitimate foreign money, and statutory power to do so to a limited extent.

Legal Tender cases (1871), 12 Wallace 457. After the American Civil War the federal government was compelled to issue paper money unbacked by gold and to make the notes legal tender. The Constitution gave no power to issue legal tender currency and a divided Supreme Court held the notes unconstitutional in *Hepburn* v. *Griswold* (1870), 8 Wall. 603. Later a differently composed court held that Congress had the power in question, by reason of the necessities of the war.

Legal theory or theory of law or theory of legal science. Phrases sometimes used as generally equivalent to jurisprudence (q.v.) in its narrower sense, or to legal philosophy (q.v.) or philosophy of law, but sometimes distinguished from legal philosophy as being more concerned with structure, classification, and organization of bodies of law in various societies than with the more philosophic problems of law.

Legatee. The person to whom a legacy is payable.

Legates. In the Roman Republic senators or experienced officers chosen by provincial governors and generals to act as staff officers and latterly as commanders of individual legions. Under the principate each legion was normally commanded by a *legatus*. *Legati Augusti pro praetore* were officials appointed by the emperor to undertake special missions or tasks.

Legates, Papal. Clerics sent by the Pope to treat on his behalf of church affairs. They may or may not have ecclesiastical jurisdiction, and may or may not have diplomatic character. The practice of sending legates to councils and provinces of the Roman Church is known from the fourth century. Diplomatic legates, *legati missi*, became common from the eleventh century, particularly when cardinals were given greater power and sent as *legati a latere*. There were also *legati nati*, persons giving a jurisdictional role as incumbents of an important see, such as Canterbury.

Legatine constitutions. Ecclesiastical laws enacted in national synods held by the Cardinals Otho and Othobon, papal legates to England, about 1220 and 1268, which are material parts of the law of the Church in England.

Legation. A diplomatic mission of the second rank, less important than an embassy. An ambassador's residence is commonly called a legation.

Leges barbarorum. The bodies of law of the main Germanic peoples who entered the Roman Empire from the fifth century onwards, such as the Visigoths, Ostrogoths, Burgundians, Franks, Saxons, Frisians, and others. Though sometimes described as codes or codifications they were compilations of existing tribal custom, not legislation, and were the laws of tribes or peoples, not of territories. They were not complete nor systematic statements of law but statements of agreed custom on various matters, notably compensation for homicide, assault, and theft of cattle. Their dates range from the latter fifth century to the ninth century and are in Latin, save that the laws of the Anglo-Saxons are in their vernacular. There are three major groups, the laws of the Visigoths, Burgundians, Alamannians, and Bavarians; those of the Salic, Ribuarian, and Chamavian Franks; and those of the Frisians, Saxons, Anglo-Saxons, and Langobards.

Leges Edwardi Confessoris. A compilation in Latin, of about 1130–35, probably by a secular clerk of French origin, made in the interest of the churches, giving a very unreliable account of English law before the Conquest and containing many invented stories.

Leges Henrici Primi. An anonymous work, possibly by the author of the *Quadripartitus* (q.v.), probably written between 1100 and 1118 and the earliest legal textbook of mediaeval Europe. It aimed to give an account of Anglo-Saxon law as amended by William the Conqueror and Henry I. It draws on the *Quadripartitus*, the *Lex Salica*, and Frankish capitularies and other sources, and Latin and Anglo-Saxon proverbs. Its statements of English law are probably substantially correct.

L. J. Downer, *Leges Henrici Primi.*

Leges Langobardorum. The laws of the Lombards, comprising the Edict of Rothair (676–656), the supplements of Luitprand (712–44) and some additional laws issued by Grunwald, Rachis, and Aistulf. They show that the law of the Lombards was basically Germanic, relying on wergelds and compurgation, but, particularly in public law and property law, influenced by Roman law.

Leges Marchiarum. An ancient code of law, said to have been written down by 12 English knights and 12 Scottish stating the laws and customs of the marches or border between England and Scotland and decisive of all actions between Englishmen and Scotsmen. They were administered at meetings held by Borderers of both countries held at fixed places and times.

Leges Quatuor Burgorum. The earliest collected body of law in Scotland of which records survive, dating from the thirteenth century. It is associated with the burghs of Edinburgh, Stirling, Roxburgh, and Berwick and resembles English codes. It contains regulations for control of burghal life and some rules of general law and practice, and casts light on early law of heritable and moveable rights, succession and the law merchant. It appears to have been recognized by the courts as the equivalent of statute law.

Leges Romanae Barbarorum. The general terms for the law codes issued by Germanic kings, from the mid-fifth century, for their subjects of Roman origin. They comprise the *Lex Romana Burgundionum*, *Lex Romana Curiensis*, and *Lex Romana Visigothorum.* Initially these codes regulated the Roman subjects, with a distinct Code for the Germanic subjects, but later they tended to coalesce. The Roman law is vulgar law, primitive and rough as compared with classical Roman law. These bodies of law were drafted by clergy and laymen with some knowledge of Roman law but influenced by the royal chancery.

Leges Visigothorum. The second Visigothic code (the first being the *Antiqua*) containing the laws of Euric, Leovigild, and Recared (fifth–sixth centuries), necessitated by the increasing power of the Church and the increasing fusion of the Roman and Germanic populations of the Visigothic kingdoms. It consists of a fusion of parts of the *Antiqua* with some rules drawn from the *Lex Romana Visigothorum* (*Breviarium Alaricianum*), which was abolished. It was strongly influenced by Roman law, including Justinian's codification. This was effected during the reigns of Chindaswinth (A.D. 641–52) and Receswinth (A.D. 649–72). It was promulgated in A.D. 654 and made binding on Roman and Visigothic subjects alike throughout the Visigothic Kingdom of Spain and Southern France.

Legis actio. In Roman law the older of the two systems of proceedings *in iure*, before the magistrate (as distinct from the trial stage, *apud iudicem*). *Legis actio* was solemn procedure and rigidly formal in character, in which the plaintiff (*actor*) and defendant (*reus*) made their assertions orally in forms settled by law and custom. There were five types of *legis actiones*; (1) *legis actio per sacramentum*, applicable to claims originating from obligations and claims to ownership, constituted in a kind of wager between the parties who each deposited a sum of money with the court. The successful party got his deposit back and the unsuccessful party's deposit was forfeited to the State. (2) *per iudicis postulationem*, for claims based on *stipulatio* or disputes about division of property among several persons. (3) *per condictionem*, for recovery of *certa pecunia* or *certa res*; neither of these required *sacramentum*. (4) *per manus iniectionem*, a method of execution against the person of a debtor; and (5) *per pignoris capionem*, a method of execution against a debtor's property. The *legis actio* system was superseded by the formulary system (q.v.) officially introduced by the *Lex Aebutia* (after 150 B.C.) but obligatory only from the time of Augustus. In trials assigned to the centumviral court, however, procedure by *legis actio* continued unaffected.

Legislation. The process of deliberate law-making or law-changing by an expression of the will of a person or body recognized by a particular legal system as having power and authority validly to declare law. The term is also used of the products of the legislative process, of the body of rules laid down thereby. In this sense it is equivalent to statute-law.

By no means everything that passes for legislation in ancient law is truly legislative, being frequently merely written formulation of accepted custom, but the possibility of a sovereign person or body consciously laying down law seems to have been realized early.

At Rome legislation has been attributed to the early kings and the Twelve Tables were legislation, while during the Republic *leges* passed by the *Comitia* and *plebiscita* passed by the *concilium plebis* were common, though more in the field of public law than that of private law. The practice of popular legislation disappeared by the end of the first century A.D. The powers of the magistrates to issue *edicta*, very important in their development of the *jus honorarium*, may be considered a kind of delegated legislative power. This came to an end when Salvius Julianus settled the Edict in about 130 A.D. From the beginning of the Empire *senatus consulta* at least indirectly altering the law became common, and by the middle of the second century it was recognized that the Emperor made law by issuing *edicta*, *decreta*, and *epistulae*. This was more clearly recognized later, whatever the form in which the imperial will was signified. A large amount of legislation was made by Justinian, notably in his Fifty Decisions, and later in his Novels.

In the West the Carolingian emperors produced a considerable amount of legislation called capitularies, but thereafter the idea of legislation declined.

In the United Kingdom legislation has been known in England since Anglo-Saxon times and in Scotland since at least the twelfth century. Since then it has been continuous, though it was long considered mainly as an appendix of amendments to the common law, and only since the early nineteenth century, largely under Bentham's influence, has it become increasingly important and substantial, in some branches of the law, e.g. taxation, social welfare law, amounting to the major part, or even the whole, of the law on the subject. In some topics, e.g. sale of goods, the existing common law has been largely restated in statutory form.

After Henry I's Charter of Liberties, charters frequently establish new rules and there was an outburst of legislation under Henry II. Down to the time of Edward I legislation came from the King in Council, or from the King in a Parliament of nobles. Parliamentary legislation became more general in the fourteenth century; the King got Parliament to lend authority to his decrees, but Parliament came to request the King to legislate on some matter, by petitioning him. By the fifteenth century, however, the practice was established of Parliament presenting a bill containing the words which it desired to have enacted. In Henry VIII's time legislation was voluminous, and important; it dealt with the most fundamental subjects, the supremacy of the Pope and the headship of the English Church. As early as 1504 there is delegated legislation, and in 1563 there is found a 'consolidating Act'.

A question of some importance was whether statute could validly innovate on what was deemed to be a fundamental law. It was accepted in the Middle Ages that statute could not operate within the Church's sphere of competence. In the seventeenth century, under Coke's influence, there is some support, but only *obiter*, for the view that statute contrary to natural law or fundamental law was void, but this was quite inconsistent with the fundamental changes made at the Reformation, establishing the headship of the Church, a new form of religion, and so on, and by the eighteenth century the modern doctrine, of the unlimited competence of parliament, was recognized.

Problems of interpretation arose early, but in the thirteenth and fourteenth centuries the view was that doubts as to meaning should be settled by the Council, or Parliament, which made the Act and should know what they meant. From the fourteenth century it was accepted that the courts must interpret, but they initially adopted a very liberal approach and did not take the text of the Act too seriously, then in the fifteenth and sixteenth centuries they began to work out rules of construction; the eighteenth century saw the adoption of the principle that preparatory materials and debates in Parliament were inadmissible to influence interpretation.

In modern practice in the U.K. a distinction must be drawn between prerogative legislation, the laying down of law, usually in the form of Orders in Council, by virtue of the royal prerogative, which is now practically confined to law making for colonies, and parliamentary legislation, the laying down of law by Acts of Parliament (q.v.)

A distinction must also be taken between superior or supreme legislation, which includes both of the previous classes, and subordinate or delegated legislation, which is legislation made by persons and bodies to whom restricted powers have been delegated by Parliament, within stated limits, to legislate. Subordinate or delegated legislation takes many forms, rules, regulations, and orders made by Ministers of the Crown, frequently in the form of statutory instruments; by-laws made by local authorities; statutes, ordinances, and regulations made by universities; rules of procedure made by courts, and so on.

As a source of law legislation is commonly contrasted with case-law, *jus scriptum* or written law, as contrasted with *jus non scriptum* or unwritten law.

Legislation as a source of law has long been the primary source in civil-law countries, where the main statement of law is commonly in the form of codes, supplemented by legislation on particular topics. In the U.K. and common law countries on the other hand case-law was for long regarded as the predominant source of law and legislation a source of amendments only. The great volume of legislation of the late nineteenth and twentieth centuries is, however, changing this position and, though codes are uncommon in common law countries, the bulk and practical importance of legislation in practically every area of law is increasing.

In theory legislation is a preferable source of law to case-law, custom, or other source, in that it can lay down clear, easily discoverable, general propositions of law, as contrasted with the difficulty of discovering propositions of law inherent in the *rationes* of various cases. In practice legislation is frequently obscure, difficult to discover by reason of inclusion of dissimilar topics in one statute, and by reason of the accretion of layers of amendments, frequently leaves lacunae and *casus omissi*, frequently is inflexible and unadaptable, and frequently is difficult to understand in its application to a specified case. Legislation does undoubtedly have superior power over all other sources in respect of power to supersede and abrogate existing contrary rules, power to create a body of rules on a matter on which there had been hitherto little or no law, and ability to provide beforehand for circumstances which have not yet arisen.

Though the function of the courts is primarily to apply the law and to decide controversies in accordance with it, it must be admitted that some decisions, and the statements by the judges of the principles on which they acted, have by reason of the doctrines of precedent and *stare decisis*, virtually had the effect of laying down new rules of law. Thus *Rylands* v. *Fletcher* and *Donoghue* v. *Stevenson* (qq.v.) were as much new formulations of law as many statutes. Such substantial reformulations of law have sometimes been called judicial legislation.

Similarly the function of the executive or administrative departments is to administer the law, not to make it, but departments of state not infrequently issue notes, circulars, directions, rulings, and the like, stating what view they take of the law as to certain cases and it is difficult to resist the conclusion that some of these are making law as much as explaining it or stating its application.

Legislation—U.S. The U.S. Constitution I, ss. 6 and 7 and VI restricts the designation 'law' to Acts, joint resolutions of both Houses of Congress, and treaties. Of these, Acts of Congress are the only true legislation.

Legislation, Judicial. The name sometimes given to the fact that, by reason of the doctrine of precedent and the rule of *stare decisis*, a decision by a superior court in a case of a particular kind is in effect laying down a general rule which governs all subsequent cases of the same general kind, and may be making new law just as much as does an Act of Parliament. Thus the House of Lords in laying down the 'rule of common employment' (q.v.) and in deciding *Rylands* v. *Fletcher* (q.v.) in effect legislated.

Legislative function. One of the major functions of government, involving the formulation, making, and promulgation of rules of law, normally of some generality of application, in furtherance of some policy. The legislative function is normally exercised by an assembly, such as Parliament or Congress, but may also, by delegation, be exercised by Ministers of the Crown, courts, local authorities, corporate bodies, and many other persons and bodies. Thus in the U.K. the Queen in Council has a very limited residual legislative function, but the legislative function is mainly exercised by parliament enacting statutes, ministers issuing statutory instruments, local authorities promulgating by-laws, courts making rules of court, and so on. Courts in deciding cases do not exercise a legislative function though the results of decisions may have an effect not dissimilar in changing the law.

Legislature. The part of the supreme governing body in a country which has the function of making and changing law by legislation, as contrasted with the executive, whose function is to execute and apply the law, and the judiciary, whose function is to interpret the law and determine particular cases according to it. It is a part of the constitution of every country to determine what persons compose the legislature or legislative body.

In the United Kingdom the legislature consists of the Queen, the House of Lords, and the House of Commons, but some legislative power in relation to overseas territories may in the exercise of the royal prerogative be exercised by the Queen in Council;

and considerable legislative power is delegated by Parliament to the Queen in Council, Ministers and Department of State, public corporations, local authorities, courts, and other bodies, to make and change subordinate rules of law within the limits of the powers conferred on them.

In the United States the federal legislature consists of the President, the Senate, and the House of Representatives, but considerable legislative power is delegated to the President. In each state the legislature consists of the Governor and the house or houses of the State legislature.

According to the principle of the separation of powers, the legislature should be distinct from the executive and the judiciary, but this is nowhere complete. In the United Kingdom the Prime Minister and all his ministers are members of one or other of the Houses of Parliament and participate fully in the work thereof. In the U.S. the President and his Cabinet are not members of either House (though the Vice-President presides over the Senate), but the President's assent is part of the legislative process.

Legitim. In Scots law *legitima portio* or bairn's part, the right of children and, since 1968, of more remote descendants representing their deceased parent, to share in the moveable estate of either parent of the children. The share is one-third of the estate if the deceased parent also left a spouse, or one half if not. The right exists by virtue of survivance and cannot be excluded by will or otherwise, nor can it be apportioned unequally among the children. It may, however, be extinguished by discharge by the persons entitled, or by satisfaction, i.e. by making a testamentary provision in lieu, in which case the child must elect between the legal right to *legitim* and the testamentary provision.

Legitimacy and legitimation. Legitimacy is the status of a person born to parents who are lawfully married. A child born to a married woman is presumed begotten by her husband and accordingly legitimate though the contrary may be proved. The same presumption arises where the child is born within a possible period of gestation after the death of the husband or dissolution of the marriage. By statute the child of a void or voidable marriage is deemed legitimate.

At common law in England a child born a bastard could not be legitimated, save by statute. The Legitimacy Act, 1926, introduced the rule of legitimation by the subsequent marriage of the natural parents with effect from the date of the Act or of the marriage, whichever is later, provided that each parent was free to marry at the time of the birth. The Legitimacy Act, 1959, removed this last proviso. Persons legitimated by these statutes may acquire rights in property like legitimate children, if no contrary intention is apparent.

In Scotland legitimation *per subsequens matrimonium* was recognized at common law, following Roman and canon law, provided that each parent was free to marry the other at the time of the conception, a proviso abolished in 1968. Persons legitimated are entitled to the same rights as legitimate in respect of deeds operative and successions opening after the Act.

At least in Scotland and so far as the rights of the Crown are concerned bastards may also be legitimated by letters of legitimation from the Crown, and probably also by statute.

Legnano, Giovanni da (Johannes de Lignano) (*c.* 1310–82). Teacher of canon law at Bologna and diplomat. He wrote on theology, moral and political philosophy, canon and civil law, and a *De Civitate Bononiae et de Bello* (1360) also a *Tractatus de Bello, De Represaliis et de Duello*, possibly the first attempt to deal with the international law of war as a whole.

Leibnitz (Leibniz), Gottfried Wilhelm (1646–1716). German philosopher, mathematician, and man of affairs, sprang from a legal family, studied philosophy, mathematics, and law, entered the service of the elector of Mainz, and in 1670 became a member of the Court of Appeal of Mainz. His earliest thinking and writings were on law. He was employed on diplomatic business, visited Paris and London, but in 1676 exchanged into the service of the Brunswick family and was based in Hanover where he had charge of the ducal library. He researched on genealogy and wrote a history of Brunswick. He wrote political pamphlets, devised a calculating machine, and discovered the differential and integral calculus, and in the last twenty-five years of his life composed his chief philosophical works, *Essais de Théodicée sur la bonté de Dieu* (1710) and *La Monadologie* (1714). In 1700 there was founded in Berlin the*Akademie der Wissenschaften*, which he had planned and of which he was made life president. His chief legal works are *Specimen Difficultatis in Jure* (1664); *De Casibus Perplexis* (1667); *De Nova Methodo discendae docendaeque Jurisprudentiae* (1667), his principal work on legal theory, which contains the first clear recognition of the importance of historical method in law, and essayed a reconstruction of the *Corpus Juris*, to adapt it better for introductory legal instruction; *Codex Juris Gentium Diplomaticus* (1693) a collection of treatises and state papers, with an introduction giving his views on international law, and a supplement ('Mantissa') in 1700; *Observationes de Principio Juris*; and *Monita quaedam ad Samuelis Pufendorfii Principia*, a criticism of Pufendorf's work. He developed a theory of law in society, conceiving justice as a communal virtue. He was one of the founders of international law and did much to give form and definiteness to that subject, but this was only one aspect of a man of almost universal genius, more famous for his other achievements.

G. Guhraner, *G. von Leibnitz*; L. Hartmann, *Leibniz als Jurist und Rechtsphilosoph.*

Leigh, Pemberton, Lord Kingsdown (1793–1867). Called to the Bar in 1816 and rapidly became a leader of the Chancery Bar. He was briefly in Parliament but in 1843, having inherited a fortune, left the Bar and Parliament. He was, however, made a Privy Councillor and Chancellor of the Duchy of Cornwall. He accepted a peerage in 1858. In the Privy Council he played a leading part in the reform of its practice, in expediting the hearing of cases, in reducing costs, and in framing many important decisions. Many of his judgments have been deemed both models of judicial expression and standard decisions. He gave his name to Lord Kingsdown's Act (the Wills Act, 1861, now repealed).

Leigh, *Lord Kingsdown's Recollections.*

Leist, Burkard Wilhelm (1819–1906). German jurist. Beginning as a Romanist he went back to the Aryan sources of Roman law in his major works *Graeco-italische Rechtsgeschichte* (1884), *Alt-arisches ius gentium* (1889), and *Alt-arisches ius civile* (1892–96). He also continued Glück's commentary on the Pandects.

Lenel, Otto (1849–1935). German jurist, author of *Das Edictum Perpetuum* (1883), a valuable reconstruction of the Roman Edict, and *Palingenesia juris civilis* (1887).

Lenin, Vladimir Ilyich Ulyanov (1870–1924). Russian statesman, organizer of the Russian Revolution in 1917 and leader of the Soviet republics till his death. He studied law at Kazan and St. Petersburg, organized revolutionary organizations on Marxist lines and was exiled to Siberia. From outside Russia he fomented revolution and when it broke out there in 1917 he returned to Russia and seized power, and over the next few years steered the new state to a position of relative safety. He had an extensive knowledge of Marxism and of relevant literature and wrote extensively. In *The State and Revolution* (1917) he stressed the replacement of the bourgeois state by the proletarian state by revolution, that it would ultimately wither away and that in the interval the proletarian state must be dictatorial. During this period there must still be law, to a certain extent bourgeois law but purely socialist law. In the ultimate communist society the State will disappear, but though there will be justice and equality it is uncertain whether the law will disappear. His views on law remain of importance in Russia because of the elevation of all his thoughts and writings to authoritative pronouncements.

Lenocinium. In Scots law a husband's connivance at his wife's adultery and his participation in the profits of her prostitution, or his lending himself in any way, directly or indirectly, to her dishonour. It is accordingly a defence competent against an action of divorce for adultery at the husband's instance.

Leonhard, Rudolf Karl Georg (1851–1921). German jurist, expositor of the German Civil Code, author of *Der allgemeine Teil des bürgerlichen Gesetzbuchs* (1900) and *Der Irrtum als Ursache nichtiger Vertrage* (1907).

Le Plat, Jodocus (1732–1810). Belgian canonist, author of *Institutionum iurisprudentiae ecclesiasticae* (1780), *Dissertatio de Sponsalibus* (1787), and *Monumentum Concilii Tridentini* (1781–87).

Lerminier, Jean Louis Eugène (1803–57). French public lawyer, author of an analysis of Savigny on Possession, *Philosophie du droit* (1831), *Cours d'histoire de la legislation romaine* (1837), and other works.

Lesbianism. Unnatural sexual practices between women, so-called from the poetess Sappho of Lesbos, who is said to have gathered a group of women about her. It has probably never been criminal in the U.K. unless probably in circumstances affronting public decency, but an imputation of lesbianism is actionable as defamatory.

Lèse majesté. In mediaeval and early modern French law, a crime, borrowed from Roman law, comprising all evil-intentioned deeds directed against the King, his councillors, or gendarmerie, or which injure the State, create public disturbances, or foment conspiracies. It was later distinguished into two degrees, *lèse majesté* proper, comprising attempts on the life of the King or any of his house, and high treason, offences against the royal authority or the public weal. The former involved cruel and dreadful punishments such as being torn asunder by horses.

Lesion. In Scots law the degree of loss, harm, or injury sustained by a minor, or person of weak capacity, necessary to entitle him to challenge the transaction by which he suffered the loss.

Les Leis Williame (or *Willelme*, or *Leges Willelmi Conquestoris*). Probably compiled in Mercia between 1100 and 1120 A.D. by an unknown person. It is preserved in both Latin and French versions, and contains certain rules of old English law as understood under the Norman Kings, a few articles borrowed from Justinian, and a translation of parts of Cnut's Code.

Leslie, Harald, Lord Birsay (1905–). Called to the Scottish Bar in 1937, he took silk in 1949, and became Chairman of the Scottish Land Court in 1965.

Leslie, John, 7th Earl and 1st Duke of Rothes (1630–81). Served Charles II, fought for him at Worcester and was taken prisoner. At the Restoration he was made Lord President of the Council, and later an Extraordinary Lord of Session and a commissioner of the exchequer, and in 1663 Lord High Treasurer. He enforced persecution of the covenanters and in 1667 was deprived of all his offices, but in consolation made Lord Chancellor of Scotland for life. In 1680 he was made Duke of Rothes. He had a reputation for drunkenness and debauchery.

Lessius, Leonard (1554–1623). Belgian theologian and jurist, author of a famous *De justitia et jure* (1645).

L'Estoile (or L'Etoile), Pierre de (1546–1611). French chronicler and pioneer of humanist jurisprudence in France.

Letter missive. Formerly when a peer was defendant in proceedings in the Court of Chancery the Lord Chancellor sent him a letter missive to request his appearance, along with the bill, petition, and order. If he failed to appear, he was sent a copy of the bill and a citation to appear and answer; if he still failed to appear a sequestration nisi was issued against his lands and goods and made absolute in the usual way.

In ecclesiastical law a letter missive is sent with the *congé d'élire* to the dean and chapter of a cathedral; it contains the name of the person whom they are to elect as bishop.

Letter of credit. A letter written by one banker, merchant, or correspondent to another, requiring him to give credit to the bearer or to a person named therein, or to furnish him with goods. If acted on, it constitutes an undertaking by the writer to meet bills drawn under the credit by the beneficiary of the credit in accordance with the conditions laid down therein. If designed to facilitate trade, it is called a commercial letter of credit and is normally addressed to a specified correspondent banker only. Letters of credit are personal and not negotiable, but may be assigned, in the case of a commercial letter of credit only with the consent of the buyer.

Letters close. These were private instructions by kings to individuals, closed and sealed on the outside (unlike letters patent). The Chancery enrolments form the Close Rolls, some of which have been indexed and calendared.

Letters of administration. The authority, granted now by the High Court in the exercise of its probate jurisdiction, to a person to act as an administrator of the estate of a person who has died intestate or without naming an executor, and giving him powers and duties similar to those of an executor. If the deceased died intestate, letters of administration are granted to some of the next of kin, or to a creditor or other person. If the deceased died testate but without appointing an executor or the executor is unable to act, letters of administration are granted with the will annexed (*cum testamento annexo*), a grant similar to probate of a will.

Letters of business. Instruments issued under the Sign Manual addressed by the Crown to the Convocation of the Province of Canterbury or of York, conveying the Crown's desire that Convocation should consider some particular business therein mentioned.

Letters of marque. The authority formerly given by a belligerent state to a private shipowner authorizing him to use his ship as a ship of war or privateer, and to wage sea-warfare on enemy shipping. They might be special, authorizing hostile acts as a means of securing reparation for individuals; or general, when issued as a means of waging warfare. Both are now obsolete.

Letters of safe conduct. A protection granted under the Great Seal and enrolled in Chancery which might be granted to an enemy alien to travel on the high seas, enter the realm, or have his goods transported free from liability to seizure. In modern practice licences from ambassadors abroad are more usually obtained.

Letters patent. Letters, delivered open, containing public directions from kings, and having the Great Seal attached, but differing from charters by their mode of address ('To all to whom these present come') and in being witnessed usually by the King himself (*teste rege* or *teste me ipso*). Copies enrolled in Chancery form the Patent Rolls, some of which have been edited or calendared. They were used for various purposes, conferring powers or privileges on persons or companies which they would not otherwise have had, sometimes for the creation of peerages and sometimes for the granting of precedence to barristers.

Formerly they were used to grant monopoly privileges to the first inventors of new methods of manufacture, but this variety of letters patent is now replaced by patents under seal of the Patent Office controlled by legislation. See PATENT.

Lettres de cachet (or *lettres du petit signet* or *lettres du roi*). Instruments of administration in France from about 1550 till 1790, signed by the King and countersigned by a secretary of state, primarily used to authorize a person's imprisonment. Such a *lettre* commanded the recipient to obey the orders therein without delay and gave no explanation. Some were used for particular police purposes, and some issued at the request of private persons or families for use against other persons. They might be used to maintain public order, to secure a person's detention for a period, or to obtain the arrest of the accused for trial, or to arrest persons suspected of spying, or to secure the functioning of political institutions such as to summon a political assembly, to have it discuss a particular matter or exclude someone from their discussions.

They could be, and were, abused so as to give rise to many complaints, though it appears that they were not issued without inquiry into the justification for the request, particularly if it came from a private person. The effect of a *lettre* was to require imprisonment in a state fortress; this might be for an unspecified period of time. There was no mode of appeal against the issue of a *lettre* and release was in the King's discretion, the whole machinery being a matter reserved to the King. *Lettres de cachet* figured frequently in the lists of grievances presented to the States General in 1789 and they were abolished in 1790.

Letzeburgish Law. See LUXEMBOURG LAW.

Leu, Johann Jacob (1689–1768). Burgomaster of Zurich, author of an important collection of statutory provisions on private law from all portions of Switzerland, arranged systematically, the *Confederate Town and Rural Law* (1724–44). He edited a *Commentaire sur la république de Suisse* and wrote a *Sur les lois des différents cantons suisses* (1727). His biggest work was a *Dictionnaire universel de la Suisse* (1746).

Levari facias. A writ of execution which commanded the sheriff to levy a judgment debt on the lands and goods of the debtor by seizing and selling the goods and receiving the rents and profits of the lands until the debt was satisfied. It was abolished in 1883.

Levinge, Sir Richard (1656–24). Called to the English Bar 1677 but went to Ireland as Solicitor-General in 1690 and sat in the Irish Parliament and became Chief Justice of the Common Pleas in Ireland in 1720, but died in 1724.

Levirate marriage. In ancient Hebrew usage, marriage between a widow and her late husband's brother, if he had died without a son. It applied only where brothers had lived together and worked common property and was intended to prevent loss of family property by the widow remarrying outside the clan. It was a disgrace for the brother not to do so. In some references other relatives had the duty, in default of the brother-in-law, to marry the widow in order of nearness of kinship, but widow and relative could refuse to marry.

Levita, Benedict (9th cent. A.D.). A deacon of the Church at Mainz who about A.D. 850 continued the work of Ansigesus (q.v.) by adding three further books of capitularies. The compilation is, however, a fraud in that the alleged capitularies are false and put together from other sources.

Levitical degrees. The degrees of affinity listed in Leviticus, c. 18, and referred to in older English and Scottish legislation relating to forbidden degrees of marriage. Modern legislation has, however, relaxed some of the prohibitions, as by permitting marrriage with a deceased wife's sister and a deceased husband's brother.

Lex. In Roman law an individual statute, usually known by the name of the magistrate who introduced it. Many are well known under these names; for some of these see following paragraphs, and see Rotondi, *Leges Publicae populi Romani.*

Lex Aebutia (after 150 B.C.). In Roman law a statute which authorized or officially introduced the formulary procedure (q.v.) as an alternative to the older procedure by *legis actio* (q.v.) and thereby greatly enhanced the power of the praetors to allow new forms of action.

Lex Aelia Sentia (A.D. 4). In Roman law a comprehensive statute regulating the manumission of slaves, completing the work of the *Lex Fufia Caninia* (q.v.). The manumitter had to be 20, the slave over 30 or he did not become a *civis*, and manumission in fraud of creditors was void.

Lex Alamannorum. The second part (the first being called the *Pactus Alamannorum*) of the law of the Alamanni, a Swabian tribe of south Germany. It dates from about A.D. 725 and comprises three parts, 23 chapters dealing with ecclesiastical law, 21 dealing with public law, and 54 dealing with private law.

Lex Anglorum et Werinorum. A compilation of Thuringian laws based on the Ripuarian law and influenced by the Saxon law.

Lex Aquilia. A Roman statute of uncertain date (probably third century B.C.) dealing with damage unlawfully caused (*damnum injuria datum*) and giving a civil remedy for damage to property. It enacted (chap. 1) that anyone who unlawfully killed another's slave or beast of the class *pecus* had to pay the owner the highest value the thing had had within the previous year, and (chap. 3) that anyone who unlawfully damaged another's property in other respects, by burning, breaking, or destroying, was liable to pay him the value the thing had had within the preceding 30 days. The damage had to be unlawful, but did not have to be wilful. The *lex* gives rise to difficult problems of interpretation. It is important as the basis in later Roman law of the general principle of liability for damage to person or property.

E. Grueber, *Lex Aquilia*; J. Monro, *ad legem Aquiliam*; J. Thayer, *Lex Aquilia*; F. H. Lawson, *Negligence in the Civil Law*.

Lex Baiuvariorum. A body of customary law comprising the native rules of several Bavarian tribes with an admixture of foreign rules, Alamannic and Frankish; it dates from about A.D. 700.

Lex Burgundionum (or *Liber Constitutionum* or *Lex Gundobada*). See Lex Gundobada.

Lex Calpurnia de repetundis (149 B.C.). A Roman statute which established a permanent court to try cases of extortion by public officials, but allowing in substance a civil claim, the trial being directed to the return of the property extorted. It was extended by later statutes.

Lex Canuleia (445 B.C.). A Roman statute which allowed intermarriage between patricians and plebeians, probably by recognizing the legitimacy of children of plebeian mothers and allowing them to be admitted to patrician *gentes*.

Lex Cincia de donis et muneribus (204 B.C.). A Roman statute which forbade gifts which might impede justice and also donations above a given amount.

Lex commissoria. In Roman law a provision releasing a seller from his obligation to sell if the price were not paid within a specified time.

Lex curiata de Imperio. In Roman law a measure by which the comitia curiata conferred *imperium* on magistrates who could not take up office unless and until their election was ratified in this way. When *imperium proconsulare* and *imperium propraetore* were developed the *lex curiata* became a formality, and in the later Republic it was frequently suspended or ignored and in the principate it fell into desuetude.

lex domicilii. The law of the country of the domicile or permanent home of a person.

Lex Falcidia (40 B.C.). A Roman statute directed against excessive legacies and intestacy; it provided that if legacies exceeded three-quarters of the estate, they might be cut down *pro rata*, so as to ensure a quarter of the net total for the *heredes* (the *quarta Falcidia*).

lex fori. In private international law the law of the country of the place of litigation, which regulates evidence, procedure, and execution.

Lex Francorum Chamavorum (or *Eura*). The compilation of laws of the Chamavi, who lived on the lower Rhine, of the early ninth century. It comprises 48 articles, and is merely a memorandum made by royal justices who were sent on circuit there by Charlemagne.

Lex Frisionum (*c.* 802). A compilation of the laws of the Frisians comprising rules from both pagan and Christian periods. It falls into three parts, and additions giving variant rules for the north and south districts.

Lex Fufia Caninia de Manumissione (2 B.C.). A Roman statute which limited the proportion and number of his slaves which a master might liberate by will. It was extended by the *Lex Aelia Sentia* (q.v.) and repealed by Justinian.

Lex Hortensia (287 B.C.). A Roman statute which enacted that *plebiscita*, i.e. resolutions of the *comitia tributa* should have full legislative force and effect.

Lex Gundobada. A collection of royal ordinances promulgated about A.D. 495 by Gundobad, king of the Burgundians, applicable to his Burgundian subjects and to relations between Burgundians and Romans. Gundobad subsequently issued for the Roman subjects a code of their own, the *Lex Romana Burgundionum* (q.v.). Gundobad and his successors issued decrees supplementary to the *Lex Gundobada* which, following Roman practice, were known as *novellae*. Even after the fall of the Burgundian kingdom in 534 the code remained valid as the personal law of the Burgundians under Frankish rule.

Lex Iulia. Many *leges Iuliae* were enacted in Roman law, some by Julius Caesar and some by Augustus.

Lex Iunia de manumissione (*c.* 17 B.C.). See *Latini Iuniani.*

lex loci. The law, particularly in private international law, of the place in which some act or event material to the case has taken place; thus *lex loci actus*, the law of the place where a formal act such as a transfer of property was made; *lex loci celebrationis*, the law of the place of celebration of marriage; *lex loci contractus*, the law of the place where a contract was made; *lex loci delicti*, the law of the place where a delict was committed; *lex loci rei sitae* or *lex situs*, the law of the place where a thing is situated; *lex loci solutionis*, the law of the place where payment or performance is made.

Lex Papia Poppaea (9 A.D.). A Roman statute designed to encourage marriage and improve the birth-rate by preventing unmarried or childless married persons from benefiting under wills.

Lex Plaetoria (or *Laetoria*) (third century B.C.). A Roman statute imposing a penalty on anyone who took advantage of a person under 25 and rendering the transaction voidable, a situation which gave rise to the practice of appointing a curator to a person between 12 and 25 to approve his transactions.

Lex Rhodia. A body of maritime law which grew up in the second or third century B.C. in the island of Rhodes, a fragment of which survives in the Digest as the *Lex Rhodia de iactu* (q.v.). Another body of maritime law, usually known as the *Rhodian Sea Law*, became current in the Mediterranean in about the eighth century A.D.

Lex Rhodia de iactu. A rule adopted in Roman law from the *Lex Rhodia* (q.v.) that where goods were thrown overboard from a ship in peril and the ship was saved, the loss was to be shared by all concerned in the adventure, shipowner and all who had goods on board. It is the basis of the modern claim for general average (q.v.) contribution.

Lex Ripuaria (or *Ribuaria*). A collection of laws of the Ripuarian Franks, probably of the sixth century. It comprises an original part, chapters 1–31 and 65–79, and one part, chapters 32–64, borrowed largely from the Salic Code. Interspersed among the earlier text is royal legislation from the seventh to ninth centuries.

Lex Romana Burgundionum. A code enacted for his Roman subjects by King Gundobad of the Burgundians about A.D. 535. It consists not of extracts from legal works but of independent statements of legal rules systematically arranged in 47 titles and drawn from the *Codices Gregorianus*, *Hermogenianus*, and *Theodosianus*, post-Theodosian Novels, Paul's *Sententiae*, and a work of Gaius, possibly an abridgement of the Institutes. There is also a paraphrase taken from older sources. It was intended not as a codification of Roman law but as a handbook for judges, and Roman law not included in the *Lex* continued to have validity. After the Frankish conquest it was enlarged or supplemented by the *Breviarium Alaricianum.* A passage from Papinian concluded the *Breviarium* and it was thought that the *Lex* was merely a continuation of the quotation from Papinian. It was accordingly sometimes called the Papian or *Responsum Papiani*, from a mistaken reading of the name Papinian which ended the previous work in a manuscript. See also *Lex Gundobada.*

Lex Romana canonice compta. A collection drawn from Justinian's *Institutes* and *Codex* and from the *Juliani Epitome Novellarum*, arranged generally according to subject-matter, prepared for ecclesiastical use and probably originating in Italy; it dates from the ninth century. It provided ecclesiastics with rules of Roman law useful to them and helped to preserve Justinianean texts for the future. This was one of the Roman sources which influenced the growth of the canon law and was drawn upon by early canonist compilers.

Lex Romana Curiensis (*c.* A.D. 900). A statement of legal custom prepared for the peoples of Eastern Switzerland, the Tyrol, and North Italy, based on an imperfect and misunderstood abstract of the *Lex Romana Visigothorum*, copying Roman ideas but representing a very decadent state of law.

Lex Romana Visigothorum. A code issued in A.D. 506 by Alaric II, king of the Visigoths, for his Roman subjects. It contains extracts from the *Codex Theodosianus*, imperial constitutions, Gaius' *Institutes*, parts of the *Codices Gregorianus* and *Hermogenianus*, Paul's *Sententiae*, and a citation from Papinian. There are also interpretations explaining the extracts and adapting them to the changed circumstances. It was prepared by legal experts and approved by an assembly. Though far inferior to Justinian's codification, it is a valuable contribution to legal science in the Western Empire, and proved by far the most important of the bodies of law prepared for Roman subjects of Germanic realms. From the sixteenth century it was known as the *Breviarium Alaricianum.*

Lex Salica. The oldest surviving Germanic code, a compilation of the laws and ancient customs of the Salian Franks, later revised, amended, and reduced to writing. It dates from the fifth century. It had great authority and wide influence among the Frankish people, and became an element in many other compilations. It influenced the development of the law in France and the Low Countries and through the Norman Conquest had influence on early English common law.

Lex Saxonum. A collection in 20 chapters of ancient Saxon law, concerned with penal matters, dating from about A.D. 800. There is a second part, known as the *Lex Francorum*, influenced by Frankish law.

Lex talionis. The principle of retaliation in kind, of an eye for an eye and tooth for a tooth, known in Old Testament society and competent at Rome under the Twelve Tables in the case of severe bodily harm. It never seems to have been established in any civilized state, though an attempt was made to introduce it into England in the case of malicious accusations, it being enacted by an Act of 1363 that such as preferred any suggestions to the King's Great Council should put in sureties of taliation, but this punishment was abandoned after one year.

Lex Theodosii. A variant name for the *Lex Romana Visigothorum* (q.v.).

Lex Thuringorum. The compilation of laws of the Thuringi, dating from about A.D. 802 comprising 25 articles on penal topics and six articles on land, theft, and miscellaneous matters.

Lex Visigothorum Ervigiana. A late compilation of law for the Visigothic kingdom, much influenced in arrangement and substance by Justinian's codification and church canons attributed to King Erwig (A.D. 680–87). It was superior to most of the other Germanic codes of law and was taken as a model in other Germanic kingdoms.

Leyden, Philip of (*c.* 1328–82). A Dutch jurist, author of *Tractatus de cura reipublicae et sorte principantis* (*c.* 1360) an important vindication of strong princely powers but contending that such powers are established for the public benefit, *Tractatus de formis et semitis rei publicae*, and other works.

Leyes de Toro. An ordinance first promulgated by the Cortes of Toro in 1505 in an attempt to resolve the doubts and diversities in the understanding of the *Fuero Real*, the *Siete Partidas* (qq.v.), and other ordinances, usually preferring the Roman or canon law solution but sometimes the native Spanish and at times adopting a compromise situation. It also introduced some new principles, notably the entailing of estates.

Leys, Leendert (Lessius, Leonardus) (1554–1623). Professor of philosophy at Douai and of theology at Louvain, author of *De Justitia et jure* (1605) and theological works.

Leyser, August von (1683–1752). German jurist, highly regarded in his time, author of a classic *Meditationes ad Pandectas* (1717). Muller published in 1786 *Observationes practiciae ad Leyserii meditationum ad Digesta opus.*

L'Hôpital, Michel de (1507–73). Lawyer, humanist, and Chancellor of France. He became first president of the Chambre des Comptes and in 1560 Chancellor of France. He advocated toleration of the Huguenots and, in effect, constitutional monarchy. He strove for judicial reform, issued the *Ordonnance de Moulins* which improved many aspects of judicial administration, and wrote a *Traité de la Réformation de la justice* and many other works.

Liability. The legal concept of being subject to the power of another, to a rule of law requiring something to be done or not done. Thus, a person who contracts to sell goods is liable to deliver them and the buyer is liable to pay the price. Each is required by law to do something, and can be compelled by legal process at the other's instance to do it; the other is empowered to exact the performance or payment. It is sometimes called subjection. The correlative concept is power.

A person is said to be under a liability when he is, or at least may be, legally obliged to do or suffer something. Thus, one may be said to be liable to perform, to pay, to be sued, to be imprisoned, or otherwise to be subject to some legal duty or legal consequence. In general, liability attaches only to persons who are legally responsible; an insane person does not generally incur any liability.

Liability may arise either from voluntary act or by force of some rule of law. Thus, a person who enters into a contract thereby becomes liable to perform what he has undertaken, or to pay for the counterpart performance, or otherwise to implement his part of the contract. If he acts in breach of contract, he becomes liable by law to pay damages in compensation for the breach. Similarly, if a man acts in breach of any of the general duties made incumbent on him by statute or common law, such as to refrain from injuring his neighbour, or to maintain his tenant's house in reasonable repair, or to exercise diligence in administering property of which he is trustee, he incurs legal liability to make good his omission or default.

Liability is commonly distinguished according to its legal grounds into civil liability, whereby one is subject to the requirement to pay or perform something by virtue of rules of civil law, and criminal liability, whereby one is subject to being fined, imprisoned, or otherwise treated by virtue of rules of criminal law. Civil liability may arise from many grounds, from the natural relations of the family, from undertaking or contract, the commission of a harm, from trust, statute, or decree of court. In respect of liability arising from harm done, it arises from intentional harm, harm brought about in breach of duty, and in some cases there is strict liability if harm befalls despite care taken. Criminal liability arises from the admitted or proved commission of some kind of conduct declared by the rules of criminal law to be a crime inferring punishment. At common law criminal liability normally also requires that the conduct has been done intentionally or recklessly but not merely negligently or accidentally, and sometimes proof of a particular intent is a necessary ingredient of a crime, but many cases under statute have been held to impose strict or absolute liability, i.e. liability irrespective of the actor's state of mind.

The term liability is sometimes used of the sanction or penalty itself. Thus a debt is spoken of as a liability.

The term is also used in the analysis of rights and duties (q.v.) as a synonym for 'subjection' and the correlative of 'power'.

Liability without fault. In general in Anglo-American law, liability in damages for personal injuries or loss depends on proof that the harm was done or allowed to happen in circumstances implying fault on the part of the doer, i.e. with intention or negligence, fault by commission or by omission. In some cases, however, usch as workmen's compensation, it was provided that liability should be incurred if there were personal injury arising out of and in the course of employment, without need for proof of fault. Subsequently it has been urged that in many other cases also, notably road accidents, there should be liability without fault to ensure that injured persons are compensated even if they cannot prove fault. In criminal law, while the general rule is that liability to penalty requires the commission of crime intentionally or recklessly, in many cases liability is incurred without either of these mental elements or anything which could be called fault.

Libel and slander. In English law libel and slander are the two varieties of defamation (q.v.), or two main modes in which defamation, or unjustified disparagement of the good name and reputation of an individual, may be committed. A defamatory statement is a libel if made in writing, film, broadcasting, or other permanent form; it is then actionable *per se.* A defamatory statement is slander if made orally or in other transient form, and, save in certain special cases, is not actionable without proof of special, i.e. actual, damage. The object of both actions is to obtain compensation for harm resulting from unjustified publication of statements defamatory of the plaintiff. Libel is, but slander is not, also a crime in certain circumstances.

Slander was originally a matter of ecclesiastical jurisdiction but early developed in the royal courts. In 1275 statute made the publication of false news tending to produce discord between king and people a crime, and thereafter several statutes on *scandalum magnatum* were passed. In the sixteenth century the civil courts developed claims of damages for breaches of these statutes. The Council and the Star Chamber dealt with slander of peers, seditious words and writings, and extended the concept of criminal libel very widely. Libel developed from *scandalum magnatum.* Common law courts early entertained actions for slander by way of actions on the case for words.

A defamatory statement is one calculated to degrade a person in the estimation of right-thinking men, or to cause him to be shunned or avoided, or to expose him to hatred, contempt or ridicule, or to convey an imputation disparaging to him in his profession, trade, or business. The statement must, in civil cases, have been published of and concerning the plaintiff to a third party. If defamatory in character and false in fact, the law implies malice. The defendant's motive is immaterial. The statement, however injurious, is not defamatory if true in substance and in fact. If the words used are not *ex facie*, i.e. obviously, defamatory the plaintiff must plead an innuendo, i.e. circumstances by virtue of which a secondary, defamatory, meaning was conveyed. It is for the trial judge to rule whether the words are capable of referring to the plaintiff and of conveying a defamatory meaning in the minds of reasonable persons, and for the jury to say whether the words complained of were in the circumstances defamatory.

In certain circumstances the publication of matter prima facie defamatory is justifiable and not actionable. This is so where the words were published on an occasion of absolute privilege, i.e. where the speaker has legal entitlement to make the statement, true or false, and immunity from liability to action therefor, e.g. by an M.P. in the House of Commons; or on an occasion of qualified privilege, i.e. where the speaker was entitled to make the statements provided they were communicated bona fide and without malice, but not if communicated with actual malice; or were fair, i.e. honest, comment made bona fide and without malice towards the plaintiff on facts truly stated which were a matter of public interest, e.g. a criticism of a book or play. In a case of a libel in a newspaper or periodical, there is a defence of apology and payment into court, and in a case of unintentional or accidental defamation, of a tender of amends, i.e. of a sum in solace of hurt feelings.

A libel (but not a slander) may also be prosecuted at common law as a crime, where it is blasphemous, obscene, or seditious. It should be likely to disturb the peace or seriously to affect the reputation of the person defamed. Publication to the person defamed alone is sufficient in crime though not in tort. Truth by itself is no defence, but truth and that publication was for the public benefit is a defence. A libel on a class of persons may be indictable though not actionable. On the other hand the crime of libel may be narrower than the tort in respect of vicarious liability. For blasphemous, obscene, and seditious libel see BLASPHEMY; OBSCENITY; SEDITION.

In Scotland the terms libel and slander are both used loosely and without the defined legal connotations of English law as synonyms for the delict of defamation (q.v.), itself a species of the genus verbal injury (q.v.).

The term libel is also used of the initial pleading in a plenary cause in the ecclesiastical courts.

Liber Assisarum (or *Book of Assizes*). The name given to Part V of the 1679 standard edition of the Year Books, containing reports of cases of the years 1–50 Edward III according to regnal years.

Liber Augustalis. The first mediaeval code of laws based on Roman jurisprudence, promulgated for the Kingdom of Sicily by Frederick II in 1231.

Liber decretalium Gregorii IX. See DECRETALS.

Liber extra. See DECRETALS.

Liber Judicioriam. See *Fuero Juzgo.*

Liber Papiensis (or *Liber legis Langobardorum*). A treatise of a practical character, compiled at the Lombard Law School at Pavia about A.D. 1040, commenting on the texts of the Lombard law. It rapidly attained a reputation throughout Italy and was the subject of an *Expositio ad librum Papiensem* of about A.D. 1090.

Liberate. A writ granted under the Great Seal and addressed to the treasurer and chamberlain of the Exchequer, for the payment of an annual pension or other sum. Also a writ addressed to the sheriff ordering him to deliver possession of lands and goods taken on the forfeiture of a recognizance. Also a writ issued from Chancery to a gaoler, directing release of a prisoner who had found bail for his appearance.

Liberate rolls. Rolls in the Public Record Office covering 1201–1436 containing the list of writs of Liberate, Allocate, and Computate issued by the Chancery.

Liberties, Charters of. The general name for a series of twelfth- and thirteenth-century deeds whereby English kings sought to define the rights and obligations of persons under feudal law. They include the coronation charter of Henry I, charters issued by Stephen and Henry II, Magna Carta (q.v.) granted in 1215 and reissued in 1216, 1217, and 1225, and confirmed by Edward I in 1297. Most of these have been superseded or repealed.

Liberties of the subject. In the United Kingdom there are no guaranteed liberties of the subject or individual and no authoritative and exhaustive enumeration of them. They are implications from the general principle that an individual may say or do whatever he pleases, provided he does not act in breach of any legal prohibition, or thereby infringe the legal rights of other persons. In general they have developed over a long period and been formulated in various ways in judicial decisions. Some have been defined and sought to be better secured, by major charters, namely Magna Carta (1215), the Petition of Right (1628), the Bill of Rights (1688), and the Act of Settlement (1700) (qq.v.). Several would be better described as liberties than as rights. In consequence of the supremacy of Parliament, any liberty may be abridged, defined, qualified, modified, or abolished by statute, though only clear expression of intention to do so will be adequate, and only in emergency is Parliament likely seriously to restrict such liberties. Few, if any, liberties are in fact unqualified, and the problems are mainly of the extent of, and authority for, restrictions and qualifications of liberties.

The main recognized liberties are: (1) the right to personal liberty and freedom from detention or restriction on movement, protected by the writ of habeas corpus, and the right of action for false imprisonment; (2) the right of freedom of conduct, to do anything not expressly prohibited and not likely to cause breach of the peace or other disturbance; (3) the right to freedom of speech, writing, publication, and discussion, limited only by liability to action for libel or slander, or to prosecution for sedition, disclosure of official secrets, or the like; (4) the right of freedom of conscience and of belief, the manifestation of religious and political toleration; (5) the right of public meeting, limited by liability for trespassing or causing nuisance, for causing breach of the peace, riot, or unlawful assembly; (6) the right of association in clubs, trade unions, partnerships, companies, and the like for any lawful purpose; (7) the right of private property in land, moveable goods, and incorporeal rights, protected by various proprietary and delictual actions, but subject to many qualifications imposed by statutes providing for taxation, compulsory purchase, expropriation, and by common law and statutory rules controlling and limiting uses in the interests of the public and of other persons; (8) the right to participate in elections to, and to stand as a candidate for election to, Parliament and elected local authorities and thus to have a say in the government of the country and the locality; and (9) the right to have a claim or grievance decided by an independent judicial person or tribunal in accordance with principles and rules generally known, according to the principles of natural justice and established rules procedure.

Some claims occasionally asserted, such as the right to work and the right to strike, are not rights. The alleged right to work is a liberty to work if one chooses, but there is no right to work in the sense of an enforceable claim to be found or given a job. The alleged right to strike is a liberty to decline to work, frequently in circumstances amounting to breach of contract, but inferring no claim against anyone. Neither of these alleged rights imposes duties on anyone.

The liberties of the subject are secured by various forms of action, by the principle that statutes will be interpreted in favour of freedom and so as not to interfere with the rights of individuals unless clearly expressed to do so, by the rule that interference with liberty, even by public authorities, must be justified by law, and by the force of public opinion.

More general formulations of liberties of the individual have been made in the U.N. Universal Declaration of Human Rights of 1948 and the Council of Europe's Convention on Human Rights of 1950 (q.v.) but neither of these is a part of United Kingdom law.

Liberty. (1) A district within a hundred in Anglo-Saxon England, detached from its jurisdiction and subject to the jurisdiction of the Church or secular lord, who was entitled by royal grant to dispense justice in the hall of his manor-house, whence it was called a hall-mote. The jurisdiction of a whole hundred, or several hundreds, was frequently granted in this way, which weakened the administration of justice.

(2) The legal concept of not being subject to duties or restrictions, but being free to do or not to do according as one chooses. Thus, an individual has a liberty to express his opinion on public affairs or the performance of an actor, but no liberty to publish seditious opinions or defamatory imputations. It may be equated with privilege. The correlative concept is absence of right, or no-right, or inability.

The term liberty is also used for a privilege or branch of the Crown's prerogative granted to a subject, and consequently for the area over which that liberty is enjoyed. It was also used in England prior to 1850 for a district, sometimes also called a franchise, within the limits of a county but exempt from the jurisdiction of the sheriff and having a separate commission of the peace. Each of the Inns of Court was originally a liberty, and the Liberty of the Savoy is a place subject to a franchise. In some districts commons are called 'liberties'.

(3) (or freedom). The quality of not being subject to control or restriction. In its most general sense the principle of liberty involves that the State or community treats each and every moral person as a free agent, capable of thinking and acting on his own behalf, of developing his own capacities in his own way, and of exercising and enjoying the rights which are the conditions of such development. This implies responsibility for his conduct, for controlling it, and for the consequences of misconduct. Liberty in society for each is qualified by the need of liberty for others and for the society as a whole; liberty can be only such as can be reconciled with similar liberties of others, and must accordingly always be regulated and restrained.

Liberty may be secured by law partly by absence of regulation and control of particular activities, as in the case of liberty of the Press, partly by express declaration of rights of free exercise of stated kinds of conduct, but largely by the imposition of restrictions on the liberties of persons to prevent any other from exercising particular liberties. Thus, the liberty of the subject to hold public meetings is secured by restrictions on the liberties of the police to prevent public meetings.

In society liberty secured by law is distinguishable into civil or personal liberty, involving physical freedom from injury to life, health, reputation, or movement; intellectual freedom for the holding and expression of thought and belief (including religious belief), and practical freedom for the exercise of will and choice in personal, contractual, and other relations with other people and in the acquisition and holding of property; political liberty to participate in government, as voter, candidate, or member; and economic liberty to work, trade, earn, spend, invest, and so on. The 'liberty of the subject' is accordingly not a single liberty but a group of distinct liberties. These are commonly the subjects of bills of rights, and liberty as a whole is frequently identified with constitutional, democratic government.

These different liberties may conflict with one another, and the liberty of one man in one sphere may conflict with the liberty of another or others in that or a related sphere. The liberty to strike frequently involves denial of the liberty of other persons to work or to earn.

Furthermore no distinguishable liberty can be allowed by society to be exercised entirely without restriction imposed by society in the interest of others. Complete liberty is anarchy, and every liberty must be subject to certain restraints imposed by law for the common good. Thus freedom of speech is limited by rules as to libel, slander, and sedition, imposed to preserve the freedom of others from unjustified damage to reputation which would follow if freedom of speech were completely uncontrolled.

Library lending right. The right of authors of books to a payment for each occasion on which one of their books is borrowed from a library, analogous to the performing right (q.v.) recognized for composers. It has long been claimed by authors and it was recognized by statute in the U.K. in 1979.

Libri feudorum. A book on feudal law compiled by lawyers of Milan and Pavia from the feudal laws of early German emperors, and probably put into final form by Bolognese jurists. The oldest version is known as the *Obertian Recension* because it concludes with a letter from Obertus de Orto, a judicial officer, to his son Anselm. It was also annotated by one Gerardus Niger. It was accepted as authoritative all over Europe, was glossed and in its final version was appended to the *Corpus Iuris* following the Novels as *Decima Collatio*. In most countries it was supplemented by local sources of feudal law.

Libro de Montalvo (or *Ordenanzas Reales de Castilla*). A compilation of mediaeval Castilian law made by Alfonso Diaz de Montalvo in 1485 and containing 1,163 laws, about a fifth of them promulgated by the then reigning Ferdinand and Isabella. Many editions were published in the next 30 years.

Licence. A permission to do what is otherwise restricted, prohibited, or illegal. Thus a licence is an authority to enter on and be on land, which would otherwise be a trespass or prohibited. The word is also used of the document expressing the permission. Thus there are dog licences, gun licences, and wireless licences. In penology a licence is a release of a prisoner on conditions, normally to obey directions of and to report to a supervisor.

The commonest use of the word is for the permission to sell intoxicating liquor. This is of two kinds, the excise licence necessary to manufacture, deal wholesale in, or sell by retail, any intoxicating liquor, and licences granted in England by justices sitting in divisional licensing committee or borough licensing committee, and in Scotland by district licensing boards. Licences are also required for the provision of music, dancing, and entertainment and for a wide variety of trades, such as bookmakers, hawkers, and the provision of loans, credit, and related financial services.

Licence Cases (1847), 5 Howard 504 (U.S. Supreme Ct.). The Supreme Court held, slightly modifying the principle of *Gibbons* v. *Ogden* (q.v.), that a tax imposed in exercise of the State policy is valid, even though it impinges on interstate commerce. It reflects a movement of opinion in favour of state rights.

Lie, Trygve Halvdan (1896–1968). Norwegian statesman and first Secretary-General of the United Nations (1946–52) who had to contend with war in Palestine resulting from the establishment of the State of Israel, disputes between India and Pakistan over Kashmir, and the Russian attack on Korea. He wrote *In the Cause of Peace* (1954).

Lieber, Francis (1798–1872). Born and educated in Germany, he moved to the U.S. in 1827 where he produced the *Encyclopaedia Americana* (13 vols., 1829–33). In 1835 he was elected to the chair of history and political economy at South Carolina and there wrote his major works *Manual of Political Ethics* (1838–39), *Legal and Political Hermeneutics* (1839), and *On Civil Liberty and Self-Government* (1853). In 1857 he moved to Columbia College and in 1865 to the Law School there. He had a lifelong interest in penal law and latterly devoted much attention to international law, and wrote a *Code for the Government of the Armies of the U.S.* (1863) which was widely accepted and has continued ever since to be the basis of international understanding on the conduct of war.

Liebermann, Felix (1851–1925). German scholar who worked particularly on early English law. His great work was *Die Gesetze der Angelsachsen*, 3 vols. (1898), and he also wrote *The National Assembly in the Anglo-Saxon Period* (1913), and worked on the *Monumenta Germaniae Historica*.

Liege. An adjective denoting in feudal law an unconditional bond between a man and his superior or overlord. The concept of ligeance is found in France in the eleventh century. If a man held lands of several lords, his obligations to his liege lord were greater than and took precedence over obligations to other lords; to the first he had done liege homage, but to others only simple homage. By the thirteenth century ligeance determined which lord a man must follow and also which lord was entitled to the profits of overlordship of that man. The King was always held to be every subject's liege lord and all feudal contracts came to reserve the allegiance due by a subject to him.

Lien (from *ligamen*, a binding) is the right in English law of one person to retain in his possession property belonging to another until the claims of the person in possession against the other are satisfied. Actual possession, obtained rightfully, is normally necessary, but exceptions exist, as in the case of maritime liens, under statute, and in equity, where possession is not required.

Liens are distinguished into legal liens, which may be general, arising by common law or express agreement, entitling the possessor to retain until all claims of the possessor are satisfied; or particular, entitling him to retain in respect of charges incurred in respect of the thing possessed only and not for a general balance due; and equitable liens, which are equitable rights normally founded on contract to a charge on the real or personal property of another until certain specific claims have been satisfied. The legal right is a right in security only, not of property nor implying any right to use or sell the thing held under lien, but the equitable right exists irrespective of possession and confers on the holder the right to a judicial sale.

The term lien is frequently used in Scots law for one of the varieties of retention (q.v.).

Lieutenancy, Commission of. Commissions, in form similar to the earlier commissions of array, issued first in the reign of Edward VI and addressed to the then newly created lieutenants or lords lieutenant of the counties. Down to the time of the Civil War they were issued from time to time, but fell into disuse after 1660 with the development of the militia system.

Life estates. In English law estates of freeholds not of inheritance, existing for the life of the grantee or of another.

Life peerages. A few life peerages were created prior to the reign of Henry VI, and even women sometimes received life peerages in the seventeenth and early eighteenth centuries, but apart from that it had become settled that peerages were hereditary. In 1856 it was thought desirable to strengthen the House of Lords in its judicial capacity and Sir James Parke (q.v.), a Baron of Exchequer, was created a life peer. But the Committee for Privileges of the House of Lords held that this did not entitle him to sit and vote in the House and an hereditary peerage was conferred on him as Lord Wensleydale. In 1876 the Appellate Jurisdiction Act authorized the creation of four life peers as Lords of Appeal in Ordinary to assist in the exercise of the House's judicial function. The Life Peerages Act, 1958, made general provision for the creation of life peerages, with the rank of baron, and also removed the disqualification on women as members of the House of Lords. Since then only life peers have, in practice, been created. Life peers are entitled to a writ of summons for life, notwithstanding retiral from active life.

Liferent and fee. In Scots law the consecutive interests of persons entitled respectively, one, to the use and profits of certain subjects for his lifetime, and the other, to the absolute ownership thereof after the expiry of the liferent. In the case of heritable property, a liferent may be a proper liferent under which the liferenter has sasine of the subjects, but on his death they pass to the fiar. While he possesses the subjects he is entitled to all rents and profits, but not to do anything which will permanently diminish the value of the lands. In the cases of moveable property and mixed estate, and in modern practice increasingly in heritage also, a liferent is normally an improper liferent only in which the estate is vested in trustees who allow the liferenter the use and profits for his lifetime and preserve the fee for the fiar. The trustees have to balance as fairly as they can the interests of liferenter and fiar, apportioning liabilities and outlays between income and capital.

Ligan. A variant of lagan (q.v.).

Light. At common law there is no right to have the access of light to one's windows not interrupted by any obstruction, but the Prescription Act, 1832, made uninterrupted enjoyment of light for 20 years an indefeasible right thereto, commonly called 'ancient lights', provided it was not enjoyed under any deed, written consent, or agreement. The right is to uninterrupted access of such light only as is required for ordinary purposes, and not to light appropriate for any particular purpose for which the light has been used.

Limitation. A clause in a conveyance fixing the extreme period during which an estate or interest may continue. A limitation is distinguished from a condition or defeasance, which may bring an interest to an end before the period of its extreme duration. A limitation may be conditional, where an interest is limited to end and another to commence on the doing or the happening of something; or collateral, where the interest is limited to end but may be ended sooner by the occurrence of an uncertain event. A conditional or collateral limitation by way of use was sometimes called a springing or shifting use, and if by way of will was called an executory devise.

Limitation of actions. The principle in English law that after the lapse of a fixed period of time an action is not maintainable. In general the running of the limitation period excludes a remedy by action or set-off but does not affect the right which subsists for other purposes, such as for a lien, but in the case of an action to recover land or to enforce an advowson, the title is extinguished by the running of the period of limitation.

In general the period of limitation in English law is 12 years for actions upon a specialty (q.v.), or a judgment, six years for actions founded on simple contract or tort, but three years for actions founded on personal injuries or death. Particular statutes fix other periods for various particular cases. Actions for recovery of land are in general barred after 12 years and the owner's title extinguished, and actions to recover a principal sum secured by a mortgage are barred after 12 years. No period of limitation applies in the case of an action by a trust beneficiary, in respect of any fraud or fraudulent breach of trust, but in other cases trust property may be recovered only within six years. All the limitation periods are subject to provisions for extension in the case of disability, and the postponement in cases of acknowledgment, or part payment, or fraud, or mistake.

See also PRESCRIPTION.

Limitation of liability. A legal device setting a limit on the extent of the liability of a party in certain circumstances. It exists in relation to companies (see LIMITED LIABILITY) and by statute in respect of the liability of innkeepers, carriers, and of shipowners for loss of or damage to goods in the ship in certain circumstances, or for loss of life or injury to persons or things caused by or on board the ship beyond a fixed amount per ton of the ship's tonnage in certain other cases.

Limited liability. A condition in which the liability of a participant in a business enterprise is limited to the amount of capital invested by him in it and does not extend in case of loss to his other assets. It has been a very important factor in the development of large enterprises because many persons would not be willing on other terms to risk their capital, and acceptance of the principle has enabled enterprises to secure investments from large numbers of people. The first form of enterprise having this characteristic was the limited partnership, in which partners were liable only for the amounts invested by them so long as at least one partner was liable, in the ordinary way, to the full extent of his assets for the firm's debts. In Britain the possibility of forming an incorporated company with the liability of all shareholders limited to the nominal value of their shares was recognized in 1862, but unlimited companies are still competent, as are partnerships and limited partnerships. In Britain a company with the liability of its shareholders limited must append the word Limited to its name.

Limited owner. A tenant of land for life, in tail or by the curtesy, or otherwise not having a fee simple in his absolute disposition. Since 1925 the interests of limited owners are equitable interests only, not legal estates in the land.

Limited partnership. A partnership in which, so long as one partner has unlimited liability, the other partners may have their liability limited to the extent of the sums severally contributed to the firm's capital.

Lincoln's Inn. See INNS OF COURT.

Linden, Johannes van der (1756–1835). An advocate and judge at Amsterdam, author of many works, notably a *Rechtsgeleerd Practicaal en Koopmans Handboek* (1806), giving a practical survey of law in the Netherlands with extensive reference to older traditional law, which was made the official law book of the South African Republic in 1859, and of a supplement to Voet's *Commentaries*, Bks. I–XI, and translator of various works. He translated Grotius into Latin and much of Pothier into Dutch, and drafted a Civil Code for Holland, some of which was incorporated in the version of the *Code Napoléon* translated and adapted to Holland.

Lindley, Nathaniel, Lord (1828–1921). Called to the Bar in 1850, but for some years had only a modest practice; from about 1865 it developed rapidly and he became Q.C. in 1872 and a judge of the Common Pleas in 1875. He was the last judge to have become a serjeant at law (q.v.), and eventually the last English survivor of the order of serjeants. In 1881 he was promoted to the Court of Appeal, made Master of the Rolls in 1897 and was a Lord of Appeal in Ordinary from 1900 to 1905. He had a very acute intellect and was very highly regarded as a judge; not least of his merits was his ability in every field of law. Lindley was a scholar of considerable attainments; in 1855 he published a translation of part of Thibaut's *System des Pandektenrecht* as *An Introduction to the Study of Jurisprudence*, and in 1860 his *Treatise on the Law of Partnership including its application to Companies*, a book which made his name and has been a standard book ever since, repeatedly re-edited.

Lipit-Ishtar, Code of. See BABYLONIAN LAW.

Liquidated damages and penalty. The parties to a contract may include therein provision for payment of a sum of money by either party, if in breach of contract in stated, or any, ways, to the other, in compensation for loss thereby caused. Such a provision is deemed a liquidated damages provision if it is regarded as a genuine pre-estimate of the loss which the parties contemplated would probably flow from the breach; but as a penalty provision if it does not attempt to assess the loss but is imposed as a threat, and as a punishment in the event of breach. Whether a provision is one or the other is a question of interpretation, the words used not being conclusive. If a provision is held to be one for liquidated damages, that sum is payable on breach without proof of loss; if it is held to be one for a penalty, the innocent party can only recover for loss proved, not exceeding the penal sum, or may ignore the penalty provision, sue for damages and recover for proved losses, either more or less than the penalty stated. Parties may also provide that either or both may terminate the contract, or do some other act affecting performance, on payment of a specified sum, in which case the sum is payable, without question whether it is liquidated damages or penalty.

Liquidation. A synonym for winding-up or the process whereby a company incorporated under the Companies Acts is brought to an end and, once its affairs have been settled, dissolved.

Liquidator. The person appointed by the court or the company itself to conduct the winding-up of a company. His duties are to get in and realize the company's assets, pay its debts and distribute any surplus among the creditors having regard to any preferential claims which exist among them. A liquidator in a winding-up by the court is restricted in that in general he requires the authority of the court to take any important step.

Lis alibi pendens. A lawsuit depending elsewhere, a plea sometimes alleged in defence to a claim, alleging that the same matter is in dependence between the same parties in another court.

Liszt, Franz von (1851–1919). German jurist, author of *Lehrbuch des deutschen Strafrechts* (1881) and *Das Völkerrecht* (1898). His life-work was the scientific foundation and reform of the criminal law of which he stated the basic principles in his *Der Zweckgedanke im Strafrecht* (1882). He was a founder of the *Internationale Kriminalistische Vereinigung* in 1889.

Lit de justice. In old French law a solemn judicial occasion when the King, accompanied by the princes of the blood, the Chancellor, peers of France, and high officials, seated on his throne, exercised his powers as high justiciar of France and made law by decree. The power was sometimes exercised to overcome the resistance of the Parlement of Paris to register a royal decree. The *lit de justice* was most often used in the eighteenth century in the face of opposition by the Parlement. The legality of the practice was not questioned until shortly before the Revolution when it was criticized as an exercise of arbitrary power.

Literal contract (or *literarum obligatio*). In Roman law a contract constituted by formal writings, possibly by fictitious entries by the creditor in the traditional Roman kind of account book.

Literature, legal. See LEGAL LITERATURE.

Litigant. A party to a civil lawsuit.

Litigiosity. In Scots law a legal prohibition on a debtor's alienating heritable property to the effect of defeating an action or diligence commenced or inchoate. Lands are rendered litigious by inhibition (q.v.) or by the service of any of a number of kinds of summonses of actions affecting the land.

Litiscontestation. Was, in Roman law, the stage reached by a litigation when both parties had stated their pleas to the praetor. It was deemed a kind of contract, that both parties would abide by the decision of the judge. In Scots law it is constituted by lodging defences, and any decree granted thereafter is not a decree in absence.

Littleton, Sir Thomas (1402–81). Became a member of Inner Temple, King's serjeant in 1455, and went the northern circuit as a judge of assize. He became a justice of the common pleas in 1466. His fame rests on a short treatise on *Tenures* in three books which was published about 1481, the earliest printed treatise on English law, and early recognized as a work of authority, unquestionably the most important book between Bracton and Coke. It owed nothing to Roman law, a little to *The Olde Tenures* of about 1375, printed in the sixteenth century, but was chiefly founded on the Year Books. Littleton's achievement consisted in giving a scientific account of the tenure of land. It is the first great book on English law not in Latin (in law French) and wholly un-influenced by Roman law, and sums up the development of what was then the most important branch of the common law. It was repeatedly reprinted in both French and English and taken by Coke (q.v.) as the foundation of his First Institute—*Coke upon Littleton*, 1628. A modern edition by Wambaugh, *Littleton's Tenures in English*, was published in 1903.

Liutprand (713–35 or 744). The greatest king of the Lombards, notable legally for the large volume of legislation which he added to the *Edict of Rothar*, the main statement of Lombard law. It comprised 153 chapters in 15 volumina. Many of his enactments introduced principles from the Christian Church and others show the influence of Roman law, while others again are conscious law reform, laying down a new principle to deal with a difficulty raised by litigation.

Liverpool Court of Passage. A court granted by charter in the seventeenth century with jurisdiction in personal actions to any value if the defendant resided or carried on business within the City of Liverpool and its port. It had the same admiralty jurisdiction as the Liverpool county court. The judge was a barrister of at least seven years' standing and had the powers of a High Court judge. In his absence the Recorder of Liverpool might act. Appeal lay to the Court of Appeal. The court was abolished in 1971.

***Liversidge* v. *Anderson*,** [1942] A.C. 206. The plaintiff was detained in the Isle of Man under Statutory Defence Regulations during World War II. He sought release. The defendant, the Home Secretary, had power to order his detention if he had reasonable cause to believe that the detainee was of hostile origin or associations and his detention necessary. It was held by the House of Lords that the grounds for the Minister's belief could not be questioned and his statement that he had reasonable cause was final, unless there were evidence of bad faith on his part, which there was not. The case has subsequently been treated as justifiable by wartime conditions and not laying down a general principle.

Livery of seisin (i.e. delivery of possession). From mediaeval times the essential part of the feoffment by which a freehold interest in land was created or transferred. It consisted in putting the donee in possession of the land, evidenced by handing over some symbols thereof, and leaving the land vacant. Later a written charter was drawn up as a record of the transaction, and in time the delivery of the charter came to be in place of the symbols and to evidence a feoffment with livery of seisin.

Livery stable. A stable where horses are kept and let out on hire.

Lives of the Chief Justices. A not wholly reliable series of lives of the Chief Justices of England from the Norman Conquest to 1832 by John Campbell, Lord Chancellor, 1859–61 (q.v.).

Lives of the Lord Chancellors. An entertaining but not entirely accurate series of biographies of the Lords Chancellors of England down to 1837, published in 1845–57 by John Campbell, Lord Chancellor, 1859–61 (q.v.).

Lives, Lease for. Formerly a common kind of lease in the Inns of Court and in parts of England.

Such a lease is now under statute a converted copyhold lease for life without right of permanent renewal, or a lease for 90 years determinable by notice after any event determining the term under the original demise. In all other cases it is an equitable interest.

Livingston, Edward (1764–1836). American statesman and jurist, practised law first in New York. He entered Congress in 1794, but moved to New Orleans in 1803 and practised there. In 1820 he was elected to the Louisiana legislature and prepared a provisional code of procedure which was in force from 1805 to 1825. He was commissioned to revise the penal law of that state, and in 1825 he published his penal code for Louisiana which, though not adopted there, brought him international renown. He re-entered Congress and in 1828 presented to the U.S. Senate his *System of Penal Laws for the U.S.A.*, but no action was taken on it, though it became a model for penal codes in other states of the U.S. and elsewhere. Under President Jackson, who regarded him as a trusted adviser, he was Secretary of State, 1831–33, and Minister to France, 1833–35. His writings include *Reports of the Plan of the Penal Code* (1822); *System of Penal Law for the State of Louisiana* (1826); *System of Penal Law for the United States* (1828); and *Complete works on Criminal Jurisprudence* (1873).

C. H. Hunt, *Life of Edward Livingston* (1864).

Livingston, George, 3rd Earl of Linlithgow (1616–90). Sat in the Scottish Parliament from 1650 and in the British Parliament 1654–55. He became a Privy Councillor in 1661. He was Lord Justice-General of Scotland 1684–88, being removed at the Revolution.

Livingston, Robert R. (1746–1813). U.S. lawyer, brother of Edward Livingston (q.v.), was a delegate to the Continental Congress and a draftsman of the Declaration of Independence. He also helped draft the original constitution of New York State and was first chancellor of that state (1777–1801), hence being called 'Chancellor Livingston'. From 1781 he organized the conduct of foreign affairs under the government established by the Articles of Confederation and from 1801 was minister to France, in which capacity he negotiated the Louisiana Purchase. In retirement he was involved in running the first steamboat on the Hudson River.

J. L. Delafield, *Chancellor Robert R. Livingston and his family* (1911).

Livingston, William (1723–90). American politician. He prepared a digest of the laws of New York for 1691–1756 (1752–62) and became the first revolutionary governor of New Jersey, 1776–90. His son, Brockholst Livingston, became a justice of the Supreme Court.

Livre de Justice et de Plet. A law book, borrowing heavily from Roman law, probably written about 1260 by one of the teachers of the Law School of Orleans and well-known in Northern France.

Llewellyn, Karl Nickerson (1893–1962). Became a member of the Connecticut Bar and taught law at Yale, Columbia, and Chicago. He was a life member of the Commissioners on Uniform State Laws and chief draftsman of the Uniform Commercial Code. His writings included *The Bramble Bush*, *The Cheyenne Way*, and *The Common Law Tradition in Deciding Appeals*, and many papers in journals. His major contribution was in jurisprudence, in the theory of realism; he was a 'law skeptic', insisting that law was not a logical body of principles, but the reactions of courts to problems coming before them.

W. Twining, *Karl Llewellyn and the Realist Movement.*

Lloyd's. An international insurance market in London, originating in the practice of underwriters and merchants congregating in the coffee-house of one Edward Lloyd (*c.* 1650–1713). They gradually developed into an association through which marine insurance is chiefly effected. It was incorporated by statute in 1871. The Lloyd's policy of marine insurance in substance goes back to the seventeenth century; the policy as revised and confirmed in 1729 is still in use and practically every phrase has been judicially interpreted. It is supplemented by clauses devised to meet the requirements of each class of insurance. Lloyd's publishes *Lloyd's List*, a record of events in shipping, including sailings and casualties.

A subsidiary product is the publication of *Lloyd's Law Reports*, reports of cases of commercial and maritime interest.

Lloyd's Register of British and Foreign Shipping. A publication issued annually by the Society of Lloyd's Register, an unincorporated body distinct from Lloyd's (q.v.) which has existed since 1834. It gives a list of ships, classified according to materials, construction, state of repair, etc. The society publishes construction standards to which vessels must comply if they are to be registered and classified.

Lo Codi. A law book written in Portugal about the mid-twelfth century, summarizing Justinian's *Code* for practical use and showing clear traces of the influence of the glossators. It indicates the extent to which Roman law was being studied in the south of France at that time.

Loan. A contract whereby one party gives another gratuitously the use for a time of some property. It is distinguishable into proper loan (*commodatum* in Roman law) and improper loan (*mutuum* in Roman law). In proper loan the property in the thing lent remains with the lender, and the borrower must, at the end of the period of loan or on demand, restore the specific thing lent in as good condition as he got it, fair wear and tear excepted. In English law this is a species of bailment (q.v.). In improper loan the property in the thing lent is transferred to the borrower, who may consume or use it up, but must at the end of the period of loan or on demand, restore an equal amount of the thing lent him, and of the same quality. A special case of improper loan is the loan of money, in which it is presumed that interest is due. Special statutory rules apply to the loan of money by persons and bodies lending money at interest as a business.

Loan societies. Institutions for establishing funds from which to make loans to the industrious classes, repayable by instalment with interest. A loan society may be registered as a company, or under the Industrial and Provident Societies, or Friendly Societies, legislation, or certified under the Loan Societies Act, 1840. A certified society enjoys certain legal privileges. Such a society is not a corporation and its property is vested in trustees. Its rules are binding on officers and members of, and persons receiving loans from, the society. The amount which may be lent is strictly limited, as is the interest which may be charged.

Local Act. An Act or statute is deemed local if limited in application in respect of area. Such statutes include those conferring on particular local authorities powers not enjoyed by local authorities generally or establishing bodies for the performance of purely local functions. They normally originate in private bills, or in provisional orders later confirmed by Act, and in the lists of statutes they are numbered by chapter numbers in small Roman numerals.

Local authority. A term for the persons or body entrusted with the execution of particular statutory functions in particular areas. Accordingly it has meant different things for the purposes of different statutes in the past. Since the development of multipurpose elected bodies for the exercise of numerous functions of local government, local authority means a county or district council, the Greater London Council, a London borough council, a parish council, or in Wales a community council. In Scotland it is a regional, islands, or district authority. In particular contexts there are local education authorities and local planning authorities.

Local courts. In England there were formerly a great number of local courts established by royal charters, by local and personal Acts, or by prescription. Most of these had fallen into decay by the early nineteenth century and the creation of the modern system of county courts (q.v.) in 1846 largely completed the process. Some survived till 1886, and 141 borough civil courts were abolished when local government was reorganized in 1972. In 1971 seven local courts still regularly exercising civil jurisdiction were abolished, namely the Chancery Courts of the Counties Palatine of Lancaster and Durham, the Mayor's and City of London Court (replaced by a county court of the same name), Bristol Tolzey and Pie Poudre Court, Liverpool Court of Passage, Norwich Guildhall Court, and Salford Hundred Court. Other Acts abolished or removed the judicial bases of certain other courts. Those still surviving are the Chancellor's Courts of Oxford and Cambridge, the Court of the Duchy Chamber of Lancaster, the Court of Husting of the City of London, the two sheriffs' courts of the City of London, the court of the Chamberlain of the City of London, and the Great and Small Barmote Courts of the High Peak, and the Barmote Courts of Wirksworth, in Derbyshire.

Local government. In the United Kingdom, local government is the administration locally by elected bodies of functions conferred on them by the central government. There are different structures of local authorities in England and Wales, Scotland, and Northern Ireland with different powers and functions, while London has a structure peculiar to itself.

The oldest scheme of local government in England consisted of the ancient communities of the county, the hundred, the township and the borough, all having common law powers. Later the justices of the peace had many local government functions conferred on them.

In England, prior to the nineteenth century, the only subjects locally administered were poor law and highways, both administered on the basis of the parish, originally the ecclesiastical parish and later adopted as the area for local civil administration, chiefly by the church wardens. The Poor Law Amendment Act, 1834, formed unions of parishes for poor law purposes, managed by locally created boards of poor law guardians which were subject to the control of the Poor Law Commissioners. The Municipal Corporations Act, 1835, made the corporations of boroughs into local authorities and directed them to set up police forces and watch committees. Subsequent legislation provided for the establishment of local boards of health, of the Local Government Board in 1871, and for the creation of sanitary districts all over England under sanitary authorities. The Local Government Act, 1888, set up county councils, which assumed most of the administrative business of county justices in quarter sessions and certain functions as to highways. Certain towns were made county boroughs. In 1894 there were established parish councils and parish meetings, and urban and rural district councils. The Local Government Act, 1929, made county councils poor law and highway authorities. After 1945 many services provided wholly or partly by local authorities were transferred to public corporations, such as national assistance (formerly poor law), electricity, and gas.

From 1933 to 1974 the structure of local government was that the whole of England and Wales, apart from London, was divided into administrative counties, subdivided into county districts, which were non-county boroughs, urban districts or rural districts, and county boroughs. Each had an elected council as authority. County councils comprised elected county councillors, with county aldermen and a chairman elected from their own number. Boroughs elected councillors, aldermen, and a mayor (or Lord Mayor); county boroughs had most of the powers and functions of counties. Urban and rural districts had councils composed of elected councillors with a chairman elected from among them. Rural parishes had parish councils and parish meetings.

From 1974 England, apart from London, is divided into counties (six metropolitan counties and 39 non-metropolitan counties) each divided into metropolitan and non-metropolitan districts, and Wales into eight counties each divided into districts. Districts are divided in England into parishes, in Wales into communities. Counties and districts have as their authorities elected councils, which elect their own chairmen. Municipal corporations of boroughs have ceased to exist.

Local authorities have wide powers of legislation for minor matters by by-laws.

In London the City has had a county organization since Anglo-Saxon times. The Corporation of the City of London discharges functions through three assemblies, the Court of Aldermen, the Court of Common Council, and the Common Hall.

In the nineteenth century special legislation was passed for London dealing with most of the subjects of local government and numerous *ad hoc* bodies of trustees and commissioners were established to provide various services. In 1856 most of these were replaced by elective vestries and district boards, and the Metropolitan Board of Works was constituted. In 1889 the metropolis became the administrative county of London and the London County Council took over from the Metropolitan Board of Works the functions of that Board and the administrative functions of quarter sessions. In 1900 the vestries and district boards were replaced by 28 metropolitan borough councils. In 1963 the administrative county of London was abolished, being replaced by the administrative area known as Greater London with a Greater London Council. The metropolitan boroughs have become 12 inner London boroughs, and there are 20 further outer London boroughs.

Certain services, police, water supply, passenger transport, and port administration are provided by special authorities.

In Scotland local government developed from burghs, counties, and parishes. Burghs acquired a large measure of self-government by charters and statutes and their councils became elective only in 1833. Elected county councils were established in 1889 and acquired powers previously exercised by various other bodies, notably the commissioners of supply.

Under the Local Government Act of 1929 functions were exercised by the councils of four counties of cities, large burghs, small burghs, and counties. Within counties district councils had minor powers. In 1975 a new structure came into being comprising nine regions, each divided into districts, and three island areas. Burghs have lost all local government functions.

There is much public dissatisfaction with the two-tier systems (counties and districts, or regions and districts) of local government which causes great proliferation of bureaucracy, with the poor quality of councillors, and with the inefficiency and profligate expenditure of local authorities, while there is suspicion of corruption from time to time and sometimes proof of it.

S. & B. Webb, *English Local Government from the Revolution to the Municipal Corporations Act* (1906–29); J. Redlich and F. W. Hirst, *History of Local Government in England* (1958); W. O. Hart, *Introduction to the Law of Local Government and Administration* (1975).

Local Government Board. A board established in 1871 to bring together in one department all the central government supervision of the various kinds of local authorities entrusted in their districts with the law of local government, health, poor relief, and similar functions. It replaced the Poor Law Board and the General Board of Health and was itself transformed into the Ministry of Health in 1919, then into the Ministry of Housing and Local Government and now the Department of the Environment.

***Local Government Board* v. *Arlidge*,** [1915] A.C. 120. A metropolitan borough council prohibited the use for habitation of a house owned by Arlidge until it had been made fit for habitation. Arlidge appealed to the Local Government Board which held a public inquiry and, having considered the report of the inquiry, confirmed the closing order. Arlidge sought to have the Board's decision quashed as not having been decided in accordance with natural justice. The House of Lords held that there had been nothing contrary to natural justice in the procedure, even though the report of the inspector who had held the inquiry was not disclosed and the Board had not heard the complainer orally, provided it had given him the opportunity it had to present his case. The procedure was usual and complied with the relevant legislation.

Local inquiry. An inquiry held in the locality concerned into such a matter as the grant or refusal of planning permission or into objections to a proposed development.

Local law. Law may exist in particular districts, areas, or localities which constitutes an exception to or derogates from the general law of the land. This may take the form of local legislation, by-laws made by a local authority in the exercise of power to do so conferred on it by charter or statute; such local legislation is recognized and enforced by the courts so long as *intra vires*, validly made, not unreasonable and not contrary to any general rule of law. Or it may take the form of local customary law, based on long-standing and settled custom, so notorious as to be deemed to have the force of law. Most local customs in England and Scotland disappeared with the growth of systems of common law general to the whole country, but some, e.g. gavelkind, borough English, survived until recently.

Locatio conductio. The letting to hire of goods or services in Roman and Scots law. It is distinguishable into *locatio rei*, the hiring of a thing; *locatio custodiae*, the hiring of custody of a thing; *locatio operarum*, the hiring of service or the employment of workpeople; *locatio operis faciendi*, the hiring of services to do a job or the employment of a contractor; and *locatio operis mercium vehendarum*, the hiring of the carriage of goods. The first two and the last of these categories fall within bailment (q.v.) in English law.

Loccenius, Johannes (1599–?). Professor at Uppsala, author of a history of Sweden, and the first methodical works on the civil law of Sweden, *Synopsis juris sueco-gothici* (1648), *Lexicon juris sueco-gothici* (1650), *Sueciae leges provinciales et civiles latine versae* (1672), *Synopsis iuris privati ad leges Suecanas accommodata* (1653).

Lochner* v. *New York (1905), 198 U.S. 45. A New York statute forbade employment of anyone in a bakery or confectionery shop for more than 60 hours a week. The Supreme Court found the statute unconstitutional as a violation of the right of an individual to make contracts in relation to his business, which was part of the liberty of the individual protected by the Fourteenth Amendment. The case provoked a notable dissent by Justice Holmes and illustrates the extreme *laissez-faire* attitude of a majority of the courts at its date.

Loci communes. Collections of passages in the civil (Roman) law which came to form the starting point of discussions on subjects of use to legal practitioners. Later they came to mean useful general principles or maxims on which controversy had centred, and *loci* superseded the older word *brocarda.* Such collections began to appear in the fourteenth century. They had an eminently practical aim and were well-known to practical civilians. The best-known collection is that of Nicolaus Everardus, *Topicorum seu Locorum legalium Opus* (1544), containing 100 *loci*, enlarged in later editions to 131. By the end of the fifteenth century classification had settled the number of *loci* at about 100.

Locke, John (1632–1704). English philosopher, was a tutor at Oxford and later travelled in Europe. He was secretary to the Earl of Shaftsbury (1666) and when the latter fell from power Locke had to leave the country. He was an exile in Holland in 1683–89, returned with William of Orange, and was a commissioner of appeals (1689–1704) and a member of the Board of Trade (1696–1700). The events of his lifetime led him to consider the role of the individual in society in terms of personal freedom, religion, and government. His writings include *Essays on the Law of Nature* (published 1954), *Letters on Toleration* (1689, 1690, and 1692), *Two Treatises on Civil Government*, in defence of the ultimate sovereignty of the people, and an *Essay concerning Human Understanding* (1690). The first *Letter* involved him in controversy and was followed by second and third *Letters.* He wrote also on theology, economics, and education. His *Treatises on Civil Government* were meant to refute the ideas of absolute monarchy of Hobbes and Filmer. They assert that natural rights of freedom and property are not abrogated by the change from a state of nature to society. The State rests on mutual contract by the people themselves to be thus governed; civil rulers hold their powers not absolutely but conditionally and subject to forfeiture if they did not obey their moral trust. The *Treatises*, though possibly written earlier, are the classic defence of the Revolution of 1689, and state principles later influential in the American War of Independence and the French Revolution and influenced Adam Smith, Berkeley, and Kant.

M. W. Cranston, *John Locke* (1957); R. Aaron, *John Locke* (1971).

Lockhart, Sir George, of Carnwath (?1630–89). Became an advocate in 1656 and in 1658 was named advocate to the Lord Protector. After the Restoration he had a large practice and became Dean of the Faculty of Advocates, being reputed the most skilful and eloquent pleader of the time. He also sat in the Scottish Parliament. During the covenanting persecutions he frequently defended political prisoners, and in 1679 was employed to impeach Lauderdale's administration before the King. In 1685 he became Lord President, but in 1689 was murdered by a disappointed litigant, Chiesly of Dalry. He has been characterized as proud and avaricious. His son, George, was author of *Memoirs of the Affairs of Scotland*, 1702–07, an important source for the history of the period.

Lockhart, Sir James, of Lee (?–1674). He sat in the Scottish Parliament, and in 1646 became a Lord of Session but was deprived of office in 1649. He was restored in 1661, again became a Lord of the Articles and a commissioner of the exchequer. He was Lord Justice-Clerk 1671–74. His son, George (q.v.), became Lord President.

Lockhart, Sir William, of Lee (1621–76). Elder brother of George, Lord President (q.v.), fought in the civil wars and became one of Cromwell's commissioners for the administration of justice in Scotland in 1652 and a member of the Scottish Privy Council. He was Lord Justice-Clerk 1674–75 and later English Ambassador in Paris.

Lock-out. A refusal by an employer to allow his employees to work, as by excluding them from the workplace, a measure sometimes employed as an element in an industrial dispute, the counterpart of a strike (q.v.).

Locus. A place, a word frequently used to denote the place in or at which some material act or event, such as a crime, delict, or breach of contract, took place. Hence it appears in such phrases as *lex loci delicti.*

Locus poenitentiae. The opportunity to withdraw from a bargain before it has been fully constituted and become binding. Thus in consensual contracts *locus poenitentiae* exists so long as an offer has not been accepted; in contracts where formally executed writing is required *locus poenitentiae* exists so long as no formal writing has been executed or no part performance supplying the lack thereof has taken place.

Locus regit actum. The place governs the act, a phrase sometimes used in the context of international private law to denote the principle that the place where an act is done supplies the law to regulate the validity of the act. The principle is, however, subject to many exceptions.

Locus sigilli (or L.S.). The place of the seal, sometimes marked by a circle at the end of a document requiring to be sealed.

Locus standi. A place to stand, or the right of a party to appear before and be heard by a tribunal, especially in applications for prerogative writs and orders. The right of a petitioner to be heard in opposition to a Private Bill, Hybrid Bill, Provisional Order, or Special Procedure Order depends on his having *locus standi*, which depends on whether his property or interests will be adversely affected if the proposed Bill or Order is passed. Questions of *locus standi* are determined by the Court of Referees in the House of Commons and by the committee dealing with the Bill in the House of Lords, by the committee hearing objections to the Provisional Order, and the Lord Chairman and the Chairman of Ways and Means in respect of Special Procedure Orders.

Lodemanage (or Loadmanage). A charge for piloting a ship. Formerly the pilots of the Cinque Ports were known as lodesmen or leadsmen (from the lead used to sound the depth of water) and their society was governed by the Court of Lodemanage, a branch of the Court of Admiralty of the Cinque Ports. The society's property was transferred to the Trinity House in 1853.

Loder, Bernard Cornelius Johannes (1849–1935). Dutch jurist, judge of the high court of the Netherlands, 1908–21, and first President of the Permanent Court of International Justice, 1922–25.

Lodger. A person occupying one or more rooms in a house by the licence of the landlord. He is not a tenant but a licensee only.

Log. A record maintained by every ship, save in certain excepted cases, in which are recorded all happenings in and to the ship as soon as possible after the happening of the event. The log is accordingly good evidence of facts and events in many kinds of legal proceedings.

Logic in legal contexts. Both legal study and the application of law in practice rely extensively on logic. In legal study aspects of logic which are utilized are analysis and classification of legal systems and of the principles and rules of each individual legal system and of each branch thereof, and analysis of legal terms, concepts and their connotations and implications and their logical relationships. The most notable work of this kind was Hohfeld's (q.v.) *Fundamental Legal Conceptions*. To some extent every good treatise on any legal system or branch of law utilizes logical techniques.

In the application of law in practice, whether ascertaining the law applicable to a problem, seeking to persuade by argument, or deciding a dispute, logic is relevant. In the simple cases of ascertaining applicable law or deciding a dispute, application of the syllogism and deduction is common. Thus: X is a crime carrying the penalty Y; AB has done X; therefore AB is guilty of the crime of X and liable to the penalty Y. This mode of reasoning is, however, subject to formal fallacies, and to verbal fallacies, or fallacies of ambiguity, arising from improper use of words, e.g. Negligence (i.e. breach of legal duty) is a tort inferring liability in damages; AB has been negligent (i.e. careless) and has injured XY. In this case no conclusion follows because carelessness is not equivalent to breach of legal duty. Reasoning is also subject to material fallacies, such as the argument *ad hominem* (attacking the person), the argument *ad misericordiam* (appealing to pity) and the fallacy of accident (applying a general rule to a particular case though it may be inapplicable because of some speciality in the particular case). In more complicated cases determination of the existence and proper formulation of the major premiss of the legal syllogism is the major enquiry and may involve analysis of many cases and inductive discovery from them of a general principle, or a difficult problem of interpretation of a statute. The minor premiss is a question of fact and of evidence. Thus in the case: Murder is criminal and punishable by life imprisonment; AB has committed murder; therefore AB is punishable by life imprisonment; the syllogism is simple and the conclusion undoubted, but what AB did and whether in the circumstances it amounted to murder may raise difficult issues of fact and evidence (what he did), and issues of mixed fact and law (whether what he did was legally murder or only assault or justifiable homicide).

It would be wrong to think that advising on law or deciding controversies is always or only an application of logic. That is not so; frequently personal attitudes, questions of policy, the exercise of discretion and considerations of what is right and fair enter into the ultimate determination.

E. Levi, *Introduction to Legal Reasoning* (1948).

Lombarda. The last collection of Lombard legislation, made at Bologna about A.D. 1100, in which the material is arranged topically in books and titles, and which was long the basis of study of Lombard law. There are two texts, the *Lombarda cassinense*, found at Monte Cassino, and the *Lombarda vulgata*, used in the Bologna law school.

Lombards, Laws of the. The edicts of the Lombard kings after their conquest of Italy in A.D. 568 were collected and promulgated as the *Edictus Langobardorum* by King Rothari in A.D. 643. This work exhibits order and system, draws on Roman law, but represents pure Germanic law. It was supplemented by the legislation of successors, of Grimoald, Liutprand (713–35) who added 15 volumes, largely of ecclesiastical enactments, Ratchis, and Aistulf. After the union of the Lombards with the Frankish kingdom, capitularies made for the whole kingdom were applied to Italy, and there were special capitularies for Italy, *Capitula Italica*, some of which were appended to Rothari's Edict.

This body of law substantially influenced Italian law and the study of the Edict was the basis of substantial legal literature, which had considerable influence on the revival of legal learning in the eleventh and twelfth centuries. The Lombard Edict was edited in the early ninth century under the title *Capitula legis regum Langobardorum* or *Concordia de singulis causis*. A later compilation prepared at Pavia of the capitularies for use in Italy was known as the *Capitulare Italicum* or *Langobardorum* (*c.* 1090). At the law school of Pavia the *Liber Papiensis* or *Liber legis Langobardorum* was compiled, largely for teaching purposes, in the eleventh century. It was later supplemented, between 1019 and 1037, by the *Walcausina*, a collection of annotations and formularies, and about 1070 by the *Expositio ad librum papiensem*, a commentary utilizing the works of Justinian, the whole being systematically revised as the *Liber Lombardae* of which there are two forms, the *Lombarda Cassinensis* and *Lombarda Vulgata*. In addition to the statutes in the Edict there was a considerable volume of subordinate royal ordinances and of popular customary law.

The Lombard law, unlike that of other Germanic systems, which disappeared from Italy as soon as the power which brought them waned, penetrated deeply and survived, becoming intermixed with Roman law, giving rise to schools of jurists, particularly those of Pavia. It survived in places in Italy until the seventeenth century and did not wholly disappear until the introduction of the French-based Codes in the early nineteenth century.

E. Merkel, *Geschichte des Langobardenrechts*.

Lombroso, Cesare (1836–1909). Italian criminologist, who propounded the view that the criminal population included a higher proportion than the non-criminal of persons exhibiting physical and mental abnormalities due partly to atavism, partly to degeneration. The theory is now regarded as unproved. His major works were *L'Uomo delinquente* (1889), *La Donna delinquente* (1893), and *Le Crime, causes et remèdes* (1899).

London Court of Bankruptcy. In 1831 bankruptcy jurisdiction was taken away from the Lord Chancellor and assigned to a Chief Judge in Bankruptcy assisted by three other judges and commissioners. Appeals lay at first to a Court of Review, and from 1847 to a Vice-Chancellor, and then to the Chancellor. In 1869 the London Court of Bankruptcy was established under the Chief Judge in Bankruptcy assisted by such other judges as the Chancellor might appoint, appeal lying to the Court of Appeal in Chancery. In 1875 the Court was embodied in the new High Court.

London Sessions. A name sometimes given to the district sittings of the Central Criminal Court or Old Bailey in London.

Long Parliament. The fifth and last Parliament of Charles I, which met in 1640, took up arms against the King in 1642, was purged of members well-disposed to the King by Pride's Purge in 1648, abolished the monarchy and the House of Lords in 1649, was expelled by Cromwell in 1653, was recalled and expelled again in 1659, and was restored at the end of 1659 and dissolved in 1660. It was the major instrument of opposition to Charles I and was the organiser of resistance until the army took over political power in 1645.

Long quinto. The longer of the two Year Books for the year 5 Edn. 4 (1465–66), sometimes called the Long Report. It is of a better standard than most Year Book reporting and some of the cases seem to be eclectic compilations.

Long vacation. The period of August–September during which the courts do not sit regularly and only urgent business is done.

Longchamp, William (?–1197). A Norman of low birth who became chancellor of Richard I, bishop of Ely, and one of the chief justiciars of the kingdom, and then papal legate in England. When Richard was absent from the kingdom Longchamp acted as viceroy, ruling arrogantly and oppressively. He was dismissed in 1191 by a council of the bishops, earls, and barons, and had to flee the country, but towards the end of Richard's reign again exercised the office of chancellor and acted in state affairs.

Longi temporis praescriptio. A doctrine of Roman law first found in the later second century A.D., of the nature of limitation of actions (q.v.), but by Justinian's time being a mode of acquisitive prescription, applicable to land. The period was 10 years if the parties were in the same district, of 20 years if they were not. *Longissimi temporis praescriptio* was a similar acquisitive prescription established by Justinian, applicable to moveables, whereby if a thing had been acquired in good faith and held for 30 years, it became absolute property.

Longinus, Caius Cassius (1st cent. A.D.). A famous Roman jurist and head of the Sabinian school, consul, governor of Asia and Syria. He wrote an important work on the *ius civile* and *libri ad Vitellium*.

Lopez de Viveros, Juan (Palacios Rubios) (1447–1523). A Spanish jurist, professor at Salamanca, editor of the *Leyes de Toro* and author of a commentary thereon, compiler of the liberties of the Mesta, and author of treatises on husband and wife and political works.

Lord. The title of honour accorded to barons, and by courtesy to the sons of a duke and the eldest son of an earl. Judges of the High Court and Court of Appeal are, in court, addressed as 'My lord'. By custom and courtesy it is the title given to all judges of the Court of Session in Scotland, though they are usually not peers, the name being either the judge's family name, or one taken from his estates. They also are addressed in court as 'My lord'. In feudal law the term is frequently used of a superior in the feudal hierarchy, whether or not he is of the peerage, as contrasted with a man or vassal, and this usage survives in the word landlord, used of the lessor of lands to a tenant, and the phrase 'lord of the manor'. The interest held by a lord in the vassal's lands is sometimes called lordship.

Lord Advocate. The King of Scots is known to have had his Advocate from the fifteenth century and from the sixteenth century the Advocate was a prominent person with legal, political, and diplomatic functions. He had a seat in parliament *ex officio*, became public prosecutor, and was recognized as a great Officer of State. In the seventeenth century the office was sometimes combined with a judgeship in the Court of Session.

After 1746 there was no Secretary of State for Scotland and the whole management of Scottish affairs fell on his shoulders. This lasted until, in 1885, the Department of the Secretary (since 1926 the Secretary of State) for Scotland was established.

In modern practice the Lord Advocate is, in the first place, the senior adviser of the Government of the day on all matters of Scots law. As a member of the Government, he demits office on a change of Government. The office is sometimes held on a non-political basis. He is frequently an M.P., and is sworn of the Privy Council. His legal secretaries act as parliamentary draftsmen for Scottish bills.

Secondly, he is the person who sues and is sued on behalf of the Crown and all Government departments and agencies, responsibilities in which he is assisted by members of the Bar appointed as Standing Junior Counsel to the various departments.

Thirdly, he is head of the system of public criminal prosecution in Scotland and all criminal cases of importance are reported to the Crown Office for instructions on prosecution, and he frequently appears personally. He has full power to prosecute all crimes committed in Scotland, and cannot be compelled, even by the High Court, to prosecute. When proceedings have been instituted he remains master of the instance and may abandon the charge or restrict it or restrict the pains of law by not moving for sentence. In these functions he is assisted by the Solicitor-General for Scotland, and by several Advocates-Depute, appointed by himself, who formerly also demitted office on a change of Government.

In court he is entitled to sit within the Bar on the right-hand side of the Bench. His position corresponds roughly to that of Attorney-General in England, but there are many points of difference. The office is commonly a stepping-stone to the Bench, frequently to the chair of one of the Divisions of the Court of Session.

A list of holders of the office is in the Appendix.

Lord Chamberlain. An officer of the royal household from Norman times. Today he is appointed by the government of the day, nominally as head of the Royal Household, but does not go out of office on a change of government. As such, he has duties to protect the royal arms from misuse. Formerly he had the function of licensing stage plays for public performance.

Lord Chancellor (of England and, since 1707, of Great Britain). This is an ancient and dignified office of state known since the eleventh century, which evolved out of the headship of the Chancery, or royal secretariat. He also holds the office of Keeper of the Great Seal, and formerly a person was sometimes appointed to the latter office and not also, or only later, to the office of Lord Chancellor.

The name chancellor is derived from the *cancelli* or screen behind which the mediaeval royal secretaries worked. Edward the Confessor was the first king to have a Chancellor to keep the King's seal. He was always an ecclesiastic, chief of the royal chaplains, secretary of state for all departments, and he drew and sealed royal writs. He was a member of the Exchequer branch of the *Curia Regis*, assisted in the judicial business of it and of the Exchequer, and as the business of his office developed, he became head of the Chancery. In 1199 the departments of the Exchequer and the Chancery were separated.

The office was originally always held by an ecclesiastic, a class which included such persons as Thomas Becket, Hubert Walter, and Thomas Wolsey, but lay chancellors were appointed sometimes from the fourteenth century. The last clerical chancellor was Archbishop Williams, 1621–25, but after the Reformation chancellors were normally, and after 1625 always, lawyers. The great legal importance of the chancellor sprang from his headship of the Chancery which granted or refused writs to persons seeking to initiate actions, and from the practice from the fifteenth century of the Council referring to him petitions to the King and Council for justice in special circumstances, which enabled successive Chancellors to create, develop, and elaborate, the whole system of equity. In this task of developing equity the most notable work was that of Lord Nottingham, Lord Hardwicke, and Lord Eldon.

The Chancellor is appointed by the Crown on the nomination of the Prime Minister from among his supporters and is today always a member of the Cabinet. The appointment is made by the Queen delivering the Great Seal of the United Kingdom to the appointee and addressing him by the title. He holds office during the royal pleasure, but by convention retires from office if his party is defeated. He is not necessarily a peer but is ennobled on appointment and not infrequently has been promoted in the peerage on retiral from office. As historically he is deemed Keeper of the King's Conscience it was long thought that he must belong to the Church of England, but statute in 1974 declared that the office was open to Roman Catholics.

The Lord Chancellor has many functions. In the Cabinet he is the nearest approach to a Minister of Justice yet developed in the British Constitution, makes recommendations for the appointment of Lords Justices of Appeal, appoints judges of the High Court, circuit judges and many lesser legal officers, and exercises huge ecclesiastical patronage.

In the House of Lords he is Speaker but may participate in debates and divisions. He is the means of communication between the Queen and Parliament and, if she is absent, reads the Speech from the Throne at an opening of Parliament. He reports the Royal Assent to Bills.

In the House of Lords sitting as a final Court of Appeal he sometimes, though less frequently than before 1939, takes part in the hearing of appeals and when doing so presides and usually delivers the leading judgment.

He is also President of the Supreme Court of Judicature, an *ex officio* member of the Court of Appeal, and nominal President of the Chancery Division of the High Court, but in modern times has very rarely sat in any of these capacities.

The Chancellor now ranks as the leading subject in the land, yielding precedence only to the Royal Family and the Archbishop of Canterbury.

After retirement, when he is entitled to a pension, there is a customary obligation to assist in the hearing of appeals in the House of Lords if he is requested to do so, and an ex-chancellor may be appointed a Lord of Appeal, as in the case of Viscount Dilhorne (q.v.).

For list of holders of the office, see Appendix.

Lord Chancellor of Ireland. An Irish chancery was established in 1232 and granted to the Chancellor of England, who executed the duties by deputy. Later it was normally held by English lawyers, few of whom were distinguished, though notable exceptions were Lord Redesdale (1802–6), Lord Plunket (1830–34, 1835–41), and Lord Ashbourne (1885–86, 1886–92, 1895–1905), as well as Sugden (1834–35, 1841–46) and Campbell (1841), both later Lord Chancellors of Great Britain.

For holders of the office see Appendix.

Lord Chancellor of Scotland. In Scotland a chancellor is met with from the early twelfth century as witnessing charters and, by the King's writ, granting redress to persons wronged by the Courts or persons in power. By the sixteenth century he was a high officer of state, head of the law, presided in the Court of Session, presided over the Privy Council and the Lords of the Articles, was entitled to have the Great Seal carried before him and to have first place in all public meetings, sat thrice annually with certain members of the Three Estates to hear causes remitted to him by the King's Council, and had numerous other functions.

A salient difference from the office in England was that the law in Scotland did not develop the premature rigidity of the common law in England, so that there never arose the need for any equitable jurisdiction, independent of, parallel to, and sometimes in conflict with, common law jurisdiction as in England. In Scotland equity could be, and was, administered within legal forms and principles by the ordinary courts.

Since the Union of 1707 there has been one Lord Chancellor for Great Britain (and since 1922 for the United Kingdom), who is always chosen from the English bench or Bar. The last Chancellor of Scotland, Lord Seafield, was in 1708 appointed Lord Chief Baron of Exchequer in Scotland and in 1713 was appointed Keeper of the Great Seal of Scotland and presided as Chancellor in the Court of Session; he died in 1730.

For list of holders of the office see Appendix.

Lord Chief Justice of England. The Courts of King's Bench and Common Pleas had Chief Justices from early times and the C.J.K.B. sometimes claimed the title of Lord Chief Justice. The office officially came into being in 1859 when Sir A. Cockburn, Chief Justice of the Common Pleas since 1852 was given the title. (He was replaced by Sir W. Erle as C.J.C.P. and Sir J. F. Pollock was C.B. The titles of C.J., C.P., and C.B. were abolished in 1880 when these divisions were merged with the Q.B. Division, Lord Coleridge, C.J.C.P., succeeding Lord Cockburn as L.C.J.). The Lord Chief Justice of England is now the head of the Queen's Bench Division of the High Court and ranks second only to the Lord Chancellor in the judicial hierarchy. He is appointed by the Crown on the nomination of the Prime Minister; if not already a peer he is given a peerage and may sit in the House of Lords and Privy Council, and is an *ex officio* judge of the Court of Appeal. In practice he spends most of his judicial time presiding in the Court of Appeal, criminal division, and in the Divisional Courts of the Q.B. Division.

For holders of the office see Appendix.

Lord Clerk Register. One of the Officers of State in Scotland from early times. The holder of the office was originally Clerk of the Scottish Parliament and of the other great Courts and Councils of the realm, Chief Clerk of the Court of Session and custodier of all its records. Latterly he was *ex officio* a member of the Privy Council and Parliament. He frequently acted as a Lord of Session, and in the course of time a great variety of records and registers were entrusted to his care. After 1707 the office became largely honorary. In 1817 the office of Principal Keeper of the Signet was conjoined by statute.

Since 1879 his office has been a titular one only, the Deputy Clerk-Register having charge of the public registers and records of Scotland until 1928 when that office was abolished and replaced by those of Keepers of the Records and of the Registers of Scotland, offices separated in 1948. The last function preserved was that of conducting the election of representative peers of Scotland.

A list of holders of the office is in *Handbook of British Chronology*, 189.

Lord Great Chamberlain. One of the great Officers of State of England. The office is hereditary. The holder was originally the financial officer of the King's household and formerly had responsibility for the whole Palace of Westminster, but this has now been limited to parts thereof. He has various responsibilities connected with the Opening of Parliament, the homage of bishops after consecration, and the introduction of new peers into the House of Lords. At coronations he participates in presenting the sovereign to the people, which ceremony is a relic of the ancient election, and girds the sovereign with his sword, signifying justice and protection. The office is distinct from that of Lord Chamberlain.

Lord High Admiral. Formerly the officer having supreme command of the Navy, a function entrusted from 1708 to the Board of Admiralty and now to the Admiralty Board acting under the Defence Council. Queen Elizabeth II reassumed the style of Lord High Admiral.

Lord High Chancellor. See LORD CHANCELLOR.

Lord High Commissioner. Between 1603 and 1707 a Lord High Commissioner to the Parliament of Scotland frequently maintained a vice-regal court at Holyroodhouse, Edinburgh. Since 1707 this vice-regal ceremonial has attached to the Crown's Commissioner to the annual General Assembly of the Church of Scotland (which meets annually in Edinburgh for 10 days in May). A person, a prominent and respected layman, is appointed by the Queen on the advice of the Government and for the duration of the Assembly resides in Holyrood, is styled 'Your Grace', the ancient Scots form of addressing the sovereign, and has the precedence of and is attended with all the ceremony proper to a king or queen. He is frequently re-appointed for a second time, but rarely again. He is attended by a purse-bearer, a master of the horse, a chaplain, and by aides-de-camp, while Her Grace has a lady-in-waiting and maids-of-honour. In 1969 the Queen attended the Assembly personally and there was no Lord High Commissioner.

Lord High Constable. Formerly a great office of state, who like the Earl Marshal had duties largely connected with the army in the field in the Middle Ages. From about 1348 they jointly held a *Curia Militaris* or Court of Chivalry (q.v.) in which were triable offences committed out of the realm and matters relating to arms. Since the early sixteenth century appointments to the office of Lord High Constable have been made only on special occasions, usually coronations. There is also a Lord High Constable of Scotland, an office hereditary in the family of the Earls of Erroll.

Lord High Steward. So long as peers retained the right to be tried by their peers for treason or felony, which was the case till 1948, the presiding officer was the Lord High Steward, who was appointed for the occasion and was usually the Lord Chancellor. All the peers present were judges of fact and law and the Lord High Steward had only one vote like the others. If the trial took place while Parliament was in recess, the Lord High Steward was the sole judge on matters of law, being assisted by a jury of peers. This court was quite distinct from that of the Lord Steward of the Queen's Household. The office of Lord High Steward early became hereditary in the family of the Earls of Leicester, but was forfeited to the Crown in 1265 and since 1407 appointments have been made *ad hoc* when necessary, by commission under the Great

Seal in response to an address by the House of Lords. The holding of the commission was signified by a staff of office and at the end of the proceedings the Lord High Steward broke the staff and declared the commission dissolved.

Lord High Treasurer. In England the office of treasurer may have evolved out of that of chamberlain or have been instituted as a new office by Henry I. Under him the exchequer developed. The titles Lord Treasurer and Lord High Treasurer came into use in the sixteenth century and were used regularly from 1612. The office was put into commission first in 1612 and always since 1714. One of the lords commissioners was the chancellor of the exchequer and under-treasurer and from 1730 the post of First Lord of the Treasury has always been held by the leader of the ministry, now called the Prime Minister.

The revenues of the Scottish kings appear to have been under the administration of the chamberlain. Probably in 1424 new officers, the comptroller and the treasurer, were appointed and came to supersede the chamberlain. The office was in commission continually from 1686 till the Union after which the exchequer was reorganized under the nominal presidency of the Lord High Treasurer of Great Britain.

Lord Justice of Appeal. The title of any of the ordinary judges of the Court of Appeal in England. Such judges are usually referred to in legal books as, e.g. Smith, L.J., and in the plural (Lords Justices), as, e.g. Smith and Jones, L.JJ. From 1851 to 1875 the title was borne by the two Lords Justices of Appeal in Chancery who heard appeals from a vice-chancellor (q.v.). See also LORDS JUSTICES.

Lord Justice-Clerk. The Justice-Clerk of Scotland was originally clerk to the Justiciar's Court. The post in time became one of considerable standing and by the late sixteenth century the holder was always a member of the Privy Council. In the seventeenth century, from his familiarity with the proceedings of the court, he was regularly named an assessor to the Justice-General or Justice-Depute. Later in the seventeenth century he came to be recognized as one of the judges of the court and assessors were appointed to him. When the High Court of Justiciary was reorganized in 1672, the Justice-Clerk ranked next to the Justice-General and, as the latter was until 1836 a layman, he was the normal working head of the Court. In the Court of Session he ranked equally with the other judges until in 1808 he was made President of the Second Division of the Inner House of the Court of Session. Since then he has ranked as holder of the second highest judicial office in Scotland in both the civil and criminal courts.

A list of holders of the office is in the Appendix.

Lord Justice-General. The term Justice-General is used in Scotland from the fifteenth century for the supreme judge in criminal causes, superseding the older term Justiciar. The office became hereditary in the Argyll family about 1514, but was resigned to the King in 1628, under reservation of Argyll and the Western Isles. Thereafter the office was conferred on various nobles for life or for shorter periods. In 1672 statute created the High Court of Justiciary as supreme criminal court in Scotland, and directed that the Justice-General when present was to preside. From 1746 it was not competent for him to sit in circuit courts alone or with one colleague, as his vote might have overruled the opinion of a legally qualified colleague, but this restriction was abolished in 1830.

The Court of Session Act, 1830, provided that on the death of the then-living noble holder of the office (James, 3rd Duke of Montrose, who died in 1836), the office should be united with that of Lord President of the Court of Session, and since then both offices are held by the one person. Since then the Lord Justice-General has been empowered to hold circuit courts whether or not any other judge were present, but in modern practice the Lord Justice-General does not go on circuit.

A list of holders of the office is in the Appendix.

Lord Keeper of the Great Seal. The Great Seal of England was used in sealing deeds issued from the royal chancery from the eleventh century and the Chancellor was the natural and normal Keeper of the great seal. The offices of Chancellor and Keeper of the Great Seal were, however, distinct and a person was frequently appointed to the latter office and only subsequently, and sometimes not, e.g. Bishop John Williams, L.K. 1621–25, promoted to the Chancellorship. This last happened in the case of Sir Robert Henley (q.v.) (Lord Keeper, 1757; Lord Henley, 1760; Lord Chancellor, 1761). A Lord Keeper has since 1562 had the same powers and jurisdiction as a Lord Chancellor. In the time of Elizabeth the seal sometimes remained with the Queen. Frequently during intervals the seal was in the hands of commissioners; this last happened in 1835. During the Commonwealth and Protectorate, 1642–60, the seal was in the hands of commissioners who were neither chancellors nor keepers.

Lord Lieutenant. Officials first appointed, or at least first legalized, under Edward VI (1551), to suppress rebellions but also with power to repel invasions. They had supreme military authority over one or more counties each, authorized to muster and exercise all persons, to lead them against enemies, domestic or foreign, and to use all the severity of martial law. They were assisted by Deputy-Lieutenants and indemnified for any actions done by authority of their commissions.

The appointment is now made by the Crown, for life, and the holder has particular connection with the local units of the territorial and reserve forces. The holder formerly appointed, with the consent of the Crown, 20 persons who have served with the armed forces as Deputy-Lieutenants, but since 1972 these are appointed directly by the Crown.

The title Lord Lieutenant was also the title borne by the King's governor of Ireland.

Lord Lyon King of Arms. Her Majesty's Principal Officer of Arms in Scotland. The office is very ancient, and descended from the Celtic Seannachie, and Lyon has always enjoyed high rank, precedence, and distinction. He is appointed by the Queen on the recommendation of the Secretary of State for Scotland and must be a member of the Scottish legal profession. He has regalia, a crown, tabard, collar, and baton.

Under the Crown he has control of all arms, badges and signs armorial, both military and civil, and exercises the function of granting armorial bearings to persons, granting supporters, authorizing additions or alterations, differencing the arms of cadets of families, advising the Crown and Government departments on matters of heraldry, genealogy or family representation, executing Royal Proclamations, and arranging State and public ceremonial, and enforcing the Royal Warrant of Precedence.

Lyon is controller of all messengers-at-arms and hence at the head of the executive branch of Scots law.

Lyon, sitting in Lyon Court, exercises both a civil and criminal jurisdiction, both under common law and statute; he has two hereditary assessors, the Lord High Constable and the Earl of Angus. To deal with genealogical questions Lyon holds two peremptory courts on 6 May and 6 November, the old druidic feasts of Beltane and Samhain, and other sittings as required.

He has jurisdiction to protect the rights of the Crown in Scotland and of private individuals in Scottish armorial bearings; the misappropriation or unauthorized display of another's coat of arms is a real injury at common law, entitling the owner to obtain interdict against the wrongdoer.

The Lyon Court is a recognized court of the realm, in which counsel may appear, and from which appeal lies to the Court of Session and House of Lords. Important decisions are reported. Its decrees are enforceable both summarily and by fine, expenses, escheat of moveables, and imprisonment. The Lyon Court early sat at Holyrood, and now in H.M. Register House, Edinburgh.

Lyon is assisted by three Heralds—Marchmont, Albany, and Rothesay; and by three Pursuivants—Carrick, Kintyre, and Unicorn, and sometimes by two additional Pursuivants Extraordinary. Collectively they constitute Her Majesty's Officers of Arms in Scotland, are members of the Royal Household in Scotland, and have numerous ceremonial duties to perform. Lyon Clerk maintains the Public Register of all Arms and Bearings in Scotland, matriculates therein arms granted, confirmed or differenced, the Public Register of Genealogies, and prepares letters patent and other writs under Lyon's Warrant. There is also a Procurator-Fiscal of the Lyon Court, a Macer, and a herald painter.

Lord Mayor. The style and title of the chief magistrate of a city in cases where the right has been conferred by letters patent. In the cases of London and York there is right also to the prefix 'Right Honourable'.

Lord Mayor's Court in London. An inferior court formerly held before the Lord Mayor and aldermen. It was amalgamated in 1920 with the City of London Court as the Mayor's and City of London Court.

Lord of Appeal in Ordinary. The designation of the persons given life peerages and salaried appointments under the Appellate Jurisdiction Act, 1876, as amended, to exercise the judicial functions of the House of Lords. See LORDS OF APPEAL IN ORDINARY.

For holders of the office see Appendix.

Lord of the manor. A person holding land of and under the Crown and from and under whom other persons held lesser tracts of land, either freemen holding land under feudal tenure or villeins (later copyholders) holding land on copyhold tenure. An incident of a manor was the right to hold a court baron in which the feudatories were judges. By the statute *Quia Emptores* (1290) the creation of new manors became impossible save by express statutory enactment or by the Crown under a custom prior to the Act. The lordship of a manor may still carry some or all of the manorial rights expressly preserved by the property legislation of 1925, notably minerals and sporting rights, and the theoretical right to hold courts. A manor may be part of or appendant to a bishop's see, rectory, or parsonage, or belong to the Crown or the Duchy of Lancaster or of Cornwall. The rights, powers, and duties of lord and tenants were governed to a considerable extent by ancient laws or special customs.

Lord Ordinary. A single judge of the Court of Session in Scotland, sitting at first instance. The 14 junior judges of that court normally sit as Lords Ordinary but any judges, including the Lord President and Lord Justice-Clerk, may, and sometimes do, do so.

Lord President of the Council. One of the great offices of State which probably originated about the beginning of the sixteenth century but became permanent only from 1679. The Lord President has charge of the Privy Council office, and at the formal meetings of the Privy Council he puts the business before the Queen and the Privy Councillors summoned. The appointment is a political one, but the departmental duties are not onerous and the holder of the office is available for other extra-departmental duties.

Lord President of the Court of Session. When the Court of Session in Scotland was established in 1532 it comprised a clerical president and 14 judges, half lay and half clerical. The Chancellor was to be president if he should attend. After 1579 the president was chosen by his colleagues, but in 1633 the president was appointed by royal nomination. Between 1641 and 1652 presidents were chosen for one session only, but from 1661 royal appointment revived.

Not only is he head of the Scottish judiciary but he has many administrative powers and functions, many corresponding to powers vested in England in the Lord Chancellor. As the Court is now constituted he presides in the First Division of the Inner House, hearing appeals, and when a larger bench is constituted to hear difficult cases, but he may also sometimes sit alone at first instance. Since 1836 the Lord President has also held the office of Lord Justice-General of Scotland (q.v.) and, as such, been head of the High Court of Justiciary and also the supreme criminal judge.

A list of Lords President is in the Appendix.

Lord Privy Seal. The private or privy seal was used by the mediaeval kings of England when the Great Seal was unavailable or its use inconvenient. From about 1300 there was a Keeper of the Privy Seal. Later the use of the Privy Seal came to be connected with financial business as a warrant for payments from the Exchequer and for the affixing of the Great Seal. The Privy Seal office was abolished in 1884, but the Lord Keeper was retained and the post is now used as an appointment for a non-departmental minister, frequently with some special responsibility.

Lord Steward of the Queen's Household. An official of the royal household, distinct from the Lord High Steward (q.v.), though both developed from that of Seneschal or steward in the thirteenth century. The duties were originally domestic, but later developed political importance and became a cabinet office. In 1924 it ceased to be a political appointment. In theory he is responsible for the everyday management of the royal household, but the duties are now partly ceremonial though he remains the first official of the court and is always a peer and a privy councillor. Formerly he had judicial authority and presided over the Board of Green Cloth or counting house which controlled the expenditure of the household, but also had power to maintain peace within a 12-mile radius of the court. From the twelfth century he also presided over the Lord Steward's Court and exercised jurisdiction over offences committed by the king's servants, and along with the Marshal of the household over the Marshalsea Court which had civil and criminal jurisdiction over actions within that radius if at least one party was a member of the royal household, and in the seventeenth century he had control over the Court of the Steward and Marshal. Both these courts were abolished in the nineteenth century.

Lord Warden. See CINQUE PORTS.

Lords Commissioners of Justiciary. The ordinary judges of the High Court of Justiciary, the supreme criminal court in Scotland. They also hold the offices of Lords of Council and Session, as judges of the Court of Session. Till 1887 only five of the Lords of Council and Session were also Lords Commissioners of Justiciary and since then all have the dual role.

Lords Commissioners of the Great Seal. Persons, normally three in number, to whom in the past the authority and powers of the Lord Chancellor was sometimes conferred. The Seal was then said to be 'in commission'. They had the same authority and powers as a Lord Chancellor or Lord Keeper.

Lords, House of. See HOUSE OF LORDS.

Lords Justices. (1) In England persons, normally barons, were appointed under this title at various times from 1253 onwards, to exercise the royal power during a minority or absence. From 1688 commissioners were chosen from the Privy Council to exercise these functions. From the time of William III onwards, in case of the King's absence from the realm, a regency has been exercised by a member of the royal family or by Lords Justices. Under the modern Regency Acts Counsellors of State, being members of the royal family, are appointed in such circumstances. In Ireland between the sixteenth and nineteenth centuries Lords Justices were persons, normally high officers such as the Lord Chancellor of Ireland and the Archbishop of Armagh, sworn in to execute the office of Lord Lieutenant during a vacancy in that office or the absence of the holder thereof from Ireland.

(2) Two judges appointed first in 1851 to form the Court of Appeal in Chancery and to hear appeals from any of the vice-chancellors (q.v.). In 1875 they became judges of the new Court of Appeal.

For holders of the office see Appendix.

Lords Justices of Appeal. The name given since 1877 to the members, other than the Master of the Rolls, of the new Court of Appeal created in 1875. From 1875 to 1877 their title was Judges of Appeal.

For holders of the office see Appendix.

Lords Marchers. Between 1066 and 1284 the English constantly encroached on Wales, and barons who gradually conquered Wales from Hereford and Gloucester received almost royal or palatine powers. These Lords Marchers were semi-independent. After Edward I subdued Wales and annexed it to the English Crown he divided north and west Wales into six counties placed under the jurisdiction of the justices of Snowdon, Chester, and South Wales, who were to administer justice mostly according to English law. The rest of Wales was governed by the Lords Marchers, who had very wide jurisdiction, extending to life and limb, powers to try all real and personal actions, to issue their own writs and to appoint justices. There developed in their jurisdictions a mixture of English law and Welsh custom known as the custom of the Marches. Edward I tried to bring the Lords Marchers to order by writs of *Quo Warranto*, but then and for long after they were in regular opposition to the Crown. In 1536 England and Wales were united, five new counties created, and 137 Marcher lordships annexed to the old and new Welsh counties or to adjacent English counties. Only a few franchises of Lords Marchers, notably the rights to hold courts baron and courts leet, survived this and an Act of 1543.

Lords of Appeal in Ordinary. Prior to 1876 appeals to the House of Lords could in theory be heard by any three Lords and the only peers with legal qualifications were the Lord Chancellor, any ex-Lord Chancellors, and any peers who happened to have legal qualifications. The judicial strength of the House was reinforced by creation of Parke, B., as Lord Wensleydale in 1856 and of Lord President McNeill as Lord Colonsay in 1867. The Appellate Jurisdiction Act, 1876, provided for the appointment of two Lords of Appeal in Ordinary, who must have held high judicial office for two years or have been practising barristers of advocates for 15 years. There was provision for the appointment of a third and a fourth Lord of Appeal when two and then all four of the then paid members of the Judicial Committee of the Privy Council died or retired. Such persons received salaries, pensions, and life peerages (originally peerages during office only), but held office during good behaviour and might be removed on the address of both Houses of Parliament; they might sit and vote in the House of Lords. They also became members of the Judicial Committee of the Privy Council and entitled to sit also in Privy Council appeals. They are usually appointed from the Court of Appeal or the Court of Session, but may be appointed direct from the Bar. They may resign office or be appointed to another judicial office such as Lord Chancellor, Lord Chief Justice, or Master of the Rolls.

The number of Lords of Appeal in Ordinary was increased to six in 1913, seven in 1929, nine in 1947 and 11 in 1968. At first one, and now two, of the Lords of Appeal are normally drawn from the Scottish bench, and from time to time an Irish judge has been a Lord of Appeal.

A list of holders of the office is in the Appendix.

Lords of Council and Session. The full title of the judges of the Court of Session in Scotland, derived from the dual origin of the court in the sixteenth century, partly from judicial committees of the Privy Council and partly from other committees appointed to hold judicial sessions.

Lords of Erection. Persons to whom, after the Reformation in Scotland, the King gave benefices formerly held by ecclesiastics to be held as temporal lordships.

Lords of regality. Persons who in Scotland had, by grant from the Crown, within their own lordships civil and criminal jurisdiction as extensive as that of the Crown.

Lords of Session. An abbreviated version of Lords of Council and Session (q.v.).

Lords of the Articles. See ARTICLES.

Lords Spiritual. In the mediaeval English Parliament bishops, abbots, and priors sat, originally as landowners, though taxed separately from temporal lords, and by their literacy and abilities formed an important element in Parliament. After the Reformation only the bishops sat, but though bishoprics have multiplied lay peerages have multiplied much more and the Lords Spiritual now form only a small part of the House of Lords. Moreover, since 1847 it has been enacted that the number of bishops in Parliament should not exceed 26; the Lords Spiritual accordingly consist of the (Anglican) Archbishops of Canterbury and York, the Bishops of London, Durham, and Winchester, and the next 21 in order of seniority without regard to their sees. The Bishop of Sodor and Man sits in the House of Keys of the Isle of Man and is not eligible to sit in the House of Lords. On resignation of his bishopric a bishop ceases to be a lord of Parliament. When present in the House the lords spiritual wear episcopal robes and sit together on the benches on the right of the woolsack. It is customary for one of the lords spiritual to attend the House at the beginning of each day's proceedings according to a rota in order to read prayers. Other churches have no place among the Lords Spiritual though a clergyman of another denomination may be a peer and sit as such, e.g. the Rev. Lord Macleod of Fiunary.

Lords Temporal. The secular or lay members of the House of Lords, comprising the holders of hereditary peerages of England, Scotland, Great Britain, and the United Kingdom, and of life peerages (including Lords of Appeal in Ordinary). They are ranked as dukes, marquesses, earls, viscounts, and barons, and a definite precedence and place in the House is assigned to each rank of the peerage, but this is now applied only on formal occasions. The category now includes hereditary peeresses in their own right and women given life peerages. All such persons are entitled to a writ of summons, provided they have attained 21, are not aliens, not bankrupt, nor serving imprisonment exceeding 12 months.

Lordship of Ireland. A sovereignty over Ireland east of Lough Neagh to Limerick created by Henry II in 1177 and conferred on his son John. It became merged in the English Crown in 1199 when John succeeded and remained vested in the English Crown until 1540. The titular lordship was supported initially by no central authority or administration, but in the early thirteenth century Ireland was granted the laws of England and the king's court in Dublin was to be the supreme court of the realm. Magna Carta was published in Ireland in 1217.

The Lordship was at its height during the reign of Edward I, who aimed to make the Lordship the second jewel in his crown and extended many of the great public statutes such as the Statute of Gloucester to Ireland in 1285. A Parliament was summoned in 1297 and from 1310 was well-established.

From about 1320 to 1485 the native people recovered two-thirds of Ireland and the English Lordship shrank and declined. In 1540 a Parliament at Dublin extinguished the Lordship and confirmed Henry VIII as King of Ireland, but of Ireland as united, annexed, and knit for ever to the Imperial Crown of England.

Loreburn, Lord. See REID, R. T.

Lorimer, James (1818–90). Became an advocate in 1845 but practised little and acquired some reputation as a writer on political science. In 1865 he was appointed to the Chair to the Law of Nature and Nations at Edinburgh and did much to develop legal studies in Edinburgh. He also advocated many political reforms later adopted. His teachings were published as the *Institutes of Law* (1872), in which he combated the prevalent utilitarian views, and the *Institutes of the Law of Nations* (1883–84), in which he bases public international law on the law of nature. His theories had more influence on the Continent than in Britain. He was one of the founders in 1873 of the Institute of International Law and produced a draft scheme for a permanent congress of nations and an international court. Some of his other writings are collected in *Studies, National and International* (1890).

Loss. Financial detriment, which may arise from one or more of various causes, notably breach of contract, tort or other breach of legal duty, or pure accident. Particularly in the context of insurance a distinction is drawn between a partial loss, where the thing insured is damaged; an actual total loss, where it is wholly or substantially wholly destroyed; and a constructive total loss, where the thing is so seriously damaged that it is impracticable or financially unjustifiable to repair it and it is proper to abandon it to the insurers, recovering as for a total loss.

Loss leaders. Goods sold substantially below the manufacturer's recommended resale price to attract custom to the seller's premises.

Loss of consortium. The loss by reason of the tort of another party of the companionship, affection, mutual assistance, sexual relations, and other elements which cumulatively make up the substance of matrimonial consortium or cohabitation. In English law, if the loss is total, it is a ground for a claim of damages by a husband.

Loss of earnings. In claims of damages for torts causing personal injuries, a material head of damages. In assessing damages under this head the court must have regard to earnings lost down to the date of trial, and probable future losses during the rest of the plaintiff's normal working life, or expected lifetime if less.

Loss of expectation of life. Where a person has been injured or killed by the wrongful act of another, an element to be considered in assessing damages is the loss to him, if it be established, of having his normal expectation of healthy lifetime materially shortened.

Loss of profits. In claims of damages for breach of contract damages may be measured by the profits which the plaintiff would have made in the ordinary course if the contract had been implemented e.g. where he was buying goods for resale and resale was within the defendant's contemplation.

Loss of services. The old common law action *per quod servitium amisit* gave an employer a claim where, by reason of a third party's fault, his employee was injured and he lost the latter's services for a time. The principle was extended to give a remedy to a father whose daughter was seduced, on the fiction that he had thereby lost her services.

Lost document. In litigation a document must normally be proved by production of the original, but if it be adequately proved that it has been lost or destroyed, it may be proved by secondary evidence in the form of a copy or oral evidence of its contents, or in Scotland a copy can be set up as equivalent to the original by an action of proving of the tenor (q.v.).

Lost grant. After the Limitation Act, 1623, limited the time for bringing actions for possession of land to 20 years, the courts developed the principle that the lapse of a similar time was sufficient to create an easement on the hypothesis that it had been created by a deed of grant now lost, and this remains one of the modes in which an easement may be held created though it is a patent fiction.

Lost or not lost. Words which may appear in a marine insurance policy to ensure that the contract is operative notwithstanding that the vessel may have already been lost (which would otherwise defeat the contract). The words make the underwriter liable even where the subject-matter of the insurance had not vested in the assured at the time of the occurrence of the loss, but not where, when the insurance was effected, the assured knew of the loss but the underwriter did not.

Lost property. Articles of personal property can be said to be lost when the person entitled to possession thereof either does not know where the thing is, though it may still be in his premises and be discovered; or it has by theft, escape, abandonment, or mistake passed out of his possession and he does not know where it is and cannot recover it. The true possessor continues to have the best title to the thing if it is found and he can establish his possessory right. The finder has a good title to possession as against everyone other than the true possessor. If he knows who the true possessor is, he should return the property to the true possessor, and if he does not, he should hand it to the police or a lost property office.

Lottery. An arrangement for distributing prizes by lot or chance. Persons participating acquire chances by, e.g., buying numbered tickets on the undertaking that the distribution of the fund, or prizes, or some other advantage, shall be determined by drawing lots. The distribution of things by lot is known from both biblical and classical times, and in modern times lotteries have frequently been utilized to raise money for public purposes, e.g. the Sydney Opera House, hospitals, and the like. In Britain Premium Bonds, a form of National Savings, are truly a lottery in that no interest is paid, but prizes are given to holders of bonds chosen by an electronic random number machine. The conduct of lotteries is extensively regulated by statute. In general lotteries for profit are illegal, but lotteries as a means of raising money for charitable purposes are permitted within limits.

Loughborough, Lord. See WEDDERBURN, ALEX.

Louter, Jan de (1847–1932). Dutch jurist author of *Het Stellig Volkerrecht* (1910), *Le droit international public positif* (1930), and other works.

Low-water mark. The line along the seashore to which the tide recedes furthest. Between this line and the line of the high tide medium between spring tide and neap tide is the foreshore, property in which, unless acquired by a subject by grant from the Crown or by prescription, belongs to the Crown for the benefit of the community generally, so that it cannot be used so as to interfere with public rights of navigation and fishing.

Lowell, Abbott Lawrence (1856–1943). American political scientist, president of Harvard, author of *The Governments of France, Italy and Germany* (1914), *Greater European Governments* (1925), and *Public Opinion and Popular Government* (1926).

Lowndes, Sir George Rivers (1862–1943). Called to the Bar in 1892, he was legal member of the Viceroy of India's Executive Council, and from 1929 a member of the Judicial Committee of the Privy Council.

Lowry, Robert Lynd Erskine (1919–). Son of a judge, he was called to the Northern Ireland Bar in 1947, took silk in 1956, became a puisne judge of Northern Ireland in 1964 and Lord Chief Justice there in 1971.

Loyseau, Charles (1566–1627). French jurist, author of many works on customary law, notably *Des Offices, Des Seigneuries, Des ordres et simples dignités, Déguerpissement et délaissement par hypothèques*, and others.

Loysel (or Loisel), Antoine (1536–1617). French jurist, author of *Institutions Coutumières* (1607) in which the essential rules of the customary law of France were set out in the form of proverbs, *Pasquier, ou Dialogue des avocats*, and other historical works.

Lucas de Penna (*c.* 1320–90). Italian commentator, acted as a judge in Apulia and was latterly in the service of the Roman Curia. He was author of *Commentaria in Tres Libros Codicis* which was truly a complete exposition of legal principles and rules relating to all departments of law, and acquired great authority in Italy.

W. Ullmann, *The Medieval Idea of Law.*

Lucerna juris. A mediaeval title applied both to Irnerius and Baldus de Ubaldis (qq.v.).

Lucerna juris pontificii. A mediaeval title sometimes given to Nicolaus dei Tudeschis.

Lumen juris. A mediaeval title sometimes used of Pope Clement IV.

Lumen legum. A title sometimes given to Irnerius (q.v.).

Lucid interval. A period of sanity between periods of insanity, when the person is capable of understanding what he is doing and of participating in legal transactions. Acts done during a lucid interval are accordingly legally valid and effective.

Lucrum cessans. Loss of continuing profits, as distinct from *damnum emergens* (q.v.).

Lunacy. See UNSOUND MIND; MENTAL ILLNESS.

Lundstedt, Anders Vilhelm (1882–1955). Swedish jurist, professor at Uppsala, author of *A Critique of Jurisprudence* (1925), *Legal Thinking Revised*, and other works. Possibly the most extreme of the Scandinavian realist school, he viewed law as simply the fact of social existence in organized groups and the conditions which make possible the coexistence of numbers of people. He attacked every abstract concept as nonsense and substituted for justice social welfare, which is a guiding motive for legal activities.

Lupold of Bebenburg (1297–1363). Canon lawyer and political theorist, bishop of Bamberg, author of *Tractatus de iuribus regni et imperii*, dealing with the relationship between the German Kingship and the imperial dignity, and seeking to define the roles of the spiritual and secular powers in Christian society. Some of his ideas appear in the Golden Bull of 1356 which in outline regulated relations between the emperor and German princes down to 1806.

Lushington, Stephen (1782–1873). Called to the Bar in 1806 and admitted a member of Doctors' Commons in 1808. He was an M.P. for a time and supported Parliamentary reform and the suppression of the slave trade, and was one of the counsel for the defence of Queen Caroline. He became judge of the Court of Admiralty and a P.C. in 1838, and Dean of the Arches in 1858, holding both offices till 1867 and being also Master of the Faculties from 1858 to 1873. He also served as a member of royal commissions. He was a brilliant lawyer and made notable contributions to maritime, prize, and ecclesiastical law.

Luxembourg Law. Luxembourg came into existence as an independent Grand Duchy in 1839 when Belgium and Holland finally split. The territory formed part of the Frankish Kingdom of Austrasia and of Charlemagne's empire, passed to the Habsburgs but after 1815 was united in a personal union with the Netherlands, which lasted till 1890. From 1815 to 1866 it was part of the German federation, but declined to join the North German Confederation in the latter year. Since 1921 it has been economically united with Belgium and since 1960 with Belgium and the Netherlands in the Benelux economic Union.

The present constitution dates from 1868, amended in 1919, and provides for an hereditary Head of State, the Grand Duke, a Council of State, and an elected Chamber of Deputies. The Grand Duke chooses the ministers, may intervene in legislation, and has certain judicial powers. The Council of State is the supreme administrative tribunal and has certain legislative functions: it is nominated by the Grand Duke.

The judiciary comprises the Superior Court of Justice, which acts as Court of Cessation and as Court of Appeal, the Court of Assizes for serious criminal cases, two regional (also called correctional) courts, and three magistrates' or police courts, which deal with minor civil, criminal, and commercial cases. The regional courts also hear appeals from the magistrates' courts. There are special tribunals for social administration cases and a military court. There is no jury trial. Appeal lies from administrative authorities to the Council of State.

Judges are appointed for life by the Grand Duke. The Department of the Attorney-General is responsible for the administration of the judiciary and the supervision of police investigations.

The substantive law is substantially the adoption of that of the countries with which the Duchy has been connected. Thus the Criminal Code of 1879 reproduced the Belgian Penal Code of 1867 and the code of criminal procedure followed that of France.

C. Sladitz, *Guide to Foreign Legal Materials—Belgium, Luxembourg, Netherlands*; P. Majerus, *L'état luxembourgeois.*

Lynching. See LYNCH LAW.

Lynch law. In the U.S. capital punishment of alleged criminals by mobs or private persons without legal trial. Sometimes there is also torture or mutilation. The name is possibly derived from Charles Lynch (1736–96), a Virginia justice of the peace, who employed such methods to suppress outlaws and loyalist activities during the Revolution. In frontier conditions, where legal institutions were embryonic or unorganized, lynch law was frequently executed by vigilantes. In the South it became an almost traditional method of dealing with negroes suspected of crimes against whites. Public opinion has now almost completely eliminated lynching.

Similar customs have been found in the U.K. and Europe, under such names as Lydford law, gibbet law or Halifax law, Cowper justice or Jeddart justice, and Fehmgerichte in Germany (qq.v.).

J. Cutler, *Lynch law* (1908).

Lyndhurst, Lord. See COPLEY, J. S.

Lyndwood, William (?1375–1446). Took holy orders about 1407 and became official of the court of Canterbury in 1414. In 1425 he became Dean of the Arches and also served as a diplomat. He became Bishop of St. Davids in 1442 and was also Keeper of the privy seal. He took an active part in promoting the foundation of Eton College. He is best remembered for the *Provinciale* (*c.* 1470–80, and later editions) a digest in five books of the synodal constitutions of the province of Canterbury from the time of Stephen Langton to that of Henry Chichele, with an explanatory gloss. It is a principal authority for English canon law. Some later editions have supplements containing the constitutions of the cardinals Otho and Othobon, papal legates to England.

Lyon. See LORD LYON KING OF ARMS.

Lyon court. The Lord Lyon Kings of Arms as the Crown's officer of arms in Scotland early acquired authority over the bearing of arms, matters of precedency, and over heralds and messengers-at-arms. From about the fifteenth century Lyon has held a court having jurisdiction over the granting, registration, verification and bearing of arms, and over matters of heraldry. It has a civil jurisdiction to deal with applications for the grant of rights to armorial bearings, and for rematriculations. When granted, these are recorded in the Register of all Arms and Bearings in Scotland.

It has also a criminal jurisdiction, dealing with usurpations of Crown and private rights in armorial bearings, proceedings being raised by the Procurator-fiscal. The court has power to fine and imprison.

Decisions of Lyon Court may be appealed to the Court of Session and House of Lords.

Lyon's Inn. One of the Inns of Chancery (q.v.).

M

M. The mark formerly branded on the left thumb of a person convicted of manslaughter but allowed benefit of clergy (q.v.).

M.P. Member of the House of Commons of the U.K. Parliament.

M.R. The abbreviation of Master of the Rolls (q.v.).

Maasdorp, Sir Andries Ferdinand Stockenstrom (1847–1931). South African judge, chief justice of the Orange Free State division of the Supreme Court of South Africa, 1902–19. He published a translation of Grotius' *Inleidinge*, as *Introduction to Dutch Jurisprudence*, and wrote a large *Institutes of South African Law.*

Macaulay, Thomas Babington, Lord (1800–59). Called to the Bar in 1826, he did not practise seriously but entered politics and became a member of Parliament and, later, Secretary of the Board of Control, which supervised the East India Company's administration of India and, in 1833, legal adviser to the Council of India. In 1834, he went to India where he drafted the Indian Penal Code and inaugurated a system of education on Western lines. He returned to England in 1838 and was Secretary for War in 1839–41. He had already published many important essays, collected as *Critical and Historical Essays*, and, in retirement he wrote his famous and influential *History of England*, 1688–1702, a view of these years from the Whig standpoint. He became a peer in 1859 as Baron Macaulay of Rothley.

G. O. Trevelyan, *Life and Letters of Macaulay* (1876).

Macclesfield, Lord. See PARKER, THOMAS.

Mace. Originally a heavy-headed club used in warfare, now a staff, usually with a large ornamental head, borne as a symbol of authority on State or ceremonial occasions before such persons as the Lord Chancellor, the Speaker of the House of Commons, Chancellors of Universities, and Lord Mayors. In the House of Commons, the Mace is laid on the table of the House while the Speaker is in the chair. When the House is in committee and the Chairman of Committees occupies the chair, the mace is placed on hooks below the table.

Macers. Officers of the High Court of Justiciary and of the Court of Session in Scotland who carry the maces of these courts on ceremonial occasions and, at other times, perform the functions of ushers.

McCulloch* v. *Maryland (1819), 4 Wheaton 316. As a move in the political and economic struggle between the Bank of the United States and smaller banks, chartered by various states, the Maryland legislature tried to tax the note issue of the Baltimore branch of the Bank of the United States and, on its failure to pay, sued McCulloch, the treasurer. The Bank had to satisfy the Supreme Court of its right to exist and of the unconstitutionality of the tax. The Court held that though the Constitution made no mention of any power of Congress to charter a bank or any other business corporation, the federal government had implied power to charter a bank, as necessary and proper, for the furtherance of government and commerce, and further, that no state may tax an instrument of the federal government, because the power to tax implies the power to destroy. The case is an important assertion of national power and of the doctrine of implied powers, and usually deemed Chief Justice Marshall's most brilliant constitutional opinion.

MacDermott, John Clarke, Baron, of Belmont (1896–). Was called to the Irish Bar in 1921 and was Minister of Public Security for Northern Ireland and Attorney-General for Northern Ireland before becoming a judge of the High Court of Northern Ireland in 1944. From 1947 to 1951, he was a Lord of Appeal in Ordinary before returning to Northern Ireland as Lord Chief Justice, 1951–71.

Macdonald, Sir Archibald, Bt. (1747–1826). Was called to the Bar in 1770 and soon gained a reputation as a sound lawyer. In 1778, he entered Parliament and was successively Solicitor-General (1784) and Attorney-General (1788). In 1793, he became Chief Baron of Exchequer and held that office till he retired in 1818, proving himself a patient and competent, though not very distinguished, judge.

Macdonald, Sir John Hay Athole, Lord Kingsburgh (1836–1919). Was called to the Scottish Bar in 1859 and was sheriff of Ross, Cromarty, and Sutherland (1874-76), Solicitor-General for Scotland (1876–80), Sheriff of Perthshire (1880–85), Dean of the Faculty of Advocates (1882–85), M.P. (1885–88), and Lord Advocate (1885–86 and 1886–88). Thereafter, he was Lord Justice-Clerk of Scotland from 1888 to 1915. He became P.C. in 1885, K.C.B. 1900, and G.C.B. in 1916. He was a sound, commonsense judge but not a great scholarly lawyer. He specialized in criminal law, wrote a *Practical Treatise on the Criminal Law of Scotland* (1867), more a collection of notes than a systematic treatise, but which nevertheless survived till 1967, secured the passage of the very important Criminal Procedure (Scotland) Act, 1887, and presided at many famous criminal trials. All his life he was an ardent volunteer soldier and was a recognized authority on military matters; he was also devoted to all kinds of athletics. He wrote *Fifty Years of It* (1909) and *Jottings of an old Edinburgh Citizen* (1915).

Macdonell, Sir John (1846–1921). British jurist, professor of comparative law at London, and editor of the *Journal of Comparative Legislation* and of various legal works.

McDouall, Andrew, Lord Bankton (?1685–1760). Passed advocate in 1708 and became a Lord of Session in 1755, taking the title of Lord Bankton from his estate.

He published in three volumes in 1751–53 an *Institute of the Laws of Scotland in Civil Rights: with Observations upon the Agreement or Diversity between them and the Laws of England* in four books, after the general method of Stair's Institutions, and taking account of the developments in the law since Stair's time. This is a work of much value and high authority, though it has never enjoyed quite the regard shown to the writings of Stair, Erskine and Bell, but it is clear that Bankton was a diligent and scholarly lawyer.

Macgregor Mitchell, Robert, Lord (1875–1938). Was called to the Scottish Bar in 1914, took silk in 1924, and was chairman of the Scottish Land Court, 1934–38.

Mackay, Aeneas James George (1839–1911). Passed advocate of the Scottish Bar in 1864, but devoted his energies mainly to legal and historical studies, and, in 1874, became Professor of Constitutional Law and History at Edinburgh. He resigned in 1881, on becoming an advocate-depute, and was Sheriff of Fife, 1886–1901. He made many notable contributions to Scottish history and literature. His main legal works were *A memoir of Sir James Dalrymple, First Viscount Stair* (1873), *The Practice of the Court of Session* (2 vols., 1877-79), and a *Manual of Practice in the Court of Session* (1893). The latter two works are still referred to.

Mackenzie, Sir George, of Rosehaugh (1636–91). Studied in Scotland, and in France, before being called to the Scottish Bar in 1659. Shortly after the Restoration, he became a justice-depute, or assistant to the Justice-General. He entered Parliament in 1669 and became notable for resistance to the pretensions of the Crown, but, in 1677, he was made Lord Advocate and in the next few years prosecuted and persecuted Covenanters with such zeal as to earn the title 'The Bloody Mackenzie'. In many cases, he strained the law so as to obtain a conviction. In 1686, he was dismissed for disagreement with James II's favour to Catholicism, but was restored in 1688. His political career was terminated in the following year by the Revolution and he retired to Oxford. The redeeming feature of his character was his devotion to literature and learning.

He wrote extensively, *Aretina, or the Serious Romance* (1661), which was probably the first novel written in Scotland, moral essays, poetry, a short *Institutions of the Law of Scotland* (1684)—long used as a text-book—a much more valuable *Laws and Customs of Scotland in Matters Criminal* (1674), *Observations on the Acts of Parliament* (1686), *Observations on the Laws and Customs of Nations as to Precedency, with the Science of Heraldry as Part of the Law of Nations* (1690), and other works on law and constitutional theory, notably *Jus Regium or the Just and Solid Foundations of Monarchy* (1684) and *Vindication of the Government of Scotland during the reign of Charles II* (1691). His *Collected Works* were published in 1716–22.

He is also famous for having, in 1682, founded the Advocates Library, now the National Library of Scotland. Mackenzie was clearly a considerable scholar, a patriot, and a very competent lawyer but one whose views verged on fanaticism and who overstrained legal forms to obtain the conviction of those whom he prosecuted.

A. Lang, *Sir George Mackenzie of Rosehaugh* (1909).

Mackenzie, Sir George, Viscount Tarbat and 1st Earl of Cromarty (1630–1714). A supporter of the last two Stuart kings, he was a confidant of Middleton in the direction of Scottish affairs after the Restoration. He became a Lord of Session (Lord Tarbat) in 1661, but was deprived of office in 1664. He was later restored to favour and held the offices of Lord Justice-General of Scotland, 1678–80 and 1703–10, an ordinary Lord of Session, 1681–89 and Lord Clerk Register, 1681–89 and 1692–96, and Secretary for Scotland, 1702–04. In 1703, he became first Earl of Cromarty. He published a number of historical and controversial works, and was one of the original Fellows of the Royal Society. His fourth son, James, was a Lord of Session and of Justiciary (Lord Royston) 1710–44.

Mackenzie, Roderick, of Prestonhall (?–1712). Passed advocate in 1666. In 1678, he became a Clerk of Session and, in 1703, a Lord of Session. He resigned in 1710. He was also Lord Justice-Clerk, 1702–04.

Mackenzie Stuart, Alexander John, Lord (1924–). Called to the Scottish Bar in 1951, he took silk in 1963, became a judge of the Court of Session in 1972, and the first British judge of the Court of Justice of the European Communities in 1973.

Mackintosh, James (1765–1832). Graduated in medicine but abandoned it and was called to the Bar in 1795. In 1791, he published *Vindiciae Gallicae*, a reply to Burke's *Reflections on the Revolution in France.* In 1799, he gave an important course of lectures on the law of nature and nations, which was attended by many distinguished men. From 1803–11, he was in India, as Recorder of Bombay. Thereafter, he entered Parliament and wrote much. In Parliament, he succeeded Romilly as leader of the group urging legal reforms, and succeeding in getting a committee appointed to consider the question of capital punishment for felony. His efforts eventually moved Peel to take up the reform of criminal law, and the great diminution in the number of crimes carrying the death penalty was mainly due to his work. He supported Catholic emancipation and the Reform Bill and wrote a *History of the Revolution in England.*

McLaren, John, Lord (1831–1910). Was called to the Scottish Bar in 1846, but his progress was hindered by poor health. In 1880, he became Lord Advocate, but held office little more than a year, being promoted to the bench against his will in 1881. In that capacity he was very successful and his judgments are highly regarded. He edited More's *Lectures on the Law of Scotland* (1864), Bell's *Commentaries on the Law of Scotland* (1870), and wrote a *Treatise on the Law of Trusts and Trust Settlements* (1863), later rewritten as *The Law of*

Scotland relating to Wills—a book which was the leading authority in Scotland for a century.

MacMahon, Sir William (1776-1837). Was called to the Irish Bar in 1799 and was Master of the Rolls in Ireland from 1814 till his death.

Macmillan, Hugh Pattison, Baron (1873–1952). Was called to the Scottish Bar in 1897 and quickly attained a good practice, particularly at the Parliamentary Bar. He was non-political Lord Advocate in 1924, a Lord of Appeal in Ordinary, 1930–39, and 1941–47, and Minister of Information, 1939–40. As a judge his opinions are rated highly, especially on Scots law.

He was chairman or member of a large number of committees and commissions and received many honours. He was joint-author of a book on *Provisional Orders*, and of Private Legislation Reports, and wrote many articles, chiefly on conveyancing, and an autobiography, *A Man of Law's Tale*.

Macnaghten, Edward, Baron (1830–1913). Was called to the Bar in 1857 and practised in chancery; he took silk and also became an M.P. in 1880. He refused several offers of both political and judicial offices in the next few years. In 1887, he became a Lord of Appeal in Ordinary, but was also politically active in the House of Lords. As a judge, he was erudite, but with the knack of speaking concisely, yet authoritatively, and some of his judgments are literary classics. As chairman of the Council of Legal Education he played a large part in improving professional education.

M'Naghten (or McNaghten or Macnaughton) Rules. Daniel M'Naghten killed the Prime Minister's Secretary by mistake for the Prime Minister under an insane belief that the Government was persecuting him. He was tried and acquitted on the ground of insanity and committed to Bethlem Hospital ((1843), 4 St. Tr. (N.S.) 847). The House of Lords then set a number of questions to the judges of England, who returned answers in the form of rules: these are reported in 10 Cl. & F. 200. The Rules, stated shortly, are:

(1) Persons acting under the influence of an insane delusion are punishable if they knew at the time of committing the crime that they were acting contrary to law.
(2) Every man is presumed sane and to have sufficient reason to be held responsible for his crimes.
(3) To establish a defence on the ground of insanity it must be clearly proved that, at the time of committing the act, the accused was labouring under such a defect of reason, from disease of the mind, as not to know the nature and quality of the act he was doing or, if he did know it, that he did not know he was doing what was wrong. If the accused was conscious that the act was one that he ought not to do, and if that act was at the same time contrary to the law of the land, he is punishable.
(4) A person under a partial delusion only must be considered as if the facts with respect to which the delusion exists were real.

The Rules have been treated in England almost as if of statutory force, and have been followed in many Commonwealth countries and parts of the U.S.A., but not in Scotland. They have been criticized for concentrating on knowledge of right and wrong and isolating knowledge from the accused's mental disease, but have not been superseded.

N. Walker, *Crime and Insanity in England* (1968).

McNair, Arnold Duncan, 1st Baron, of Gleniffer (1885–1975). Became successively Whewell Professor of International Law at Cambridge, 1935–37, Vice-Chancellor of Liverpool University, 1937–45, Professor of Comparative Law at Cambridge, 1945–46, member of the Permanent Court of Arbitration at The Hague, 1945–65, a judge of the International Court of Justice, 1946–55 (President of that Court, 1952–55), and President of the European Court of Human Rights, 1959–65. He published *The Legal Effects of War* (1920), *Law of the Air* (1932), *Roman Law and Common Law* (with W. W. Buckland) (1936), *Law of Treaties* (1938, revised 1961), *International Law Opinions* (3 vols., 1956).

McNeill, Duncan, 1st Baron Colonsay and Oronsay (1793–1874). Was called to the Scottish Bar in 1816 and rapidly acquired a practice. He was Solicitor-General, 1834–35 and again in 1841–42. In 1842–46, he was Lord Advocate and was Dean of Faculty from 1843 to 1851 and also sat in Parliament from 1843. While an M.P., he promoted legislation on public companies, conveyancing, and poor law. He ceased to be Lord Advocate in 1846 and, in 1851, became a judge of the Court of Session as Lord Colonsay. In 1852, he succeeded Boyle as Lord President and, in 1867, was given a peerage as Lord Colonsay to assist the House of Lords in disposing of Scots appeals. He was, however, too old to make much impact in the House of Lords but, as Lord President, showed himself a sound, though not outstanding lawyer.

Macqueen, Robert, Lord Braxfield (1722–99). Was called to the Scottish Bar in 1744 and attained great professional eminence, particularly in matters of feudal conveyancing. He was appointed a Lord of Session in 1776, a Lord of Justiciary in 1780, and Lord Justice-Clerk in 1788. His reputation for coarseness, brutality, and unfairness gained in some of the treason trials towards the end of his life, have obscured his reputation for learning, particularly in land law, and a generally high reputation as a judge.

Maconochie, Alexander (1787–1860). Went to Van Dieman's Land as secretary to the governor, was horrified at conditions in the penal settlement there and, in 1839, was appointed superintendent of Norfolk Island. He introduced a humane system, particularly that of marks for work and good conduct. The hostility of officials brought about his recall in 1844, after which he campaigned in England for penal reform. He was Governor of Birmingham Borough Prison, 1849–51, but again was dismissed for leniency. He wrote a number of works on penal reform, such as *Thoughts on Convict Management* (1838) and *Norfolk Island* (1847), and greatly influenced the Irish Prisons Board. His main ideas were: sentences for the performance of a quantity of labour, not for a time; marks to be earned by good conduct; association in groups; and diminished degree of discipline to prepare a man for release.

J. Barry, *Alexander Maconochie of Norfolk Island.*

Madox, Thomas (1666–1727). Though a law student, was never called to the Bar. He was encouraged in the study of the records by Lord Somers. His publications include the *Formulare Anglicanum* (1702), a collection of charters and documents from the Conquest to 1550, and the great *History and Antiquities of the Exchequer* (1711) which, though primarily concerned with finance, touches upon most branches of government and law, and in which he published the *Dialogus de Scaccario* (q.v.) and settled its authorship. He also published the *Firma Burgi* (1722) of importance for municipal history, and a *Baronia Anglica*, which was part of a projected Feudal History and Costumier of England. He also projected a History of Parliament. All his work was based on original authorities and he sought the strictest accuracy. His work has proved of lasting value and importance, not least for legal history. He was historiographer to the King from 1714 to his death.

Maecianus, Lucius Volusius (second century A.D.). A Roman jurist, he held various high offices, tutored Marcus Aurelius, and was a member of the imperial consilium under Antoninus Pius and Marcus Aurelius. He wrote treatises on *Fideicommissa* (16 books); *De Judiciis Publicis* (14 books), and a monograph in Greek on the *Lex Rhodia*, and is cited in the *Digest.*

Magister sententiarum. A mediaeval title sometimes applied to Peter Lombard.

Magistrate. In its broadest sense, a magistrate is any person, even the sovereign, exercising judicial authority. It is normally, however, limited to subordinate magistrates appointed by the sovereign, particularly in England to a person holding the office of justice of the peace, or of stipendiary magistrate or metropolitan magistrate.

In Scotland, the term is used chiefly of the provosts and bailies of burghs, and also of stipendiary magistrates, but not of justices of the peace.

Magistrates' Courts. In England, a magistrates' court consists of justices of the peace acting under statute, or at common law, or by virtue of their commission, or of a stipendiary magistrate. Justices sitting as a court or committee of quarter sessions, or as special sessions for the purpose of executing administrative functions under statutory authority, such as licensing, are excluded from the definition of magistrates' court. Such courts are inferior courts but, as least for some purposes, courts of record. A single lay justice may act only if statute authorizes it in that kind of matter and his powers are strictly limited. The normal case is acting as examining justice to determine whether or not to commit for trial on indictment, or when granting a warrant. A magistrates' court must normally consist of at least two lay justices.

Magistrates' courts have original civil jurisdiction as to the summary recovery of certain kinds of debts, and as to domestic proceedings, including separation and maintenance, guardianship, adoption, affiliation and under various statutes.

They have original criminal jurisdiction to inquire as examining justices into all matters triable on indictment and, if they deem the evidence sufficient, to commit the person charged for trial at the Crown Court. They also have original criminal jurisdiction to try all offences triable summarily, i.e. without a jury, and offences triable either on indictment or summarily, if the court considers summary trial more suitable in the circumstances. If it considers trial on indictment more suitable the court proceeds to inquire into the information as examining justices. It may also do so, even after summary trial has begun, so long as it discontinues the summary trial before the conclusion of the evidence for the prosecution.

Magistrates, Resident. Judicial officers in Northern Ireland having powers similar to those of stipendiary magistrates in England, and sitting both as examining magistrates and as courts of petty sessions trying summary offences and minor civil cases.

Magistrates, stipendiary. In England and Wales, a stipendiary (i.e. legally qualified and salaried) magistrate may be appointed by the Crown on the petition of the council of a county having a separate commission of the peace, or a joint district of two or more areas for which separate appointments might be made. He must be a barrister or solicitor of seven years' standing. He may sit alone, or along with any J.P., and has power to do alone what any two J.P.s may do. Accordingly, he normally acts alone as a magistrates' court. Within the Greater London area, metropolitan stipendiary magistrates sit at courts throughout the metropolitan police district.

In Scotland, stipendiary magistrates may be appointed in lieu of, and to exercise the jurisdiction of, formerly, a burgh magistrate or Judge of Police in the burgh police court, and now of a district court, trying summarily minor criminal offences. Appointments have been made only in Glasgow.

Magna assisa eligenda, de. The writ issued to the sheriff to return four knights, who were to choose 12 others to be associated with them to form the grand assize or grand jury to try the writ of right.

Magna Carta. John became King of England in 1199 and roused much opposition by his tyrannical rule. In 1205, John became involved in a struggle with Pope Innocent III over the election of an archbishop of Canterbury. The Pope procured the election of Stephen Langton in 1206. John refused to receive him and England was placed under interdict (1208), John excommunicated (1209), and his kingdom declared forfeit (1212). John then surrendered, accepted Langton, and surrendered the kingdom to the Pope, receiving it back as a fief of the Holy See, subject to an annual tribute. This surrender alienated the people and the Pope, having secured submission, began to support John. Accordingly, Langton and the barons were driven to resist. In 1213, the northern barons refused to follow John to France to recover Normandy, and two councils of bishops and barons, at St. Albans and London, discussed the misgovernment of the realm. Langton produced the charter of liberties of Henry I (q.v.) and this was adopted as the basis of the barons' demands. In 1214, the barons met at Bury St. Edmunds and made a confederacy binding them, if the King would not acknowledge their rights, to make war on him until he confirmed the people's liberties by a charter. In 1215, they presented their demands to the King, who took various steps to break their unity, but they continued united, assembled in arms at Stamford, denounced their homage to the king, and marched on London. John was deserted by all but a few adherents and accepted the Articles of the Barons which were embodied in the Great Charter, to which he signified his assent at Runnymede on 12 June, 1215.

Magna Carta contains a preamble and 63 clauses, not entirely logically arranged. (1) The King declared that the Church should be free with all rights and liberties inviolate, and granted to all freemen of the kingdom certain liberties; (2, 3) heirs should pay only the ancient reliefs on succeeding; (4, 5) guardians should take only reasonable fruits and profits of wardship, without destruction or waste; (6) heirs should be married without disparagement; (7, 8) a widow should receive her dower and inheritance within 40 days of her husband's death, and should not be forced to remarry; (9) land should not be seized for Crown debt if the debtor's goods would suffice; (10, 11) debts to the Jews were not to bear interest during the minority of a deceased debtor's heir; (12) no scutage or aid should be imposed, save by common counsel of the kingdom, except in the three recognized cases, and even then, should be reasonable; (13) the city of London should have all its ancient liberties and free customs, and so of all other cities, boroughs, towns and ports; (14) to obtain the common counsel of the kingdom the King should summon clergy and barons by writ and tenants-in-chief by general writ, to a stated place, at 40 days notice; (15) the King should not empower mesne lords to exact other than the three ordinary aids, and these of reasonable amount; (16) no one should be compelled to render more than the due service for a knight's fee or other free tenement; (17) Common Pleas should not follow the King's court, but be held in some certain place; (18, 19) the assizes of *novel disseisin*, *mort d'ancestor*, and *darrein presentment* should be held only in the county where the lands lay, and two justices are to be sent into each county four times a year to hold such assizes; (20) a freeman should only be amerced for a small offence after its manner, and for a great crime according to its heinousness; (21) earls and barons should be amerced only by their peers, according to the degree of the offence; (22) no cleric should be amerced for his lay tenement but according to the said proportions, and not according to the value of his ecclesiastical benefice; (23) no town or man should be distrained to make bridges at river banks, unless anciently and of right bound to do so; (24) no sheriff, constable, coroner or bailiff of the king should hold pleas of the Crown; (25) counties, hundreds, and wapentakes should stand at the ancient ferms; (26) on the death of a tenant-in-chief of a lay fee, indebted to the Crown, the sheriff might attach the chattels thereon to the value of the debt; (27) on the death of a freeman intestate, his chattels should be distributed by the hands of his nearest kinsfolk and friends under supervision of the Church; (28) no constable or royal bailiff should take a man's corn or other chattels without immediate payment, unless the seller give credit; (29) no knight should be compelled to pay for castle-guard if he be willing to perform the service in person or by deputy, and he should be free from duty while with the army; (30, 31) the King should not take horses or carriages of freemen for carriage, or any man's timber for castles or other uses, unless by the owner's consent; (32) the King should hold the lands of convicted felons only for a year and a day, and then surrender them to the mesne lords; (33) all weirs in the Thames and Medway and throughout England should be put down, except on the sea-coast; (34) the writ of *Praecipe in capite* should not, in future, be issued to anyone so as to cause a freeman to lose his court; (35) there should be one standard of weights and measures throughout the kingdom; (36) the writ of inquest of life and limb (*de odio et atia*) should be given gratis and not denied; (37) the King should not have the wardship of land held of mesne lords by reason of the sub-vassal having other lands held of the King; (38) no bailiff for the future should put anyone to his law on his own bare statement without credible witnesses; (39) no freeman should be taken or imprisoned, or disseised, or exiled, or outlawed, or anyways destroyed; nor would the King go against him or send on him, unless by the lawful judgment of his peers or by the law of the land; (40) to none would the King sell, deny, or delay right or justice; (41) all merchants should have liberty, safety to enter, dwell in, and leave England without being subjected to any evil tolls, but only the ancient customs, save in wartime; (42) in future, any one might leave the kingdom and return at will, save in wartime, with the exception of prisoners, outlaws, and alien enemies; (43) the tenants of baronies escheated to the Crown should pay only the same relief and perform the same services as if the land were still held of a mesne lord; (44) persons dwelling outwith a forest should not, in future, be compelled to attend the King's forest courts on common summons, unless they be impleaded; (45) the King would appoint as justices, constables, sheriffs, and bailiffs only such as know the law and mean duly to observe it; (46) barons who had founded abbeys should have the custody of them when vacant; (47–59) were special and temporary, save (54) providing that no one should be taken or imprisoned on the appeal of a woman save for the death of her husband.

Clause 60 extended all the foregoing rights and liberties granted to the king's vassals to the tenants of mesne lords. Clause 61 provided means for enforcing observance of the Charter; the baronage was to elect a Council of 25 to ensure that the provisions of the Charter were carried into effect; if the King broke a provision, any four barons might complain and demand redress and, if redress were not given within 40 days, the barons, together with the commonalty of the land, might seize castles and lands and do distress until the grievance was redressed. Clause 62 was temporary, and clause 63 enjoined the freedom of the Church of England and that all men in the kingdom should enjoy these liberties and rights freely and quietly for ever, and prescribed the oath to be taken by the king and barons, on both sides to observe the Charter.

John soon tried to evade the Charter. The Pope declared the Charter void, excommunicated the barons and suspended Langton. John recruited mercenaries and renewed the struggle. The barons then renounced their allegiance and offered the crown to Louis, eldest son of the King of France. John, however, died late in 1216, and the Earl of Pembroke, regent for the child Henry III, succeeded in winning over the barons to the new King's side and in persuading Louis to abandon his claim to the throne. He also renewed the Great Charter with numerous alterations and omissions, and it was again reissued in 1217 with further alterations, additions, and omissions, the forest clauses being embodied in the Charter of the Forest.

In 1225, Henry III reissued Magna Carta and the Charter of the Forest, with only small alterations and it is in the form of 1225 that Magna Carta was confirmed repeatedly thereafter, normally in return for the grant of a subsidy to the Crown. It was thus confirmed 37 times down to the time of Henry VI.

The interpretation of Magna Carta has given rise to much controversy. At the time it was not unique, similar charters defining people's rights being granted by many rulers and lords throughout Europe round about the same time. It was a practical agreement designed to limit abuses and extortion and to define reasonable imposts, and to state the standard or desirable practice. It was not exacted solely in the interests of the barons because the definition of feudal duties benefited feudal tenants of all ranks of society. It was an attempt to put in writing acceptable principles regulating the relations of King and people, not a statement of political theory, but a practical assertion of rights. Much of it was intended to be declaratory, not innovatory.

Subsequently, however, myths developed round Magna Carta and, from Coke to Hallam, many thought that all the rights and liberties of Englishmen had been created by Magna Carta. Thus, clauses 39 and 40 were hailed as the origins of *habeas corpus*, trial by jury, and the liberties of the subject generally, whereas the clauses were merely intended to protect freemen against the kinds of arbitrary conduct which John had been committing.

Even discounting the myths and exaggerated claims, however, Magna Carta was of permanent significance as the first great instance of general dissatisfaction with the King's government provoking resistance and the setting down, in writing, of limitations on the royal power.

Most of it has now been repealed as superseded.

A. L. Poole, *Domesday Book to Magna Carta* (1951); W. S. McKechnie, *Magna Carta* (1914); H. Malden (ed.), *Magna Carta Commemoration Essays*; W. S. Holdsworth, *History of English Law*; J. C. Holt, *Magna Carta* (1965); F. Thompson, *Magna Carta, Its Role in the Making of the English Constitution, 1300–1629* (1948).

Magnus VI Lagabøter (the Lawmender) (1238–80) (King of Norway, 1263–80). Ceded the Hebrides and Isle of Man to Scotland in exchange for a payment and an annual rent. In 1274, he introduced a new national legal code based on the existing system but replacing the former provincial laws with a common national law. One change was to replace private vengeance by public administration of criminal justice. In 1277, he instituted a new municipal code which created a city-council form of government for the Norwegian towns. In 1277, he also concluded a concordat with the Church, which made the Church essentially independent but increased its prestige. These changes wholly transformed the Norwegian legal system, and served as models for Norwegian colonies. The national code continued in force for over 400 years.

Maguire, Conor Alexander (1889–). Chief Justice of the Supreme Court of the Republic of Ireland, 1946–61 and judge of the European Court of Human Rights, 1965–

Maiden. An early form of guillotine used in Scotland in the sixteenth century. It consisted of a heavy knife running in grooves in a vertical frame, raised by a rope and allowed to crash down, severing the neck of a person pinioned with his head in the appropriate place.

Maiden assize (or circuit, or sessions). Formerly a sitting of a court at which no prisoner was sentenced to death and, latterly, a sitting at which there were no prisoners to be tried. See also GLOVES.

Mail. An old Scottish law term signifying rent. Hence, blackmail is a kind of illegal rent.

Maills and duties. In Scots law, the rents of land and duties or services due from it. The action of maills and duties, originally competent to any proprietor, came to be restricted to heritable creditors seeking to recover rent from tenants in preference to landlords.

Maiming. Mutilation or physical injury involving incapacity to use, or loss of, a bodily member. It is now used mainly of injuring animals. Mayhem (q.v.) is an old word for maiming of persons.

Maimonides, Moses (1135–1204). A mediaeval Jewish philosopher and jurist, who lived in Cordoba and Egypt, and whose work influenced the great mediaeval scholastic writers. His major works were a commentary on the *Mishna*, a compendium of decisions in Jewish law from the earliest times to the third century, the *Mishna Torah*, a systematization of all Jewish law and doctrine, as well as other works of lesser scope on Jewish law, works on medicine, and a great classic of religious philosophy, the *More nevukhim* or *Guide for the Perplexed*, which sought to harmonize Biblical and Rabbinic teaching with philosophy and strongly influenced Arab, Jewish, and Christian successors. He has been recognized as the greatest of the Jewish philosophers and a pillar of the traditional faith.

S. Baron (ed.) *Essays on Maimonides*, B. Bokser, *Legacy of Maimonides.*

Maine, Sir Henry James Sumner (1822–88). An English jurist, who was a fine classical scholar. From 1847 to 1854 he was Regius Professor of Civil Law at Cambridge and, in 1850, was called to the Bar. From 1852, he was the first reader in Roman law and jurisprudence at the Inns of Court and his lectures attracted much attention. In 1861, the publication of *Ancient Law* brought him wider recognition. It contains the famous and brilliant generalization that the movement of progressive societies has hitherto been from status to contract. From 1862 to 1869, he was legal member of the council in India and carried many important measures. On his return in 1869, he was appointed Corpus Professor of Jurisprudence at Oxford. Some of his lectures were later published as *Village Communities* (1871) and *Early History of Institutions* (1875); in both he draws on his Indian experience. In 1877, he became Master of Trinity Hall, Cambridge, resigning his Oxford chair in the following year. *Dissertations on Early Law and Custom* (1883) was the last work founded on his Indian knowledge. In 1887, he became Whewell Professor of international law at Cambridge, but his posthumously published lectures on that topic, *International Law* (1888), had not been fully worked out. He also wrote in many journals. His fame rests on *Ancient Law* and the succeeding volumes, which were the first application to legal studies of historical and evolutionary views, though the generalizations were frequently founded on inadequate bases and are not wholly justifiable.

A. Grant Duff, *Maine: A Brief Memoir of his Life* (1892); G. Feaver, *From Status to Contract* (1969).

Mainprise (Fr. *main*, hand; *pris*, taken). A word used originally of the process of delivering a person to sureties or pledges (mainpernors) who undertook to produce him again at a future time. A man could be mainperned in cases where trial was inapplicable.

The writ of mainprise was a writ to the sheriff requiring him to take sureties for a prisoner's appearance.

Maintenance. Money payable to a spouse or child for that person's living expenses by the other spouse or a parent under an order of a court, corresponding to U.S. alimony.

Maintenance and champerty. At common law in England, maintenance was both a crime and tort, consisting of, without lawful excuse, lending assistance to a party to a civil action, without having a legally recognized interest in the subject matter of the action. Champerty (*campum partire*) was the variety of maintenance in which the party maintaining was to receive parts of the fruits of the action if it were successful. The crimes and torts were both abolished in 1967.

Maitland, Richard, 4th Earl of Lauderdale (1653–95). Sat in the Scottish Parliament and became a member of the Scottish Privy Council in 1678. He was Lord Justice-Clerk, 1681–84, General of the Mint, 1685–89, and a Commissioner of the Treasury, 1687–89. He accompanied James VII to France in 1689 and was outlawed. He was succeeded by his brother John, who was a Lord of Session (Lord Ravelrig, later Lord Halton, 1689–1710).

Maitland, Frederic William (1850–1906). Was called to the Bar in 1876 and became a very competent equity lawyer and conveyancer, but soon devoted himself to history and the overlapping spheres of law and history. He became Downing Professor of the Laws of England at Cambridge in 1888. Though his qualities were not recognized at the time, it is clear that he was a very great historian. He wrote extensively, in a vigorous and vital style and, though a meticulous scholar, had little sympathy with pedantry. Of his mainly historical works the chief are *Pleas of the Crown for Gloucester* (1884), *Justice and Police* (1885), *Memoranda de Parliamento* (1893), *Domesday Book and Beyond* (1897), *Township and Borough* (1898), and, jointly with Sir Frederick Pollock (q.v.), *The History of English Law before the Time of Edward I* (1895, revised edn., 1898). Of his mainly legal works, the chief are *Bracton's Note-Book* (1887), *Roman Canon Law in the Church of England* (1898), lectures on the *Constitutional History of England* (1908) and on *Equity*, and *The Forms of Action* (1909, later republished separately). He also wrote many essays and reviews. It was largely through his efforts that the Selden Society was founded and he edited for it the *Select Pleas of the Crown, Select Pleas in Manorial and other Seignorial Courts, Select Passages from Bracton and Azo*, three volumes in the Society's *Year Books* series, and collaborated in editing *The Court Baron, The Mirror of Justices*, and other volumes of the Society's Year Book series. He studied the language of the yearbooks and his introductions to these volumes contain much valuable matter on mediaeval law-French. Many of his shorter writings are in his *Collected Papers* (3 vols., 1911), *Selected Essays* (1936), and *Selected Historical Essays* (1957).

H. A. L. Fisher, *F. W. Maitland* (1910); C. H. S. Fifoot, *Maitland* (1971).

Majalla. An Ottoman code of civil law, mainly the work of Ahmed Cevdet Pasha (1822–95), and promulgated in 1869–76. Though based on the Sharia, its was influenced by European codes and was the basis of the civil law of the Ottoman Empire and its successor states for many years.

Majestas. In Roman law, sovereignty or supreme power. Under the Republic it was an attribute of the State but came to be attached to the *princeps* himself. Any derogation of *majestas* was treason and Sulla established a permanent *quaestio de majestate*. Under the early Empire, many kinds of conduct were charged as infringing *majestas*, and most trials seem to have been decided by political factors.

From being an attribute of the *princeps*, the word 'majesty' came to be an honorific title confined, at first, to the Roman emperors of the West but later extended to all kings. From the time of Henry II, it has been used in England, the full form being 'Her Most Gracious Majesty'. The usual form is 'Her Majesty' and the abbreviation H.M. appears in many contexts such as H.M.S., H.M.S.O. and O.H.M.S.

Majority. (1) The status of being a person of full age, entitled to enjoy all civil rights and to act on one's own account, as contrasted with minority (q.v.). At common law the age of majority in the U.K. was 21, but this was reduced by legislation in 1970 to 18 for most purposes.

(2) The major or greater number of a group of persons. The justification for the principle that the view of a majority rules and that of the minority is ignored depends in some cases, such as Parliament and lower governing bodies, courts, and the like, on long-standing custom, in other cases, such as companies, partnerships, trustees, and the like, on statutory provision. A majority is, however, constantly in danger of arrogantly assuming that its views are perfect wisdom and of wholly disregarding the rights and interests and views of the minority.

Make his law. A man was said to make his law when he performed the law which he had bound himself to, that is, cleared himself by his oath, and the oaths of his helpers, of an action commenced against him.

Mala fides. Bad faith, a term for the state of mind of a person who acts dishonestly, in the knowledge of lack of title or otherwise without belief in the legal justification for his action.

Male. Of the masculine sex. To which sex a human being belongs is determined by his genetic constitution, not by appearance, manner, psychological traits, or other factors. It is a material fact for marriage and certain other legal purposes.

Maleville, Jacques, Marquis de (1741–1824). Member of the French Tribunal de Cassation (1791) and one of the commission which drafted the Code Napoleon and its defender from the Roman law standpoint. He later became a senator and count of the Empire.

Malice. In its general use in British law, malice means no more than the deliberate or intentional doing of wrongful conduct. In libel and slander cases, it is common for the plaintiff to plead that the words were published maliciously, but malice is presumed if words of a defamatory character were deliberately communicated. But in some contexts, where it is sometimes called actual or express malice, malice means malevolence, spite, or ill-will. Thus the plea that defamatory matter is not actionable because published in circumstances protected by qualified privilege may be overcome by evidence that the publication was actuated by malice, i.e. by malevolence or evil motive.

In the phrase 'malicious damage', a man acts maliciously when he intended to do the unlawful act with which he is charged, or if that is the necessary consequence of some other criminal act in which he was engaged, or if he should have foreseen that as the consequence of unlawful conduct in which he was engaged, or the act was done with reckless indifference to the consequences.

In murder, the mental element is traditionally defined as malice aforethought (q.v.), which may be express, where there was deliberate purpose to kill or at least to do serious bodily harm to someone, not necessarily the victim, or implied by law, where the person killed was an officer of law legally arresting or imprisoning the accused or otherwise executing legal process, or where there was provocation but not enough to reduce the crime to manslaughter, or where the killing was done while committing another felony involving violence or an act dangerous to life.

English criminal law, prior to 1957, also recognized constructive malice, which was to the effect that where a person killed another in the course or furtherance of committing some other felony, e.g. rape, robbery, he was guilty of murder, even though he might not actually have intended to kill. This category of case included killing in the course of or for the purpose of resisting an officer of justice.

Malice aforethought. In English law, the traditional term for the mental element in murder which distinguishes it from manslaughter. Unfortunately, despite the adjective it does not mean that the conduct must have proceeded from evil motive, nor that it was thought out beforehand. It covers a number of different mental attitudes, any one of which has been held to render a homicide so heinous as to make it murder. These include intention to kill anyone, or intention to cause grievous bodily harm to anyone, or intention to do an act knowing that it will probably cause death or grievous bodily harm, or possibly even intention to do something unlawful from which death results as the natural and probable consequence. Distinct forms of malice aforethought were constructive malice, abolished in 1957, where the malice aforethought to commit murder was attributed to one who caused death in the course of committing another felony such as rape, or who caused death in the course of resisting lawful arrest by an officer of justice, and implied malice, where death is caused in the course of furtherance of some other crime, particularly where there was intent to cause grievous bodily harm.

Malicious abuse of civil proceedings. In general, a person may utilize any form of legal process without any liability, save liability to pay the costs of proceedings if unsuccessful. But an action lies for initiating civil proceedings, such as action, presentation of a bankruptcy or winding up petition, an unfounded claim to property, not only unsuccessfully but maliciously and without reasonable and probable cause and resulting in damage to the plaintiff.

Malicious arrest. An arrest, mainly in civil cases and accordingly now rare, procured by a person maliciously and without probable cause.

Malicious damage. Deliberate damage or injury to property, including many kinds of vandalism.

Malicious falsehood. A right of action exists where one person has, orally or in writing, with actual malice communicated an untrue statement disparaging the plaintiff's title, goods, or other property and calculated, in the ordinary course, to produce and, in fact, producing actual damage. The main instances are slander of goods and slander of title (qq.v.).

Malicious injuries. Any deliberate and evilly-motivated harm to the person of another, as by wounding, maiming or disfiguring.

Malicious mischief. In Scots law, the deliberate destruction or damage of the property of another.

Malicious prosecution. The tort of initiating a criminal process by making a charge against a person before a judicial officer or tribunal not only unjustifiably but maliciously and without reasonable or probable cause, resulting in actual damage to the complainer, save where the charge endangers his reputation or person, when damage is implied. The prosecution must have failed, though the initial success of the prosecution, even though reversed on appeal, would be evidence relevant to the existence or not of reasonable cause. A person who fairly and honestly states facts to a magistrate is not responsible to one against whom the magistrate issues a warrant of arrest.

Malins, Sir Richard (1805–82). Was called to the Bar in 1830 and practised as an equity draftsman and conveyancer. He took silk in 1849 and sat in Parliament, 1852–65. In 1866, he became a vice-chancellor and sat till his retirement in 1881. He was a competent but not outstanding judge.

Malitia supplet aetatem (malice supplies age). An old maxim applicable to children above the age of criminal responsibility, but under 14, to the effect that they can be convicted only if it be shown that they had mental capacity to have intention to do what they knew was wrong.

Malleus Maleficarum (The Witches' Hammer). A work produced, in 1486, by two Dominican friars, Heinrich Kraemer and Johann Sprenger, who had been instructed by Pope Innocent VIII to extirpate witchcraft in Germany. It became the authoritative encyclopaedia of daemonology and witchcraft throughout Christendom and remained of authority for three centuries. Based on folk beliefs, it became a systematized theory which attributed witches' special powers to their relations with the devil, especially sexual relations, and this led to suspected witches being persecuted by both Catholic and Protestant churches.

Malta, law of. Malta and adjacent islands have been occupied since prehistoric times, being part of the Roman and Byzantine empires and occupied by Arabs, Normans, and feudal lords. In 1530, Malta was ceded to the Order of St. John of Jerusalem, passed under British rule in 1802 and, finally, became independent in 1964. It is now a republic within the British Commonwealth. The head of state is the President. He is advised by a Prime Minister and Cabinet. A unicameral Parliament with a life of five years, comprises members elected by proportional representation. The chief justice and judges are appointed by the President on the advice of the Prime Minister and exercise civil and criminal jurisdiction in the superior courts and the courts of magistrates. One of the superior courts is known as the Constitutional Court and hears appeals on constitutional matters. Appeal no longer lies to the Privy Council. There are a Court of Appeal and a Court of Criminal Appeal, a Civil Court, Commercial Court and Criminal Court. Qualified magistrates sit in the Courts of magistrates of Judicial Police. There are also various special courts and tribunals. The language of the courts is Maltese.

Maltese law is based largely on the Code Napoléon and other French codes and was codified, mainly between 1854 and 1873. Some commercial and maritime matters are regulated by English principles and procedure is on English lines. The principle of *stare decisis* does not hold. Law is taught in the University of Malta.

J. Cremona, *Outline of the Constitutional Development of Malta under British Rule* (1963).

***Malum in se* and *malum prohibitum*.** A distinction drawn, from at least the fifteenth century, between criminal conduct, which was essentially evil, and criminal conduct which was merely prohibited by statute but was not inherently wrong. Thus, it was held that the King could dispense with penalties in the latter class of case only. The distinction is unscientific and of no legal significance in modern law, though members of the public commonly draw the distinction and regard, e.g. assault or theft, as different from and worse than, e.g. road traffic offences.

Malynes, Gerard (?–1641). About 1586, he was appointed one of the commissioners of trade in the Low Countries and was frequently consulted by the Privy Council in the reigns of Elizabeth and James I. Under James I, he took part in many schemes for developing the natural resources of the country and was much concerned with monetary affairs.

He was one of the first writers on economic affairs, particularly on foreign exchange, and his numerous works are full of information on commercial matters and are important for economic history. Most relevant for legal purposes are *Consuetudo vel Lex Mercatoria*, or the Ancient Law Merchant (1622), later published with Marius' (q.v.) *Collection of Sea Laws: Advice concerning Bills.*

Man, Isle of. One of the British Islands (q.v.) but not a part of the United Kingdom and a separate feudal lordship. From 1290, the island was under the protection of the English kings who were known from 1406 to 1649 as Kings of Man and then as Lord of Man. It became vested in the Crown by statute in 1765 when it was acquired from the Duke of Atholl.

The Crown appoints the Lieutenant-Governor, who controls the executive in consultation with an Executive Council of seven, the majority being chairmen of the principal spending boards of

Tynwald. These boards exercise administrative powers in relation to various services.

The legislature is Tynwald, or the Tynwald Court, consisting of the Lieutenant-Governor, the Legislative Council, and the House of Keys. The Legislative Council consists of the Lieutenant-Governor, the Bishop of Sodor and Man, the First Deemster and Clerk of the Rolls, the Second Deemster, the Attorney-General, two persons nominated by the Lieutenant Governor, four persons nominated by the House of Keys. The House of Keys is composed of 24 elected members. The Houses have equal powers and sit together but vote separately.

Legislation requires the assent of the Queen in Council and, save in emergencies, formal promulgation on Tynwald Hill by reading in Manx and English of a summary of the law and of the royal assent; the Lieutenant Governor and the Speaker of the House of Keys sign an attestation of the promulgation. To a considerable extent, legislation of Tynwald corresponds with that of the U.K. Parliament.

The United Kingdom Parliament legislates for the Isle of Man in important and non-local matters, which are made applicable after consultation with Tynwald, and is responsible for foreign affairs.

The High Court consists of the Common Law Division, of which the First and Second Deemsters are judges, the Chancery Division, of which the First Deemster and Clerk of the Rolls is judge, and the Staff of Government Division, which is the court of appeal from the other divisions, and of which one Deemster and the Judge of Appeal are the judges. The criminal court is the Court of General Gaol Delivery; one judge of the High Court is judge and the three judges of the High Court constitute a court of criminal appeal. A further appeal lies to the Judicial Committee of the Privy Council.

The law of the island is not English common law, but a system based on Norse customary law and developed by legislation. For long, it rested largely on tradition and was known as 'breast law', i.e. law contained in the judge's breast. In most essential matters it is similar to English law. Law and equity were merged in 1883.

Man of straw. A person of no financial substance, who is not worth suing.

Manager. One who directs, controls, and secures the doing of things, a title frequently given to the person in charge of a shop, branch of a bank, factory, or other establishment. In Parliament, when representatives of both Houses meet to settle differences between the Houses, they are designated managers. In bankruptcy, a special manager of the debtor's business may be appointed pending the appointment of a trustee. In Chancery practice, if a receiver is appointed and requires to carry on a business, he may be designated a manager or receiver and manager, and in the winding-up of a company in England, a special manager may be appointed.

Managing clerk. A person employed in a solicitor's office, who conducts business under supervision of the principals. He may be admitted as a solicitor, or be unadmitted, in which case, he is customarily entitled to be called a managing clerk after 10 years' experience. The term legal executive (q.v.) is now commonly used of managing clerks.

Manbote. Compensation paid for homicide, particularly to the lord of a villein killed.

Mancini, Pasquale Stanislas (1817–88). Italian jurist and statesman, Professor of Law in Turin. He was strong advocate of nationality and secured its acceptance, rather than domicile, in Italy as the major fact in personal matters. He was one of the founders of the Institute of International Law and its president, and held many public offices in the new kingdom of Italy.

Mancipatio. In Roman law, a solemn transaction found in the XII Tables involving the use of copper and scales (*per aes et libram*) and the presence of a *libripens* holding the scales and five witnesses. It was used for the sale of *res mancipi*, which were the major items in a primitive agricultural economy, including Italic land, slaves and beasts of draught and of burden. Applications of this means of transfer were gradually extended to the making of a gift, the constitution of a dowry or mortgage, and such acts as *adoptio*, *emancipatio*, *coemptio* and, for the making of a will, *per aes et libram*, by which the testator transferred his whole property.

Mancipium. See *Patria potestas*; *Emancipatio.*

Mandamus (Lat. we command). formerly a prerogative writ, now an order which may be granted by a Divisional Court of the Queen's Bench Division of the High Court in England, in the form of a command from the Court to any person, corporation, or inferior tribunal requiring him or it to do something which is a public duty and pertains to his or its office to do. It issues in all cases where there is a specific legal right to have a function exercised and no specific legal remedy for enforcing that right, and also in cases where there is an alternative legal remedy but one less convenient and effectual. The grant of the order is, in general, in the discretion of the court.

Like the orders of prohibition and *certiorari* (qq.v.), the chief use of the order is as a means of controlling inferior courts and tribunals which have authority to determine issues and a duty to act judicially. Thus, it lies where a tribunal has declined to exercise jurisdiction, or has been influenced by extraneous considerations, but not where the tribunal has acted honestly, though wrongly, on the facts or the law. It is not an appeal against the merits of a decision.

It lies, also, to compel the restoration of a person to an office of a public nature from which he has been wrongfully evicted, or to compel his admission thereto, to compel the delivery of public books and papers, to compel the performance of statutory duties, to compel public officials and bodies to carry out their duties, and to command inferior courts and tribunals to exercise their jurisdiction.

No order lies against the Crown, or Crown servants, unless they owe a duty to the applicant as well as to the Crown, the superior courts, or ministerial officers subject to authority.

As a rule, the court will not make an order where there is an alternative legal remedy, or when it would be futile in its result.

Mandata. Under the Roman Empire, administrative instructions to officials, particularly provincial governors, which in time came to resemble standing instructions. They were not of great importance for the development of private law.

Mandatary. In Scots law in person to whom is given a mandate to act in some way on behalf of another. A person outwith the jurisdiction may be required to sist a mandatary, i.e. to appoint a person within the jurisdiction as his substitute to act in his stead as a party to the cause. The mandatary is personally liable for expenses and to implement the orders of the court.

Mandate. In Roman and Scots law, a contract by which one person undertakes to act gratuitously for another or manage the affairs of another. In modern practice, a formal grant of authority to act may take the form of a factory and commission or power of attorney, but informal or presumed grants will suffice. The mandatary must perform what he has undertaken, adhere to instructions, exercise such care and foresight as a prudent man would in his own affairs, and account for his intromissions. He is entitled to be reimbursed for expenses *bona fide* incurred but not to be paid, though he may receive an honorarium. He is liable for neglect of duty or lack or ordinary care. In English law, the name is sometimes given to a species of gratuitous bailment, transfer of a specific chattel as to which the bailee engages to do some act without reward.

Mandated territories. After World War I, certain territories and colonies of Germany and Turkey were detached from them and entrusted by the League of Nations to 'mandatory states' to administer on behalf of the League on conditions laid down in mandates between the League and each mandatory. The dominant concept was that of guardianship in the interests of the inhabitants of the mandated areas, and the system was supervised by the council of the League assisted by the Permanent Mandates Commission. There were three classes of mandated territories designated A, B, and C, according to their proximity to attaining independent statehood. Class A comprised former Turkish provinces north of Arabia, Iraq, Palestine and Transjordan being assigned to Britain, Syria and Lebanon to France: Class B comprised former German colonies in Africa, and Class C comprised former German Pacific islands territories. The scheme was reasonably successful and the mandated territories began to achieve independence from 1932. After World War II, the system was replaced under the United Nations by the trusteeship system (q.v.).

Mandatory injunction. The form of injunction (q.v.) in which a court directs the performance of a positive act, as by directing a building to be demolished.

Mandatory and directory provisions. A distinction drawn by courts when considering whether a defect in formal or procedural requirements invalidates a subordinate legislative instrument. Mandatory requirements must be complied with and non-compliance is a potentially vitiating defect; merely directory requirements should be complied with, but the courts may hold that substantial compliance is enough, or even that total non-compliance is a mere error not vitiating what has been done. The main criterion is the practical importance of non-compliance.

Manilius, Manius (second century B.C.). A famous Roman jurist, one of the first to write on the civil law, and author of *Monumenta* and forms for contract of sale. Some of his *actiones* and *responsa* survive.

Manners-Sutton, Thomas, Lord Manners (1756–1842). Was called to the English Bar in 1780, became Solicitor-General in 1802, and a baron of Exchequer in 1805. He was made Lord Chancellor of Ireland and a peer in 1807 and held that office till he resigned in 1827.

Mannheim, Hermann (1889–1974). German judge and jurist who came to England in 1934 and taught criminology. He wrote *Criminal Justice and Social Reconstruction* (1946), *Comparative Criminology* (1965), and many other books and did much to advance the study of criminology in Britain.

Manning, William Oke (1809–78). English jurist and author of *Commentaries on the Law of Nations* (1839), one of the first English treatises on international law.

Manningham-Buller, Reginald Edward, 4th baronet and 1st Viscount Dilhorne (1905–). He was called to the Bar in 1927, became an M.P. in 1943, took silk in 1946, and was Solicitor-General, 1951–54. He became Attorney-General in 1954–62, and then Lord Chancellor (1962–64). In 1969 he was made a Lord of Appeal.

Manor. The term applied, after the Norman conquest, to estates organized under knights, ecclesiastical corporations, or otherwise, and managed and cultivated as units. By the end of the 11th century, the main element was the feudal lord, and soon he came to be regarded as the owner of the manor, and to have authority over the tenants, and the right to hold a court for them. The typical manor comprised demesne lands in the hands of the lord and lands let to free and villein tenants, all having certain rights of common in the waste or non-cultivated lands.

In the thirteenth and fourteenth centuries, a manor also implied a right of jurisdiction exercised through a court baron, attended by both freeholders and villein tenants. By the sixteenth century the freeholders owed suit of court to and were subject to the court baron, while villein tenants or copyholders were subject to the court customary in which the lord or his steward was judge, which were two aspects of the court rather than two courts.

The court baron tried all personal actions where the cause of action did not exceed 40 shillings, and till the end of the fifteenth century, all actions connected with copyhold tenements could be heard only in the court customary.

In the eighteenth century the manorial court decayed rapidly, cases being generally brought in the King's courts, the only surviving business being copyhold conveyancing.

F. Seebohm, *The English Village Community* (1883); P. Vinogradoff, *Villeinage in England* (1892); *The Growth of the Manor* (1905); F. W. Maitland, *Select Pleas in Manorial Courts*; *The Court Baron* (1888); *Domesday Book and Beyond* (1897).

Manor (U.S.). The creation of manors on the English model, in the shape of tracts of land for which the tenant paid an annual feu-duty to the proprietor, took place, chiefly, in the proprietary colonies, especially New York, Maryland, and South Carolina. The patroons, or wealthy landowners, of the colonial New Netherlands, obtained grants of land from the Dutch West India Company with perpetual ownership, the rights to exact payments in money, goods, or services from tenants, and rights of holding courts and administering civil and criminal justice. This system continued till 1775, when patroons became merely landed proprietors, but was not abolished till 1846. Maryland, especially among the English colonies, was a land of manors, where the lords of manors could hold baron courts and their stewards courts leet. The eight proprietors who obtained the grant of Carolina in 1663, planned an elaborate feudal society based on the ownership of land, in which manors were to be recognized, but the scheme never became effective.

Manpower Services Commission. A body established in 1973 to make arrangements to assist persons to select, train for, obtain, and retain employment suitable for their ages and capacities, and to obtain suitable employees, partners, and business associates. It has wide powers for the performance of its functions. It may direct the Employment Service Agency or the Training Services Agency to perform certain of its functions, and may make proposals for the establishment of industrial training boards and give such boards directions.

Manrent. Personal service or attendance. It was in Scotland the token of a kind of bondage, whereby free persons bound themselves by bonds of manrent to become bondmen or followers of those who were their patrons or defenders.

Manse. In Scotland, the dwelling-house of the clergyman. Each rural parish minister was entitled to have a manse and have it maintained by the heritors. This responsibility has now been devolved on congregations and on the General Trustees of the Church.

Mansfield, Lord. See MURRAY, WILLIAM.

Mansfield, Sir James (1733–1821). Was called to the Bar in 1758, took silk in 1772, entered Parliament in 1779, and was Solicitor-General in 1780–82 and 1783. He was not a success in Parliament and he lost his seat in 1784. He is held to have been a competent, but not brilliant advocate. In 1804, he became Chief Justice of the Common Pleas and presided in that court till 1814, when he resigned. He was a very competent lawyer and, though his decisions are sound he pronounced no decisions of outstanding importance.

Mansion (or mansionhouse). Originally a lord's house on a manor, but later used of any dwelling-house, particularly a large one.

Manslaughter. In English law, the crime of unlawfully killing another person in circumstances not amounting to murder; it covers killing under provocation, such as temporarily to deprive a reasonable man of self-control and in the heat of fury, unintentionally killing another while committing an unlawful act not likely to cause danger to others, or killing by serious and culpable neglect of a legal duty incumbent in the person causing the death, e.g. by reckless driving. It covers cases of death caused by negligence much more serious than merely carelessness, though the latter may be tortious negligence. On an indictment for murder a jury may bring in a verdict of manslaughter if they hold malice aforethought not proved.

Mansuetae naturae (Lat. of a tame nature). Domesticated, used of animals of species generally domesticated in Britain.

Manu. In Indian mythology, the first man and legendary author of a Sanskrit code of law, the *Manu-smrti*, or Tradition of Manu. This is traditionally the most authoritative of the Hindu law-books (*Dharma-sastra*) and in its present form probably dates from the first century B.C. It comprises 12 chapters, containing 2,694 stanzas and prescribes, for Hindus, the obligations incumbent on them as members of one of the four social classes and passing through one of the four stages of life. It contains a mixture of religious and secular rules and has had immense practical significance as providing ordinary Hindus with a practical code of conduct.

Manumission. The freeing of a slave. In Roman law, this could be effected in various ways, by will and, later by *fideicommissum* (q.v.) by *vindicta*, a solemn procedure before a magistrate, originally a fictitious claim of liberty, or by the entry of the slave's name with the master's consent, by the censors in the list of citizens, and later, by informal declarations such as by letter of enfranchisement or declaration before witnesses. Under Christianity, the Church added a new procedure of manumission *in ecclesia*. The effects of manumission were very varied and were subject to many changes as time passed.

Manus (Lat. hand). In Roman law, the term for the power of a husband over his wife. Marriage might be constituted *cum manu* by either *confarreatio*, the earliest form, obligatory for patricians, requiring the presence of a pontiff and 10 witnesses, or by *coemptio*, a fictitious sale of the bride to her husband by *mancipatio* (q.v.); in both cases the bride passed from being subject to the *patria potestas* (q.v.) of her father to being under the *manus* of her husband. *Manus* was acquired by marriage *usu*, constituted by actual cohabitation at the husband's house for a year without interruption. Marriage *sine manu* was, however, increasingly common as time went on, in which case the wife remained under her father's *potestas* and divorce was competent at the instance of either spouse.

Manwood, John (*d.* 1610). A barrister and gamekeeper of Waltham Forest who compiled *A Brefe Collection of the Laws of the Forest* (1598), enlarged into *A Treatise of the Lawes of the Forest* (1615).

***Marais* v. *G.O.C. Lines of Communication*,** [1902] A.C. 109. Marais was arrested and detained by the military forces of the Crown in South Africa for having contravened martial law regulations. It was held that he could not appeal as, in a state of war, the civil courts have no jurisdiction to question the actions of the military authorities. The courts can, however, decide whether a state of war existed or not.

Marbury* v. *Madison (1803), 1 Cranch 137. Marbury had been appointed a justice of the peace for the District of Columbia by President John Adams hours before his presidency terminated. Madison, the new Secretary of State, on President Jefferson's orders, refused to deliver the commission to Marbury, who applied to the Supreme Court for a writ of mandamus. The Court, in a unanimous opinion delivered by C. J. Marshall, held that Madison had no right to withhold the commission, but that the Judiciary Act, 1789, Section 13, which empowered the court to issue such a writ, was contrary to the Constitution and therefore invalid. The decision thereby asserted the Court's power of judicial review (q.v.), though by declining to issue the writ of mandamus it avoided a conflict with the President, the decision created a self-imposed limitation on the Court's own powers.

Marcellus, Ulpius (second century A.D.). Roman jurist and member of the Council under Marcus Aurelius, who wrote a *Digesta* in 31 books, extensively copied by later jurists, notably by Ulpian in his commentary on the Edict, and the subject of commentaries by Ulpian and Scaevola, notes to Julian's *Digesta* and to Pomponius' *Regulae*, *De officio consulis* (five books), a collection of *Responsa*, and a commentary *Ad legem Juliam et Papiam*. He gives the impression of being an acute and independent thinker and not a mere compiler, taking issue with earlier jurists, and he is frequently cited by subsequent jurists.

Marchers, Lords. The name given to lords permitted, by the kings of England, to conquer and possess such parts of Wales as they could. Each such lord was given sovereign rights in the areas conquered by him, which, in aggregate, comprised most of what are now the border counties. This state of affairs continued until the conquest of Wales in 1282, but the marcher lordships were annexed to the Crown by statute in 1354 and the sovereign rights of the lords marchers were abolished by statute in 1535.

Marches. The border lands between England and Wales and England and Scotland. The term is still used in Scotland for the boundaries between one estate and another.

Marches, Court of. A former court in Wales with jurisdiction in actions of debt or damages not exceeding £50.

Marches, custom of the. A mixture of English law and Welsh custom which developed in the parts of Wales governed by the Lords Marchers (q.v.).

Marchet (or merchet). A pecuniary fine payable in feudal law by a tenant to his lord on the marriage of one of his daughters in respect of any loss of service which the lord might sustain. It may be the origin of the alleged *ius primae noctis* (q.v.).

Marcianus, Aelius (third century A.D.). One of the last classical Roman jurists, probably an office-holder in the imperial chancery, and author of manuals (*Institutiones*, *Regulae*) and monographs, chiefly on criminal procedure, which appear to be addressed to persons who obtained Roman citizenship under Caracalla, instructing them in Roman law. His writings are not entirely lacking in originality.

Marculfus (seventh century A.D.). A French jurist, author of a famous Formulary, *Marculfi monachi formularum libri duo*, or collection of legal forms based on Salic law for the use of clerics, in two books dealing with public and private documents respectively. It is very valuable for French legal history.

Mare clausum. The name given, in the sixteenth–seventeenth century controversy over the control of the high seas, to the doctrine that a country could assert sovereignty over the high seas. Spain and Portugal claimed whole oceans; England claimed the Channel and the North Sea. It is also the title for a book by John Selden (q.v.) published in 1635, asserting that the sea was a subject of private property as much as is the land.

Mare liberum. The name given, in the sixteenth–seventeenth century controversy over the control of the high seas, to the doctrine that the seas were open. It is also the title of Grotius' (q.v.) book, published in 1609, maintaining that position.

Marginal notes. Indications in the margins of Acts of Parliament, beside the commencement of sections, of the subject-matter of the sections. They are not discussed in Parliament, are not deemed part of the Act, and may not be invoked in court to assist interpretation. Since marginal notes are not debated, nor amended, in Parliament they may sometimes not accurately summarize the substance of the section.

Mariana, Juan de (1536–1624). A Spanish jurist, humanist and historian, author of a valuable *Historiae de rebus Hispaniae* (1592), and *De Rege et regis institutione* (1599) in which he attacks absolutism, and concludes that it is lawful to overthrow a tyrant, and many other works.

Marine insurance. The contract whereby an insurer undertakes to indemnify the assured to the extent thereby agreed against losses incident to a marine adventure. Insurances are usually on ship, goods, freight, or profits, against maritime perils, such as perils of the seas, fire, war perils, pirates, thieves, capture, seizure, restraint of princes and peoples, jettison, barratry, and other like perils. The contract is contained in a policy of sea insurance, which may be for a voyage, or time, or for both a voyage and for time, or may be a floating policy, which leaves the name of the ship or other particulars to be defined by subsequent declaration. Almost all policies effected in the U.K. are modelled on a policy called Lloyd's policy, recognized and scheduled to the Marine Insurance Act, 1906. A large volume of case-law has grown up round the interpretation of the common-form policy.

Marischal. A Scottish Office of State who had jurisdiction through military courts over trespasses done in the army in time of war, and the release and ransom of prisoners. Along with the King and the Constable, he constituted a court of chivalry for treason-duels. The office seems to have disappeared after 1707.

Maritagium. A customary kind of gift well recognized in early mediaeval England. A donor gave lands together with his daughter to the donee in free marriage, feudal services not being due until the third heir enters, the presumption being that the family will by then have become established and capable of performing feudal services. Until the third heir had entered, the maritagium probably could not be alienated, and if he failed, it reverted to the donor and his heirs. Later, the gift was embodied in a deed and the descent limited to a specified series of heirs. It is accordingly the ancestor of the feetail or entail.

Marital rights. Formerly used of the rights of a husband to property to which his wife was entitled during the marriage. A voluntary conveyance by the wife after the engagement to marry and without notice to the husband was deemed a fraud on his marital rights. This use of the phrase is now obsolete, and the phrase is now used as synonymous with conjugal rights, for the husband's right to the society of and to exclusive intercourse with his wife.

Maritime courts. The pre-eminent maritime court has long been the High Court of Admiralty and its successors, the Probate, Divorce and Admiralty Division and, since 1971, the Admiralty court within the Queen's Bench Division.

In the Middle Ages, many English seaport towns, such as Bristol, Newcastle, and Yarmouth, had jurisdiction in maritime cases. When the court of Admiralty was established, many towns secured exemption from its jurisdiction, but the Admiralty challenged the jurisdiction of some of them. All these local jurisdictions disappeared by the early nineteenth century, except that of the Cinque Ports (q.v.).

There were formerly courts of the vice-admirals of the 19 districts of the coast of England and Wales with a jurisdiction up to high-water mark and the first bridges on rivers.

Among maritime courts with special jurisdiction are courts of survey, which hear appeals by owners of ships detained as unsafe or unseaworthy, wreck inquiries to investigate the cause of a shipping casualty, naval courts which hear complaints by officers or crew or investigate wrecks abroad.

In wartime, the Admiralty Court exercises a jurisdiction in prize (q.v.), with appeal to the Judicial Committee of the Privy Council.

Maritime law. The body of law concerned with ships, shipping, and carriage by sea. 'Admiralty law' covers much the same range of topics but is, properly and strictly, the law within the jurisdiction of, and administered by, Admiralty courts, while 'law of the sea' is mainly appropriated to the branches of international law concerned with territorial waters, restricted areas for fishing and other matters which are or should be matters of international agreements.

In ancient and mediaeval times, maritime adventures and the conduct of masters and mariners were regulated by various collections of usages connected with particular ports, recognized as having force of customary law and having an international character. Most of them probably had their origins in the Rhodian sea law of the third or second century B.C., which had great authority for centuries in the Mediterranean, supported by other developments of late Roman law and customs of the seafaring peoples of the Mediterranean. The Rhodian sea law, in effect the maritime code of the later Eastern Roman Empire, was codified about the eighth century A.D. Maritime law was closely connected with commercial law and the principles were developed in a generally similar manner.

The *Basilica* lays down law on many maritime matters, the general principles being supplemented by instances derived from the Rhodian sea law. Then the *Tabula Amalfitana* of the tenth century stated the maritime usages of Amalfi, while the decisions of the consuls of the Corporation of Navigators at Trani were compiled in the eleventh century. Venice and other Italian ports had their own compilations of maritime law.

The Crusaders set up tribunals of their own to try disputes arising among traders in the Eastern Mediterranean, and codes were formulated, known as the Assizes of Jerusalem, which included provisions on maritime law, which were applied by consuls or sea magistrates. The Latin kingdom of Jerusalem had both a mercantile court and a maritime court.

The Mediterranean ports, generally, followed the customs embodied in the *Consolato del Mare* or *Book of the Consulate of the Sea*, written in Catalan and published at Barcelona in the later fifteenth century. It takes its name from the fact that it embodies rules of law followed in the maritime towns of the Mediterranean and applied by commercial judges known as consuls. They were introduced from Barcelona to Valencia, Majorca, Sicily, and Rousillon. The earliest edition contains a code of procedure issued by the kings of Aragon for the guidance of the courts of the consuls of the sea at Valencia, a collection of ancient customs of the sea, and a body of ordinances for the government of ships of war. An appendix contains maritime ordinances of the kings of Aragon and the councillors of Barcelona. It was a most valuable collection, was translated into many languages, reprinted repeatedly, and was recognized by common consent as law throughout the Mediterranean.

Ports on the Channel, North Sea, and Baltic coasts, generally, adopted the Laws of Oleron, of Wisby or of the Hanse towns. It is said that Eleanor of Aquitaine, having observed the repute in which the *Consolato del Mare* was held in the Levant, directed the making of a record of the judgments of the maritime court of the island of Oleron (then in the duchy of Guienne and an important commercial centre) to serve as law to mariners in the West, and that Richard I brought a roll of these judgments back from the Holy Land, which he ordained were to be observed as law in England. The earliest version of the Oleron laws comprised customs observed in the wine and oil trades carried on between the ports of England, Flanders, Normandy, Brittany, and Guienne. A Flemish text of the fourteenth century, the *Purple Book of Bruges*, and a French text entitled *La Charte d'Oleroun des judgementz de la mier*, of the same period, in the *Liber Memorandorum* of the City of London Corporation, and in the *Red Book of Bristol*, are known, and an English translation, under the name of the *Rutter of the Sea*, was published in 1536. A text is also found in the *Judgments of Damme*, the sea-laws in use among the merchants of Flanders, expecially those of Damme and Bruges, in the fourteenth century. Dutch versions are also known, one accompanied by certain customs of Amsterdam.

An enlarged collection of sea-laws, which purported to be an extract from the laws of Oleron, was published at Poitiers in 1483 as *Le Grant routier de la mer* and an English translation was appended to Godolphin's *A View of the Admiral Jurisdiction* (1661) but this was superseded by Cleirac's *Us et coustumes de la mer* (1647) an English translation of which was incorporated in Justice's *General Treatise of the Dominion of the Sea, and a Compleat Body of the Sea Laws* (1705), and in the English translation of the *Black Book of the Admiralty*. Cleirac's version was also received in the English High Court of Admiralty.

In the North Sea and Baltic Sea, greater reliance was placed on the Visby sea laws, which may have originated in a code in Old Saxon dated 1240, and preserved at Lubeck, or in a code drawn up in Visby. It contains a copy of the Laws of Oleron and extracts from the laws of Amsterdam and the statutes of Lubeck. Extracts from them are quoted in Balfour's (q.v.) *Practicks*.

Another code which emanated from the Baltic is the statutory code enacted by the Hanseatic League, published as *Jus Maritimum Hanseaticum* in 1667. It was based largely on the traditional Roman Law.

In the trading centres of southern Europe, distinct magistrates and courts administering maritime law arose, and this mode of administering the law was followed in northern Europe. But in England and France, maritime jurisdiction was vested in an Admiralty court; in England, ultimately, the Admiralty lost all its jurisdiction over commercial cases but, in France, it retained jurisdiction over such cases involving carriage by sea.

The growing similarity between commercial laws of different states intensified similarity of maritime codes.

In England, the *Black Book of the Admiralty* was probably begun under Edward III and continued later, as a practical manual of the rules of maritime law for use in the Admiral's court. The original Norman French documents were later translated. Many of the rules probably originated in the decisions of the English ports which, from a very early period, had courts administering the customary law of the sea but the judgments of Oleron were adopted and later inserted into it. The Domesday of Ipswich shows that that town had a court in which maritime cases were dealt with rapidly, and in the early Middle Ages, many seaport towns had jurisdiction in maritime cases. The Court of Admiralty later questioned these jurisdictions and all had lost their jurisdictions by the nineteenth century, save only the Lord Warden of the Cinque Ports. Welwod, Professor of Civil Law at St. Andrews, published, the *Sea Law of Scotland* (1590) and followed this with his *Abridgment of all Sea Lawes* (1613). It records an inquisition conducted by Edward III at Queenborough in 1375, as a result of which certain articles concerning the Admiralty and seafaring were recorded.

The *Guidon de la Mer*, a collection which appeared at Rouen in the seventeenth century, was originally a private collection of principles of maritime law fixed by custom, and was published to assist the consuls of Rouen to administer mercantile business. Its rules formed the basis of the French Ordinance of Louis XIV of 1681.

Till the mid-fourteenth century in England, maritime law was mainly administered in the local courts of seaports. They mainly administered the laws of Oleron.

The rising court of Admiralty began, from the fourteenth century, to attract most of the maritime business of the country. Its procedure was based on the principles of the civil law and was more understandable by foreign merchants than was that of the common law courts.

The reception of Continental maritime law into English law was effected, mainly in the sixteenth and seventeenth centuries, by the court of Admiralty.

As the European nation states developed, some of them enacted distinctive codes of maritime law. These included the Maritime Code of Christian XI of Sweden (1667), the Marine Ordinance of Louis XIV of France (1681), and the Code of Christian V of Denmark (1683). The French Marine Ordinances, the work of Colbert, incorporated much of the settled custom of the sea into French national law and the French Admiralty Court was given jurisdiction in place of the former consular courts. The substance of the Ordinances was taken over into the French *Code de Commerce*, of 1807 and maritime law became merely a branch of commercial law.

In England, maritime law, in the form of the Rolls of Oleron, was first administered by the Court of the Cinque Ports. The Admiralty Court had, by the late sixteenth century, developed a wide jurisdiction, but much of this was later taken from it by the common law courts. The Court of Admiralty was amalgamated with others into the Probate, Divorce and Admiralty Division of the High Court in 1875 and, in 1971, became a distinct function of the Queen's Bench Division, dealing mainly with collision and salvage cases.

The Contract of affreightment, or contract for the carriage of goods by sea, is generally evidenced by a charter-party (q.v.) by which a ship or major part thereof is taken on hire for a time or for one or more voyages, or by bills of lading (q.v.) by which the owner or charterer of the ship carries the consignor's goods, issuing a bill of lading as a receipt for the goods and as a document of title to the goods. Marine insurance (q.v.) is important in carriage by sea.

Distinctive features of maritime law are the principles of salvage and general average. The principle of salvage (q.v.) is that persons who, by strenuous efforts and taking risks, save maritime property from loss or damage from perils of the sea, are entitled to a reward for their efforts, and have a maritime lien over the property salved therefor. The principle of general average (q.v.) is that, where one element of the maritime adventure, ship, freight or cargo, is deliberately sacrificed to save the others, the loss may be recovered from the parties in right of the other interests saved in proportion to the values involved. A great deal of international uniformity has been introduced by the York-Antwerp Rules of General Average, first promulgated in 1890 and several times revised since.

Modern maritime law has been much developed by international conventions. In many fields the initiative has been taken by the International Maritime Committee (Comité Maritime International). Thus, there have been international regulations for the prevention of collisions at sea, first adopted in 1889 and frequently revised since, the York-Antwerp Rules of General Average, conventions accepting the principle of limiting the liability of a ship in the event of collision to a stated amount per ton of the ship's tonnage and in other respects.

W. Ashburner, *The Rhodian Sea Law*; J. Pardessus, *Collection des lois maritimes* (1828); C. Abbott, *Law Relative to Merchant Ships and Seamen*, F. Sanborn, *Origins of the Early English Maritime and Commercial Law*; T. Carver, *Carriage by Sea*; T. E. Scrutton, *Charter-parties and Bills of Lading*; J. Arnould, *Marine Insurance*; R. Marsden, *Collisions at Sea*; A. Benedict, *The Law of American Admiralty*; G. Gilmore and C. Black, *Law of Admiralty*; G. Ripert, *Droit Maritime.*

Maritime lien (properly hypothec). A claim on a maritime subject, a ship, her apparel, or cargo, in respect of service done to it or for injury caused by it, not requiring possession of the subject. It is recognized in respect of the claims under bottomry and respondentia bonds, for salvage of property, seamen's wages, and damage done by the ship. Statute has created various claims similar to and enforceable in the same way as those enforceable by maritime lien. Maritime liens are enforced by proceedings *in rem* whereby the court will seize and, if necessary, sell the ship or cargo to satisfy the holder's claim.

Marius, John (17th century). Author of a book on *Bills of Exchange* (1651) regarded as of high authority.

Mark. A weight, especially of silver later a unit of account (a mark weight) valued at 13s. 4d. and later a coin of the seventeenth century valued at 13s. 4d. sterling, or a Scottish coin valued at 1s. 2d. sterling; also a cross marked on a document by a person who cannot write his name. See also TRADE MARK.

Mark system. A system of encouraging convicts, developed about 1840, by Alexander Maconochie (q.v.) at Norfolk Island, Australia, whereby prisoners were held not for a fixed time but until they had earned a fixed number of marks, the number being fixed by reference to the crime. Marks were earned for good conduct, hard work, and diligence and might be deducted for idleness or misconduct, and a prisoner became eligible for release when he attained the appropriate number of marks. The system was the forerunner of indeterminate sentences, individualized treatment, and remission for good conduct.

Market. A public place to which buyers and sellers of particular commodities resort for the purpose of doing business. A market can be held only by virtue of a royal grant of a franchise of market, or by long usage, such as raises a presumption of a lost grant.

The term is also sometimes used, even when there is no place to which buyers and sellers go, for the notional markets arising from the existence of several persons willing to buy or sell, as the case may be.

Market, Court of the clerk of the. In late mediaeval times, an incident of every market and fair, in which misdemeanours, particularly in relation to weights and measures, might be punished. The clerk of the market was a person appointed, frequently by the bishop, as the person responsible for suppressing abuses.

Market overt. The principle of English law that where goods are sold in market overt, i.e. in a public and legally constituted market or fair, including a modern statutory market, according to the usage of the market, and also including, by the custom of London, that part of every shop within the City of London to which the public is admitted without special invitation, between sunrise and sunset any day except Sundays and holidays, for the sale by the shopkeeper of such goods as he professes to trade in, the buyer acquires a good title to the goods if he buys them in good faith and without any notice of defect or want of title on the part of the seller. But if goods have been stolen and the offender is convicted, the property therein reverts in the owner notwithstanding any dealing with them even in market overt. The principle is not applicable in Wales or Scotland.

Markets and fairs. Markets and fairs are franchises conferring rights to hold an assembly of persons to buy and sell the kinds of commodities for which the franchise was given. The term is also used for the place, the time, or the gathering itself. There is no legal distinction between a market and a fair, but the former is held weekly, or more often, and the latter only once or twice a year. The right of holding a market or fair is one which may be granted to a subject by the Crown by virtue of the royal prerogative and confers a monopoly right to hold a market or fair for the benefit of the public, or it may be established by prescription based on immemorial user, raising a presumption of a lost grant, or be created by local Act. In modern times, local authorities may establish markets for various kinds of products under various public-general statutes. The person having the right of market is entitled to take action against persons who unlawfully dispute or interfere with the holding of the market, to exercise general direction of the market, and to cause an assembly of persons which might otherwise be a nuisance; he must provide an adequate place for the market, and may move it to a new place. Every member of the public has liberty to attend the market and deal there. By grant, statute or custom tolls or dues may be levied on goods brought into the market. Stallage or pennage may be charged for the liberty to erect stalls or pens for cattle. The franchise may be forfeited by the Crown for non-use, disregard of the conditions of the grant, or the taking of excessive toll.

Marlay, Thomas (?–1756). Was Solicitor-General for Ireland, 1720, and Attorney-General, 1727. In 1741, he became Chief Justice of the King's Bench in Ireland and retired in 1751.

Marowe, T. (fifteenth century). A member of the Inner Temple whose reading on the practice of Justices of the Peace was a major source of Lambard's *Eirenarcha* (q.v.). It is printed in B. H. Putnam's *Early Treatises on the Practice of the Justices of the Peace* (1924).

Marquess (or Marquis). The second grade in the peerage of the United Kingdom, ranking next to dukes. In origin, it was the title in various European countries (Old Fr. Marchis, Germ. Markgraf (margrave)) of the ruler of a mark, march, or border territory, but later became a title only. The first creation in England was the Marquess of Dublin in 1385.

Marriage. A term both for the relationship of husband and wife, and for the ceremony which creates such a relationship. In English and Scots law, the basic legal concept of marriage is that generally recognized in Christian communities, namely the voluntary and permanent relationship of one man and one woman to the exclusion of all others, for life or until earlier legal dissolution. The concepts and rules have been materially influenced by the pre-Reformation canon law of marriage. The requisites of marriage are that each party should be capable in respect of sex, age, mental capacity, and single status of contracting marriage, that they should not be within degrees of relationship prohibited from marrying one another, they they should freely consent to take each other in marriage, and that prescribed notices and formalities be complied with. Marriage is normally preceded by an exchange of promises to marry in future (*sponsalia de futuro*) after which the intending spouses are said to be engaged or betrothed. See PROMISE OF MARRIAGE. But this is not a necessary or legal requisite.

In England and Scotland, the intending spouses must be of opposite sexes, each must have attained the age of 16, have sufficient mental capacity to understand the nature of the contract of marriage and what he or she is undertaking in agreeing to marry, and be of single status (unmarried, widowed or divorced). Incapacity to have issue is irrelevant. Sexual impotency at and after the time of marriage is a ground for having the marriage avoided.

The parties must not be within the prohibited degrees of consanguinity or affinity, as laid down for England in the Marriage Act, 1949, and for Scotland by the Marriage (Sc.) Act, 1977. Half-blood relationship, illegitimate relationship and adoptive relationship are within the prohibited degrees.

There must be free and genuine consent to marry and consent obtained by duress, fraud or personation, or while a party is incapacitated by drink or drugs, is invalid.

At common law in England, the parties had to take each other as husband and wife in the presence of an episcopally ordained clergyman, before the Reformation a priest, and after the Reformation a priest or deacon. In 1753, statute required all marriages to be solemnized in the parish church or a public chapel of the Church of England, by licence, or after due proclamation of banns, and since 1823, the formalities have been determined by statute. The modern formalities are that every marriage shall be by publication of banns, common or special licence, or superintendent registrar's certificate, or certificate and licence, or naval officer's licence and, save in the cases of marriages according to the usages of Jews or Quakers and in the case of a marriage by special licence, solemnized in a church or chapel of the Church of England in which marriages may lawfully be solemnized, or in a non-conformist church or building duly registered for the solemnization of marriages, or in a naval, military, or air force chapel, or in a superintendent registrar's office. Where a party to a marriage is under 18, the consent of both, or sometimes one of them, is generally necessary or, alternatively, the consent of the High Court, a county court, or a magistrates' court. The only essential of the ceremony is the declaration that each takes the other in marriage, the remaining formalities being matters of choice or of religious practice.

At common law in Scotland, following canon law, parties might marry irregularly, by exchange of consent in writing or in the presence of two witnesses (declaration *de praesenti*), or by promise, proved by writing or admission on oath, followed by sexual relations on the faith of the promise (promise *subsequente copula*), or by cohabitation with acquisition of the reputation of being regarded as married persons, or regularly, by banns and solemnization by a clergyman of the Church of Scotland. After 1852, to stop runaway marriages, at least one party had to be usually resident in Scotland for 21 days before the marriage. Since 1940, the first two modes of contracting irregular marriage are incompetent and, apart from cohabitation with habit and repute, marriage requires notice to a registrar, and must be solemnized by a clergyman or by an authorized registrar.

No consent of parents or guardians is required for marriage in Scotland of a person aged 16 but under full age.

A marriage is void *ab initio* if any of the legal requisites are not satisfied, or if there is not genuine consent to the marriage.

A marriage is voidable and may be avoided by the court if either party is physically or psychologically impotent and incapable of consummating the marriage or, in England only, if either party is of unsound mind or a mental defective, or subject to recurrent fits of insanity or epilepsy, or the respondent was suffering from a venereal disease in a communicable form, or the respondent was pregnant by some person other than the petitioner. A voidable marriage which is approbated by the parties or not challenged is legally equivalent to a valid one.

Once a valid marriage has been contracted the parties have the status, rights, and liabilities of married persons (see HUSBAND AND WIFE) and the marriage is dissoluble on the death of one or by legal divorce (see DIVORCE).

J. Jackson, *Formation and Annulment of Marriage*; P. Bromley, *Family Law*; J. Wilson and E. Clive, *Husband and Wife in Scotland.*

Marriage, in feudal law. The lord had an interest in the marriage of the widows and daughters of tenants to control who by marriage came to hold their lands. By the late twelfth century the King was known to sell the marriage of male heirs.

Marriage Act. Lord Hardwicke's Marriage Act (1753) provided that no marriage was valid unless solemnized by an Anglican clergyman after banns called on three successive Sundays in the parish church. It applied to Protestant dissenters and Roman Catholics, who felt aggrieved, but not to the royal family, Jews, or Quakers. It put an end to Fleet marriages and other hasty and disreputable unions and ensured that marriage records would be more reliable in future.

Marriage articles. A contract in consideration of marriage to settle property on terms intended to be embodied subsequently in a formal marriage settlement.

Marriage broking. Arranging marriages as a business. The consideration for arranging a marriage, marriage brokage, is irrecoverable, the transaction being deemed contrary to public policy.

Marriage counselling. Advice and guidance by a third party to married persons whose relationship has become disharmonious, seeking to enable them to resolve their conflicts and difficulties.

Marriage, promise of. A promise by one person that he or she will contract a valid marriage with the promisee within a stated, or a reasonable time. To refuse or fail to do so was formerly a breach of contract in England. In early times, spiritual courts enforced specific performance of such a contract and jurisdiction to do so survived till 1753, but since long before then only damages were given. The right to recover damages was abolished in England in 1970.

Marriage settlement. An agreement made prior to a marriage but in contemplation and in consideration of it, under which property, real or personal, is settled for the benefit of the prospective husband and wife and any issue of the marriage. They are effected by vesting the settled property in trustees, with a trust instrument declaring the terms of the trusts created. A post-nuptial settlement may be made containing similar provisions.

In Scotland, the corresponding concepts are antenuptial and post-nuptial marriage contracts.

Married woman. At common law in England and Scotland, husband and wife were one person in law, the legal personality of the wife being submerged in that of the husband. Accordingly, a married woman was in general incapable of acquiring, holding, or alienating any property. In England, the husband acquired a freehold interest in all estates of inheritance and life estates of which the wife was or became seised, and on the birth of issue he became entitled to an estate for life in her estate of inheritance on her death (see CURTESY). Money and personal chattels belonging to the wife at the time of her marriage or acquired by her during it, were vested in the husband absolutely. In equity, however, property might be given to a married woman for her separate use and was then protected during coverture, while a married woman carrying on a separate trade might, by agreement with her husband, become entitled to the business and its profits for her separate use. She could dispose of the equitable interest in any property held to her separate use and not subject to a restraint on anticipation without her husband's concurrence. In Scotland, the law was to the same general effect, a married woman's property being often protected by the creation of an ante-nuptial marriage-contract trust. In both countries, the law has been completely altered by a series of statutes dating from 1857 and a married woman is now a separate legal person with full power of acquiring, holding, and dealing with any kind of property. A few traces of the former rules long remained, such as that a wife's domicile was determined by her husband; some still do, such as that the incomes of the spouses are aggregated for the purposes of income tax.

Marriott, Sir James (?1730–1803). Became an advocate of Doctors' Commons in 1759 and, in 1764, King's Advocate. He was judge of the Court of Admiralty, 1778–98, and an M.P. from 1781. He seems to have been able and industrious, and an effective advocate, but as a judge he lacked dignity and manners, though some of his opinions are valuable.

Marshal. Originally a person having charge of horses. A marshal is a person, usually a recently-called barrister, whom a judge on circuit takes with him to act as his secretary and personal assistant. The Marshal of the Admiralty Court has duties in connection with executing warrants issued by the court, the custody of embargoed ships, the appraisement and sale of condemned ships and cargoes, and other similar matters.

Marshal, Earl. See EARL MARSHAL.

Marshal of the King's Household. An ancient office, distinct from that of Earl Marshal, but now obsolete. The Marshal had duties in the management of the royal household and, along with the Steward of the King's Household, sat in judgment in the Court of the Marshalsea (q.v.) These officers were also nominally judges of the Palace Court.

Marshal of the Queen's Bench. An officer who, prior to 1841, had custody of the Queen's Bench Prison.

Marshall, John (1775–1835). Served in the Continental Army, attended William and Mary College for a time, and was admitted to the Bar in 1780. He became a member of the House of Delegates of Virginia from 1782–86, and of the Council of State of Virginia, also in 1782. He became a member of the Virginian Convention to ratify the Federal Constitution in 1788. During his early years at the Bar he studied in rather desultory fashion. He was without political ambition but served several terms in the House of Delegates of Virginia from a desire to serve the public interest. Thereafter, he served as a special envoy to France in 1797–98, a member of Congress in 1799–1800, and Secretary of State 1800–01, before being appointed Chief Justice of the Supreme Court, an office he held till his death. With the appointment of Marshall the Supreme Court developed a strong independence. At the very outset of his headship, *Marbury* v. *Madison* came before the Court and, in his judgment, Marshall declared as unconstitutional a section of the Judiciary Act, 1801. Shortly thereafter, following on the acquittal of ex-Vice-President Burr at a trial for treason at which Marshall presided, President Jefferson quarrelled with Marshall.

For 34 years, Marshall presided over, and dominated the Supreme Court. He introduced the practice of the Court, or at least the majority, speaking as a body. He created the doctrine of judicial review, which was enunciated in *Marbury* v. *Madison* (1801) and in *McCulloch* v. *Maryland* (1819) gave judicial sanction to the doctrine of implied powers of Congress; in the *Dartmouth College* case (1819) the Court held a college charter irrevocable by a state legislature; and in these and many other cases he formulated fundamental principles of American constitutional law. His view of the Constitution was systematized and put in its historical setting by Story's *Commentaries on the Constitution* (1833) dedicated to Marshall.

His fame rests on his opinions on constitutional law and, to a lesser extent, on international law; in private law, he was a strong but hardly outstanding judge. Above all he placed the Supreme Court in a position of independence and of power in the government. He wrote a *Life of George Washington* (5 vols., 1804–07), which was hastily-written and disappointing.

A. J. Beveridge, *Life of John Marshall* (1919); J.

Dillon, *John Marshall, Life, Character and Judicial Services* (1903); J. B. Thayer, *John Marshall* (1904); C. Warren, *The Supreme Court in U.S. History*; E. Corwin, *John Marshall and the Constitution.*

Marshalling. A doctrine having several applications. In the administration of the estate of a deceased, marshalling consists of arranging the assets so as to give effect to priority of debts. In distribution of an estate, marshalling arises where there are several claimants, and some funds are available only for particular claimants and others generally available; marshalling consists in so allocating the assets as to satisfy all the claimants so far as possible. In relation to mortgages marshalling involves distributing the assests so that each mortgagee, so far as possible, has his debt satisfied.

Marshalsea, Court of the. A court, originally held by the Steward and Marshal of the Royal Household, to administer justice between the sovereign's domestic servants. It dealt with cases of trespass committed within the verge of the court, fixed at 12 miles round the sovereign's residence, if only one party was in the sovereign's service, and with debts, contracts and covenants, where both parties belonged to the royal household, in which case the inquest was composed of men from the royal household only.

Charles I created a new court called the Palace Court to be held by Steward of the Household and Knight Marshal, and the steward of the court or his deputy, and having jurisdiction to hear all kinds of personal actions between parties within twelve miles of Whitehall Place. This was held weekly together with the ancient Court of Marshalsea. It was abolished in 1849.

Marshalsea Prison. Originally the prison of the Court of the Marshalsea and known from about 1300. It was on a site in Borough High St., Southwark, and was used as a debtor's prison down to 1842 when, like the Queen's Bench Prison and the Fleet Prison, it was replaced by the Queen's Prison.

Marsilius of Padua (1270–1342). Italian scholar, author, with John of Jandun, of *Defensor Pacis* (1324) an obscure and difficult work urging that the people are the source of the law and should themselves choose the head of the government, which is responsible to them. The major cause of trouble among people is the papacy, and Pope, bishops and clergy should have no jurisdiction, wealth, privileges, or even independence, but should be subordinate to the civil power. These ideas were novel and audacious in their time but the seed of some of the doctrines of the Reformation and the French Revolution. A later work, *Defensor Minor* elaborated certain matters. He wrote also a *De translatione imperii Romani*, intended to prove the emperor's exclusive jurisdiction in matrimonial affairs. His work was disowned, condemned, and censured by many authorities, but had considerable subsequent influence.

Martens, Fyodor Fyodorovitch (*or Friedrich*). See MARTENS, THEODOR THEODOROVICH.

Martens, Georg Friedrich von (1756–1821). German diplomat and jurist, author of *Primae lineae juris gentium Europaearum practici* (1785) revised as *Précis du droit des gens modernes de l'Europe* (1789). He was also responsible for the great *Recueil des traités* from 1761 (1817–35), *Nouveau recueil des traités depuis 1808* (1817–42) and *Nouveau recueil général des traités* (1843–75), works supplemented by others, and continued in later series by other scholars down to the present time.

Martens, Baron Karl von (or Baron Charles de) (1781–1862). German diplomat, nephew of the proceding, author of *Causes célèbres du droit des gens* (1827), *Manuel diplomatique* (1882), *Guide diplomatique* (1832) and *Nouvelle causes célèbres du droit des gens* (1843). He also continued some of his uncle's work in the collection of treaties.

Martens, Theodor Theodorovich (Fedor Fedorovich or Fedor de or Friedrich von) (1845–1909). Russian jurist and diplomat, Nobel peace prizeman in 1902, Russian representative to the second Hague Convention (1907), author of *The Right of Private Property in War* (1869), *The Office of Consul in the East* (1873), a work translated as *Traité de droit international* (1883-87), and responsible for a *Recueil des traités conclues par la Russie* (1874–1909).

Martial law. In the Middle Ages, martial law meant the law administered by the Court of the Constable and the Marshal, and from that court originated both the martial and the military law of today. It now means the law applicable, by virtue of the royal prerogative, to foreign territory occupied for the time being by the armed forces of the Crown. Except in so far as the ordinary courts of the occupied territory are permitted to continue to exist and administer justice, justice is administered by courts-martial or military tribunals according to rules established by the military authorities of the occupying forces.

Martial law may, exceptionally, be established within the State itself, in substitution for the ordinary government and administration of justice, when a state of war, or rebellion, or invasion, or other serious disturbance exists. In that event justice is administered by courts-martial and military tribunals. This has not happened in Britain since the seventeenth century. In the United Kingdom, it is uncertain whether martial law may be declared in emergency by act of the prerogative, or always requires the approval of parliament. It cannot be declared in time of peace. In either case, the civil courts cannot interfere with acts done by military authorities in pursuance of their powers, and after the emergency acts done may be justified so far as done in good faith and with reasonable justification, having regard to the emergency and necessity for their being done.

The term is also used, inaccurately today, as a synonym for military law (q.v.), though that was its original connotation.

Martin, François Xavier (1762–1846). American judge and jurist, a judge in Louisiana 1815–46, and author of *A General Digest of the Acts of Legislature of the Late Territory of Orleans and of the State of Louisiana* (1816), and of valuable reports of the superior courts of the state. He is sometimes called the father of Louisiana. He wrote also histories of Louisiana and North Carolina.

Martin, Sir William (1807–1880). Chief justice of New Zealand, 1841–57 and founder of that country's judicial system.

Martin* v. *Hunter's Lessee (1816), 1 Wheaton 304 (U.S. Supreme Ct.). A Virginia Court of Appeals refused to accept an earlier reversal by the Supreme Court of their decision to uphold a state law denying the right of an alien to inherit real property. The Supreme Court held that it could review the decisions of State Courts and upheld the constitutionality of the Judiciary Act of 1789, s. 25, which gave that power.

Martin* v. *Mott (1827), 12 Wheaton 19 (U.S. Supreme Ct.). The Supreme Court held that when the President, acting under authority of Congress, calls out state militia, his decision is conclusive upon all persons and not subject to be blocked by the states.

Martinez, Fernando (thirteenth century). A famous Spanish jurist, Bishop of Oviedo, and author of *Margarita de los Pleitos* and, in Latin, *De Orden de los Juicos.*

Martinmas. The feast of St. Martin of Tours, 11th November, one of the cross quarter-days and, in Scotland, one of the term or half-year days, the other being Whitsunday (q.v.).

Martinus de Fano (?–1275). Italian jurist, a pupil of Azo, and author of *Tractatus de positionibus* (1234), *Notabilia super Decreto* and *Notabilia Decretalium* (*c.* 1240.).

Martinus Gosia (?–1166). One of the 'four doctors' who succeeded Irnerius in the law school at Bologna, founding the school of glossators, who commented on the rediscovered Roman law. One of his pupils was Placentinus, who taught at Mantua, Bologna, and Montpellier.

Martinus Syllimanus (sixteenth century) a commentator, and author of *Tractatus super Usibus Fendorum.*

Marx, Karl Heinrich (1818–83). German political philosopher, studied law at Bonn and Berlin, and early wrote articles stressing the political struggle as the means of social advance of the working class. He became friendly with, and collaborated with Friedrich Engels and, in 1847, they issued the *Communist Manifesto.* In 1849, following the collapse of the risings of 1848 in Germany and France, Marx settled in London and engaged in writing. In 1859, he published *The Critique of Political Economy*, later rewritten and published in 1867 as *Capital*, Vol. I (Vols. 2 and 3 were posthumous). In 1864, the International Working Men's Association, later known as the First Communist International, was founded and Marx was, in fact, head of its general council until it was removed to New York in 1872 and dissolved in 1876. The major points in his theory of Socialism are the materialist conception of history, that in any society there are certain material forces of production which, to be exploited, require a power able to enforce them, namely the State, which takes its form from the character of the economic institutions it exists to uphold, and the class-struggle, whereby the proletariat, created by industrial capitalism, will turn on its masters and set up a new social and political order by violent revolution. The main theme of *Capital* is the doctrine of surplus value, to the effect that the capitalist class exploits the labouring class by getting more value from a day's work than the worker gets for his subsistence. The right to the whole produce of labour belongs to the workers. His economic views are unsound but have been influential. The State is part of the political and legal superstructure with the function of upholding the dominant productive system and the social relationships necessary for its working. When the working class takes over it must remake the State: this is the doctrine of the dictatorship of the proletariat. But when the proletarian dictatorship has established the new order all exploitation will have been abolished and the State will wither away; government will be replaced by social administration in a classless society in which all economic antagonisms will have been resolved. Law is an instrument of the capitalist state, the coercive machinery for the maintenance of exploitation of the proletariat by the bourgeoisie. The acceptance of Marxism-Leninism in Russia makes the Marxist view of the state and law underlie modern Russian jurisprudence, but developments there lend no support to the theory that State and law will wither away; on the contrary they are stronger and more oppressive than ever before. But Marx did not develop a legal theory of socialism or communism. Later followers have to some extent done so.

I. Berlin, *Karl Marx* (1963); R. N. Hunt; *Marxism, Past & Present* (1954); G. Lichtheim, *Marxism; An Historical & Critical Study* (1961).

Marxist legal theory. Marx himself worked out no theory of law, though his attitude to it is discoverable from his writings on politics and economics. Since his time, Marxism has been understood in various forms and practised in various ways by various socialist writers and thinkers. The major contribution to thinking about the relation between legal institutions and economic conditions from the Marxist standpoint has been that of Karl Renner in *The Institutions of Private Law and Their Social Function* (1949), in which he examines the transformation of ownership from an institution of private law into one of public law with a changed function. The theory that law, like the State, is an instrument of capitalist oppression and disappears in a society, where ownership of the means of production has passed into the hands of the community, has been developed by the Soviet jurist Pashukanis. In Soviet Russia, what has developed in practice is that law is a weapon used by the new ruling class for the suppression of their opponents and an authoritative instrument of the coercive power of the State. Law is wholly subservient to politics; it does not exist to protect the individual or give him rights, but to keep him in subjection. Traditional legal institutions, such as contract and property, survive mainly in relation to State organizations and only in limited spheres in relation to individual persons.

R. Schlesinger, *Soviet Legal Theory* (1950); J. Hazard, *Soviet Legal Philosophy* (1951).

Mascardi, Joseph (seventeenth century). Italian jurist, humanist, and author of a famous *Conclusiones probationum omnium* (1624) on the theory of evidence.

Mason, Charles (1804–82). Was admitted to the Bar in 1832 and became Chief Justice when the territory of Iowa was organized in 1838. As a member of the commission which drafted *The Code of Iowa*, 1851, he greatly influenced the law of that state and, subsequently, of others.

Massachusetts Body of Liberties. A statement based mainly on English common law, drawn up by Nathanial Ward and adopted by the General Court of Massachusetts in 1641 as the basic law of Massachusetts Bay Colony. It gave too much authority to magistrates and was superseded by the Massachusetts General Lawes and Libertyes (q.v.) in 1648, but had influence on later assemblies in leading them to think of bills of rights.

Massachusetts General Lawes and Libertyes. A statement of law, built on and superseding the Massachusetts Body of Liberties (q.v.), adopted in 1648. It adopted features from English common law but rejected others, and gave a lead to the world in declaring basic human rights and freedoms, such as the right of all to equal protection from the law, the right to bail, the right of appeal, freedom from double jeopardy, the privilege against self-incrimination, freedom of speech, of travel, of judicial dissent, and the like. In time, most of these principles were incorporated in the Bill of Rights or elsewhere in the legal system.

Master. A term for a person holding authority or having control over persons or things. The title was frequent in mediaeval times but many of the uses are now obsolete. Thus, the Master of the Temple is the priest in charge of the Temple Church, London. The Benchers (q.v.) of an Inn of Court are properly called the Masters of the Bench. Other instances are Masters of the Supreme Court, Master of the Rolls, and Master of the Faculties.

Master, particularly in the proper sense of one being a Master of Arts of a university, was the usual prefix to a man's name if he were of some standing. It became corrupted to 'mister' and, prefixed to a title, is the proper form of address in many cases, such as Mr. Speaker, Mr. Justice, and in the U.S. Mr. President.

The term master is also applied to the chief officer of a ship, commonly called the captain. A master has extensive authority as agent for the owners.

Master and servant (or employer and employee). The legal relationship under which one person engages another to serve him, and to do certain work for him, having power to direct and control what work is to be done and when, where and how it is to be done. It is therefore distinguishable from the relationship between employer and independent contractor, under which one person engages another to bring about a given result, using his own knowledge, skill, and discretion in respect of how he does it. This distinction is also referred to as that between *lacatio operarum* and *locatio operis faciendi*, between a contract of service, and a contract for services. Distinguishable relationships are those between principal and agent, and between partners. Whether a relationship is that of master and servant or not is a question of fact, regard being paid to the terms of the engagement, the time and place of doing the work, the method of remuneration, the master's power of direction and control, the power to suspend or dismiss, but none of these factors is, by itself, conclusive and persons may be servants though they have to, and are possibly deliberately employed to, exercise a high degree of professional skill and independent judgment and discretion in doing the work. In certain cases, such as Crown servants, constables, and clergymen, persons hold office rather than are employees.

The relationship is formed by contract but there are many statutory restrictions on employment of particular kinds of persons, on hours of work, modes of payment of wages, permissible deductions therefrom, minimum wages payable, allowance of holidays, and, in many other circumstances, terms are negotiated by trade unions acting locally or even nationally on behalf of employees of particular classes.

During the relationship it is the duty of the master or employer to accept the servant into his employment, to provide board and lodging if that be a term of the contract, to provide work, if that be an express or implied term of the contract, to pay the wages agreed upon, to take reasonable care for the safety of the employee, that he is not injured by reason of the master's personal negligence or of the negligence of persons for whom the master is responsible, to indemnify the employee against expenses, losses, and liabilities incurred in the performance of the duties of the post, and to comply with all statutory duties incumbent on him relative to the kind of employment in question. If by reason of failure to take reasonable care for the servants' safety, the latter sustains injury or death, a claim of damages lies, and claims are also competent under the National Insurance (Industrial Injuries) Acts. If the servant is in breach of contract the master has a claim for damages.

During the relationship, it is the duty of the servant or employee to serve faithfully and honestly, attend regularly at the time and place set for the work, carry out instructions, observe good faith and confidentiality, take reasonable care for others in the performance of the duties, and obey any relevant statutory duties incumbent on employees. If the master is in breach of contract, the servant has a claim for damages.

The master is vicariously liable in tort or delict to a third party injured by fault of the servant if the act causing the injury were expressly authorized, or were done in the course of the employment, but not if, in doing it, the servant was acting outside the course of the employment. The servant remains personally liable, i.e. master and servant are jointly and severally liable, and the master has a right of indemnity against the servant.

A contract of employment may be permanent, till a retirement date, or be for a stated period, but otherwise there is some presumption for a hiring for a year and in industrial employment for employment by the week. The intervals at which remuneration is paid are not conclusive of the duration of employment. Persons in the employment of the Crown normally hold office during the pleasure of the Crown but civilian employees are protected by modern legislation against unfair dismissal.

The contract may be terminated by agreement, performance, expiry, or the death of either party. An employee may be summarily dismissed for wilful disobedience to a reasonable order, substantial misconduct, incompetence, neglect of duty or conduct prejudicial to the master's business, and may refuse to continue in the employment if the master is materially in breach of contract or he has reasonable apprehension of injury if he continues in the employment, or the contract may be terminated by either party giving reasonable notice to the other. If the contract is terminated by reason of redundancy, the employee is entitled to a statutory redundancy payment.

Third parties may be liable for interfering with the contractual relation of master and servant as by inducing either party to act in breach of contract, or by harbouring or employing a servant during the subsistence of his contract with a previous master. In English law a master may recover damages in an action *per quod servitium amisit* for loss of services by reason of the injury of a member of his family or domestic servant by the wrongful act of a third party.

C. Smith, *Master and Servant*; F. Batt, *Master and Servant*; W. Fridman, *Modern Law of Employment*; B. Hepple and P. O'Higgins, *Employment Law* (1976); J. Munkman, *Employer's Liability*.

Master of the Bench. The proper title of a bencher (q.v.) of any of the Inns of court.

Master of the Court of Protection. An officer of the Court charged with administration of the property of persons of unsound mind.

Master of the Court of Wards and Liveries. The chief officer of the court of Wards and Liveries. It was abolished with that court in 1600.

Master of the Crown Office. The title of the Queen's Coroner and attorney in the criminal department of the Court of Queen's Bench who prosecuted, nominally on behalf of the Crown, on the relation of a common informer or private person. He is now an officer of the Supreme Court appointed by the Lord Chief Justice, registrar of the

Court of Appeal, Criminal Division, and *ex officio* a master of the Queen's Bench Division.

Master of the Faculties. An officer who, subject to the authority of one of the archbishops, grants faculties to notaries and grants licences and dispensations. Since 1874, the office has been conjoined with that of judge of the provincial courts of Canterbury and of York.

Master of the Horse. Originally a person in charge of the royal stables, now a peer and one of the three principal officers of the household (with the Lord Steward and the Lord Chamberlain). It is a political office, changing with changes in government, and his main functions are to attend the sovereign on State occasions. The Crown Equerry and Secretary of the Master of the Horse now has responsibilities for the royal stables.

Master of the Household. A title formerly sometimes given to the Lord Steward, but now to a permanent officer of his department having managerial functions in relation to the royal household.

Master of the King's Household. One of the Officers of State in Scotland, who seems to have shared financial responsibility with the Comptroller. The Master was also the King's Carver. In 1707, the heritable office of Master of the King's Household in Scotland was conferred on the Duke of Argyll, by whose successor it is still held.

Master of the King's Wardrobe. One of the Officers of State of Scotland. The office has been discontinued since 1603.

Master of the Mint. Prior to 1870, an official having charge of the Royal Mint. The title of Master Worker and Warden of Her Majesty's Royal Mint is now held by the Chancellor of the Exchequer, the actual management of the Mint being in the hands of permanent officers.

Master of the Ordnance. Formerly an officer having charge of the king's artillery and munitions of war. The modern counterpart is the Master-General of the Ordnance, a senior military officer holding that post in the Ministry of Defence.

Master of the Rolls. The office probably originated in the Chancery clerk whose duty was to oversee the scribes who composed the Chancellor's Roll. In the Middle Ages, the title was Clerk or Curator of the Rolls, and he was one of the Masters in Chancery who assisted and acted for the Chancellor.

With the development of Chancery jurisdiction, the judicial duties of the Masters in Chancery increased and a large part of these fell on the Master of the Rolls; as early as the mid-sixteenth century he was sometimes called vice-chancellor, and some jurisdiction was delegated to him by special commission. By the early seventeenth century, he heard causes in the absence of the Chancellor and became the Chancellor's general deputy.

In 1729, it was provided that all orders made by the Master of the Rolls should be valid, subject to an appeal to the Lord Chancellor, and thereafter he sat regularly as a Chancery judge. In 1833, his jurisdiction was extended and it was enacted that he should sit continuously, but his orders were subject to appeal to the Lord Chancellor and after 1851 to the Court of Appeal in Chancery. He could also, and frequently did, sit in the Court of Appeal in Chancery.

Since 1875, the Master of the Rolls has been president of the Court of Appeal. Until 1958 he had the general responsibility for the public records (a responsibility then transferred to the Lord Chancellor) and is still responsible for the records of the Chancery of England. He admits persons as solicitors of the Supreme Court.

Master of the Rolls in Ireland. The office originated in the office of the keeper of the Rolls in the Irish Chancery and became an office granted by letters patent in 1333. It survived till 1921.

Master of the Temple. The clergyman of the Temple Church. He is nominated by the Crown and has an official residence in the Temple.

Master of the Wards. The judge of the Court of Wards and Liveries (q.v.).

Masters Extraordinary (or Masters Extraordinary in Chancery). Officers known from the early seventeenth century but abolished in 1853. Their main functions were to administer oaths in Chancery proceedings.

Masters in Chancery (or Masters in Ordinary of the High Court of Chancery). Senior officials of the Court of Chancery who assisted the Chancellor in dealing with petitions, issuing original writs, hearing witnesses and, in time, increasingly undertaking delegated judicial functions, hearing and reporting on cases to the Chancellor. The chief among them was the Master of the Rolls who, from the sixteenth century, was the chief assistant and deputy of the Chancellor. The office of Master was a saleable one and large sums were paid for the appointment; such sales were prohibited after the scandals which resulted in the impeachment of Lord Chancellor Macclesfield. The offices were abolished in 1852 but the chief clerks appointed in their place have since 1897 been called Masters of the Supreme Court (q.v.).

Masters of Requests. At first the legal assessors to, and later the judges of, the Court of Requests (q.v.), a standing committee of the Council to deal with the suits of poor men and of the King's servants. Also, one of the Officers of State of Scotland. In the sixteenth century, his duties were to receive petitions from subjects and to remind the Lord Chancellor of affairs of state, pending business and petitions, that they might be put before the King for consideration. The office seems to have ended with the Union of 1707.

Masters of the Supreme Court. Officers of the Queen's Bench and Chancery Divisions of the Supreme Court in England. In the Queen's Bench Division, masters are barristers and they perform judicial work in chambers, subject to appeal to a judge, and issue directions on matters of practice and procedure. The Senior Master is also Queen's Remembrancer and Registrar of Judgments. The Master of the Crown Office and Queen's Coroner and Attorney is registrar of criminal appeals and also a master of the Q.B.D. In the Chancery Division, masters are solicitors and may exercise the jurisdiction of a judge in chambers, with certain exceptions, subject to adjournment to a judge. They take accounts, supervise sales, settle conveyancing documents, give directions to trustees and perform other similar functions. There are also taxing masters who scrutinize solicitors' bills of costs.

Materiality. The issue of the importance of something, which arises in several different contexts. A misrepresentation is material if substantial and important and actually influencing the person imposed upon. An alteration in a bill or cheque is material in some cases, such as of alteration of the date. A material change in the use of a building is a development for the purposes of town and country planning law.

Matienzo, Juan de (sixteenth century). Spanish jurist, sent to Peru as a judge, and author of *Gobierno del Peru* and *In librum V collectionis legum hispaniae commentaria.*

Matrimonial causes. Actions of divorce, nullity of marriage, judicial separation, jactitation of marriage, or restitution of conjugal rights, jurisdiction in which belonged originally to the ecclesiastical courts, was transferred to the Court for Divorce and Matrimonial Causes in 1857, to the Probate, Divorce and Admiralty Division of the High Court in 1875, and to the Family Division of the High Court in 1972, though certain county courts have jurisdiction in undefended divorce cases.

Matthaeus, Antonius (1564–1637). Dutch jurist, professor at Marburg and Groningen, and author of *Notae et animadversiones in libros IV Institutionum* (1600).

Matthaeus, Antonius (1601–54). A Dutch jurist, professor at Utrecht, and author of *De Criminibus* (1644), which had wide influence in Europe, *De auctionibus*, and of *Paroemiae Belgarum jurisconsultis usitatissimae* (1667). His son, Antonius (1635–1710), was Professor of Law at Leiden and a learned jurist, and author of *Manuducatio ad jus canonicum* (1706), and his grandson, Antonius (1672–1719), was also a Professor of Law.

Matthaeus de Afflictis (1448–1528). An Italian commentator and author of *Super tres Fendorum libros commentaria* (1159) and *Decisiones sacri regii consilii Neapolitani* (1575).

Maudsley, Henry (1835–1918). A forensic psychiatrist who devoted much attention to the connection between insanity and crime and investigated the causation of crime. He was Professor of Medical Jurisprudence at University College, London, and very influential on thinking on crime. His writings included *Responsibility in Mental Disease* (1874), *Physiology of Mind* (1876), and *Pathology of Mind* (1895).

Maugham, Frederic Herbert, 1st Viscount (1866–1958). Brother of Somerset Maugham, the author, was called to the Bar in 1890, and practised in Chancery. He took silk in 1913 and was engaged in much heavy business. In 1922, he was offered the Solicitor-Generalship but had to decline as he had no chance of getting into Parliament. In 1928 he became a judge of the Chancery division and proved himself a good judge; in 1934, he was promoted to the Court of Appeal; and, in 1935, to the House of Lords. In 1938, to general surprise, he was named Chancellor, though he had never been in the House of Commons, and took a leading part in putting through a great deal of important legislation, notably the Evidence Act, 1938. He gave up office when the government was reorganized on the outbreak of war in 1939 and was reappointed a Lord of Appeal, sitting until he retired in 1941 and occasionally thereafter. As a judge, he displayed both breadth and depth of learning and was notable for the clarity and style of his judgments.

F. Maugham, *At the End of the Day* (1954).

Maurer, Konrad von (1823–1902). German legal historian, mainly interested in philosophy of religion and folklore, in Scandinavian legal origins, and in ecclesiastical and canon law.

Maxims, legal. Short, pithy formulations of broad and general principles of common sense and justice, many being derived directly or indirectly from Roman law and hence being in Latin. They include, e.g. *vigilantibus, non dormientibus, jura subveniunt* (the laws help those who are vigilant, not those asleep), *falsa demonstratio non nocet* (false designation does not harm), and *caveat emptor* (let the buyer beware). Francis Bacon made a collection of Latin maxims of the common law, with elaborate English commentaries and Coke's writings contain many maxims, some drawn from Roman law and some invented by Coke himself. With the increasing complexity of modern law exceptions are increasingly developing and maxims cannot always be relied on as stating principles of law completely or accurately.

F. Bacon, *Collection of some Principal Rules and Maxims of the Common Law* (1630) H. Broom, *Legal Maxims*; J. Trayner, *Latin Maxims and Phrases*; J. Wharton's *Law Lexicon*; W. Jowitt, *Dictionary of English Law*.

Maxwell, Sir John, of Pollock (1643–1732). Sat in the Scottish Parliament from 1689 and was a Privy Councillor and a Lord of the Treasury. He became a Lord of Session in 1699, holding that office till his death, and also Lord Justice-Clerk from 1699–1702.

Maxwell Fyfe, D. P. See FYFE, D. P. MAXWELL.

Mayer, Max Ernst (1875–1923). German jurist, primarily in the fields of criminal and legal philosophy. He wrote a *Lehrbuch des deutschen Strafrechts* (1915) and a *Rechtsphilosophie* (1922), which has a sociological standpoint.

Mayer, Otto (1846–1924). German jurist, notable for having developed the study of administrative law, particularly its general concepts. His major works are *Theorie des franzosischen Verwaltungsrecht* (1886) and *Deutsches Verwaltungsrecht* (1895–96).

May, George Augustus Chichester (1815–92). Was called to the Irish Bar in 1844, edited the *Irish Law Reports*, became Attorney-General in 1875, Chief Justice of the Queen's Bench in Ireland in 1877 and, in 1878, Chief Justice of the Queen's Bench Division of the High Court of Ireland. He resigned in 1887.

May, Thomas Erskine (Lord Farnborough). See ERSKINE MAY, THOMAS.

Mayflower Compact (November 1620). An agreement drawn up and signed by the 41 male adult Pilgrims on the *Mayflower*, anchored off Cape Cod. It joined them into a civil body with powers to make laws and appoint officers for the general good of the colony. This remained Plymouth colony's basic charter until 1691.

Mayhem. In mediaeval English law, an injury to the person which amounted to the deprivation of a member useful for fighting. It could originally be prosecuted by an appeal of felony, but usually by proceedings for trespass. It is now treated as malicious injury to the person.

Maynard, Sir John (1602–90). Was called to the Bar in 1626 and rapidly acquired a large practice. He was one of the managers of Strafford's and Laud's impeachments, sat in Cromwell's Parliament and in those of Charles II and James II, and took a leading part in the debates on the Bill of Rights. He was also involved in most of the State trials of the period. He was a great student of the Year Books and old records as well as an accomplished pleader and an eminent lawyer though somewhat lacking in professional honour. To him is owed the printed edition of the Year Books of Edward II, covering most of that King's reign, published in 1678.

Mayno, Jason de (Maynus) (?–1519). An Italian jurist and commentator, professor at Pavia, where he taught Alciatus, and author of *In Digestum Commentaria* (1573) and *Consiliorum sive Responsorum Juris quattuor volumina* (1611).

Mayor. From the twelfth century, the title of the chief magistrate of a municipal borough in England and Wales, who was latterly elected by and acted as chairman of the elected council. The mayors of certain cities have acquired, by prescription or letters patent, the title of Lord Mayor.

In the U.S., the office of mayor was appointive in colonial time but by the mid-nineteenth century was elective practically everywhere. Save where commission or city-manager government is in force, he is the chief executive of the city government.

Mayor's and City of London Court. The Mayor's Court of London was an inferior court of record originally held before the Mayor and Aldermen of the City. It had jurisdiction in all actions, at law and in equity, concurrent with that of the High Court, in which the whole cause of action arose within the City, but not in divorce, admiralty, probate, or bankruptcy.

The City of London Court was the Sheriff's Court of the City of London and was a county court.

The Mayor's and City of London Court was an amalgamation of these courts, which took effect in 1921, and was a court of record, which exercised the jurisdiction formerly exercisable by the Mayor's court and was also deemed to be a county court for the purpose of all proceedings within the jurisdiction of a county court. In 1925, the amalgamated court's powers over matters within its jurisdiction was equated to those of the High Court. In 1971, it was abolished and the county court for the district of the City of London was entitled the Mayor's and City of London Court. It sits in the Guildhall. The Lord Mayor and Aldermen were nominally judges but, in practice, took no part in the work of the court. The real judges were the Recorder of London and the Common Serjeant. Since 1971, the judge assigned to the county court district for the City is known as the judge of the Mayor's and City of London Court.

Measure. An enactment of the National Assembly, and now of the General Synod, of the Church of England. Also a standard or rule by which anything is quantified. By statute imperial standard measures and metric measures are both competent, but the latter are increasingly being adopted.

Measure of damages. The determination of how much money is to be ordered to be paid by defendant to plaintiff in compensation for loss or harm caused by the former's breach of duty, whether breach of contract, tort, or other breach of legal duty. The question of measure of damages arises only if and when it has been admitted, or established, that the defendant is liable, and after any questions of remoteness of damage (q.v.) have been decided. What factors have to be taken into account in measuring damages in particular cases are, in many cases, determined by previous decisions, but the general principles are that in a case of breach of contract the plaintiff should be awarded such sum as will restore him to the position he would have been in if the contract had been duly performed in the first place, or such sum as must reasonably be supposed to have been in the contemplation of the parties as the probable result of a breach. This will cover such factors as the extra cost of obtaining goods elsewhere, or loss on their resale to another buyer, extra cost of carrying them by another means, and loss on a sub-sale if such were in contemplation. In a case of tort the basic principle is that the plaintiff should be restored, so far as money can do so, to the position he would have been in if the tort had not been done to him. In pursuance of this principle in a case of personal injuries, general damages are given for pain and suffering, loss of limbs or bodily function and other non-pecuniary losses, and special damages for loss of earnings suffered or likely to be suffered in consequence of the injuries. In a case of wrongful death, damages are given for the loss of the pecuniary benefits which the plaintiffs had been enjoying and would probably have continued to enjoy from the deceased's activities. In a case of damage to property damages are given for the diminution in value of the thing by reason of the damage done to it. In general, damages are measured by fair compensation, not punishment, but in a few cases English law allows exemplary or punitive damages.

Media (or media of communication). A compendious term for the daily and periodical press, films, radio, and television. Each and all have considerable power for communicating information, expressing opinions, and influencing ideas and opinions. All forms of media are subject to the restrictions of liability for libel, sedition, and infringement of copyright. See PRESS, FREEDOM OF.

Mediaeval law. The mediaeval period of western law covers the period from the rise of the Eastern Roman Empire in the fourth century and the decline and final extinction of the Roman Empire in the West in A.D. 476 up to the late fifteenth century, which saw the fall of Constantinople and the extinction of the Byzantine or East Roman Empire in 1453, the quickening of intellectual development which is called the Renaissance, the beginnings of the Reformation in religion, the Reception of Roman law, the rise of the nation states, the beginnings of the great explorations of the world, and the complex of changes which inaugurated modern times. The period has a certain unity and coherence and in that time developments in legal institutions and ideas took place which have profoundly influenced modern law.

From the time when Diocletian divided the Roman Empire between himself and Maximian and when (A.D. 330) Constantine established his new capital at Byzantium, the Roman Empire was under increasing pressure of barbarian invaders from the east and north. In 376, the Visigoths, fleeing from the Huns, crossed the Danube and in 378, they destroyed a Roman army at Adrianople. The Goths were now settled within the empire and the Franks in Gaul. In 390, the Visigoths devastated Thrace and years of confused attacks and attempts to buy off invasions followed.

In 401, the Visigoths under Alaric entered Italy and, in 410, they captured and sacked Rome. Then they moved on into Gaul and their kings established themselves in Toulouse. By 418 they had settled from the Loire to Bordeaux. In 466, Euric became King of the Visigoths and largely succeeded in conquering the derelict Roman provinces in Gaul and Spain; by 478 he ruled from the Loire to Gibraltar.

In 476, Odovacar deposed the last titular Emperor at Rome and brought to an end the Roman Empire of the West. But in 488, the Ostrogoths migrated across the Danube, entered Italy and defeated

Odovacar, who was besieged in Ravenna (491–3), and Theodoric became King of Italy.

The Vandals entered the Empire across the Rhine in 406–7, passed into Spain (409), established themselves in Andalusia, and then passed over into Africa (429), where their king established their capital at Carthage (439). In 455, the Vandals under Gaiseric mastered the western Mediterranean and plundered Rome. Over the next 20 years, they ravaged the East and, in 476, a treaty recognized the Vandal dominion in Africa, Sicily, Sardinia, Corsica, and the Balearic Islands. But after Gaiseric's death in 477, the Vandal danger was at an end.

The Huns, under Attila, first devastated the East (441–47) then turned on the West and entered Gaul until they were checked at Mauriacus (451). Then they turned to Italy but made little headway against famine and pestilence, only leaving the spoils for the Vandals four years later. The Burgundians, established in Savoy, extended their power east of the Rhone in 435. By 476, Gundobad had consolidated the Burgundian Kingdom in the valleys of the Rhone and the Saône. Further north, the various tribes of Franks overran north-eastern Gaul and founded the most enduring of all the barbarian kingdoms. Other German tribes (Jutes, Angles and Saxons) settled in the south and east of Britain, bringing with them the customs of the north-western European mainland.

Though the *Codex Theodosianus* (q.v.), collected and promulgated in the East (438), was received and accepted by the Roman Senate, the real separation of the Latin and Greek halves of the Empire is shown by the declaration at the same time that future edicts should be valid only in the dominions of the Emperor who decreed them. When Valentinian, in 445, confirmed and extended the appellate jurisdiction of the Pope over all bishops, this edict affected the west only. The *Codex Theodosianus* was the major repository of law for the Western provinces.

When Justinian's troops recaptured Italy he promulgated an *Edictale Programma* (*c.* 540) and, in 554, the *Pragmatic Sanction,* annulling all laws of Ostrogothic conquerors and putting into force for everyone the Roman law in its revised form of *Digest, Code*, and *Novels*. It did not in fact, wholly supersede the Theodosian Code and, in any event, by 568 the Lombards were pouring into Italy.

The barbarian kingdoms of the West

Accordingly, by 476 the European territory of the Roman Empire was divided up into Germanic kingdoms, principally those of the Vandals in Africa, the Visigoths in Spain, the Ostrogoths in Italy, the Franks in Gaul, the Burgundians on the Rhone, and the Jutes, Angles, and Saxons in south Britain. A few remnants of the old Empire survived, the Celtic kingdoms in Britain, a Celtic confederacy in Armorica, A Romano-Gallic kingdom on the Seine and Loire, the Cantabrians and Basques in north Iberia, and, for a short time, a small Roman kingdom in Dalmatia.

The Germanic kingdoms varied in institutions and character. Common characteristics included the importance of the kindred or group of kinsmen, the basis of the duty of blood-revenge if one of the kindred were killed or injured, and of the payment of wergeld (q.v.), the habit of assemblies of freemen to declare customs in dispute and make decisions, meetings which had a long history as local courts of justice, the institution of kingship, of a leader with a band of followers under his protection and bound to his service in peace and war, which contained the germ of feudalism, and the fact that the barbarian settlers became landlords over a numerically superior Roman population, and tended to be influenced by that majority and to adopt its language and institutions, and to be absorbed thereby. As the tribal kings became masters of powerful countries, they became administrators and legislators and almost unwittingly stepped into the Roman heritage.

The Vandal state, founded by Gaiseric, was the first barbarian kingdom to be formed and the first to fall. He conquered the central African Roman provinces and became the terror of both Eastern and Western empires. He died in 477 leaving the kingdom strong. The old Roman civil administration was maintained and the Romans kept their own courts and law, save when a Vandal was a party, and filled many offices under the indolent lords. The Moors captured much of Mauretania and Numidia, and Justinian's armies completed the extinction of the Vandal state in 534.

The Visigothic power was consolidated in Gaul and established in Spain under Euric (466–84). He began the conquest of Spain in 469 and, by the time of his death, ruled most of the centre and east down to the Straits of Gibraltar, though the Basques held out in the north and the Suevi in the north-west. Euric sought to establish a Gothic Empire in the west. He was the first of the Germanic kings to issue a code of law additional to the Visigothic customs and adapted to their new relations with the Romans and among themselves (*Codex Eurici* or *Euricianus* (*c.* 475)). This was the earliest of the *leges barbarorum*. But the two peoples lived each under their own law, though the central administration was organized on simplified Roman lines with Gothic dukes and counts. The Goths lived as landlords whose land was tilled by the Romans.

Euric's successor Alaric II was feeble but more favourable to the Church. He issued the *Breviarium Alaricianum* or *Lex Romana Visigothorum* (506), a compendium of Roman law, simplified and approved by a Gallo-Roman assembly for the use of Gallo-Romans, which became the standard source of Roman law in Western Europe throughout the Middle Ages. Alaric was overthrown by Clovis, King of the Franks, and the Goths lost Gaul except Provence. A period of struggle ensued until Leovigild (568–86) succeeded and tried to unify the peninsula and to consolidate royal authority. Struggles continued but the kings remained strong; Chindaswinth (642–53) was a legislator and his son Receswinth, having regard to the continuous Romanization of Visigothic law, abolished the *Breviarium Alaricianum* and issued a code for both Visigoths and Romans, the *Lex Visigothorum* or *Liber Judiciorum* (654), under which personal law was replaced in Spain by the territorial law of the kingdom. Its unofficial revision, the *Forum Judicum* or *Fuero Juzgo*, was the law-book in the revival of Christian Spain. The Visigoths maintained much of the late Roman social structure, but their customs reappear in the early Fueros, or town charters, of mediaeval Spain. Law, however, became more and more Romanized; Euric's code shows Roman influence, and subsequent changes showed more and more Roman influence. But the kingdom remained disunited and weak, the nobles unruly, the peasantry oppressed. In 711, the Moslems landed from Africa and by 720 had occupied southern Spain.

In Italy, Theodoric ruled Ostrogoths, who lived under their own courts and customs, and Romans, whose lives were still regulated by Roman law. Where both nationalities were involved in a case, a Roman jurist sat with Ostrogothic magistrates.

Roman civil administration of the late Empire continued and, about A.D. 500, the King, as a Roman magistrate, issued an *Edict* (*Edictum Theoderici*) derived from Roman sources, dealing chiefly with criminal law and procedure and binding on Goths and Romans alike. Theodoric was also responsible for some lesser edicts, and his successor Atalaric also promulgated edicts. But Ostrogothic law ended with the reconquest of Italy by the Byzantines and the *Edictum* was replaced by Justinian's *Corpus Juris*.

In the third century, the Franks emerged as a loose grouping of tribes; by the fifth century they had developed into two main groups, the Salian Franks, in what is now Belgium, and the Ripuarian Franks around Cologne. The code of the former group, the Salic Law, was a compilation of ancient customs made in the fifth century, which had wide influence and influenced many later compilations.

The Burgundian kingdom was fairly tolerant. In about 495, King Gundobad issued, in Latin, a *Lex Gundobada* (or *Lex Burgundionum* or *Liber Constitutionum*) for both Roman and Burgundian subjects and a *Lex Romana Burgundionum* (506) for Romans only. The tribal kingdom came to an end in 534 when it fell to the Franks and the *Lex Romana* was replaced by the *Breviarum*.

The Arabs

In the early seventh century, Muhammad arose as a prophet and extended his influence by conversions and war and began to unite the Arab tribes. He achieved immense moral and religious influence and launched a powerful religious and national movement and, in the Koran, laid down what was at once a moral and religious guide and a system of legal principles. Under his successors the Arabs conquered Palestine, Syria (634–36), Mesopotamia (639–46) and Persia (649), Egypt (642) and Cyprus (649), Cyrenaica (642), Morocco, and, in 711, they entered Spain. Shortly after, most of Spain had been occupied and, despite constant struggles with the Christians to the north and among themselves, the Moors became the most cultured people of the West. Rules of their law seem to have entered into many Spanish customs.

The Franks

The invasion of Gaul by the Franks had been a slower and more gradual expansion than the barbarian migrations. In 476, there were two main groups, the Salian Franks in Flanders, and the Ripuarian Franks around Cologne. Clovis became king of the Merovingian part of the Salians in 481. In a few years, he became monarch of a state stretching south to the Loire, and defeated the Alemanni near Strasbourg in 496. He continued to wage war, occupied Alsace and conquered Aquitaine to the Pyrenees. He fixed his capital at Paris and united the Frankish tribes. Shortly before his death, under episcopal influence, he issued a written Latin codification of the customary law of the Salian Franks, the *Salic Law*. Revised texts of this were later issued, the last in about A.D. 820. The *Lex Salica* considerably influenced the laws of France, Belgium, and the Netherlands. Closely related to it is the *Lex Ribuaria* of the sixth century, in part a modernized version of the *Lex Salica.*

There followed a period of struggle among his sons and grandsons until Chlotar II became the only King of the Franks (584–629). After him, further civil war followed before Pepin II (680–714) was able to assert general authority. His son Charles Martel decisively repelled the Moors from France at Tours (732) and issued decrees, and acted as king, the first of the Carolingians. In 751, his son Pepin was recognized as King of all the Franks.

The Merovingians were absolute monarchs but copied the Empire in the matter of central administration. The Kingdom was divided into districts under counts who held the courts; several counties might be combined under a duke. Personal laws prevailed, the most widespread being the Salic Law for Salian Franks and the *Breviarium Alaricianum* for Gallo-Romans. The laws of the Ripuarian Franks, the Alemanni, and the Bavarians were partly written down in about 700. Rather later written law appears among the Frisians, Saxons, and the peoples of Thuringia and Hamaland. Justice was administered by vicars, counts, and the King's court but justice cannot have been certain, speedy, or just. The Church grew wealthy and powerful but was decadent. Slavery was common.

England

The main Germanic peoples who entered Britain were Jutes, Angles, and Saxons; from the early fifth century they spread over England and, by 615, had reached the Irish Sea, splitting the Welsh from the Celts of Scotland. The Anglo-Saxons retained and imposed their own language, kept their barbaric institutions and customary laws, and eradicated most of the Roman civilization. They were grouped in tribes under petty kings, and met in folk-moots to administer justice. In the seventh century, Christianity was widely adopted, literacy developed and civilization spread. An early result was the promulgation of written amendments to the tribal customs; the earliest legislating king was Aethelberht of King (*c.* 600) but there were others. His are the first known Germanic laws in a Germanic tongue. Additions were made to the Kentish laws by Hlothaer, Eadric, and Wihtraed (690–6).

Ine of Sussex and Wessex (688–726) issued a code of laws supplementary to customs which evidence a developing society. Offa of Mercia conquered up to the borders of Wales and united central and south-eastern England into a single state by the end of the eighth century.

There was also great development of grants of charters of land.

Scotland

Only the South of Scotland was occupied by the Romans and even that was held as an armed frontier territory. After their withdrawal in the fifth century the main Pictish kingdoms were Dalriada in the west, founded by immigrants from Ireland, and Strathclyde with a capital at Dumbarton. The north-east was held by heathen Picts.

The Lombards

The Lombards first appear as a tribe on the west of the Elbe and are found in Pannonia. In 568, they invaded Italy, where they settled in the northern plains, though some groups ranged further south and set up independent principalities, such as of Benevento and Spoleto. By 605, the Lombard kingdom covered most of Italy, as conquering power. Justinian's reconquest of Italy was reversed. The Lombards became the landowners and the free subjects lived by Lombard law, Roman, chiefly Theodosian, law being practised by free and unfree Romans within their lands.

Rothari, Duke of Brescia, was elected King in 636, and conquered Genoa, but earned his chief fame by publishing the *Edictum Rotharis* in 643, the contents of which were Germanic but showed Roman influence in being in Latin and in being a systematic codification of the law rather than a restatement, or amendment, of custom. Where silent, Romans were left to be ruled by Roman law. It was supplemented by edicts of later kings such as Liutprand.

The Lombards gradually adopted the romance language of the natives, intermarried with them and, by 730, Roman law was recognized by the kings for those who had inherited it, though Lombard law was more prevalent and the kinship blood-feud was still widespread. Roman culture took hold of the capturers. Liutprand (712–44) aspired to unite all Italy in the Lombard kingdom and was a conqueror, but also a legislator. The recording of the laws was done in the *Liber Edictus* or *Edicta Langobardorum.* The Lombard law took root in Italy, became a subject of scientific study and treatment, competed with Roman law for domination, and persisted in use until well into modern times. Lombard law gave rise to the law school of Pavia where, in the eleventh century, was compiled the *Liber Papiensis.* To this there was added the *Walcausina*, a collection of annotations and formulae and, about 1070, the *Expositio*, a legal commentary, the whole material being later revised in the *Liber Lombardae.*

Both Roman law and Church law were influential in Italy in Lombard times, and schools of Lombard law arose, notably at their capital, Pavia.

The Lombard Dynasty fell in the later eighth century and their interests centred thereafter in Benevento. The Dukes of Benevento continued to legislate and made amendments to the Lombard edict and issued ordinances of their own. The south of Italy later fell to the Normans.

The Carolingian empire

In 751, Pepin, son of Charles Martel, was elected King of the Franks, and secured papal anointing and confirmation. His son Charles (Charlemagne) captured north Italy and styled himself King of the Franks and Lombards, and in effect controlled the Pope, then waged war on the Saxons and, after savage warfare, subdued them (785). Then he annexed Bavaria, overthrew the Avars and made the Slovenes and Croats submit to Frankish supremacy; in Spain he pushed the frontier south of the Pyrenees.

He personally administered his vast empire which stretched from the Elbe to the Pyrenees and south of Rome and from the Atlantic to east of the Alps. In 800, the Pope crowned him as Emperor of the Romans and in 812 the Byzantine Emperor recognized him as Emperor of the West.

He legislated perpetually; he commanded the unwritten customary law of various tribes to be reduced to writing and that of others, such as the Salic law, was edited authoritatively. This gave rise to many of the surviving tribal codes.

He issued numerous capitularies (q.v.), not part of a systematic code but each a collection of laws, regulations, and temporary or permanent commands. These tended to produce greater uniformity of law over wider areas. In 827, Ansegisus collected many capitularies in four books, and subsequently Benedict Levita added three books of false capitularies. He instituted a body of experts in law for each county, *scabini* in the north and *iudices* in the south, who acted as assessors to the court of the district. In each district a local lord was judge, leader of the forces and responsible for law and order. Charlemagne sent out *missi dominici* to all the provinces to inspect, check, report and regulate. They heard complaints locally, and punished the misdeeds of officials.

In north Italy, the Franks did not abolish the Lombard legislation and Lombard Italy developed on its own lines, though influenced by Frankish legislation. A notable work was the *Exposition to the Book of Pavia*, a commentary on a compilation of laws in force in Lombardy. In 832, Lothar I submitted to the Assembly at Pavia, a revision of the entire body of capitularies from the time of Charlemagne. A selection, the *Constitutiones Papienses* was enacted and became a local law for Italy.

Under the Carolingian Empire, the system of personality of laws developed fully. The Roman law and the native customs of the Germanic settlers, in what had been the Roman Empire, came together in an area politically united, but as tribal distinction continued adherence to their own laws and customs remained. The capitularies, however, exercised a unifying influence.

The Carolingian empire produced a great revival of learning and culture and a growth of scholarship. There developed schools of Lombard law, at Milan, Mantua, and Verona, but above all at Pavia. Among the works of the Pavia school were the editing of the Lombard Edict, the compilation of the Capitulare Italicum at the end of the eleventh century, and the production of a commentary, the *Liber legis Langobardorum* or *Liber Papiensis*. Similarly, the schools of Roman law produced Summaries, Glosses, and Compendiums, notably the *Exceptiones Legum Romanarum*, the *Brachylogus*, the *Summa Codicis*, and the *Quaestiones de Juris Subtilitatibus*. In the eleventh century, Pavia declined and the centre of legal studies moved to Bologna.

The Franks were converted to Christianity by the end of the eighth century and a network of dioceses spread through the Frankish Empire.

The Carolingian Empire decayed after Charlemagne's death. By the treaty of Verdun (843) the empire was divided among his three sons into the lands later known as France, Lorraine and north Italy, and Germany. The partition of Meersen (870) eliminated Lorraine.

In the decay of the Frankish Empire, a factor was the raids of the Northmen or Vikings on Britain, Ireland, and France. In the ninth century they began to settle in the river valleys. In 911, the lower Seine basin was ceded to them as Normandy and they settled in large numbers, rapidly adopting the language and institutions of France. At the same time Magyar hordes devastated Germany and Lombardy but after their crushing defeat at the Lechfeld (955) the Magyars settled in Hungary.

France disintegrated after 888, under a series of weak kings, into a collection of provinces, as the great vassals of the Crown fought among themselves. The Duchy of Normandy was notable for the vigour of its ruling dukes, and a regular organization of feudal service developed there in the eleventh century. In 1066, the Duke of Normandy invaded England, conquered it and became King of England, and by the end of that century the Capet kings were established in the kingship.

In France, the rough division long persisted between northern France, where customary law compounded of remnants of Roman and Germanic law, capitularies, canon law, and local usages, was dominant and southern France, where written law was dominant and Roman legal influence persisted, being in substance the custom of the south. At an early date, the texts of the Digest and the writings of the jurists were translated into French, and the Roman law was looked to for aid in the interpretation of the customs. Only the Roman and the canon law were studied in the universities. Ordonnances and royal legislation helped to produced uniformity and there was a movement towards the codification of custom.

England—ninth and tenth centuries

In the early ninth century, Wessex became the leading kingdom but, in 835, the Danes began to raid the east coast and, after years of raiding, began to settle, overthrowing Northumbria and forcing Mercia to pay Danegeld. In 869, they occupied East Anglia and for a time, only Wessex barely survived. Wessex under Alfred fought back, gained London in 886, and, in effect, he became king of all the English. The boundary was Watling Street, east of which lay the Danelaw, densely settled by Danes with Scandinavian institutions and law.

Alfred was not only a leader but a reformer; he strove to revive religion and learning, and worked on revision of the West Saxon, Kentish, and Mercian laws. Grants of land and rights by charter became common.

Alfred's son Edward and grandson Athelstan began the reconquest of the Danelaw; by 918, Edward had conquered as far as the Humber and, in 927, Athelstan annexed the kingdom of York. By

975, England was substantially united. But from 991, fresh Danish raids exacted heavy Danegelds and, by 1016, Canute was King of England. All of these kings legislated, and Canute published a substantial code, of which two versions are known, the *Consiliatio Cnuti* and the *Instituta Cnuti*.

The substantive law mostly comprised wrongs and offences and tariffs of compensation were recognized.

In the Danelaw small freeholders or socmen were numerous. The Danes kept to their own law.

Scotland

By 844, the King of the Scots of Dalriada was able to make himself King of the Picts as well and, in 945, the Scottish king obtained Cumbria and, in 954–975, Lothian and the Tweed valley. By 1035, the Scottish kingdom was united within its present frontiers. Law was probably local and customary, traditional and unwritten.

Ireland

In Ireland, a hierarchical pattern evolved. The high-king had a struggle to maintain authority over the five kings of the provinces, who in turn dominated petty tribal-kings. Despite petty wars, art and learning flourished in courts and monasteries. A second period of Viking invasion began in 914, but Brian Borumha came near to uniting Ireland and at Clontarf (1014) heavily defeated the Vikings.

Scandinavia

In the northern countries, the earliest laws were customary, preserved and declared orally by law-men, who recited the laws at annual meetings of the people. In Sweden, the law-men early became elected public officers. Popular assemblies early acquired lawmaking power and written laws appeared in the twelfth century—in Iceland from 1118, in Norway the Frostathing-lov about 1190, and in Sweden and Denmark in the next century.

Official codes of law appeared in Denmark (King Waldemar's Lov-bog) in 1241 and in Norway in 1267–68 when King Magnus Lagaboter promulgated a code common all districts, the Landslov, which was introduced also in Orkney and Shetland. This was extended to Iceland in 1271–73 under the name of Ironside and superseded the ancient law of Iceland. Swedish law was slower in being unified but King Magnus Eriksson caused a well-developed Code to be compiled in about 1347. These bodies of law lasted till well after the mediaeval period.

The growth of feudalism

All over Europe by the ninth century, there had developed a landed aristocracy who frequently acted as officers of government and had devolved on them functions over their own and neighbouring lands. Feudalism (q.v.) developed from the breakdown of earlier structures of law and government but created new institutions and rules. Land tenure determined political rights and duties and a person's status depended on his place in a hierarchy of landholders. Each vassal owed duties to his superior, frequently onerous, the major one of which was military service. In return he was entitled to protection and to hold the lord's land. Ownership of land was divided, each person in a feudal chain holding some of the rights in the land. The vassal had to attend the lord's court and to be judged by his fellow-vassals there. The right to give counsel to the lord became a right to consent, a principle containing the seed of acting by majority opinion only.

In France, feudalism led to disruption of sovereignty as tenants-in-chief acquired the greater part of the royal rights and powers, and many of these they transmitted to their own vassals, so that a host of seigneurs had rights of government and jurisdiction. The class of knights developed from a clan of horsed warriors, and knighthood became a privilege of rank.

Feudalization was never complete. In England and France, all land was deemed held of the Crown and under feudal lords, but in southern France and Germany allodial ownership of land persisted.

Feudalism was a legalistic organization of society and gave rise to much legal literature. The earliest known feudal code was the *Usatges* of Raymond Berengar I, Count of Barcelona (*c.* 1065). The *Books of the Fiefs* or Feus were first compiled in Milan and were later revised, translated and obtained vogue in France and Germany as well. The *Assizes of Jerusalem* was a compilation of feudal law for the Latin Kingdom of Jerusalem established by the Crusaders.

The Holy Roman Empire

The revival of the German monarchy began with the election of Henry the Fowler as King of the Saxons and Franconians in 919. He quickly made himself chief of the confederation of German duchies and his son Otto the Great (936–73) rescued Germany and the West from the Magyars, and aimed at the headship of Christendom. He conquered north Italy and, in 962, received the imperial crown from the Pope. Thus, he recreated the Western, or as it was later called, the Holy Roman Empire. In 1034, Burgundy was added to the Holy Roman Empire.

In this time, the customs of the various Germanic tribes and also the capitularies of Charlemagne and his successors were forgotten. Down to the thirteenth century, law was normally unwritten, but from that time written law appears, both statutes and law-books such as the *Sachsenspiegel*, the *Spiegel der deutschen Leute*, the *Schwabenspiegel*, and similar books, as well as enactments of manorial law, privileges conferred on towns, town-law books.

In the early thirteenth century Frederick II (1220–50) attempted to subjugate Italy. He developed the administrative system and, in 1231, issued in both Latin and Greek the *Liber Augustalis*, an imitation of Justinian and supplemented by *Novellae*, the first code of law to be promulgated by a mediaeval secular monarch. Eleven provinces each had a justiciar and chamberlain for civil and criminal cases. But conflict with the Pope kept strife active and his attempt to unite Italy to the Empire failed. Henceforth, it was abandoned to communes and tyrannies.

In the fourteenth century, Charles IV issued, in 1356, his Golden Bull which became a kind of fundamental law of the Empire, an attempt to define the body of electors, and to give some solidarity and functions other than election. Throughout that century, Germany was in chaos, a quarrelling mass of feudal lords, self-governing towns, lay and ecclesiastical magnates. Might was supreme and law unenforced. The Empire was unworkable with no capital, administration, or revenues. The Reichstag was powerless. In 1438, Albert V of Hapsburg, ruler of Austria, was elected and henceforth the Hapsburgs were always elected Emperors. Feudalism grew stronger and landless serfdom increased.

In the Empire, every principality, lordship and town had its own law; there was municipal law, manorial law, guild law, knights' law, and peasants' law. A beginning of systematization was made by the spread of authoritative treatises on customary law, of which the most important was the *Sachsenspiegel* (q.v.), which gave German courts a basis, widely accepted and maintained.

When the central Imperial court, the *Reichskammergericht* was constituted a standing tribunal in 1495, it adopted Roman law for its guidance as the common law of the Empire, and its example was followed by lower courts. This gave a great impetus to the reception of Roman law in Germany, and towns and principalities began to remodel their laws on Roman principles.

The 'Reception' of Roman law in Germany started about 1400 but went on for several centuries. Roman legal learning had been penetrating Germany since the thirteenth century and the idea of a common law for the Empire became powerful. Then there developed the idea that Roman law had a claim to validity in Germany, based on the idea that the German Roman Empire was a continuation of the ancient Roman Empire. Knowledge of Roman law developed, it was taught in universities, and jurists skilled in Roman law began to be employed in Germany. The decisive reasons for the Reception were the weakness of German law, the lack of a strong civil court and the absence of a strong legislative authority to develop it, and the lack of scientific cultivation of German law. The Reception fundamentally altered German private law for the future; it postponed legal unity and prevented German private law becoming national.

From their foundation, Roman and canon law were important studies in the German universities, the latter being more important till the end of the fifteenth century. Thereafter, the Roman law was dominant and its superior technical apparatus led to its widespread influence in practice and to a transformation of the administration of justice.

The Netherlands

In the Netherlands, the power of the Carolingian central administration was weak and law, for long, remained customary common law, developed by local traditions and the precedents of local courts, influenced by the ancient Frankish compilations of law. Towns and local districts formulated their own statutes and, at the end of the Middle Ages, there was a movement for the framing of a general common law. The Roman law was, from an early date, an important subsidiary source, though not equally so in all the provinces, and down to the Reformation regard was had to the canon law. The Roman influence was such in the seventeenth and eighteenth centuries to coalesce with native law and to give rise to the Roman-Dutch law.

Switzerland

In the districts which, from the thirteenth century, formed the Old Swiss Confederation, Germanic folk codes and Roman law were replaced by local customary laws on the basis of the older racial laws. In the Middle Ages, there was acquaintance with and some reliance on Roman law, but never a wholesale reception. By the end of that time the Confederation had detached itself from the Empire and made alliances with France. Individual cities had charters, town or territorial laws, and law-books.

Italy, 1100–1500

In Italy, the Roman law continued to be the general law, though Lombard law survived in many parts of the north. The centre of Lombard, as of Roman legal studies was Bologna and the Lombardists developed their system almost as highly as did the Romanists and there was rivalry between the two systems.

Italy, particularly Bologna, became the centre of legal study and of the revival of Roman law, which inspired a legal renaissance all over Europe. In the thirteenth century, Popes ineffectually tried to stifle the study of Roman law. Students flocked from all over Europe to study the Roman law at the feet of the Glossators, and after about 1250, their successors, the Commentators, who applied logical discussion and reasoning to legal studies. From about 1400, the growth of humanism affected legal studies also and the Roman law began to be studied in relation to ancient civilization generally.

About the same time, the growing independence of cities and merchant gilds gave rise to bodies of mercantile and maritime law, such as the sea laws of Amalfi, Trani, Pisa, Genoa, and other ports.

From 1100, onwards, there were substantial bodies of legislation being enacted in the various Italian states, trying to relate such diverse sources of law as city statutes, feudal customs, Roman law, Lombard law, canon law. In Savoy, a revision by a commission of jurists was published in 1430, in Sardinia a code was published in 1395, and in Venice various codes were adopted in the thirteenth century.

In the south of Italy, the Lombards were succeeded by the Normans to whom is owed the *Assisae regum regni Siciliae* of the twelfth century, and then by the Swabians. In Naples, under the Angevins, there was much legislation and there was Aragonese legislation in Sicily. In the papal states Roman and canon law survived but there was important legislation in each of the provinces such as Cardinal Albornoz's *Constitutiones Marchiae Anconitae*, a revision of the law of the Marches (1357).

The Papacy and the Canon law

Despite the troubles of the Dark Ages, Christianity survived and spread and the Church preserved many remnants of ancient civilization, giving the impetus to many developments which assisted in reviving civilization. The *foci* of Christian influence were the monasteries. They rapidly accumulated lands and endowments. Under Pope Gregory, the Papacy achieved pre-eminence throughout the Christian world and the ideal of unity in government and principles of action was better exemplified by the Church, in having an accepted head in the Papacy, than by any state.

The bishops alone in the Dark Ages retained an ideal of the Christian state as an institution for internal peace and good government. They developed a theory of the Church as the kingdom of God on earth, which made kings to act as its instruments. Bishops were confronted by the force and greed of laymen, and were not always united among themselves nor disinterested. The growing power of the Papacy was a safeguard against both.

The sources of the canon law were the Bible, decrees of General Councils of the Church and ordinances of the Popes, which took various forms, decrees, rescripts, epistles, bulls, and writs. In some instances, state law confirmed and reinforced canon law. From an early date, compilations were made, notably the *Prisca* (fifth century), the *Dionysiana* (fifth–sixth century) made by the monk Dionysius Exiguus, and a version of it, the *Dionysio-Adriana*, presented by Pope Hadrian to Charlemagne in 744.

About 850, a Frankish cleric, assuming the name of Isidorus Mercator and taking as a basis the *Hispana*, the collection of canon law materials collected by St. Isidore of Seville, forged, altered, and rearranged a mass of material, partly genuine papal decretals, partly spurious ones, partly altered ones, so as to make the bishops safe against their archbishops, the clergy and the laity. The False Decretals laid down the universal and absolute supremacy of the Papacy. The Pope was sovereign legislator and supreme judge, and the episcopate was free from their own metropolitans and from lay judgment. The papal power benefited from these documents, though from the first some scholars doubted the veracity of the collection.

When Charlemagne's empire declined into anarchy the Western Church degenerated. Bishops were nominated and invested by lay rulers and the whole Church was rotten with simony. Monasteries were corrupt and the rules of the founders disregarded. A revival set in with the foundation of Cluny (910). In the struggle against corruption the renewed study of the canon law was a guide to the path of reform. A new kind of compilation of canon law began to come into vogue with the publication, in about 915, of Regino of Prum's *De synodalibus causis*, a treatise of episcopal procedure, and about 1012, of the *Decretum* of Burchard, Bishop of Worms, containing classified excerpts from the canon law.

After a period of chaos Pope Leo IX gave impetus to reform by active leadership and intervention; he began to transform the Papal Chancery and sent legates *a latere* to keep in touch with remote provinces. He attempted, however, to assert papal power over the Eastern Church and this resulted in the division of the Eastern and Western Churches (1054). With the accession of Gregory VII (Hildebrand) the principle of Papal supremacy over Church and Christendom was carried far, and also the principle that the Church was supreme over the State. He forbade the investiture of ecclesiastical office by laymen. This provoked a quarrel with the Emperor in which the latter was humbled at Canossa (1077) but subsequently got his revenge. The Concordat of Worms (1122) marks an epoch in the reform of the clergy and the government of the Church and, increasingly, the Church acted as a united, disciplined body through the canon law and the courts of the Church.

As the power of the Popes and the Curia grew the canon law, comprising canons passed by councils and papal decretals, were recognized as binding in the whole Church. Canon lawyers began to develop their science, following the lines of Burchard's *Decretum*. The works of Ivo of Chartres (q.v.) were

MEDIAEVAL LAW

followed by Gratian's (q.v.) *Decretum* (1141) which rapidly became official. Decretals increased in number as appeals to Rome multiplied, and were collected in the Decretals of Gregory IX (1234), while the canon law was glossed and expounded in numerous texts, such as the *Summa Aurea* of Hostiensis (q.v.). The canon law had become a vital formative and widely persuasive influence in mediaeval civilization. Under Innocent III (1198–1216), the Roman Chancery was systematically organized. The method of hearing appeals was elaborated and later developed into a rota of expert auditors, who heard evidence and prepared cases for papal decision, and the Fourth Lateran Council (1215) passed much important legislation.

Additions to the official corpus of canon law were made in the *Sextus* of Boniface VIII (1298), the *Clementines* of Clement V (1317), and the *Extravagantes* of John XXII (1325) and finally, at the end of the fifteenth century, the *Extravagantes Communes*. A definitive official text was given to the whole Corpus by a commission, and their official edition was published in 1582.

Subsidiary sources of canon law include concordats and judicial decisions, particularly of the Rota.

The authority of the canon law was greatest in the Middle Ages, and in regions coming directly under the government of the Church, but throughout Christendom, it was important in relation to marriage and divorce, the status of and jurisdiction over clergy, and certain kinds of misconduct.

From their earliest years canon law was studied in the European universities as a branch of theology, and a vast literature of commentaries, summaries, questions, and other works on it developed. Among great canonists were Rufinus and Huguccio in the twelfth century, Tancred, Hostiensis and Durandus in the thirteenth, Johannes Andreae in the fourteenth, and John of Torquemada and Panormitanus in the fifteenth centuries.

From about 1270 the Papacy declined from predominance in European politics and was one competitor among others for secular power. During the years at Avignon, the centralizing tendencies of the mediaeval Papacy reached their culmination. Judicial business had grown enormously and the *Audientia Sacri Palatii*, known from 1336 as the *Rota*, was busy dealing with cases reserved to the Papacy. A vast fiscal system was developed.

Though the Reformation destroyed the formal authority of the Roman Church and the canon law in many countries, it frequently remained the basis of national law in the spheres in which it had formerly been authoritative and was modified and adapted to a different ecclesiastical structure.

The revival of Roman law

Parallel with the growth of canon law, there was a rediscovery of the Roman law of Justinian, which initially was accepted and even known only in the East, and which had hardly penetrated the West. This began in Italy, at Pavia and then at Bologna. Irnerius (q.v.) began the school of Glossators (q.v.) or textual commentators. The *Summa* of Azo (q.v.) became indispensable for an Italian lawyer and the Gloss of Accursius (q.v.) the standard accompaniment to the text. The Roman law, though not in force, became a guide to those who administered local and customary laws, a quarry for ideas and a powerful influence for change. In the fourteenth century, the Glossators were replaced by the Commentators (q.v.) who sought to make the civil law applicable to contemporary conditions and to harmonize it with Germanic, feudal, and canon law. In the fifteenth century the humanists such as Valla, Politian, and Leto, and the jurists such as Alciatus, tried to restore the classical law and set it in its context.

In southern France, Justinian's Code, modified by local custom, gradually superseded the earlier Roman law-books as the law of the land, and, in northern France, it mingled with traditional customary law. In Germany, it began to oust native law and became authoritative in the Imperial Court of Justice set up by the Emperor Maximilian I in 1495. In France, Italy, Sicily, and Spain it had a vital and permanent effect on the law. But in England, it had little influence as royal judges were already developing a common law.

The revived Roman law was very influential in the development of mediaeval procedure, a constant guide in unprecedented cases and a major contribution to mediaeval thought as well as to law and order.

The revived study of Roman and canon law spread rapidly all over Europe and students carried these studies from Italy to their home countries. In the twelfth century, there was a need everywhere for judges, notaries, advocates, and administrators, and this could only be met by jurists trained in the methods developed at Bologna. Centres of legal scholarship multiplied. Vacarius carried civilian scholarship to Oxford, Placentinus to Montpellier, and new centres of canon law scholarship developed in many places.

Law in the Eastern Roman Empire

From the division of the Roman Empire into Eastern and Western portions in 395, Roman law continued to apply in the east. The *Codex Theodosianus* (q.v.) of A.D. 438 collected the imperial constitutions since Constantine and was supplemented by later constitutions. Justinian, in the years 528–34, superseded all earlier law by having prepared and enacting the *Digest* and *Code* (qq.v.), rearranged and more or less revised collections of older law and legislation respectively. These were supplemented by later constitutions, the *Novels* (q.v.), of which there are three unofficial collections, and various works were written to make his compilations more accessible to the Greek populace. A selection of matter from the Justinianean literature, modified, appeared as the *Ecloga* (q.v.) about 740, but was replaced by the *Procheiron* (q.v.) of about 879 and the *Epanagoge* (*c.* 886) and, about 900, there appeared the *Basilica* (q.v.), a major compilation of materials from Justinian's books, with changes to adapt them to changed circumstances. Thereafter ensued a very decadent period productive of nothing important. About 1345, appeared the *Hexabiblos*, a poor compilation, which however, survived the overwhelming of the empire by the Turks in 1453 and was, in 1835, decreed to be in force in the kingdom of Greece, subject to contrary custom or practice, pending the promulgation of a civil code.

The Norman Conquest and settlement in England

In 1066, Duke William of Normandy landed in England, defeated the Saxons, and became King. This substantially determined the whole subsequent history of English law. He introduced feudalism, built castles, and redistributed lands to his followers, but claimed to maintain the laws and rights previously established. In succeeding years, he extended his sovereignty and was supreme by 1072. The Normans did not bring any written law or legal science with them, but some of William's men, such as Lanfranc, were familiar with the scholarship of the Pavia law school. William did not attempt to replace English by Norman law. The ruling aristocracy became almost entirely Norman and French, and land and power were transferred to Normans. The Witan became the court of the Crown tenants-in-chief. Serfdom became dominant among the peasantry and the country was thoroughly feudalized. The forest law was introduced from France to protect his hunting rights. The Crown was represented by sheriffs in the shires who held shire courts. Legislation was scanty but far-reaching; ordeal by combat came in. The King dominated both the baronage and the Church and foreigners, such as Lanfranc, were imported as bishops.

The reign of Henry I (1100–35) was notable for the development of a system of government and law. There developed, within the *Curia Regis*, a group of barons who audited the sheriffs' accounts and tried cases concerning the revenue, the Exchequer. A group of lesser men began to be employed as royal judges and sent on circuit to hear Pleas of the Crown in county courts.

Civil war broke out on Henry's death, but Henry II (1154–89) strove to reimpose order. A major dispute with the Church arose over whether clerics charged with offences should be tried in lay or church courts. This gave rise to the murder of Becket, Archbishop of Canterbury, which resulted in the surrender of the punishment of criminous clerics to the Church courts and acceptance of appeals from English Church courts to the Pope. He effected major reforms in finance, law, and administration, and in his reign, Glanville (q.v.) wrote the *De Legibus Angliae*, the first major text on English law. From 1166, he sent justices to tour the shires, inspecting, levying taxes, and doing justice, and this led to the gradual replacement of provincial and customary rules by the common law of England, developed largely by decisions of royal judges into a national system.

After Henry II, Richard I was long absent from the realm but this allowed scope to great administrators such as Hubert Walter (q.v.), under whom Exchequer and Chancery became important departments of State, while regular judges sat at Westminster to hear Common Pleas, and moved with the King as his Court of King's Bench. John antagonized the Pope and the barons and had to submit to sealing Magna Carta (1215) which, in substance, accepted the principle that the King was not a despot but was bound by feudal custom; he must observe feudal rights and not overstep old and reasonable customs. Henry III also warred with the barons and, when he was their prisoner, their leader Simon de Montfort took the crucial step of summoning a Parliament to which not only barons and prelates and knights of shires were summoned, but burgesses from some boroughs. In his time, Bracton wrote the second great treatise on the laws of England. The reign of Edward I (1272–1307) was of outstanding importance. Legislation was regular and many statutes set out to regularise and protect the rights of the king and feudal lords. The three regular courts, King's Bench, Common Pleas, and Exchequer were regularly at work, competing with one another, and steadily developing the common law by their decisions. Above these was the jurisdiction of the King in his council, which was closely linked with the rise of Parliament. Two functions of the *Curia* had become departments of State, the Exchequer and the Chancery. In 1295 the so-called Model Parliament was held, in which barons, clergy, knights, and burgesses were all represented and, since then, the idea has been accepted that a representative Parliament could bind all England by its decisions, legislative or judicial.

Edward subjugated Wales but did not abolish Welsh customary law, though he imposed English county organization. In Ireland, he extended the area of English domination, extended his major English legislation to Ireland, and summoned Irish parliaments on the English model. In Scotland, he took advantage of a break in the royal succession and competition for the throne to claim sovereignty, but provoked only resistance which outlived him and extorted a recognition of independence (1328). The costs of his wars forced reliance on Parliaments to vote money. He had to concede confirmation of Magna Carta and the Charter of the Forest, and in 1300 accepted 20 *Articuli super Cartas* for the relief of grievances.

Parliament achieved more power under the weak Edward II and the warring Edward III. In 1340 it was agreed that future taxes should be granted only in Parliament.

Scotland

The Norman Conquest did not affect Scotland directly but feudal institutions were copied, which progressed until the failure of royal heirs in 1290 gave rise to the long fight for independence from England, not settled till 1328. From 1326, burgh representatives attended Parliament though it remained local, not least because it delegated its powers to a committee. The reign of James I (1424–37 (in captivity 1406–24)) marked the start of a century in which kings sought, with some success, to revive royal power and subdue anarchic nobles, and in which law and justice began to make some impact.

Ireland

The Anglo-Norman dominion of Ireland was shaken by the English defeats in Scotland, and English authority was restricted to the Pale round Dublin. The Statute of Kilkenny (1366) sought to forbid intermarriage and the use of the Irish Brehon laws in English districts. The Pale shrank and the Anglo-Irish became more Hibernized.

Rise of the French monarchy

As the Holy Roman Empire declined in the twelfth and thirteenth centuries, a strong monarchy was built up in France, mainly by Philip II Augustus (1180–1223). He recovered the Angevin possessions in northern France from John of England, established in Paris an efficient bureaucracy and developed strong control from the centre. The *Curia Regis* became more organized and specialized, and its legal functions were becoming a department to which the name *Parlement*, once the general name for an assembly, was tending to be restricted. Its procedure and system of recording cases was influenced by that of Normandy. In the early fourteenth century, the King called all the three Estates, later known as the States General, to Paris, for support and financial aid.

A hierarchy of courts developed from the *Parlement* of Paris, which was organized as a court in 1345. The final court of appeal was *La Grande Chambre*. Provincial *Parlements* were set up in Toulouse, Grenoble, and elsewhere in the fifteenth century.

Law remained largely customary, founded on charters of privileges or municipal statutes, collections of customs, decisions, chartularies, deeds and formularies. In the north (*pays de droit coutumier*) The customs were inspired by Germanic, mainly Frankish, traditions. In the south (*pays de droit écrit*) Roman law prevailed almost everywhere. Among the collections of customs the most notable was the *Très Ancien Coutumier de Normandie*, the *Grand Coutumier de Normandie*, the *Établissements de Saint Louis*, Beaumanoir's *Customs of Beauvaisis* and the *Grand Coutumier de France*. By the Ordinances of Montil-les-Tours in 1454, Charles VII ordered official compilations of the customs of the country.

Iberia—the reconquest and the building of Spain

From 1031, the Christian reconquest of the Iberian peninsula from the Moors proceeded slowly. In 1139, the Count of Portugal assumed the title of king, and there was as much civil war as ever against the Moors. About 1068, there is attributed to Raymond Berengar I of Catalonia a code, the *Usaticos* (or *Usatges*), with constitutional, civil and criminal provisions. From the thirteenth century, Roman influence appeared and came to dominate legislation and the grant of municipal *fueros* (charters) made much new law. In 1254, Alfonso X of Castile promulgated a charter, the *Fuero Real*, based on the ancient Visigothic law, the *Fuero Juzgo*, systematized and modernized. Between 1254 and 1265, jurists compiled the *Leyes de las Siete Partidas*, a compilation comprising, as well as these materials, local *Fueros* of Castile, Leon, and Cordoba and showing the influence of Roman and canon law. In the Christian states of Spain, social conditions were less feudal and more developed than in France. Towns were gaining *fueros* (charters) which Alfonso did much to unify by granting a model *fuero* to certain cities. The Cortes or assemblies of magnates were as highly developed as those of England or France; by the thirteenth century they were found in every state except Navarre and everywhere exercised some check on monarchical absolutism. From 1469, Castile and Aragon were united, Spain was, for the most part, unified, and the baronial turbulence checked. Though Roman law was influential, each state also had substantial bodies of legislation of the Cortes and royal ordinances; of these a great collection, the *Ordenamiento de Montalvo*, was made in the fifteenth century.

In 1492, Granada was captured and the last of the Moorish power in Spain extinguished, in 1494 Rousillon recovered and, in 1512, Navarre was conquered.

By 1500, the formation of native law had ended everywhere in the Peninsula.

From 1200, there were numerous Spanish jurists of note, including the great canonist Raymond of Penaforte.

Commercial and maritime law

In the eleventh and twelfth centuries, several towns in northern Italy won the right to appoint 'consuls' who were to be elected annually to supervise administration and to dispense justice, and there was also great growth of statutes of cities, departments of city administration, and of trades and crafts, particularly the merchant gild. The bodies of mercantile statutes were influential in forming a commercial law for Italy. In France, commercial rules were found in municipal charters, by-laws of trading gilds and books of customs.

The maritime ports of Italy developed codes of maritime law which had enduring influence, such as the *Tabula Amalfitana* of the twelfth or thirteenth century, the ordinances of Trani, and the codes of Pisa and Genoa. The *Consolato del Mare* was reduced to writing at Barcelona in the fourteenth century and became practically the common sea law of the Mediterranean. It contained the rules to be followed by commercial judges or judge-consuls and was the Book of Custom for the whole Mediterranean. The Rules of Oleron, a compilation of decisions said to have been rendered by the sea-judges of the island of Oleron, and dating from about the eleventh century, became the common law of the Biscay coast. From about the same period can be traced the origins of banking, bills of exchange, and insurance policies.

Growth of royal power in Western Europe

In the later Middle Ages, kings greatly increased their prestige, power, and domains. Thus the King of England subdued Wales and secured the cession of large parts of France. In Spain, the King of Aragon conquered Sicily and Sardinia, and the marriage of Ferdinand of Aragon and Isabella of Castile, in 1474, led to the establishment of a monarchy ruling all Spain. By the end of the fifteenth century, the King of France had united to his domain nearly all the provinces and most of the country between Rhone and Alps.

Royal power mattered very much; if the King were weak, anarchy would result; if he were strong, he had an enormous task to perform. He was the interpreter of law and custom, the first Minister of State, the inspiration of most of the legislation passed by his Parliament. The royal power came to be exercised by the household officials, who became, in time, the heads of a series of departments, such as Exchequer, Chancery and the like.

Growth of national assemblies

In nearly every mediaeval state there was some sort of assembly which the King consulted from time to time, *thing, witan, curia*, council. In England, it began to include representatives from the shires and boroughs, developed into a consultative body of two Houses, acquired the right to vote taxes and then began to assert the claim to say how the taxes might be spent. Parliamentary privilege and procedure developed. In France, on the other hand, meetings of the States General were sporadic, they never developed corporate feeling nor achieved control of taxation; they had no part in legislation or the administration of justice. In Spain, all the Cortes were fairly representative and exercised considerable constitutional checks. The kings employed these assemblies chiefly for securing consent to taxation which supplemented the Kings' revenues from their own estates.

Round about the thirteenth century, kings were, in many countries, persuaded to concede charters of liberties, regulating and limiting the royal power. Thus, the Emperor Frederick Barbarossa ceded a charter of liberties to the northern Italian towns of the Lombard League in 1183. Alfonso VIII of Leon affirmed feudal privileges in 1188. In England, John granted Magna Carta in 1215. Andrew II of Hungary granted the Golden Bull which recognized the customary liberties of his vassals in 1222. The Emperor Frederick's son, Henry, made concessions to the aristocracy by the *Constitutio in Favorem Principum* in 1232. Peter III of Aragon assented to a *Privilegio General* in 1283, and Louis X of France published ordinances to a similar effect in 1215.

Revival of towns

The towns which, in Roman times, had been capitals of provinces, were reduced, in the Dark Ages, to small fortified strongholds. As the Dark Ages passed they revived and again became centres of commerce and industry. This began in the Italian Mediterranean ports which had commercial relations with Byzantium and the Arab countries. Only later did towns develop again in Western Europe. Everywhere, they began to acquire charters and privileges and burgesses became a distinct class in the community.

Other developments in the Middle Ages

There were many developments in the millennium of the Middle Ages which were common to most countries. Among these were the development of orders of knighthood, based on the mounted warrior of feudalism, the idea of chivalry, and the science and law of heraldry. Among merchants there developed, in many countries, guilds of associations of merchants, craftsmen, and others, with their own rules, privileges and machinery for the resolution of disputes.

From the thirteenth century onwards, there was a great spread of literacy, and a growth of clergy, bureaucrats, lawyers, officials, merchants, and traders, all utilizing literacy in their work. Education spread generally and, from 1450, the diffusion of ideas, and of the written word, was enormously facilitated by the development of printing.

In the fourteenth century began the great intellectual flowering known as the Renaissance. The study of Greek and Latin civilizations, literatures, and remains developed rapidly. This influenced the study of Roman law, and caused it to be examined more critically, as an aspect of ancient civilization.

Law in the universities

The earliest *studium generale*, which can be regarded as a university, is that of Salerno. Its fame was chiefly in medicine. By the thirteenth century, the colleges of doctors of civil and canon law (soon to be accompanied by colleges of arts and of medicine) existed at Bologna, and students of law came from all over Europe to the Law School there. In the north, universities developed from the cathedral schools and, by the end of the Middle Ages, there were universities throughout Western Christendom. Paris was the archetype of northern universities but, as a school of law, it never rivalled Bologna and stood below the later French foundations of Orleans, Angers, and Toulouse. Its students were confined to the canon law.

From Bologna, the scientific study of law, particularly Roman law, spread all over Europe, from Oxford, Cambridge, and Louvain to Salamanca, Coimbra, and Naples, from Vienna to Orleans. The courses of study varied from university to university. The full course of study might extend to 10 years for the doctorate in both civil and canon law.

Conclusion

In the millennium which comprises the mediaeval period, most of the countries of Europe had proceeded a considerable way to becoming organized states and most of them had developed the beginnings, at least, of bodies of national law. The Roman law had been rediscovered and become a subject of intense study, the canon law had been largely created, been reduced to order and also become a subject of study, feudalism had developed and many of the customary bodies of rules and practices, such as of merchants, were becoming settled bodies of principles. In every country, there were professional judges, legal scholars, advocates, notaries, and working lawyers. There was legislation and books and bodies of legal learning, much of it common to most of western Europe. From the fifteenth century onwards, there were accelerating developments, of which, in the legal sphere, the most noteworthy are the growth of international law, the Reception of Roman Law as a body, particularly in Germany, and the further development of legal ideas and studies as a by-product of the Renaissance.

J. Gilissen (ed.) *Bibliographical Introduction to Legal History and Ethnology* (1963–); *Cambridge Mediaeval History* (8 vols. 1911–67); *Continental Legal History Series* (10 vols. 1912–18); E. Jenks, *Law and Politics in the Middle Ages* (1919); Gaines Post, *Studies in Mediaeval Legal Thought*; F. Kern, *Kingship and Law in the Middle Ages*; (1939); W. Ullman, *Mediaeval Idea of Law*; (1946); W. Carlyle, *History of Medieval Political Theory in the West* (1903–36); F. C. Savigny, *Geschichte des römischen Rechts im Mittelalter* (1834–51); P. Vinogradoff, *Roman Law in Medieval Europe* (1929); P. Koschaker, *Europa und das römische Recht* (1958); P. Fournier et G. Le Bras, *Histoire des collections canoniques en Occident* (1931–32); J. F. Von Schulte, *Die Geschichte der Quellen under Literatur des Canonischen Rechts* (1875–77); S. Kuttner, *Repertorium der Kanonistik* (1937); M. Bloch, *Le Société féodale* (1939–40); R. Boutruche, *Seigneurie et feodalité* (1968–70); H. Mitteis, *Der Staat des hohen Mittelalters* (1968); F. Pollock and F. W. Maitland, *History of English Law before the time of Edward I* (1898); T. Plucknett, *Concise History of the Common Law* (1956); W. S. Holdsworth, *History of English Law*; P. Olivier-Martin, *Histoire du droit français des origines à la Revolution* (1951); P. Ourliac et de Malafosse, *Histoire du droit privé* (1961–68); H. Brunner, *Deutsche Rechtsgeschichte* (1906–28); H. Conrad, *Deutsche Rechtsgeschichte* (1962); F. Calasso, *Medioevo del diritto* (1954); E. Besta, *Fonti: legislazione et Scienza juridica* (1923–25); E. Solmi, *Storia del diritto italiano* (1930); A. Garcia Gallo, *Manual de Historia del Derecho español* (1967); L. G. de Valdeavellano, *Curso de historia de las instituciones españolas* (1968).

Mediation. The intervention of a third party attempting to resolve a conflict between two others. It sometimes arises in international relations. The Hague Conventions of 1899 and 1907 and the Covenant of the League of Nations sought to create machinery for mediation. Under the Charter of the United Nations, members in dispute must first seek a solution by, *inter alia*, mediation, and a dispute reported to the Council or the Assembly may be submitted to mediation. The Council or the Assembly have appointed mediators several times, as between Israel and Arab states, or between India and Pakistan over Kashmir. It differs from arbitration in that there is no agreement to accept any award or suggestion made, and from 'good offices' in that a mediator usually takes a more active part in trying to settle the dispute.

The term mediation is also used in some countries in relation to resolving industrial disputes. Thus, the U.S. Government maintains a labour mediation or conciliation service.

Medical jurisprudence. An alternative name for forensic medicine (q.v.) or legal medicine.

Meditatio fugae. An old principle of Scots law that if a creditor could make an oath that his debtor was *in meditatione fugae* to avoid payment of his debt, he may apply to a magistrate who may grant warrant for apprehending the debtor for examination, and subsequently, grant warrant to imprison the debtor until he finds caution *judicio sisti.* In practice, such applications are obsolete.

Meeting. A gathering of two or more persons called to receive a report, take a decision or otherwise take some lawful action. The transaction of the business of many kinds of bodies, directors of companies, trustees, members of local authorities, members of societies, is effected by calling meetings and taking decisions thereat.

Under many statutes, meetings must be summoned for particular purposes, e.g., annual meetings of members of companies and, in certain circumstances, meetings of holders of particular classes of shares or meetings of creditors, or meetings of the creditors of a bankrupt. The regulations of the body concerned, or the relevant statute, may prescribe the majority required for particular kinds of action.

The liberty of persons to meet for any purpose not unlawful is one of the important liberties of the subject. Meetings in certain places, e.g. Trafalgar Square, require ministerial authority, and meetings cannnot be held in open spaces merely because the public have access thereto. Meeting on highways, unless permitted, are trespass and may be obstruction or breach of local byelaw. If the police reasonably fear that the holding of a meeting will give rise to a breach of the peace they are bound to take reasonable steps to prevent it taking place, and this is so even if the meeting is a public one held on private premises.

Meijers, Eduard Maurits (1880–1954). Dutch jurist and professor at Leiden who wrote extensively on legal history, international law, customary law and many branches of civil law and edited Bynkershoek. He was also draftsman of the new Dutch Civil Code, adopted in 1954–70.

Melfi, Constitutions of (1217) (or *Liber Augustalis*). The first major code of law issued by a western European ruler in the Middle Ages, issued by the Emperor Frederick II in an attempt to codify the mass of feudal Byzantine and Islamic customs current in the kingdom of Sicily, and make it a coherent and rational system. It emphasized the imperial authority, but also the equality of persons and their common subjection to the law.

Mellish, Sir George (1814–77). Practised as a special pleader before being called to the Bar in 1848. He developed the reputation of a good commercial lawyer and took silk in 1861. Though a common law practitioner, he was appointed a Lord Justice of Appeal in Chancery in 1870 but had inadequate knowledge of equity and its procedure and was less successful than at the Bar. His judgments are lucid and learned and the court was well regarded in his time.

Melville, Viscount. See DUNDAS, HENRY.

Memoranda Rolls. Records covering 1199–1848, complied by the King's Remembrancer and the Treasurer's Remembrancer, who noted the business to come before the barons of the Exchequer.

Memorandum. A brief note of a matter to be remembered. In a policy of marine insurance it is a clause inserted to avoid the danger of the underwriters being liable for damage to perishable goods and for minor damage. In a sale of land by auction, a memorandum is a note appended to the conditions of sale to the effect that a particular person was declared the purchaser at a stated price, and signed by the purchaser and auctioneer. A memorandum in writing was formerly necessary in the case of sale of goods over £10 in value and still is required for a contract for the sale of land.

Memorandum of Association. See ASSOCIATION, MEMORANDUM OF.

Memorial. A document prepared by an instructing solicitor and sent to counsel, narrating the facts and circumstances and setting out the questions on which counsel's opinion is sought. Also an abstract of the material parts of a deed which must be registered, with a request that the deed be registered; it is this instrument which is registered.

Memory, legal, time of. The period commencing with the accession of Richard I (July 6 or September 3, 1189), a date chosen because the Statute of Westminster I (1275), fixed that as the period of limitation for bringing certain real actions.

Memory, living, time of. The period within which no man alive has heard any proof to the contrary, a period material where it is alleged that a custom or prescription has existed from time whereof the memory of man runneth not to the contrary.

Menger, Anton (1841–1906). Austrian jurist, influential on the draft German Civil Code, and notable for his work on the juridical theory of socialism. His major work is *Neue Staatslehre* (1903), which seeks greatly to enlarge State influence in legal relations.

Menochius, Jacobus (1532–1607). Professor at Padua, highly regarded in his time, author of *De*

Recuperanda Possessione Commentaria (1565), *De Praesumptionbus* (1575), and *Consilia sive Responsa* (1572), and other works.

Mens rea. In Anglo-American law, along with *actus reus* (q.v.), the essential mental element of a completed crime. The definition of every crime, expressly of impliedly, states what *actus reus*, and what concomitant *mens rea*, are necessary to constitute commission of that crime. It is neither general criminal depravity, nor moral wickedness, but the guilty mind or mental state required for the commission of a particular crime. It is part of the definition of every crime to say what kind of guilty mind or mental state is necessary for the commission of that particular crime. Commonly, it is intention or recklessness but it may be mere inattention or negligence. In crimes of absolute liability, it consists merely in allowing the prohibited result to happen. Sometimes the intention may be merely general, no more than deliberate conduct, but in some crimes, the requisite *mens rea* is an intention to bring about some specific result. Thus theft is taking 'with the intention of permanently depriving' the owner of the thing, and the intention of personal gain is no part of the *mens rea* of theft. There is a general presumption in criminal law that *mens rea* in the sense of intention or recklessness is necessary for the commission of crime but many statutory offences are crimes of absolute liability and are committed if the prohibited event is allowed to take place, the justifications being difficulties of proof of intent, and the overriding need to discourage conduct of the kind prohibited, whether intended or not.

Where *mens rea* of any kind is required for conviction it is a defence to show that the accused was incapable of forming or holding the requisite *mens rea*, as by reason of infancy, unsoundness of mind, or diminished responsibility.

Mental disorder and illness. Subnormal mental capacity, whether arising from innate mental defect or deficiency or from mental disorder or illness, has long presented legal problems, of care and custody, of responsibility for such persons' rights, of their capacity for legal acts, of responsibility for wrongs and crimes, and of liability to punishment. At various times, a wide variety of terms have been used, such as frenzy, madness, *non compos mentis*, idiocy, lunacy, insanity, and others, but none have any clear or fixed connotations. The modern British classification of mental disorders is into severe subnormality, subnormality and psychopathic disorders.

At common law, if a person was found by inquisition to be lunatic, detention and care was secured through the appointment of a committee. From 1774, reception of such persons in institutions was regulated by statute, a justice's order being normally required. Since 1959, a person may be admitted to a mental hospital voluntarily, or detained compulsorily on medical recommendation, or received into guardianship on medical recommendation. In Scotland at common law, a person might be found lunatic in proceedings on a brieve of cognition, and reception required a sheriff's warrant. Since 1960, similar provisions apply to those in England.

Under the royal prerogative, the King had the custody of the lands of natural fools, and the estates of idiots were a source of revenue until the end of the eighteenth century. Until the Mental Health Act, 1959, the jurisdiction in lunacy under which a patient's property was administered was partly the inherent jurisdiction under the royal prerogative and partly statutory and exercised by the Lord Chancellor and certain judges; the jurisdiction is now entirely statutory, exercised by nominated judges and the master of the Court of Protection. A committee of the person or estate of a mental patient is no longer appointed. In Scotland, the practice is for the court to appoint a curator bonis to look after the patient's affairs.

In relation to responsibility for crime, it has long been settled that a man cannot be tried if he is so disordered mentally that he cannot instruct his defence and that it would not be just to proceed with his trial (insanity on arraignment). In Scotland, the plea of insanity in bar of trial became so widely interpreted that the defence of insanity at the trial is very rare. In theory, a person unfit to plead and transferred to mental hospital may later be tried, but this is uncommon.

In respect of liability to punishment, early English cases indicate that in pre-Norman times the offender's family had to pay and to look after him, that when jury trial developed a criminal regarded as insane was remitted to be dealt with by the King. At least from 1505, it seems to have been the practice to acquit accused found to be insane. Hale, in his *Pleas of the Crown*, distinguished native idiocy, severe mental illness and induced witlessness, and total and partial insanity, and stated that an insane accused might be acquitted. The test was the ability to distinguish between right and wrong. The insane was returned to his relatives or confined, frequently in gaol by reason of the shortage of madhouses. Some became vagrants.

In the latter part of the eighteenth century, some judges appeared willing to allow a defence of insanity. In 1800, Hadfield, who tried to murder George III, was acquitted ((1800), 27 St. Tr. (N.S.) 1281) and returned to prison. In effect the test applied was whether the accused suffered from a delusion which prompted his act, even though he did not lack all understanding. This case prompted the Criminal Lunatics Act, 1800, which permitted only the special verdict of acquittal on the ground of insanity, in which case the court must order the accused to be kept in strict custody until His Majesty's pleasure be known. The Act, however, applied to treason, murder and felonies only.

Thereafter, there was increased interest in the effect of insanity both in literature and in the courts.

In 1843, Daniel M'Naghten shot Sir Robert Peel's private secretary in mistake for the Prime Minister himself. He had a history of insane delusions. The jury returned a special verdict of insanity and M'Naghten was confined in Broadmoor. Public concern, notably in Parliament, led to the House of Lords putting certain questions to the judges who, by a majority returned the answers ever since known as the M'Naghten Rules (q.v.), notably that a person is punishable 'if he knew at the time of committing such crime that he was acting contrary to law' and that to establish a defence of insanity it must be proved that 'the party accused was labouring under such a defect of reason, from disease of the mind, as not to know the nature and quality of the act he was doing, or, if he did know it, that he did not know he was doing what was wrong'. This was the test long utilized in textbooks and by judges.

The Royal Commission on Capital Punishment (Report, 1866) condemned the right-wrong test and seem to have wished to include uncontrollable impulses, and Fitzjames Stephen, in his *General View of the Criminal Law and History of the Criminal Law of England* moved in the same direction.

The Atkin Committee in Insanity and Crime (Report Cmd. 2005, 1924) appointed following the acquittal of True on the ground of insanity, recommended that irresistible impulse should be included in the category of irresponsibility, and as time went on, the Rules were stretched considerably.

The (Gowers) Royal Commission on Capital Punishment (Report, Cmnd. 8932, 1953) recommended that the Rules be abrogated, or at least that irresistible impulse be added to the Rules. No action was taken on the Report.

In 1883, after an attack on Queen Victoria, and at her instance, the Trial of Lunatics Act provided that insane accused be convicted as 'guilty but insane', they being detained in an asylum during Her Majesty's pleasure. Since then, the Atkin Committee and the Gowers Commission recommended that the verdict be not guilty by reason of insanity and this was effected by the Criminal Procedure (Insanity) Act, 1964.

In Scotland, the M'Naghten Rules were never treated as authoritative and there developed a recognition of 'diminished responsibility', that weakness of mind, not amounting to insanity, might be treated as reducing the crime from murder to culpable homicide, i.e. manslaughter, thereby avoiding the capital penalty. First recognized in *Dingwall* (1867), 5 Irv. 466, it was extensively used later. It was introduced into English law by the Homicide Act 1957, and was interpreted more favourably to the accused than was the original doctrine in Scotland. In the result, the number found insane on arraignment and unfit to plead, and those found guilty but insane both diminished substantially.

The royal prerogative of mercy has been exercised since 1837 by the Home Secretary or the Secretary of State for Scotland and, from 1840, statute provided for an inquiry into the sanity of convicted prisoners and, if they were found to be insane, their transfer to an asylum. From 1882, at the latest, all capital cases were reported to the Home Office and the question of reprieve was accordingly considered and, from 1884, it was obligatory to reprieve if certificates were given that a prisoner, though convicted, was insane. These statutory inquiries were abolished in 1959.

N. Walker, *Crime and Insanity in England.*

Mentally disordered persons. Mental incapacity or disorder has been named and classified in many different ways at different times. At common law in England, a distinction was drawn between idiots, who from birth had defective mental capacity, and lunatics, who became insane thereafter and whose incapacity might be or was temporary or intermittent. In modern statutes, a wide variety of terms has been employed and the same terminology is not used in all branches of the law.

In all branches of the law, every person is presumed sane and capable of acting on his own behalf until the contrary is proved, but once proved to be mentally disordered he is presumed to continue so, unless it is established that he has recovered or had a lucid interval at the time material to the investigation.

In respect of civil capacity, to make a contract or a will, to dispose of property, to incur liability for tort, and in matrimonial relations, the tests vary. To make a valid contract a person must be capable of understanding, of volition or choice and of giving or withholding assent. Similarly, to make a will or dispose of property a person must be capable of understanding the effect of the proposed disposition and of giving or withholding assent. In tort and matrimonial relations, capacity to be held liable is tested by standards similar to those invoked in respect of criminal liability.

A mentally disordered person is represented in litigation in England by his next friend or guardian *ad litem* and in Scotland by his curator bonis or a curator *ad litem*.

In respect of capacity to commit crime, to stand trial therefor, and to be held legally responsible therefor the question is whether the presumption of sanity has been overcome and it been proved that the accused was subject to such a defect of reason as not to know the nature and quality of his act or, if he did so, not to know that he was doing legal wrong.

Statute defines the circumstances in which a mentally disordered person may or must be taken into care and relieved of capacity to transact legally on his own. A person convicted of an imprisonable offence for which a penalty is not fixed may be compulsorily admitted to hospital for observation or treatment on medical recommendation, or received into guardianship, or become a voluntary patient.

Mercantile agent. A person having in the customary course of his business as agent, authority to sell goods or consign goods for sale, or to buy goods, or raise money on the security of goods. The category include brokers and factors (qq.v.).

Mercantile law. A term older than but substantially synonymous with business law and commercial law. Like them, it is a common compendious term for those topics of law relevant to persons engaged in business or commerce, such as sale, carriage, insurance, bills and cheques and bankruptcy. Like them, from the analytical point of view, it is not in common-law countries a distinct branch of the legal system because rules of law based on mercantile customs and the Law Merchant (q.v.) have, in those countries, long since been fully incorporated into the private law, whereas in countries with legal systems based on Roman law, mercantile or commercial law is usually still a body of principles applicable to distinct kinds of persons or transactions, distinct from the civil law applicable to ordinary persons not engaged in trade.

Mercantilism. A general term for an economic theory and practice commonly adopted by European nations from about 1600 to 1800. It sought to make individual nations wealthy by encouraging exports and limiting imports, and by accumulation, the holding of bullion. The system was identified with the encouragement of British shipping by Navigation Acts and with the way of dealing with colonies which led to the American Revolution. The first major attack on mercantilist principles was Adam Smith's *Wealth of Nations*, and the system declined as changes were brought about by the Industrial Revolution.

Mercenary troops. Professional soldiers willing to serve any country or organization for pay. They were common in the ancient world and, from the thirteenth century, in Europe, as the feudal system of providing troops broke down and was commuted into money payments. Their legal position has varied. The Foreign Enlistment Acts now seeks to prevent British nationals enlisting in foreign forces.

Merchandise marks. The Merchandise Marks Acts, 1887 to 1953, prohibited the fraudulent marking of goods and the fraudulent sale of merchandise falsely marked or not bearing the proper marks. They have been replaced by the Trade Descriptions Act, 1968.

Merchant Adventurers. An association of English merchants who became rivals of the Hanse (q.v.). Henry VII, in 1505, conferred on them the title of the Fellowship of Merchant Adventurers and, in 1564, Elizabeth chartered them as the Governor, Assistants, and Fellowship of the Merchant Adventurers of England with liberty to trade in the north of Germany. Their main centre was originally Calais, later Antwerp, and finally Hamburg, and they were sometimes known as the Hamburg Company. They seem to have ceased to trade about the end of the seventeenth century.

G. Cawston and A. Keane, *Early Chartered Companies*, 1296–1858; W. E. Lingelbach, *Merchant Adventurers of England*.

Merchant shipping. The compendious term for seagoing ships, not being warships nor engaged in fishing, engaged entirely or mainly in carrying goods. They are regulated as to ownership, registration, safety, equipment, crew, navigation, pilotage, and many other matters by a large body of legislation contained in the Merchant Shipping Acts and regulations made thereunder. Parts of the legislation apply to fishing vessels and parts to passenger steamers. Contracts for the carriage of goods by sea are regulated mainly by charter-parties and bills of lading (qq.v.) modified by international agreements embodied in the Carriage of Goods by Sea Act 1971. Much important shipping law is the product of international conventions by legislation made part of the municipal law of the U.K.

Mercheta mulierum. An alleged, but probably mythical right in Scots law that the lord of lands might have the first night of each married woman in his lands, a right later commuted to a payment by the bridegroom. There is no evidence of such a right ever being exacted. It has also been explained as a fine paid to the lord by a villein when his unmarried daughter was deflowered, or a composition payable by him for the lord's permission to give her in marriage to a man not subject to the lord's jurisdiction. Compare *ius primae noctis*.

Mercy, prerogative of. The exercise of the royal prerogative (q.v.) to pardon a convicted criminal or to grant a reprieve and commutation of sentence, particularly, when capital punishment was competent, to commute a death sentence to life imprisonment.

Mercy, recommendation to. When capital punishment was in force, a jury might make a recommendation to mercy, which was a factor which the Home Secretary or Secretary of State for Scotland might have in mind when deciding whether or not to recommend a reprieve and commutation of sentence.

Meredith, Richard Edmund (1856–1916) Called to the Irish Bar in 1879, he became judge of the Land Commission in 1898 and Master of the Rolls in Ireland, 1906–12.

Merger. The incorporation of one right in another, a principle of wide application. In real property, it arises where two estates in land become vested in the one person, so long as there is no contrary intention, express or implied, of the parties. In relation to charges on land, if a charge on land and the ownership of the land become united in the same person, merger results if that is the intention. Two or more charges also may be merged if that is the intention. Business mergers arise where two or more companies unite or one buys over another. This is now regulated by the Fair Trading Act, 1973.

Merger treaty. The Brussels Treaty of 1965, establishing a single Council and a single Commission for the European Communities in place of the separate ones for the Coal and Steel Community, Economic Community and Atomic Energy Authority established by their respective constitutive treaties. A convention of 1957, on certain institutions common to the European Communities, had already provided for a single European Parliamentary Assembly and a single Court of Justice to serve the three Communities. The merger treaty also makes uniform some divergent provisions in the three Community treaties.

Merignhac, Alexandre Giraud Jacques Antoine (1857–1927). Author of *Traité de droit public international* (1905–12), *Le droit des gens et la guerre 1914–18* (1921), and other works on international law.

Merits, on the. A person has a good cause of action or defence on the merits where he has a claim or defence on the real matter in issue, and not merely on a technical plea, such as absence of jurisdiction. 'The merits', however, include legal pleas or defences, such as that a limitation period has run, and are not confined to pleas morally good.

Merkel, Adolf Josef (1836–96). German jurist, concerned mainly with crime and punishment, author of *Kriminalistische Abhandlungen* (1867), *Juristische Enzyklopädie* (1885), and other works.

Merlin, Philippe Antoine, Comte (Merlin de Douai) (1754–1838). Leading French jurist of the Napoleonic period. In the Constituent Assembly, he secured the abolition of feudal and seignorial rights. He drafted the code of crimes and penalties enacted by the Convention in 1795, became Minister of Justice under the Directory and, later, one of the directors. Under Napoleon he was *procureur general* (1804) and did much to establish the interpretation of the Code Civil. He became a Councillor of State in 1806 and a count in 1810. Later he was Minister of State during the Hundred Days but was in exile from 1815 to 1830. While in exile, he compiled a *Répertoire de jurisprudence* (1827–28) and *Recueil alphabétique des questions de droit* (1827–28).

Merriman, Frank Boyd, Lord (1880–1962). Called to the Bar in 1904, he took silk in 1919, was Solicitor-General in 1928–29 and 1932–33, and President of the Probate, Divorce and Admiralty Division from 1933 to 1962.

Merrivale, Lord. See DUKE, H. E.

Merryman, *ex parte* (1861), 17 Fed. Cas. 9487. Merryman was imprisoned by Presidential authority in Fort McHenry. Chief Justice Taney, sitting as circuit judge, held that he was wrongfully detained and issued a writ of *habeas corpus*. The commanding officer refused to comply, and Taney cited him for contempt and issued an opinion that Congress alone has power to suspend the writ of *habeas corpus*. President Lincoln ignored the judgment taking the view that in the absence of legislation he had the power, but there is considerable support for Taney's ruling.

Mersey, Viscount. See BIGHAM, JOHN CHARLES.

Merton, Statute of (1235). Generally deemed the earliest statute of English law, the first listed in the *Chronological Table of the Statutes*, so called because it was enacted at the priory of Merton in Surrey. It allowed the enclosure of common by lords of manors and declared the illegitimacy of children born before marriage. On the latter point, the bishops wished the law altered, in conformity with the canon law, but the barons unanimously asserted that they did not wish the laws of England altered, and it was not altered unitl 1926.

Merula (van Merel) Paul (1558–1607). Professor of history at Leiden, a teacher of Grotius, and author of *Manier van procederen* (1592), the leading text on procedure in Holland.

Mesne (Lat. *medius*) (pronounced 'mean'). Middle, intermediate or intervening. In feudal contexts a mesne lord was one intermediate between the Crown and the tenant actually holding the land, himself holding of a superior lord and having tenants holding of him. Mesne process were writs intervening between the beginning and end of a case. Mesne profits are profits derived from land while wrongfully possessed by another and include the rental value of the land and elements for deterioration.

Mesne process. Writs arising at intermediate stages in an action, as distinguished from primary and final process, e.g. the *capias* issued after a writ of summons (the primary process) and prior to the *capias ad satisfaciendum* or writ of execution (the final process). Power of arrest on mesne process is now greatly limited and confined to the case where the defendant is likely to quit the country and his absence would prejudice the plaintiff.

Mesne profits. Profits derived from land by a person wrongfully withholding the possession of it from another during the period of his occupation, and which accordingly he must pay over to the true possessor as part of the compensation for his wrongful possession. The sum is the yearly value of the premises.

Messenger. A person who carries messages or otherwise communicates instructions or requests. Queen's Messengers are persons employed under the instructions of Secretaries of State to convey both domestic or foreign dispatches. They were formerly employed to arrest persons for treason or other offences not properly falling within the common law. There was formerly a messenger or pursuivant of the Great Seal who had duties in connection with writs for the election of members of Parliament. His functions were, in 1874, transferred to an officer selected by the Lord Chancellor and known as the Clerk of the Chamber and Messenger of the Great Seal. The Messengers of the Exchequer were officers attached to that court who acted as pursuivants to the Lord Treasurer. The Stationers' Company had, in the eighteenth century, an officer called the Messenger of the Press with the function of searching for unlicensed and later for seditious printing.

Messenger-at-arms. In Scots law, junior members of the corps of Officers-of-Arms, appointed by and subject to the control of the Lord Lyon King of Arms. The office may be held by persons who are also sheriff-officers (q.v.) and a messenger-at-arms may execute sheriff court writs. Their function is to execute all writs proceeding from the Court of Session and High Court of Justiciary. Every messenger is bound to serve any of the lieges on payment of his fees and expenses.

Messuage (Late Lat. *messuagium*; Fr. *maison*). A dwelling-house together with its own buildings, attached garden ground and the surrounding land. A capital messuage is the chief mansion house of an estate. In Scotland, a messuage is a principal dwelling-house without a barony.

Metage. Measurement. The City of London Corporation had, by charter and prescription, a right of compulsory metage, first of all merchandise and, ultimately, only of grain brought into the Port of London. The dues were known as fillage and lastage. Compulsory metage and fillage and lastage were abolished in 1872.

Metaphysical jurisprudence. The general term for the body of thinking about law founded on metaphysical speculation, particularly on Kant's doctrine of the free-willing individual, able to use his reason to choose one course of action rather than another. To Kant, law was concerned with the external and practical relations of persons; it was the sum total of the conditions under which the will of one could be reconciled with the will of others, according to a general rule of freedom. The general characteristic of a theory of law and justice of this class is that it claims to impose obligation by virtue of its deduction from a pure idea and not by derivation from the social facts to which it is to be applied. It normally also seeks to establish absolute standards valid for all times and places, as contrasted with a relativist theory.

Metayer system. A system under which land is divided into small farms, the landlord generally supplying the stock and receiving in return a fixed proportion of the produce, in kind not by way of rent, a proportion frequently fixed at one-half.

Metes and bounds. The limits or boundaries of a tract of land, as marked by natural features, man-made structures, or markers. Where a widow was entitled, at common law, to dower, her share was set apart to be held in severalty and she was said to hold it by metes and bounds. In the U.S., metes and bounds descriptions of land are common where survey areas are irregularly shaped and of varied sizes. The boundaries are stated by directions and distances, and identifiable features or marks at corners.

Methodology, legal. The body of knowledge of the methods which may be utilized to discover the principles and rules relevant, within a particular legal system or body of law, to the determination of a particular problem or controversy. Application of legal methodology depends, in the first place, on determination of the facts raising the problem and discovery of what is truly in issue. The facts may be agreed or admitted, but if not admitted, must be resolved by receiving and interpreting and evaluating evidence of the relevant facts. Once the facts have been determined, they must be classified or categorized to discover what legal issues or points have to be investigated. This must be done by reference to the analytical classification of the particular legal system, to the major divisions or branches of that system, the main heads thereof, sub-heads, sub-sub-heads, and so on, under which the principles and rules are grouped in the encyclopaedias, standard text-books, indices, digests and other reference works of the particular system. Thus, a set of facts may be analysed as raising an issue of Private law—obligations—contract—sale of goods—seller's liability for defective goods. Such analysis and reference to the appropriate books, and other sources of information, leads to the sections of statutes, interpretative case-law, academic commentary, and other sources (q.v.) of principles and rules relevant to factual issues of the particular kind being investigated. It is essential that the classification or categorisation of the facts be done by reference to legal heads and concepts, and not to the everyday or commonsense heads to which a layman might refer a problem. Thus, the sale of a defective car is properly analysed and classified as sale of goods – liability for defective goods, and not under such heads as Automobile, Road Traffic or Transport.

The application of legal methodology in this way is a skill much developed by experience and extensive acquaintance with the legal system in question, so that an experienced lawyer can frequently omit several stages in the process of scientific enquiry and go at once to the statute or case which provides an answer to his problem, or even answer it without needing to look at the legal sources at all.

Complications arise from the facts that a given set of facts may involve a number of distinct legal issues, such as of the jurisdiction of the court, of the system of law to be applied, of the principle of law involved, of the significance of the facts for the principle (e.g. whether, where goods were really 'defective' or, in fact, adequate) and of the remedy competent and appropriate, and factors affecting it. Only extensive legal knowledge and experience can suggest to a lawyer all the points which require investigation before he can advise what law is applicable to a given set of facts and what the rights and duties of the parties are in the circumstances.

Methuen, John (1650–1706). Was called to the English Bar in 1674, became a Master in Chancery in 1685, and Chancellor of Ireland, 1699–1703. He was also twice ambassador to Portugal and, in 1703, concluded the Methuen Treaty with that country which reduced the duty on Portuguese wines and stimulated the consumption of port in Britain.

Metric system. The system of weights and measures utilizing multiples and sub-multiples of 10. The different metric units are defined by statutory instrument on the basis of internationally agreed definitions of these units. Metric units of measurement have long been equally valid with imperial units and are now being increasingly utilized in practice.

Metropolis (Gk. mother city). The principal city, generally also the seat of government in any country. In the U.K., it denotes London, not merely the City of London, but the whole of Greater London.

Metropolitan. An archbishop. The Archbishop of Canterbury is the metropolitan, and the primate of all England.

Metropolitan Board of Works. A board created, in 1855, for the local government of London and replaced by the London County Council in 1888.

Metropolitan police. The police force of Greater London, but excluding the City.

Metropolitan stipendiary (formerly police) magistrates. Legally qualified persons appointed to act as stipendiary (salaried) magistrates in the Greater London area, holding courts at the various metropolitan magistrates' courts.

Mevius, Davidus (1609–70). German jurist, professor at Greifswald, syndic of Stralsund, and author of *Commentarius in jus Lubecense* (1642), *Prodromus jurisprudentiae gentium communis* (1671), and *De Arrestis* (1674).

Michaelmas. The feast of the Archangel Michael and All Angels, September 29th. It is one of the quarter-days in England and Wales, and the name is given to the autumn sittings of the High Court and to the autumn term of the Inns of Court. In Scotland, the Michaelmas head court was the meeting of the heritors at which the roll of freeholders was revised.

Middle Temple. One of the Inns of Court (q.v.).

Middlesex, Bill of. A fiction whereby, from the fifteenth century, the court of King's Bench extended its jurisdiction, nominally limited to causes of action arising in the county where the court happened to be sitting at that time. A real or fictitious bill of Middlesex was proffered against a person, stating that he had committed within that county a trespass with force and arms. If the defendant surrendered or was committed or bailed, he was within the court's jurisdiction and could be called to answer any other bill proffered against him while in custody. If he was not within the jurisdiction process called *latitat* (q.v.) issued. Statute in 1661 enacted that arrest and bail could follow only if the process disclosed the true cause of action. This nullified the bill of Middlesex until the King's Bench added to the fictitious trespass a clause '*ac etiam*' (and also) which set out the real cause of action. Later, the bill of Middlesex was not issued, and process began with the *latitat*. The device was used to capture business for the King's Bench from the Common Pleas, but the latter court retaliated with similar fictions.

Midsummer day. The summer solstice and the feast of St. John the Baptist, June 24th, one of the usual quarter-days in England and Wales.

Milan, Edict of. The proclamation, product of a political agreement made at Milan between the Roman Emperors Constantine I and Licinius in 313, and made by Licinius for the East in June 313, which granted all persons freedom of worship, gave Christians legal rights, including the right to establish churches and directed return to Christians of confiscated property and, accordingly, established toleration for Christianity throughout the Roman Empire.

Military courts. See COURT OF CHIVALRY; COURT MARTIAL.

Military government. In international law, the administration of government in territory occupied by an occupying power, arising from the collapse or supersession of the former government. The government itself may be military, civil, or mixed, and its functions comprise all the functions of government, legislative, executive, and judicial functions. It falls to be distinguished from both martial law and military law (qq.v.). Also, it does not cover cases of military forces stationed in neutral or friendly territories and sharing some responsibilities of administration with the established government.

Military law. Like martial law (q.v.) military law originated from the law administered in the Middle Ages by the Court of the Constable and the Marshal. It has now come to be the special body of rules applicable to, first the army and now to all the armed forces of a state, governing their terms of service, discipline, and the punishment of specifically 'service' offences which are, in many cases, not criminal by the ordinary law of the land. Military law is additional to the ordinary law in its application to members of the forces, and not in substitution therefor. It is applied by the officers commanding units and higher formations, and by courts-martial. In Rome, military law derived from the imperium of magistrates in their capacity as commanders of the forces. Military justice became more formalized and was included in Justinian's *Digest* and *Code*. In the Middle Ages, discipline was enforced by ordinances or articles of war issued by the King or other commander for each campaign. An early instance are those of King Richard I, in A.D. 1189, for the regulation of forces going to the Holy Land.

In England, it originated in the law applied in the Courts of the Constable and the Marshal, but codes of rules were drawn up for the government of the army from 1642 onwards. It was recognized, by 1687, that the jurisdiction of the Constable's and the Marshal's court was obsolete and that jurisdiction was being exercised by officers of the army. Military law was legalized by the first Mutiny Act of 1689 and annual re-enactments thereafter. By statute of 1717, the power to make articles of war was embodied in the Act. Since 1879, it is governed by the Army Act. Since 1955, there is a quinquennial Army and Air Force Act continued annually by Order in Council.

C. Clode, *Military and Martial Law* (1874); C. Clode, *Military Forces of the Crown* (1869).

Military Law (U.S.). The first Congress of the U.S. established courts-martial to enforce the discipline necessary for the efficiency of the army and navy. In 1775, and again in 1806, Articles of War were adopted, modelled on the Mutiny Act and Articles in force in Great Britain at the time. Military law is now administered under the Uniform Code of Military Justice adopted in 1951.

Military tenures. In feudal law, the tenures (q.v.) in which the duty owed was some form of military service, including grand serjeanty, knight service, cornage, and the like. See also TENURE.

Military testament. A will which, following the Roman law, if made on active service is valid though by a person under 18 and not having all the normal formalities. In England, it is recognized by statute and extends to members of all the armed forces.

Militia. A citizen force of soldiers as distinct from the regular army. The name was used during the Civil War and a force was authorized by statute in 1662 to be raised and trained by lords lieutenant of counties. In 1757, each county had to find a quota and men were to be chosen by ballot, but after 1815, voluntary enlistment was substituted, though ballot was not officially suspended until 1865. In 1907, militia units became part of the special reserve. In 1921, the portion of the army reserve previously called the special reserve was called the militia and, when conscription was introduced in 1939, men called up were called militiamen. Since, 1950, the term has again been used for a category of reserve forces.

Militia (U.S.). Now the National Guard. In colonial times, each community had its militia, composed of all able-bodied male citizens, but provincial militia were very ill-disciplined. The Federal Constitution gave Congress power to raise a regular army but the Second Amendment authorized state militias and state-organized bodies developed in the fifteenth century. The name National Guard was taken by N.Y. state militia in 1824 and an Act of 1903 made the National Guard the country's reserve force, federally equipped but state controlled. The Reserve Force Act, 1955, revised the requirements and training of the National Guard which has frequently been called on by state and federal officials to maintain order in civil troubles.

Mill, John Stuart (1806–73). English philosopher, official of the East India Company, and M.P., 1865–68. Philosophically he was influenced by Bentham and, at one time, studied Roman law with Austin. His extensive writings include a *System of Logic* (1843), *Principles of Political Economy* (1848), *Liberty* (1859), *Representative Government* (1861), *Utilitarianism* (1863), *The Subjection of Women* (1869) and essays on many political and related subjects. He systematized and expounded utilitarianism and did much to promote the status of women and of the lower classes in society. His *Autobiography* (1873) tells much of the development of utilitarian and liberal ideas.

M. Packe, *Life of John Stuart Mill*; Leslie Stephen, *The English Utilitarians* (1900).

Millar, John (1735–1801). Having studied under Adam Smith, been tutor to Lord Kames' children, and become a close friend of David Hume and tutor to David Hume (later Baron Hume (q.v.)), John Millar was called to the Scottish Bar in 1760 and, in the following year, appointed Professor of Law at Glasgow. For 40 years he taught with great success, acquiring a great reputation as a teacher. He lectured on civil law, jurisprudence, and the whole range of Scots law, government, and English law, and had a great reputation as a lecturer. He also distinguished himself for his liberal opinions in favour of the American colonists and of parliamentary reform. He was a zealous member of the Society of Friends of the People.

In 1771, he published *The Origin of the Distinction of Ranks*, valuable as an early sociological study, influenced by Montesquieu and Hume, and translated into several languages. His other major work is a *Historical View of The English Government* (1787), probably the first attempt to write a constitutional history, which attained a high reputation. His son, John, published a book on Insurance in 1787.

Millbank penitentiary. A prison erected near Westminster after 1812 on the site bought for the erection of a Benthamite panopticon (q.v.), and now occupied by the Tate Gallery. It was used for convicts awaiting transportation or for those not to be transported. It consisted of six pentagonal blocks around a hexagonal central space. From 1846, convicts spent the first 15–18 months of their sentence in separate confinement in Millbank or Pentonville with hard labour before being transferred to employment on public works. It was demolished in the 1880s.

Millenary Petition. A petition presented to King James I in 1603 by certain Anglican clergy desiring reformation of certain ceremonies and abuses of the church, in its services and discipline, the withdrawal of regulations, which excluded many conscientious ministers from the Church, and a reform of the ecclesiastical courts. It gave rise to the Hampton Court Conference and some modifications were made in the Book of Common Prayer, but the King rejected the requests and some of those who presented the petition were committed to prison.

Miller, David Hunter (1875–1961). American international lawyer, legal adviser to the American delegation to the Paris Peace Conference (1918–19) and who participated in drafting the Covenant of the League of Nations. He later led the U.S. delegation to the 1930 Hague Conference for the codification of international law. His writings include *Treaties and Other International Acts of the United States* (8 vols. 1931–48).

Miller, Sir Thomas, of Barskimming and Glenlee (1717–89). Called to the Scottish Bar in 1742, he was shortly after appointed Sheriff of Kirkcudbright and acted as one of the Faculty Reporters of decisions for volume 1 of the Faculty Collection. In 1759, he became Solicitor-General for Scotland and, in 1760, Lord Advocate, in which capacity he took an important part in the government of Scotland. He entered Parliament in 1761. In 1766, he became Lord Justice-Clerk, and, in 1788, Lord President of the Court of Session. He was made a baronet in 1789. He had a high reputation as a lawyer and as a member of the group of scholars and men of letters who lived in Edinburgh at that time. His son, William (1755–1846) became a judge as Lord Glenlee (1795–1846).

Miller Arnold's Case (1779). A famous Prussian case, which gave rise to great controversy and vast literature. Miller Arnold owned a water-mill powered by a stream. A noble neighbour diverted water from the stream to fill fish-ponds and Arnold's mill could not be used, his revenues fell and, by reason of his inability to pay his rent, his landlord took judgment and had the mill sold in execution. Arnold's actions for recovery of his mill and for damage for diversion of the water were dismissed. Frederick the Great intervened, ordered prosecution of the judges who had decided against Arnold and, on their acquittal, removed from office and fined the judges who acquitted them, ordered the purchasers to restore Arnold his mill, reimbursing them out of money paid by the convicted judges, and ordered destruction of the diversion channel and the restoration of Arnold's water-supply. The case is notable as an instance of royal interference to do justice where injustice resulted from fixed rules of law. On Frederick's death in 1792, the convicted judges were exonerated by royal decree and indemnified, but Arnold kept his mill.

Milligan, *Ex parte* (1866), 4 Wallace 2 (U.S. Supreme Ct.). Milligan, a southern sympathizer in the Civil War, was arrested in northern territory for subversive activities, tried by a military commission, and sentenced to death. He sought a writ of *habeas corpus* on the ground that the military had no right to try him. The Supreme Court held his detention and trial illegal, since the trial of civilians by military commissions in non-military areas had been provided for by executive orders only and not by act of Congress, and carefully delimited the line between civil and military power over civilians in wartime.

Miltiz, Alexander von (1785–1843). German jurist and author of *Manuel des consuls* (1837).

Mines and minerals. Minerals, and mines and quarries opened to enable them to be extracted, are parts of land and the owner of the surface of land is *prima facie* entitled to everything beneath the surface of his land down to the centre of the earth, subject to the exceptions that at common law mines of gold and silver belong to the Crown and, by statute, the property in petroleum is vested in the Crown and in coal in the National Coal Board. The ownership of minerals under land may be severed from that of the surface and held separately, and the right to open mines and quarries and extract minerals may be granted by way of lease, though such a lease is truly a sale of the minerals with liberty to extract and remove the minerals.

The owner of the surface has a natural right to have the surface supported and accordingly, unless the right to let down the surface has been granted, he has a claim against the owner of the minerals if the latter's operations cause subsidence of the surface and damage thereto. A right of support for buildings on the surface may be acquired as an easement.

The conduct of mines and quarries is regulated in great detail by statute, particularly in the interests of safety, health and welfare of workmen.

Minister. A servant; in nonconformist churches, a person in holy orders. In public law, a Minister is one placed by the Crown in charge of a department of State, comprising persons with special titles, e.g. Chancellor of the Exchequer, persons holding the office of Secretary of State, and persons holding the offices of Minister of or for X or Y, e.g. Defence. By convention a Minister must be a member of one of the Houses of Parliament, usually of the Commons. The Ministers in charge of the major departments form the Cabinet (q.v.) and the chief Minister is the Prime Minister (q.v.). See also MINISTER OF STATE; MINISTER OF THE CROWN.

Minister of State. A grade of Minister in the United Kingdom government below that of a head of a department but above that of Parliamentary Secretary and, as a rule, not carrying membership of the Cabinet. Ministers of State are usually appointed in departments of wide-ranging and heavy responsibility to assist and deputize for the Minister and to assume semi-independent charge of major branches of the departments' responsibilities.

Minister of the Crown. The holder of one of the chief political offices in Her Majesty's Government in the U.K. Some Ministers, usually including the Prime Minister, the Lord President of the Council, the Lord Privy Seal, the Chancellor of the Duchy of Lancaster, and the Paymaster-General have no, or only nominal, departmental duties and can concern themselves, in the Prime Minister's case, with the general oversight of government, and in the other cases, with particular tasks assigned to them. But the majority of Ministers are placed in charge of particular departments of State, with the title of Secretary of State for X, or Minister of Y. Ministers are appointed by the Crown on the nomination of the Prime Minister. They may or may not be members of the Cabinet, but must by convention be members of one or other House of Parliament, mainly of the House of Commons. The main task of the Minister is to direct the policy of the department in accordance with the general policy of the government, to justify that policy to Parliament, pilot through Parliament legislation affecting his department, and to answer to Parliament for it. Each Minister is responsible for the whole work of his department and if there is a failure or a scandal he must answer for it and may have to resign.

The House of Commons Disqualification Act, 1975, limits to 95 the number of persons who may sit and vote in the House of Commons, being the holders of specified offices, both senior (Secretaries of State and Ministers) and junior (Parliamentary Secretaries). Apart from the Lord Chancellor, there need be no members of the House of Lords in the government.

Minister plenipotentiary and envoy extraordinary. A representative from a foreign power of a class lower than ambassador, but higher than minister resident.

Minister without Portfolio. A Minister in the U.K. Government not having a department but available for some special or temporary assignment. He may rank as a member of the Cabinet or as a Minister of State.

Ministère public. The office of public prosecutor in France, having the function of prosecuting criminal cases and representing the public interest in civil cases. Members of the *ministère public* have the same qualifications as members of the career judiciary, are considered as magistrates, and referred to as *magistrature debout*, and many later become judges. They are found in the highest courts (*procureurs généraux*) and in courts of first instance and superior courts (*procureurs de la république*). In criminal prosecutions, the *procureur* has the decision whether prosecution should be initiated or not, though he is subject to the ultimate control of the Ministry of Justice. In civil cases, the *procureur* may act in his own name or in that of a government agency, or may be merely an observer or *amicus curiae*, but, in all cases, his major interest is that the law should be correctly interpreted and applied. In the *Cour de Cassation* the *procureur général* may bring to the court's attention decisions of lower courts which may require review by reasonable if questionable interpretation of the law. In all courts the interpretations suggested by the *procureurs* are persuasive on the court.

See also PROCUREUR.

Ministerial functions. Functions involving the application or execution of a rule of law in a particular case, when there is little or no discretion as to the mode of exercise, as in the execution of a warrant or repaying overpaid tax. They are distinguished from legislative functions and judicial functions and also from discretionary executive or administrative functions, which involve the exercise of a substantial measure of individual discretion and judgment.

Ministers' Powers, Committee on. A government committee, established under Lord Donoughmore in 1929, to enquire into Ministers' powers following Lord Hewart's (q.v.) attack on delegated legislation in *The New Despotism* (1929), criticizing the use of statutory instruments by departments. The committee recognized the need for delegating legislative powers in certain cases but recommended greater Parliamentary control of delegated legislation, including a standing committee to consider statutory instruments. It also considered administrative adjudication and, while rejecting a system of administrative courts, recommended various safeguards against abuse, including that the reports of inspectors holding inquiries should be published.

See also TRIBUNALS AND INQUIRIES.

Ministry. A term sometimes used of a department of State for which a Minister is responsible to Parliament, e.g. the Ministry of Defence, and sometimes of the whole group of Ministers who formed a government, e.g. the Second Gladstone Ministry. In the latter sense the term is wider than the Cabinet.

Minor, John Barbee (1813–95). Became Professor of Law in the University of Virginia in 1845 and became an outstanding law teacher, doing much to raise the standard and reputation of the Virginia Law School. He published reports, *Institutes of Common and Statute Law* (1875–95), a monumental contribution to American jurisprudence, and long deemed a high authority, and an *Exposition of the Law of Crimes and Punishments* (1894). His son, Raleigh Colston Minor (1869–1923) succeeded his father. He wrote a *Law of Real Property* (1908) and was a pioneer in private international law, his book on *Conflict of Laws* (1901) attaining international recognition.

Minor. In Scots law, a person of either sex above the age of pupillarity (14 for boys, 12 for girls) and below the age of majority (18). A minor has substantial legal capacity. He may marry at 16 without anyone's consent. If he has no curator (normally his father) he has full contractual powers; if he has a curator and contracts with the latter's consent, he has full powers; if he has a curator but contracts without the curator's consent, his contracts are void against him but enforceable by him. In the first two cases, his contracts are voidable at his instance until four years after majority (*quadriennium utile*) on proof of his minority at the time and *enorm lesion*, i.e. serious inadequacy of consideration. In certain cases *enorm lesion* is presumed. He has title to sue for delict and may be liable for harm caused. He may hold property, be made bankrupt, act as a trustee, make a will, and sue and be sued, the curator, if any, being called for his interest. The term minor is also being adopted in English law as a synonym for infant.

Minorities. Groups of individuals within a state, belonging to a race or having customs, language, religious beliefs, or other practices materially different from those of the majority of the individuals in that state, but not so objectionable that they should not be tolerated. The problem of protecting minorities in the enjoyment of their distinctive practices arises nationally and internationally. The practice of making treaty stipulations to ensure certain rights to minorities, particularly in matters of religion, has been found since the Reformation, and protective clauses were included in many treaties after the First World War, and placed under the guarantee of the League of Nations. After the Second World War treaties abandoned protection of minorities in favour of the principle of basic equality for all human beings in a country, but international means for enforcing equality is lacking.

Minute. A note or record of some transaction. Thus, minutes must be kept of proceedings and decisions at meetings of directors or shareholders of a company. In Chancery, minutes of the order of judgment are frequently drawn up by the parties settling in outline the order to be made which is then put into final form by the registrar. In Scottish procedure, a minute is an incidental intimating in the procedure, e.g. proposing an amendment.

Minute Book. In Scotland, a book kept in the office of the General Register of Sasines in which a brief entry is made of deeds received for registration; and also a book maintained in the Court of Session recording brief entries of judgments issued.

Mirror. A generic term for a kind of literature popular in late mediaeval and Renaissance times, works setting out to exhibit an idealized kind of person whom lesser mortals might look upon and model themselves on. The literature goes back to Plato's description of the philosopher-kings in the Republic, to Seneca's Moral Essays, and other classical models. An early mediaeval example is Augustine's *City of God.* It is a common kind of literature in relation to manuals for the guidance of young kings and princes, the most renowned being Machiavelli's *The Prince.*

In law, examples of this genre include the *Sachsenspiegel*, the *Schwabenspiegel*, the *Mirror of Justices*, Fortescue's *De Laudibus Legum Angliae* (qq.v.) and other similar books.

Mirror of Justices. A book in law French, dating from 1285–90, possibly composed by one Andrew Horn, a fishmonger and chamberlain of the City of London, and dealing in five books with sins against the Holy Peace, Actions, Exceptions, Judgment, and Abuses. It is a compound of fable, skit, criticism, and legal textbook. Coke, however, treated it as a learned treatise, and some of it may have passed into his *Institutes.* It was printed in 1642, and edited by Whittaker and Maitland for the Selden Society in 1893.

Mirror of Parliament, The. A record of parliamentary debates inaugurated by one John Henry Barrow, which ran from 1828 to 1841 and was then a serious rival to *Hansard* (q.v.).

Misadventure, death by (or homicide *per infortunium*). A death caused by another unintentionally and in the doing of something lawful. It subjects to no criminal liability.

Misappropriation. The crime of a person dealing dishonestly with property entrusted to him for safe custody or other specified purposes.

Miscarriage. Medically an abortion. A miscarriage of justice is a term of indefinite meaning for a trial of proceedings which appears to have worked injustice.

Mischief. Sometimes used with reference to the defect the object of which a statute was intended to remedy, it being said that the judicial function is to interpret the statute so as to suppress the mischief and advance the remedy.

Misconduct. A very general term for conduct falling short of what it should be. In matrimonial law, it is a euphemism for adultery.

Misdemeanours. At common law in England, a crime which was neither a treason nor a felony (qq.v.) but a lesser offence. Some crimes which were technically misdemeanours were serious, e.g. conspiracy, riot, assault, but many were trivial offences. In 1967, all distinctions between felony and misdemeanour were abolished, the rules applicable to misdemeanour being made applicable to both categories. The distinction has been replaced by that between arrestable and non-arrestable offences, the former being those for which the sentence is fixed by law or for which a person may be sentenced to five years' imprisonment.

Misdirection. An error in law by a judge in charging or directing a jury, frequently alleged as a ground for an appellate court quashing a conviction.

Mise. An Anglo-French legal term for any outlay of money, but particularly the payment of taxation. Thus, in John's reign, the mise rolls (*rotuli misae*) record payments from the Exchequer to departments of the royal household. In the *Confirmatio Cartarum* of Edward I of 1297, it is used as meaning taxation. It also meant the settlement of a dispute by arbitration. Thus the Mise of Lewes was an agreement between the parties after Henry III's defeat at Lewes in 1264.

Misfeasance. The improper performance of something which a person should have done properly. It is used particularly of improper conduct in the management of a company. See also MALFEASANCE.

Mishna. The oldest authoritative post-Biblical collection of Jewish oral laws, compiled by various scholars over some two centuries and presenting different interpretations of traditions preserved from the time of Ezra (*c.* 450 B.C.). It reached its final form in the third century A.D. Study by later scholars gave rise to an extensive apparatus of notes and commentary, the Gemara. The Mishna and Gemara were then combined into the Talmud. As scholars in Palestine and Babylon produced different annotations the versions of the Talmud differ in respect of Gemara.

The Mishna comprises six major sections or orders, including in all 63 tractates, each divided into chapters. The six orders deal respectively with religious laws concerning agriculture, with ceremonies and religious festivals, with religious laws concerning marriage and divorce, with civil and criminal duties and penalties, with laws as to temple sacrifices and offerings, and with ritual purity and purification.

Mishna Torah. A twelfth-century commentary on the Talmud in Arabic composed by Moses Maimonides (q.v.).

Misprision. A neglect or oversight. The word was formerly used of every substantial misdemeanour which did not have a certain name assigned to it by law. It was sometimes also used to denote contempts or high misdemeanours, such as undutiful behaviour towards the sovereign. The word is now almost confined to the phrases misprision of felony and misprision of treason (q.v.).

Misprision of felony. The crime of knowing that another person had committed a felony but concealing that knowledge or procuring the concealment thereof. At common law, it was a misdemeanour. In 1967, it was replaced by 'compounding an arrestable offence'.

Misprision of treason. The crime of knowing that another person has committed high treason and not within a reasonable time giving information thereof. At common law, it was punishable by life imprisonment and forfeiture of property. Statute of 1534 made concealment of treason a misprision and not, as at common law, amounting to aiding and abetting the principal offence.

Misrepresentation. An incorrect statement of fact, or of mixed fact and law, made by one party to another with the object, and having the result, of inducing the other to enter into a contract or similar relationship with the representor. It may be made by statement or other actings, or by concealment but not by mere omission, silence or inaction save where such would distort the natural inference from other facts or where, exceptionally, there is a positive duty to disclose all relevant facts.

A representation of fact must be distinguished from mere words of advertisement, from a promise, from a statement of intention, belief or of opinion, and from a statement intended to have contractual force and to be a condition of a warranty of the contract.

A misrepresentation, having the result of inducing the representee to enter into a contract with the representor, renders the contract void, or at least voidable by the representee, so that he is not bound thereby and may recover any money or property transferred in pursuance of the contract. He may further recover damages if the representation were made fraudulently or, probably, negligently, but not if made innocently, though in this case the representee may recover an indemnity against any consequence of the obligation set aside.

Missi dominici. Officers sent by Frankish kings and emperors to supervise provincial administration. They were used occasionally by Merovingian and early Carolingian kings and extensively by Charlemagne. From about 802, most of his empire was from time to time divided into areas or circuits for inspection which were visited by at least two *missi*, one a bishop or abbot, one a layman, usually a count. They had extensive investigatory powers and a mandate to rectify all injustices and errors of government. They reported on local conditions and problems and the state of the district, promulgated imperial decrees in the locality and, on the accession of a new emperor, received the oath of allegiance from freemen. They had great prestige and enjoyed great respect, having a wergeld equal to that of the emperor's own family. The institution fell into abeyance when disputes rent the empire after Charlemagne's death. It has some similarities to the English institution of the justices in eyre (q.v.).

Mississippi* v. *Johnson (1867), 4 Wallace 475. In this case, an attempt was made to have the Supreme Court grant an injunction forbidding the President to enforce the Reconstruction Acts, passed for the government of the defeated South after the Civil War, as being unconstitutional. The Court held that it had no power to grant such a remedy against the President and, in doing so, it made an important exposition of the relations of the powers of government; the Court could restrain neither President nor Congress, though it could scrutinize the acts of both once they were performed.

Missives. Letters; in Scots law documents in the form of letters exchanged between parties, particularly for the sale or lease of land or buildings.

Missouri Plan. A scheme, originating in the State of Missouri and later adopted elsewhere, to limit political considerations in the choice of judges. A nominating commission considers candidates for judicial office and submits a short list of those considered qualified and suitable to the governor or other person making the appointment. The appointing authority chooses a person who takes office and, after serving a probationary period, he stands for election for a longer term on the strength of his record, and not against other candidates, the question being solely whether he should be continued in office or rejected. If rejected the nominating commission puts forward a further suggested list.

Mistake. In English law, an erroneous belief as to some matter relevant to legal liability. It is akin to ignorance. Mistake may exist as to some matter of fact, or as to private rights, or as to some matter of general law. In general, no relief can be claimed in case of mistake of general law. In criminal law mistake of law is generally no defence to a criminal charge. A mistake of fact is, however, generally a defence if the mistake was reasonable and led to a belief in the existence of facts which would constitute a defence if they existed.

In civil law, mistake is a basis of relief at common law in an action of deceit where fraudulent misrepresentation has induced mistake, in an action for money had and received, in order to recover money paid under a mistake of fact, and as a defence to an action on contract where the mistake has been so fundamental as to preclude agreement ever truly having been reached. Equity gives relief in a wider range of cases of mistake. Mistake is also relevant where it has arisen in the expression of an intention. If there has been mistake in the expression of a common intention the court may grant relief by rectification or rescission of the deed or instrument.

Mistress. The style of the wife of an esquire or gentleman, but also used of a kept woman.

Mistress of the Robes. A lady of high rank in the peerage who is nominally the Queen's chief personal attendant, but who, in practice, attends on great state occasions only. At the beginning of Queen Victoria's reign, the principle was accepted that the Mistress of the Robes was a political appointment which changed with changes in the ministry.

Mistrial. A trial which is vitiated by a fundamental defect, such as lack of jurisdiction, and which accordingly is legally ineffective.

Mitford, John, Lord Redesdale (1748–1830). Was called to the English Bar in 1777, and became Solicitor-General in 1793, Attorney-General in 1799 and Speaker of the House of Commons in 1801; he became Lord Chancellor of Ireland and Lord Redesdale in 1802, but resigned in 1806.

Mitigation. Evidence led or statement made to lessen or minimize the effect of other evidence or facts. Thus, in an action for defamation, an apology may be pleaded in mitigation of damages. In a criminal trial, a plea in mitigation is a statement made after conviction by or on behalf of the accused suggesting why the penalty should be moderated.

Mitteis, Ludwig (1859–1921). German jurist and papyrologist, whose major work was in ascertaining the relations of Roman and local customary law, in the Hellenistic provinces. His major writing was *Reichsrecht und Volksrecht in den östlichen Provinzen des römischen Kaisserrechts* (1891) and *Grundzüge und Chrestomathie der Papyruskunde* (1912).

Mitter, Sir Binod Chandra (1872–1930). Was standing counsel to the Government of India, 1910–16, a member of the Council of the Governor of Bengal, 1910–17, and a member of the Judicial Committee of the Privy Council, 1929–30.

Mittermaier, Carl Joseph Anton (1787–1867). German jurist, professor at Landshut, Bonn and Heidelberg, author of numerous works notably on German private law, criminal law and procedure. He travelled extensively and observed procedural methods in the U.K. and U.S. and edited leading periodicals, and was also president of the first German Parliament in 1848.

Mittimus. A writ formerly used in sending a record or its tenor from one court to another. Also a written command to the keeper of a prison to receive and keep safe, until released in due course of law, a person charged with any crime.

Mixed action. An action at common law partaking of the natures of both real and personal actions (qq.v.), both demanding property and seeking damages for wrong.

Mixed questions of fact and law. Questions which involve both the determination of matters of fact and the decision of the effect in law of those factors. Thus, whether a person is guilty of manslaughter depends on what he did and in what circumstances (fact), and whether these actings fell within the category of manslaughter as defined by law (law). Many questions put to juries and answered by their verdicts are accordingly mixed questions of fact and law.

Mobbing. In Scots law, the crime of participating in a group of persons acting together for a common illegal purpose which they effect or attempt to effect by violence, intimidation or a demonstration of force, and in breach of the peace and to the alarm of the lieges.

Mobilia sequuntur personam. The principle in international private law that ownership, transmission, and inheritance of moveable property are determined by the personal law (law of the domicile) of the person having right thereto.

Model clauses. Clauses appropriate for inclusion in private bills. A committee was appointed in 1948, consisting of counsel to the Lord Chairman in the House of Lords and including representatives of government departments and parliamentary agents and presided over by Counsel to the Speaker, to revise and redraft model clauses and a volume of these, revised each session, has been issued.

Model Parliament. The parliament summoned by Edward I in 1295 and so-called because it was the first broadly representative assembly and provided the model for future Parliaments. In addition to the bishops, abbots, earls, and barons, it included two knights elected from each shire and two burgesses elected from each city and borough; this was not the first time that commoners had been summoned (see PARLIAMENT) but only from the Model Parliament onwards were they summoned in the normal course.

Moderator. A person who presides at an assembly. In the Church of Scotland, the persons elected to act as chairmen of the General Assembly, of synods and of presbyteries are styled Moderators.

Modern law. In legal history the modern period can be said to date from the watersheds of the Renaissance and the Reformation. It is impossible to give a chronological account of legal development in this time but it is possible to point to a number of important general features. The main lines of development of, and features of each distinct legal system are mentioned under the heads thereof.

In the first place, nation-states had already been established in some European countries (notably England, France, and Spain) by the opening of the modern period and this trend continued, so that by the later twentieth century, the whole of the habitable lands in the world were occupied by distinct nation-states. As this movement has progressed, the principle of the territoriality of law has almost completely superseded the principle of personality of law; a man, that is, is normally subject to the law of X if he lives there, not to the law of Y because he is by birth or nationality or domicile a man of Y.

From the sixteenth century, the major European nation-states sent out explorers all over the world, followed, in many cases, by settlers who established settlements in areas hitherto unoccupied or occupied only by a native population at a lesser stage of material development. Thus, French and English settlers colonized the eastern seaboard of North America, Spanish settlers parts of North and all of Central and South America, Dutch the southern tip of Africa, French and English parts of India and, later, parts of eastern, western and northern Africa. These settlers and traders took their own legal ideas with them and established legal systems on the lines familiar to them where there were none, and where there were, overlaid and influenced the existing systems with institutions and ideas from their home countries. As colonized areas passed from one European power to another by reason of war and treaty, as in Canada and South Africa, one European system sometimes superseded and sometimes overlaid another.

As these colonized areas developed they, in some cases, broke away from the colonizing country, e.g. the U.S.A., and, in other cases, developed increasingly independently until they attained independent status by agreement, e.g. Canada and Australia. Political independence resulted in legal independence, but frequently in the form of independent development of principles growing from common roots. Thus, the legal systems of the U.S.A. and the major countries of the Commonwealth all stem from English common law, and though, today, in many respects very different, they all have family resemblances and affinities.

Influence of the Renaissance

The renaissance was a great intellectual and cultural transformation based on the rediscovery of classical antiquity and revival of the study of everything connected with it, which began to be first noticeable in Italy in the fourteenth century and developed in the fifteenth century, and which affected art, literature, and scholarship. Legal science was later in feeling the influence of humanism. In law, the major manifestations were the application of the methods and standards of humanism, looking at texts critically and historically, to the study of the Roman law, and the expansion of these studies from Bologna and Pavia to the rest of Europe. Thus, Budaeus produced his *Annotationes in Pandectas* (1508), showing contempt for the mediaeval interpreters of Roman law, and he, Alciatus, and Zasius were the founders of modern legal science. The critique of Roman law started at the instance of Politian, and Valla exposed the False Decretals to ridicule.

From 1500, there developed a great outburst of legal scholarship in Western countries, scholars producing texts of Roman law (Torelli, Cujas, and others), legal historians (Gravina, Vico, Conringius, and others), and practical jurists (Clarus, Farinacius, Hotomanus, the two Gothfredi, Thomasius, and many others).

The Reception

The revival of interest in, and study of Roman law from the eleventh century foreshadowed the Reception of Roman law, particularly in the Holy Roman Empire, from about 1400 to about 1700. In its earlier stages it was a theoretical and academic reception, but from 1500 it was a practical reception. Save where abrogated, the Roman law was the common law of the Holy Roman Empire and accepted as applicable common law. Princes and their legal advisers found in it legal justification for building powerful states. A significant factor was the establishment of the Imperial Chamber of Justice (the *Reichskammergericht*) in 1495, with the rule that its members had to judge according to the common laws of the Empire, which included the Roman law. It had a unifying influence and tended to produce uniformity through the Empire. By the Reception, Roman law became a major element in modern law and in civilization.

This particular factor had lesser influence in England, partly because the English monarchy never regarded itself as a successor to the Roman emperors and particularly because, by 1400, an indigenous system of common law, with a strong legal profession, had already developed; it was building a system of its own. Roman law was not without influence, particularly in ecclesiastical and admiralty courts, but it was an influence only and there was no reception. There was rather more Roman influence in Scotland, though no reception in nearly as full measure as in Europe.

Influence of the Reformation

The split in the Church in the sixteenth century, known as the Reformation, had profound legal consequences. The authority of the Pope, ecclesiastical courts and the canon law were repudiated over roughly half of Europe, and their jurisdictions transferred to other authorities, though in many cases, the principles of the canon law continued to be applied, or at least to be influential, in some of the spheres in which it had formerly been authoritative. In those countries too, the divided allegiance to Pope and King was converted into one to national Church and King and, in some cases, as in England, the King was head of the national Church also. Protestant ecclesiastical and matrimonial law now developed in Protestant countries.

The Reformation gave an impetus to social and political thinking which, in England, contributed to the resistance to the King, the civil war which led to the Revolution of 1689, and, ultimately, to the establishment of the supremacy of Parliament. It led to the work of Hobbes and Locke. In France, the religious wars gave rise to the writings of Brutus, Hotman, and others.

The study of national laws

Down to about the eighteenth century, the main academic study of law was the study of Roman and canon law. National law was studied and learned in England in the Inns of Court and elsewhere by apprenticeship, and by practice and experience. But the study of national laws developed.

Swedish law was taught at Uppsala from 1620 and French law in Paris from 1679. German law was taught at Wittenberg only from 1707, Scots law at Glasgow and Edinburgh from about 1715, and English law at Oxford from 1753 and at Cambridge from 1800.

Influence of the Enlightenment and natural law

The eighteenth century was the time of the Enlightenment, the Age of Reason, which was marked by a challenge to accepted ideas in religion and every other sphere, led by Diderot, Voltaire, and others. Man, and everything else, should be studied fresh with regard to observation and experiment.

In social science, Montesquieu tried to analyse the laws operating in social life and to synthesize law, politics, economics, morals, religion, and their inter-relations. He summed up the tradition of power checking power and authority limiting authority. Rousseau founded government on a social contract.

The concept of natural law had long been intermixed with Christian theology and morals. After the Reformation, it was transformed and gave a new philosophic basis to law. The leading thinkers in this development were Vico, Gentilis, and Grotius. They were followed by many more, such as Pufendorf, Thomasius, Wolff, and others. In more practical terms, this body of thought gave rise to the Declaration of the Rights of Man, and the Napoleonic Codes. From natural law developed rationalism, and the legal thought of Kant. As a reaction to natural law there arose the historical school of jurists led by Savigny, urging that law was a manifestation of the attitudes, beliefs, and interests of the people.

Utilitarianism developed later and became the prevailing attitude of the moralist and of the legislator, though in Germany, the natural law school of thought continued to flourish. In Britain, Bentham and his followers had immense influence on legal thinking and practical reform, on the rational reconstruction of courts, procedure and evidence, and on substantive law.

The Industrial Revolution

From 1760 in England, and later in Europe, there began the development of heavy industry and the movement of population from the countryside into towns, which gave rise, in due course, to trade unionism, industrial legislation, and eventually to workmen's compensation and social security. It created railways, with new legal concepts and problems.

Utilitarianism

In England, natural law thinking did not make great headway and more influence was had by the utilitarianism of Bentham, Mill, and others, which inspired the English philosophical radicals of the early nineteenth century and resulted in much practical law reform, including mitigation of the worst excesses of the criminal law and punishment.

Codification

Codification in modern law has sought to replace a mass of older, frequently inconsistent provisions of law by coherent statements of principle covering large tracts of law, even the whole of civil law or of criminal law. The ideal can never be completely realized, as a legal system must evolve and develop. Codification has been heavily influenced by Roman law and all the major codes have been Romanistic in concept and structure.

Notable modern codes include the Carolina (1532), the Ordinances carried through by Louis XIV and Colbert in France (1667–70), and the Prussian code (*Allgemeines Landrecht*), undertaken in 1749 but not completed till 1794. The Bavarian Code (*Codex Maximilianeus Bavaricus civilis*) was published in 1756. The major codifications were, however, those of Napoleon, the Civil Code (1804), Civil Procedure Code (1807), Commercial Code (1808), Code of Criminal Procedure (1811), and Penal Code (1811).

These codes were imposed on countries conquered by the French (Belgium, Luxembourg, Italy, Holland, and some German states) and were voluntarily adopted in other countries, but after the fall of Napoleon, some countries retained them, while others repudiated them. The codes also spread, and were taken as models in other countries, notably in most of the states of Central and South America.

The Austrian code of 1811, initiated by Maria Theresa in about 1750, was adopted in parts of central and eastern Europe.

Though codification made little headway in Britain, codes were enacted for British India in 1837–82 and, in the U.S.A., various codes were projected and, in some states, adopted—the most widespread being the New York Code of Civil Procedure.

At the end of the nineteenth century, the German Civil Code provided another major specimen, which also has subsequently been extensively copied, and which influenced the Swiss Civil Code of 1907. The unification of Italy prompted a Civil Code there in 1868, while the German Commercial Code of 1861 was a notable landmark in that sphere of law.

Bulk and complexity

In all developed countries there has, since 1800, been enormous growth in the bulk, detail, and complexity of the legal regulation of society. This arises from increasing concern with the welfare of society, the protection of underprivileged, increased regulation of political and economic power to check monopolies.

This has been achieved largely by an enormous growth of statute law, creating whole new tracts of legal science.

In common law countries, where legal development was traditionally effected by judicial initiative, restating principles, extending them to new cases and applying them to phenomena which have recently developed, this development has continued and been marked, in many cases, by bold empiricism. Yet even so, statute everywhere has added more chapters to the law and rewritten more passages than has judicial work.

Law reform

Conscious law reform has been provoked not only by the increasing bulk and complexity of the law, but by realization that many of its rules had become outmoded and unsuitable. In France, the Revolution swept away much old law and created a new structure of society and a new system of law. In England and Scotland in the nineteenth century, there was a steady flood of commissions and reports culminating in the reconstruction of the system of courts and the merger of law and equity in 1875. Statute law revision and consolidation were also undertaken. In 1965, the two Law Commissions were established as permanent agencies for revision and reform of the law.

International law

The steady increase in international intercourse has brought problems of choice of jurisdiction and of the legal system suitable to be invoked. Thus, conflict of laws or international private law, though foreshadowed by Bartolus, Baldus, and others, has grown up almost entirely since 1800, or even since 1850. For a century it was developed by jurists and courts but, in the twentieth century, has increasingly been developed by international conventions.

Public international law became important as the nation-states developed. It has been developed and substantially influenced by text-writings by such as Ayala, Gentilis, Suarez, Grotius, and others, by custom, judicial decisions, and analogies with national private law and, latterly, by bilateral and multilateral treaty-making, dealing with a wide variety of subjects. The first of these is sometimes said to be the Declaration of Paris of 1856.

In the twentieth century, international institutions have become of increased importance and the League of Nations, the U.N.O., the Permanent Court of International Justice, and the International Court of Justice have created much new law. There has been a tendency to codify rules of international law.

From the mid-nineteenth century there was proliferation of international institutes and organizations concerned generally with reducing conflicts between national legal systems, such as the Institute of International Law (1873).

Similarly, there were many moves from that time onwards by multilateral or international conventions to settle uniform regulation of various problems or to limit the inconvenience of disparate national rules. Uniform legislation was sought on railway freights, negotiable instruments, and the maritime law of general average, the last resulting in the York (later York–Antwerp) Rules, the International Law Association pressed for agreements on maritime law and numerous international administrative agencies were established, such as the International Telegraphic Union, the Universal Postal Union, and the Metric Union, while international conferences increased knowledge of and borrowing from other countries. The Hague Conferences of 1893 to 1904 sought to achieve some uniformity in matters of conflict of laws.

Supranational law

More recently, the movement towards closer association of various European countries has brought into existence a body of supranational law, that of the European Community, with its own legislature and judiciary. Independently thereof, the Council of Europe has developed and created the European Commission on Human Rights with its own Court.

New concepts and rules

In the modern period, a host of new concepts and rules of law have been developed in response to the problems created by industrialization, large-scale capitalism, democracy, and international trade. Thus, there have developed large bodies of law on patents and trade-marks, monopoly and competition, social security, town and country planning, control of rents, consumer protection, aerial navigation, broadcasting, and a hundred other branches of law unknown to the jurist of A.D. 1500. Corporate bodies, companies, public corporations, local authorities, and others have steadily increased their importance both as subjects of law and as participants in every kind of legal activity.

Growing importance of public law

There has been a steady growth in the importance of public law, in multiplication and proliferation of ministries, public corporations, local authorities, boards, administrative agencies, officials and the like, bringing into the forefront problems of their status and powers, the question of checking and controlling them, and of how and how far they are amenable to the rule of law. Such authorities, even in western democracies, carry on almost every kind of social, industrial, and commercial activity and regulate practically every kind of individual action.

Legal scholarship

The foundations of modern legal science were laid by the humanist-jurists Alciatus, Budaeus, and Zasius, but there followed, in their footsteps in every developed country, a host of scholars. Initially, and for long, they devoted most attention to the Roman law (Cujas, Doneau) but increasingly became emancipated therefrom and concerned themselves with national law, striving to reduce the rules to a series of ordered principles and to state them lucidly, and in order. Thus, in France, Domat and Pothier prepared the way for codification, in England Blackstone, in Scotland Stair and Erskine, and in the U.S. Kent stated their national systems as orderly bodies of knowledge.

In the countries of the Reception, work of immense importance in developing the received Roman law, adapting it to modern conditions, systematizing it and preparing the way for the German Civil Code was done by such academic jurists as Gluck, Thibaut, Savigny, Puchta, Vangerow, Jhering, Windscheid, Brinz, Dernburg, and Kipp.

While in the civil law countries the tradition has continued of major treatises on the civil law as a whole, e.g. Colin and Capitant, *Traité de droit civil français*, in common law countries the tendency has been to treat of branches of the law only. Many of these books have had great formative influence on the law. The works of Story, Anson, Pollock, and Salmond have been very influential.

The development of international law has been both materially assisted by and produced much juristic thought, analysis, and speculation. Among the most outstanding names are Suarez, Grotius, Puffendorf, Vattel, Bynkershoek, and among the moderns, Wheaton, Oppenheim, Calvo, von Martens, and others.

Thinking about law as contrasted with examination and exposition of rules of law had, from early times, been in the hands of philosophers and moral and political thinkers. In the nineteenth century, it began to fall more into the hands of lawyer-thinkers. The critical scrutiny of Bentham was followed by Austin, whose analytical approach was followed by a long series of English jurists (Markby, Holland, Salmond). Natural law thinking, previously dominant, continued (Ahrens, T. H. Green, Lorimer) and the historical school flourished for a time (Savigny, Maine, Vinogradoff) while, more recently, utilitarianism and a sociological standpoint have underlain the thinking of Ihering, Duguit, Pound, and Stone.

Socialism and Marxism have also produced jurists looking at law from these standpoints.

See also under individual legal systems.

R. David and J. Brierly, *Major Legal Systems in the World Today*; A. V. Dicey, *Law and Opinion in Britain*; *Progress of Continental Law in 19th century*.

Modestinus, Herennius (3rd Cent. A.D.). Roman jurist, a pupil of Ulpian, and the last of the classical jurists. A prolific writer on law, he was the author of *Differentiae* (nine books), *Pandectae* (12 books), and *Regulae* (12 books). *Responsa* (19 books) notes on such subjects as *de inofficioso testamento*, *de legatis et fideicommissis*, *de manumissionibus*, *de praescriptionibus*, and other topics.

Modification. A decree of the Teind Court in Scotland awarding an appropriate stipend to the minister of a parish.

Modo et forma (in manner and form). A form of words used in older pleadings by way of traverse to deny the fact alleged in other party's pleadings in the manner and form in which it was alleged and, accordingly, to throw on the other party the onus of proving the fact, not only in general, but in the manner and form alleged. The only traverses which were not pleaded *modo et forma* were the pleas of *non est factum* and *de injuria*.

Modus et conventio vincunt legem (custom and agreement overcome law). A maxim expressing the general principle that parties may, generally, by their practice or by express agreement, overrule a legal provision. Increasingly, however, it is now provided that parties may not exclude or modify provisions made by law, such as provisions implying terms or imposing standards for the general public benefit.

Modus operandi. In criminology, the characteristic manner of working which is associated with a particular criminal. It has been observed that professional criminals generally use a particular pattern and method of committing their habitual kind of crime, so that the occurrence of a crime committed in a particular manner points to (but does not prove) its having been committed by one who habitually uses this method.

Modus tenendi Parliamentum. A tract, probably dating from the early fourteenth century, describing the constitution and powers of Parliament and the ancestor of works on parliamentary practice. It shows that the constitution and powers were, even then, becoming fixed. It was translated in the seventeenth century and edited by T. D. Hardy in 1846.

***Mogul Steamship Co.* v. *McGregor, Gow & Co.*,** [1892] A.C. 25. Certain shipowners, to maximize their share of a particular trade, formed an association to regulate the division of cargoes between them, rates of freight and other conditions, and offered rebates to shippers who shipped only with them. The plaintiffs, having lost much business, alleged conspiracy on the defendants' part to injure them. It was held that the defendants' acts were not unlawful, nor had they employed any unlawful means, and that the action failed.

Mohammedan law: See ISLAMIC LAW.

Molestation. In the context of industrial disputes, troubling and annoying a person, particularly by following him in a persistent or disorderly manner, hiding his property, or besetting his place of work or residence. In the context of separation of husband and wife it is intentionally annoying the other spouse. In Scots law, it is the troubling of a person in the possession of his lands.

Molina, Luis de (1535–1600). A Spanish Jesuit, who taught at Coimbra and Evora and wrote a famous *De Justitia et Jure* (1599), numerous theological works and a commentary on Aquinas which provoked reaction from other thinkers. His theological system is termed Molinism.

Molinaeus. See DUMOULIN, CHARLES.

Molina y Morales, Luis (sixteenth century). Spanish jurist, councillor of the Indies and later of Castile, and author of *Pro successione regni portugalliae allegatio* and *De hispaniarum primogeniorum origine ac natura*.

Molliter manus imposuit. At common law, a plea in defence to an action for battery (q.v.) to the effect that the conduct alleged as battery was lawful and that the defendant laid hands on the plaintiff gently and used no more force than was necessary, as in the case of a constable restraining a disorderly person.

Molloy, Charles (1646–90). Was a member of the Bar, but made no reputation in practice. He was the author of *De Jure Maritimo et Navali* which deals with international, maritime and commercial law, with naval and military discipline, and with the prize jurisdiction, and contains information on many related matters, particularly those relevant to maritime and commercial business. Published in 1676, it went through many editions and was a standard authority till the nineteenth century, though containing little not also in the works of Malynes and Marius.

Molony, Sir Thomas Francis (1865–1949). Was called to the Irish Bar in 1887 and to the English Bar in 1900. He became Solicitor-General for Ireland in 1912, Attorney-General in 1913, and, almost at once, a judge. In 1915, he went to the Irish Court of Appeal and, in 1918, he became Lord Chief Justice of Ireland, the last holder of the office. He resigned in 1924 when the Irish Free State judicature was established. His years as a judge were difficult politically, but he sought to maintain justice and law notwithstanding.

Mommsen, Theodor (1817–1903). Distinguished German historian and jurist. As a student, he was much influenced by the writings of Savigny. He was Professor of Civil Law at Leipzig and later at Zurich and wrote a classic *Roman History* (1854–56) which won him a Nobel prize for literature in 1902. At Berlin, where he moved in 1858, he was made editor of the *Corpus Inscriptionum Latinarum*, but also wrote a history of the Roman coinage, a work on the Roman provinces, *Römisches Staatsrecht* (1871–88) an analysis of Roman constitutional law, and *Römisches Strafrecht* (1899) on Roman criminal law. He also edited Justinian's *Digest* in the modern standard Berlin edition, edited other works, and sat in the Prussian Parliament, 1873–79, and in the German Reichstag, 1881–84.

Monahan, James Henry (1803–78). Was called to the Irish Bar in 1828, became Solicitor-General in 1846, and Attorney-General in 1847. He was Chief Justice of the Common Pleas in Ireland from 1850 to 1876.

Monarcha juris. A title sometimes given, in mediaeval times, to Bartholomew of Saliceto (q.v.).

Monarchy. The structure of government in which the office of Head of State is held by a single person for life, under the title of King or Queen. Originally, probably in every monarchy, the king not only reigned but ruled and governed personally, but in the United Kingdom the significant growth in the power of Parliament, particularly of the elected House of Commons since about 1500, and the development of the Cabinet system of government since about 1700, have given rise to the modern British system of constitutional or limited monarchy, under which the king or queen reigns but all significant acts of government are done on the advice of ministers who are responsible to Parliament, or by such ministers in the name of the monarch. The monarch does not participate in political controversies and is neutral, co-operating with whatever group of leaders have the confidence, for the time being, of a majority of the House of Commons and of the country. Generally, similar conditions apply in the other monarchies of western Europe.

In modern Britain, the monarchy is valuable as the major, if not the only factor unifying kingdoms and regions which sometimes show diverging tendencies. The strength of the monarchy in modern Britain rests largely on its reputation, developed particularly since the middle years of Queen Victoria's reign (about 1870), for honourable character and dedication to duty and the good of the nation, and on the enormous strength of personal loyalty, devotion and affection generally felt for the Queen and the Royal Family. This is regularly and spontaneously expressed on such occasions as royal jubilees, weddings, and funerals.

The system has great advantages in avoiding the regular disturbances necessitated by presidential election, and competitions for the office of Head of State, in maintaining the neutrality of the office, and in providing a focus for loyalty, admiration and respect.

H. Nicolson, *Monarchy*; C. Petrie, *The Modern British Monarchy*; E. Barker, *British constitutional monarchy*.

Monasticon Anglicanum. The major work of Dugdale (q.v.) containing historical accounts of the abbeys, monasteries, friaries, hospitals, cathedrals, and collegiate churches in England and Wales, together with Scottish, Irish, and French houses which were their dependencies. Originally published in 1655–73, a modern edition appeared in 1817–30.

Monboddo, Lord. See BURNETT, JAMES.

Moncreiff, Sir James (1811–95). Grandson of the Rev. Sir Henry Moncreiff, later Moncreiff Wellwood, the outstanding Scottish clergyman of his time, and son of Sir James Wellwood Moncreiff, Bt. of Tullibole, a judge of the Court of Session

(1829–51) as Lord Moncreiff. His elder brother was the Rev. Sir Henry Wellwood Moncreiff, who became a leader of the Free Church of Scotland. He was called to the Scottish Bar in 1833, became Solicitor-General for Scotland in 1850 and was Lord Advocate in 1851–52, 1852–58, 1859–66, and 1868–69. He was also Dean of the Faculty of Advocates, 1858–69. As Lord Advocate he conducted the prosecution of Madeleine Smith, and in Parliament promoted much beneficial Scottish legislation. In 1869, he became Lord Justice-Clerk and proved himself a competent judge. In 1871, he became a baronet and, in 1874, a peer, as Baron Moncreiff of Tullibole. He resigned in 1888. His eldest son, Henry James Moncreiff (1840–1909) second baron, became a judge, as Lord Wellwood (1888–1905), and his son-in-law J. B. Balfour (q.v.) became Lord President Kinross.

Moncreiff, Alexander (1870–1949). Called to the Scottish Bar in 1894, he took silk in 1912, and became a Judge of the Court of Session in 1926. He was promoted Lord Justice-Clerk in 1947 to hold the place until the Lord Advocate could be released to succeed him, as he did later in the same year.

Money. Those tokens of value which pass from hand to hand in discharge of obligations to pay, particularly coin of the realm and banknotes (or at least those banknotes which are legal tender). Cheques and bills of exchange are not money but frequently treated as equivalent thereto. In particular contexts, such as wills, the word money may have an extended meaning, including an account at bank, investments and even, occasionally, all assets, even including real property.

Money Bill. In the U.K. Parliament, a Public Bill the main object of which is to impose or alter taxation or authorize expenditure. By the Parliament Act, 1911, a Money Bill is defined as one which in the opinion of the Speaker contains only provisions dealing with taxation, the imposition of charges on the Consolidated Fund, and related matters. Money Bills must be endorsed as such by the Speaker before being sent to the House of Lords. In deciding whether or not to endorse a Bill as a Money Bill the Speaker, if practicable, consults two members to be appointed from the Chairman's Panel each session. The Lords may not initiate or amend a Money Bill and, under the Parliament Acts, 1911 and 1949, may delay it for only a month, but if it is not endorsed as a Money Bill it can be delayed for a year by the Lords withholding its consent. Even the annual Finance Bill is frequently not a Money Bill as defined by the 1911 Act because it contains provisions other than relative to taxation.

Money counts. In common law pleadings, the counts or grounds of action usually founded on simple contracts, giving rise to claims for payment of money.

Money had and received. A ground of action arising where one person received money which, in justice and equity, belonged to another, entitling the other to recover the money as money had and received to the use of the plaintiff. It arose, e.g. against agents, to make them pay their principals.

Moneylending. The lending of money at interest as a business has been controlled, first by the Moneylenders Acts, 1900–27, and now by the Consumer Credit Act, 1974, which requires persons carrying on such a business to be licensed by the Director-General of Fair Trading, imposes stringent controls on the terms and conditions which may attach to a transaction, and gives the courts wide power to re-open extortionate transactions and order repayment of excessive interest. Attempts to control excessive interest by legislation were made from the sixteenth century and the Church had, from earlier times, forbidden usury, and many devices were resorted to to avoid these prohibitions. All these controls were repealed by the Usury Laws Repeal Act, 1854.

Monism. In international law, the theory that neither international law nor any system of municipal law is superior to the other but, that all form parts of one legal order or system of norms binding states and individuals alike, their rules being interrelated. Some jurists, such as Kelsen, however, understand monism in the sense that international law has ultimate primacy and systems of municipal law are subordinate forms of law. See also DUALISM.

Monition. A summons or order. In admiralty practice, a monition was the process, similar to a writ of summons, whereby an action was commenced, and also an order of the court that something be done. In ecclesiastical procedure, it was an order warning the party complained against to do or not to do some act, to lodge answers to pay costs, to refrain from repeating an offence. In ecclesiastical appeals to the Privy Council, a monition calls for the officers of the court below to transmit the proceedings for the use of the Board.

Monkswell, Lord. See COLLIER, R. P.

Monnet, Jean (1888–1979). French businessman and economist, served as a deputy secretary of the League of Nations, 1919–23, and entered international banking, advising countries on economic development. In France after 1945, as head of the French planning office, he initiated the Monnet

Plan which aimed at the reconstruction of French industry on a modern basis, and developed the idea of the European Coal and Steel Community, was first chairman of its High Authority (1952–55), and played a leading part in the economic integration of Europe.

Monopolies. A monopoly, literally a sole seller, exists where only one, or a limited number, of sellers exist of particular goods, or services, and barriers, either legal or economic, exist to the entry of competitors into the market for the supply of the goods or services in question. The term was originally given to grants from the Crown, usually to a favourite, of the exclusive right to manufacture or sell particular kinds of goods. The practice became so common that Parliament raised it as a grievance and, in 1624, the Statute of Monopolies made all monopolies illegal, save those authorized by Parliament, or granted in respect of the working or making of new manufactures or inventions, an exception which gave rise to the modern patent (q.v.) system. Some monopolies, such as that enjoyed by a village store, exist by force of situation, circumstances, and economics, not by law. Others exist by law. In modern practice, Parliament has created numerous monopolies such as in the supply of gas, electricity, carriage by railway, postal, telephone and telegraph services, and many more, and these are accordingly supposed not to be against the public interest. In the professions, monopolies exist in that only enrolled members of various professions, e.g. barristers, solicitors, physicians, may act in certain contexts or take fees for certain services. There is however no legal limit on the number of persons who may qualify to provide these services and the restriction of providing various professional services to such persons is in the public interest, to ensure that those providing the services are competent, and to protect the public against incompetent persons or charlatans.

Monopolies and Mergers Commission. A body established as the Monopolies and Restrictive Practices Commission by the Monopolies and Restrictive Practice (Inquiry and Control) Act, 1948, to investigate and report on any trade in which the supply, processing, or export of any goods is subject to monopoly conditions as they are defined. It later became the Monopolies Commission and, in 1973, the Monopolies and Mergers Commission. Its functions are to investigate cases referred to it and report on the existence or possible existence of a monopoly situation in respect of the supply of goods or services, or in relation to exports, with respect to the transfer of a newspaper or newspaper assets or to a merger situation requiring investigation. Its composition and powers have subsequently been altered several times and extended to the supply of services, commercial and professional. It has produced numerous reports on particular cases referred to it and also general reports on monopoly practices in industry and commerce. If it finds on a reference that monopoly conditions prevail in the subject-matter under investigation certain Ministers may make orders for remedying or preventing any mischief which may result from the monopoly conditions, requiring the parties to determine the arrangements in whole or in part.

Monopolies and restrictive trade practices. It is against legal policy for any person or body to have the sole exercise of any business or trade in the country, and the Crown cannot grant a monopoly without statutory authority. Various statutory monopolies exist, e.g. that of the National Coal Board in respect of coal-mining.

Investigations by the Monopolies (now Monopolies and Mergers) Commission (q.v.) after 1948, revealed the existence of monopolies in certain industries and the Restrictive Trade Practices Act, 1956, required particulars of most restrictive agreements affecting the production, supply or processing of goods (but not the provision of services) to be registered with the Registrar of Restrictive Trading Agreements (now Director General of Fair Trading) and then considered by the Restrictive Practices Court (q.v.), to determine whether they were or were not in the public interest. The statutory presumption is that a registered agreement is void as contrary to the public interest and the onus is on the parties propounding the agreement to establish that the agreement was not contrary to the public interest by reference to one or more of stated criteria. The law was revised but not substantially altered by the Fair Trading Act, 1973.

R. Wilberforce, A. Campbell, and N. Elles, *Restrictive Trade Practices and Monopolies.*

Monroe Doctrine. In 1823, it was rumoured that France and Spain would jointly send an expedition to South America and there were discussions on a joint British-U.S. warning to those countries to keep out of America entirely. President Monroe, in his annual message to Congress, reaffirmed Washington's policy of non-intervention by the U.S. in European affairs and enunciated the principle that the U.S. could countenance no intervention in the American continent by a foreign power. The declaration had little immediate result but by 1860 had become widely known abroad. The Monroe Doctrine was invoked to eject the French regime of Maximilian in Mexico in 1864–67 and to persuade Britain to accept arbitration in the Venezuela boundary dispute in 1899. Later, the Doctrine was interpreted to mean not only the exclusion of European powers from Latin America but U.S. power there, but under President F. D. Roosevelt it was generalized and liberalized to mean that its application would require consultation with and the co-operation of Latin American states.

D. Perkins: *Hand Off: A History of the Monroe Doctrine.*

Monstrans de droit (manifestation of right). At common law, a method of obtaining possession or restitution from the Crown of real or personal property, brought on the common law side of the Chancery or in the Exchequer. It was superseded by the petition of right which, in turn, has been replaced under the Crown Proceedings Act, 1947, by an ordinary action against a government department.

Montagu, William (?1619–1707). Was called to the Bar in 1641 and appointed Chief Baron of Exchequer in 1676, but was removed in 1686 for denying the royal dispensing power.

Montague, Sir James (1666–1723). Became Solicitor-General in 1707 and Attorney-General in 1708, but was removed in 1710. He became a baron of Exchequer in 1714, a commissioner of the Great Seal in 1718, and was Chief Baron of Exchequer, 1722–23.

Montemayor, Juan Francesco (1620–1685). Spanish jurist, judge in Mexico, and one of the Council of the Indies. He was author of a *Discorso del derecho y repartimento de presas* (1658) on sea warfare, and a summary of the law applicable to Mexico which was in use until Mexico attained independence.

Montesquieu, Charles Louis de Secondat, Baron de la Brede et de (1689–1755). French thinker, president of the *Parlement* of Bordeaux (1716–28). He wrote *Lettres Persanes*, criticizing the social, political, and other follies of the time in France, and travelled widely in Europe. In *Considerations sur les causes de la grandeur et de la decadence des Romains* (1734), he wrote an important early essay in the philosophy of history. His great work, however, is *L'Esprit des Lois* (1748) in 31 books, eight on law in general and forms of government, five on military matters and taxation, six on mariners and customs and their relation to climatic conditions, four on economic matters, three on religion and five on Roman, French and feudal law. It is based on wide reading and study, and contains many stimulating and critical ideas. It at once provoked enemies, but opposition gradually diminished and it became recognized as a classic, and it was very influential on later political thought. It may be regarded as the first work on comparative or sociological jurisprudence. It is also noteworthy as having enunciated, on the basis of his observations of English government, the doctrine of the separation of powers (q.v.), and having stressed the effect of climate on individuals and consequently on the intellectual outlook of society, though he accepted that this was only one factor, along with laws, religion and maxims of government.

R. Shackleton, *Montesquieu: A Critical Biography.*

Montfort, Simon de, Earl of Leicester (*c.* 1200-65). Served and opposed Henry III and, in 1258, headed the opposition to the King and became one of the 15 who, under the Provisions of Oxford, were to control the administration. In 1263, being convinced of the King's hostility to all reform, he raised a rebellion and defeated and captured the King at Lewes. He then established a government of three and summoned a parliament in 1265, remembered for having included representation of the towns. He was then, however, defeated at Evesham. He was masterful and ambitious but the attempt to base parliamentary support on wider representation contained the seeds of important later developments.

Montgomery, Sir James (1721–1803). Was called to the Scottish Bar in 1743, appointed joint Solicitor-General for Scotland in 1760, sole holder of the office in 1764, and Lord Advocate in 1766. As such, he was responsible for the Entail Act of 1770, commonly called by his name. He was a well known improving landlord. In 1775, he became Chief Baron of Exchequer in Scotland; he resigned in 1801, being made a baronet. He was an assiduous agriculturist and his main interests were his estates.

Month. A lunar month is the period of 28 days from the rising of one new moon to the rising of the next; a solar month is the time the sun takes to pass through one of the 12 signs of the zodiac, a little over 30 days; a calendar month is the period of 30 or 31 days (in February 28, or in a leap year, 29 days) between a day bearing a particular number and the day of the same number in the next month. At common law, a month is a lunar month unless there are indications that a calendar month was intended but in statutes since 1850 and deeds since 1925 a month means a calendar month. In commercial relations, a month is a calendar month.

Monumenta Germaniae Historica. An extensive and critically edited collection of source-materials relative to German history from A.D. 500 to A.D. 1500. It was undertaken in response to nationalist feeling and initiated by Karl Freiherr von Stein, the statesman who, in his retirement after 1816, aroused interest in German history. The success of the project was due largely to the energy and ability of Georg Heinrich Pertz, who was editor in charge from 1823. The project eventually ran to 120 volumes, comprising series containing *scriptores*, *leges*, *diplomata*, *epistolae*, and *antiquitates*. The series also had great importance as giving an impetus to work by historical scholars in other European countries and setting standards for other similar collections.

Moore, John Bassett (1860–1947). Was admitted to the Delaware Bar in 1883, served in the Department of State, 1885–91, and then became Professor of International Law and Diplomacy at Columbia, 1891–1924. There, he made the subject a matter of serious legal study. He was a great scholar and a recognized authority on all questions of international law. In 1898, he was temporarily re-employed in the Department of State and drafted the peace treaty with Spain, and was again employed in 1913–14. He published *A Report on Extraterritorial Crime* (1887), *A Report on Extradition to the International American Conference* (1891), and *A Treatise on Extradition and Interstate Rendition* (1891). His great *History and Digest of International Arbitrations*, commissioned by the State Department, was published in six volumes in 1898 and he also published *A Digest of International Law* (8 vols., 1906) and *International Adjudications, Ancient and Modern* (7 vols., 1929–36). He served as a judge of the Permanent Court of International Justice, 1921–28.

Moore, Sir William, Bart (1864–1944). Was called to the Irish Bar, 1887, and the English Bar, 1899, and became a judge of the King's Bench Division in Ireland, 1917–21, a Lord Justice of Appeal in the Supreme Court of Northern Ireland, 1921–25, and Lord Chief Justice of Northern Ireland, 1925–37.

Moore, William Underhill (1879–1949). Admitted to the New York Bar in 1902, he taught law at Kansas, 1908–09, Wisconsin, 1909–14, Chicago, 1914–16, and Columbia. 1916–45. His work was in commercial law and he edited *Cases on the Law of Bills and Notes* (1910) and annotated Norton's *Handbook of the Law of Bills and Notes* (1910). He was a man of wide erudition and interest and sought to apply strictly scientific methods to legal study.

Moots. A moot was an assembly in Anglo-Saxon England, and is found in such compounds as folkmoot, and in the word *witanagemot* for the national council. Moots were also mock hearings held in the mediaeval Inns of Court as part of the education of young men for practice at the Bar. Procedure at the moots was modelled on that of the Court of Common Bench or Common Pleas. The senior members of the Inn, the benchers, formed the bench and those who, in imitation of the serjeants, argued at the Bar were called Masters of the Utter or Outer Bar or Utter Barristers, because they sat uttermost on the forms which constituted the Bar in the library or hall of the Inn where the moots were held. The members of the Inn who sat inside on the forms and recited the pleadings were called inner barristers and later students. It was a condition of admission to audience in the courts that a man had performed satisfactorily in a number of moots and even after being called to the utter bar that he had for three years participated in ordinary mootings. The term is now used for similar moot arguments held in university law schools and the Inns of Court today as part of the training of students.

Moot case or moot point. A case or point which is doubtful or arguable, and suitable for argument in a moot.

Mora. In Scots law, delay, a plea which, by itself, has no effect, the plea properly being that the other party's mora, taciturnity and acquiescence bars him from now objecting to what he complains of.

Morality, Morals and moral law. Morality or morals overlap with manners, folkways, religion, law, customs, and public opinion. Generally, one may say that morals are those standards of conduct which are accepted in the society and stratum of society in which the person lives. Such accepted standards are studied from the standpoint of what should be accepted and allowed or disapproved by religion, moral theology, and ethics or moral philosophy, and from the standpoint of what is, in practice, accepted in particular societies by social anthropology and sociology. Law generally approves and reinforces what is generally accepted as good moral behaviour in the society in which it operates and disapproves and penalizes what is regarded as bad moral behaviour, such as sexual immorality, dishonesty and unfair dealing. But there is no exact correspondence between law and accepted morality.

The moral law so called is not a body or principle or rule of law, strictly so called, at all but a name for the prescriptions of conscience and the socially controlling force of approved conduct. It is sometimes identified with natural law (q.v.).

A. L. Goodhart, *English Law and the Moral Law* (1952).

Morden, Lord. See YORKE, CHARLES.

More, Sir Thomas (1477–1535). English lawyer, humanist, scholar and martyr. Son of a judge, More was called to the Bar in 1501. From 1518, he was in the royal service as diplomat, scholarly secretary, and controversialist, and assisted Henry VIII in his book against Luther. In 1523, he became Speaker of the House of Commons and, in 1529, Lord Chancellor, the first layman to hold the office. He resigned in 1532, declined to attend the coronation of Anne Boleyn as Queen, and was lodged in the Tower. In 1535, he was tried and beheaded for refusing to accept Henry's alteration of the royal succession in favour of his children by Anne. In 1935, he was canonized. He was a great scholar, an intimate of Erasmus and the learned men of Europe. His writings include *Utopia* (1516), various works attacking Luther, and numerous polemics, notably *A Dialogue Concerning Herecies* and *The Confutation of Tyndal's Answere* (1532–33).

R. Chambers, *Thomas More* (1935); E. Reynolds, *The Field is Won: The Life and Death of Sir Thomas More* (1968).

More or less. Words used in contracts and conveyances to cover minor inaccuracies in description or extent. Whether a particular discrepancy is too large to be covered by 'more or less' depends on the circumstances.

Moreau-Lislet, Louis Casimir Elisabeth (1767–1832). Practised law in Louisiana and, in 1806, published an *Explication des Lois Criminelles du Territoire d'Orleans.* He was then commissioned to prepare a code which was published in French and English in 1808, and, in 1820, published *The Laws of Las Siete Partidas Which are Still in Force in the State of Louisiana.* He was then selected with Derbigny and Livingston to prepare a revised code which appeared as the *Civil Code of the State of Louisiana* in 1825. He was a good legal scholar and did much to preserve French interests in Louisiana.

Moreno Quitana, Lucio Manuel (1898–). Argentinian jurist, judge of the International Court of Justice, and author of *Tratado de derecho international* (1963) and other works on international law.

Morganatic marriage. A legally valid marriage between a male member of royal, or princely, family, and a woman of lower birth or rank with the conditions that she does not acquire his rank and that any children are not to succeed to his rank, dignity or hereditary property. It is essentially a German institution, based on a German idea of equality of birth which, in the Middle Ages, had numerous applications in German law, requiring parties to many kinds of transactions to be of the same rank or standing. The name is from late Latin *matrimonium ad morganaticum*, meaning either marriage on the morning gift (the latter to be all that the bride could expect), or restricted marriage, or early morning marriage. The institution was adopted and utilized in some East European royal houses, but never in France or Britain. It was suggested, in 1936, that King Edward VIII might contract a morganatic marriage with Mrs Simpson, but it was decided that the institution was not known in British law and would have required legislative sanction.

Morris, Michael, Baron Morris and Killanin (1826–1901). Called to the Irish Bar in 1849, he took silk in 1863, and entered Parliament in 1865. In 1866, he became successively Solicitor-General and Attorney-General for Ireland and, in 1867, he became a judge of the Irish Common Pleas, Chief Justice in 1876, and Lord Chief Justice of Ireland in 1887. In 1889, he became a Lord of Appeal in Ordinary. When he retired in 1900 he received a barony as Lord Morris and Killanin. He is said to have made no claim to legal erudition and to have scorned precedent, but was nevertheless a sound and popular judge. He took a great interest in Irish education but was opposed to Home Rule.

Morris, Sir John William, Baron, of Borth-y-Gest (1896–1979). Called to the Bar in 1921, he was a judge of the King's Bench Division 1945–51, a Lord Justice of Appeal, 1951–60, and a Lord of Appeal in Ordinary, 1960–75. He was chairman of the Committee on Jury Service (1965) and was made a Companion of Honour in 1975.

Mort d'ancestor, assize of. A procedure instituted by the Assize of Northampton in 1176, whereby if a man had died in possession of a tenement of land, otherwise than as a tenant for life, his heir was entitled to obtain and hold possession against anyone else even though that other had a better right to the land than the deceased had. The other's right had to be asserted by action, not by seizure. The question whether the deceased had died seised of the lands and whether the claimant was his heir had to be decided by the verdict of a jury of 12 from the neighbourhood. The procedure survived till 1833.

Mortgage. In English law, a disposition of property to another in security of a debt, in supplement of a personal contract for payment of the debt. Mortgage (*mortuum vadium*, dead pledge) is so-called because, in many cases, the mortgagor does not perform the condition in the provision for redemption and the pledge is forfeited. It is distinguished from vifgage (*vivum vadium*) where neither loan nor pledge is lost. Practically any kind of property may be mortgaged. It may be effected by a demise i.e. lease, or sub-demise of land, the transfer of a chattel, the assignment of a chose in action, a charge on any interest in real or personal property or an agreement to create a charge, the condition being defeasible and the security being redeemable on payment or discharge of the debt or other obligation.

A legal mortgage of land is created by demise, or sub-demise of the land, or creating a legal charge over it. A legal mortgage of chattels may be made by pledge or bill of sale, and of a legal chose in action, such as a debt, by written assignment subject to statutory requirements. Mortgage of ships is regulated by statute.

An equitable mortgage of land may be created by the deposit of the title deeds of the land or by a written contract for the mortgage of land or an interest in land. An equitable mortgage of personal property may be made by an agreement to charge personal estate, an assignment of debts or funds in the hands of trustees, or a deposit of share certificates with a blank transfer signed by the holder.

A mortgagor (borrower) continues to have an equity of redemption (q.v.) or equitable right to redeem the mortgaged property at any time on payment of principal, interest and costs, notwithstanding expiry of the time by which, in law, he was bound to repay, and this equitable right is an interest in the land which may itself be sold, mortgaged, or disposed of by will.

The mortgagee (lender) may transfer his mortgage absolutely or by way of sub-mortgage, and it devolves on his death as personal estate. So long as he holds it he is entitled to possession of the mortgaged property or receipt of the rents and profits thereof but, so long as the equity of redemption is vested in the mortgagor, cannot sell the property save under an express or implied power of sale or a statutory power or the mortgagor's concurrence.

A mortgagee may make his ownership absolute by foreclosure, by bringing an action for foreclosure under which a further date is appointed for payment and, if the money is not then paid, the property belongs to the mortgage absolutely. Alternatively, a sale may be ordered to enable the mortgagee to be paid out of the proceeds.

The mortgagor may redeem his property free from the mortgagee's security on payment of all money secured by the mortgage, with interest and costs. On such payment the mortgagee must execute any necessary reassignment and reconvey any chattels.

Mortification. In old Scots land law, a kind of tenure whereby lands might be held in perpetuity in return for prayers and masses. It was abolished at the Reformation. The term is also sometimes applied to a charitable foundation.

Mortmain (dead hand). Under the feudal system of land tenure in England, the valuable incidents of wardship, marriage, relief, and others became exigible on the death of the vassal, but did not if the land was vested in a dead hand, or corporation which never married or died. Hence, from early times, mortmain statutes required the authority of a statute or a Crown licence before land could be vested in a corporation without it being subjected to forfeiture to the Crown. These included Magna Carta, 1215, the Provisions of Westminster, 1252, the statute *de viris religiosis*, 1279, the Statute of Westminister II, 1285, and the statute of 1392 which extended the mortmain statutes to all corporations. In modern times, the principle of forfeiture remained but many statutory exceptions limited its operation. The Mortmain Acts were repealed in 1960. These principles had no application or counterpart in Scotland.

Morton of Henryton, Fergus Dunlop, Baron (1887–1973). Was called to the Bar in 1912, took silk in 1929, became a judge of the Chancery Division in 1938, a Lord Justice of Appeal in 1944, and a Lord of Appeal in Ordinary, 1947–59. He was chairman of the Royal Commission on Marriage and Divorce, 1951–56.

Mos Gallicus. The name given in the sixteenth century to the new style, particularly associated with Alciatus at Bourges, of studying the Roman law according to the methods of the humanist jurists (q.v.), as contrasted with the older *Mos Italicus* (q.v.).

Mos Italicus. The name given to the style of discussing the Roman law adopted in Italy by the school of commentators (q.v.), who wrote systematic commentaries on the texts.

Mosaic Code. A name sometimes given to the earliest compilation of New England legislation, prepared by John Cotton in 1636 at the request of the Massachusetts General Court. John Winthrop called it 'Moses His Judicials', as it sets out a frame of government consonant with the Pentateuch. It was never adopted by Massachusetts but many clauses were included in the Fundamental Orders of New Haven Colony (1639), which was more strictly theocratically organized than Massachusetts.

Mosaic law. See TORAH.

Mosca, Gaetano (1858–1941). Italian jurist and political theorist, author of *Teorica dei governi e governo parlamentare* (1884), *Elementi di scienza politica* (1896), and other works, stressing that societies are governed by minorities, hereditary, military or priestly or aristocracies of wealth or merit.

Moser, Johann Jacob (1701–85). German jurist, author of many works on public and international law, and particularly of *Deutsches Staatsrecht* (1737–54), *Deutsches Staatsarchiv* (1751–57), *Grundsätze des jetzt üblichen Völkerrechts in Friedenzeiten* (1750), *Grundsätze des jetzt üblichen Völkerrechts in Kriegzeiten* (1752), *Neues deutsches Staatsrecht* (1766–75) and *Versuch der neuesten europäischen Völkerrechts in Frieden und Kriegszeiten* (1777–80). He was founder of the positivist treatment of international law in Germany. His son Freidrich Karl (1723–98) wrote mainly on politics. Both are to be distinguished from Justus Moser (1720–94), German historian, one of the fathers of the historical school of law, economics and ethnology.

Most-favoured-nation clause. A clause sometimes included in commercial treaties, to the effect that each party must grant to the other all benefits which either has granted in the past, or will grant in the future, to any third state, i.e. treat the other party equally with the nation most-favoured by any other treaty, past or future. The U.S. has frequently employed a qualified form of the clause, that all favours granted to third states accrue to the other party only subject to the same compensation as granted by the third state concerned.

Motion. An incidental application or request made to a court, generally made orally in open court. Normally, a motion is made only after notice to the other party, but in some cases, it may be made *ex parte*, without the other party having had notice of it. A motion of course is one which is granted by the court as a matter of course. In Parliament, a motion is a proposal to the House made by a member for the purpose of getting a decision, and is utilized in many circumstances; thus, there are motions for the adjournment of the House or for leave to bring in a Bill or that a Bill be read a second time. Every matter in both Houses is determined on questions put from the Chair on a motion made by a member. Some matters cannot be debated save on a substantive motion which admits of a distinct vote of the House.

Motive. That which moves or prompts a person to a particular course of action, or is seen by him as the ultimate purpose or end he seeks to achieve by that action. It is distinguishable from intention, which is the more immediate foreseen end to which he directs his acts. Thus, A may intend to kill B, with the motive of getting B's money, or of ridding his country of a tyrant, or of gratifying his hatred of a heretic. Motive accordingly supplies the reason for an intended act.

Investigation of a person's motive for particular conduct is truly a psychological inquiry and it may reveal mixed, irrational, or unsuspected motives.

The motive for an act or abstention is normally legally irrelevant; a good motive will not exculpate from the consequences of an intentional harm or crime, though it may be relevant to penalty; a bad motive will only exceptionally make lawful conduct tortious or criminal. Motive accordingly need not usually be investigated or proved.

Motor Insurers' Bureau. A body established by the general body of insurance companies transacting motor insurance business, which, by agreement with the Ministry of Transport, is obliged to satisfy any unsatisfied judgments arising from a liability which is required by statute to be covered by insurance but which is, in fact, not covered.

Moulins, Ordinance of (1566). A programme of legal reforms in France in the articles, presented to an assembly of notables convened at Moulins by Charles IX. It remained an influential element in French law down to the Revolution.

Moulton, John Fletcher, Baron Moulton of Bank (1844–1921). Won fame as a mathematician before being called to the Bar in 1874, where he specialized in patent work. He later became a F.R.S. Between 1885 and 1906 he sat in Parliament for various periods, and in the latter year was appointed a Lord Justice of Appeal, where he proved an outstanding judge. In 1912, he became a Lord of Appeal, where he sat till his death, save for the war years when he organized the supply of explosives for the War Office with great success.

Movables and immovables. The main distinction drawn in later Roman law and modern systems based thereon between kinds of things subject to ownership and possession. While basically the distinction corresponds to everyday conceptions, assigning animals and vehicles to the former and land and buildings to the latter category, particular things may be assigned to one category rather than the other for reasons of convenience. Thus, in French law, farm implements and animals are immovables. The distinction is also important in international private law, more so than that between real and personal (q.v.) or other distinctions peculiar to one legal system and largely determined by its history. Thus, land held on lease is personal property by English law for historical reason, but in international private law it is a right in immovable property.

Moveable property. In Scots law, those kinds of property which formerly passed on death to the executor for division among the next-of-kin, as contrasted with heritable property (q.v.), corresponding generally to personal property in English law. It is distinguished into corporeal moveable property, comprising goods, chattels, animals, vehicles, and the like, and incorporeal moveable property, comprising debts, claims of damages, shares in companies, goodwill, patents, copyrights, and the like. In relation to corporeal moveables, great importance attaches to the rule that for the transfer of right of poverty or in security delivery is essential, save under the Sale of Goods Act, actual, symbolical or constructive delivery.

Mudaeus (van der Munden) Gabriel (1500–?). Belgian jurist and humanist, who revived the study of law in Belgium, author of *De Contractibus* (1586), and commentaries on various titles of the Digest.

Mufti. An Islamic canon lawyer who, in response to enquiries by judges or private persons, gives opinions based on the Koran, tradition, and previous practices and decisions. Under the Ottoman Empire, the mufti of Istanbul was regarded as the supreme legal authority in Islam and, theoretically, had precedence over the whole judicial and theological hierarchy. As the sphere of Islamic law has declined, however, even in Islamic countries with the adoption of modern civil codes, the sphere of special knowledge of the mufti is being limited to matters of personal status, marriage, divorce, and succession.

Muir of Huntershill, Thomas (1765–99). A Scottish advocate who advocated parliamentary reform and joined the Society of the Friends of the People, a body with reformist aims along constitutional lines. He was a delegate to a General Convention of the Friends of the People in Scotland at Edinburgh in December, 1792, and the government, being afraid of the proceedings of the reformist and revolutionary societies in the United Kingdom, took steps to charge some of the more prominent among the Friends of the People. Muir was arrested and tried for sedition in 1793, the proceedings being a travesty of justice. He was convicted by a packed jury and sentenced to 14 years' transportation. He escaped from Botany Bay in 1796 and died in France. His trial was followed by others, but from 1795, the fears of the government receded. Muir is remembered as the martyr in the cause of reform.

G. Pratt-Insh, *Muir of Huntershill.*

Mulct. A fine or penalty; as a verb, to extract money.

Muller* v. *Oregon (1908), 208 U.S. 412. In this case, the Supreme Court upheld an Oregon law limiting the hours of employment for women, and established that the courts might recognize the special circumstances justifying the exercise of the police power. It is also notable because the Court admitted in evidence the 'Brandeis Brief' introduced by Brandeis (q.v.) as counsel for Oregon, in the shape of philosophical, sociological, statistical, and economic data relevant to the issue and the impact of long working hours.

Multiplepoinding (pron. multiplepinding). In Scots law, an action to determine which of several claimants on a thing or fund held by another should be preferred, originating from a process utilized where double or multiple poindings, i.e. seizures with a view to execution, had been effected. It is comparable to but rather wider than the English interpleader.

Multital rights. A term in the writings of Hohfeld (q.v.) for a right *in rem*, meaning one of a large class of fundamentally similar yet separate rights, actual and potential, residing in a single person or single group of persons but availing against persons constituting a very large and indefinite class of people. See also PAUCITAL RIGHTS.

Multures. In older Scots law, the quantity of grain deliverable to a miller in return for grinding corn. Multures due by persons astricted to, or bound to use, a particular mill were insucken multures, those due by persons who voluntarily used the mill outsucken multures. Dry multures were duties in grain or money, paid whether grain be ground or not.

Mund. In Germanic law, the rights of protection and guardianship over one's family, household and property.

Mundbryce. In Anglo-Saxon law, compensation for breach of the peace.

Municipal corporations. Bodies corporate constituted by the incorporation, usually by charter, of the inhabitants of a borough in England or Wales and composed of the mayor, aldermen, and burgesses or citizens. Since the Local Government Act, 1972, municipal corporations have ceased to be local government authorities and save in London, aldermen are abolished. The burgesses are now the ratepayers.

Municipal Corporations Act (1835). An Act which dissolved 178 variously constituted English municipal corporations and constituted, in their places, uniformly constituted elected bodies, thus largely getting rid of many of the defects of the existing authorities, such as intolerence, co-optation, and corruption. The new authorities comprised councillors democratically elected for three-year terms, who themselves chose aldermen for six-year terms. The council elected a mayor and had to have a clerk, treasurer, public meetings, and annual audit. Each council was given such functions as police, street lighting and power to administer local revenues and make local bye-laws. The system thus established has been drastically altered by the Local Government Act, 1972.

Municipal courts. The courts formerly held in boroughs by virtue of their charters. A notable instance is the Mayor's Court of the City of London now amalgamated with the City of London Court as the Mayor's and City of London Court.

Municipal law. In Roman law, a *municipium* was a self-governing political community within the Roman Empire; like *civitas* a *municipium* was a body politic or civil society, and municipal law is sometimes used by jurists as a synonym for the national or state law of a country as distinct from international law. In modern times, a municipality is a city or borough with some power of making bye-laws and municipal law is sometimes used to cover rules applicable to or made by such a local authority. See STATE, NATIONAL, OR MUNICIPAL LAW.

Municipality. In the United Kingdom, a term usually applied to large boroughs or cities. In the U.S., it is a unit of local government, an urban subdivision of a state having a municipal corporation to exercise local government functions therein. It may be variously designated a city, town, borough, or village. In European countries, the term may be applied to various types of smaller units of local government.

Municipium. In Roman times, an Italian community admitted to a kind of alliance with Rome, exchanging rights of intermarriage and commerce. *Municipia* had local autonomy, save in matters of foreign policy, but their residents acquired Roman citizenship only by settling at Rome. The first *municipia* were allies but the status was later extended to conquered peoples, though some lacked the right to vote down to 90 B.C. After the Social War of 90–89 B.C., many communities became *municipia civium Romanorum*, and, after 44 B.C., *municipium* meant any self-governing Italian borough, irrespective of origin. Under the principate, citizen rights were extended to provincial communities in the same way. *Municipia* spread rapidly in the West, but more slowly in the North, and in the East. In and after the second century, ambition led provincial municipalities to seek colonial rights. The whole system declined in the third and fourth centuries.

Munn* v. *Illinois (1877), 94 U.S. 113. An Illinois statute fixed the maximum rates which privately owed grain elevators in Chicago might charge for the storage of grain. The owners were in a position to charge high rates for storage, and contended that since their property was private, the regulation of rates violated the Fourteenth Amendment. The Supreme Court held that the business was so affected with a public interest that regulation was not unconstitutional, and the case is an important initial statement of the power of a state to regulate business in the public interest.

Munro, Robert Munro, Baron Alness (1868–1955). Was called to the Scottish Bar in 1893, sat in Parliament, 1910–22, and became Lord Advocate, 1913–16, thereafter becoming Secretary for Scotland, 1916–22. He was Lord Justice-Clerk, 1922–33, as Lord Alness, receiving a barony in 1934 and the G.B.E. in 1946.

Munus. Originally the duty of a Roman citizen to the State. Under the Republic, the main *munus* was military service, but later *munus* was merely the duties of a person to his municipality to pay tax, maintain roads and buildings and the like, it being the duty of the municipal magistrates to ensure that all performed their *munera* according to their liabilities as determined by standing and wealth. Under the Empire, *munus publicum* came to mean the liability of municipal councillors themselves for imperial taxation, which made municipal office a ruinous burden. In Scots law, the term *munus publicum* is given to certain offices, the holders of which were deemed to owe duties to the public and to hold *ad vitam aut culpam*.

Murder. In English law, it is murder if a person of sound mind unlawfully kills any human creature in being and within the Queen's peace, with malice aforethought, either express or implied by law, the death occurring within a year and a day after the injury. The accused must be sound mind; if not, he is entitled to be excused on the ground of insanity. The dead one must be a human creature, and one in being, not an unborn child nor one being born, and one within the Queen's peace, which extends to all persons in Her Majesty's dominions except rebels and alien enemies actually engaged in hostile operations against the Crown. The accused's conduct must have been the main direct cause of the death and must have been done with express malice, where the accused's deliberate purpose is to kill or, at very least, cause grievous bodily harm, or malice implied in law, where the malice is implied from the voluntary act of inflicting grievous bodily harm. If death does not ensue within a year and a day, the death is presumed attributable to some other cause. The former doctrine of constructive malice, whereby a person who killed another in the course of committing another crime, e.g. rape, was deemed guilty of murder, was abolished in 1967. In Scotland, murder is homicide committed intentionally or in circumstances showing complete indifference to or wicked disregard of consequences. The doctrine of constructive malice is probably not recognized but homicide in the course of committing a serious crime, if displaying recklessness, is probably murder. At common law, murder was a capital offence. Between 1957 and 1969 murder was, in both countries, divided into capital murder, punishable by death, and non-capital murder, punished by life imprisonment only, but the distinction disappeared when the death penalty was abolished in 1969.

Murdrum. In mediaeval English law, secret homicide. Legislation of William I required the hundred (q.v.) to pay a murder fine whenever a dead body was found within its limits which could not be proved to be that of an Englishman and the delinquent was not produced. In 1340, the murder fine was abolished and the term reverted to meaning the most serious form of homicide, killing with malice aforethought.

Murmuring a judge. In older Scots law, to slander or defame him.

Murray, David, 7th Viscount Stormont and 2nd Earl of Mansfield (1727–96). Was a Scottish representative peer and ambassador to Vienna and Paris, 1763–78, a Secretary of State, Lord President of the Council, and Lord Justice-General of Scotland, 1778–94.

Murray, John, 2nd Earl and 1st Marquess of Atholl (1631–1703). Became a Privy Councillor at the Restoration, and was Lord Justice-General of Scotland, 1672 and 1685–89, an Extraordinary Lord of Session, 1673–89, and Vice-Admiral of Scotland, 1680.

Murray, Sir Robert, of Craigie (?–1673). Entered the French Army but joined Charles I at Newcastle. Charles II appointed him Lord Justice-Clerk in 1651 and nominated him a Privy Councillor and a Lord of Session. At the Restoration, he was appointed a Lord of Session and a Lord of Exchequer but never sat as a judge. In 1667, he was reappointed Lord Justice-Clerk.

Murray, Sir Thomas, of Glendoick or Glendook (?1630–1684). Was called to the Scottish Bar in 1661, became Lord Clerk-Register in 1662 and a judge, as Lord Glendoick (or Glendook) in 1674. He published, in 1681 (in two editions—folio and duodecimo), the acts of the Scottish Parliament, but despite his position as Lord Clerk Register, he seems to have paid little attention to accuracy. The edition is copied from Skene's (q.v.) edition of 1597, with later additions, and includes all the mistakes of Skene. Nevertheless, the Glendook edition is that still commonly referred to in books and cases. He lost his office in 1681.

Murray, Sir Thomas David King, Lord Birnam (1884–1955). Called to the Scottish Bar in 1910, he was a sheriff-substitute 1928–33, returned to the Bar and became an advocate-depute, was chairman of the Scottish Land Court, 1938–41, as Lord Murray, returned again to the Bar and resumed his ordinary name, being knighted and made Solicitor-General for Scotland, and finally served as a judge of the Court of Session, as Lord Birnam, 1941–55.

Murray, Sir William, Lord Mansfield (1705–93). Born in Perth, he came to England in 1718 and was called to the Bar in 1730. He studied extensively, ancient and modern history, and international, Roman and Scots law as well as English law. His progress at the Bar was rapid and he was deemed one of the most effective advocates of the day. He also had a great mastery of law based on extensive knowledge. In 1742, he was made Solicitor-General and entered Parliament; in 1754, he became Attorney-General. In Parliament, he was an outstanding speaker until, in 1756, he became Chief Justice of the King's Bench and Baron Mansfield. He remained a member of the government till 1763 and even thereafter was a leading member of the House of Lords. In 1776, he was made an earl. His house was destroyed in the Gordon Riots for his having supported Catholic emancipation. He was devoted to the knowledge and administration of the law and as he aged increasingly devoted all his ability to his judicial duties. He resigned in 1788.

He exercised a very great influence on many branches of the common law by his judgments. He made some new rules of procedure and altered others to enable the courts to lay down clear rules for future similar cases, and was insistent on arriving at the real merits of the case, rather than deciding on points of pleading. His judgments affected every branch of the law, but particularly important was his settling the leading principles of law applicable to commercial cases on the basis of principles widely accepted in commercial communities, by the admission of evidence of mercantile custom to establish the rules of law. In this way, the practice as to insurance, negotiable instruments, maritime contracts, and bankruptcy was incorporated into the common law.

He had great reverence for the common law and, on the whole, did not favour legislative changes, but was very ready to develop and reformulate the law by decisions. Some of his approaches were subsequently disapproved, such as the view that moral obligation was adequate consideration for a contract, but nearly all of his work has stood the test of time, and he is unquestionably one of the makers of modern law. He also had an influence on legal studies in that he suggested to Blackstone the project of lectures on English law at Oxford and, sometimes, particularly explained a principle in court for the benefit of students present.

C.H.S. Fifoot, *Lord Mansfield.*

Muslim law. See ISLAMIC LAW.

Mute (silent). A prisoner is said to stand mute when, being arraigned on a charge of crime, he makes no answer, or no answer relevant to the question. If he makes no answer a jury must try the question whether he is mute of malice (i.e. deliberately) or by visitation of God (i.e. being dumb). In the former case, a plea of not guilty is entered, and in the latter the trial proceeds as if he had pleaded not guilty.

Mutiny. Any overt act of insubordination or defiance of and of attack on established authority in one of the armed services by persons subject to the authority and disciplinary code of one of these services, as by refusal to obey orders, seizing a ship or shore establishment, seizing officers and obeying instructions of others, or otherwise. It may have a political object and revolutionaries frequently try to incite members of the forces to mutiny, or it may be provoked by harsh treatment. It has always been regarded as a most serious breach of discipline and a major crime under military law, involving heavy penalties. Mutiny may be an element of or distinct from revolt or rebellion, the latter terms being usually used of more widespread attack on authority. Prior to 1879, the code of law for the army was continued in an annual Mutiny Act together with annual Articles of War.

Mutiny Act. The Bill of Rights, 1689, declared that to maintain a standing army within the realm in peacetime was against the law unless with the consent of Parliament. An Act was accordingly passed annually from 1689 to 1879 authorizing the maintenance of a standing force and providing for its discipline, and accordingly called the Mutiny Act. The substance of its provisions were then embodied in the Army Discipline and Regulation Act, 1879, which was later replaced by the Army Act, 1881, renewed annually, and then by the Army Act, 1955, and the Air Force Act, 1955; which are continued annually by orders empowered by the Armed Forces Act, 1971.

Mutual. A word which properly means the feelings, views, etc. of each of two persons in relation to the other, or *inter se*, but which is commonly used as meaning 'common', i.e. held or felt by both in relation to an external person or object.

Mutual contracts. Engagements by two or more persons by which each is bound to give or do, or abstain from doing, something to or for the other.

Mutual gable. A wall between two buildings, serving as gable wall to each.

Mutual mistake or error. In contract, circumstances where it appears that each party has been under a mistake as to the other party's identity or intention. It has to be distinguished from common error (q.v.).

Mutual settlement or will. A settlement or will by which each of two persons gives the whole or at least a substantial part of his estate on his death to the other. Such settlements give rise to problems of revocability, save by joint assent, being frequently deemed contractual and not purely testamentary.

Mutuality. In agreements, the requirement that each party should do something to or for the other.

Mutuum. In Roman and Scots law, the kind of contract of loan whereby one person lends to the other some of such things as are consumed by use, such as food or fuel, and wherein accordingly the property in the thing lent passes and the obligation is to restore not the thing lent but as much, and of the same kind, quality and value, as was lent. See also COMMODATUM.

Myers* v. *United States (1926), 272 U.S. 52. The question was whether the President could dismiss at will and without the intervention of the Senate, a person appointed to serve under him with the advice and consent of the Senate. Myers, a first-class postmaster, was removed in 1920 by President Wilson before the end of his term and without senatorial permission, as required by statute, being obtained. The Supreme Court held that Congress could not limit the President's power to remove executive officers. In *Humphrey's Executor* v. *U.S.* (1935), 295 U.S. 602, the Court later restricted the principles of *Myer's* case so far as applicable to officers of quasi-legislative and quasi-judicial agencies but the principle stands as to administrative officers.

Mynsinger or Münsinger, Joachim (1514–88). German jurist and author of *Observationes iudicii imperialis camerae* (1663).

N

Naeranus, Joannes (?1600–70). Dutch historian and jurist, compiler of *Sententien en gewezen Zaken van den Hoogen, en Provincialen Raad* (1662), an important collection of decisions of the superior courts of Holland in the seventeenth century, and of *Hollandsche Consultatien* (1645–66), a valuable collection of opinions of Grotius and other jurists.

Naish, John (1841–90). Was called to the Irish Bar in 1865, became Solicitor-General, 1883, Attorney-General, 1884, and Lord Chancellor of Ireland in 1885 and 1886.

Name and arms clause. A clause sometimes found in settlements of property imposing on persons succeeding to estate under the settlement an obligation to take the settlor's surname and bear his arms, on pain of forfeiture of the estate.

Names and change of name. By custom in the United Kingdom, a person takes the surname of his father, if legitimate, or his mother, if illegitimate, and is given one or more forenames or Christian names. By custom, on marriage, a woman commonly assumes the surname of her husband instead of, or in the U.S. in addition to, her maiden name but she may legitimately continue to use her maiden surname. She may continue to use her husband's surname though the marriage be dissolved and she remarries. The prefixes Mr., Miss and Mrs. are not part of a person's name.

A person may, however, assume and use any name he pleases, so that a child need not be given his parents' surname. An adopted child may be given his adoptive parents' surname.

The Roman Canon law, following a decision of the Council of Trent (1561), has adopted the principle that children baptized as Catholics must have as a baptismal name the name of a saint. This is important in mainly Catholic countries. In France in 1803, a law restricted Christian names to names known from ancient history or used in various calendars and this principle is still applied. The Council of Trent also decreed that every parish must keep registers of baptisms with the names of child, parents, and grandparents, and this rule was soon followed in Protestant countries. It did much to establish the basic rule that a child takes the family name of his parents. In the same way developed the general practice of a woman taking her husband's surname on marriage.

There is no right of property in a name, so that, in the absence of misrepresentation, fraud, or passing-off, a person may use his own name or adopt another name for himself or his property even though it causes inconvenience or annoyance to another. But, if a name has been habitually used in connection with a particular business or profession, the user may have acquired a right to prevent anyone else using that name in a manner likely to cause confusion, because the name has become part of the user's goodwill.

There are no restrictions on a person changing his name or assuming an additional name, not even if by so doing he assumes the name or names by which another person is known, though he may incur legal sanctions if he seeks to pass himself off as that other person. It is said, in England, that a first name, given at baptism, can be changed on confirmation, and it can certainly be changed by Act of Parliament or royal licence. A surname can be changed without any formality. But to avoid confusion, it is common to obtain a private Act authorizing a change of name, or to obtain a royal licence, or, in England, to execute a deed poll declaring the change and enroll it in the Central Office of the Supreme Court or the College of Arms, or, in Scotland, to obtain from the Lord Lyon King of Arms a certificate of recognition of change of name. In all cases, it is customary to make public advertisement in the press of change of name.

A person who is enrolled in any professional register must, on changing his name, take steps to have his professional registration altered.

Similarly, in the U.S., a person may use such name as he pleases, and change it as he pleases, provided he does not do so for fraudulent purposes.

Naming a Member. The procedure whereby a member of the House of Commons can be called on by the Speaker to leave the Chamber for the remainder of the day's sitting if he has refused to withdraw offensive language, and if he refuses or persists in disregarding the Speaker's rulings the Speaker may name Mr. X for disregarding the authority of the Chair. The Leader of the House moves that the Member be suspended and this motion is put at once. If the House agrees, the Speaker calls on him to withdraw and if need be summons the Serjeant-at-Arms.

Nantes, Edict of. A law promulgated in 1598 by Henry IV of France which granted a measure of religious liberty to his Protestant subjects, gave them full civil rights, and established a special court, the Chambre de l'Edit, composed of Protestants and Catholics, to decide disputes arising from the edict. It was resented by the *parlements* (q.v.) and the Catholic clergy. In 1685, Louis XIV revoked the Edict and deprived the French Protestants of all civil rights and religious liberties, which drove many of them to take refuge in Protestant countries outside France.

Napier, Sir Joseph (1804–82). Practised in London as a pleader before being called to the Irish Bar in 1831. He became Lord Chancellor of Ireland in 1858–59 and was a member of the Judicial Committee of the Privy Council from 1868.

Napier, MacVey (1776–1847). Lecturer on and first professor of conveyancing at Edinburgh University (1816–47) edited the *Edinburgh Review*, the supplements to the 4th, 5th and 6th editions, and the 7th edition of the *Encyclopaedia Britannica*.

Napoleonic codes. An unofficial term for the codes adopted in France at the instance of Napoleon, namely the Code Civil (1804), Code de Procédure civile (1806), Code de Commerce (1807), Code Pénal (1810), and Code d'instruction criminelle (1811). There are other, though less important, codes in the body of French law but these are not included in the group of Napoleonic codes.

Narratio. In old common law pleading, the oral claim of a plaintiff. It was also sometimes called the *conte* or tale.

Narrative. In Scots law, the clause of a deed stating the names of granter and grantee and the cause of granting.

Narrator. In the early common law, a pleader or sometimes a reporter. They became a group of professionals in the thirteenth century, concerned with pleading in court, and distinct from the attorneys who were concerned with procedural details of uses. In older common law, a *serviens narrator* was a serjeant-at-law (q.v.).

Narrow Seas. The seas between two coasts not far apart, and in particular the English Channel.

National Assembly. The name of various French parliamentary assemblies, including that formed by representatives of the Third Estate, June 17 to July 9, 1789, and that of the Third Republic, 1875–1940 (comprising both the Senate and the Chamber of Deputies). In the Fourth Republic (1946–58) and Fifth Republic (1958–), it is the name of the lower house only. The term has also been used of the National Constituent Assembly (1789–91), the Constituent Assembly of 1848, and that of 1945.

National Assembly of the Church of England (or Church Assembly). A body set up, in 1919, consisting of the House of Bishops (comprising the members of the Upper Houses of the Convocations of Canterbury and York), the House of Clergy (comprising the members of the Lower Houses of the two Convocations), and the House of Laity (consisting of elected representatives of the two Provinces). It had a legislative committee which submitted Measures, which the Assembly wished to become law to the Ecclesiastical Committee of Parliament (consisting of 15 peers nominated by the Lord Chancellor and 15 members of the House of Commons nominated by the Speaker) which reported thereon to Parliament. If each House resolved that the Measure be presented for the Royal Assent it had, on receiving the Royal assent, the same force as an Act of Parliament. The Church Assembly was superseded by the General Synod (q.v.) of the Church of England in 1969.

National Assistance. See SUPPLEMENTARY BENEFITS.

National Convention. The assembly which governed France from 1792 to 1795 when it was replaced by the Directory.

National Covenant. Charles I had, in 1637, attempted to thrust a new liturgy and canons, of Anglican style, on the Scottish people, who were Presbyterian. This provoked resistance and the National Covenant, a religious manifesto protesting against the King's religious policy in Scotland, was widely signed throughout Scotland and led to a meeting of the General Assembly of the Church of Scotland at Glasgow, which defied the King's Commissioner and declared Episcopacy abolished in Scotland.

National Debt Commissioners. In full, the Commission for the Reduction of the National Debt of the United Kingdom, consisting of the Speaker of the House of Commons, the Chancellor of the Exchequer, the Master of the Rolls, the Lord Chief Justice, the Accountant-General of the Supreme Court, and the Governor and Deputy-Governor of the Bank of England. Their functions are exercised by the Comptroller-General of the National Debt Office.

National Health Service. Separate legislation for each of England, Scotland, and Northern Ireland established a national health service in 1948, replacing a patchwork system of voluntary hospitals, local authority hospitals, hospitals, and institutions administered under the poor law, and of entitlement to medical services under the national health insurance scheme established in 1911. Under the legislation, all hospitals, except private ones, were transferred to the Minister of Health (now the Secretary of State for Social Services) or the Secretary of State for Scotland and administered by regional hospital boards and hospital management committees. General medical, dental, pharmaceutical, and ophthalmic services were organized nationally, administered through local executive councils and local health authorities were made responsible for ancillary services, such as health visitors and domestic help. There are also various advisory bodies. Limited charges are made, in some cases, for certain services and the provision of certain drugs or appliances. Each person may, in general, choose the medical practitioner by whom he is to be attended, and who treats him and issues any certificates required. The practitioner is remunerated by the Minister and, in general, may not demand any fee from his patients for services. The whole cost of the service is borne by monies provided by Parliament, mostly out of general taxation but, to a small extent, by contributions collected with national insurance contributions and by charges for drugs or appliances.

National Industrial Relations Court. A British superior court established in 1971 and abolished in 1974. It consisted of judges nominated by the Lord Chancellor from the High Court and Court of Appeal, at least one judge nominated from the Court of Session, and lay members with specialized knowledge of industrial relations appointed by the Queen. It might sit anywhere in Great Britain, in any number of divisions each composed of a judge and from two to four other members. Its jurisdiction covered numerous matters arising out of industrial disputes and it had certain appellate jurisdiction from industrial tribunals. It was replaced by the Employment Appeal Tribunal (q.v.).

National Insurance. Compulsory insurance against unemployment and sickness was introduced in Britain in 1912 and for contributory pensions in 1926. Both schemes were widened by later enactments and superseded, in 1948, by the national insurance scheme. For the purposes of this scheme the adult population, apart from married women, students, and persons in receipt of small incomes, was divided into employed persons, self-employed persons and non-employed persons, who all had to pay weekly contributions at specified rates into the National Insurance Fund. The Exchequer contributed a further sum and employers also paid weekly contributions for each employee. Contributions were, in general, paid by affixing stamps to cards. Since 1961, graduated contributions are paid by employed contributors earning over a stated minimum remuneration. So far as self-employed persons and employers are concerned, the contributions are a form of taxation and not related to benefits; certain contributions from self-employed persons are assessed and recovered by the Inland Revenue.

The benefits provided by the scheme comprised weekly payments as unemployment benefit, sickness benefit, maternity allowance, widow's allowance, widow's pension, widowed mother's allowance, child's special allowance, guardian's allowance, and retirement pension and lump sum payments by way of maternity grant, home confinement grant, and death grant. Each kind was paid at a specified rate and entitlement depends on satisfaction of qualifying conditions, including payment of the requisite number of contributions of the appropriate class. An increased weekly pension was payable in respect of graduated contributions.

Claims for benefit were made to insurance officers, with right of appeal to local tribunals of three and then to the National Insurance Commissioner or deputy commissioners.

The scheme was replaced in 1973 by social security (q.v.) pensions.

National Insurance (Industrial Injuries). A scheme for workmen's compensation (q.v.) was introduced in Britain in 1897 and later extended and replaced, in 1948, by the industrial injuries scheme which provided a system of compulsory insurance against personal injury caused by accident arising out of and in the course of the employment and against prescribed diseases and injuries due to the nature of a person's employment. It applied to all persons employed under a contract of service and was, accordingly, generally coextensive with the class of employed persons under the national insurance scheme. Benefits were payable from a state insurance fund to which contributions were made by employers, employees, and the Exchequer. The contributions were included in the amount of the stamps which had to be affixed under the national insurance scheme. Title to benefit did not depend on the payment of a qualifying number of contributions. The benefits payable were injury benefit for a period not exceeding 156 days, during which the employee was incapable of work, disablement benefit thereafter, so long as the employee was continuing to suffer from a loss of physical or mental faculty as a result of accident or disease, and death benefit, payable to dependants. Disablement benefit might take the form of a pension or a gratuity, and special increases might be given for unemployability, special hardship, need for constant attendance or hospital treatment.

Claims for benefit were made to insurance officers with right of appeal to local tribunals of three and then to the Industrial Injuries Commissioners or deputy commissioners. Medical boards and medical appeal tribunals decided issues of capacity for work and degree of disablement.

The scheme was replaced in 1975 by a scheme under the Social Security Act, 1975.

National judges. A judge of the International Court of Justice (q.v.) retains his right to sit, even though the country to which he belongs is a party to a case before the court, but if the other country concerned does not have one of its nationals among the judges it may nominate a national judge for the hearing of that case.

National Labour Relations Board. An independent federal agency established in the U.S. in 1935 to administer the (Wagner) National Labour Relations Act, later amended by the Taft-Hartley Act (1947) and the Landrum-Griffin Act (1959). Its functions are to prevent or correct unfair labour practices. It deals with labour disputes by investigation and informal settlement or quasi-judicial proceedings. It may not act on its own motion and has no independent power to enforce its orders, but may do so through a U.S. Court of Appeals.

National Labor Relations Board* v. *Jones & Laughlin Steel Corpn. (1937), 301 U.S. 1. The Board attempted to regulate labour relations in a steel plant, labour which was not directly concerned with shipment of goods. In a question whether this exercise of federal power was legitimate the Supreme Court held that, whether or not the processes of manufacture were interstate commerce, labour troubles at the plant would restrain interstate commerce and were accordingly subject to federal regulation. The decision evidenced an attitude of the Supreme Court more favourable to New Deal legislation, opened the way to more federal regulation by way of the commerce power, and had an influence in defeating Roosevelt's attempt to introduce new blood into the Supreme Court.

National Law. The law of any particular nation-state, as distinct from international law and supra-national law (qq.v.). See STATE, NATIONAL, OR MUNICIPAL LAW.

National Parks Commission. A body, established in 1949, for the preservation and enhancement of natural beauty in England and Wales, particularly in the areas designated as natural parks or as areas of outstanding natural beauty and the encouragement of the provision or improvement of facilities for the enjoyment of such national parks.

National Trust. Formed originally as an association and incorporated by statute in 1907, the National Trust for Places of Historic Interest or Natural Beauty exists to promote the permanent preservation for the nation of lands and buildings of beauty or historic interest and the preservation of their natural aspect, features, and animal and plant life, including the preservation of buildings of national interest, or architectural, historic or artistic interest and the protection and augmentation of the amenities of such places, the preservation of furniture, pictures, and chattels, and the access to and enjoyment of such buildings and chattels by the public. It is a private, not a state, body, with thousands of members, and is recognized as a charity. It owns many historic buildings and sites and may enter into agreements with landowners to prevent the land being used or developed disadvantageously. A separate but allied National Trust for Scotland with similar objects was established in 1931 and performs similar functions in Scotland.

Nationalization. Transfer by a State of rights of ownership or control of property from individuals or corporations to the State itself, or to a State agency created to hold and manage the property. Technically, it is distinct from expropriation, or exercise of the power of eminent domain but, in practice, it is hardly distinguishable and nationalization is normally effected by socialist or communist governments, to whom expropriation is a major object of policy, for political rather than economic reasons. This has been the case in Russia since 1917 and Britain by Labour governments since 1945. Sometimes nationalization has been resorted to to get rid of foreign control of an industry or asset, as in the case of Egyptian nationalization of the Suez Canal in 1956. Where international issues are involved, nationalization usually implies some compensation to the former owners but it is always inadequate, and nationalization is truly a euphemism for theft.

In the U.K., when an industry has been nationalized, a nominally independent board is set up to run it but, in practice, such boards are always subject to political pressures and ministerial interference on political grounds. There is also a large bureaucratic organization. Economically, nationalization of industries has been disastrous, as they have unlimited capacity for loss-making and a voracious appetite for subsidies from taxation. Legally, the corporations running industries are independent entities and not departments of state, nor local authorities. They report to the appropriate Minister and he is, in general, answerable to Parliament and certain ones have, from time to time, been inquired into by a select committee of the House of Commons. They normally have large monopoly powers, which places the individual in a worse position than before in that he cannot obtain supplies from any alternative.

Nationality. The concept of legal relationship between a person and a state. It implies correlative duties of allegiance by the individual and of protection by the State. It is commonly regarded as a fundamental right of an individual and the U.N. Declaration of Human Rights (1948) declares that everyone has a right to a nationality. It is distinct from belonging to a certain nation by race. Thus

English, Scots, and Welsh belong to different nations racially but all have British nationality in law. Nationality is distinct from domicile, which is the territory of a person's permanent home. Each state may primarily determine in what circumstances and on what bases a person may claim nationality of that state, and what the incidents of that relationship are, but nationality is also affected by principles of international law. Many legal consequences both internally and in international law and relations flow from a person being a national of a particular state.

Nationality is commonly acquired by birth of a father, or parents, or a mother, who is a national of the particular state, or by birth in the territory of that state. Where the factors of blood and place of birth differ, the child or parents may sometimes elect which nationality to claim. It frequently is, or may be, acquired by marriage with a national and regularly may be acquired by naturalization in a particular state and relinquishment of former nationality. Nationality lost, e.g. by marriage, may sometimes be resumed if the marriage is ended. So too, a person naturalized in another state may be allowed to re-acquire his original nationality by redintegration or resumption. In general, the nationality of persons changes with a change of sovereignty, as where territory is transferred to another state, having been subjugated or ceded.

Within the Commonwealth, a person may acquire citizenship of the U.K. and Colonies or of any one of the Commonwealth countries and, if so qualified, he acquires the status of British subject or Commonwealth citizen. The latter status probably has significance in other Commonwealth countries only.

A distinction may be drawn by national laws between citizens and other grades of nationals, such as naturalized aliens.

Among the consequences of nationality are that a state may claim diplomatic protection only for its own nationals. Persons not enjoying the protection of the state of their nationality by its internal law are, *de facto*, stateless and there may be a rule that neither state of a dual national may exercise diplomatic protection against the other. A state is also bound to receive on its territory such of its citizens as are not allowed to remain on the territory of other states.

There is no general rule that a person may voluntarily renounce his nationality and sever his connection with his state, even by acquiring the nationality of another state under its law. A person does not necessarily renounce his nationality by going abroad, even permanently, as he may do this for reasons of health, business, to avoid taxation, or otherwise, but some states provide for expiry of citizenship on long residence abroad. These are matters of municipal law.

There is no general rule of international law that deprivation of nationality by municipal law is illegal. Sometimes nationals of certain categories may lose their nationality if they acquire a domicile abroad. The Universal Declaration of Human Rights prescribes that no one shall be arbitrarily deprived of his nationality nor denied the right to change his nationality.

In view of differences between systems of municipal law, an individual may have dual nationality, or may have to elect between one nationality and another, or may not be a national of any state, i.e. stateless. This has become an increasing problem and a U.N. Convention on the Reduction of Statelessness was adopted in 1961.

The concept of nationality has been extended from natural persons to corporations, ships, which have nationality by registration and acquisition of the right to fly a flag, aircraft, whose nationality is governed by the state of registration, and property in general.

Nations, law of. The older term for public international law.

Nativo habendo, de. A writ available, in mediaeval times, to a lord whose villein had run away, requiring the sheriff to apprehend the villein and return him to the lord. Trial of the writ determined the question whether an alleged villein were a villein or a free man and, if a villein, a villein of the lord claiming him.

Natura brevium. A book dating from the reign of Edward III, containing a selection of writs, together with a commentary thereon. The writs tallied with those in the *Registrum brevium*, and it was said to have helped very much with the understanding of the law and of the Register. After Fitzherbert published his *New Natura Brevium* in 1534, it was called the *Old Natura Brevium*. There were many printed editions in the sixteenth century.

Natural allegiance. The allegiance permanently due by natural-born subjects to their sovereign, as contrasted with the temporary allegiance due by an alien or stranger so long as within a sovereign's dominions and within his protection.

Natural child. An illegitimate child or bastard.

Natural equity. Generally, a synonym for natural justice.

Natural justice. Until the eighteenth century, the term natural justice was often used interchangeably with natural law, equity, *summum jus* and other similar expressions, and the idea undoubtedly originated from the concept of natural law.

In modern practice the term usually denotes the general principles and minimum standards of fairness in adjudicating on a dispute, embodying the specific requirements (a) that no man be judge in his own cause—*nemo judex in parte sua*, and (b) that each side be heard and no man be condemned unheard—*audi alteram partem*. Both principles have been stated since very early times. The first requires a judge to decline jurisdiction if he has any legal or pecuniary interest in the outcome of the decision, or if there is any real likelihood or appearance of bias, such as arising from relationship to a party or membership of a body concerned. An objection may be waived or statute may authorize what might be objectionable under this principle. The second principle requires a judge to give both or all parties to a cause equal opportunities to be heard, and to give equal attention to their submissions.

The principles are most commonly invoked today when non-observance of either or both is made the ground for challenge of the validity of decisions. Thus, challenge may be made of arbitrators' awards, decisions of domestic tribunals, committees of clubs and societies, and statutory disciplinary tribunals on these grounds. Where, however, a minister or department has to decide or determine some issue as a matter of policy parties are not entitled to a hearing if the act is an administrative act, but if he has to hear objections or an appeal the hearing is quasi-judicial and the requirements of natural justice apply to them. Where the principles apply and have not been observed a court will annul the decision and require the matter to be considered afresh.

Natural justice is also invoked where the question arises of the enforcement in Britain of judgments obtained in foreign courts. A foreign judgment is unenforceable if obtained in circumstances contrary to natural justice, as where granted without jurisdiction, or without service on the complainer, or opportunity to the complainer to state his case, but British courts are slow to stigmatize as contrary to natural justice proceedings which have counterparts in British law.

Natural law (or law of nature). The belief in natural law and natural law theories has been one of the most ancient and persistent themes in philosophy, and in legal history and theory. In general, it denotes belief in a system of right or justice common to all men prescribed by the supreme controlling force in the universe and distinct from positive law, law laid down by any particular state or other human organization. The starting points of all natural law thinking have been 'reason' and 'the nature of man'. But there have always been disagreements about the meaning of natural law and its relation to positive law.

Greek thinkers laid the basis of natural law and developed its essential features. Observation of the uniformities of physical nature and the rhythm of natural events gave rise to the belief in a force which directed those happenings. Thinkers, particularly the sophists, began to reflect on the problems of social and political life in the Greek city-states and about the reason and validity of laws, and whether some permanent feature underlay the varied and changing particular laws. The concept of Nature as the underlying, ordering force came to be invoked. Natural law as the just, permanent element was thought of in contrast to changeable, sometimes oppressive, sometimes unjust, human law. Aristotle's *Logic* substantially adhered to the sophist point of view; man is a natural creature, but is also endowed with reason. Later, in the *Ethics*, he distinguished natural and legal or conventional justice. Natural law had authority everywhere and was discoverable by the use of reason.

With the Stoics a century later, nature was not only the order of things but man's reason, and man's reason was part of nature. Reason governed the universe, and man, a part of universal nature, was governed by reason; reason orders his faculties in such a way that he can fulfil his true nature; when man lives according to reason, he lives naturally. The Stoics rejected the differences of city-states, of Greeks and barbarians and postulated a world-society in which men lived as equals; with this ideal they coupled the distinction between absolute and relative ideals of natural law. Relative natural law in an imperfect, human world required from legislators law, guided by reason, which approximated as closely as possible to absolute natural law.

At Rome, long before natural law was a part of philosophic or juristic teaching, Roman jurists used the concept of the law of nature to justify the taking, from materials supplied by the foreign laws and customs of non-Roman peoples, principles which seemed capable of general application and developed them into general legal principles. The Roman jurists and magistrates developed these principles of reason and justice empirically rather than by deduction from a general idea. What they created was not a body of principles of natural law but a *ius gentium* as the embodiment of laws and usages found among various peoples and representing good sense. Gradually, the sphere of application and the meaning of *ius gentium* changed. The *ius naturale* exercised its creative function through the *ius gentium* and was recognized independently when Stoic philosophy was received in the time of the Empire.

Natural law had been influential as a standard for criticism of the *ius civile* and, when Caracalla extended Roman citizenship to practically all persons living in the Empire, *ius civile* and *ius gentium* coalesced, though as slavery was still recognized, the joint system still diverged from the ideal of the *ius naturale*. But in many respects the law, through the *ius gentium*, was based on *ius naturale*; many natural law precepts were part of positive law, and judges and magistrates introduced new developments by invoking the concepts of *iustum*, *aequum et bonum*, *natura*, and others. For Cicero and some other philosophers, moreover, *ius naturale* was a higher law by which the validity of positive law could be and was measured.

In the Dark Ages, the Fathers of the Church, notably Ambrose, Augustine and Gregory, preserved the continuity of natural law and also began to change its basis and meaning. It was the task of the Christian Church, they held, to require the best possible approximation of human laws to Christian principles. For this purpose the Church was given supremacy over the State, which was a necessary evil, justifiable only so far as it protected peace and the Church and tried to implement the requirements of natural law.

When Europe emerged from the Dark Ages in about the ninth century, it was based on the dualism of Church and State, the Pope and the Emperor, each having an elaborate hierarchy and system of rules. Both professed Christianity, which pointed to a theory of law in which Christianity was the supreme spiritual and legal force and value, superior to all other laws. The Fathers of the Church had already begun to substitute Christianity for the Stoic reason as the supreme force in the universe, and the Church continued this trend. Thus, Occam postulated a hierarchy of laws comprising universal rules of conduct, dictated by natural reason, rules which would be deemed reasonable and binding in a society governed by natural equity without positive law, and rules deduced from general principles of the law of nature but not fundamental and accordingly liable to modification by authority. In Gratian's *Decretum*, the law of nature was treated as part of the law of God, as immutable and prevailing over contrary custom or positive law; the Church, moreover, was the authentic expositor of the law of nature. The investiture conflict gave a powerful incentive for the development of the doctrine of natural law, and produced the scholastic system of law which supported the claim of the Church to legal supremacy.

In St. Thomas Aquinas, natural law was derived from the law of God. The world was ruled by divine providence and the whole community of the world governed by divine reason. Divine law was supreme; such of it as was intelligible to men was revealed through Church law as the incorporation of divine wisdom, which gave direction to all actions and movements. Natural law was the part of divine law which revealed itself in natural reason, and man as a reasonable being applied this part of divine law to human affairs. From the principles of external law as revealed in natural law all human law was derived. There was also divine positive law enacted for men by God, contained in the scriptures, and all laws enacted by human authorities had to keep within these categories. The hierarchy of legal values was accordingly: divine law; eternal law; natural law; divine positive (scriptural) law; and human positive law, which was valid only so far as compatible with natural and eternal law. It was variable according to time and circumstances and served to further the interests of society and to be useful to men. St. Thomas took the view that the State had a legitimate sphere and function, to regulate social life justly: State laws were unjust and contrary to natural and divine law if directed to a false end, or beyond the power of the legislator, and thus invalid. St. Thomas' system, based on Aristotle and the Church Fathers, upheld the supreme authority of the Church, including its rights in the investiture controversy, but had room for the authority of the Emperor. Dante's *De Monarchia* championed the supremacy of the Holy Roman Empire, envisaging a universal order without the Church. The Emperor was the legitimate successor of the Roman people and was chosen by God to rule the world. In succeeding years, the authority of the law of nature was repeatedly invoked in the struggle for supremacy between Pope and General Council, and between Pope and Emperor; later it was invoked again in the struggles between Catholic and Protestants, both sides finding in it the true interpretation of scripture, or the right of the state in relation to spiritual jurisdiction.

Bracton recognized natural law as that which nature, that is to say, God, teaches to all animals, and he tried to reconcile natural and human law, but recognized the difficulties; there were rules of positive law which could hardly be reconciled.

The collapse of the mediaeval order, the Renaissance, the Reformation, widening intellectual horizons, the rise of nation-states, and the development of commerce posed new problems. The intellectual authority of reason began to be substituted for the spiritual authority of divine law as the basis of principles of natural law. Grotius gave expression to the new foundations of natural law in his *De Jure Belli ac Pacis*; men, he urged, desired life in society spent in peace; natural law was derived from the nature of the human intellect which desired a peaceful society, and was independent of divine command. These principles of reason could be deduced *a priori*, by examining anything in relation to the rational and social nature of man, or *a posteriori*, from examination of the acceptance of these principles among nations. Natural law was still the superior law, and on the principles of natural law, Grotius founded his system of international law, on such principles as *pacta sunt servanda*, the obligation to repair damage done by one's fault and others. Natural law also figured in political controversy: Grotius founded on natural law the plea that the seas were free to all (*Mare liberum*, 1609) and Selden founded on natural law the plea that public and private dominion over the seas was permitted (*Mare clausum seu de dominio maris*, 1635).

The origins of the social contract theory can be traced in Plato, John of Salisbury, John of Paris, St. Thomas Aquinas and it was developed by Marsiglio di Padua. Grotius considered a social contract as an historical fact but also thought that the ruler chosen by the social contract was bound by natural law, apart from any undertakings to his subjects. Hobbes developed the social contract theory but acknowledged the authority of natural law which, however, he understood not as an objective order superior to positive law but as a body of natural rights, subjective claims based on the nature of man. For him the chief principle of natural law was the natural right of self-preservation, in pursuance of which men sought to find security by transferring all his natural rights to a ruler and promising to obey him; there is to him only one kind of pact, an unconditional subjection of men to a ruler who obtains absolute power. Natural law to him was not superior law. Locke, however, restored the older concept of natural law in that he made it superior to and immutable by positive law. He used the social contract theory not to prove the transfer of all rights to a ruler, but to justify government by the majority, having the duty of preserving the individual rights entrusted to them by individuals. Locke, accordingly, opposed to Hobbes' absolutism the theory of the individual's inalienable natural rights, notably the right of private property.

The codification movement of the eighteenth century was commonly based on natural law, and embodied natural law ideas crystallized into detailed and formulated rules.

Hume destroyed the theoretical basis of natural law, rejected reason as the guide to action and constituted the moral sense, guided by pleasure or pain, as the force which established moral distinctions. The social contract theory, which was connected with natural law, was also pushed aside by utilitarianism and positivism. The nineteenth century was a period of eclipse for natural law, though it was continued by Ahrens, Krause, and Lorimer's *Institutes of Law* but, at about the end of the nineteenth century, there was a revival of natural law theories.

One variety of revived natural law is a modern version of scholastic and Catholic doctrines of natural law exemplified by the thinking of Renard, Delos, Maritain, Dabin, and Rommen. Another is exemplified by Stammler's theory of 'natural law with changing content', and another again by the theory of natural law as an evolutionary ideal, represented by the theories of Geny and Del Vecchio. In America, Pound, to some extent, and more frankly, Fuller and Hall, have been influenced by natural law thinking.

Natural law ideas were extensively utilized in England in several contexts. The first is the struggle between the judges and Parliament for legal supremacy in the seventeenth century. Coke, in *Bonham's case* (1610) 8 Co. Rep. 114a, observed *obiter* that the common law could adjudge statutes against common right or reason or repugnant or impossible to be performed to be void, and there are some later assertions to the same effect, but after 1688, the supremacy of Parliament and legislation was admitted. Natural law ideas and natural justice have also been invoked in such contexts as the recognition or otherwise of a custom, the applicability in a case before an English court of foreign law or foreign judgment and, most importantly, the challenge of certain administrative and quasi-judicial decisions on the ground of non-compliance with the rules of natural justice. Probably the most important context in which natural law ideas have been important is the development of equity; in the early stages natural law ideas drawn from civilians and canonists were probably of considerable importance; in later stages natural law and such ideas as *aequum et bonum* have contributed to the ideas of restitution for the avoidance of unjust benefit.

In the U.S.A., the Federal and State constitutions were drafted under the influence of natural law and natural rights concepts and judicial review of legislation by the Supreme Court has been largely influenced by ideas of natural law and natural rights.

After Grotius' time, natural law continued in the works of Pufendorf, Thomasius, Wolff, and Selden to provide a foundation for general theories of law and for international law. Vattel, however, though paying deference to natural law distinguished internal law (of conscience) and external law (of action). In practice, only the latter is material and natural law is law of conscience only. Thereafter international law has been primarily positive law.

In the nineteenth and twentieth centuries, there was a reaction against natural law theories; scientific investigation did not favour vague and unprovable hypotheses; the reaction against the 'reason' of the French Revolution led to a reaction against reason as a basis for law; and a utilitarian view suggested that the basis for law was the practical inquiry as to what would most conduce to the general benefit. But in the twentieth century, disillusionment with State power and positive law has stimulated a revival of interest in natural law.

The theory of natural law has thus at various times fulfilled various functions. It helped to transform the rigid Roman *ius civile* into a broader and more equitable system; it provided arguments in mediaeval struggles between Pope and Emperor; it has provided a basis for much of modern international law; it has underlain interpretations of the U.S. Constitution which resisted statutory attempts to restrict the economic freedom of the individual; it gave rise to the concept of natural rights (q.v.), which developed into fundamental rights (q.v.) and in modern terms, human rights (q.v.); it has in different ways provided many jurists with a basis from which to deduce an ideal or a right system of law, or with a standard by which to judge positive law. The idea has never died and seems immortal.

J. Bryce, in *Studies in History and Jurisprudence* (1901); F. Pollock, in *Essays in the Law* (1922); R. Pound, *Introduction to the Philosophy of Law* (1954), J. W. Jones, *Historical Introduction to the Theory of Law* (1940); G. Sabine, *History of Political Theory* (1937); C. Haines, *Revival of Natural Law Concepts* (1930); A. P. D'Entreves, *Natural Law* (1951); J. Stone, *Human Law and Human Justice* (1965).

Natural obligations. A term sometimes applied to obligations said to arise from the law of nature only or from natural justice and equity, such as the obligations between parents and children. The term is sometimes applied also to obligations binding morally but, for technical legal reasons, such as defective formalities, not legally enforceable, though having some effect, such as barring a claim for repayment of a debt due but not enforceable.

Natural persons. Those legal persons who are also living human beings as distinct from legal persons who are artificial or juristic persons such as corporations. Not every living human being is in all legal systems a natural person; thus slaves have, in certain systems, either been non-persons in law, or persons with very restricted capacities.

Natural rights. Claims asserted as being rights inherent in an individual by virtue of natural law or the nature of man rather than by the positive law of his State. The origins of these asserted rights go back to the mediaeval idea of the law of nature and to the Puritan revolution of the seventeenth century when the appeal to historic right, anciently enjoyed in the golden age of England, was replaced by an appeal to natural rights.

The theory of natural rights is the logical outgrowth of the Protestant revolt against the authority of tradition and the Protestant appeal to the reason and conscience of the individual. Puritan England produced the theory of natural rights, and the Bill of Rights of 1689 sought to formulate them.

Locke's *Second Treatise on Civil Government* restored the mediaeval concept of natural law and held it superior to and unalterable by positive law. He also put the individual in the forefront and in the name of supreme ethical principles invested him with inalienable natural rights, notably to life, liberty, and estate or private property. The duty of governments is to preserve those rights, the protection of which has been entrusted to them by individuals by the social contract. Tom Paine (q.v.) even asserted the divine origin of the rights of man at the creation.

Locke's doctrine of natural rights was adopted verbatim in the Virginia Declaration of Rights, 1776, and with verbal modification in the American Declaration of Independence, 1776, and appeared further modified in the French Declaration of the Rights of Man and the Citizen of 1791 and 1793, and in the French Declaration of the Rights and Duties of Man and the Citizen prefixed to the Constitution of 1795.

The individual natural rights normally asserted are life, liberty, property, and happiness. All of these require interpretation and, frequently, qualification, before the assertion can have substance or meaning.

Apart from rhetorical appeal the principal value of such claims to natural rights has been an attempt to begin formulating the protections which ideally and theoretically many individuals would like to see secured and protected in a well-conducted society. In modern times, more attention is focused on civil rights or liberties (q.v.) which positive law should secure, or on even more broadly formulated human rights (q.v.).

D. G. Ritchie, *Natural Rights* (1916); H. Lauterpacht, *International Law and Human Rights* (1950).

Natural use (or user). Natural use, in the context of the rule of strict liability for harm done by the escape of something dangerous from one's land, is such use as is ordinary, normal, and natural (such as agriculture) as compared with some substantial innovation on or alteration of the state of nature such as damming up a stream.

Naturalization. The process of making a person a subject of a state of which he was not a natural-born subject but hitherto an alien. The effect is to make the naturalized person have, in general, the same rights and privileges and to make him subject to the same duties and obligations as natural-born members of the State in which he is naturalized, save that he cannot without the consent of the state in which he was a natural-born subject, divest himself of his obligations to it.

Prior to 1844, an alien could become a British subject only by private statute, and this is still competent but since, and by virtue of the Naturalisation Act, 1870, an alien could become a British subject by obtaining a certificate from the Home Secretary. Under the British Nationality Act, 1948, as amended, conditions and qualifications are imposed for obtaining the certificate, including periods of residence, good character, and sufficient knowledge of the English language.

Nature Conservancy. A body incorporated in 1949 to give advice on the conservation and control of natural flora and fauna in Great Britain, and to manage nature reserves and organize research and scientific services in relation to natural flora and fauna. It may enter into agreements with landowners to secure that their land should be managed as a nature reserve where it appears expedient that the land should be so managed.

Nature, law of. See NATURAL LAW.

Nautae, caupones, stabularii. The title of the praetorian edict in the Roman law, contained in Dig. IV, 9, 1, which imposed strict liability on shipmasters, innkeepers, and stablekeepers for the safety of property brought on to their premises and entrusted to them. See also CARRIERS; INNKEEPERS.

Nautical assessor. A person of experience in maritime matters sometimes summoned to assist a court dealing with a case relating to collision at sea, salvage, towage, or other maritime matter. His or their function is to advise the court on such matters of nautical skill and practice as the court may submit to them in the form of written questions. They may not question witnesses and do not give judgment, the decision being one for the court though it will normally be hesitant to disregard the view of the assessors on the matters submitted to them.

Naval court. A court which may be summoned by the captain of one of H.M. ships on a foreign station, or by a consular officer. It consists of three, four or five members, including, so far as possible, a naval officer, a consular officer, and the master of a British merchant ship. Its jurisdiction is to hear a complaint by the master, mate, or any seaman, or to investigate if the interest of an owner or cargo-owner required it, or if a British ship is wrecked, abandoned, or lost. The court must report to the Board of Trade and has wide powers to cancel or suspend officers' certificates, discharge, imprison or fine seamen, or order a ship to be surveyed. It is entirely different from a naval court-martial.

Naval court-martial. See COURT-MARTIAL.

Navigation, right of. The right of members of the public to use a river, arm of the sea, or other piece of water for passage in boats, including the right to anchor. It is a right of way, not of ownership, so that the owner of the bed of a navigable river retains his right of ownership but must not interfere with the passage of boats.

A river subject to a right of navigation is designated 'navigable'; whether a particular river at a particular place is navigable or not depends on its breadth, depth and similar factors and on whether it has customarily been used by boats belonging to members of the public.

Navigation Acts. A series of statutes of 1381, 1485, 1489, and 1532 designed to protect British shipping, promote trade, and naval defence. They proved unenforceable and ineffective. Also a series of statutes passed by the English Parliament in the seventeenth century to foster mercantilism by making England and its colonies self-sufficient. The 1650 Act forbade foreign ships to trade with English colonies without special licence, and that of 1651 provided that all goods imported into England from Asia, Africa, or America should be carried in English ships manned by English crews. The Act of 1660 enumerated commodities, including sugar, cotton and tobacco, which could be exported from English colonies only to English ports. The Act of 1663 imposed a duty on all European goods entering English colonies unless trans-shipped via England in English or colonial vessels, and the Act of 1696 required royal approval for the acts of colonial governors and provided for the establishment of customs houses and admiralty courts subject to direction from London. The effect was to give Englishmen and English colonists a legal monopoly of all trade between England and its colonies. The Acts operated against trade by outsiders, including Scots, Dutch, and Europeans generally, and probably stimulated English and colonial merchant navies. All the Navigation Acts were repealed in 1849.

See also MERCANTILISM; TRADE ACTS.

Ne exeat regno. At common law in England, the King could forbid his subjects going outside the realm, and this writ was issued by the Chancellor for reasons of State on the application, in the sixteenth and seventeenth centuries, of a Secretary of State and, in the next century, on the application of a private person, to prevent the evasion of legal or equitable liabilities. It is probably not wholly obsolete.

Near* v. *Minnesota (1931), 283 U.S. 697 (U.S. Supreme Ct.). the Supreme Court held unconstitutional a Minnesota law forbidding publication of malicious, scandalous, and defamatory articles, declaring that the right to criticize public officials was a fundamental principle of democratic government and that the guarantee of personal liberties under the First Amendment applied to state as well as federal laws.

Nebbia* v. *New York (1934), 291 U.S. 502. A state law had fixed retail milk prices and was challenged as a violation of the due process clause of the Fourteenth Amendment. The Supreme Court extended the police power of states, by denying that there was any closed category of business affected with a public interest.

Nebrija, Antonio de (Nebrissensis) (1444–1532). A leading Spanish jurist and grammarian, reviser of the glosses of Accursius, and author of *Observaciones sobre las Pandectas* and of a *Lexicon Juris Civilis* (1486).

Necessaries. Those things which law deems obviously necessary for the maintenance of life, such as food, clothing, and shelter. If a wife is deserted she has the legal right to pledge her husband's credit for necessaries. Minors and persons of unsound mind do not have capacity to contract but must pay a reasonable price for necessaries sold and delivered to them, such as are suitable to their condition in life and to their actual requirements at the time of the sale and delivery. Wilful neglect to provide necessaries for children is criminal at common law. In the context of shipping, necessaries are such things as are fit and proper for the service in which the ship is engaged and such as a reasonably prudent owner would furnish. A master may pledge the ship to the effect binding the owner for necessaries supplied abroad.

Necessity. It is uncertain in British law to what extent, if at all, necessity is a defence to a criminal charge. It may be a defence if the necessity is to preserve an obviously greater moral value than that upheld by refraining from the crime, e.g. breaking the speed limit to save life, causing injury or even death in self-defence, but in *R.* v. *Dudley and Stephens* (1884), 14 Q.B.D. 273, where two men drifting in an open boat killed and ate the body of a companion to save themselves, their conduct was held murder. It is pretty certainly not a generally available defence, but only allowed in very exceptional cases where the crime is obviously justifiable in the circumstances.

Necessity, agent of. See AGENT.

Neck verse. When a prisoner claimed benefit of clergy (q.v.) he was given to read a passage of scripture to prove his literacy and his entitlement to the privilege. The passage set came to be known as the neck verse; it was the opening verse of Psalm 51: Have mercy upon me, O God, according to thy loving kindness, according unto the multitude of thy tender mercies blot out my transgressions. Criminals frequently memorized the verse and while pretending to read it, in fact recited it.

Negative pregnant. In common law pleading, a reply to an allegation which denies the allegation literally, but does not answer the substance of it or contains in itself an implied admission of part of what is alleged. Thus, if a gift by deed were alleged the reply that 'he did not give by deed' is pregnant with the affirmative that he gave by parole.

Neglect. Failure to take some action which the person should have taken. It may be wilful or deliberate, or inadvertent or careless, and either variety may give rise to civil or criminal liability, if there should have been action and not neglect. Bare neglect without duty to act, e.g. neglect to clean one's shoes, infers no legal liability.

Negligence. In everyday usage, the word negligence denotes mere carelessness. Secondly, in legal usage it signifies failure to exercise the standard of care which the doer as a reasonable man should, by law, have exercised in the circumstances; if there is no legal duty to take care, lack of care has no legal consequences. In general, there is a legal duty to take care where it was or should have been reasonably foreseeable that failure to do so was likely to cause injury. Negligence is, accordingly, a mode in which many kinds of harms may be caused, by not taking such adequate precautions as should have been taken in the circumstances to avoid or prevent that harm, as contrasted with causing such harm intentionally or deliberately. A man may, accordingly, cause harm negligently though he was not careless but tried to be careful, if the care taken was such as the court deems inadequate in the circumstances.

Thirdly, in English law, the name negligence is given to a specific kind of tort, the tort of failing in particular circumstances to exercise the care which should have been shown in these circumstances, the care of the reasonable man, and of thereby causing harm to another in person or property. It implies the existence of a legal duty to take care, owed to the complainer, which duty exists, in general, where there is such proximity between two persons that a want of care on the part of the one is likely to affect the other injuriously, a failure to exercise the standard of care deemed right in the circumstances, which is normally defined as reasonable care, but which may be higher in particular circumstances, e.g. the airline pilot, the surgeon operating, causal connection between the failure to take care and injury suffered, not interrupted by the intervention of some other causal factor, and not too remotely connected with the ultimate harm, and actual loss, injury or damage to the complainer. Negligence takes innumerable forms, but the commonest forms are negligence causing personal injuries or death, of which species are employers' liability to an employee, the liability of occupiers of land to visitors thereon, the liability of suppliers to consumers, of persons doing work to their clients, of persons handling vehicles to other road-users, and so on.

The categories of negligence are not closed and new varieties such as negligence causing economic loss may be recognized.

This third sense of the term is derived from the second and developed probably from English law's continuing classification of torts by their origins as distinct forms of action. English law does not distinguish harms logically into harms caused intentionally and harms caused negligently but into trespass, conversion, slander, and the like (see TORT), so that physical harm caused negligently has been regarded as another nominate tort, negligence. In Scots law, which never recognized forms of actions, any kind of delict (q.v.) may be done either intentionally or negligently (in the second usage of the term) and wrongs of negligence are wrongs committed negligently.

In the context of contributory negligence (q.v.), negligence has its everyday meaning, of ordinary carelessness.

In criminal law, negligence bears its ordinary meaning and is not generally a ground of criminal liability but only in particular cases, such as the offence of driving a vehicle without due care and attention, or where the negligence is serious, such as to amount to recklessness or indifference to consequences.

Negotiable instruments. A class of undertakings to pay, evidenced in writing, which by long-settled mercantile custom have the characteristics that valuable consideration is presumed and need not be stated or proved, that they may be transferred from one person to another by delivery, or endorsement and delivery, without more formal assignment, enabling the transferee to sue thereon in his own name, and that a transferee taking such an instrument in good faith, for value, and without notice of any defect in the title of the transferor obtains a good title notwithstanding any defect in the title of the transferor. Negotiable instruments include bills of exchange, cheques, and promissory notes (including bank notes), dividend warrants, bankers' drafts, and scrip, debentures or share warrants payable to bearer, but do not include postal orders or share certificates. Bills of lading are not truly negotiable instruments.

Negotiation (1) A means for, and usually the first step towards, the settlement of international disputes. The Charter of the U.N. makes negotiation the first means of settlement incumbent on parties prior to invoking the jurisdiction of the Security Council. It is used in a similar sense in relation to industrial disputes.

(2) The transfer of a bill of exchange or other negotiable instrument (q.v.) from one person to another in such a manner as to make the transferee the holder of the bill.

Negotiorum gestio. One of the heads of quasi-contractual liability in Roman law under which, if one person intervenes voluntarily, unasked, and unauthorized, in the affairs of another in his absence, during his incapacity, or in other circumstances of emergency, he is entitled to be relieved of liabilities reasonably undertaken and to be compensated for expenditure reasonably incurred in the interest of and for the benefit of the absent or incapacitated party. The *gestor* is bound to take reasonable care and exercise reasonable diligence in what he does and is liable if, by lack of diligence or care, the other party's property or affairs be damaged or prejudiced. This class of obligation is not recognized by English common law though a very similar obligation is known under the principle of agency of necessity, but the obligation is recognized by Scots law and other systems founded on Roman law as one of the heads of obligation arising from unjust enrichments.

Nelson, Leonard (1882–1927). German philosopher, initially a Kantian interested in natural science but subsequently in ethics and jurisprudence. To him, rational jurisprudence was the science of the social conditions which would give each person equal external opportunities for attaining self-determination and spiritual freedom. His main writing on legal matters was *Die Rechtswissenschaft ohne Recht* (1917).

Nelson, William (eighteenth century). Compiler of an *Abridgment of the Common Law of England* (3 vols. 1725–26), said to be inaccurate and only a bad copy of Hughes' *Abridgment* (q.v.).

Nemo plus juris ad alium transferre potest quam ipse habet (No man can transfer to another a higher right than he has himself). The general principle that a possessor with a limited title, such as a borrower, cannot by selling or transferring property, convey to another a higher right than he has, such as ownership. Law has, however, out of regard for commercial convenience and practice, recognized that, in some cases, where a holder of goods is customarily authorized to deal with them as his own, a person in possession though not himself owner, may transfer a title of ownership. This applies particularly to factors (q.v.), i.e. mercantile agents. The same principle is expressed by the maxim *nemo dat quod non habet* (no one gives what he does not have).

Neo-Hegelian theories of justice. Neo-Hegelianism covers all revived interest in idealistic ways of thinking since the time of Hegel, and fresh attempts to build theories on Hegelian foundations.

The most notable jurist in this group is Josef Kohler (q.v.), who accepted the Hegelian view that the story of mankind, if its true significance were perceived, would show a persistent trend in the unfolding of the Idea, though he did not accept that the development of world history moved logically in accordance with the Hegelian dialectic. For legal purposes, Kohler translated the Idea into the idea of civilization or the raising of man's powers over nature and his own nature to the highest possible level. He developed the concept of the jural postulates of a civilization as the criterion of its just law; every civilization has its definite postulates and each society must shape the law according to its requirements. Kohler's work and thinking has been influential on later jurists, such as Roscoe Pound, though these successors have not all accepted his philosophical presuppositions. More modern neo-Hegelians, such as Binder, seek a strong state and the subordination of the individual to the State, and neo-Hegelianism underlay much of the juristic thinking of Nazi Germany and Fascist Italy.

In England, neo-Hegelianism had some influence on political philosophy, as on Bosanquet and T. H. Green, but though to the former the state embodied the highest intellectual and moral conduct of which the average man was capable, the latter saw the line of development as leading to Fabianism and social democracy.

Neo-Kantian theories of justice. Neo-Kantianism developed from the mid-nineteenth century. In its application to theory of justice and legal problems, the most notable figure is Rudolf Stammler (q.v.), who accepted Kant's major distinctions in the field of law and justice, but turned attention from the relation of morals and ethics to abstract rules and directed it to the relation of these matters to the administration of justice through rules. He deduced the principles of just law but had difficulty in applying them to particular cases. Nelson (q.v.) offered a reconstructed ethics as a framework within which justice is part of the material theory of duties. Binder (q.v.) was also, at least initially, neo-Kantian but later turned to a neo-Hegelian philosophy and the justification of Fascism. More importantly, Kelsen (q.v.) also started from a neo-Kantian standpoint, in his case from the neo-Kantian theory of the logic of the law, in developing his theory of the hierarchy of legal norms.

Neo-scholastic theories of justice. The scholastic theory of justice embodied in the work of St. Thomas Aquinas is based on the concept of divine law, of which natural law is the part which reveals itself in natural reason and from which all human laws are derived. Catholic doctrine was developed from these scholastic premises and modern natural law doctrines of Catholic origin are derived from the same bases. A number of modern jurists have built theories on these foundations, notably Renard, Delos, Maritain, Dabin, and Rommen. In general, they oppose the omnipotence of the State but, as with all natural law thinking, conclusions tend to be drawn equally justifiable in other premises.

Neostadius, Cornelius Matthaeus (Nieustadt) (1549–1606). Dutch jurist and compiler of *Utriusque Hollandiae Zelandiae Frisiaeque Curiae Decisiones* (1667), an important source of Roman-Dutch law, and other works.

Neratius Priscus, Lucius (first century A.D.). An important Roman jurist, consul and member of the council under Trajan and Hadrian, and with Celsus, last head of the Proculian school. He wrote various works, such as *Regulae, Epistulae, Membranae*, and a dissertation *De nuptiis*, which were extensively excerpted by later jurists.

Nerva, Marcus Cocceius (first century A.D.). A Roman jurist, whose works are frequently quoted by later jurists, said to have been head of the Proculian school before Proculus. His son was also a jurist, is said to have given responsa at 18, and wrote a *de usucapionibus*. His grandson was the Emperor Nerva.

Nervous shock. It is now recognized that nervous shock is as much a cause of action in tort as physical injury. Nervous shock is more than a mere fright but distinct and substantial lesion to the mind suffered through the medium of one or more of the senses and having discernible manifestations.

Netherlands law. The Netherlands were within the Roman Empire and, after its authority collapsed, that part of Europe came under the rule of Frankish tribes and within Charlemagne's Empire. Law was mainly customary, as restated in part in the Salic and Frisian laws. After the dismemberment of Charlemagne's Empire the northern Netherlands belonged to Lorraine, but government was in the hands of local rulers, all feudal dependents of the German Emperor. Each territory had its own law and judicial organization and towns shaped their own law in their own statutes. Attempts were made to create a unified state. The *Groote Raad* was established as an executive and judicial council for the whole of the Netherlands in 1446 and from 1473 to 1477 the judicial section, as the *Parlement van Mechelen*, was a central court of appeal.

By the mid-sixteenth century, all the Low Countries were under Charles V, King of Spain and Holy Roman Emperor. In 1548, they were formed into the Burgundian Circle within the Empire. In 1568, the Netherlands revolted and their independence of Spain was recognized by the Peace of Westphalia in 1648.

By this time there was a substantial amount of imperial legislation, administrative regulations, and laws for the preservation of the public peace, the precursors of criminal codes. Law books included von Repkow's *Sachsenspiegel*, of which a Dutch version, the *Hollandscher Saksenspiegel* was published in the fourteenth century, the *Libri Feudorum* supplemented by local sources of feudal law, and the *Somme rural* of Jean Boutillier (*c.* 1380) which borrowed much from Roman law. There were numerous compilations of the customary law of a territory and of local legislation. In 1531, Charles V called on all provincial and local authorities to submit their local laws to the government for approval, but this was largely ineffective. There were also compendia of municipal statutes and privileges from the twelfth century onwards. An important one was Jean Matthijsen's *Rechtboek van den Briel* (fifteenth century) and, by the late sixteenth century, the municipal laws of the major towns were voluminous compilations. Courts were local and territorial, the imperial court, the *Reichskammergericht* (1495), having little influence on the law of the Netherlands.

During the thirteenth and fourteenth centuries, Roman law exercised a growing influence on both legal theory and practice. The Reception of Roman law in Germany, and the view that the Holy Roman Empire was the successor of the ancient Roman Empire affected the Netherlands also, as being a feudal dependency of the German Empire, and was supported by the princes and their legal advisers, and by the rising universities. It became a subsidiary common law. There was also a reception of canon and feudal law. The speed and extent of the reception in the Netherlands varied in the different provinces, but was not so complete as in Germany. Thus there came into being the Roman-Dutch law.

The independence of the Netherlands from Spain and the German Empire was recognized by the Treaty of Westphalia (1648). In the seventeenth and eighteenth centuries, the Roman-Dutch law flourished. Legislation was collected in the *Groot Placaet Boeck*, decisions by various eminent jurists, including Bynkershoek's *Observationes tumultuariae*, Pauw's *Observationes tumultuariae novae*, and opinions by such jurists as Naeranus. Systematic treatises were written, the most notable being Everardus' *Topica* (1516) Wielant and Damhouder's works on civil and criminal procedure, Peck's books on arrest and testaments, Andreas Gail's *Practicarum observationum* (1580), and works of Merula and Goudelin.

Of the first importance is Grotius' *Inleiding tot de Hollandsche Rechtsgeleertheyd* (1631); quite apart from his writings on legal philosophy and international law, it treated the Roman-Dutch law as a system on its own and is still the basic work thereon. Later important works include the writings of Antonius Matthaeus, à Sande, Rodenburg, Groenewegen, and Decker. Vinnius wrote a standard commentary on Justinian's *Institutes*, later edited by Heineccius, and Noodt a Commentary on Digest 1–27 and various monographs. Both acquired European reputations. Van Leeuwen's *Het Roomsch Hollandsch Recht* (1664), Huber's *Praelectiones juris civilis* (1687) and Voet's *Commentarius ad Pandectas* (1698–1704) are all important. Bynkershoek collected decisions of the *Hooge Raad*, a series continued by Pauw, and wrote books on Roman and Roman-Dutch law, and is also one of the fundamental authorities in public international law. Van der Keessel published lectures on Roman law and on Grotius' *Inleiding*, and Van der Linden wrote a *Rechtgeleerd Practicaal en Koopmans Handboek* (1806). There were many other notable jurists and at this time the universities of the Netherlands were the leading law schools of Europe. The Roman Dutch law (q.v.) was carried overseas by Dutch colonists and thus became the basis of the law of South Africa, Ceylon, and till 1917, Guiana.

In 1795, the Netherlands was subjugated by France and the Batavian Republic established. In 1815, the kingdom of the Netherlands was established but in 1830 the southern province became the Kingdom of Belgium. The Netherlands is now a parliamentary democracy under an hereditary constitutional monarch. The constitution dates from 1814 and embodies the principle of ministerial responsibility. The legislature comprises the States-General, consisting of two chambers, a First Chamber elected by the councils of the 11 provinces and a larger Second Chamber, directly elected, the latter having the dominant power but both sharing legislative power with the government. Local government is effected by 11 provinces with directly elected councils, and nearly 900 municipalities with directly elected councils, but the chairmen of provincial executives and burgermeisters of municipalities are appointed by the government for six years. In 1809, the Roman-Dutch law was abolished in the Netherlands and replaced by a Napoleonic Dutch code and, in 1811, by the Code civil. In 1838, the *Burgerlijk Wetboek*, substantially modelled on the French Code Civil, was adopted and in 1886 the Criminal Code. Other codes, commercial, civil procedure, were adopted in 1838. From 1947, work proceeded on a new civil code drafted by Prof. E. M. Meijers and it was adopted in 1954–70.

The judicial system comprises a Supreme Court (*Hoge Raad*), five Courts of Appeal which hear civil and criminal appeals, 19 regional courts with original jurisdiction in nearly all cases and appellate jurisdiction from cantonal courts, and 62 cantonal or district courts, which deal with minor civil and criminal cases. There is no jury trial. There is a great variety of administrative courts and organs of the administration having quasi-judicial powers. There is a Public Prosecutor's Department, comprising Procurators General, Advocates General, and Public Prosecutors. Their functions extend to some civil cases also.

A. de Goede, *Nederlandse rechtsgeschiedenis*; R. Kranenburg, *Het Nederlands Staatsrecht*; W. Pompe, *Handboek van het Nederlandse Strafrecht*; C. Asser, *Handleiding van het Nederlandsch Burgerlijk Recht*; W. Molengraaff, *Inleiding tot het Nederlandsche Handelsrecht.*

Nettelbladt, Daniel (1719–91). One of the greatest law teachers in Germany, professor at Halle, and author of *Systema elementare universae jurisprudentiae naturalis* (1748), *Systema elementare jurisprudentiae positivae Germanorum generalis* (1781), and other works.

Neutralization. The making of a piece of land or water neutral, by treaty, e.g. the Kiel Canal, or declaration of neutrality, e.g. Switzerland.

Neutrality. The position of a state not a party to a war between other states. The essence of neutrality is that the neutral state stands aloof from the conflict and treats both sides impartially. In antiquity neutrality was unknown. Mediaeval international law distinguished between the friend not allied in the war and the enemy with whom one was at war. The *Consolato del Mare* provided that a friend's goods must be respected wherever they are found, and a friend's ship, even if carrying enemy's goods, was to be respected. Neutrality was recognized only in the eighteenth century. Practice thereafter varied but the law was much advanced by the action of neutral countries in later wars to assert their freedom. In 1780, the Armed Neutrality of Russia, Sweden, and Denmark, to which other countries later adhered, adopted an advanced neutral position and, in 1800, Russia formed a second Armed Neutrality against Britain. During the nineteenth century, neutrality was developed by the attitude of the U.S. during the French Revolutionary Wars, the permanent neutralization of Switzerland and Belgium and the Declaration of Paris of 1856, which created the rules that neutral goods on enemy ships must not be confiscated and that blockades must be effective.

Under the U.N. Charter, no member of U.N. is entitled to remain neutral in a war in which the Security Council has found a particular state guilty of aggression or breach of the peace and in which it has called on members to take military action against that state. But in other cases there is still room for neutrality.

Unless it is bound by treaty, no state is bound to remain neutral in a war, but a state is presumed to be neutral unless it declares by word or act its adherence to one side or the other. It may make a declaration of neutrality. The consequences of neutrality are that a neutral state must neither help nor harm either belligerent, nor allow either party to make use of its territory or resources for military purposes. It may not provide money or munitions to belligerents, nor allow its territory to become a theatre of war, nor allow its citizens to make neutral territory a base of war preparations or operations. It must not allow passage of troops through its territory, but may allow passage of wounded, and of warships through its territorial sea. Intercourse with each belligerent may continue as before. A neutral state must, however, acquiesce in the claim of belligerents to visit and search its ships and to capture them if carrying contraband, breaking blockade and otherwise acting un-neutrally. Belligerent states must avoid violation of neutral territory, interference in legitimate trade between neutral and the enemy, and seizure of neutral goods, other than contraband, on enemy vessels.

Neutral territory is an asylum for prisoners of either belligerent, who become free if they reach neutral territory; they may be detained or be allowed to rejoin their own forces. Bodies of troops may enter neutral territory to avoid capture, and may be rejected or granted asylum and disarmed. Belligerent aircraft may not enter the airspace of a neutral state and if they land they must be interned. Belligerent warships may be granted temporary asylum, particularly for repairs, or in distress, but abuse of this may result in the vessel being interned or ordered to leave.

Neutrality may be violated by either neutral or belligerent doing what is a breach of duty incumbent on him flowing from the state of neutrality. Such conduct is an international delinquency and the offender may be required to pay compensation.

Neuva Recopilacion. A compilation of nine books of ordinances of Cortes of Castile, royal orders, and other materials made by one Bartolome de Arrieta on the instructions of Philip II and promulgated in 1564. The work was very incomplete yet it was several times re-issued with later statutes added and the material in it with the supplements was re-arranged in 12 books in the *Novisima Recopilacion de las Leyes de España* in 1805.

New Abridgment. A name sometimes given to Matthew Bacon's (q.v.) *Abridgment of Law and Equity* (1736). There was also *A New Abridgment of Cases in Equity* by Josiah Brown (1793).

New Assignment. In common law pleading, if the declaration in an action was ambiguous and the defendant pleaded facts which literally answered the declaration but did not answer his real claim, the plaintiff had to proceed by new assignment, by pleading in reply that his action was not for the cause the defendant thought but for another to which the plea was inapplicable. It also arose where the plaintiff's declaration was ambiguous and it was necessary to reassign the cause of action with fresh particulars.

New Book of Entries. A name sometimes given to William Rastell's (q.v.) *Collection of Entries* (1566) and sometimes to Coke's (q.v.) *A Booke of Entries* (1614).

New Entries. A name sometimes given to Coke's (q.v.) *A Book of Entries* (1614).

New Inn. One of the Inns of Chancery (q.v.).

New England, Dominion of. A consolidation of eight English colonies in North America in 1688–89. Colonial assemblies were abolished, taxes and quit rents levied, and the Church of England was encouraged. The Dominion collapsed after the revolution in England in 1688 and Andros, the leading figure, was imprisoned.

New England Confederation (1643). A colonial union of Massachusetts Bay, Plymouth, Connecticut, and New Haven for defence against Dutch, French, and Indians, and the first federation in North America. A board of eight commissioners, chosen annually by the colony legislatures, was empowered to declare defensive and offensive war and given jurisdiction over inter-colonial disputes and Indian affairs. It directed operations during King Philip's War (1675) but was dissolved when the Charter of Massachusetts Bay was revoked in 1684. Nevertheless, it was a precedent for joint action and federation.

New Laws of 1542. The first comprehensive code for the government of Spanish America, in 54 articles, including a virtual abolition of Indian slavery. Spain sent out special commissioners to enforce the New Laws, which were unwelcome to the Spanish colonists. In Peru, there was a rebellion against the laws and the Spanish government had to make substantial concessions to make the laws enforceable. Nevertheless, they were an important stage in bringing the conquistadors into subjection to the Spanish Crown.

New Natura Brevium. A work published as *Natura Brevium* by Sir Anthony Fitzherbert in law French in 1534, and many later editions to 1794, and later called the *New Natura Brevium* to distinguish it from the earlier (Old) *Natura Brevium.* It has always been deemed a work of great authority and value on pleading and procedure.

New Style. The Calendar (New Style) Act 1750, introduced the Gregorian Calendar into England in place of the Julian Calendar in 1752 and to bring the calendar into line with that in use in countries already using the Gregorian Calendar, the day after September 3rd was numbered September 14th.

See also NEW YEAR'S DAY.

New trial. The retrial of an issue of fact. This is permitted only in limited circumstances. Where an issue has been tried by civil jury trial, the only way of having the jury's verdict reviewed is by moving for a new trial, which is allowed only if substantial miscarriage has been occasioned by the judge misdirecting the jury, improperly admitting, or rejecting, evidence or by the jury giving a verdict against the weight of the evidence, or awarding damages so excessive or inadequate as to be wholly disproportionate to the facts of the case or on various other grounds. A new trial is not allowed merely because the appellate court does not agree with the jury's verdict. If a new trial is allowed, the evidence is re-heard by a fresh jury. New trial is not the normal way of reviewing the decision of a judge sitting without a jury.

In criminal cases, the Court of Appeal may order a new trial if this is required in the interests of justice but only when an appeal against conviction is allowed in the light of fresh evidence. It proceeds on a fresh indictment and can be only for the offence in respect of which the appeal has been allowed, and not for any fresh offence, or offence of which the accused was found not guilty.

New Year's Day. At Rome, according to the Julian Calendar, established by Julius Caesar the year began on January 1st. From the seventh to the twelfth century, the year was deemed by chroniclers to begin on Christmas Day. Thereafter, in ecclesiastical matters, and from the fourteenth century for legal purposes, it commenced on March 25th (Lady Day). Consequently, a document dated in England between January 1st and March 24th truly belongs, according to modern views, to the year after that whose date it bears. The Calendar Act, 1751, made it commence thereafter on January 1st. In Scotland, from 1600, the year was ordered to begin on January 1st, though in other respects Scotland also adhered to the Old Style (Julian) Calendar.

New Year's Day has traditionally been a public holiday in Scotland and the Bank Holidays Act, 1871 (now Banking and Financial Dealings Act, 1971), made 1st and 2nd January bank holidays in Scotland.

New Zealand law. Captain Cook took possession of New Zealand in 1769 and, in 1839, letters patent extended the colony of New South Wales to include any New Zealand territory acquired by the Queen. In 1840, the Treaty of Waitangi was signed with

Maori chiefs and British sovereignty proclaimed and, in 1841, New Zealand was constituted a separate colony. Courts of law were established in the same year. When British sovereignty was proclaimed, New Zealand became subject to the laws of England, and when it became a separate colony, many Ordinances were passed modelled on Ordinances in force in Australian colonies. In 1858, the New Zealand Parliament provided that the laws of England, so far as applicable to the circumstances of the colony, should be deemed to have been in force since 14th January 1840.

In 1852, the United Kingdom Parliament created a constitution with a Governor, an appointed Legislative Council, and an elected House of Representatives, the three forming the General Assembly, and six provinces each with an elected Superintendent and elected Council having wide powers. The first Parliament met in 1854. The provinces were abolished in 1876. Under the 1852 constitution, responsible government developed and the Governor's responsibility to the British Colonial Office was gradually eroded.

Courts were created on the English model and the general ideas and conceptions of English law were adopted from the start, but the early judges had great difficulties. In 1857 there were only two judges of the Supreme Court for the colony, and persons appointed as Resident Magistrates were often legally unqualified.

In the latter part of the nineteenth century, there was increasing regard for U.S. decisions, though this subsequently declined.

The principle has been accepted in New Zealand that precedents from any part of the world, where the common law system prevails, are authoritative, to maintain the interpretation of that law as uniform as possible. Every court in New Zealand is bound by the decisions of all courts superior to itself. The decisions of superior courts in other British jurisdictions are persuasive precedents.

From an early stage, the New Zealand Parliament has shown itself bold in legislation on matters frequently not yet covered by legislation in other common law jurisdictions.

Constitution

The core of the constitution is contained in the U.K. Act to Grant a Representative Constitution to the Colony of New Zealand of 1852. The constitution is a unitary one.

The Queen is represented by the Governor General, appointed on the advice of the New Zealand Government, and he legally has a power of veto on legislation.

The upper house, the Legislative Council, which was appointive, was abolished in 1950. The lower house, now the sole house, is the House of Representatives. It includes four Maori members, and in organization, ceremonial, and procedure generally follows the House of Commons.

Since 1947, the New Zealand Parliament has had plenary powers. It is not a High Court, not having inherited the *lex et consuetudo Parliamenti*. Its powers are derived from the 1852 Act, as later amended, and the Statute of Westminster 1931.

Judicial system

The power to establish courts of justice was conferred in 1840, and was exercised on the English model. The Supreme Court of New Zealand was established in 1841 and has jurisdiction in law and equity, which have always been administered concurrently. It also has criminal jurisdiction, original and appellate.

New Zealand is divided into 10 judicial districts. Supreme Court judges are normally resident in each of the main cities and attend to the judicial work there and in the adjacent districts. There is an Administrative Division of the Supreme Court, which hears appeals from administrative tribunals. The Court of Appeal was created in 1862 and consists of the Chief Justice *ex officio*, the president, and two judges; the quorum is three. It hears appeals from the Supreme Court in civil and criminal cases. Further appeal lies to the Judicial Committee of the Privy Council.

The inferior courts are the magistrates' courts, held by stipendiary magistrates, which exercise wide and varied jurisdiction, civil and criminal, generally concurrent with the Supreme Court. Lay justices sit occasionally in minor criminal cases.

There are also courts of special jurisdiction, notably a Maori Land Court and Maori Appellate Court, coroner's courts, Court of Arbitration for the Settlement of industrial disputes, Compensation court to deal with workmen's compensation claims, and warden's courts.

Public Law

In general, the public law of England has been accepted and followed. The Crown has long been liable for breach of contract or of duty in the same way as a private individual.

Public administration has been notable for the large number of government departments and agencies which have been set up, many much earlier than similar bodies in other comparable countries. There are also a number of public corporations. There is a very extensive system of social security.

Local government is organized under statute; there are county councils; borough councils; and harbour, road, and hospital boards, all with powers and duties defined by statute, and subject to judicial control to keep them within their jurisdiction.

Delegated legislative power is extensively organized, but is subject to control by the courts deciding whether regulations are *intra vires* or *ultra vires*.

Numerous administrative tribunals exist under particular statutes. The Supreme Court exercises control by the writs of mandamus, prohibition, and *certiorari* and New Zealand was the first country in the Commonwealth to have an Ombudsman.

Criminal law and procedure

Initially, English criminal law was adopted so far as suitable to the circumstances of the colony. The law relating to indictable offences was codified in 1893, great simplification being effected.

Criminal jurisdiction is exercised by the Supreme Court, stipendiary magistrates, or at least two justices of the peace sitting together.

In criminal trials, the grand jury remains and considers the bill of indictment, normally after preliminary investigation by, and committal for, trial from a magistrates' court. The petty jury remains but a large number of indictable cases may be tried summarily unless the defendant insists on jury trial, and persons charged with summary offences may, in certain cases, claim jury trial. Appeal lies to the Court of Appeal.

The penal system has all the customary features of a modern one—probation, parole, special institutions for young offenders.

Private law

The major branches of private law are based on and, in general, are the same as in English law.

Family law has followed clearly the English model. Married women have full rights to own property, as if single. Divorce by judicial process was introduced in 1867, the grounds being several times subsequently widened.

The law of contract and commercial transactions has not deviated in large measure from English law and most of the English legislation on these topics has been copied in New Zealand.

In relation to employment, industrial disputes are dealt with first by a Council of Conciliation, but a party may apply to the Court of Arbitration for exemption from the terms agreed, which may make an award as to the mode, term, and conditions of employment. The Court of Arbitration has become a major force in formulating industrial codes and making awards covering large sectors of the national economy.

The law of torts, similarly, follows very closely the corresponding law of England.

In the sphere of real property, the English principles were, from the first, statutorily simplified in the application in New Zealand. A system of registration of deeds has been in operation from the beginnings and, in 1860, there was introduced the Torrens system of registration of title to land and this has almost completely superseded the system of registration of deeds.

Since 1879, all a deceased's property descends on intestacy in the same manner and, since 1945, there has been a code regulating the succession to all property on intestacy. Since 1900, it has been possible for the court to overrule a will to make provision for members of the family of a deceased who has not adequately provided for them.

Procedure

In civil cases, trial may be before one or more judges, or before a judge and jury, either special or common, depending on the class of action. Stipendiary magistrates have a jurisdiction limited only by the value of the cause, but otherwise concurrent with the Supreme Court.

Legal profession

The legal profession is a unified one, but a barrister may practise exclusively as such and, if he is a Queen's Counsel, must do so, and a solicitor is precluded from practising as a barrister until he has fulfilled the requirements for admission to the Bar. Barristers have the privileges and responsibilities of English barristers. The New Zealand Law Society was constituted in 1869.

In 1931, a Council of Legal Education was constituted and the examination of candidates for admission became the province of the University of New Zealand (now the separate universities of New Zealand) and, in practical subjects, of the Council of Legal Education.

A Law Revision Committee was established in 1936 and has been responsible for much valuable reform of private law.

Extensive reliance is placed in New Zealand on English textbooks, though there is a good number of books on local developments, and in Sir John Salmond (q.v.), New Zealand has produced a jurist renowned throughout the common law world.

J. Hight and H. Bamford, *The Constitutional History and Law of New Zealand*; J. L. Robson, *New Zealand, The Development of its Laws and Constitution*; K. J. Scott, *The New Zealand Constitution*; F. B. Adams, *Criminal Law in New Zealand*; J. Garrow and S. Goodall, *Law of Real Property in New Zealand*; J. Garrow and H. R. Gray, *Law of Personal Property in New Zealand*; *New Zealand Official Yearbook*.

Newcastle, Propositions of (1646). One of a series of proposals put to Charlies I by Parliament during the Civil War as a basis for a settlement. They were framed on the Propositions of Uxbridge, 1644. The main points were the abolition of the ecclesiastical hierarchy, the enforcement of Presbyterian Church government, penalties against Catholics, power to Parliament to control the armed forces for 20 years, and many other restrictions on the royal power. Charles made three evasive replies, indicating acceptance of some only of the propositions, and no settlement was reached.

Newgate. The New Gate of the City of London, the principal west gate at the point where Watling Street reached London, roughly along the line of Oxford Street and Holborn, built in the reign of Henry I, used as a prison from at least 1188, and rebuilt as such in 1420. It was destroyed in the Gordon riots in 1780 but rebuilt in 1783 and used for both civil debtors and criminals until 1815. Thereafter, it was used for criminals only and, from 1881, only during the sittings of the Central Criminal Court. It was finally demolished in 1902, part of the site being occupied by the Central Criminal Court then built.

Newgate Calendar (or *Malefactors' Bloody Register*). The original series of this work, by R. Sanders, was published in five volumes in 1760 and narrated notorious crimes from 1700 till then. There were many later editions. Later series were issued from about 1820 as the *Newgate Calendar*, and the *New Newgate Calendar* appeared weekly in 1863–65. There was also an *Annals of Newgate* by the Rev. M. Villette and others (1776).

Newton, Sir Henry (1651–1715). Became an advocate of Doctors' Commons in 1678, Judge-Advocate in 1694, and was envoy to various Italian states. In 1714, he was made judge of the Court of Admiralty but died the following year.

Next friend. A person of full age who is joined with a minor as plaintiff to enable the latter to sue in an English court. The same rule applies to a person of unsound mind, in certain cases, and formerly applied to a married woman.

Next-of-kin. Those who are next in degree of kinship to a deceased person, according to the civil law mode of reckoning degrees. Prior to 1926, the next-of-kin were those relatives of the deceased who, under the Statute of Distributions, 1670, were entitled to his personal estate. Since 1925, a statutory table (revised in 1952) defines what relatives are entitled to his estate on his death intestate. Similarly in Scots law, the next-of-kin were those relatives who prior to 1964 were entitled at common law to the deceased's moveable estate. The class was extended in 1855 and 1919.

Nexum. A transaction in early Roman law, probably a form of loan contracted *per aes et libram* in the presence of a *libripens* and five witnesses and the counterpart in obligations to *mancipatio* (q.v.) in the law of property. The lender probably originally weighed out the money lent to the borrower. The borrower probably bound himself to repay, the obligation to do so being enforceable by immediate execution against the person, so that if he did not repay the creditor could seize him and kill him or sell him as a slave. An alternative view is that the transaction was a kind of self-sale of the debtor in security of repayment. The possibility of this bondage resulting from debt was ended by the *Lex Poetilia*, probably of 326 B.C.

Nicholas of Cusa (1401–64). German cardinal, philosopher, leader in the conciliar movement in the Catholic Church, and the leading German thinker of the fifteenth century. His views on social theory are in his *De concordantia catholica* (1431). He was the first to raise doubts about the Donation of Constantine and the Pseudo-Isidorian Decretals, and his inquiring attitude had great subsequent influence.

Nicholl, John (1759–1838). He was admitted to Doctors' Commons in 1785, became King's Advocate in 1798, and was a founder of King's College, London. He was in Parliament 1802–33. He became Dean of the Arches (1809), judge of the Court of Admiralty (1833), and vicar-general to the Archbishop of Canterbury (1834), resigning the office of Dean of the Arches. He was later a member of the judicial committee of the Privy Council. His decisions are based on extensive knowledge of the authorities and great ability to deduce the relevant principles and to apply them.

Nightwalker. Persons who go about at night and often disturb the public peace. At common law, they might be taken into custody till the morning but the offence of being a common nightwalker was abolished in 1967. The word is used in some older statutes to denote a prostitute.

Nil Debet (he owes nothing). The plea of the general issue in actions of debt not founded on a specialty debt. In 1853, the plea of *nunquam indebitatus*, that the defendant was never indebted as alleged, was substituted, but this in turn has been replaced by the rule that the defendant must deal specifically with every allegation of fact not admitted.

Nineteen Propositions, The. Terms which Parliament intimated to Charles I in June 1642, before the Civil War broke out. They amounted to a demand that Parliament should be the sovereign in the country and would have established government by persons appointed by Parliament in lieu of government by the King. The points were that (1) the King's officers and Ministers of State should be approved by Parliament; (2) the great affairs of the kingdom be transacted only in Parliament; (3) the major officers of State be chosen with the approbation of Parliament; (4) the persons to whom the education of the King's children should be committed be approved by Parliament; (5) no marriage of the King's children be concluded without consent of Parliament; (6) the law against Catholics be strictly put into execution; (7) the votes of Catholic lords be taken away; (8) reformation be made of the Church government and liturgy; (9) the King should accept Parliament's plans for the ordering of the militia; (10) members evicted from office be restored; (11) all Privy Councillors and judges should take an oath to maintain the Petition of Right; (12) all judges and officers appointed with the approval of Parliament to hold office *quamdiu bene se gesserint*; (13) the justice of Parliament to pass on all delinquents; (14) the King's general pardon be granted; (15) the forts and castles of the kingdom to be put under the command of persons appointed by the King with the approval of Parliament; (16) the King's extraordinary guards and military forces to be removed and discharged; (17) an alliance be concluded with the United Provinces; (18) to clear Lord Kimbolton and the five members of the House of Commons; and (19) a Bill be passed restraining peers made hereafter from sitting or voting in Parliament unless admitted with the consent of both Houses. The King rejected the propositions and the Civil War was inevitable.

Nisbet, Sir John, Lord Dirleton (1609–87). Called to the Scottish Bar in 1633, he became Lord Advocate in 1664 and also a judge of the Court of Session as Lord Dirleton, being the last person to combine these offices. His severity as a persecutor of covenanters almost equalled that of his successor Sir George Mackenzie (q.v.). In 1670, he was a member of the Commission to treat of union with England. He had to resign in 1677. At a time when bad men were common, he was one of the worst. Nevertheless, he was a distinguished pleader, and was deemed learned. He is, today, best remembered for his book *Some Doubts and Questions in the Law especially of Scotland*, a collection, under alphabetical heads, of difficult points, commonly known as *Dirleton's Doubts*, and also a *Collection of Decisions of the Lords of Council and Session*, 1665–77. These works are usually bound together. Many years later his *Doubts* were answered by Stewart (q.v.).

Nisi (unless). A decree, rule, or order is made *nisi* when it is not to take effect unless the person affected fails within a stated time to appear and show cause why it should not take effect. Orders and rules *nisi* were frequently made by common law courts prior to the Judicature Acts but they are now much less common. In divorce cases, the decree is decree *nisi*, to enable the Queen's Proctor (q.v.) to determine whether or not he need intervene. Only later is decree made absolute.

Nisi prius (unless before). The forerunner of the *nisi prius* system was that justices in eyre ordered all cases in the Court of Common Pleas, originating in the county in which they then were, should be adjourned before them, but that time in the Common Pleas was always reserved for those cases in case the itinerant justices had not previously come to those parts. The Statute of Westminster II (1285) provided that, for holding inquisitions into minor trespass and the trial of pleas commenced before either King's Bench or Common Pleas, a day and place should be set aside in the circuits of the justices of assize, and that such cases should not be determined in the Benches at Westminster unless the judges of assize had not previously (*nisi prius*) come into the county to try them. This system was extended by later statutes to common pleas begun in the Exchequer, and to criminal cases. The system proved popular and satisfactory and was the basis of the modern system of assizes, now superseded, since 1971, by the system of Crown Courts.

No case. A submission which may be made by a defendant at the close of a plaintiff's case or the case for the Crown. The plea is to the effect that the plaintiff's case or the case for the Crown is insufficient in law so that, even if the evidence is accepted, a finding for the plaintiff or for the Crown would be incompetent and unjustified in law. When such a submission is made, it is for the trial judge to decide whether there is a case to go to the jury or not. Such a submission may also be made in a magistrates' court.

No-right or no-claim. In Hohfeld's (q.v.) scheme of legal relationships, the opposite concept to a claim and the correlative concept to a liberty or privilege, in that if A has a liberty or privilege, B, the other party to the relationship, has no-right or no-claim, i.e. cannot stop the exercise of liberty of privilege. The phrases are unsatisfactory and 'inability' would be better.

Noahide laws. The designation in the Jewish Talmud of seven Biblical laws given to Adam and Noah before the revelation on Mount Sinai and, hence, deemed binding on all mankind. They include prohibitions on worship of idols, blasphemy, murder, adultery, and robbery and the command to establish courts of justice. Others were added later. Later scholars have regarded these precepts as universal norms of conduct, guarantees of fundamental human rights, and as fundamental concepts of international law.

Nobel Prizes. Five prizes awarded annually by the Nobel Foundation, established under the will of the Swedish industrialist Alfred Nobel. They are for the most significant discoveries or advances in physics, chemistry, physiology or medicine, eminent literary work, and outstanding work in promoting international good will and in organizing peace conferences. The peace prize is awarded by the Norwegian Nobel Committee. Among winners of the peace prize have been such distinguished jurists as T. M. C. Asser (1911), Elihu Root (1912), Viscount Cecil (1937), Dag Hammarskjold (1961), and Rene Cassin (1968). The Institute of International Law was given the award in 1904.

Nobile Officium. In Scots law, an extraordinary equitable power vested in the Court of Session, inherent in it as a Supreme Court, and exercised by the Inner House of that Court. Many of the specific powers within this general power have, in time, been extended also to the Outer House, e.g. granting special powers to trustees, and even to the Sheriff Court, e.g. appointing a judicial factor. The main categories within which this jurisdiction is now exercised are the supplying of defects in statutory procedure, the supplying of omissions or defects in deeds, particularly in relation to trusts, and affording remedies in cases for which there is no other remedy apparent. In modern practice, the Court will rarely exercise this power save where there is precedent or analogy. The High Court of Justiciary has a similar but more restricted, and only rarely invoked, power.

Nobility. Persons deemed of noble birth, or ennobled by the sovereign by the grant of any dignity or degree of the peerage, i.e. being made a duke, marquess, earl, viscount, or baron, or their female counterparts. Corresponding grades of nobility exist or existed in all the older European countries. In all cases the concept of nobility originated in feudalism and in rewards by kings for outstanding services. Ennoblement was connected with endowment with large estates and jurisdiction; thus dukedoms and earldoms are frequently of the same names as counties. Since 1963, no new hereditary ennoblements have been made in the U.K.

Noblesse de robe. In seventeenth- and eighteenth-century France, a class of hereditary nobles who acquired their rank from holding high state office. They developed in the early seventeenth century by the grant of charters of nobility to judicial offices to secure their support. At the apex of this privileged class were officers of such courts as the Parlement of Paris. At first, they were disliked and opposed by the older nobility and by those who rose by way of military service, but the groups tended to coalesce in the eighteenth century in defence of their privileges.

No-fault liability. The doctrine that a person who has sustained any of stated injuries or losses should have a claim against an injurer or a fund, frequently managed by the State, without having to establish that another person was liable for fault in causing him the harm complained of. This is the principle underlying workmen's compensation and industrial injuries liability in the U.K. and automobile insurance schemes in certain Canadian provinces and U.S. states. No-fault liability may be complete, in excluding tort claims completely, or partial, if permitting the injured person to bring a claim in tort against the person responsible and to recover in addition to, or so far as in excess of, his insurance claim.

Nolle prosequi. In civil proceedings, an undertaking by the plaintiff not to proceed with his action at all or as to part of it, or as to certain defendants. The Attorney-General of England has power in any criminal proceedings on indictment at any time to enter a *nolle prosequi* and thereby to stay proceedings. The origin of the power is uncertain but the basis appears to be that the Crown, in whose name criminal proceedings are taken, may discontinue them. The first instance was in 1555. The court will not thereafter allow any further proceedings to be taken in the case, nor inquire into the reasons or justification for the Attorney-General's decision. It is not equivalent to an acquittal and does not bar a fresh indictment for the same offence.

In the U.S., the discretion is vested in the prosecutor such as the district attorney and may be used if the accused agrees to make restitution or to plead guilty to a lesser charge.

Nolumus leges Angliae mutare. Traditionally, the reply given by the barons of England at the Parliament of Merton (1235) when it was proposed, in conformity with the canon law, to introduce the principle of legitimation *per subsequens matrimonium* into English law, a change which was not made until 1926.

Nominal damages. An award of a trivial sum of money, such as ½p., as damages in recognition of the fact that technically there has been an infringement of a right vested in the plaintiff but also indicating that no discernible loss had been established. Nominal damages are frequently awarded where the court or jury takes the view that, technically, the plaintiff was wronged but it has formed a very low opinion of his conduct and of the real merits of his claim.

Nominate and innominate. A nominate right or obligation is one having a *nomen iuris* which defines its limits and settles its consequences. In Roman law, there were 12 nominate contracts, to each of which was appropriate a particular action, such as the *actiones commodati directa et contraria*. In modern law, the term is applied to recognized transactions, such as sale or lease, the term innominate being applied to anomalous transactions not falling into a recognized category and not having defined consequences.

Nominating and reducing. A mode formerly used to obtain a panel of special jurors in London and Middlesex from which to obtain the jury to try a particular action. Numbers denoting the persons on the sheriff's list were drawn from a box until 48 unchallenged persons had been nominated. Each party struck out 12 and the rest formed the panel. This mode is now obsolete.

Nomination. The naming of a person to an office or post; the proposing of a person as a candidate for election. Also, a means whereby a person, entitled to limited sums of money from a friendly society or other similar body, may name a person to whom the money is to be paid on his death, without need for making a formal will.

Nomocanon. A word used from the eleventh century to designate canonical collections composed of both ecclesiastical and civil laws dealing with ecclesiastical matters and canonical collections containing both secular and ecclesiastical laws, a type proper to the Oriental Churches from the early Middle Ages and having an important role in the history of Oriental canon law. From the fourth century, ecclesiastical matters had an important place in imperial law, as is seen in the *Codex Theodosianus* and the *Basilica*. Collections of nomocanons have been a principal source of Oriental canon law since the Middle Ages. The earliest is the *Nomocanon L titulorum* (sixth century) falsely ascribed to Joannes Scholasticus and in use till the twelfth century. The most important is the *Nomocanon XIV titulorum*, compiled about 629, probably by Enantiophanes, comprising decrees of councils, texts of letters of the Father, and imperial constitutions. In 982, it was accepted as the universal law of the Oriental Church. It was revised several times, notably in 1198 by Theodore Balsamon.

Sava, Archbishop of Serbia, made a Slavic adaptation of the Byzantine nomocanons in 1219 under the name *Kormchaya Kniga* which was adopted by all the Slavic Orthodox Churches. In the eighteenth century, new compilations of excerpts from the nomocanons appeared, notably the *Pedalion* for the Greeks and the *Kniga pravil* for the Russians.

Nomotechnia (*un description del Common Leys d'angleterre solonque les rules des art*). A work published in law French by Sir Henry Finch in 1613. A translation by Finch himself was published in 1627 under the title *Law, or a Discourse Thereof, in Foure Bookes*, another translation as *A description of the Common Laws of England according to the Rules of Art compared with the Prerogatives of the King* in 1759, and an abridgment called *A Summary of the Common Law of England* in 1654. The first book is, in some respects, a forerunner of modern books on jurisprudence, the second deals with property, the third with trespasses and crimes, and the fourth with procedure, an arrangement which may owe something to Justinian. It was a useful book and a predecessor of Blackstone's *Commentaries* as a literary exposition, and till Blackstone was regarded as the best elementary book for law students, but it was from the first overshadowed as a practitioner's work by Coke's *Institutes* (1628).

Non-access. In matrimonial and affiliation proceedings, the impossibility of a husband having had sexual relations with his wife at or about a particular time, a fact relevant to rebut the presumption that a married woman's child was begotten by her husband.

Non-age. As opposed to 'full age', the period during which a person is in minority or below the age of being an adult, formerly 21 and now, since 1970, 18.

Non-assumpsit. A plea by way of traverse in the old action of assumpsit, which denied the existence of any undertaking to the effect alleged, or of the matters of fact from which the alleged promise would be implied.

Non-belligerency. Illegal discrimination by a neutral between belligerents.

Non-entry. In Scottish feudal law, a casualty, now obsolete, payable to the superior of lands where the heir of a vassal neglected to have himself entered with the superior, i.e. neglected to renew his investiture. By virtue of the casualty the superior was entitled to the rents of the feu.

Non est factum (it is not his deed). A plea denying that a deed mentioned in the plaintiff's pleading was truly that of the defendant, allowing the latter to plead that the deed was never executed by him, and also giving an opening for the plea that there had been a mistake by him as to the nature of the deed, not merely as to its terms or effect.

Nonfeasance. The neglect or failure of a person to do what he ought to do, particularly to perform a duty owed to the public, whereby an individual sustains special or particular damage to himself. See also MALFEASANCE; MISFEASANCE.

Non-jurors. Those who, by reason of attachment to the Stuart dynasty, refused to take the oaths of allegiance to the government established by the Revolution of 1688, particularly those who refused to take the oath which, by statute of 1714, might be tendered to anyone suspected of being disaffected. A person refusing was adjudged a popish recusant convict and subject to heavy penalties. In Scotland, the entire episcopate and most of the episcopal clergy were non-jurors and, accordingly, the presbyterian Church became finally established as the Church of Scotland in 1690.

Non-justiciable (disputes). In international law disputes which are deemed not to be capable of resolution by a judicial tribunal. Some constitutional provisions may also be non-justiciable.

Non liquet (It is not clear). In civil law, *non liquet* means the failure of a party to discharge a material burden of proof and is limited to matters of fact. In international law, it is used of the question whether an international tribunal is under a duty to decide or not to decide a question when it finds that appropriate rules do not exist for solving the problem before the tribunal. It is questionable whether this is ever proper, because in default of recognized rules on a subject, a tribunal must either find for the respondent, as not having infringed any rule of law, or develop rules by analogy or by resort to equity.

Non obstante (notwithstanding). It became the practice of the Crown, from an early date, to permit the doing of various things *non obstante* any law to the contrary. Later, more important examples were the attempts of James II to dispense with the provisions of the Test Act, 1672, and after the Revolution the Bill of Rights (1688) enacted that any dispensation *non obstante* should be void and of no effect.

Non-suit. The conclusion ordered by a judge when the plaintiff fails to establish a legal cause of action or to support his pleadings by any admissible evidence. The term is technically obsolete since 1883, but is still colloquially used where a judge withdraws a case from a jury and directs a verdict for the defendant.

Non-user. The failure of a person to utilize or exercise a right, particularly with reference to easements, profits, and similar rights which can be extinguished by non-user for at least 20 years, of such a nature as indicates intention to abandon the right.

Noodt, Gerard (1647–1725). A famous Dutch jurist with an international reputation, Professor of Law at Leiden, with such a reputation for exposition that scholars talked of *methodus Noodti*, author of *Dissertatio de jure summi imperii et lege regia* (1699), *Probabilium juris civilis libri IV*, a commentary on Digest I–XXVII (1713) and monographs *Ad Legem Aquiliam, De Pactis et transactionibus* and *De Usufructu*, and other works.

Norddeutsches Strafgesetzbuch. The criminal code of the North German Confederation, enacted in 1870. Though not a radically reformist code, it marked an advance on the law of the confederated states. From 1871, it became the law of the German Empire.

Nordenfelt* v. *Maxim-Nordenfelt Guns and Ammunition Co. [1894] A.C. 535. N. sold his world-wide armaments business and undertook not to compete with the purchaser anywhere in the world for 25 years. It was held that restraint of an individual's future liberty of action might be justified in special circumstances, if reasonable in reference to the interests of the parties and to the interests of the public, and that in the present case it was reasonable notwithstanding the unlimited area and the length of time.

Nordic Council. An organization of Norway, Sweden, Finland, Denmark, and Iceland, formed in 1952 for consultation and co-operation on matters of common interest. It has sought to make uniform provisions on legal, economic, and social matters. The executive is the annual Council, of representatives from the national legislatures.

Norfolk Island. An island in the south-west Pacific discovered in 1774 and now part of Australia. It was used as a penal colony from 1788 to 1813 and 1825 to 1855. The prisoners sent there were the most depraved and dissolute, removed there from Botany Bay and Van Diemen's Land, and treatment was usually harsh and sadistic. It was, however, the scene of Maconochie's (q.v.) enlightened experiments in rewarding good conduct.

Norm. A rule or standard of behaviour accepted by members of a group. Norms are more specific than values or ideals but usually less so than rules of law. They are essentially social, created by society and stating what it will allow or condemn. They may influence law or be influenced by law. They may relate to any human activity or aspect of social conduct, and may be accepted by individuals voluntarily, or be enforced by external sanctions. Kelsen uses the phrase 'legal norm' as meaning a legal rule, and in his philosophy the Grundnorm or fundamental norm is the basis of a legal system. The word is also used to denote a statistically settled standard of average behaviour, attitude or opinion of a group. In this sense, it concerns actual rather than expected behaviour.

Norman* v. *Baltimore & Ohio Railroad Co. (1935), 294 U.S. 240. As a measure to deal with the great depression of the 1930s, Congress withdrew gold from circulation and devalued the dollar by reducing the amount of gold measuring the value of each dollar in circulation, and also required that existing contracts requiring payment in gold or in dollars of the former gold value were to be satisfied by payment of the same number of dollars though of the reduced gold value. Creditors challenged this as an impairment of the obligation of contracts and a taking of property without due process of law. The Supreme Court, by a majority, upheld the action of Congress, which minimized the confusion resulting from devaluation.

Norman Conquest. The conquest of England by the Normans under Duke William in and after 1066. The legal consequences of the conquest were very important and far-reaching. The major consequences were the replacement of the Anglo-Saxon kings and nobility by a Norman-French series of kings and nobility, the latter holding large tracts of land by knight-service, the enforced adoption of Norman-French as the language of the court and the King's courts, the conversion of the Witanagemot into the *Curia Regis*, which later threw out shoots which became the Exchequer, King's Bench, and Common Pleas, the reorganization and strengthening of the Church while restricting its political power, the separation of the courts of the State and of the Church, a new structure of social organization, and of jurisdiction, and the establishment of the unquestioned overlordship and supremacy of the King. In short, the consequences can be said to be systematization, the imposition of orderly government, and the ascertainment and strict enforcement of royal rights. Thus were laid the essential foundations of the common law which Henry I and Henry II began to build. If the Norman Conquest had not taken place, the cultural, political and economic links would not have been with France and the south, but with Scandinavia and the north.

G. O. Sayles, *The Mediaeval Foundations of England*; F. M. Stenton, *The First Century of English Feudalism*; F. M. Stenton, *English Society in the Early Middle Ages, 1066–1307*; A. L. Poole, *Domesday Book to Magna Carta, 1087–1216*; H. W. C. Davis, *England under the Normans and Angevins 1066–1272*; F. Pollock and F. W. Maitland, *History of English Law before the time of Edward I.*

Norman-French. The language introduced into England by the Normans after the Norman Conquest, which was used at court and became the official language of English law down to 1362, though charters and land grants were commonly expressed in Latin and treatises were in Latin. The Parliament Roll and the Statute Roll were in French. As time went on, it retained its archaic character and increasingly diverged from the French used in France; it degenerated into law French which long continued and, though abolished by Cromwell in 1650, revived a little after the Restoration. English was made the language of the law in 1731.

Normand, Wilfrid Guild, Baron, of Aberdour (1884–1962). Became an advocate of the Scottish Bar in 1910, was an M.P., 1931–35, Solicitor-General for Scotland, 1929 and 1931–33, and Lord Advocate, 1933–35. He was Lord Justice-General of Scotland and Lord President of the Court of Session, 1935–47, and a Lord of Appeal, 1947–53.

Normandie, Grand Coutumier de. An important mediaeval legal work in three versions, in Latin, entitled *Summa de legibus Normannie* (*c.* 1250), in French prose, entitled *Grand Coutumier de Normandie* (*c.* 1260), and in French verse (*c.* 1280). It codifies the customs and shows scientific arrangement. There is not much evidence of Roman or canon law. It held the field until the official redaction of the Coutume de Normandie in 1583 and is still a source of common law in Jersey.

Normandie, Très Ancien Coutumier de. A private collection of the customs of Normandy contained in two treatises dealing respectively with private law, procedure, penal law, and the seigniorial courts, and with the progress of an action involving title to property, dating from *c.* 1200 and *c.* 1220 respectively.

North, Francis, Lord Guildford (1637–85). He was called to the Bar in 1661 and quickly established a reputation. He became Solicitor-General in 1671 and Attorney-General in 1693 and succeeded Vaughan as Chief Justice of the Common Pleas in 1674. There he actively reformed the abuses of the court and restored its jurisdiction, encroached on by the King's Bench. He was a learned lawyer and advocated legal reforms, and a man of upright character. As a politician, and in political cases he was less at home. In 1682, he succeeded as Lord Keeper and, in 1683, became

Baron Guildford. As Lord Keeper, though fitted for the post and desirous of making reforms, he achieved little, and Jeffreys undermined his influence. His brother was Sir Dudley North, the economist, and his great-grandson was Prime Minister as Lord North, in 1770–82.

Roger North, *Lives of the Norths* (1735).

North, Council of the. This probably originated in the council of Richard, Duke of Gloucester, who administered the north of England for his brother, Edward IV. When he became Richard III he appointed his nephew Lincoln with a council to govern the northern parts. Thereafter, the council was intermittent and it had lapsed by 1509. By 1525, the Duke of Richmond, Henry VIII's lieutenant in the north, had been given a council staffed by lawyers and administrators which administered the Tudor lands there and had also wide civil and criminal jurisdiction. Its jurisdiction waxed and waned and, after the Pilgrimage of Grace in 1536, it was reorganized on a permanent basis, as a body of administrators and judges under a Lord President, responsible for the whole of the north—an instrument of central government.

It had jurisdiction over the five northern counties of England and comprised a large number of persons, of whom five, other than the President and Vice-President, were salaried and bound to continuous attendance. The special object of the court seems to have been the suppression of riots and disturbances but its jurisdiction extended to aiding the bishops and the High Commission in the repression of recusants and the maintenance of uniformity and good morals, supervising the justices of the peace, providing for the defence of the border and generally maintaining the laws. In cases of difficulty the President and Council might apply to the justices of the courts or the Privy Council and act according to their advice. They had power to determine real and personal actions, to hold ordinary sessions, execute commissions of Oyer and Terminer and gaol delivery, to hear indictments for murder and felony and execute felons. It developed a jurisdiction in equity and had Star Chamber powers in relation to riot and the supervision of administration. Its most important years were 1572–95 under the Earl of Huntingdon. It came to an end in 1641.

R. Reid, *The King's Council in the North* (1921).

North Atlantic Assembly. The forum for parliamentarians of N.A.T.O. member-states; it has no statutory basis but meets annually and makes recommendations and resolutions to the North Atlantic Council (the chief authority of N.A.T.O.). It is composed of delegates from the national parliaments of the member-states and works through various committees.

North Atlantic Treaty Organisation (N.A.T.O.). An international organization formed by treaty in 1949 for the defence of any member attacked by an outsider and the development of friendly and peaceful relations. The chief authority is the North Atlantic Council, assisted by a Secretary-General, and a Military Committee. The headquarters were in Paris until 1967 and since then in Brussels. Since 1949, it has done much to develop a common strategy and common military installations, and to improve the ability of the group of resist armed attack.

North Briton. A newspaper, begun in 1762 by John Wilkes (q.v.), in which he asserted the freedom of the press. He attacked George III and his favourites and, in No. 45, described the Speech from the Throne to Parliament as a falsehood. General warrants were issued against all concerned in the production of the paper and Wilkes was arrested. He asserted his privilege as an M.P. but, in 1763, the House of Commons declared No. 45 to be a seditious libel, denied that the privilege of members protected them in such a case, and banned the publication. Wilkes fled from the country and the *North Briton* ceased publication though a few issues were produced some years later by a supporter.

North German Confederation. Established after the Austro-Prussian War (1867) replacing the German Confederation of 1815. The King of Prussia was president of this confederation of all states north of the Main, and responsible for the army, foreign policy, and the appointment of the Chancellor. The Bundersrat, composed of delegates from the states, but dominated by the Prussian delegation, was responsible for legislation. The Reichstag, elected by manhood suffrage, shared control of the Budget with the Bundestag. Austria, Bavaria, Baden, Wurtemberg, and Hesse-Darmstadt were excluded from the confederation but all other than Austria were linked to Prussia by secret military treaties and linked economically to the Confederation by the Zollverein. In 1871, the Confederation was superseded by the German Empire.

Northampton, Assize of. Decisions reached at Henry II's council held at Northampton in 1176. They deal with administration of criminal law, appearing to repeat and amend, provisions of the Assize of Clarendon (1166); land law, including reference to the assize of mort d'ancestor (q.v.); and instructions for the justices in eyre, requiring them to exact oaths of fealty from everyone and to see that castles ordered to be demolished were in fact pulled down.

Northern Ireland Law. Northern Ireland, as a distinct area of government and law, was created by the Government of Ireland Act, 1920, which provided for the setting up of a separate Parliament for Northern Ireland within the United Kingdom. Northern Ireland voted not to accept the dominion status for Ireland envisaged in 1922. At its creation, it inherited what had hitherto been the law of Ireland (see IRISH LAW).

Though having an executive, legislature, and judiciary of its own, Northern Ireland was not independent of the United Kingdom, not was the relationship truly federal in that sole legislative power on certain matters was reserved to the Westminster Parliament and it retained full legislative power even in delegated matters. Northern Ireland continues to return members to the Westminster Parliament. Any court in Northern Ireland may hold to be *ultra vires* and void any statute of the Northern Ireland Parliament. The Northern Ireland Parliament showed a strong tendency to follow the lead of the Westminster Parliament in legislation, both as to the subjects dealt with, and as to the substance of the legislation.

Executive

The executive power is vested in the British Crown and exercised down to 1972 by the Governor of Northern Ireland. His executive committee or Cabinet comprised a Prime Minister and Ministers administering departments. Only members of the Privy Council of Northern Ireland were eligible for appointment as Ministers and they must be or become members of the Northern Ireland Parliament. A Minister might sit and speak in either House but might vote only in the House of which he was a member. In 1972, on account of the disturbances and terrorism, the Secretary of State for Northern Ireland assumed the position of chief executive officer and administered all departments.

Provision was made in 1973 for a Northern Ireland Executive consisting of a chief executive, the heads of the Northern Ireland departments, and other persons all appointed by the Secretary of State, but this did not come into being. The Attorney-General for England and Wales superseded the Attorney-General for Northern Ireland.

Legislature

The Northern Ireland Parliament consisted of two Houses—the Senate, comprising the Lord Mayor of Belfast, the Mayor of Londonderry and 24 senators elected for eight years by members of the House of Commons, half retiring every four years, and the House of Commons comprising 52 members elected until dissolution or for five years. The privileges of members were similar to those of the Westminster House of Commons. The Parliament had general power to make laws for the peace, order and good government of Northern Ireland, and to legislate in matters not reserved to the Westminster Parliament. But it was not a sovereign legislature because some subjects were reserved, and because the Westminster Parliament retained overriding power to legislate even on subjects delegated to the Northern Ireland Parliament.

The Parliament was suspended in 1972 and direct rule from London substituted. In 1973, provision was made for a Northern Ireland Assembly with 78 members elected by single transferable vote, with power to legislate by Measure subject to approval by the Queen in Council, and the Northern Ireland Parliament was abolished. The Assembly might not legislate on excepted and reserved matters without the consent of the Secretary of State and the U.K. Parliament might still legislate for Northern Ireland. The Assembly was prorogued in 1974 and dissolved in 1975 and direct rule continued. A Northern Ireland Constitutional Convention was elected later in 1975 but did not come into being.

Judicial organisation

The superior court is the Supreme Court of Judicature, consisting of the High Court of Justice, comprising the Lord Chief Justice of Northern Ireland and six puisne judges, and divided into Queen's Bench, Chancery, and Family Divisions, the Court of Appeal, consisting of the Lord Chief Justice and three Lords Justices of Appeal, and the Crown Court, consisting of the Lord Chief Justice, the judges of the Court of Appeal and High Court and the county court judges. All the judges of the Supreme Court are members of the Court of Appeal when it sits in criminal cases. Appeal lies with leave from the Court of Appeal in both civil and criminal cases to the House of Lords. The Judicial Committee of the Privy Council has jurisdiction to determine questions of constitutional law referred to it, but if a question of the constitutional legislation arises, any court must decide it, even if the decision requires it to hold *ultra vires* legislation of the Northern Ireland Parliament.

Judges of the Supreme Court and county court judges hold Crown Courts for trial of cases on indictment.

County court judges (or in Belfast and Londonderry, Recorders) sit as county court, without a jury, trying civil cases of intermediate importance, and as Crown Court, with a jury, trying all but the most serious criminal offences. They also act as courts of appeal from Resident Magistrates. Appeal lies to the Court of Appeal.

Resident magistrates sit as examining magistrates and as courts of petty sessions trying summary offences and minor civil cases. Justices of the peace may act similarly. There is a separate commission of the peace for each county court division.

In 1973, in consequence of the amount of terrorism, numerous offences commonly committed by terrorists were made triable without a jury at Belfast City Commission or Belfast Recorder's Court.

Coroners' courts are held as in England to enquire into sudden and unexplained deaths.

There is a lands Tribunal to assess claims for compensation for the taking of land.

Sources of the law

In legislation, the Irish courts must have regard to legislation of the former Parliament of Ireland, of the United Kingdom Parliament affecting Ireland and of the former Parliament of Northern Ireland.

Among precedents, they have regard to precedents of the Northern Irish courts, English courts, Irish courts before 1921, courts of the rest of Ireland since 1921 and, as persuasive authority, to precedents of Scottish and Commonwealth courts. The principle of *stare decisis* and the doctrine of binding precedents is accepted in Northern Ireland. Judicial notice is taken of both English law and the law of the Republic of Ireland.

There are separate series of Northern Ireland reports.

Public law

The main execution of government policy is through the government departments, each headed by a Minister who is a member of the Cabinet. Ministers also have considerable powers of delegated legislation. An Act of 1958 makes comprehensive provision for the numbering, publication and sale of subordinate legislation.

Most of the taxes are imposed and collected by the Westminster Parliament, part being paid into the Northern Ireland Exchequer, but a small proportion of the revenue is raised from taxes levied by the Northern Ireland government. There is a Joint Exchequer Board which determines every year what part of the taxes reserved to Westminster is attributable to Northern Ireland, which is paid to the Northern Ireland Exchequer, and the contribution from Northern Ireland to United Kingdom liabilities and expenditure.

Local government is through elected county borough councils, county councils, and urban and rural district councils, all exercising delegated statutory powers.

Social security legislation and legislation protecting workpeople are on the same lines as in England.

The police, the Royal Ulster Constabulary, come under the Minister for Home Affairs and are not organized on a county or regional basis.

The liberty of the subject has been less protected in Ireland than in England. Habeas Corpus was not enacted in Ireland till 1781 and, in the nineteenth century, Coercion Acts and other measures frequently enabled the government to interfere arbitrarily with the liberty of the subject, and there remains considerable power to curb unrest.

Criminal law

In general principles and in respect of most statutes, the criminal law is the same as in England. The distinction between felony and misdemeanour was abolished in 1967.

In procedure and treatment of offenders, most English reforms have been copied in Northern Ireland. The grand jury still exists for cases tried at assizes. Certain indictable offences may, with the accused's consent, be tried summarily, and in certain summary offences the accused may claim jury trial.

The Court of Criminal Appeal may, in particular cases, order a retrial. The Minister of Home Affairs may, in certain circumstances, refer a case to the Court of Criminal Appeal or refer to that court a point on which he desires the assistance of the court.

Since 1957 there has been a statutory scheme of compensation for damage caused as a result of an act committed by a malicious person acting for an unlawful association.

Private law

There are no significant variations from English law in the sphere of family law, though it was only in 1939 that the courts acquired jurisdiction to grant divorce.

The general principles of contract and of mercantile contracts are the same as in English law. There is distinct companies legislation which is, however, similar in principle to English legislation. In torts there is little variation from English law.

There are significant differences from English law in the sphere of real property, though broadly the common law of tenures and estates was adopted, and there developed types of landholding unusual or unknown in England. Statutory changes were made in the nineteenth century, particularly modifying the law of landlord and tenant in favour of the tenant, securing them fair rents, fixity of tenure and free sale, and permitting purchase of land by tenants. The great modifications of English real property law effected in 1925 were not copied in Northern Ireland. There is registry of deeds affecting land, established in 1707, and local registration of title, established in 1891.

The law of trusts is generally the same as that of England before the reforms of 1925, but was substantially revised in 1958, and the general doctrines of equity are similar to those accepted in England.

The rules determining succession on intestacy are similar to those applicable in England before

1926, realty and personalty descending separately; the rules as to the administration of estates are substantially the same as in England. The court may make provision out of an estate for dependants not reasonably provided for by a deceased.

Civil procedure

Law and equity are administered concurrently.

The principles of civil procedure are generally similar to those in force in the English courts. A civil jury numbers seven.

Legal profession

There is a separate Northern Ireland Bar, members of which belong to the Inn of Court of Northern Ireland, with the ranks of Queen's Counsel and of barrister, and a separate Incorporated Law Society of Northern Ireland. The great majority of lawyers in Northern Ireland obtain the main part of their legal training in the Faculty of Law of the Queen's University, Belfast.

There was an Attorney-General for Northern Ireland, who was a member of the Privy Council and of the Parliament of Northern Ireland, who advised and represented the Crown in matters affecting Northern Ireland. There was no Solicitor-General but a deputy might be appointed to act for the Attorney-General. Since the Northern Irish Parliament was abolished the office of Attorney-General for Northern Ireland has been held by the Attorney-General (for England).

As in England, there is a legal aid and advice scheme to assist persons of limited means to have legal advice or litigate, administered by the Law Society.

Chart, *History of Northern Ireland*; A. S. Quekett, *Constitution of Northern Ireland*; P. N. Mansergh, *Government of Northern Ireland*; F. H. Newark, *Devolution of government: the experiment in Northern Ireland*; *Ulster Year Book*.

Northern Ireland, Supreme Court of Judicature of. The Supreme Court of Northern Ireland, created in 1920 by adaptation of the Judicature (Ireland) Act, 1877, and restructured by the Judicature (Northern Ireland) Act, 1978. It consists of the High Court of Justice, the Crown Court, and the Court of Appeal. Rules of Court distinguish the High Court into a Queen's Bench Division (which has common law, and admiralty jurisdiction) Chancery Division and Family Division.

The Crown Court exclusive jurisdiction in criminal trials on indictment.

All the judges of the Supreme Court are also judges of the Court of Appeal when hearing criminal appeals; this replaces the Court of Criminal Appeal, established in 1930.

Further appeal lies from both Court of Appeal and Court of Criminal Appeal to the House of Lords. There is a Supreme Court Rules Committee and a unified civil service, the Northern Ireland Court Service, serving the courts.

Northern Securities Company* v. *U.S. (1904), 193 U.S. 197. An anti-trust suit brought by the Department of Justice to dissolve the company which was a company holding railroad stock. The Supreme Court held that the Sherman Act was aimed at all combinations in restraint of trade which directly operated in restraint of commerce and could apply to corporations lawfully organised under state statutes.

Northington, Earl of. See HENLEY, ROBERT.

Northstead, Manor of. See CHILTERN HUNDREDS.

Norwegian law. in Norway, law was originally customary but from about 1000 there is record of Law-men who appeared at the local and general Things or assemblies, recited the law orally and interpreted it in particular cases brought before the Thing. About 1200, the Law-men obtained an official character, though their decrees were not regarded as binding. Groups of districts meeting at Law-Things began to exercise both legislative and judicial functions but when the monarchy achieved general power the kings quickly assumed the lead in legislation.

Norwegian laws were reduced to writing in the various regions in the twelfth century; the earliest written texts were probably the Gulathing-lov of about 1150 and the Frostathing-lov of 1190, containing what was generally known as St. Olaf's laws. These were law-books rather than collections having official authority.

Under King Magnus Lagaboter (Lawmender) (1263–80), there was finally created 1274–76, a common national code, the Landslov, based on regional codifications, such as the Frosta-Thing Law, which deals with church regulations, criminal law, land law, and commercial law. It was introduced in Norwegian dependencies, Greenland, Faroes, Orkney, and Shetland. A code based on this was enacted in Iceland under the name of Ironside (Jarnsida) in 1271. In 1276, a new Town-law common to the whole country was prepared on the orders of King Magnus based on the National Law Codification and superseding distinct town-laws dealing with commerce and shipping. King Magnus' codifications were officially in force for about 400 years, though increasingly superseded by legislation.

In 1602–4, King Christian IV had a new code prepared, known as the Code of Christian IV; it was mainly a revision of the laws of Magnus Lagaboter. It was not long satisfactory, and revision was undertaken after 1666, but little progress was made until the 1680 draft of the Danish code appeared.

In 1688, Norway was given a new complete codification of law, King Christian V's Norwegian Law, in six books, which to a considerable extent introduced Danish provisions and was, in lesser degree, influenced by conceptions of Roman law. It sought to supersede prior legislation and some of it remained in force into the twentieth century.

Modern Norway is a constitutional hereditary monarchy. The constitution, framed in 1814, was influenced by the U.S. Constitution, French revolutionary ideas and British political traditions, and provided for a limited monarchy and separation of powers.

The legislature is the Storting which cannot be dissolved during its four-year term. The King has a suspensive veto on legislation but this has not been exercised for many years.

The King is administrative head of the government and selects the Cabinet. If the Storting votes against Cabinet proposals on a major issue the Minister or even the whole Cabinet resigns. For local government purposes the country is divided into 20 regions, which comprise about 450 municipalities, urban and rural, having councils elected every fourth year, which elect mayors and aldermen.

A Supreme Court was established in the mid-seventeenth century and re-established in 1815 under the Constitution of 1814. There are five courts of intermediate appeal and 100 town courts and district courts throughout the country exercising civil jurisdiction. For criminal cases there are the town courts and county courts which, in certain cases, function as examining and summary courts, the intermediate appeal courts, which also act as trial courts in serious cases, and the Supreme Court. Special ecclesiastical courts were established during the Middle Ages, the procedure in which was regulated by canon law, and, despite the Reformation, existed till 1887.

There are a General Attorney, who acts for the state in civil cases, and a State Director of Public Prosecutions.

The Norwegian constitution of 1814, enacted after dissolution of the union with Denmark, provided for new general codes. A general Penal Code was enacted in 1842 and replaced in 1902. Some parts of public, civil, and procedural law have been codified but there has been no general civil codification, though a commission was appointed in 1953 to undertake this. Custom has been and still is a major source. Parts of the 1688 code still stand and there is a great deal of legislation on various topics. Since 1880 there has been an increasing amount of uniform Scandinavian law. Judicial precedents and legal writings contribute to the development of customary law rather than are independent sources.

S. Juul, A. Malmström and J. Søndergaard, *Scandinavian Legal Bibliography*; T. Andenaes, *The Constitution of Norway*.

Norwich Guildhall Court. This court may have originated in a court of husting like that of London. It had jurisdiction in personal and mixed actions to any value and was held six times a year before the steward, who was a barrister of at least five years' standing. The Recorder of Norwich was not *ex officio* judge of the court. The court was active until its abolition in 1971.

Noscitur a sociis (a man is known from his associates). A principle in the interpretation of deeds and statutes, that a word of uncertain denotation may be assigned a meaning by reference to the accompanying words and the contract. See also *Ejusdem generis*.

Not guilty. The plea in reply to a criminal charge where the accused wishes to put the Crown to the task of proving his guilt, and also the proper verdict of a jury or court of summary jurisdiction where the prosecution has not established the accused's guilt beyond reasonable doubt. It does not mean that the accused was wholly innocent, but merely that the prosecution has not established his guilt. In older practice, the plea was used in common law actions of tort, where the defendant was raising the general issue (q.v.) and denied that he had committed the wrong complained of.

Not negotiable. Words which may be added to the crossing of a cheque which have the effect that no one who takes the cheque can have or can give a better title than the person from whom he took it. The words do not have the effect of rendering the cheque incapable of being negotiated, i.e. transferred from one person to another.

Not proven (past participle of preve, an old variant of verb prove). In Scottish criminal procedure, a court or jury may find an accused Guilty, Not proven, or Not guilty. The intermediate verdict developed from the practice in the seventeenth century of drafting indictments in detail and remitting to the jury to find the facts proven or not proven, it being then for the judges to decide the legal effect of a finding that the facts charged were proven. But in later cases, the jury was allowed to find the accused Guilty or Not Guilty, and in the result all three verdicts have survived. The effect of a Not proven verdict is the same as that of Not Guilty, in that the accused is released and cannot be tried again, but the implication is that he has escaped conviction only because of some slight doubt or some technical inadequacy of evidence, e.g. lack of corroboration, and that serious suspicion of guilt attaches to him so that a verdict of Not Guilty is not justified, and the verdict is a valuable preventive against unjustified verdicts of Not Guilty.

Notarial instrument. In Scots law, a deed under the hand of a notary public, narrating procedure transacted by or before him in his official capacity, formerly much used in conveyancing.

Notary. An office of great antiquity known in all Western European countries. In Roman law, a *notarius* was a person who took notes of judicial proceedings. A notary's functions are to draw, attest and certify deeds and other documents, to prepare mercantile documents, to note and certify transactions relating to negotiable instruments and to draw up protests or formal documents relating to occurrences on ships' voyages and their navigation. They administer oaths and take affidavits. Before the Reformation the appointment of notaries throughout Christendom was made by the Pope. In England, he delegated his powers to the Archbishop of Canterbury. Since 1533, all faculties appointing notaries in England and Wales have been issued by the Archbishop acting through the Court of Faculties presided over by the Master of the Faculties. Since 1920, the appointment of district notaries practising wholly in Wales has been vested in the Lord Chancellor. In England, there are now ecclesiastical notaries, an office held by registrars of ecclesiastical courts and legal secretaries of bishops, general notaries, who hold faculties entitling them to practise anywhere in England and Wales, and district notaries, who may practise only in a limited area in England or Wales. Diplomatic and consular officers in foreign countries may do any notarial act which any notary public may do in the United Kingdom.

In Scotland, any solicitor may, on petition to the Court of Session, be admitted a notary public. His main functions are to authenticate wills for disabled persons, take affidavits in bankruptcy, note and protest bills, and authenticate the due execution of deeds to be used abroad. Each notary has a motto which he uses in official actings and a protocol book in which to record protests and other matters authenticated by him.

Note. An abbreviation for promissory note (q.v.), for bank note, and a term for various kinds of documents used in trade, such as contract notes, bought and sold notes, consignment notes and the like.

Note a bill. The action of a notary public in presenting again a foreign bill which has been dishonoured and if it is not paid, to make a note on the back consisting of his initials, the date, and the reason, if any given, for non-payment.

Note, judge's. The notes made by a judge of the evidence given during the trial of an action. They may include observations on witnesses' demeanour or credibility and points other than merely the evidence given. In criminal trials, the judge must make such notes and furnish them to the Court of Appeal in the event of an appeal. These notes are quite distinct from the official shorthand writer's notes of the evidence given and arguments adduced. In Scottish sheriff courts, a sheriff is obliged to append a note to all interlocutors other than merely formal ones, giving his reasons in law for his disposal of the case. This is in effect an opinion.

Notice (knowledge). Judicial notice (q.v.) comprises those facts of which a judge is deemed to have knowledge without the need for parties to lead evidence of them. To give notice to a person means bringing it to his knowledge.

Notice, generally, may be actual, imputed, or constructive. Actual or express notice is that conveyed by one person to another orally or in writing. Imputed notice is notice given to an agent and therefore attributed to the principal, it being presumed that the latter will have communicated it to his principal, but this presumption may be rebutted. Constructive notice exists where the person would have had knowledge if he had made such ordinary inquiries as would have been reasonable in the circumstances, and where he is accordingly deemed to have knowledge. Constructive notice is frequently given by registration in a public register or notification in a public way, e.g. by publication in the *Gazette*, in which case persons liable to be affected are taken to have notice. The doctrine of notice is important in many contexts in equity.

Notice of trial. In certain cases in the High Court the plaintiff must give the defendant notice of intention to lodge a certificate of readiness for trial. If the plaintiff fails to set down the action for trial within the time prescribed, the defendant may do so, or apply to dismiss the action for want of prosecution.

Notice to quit. A notice, given by either landlord or tenant of land or tenements leased to the other party as a normal prerequisite of determining the tenancy otherwise than by agreement. The notice should be clear and unambiguous.

Notice to treat. The notice which a public authority, having statutory powers to acquire land compulsorily, must give to the persons having interests in the land as the first step in acquiring the land. It intimates that the authority wishes to acquire the land and is willing to treat for the purchase of it, and requests particulars of the estate and interests of the persons to whom it is given.

Noting. The making of a note or memorandum by a notary public on a bill of exchange of the fact that he has presented the bill for acceptance or for payment, as the case may be, and that it was dishonoured. The note consists of his initials, the date, the fact of presentment, and refusal. Noting is, for business purposes, deemed as evidencing due presentment. It is not necessary for preserving recourse against the drawer or indorser of an inland bill. In the case of foreign bills, they must be noted and protested, and notice must be sent to the drawer and indorsers with a copy of the bill, if the drawer and indorsers are abroad.

Notorious. That which is matter of notoriety or public knowledge, and accordingly does not need to be proved by evidence.

Notour bankruptcy. In Scots law, a state of insolvency which is notorious, as evidenced by definite statutory indicia, and which has important effects in restraining the debtor's powers of dealing with his property and in limiting his creditors' powers of securing preferences by diligence. It is a prerequisite of sequestration, and the procedure for realizing and distributing the debtor's estate commonly known as bankruptcy.

Nottingham, Earl of. See FINCH, HENEAGE.

Nova causa interveniens. An alternative for *novus actus interveniens* (q.v.).

Nova Scotia, baronets of. In 1621, Sir William Alexander, later Earl of Stirling, was granted the province of Nova Scotia by Charles I and, in 1624, he announced that it was to be divided into 150 baronies to be held by Scottish baronets each of whom was to send six men to serve in the colony for two years or to pay 2,000 marks in lieu, and also 1,000 marks to Alexander. Charles I chartered the order in 1625. About 120 grants were made but applications declined and the baronetage became unimportant when Nova Scotia was ceded to France in 1632. But the titles remained a part of the Scottish peerage and were opened to English and Irish gentlemen in 1633, and a number of U.K. peers still derive their rank from the original creations of Nova Scotia baronets.

Nova statuta* and *vetera statuta. Two groups of English statutes. The division is based on the earliest printed copies, itself based on arrangement of manuscripts. The *Vetera Statuta* end with Edward II (1327) and were first printed in 1508. The *Nova Statuta* begin with Edward III and were printed in 1497. Nevertheless, the division corresponds to a substantial dividing line, the old statutes being really part of the common law.

Novae narrationes. A book, in law French, probably compiled in the early part of Edward III's reign, dealing chiefly with narrationes, i.e. plaintiffs' statements of claim, but including some defences. The list of writs on which pleadings are given is similar to those annotated in the (Old) *Natura Brevium.* It was printed several times in the sixteenth century and seems to have been of considerable value, and of substantial authority, even into the fifteenth century. The *Articuli ad Novas Narrationes* (q.v.) is frequently printed with this work.

E. Shanks and S. Milsom, *Novae Narrationes* (Selden Socy.).

Novation. A transaction whereby, with the consent of all parties concerned, a new contract is substituted for one in being, either substituting a new agreement or substituting a different person for creditor or debtor. It differs from assignment in that the consent of all parties is necessary, though this may be inferred from conduct and need not be expressed.

Novel disseisin, assize of. A procedure established by enactment of Henry II (the Assize of Clarendon) in 1166, whereby if one person were dispossessed (disseised) of his free tenement unjustly and without a judgment, he was to have a remedy by royal writ in that a jury of 12 free and lawful men of the neighbourhood was to be summoned and in the presence of the King's justice to answer whether the disseisin were unjustified or not. The importance of this ordinance was that lawful possession, as distinct from ownership, was to be protected by a rapid remedy and that seisin was to be protected by the King, no matter from what lord the lands were held. One man could not turn another out of possession without a judgment. The procedure survived until 1833.

Novellae (*leges novellae*, supplementary laws). Constitutions issued by Justinian after the publication of his (second) Code (q.v.) in A.D. 534, many dealing with public or ecclesiastical matters, some resolving difficulties in earlier legislation and some substantially altering earlier private law. They vary greatly in length. Most were in Greek, and some in both Latin and Greek. There are three known unofficial, or semi-official, collections, the *Epitome Juliani* (*c.* A.D. 555) compiled by Julianus, a professor at Constantinople, containing 124 constitutions in an abridged Latin version. This was probably intended for use in Italy; the *Authenticum*, a collection of 134 constitutions, the Latin in original, the Greek translated into Latin, possibly made officially (*c.* A.D. 557) for promulgation in Italy; and the Greek collection or collection of 168 constitutions (*c.* A.D. 580) arranged for the most part chronologically. It must have originated in Constantinople and cannot have been completed before the reign of Tiberius II (572–82). The customary citation is Nov. and the number. They are divided into chapters, in the case of the larger ones, and paragraphs. The standard modern edition is by Schoell and Kroll.

P. Noailles, *Les Collections de novelles*; N. van der Wal, *Manuale Novellarum.*

Novellae posttheodosianae, collections of Roman imperial constitutions issued after the *Codex Theodosianus*, made for the western Empire and containing constitutions of A.D. 438–58. They are included in Volume II of Mommsen's *Theodosiani libri XVI.*

Novelty. The element of originality in an invention which is requisite for the grant of a patent to the inventor. Accordingly, lack of novelty is an objection to the grant of a patent.

Novodamus, charter of. In Scottish feudal land law, a fresh grant of lands to the grantee usually to make some change in the incidents of tenure of land already granted, or to resolve doubts about the grant or its terms.

Novus actus interveniens (or *nova causa interveniens*). A fresh factor intervening between an initial causal factor, such as negligent act or omission by a defendant, and its consequence, in the shape of harm to the plaintiff, which may nor may not, depending on the circumstances, be deemed to have broken the causal connection between cause and consequence and to have superseded the original causal factor and replaced it by a new one, with consequences for the legal liability of the person responsible for the initial causal factor. A *novus actus* (or *nova causa*) *interveniens* may be constituted by an act or omission of a third party, or one by the plaintiff himself, or by an event of nature.

Noxal liability. In Roman law, if a son *in potestate* or a slave committed a delict the injured person could take vengeance on him, but this would have infringed the father or master's *potestas*, and the latter could satisfy his liability by making noxal surrender of the wrongdoer to the injured person or buy off the liability by paying the penalty. Roman law also recognized, under the *actio de pauperie*, noxal surrender of animals which had done harm as an alternative to compensating the person injured by the animal. The maxim *noxa caput sequitur* stated the rule that the purchaser of a slave or animal which had done harm took over the liability to make good the damages.

Nudum pactum. In the Roman law, an informal bargain not amounting to a contract and not the basis for an action. *Pacta* were, however, admitted as defences to actions and then to modify obligations. The term is sometimes used in common law in the quite different sense of an agreement lacking consideration and on which accordingly, unless made under seal, no action will lie; hence the maxim *ex nudo pacto non oritur actio.*

Nugent, Thomas, Lord Riverston (?–1715). A member of the English and Irish Bars, he became a judge of the Irish King's Bench in 1686, Chief Justice in 1687, and was given the title of Chief Justice of Ireland by James II.

Nuisance (from *nocumentum*, hurt, inconvenience, or damage). A term of not very definite meaning. It generally covers acts unwarranted by law which cause inconvenience or damage to the public in the exercise of rights common to all subjects, acts connected with the occupation of land which injure another person in his use of land or interfere with the enjoyment of land or some right connected therewith, and acts or omissions declared by statute to be nuisance. Nuisance includes escape of water or noxious fumes, noise, vibration, pollution and unreasonable injury to health or comfort. It must be a continuing, or at least not single or short, inconvenience. A public, general or common nuisance inflicts damage, injury or inconvenience on all persons, or all who come within the area of the nuisance. It may be restrained or damages recovered by civil proceedings by or in the name of the Attorney-General alone or at the relation of a local authority or private individual, or by action by a local authority or private individual. A public nuisance is also a common law crime and indictable, and no lapse of time legalizes a public nuisance.

A private nuisance does not inconvenience the public at large but does interfere with a person's use or enjoyment of land. It may be restrained or damages recovered by action at the instance of the party aggrieved, and may be abated, i.e. summarily removed or stopped, by the party injured without resort to legal proceedings at all. The right to perpetrate a nuisance may be acquired by statute or prescriptive right. The distinction between nuisance, trespass, negligence and liability under the principle of *Rylands* v. *Fletcher* (qq.v.) may be narrow and difficult.

In Scotland, the distinction between public and private nuisances is not recognized and, apart from nuisances declared to be so by statute, nuisance is not a ground for criminal proceedings. Only the party materially inconvenienced by actings may complain of them as a nuisance.

Nullification. In the U.S., the doctrine which asserted the right of a state to declare void within its boundaries federal government actions judged by that state to be unconstitutional. It is found in Jefferson's resolutions to the Kentucky legislature (1798) and Madison's resolutions to that of Virginia (1798), which declared that a state could lawfully oppose decisions reached by the federal courts. It was formulated by Calhoun in his *Exposition and Protest* (1829), in which he contended that the states were sovereign bodies. The South Carolina state legislature accepted this argument and, in 1832, adopted an ordinance nullifying the U.S. tariffs of 1828 and 1832. President Jackson denounced the doctrine of nullification and Congress passed a Bill empowering him to collect the tariffs by force. A compromise was, however, arrived at and South Carolina withdrew its nullification ordinance.

Nullity of marriage. A marriage is null and void when it has, from the outset, been vitiated by an inherent defect, rendering it in law a non-marriage or only an apparent marriage. Such a marriage does not produce any of the legal consequences of a valid marriage, even though the parties cohabit for some time, or even have issue. Thus children are, in general, illegitimate. Nullity, save in the most obvious cases, needs to be declared by the appropriate court.

The grounds on which nullity may be declared include any legal incapacity to marry, such as non-age, unsoundness of mind, or prior existing marriage, relationship, within the prohibited degree of consanguinity or affinity, non-compliance with the prescribed preliminaries or formalities of marriage and absence of free and genuine consent to take each other as spouses.

Where a marriage is not void but only voidable, e.g. for non-consummation (see MARRIAGE) proceedings must be taken to have the marriage declared null, failing which it must be treated as valid.

When a marriage is declared null, parties are entitled to mutual restitution of any money or property transferred on the faith of the marriage, but not to such provisions as may be demanded in case of divorce.

J. Jackson, *Formation and Annulment of Marriage*; P. Bromley, *Family Law*; J. Wilson and E. Clive, *Husband and Wife in Scotland.*

Nullum crimen* (or *nulla poena*) *sine lege (no crime (or penalty) without law making it so). A maxim embodying the basic principle of the criminal law that conduct cannot be punished as criminal unless some rule of law has already declared conduct of that kind to be criminal and punishable as such. This principle, accordingly, denies the validity of retrospective declaration of the criminality of any kind of conduct and also the justifiability of a court or judge declaring to be criminal when a case arises anything not previously declared criminal. The U.S. Constitution and the European Convention on Human Rights prohibit retrospective creation of criminality.

Nullum tempus occurrit regi (time never runs against the Crown). The common law principle that Crown rights are not affected by rules of limitation or prescription, nor can laches (q.v.) be imputed to the Crown. But many modern statutes expressly prescribe the contrary, though with exceptions and qualifications.

Nuncio. A representative of the Vatican accredited as an ambassador to a government maintaining diplomatic relations with the Holy See. Like an ambassador, he represents his state, seeks to maintain good relations with the host country, but also reports on the state of the Catholic Church in that country. A full nuncio is accredited only to those countries which observe the decision of the Congress of Vienna (1815) that the papal nuncio is automatically doyen of the diplomatic corps in the host country. If a country does not observe that practice the title pro-nuncio has, since 1965, been given to the papal representative. An internuncio is a Vatican representative having the rank of minister plenipotentiary, but his duties are similar to those of a nuncio.

Nuncupative will. A will not reduced to writing but declared orally to and proved by the evidence of witnesses. They have been abolished since 1837, save for wills of personal estate by soldiers in actual military service, or marines or seamen at sea, and this privilege was extended, in 1918, to real property. Nuncupative wills are valid under Scots law to the value of £100 Scots (£8.33) only.

Nuremberg trials. A series of trials, held at Nuremberg in 1945–46, in which leaders of Nazi Germany were tried for war crimes by an International Military Tribunal composed of two judges from each of the U.K., U.S., France, and Russia. The tribunal was set up pursuant to a London Agreement of 1945, to which many states later acceded. The indictment charged the Nazi leaders with crimes against peace, i.e. waging war; crimes against humanity, i.e. deportation, genocide; war crimes, i.e. violations of the laws of war; and conspiracy to commit these three kinds of criminal acts. Twelve defendants were sentenced to be hanged, three to life imprisonment, four to imprisonment for from 10 to 20 years, and three were acquitted. The significance of the trials was that the Tribunal affirmed that individuals as well as states could be guilty of war crimes, that individuals were punishable for crimes against international law, that the crimes charged had long been deemed criminal, and that the trials were not covering vengeance with a veneer of legality. Some writers attack the legality of the trials as proceeding on *ex post facto* declaration of the criminality of the conduct charged. But there is little doubt of the inherently criminal character by international common law of the conduct in question.

Nys, Ernest (1851–1920). Belgian jurist, member of the Permanent Court of Arbitration, and author of *La Guerre Maritime* (1881), *Le Droit de la Guerre et les précurseurs de Grotius* (1882), *Les Origines du Droit International* (1894), and *Le Droit International et Les Principes, Les Théories, Les Faits* (1904).

O

Oath. An assertion or promise made in the belief that supernatural retribution will fall on the taker if he violates what he swears to do. The concept of an oath as sacred and binding is ancient and widespread, but some ancient peoples swore by their swords or other weapons.

Many forms of oath are taken, among the most notable being the coronation oath, the oath of allegiance, the judicial oath, oaths taken by privy councillors, by peers on their creation, by M.P.s before sitting or voting, by archbishops and bishops, by aliens on naturalization, by jurors, and by witnesses before giving evidence. In court in England, an oath is normally taken by the person taking a New Testament in his hand and repeating the words of the oath after the officer administering it. Jews swear on the Old Testament and children under 14 'promise' but do not 'swear'. In Scotland, a person does not take any book, but holds up his right hand at head height, palm to the front, and repeats the words of the oath after the presiding judge, who stands up, holds up his right hand similarly and says the words to be repeated. In many cases, persons who object to being sworn, having no religious belief, are entitled to make a solemn affirmation instead of taking an oath, with the same force and effect. The administering of unlawful oaths is unlawful as tending to sedition. As religious belief diminishes there is increasing doubt nowadays whether taking an oath has any effect on many witnesses, and whether the practice should be continued.

Oath *ex officio*. The oath by which a person in holy orders could swear himself innocent of a charge, and also the oath of his compurgators.

Oath of calumny. An oath administered to the pursuer in consistorial actions in Scotland that he believed that the facts stated in his summons were true and that there had been no agreement to put forward a false case or to withhold a just defence. The oath was prescribed to prevent collusion. An oath of calumny was also formerly used in other kinds of cases.

Oath of party. In Scots law, a form of proof whereby a litigant may require his opponent to answer on oath as to the truth of his case or some specific part of it, by referring the matter to his opponent's oath. Challenge, or contradiction of the answers given is incompetent. It is in effect an appeal to the opponent's conscience. It is now uncommon.

Oath-helpers. The persons who, in trial by wager of law (q.v.) or compurgation, took an oath that they believed the oath the party they were assisting was truly sworn.

Ob non solutum canonem (on account of annual duty not paid). In Scots law, the term used of an action to terminate a feudal relationship for non-payment of feuduty. See also EMPHYTEUSIS.

Obedience. Conformity to a rule for conduct accepted as authoritative as emanating from God, moral principles, law, superior officer, or other source of obligation. The existence or not of a duty to obey law is deemed obligatory and binding persons to obedience.

The question of obedience to law raises at least two issues, why people should obey law, and why they do, or some times do not, obey law. The first is a philosophical, the second a psychological, question. They should obey law because law is made by or under the authority of the superior authorities in the community, in the policy of whom all adults have their part, and because the law on any matter must, in theory, be taken to have the assent of at least a majority of the adults in the community. Moreover, widespread disobedience to law is creative of social disorder and operates unfairly on those who continue to obey law and discharge their obligations. This leads to the difficult issue of whether it is ever, and, if so, when it is, permissible to disobey the law. It is never legally permissible, but it may sometimes be morally possible. Persons do obey law for various reasons, because they accept and approve of the rules in question, because they have been brought up and educated to obey the law, because they believe that there is a social and moral obligation to obey the law even if one disagrees with it, and because of fear of the sanctions and penalties for disobedience. Persons do not obey law sometimes because of religious scruples or conscientious dissent from the provisions, sometimes from ignorance or carelessness, but most commonly from desire to evade duties and responsibilities and to reap unjustifiable rewards. It is much more difficult to persuade people to obey, and to make them obey, unjust, biased, or stupid legislation or legislation which conflicts with widely-held beliefs than to enforce obedience of reasonable laws.

Obediential obligations. A term sometimes used in Scots law for obligations imposed by force of law, particularly those arising from family relationships.

Obiter dictum. In the terminology of the principles of binding precedent or *stare decisis* (q.v.) recognized in Anglo-American systems of law, is any statement on a point of law in a judge's opinion which is not, or not part of, the *ratio decidendi* (q.v.) or principle of the decision. The category includes propositions stated as illustrations, or converse cases, or based on hypothetical facts, or on facts held not to be material, or not to be proved. As it is the function of the court deciding a later case to determine what the *ratio* of a precedent is, it must inferentially decide what are mere *obiter dicta*, and a superior court may ultimately decide that some statement long thought to have been part of the *ratio* of a precedent is *obiter dictum* only, or conversely may hold that a proposition hitherto regarded as *dictum* is part of the *ratio.*

Though never binding on a court in a subsequent case *obiter dicta* may be of weight and persuasive value. A casual expression carries little or no weight; other *dicta* may be carefully considered statements and may be acted upon and acquire substantial authority. The value of *dicta* also depends on the status of the court and the reputation of the judge.

A distinction has also been suggested between *obiter dicta*, irrelevant to the case, and judicial dicta, not part of the *ratio* but relevant to a collateral matter.

Object of a power. The person in whose favour a power of appointment (q.v.) may be exercised. A person in whose favour it may not be exercised is a non-object.

Object of a legal right. The thing in respect of which it exists, or sometimes the act or omission to which a person is bound. Thus, when it is said that A owns a house A is the subject of the right of ownership to whom it attaches, and the house is the object in respect of which it is held; when it is said that A is entitled not to be injured by A, A is the subject and the object is the abstention from wrongful conduct which attaches to B.

Objection. An intervention by counsel for one party, contending that a question by another counsel to a witness is improper or objectionable and should be disallowed, or that a document sought to be tendered by another party should not be received.

Objective. A term frequently used in law, contrasted with subjective, particularly with reference to the standard by which some action is judged. An objective standard is that which is deemed right, fair, or otherwise correct independently of whether the doer or person concerned himself regarded the action as right, fair or correct, whereas a subjective standard has reference to the belief and opinion of the doer himself. As a noun, the word also means an aim or motive.

Obligation. A legal concept signifying a bond or tie linking two legal persons, conferring on each mutual legally enforceable rights and duties. Thus, where one person sells to another they are deemed to create an obligation which links them and gives rise to the rights to delivery or to payment, and to the correlative duties to pay or to deliver respectively. Where one person injures another, the injury creates an obligation which gives rise to the right to recover damages and the duty to pay damages.

The sources of obligation may be distinguished into voluntary conduct of kinds which, according to the particular legal system, give rise to obligation, and acts or omissions which, under the legal system, give rise to obligation independently of the wills of the parties. In the first category are voluntary undertakings or promises and bilateral consensual agreements or contracts. In the second category are circumstances giving rise to an obligation of restitution (obligations *quasi ex contractu* in Roman law), circumstances giving rise to an obligation to make reparation or pay compensation for harm done (obligations *ex delicto* and *quasi ex delicto*), natural obligations springing from family relations, obligations arising from the relationship of trust and other fiduciary relations, certain relationships imposed by statute, and relationships imposed by judgments of courts.

The rights arising from an obligation are *in personam* only, enforceable only against a determinate person though, in some cases, such as the commission of a tort, an obligation arises from the breach of a right *in rem* and a duty *in rem*.

Obligation once created subsists until discharged, normally by performance or payment, but sometimes by discharge, novation, delegation, frustration, or prescription.

The law of obligations is accordingly in legal systems based on the civil law, the term for the bodies of law which deal with relationships creative of obligations, particularly with the law of contract, tort or delict, and restitution or quasi-contract.

The term obligation is also sometimes used as meaning a debt, or a legal duty or liability, or a deed evidencing a debt or liability. Sometimes an obligation, e.g. to save life, is entirely or mainly a moral one, created by principles of morality, or a religious one, created by religious belief.

Obligations, law of. In Roman law and systems based thereon, the law of obligations covers the whole mass of relations whereby two persons become mutually linked by a legal bond or tie, binding the one to do something to or for the other. In Roman law, obligations were said to arise *ex contractu*, *quasi ex contractu*, *ex delicto*, or *quasi ex delicto*. In modern law, obligation can arise from at least six main grounds, from an undertaking or contract with another, from circumstances where a rule of law on equitable grounds imposes a duty to make repayment, or recompense, or restitution to another for the avoidance of unjust benefit, from the legally unjustifiable causing of harm to another, imposing a duty to pay compensation, from the provisions of a statute, from the trust or fiduciary relationship, and from a judgment of a court.

Common law systems have never regarded the law of obligations as a distinct branch of law, but have recognized the various distinct grounds of obligation, contract, tort, quasi-contract or restitution, trust and so on. Civil-law systems, on the other hand, maintain the Roman view of obligations as a distinct branch, subdividing it by reference to the basis of obligation, in promise or contract, in circumstances requiring restitution, in delict, in statute, trust, judgment or otherwise. Logically the distinction is between obligations created by the parties' wills (contractual obligations) and obligations between parties created in particular circumstances by the operation of a principle of law, independent of the parties' volition. In the former of these categories, the acts of the parties in creating the obligation create mutual rights *in personam*, whereas in the latter categories the law imposes duties *in rem*, owed to persons generally, and rights *in personam* arise only if the legal duty is not implemented in relation to some person to whom the duty was owed, e.g. if the statutory duty to take precautions for employees' safety in a factory is not fulfilled and an employee is injured, he acquires a right *in personam* against the factory occupier for compensation.

O'Brien, Ignatius John, Baron Shandon (1857–1930). Was called to the Irish Bar in 1881 and became Solicitor-General 1911, Attorney-General, 1912, and Lord Chancellor of Ireland, 1913–18. He then became a peer as Baron Shandon.

O'Brien, Peter, Lord (1842–1914). Called to the Irish Bar in 1865, he became Solicitor-General in 1887 and Attorney-General in 1888. In 1889, he became Chief Justice of Ireland. He was made a peer in 1900 and retired in 1913.

Obscene libel. A common law crime consisting of the publication with intent to corrupt, of any matter which is obscene, that is, having a tendency to deprave or corrupt those whose minds are open to immoral influences, particularly but not exclusively in relation to sex. The common law crime has been largely superseded by a statutory offence under the Obscene Publications Act, 1959.

Obscenity. The quality of offending decency, and of tending to deprave and corrupt persons whose minds are open to sexually immoral influences and into whose hands the matter in question may come. It, accordingly, generally covers language, literature, or representations dealing with erotic, pornographic, and sexually perverted subjects. But the obscenity of any matter lies in its effect on the mind of the reader or viewer more than in any definable quality of the matter itself.

The importance and interest of obscenity lies in the attempts made to define it and to suppress it, particularly obscene books and pictures. Plato's *Republic* called for censorship and the Twelve Tables penalized the singing of bawdy songs. In England, it was recognized from 1727 that it was criminal at common law to publish indecent matter, and the Obscene Publications Act, 1857, prohibited the publication of obscene matter but did not define obscene.

It is difficult, if not impossible, to define obscene satisfactorily, and to fix the person by whom and the standard by which the obscenity or otherwise of matter is to be judged.

At common law, it was an offence to publish any matter tending to deprave and corrupt those whose minds were open to immoral influences and into whose hands the publication might come: *R.* v. *Hicklin* (1868), L.R. 3 Q.B. 360, 371. Under statute in England (Obscene Publications Acts, 1959–64), matter is deemed obscene if, in its effect, it is such as to tend to deprave or corrupt persons who are likely, having regard to all relevant circumstances, to read, see or hear the matter contained or embodied in it. A person cannot be convicted if it is proved that the publication is justified as being for the public good on the ground that it is in the interests of science, literature, art or learning, or of other objects of general concern. The opinion of experts as to literary, artistic, scientific or other merits of matter is admissible to establish or negative this defence.

Obscene performances of plays are distinct offences.

In the U.S., the Comstock Act (1873) penalized the sending or receiving of obscene, lewd or lascivious publications. Until 1934, the English text of *Hicklin*'s case supplied the test of obscenity, but, in *U.S.* v. *Book entitled Ulysses* (1934), 72 N.Y. 705, it was held that the test was whether the book as a whole had a libidinous effect. In *Roth* v. *U.S.* (1957), the Supreme Court considered that the standard was the reasonable adult and that a work was obscene if it appealed to a prurient interest in sex.

Many countries are parties to a U.N. Convention for the Suppression of the Circulation of and Traffic in Obscene Publications, which does not define 'obscene', it being accepted that this would vary from country to country.

Obstruction. Wilful obstruction of a constable in the execution of his duty is criminal. This certainly covers interference with him by force, threats, or otherwise, but the act amounting to obstruction need not itself be criminal. It is sufficient that it is something the person obstructing should not have done in the circumstances, though it is probably too wide to say that it covers making it more difficult for the police to carry out their duties. Wilful obstruction of the highway is also an offence and obstruction may be a public nuisance.

Occupancy. Actual possession and use of real property by permission or agreement without any necessary claim to ownership or tenancy, or sometimes actual possession without the owner's permission, as by squatters. Occupancy is a mode of acquiring property in certain kinds of goods, notably fish and fowl, always except royal fish and fish in a fishpond, and animals in a forest or protected park.

Occupation. In international law, it is a mode of acquiring sovereignty over territory not already under the dominion of a recognized state. It has been recognized that discovery is not enough and that, particularly in habitable areas, occupation must be effective, and must amount to actual settlement and administration and the claim to exclude others. But it is a question over what area occupation is effective, and whether a single act of sovereignty or only a continuing exercise of jurisdiction will suffice is a matter of circumstances. If the basis of title is the exercise of sovereign jurisdiction, acquisition by occupation can be effected only by authorized persons in name of the claiming state and not by private persons, though the actings of private persons may afford evidence of possession.

In municipal law, it is a mode in which title to personal property may be acquired, in cases of goods left ownerless or game or fish taken by the finder or captor, with certain exceptions. It is not a mode in which title to real property can be acquired.

In election law and rating law occupation means actual use and enjoyment.

An occupation also denotes an employment or trade.

Occupied territory. In wartime, territory hitherto belonging to one state which has been in fact occupied by the armed forces of another state. Occupation has sometimes been a prelude to cession but frequently occupied territory has subsequently been evacuated.

Occupier. A person having actual possession and control of premises and using them for his own purposes. In general, it is the occupier rather than the owner of land who can protect his occupation by proceedings for nuisance or trespass. Occupiers of premises owe a duty to take such care as in all the circumstances of the case is reasonable to see that visitors will be reasonably safe in using their premises for the purpose for which they are invited or permitted by the occupier to be there.

Ochs, Peter (1752–1821). Swiss revolutionary who drafted the constitution of the Helvetian Republic (1798), in which he served as president of the Senate and later president of the executive, the directory. Later, he produced new constitutional and penal codes for the state of Basel.

O'Connor, Charles Andrew (1854–1928). Called to the Irish Bar in 1878, he took silk in 1874, became successively Solicitor-General (1909–11) and Attorney-General for Ireland (1911–12), and was then Master of the Rolls in Ireland (1912–14) and a judge of the Supreme Court of the Irish Free State (1924–25).

O'Dalaigh, Cearbhall (1911–). Was Attorney-General of Ireland, 1946–48 and 1951–53, a judge of the Supreme Court of Ireland, 1953–61, Chief Justice of the Supreme Court of the Republic of Ireland, 1961–73, later a judge of the European Court, 1973–74, and President of the Republic of Ireland, 1974–76.

Odofredus, Denari (?–1265). Italian jurist, professor at Bologna, and author of *Commentaria in Codicem*, *Lecturae in Digestum Vetus*, *Summa de libellis* *formandis*, *Lecturae in tres libros* and *Lecturae in digestum novum*.

Of course. That which necessarily takes place. An order of course is one to which the applicant is entitled on his own application and at his own risk, the court and its officers having no discretion to refuse it so long as the application is made in the proper form.

Offence. A word with no fixed or technical meaning, denoting an act contrary to, offending against, and punishable by, law, but particularly one made to by statute rather than by common law, the latter being usually called crimes, and also particularly one punishable on summary conviction. The distinction between crimes and offences is a matter of common usage of words rather than of legal definition. Thus, the usual terminology is of 'road traffic offences'.

Offender. Strictly, a person who has committed an offence but also used for one who has contravened the criminal law, whether a crime or an offence.

It is particularly used in a number of common phrases such as first offender, fugitive offender, juvenile offender, and young offender.

Offensive weapon. Any article made, or adapted for use, for causing injury to the person, or intended to be used for such a purpose. If carried in a public place it is an offence. The expression includes anything not generally carried for any purpose other than offence, e.g. a bayonet, but may, in particular circumstances, include things not generally so carried, e.g. a Boy Scout knife, if in the circumstances the apparent intention of carrying it was offence and there was no lawful authority or reasonable excuse for carrying the thing.

Offer. An indication or expression of willingness to become bound by some obligation if another party, the offeree, accepts the offer. An offer is, unless declared irrevocable, revocable until accepted, and it lapses if refused, or after the lapse of any period for which it has been stated to be open for acceptance or, in any event, after the lapse of a reasonable time. It is probably impliedly revoked by the offeror's death or incapacity. Acceptance of an offer creates an obligation or contract between the parties.

Office. A post carrying rights and duties. It may be of the nature of an employment and be remunerated, e.g. the office of judge or professor or company secretary, or be gratuitous, e.g. the office of trustee. Offices may be public or private. Public offices entitle and even require the holder to intervene in the affairs of others without their request or permission, and may be granted to a person for a period, or, as in the case of most Crown appointments, during the Crown's please (*durante bene placito*) or, as in the case of judges (*quamdiu bene se gesserit*) or occasionally be heritable, granted to the grantee and his heirs, e.g. the office of Earl Marshal. Private offices, such as of executor or trustee, have limited ambits of intervention and differ from employments in that the holders have duties and responsibilities by force of law and notwithstanding any contractual provisions.

Office of Profit under the Crown. From the time of James I, it was a principle of the English Parliament that to hold an office requiring prolonged absence was inconsistent with Membership of the House of Commons. The House also disliked members being in the pay of, and under the influence of, the Crown. The Act of Settlement, 1701, provided that no person having an office or place of profit under the Crown could be a member of the House of Commons, and, in the eighteenth century, various statutes excluded various classes of office-holders from membership. The law now rests on the House of Commons Disqualification Act, 1975, which lists the offices disqualifying from membership (judicial and similar offices, membership of numerous boards, authorities and commissions, and a large number of miscellaneous offices) and includes among them 'the office of Stewart or Bailiff of Her Majesty's three Chiltern Hundreds of Stoke, Desborough, and Burnham, or of the Manor of Northstead' and appointment to either of these offices is, accordingly, still the means whereby an M.P. may resign his seat. The basic distinction is between political and non-political offices; if the former had been, or were now, excluded from the House of Commons, it would have prevented, or now prevent, the development of the principle of ministerial responsibility to Parliament.

Officers of State. A term of inexact connotation in Scots law, generally equated with Officers of the Crown.

Great Officers of the Crown appeared from the time of Alexander I onwards, first the Chancellor, Constable, and Justiciar. In the time of David I, there appeared the Great Steward, High Constable, and Great Chamberlain and, in the fifteenth century, the Secretary, Keeper of the Privy Seal, Clerk of the Rolls and Register; others are mentioned in that and the following centuries. There were frequent rivalries for precedence and in 1617 King James VI declared in Privy Council that only eight Officers of State should sit as such in Parliament: Treasurer, Privy Seal, Secretary, Register, Advocate, Justice-Clerk, Treasurer-Depute, Master of Requests, but nevertheless the Chancellor, Collector and Comptroller sat as Officers of State.

By the Officers of State Act, 1685, the Officers of the Crown were declared to be the Treasurer, the Secretary, the Collector (an office later joined to that of Treasurer), the Justice-General, the Justice-Clerk, the Advocate, the Master of Requests and the Clerk of Register.

The full list is probably the Admiral, Advocate, Almoner, Chancellor, Chamberlain, Clerk of Register, Collector-General, Comptroller, High Constable, Justice-Clerk, Justiciar, Master of the King's Household, Master of the King's Wardrobe, Master of Requests, Marischall, Keeper of the Privy Seal, Secretary and Treasurer. Most of these lapsed after 1707, some being superseded by their English equivalents but there remain the Lord Lyon King of Arms, Lord Justice-General (in place of the Justiciar) Lord Justice-Clerk, Lord Advocate, Solicitor-General, Keeper of the Great Seal, Keeper of the Privy Seal, Lord Clerk Register and Lord High Constable.

Official. In canon law, a person designated by a bishop to exercise the latter's jurisdiction in non-criminal cases; in mediaeval Scots law, a priest, normally a canonist who exercised the bishop's judicial powers in his diocese or an archdeaconry thereof. Sometimes the official was designated commissary general, or both might be found acting concurrently. Jurisdiction extended to matters of status, wills, and intestacy, the care of widows and orphans, and many other causes, and the law applied was the canon law. Appeal lay to the official Principal of St. Andrews or Glasgow and further to the Rota of Rome. At the Reformation the courts of the officials were superseded by the Commissary (q.v.) court. In modern English ecclesiastical law the official is the person appointed by an archdeacon as his substitute.

Official assignees. Persons appointed by the Lord Chancellor under the pre-1869 bankruptcy system from among merchants or accountants to be assignee of a bankrupt's estate together with the assignee or assignees chosen by the creditors.

Official Journal. The official means of publication by the Council and Commission of the European Communities. *The Official Journal of the E.C.S.C.* appeared from 1952 to 1958 and the *Official Journal of the European Communities* has appeared since then. It appears almost daily in each of the official languages of the Communities. Since 1968, there have been two series, one devoted to secondary legislation, designated L, and the other to communications and information, designated C. Matter published in L part includes regulations made by the Council or Commission, which become binding on the date specified or on the 20th day following publication, directives, and decisions, and decisions and recommendations of the E.C.S.C. The C part contains Commission proposals and opinions, reports of the sessions of the European Parliament and notices of the proceedings of the Court of Justice, and various other kinds of notices.

Official liquidators. Persons formerly appointed under the Companies Act, 1862, to carry through the winding up of companies under the direction of the court.

Official managers. Persons formerly appointed to superintend the winding up of insolvent companies under the control of the Court of Chancery.

Official Principal. A person appointed by an archbishop or bishop of the Church of England as his delegate to hear causes between two parties. He also holds the offices of Chancellor and of Vicar-General. The Official Principal of the Province of Canterbury is the Dean of the Arches (q.v.) and of the Province of York, the Auditor of the Chancery Court of York (q.v.).

Official receivers. Persons appointed by the Department of Trade to act as interim receivers and managers of the estates of bankrupts, pending the appointment of trustees in bankruptcy.

Official referee. Officers of the High Court to whom a judge might formerly refer any civil matter for inquiry or report on any question of fact arising therein. For the purposes of his inquiry the official referee had the same powers as a judge. Following a reference the official referee reported to the court which might adopt or vary his report, refer the matter back or decide the question referred. The office was abolished in 1971, the functions being discharged by certain circuit judges.

Official secrets. The Official Secrets Acts, 1911 and 1920 make it an offence to spy, or wrongfully to communicate plans, documents, or information about matters of defence, military, or naval matters, and other confidential information, to use such materials for the benefit of any foreign power or in a way prejudicial to the interests of the State, or not to take proper care of such materials. The Act has been criticized as enabling a government to suppress any criticism it does not like and to keep secret matters which should be publicly known.

Official Solicitor. Formerly an officer of the Court of Chancery with the function of protecting the Suitors Fund and administering it, so far as within the spending power of the court, and now an officer of the Supreme Court. When directed, he acts as guardian *ad litem* to persons under disability, visits persons committed for contempt, acts as judicial trustee, and generally acts as solicitor in cases in which the Court requires his services in that capacity.

Official Trustees of Charitable Funds. Officials of the Charity Commission who hold funds invested and pay the income to the administering trustees of charities, and when appropriate transfer capital according to the directions of the court or the charity commissioners.

Official Trustee of Charity Lands. The secretary of the Charity Commissioners, who may hold land or other hereditaments, held on charitable trusts by order of the court of the Charity Commissioners if it is vested in persons other than the administering trustee, or there are no trustees, or they are unwilling or unable to act and in certain other cases. Land held by him may be transferred to the trustees of the charity by the court or the Commissioners.

Officialité. A French ecclesiastical tribunal active from mediaeval times till the Revolution. Presided over by a judge or official, it dealt with cases involving breaches of Church discipline and civil actions involving marriage, legitimacy, heresy but, from the fourteenth century, the *parlement* of Paris began to undermine its authority over such matters as heresy and blasphemy.

Officio, ex. By virtue of his office. Thus the Lord Chancellor is *ex officio* a judge of the Court of Appeal and the President of the Chancery Division of the High Court and the Lord Chief Justice is *ex officio* a judge of the Court of Appeal.

Ogilvy, James, 4th Earl of Findlater and 1st Earl of Seafield (1664–1730). Was called to the Scottish Bar in 1685 and sat in the Scottish Parliaments. He was Solicitor-General in 1693, Secretary of State in 1696 and, as Viscount Seafield, presided over the Parliament which met in 1698. In 1701, he became Earl of Seafield. In 1702, he became Lord Chancellor of Scotland but was superseded in 1704, although he was reappointed in the next year and held office till 1708. He was one of the commissioners who, in 1706, negotiated the Union and actively supported it. Thereafter, he sat in the British Parliament as a Scottish representative peer. In 1707, he became a member of the English Privy Council and, on returning to Edinburgh, produced a fresh commission appointing him Chancellor of Scotland, but the validity of this office having been doubted, he was appointed Lord Chief Baron of the Exchequer, 1708–9. In 1713, he brought in a bill to repeal the Act of Union of 1707 which was lost by only four votes. Shortly thereafter he was appointed Keeper of the Great Seal of Scotland, 1713–14, and presided as Chancellor in the Court of Session. He was reputed a learned lawyer and a courteous and just judge.

O'Grady, Standish, Baron O'Grady and Viscount Guillamore, (1766–1840). Was called to the Irish Bar in 1787, became Attorney-General in 1803, and Chief Baron of Exchequer in Ireland 1805–31. When he retired he became Baron O'Grady and Viscount Guillamore.

O'Hagan, Thomas, 1st Lord (1812–85). Was called to the Irish Bar in 1836 and, for a time, was associated with the Irish nationalist movement. He became Solicitor-General for Ireland in 1861 and Attorney-General for Ireland in the next year. He entered Parliament in 1863. In 1865, he became a judge of the court of Common Pleas in Ireland and, in 1868, Lord Chancellor of Ireland. He was made a peer in 1870. He lost office in 1874 but resumed it in 1880, resigning in 1881. In the intervals, he sat regularly in the House of Lords and assisted in the passage of much useful Irish legislation.

Oidores. Judge of the *audiencia*, the highest court of appeal in colonial Spanish America, who also had extensive political power. They lived a semi-monastic life, being forbidden to take fees or gifts, to hold lands or Indians, or to engage in trade; their dress was prescribed and their only relaxation was attendance at religious festivals. They were usually peninsular Spaniards but, from the eighteenth century, were frequently Creoles. The Spanish Crown relied heavily on them and they were a consistently loyal department of the colonial administration.

Oil and gas law. The term for the body of law concerning the discovery, ownership, extraction, transport, and disposal of deposits of natural oil or gas under the surface of the land or under territorial waters or the continental shelf.

Old age pensions. Pensions of 5*s.* weekly payable to persons aged 70 or over and having a yearly income not exceeding £21, and of lesser sums to persons having a yearly income between £21 and £31.10*s.*, under the Old Age Pensions Act, 1908. The provision was later amended and is now represented by retirement pensions. See SOCIAL SECURITY.

Old Bailey. See CENTRAL CRIMINAL COURT.

Old Booke of Entries. A name sometimes given to William Rastell's (q.v.) *Collection of Entries* (1566) It is sometimes also called the *New Book of Entries.*

Old *Natura Brevium*. Originally called *Natura brevium* and renamed after Fitzherbert (q.v.) published his (New) *Natura Brevium* in 1534. It is a selection of writs together with a commentary and originated in a law-French manuscript of the mid-fourteenth century. The writs correspond with the writs in the printed *Registrum brevium*. There are many editions of it, alone or bound up with other short treatises.

Old Tenures. A short tract in law-French by an unknown author dating from the late fourteenth century. It was printed several times in the sixteenth century, translated by Rastell and printed with his *Olde Termes de la Ley*, and later. It deals briefly with the various kinds of tenures, and their incidents but is a connected exposition, and provided some of the material for Littleton's (q.v.) *Tenures*.

***Olde Terms de la Ley*.** A legal dictionary by Rastell, published in 1624 and many later editions.

Oldendorp, Johann (1480–1567). Syndic at Rostock and Lubeck and professor at Cologne, one of the most important jurists of the sixteenth century, author of *Juris naturalis, gentium et civilis isagoge* (1539) in which he developed the outlines of a system of legal philosophy or natural law, *Practica actionum forensium* (1540), *Sylloge sive Enchiridion actionum forensium* (1552) and other works.

Oldfield, Thomas Hinton Burley (1755–1822). Constitutional historian and Parliamentary reformer, author of *History of the Boroughs* (1792) and *Representative History of Great Britain and Ireland* (1816), which gave theoretical foundations for attempts by the House of Commons to limit royal influence and amend its own composition.

Oldradus de Ponte (?–1335). A distinguished Italian commentator jurist who taught at Padua, Siena, Montpellier, Perugia, and Bologna, teaching Bartolus among others, author of *Consilia* (1515) and commentaries on the Code and Digest.

Oléron, Rules of. A compilation of decisions of the eleventh century said to have been rendered by the maritime judges of Oléron (an island off the west coast of France) and accepted as stating the common maritime law of the Atlantic coast corresponding to the Consolato del Mare of the Mediterranean and, as such, an element in later books of maritime law (e.g. the maritime law of Wisby in the Baltic). It is said to have been made about 1150 for Eleanor of Acquitaine, queen successively of Louis VII of France and of Henry II of England. In the reign of Edward III, it was declared to be the law according to which the Admiralty determined disputes in England. Various versions exist, as *La Charte D'Oleroun des Juggements de la Mer* in the *Liber Memorandum* in the London Guildhall, in Garcie de Ferrande's *Grand Routier de la Mer* (1542), as *Lawes of ye Yle of Oleron and Ye Judgements of Ye See* in R. Copland's *The Rutter of the Sea*, in Cleirac's *Us et Coutumes de la mer* (1647), Malynes' *Lex Mercatoria*, and elsewhere. The most complete is in Pardessus' (q.v.) *Collection des Lois Maritimes* (1828–45).

Oligarchy. Government by the few, is difficult to distinguish from monarchy (q.v.) because the sole ruler (monarch) must operate through a small group of agents and advisors, and from democracy because the people as a whole can only, save in small city-states, govern through representatives. It is also difficult to distinguish oligarchy from aristocracy, government by a a minority and even from the tyranny of a group of party leaders. Historically, oligarchy has generally implied a greater identification of political with economic power than either monarchy or democracy. In modern times, there is a strong oligarchic trend both in democratic and totalitarian countries, in that in both the whole effective power has tended to fall into the hands of a small group (the Cabinet, the Praesidium) and the general body of the representatives of the people have individually no power and even collectively not much influence.

Olivecrona, Knut Hans Karl (1897–). Swedish jurist, a realist and follower of Hagerstrom, author of *Law as Fact* (1939 and 1971), and books on a variety of other legal subjects. He contends the law is nothing but a set of social facts, though he does not clarify 'facts'. Law has binding force so far as it is valid. He has attacked some traditional legal myths and his work is a valuable criticism of some older views.

O'Loghlen, Sir Michael (1789–1842). Was called to the Irish Bar in 1811 and was successively Solicitor-General, Attorney-General, a baron of Exchequer in Ireland 1836, and Master of the Rolls in Ireland 1837.

Ombudsman. A person appointed by Parliament to investigate citizens' complaints of executive or bureaucratic incompetence or injustice, but not illegality. The first such officer was appointed in Sweden in 1809–10 with wide powers of enquiry. The institution was adopted in other Scandinavian countries—by Finland (1919), Denmark (1954), and Norway (1962). He is an independent investigator and must act impartially between government and individual, but the scope of his investigations usually excludes legislative and judicial conduct.

The office supplied the model for the office of Parliamentary Commissioner (q.v.) in the U.K. and the similar officer in New Zealand (1962) and the Scandinavian word is often used colloquially for the U.K. officer.

H. Rowat, *The Ombudsman.*

Omission. A failure to do an act, or the not-doing of it. An omission may be deliberate, i.e. an abstention, or be inadvertent, as by forgetfulness. In general, an omission is legally significant only if there was a legal duty to act and not to omit, so that the omission is a breach of legal duty to have acted. A bare omission where there was no duty to act, e.g. the omission to wash in the morning, is normally legally irrelevant.

Ompteda, Dietrich Henrich Ludwig von (1746–1803). German diplomat and author of *Litteratur des gesammelten sowohl natürlichen als positiven Völkerrechtes* (1785).

Onerous. The opposite of gratuitous, implying burdens and liabilities which more or less balance the advantage anticipated.

Onus of proof (*onus probandi*, or burden of proof). The question of which party to a dispute on fact must undertake to try to establish the disputed fact, a question of much importance in the law of evidence. The basic rules are that a party asserting an affirmative must prove it, and that a party relying on a fact peculiarly within his own knowledge must establish it. The onus of proof may shift at various stages in a case.

Open cheque. One not crossed but payable to bearer or to order.

Open contract. A contract which does not seek to modify or exclude the implications attaching by law to a contract of that category.

Open court. A court-room open to any member of the public, as contrasted with proceedings in chambers (q.v.). The general rule is that all courts in which judicial proceedings are heard are open courts, but exceptionally the public may be excluded, if statute or the interests of justice require it.

Open cover. An informal agreement whereby an underwriter agrees to issue a marine insurance policy on cargo not yet shipped once it is shipped and the premium has been paid.

Opening the case. Commencing the exposition and discussion of a case before a judge. The onus of opening, in general, lies on the party asserting the claim.

Opening the pleadings. In a case heard by jury, stating briefly the substance of the case made in the pleadings, normally done by junior counsel for the plaintiff. In the Chancery Division, the opening is the speech of the plaintiff's senior counsel.

Operation of law. A mode by which certain legal consequences ensue, as contrasted with their happening by an act of a party. Thus, agency may be created by parties in various ways and also arises in other circumstances, e.g. necessity, by operation of law.

Operative words. Words in a conveyance, such as grant, demise or release, which have an operation or effect in creating or transferring an estate.

Opinion. (1) What a person thinks about a particular thing or matter, particularly a view based on grounds not accurately established or even incapable of proof or demonstration.

(2) A written statement by counsel of what he believes to be the legal position on some matter referred to him in a case or Memorial.

(3) A statement by a court or judge of the law relative to a case and of his reasons for reaching a particular decision of the case, as distinct from the order made by him. In the House of Lords, the judgments of the Lords of Appeal are called opinions or speeches.

Opinion evidence. Evidence by an expert of the opinion he has formed of some matter in issue, e.g. the cause of a death, based on his professional or practical knowledge and experience, applied to facts observed or ascertained by or reported to him.

Oppenheim, Heinrich Bernhard (1819–80). German jurist and author of *System des Völkerrechts* (1845), *Philosophie des Rechts und des Gesellschaft und das offentliche Recht Deutschlands* (1850), and *Praktisches Handbuch der Consulate aller Länder* (1854).

Oppenheim, Lassa Francis Lawrence (1858–1919). Was born, educated, and taught law in Germany but settled in London in 1895, became naturalized, and devoted himself to international law. In 1908, he succeeded Westlake in the Whewell Chair of International Law at Cambridge. His major work is his *International Law: A Treatise* in two volumes (1905–6), both repeatedly re-edited, and still highly authoritative; he was a positivist who found international law in custom, international decisions and quasi-legislation, and a strong advocate of a league of nations. He also contributed extensively to periodicals.

Option. The privilege of doing or not doing something as one chooses, frequently exercisable for a determinate time only or subject to stated conditions, such as to purchase land or goods.

Optional clause. The Statute of the Permanent Court of International Justice of 1920 proposed that the jurisdiction of that Court should be compulsory but this was not accepted and was replaced by Art. 36—the Optional Clause—which enabled members of the League of Nations to declare that they recognized as compulsory *ipso facto* and without any special agreement, in relation to any other member or state accepting the same obligation, the jurisdiction of the Court in any or all of the classes of legal disputes concerning (a) the interpretation of a treaty, (b) any question of international law, (c) the existence of any fact which, if established, would constitute the breach of an international obligation, (d) the nature and extent of the reparation to be made for the breach of an international obligation. This obligation might be undertaken for a period of time or permanently and other conditions or reservations might be and were attached. The latter sometimes reduce the value of accepting the Optional Clause to something purely nominal. The same provision is contained in the Statute of the I.C.J. Despite reservations the Optional Clause has permitted a large measure of obligatory jurisdiction by the I.C.J.

Oral. That which is done by the mouth, or conveyed by speech as contrasted with writing. 'Verbal' (q.v.) is frequently used when 'oral' is meant.

Orde or Orde, Robert (*d.* 1778). Became a baron and latterly (1755–75) Chief Baron of the Court of Exchequer in Scotland. His daughter married Lord Braxfield (q.v.).

Ordainers, Lords. A group of barons opposed to Edward II who demanded extensive reforms in the Parliament of 1310. Edward had to agree to their control of the council and the preparation of reforming ordinances, 41 in number, which were promulgated in 1311 and imposed substantial limitations on the King's actings. Over the years 1311–21 there were several attempts to enforce ordinances on Edward but, in 1322, he defeated the Lords and executed several of them but this, in turn, led to his downfall in 1326.

Ordeal. The mediaeval practice of trying a man's guilt by, it was believed, obtaining the verdict of God. Trial by ordeal has been found in various forms in many parts of the world. A man was obliged to submit to the ordeal where he was taken redhanded, or was unable to produce a sufficient number of compurgators, or had been guilty of perjury on a previous occasion, or was not a freeman.

It took one of several forms, carrying red-hot iron, the burns healing within three days if the man were innocent; plunging the hand or arm into boiling water, the scalds healing in case of innocence; being thrown into a pool of water, the guilty floating (being rejected by the water); or being set to swallow the consecrated morsel of bread, the guilty choking. The ordeal decayed early because there was ample scope for corruption, and it seems to have been hard to obtain a conviction. All, save the cold water ordeal, had to be undergone in Church under the superintendence of priests. The fourth Lateran Council of 1215 forbade priests to take part and ordeal quickly vanished thereafter in England.

H. C. Lea, *Superstition and Force* (1866).

Ordenancas Reales. See *Libro de Montalvo.*

Ordenanzas sobre descumbrimientos. A legal code promulgated by Philip II of Spain in 1573 to govern the conduct of Spanish settlers in America, on lines generally similar to those of the New Laws (q.v.) of 1542.

Order. A mandate or direction, as when a cheque is drawn in favour of AB 'or order', i.e. or the person named by AB; also a command or direction issued by an English court, particularly in summary proceedings or in interlocutory applications, as contrasted with final judgments. The Rules of the Supreme Court (q.v.) are divided into numbered Orders dealing with distinct topics.

Order in Council. A decree or order made by the Queen by and with the advice of the Privy Council. The Queen had power at common law to legislate by Order in Council, a survival of the time when legislation was part of the royal prerogative but, in modern practice, this form of legislation is limited to legislation for newly ceded or conquered territories, though it has been used for other purposes, as when in 1870, Queen Victoria, on Gladstone's advice, abolished the purchase of army commissions, a reform which he could not then have achieved by legislation.

Orders in Council are also used as a means of delegated legislation, giving effect in some respect to an Act of Parliament which empowered the issuing of Orders in Council. Such Orders in Council are published along with other delegated legislation among the Statutory Instruments.

Order of Council. An order made by the Privy Council without the concurrence of the Queen.

Order, payable to. A bill of exchange or cheque drawn payable to order is payable to the person named or to another person as ordered by the payee by his indorsement.

Orders of knighthood. See KNIGHTHOOD.

Orders other than of knighthood. These are societies created in modern times, similar to the orders of knighthood but not conferring the status of knighthood on those appointed to them. They include the Order of Merit (1902), the Order of Companions of Honour (1917), the Order of Victoria and Albert (1862), the Order of the Crown of India (1878), the Distinguished Service Order (1886), and the Imperial Service Order (1902).

Ordinances. According to old authorities, a declaration by the Crown in answer to a petition by the Commons enquiring as to the law relative to a particular matter, as contrasted with a statute, laying down new law. More particularly, it is an enactment of Parliament not having the consent of Queen, Lords, and Commons but of only one or two of these. It is particularly applied to the legislation of the period between the split between Crown and Parliament in 1642 and the Restoration in 1660. But some legislation passed in normal form has been called ordinances, e.g. the Ordinances of the Forest of 1305 and 1306.

The term is also sometimes given to various kinds of prerogative orders issued by English kings prior to the Bill of Rights of 1688, such as those suspending or dispensing with laws.

In French law, an ordinance was an exercise of the King's legislative power, notable instances being Colbert's ordinances, issued by Louis XIV, on civil procedure (1667), criminal procedure (1670), commerce (1675), and maritime commerce (1681).

In modern times, an ordinance is an order issued by the President in execution of laws, dealing with the organization of public services, as delegated legislation, colonial ordinances, and emergency legislation.

Ordinary. In canon law, any cleric having ordinary jurisdiction, i.e. jurisdiction following automatically from an office that he holds, as distinguished from delegated jurisdiction, in both the internal, i.e. spiritual and moral, and external, i.e. in the world, fora. It belongs to the Pope, bishops, and certain lesser ranks of clergy. In English ecclesiastical law, it is applied to a bishop when exercising the jurisdiction belonging to his office as *iudex ordinarius* of the diocese, and to an official of a bishop having judicial power. In England, the office of judge ordinary was created by the Matrimonial Causes Act, 1857, as the judge of the Court of Matrimonial Causes. It disappeared with the Judicature Act, 1875.

Ordinary Lord. A judge of the Court of Session other than the Lord President and Lord Justice-Clerk, particularly one sitting in the Outer House hearing matters at first instance.

Ordinary of assize and sessions. A deputy of the diocesan bishop appointed to offer criminals the neck-verse (q.v.) and judge whether they could read, and to give them spiritual comfort before execution.

Ordinary of Newgate. A clergyman who attended on condemned criminals in Newgate (q.v.). He customarily noted their behaviour and frequently published a pamphlet on a noteworthy execution.

Ordines judiciarii. A variety of legal literature, common from the twelfh to the sixteenth century, comprising practical treatises describing procedure in court, more commonly civil procedure according to Roman and Canon law but sometimes dealing with procedure in ecclesiastical courts. The earliest extant example is probably the *Excerpta legum edita a Bulgarino causidico* (*c.* 1140), followed by the *Ordo judicarius Causa II, quaestio 1* (*c.* 1171). Other notable ones are the *Ordo judiciarius Bambergensis* (*c.* 1185), the *Rhetorica ecclesiastica* (*c.*1190), and *Ordo judiciarius* of Tancred (*c.* 1215). Those composed after 1234 had regard to the titles in the Decretals dealing with procedure; these include *ordines* by Gratia of Arezzo (*c.* 1240), Peter Penerchio's *Scientiam* (*c.* 1235–40), William of Drogheda's *Summa Aurea* (1239), Master Arnulph's *Summa Minorum* (*c.* 1250). The most famous of all treatises of this class is William Duranti's *Speculum judiciale* (1272) which became the standard work on procedure in the Middle Ages and influenced later works.

Ordo Judiciarius. A work composed in Germany in the early twelfth century as a text book of Roman canon procedure, which went through many editions and was extensively commented on.

Oregon Question. A dispute involving Great Britain, the U.S., Spain, and Russia over ownership of the Oregon country. All four countries had claims based on exploration or settlement. Spain abandoned her claims under the Nootka Sound Convention with Britain (1790) and the Transcontinental Treaty with the U.S. (1819). Russia abandoned its claims by treaties of 1824–25. Eventually by the Oregon Treaty (1846), the British and U.S. claims were compromised by drawing the boundary along latitude 49°N. The marine boundary south of Vancouver Island was settled in 1872.

Oregon school-case (*Pierce* v. *Society of Sisters*) (1925), 268 U.S. 510. A decision of the Supreme Court upholding the right of parents to control the education of their children and declaring unconstitutional an Oregon state law which would have made compulsory attendance at state schools.

Organization for Economic Co-operation and Development (O.E.C.D.). In 1960 the members and associate members of O.E.E.C. found that its objects had been mainly realized and by convention created O.E.C.D. It has three major aims, to promote the highest sustainable standard of economic growth and employment and a rising standard of living while maintaining financial stability, to contribute to the sound economic expansion of member and non-member countries, and to further world trade of multilateral, non-discriminating kind in accordance with international obligations. It is essentially a consultative organization, which is controlled by a Council comprising one representative of each member country and its decisions are binding on member countries unless they abstain from the decision. The Council works through an executive committee and has various subordinate committees. It maintains relations with the Commission of the European Communities.

Organization for European Economic Co-operation (O.E.E.C.). An organization created in 1948, following General George C. Marshall's announcement of the Marshall Plan for economic aid to Europe. It endeavoured to coordinate national economic policies and provide advisory experts. Its usefulness became exhausted and, in 1961, it was transformed into the Organization for Economic Co-operation and Development (O.E.C.D.) for the purpose of assisting developing countries.

Organization of American States. A regional system of international co-operation originating in a conference held at Washington in 1889. In 1936, the U.S. accepted the principle of non-intervention in Latin America in return for an assumption by Latin American states of a collective obligation to meet threats to peace. By 1940, the states concerned had established a regional security system which they secured should not be superseded by the U.N. system established in 1945. In 1947, they entered into a treaty of reciprocal assistance and, in 1948, signed the Charter of the Organization of American States. It has proved an effective agency for promotion of common interests as well as an organization for peace and security. There has been extensive co-operation in economic affairs and attempts to raise the general standard of living, notably through the Alliance for Progress established in 1961. It has a large number of specialized agencies attached to it, notably the Inter-American Peace Committee and the Special Consultative Committee on Security.

C. G. Fenwick, *The Organisation of American States.*

Organization of Central American States. A body established in 1951 which, by 1961, had very largely established a Central American common market and had begun to develop a more formal political and judicial structure.

Original charter. In Scots law, one making a first grant of a particular piece of land as contrasted with a charter by progress, renewing the grant to an heir or successor of the first or a succeeding vassal.

Original estate. The first of several estates related to each other as particular estate and reversion, as distinct from a derivative estate which is a particular estate carved out of another and larger estate.

Original writ. The beginning of a real action at common law. It took the form of a mandatory letter from Chancery, running in name of the sovereign and under the Great Seal, addressed to the sheriff of the county, setting out briefly the cause of complaint and requiring the sheriff to command the defendant to satisfy the claim and if he did not, to summon him to appear in one of the superior common law courts. The original writ was not the assertion of jurisdiction, but a royal command conferring on the judges power to try the matters contained in it. There were various original writs varying according to the plaintiff's complaint and generally stated in settled verbal form.

In the course of time, new kinds of wrongs were recognized as justifying inquiry and remedy, and the Statute of Westminster II (1285) provided that if in one case a writ was found and in a similar case involving the same law and requiring the same remedy no writ could be found the clerks of the Chancery should agree in framing a writ or adjourn the plaintiffs to the next Parliament, and so a writ shall be framed by the consent of the learned in the law.

Originalia rolls. Records covering 1236–1837, compiled in Chancery, recording the extracts transmitted from Chancery to the Exchequer.

Originating summons. A summons without writ, returnable in the chambers of a judge of the High Court, frequently used in the Chancery Division to determine questions arising in the course of the administration of an estate or trust, or as to the interpretation of a deed, for settling questions between vendors and purchasers and other purposes. It may be taken out by any person interested. In simple cases the judge will determine the matter in chambers; in cases requiring argument it will be adjourned into court for hearing.

Organized crime. The commission of crime by enterprises set up for the purpose of achieving ends by criminal means. It includes smuggling, drug trafficking, large-scale thefts and robberies, and kidnapping by terrorists, and may exist anywhere there are substantial gains to be made. It has, however, had its largest development in the U.S. In the Prohibition era (1920–33), there was an enormous increase in illicit import, manufacture, transport, and sale of liquor (generally known as bootlegging) and, after the end of prohibition, the organizations which had grown up to further the liquor traffic turned to other ends, and seem to have become even more organized than before. A by-product of organized crime was gangsterism and gang warfare.

Activities commonly conducted by criminal organizations also include gambling dens, drug peddling, and loan companies charging extortionate interest and using threats and violence against defaulting debtors. They also frequently operate legitimate businesses, though often as a cover for illegal activities and frequently in illegitimate ways.

Organized crime has always invested heavily in protection from interference by law and in securing privileges and concessions. Accordingly, bribery and corruption of politicians, officials, prosecutors, and police have been frequent and the prevalence and success of organized crime and the impotence of honest men against it frequently stems from the weakness or corruption of those who should be enforcing the law. Even when a leader of organized crime is caught he is frequently charged only with some petty technical offence and given only nominal punishment.

Ørsted, Anders Sandoe (1778–1860). Danish jurist and politician and founder of Danish–Norwegian jurisprudence. He was a productive writer and author of a *Handbog over den Danske og Norske Lovkyndighed* (Manual of Danish and Norwegian Jurisprudence) (1822–35). Latterly, he was politically unpopular as a reactionary.

Orto, Anselmus de (twelfth century). Italian jurist, son of the following, and author of an *De Instrumento actionum* and works on contracts.

Orto, Obertus de (twelfth century). Italian jurist and author of a *Libri Feudorum*.

Ortolan, Joseph-Louis-Elzéar (1802–73). French jurist and author of many works, particularly on Roman law, public law, and comparative criminal law, notably an *Explication historique des Institutes de Justinien* (1827) and *Histoire de la legislation romaine* (1828).

Ortolan, Jean-Félicité-Theodore (1808–74). Brother of the preceding, a sea-captain, and author of *Règles internationales et Diplomatie de la mer* (1844–45).

Osborne, Thomas Mott (1859–1926). Became chairman of the New York state commission on prison reform, in 1913, warden of Sing-Sing prison, 1914–16, but had to resign owing to public hostility to his introduction of a system of self-government for the prisoners there, and later (1917–20) was in charge of the U.S. naval prison at Portsmouth, N.H. In his writings *Within Prison Walls* (1914), *Society and Prisons* (1916), and *Prisons and Common Sense* (1924), he advocated individual treatment of prisoners and held the view that prison should seek to rehabilitate offenders.

Osborne judgment (i.e., *Amalgamated Society of Railway Servants* v. *Osborne*, [1910] A.C. 87). The union had made the furtherance of parliamentary representation one of their objects and provided for a compulsory annual levy on members to maintain members in the House of Commons who should support the Labour Party. Osborne challenged this object as *ultra vires* and this view was upheld by the courts. Statute subsequently (1913) permitted unions to have political objects but provided that any dissenting member might contract out from paying the political levy.

Osgoode, William (1754–1824). Was born in England and called to the English Bar in 1779. In 1791, he was appointed Chief Justice of Upper Canada and, in 1794, Chief Justice of Lower Canada but, after being involved in disputes over the management of the public lands, resigned in 1801 and returned home, thereafter taking no part in law or politics. The building occupied by the superior courts at Toronto is named Osgoode Hall after him and his name is also remembered in Osgoode Hall Law School of York University, Ontario.

O'Shea, Kitty, (née Katharine Wood) (1846–1921). Wife of William Henry O'Shea, M.P. for County Clare from 1880, and a supporter of Charles Stewart Parnell, leader of the Irish Nationalist Party. Mrs. O'Shea lived with Parnell from 1881 and acted as intermediary between Parnell and Gladstone on Irish problems. In 1890, O'Shea divorced her, naming Parnell as co-respondent. This resulted in Parnell losing the leadership of his party and being ostracized by the Irish Catholic hierarchy and large sections of English opinion.

Parnell married Mrs. O'Shea in 1891, not long before his death.

Ostensible. That which is shown or disclosed by or apparent from circumstances, whatever the reality may be. Thus, ostensible agency is the power to act as agent indicated by circumstances, though the person may not truly have had such power at all.

Ostracism. The ancient Athenian political device whereby a citizen could be banished without charges or trial if thought to be threatening the stability of the state. The populace decided, at a meeting, whether to have a vote on ostracism. On the date of the vote, any citizen could write the name of any other on a potsherd (ostrakon) and, if a sufficient number wrote the same name, the person named had to leave Attica within 10 days and stay away for 10 years, although he retained his property and might be recalled early. It accordingly differed from exile in Roman law, in which the exiled person was deprived of his property and exiled for an indefinite period. Ostracism was resorted to several times in the fifth century B.C. against Hipparchus (487), Megacles (486), Xanthippus (485), Aristides (483), Themistocles (?471), and some others. Similar machinery existed in some other Greek cities.

Ostrogothic laws. Edicts promulgated by the Ostrogothic conquerors of Italy in the seventh century. The principal is the Edict of Theodoric (q.v.), but others include the Edict of Atalaric.

The sway of these enactments ended with the reconquest of Italy by Justinian. The Pragmatic Sanction of Justinian confirmed the edicts of the earlier Ostrogothic Kings, so far as consistent with his Code, and annulled later ones.

Ottoman constitution. The first major constitution of the Middle East, modelled on the 1821 Belgian constitution and promulgated in 1876. It was suspended in 1878 and its restoration was a principal object of the Young Turk revolutionaries of 1908. It was superseded by the Turkish constitution of 1922.

Ouster. Dispossession, a kind of wrong which may be committed in respect of any kind of hereditament. Ouster of freehold could be effected by abatement, deforcement, discontinuance, disseisin, or intrusion. Ouster of chattels real could consist in amotion of possession from estates held by statute, recognizance or *elegit*, or by amotion of possession from an estate for years taking place by a kind of disseisin or ejection or turning the tenant out.

Ousterlemain. A writ directing that seisin or possession of land be delivered from the hands of the Crown into the hands of a person entitled to it, as where an heir in ward of lands held by the crown obtained possession of the lands when he attained majority. It was also the name of a judgment on a *monstrans de droit*, deciding that the Crown had no right to a thing seized by it.

Outer bar. A collective term for junior barristers because they plead from outside the bar (q.v.) of the court, while Queen's Counsel plead from within it. On being called to the bar (q.v.), a person is 'called to the degree of the outer bar'.

Outer House. See COURT OF SESSION.

Outer space. Generally understood to be the parts of the universe outside the atmosphere and stratosphere of the world, but there is no settled definition. International law applies to outer space and celestial bodies, and neither may be appropriated by any earthly state, although all states may freely explore and use them on a basis of equality, but in accordance with international law. States are liable for national activities in outer space and retain ownership of and jurisdiction over objects launched into space. The law, particularly on liability for damage caused by space vehicles, is in a very early stage of development.

Outfangthief. The right of a feudal lord to bring within his own jurisdiction and hang there, one of his own men who had been condemned for theft by another lord, or the right to try thieves who had come within his jurisdiction having stolen elsewhere, or the right of a lord to bring back and try one of his own men who had stolen outwith his jurisdiction. See also INFANGTHIEF.

Outlaw. A person put outside the protection of the law. Outlawry was originally the punishment of one who could not pay the wergeld or blood-money to the relatives of one whom he had killed, and it later became a punishment for refusing to appear in court or for evading justice by disappearing. An outlaw was *civiliter mortuus*, could hold no property, could not sue in any court and had no enforceable legal rights. The state was abolished in civil proceedings in 1879 and in criminal in 1938 but had been in disuse long before these dates.

Overdue. When the time for payment or performance is past. A bill of exchange is overdue when it appears to have been in circulation for an unreasonable length of time. If it is then negotiated it can be done so subject only to any defect of title affecting it at its maturity and, thereafter, no person taking it can acquire or give a better title than the person from whom he took it had.

Overreaching. The principle whereby equitable interests in land held by an estate owner, or on trust for sale, attach from the moment of sale to the purchase money and no longer to the land itself, the purchaser taking the land free of the equitable interests. Overreaching applies to many equitable interests but also to some legal ones.

Overriding interests. Certain kinds of rights, e.g. leases, which bind the proprietor of registered land in England, even though he has no knowledge of them and no reference is made to them in the register.

Overruling. The action of a superior court in stating that a precedent or previous decision of a lower court, on a point before it, was a wrong statement of the law and should no longer be regarded as authoritative. The Supreme Court of the U.S. has long, and the House of Lords has since 1966, had the power to overrule their own previous decisions, but in other courts, only a superior court can overrule a decision and a court of equivalent standing cannot do so; at most it can distinguish the precedent case. Hence, the Court of Appeal cannot overrule a previous decision of the Court of Appeal, but it can overrule a previous decision of a single judge of the High Court.

Overseers. An office, created in 1572, as assistants to the churchwardens in the care of the poor. In 1601, it was provided that in every parish the justices should nominate the churchwardens and from two to four householders to be overseers of the poor. The churchwardens usually left the task to the overseers, who originally levied and disbursed the poor rate. The office was abolished in 1925.

Oversman. In Scottish arbitrations, an umpire appointed to decide where two arbiters have differed in opinion. The oversman was either appointed by the deed of submission to arbitration or named by the arbiters under powers given them by the submission.

Overt act. An act openly done, as distinct from private intention or even words spoken. The phrase is met with in the law of treason, where treasonable intention is not punishable unless evidenced by an overt act, and in criminal conspiracy, where the unlawful agreement is an overt act.

Ownership. The legal relationship between a person and an object of rights, whether corporeal (immovable, e.g. land, or movable, e.g. animals) or incorporeal, e.g. shares. A person having ownership has the fullest group of rights which a person can legally have in relation to things of that kind, including at least some of the rights to occupy, possess, use, abuse, use up, let out, lend, transfer in security, sell, exchange, gift, bequeath, and destroy. The precise rights vary as between immovables, movables, and incorporeals. Ownership is always, accordingly, ownership of a group of rights in and to some object of these rights. The owner may, accordingly, transfer to others some or many of the rights but he remains owner so long as he retains the radical or reversionary right of getting the thing back when the other party's right has terminated; he ceases to be owner if he alienates the reversionary right and can no longer recover the thing.

Rights of ownership may be abridged by agreement, as by grant of an easement over land or entering into a restrictive covenant, or by common law, such as the restrictions on committing nuisance, or by statute, as by building or planning legislation, compulsory purchase powers or taxation. Ownership is accordingly the entirety of the powers of use and disposal permitted by law.

The rights which amount to ownership are generally rights *in rem* though when a particular right is transferred to another, as by loan or lease, the right to recover it is *in personam* against the temporary possessor as well as *in rem* against all third parties.

The rights of ownership may be vested in a single person, or in two or more either as joint owners, or as owners in common. Joint owners possess the thing undivided but have only one title to the thing, so that on the death of any one the title accresces to the other or others until it is vested in one, who is then sole owner; the best case is the ownership of trustees. Owners in common possess the thing undivided but each has a distinct title to a determinate share which on his death passes to his representatives.

Ownership may again be vested simultaneously in one or more persons as trustees and one or more persons as beneficiaries, the former having legal title and legal ownership as against third parties, the latter having a claim against the trustees only but being recognized in equity as the owners.

In feudal land law, such as applied in England down to 1925, several parties, the Crown, one or more mesne lords, and the tenant, concurrently held distinct estates in the land and were linked to one another by a relation of tenure, whereby each owed to the one superior to him in the tenurial chain certain duties, frequently originally military service, or money payments in lieu thereof. There was no outright ownership of land save by the Crown and the main remedies were to protect possession, by writ of trespass, and to secure possession, by proof of a better right than the posssessor had, by ejectment.

Oxford Law School. Law was being studied at Oxford from the twelfth century. Vacarius, the founder of the study of civil and canon law in England, taught there in the latter part of that century.

Henry VIII founded the Regius Chair of Civil Law and its study began to revive from the later sixteenth century; a notable holder of that chair was Albericus Gentilis (1587–1611) while other notable holders were John Budden (1611–20) and Richard Zouche (1620–61). Henry's break with Rome put an end to the study of the canon law and English law had, by then, for centuries been studied in the Inns of Court. In 1753, William Blackstone (q.v.) offered a course of lectures on the Laws of England and, in 1758, he was elected first holder of the Vinerian Chair, established in 1758, pursuant to the will of Charles Viner (q.v.) compiler of Viner's *Abridgment*. He held the chair only till 1766 and subsequently published his famous *Commentaries* (1765–69) based on his lectures—the first modern exposition of the law of England having literary form. It was immensely popular and influential, even more so in America than in England. Thereafter, legal studies stagnated until the establishment of the School of Law and Modern History in 1850; they were divided in 1876. Since then, Oxford legal studies have always been concerned with the development of intellectual capacity in law rather than with inculcating professional expertise and attention has always been given to jurisprudence, legal history, Roman law, and other topics as well as to the major branches of English law.

The greatness of the Oxford Law School dates from the election as Vinerian professor, in 1882, of A. V. Dicey (1882–1909) (q.v.), whose most distinguished successors have been Sir William S. Holdsworth (1922–44) (q.v.) and G. C. Cheshire (1944–49). Among the other major scholars who have taught in the school have been Sir W. R. Anson warden of All Souls (1881–1914), James Bryce (1870–93), Henry Goudy (1873–1919), and Francis de Zulueta (1919–48) in the Regius Chair of Civil Law; T. E. Holland (1874–1911) and J. L. Brierley (1922–47) in the Chichele Chair of International Law; Sir Henry Maine (1868–83), Sir Frederick Pollock (1883–1903), and Sir Paul Vinogradoff (1903–26) in the Corpus Chair of Jurisprudence. These and many other teachers in the school have contributed very substantially to legal literature.

H. G. Hanbury, *The Vinerian Chair and Legal Education* (1958); F. H. Lawson, *The Oxford Law School, 1850–1965* (1968).

Oxford Propositions (1643). Proposals made early in the Civil War by Parliament for peace negotiations with Charles I, which were rejected by the King.

Oxford, Provisions of. A plan of reform imposed in 1258 on Henry III by his barons. In return for a grant of revenue the King agreed to abide by a programme to be formulated by a commission of 24, half nominated by the King and half by the barons. Their recommendations constitute the Provisions and are the first constitutional provisions in English history. They were to remain in effect for 12 years. The government of the realm was placed under the King and a 15-member baronial advisory council. All high officers of the realm were to swear allegiance to the King and council, and Parliament was to meet three times a year to discuss further reforms. A justiciar was appointed to oversee local administration and most of the sheriffs were replaced by knights, who held land in the areas which they administered. The Provisions were annulled by Papal Bulls in 1261 and 1262 and by Louis IX of France in the Mise of Amiens in 1264 but restored by baronial action in 1263, and, with modifications in 1264, but were finally annulled by the Dictum of Kenilworth in 1266.

Oxgang (or oxgate). As much land as a single ox could, theoretically, plough. It was one-eighth of a carucate, or the area which the ordinary team of eight oxen could plough in a year, and amounted to from 12 to 15 acres.

Oyer and Terminer, commission of (to hear and determine). A commission issued to certain of the King's justices, and others directing them, to enquire concerning certain crimes committed in certain counties, either special, as to a particular case, or particular place, or certain specified crimes, or general, as to all crimes within the commission's area. Till 1971 it was one of the commissions issued to judges of assize.

In the U.S., the term is given to some state courts having criminal jurisdiction.

P

P.C. Privy Council, or Privy Counsellor.

Pacific blockade. A measure of coercion short of war exercised by a more powerful state on a less powerful one. It is an act of violence and of the nature of war, but as its object is to avoid war, rights of warfare cannot be exercised against ships of other states. The first recorded instances are the Anglo-

Swedish blockade of Norwegian ports in 1814 and the intervention of Britain, France, and Russia in Turkish affairs in 1827. The Declaration of Paris of 1856 required that blockades must be effective and this gave rise to a question of the validity of the French blockade of Chinese ports in 1884. Since then pacific blockades have been used by the major powers as a joint measure in the interests of peace, which alters the meaning of 'pacific'. By the Hague Convention of 1907, No. 2, the contracting powers agreed not to use armed force for the recovery of contract debts unless arbitration is unsuccessful. This in practice supersedes some of the cases in which pacific blockade had been employed.

Pacta sunt servanda (agreements are to be kept). An abbreviated form of the rule stated in Justinian's Code 2, 3, 29, an expression of the principle that undertakings and contracts must be observed and implemented.

Pactio. An agreement or contract in Roman law, synonymous with *conventio* and distinct from *pactum* (q.v.).

Pactional damages. Agreed pre-estimate of damages, or liquidate damages (q.v.).

Pactum. Originally, in Roman law, a compromise or agreement not to sue, but later any agreement not falling within the defined terms of any of the recognized contracts. Originally *pacta* might be pleaded as a defence only but did not give rise to a claim, but latterly in late law they might in some cases give rise to an an action. *Nudum pactum* was a *pactum* not giving rise to an action, but the phrase is sometimes used in common law for an agreement without consideration. *Pacta vestita* were agreements which praetors, and later emperors, accepted as giving rise to action though not falling within the list of contracts. The principal example was *constitutum debiti*, an agreement to pay an existing debt at a fixed time.

Pactum de non petendo. An agreement by which the creditor binds himself not to insist for payment or performance, for a time or absolutely.

Pactum de retrovendendo. A stipulation that the seller should be entitled to buy back the property sold.

Pactum legis commissoriae. A provision that a sale became void if the price were not paid before a certain day. The *pactum legis commissoriae in pignoribus* was a provision sometimes attached to a redeemable right cutting off the right of redemption if it had not been exercised by a stated date. In Roman law such a stipulation was *contra bonos mores*.

Pactum illicitum. In Scots law a general term for contracts which are objectionable as contrary to positive law or to public policy.

Pactus. An agreement in Germanic law between a king and his people.

Page Wood, William, 1st Lord Hatherley (1801–81). Studied abroad and became acquainted with all the literary idols of London, and was called to the Bar in 1827. His practice was at the Parliamentary and chancery bars. In 1846 he entered Parliament and became Solicitor-General in 1851 and in 1852 a Vice-Chancellor, in which capacity he was a very sound judge, though his judgments were long and rambling. In 1868 he became a Lord Justice of Appeal in Chancery and, later in the same year, Lord Chancellor as Lord Hatherley. He resigned in 1872 but sat in the House of Lords thereafter. As Chancellor he put through some important government measures, but failed to carry a bill which anticipated the Judicature Act, and during his earlier judicial career was constantly engaged on commissions on various topics. He was a very sincere churchman but not a vigorous law reformer. He was not a great Chancellor, not being intellectually outstanding nor a profound master of law, though he was a sound and industrious lawyer.

W. R. Stephens, *Memoir of Lord Hatherley*.

Paine, Thomas (1737–1809). English political writer, worked in the excise, went to America in 1774 and in 1776 published *Common Sense*, a powerful republican pamphlet. In 1787 he returned to Europe and in 1790 wrote *The Rights of Man* (1791, Part II, 1792) defending the French Revolution, as an answer to Burke's *Reflections on the Revolution in France*. For this he was found guilty, in absence, of seditious libel and declared an outlaw. He took refuge in France where he sat in the French Convention and was nearly guillotined. His final work was *The Age of Reason*, urging pure morality founded on natural religion. In 1802 he returned to America where he spent his last years, but by this time he had lost his popularity.

M. D. Conway, *Life of Thomas Paine* (1892).

Pains and penalties, Bill of. Bills introduced in Parliament, usually in the House of Lords, to condemn persons as guilty of treason or serious crime, or for special purposes to impose pains and penalties contrary to or different from those allowed by common law. Persons affected are entitled to be heard in person or by counsel at the Bar of the House. Such bills were competent in cases other than treason or felony, including cases of conduct tending to grave public mischief. The last one was that against Queen Caroline in 1820 but the Titles

Deprivation Act, 1917, was really of the same nature.

Pairing. The practice in the House of Commons whereby two members of the House of Commons of opposite parties are allowed by the Whips to absent themselves from a particular division or for an agreed period. This prevents the government majority being lost by reason of necessary absence.

Pais. The country, the people from whom a jury may be drawn; thus trial by pais is trial by jury.

Palace. An official royal residence. There is a privilege from the execution of legal process within the precincts, but this seems not to extend to a royal palace which is not a royal residence.

Palace court. A court established by James I by letters patent and having jurisdiction in personal actions to any amount arising between any parties within 12 miles of Whitehall, except actions arising within the City of London, or within the jurisdiction of the Court of the Marshalsea, or the courts at Westminster. Important cases were usually removed into the King's Bench or Common Pleas. The judges nominally were the Lord Steward and the Marshal of the King's Household and the steward of the court or his deputy. The deputy was the real judge. Only four barristers and six attorneys could practise in the court. It was abolished in 1849.

Palace of Westminster. The chief residence of the kings of England from Edward the Confessor to Henry VIII. It remained a Royal Palace though it ceased to be a royal residence after St. James's Palace was built in 1532. In 1965 the Queen transferred control of it to Parliament. Parliaments were held there from early times. It includes Westminster Hall, completed in 1099, in which, from the middle of the thirteenth century till 1882 the courts of law were held. In 1834 a fire destroyed all the Palace except Westminster Hall, the Crypt, and the Cloisters, and the present building was built 1837–52, and extensively damaged by bombs during the Second World War and reconstructed thereafter.

Palatine (*palatinus*, belonging to the palace). Originally the chamberlains and troops guarding the palace of the Roman emperors, and later the force of the army which accompanied the emperor. Later the term was applied to various officials in many countries of mediaeval Europe. Among these were the counts palatine in Carolingian times who were officials of the sovereign's household and in particular of his law courts. The count palatine was present officially at such ceremonies as the taking of oaths, judicial sentencing, and had charge of the records of such happenings. He examined cases in the King's courts and later exercised judicial authority in his own court. Under the German kings of the tenth to twelfth centuries counts palatine had functions like those of the *missi dominici*. The count palatine of Lotharingia became the real successor to the Carolingian count palatine and this office developed into the Countship Palatine of the Rhine or Palatinate, which became a great territorial office, existing till the end of the Holy Roman Empire and still a territorial designation, as a *land* of the Federal German Republic. The term palatine was also used of a body of household counts palatine at the court of Charles IV, but they had only honorary functions, and, in the form paladin, was used in literature of the companions of Charlemagne.

In the Middle Ages the counties palatine of Durham, Lancaster, and Chester were independent principalities of the continental type in which the King's writs did not run, though they were bound by statutes of the Westminster Parliament, their judicial system copied that of the rest of England, and the rules of the common law were applied. In palatinates or counties palatine the lord had royal powers, to create courts, pardon crimes, issue writs and indictments, and rule like a petty king.

In Durham powerful bishops with large grants of lands developed a Council and Chancery, and a central court of pleas, judicial developments closely paralleling that of the rest of England, though there was a tendency to emphasize the control of the King's courts. A statute of 1536, though retaining an independent judicial system, made it immediately dependent on the King. The establishment of the Council of the North in 1537 made dependence more evident. In the nineteenth century the temporal powers of the bishoprics were vested in the Crown, first as a franchise separate from the Crown, but since 1858 by right of the Crown. The Court of Pleas was reorganized in 1839 and merged in 1873 in the High Court of Justice. The Court of Chancery survived till 1971. Since 1873 the county was not independent concerning commissions of assize.

Lancaster was made a county palatine in the fourteenth century and developed a separate judicial system with court of pleas, justices of assize, justices of the peace, and court of chancery. Though it remained a separate franchise it has, since 1399, always been in the hands of the Crown. The Duchy was managed by a Chancellor and council, and from the late fifteenth century the Chancellor held a court of Duchy Chamber with equitable jurisdiction, distinct from the palatine court of Chancery. In 1873 the Lancaster Court of Pleas was merged in the High Court, but the palatine court of Chancery survived with generally the same jurisdiction as the Chancery Division till 1971.

The Duchy of Lancaster formerly had its own

Q.C.s, a famous one being J. P. Benjamin, appointed in 1867. They received no patents of appointment, being created by merely being called within the Bar by the senior judge on circuit sitting in the Court of Common Pleas at Lancaster. A Q.C. of the Duchy could revert to the rank of barrister, which an ordinary Q.C. could not do unless he were also being disbarred.

Chester and Shrewsbury were made counties palatine by William I, and Pembroke also became a county palatine later. Chester reverted to the Crown in 1237, was made a principality and settled on the King's eldest son. The justice of Chester held a court for common pleas and pleas of the Crown, and the chamberlain had chancery and exchequer jurisdiction, error lying to the King's Bench. In 1536 it too was made more dependent on the Crown. When the courts of Great Sessions for Wales were constituted in 1543 the justice of Chester became judge of a group of three Welsh counties, and later he was usually a member of the Council of Wales. In 1830 the palatine court of Chester was abolished and the English circuit system extended to Wales and Chester.

There were palatine provinces among the English colonies in North America. Lord Baltimore was granted palatine rights in Maryland in 1632. The lord of Maine obtained in 1639 a charter granting him as large and ample prerogatives as were enjoyed by the Bishop of Durham, and the proprietors of Carolina were given such rights in 1663.

Pale. A district marked off from the surrounding country by boundaries or by a distinct system of government, particularly, in Ireland, the area around Dublin which from about 1395 was to be kept definitely English. In 1465 it extended to the counties of Meath, Louth, Dublin, and Kildare, and Poyning's Parliament of 1495 provided that a double ditch should be built round the Pale to prevent cattle-raiding, and that every subject therein must be armed after the English fashion for the defence of this land, and ready to do military duty when called on. Only within the Pale were English law, speech, and loyalty regarded as secure. As time went on the Pale shrank and became more Irish, but was extended by the later Tudors to include most of Leinster and Meath. There was also an English Pale round Calais in France from 1347 until 1558, and for a short time under the Tudors an English Pale in Scotland.

Palgrave, Sir Francis (1788–1861). British archivist and historian, head of the Public Record Office 1838–61. He edited various series of records, notably *Rotuli Curiae Regis* (1835), and set new standards of scholarship in the editorial work. He also wrote a series of books on early English history, the *History of Normandy and England* (1851–64) and the *Rise and Progress of the English Commonwealth, Anglo-Saxon Period* (1832), which emphasized the importance of law and legal institutions for understanding political history.

Palgrave, Sir Reginald Francis Douce (1829–1904). Clerk of the House of Commons 1886–1900 in succession to Erskine May and a great authority on Parliamentary procedure. He edited Erskine May's *Parliamentary Practice* and several editions of the *Rules, Orders and Forms of Procedure of the House of Commons*, and wrote *The House of Commons* (1868) and *The Chairman's Handbook* (1877).

Palles, Christopher (1831–1920). Called to the Irish Bar in 1853, became Solicitor-General and Attorney-General in 1872, and Chief Baron of Exchequer in Ireland in 1874, but allowed his court to be merged with the Irish Queen's Bench Division in 1898. He retired in 1916 having received many academic and other honours, and been recognized as an outstanding exponent of the common law.

V. Delaney, *Christopher Palles*.

Palm tree justice (or the justice of the cadi (q.v.) sitting under the palm tree). A term for justice without law, descriptive of the situation where a person has to decide with no principles or rules of law to guide him on how to decide. Such a system may result in very fair decisions in individual cases, but every case is decided on its own and there may be no consistency or uniformity between similar cases.

Palmer, Roundell, 1st Baron and 1st Earl Selborne (1812–95). After a brilliant career at Oxford he made a slow start at the Bar, but his practice developed and he entered Parliament in 1847 but lost his seat in 1857. In 1861 he was made Solicitor-General and re-entered the House, and in 1863 Attorney-General. His party was out of office from 1866 to 1868 and in 1872 he became Lord Chancellor. During the years 1868–72 he was involved in many schemes for legal reform: he proposed the Judicature Commission and as Chancellor carried the Act of 1873, founded on its recommendations, which effected the fusion of law and equity, swept away the existing courts and established the Supreme Court of Judicature. He founded an Association for the Improvement of Legal Education and tried unsuccessfully to found a school of law in London. He ceased to be Chancellor in 1874 but held the office again in 1880–85, during which time the new Law Courts in the Strand were opened and much important legislation was passed. He declined the Chancellorship in 1886. All his life he had a reputation for energy and industry and he proved himself a master of equity, an outstanding judge, and initiator of much important legislation.

Selborne, *Memorials, Family and Personal; Memorials, Personal and Political.*

Palsgraf* v. *Long Island Railroad Co. (1928), 248 N.Y. 339. A leading U.S. case on negligence. A woman on a railroad platform was injured by the explosion of a parcel dropped by a person who boarded a train as it departed. It was held that she had no right of action because there was nothing in the circumstances to affect the defendant railroad with knowledge that anything done by it might result in harm to her.

Pan-American Union. An organization founded in 1890 and known till 1910 as the International Bureau of American Republics. It serves as a permanent agency for exchanging information about all the American republics. Its headquarters are the Pan-American Union in Washington, D.C., and its governing board comprises the U.S. Secretary of State and the chiefs of mission in Washington from the other countries in the American continent. In 1948 it became the general secretariat for the Organization of American States (q.v.).

Pandectists. A name given to the group of German jurists, notably Savigny, Dernburg, Puchta, Vangerow, and Windscheid (qq.v.), who devoted themselves to the study of the Pandects or Digest of the Roman law, seeking to discover the original meaning in Roman times of the legal propositions in the Corpus Juris, and to apply analytical jurisprudence to that system, seeking to organize the materials of that system into a coherent system. Classic works of the group are Savigny's work on *Possession* and his *System of Modern Roman Law*. Their writings, however, tended to become elaborate structures of legal concepts and logical deductions, sometimes ill-related to the needs of the countries which in the nineteenth century had based their law on Roman law. They sometimes deduced logical consequences with scant regard to the possibility or practicability of the result. Their methods and work were satirized by von Jhering in his *Scherz und Ernst in der Jurisprudenz* (1885). The group's work did have considerable influence in adapting Roman law to their contemporary society and in introducing the Roman element in the German Civil Code of 1900. But reaction against the over-emphasis on logic in German legal thinking gave rise thereafter to the 'free law' movement and the 'jurisprudence of interests'.

Pandects (*Pandectae*). An alternative name for the *Digest* (q.v.) of the emperor Justinian.

Pandulf (?–1226). Papal legate to England in the reign of King John. He received John's submission, absolved him from excommunication, lifted the interdict, and received England as a fief, which he regranted to John to hold of the Pope. He later supported John during the baronial rebellion and became Bishop of Norwich. During the minority of Henry III he was a leading member of the regency, but was recalled to Rome in 1221.

Panel (or pannel). In Scottish criminal law, the name for the accused, from the time of his first appearance in court. Also the schedule or page containing the names of the jurors summoned to attend for a trial. Also any list of persons, such as those legal practitioners willing to act for persons entitled to legal aid, or the list of J.P.s from whom juvenile courts are composed in England.

Pannage. Acorns, beech-mast, and other food which swine feed upon in woods. Common of pannage is the right of feeding swine at certain times of year in a commonable wood or forest.

Pannier (or pannierman). Servants of the Inner Temple who serve at dinner are sometimes colloquially called pannier. In the Inns a servant formerly used a horse with panniers to bring in provisions required for meals in hall, and was accordingly called a pannier-man.

Panopticon. The name given by Jeremy Bentham to his planned penitentiary in which cells in a circular building would all constantly be under observation from a central tower. The idea was developed in his *Panopticon or the Inspection House* (*Works*, ed. Bowring, IV, 37).

In 1794 Parliament agreed to take up his idea and a site was bought at Millbank, London, but it was not built, and when a prison was built on the site in 1812 it was not to Bentham's plan. A version of the Panopticon was built as the Western Penitentiary, Pittsburgh, U.S.A. in 1818 and another version was built at Breda, Holland.

Panormitanus (Nicola de Tudisco or Tedeschi) (1389–?). Famous Italian commentator and canonist, professor at Bologna and Florence, later Archbishop of Palermo, author of *In quinque decretalium libros commentaria* (1475), *Glossae in Clementinas* (1474), *Disputationes et allegationes subtilissimae* (1474), and other works.

Papal arbitration. An aspect of international arbitration. The Theodosian Code recognized the validity of decisions by bishops as judge or arbitrator and the obligation of the State to enforce such decisions. The great popes of the twelfth and thirteenth centuries appointed Roman law forms of arbitration for many controversies, and in the feudal period the popes were frequently called on to act as arbitrators in international disputes. This tradition was revived in modern times and Pope Leo XIII was selected to arbitrate between Germany and Spain over the Caroline Islands, and the idea provoked a considerable literature. But in neither World War did the Pope succeed in acting as arbitrator or mediator.

Papal documents. The general term for formal communications from the Pope to one of his authorized representatives, to Roman Catholics generally, or groups, or individuals. They may have many uses, doctrinal, pastoral, disciplinary, or administrative. They are of various kinds though the names are not always used consistently. A bull is a circular leaden seal having on one side the images of SS. Peter and Paul and on the other the name of the reigning pope and the date of his accession. A bull is affixed to certain documents issued by a Pope and these documents themselves have come to be called Bulls. A Bull is the most solemn form of papal document used, e.g. the definition of a doctrine as a matter of Catholic belief. An apostolic or papal constitution is a solemn pronouncement concerning serious doctrinal or disciplinary matters. A papal brief is also important and sealed with red wax impressed by the Pope's fisherman's ring. An encyclical letter is a formal pastoral letter to the entire Church relating to moral, doctrinal, or disciplinary matters; it is authoritative but not necessarily infallible. An encyclical epistle is addressed to part of the Church.

A *motu proprio* is an informal document issued by the Pope in his own name dealing with important doctrinal or disciplinary matters. A *chirograph* is an autograph letter from the Pope, usually to a cardinal.

Decrees are extensive orders of the Pope, a council or one of the departments of the Roman Curia having binding force for the Church or for the parties. A rescript is a reply by the Pope or a department of the Curia to a request or question and generally affects only the addressee.

Papal line of demarcation. A line defining the spheres of influence of Spain and Portugal in the New World, defined by a Bull of Alexander VI of 3 May 1493, modified by the Treaty of Marseilles (1494) between Spain and Portugal, assigning to the Portuguese the sphere of influence in discoveries east of, and to the Spanish in discoveries west of, a line 370 leagues west of the Cape Verde Islands.

Paper, Revenue. A list maintained in the Central Office of the Supreme Court of appeals from decisions of the Commissioners of Income Tax which are awaiting hearing.

Papinianus, Aemilius (*c.* A.D. 150–212). One of the greatest classical Roman jurists. His origin is uncertain but may have been Syria or Africa. He had a distinguished official career and became *praefectus praetorio* in 203 but was executed by order of Caracalla in 212. He wrote no comprehensive or systematic text but various collections, including *Quaestiones* (37 books) and *Respona* (19 books), both of which date from A.D. 190–210 and contain citations from earlier jurists and doctrinal discussions. Paulus wrote *Notae* to both works and Ulpian *Notae* to the *Responsa*. He also wrote *Definitiones* and *De adulteriis*. His work came to be highly regarded. In the Valentinian Law of *Citations* (426) it was provided not only that his works should be authoritative but that, in case of equality of opinions on each side, Papinian's view should be decisive. There are 601 excerpts from his works in Justinian's *Digest* and they indicate clear and logical thinking, independence of judgment, and readiness to reconsider his view.

E. Costa, *Papiniano.*

Papinianistae. In the law schools of Justinian's time, a term for third year students, given them because they were mainly studying Papinian's works. To reinforce this connection Justinian put a quotation from Papinian at the beginning of each title of *Digest*, Book 20, the first of the books (20–22) which they had to study in addition to the topics not covered in the second year.

Paraphernalia. In older law, jewellery and ornaments given by a husband to his wife before or during marriage, disposable by the husband during his life but not by will, falling to the wife on the husband's death. In older Scots law, they were those goods which remained the sole property of a woman, notwithstanding marriage, and did not become her husband's property. They comprised clothing and jewellery. The concept is now obsolete.

Paraphrase of Theophilus. An almost complete Greek paraphrase of Justinian's *Institutes*, but including elements from Gaius' *Institutes*, probably by the Theophilus who collaborated in Justinian's codification.

Parcels. A part or portion of land, particularly the property described as dealt with by a deed of conveyance or mortgage.

Pardessus, Jean Marie (1772–1853). French jurist and judge, author of *Cours de droit commercial* (1813), *Collection des lois maritimes anterieures au XVIIe siècle* (1828), *Us et coutumes de la Mer* (1847), and other works.

Pardon. A release from guilt or remission of punishment, which generally may be granted only by a head of state, by executive act, but may be granted by legislation, or by an Act of Indemnity. Pardon may be either a release from liability to penalty, or discharge the liability which has been incurred by conviction. It may also be complete and unconditional, or qualified and conditional. In modern Britain a pardon is granted by the Crown only on the advice of the Home Secretary or the Secretary of State for Scotland, and is effected by warrant under the Sign Manual. At common law the Crown could not pardon an offence against a penal statute after information had been brought because the informer had acquired a property in part of the penalty. By the Act of Settlement, 1701, a pardon cannot be pleaded to prevent an impeachment, but it may be granted to the person condemned after an impeachment has been determined. The effect of a full pardon is not settled. It is said by some authorities that in England a full pardon totally obliterates the individual's crime. In the U.S. the President may pardon all offences against the government except in case of impeachment. The Supreme Court has held that pardon wholly extinguished guilt and the offender is deemed innocent, but in some states pardon merely extinguishes the sentence and does not wholly remove disqualifications consequent on conviction.

Parent and child. The natural relationship of parent and child gives rise to various legal duties. At common law in England a parent is under a duty to maintain a child only if neglect to do so would be criminal, but under statute each spouse is liable to maintain his or her children until 16.

At common law the father has a right prior to the mother to the custody of a legitimate child during infancy, to control, correct and chastise the child but under statutes both parents have equal rights. By statute the court may regulate custody and access on the application of either parent, the welfare of the child being the first consideration. On the father's death the mother has a common law right to custody, unless she has forfeited it by misconduct. On divorce the court has power to regulate custody and access, having regard to the welfare of the child.

A parent of a child of compulsory school age must cause the child to have efficient full-time education suitable to the child's age, ability, and aptitude, by attendance at school or otherwise.

A father has the right to determine in what religion his child shall be brought up; the child cannot finally decide for himself until he attains full age. The right continues even after the father's death. But in legal proceedings the court will regard the infant's welfare are paramount.

Though a young person may marry at 16 the consent of the parents or surviving parent is generally requisite until 18, failing which the consent of the court.

In Scotland the father has the right of custody of a pupil (q.v.) child and authority to regulate its upbringing.

By statute each parent has equal rights in respect of custody, upbringing, or administration of property and the child's welfare is the paramount consideration. The father has a lesser right of custody and authority over a minor child.

The court has power to deal with custody and access up to 16 in matrimonial cases.

The father is under a natural obligation to aliment his child until it can maintain itself; failing him the mother and other ascendants are liable. Parents are obliged to provide education, and now have equal rights in determining the child's religious upbringing.

The mother is the natural custodier of her bastard child and alone liable to aliment it unless she can establish paternity.

Parentelic system. In the law of succession, the system of kinship groups within which the heir of a given person must be sought. The primary parentelic group is the deceased and his own descendants, the next the deceased's father and his descendants, the next the deceased's paternal grandfather and his descendants, and so on. Each successive group is larger and is resorted to, failing an heir from the previous group. The choice of heirs within each *parentela* has developed differently in different systems. In mediaeval English and Scots law the first distinction was that land devolved on a single heir, but chattels devolved on the other members of the *parentela*. Once feudalism had become accepted the heir of land was found by the twin principles of the preference for males over females and of primogeniture. Subsidiary rules developed largely by custom.

Pares. A person's peers or equals. Magna Carta *c.* 39 provided for condemnation only by judgment of a man's peers, i.e. other vassals of the same lord as the defendant, but this was interpreted in the seventeenth century as a declaration of the right to trial by jury. The *pares curiae* are the vassals who attend a superior's court.

Pari-mutuel. A method of wagering introduced in France in 1870 by a Parisian, Pierre Oller, and widely adopted since. The gambler buys a ticket on the horse he wishes to back and the pay-out to winners is made, after deduction of the operator's commission, from the pool formed by all the bets on all the runners in the race. More than one person may accordingly collect on a winning ticket. The method was developed further by the introduction of the totalizator, a machine which calculates odds and totals bets and displays these to persons intending to bet.

Paris a Puteo (Paris del Pozzo) (?–1493). Author of a *Tractatus de re militari*.

Paris, Treaty of (1763). Concluded the Seven Years' War. Great Britain acquired the whole of French Canada and Spanish Florida, and France gave up to Spain all American territory east of the Mississippi except the city of New Orleans. This completed the creation of the British Empire.

Paris, Treaty of (1783). Ended the American War of Independence. It provided for British recognition of U.S. independence, the fixing of boundaries between the U.S. and Canada, and for U.S. fishing privileges in the territorial water of British North America.

Paris, Treaty of (1814). Marked the end of Napoleon's conquest of Europe, the Bourbon dynasty being restored in France.

Paris, Treaty of (1815). After Waterloo and the collapse of Napoleon's attempted return, reduced France to its 1790 boundaries and imposed on France an army of occupation.

Paris, Treaty of (1856). Ended the Crimean War, guaranteed the integrity of Ottoman Turkey, and neutralized the Black Sea.

Paris, Treaty of (1898). Ended the Spanish American war. Spain granted Cuba independence and ceded Puerto Rico, Guam, and the Philippines to the U.S., receiving $20m. for the Philippines.

Paris, Treaties of (1947). Between the allied powers and the allies of Germany during World War II.

Paris, Treaty or Pact of (or Kellogg–Briand Pact). The General Treaty for the Renunciation of War of 1928, by which the signatories condemned recourse to war for the solution of international controversies, renounced it as an instrument of national policy in their relations with one another, and agreed that the settlement of all disputes or conflicts should never be sought save by pacific means. The Treaty does not, however, exclude resort to war as a means of permissible self-defence, as between signatories and non-signatories, as against a signatory who has resorted to war in breach of his obligation, and as a measure of collective action for the enforcement of international obligations by virtue of such instruments as the U.N. Charter. It is doubtful whether the Treaty has prohibited resort to force short of war. It contains no specific obligation to submit controversies to settlement, judicial or otherwise, and no means of coercing signatories who contravene the Treaty, though a contravention is an international wrong against all the other signatories.

Parish. Originally a subdivision of an ecclesiastical diocese, comprising the district served by one church and clergyman. Local administration, chiefly relating to poor law and highways, was committed to parishes and churchwardens from the seventeenth century and the areas of civil and ecclesiastical parishes in many cases ceased to be the same.

The urban parish ceased to be a local government unit in 1933, but rural parishes continued. Since 1974 parishes are sub-divisions of local government districts, having a parish council and parish meetings, and charged with duties in relation to village greens, recreation grounds, and public footpaths. In Wales the equivalent is the community council and community meetings.

In Scotland the parish, originally a purely ecclesiastical division of the country, became a civil administrative district by becoming the area for the administration by parochial boards of the poor law; but later nineteenth-century legislation gave power to form new parishes for civil purposes, and the civil parish was adopted as the area for various administrative purposes though actual areas differed for different purposes. Parish councils existed from 1894 to 1930. In the twentieth century the parish has ceased to be an area or unit for administrative purposes, all functions having been transferred to units dealing with larger areas.

Parke, James, Lord Wensleydale (1782–1868). After practising as a special pleader he was called to the Bar in 1813 and developed a large common law practice. In 1828 he became a judge of the King's Bench, in 1833 a member of the Judicial Committee of the Privy Council, and in 1834 a Baron of Exchequer, in which court he became the dominant personality. He resigned in 1855 and was given a life peerage to enable his great abilities to be used in the House of Lords. But that House resolved in the Wensleydale Peerage Case that a life peerage gave no right to sit and vote in the House, and his peerage was made hereditary. He continued to sit in the House of Lords and Privy Council till his death. He was a very fine judge, but his defect was his attachment to the strict rules and technicalities of pleading. Many of his judgments are important, and all are carefully and lucidly expressed.

Parker, Sir James (1803–52). Called to the Bar in 1829 and practised as an equity draftsman and conveyancer as well as on circuit. He took silk in 1844. He was appointed a Vice-Chancellor in 1851 and proved himself an excellent judge, but died in the next year.

Parker, Hubert Lister, Baron Parker of Waddington (1900–72). Third son of Lord Parker of Waddington (q.v.). Called to the Bar in 1924, he became junior counsel to the Admiralty and to the Treasury, a judge of the King's Bench Division, 1950–54, a Lord Justice of appeal, 1954–58, and Lord Chief Justice of England and a life peer, 1958–71. He was a firm but not harsh judge, interested in the reform of the criminal courts to cope with increases in crime, and sought to improve sentencing and to promote its uniformity over the country.

Parker, Robert John, Baron Parker of Waddington (1857–1918). Called to the Bar in 1883 he soon developed an equity practice, and in 1906 became a chancery judge. In 1913 he was promoted direct to the office of Lord of Appeal, promotion of unprecedented rapidity. He was recognized as a judge of high quality, not least when sitting on the board hearing prize appeals during the World War, cases involving intricate practice and unfamiliar tracts of law.

Parker, Thomas, 1st Earl of Macclesfield (1666–1732). Called to the Bar in 1691, he became an M.P., Queen's serjeant, and was knighted in 1705, and was appointed Chief Justice of the King's Bench in 1710. He became a peer in 1716 and an earl in 1721. He was appointed Lord Chancellor in 1718 and held office till 1725 when he resigned on account of scandals arising out of the uses made by the Masters in Chancery of funds in court. He was impeached and charged with selling offices in the Court of Chancery, and fined £30,000. Thereafter he lived in retirement. Nevertheless, he was a great lawyer, both in common law and equity, and in the latter sphere stated principles clearly and logically and gave them fresh precision, and thereby contributed substantially to the development of equity.

Parker, Sir Thomas (1695–1784). Called to the Bar in 1724, became a judge of Common Pleas, 1740, and Chief Baron of Exchequer, 1742–72. He compiled a volume of reports of revenue cases in the Exchequer (1776) which included also cases of 1678–1718.

Parkhurst. A prison established in 1838, in a former military hospital on the Isle of Wight, as a convict prison for young offenders awaiting transportation. For some time it was conducted more on the lines of a reformatory than of a prison. Subsequently it has been used, like Dartmoor, as a central prison for long-term prisoners.

Parlement. In pre-Revolution France a court of appeal with jurisdiction over a defined territory and having political and administrative privileges. They originated from occasional assemblies at which the legal members of the French *curia regis* and prelates considered appeals from decisions of royal courts of justice. A settled tribunal became established on the island of the city of Paris in the early fourteenth century, sitting from November to July or August. Lay and ecclesiastical peers of France had the right to sit in the *parlement* of Paris. Bishops and abbots ceased to sit after 1319, but other clerics continued to participate. The *parlement* of Paris was the oldest and the most prestigious but other *parlements* were established at Toulouse (1420), Grenoble (1456), Bordeaux (1462), Dijon (1477), Rouen (1499), Aix-en-Provence (1501), Rennes (1553), Pau (1620), Metz (1633), Douai (1668), Besançon (1676), and Nancy (1775); while sovereign councils not called *parlements* but exercising similar functions existed at Arras, Bastia, Colmar, and Perpignan. Until about 1515 *parlements* were mixed tribunals with both lay and ecclesiastical judges, but the composition of a *parlement* depended on the extent of its jurisdiction. Each had a first president named by the King, presidents, counsellors, a public minister, a procurator general, and advocates general. It had a Great Chamber, open to the senior counsellors, which dealt with the chief appeal cases, and, in Paris, matters relative to the royal regalian rights. Each had also one or more chambers of enquiry which handled most appeals, and a chamber of requests which dealt at first instance with questions concerning privileges provided with royal letters. New legislation had to be presented, verified, and registered before the assembled chambers before it could fall within the *parlement*'s jurisdiction, and a *parlement* could defer registration by using its right of remonstrance. Similarly they refused to receive the decrees of the Council of Trent.

Members of *parlements* frequently were promoted to high offices in the State or to bishoprics. They tended to consider themselves fathers of their country and tutors to the King and latterly liked to appear as representatives of the people. All the *parlements* were abolished at the Revolution.

H. B. de Bastarde d'Estang, *Les parlements de France*; F. Lot et R. Fawtier, *Histoire des institutions françaises au moyen âge*.

Parlement de Paris. A body which in France developed from the mediaeval *Curia Regis* and from about 1250 began to hold more regular sessions primarily devoted to judicial business; judicial procedure developed rapidly, and by 1278 a royal ordinance regulated the organization and procedure of this special group. By 1296 its regular membership was fixed at 51 judges, and, after rising to 180 about 1340, was reduced to about 80 in 1345. The large numbers were required by the inquisitorial procedure. It was primarily an appellate court but had original jurisdiction over cases involving high nobility or specially privileged persons. These cases came to be dealt with by the *Chambre des Requêtes*. Most of the work came by appeal from royal courts or seignorial courts, but involved full rehearing of the case. Evidence was commonly taken by an examiner, recorded and transmitted to Paris where it was examined by a *rapporteur* who summarized the facts for the court and recommended a decision.

In the early period, at least half the judges were drawn from the clergy.

A series of selected decrees of the period 1254–1318 was made for the use of the court, and lawyers compiled reports of decisions, while a manual on the *Parlement*'s procedure was written about 1330.

The *Parlement* also had considerable legislative and administrative powers and these expanded in consequence of the breakdowns of government caused by the Hundred Years' War, and by the end of the fifteenth century there had been established the practice of registering royal legislation with the *Parlement*, with opportunity for it to comment thereon or protest. This led to disputes with Crown officials on political issues of many kinds.

The *Parlement* was reconstructed after the Hundred Years' War and from the fifteenth century onwards 12 provincial *Parlements* were established, each being a superior court within its jurisdictional boundaries with legislative power. Their rule-making powers contributed much, particularly in the field of private law. From about 1500 judges could normally count on life tenure, but from the fifteenth century the sale of judgeships became established. By about 1600 saleability and heritability of offices were conceded. In the seventeenth and eighteenth centuries judgeships in the *Parlements* came to confer automatically the status of nobility. By 1715 there were 240 judges in nine chambers in the *Parlement* of Paris. In the eighteenth century reports by judges of decisions were published and numerous collections of decisions were issued by practising advocates.

The *Parlement* had long had the function of registering royal edicts and ordinances and sometimes delayed to do so, so that two or more royal commands were needed to obtain registration, or even the ultimate solemnity of a *lit de justice*. Initially the kings did not concede that registration was necessary to the validity of royal legislation, but *Parlements* increasingly refused registration to legislation disapproved of, or sometimes introduced amendments. In so doing they, to some extent, filled the place vacated by the States-General. Louis XIV silenced the *Parlement*, but its powers revived in 1715 and from then to 1789 it was the main opposition to autocracy, while the provincial *Parlements* frequently repeated the claims of that of Paris. Finally in 1789 it demanded the recall of the States-General which, when they assembled, promptly abolished the *Parlements*.

While in existence the *Parlement* contributed greatly to the evolution of French law and developed techniques which were influential later.

Parlement Belge, The (1880), 5 P.D. 197. The P.B. was a mail steamer owned by the King of the Belgians and manned by the Belgian navy, and hence a state or public ship, and conveyed mails, passengers, and luggage between Ostend and Dover. It was held that, as a public vessel, it was not liable to be seized in an action *in rem* for collision damage to another ship.

Parliament Acts, 1911 and 1949. The 1911 Act had its origin in the rejection by the House of Lords of the Budget of 1909 and the determination of the government to restrict the delaying powers of that House. The 1911 Act provides that (1) if a money bill, endorsed as such by the Speaker, has been passed by the Commons and sent to the Lords and is not passed by the Lords without amendment within one month, it shall (unless the Commons direct to the contrary) be presented for Royal Assent without the Lords' consent; (2) if any public bill (other than a money bill, a bill extending the maximum duration of Parliament beyond five years, or a provisional order bill) was passed by the Commons in three successive sessions (whether of the same parliament or not) and was rejected by the Lords in each of these sessions, it should on the third rejection, unless the Commons directed to the contrary, be presented for the Royal assent without the Lords' consent, provided that two years had elapsed between its second reading in the first session and its passing the Commons in the third session. The 1911 Act also reduced the maximum duration of a Parliament from seven years (Septennial Act, 1715) to five years.

The 1949 Act had its genesis in a Bill of 1947 introduced to ensure the passage of nationalization statutes, particularly the Iron and Steel Bill, which, it was feared, might be frustrated by the Lords. The Bill was eventually passed in 1949 under the terms of the 1911 Act. It reduced the delaying power on public bills, with the exception already stated, to one year by providing that a bill might be presented for Royal Assent if passed by the Commons in two successive sessions, provided that one year had elapsed between its second reading in the first session and its passing the Commons in the second session.

The only Acts passed under the 1911 Act powers have been the Government of Ireland Act, 1914, and the Act of 1949 itself.

Parliament of Australia. The Federal Parliament of the Commonwealth of Australia first met in 1901, until 1927 in Melbourne and since then in Canberra. It comprises a Senate, consisting of members elected from and representing the States; and a House of Representatives, consisting of members elected by proportionate representation representing constituencies into which the populace is divided. Senators are elected for six years, half retiring every three years, members of the House for the duration of a Parliament, the limit of which is three years. The Federal Parliament's powers are enumerated in the Constitution and limited, all residuary powers being vested in the State Parliaments. Where federal law is inconsistent with state law, the former prevails. Parliamentary procedure is based on that of the U.K. Financial initiative is confined to the House and the Senate may not originate nor amend, but may reject, a money bill. In the event of disagreement between the Houses there is provision for conferences of managers appointed by each house and for joint sittings. There are, however, differences from U.K. Parliamentary practice. Thus the Speaker is the nominee of the majority party and does not always stand apart from political conflict. Ministers are virtually nominated by the party caucus. Private bills have no place in procedure and private members' bills are rare. The Cabinet system exists in form similar to that in the U.K.

Parliament of Canada. The Dominion Parliament of Canada was established in 1867 and sits in Ottawa. It comprises the Governor-General representing the Queen, the Senate, consisting of members appointed till age 75, by the Governor-General on the advice of the Cabinet to represent the Provinces in proportion to population, and the House of Commons directly elected by the adult population for five years, unless dissolved sooner. The Houses have equal powers and the Senate has legal power to reject legislation initiated in the Commons and to amend money bills. The Speaker of the Senate is a senator appointed by the Governor-General, and the Speaker of the House elected by the House, normally from the party in power and normally changing with each Parliament. The procedure of the House is based on that of the U.K. Parliament, though there are many differences of detail. Under the constitution specified legislative powers are assigned to the provinces, each of which has a legislative assembly, and the residue is vested in the Dominion Parliament. The Dominion Parliament cannot change the federal distribution of powers which still requires an Act of the Westminster Parliament, one which the latter would not refuse to pass if requested to do so.

Parliament of England. Prior to the Conquest of 1066 the witenagemot (q.v.), comprising the King, King's thegns, earldormen, bishops, and abbots, met periodically and participated in acts of government, and acted sometimes as a supreme court in both civil and criminal cases. After the Conquest this was superseded by the *Curia Regis* or court of the King's feudal vassals. It comprised the earls, bishops, and abbots and, theoretically, all of the King's tenants-in-chief, though the King seems early to have assumed the power to select those who were summoned. Magna Carta recognized the right of all tenants-in-chief to be summoned when extraordinary aids were to be granted. But the national assembly gradually became a gathering of only the greater lords. It became recognized that baronial fiefs, the right to which carried right to summons to the national council, were hereditary and consequently the right of membership became hereditary. In addition, in the fourteenth century, the King asserted the right to summon persons who did not hold baronies. Such a summons did not at first create an hereditary right, but this was later recognized, even in the case of the creation of baronies by letters patent. Hence the hereditary character of the temporal Lords in Parliament was early settled.

The word *parliamentum* as meaning a deliberative body was used in 1247, but for a while thereafter was sometimes used of sessions of the King's council and of assemblies of the *magnum Concilium.*

The idea of representation had long been familiar, and to the council of St. Albans in 1213 were summoned bishops and barons and also the reeve and four men from each township on the royal demesne, and representatives of the shires were summoned to the national council in 1213, 1254, 1261, and 1264. After his victory over Henry III at Lewes, Simon de Montfort appointed certain persons to be guardians of the peace in every county and summoned four lawful and discreet knights chosen by assent of the county to attend the King in Parliament at London to treat of business with him. For 1265 he summoned not only bishops and barons but directed the sheriffs to return two knights from each shire, two citizens from each city, and two burgesses from each borough. This assembly was called to approve the newly established form of government, but the precedent was followed by Edward I in 1275. Only from 1295, however, was representation of counties and boroughs regular, and only thereafter can there be said to be representation of the commons. Even under Edward I great councils are common while representation is sporadic.

In 1295 Edward I, being oppressed with a Welsh rebellion, war with Scotland, and war with France, summoned an assembly of greater clergy, barons, two knights from each county, two citizens from each city, and two burgesses from each borough. The inferior clergy were also represented by priors or cathedral deans, archdeacons, and proctors representing the cathedral and parochial clergy. This assembly sat in three groups and voted an aid separately, each giving a different grant. The lesser clergy preferred to grant aids through convocation and ceased to attend Parliament in the fourteenth century, but for two further centuries (till 1664) retained the right to tax themselves in convocation.

In the fourteenth century the function of parliaments was unclear; they were summoned to counsel and advise the King, to vote finance, but their functions might include adjudication on disputes, or supervision of the abuses of sheriffs and local officials, and they asserted the right to air grievances. The representatives of the Commons were sometimes summoned merely to consent. By the mid-fourteenth century the knights, citizens, and burgesses usually met together.

The growth of the power of the Commons was slow and gradual. As early as 1309 the Commons granted a subsidy on condition that the King granted redress of grievances according to certain articles, the Articles of Stamford.

In the time of Edward III the Commons began to sit separately and asserted three important claims, that all taxation without the consent of Parliament was illegal, that both Houses must concur in legislation, and that the Commons had the right to inquire into and amend the abuses of the administration; and two consequential claims, to examine public accounts and appropriate money, and to impeach the King's ministers for maladministration. By repeated remonstrances and conditional grants the Commons at length managed to assert some control over taxation.

In 1353 appropriation of supplies occurred, when a subsidy on wool was granted, but to be applied solely to the purposes of the war, and in 1340 a committee was appointed to examine the accounts of the collectors of the last subsidy.

At this time statutes were normally based on petitions from the Commons praying for remedies for particular grievances. In 1340 a committee was appointed to convert such petitions and answers as should be permanent into statutes. Temporary matters were regulated by ordinances rather than statutes, and were sometimes made in a great council, not a Parliament.

On two occasions in the mid-fourteenth century the Commons boldly interferred in administration; in 1341 they tried along with the Lords to establish the responsibility of the King's ministers to Parliament, by petitioning that commissioners should be appointed to enquire into the accounts of those who had received aids and public money, and that ministers and judges should be appointed in Parliament and be sworn to observe Magna Carta and other statutes. The King conceded the demands, passing a statute, but declared it null and void after Parliament was dissolved. In 1376 the Commons initiated an impeachment of Lords Latimer and Nevill and four commoners. The House of Lords tried and convicted all six accused. In 1386 the Commons impeached de la Pole, the Chancellor.

Under the Lancastrian kings Parliament continued to assert the rights it had secured during the fourteenth century, but also made headway in settling its own constitution, privileges, and forms of procedure.

The Wars of the Roses increased the power of the Commons by the large-scale destruction of the baronage, but left it almost alone to face the increased power of the Tudor monarchs.

Freedom of speech in Parliament was slowly recognized in the fifteenth and sixteenth centuries; freedom from civil arrest had been occasionally recognized from the thirteenth century, and though denied in *Thorpe's case* (q.v.) the privilege became regularly recognized in the fifteenth century, though only in 1543 did the Commons deliver a member from custody, or commit to prison, by their own sole authority; and the Commons gave much attention to their right to determine contested elections.

The office of Speaker developed markedly in the fourteenth and fifteenth centuries. An official by that name is mentioned in 1377. The kings took care to have a trusted councillor elected Speaker, and he was not only spokesman of the Commons but a king's officer or government representative.

Down to the early fifteenth century statutes were drawn up from petitions presented by the Commons and the King's answers thereto, but the King's officers sometimes enrolled as statutes provisions differing materially from what the Commons had sought. In 1414 they petitioned the King that he enact precisely what they sought by petition, but the King for some time longer exercised a wide discretion in legislation. In the mid-fifteenth century proposed complete statutes began to be introduced as bills, possibly by the Crown as an improved method of business, and under Henry VIII it had become established that ordinary legislation originated in draft bills, that bills might originate in the House of Lords and be sent to the Commons, but that money bills must come from the Commons.

Parliament's legislative authority was, however, subject to the King's rather vague suspending and dispensing powers (qq.v.).

Under the Tudors, despite the great powers of the monarch, Parliament continued to be important, though the House of Lords was small and the Commons was strongly influenced by the group of King's servants, some members of the King's council, therein, but in many cases it strongly opposed the monarch. The Reformation Parliament of 1529–36 passed a mass of important legislation, restraining the personal privileges and emoluments of the clergy, depriving the Pope of annates or first-fruits of benefices, restraining appeals to Rome, requiring the submission of the clergy, providing for the election of archbishops and bishops, terminating payment of Peter-pence and other papal exactions, settling the royal succession, the Act of Supremacy (1534) making the King the supreme head of the Church of England, annexing first-fruits and tenths to the Crown, and suppressing the religious houses. Statute imposed the use of the First Book of Common Prayer (1549). In Mary's reign all the anti-papal legislation was repealed and the Catholic religion completely re-established, and this reversal was itself reversed with some modifications under Elizabeth by the Acts of Supremacy and Uniformity.

Under Elizabeth conflicts arose between Crown and Parliament over the further reformation of the Church and the settlement of the succession.

After the accession of James I the King's belief in the divine right of kings inevitably provoked conflict with the Commons. The Commons had to vindicate their right to determine contested elections against the new King, who wished the returns to be examined in Chancery, and established it against the King. After *Bate's Case* (q.v.) the Commons, in face of royal veto, insisted on debating the illegality of impositions, and complained of the Court of High Commission. James tried to rule without Parliament, 1611–14, but when Parliament was summoned it immediately took up again the grievance over impositions, was dissolved and some members were sent to the Tower. After six years of arbitrary government James had to summon Parliament again. It promptly impeached, among others, Lord Chancellor Bacon and Lord Treasurer Cranfield and took up the redress of its grievances and, being told by the King not to deal with affairs of state, recorded a vigorous protestation in their Journal, which the King tore out with his own hand. By the time of James's death the Commons had shown vigour in opposing the royal policy and made a claim to debate all matters of public concern.

Charles I's second Parliament impeached Buckingham and he dissolved it, resorting to illegal methods of raising money, as by forced loans, the legality of which was challenged in Darnel's or the Five Knights' case (q.v.). The third parliament presented the Petition of Right (q.v.) but was dissolved in 1629 after which Charles governed without Parliament till 1640, raising money by such expedients as ship-money, which gave rise to Hampden's case (q.v.). Parliament had to be summoned again in 1640 but, insisting on grievances, was dissolved. Later in the year, however, there had to be summoned what became the Long Parliament. It impeached Strafford and others, abolished ship-money, the Court of Star Chamber and the Court of High Commission and presented the Grant Remonstrance denouncing the King's illegalities of recent years. Charles then sought to impeach and to arrest personally five members of the Commons. War was then inevitable. Power passed into the hands of the leader of the army and, after Pride's Purge (q.v.) of the House of Commons in 1648, the Rump became merely the tool of the Army.

Early in 1649 the King was executed, the House of Lords was abolished and the old form of government replaced by the Commonwealth of England. In 1653 Cromwell dismissed the Rump and under the Instrument of Government ruled as Lord Protector with a unicameral Parliament for England, Scotland, and Ireland. After Cromwell's death the army restored the Rump and then the rest of the Long Parliament. A Convention met, declared itself a Parliament, and restored the monarchy and the old constitution. The major development of the period since 1603 had been the permanent establishment of the powerful influence of the House of Commons in the government of the nation.

After the Restoration Parliament again imposed disabilities on protestant dissenters and catholics by the Act of Uniformity, 1662, the Conventicle Act, 1664, the Five-Mile Act, 1665, and finally the Test Act, 1673, but a beneficial piece of legislation was the Habeas Corpus Amendment Act, 1679.

Political parties had been foreshadowed by the groups supporting and denouncing the royal prerogative from the beginning of the seventeenth century, and they become more distinct and more like modern political parties by the latter part of that century.

James II intended to make himself an absolute monarch but soon had to summon Parliament, which initially was compliant. But James by extensive exercise of the dispensing power (q.v.), creation of a new Court of High Commission for Ecclesiastical Causes, and use of the suspending power (q.v.) in the two Declarations of Indulgence, provoked leaders of both parties to invite jointly William of Orange and his wife Mary (daughter of James II) who came to England, whereon James fled. A Convention Parliament met, declared itself a legal Parliament, and resolved that James had abdicated the Government and that the throne was thereby vacant and offered the throne to the Prince and Princess of Orange jointly. The throne was accordingly offered, and accepted, on the conditions in the Declaration of Right (q.v.), later confirmed as the Bill of Rights. In 1701 the Act of Settlement (1701) settled the succession on the Electress Sophia of Hanover and her descendants, being Protestants.

The reign of Anne was important for the growth of parties, an advance towards cabinet government and several great cases on parliamentary privilege.

The decline in the royal power is marked by the fact that in 1708 the monarch for the last time vetoed a bill passed by Parliament.

Finally in 1707 after disputes between the Parliaments of England and of Scotland, and negotiations between commissioners, each Parliament passed an Act approving the Treaty of Union (q.v.) and each Parliament came to an end on 1 May 1707, being replaced by the Parliament of Great Britain (q.v.).

Parliament of Great Britain. The Parliament of Great Britain was created by the Treaty of Union of 1707, approved by Acts of the English and Scottish Parliaments, which came into force on 1 May 1707. From the first it sat at Westminster and assumed all the powers, privileges, and procedure of the former Parliament of England, so much so as to be commonly regarded as a continuation of the Parliament of England. Each country retained its own electoral system, but not all Scottish peers but only 16 representatives of them, elected at the beginning of each Parliament, were admitted to the House of Lords and the representation of Scotland in the Commons (45 members) was less than in proper proportion to population; England had 513. Scotland was under-represented till 1882. While in both Houses the Scottish representatives sometimes held the balance of power, they were repeatedly voted down on Scottish matters, as on the Scottish Treason Bill of 1709 and on the Malt Tax of 1713, and Scottish interests were neglected or overruled as by the abolition of the Scottish Privy Council and the Patronage Act of 1712 which restored lay patronage of vacant charges in the Church of Scotland. The Union was for some time unpopular in Scotland, but the freedom of trade came to benefit both countries.

The eighteenth century was noteworthy for the last exercise of the royal veto on a bill (the Scottish Militia Bill, 1708), a great multiplication of peers, the steady development of parties and the rise of the cabinet system. It was the era of political corruption, elections being carried in small electorates by promises, threats, and cash. It is said that in 1793 354 of the 558 members were returned nominally by less than 15,000 voters and truly by the government and 197 private patrons. There was a strong family element in the Commons.

Similarly ministers and George III managed the House of Commons by rewards and honours to members, or displeasure. During Lord North's administration (1770–82) the King's personal influence was at its greatest, and the King exercised most of the functions of Prime Minister himself.

In the eighteenth century many issues of parliamentary privilege were determined, such as the right to expel a member, as in the conflicts with John Wilkes and the recognition of the liberty of the Press to publish accounts of debates. The office of Speaker, particularly under Onslow (1727–61), became more settled in position, and tended to become more neutral.

The defects of the representative system, seats at the disposal of patrons, rotten boroughs, developing cities without a representative, provoked Chatham in 1766 to advocate parliamentary reform and John Wilkes produced a comprehensive scheme in 1776. The younger Pitt raised the matter in 1782–85 but could make no headway, and nothing had been done, when in 1801 the Parliament of Great Britain and the Parliament of Ireland both gave way to the Parliament of the United Kingdom (q.v.).

Parliament of Ireland. Norman expeditions established an uneasy supremacy over much of Ireland in the twelfth century and John established, at least in the settled districts, English law and justice. The germ of the Irish Parliament may be in a meeting of the Great Council held at Kilkenny in 1297 to which two knights from each county and liberty were summoned, but only in 1310 was a representative Parliament established. Under Edward I the Irish Parliament was rather a solemn and special session of a court held before justiciar and council; there is little evidence of the presence of Commons before 1370, but in the fifteenth century the Commons were better represented.

In 1366 Lionel of Clarence, the King's Lieutenant in Ireland, secured the passing of the Statutes of Kilkenny, which represent an attempt to save the English colony within the Pale and exclude the Irish outwith that area from government; they were excluded from ecclesiastical office and the English were to have no business with them. These were repeatedly confirmed and not repealed till 1613. In the fifteenth century a feature of Parliament was the concentration of administrative and judicial business introduced by private bills.

By the time of Henry VII English authority in Ireland was limited to a small part of the Pale round Dublin. In 1494 Parliament provided a complete scheme for the reform of the English government in Ireland. It passed Poyning's Acts, which provided that no Parliament should be held in Ireland without the sanction of the King and Council, and that they should be entitled to disallow statutes of the Irish Parliament. It also provided that all laws made in England should apply also to Ireland.

It was only in the mid-sixteenth century that Parliament became fixed in Dublin; previously it met wherever was convenient. It represented only the great landed proprietors, the Protestant Church dignitaries, and areas under English control. It was an organ of English rather than Irish opinion. It had small influence and its freedom of action was curtailed by Poyning's laws and the dominant position of the Lord Deputy, who was subject to the English Privy Council.

A Parliament at Dublin in 1536 imposed the Reformation on Ireland, Henry VIII being declared Supreme Head of the Church in Ireland. But this had little effect outside the Pale and the destruction of monasteries and the plunder of the Church antagonized the people.

Wentworth (Strafford), as Lord Deputy in Ireland after 1630, took care not to allow Parliament to have any power and it was dissolved before it could discuss any grievances. After his departure in 1640 the independence of the Irish Parliament revived and the members queried the legality of many of Wentworth's acts. In 1649 Cromwell suppressed an Irish insurrection and the English Parliament annexed some two-thirds of the land in Ireland and allotted it to settlers from England. But on the other hand he gave Ireland representation in the Parliament at Westminster. Many of the Cromwellian settlers, however, became Irish in sentiment. At the Restoration the Irish Parliament was restored with its former subordination to that of England. It was entirely Protestant, largely impotent, and did not meet at all between 1666 and 1685. An almost entirely Catholic Parliament met in 1688 and repealed Poyning's Law, abolished all judicial appeals to England, and passed an act establishing religious equality. But under William III and Anne very severe legislation was passed against Catholics, including their exclusion from Parliament, and in 1696 all the legislation of the Catholic Parliament was repealed. In 1707 when the Parliaments of England and Scotland united the Irish House of Commons petitioned for union, but Ireland was deemed so effectively subordinated to England that the request was ignored. A British Act of 1719 declared the full power of the Westminster Parliament to make laws for Ireland. In 1727 Catholics were excluded from the parliamentary franchise and the legal profession.

By 1760 the Irish Parliament had sunk low, but thereafter the government in London made concessions and Grattan was able to secure legislative independence for it in 1782 but without responsible government; the franchise and other civil liberties were conferred on Catholics in 1792–93, but enmity led to unsuccessful rebellion in 1798. Thereafter a plan for Union of the Parliaments of Great Britain and Ireland was prepared and carried by bribery. The Union was accepted by both British and Irish Parliaments in 1800 and came into operation in 1801. By the Act of Union (Ireland), 1800 and the Union with Ireland Act, 1800 (of the British Parliament) Ireland was to be represented in the United Kingdom Parliament by four bishops sitting in rotation, 28 peers of Ireland elected for life by the Irish peers, and 100 members of the Commons. Till 1921 Ireland was represented in the United Kingdom Parliament. For development thereafter see DAIL; PARLIAMENT OF NORTHERN IRELAND.

E. Curtis, *History of Mediaeval Ireland*; R. Bagwell, *Ireland under the Tudors*; *Ireland under the Stuarts*; W. Lecky, *History of Ireland in the 18th century*; H. Richardson and G. O. Sayles, *Parliaments and Councils of Mediaeval Ireland*; *Irish Parliament in the Middle Ages*; J. G. S. MacNeill, *Constitutional and Parliamentary History of Ireland*; J. T. Ball, *Historical Review of the Legislative systems in Ireland, 1172–1800*; R. Berry (ed.), *Statutes and Ordinances of the Parliaments of Ireland.*

Parliament of New Zealand. The Parliament of New Zealand first met in 1854. It comprises the Governor-General, representing the Queen, and the House of Representatives, elected by adult suffrage. The Legislative Council or upper house was abolished in 1950. The duration of a Parliament is three years. It follows the procedure of the U.K. Parliament fairly closely but with variations, particularly in respect of financial procedure. The Speaker does not remove himself entirely from party politics. Its powers are not legally limited, but it has passed legislation entrenching certain provisions of the constitution against repeal save by large majorities.

Parliament of Northern Ireland. The Union of 1800 of the British and Irish Parliaments was very unpopular in Ireland and the Irish members at Westminster maintained pressure for Catholic Emancipation (achieved in 1829), the disestablishment of the Protestant Established Church of Ireland (secured in 1869), and Home Rule. Under Butt and Parnell the Irish party was a continual thorn in the side of the government at Westminster. Gladstone became converted to Home Rule and introduced Home Rule Bills in 1886 and 1893. A Government of Ireland Bill was introduced in 1912 and was acceptable to the Irish Nationalists but not to Sinn Fein, nor to Ulster where loyalty to the United Kingdom was strong. The government then proposed an amendment whereby any county in Ireland might vote to be excluded from Home Rule for six years; this was unacceptable to the Ulster leader, Carson (q.v.). In 1914 the Bill was passed and would have become law under the Parliament Act, 1911, but, owing to the outbreak of the First World War it was passed but its operation postponed. In Ireland a rebellion broke out in 1916 and in 1918 Sinn Fein candidates elected to the U.K. Parliament boycotted it, met in Dublin and called themselves the Dail. In 1920 the Government of Ireland Act repealed the 1914 Act and provided for what was originally intended to be a transitional stage only, separate Parliaments in Northern and Southern Ireland, with power to legislate on all transferred matters and, as a symbol of unity, a

Council of Ireland, composed equally of members from North and South of Ireland with minimal powers. Representation at Westminster was maintained on a reduced scale.

The Sinn Fein party repudiated this settlement and civil disorder ensued until the conclusion of a treaty of 1921, enshrined in the Irish Free State (Agreement) Act, 1922, under which the 1920 Act continued to apply with modifications to Northern Ireland while Southern Ireland, as the Irish Free State, attained dominion status, equivalent to that then enjoyed by Canada. Northern Ireland was settled as six counties of the nine in the historical province of Ulster.

The Parliament of Northern Ireland consisted of the Queen (who was represented in Northern Ireland by the Governor), the Senate, consisting of the Lord Mayor of Belfast and the Mayor of Londonderry, and of 24 senators elected by the House of Commons on a system of proportional representation for eight years, half retiring every four years, and the House of Commons, consisting of 52 members, elected by a system of parliamentary franchise similar to that in Great Britain. Unless sooner dissolved a Parliament continued for five years.

It had power to legislate for the peace, order, and good government of Northern Ireland in all matters except those reserved to the Parliament of the United Kingdom (the Crown, defence, foreign relations, customs and excise, and various other topics) and the convention was followed that, save on reserved subjects, the U.K. Parliament did not legislate for Northern Ireland. But it was not a sovereign legislature and the United Kingdom Parliament might legislate for Northern Ireland not only on reserved topics but on any topic, and, as a matter of law, could abolish the Northern Ireland Parliament. Nor was the relationship truly that between federal government and provincial government, as the sovereignty of the Westminster Parliament in all matters was expressly preserved.

The Parliament had the powers, privileges, and immunities of the United Kingdom Parliament and exercised the same checks on the Northern Ireland executive as does the U.K. Parliament. Money bills could originate only in the House of Commons and could not be amended by the Senate. If the Senate rejected, or failed to pass, or passed with unacceptable amendments a public bill sent up at least a month before the end of the session and the House passed the bill again in the following session, provision was made for a joint session.

In 1972 Northern Ireland was placed under temporary government from Whitehall. In 1973 a new legislature was established under the name of the Northern Ireland Assembly, and the office of Governor of Northern Ireland and the Parliament of Northern Ireland were abolished. The Northern Ireland Assembly was prorogued indefinitely in 1974.

N. Mansergh, *The Irish Question*; A. Quekett, *The Constitution of Northern Ireland*; N. Mansergh, *The Government of Northern Ireland*; G. Lawrence, *The Government of Northern Ireland*; H. Calvert, *Constitutional Law in Northern Ireland.*

Parliament of Scotland. In Scotland Parliament originated from the King's council, which sat in various capacities, as a court of justice or of enquiry, handling financial business or general administration. In 1293 there is mention of pleas being heard by King and council in Parliament. By the end of the thirteenth century Parliament was primarily a sitting of the council to declare the law in a particular case or for all like cases. Distinct from Parliaments were councils reinforced by additional persons summoned, General Councils; these could be summoned less formally and more easily, and the King could summon whom he liked. Representatives from burghs may have been summoned to Parliaments in 1326 and 1328, but from 1357 they attended fairly regularly, three or four from each burgh but later one only, Edinburgh sending two. Till the end of the sixteenth century there were no representatives from the shires. Parliament was unicameral, the three estates of nobles, clergy, and burgesses sitting together. Though in the fourteenth century a General Council had legislative and taxing powers, burgh representatives were not summoned to a General Council till the sixteenth century. In the sixteenth century a General Council began to be called a Convention of Estates and by the seventeenth century there was little difference in membership between a Convention of Estates and a Parliament.

Down to the sixteenth century Parliament was the recipient of pleas and complaints and in the fifteenth century various committees were established to hear causes, a practice which persisted until the Court of Session was established, superseding these committees, in 1532.

From 1424 it became the practice for Parliament to appoint a committee, the Lords of the Articles, to prepare legislation which Parliament later passed without discussion. Control of the Lords of the Articles accordingly meant control of legislation, and Parliament failed to develop as a debating body, checking the power of the Crown.

In 1560 Parliament abolished the authority and jurisdiction of the Pope, forbade the mass, and approved a Confession of Faith. Thereafter the General Assembly of the Church of Scotland was a powerful force, usually in opposition to the Crown.

In 1603 James VI of Scotland became also James I of England, and Scotland ceased to have a resident king, and in the century before 1707 several attempts were made to unite the two Parliaments.

He ruled Scotland through the Privy Council and control of the Committee of the Articles. But he and his son, Charles I, built up antagonism by their policy of enforcing episcopacy, so that Scottish forces assisted the English parliamentary force in the Civil War and from 1638 Parliament was more active than ever before. In 1640, in defiance of the King, it elected its own president, abolished the clerical estate and the Lords of the Articles and passed a Triennial Act to ensure that no longer than three years would elapse between sessions of Parliament. In 1641 it insisted that officers of state, privy councillors, and judges should be chosen only with its advice.

From 1651 Cromwell treated Scotland as a conquered country and made it part of the Commonwealth with, from 1654, 30 members in the Protectorate Parliaments.

After the Restoration the revived Scottish Parliament passed a Rescissory Act which annulled all legislation since 1633. The Lords of the Articles returned, and control was exercised from Whitehall through the Scottish Privy Council. Episcopacy was restored and the Privy Council tried repression of presbyterians and conciliation alternately.

James VII and II governed arbitrarily through the Privy Council and alienated sympathy in Scotland as much as in England. After William III had become King of England he summoned a Convention of Estates which resolved that James had forfeited the Crown and offered the Scottish Crown to William and Mary. It set out in the Claim of Right and the Articles of Grievances the illegalities of James VII's regime and the terms of the settlement with William III. The Convention was turned into a Parliament in 1690 and finally abolished episcopacy, the clerical estate, and the Lords of the Articles. William III alienated much support in Scotland by the massacre of Glencoe and the conflict between Scottish and English commercial interests, culminating in the failure of the Darien scheme. In 1701 the English Parliament passed the Act of Settlement entailing the English crown on Anne, the Electress Sophia of Hanover and her heirs, being Protestants, but in 1704 the Scottish Parliament declared in the Act of Security that it would not adopt the Hanoverian succession save on satisfactory terms. The English Parliament replied with the Alien Act, 1705, which would have made Scots aliens in England. Out of a rapidly worsening situation a project for closer union emerged and commissioners negotiated a Treaty of Union which was passed by both Parliaments and came into force in 1707. The Scottish Parliament then came to an end, being replaced by the Parliament of Great Britain (q.v.).

The Scottish Parliament did not, until the seventeenth century, show the independence or resistance to royal pressure which the English Parliament had done, but in its last century it made great strides. In particular, after the Restoration it had acted vigorously and sensibly, adjusted its own composition, settled rules of debate, established the Bank of Scotland and promoted economic development, and the clash of an expanding commercial community with English restrictions in the shape of the Navigation Acts.

Moreover, between 1424 and 1707 the Scottish Parliament passed a great quantity of legislation, many of the Acts laying down fundamental principles of Scots law, and a substantial number of which have continued in force down to the present time.

J. Mackinnon, *Constitutional History of Scotland*; R. S. Rait, *The Parliaments of Scotland*; C. Terry, *The Scottish Parliament: Constitution and Procedure, 1603–1707*; A. V. Dicey and R. S. Rait, *Thoughts on the Union; Acts of the Parliaments of Scotland, 1424–1707*.

Parliament of the United Kingdom. The Parliament of the United Kingdom of Great Britain and Ireland was formed by the amalgamation in 1801 of the Parliaments of Great Britain (q.v.) and of Ireland (q.v.). It comprised initially 513 members for England, 45 for Scotland and 100 for Ireland, about 310 English and British peers, 16 Scottish representative peers, and 28 Irish representative peers.

The pressure for parliamentary reform was revived after Waterloo and after mounting pressure and public excitement, a dissolution of Parliament, the resignation and recall of the government and a threatened large creation of peers, the Reform Act of 1832 was passed. It disfranchised 56 boroughs entirely and 30 others partially, created 43 new boroughs and increased the county representation, and extended the franchise. A similar Act for Scotland increased representation, widened the franchise, and an Act for Ireland added five members and widened the franchise. The property qualification for members was modified in 1838 and abolished in 1858; a further extension of the franchise was effected in 1867 and, with differences of detail, for Ireland and Wales in 1868. In 1872 voting by secret ballot was introduced.

The franchise was again widened in 1884 and a redistribution of seats effected in 1885. In 1918 the franchise was extended to all males of 21 and to women of 30. In 1928 the age for women voters was reduced to 21 and in 1969 the age was reduced for both sexes to 18.

The power and influence of the sovereign has declined considerably, particularly since 1820. So has the power and influence of the House of Lords; the House of Commons is undoubtedly the major element in Parliament and has strong dictatorial tendencies, particularly if a government has a substantial majority.

PARLIAMENT OF THE U.K.

By 1800 the cabinet system was well established and since the end of George III's personal intervention in government the power of the cabinet and particularly of the Prime Minister has greatly developed. The enormous increase of governmental interference in every aspect of public affairs, particularly the increase of patronage which developed during and after two world wars, have increased its powers.

The growth of the party system has been notable, and since 1832 most candidates belong to and stand as local representatives of one of the major national political parties. During the nineteenth century the parties were Conservative and Liberal. In the early years of the twentieth century the Labour party made its appearance; after the First World War the Liberal party split and collapsed and the Labour party superseded it. Since 1945 the Conservative and Labour parties have shared 95% of the votes cast.

Parliament now comprises; (a) in the House of Lords, all holders of either sex of hereditary English, Scottish, Irish, British or United Kingdom peerages, the Lords of Appeal in Ordinary, life peers, and the 26 spiritual Lords, and (b) in the House of Commons 630 members elected each to represent a single-member constituency, 511 for England, 36 for Wales, 71 for Scotland, and 12 for Northern Ireland.

Parliaments are elected when the Prime Minister advises the Queen to dissolve Parliament and summon a fresh one. The maximum duration of Parliament is five years but in practice Parliaments average about four years. By-elections are held to fill casual vacancies.

Any British subject aged at least 21 may stand for election, except peers, most clergymen, undischarged bankrupts and persons, such as judges, civil servants, and holders of public offices excluded by the House of Commons Disqualification Act, 1975. Most candidates stand as the approved representative in the constituency of one of the major political parties, and wholly independent candidates have in practice practically no chance of election.

Election is by secret ballot by British subjects or citizens of the Irish Republic, if aged 18 or over, resident in the constituency and on the register of electors thereof. Provision is made for postal and proxy voting in the case of persons unable to vote personally.

The leader of the party which wins the largest number of seats becomes Prime Minister and forms the government, his major supporters being appointed to the chief ministerial offices. The largest minority party becomes the official Opposition.

In association with the Queen Parliament is the supreme legislature for the whole United Kingdom and for each of the constituent countries singly or in any combination, though in Northern Ireland the subordinate Parliament (now abolished) might legislate on most domestic subjects and the Scottish and Welsh Assemblies (qq.v.) may do so.

Till 1911 the Lords and the Commons had co-ordinate legislative authority except in money matters which, by convention and according to privileges claimed by the Commons, had to be introduced in the Commons and their money clauses could not be amended by the Lords. Since 1911 by virtue of the Parliament Act, 1911, a Bill certified by the Speaker to be a money Bill may be presented for royal assent notwithstanding that the Lords have failed to pass it without amendment after it has been before them for one month. Other Bills, except Bills to prolong the duration of Parliament, provisional order confirmation Bills and private Bills might be presented for royal assent if passed by the Commons in three successive sessions and two years had elapsed between second reading in the first session and third reading in the third session, and it had been before the Lords for at least a month and the Speaker had certified that the 1911 Act had been complied with. By the Parliament Act, 1949, the number of sessions was reduced to two and the interval from two years to one. In effect the Lords have a delaying power of thirteen months. The Act of 1949 and most controversy about the Lords' power since 1945 have arisen from nationalization Bills. The Lords' function is not to rubber-stamp the Commons' Bills but to re-examine them in principle and in detail, and the refusal by the Commons to accept Lords' amendments is commonly arrogant and dictatorial.

One Parliament cannot bind itself or any subsequent Parliament, and in consequence after an election resulting in a change of government it commonly modifies some of the legislation of the previous government. The accepted view is that there are no legal (though there are political and moral) limitations on Parliament's power; it can overrule decisions of the highest court, alter any rule of law to any effect whatever, prolong its own life, alter its own composition. It is not bound by any constitutional document except perhaps, in theory, some provisions of the Union with Scotland Act and its decisions are not challengeable nor reviewable in any way. Nor do any particular kinds of legislation require particular majorities or special procedure.

The life of a Parliament is divided into sessions, each usually of about a year, and each terminated by a prorogation, which has the effect of terminating all business and causing all legislation in progress to lapse. After a short interval a new session is opened by the Queen. During each session, either House may adjourn to such date as it pleases; adjournment does not affect business in progress. The life of a Parliament is ended by dissolution, effected by royal proclamation on the advice of the Privy Council, in fact at the request of the Prime Minister, which proclamation in modern practice also orders the issue of writs for the election of a new Parliament. On average the House of Commons sits for about 160 days a year, and the House of Lords for about 110 days.

The main functions of Parliament are to control, question, and criticize Her Majesty's government, both general policy and detailed execution thereof, to make available finance for the needs of the community and to appropriate to the funds to the different services of the State, to represent the general opinion of the community, and by legislating to make, repeal, or modify the law on any topic.

In all its roles it is an inefficient and ineffective agency. In particular as a legislative body, it increasingly enacts a large volume of confused, and much of it practically incomprehensible, legislation, much of it not properly discussed and much of it thrust through by a government majority which may have no regard for the wishes or needs of the community.

Since 1945 and particularly since 1965 there has been a very widespread fall in the qualities of M.P.s and in public regard for Parliament. The sayings and doings of most politicians are regarded with considerable cynicism and scepticism by the public. Parliamentary debates are largely a farce when members vote solidly according to party groupings and the outcome of divisions is a foregone conclusion.

A. Pollard, *Evolution of Parliament*; E. and A. Porritt, *The Unreformed House of Commons*; L. O. Pike, *Constitutional History of the House of Lords*; W. I. Jennings, *Parliament*; E. C. S. Wade and G. G. Phillips, *Constitutional and Administrative Law*; T. Erskine May, *Treatise on the Law, Privileges, Proceedings and Usage of Parliament.*

Parliament House. A building behind St. Giles Cathedral in Edinburgh, built in 1632–39 and extensively altered in the early nineteenth century. It contains the great Hall in which the Parliament of Scotland sat until the Union of 1707 and the courts and some of the offices of the Court of Session and High Court of Justiciary. Connecting with it are the libraries of the Faculty of Advocates, Society of Writers to the Signet, and Society of Solicitors in the Supreme Courts of Scotland. The buildings have been much altered at various times and courts have sat in various rooms; in the nineteenth century Outer House judges sat in the Great Hall itself.

Parliament Rolls. The record of the proceedings of Parliament from the early fourteenth century, containing a record of the proceedings in Parliament, with private petitions and petitions on which statutes were framed. A selection from the Rolls of the times of Edward I and II is known as the *Vetus Codex* or *Placita Parliamentaria*, and other abridgements were published in the seventeenth century. An official edition was ordered in 1765 and two volumes of a new edition were published in 1827–34. These rolls end in 1503 being replaced by the *Journal of the House of Lords* and *Journal of the House of Commons.*

Parliamentary agents. Persons, normally solicitors, who act for the promoters of, or the petitioners opposing, a Private Bill before Parliament. A register of parliamentary agents is maintained in the Private Bill Offices of both Houses and only persons registered may describe themselves or act as parliamentary agents. To be eligible for registration, persons must satisfy the Speaker and the Lord Chairman of their practical knowledge of the Standing Orders and procedure of the House of Commons regulating Private Business.

Parliamentary Bar. The group of barristers who specialize in appearing for parties promoting or opposing Private Bills. They were important in the nineteenth and early twentieth centuries, when there was much work, particularly with railway Bills, municipal undertakings, and extensions of the powers of local authorities. Much of this work has declined since 1945.

Parliamentary Commissioner for Administration. This office was created in the United Kingdom by the Parliamentary Commissioner Act, 1967, largely on the model of the 'Ombudsman' (q.v.) of the Scandinavian countries, and of the similar officer in New Zealand appointed in 1962. He is appointed by Letters Patent and holds office during good behaviour, but may be removed by the Queen in consequence of addresses from both Houses of Parliament. He may investigate any action taken by or on behalf of a government department, or other authority to which the Act applies, in the exercise of its administrative functions, where written complaint is made to an M.P. by a member of the public who claims to have sustained injustice in consequence of maladministration, and the complaint is referred to the Commissioner by the M.P. with a request to conduct an investigation, but he may not act on his own initiative nor conduct an investigation where a right of appeal or review exists. Many complaints are outside his terms of reference. Where he investigates or decides not to investigate he sends a report to the M.P. and, if he investigates, reports the result to the principal officer of the department concerned. If he thinks injustice has been caused by maladministration and has not been, or will not be, remedied, he may make a special report to each House of Parliament. 'Injustice' and 'maladministration' are undefined, and both are very broad terms. He makes a general report annually to Parliament on his functions. The list of authorities subject to investigation does not include all government departments, nor public corporations, nor local authorities, and certain matters (including court proceedings) are excluded from investigation. There are now separate Commissioners for Northern Ireland (1969), for Local Administration in England and Wales (1973), and in Scotland (1975), and for the National Health Services of England and Wales (1973) and Scotland (1972).

Parliamentary committees. There are many kinds of committees of each House, namely committees of the whole House, which are the whole House sitting as a committee presided over by the Chairman rather than by the Lord Chancellor or the Speaker, standing or sessional committees, select committees, joint committees of the two Houses, and committees appointed *ad hoc.*

Parliamentary Counsel to the Treasury. The chief officer responsible for the drafting of Bills for consideration by the U.K. Parliament. Prior to 1869 Bills were drafted by judges, by committees of the Privy Council, by distinguished lawyers, but about 1790 a special pleader was appointed to draw or settle all Bills for the Treasury, though he sometimes acted for other departments also. From 1837 the Home Secretary had a regular assistant, charged with the duty of preparing Bills. As the century went on different departments employed different counsel to draft their Bills. In 1869 the office of Parliamentary Counsel was established with a small permanent staff, originally only of barristers, but including some solicitors also, charged to draw Bills for all departments. The office of First Parliamentary Counsel has been held by some very distinguished lawyers, notably Lord Thring, Sir Henry Jenkyns, and Sir Courtenay Ilbert. The counsel are also responsible for statute law revision, and consolidation Bills and for much subordinate legislation. Scottish Bills are drafted by the Lord Advocate's Legal Secretaries and Parliamentary Draftsmen, and private members are responsible for drafting their own Bills.

Parliamentary Debates. The reports of the debates in the U.K. Parliaments. The first continuous record of contemporary debates, compiled by Anchitell Grey, covering 1667–94, was published as *Grey's Debates* in 1769. Debates were also published by Richard Chandler (1660–1742), Ebenezer Timberland (H.L., 1660–1742), John Almon (1742–74), Dr. Johnson (1741–44), John Debrett (1743–74), and William Woodfall (1789–1803). Various diaries and collections of materials contain records of proceedings in Parliament between 1550 and 1688, but are incomplete. Parliament long regarded the publication of reports of its debates as a breach of privilege, twice (1738 and 1762) declared so, and, strictly speaking, still does so, but since the early nineteenth century has acquiesced in all fair reports. The modern Parliamentary Debates commenced when Cobbett, having completed his *Parliamentary History, 1666–1803* (36 vols.) included reports in his *Political Register.* In 1811 T. C. Hansard (q.v.) bought Cobbett's interest and issued reports of debates at monthly intervals, compiled from newspaper reports. From 1878 the Hansard business received a small subsidy, but in 1890 had to sell the business to the Hansard Publishing Union which became bankrupt. In 1908, after several firms had sought with little success to perform the function and a Select Committee had recommended that the reporting be undertaken by the government, a permanent staff was appointed and the *Official Reports* commenced in the House of Commons in 1909 and in the House of Lords in 1920. The name Hansard, missing from 1892, was restored, in brackets, to the title page in 1943, and is colloquially used for the whole body of debates.

The Debates comprise five series, viz: First Series: 41 volumes, 1803–20; Second Series: 25 volumes, 1820–30; Third Series: 356 volumes, 1830–91; Fourth Series: 199 volumes, 1892–1908; Fifth Series: separate runs of volumes for Lords and Commons Debates, from 1909. They are cited by volume, series and column, e.g. 456 H.C. Deb. 5s. 789.

The Debates may not be taken into account by courts seeking to interpret legislation because of the impossibility of knowing whether Parliament really intended to enact what any member said was the intention of a clause in a Bill.

Parliamentary draftsmen. Barristers and solicitors comprising the staff of the office of Parliamentary Counsel to H.M. Treasury (q.v.), who are responsible for the drafting of Bills for consideration by Parliament. When a decision has been taken to legislate on a particular topic, the Treasury instructs Parliamentary Counsel to draft a Bill on the lines proposed. The drafting is normally done by a senior and a junior draftsman working together and consulting the Minister concerned and his senior officials. Frequently several, or many, drafts are needed.

While a Bill is under discussion in either House the draftsmen are in attendance and they attend to the drafting of amendments accepted and of consequential amendments.

Parliamentary government. The system of government which centres on a Parliament, or assembly in which an elected house of representatives is the major element, from which the chief executives are drawn and to which, and through which to the general body of the community, the chief executives are, individually and collectively, responsible. It differs from presidential government in which the chief executives are not members of and responsible to the assembly. It is exemplified chiefly in the United Kingdom and the older countries of the Commonwealth.

Parliamentary History. A work in 36 volumes compiled by William Cobbett and completed by J. Wright, which constitutes a legislative history of England, Great Britain and the United Kingdom from the Norman Conquest to 1803. It is founded mainly on the *Old Parliamentary or Constitutional History of England to the Restoration* (24 vols.), Sir Simonds D'Ewes' *Journals*, Rushworth's *Historical Collections*, 1618–48, Somers' *State Tracts*, Hardwicke's *State Papers*, Chandler and Timberland's *Debates*, Grey's *Debates*, Almon's *Debates*, Dr. Johnson's *Debates in Parliament*, Debrett's *Debates*, and Woodfall's *Debates*. For the period prior to Cobbett's *Political Register* it is the principal source of information on the course of debate and activity in Parliament.

Parliamentary papers. These include Bills, reports of Parliamentary Committees, Royal Commissions and Departmental Committees, Accounts and Papers. Some papers are required by Act to be laid before Parliament ('Act Papers'), others ordered to be printed by one House or the other ('House of Lords Papers', 'House of Commons Papers'), and others ordered to be printed by the departments concerned ('Command Papers'). From 1801 for the House of Commons and from 1804 for the House of Lords all papers presented to Parliament have been bound in sessional volumes, classified and indexed.

See also BLUE BOOKS; GREEN PAPERS; WHITE PAPERS.

Parliamentary Private Secretary. An ordinary, but frequently a young and promising, Member of Parliament appointed by a Minister, to assist him, maintain contact with back-benchers and advise his Minister on feeling in the House and in the party generally on the Minister's field of responsibility. There are no defined duties, and the appointment carries no salary but it gives the young member an introduction to the ministerial function.

Parliamentary privilege. See PRIVILEGE, PARLIAMENTARY.

Parliamentary Secretary. A junior minister ranking below Ministers and Ministers of State, not appointed by the Queen but by the Prime Minister as assistant and deputy to a Minister of the Crown. If the Minister is a Secretary of State, the junior minister is designated Parliamentary Under-Secretary of State. Each major department has at least one; the Ministers of the Crown Act, 1964, allows the appointment of not more than 36. If the Minister is in the Lords, the Parliamentary Secretary is usually in the Commons, and a Minister in the Commons may have a Parliamentary Secretary in the Lords. As junior ministers Parliamentary Secretaries are subject to the principle of collective responsibility though not to the same degree as Cabinet ministers.

Parmoor, Lord. See CRIPPS, C.A.

Parol (or parole). By word of mouth, but including writings not under seal.

Parol or parole evidence. Evidence given by oral testimony by witnesses. It is a general rule that parole evidence cannot be used in substitution for a written instrument where that is required by law, nor to qualify or contradict a written instrument, unless there are allegations of fraud or mistake.

Parole. A form of conditional release from prison sometimes granted before the date when a prisoner is ordinarily due for release. Its use follows from the belief that detention should be reformative not punitive and that prisoners should be reintroduced gradually to living freely in society. The conditions for release on parole differ from one jurisdiction to another but generally take account of the nature of the offence, the period spent in detention, the reaction of the offender to the training given him in prison, and the danger to the public if he is released. Parole schemes are administered in various ways, usually by the ministry of justice but sometimes by the judiciary. In England and Scotland cases are considered by Parole Boards, which include legal, clerical, and social work members. A prisoner granted parole is discharged and may be recalled to prison if his reaction to parole is unsatisfactory. The supervision may take the form of periodical reporting to the police or regular contact with a social worker.

In international law, parole is the promise of a prisoner of war, if released, to return at an appointed time, or the practice of releasing prisoners of war on their undertaking that they will not again serve during the war. If their government does not honour their undertaking, it should return them to captivity.

Parquet. The department of the public prosecutor in France.

Parson. In English law the parish priest, sometimes called the rector, who represents the Church and possesses all the rights of a parochial church. He is in law a corporation sole having the freehold for life of the parsonage, glebe and, formerly, tithes and other dues, unless these are appropriated to a spiritual corporation or a lay person which makes provision for the spiritual functions by appointing a vicar and assigning to him a portion of the income.

Parsons, Theophilus (1750–1813). American jurist, was a member of the State constitutional convention of 1779–80 and a draftsman of the State constitution. He was a delegate to the State convention which ratified the federal constitution in 1788 and was chief justice of the Supreme Court of Massachusetts, 1806–13. He wrote a *Commentaries on the Laws of the United States* (1836).

Parsons, Theophilus (1797–1882). Son of the chief justice of Massachusetts of the same name (q.v.), practised law and journalism, and in 1848 became a professor in the Harvard Law School. He was a prolific writer, on *Contracts* (1853–55), *Mercantile Law* (1856), *Maritime Law* (1859), *Bills and Notes* (1863–76), *Shipping* (1869), and other legal topics. He retired in 1869, when Langdell was appointed and introduced the case-book method.

Part performance. The equitable doctrine that where a contract, e.g. for the sale of land, requires by law to be evidenced in writing but is not so evidenced, if the contract has been partly performed, e.g. by the buyer taking possession or paying a substantial part of the price, the court will enforce the contract by decree of specific performance.

Particular average loss. A loss fortuitously caused by a maritime peril which has to be borne by the party on whom the loss falls, as distinct from general average losses (q.v.) where the loss is voluntarily incurred and is made good by a rateable contribution from all the parties concerned in the adventure.

Particular estate. In pre-1926 English real property law, the first of the two or more successive estates into which the legal estate in land could be divided, the estate in possession, as distinct from the subsequent estate or remainder. The estate in possession or particular estate might be an estate for years, an estate for life, or an estate tail; the ultimate estate was always an estate in fee simple.

Particular lien. A right to retain goods in respect of which charges have been incurred until those charges have been paid, but not to retain in respect of a general balance of charges incurred.

Particulars. In pleading, details of a claim or defence which need be known to enable the other party to know what case he has to meet. The court may always order that further and better particulars be given of any matter stated in a pleading, or notice, or of the nature of the defence.

Parties. Those persons who, and whose interests, are involved in any act, deed, or legal proceeding, as distinct from third parties, witnesses, and others. Much procedural law is concerned with who may, or must, or may not be parties in various circumstances.

Partnership. The relation which subsists between persons carrying on a business, including every trade, occupation, or profession, in common with a view of profit. Associations for objects in common which are not partnerships include clubs, building and friendly or other benefit societies, and societies for religious, charitable, literary, or scientific purposes. The individuals who constitute the partnerships are collectively called the 'firm' and they trade under the firm name. In some professions, e.g. that of advocate or barrister, partnership is contrary to etiquette, and partnerships for illegal purposes are void. In Scotland, but not in England, the firm is a legal *persona* distinct from that of the members.

Whether a partnership exists or not is a question of fact; it is not necessarily created by co-ownership of property, still less by sharing of gross returns, though sharing of profits and losses is indicative of partnership, and a person may be held to be a partner if he has represented himself, or allowed himself to be represented, as a partner.

Partnership is constituted by agreement, usually in writing. It is presumed based on mutual trust and confidence, and the utmost good faith is required in the relations of partners. Each has power to bind the firm. Each must account to the firm for any benefit obtained from use of the firm name or business connection, and for profits made in a competing business, and not take any exclusive profit for himself. Every partner may take part in the management of the partnership business and the majority view must rule in case of difference. A new partner may be admitted only by consent of all, but, unless the agreement provides for it, no majority can expel a partner. Partnership property must be held and applied by the partners exclusively for partnership purposes and they are tenants in common as regards the assets of the partnership.

Prima facie all share equally in capital and profits and contribute equally to losses, but this presumption may be varied by their agreement. A partner may assign his share in the partnership although such assignment does not make the assignee a partner, but only a creditor for the share of profits; provision can be made for an assignee to be a partner.

Among themselves each partner is an agent of the firm and of the other partners for the purposes of the partnership business, and the acts of any one doing an act for carrying on in the usual way the business of the firm binds the firm and the partners. If the authority of one partner is restricted, an act done by him in contravention of the restriction is not binding on the firm as regards persons having notice of the agreement but is as regards persons ignorant of the restriction.

As regards third parties, every partner is jointly liable for the firm's debts and obligations and jointly and severally liable for the wrongful acts of any one done to a third party. A partner is not liable for anything done before he became a partner, but when he retires he does not cease to be liable for partnership obligations incurred before his retirement.

Where no fixed term has been agreed a partnership may be dissolved by any partner giving notice to the others, and it is dissolved on the expiry of the term or the completion of the enterprise for which it was established or on the death or bankruptcy of any partner. It may also be dissolved by the court on the ground of the insanity or incapacity of a partner, conduct prejudicial to the partnership, breach of the partnership agreement, or unreasonable conduct by a partner, where the business can only be carried on at a loss or where it is just and equitable to do so. On dissolution the assets are applied in payment of the firm's debts and liabilities, in payment of what is due to the partners, and any surplus is distributed among the partners in the proportion in which they are entitled to share profits.

Since 1907 it has been permissible to form partnerships the liability of the partners of which are limited to the amounts of capital contributed, but at least one partner must be a general partner with unlimited liability. Such a partnership must be registered with the registrar of companies. The general principles of partnership law apply, but each limited partner may not take part in the management of the business and he cannot bind the firm nor withdraw his capital.

Party wall. A wall separating adjoining lands of different proprietors, who are at common law presumed tenants in common of the wall. Under statute it is divided longitudinally into two halves, each being subject to a cross easement in favour of the owner of the other half.

Paschal, Carlo (Paschalius, Carolus) (1547–1625). Dutch jurist, author of *Legatus* (1648).

Pashukanis, Evgenii Bronislavovich (1891–). Russian legal philosopher, prominent in the 1930s. He was Vice-Commissar of Justice in 1936 charged with supervising the drafting of the new codes required by the new Constitution of 1936, but was repudiated in 1937. He was author of *General Theory of Law and Marxism* (1924), *Essays in International Law* (1935), and *Text-Book of Soviet Economic Law* (1935). He communicated a commodity exchange theory of the origin of law. Law was essentially bourgeois in character and had little future in a socialist society; it would disappear as the capitalist system was replaced.

Pasley* v. *Freeman (1789), 3 T.R. 51. The defendant, by falsely asserting that a third party was trustworthy, induced the plaintiffs to deliver goods to him on credit. It was held that they might recover damages in an action for deceit.

Pasquier, Étienne (1529–1615). A French humanist jurist, later Advocate-General at the Paris cour de comptes, author of an encyclopaedic history, *Recherches de la France* (1560–1621), *Interpretation des Institutes de Justinien*, and works on government.

Pasquier Étienne-Denis, Duc (1767–1862). A descendant of Etienne Pasquier (q.v.) he was a counsellor of the Paris *Parlement* before the Revolution, and later a counsellor of state and prefect of police under Napoleon, president of the chamber of peers in 1830, and chancellor of France 1837–48.

Passing off. An action brought against a person for having represented the marks on, set-up, packaging, or other features of his goods, as being those of the plaintiff, and having caused the plaintiff actual or probable loss thereby. It is unnecessary to prove fraudulent motive or misrepresentation. The remedies granted are injunction against repetition or continuance, and damages or an account of profits, and, in some cases, orders may be made for delivery up of infringing articles or labels or the erasure of offending marks. Such an action also lies if one trader represents his business as being that of, or connected with that of, the plaintiff.

Passport. A certificate issued by the Foreign Office to citizens of the United Kingdom and Colonies, British subjects without citizenship and British protected persons, certifying their identity and citizenship and requesting British consuls and officials abroad to give them protection and any necessary assistance when travelling abroad.

Passy, Frédéric (1822–1912). French lawyer and advocate of peace. In 1867 he founded the Ligue Internationale de la Paix, later the Société Française pour l'Arbitrage entre les Nations, and successfully urged arbitration in various international disputes. In 1901 he was, jointly with Dunant, founder of the Red Cross, awarded the Nobel Peace Prize.

Pasture. Ground covered by grass cropped by sheep, cattle, and horses. Common of pasture is the right to pasture animals of these kinds on common land. The right may be appendant, i.e. an incident to the grant to certain tenants of arable land prior to the statute of *Quia Emptores* (1290); or appurtenant, i.e. depending on a grant annexing to particular lands; or in gross, i.e. depending on a grant without reference to a particular land; or by reason of vicinage, i.e. having to allow the cattle of commoners of adjoining land to come into the common.

Patent (pronounced pătent, not pātent). A grant from the Crown of a monopoly right in respect of an invention. The term is derived from the fact that the grant was made by *literae patentes*, open letters addressed to all to whom they may come. The origin of the modern patent system is in the common grants of monopolies to favourites of Elizabeth and James I. This was checked by the Statute of Monopolies, 1623, which declared all grants of monopolies void, but saved future patents and grants of privilege for 14 years or less of the sole working or making of any manner of new manufactures to the true and first inventor of such manufactures.

There is now, accordingly, no prerogative power to grant a monopoly save where permitted by statute, but there is a prerogative to refuse a patent. A patent is now granted by and under the seal of the Patent Office.

The granting of patents is now regulated by statute, particularly the Patents Act, 1977, but the interpretation of many terms depends on common law and earlier statutes. To be patentable there must be an invention which is new, involves an inventive step, and is capable of industrial application. The legislation has effect throughout the U.K. and Isle of Man, subject to modifications for Scotland, Northern Ireland, and the Isle of Man, but the Patent Office is in London and is controlled under the Department for Trade by the Comptroller-General of Patents, Designs, and Trade Marks. It maintains a public register of patents, in which transmissions and assignments are registered. The Channel Islands and other parts of the Commonwealth have their own patent systems.

A patent is granted by the Patent Office after examination of a complete specification (which may be preceded by a provisional specification) of the patent sought to be patented. It must particularly describe the invention and the method by which it is to be performed, disclosing the best method of performing the invention known to the applicant, and have a claim or claims defining the scope of the invention claimed. Every claim has effect from a priority date, which is the date of the filing of the application. The patent will not be granted if the invention has been anticipated by prior publication of another patent specification.

A patent may be refused, or, if granted, be challenged as invalid if it was not for an invention within the meaning of the Act, or the invention was not new, or was obvious, or useless, or the claims of the specification ambiguous, or was not explicit. The court may revoke a patent on a petition or on a counter-claim in an action for infringement.

If granted, a patent is valid for 20 years from the filing of the complete specification, but subject to revocation, surrender, and lapse for non-payment of fees.

Patents are choses in action and assignable outright or in security, and the patentee may grant licences to exploit the patented invention. In certain cases the Comptroller may grant a compulsory licence, where the patented invention is not being worked or the monopoly being abused, and in certain other cases. If a patent is endorsed 'licences of right', any person is entitled to a licence under the patent as of right on terms settled, failing agreement, by the Comptroller. Government departments may use patented inventions for the service of the Crown on payment of compensation.

Whether conduct by a competitor amounts to infringement depends on whether his article or process falls within the scope of the monopoly granted by the patent and whether the alleged infringer has done anything the monopoly in which is given to the patentee. A patentee is entitled to an injunction to stop infringement and to damages or an account of profits. If the validity of a claim has been challenged and found valid the court may certify that the validity was contested and upheld. A person aggrieved by groundless threats of proceedings for infringement of a patent may recover damages therefor and obtain an injunction against continuance of the threats.

By the Act of 1977 U.K. courts have jurisdiction over infringements of a Community patent, granted by the European Patent Office in Munich and having a common character throughout the E.E.C., though it is governed primarily by the E.E.C. Patent Convention. There is also for a transitional period a 'European patent (U.K.)' granted by the European Patent Office under the European Patent Convention which is equivalent to a British patent.

Patent (U.S.). Patents were granted in the American colonies to encourage new industries, the first in Massachusetts in 1641. The Constitution of the U.S. (Art. I, sec. 8) authorized Congress to secure to inventors the exclusive right to their discoveries, and the U.S. Patent Office began work in 1790. In 1802 it became a Bureau, and in 1809 it was transferred to the Department of the Interior, and in 1925 to the Department of Commerce. Patents give a monopoly right for 17 years.

Patent of precedence. Letters patent formerly sometimes granted to particular barristers, normally serjeants (q.v.), as a mark of distinction and entitling them to have the precedence of Queen's Counsel. They are said to have originated with William Murray, later Lord Mansfield, and were at first confined to M.P.'s who, if appointed Q.C., would have had to vacate their seats as holding an office of profit under the Crown. Patents expired on the demise of the Crown. They are now obsolete.

Patent Rolls. Chancery rolls from 1202 to the present, containing documents relating to the Crown, the revenue, the judicature, grants and confirmations of offices, and other matters.

Patents Appeal Tribunal. Appeals from the Comptroller of Patents, Designs, and Trade Marks lay to this tribunal, which is not a part of the High Court but consisted of a judge of the High Court, nominated by the Lord Chancellor. The tribunal was not bound by its own previous decisions. Appeals from Scottish cases went to the Scottish Appeal Tribunal which consisted of a judge of the Court of Session nominated by the Lord President. Further appeal lays to the Court of Appeal or the Court of Session. In 1977 the functions of this tribunal were transferred to the Patents Court (q.v.).

Patents Court. A court, part of the Chancery Division of the High Court, constituted by the Patents Act, 1977, consisting of one or more nominated judges of the High Court, having the function of hearing appeals on patent matters from the Comptroller-General of patents, designs, and trade marks. Further appeal lies to the Court of Appeal in certain cases only.

Pater decretalium. A title sometimes given in mediaeval times to Pope Gregory IX.

Pater et tuba juris canonici. A mediaeval title sometimes given to Joannes Andreae.

Pater juris. A title sometimes given to Pope Innocent III.

Paterfamilias. In Roman law, the male head of a family and the only full person known to the law. He alone owned property and all his children, of whatever age, were in his power (*patria potestas*) and subject to his power of life and death. Particularly in earlier private law there was little difference between a son and a slave in that the *paterfamilias* could dispose of either as he wished.

Paterson, Sir Alexander Henry (1884–1947). Early on he became interested in criminals and social outcasts, and in 1911 became assistant director of the newly-formed central Association for the Aid of Discharged Prisoners. In 1922 he became a member of the Prison Commission; as such his purpose was to counter the deterioration caused by imprisonment, and he stressed rehabilitation and training of prisoners, introduced evening classes and lectures, open prisons, and prison visitors. He developed the Borstal system and introduced the house system therein. He visited prisons abroad and advised the colonial office on their prisons and maintained contact with many ex-prisoners. He wrote a handbook on *The Principles of the Borstal System* (1932) and various papers by him were published as *Paterson on Prisons* (1950).

Paterson, Marcus (1712–87). Called to the Irish Bar in 1742, became Solicitor-General in 1764 and Chief Justice of the Common Pleas in Ireland 1770.

Pateshull, Martin of (?–1229). A royal clerk, Dean of St. Paul's, a justice of the bench from 1217 and one constantly employed on judicial eyres. He was reputed by Bracton the most able lawyer in England.

Patria potestas. In Roman law, the power which a father exercised over his male descendants, including those adopted into his family. Under it he had control of their persons, including in early law the right to inflict capital punishment, and owned all the family property, though he might allow a son or a slave to treat some property as if it were his own. It might be ended, short of the father's death, by emancipation of a son, and daughters were released from *patria potestas* if married *cum manu*, when they passed into the husband's family. By classical times, the *patria potestas* was much reduced, the power of punishment being limited to modest punishment, while sons could keep as their own what they earned in military service (*peculium castrense*). In later law this exception was extended to earnings from other sources, while the father's right in other kinds of property, such as property inherited from the mother, was limited to a life interest.

Patrial. Under the Immigration Act, 1971, a person having the right of abode in the United Kingdom. This right is enjoyed by citizens of the United Kingdom and Colonies by birth, adoption, naturalization or registration in the U.K., by children or grandchildren born to or legally adopted by a person having that citizenship, by such a citizen settled in the U.K. and ordinarily resident there for at least five years, and the child or wife of such a citizen. Non-patrials may live, work, and settle in the U.K. by permission and subject to control of their entry into and departure from the U.K.

Patrician. In early Rome, the privileged class of citizens who prior to about 400 B.C. monopolized the magistracies and priesthoods. Their privileges were reduced by the plebeians, and by the end of the Republic the distinctions were politically unimportant. Under the Empire only the empress could create patricians but the rank carried few privileges. After Constantine in the fourth century *patricius* was a non-hereditary title of honour, ranking after the emperor and the consuls. Later the title patricius Romanorum was the title of the patron of the Catholic Church, conferred between the ninth and twelfth century by the Pope on the emperors from Charlemagne to Frederick Barbarossa.

Patrimonial rights. Rights pertaining to a person's patrimony or estate, comprising rights of property and rights secured by obligation, as distinct from personal rights, pertaining to his person only.

Patton, George, Lord Glenalmond (1803–69). Called to the Scottish Bar in 1828. He was Solicitor-General briefly in 1859 and went into parliament in 1866 and was then appointed Lord Advocate. In 1867 he became Lord Justice-Clerk but in 1869 committed suicide, when nervous by reason of a pending investigation into bribery at the election for Bridgewater, where he had both won and lost elections which, he feared, probably unnecessarily, might compromise him. He was a man of considerable ability but had no chance to display it.

Paty's case, (or Case of the men of Aylesbury) (1704), 2 Ld. Raym. 1105. Following the decision of *Ashby* v. *White* (q.v.) and the Lord's denial that the Commons had the sole right of deciding all matters relating to elections, Paty and other electors brought actions against the constables of Aylesbury. The Commons committed them to prison for breach of privilege and the Queen's Bench, Holt, C. J., dissenting, refused to grant habeas corpus. The matter was resolved by the prorogation of Parliament but Holt, C. J.'s, dissenting view is probably the accepted one today.

Paulette. A royal edict of 1604 in France which provided that an office could be sold or bequeathed free of the customary rule that, if not so transferred, more than 40 days before the holder's death, the office would revert to the Crown, in consideration of an annual payment to the crown of one sixtieth of its value. This resulted in offices becoming hereditary and in the creation of a permanent class of judicial officers, the *noblesse de la robe*, as well as diminishing the Crown's liberty of appointment.

Paucapalea (twelfth century). A canonist, possibly a student of Gratian (q.v.), probably responsible for some of the earliest additions to Gratian's *Decretum*, and author of early glosses and a *Summa* on the *Decretum* which is frequently referred to by later canonists.

Paucital rights. A term in the writings of Hohfeld (q.v.) for a right *in personam*, meaning either a unique right residing in a person or group and availing against a person or group, or one of a few fundamentally similar yet separate rights availing respectively against a few definite persons. See also MULTITAL RIGHT.

Paul, Sir George Onesiphorus (1746–1820). Travelled abroad and became high sheriff of Gloucestershire, in which capacity he became interested in prison reform. He secured the building of a new county gaol at Gloucester, opened in 1791, which contained rooms where debtors could work; they were allowed to retain two-thirds of what they earned. Five new bridewells were also erected in various parts of Gloucester. He was intimately acquainted with the writings of Howard (q.v.) though probably did not know him. He was also involved in a society for providing free medical assistance for the poor and other philanthropic works.

Paulo di Castro (Paulus Castrensis) (?–1441). A famous Italian jurist and commentator, author of *In Digestum praelectiones* (1546), *Consiliorum sive responsorum* (1571), and *In Digestum Commentaria* (1582).

Paulus, Julius (fl. *c.* A.D. 200). One of the most famous classical Roman jurists. He practised as an advocate, was assessor to Papinian as *praefectus praetorio*, *magister memoriae*, and a member of the imperial *consilium* of Severus and Caracalla. Under Alexander Severus he was *praefectus praetorio*. He is known to have been in great demand for *responsa* and was a voluminous writer, including educational works, *Institutiones, Manualia, Regulae, Sententiae, Quaestiones, Responsa*, commentaries on the works of older jurists (Plautius, Neratius, Vitellius, Labeo, Alfenus), and on Papinian, monographs on *leges, senatusconsulta*, constitutions and various topics of law, a commentary on the Edict in 80 books and an exposition of *ius civile* (*Ad Sabinum*, in 16 books). He was highly regarded by contemporaries and successors and was named in the Valentinian Law of Citations (426) as one of the jurists whose works were authoritative. There are 2,081 excerpts from his works in Justinian's Digest, nearly one sixth of the whole. He was clearly a masterly jurist, clear, logical, and critical, cogent in argument and accurate in exposition. The surviving *Sententiae* may not be authentic but a late anthology compiled from Paulus' works.

Paulus de Liazariis (*c.* 1295–1356). Canonist, author of a *Lectura super Clementinas* and a *Casus summarii* or *Epitome Clementinarum.*

Pauw, Willem (1712–87). Dutch jurist, son-in-law of Bynkershoek (q.v.), president of the Hooge Raad van Holland from 1784, editor of some of Bynkershoek's works and continuator of his *Observationes tumultuarie* with his own *Observationes tumultuariae novae*, containing decisions of the Hooge Raad, 1743–87 (1964–67).

Pavia. Capital of Lombard Italy and the site of a famous law-school, the principal school of Lombard law. Its history is recorded in the *Expositio ad Librum Papiensem* (*c.* 1090) which is an extended commentary on the laws of Lombardy and mentions the major Lombard jurists, Valcausus, and his successors Bonifilius and Guglielmius, Lanfranc, who went to Normandy and died Archbishop of Canterbury in 1089, Sigifred, Bagelard, Ugo, and others. The great years of the Pavia school were the eleventh century, after which Roman law influence became important and the school was transferred to Bologna.

Pawn (or pledge). The delivery of personal property to another in security for some debt or engagement. Various statutes prohibit the pawning of certain kinds of property. It is essential that the property be actually or constructively delivered to the pawnee, who acquires a special property or special interest in the property pawned, the general property remaining in the owner. The pawnee must take ordinary care of the thing pawned, but not use it if that would deteriorate or depreciate it. The pawnee may assign his rights by assignment or sub-pledge. At common law the pawner has an absolute right to redeem the thing pledged at any time on repayment of the money lent. The pawnee has a power of sale on default in payment by the time fixed for repayment, or on default, after request for payment and notice given of intention to sell, the pawner being entitled to redeem until the sale is effected. If the sale makes a surplus it belongs to the pawner, if a deficiency the pawnee can sue for the balance. The pawnee has no right of foreclosure.

The contract is extinguished by satisfaction of the debt or engagement, with interest, and redelivery of the property pawned to the pawner.

The pawning of goods with pawnbrokers is now regulated by the Consumer Credit Act, 1974.

Pawnbroking. The business of making advances of money to persons who pledge goods in security of repayment. The trade is a very ancient one. In the Middle Ages because of the usury laws pawnbroking was usually carried on by groups exempted by religion or custom from the prohibitions on usury. Public pawnshops were occasionally organized, e.g. in 1198 at Freisung in Bavaria. The Franciscan Friars in Italy in 1462 established *montes pietatis* or charitable funds from which they might make interest-free loans to poor persons, secured by deposits of pledges, but these funds came to have to charge interest and sell unredeemed pledges. Public pawnshops were later tried in some European countries as a means of avoiding exorbitant interest.

In the U.K. pawnbroking has been regulated by statute since 1872 and is now subject to the control of the Director-General of Fair Trading, who must licence pawnbrokers and under statute controls the terms and conditions of transactions, rates of interest, and the terms of redemption or of sale of forfeited pledges.

Paymaster-General. An office in the United Kingdom created in 1835 by the merger of the offices of Paymaster-General of the Forces and Treasurer of the Navy, In modern practice the office of Paymaster-General is a political one, with negligible financial functions, the functions in connection with the disbursement of public money being handled by the permanent staff. Accordingly, the office is frequently held by a minister, whose position may vary from a full Cabinet Minister to a junior minister, charged with some special and non-departmental function.

Payment into court. The deposit of money with an official of, or the banker to, the court by way of satisfaction of a claim made against the party paying. It is of the nature of an offer of compromise. In the Chancery Division it is a common practice to pay into court money in the hands of a trustee or executor who is unable to obtain a good discharge from the person beneficially entitled to the money. The persons who claim to be entitled can apply to have the money paid out to them.

***Pays du droit coutoumier* and *pays du droit écrit*.** In France from the time of the Franks a division developed between the north, where mixed remnants of Germanic and Roman law, capitularies, canon law, and local usages made up the customary law, which varied in details in different districts though it had a certain amount of unity in general principles; and the south where the Roman law, which was the personal law of the majority of the population, was applied to everyone as the custom of the region. The line of demarcation coincided approximately with that separating the language of Oïl from that of Oc, and ran along an irregular line from the island of Oleron to the lake of Geneva, more often above than below the central mountain mass. The distinction was not sharp or absolute;

there were local customs in the regions of written law and Roman law had some authority in the regions of customs.

Peace, Charles (1832–79), English criminal, committed several murders, for one of which another man was convicted but later pardoned, and many robberies and housebreakings. After his execution for murder he became something of a legend and fantastic exploits were ascribed to him.

Peace. In international law, the normal state of relations in international society, marked by absence of overt hostilities between states. International law is generally stated by reference to the state of peace, and the state of war treated as abnormal, a more or less localized interruption of that normal state.

As war has become increasingly costly and destructive, efforts have been made to try to maintain peace. Three plans deserve particular attention. The Abbé de Saint-Pierre in 1713 published a project whereby the 24 Christian states of Europe should form a perpetual alliance for common security and mutual guarantee of their possessions and treaties. Disputes were to be resolved by mediation or arbitration by a Senate of Peace. Jeremy Bentham's plan of 1789 urged the establishment of an International Court of Judicature and a common legislature, and the banning of any state which should not conform to the resolutions agreed on by the states. Immanuel Kant's project for a perpetual peace of 1795 required that all states be republican and that the public law of Europe should rest on a federation of free states; he developed the idea of an organization of all states, not merely Christian or European ones.

The Hague Peace Conference of 1899 and 1907 adopted and confirmed a Convention for the Pacific Settlement of International Disputes, by use of mediation and arbitration, and it set up the Permanent Court of Arbitration. For investigations the Hague Conference recommended the setting up of International Commissions of Inquiry.

A determined effort to prevent war was made after World War I by the establishment of the League of Nations (q.v.) with its organs, the Council, the General Assembly, and the Permanent Court of International Justice (q.v.). The League's peace machinery broke down, however, in face of the Japanese aggression in Manchukuo in 1931 and the Italian attack on Abyssinia in 1935. The League of Nations did not abolish war, but in the Paris Peace Pact of 1928 practically every sovereign state renounced war as an instrument of national policy in their relations with one another.

After World War II fresh efforts were made by the establishment of the U.N. Organization and the International Court of Justice, but nevertheless Communist powers have sought the fruits of war by aggression in Korea and Vietnam which forced other states to resort to war. The 'cold war' which has existed since 1945 has made the threat of war as a means of Communist aggression, aggrandisement, and imperialism, a constant worry for Western states. Given that attitude by Communist states there can be no certainty of peace.

Peace, breach of the. Any disturbance of the quiet and security sought to be maintained by law, and promised by the sovereign to be maintained, including a riot, an affray, and an assault or battery. Indictments formerly concluded with the statement that the offence was committed 'against the peace of our lord and king'. In Scotland it is a common summary offence, covering shouting, swearing, fighting, and generally creating a disturbance.

Peace, Commission of the. A commission or authority issued under the Wafer Great Seal under which justices of the peace are appointed by the Crown and exercise their jurisdiction. A separate commission is issued for every county, county borough, and borough which has a separate commission of the peace. It is kept by the clerk of the peace and a person is appointed a justice by having his name inscribed in the schedule to the commission. The names are entered by authority of the Lord Chancellor (or Chancellor of the Duchy of Lancaster) on the recommendation of advisory committees. The Lord Chancellor, Lord Chief Justice, Master of the Rolls, and judges of the High Court and Court of Appeal, the Lord President of the Council, Lord Privy Seal and members of the Privy Council, and certain others are included in the commission for every county. Local officials such as mayors were formerly justices *ex officio*. Persons holding lower judicial appointments such as Recorder are also *ex officio* justices.

On appointment justices must take the oath of allegiance and the judicial oath.

On attaining the age of 70 a person, with certain exceptions, is placed on the supplemental list and he does not then exercise judicial functions. Justices may be removed from the commission by the Lord Chancellor without need to show cause. This has occasionally been done where justices have refused to administer some statute of which they disapproved.

Peace, Justices of the. The origin of justices of the peace is to be found in many appointments from the end of the twelfth century to the middle of the fourteenth of knights to take oaths from persons to keep the peace (1195), to act as custodians of the peace (1264), to receive the presentment of constables as to infringements of the Statute of Winchester (1285), and so on. From 1327 statute provided for the appointment of conservators of the peace in each county and their powers were repeatedly enlarged. After the Black Death justices of labourers were appointed to enforce the Ordinance of Labourers. Their commissions were later united with those of the keepers of the peace. In 1361 by the Justices of the Peace Act, in every county in England a lord and three or four of the most worthy in the county were assigned to keep the peace and to hear and determine felonies and trespasses, and in 1363 they were ordered to hold sessions four times a year and from this time they were known as justices of the peace. Thereafter their duties were repeatedly multiplied and extended in scope, and their numbers increased. The practice developed of making all the justices members of the quorum to enquire into the breaches of the peace, though not learned in the law. Statute from the sixteenth century onwards empowered them to deal summarily with certain offences, and these sittings came to be known as petty sessions.

In the early eighteenth century it came to be the practice to appoint justices on the nomination of the Lord Lieutenant, though the Lord Chancellor may appoint on his own initiative. In the eighteenth and nineteenth centuries they were usually country gentlemen, latterly with an admixture of manufacturers and other men of substance. In the twentieth century the field from which they have been chosen has greatly widened, and advisory committees assist the Lord Lieutenants to secure representatives from all classes in the community. They have always been unpaid, though they are nowadays entitled to allowances for loss of earnings and expenses incurred.

In the nineteenth century many of their administrative duties were transferred to elective local authorities, but they have retained their judicial functions. The judicial functions are both civil and criminal and are exercised in courts of Petty Sessions now called magistrates' courts, and formerly in courts of General or Quarter Sessions.

Petty sessions developed from the practice of statutes conferring summary, i.e. without a jury, jurisdiction on two or more justices to inflict penalties. In 1848 statute codified the rules of procedure to be observed by justices in the exercise of their summary jurisdiction, and since then such courts have commonly been called petty sessions but officially magistrates' courts.

A magistrates' court of from two to seven justices has a civil jurisdiction in respect of certain kinds of civil debts, particularly affiliation and matrimonial orders, and a criminal jurisdiction over offences triable summarily, including offences triable summarily or on indictment which are more suitable for summary trial.

In the case of indictable offences justices must conduct a preliminary enquiry, normally in open court, to ascertain whether there is sufficient evidence to justify committing the accused for trial. Evidence is heard for the prosecution, but the defence usually reserves its evidence.

A single justice has very limited powers, e.g. to sign warrants, but may act as examining judge.

Justices have also important functions in issuing summonses, warrants, and granting bail, and authenticating various kinds of documents.

The justices have always been subject to the supervisory jurisdiction of the High Court and appeal lies, in civil matrimonial cases, to a Divisional Court of the Family Division of the High Court; and in criminal cases from the magistrates' court to the Crown Court, or by way of case stated to a divisional court of the Queen's Bench Division, and then to the House of Lords, and from the Crown Court to the Court of Appeal, Criminal Division, and the House of Lords.

In Scotland the office of justice of the peace was envisaged in 1587 and was introduced on the English model in 1609. They were to bind evil-doers to keep the peace and present them to the King's Justices for punishment. In 1611 jurisdiction to try breaches of the peace was conferred on them. Many later Acts sought to increase their effectiveness but their powers were inadequate and the institution never took root or flourished in Scotland, particularly in face of the important heritable jurisdictions and the powers of the sheriff courts.

The justices were for long appointed in Scotland also by the Lord Chancellor, but this function has been transferred to the Secretary of State for Scotland acting on the advice of local advisory committees.

There is now a separate commission for each district and islands area, and it includes certain *ex officio* justices.

The justices have certain administrative powers and duties, many of their former ones have been transferred to elected local authorities, the most important being in relation to licensing, while individual justices have powers in respect of authenticating documents.

They had judicial functions both civil and criminal. The civil jurisdiction (the J.P. Small Debt Court) was limited to actions for small debts, was very unimportant and was abolished in 1975. The criminal jurisdiction (the district court, formerly the J.P. Court) is confined to trial of petty offences, mainly breaches of the peace, the powers of punishment being limited, all cases of any consequence being dealt with by the sheriffs.

Quarter sessions nominally existed in Scotland but did not function as a court of trial and in many counties never met.

The justices and their courts are in fact an alien excrescence on the Scottish legal system and wholly unnecessary, the sheriff court being a far more important lower court in both civil and criminal respects.

Peace of God. A movement of ecclesiastical inspiration originating in Aquitaine in the late tenth century and spreading over Europe in the next two centuries to arrest anarchy by censures and pacts of peace. By about 1000 the Carolingian organization of France had fallen into anarchy and churches began to take measures to protect life and property. It started when bishops pronounced excommunications on robbers and aggressors and the movement spread. Baldwin IV of Flanders organized a great collective oath of peace in 1030 and in Normandy the State assisted the Church against violators of peace pledges. Pope Urban II and Norman barons proclaimed it in Southern Italy in 1089. The Peace was only a name in England because the monarchy was generally strong from 1066 onwards. The movement generally had moderate success but sometimes got out of hand.

Akin to the Peace of God but distinct was the Truce of God during which hostilities were suspended. It possibly originated in 1027 when days of rest from fighting were proclaimed.

Peace, The King's (or Queen's). In the eleventh century the King's peace was personal and confined to places, notably his household and the great highways, but it existed throughout the land during the festivals of the Church. Until the time of Edward I the King's peace died with him. It coexisted with the peace of other lords and conduct, though not a breach of the King's peace, might be a breach of the peace of the lord of the manor. Later the King's peace extended over all England at all times, so that a breach of the peace, though not defined by law, was always a crime against the King.

C. K. Allen, *The Queen's Peace.*

Peace treaty. The normal mode of terminating a war. Hostilities are normally suspended by an armistice (q.v.) and negotiations take place. Hostilities are usually brought to an end by preliminaries of peace, which are a treaty in themselves, but superseded by the definitive peace treaty. Unless otherwise specified, peace is restored by the signing of the peace treaty. All conduct legitimate in warfare then ceases to be legitimate and all rights and duties attaching to the state of peace *ipso facto* revive. Unless stipulated to the contrary, all conditions remain as at the conclusion of hostilities. Accordingly, territory occupied and ships and property seized remain the property of the possessor. It is not legally necessary, though usual, for conquered territory to be ceded to the occupier. Unless there is reservation the treaty operates an amnesty for all troops and individuals for war crimes and wrongful acts. Prisoners of war have to be repatriated as soon as possible. Parties are bound to perform the treaty in good faith, but either party may denounce the treaty if the other acts in fundamental breach thereof.

Peacekeeping forces. Forces created by the United Nations in the exercise of its competence to keep world peace. Such forces have been established by resolutions of the Security Council or of the General Assembly. They have been composed of contingents from various member states, which have remained subject to the jurisdiction of their home states and have generally been regarded as a subsidiary organ of the U.N., a body of international personnel under the authority of the U.N. and having immunities and privileges as such. Their functions have been specified in each case with reference to the particular tasks to be performed. The major instances of their establishment have been in the Middle East, in the Congo, and in Cyprus.

Peacock, Sir Barnes (1810–90). Practised as a special pleader before being called to the Bar in 1836. He became Q.C. in 1850 and in 1852 legal member of the Council in India, where he was responsible for various codifying acts. From 1859 to 1870 he was Chief Justice of the Supreme Court at Calcutta. In 1872 he was appointed a paid member of the judicial committee of the Privy Council and sat in that capacity till shortly before his death.

Pearce, Edward Holroyd Pearce, Lord (1901–). Called to the Bar in 1925 he took silk in 1945 and was a puisine judge of the P.D.A. Division, 1948–54, and of the Queen's Bench Division, 1954–57, a Lord Justice of Appeal, 1957–62, and a Lord of Appeal in Ordinary, 1962–69.

Pearson, Colin Hargreaves Pearson, Lord (1899–). Called to the Bar in 1924, took silk in 1949, was a judge of the Queen's Bench Division, 1951–61, a Lord Justice of Appeal, 1961–65, and a Lord of Appeal in Ordinary, 1965–74. He was also a judge of the Restrictive Practices Court, 1957–61, and its President, 1960–61, and Chairman of the Royal Commission on Civil Liability, 1974–78.

Pearson, Sir John (1819–86). Called to the Bar in 1844 but made slow headway. In 1864 he took silk and in 1882 was appointed to succeed Vice-Chancellor Hall but without the title of Vice-Chancellor. In his short time on the bench he showed himself a very competent judge.

Peck, Peter (Peckius) (1529–1589). Counsellor of the High Court of Mechlin, author of *Commentaria* *in juris civilis titulos ad rem nauticam pertinentes* (1556), *Tractatus de testamentis conjugum* (1556), and *Tractatus de jure sistendi et manuum injectione* (1564), later translated into Dutch.

Peculiar. In ecclesiastical law, a place which is a jurisdiction by itself, not subject to the jurisdiction of the bishop of the diocese. They include royal peculiars such as Westminster Abbey, or the sovereign's chapel, such as St. George's Chapel, Windsor, peculiars of archbishops, of bishops, of deans, deans and chapters, and the like. Formerly numerous, they are now much reduced in numbers.

Peculiar courts. These were courts which developed from the exemption of the greater abbeys and of the King's chapels royal from episcopal jurisdiction and from the conflict in some cases between bishops and their chapters, resulting in the apportionment of lands and in each exercising jurisdiction. Peculiars had a great variety of origins, some being exempt from all episcopal and even archiepiscopal jurisdiction. Appeal lay originally to the Pope and later to the High Court of Delegates. Most peculiars have subsequently been abolished. A court of Peculiars was held by the Dean of the Arches at Bow Church for those London parishes which were exempt from the diocesan jurisdiction of the Bishop of London.

Peculium. In Roman law, property assigned to be managed, used and, within limits, disposed of by a person lacking full legal right of property, particularly a *filius familias* or a slave. It became increasingly important as time went on. Legally it was a voluntary grant and revocable, and involved liabilities to third parties up to the amount of the *peculium*, but in practice the possessor had a very free hand and a slave might buy his freedom with the profits of his *peculium*. From the time of Augustus, anything a soldier acquired on military service was automatically his *peculium* (*peculium castrense*) and this was extended to the profits of state employment other than military service (*peculium quasi-castrense*).

Pedius, Sextus (*c.* A.D. 50–*c.* 120). Roman jurist, who wrote a commentary on the praetorian and aedilician Edict, frequently quoted by later jurists, and a treatise *De Stipulationibus*. He appears to have been a jurist of some originality of thought.

Peel, Sir Lawrence (1799–1884). Cousin of the statesman, Sir Robert Peel, was called to the Bar in 1824, and served as Advocate-General at Calcutta 1840–42, before being promoted Chief Justice of the Supreme Court at Calcutta. He retired in 1855. After returning to England he served as a paid member of the Judicial Committee of the Privy Council and was legal correspondent of *The Times*.

Peeping Tom. A voyeur or person who obtains sexual pleasure from secretly watching other persons undress or engage in sexual acts. The name is taken from one who is said to have been struck blind or dead for watching Lady Godiva as she rode naked through Coventry to protest against heavy taxation, but this is a late addition to the story of Lady Godiva.

Peerage. A term designating both a rank of nobility which may be granted by the Crown to a person, or the whole group of persons possessing any of the ranks of nobility. The term peers is from the Latin *pares*; it was a principle early recognized that a man should be tried by his equals (*pares*), and the barons early succeeded in establishing that lords should be tried by lords for treason or felony, and this privilege defined the class of *pares* or peers as persons who were tenants-in-chief of the Crown and legally equals of each other. These were also the persons on whom lands and titles were conferred and who were commonly summoned to the King's Council. The doctrine of ennobled blood was imported under Henry VIII and has given rise to much confusion. Bishops and life peers are not deemed ennobled in blood and were never triable by peers only.

In the United Kingdom peerage there are five degrees, viz. (in descending order) dukes, marquesses, earls, viscounts, and barons. Earls and barons were the earliest to appear; dukes were created from 1337, marquesses from 1385, and viscounts from 1440. There are five classes of peers, viz. peers of England, peers of Scotland, peers of Ireland, peers of Great Britain, and peers of the United Kingdom. No English or Scottish peerages have been created since the Act of Union, 1707; no peers of Ireland have been created since 1898; no peers of Great Britain have been created save between 1707 and 1800, and the only peers created since 1800 have been peers of the United Kingdom. Baronetcies and knighthoods (q.v.) are dignities but not peerages.

A peerage is an incorporeal heritable right which descends according to the words of limitation in the grant. It is impartible and inalienable. It may be conferred on a woman, and, where a peerage may descend in the female line, women have on various occasions inherited peerages. Such are peeresses in their own right as distinct from peeresses by marriage.

The creation of peerages for life, long obsolete, was authorized by the Appellate Jurisdiction Acts, 1876, which permitted the conferment of baronies for life on persons appointed Lords of Appeal in Ordinary (q.v.). The Life Peerages Act, 1958, authorizes the creation of life peers generally, male or female. The power has hitherto been exercised to create life barons only.

The Crown has power to create any number of peers and of any degree. In modern practice the power is exercised on the advice of the Prime Minister and the honour is most commonly a reward for political services. Peerages can be, and have been, conferred for party political reasons; 12 were created in 1712 to save the government, and 16 to help pass the Reform Bill in 1832. In 1832 and 1911 the Opposition of the House of Lords was overcome by the threat to create enough peers to secure a majority. Peers have at various times been created by Act of Parliament, by charter, and by letters patent. A writ of summons to Parliament followed by sitting in Parliament creates a barony by writ which descends to the heirs general of his body. In modern practice the grant is normally by letters patent under the Great Seal, specifying the grantee, the name of the dignity, and the limitation thereof to future heirs. The name of a place is not essential to the creation of a peerage, though now customary, and necessary if there be more than one peerage having the same personal name.

A person may simultaneously hold several peerages of different degrees and creations, e.g. the Duke of Devonshire (1694) is also Marquess of Hartington (1694), Earl of Devonshire (1618), Earl of Burlington (1831), and Baron Cavendish (1605). In such a case the second title is by courtesy customarily borne by the heir to the highest dignity and such a person may sit in the House of Commons. A peer may be promoted to a higher degree of peerage, e.g. Baron Jowitt (1945) became Viscount Jowitt in 1947 and Earl Jowitt in 1951.

The main privilege of a peer is to sit and vote in the House of Lords. Between 1707 and the Peerage Act, 1963, Scottish peers were represented by 16 of their number, elected for each Parliament; now all sit, as do peers of England, Great Britain, and the United Kingdom. An Irish peer has all the privileges of peerage, unless he chooses to waive them to enter the House of Commons, e.g. Viscount Palmerston, but not that of sitting and voting in Parliament. The Act of Union, 1800, provided for the representation of Irish peers in the House of Lords by 28 elected representatives, but no election has been held since 1922 and none now survive. Peeresses in their own right had no right to sit and vote in Parliament until the Peerage Act, 1963. Peeresses by marriage do not sit or vote in Parliament.

Peers have certain minor privileges including freedom from arrest in civil causes. The privilege of trial by their peers in cases of treason or felony was abolished in 1948. Peers are disqualified from election to the House of Commons.

A claim to a peerage which was thought to be extinguished or is in abeyance is made by petition to the sovereign. All claims of any complexity are referred to the House of Lords which refers them to the Committee for Privileges, which hears the claim (including hearing counsel, ordering production of documents, and examining witnesses on oath) and reports to the House.

A life peerage terminates on the recipient's death. An hereditary peerage is extinguished if there is no heir within the classes of persons to whom the peerage is destined. An hereditary peerage may fall into abeyance if the succession opens to co-heirs such as daughters. The Crown may terminate an abeyance in favour of a co-heir.

An hereditary peerage may by the Peerage Act, 1963, under certain conditions and within stated timelimits, be disclaimed by delivery to the Lord Chancellor of an instrument of disclaimer. Thus Mr. Wedgwood Benn renounced the Viscountcy of Stansgate and became again Mr. Benn. Disclaimer is irrevocable during the lifetime of the individual concerned, but does not extinguish the peerage which remains dormant until the next heir succeeds. The person disclaiming renounces all titles, rights, and privileges of peerage and becomes a commoner; he is ineligible for the grant of any other hereditary peerage but not precluded from accepting a life peerage. Thus the Earl of Home renounced his peerage, becoming Sir Alec Douglas Home, K.T., and later received a life peerage as Lord Home of the Hirsel.

In France the origins of the peerage are obscure; it emerged in the fourteenth century. A French peerage was an honorary distinction, usually associated with a duchy, but without political significance. The old peerage disappeared at the revolution, but a revived peerage existed between 1815 and 1848 and there was a chamber of peers with a voice in legislation.

W. Cruise, *Origin and Nature of Dignities or Titles of Honour* (1810); F. B. Palmer, *Peerage Law in England* (1907); J. H. Round, *Peerage Law and History*; Debrett, *Complete Peerage*.

Peerage Bill (1719). A proposal to limit the House of Lords to existing peerages, apart from the princes of the blood, replacement of extinct lines and six new creations, and to replace the 16 Scottish representatives peers by 25 hereditary peers. The effect would have been to create a House not subject to the influence of either king or Commons. The proposal was defeated in the Commons, but thereafter the royal prerogative of creation of peers was more sparingly used.

Peers, Threat to create. A device twice used and several times contemplated in British history. In 1832 and 1911 the governments of the day secured the King's promise to create, if necessary, enough peers to pass in the Lords a Bill which the government was determined to pass, the Reform Bill in 1832 and the Parliament Bill in 1911. Forty creations would have been needed in 1832 and nearly 500 in 1911. In each case the threat persuaded the House to allow the measure to pass, so that on neither occasion was there an actual creation. A material difference is that in 1832 the threat was made after the Lords had rejected the Bill, but in 1911 the King's promise was secured before the Bill had been discussed in Parliament, but was kept secret, and this use is of questionable constitutional propriety.

Pegasus (first century A.D.). A Roman jurist, and successor of Proculus as head of the Proculian school, considered to have been very erudite. His works are frequently quoted.

Peine forte et dure (strong and hard punishment). In old English criminal law a prisoner could not be compelled to accept jury trial. The Statute of Westminster I (1275) provided that felons who refused jury trial should be committed to a hard and strong prison. The words *prison forte et dure* became transformed into *peine forte et dure*, into a form of torture whereby, in the sixteenth century, the prisoner was placed between two boards and weights piled on him until he accepted jury trial or died. Some chose to die in this way as such a person had never been convicted and his goods were not forfeited to the Crown. It was inapplicable to treason, where silence was held to imply guilt. The last recorded case of crushing a prisoner to death is said to have been at Cambridge in 1741. It was abolished in 1772 and a conviction substituted. In 1827 a plea of not guilty was ordered to be entered if an accused declined to plead. It was rarely used in America and the few instances prompted the constitutional prohibition against cruel and unusual punishments. In 1692 in a Salem witchcraft trial an accused was pressed to death.

Pemberton, Sir Francis (1625–97). Called to the Bar in 1654, he is said to have been long imprisoned for debt, but developed a good practice in the time of Charles II and acquired a good reputation as a lawyer. He became a judge of the King's bench in 1679, but was removed in the next year at the instance of Scroggs, but in 1681 became Chief Justice of the King's Bench, being removed to the same office in the Common Pleas in 1683 and dismissed in the same year for impartiality in a treason trial, which was unwelcome to the Court. He then practised at the Bar till his death, appearing for the defence at the trial of the Seven Bishops (q.v.).

Penal action. An action for a penalty imposed as punishment by a statute. It might be recoverable by anyone who could sue for it, or go partly to the Crown and partly to the informer. Proceedings by common informers were abolished in 1951.

Penal code. A compendious term for the legislation of the (Protestant) Irish Parliament of 1695–1727, in violation of the Treaty of Limerick of 1691, which excluded Catholics from Parliament, disarmed them, made it illegal for them to go abroad for education, closed Dublin University to them, excluded them from professions and legal practice, and otherwise set out to humble and restrict them. Acts forbade landowners to acquire freehold other than by inheritance or to take leases of more than 31 years at heavy rent. A 'Gravelling Act' required partition of lands among sons, unless the eldest son conformed in religion, when he inherited the whole, and this impoverished and largely extinguished landowners. The final Act of 1727 debarred Catholics from voting for M.P.s. The penal laws lasted in their entirety for about 70 years and were not repealed until 1829; they crushed, humbled, and antagonised the upper classes.

Penal colony. An overseas settlement established for isolation of criminals from society and their punishment by hard labour. England transported criminals to America until the Revolution and to Australia until the mid-nineteenth century. France had penal colonies in Africa, New Caledonia, and French Guiana, the last including the notorious Devil's Island. Penal colonies were frequently associated with cruelty, torture, and brutalizing and degrading conditions of life, and have generally now been abolished. But correctional camps in Siberia maintained by Russia, from the times of the tsars onwards, maintain the practice.

See also PENOLOGY; TRANSPORTATION.

Penal laws. Bodies of legislation in England, Scotland, and Ireland from the sixteenth to the nineteenth centuries designed to discriminate against and oppress Catholics. They exhibit no settled scheme or consistency, but were products of prejudice and reactions to particular political conditions.

In England, Elizabeth's Acts of Supremacy and Uniformity (1559) required all public and ecclesiastical officers to take an oath of supremacy and penalized any who should maintain any spiritual jurisdiction within the realm. It was made an offence to say or hear mass and failure to attend Anglican service was punishable. The Northern Rebellion of 1570 resulted in three penal laws in 1571. In 1581, it was made treason to do anything to reconcile or be reconciled to the Romish religion while, in 1585, an Act made it treason for a priest to enter or remain in England and, in 1593, it was punishable to aid recusants.

Oppressive measures were passed under James I, particularly after the Gunpowder Plot. After some relaxation under the Commonwealth, the penal legislation was restored in 1660 and Charles II was unsuccessful in seeking to relax it. There were no additions to the penal laws after 1727. The Catholic Relief Acts, 1778 and 1791 released some of the disabilities and the Catholic Emancipation Act, 1829 removed the bars on Catholics voting for or sitting in Parliament.

In Scotland, the Reformation Parliament of 1560 made the saying or hearing of the Mass punishable by death and down to the Restoration penal statutes were enforced. In 1681, a Test Act imposed an oath on all office-holders obliging them to disown all popish principles. Fresh statutes were passed in the eighteenth century and were enforced after the Union of 1707. There was no relaxation until 1793.

In Ireland, the Irish Parliament passed Acts of Supremacy and Uniformity in 1560 requiring adherence to the State Church. There was some relaxation under the Stuarts but the Irish Parliaments refused to give effect to the Treaty of Limerick of 1691, which gave limited toleration. The 'penal laws' are particularly applicable to the legislation of the Irish Parliament from 1692 to 1727 and the legislation of the British Parliament against Irish industry and commerce. Together, these bodies of legislation constituted a harsh complicated body of law effecting heavy cultural, political, economic, and religious oppression. Relaxation began in 1778 and much of the oppressive legislation was repealed in 1782, but not till 1829 and later was toleration accepted.

The history of the topic in all three countries is a sad commentary on bigotry, hatred, and prejudice, which has done much to poison relations between groups down to the present.

C. Butler, *Historical Account of the Laws respecting Roman Catholics and of the Laws passed for their Relief*; J. Brown, *Historical Account of the Laws enacted against the Catholics*; A. Bellesheim, *History of the Catholic Church of Scotland*; E. Curtis, *History of Ireland.*

Penal Servitude. A sentence substituted in England in 1853 for sentences of transportation of less than 14 years. The duration of sentences was related to, but shorter than, those of transportation, the minimum being three years with no remission. In 1857, an Act restored correspondence with the lengths of transportation sentences together with remission on what was called a licence to be at large. The system involved three stages, separate confinement in Pentonville or one of the local prisons, associated labour in a public works prison, itself divided into three progressive stages carrying increasing privileges and gratuities, and finally release on licence. Penal servitude was abolished in 1948.

Penalty. A sum of money payable as compensation or as punishment. Many statutes impose penalties for non-implement of a public duty. Penalties may be agreed upon by the parties, as in a bond subject to a condition, where a party binds himself to pay a sum, frequently double the amount secured, if the condition is not complied with. Parties to a contract may agree that, in the event of a breach, the one in breach will pay to the other an agreed sum; if this sum cannot be regarded as a genuine pre-estimate of the damage likely to be sustained by a breach of contract, but is rather a sum stipulated *in terrorem* of the party in breach, it is deemed a penalty and is irrecoverable so far as in excess of the damage actually sustained.

Pengelly, Sir Thomas (1675–1730). Was called to the Bar in 1700, attained a good practice, and later sat in Parliament. He was one of the managers of the Earl of Macclesfield's impeachment. A very able, firm and conscientious lawyer, he was Chief Baron of Exchequer, 1726–30, but died of gaol fever contracted at assizes.

Penitentials. Manuals for confessors, setting out prescribed penances for specified sums. They originated in the Celtic Church. Some of the later penitentials drew on collections of canon law as sources, and there are traces of penitentials in Burchard's *Decretum.* They had some influence on the development of the law of tort in that, by the seventh and eighth centuries, they were abandoning the idea of fixed tariffs of penalties and this influenced law to abandon settled penalties in favour of discretionary modification of penalties according to the circumstances.

Penitentiary. An older English and a modern American term for a prison, so called because the original intention and hope was that convicted persons would reflect on their misdeeds and repent of them.

Pennefather, Edward (1775–1847). Was called to the Irish Bar in 1795. He became Solicitor-General in 1835 and 1841 and held the office of Chief Justice of the King's Bench in Ireland, 1841–46.

Penn's Charter of Liberties. A charter which, in 1701, became the constitution of Philadelphia, replacing several other earlier Frames of Government, of 1683 and 1696. It granted liberty of conscience, the right to choose representatives for the Assembly, which had legislative power, to nominate for the offices of sheriff and coroner and other liberties customary in the colonies. It dealt also with the preservation and recording of laws, the liberty of criminals to have counsel and give evidence, the safeguarding of property, and the licensing of taverns.

Pennsylvania system. A penal method based on the idea that solitary confinement encouraged penitence and reformation. Advocated by a Philadelphia Society, it was tried first in the Eastern State Penitentiary, Cherry Hill, Philadelphia, in 1829. Prisoners were kept in separate cells each having an enclosed exercise yard attached to it and saw no one apart from guards and occasional visitors. The solitary régime was, however, soon modified to permit the doing of some work. The system spread and became the predominant one in European prisons, but it fell into disfavour and was superseded by the Auburn system (q.v.).

Penology. The study of the philosophy and the methods of punishment and treatment of persons found guilty of crime, the methods which have been used at various times in various countries, their merits and defects, those which are currently in use, and those which might with benefit be introduced. It also includes evaluation of particular methods, their difficulty, cost and effectiveness, or success-rates. Modern penology originated with the publication of Beccaria's (q.v.) *Crimes and Punishments* in 1764, which founded the classical school which accepted that criminal acts were based on a rational calculation of the prospective pleasures and pains of the act. What was needed in each case accordingly was a penalty sufficient to counterbalance the advantages of the crime; excessive penalties were unjust and unnecessary. The succeeding neo-classical school insisted on recognition of varying degrees of moral and legal responsibility, and of mitigating circumstances. From their views developed the modern idea of individualization of punishment. The Italian, Lombrosian, or positive school developed in the latter nineteenth century and held that the criminal is predestined by his inherited traits to criminality; to punish him is both irrational and pointless, because he is irresponsible. This view is now wholly discredited. Modern investigation stresses the wide variety of causal factors in crime.

Views of the purposes of punishment and treatment have varied. The most ancient is retribution, an outgrowth of the sentiment of vengeance; resentment at injury can be solaced only by expiation or retribution. In early society this took the form of talion, an eye for an eye and a tooth for a tooth. Retribution is justified by the philosophy which regards punishment as a moral necessity in wrongdoing, a view held by idealistic philosophers from Plato to Aquinas and Kant to T. H. Green. Deterrence has been favoured by utilitarians from Bentham and Mill to Spencer. In modern practice, reform is the predominant declared aim, but one which is difficult, probably quite impossible, to achieve. Too little regard is frequently paid to the interests of the general public, the need to protect them and to make the way of the lawbreaker hard and unpleasant by firm discipline, and to punish wrongdoing, whether it reforms or not.

L. Radzinowicz, *History of English Criminal Law*; W. Moberley, *The Ethics of Punishment*; N. D. Walker, *Crime and Punishment in Britain*; Lord Longford, *The Idea of Punishment*; L. Fox, *The English Prison and Borstal System*; G. B. Vold, *Theoretical Criminology*; P. Tappan, *Crime, Justice and Correction*.

Penrice, Sir Henry (c. 1680–1752). Was judge of the court of Admiralty, 1715–51, and seems to have been a capable civilian, but little is known of him.

Pentonville. A prison in the Clerkenwell district of London, established in 1842 to hold convicts awaiting transportation and to provide local authorities with a model prison showing how the 'separate system' ought to work. Many new prisons were, in fact, built on this model. After the substitution of penal servitude for transportation it was used for prisoners serving sentences of penal servitude. It was closed for a time but reopened in the 1940s and is one of London's principal prisons.

Penzance, Lord. See WILDE, J. P.

Pepper Arden, Richard, Baron Alvanley (1745–1804). Was called to the Bar in 1769 and took silk in 1780. In 1782 he entered Parliament and became Solicitor-General and Attorney-General in 1784. He succeeded Kenyon as Master of the Rolls in 1788 and was knighted, and in 1801 became Chief Justice of the Common Pleas and Baron Alvanley. In both of his judicial capacities he acquired the reputation of being a sound and learned lawyer.

Peppercorn rent. A purely nominal rent, stated in the lease as 'a peppercorn, if demanded'.

Pepys, Charles Christopher, 1st Lord Cottenham (1781–1851). Studied in Tidd's chambers and was called to the Bar in 1804, but made rather slow progress. He took silk in 1826 and became an M.P. in 1831, Solicitor-General in 1834 and, in the same year, Master of the Rolls. In 1835 he was made one of the commissioners of the Great Seal and, in 1836, Chancellor as Lord Cottenham. As a politician he was a failure but he was, at first, a useful member of the Cabinet and a very successful judge. He was out of office 1841 to 1846 but Chancellor again, 1846 till 1850. His despatch of judicial business became increasingly unsatisfactory and heavy arrears of business built up. Nevertheless, while in office he introduced legislative reforms, particularly in the practice and procedure of Chancery and made the largest contribution to equity of the Chancellors since Eldon. He was a very sound equity lawyer and made developments particularly in relation to injunctions and the rules of company law.

***Per capita* and *per stirpes*.** In the law of succession, terms fixing the shares descendants or other beneficiaries are to take. If *per capita* all in the designated class take equally, by heads; if *per stirpes*, they take by the branches of the family, e.g. grandchildren jointly taking their parent's share.

Per incuriam (by carelessness). A decision rendered *per incuriam*, in ignorance of a relevant statute or precedent, is not a precedent which need be followed.

Per quod consortium amisit (whereby he lost the benefit of her society). The allegation in the old action of damages by a husband against a third party for injuring his wife.

Per quod servitium amisit (whereby he lost his services). The allegation in an action by a person against a third party who, by his wrongful act, had deprived him of the injured person's services; it applies to household servants only.

Per subsequens matrimonium (by subsequent marriage). The prime mode of legitimation of an illegitimate child.

Perambulation. The process of walking round the boundaries of a piece of land or estate, to fix them or to preserve evidence of them.

Perceval, Spencer (1762–1811). Was called to the Bar about 1785 and became K.C. in 1796. He enjoyed a high public reputation, entered Parliament in 1796, became Solicitor-General in 1801, and was Attorney-General from 1802 to 1806. He refused the Chief Justiceship of the Common Pleas in 1804 and his interest turned more to politics than to law. He became Chancellor of the Exchequer (1807–9) and thereafter was Prime Minister till his assassination. He was one of the few men to be as successful in the House of Commons as in the courts.

Perduellio. In Roman law, hostile activity against the State, one of the earliest crimes punishable, under the XII Tables, by death and, in early times, subject to the jurisdiction of *duoviri perduellionis*. In the late Republic, it became merged in *maiestas*.

Peregrini. In Roman law, citizens of any state other than Rome. *Socii Latini* were *peregrini* down to 89 B.C. and all provincials, though having *conubium* or *commercium*, were *peregrini*. In practice, the grant of Roman status to a *peregrinus* freed him from liabilities to his original state though Augustus affirmed the liability of such citizens to the *munera* of their own state.

Peregrinus, M. A. (sixteenth century). A humanist and author of *Tractatus de possessione et proprietate*.

Perezius, Antonius (1583–1672 (?1674)). A Spaniard who became Professor of Law at Louvain and a notable writer on Roman law. He was author of *Institutiones Distinctae* (1634), *Commentarii in libros novem Codicis* (1626), *Praelectiones in tres posteriores libros Codicis* (1613), *Ius Publicum* (1657), *Commentaria in XXV Digestorum libros* (1669), and other works.

Performance. The doing of what is required by an undertaking or contract or statute. It is always a question of the interpretation of the undertaking, contract or statute to determine what is required for due performance of the obligation, and a question of fact whether due performance has been made or not. Formerly, the tendency was to require strict and literal performance but in modern times, the tendency is to require satisfaction of the real substance and intent of the obligation. If performance is not made, the default may give rise in civil cases to a claim for specific performance (q.v.) or damages (q.v.) in compensation, and in statutory cases to prosecution for neglect of duty.

Performing right. One of the rights comprised in copyright (q.v.) being, in the case of musical or dramatic work, the right to perform the work or part thereof in public or to license others to do so. The Performing Right Society exists to grant, on behalf of composers and dramatists, licenses to perform works, to recover fees for doing so, and to seek remedies against persons infringing composers' or dramatists' copyright. The Performing Right Tribunal exists to resolve disputes as to the terms on which such licences are given.

Perils of the sea. Every kind of natural misfortune which may affect a maritime adventure, such as hurricane, shipwreck, stranding, and damage by heavy weather beyond the ordinary damage of wind and waves. The term extends also to acts attributable to the acts or fault of other parties such as loss or damage by collision, capture by pirates, or damage by rats.

Periodicals, legal. Legal periodicals serve different functions, to give information about developments in law and practice, to supply professional information to legal practitioners, to serve as a forum for the critical analysis and discussion of legal topics, and to be an outlet for work of research. They vary accordingly from highly academic and learned to the severely practical and professional. Some again are general, dealing with any branch of law, and some are specialized, concerned only with a particular field. Some contain notes on recent statutes and cases, some contain full reports of cases.

In the United Kingdom, the principal general academic and scholarly journals are the *Law Quarterly Review* (1885–), the *Cambridge Law Journal* (1921–) and the *Modern Law Review* (1937–). In Scotland there is the *Juridical Review* (1889–), in Northern Ireland the *Northern Ireland Legal Quarterly* (1936–), and in Ireland the *Irish Jurist* (1966–).

Among those concerned with particular fields of law may be mentioned the *International and Comparative Law Quarterly* (1952–), superseding the *Journal of Comparative Legislation and International Law* (1896–1950) and the *International Law Quarterly* (1947–51); the *Common Market Law Review* (1963–); *Public Law* (1956–), incorporating the *British Journal of Administrative Law* (1954–58); the *Journal of Criminal Law* (1937–), the *Criminal Law Review* (1954–) and the *British Journal of Criminology* (1960–); the *Industrial Law Journal* (1972–) and the *British Journal of Industrial Relations* (1963–); the *British Tax Review* (1956–); the *Business Law Review* (1954–58) and the *Journal of Business Law* (1957–); the *Conveyancer* (now *Conveyancer and Property Lawyer*) (1915–); the *Journal of Planning and Environment Law* (1948–); the *British Journal of Law and Society* (1974–); the *Journal of the Society of Public Teachers of Law* (1924–).

The principal practitioners' journals are the *Solicitors' Journal* (1857–) and the *New Law Journal* (1956–) (superseding the *Law Journal* (1866–1965), and the *Law Times* (1843–1965)). In Scotland there is the *Scots Law Times* (1893–) and in Ireland the *Irish Law Times and Solicitors' Journal* (1867–).

In the U.S., the major academic and scholarly journals are those published by, and edited by senior students of the major law schools. Practically every law school publishes a journal. The major ones are the *Harvard Law Review* (1887–), the *Yale Law Journal* (1891–) and the *Columbia Law Review* (1901–). Also important are the *American Journal of International Law* (1907–) and the *American Journal of Comparative Law* (1952–). Similarly in Canada there is the *Canadian Bar Review* (1923–), in Australia the *Australian Law Journal* (1927–), in New Zealand the *New Zealand Universities Law Review* (1963–), and in South Africa *Acta Juridica* (1959–).

Apart from periodicals published at intervals there are various annuals, such as *Current Legal Problems* (1948–), the *Annual Survey of Commonwealth Law* (1965–), the *British Yearbook of International Law* (1920–), the *Annual Survey of American Law* (1942–), the *Yearbook of World Affairs* (1947–), and others.

Similarly, in all the major European countries there are numerous periodicals, general and particular, some of which have interest and importance far outside their own jurisdictions.

Perjury. The crime of, having been lawfully sworn as a witness, or interpreter, in a judicial proceeding, wilfully making a statement in that proceeding which the maker knows to be false or does not believe to be true. The false statement must be definite and unequivocal. Subornation of perjury is procuring another to commit perjury, by persuasion or, more often, by bribery or threats.

Perkins (or Parkins) John (?–1545). Author of *A Profitable Booke treating of the Lawes of England* in Law French (1528), with later English translations, dealing largely with conveyancing. It was a popular text-book for law students and was regarded as witty and learned.

Permanent Court of Arbitration. The first Hague Peace Conference (1899) laid down provisions, later enlarged at the Second Hague Peace Conference (1907), setting up a Permanent Administrative Council of the Court at the Hague, and consisting of the diplomatic envoys of the contracting powers accredited to Holland, to control the International Bureau of the Court. It serves as registry for and conducts the administrative business of the Court. The Court of Arbitration itself consists of individuals of recognized competence in questions of international law selected and appointed by the contracting powers, each being entitled to appoint not more than four members for renewable six-year terms. The Court does not itself decide cases but, in each individual case, parties electing arbitration each select two members from the panel of arbitrators, and the four then select an umpire.

The Court made awards in some 20 cases prior to 1932, but has not been invoked since then. It still exists and the national groups of arbitrators form the basis of the lists of candidates for election as judges of the International Court of Justice (q.v.).

Permanent Court of International Justice. In 1920, the League of Nations (q.v.) appointed an advisory committee of jurists to prepare a draft statute for a P.C.I.J. This developed from three sources, previous experience of international arbitration (q.v.) and a draft convention relative to the creation of a Permanent Court of Arbitral Justice discussed at the Second Hague Peace Conference in 1907; a proposal of neutral states for compulsory jurisdiction; and a plan put forward by Elihu Root and Lord Phillimore for the election of judges. The draft Statute proposed compulsory jurisdiction but this provision failed to secure support. As amended, the Statute came into force in 1921 and the Court began to function in 1922.

The Court consisted of 11 judges and four deputy-judges elected from members of the Permanent Court of Arbitration by ballots held simultaneously but independently by the Council and Assembly of the League; the Court elected its own President and Vice-President for terms of three years. Judges were elected for nine years and eligible for re-election. Casual vacancies were filled in the same manner as at a general election and judges so appointed held office for the rest of their predecessors' terms only. The deputy-judges were to replace judges if one or more was unable to sit. The quorum of judges was nine. The seat of the Court was at The Hague. In 1930, the deputy-judges were abolished and the Court increased to 15.

The jurisdiction of the Court extended to states or members of the League of Nations only, and to non-member states if they accepted the jurisdiction of the Court and undertook to carry out its decisions in full good faith and not to resort to war with a state complying therewith. It was not compulsory but depended on voluntary submission by parties of a dispute, but states adhering to the Statute of the Court might under the 'optional clause' (q.v.) declare that they recognized as compulsory, in relation to any member or state accepting the same obligation, the jurisdiction of the Court in all or any of the classes of legal disputes concerning (a) the interpretation of a treaty; (b) any question of international law; (c) the existence of any fact which, if established, would constitute the breach of an international obligation; and (d) the nature or extent of the reparation to be made for the breach of an international obligation.

The Court was directed to apply (1) international conventions, whether general or particular, establishing rules expressly recognized by the contesting states; (2) international custom, as evidence of a general practice accepted as law; (3) the general principles of law recognized by civilized nations; (4) subject to the provisions of Art. 59, judicial decisions and the teachings of the most highly qualified publicists of the various nations as subsidiary means for the determination of rules of law. This provision was not to prejudice the power of the Court to decide a case *ex acquo et bono* if the parties agreed thereto. (Art. 59 provided that the decision of the Court had no binding force except between the parties and in respect of the particular case.)

The official languages of the Court were French and English, but it might authorize another language to be used. The procedure of the Court was laid down in the Statute of the Court and in Rules of Court made thereunder; it provided for both written cases, evidence and oral submissions.

Cases were decided by a majority, the President or his deputy having a casting vote. There was one judgment only, stating the view of the judges or the majority. Dissenting judges might deliver separate opinions.

Provision was made for a special chamber of three judges to hear and determine cases by summary procedure at the request of the parties, and for special chambers for labour cases and transit and communication cases.

The Court was also given power to give advisory judgments, and this proved of great value and importance.

The Court was in suspense between 1940 and 1946 when it was reconstituted as the International Court of Justice.

M. O. Hudson, *The Permanent Court of International Justice*; A. P. Fachiri, *The Permanent Court of International Justice.*

Permanent Representatives of the Member States, Committee of. A body not set up by any of the treaties establishing the European Communities but recognized by them. Its functions are to prepare the work of the Council, to narrow and clarify areas of dispute and difficulty and to prepare issues for eventual political decision. The members are diplomats of ambassadorial rank rather than politicians or European bureaucrats and are an integral part of the Community decision-making machinery. They may also carry out tasks entrusted to them by the Council, though no substantial delegation of power to them is possible. They are not alternates for the members of Council and the committee has no power of decision, but it provides continuity between Council meetings. It may itself set up special committees or working groups and its conclusions have considerable value and force.

Permutation. Barter, or the exchange of one movable object for another.

Perpetual injunction. An injunction (q.v.) which is granted without limit of time as contrasted with an *interim* injunction.

Perpetuating testimony. The Chancery process whereby steps could be taken to preserve and perpetuate evidence when it was likely to be lost by reason of the witness' age, health, or intention to go abroad before his evidence could be taken in the normal course.

Perpetuities. Dispositions of property which infringe a group of rules which, on grounds of public policy, discourage dispositions imposing restrictions on future alienations of that property or fettering to an unreasonable extent the future devolution or enjoyment of that property. The main rules invalidate limitations and trusts which impose restrictions on the future alienation of property or establish trusts of very long duration, invalidate limitations of property so that it may vest outwith certain periods permitted by statute, and prohibit the accumulation of income for longer than one of certain periods permitted by statute. The English law is now contained in the Perpetuities and Accumulations Act, 1964.

J. C. Gray, *Rule Against Perpetuities*; J. H. C. Morris and W. Leach, *The Rule against Perpetuities.*

Persistent offender. A person who has repeatedly offended against the criminal law. Such an offender is normally punished more heavily than one who has not previously or has occasionally offended, and statutes sometimes expressly provide greater maximum penalties or special regimes, such as preventive detention, for persistent offenders.

Person. In law, a person is any entity recognized as having an existence in law capable of suing and being sued, and otherwise having rights and being subject to duties and liabilities. The category of legal persons includes most living natural persons and also legal or juristic persons, such as corporations, which have legal existence quite independently of the persons who, for the time being, are members of the corporation. Among living human beings, it is sometimes denied that young children or persons mentally incapacitated are legal persons; but probably they are, though all legal rights and duties must be discharged by a guardian or representative. But at various times and in various legal systems outlaws, persons deemed civilly dead, and other categories of individuals have been, though alive, not deemed to be legal persons. Among juristic persons, it is for each legal system to determine what things or groups of individuals are to be deemed legal persons, and personality may attach to idols, funds, institutions, and groups of individuals associated in particular ways. In Britain, building societies and companies are incorporated and therefore persons, but partnerships, trade unions, and many clubs, societies, and associations are not, and are not legal persons, but only groups of individuals.

Personal Act. An Act of Parliament dealing with the status, powers, property, or other rights of a particular person only, e.g. an Act decreeing a divorce or dealing with a person's estates.

Personal action. At common law in England, personal actions were distinguished from real actions (q.v.) and mixed actions (q.v.). They were claims against persons arising out of contracts or out of torts, the former comprising the actions of account, assumpsit, covenant, debt, and certain others, the latter comprising attaint, case, deceit, champerty, conspiracy, detinue, replevin, trespass, trover, and certain others. All these were abolished in the nineteenth century together with their individual original writs and distinct forms of procedure. The term is now frequently given to an action *in personam*, where the judgement of the court is a personal one, normally for payment of money, as contrasted with an action *in rem*, where the plaintiff seeks to make good a claim to or against certain property in respect of which, or in respect of damage done by which, he alleges that he has an actionable demand. The term is also used as translation for *actio personalis* as in the maxim *actio personalis moritur cum persona.* A personal action in this sense is a claim which falls on the death of the claimant and does not pass to his personal representative.

Personal bar. In Scots law, the general doctrine that a person may not be heard to deny what he has previously, expressly, or by conduct, affirmed, corresponding generally to estoppel in English law.

Personal estate. The part of the estate left by a deceased which consists of personal property, i.e., leasehold estates, money, securities for money, debts, goods and chattels.

Personal injury. In the law of torts, injury sustained by an individual in his person, as by broken limbs, as distinct from injury to his reputation or property.

Personal law. A system of personal law exists where the law applicable to a person and his transactions is determined by the law of his tribe, religious group, caste, or other personal factor, as distinct from the territorial law of the country to which he belongs, in which he finds himself, or in which the transaction takes place. After the fall of the Roman Empire and barbarian settlements in former Roman territory, a system of personal laws developed, and each people and tribe retained its own tribal law while Roman citizens still living in the conquered territories still lived by Roman law. Franks, Burgundians, Goths, and others each lived by their own personal law. From about the eleventh century, this system broke down with the development of feudalism and the growth of the city-states, and the principle of territoriality of law became the general one.

There remain, however, instances in which personal law still has relevance. Thus in India, Hindus and Mohammedans have long been governed in family matters by their religious laws. In England, divorces effected in England not by judicial process but according to the personal religious laws of the parties, and valid according to the law of their domicile, have been held valid, though it would be otherwise if the parties had been domiciled in England.

The term is also sometimes used for the aggregate of the elements affecting the legal status of a person.

Personal liberty laws. Laws passed by certain northern states before the American civil law to counteract the Fugitive Slave Acts and to protect negroes and escaped slaves settled in the North. Thus, Indiana (1824) and Connecticut (1828) allowed jury trial on appeal. Vermont and New York (1840) allowed jury trial and allowed fugitives attorneys. After the Supreme Court ruled in 1842, that enforcement of the Fugitive Slave Act was a federal responsibility, some northern States forbade their authorities to co-operate in capturing and returning slaves. After 1850, most northern states allowed jury trial, allowed punishment for illegal seizure of alleged fugitives and declined to allow their authorities to recognize claims to fugitives.

Personal Property (or personalty). A general term in English law for all forms of property, other than freehold estate and interests in land, a term used because no action lay to recover these kinds of property from a wrongful taker but only a personal claim against him for damages. It is contrasted with real property (q.v.). Personal property includes both corporeal and incorporeal, and movable and immovable elements. At common law, rules applicable to personal property differed in many respects from those applicable to real property, e.g. in respect of formalities of wills, of descent on intestacy, but statutes have removed most of these differences.

Personal property is divided into chattels real, which are interests concerning realty, such as a term of years in land, or the next presentation to a church, and chattels personal, which comprises choses in possession where a person can have actual possession of corporeal chattels, such as animals, books, money, and everything else physically transferable, and choses in action, where a person has right to recover by suit or action at law things incapable of physical transfer, such as debts, claims, patent rights, and other legal rights. Personal property may be partly in possession and partly in action; thus, a bill of exchange is a chose in possession but the debt thereby secured is a chose in action.

The rights which may subsist in personal property are the rights of ownership and of possession (q.v.). Ownership includes the rights to enjoy, use, alter, alienate, use up or destroy, and to maintain, or recover, possession of the property against all persons. Ownership of personal chattels may be obtained by taking original possession, as by catching fish, by invention or creation, by accession (q.v.) or confusion (q.v.), by gift, barter or sale by a previous owner, by succession under the will or intestacy of a previous owner, or by bare acquisition of possession in the case of negotiable instruments and coin, so long as in good faith and for value. The finder of lost goods in a public place acquires ownership as against everyone except the true owner. By royal prerogative, the Crown acquires ownership of treasure trove, waifs, strays, wreck, flotsam, jetsam and ligan, and by statute the goods of a bankrupt are transferred to the trustee. Some of the rights which make up ownership may be separated, as where possession is transferred when lending, hiring out, or pledging the property. Since 1925, successive interests in leasehold property cannot exist at law but may exist as equitable interests under a will or trust, and successive interests in personal chattels may be created by will with or without trustees. Concurrent ownership may be joint or in common.

Personal chattels may be alienated and ownership transferred by delivery in pursuance of gift, barter or sale, or under a will or on intestacy. They may also be settled to devolve with settled land or with a title of honour. Involuntary alienation arises under distress, execution and bankruptcy.

Personal representative. A term used to describe one who is either the executor (q.v.), original or by representation, or the administrator (q.v.) of the estate of a deceased. He represents the deceased as regards both real and personal estate and is deemed in law to be his heir within the meaning of all trusts and powers. He is within the statutory definition of trustee and the administration of the property of a deceased is a trust for certain purposes.

Personal right (*jus in personam*). A claim against another determinate legal person only, enforceable by action against him, as distinct from a real right against everyone, enforceable by possession of property. The commonest kind of *jus in personam* is the *jus ad rem*, the right to have some thing (e.g. a thing bought) transferred to the holder of the right.

Personal security. Security for the performance of an obligation by one party to another constituted by the personal undertaking of a third party that if the first does not perform his undertaking, he can be called on to do so and binds himself to do so. In Roman law, personal security was constituted originally by *sponsio* and *fidepromissio* and later by *fideiussio*, all created by *stipulatio*. Of these, only the third survived to the time of Justinian. In modern English law, the corresponding concept is the undertaking of surety (q.v.) for his principal and, in Scots law, that of a cautioner (q.v.) for his principal. Where security involves transfer of some thing or valuable right to be held by the creditor and, in case of need, realized by him, it is real security (q.v.).

Personal union. A link between two states established by the fact of having a head of State in common. Between 1603–49 and 1660–1707 England and Scotland were joined in a personal union only.

Personality. In law, personality is the legal quality of being a legal person, i.e. a unit capable of sustaining and exercising legal rights and being subject to legal duties and liabilities. It is for each legal system to prescribe whether and in what circumstances it will attribute legal personality to particular kinds of persons, groups, entities, institutions, and the like. In earlier systems of law, some persons, notably slaves, had no legal personality; they were chattels. Persons outlawed lost legal personality. The major categories to which modern systems attribute legal personality are natural persons and corporations, but there is no logical impossibility in attributing personality to animals, groups of persons, institutions, foundations, funds, societies, idols, and other entities.

Personality, particularly in relation to natural persons, has two attributes, status and capacity (qq.v.) which together give legal content to the idea of being a legal person by defining what legal rights and duties attach to particular kinds of persons. Though all natural persons may have personality not all have the same status or capacity. Thus, aliens, infants, and mentally afflicted may have limited capacity.

A person may and frequently does act at the same time in various guises, as an individual, as executor of A, trustee of B, agent for C, and so on. It is doubtful whether in such cases one should say that the person simultaneously sustains several legal personalities, or merely that he simultaneously acts in several capacities. The former seems the better view; A's executor may have to deal with B's trustee, notwithstanding that one individual sustains both personalities.

In modern systems, personality generally commences with birth; the unborn child in the womb has no legal personality but sometimes the maxim *nasciturus pro iam nato habetur* is applied, so that in matters affecting its interests the unborn child, provided it is later born alive, is treated as if it were already born. In some jurisdictions this principle extends to enable the child, once born, to claim for injuries inflicted on it before birth, though the difficulties of establishing that any duty was owed to the child, and of proving causal connection between the alleged wrong and any harm sustained, are very great.

Similarly, personality in general ends with natural death, though personality continues to the extent that the deceased's estate may, in some cases, recover assets due to the now deceased, and may be liable for debts or damages. Also if the deceased has expressed his wishes as to the disposal of his estate in a way recognized as valid, many legal systems give effect to these wishes, enforcing the duties of the executor or, if necessary, appointing an administrator. The executor may be said to be *eadem persona cum defuncto* and to continue the deceased's personality until the administration of his estate has been completed.

In the case of corporations, personality commences only when the constituting formalities have been completed and the appropriate authority has granted legal existence, and personality terminates when the grant of personality expires or is rescinded in appropriate manner.

In international law personality is attributed to sovereign states and not to dependent states, provinces or parts of states, and to major international organizations, and only to a limited extent to natural persons and associations.

Personality, international. The quality of being recognized as a right- and duty-bearing entity in international law. While the concept generally corresponds to that of personality in municipal law, the entities having international personality are different from those having personality under municipal law. It is clear that every independent state has international personality but difficulties exist in other cases. Colonies, protectorates and protected states, and trust territories are generally not deemed states having international personality, but may develop into states. The United Nations has been held to have international personality by reason of its purposes and functions as specified in its constituent documents. In the cases of other international organizations, whether a particular institution possesses international personality or not depends on its constitution, functions, powers, privileges and immunities and other indicia of personality, such as power to make treaties, to make claims at the international level.

Though many rules of international law exist for the benefit of individual natural persons and many others impose responsibility on natural persons, e.g. for war crimes, this does not involve that natural persons are persons in international law, and there would appear to be no circumstances in which individuals have recognized status under international law and the right to operate on the international plane.

Personality disorders. Persistent disturbances in individuals' patterns of adjustment to life and responses in situations with which they are faced. They comprise a group of behaviour patterns which define an individual's relationship to society, to his work, and to himself. The main types are the inadequate, the schizoid, the cyclothymic, and the paranoid. The inadequate make inadequate responses to intellectual, social, emotional, and other demands. They show lack of adaptability, judgement, and ability to co-operate. The schizoid is quiet, shy, and withdrawn, avoids competition and society and is introverted. The cyclothymic is alternately elated and depressed; the paranoid suffers from uncertainty, suspicion, and inability to trust others as being fair and reasonable beings.

Some persons also suffer from personality trait disturbance where the basic personality maldevelopment is a distinctive factor. This group includes persons regarded as emotionally unstable, passive-aggressive and compulsive personalities.

Again there are persons suffering from sociopathic or psychopathic personality disturbance, who frequently act antisocially, a category which includes sexual deviants.

Understanding of personality disorders is important legally because some of them may account for conduct of legal significance, such as sexual offences.

Personality of Laws. In the Frankish Empire, the legal system regulating the rights and duties of a person was that of the people or tribe to which he belonged, which followed him whenever he went, rather than that of the territory in which he was resident. The personal law was usually that of the nation or tribe to which he belonged, though it might exceptionally be a law adopted by him for a particular reason. This fact developed from the co-existence in the Empire of a number of systems of law, all of equal standing and each recognized and sanctioned by the others.

The principle of personality became universal in Western Europe when different peoples were loosely united under Charlemagne, each preserving its own customs and legal autonomy as well as substantial political independence. It would not, in any event, have been feasible for the Carolingians to impose a single legal system over such a large area and such diverse peoples.

As an individual's rights and duties were determined by his personal law it was necessary for everyone to make *professio juris* or declaration of law, when called upon, according to prescribed conditions. Thus, *missi* or itinerant royal judges would, when on circuit, enquire of persons by what law they lived. Thus, at an inquest held at Rome in 824 by order of the Emperor Lothar, the judges enquired as to the law professed by every subject. More commonly a person declared his law as a preliminary to the determination of how some solemn act be done, such as swearing loyalty to the Emperor, contracting marriage or a binding obligation, or being a litigant or a witness. Notaries' documents recited these declarations as part of the record of the transaction. Individuals probably did not have a free choice of law but had to abide by the law of the nation to which they belonged, though exceptions were recognized to this principle where a person's nationality was in doubt, as in the case of the bastard or the manumitted slave, and where a new status required a change, as on a woman's marriage or a person's entry into holy orders; a wife took her husband's law, and churchmen normally took the Roman law. A parcel of land was governed by the law of the first possessor thereof, not of any later owner.

Exceptions to the principle of personality of law were recognized, in two main classes of cases, where the individual's personal law could not be recognized, as in the cases of aliens, i.e. persons not within the Frankish Empire, and non-Christians, and where there was conflict between the personal law and the public interest, as in the sphere of criminal law, where it was early determined that a penalty should be determined by the law of the place of the offence, not by the offender's personal law, and in the sphere regulated by ecclesiastical law, as in the case of marriage.

Moreover, as the power of the Emperor increased and superseded popular government, the Imperial law increased its dominance.

The mingling of persons subject to different personal laws gave rise to problems of conflict of laws and to the need for rules to determine such conflicts. So far as possible, all the systems of law involved were given effect to, failing which reliance was placed on the law of the person whose interest was the dominant one. Thus, contracts were determined by the law of the obligor or defendant, succession by the law of the deceased.

In the feudal period, various factors combined to lead to the supersession of the principle of personality by that of territoriality. Tribes and races had merged regionally and customs were developing on national lines. The feudal political system weighed in favour of territoriality as differences of race disappeared within each fief, which themselves in any event depended on a common feudal law with common general principles, and the major classes of population, ecclesiastics, agriculturalists, and towndwellers had strong common interests in a local (or even wider) basis for law than in personal laws. Different personal systems tended to overlap and mix, and in areas rules were adopted from one into another personal system, while legal manuals tended to mingle and confuse conflicting rules. Accordingly, the tendency was for a system of law to become dominant and uniform in an area. Thus, in the southern province of France Romanesque law became dominant and in the northern provinces Germanic law dominated. In northern Italy Lombard law prevailed over Germanic systems and even over Roman law. In central Italy Roman law was dominant, but in southern Italy the principle of personality lasted longer than elsewhere. The territorial systems of course incorporated much of the prevailing personal systems in their areas. Moreover, initially the ambit of a territorial system was small and only much later when large states were established, did territorial law expand and become national.

Personation. The act of a person representing himself to be someone else, whether real or fictitious, living or dead. If done with fraudulent intent it is criminal at common law. Personation in various circumstances, e.g. of a policeman, is made criminal by various statutes. Personation of a voter is a corrupt practice.

Persons. In law, persons are those units or entities which under a particular legal system have legal personality and are the subjects or holders of legal rights, to whom legal rights attach, who exercise them, and are subject to legal duties and liabilities. It is for each legal system to define what kinds of individuals, groups, and entities are regarded as legal persons.

In English law, persons comprises natural persons, i.e. living human beings, and legal or juristic persons, i.e. bodies of human beings treated collectively as legal entities.

Among natural persons, all living human beings are legal persons. It is questionable whether a living but unborn child is a legal person. Animals are not legal persons. Some legal persons, particularly infants or minors and persons mentally incapacitated, are legal persons but have restricted capacity.

Other legal systems take different attitudes on various matters; in Scots law a partnership or firm is a legal person distinct from the partners, but is not incorporated; in Hindu law an idol may be a legal person; in other systems foundations of institutions may be legal persons. Various legal systems have, in earlier times, denied the quality of being a legal person to slaves, aliens, outlaws, persons civilly dead, and some other categories.

The attributes of natural persons depend on their status and capacity (qq.v.).

Legal or juristic persons include groups of people who have been incorporated in an appropriate way and made into a corporation (q.v.) or corporate body. Not all groups can be, or are, incorporated. Thus partners, trustees, and members of an Inn of Court are not incorporated. Trade unions are in English law not incorporated.

Persons, law of. The branch of law dealing with the personal status and capacities of individuals, and sometimes including also the law of domestic or family relations. Thus, it deals with the different status, powers, capacities, liabilities, and disabilities of legitimate, illegitimate, legitimated, and adopted persons, of those of sound and unsound mind, of nobles and commoners, of married women, prodigals, and the like, and sometimes includes the law of the relations of husband and wife, parent and child and guardian and ward.

Persuasive authority. A precedent or statement of law from another source, e.g. statement in a textbook, decision of a foreign court, which is not in the circumstances, particularly under the rules as to the authority of precedents (q.v.), binding on a court and bound to be applied, but only one having a persuasive effect depending on the court's estimate of its value and worth in the circumstances.

Perverse verdict. A verdict in which the jury has refused to obey the direction of the prescribing judge on a matter of law. Justices of the peace may make a perverse decision if they have declined to draw an obvious conclusion or have taken into account any matter which they should have ignored.

Perverting the course of justice. A generic term for various crimes all having that tendency, such as fabricating false evidence, interfering with witnesses, embracery, maintenance, champerty, barratry, prison breaking, escape and rescue, and public mischief.

Peter of Sampson (thirteenth century). French canonist and author of a *Summa* on the Decretals and of glosses on some Novels of Innocent IV.

Peter of Spain (*Petrus Hispanus*) (twelfth century). Spanish canonist, author of glosses on Gratian's *Decretum*, and compiler of an apparatus of glosses on the *Compilatio Prima* (*c.* 1195).

Peter of Spain (*Petrus Hispanus*). Portuguese canonist, author of a *notabilia* on the *Compilatio Quarta*, and possibly author of some treatises on civil law.

Peterloo massacre. The dispersal by volunteer cavalry of a radical meeting held on 16 August 1819, at St. Peter's Fields, Manchester, later commonly referred to as symbolizing Tory reaction to reform and harshness. The myth is of a peaceful crowd calling for Parliamentary reform being dispersed by savage cavalry with many casualties. The truth is of worried local residents appealing for protection from the mob and of militants, in a crowd of about 50,000, attacking local Yeomanry before Hussars were ordered to disperse them, and that most of the casualties, many fewer than claimed in local accounts at the time, were caused by panic and by being trampled down in the crowd. The magistrates had no option but to disperse the mob.

L. Walmsley, *Peterloo: The Case reopened.*

Petersdorff, Charles Erdman (1800–86). Was called to the Bar in 1833 and was a judge of county courts, 1863–85. He was the author of various legal books, of which the principal are *A Practical and Elementary Abridgment of Cases in the King's Bench, Common Pleas, Exchequer and at Nisi Prius since the Restoration*, 15 vols. (1825–30) and *A Practical and Elementary Abridgment of the Common Law as altered and established by the Recent Statutes*, 5 vols. (1841–44 and later editions). Though called an *Abridgment* this work approaches closely in arrangement to modern Digests of case-law, abandoning any attempt to give an exposition of legal principles.

Petit or petty jury. See JURY.

Petit treason. See PETTY TREASON.

Petition. A written request made to a person, official, legislature or court for the grant of a favour, or authority, or the redress of some grievance. In England, the right of petitioning the Crown was recognized as early as Magna Carta. In Parliament, petitions to the Crown came in the fifteenth century to take the form of statements of how the petitioners desired the law to be changed and, on being granted, became statutes. In the sixteenth century, petitions came to contain the exact words to be enacted, in short to be Bills. The right to petition the Crown was fully recognized in the Bill of Rights of 1689, and by convention the right to petition Parliament has also been recognized. Petitions to the House of Commons became so common that, in 1839, the House resolved that in future they should not be debated.

In court procedure, petition is the appropriate mode of commencing certain proceedings, such as for divorce, of seeking authority for or ratification of various kinds of action, permission to do something, such as to present an appeal, or some administrative act such as to order a company to be wound up. In the House of Lords and Privy Council, appeal is commenced by petition of appeal praying that the judgment appealed from may be reversed or varied.

In the U.S. the right to petition was recognized during the Revolution and embodied in the First Amendment, and now both Congress and state legislatures have settled procedures for dealing with petitions.

Petition of right. At common law, a subject could not sue the Crown in the King's courts, as the King was not subject to their jurisdiction. Hence, when a subject's property was alleged to be in the possession of the King or servants of the Crown and should be transferred to the subject, it was necessary to present a petition of right, which the King might remit to one of the courts with a direction (*fiat justitia*) to the judges to do justice. In time this came to be the means of securing a remedy for breach of contract. In the nineteenth century, the functions of deciding on whether a cause of action had been disclosed and of issuing the *fiat* were devolved on the Attorney-General and the Home Secretary. This machinery was abolished in 1947 when it was made competent for a subject to sue the Crown by way of action against the relevant government department.

Petition of Right, The. A claim declaratory of rights, and a stage in the struggle between Parliament and Charles I and the first statutory restriction of the powers of the Crown since the accession of the Tudors. It was directed against the two chief abuses of power by the King, arbitrary imprisonment and taxation without Parliamentary authority. It recites the tenor of various older statutes and conduct by or under authority of the King in contravention thereof, and prays the King that no man be compelled to make any payment without consent by Act of Parliament, and no one be molested concerning these matters or for refusal thereof; that no freeman be imprisoned or detained as previously narrated; that no soldiers or mariners be billeted on persons against their will; that commissions for proceeding by martial law be revoked and annulled; and that no commissions of the like nature issue in future. It was passed by both Houses in 1628. The King at first prevaricated but finally assented. The Petition is memorable as the first restriction of the powers of the Crown since the accession of the Tudors.

Petition of Thirty-One Articles. A petition presented by the House of Commons to Henry IV in 1406 setting out in detail the canons of good government to which politicians of that time regularly appealed when things were going badly. The King accepted the articles, without reserve.

The major points were (i) the King should name 16 counsellors and officers to advise him; (ii) the Chancellor and Privy Seal should pass no grant or other matter to the seals contrary to law; (x) the council should determine nothing cognizable at common law unless for a reasonable cause and by the advice of the judges; (xxiv) the council and chief officers of state were to be sworn in Parliament to observe the common law of the land, and the statutes and ordinances before made and ordained, as well for the King's household as for the good government of the realm of England; (xxvi–xxx) the administration of the courts of law and of the household was to be regulated, and full enquiry made by the Chancellor and Treasurer of England and others, the justices of either bench and the barons of the Exchequer, into all torts, oppressions, and defaults done to the people who had had business therein and a report made to the council in order that full and due correction might be made. The articles were to be in force only from the beginning of the then parliament to the end of the next one.

Petrazhitsky, Leo Josifovich (1867–1931). Russian jurist, studied under Dernburg and became professor at St. Petersburg. His early work criticized the Roman basis of the projected German Civil Code as being opposed to the gospel of love in its legal sphere. He later developed a theory of law based on an emotional psychology. He was author of *Introduction to Legal Policy* (1900), *Introduction to the Study of Law and Morals* (1905), and *Theory of Law and the State in connection with a Theory of Morals* (1907). In 1921 he moved to Warsaw and contributed to the project of unifying the civil code of the revived Polish state.

Petri Exceptiones Legum Romanorum. A small legal manual in four books, probably composed in southern France in the twelfth century, drawing on Justinian, intended for the use of local judges.

Petrus (twelfth century). A Provençal jurist and author of a famous *Exceptiones legum Romanorum*, distinguished from earlier works by its systematic character and equitable tendency, setting out Roman law as a Provençal jurist wished it to obtain in his country. It was dedicated to a judge of Valence and designed to help him in his work.

Petrus Collivaccinus of Benevento (?–1220). Italian canonist, and cardinal, compiler of the *Compilatio III Antiqua* at the instance of Innocent III in 1210, and probably author of the *Summa Reginensis* on Gratian's *Decretum*.

Petrus Ravennas (or Petrus Thomas) (1448–1518). Civilian and canonist, teacher in German universities, and author of *Compendium juris civilis* (1503) and *Alphabetum aureum* (1508).

Petrus de Stagno (Pierre L'Estaing) (?–1377). Canonist and cardinal, papal legate to Bologna, and author of unpublished *Reportationes super Clementinis.*

Petty Bag Office. The principal office on the common law side of the old Court of Chancery, from which issued all original writs, various kinds of writs of error and of certiorari, commissions of charitable uses, commissions to seize escheated and forfeited lands, writs calling a new Parliament and various other kinds of writs. It was so-called because the proceedings in it, in which the Crown was concerned, were preserved in a little sack or bag instead of being enrolled on rolls as were other proceedings. The office of Clerk of the Petty Bag was abolished in 1888 and the powers attaching thereto are now vested in the Crown Office Department of the Central Office of the High Court. See also CHANCERY; HANAPER OFFICE.

Petty Jury. See JURY.

Petty larceny. The stealing of goods of a value not exceeding one shilling. The distinction between grand and petty larceny, which affected liability to the death penalty, was abolished in 1826.

Petty serjeanty. The former tenure of holding lands of the Crown by the annual service of some small article of warfare such as a sword, bow, arrow, or the like. The services incident to grand (q.v.) or petty serjeanty have not been abolished, but the tenure has, in other respects, been assimilated to socage or copyhold.

Petty sessions. A meeting of two or more justices of the peace, as distinct from a general or quarter sessions. Every sitting of justices of the peace or a stipendiary magistrate is deemed a petty sessions of the peace and is now known as a sitting of a magistrates' court.

Petty treason. Treason of a less serious kind than high treason, and including a servant killing his master, a wife killing her husband, or a secular person his prelate. Since 1828, offences hitherto deemed petty treason are deemed murder.

Pharmaceutical Society. The body in the United Kingdom which, by statute, regulates membership and discipline of the profession of pharmaceutical chemist. Through its statutory committee it exercises disciplinary powers and may direct removal from the register of the name of a member convicted of a criminal offence or guilty of misconduct which renders him unfit to have his name on the register. An appeal lies to the Divisional Court of the Queen's Bench Division.

Phenomenological jurisprudence. The term for the attitude of law of a group of thinkers, principally Edmund Husserl and Max Scheler, who are opposed to the rigid separation of form and matter in the Kantian and neo-Kantian philosophy. They believe that objects have an essence of their own which can be apprehended by intellectual intuition or emotional intuition. On this basis, some legal philosophers have developed various legal principles based, in their view, on the essence or nature of legal concepts and institutions. All particular instances, e.g. of contracts, have certain characteristic elements valid irrespective of time and place. This approach has value in allowing for divergences of detail between particular legal systems though recognizing underlying unity.

Philadelphia Convention. A meeting, proposed by the Annapolis Convention of 1786 and summoned by the Continental Congress to find ways of strengthening the federal government through revision of the Articles of Confederation. It met at Philadelphia from May to September 1787. The delegates included a large number of the leading men of the states. It was generally agreed that the central government needed to be strengthened and that a balanced government of legislature, executive and judiciary was needed, but there was much debate on the relative powers of federal and state governments and large and small states. The Convention finally adopted a draft constitution which was signed by only 39 of the 55 delegates but accepted by the Continental Congress. Thereafter ensued a struggle between the Federalists, who advocated a strong central government, and their opponents, but the draft was ratified by the state conventions and came into effect in 1789.

Phillimore, John George (1808–65). Eldest son of Joseph Phillimore (q.v.) was called to the Bar in 1832 and was later Reader in Civil Law and Jurisprudence at Middle Temple and Reader in Constitutional Law and Legal History to the Inns of Court. He was a jurist of some repute and published, *inter alia*, an *Introduction to the Study and History of the Roman Law* (1848), *Principles and Maxims of Jurisprudence* (1856), and *Private Law among the Romans* (1863).

Phillimore, Joseph (1775–1855). Became an advocate of Doctors' Commons in 1804 and practised successfully in the Admiralty and Ecclesiastical courts. In 1809, he became Regius Professor of Civil Law at Oxford and subsequently sat in Parliament. He served on the board of control for India from 1822 and became King's Advocate in 1834. He edited several volumes of admiralty and ecclesiastical reports.

Phillimore, Sir Robert Joseph (1810–85). Son of Joseph Phillimore (q.v.), became an advocate of Doctor's Commons in 1839 and a barrister in 1841. He became Admiralty advocate in 1855 and Queen's Advocate in 1862. He was also an M.P., 1852–57. He became judge of the Court of Admiralty and Dean of the Arches 1867–75, and thereafter sat as judge of the Probate, Divorce and Admiralty Division of the High Court till 1883. He was a great scholar and jurist, and pronounced many important judgments in maritime, ecclesiastical and probate and divorce law. He also sat on many royal commissions, edited Burn's *Ecclesiastical Law* and was the author of *Ecclesiastical Law of the Church of England* (1873), the important *Commentaries on International Law* (4 vols., 1854–61 and later editions), and of several other books.

Phillimore, Walter George Frank, Baron (1845–1929). Grandson of Joseph Phillimore (q.v.) and son of Sir Robert Joseph Phillimore (q.v.), last judge of the High Court of Admiralty, was called to the Bar in 1868, and specialized in ecclesiastical and admiralty cases. He became a judge of the Queen's Bench Division in 1897; he was promoted to the Court of Appeal in 1913. He retired in 1916 but as a peer (1918) sat in the House of Lords and the Judicial Committee, where his wide knowledge of law was seen to best advantage. As an international lawyer he acquired a European reputation, drafted the scheme for the League of Nations and was a member of the committee which framed the statute establishing the Permanent Court of International Justice. He edited his father's *Ecclesiastical Law of the Church of England* (2 vols., 1895), part of his father's *Commentaries on International Law* and many editions of Blunt's *Book of Church Law*. He has been characterized as one of the great lawyers of his generation.

Philosophical jurisprudence. The general name for theories of and about law propounded mainly, if not entirely, by scholars who were primarily philosophers and only subsidiarily, if at all, jurists, and which are characterized by abstractness and generality as compared with on the one hand historical jurisprudence and on the other the more down-to-earth and practical jurisprudence of analytical or positivist jurists and sociological jurisprudence. Philosophical jurisprudence, in general, is concerned to understand, organize, and criticize the ideal element in law, and to seek ideal standards by which to criticise law and direct lawmaking. It includes a great variety of mutually inconsistent and contradictory bodies of theory. The most notable single theory is the natural law (q.v.) theory, though this itself includes many divergent varieties of thinking. Other important theories are German transcendental idealism, particularly the legal theory of Kant, Fichte, and Hegel, followed by neo-Kantians such as Stammler and Del Vecchio, and neo-Hegelians such as Binder, who are all social-philosophical rather than metaphysical jurists; utilitarianism, developed by Bentham, Mill and their followers in nineteenth century Britain; positivism, developed by Comte, Spencer, Durkheim and Duguit; and Marxism and its derivatives in modern Communist legal theory. In the twentieth century, there has been a revival of forms of natural-law thinking, both neo-scholastic, as in the work of Geny and positivist-sociological, as in Duguit.

Philosophy of law. A phrase sometimes used as a synonym for jurisprudence (q.v.) in its narrower sense, but more commonly as a synonym for legal philosophy, for thinking and writing of a theoretical character on the basic principles of law and major problems of legal theory. It may be described as a branch of philosophy, related to the philosophies of morals, politics and society, concerned with examining the most general and fundamental problems of law, as distinct from an historical or practical inquiry as to what the law of a particular state is on some subject. It is concerned with such questions as the nature and purpose of law, the relations between law, religion, morals and politics, the values which are, or should be, promoted by law, the need for a sanction, why people should obey law, and what is meant by such fundamental terms as law, obligation, liability and the like.

Many of the great philosophers have discussed legal issues as part of and in relation to their general philosophic system but much legal philosophy has sprung from thoughtful lawyers enquiring and thinking about what underlies all the particular principles and rules and what courts and judges and legal systems exist to do.

H. Cairns, *Legal Philosophy from Plato to Hegel* (1949); C. J. Friedrich, *Philosophy of Law in Historical Perspective* (1958); R. Pound, *Introduction to the Philosophy of Law* (1954).

Phipps, Sir Constantine (1656–1723). Was called to the English Bar in 1684 and went to Ireland as Chancellor in 1710 but was removed in 1714.

Picketing. Posting men so as to communicate to workmen the existence of a trade dispute and to attempt to dissuade them from going to a place to do work or anything in connection with the work normally done there. If confined to communicating information or seeking to persuade or dissuade, it is legal as being 'peaceful picketing' but if the numbers of men are such or their conduct is such as to be calculated to intimidate or amount to watching, besetting, or persistent following, or actually causes injury or damage, it is illegal. In practice, picketing is frequently mass intimidation and hooliganism and is always an infringement of the liberty of workmen to work if they wish.

Pickford, William, Baron Sterndale (1848–1923). Was called to the Bar in 1874. His practice was chiefly in commercial cases and he took silk in 1893. In 1907, he became a judge of the King's Bench Division and, in 1914, a Lord Justice of Appeal. In 1918, he became President of the P.D.A. Division and received a barony, but, in 1919, returned to the Court of Appeal as Master of the Rolls, where he presided till his death. Though not a very learned lawyer he was a very sound judge. He was active in the movement for the unification of maritime law, which gave rise to the Maritime Conventions Act, 1911, and assisted the passage of the Carriage of Goods by Sea Act, 1924. He was a member and later chairman of the Dardanelles Commission of 1916–17.

Piepowder, courts of. See FAIRS AND BOROUGHS, COURT OF.

Pignus. In Roman law, the contract of pledge or the giving of a thing in security for a debt or the performance of some act. The property remained in the pledgor and the pledge was returnable when the debt was paid or the act performed.

Pigot, David Richard (1796–1874). Was called to the Irish Bar in 1826. He became solicitor-general in 1839 and Attorney-General in 1840, and was Chief Baron of Exchequer in Ireland, 1846-73.

Pilate, Pontius (?–*c*. A.D. 39). A Roman equestrian, appointed procurator of Judaea in A.D. 26. He incurred the enmity of the Jews by insulting their religion. He condemned Jesus Christ to be crucified. He was recalled to Rome in A.D. 36 to be tried for cruelty and oppression, particularly on the charge that he executed persons without proper trial. According to a tradition he killed himself in A.D. 39 on orders from the Emperor Caligula. His part in the crucifixion is variously explained. Some take the view that he reluctantly ordered it, despite a belief in Jesus' innocence, in order to avoid being accused of disloyalty to the Emperor, others that the Jews took advantage of his weak position in relation to the Emperor to secure the condemnation of Jesus.

Pilius or Pileus or Pillius Medicinensis (twelfth century). Italian jurist, author of glosses on the *Corpus Juris*, *Distinctiones*, a *Summa* to the *Tres*

Libri Codicis, and said to be the first glossator of the *Libri feudorum*.

Pillet, Antoine (1857–1926). French jurist and one of the chief figures in the theory of private international law after Savigny. He took as fundamental the unity of public and private international law; the choice of law was a question of securing the respect of sovereignties, the law of the sovereign to be preferred being the one having the most direct interest in the solution, though public policy might be an overriding factor. His major works were *Principes de droit international privé* (1903), *Traité pratique de droit international privé* (1923–24), and with Niboyet, *Manual de droit international privé* (1924).

Pillory. A form of punishment in which an offender was made to stand in a public place with his head and hands secured through holes in a frame, sometimes to be derided and pelted by the public and held up to them as an infamous person. It is very ancient and was used at common law, particularly for frauds on the public, who might have their revenge by pelting the offender, sometimes even with fatal results. In the seventeenth and eighteenth centuries, it was used as punishment for persons deemed guilty of minor political offences, libel, and sedition. Thus, it was inflicted on Prynne. It was finally abolished in 1837. See also STOCKS.

Pilotage. The supervision of the navigation of a ship by a pilot. It is a requirement that in waters to which orders made under the Pilotage Act, 1913, apply, most vessels may or must proceed only in the charge of a pilot. Pilots are provided by the pilotage authorities for particular districts. Where a ship is under pilotage the owner or master is not relieved of any liability for loss or damage caused by her wrongful navigation.

Pinkerton, Allan (1819–84). Born in Scotland, he became deputy sheriff of Kane County, Illinois, and then of Cook County (1846–50), and founder of the notably successful Pinkerton National Detective Agency, which specialized in railway theft cases. He worked in intelligence during the Civil War and after it helped to break up the Molly Maguires, a coal miners' terrorist organization, and to oppose labour unions during strikes in 1877.

Pin money. An allowance made by a husband to his wife, for her dress and other personal expenses. It might be secured by a settlement.

Pinto, Aaron Adolf de (1828–1908). Dutch jurist and judge, largely responsible for the adoption of the Netherlands penal code (1886) and the revision of the Netherlands code of civil procedure. Later, he was a judge and vice-president of the Supreme Court, and he also wrote extensively.

Pipe Rolls (or Great Rolls of the Exchequer). The record of accounts of the English Exchequer, comprising 676 rolls, covering the years 1131 and 1156 to 1833 complete, save for gaps in 1216 and 1403. They contain records of accounts rendered by those responsible for the royal revenues, particularly sheriffs, but also stewards and bailiffs of honours and other similar persons, and show feudal revenue from feudal dues, judicial fees and other sources, and also expenditure on royal administration. The rolls are accordingly important as sources for legal and administrative history. They are also authority for the descent of many of the great families of mediaeval England and for the history of English towns. The origin of the name is uncertain. Portions of the Pipe Rolls have been printed, particularly by the Pipe Roll Society, which was established in 1883.

Piracy. In international law (piracy *iure gentium*), any illegal act of robbery, detention, or violence committed against persons or property on the high sea for private ends by one private vessel against another, or by a mutinous crew or passengers against their own vessel. Piracy has been known from the earliest times and was practised by Mediterranean peoples, Vikings and Moors. Ships commissioned to act as privateers (q.v.) sometimes engaged in piracy after the end of hostilities, and the line between privateering and piracy was sometimes narrow. Piracy was tolerated by the states of North Africa from the sixteenth to the eighteenth centuries and was suppressed only in the early nineteenth century. It has occasionally occurred in the China Sea in the twentieth century. By international law, an act of piracy on the open sea is an international crime, the pirate being considered the enemy of every state, and liable to be captured, tried and punished by any state. By committing piracy a vessel loses the protection of its flag. A public vessel is not a pirate but if she commits unjustified violence redress may be demanded from her flag state. A privateer was not a pirate so long as her acts were authorized and confined to enemy vessels, and warships siding with insurgents should not be treated as pirates. Piracy has now been extended to similar acts on board, or against, aircraft. There is some authority for saying that insurgents or even states may commit piracy.

Piracy may be defined more widely or more narrowly in municipal law or by treaty, and may include, e.g. transporting slaves, insurgency, mutiny. In territorial waters, a ship is amenable to municipal law for these purposes.

The term piracy is also applied in municipal law to deliberate and large-scale infringement of copyright.

Pirate radio station. An independent radio station, broadcasting, usually from outside territorial waters, without authority, and on unauthorized wavelengths.

Pirhing, Ehrenreich (1606–79). German canonist and author of a classic *Universum jus canonicum secundum titulos Decretalium distributum novo methodo explicatum*, later abridged as *Synopsis Pirhingana* (1693), in which he sought to co-ordinate the principal elements of the canon law in a logical system.

Piscary, common of. A right of fishing with other persons in another man's pond, lake, or non-navigable river. It may be appurtenant to certain lands, or be in gross, i.e. distinct from any landed property.

Pit and gallows. A clause sometimes included in a grant of jurisdiction to a feudal lord, entitling him to maintain a dungeon and a gallows on which to hang offenders.

Pitcairn, Robert (1793–1855). Was admitted a Writer to the Signet in 1815 and was, for long, an assistant to Thomas Thomson in the Register House, Edinburgh. He collected from the records and published *Trials and Other Proceedings in Matters Criminal before the High Court of Justiciary in Scotland* (3 vols., 1833) covering cases from 1488 to 1625, and the principal source of information on older Scottish criminal law, and various other works.

Pithou, François (1543–1621). A French humanist jurist, editor of Julian's *Epitome* and of the laws of the Visigoths, author of a *De la grandeur des droits du royaume de France* (1587).

Pithou, Pierre (1539–96). A French humanist jurist, brother of the foregoing, editor of *Lex Dei* and the post-Theodosian *Novellae*, the *Leges Visigothorum* (1579), and author of an important *Recueil des libertés de l'Eglise gallicane* (1594), of an *Observations sur le Code Justinien*, and of works on canon law.

Place. The precise locus at which an act or event of legal significance takes place may be legally important, and in a minority of cases also be difficult to determine. Thus, in criminal law, the place of a crime may affect the questions which State or authority has jurisdiction to arrest or prosecute or punish; in civil law the place of a happening may affect the questions which state has jurisdiction, what system of law falls to be applied, and related questions of remedies, evidence and procedure. Place is accordingly frequently material in cases of conflict of laws.

Determination of the place of an act or event is difficult where human agency in one place has legally significant consequences elsewhere; if A in Scotland posts a defamatory letter to B in England, in which country is the wrong of libel committed? If P in Maryland fires a gun and the bullet kills Q in Virginia where was the tort, or crime, committed? There is much to be said for the view that the important place is the place where the consequences took place. Further complications arise in cases of events on a ship at sea, or in the territorial waters of another state, or in an aircraft passing through the airspace of another state.

In many statutory contexts, questions of interpretation of such phrases as 'public place' have arisen.

Place bills. Bills excluding Crown Office-holders from, or limiting their numbers in, the House of Commons, and intended to secure the independence of the Commons from Crown influence. The Act of Settlement of 1701 imposed total exclusion but the Succession to the Crown Act (1705) allowed the holders of pre-1705 offices to sit, provided they stood for re-election on appointment, a rule which lasted, with modifications, till 1926. Revenue officers were excluded in the eighteenth century and clerks of the main government departments in 1742. The administrative reforms of 1780–1830 abolished many places. The principle of place bills, however, played a part in excluding the civil service from politics. The law is now regulated by the House of Commons (Disqualification) Act, 1975, which lists the offices the holding of which disqualifies for membership of the House of Commons.

Placemen. The term in political history for M.P.s who hold office under the Crown. They were, in fact, more important as a link between Ministers and ordinary members than as agents of royal influence. They never constituted a party and commonly regarded their offices, which were frequently court or royal household posts and usually sinecures as rewards rather than as bribes for future support. Moreover, as most of the posts were permanent no single government could appoint to more than a few posts. In the mid-eighteenth century there were about 160, together with some 40 holders of administrative offices. Thereafter numbers declined to about 20 in 1830.

Placentinus (?–1192). An eminent glossator, founder of the law school of Montpellier, and author of *Summa Institutionum* (1535), *In Codicis Libros Novem Summa* (1536), and a treatise on actions, *De varietate actionum*.

Plaint. A written statement of a cause of action. Before real actions (q.v.) were abolished, a copyholder could only plead in respect of his copyhold land by so-called customary plaints, analogous to the writs used in real actions at common law, called, e.g. plaints in the nature of writs of right. The term is now used for the initial writ in a county court on which a summons is issued.

Plaintiff. In the common law, the party who complains and brings an action. In older procedure, the appellant was sometimes called plaintiff in error.

Planck, Gottlieb (1824–1910). German jurist and statesman, took a major part in the drafting of the German Criminal Code, Code of Civil Procedure, and of the German Civil Code, being a member of both the first and second drafting commissions. He wrote a major commentary on the Civil Code (7 vols., 1897–1902).

Planck, Johann Julius Wilhelm von (1817–1900). German jurist, most famous for the comparative study of civil and criminal procedure. His great works were *Das deutsche Gerichtsverfahren im Mittelalter* (1879) and *Lehrbuch des deutschen Civilprozessrechts* (1887–96).

Planiol, Marcel (1853–1931). French jurist and author with Ripert of a famous *Traité élémentaire de droit civil* (1899), many times re-edited.

Plant breeder's rights. Proprietary rights which may be obtained by breeders and discoverers of new varieties of plants from the Controller of Plant Variety Rights conferring on them a monopoly right of selling the reproductive material of the plant variety, of producing it for sale, and of certain other rights, for from 15 to 25 years. During the period of protection, a holder of plant breeders' rights must be in a position to produce to the Controller reproductive material which is capable of producing the variety to which the rights relate. Infringement is actionable giving rise to claims for damages, injunction, account of profits, or other appropriate remedy.

A Plant Variety Rights Tribunal exists to determine disputes and it may arbitrate in infringement disputes.

Plato (*c.* 428–348 B.C.). Greek philosopher, disciple of Socrates until the latter's death in 399 B.C., and subsequently, founder and head of the Academy, an institute for the pursuit of philosophy and scientific research. He was author of a large number of dramatic dialogues, an early group including *Apology*, *Symposium*, *Phaedo*, and *Republic*, and a late group including *Sophistes*, *Politicus*, *Philebus*, *Timaeus*, and *Laws*, in many of which Socrates appears as a leading character. The later dialogues show less dramatic power but more developed critical acumen and more mature thought. Most are concerned with the discussion of particular ethical problems and Plato is the earliest, and one of the greatest, philosophical writers.

The Academy was active in mathematics and science, but also in jurisprudence and practical legislation, and sent out legislators to many states and, in his *Laws*, Plato provided a model for constitution-making and legislation by members of the Academy who might be called on to assist in these activities. The work is rich in legal and political wisdom and may have influenced later Greek law and, through it, Roman law.

Plato regarded law as a product of reason, and identified it with nature itself. It sought to be the discovery of true reality and the philosopher-king will prescribe laws based on the ideal laws which he has perceived in the world of reality; the end of law was to produce men who were completely good, and to assist in the maintenance of an ordered society, and the method of attaining this was a benevolent dictatorship.

In relation to legislation, Plato's theory was that the legislator by reason alone could formulate rules adequate for the needs of the community; he accepted the traditional distinction between written and unwritten law, between human and divine law. Some of his specific ideas anticipate Bentham. The legislator had to have in view freedom, the unity of the state, and temperance among the citizens. He disapproved of the Athenian jury courts and proposed a much better system, with provision for appeals.

To Plato, Justice was a general principle of human conduct and not merely a rule for the courts. The idea of rendering each man his due was to him inadequate as a guide for human behaviour. Justice was a harmony of the state, in which each individual performed the necessary work which fell to his lot in the properly organized state. Plato's discussion of the nature of justice in the *Republic* may be taken as the beginning of legal philosophy. Quite apart from all detailed and practical discussion Plato's examination of fundamental legal issues is, everywhere, stimulating and thought-provoking.

Plato's Academy continued to exist down to A.D. 529 when, like other Athenian schools of philosophy, it was closed by Justinian. Platonism was revived by Plotinus in the third century, and it influenced early Christian philosophy and, through Augustine, mediaeval western Christian thoughts; indeed, down to the thirteenth century, Plato dominated European thinking, and thereafter his influence in ethics, religion, and philosophic thought has never been extinct.

A. E. Taylor, *Plato: The Man and His Work*; G.

C. Field, *The Philosophy of Plato*; N. R. Murray, *The Interpretation of Plato's Republic.*

Plautius (first century A.D.). Roman jurist, author of works on praetorian law, now lost, but highly esteemed by later jurists, and the subject of commentaries by Neratius, Javolenus, Paulus, and Pomponius.

Plea. Originally any legal proceeding. Hence, civil actions between subjects were called common pleas and criminal cases were called pleas of the Crown. Similarly, the Court of Queen's bench had a plea side, dealing with ordinary actions, and a Crown side, concerned with criminal cases, and the Court of Exchequer had a plea side as distinct from its revenue side. The term was also used in civil cases as equivalent to the modern statement of defence, a statement of facts put forward as an answer to the plaintiff's declaration. Pleas might be dilatory, comprising pleas to the jurisdiction, pleas in suspension and pleas in abatement, or peremptory, comprising pleas in bar or pleas which offered a defence of substance to the claim, whether by traverse or by confession and avoidance.

In modern usage, a plea is any contention put forward, particularly by a defendant by way of answer to the plaintiff's declaration, or in criminal cases a statement of being guilty or not guilty, or a special plea such as *autrefois acquit* (q.v.).

Plea-bargaining. The practice accepted in criminal courts in some jurisdictions, whereby the prosecutor, in return for an admission of guilt of the crime charged or a lesser variety of the crime charged or some part of the crime charged, will modify the charge by dropping other crimes charged or offering no evidence on them, or use his influence to secure a lighter sentence than might otherwise have been imposed. Theoretically undesirable, the practice has some advantages in securing admissions in cases where it might be difficult to prove the charge laid against the accused. In U.K. courts, it is regarded as improper and the trial judge should never be implicated in any such agreement. It is regarded in the U.S.A. as permissible.

Plea in bar of trial. In criminal procedure, a preliminary plea, such as of insanity at the time of trial, which, if upheld, prevents further proceedings.

Plea in law. In Scottish civil procedure, concise propositions of law, without reference to authorities therefor, appended by parties to their summons, defences or other pleading, indicating the points of law on which the claim or defence is being maintained. Later argument and evidence is confined to matters relevant to pleas tabled.

Pleas Roll. The records of the proceedings of mediaeval English courts. They comprise a number of pieces of vellum, attached together at the top, not sewn into a continuous strip. From probably 1234, there were separate sets of rolls for the two different courts, the justices at Westminster—*de banco* rolls, and the justices following the King—*coram rege* rolls. Later classification of plea rolls has been rather arbitrary, depending partly on the places where the rolls happened to be kept. *Curia regis* rolls, consisting of cases heard before the Bench or *coram rege* date from about 1181; Eyre rolls and Assize rolls from about 1189; *Coram rege* and *De banco* rolls from 1234; Exchequer plea rolls from 1236; Chancery rolls (non-judicial) from 1199; *Rotuli Parliamentorum* from 1290. Various plea rolls have been edited. They give first-hand information of the working of the *Curia regis* in the twelfth and thirteenth centuries. On the Plea Roll were entered the pleas put in by counsel for the parties and the procedural steps, with the decisions on the points raised. By Bracton's time, the lawyers settled the pleadings in the form in which they entered the plea roll, the enrolment being formal. In many cases it is possible to collate the entries in the Plea Rolls with the reports of the same cases in the Year Books (q.v.).

Plead. To make an allegation in a case, particularly one in answer to the previous pleading of the other party, as by denying facts stated therein; also to tender a plea of guilty or not guilty in a criminal prosecution; also, generally, to argue a case in court.

Pleader. In general, one who pleads in court on behalf of another; more correctly, one who draws pleadings. When common law pleading was very technical special pleaders (q.v.), who were neither counsel nor solicitors, drew pleadings.

Pleading. A document in which a party to a civil proceeding formulates his claim as a basis for a hearing by a court. The word is also used of the art of framing any such document and of the art of arguing orally in court on the basis of such pleadings. Originally in England, pleadings were oral. When the case was called each party made an oral statement of the facts as he saw them, and the judge regulated this oral discussion so as to ascertain the real points of controversy. The parties were then said to be 'at issue' and the issue was decided, if a matter of law, by the court, and if a matter of fact, tried according to the mode then competent. The allegations were recorded by an officer of the court on a roll, the record, which was the official record of the pleadings. Subsequently the pleader entered his statement directly on the roll, and allowed his adversary to see it to frame his reply. Later still pleaders delivered their pleas already written, but this practice was still governed by the former rules and they had to come to an issue in written proceedings as formerly they had done orally.

Prior to the Judicature Act, 1873, the plaintiff delivered his statement of claim and the defendant a defence, to which the plaintiff might reply, the defendant deliver a rejoinder, the plaintiff a surrejoinder, the defendant a rebutter, and the plaintiff a surrebutter. In modern practice, the pleadings normally consist only of statement of claim, defence, and sometimes reply.

The function of pleadings is to give fair notice of the claim and to define the issues on which the court has to decide, of law or fact or both, thus enabling the court to determine the mode of trial. Pleadings must contain a summary statement of the material facts on which the party founds his claim or defence, but not of the evidence whereby those facts are proposed to be proved, nor of rules or inferences of law.

Every allegation of fact in a pleading which is not expressly or by necessary implication denied or not admitted by the other party's pleading must be taken to be admitted. A general denial of the other party's case is insufficient but the allegations must be dealt with specifically. Particulars must be given of important facts such as allegations of adultery or fraud or negligence.

The detailed rules regarding pleadings are determined by the rules of procedure of each particular court supplemented by the practice of that court.

In Scotland, the initial pleadings in most cases comprise a summons, containing conclusions (requests for the remedies claimed), a condescendence or statement in numbered paragraphs of the facts alleged by the pursuer to justify his claim for the remedy, and pleas-in-law, and defences, containing the defender's answers and his pleas-in-law. Summons and defences are combined in a single document, the record, on which argument and hearings take place. The function of pleadings is the same as in England.

Oral pleading today is an application of the art of rhetoric and the principles of logic to stating the contentions of a party, in fact and law, most clearly and effectively, and thereby seeking to persuade the court or tribunal of the rightness of the party's case.

Pleas of the Crown. In English law, pleas in which the Crown had a financial interest, as by exacting a fine, as distinct from common pleas or claims between subjects, a distinction similar to but not quite coincident with that between criminal and civil cases. The older textbooks on criminal law were all entitled Pleas of the Crown. Originally, most wrongs came to be emendable in money, but the most serious were corporally punished and not emendable save by the King's special grace. By Norman times, there were lists of royal pleas, which differed from county to county, but the Normans gave a more precise meaning to the idea of pleas of the Crown: they were not merely pleas in which the Crown had a pecuniary interest but conduct held to be committed against the Crown. Glanvill listed as pleas of the Crown treason, breach of the peace, homicide, arson, robbery, rape and counterfeiting the King's seal or coinage. Larceny was a royal plea only in some places, elsewhere a plea of the sheriff. As the essential concern of the King's Bench was with matters affecting the King, the Pleas of the Crown fell within the jurisdiction of this court, and for that reason the King's (or Queen's) Bench has always been concerned with criminal law in a way the other courts, or Divisions of the High Court, have not been.

In Scotland, the pleas of the Crown are said to be murder, robbery, rape, and wilful fire-raising, probably because all were at one time within the exclusive jurisdiction of the High Court and were punishable capitally, but the phrase is not native to Scotland and is of no significance today.

F. W. Maitland, *Select Pleas of the Crown*; Sir M. Hale, *Summary of the Pleas of the Crown* (1678); *History of the Pleas of the Crown* (1736); W. Hawkins, *Pleas of the Crown* (1716–21); Sir E. H. East, *Pleas of the Crown* (1803).

Pleas of the sword (*placita spatae vel gladii*). Rights over all men enjoyed in certain circumstances by the Dukes of Normandy, comparable in many ways to the pleas of the Crown enjoyed by the Kings of England. They seem originally, to have been few in number—assault on a highway leading to a city or ducal castle, hamsoken (q.v), and an attack on a man while at the plough, offences against the duke's money and offences against his writs of protection—but gradually the list expanded.

Plebeians. In Rome, the general body of citizens, not belonging to the privileged patrician class. Originally plebeians were excluded from the Senate and all public offices, and until the *Lex Canuleia* (445 B.C.) could not marry patricians. The plebeians elected their own officers (tribunes and plebeian aediles) and had their own assembly (*concilium plebis*). In 287 B.C., a *Lex Hortensia* made *plebiscita*, measures passed by the *concilum plebis*, binding on the whole community.

Plebiscite. The means of securing an expression of view on a question from the general community of a territory by having them vote for or against a question. It differs from the expression of view at a general election made by electing representatives to a legislative assembly. In the Middle Ages, some inhabitants of ceded territory sometimes protested against the cession. Later there are instances of including in treaties of cession a clause giving a right of withdrawal from the ceded territory to all who were unwilling to accept allegiance to the new lord. The French revolutionary National Assembly seems first to have proclaimed that annexation required the consent of the population transferred, and this principle was applied in several cases in the 1790's and again in 1848 and 1860, and later in Italian states. The device of the plebiscite was later applied to other circumstances as that in Norway in 1904 on whether to break with Sweden or not. Plebiscites have several times been resorted to in France to secure popular approval for constitutional changes. See also REFERENDUM.

Plebiscitum. In Roman law, a resolution passed by the plebs, originally binding only the group which had issued it in the *concilia tributa plebis*. But from the fifth century B.C. many *plebiscita* must have been recognized by the whole community and the *Lex Hortensia* (*c.* 287 B.C.) gave *plebiscita* the force of law. By the later Republic, this was the normal form of legislation.

Pledge. The transfer by one person to another of the possession of certain goods to be held by the latter as security for the performance by the former of some obligation to pay or perform, which being performed, the pledge must be restored. Negotiable instruments and documents of title may also be pledged. The term is also used of the thing pledged. The commonest case of pledge is the transaction with a pawnbroker by which some article is pledged as security for a loan of money and either later redeemed or sold under statutory power.

Pleine jurisdiction (plenary jurisdiction). An aspect of the jurisdiction of the French Conseil d'Etat and of the Court of the European Communities, which enables the court, in certain cases, to go into the merits of the parties' cases and to substitute its own judgments for those of the Communities' institutions. The main instances are in connection with violations of the treaties by member states, cases of non-contractual liability amounting to *faute de service*, the settlement of disputes between the Communities and their employees over contracts of employment, and appeals against fines and other pecuniary penalties.

Plenipotentiary. A person having full powers, a term usually used of diplomatic representatives. A minister plenipotentiary and envoy extraordinary is a diplomatic agent of the second class, ranking after ambassadors and above ministers resident.

Plessey* v. *Ferguson (1896), 163 U.S. 537. As part of the continuing discrimination in the southern states agains negroes, a Louisiana statute required railroads to provide 'equal but separate' accommodation for white and black passengers and forbade intermingling. The Supreme Court, with one dissent, held that provision of separate but equal facilities did not conflict with the Thirteenth or Fourteenth Amendment. The decision represented the law until *Brown* v. *Board of Education* (1954) (q.v.).

Plimsoll, Samuel (1824–98). An M.P. and social reformer who attacked the practice of shipowners sending unseaworthy and overloaded ships to sea. He secured the passage of legislation providing for the marking of a load line on the hulls of ships, indicating the maximum depth to which the ship can safely be loaded. This has now been adopted by international convention.

Plowden, Edmund (1518–85). Was called to the Bar about 1540, sat in Parliament, and acted as a justice of gaol delivery. His professional advance was prevented by his adherence to Roman Catholicism, but he attained a great professional reputation, was esteemed possibly the most learned lawyer of his time, and published, in law French, *Commentaries* or Reports of cases (1571 and 1579). These were notable for their accuracy and concentration on the decisive issues, were repeatedly reprinted, and translated and have always been highly esteemed. Among other works was a Treatise on Succession by Mary Queen of Scots. He is said to have declined an offer of the Chancellorship from Elizabeth I.

Plucknett, Theodore Frank Thomas (1897–1965). Legal historian, who taught at Harvard and London. His major work was the *Concise History of the Common Law* (1929) and he edited or supervised many volumes for the Selden Society. Smaller works include *Statutes and Their Interpretation in the Fourteenth Century*, *Edward I and Criminal Law*, *Early English Legal Literature* and the *Legislation of Edward I.*

Plumer, Sir Thomas (1753–1824). Called to the Bar in 1778, he was of counsel for the defence of Warren Hastings and appeared in many of the State trials at the turn of the century. He became Solicitor-General in 1807, Attorney-General in 1812 and, in 1813, the first Vice-Chancellor of England. In 1818, he became Master of the Rolls. He was an able lawyer and an outstanding judge, his judgments being noted for their length.

Plunket, William Conyngham, Lord (1764–1854). Studied law in London and Dublin, was called to the Irish Bar in 1787, and sat in the Irish Parliament and opposed the union of the Irish and British Parliaments. He became Master of the Rolls in England in 1827 but was badly received and resigned. He was then appointed Chief Justice of the Common Pleas in Ireland and in 1830 Lord Chancellor of Ireland but retire in 1834. He held the office again 1835–41.

Pluralism. A term for the various doctrines which in opposition to the traditional theory of State sovereignty emphasize the importance of associations and groups smaller and more specialised than the State, such as churches and trade unions. Pluralist thought owes much to Mill's individualism and T. H. Green's idealist doctrine of self-realization, and, rather later, to Gierke, Maitland, Durkheim and others, who investigated the nature and development of corporations. It was developed, in the twentieth century, extensively by J. N. Figgis, Harold Laski, Ernest Barker, A. D. Lindsay, and others in England and, in a special way, by Duguit in France. The English scholars contend that the state should be looked on as an organization of groups and that government should not trench on the spheres of these groups. The more extreme pluralist views deny the sovereign power of the State or would give it only residual functions. This conflicts with the traditional view of state sovereignty held by the analytical jurists from Bodin and Hobbes onwards. There is much of value and importance in pluralist doctrine though no theorist has been able convincingly to eliminate the State from his theory.

The term is sometimes used, alternatively to dualism (q.v.), for the theory that international law and systems of municipal law belong to distinct legal orders, neither being subordinate to the other.

Poaching. The unauthorized taking of game or fish from private property or from a place where trapping, shooting, or fishing are restricted. In the past, poaching was frequently done by poor people for food, but today it is frequently done for sport or profit. It was long regarded as a major crime and heavily punishable, and sometimes, in the eighteenth and nineteenth centuries, gangs of poachers fought with gamekeepers. The most objectionable features of some modern poaching are the cruel and destructive methods employed, such as poisoning rivers.

Pocket boroughs. The general name, under the unreformed British parliamentary system, for boroughs where a landowner owned all, or at least a majority, of the lands having a vote by virtue of being held by burgage tenure, and accordingly could nominate the member to sit for the borough.

Pocket veto. A method whereby the President of the U.S. may prevent the enactment of legislation of which he disapproves, competent under Art. I, sec. 7 of the Constitution, whereby any bill presented to the President within 10 days before Congress adjourns shall not be a law, unless the President signs it. It was first used by Madison and most frequently by F. D. Roosevelt.

Podesta. The chief judicial and military officer of mediaeval Italian communes, an office established by the Emperor Frederick Barbarossa to control turbulent Lombard cities. Later, however, communes began to elect their own podesta, frequently from noblemen from another city. After the thirteenth century, the office declined in importance and in Florence, by the fifteenth century, the functions were judicial. The title was used in the nineteenth and twentieth centuries for the mayors of Italian towns.

Poinding (pronounced 'pinding'). In Scots law, the form of diligence (q.v.) by which a creditor holding a decree of court against a debtor can make the debtor's goods available for the satisfaction of his decree. The debtor is first charged, i.e. judicially called upon to pay. If he does not do so a messenger-at-arms or sheriff officer proceeds to the debtor's home or business, inventories the goods there and sets a value on each, and reports to the court, which may then grant warrant to sell the goods by public auction or adjudge them to the creditor.

Poinding of the ground. In Scots law, the form of action by which a creditor having a debt secured over land may attach all the goods on the lands burdened in order to secure payment, decree in which authorizes the poinding (q.v.) and sale of goods on the debtor's ground.

Point of law. An issue or matter of the true interpretation of some principle or rule of law in the circumstances alleged or established in evidence. Thus, it is sometimes provided that objection may be taken in point of law, or that appeal lies on a point of law.

Point of order. In parliamentary procedure, an objection to the proceedings raised while another member is speaking.

Police. The modern British police forces originated in the petty or parish constable appointed to preserve the peace in his area and to execute the orders and warrants of the justices of the peace. It became customary for the justices to swear in a constable and he came to be recognized as a ministerial officer of the Crown. The office of constable was unpaid and unpopular and persons nominated to the office frequently paid substitutes to act for them. There also existed in towns the town watch, though watching was not everywhere enforced. By the eighteenth century, the system, particularly in urban districts, was in disorder. In some towns, improvement commissioners appointed under local Acts were responsible for police. In London, some of the justices, notably the Fielding brothers, organized bodies of paid constables of which the Bow Street runners were one, and at the end of the eighteenth century, seven police offices were established near London each employing a few constables. Police forces were established for the metropolitan police district in 1829 and the City of London in 1839. In 1833, the Lighting and Watching Act permitted the appointment in parishes of inspectors responsible for establishing a parish watch, while in 1847 the Town Police Clauses Act allowed towns to make provision for police in local Acts, and later in the century various statutes provided for the establishment of police forces but these preserved the relationship between the constable and the justices.

The modern police force is not a national one but a number of forces exist for defined areas, counties and combined areas, maintained by police authorities. The number of police authorities has been much reduced by amalgamations over the last century. The police authority for the metropolitan police district is the Secretary of State and for the City of London it is the Common Council of the City. The chief officer is the Commissioner of Police of the Metropolis. Elsewhere, police authorities are the police committees of county councils, one-third of the members of each committee being magistrates and the rest members of the council. The chief officers are the chief constables. The Home Secretary has certain powers such as the appointment of chief constables, the prescription of pay, uniforms and other general regulations.

A policeman's authority arises from his having sworn to serve the Crown in the office of constable and when on duty he acts as an officer of the Crown and a public servant. His powers are exercised by virtue of his office and on his own responsibility, save when acting in execution of a warrant. Accordingly, chief constables and other superior officers are not, save under statute, liable for the actions of a subordinate, and constables are not employees of the police authority, nor of the Crown.

A constable of any police force in England or Wales has the jurisdiction, powers, privileges, and duties attaching to his office throughout England and Wales.

The main functions of constables continue to be the maintenance of the Queen's Peace and good order, the prevention of crime and the protection of life and property. A constable has the same powers of arrest as a private person but with additional statutory powers, exercisable in many cases without warrant. To effect arrest or execute search warrants he has powers of entry but no general statutory power to search a person arrested but he may be authorized to search premises for stolen goods.

Apart from constables appointed by police authorities, constables may be appointed by the British Railways Board, harbour authorities, and certain other bodies.

L. Radzinowicz, *History of English Criminal Law.*

Police Complaints Board. A body of not less than nine persons appointed by the Prime Minister for periods of three years, set up in 1976, to deal with complaints by members of the public against police officers not above the rank of superintendent. It is not concerned with matters of internal discipline nor with complaints that criminal offences have been committed by a police officer. A police officer convicted of an offence may appeal to the Home Secretary.

Police court. The former name for a court held by a stipendiary or metropolitan stipendiary magistrate and now known as a magistrates' court. In Scotland, police courts are now known as district courts.

Police magistrate. A stipendiary, or metropolitan stipendiary, magistrate.

Police power. The title given in American constitutional law to the governmental power to regulate matters of safety, health, welfare, and morals for the general public good. The Supreme Court held, in the *Charles River Bridge* case (1837) (q.v.), that, though Congress had the power to regulate interstate commerce, a state's police power entitled it to make reasonable laws for regulatory purposes even though such might not be expressly authorized by the Constitution. After the passage of the Fourteenth Amendment the Supreme Court frequently, under the 'due process' clause, struck down legislation creating regulatory functions, but latterly the Court has taken a much more liberal view of such regulation. Moreover more recently the police power has been extended beyond safety, health and morals to the conservation of natural resources, the protection of the public against fraud and waste, and other purposes not so narrowly limited as in earlier interpretations.

Policy (Greek politeia, government). The general considerations which a governing body has in mind in legislating, deciding on a course of action or otherwise acting. The policy of a statute, rule of law, decision or other conduct is its intended or probable effects. Many statutes and administrative acts are done in pursuance of a particular social, political, or economic policy.

Policies of Insurance, Court of. A court consisting of the judge of the Admiralty, the recorder of London, two doctors of the civil law, two common lawyers and eight merchants, all appointed by the Lord Chancellor under an Act of 1601 to decide summarily disputes between merchants arising out of insurance policies. Appeal lay to the Court of Chancery. The court fell into disuse in the eighteenth century and was formally abolished in 1863.

Policy of insurance. An instrument evidencing a contract between two parties, one of whom is normally a company engaging in insuring risks as a business, that that one will pay to the other a determinate sum of money or will indemnify him in the event of the occurrence of a harm of a stated kind, in return for a payment by the other of a sum, the premium, calculated by reference to the chance of the occurrence of such a harm. According to the kind of risk run contracts of insurance are distinguished into marine, accident, fire, motor, aircraft and other kinds of insurance.

Policy, public. A general principle, amounting to taking into account the general public interest or well-being of the community, which courts have held justify their refusing to give effect to certain kinds of transactions, such as agreements in undue restraint of trade, or trading with the enemy in wartime, even though such agreement be not in contravention of any rule of positive law.

Political asylum. Taking refuge in a foreign embassy or consulate or on board a foreign ship in order to escape from the jurisdiction of the local state because of fear on the refugee's part that he is going to be imprisoned or oppressed because of his political beliefs or activities. Such action frequently gives rise to a claim for surrender or for extradition, which is met by the defence of 'political offence' (q.v.). It is of particular importance in localities where the state authority is intolerant of political dissent or subversive or counter-revolutionary activities. In Eastern Europe, much which in Western Europe would be permissible dissent from government policy is treated as subversive or criminal. It is doubtful if there is a general international law concept of asylum. At least two questions are involved, the immunity which attaches to a foreign embassy or ship and prevents a refugee from being removed by force, and the possible responsibility which the sheltering state may incur to the territorial state by granting refuge. On the latter point it seems that asylum may be granted on humanitarian grounds to protect political offenders from mob violence or an administration of justice corrupted by political aims.

Political offences. The right of one state to claim the extradition (q.v.) of an offender against its laws, who has taken refuge in another country, suffers exception in the case of 'political offences', but this concept is an indefinite one and not authoritatively defined. One approach to the question is to consider whether the facts alleged amount to a known crime ignoring political motivation; another is to exempt from extradition persons politically motivated, whatever the crime charged. On the former view the assassination of a tyrant is an extraditable offence; on the latter view it is not. The view of U.K. courts is that a political offence is an offence committed in connection with a political disturbance.

Since 1945, it has been held that refugees could contend that their extradition was sought for a political purpose, even though they had not been involved in any political disturbance or rising. A refugee may not argue that the state requesting extradition is not acting in good faith nor that, if surrendered for an ordinary offence, he will be tried and dealt with for a different offence of a political character, since by the principle of speciality, to which effect is given in extradition treaties to which the U.K. is a party, the state requesting extradition may not try a person extradited save for the offence for which he was extradited, unless with the consent of the extraditing state.

Poll. Originally a head, hence a counting of heads. Thus a taking of the votes of persons entitled to vote, by ballot or by proxy, as compared with counting the votes of persons present. Hence, U.K. electoral laws speak of 'polling stations'. When a person has to be chosen, or a course of action adopted, by a majority of those entitled to vote, there is a right at common law to demand a poll. The articles of association of companies normally provide for this.

Poll tax. A tax of uniform amount on each individual or head. It has been levied several times in England, the first in 1377 and the most famous being that of 1380, which gave rise to Wat Tyler's peasants' revolt in 1381. At other times the amount varied according to the rank of the party assessed. Money was raised in this way several times in the seventeenth century. Its use caused discontent in the United States and was declared unconstitutional by the 24th Amendment (1964).

Pollexfen, Sir Henry. (1631–91). Called to the Bar in 1658, he early became prominent. He appeared on the constitutional side against the Crown in all the great trials of James II's reign and was deemed an honest and learned lawyer. On the Revolution Settlement, he became Attorney-General (1689) and Chief Justice of the Common Pleas (1689–91). His reputation is of an honest and learned lawyer, but not a great one. He is better remembered for his reports, though they are very defective.

Pollicitation. In Roman law, a promise, particularly in classical law, an undertaking to a municipality to make it a gift, in respect of an honour conferred or to be conferred, usually considered irrevocable even without acceptance.

Pollock, Sir Ernest Murray, Viscount Hanworth (1861–1936). Grandson of Sir J. F. Pollock (q.v.), Chief Baron of Exchequer, and cousin of Sir Frederick Pollock (q.v.) was called to the Bar in 1885. He acquired a fair practice, took silk in 1905, and became an M.P. in 1910. In 1919, he became Solicitor-General and, in 1922, Attorney-General. He had never been in the first rank at the Bar and was deemed fortunate to become a Law Officer, yet in 1923 he unexpectedly became Master of the Rolls. He was satisfactory as a judge. In 1926, he became a peer. His greatest service was in the collection and preservation of mediaeval documents, in his capacity as keeper of the public records. He published a life of his grandfather, Lord Chief Baron Pollock, in 1929.

Pollock, Sir Frederick, Bart. (1845–1937). Grandson of Chief Baron Sir Jonathan Pollock (q.v.), was called to the Bar in 1871. He became Corpus Professor of Jurisprudence at Oxford 1883–1903. He drafted what became the Partnership Act, 1890, and from 1895 to 1937, was Editor in Chief of the Law Reports. He was one of the founders of the *Law Quarterly Review* and edited it from 1884 to 1919. He became a Privy Councillor in 1911 and a K.C. in 1921, Judge of the Admiralty Court of the Cinque Ports and Chairman of the Royal Commission on the Public Records.

He was a poor lecturer who drew small audiences, but was outstandingly successful as a writer, in systematizing and presenting in readable form branches of the law. His works were among the first scientific legal treatises. His writings included *Principles of Contract* (1876), *Digest of The Law of Partnership* (1877), *Essays in Jurisprudence and Ethics* (1882), *The Law of Torts* (1887), *A First Book of Jurisprudence* (1896), *The Expansion of the Common Law* (1904), *The Genius of the Common Law* (1912), *The League of Nations* (1920) and, with R. S. Wright, *Possession in the Common Law* (1888) and, with F. W. Maitland (q.v.) *The History of English Law before the Time of Edward I* (1895). *Contract* and *Torts* in particular were standard books and went through many editions. The correspondence of many years with Oliver Wendell Holmes (q.v.) was published as the *Holmes-Pollock Letters.*

Pollock, Sir Jonathan Frederick (1783–1870. Was called to the Bar in 1807 and attained the leadership of the Northern circuit. In 1831, he became an M.P. and also chairman of the common law commission appointed in 1828. He was Attorney-General in 1834 and again 1841–44, when he became Chief Baron of the Exchequer, an office which he held till 1866. On his retirement he was made a baronet. His opinions were scholarly and felicitious but he was not an outstanding judge. His brother George (1786–1872) became a field marshal, his eldest son William Frederick (1815–88) Queen's Remembrancer and Senior Master of the High Court, and his grandson was Sir Frederick Pollock (q.v.).

Lord Hanworth, *Lord Chief Baron Pollock.*

Pollock* v. *Farmers' Loan & Trust Company (1895), 158 U. S. 601. Congress, in 1894, enacted a federal income tax law, progressive in its incidence, and this was challenged as unconstitutional, as being a direct tax, which would have to be collected by apportioning the amount to be collected among the states according to population. In this case, the Supreme Court held that a tax on rents was the same as a tax on the property from which the rents were collected, and was therefore a direct tax, and that a tax on income was also a direct tax. The decision resulted eventually in adoption of the Sixteenth Amendment authorizing collection of income taxes without apportionment.

Pollution. The discharge, into water or the atmosphere, of something which detrimentally affects the natural state of either. Pollution of a river may be actionable as a nuisance, but a legal right to pollute the water may be acquired by grant, statute, or prescription though such right does not permit anything prejudicial to public health, amounting to a public nuisance, or in contravention of statute. Pollution may be punishable criminally under various statutes. Pollution of the atmosphere is similarly actionable and criminal. Noise may also be punishable as pollution. Statute makes pollution of territorial waters by the escape of oil an offence.

Polyandry. The system under which a woman is married at the same time to two or more men. It is very rare and distinguishable from the practice obtaining in some societies whereby several men have privileged sexual access to a woman, a custom often associated with kinship and hospitality.

Polygamy. The practice of a person having more than one spouse at one time, including both polyandry and polygyny (qq.v.). European legal systems, influenced by Christianity, have not permitted polygamy, but have increasingly recognized the legal validity of polygamous marriages entered into in countries which recognize such practices.

Polygyny. The system under which a male is married to more than one woman at the same time. It has been found in many societies, being permitted by Islamic law (maximum of four wives) and among Mormons.

Pomeroy, John Norton (1828–85). Was admitted to the Bar in 1851 but had little practice and turned to legal writing, joining New York University Law Faculty in 1864. He published *An Introduction to Municipal Law* (1864), *An Introduction to the Constitutional Law of the United States* (1868), and *Remedies and Remedial Rights* (1876), which was later republished as *Code Remedies.* In 1878, he moved to Hastings College of Law, where he wrote his *Treatise on Specific Performance of Contracts* (1879) and his greatest work, *Treatise on Equity Jurisprudence* (1881–83). His writings had great influence on bench and bar, and on the development of the law.

Pomponius, Sextus (second century. A.D.). A Roman jurist and one of the most prolific writers—a compiler rather than an independent thinker. His major work was a commentary on the Edict (*c.* 150 books) instructed by Hadrian, but known only by quotations in later writers. He also wrote two text-books on the *ius civile, Ad Sabinum* (36 books) and *Ad Q. Mucium* (30 books), *Variae lectiones, Epistulae,* and monographs on *Senatus consulta, Fideicommissa* and *Stipulationes,* and a valuable *Liber singularis enchiridii* on the history of legal sources and legal science. His works are extensively quoted in the *Digest.*

Ponsonby, George (1755–1817). Was called to the Irish Bar in 1780 and sat in the Irish and United Kingdom Parliaments. He was Lord Chancellor of Ireland 1806–7 and later led the opposition in the U.K. Parliament.

Poor law. Statutes to repress begging were passed in England from 1388 but the 'old poor law' was created by the Poor Relief Act, 1601. The relief of the poor was the charge of each parish which had to appoint overseers of the poor, with the task of setting to work all paupers and their families and of taxing all occupiers to raise a parish fund for the relief of the parish poor. The system gave rise to the rules about settlement and removal under which all persons likely to resort to a parish were compellable to return to the parish where they originally belonged so as to relieve each parish of the duty of maintaining other persons than their own. Parishes tended to unite and combine for the administration of poor relief and statute grouped parishes into poor law unions each under the control of elected boards of guardians. In 1782, parishes were authorized to appoint guardians in lieu of overseers and could enter into unions for the better accommodation and maintainance of the poor. The 'new Poor Law' of 1834 placed the general management of the poor and of relief funds under the control of the Poor Law Commissioners. Their functions were transferred successively to the Poor Law Board (1847), the Local Government Board (1871) and the Ministry of Health (1914). From 1925, the poor-rate became part of the consolidated general rate for each area. In 1930, the functions of overseers and boards of guardians were transferred to counties and county boroughs which were required to set up public assistance committees, subject to the supervisory powers of the Ministry of Health. In 1948, the poor law was superseded by national assistance, itself replaced, in 1966, by supplementary benefits.

G. Nicholls, *History of the English Poor Law* (1854–99); S. & B. Webb *English Poor Law History* (1927–28).

Pornography. Books, pictures, films, and similar matter dealing with erotic subjects, and particularly with sexual behaviour, and designed to stimulate sexual interest and excitement. The import, possession or sale of such materials is frequently treated as criminal, as being likely to deprave or corrupt the morals of young persons and conducive to crime, but it is uncertain how far these fears are justified, and frequently difficult to determine whether materials are pornographic or not. See also OBSCENITY.

Portalis, Jean Etienne Marie (1746–1807). French jurist who was made a *conseiller d'état* by Napoleon, was a leading member of the commission which drafted the French Civil Code, an inspirer of its political doctrines, and was largely responsible for the draft Code, particularly the sections on marriage and succession. Later he was in charge of the department of public worship and framed the concordat between Napoleon and the Pope.

Port Arthur, Tasmania. Named from Governor Sir George Arthur, site of a penal settlement from 1830 till 1873, and of a model reformatory for boys, 1835–49.

Porteous riot. On the occasion of the execution, in Edinburgh in 1736, of a smuggler named Wilson, who had overpowered his guards and allowed his co-accused to escape, Porteous, captain of the Edinburgh City Guard, fearing a rescue, ordered the guard to fire on the crowd, whereby several persons were killed. Porteous was tried, convicted, and sentenced to death but on the eve of execution was reprieved. He was unpopular with the general populace who, when the news was received, stormed the Tolbooth in which he was confined and hanged him in the Grassmarket. The incident is incorporated in Scott's *Heart of Midlothian*.

W. Roughead, *Trial of Captain Porteous* (1909).

Porter, Andrew Marshall (1837–1909). Was called to Irish Bar in 1860, became Solicitor-General, 1881, and Attorney-General 1883, and was Master of the Rolls in Ireland, 1883–1906. He became a baronet in 1902.

Porter, Sir Charles (?–1696). Was called to the English Bar in 1663 and was Chancellor of Ireland, 1686–87 and 1690–96. He was several times charged with misconduct. He also acted as a Lord Justice.

Porter, Samuel Lowry, Lord (1877–1956). Was called to the Bar in 1905 and practised mainly in the Commercial Court. He became a judge of the King's Bench Division in 1934, and a Lord of Appeal in Ordinary in 1938. He retired in 1954. He was a sound and learned lawyer but never courted publicity. He was interested in international law and presided over several committees and inquiries.

Portions. Sums of money provided under a strict settlement of settled land for the benefit of the children of the settlor, other than the one who succeeds to the land.

Portius, Christopherus (Porgio) (fifteenth century). A commentator and author of *Super Institutionum Commentaria* (1522).

Porto Alexandre, The. [1920] P. 30. The *P.A.* a ship, owned by the Portuguese government, was used for revenue-earning trading. It was held, in an action for salvage services, that as the *P.A.* was the property of a sovereign state, the action could not proceed.

Ports and Harbours. At common law, the right of creating ports and harbours is vested in the Crown under the prerogative, but a right of port may be granted to a subject as a liberty or franchise, or acquired by prescription with the right to take tolls and dues, or be created by statute. The owner of a natural harbour may not make it a port without the authority of charter, statute, or prescription. Such authorization normally creates a harbour authority to manage and control the harbour, with power to charge tolls.

Portuguese law. Portugal has, at different times, been subject to Celtic, Phoenician, Roman, Visigothic, and Muslim domination, but became an independent monarchy in the twelfth century. It was united to Spain during 1580–1640 and established major colonial empires in Brazil and Africa.

Since 1910, Portugal has been a republic and, since 1928, has been in effect a dictatorship, the 'New State'. Under the constitution of 1933 revised in 1971, the Chief of State is the President elected for a seven-year term, who is advised by a Council of State. He appoints the Prime minister who forms a Cabinet. The National Assembly is elected every four years. A Corporative Chamber acts as an advisory and technical body; it is composed of representatives of corporations. The country is divided into 18 administrative districts with civil governors assisted by boards. Since 1974, there have been provisional governments, following a revolution, and the Constitution has been partly suspended.

The courts comprise a Supreme Court, four courts of appeal and numerous courts of first instance. Special courts handle juveniles, domestic affairs, and decisions of administrative authorities. Judges are appointed for life after competitive examination.

Commercial law was codified in 1833, revised in 1888, and a civil code adopted in 1867 and a new one in 1967.

A. S Correia, *Codigo penal portugues*; D. A. Fernandes, *Lições de dereito civil.*

Positive law. In Austinian jurisprudence, law posited or laid down within a particular political community by men, as political superiors, to other men, as distinct from moral law or law existing in an ideal community or by virtue of the law of nature or other non-political authority. Positive laws, or laws strictly so-called, were established directly by monarchs or sovereign bodies, as supreme political superiors, by men in a state of subjection, as subordinate political superiors, and by private persons, in pursuance of legal rights. Every positive law was a direct or indirect command.

Positive morality. In Austinian jurisprudence, a term for laws set by men, not being political superiors, to other men, and for rules set and enforced by mere opinion, or by the opinions or sentiments held or felt by men in regard to human conduct. These two classes were to be grouped as positive morality, as distinguished from divine law and from positive law (q.v.).

Positivism. The term for a general approach or attitude to thinking about law based on observation rather than speculation or reasoning. Though the word has several meanings, the common core of these is attention to *jus positum*, law actually set by men to men in a community, as contrasted with any ideal system or concept of what law ought to be. The word is accordingly used to designate the general attitude of those jurists whose main concern is the examination and clarification of the legal concepts of systems of law which have been or are actually in force. But positivism in jurisprudence comprises approaches to law widely divergent among themselves though all opposed to metaphysical and natural law theories.

In jurisprudence, positivism is an attitude to law concerned primarily with the identification of law and the expounding of the meaning of law thus identified. On identification of law, the leading positivists Austin and Kelsen sought to identify law by total separation from moral justice. Austin saw law as the command of a superior who had power to impose a sanction on one who did not comply, Kelsen as directives to officials to apply coercion when these directives were issued in accordance with a basic norm. Hart has proposed a secondary rule of recognition whose function is to provide procedural criteria by which to determine the validity of primary rules of conduct, and other secondary rules which govern the application and modification of the primary rules.

On exposition of the meaning of law, positivists have largely or entirely excluded the concept of justice. Austin contended that expressions of law were to be viewed as having objective meaning drawn from the provisions themselves, and Kelsen conceived of an objective meaning of law.

Positivism has been heavily criticized for failing to recognize the law must serve the function of a social institution, and for failing to take account of the problems of justice involved in the operation of a legal system.

Positivism is associated firstly with Bentham, Austin (qq.v.) and their followers, the analytical positivists, whose views embodied a reaction against the natural law theories of the previous centuries. They concentrated on the formal analysis of the terms and concepts of law in force, in fact almost entirely English law with some reference to Roman law and international law. Austin's analysis of the notion of law led him to the conclusion that laws properly so-called were general commands; laws set by men as political superiors, or in pursuance of rights conferred by such superiors, to other men were positive law or law strictly so called; other laws properly so-called were laws of God, or laws not set by men as political superiors, which he designated mere positive morality. Laws improperly so-called included international law, laws of fashion, and the like, which also were mere positive morality, and laws by metaphor, such as the laws of natural phenomena. Positive law was characterized by command, sanction, duty, and sovereignty, and must be kept quite distinct from ethics. Austin's command theory was refined and developed by Markby, Holland, Gray and Salmond. His major failing was to suppose that his view was the only proper approach to jurisprudence.

Various aspects of Austin's teaching were influential on German positivism, notably the work of Bergbohm, Jhering and Somlo. In France positivism was influenced by the authority and standing of the Civil Code, the most notable work being that of Saleilles, Ripert, and Roguin.

Comte developed a theory of scientific positivism on empirical rather then metaphysical bases and, in turn, influenced Durkheim and Duguit, who based their legal philosophy on the concept of social solidarity; law was an aspect and requisite of social solidarity and there was a duty to maintain social solidarity.

The most important modern positivist legal theory is Kelsen's Pure Theory (q.v.), a theory which seeks to be entirely formal, and free from all taint of history, ethics, politics, sociology, idealism, and other external influences; he sees law as a hierarchy of normative relations. Also important has been Hart's *The Concept of Law* which emphasises rules of obligation and distinguishes primary rules which regulate conduct and secondary rules which serve to identify primary rules. Unlike Kelsen he regards some shared morality as essential to the existence of any society though he excludes morality from his concept of law.

Positivism in various forms has been and is a very widely accepted approach to law, particularly in Anglo-American legal systems and has contributed very greatly to the terminology and general understanding of these systems.

Posse comitatus (the force of the county). In early English law, the force of able-bodied citizens of the county summoned and commanded by the sheriff to assist in maintaining public order, to pursue felons, or to participate in the military defence of the county. Attendance was enforced by the penalty of culvertage or turntail, which implied forfeiture of property and perpetual servitude. As the sheriff's authority declined, the *posse* became a purely civil body and, in time, the authority to call out such assistance was entrusted to justices and magistrates.

In the U.S., the *posse* has been preserved in many states and was important on the frontier. Sheriffs have the authority to summon it and it is sometimes criminal to fail to assist. Members of the *posse* have generally been held authorized to use force if necessary in performance of their function, but the question of liability where the officer summoning the *posse* is acting outwith his authority is uncertain.

Possession. A legal concept of variable meaning, the word being used in different contexts with different meanings. It denotes a kind of relationship between a person and some object of property, real or personal. It is sometimes used as meaning merely physical control or detention, without any question of legal right. Secondly, it may mean legal possession which will be recognized and protected by law, and in this sense it is usually said that the elements are *animus possidendi*, intent to hold against others, and *factum possidendi*, the amount of occupation or control of which the thing is capable and which excludes strangers. The elements may be separated in that an owner has *animus possidendi* while away from home or while a borrower from him has *factum possidendi*, the latter having *animus* as against all except the true owner. The possession of the servant, hirer, borrower, etc. is sometimes distinguished as custody or *de facto* possession rather than possession. A finder of lost property has *factum* but may have no *animus*. A thief may have legal possession, having both *animus et factum possidendi*, even though he obtained possession illegally.

Possession is acquired originally, as by catching fish, or cutting growing crops, by making or creating a thing, by accession, as by the birth of lambs to a farmer's sheep, or derivatively, by transfer from another in pursuance of a gift, sale, or other contract. The right of property in a chattel may be transferred without change of possession, as where goods are sold but not yet delivered, but the possessor thereafter holds not for himself but for the new proprietor.

A person having actual possession is presumed to have the right of property and is accordingly entitled to have his possession protected unless anyone can prove a superior title. Even if he has no right to possession, he is entitled as against a stranger or a wrongdoer to have the rights and remedies of a person entitled to possession. A stranger or wrongdoer cannot merely point to the defect of the possessor's right or rely on the right of a third party, unless he is acting on behalf of the third party.

De facto possession may in some circumstances be lost by loss of physical control, but loss of physical control does not necessarily involve loss of legal possession, and a person who has relinquished *de facto* possession has the rights of a possessor.

Possession is lost when a chattel is delivered outright to another, or used up or destroyed, or abandoned without intention to transfer it to another, or taken under process of law, as under distress, execution or bankruptcy but not when a chattel is taken unlawfully.

Possession is one of the rights which made up property or ownership, but it may be granted separately by the owner to another, as when he leases or lends the object of his property, or may be taken from the owner as when a thief steals the object from the owner, or from one who possesses legitimately under and for the owner, such as a borrower.

The concept of possession in Roman law has been extensively analysed, particularly by Savigny, as comprising two elements, *animus* or intent to exercise control, and *corpus* or fact of physical control; both were required for the acquisition of possession. But neither the theory of Savigny nor that of Jhering can account satisfactorily for all the cases.

Similarly in the common law, attempts to apply Roman theories are not wholly successful and it seems that recognition of possession varies in different contexts in response to different pressures and with regard to the different purposes pursued by different branches of the law. It is only useful to examine what possession means in particular contexts, criminal law, land law, and so on, and there is no certainty that usage of the concept is uniform even within one branch of law.

A distinction is frequently drawn between possession and mere custody, the custodier having physical control but holding for the possessor not for himself; thus employees are frequently said to have custody only. Sometimes possession, custody and detention are distinguished, the latter terms generally denoting physical control which did not carry the legal advantages given to the possessor.

Possession may be exercised immediately, by the possessor himself, or mediately, by an agent, or employee, or borrower, or hirer, or other person who has custody but recognizes the possessor's title and is willing to return it on demand or when the purpose of his custody has been served.

Possession is a relationship of great legal importance; it may create ownership, as by *occupatio*; it is one of the elements in, or rights which collectively constitute, ownership; it raises a presumption of right, such that a claimant from the possessor must show that he has a better title than the possessor; it frequently evidences title of ownership, though by no means always; in certain cases, notably those of money and negotiable instruments, a possessor can confer on one who takes from him a good title, though he had none himself, provided the taker takes in good faith, for value and without notice of the defect in the other's title.

Because possession is usually justifiable, law frequently protects possession, until or unless another claimant can establish a better title to possess than the possessor has. Thus, the true owner can recover from the mere possessor, the lender from the borrower, and so on.

In feudal land law, actual possession of land was less important than seisin (q.v.), legally recognized possession. The person seised, if dispossessed, could recover possession by the writ of right if he could show a superior title to that of the possessor. A clear distinction between seisin and possession emerged with the concept of the term for years. The termor's interest was not seisin, but possession and was protected by a form of trespass, *de ejectione firmae*.

In relation to some incorporeals, e.g. debts, claims of damages, the concept of possession is inappropriate. A person has a right to a debt or a patent or to damages only if by law he has such a right, and there is no means of possessing such a right. It may be asserted or claimed or exercised but not possessed. In relation to other incorporeals, e.g. an easement, a right of fishing, they may be possessed by actual exercise and use and the exclusion of trespassers.

F. C. v. Savigny, *Das Recht des Besitzes*; R v. Jhering, *Grund des Besitzesschutzes*; *Der Besitzwille*; F. Pollock and R. S. Wright, *Possession in The Common Law;* J. Salmond, *Jurisprudence;* A. L. Goodhart, *Essays.*

Possessory action. An action which determines which of two or more claimants is entitled to lawful possession of a particular thing but does not determine proprietary title. In Roman law, the possessory remedies were *interdicta, uti possidetis* for immoveables and *utrubi* for moveables. At common law in England, the principal possessory actions were ejectment (*ejectione firmae*) for the recovery of land where the lessee for years, had been ousted from his possession; replevin, to recover specific goods which had been wrongfully taken out of the plaintiff's possession; detinue, to recover specific goods of the plaintiff; and trover, to recover lost property from the finder thereof.

Possessory title. The title to land acquired in England by occupying the land for 12 years (30 years in the case of the Crown) without paying rent or otherwise acknowledging the title of the true owner.

Possibility. A future event which may or may not happen; also, in the law of real property, an interest in land which depends on the happening of such an event. In the latter sense a possibility may be a bare possibility, such as that a son will succeed to his father's land, or a possibility coupled with an interest, such as that a remainderman (q.v.) will succeed when a life interest falls in, if a contingency is satisfied.

Post, Albert Hermann (1839–95). German jurist, who wrote on primitive forms of law and society, particularly the family, and developed a system of ethnological jurisprudence, aiming to construct a universal history of law, to discover the natural evolution of law and society. The attempt was unsuccessful since divergent lines of evolution are found. His major works were *Einleitung in das Studium der ethnologischen Jurisprudenz* (1886) and *Grundriss der ethnologischen Jurisprudenz* (1894–95).

Postglossators (or commentators). The second group of jurists who studied the Roman law after the revival of Roman law studies in the twelfth century, following the glossators. Unlike the glossators, they had regard to the requirements of legal practice and sought to relate the principles of Roman law to feudal and Germanic customs, canon law, city statutes, and other law of the times. The jurists of this school wrote substantial, frequently vast, systematic commentaries, tracts and treatises. Among the most notable members of the school were Cino of Pistoia, Bartolus of Sassoferrato, Baldus degli Ubaldi, Lucas de Penna, Bartolomaeus Salicetus, and others.

Postliminium. In Roman law, the principle that if a Roman entered a foreign state with which Rome had no treaty of friendship, he would be enslaved and Roman goods taken there could be appropriated, but if Roman or property returned to the territory of the Roman Empire, he became a Roman citizen again and the property reverted to its former owner. In international law, it is the principle that individuals, property and territory which have come in time of war under the authority of the enemy come again under their original sovereign if abandoned by the enemy, reconquered or restored by a third state. The principle does not affect acts which a former military occupant was entitled to perform while in occupation, but it does affect illegitimate acts by the occupying state. Postliminium does not operate where territory ceded or annexed later reverts to a former owner state, or where a territory again becomes independent.

Post-mortem (or autopsy). An examination of the body of a person after his death to determine the cause of the death and, if the death is not due to natural causes, to discover any information which may be relevant to either a civil claim or a criminal prosecution arising out of the death. Post-mortems are accordingly frequently called for by coroners, procurators fiscal and others concerned to know whether any legal liability, civil or criminal, on anyone's part may arise. Post-mortem techniques are basically those of pathology but may involve chemical analysis, ballistic problems, and require the assistance of other experts.

Post-obit bond. A bond executed by a person in return for money lent whereby the borrower binds himself to pay the lender a larger sum, with interest, on the death of a third party on whose death the borrower expects to become entitled to property.

Postman. In the old Court of Exchequer, a junior barrister who, except in Crown business, had precedence in motions. The name arose from the fact that the postman's place in Court was by the post formerly used as a measure of length in excise cases. The office ceased to exist when the Court disappeared in 1875. See also TUBMAN.

Pothier, Robert Joseph (1699–1772). French jurist. At 21 he became Conseiller au Présidial d'Orléans and sat on the bench there for 50 years, combining practice with study of the theory and history of law. He also long maintained study of theology. For years he studied Roman law, particularly the problem of the order of the Digest and, in 1748, published his *Pandectae Justinianae in novum ordinem digestae*, in which he rearranged the texts in methodical order though preserving the old arrangements of the titles. From 1749, he was also Professor of French Law at Orléans and turned to his own law, producing, from 1761, a great series of treatises on particular branches of that law, obligations, sale, hiring, maritime hiring, partnership, loan, deposit, mandate, negotiorum gestio, pledge, contingent contracts, insurance, bottomry bonds, gaming contracts, marriage, community of property, dower, property and possession. Certain others were published posthumously.

The Treatise on Obligations was soon recognized as a major contribution to legal science, translated by Evans and frequently cited in British courts.

Potwalloper boroughs. Borough constituencies in the unreformed British parliament where the vote was held by those who kept their own house and dressed their own meat.

Pound, Roscoe (1870–1964). Majored in botany at Nebraska, conducted the botanical survey of that state in addition to conducting a legal practice, 1892–1903, and was successively an Assistant Professor of Law at Nebraska, 1899–1901, and a Commissioner of Appeals of the State Supreme Court, 1901–03. He then became Dean of the University of Nebraska College of Law, 1903–07, a professor at North-western University Law School, 1907–09, and Chicago, 1909–10. He then became Story Professor of Law at Harvard, 1910–13, Carter Professor of Jurisprudence, 1913–37, Dean, 1916–36, and, thereafter, University Professor, 1937–47. His major interest was jurisprudence and he was the leader of the group which regarded law as social engineering, the means of reconciling men's interests with the minimum of heat and friction, a view which came to be widely accepted under the name of sociological jurisprudence.

A most prolific writer and expositor, his major work is *Jurisprudence* (5 vols., 1959) but other important books are *Introduction to the Philosophy of Law* (1922), *Law and Morals* (1926), *Criminal Justice in America* (1930), *Readings on the History and System of the Common Law* (1927), *The History and System of the Common Law* (1939), *Social Control Through Law* (1942), and *New Paths of the Law* (1950). *Outlines of Jurisprudence* (1942) is a useful outline and bibliography. He gave many addresses and wrote many articles.

F. B. Sayre, *Life of Roscoe Pound.*

Pound. A unit of weight in the imperial standard system of weights and measures; also the main unit in the British monetary system. The pound Scots was valued at one-twelfth of the pound sterling. A pound is also an enclosure, building, or place where goods, taken by distress (q.v.), are placed by the distrainer. A pound overt (open) is used for cattle impounded by way of distress, straying cattle, or the confinement of cattle *damage feasant.*

Pound-breach. The act of taking goods distrained out of pound (q.v.) before the distrainor's claim has been satisfied. The distrainor may bring an action for treble damages or proceed for a penalty. Pound-breach to rescue animals distrained or taken *damage feasant* is criminal at common law.

Poverty law. The branches of law, particularly of the law of social security, concerned with persons who lack the means to supply themselves with a reasonable minimum standard of housing, clothing, food, and other essentials of life.

Povey, Sir John (? 1621–79). Was called to the English (1645) and Irish (1658) Bars, became a baron of the Irish Exchequer in 1663, and Chief Justice of the Irish King's Bench, 1673–79.

Powell, Richard Roy Beldon (1890–). Was admitted to the New York Bar in 1914, but turned to law teaching at Columbia, 1921–25, and Harvard 1925–60, where he became the leading scholar on the law of property. He was reporter for the *Restatement of the Law of Property* and published *Cases and Materials on Trusts and Estates* (1923–33), *On the Law of Possessory Estates* (1933), and *On the Law of Trusts* (1940).

Power. The legal concept of entitlement to do something of legal force and effect, such as to make a will, to sell property mortgaged if the loan is not repaid, to rescind a contract if materially breached by the other party, and to exercise many other rights. It is, accordingly, one sense of the more general term 'right'. The correlative is subjection or liability (qq.v.). A public power is one conferred by the Crown or Parliament on an authorized delegate of either, for a public purpose, such as to take land compulsorily. Private powers are conferred partly by general law (common law or statute or both) and partly by authority conferred by a person entitled to confer the power. Such powers include those of guardians, trustees, executors, directors, etc.

The term is also used in particular of an authority conferred on a person to deal with or dispose of property not his own. Powers may be administrative or managerial, such as the powers conferred on a tenant for life or trustees for sale, or dispositive powers, or powers of appointment, authorizing the donee of the power to dispose of beneficial interests in property; this latter category may be general, if exercisable in favour of any person at all, or special, if exercisable in favour only of certain limited persons or classes. In the category of administrative powers, many kinds of powers are conferred by statute on, e.g. trustees for sale, but express powers may be conferred in addition by the person creating the trust.

The exercise of a power of appointment must be made in strict accordance with the conditions stipulated by the donor of the power and it is a fraud on the power to exercise it for other than the purposes intended by the creator of the power.

See also SEPARATION OF POWERS.

Power of appointment. A power given by a deed or will to a person to direct that an interest in property is to devolve on a person or persons, or on one or more of those within a designated and limited group of persons.

Power of attorney. Formal authority conferred by deed by one person on another to act as his attorney, or agent, or legal representative. The powers conferred are set out in the deed and may be limited to one transaction or general and unlimited. Even in the latter case, authority to exercise important or unusual powers, such as to sell land, should be conferred expressly.

In England, powers of attorney, the execution of which had been verified, formerly in some cases had to be filed in the Central Office of the Supreme Court and/or at the Land Registry. In Scotland, powers of attorney (properly called factories and commissions) may be recorded in the Books of Council and Session.

Powle, Henry (*c.* 1629–92). Though called to the Bar in 1654, was a politician rather than a lawyer. He was Speaker of the Convention Parliament which offered the Crown to William of Orange. He was then (1690) appointed Master of the Rolls but left no mark as a judge.

Poyning's law. An Act passed in 1495 by an Irish Parliament of 1494–95 summoned, and packed, by Sir Edward Poynings, Deputy to Prince Henry (Henry VII's Lieutenant), designed to restore royal power in Ireland and render the Irish Parliament harmless. No Parliament was to be held in Ireland until the Lieutenant and Council of Ireland certified to the King what causes and acts they should pass and the King should send his licence to affirm the causes and to summon the Parliament under the Great Seal of England. Any Parliament held contrary to these forms was to be void and of no effect. As explained by an Act of Philip and Mary, any Bill the Irish Parliament might pass must first be approved by the Irish Privy Council and then by the English Privy Council, and had to be accepted or rejected as modified by the latter. This Act was in force until 1783 and marked the complete subjection of the Irish Parliament to the English Crown. In the seventeenth century, attempts were made several times to suspend the Act, but thwarted by opposition, the members of which saw advantage in having a reference to Westminster.

Practice. A term sometimes used as distinguished from legal study, for holding oneself out as 'practising' or 'in practice' or willing to act for others in legal affairs. Practice is also distinguished from legal theory as being what actually happens as distinct from what should happen according to the rules. The term is also used of the customary, usual, and appropriate mode of dealing with particular situations. In relation to chamber business and conveyancing, much of the course of business is dictated by practice, by usages based on experience, prudence, and convenience rather than by requirements of law. Thus, examination of title is largely a matter of practice rather than law.

In relation to court business, practice is a slightly less formal regulation of proceedings, concerned with what is usual and customary and deemed proper rather than with what is prescribed by procedural rules though the distinction between practice and procedure is indefinite. Thus the law is that the award of costs is entirely in the court's discretion; the practice is to award a successful plaintiff his costs. Practice is, to some extent, stated or altered by Practice Directions and by practice masters' rules, and to some extent it is determined by the customary habits and usages of the courts and the profession.

Parliamentary practice includes the custom and usage of Parliament as well as the rules of law and procedure.

Practice directions. In the High Court, directions issued with the authority of the court or of the judges of the Division to which the particular kind of business in question is assigned. Practice directions and notes are usually published in law reports and legal practitioners' journals for the information of the profession.

Practice masters' rules. In the High Court, directions for the regulation of business concerning the Central Office of the Supreme Court. They are printed in the *Annual Practice.*

Practicks. (from *practicae observationes*, practique, practice). Initially, a brief record kept by one or other of the judges of the Court of Session in Scotland, from its foundation in 1532, of the decisions of that court. The early collections were chronological and were designated by the name of the compiler, *e.g.* Sinclair's *Practicks*, 1540–49. Later collections included abstracts of statutes and other materials, and the notes were arranged under subject-heads, *e.g.* Balfour's *Practicks*, 1469–1579 (pub. 1754), founded on older collections now lost. The two kinds, sometimes distinguished as decision-practicks and digest-practicks, were latterly concurrent and from the mid-seventeenth century were superseded, the former by collections of reports of cases and the latter by systematic treatises and, later, by digests of cases, commencing with Kames' *Folio Dictionary* (1741). Many volumes of *Practicks* are extant and some have been published. They are of great value as an authentic record of mediaeval Scottish customary law and of early decisions and the published volumes have frequently been referred to as authorities.

Pradier-Fodéré, Paul Louis Ernest (1827–1904). French jurist, author of a major *Traité de droit international public (1885–1906)*, of *Cours de droit diplomatique* (1881), *Précis de droit administratif* (1872), and other works.

Praecipe. Formerly a kind of original writ, instructing the sheriff to command the defendant to do something or to show cause why he had not done it. Among such was the *praecipe quod reddat*, the writ by which a common recovery was commenced, to command the defendant to surrender the land to the claimant. The term is now given to a slip of paper containing particulars of a document which the party wishes to have prepared or issued, such as a writ of execution, given to the appropriate officer of the court to enable him to prepare it.

Praefectus urbi. At Rome, from the time of Augustus, an official in charge of the city and its police and having jurisdiction as far as one hundred miles from Rome and power to decide both civil and criminal cases. He was held to be the Emperor's representative and acquired much of the jurisdiction which had formerly belonged to the *praetor urbanus* (q.v.).

Praemunire. The name of a royal writ designed to protect the royal rights of jurisdiction, requiring the sheriff to warn an accused to appear to answer for contempt, which then came to denote a wide but indefinite complex of offences prosecuted by the writ, and also the penalties incurred thereby. The foundation statute was the Statute of Provisors of 1352. The writ was developed in the fourteenth century against papal attempts to invade lay rights in appointments in the Church, and a statute of 1533 on the matter could be interpreted as punishing all invasions of the royal power, particularly any activity of spiritual judges to the detriment of the royal courts, with loss of property and imprisonment at pleasure. It was used against Wolsey and clergy who had participated in the illegal exercise of his legatine power. In 1531, all the clergy in England was charged under it with breaking the King's law by exercising their spiritual jurisdiction, and the Church bought a pardon. In later legislation, the word was transferred to both the offence of suing at Rome or elsewhere in contempt of the King, and to the procedure and penalties contained in the Act of 1533.

The penalty was made applicable by various statutes to the offences of appealing to or obtaining of bulls and other matters from the See of Rome, refusing to confirm a bishop when nominated by the Crown, affirming that the writer, or speaker, or another, is under obligation to change the government of Church or State or that Parliament may legislate without the Crown, affirming the right to the Crown of any persons other than those in the lawful succession or that Parliament may not make laws binding the Crown and the descent thereof, and knowingly and wilfully solemnizing, or assisting, or being present at a marriage, or the making of a matrimonial contract forbidden by the Royal Marriages Act, 1772. Some of these offences might now be treated as treason or treason felony. In the only reported case of praemunire (*R.* v. *Cook* (1662), 6 St. Tr. 202) the prisoner was put out of the King's protection, his real property was forfeited for life, his personal property forfeited absolutely, and he was imprisoned at the King's pleasure. It is now abolished.

Praepositura. The implied authority of a wife or other housekeeper to contract to the effect of rendering the husband liable for goods necessary for the household and suitable to its style of living.

Praepositus (Joannes de San Georgio) (?–1378). Italian canonist and author of a commentary on the Clementines (*Reportationes super Clementinas*), supplementary to Joannes Andreae's apparatus on the Clementines, and a *Quaestiones.*

Praestita rolls. Rolls covering 1199–1603 containing the list of payments advanced by the Exchequer to royal officials.

Praetor. A Roman magistrate, originally the highest one and connected with military affairs. In about 366 B.C., a praetor was elected to supervise the administration of justice at Rome (*praetor urbanus*), the earlier praetors having been now entitled consuls, and in about 242 B.C. a second praetor (*praetor peregrinus*) was appointed to deal with causes in which one or both parties were foreigners. Other praetors were later added, with responsibility for provinces. At about 100 B.C., it was provided that all the (eight) praetors should remain in Rome during their year of office, six as judges or presidents of jury courts (*quaestiones*), (q.v.) thereafter going to provincial governships.

It early became custom for praetors to announce the principles on which they would act, grant remedies and so on, and the Edict became an important source of law supplementary to and corrective of the older *ius civile*. The bulk of the Edict was carried on from year to year, amended and modified from time to time. About A.D. 130, Hadrian caused the jurist Salvius Julianus to revise and restate the Edict which thereafter was unalterable (*Edictum Perpetuum*).

The office declined in importance under the Empire, special functions being assigned to particular praetors, and eventually became a merely honorary appointment though some of the praetorships with special jurisdiction lasted into the fourth century A.D. and were copied in the constitution of Constantinople.

Pragmatic sanction. A fundamental law of the later Roman State regulating important matters particularly of Church and State. This title was given to the series of enactments promulgated by Justinian in A.D. 554 after his reconquest of Italy, to restore Roman law, and putting into force for all in Italy the revised Roman law contained in the *Digest*, *Code*, and *Novels*. It reinforced an earlier *Edictale Programma* (*c.* 540) but, though very influential, did not completely supersede the Theodosian Code of A.D. 438. Later, the term was used in countries influenced by Roman law for an expression of the will of the sovereign defining the limits of this own power or regulating the succession. In France the term was applied to regulations of general councils which were enforced by the King on legal advice, a notable one being the Pragmatic Sanction of Bourges of 1438 which asserted the supremacy of a council over the Pope and went far to establish an independent church in France and to develop Gallicanism. It was ultimately superseded by the Concordat of Bologna in 1516. Pragmatic Sanctions dealing with dynastic succession include those of 1713 which settled the succession to the Emperor Charles VI of his daughter Maria Theresa as sovereign of the Austrian dominions, and of 1830 in Spain.

Pragmatism. A philosophical approach of which the principal exponent was William James. In the sphere of legal history, it emphasizes an experimental, flexible attitude rather than one relying on strict logic. Law is an experimental process in which logic is only one of many factors leading to a conclusion. The pragmatic approach concerns itself not with what legal rules in the books are, but with how rules work. This attitude underlies and was a forerunner of realism (q.v.) in jurisprudence, also an American development.

Pratt, Charles, 1st Lord and 1st Earl Camden (1714–94). A son of Sir John Pratt, C.J.K.B., he was called to the Bar in 1738, and was initially unsuccessful, but after some time succeeded. He became Attorney-General in 1757 and an M.P. In both capacities he showed liberal views, In 1761, he was promoted Chief Justice of the Common Pleas and, as such, pronounced the decisions in 'general warrant' cases which made him as popular as Wilkes. In 1765, he became a peer and, in the following year, Lord Chancellor. He was removed in 1770 but later served as President of the Council, 1782–83 and 1784–94. He was made an earl in 1786. As a judge he carefully followed precedent but was not dominated by it; he knew equity practice thoroughly and was respected by the Bar and suitors in the court. He gave the leading judgment in the great *Douglas cause* (q.v.) but his greatest fame was won in the field of public law. He wrote a *Discourse against the Jurisdiction of the King's Bench by Process of Latitat* (1745) which displays considerable learning.

H. S. Eeles, *Lord Chancellor Camden and his Family*.

Pratt, Sir John. (1657–1725). Called to the Bar in 1682, he acquired a large practice after the Revolution and was deemed one of the ablest counsel of his time. He became a judge of the King's Bench in 1714 and Chief Justice in 1718, and was esteemed a strong and impartial judge of great legal ability. Many of his decisions are still of importance. His son Charles (q.v.) became Lord Chancellor Camden.

Preamble. The part of a bill presented to Parliament which states the need for, the purposes of, and the intended effect of the legislation. It is not an essential part of a Public Bill and is now regularly omitted. If, however, there is a preamble it may be referred to in order to resolve doubts as to the interpretation of a section. It is always, however, found in a private bill and is considered first. If the bill is opposed petitioners against the bill are heard and the Committee on Opposed Bills must resolve whether the preamble is or is not proved. Only in the former case are the clauses of the bill considered in detail. Even in an unopposed bill the preamble has to be proved.

Preaudience. The right which certain counsel have in courts of being heard before others in business which is not set down to be heard in a particular order. The Attorney-General and the Solicitor-General have priority in most matters, followed by Queen's Counsel and by junior counsel according to the dates of their call to the Bar.

Precarium. In early mediaeval times, a grant of land for life in return for military services, and accordingly the basis of feudal tenure. Also in Scots law, a gratuitous loan of goods, revocable at pleasure.

Precatory trust. Words in a will praying, requesting, or recommending that something be done. The question which arises is whether such an expression is to be treated as a mere expression of hope or desire or as creating a trust. The words will generally be so treated only if they can be interpreted as truly imperative.

Precedence. The order in which persons rank on ceremonial occasions. In England, the basis of precedence is the House of Lords Precedence Act, 1539, modified by later statutes, Royal Warrants, Orders in Council, and ancient usage. In Scotland, attempts were made to settle the precedence of nobles in the sixteenth century, but modern precedence is regulated by a Royal Warrant of 1905. Precedence may be determined by holding, or being the wife, or son, or daughter of the holder of, an hereditary dignity, (see PEERAGE) or by holding an office listed in the Table of Precedence, e.g. judge of the High Court, or by membership of a class of one of the Orders of Knighthood (see KNIGHTHOOD).

In the United States, precedence depends on official position held and is determined by custom.

Precedence, patent of. A mark of distinction formerly sometimes conferred by letters patent on a barrister whom the Crown wished to honour. It conferred such rank and precedence as was assigned in the patent, sometimes after the King's Attorney-General, but usually after His Majesty's Counsel. Patentees sat within the Bar but received no salary and were not sworn, and might be retained in cases against the Crown, unlike King's Counsel.

Precedents. Things which have gone before. The term is used, legally in three senses: firstly, of court precedents, which are forms of writs and other pleadings which have been found valid and effective and useful in the past and may be used or copied again; secondly, of conveyancing precedents, which are forms of conveyances, settlements, and other deeds and of clauses thereof, used in conveyancing, which may usefully be copied or adapted. Many collections have been published of both kinds of precedents to assist in drafting pleadings or conveyancing deeds, e.g. Bullen and Leake's *Precedents of Pleading in Actions in the Queen's Bench Division*, Key and Elphinstone's *Compendium of Precedents in Conveyancing*. Thirdly, the term is used of judicial precedents, which are previous decisions of the superior courts deemed to embody a principle which in a subsequent case raising the same, or a closely related, point of law, may be referred to as stating or containing the principle which may be at least influential on the court's decision of it, or even, under the principle of *stare decisis* (q.v.), decisive of it. A precedent, that is, is a previous judicial decision considered as a source of law in a later case.

The use of precedents as a source for a rule of law to be applied to the decision of a later case depends on the long-standing custom of judges of the superior courts, when deciding cases, giving judgments which review the relevant law, including any relevant previous decisions, and give their legal reasons for the decision reached; on the equally long-standing custom of reports being published of interesting and important cases, setting out the judgments in full; on the highly centralized system of the English courts; and on the natural tendency to seek to rely on the wisdom of previous judges and to decide a case in a way consistent with the decision of previous similar cases.

While reference to precedents has long been common, it is only since 1800, and particularly since the late nineteenth century, that the full rigidity of *stare decisis* has been accepted in England, namely that appellate courts are bound by their own previous decisions, and that each court is bound by a decision of a court above it in the hierarchy of courts. This is more rigid than the practice in Scotland where, until the nineteenth century, only a consistent line of decisions had binding force, where a larger court may still be convened to reconsider a particular precedent, and a precedent may cease to be binding if superseded by changed social and other conditions, more rigid than the practice in the U.S.A., where superior courts, including the Supreme Court, may reconsider their own previous decisions, and more rigid than practice in France where only a consistent course of practice of the courts has binding force. While the Judicial Committee of the Privy Council is slow to dissent from its own precedents, it may do so, and the supreme courts of Commonwealth countries generally do not hold themselves bound by their own previous decisions.

The older theory in England was that, in following and applying precedents, judges were simply declaring what has always been the law and was implicit in customary usages and previous decisions, but it is more correct to recognize, as is now commonly done, that judges in deciding cases, applying, extending, or restricting precedents, are in truth making new law. A great deal of common law, and most of the principles of equity, has been created in particular cases by judges.

Moreover, when a superior court overrules a precedent or distinguishes it restrictively, the older theory was that the court was merely declaring what had always been the law and declaring that it had been misunderstood or misstated in the precedent, whereas, in fact, the court is truly changing the law or at least restating the rule in modified terms.

Precedents may be classified firstly as being 'in point' in, or 'on all fours' with, a subsequent case, or as not so being. A precedent is in point if there was raised, argued, and decided in it in relation to one set of facts, some issue of law which is the same as has arisen in the case now before the court in relation to a different set of facts of the same general kind, e.g. is *coitus interruptus* adultery? Precisely the same set of facts as were before the court in a precedent never recur, and it is always a question whether or not the different facts which have arisen in a later case are sufficiently similar to those of the precedent to make the decision in the precedent a guide to the decision in the present case. If a precedent is not in point it may yet, by analogy, yield a principle relevant to the case for decision, or may be so different as not to be helpful at all. Much argument frequently arises on whether a particular precedent is in point or, on the other hand, can be 'distinguished' as dealing with a different kind of situation. If a judge, or court, wishes not to be bound by a precedent he will seek to 'distinguish' the precedent by pointing to differences in facts which, in his view, suffice to render the precedent not 'in point'.

Precedents must be classified, secondly, as binding or persuasive. A precedent, if in point, may be binding on a subsequent court, in which case the subsequent court *must* apply the principle of the precedent to the facts of the subsequent case and decide accordingly, whether the judges of the subsequent court agree with the principle or not, or it may be merely persuasive, in which case the subsequent court must take the precedent into consideration and give it such weight as they think proper; they are not bound to follow it but may do so; if they do it is because the precedent has persuaded them by its merits, by its reasoning, not because it is authoritative.

Whether a precedent is binding or only persuasive depends primarily on the place if the court which decided it in the hierarchy of courts in relation to the place of the court now considering the precedent, irrespective of whether appeal lies to the court which decided the precedent or not. For this purpose the regular civil and criminal courts of England and Wales, Scotland and Northern Ireland stand as shown in table below.

Decisions of the House of Lords, on appeal from any one country of the U.K., are binding on the House itself in later appeals from that same country, unless decided *per incuriam* (i.e. in ignorance of a relevant precedent or statute) or, unless, since 1966, the House thinks it right, exceptionally, to reconsider and not follow its own precedent, and binding on all lower courts in the same vertical line and strongly persuasive on all other courts in all other lines, particularly on matters of general principle and on the interpretation of statutes common to the

England and Wales		*Scotland*		N. Ireland	
Civil	*Criminal*	*Civil*	*Criminal*	*Civil*	*Criminal*
House of Lords	House of Lords	House of Lords	—	House of Lords	House of Lords
Court of Appeal (Civil Div.) (*a*)	Court of Appeal (Criminal Div.) (*b*)	Court of Session (Inner House)	High Court of Justiciary (*c*)	Court of Appeal	Court of Appeal
Divisional Court.; High Court (*d*)	Divisional Court.; Crown Court (*e*)	Court of Session (Outer House)	High Court of Justiciary (*f*)	High Court	Crown Court
	Crown Court	Sheriff Court (Sheriff Principal)	Sheriff Court (On indictment)		Crown Court
County Court	Magistrates' Court	Sheriff Court	Sheriff Court (summary)	County Court	Magistrates' Courts

(*a*) Including the former Court of Exchequer Chamber and Court of Appeal in Chancery.
(*b*) Including the former Court for Crown Cases Reserved and Court of Criminal Appeal.
(*c*) Sitting as Court of Criminal Appeal.
(*d*) Including courts with the standing of the High Court, and the former separate courts of Queen's Bench, Common Pleas, Exchequer, Probate, Matrimonial Causes, and Admiralty.
(*e*) Including the Central Criminal Courts and the former assize courts.
(*f*) Sitting as court of trial.

precedent line and to the line later considering the statute. Decisions on appeals from one country do not bind the House in later appeals from another country, unless the point in issue is common to both systems, such as the interpretation of a statute applicable to both countries.

Decisions of the Court of Appeal, and of older courts of equivalent rank, are binding on all divisions of that court in later cases, and on all courts below in the same vertical line, unless decided *per incuriam*, or inconsistent with a decision of the House of Lords, or there is inconsistency between two prior cases and one must be followed, and persuasive, in greater or lesser degree depending on the subject-matter in question and its similarity or dissimilarity to the law of the later court, on all courts of lower rank in all vertical lines. This rule is less rigidly applied in criminal cases, where the liberty of the individual is at stake. A decision of a full court of five judges has no more weight than that of a quorum, but in the Court of Session a decision of a larger court, usually of seven judges, has greater weight that that of a quorum of either Division of the Inner House and may overrule a previous decision of a Division. The Court of Appeal regards itself as bound by the Court of Session in matters on which the legal systems are the same, and conversely.

The former Court of Criminal Appeal adopted a slightly more liberal attitude to precedent in cases where the liberty of the subject was involved.

Decisions of a Divisional Court, whether civil or criminal, are probably binding on that court in later cases, unless the decision was reached by mistake, and on lower courts; they possibly do not bind single judges of the High Court, though they will be very slow to differ from such decisions. Such decisions are of some persuasive weight on courts in other vertical lines.

Decisions of single High Court judges are not regarded as binding on other single judges, though they will normally be followed, but they are probably binding on lower courts, and of some persuasive weight in courts of other vertical lines. The same is true of single judges in other vertical lines.

In the inferior courts, one court is not bound by a decision of another though the practice is to follow such decisions where reported unless there is reason for not doing so.

Judgments of the Judicial Committee of the Privy Council are not binding on the Board itself, nor on any court in the U.K., save probably in those few cases where it sits on appeal from a U.K. court, but because of the eminence of the judges who sit in the Judicial Committee and of the desire to maintain harmony of decisions throughout the English-speaking world, many decisions of the Judicial Committee are persuasive in U.K. courts.

Persuasive value attaches to decisions of the Supreme Court and other superior courts of the U.S.A., of the superior courts of the older Commonwealth countries, and of foreign superior courts, in all cases depending on the point of law in question and the similarity or dissimilarity of the systems of law concerned on the point in issue.

The binding or persuasive element in a precedent

A precedent is binding or persuasive in a later case only in respect of the principle of law, if any, on the basis of which it was decided, and which can be extracted from it for subsequent use. This is termed the *ratio decidendi* (q.v.).

It is for the subsequent court to determine what the *ratio* of a precedent is. A case may yield no useful *ratio*, if decided on a point of pleading or of fact, or entirely on its own facts, or it may be so ill-reported, or the judgments may be so obscure, that the later court cannot discover on what basis the previous court decided the precedent.

Everything else in a precedent is *obiter dictum* (q.v.) and not binding, though it may be highly persuasive.

Evaluation of precedent

The weight to be attached to a precedent is affected by various factors. It gains in weight and authority the older it is, particularly if it has been considered as settling the law and approved, followed and applied, and if transactions have been entered into on the basis that it represents the law. Conversely, it lacks weight and authority if it has become obsolete or been superseded by changed conditions or legislation, or has been criticized, distinguished, not followed or explained away. The personal eminence of the judge or judges who decided the precedent is of importance. Whether the court which decided the precedent was unanimous or divided, and in the latter case the grounds for the dissents, have to be considered. Whether the decision has been approved or disapproved by later courts, by academic writers and textbooks has all to be taken into account. In the case of old precedents, the quality of the report must be considered and a case can be disregarded if so inadequately reported that the ground of decision is uncertain.

The handling of precedents

A court, when dealing with one or more precedents, has, accordingly to decide whether each precedent is binding or persuasive and, in the later case, the degree of persuasive force, and to decide what *ratio* or principle each precedent yields. If the precedent is, in the circumstances, binding the later court must apply it, or follow it, whatever it may think of the rightness of the *ratio* of the precedent. Thus a judge of the Q.B.D. must follow precedent of the House of Lords and Court of Appeal.

A court may 'distinguish' a precedent on the ground that its facts differed from those of the case before it in a respect which makes it raise a different issue of law. Thus, a judge of the Q.B.D. may distinguish precedents of the House of Lords and Court of Appeal and say: They deal with a different point and accordingly do not bind me. Whether a precedent is reasonably distinguishable or not may be a very difficult question.

A court superior to the court which decided the precedent may overrule it, i.e. declare it to be bad law, in which case the *ratio* is disapproved and no longer has any binding force, at least on the point on which it has been overruled. It may be good law on other points. A court may impliedly overrule one of its own precedents by choosing to follow a later decision of a higher court which is inconsistent with its own precedent.

Precedent as a source of law

Any rule based on a precedent is inferior to a rule laid down by statute or subordinate legislation, in that statute or subordinate legislation may overrule or cancel any rule established by precedent and, in case of conflict or dispute, the statutory rule must always have precedence. But decisions on the interpretation of statutory provisions are precedents as much as are decisions on any other point of law, and may be binding or persuasive, be distinguished, followed or overruled in the same way as other precedents.

R. Cross, *Precedent in English Law*; C. K. Allen, *Law in the Making*; T. B. Smith, *Doctrines of Judicial Precedent in Scots Law*; and books on *Jurisprudence* generally.

Precept. Strictly speaking, a command by a legitimate authority which binds in conscience. More generally, a command, instruction, order, or direction.

Precept of sasine. In older Scottish conveyancing, an order by a superior to his bailie to give infeftment of certain lands to a vassal. Originally a separate deed, it later became a clause in the charter making the grant of the land, and is now unnecessary.

Precognition. In Scotland, a written statement, not on oath, of what a witness knows about a matter under investigation and consequently of the probable substance of his testimony in court on the matter, corresponding to the English proof of evidence. In criminal cases, precognitions of witnesses are usually taken by a procurator-fiscal or one of his staff.

Prediction studies. A branch of criminology which, by comprehensive investigation of the personal conditions, antecedents, and careers of prisoners, seeks to correlate certain factors with success and others with failure in social readjustment, and to make it possible to predict how the offenders might behave after a corresponding penal treatment. Prediction tables have been formulated which, on the basis of given facts, indicate the probability of an offender's success or failure during and after particular treatment. Checking by follow-up studies indicates a substantial correlation between prediction and actual subsequent career, at least as high as the correlation between educational prognoses based on intelligence tests and actual achievements. Further work on these lines offers hope of being able to offer guidance to courts in sentencing and to institutions in dealing with offenders.

Pre-emption. A right of purchasing property in preference to another or others. It exists in some cases by statute and may be given by contract.

The term was also given to a prerogative right by which the royal purveyor might buy up goods for the use of the royal household at an appraised price, in preference to other prospective buyers and without the owner's consent. This right was abolished in 1660.

In international law, pre-emption is the right of government to purchase for their own purposes the property of subjects of another state in transit through their states. The right, formerly general, is now limited to times of war and the circumstances where the goods, though not contraband, would be harmful to the pre-empting state if allowed to be delivered to their destination.

Prefect. In ancient Rome, an officer appointed by a superior magistrate as his deputy or subordinate. Under the early republic, a prefect of the city (*praefectus urbi*) was appointed by the consuls if they had to be absent from Rome. After 367 B.C. the praetors exercised this function. The office of prefect was revived by Augustus and the holder was placed in charge of law and order in Rome. Two praetorian prefects (*praefecti praetorio*) were appointed to command the imperial bodyguard, the praetorian guard. They attained great power, acting as prime ministers or, in the case of weak emperors, virtual rulers; they sometimes commanded armies or exercised judicial functions. From Diocletian's time there were praetorian prefects in both Eastern and Western Empires, with financial powers and powers of supervision of governors. Under Constantine they became civil officials only, but the highest ones with judicial and financial functions.

There were also prefects drawn from the equestrian order; these included prefects charged with the Roman fire brigade and with managing the corn supply.

Under the Republic, consuls and praetors appointed an equestrian *praefectus fabrum*, initially a chief engineer to the army and later a kind of chief of staff, and the urban praetors appointed judicial prefects (*praefecti juri dicundo*) to administer justice, while later the title was given to commanders of subordinate task-forces, and to officials dealing with finance.

The name was revived in 1800 by Napoleon who created a corps of prefects to replace the older *intendants*, to act as administrators of the *départments* and ensure that government policy was effectively executed throughout the country. In the nineteenth century, they were mainly concerned with police and elections and with maintaining the government's parliamentary majority.

In time, they were increasingly concerned with social and economic matters and promoting government policy on these matters in their areas. Prefects are appointed by the President on the recommendation of the Minister of the Interior, to whom they are responsible. The prefect is now the chief executive officer of the *conseil général* or elected authority for the *département*, the police authority and supervisor of the *communes*, and his sanction is required for many administrative acts of local authorities. Within the *département* he is the sole legal representative of the central government, delegate for the exercise of ministerial powers, and channel for the communication of power from the central government to subordinate authorities.

The prefectoral corps includes as well as prefects, secretaries-general, who are deputies of prefects, sub-prefects, responsible for *arrondissements* and chefs du cabinet, who are personal assistants of prefects. In Paris the prefect of the *département* exercises the powers usually exercised by the mayor of a commune, and there is a special prefect of police.

Préfecture de Police. The police force for Paris and the *Département de la Seime*, controlled by the Minister of the Interior and headed by the prefect of police. It comprises uniformed police (*gardiens de la paix*) responsible for crowd and traffic control, and plain-clothes police (the directorate of civil police) who investigate serious crime anywhere in France. See also *Gendarmerie Nationale*; *Sûrété Nationale*.

Preference shares or stock. In British company law, shares in or stock of a company carrying entitlement to dividend at a stated rate in preference to the claims of holders of ordinary shares to any dividend. Preference shares may or may not be cumulative; if they are, any preference dividend not paid in any year must be made up in a future year also in preference to the claims of ordinary shareholders. Preference shares may or may not be participating; if they are, preference shareholders have claims, in addition to their preference, to participate in the funds available for distribution as dividend. Preference shares, generally, offer greater security then do ordinary shares, but have less prospects in times of expansion. Preference shares usually also have priority of repayment in the event of winding up.

Preferential payments. In the administration of the estates of deceased persons, in bankruptcy, and in the winding up of companies, claims which, by statute, have to be paid in preference to other claims against the assets. *Inter se*, they rank equally and if the assets are insufficient to satisfy them all they abate in equal proportions.

Prejudice, without. A phrase frequently employed in negotiations for settling a dispute, meaning 'without prejudice to the writer's claims, pleas and arguments', conveying the meaning that if the negotiations break down, nothing communicated in the negotiations may subsequently be taken advantage of by the other party. Accordingly, an offer made to settle the dispute, if made 'without prejudice' cannot be treated as an admission of liability. Nothing said or written 'without prejudice' may be put in evidence or considered at a trial or hearing unless by consent of both parties. Nor can it be used as cover for a libellous statement.

The phrase 'without prejudice to ... ' is used in statutes and deeds signifying that what is being enacted or provided is not to affect the instances excepted by the 'without prejudice' clause.

Preliminary act. In actions of damages arising from ship collisions, a questionnaire and answers normally required to be filed, sealed-up by each party before any pleading is lodged, stating the party's view on various facts, such as speeds, lights, bearings, wind, tide, weather, and other particulars. The object is to get a statement of these facts from each party as soon as possible, and they cannot normally be later amended. They are not opened until ordered by the court, when the court may order trial on them without any further pleadings.

Preliminary criminal proceedings. All the proceedings prior to trial, comprising bringing the person charged before a justice or justices by summons or by arrest, with or without a warrant, preliminary examination by justices to determine whether the accused should or should not be committed for trial and discharge or committal for trial. Incidental matters may include application for bail and application for legal aid.

Preliminary examination. (1) A private examination by the official receiver of a person against whom a receiving order has been made, so as to discover his assets and liabilities. The debtor's answers are taken down in writing.

(2) A stage in criminal proceedings, introduced in the French Code of 1808 and adopted over most of Europe, intended to ensure that only accusations which have foundation in fact and in law, are submitted to the trial court and to dismiss the unfounded charges before trial. It applies only to serious offences. It is conducted by an examining magistrate who has extensive powers to interrogate the accused, make inspections, searches and seizure, hear witnesses, and discover the truth. Under various procedural codes, provision is made for controlling the proceedings and ensuring fairness for the accused.

Preliminary expenses. The expenses incurred in promoting, forming, and registering a company.

Preliminary inquiry. In English criminal procedure, where a person is brought before a magistrates' court charged with an indictable offence, a single magistrate or justice holds a preliminary examination to determine whether or not the accused should be committed for trial. Evidence is led for the prosecution and the accused may be represented and may cross-examine prosecution witnesses. The evidence is taken down and the written version, or deposition, is read over to the witness in the presence of the accused and signed by the witness and the presiding magistrate. The accused is then cautioned and may give evidence and call witnesses or may reserve his defence. The court must then decide whether there is evidence justifying the accused being tried, and if so it commits the accused for trial, and if not discharges the accused. This decision must be taken in respect of each count of the indictment separately.

Preliminary rulings. The European Court has jurisdiction to give preliminary rulings on the interpretation of the E.E.C. and related treaties, the validity and interpretation of acts of the institutions of the Community, and the interpretation of the statutes of bodies established by an act of the E.E.C. Council, if the statutes so provide. Any court or tribunal in a member state may, if it considers that a decision on a question under any of these heads is necessary to enable it to give judgment, request the Court to give its ruling thereon. Such a reference is obligatory in the case of matters before courts against whose decisions there is no judicial remedy under national law. 'Judicial remedy' covers appeal, cassation, and probably review of administrative action under the prerogative writs (q.v.). A court low in the judicial hierarchy must refer a question if no appeal lies from it. The reference to, and the ruling by, the Court do not decide the case before the national court but give that court an authoritative ruling on the issue of Community law raised and enable it then to decide the case, using the ruling. If a judge of a national court finds a question one of the kinds which is referable in a case before him he may, or must, refer it even though he sees no difficulty of interpretation. If the Court has previously ruled on the matter there is no need to refer a question again, but a national judge is not prevented from doing so.

Premises (praemissa). Things set out before, and consequently, in deeds, things previously mentioned. In conveyances, the word frequently refers back to subjects fully described earlier in the deed, and has accordingly come to be used generally as meaning 'land' or 'buildings'. It is also used as meaning all the parts of a deed prior to the *habendum* (q.v.), comprising the introduction, date, parties, recitals, consideration, and operative word of gift, grant, release, or otherwise.

In logic, the premises (or premisses) of a syllogism are the two propositions, major and minor, from which a conclusion can be drawn.

Premium. A lump sum consideration; the annual or other periodical payment for the renewal of a policy of insurance; the value of shares or other securities so far as exceeding their par or nominal value.

Prepense. Preconceived, previously thought out. Malice prepense (*malitia praecogitata*) or malice aforethought is the mental element requisite for a killing to be murder.

Prerogative, royal. The pre-eminence which the sovereign enjoys at common law over and above all other persons in right of her royal dignity, including all the special dignities, powers, privileges, and liberties allowed by law to the person in right of the Crown. It is created and controlled by common law, modified in some cases by statute. The special privileges comprise (1) those concerned with the qualities of dignity and pre-eminence ascribed to the sovereign, such as personal inviolability, immunity from all civil and criminal proceedings, exemption from the operation of statutes and of custom; (2) prerogatives of power and authority, including being titular head of the Church of England and the armed forces, chief executive of the State, controller of foreign relations, maker of peace and war, fountain of justice and mercy, source of all honours and dignities, and protector of all subjects; and (3) special privileges in relation to rights of property. The extent of the prerogative has been defined, and in some respects limited, notably by Magna Carta (1215), the Petition of Right (1628), the Bill of Rights (1688), and the Act of Settlement (1700). Thus the powers of suspending and dispensing with laws, save by consent of Parliament, have been abolished. Taxation can be imposed only with the sanction of Parliament. The growth of the principle of the responsibility of the Queen's Ministers to Parliament has meant that in practice many prerogative powers are exercised by Ministers in accordance with the policy of the majority in Parliament.

Since the prerogative is created and defined by the common law, the courts can inquire into the existence or extent of any alleged prerogative, but must take judicial notice of it if it exists.

Law and custom impose various restraints on the arbitrary and improper exercise of the prerogative. Thus, by convention, the sovereign acts only on the advice of her constitutional advisers, her Ministers, and one or other Minister is responsible to Parliament for everything of an executive nature done in her name. The relationship between statutory and prerogative powers remains very obscure.

By virtue of the royal prerogative, the sovereign is the source and fountain of all justice and all jurisdiction is derived from her. All commissions to administer the law are granted by the sovereign; all writs run in her name and are executed by officers of the Crown; she may grant pardons and reprieves.

In foreign relations, the Crown appoints ambassadors and other diplomatic agents and conducts foreign policy, making peace and war, and concluding treaties. But, by convention, these relations are effected in conformity with the wishes of Parliament. In general, treaties are binding without parliamentary sanction, but in important cases the agreement of Parliament is made a condition of the treaty and is accordingly necessary. In wartime, the Crown has wider powers in the exercise of the prerogative than in peacetime; thus power to requisition ships, to require the services of persons capable of bearing arms and others apply in wartime.

Among surviving prerogative rights in domestic affairs are the rules that the Crown can do no wrong and accordingly, save as permitted by the Crown Proceedings Act, 1947, cannot be sued for breach of contract or tort, that the Crown has a common law priority as a creditor on the estate of an insolvent debtor, that save as provided by statute, time does not run against the Crown in litigation, that the Crown is not bound by statute save by express words or necessary implication, and that the Crown has a duty to protect its subjects, including the welfare of minors and persons of unsound mind. The Crown also has prerogative rights and powers in respect of coinage, the granting of charters of incorporation, the grant of franchises, including the rights to hold markets and fairs and to take tolls from bridges or ferries, the construction of ports and harbours, the printing and publishing of statutes and other legislation and of the authorized version of the Bible and the Book of Common Prayer.

Prerogative rights can be abridged or modified by Act of Parliament, but only by express words.

Prerogative courts. Courts through which the English kings exercised their prerogative and discretionary powers as distinct from the courts exercising the ordinary jurisdiction. They included the Star Chamber, dealing with offences against public order, the High Commission, to deal with ecclesiastical affairs arising from the Reformation, the Court of Requests which handled small claims, and the Court of Chancery which granted equitable relief in supplement of remedies available at common law (qq.v.). All had aroused strong opposition from the common law courts by the seventeenth century and suffered by association with the royal prerogative when all the royal powers were attacked in the reign of Charles I, particularly when used, as the Star Chamber and High Commission were, to enforce the King's policies which did not have Parliamentary authority or support. All these prerogative courts, except the Chancery, were accordingly abolished before 1660:

Prerogative Court of Canterbury. This court originally sat in the archbishop's palace, but was moved about the Reformation to Doctors' Commons. It exercised the archbishop's testamentary jurisdiction and most of the testamentary business of the court flowed to this court, particularly by claiming to oust the bishops' jurisdiction when a man left property in more than one diocese, a claim recognized in 1604. The court was sometimes presided over by the Official Principal of the Archdiocese, sometimes by a special commissary.

Prerogative Court of York. A similar court to the Prerogative Court of Canterbury (q.v.) was held for the province of York.

Prerogative of mercy. In early times, an exercise of the royal prerogative to pardon offenders was more common than it later was, and few offences could not be pardoned. Later this prerogative was exercised to change the mode of execution e.g. from hanging to decapitation, or to avoid the capital penalty altogether, as where felons were pardoned conditionally on going abroad for a period of years. More recently, it has normally been exercised, on the advice of the Home Secretary or the Secretary of State for Scotland, by way of commuting a death sentence to one of life imprisonment, but since the abolition of capital punishment the prerogative has not required to be exercised in this way. See also PARDON.

Prerogative writs and orders. Processes issued by English courts in special circumstances; they are the writ of *habeas corpus* (q.v.), the orders of mandamus, prohibition and certiorari (qq.v.) and formerly the writ of Quo Warranto (q.v.).

Presbytery. In the Church of Scotland and other presbyterian churches, a meeting, constituted by the minister of and an elder from each of the churches in a district, which has certain supervisory functions and power to determine complaints and appeals from kirks within its bounds, and to investigate charges against ministers. A presbytery is presided over by an elected moderator.

Prescription. The term for the principle whereby a right or obligation is created by lapse of time, or a right or liability is extinguished by lapse of time. Positive or creative prescription originated in Roman law in relation to time fortifying title to property in fact possessed for stated periods of time. In common law, it has many applications in relation to incorporeal hereditaments, which if actually used and enjoyed, continuously and peaceably, from time immemorial or from time whereof the memory of man runneth not to the contrary (which is held to be from the beginning of the reign of Richard I and is presumed if the right has been enjoyed for 20 years on the basis, according to some, of the fiction of a presumed lost grant) are indefeasible. The duration of prescription is fixed in certain particular cases by statute. Negative or extinctive prescription applies to realty or corporeal hereditaments and has the effect of giving one who has occupied without interruption for a stated time a valid and unassailable title by extinguishing the claims of all possible competing claimants. It is now mainly regulated by the Limitation Act, 1939.

In Scots law, positive prescription fortifies the title of one who has possessed land or buildings, or exercised servitude rights, for stated periods, and negative prescription extinguishes rights and obligations, and servitude rights, after stated periods if the rights have not been claimed or acknowledged during the periods.

See also LIMITATION.

Presence, hearing in. A rehearing before a larger number of judges of a case heard by either Division of the Court of Session and which has been ordered to be re-argued.

Present, to. To put a bill of exchange before the drawee for acceptance by him, or for payment by him. See also BILL OF EXCHANGE; DISHONOUR; NOTING.

Presentation. In ecclesiastical law, the offering by the patron of a benefice to the bishop of the diocese of a person to be instituted to the benefice. The branch of the Lord Chancellor's office concerned with the exercise of his ecclesiastical patronage was formerly called the Presentation Office; it was abolished in 1832.

Presentment. A report or return made by a jury or other similar group. Thus, at common law, a jury made presentment, as when a grand jury named a person to be indicted, or after holding an inquisition of office, or more commonly reported any offence or other matter from their own knowledge, such as a libel or nuisance, on which an indictment might follow.

Presentment of a bill of exchange. See PRESENT.

Presents. In a deed, 'these presents' means the deed itself. It is probably a translation and abbreviation of '*has praesentes litteras*'.

President. The usual name for the head of state in a republic. His position and powers vary from state to state according to the constitution and practice. There are two main kinds of Presidents, those who are heads of state with limited executive powers, and those who are both heads of state and chief executives.

In the U.S. the President is not only the titular and ceremonial head of state, but the chief executive, charged by the Constitution to take care that the laws be faithfully executed; Commander in Chief of the armed forces; and the leader of his party and, as such, controller of party patronage. He has wide powers of nominating persons to office, including to the seats on the Supreme Court, and of appointing to other offices. Increasingly, Congress has delegated power to the executive by administrative regulations.

The President is assisted by a growing staff in the Office of the President, by his Cabinet of heads of the Departments of State, and sometimes by personal advisers or special assistants. Increasingly, Congress and the nation looks to the President for a lead, a trend enhanced by the crises of the twentieth century, the World Wars, the Depression, the Cold War, and by the growing importance of the U.S. as a world power and the leader of the non-Communist nations.

In some other countries, on the other hand, such as West Germany, the President is a less important and largely ceremonial officer, the effective political leader being the Chancellor.

The title is also given to various other persons who preside, formerly to persons doing so by delegation for the Crown and now in some cases to heads of universities, colleges, societies and commercial organizations.

Press and printing. In British law, there is no distinct body of law affecting the Press save for a few rules applying to newspapers only. There is no system of prior censorship or public control of the Press or of printing. In 1538, a requirement that books be licensed for printing by the Privy Council or other royal nominees was introduced. After 1557, when the Stationers' Company was incorporated, only members thereof or other persons holding a special patent might print any work for sale in England. In 1559, Elizabeth issued Injunctions requiring that no book be printed unless licensed by her, the privy council, or certain Church dignitaries, but this seems to have been largely ineffective and, in 1586, an Ordinance of the Star Chamber forbade printing presses outside London, apart from Oxford and Cambridge. Throughout the seventeenth century, control and licensing of the Press were continued, but licensing of the Press came to an end in 1694 when the controlling Act expired and was not renewed. Thereafter control of the Press was effected by prosecutions for treason, seditious libel, and similar offences, and governments relied more on meeting attacks and criticisms by publishing their own journals and pamphlets.

Writers, printers, and publishers are, however, subject to general legal rules of wide application, notably to the law of copyright, libel and slander, obscenity, official secrets, contempt of court and the like, which impose control by way of possible civil or criminal liability after publication. Relations between publisher and printer and between publisher and author are regulated by contract.

Papers printed by order of either House of Parliament are absolutely privileged, while fair and accurate reports of proceedings before courts and recognized legal tribunals and, by statute, fair and accurate reports of many other kinds of proceedings, enjoy qualified privilege, in some cases subject to explanation or contradiction. The Press may also claim the defence of fair comment on matters of public interest. The printing of certain kinds of information is criminal. In general, representatives of the Press are entitled to be present at meetings of local authorities unless expressly excluded, and to be present at the sittings of any court, save when the public is excluded.

W. A. Copinger, *Copyright*; C. Gatley, *Libel and Slander*.

Press Council (or General Council of the Press). A body, formed in 1953 and reorganized as the Press Council in 1963, with the objects of preserving the freedom of the Press, maintaining high professional and commercial standards, considering complaints, watching for restrictions on the supply of information of public interest and importance and reporting on developments tending towards concentration or monopoly in the Press. It issues annual reports.

Press gang. A naval party which, down to the beginning of the nineteenth century, was used forcibly to impress or take men for service in the Navy. From mediaeval times the Crown claimed the power to impress men for the defence of the realm. Only at the end of the sixteenth century were seamen exempted from being pressed as soldiers, but they remained liable to impressment for naval service. Under Elizabeth, a Vagrancy Act rendered 'disreputable persons' liable to impressment for naval service. In the seventeenth and eighteenth centuries, the quotas of men needed were frequently made up from the gaols. During the Napoleonic wars, the press gang was the main means of recruiting naval seamen. In 1795, each county was directed to find a quota of men, who were, of course, all the troublemakers and nuisances of the district. Insistence on the right to press British subjects was a contributory cause of the American War of Independence and the war between Britain and the U.S. in 1812. After 1815, the press gang was not used, and need for it disappeared in the nineteenth century though the right to impress probably survives in theory. In both World Wars, conscripted men served in the Navy.

Pressing to death. See *Peine forte et dure.*

Prestable. In Scots law, payable, exigible, enforceable.

Prestation. In Scots law, that which is due or payable or to be performed.

Pressure group. Any group or organization which exerts continuous influence and pressure on government to take account of its views and to enact legislation and take executive decisions agreeable to its interests and views. Notable examples are the T.U.C. and the trade unions, the C.B.I., the B.M.A., the motoring organizations, the licensed trade and many others, which may be called sectional or interest groups, because their interests are involved, and bodies such as the Howard League (q.v.) and the R.S.P.C.A. which may be called cause or promotional groups, because they seek to promote causes rather than promote their own interests.

Presumption. In the law of evidence, an inference or conclusion of fact which must or may be drawn from certain other established facts. Presumptions are commonly grouped into three classes; firstly, conclusive or irrebuttable presumptions of law, *praesumptiones juris et de jure*, which are conclusions which by law must be drawn from certain facts and which cannot be rebutted by any contrary evidence, and which are consequently not truly matters of evidence but rules of law. These include e.g. the presumption that no child under 10 can be guilty of any criminal offence.

Secondly, there are rebuttable presumptions of law (*praesumptiones juris*), which are conclusions which, by law, are required to be drawn in the absence of evidence to the contrary, such as that a person accused is of sound mind and responsible for his conduct.

Thirdly, there are presumptions of fact (*praesumptiones facti*), which are conclusions which a judge of fact or jury may draw from other proved facts. The strength or weakness of a particular presumption of fact depends on the circumstances and the presence or absence of alternative explanation. Thus possession of stolen property very shortly after it was stolen, in the absence of explanation, raises an inference that the possessor was connected with the theft. Possession of property, particularly if for a substantial time, raises a presumption of ownership. Many particular presumptions arise in particular contexts, such as the presumption, based on common medical knowledge, that a woman in her fifties or older is beyond the age of child-bearing.

In the interpretation of statutes certain so-called presumptions exist, which are not truly concerned with evidence at all, but are canons of interpretation.

Presumption of innocence. A fundamental principle of criminal law in common-law countries. It is to the effect that a person charged with crime must be presumed innocent until or unless the contrary is admitted by him or proved beyond reasonable doubt by competent evidence. There follow from this various consequences; if no evidence is led the accused is entitled to be discharged; the onus of proof is on the prosecution, which must rebut the presumption of innocence by establishing by competent evidence, beyond reasonable doubt, that the accused was guilty of the crime charged; there is no onus on the accused to explain his conduct, nor to exculpate himself unless he is relying on a substantive defence such as alibi, mental irresponsibility at the time, or self-defence, in which case the accused must give evidence of facts supporting his plea in defence. Statutes have, however, in some cases made particular kinds of conduct an offence, making certain facts, such as possession of drugs, presumptive of guilt, and throwing the onus of exculpation on the accused.

Presumption of life. Where a person is known to have been alive at a particular time, there is a rebuttable presumption that he continued in life for some time beyond the ordinary and normal duration of life for a person born at the time he was born. If, however, he has not been seen, reported or heard from or about, for at least seven years continuously he will be presumed to have died not later than the expiry of the seven years. If it is necessary to establish the precise date of death within the seven-year period, it must be done by evidence and is not established by presumption.

In Scotland, the common law presumption is that a person lives 80 to 100 years, but this may be overcome by evidence of facts pointing to the high probability of earlier death or, under statute, by evidence of disappearance for at least seven years.

Presumptive heir. A person who is the heir of another person, but subject to the contingency that a child may be born who will displace him and be a nearer heir.

Presumptive title. A title arising from the bare possession or occupation of some property but without apparent right or claim of right to maintain the possession. Such a title is sufficient to ground an action against a trespasser and may by lapse of time ripen into an indefeasible title.

Preuss, Hugo (1860–1925). German political theorist. A follower of Gierke and an expert on public law, he was the chief draftsman of the German post-1918 constitution. His chief work was *Die Entwicklung des deutschen Städtwesens* (1906).

Prevarication. The conduct of a witness who seeks to avoid answering lawful questions or gives evasive answers. It is punishable as a contempt of court.

Preventive detention. A mode of imprisonment resorted to where there appears good grounds for believing that the individual must be kept in custody to prevent him continuing or repeating his offences, or occasionally, to ensure that an individual does not interfere with evidence before his trial. The Prevention of Crime Act, 1908, authorized an additional sentence of preventive detention where an accused was charged with being an habitual criminal, and the Criminal Justice Act, 1948, replaced this by a provision permitting, in certain cases, a sentence of from five to 14 years' preventive detention. In 1973, this was superseded by power to impose an extended term of imprisonment if it is expedient to protect the public from the offender for a substantial time.

Previous conviction. The fact that an accused had previously been convicted of the same or an analogous crime was formerly charged as an aggravation of the crime. It is now deemed a fact relevant to penalty only and not disclosed to jury, or magistrates prior to conviction. Some statutes provide for increased penalties where the accused has previously been convicted. An accused, if giving evidence, may not be questioned as to whether he has previously been convicted unless evidence of a previous conviction is admissible to show whether he is or is not guilty of the crime charged, or unless he has sought to establish his good character or the bad character of the prosecutor or any of the witnesses for the prosecution or given evidence against anyone charged with the same offence.

Prevôt. In pre-revolutionary France, an inferior royal judge. The office dated from the eleventh century and the prevôts served as general administrators, military commanders, and lower court judges, and also collected fines and taxes. They held their offices as feudal fiefs and, gradually, achieved considerable independence. By the thirteenth century, however, they were brought under the control of the *baillis* and, from 1493, were salaried officials. By the seventeenth century, they were judicial officers only.

Price, Richard (1723–91). English moral and political philosopher, supported American independence in *Observations on Civil Liberty and the Justice and Policy of the War with America* (1776). His favourable views on the French Revolution were attacked by Burke. He wrote also a *Review of the Principal Questions in Morals* (1757) which won some reputation at the time.

Price maintenance agreements. Agreements between manufacturers and retailers whereby the latter undertake not to resell the manufacturers' products at less than prices fixed by them. Under the Resale Prices Act, 1964 (now 1976), such contractual provisions are void unless they can be shown to the satisfaction of the Restrictive Practices Court to be in the public interest, as was held proved in the case of the Net Book Agreement.

Pricking for sheriffs. The annual ceremony at which the Queen marks with a silver bodkin one of the three names of persons nominated for each county to serve as sheriff for the ensuing year. In practice the names have been chosen beforehand. In the Duchy of Cornwall the sheriff is appointed by the Prince of Wales as Duke of Cornwall and in Lancashire by the Crown in right of the Duchy of Lancaster.

Pride's Purge. The unconstitutional exclusion by force, on 6 December 1648, of certain members of the House of Commons, effected by Colonel Pride, with a large force of soldiers, at the instance of the leaders of the Army and with the connivance of certain members of the House. It was done to ensure a majority for the policy of bringing Charles I to trial and was successful in this, thus leading up to his trial and execution.

Priestly Code. A collection of ancient Hebrew law, found in the books of Exodus, Leviticus, and Numbers, and containing in Leviticus 17–26 a major section known as the Code of Holiness. It belongs to the post-exilic period, afer 538 B.C. and emphasizes institutional, ceremonial, and ritualistic practices. The main emphasis is on purity of worship of Jahweh. Most of the ordinances of the Code of Holiness come from the pre-exilic period, but show a reinterpretation in the light of the experience of exile in Babylon.

Prima facie. (at first appearance). Thus a party must frequently make a *prima facie* case, that is, a case sufficient to call for an answer. *Prima facie* evidence is evidence which is sufficient to establish a fact in the absence of any evidence to the contrary, but is not conclusive.

Primae impressionis (of first impression). A case of first impression is one on which there is no clear authority and which must be decided by reason.

Primage. A small sum payable by the owner or consignee of goods to the master of the ship by which they have been carried for his care and trouble. It varies in amount according to the custom of the port or of the particular trade and is usually stipulated for in bills of lading. It is sometimes called hat-money.

Primate. A chief ecclesiastical person. The Archbishop of Canterbury is Primate of All England and the Archbishop of York is Primate of England. In both the Roman Catholic Church in Ireland and the (Anglican) Church of Ireland the Archbishops of Armagh are Primates of All Ireland and the Archbishops of Dublin are Primates of Ireland. In Scotland there is no primate, but one of the bishops of the Scottish Episcopal Church is elected Primus.

Prime Minister. The leader of the Queen's Ministers in the United Kingdom Parliament. The first Prime Minister is usually held to be Sir Robert Walpole who became head of the administration as First Lord of the Treasury and Chancellor of the Exchequer in 1721. Since his time (1721–42) there has always been one statesman who has been deemed head of the government and who has usually held the office of First Lord of the Treasury, though occasionally he has held other offices. He has always been a Member of one of the Houses of Parliament and in modern practice (since Lord Salisbury in 1895–1902) always a member of the House of Commons.

In law the office probably does not exist, though it is recognized by statute for purposes of salary and pension. He is merely the person invited to head the government. By modern convention the Queen must invite the leader of the largest party in the House of Commons to be Prime Minister. He is accordingly both the Queen's chief minister and the leader of the largest party, owing allegiance as chief minister to the Queen and to the country, and as party leader to his party and its policy and interests. In consequence a General Election is in effect a national choice of a Prime Minister. He chooses persons from among his chief supporters to hold the chief and subordinate offices, and may dismiss them or move from one post to another at will. He also has enormous powers of patronage, and the power to advise a dissolution of Parliament when he thinks it will suit his political advantage.

He is the person mainly responsible legally to the Crown and practically to the community as a whole for the whole administration of government, and the leader and co-ordinator of all government policies. Constitutionally he is the Queen's chief adviser on all matters of government.

By Royal Warrant he has precedence in England after princes of the blood royal, the Archbishop of Canterbury, the Lord Chancellor and the Archbishop of York, and in Scotland after the royal family, the Lord High Commissioner to the General Assembly of the Church of Scotland, while it is in session, the Lord Chancellor and the Moderator of the General Assembly.

While he holds the office he has the use of the town house, No. 10 Downing Street, and the country estate of Chequers, Bucks. On relinquishing office or subsequently he is normally offered an earldom, e.g. Earl of Avon (Mr Anthony Eden) or other high honour such as membership of the Order of the Garter, e.g. Sir Winston Churchill.

Primer seisin. In feudal law, on the death of a tenant in chief the King was entitled to primer seisin, or a year's profits of the land, and only when the heir had done homage and paid a relief (q.v.) might he recover the lands from the King and get seisin thereof. Primer seisin was abolished in 1660.

Primitive law. Primitive law is not synonymous with ancient law. It comprehends systems, institutions and rules regulative of conduct in societies which are immature and undeveloped and reflect an early stage of civilization, and is exemplified by some ancient and mediaeval legal systems, as evidenced by surviving literary fragments and as reported by later writers, and by social anthropological observations of peoples living at primitive stages of development at the present time or in the recent past. Thus such distinctions as civil and criminal law, and concepts such as rights and interests, duty and sanction, are unknown, though relationships such as marriage are known.

Preliterate peoples have no law in the sense of statutes and codes, nor legal institutions of courts and judges, but frequently recognized customs as having obligatory force and there is frequently collective enforcement by the elders or the collective opinion of the group. Such obligatory customs are partly social, partly moral, partly religious and it is difficult to distinguish custom, morality, taboo, ritual, and belief in the supernatural. Whatever their nature regulation of conduct is common and effective, and societies with only primitive law are not anarchic.

There are probably no customs or practices found universally, but the binding practices of different peoples vary considerably. Common characteristics of primitive law are collective enforcement of conformity by the tribe, collective responsibility of the family or tribe for misdeeds, blood-feuds, frequently averted by compensation according to recognized tariffs, and the importance of relationships, in marriage, property, responsibilities, and succession.

As societies become literate unwritten primitive law merges into the primitive law evidenced by such immature codes as the barbarian laws of Western Europe, and the earliest laws of English Kings, though even these represent an advance in that legal rules and sanctions are becoming differentiated from social, moral, and religious rules and that customs are being written down and sometimes consciously amended and systematized. Maturity of codes is not correlated with antiquity. The most immature and 'early' codes include the barbarian law and the earliest English laws down to those of Alfred the Great, in most of which the sanctions for wrongs are pecuniary.

Moderately mature (middle Codes) include the Hittite Code, the Twelve Tables, the Hebrew Code, and the Lombard laws (qq.v.), in which crimes are more important and commercial transactions are found, and substantially mature (late Codes) include those of Hammurabi, the Visigoths, Burgundians and Ostrogoths, and the Norman kings, related to conditions of substantial material culture in which commercial transactions are important.

H. S. Maine, *Ancient Law; Early Law and Custom*; R. Lowie, *Primitive Society*; L. T. Hobhouse, *Morals in Evolution*; A. Kocourek and J. Wigmore, *Evolution of Law*, 3 vols; E. S Hartland, *Primitive Law*; B. Malinowski, *Crime and Custom in Savage Society*; H. Hogbin, *Law and Order in Polynesia*; A. S. Diamond, *Primitive Law, Past, and Present*; E. A. Hoebel, *Law of Primitive Man*; Radcliffe Brown, *Encyc. Soc. Sc. IX 202.*

Primogeniture. Seniority of birth, the rule that the whole of an inheritance descends entire to the eldest son to the total exclusion of the younger sons. It developed as custom in Normandy and England in relation to fiefs of land held by military tenure, and it kept the fief intact and ensured that one man and one piece of land was liable for the feudal services, but later became common and extended to all lands, save in burghs, the county of Kent, which retained the custom of Gavelkind (q.v.), and in places where division of land among sons continued as a custom. The rule long outlived its original rationale and survived till 1926 in England and 1964 in Scotland. But where land descended to females, the rule did not apply and all took equally. It never applied in relation to personal or moveable property. It continues to apply in relation to inheritance of the royal dignity and, generally, of peerages and other dignities. The converse principle, ultimogeniture (preference for the youngest son) is occasionally found in other societies.

Primrose, Sir Archibald, Lord Carrington (1616–79). Succeeded his father as clerk to the Scottish Privy Council. At the Restoration he became Lord Clerk Register, then a Lord of Session as Lord Carrington and in 1676 Lord Justice-General, but was deprived of this office in 1678. From him is descended the family of Earls of Rosebery.

Prince (from *princeps*, leader). A title of high rank, used sometimes to designate a territorial sovereign of a territory less than a kingdom, as in the case of many pre-1815 states in Europe, and sometimes to designate junior members of a kingly house. Thus, in the United Kingdom the title given to all the sovereign's sons and that of princess to his daughters. The title of Prince of Wales was originated by Edward I and conferred on his son, Edward, on the conquest of Wales in 1284. It has since 1301 regularly been conferred (it does not pass automatically) on the eldest son and heir of the Kings and Queens of England and now of the U.K. The title of Prince of Scotland was borne by the eldest son of the King of Scots. The title of Prince Consort was bestowed by Queen Victoria on her husband Prince Albert in 1857. Prince Philip was granted the style and titular dignity of a Prince of the United Kingdom in 1957.

Princess Royal. The title which may be conferred by the sovereign on his or her eldest daughter.

Principality of Scotland. Lands in Ayr (now Kyle and Carrick), Renfrew, and Ross and Cromarty erected into a principality in the fourteenth century and vested in the eldest son of the sovereign or, until a son is born, in the sovereign as Prince and Steward in right of the Crown.

Principality of Wales. The whole of Wales, erected into a principality by Edward I in 1301 after his conquest of Wales. The title of Prince of Wales is not automatically inherited by, but is normally conferred on, the eldest son of English, and now British, sovereigns. When not so held the principality is vested in the sovereign.

Principal. A person who appoints and authorizes another to act as his agent. See AGENCY. Also a capital sum of money to which interest accrues.

Principal or principal debtor. A person who is indebted to another and whose liability, if not discharged, has been undertaken to be discharged by another, the surety or guarantor or, in Scotland, cautioner.

Principals and accessories. At common law in England participants in criminal actings were distinguished into principals in the first degree, being those whose acts were the most immediate cause of the criminal act; principals in the second degree, those who were present, aiding and abetting the principal in the first degree at the time of the commission of the offence; accessories before the fact, being those who before the commission of the crime, advised or encouraged its commission or knowingly gave assistance to one of the principals; and accessories after the fact, those who gave a principal or an accessory any assistance tending to or enabling him to evade arrest, trial, and punishment. In treasons (q.v.) participants of all four categories may be indicted as principals; in felonies (q.v.) the quadruple distinction was drawn, and in misdemeanours (q.v.) participants of the first three categories might be charged as principals and participants of the fourth category were not chargeable at all. By reason of the abolition in 1967 of the distinction between felony and misdemeanour the rule as to misdemeanours now applies to all crimes other than treasons.

Principate. A term for the period of Roman history from 27 B.C. when Augustus resigned his powers and retained the office of *princeps senatus*, to the accession of Diocletian in A.D. 284. During the principate the emperors were in fact supreme but their powers continued to be disguised under republican forms. From A.D. 284 the emperors were openly autocrats.

Principles of law. General precepts which serve as authoritative starting points for legal reasoning, employed continually and legitimately where cases for decision are not covered or are not fully or obviously covered by more specific and particular rules. Examples are the principle that an employer must take reasonable care for the safety of an employee, that a person cannot be held guilty of crime unless he intended the conduct deemed criminal. Such principles comprehend and explain numerous particular rules and specific decisions. In common law systems principles are developed inductively by judges on the basis of several particular decisions or worked out by academic writers in legal treatises. Principles may suffer exceptions and may conflict with one another, in which case the conflict must be resolved by resorting to a more general principle or to a higher ideal or moral standard.

The difference between a principle and a doctrine of law is frequently a matter of terminology or a choice of words rather than a difference made on any logical or legal ground. A principle is generally regarded as much more general than a rule of law, which is generally regarded as more particular and specific than a principle and may be very particular and specific.

Prins, Adolphe (1845–1919). Belgian penologist, for long Inspector-General of prisons, and very influential on Belgian criminal legislation. He was a leader in penal law reform and believed in the individualization of punishment. He also wrote extensively on penal and public law.

Priority. The quality of being before another in point of time or of having a higher right than another. Priority is frequently determined by such factors as date of intimation or of registration. In administration of estates, bankruptcy, winding-up of companies, and certain other cases, statutes confer priority on particular creditors' claims.

Prison. Originally imprisonment was not a place for or mode of punishment by itself but only a place for holding offenders. The dungeon of a castle was used to hold prisoners pending trial or sentence. In 1166 the Assize of Clarendon directed the building of gaols in counties and boroughs. The King's courts at Westminster had long maintained their own prisons, the Marshalsea and the Fleet, chiefly for debtors. Latterly these were the prisons of the Court of Chancery and of the Star Chamber respectively. By the mid-sixteenth century Workhouses, or Houses of Correction, modelled on and named after the Bridewell organized in London in 1552, were established under the local justices. Their original functions, of providing work for vagrants and unemployed, were superseded by the function of imprisoning petty offenders, and in 1720 statute recognized this change. In 1729 a report to the House of Commons disclosed the cruelties practised in the Marshalsea and Fleet prisons, in which fees for the safe custody of the prisoners were exacted from them by many kinds of oppression and cruelty. All kinds of prisoners were kept without segregation and there was no attempt to help or reform them.

In the 1770s, John Howard (q.v.) began his crusade to expose the misery and degradation to which prisoners were subjected, and started the movement for prison reform. This was indeed only one aspect of a European movement for the reform of penal systems which was a characteristic of the Age of Enlightenment, a movement which started with Voltaire and Montesquieu and was stirred by Beccaria's *Crimes and Punishments* (1764). In 1774 Howard received the thanks of the House of Commons. The Prison Act, 1778, began the modern British system, on the basis of separate confinement with labour and with religious and moral instruction, but this principle was not generally adopted by the justices, who then controlled the county prisons. For the next half-century, moreover, transportation to Botany Bay offered an alternative to imprisonment, until condemned by a parliamentary inquiry in 1837, while another inquiry also condemned the 'hulks' or floating prisons which were another alternative to imprisonment.

The adoption of the principle of separate confinement was stimulated in the early nineteenth century by Bentham's ideas, of the 'panopticon', a kind of prison in which the regime was of safe custody, separate cellular confinement, and hard labour, all cells being radially arranged and under constant surveillance from a central observation point. Bentham also, however, was a strong advocate of classifying and training prisoners, fitting them for discharge, and of ascertaining the causes of crime. His ideas, based on the ideas of reformation by seclusion, instruction and employment, and foreign (especially French) jurists, influenced the building of Millbank penitentiary (1813–23).

About the same time the Prison Discipline Society were working hard to remedy the state of neglect, filth, and privation which was the lot of most prisoners, and in 1823 a prison act was passed, as well as to ensure safe custody, to preserve health, improve morality, and enforce hard labour. Under this Act many local prisons which still survive were erected, based on classification rather than separation of prisoners.

In 1831 a parliamentary committee recommended that all prisoners should be confined in separate cells by day and night. This view was reinforced by those of the experts who inspected the rival systems in the United States, where the showpieces were the Walnut St. jail in Philadelphia, which enforced strict cellular confinement by night and day (Pennsylvania system), and the State penitentiary at Auburn, N.Y., where there was cellular confinement by night and labour in association by day under a strict rule of silence (Auburn system). In 1842 Pentonville was built as a model of cellular construction and as a test of the merits of the separate system. Those sent to Pentonville suffered 18 months isolation and masks were worn to ensure separation and non-recognition.

The Pentonville plan led to the establishment of that system throughout the country. Wormwood Scrubs was later built as an improvement on Pentonville.

The essential features of the English prison system down to 1857 were accordingly separate confinement and hard labour and much ingenuity was expended in devising forms of labour compatible with separation; these included cranks, treadmill, shot-drill, and stone-breaking. In 1857 the refusal of colonial territories to continue to accept transported convicts forced the introduction of the system of penal servitude. Convicts has to be kept under special discipline in prisons where there were opportunities for their employment, and a feature was the progressive stage system; initially there was nine months' separate confinement but the remainder of the sentence was divided into three stages and remission of sentence, not exceeding one fourth, could be earned.

In 1878 the Government assumed control of all prisons, local and convict, and the Prison Commission was established with Sir Edmund du Cane as Chairman. Many of the features of the penal servitude system were then introduced into local prisons, which dealt with short-term prisoners, the male and progressive stage systems, cellular confinement in the early stages followed by industry in association. Thinking about prisons was affected then and thereafter by the developing body of penological thought, particularly stressing the individualism of punishment and criticizing imprisonment as not suitable for all cases.

Another important stage was the report of the Gladstone Committee in 1894 which was followed by the Prison Act, 1898; this abolished hard labour, provided for division of prisoners according to degree and character of offence, empowered remission of sentence not exceeding one-sixth, and provided against excessive or unnecessary exercise of powers of corporal punishment. The general administration of prisons is in the hands of the Home Secretary. Each prison has not only a governor but a board of visitors appointed by the Home Secretary.

In the early twentieth century notable features were the removal of adolescent offenders from prison, they being committed instead to Borstal institutions (q.v.) the institution of preventive detention for habitual criminals, and the prohibition of the imprisonment of children. Probation (q.v) became a workable alternative in many cases. 'Open' prisons were developed for offenders who were not dangerous to the public and did not require close custody. In 1948 penal servitude, hard labour, and the three divisions of imprisonment were abolished. In 1963 it became competent to release prisoners on parole when they had completed one-third of their sentence.

Throughout the twentieth century the major obstacles to developments in penology have been the rise in the crime-rate, with resultant serious crowding in prisons, and the slow rate of replacement of out-of-date buildings. But modern penologists are frequently sceptical about the deterrent and reformative value of imprisonment, particularly of long sentences.

In Scotland a General Board of Directors of Prisons in Scotland was established in 1839 to supervise both the general prison at Perth and the local prisons, to manage which the Act set up County boards. The Board was abolished in 1860 and its powers transferred to the Home Secretary. In 1877 an Act established the Prison Commission for Scotland, the functions of which passed in 1929 to the Secretary of State for Scotland, acting first through the Prison Department and since 1939 the Scottish Home and Health Department. Though regulated by different legislation and with differences in detail, the general function and operation of prisons in Scotland is much as in England.

J. Howard, *State of the Prisons in England and Wales*; T. F. Buxton, *An inquiry whether crime and misery are produced or prevented by our present system of Prison Discipline*; J. Jebb, *Thoughts on The Construction and Policy of Prisons*; S. & B. Webb, *English Prisons under Local Government*; E. du Cane, *The Punishment and Prevention of Crime*; E. Ruggles-Brise, *English Prison System*; L. Fox, *The English Prison and Borstal Systems*; R. S. E. Hinde, *The British Penal System*, 1773–1950; M. Grunhut, *Penal Reform*; L. Radzinowicz, *History of English Criminal Law from 1750*.

Prison Commission. A body established in England by the Prison Act, 1877, charged under the Home Secretary with the general superintendence of all local prisons. It was a distinct corporate body, comprising not more than five Commissioners who were civil servants, and was responsible to the Home Secretary. There were also various specialist Directors and Assistant Commissioners. The successive chairmen were Col. Sir Edmund du Cane (1878–94), Sir Evelyn Ruggles-Brise (1894–1921), Sir Maurice Waller (1921–28), Sir Alexander Maxwell (1928–1932), Sir Harold Scott (1932–39), C. D. Carew Robinson (1939–42), and Sir Lionel Fox (1942–60).

Its functions were transferred to the Prison Department of the Home Office in 1963 and the Commission dissolved.

Prison reform. Prison reform is only one aspect of the wider topic of penal reform. It originated with the investigations and writings of John Howard (q.v.) from 1773 onwards and with the thinking of Beccaria and Bentham. Initially it was concerned to improve physical conditions in prison and to mitigate corruption and contagion of prisoners by more experienced ones, but quite soon developed into the positive task of developing reasonable methods of discipline and treatment for the re-adjustment of criminals. The Penitentiary Act, 1779, prescribed provision for three classes of prisoners with a gradual relaxation in confinement and labour. From 1787, however, a new period of transportation began and interest in prisons at home declined.

After Bentham's death his followers began a new era of penal reform. Peel's Prison Act, 1823, prescribed standards of space, order, and discipline for prisons and relied on classification of prisoners, but provided no central controlling authority. The reduction in the number of capital offences and the reduction and final ending of transportation increased the need for prisons as places of detention; Milbank was opened in 1821. In 1835 prison inspectors were appointed and they influenced acceptance of the principle of separate confinement, which was the dominating feature of nineteenth-century prison development in Europe.

New reformative influences came in with the work of Elizabeth Fry, who emphasized religious instruction, classification, and employment in association. She and her supporters and followers emphasized moral authority, re-education, and reformation.

In 1877 the prison administration was transferred to the Prison Commissioners and a rationalized system of prison management introduced. The Gladstone Committee Report of 1895, and the 1898 Act based thereon, treated reformation as the major object, to be achieved by individualized treatment, productive work, educational training of young offenders, but long-term detention of habitual criminals.

Twentieth-century developments included the Borstal system for young offenders, and later the introduction of corrective training and preventive detention, but increasing attention was being devoted to non-custodial methods of treatment, probation, attendance centres, and community service orders. Economic conditions have frustrated many ideals of prison reform in the U.K. and to far too great an extent, imprisonment was still having to be served in old buildings overcrowded and unsuitable for constructive training.

Prison reform (U.S.). In the U.S. the Philadelphia system was based on the idea of solitary confinement leading to meditation and consequent moral regeneration. The Auburn system of 1816 permitted association of prisoners during the day.

In 1870 the American Prison Association was formed to propagate the idea of rehabilitation, and subsequently there were introduced segregation of young offenders from older criminals, parole, probation, suspended sentences, and post-release supervision.

Prison visitors. Prison visiting started with Elizabeth Fry and her Quaker ladies, but after her time the practice died. There seems to have been some revival in the later nineteenth century and the practice was developed in the twentieth century, initially with women and, from 1922, with men. A National Association of Prison Visitors was established in 1944, combining former distinct associations for men and women prisoners. It now has a branch in each prison, which arranges visits to prisoners. Visitors are volunteers but act by invitation. The function of visitors is not to deal with complaints or grievances, nor to preach or deliver moral homilies, but to keep the prisoner in touch with the outside world, and give him some change of companionship and conversation. The prison visitors maintain close co-operation with after-care agencies. They are wholly distinct from Visiting Committees (q.v.).

Prisoners of war. Members of the forces of one country captured by or surrendering to the forces of the enemy. In early times prisoners were enslaved and in many cases mutilated or put to death, the more valuable being held to ransom. From the seventeenth century negotiations were sometimes made for an official exchange of prisoners.

The custom of releasing prisoners on parole, on their giving their word not to serve against their captors again for a period, was known from the eighteenth century. At the same time the principle of decent treatment of prisoners was developing.

In the twentieth-century wars exchange of prisoners has been limited to the sick and disabled, and other prisoners have been confined in camps and prisons, frequently underfed, bullied, and ill-used. Protection of the rights of prisoners has been sought to be given by a Hague Convention, 1899, and the Geneva Conventions, 1949.

Privacy. Sometimes explained as the right to be left alone, the claim not to have one's private life intruded on or unjustifiably brought into public consideration, a legal interest imperfectly recognized in many legal systems. A remedy can sometimes be given for infringement of privacy on grounds of breach of confidence or trust, breach of contract, libel, nuisance, or other grounds. The Universal Declaration of Human Rights denounces interference with privacy, family, home, or correspondence.

Private Acts. A term used at different times meaning different things. Prior to 1798 private Acts of Parliament were not printed or numbered. From 1798–1813 private and personal Acts were not printed or numbered. From 1814–1868 private Acts were distinguished into those printed and those not printed. Fron 1868 private Acts were confined to personal statutes, and in 1876 they ceased to be numbered. In 1948 the remaining private Acts came to be called personal Acts. Private Acts are now Acts of Parliament of which judicial notice is not taken, but which require in each case to be pleaded and proved by the party seeking to rely on the Act. From the end of the seventeenth century, however, the practice developed of inserting in local and personal Acts provisions that they were to be deemed public, and to be judicially noticed, and since 1850 every statute is public in this sense unless it contains an express provision to the contrary. The distinction between public and private Acts is accordingly quite different from that between public and private Bills, the former dealing with matters of public policy and general interest and the latter affecting the interests of individuals or particular localities. A public Bill may be general, local, or personal in nature, but becomes always a public statute; a private Bill may be local or personal in nature and will result in a public statute in the former case and a private statute in the latter.

Private Bills. Bills submitted to Parliament to secure Parliamentary authority, special powers or benefits for particular persons, local authorities, corporations, or statutory companies. They are quite distinct from Private Members' Bills (q.v.). Unlike a public Bill (q.v.) a private Bill is based on a petition presented by the person or body desirous of obtaining the authority, powers, or benefit desired. It may be local in character, seeking powers for a local authority in relation to its own area; or personal, relating to the affairs of a particular person, usually in relation to such matters as change of name, naturalization, divorce, or succession to estates, but these are all uncommon today. There was a large amount of private legislation in the mid-nineteenth century, particularly in connection with the grant of powers to railway companies. The need for private Bills has now been largely superseded by Provisional Orders and Special Procedure Orders (qq.v.).

Special procedure exists for dealing with private Bills and there are separate Standing Orders in both Houses of Parliament relating to Private Business. The work of carrying through the proceedings on behalf of petitioners is carried out by parliamentary agents (q.v.).

Private Bills originate in petitions deposited by the promoters in the Private Bill Office, 27 November being the last date in each year for the deposit of the petition and Bill and 30 January the last date for the presentation of petitions opposing the Bill. Public notices of intention to promote a Bill for stated purposes must be given by the promoters by advertisement in newspapers and in the Gazette, and individual notices to owners and occupiers of land or buildings liable to be affected by the proposals. The petition and Bill, once deposited, are open for inspection and copies must be furnished for the officers of Parliament and government departments concerned. The two Houses each appoint an examiner and on 18 December they separately commence to examine all petitions to ensure that Standing Orders have been fully complied with. If an examiner considers that a petition has not complied with Standing Orders, he refers it to the Standing Orders Committee, which considers it and makes a recommendation to the House.

Thereafter the Chairman of Ways and Means in the House of Commons and the Lord Chairman of Committees in the House of Lords decide in which House each Bill shall be introduced. The Bill is deemed read a first time once it has been laid on the Table of either House.

If petitions have been deposited opposing a Bill, and the promoters challenge the *locus standi* of the petitioners, the matter is decided by the Court of Referees (q.v.).

Private Bills must pass through the same stages as public Bills. Many are unopposed and the Second Reading is taken as unopposed Private Business before Question Time, but if opposed in the House the matter must be debated. Apart from opposition in the House a private Bill is deemed opposed if a petition has been presented against it, of if the Chairman of Ways and Means reports to the House that in his opinion it should be treated as opposed. After Second Reading every Bill is referred to the Committee of Selection which allocates the Bill to an appropriate committee for its committee stage. Unopposed Bills are referred to the Committee on Unopposed Bills, consisting of the Chairman of Ways and Means, the Deputy Chairman, and three members, assisted by Counsel to the Speaker or, in the House of Lords, of the Lord Chairman of Committees. Opposed Bills are referred, sometime in groups, to select committees consisting of a Chairman and three members unconnected with the locality and having no interest in the subject-matter. Their function is to determine whether or not it is in the public interest that the special powers sought should be allowed, and they sit as a semi-judicial body, and hear counsel for the promoters and objectors. A Bill may be amended in committee, in which case it must lie on the Table of the House so that amendments may be moved in the House. Report stage, third reading, and all stages in the other House are normally purely formal.

F. Clifford, *History of Private Bill Legislation*; O. C. Williams, *Historical Development of Private Bill Procedure and Standing Orders in the House of Commons.*

Private company. A company incorporated under the Companies Acts which by its Articles of Association limits the number of its members to 50, restricts the right to transfer its shares, and prohibits any invitation to the public to subscribe for any of its shares or debentures. Various provisions of the Companies Acts do not apply to private companies. The great majority of companies incorporated are private companies since incorporation confers on sole traders and small groups the privilege of limited liability.

Private international law. See INTERNATIONAL PRIVATE LAW.

Private law. While the division of any one state legal system into public law and private law dates back at least to Roman law, is fundamental in all countries with legal systems based on civil or Romano-Germanic law, and has been mentioned repeatedly since Roman times as the fundamental division of the law of a state, it is difficult to find a clear principle on which to make the division. In general, however, private law may be defined as the part of the whole body of principles and rules included in a legal system which comprises the principles and rules dealing with the relations of ordinary individuals with one another, and also those dealing with the relations of the State or an agency thereof with an individual in circumstances where the State or its agency does not have any special position or privilege by virtue of being a department of state.

The distinction must not obscure the fact that in many circumstances both public and private law may be relevant; conduct may be both a crime and a tort, or may give rise to both a claim for damages and for social security benefit. A local authority's contractual powers are fixed by public law, but the substance of most of its contracts is fixed by private law.

In all countries of the civil law family the private law has a generally similar structure, based on the Roman law. Sometimes such countries distinguish civil and commercial law, sometimes not. In countries of the common law family the concept of private law is almost unknown, though it corresponds generally to the spheres of common law and equity combined It includes both substantive law and civil procedure. Private law is entirely civil in character, administrative law and criminal law belonging entirely to the sphere of public law. Bodies of procedural law are appendages to private, administrative, and criminal law respectively.

Within the category of private law the main branches were defined by the Roman jurists as the law of persons, of things, and of actions; but more modern, and more detailed division of this category would distinguish the law of persons and family law, obligations arising from contract, delict or tort or on other grounds, property, trusts, succession on death, and, remedies, and also international private law. Commercial or mercantile law, industrial or labour law, and maritime law also fall into the category of private law, though in some modern civil-law systems these bodies of principles are contained in codes or bodies of legislation distinct from the civil code.

See also PUBLIC LAW.

Private Members' Bills. Public Bills introduced not by a Minister in implement of government policy but by a back-bencher or private member. They are quite distinct from private Bills (q.v.). At the beginning of each Parliamentary session a ballot is held for private members wishing to introduce Bills, and this determines the order in which those who are successful may do so. Many members have pet ideas for legislation and any successful in the ballot who have not will be provided with Bills by other members, pressure groups, or small Bills which the Government wishes to introduce but cannot find time for in its programme. A private member must himself get his Bill drafted though a grant is now sometimes made to assist in getting this done. A very limited amount of time is available each session for this purpose and very few private members' Bills pass. If the Government approves of the principle of a private members' Bill, it may assist its passage by allowing government time for debates. A private member's Bill may not deal with any object primarily financial. It may contain money clauses in the hope that, if given a second reading, the Government will sponsor a financial resolution. Private members' Bills pass through the same stages as all other public Bills, save that at committee stage they are referred to a Standing Committee which gives priority to private members' Bills, and emerge as Public Acts.

Private members' Bills have given rise to a number of useful Acts on such topics as superannuation and protection of animals, and to some notable statutes such as the extension of the grounds of divorce, and the abolition of the death penalty for murder.

Privateering The practice of naval belligerents licensing private individuals, not being members of their own naval forces, to operate against the enemy's commerce on their own account. In some cases government ships were lent to individuals for privateering. There is no doubt that in some cases privateering was no better than licensed piracy.

In the sixteenth century such sailors as Hawkins and Drake sometimes sailed as privateers and in the seventeenth century buccaneers in the West Indies such as Morgan sometimes sailed as privateers. Extensive use was made of privateers in the eighteenth century and in the war of 1812 between Britain and the U.S.

In the seventeenth and eighteenth centuries some countries were bound by treaties to require security from privateers before they set out, and to hold the money in readiness to indemnify any who could prove illegal treatment at the hands of the privateer.

The Declaration of Paris, 1856, declared privateering abolished, but it was many years before some nations accepted that view and the disappearance of the practice is really a consequence of the growth of organized navies. The U.S. did not accept the abolition of privateering until the end of the nineteenth century. In modern practice if merchant ships are armed, they are treated as warships by all parties.

Privilege. In general the legal concept of being entitled or authorized to do or not to do something, as one pleases, free from legal restriction by another. Thus, a person in circumstances of absolute privilege may communicate defamatory matter with immunity from action for damages. It is frequently equated with liberty, the correlative concept attaching to the other party being absence of right, or no-right, or inability.

Many kinds of privileges are recognized in law. Certain categories of persons have special privileges, notably members of the royal family, peers, members of Parliament, ambassadors, barristers, and others. Trade unions have wide privileges. Privilege also attaches to particular circumstances, notably the privilege protecting certain kinds of goods from legal distress, and the privilege which attaches to certain kinds of statements in defamation. This is of two kinds, absolute privilege, which attaches to parliamentary and official communications, communications in the course of administration of justice, and certain other cases and wholly protects the maker from liability for defamation; and qualified privilege, which attaches to communications made by persons having a duty to make them to persons having an interest to receive them, which protects the maker provided the communication was not made maliciously or spitefully. Another kind of privilege is the claim of a professional man not to be bound to disclose communications made to him by his client or patient in the course of their professional relationship.

Privilege, Parliamentary. An important part of the law and custom of Parliament, comprising the freedom of each House to conduct its proceedings without interference by the Crown, the courts, bodies outside Parliament, or the public. Its existence is justified as essential for the conduct of Parliament's business and the maintenance of its authority. Privilege may exist by common law or by statute; neither House can by its own resolution create new privileges. The courts will not examine the exercise of privilege nor the internal proceedings of either House, but will not allow either House to extend its privileges at the expense of the rights of the subject.

Certain privileges attach to the House of Commons and its members. At the opening of each Parliament, the Speaker claims for the Commons from the Crown their ancient and undoubted rights and privileges, particularly freedom of speech in debate, freedom from arrest, and freedom of access to Her Majesty whenever occasion shall require; and that the most favourable construction shall be placed upon all their proceedings.

The privileges of individual members are: (1) Freedom from arrest in civil proceedings from 40 days before to 40 days after a meeting of Parliament, but not from arrest on a criminal charge, nor from preventive detention by order of the executive under statutory authority, nor from proceedings in bankruptcy, nor against committal for contempt of court. Members have no general immunity from civil actions being brought against them. (2) Freedom of speech is an important privilege. It was challenged in the seventeenth century, but the Bill of Rights 1688 declared that 'the freedom of specch and debates or proceedings in Parliament ought not to be impeached or questioned in any court or place out of Parliament'. Hence, members are immune from actions for defamation for anything said in the House. Protection for members extends to both civil and criminal liability, questions and possibly even discussions with Ministers outside the House. (3) The House has always maintained the right to privacy of debates in the House, and formerly maintained the right to control publications of its debates. It still maintains the power to permit select committees to sit in private. At common law a fair and accurate report of proceedings in Parliament is privileged provided it is an honest and fair comment on the facts.

The House as a whole has the exclusive right to control its own proceedings and regulate its internal affairs and to maintain order and discipline during debates. Accordingly the courts cannot investigate alleged irregular procedure in passing a Bill. It may also control its own constitution. Formerly this extended to determining disputed elections, but now election disputes are decided by the courts. The House, however, still retains the right to regulate the filling of vacancies, the right to determine whether a member is qualified to sit in the House, and the right to expel a member whom it considers unfit to continue as a member, though it may not create disqualifications unknown to the law.

By virtue of its power to control its own proceedings the House may protect its privileges and punish persons who violate them or commit contempt of the House, as by creating a disturbance in the House, molesting a member on account of his conduct in the House, or publishing material derogatory of the House. The courts may decide the existence and extent of the privileges of the House, but only the House can decide whether conduct is a contempt of the House. A person may commit a contempt without infringing any of the privileges of the House. The House's powers include power to order the offender to be reprimanded or admonished by the Speaker, and to commit any person to the custody of its own officers or to prison for breach of its privileges or contempt of the House. It may suspend or expel members, dismiss its officials, withdraw facilities from lobby correspondents, and so on. It probably cannot fine.

The House's privilege of access to the sovereign is a collective one of the members exercised through the Speaker.

In the past the House claimed to be the sole and absolute judge of its own privileges, but the court in *Stockdale* v. *Hansard* (q.v.) maintained the right to determine the nature and limit of parliamentary privilege if it is necessary to do so in determining the rights of individuals outside the House. There may accordingly be conflict if the House should seek to extend its privileges or abuse its powers for political ends.

When a complaint of breach of privilege or of contempt is raised the Speaker has 24 hours in which to consider whether there has been a prima facie breach of privilege or not. If he holds that it does, a motion is moved that the matter be referred to the Committee of Privileges; if this is carried the Committee investigates the matter, hears witnesses, reports to the House, and may recommend what action should be taken. The individual against whom the complaint is made is not entitled to be heard or legally represented, to call evidence, nor to cross-examine witnesses, and the procedure is unsatisfactory.

In the House of Lords peers have freedom from civil arrest; an individual peer may claim it at any time, but the House claims it only within the usual times of privilege of Parliament. They also have freedom of speech, and freedom of access to the sovereign which exists for the House collectively and, by virtue of peerage, for each peer individually. The House can, through its Committee for Privileges decide the right of newly created peers to sit and vote, and decide claims to old peerages, referred by the Crown to the House. The House can commit a person to prison for an indefinite time, the period not being terminated by prorogation of Parliament, or fine, or order security to be found for good conduct.

Erskine May, *Law, Privileges, Proceedings, and Usage of Parliament.*

Privileged communication. One which the recipient cannot be required by a court to disclose in evidence. The category includes communications by a client to his legal adviser, but not to clergymen, bankers, or other advisers, communications between spouses during marriage, and, at least formerly a witness's title-deeds or equivalent documents.

Privileged debts (or preferential debts). Those debts, such as funeral expenses, which an executor may pay in preference to all others.

Privileged places. Places in or near the city of London where a right of sanctuary (q.v.) was believed to exist. They comprised Whitefriars, The Savoy, Salisbury Court, Ram Alley, Mitre Court, Fallers Rents, Mintage Close, the Minories, the Mint, and Clink, or Deadman's Place. The right, if any, was abolished in 1697.

Privileges, Committee for. See COMMITTEE FOR PRIVILEGES.

Privileges, Committee of. See COMMITTEE OF PRIVILEGES.

Privity. Originally knowledge, and later the relation between two parties connected by blood relationship, tenure, or contract. Thus privity of blood exists between an heir and his ancestor, privity in representation between a testator and his executors, privity in tenure between a lord and a tenant who holds of him by service, privity of estate between lessor and lessee, tenant for life, and remainderman. Privity of contract is the relation between the parties to a contract which entitles one to sue the other, but in general disentitles a third party from suing on the contract.

Privy Council. The *Curia Regis* of the Norman kings was the feudal version of the Anglo-Saxon witenagemot, but from the first it took two forms—a large, great council of the realm, and a small select body of officials who met regularly to carry on everyday government. The *Curia* inherited the old appellate jurisdiction of the witenagemot and also exercised jurisdiction over the King's tenants-in-chief.

In the reign of Henry III much of the judicial functions of the *Curia Regis* became permanently divided among three judicial bodies, the Exchequer, dealing with fiscal matters, the Common Pleas dealing with civil disputes between commoners, and the King's Bench, dealing with criminal cases and other cases involving royal powers. These three courts exercised most of the judicial business of the State.

Slightly later the large Councils of the nobility which joined with the King and representatives of the shires and boroughs in a Parliament became the House of Lords, and retained some judicial powers from earlier times. The smaller Council was coming to be regarded as a mainly executive body, but still retained some judicial functions.

After the courts of King's Bench, Common Pleas and Exchequer split off from the *Curia Regis* the King retained, and exercised in his 'continual' Council, a personal jurisdiction both as a court of royal justice in all cases not delegated to the courts of common law and as a superior court of appeal. The King's continual or permanent Council became especially important from the early thirteenth century. It comprised officers of state, judges, some bishops and barons, and certain others. As developed by Edward I it became the main instrument of royal power, sitting regularly for the despatch of business, advising the King, exercising a wide jurisdiction, civil and criminal, and dealing with a vast number of petitions, seeking relief of various kinds, many of which were remitted to common law courts. The council exercised jurisdiction where it was unlikely that the ordinary courts could manage, as where they might be overawed or coerced or great nobles were parties.

The Chancellor usually presided when the Council sat judicially; apart from his connection with the Council he had an important ordinary legal jurisdiction. In the reign of Edward I appeared the first signs of the chancellor's extraordinary or equitable jurisdiction, when petitions for extraordinary remedies were remitted to the chancellor to give such remedy as seemed to him consistent with justice. In 1349 all matters which were 'of grace' were directed to be dealt with by the chancellor or keeper of the privy seal.

In the fourteenth century the Council, without legal warrant, assumed jurisdiction in many cases within the jurisdiction of the common law courts and acted in violation of Magna Carta. Various statutes were passed on the solicitation of the House of Commons seeking to restrain these illegalities, but had little effect.

Apart from the continual Council, the barons retained certain powers and, under the name of Magnum Concilium, continued to meet down to the fifteenth century, sometimes along with the continual Council and sometimes in an assembly containing the main elements of a parliament. The lords together with the continual council exercised the judicial function of the national assembly, and from this the judicial function of the House of Lords is derived. In the time of Edward III the lords took jurisdiction into their own hands and reduced the position of the judges and members of the council who were not peers to that of mere advisers.

By the end of the mediaeval period the Council was becoming mainly an executive body, but it still retained wide but vague judicial powers, because there were so many kinds of cases with which the common law could not deal effectively. Cases outside the jurisdiction of the ordinary courts, in which the King's interest were affected, in which the law was at fault or its common law process ineffective all came before the Council. By this time one branch of this jurisdiction began to separate itself and to become recognized as the court of Chancery.

When the Tudors attained power they were able to utilize the wide jurisdiction of the Council to aid them in developing the modern state, and the Council divided into the administrative aspect of the Council and the judicial aspect which in 1487 became the Court of Star Chamber (q.v.). In the early seventeenth century there was strong feeling by the common lawyers against the jurisdiction of the Council.

In 1641 the Parliament abolished the jurisdiction of the Council in England, the Star Chamber, and the other prerogative courts, but this left untouched jurisdiction to deal with appeals from places beyond the jurisdiction of the ordinary courts.

In the late seventeenth and eighteenth centuries there were committees of the Council for trade and to hear appeals from the plantations. As colonies increased there was an increase in the number of appeals and rules of procedure were gradually formulated, and there was also an awareness of the need to lay down uniform rules throughout the territories overseas. The quantity of business gave the committee many of the characteristics of a court. Reports of decisions began in 1829 and in that year statute abolished the Court of Delegates and transferred its jurisdiction to hear appeals from the ecclesiastical courts and court of Admiralty to this Committee of the Privy Council. In consequence statute in 1833 created the Judicial Committee of the Privy Council (q.v.).

Today the Privy Council numbers about 300 persons who hold or have held high political, legal, or ecclesiastical office in the U.K. or Commonwealth. Cabinet Ministers are invariably made members, the two archbishops and such others as the Lord Chief Justice, Master of the Rolls, Lords Justices of Appeal, Attorney-General, Lord Advocate, and others. They are appointed by letters patent, are entitled to the style 'The Right Honourable . . .' and the designation P.C., and must take the oath of allegiance and a special Privy Councillor's oath. A Privy Councillor may be dismissed from the Council by the sovereign, or be removed at the person's own request.

The Privy Council is summoned as a body only to sign the Proclamation of the accession of a new sovereign and when the sovereign announces an intention to marry. The normal functions of the Privy Council are to give formal effect to Proclamations or Orders in Council, issued by the Crown under powers of the prerogative or statutory authority, in either case acting on the advice of the government. At a meeting of the Privy Council the quorum is three and four are normally summoned to attend, the Lord President and the Ministers chiefly concerned with the business being dealt with. The sovereign presides and the Clerk of the Council attends, records the names of those present, and authenticates by his signature the sovereign's assent to a measure. The fact that Her Majesty held a Council, and those present is reported in the daily Court Circular.

There are various standing committees of the Privy Council dealing with such matters as the universities, the Channel Islands, and research, and committees are established to deal with particular matters such as petitions for the grant of a royal charter. An important committee is the Judicial Committee (see PRIVY COUNCIL, JUDICIAL COMMITTEE OF THE).

J. Jolliffe, *Constitutional History of Mediaeval England*; D. L. Keir, *Constitutional History of Modern Britain*; A. V. Dicey, *The Privy Council.*

Privy Council, Judicial Committee of the. The Privy Council developed distinct administrative and judicial capacities under the Tudors. An Act of the Long Parliament deprived the Council of jurisdiction over all English Bills or petitions, but left it able to deal with appeals initiated by Bill or petition from places outside the jurisdiction of the ordinary courts of common law and equity. From the first these appeals were dealt with by an appeal committee. In 1832 its jurisdiction was increased when the Court of Delegates was abolished and jurisdiction to hear appeals from the Court of Admiralty and the ecclesiastical courts was transferred to the appeal committee. In 1833 statute constituted the Judicial Committee, comprising the Lord President of the Council, the Lord Chancellor, such members of the Privy Council as hold or have held the office of Lord Keeper or First Commissioner of the Great Seal, Chief Justice of any of the three common law courts, Master of the Rolls, Vice-Chancellor, Judge of the Prerogative Court of the Archbishop of Canterbury, Judge of the Court of Admiralty, Chief Judge in Bankruptcy and, in ecclesiastical cases, every archbishop or bishop who was a privy councillor. The Crown might appoint two other persons to be members, summon other members, or direct the attendance of the judges.

By the Judicial Committee Act, 1871, provision was made for the appointment of four persons to act as members of the Judicial Committee, who must be or have been judges of one of the superior courts at Westminster or a Chief Justice of the High Court at Fort William in Bengal, or Madras, or Bombay, and who were to be paid. By a shabby trick Sir R. P. Collier was appointed a judge of the Common Pleas for a few days to give him the qualification for appointment to the Judicial Committee. Under this Act were appointed Collier, Montague Smith, Barnes Peacock, and Colville (qq.v.).

The Appellate Jurisdiction Act, 1876, which established the modern appellate jurisdiction of the House of Lords, provided for the appointment of two Lords of Appeal in Ordinary, who had been for at least 15 years practising barristers in England or Ireland or practising advocates in Scotland, or holder for at least two years of a high judicial office as defined in that Act, and for the appointment of a third Lord of Appeal when two of the four salaried members of the Judicial Committee had died or retired and of a fourth Lord of Appeal when the remaining two had died or retired. Lords of Appeal, if Privy Councillors, were to be members of the Judicial Committee.

By the Appellate Jurisdiction Act, 1887, the committee included such members of the Privy Council as hold or had held any of the offices described as 'high judicial offices' in the Appellate Jurisdiction Acts, 1876 and 1887, and also those of Lords of Appeal in Ordinary and of members of the Judicial Committee of the Privy Council. Only under this Act was Lord Hobhouse, who was a peer and member of the Judicial Committee but had not held high judicial office in the United Kingdom, enabled to sit in the House of Lords also.

The Judicial Committee Amendment Act, 1895, added any person who was a P.C. and was or had been Chief Justice or a judge of the Supreme Court of Canada or of a superior court in a province of Canada, or of one of the Australasian colonies to the number of not more than five. The Appellate Jurisdiction Act, 1908, added the Chief Justice or justices of the High Court of Australia, or of the Supreme Court of Newfoundland, and provided that not more than two persons who were or had been chief justices or judges of the Federal Court in India or a High Court in British India and were members of the P.C. should be members of the Judicial Committee. The Appellate Jurisdiction Act, 1913, added persons who were members of the P.C. and were or had been Chief Justice or a judge of the Supreme Court of South Africa. The Administration of Justice Act, 1928, repealed the limit in the 1895 Act on the number of dominion judges who might be members of the Judicial Committee (but the limit of numbers in the 1908 Act is unrepealed). The quorum was fixed at four, later reduced to three. Since 1945 the composition has been varied to take account of the changing structure of the Commonwealth and the list of countries from which appeal lies to the Privy Council.

The jurisdiction of the Judicial Committee is varied. Within the United Kingdom appeals from the ecclesiastical and prize courts go, and till 1875 appeals from the Court of Admiralty went, to the Privy Council. Constitutional questions affecting Northern Ireland and appeals from the Joint Exchequer Board also go to the Privy Council, as do appeals from the professional discipline tribunals of the General Medical and similar Councils.

The main jurisdiction has always been to hear appeals from the highest courts of Commonwealth countries and colonies, but since the Statute of Westminster, 1931, many Commonwealth countries have restricted or abolished such appeals.

The Judicial Committee may sit to hear appeals in more than one division at the same time, each consisting normally of five members. For the hearing of ecclesiastical appeals provision is made for the attendance of bishops as assessors according to a rota.

The Queen may also through the Attorney-General refer to the Judicial Committee any matters whatever which she may think fit for opinion, and the Committee has considered and reported on a variety of miscellaneous matters under this head.

The judgment takes the form of a report to Her Majesty, advising her as to the appeal. For long only one judgment was given, but since 1966 dissenting judgments have been permitted. The result of the appeal is delivered orally, in open court; the reasons are not normally read out but printed copies are distributed. The committee's report is submitted to Her Majesty at the next Council and, after approval, is embodied in an Order in Council, thereby becoming the final decree.

The Judicial Committee does not regard itself as bound by its own previous decisions, and its decisions are not binding on any United Kingdom court, though, on points where the law is similar in the U.K., they may be very persuasive.

The jurisdiction of the Judicial Committee extends to appeals from the Channel Islands and the Isle of Man, appeals from Commonwealth countries and colonies, save in so far as independent countries of the Commonwealth are entitled to, and have, abolished appeals to the Privy Council, appeals in ecclesiastical cases, and appeals from Vice-admiralty courts in the colonies, and appeals in Prize Cases. In addition the Judicial Committee has by statute certain quasi-judicial powers, and a general power to determine miscellaneous disputes referred to it by the Crown.

Privy Council, Scottish. The Privy or Secret Council seems to date from the reign of David II but its character and functions were very indefinite until the late fifteenth century, and the distinctions between the Privy Council, a Convention of Estates, and Parliament are uncertain. It originally had functions of advising and administering justice, but later came to exercise both legislative and executive powers. In 1489 it was given a statutory constitution for the administration of justice.

In the fifteenth and early sixteenth centuries committees of the Lords of Council were repeatedly established to administer justice and from these, and the court known as the Session established in the mid-fifteenth century, there developed the Court of Session, established as the permanent superior civil court in 1532.

By the time of James VI it was a legislative, executive, and judicial body and its officials were the King's nominees. After James moved to England in 1603 he ruled Scotland through the Privy Council. It supervised the laws, the courts and the justices of the peace, regulated trade and wages, restricted the Press, punished recusants, and generally managed the country. Charles I reconstructed the Privy Council adding bishops to its membership. During the Commonwealth it was in abeyance, but was revived after the Restoration and from 1662 the Privy Council governed the country, principally enforcing penal religious statutes on recusants. In 1686 James VII and II purged the membership to permit the practice of Catholicism. William III reconstituted the Council but it was less important as an agency of government. Under the Treaty of Union the Scottish Privy Council was abolished. It had for much of its existence been an agency of oppression of Parliament and people, and its demise was unlamented.

Privy seal (or private seal). A seal intermediate between the great seal (q.v.) and the signet (q.v.) under which charters and grants signed by the sovereign passed before passing the great seal, and also used as an authority to the Lord Chancellor to affix the great seal to certain documents.

See also LORD PRIVY SEAL.

Privy signet. One of the sovereign's seals used in sealing private letters and grants which pass her hand by bill signed.

Prize. The term for a ship or goods captured at sea or seized in port by the maritime forces of a belligerent, and now extended by statute to aircraft and goods carried thereon. A ship or goods may be lawfully taken as prize at sea or in port, at any time after a declaration of war and before its final termination. Enemy ships, and neutral ships acting in an unneutral manner, such as attempting to break a blockade, are lawful prize, but hospital ships and certain other categories are not lawful prize. By the Declaration of Paris, 1856, enemy goods carried in a neutral vessel and neutral goods carried in an enemy vessel are not lawful prize, but neutral goods declared by a belligerent to be contraband are liable to seizure if destined for an enemy country.

The commander of a belligerent warship may

visit a ship at sea or in the belligerent's port and examine the ship's papers, and if not satisfied search the ship. If not satisfied, he may seize the ship and send it in to a convenient port for adjudication by the Prize Court whether it has to be condemned as lawful prize or must be released. A sentence of condemnation is necessary to vest the property in the ship and goods in the captors. The Crown may release a captured ship or goods before condemnation.

In the United Kindom jurisdiction in prize was formerly exercised by the High Court of Admiralty under a commission from the Crown, and later under statute, and is now exercised by the Admiralty Court of the Queen's Bench Division of the High Court with appeal to the Judicial Committee of the Privy Council. A Vice-Admiralty Court or a Colonial Court of Admiralty may be at any time commissioned as prize court in a British possession.

The law administered by the Prize Court is international law and the British Prize Court is bound by U.K. statutes but not by Orders in Council or proclamations if not in accordance with international law, unless issued under authority of a statute or waiving Crown rights in favour of enemies or neutrals.

Enemy vessels or goods seized at sea by one of H.M. ships will be condemned as droits of the Crown, and if seized in a port as droits of Admiralty, provided the ship entered the port voluntarily and not be reason of war or operations of war. Ships or goods captured by a ship other than a ship of war or by land forces are also droits of Admiralty. Formerly a grant of droits of the Crown or of Admiralty or part thereof were made to the actual captors; provision is made instead for the proceeds of droits of the Crown to be paid to Naval and R.A.F. Prize Funds. Droits of Admiralty are paid to the Exchequer.

Prize Cases (1863), 2 Black 635. When in April 1861 the Confederacy began hostilities against the Union the President, without waiting for Congress to declare war, declared a blockade of Southern ports. The owners of ships seized when running the blockade challenged the legality of their seizure. The Supreme Court held that the blockade was legitimate and enforceable from the date when it was proclaimed, and certainly from the date when Congress enacted legislation supporting the blockade.

Prize court. In the Middle Ages when a ship captured an enemy ship or enemy goods the captors kept the ship as booty without any judicial decision. In England in 1426 a royal proclamation provided that captured goods were not to be dispersed until the Council, the Chancellor, the Admiral, or his deputy had certified whether the goods were of enemy character or not. But not till 1589 was it laid down that the Council required that prizes be kept safely until judgment in the High Court of Admiralty that the goods be lawful prize. In the eighteenth century the judge of Admiralty was given a special warrant, based on a commission by the Crown to the Admiralty on the outbreak of war, to deal with trial of prize cases. The law administered was found in the collections of customary usages, such as of Rhodes or the Consolato del Mare, in treaties, and in the books of the commentators on the civil law or the law of nations, but the Privy Council sometimes gave instructions to the court on both law and procedure. The cases of the Napoleonic and World Wars have subsequently developed a considerable volume of case-law. The materials on which the court exercises judgment are the ship's papers, the ladings of the cargo, the proprietorship thereof, their destination, and examination of the master and crew. The judgment might be of condemnation as lawful prize, or acquittal with costs against the captor, or even with damages in case of grossly improper capture.

In earlier times the court also estreated bonds deposited by owners of privateers if they had gone beyond their letters of marque, decided disputes as to prize salvage and disputes between the government and captors over droits of Admiralty. In the eighteenth century there were also Vice-Admiralty courts acting in overseas territories under special warrants to act in prize from the Lords of the Admiralty, and this was done again in the First World War.

Appeals were directed in 1648 to lie to seven members of the Privy Council and still lie to the Judicial Committee of the Privy Council.

Prize courts are national courts constituted by municipal law, but states are bound by international law to enact for their prize courts only such statutes and regulations as are in conformity with international law.

E. S. Roscoe, *History of the English Prize Courts.*

Prize law. The body of law administered in wartime by maritime nations, through prize courts, regulating the practice of the capture of ships and cargoes at sea in time of war. The prize courts have to investigate the legality of all captures and administer justice as between the captors and the persons interested in the ships seized and their cargoes. If the seizure is lawful, the property is condemned as lawful prize of war; if unlawful, restitution or redress is ordered.

Prize courts are not international tribunals; in the United Kingdom prize jurisdiction was exercised by the Court of Admiralty and subsequently by the Probate, Divorce, and Admiralty Division of the High Court, and now by the Admiralty Court of the Queen's Bench Division, appeal lying to the

Judicial Committee of the Privy Council, and that court is bound by relevant law laid down by the United Kingdom Parliament, save in so far as such legislation is inconsistent with international law. For, though a prize court is a municipal rather than international court, the law it administers is a branch of international law or the law of nations, and it must ascertain and give effect to rules of law laid down, not by its own or any other state but which originated in the practices and usage long observed by civilized nations in their relations towards each other, or in express international agreement. There is a presumption that neither Act of Parliament, nor Orders in Council in execution of the royal prerogative, are intended to make law for the prize court in substitution for, nor in derogation of, the relevant rules of international law: *The Zamora*, [1916] 2 A.C. 77.

Probate. The certificate granted by the court that the will of a deceased person has been proved and registered in the court and that a right to administer his effects has been granted to the executor proving the will. Probate may be granted in common form, which is done in ordinary cases where the executor swears and files an affidavit that the will is the true and original last will of the deceased and that he will faithfully administer the estate; or in solemn form, which is done where there is or is likely to be a dispute as to the validity of the will, in which case the person propounding the will brings an action against any person disputing it. If the court is satisfied, it pronounces for the validity of the will and grants probate of it. Probate business includes all business relating to the grant or revocation of probates and letters of administration. Non-contentious business is dealt with in the Principal Probate Registry of the Family Division or a district registry and contentious business in the Chancery Division or, in small cases, the county court.

Probate, Court of. Established in 1857, taking over from ecclesiastical courts jurisdiction in respect of the grant of probate of wills, and of letters of administration in cases of intestacy, and all testamentary matters. In 1875 it became part of the Probate, Divorce, and Admiralty Division (now the Family Division) of the High Court.

Probate, Divorce and Admiralty Division. A division of the High Court in England created in 1875 by the amalgamation of the former separate Court of Probate, Court for Matrimonial Causes, and High Court of Admiralty. In 1971 it was renamed the Family Division, admiralty jurisdiction being transferred to the Queen's Bench Division and contentious probate business being transferred to the Chancery Division.

Probation. A mode of treatment of offenders introduced in Britain in 1907 as an alternative to punishment. The offender is placed by a court for a period of from one to three years under the supervision of a social worker, must keep in touch with him and conform to his instructions and generally be of good behaviour. Conditions, such as to undergo treatment, may be attached to a probation order. If the offender again offends during his period of probation, or does not comply with the probation order, he is brought back to court and may be punished for breach of probation by fine, community service order, or requirement to attend at an attendance centre, as well as punished for the subsequent offence.

Probationer, Lord. Prior to 1933, a newly-appointed judge of the Court of Session while undergoing his trials, i.e. hearing his trial cases.

Probative. In Scots law, the quality attaching to deeds solemnly attested which renders such a deed proof of its own authenticity, without need for proof of signature or witnessing and which can only be overcome by judicial reduction of the deed.

Proby, John Joshua, Earl of Carysfort (1751–1828). Was joint Master of the Rolls in Ireland, 1789–1801. He was later a peer and Privy Councillor of the United Kingdom.

Probyn, Sir Edward (1678–1742). Called to the Bar in 1702 he defended the Earl of Macclesfield and, in 1726, was promoted a judge of the King's Bench and was thereafter Chief Baron of Exchequer (1740–42).

Procedure. A term used in two senses: in the wider sense it is contrasted with substantive law (q.v.), is probably co-extensive with adjective law (q.v.), and includes the whole of the legal machinery whereby the rights and duties conferred and imposed on persons by rules of substantive law are declared and enforced. The distinction in this sense is important in cases involving issues of international private law, where the general rule is that the rights of parties are determined by the system of law under which their relationship (e.g. buyer–seller, wrong-doer–victim) existed, but all matters of procedure are regulated by the *lex fori*, the law of the court in which the claim is brought. In this sense procedure includes the rules of jurisdiction, procedure in the narrower sense, pleading, evidence, and execution.

In the narrower and more exact sense, it comprises only a part of adjective law, and includes the principles and rules governing the steps to be taken in initiating and carrying through a legal claim or other proceeding, from framing the writ, petition, or other means of initiating the claim, to final judgment. It has been judicially defined as 'The mode of proceeding by which a legal right is enforced, as distinguished from the law which gives or defines the right, and which by means of the proceeding the court is to administer; the machinery as distinguished from the product' (*Poyser* v. *Minors* (1881), 7 Q.B.D. 329, 333).

It is normal for legal systems to have distinct codes of procedure for civil and criminal, and frequently also for specialized kinds of, cases. The essential stages in civil procedure are the statement of a claim by one person alleging that another has infringed his rights, the lodging of an answer or defence by that other, investigation of the disputed facts and any dispute on the relevant law by court or jury, and decision on the issues and, where appropriate, the grant of a remedy. In criminal procedure the essential stages are investigation, the levelling of a charge against a person, his admission or denial, investigation of the disputed facts and any dispute on the relevant law, decision on guilt or innocence and, in the former case, determination of the appropriate penalty. Procedure in administrative and other tribunals follows the same general principles but is frequently less formalized or regulated. Further rules regulate appeals. In developed systems of law, procedure is regulated by bodies of rules applicable to the different courts, but many details are supplied by practice, the custom, and usage of the particular court.

In the United Kingdom the procedure of the High Court and Court of Appeal is mainly regulated by the Rules of the Supreme Court, and subsidiarily by Practice Directions issued by the judges and by the Practice Masters' Rules, which are directives for the regulation of business concerning the Central Office. In the Court of Session it is mainly regulated by the Rules of the Court of Session. Other courts have similarly their own bodies of procedural rules. Rules of procedure are in many cases supplemented by the practice and custom of the court and the profession.

In the U.S. procedure was generally based on the English common law system, but many states have adopted the code of procedure, mainly the work of David Dudley Field (q.v.) adopted in New York in 1848. The chief characteristic of this is the adoption of a single form of civil action for all cases, though in some jurisdictions the underlying basis of the action, legal or equitable, is still relevant and in some the requirement of jury trial in actions at law has perpetuated a distinction between actions founded on law, in which jury trial is required, and actions founded in equity, where there is no jury. Even non-code states have moved away from the original modes of procedure to modified ones. In the Federal courts there is uniform simplified procedure in equity, but procedure in law has for long been tied to conformity with the procedure of the State courts.

Procedure, parliamentary. The rules regulating the conduct of business in the two Houses of Parliament. In the House of Lords the procedure of the House in general is regulated by its standing orders, by resolutions and orders agreed from time to time, and by the established practice of the House. Business is distinguished into business transacted in the House itself, itself divided into public business, including the first, second, and third readings of all Bills, amendments to Bills made in committee, and consideration of reports from select committees, and private business, including the consideration of private Bills, business transacted in committee of the whole House, business dealt with by select committees, committees appointed for a session to perform certain definite duties connected with the business, administration or procedure of the House, business dealt with in private Bill committees, and business dealt with in joint committees of the two Houses.

In the House of Commons business is similarly distinguished. The conduct of business and the procedure of the House are regulated by the standing orders of the House, by resolutions and orders agreed to from time to time by the House, and by a series of customs and precedents based partly on tradition and partly on rulings by Speakers on matters of practice or points of order. Business transacted by the House is divided into public business, which includes motions, financial business, and all the stages of public Bills; and private business which includes the readings and report stages of private Bills and provisional order confirmation Bills. In committee of the whole House the House considers some public bills already read a second time, any Bill recommitted after report from a select committee, or any Bill specially recommitted, and any other matter the House resolves to discuss there. Most Bills, however, after second reading, are referred for detailed consideration to a standing committee. Select committees are appointed to consider or inquire into public matters or Bills. Sessional committees are appointed each session to consider particular matters. Private Bills committees hear evidence in support of private Bills.

Decisions in the House and committees are reached on a question put from the chair on a motion made by a member of the House and, with certain exceptions, seconded. Voting is by preponderance of voices or by counting the members who go through the division lobbies and are there counted by tellers.

Proceedings (or judicial proceedings). A term sometimes used as including, or meaning, an action or prosecution, and sometimes as meaning a step in an action.

Procès-verbal. In French law, a statement of the facts of a case in a criminal charge. Also a detailed account by a magistrate, police officer, or person of similar authority of acts done in the execution of his duty. The term is also given to the minutes of a meeting or assembly.

Process. The proceedings in any action or prosecution, and particularly the summons by which a person is cited to court. At common law the process of the old superior courts of King's Bench, Common Pleas, Exchequer, and Chancery differed materially prior to the Uniformity of Process Act, 1832. It consists now of writs and originating summonses. Process includes process against a defendant and process against third parties, e.g. to summon witnesses. In criminal matters process means the proceedings issued to bring in a person against whom an indictment has been framed.

In Scotland a process is a civil action, and also the collection of pleadings and documents which are put before the court in such an action.

Procession. A more or less organized group of persons moving along a highway or other defined path. While passage by persons along a highway is merely the use for which highways exist, a procession may be a serious obstruction to the use of the highway by non-participants. Public processions are subject to many statutory powers of regulation, particularly if the police reasonably apprehend that the procession may give rise to serious public disorder. Conditions may be imposed and local authorities may regulate certain kinds of public processions. A procession may be a public nuisance if it involves unreasonable use of the highway, causing obstruction or excessive noise. If the purpose of the procession is to commit a crime of violence or to achieve any other purpose in such a way as to give persons of ordinary courage in the vicinity reasonable grounds for apprehending a breach of the peace, participants commit the common law crime of unlawful assembly.

Procheiron. A manual published by the Eastern Emperor, Basil I, *c.* A.D. 879, intended to supersede *The Ecloga* (q.v.) but retaining many of its innovations.

Proclamation. A royal proclamation is a formal announcement, under the great seal, of a matter of public concern such as a dissolution of Parliament, a declaration of war, or the like.

In the sixteenth and early seventeenth centuries in England legislation might be made by proclamation as well as by statute, though the latter was more important. The boundary between royal and parliamentary legislative power was uncertain, and proclamations were used in many cases which would in later law have been illegal. Henry VIII's Statute of Proclamations (1539) was intended to define the sphere within which they should have the force of law and the forms under which they should be issued. It enacted that the King with the advice of the Council might issue proclamations which should have the force of an Act of Parliament, but the common law, existing statutes, and rights of property were safeguarded, and the death penalty for breach of proclamation prohibited. The Act was repealed in 1548, but proclamations were extensively used thereafter, but so tactfully that the limits of their legal use were never finally settled. But this issue came to the fore under James I and a protest from the Commons evoked the resolution of the judges in the Case of Proclamations (1611) (q.v.) to the effect that the King cannot change the common law nor create any offence by proclamation which was not an offence before, without parliament, and after the Great Rebellion this was taken as settling the law. During the periods down to 1640 proclamations were extensively used to put law into effect and also to make new law, particularly in relation to administration, defence and foreign affairs, commercial and industrial regulation, social and religious life, the Press, and other topics. After the Restoration proclamations were less important and restricted to executive rather than legislative acts, and after the Revolution Settlement became mere announcements or statutory declarations.

The Crown has power under the prerogative to legislate by Order in Council for newly conquered territory and in protectorates a high commissioner may be empowered to legislate by proclamation. Proclamations are also used for the purposes to bring Acts or ordinances into force, or to give effect to executive acts.

Statute may also empower the crown to make a proclamation, as of a day as a bank holiday, though Orders in Council or ministerial orders are now more commonly authorized.

Proclamations, Case of (1611), 12 Co. Rep. 74. King James I raised the question whether he could by proclamation prohibit new buildings in and about London and the making of starch of wheat. The Chief Justices and Chief Baron held that the King by proclamation or otherwise could not change common law, or statute, or the customs of the realm, that the King by proclamation could not create any offence which was not an offence before, and had no prerogative but what the law of the land allowed him.

Proclamations, Statute of. In 1539 Henry VIII issued certain royal proclamations and the judges held that those who disobeyed them could not be punished by the Council, whereupon the King appealed to Parliament to give his proclamations the force of statutes. Parliament did so, enacting that proclamations made by him with the advice of a majority of his Council should have the force of statutes but so as not to be prejudicial to any person's inheritance, offices, liberties, goods and chattels, or to infringe the established laws. The fact that the King had to obtain the statute, and the limitations under which it was passed, show the limitations on the royal power. It was repealed in 1547, but proclamations continued to be issued none the less, as by Mary Tudor declaring the importation of heretical books punishable by martial law as rebellion.

Proclamations, statutory. Proclamations issued by the Crown not under the prerogative but under statutory authority.

Proctor (from procurator (q.v.)). Persons who had studied Roman and canon law at Oxford or Cambridge and obtained from the bishop of a diocese a patent to practise in his diocesan consistory court. They practised rather like solicitors, drafting the pleadings in matrimonial cases in ecclesiastical courts and appearing to argue the case. They also practised in admiralty courts.

After the Judicature Acts, 1873–75, the title became obsolete, the class of proctors being absorbed by that of solicitors. The title, however, survives in the office of Queen's Proctor, who represents the interest of the Crown in divorce cases. In modern English ecclesiastical law a proctor is a representative of the clergy in Convocation, and in Oxford and Cambridge universities the proctors are persons charged with supervision of student discipline. In the U.S. the word is sometimes still used of a legal practitioner in probate and admiralty courts.

Proculians and Sabinians. Most of the Roman jurists of the imperial period adhered to one or other of the two schools of jurists, the Proculians, called from their leader Proculus, and the Sabinians, or Cassians, from their leader Sabinus (q.v.) or his disciple Cassius. The antagonism originated with their predecessors M. Antistius Labeo and G. Ateius Capito. The principal adherents of the Proculians were Proculus, the Nervae, father and son, Longinus, Pegasus, the Celsi, father and son, and Neratius; and of the Sabinians, Sabinus, Cassius, Caelius Sabinus, Javolenus, Aburnius Valens, Tuscianus, Julianus, and Gaius. No jurist after Gaius is known to have belonged to either school, so that the schools probably ended about that time. The basis of the differences of view is obscure though different views are known on many points. It appears not to have been the difference between *ius strictum* and *aequitas*, nor between *ius civile* and *ius gentium*, nor between philosophical views nor in methods of exposition, though the Proculians are known to have been casuistic and the Sabinians to have favoured rather systematic description; Labeo was an innovator, and Capito a traditionalist. It may be that the 'schools' were clubs or discussion-groups centring on two series of distinguished jurists.

Proculus (First century A.D.). A notable Roman jurist and teacher, who gave his name to the Proculian school (see PROCULIANS AND SABINIANS). He was author of *Epistulae* in 11 books, a collection of opinions and discussions from practice, and of *Notae* to Labeo, and was frequently cited by later jurists.

Procurator. Originally at Rome one who acted for another, particularly a government financial agent. From the time of Augustus procurators were frequently appointed to offices in the imperial government or the administration of the provinces. In imperial provinces the procurator was subordinate to a legate, but in senatorial provinces he cooperated with the governor and his quaestor. Procurators were also appointed to govern smaller provinces; thus Pontius Pilate was procurator of Judaea. Such procurators had judicial and financial authority, but usually subject to the overriding authority of the governor of a major province.

In the Middle Ages it signified various kinds of officials. In older books the term is sometimes given to proxies of peers, and bishops are described as *procuratores ecclesiarum*. The term has long been used in Scotland for a lawyer who represents clients in lower courts. In Glasgow the practitioners of the Commissary Courts of the Archdiocese of Glasgow formed a society, acquired privileges, and came to be allowed to practise in all Glasgow courts. The Faculty of Procurators obtained a royal charter in 1796 and the title Royal in 1950. There are similar but lesser Faculties of procurators in many counties in Scotland. The Procurator of the Church of Scotland is a senior member of the Scottish Bar who acts as legal adviser to the Church. The word also survives in procurator-fiscal (q.v.) and in its English form of proctor (q.v.).

The term in various forms is used in countries of the civil law family for officials who act as representatives in courts. In Germany prior to 1879 the procurator represented the client in court whereas the advocate advised him.

Procurator-fiscal. The procurator of the fiscus or treasury, an ancient legal officer in Scotland, recognized in the seventeenth century as the prosecutor in the Sheriff Courts. Though until 1907 they were appointed by the Sheriff they gradually, in the eighteenth and nineteenth centuries, became increasingly officers of the Lord Advocate. They are now persons qualified as advocate or solicitor, appointed by the Lord Advocate and, except in some remote districts, are civil servants and debarred from private legal practice.

There is now a procurator-fiscal, with assistants and deputes, in each Sheriff Court district in Scotland. Their main functions are to make the investigations on which criminal charges may be founded, to take the precognitions of witnesses, to take the instructions of the Lord Advocate and Crown counsel on the mode of prosecution, and to conduct the prosecution case in Sheriff Court prosecutions, both on indictment and summary, in district court prosecutions, and in inquiries into cases of sudden and suspicious deaths. In summary cases the procurator fiscal prosecutes in his own name, in the public interest.

They also collect for the Exchequer fines imposed in proceedings instituted by them. As public prosecutors they enjoy considerable privileges and immunities from action for malicious prosecution.

Procurator-General and Treasury Solicitor. See TREASURY SOLICITOR.

Procuratory. In Scots law a mandate or commission.

Procuratory of resignation. In Scottish conveyancing a mandate whereby a vassal authorizes the fee of his land to be resigned into the hands of the superior, either *ad remanentiam*, to remain with him, or *in favorem*, to enable him to make a fresh grant in favour of another vassal, or to the present vassal and a different series of heirs.

Product liability. The issue of whether, on what basis, and to what extent a manufacturer or supplier of some product should be liable to the ultimate consumer or user for harm done by reason of a defect in design or manufacture. In general, a manufacturer is liable in contract only to the purchaser from him, but he is liable in tort to the ultimate consumer if he failed to take reasonable care in the design or manufacture and that failure caused the harm. The duty is not strict, nor absolute (qq.v.), but to take reasonable care only. Frequently the occurrence of harm sufficiently evidences lack of reasonable care or gives rise to a rebuttable inference of fault. Difficulties may arise where the defect was in a component obtained by the manufacturer from a supplier. Proposals have been made for imposing strict liability on manufacturers for defective products. In certain cases statutes have imposed minimum safety standards for particular products.

Production. In England the High Court has power to order a party to produce any documents relating to the matter in issue, or to produce any documents for the inspection of any other party.

In Scottish criminal procedure, productions are things produced at the trial and incorporated in the evidence if authenticated by the testimony of witnesses, e.g. weapons, property stolen and recovered.

Profanity. The irreverent use in everyday speech of the name of God or Christ, distinguishable from blasphemy. It was originally considered an offence against God and religion and punishable in the ecclesiastical courts, but made a temporal offence in the seventeenth century. In Scotland profanation of the Sabbath by any occupation, business or sport, was at one time criminal, as was the disturbing of divine service.

Profit. The financial gain from a transaction or series of transactions. In an action for infringement of patent or copyright a plaintiff may elect to claim damages or an accounting by the defendant for the profit made by him by the infringement.

In real property a profit is a product of the land or a portion of the soil of the land. Hence a profit *à prendre* is the right of a person to take some profit from the land of another, such as sand from a seashore, ice from a canal, or animals such as fish and fowl, or grass growing on or minerals under the surface, or rights of common (of pasture, of piscary, or turbary, of estovers). Profits *à prendre* may be acquired by express grant, implied grant, statute, prescription at common law, or under statute.

Progressive stage system. A system of managing prisoners in prison, by giving prisoners the stimulus of hope that, by good conduct and industry, they can make quicker progress towards release. The idea owes much to penal settlements in Australia, particularly to the ideas and 'mark system' of Maconochie (q.v.) as superintendent at Norfolk Island. In the later nineteenth century the stages were, initially penal labour, then useful labour, then a measure of material comfort. When earnings at work were introduced the right to earn became a stage privilege. In modern practice the stage systems vary according to the prisoner's classification and his length of sentence. In addition there are certain privileges (letters, visits, books, smoking) which may be forfeited if abused or as punishment for misbehaviour.

Prohibited degrees. See AFFINITY; MARRIAGE.

Prohibition. The term for the legal prohibition of the manufacture, sale, or transportation of alcoholic drink. State prohibition laws were passed in several states, commencing with Maine in 1846 and by 1920 it was in force in 33 states. The 18th amendment to the Constitution was ratified in 1919 and became effective in 1920. Enforcement varied from district to district and the existence of prohibition gave rise to illegal manufacture, importation, and sale of spirits. In 1933 the 21st amendment repealed the 18th and national prohibition was ended. A few states continued state prohibition but this disappeared in time. Finland also tried national prohibition from 1919 to 1931.

Prohibition, Order of. Formerly a prerogative writ, now an order which may be granted by the High Court in England, directed to an inferior court or tribunal, forbidding it to commence or continue proceedings in absence of or excess of jurisdiction or in contravention of law. It lies to all inferior courts with limited jurisdiction, lay and ecclesiastical, civil, criminal, military, maritime, or otherwise having special jurisdiction, but not to legislative bodies. It may be sought alone or along with certiorari (q.v.).

Like the orders of mandamus and certiorari (qq.v.) the chief use of the order is as a means of controlling inferior courts and persons having the power to decide issues and the duty to act judicially. Like certiorari prohibition lies primarily in respect of judicial acts, but not for error in law, unless there is excess of jurisdiction, nor for error in fact. It lies for departure from the dictates of natural justice. In modern practice it had been extended to the control of all kinds of administrative as well as judicial acts.

It matters not that there is an alternative remedy to correct the defect in jurisdiction, or that an appeal lies against the absence or excess. If the objection to jurisdiction appears on the face of the proceedings, prohibition lies at any time, even after judgment.

Prohibitions del Roy (1607), 12 Co. Rep. 63. In this case Coke, C.J., with the consent of all the judges, stated that the King could not personally judge any case, but that any case affecting his interests should be decided by the judges.

Prolocutor. A term formerly sometimes applied to the Keeper of the Great Seal as Speaker of the House of Lords. It is also given to the member of the Upper House of Convocation who presides at its sittings, and to a member of the Lower House of Convocation selected by the members thereof to present its resolutions and opinions to the Upper House. The person appointed and his deputy require confirmation by the archbishop.

Promise. A voluntary undertaking by one person to do something to or for another or forbear from something. Such a promise is not binding in English law unless under seal or for consideration, and in Scots law not unless made in writing or proved by the promisor's admission on oath. The term promise is also applied to each of the mutual undertakings of parties to a bilateral contract, to do something to or for each other.

Promise of marriage. An exchange of promises by a man and a woman to marry each other. The mutual promises may be express, or inferred from conduct evidencing intention to marry. It is void if either party is, to the knowledge of the other, already married, or if made in consideration of the promisee permitting sexual intercourse or for other immoral consideration. No parental consent, or writing, or other formalities, are necessary. Exchange of rings or other gifts is not essential, but evidential. A promise by a infant in England was voidable, but a Scottish minor may validly promise, even before attaining the age requisite for marriage. Breach of promise was until 1971 in England, and still is in Scotland, actionable by either party for damages, unless there is justification for breaking off the engagement, on such grounds as bad character, supervening illness or insanity, or that the promise was induced by material misrepresentation.

Promissory note. An unconditional promise in writing, made by one person to another, signed by the maker, engaging to pay on demand or at a fixed or determinable future time a sum certain in money to or to the order of a specified person or to bearer. It is a kind of negotiable instrument, and the law applicable to bills of exchange applies with necessary modifications to a promissory note, the maker corresponding with the acceptor of a bill, but the provisions relating to presentment for acceptance, acceptance, and other related matters do not apply. Promissory notes have been known in England since the seventeenth century but were at first not admitted to be negotiable.

Promoter. A person who arranges for the formation of a company, the raising of necessary finance, and other matters necessary to bring it into being. The Companies Acts impose heavy liabilities on promoters for misstatements in prospectuses, and at common law promoters were regarded as being in a fiduciary relationship towards the company which they are promoting. The term is also sometimes applied to a person who procures the passing of a private Act of Parliament, or to a person seeking to further some undertaking for which statutory powers, e.g. of compulsory purchase, are being sought.

Promulgation. The act of making known to the general public. Acts of the Scottish Parliament were promulgated by proclamation in all the county towns, and, later, by publication at the market cross of Edinburgh. Modern U.K. statutes require no promulgation for their coming into force or validity.

Proof. (1) That which establishes the existence or non-existence of a fact unknown or disputed, namely admission or evidence; much of the law of evidence is concerned with onus of proof, modes of proof, manner of proof of various kinds of facts or facts in various circumstances, standard of proof, and similar matters; (2) also in England, a written statement of the facts on which a witness is able to give evidence, taken before a trial, a copy of which is available to counsel to assist him in examining the witness orally; (3) also, in Scotland, a hearing of evidence by a judge alone, as contrasted with jury trial.

In bankruptcy a proof is an affidavit, declaration, or other means whereby a claim against the estate is proved. A creditor is said to 'prove against the estate'.

Proper law. In international private law problems a question of importance and difficulty is: by what system of law are the rights and duties, liabilities and remedies, of parties to be determined? The answer frequently is: the proper law of the transaction. The proper law is defined as the system of law with which the transaction, or the part of it in question, has the most substantial relationship. In contract cases one legal system, e.g. the *lex loci contractus*, may be the proper law in respect of certain matters, such as of formation, but another, e.g. the *lex loci solutionis*, may be the proper law in respect of other matters, such as performance or breach. Attempts have been made to impose the 'proper law' as the best solution in questions of tortious liability, but in this sphere the 'proper law' has not yet been accepted in the U.K. In other branches of law the choice of the proper law is more settled; thus in relation to real property the proper law is the *lex situs*.

Property. The term property is used properly as denoting the right of ownership, as where a rule of law provides for the passing of the property in a thing, but is also, and more commonly, used in a transferred sense, of the object of the right of property, i.e. for the thing owned, as in such phrases as 'the property market', 'a desirable property'. In the former sense a right of property is the fullest right which may exist in and over any subject, including the rights to possess, use, lend, alienate, use up, consume, and otherwise deal with it. In the latter sense the term property is often used as including incorporeal assets such as shares, though it seems an excessively wide usage to use the word as including all a man's rights *in rem*, including such as the right not to be defamed.

Property in the sense of right of ownership may exist in respect of both corporeal things (e.g. buildings, animals) and incorporeal objects (e.g. copyrights, claims of damages). These categories are cross-divided into immovable objects (e.g. land) and movable objects (e.g. animals, claims). Particular systems sometimes treat as immovable in law certain things movable in fact. Particular legal systems may, however, proceed also, or indeed primarily, on other distinctions. Thus Roman law distinguished *res mancipi* from *res nec mancipi*. English law has divided property into real and personal property, according as the result of a claim was real, for the recovery of a specific thing, or personal, for the recovery of damages. By an historical accident land was the only thing which could be specifically recovered, so land came to be described as real property. In the case of movables the defendant could either return the thing or pay damages, and accordingly movables came to be described as personal property. Also for historical reasons it became settled that leaseholds were personal property. Down to 1925 the rules for descent on intestacy were different and there were, and are, other important differences. Personal property is distinguished into choses in action, which can be claimed and enforced by action only, and choses in possession, which can be secured by taking possession.

In Scots law the roughly corresponding distinction is between heritable property (land and rights akin to land) and moveable property (movable things and most kinds of claims) and these bodies of objects of property formerly descended on intestacy to the heir and to the executor respectively. There are other important differences. Both heritable and moveable property are subdivided into corporeal and incorporeal branches.

The term 'intellectual property' is simply a convenient label for such kinds of property as copyrights, patents, trade-marks, which fall under the heading of choses in action or incorporeal moveable property.

A right of property may be acquired in various ways. The main distinction is between original and derivative acquisition. The days are long past in settled countries when land could be acquired originally, by occupation or settlement, but movables can still be acquired in this way, as when a man catches fish or makes a table from wood. Much commoner and more important is derivative acquisition from another, by gift, purchase, or bequest.

Not everything may be acquired or made an object of property. Things such as the sea or the air cannot be appropriated and property held by the

Crown in the public interest, or owned by public authorities, are deemed *res communes* or *extra commercium.*

The right of property is best conceived not as a single right but as a bundle of distinct rights, some or even many of which may be relinquished temporarily without loss of ownership. The kinds of rights which a right of property confers over the objects of that right vary according to the nature of the object, but they normally include the rights to possess, use, use up, abuse, lend, let on hire, grant as security, gift, sell and bequeath the object. The owner loses his ownership only if and when he uses up the object, or transfers it without retaining any reversionary right. In short, ownership is the most absolute power permitted by the law to enjoy or dispose of things of the kind in question. But increasingly modern systems place restrictions in the public interest on the ways in which objects of property may be used, so that no ownership implies absolute rights.

The extent to which, and the ways in which, an owner may, while retaining ownership, grant to others some of the bundle of rights which he has by ownership, give rise to important bodies of rules. Thus, an owner may surrender possession by granting a licence to use land or lending a chattel, he may grant a neighbour an easement over his land, let his land on lease or his chattels on hire, create a security over his land or chattels in various ways, and so on.

The social, moral, political, and economic justifications for the legal recognition and protection of private ownership of property are diverse. There is much substance in the view that invention and industry should be rewarded by letting the creator keep what he has created or the proceeds thereof. Again, property provides an incentive to work and private property is normally better cared for and managed than public property. The Hegelians regarded some holding of property as essential for the development of personality; it is the control of some property which makes a man free and responsible, and public policy should facilitate the widest acquisition and holding of property, while also preventing excessive concentration of private property or the misuse of property for anti-social ends. A society in which all property was owned by the State would be a society of slaves. Socialists on the other hand seek to vest much, or even nearly all, property, in the means of production and exchange in the State and its agencies, nominally to be enjoyed and used for the benefit of all.

The right of property may be lost in various ways, particularly by transfer by sale, gift or bequest to another, by the operation of law such as rules of bankruptcy, execution of process or expropriation in the public interest, or by consumption or use, or total destruction of the object of property.

R. Noyes. *The Institution of Property*; F. Vinding Kruse, *Right of Property.*

Proprietary rights. The rights which a person has against persons generally to or over some object of property by virtue of owning or being entitled to it, as distinct from personal rights which he has against another person or persons only.

Prorogation. (1) An exercise of the royal prerogative by which a session of Parliament is terminated, the Parliament itself continuing in being until dissolved. By convention the sovereign acts on the Prime Minister's advice. It is effected by an announcement made in the House of Lords by commissioners appointed for the purpose by the sovereign. The Lord Chancellor reads to both Houses a speech from the Throne reviewing the proceedings of the session. The Speaker, on returning to the Commons Chamber with those who with him attended in the House of Lords, again reads the Speech from the Throne. Prorogation terminates sittings and suspends all business and terminates all proceedings pending, except impeachments and appeals before the House of Lords. Committees cease to exist and have to be appointed afresh for the next session. All Bills which have not passed through all their stages lapse and have to be reintroduced and debated afresh.

(2) Conferment by consent of parties of jurisdiction on a judge not otherwise having jurisdiction, provided the jurisdiction is not statutorily excluded. A defendant may prorogate jurisdiction by appearing and lodging a defence.

(3) An extension of time, or the extension of a lease.

Prosecution. The process of preparing and presenting the case against a person accused of having committed crime. It is technically distinct from, but closely connected with, the investigation of a crime, the search for and arrest of a person suspected of having committed the crime, and any preliminary hearing to determine whether or not there is evidence sufficient to justify putting the suspect on trial. In England the investigation is done by the police, the prosecution by an authority, the police, or a private person. In the U.S. the prosecutor is largely responsible for the police investigation and presents the evidence at a hearing before a grand jury which determines whether or not to allow indictment. In Scotland the investigation is by the police but subject to the general supervision of the Lord Advocate and the local procurator-fiscal and the Lord Advocate or procurator-fiscal prosecutes in court. In France the police conduct the investigation but an investigating magistrate has to conduct the preliminary hearing. The word is also used of a proceeding against a person in the criminal courts on indictment, and even, though inaccurately, of taking civil proceedings, e.g. prosecuting a claim.

Prosecutor. The person who, in a criminal court, presents the case against a person accused of crime, In some countries, such as France, prosecution is a public function conducted by officers of a nationwide service, a branch of the civil service. In Scotland prosecutions in the superior courts are conducted by the Lord Advocate and the Advocate-deputes, and in the inferior courts by the procurators-fiscal, who are civil servants. In other countries, such as the U.S., states and counties have their own prosecutors, mostly elected to office. At the federal level, a district attorney is appointed for each federal district by the U.S. Attorney-General's office, and as belonging to the executive branch of government may be replaced on a change of administration. In England on the other hand there is no general system of public prosecution. The Director of Public Prosecutions (q.v.) undertakes only a small proportion of prosecutions and most prosecutions are conducted by barristers instructed by local authorities, the police, and other authorities. In summary cases police officers frequently prosecute, in theory as private individuals.

Prospectus. The statement issued by the promoters of a nascent company, or by the directors of a company in being, when inviting members of the public to purchase a new issue of shares in the company or to lend money on debentures. Statute prescribes in detail what information must be given in a prospectus and provides remedies for misrepresentations therein causing loss.

Prostitution. Promiscuous sexual intercourse or sexual practices, done or permitted by women usually for reward. It is distinguished from adultery, in which the emphasis is on infidelity to the spouse and which may not be promiscuous or for gain, from fornication which is extra-marital intercourse not necessarily adulterous, promiscuous, nor for reward, and from concubinage which is an inferior form of marriage. In the ancient East, save among the Jews, prostitution was associated with religion and at Babylon was even compulsory in honour of a goddess. Among the Jews the Mosaic law strongly discouraged the practice. In Ancient Greece substantial laxity existed and houses of prostitution were regulated, but women of the highest grade of prostitute were socially recognized and wielded great and open influence. At Rome prostitution was severely regulated but latterly failed of effect and certainly did not replace widespread immorality. The early Christian Church laid great stress on chastity but had sympathy for prostitutes. Nevertheless, the practice flourished everywhere in the Middle Ages and there were licensed brothels in London. A change took place in the sixteenth century by reason of the Reformation and widespread prevalence of venereal diseases. Since then there have been alternations of licence and of repression.

The causes of prostitution are various, including male lust, separation from home and wife, marital unhappiness, female unemployment or need for money or desire for more luxurious life. Many of the women who resort to it are of inferior mental attainments.

Legally, as in the cases of drink and gambling, the choice is between repression, which can never be wholly successful and generates other evils, and regulation and control, though the latter cannot eliminate enthusiastic amateurs. Many European countries regulate brothels in various ways, and in Great Britain between 1864 and 1886 statute required compulsory medical examination of prostitutes in garrison towns.

British law makes procuring, causing, or encouraging prostitution, using premises for prostitution, detention of women in a brothel, and living on earnings of prostitution offences; and it is an offence in England to loiter or solicit in a street, but prostitution is not criminal in itself. Prostitution is often associated with other criminal activities, women decoying men to places where they are robbed, blackmailed, and the like.

Protection, Court of. This is not a court in the ordinary sense but an office of the Supreme Court of Judicature in England having charge of the protection and management of the property of persons under disability, and under the charge of a Master and Deputy Master appointed by the Lord Chancellor. A nominated judge hears appeals from decisions of the Master or Deputy Master, and exercises functions reserved to a judge.

Protector of the settlement. In English real property law, a person whose consent is required to enable a remainderman in tail to bar the entail, and without whose consent the remainderman can only bar his own issue and create a base fee (q.v.).

Protectorate. (1) In international relations and constitutional law, a relationship between one state and another under which the one exercises substantial control over the other. Theoretically and ideally a protectorate is a form of guardianship and the emphasis is on the duty of the protecting state to foster the social, economic, and political development of the protectorate with a view to it becoming independent, but in some cases, as of the German protectorate over Czechoslovakia in 1939, it is a euphemism for annexation. In the history of the British Empire and Commonwealth a protectorate or protected state frequently existed by treaty with a native ruler or chief under which the territory was administered in a way generally similar to that in Crown colonies.

(2) In British history the period (1653–59) during which Oliver Cromwell and Richard Cromwell successively governed under the title of Lord Protector of the Commonwealth of England, Scotland, and Ireland, assisted by a Council of State and Parliament, as provided by the Instrument of Government (q.v.). The regime was based on the army, with Major-Generals co-ordinating military and local government. It was modified by the Humble Petition and Advice (1657) (q.v.) which established a more nearly monarchical form of government with an upper house of parliament. The resignation of Richard Cromwell in May 1659 opened the way for the Restoration.

Protest. A declaration of opinion, usually of dissents; a notarial instrument in which the notary protests that the party against whom the instrument is directed shall be liable to certain consequences stated in the instrument. In the case of a bill of exchange it is notarial evidence of a demand for payment having been made and not having been met. Foreign bills must be protested in case of dishonour and inland bills must be protested in various cases. In maritime law it is a writing drawn up by the master of a ship and attested by a justice of the peace or consul setting down the circumstances in which an injury has befallen ship or cargo, or other circumstances have arisen affecting the liability of shipowner or charterer.

In the House of Lords the name is given to the dissent which may be entered in the *Journals* of the House by any peer who disagrees with the majority view of the House.

Protestant. A term originally applied to those who protested against a decree of the Emperor Charles V and the Diet of Spires of 1529, and coming to be applied to all Christians not adhering to the Roman Catholic or Eastern Orthodox branches of the Christian Church. It includes Anglicans, Methodists, Presbyterians, and members of other denominations.

The term is used in many statutes of the seventeenth century, notably in the Bill of Rights of 1688 and the Act of Settlement of 1701 which settled the succession to the Crown on Princess Sophia and her descendants, being Protestants. This was confirmed by the Union with Scotland Act, 1706, and the Union with Ireland Act, 1800.

Protocol. (1) In diplomacy, the original draft or record of a treaty or other diplomatic document or instrument; an addendum to a treaty; or the minutes of a deliberative assembly of the representatives of nations.

(2) In diplomatic circles, the official formulae used at the beginning and end of charters, papal bulls, etc., as distinct from the text.

(3) The etiquette and ceremonial to be observed in official ceremonies, particularly by a head of state.

Protocol book. A book kept by a person admitted a notary public in Scotland in which he was formerly bound to enter a note of all notarial instruments and deeds executed by him. Some such books survive from the fifteenth century onwards and are valuable sources of information on mediaeval law and practice.

Protonotary (or prothonotory). A principal notary or chief clerk, originally the holder at the Byzantine court of that office; formerly in England the chief clerk or registrar of the Courts of Chancery, Common Pleas and King's Bench; in the Catholic Church, one of a college of prelates whose function is to register papal acts and keep records of beatifications and canonizations. In some Australian states the title survives for a chief administrative officer of a Court.

Province. The area over which an archbishop exercises jurisdiction. In England there are two provinces, Canterbury, comprising all England south of the Humber, and York, comprising all north of that river. Each includes a number of dioceses, each with a bishop. Also in Canada the ten parts which since 1867 have formed the Dominion of Canada.

Province of Jurisprudence Determined, The. See AUSTIN, JOHN.

Provincial constitutions. Decrees of provincial synods held under various Archbishops of Canterbury, sometimes adopted later in the Province of York. Lyndwood's (q.v.) *Provinciale* is a valuable compilation of these constitutions from the time of Stephen Langton to Archbishop Chichele. See also R. Sharrock's *Provinciale* (1664) and M. E. Walcott's *Constitutions of the Church of England* (1874).

Provincial courts. The ecclesiastical courts of the two provinces, namely, in the province of Canterbury, the Court of Arches, the Court of the Vicar-General, the Court of the Master of the Faculties, the Court of the Commissary of the Archbishop, and the Courts of Audience and of Peculiars, both now in disuse; and in the province of York, the Chancery Court of York, the Consistory Court, and the obsolete Court of Audience.

Provincial estates. Representative assemblies in the provinces of the old French monarchy. They emerged in the fourteenth century as general assemblies of the three estates of the provinces to provide for consultation and defence, and multiplied in the following century. Their powers varied greatly from one place to another; some merely voiced grievances, but others had responsibility for collection of taxes. Though they existed mainly to vote subsidies, they were allowed to petition the Crown and to publish edicts, and were consulted on diplomatic issues. They were mainly aristocratic and feudal in composition and not fully representative. In the fifteenth century there were over 30 provincial estates, but in the sixteenth century some smaller ones were absorbed by larger ones while others were discarded by cardinals. By the eighteenth century only 10 remained with attenuated powers, these having been reduced by their own incompetence, lack of public support, and the growth in the power of the Crown and its determination to develop a centralized administrative and financial structure, subject to its own control.

Provinciale. See LYNDWOOD, WILLIAM.

Proving a will. Obtaining probate (q.v.) of the will.

Proving of the tenor. In Scotland the action whereby terms of a deed lost or destroyed may be established from drafts, copies, or other evidence to the same effect as the original deed.

Provisional orders. In 1848 provisional order procedure was introduced as a simplified means of obtaining legislative powers, particularly for local authorities, which would otherwise have required a private Bill. A Minister of the Crown under the authority of a statute makes a provisional order and then introduces a public Bill to confirm it or several such orders. It is alternative to and does not prevent the promotion of a private Bill. The provisional order requires publicity and, if need be, inquiry, in much the same way as a private Bill.

In relation to Scotland the Secretary of State for Scotland has power under the Private Legislation Procedure (Scotland) Act, 1936, to issue a provisional order in relation to any matter which could be a subject for a private Bill, and subsequently introduces a Provisional Order Confirmation Bill to confirm it.

In England and Wales provisional orders have been in practice superseded by Special Procedure Orders under the Statutory Orders (Special Procedure) Acts, 1945 and 1965.

Provisions of Oxford. A constitution drawn up by a commission of 24 drawn from the Mad Parliament summoned to meet at Oxford in 1258 when Henry III and Parliament were at loggerheads. It established a standing council of state of 15 with whose advice and consent Henry III was to exercise the royal authority. Three times a year the 15 were to meet with another 12 to treat of the common affairs of the realm, while another 24 members were to make grants of money to the Crown.

Proviso. A clause in a deed or statute, beginning 'Provided that' and operating as a condition or qualification, frequently inserted to save or except from the effect of the preceding words some rights, or instances, or cases. In criminal procedure 'the proviso' is the concluding words of a section of the Criminal Appeal Act, 1968, to the effect that, even though there was a technical irregularity in the trial the Court of Appeal may dismiss the appeal if no miscarriage of justice actually occurred.

Provisors. Persons who sought from the Pope a provision or undertaking that they should secure particular benefices when they fell vacant in return for an immediate payment by the provisor. Various statutes, Statutes of Provisors, were passed at various times to counter the practice, notably an Act of 1351.

Provocation. An act or series of acts done by one person to another which could cause in a reasonable person, and actually causes in another, a sudden and temporary loss of self-control, rendering him so subject to passion as to make him for the moment not master of his conduct. Killing subject to provocation by the deceased and done in the heat of the moment is not murder but manslaughter. Provocation accordingly is not a defence or an excuse but diminishes the quality of the crime and consequently the liability to punishment. Acts may readily amount to provocation, or a series of affronts, or even words.

Provost. The title of the chief officer of some colleges, and, in Scotland, the chief magistrate of a burgh. The Provosts of Edinburgh, Glasgow, Dundee, and Aberdeen are entitled to the designation Lord Provost, and this title was formerly claimed by some other towns also.

Provost-Marshal. The officer of the royal forces who is responsible for good order among the forces, the suppression of offences, and the execution of the sentences of courts-martial.

Proximate cause. The causal factor which is closest, not necessarily in time or space, but in efficacy to some harmful consequence. The general principle is *causa proxima non remota spectatur*, so that loss or harm will be attributed to the most immediate effective causal factor rather than to something earlier, though that may have been an essential prerequisite of the loss or harm.

Proxy (from procuracy). A person appointed to act and vote for another, such as for a member of a company at meetings of shareholders; also the instrument appointing a person to act as proxy.

Voting by proxy is permitted in parliamentary and local government elections in certain circumstances. Until 1868 members of the House of Lords could vote by proxy.

Marriage by proxy is permitted by some legal systems and, if valid by the law of the country of celebration, will be recognized as valid by British law.

Prussian Civil Code (Allgemeines Landrecht). The general code of the Prussian states. It sprang from the reform movement of Frederick the Great (1740–86) and was influenced by the eighteenth-century Enlightenment. It was begun under Frederick but not promulgated until 1794. It was to be enforced whenever local customs were not in conflict with it. During the nineteenth century it was adopted by other German states and it remained in force until superseded by the German Civil Code in 1900. Though predominantly a civil code, it included elements of constitutional and administrative law and criminal law. It was large, extending to 17,000 paragraphs and sought to be comprehensive and to provide a solution for every difficulty, thereby excluding judicial empiricism. It granted freedom of conscience and of religion, but censorship was imposed and political dissenters were subject to penalties. The criminal provisions were liberal, aiming at prevention rather than punishment, and torture was abolished and capital punishment limited. On the other hand class distinctions were strictly preserved; only the nobility could own manorial estates and they might not engage in trade or industry.

Pseudohermaphroditism. A condition in which a person combines features of both sexes in external genitalia, rendering determination of sex difficult, yet has a single chromosomal and gonadal sex. In male pseudohermaphroditism the gonads are testes and the chromosomal pattern is male but the external genitalia and secondary sexual characteristics resemble those of a female. In female pseudohermaphroditism the individual has ovaries but external genitalia or secondary sexual characteristics resemble those of a male. At puberty secondary female characteristics usually appear.

Pseudo-Isidore. See FALSE DECRETALS.

Psychological treatment of prisoners. An idea originating in psycho-analysis, is treatment by mental analysis, suggestion, persuasion, or re-education. It aims to reveal to the patient the true working of his conscious and unconscious motives to help him meet the difficulties in his own life, understand himself better and adapt better to changing circumstances. Psychological treatment requires the patient's confidence and co-operation. It also provides an argument against fixed prison terms and in favour of indeterminate sentences, but is not a general alternative to prison. Hitherto psychological treatment is of unproven and uncertain efficacy.

Psychopathy. A kind of mental illness displaying less specific symptoms than most mental illnesses and less easily measurable than mental subnormality. It is not a specific illness or disorder, but a term for many kinds of abnormal and anti-social conduct, including inadequacy, aggressiveness, and sexual promiscuity. It has been statutorily defined as a persistent disorder or disability of mind which results in abnormally aggressive or seriously irresponsible conduct which requires or is susceptible to medical treatment. It may or may not co-exist with subnormality or mental illness. Cases raise in acutely difficult form the problem of distinguishing between persons who should be held responsible and punishable and those who should not. In England it has been held several times that psychopathy can justify a defence of diminished responsibility. In Scotland the view has been accepted that it does not amount to that state.

Puberty. The stage of life at which a young person becomes capable of begetting or bearing children. For legal purposes this is fixed at 14 for males and 12 for females and at common law young persons might marry at these ages, but in Britain the age has been raised to 16 for both sexes.

Public Act. An Act of Parliament which is judicially noticed, without need for proof of its terms, as contrasted with a private Act (q.v.). Every Act passed since 1850 is deemed public in this sense unless the contrary is expressly provided in the Act. This significance accordingly has no connection with the classification of Bills into public and private. Local Acts may be either public or private in this sense.

Public administration. The implementing and application of the policies of governments, particularly by departments of state, by public corporations, such as nationalized industries, and by local authorities. In general the policies are determined by the Cabinet and Parliament, but the details are frequently worked out by the departments and other authorities concerned and the detailed policies applied to particular cases.

Public Bill. A Bill presented to Parliament which deals with matters of national policy and is generally applicable to the United Kingdom, or to England and Wales, or to Scotland, or to a major place or area, such as London.

It originated in the petition presented to the King for legislation. In 1414 the Commons requested the King that no law be made save in the terms requested, which the King conceded, and thereafter a draft statute was sent up, rather than a petition for legislation.

A public Bill may be introduced by a Minister of the Crown in implement of government policy, or by a private member who has won a place in the ballot to introduce private members' Bills. In either case, if passed, it results in a public general statute.

Public company. A company which does not by its Articles of Association impose on itself the restrictions required of a private company (q.v.), and which can accordingly be of any size and solicit public subscription for its shares. Only a small minority of companies are public companies, but these include nearly all the major commercial and industrial businesses.

Public corporations. The name commonly given to corporate bodies established by statute to own, manage, and operate utilities and industries in the public interest. As compared with private corporations they have no shareholders, no equity capital, and no annual meetings. They frequently have no competition and no stimulus to efficiency. Early examples of such public corporations were the Mersey Docks and Harbour Board, the Port of London Authority, and the Metropolitan Water Board. Since 1918 and more particularly since 1945 many more have been established to manage nationalized industries. In theory the public corporation is accountable through the responsible minister to Parliament and thus to the community, is financially self-supporting, and is free from political interference. In practice such bodies have no regard for the community, are constantly losing money and having to be subsidized by taxation, and are subject to constant political interference, and reversals of policy. They are run by Boards appointed and renewable by Ministers, and the qualifications of many of the members of the Boards are not apparent; frequently the appointments are political perquisites, rewards for failed or superannuated politicians, and it has constantly proved difficult to get men knowledgeable about the industry in question to accept appointment.

Public defender. The name given, particularly in the U.S. to a lawyer employed at public expense to defend persons accused of crimes. The U.S. Supreme Court has ruled since 1963 that poor persons are entitled to the aid of counsel during police interrogation, trial, and appeal, at least in case of felonies, and some jurisdictions have extended the principle to misdemeanours.

See also LEGAL AID.

Public domain. In the U.S., land owned by the federal government. It included land extending west to the Mississippi ceded to the federal government by seven of the original states in 1781–1802, land acquired by the Louisiana Purchase of 1803, the Florida purchase of 1819, the acquisition of the Oregon region in 1846, the Mexico cession of 1848, the Gadsden purchase of land from Mexico in 1853, and the purchase of Alaska from Russia in 1867. Much of this land was sold off by public land sales or granted to states or railroad corporations, but in the twentieth century much of the public land has been set aside as National Parks and National Forests.

Public General Statutes. The body of statutes which are both public, i.e. to be judicially noticed by courts without further evidence of their terms, and of general application, to the United Kingdom, or to England and Wales, or to Scotland, or to a major area such as London. The presumption is that they apply throughout the U.K. Since 1798 Public General Statutes have been published separately from Local and Personal Acts in annual volumes.

Public international law. See INTERNATIONAL LAW.

Public law. The distinction between public law and private law (q.v.) has been recognized since at least the Roman law, is fundamental in all countries with civil or Romano-Germanic law, and been regarded as the fundamental division of the principles and rules of any legal system, but it is hard to state the precise basis of the distinction. In general it may be said that public law comprises the principles and rules which relate to the structure, activities, rights, powers and immunities, duties and liabilities, of the State, of the organized political community, the government and its departments and agencies, save that in circumstances where the State, or a department or agency, enjoys no special rights or powers its relations with individuals are regulated by private law. Public law is accordingly the part of the whole legal system which is applicable to the State and its relations with ordinary individuals, which are different from the private law concerning the subjects of the State and their relations with each other.

The distinction must not be allowed to conceal the fact that in many sets of circumstances both public and private law may be relevant, or may interact: conduct may be both a crime and a tort, or give rise to claims for damages and also for social security benefit.

Even in civil law countries, public law has attained a standard of development very inferior to that of private law, and in the U.K. and common law countries it hardly exhibits any unity or system at all.

Public law is generally considered to comprise constitutional law, administrative law and procedure, local government law, social security law, revenue law, ecclesiastical law, and military law, Criminal law and procedure are sometimes included in public law, and are at least akin to public rather than to private law, but are sometimes considered distinct from both.

The category comprises rights and duties enforced by civil, by administrative, by ecclesiastical, by other special and by criminal courts and procedure.

Public meeting. A meeting which any person may attend. Persons may lawfully meet in any place with the consent of the owner or occupier for any lawful purpose, but not in any place without the consent of the owner or occupier, nor in any public street, which may be used only for passage and purposes incidental thereto. By custom public meetings may be held at various particular places.

Public mischief. A concept of very indeterminate connotation. It has been held in England to be criminal to conspire to commit, or to commit, a public mischief. This probably includes any criminal conduct, but also conduct to the prejudice of the State if done by dishonest, tortious, or criminal devices. Thus a false report to the police which causes them to waste time investigating a crime never committed has been held a public mischief. It has been said to include corrupting public morals (q.v.). The vagueness of this crime makes it a dangerous and undesirable one, which would probably be better expunged from the criminal law. Some specific kinds of public mischief, such as making hoax telephone calls, are statutory offences.

Public morals. Also a concept of very uncertain connotation. It has been held in England to be criminal to conspire to corrupt public morals, as by publishing a directory of prostitutes or advertising opportunities for committing homosexual acts, but the vagueness of the concept has given rise to criticism of the recognition of the crime.

Public nuisance. At common law in England the crime of doing an act not warranted by law or omitting to discharge a legal duty, which obstructs or causes inconvenience to the public in the exercise of rights common to all persons. It includes obstructing the highway, polluting the atmosphere or a river, sending to market food unfit for human consumption, causing excessive noise and dirt in the conduct of quarrying, and other similar conduct. An isolated act may be a public nuisance and the interference with the public's rights must be substantial and unreasonable. It is not necessary that there was any intention to create a nuisance; it is enough if it was the natural consequence of what was done. Many kinds of public nuisance are also made criminal by particular statutes.

Public opinion. The term for the predominant sense of the whole body of individual opinions, views, attitudes and opinions about particular topics held and expressed by a substantial proportion of a community. Public opinion is not necessarily majority opinion, but the opinion held by a substantial, though possibly minority, group. It must be distinguished from professional opinion, e.g. the views of doctors on a method of treatment. On some topics there may be no discoverable public opinion, but only the general opinion of a particular section of the community. Some topics may give rise to strong and widely held views at one time but to little or no general opinion at another time. On many topics there may be simultaneously held two or more contradictory public opinions.

The existence and state of general public opinion has long been regarded as important by rulers, who have long sought to ascertain it and influence it, at least among the classes whose views mattered. In the past proclamations, sermons, and other means of influencing opinion were used. The existence of a general public opinion has been much more significant since the development of printing and from the sixteenth century exchanges of information between financial centres, propaganda tracts, broadsheets, and newspapers became important. In recent times newspapers, magazines, radio, and television have assumed great importance as moulders of public opinion both by what they state and what they omit, and by the slant they impart to the facts. In modern times politicians, parties, pressure groups, public relations officers, and advertisers spend much time trying to create a public opinion favourable to what they are trying to promote.

Public opinion is expressed by books, pamphlets, speeches, advertisements, petitions, letters to the editors of newspapers, demonstrations, strikes, and in the last resort insurrection. Bentham believed that public opinion was a useful check on the authority of rulers. But there is great danger of government policy being influenced by public opinion artificially whipped up, or opinion based on incomplete or inaccurate information, or by opinion on a matter which it is very difficult properly to understand and evaluate, and also danger that politicians may in deference to public opinion seek to do what will be popular rather than what is right or really necessary. Indeed deference to ignorant public opinion for the sake of votes is the greatest defect of democracy.

Public opinion on matters of law is frequently non-existent, e.g. what is the present public opinion as to the law of future interests in land? or so nebulous and ill-informed as to be valueless, e.g. the opinion that the law is a complicated mess and lawyers just crooks, but it may be focused sharply on a topic and crystallized by a particular case or scandal, the report of an investigation, the campaign of a pressure group, or a speech by a public figure. There may be a quite spontaneous outburst of feeling on an issue. There may consequently be a surge of public feeling such that a government feels bound to make a change in the law.

Public opinion influences lawmaking and law reform principally through elections and the resulting composition of legislatures and executives, but also does so through the Press and other comments on the law and its administration, on proposals, reports, and bills, petitions and representations, delegations to ministers, and the like.

The ascertainment of what public opinion is on any matter is usually made by public-opinion polling, i.e. ascertaining by questions the views of a sample of the general population. It is commonly done in relation to major political issues. The reliability of the results depends largely on the choice of the sample and the framing of the questions.

Bryce, *American Commonwealth*; Dicey, *Lectures on the Relation of Law and Public Opinion in England during the Nineteenth Century*; Lowell, *Public Opinion and Popular Government*; Lippmann, *Public Opinion*; Ginsberg (ed.) *Law and Opinion in England in the Twentieth Century.*

Public order. The state of peaceful co-existence among members of the public generally in which there is an absence of breach of the peace, fighting, rioting, disturbance, or conduct which causes unreasonable interference with or disturbance to quiet living. Infringements of public order have in particular circumstances been defined by statutes as including going armed, the wearing of para-military uniforms or uniforms indicating membership of a political or other group, inciting mutiny or disaffection in the police or the armed forces, sedition, and other offences.

Public policy. A very indefinite moral value sometimes appealed to by Anglo-American courts as justifying a decision. It has been said to be a principle of judicial legislation or interpretation founded on the current needs of the community. It normally prohibits and rarely creates: the standard phrase is 'contrary to public policy'. It depends not on evidence but on judicial impression of what is or is not in the general public interest. For that reason judges have criticized it as providing an uncertain, even dangerous, standard, an 'unruly horse', and have been reluctant to invoke it in unprecedented circumstances.

It has been referred to as upholding the sanctity of marriage, justifying religious tolerance, maintaining the integrity of political life, justifying the principle that restraints of trade are void unless shown to be reasonable in the interests of the parties and the public, the principle which treats many kinds of contract as illegal and unenforceable at common law, the revival of an old or uncertain crime to justify punishment of what is obviously undesirable, and in other contexts. In essence to declare something contrary to public policy is for the judge to declare that he thinks it wrong to allow it.

In many contexts rules originally based on public policy have become so settled that the justification is forgotten, and only statute could alter the rule.

The continental counterpart is the concept of *ordre publique* which essentially depends on ideas of natural law.

Public Prosecutions, Director of. An English official, first appointed in 1878, though appointment had been called for repeatedly in the nineteenth century, with the function, under the superintendence of the Attorney-General, of instituting and carry on criminal proceedings and of advising others doing so. He is assisted by a number of Assistant Directors. It is, however, far from the case that the Director and his staff form a corps of state prosecutors or that anything like all prosecutions in England and Wales are conducted by, or under instructions of, the Director and his staff. The great majority of prosecutions are initiated by the police or other private persons. From 1883 to 1908 the office was combined with that of Treasury Solicitor. Counsel representing the Crown in prosecutions conducted by the Director are selected by the Attorney-General. At the Central Criminal Court they are known as Treasury Counsel (q.v.). The police must report all instances of various categories of crimes to the Director and the Director may take over the conduct of proceedings initiated by a private prosecutor.

Public prosecutor. See DIRECTOR OF PUBLIC PROSECUTIONS; LORD ADVOCATE; PROCURATOR-FISCAL.

Public Records. Defined in England by the Public Record Act, 1837, as all rolls, records, writs, books, proceedings, decrees, warrants, accounts, papers, and documents whatsoever of a public nature belonging to Her Majesty. Most of the Public Records, after many years of neglect, are now housed in the Public Record Office, Chancery Lane, London. Such records are available for public inspection only 30 years after they were created. The Act of 1837 made the Master of the Rolls titular custodian of all the muniments of the superior courts of law and of special or abolished jurisdictions. In 1958 responsibility was transferred to the Lord Chancellor.

See further M. S. Giuseppi, *Guide to the Manuscripts preserved in the Public Record Office*.

See also Domesday Book; Exchequer Records.

Public Record Office. A national repository established by and maintained under statute of 1838, the main repository being in Chancery Lane, London. Previously records were kept in the Tower of London and the Rolls House in Chancery Lane which, originally a house of maintenance for converted Jews, was annexed by Edward III to the office of the Master of the Rolls, who had his official residence there. Distinguished Keepers of the records have included Selden, Prynne, and Palgrave (Deputy Keeper, 1838–61) who did much to establish the modern office and promote the critical study of English history from record evidence. Sir John Romilly, M.R. also played a great part in developing the modern care of records. It was under the general direction of the Master of the Rolls until 1958 when the general responsibility was transferred to the Lord Chancellor, but the working head of the Office is the Deputy Keeper of the Records, appointed with the approval of the Queen and assisted by a large staff. The records originally put in the charge of the M.R. to be deposited in the P.R.O. were mainly Court records, but all records belonging to Her Majesty are now directed to be placed there, and it contains non-current records of the courts and of most of the government departments. An annual report is made to Parliament and calendars, catalogues and indexes are produced from time to time. The National Register of Archives, containing particulars of numerous local and private records, is maintained at the P.R.O.

Public right. A right at common law or under statute pertaining to the public generally and exercisable by any member thereof, such as the right to pass along a highway, as contrasted with a private right attaching to an individual. The remedy for infringement is an indictment or an action by the Attorney-General for an injunction, either brought directly or on the relation of an individual aggrieved by the infringement.

Public ships. Vessels publicly owned, including warships, unarmed ships reserved for governmental functions, and state trading vessels. The first two categories are immune from the jurisdiction of courts in ports which they visit, unless the immunity is waived. Publicly owned merchant ships may not be entitled to the same immunity from the jurisdiction of U.K. courts, but in U.S. law the general principle that they were entitled to the same immunities has been narrowed by exceptions.

Public trust. A trust for public purposes being either a charitable trust or one for public objects which are not of a charitable character. The latter group are, with certain exceptions, invalid if they infringe the law against perpetuities (q.v.), and will in any case not be recognized by the court except in so far as they are for the benefit of ascertained or ascertainable beneficiaries. The distinction between purposes of a public and those of a private nature is a fine one and almost impossible to define in general terms.

Public Trustee. A public officer, established by statute of 1906 as a corporation sole, having a central office in London, who may act as an ordinary trustee, or be custodian trustee, or be appointed a judicial trustee. He may act alone or jointly with any person or body of persons, but cannot accept a trust exclusively for religious or charitable purposes, or unless to a limited extent one which involves managing or carrying on a business, or one under a deed of arrangement for the benefit of creditors, or one for the administration of an estate known to be insolvent, or one to be construed and governed by the law of any foreign country. He may hold any of the funds of a friendly society or trade union. He has a duty to administer small estates unless there are good reasons for refusing to do so.

Public Works Loan Commissioners. A body constituted in 1875 for the purpose of making loans to local authorities and certain other bodies. The money is provided by the National Debt Commissioners out of the National Loans Fund to an extent provided by Public Works Loans Acts. In making loans they must consider whether the work or purchase of land for which the loan is asked would be such a benefit to the public as to justify a loan out of public money, and have regard to the sufficiency of the security. The main purposes for which money may be lent are housing, public buildings and works for health and sanitary purposes. Loans are repayable by instalments with interest, and security must be taken for the loan.

Publication. Communicating or making available to the public. In defamation it is essential in English civil law that the matter complained of have been published to some person other than the plaintiff. In patent law publication before protection is obtained defeats an inventor's claim to protection by patent because only a new and original invention can be patented. In copyright, publication means the issue of copies to the public and has the consequence that copyright exists in the work as a published and not as an unpublished work.

Publici juris. Of public right, available to any person. The term is used of those things such as light, air, and the open sea which may be used by anyone, and of rights, formerly exclusive to a person by virtue of copyright, patent, or similar right but where the exclusive right has expired and the right of publication, manufacture, or otherwise has become public.

Publicists. The statute of the International Court of Justice lays down as a subsidiary source of law for judicial decision the 'teachings of the most highly qualified publicists'. These 'publicists' are the leading text writers on public international law. There is no settled list of who are or are not within the class, and reference to the teachings of publicists has normally been made only in dissenting or separate opinions of individual judges, where such reference has been frequent.

In French law jurists are generally known as *publicistes* or *privatistes* according as their fields of interest concentrate on topics of public or of private law.

Puchta, Georg Friedrich (1798–1846). German jurist, Savigny's successor at Berlin, became the leader of the German historical school. He gave a profound basis for the doctrines of the historical school in his book on customary law, which influenced Savigny, though some of his other books had too little regard for modern developments or just results. His main works were *Das Gewohnheitsrecht* (1828–37), *Cursus der Institutionen* (1841–47, and later editions), *Lehrbuch der Pandekten* (1838 and later editions), and *Vorlesungen über das heutige römische Recht* (1847).

Pufendorf, Samuel von (1632–94). German jurist, while in prison in Denmark thought out a system of universal law which became his *Elementa jurisprudentiae universalis Libri duo* (1661). He became professor of the law of nature and nations at Heidelberg in 1662. His book *De statu imperii germanici* (1667) was an attack on the Holy Roman Empire. In 1670 he moved to Lund where he published *De jure Naturae et Gentium libri octo* (1672) and in 1675 *De officio hominis et civis juxta Legem Naturalem Libri Duo*. In 1677 he went to Stockholm as historiographer-royal and in 1688 to Brandenburg. In the *De Jure Naturae et Gentium libri octo* Pufendorf built on the work of Grotius and extended and developed it. It is a complete system of law public, private, and international, based on natural law. Against Hobbes's view he contended that the state of nature was one of peace, not war, and he urged the view that international law was not confined to Christendom but existed between all nations. His work, though for long inadequately appreciated, is of great importance. He presented the basis of a universal legal science on the basis of natural law. His most original work was in German public law, on the constitution of the German Empire.

Puisne judge (pronounced 'puny'). Any judge of the English High Court, other than the Chief Justice, or, in the former Court of Exchequer, Chief Baron. The Master of the Rolls is not generally regarded as a puisne judge.

Puisne mortgage. Formerly a mortgage subsequent to a mortgage of a legal estate, but since 1925 it is any legal mortgage of the legal estate not being one protected by a deposit of documents relating to the legal estate affected and, in appropriate cases, registered in a local deeds registry.

Pulszky, Agost (1846–1901). Hungarian jurist and statesman, sought in his writings to synthesize the historical school and sociological jurisprudence. His writings include *The basis of a philosophy of law and the State* (in Hungarian, 1885), and *The Theory of Law and Civil Society* (1888).

Pulton, Ferdinando (16th–17th cent.). Devoted his life to editing the statutes, publishing in 1560 *An Abstract of all the Penal Statutes which are General*, a practical and useful book, and in 1606 *A Kalendar or Table of all the Statutes ... and An Abridgement of All the Statutes*, comprising an edition of the statutes arranged chronologically and a part in which their contents were summarized under alphabetical headings. He also published in 1609 a *De Pace Regis et Regni*, a comprehensive work on the criminal law.

Punishment. The infliction of some pain, suffering, loss, disability, or other disadvantage on a person by another having legal authority to impose punishment. Punishment must be legally authorized, otherwise it is prima facie tortious or itself criminal. In modern societies punishment is generally confined to the consequences of infraction of the criminal law, and in general one person may not punish another for infractions of civil law. But punishments may be inflicted by parents or teachers on children as a means of discipline.

Similarly punishment is normally personal, attaching to an individual for his own transgressions;

collective punishments are generally of questionable legality and fairness.

Punishment may be imposed in various contexts: in the family, the school, the armed forces, the professional association, the trade union, and in the political society generally. Within the family the head of the family has traditionally had limited powers of punishment of the subordinate, and particularly younger, members of the family; this reached heights in the Roman paterfamilias, the mediaeval master craftsman, and the Victorian father. Schools and colleges have frequently exercised a quasi-paternal jurisdiction, the armed forces to maintain discipline, professional associations to maintain professional standards and regulate what they consider unethical practices, and trade unions to ensure solidarity and cohesion. The legal basis of all these kinds of jurisdiction is either delegated state authority or contractual acceptance of the rules and standards of the group or association. But the most important are those imposed by the political society generally on law-breakers.

In primitive societies punishment is often conceived of as the vengeance of the gods against one who has offended against the customs and rules of the society. In more developed societies a distinction is drawn between religion-based punishments and ethics-based punishments imposed by senior men in the society. In Jewish and Christian thought an essential element of punishment is the guilt of the wrongdoer and this relationship was extensively worked out in mediaeval theology. The ethical approach emphasized the importance of the moral responsibility of the offender and the immunity from punishment of offenders not responsible for their actings, and also the element of intention in criminal conduct.

There was not much serious thought about punishment until the age of the Enlightenment. The first, and very influential, writer was Beccaria whose *Dei delitti e delle pene* was published in 1764, whose view was that the sole object of punishment was prevention. Bentham's ideas in his *Introduction to the Principles of Morals and Legislation* (1789) were also very influential; he believed that punishment should be sufficient to prevent crime and no more, and argued that cruel and excessive punishments were valueless and evil. These ideas were enormously influential on nineteenth-century reformers. Since the mid-nineteenth century penal theory has been largely based on observation of the needs of individual criminals, on firm but humane treatment, on attempts at constructive reform and training and on bold experiments.

Purposes of punishment

Various purposes and functions have been suggested. The first is retribution, the idea that the offender should be made to suffer in return for the wrong done and in proportion to the evil. This has support from a wide range of theological and philosophical sources and accords with many popular views about crime and punishment; it may afford some psychological satisfaction to the victim. But the principle of the *lex talionis*, 'an eye for an eye' is difficult, sometimes impossible, to apply, and does not take much account of the wickedness or malice of the offender. In modern penology logical retribution does not bulk largely.

A semi-theological development is the theory of expiation, that an offender must, by undergoing suffering corresponding to that which he has caused, work out the evil of his conduct.

Prevention is a material function, though by itself rather negative; it is properly justifiable in cases where an offender must be kept out of society for a substantial period because of the danger of his repeating his offences.

A purpose frequently invoked is deterrence which has two aspects, deterring the offender from offending again, and by example deterring others from offending in a similar way. This purpose assumes that criminals and potential criminals are rational and will seek to avoid punishment and will remember past experiences and weigh the possible consequences of committing crime and being caught. Belief in deterrence has often underlain the former public infliction of punishments, and the cruel and often brutal nature of many punishments. The deterrent effect of punishment in fact depends largely on certainty of detection and conviction, but even if conviction were certain there would still be irrational and optimistic criminals who would not be deterred thereby.

Modern penology stresses the reformation of criminals, but some thinkers contend that punishment is inconsistent with reformation, and constructive treatment of offenders cannot be punitive, and any kind of punishment cannot be reformative. Moreover, reformation assumes that criminals are capable of being reformed; in practice many custodial sentences are too short for any reformation and many criminals are too corrupted for any reformation. Again, many courts and members of the public are sceptical about the "soft" regime of reformation and retraining.

Protection of the public is an important purpose of punishment and in some cases, e.g. sex offences, can only be achieved by removing the offender from society for long periods. Similarly, persistent offenders are commonly thought to require longer sentences or special sentences of preventive detention.

Modes of punishment

Many kinds of punishments have been used at different times in different societies. Death has been frequently employed, the commonest forms having been hanging or decapitation by sword, axe, or guillotine. In biblical times persons were often stoned to death. Under imperial Rome crucifixion was used for thieves and slaves, while in Rome being hurled alive from the Tarpeian rock was utilized as was being fed to the wild animals in the Colosseum. In mediaeval Europe decapitation was reserved for persons of better status, and hanging for those of lower rank. Heretics and witches were burned alive. In England the penalty of hanging, drawing, and quartering was normally reserved for treason; the victim was hanged till partly strangled and while still alive disembowelled; his entrails were burned and his head was severed and his body divided into four pieces. In France traitors might be pulled apart by horses or tied to a large wheel and their bones broken by hammer-blows.

Maiming has also frequently been used, methods including branding with hot irons, blinding, amputation of hands, ears, tongues, and flogging. The offender might be placed in the stocks or the pillory and pelted by the general populace, frequently to severe injury.

Until recent times members of the armed forces were subject to death or to substantial corporal punishment, as by flogging, normally inflicted publicly, as well as to various forms of detention and extra duties, particularly of heavy and dirty kind, e.g. carrying coal.

In various cases punishment of the offender from the community have been used, by way of exile, or transportation to a penal colony.

In modern times the main kinds of punishment utilized are custodial, comprising imprisonment and other varieties of detention (including Borstal training and detention in young offenders' institutions); pecuniary, by imposition of pecuniary fines; supervision, as by putting offenders on probation; disabling, as by disqualification from holding a driving licence; and miscellaneous other penalties. In custodial institutions, particularly for young offenders, there is considerable reliance on training and reformation.

One major difficulty of penal policy is the choice between the determinate sentence, fixed at the time of imposition, which may not be long enough for constructive treatment, or training or reformation in the particular case; and the indeterminate sentence, which leaves it to a board or an expert to determine when the prisoner may safely be released; this may be open to abuse, is much disliked and implies that a prisoner might be detained for years for a comparatively slight offence.

Another difficulty is the fact that public attitudes demand death or long detention for murderers, though some of them could safely be released after comparatively short periods, whereas in other cases long sentences may be justifiable for offenders whose prognosis is poor though they may not have committed the most serious offences.

Effectiveness of punishment

Judged by reconviction records, no form of punishment is completely effective in deterring or reforming criminals or protecting society. The best course would appear to be the development of better techniques of prediction of likelihood to offend again and the adaptation of a greater variety of methods suitable for different kinds of criminals.

A. Ewing, *Morality of Punishment*; Lord Longford, *The Idea of Punishment*; H. L. A. Hart, *Punishment and Responsibility*; N. Walker, *Crime and Punishment in Britain*; L. Fox, *The English Prison and Borstal Systems*; M. Grunhut, *Penal Reform*; D. Glaser, *The Effectiveness of a Prison and Parole System.*

Pupil. In Scots law, a boy aged under 14 or a girl aged under 12, i.e., children who have not attained the presumed ages of puberty. Pupils have very limited legal capacity. Contracts must in general be entered into by the pupil's tutor, but a contract by a pupil is probably enforceable by but not against him, and is voidable for 20 years. He has no title to sue for wrong done nor liability to be sued, and property cannot in general be held by him. Legal proceedings must be taken by his tutor as such. If under eight, he is deemed incapable of committing crime, and over that age is subject to special procedure.

Purchase. In its most general sense, buying. But the word is used legally also in England in the technical sense of acquiring land by lawful act of himself or another, e.g. by conveyance, devise, or gift, as distinct from acquiring by act of the law, e.g. inheritance, or by wrong, e.g. disseisin. Hence, in real property 'words of purchase' were the words by which an estate in land is deemed to commence in point of title in the person described by them.

Pure theory of law. The theory of law developed by Kelsen, who attempted to arrive at a theory of law uncontaminated by history, ethics, politics, sociology, and other external factors. He built on Kant but modified Kantian theory and introduced elements from other writers, and was at one with analytical jurists in seeking to separate law from other sciences.

The science of law is not a sequence of causes and effects like natural science, but a hierarchy of normative relations. The aim of a theory of law is to reduce chaos and multiplicity to order, and such a theory is formal, a theory of the way of ordering or changing contents in a specific way. It stands to a particular system of positive law as does possible to actual law. The sole object of study for jurisprudence is the nature of the norms or standards established by law. Law and the State are the same thing viewed from different angles. Rules of law cannot be defined in terms of the State, nor by reference to justice, nor by their content, but only by the way in which the rule is created, by inquiring what norm attaches legal validity to particular rules or legal behaviour, such as a court awarding damages. One must trace back to the ultimate source of validity, such as the norm: the will of the Queen in Parliament (or the Constitution) must be obeyed; this is the *Grundnorm* or fundamental axiom of the particular legal system; the validity of the *Grundnorm* can itself be tested only by non-legal factors. It is the product of historical development or revolution, but not of law. The theory is of great logical value and critical force, but many critics feel that the theory has no room for morality, social factors, or other influences on law.

H. Kelsen, *General Theory of Law and State; Pure Theory of Law; What is Justice?*

Purgation. A person's clearing himself of a crime of which he was accused. It might be either canonical, the form used in the spiritual courts in which mode the accused took an oath that he was clear of guilt and brought neighbours to swear that they believed that he swore truthfully; or vulgar, by combat or by ordeal of fire or water.

Purging an irritancy. In Scots law, when liability is incurred by non-implement of some obligation to have the obligation irritated or cancelled, the person liable may in certain cases, by paying or performing in terms of his obligation at any time before decree declaring the irritancy to have been incurred passes against him, purge the irritancy.

Purpresture (or purprision). In mediaeval law, any infringement of the rights of the Crown, in the Royal demesne, or by obstructing the public highway, or impeding a public water course. It extended to encroachment on the interests of the feudal lord, other than the Crown, and was a feudal delinquency inferring forfeiture of the fief.

Pursuer. In Scots law, the person who claims in a civil action, equivalent to the English plaintiff.

Purveyance. The prerogative of the sovereign to compel the sale to himself at a reduced price of goods necessary to maintain the royal household as it travelled about the country. It included the hiring of carriage of the goods and often the obtaining of forced labour also. It was a grievance from the Middle Ages until the seventeenth century and was limited by Magna Carta and by various statutes thereafter, but it persisted until it fell into disuse during the Commonwealth and was abolished in 1660.

Pursuivants (pronounced 'pirswivant'). Junior members of the College of Arms. They are entitled Rouge Croix, Bluemantle, Rouge Dragon, and Portcullis. Pursuivants extraordinary are occasionally appointed, with ceremonial duties only. In Scotland there are three, Carrick, Kintyre, and Unicorn, and there may be pursuivants extraordinary.

Putative. Supposed, believed, or reputed, used in the phrases 'putative marriage' (believed to be valid) and 'putative father' of an illegitimate child.

Puture (or pulture, or serjeants' food, or fold). A customary exaction demanded by serjeants of the peace, keepers in forests, and bailiffs of hundreds, of provisions for themselves and their horses and dogs, without payment, from the tenants and inhabitants of the forest or hundred. By the fourteenth century it has been commuted to a small sum of money and statute of 1351 prohibited the taking of anything under the name of puture beyond the customary payments.

Putter, Johann Stephan (1725–1807). German jurist, the leading expositor of the public law of the old Empire, author of *Institutiones iuris publici germanici* (1770); *Teutsche Reichsgeschichte in ihren Hauptfächen entwickelt* (1778); *Historische Entwickelung der heutigen Staatsverfassung des Teutschen Reichs* (1786–87), and *Literatur des teutschen Staatsrechts* (1776–83).

Pyne, Sir Richard (1644–1709). Called to the English Bar in 1669 and the Irish Bar in 1674. In 1691 he became Chief Justice of the Common Pleas in Ireland and in 1695 Chief Justice of the Irish King's Bench.

Pynson, Richard (d. 1530). A Norman by birth, probably came to England in the time of Caxton and learned printing from one of Caxton's apprentices. Some time between 1490 and 1493 he began to print on his own account, and about 1498 he succeeded William de Machlinia as principal printer of law books in London. From 1508 to 1530 he was King's Printer to Henry VIII. Initially he printed mostly religious books, but from about 1500 he printed mainly legal works, and printed more of these than any other printer before 1557. During his career he printed over 300 different books including over 70 editions of the Year Books. His work was superior to that of contemporary English printers, both in letterpress and in illustrations.

Pyx. A box in which sample coins are kept. By the Coinage Act, 1971, a trial of the pyx is held at least once in each year in which coins have been issued to ascertain whether coins issued by the Royal Mint are in accordance with law. It is held by the Queen's Remembrancer with a jury of not less than six freemen of the Goldsmiths' Company of London, and attended by certain officers of the Treasury, the Department of Trade and Industry, and the Mint. Sample coins are weighed, melted, and assayed to determine their weight and fineness and the jury returns a verdict on these matters. The standards of weight and fineness of coins are prescribed by the same statute.

Q

Q.B. Queen's Bench.

Q.B.D. Queen's Bench Division.

Q.C. Queen's Counsel.

Qadi (or cadi). A Muslim judge who hears religious cases, such as concerning inheritance, pious bequests, marriage and divorce, and decides according to the Shariah, or Islamic canon law. From early Islamic culture such judges were appointed and detailed requirements specified for persons appointed. Originally their functions were judicial and arbitral, but later they assumed the management of pious funds, guardianship of property, and control of the marriages of women without guardians.

Qaro, Joseph. (1488–1575). Author of the *Shulchan 'Arukh* (The Set Table) the most authoritative code of rules of Jewish law of practical application, still used by modern judges and rabbis, though subject to later commentaries and rabbinical opinions.

Quadragesimo Anno. An encyclical of Pope Pius XI of 1931, 40 years after Leo XIV's *Rerum Novarum* (q.v.), on changes in the social order. It advocated social justice, fairer distribution of the fruits of industry, better relations between capital and labour, and clarified Catholic teaching on private property and related matters. It condemned excessive concentration of economic power, communism and socialism, and gave a strong impetus to Catholic social action.

Quadragesms. The common name for the volume of the Year Books (q.v.) containing the cases of 40–50 Edward III (1367–77).

Quadriennium utile. In Scots law, the period of four years after attaining majority within which a person may seek to reduce any transaction entered into during minority and seriously prejudicial to him.

Quadripartitus. A small treatise compiled between 1113 and 1118 by an anonymous scribe at Winchester employed in the King's court or Exchequer, in four parts, of which only two exist. The first gives a Latin translation of the Anglo-Saxon laws and the second some important state papers of the writer's own time. The third, on legal procedure, and the fourth, on theft, are lost.

F. Liebermann, *Quadripartitus* (1892).

Quaestiones. (1) In early Roman law alleged crimes against the State were tried before the Assembly but, if particularly serious or requiring detailed investigation, were often assigned to a *quaestio* or *ad hoc* tribunal under a magistrate appointed by Senate or people or both. In 149 B.C. a standing commission (*quaestio perpetua*) of the Senate on extortion or abuse of power (*repetundae*) was constituted. At first civil in character it was converted into a criminal court. Thereafter both special commissions and standing commissions were established, for *veneficia* and *majestas*, and probably for *ambitus* and *peculatus*. The juries were originally of *equites* but later of senators and *equites*. Sulla increased the standing *quaestiones* to at least seven and confined the juries to senators but later jurors were chosen from panels of senators, *equites* and *tribuni aerarii*. The size of jury varied from one case to another and procedural details also varied. Any private person could with the magistrate's leave prosecute; a preliminary inquiry (*divinatio*) might be needed to determine which of several should prosecute. The indictment was in writing. The jury decided by majority vote and the presiding magistrate pronounced judgment and sentence, from which there was no appeal. The system continued unchanged in principle from 70 B.C. till the early Empire, but *quaestiones* became of less importance with the development of imperial jurisdiction in important cases and of the *praefectus urbi* and *praefectus praetorio* over lesser cases.

(2) Legal questions, a literary form employed by scholars on Roman or canon law, investigating the subject discussed by dialectical methods. Many of the great classical jurists wrote *quaestiones*. In the canon law *quaestiones* owe something to the Roman

law *quaestiones*, but were probably more immediately prompted by the dialectical framework of the *Decretum*. The mode was popularized by Sicardus and Everard of Ypres. It might take the form of a dialectical examination of a problem suggested by or arising out of a *decretum*, or of a fictitious problem or classroom exercise.

Qualification. That which is required before a person is legally entitled to do something. Thus a director of a company is frequently required to hold certain shares as qualification for office. At various times statutes have imposed various property qualifications for exercising the franchise, being a juryman or a justice of the peace.

A qualification is also a restriction or diminution of a right of grant. Thus, an offer may be accepted with or subject to qualification.

Qualified privilege. See PRIVILEGE.

Qualified property. A right of ownership of a special and restricted kind, arising from the circumstances of the possessor, as in the case of a bailment, or from the circumstances of the subject-matter.

Qualified title. A title to land registered subject to an excepted estate, right or interest arising before a specified date, or under a specified instrument, or otherwise particularly described in the Land Register.

Quality of estate. In real property, the period when, and the manner in which, the right of enjoying an estate or interest in land is exercised, namely, whether it is given at present or in the future, and whether it is to be exercised solely, jointly, or in common.

Quamdiu se bene gesserit (so long as he shall have behaved himself well). A clause found in the grant of certain offices securing the office to the grantee so long as the tenure is not abused, as distinct from tenure during the pleasure of the granter (*durante bene placito*) and consequently liable to revocation at any time, or tenure for life.

The clause is associated particularly since the Act of Settlement with the tenure of judges of the superior courts, protecting them from liability to dismissal if displeasing the King or Parliament.

Quangos (quasi autonomous national government organizations). A term for the Authorities, Boards, Commissions, Corporations, Organizations, and other bodies freely created by modern government which are neither departments of state nor local authorities. They comprise such bodies as the B.B.C. New Town Development Corporations, Regional Water Authorities, and many more of various kinds, all administering, managing, controlling, and frequently doing work the need for which is not apparent. Much criticism has been directed at the shameless way in which politicians have used their powers to appoint as members of these authorities their political friends, usually at excessive salaries. They are among the worst modern instances of political nepotism.

Quanti minoris, actio. In Roman law an action competent to a buyer who discovered a non-fundamental defect in the thing bought, to claim repayment of part of the price proportional to the extent of defect in the thing.

Quantity of an estate. In real property, the time of continuance of an estate or interest in land, as in fee, for life or for a term of years.

Quantum lucratus (as much as he has gained or benefited). A possible measure of restitution in case of unjust enrichment (q.v.).

Quantum meruit (as much as he has deserved). Another possible measure of restitution in case of unjust enrichment (q.v.), or measure of payment where a contract has not fixed a price.

Quantum of damages (or measure of damages). The amount fixed in money as proper compensation for particular proved losses.

Quarantine. The detention of persons or animals who or which may have been in contact with communicable disease until it is deemed certain that they have escaped infection and will not communicate the disease further. The name is drawn from the fact that the period was formerly commonly 40 (Ital. *quaranta*) days. In 1423 Venice established a lazaretto or quarantine station on an island to check the growth of disease brought in by ships. In the sixteenth century quarantine became widespread and there developed the system of bills of health, certificates that the last port was free from disease; a clean bill entitled a ship to use the port without subjection to quarantine. From 1850 international conferences sought to standardize practice. Under the World Health Organization only a few diseases justify quarantining of humans. In modern practice quarantine is applied chiefly to animals, e.g. to prevent the spread of rabies.

In Magna Carta it was provided that a widow should be allowed to remain 40 days after her husband's death in his mansion-house, called the widow's quarantine, during which time her dower was to be assigned; this right was enforceable by a writ of *quarentina habenda*.

Quare actions. A large class of mediaeval writs, of which trespass is the most important variety, calling on the defendant to come before the justices *ostensurus quare*, to show why he caused damage to the plaintiff. The principal specimens were trespass *quare clausum fregit* (why he broke the close), and *quare ejecit infra terminum* (why he ejected within the term).

Quare clausum fregit (why he broke the (plaintiff's) close). The standard phrase in the common law writ of trespass to land.

Quare ejecit infra terminum (why he ejected within the term). A writ competent where an ejector or wrongdoer was not himself in possession of lands, but another who claimed under him was. It later fell into disuse and was abolished in 1833.

Quare impedit (why he hindered). A real possessory action, competent originally only in the Court of Common Pleas, to recover a right of presentation when the patron's right is disturbed, as where another tries to present in his place, or to try a disputed title to an advowson.

Quarter Seal. A seal, so-called from having been originally the quarter of the Great Seal, kept by the director of the Chancery in Scotland. It is used to seal certain kinds of Crown grants and there is a register of grants which pass it.

Quarter sessions. Quarter sessions were the quarterly meetings of the whole body of the justices of the peace for a county for the transaction of business. In counties quarter sessions consisted of the justices who sat under the commission of the peace for the county. They were presided over by a chairman, and there were usually one or more deputy-chairmen. Legally qualified chairmen included the holders of such offices as High Court judges, county court judges, recorders of certain boroughs, and the like, and the jurisdiction of the court was greater when the chairman was legally qualified.

In boroughs having a separate court of quarter sessions, quarter sessions were held by the Recorder, a barrister who visited the borough regularly to hold the sessions, but might still continue in practice at the Bar.

Quarter sessions had original criminal jurisdiction to try all indictable offences, except those excluded by statute, and to sentence an offender convicted by a magistrates' court and committed for sentence because the justices considered a greater sentence was deserved than they could pass. They also had appellate jurisdiction from magistrates courts, as to conviction and sentence, unless the appellant pleads guilty when appeal lay against sentence only. Quarter sessions had original civil jurisdiction only in a few matters under statute. They had extensive appellate civil jurisdiction in respect of bastardy orders, and in certain other matters under statute. In 1971 quarter sessions were abolished and replaced by the Crown Court system.

In Scotland quarter sessions were sessions of the justices appointed to be held quarterly to review sentences passed at special or petty sessions. But these sessions have long been practically obsolete and have now been abolished.

Quarter-days. The days which begin the four quarters of the year, namely, in England, Lady Day (25th March), Midsummer Day (24th June), Michaelmas Day (29th September), and Christmas Day (25th December). In Northern Ireland the same are the quarter days. In Scotland they are Candlemas (2nd February), Whitsunday (15th May), Lammas (1st August), and Martinmas (11th November). These days are important as the common law times for payment of rent.

Quartering. Originally part of the punishment for high treason, to the effect that after hanging and disembowelling, the head should be severed from the body and the body divided into four quarters. This was, however, abolished in 1870.

In heraldry quartering is the division of a shield into four so as to show two, three, or four coats of arms on the same shield, as exemplified in the royal arms.

Quash (cf. Fr. *casser*). To annul or cut down, as of an indictment; to set aside or cancel, as of a conviction in an inferior court.

Quasi contract. In Roman law the branch of the law of obligations said to arise *quasi ex contractu*, dealing with cases where a duty arose to make payment or transfer property as if there were a contract to do so. This covered the cases of the recovery of money paid when not due, property delivered by mistake and the like. In modern practice, though the name quasi-contract is often retained, it is recognized that no fictional contract need be invoked as the obligation arises by force of law alone for the avoidance of what would otherwise be unjustified enrichment. In Scots law the category of quasi-contract covers the cases of the recovery of money unjustifiably retained, the return of property, the obligation to compensate a voluntary benefactor (*negotiorum gestio*), the duty to reward a maritime salvor, and the duty to apportion loss resulting from general average (q.v.) loss, and corresponds generally to obligations of restitution in common law systems. In English law the term is not fully recognized though it is sometimes used to cover cases of money paid at the request of another, money had and received by the defendant for the use of the plaintiff, *quantum meruit* claims, account stated, salvage, and necessaries. The term restitution is favoured instead.

Quasi-delict. In Roman law a category of obligation said to arise *quasi ex delicto*, imposing liability to compensate for harm, but comprising cases with no clear common factor, namely the cases of the judge who misconducted a case, of the occupier of a building for anything thrown or poured down from the building on to a public place, of the occupier of an object suspended from the building which falls and does damage, and of the shipowner or keeper of an inn or stable for theft or damage done by residents, slaves, or employees. The common feature may have been vicarious liability or strict liability, but the category of quasi-delicts was later understood as distinguishable from the category of delicts as comprehending cases of negligence as distinct from cases of intentional harm. This does not agree with the cases included. In truth the category was probably invented by Justinian's jurists to have four categories of obligation with four cases in each. The category is met as a name in modern systems based on Roman law, but is unnecessary and unhelpful.

Quasi-judicial. Akin to judicial, substantively or procedurally, an adjective used in many contexts with various connotations. An older view was that a quasi-judicial decision differed from a judicial one in that the former involved a discretionary element whereas the latter did not. This is inadequate because judicial decisions frequently involve discretion. The term is often used of the functions and acts of persons and bodies not strictly called judicial, not being courts or judges, but similar thereto in having authority or discretion to decide issues involving other persons, such as justices exercising their licensing jurisdiction. On this view a decision is quasi-judicial if it emanates from a body which is not a court, properly so-called. The term is commonly used of decisions involving discretion, but following on a judicial-type investigation, as where a minister, after local inquiry, decides to confirm a compulsory purchase order; but this can equally well, or better, be called an administrative function.

Queen. The term applied to a female holding the Crown in her own right, who had the same powers, prerogatives, and rights as a king has (Queen regnant), to the spouse of a king (Queen consort), and to the widow of a deceased king (Queen Dowager or Queen mother). Queens of the latter two kinds have no royal or political powers save any specially conferred, as a Counsellor of State or under a Regency Act, but are protected as to person and chastity by the Treason Act, 1351. A queen consort might formerly appoint her own Attorney-General and Solicitor-General, but this has not been done since 1837.

Under British law a Queen regnant has all the same powers as a king. Queen Elizabeth II's title to the Crown is based on the Act of Settlement, 1700, the Union with Scotland Act, 1706, and certain other statutes. She is an essential element of Parliament, head of the armed forces, supreme governor on earth of the Church of England, and Head of the Commonwealth.

Queen Anne's Bounty. Duties called first-fruits and tenths payable originally to the Pope but made payable to the Crown by Henry VIII, and directed by Queen Anne to be payable to commissioners called the Governors of the Bounty of Queen Anne, for the augmentation of the livings of the poorer clergy. The commissioners were, in 1947, amalgamated with the Ecclesiastical Commissioners for England and renamed the Church Commissioners.

Queen's and Lord Treasurer's Remembrancer. An officer in Scotland holding two offices both instituted at the foundation of the Court of Exchequer in Scotland in 1707 and conjoined in 1836. The Queen's Remembrancer was the chief executive officer of the Exchequer under the Barons of Exchequer. The Lord Treasurer's Remembrancer's principal duty was the examination and audit of the criminal accounts for Scotland. In modern practice this officer is Treasury representative on various Scottish boards and acts as Paymaster-General in Scotland. He is also Registrar of Companies, Limited Partnerships, and Business Names, auditor of the accounts of sheriff clerks and procurators-fiscal, responsible for the collection of fines and penalties imposed in Scottish courts, Keeper of the *Edinburgh Gazette*, administrator of treasure trove and of estates of deceased persons which fall to the Crown as *ultimus haeres*, and responsible for the custody of the Regalia of Scotland kept in Edinburgh Castle.

Queen's (or King's) Bench. One of the three English superior courts of common law which, like the Exchequer and the Common Pleas, developed by differentiation of function and specialization out of the King's Council. Though like these other courts it had jurisdiction in civil actions it was considered superior to the others in power and importance, probably because of its original association with the King himself. Also by the thirteenth century it had jurisdiction in error (q.v.), including, till 1830, the errors of the Common Pleas and, till 1783, error of the King's Bench in Ireland; it also had jurisdiction to control inferior courts, magistrates, and corporations. It alone (apart from magistrates' summary jurisdiction) had criminal jurisdiction. Accordingly it was said to have 'two sides' or sets of offices, the plea side, concerned with civil business, and the Crown side dealing with criminal and supervisory jurisdiction.

By the Judicature Acts, 1873–75, its jurisdiction was transferred to the new High Court, but the jurisdiction it had exercised continued to be the main work of the Queen's Bench Division of that Court.

Queen's (or King's) Bench Division. One of the five Divisions into which the High Court, established by the Judicature Act, 1873–75, was divided. It exercised the jurisdiction formerly exercised by the Court of Queen's Bench. In 1881 the Common Pleas Division and the Exchequer Division were merged in the Queen's Bench Division. From 1901 to 1952 it was styled the King's Bench Division. In 1971 it assumed also the Admiralty jurisdiction of the Probate, Divorce, and Admiralty Division (which was then renamed the Family Division).

Queen's Bench Prison. A debtors' prison in Southwark, replaced in 1842 by the Queen's Prison (q.v.).

Queen's chambers. The portions of the seas adjacent to the coast of Great Britain enclosed by imaginary straight lines drawn from one promontory to another and always deemed to have formed part of the territorial waters.

Queen's consent. The Queen's consent, which is quite different from the Royal Assent (q.v.) to Bills, must be signified by a Privy Councillor, normally on Third Reading, to a Bill which affects local or personal interests concerning the royal prerogative, the personal property or interests, or the hereditary revenues of the Crown or of the Duchy of Cornwall. A Bill is null and void if the Queen's Consent is required but not obtained. If the Prince of Wales is of full age, consent in matters relating to the Duchy of Cornwall must be signified on his behalf.

Queen's Coroner and Attorney. An officer distinct from the Coroner of the Queen's Household or the Coroner of the Verge. He was an officer on the Crown side of the Queen's Bench and is now one of the Masters of the Supreme Court appointed to the office by the Lord Chancellor, the office being conjoined with that of Master of the Crown Office. The only inquests he ever held were those held upon persons dying within the old Queen's Bench prison (q.v.) or the rules (q.v.) thereof, and he had certain duties as to extracting amerciaments and fines due to the Crown.

Queen's (or King's) Counsel. The senior grade of counsel at the Bars of United Kingdom countries. The rank was created in England in the sixteenth century, one of the first to receive it being Sir Francis Bacon, originally as a group of counsel retained by the Crown, ranking next after serjeants, and acting as assistants to the Attorney-General. It is now an honour conferred on barristers of standing and experience by the Queen on the recommendation of the Lord Chancellor, by letters patent appointing them to be 'one of Her Majesty's Counsel learned in the law'. They thus become leading counsel, leaders or 'silks' and take precedence of all counsel not having that rank. Queen's Counsel wear silk gowns and in the Supreme Court sit within the Bar. Originally Queen's Counsel were created to assist the Crown and might not appear against the Crown without a licence from the Crown, but there has now for long been a general dispensation from this rule.

Until 1977 the rule was that a Q.C. should not appear in a court unless accompanied by a junior counsel, but this rule does not apply to other appearances. He does not draft pleadings or write opinions on evidence.

Separate rolls of Queen's Counsel are maintained in England and Wales, Scotland (since 1898), and Northern Ireland (since 1921). In the Republic of Ireland, the order has been replaced by that of Senior Counsel. There used also to be Queen's Counsel in the County Palatine of Lancaster.

In Commonwealth countries and British colonies separate rolls of Queen's Counsel are maintained. In countries, e.g. Canada, where the legal profession is not divided into barristers and solicitors, rolls are maintained in each province.

Queen's enemies. The traditional term in contracts, particularly of affreightment, for enemies of the State.

Queen's evidence. An accused person who gives evidence against persons associated with him in the alleged crime and who is, in consideration thereof, not put on trial.

Queen's Peace. See KING'S PEACE.

Queen's Printer. The person appointed to exercise the Crown right of printing, under royal authority, royal proclamations, statutes, and similar public documents.

Acts of Parliament purporting to be printed by the Queen's Printer are admissible in evidence without proof that they were so printed or evidence as to the terms thereof and are prima facie evidence of the correctness of the statute so printed.

Queen's Prison. A prison established in Southwark in 1842 to replace the Queen's Bench Prison, the Fleet Prison, and the Marshalsea Prison. It was abolished in 1862 and prisoners transferred to Whitecross Street Prison and from there in 1870 to the City Prison at Holloway.

Queen's Proctor. An office now held by the Treasury Solicitor. He represents the Crown in probate, matrimonial, and admiralty cases. In matrimonial causes he might intervene at any stage if he desired to allege collusion, or to show cause why a decree nisi of divorce should not be made absolute, on the ground the material facts have not been disclosed to the court. In such causes the court may direct that the papers be sent to the Queen's Proctor who is required, under the direction of the Attorney-General, to have argued any matter which the court thinks should be fully argued. In admiralty cases he acts for the Crown in proceedings for the recovery of droits of Admiralty (q.v.).

Queen's Recommendation. No taxation or expenditure may be proposed without the Queen's Recommendation, because the sovereign is the executive power and charged with responsibility for all revenue and all payments. In practice this means that such proposals are reserved to the Government. Hence, if a private member's Bill requires the expenditure of public money the sponsoring member must be able to persuade a minister to move a financial resolution in the House, failing which the Bill cannot be passed. The Recommendation is made in writing to the Clerk of the House of Commons, though Recommendation of the Annual Estimates is made in the Speech from the Throne at the beginning of each session of Parliament.

Queen's Regulations. Sets of general regulations for the good government of the Army, Royal Navy, and Royal Air Force respectively, and supplemented in each case by Instructions issued by the relevant branch of the Ministry of Defence.

Queen's Remembrancer. Formerly an officer on the revenue side of the Court of Exchequer and now an officer of the Supreme Court, held by the Senior Master of the Queen's Bench Division.

Queen's serjeant. If a serjeant-at-law (q.v.) were made a Q.C. he was called a Queen's serjeant, and there were superior offices of serjeanty such as the Queen's Premier Serjeant and the Queen's Ancient Serjeant, to which appointments were sometimes made. James Manning, made serjeant in 1840, Queen's serjeant in 1846, and Queen's ancient serjeant in 1849, and who died in 1866, is believed to have been the last holder of these offices.

Queen's shilling. Money originally given ('press money') both to men impressed into the naval service of the Crown and to soldiers and sailors who enlisted voluntarily as an earnest of their contract to serve.

Queen's ship. Strictly speaking 'Her Majesty's Ship' means a ship commissioned for naval service and flying the white ensign, whereas 'Her Majesty's vessel' means a ship or vessel engaged in the naval service of Her Majesty, whether belonging to her or not, and includes 'fleet auxiliaries', hospital ships, store ships, and the like.

Queen's speech (or speech from the throne). The speech read by the Queen at the State opening of a session of Parliament, but in fact prepared by the Cabinet, outlining the policies and legislation proposed for the session. A similar speech is read at the conclusion of a session in which the Cabinet congratulate themselves on what they did during the session.

Querela inofficiosi testamenti. An action allowed in Roman law to an heir who contended that he had been disinherited on grounds falsely alleged, on the hypothesis that the testator must have been of unsound mind at the date of the testament thus to have excluded the heir.

Quesnelliana. A collection of canon law materials published by Pasquier Quesnel in his compilation of the works of Pope Leo the Great. It is believed to have been compiled between A.D. 494 and 523 possibly in Gaul but probably in Rome. It consists of excerpts from oriental and African councils of the Church, writings, decretals of Popes, and many letters of Pope Leo I. It seems to have been designed to replace earlier collections but has neither logical nor chronological order. It is important for the history of the sources of canon law and was the principal collection used in Gaul till the mid-eighth century, and some of its materials passed into later collections, such as the *Hadriana*, the *Hispana* (qq.v.), and the *Capitularies* of Benedict Levita.

Questions in Parliament. The practice whereby members of Parliament can question a Minister of the Crown or sometimes the chairman of a committee, on any matter for which he is departmentally responsible seems to have started in 1721. It is a valuable means of obtaining information, uncovering abuses or maladministration, stimulating action, and investigating grievances. The number of oral questions is limited to two per day, but any number of questions requiring written answers may be asked. Prior notice must be given to enable the minister and department to ascertain the facts necessary for the answer. The Clerks at the

Table may disallow proposed questions as inadmissible subject to the Speaker's final ruling. Question time in the House of Commons is a period of three-quarters of an hour shortly after the commencement of the sitting. The order of questions is so arranged that each Minister has his turn of answering questions orally. Answers to questions not required to be answered orally and to questions not reached by the end of Question Time are printed in *Hansard.* The member asking the question, or another member, if not content with the answer, may put supplementary questions. In the House of Lords there is a similar practice, though there is greater latitude than in the House of Commons.

D. N. Chester and N. Bowring, *Questions in Parliament* (1962).

Questions of fact and questions of law. The two major kinds of matters which arise for consideration and decision in legal disputes. A question of fact is any matter in issue which if not admitted, has to be determined by hearing and evaluating evidence.

A question of law has usually to be determined by considering and weighing legal authorities and arguments. It may mean a question as to what the true rule of law is on a certain matter, or one which has to be determined according to a rule of law, or one which has to be determined by the trial judge rather than by the jury, if there is one. It is by no means always easy to decide whether a question is one of fact or one of law.

Questions of mixed fact and law also arise, such as of X's guilt of manslaughter which depends on what he did (fact) and whether these facts amount to manslaughter as legally defined (law). Also distinct from both questions of fact and questions of law are questions of judicial discretion which must be determined by the application of judicial discretion, the exercise of moral judgment, the decision of what seems right and fair and best in the circumstances.

Appeal procedures frequently allow the raising only of questions of law, or of questions of fact only subject to conditions and limitations.

Quetelet, Lambert Adolphe Jacques (1796–1874). Belgian statistician and sociologist, notable for his applications of statistics to social phenomena, mortality, crime, and related subjects. He sought through statistics to discover the causes of conduct of various kinds in society and developed the theory of the relative propensity of various age groups to crimes and other conduct, a theory which gave rise to acute controversy among social scientists.

Qui facit per alium, facit per se (one who does something through the agency of another, does it by himself). An ancient maxim expressing but not explaining or justifying the rule of vicarious liability, whereby a principal is normally liable for harm done by his agent and an employer for harm done by his employee, provided in each case that the agent or employee was acting within the scope of his authority or of his employment.

Quia Emptores. A statute of 1290, so-called from its introductory words, which gave the tenant in fee simple of land held by free tenure, other than a tenant in chief of the Crown, the power of free alienation of land and defined the effect of such alienation as being that, if the tenant alienated in fee simple, the alienee should hold not of the alienor but directly of his lord. It accordingly prohibited subinfeudation and contributed to the extinction of the mesne tenures and to the concentration of all feudal lordships in the Crown.

Quia timet (because he fears). A claim, particularly for an injunction, to quieten fears of a probable future injury to property. It must be shown that there is imminent danger of a substantial kind, or that, if the injury apprehended takes place, it will be irreparable.

Quiet enjoyment. A covenant implied in leases, but frequently superseded by an express and qualified covenant. The implied covenant may protect the lessee against any lawful entry whatever, but an express covenant will frequently guarantee only against entry by the lessor or persons claiming through or under him. A covenant is statutorily implied in any conveyance of value as beneficial owner and for value.

Quietus. Originally a writ addressed to the Barons of the Exchequer directing that a debtor should be discharged of claims against him, and later a term used by the Clerk of the Pipe (q.v.) in the certificate given to a sheriff after the latter had paid any balance due. It accordingly came to be a general term for a discharge by the Crown to a person indebted to the Crown, and was a defence to a writ of extent and might be recorded in a register of judgments in discharge of an execution issued.

Quinquaginta decisiones. A collection of decisions on particular points which had arisen during the work of Justinian's codifying commissions, particularly between the publication of the *Codex* of A.D. 529 and the commencement of work on the *Digest*, made probably in A.D. 530. It has not survived. These decisions were incorporated in the revised *Codex* published in A.D. 534.

Quinque Compilationes Antiquae. Between the *Decretum Gratiani* (*c.* 1140) and the Decretals of

Gregory IX (1234) the attention of canon lawyers was devoted to literary commentary on the *Decretum* and systematization of the newer canon law expressed in decretals. The five most important, though not the only, decretal collections are the *Quinque compilationes. Compilatio prima* or *Breviarium Extravagantium* of Bernard of Pavia (1187–91), which was itself the culmination of a tradition of codification dating from the 1170s, was composed mainly of post-Gratian decretals, and included the canons of the Lateran council of 1179, and was arranged in five books entitled *judex, judicium, clerus, connubium*, and *crimen*, according to subject-matter, a scheme later adopted by the other four compilations and taken over in the Decretals. *Compilatio tertia* was composed by Peter of Benevento for Innocent III, comprised decretals of the first 12 years of his pontificate, and was promulgated in 1210. *Compilatio secunda* was composed by John of Wales (1210–15) and is called *secunda* because it deals with decretals of an intermediate period between *prima* and *tertia. Compilatio quarta* is attributed to Joannes Teutonicus or sometimes to Alan and contains decretals of the later years of Innocent III, and *compilatio quinta* may be attributed to Tancred and was promulgated in 1226. Other collections were made contemporaneously, but those five attained a pre-eminent reputation, were the basis of canonical study and became the basis of reference for decretalists, and were the subject of important glosses. They had great influence on the *Decretals* of Gregory IX in that Raymund of Penaforte accepted Bernard of Pavia's scheme of arrangement and most of the materials of the five collections are incorporated therein. All five compilations were glossed and studied in detail.

Quit-rent. A rent payable by a freeholder or ancient copyholder of a manor to the lord, payment of which made him quit or free of other services and dues. As no manor has been created since *Quia Emptores* (q.v.) in 1290, any quit rent must have become first payable before then. By statute any still remaining may be redeemed.

Quo Warranto. A mediaeval royal writ enquiring by what warrant a party claimed to exercise a particular right, especially in jurisdiction. In 1274 Edward I sent commissioners to enquire into usurpations of royal rights, and on the basis of the information obtained the King sent commissioners to enquire by what warrant particular landowners exercised royal rights. Initially, if no charter could be produced the commissioners claimed the rights for the Crown. The commissioners' conduct gave rise to much resentment, so that in 1290 two statutes of Gloucester allowed that uninterrupted possession since 1189 should give a good title and that owners of certain lost franchises should have restitution. In the sixteenth century the writ was replaced by an information in the nature of a *Quo Warranto* filed by the Attorney-General, of which the most famous instance is the information filed against the corporation of London in 1681–83. It was frequently used to try the right to be elected to municipal offices. This was in turn abolished in 1938.

Quominus. A writ whereby the mediaeval court of Exchequer obtained a general jurisdiction over common pleas and a notable example of the use of a legal fiction. The plaintiff who wished to sue was supposed to be a debtor to the Crown and by reason of inability to recover his debt or damages from the defendant was thereby the less able (*quo minus sufficiens existit*) to satisfy the Crown. The device was generally used from the fourteenth century but was abolished in the nineteenth century.

Quoniam Attachiamenta. A Scottish legal treatise, so called from its opening words. It dates from the fourteenth century, is purely Scottish, probably the oldest purely Scottish law book, and is a manual of procedure in feudal courts based on experience and including forms and styles of writs, and utilizing a technical vocabulary. The author is unknown, but obviously had practical experience of what he wrote about. Like *Regiam Majestatem* it was referred to several times in later statutes as one of the books of law of the realm to be taken as a basis for a restatement of Scots law, and was frequently referred to by later writers. It was first printed, along with *Regiam Majestatem*, by Skene in 1609, and remains a valuable historical source of information on procedure.

Quorum (of whom). An expression taken from the form of a commission of the peace used down to 1878—*quorum aliquem vestrum A.B. unum esse volumus*—specifying named persons to be among the justices (though later every justice was named as of the quorum). Hence it now means the necessary persons or number of persons who must be present for a meeting of members, shareholders, directors, creditors, or the like, to be entitled to do business. Statutes, Articles of Association, club constitutions, and the like normally state the quorum requisite for particular kinds of meetings. If a quorum is not present no business can be validly transacted.

Quot. In old Scots law, the proportion of the moveable estate of a deceased due to the bishop of the diocese. It was a twentieth part of the moveables before deduction of debts, even where the assets were no more than sufficient to satisfy the debts, and was accordingly frequently prejudicial to the creditors. In 1669 it was made exigible from the net estate only, and was finally prohibited in 1701.

R

R. *rex* (the king) or *regina* (the queen).

R.S.C. Rules of the Supreme Court of England and Wales.

Rabbinical courts. Courts in modern Israel having jurisdiction in matters of marriage and divorce of Jewish citizens or residents and in all other questions having concurrent jurisdiction with ordinary civil courts. Appeal does not lie to civil courts but the Supreme Court may exercise supervisory jurisdiction over rabbinical tribunals and rabbinical judgments can be executed only through the offices of the civil courts. In the U.K. the Bet (or Beth) Din is a rabbinical court but its decisions are ineffective in civil law.

Race relations. Mankind is commonly classified into three or five races, Caucasian, Mongolian, Negroid (and sometimes also American and Malayan), race designating a group of people who have the majority of their physical characteristics in common. The important distinguishing factors are colour of skin, cultural factors, and language. Racism is the attitude and doctrine that certain racially different people are necessarily inferior to others. It originated in the expansion of European peoples overseas from the sixteenth century, and the temptation and tendency to treat coloured people in countries occupied as inferior, and to deny them human rights. Popes and theologians fought against this attitude and Spain passed much enlightened legislation to protect natives in Latin America.

In the U.S., however, black slaves were imported from Africa from an early date and slave owners became convinced that the racial differences were so fundamental that no black could be fit to be other than a slave. The Civil War, emancipation, and a century of agitation for civil rights have convinced most people that any subordination of coloured persons as a race or group is unfounded and socially dangerous.

Another manifestation of racial prejudice is anti-Semitism, which has been practised in many countries at various times, but nowhere more extremely than in Nazi Germany where the extermination of the Jewish population was a major object of policy.

In South Africa racial subordination has taken the form of the principle of apartheid under which racial segregation is enforced.

In Britain the large influx of coloured persons from various parts of the Commonwealth in the 1960s and 1970s caused the passing of Race Relations Acts, 1965 and 1968, which made discrimination in housing, employment, and other respects illegal. A Race Relations Board was established to act as a supervisory and conciliatory body with power to bring cases before specified courts. In 1976 a further Act sought to outlaw discrimination on racial grounds and to promote good relations between people of different racial groups. It established a Community Relations Commission in place of the Race Relations Board with wide powers to promote good racial relations, Incitement to racial hatred is a statutory offence. There nevertheless continues much resentment of the presence of so many coloured people in Britain.

Rachel, Samuel (1628–91). German jurist, became professor of moral philosophy at Helmstedt, and in 1665–78 professor of natural and international law at Kiel. Later he served as a diplomat at various German courts. He was author of *De Jure naturae et gentium dissertationes* (1676), *Introductio in jus Germanicum* (1680), and *Institutionum jurisprudentiae libri IV* (1681). He opposed Grotius in distinguishing the law of nations from natural law and basing it on agreements or custom, and sought to free his system from theological and moral principles and to introduce utilitarian ideas. He was one of the first to establish international law as a separate science and to stress its legally binding character. He was the forerunner of the nineteenth-century positivist movement in international law.

Rack. An instrument of torture, in the form of a frame in which an individual was stretched prone, arms and legs attached to cords on pulleys at the corners. By means of levers the cords could be pulled, stretching the victim's arms and legs, and in extreme cases dislocating them or pulling the victim apart.

Rack-rent. The highest rent obtainable, the full annual value of the property.

Radbruch, Gustav (1878–1949). German jurist, professor at Konigsberg, Kiel and Heidelberg and Minister of Justice of the Reich (1921–22 and 1923) under the Weimar Republic. He was author of *Introduction to Jurisprudence* (1928), *Philosophy of Law* (1932), and *The Spirit of English Law* (1947). His legal philosophy is based on Kant. All law in the final analysis depends on value notions and philosophy of law can only seek to discover these values and indicate them more clearly, so that the positions of parties are always relative to each other.

Radcliffe, Cyril John, Viscount of Werneth (1899–1977). Called to the Bar in 1924, took silk in 1935 and moved straight from the Bar to become a Lord of Appeal in Ordinary in 1949, sitting till 1964. He was Chairman of many Commissions and committees including the Royal Commission on the Taxation of Profits and Income and the Royal Commission on the Monetary System, as well as of the commission which proposed the frontier between India and Pakistan in the Punjab and Bengal.

Radzinowicz, Sir Leon (1906–). Born in Poland, he taught law there and later came to England and became Assistant Director and Director of the Department of Criminal Science at Cambridge, 1946–59. He became Wolfson Professor of Criminology, 1959, and Director of the Institute of Criminology at Cambridge, 1960, and a member of numerous societies and commissions concerned with crime, and lectured on criminology in many centres. He edited the series of *English Studies in Criminal Science*, (later *Cambridge Studies in Criminology*) and wrote a major *History of English Criminal Law* (4 vols., 1948–68). He was knighted in 1970.

Rae, Sir David, of Eskgrove (1725–1804). Passed advocate in 1751, worked as a reporter of decisions and early obtained a reputation as an able lawyer, being for many years the leader in the Scottish Court of Exchequer. He became a Lord of Session in 1782, a Lord of Justiciary in 1785, and Lord Justice-Clerk in 1799. In 1804 he was made a baronet. He was highly esteemed as a lawyer and judge. His younger son William (1769–1842) was Lord Advocate 1819–30, 1834–35, and 1841–42, but declined promotion to the bench.

Raffle. A kind of lottery, the sale of a thing by selling tickets for small sums of money and giving the thing to the holder of whichever ticket is picked at random. Raffles are in general illegal, but are permitted under conditions for charitable, cultural, or similar purposes.

Rageman, Statute of. An Act of 1276 under which justices were appointed to hear and determine complaints which had arisen during the 25 years prior to Michaelmas 1276. The justices were sometimes called *juratores de ragemen*. Under this statute the commission of trailbaston (q.v.) was issued.

Ragman roll. The name given to the roll of barons and other leading persons of Scotland who subscribed the submission to Edward I in 1296.

Railway and Canal Commission. A court consisting of a judge of the High Court, or Court of Session, or High Court in Ireland, and two technically skilled members, established as the Railway Commission in 1873 to enforce the statutory obligations of the railway companies to grant reasonable facilities to the public and to refrain from undue preference. It was remodelled, renamed the Railway and Canal Commission, and its jurisdiction extended in 1888, but it lost its jurisdiction over rates and other charges in 1921 though still having functions in connection with railway companies' duties to the public. Its jurisdiction was further abridged in 1947 and it was abolished in 1949.

Railway Rates Tribunal. A court established in 1921 comprising a judge of the High Court or Court of Session and two technically experienced members, having complete jurisdiction over the charges to be made for the carriage of passengers and goods by railway, the classification of merchandise, the fixing and reviewing of standard charges, and other similar matters. In 1947 it was replaced by the Transport Tribunal.

Railways. The successful operation of the Liverpool and Manchester Railway in 1830, and after, aroused public enthusiasm for railway construction which reached its peak in the railway manias of 1837–40 and 1845–49. By 1870 most of the British railway network was complete. This construction has important legal consequences. It gave rise to much private legislation authorizing compulsory acquisition of land, largely reduced to order by the Railways Clauses Consolidation Acts, 1845, to the development of companies and the principle of limited liability, to developments in contract and employment law, and the law of liability for accidents, as well as to changes in the law of carriage. The growth of the railways rapidly put an end to stage-coach and other long-distance road transport, and consequently to turnpike trusts.

The operations of the early railways, the dangers of monopoly and other factors early led to government intervention. The Carriers Act, 1830, modified the liability of carriers by rail for articles of great value in small bulk, and allowed them to make extra charges. Under the Railway Regulation Act, 1844, fares were controlled and the Board of Trade began to enforce standard methods of auditing and safety precautions. This Act and its replacement, the Cheap Trains Act, 1883, imposed on railways duties to run certain trains at fares not exceeding one penny per mile, particularly for workmen ("Parliamentary trains"). The Railway and Canal Traffic Act, 1854, required every railway to afford "all reasonable facilities for the receiving and forwarding and delivery of traffic" and not to give any undue or unreasonable preference or advantage to any particular person. Jurisdiction as to facilities was entrusted to the Railway and Canal Commission, an early example of a regulatory authority for an industry.

Down to 1888 a special Act for each company fixed the maximum rates it might charge for each description of goods. By 1888 there were more than 900 such Acts in force. Only in 1873 were companies obliged to have rate books available to the public. The Railway and Canal Traffic Act, 1888, provided for the creation and enforcement by the Board of Trade of a code of rates and charges for merchandise traffic.

During the First World War the railways were placed under governmental control and compulsorily grouped into four large groups in 1921. Standard charges for various classes of goods were fixed by the Railway Rates Tribunal. In 1949, following a further period of state control during the Second World War, the railways were nationalized and placed under the British Transport Commission, later renamed the British Railways Board.

Rainsford, Sir Richard (1605–79). Called to the Bar in 1632 and sat in the Convention Parliament. A respectable but mediocre lawyer, he became a Baron of Exchequer in 1663, a judge of the King's Bench in 1668, and Chief Justice of the King's Bench in 1676, but was removed in 1678 to make way for Scroggs. He left no mark in the law.

Raleigh, William of (?–1250). A justice of the bench in 1228, later treasurer of Exeter Cathedral, Bishop of Winchester twice and of Norwich, and a pre-eminent lawyer of his time.

Ramistic method. A method of legal exegesis and teaching so-named from the movement associated in all fields of science with Pierre de la Ramée (Petrus Ramus) (1515–72), who revolted against the Aristotelian-scholastic dialectic and developed a distinct method. It was applied to legal study by Connanus, Duarenus, Donellus, and others though, particularly in the seventeenth century, it was practised in pedantic fashion.

Rampton. A special hospital near Retford for persons of unsound mind, convicted of crimes, or having dangerous or criminal propensities. It specializes in patients with some degree of subnormality or brain damage.

Ranconcet, Aimar de (?–1559). An eminent French humanist jurist.

Rankin, Sir George Claus (1877–1946). Called to the Bar in 1904, but after the war became a judge of the High Court of Calcutta and in 1925 Chief Justice of Bengal. His health failed and he resigned in 1934 but in 1935 became a paid member of the Judicial Committee of the Privy Council; he retired in 1944. He had made an exhaustive study of Hindu and Mohammedan law and was recognized in the Judicial Committee as the greatest judicial authority of the time on Indian law.

Ranking and sale. In Scots law an action, now obsolete, whereby an insolvent person's heritable property is judicially sold and the proceeds distributed among his creditors according to their due rights and preferences.

Ransom. A sum of money demanded by persons holding captive hostages, a ship or aircraft, or prisoners of war. In feudal times one of the obligations of persons holding by knights' tenure was to contribute to the ransom of their lord if he were taken prisoner in war. It is now most commonly either simply a means of extortion, or a means of trying to enforce political demands.

Rape. (1) The crime of a man having sexual intercourse *per vaginam* with a female not being his wife (unless they are separated) knowing that it is done without her consent or reckless as to whether she consents or not. The requisite intercourse requires penetration but not necessarily emission. It is a defence that the woman consented, but consent obtained by force or threats, or fraud, or personation of her husband or lover, is not consent; and there is no consent if the woman was asleep, drunk, or drugged. Proof of absence of consent is difficult. By statute it is rape also to have intercourse with a female who is mentally defective. Sexual relations with females under 16, even with their consent, though not rape, is criminal. In Scotland rape is the crime of having sexual relations with a female against her will. The overcoming of the will may be effected by force, threats, drugging, or personation. By statute it is rape to have intercourse with a female mentally defective or by personating her husband.

(2) The term is also given to the ancient divisions of Sussex, which appear to have been areas of military government in Norman times. They are Arundel, Bramber, Chichester, Hastings, Lewes, and Pevensey. See also LATHE; WAPENTAKE.

Rapine. Robbery, or the taking of a thing from a person against his will.

Rapporteur. In French law, a judge who examines evidence collected by investigators and reports his conclusions and recommendations to the whole court. The function originated in mediaeval ecclesiastical courts and was taken over by the Parlement of Paris in the thirteenth century. It sprang from the need to have one person analyse the material gathered by investigators from witnesses and documents. The position of rapporteur was held by different judges in turn. At first rapporteurs were not members of the court, but by 1336 they were allowed to participate as judges in arriving at the decision.

After the Revolution the rapporteur had to present his analysis in open court and it was published as part of the report of the case. His recommendations included examination of the relevant law with references to authorities. In the *Cour de Cassation* the rapporteur is a member of the court, examines the record, determines matters in issue and makes recommendations to the full court for disposal of the case. Compare ADVOCATE-GENERAL; REFERENT.

Rastell, John (d. 1536). Was trained as a lawyer, and also sat in Parliament, but before 1516 began work as a printer, editing and printing the *Liber Assisarum* and Fitzherbert's *Graunde Abridgment.* Most of the works he issued were legal, but he published others and took part in the religious controversies of the times. He also compiled *Expositiones Terminorum Legum Anglorum* (1527 and later editions). He married the sister of Sir Thomas More, and a son, William (?1508–65) (q.v.), became a judge, and also edited legal works.

Rastell, William (?1508–1565). Became a printer and rescued the writings of Sir Thomas More, his uncle, which he later edited and published. He was called to the Bar in 1539 and became a judge in 1558 but, being a Catholic, went into exile in 1563. He also published *A Collection of All the Statutes* (1557), *A Collection of Entrees* (1566) (called by Rolle *Antient Entries*, by some authors the *New Book of Entries*, and occasionally the *Old Book of Entries*), and wrote a biography of More of which only a fragment survives.

Rate. A sum assessed on and made payable by a person in respect of his occupation of land and buildings. Rates can be imposed only by public authorities under statutory powers. Rates have been known since at least the Poor Relief Act, 1601, and later statutes imposed rates for other purposes. Since 1925 one general rate has replaced nearly all the rates imposed by various authorities under different Acts.

Ratification. In international law, the confirmation of a treaty or other international agreement. Formerly, if an agent concluded a treaty within the authority of his Full Powers (q.v.), his king was expected to ratify the agent's signature, as final acknowledgment of the agent's act. In modern practice ratification is discretionary, unless the treaty specifies that it is to come into force on signature. It is only required when the treaty specifies or implies this. In U.K. practice ratification is not done if the treaty does not contain a ratification clause, and this is not included unless the other party is required by its constitution to ratify. Ratification is by the Crown. U.S. practice is to omit ratification in the case of executive agreements, which are made only when the Constitution permits it. Ratification is by the Senate.

An unratified treaty is not devoid of legal significance. If a state commits itself to a course of action on the faith of a treaty which the other state fails to ratify, and suffers loss, it may have a claim founded on abuse of rights.

In bilateral treaties an exchange of ratification is made, and in the case of multilateral treaties ratifications are deposited with the Secretary-General of the U.N. or in the archives of a designated government. Unless otherwise provided, the treaty does not come into force until all, or a stated number of, ratifications have been deposited.

Formerly ratification was retrospective to the date of signature, but in modern practice this is not so, and the treaty is effective from ratification only, though the U.S. adheres to the older view.

In municipal law ratification is confirmation of acts by a person who might have repudiated those acts. Thus actings by one purporting to act as agent for another may be ratified by the beneficiary, provided he could have authorized the acts in the beginning.

Rating and valuation. In the United Kingdom rates, or a levy on the occupiers, or in certain circumstances, owners, of land and buildings, are a major means whereby local government authorities raise money for the finance of their functions. They also receive substantial block grants from the Exchequer for particular functions. Liability to rates originated in the Poor Relief Act, 1601, though the purposes for which rates are imposed have been greatly extended and altered and now extend to education, police, public health, social work, lighting, cleansing, public parks and libraries, and other local authority functions.

The general rate is a consolidated rate levied annually by each rating authority for its area on the basis of an amount per pound on the rateable value of each hereditament in the rating area. County councils issue precepts to other rating authorities, viz. district councils, for money to discharge liabilities falling on them in the latter's area and the amount precepted is taken account of when the other rating authorities are fixing their own rates.

The value of land or buildings to an occupier for the purposes of liability to rates is fixed by valuation officers of the Commissioners of Inland Revenue, new valuation lists being prepared every five years, on the basis of the rent at which the premises might reasonably be expected to let from year to year. In the case of houses and non-industrial buildings, a gross annual value is fixed on the basis that the landlord undertakes the cost of insurance and repairs necessary to maintain the premises in a state to command the rent, and certain deductions are allowed to arrive at the net annual value. In the case of other kinds of buildings the net annual value is assessed direct. Various methods of settling valuations of various kinds of premises have been approved by the courts. The new annual value is in general the rateable value, but in certain cases deductions from net annual value are permitted in arriving at rateable value. There are many decisions on whether a person is or is not in rateable occupation of premises. Crown property is in general not rateable and certain other kinds of property are exempt. Special provisions apply to premises occupied by gas and electricity boards and railway and canal premises. Proposed valuations may be challenged by appeal to a local valuation court, composed of laymen, or referred to arbitration. A further appeal from a local valuation court lies to the Land Tribunal with an appeal therefrom on a point of law to the Court of Appeal.

In Scotland valuation is done by county assessors and appeal lies to a local valuation appeal committee and from there to the Lands Valuation Appeal Court, whose decision is final.

Demand notes are issued to occupiers requiring payment of the sum due in respect of each premises and payment is enforceable by distress, not by action.

The rating system is very unsatisfactory in that liability for rates is related neither to use of the services provided nor to ability to pay nor to the profits earned by the occupation of the premises, and rates are imposed by local authorities without any regard to the ratepayer's concurrent liability to national taxation. The incidence of rates varies substantially from one part of the country to another. The power to raise some revenue locally is supposed to engender responsibility in local authorities and interest by residents in their membership, powers, and operation. It does none of these things and utter irresponsibility is the hallmark of most local authorities.

Ratio decidendi. In the Anglo-American doctrine of judicial precedents or *stare decisis* (q.v.) the principle of law which is the basis of the actual decision and therefore, by virtue of the doctrine of *stare decisis*, the principle which subsequent courts, faced with a set of facts indistinguishable in any material particular from those in the precedent case, must apply to the decision of the subsequent case also. *Ratio decidendi* is unsatisfactorily translated as 'the reason for the decision' because the reason may in fact be something other, such as the judge's dislike of the defendant. Nor is the *ratio* the decision itself, for this binds only the parties whereas the *ratio* is the principle which is of application to subsequent cases and states the law for all parties.

The *ratio* of a decision is not necessarily enunciated expressly, nor contained in any single phrase or sentence of the judge's opinion, still less is it necessarily contained in any sentence of the reporter's headnote to the published report of the decision, as the reporter may misinterpret the decision. The *ratio* must be discovered by determining what facts were deemed material to the decision and what proposition of law justified that decision on these material facts, ignoring all facts disregarded as immaterial. The more facts are deemed material, and the more detailed they are, the narrower and more particular the *ratio* of the decision will be. If the fact that the plaintiff was female be deemed material, the *ratio* is narrower than if that fact was immaterial. Thus in *Rylands* v. *Fletcher* (1868), L.R. 3 H.L. 330 the facts were (i) Fletcher got a reservoir built; (ii) Fletcher employed contractors to build it; (iii) the contractors were negligent; (iv) Fletcher was not negligent; (v) by reason of negligent construction water escaped and flooded Ryland's mine. The House of Lords held facts (i) and (v) only to be material, and thus the *ratio* of the decision is one of strict liability for the escape of water, irrespective of whether the work was done personally or by a contractor and irrespective of negligence. Moreover, in laying down this *ratio* the House stated the material facts in generalized form, stating that strict liability applied where a person accumulated a 'dangerous substance' on his land and it escaped. The *ratio* was not confined to escape of water.

The ascertainment of the *ratio* of a case is the task of a later court trying to discover whether the case is a precedent in point or one distinguishable, and if in point and binding, what proposition of law it lays down which the later court must or may accept and apply.

The discovery of the *ratio decidendi* of a case may be a matter of great difficulty because it is not necessarily expressly or accurately formulated in the, or a, judicial opinion, still less in the reporter's headnote to the report of the case. It is a generalized statement of legal principle which, applied to the material facts of the particular case, has produced, and justifies, the result reached by the court. The more numerous and detailed the facts held by the court to have been material and necessary for the decision, the more narrow and specialized the *ratio* will be; the fewer the facts held material the more general and widely applicable the *ratio* will be. A case may have been decided on its own facts, or the particular terms of its pleadings, or on the basis of some admission or concession, and may yield no *ratio* or general principle at all. Or the later court may find itself unable to discover the basis on which the previous court decided the precedent case; this greatly weakens the value of the case as a precedent.

Even if no reasons are given, or no opinion is delivered, a decision must have some legal justification, and the *ratio* may usually be inferred from the facts and the decision thereon. But the *ratio* of such a case cannot be any more general than the circumstances of that case. Similarly, where the judicial opinion is entirely concerned with stating the facts found, the *ratio* may be inferred from those facts and the decision thereon.

The judgment in a precedent must, moreover, be read *secundum subjectam materiem*; it is a judgment in relation to the facts of a particular case and the judge may not be laying down a rule for any case other than the one before him and precisely similar cases. If he does lay down a rule for kinds of cases other than that before him, the validity of the propositions for other cases must be considered if and when those other cases arise, when the rule may be regarded as too widely and generally stated.

There is no fixed way of ascertaining the *ratio* of a decision. The test propounded by Wambaugh (q.v.) for discovering the *ratio* was that the *ratio* must be a general rule which, if reversed, must have resulted in the case being decided otherwise. But in many cases various *rationes* can be formulated which would all result in the same decision.

Goodhart (q.v.) suggested that the *ratio* is the principle derived from the judge's decision on the basis of the facts treated by him as material and ignoring the facts treated by him as immaterial. A person considering a precedent must seek to discover what facts the judge treated as material and formulate the proposition by reference to the actual decision on those facts only.

Where a judge states more than one principle as justifying his decision each may be a distinct and alternative *ratio*, or one may be *ratio* and one *obiter dictum*. Each *ratio* is as binding as any other, but a later court may treat one *ratio* as the main one, and the other or others as mere *dictum*.

Different *rationes* can be derived from a decision by taking different combinations of material facts; there may accordingly be subsequent doubt and dispute as to what the *ratio* of a particular decision is.

A *ratio* may be held inherent in a decision even though no opinion has been delivered, but that *ratio* may be of no future utility if the decision is one entirely on its own facts, such as on a matter of pleading or the interpretation of the particular contract or will in issue. A later court is not obliged to find a *ratio* for a precedent and may in despair abandon the quest, treating the precedent as a decision on its facts only.

Determination of the *ratio* may be difficult in the case of decisions of multi-judge appellate courts, where each judge, even if concurring in the result, may enunciate a different legal proposition. If the court is divided on the result, it may be even more difficult to find a *ratio*.

Difficulties also arise where a case involves two or more points and a decision on either would suffice for the decision of the case, e.g. if the judge holds that he had no jurisdiction, and also that the plaintiff had failed to establish his claim, or where the judge gives two reasons for his conclusion.

Determination of what the *ratio* of a decision is is a matter for a subsequent court, to which that decision has been cited as a precedent, to determine. As the subsequent case is unlikely to coincide with the precedent in respect of all material facts, the judge in the subsequent case must normally either restrict the *ratio* of the precedent and say that it does not extend to the different material facts before him, or enlarge the *ratio* and say that it covers also the case before him.

Any two or more decisions, that enunciating the *ratio* and any in which the same *ratio* is applied, are the basis for judicial induction of a more generalized principle, such as of manufacturer's liability for packaged food and drink, or even more widely, for packaged manufactured goods. Thus the decision in *Donoghue* v. *Stevenson*, [1932] A.C. 562 related to the liability of a manufacturer to a consumer for defective bottled ginger beer; if in a later case the liability for defective tinned meat arose the later court might say that the *ratio* of *Donoghue* applied to ginger beer only, or to bottled drink only, or it might enlarge the *ratio* by saying that it applied equally to tinned meat, to packaged food and drink. If, in a still later case, liability for defective clothing or mechanical parts were in issue, the *ratio* might be confined to food or to goods causing personal injuries, or be extended and generalized further to cover manufactured goods.

Everything contained in a judgment which is not part of the *ratio decidendi* is classified as *obiter dicta* (q.v.). This clearly covers hypothetical or unproved facts, illustrations, and casual expressions. But a subsequent court, in determining what the *ratio* of a precedent was, may interpret the decision so as to downgrade certain parts of what had hitherto been thought to be the *ratio* to mere *dicta*.

The importance of determining the *ratio decidendi* of a decision is that, under the rules of *stare decisis* (q.v.) a subsequent court is bound, if at all, only by what it holds to have been the *ratio decidendi* of the precedent. If a later court is willing to be bound by a particular proposition, it will treat that as the *ratio*; if not, it will seek to interpret that proposition in whole or in part as mere *dicta* and to treat another proposition as the *ratio* of the precedent.

C. K. Allen, *Law in the Making*; R. Cross, *Precedent in English Law* (1976).

See also PRECEDENT; *Stare Decisis*.

Ratio legis. The underlying principle or objective of the law, particularly of legislation, a factor sometimes taken into account in civil law systems in interpreting legislation, on the basis that the legislators' intention should be considered and not merely the verbal forms in which the law has been expressed.

Rationalism. The movement of thinking produced by the Renaissance, the Reformation, and the rise of nation-states which substituted for the spiritual authority of divine and ecclesiastical law the intellectual authority of reason. Grotius found a new foundation for natural law as well as for international law in the nature of the human intellect which desires a peaceful society. The principles of reason could be deduced both *a priori*, by examining anything in relation to the rational and social nature of man, and *a posteriori*, by examining the acceptance of these principles among people. Pufendorf similarly based natural law on the two sides of human nature, and he, Thomasius, Wolff, and Selden all asserted in their general theory and particularly for the law of nations the supremacy of the law of nature. Vattel on the other hand relegated natural law to the recesses of a state's conscience.

Ray, Isaac (1807–81). Forensic psychiatrist, very important in developing and influencing thinking about his subject; he wrote a *Medical Jurisprudence of Insanity* (1838), much influenced by French psychiatrists and French law. His book was one of the most influential in the nineteenth century on the subject.

Raymond, Sir Robert, Lord (1673–1733). Son of a judge of all three common-law courts, was called to the Bar in 1697 and soon had a practice, acquiring a reputation for ability. He was made Solicitor-General (1710) and Attorney-General (1720) before becoming a judge (1724) and then Chief Justice of the King's Bench (1725–33). In 1731 he became Lord Raymond. Like his father he compiled a volume of common law decisions for the period 1694–1732, which had a good reputation. His decisions were of importance principally in the spheres of obligations and criminal law and he was an impartial, careful, and discriminating judge.

Raymond de Peñaforté, St. (*c.* 1175–*c.* 1240). A Catalan canonist, third master-general of the Dominican order, professor at Bologna, and compiler, on the orders of Pope Gregory IX, of the collection of papal decretals and constitutions since the *Decretum* of Gratian, known as the *Decretales Gregorii IX* (1234) which form Book V of the *Corpus Iuris Canonici* and was the first official collection included therein. It comprises five books and a gloss was written on it by Bernard of Parma. He also wrote a *Summa de poenitentia et matrimonio*, a lost *Tractatus de bello et duello* and many other works.

Raynerius de Forlivio (?–1358). Author of *Lectura super Digesto Novo* (1523).

Reader. At the Inns of Court in the fifteenth to seventeenth centuries a reader was a senior barrister who gave a reading or series of lectures to student members of the Inns in Lent and August vacations. Readers included many who became famous later. The Lent reading was reserved for double readers, who had read some years before and were benchers. To be reader was a prerequisite of becoming a bencher, but in the seventeenth century benchers began to be elected without having first delivered a reading. In the fifteenth century readings consisted of short standard lectures which over a period covered a cycle of the most important old statutes. By mid-sixteenth century readings were expositions of a statute or branch of law, particularly on land law. They provided themes for the case discussions which followed. Thus Bacon read on the Statute of Uses. Some readings, such as Callis's on the Statute of Sewers (1622), were later published as authoritative works. The system broke down in the seventeenth century. The title is also given to an official in the Temple Church who reads the lessons and preaches on Sunday afternoons, and is used in the Inns of Court School of Law and in universities for teachers immediately below professorial rank.

S. E. Thorne, *Readings and Moots at the Inns of Court.*

Reading, Lord. See ISAACS, RUFUS.

Readings of Bills. The main stages through which a Bill must go in each of the Houses of the U.K. Parliament. The first reading is formal, after which the Bill is printed. The debate on second reading is the main discussion on the principles of the Bill; if passed, it is then examined in detail in committee and reported to the House. Third reading is final approval. The name comes from the fact that till late in the eighteenth century Bills were read out by the Clerk to inform the Members of the proposals.

Real actions. In civil law, actions in which a party sought to vindicate property as his own. In mediaeval English law a real action was an action in which the specific thing demanded could be recovered, in particular title to land. Accordingly, learning as to real actions is the major part of the land law of the Middle Ages. The actions by which land held by free tenure might be asserted were the writ of right group of actions, the assize of novel disseisin, and the writs of entry *sur* disseisin. There were numerous other real actions dealing with the various interests which persons might have in land, as between landlord and tenant, for the protection of incorporeal rights, and so on. Land held by unfree tenure was not protected by the royal courts, nor by the real actions, nor were chattels real, such as the interest of the tenant for a term of years. Real actions were abolished in 1833 with a few exceptions which were later assimilated to ordinary actions.

Real burden. In Scots law, a burden, charge, or incumbrance in the form of a liability to pay, which is charged on land or buildings and which attaches to them as well as to the debtor personally and, in default of payment, justifies remedies against the lands themselves.

Real contracts. In Roman law the contracts of *mutuum* (loan), *commodatum* (loan for use), *depositum* (deposit), and *pignus* (pledge), in all of which obligation arises not merely from agreement but from the delivery of the thing, which is the subject-matter of the contract, and called "real" because the obligation arose *re*, by the fact of delivery.

Real estate. Prior to 1926, land and estates and interests therein, of any tenure, except terms of years. After 1925 it includes chattels real and all interests in land but not money to arise under trust for sale, not money secured, or charged on land.

Real property. Historically in English law real property comprised things, or rights in things which could be recovered specifically by a real action, and since originally specific recovery was competent only where the claimant was entitled to a freehold interest, i.e. an estate for life or a greater estate, real property came to mean land and things so attached thereto as to be part thereof, and rights in lands which were heritable or enduring for life, whether they involved full ownership or only partial enjoyment of the land or the profits thereof. Rights in land for terms of years only were not real property but chattels real; real property and chattels real differed in mode of devolution on death, right of succession on intestacy, and legal remedies for recovery thereof.

At common law under the feudal system no person owned any land outright, but held it under the sovereign directly or under one, or a succession of, mesne lords, who held in a chain from the sovereign. The relationship was that of tenure, each holding from and under another in return for a service or money, and each owned not the land but an estate, or group of rights, in the land. Tenure imported reciprocal rights and duties; the lord was bound to defend his tenant's title, and the tenant was bound to render certain services to the lord. The major kinds of tenure were knight-service, the provision of one or more knights for military service, later generally commuted for a money payment; tenure in serjeanty where the land was held by personal service, later distinguished into grand serjeanty (where the services were honourable and important), and petty serjeanty, socage (where the tenant usually rendered services on the lord's land or paid a money rent), and frankalmoin, or free alms (where land was held by ecclesiastics and no definite service, spiritual or secular was due). Tenures in chivalry and free socage were the tenures of freemen and the tenant had a free holding. Only freeholders could utilize the proprietary and possessory actions allowed in the King's courts, the writ of right and the assizes of novel disseisin and mort d'ancestor respectively. Tenure in villeinage was the tenure of an unfree man, though a free man might hold in this way, and had only the remedies he could obtain in the manorial court. Subsequently villein tenure became recognized as copyholding, and as well protected as freehold tenure. Leases for terms of years conferred no freehold and did not entitle the termor to freehold remedies; disseisin was confined to freeholders and a dispossessed termor could seek to recover possession by an action of ejectment.

While tenure in common socage was subject to general rules of the common law and statute, special incidents might be attached to it by the custom of particular manors or districts. The main special tenures were ancient freeholds, where land held of a manor was subject to customary incidents, mainly regarding alienation, but the tenure was not at the will of the lord and admittance was not necessary to complete an alienee's title; tenure in ancient demesne, of manors which had belonged to the Crown in the time of Edward the Confessor or William I; gavelkind, where lands of socage tenure, particularly in Kent, were subject to special customs mainly relating to descent, notably that there was equal division of inheritable estates on intestacy among all sons or male heirs; and borough English, sometimes associated with burgage tenure but also found in rural manors, particularly in the east midlands, whereby the youngest son inherited the lands.

Apart from the common law systems of tenures of and estates in land, a system of equitable interests in land developed. Chancery recognized and enforced uses, whereby land was conveyed by one person to another and his heirs to the use of a third. The Statute of Uses, 1535, provided that the feofee to uses should be deemed the absolute owner of the lands. But this was circumvented by the practice of tacking a trust to the use which was executed by the Statute, and the limitation in trust was enforced by the Chancery. Similarly Chancery recognized the equity of redemption (q.v.) as an equitable estate in the land surviving in a mortgagor, notwithstanding that the period for redemption of the mortgage was past. Tenure in chivalry was obsolete by the fourteenth century, but was kept alive by the casual profits, notably wardship and marriage, but knight service and its incidents were abolished in 1660 and converted into common socage. In the nineteenth century there was much reform of conveyancing and major changes in real property were effected by a series of statutes in 1925. These effected the abolition of copyhold; the reduction of tenures of land to two, namely freehold and leasehold; the reduction of the legal estates in land to two, namely the fee simple absolute in possession and the term of years absolute; the reduction of legal interests or charges in or over land, all other estates, interests or charges, such as life interests, taking effect as equitable interests only; a great extension of the systems of registration of charges over land and of title to land, and a considerable measure of assimilation of the law of real property to that of personal property.

Under the modern law a fee simple absolute in possession is as nearly as possible complete ownership, and the owner may use the land as he wishes, subject to overriding statutory controls, alienate it, dispose of it by will, and on his death intestate it devolves on his personal representatives, subject to a trust for sale, the proceeds of sale being distributed according to the rules of intestate succession. A term of years absolute must have a fixed beginning and a fixed ending.

Equitable interests include estates upon condition and determinable fees, estates tail or entailed interests, estates for life of the tenant or *pur autre vie*, future interests, namely remainders and reversions.

G. C. Cheshire, *Modern Real Property*; R. E. Megarry and H. W. R. Wade, *Law of Real Property*.

Real representative. A person on whom, after 1897, real estate devolved on death. He held the estate as trustee for the person legally beneficially entitled thereto. He is now the executor or administrator, as real estate now devolves on the deceased's personal representative in the same way as chattels real formerly did.

Real right (or *jus in re*). A right of property in an object of the right by virtue of which the person vested with the right may secure possession of the object and exercise the right of an owner in relation thereto, as contrasted with a personal right entitling the person only to a claim for performance of an obligation or compensation for non-performance.

Real security. Security for the performance of an obligation constituted by mortgage or other transfer of a real right in property as contrasted with merely personal security constituted by an individual's guarantee.

Realism. The general name for the basic approach to law of two groups of jurists, distinct but sharing a commonsense approach. The American realists stem from the judge-centred approach of J. C. Gray and the scepticism of Holmes, J., and emphasize the uncertainty of the law and the importance of the attitude of the judges; what these officials do about disputes is the law. Realism had some affinities with the pragmatism of William James and the work of John Dewey. After Holmes the main figures were Llewellyn, Cook, and Jerome Frank (qq.v.). Their approach is a combination of analytical, positivist, and sociological. The importance of their work was to concentrate attention on litigation, judges and judicial attitudes, empiricism, the influences which affect judges (of which the words of statutes and cases are only some), and on the operation of law in society. Law is a means to social ends, but must be evaluated in terms of its effects.

The Scandinavian realists have all been influenced by Hägerström; the most notable were Olivecrona, Lundstedt, and Ross (qq.v.) who, though they diverge *inter se*, had in common that they wanted to see how law operated as social fact. Unlike the Americans they sought to develop a philosophy of law. To Olivecrona law must be studied as a social fact; though rules of law do not act as imperatives issuing from a formal or human source carrying any binding force, they do influence the behaviour of judges and officials. To Lundstedt law needs no higher basis than that it is indispensable for the maintenance of human society and its content should be determined by the requirements of social welfare. Ross sees rules of law as directing the ways in which official agencies of the legal system should decide issues, not as controlling the behaviour of individuals.

Both groups of thinkers have made useful contributions to thinking about law by their insistence on how it operates as distinct from its logical structure.

Reason. The intellectual faculty of being able to discriminate, judge and evaluate, to know truth, and to adapt one's action to a particular end. It is distinguishable from, though it operates along with, the faculties of imagination, memory, instinct, emotions, sensations, and will. It is distinguishable from faith in that reason finds truth by arguments and evidence which persuade and carry conviction, whereas faith relies on belief in authority, divine or human, and from sense-perception in that it arrives at knowledge by application of logic, not solely by appearances. The divergence of reason from faith was clearly established by Bacon, and Descartes and Kant established the critical examination of reason as the central subject of philosophy. Hegel saw the function of philosophy as being to discover the place of reason in nature, in experience, and in reality.

Reason was, with "the nature of man", the starting point of natural law thinking because reason was the prime quality which distinguished man from other things in the world and the application of reason would disclose what conforms or conflicts with man's ideal nature. Natural law was derived from the application of reason to man's nature. Reason has accordingly been a factor of great importance in the history of legal thought because of the perennial prevalence of natural law thinking.

Reason, Age of. The stage of life, when a child begins to be able to reason, to choose, to will, and to be responsible for his conduct. It varies in different children and is affected by education and environment.

Reason of state (*raison d'état* or *realpolitik*). The doctrine developed during the sixteenth century, foreshadowed by Machiavelli's *Il Principe*, expounded more frankly by Giovanni Botero's *Della Ragione di Stato* and implicit in Grotius' *De Jure Belli ac Pacis*, that the new sovereign states had replaced all mediaeval ideas of the unity of Christendom and moral considerations in international relations. It entered international law from the time of the Peace of Westphalia (1648) and has remained a dominant factor in international relations. In substance national interests must be regarded and justify disregard of other obligations.

Reasonable. An adjective used in many phrases in legal contexts, generally signifying that which is agreeable to reason, not irrational, extravagant or excessive, nor yet trivial, and hence coming to mean moderate.

Reasonable care. A standard of care and precautions desiderated by many rules of common law, which can be described only as being the care which a person possessed of reason would take. Such a person will guard against the obvious, the possible, and the foreseeable, but not against the bare possibility, the highly unusual, and the completely unexpected. He will take at least the precautions customary and normal in the circumstances. Whether care and precautions taken in a particular case measure up to reasonable care depends on the circumstances of the particular case. What is customary and normal may not be adequate and cannot be equated with reasonable care. There are some circumstances in which it would be folly to neglect certain precautions.

Reasonable cause. A ground or justification for doing something which commends itself to an individual's reason. In legal contexts it is frequently combined in the phrase reasonable and probable cause, probable cause being grounds which are capable of proof. The question arises in such problems as an action for malicious prosecution or for false imprisonment, where the absence of reasonable and probable cause is an essential of success in the former and its existence is a defence in the latter. Whether there is reasonable and probable cause is a question of law.

Reasonable man. A hypothetical creature whose imaginary characteristics and conduct by way of foresight, care, precautions against harm, susceptibility to harm, and the like are frequently referred to as the standard for judging the actual foresight, care, etc. of a particular defendant. He is a man of ordinary prudence (*Vaughan* v. *Menlove* (1837) 3 Bing. N.C. 468), a man using ordinary care and skill (*Heaven* v. *Pender* (1883), 11 Q.B.D. 503). The reasonable man is sometimes described as "the man in the street", "the man on the Clapham omnibus", or "the man who takes the magazines at home and in the evening pushes the lawn mower in his shirt sleeves". "Negligence is the omission to do something which a reasonable man ... would do, or doing something which a prudent and reasonable man would not do": *Blyth* v. *Birmingham Waterworks* (1856), 11 Ex. 781. "The standard of foresight of the reasonable man eliminates the personal equation and is independent of the idiosyncrasies of the particular person whose conduct is in question.... The reasonable man is presumed to be free both from over-apprehension and from over-confidence"; *Glasgow Corpn. v. Muir*, [1943] A.C. 448. The standard provided by consideration of the "reasonable man" is objective and impersonal, but varies with circumstances, the known characteristics of the thing involved, the magnitude of any known risk, the practicability of precautions, the customary practice in the circumstances, the existence of emergency, and the like.

It has been observed that Lord Bramwell occasionally attributed to the reasonable man the agility of an acrobat and the foresight of a Hebrew prophet, but the reasonable man has not the courage of Achilles, the wisdom of Ulysses, or the strength of Hercules, nor has he the prophetic vision of a clairvoyant. In truth the reasonable man is a personification of the court or jury's social judgment. There is, however, apparently no "reasonable woman" known to the common law.

Reasonable parts (*rationabiles partes*). The two-thirds of a deceased's personal property which at common law fell respectively to his widow and to his children, the remaining third only being capable of being bequeathed freely. Magna Carta preserved these rights of the survivors, but in the course of time they vanished. Similar rights survived in Scotland under the names of *jus relictae* and *legitim* (qq.v.), and similar rights are found in most European systems of law.

Reasoning, Legal. Legal reasoning is in large measure the application of ordinary processes of logical reasoning to legal propositions. Many kinds of reasoning may be utilized in different circumstances. In interpreting legal rules and deeds much reliance is placed on immediate inference of propositions implied by the terms of the rule or deed in question. One of the common kinds of reasoning is by analogy, e.g. rule A applies to case X; the present case is in all material respects similar to case X; therefore rule A is applicable to the present case. Sometimes, however, there are competing analogies, and there is frequently a question whether a case is sufficiently similar to a model to be properly governed by the same rule as applied to it.

Inductive reasoning seeks to derive a general proposition from two or more particular propositions of the same kind; thus, in case X a manufacturer has been held liable for harm done by negligently manufactured bottled drink; in case Y a manufacturer has been held liable for harm done by negligently manufactured packaged clothing; from these propositions one may inductively derive the more general proposition: a manufacturer may be held liable for harm done by negligently manufactured packaged products for human use. Such a more general proposition may then be taken as a basis for deductive reasoning. The difficulty is that several propositions of different degrees of generality may be derived inductively from two or more particular propositions.

Deductive reasoning is best illustrated by the syllogism; thus, e.g. circumstances of type A are governed by rule X; the circumstances of this case are within type A; therefore the circumstances are governed by rule X. In legal syllogisms the difficulties are determination of how the type or category of case is defined or delimited, and determination of how rule X is stated, and of whether the circumstances of a case bring it within type A or another type (a problem of ascertaining and evaluating facts). There may, that is, be several competing major premisses. A major premiss may be supplied by a proposition of statute, or by the *ratio* of a precedent case, or by a general proposition derived inductively from the *rationes* of several precedents, or by a statement in a book of authority. An example would be: legitimate visitors on real property are owed a duty of reasonable care for their safety by the occupier; A.B. was in the circumstances a legitimate visitor; therefore A.B. was owed a duty of care by the occupier.

The *a fortiori* agreement is commonly employed, e.g. a duty of care is owed to normal visitors; *a fortiori* it should be owed to blind and deaf visitors, as less able to observe the dangers.

Legal reasoning is liable to logical fallacies or defective argument forms. These include fallacies in material content, verbal fallacies and formal fallacies. Material fallacies include: (1) The fallacy of accident, applying a general rule to a particular case in which a special circumstance makes the general rule inapplicable, e.g. the rule that persons who commit crime are subject to punishment does not warrant that conclusion where the person is a child below the age of criminal responsibility. (2) The converse fallacy of accident argues improperly from a special case, e.g. strict liability for flooding, to a general proposition. (3) The fallacy of irrelevant conclusion involves the conclusion changing the point in issue in the premisses; specimens of this are the argument *ad hominem* (with reference to the man), e.g. X is an honourable man; therefore he is not guilty of this offence; the argument *ad populum* (to the people); The argument *ad misericordiam* (to pity), e.g. think of my client's poor wife and children; the argument *ad verecundiam* (to respect), e.g. Lord Chancellor X said ...; the argument *ad ignorantiam* (to ignorance), e.g. no one can explain this, therefore my client is innocent; the argument *ad baculum* (to force), e.g. if the court does X, there will be a strike. (4) The fallacy of circular argument, begging the question or *petitio principii*, where the premisses assume the conclusion, e.g. A was present as of right, therefore a duty of care was owed to him; but what right? (5) The fallacy of false cause, of which the commonest form is *post hoc, ergo propter hoc* (after this, hence because of this), e.g. Q came on Monday; on Monday night the jewels were lost; therefore Q stole them. Another form is *reductio ad absurdum* which declares that a hypothesis is unfounded if a contradiction is deduced from it. (6) The fallacy of many questions, demanding a single answer to a question which involves two or more distinct questions, or presupposes something, e.g. when did X stop committing adultery? (7) The fallacy of *non sequitur* (it does not follow), where the conclusion does not follow logically from the premisses, e.g. cricket is lawful, and therefore nobody can be liable in damages for harm done by playing it.

Verbal fallacies or fallacies of ambiguity arise from improper use of words: notably equivocation, using a word in different senses in a premiss and in the conclusion; amphiboly, where the grammar permits of more than one meaning of a sentence; accent, where the meaning depends on the stress or placing of a word, e.g. the word 'only'; composition, where the premiss that the parts are of a certain kind is taken to justify the conclusion that the whole is of the same kind; and division, where the premiss that the whole is of a kind is used to infer that the parts are of the same kind.

Formal fallacies consist in defects in the logical or structural pattern of reasoning; most of them consist of fallacies in the syllogism and are treated in books on formal logic.

M. R. Cohen and E. Nagel, *Introduction to Logic and Scientific Method*; R. H. Thouless, *Straight and Crooked Thinking*.

Reasons for decisions. The giving of legal reasons for their decisions by judicial and quasi-judicial persons and bodies is a matter of great importance. Parties are more likely to accept and be satisfied with a decision if the reasons are set out. Where appeal is competent it is much easier if there is not only a decision but reasons to be challenged. The practice in courts varies. In the superior courts there is no general requirement of giving reasons, but it is the practice in all save the simplest or most formal cases for the judge or judges to give their reasons, pointing to the factors which in their view justify the actual decision. In inferior courts there is sometimes a requirement of giving reasons. Juries are never required to give reasons for their verdicts. To do so would, in the view of some critics, merely make plain the haphazard and irrational working of juries.

In administrative law civil servants have stubbornly resisted any idea of giving reasons for decisions, though statutes have increasingly imposed a duty to give reasons for certain types of decisions, usually those made after a formal hearing or inquiry. The Franks Committee recommended full disclosure of the reports of inspectors who held public local inquiries along with the Minister's reasoned letter of decision on the dispute, and subsequently minor appeals could be decided by inspectors issuing a letter of decision containing reasons. Tribunals under the supervision of the Council on Tribunals are required, when requested (unless exempted by order), to give oral or written reasons when giving or notifying their decision.

Rebellion. Violent opposition by a substantial group of persons against the lawfully constituted authority in a state, so substantial as to amount to an attempt to overthrow that authority. The classic case in British history is the Great Rebellion or Civil War of 1642–46.

Under older Scots law a debtor who disobeyed a charge on letters of horning (q.v.) to pay or perform in terms of his obligation was, in respect of his disobedience to the sovereign's command, accounted a rebel, and denunciation on letters of horning was formerly followed with the penal consequences of actual rebellion and termed civil rebellion.

Rebuffus, Petrus (1487–1557). French canonist, later an auditor of the Rota at Rome, author of *Tractatus Varii* (1581), *Praxis beneficiorum* (1584), a commentary on *De Verborum et rerum significatione* (1586), and other works.

Rebus sic stantibus. The doctrine in relation to international treaties that the treaty is subject to an implied condition that, if there has been a material change in the state of things which was at the foundation of the agreement, and if by an unforeseen change of circumstances an obligation provided for in the treaty should imperil the existence or vital development of one of the parties, it should be entitled to claim to be released from the obligation concerned. The doctrine does not provide that a state is automatically released from its treaty obligations, but only that it may request the other parties to agree to abrogation of the treaty. It does permit suspension rather than termination of a treaty. The doctrine was much asserted in the period of insistence on state sovereignty in the nineteenth century when some jurists would have released states from obligations which conflicted with the development of vital interests, and this led it to be too readily invoked as an excuse for non-compliance with an obligation. International courts have recognized the doctrine but rarely found it applicable.

Rebut. To adduce evidence or arguments countering that adduced by the other party.

Rebutter. In old practice, the pleading which followed the rejoinder and the surrejoinder, and which might in turn be answered by the surrebutter.

Rebutting evidence. Evidence which counteracts or disproves evidence adduced by the other party.

Recaption. The act of claiming and retaking goods wrongfully taken, or a wife, child, or servant wrongfully detained, by another. It is permissible so long as not done in a manner causing a breach of the peace.

Receipt rolls. In use in the times of Henry II and III, contained an account of money received, and were superseded by Pells of Issue and Receipt, which were journals of daily expenditure and receipt.

Receiver. A person appointed in England by the court, or by individuals or corporations, for the protection or collection of property. If appointed by the court, he is an officer of the court and his authority is derived from the court's order. His main function is the protection or preservation of property for the benefit of persons who have an interest in it, and appointments may be made pending an action or arbitration, or pending the appointment of the legal representative of a deceased, in cases of disputed title to land, if no one is in possession of property, if a trustee has been neglecting the trust property, and in other cases. The property in respect of which he may be appointed may be of any kind. He must find security duly to account for his receipts and to apply them as the court directs, and is entitled to remuneration. He is bound to account for all monies received by him.

If appointed out of court, a receiver is an agent and his duties and liabilities depend on the general law of agency and the instrument under which he is appointed. Instances are the appointment by holders of debentures issued by a company, or by partners to realize the firm assets.

In bankruptcy the official receiver, an officer of the court, is appointed receiver of the debtor's property until a trustee is appointed.

In Scotland a receiver may be appointed only by or on behalf of creditors having a security over a company by way of floating charge. There is no office of official receiver in Scotland.

Receiver of the Metropolitan Police. An official, incorporated as a corporation sole, in whom all Metropolitan police property is vested. He is statutorily liable for property damaged in a riot.

Receiver of wreck. Officers appointed by the Department of Trade to take steps for the preservation of vessels stranded or in distress, to take possession of articles washed ashore, to suppress plundering, to conduct an inquiry as to the stranding and, if necessary, to sell the vessel, wreck, or cargo.

Receiving order. An order of a court for the protection of the estate when an act of bankruptcy has been established, and making the official receiver custodian of the debtor's property. Thereafter, unless a scheme is agreed or a composition accepted, the debtor may be adjudged bankrupt.

Receiving stolen goods. At common law and by statute in England, the crime of receiving any property, knowing it to have been stolen or obtained in any way under circumstances which amount to felony or misdemeanour. Since 1968 it has been replaced by the offence of handling stolen goods, namely, of knowing or believing them to be stolen goods, dishonestly receiving the goods, or dishonestly undertaking or assisting in their retention, removal, disposal, or realization by or for the benefit of another person, or arranging to do so.

Reception, The. The name given to the process whereby the Roman and to a lesser extent the canon and feudal laws were taken into the legal systems of Germanic countries, particularly between about A.D. 1400 and 1700. Down to roughly 1400 the laws of the Germanic peoples of what are now Germany, the Low countries, and Austria were basically tribal customs, overlaid by legislation of Charlemagne and his successors and of German emperors. Later distinct provinces with variant legal systems developed within the domain of Germanic law.

With the spread of the Christian Church the canon law penetrated these countries and was authoritative in its own sphere. From the thirteenth century Roman legal learning was carried into Germany and increasing attention was given to it in legal literature and practice. From the start, German Universities studied canon law and Roman law, and popular legal literature made accessible canon and Roman legal ideas. The theory underlying the reception was the growing conviction that the Roman Empire of the German nation was a continuation of the ancient Roman Empire, so that Roman Emperors were the predecessors of the German kings and their laws had a claim to validity in Germany. Its practical manifestation became apparent in legal literature as jurists trained in Roman law came to be employed in Germany. The decisive causes of the reception lay in the absence of a powerful supreme court in Germany, the lack of a strong legislative body, and the lack of scientific cultivation of the law as compared with Italy, France, and England.

When the Reichskammergericht (Imperial Chamber of Justice) was established in 1495 half of its members had to be learned in law and all had to swear to judge according to the common laws of the Empire, which meant the *Corpus Juris Civilis*. This lead was followed in lower courts. Thus, reference was made increasingly to Roman law to the exclusion of Germanic law, save in Saxony where traditional legal ideas survived more, and Switzerland and Schleswig which were outwith the jurisdiction of the Imperial Chamber. But native legal development was provincial and the Roman law offered a developed system, scientifically treated, capable of use as a common law.

By the end of the fifteenth century Roman law had secured an important place in the German universities alongside the study of the canon law. The courts of the lay-judges declined as litigants were referred to princes and their councils, or to the law faculties of universities, or to superior territorial courts, composed partly of learned assessors, for justice.

The reception was a decisive fact in German legal history and fundamentally altered German private law. It postponed legal unity, made private law a composite rather than a coherent system, and made the study of Roman law and its application to modern German conditions the major subject of study. It decisively influenced the character of the German Civil Code of 1900.

The term "reception" is sometimes used more generally connoting the taking of substantial quantities of principles, ideas, rules, and techniques from Western legal systems, particularly English, French, and Dutch law, into the legal systems developed by colonists from these countries in territories in Africa, India, and South-east Asia colonized from these countries.

Recess. The period between the prorogation, which terminates a session of Parliament, and the opening of the new session. The term is sometimes applied also to the parliamentary vacations, when the Houses have adjourned, e.g. over Christmas.

Recidivism. The term for the relapsing of discharged convicts, their commission of further offences, and reconviction. The extent to which it occurs, which is considerable, is a measure of the lack of success of the system of punishment and treatment, which has neither reformed nor deterred them, nor prevented repetition of crime. The probability of relapsing into crime, reconviction, and becoming a confirmed criminal becomes steadily greater with each new conviction, particularly after the first. This justifies more lenient treatment of first offenders, heavier penalties for those previously convicted, corrective training, and ultimately preventive detention.

Recitals. Statements in a deed normally commencing with the word "Whereas...", introductory of the operative clauses. They include narrative recitals, setting out the facts on which the deed is based, and introductory recitals explaining the motives for the operative clauses. Operative clauses prevail over recitals unless ambiguous, in which case the recitals may be looked to to aid the interpretation of the operative clauses.

Recklessness. A state of mind in which a person may do certain acts and which may be relevant to his legal liability for those acts and their consequences. Recklessness, like intention (q.v.), requires foresight of certain consequences of his acts as inevitable, or probable, or sometimes even as possible but, unlike intention, involves no desire that these consequences should result or will to bring them about. It is the state of mind of the man who takes a chance or a risk, knowing that there is a risk. As in the case of intentional conduct, foresight may be imputed to a person if a reasonable man would have foreseen the consequences as inevitable, or probable, or possible, but absence of desire may appear from the circumstances. If the actor foresees, or must be taken to have foreseen, it does not matter whether he was willing to run the risk or indifferent to it. Recklessness is frequently a requisite of particular crimes, or is specified as the mental element of a crime as a weaker alternative to the intentional doing of the same act.

Reclaiming. In Court of Session procedure, to appeal from the Outer House to the Inner House. It is done, formerly by lodging a reclaiming note, and now by reclaiming motion. Also, formerly, the act of a lord recalling one of his vassals who had without his permission gone to live elsewhere.

Recognizance. An obligation or bond acknowledged before a court of record or authorized magistrate and later enrolled in a court of record, whereby the person bound (cognizer or conusor) is bound to secure the performance of some act such as to pay a debt, keep the peace and be of good behaviour, appear to stand trial, or otherwise. A recognizance is usually enforced by estreat (q.v.).

Recognition. (1) In international law, any act on the part of one state which expresses or implies a legally significant reaction by it to the act of another state which does or may affect the legal rights or political interests of the first state. It is a political and executive act having legal consequences. It may be done by unilateral act or acquiescence, or by bilateral act such as treaty or informal agreement, or may be effected collectively, by admitting a new state to a multilateral treaty. The most important instances of recognition are recognition of new states, of governments, of territorial changes, and of belligerency, and of insurgency. It is debated whether recognition of states or governments is declaratory, acknowledgement of an existing state of law and fact, or constitutive, recognition being creative of the legal personality of the state and legal rights and duties. A typical act of recognition has the function of determining statehood, in such cases as where a territory has revolted from the mother-state, and of providing a basis for formal relations, including diplomatic representation. It is only possible if the new state is organized and exercising independent public authority over a territory.

A distinction is commonly drawn in relation to recognition of governments between *de jure* and *de facto* recognition; the difference is that *de facto* recognition takes place when a new authority has attained power, but has not attained such stability as to be apparently permanent or does not give prospects of being able to fulfil international obligations, whereas *de jure* recognition is recognition of the government which is lawfully entitled to exercise sovereignty, though it may not in fact do so, but the effect of either depends very much on the facts and the intention of the recognizing government. *De facto* recognition may not bring about diplomatic intercourse or conferment of diplomatic immunities on representatives. *De jure* recognition may follow *de facto* recognition if the new government appears to be firmly established and such that it should be acknowledged as the lawful authority.

The consequences of recognition of a new state or government are that it acquires capacity to enter into diplomatic relations, to sue in the courts of the recognizing State, has immunity from the jurisdiction of these courts and the right to recover property belonging to the preceding government. A wholly unrecognized state or government has no legal existence in the eyes of the state refusing recognition.

In British and American practice recognition of a state or government is retroactive to the commencement of the activities of the authority recognized.

Recognition may lapse if a state ceases to be independent, or a government ceases to be effective, or belligerents in a civil war are defeated, and it may be withdrawn by express notification or by recognition *de iure* of the sovereignty of another state or of the establishment of a rival government.

(2) In old Scottish land law, a casualty by which if a vassal alienated his lands without his superior's leave, consent, or confirmation, the lands fell to the superior in satisfaction for the contempt shown him.

Recompense. In Scots law, a quasi-contractual obligation incumbent on a party who has benefited by the act of another done without intention of gift, to compensate that other to the extent of the benefit received. It does not arise where one reaps a merely incidental benefit from expenditure by the other for his own purposes or benefit.

Reconvention. The principle that if a party, not subject to the jurisdiction of a country's courts, brings an action therein against one subject thereto, he thereby submits himself to the jurisdiction of these courts in any counter-claim or cross-action arising out of the same circumstances.

Reconversion. The principle by which a prior constructive conversion (q.v.) of property is cancelled and the property deemed restored to its original form. Thus, if real property is devised to one on trust to sell it and pay the proceeds to another, the property is deemed constructively converted into personalty; but if the other, before the property is sold, elects to take it as land, it is constructively reconverted and the trustee must convey it rather than sell it. Reconversion may also operate by force of law, where the property directed to be converted comes into possession of a party absolutely entitled to dispose of it but he does not convert it nor indicate any intention regarding it.

Reconciliation Parliament. The Parliament of Mary Tudor which met in 1554 to reunite England to the Catholic Church. It repealed all the ecclesiastical legislation of the reign of Henry VIII, and both Houses made supplication to the Pope seeking pardon for their past defection and received absolution from Cardinal Pole.

Recopilacion de las leyes de los reinos de Indias. A codification begun in 1624 under the supervision of Rodrigo de Aguiar y Acuña and Juan Solorzano Pereira of the legislation passed by Spain for its American colonies. It was finally edited by Fernando Jimenez Paniagua and promulgated in 1681. It comprises nine books, divided into 218 *titulos* and containing 6,377 enactments dealing with: (1) Church government and education, (2) the Council of the Indies and the courts, (3) political and military administration, (4) discoveries, colonization, and municipal government, (5) provincial government and lower courts, (6) Indians, (7) criminal law, (8) public finance, and (9) commerce and navigation. It became obsolete with the passage of new legislation in the eighteenth century and recodification was begun in 1805 but never finished. The last editions contained supplementary materials; in this form it applied to Cuba, Puerto Rico, and the Philippines until 1898. Though defective in many respects it was the most comprehensive body of law ever passed for a colonial empire and laid down humane principles for the treatment of Indians.

Record. That which can be brought from an inferior court before the High Court under an order of *certiorari* (q.v.) including the document initiating the proceedings, the pleadings, if any, and the adjudication. In Scottish civil procedure it is the document made up after a defender has lodged defences, comprising the summons, conclusions (q.v.), the pursuer's condescendence (q.v.), or statement of facts with the defender's defences fitted therein paragraph by paragraph, and the pleas in law for pursuer and for defender. The first version is the open record, which may thereafter be modified by each party by adjusting his pleadings. When this is done the record is closed and further procedure determined, usually by way of debate on the relevant law or proof of facts in dispute. The record controls the debate in that each party's contentions are limited to the points raised in his pleadings, and controls the proof in that a party is allowed to lead evidence relevant only to those allegations and pleas of which he has given notice in his condescendence or defences.

The name is also given to the volume containing the cases, evidence, and other materials submitted by appellants and respondents in appeals to the House of Lords or Privy Council.

Record, Court of. Originally one whose acts and proceedings were enrolled in parchment. Accordingly the Council, Star Chamber, Chancery as a court of equity, Admiralty and ecclesiastical courts were not courts of record. In about the seventeenth century the common law courts developed the principle that only a court of record could fine and imprison for contempt of court. This gave rise to the modern principle that a court of record is one which has power to fine or imprison for contempt of itself. This is the only test for criminal courts, but in the case of civil courts it is said that courts of record are such as have power to hear and determine actions in which debt, damages, or the value of the property claimed is £2 or over. Also, in the case of civil courts, a writ of error lay where a judgment was alleged to be wrong whereas the remedy in a court not of record was by writ of false judgment. The proceedings of a court of record are called records and are conclusive evidence of matters recorded therein.

Apart from common law criteria, some courts have by statute been declared to be courts of record, e.g. the Restrictive Practices Court, the Transport Tribunal, the Courts-Martial Appeal Court.

Save under statute the concept is unknown in Scotland.

Record Commission. A body appointed in 1800 on the recommendation of a Select Committee of the House of Commons. Between 1810 and 1825 it published the nine folio volumes of the *Statutes of the Realm, 1235–1713*. Though containing mistakes in transcription and translation, inaccurate dating, and other errors, the work is of great value. It published also other materials for legal history such as Domesday Book.

Record, Debts of. Debts which are shown to be due by the evidence, such as a judgment, of a court of record.

Record Office. See PUBLIC RECORD OFFICE; SCOTTISH RECORD OFFICE.

Record, Trial by. Prior to the Judicature Acts, 1873–75, a trial by inspection and examination of the record, where it appeared from the pleadings that one party alleged and the other denied the existence of a judgment or other record decisive of the issue.

Recorder. In England and Wales a recorder was a barrister of at least five years' standing, appointed by the Lord Chancellor to a borough having a separate court of quarter sessions. He acted part-time only, visiting the borough at least four times a year, and there acting as sole judge of the court of borough quarter sessions. In Liverpool and Manchester the Recorders were full-time judges of the Crown Court (q.v.) there, which were in continuous session.

Since the Courts Act, 1971, a recorder is a barrister or solicitor of 10 years' standing, appointed to act for a term as a part-time judge of the Crown Court (q.v.). When sitting as such judge he is addressed as "Sir" (or "Madam"). By virtue of his office a recorder is capable, with his consent, of sitting as a judge for any county court district and may be requested to sit as a judge of the High Court.

In Northern Ireland Recorders of county boroughs hold both county courts, without a jury, trying civil cases of intermediate importance, and the Crown Court, with a jury, trying all except the most serious indictable criminal offences.

In Scotland there are no recorders.

Recorder of London. The Recorder of London is elected for life by the Court of Aldermen of the City. He is chief adviser and advocate of the City corporation and may not exercise any judicial functions unless appointed by the Crown to exercise them, as happens when as an *ex officio* circuit judge he acts as a judge of the Mayor's and City of London Court and of the Central Criminal Court.

Records. The rolls of courts which are courts of record (q.v.).

Recovery. In older English law, the effect of a judgment that the plaintiff may recover his lands in terms of his claim. Formerly in order to get rid of the fetters of an entail, a fictitious and collusive process was instituted in the Court of Common Pleas in which a claimant or recoverer obtained judgment for the lands against the tenant in tail, under a secret trust that, on the recovery being completed, he would reconvey the lands to the defendant in fee simple. The process was abolished in 1833.

In modern practice an action lies for recovery of land from a person wrongfully in possession of it; it replaces the old action for ejectment (q.v.).

Rectification (correction). English courts have equitable power to rectify a deed which by mistake does not correctly express the common intention of the parties to it, and statutory powers, on application by a party affected, to rectify entries in various registers, e.g. of members of a company, which are shown to be incorrect.

Rectitudines singularum personarum. A small treatise dating from between 960 and 1060, setting out the services rendered to a lord by the various classes of persons on a manor.

Recto, breve de. A writ of right by which a person sought to recover both seisin and possession of land. There were two kinds, the writ of right patent, which was appropriate to the holder of a fee simple, against the tenant of the freehold, and the writ of right close, competent where a person held lands by charter in ancient demesne, in fee simple, fee tail, for a term of life, or in dower, and was disseised by another. Writs of right were abolished in 1833.

Rector. In English ecclesiastical law, a person in holy orders having the cure of souls together with the revenues of the church, or a layman who has the part of the revenues of a church which, before the dissolution of the monasteries, was appropriated to a monastery, the spiritual offices being performed by a vicar. A rector holds his office by ordination, presentation by the patron, institution, and induction. He has spiritual charge of the parishioners and the right and duty to celebrate services in the church. He has exclusive title to the revenues of the living, the rectory or parsonage house, and the glebe. A lay rector is sometimes called an impropriator.

Recueil de la Jurisprudence de la Cour. The French version of the reports of the case-law of the European Court, comprising the submissions of the Advocates-General, and the judgments and opinions of the court. Since 1973 there has been an English version. The version in the procedural language of the particular case is the authentic version.

Recuperatores. In Roman civil procedure, jurymen who acted in the second stage of proceedings in place of the single *iudex*. They were first used in cases involving foreigners, but later also in cases involving citizens only. It is uncertain what kinds of cases were dealt with by them. They are not found in post-classical procedure.

Recusants. A general term for those who after the Reformation in England denied the royal ecclesiastical supremacy and refused to attend the services of the Church of England. The Act of Uniformity, 1558, imposed penalties on all such persons. Later statutes distinguished between Roman Catholics (and the name of recusants came to be applied particularly to them) and Protestant dissenters and non-conformists.

Red Book of the Exchequer. A book compiled, at least in parts, by Alexander de Swereford, a Baron of the Exchequer and Treasury Official, about 1230. Its main purpose was to create a permanent record of the liability of royal tenants for scutage, but it includes also a register of surrenders to the Crown and other matters relative to Crown property, some charters and the *Dialogus de Scaccario* and the *Leges Henrici Primi* (qq.v.). It is inaccurate and sometimes erroneous, but preserves matter unobtainable elsewhere. It was edited by H. Hall and printed in the Rolls Series in 3 volumes in 1896. See also J. H. Round's *Studies on the Red Book of the Exchequer* (1898).

Red Cross, International. An international humanitarian agency organized after the battle of Solferino (1859), concerned in peacetime with health and safety and in wartime with caring for wounded and the prevention and alleviation of human suffering generally. It has now a world-wide organization, based in Geneva, together with national societies. The first multilateral agreement on the Red Cross, the Geneva Convention of 1864, committed signatory governments to care for wounded, both friend and foe, and this was later revised and extended to warfare at sea (1907), prisoners of war (1929), and civilians in wartime (1949).

Red Mass. The solemn votive Mass in honour of the Holy Spirit, commonly celebrated at the opening of a judicial year and attended by judges and lawyers. It was so-called because the celebrants wore red and the judges wore their scarlet. The custom originated in Europe in the thirteenth century and from the time of Edward I was offered in Westminster Abbey at the opening of Michaelmas term. It is offered to invoke divine guidance for the new court term, and is celebrated in honour of the Holy Spirit as the source of wisdom, understanding, and counsel, qualities which should be utilized in the administration of justice. It is celebrated now in Westminster Cathedral, Protestants attending a service in Westminster Abbey. It has also long been associated with the opening of the Rota (q.v.).

Redlich, Joseph (1869–1936). Austrian historian and statesman, author of an important *Local Government in England* (1903) and of an unfinished history of Austrian governmental policy from 1848. He sat in the Austro-Hungarian Reichrat from 1907 to 1918 and consistently worked for liberal reforms.

Reddendo. The name in Scottish conveyancing for the clause in a charter specifying the feuduty or other services stipulated to be paid or performed by the vassal to the superior for the lands, so-called from the Latin initial words of the clause, *reddendo inde annuatim*. In English conveyancing the *reddendum* is the clause in a lease which states the rent and the time when it is payable.

Redeemable rights. Rights conveyed by a proprietor to another which are of the nature of rights in security only and not absolute rights, and which consequently may be redeemed on repayment of money lent or satisfaction of other specified conditions.

Reduction. In Scots law, the form of action used where it is desired to annul or cancel a deed or writing on such grounds as fraud, or a decree of court on grounds of inherent nullity.

Reduction into possession. The act of exercising a right conferred by a chose in action (q.v.) so as to convert it into a chose in possession (q.v.). It was particularly used of the former right of a husband to secure rights belonging to his wife, e.g. debts due to her.

Reduction of capital. The act of a company limited by shares diminishing its nominal capital. It requires a special resolution of the members and confirmation by the court which must be satisfied that the company's creditors have consented or their claims been discharged or secured. A company limited by shares or by guarantee may, if authorized by its articles, without the need for confirmation by the court, reduce its capital by cancelling unissued shares.

Redundancy. An employee's being surplus to the needs of his employer and losing his employment accordingly. Since 1965 employees continuously employed for certain periods and dismissed by reason of redundancy, laid off, or kept on short time, are entitled to payments, called redundancy payments, from their employers who in turn, are entitled to rebates from the redundancy fund managed by the Secretary of State for Employment. The qualifying period is two years' employment after the age of 18. Certain categories of employees are excluded. Disputes as to entitlement to redundancy payments are determined by industrial tribunals.

Re-entry. The resuming of possession of land surrendered to another. Provision for re-entry is normally made in leases, in such events as non-payment of rent.

Re-examination. The further examination of a witness by counsel for the party calling him, after he has been cross-examined by the opposing party. It is limited to clearing up points left in doubt. If re-examination discloses new matter, the court may permit further cross-examination on it.

Re-exchange. The damages which the holder of a bill of exchange suffers if the bill is dishonoured in a foreign country where it was payable. It comprises the amount of sterling which will purchase the amount of foreign money due, at the rate of exchange on the day of dishonour, together with interest and necessary expenses, and is claimed by drawing a cross-bill on an indorser in this country of the bill dishonoured.

Reeve. The official of the mediaeval hundred, sometimes called hundred-reeve or hundred-man. He was a bailiff or deputy of the sheriff and held the hundred court, along with the suitors of that court.

Reeve, Tapping (1744–1823). Practised law in Connecticut from 1772 but is noteworthy for founding the Litchfield Law School in 1784, at a time when only the college of William and Mary offered systematic instruction in law, which was for long the most important one in the country. He was sole teacher for 14 years. Reeve was appointed a judge of the Connecticut Superior Court in 1798 and thereafter most of the teaching was done by colleagues. The school closed in 1833. Among its alumni were Aaron Burr and John C. Calhoun. In 1814 Reeve was made chief justice of the Supreme Court of errors; he retired in 1816. He wrote *The Law of Baron and Femme; of Parent and Child; of Guardian and Ward; of Master and Servant; and of the Powers of Courts of Chancery* (1816 and later editions).

Fisher, *The Litchfield Law School.*

Reeve, Sir Thomas (?–1737). Called to the Bar in 1713; an able lawyer, he became a judge of the Common Pleas in 1733 and Chief Justice in 1736.

Reeves, John (1752 or 53–1829). The first historian of English law, was chief justice of Newfoundland 1792–93 and later became King's printer. He was also F.R.S. and F.S.A., and wrote many books and pamphlets. His *History of English Law* down to 1509 appeared in two volumes in 1783–84, later editions carried on the tale to 1601. It was the first general history and, though dull and failing to relate the law to its social and political context, courageously challenged some traditional beliefs and represented a considerable achievement having regard to the materials then available. He later (1792) published a *History of the Law of Shipping and Navigation*, in effect a history of the Navigation Acts, and various other works.

F. W. Maitland, *Coll. Papers*, II, 6.

Referee. A person to whom a matter is referred for his opinion or decision. The term may accordingly include an arbitrator, but it is applied particularly to judicial references. The Judicature Acts, 1873–75, provided for the appointment in the Supreme Court of official referees, to one of whom questions arising in an action might be referred for inquiry and report, or, with the consent of parties, for trial, or, without consent, in any cause requiring prolonged examination of documents or accounts or any scientific or local investigation which, in the court's opinion, could not conveniently be made before a jury or by other officers. Since 1971 designated circuit judges exercise the functions conferred on official referees and distinct appointments will not thereafter be made.

In parliamentary business referees on private Bills are members appointed by the House of Commons to decide question of *locus standi* and, prior to 1868, engineering questions. They are now the Chairman and Deputy Chairman of the Committee of Ways and Means and at least seven other members (three being a quorum), and known as the Court of Referees.

Reference, Incorporation by. The practice, in statutes, of incorporating sections or larger parts of a previous Act into a later Act by reference thereto and not by express inclusion or repetition.

Reference to the European Court. See PRELIMINARY RULINGS.

Referendum. A piece of constitutional machinery for obtaining from the electorate an expression of view on a specific question. It has been used in Canada and New Zealand on the question of prohibition, in Australia on amendments to the constitution, and in South Africa on whether to become a republic. In the United Kingdom it was used in 1975 on the question of whether or not to remain a member of the European Communities and in 1979 on whether or not the Scottish and Welsh Assemblies (qq.v.) were to be established.

Referent. An office in the Reichskammergericht (q.v.), the duties of which were to analyse the evidence in a case, bring out the issues at stake and make recommendations to the court for disposal of the case. These discussions and reports were kept secret and not disclosed in the decisions, but from the seventeenth century reports of cases began to give the referents' conclusions and recommendations as well as the opinions of the judges. See also ADVOCATE-GENERAL; RAPPORTEUR.

Reform Act. A common term for the Representation of the People Act, 1832, which abolished "rotten boroughs", increased county representation and gave representation to many towns hitherto unrepresented, and extended the franchise in counties to possessors of £10 copyholds, £10 long leaseholders, £50 short lease holders, and occupiers paying a minimum rent of £50 p.a., and in counties to householders paying £10 annual rent. The term is also sometimes given to the later Representation of the People Acts of 1867 which extended the franchise to many working men in towns, of 1884, which extended it to agricultural labourers, and 1918, which conferred it on women.

Reformation. (1) The religious movement which swept over Europe in the sixteenth century challenging the authority and doctrine of the Roman Catholic Church, and resulting in many countries repudiating papal authority and Catholic doctrine and adopting Lutheranism, Calvinism, and other forms of Protestantism. In England the separation of the Church of England from that of Rome effected under Henry VIII was a political and legal rather than a religious reformation. Only later, under Edward VI and, after a reaction under Mary, under Elizabeth, were there doctrinal changes. The break with Rome was prompted by the quarrel between Henry VIII and the Pope over the King's application for a bull annulling his marriage to Catherine of Aragon. The Reformation Parliament of 1529–36 abolished the papal supremacy in England, reformed the constitution and administrative system of the Anglican Church, established the royal supremacy, making the King supreme head on earth of the Church of England, and commenced the dissolution of the monasteries. The legal implications of this included the assertion of the supremacy and omnipotence of the State over the Church, and major changes in land law and in the distribution of wealth.

In Scotland the Reformation was primarily a religious one but had profound legal consequences, terminating the papal supremacy, appeals to Rome and ecclesiastical jurisdiction, altering the courts and their powers, and changing much substantive law.

(2) The object increasingly of penal systems in dealing with offenders. In Britain it was accepted by the Gladstone Committee of 1895 and has been increasingly adopted since then. Various techniques have been used in prisons to seek to achieve reformation, particularly solitude to induce meditation and repentance, a theory which gave rise to the separate system in the English convict prisons of the nineteenth century; secondly, moral and religious instruction; thirdly, hard, unproductive labour, latterly replaced by productive work, such as farming; fourthly, particularly in Borstals and detention centres, hard physical activity, strenuous games, and physical training; fifthly, altering the offender's environment, as by rehousing or finding employment; and sixthly, psychology and group counselling. However laudable the ideal and the purpose, none of these methods have, in a large number of cases, more than limited or temporary success.

Reformation (or Black) Parliament. The fifth Parliament of Henry VIII, which sat 1529–36 and enacted statutes to give effect to the religious reformation made by the King. It is also significant as the Parliament in which the Commons took the initiative and became the dominant House, the Lords finding themselves unable to resist the combined wills of the King and the Commons.

Reformatory. A kind of penal institution developed in the U.S., arising out of the discussions at the First National Prison Congress at Cincinnati in 1870. The first was the N.Y. State Reformatory opened at Elmira in 1877. The system was a special form of prison discipline for young adults between 16 and 30, committed for indeterminate sentences. From this, the movement for the erection of reformatories spread rapidly over the U.S. and the system remains a lasting American contribution to penal reform.

Reformatory schools. Places established by statute from 1854 which took young offenders who had committed crimes punishable by imprisonment. They were modelled on voluntary establishments founded from 1818 onwards, while from 1838 Parkhurst was used as a penal establishment for young offenders. Committal to reformatory became routine on second conviction and this removed many young offenders from criminal surroundings before they became confirmed criminals.

Their establishment is believed to have produced a great reduction in juvenile delinquency. From the early years of the twentieth century the idea of a penal reformatory was under consideration, and from it there developed the first Borstals.

Refresher. The fee payable to counsel, in addition to the fee marked on his brief, for each hearing after the first in magistrates' courts and for the second and each subsequent day's hearing in other courts. The refresher fee should be agreed and marked on the brief at the same time as the brief fee, but it is frequently two-thirds of the brief fee. Strictly speaking refreshers are payable only where the cause is heard on oral evidence in open court, but fees paid for continued hearings where no evidence is being heard are commonly but inaccurately called refreshers.

Refreshing his memory. A witness may be allowed, while giving oral evidence, to refresh his memory by referring to a note made by him at the material time, where the writing is not by itself admissible, e.g. a police constable may refer to his note-book to check his evidence of vehicle numbers, road widths, etc.

Refugees. Persons who, owing to fear of persecution on account of race, nationality, political beliefs, or otherwise, is unable or unwilling to remain in the country in which they are and are seeking to settle elsewhere. The existence of refugees is due to wars and natural forces such as droughts, but mainly to intolerance or hatred of the authorities of particular countries for those not belonging to their group or conforming to their regime. The League of Nations had an International Office for Refugees and after 1945 U.N.R.R.A. did much to repatriate and resettle refugees. Its work was continued by the International Refugees Organization and the office of the U.N. High Commissioner for Refugees and many voluntary organizations. In 1951 the U.N. Convention relating to the Status of Refugees granted certain minimum legal rights to refugees in all signatory countries.

In the U.K. the admission of refugees expelled from East African countries has caused serious problems and given rise to much racial tension.

Regalia (or *jura regalia*). (1) Rights which under feudal law attach to the sovereign. They are divided into *regalia majora*, which include the various heads of the royal prerogative, and which cannot be communicated to a subject; and *regalia minora*, which are rights in land held in trust for the community generally, and rights in land, such as salmon-fishings, which may be communicated to a subject, but only by express grant.

(2) Regalia also means the articles given to the sovereign at his or her coronation, namely the crown, the orb with the cross, sceptre with the cross, rod with the dove, the sword of justice, the ring. In normal times they are preserved in the Tower of London.

The regalia of Scotland comprising the crown of the Scottish kings, the sceptre, the sword of state, and the mace are preserved in Edinburgh Castle.

Regality. A territorial jurisdiction which in old Scots law might be created by the King only, by granting lands to a subject *in liberam regalitatem*, and also the tract of land over which such a right extended. A lord of regality had a civil jurisdiction equal to that of the King's sheriff, and more extensive criminal jurisdiction, equivalent to that of the Court of Justiciary, excepting only treason. A regality was a superior jurisdiction to that of a barony (q.v.) and might be exercised over baronies within the regality. The jurisdiction was exercised by the regality court, usually presided over by the bailie or his depute, and composed of the suitors of court, who held lands by suit of court. Initially regalities were a part of the system of government, delegated jurisdiction, but from the fourteenth century the lords of regality frequently sought to usurp royal authority and establish semi-independent domains. But in the fifteenth century regalities again became a means of governing by delegated authority. Regalities and regality jurisdiction were abolished in 1747.

Regality court. In mediaeval Scotland the King sometimes granted to feudal lords not only the jurisdictional rights of a baron (see BARON COURT) but also the higher jurisdictional rights belonging to the Crown and normally exercised by the sheriffs and justiciars. Hence a regality was an area where the lord had royal rights of justice and the King's sheriff and justiciar had neither need nor right to exercise their jurisdiction. All lands within the regality, though discontiguous, formed one jurisdictional unit. The court was generally held by the lord's bailie or his depute or deputes, but the regality justiciar presided when it sat as a criminal court, and the regality chamberlain when it sat as chamberlain ayre. The body of the court comprised the tenants who owed suit of court. Regality civil jurisdiction was very extensive, and its criminal jurisdiction included all crimes except treason. It had also jurisdiction over a great range of miscellaneous disputes and acts of misconduct by tenants. Regality jurisdiction declined from the fifteenth century and was abolished in 1747.

Regard. Money paid to a serjeant at law on his elevation to that rank.

Regard, Court of. A forest court, held triennially to investigate waste and other matters affecting the forest. The area which constituted the forest was known as the regard, and it was inspected by the regarder.

Regency. The situation in which the headship of state is exercised by one or more persons in lieu of and on behalf of a king who is out of the country, a child, ill, or of unsound mind. With the development of hereditary succession to the throne, expedients were necessary to cope with such difficulties. In the Middle Ages practice varied. Henry VIII provided by statute of 1536 that if the successor to the throne were under age (18 if male, 16 if female) he should be, till of full age, in the government of his natural mother and of such other counsellors as the King by will should appoint. On the death of Henry VIII the King's executors appointed the Earl of Hertford as protector of the realm and guardian of the new king's (Edward VI's) person. From 1689 to 1760 Lords Justices were frequently appointed to exercise the royal powers when the King was out of the country.

In 1751 when Frederick, Prince of Wales, died, a Regency Act appointed the Princess Dowager of Wales to be guardian and regent, and a Council of Regency was nominated. In 1765 statute empowered George III to nominate by sign-manual the Queen, the Princess Dowager of Wales, or any descendant of George II residing in the kingdom. A Council of Regency was appointed by the Act. In 1830 the Duchess of Kent was appointed guardian and regent in the event of Queen Victoria's succession while under 18, and in 1837 on her accession, it was provided that, in the event of her death, the government should be carried on by lords justices till the heir presumptive (the King of Hanover) arrived. When Queen Victoria married in 1840 an Act provided that if she died before any of her children attained 18, Prince Albert should be regent.

The Regency Act, 1937, provided that in the case of a severeign succeeding while under 18, or being totally incapable of acting, regency should be exercised by the person who is next in succession to the Crown and in the case of illness or absence from the U.K. the sovereign might by letters patent delegate to Counsellors of State such of the royal functions as might be specified in the letters patent. The Counsellors of State are to be the wife or husband of the sovereign and the four persons who are next in line of succession to the Crown. This is modified by the Regency Acts 1943 and 1953, which provided for the Duke of Edinburgh being regent unless or until there is a child of Queen Elizabeth who can be regent, and add Queen Elizabeth, the Queen Mother, to the category of Counsellors of State.

The King's temporary absences from the country have occasionally been dealt with by conferment of a regency on a queen or prince of Wales. When William III was out of the kingdom statute conferred the office on Mary, and after her death he appointed a commission of Lords Justices and this precedent was frequently followed down to 1760 and between 1820 and 1837. Counsellors of State were appointed when George V went to India in 1911. It is now covered by the provisions of the Regency Act, 1937.

Regency Bill (1788). A proposal by the younger Pitt to make the Prince of Wales regent but without full royal powers. In 1765 Parliament had legislated for a regency on George III's own suggestion, but in 1788 he was ill and unable to assent to the Bill. Pitt proposed that Royal Assent be signified by a commission. Fox and Burke contended that the heir should immediately and automatically assume full royal power as regent, and mutual accusations arose between Pitt and them. In 1789 the King recovered and the Bill was withdrawn, but a similar Bill was introduced when the King became ill again in 1810 and was passed in 1811, the royal assent being given by commission.

Regiam Majestatem. A Scottish legal treatise, so-called from its opening words. It appears to have been unofficially written by an unknown author as a handy manual of law and practice. As was observed in the seventeenth century it draws heavily on Glanvill's *De Legibus et Consuetudinibus Angliae* (1187), but interpolates material from native sources, from early Scottish statutes, from Roman and canon law, and from the primitive laws of the Brets and Scots. It is in four books, the earlier parts being more thoroughly revised than the later and showing more evidence of Glanvill's work being edited and modified to suit Scottish conditions. The date may be about 1230–60 or considerably later, about 1320–50.

Among theories as to its origin one is that it was produced at the instance of Edward I, the outcome of his *Ordinacio super stabilitate terre Scocie* of 1305 to prepare a restatement of Scots law, and another that David I of Scotland caused it to be compiled. These are unfounded; it may be that it represents an attempt, after the War of Independence from England (1290–1329), to rebuild the legal system shattered during the war.

After the Restoration opinion was divided as to its authenticity, though most legal scholars accepted it as genuine.

The book was referred to in statute in 1425, when an Act set up a commission to examine the books of law and amend the law where necessary. The project of revising the laws, commencing with *Regiam Majestatem*, is mentioned several times later in statute. It was in common use by practitioners in the sixteenth century, and was repeatedly cited in Balfour's *Practicks*, and frequently in Craig's *Ius Feudale*. It was first printed by Sir John Skene, both in Latin and in English, in 1609 and this has been regarded as the authorized version. Later texts are by Thomas Thomson in the *Acts of the Parliaments of Scotland* (1844) and by Lord Cooper (Stair Society, 1947).

Regino of Prum (?–915). Abbot of the monastery of Prum, author of *De synodalibus causis et disciplinis ecclesiasticis*, a collection of materials on church councils, the Theodosian Code and utilized as a source in later compilations on canon law.

Region. In Scotland since 1975 the major areas exercising local government functions. They are divided into districts.

Regional organizations. Groupings of states linked by common interests or policies, such as the Organization of American States or the League of Arab States. Such regional groupings were not in conflict with the Covenant of the League of Nations and do not conflict with the Charter of the U.N., though "regional arrangements" for peace-keeping are subordinate to the overriding responsibility of the Security Council. What is or is not a "regional arrangement" is undefined and N.A.T.O. describes itself as an organization for collective self-defence, because self-defence does not require authorization from the Security Council. Regional organizations are probably, however, wider than regional arrangements, and there are numerous regional economic commissions. Notable regional organizations are the Council of Europe, N.A.T.O. and the European Communities, the Organization of American States, the Arab League, and the Warsaw Treaty Organization.

Register of Debates. An unofficial record of speeches in Congress, 1825–37.

Register of M.P.s' interests. By an old rule of the House of Commons no member with a direct pecuniary interest in a question may vote on it. By custom members are required to declare their interest when speaking in a debate in the House or in Standing Committee. In 1975 the House resolved to establish a compulsory register of members' interests which might be thought to affect his conduct as a member or influence his actions in Parliament. The register comprises nine kinds of interests, employment or vocation, and company directorships, clients for whom professional services are provided related to membership of the House, financial sponsorship, overseas visits, payments or benefits from foreign governments or organizations, land or property of substantial value, and companies in which the member owns more than 1% of the issued share capital. It is maintained by a senior clerk of the House under the supervision of a Select Committee. The only sanction for non-disclosure is the power of the House to treat it as contempt. A member must still declare an interest in debate, even though it is registered, but the register is sufficient disclosure for the purpose of voting.

Register of Sasines. General and particular registers of Sasines were established in Scotland in 1617, the particular registers being abolished in 1868. In this register must be registered all deeds creating, burdening, transferring, and extinguishing rights and interests in land. The register is accordingly of great historical value and fundamental to conveyancing in that the validity and priority of many rights depends on registration. Moreover, being public and open to inspection it secures the right of the party registered against all claimants.

Register of Writs (or *Registrum Brevium*). A book recording the forms of the original writs. In mediaeval English law a writ from the King was necessary to initiate any application to court for redress. Each action had its appropriate writ, which could not be used for other purposes; the clerks of the royal chancery probably had discretion to devise new writs to suit new claims. The Statute of Westminster II (q.v.) of 1285 may have restricted the creation of new writs but not the power to adapt old forms to new circumstances. Accordingly later manuscripts of the *Registrum Brevium* included more writs than earlier ones, and more variants than entirely new writs. Manuscript collections of writs are known from the thirteenth century, but there probably were earlier ones; all are individual, and the later ones contain more; there was no single, still less official, Register. But when the Register was printed, as it was by Rastell in 1531, it ceased to develop. In its final form the writs fell into a number of recognized groups, though there was no definite arrangement. The availability of a writ governed the practice of the courts, as each writ had its own appropriate process, mode of trial, and mode of execution. The existence of the writs gave rise to the tracts dealing with them and with pleading. In Edward III's time a selection of writs was published with commentary under the title *Natura Brevium*, later known as the *Old Natura Brevium*.

Registered designs. Designs are those features of shape, configuration, pattern, or ornament applied to any article by any industrial means or process which in the finished article appeal to and are judged solely by the eye, and excluding a method or principle of construction or features of shape or configuration dictated solely by function. The protection of designs has been effected by statute since 1842, but now depends on the Registered Designs Acts, 1949–61. The Comptroller-General of Patents, Designs, and Trade Marks is registrar of designs and the author of a design may apply for registration thereof.

A design may be registered for any article of manufacture other than a building, works of sculpture, and printed matter. To be registrable a design must be new or original, whether the purpose is beauty or utility, or both, though parts of the design may be old.

Once registered the design is copyright for five years, which may be extended for two further periods of five years. Registration gives the proprietor the exclusive right to make or import for sale or to sell, hire or offer for sale, or hire any article in respect of which the design is registered.

The register may be corrected or rectified if proprietorship of the design or its validity is challenged. An action for infringement lies against a copier, for an injunction, damages or an account of profits, and possibly for an order for delivery up of infringing articles.

A registered design is assignable and licences may be granted to manufacture articles of the design.

Registrar. An officer whose function is to maintain a register. Registrars in the Chancery Division draw up the orders of the court. Registrars of county courts have limited judicial functions as well as executive ones. Under many statutes persons are appointed to act as registrars, e.g. of friendly societies, of companies, of solicitors.

Registrar and merchants. In Admiralty actions matters of accounts or assessment of damages are usually referred to the registrar, assisted by, usually two, merchants, for investigation. The registrar's report is prepared after they have heard evidence.

Registration for execution. In Scots law, registration of an obligatory deed pursuant to a clause therein authorizing registration, which makes it equivalent for purposes of enforcement to a decree of court.

Registration for preservation. In Scots law the registration of any deed in the Books of Council and Session or of any sheriff court, to secure against possibility of loss. Official extracts (photocopies) are obtainable which are equivalent to the original.

Registration for publication. In Scots law, the registration of deeds relating to land in the General Register of Sasines, which makes public to any enquirer the title to every piece of land, the burdens affecting it, and what persons have interests in it.

Registration of births, deaths and marriages. In England from early times baptisms, marriages, and burials have been registered in the parish church.

National registration of births, deaths, and marriages was introduced in England in 1837, when there was established the General Register Office under a Registrar-General and a network of local offices. Registration is compulsory under penalty of live births, still births, marriages and deaths, and adoptions of children are directed by the adoption order made by the court to be registered. The indexes of certified copies of entries are kept in the General Register Office and may be searched by anyone on payment of a fee. Indexes kept by superintendent registrars and registrars may also be searched and copies obtained. A person legitimated *per subsequens matrimonium* may have his birth re-registered. Birth, deaths, and marriages at sea, in civil or service aircraft, of service personnel overseas, and in British consular districts must similarly be registered. In Scotland registration of births, deaths, and marriages was done parochially from 1551, but a statutory system was introduced in 1854 under the direction of the Registrar-General for Scotland.

Registration of charges on land. By the Land Charges Act, 1925, charges affecting land must be registered, purchasers of the land for value being protected in the case of non-registration.

Registration of deeds. In England, in Yorkshire and formerly in Middlesex, registries exist where certain instruments may be registered for the protection of purchasers and mortgagers, while various land charges may be registered under the Land Charges Act as a means of making public and protecting rights or claims affecting land, but not necessarily discovered by inspection of property or investigation of title. Bills of sale must be registered. In those parts of England where title to land must be registered at the Land Registry, dispositions of land must be registered to effect transfer of the legal title to the land.

In Scotland deeds may be registered for one or more of three purposes, execution, preservation, and publication. Registration for execution is the recording of a deed in the books of a competent court so that the person in right of the deed may obtain an extract (official copy) with a warrant for the enforcement by the appropriate diligence of the obligation undertaken by the granter of the deed. Registration for preservation is the recording of a deed in a public register and its retention in a public office to guard against its being lost or destroyed. Registration for publication concerns mainly deeds affecting rights in land which since 1617 have been recorded in the General Register of Sasines and thereby are open to public knowledge, but also concerns the Register of Entails, and the Register of Inhibitions and Adjudications which contains public intimation of personal restrictions on dealing with heritable property.

Registration of title. The system of conveyancing whereby the entitlement of named persons to legal estates in land is shown by their being recorded in a national register. Under the system of registration of title adopted in England since 1862, originally voluntary, but later made compulsory in designated areas, the provisions being applied initially to London and extended to other areas at various times, and eventually going to cover the whole country, on first registration there is obtained a state insured record of entitlement to legal estates in identified pieces of freehold and leasehold land, which is kept up to date by registering each transfer and transmission. The system centres on the Land Registry in London and certain branch offices for various parts of the country. The entitlement of the registered proprietor is subject to such mortgages and other burdens, if any, as are set out on the register and to other rights, overriding interests, not mentioned on the register. A registered proprietor receives a land certificate containing a facsimile of the entry in the register and relevant plan.

The advantages of registered title are that purchasers need not investigate the title further than shown on the register, that conveyance is simple and cheap, that it guarantees the title to purchasers and mortgagees, and that equitable claims not disclosed by the register do not affect purchasers for value.

Regius Chairs of law. Professorships created in several of the older British universities by various kings and queens and by reason of their antiquity and royal foundation deemed prestigious. Appointment is made by the Crown on advice taken. Regius Chairs of Civil (i.e. Roman) Law at Oxford and Cambridge were founded by Henry VIII in 1546 and 1540 respectively. Distinguished holders at Oxford have included Albericus Gentilis, Joseph Phillimore, Sir Travers Twiss, James Bryce, Henry Goudy, Francis de Zulueta, and H. F. Jolowicz; and at Cambridge Sir Thomas Smith, Sir Henry Maine, E. C. Clark, and W. W. Buckland. There is a Regius Chair of Laws at Dublin. The Regius Chair of Public Law at Edinburgh was founded in 1707; it is devoted to jurisprudence and a distinguished holder was James Lorimer. The Regius Chair of Law at Glasgow was founded in 1712 and, originally devoted to Roman Law, has since about 1800 been devoted to Scots law; distinguished holders include William Forbes, John Millar, and W. M. Gloag. In these ancient universities there are also Regius Chairs in some other subjects, e.g. divinity, medicine.

Regnal years. The years marked by the successive anniversaries of a particular sovereign's accession to the throne, which consequently do not coincide with calendar years. Prior to 1963 Acts of Parliament were numbered serially within each Parliamentary session, which itself was described by the regnal year or years of the sovereign during which it was held. A session starting in one regnal year, e.g. 9 Geo. VI, became renumbered 9 & 10 Geo VI when it passed the ninth anniversary of the accession. A parliamentary session might and usually did extend over parts of two regnal years, such as 9 and 10 Geo. VI (1945–46), or even over parts of two regnal years and the whole of an intervening regnal year, such as 12, 13 & 14 Geo. VI (1948–49), or parts of the regnal years of two sovereigns, such as 15 & 16 Geo. VI and 1 Eliz. II (1951–52). The system was confusing: thus there were the Finance Act 1955 (3 & 4 Eliz. 2, c. 15); the Finance (No. 2) Act, 1955 (4 Eliz. 2, c. 17); and the Finance Act 1956 (4 & 5 Eliz. 2, c. 54), the latter two of which belong to the same parliamentary session, and the former of which was later renumbered as 4 & 5 Eliz. 2, c. 17.

Since 1963 Acts of Parliaments have by statute been numbered by the calendar year in which they were passed and by the chapter number, without reference to regnal year, e.g. New Towns Act, 1965, c. 59. Regnal years are listed in Cheney (ed.) *Handbook of Dates*; Sweet and Maxwell's *Guide to Law Reports and Statutes*; and *Where to Look for Your Law.*

Regrating. The former offence of buying corn or other goods in a market and later reselling it in the same place, so as to raise the price by causing a shortage. It ceased to be an offence in 1847.

Regulation. In a broad sense any prescription intended to govern conduct. More specifically the term is commonly used of subordinate legislation issued by Departments of State in pursuance of statutory powers.

In the law of the European Communities regulations are one kind of administrative act made by the Council or the Commission of the Communities in order to carry out their task in accordance with the treaties. Regulations, which have to be distinguished from directives and decisions, have a general scope, are binding in their entirety and are directly applicable in all member states. Both the Council and the Commission can make regulations. They are meant to be an instrument of uniformity within the Community, though they may have to leave implementation to the member states. They must rest upon the authority of the treaty and must indicate in general terms their purpose, the reasons which justify them, and the outlines of the system proposed.

Regulatory agency. In the U.S., a quasi-judicial governmental commission established to supervise a specified economic field, having power to make regulations, enforce them, and decide cases of alleged violations of its regulations. The first was the Interstate Commerce Commission (1887) to regulate the railroads, which fixes reasonable rates, grants licences, stops discriminatory practices, promulgates safety rules, and gives substance to the policies of Congress on its subject-matter. Others included the Federal Trade Commission (1914), Federal Power Commission (1930), Federal Communications Commission (1934), Securities and Exchange Commission (1934), and Civil Aeronautics Board (1940).

Rehabilitation of offenders. The principle that a person convicted but who has not been reconvicted of any serious offence for a period of years should be treated as if he had never been convicted or sentenced. In Britain it is accepted by the Rehabilitation of Offenders Act, 1974. His convictions are treated as spent. The person must have undergone the sentence imposed. There are excluded from rehabilitation sentences of life imprisonment, of imprisonment or corrective training for more than 30 months, of preventive detention, of detention during Her Majesty's pleasure or for life, or in the case of young offenders for longer than 30 months. The legislation restricts information and evidence which may be given subsequently about his previous convictions. After the rehabilitation period the convicted person need not disclose his conviction even in answer to questions, unless informed that spent convictions are to be disclosed. The rehabilitation periods vary according to the severity of the sentence from 5 years for a fine, to 10 years for a sentence of imprisonment for between 6 and 30 months, and in the case of young offenders vary from 3 to 7 years, and in the case of minor penalties are generally 1 year.

Re-hearing. The re-arguing of a cause which has already been heard and decided. In the old Court of Chancery an appeal from the Vice-Chancellor or Master of the Rolls to the Lord Chancellor was deemed a re-hearing rather than an appeal, as the former judges were considered delegates of the Chancellor rather than independent judges. Since the Judicature Acts, 1873–75, every appeal to the Court of Appeal is by way of re-hearing. That Court may accordingly reconsider the whole case and, if need be, hear further evidence.

Rei interventus. In Scots law, the name for actings permitted by one party to be done by the other on the faith of a contract which should have been, but is not, constituted by writings probative of both parties, and unequivocally referable thereto, in consequence of which the law holds that neither party is entitled thereafter to resile from the bargain. Thus, if a seller permits the buyer to take possession, he cannot thereafter deny the contract. It is an instance of the doctrine of personal bar and comparable to the English equitable doctrine of part performance.

Reichskammergericht. The permanent high court of the Holy Roman Empire created by Maximilian I in 1495. It was to consist of a president nominated by the Emperor, eight nobles, and eight doctors of law approved by the Reichstag and to be paid for by a general tax, the Common Penny, voted by the estates. The court sat first at Frankfurt, later at Speyer and early adopted Roman law principles, thus materially influencing the Reception of Roman Law in Germany. It was reorganized by Ferdinand I in the mid-sixteenth century with 24 judges, but its proceedings were reviewed annually by an appellate committee of the Reichstag, though this provision lapsed after 1588. By the latter part of the sixteenth century the court was overwhelmed with litigation and bogged down by its own slow and complicated procedure. Increasingly accordingly litigation went to the Reichhofrat or Aulic Council.

Reichstag. The diet or assembly of the Holy Roman Empire, a body of mediaeval origin and meeting only when summoned by the Emperor, for short periods. Nevertheless, by the end of the fifteenth century it had become the chief executive power in the empire. It purported to represent the estates and consisted of three curias, the electors, the princes, and the free imperial cities of Germany. The Emperor summoned it and put proposals before it and withdrew while they debated separately. They then consulted together and conveyed their unanimous decision to the Emperor. At the Diet of Worms (1495) the Reichstag's functions were extended to include the maintenance of peace, determining the modes of tax collection, foreign policy, and the right to declare war. In the sixteenth century it met frequently to take decisions on military and financial matters in face of the attacks from the Turks and the estates frequently voiced complaints and grievances, and in 1555 it appointed a business committee to prepare legislation for the full diet. But from about the same time it became increasingly split on religion and its efficiency impaired. By the Peace of Westphalia (1648), though it participated in the negotiations, it had ceased to be an effective agency of government and power had passed to the individual princes. From 1663 it became a permanent congress of ambassadors at Regensburg without executive authority. It was abolished in 1806.

The lower house of the German Parliament down to 1933 was also called the Reichstag.

Reid, James Scott Cumberland, Baron Reid of Drem (1890–1974). Became an advocate of the Scottish Bar in 1914 and a K.C. in 1932. He was an expert on agricultural law. He was an M.P., 1931–35 and 1937–48, Solicitor-General for Scotland, 1936–41, Lord Advocate, 1941–45, and Dean of the Faculty of Advocates, 1945–48, before becoming a Lord of Appeal in Ordinary, 1948–74. He was made a Companion of Honour in 1967. He delivered many leading judgments and won very high regard for his judicial qualities, and the older he grew the more generally he was regarded as having improved in judicial qualities!

Reid, Robert Threshie, 1st Earl of Loreburn (1846–1923). After a brilliant record at Oxford, Reid was called to the Bar in 1871 and developed a practice in commercial cases. He took silk in 1882. He entered Parliament in 1880 and supported the cause of Home Rule, in 1894 became Solicitor-General and, a few months later, Attorney-General. As such he aided Harcourt in putting through the Bill which established death duties. In 1895 his party lost power but he assisted in the Venezuelan arbitration and increasingly in cases in the House of Lords and Privy Council. In 1905 he became Lord Chancellor when the Liberals regained power. In the House of Lords he was a greater success than in the Commons, and made some good judicial appointments. In 1911 he became an Earl and resigned in 1912.

He was responsible for putting through a great deal of important legislation, including the establishment of the Court of Criminal Appeal (1907), and the Parliament Act, 1911, and after his retiral participated in negotiations seeking a solution for the Irish question. As a judge he gave many important judgments, showing a strong partiality for principle.

Reiffenstuel, Anacletus Johann Georg (1642–1703). A Franciscan theologian and canonist, author of a *Ius canonicum universum* (1700), treating the entire field of canon law, including the decrees of the Council of Trent, the papal constitutions issued thereafter, and the practice of the Roman Curia, which was thereafter a standard authority. He wrote also on moral theology.

Reinsurance. The action of an insurer or underwriter protecting himself by insurance against the risk he has undertaken by insuring an individual against a particular risk. If the original insurer fails to pay, the insured has no claim against the reinsurer; if the insurer has to pay, he has a claim against his reinsurer.

Rejoinder. In common law pleading, the defendant's answer to the plaintiff's reply. It might be followed by surrejoinder, rebutter, and surrebutter.

Relation. (1) A general term for a blood-relative or one of the next-of-kin.

(2) A term connecting something done at one time to something done or happening at an earlier time. Thus an adjudication in bankruptcy relates back to the act of bankruptcy on which the receiving order was made, to the effect of invalidating alienations made in that period fraudulently or in favour of anyone having notice of the act of bankruptcy.

Relations. Dealings by one person with another in particular contexts, such as domestic, commercial, and industrial. Much law can be stated by reference to the particular relationships involved, such as husband and wife, buyer and seller, employer and employee, landlord and tenant, company and shareholder, and the rights and duties which attach to persons in particular relationships are frequently of more importance than any which they can or do create for themselves when entering into such a relationship.

Relator. An informer, the name formerly given to the plaintiff in an information in Chancery, and now given to the person who suggests that an action be brought by the Attorney-General. Thus, a private person cannot sue to compel the performance of a statutory duty by a public body unless he has suffered special damage, but must obtain the consent of the Attorney-General, and such an action is designated, e.g. *Att-Gen ex rel. (i.e. ex relatione) X.* v. *Y.*

Relativism. The rejection of absolute standards, principles, or values and assertion of standards, principles, and values which vary with the time and place. In some modern theories of justice the moral values asserted are not absolute but determined by variables. Jurists who have adopted such a relativist standpoint have included such diverse thinkers as Bentham, whose hedonistic criterion was dependent on individually felt pains and pleasures; Holmes, who contended that men could not get away from their own beliefs and preferences; Max Weber and Roscoe Pound. Pound viewed justice as expressible in terms of a scheme of interests recognized in a society at a time and place; justice was an ideal relation between men, not the ideal or the only one. The most elaborate modern relativist theory of justice is that worked out by Gustav Radbruch.

Release. A discharge or renunciation of a claim or right of action. Also at common law the conveyance of a larger estate, or a remainder or reversion, to a party already in possession.

Relegation. In Roman law a punishment similar to deportation or exile but less severe in that the person relegated did not lose his rights as a citizen; in English law it is used as meaning exile by special Act of Parliament, or judicial banishment for a time only, as distinct from abjuration.

Relevancy. In Scots law a plea of irrelevancy, or plea to the relevancy, is a plea that, even if the other party's statements in his pleadings are accepted as true, he is not entitled to the remedy he seeks or to be absolved from liability, as the case may be. It is similar to the former English demurrer (q.v.). The relevancy of a party's pleadings may be challenged, debated, and decided before hearing evidence, and an action may be dismissed or defences repelled and decree granted at that stage, or the relevancy may be reserved and debated after evidence has been heard. In *Donoghue* v. *Stevenson*, [1932] A.C. 562 (q.v.), the decision at all stages was on a defender's plea to the relevancy of the pursuer's pleadings, and the decision that the action was relevant merely opened the door to allow the hearing of evidence to prove the facts alleged, which hearing never actually took place, as the action was settled. Whether there ever was a snail in the bottle remains unknown.

Similarly in criminal trials the relevancy of the indictment or complaint, i.e. whether it charges anything criminal, may be challenged before the trial proceeds.

The term relevancy is also used for the pertinency of evidence, evidence being relevant only if it tends to establish the state of some fact in issue. Logical relevancy and legal relevancy or admissibility are not coterminous. Facts relevant in a judicial inquiry include facts tending to prove or disprove a fact in issue, facts showing identity or connection of parties, facts showing states of mind and, sometimes, similar facts and evidence of character.

Relief. In feudal law, an incident of tenure, the sum which the heir of a tenant had to pay to the lord so that he might succeed to his ancestor's property, in recognition of the lord's seignory. Hereditary right was recognized in return for the relief. It was at first doubtful whether socage tenants paid relief, but by the fourteenth century this was settled and only in the fifteenth century were tenants by serjeanty deemed liable. Originally fixed by bargaining it came to be fixed at one year's additional rent. On the death of a tenant in chief the King was always entitled to *primer seisin,* and the heir might sue the lands out of the King's hands and take seisin only when he had done homage and paid his relief. Primer seisin was abolished in 1660.

In modern law relief is the remedy or satisfaction sought in an action, particularly in Chancery, where a plaintiff frequently claims specific relief and also general relief, such further or other relief as the nature of the case may require, or claims one or another kind of relief is the alternative.

The term relief is also given in taxation law to certain allowances granted in computing tax payable, e.g. earned income relief.

Religion. See LAW AND RELIGION.

Religious law. The generic name for legal systems which attach to an individual not by reason of his nationality or domicile, or presence in a particular territory, but by reason of his religion. The main bodies of law within the group are Buddhist law, Hindu law, Islamic or Muslim law, and Jewish law (qq.v.).

Relocation, Tacit. In Scots law, the implied renewal of a lease inferred from neither party having given notice of intention to terminate the lease at the stipulated date of expiry and having continued the relationship though without express agreement. The duration of renewal is one year or, if the lease were originally for a shorter period, for the same period again. The same principle has been applied by analogy to certain categories of contracts of employment.

Rem, Action in. An action directed against particular property.

Rem, Information in. An information filed in the Exchequer where goods were supposed to have become the property of the Crown and no claimant appeared to claim them, as in the case of treasure-trove or wreck. After this proclamation was made for the owner, if any, to come forward to claim his property, a commission of appraisement was issued to value the goods, and a further proclamation that the goods were supposed derelict and condemned to the use of the Crown. The process was later used when forfeiture of goods was imposed by statute for transgression of the customs and excise law, even though the offender escaped punishment.

Rem, Judgment in. A legal determination binding not only the parties but all persons. It applies particularly to judgments in Admiralty, declaring the status of a ship, matrimonial causes, grants of probate and administration, and condemnation of goods by a competent court.

Remainder. An estate in land in English law which fell to the remainderman after the particular estate, i.e. the first estate which entitled its owner to actual possession, had determined. Thus under a grant to A for life, remainder to B in fee simple, B had a remainder. A remainder was vested if the remainderman or his representatives were continually entitled and ready to take actual possession of the land whenever the particular estate ended. A remainder was contingent if the grantee were not an ascertained person, or his title depended on the occurrence of some event, so that there was no person ready to enter the land as soon as it became vacant. A contingent remainder became a vested remainder when the person to whom it was limited was ascertained, or when the event on which it was dependent happened. Vested remainders are classified as present, contingent remainders as future interests in land.

Remand. To adjourn a hearing before a magistrates' court to another date, ordering the defendant, unless he is admitted to bail, to be held in custody in the meantime.

Remand centres. Places for the detention of young persons between 14 and 21 pending trial or sentence and the unruly and depraved juveniles from remand homes (q.v.).

Remand homes. Places provided by local authorities for the detention in custody of young offenders, under 17, and for the reception of young persons awaiting trial, awaiting transfer, or awaiting examination and final disposal. They have since 1967 been renamed community homes.

Remedies. A remedy is that which redresses, rectifies, or corrects that which has been done wrongly, or has caused injury, harm, loss, or damage. Remedies may take many forms, notably acts of grace or clemency, such as discharge of claim or *ex gratia* payment; political remedies, such as complaints to an M.P. or councillor; petition to Parliament or other representations to a political authority; and legal remedies, obtainable in accordance with rules of law. Legal remedies in turn may be administrative, obtainable by application to a higher administrative officer or Minister, to a special administrative body or court, or tribunal (which itself may or may not be subject to control by proceedings in the ordinary courts), or civil, by way of proceedings in civil courts, or by negotiation backed by the possibility or threat of taking legal proceedings. Criminal proceedings do not in general afford remedies to aggrieved individuals, but enforce the social policy of the community; they may, however, indirectly afford a remedy. Within the system of legal remedies appeal to a superior court or authority may itself be called a kind of remedy.

An antithesis is implied in the common phrases law and remedies, or rights and remedies, supported by such alleged principles as that in private international law the rights of parties are determined by the *lex causae* and the remedies by the *lex fori*. A more accurate analysis is to say that a legal system confers on parties to a particular relationship both primary rights and duties, such as to obtain delivery of goods bought and to have payment of the price, and secondary rights and duties, such as to enforce delivery or obtain compensation for non-delivery, or on the other part to compel payment of the price or compensation for non-acceptance. Rights to performance or to a remedy for non-performance or inadequate performance are both truly kinds of legal rights, though the secondary or remedial rights came into play and can be invoked only if the primary rights are not voluntarily or satisfactorily implemented. A remedy is accordingly a right to redress or relief which will, so far as possible, rectify the consequences of a breach of duty by another party to a legal relationship. In a particular legal system civil remedies may be distinguished according to subject matter, and the courts which award them into admiralty, common law, equitable, matrimonial, and other remedies. Particular remedies include divorce, payment of debt, injunction, specific performance and, most commonly, pecuniary damages.

Remedies, law of. The branch of law dealing with the remedies which may competently be exercised or sought where various kinds of rights have been infringed. In English law, certainly until the nineteenth century, the law was dominated by procedural considerations and remedies preceded rights; a man could be said to have a right only if there existed recognized procedure which allowed him a remedy in the circumstances. In civil law systems, on the other hand, a remedy will be made available if there has been an infringement of a recognized legal right. The distinction between rights and remedies is frequently equated to that between substantive and adjective law or to that between substantive law and procedure. This is inaccurate because a remedial right, arising on breach of a duty or infringement of a primary right, is as much a legal right as is the primary right, and adjective law or procedure merely deal with the machinery, the means, whereby the remedial right can be declared or made effective. The right to delivery of goods bought and the right to damages in the event of non-delivery are both substantive legal rights, though the steps to obtain a judgment for damages and the modes of execution belong to adjective law. Remedies are rarely regarded as a distinct branch of law, the different remedies being usually considered in relation to the kinds of primary rights and the ways in which they may be infringed.

The first kind of remedy is self-help, which is competent without resort to a court but only in a few restricted classes of cases. Much more important and common are remedies granted by courts; these include judicial declarations of rights, the stopping of continuing or threatened infringements of rights (as by injunction), the rescission or setting aside of transactions, securing or protection of possession of property, the specific enforcement of a primary right (such as payment of a debt), and, most common of all, the payment of damages as pecuniary compensation for loss caused.

In addition there are various remedies appropriate to particular kinds of cases, such as dissolution of marriage, taking of accounts, applications for a prerogative order, orders for the administration of trusts, the liquidation of a company, or the realization of an insolvent estate, and the like.

Remedies, which are all intended in some way to make good the default of one party, fall to be sharply distinguished from penalties, punishments, and sanctions designed to punish or deter, and which in general are proper to the criminal courts. In a few exceptional instances penal considerations arise in civil courts, as in the former actions by a common informer for a statutory penalty.

Remembrancer. See LORD TREASURER'S REMEMBRANCER; QUEEN'S REMEMBRANCER; QUEEN'S AND LORD TREASURER'S REMEMBRANCER.

Re mercatoria, in. In Scots law transactions *in re mercatoria* are those commonly entered into between persons engaged in commerce. Out of favour for trade documents used in such transactions, e.g. contract notes, orders, invoices, cheques, etc. are valid though not formally authenticated but merely signed, or even initialled or stamped.

Remissio injuriae. In older Scots law the plea in an action of divorce for adultery that the pursuer had forgiven the wrong. It might be established by express pardon, or by facts and circumstances implying condonation (q.v.). In either case it was a complete defence.

Remission. A pardon under the Great Seal. Also a part of a convict's prison sentence which is let off. Originally convicts were eligible to earn remission by good conduct and industry. It has been gradually extended and remission of one-third of the prison sentence is now given automatically (except in short sentences) subject to forfeiture of a period of this remission for bad conduct or offences against prison discipline.

The term is also used of the action of a superior court, when altering the determination of an inferior court, sending the case back to the inferior court to enable it to take steps to put the decision into effect.

Remoteness. A concept relating to liability for the consequences of wrongful acts or omissions and operating as a limitation on liability. It arises in at least two distinct ways, sometimes confused, which are governed by distinct principles. The first is the principle that, if consequences which have resulted are deemed so distantly related to the act or omission that they could not reasonably have been foreseen as liable to result from it, there is no liability at all on the doer for those consequences, even though a casual connection can be traced. The person responsible for an act or omission, that is, is not liable for wholly unlikely consequences that can be traced as having been caused by his act or omission; consequences not reasonably foreseeable are deemed too remote and give rise to no liability.

The second is the principle that, if an act or omission has brought about certain immediate consequences which in turn bring about a chain of further consequences, the person responsible for the act or omission is responsible only for certain of the consequences. At some point the court must draw a line and all further consequences are deemed too remote. The second principle is called the principle of remoteness of damage; the first principle is properly called the principle of remoteness of injury, but sometimes also called remoteness of damage.

The concept is important in cases of damages for breach of contract or for tort. A party in breach of contract is liable only for the consequences of the breach which he did or must be taken to have contemplated, having regard to the knowledge he had at the time the contract was made, both knowledge he must have had as a reasonable man and knowledge he had actually been given when the contract was made. A party causing a tort is liable only if some harmful consequences of the kind which happened should have been foreseen as liable to result, and he is liable only for the reasonably foreseeable, or possibly for the direct, consequences of his wrongful conduct, not necessarily for all that has actually followed.

Removal. The transfer of an action from the High Court to the county court or conversely, or the bringing of a cause from an inferior court into the High Court by certiorari (q.v.). Under the poor law (q.v.) casual poor of a parish were liable to be removed into the parish of their birth under an order of removal made by two justices, unless they had acquired the privilege of irremovability.

Removing. In Scots law, the action brought by a landlord to have ejected from premises a tenant whose lease has expired (ordinary removing) or who has incurred an irritancy (q.v.) (extraordinary removing), or, under statute, where the letting was for less than a year (summary removing).

Renard, Georges (1876–1944). French jurist, professor at Nancy, and later a Dominican monk, author of *Notions très sommaires de droit public français* (1920), *La Théorie de l'institution* (1930), *La Philosophie de l'institution* (1929), and other works, and a leading writer on the theory of institutions.

Renault, Louis (1843–1918). French jurist, the founder of international law as a legal science in France. He took up a position intermediate between the natural law and the positivist standpoints. He served as arbitrator in international disputes and wrote extensively, notably a *Manuel de droit commercial* (1887) and a *Traité de droit commercial* (1889–99).

Renner, Karl (1870–1950). Chancellor (1918–20) and later president of Austria (1945–50), author of *The Institutions of Private Law and their Social Function*, an attempt to utilize Marxist sociology for the construction of a theory of law, concerned with the impact of economic forces and social changes on the functioning of legal institutions.

Rent. The consideration, usually in money, paid to a landlord by a tenant of land, buildings, or other corporeal hereditaments for exclusive possession of them for a period. Prima facie rent is fixed by negotiation and it is usually payable weekly, monthly, quarterly, or half-yearly. The landlord may bring an action for rent unpaid or distrain for arrears of rent.

The extent to which the rents of dwelling-houses may be increased has been controlled by a complex body of statute since 1914 in a way which, in view of the greatly increased cost of repairs and maintenance since then, has materially contributed to the decline in maintenance. Rents may be fixed and registered by the local rent officer subject to appeal to a rent assessment committee.

In the case of furnished dwellings contracts may be referred to rent tribunals which fix a reasonable rent.

Rent-charge. A right created by statute, deed, or will to receive a definite annual sum of money payable exclusively from land. It is real property if charged on freehold land or on another perpetual rent-charge, but, if charged on leasehold land, it is a chattel real in the nature of personal property. It is not an incident in the tenure of land but a burden on the ownership of land. It may be perpetual, as where land is sold in consideration of a perpetual rent-charge, or for a term of years, or for life. A rent-charge is recoverable by distress or appointment of a receiver as well as by actions of covenant or debt.

Renunciation Act, 1783. This Act of the British Parliament, accepted the claim of the people of Ireland to be bound only by their own laws and courts. It completed the Irish constitution of 1782, though this still fell short of independence of Westminster.

Renvoi. In international private law, the problem, where conflict of law rules of one system refer the decision of a point to a foreign system of law, whether the 'foreign system' means the internal law of the foreign system only or the whole law of the foreign system, including its conflict of laws rules, since the latter may remit the decision of the point back to the referring system, or transmit it to a third system. The problem arises because some systems refer certain matters, e.g. of succession, to the law of the domicile and others to the law of the nationality of the party concerned, or otherwise treat particular matters differently.

Reparation. In Scots law, properly a synonym for damages or other compensation, but a term formerly commonly used for the whole branch of law concerned with civil liability for harm done whether intentionally, negligently, in breach of a duty of strict liability, or in breach of a statutory duty, and hence generally corresponding to tort in common law. This branch is now more usually called delict (q.v.). Alternatively reparation was sometimes used as meaning the part of this branch concerned with liability for intentional harms, as contrasted with that concerned with negligence.

Repeal. To revoke, abrogate, or cancel, particularly of a statute. Any statute may repeal any prior Act in whole or in part, either expressly, or impliedly by enacting matter contrary to and inconsistent with the prior legislation. Thus a statute frequently states that certain prior statutory provisions are thereby repealed. The courts will treat matter as repealed by implication only if the earlier and later statutory provisions are clearly inconsistent. When a repealing provision is itself repealed, this does not revive any provision previously repealed by it, unless intent to revive is apparent, but it may allow common law principles again to apply.

Repentance. A change of mind and feeling of regret for what one has done or not done. In former times reformers of the prison regime placed much reliance on giving an opportunity for and stimulating repentance. Hence early nineteenth-century prisons were called penitentiaries. The ideal of repentance led to the belief that the compulsory seclusion would be beneficial and the "separate system", a modified form of solitary confinement in which the prisoner did not communicate with other convicts, was enforced and lasted well into the twentieth century. Evidence of genuine repentance by an offender for the crime he has committed has sometimes resulted in a lesser sentence.

Repetition. In Roman and Scots law, a claim for recovery of money unjustifiably paid, e.g. by mistake.

Repgow or Repkow, Eike von (*c.* 1180–1233). The Saxon author of the *Sachsenspiegel* (q.v.), the first substantial account of German law and the beginning of German legal literature, in which he recorded the ordinary and feudal law of Saxony. It attained almost the force of law in Lower Germany and was deemed common law in parts of Germany until 1900. In Upper Germany it was the basis of the *Schwabenspiegel* and *Deutschenspiegel* and it was translated into other languages. He also wrote the first book of history in German.

Repledging. In old law, the power competent to certain private jurisdictions to demand the transfer of an offender charged before another tribunal, on the allegation that the offence had been committed within the repledger's jurisdiction.

Replevin. A process at common law to obtain the redelivery to an owner of chattels wrongfully distrained or taken from him, normally in satisfaction of rent allegedly due and unpaid, on his finding sufficient security for the rent and costs of action, and undertaking that he will pursue an action against the distrainer to determine the latter's right to distrain.

Replication. The former term for the plaintiff's or petitioner's pleading delivered in answer to the defendant's or respondent's plea or answer, now called a reply.

Reply. In the High Court, the pleading delivered by the plaintiff in answer to the defendant's statement of defence. When a case is being tried or argued in court the reply is the plaintiff's counsel's speech or argument in answer to that of defendant's counsel.

Repone. In Scots law, to restore a defender and allow him to defend notwithstanding that he had allowed decree to pass against him in absence.

Reporter. A person instructed to hold a public local inquiry, e.g. into a planning application, and to report to the Minister; a barrister who observes a case in court for the purpose of preparing a report for one of the series of law reports (q.v.); in the Scottish system of dealing with children, the official who brings a child before a children's hearing (q.v.) for its consideration.

Reports. See LAW REPORTS; REVISED REPORTS.

Reports, The The first 11 of the 13 parts of *Les Reports de E.C.* published in French in 1600–15 by Sir Edward Coke (q.v.), republished in translation in 1658–59, and in later editions, which were influential in extending understanding of the common law. There are various Tables and Abridgments of Coke's Reports, and even a version in verse! There was another series, called simply *The Reports*, edited by Mews, in 15 vols., covering cases in all the courts in 1893–95.

Representation. (1) An oral or written statement made by one party contemplating entering into a contract with another, to that other, which is intended to influence the other to contract and which is a material factor in the latter's decision to contract on the terms on which he does contract, but which is not a term of the contract itself. If inaccurate, it is a misrepresentation (q.v.) and may give rise to a claim to rescind the contract.

(2) The substitution of one person to come in place of another, particularly in the law of succession where the children of a person are entitled to stand in his place if he has predeceased.

(3) The entitlement of a person to have another stand in his place and act for him legally, presenting his case in court.

(4) The principle underlying democratic government, that each person of full age and sound mind

shares in sending a representative to Parliament and through his representative participates in government and legislation. In theory the elected representative is a representative, one chosen from the group who should act as he thinks best, and not a delegate, one sent to maintain a point of view. In practice, with the growth of political parties, the contest in a constituency at an election is between the parties and the person elected normally goes pledged to maintain the party view, right or wrong, and he represents the views of his party supporters only. In many constituencies selection of the candidate is in the hands of a group, such as a trade union, and when elected the member is not truly a representative of the views of the voters but regards himself as a delegate of the group.

The question of who are represented is determined by the franchise and the extension of representation. See REPRESENTATION OF THE PEOPLE ACTS. Originally the franchise was determined generally by the ownership of property on the basis that property-owners had "a stake in the country", but this has gradually been extended to all adults of full age and nominally sound mind.

Representation is very defective in that in the British system of single-member constituencies, the person obtaining the most votes wins the seat, but there may be, and frequently are, more votes cast against the winner than for him, so that he may in fact represent the views of only a minority of the electorate. Proportional representation in various forms is arithmetically more complicated but does in large measure, by utilizing multi-member constituencies, produce numbers of members proportional to the votes cast for the different parties. Representation is also based on the principle of one man, one vote, which takes no account of the different contributions different persons make to the community, financially and otherwise. There is no good reason why persons should not have votes in proportion to tax paid. Similarly, in the eighteenth century much play was made with the slogan "No taxation without representation", but this is exactly what is done to industry and commerce where large organizations contribute enormously to the national economic well-being and pay millions of pounds annually in taxation but are totally denied any say in the governing process, and in particular in the imposition of taxation.

Representation of the People Acts. The series of statutes extending the Parliamentary franchise and making Parliament more representative. The 1832 Reform Act made a redistribution of seats by disfranchising rotten and pocket boroughs and enfranchising expanding towns and standardized the qualification, giving the vote to a £10 householder in boroughs, to copyholders and long leaseholders in counties, and to tenants-at-will paying £50 p.a. The 1867 Reform Act extended the borough franchise and introduced the £10 lodger franchise, and gave the vote in counties to £12 occupiers, £5 owners, and long leaseholders. This Act generally enfranchised the urban working man. It also redistributed seats and provided that Parliament should not be dissolved automatically on the death of the sovereign. The 1884 Reform Act extended householder, lodger, and service franchises to counties and gave the vote to most agricultural workers. In 1918 the vote was extended to women aged 30 and in 1928 to all women equally with men.

The Scottish Reform Act of 1832 increased the number of Scottish members from 45 to 53 and extended the burgh franchise to £10 householders and the county franchise to owners of property worth £10 a year. The 1868 Act increased the Scottish members to 60 but in other respects reproduced the 1867 English Act.

Later statutes have modified the system only in details.

Representative. A person who represents or stands in the place of another. Thus the executor or administrator of a deceased is called his personal representative, as representing him in respect of his estate. Prior to 1925 there might be a real representative.

A representative action is one brought by or against a member of a class of persons on behalf of the whole class or group, including himself. Such a fact must be stated on the writ of summons and in the statement of claim.

Representative peers. After the Treaty of Union of 1707 the Scottish peers were represented in the House of Lords by 16 of themselves, elected by themselves for the duration of a Parliament. Those not elected were not eligible for the House of Commons. Scots holding a peerage of Great Britain or of the United Kingdom sat in the Lords as of right. Under the Peerage Act, 1963, all Scottish peers sit in the Lords, unless they disclaim their peerage.

From the Union with Ireland in 1800, 28 Irish peers were to be elected for life and the rest might be elected to the House of Commons, though not for Irish seats. The total number of Irish peerages was to be limited to 100. No election of Irish representative peers took place after 1919, the machinery for holding an election having been abolished when the Irish Free State was established in 1922. The last Irish representative peer died in 1961. Twelve Irish peers claimed a right to be represented in the House of Lords in 1966, but their claim was rejected. The provisions were repealed in 1971.

J. Fergusson, *The Sixteen Peers of Scotland.*

Reprieve. The suspending for a time of the execution of a sentence of a criminal court. When capital punishment was competent the court could grant a reprieve if a woman were convicted and were pregnant, or if a person convicted became insane after conviction, and might do so if the judge was not satisfied with the verdict and thought that time should be allowed to apply to the Crown for a pardon. In all other cases only the Crown could reprieve, which might be done on the advice of the Home Secretary or the Secretary of State for Scotland. Reprieve was normally associated with commutation of sentence to life imprisonment, but reprieve and commutation were quite distinct.

Reprimand. A formal and public reproof of his conduct, delivered to an offender by a judge, or by a superior to an inferior, as in the armed forces, or by the Speaker of the House of Commons to a person guilty of breach of privilege or contempt of the House.

Reprisals. A term used from the thirteenth century, originally of a kind of measure of self-redress or forcible seizure of goods. It was used also for redress of injuries inflicted by the citizen of one state on the citizen of another on the basis that the latter was vicariously liable for the injustice done by his state or fellow-citizens. This was, however, very rough justice and English kings later provided that reprisals were to take place only after letters of request for redress had been sent out under the Privy Seal but had been disregarded. The aggrieved party might then obtain letters of marque under the Great Seal which authorized him to make reprisals until adequately compensated for what he had lost. Peace treaties from the fifteenth century frequently contain the condition that no reprisals shall be authorized until the one monarch has applied officially to the other for redress. By the end of the seventeenth century reprisal under letters of marque and reprisal was obsolete, though occasional instances occurred later. But reprisal as between states has continued to be used for the redress of grievances down to modern times. When a state by diplomatic means has failed to secure redress for a legal wrong inflicted by another state, it may take exceptional measures short of war called reprisals against the other state, designed to compel consent to a just settlement and inflicting not disproportionate injury. These are injurious and otherwise illegal acts for the purpose of compelling the other state to consent to settlement of a dispute created by its own international delinquency. Frequently, however, what one state calls reprisals the other will call aggression. In fact reprisal has been confined in modern times to action by greater powers against smaller, as by Britain against Greece in the Don Pacifico affair (1847). Reprisals are permitted in all cases of international delinquency when the aggrieved state cannot get reparation by negotiation or other amicable means, but not for internationally injurious acts on the part of officials, armed forces, and private individuals if their state discharges its obligations of vicarious responsibility, and British action in the Don Pacifico affair was unjustified.

Acts of reprisal may be performed only by state officials or forces, by authority, and against anything and everything belonging to or due to the delinquent state or its citizens, as by seizing ships or goods, or occupying territory, or seizing hostages. Negative reprisals are a refusal to perform such acts as are ordinarily obligatory, such as to pay a debt. Reprisals should be proportionate to the wrong done and no more, and limited to the compulsion necessary to obtain reparation, and are admissible only after negotiations to obtain reparation have been in vain.

In time of war reprisals are retaliation by one belligerent against another, by otherwise illegitimate acts of warfare, to compel the latter to comply with the rules of legitimate warfare. They may be resorted to for any and every act of illegitimate warfare. In practice alleged reprisals are frequently made the cloak for illegal conduct of many kinds.

Republication. The republication of a will or codicil arises where it is re-executed by the testator, as where it has been revoked and he wishes to revive it.

Repudiation. The act of one party to a contract expressing or implying that he is not going to perform it, conduct which entitles the other party to treat the contract as at an end and claim damages. Also the putting away of a wife, or woman betrothed, and the renunciation of an ecclesiastical benefice.

Repugnant. Inconsistent with or contradictory of. Where in a deed one provision is repugnant to another the earlier is deemed to prevail, except in wills where the later prevails as the latest word of the testator.

Reputation. The general estimation in which a person is held with reference to moral character, ability, and other qualities (his good name, honour, credit or standing). Unjustifiably to cast aspersions on a person's reputation may be defamation (or libel or slander) and be actionable by him. In the law of evidence general reputation, i.e. that which is generally believed, may be adduced to prove matters of public and general interest such as parish boundaries, rights of highway or common, etc.

Reputed ownership. A principle introduced in bankruptcy law by statute in 1623, whereby the creditors of a trader could claim to have sold and divided property not belonging to the debtor but which, having been in his possession, had given rise to the inference that he owned it. It is now applicable to all goods possessed by a bankrupt in business. In older Scots law it is the principle that ownership of moveables was presumed from possession.

Requests, Court of. With the great increase in the judicial business of Council and Chancery under the Tudors, there was created a court, more closely related to the Council and the Court of Star Chamber than to the Chancery. It was presided over by the Lord Privy Seal and exercised an equitable jurisdiction in civil cases similar to that of the Chancery, and also occasionally a quasi-criminal jurisdiction in riots, forgery, and similar offences, similar to that of the Star Chamber. Originally it exercised jurisdiction at the suit of poor men or of the King's servants. From the mid-sixteenth century the legal assessors, known as Masters of Requests, had entire control of the court so that it became quite separate from the Star Chamber.

It seems to have been a useful and popular court, but at the end of the sixteenth century was attacked by the common law courts although it survived these attacks. It was killed, if not directly abolished, by an Act of 1641 which abolished the jurisdiction of the Council in England.

I. S. Leadam, *Select Cases in the Court of Requests* (Selden Socy.)

Requisitions on title. In conveyancing, inquiries and requests arising on examination of the abstract of title, made on behalf of the prospective purchaser or mortgagee, which the intending vendor or mortgagor must satisfy and comply with. A contract of sale may stipulate that the title is to commence with a specified document and that no requisitions shall be made in respect of the earlier title, and usually also stipulates that requisitions shall be made within a stated time after delivery of the abstract of title.

Rerum Novarum. An encyclical letter of Pope Leo XIII issued in 1891. It examines the problem posed by socialism, then attracting significant support, and develops arguments establishing the natural right to private property, outlines the role of the Church in social affairs, expounds the social role of the Church and of the State, including a rejection of *laissez-faire*, and emphasizes the beneficial powers of voluntary organizations. It was the first of the great social encyclicals and had great influence on social thinking throughout Europe.

See also *Quadragesimo anno.*

Res communes. Those things which are naturally common property, such as light and air, and incapable of appropriation by any person.

Res furtivae. Things stolen and accordingly tainted so that they can still be reclaimed by their true owner even from a bona fide purchaser for value. This is subject to certain exceptions, particularly in England, by reason of the rule of market overt (q.v.).

Res gestae (things done). In the law of evidence, the facts constituting and immediately accompanying the matter which is in issue in the proceedings. It includes facts leading up to, explaining and following continuously from the facts in issue. Thus evidence by a hearer of what the victim shouted when assaulted is admissible as part of the *res gestae*, whereas otherwise it would be inadmissible as hearsay.

Res ipsa loquitur (the thing speaks for itself). A principle frequently invoked in the law of tort, to the effect that the event itself is indicative of negligence. It has been said to apply where the thing which caused the accident, e.g. an aircraft, was under the control and management of the defendant or his servants and the accident was such as would not have happened in the ordinary course if due care had been exercised by him or them. The happening then by itself affords prima facie evidence of negligence, which may, however, be rebutted by the defendant. The principle does not apply to, e.g. a car hitting a pedestrian, where the fault of either or both may have contributed to the harm.

Res judicata. The principle that a matter, finally adjudicated on by a competent court, may not subsequently be reopened or challenged as to the matter or point decided, by the original parties or successors in their interests. It does not preclude any appeal which may be competent in the circumstances, nor prevent other parties raising the same controversy. The basis of the principle is the need for finality and the inexpediency of allowing a matter once decided to be litigated again. It applies only where the parties are the same, the issue the same, and the means of raising the question the same as has already been considered by a court.

It differs from the doctrine of *stare decisis* (q.v.) or binding precedent in that the latter relates to a principle or rule of law being treated as settled, whereas *res judicata* relates to the finality of a decision of a particular dispute, as between particular parties. *Stare decisis* binds everyone, *res judicata* binds the parties only and in relation to a particular controversy only.

Res mancipi. In Roman law, the basic forms of property, Italic land, slaves and beasts of burden, horses and oxen. They could be transferred only by *mancipium* or mancipation. Other things were *res nec mancipi* and could be passed by delivery.

Res merae facultatis. In Scots law, a right which may be exercised or not at pleasure, but which is not extinguished by non-use for a period, however long.

Res nullius. A thing not belonging to anyone, such as fish and fowl not caught or captured.

Resale price maintenance. At common law agreements between manufacturers or traders in particular fields of enterprise were subject to the common law rules of restraint of trade (q.v.) and of conspiracy, and accordingly agreements fixing prices, regulating output and the like were prima facie void, unless they could be shown to be reasonable in the circumstances, while conspiracy, as by the publication of 'black lists' was actionable unless the predominant purpose were the furtherance of legitimate interests of the parties combining. Individual agreements between supplier and retailer for the maintenance of resale prices were valid and enforceable, but not generally enforceable against third parties, and collective agreements were in general valid.

Under the Restrictive Trade Practices Act, 1956, individual agreements were made enforceable against third parties who took the goods with notice of the condition maintaining resale prices, but collective agreements were required to be registered, adjudicated on by the Restrictive Practices Court and, unless allowed by that Court, enforcement thereof by withholding of supplies, discriminatory terms, penalties, and the like was prohibited. Under the Resale Prices Act, 1964, a term in a contract for the sale of goods providing for minimum prices to be charged on the resale of these goods is void, but the Restrictive Practices Court may on the application of suppliers and trade associations exempt certain categories of goods from the rule on specified and limited grounds. Indirect means of enforcing resale price maintenance is in general prohibited.

Rescission. The cutting down or terminating of a contract by the parties or one of them. It may be done by agreement, or by one party who is entitled to do so by reason of the repudiation or material default of another, or by reason of the contract having been induced by fraud or misrepresentation by the other party. It is effected by taking proceedings to have the contract judicially set aside, or by giving notice to the other party of intention to treat the contract as at an end.

Rescript. In later Roman law, an answer by the emperor to a problem put to him by a person, such as a provincial governor. It was permissible for a judge or litigant to seek an opinion on a point of law; the imperial ruling would determine what the decision should be if certain facts were established. This practice became common from the time of Hadrian onwards. Such answers were deemed equivalent to edicts and of general application and many are preserved in the *Codex Theodosianus* and in Justinian's *Codex.*

Rescue. Voluntary intervention by an individual in an attempt to save life or property from being lost. There is in general no legal duty to seek to rescue, but if a person does seek to rescue and is himself injured in so doing, it may be held that a party who created the danger calling for rescue owed a duty of care to a possible rescuer, or that the rescuer's injuries were a foreseeable and not too remote consequence of the harm done to the person primarily endangered and that accordingly he can recover from the person who created the danger. To act in an attempt to rescue will only rarely be held to make the rescuer subject to the plea of *volenti non fit injuria* (q.v.) or of contributory negligence. In some cases, under statute, e.g. at sea, there is a positive duty to seek to rescue.

Rescue may also mean knowingly and forcibly releasing a person from lawful custody. This is criminal under various statutes.

Resealing. The process formerly necessary before probate (q.v.) granted in England could be used to deal with assets in Scotland, and conversely.

Reservation. A clause in a deed whereby a granter reserves something to himself from the grant made, such as the rent in an ordinary lease.

Reservations. In international law, statements made by one party at the time of signature or of ratification of a treaty withholding assent from one or more provisions. The extent to which this can be done while preserving adherence to the treaty depends very much on the circumstances of the case. In a bilateral treaty the one party may decline to agree to the other's reservations, in which case the treaty remains unratified. The practice of entering reservations developed only since the late nineteenth century and initially the general view was that all signatory states must assent to reservations by any one before they could be admitted, but since 1945 the tendency to attach reservations to acceptance of multilateral conventions has grown. The right to enter reservations may be, but need not be, expressed in the treaty but, if it is not, it is a question of circumstances whether reservation is permitted. If reservation is entered by one state, it is doubtful whether the treaty is binding on other states *vis-a-vis* the reserving state unless they have consented to the reservation. If a reservation is incompatible with the predominant object of the treaty, any other party may object to it in which case no obligation arises between reserving and objecting states, but if no objection is made, obligation probably arises, whether the reservation is incompatible with the object of the treaty or not.

Reserved judgment. A judgment which had not been delivered at the conclusion of a hearing but has been withheld and delivered later, after full consideration of the evidence and arguments. It is normal in the House of Lords and Privy Council and may be done by other courts in cases of difficulty. Where the court reserves judgment the report frequently states *c.a.v.*, i.e. *curia advisari vult.*

Reserving points of law. At common law it was regular practice for judges sitting at assizes to reserve points of law for consideration by the full court at Westminster. This practice was maintained until the Appellate Jurisdiction Act, 1876, substituted argument on further consideration before the judge himself.

In criminal cases judges reserved points for argument before the Court for Crown Cases Reserved.

Reset. In Scots law, the crime of the retention of goods obtained, to the resetter's knowledge, by theft, robbery, fraud, or embezzlement, with the intention of keeping them from their true owner. A person charged with theft may be convicted of reset. Compare RECEIVING (q.v.).

Residence. The place in which an individual actually lives, or where the management of a corporation is carried on. It is distinct from domicile (q.v.), but is a factor relevant to the determination of domicile. It is frequently relevant for determining the jurisdiction of a court, for taxation, voting at elections, and other purposes. The concept is sometimes qualified by 'actual' residence or 'ordinary residence'.

Resident. An agent or officer of a government residing in another state, particularly a small state or one less than a kingdom. A resident ranks lower than ambassadors or envoys.

Residue. That which is left over, particularly, in a case of testate succession, what is left over from an estate, not having been needed to discharge debts, nor specifically bequeathed. The person to whom residue is bequeathed is the residuary legatee. If no person be so named, any residue falls into intestacy and is distributed according to the rules of intestate succession.

Resignation. (1) The giving up of an office, claim, or possession. Most offices can be resigned, normally by intimation in writing, but a member of the House of Commons cannot resign his seat; he may, however, obtain appointment to an office of profit under the Crown, notably the Stewardship of the Chiltern Hundreds (q.v.), which has the effect of vacating his seat.

(2) In older Scottish conveyancing, the mode in which a vassal returned the lands to the superior. This might be permanent, resignation *ad remanentiam*, or in favour of another, resignation *in favorem*, when the object was to transfer the lands to a third party and the resignation was made to allow the superior to grant the lands afresh to the new vassal. It is now obsolete.

Resolution. The decision of a meeting. Thus the members of companies take decisions at meetings by passing ordinary, special, or extraordinary resolutions. The House of Commons may under statutory powers vary or renew taxation for limited periods by passing resolutions.

Resolutive condition. A condition in a contract providing for the contract being ended if the condition be satisfied, e.g. a lease granted only for so long as the tenant is employed by the landlord.

Respite. A reprieve, suspension of sentence, or extension of time. As a verb to respite is to discharge or dispense with, e.g. to respite homage, or to delay, e.g. to respite an appeal.

Respondeat superior (let the superior person answer). A maxim embodying but not explaining or justifying the rule of vicarious liability (q.v.) that if an agent or employee causes harm while acting within the scope of his authority or employment, the principal or employer is responsible to the person harmed. Another maxim to the same effect is *qui facit per alium facit per se.*

Respondent. The person called to answer in proceedings by petition. Compare DEFENDANT.

Respondentia bond. A bond which may be granted by the master of a ship, granting security over the cargo of his ship without surrendering possession thereof, for the purpose of raising money to enable him to complete the voyage of his ship, or to forward the goods to their destination. The bond gives the lender a maritime lien (or hypothec) enforceable against the cargo, and the lender is entitled to repayment of his advance only if the whole of the cargo hypothecated arrives; if only a proportion arrives, he is entitled only to a proportionate part of his advance.

See also BOTTOMRY.

Responsa. In Roman law, the answers of learned jurists to difficult legal problems put to them. It is uncertain how far these opinions were deemed authoritative or binding on a judge. It seems that during the Republic a judge could evaluate a *responsum* for himself. Augustus, however, conferred on certain jurists the *ius respondendi* and later emperors continued the practice. This probably did not make the *responsa* of the jurists so honoured formally binding but it was bound to make them be more highly regarded. By the second century it seems that this authority attached not only to *responsa* given for the particular case but to all writings of the recognized jurists, whether living or dead, which must have produced conflicts of opinion in many cases.

Responsibility. A word used in several senses. A person may be said (1) to be responsible if he generally displays care and forethought and considers the possible results of his actings. He may also be said (2) to be responsible for certain events if his conduct has been a material factor in bringing them about; thus a reckless driver may be said to be responsible for an accident. In this sense the word means little more than that he has caused the events and does not necessarily imply accountability. An animal or a snowfall may be said to be responsible for causing a happening. A person may also be said (3) to be legally responsible when of such an age and in such a state of mind and body that he is deemed to be capable of controlling his conduct rationally and such that he can fairly be held accountable and legally liable for the consequences of what he does. Conversely, a person mentally ill or under the influence of drink or drugs may be held legally irresponsible. Responsibility in this sense is fundamental to liability to punishment.

Responsibility in the third sense has a substantial moral flavour, but moral responsibility or blameworthiness and legal responsibility are not wholly equivalent. A person may by law be held responsible in cases where he has not been personally blameworthy at all. Thus under the principle of vicarious liability a person is held responsible for, and legally liable for the wrongs of, his employees, though not personally in fault at all.

The concept of diminished responsibility (q.v.) is based on the appreciation that some persons may be less than fully capable of appreciating the possible consequences of their conduct and of controlling it, yet not wholly irresponsible, and that they should be treated differently from both categories.

H. L. A. Hart, *Punishment and Responsibility*.

Responsibility, Collective. In Cabinet government the principle that each and every member of the Cabinet is responsible for every Cabinet decision, must accept it, defend it, and implement it, and not disclose any doubts he has or opposition to it. If he is not prepared to accept it, he should resign. The principle has grown up with the growth of the Cabinet and was an accepted principle by the mid-nineteenth century. Exceptionally Cabinets have allowed certain Ministers openly to differ from their colleagues and oppose Cabinet policy, e.g. in 1975 over the issue of Britain's continued membership of the European Communities.

Responsibility, Ministerial. The principle that the minister in charge of a department of state is answerable to Parliament and to the community for everything done by his department, whether he personally instructed or authorized it, or even knew of it, or not. Hence a minister must reply to criticisms of the department, and cannot plead ignorance, and if there is inefficiency, muddle, dishonesty, or other mismanagement he may have to resign.

Responsible government. A term chiefly used in relation to the development of self-government in British colonies in the nineteenth century, indicating that the senior executive were collectively and individually answerable to the elected legislature, as distinct from being responsible only to the Governor. The term is also used more generally to denote a form of government which is responsive to public opinion and answerable to the electorate.

Restatement of the Law. An attempt made by the American Law Institute to have formulated in propositions rather like the articles of a code what are deemed to be the best doctrines and principles on the main branches of the law of the U.S., particularly the branches still mainly dependent on case-law. The first edition appeared in 1932–57. The volumes are legally unauthoritative and a purely private compilation, though having the substantial authority of the very eminent jurists who acted as Reporters for the various subjects and their committees, and differ from text-books in not citing case-law and in putting forward not the settled or predominant view on any point but what seemed the most rational view. The volumes of the *Restatement*, though unauthoritative, have been frequently referred to in the courts and have had persuasive influence on judicial decisions. The volume *The Restatement in the Courts* shows in what judicial decisions of which states an article of the *Restatement* has been cited and adopted, distinguished or rejected. A new edition, the *Restatement Second*, was undertaken in 1952.

Restitutio in integrum. In Roman law the act of a court in restoring to his former position a person who, by strict law, had lost a right or position. In English and Scottish law it is used of the restoration to his former position of a person prejudiced by a contract entered into in circumstances of coercion or duress, fraud, error or mistake, by rescission of the contract and requiring return or repayment of anything transferred or paid under it.

Restitution (1). The branch of obligations dealing with the circumstances where one person must restore property or money to another in order that he may not be unjustly enriched. It covers cases where goods or money have been transferred under a mistake, under compulsion, under transaction vitiated by illegality, informality, lack of authority or otherwise, or where the defendant has acquired a benefit through his own wrongful act or otherwise unjustifiably. It corresponds generally to the branch of law in civil law systems known as quasi-contract (q.v.).

R. A. Goff and G. Jones, *Restitution.*

(2) In Scots law, an obligation within the group of quasi-contractual obligations, incumbent on a person in possession of property truly belonging to another, to return it to the true owner.

Restitution of conjugal rights. Either spouse might in England, prior to 1970, seek a decree of restitution of conjugal rights if the respondent had refused or ceased to render conjugal duties to the petitioner. A spouse who had just cause for withdrawing from cohabitation might petition when the cause terminated and a spouse who had wrongfully withdrawn from cohabitation might similarly petition on refusal of a bona fide offer to return. Non-compliance with a decree entitled the petitioner to seek a decree of judicial separation.

Restitution order. An order which an English criminal court may make on the conviction of a person of an offence having reference to the theft of property, for the restoration of the property by the person having possession or control thereof (who may be a person other than the thief) to the person entitled to recover them.

Restraint of marriage. The law favours freedom of marriage and consequently treats as void any contract or provision in a contract the effect of which is to restrain a person from marrying at all, or is to impose excessive or unreasonable restriction on freedom of choice in marriage.

Restraint of trade. An individual is prima facie entitled to exercise any profession, trade, or calling as and where he likes and it is generally deemed contrary to public policy to enforce any restraint on his liberty of action. But many restraints are in fact recognized: firstly, those imposed by statute or common law to ensure a proper standard of competence in the practitioners of a profession, or to protect public order, the revenue, the trade, public health and safety, temperance, and the like; and secondly, those undertaken by agreement, provided they are deemed reasonable in the circumstances, having regard to the interests both of the parties and of the general public. Restraints under the second head include restraints imposed by the buyers of businesses on the sellers thereof, to prevent the latter recommencing business in competition, restraints accepted by employees or partners on their future liberty of practising or trading after they leave the employment or partnership, restraints undertaken by associations of manufacturers or traders to regulate prices, output, and other terms of trade, restraints connected with the use of land mortgaged, and restraints between employees as to the terms on which they will be employed. If any such restraint is challenged it is unenforceable unless the court considers it reasonable in the circumstances, having regard to the interests of the parties and of the general public. By reason of the principle of restraint of trade, trade unions were unlawful at common law because they restricted the liberty of persons to take employment on whatever terms they chose, and became lawful only by statute.

Restraint on alienation. A restriction on the powers of a person to alienate some property. Conditions in restraint of alienation of an absolute interest in possession, in either real or personal property, are generally void as being repugnant to the nature of the interest, but such conditions restraining alienation of lesser interests are valid, if there is a gift over, to a person who is to take, if there is alienation and consequent forfeiture of the interest.

Restraint on anticipation. A restraint imposed on a woman to whom any property was given, or limited so as to become her separate property, to prevent her from alienating or anticipating the income from it during her marriage. Such restraints are not competent since 1949.

Restraints of princes. An expression common in bills of lading, marine insurance policies, and similar deeds, indicating one of the contingencies provided against, namely any forcible interference with the voyage by the government of another country.

Restrictive covenant. An agreement restrictive of the use of particular land for the benefit of other land and binding on every owner having notice of the covenant. It must be registered as a land charge and, if not so registered, is void against a purchaser for value.

Restrictive practices. Arrangements, understandings, rules, agreements, and the like, whereby persons deliberately restrict intake into trades and professions, restrict output and otherwise control working so as to raise prices or earnings, or exclude outsiders from the trade. The Monopolies and Restrictive Practices (Inquiry and Control) Act, 1948, provided for inquiry into the existence and effects of monopolies or restrictions in certain contexts. The Restrictive Trade Practices Act, 1956 (now 1976), provided for the registration and subsequent judicial investigation of certain kinds of restrictive trading agreements by the Restrictive Practices Court (q.v.), it being presumed that a restriction was contrary to the public interest and the onus being on the parties thereto to justify it to the satisfaction of the court on one or other of specified and limited grounds. The European Communities also have rules against restrictive practices.

Restrictive Practices Court. A United Kingdom superior court, established in 1956, consisting of three puisne judges of the High Court nominated by the Lord Chancellor, one judge of the Court of Session nominated by the Lord President and one judge of the Supreme Court of Northern Ireland nominated by the Lord Chief Justice of Northern Ireland, and not more than 10 persons qualified by their knowledge or experience of industry, commerce, or public affairs, and appointed for periods of at least three years. One of the judges is president of the court. It may sit in more than one division, and in any country of the United Kingdom; a division must consist of at least one judicial member and at least two appointed members. On matters of law the judicial members' opinions must rule; on other matters the majority opinion rules. Judgment is given by the presiding judge.

The court has jurisdiction to declare whether or not any restrictions, which have rendered an agreement containing them registrable with the Registrar of Restrictive Trade Practices, are or are not contrary to the public interest, there being a statutory presumption that they are contrary thereto. The court may make orders for restraining a party to the agreement from giving effect to it or making another agreement to the like effect.

It also has jurisdiction to determine whether goods of any class should be exempted from the statutory prohibition on resale price maintenance.

Result. A thing results if it has been partially or ineffectually disposed of and returns to the former owner. A resulting trust arises where an owner of property transfers the property but does not provide for the equitable or beneficial interest, or where the object of the trust fails, in which case a trust results in favour of the owner of the property.

Resumption. A taking back, particularly by the landlord of some part of the premises leased, under a power reserved in the lease to do so.

Retainer. The engagement, for a fee, of a barrister to give his services to a client. Acceptance of a retainer binds counsel to accept a brief for the client and not to accept a brief against him. Retainers may be general or special. A general retainer is ordinarily for the Supreme Court and House of Lords, or otherwise must specify the courts, tribunals, or matters to which it applies. It may be limited in duration but, if not, it lasts for the joint lives of counsel and client. The retainer by itself confers no authority on counsel; a brief must be delivered to authorize him to take any step in a particular proceeding. A special retainer can be given only after the commencement of the proceeding to which it relates; subject to any general retainer it is binding when tendered. Counsel specially retained is entitled to the delivery of the brief on every occasion on which counsel is briefed in a proceeding to which the special retainer applies. The brief to which counsel is entitled is a regular brief and not a complimentary one or one marked with an improper fee.

The term is also used of retainer to a solicitor, which equally may be general or special.

Retention. In Scots law, the right of retaining moveable property belonging to another of which the party is in possession until a debt due to the party claiming the right of retention is paid, corresponding to the right of lien in English law. Also the right of withholding a debt until due performance has been made of the counterpart obligation or compensation paid for non-performance of the counterpart, e.g. withholding the price of goods until they are supplied or damages agreed for non-supply. Also the right of withholding the retainer's own property which he should transfer to another, such as land sold but possession of which has not been given, until payment has been made.

Retorsion. In international law, retaliation for discourteous, unkind, unfair, or inequitable, but not necessarily illegal, conduct by one state by means of acts of the same or a similar kind. Whether and when retorsion is justified depends very much on circumstances, and all acts of retorsion should cease when the original unfriendly acts are terminated.

Retrial. A fresh trial of a case. In civil cases tried by jury, if the verdict is set aside on appeal as unjustifiable, a retrial is ordered. In criminal cases a retrial may be ordered where the first trial has in truth been only an apparent trial, having been vitiated by a defect so fundamental as to render it a nullity, and also where the Court of Appeal allows an appeal against conviction only by reason of fresh evidence received or available and it appears to the court that the interests of justice require it, it may order a retrial.

Retour. In old Scottish procedure proceedings on a brieve issued from Chancery, the extract or copy of the return made to Chancery by the court stating the finding of the jury on the matter ordered to be inquired into.

Retrait lignager* and *retrait féodal. At common law land in towns might be held by burgage tenure. Frequently by local custom such land was devisable, and if a tenant wished to alienate, his relatives often had the right to the first option (*retrait lignager*) and the lord a second option (*retrait féodal*).

Retribution. Return or recompense for, or requital of, harm done. It is sometimes deemed one of the purposes of punishment, as satisfying the instinct of retaliation or revenge which naturally arises in a victim, but also to a considerable extent in society generally. It may be deemed controlled and regularized vengeance, exacted by society. Many modern thinkers criticize retribution on the basis that one can never produce good by repaying evil with evil, but prevention, deterrence, and reformation as purposes of punishment are conspicuously unsuccessful and the fear of unpleasant consequences by way of retribution may have more powerful an effect on the mind of the wrongdoer than any other. Retribution underlies the ancient principle of talion, of an eye for an eye, but it has powerful support from philosophers such as Kant and theologians. Utilitarians, however, regard retribution as inadmissible, as multiplying pain and not producing happiness or benefit. A variant of retribution is expiation, the theory that evil must be cancelled or blotted out by the suffering of penalty. In modern practice, a distinction can be drawn between retaliatory retribution, where suffering is intentionally inflicted according to some scale on an offender capable of understanding that this is intentionally inflicted because of his offence; distributive retribution, the restriction of penal measures to persons who have committed offences; and quantitative retribution, the limitation of penal measures having aims other than retribution so that they do not exceed a degree of severity considered appropriate to the offence. There are great difficulties in assessing the retributive appropriateness of particular penalties by any objective standard.

Retrocession. In Scots law the reconveyance of a right by an assignee thereof to the party who originally assigned it to him.

Return. A report by an officer of a court indicating how he has performed a duty imposed on him. At common law returns to writs of execution were known by the Latin words indicating what the outcome of the execution was, e.g. *non est inventus, nulla bona.*

The term is also used of reports required periodically, usually annually, from persons, e.g. of income for tax purposes, from companies—e.g. of capital, directors, etc.

Returna brevium. A short tract in law French giving an account of the returns to writs. There are many sixteenth-century editions.

Revenue. The public revenue of the United Kingdom is vested in the Crown, being originally derived from the lands, prerogative rights and privileges of the Crown, supplemented by aids granted by Parliament, in effect by the House of Commons. The sovereign now customarily on accession surrenders to Parliament the revenue from Crown lands and prerogative rights in return for an annual grant (the Civil List) and Parliament annually grants aids to provide for the expenses of Her Majesty's Government.

The necessary money is raised by taxation of various kinds as authorized and controlled by Parliament.

Revenue is produced partly by impositions which are permanent in form but subject to annual renewal and annual fixing of rates by Parliament; these include income tax, corporation tax, and capital gains tax. It is also produced partly by permanent impositions, the rates of which are, however, varied from time to time; these include customs and excise duties on, e.g. tobacco, wines and spirits, petrol, stamp duties, estate duty, and numerous other lesser sources of income.

Most of the revenue is collected by the Commissioners of Inland Revenue (income tax, corporation tax, capital transfer tax) and by the Commissioners of Customs and Excise (customs and excise duties), and the remainder by the Post Office, the Crown Estates Commissioners, and some other authorities.

The revenue of local authorities is by way of grants from the central government and by imposing and collecting rates.

Revenue paper. See PAPER, REVENUE.

Reversal. The decision of an appellate court that the judgment of a lower court appealed against is erroneous and must be altered to the opposite effect, e.g. that the plaintiff is not entitled to the damages awarded him by the lower court. Reversal of a judgment is different from overruling; reversal says that the decision between the parties was wrong; overruling is to the effect that a precedent or prior decision of a lower court, usually between other parties, proceeded on an incorrect statement of law. Both may, however, occur in one case; if, e.g. a judge of the High Court proceeding on principle A gives judgment in favour of the plaintiff, the Court of Appeal may both overrule his statement of the law, and reverse his actual judgment.

Reversion. In English land law, an interest in land arising by operation of law whenever the owner of an estate grants to another a particular estate, e.g. a life interest, or term of years, but does not dispose of the whole of his interest. The reversion is the interest the owner has during the duration of the particular interest.

In Scottish land law it is a right of redemption, and may be either legal, provided by the operation of law itself, or conventional, provided in a security deed.

Revigny, Jacques de (Jacobus a Ravanis) (*c.* 1235–96). An early commentator in France, professor at Toulouse and Orleans, applied to law the dialectics of scholasticism, and was author of *Lectura super Codicis.*

Review. To reconsider on appeal.

See also JUDICIAL REVIEW.

Revised Reports. A reprint of such of the reports of cases in the English Common Law and Equity courts, from 1785 to 1865, as were deemed still of practical utility, edited by Sir Frederick Pollock, and published in 152 volumes in 1891–1920. (*The Law Reports* (q.v.) commenced in 1865 and follow on this series). It is not as complete as the *English Reports.* A table in Sweet & Maxwell's *Guide to Law Reports and Statutes* shows in which volume of the *Revised Reports* older reports, e.g. Adolphus and Ellis's reports, can be found.

Revising barristers. Barristers, who from 1843, held courts throughout the country each autumn to revise the list of persons entitled to vote in Parliamentary and local government elections. From 1888 they also performed the functions conferred on revising assessors in 1882 of revising the parish burgess lists. Appointments were annual, but in practice the same barristers were appointed regularly. In 1918 their functions were transferred to town clerks and county clerks.

Revocation. The recalling of an act formerly done. It may be done by an act of the party, as by revoking an offer, or an agent's authority, or a will, or by operation of law, as where a will is revoked by law on subsequent marriage, or an offer by the death of the offeror, or by the court, as where it revokes a grant of probate wrongfully obtained.

Reynell, Sir Richard (?1625–99). Called to the English Bar in 1653 and the Irish Bar in 1658, he became a judge of the Irish King's Bench in 1674 but was superseded in 1686; he returned to Ireland as Chief Justice of the King's Bench in 1691 and was superseded in 1695.

Reynolds, Sir James (1684–1747). Probably a nephew of Chief Baron Sir James Reynolds (*infra*), was called to the English Bar in 1712 and went to Ireland as Chief Justice of the Common Pleas in 1727, but was transferred to the English Court of Exchequer in 1740.

Reynolds, Sir James (1686–1739). Called to the Bar in 1712, he became a judge of the King's Bench in 1725 and was Chief Baron of the Exchequer, 1730–38.

Rex peccare non potest (the king cannot do wrong). A principle of law in feudal times, giving rise to the rule, which obtained down to the Crown Proceedings Act, 1947, that the Crown was not liable in tort, though the actual wrongdoer, though a Crown servant, was, and the liability he incurred was normally satisfied by the Crown from public funds.

Rhodes, Sea-law of. A code of sea-laws thought to have been compiled on the island of Rhodes between A.D. 600 and 800. It includes the *lex Rhodia de iactu*, to the effect that where goods are jettisoned to save the ship and the rest of the cargo all the owners of the ship and cargo must share the loss, the basis of the modern law of general average (q.v.).

W. Ashburner, *The Rhodian Sea Law* (1909).

Rhuddlan, Statute of (1284) (also known as the Statute of Wales). An enactment of Edward I to settle the administration of Wales following his conquest of Wales in 1282–83.

Riccobono, Salvatore (1864–1958). Italian jurist and civil-law scholar, one of the editors of *Fontes iuris Romani antiquissimi* (1908), and of writings on the influence of Christianity on Roman law.

Rice, Sir Stephen (seventeenth century). Joined the English and Irish Bars, became a Baron of Exchequer in Ireland in 1686, and Chief Baron in the next year. He followed James II to France but later returned to Ireland.

Richard de Mores (Ricardus Anglicus) (*c.* 1160–1242). The first English canonist who taught at Bologna; he may also have been connected with Vacarius' civil law school at Oxford. His writings include *Summa brevis* (an introduction to the Decretum), *Distinctiones decretorum, Summa de ordine iudiciario*, on procedure, *Casus decretalium, Apparatus decretalium*, which became a source for the glosses on *Compilatio prima* and the gloss on the Decretals, and *Generalia* or *Brocarda*.

Richard of Canterbury (?–1184). Successor of Becket at Canterbury; he was an efficient administrator and, in 1175, presided over one of the earliest English provincial councils and promulgated some important canons. He served as legate, papal judge delegate, and legislator and helped to develop the body of decretal law.

Richards, Sir Richard (1752–1823). He was called to the Bar in 1780 and practised in Chancery. He sat in Parliament 1796–99. He became a baron of Exchequer in 1814 and Chief Baron in 1817, and was respected as a sound and capable but not brilliant lawyer.

Richter, Aemilius Ludwig (1808–64). German canon law scholar, author of the important *Beitrage zur Kenntnis der Quellen des Kanonischen Rechts* (1837), and editor of the *Corpus Iuris Canonici* (1833). He also wrote on Protestant doctrine and church law.

Rider. An extra clause added to a Bill passed through Parliament. Also an additional statement made by a jury along with their verdict, such as, formerly, in criminal cases, a recommendation to mercy.

Rigby, Richard (1722–88). Sat in the English Parliament and then in the Irish one and was appointed M.R. in Ireland in 1759. He was also, at different times, Vice-Treasurer and Paymaster of the Forces.

Right. A much ill-used and over-used word. In a legal context it is a legal concept denoting an advantage or benefit conferred on a person by rules of a particular legal system. In Greek philosophy and Roman law, a right seems to have been identified with what was right and just. Later, a right was sometimes deduced from the fundamental datum of free will or sometimes seen as essentially based on legal relationships between persons, determined by a rule of law and sanctioned and protected by the legal order, or sometimes as an interest recognized and protected by a rule of legal justice.

A distinction has frequently been drawn between natural or moral rights and legal rights. The former are claims which, it is asserted, should by natural justice or principles of morality, be recognized and protected, whereas the latter may or may not have any moral basis but are in fact recognized and protected by the particular legal system in question. Only the latter have substantial legal significance.

Similarly, one meets such phrases as 'social rights', 'political rights', 'economic rights' and the like. See also Civil Rights; Natural Rights; Rights of Man. In so far as these are recognized by the rules of a particular legal system, they are legal rights; in so far as they are not, they may be moral claims or mere wishes or aspirations or assertions; thus, the 'right to a living wage' may be a legal right if there is relevant legislation defining and enforcing such an entitlement; otherwise it is a mere assertion or aspiration. In most political discourse, 'we have a right' really only means 'we want'. Precisely what rights any particular kind of person or particular individual has in law depend on the particular legal system in question at the time in question. It may be doubtful or disputed whether an alleged right, e.g. a right of privacy, is recognized or not. Rights may be created or taken away by legislation, or may be recognized or declared not to exist by courts.

The basis for the existence of a legal right is found in various ideas. One basis is morality: a legal right exists if it is morally right that the kind of claim ratified by the legal right should be recognized; but not every claim morally approved is recognized as a legal right, nor conversely. Another basis is interest (q.v.), but not every interest is protected by a legal right. Another basis is in duties; rights are the ability to enforce the correlative duties. But there may be lack of correspondence between rights and duties.

Generally, the counterpart and correlative of a legal right is a legal duty, in that if one person has a legal right of a particular kind some other person or persons must be subject to a legal duty. Thus, the right of a buyer to delivery of goods is correlative to the duty of the seller to deliver, and conversely with the right and duty of payment for the goods. But the presence of a duty is not always essential for the existence of a right, and a duty may exist, as under the criminal law, without conferring on any person any right to demand compliance with the duty or any right to invoke prosecution or impose punishment.

It has long been recognized that the term 'right' comprised different kinds of advantages and that the counterpart 'duty' similarly was distinguishable into different kinds of disadvantages. Salmond, founding on the work of German jurists, distinguished legal advantages into rights, liberties and powers and legal burdens into duties, disabilities, and liabilities. Hohfeld worked out the following pattern:

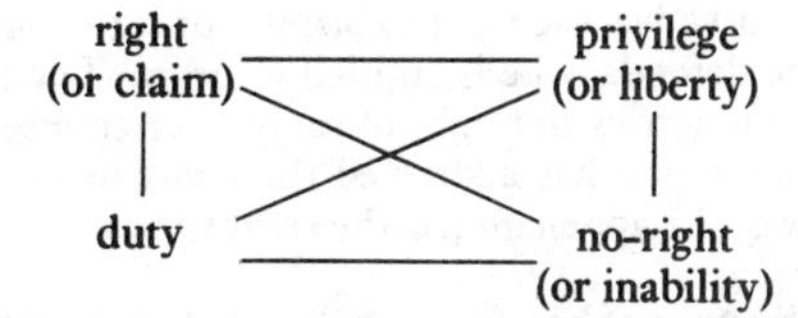

in which each concept was said to be the jural correlative of the one vertically above or below, in that the presence of one in one party to a relationship implied the presence of the other in the other party, and in which each concept in each square was the jural opposite of that diagonally opposite it, in that the presence of one in one party implied the absence of the opposite in himself. Later study revealed that the upper and lower pairs of concepts in each square, the concepts linked horizontally, could be regarded as jural contradictories in that the presence of either in one person implied the absence of the contradictory in the other party. Later scholars have also, in some cases, suggested variants on Hohfeld's terms, noted above in brackets.

In the right-duty relation, the position of parties *vis-à-vis* each other is essentially: I claim—you must: in privilege—no-right it is: I may—you cannot: in power—liability it is: I can—you must accept it; in immunity—disability it is: I can with impunity—you cannot. Examples of the four different kinds of rights would be a right (claim) to payment of a debt and a duty to pay it; a privilege or liberty to carry an umbrella and no-right in others to prevent one doing so; a power to give orders and the liability to have to obey them; and the ability to make a defamatory statement in circumstances of absolute privilege free from liability to be sued and the disability of the person defamed to sue. These eight conceptions Hohfeld called the lowest common denominators of the law.

This analysis has done much to clarify thinking and to encourage more accurate use of ideas, if not of words; it emphasizes the need to consider what kinds of advantages or disadvantages are involved in a relationship, and not to assume that all are of the same pattern. But many of the eight terms are themselves words of variable and uncertain meanings.

Every legal right attaches to one or more legal persons, sometimes called the owner of the right, the subject of it, or the party entitled; it attaches by virtue of a title, by virtue of certain facts or events by reason of which under the particular legal system the owner comes to have the right; it avails against one or more other legal persons, sometimes called the person bound, or the subject of the correlative duty; it relates to some obligation or property, which is the object of the right, and obliges or binds the person subject to the duty to some act or abstention in favour of the person entitled: this is the content or substance of the right. Thus, if A

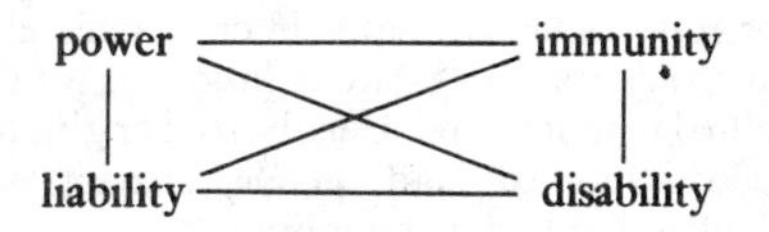

contracts with B, A acquires a right to performance, by virtue of the contract, against B, relative to the thing contracted for, obliging B to perform or deliver as may be appropriate. If X is so placed relative to Y that Y's carelessness may injure X, X acquires a right, by force of law in these circumstances, against Y, relative to their relationship, obliging Y to take care that his conduct does not injure X. If in breach of duty he does injure X, X acquires the secondary or consequential right to recover damages from Y in compensation for the harm done.

Such rights as to matrimonial fidelity, performance of a contract, abstention from injury, ownership or possession of property and the like may be called primary or substantive or principal rights. But relationships giving rise to such rights also give rise, by legal implication, to secondary or remedial or accessory rights, to demand an appropriate legal remedy if the primary right is not duly implemented. Thus, in the event of breach, rights arise to dissolution of marriage, damages for breach of contract or for tort, action to vindicate ownership or possession, and so on.

Rights may be positive or negative, such as to have performance of a duty, or not to have one's person harmed or one's reputation besmirched.

Rights may also be distinguished into rights *in personam* or personal rights, availing only against a definite legal person, as in the case of a right under a contract, and rights *in rem* or real rights, availing against persons generally or the world at large, as in the case of a right not to be personally injured or a right to the ownership of property. This distinction is based, not accurately, on the Roman distinction between *actio in rem*, to recover a specific *res*, and *actio in personam*, a claim against a particular person for enforcement of an obligation. Rights *in personam* are generally positive and rights *in rem* generally negative. A remedial right is always *in personam.* A right *in rem* need not relate to any tangible *res*; thus a right to have one's reputation not blackened is a right *in rem.* Moreover a right *in rem* is truly not a single right but a bundle of as many rights as there are persons subject to the duty not to infringe that right. While the distinction is useful it does not suit every legal system.

Another distinction is into personal and proprietary rights. Personal rights are those which attach to a person, but are not part of his assets or estate or property; they include rights to freedom, rights against spouse and children, rights to honour and reputation, to honours, dignities and official position and the like. In general, they cannot be transferred.

Proprietary rights are rights in or to some element of a person's assets or estate or property, which may be valued in money and, usually, sold or gifted; they include rights to land, goods, shares, patents, goodwill, debts and many more.

Yet another distinction is that between rights over one's own (*iura in re propria*) and over another's property (*iura in re aliena*). The former comprise all the individual rights which a man may exercise over something which is his own property, as for example to use, abuse, consume, destroy, sell or gift it; the latter comprise the individual rights which one person may exercise over something which is the property of another; thus, a mortgagee or party entitled to an easement or tenant may exercise certain rights in or over the property of the mortgagor, owner or landlord as the case may be.

English law and systems founded thereon also recognize a distinction between legal and equitable rights based on the different circumstances in which courts of law and of equity respectively would recognize claims asserted.

J. Salmond, *Jurisprudence*; W. Hohfeld, *Fundamental Legal Conceptions*; A. Kocourek, *Jural Relations*; J. Stone, *Lawyer's Systems and Legal Reasoning*.

Right of action. A right to bring an action. Whether a person has a right of action depends on whether he has an enforceable claim against another for some relief or remedy.

Right of entry. The right to take or resume possession of land by entering on it peaceably. A person has an original right of entry where he is entitled to possession of the land, as where a previous estate has determined. There is also a right of entry attaching to a reversion whereby an existing estate is defeated, as where a lease reserves to the lessor the right to determine the lease by re-entry if the lessee fails to pay the rent.

Right of way. A right of passage over the lands of another. It may be either a public right of way or a species of easement or servitude. A public right of way may be acquired by statute, grant, or long usage; the way must run from one public place to another and the use must be such as may be ascribed to public right and not to contract, permission or some other right.

An easement or servitude right of way may be created by statute or by grant, express or implied, or by express reservation in a grant of land, or by long and continuous employment and uninterrupted use for 40 years, and may be destroyed by statute, release, and certain other ways.

Right to begin. The right of first addressing the court or jury at a hearing or trial. In general, the party on whom lies the onus of proving the affirmative has the right to begin, but if the onus is on the defendant he is entitled to begin. The right to begin carries the right of reply after counsel for the other side has addressed the court or jury. In appeals, the appellant has the right to begin.

Right to strike. The so-called right to strike, i.e. to withdraw one's labour, usually collectively, is a liberty, exercise of which is legal and permissible so long as it is not used contrary to law. But unless notice of intent to terminate their contracts of employment has been given it amounts to a breach of contract by each employee.

Under the European Social Charter, states which are members of the Council of Europe recognize the right of workers to collective action including the right to strike. This seems to promote striking from a freedom or liberty to a fundamental human right, the exercise of which is not a breach of contract whatever the contract may say and which would never justify sacking. This is probably not British law.

By statute, a trade union is protected from tortious liability for a strike in contemplation or furtherance of a trade dispute. Hence action in support of a political strike is not protected, but the boundary between 'political' and 'economic' may be vague.

Right to work. A phrase used in at least two senses frequently confused. It may mean the right, i.e. liberty, to work, free from being prevented by restrictive rules, such as a trade union closed shop. This is a legal liberty frequently denied by trade unions. Or it may mean the right, i.e. claim, to work, or to have employment furnished or made available for the claimant. The latter, though frequently asserted, is not a legal right because the availability of work always depends on economic factors and cannot be secured by law. The former is, however, an important liberty of the individual. It may legitimately be excluded or restricted by provisions seeking to ensure that only adequately trained and skilled persons are permitted to work in certain capacities, e.g. qualified physicians. But it is an illegitimate infringement of individual liberty to exclude a person from work on account of sex, race, colour, or other irrelevant factor or, most commonly, because he is not a member of a, or of the specified, trade union, as where a closed shop (q.v.) is in operation.

Right, writ of. The writ initiating the principal action in early mediaeval English law to determine the title to land.

In the time of Henry II, it came to be accepted that no action to recover land could be commenced save by a Writ of Right, which, addressed to the lord of the manor to do justice between the claimant to the land and the tenant, brought the action within the control of the royal courts by the threat that, in default of justice from the lord, the King would assert the right to do justice. Proceedings could be brought before royal judges by the writs of Tolt and Pone. Later, the procedure was replaced by a writ of Right Close based on the pretence that the lord had waived his right to hold the court. The issue was determined by the Grand Assize, under which a writ was issued to the sheriff of the county to summon four knights to select 12 more, which body should say whether the demandant had truthfully alleged that he had been wrongfully deforced of the land which he claimed. The decision on a writ of right determined the claim of the parties to the land in dispute.

Rights and remedies. One of the familiar antitheses of legal discourse but, in fact, it is not clear that there is any antithesis and, if so, what it is. For a right or substantive right, such as to have a contract implemented or to be free from physical harm, and a remedy, or claim, if a contract be broken or a tort done, to recover damages in compensation, are both kinds of legally recognized rights or claims. The only valid distinction is between antecedent or primary rights, such as to performance of a contract, to be kept free from tortious harm, and many more, and consequential, secondary or remedial rights, such as to recover damages, to exact payment, to have a marriage dissolved and the like, which differ from primary rights only in that, though always existing in potentiality, they become actually claimable only if and when the antecedent right exists in the relationship in question and has not been implemented. Thus, if A, in the circumstances, owes a duty of care to B not to injure him, B has a legal right not to be injured and a potential remedial right, if he should be injured, to damages. If, in fact, he is injured by A's breach of duty he has an actual remedial right to recover damages. If A does not owe a duty of care to B but injures B, B has no right to have been kept harmless, and no consequential right to, or remedy by way of, damages. Rights for the infringement of which there is no remedy are readily conceivable and sometimes arise, but are valueless. In ancient and mediaeval law the principle was frequently *ubi remedium, ibi jus*, as under the old English system of forms of action, where no right of action lay in various circumstances, so that it was questionable whether there was any primary right, and it certainly was unenforceable. The modern principle is generally *ubi jus, ibi remedium*; if there is a right the law will find a remedy for its infringement.

Rights *in personam* and rights *in rem*. In Roman law and, derived therefrom, in modern law, the distinction between rights which can be asserted against a determinate other person only, such as a claim of debt, and those which can be asserted in respect of a thing against any person at all, or persons generally, such as to ownership of a thing. In Roman law, a contract created rights *in personam* but to create rights *in rem*, a conveyance was necessary in addition.

Rights of man. A phrase associated in the first place with the book of that name by Thomas Paine (q.v.), and with the Declaration of the Rights of Man and of the Citizen adopted by the French National Assembly in 1789 and which served as a preface to the French Constitution of 1791. It asserted that all men were born free and equal in rights, specified as being rights of liberty, private property, the inviolability of the person and resistance to oppression, that all citizens were equal before the law and were to have the right to participate in legislation directly or indirectly, that no one was to be arrested without judicial order; freedom of religion and freedom of speech were safeguarded within the limits of law and public order. The Declaration was fundamentally an attack on the pre-Revolutionary monarchy, but included many of the liberal political ideas of the eighteenth century, such as natural rights, sovereignty of the nation and the separation of powers, and reflected bourgeois ideas. It came to be regarded as a fundamental statement of liberal beliefs. Other declarations to substantially the same effect were the Constitution of Virginia of 1776 and the American Declaration of Independence.

In a broader and more general sense, the phrase is associated with and largely synonymous with the concepts of natural rights, fundamental rights and human rights (q.v.).

Riksdag. The Swedish Diet. It originated in the fourteenth century but remained unimportant until the eighteenth century. In 1617, an ordinance fixed the number of estates at four and provided that in the event of disagreement the King would take the final decision. From 1680 to 1718, its power was in decline but subsequently the council became responsible to the Riksdag.

Riot. A tumultuous disturbance of the peace by three or more persons assembled together without lawful authority, with intent to assist each other, if necessary by force, against anyone who opposes them in the execution of a common unlawful purpose, and who execute or begin to execute that purpose in a violent manner so as to alarm at least one person of reasonable firmness and courage. The purpose is commonly to destroy property. A gathering may constitute a riot though not using force because not interfered with. A peaceful assembly may become a riot if it resolves to go in a body to do any act of violence and commences to do so.

At common law, riot was a misdemeanour but under the Riot Act 1714 (now repealed) if 12 or more persons assembled to the disturbance of the public peace and did not disperse within an hour of the reading of a proclamation in statutory form calling on them to disperse, they were guilty of felony.

At common law, any citizen may suppress a riot and use weapons to do so.

By statute, where a house, shop or building, or property therein has been injured, stolen, or destroyed by rioters, the owner is entitled to compensation from public funds.

Riot Act. A statute passed in 1714 to clarify the law of riots and maintain law and order during the Jacobite rebellion. Previously, a riot which damaged property might be construed as levying war against the King and treasonable. The Act was intended to clarify the law. It enacted that if any persons to the number of 12 or more were unlawfully, riotously, and tumultuously assembled together to the disturbance of the public peace and did not disperse within one hour after being commanded to do so by a justice of the peace or similar person, by proclamation in the King's name, in the form set out in the Act, they should be guilty of felony. If they did not disperse the justices might apprehend them, and were free from liability if any of the rioters were killed or injured. The procedure gave rise to the phrase 'reading the Riot Act'. The Act was repealed in 1967.

Ripuarian Franks, Law of the. The *Lex Ripuaria* or *Pactus legis Ripuariae* is a Frankish Code stating the laws of the Ripuarian Franks. Part is copied from the Salic Code, but the other, and probably earlier, part is an original compilation. The date may be about 600.

Risk. A danger, or hazard, or possibility of loss. In sale an important question is when the risk passes from seller to buyer. Under the Sale of Goods Act, 1893, this is when the contract is concluded, notwithstanding that delivery is not then made but is delayed. In insurance, the risk covered is the event, e.g. fire, accident, the occurrence of which makes the insurer liable under his contract to compensate the insured.

Rittershausen, Konrad (Rittershusius) (1560–?). A German humanist and author of *Disputationes ad Institutiones* (1580), *Jus Justinianum* (1615), *De Differentiis juris civilis et canonici* (1616), and *Commentarius Novus in quattuor libros Institutionum* (1618).

Rivers in municipal law. Rivers are quantities of water flowing in natural and, at least, reasonably defined channels towards the sea. Distinctions must be taken for legal purposes between rivers which are both non-tidal and non-navigable, rivers which are navigable though non-tidal, and rivers which are navigable and also tidal. The same river may fall into all three classes at different places on its course. In the case of a non-tidal and non-navigable river, the bed of the river and the banks belong to the riparian proprietor. In the case of a navigable river members of the public have liberty of navigating it and of using the banks so far as incidental to navigation, e.g. for mooring, or landing goods, but the bed of the river and the banks belong to the riparian proprietor. In the case of a tidal river, the bed of the river belongs to the Crown.

Rivers in international law. Rivers may be national, if entirely within the territory of one state, or boundary, if marking a territorial boundary, or multi-national, if owned by more than one state. In all these cases, riparian states can regulate navigation within their own territories and can exclude vessels of non-riparian states. There are also international rivers which, though passing through several states, are navigable for some distance from the sea and, in which cases customary international law has, since the nineteenth century, recognized the freedom of vessels of all nations to navigate. The Congress of Vienna recognized the principle of free navigation on the international rivers of Europe, namely the Scheldt, Meuse, and Rhine and the latter's navigable tributaries. The Peace of Paris of 1856 stipulated for free navigation on the Danube. The Peace Treaties after the First World War declared various other rivers international. A Barcelona Convention of 1921 produced a Convention and Statute on the Regime of Navigable Waterways of International Concern.

The case of the flow of water of a national river is a matter for the local state, but the flow of other rivers is subject to the principle that no state may alter natural conditions to the detriment of another state, as by diverting the flow.

Rivier, Alphonse Pierre Octave (1835–98). French-Swiss jurist and author of *Lehrbuch des Völkerrechts* (1889), *Principes du droit des gens* (1896), and *Introduction historique au droit romain* (1881). He was president of the Institute of International Law, 1888–91, and editor of the *Revue de Droit International.*

Road. A determinate track from one place to another. At common law the subsoil continued vested in the proprietors of the adjacent lands, members of the public merely having a liberty of passing along the surface either by virtue of public right or by virtue of right of way. Increasingly under statute highway authorities are empowered to acquire the land along the line of which a road runs as well as to make and maintain the surface of the road and ancillary works.

Road traffic. The great increase in the use of vehicles on the roads in the twentieth century has given rise to a great volume of complicated law, which deals with many different topics. The main topics are the testing and licensing of persons to drive vehicles of different classes, the creation of offences of reckless, dangerous, and careless driving, speeding, driving while subject to drink or drugs, and civil liabilities for harm done by and while driving, the imposition of requirements of compulsory insurance in certain cases, the control of construction, licensing, registration, testing, and use of different classes of motor vehicles, the control of their speeds, parking and stopping, and use generally, the control of licensing of the use of public service vehicles and goods vehicles, and the regulation of the use of roads of various kinds.

G. Wilkinson, *Road Traffic Law; Encyclopaedia of the Law of Road Traffic.*

Robbery. In English law, an offence at common law and under statute; since 1968 it consists of stealing and, immediately before or at the time of doing so and in order to do so, using force on any person or putting or seeking to put any person in fear of being then and there subjected to force.

In Scots law, it is, at common law, theft accomplished by means of personal violence or intimidation, and the requirements for theft apply to it also.

Robertson, James Patrick Bannerman, 1st Lord Robertson of Forteviot (1845–1909). Passed advocate of the Scottish Bar in 1867, at first made slow progress, but gradually acquired a large practice. In 1885, he became a Q.C. and was Solicitor-General, 1885–86 and 1886–89. In 1889, he became Lord Advocate and, in 1891, Lord President of the Court of Session, in which office he was relatively undistinguished. In 1899, he succeeded Lord Watson in the House of Lords and in that court, and in the Privy Council, showed his abilities to better advantage. He was chairman of the Irish University Commission (1904) and always retained a keen interest in politics.

Robes, judicial. In the U.K., a wide variety of judicial robes have evolved over the years and are worn on different occasions. The Lord Chancellor wears a heavy black damask gold-embroidered robe and a full-bottomed wig. On ceremonial occasions, the Lord Chief Justice wears a robe of scarlet and ermine with a train and the Master of the Rolls, the Lords Justices of Appeal, and the President of the Family Division wear robes similar to that of the Lord Chancellor, all with full-bottomed wigs.

On ordinary occasions, for hearings, the Lords of Appeal in Ordinary do not wear wigs or robes of any kind, and the Master of the Rolls, the Lord Justices, and the judges of the Chancery and Family Division wear black waistcoat and morning coat and the same black gown as a Q.C. wears on ceremonial occasions, with short wigs. The Lord Chief Justice and judges of the Queen's Bench wear, in winter, a black and ermine robe for civil business and a scarlet and ermine robe for criminal business, and, in summer, a black and silk robe for civil business and a scarlet and silk robe for criminal business, all with short wigs. Scarlet robes are worn on Red Letter days.

A circuit judge, sitting in the county court (civil business), wears a blue robe with a purple sash, and when sitting in the Crown Court the same robe with a red sash. The Recorder of London wears a special red robe rather like that of a High Court judge; the Common Serjeant has a special black ceremonial robe. Other judges of the Central Criminal Court and Crown Courts wear the Q.C.'s morning coat and black gown. A Recorder wears a wig and barrister's or solicitor's gown.

In Scotland, the judges of the Court of Session wear dark blue worsted robes with maroon facings and scarlet crosses on the facings. When sitting as Lords Commissioners of Justiciary, they wear scarlet robes faced with white and decorated with scarlet crosses. The Lord Justice-General's robe is faced with ermine and the Lord Justice-Clerk's robe has rows of small square holes punched in the white satin facings to give the appearance of ermine. They always wear their justiciary (criminal) robes and full-bottomed wigs for State and ceremonial occasions. Sheriffs-principal and sheriffs wear the Q.C.'s morning coat and black gown.

In the U.S., black gowns are generally worn, but not wigs.

Robinson, Sir Christopher (1766–1833). Became D.C.L. and advocate of Doctors' Commons in 1796. In 1798 he initiated the first regular series of reports in the Admiralty court. In 1809 he became King's Advocate, in 1821 judge of the Consistory Court and Chancellor of the Diocese of London, and, in 1828, judge of the court of Admiralty and a Privy Councillor. He was an able and industrious judge, but made little mark on the law. He published several volumes of reports of admiralty cases. His son, William, also published reports of admiralty cases.

Robotpatent. A law governing *robot* or the labour due performed by peasants in the Austrian Empire for their lords. Various laws regulating *robot* were known from at least the sixteenth century, and the system varied very much from place to place in extent and burdensome nature. In 1771–78, Maria Theresa issued a series of patents regulating and restricting peasant labour in the German and Bohemian parts of the Empire, though not in the Hungarian lands where the nobility resisted any restriction. The object of restriction was not only to limit excessive exaction of labour but to ensure that peasants could contribute to general national expenditure. In 1774, it was decreed that an adjustment of the *robots* were to be settled between lords and serfs and, in 1775, this was made compulsory and a graduated scale of labour services related to taxes paid fixed, with maximum and minimum demands. Ultimately, most of the patents merely confirmed prior laws, though abolishing illegalities and encroachments. *Robot* was abolished in 1789 but later restored and not finally abandoned until 1848.

Robson, William Snowdon, Lord, of Jesmond (1852–1918). Though not called to the Bar till 1880, he was a success in Parliament and became Solicitor-General in 1905 and Attorney-General in 1908. He became a Lord of Appeal in 1910 but had to retire in 1912. Though a good advocate, he was never an outstanding lawyer and as judge made practically no contribution to the law.

G. Keeton: *A Liberal Attorney-General.*

Roche, Alexander Adair, Baron (1871–1956). Developed a large commercial practice and became a judge of the King's Bench Division in 1917, a Lord Justice of Appeal, 1934–35, and a Lord of Appeal in Ordinary, 1935–38.

Rochester's cook, Bishop of. A statute of 1530, having narrated that one Richard Roose or Coke had put poison in yeast in the kitchen of the Bishop of Rochester, whereby several persons were made ill and two died, enacted that the poisoning should be high treason and that 'the said Richard Roose shall be therefore boiled to death without benefit of clergy' and that all future murders by poisoning should be punishable by death by boiling. Roose was boiled to death, and there are said to have been two more such executions. In 1547, poisoning was declared to be merely murder but the 1530 Act was not repealed until 1863.

Rochfort, Robert (?1652–1727). Was recorder of Londonderry, sat in the Irish Parliament, and was later Speaker and Attorney-General. He became Chief Baron of Exchequer in Ireland in 1707, but was superseded in 1714.

Rodenburg, Christian (1608–68). Dutch jurist and judge at Utrecht, author of *Tractatus de iure coniugum* (1653), an important book in Roman-Dutch law on community of property.

Röder, Karl David August (1806–79). German jurist and author of *Grundzüge des Naturrechts oder der Rechtsphilosophie* (1860) and other works.

Roffredo da Benevento (Roffredus Beneventanus) (?–1242). Author of *Tractatus Libellorum seu de Ordine Judiciorum* (1500), *Quaestiones Sabbatinae*, and other works.

Rogerius (?–1192). A famous Italian glossator, author of a *Summa* on the Codex, which was the basis of Azo's *Summa*, a *Summa in Tres Libros, Quaestiones super Institutis, Enodationes quaestionum super Codice*, a treatise on prescription, and a collection of controversies of the glossators, *De dissensionibus dominorum.*

Rogerson, John (?–1741). Was called to the English Bar in 1698 and to the Irish Bar in 1701. He was Solicitor-General, 1714, Recorder of Dublin, 1715, and Attorney-General, 1720. He became Chief Justice of the King's Bench in Ireland in 1727 and held that office till his death.

Roldan, El maestro (thirteenth century). A Spanish jurist of high reputation, known to have edited the *Ordenamiento de las Tafurerias*, the ordinance regulative of gambling-dens, and probably participated in the redaction of the *Siete Partidas.*

Rolfe, Robert Monsey, 1st Lord Cranworth (1790–1868). Was called to the Bar in 1816 and developed a chancery practice. He became an M.P. in 1831 but was not successful in Parliament. He took silk in 1832, was appointed Solicitor-General in 1834 and, in 1839, became a Baron of Exchequer, in which capacity he was a considerable success. In 1850, he was made a Vice-Chancellor and given a peerage and, in 1851, became one of the two first Lords Justices in Chancery. In 1852, he became Lord Chancellor, holding office till 1858, and again from 1865 to 1866. He did much, as a judge, for the development of equitable principles and was a good Chancellor, but was less successful in the House of Lords, being a poor debater, though he had considerable success as a legislator, in carrying useful reforming measures.

Rolle, Henry (?1589–1656). Was called to the Bar in 1618 and practised successfully in the King's Bench, becoming a serjeant in 1640. He sat in Parliament from 1614 to 1629 and supported the Parliamentarians against the court party. In 1645, he became a judge and, in 1648, Chief Justice of the King's Bench; his reputation was high and, at the time, second only to Sir Matthew Hale. He devoted much time to compiling reports and is remembered chiefly for his *Abridgement des Plusieurs Cases et Resolutions del Commun Ley* (2 vols., 1668), edited by Hale (q.v.). In law-French, it includes matter from the Year-Books and subsequent reports, but also includes cases not elsewhere discoverable, and contains summaries of Parliamentary records and statutes, and makes a great advance on previous works in making some arrangement of the cases within each alphabetical title under sub-titles and accordingly influenced later Abridgements.

Rolls. Pieces of parchment stitched together to form a long continuous piece which was rolled up when not in use. Formerly, all the main kinds of legal records were written on rolls. At common law, the steps of every action were entered at various stages on rolls, known as the plea roll, the issue roll, and the judgment roll respectively. The records of the proceedings of Parliament, including decisions of points referred to Parliament by the judges, and the petitions to Parliament and the statutes, were entered on the Parliament Rolls. There were numerous other kinds of rolls maintained, notably patent rolls, close rolls, charter rolls, exchequer rolls, and others. These all form part of the public records kept nominally in the custody of the Lord Chancellor and, in fact, in the Public Record Office in London.

The court rolls of manors comprised records of the business of the court, including names, rents, services, admissions and surrenders, and other matters. Copyhold lands were held by copy of the court roll relating to the admission of the tenant to the lands.

Rolls Court. As the jurisdiction of the Chancery developed, many judicial duties devolved on the Master of the Rolls as chief of the Chancery officials, and from the fifteenth century, he was sometimes given a commission to hear cases generally. From 1833, when it was provided that the Master of the Rolls should sit continuously for judicial business, there was a Rolls Court, from which appeal lay to the Chancellor and after 1851 to the Court of Appeal in Chancery. There are series of reports, notably Beavan's, of cases in the Rolls Court.

Rolls, Master of the. See MASTER OF THE ROLLS.

Rolls of the Court of Session. A number of rolls, published daily during session, under authority of the Court, listing under the names of the different rolls, in order, the cases which are awaiting hearing. The rolls also list the cases which are to be called before the different courts or judges on the following days.

Rolls of Parliament. Parliamentary records, covering 1278 to 1503, consisting mainly of petitions, mostly public, but some private, and the answers thereto, and giving the proceedings at the opening of Parliaments and some reports of the actual proceedings of Parliament. They were published as *Rotuli Parliamentorum* in six volumes in 1767–77, with an index in 1832.

Rolls Series. A series of volumes, entitled *Chronicles and Memorials of Great Britain and Ireland during the Middle Ages*, which began in 1858 under the general direction of Sir John Romilly, M.R. and eventually ran to 253 volumes. It was planned to include editions of all the more significant works of the mediaeval chroniclers of Great Britain and Ireland and is indispensable for the study of mediaeval Britain. The term is also given to a series of 20 volumes, issued between 1863 and 1911, printing edited editions of certain of the Year books (q.v.), comprising a text based on a careful collation of manuscripts and a translation on the facing page. The editors were successively A. J. Horwood and L. O. Pike. The editorial introductions to many volumes are themselves of great interest and value.

Rolt, Sir John (1804–71). Worked up from serving in a shop to be called to the Bar in 1837. He practised in Chancery and took silk in 1846. He became an M.P. in 1857 and Attorney-General in 1866 and, in 1867, a Lord Justice of Appeal and Privy Councillor but had to resign for health reasons in 1868. He was a good advocate and equity lawyer but deficient in other branches of law and not a profound lawyer.

Romagnosi, Gian Domenico (1761–1835). Italian jurist and scholar, author of works on criminal and public law, natural law and other subjects.

Roman Catholic. Prior to the Reformation, the Christian Church enjoyed legal privileges in both England and Scotland and did not tolerate any divisions.

After the Reformation, in both England and Scotland, the adherent to the Catholic Church was in the minority, labelled a papist, popish recusant, and similar names, and not only not tolerated but, save for short periods, hunted and persecuted. There were numerous penal statutes in aggregate totally proscribing Catholicism. The prohibition on the saying or hearing of Mass lasted till 1791 and not only were adherents to Catholicism forbidden to practice their faith but they were compelled to attend the parish church under pain of recusancy. In 1791, the Relief Act allowed Catholics to stay away from Protestant churches and say and hear mass and to exercise certain other liberties, but only on taking an oath of allegiance.

In Scotland, a long line of statutes from 1560 prescribed penalties, including death, for such offences as hearing or saying mass, harbouring or assisting priests and possessing Catholic books. In addition, Catholics suffered serious civil disabilities, being disabled from holding office, holding land, or inheriting estates, being schoolmasters, teaching any art or science, and in many other ways. Only in 1793 did Scots Catholics secure relief, the penalties of former Acts being remitted if they accepted the terms of a prescribed oath, and while the Act secured to them peaceful possession and free disposition of their property, they continued to be excluded from almost every public office including those of professor and teacher.

The Catholic Emancipation Act, 1829, gave Catholics the rights to vote for and sit in Parliament, to hold all civil and military offices under the Crown and to exercise any other franchise or civil right, with a few exceptions, on taking an oath prescribed by the Act, but many restrictions on religious liberty remained. Further relief was granted in 1832 and 1844, which repealed many penal Acts. The penalties for recusancy were repealed in 1846, and the oath was repealed in 1871. A further relief Act in 1926 cleared away some remaining penal legislation. Remaining disabilities are from inheriting the Crown or marrying the King or Queen Regnant, holding the office of Lord High Commissioner to the Church of Scotland, and from all offices in the Church of England and Church of Scotland.

The prohibition on the assumption of territorial ecclesiastical titles by Catholic bishops was imposed in 1851 but was repealed in 1871.

Judicial decisions moved in the same direction by recognizing, in 1919, that a bequest for masses for the souls of the dead was valid. Modern legislation and decisions do not discriminate and recognize the Catholic religion equally with others.

Roman law, or civil law. The system of law developed at Rome and latterly in the Roman Empire between the foundation of the city (traditionally 753 B.C.) and the fall of the Eastern Roman Empire at the fall of Constantinople in A.D. 1453. Its history accordingly spans much of the period of ancient law (down to the fall of Rome in A.D. 476) and part of the period of mediaeval law down to the Dark Ages, while thereafter and down to the present day Roman law, though itself dead, has been an important source of, and influence on, other legal systems. The ancient period, when Roman law was made at and for Rome itself and for its colonies and empire, can be divided, according to the constitution in force at Rome, into the periods of the Monarchy (753–510 B.C.), the Republic (510–31 B.C.) and the Empire (31 B.C.–A.D. 476), though both the latter periods can be subdivided. In the early mediaeval period, the centre shifted to Constantinople and, when Europe emerged from the Dark Ages, Roman law became a factor in the law of many peoples, not a system enforced from Rome or even Constantinople. This influence continued powerfully into the early modern period (1453–1800) and, though diminished, it continues to this day.

The Monarchy

There is no certain knowledge about the constitution or the law of Rome in the times of the kings; only conjectures based on traditions reported by the historians. It may be inferred that the constitution comprised the King, the council of elders, and the assembly of the people but we have no knowledge of the distribution of power or differentiation of functions, nor of the private law between individuals.

Law under the early Republic

The period of the Republic extended from the abolition of the monarchy (510 B.C.) to the assumption by Octavian of the title of Augustus in 27 B.C. and the inauguration of the Empire. The constitution and law of the very early Republic are as conjectural as for the previous centuries but, from about 450 B.C., there is substantial information about the constitution and some knowledge of private law.

The chief magistrates were two consuls, holding office for a year and the office was originally confined to patricians, but a long struggle took place between patricians and plebeians on political and economic grounds, which resulted in concessions to the plebeians. Thus, from 367, one consul must be plebeian; in 351 a plebeian was elected censor; from 339 one censor had to be plebeian; in 300 the pontificate was opened to the plebeians; and in 287 resolutions of the plebeian *concilium plebis* were given the force of law.

After 287 there were three legislative bodies, the *comitia centuriata* and the *comitia tributa*, both consisting of the same persons, though organized differently, which passed *leges*, and the *concilium plebis* which passed *plebiscita*. The last became, increasingly, the usual legislative body—legislation being introduced by the tribunes. The senate was a consultative body only.

Among the magistrates, particular legal importance attaches to the praetorship, an office instituted in 367 B.C. At first, one praetor only was appointed but, from 242 B.C., there were two, one (*praetor urbanus*) dealing with suits between citizens, and the other (*praetor peregrinus*) with litigation between citizens and foreigner or between foreigners. Further praetors were later appointed to govern provinces, two in 227 and two more in 197.

At this time there were four distinct classes of

persons recognized: Roman citizens, including inhabitants of communities incorporated with citizenship and of Roman colonies; citizens without voting rights; Latins, i.e. inhabitants of Latin colonies; and allies.

One aspect of the struggle between patricians and plebeians was the pressure which resulted in the enactment of the Twelve Tables (q.v.) and made the law less of a mystery known only to patricians. Fragments of this code are known and it appears to have taken the form of a series of imperatives. It was almost entirely devoted to private law, almost entirely secular, and based on the customary law of Rome, though a few points may have been adopted from Greek models. This code was always highly regarded and treated as the fundamental element of the Roman system.

At the time of the Twelve Tables, the Roman family was patriarchal; the wife was *in manu* of the husband, the children and slaves *in potestate*. Marriage might be effected by *confarreatio, coemptio*, a fictitious sale, or *usus*, uninterrupted cohabitation for a year. Divorce was competent but not lightly entertained; by custom it required the consent of a family council, but no judicial process. Adoption was recognized in two forms, *adoptio*, a fictitious sale, and *adrogatio* which required investigation by the pontiffs and a legislative act by the *comitia curiata*. Guardianship was in force for children under puberty and for women of any age.

Intestate succession was provided for, estate passing to those who would pass out of *manus* or *potestas* by reason of the death, whom failing to the nearest agnates. Wills were recognized, in the forms of legislative sanction given in the *comitia curiata*, or when the army was about to set out. Later, a form of will was devised, operating by a fictitious sale of the testator's estate to a friend who was requested to deal with the estate as instructed.

In property law, the basic distinction was between *res mancipi* (land subject to Roman ownership, rural servitudes, slaves, beasts of draught and burden, and cattle) and all other things, *res nec mancipi*. The former could be transferred only by mancipation (real or fictitious sale) or *in iure cessio* (a collusive action before a magistrate), the latter by bare delivery. Ownership might also be acquired, subject to qualifications, by *usucapio*, possession for stated periods of time.

The concept of contract was very ill-developed.

The distinction between civil and criminal wrongs was not clearly drawn. There were recognized *furtum* (theft) and *injuria* (insult). The *Lex Aquilia* (possibly 280 B.C.) gave redress for damage to property.

Procedure was characterized by a division of proceedings into two stages, that *in iure* before a magistrate, at which the claimant made his claim in set words, the defendant replied in set words and the magistrate added his authority to the case being sent for trial, and that *apud iudicem* or *in iudicio* before a person appointed by agreement between the parties from a list of persons kept by the magistrate, who heard the evidence and gave judgment.

Leges, apart from the Twelve Tables, were not an important source of private law during the Republic, though some were (notably the *lex Aquilia*, which restated the law of damage to property was one). Private law was developed largely by interpretation and magisterial edict.

Interpretation of enacted law was originally in the hands of the college of pontiffs, who advised magistrates and individuals. They were also important in formulating *legis actiones* or forms of words used to make legal claims. From about 250 B.C., there came into being a class of *juris consulti* or *juris prudentes* and some of these wrote treatises on various branches of law, including *P. Mucius Scaevola, M. Junius Brutus*, and *M. Manilius*, who became known as the three founders of the civil law. Rather later, *Q. Mucius Scaevola* wrote the first systematic legal treatise yet known. The functions of the *juris prudentes* were giving opinions and advice to magistrates, framing legal documents and guidance in litigation and these legal functions were commonly ancillary to political careers.

All the higher magistrates had the right to issue edicts and the practice came to be for the praetors and other magistrates having jurisdiction to publish edicts, when they took up office, stating in what ways they would exercise their jurisdiction. It became customary for each praetor to republish the bulk of his predecessor's edict with amendments and thus there developed the *edictum tralaticium*, carried on from year to year. The praetor had no power to change the law but could give remedies in cases not covered by the strict law. This exercise of the praetorian power to give remedies in stated cases gave rise to the body of *jus honorarium* which came to be a distinct source of loss, to stand alongside the *jus civile* and be interwoven with it, having the function of aiding, supplementing, or correcting the civil law. The praetorian edict, like the civil law, became the subject of jurists' commentaries. The peregrine praetor had even greater liberty in developing his edict but, by the first century, the two edicts seem to have corresponded closely.

The later Republic

From about 150 B.C. there is fairly extensive information about public law and a fair amount of knowledge of private law. This period is important as including the foundation on which later law was developed and the beginnings of legal literature.

In the later republic there were three legislative

bodies, the *comitia centuriata*, the *comitia tributa*, both consisting of the same people though organized differently, which passed *leges*, and the *concilium plebis* which passed *plebiscita*. Any one of these might approve proposed legislation. The senate was a purely consultative body.

In the last half century of the Republic the republican constitution was not functioning regularly and Caesar, in substance, exercised royal power. In 43 B.C., Anthony Octavian and Lepidus were appointed triumvirs for settling the republic but, by 31, Octavian was sole master and, in fact, dictator. In 27 B.C., he renounced his powers and the republican constitution nominally revived but, in fact, the old system had broken down. It did not make much difference to private law. Citizenship was gradually extended to the whole of Italy.

The jurists writing on various branches of the law were numerous in the late Republic and the legal career tended to become more specialized and distinct from political advancement; their *responsa* given not only to individuals, but to lay-judges trying cases, became increasingly influential and a distinct source of law. So, too, in drawing up their edicts and making alterations thereto, magistrates must have been assisted by jurists, so that they were influential on the form of the edictal law. They also functioned in framing legal forms for contracts, wills, and other transactions. Advocacy proper was not the function of the jurists but of the orators. Though little has survived from the legal writing of the jurists, it is evident, from later work, that the writings of the republican jurists were the essential foundation for the great development in the early Empire.

In the later republic changes in substantive law had taken place from the time of the Twelve Tables. Marriage was normally effected by bare consent and divorce was free, and latterly very common. Guardians might be appointed by the praetor as well as by will, and curators were appointed to young men above puberty but under the age of 25. Testamentary succession was commonly, by transfer *inter vivos*, by fictitious sale to a *familiae emptor*, but the praetors gave possession of deceased's estates to persons nominated heirs in written wills sealed with the requisite seven seals. In cases of intestacy the rules of the Twelve Tables remained unchanged, though the praetors recognized the claims of blood-relatives. In the law of property, there came to be recognized, distinct from ownership, the right to possess, which was protected by possessory interdicts. Among obligations, there were recognized contracts by stipulation, i.e. by formal question and answer, literal contracts, by book-entries, and contracts based on bare consent, while praetorian action extended *injuria* beyond physical assault to non-physical affronts. The *lex Aquilia* of *c.* 280 B.C. gave extensive remedies for damage to property. The praetors also recognized the wrongs of threats and fraud.

Procedure was substantially developed, largely through the praetorian edict which developed the formulary system. It was introduced officially by the *Lex Aebutia* (not before 150 B.C.). Under this, the question at issue was settled at the stage *in iure* before the magistrate and then submitted to the *iudex* in a formula adapted to the particular dispute between the parties. In the request for a formula, the defendant's claim to have it altered or an exception inserted, and the magistrate's settling of the form of words to be allowed, jurists acted as advisers to the parties and the magistrate. The new system gave fresh importance to the function of the magistrate and to his advisers. The formulary system appears to have brought no material change in the proceedings *apud iudicem*. Execution, however, was changed by the introduction of execution against the judgment debtor's property. This system may have arisen in courts exercising jurisdiction over foreigners, and then been introduced at Rome. By a *lex Aebutia* (probably after 150 B.C.) this formulary process was introduced officially as an alternative to *legis actiones* and it became obligatory for all lawsuits by statute under Augustus.

Apart from the formulary system, the praetors developed the law by recognizing new forms of stipulation, authorizing possession, granting restitution *in integrum*, and granting interdicts, particularly protecting possession.

Only in the last century of the Republic was permanent provision made for special tribunals to try criminal cases. Previously though murder appears, from the earliest times, to have been punished by the State, theft and assault gave rise only to private claims. In the late Republic, criminal practice was administered by *quaestiones perpetuae* established by various statutes from 149 B.C. onwards, each of which made provision for a court consisting of a magistrate and a number of jurors to try some offence or class of offences. The number of praetors was raised to eight, to provide six presidents for such *quaestiones*. The manning of these jury courts was latterly a serious political issue.

Law at Rome under the Empire

Under the Empire (27 B.C. to A.D. 476) legal literature developed and the years A.D. 100–250 are the classical period of Roman law when it reached its fullest development in the hands of lawyers who were frequently both writers and practitioners. A post-classical period followed, marked by a decline in legal thought and literature.

In the first part of this period, commonly called the principate, from 27 B.C. to the death of Alexander Severus in A.D. 235, there was a preservation of republican forms of government though, in fact, there was a development of imperial authority throughout all departments of government. Though the *comitia*, the senate, and the authority of the magistrates continued unchanged, Octavian could, in fact, control all administration. He took the title Augustus. The powers of the people diminished and those of the senate were increased. Imperial rights came to be recognized for dispensation from the operations of any rule of law from which dispensation was possible, and to legislate. The imperial civil service almost completely superseded the republican hierarchy of magistrates, though these nominally remained. Citizenship was gradually extended to the provinces by the foundation of Roman colonies and by direct grant to the communities and individuals. In A.D. 212, by the *constitutio Antoniniana*, Caracalla extended the citizenship to most of the inhabitants of the Empire. The praetors continued to act judicially; the peregrine praetorship continued until the edict of Caracalla in A.D. 212 conferred Roman citizenship on nearly all the inhabitants of the Empire; the urban praetor remained the chief judicial officer in civil cases, while other praetors presided over the *quaestiones perpetuae* until these were superseded.

Law continued to be made by *leges* under Augustus but legislation by the *comitia* disappeared by the end of the first century A.D. The magistrates' edicts continued to be issued but changes were less frequent, and the power to alter the edict ended in the time of Hadrian. About A.D. 130–138, Salvius Julianus was commissioned to revise the praetorian edict and, as revised, it was confirmed by *senatus consultum* and was thenceforth fixed. Thereafter, the edict was no longer a source of new law. The revision probably applied to the edicts of the peregrine praetor and also the aediles. The edicts of the provincial governors were probably, in large part, common to all the provinces, and this part was also probably revised by Julianus, leaving parts for each province which were altered on imperial instructions.

Under the principate, *senatus consulta* became an important source of new law as the *senatus* came to replace the *comitia* and as it came to give advice to magistrates in their full use of the *jus edicendi*. The legislative initiative, however, normally came from the Emperor himself. Latterly, the senate merely confirmed the Emperor's proposal, and latterly the proposals merge into *edicta* as direct expressions of the Emperor's wishes. By the mid-second century, it was recognized that the Emperor himself could actually make law, by *constitutiones principum* (q.v.). These might take the forms of *edicta, decreta,* or *epistulae*, rescripta and *subscriptiones*. He might also issue *mandata*, instructions to governors and other officials, which were treated as having the force of statute.

Responsa prudentium continued to be an increasingly important source of law but Augustus conferred on certain more eminent jurists the *jus respondendi*, which, though it did not make their opinions of binding force, must have made them of higher authority than the opinions of jurists not so favoured. In fact, the *responsa* and the writings of jurists were the chief instrument of legal development under the principate. Custom was not an important source of law.

The period of the principate included the years when legal science was at the highest development, and it is to the jurists that is owed the fame and the subsequent influence of Roman law. The jurists were normally public men, many holding high offices of state, though some were purely scholars and teachers. They were chiefly concerned with practical problems and less with philosophical jurisprudence, analysis of conceptions, or systematization of the law. Their writings are generally casuistic, based on discussion of cases, rather than systematic or statements of principle. Generally, books follow the traditional order of Sabinus' work (see *infra*) or of the praetorian edict. The main kinds of books were (a) *institutiones* or *enchiridia*, introductory textbooks; (b) *regulae, definitiones,* or *sententiae*, books containing brief statements of law, designed for practitioners as well as for students; (c) general books on the *ius civile*, some following Q. Mucius Scaevola's 18 books on the civil law, others based on Massurius Sabinus' *iuris civilis libri* IIII and hence called *libri ad Sabinum* or *ex Sabino*; (d) commentaries on the praetorian edict, necessarily involving those parts of the *ins civile* which were supplemented or modified by the edict; (e) *Digesta*, or treatises on the law as a whole, dealing with the praetorian edict, the civil law, and criminal law; (f) commentaries on particular *leges* or *senatus consulta*; (g) *Responsa*, answers to problems referred to the author in his practice for his opinion; *epistulae*, dealing with cases, some from actual practice, in which the jurist's opinion was sought by letter; *quaestiones* and *disputationes*, dealing with cases normally raised not in actual practice but by discussion with pupils. The scope and contents of particular books must, however, have been partly matters of personal choice. It has been possible from the excerpts contained in Justinian's *Digest* to form in some cases a good idea of the size, scope and contents of particular books.

The jurists belonged to two schools or groups, known as the Proculians and Sabinians (q.v.), the distinction between which is rather obscure. Among the most notable jurists were *M. Antistius Labeo*, who wrote commentaries on the Twelve Tables and on the urban and peregrine edicts and other works, *C. Ateius Capito, Massurius Sabinus, Cocceius Nerva*, and his son, *Proculus, C. Cassius Longinus, Cn. Arulenus Caelius Sabinus, Pegasus, Plautius Javolenus Priscus*, a most influential jurist who wrote *epistulae* in 90 books, *Titius Aristo, P. Juventius Celsus*, who composed a *Digesta* in 39 books. *Salvius Julianus* who revised the edict and wrote a Digesta in 90 books, *Neratius Priscus, Sextus Pomponius*, a most prolific writer, *Sextus Caecilius Africanus, Gaius*, who wrote an *Institutes* in four books, commentaries on the Twelve Tables, the urban, provincial, and aedilician edicts and numerous other works. Alone of the works of this period, his *Institutes* (*c.* A.D. 161) survives almost complete. Others were *Ulpius Marcellus*, who wrote a *Digesta* in 31 books, *Aemilius Papinianus*, who wrote many books of *quaestiones* and *responsa, Julius Paulus*, a voluminous writer of commentaries and responsa, *Domitius Ulpianus*, who wrote on the edict and other works, and *Herennius Modestinus* (qq.v.). Most of these works were practical, concerned with concrete questions and discussion of problems, and frequently weak in arrangement, in theory, and in analysis of concepts, but they included many elementary textbooks as well as general works, commentaries on the edicts, commentaries on particular *leges*, and collections of *responsa*. The total bulk of the legal literature, by the end of the classical age, must have been enormous.

During the principate, the courts and procedure of the republican period were not abolished but frequently atrophied as imperial institutions developed and superseded them. The Emperor might assume jurisdiction in any matter on his own initiative or at the request of a party, and hear appeals personally or by judge-delegate. The jurisdiction of the prefects was based entirely on delegation by the Emperor. He also, sometimes, delegated to magistrates functions in connection with particular matters, such as *fideicommissa*, guardianship and claims for *alimenta*. His legislative power might be exercised through republican forms, by obtaining the passing of *senatus consulta*, or personally, by rescript.

In civil procedure, there developed in parallel with older procedure the system of *cognitio extraordinaria*, under which the judge used his power of investigation and compulsion to decide issues between private parties. It was not based on agreement of parties but on authority, and the case might be heard by the official whose authority was invoked, or by a delegate from him, or even by settling a formula as under the older procedure. The *praefectus urbi* exercised an increasing civil jurisdiction. Judgment by default was introduced, and appeal was introduced, culminating in appeal to the Emperor.

The criminal law was never formulated in anything comparable to the Edict, nor did it receive the attention of the jurists to anything like the same extent.

In criminal procedure, the *quaestiones perpetuae* continued to exist for the first two centuries A.D., but other tribunals developed, notably the Emperor and his delegates, the prefects, and provincial governors, in which the magistrates could and did investigate on their own initiative. Punishments became more severe.

The classical law was marked by a decrease in rigidity and formalism, an increased tendency towards abstract thinking, classification and recognition of the rights involved rather than merely the form of action. The influence of Greek thought and law was marked, and writing was in use much more to create obligation.

The later Empire

The later part of the Empire, commonly known as the dominate, extended from about A.D. 280 to the end of the Western Empire in A.D. 476. Diocletian (284–305) re-established peace and order in the Empire and under him and his successors the imperial power was developed into a monarchy with all the panoply of an Oriental despotism. In 286 the territorial division of the Empire was introduced, with two Augusti, each with a subordinate Caesar, ruling over East and West. From 395 to 476, though the Empire remained, in theory, a unity under joint rulers, the two halves were separate and occasionally hostile. Administration came into the hands of a hierarchy of officials, dependent solely on the Emperor, local and provincial differences largely disappearing. The Emperor's court was moved to Milan and later to Ravenna.

The distinction in status between Italy and the rest of the Empire had become blurred and, under the dominate the whole Empire, apart from the cities of Rome and Constantinople, was administered on a uniform plan. The whole Empire was divided into four great prefectures, two in the East (Oriens and Illyricum) and two in the West (the Gauls and the Italies), the latter of which ceased to exist on the fall of the Western Empire. The prefectures were divided into dioceses, each comprising a number of provinces, ultimately numbering over 120. There were some exceptions to the strict hierarchy.

Under the dominate, it was recognized that the Emperor was the legislator and he commonly laid down new general rules, but under various forms, commonly by letter or document to a praetorian prefect or other official, or to the senate, or to the people or a section thereof such as the inhabitants of a city. As administrator or judge also the Emperor could lay down general rules.

Justinian enacted that imperial judgment given in presence of the parties were to be valid as precedents in similar cases, but insisted on a general imperial power of interpretation.

Imperial legislation was variously called edicts, rescripts, *adnotationes, sanctiones pragmaticae*, which seem to have been replies to petitions usually containing matter of general interest, and *mandata*.

All constitutions were recorded in the imperial archives and signed by the emperor himself.

Jurisdiction was exercised in the two capitals by the *praefectus urbi*, elsewhere by the provincial governors and their delegates. Appeal lay to the praetorian prefect or the *vicarius* with, in the latter case, further appeal to the Emperor. The Emperor might himself hear cases in his *consistorium*. There were many special jurisdictions. Christians had, since the third century, brought cases before bishops and after 398 A.D., subject to parties consenting to the jurisdiction, a bishop had jurisdiction in all civil matters and later bishops acquired general powers of supervising the administration of justice in their dioceses.

In civil procedure, the formulary system disappeared in the third century, though *formulae* were still used until formally abolished in A.D. 342. *Cognitio extraordinaria* was the sole procedure, though the law applied was much influenced by older law, and witnesses were summoned by the court and examined by the judge.

In criminal procedure trials were conducted by public officials under a procedure *extra ordinem*. Apart from political offences and matters reserved to the Senate, jurisdiction was exercised by the Emperor and the prefects and, in the provinces, by *praesides* and *procuratores*. *Cognitio extra ordinem* superseded the *quaestiones perpetuae*. The practice of the cognitiones distinguished the states of mind of offenders, and imperial judges had discretion in grading penalties according to the circumstances of the case. Under the late Empire, discretion was abolished and penalties were more severe and mandatory.

Since the jurists had ceased to write, it became necessary to take steps to acquaint persons with the law. Two unofficial collections were made (by unknown authors, probably in the East), the *Codex Gregorianus* (probably A.D. 291) and the *Codex Hermogenianus* (probably A.D. 295). The former was the larger and divided into books and titles, the arrangement of subjects being that traditional in the *Digesta* of the classical jurists. The latter was divided into titles only. Both gave the text of the constitutions in full. Neither has survived, save in quotation. In 429, Theodosius II appointed a commission to compile an official collection of all *leges generales* enacted since Constantine's time, including obsolete matter. In 435, however, a second commission was appointed with a different remit and with powers to alter constitutions but not to omit any merely because they were obsolete. The *Codex Theodosianus* was published in 438 and effective the next year. It was divided into 16 books, sub-divided into titles which generally followed those of the *digesta* and the individual constitutions arranged chronologically. In the West, the Theodosian Code continued to be used but, in the East, it was superseded by Justinian's legislation. A collection of later constitutions was made under Majorian (A.D. 455), and there is a collection of 16 constitutions, mostly concerning ecclesiastical matters (the last dated 425), known as the *constitutiones Sirmondi*.

After the classical period, the *jus respondendi* was no longer conferred and *responsa prudentium* ceased to be a source of law. A period of decadence set in, brought on by the disturbed condition of the empire, the increased absolutism of the rulers, the shifting of the intellectual centre to the East and the development of the Christian Church and theology as an outlet for ability. Hence, under the dominate, legal literature was largely the collection or epitomizing of earlier materials.

In 321, Constantine abolished the notes of Upian and Paul on Papinian and, in 327, he confirmed the works of Paul, in particular the *Sententiae*. In 426, however, Theodosius II issued his Law of Citations by which the works of Papinian, Paul, Gaius, Ulpian, and Modestinus were affirmed as authoritative, and it was provided that the works of any jurist cited by the five authoritative writers might also be cited if confirmed by comparison of manuscripts. If there were divisions of opinion among the five writers, the majority view was to be followed; in case of equality, the view of Papinian was to prevail, and if there were equality and Papinian had expressed no view, the decision was left to the judge's discretion. These rules remained nominally in force until Justinian's time.

The surviving post-classical legal literature is all anonymous and is, very probably, mostly of Western origin. It includes Paul's *Sententiae*, which is probably a post-classical anthology from various writings by Paul, probably compiled in the third century, the *Epitome of Ulpian*, an epitome probably based on a post-classical compilation, itself based on Gaius and other classical works, the *Opiniones of Ulpian*, which are probably post-classical rather than original, the *Autun Gaius*, a paraphrase of Gaius, with illustrations, for students' use, the *Vatican Fragments* dating from between 372 and 438, of which seven titles survive, the *Collatio Legum Mosaicarum et Romanorum*, comprising 16 titles dealing with crime, delict, and succession, dating from between 390 and 438, founded on comparison between rules in the Pentateuch and those of the five authoritative jurists and of constitutions in the Gregorian and Hermogenian codes, the *veteris cujusdam jurisconsulti consultatio* of the late fifth or early sixth centuries, comprising answers to queries and short theoretical discussions, founded on Paul's *sententiae* and the three codes. From the East, there survive the *Scholia sinaitica*, a work of the fifth century, probably for teaching purposes, containing notes on Ulpian's *Libri ad Sabinum*, and the *Syro-Roman Law Book*, known in versions in several Oriental languages, dealing mainly with the family, slaves, and succession, and basically Roman law. The purpose of the book is uncertain.

The division of the Empire and the end of the Western Empire

The reign of Constantine the Great (324–337) was marked by the recognition of Christianity as the religion of the Empire and the establishment at Byzantium of a second capital, soon known as Constantinople. In 364, Valentinian became Emperor of the West and Valens of the East and this division became permanent with the accession of Honorius and Arcadius, the sons of Theodosius in 395. Though, in theory, the Empire remained a single whole it was in effect divided. The following century was marked by declining prosperity, increasing taxation, civil wars and, above all, by barbarian raids from the north-east into the Empire. The Visigoths under Alaric penetrated into Lombardy and, in 410, captured and sacked Rome. Spain became divided between the Vandals, Suevi, and Alani. In the mid-fifth century, though no province had broken away, the authority of the emperors was merely nominal and, in Gaul and Spain, barbarian states were growing up which admitted the Emperor's supremacy but, in substance, were independent. The Vandals conquered Africa and in 451 Attila and the Huns invaded Gaul but were defeated at Chalons. In 455, the Vandals under Genseric invaded Italy and sacked Rome; for 20 years thereafter barbarian leaders were the effective power behind the emperors of the West and, in 475, the barbarians deposed Romulus Augustulus and proclaimed their leader, Odoacer, King of Italy. But Odoacer recognized the supremacy of the Emperor in Constantinople and was invested with the title of 'patrician', and in Italy, as in Spain and Gaul, the language, government, and law remained Roman. The break, however, facilitated the growth of new states and nationalities in the west, and made possible the growth of a Romano-German mediaeval civilization. For the continuation see MEDIAEVAL LAW.

Law in the Eastern Empire to the time of Justinian

The Eastern Empire dates from the foundation of Constantinople in about 340 and, particularly, from the division of the Empire from 395. The Eastern Empire escaped the fate of the West because the regions were more populous, better governed, and they enjoyed peace with Persia for over a century. Theodosius II built the great land walls of Constantinople, and promulgated the *Codex Theodosianus* (439), a copy of which was received by the Roman Senate. After the fall of the Western Empire, the Roman law lived on in the East.

The work of Justinian

Justinian became Emperor at Constantinople in 528 and sought to effect the reconquest of the lost western provinces; Belisarius recovered Africa (533) and Belisarius and Narses won back Italy (554) but the conquests were lost after his death, when Italy fell to the Lombards. As soon as he came to the throne he commissioned Tribonian and others to make a new collection of imperial constitutions from the three older Codes and later enactments, with power to omit what was obsolete or unnecessary, remove repetitions and contradictions, and make changes and additions. This Code was promulgated and given the force of law in 529 and thereafter, with certain exceptions, previous enactments might be cited only as they appeared in the new collection. The next stage was the promulgation of a number of constitutions to settle outstanding controversies and abolish obsolete institutions, and 50 of these were apparently collected and published as a collection. It has not survived. Similar legislation continued to be passed as later work proceeded.

In 530, a constitution was issued to Tribonian with power to choose a commission to help him in the work, to read and make excerpts from the ancient jurists who had the *jus respondendi* and to collect them into a single work divided into 50 books and subordinate titles. The commissioners were to choose whatever they thought best, not necessarily to follow the majority opinion, and the collection was to contain no contradictions or repetitions or superfluous or obsolete matters and nothing that was already in the Code save where arrangement of the material made this advisable for the sake of clarity. They had power to alter and abbreviate texts, including the words of ancient leges or constitutions quoted by the jurists. Commentaries were forbidden. Tribonian's commission consisted of himself, a high official (Constantinus), two professors from Constantinople (Theophilus and Cratinus), two from Beyrout, (Dorotheus and Anatolius), and 11 advocates.

The labours of this commission resulted in the *Digest* or *Pandects.* They are said to have read nearly 2,000 books containing 3 million lines, many unknown even to the commissioners, and from those they selected 150,000 lines representing 39 authors. Selection was made both from the works of the jurists, mentioned in the Theodosian Law of Citations, and from authors cited by them, and even from a few earlier jurists. The bulk of the quotations come from the late classical jurists and Ulpian and Paul between them contribute about half of the total. The commissioners had to follow the pattern of the commentaries on the Edict and of the Code of 529, and seem to have taken Ulpian's commentary as their chief guide. The 50 books have no headings, but the titles have headings taken mostly from existing works. Within each title the excerpts are grouped into four masses: the Sabinian mass, beginning with the books of Ulpian, Pomponius and Paulus *ad Sabinum*; the Edictal mass beginning with some of the books from Ulpian's and Paul's commentaries on the Edict; the Papinian mass beginning with Papinian's *quaestiones* and *responsa*; and the post-Papinian mass or Appendix, containing only a few works. The order in which the masses appear in each title is not always the same and not every title contains all the masses, while sometimes a fragment is taken out of its mass and inserted elsewhere. The Digest came into force on 30 December, 533.

As well as the division into books and titles, the Digest was divided, particularly for educational purposes, into seven parts (Books 1–4, 5–11, 12–19, 20–27, 28–36, 37–44, and 45–50) while, in the Middle Ages, the Digest was commonly divided up into *Digestum vetus* (1 to 24:2), *Infortiatum* (24.3 to 38) and *Digestum Novum* (39 to 50), the latter part of the *Infortiatum* being further distinguished as the *tres partes*; This resulted, it is said, from the fact that the earliest scholars found manuscripts of only the first and third parts, and subsequently got to know the second.

The Digest was unsuitable for use in law schools and, before it was complete, Tribonian, Theophilus, and Dorótheus were instructed to compile a new students' manual, based on the classical institutional works and especially on the *Institutes* of Gaius. This was published in 533, given the validity of an imperial statute, and it came into force on the same day as the Digest. The *Institutes* of Justinian follows closely the arrangement of the *Institutes* of Gaius. A large part of the text is taken from Gaius' *Institutes* and *res cottidianae* but use was also made of the *Institutes* of Marcian, Florentinus, and Ulpian and possibly also of Paulus. These texts were, however, modified to bring the statement of law up to date, Most of what Gaius wrote on actions was omitted as obsolete, and titles were added on *officium judicis* and on criminal law. As a whole, the *Institutes* are as much a compilation from the work of long-dead authors as is the Digest.

All commentaries on the Digest were forbidden, as Justinian believed that this would only load the text with confusion, as had happened with jurists' commentaries on the praetorian edict.

By 533, the Code of 529 and the 50 Decisions had been supplemented by a large number of other constitutions and Tribonian, Dorotheus, and three members of the commission which had produced the Digest were instructed to prepare a revised Code containing the new matter, again with wide powers to alter or omit. The *Codex repetitae praelectionis* was published late in 534 and superseded the earlier Code and constitutions. The Code is in 12 books, subdivided into titles within which the constitutions are arranged chronologically. The arrangement is based on the older codes and, thus, indirectly on the Edict. It is ecclesiastical law, sources of law and duties of the higher officials (Bk. I), private law (Bks II to VIII), criminal law (Bk. IX) and administrative law (Bks. X–XII). Considerable changes were made to bring the matters up to date and to achieve consistency.

Thereafter, Justinian continued to legislate, by means of Novels (*Novellae constitutiones*), the tendency of which was to reflect current Byzantine practice rather than pure Roman law. Most are in Greek, a few in Greek and Latin, and only those addressed to Latin-speaking provinces in Latin. An apparent intention to publish an official collection of Novels was never executed, but three unofficial collections are known; the *Epitome Juliani* containing 124 constitutions in an abridged Latin version dating from about A.D. 555; the Greek collection or the collection of 168 constitutions, of about A.D. 580; and the *Authenticum*, a collection in which Latin constitutions are in the original and Greek ones in literal translations. It contains 134 constitutions extending from 535 to 556. It may date from the sixth century and may be semi-official.

The war of Justinian's reign strained the Empire and brought on financial oppression which led to a decline after his death, when the Empire was attacked on all sides. But it survived all attacks until finally Constantinople fell to the Ottoman Turks in 1453 and the Eastern Empire also came to an end. For law in the East after Justinian see MEDIAEVAL LAW.

The influence and importance of Roman law

The enterprise of compiling the *Digest* and *Institutes* is extraordinary. It is as if law commissioners in twentieth-century Britain were to make excerpts from Coke, Littleton, Bacon and authors of similar date, amend the statements of law, and publish it as current law. But that very fact makes it an invaluable repository of earlier law which would otherwise have been lost. Much attention has been given to studying the interpolations and discerning the original texts under the overlay of later alterations and revision.

Roman law was, by A.D. 535, by far the longest-lived, most mature, and fully developed system which had yet developed on earth. Moreover, it had spread over all Europe, as well as into Africa, Asia Minor, and Palestine and parts of it were incorporated in codes prepared for the use of Roman citizens living in the barbarian kingdoms of Western Europe. The original texts were rediscovered in the eleventh century and began to be studied eagerly all over Europe, partly as the sources of some of the law of the different kingdoms, partly as the true and pure law, partly as a source of rules in default of others.

History: P. Kruger, *Geschichte der Quellen des Römischen Rechts*; T. Kipp, *Geschichte der Quellen*; B Kübler, *Geschichte des römischen Rechts*; H. F. Jolowicz, *Historical Introduction to the Study of Roman Law*; W. Kunkel, *Introduction to Roman Legal and Constitutional History*; F. Schulz, *History of Roman Legal Science*. Texts: C. G. Bruns, *Fontes juris Romani*; P. Girard, *Textes de droit romain*; O. Lenel, *Edictum Perpetuum*; F. de Zulueta, *Institutes of Gaius*; J. Moyle, *Institutes of Justinian*. Substantive Law: W. A. Watson: *Law of Obligations, Persons, Property, Succession in the late Roman Republic (4 vols.)*; W. W. Buckland, *Textbook of Roman Law from Augustus to Justinian*; P. Bonfante, *Corso di diritto Romano*; V. Arangio-Ruiz, *Corso di istituzioni di diritto Romano*; F. Schulz, *Classical Roman Law*. Civil Procedure: L. Wenger, *Institutionen des römischen Zivilprozess rechts*; P. F. Girard: *Organisation judiciaire des Romains*; A. Greenidge, *Legal Procedure of Cicero's Time*. Criminal Law and Procedure: T. Mommsen, *Römisches Strafrecht*; E. Ferrini, *Diritto penale romano*; J. Strachan Davidson, *Problems of the Roman Criminal Law*; Later developments: P. Vinogradoff, *Roman Law in Mediaeval Europe*; H. L. Hahlo and E. Kahn, *The South African Legal System*; P. Koschaker, *Europa und das römische Recht*.

Roman-Dutch Law. The legal system, produced by the fusion of mediaeval Dutch law, mainly of Germanic origin, and the Roman law of Justinian as accepted in the Reception of Roman law, which applied in the province of Holland in the sixteenth, seventeenth and eighteenth centuries and was carried to South Africa and became the main constituent of modern South African law. The name seems to have been used first by Van Leeuwen as a sub-title of his *Paratitla iuris novissimi* (1652) and as the title of his *Roomsch Hollandsch Recht* (1664). But the system was first created by Grotius' *Inleidinge tot de Hollandsche Rechtsgeleerdheid* (1631), which treated Roman-Dutch law as a system on its own and is still a basic work thereon. Major contributions to its development were made by Groenewegen, Van Leeuven, Huber's *Heedensdaegse Rechtsgeleerheyt* (1686), Voet's *Commentarius ad Pandectas* (1698–1704), Bynkershoek's *Quaestiones Juris Privati* (1744), Van der Keessel's *Theses Selectae juris Hollandici et Zelandici* (1800) and Van der Linden's *Rechtsgeleerd, Practicaal, en Koopmans Handboek*. These were supplemented by many volumes of statutory decisions of the courts and opinion of jurists.

The Roman-Dutch law was carried to Dutch colonies at the Cape of Good Hope (1652), Ceylon (1656), and Guiana. These settlements passed into British hands at the end of the eighteenth and early nineteenth century. The Cape was taken in 1795, given back in 1803, and retaken in 1806, and the Roman-Dutch law subsisted, so far as not expressly repealed. It was supplemented by a steady stream of Cape legislation. But in the Netherlands, the republic fell to the French in 1795, being replaced by the Batavian Republic, and French law was introduced. The kingdom of the Netherlands was established in 1815 and subsequently a Dutch civil code, much influenced by the French one, was adopted.

After 1806, as the boundaries of European settlement in South Africa were pushed north, the area subject to the Roman-Dutch law expanded. Principles from English common law were introduced and were otherwise influential, not least because the training of many judges and officials had been in English law, and because of appeals to the Privy Council. Nevertheless, the Roman-Dutch law was the common law.

In 1910, the South African states were united in the Union of South Africa which attained sovereign independence in 1931 and, in 1961, became a republic and left the Commonwealth. Roman-Dutch law is still the common law basis of modern South Africa and since severance of the formal links with Britain and discontinuance of appeals to the Privy Council it has acquired fresh importance.

H. L. Hahlo and E. Kahn, *The Union of South Africa—The Development of its Laws* (1960), H. L. Hahlo and E. Kahn, *The South African Legal System* (1968); R. W. Lee, *Introduction to Roman-Dutch Law* (1953). A. Maasdorp, *Institutes of South African Law*, 4 vols.

Romano-Germanic family. A term for the family of legal systems, also called civil law systems, or the civil law family, comprising the systems based on Germanic tribal customs and Roman law which developed in Western Europe, in what are now France, Belgium, the Netherlands, Germany, Austria, Spain, and Portugal.

Systems based on one or another of these European systems have been carried by colonization to, or been adopted by many other countries, particularly in Latin America, parts of Africa, the Near East, Japan, and Indonesia. All of these have distinct family resemblances in structure, concepts, and language; the standard form of law is the code and judicial exposition, though important, is not the major creative force in the way it is in common law systems (q.v.).

Romanus, Ludovicus (?–1439). Author of *Consilia* (1565) and *Commentaria in Digestum* (1580).

Rome, Treaties of. Two treaties, framed by inter-governmental committees at the Messina Conference (1955), which established the European Economic Community (q.v.) and the European Atomic Energy Community (q.v.), were signed in 1957, and came into force in 1958.

Romer, Sir Mark Lemon, Baron (1866–1944). Son of Lord Justice Sir Robert Romer, was called to the Bar in 1890, became a judge of the Chancery Division in 1922, a Lord Justice of Appeal in 1929 and, in 1938, a Lord of Appeal. He was a very fair and successful judge, though he made no parade of his legal knowledge in his judgments, which are lucid but concise. His son Sir Charles R. R. Romer became a Lord Justice of Appeal.

Romilly, John, Lord (1802–74). Second son of Sir Samuel Romilly (q.v.), was called to the Bar in 1827, became Solicitor-General in 1848, Attorney-General in 1850, and Master of the Rolls in 1851. He resigned in 1873. He sat in Parliament from 1832–35 and 1847 to 1852, the last Master of the Rolls to do so. He was an industrious and conscientious judge, and some of his judgments lay down important principles of equity, but he cannot be rated outstanding. On the other hand, he performed great services by carrying on Langdale's work at the Public Record Office, and initiated the policy of printing calendars of State Papers and of publishing the Rolls Series, edited editions of chronicles and documents.

Romilly, Sir Samuel (1757–1818). Was called to Bar in 1783 and practised successfully in Chancery. He took silk in 1800 and was Solicitor-General in 1806–7. He was one of the best lawyers of the day and was deemed an outstanding advocate. He was in Parliament, 1806–18, and consistently urged measures of law reform. He had long been known for liberal opinions, and was a friend and adviser of Bentham, argued that the Lord Chancellor should devote himself to law reform, and wrote many essays urging particular reforms. In particular, he attacked the indiscriminate capital punishment then competent. He was intimate with Bentham, but drew his original inspiration from Rousseau and Beccaria. The times were against him and his efforts had little success, but he was recognized by those who followed as their leader and inspirer and he succeeded in having the death penalty abolished in certain minor cases. He was also a strong supporter of Catholic and slave emancipation and was author of *Observations on the Criminal Law of England as it relates to Capital Punishment* (1810), *Objections to the Project of creating a Vice-Chancellor of England*, and other works.

Life by himself; Medd, *Romilly, Lawyer and Reformer*; Brougham, *Statesmen of the Time of George III*, I, 290.

Rookeries. A term for the criminal districts of London, which may have grown up round the sanctuaries and older hospitals of London. In the fifteenth century, the major districts were St. Giles, south of Great Russell Street, Drury Lane, and Covent Garden, the Spitalfields-Whitechapel area, Saffron Hill to the north of Holborn, the Whitecross Street part of Clerkenwell, and Southwark. Many of these areas were broken up by new roads and redevelopment.

T. Beames, *The Rookeries of London*; C. Booth, *Life and Labour of the People of London*.

Root, Elihu (1845–1937). U.S. lawyer and politician, became Secretary of War, 1899–1904, and Secretary of State, 1905–9. He worked devotedly for world peace, becoming president of the Carnegie Endowment for International Peace (1910–25) and, in 1912, was awarded the Nobel Peace Prize. He participated in founding the League of Nations and advocated U.S. membership of the Permanent Court of International Justice.

Root and Branch Petition. A petition presented to the House of Commons in 1640 by many persons in London bitterly attacking episcopal government of the Church and demanding its extirpation, root, and branch. It gave rise to a Root and Branch Bill for transferring episcopal jurisdiction to Parliamentary Commissioners which reached the committee stage in the House of Commons. Finally in 1642, after several abortive Bills, an Act to remove bishops from the Lords was passed and, in 1643, an ordinance of Parliament declared bishops, deans and chapters abolished and another ordinance confiscated their lands.

Root of title. An instrument which purports to deal with the absolute ownership of land to be transferred, which is valid without reference to any earlier document and casts no doubts on the title of the disposing party, and is at least 15 years back from the current transaction. Thus, a conveyance on sale or a specific devise of land is a good root of title. This is the necessary basis for a good title acceptable by a purchaser.

Rose-Innes, Sir James (1855–1942). Was admitted to the Cape of Good Hope Bar in 1878 and sat in the Cape Parliament, being Rhodes' Attorney-General in 1890–93 and again in 1900–2. In 1902, he became Chief Justice of the Transvaal, a judge of the Court of Appeal in 1910 and, in 1914, Chief Justice of South Africa. His judgments displayed learning and he ranked high as an expositor of the Roman-Dutch law. He retired in 1927.

Rosenthal, H. à (sixteenth century). A humanist, author of *Tractatus et Synopsis totius juris feudalis* (1597).

Ross, Alf Niels Christian (1899–). Danish jurist, professor at Copenhagen, judge of Court of

Human Rights, and author of *A Textbook of International Law* (1948), *Towards a Realistic Jurisprudence* (1948), *The Constitution of the United Nations* (1950), *On Law and Justice* (1958) and other books.

Ross, Sir John, Bart. of Dunmoyle (1853–1935). Was called to the Irish Bar in 1879, took silk in 1891, and entered Parliament in 1892. In 1896, he became land judge in the Irish Chancery Division, a baronet in 1919, and, in 1921, Lord Chancellor of Ireland for life. The office was abolished in 1922 and he was the last holder. He was a lawyer of great ability and learning, and his judgments show much independence of thought.

Rosslyn, Earl of. See WEDDERBURN, ALEXANDER.

Rota. From the earliest period of the Christian Church, judicial issues were referred to Rome for decision. By the thirteenth century, the popes were referring them to chaplains, members of the papal household, to hear the cases. Initially, the auditors merely heard the case and remitted the case to the Pope for decision with their opinion thereon. Innocent III first gave the auditors power to decide certain cases and John XXII, in 1332, gave the auditors powers in general to render decisions. From time to time, regulations have been made to govern the procedure. The Rota reached the climax of its power in the fifteenth century, but continued to function and was reconstituted and reorganized in 1908.

Its constitution and powers are regulated by the Code of Canon Law and papal rules. It is an ordinary tribunal of the Holy See with an undefined number of auditors, drawn from many countries. The presiding officer is the Dean of the Rota. Normally judgment is delivered by a *turnus* of three auditors. There are no territorial limits of jurisdiction and it may deal with any kind of ecclesiastical case, even criminal ones, though most of its work is in the field of matrimonial law. It is primarily a court for appeals from lower tribunals, but may act as a court of first instance for cases reserved by the Pope to himself and in certain other cases. It conducts a school for men aspiring to act as rotal procurators and advocates. Since the Rota was reorganized in 1908, its decisions and sentences have been compiled and published.

The name is of obscure origin, from the fact that judges sat in a circle or from the circle on the floor of the hall in Avignon (the name dating from that time) where cases were being heard, but the most likely explanation is said to be that the bookcase which held the papers in cases was on wheels and could be moved about.

Rothar, Edict of. A written formulation of Lombard law, based on Lombard custom but with traces of Roman and common law, promulgated in Italy in 643, and intended as a written formulation of law for both Lombards and Romans. Where not applicable, each people was ruled by its own former laws. It comprises 388 chapters divided into criminal law, family law, property law and procedure. It is the best of the legislative works of the German invaders of Italy, being systematically arranged and giving reasons for rules. It was supplemented by later Lombard kings, notably by Liutprand, Rachis, and Astolf.

Rotten (or pocket) borough. The term for those parliamentary constituencies which, by reason of movements of population and other causes, had by the late eighteenth century, few voters and were accordingly entirely in the hands of wealthy landlords or patrons. Thus Sarum was only a walled mound and Dunwich had been eroded by the sea. The persons elected were the nominees of the patron and had to obey his wishes. Some such seats were openly sold. Many such boroughs were disfranchised by the Reform Act in 1832.

Rotuli Parliamentorum. Rolls now in the Public Record Office, covering the period 1290–1503 and containing entries of what occurred in Parliament in each session, particularly the petitions of aggrieved persons or of the Commons and replies thereto, and occasionally also the statutes which resulted from the petition. They include some statutes which do not appear on the Statute rolls (q.v.) and there is evidence that these rolls were used to verify the statute rolls. They also contain matter of historical interest.

The printed edition of *Rotuli Parliamentorum* (6 vols., 1765, and Index 1832) relies on these rolls and also on a book variously called *Liber Irrotulamentorum de Parliamentis*, the *Black Book of the Tower*, and *Vetus Codex*, consisting of extracts from the rolls made in the fourteenth century, but is very defective. A new edition entitled *Parliamentary writs and writs of military summons* stopped after two volumes (1827–34).

Rousseau, Charles (1902–), French jurist, member of the Permanent Court of Arbitration, and author of *Principes générales du droit international public* (1944), *Principes de droit international public* (1958), *Traité de droit international public* (1970), and other works.

Rousseau, Jean Jacques (1712–78). French philosopher, who lived a vagabond existence until, in 1749, he won the prize offered by the academy of Dijon for an essay on the effect of the progress of civilization on morals. He later published *La Nouvelle Héloise* (1760), *Du Contrat Social* (1762), *Émile* (1762), and *Confessions* (1767), as well as minor works. In the *Social Contract*, he upholds the view that a society without laws is not a true community; law in society can be deemed legitimate because the laws have been decided by the general will, the constant will of all the members of the State, by virtue of which they are citizens and free. The general will is not the sum of the wills of the individuals composing the community. True law is always general and the chief problem for government is how it can secure such laws; this is a problem which Rousseau leaves unsolved. He was an exponent rather than the originator of ideas and his work has been influential by its vigour rather than its substance; it is incomplete and logically defective.

F. C. Green, *Jean Jacques Rousseau*; R. D. Masters, *Political Philosophy of Rousseau.*

Rout. A disturbance of the peace by three or more persons who have assembled with intent to commit a crime of violence or to do anything which causes reasonable men to apprehend a breach of the peace and who actually make a move towards the execution of their purpose. It is intermediate between an unlawful assembly and a riot (qq.v.).

Royal arms. The heraldic insignia appropriate to the sovereign. In England, the royal arms are a shield bearing, in the first and fourth quarters, the arms of England (three lions passant guardant), in the second, those of Scotland (a lion rampant), and in the third, those of Ireland (a harp), encircled by a garter, bearing the words 'Honi soit qui mal y pense', with a crest (a royal helmet with a crown, surmounted by a lion) and supporters (a lion crowned and unicorn with coronet), and the motto 'Dieu et mon droit'. In Scotland, the royal arms are a shield bearing in the first and fourth quarters, the arms of Scotland, in the second, the arms of England and, in the third, those of Ireland, surrounded by the collar of the Order of the Thistle, having a crest (a crown surmounted by a lion seated holding a sword and sceptre with the motto 'In Defens') and supporters (a unicorn crowned and lion crowned, the former sustaining a banner with a saltire (St. Andrew's cross) and the latter sustaining a banner with a St. George's cross) and the motto 'Nemo me Impune Lacessit'). The arms of France were omitted from 1801 and those of Hanover from 1837. For the bearing of the royal arms by subjects see ROYAL WARRANT.

Royal Assent. The signification of the sovereign's assent to a Bill which has been finally agreed to by both Houses of Parliament. In theory, the sovereign has a discretion to grant or refuse assent, but the royal assent has not been refused since 1707 and it could not now be refused without provoking a constitutional crisis.

Prior to about 1700, the Royal Assent was normally signified in person, but the Royal Assent to the Bill of Attainder against Queen Katherine Howard was signified on Henry VIII's behalf by a Commission of Peers, and this became the normal practice after the accession of George I. Royal Assent was last signified in person in 1854. The later practice was for three or more Lords Commissioners to be empowered. The Commons were summoned to the Lords' Chamber, the commission read, the title of each Bill read by the Clerk of the Crown, and the Royal Assent signified by the Clerk of the Parliaments in Norman French. Bills granting aids and supplies, i.e. money, receive the Royal Assent first, the formula being *La Reyne remercie ses bons sujets, accepte leur benevolence, et ainsi le veult.* For Public and Local Bills the formula is *La Reyne le Veult*, and for personal Bills it is *Soit fait comme il est desiré.* The words for the refusal of the Royal Assent would be *La Reyne s'avisera.* When the Commons returned to their own Chamber, the Speaker reports that the Royal Assent has been given and this is recorded in the Votes and Proceedings and the Journal of the House.

Under the Royal Assent Act, 1967, the Royal Assent may merely be announced by the Lord Chancellor and the Speaker to their respective Houses, and this is now the normal method, though personal Assent or by Lords Commissioners, remain competent, and the latter is sometimes used.

In view of the convention, since 1707, that the Royal Assent is never refused there is no machinery for sending a Bill back for reconsideration nor for the Houses overriding a veto.

***Royal British Bank* v. *Turquand* (1856), 6 E. & B. 327, Rule in.** A company issued a bond under its common seal, signed by two directors, though the directors could borrow only with the authority of a resolution of the company and none had been passed. It was held that the bond was binding as the lenders were entitled to assume that the requisite resolution had been passed: *omnia praesumuntur rite esse acta.*

Royal burgh. In Scotland, a municipal corporation created by royal charter prior to 1707.

Royal commission. A group of persons appointed by royal warrant at the instigation of the government of the time to investigate some matter and recommend legal changes deemed necessary. They were used infrequently prior to 1830, these functions being mainly performed by Select Committees of the House of Commons. From 1830, Royal Commissions were more frequently appointed; they included persons of varied experience and took evidence and opinions from a wide variety of sources. A notable early example was the Royal

Commission on the Poor Laws (1832–34) which preceded the Poor Law Amendment Act 1834. Others were on Judicial Reform (1833), Municipal Reform (1835), and Church Reform (1835). By 1849, more than 100 Royal Commissions had been established and nearly all the major pieces of social legislation between 1832 and 1867 resulted from this kind of prior investigation. The influence of Benthamite thinking was notable in many of these, and the general tenor of the reports was towards greater efficiency, settled procedures and better administration. Later in the century, the Royal Commission had become a standard piece of the machinery of government, to investigate, analyse information and recommend action. The Reports of Royal Commissions, published as Blue Books, are invariably interesting and valuable for their collection and presentation of information and opinion, even after action has been taken on the report and the immediate occasion has passed away.

Royal Courts of Justice. The statutory name of the Law Courts on the north side of the Strand, London, in which the main civil business of the Supreme Court of Judicature is transacted. For centuries, the Queen's Bench and other courts and latterly the Probate, Divorce and Admiralty Division occupied courts at Westminster Hall and the Chancery Division courts at Lincoln's Inn. Concentration of the courts was recommended by a Royal Commission in 1858, and authorized by statute in 1865. The courts were opened formally by Queen Victoria in 1882 and opened for business in January 1883. But sittings for judicial hearings take place at many other places in London, of the House of Lords in the Palace of Westminster, the Privy Council in Downing Street, the Central Criminal Court at the Old Bailey, and inferior courts at many places.

It is also the name of the building in Chichester Street, Belfast, in which the High Court and Court of Appeal of Northern Ireland sit.

Royal declaration. A declaration against Catholic doctrine required by the Bill of Rights (1689) to be subscribed and repeated by every English monarch at his coronation or the opening of his first Parliament. It was a product of the violently anti-Catholic feeling at that time. In the nineteenth century, attempts were made to have the wording modified and a new declaration was adopted in 1910, affirming that the monarch was a Protestant and would maintain the enactments securing the Protestant succession.

Royal Family. There is no legal meaning of the term 'Royal Family'. For the purposes of the Royal Marriages Act, 1772, the Royal Family is limited to descendants of the body of George II, and a member of the Royal Family as so defined, if under 25, apart from the issue of princesses who have married, or marry, into foreign families, may not marry without the prior consent of the Queen under the Great Seal and declared in Council. If over 25, they may marry after 12 months after notice given to the Privy Council. By virtue of the ruling of King George V that the title 'Royal Highness' should be limited to the children and grandchildren of a sovereign, the 'Royal Family' may be deemed limited to the aggregate of such persons, more remote descendants of a sovereign not being royal. Descendants, not being royal, bear the family name Windsor. Descendants of Queen Elizabeth II, other than those enjoying the style or title of Royal Highness and the titular dignity of Prince or Princess, and female descendants who marry and their descendants, bear the family name Mountbatten-Windsor.

Royal fish. In English law, whales and sturgeon. When either are stranded or caught they belong to the Crown, by reason, it is said, of their superior excellence. The right to royal fish may be granted by the Crown to a subject or acquired by prescription.

Royal Forces. The existence of armed forces of the Crown is authorized by royal prerogative and by statute and the supreme command of these is vested in the Crown, but the regulation of the forces is exercised by the Ministry of Defence acting through the Defence Council.

The only authority for existence of the naval forces is the royal prerogative. Since the Bill of Rights, 1688, declared illegal the maintenance of a standing army in the realm in time of peace without the consent of Parliament, the maintenance of an army depends on the periodic renewals of the Army Act. The Air Force was authorized by the Air Force (Constitution) Act, 1917, but its continued existence depends on the periodic renewal of the Air Force Act. The Royal Marines, a separate military force raised for naval service, are subject to parts of both the Army Act, 1955 and the Naval Discipline Act, 1957, The women's services are authorized by the Army and Air Force (Women's Service) Act, 1948, and are regulated by parts of the Army Act or the Air Force Act. The Women's Royal Naval Service is not authorized by statute but is treated as part of the armed forces for the purposes of certain statutes.

The relationship between the Crown and a member of any of the armed forces is not a contract of employment and is always terminable without notice. The relations are determined not by contract but by Royal Warrants, Orders in Council, and regulations issued under prerogative authority and by statute, and these sources have gone far to replace the authority formerly based on the prerogative only.

Persons subject to service discipline are subject to

the jurisdiction of the ordinary civil and criminal courts and also to the jurisdiction of courts-martial constituted under the various codes of discipline, but a person charged with an offence cognizable both by a court-martial and a magistrates' court is not liable to be tried by both for the same offence.

Members of the armed forces continue to be subject to nearly all the rights and duties of ordinary citizens, but are subject to certain privileges and disabilities and, further, are subject to the codes of discipline contained in the Naval Discipline Act, 1957, the Army Act, 1955, or the Air Force Act, 1955, all as later amended.

Royal forests. Areas, not necessarily wooded, whether belonging to the Crown or to a subject, reserved for hunting game by the sovereign. Their areas were fixed by record or prescription and they were formerly administered by special officers with special courts and special bodies of law. These have been obsolete since the sixteenth century and the Crown has long since ceased to enforce any prerogative as to game. See FOREST LAWS.

Royal Household. Various offices, originally hereditary, but now held by persons appointed by the sovereign, some of whom are selected from members of the political party in power, either in the Lords or Commons. Until the twentieth century, it was consistent practice that the principal officers of the household changed on a change of government but, in 1924, it was decided that the only officers to change on a change of government should be the Treasurer, the Comptroller, and the Vice-Chamberlain of the Household, who are government whips and Lords in Waiting. The other officers of the household are the Mistress of the Robes, the Ladies of the Bedchamber, the Lord Steward, the Lord Chamberlain, the Master of the Horse, the Captain of the Corps of Gentlemen-at-Arms, the Captain of the Corps of Yeomen of the Guard, and the Chief Equerry or Groom-in-Waiting. The privileges of some ancient hereditary offices, subject to allowance by the Court of Claims (q.v.), are exercisable at coronations. Some other offices connected with, though not today part of the royal household, are still hereditary, e.g. the Lord Great Chamberlain, and the Earl Marshal. The Private Secretary to the Sovereign is appointed by her personally.

Royal Marriages Act, 1772. An Act occasioned by George III's fear of the effect on the dignity and honour of the royal family of members thereof contracting unsuitable marriages, two of his brothers having done so, the Duke of Cumberland to Mrs. Horton and the Duke of Gloucester to Lady Waldegrave. It provided that marriages of descendants of George II, other than the issue of princesses who marry into foreign families, should not be valid unless they had the consent of the King in Council, or, if the parties were aged over 25, they had given 12 months' notice to the Privy Council, unless during that time both Houses of Parliament expressly declare disapproval of the proposed marriage.

Royal pardon. An exercise of the royal prerogative absolving criminals from guilt and liability to punishment. There were no restraints on its use until the Act of Settlement of 1701 made illegal the use of pardon under the Great Seal to protect persons facing impeachment by the Commons. The right to pardon other offenders remains and is exercised by the Home Secretary or the Secretary of State for Scotland.

Royal prerogative. The powers which the sovereign has, by virtue of her position and royal dignity, distinct from those possessed by subjects, and particularly the residue of royal powers which the Queen or her Ministers may exercise without the authority of Act of Parliament. The royal prerogative, originally very extensive and unfettered, has, over the centuries, been in many respects abridged by statute but much remains; in other respects it has been limited by constitutional conventions.

The personal prerogatives include immunity from civil and criminal action and the rules that the King can do no wrong, is not bound by statute unless expressly stated to be bound, never dies and is never affected by non-age. These latter rules have been much modified by legislation.

The political prerogatives include, on the domestic side, in relation to the executive function, the summoning, proroguing and dissolution of Parliament, the appointment and dismissal of Ministers, the creation of peers and the conferment of honours, to the judicial function, the grant of pardon or reprieve and commutation of sentence, to the legislative function, the provision of legislatures for British colonies abroad, and various ecclesiastical prerogatives; on the foreign side they include the rights to make war and peace, to make treaties, to send and receive ambassadors and diplomatic representatives. All these political prerogatives are now exercised on the advice of Ministers of the Crown, the appropriate Minister affixing the seal of which he has custody or countersigning the relevant document, save that occasional circumstances arise where the sovereign must make a personal decision, such as the choice between two equal candidates for appointment as Prime Minister.

An exercise of the prerogative is effected in various ways, by proclamations, writs or letters patent under the Great Seal, Orders in Council, orders, commissions, or warrants under the royal sign manual.

Despite statutory abridgments and limitation by constitutional convention, the prerogative remains a valuable reservoir of power enabling action to be taken in exceptional circumstances without prior authorization.

The courts may adjudicate on whether a particular exercise of the royal prerogative is valid or not, whether, e.g., property may be taken in emergency without compensation.

Royal proclamations. Though formerly used as a means of legislation, royal proclamations have, since the seventeenth century, had no such force. The usual purposes are to assemble or dissolve Parliament, to declare peace or war, to constitute new colonies, to declare martial law or to appoint days for public thanksgiving.

Royal standard. The personal flag of the sovereign, which may only be flown by the Queen's permission when she is present. In England, it is the same as the shield in the royal arms (q.v.). In Scotland, it is the scarlet lion rampant within a double tressure flory counterflory, on a gold field. (The Union flag or Union Jack may be flown by any citizen of the Commonwealth. In Scotland the national flag is the saltire or white St. Andrew's cross on a blue ground.)

Royal style and titles. The style and titles of the King or Queen Regnant are those announced after accession by proclamation under statutory authority. The present royal style and titles in the U.K. are 'Elizabeth the Second, by the Grace of God of the United Kingdom of Great Britain and Northern Ireland and of Her other Realms and Territories, Queen, Head of the Commonwealth, Defender of the Faith'. The title of Emperor of India was abandoned in 1947. The numbering of sovereigns bearing the same name follows whichever was the highest number borne by a previous sovereign of that name in the monarchy of England or of Scotland.

The title of Queen Elizabeth in member states of the Commonwealth is determined by Her Majesty at the instance of each member concerned, and accordingly varies slightly in different Commonwealth countries.

Royal veto. The prerogative right of English and now British monarchs to refuse assent to Bills passed by the Lords and Commons. It was much used by Elizabeth I, and also asserted by her as a threat when Parliament was proving difficult. The veto was last used by Queen Anne in 1708 on a Scotch Militia Bill. It is now a convention of the constitution that it will not be used and any use of the power would bring about a constitutional crisis.

Royal warrant. A writing issued by a royal official under which a tradesman is authorized to act in a particular capacity, e.g. supplier of fishing tackle, for some member of the royal family. Such a tradesman may display the royal arms and designate himself as 'By Royal Appointment supplier of . . . to . . .'.

A royal warrant may also occasionally be used as an instrument for exercising authority under common law or statute.

Royalty. A periodical payment made by a publisher to an author who has licensed him to publish his work, usually calculated as a percentage of the sale price of each copy sold of the book. Also a periodical payment to a patentee by a person licensed to exploit the patent. Also a periodical payment to the owner of minerals by a party authorized to extract and remove the minerals.

Rubric. The rubric of a statute is its long title, so-called because originally it was written in red letters.

The rubric of a law report is the headnote indicating the main topics of law in issue and discussed in the case.

In ecclesiastical law, the rubrics of the Book of Common Prayer include all directions and instructions contained in it, and all its tables, calendars, rules, and other contents. The name is taken from the red ink in which the rubrics are sometimes printed.

Rufinus (?–*c.* 1192). Italian archbishop, canonist and theologian, and author of *Summa decretorum*, the most influential decretist work until the appearance of Huguccio's *Summa Super Decretis*, and one which introduced analytical techniques in imitation of the civilians and first gave systematic exposition of the text.

Ruggles-Brise, Sir Evelyn John (1857–1935). Entered the home civil service and, having served as private secretary to four Home Secretaries, became a prison commissioner, and, in 1895, chairman. He remained in this post till 1921. His first task was to implement the recommendations of the Gladstone committee of 1895; the changes included the building of workshops, mitigation of the rigours of separate confinement, reforms in the classification of prisoners, in medical care, and in the training of prison staffs. He introduced the Borstal system for the rehabilitation of young offenders with, in 1908, the indeterminate sentence, related not to the crime but to the youth's progress under training. His work was influential abroad also and he was long President of the International Penal and Penitentiary Commission. In 1924, he published *Prison Reform at Home and Abroad.*

Ruinus, Carolus (sixteenth century). A commentator, author of *Responsorum sive consiliorum tomi* (1571).

Ruiz, Jacome (Jacobo de la Leyes) (thirteenth century). An Italian naturalized in Spain, who wrote, for Alfonso X, the *Flores de las Leyes,* an anthology of materials on civil law, judicial organization and procedure from Italian jurists, some of which was incorporated in *Las Siete Partidas* (q.v.).

Rule. A statement of the prescription of the law on some matter, usually applied to one more specific and detailed than a doctrine or principle (qq.v.), though whether a statement is called a doctrine or principle or rule may be a matter of doubt and dispute. Rules may be prescribed by statute, e.g. a person is not to drive a vehicle on a road in a built-up area at more than X miles per hour, or by common law, e.g. if a person slanders another he is liable to pay him damages. Some rules have names of their own, e.g. the rule in *Shelley's case.*

Rules are of different kinds; a material distinction is between rules of recognition, determining whether a statement is a prescription of law, or of morals, or manners, or otherwise non-law, rules of substantive law, which state the rights and duties of persons, and rules of adjective law, which prescribe the means of enforcing rights and duties and securing remedies. In all cases, much importance attached to the precise verbal formulation of rules.

The term rule is also given to a regulation made by court or public office under the authority of statute, and formerly to rules made by ministers under statutory authority in further implement of statutes: see STATUTORY RULES AND ORDERS. It may also mean a direction made by a court in an action or proceeding. Thus, a court may make a rule absolute, or a rule *nisi,* in the latter case allowing the party to show cause why it should not be made absolute.

Rule of law. A concept of the utmost importance but having no defined, nor readily definable, content. It implies the subordination of all authorities, legislative, executive, judicial, and other to certain principles which would generally be accepted as characteristic of law, such as the ideas of the fundamental principles of justice, moral principles, fairness and due process. It implies respect for the supreme value and dignity of the individual.

In any legal system it implies limitations on legislative power, safeguards against abuse of executive power, adequate and equal opportunities of access to legal advice and assistance and protection, proper protection of individual and group rights and liberties, and equality before the law. In supranational and international communities, it implies recognition of the different traditions, aspirations and claims of different communities, and the development of means to harmonize claims, resolve conflicts and disputes and eliminate violence. It means more than that the government maintains and enforces law and order, but that the government is, itself, subject to rules of law and cannot itself disregard the law or remake it to suit itself.

It probably originated in mediaeval time in the belief that the law, whether of God or of man, ought to rule the world. One development of this was the idea that certain fundamental laws are unalterable by even the government, an idea itself connected with ideas of natural law.

In England, as a result of the constitutional struggles of the seventeenth century, in which Parliament and the common lawyers were victorious, the supremacy of the law came to mean the supremacy of Parliament. But this involves that Parliament can, at any time, and even retrospectively, alter the law to suit its purpose, or to absolve its misdeeds, or can overrule or reverse judicial decisions. Parliament is above the law, and the executive, the Cabinet, so long as it can carry a majority in the House of Commons, is in control of Parliament.

A. V. Dicey, in his *Introduction to the Study of the Law of the Constitution* (1885), reached the conclusion that the phrase had three meanings in the England of his day, namely, that no man could be made to suffer punishment or to pay damages for any conduct not definitely forbidden by law, that every man's legal rights or liabilities were almost invariably determined by the ordinary courts of the realm, and that each man's individual rights are far less the result of the constitution of the United Kingdom than the basis on which the constitution is founded. The first meaning, that liberty was conditioned by the regular rules of law and this excluded arbitrary interference by the government is now much less true than in Dicey's time.

The second meaning, that the courts alone were able to determine contract, tort, or crime, has increasingly become untrue since Dicey's time.

The third meaning, that in some countries constitutions state and create guaranteed rights but that, in Britain, these proceed from the enforcement of private rights by the courts is also less true.

Since Dicey wrote, important qualifications have developed to the doctrine. Thus, though certain Crown privileges in litigation were abolished by the Crown Proceedings Act, 1947, others, such as the privilege of non-disclosure of documents, remain. Many liberties have been curtailed by statute. Tribunals have been established to decide many issues which do not administer the common law or its procedure. Again, some groups, notably trade unions, have coerced Parliament into granting them immunity from the ordinary rules of law, such as their immunity from liability in tort. Not least, there has been a marked decline in trust in the courts and the law and a marked increase in lawlessness, not merely in crime, but in dissent, demonstrations, and violence.

An important issue in relation to maintaining the rule of law is: what law? It cannot be assumed that the positive law of a particular state or province thereof existing at a particular time is entirely, or at all, worthy of being maintained or that governing people by subjection to that rule is necessarily right. The German people may, during the Third Reich, have been subject to the rule of law, but it was a law which many of them and many outsiders found abhorrent in many respects. The law, to the rule of which governors and governed must be subjected, must be a body of principles and rules consistent with the most generally accepted bodies of human rights and liberties which have been formulated. Much value accordingly attaches to such formulations as the Universal Declaration of Human Rights. It cannot be assumed that the existence of nominally democratic forms of government will ensure this because democracy is at the mercy of ignorant and unthinking majorities, frequently jealous and biased, obsessed by class-hatred, quite capable of, and frequently willing and ready to, suppress liberties not in accordance with their way of thinking. The miserable story, in many countries since the Reformation, of religious intolerance enforced and promoted by law evidences this. Racial hatred is another example.

Intranationally, it is necessary to examine the law and practice of each state to determine whether and how the rule of law applies therein.

In the United Kingdom, it applies only to a limited extent and with large exceptions, so large as to make it doubtful whether it can be said to apply at all. In the absence of a written constitution, with no Bill of Rights or entrenched rights, and with well-established principles that the legislative power of the Queen in Parliament is unlimited and that no Parliament can bind its successors there is no legal restriction on Parliament enacting legislation in complete conflict with any or all of the most fundamental human or moral rights, and it has frequently enacted confiscatory or expropriative legislation, retrospective legislation, and much that many persons have considered in flat contradiction of human rights. Parliament has, moreover in the twentieth century, repeatedly conferred on individuals, authorities, departments of State, and others very wide and far-reaching powers, enabling such persons and bodies to do practically whatever they like with ordinary individuals and their rights and property, and has frequently excluded completely or almost completely appeal or resort to the courts, or limited appeal in a mode or to an extent which makes it worthless. Parliament is wholly incapable of controlling the exercise of the powers it has conferred on Ministers and their departments. Again Parliament, at the dictation of the major trade unions, has very largely exempted them from liability to law, conferring immunity from legal proceedings for torts arising from industrial disputes. Very significant in the twentieth century has been the increasing refusal of trade unions to be subject to the law, and their insistence that any judicial decision restrictive of their rights be reversed by legislation, which a subservient Parliament has regularly done. The trade unions are no more subject to State law than was the mediaeval Church. Parliament, moreover, has frequently indemnified individuals from even deliberate contraventions of the law if the political majority was minded to do so. The rule of law does not apply in modern Britain which, despite the forms and some appearances of liberal democracy, when a Labour government is in power, is a dictatorship headed by a group of trade union leaders and their political servants.

In the U.S., the legislative power of Congress is limited in general terms by the terms of the Constitution, including the Bill of Rights, and the ultimate power lies with the Supreme Court which says what, in particular circumstances, the Constitution means and allows or forbids. By its interpretative power it has hitherto had considerable success in upholding the rule of law substantively and procedurally. In states and municipalities, the external influence has sometimes been simple graft and corruption, sometimes the economic power of large business enterprises.

In 1958, the American Bar Association created a Special Committee on World Peace through Law to examine what could be done to advance the rule of law among nations. Later a Conference at Athens (1963) adopted a declaration of general principles for a world rule of law and resolved to create a World Peace through Law Center.

Rule of the road. Customary rules, judicially recognized and now recognized by statute. The principal ones are that animals and vehicles keep to the left hand when meeting others; when passing others going in the same direction they keep to the right and consequently must beware of oncoming vehicles; and when crossing the course of another they bear to the left of the other and pass behind it. Pedestrians may walk on either side of the road but it is prudent to walk on the side on which there is a footpath, if there is one, failing which to walk on the right so as always to be facing the traffic which is nearer when passing.

These customary rules are the basis of the modern Highway Code.

Rule of the road at sea. Since 1840, many sets of regulations of varying application have been made laying down rules to prevent collisions at sea. Those now in force are International Regulations for Preventing Collisions at Sea, 1960, which provide for lights, warning signals, and altering speed or course to avoid collision. The basic principle is that vessels meeting must keep to starboard and pass port to port and when crossing the vessel to starboard has the right of way.

Rule of the war of 1756. At the outbreak of war in 1756, British maritime supremacy threatened to end the export of colonial materials from French colonies to Europe and prevent the provisioning of these colonies from Europe. The French were prepared to allow Dutch ships to share in the West India trade under limited licence. The British, however, undertook to seize and the prize court to condemn, Dutch ships which engaged in the French colonial trade because they were using neutral character to cover continuance of the French trade.

Rules of construction or interpretation. These rules are different from rules of law in being flexible and not binding and really only guides to the meaning which should be put on phrases or words in statutes, deeds, and other writings. They are never to be applied if the application is inconsistent with what appears to be the most reasonable meaning of the phrase or words in question. Subject to that overriding principle, the chief rules or canons of construction are: (1) words and phrases are to be given their ordinary, everyday meaning unless they are technical legal terms, in which case they are given their technical meaning; (2) words and phrases are to be understood having regard to the rules of English grammar; (3) they must be construed consistently with the context in which they are used, the statute or deed being considered as a whole and consistently with other statutes or deeds on the same subject-matter; (4) in the case of statutes, regard may be had to the preamble, the long title, headings to sets of sections, and schedules, but not to marginal notes, punctuation, short title or schemes or orders made under the Act. Regard may also be had to the prior state of the law, cases interpreting the provisions, usage, standard textbooks but not to international law, Royal Commission or other reports, White Papers, or Hansard.

Rules of court. Rules, usually made by the courts under statutory authority, regulating procedure. In England, the Appellate Jurisdiction Act, 1876, gave the rule-making power to a committee of judges, barristers, and solicitors, who have made the Rules of the Supreme Court, frequently amended and periodically largely recast. In Scotland, statutes give power to a Rules Council, who have made the Rules of the Court of Session. These rules are frequently amended and, from time to time, recast.

Rules of good seamanship. Practices based on custom and experience in the handling of ships which should be adhered to over the whole range of conditions in the working of a vessel. What is good seamanship is a question of fact to be decided in each case, on consideration of all relevant circumstances. In case of conflict with the International Regulations for Preventing Collisions at Sea, the latter must be obeyed.

Rules of law. A phrase sometimes used for the whole body of prescriptions comprised in a legal system. See RULE.

Rules of law (*Regulae iuris*). A body of statements of fundamental norms, ordered to be made by Pope Boniface VIII as the completion of the *Sext* (q.v.), after the model of Justinian's *De Diversis Regulis Iuris Antiqui* in Digest 50, 17. The collection was edited by Dino Rosoni of Mugello, the distinguished jurist; there are 88 in number, mostly drawn from Roman law, but some from canon law and, frequently, from the Bible. Many are obvious or so vague as to be of little value, yet they are frequently mentioned in the modern Code of Canon Law.

Rules of the Air and Air Traffic Control. A body of regulations, made under statute, governing the management of aircraft taking off, flying, and landing in the interests of safety and so as to minimize the risk of collision or other accident. They include right-of-way rules for take-off, flight, and landing, the general principle being that aircraft keep to the right and give way to other aircraft on the right.

Rules of the Supreme Court. Rules regulating the practice and procedure of the Supreme Court of Judicature in England and Wales, made by a Rule Committee comprising of the Lord Chancellor, various judges, barristers, and solicitors. They must be laid before Parliament and are subject to annulment. There are separate rules for the Crown Court, the Criminal Division of the Court of Appeal, and for matrimonial causes. They are supplemented by practice directions and practice notes and, in the Queen's Bench Division, masters' practice directions. They are printed with copious annotations in the Annual Practice.

Ruling cases. Leading cases, the principle of which rule later cases.

Rump Parliament. The name sometimes given to what remained of the Long Parliament (q.v.) after the exclusion of many members of the Commons by Pride's Purge (q.v.) and which was expelled by Cromwell in 1653. It was also the name sometimes given to what remained of the Long Parliament which was recalled in 1659. The restored Rump was again expelled later in 1659 but recalled at the end of that year. The excluded members were recalled by Monk and, having prepared the restoration of Charles II, the Rump and the Long Parliament dissolved themselves in March 1660.

Run with the land. A condition or covenant is said to run with the land when the benefit or burden of the condition or covenant transmits with the land to a purchaser or other assignee thereof and does not continue to attach to the party originally benefited or bound personally.

Running days. In charter-parties, the whole period of time allowed for loading or unloading the cargo, including Sundays and holidays and not, unless by local custom, equivalent to working days.

Runnymede. The place on the south bank of the Thames, 20 miles west of London, where King John sealed Magna Carta (q.v.) in 1215. The area, now owned by the National Trust, contains memorials commemorating Magna Carta, erected by the American Bar Association, Commonwealth airmen killed in World War II, and President J. F. Kennedy.

Runrig lands. In Scots law, lands where successive ridges of a field or successive portions belong to different proprietors. Statute of 1695 authorized the consolidation of such lands and their redivision among the different proprietors according to their respective interests.

Russell, Charles, Baron Russell of Killowen (1832–1900). Was admitted a solicitor in Belfast in 1854 and, in 1856, he moved to London where he was called to the Bar in 1859. In 1872 he took silk and became a leader in the commercial field; he was an aggressive cross-examiner and an impassioned pleader. In 1880 he entered Parliament and supported Irish home rule, became Attorney-General in 1886 and, in 1888, was leading counsel for Parnell before the special commission and made a brilliant speech for the defence. In 1892 he became Attorney-General again and, in the following year, represented the United Kingdom at the Behring Sea arbitration. In 1894 he succeeded Lord Bowen as a Lord of Appeal and received a peerage but, a month later, succeeded Lord Coleridge as Lord Chief Justice. As such he was still masterful, but patient and courteous, and showed an excellent knowledge of law. In 1899, he acted as arbitrator in the arbitration on the Venezuela-British Guiana boundary dispute. He also took considerable interest in legal education, urging raising of the existing standards.

Russell, Charles Ritchie, Baron Russell of Killowen (1908–). Son of Francis, Lord Russell of Killowen (q.v.), he was called in 1931, took silk in 1948, and was a judge of the Chancery Division 1960–62, a Lord Justice of Appeal, 1962–67, and a Lord of Appeal in Ordinary 1975– . He was also a judge of the Restrictive Practices Court, 1960–61, and President of that court, 1961–62.

Russell, Frank (Francis) Xavier Joseph, Baron Russell of Killowen (1867–1946). Son of Charles Russell, L.C.J. (q.v.). He was called to the Bar in 1893 and practised in Chancery. In 1919 he became a Chancery judge, a Lord Justice of Appeal in 1928 and, in 1929, a Lord of Appeal. His judgments are models of concise and lucid English and of high authority. His son Charles (q.v.) also became a Lord of Appeal.

Rutherforth, Thomas (1712–71). Regius Professor of Divinity at Cambridge (1745) and archdeacon of Essex, published an *Institutes of Natural Law* (1754), based on lectures at Cambridge on Grotius' *De Jure Belli ac Pacis*, which had substantial influence. It was the first systematic exposition of natural law by an English writer. The two parts deal respectively with natural and with politic law, and it relies considerably on Old Testament history and legal rules. Though displaying considerable learning, the lectures are not always clear, and some of the criticisms are ill-founded. The theory of the relation of the law of nature to international law is unclear.

Rutledge, John (1739–1800). Was called to the English Bar in 1760, was a delegate to the first and second Continental Congresses, participated in writing the South Carolina Constitution of 1776, and became, successively, President (1776–78) and Governor (1779–82) of South Carolina. He was a judge of the Court of Chancery of South Carolina, 1784–91, and an Associate Justice of the U.S. Supreme Court, 1789–91, before becoming Chief Justice of the Common Pleas of South Carolina in 1791. In 1795 he was named Chief Justice of the Supreme Court, but the Senate rejected the nomination and he sat for only five months.

Ryder, Sir Dudley (1691–1756). Called to the Bar in 1725, he was successful and sat in Parliament from 1733. He became Solicitor-General in 1733,

Attorney-General in 1737, and Chief Justice of the King's Bench in 1754. He appears to have been efficient, honest, and sound, but not noteworthy. He was made a peer the day before he died. His descendants became barons and earls of Harrowby.

Rylands* v. *Fletcher (1868), L.R. 3 H.L. 330. The defendant employed an independent contractor to construct a reservoir on his land. When it was filled water flowed into disused mine workings underneath which communicated with, and consequently flooded, the plaintiff's mines. There was no proof of negligence. It was held that a person who used his land in a non-natural way, as where he, for his own purposes, brought on his land and collected and kept there anything likely to do mischief if it escaped, must keep it in at his peril, and was *prima facie* answerable for all the damage which was the natural consequence of its escape, unless he excused himself by showing that the escape was due to the plaintiff's default, or was the consequence of *vis major*, or act of God.

Rymer, Thomas (1641–1713). Though called to the Bar in 1673, Rymer had been a dramatist, literary critic, and poet before he became an antiquarian. In 1684 he published a tract on the *Antiquity, Power and Decay of Parliaments* and, in 1692, became historiographer-royal. In 1693 he was appointed to edit a collection, undertaken by the government, of past treaties entered into by England with foreign powers. Taking as his model Leibnitz's *Codex Juris Gentium Diplomaticus*, he put out, under the title *Foedera, Conventiones, Literae et cujuscumque Generis Acta Publica ... ab 1066 ad nostra tempora*, 15 volumes between 1704 and 1713 and his co-editor, Sanderson, brought out five further volumes after his death. The documents cover the period 1101 to 1654 and, though intended to be a collection of treaties, it contains much matter of legal importance. The third edition (10 vols., 1735–45) contained additional documents. There is a *Syllabus of Documents in Rymer's Foedera* by T. D. Hardy published in three volumes in the Rolls Series in 1869–85.

S

S. The mark at one time branded on the forehead or cheek of a person adjudged under the Statute of Vagabonds, 1547, to be a slave for two years, and who ran away and was caught. He was also to be adjudged a slave for ever. This statute had been obsolete for centuries before it was repealed in 1863. It is also the abbreviation for a section of an Act of Parliament.

S.I. Statutory instrument (q.v.).

S.R. & O. Statutory rules and orders (q.v.).

Sabinians. One of the school of jurists of the Roman Imperial period down to the time of Salvius Julianus and Gaius, taking its name from Massurius Sabinus (q.v.), though the founder of the school was his teacher G. Ateius Capito or Cassius. The rival school was the Proculians. The schools were educational establishments and also groups holding different views, but it is difficult to ascertain what were the real differences in principle between them. The major adherents of the Sabinian school were Cassius, Javolenus, Aburnius Valens, Salvius Julianus, and Gaius.

Sabinus, Massurius (first century A.D.). Roman jurist, famous for an exposition of the *ius civile* in three books, taken later as a model for other systematic treatises on private law which were consequently called *Ex Sabino* or *Ad Sabinum*, and author also of *Ad edictum praetoris urbani, De furtis*, and *Responsa*. He was the second leader of, and gave his name to, the Sabinian school (q.v.) of jurists.

Sabinus, Cnaeus Arulenus Caelius (first century A.D.). Roman jurist, head of the Sabinian school, and author of a book on the aedilician edict.

Sac and soc (or sake and soke). Words commonly included in early charters denoting a kind of jurisdictional right granted. Maitland's view was that the Anglo-Saxon sac and soc included the right to hold a petty court, to compel tenants to attend it and to take profits from it, but this view has been doubted. In the thirteenth century the words seem to describe jurisdiction such as every lord had by implication and even without a royal charter.

Sacco and Vanzetti, Trial of. Nicola Sacco and Bartolomeo Vanzetti were immigrants to the U.S. from Italy, employed in Massachusetts in a shoe factory and as a fish pedlar respectively. On 15 April 1920 the paymaster and guard of a shoe factory were shot dead in the street of South Braintree, Mass., and robbed of $16,000. Sacco and Vanzetti were arrested, tried, and convicted. Applications for rehearing were denied, but in 1925 a condemned criminal made a confession and in 1926 much evidence was found corroborative of the view that the robbery was the work of professional bandits from Providence, R.I. Large sections of public opinion demanded reconsideration and the governor of Massachusetts appointed an advisory committee to reconsider the case. The Governor and committee were unanimous that the trial was fair and the defendants guilty and they were electrocuted on 23 August 1927. Doubts as to the rightness of the conviction linger, however, as both accused had evaded military service in 1917–18 and had been connected with socialism and anarchism and there had been considerable local prejudice against them.

F. Frankfurter, *The Case of Sacco and Vanzetti*; J. Fraenkel, *The Sacco-Vanzetti Case.*

Sachsenspiegel (Mirror of the Saxons). A very important mediaeval German legal text composed probably about 1230 by a Saxon knight and judge, Eike von Repkow, first in Latin, then in German. It includes a book of territorial law (*Landrecht*) and one of feudal law (*Lehnrecht*), mainly for the Ostphalian part of Saxon territory. The book attained great prestige and was given the authority of a statute book in the Saxon courts, was translated, furnished with a gloss by Johann von Buch, later enlarged by Nikolaus Wurm and finally given a commentary. Revised and modified versions of it became the basis of other later law-books such as the *Gorlitz Lawbook*, the *Breslau Territorial Law*, the *Dutch Sachsenspiegel*, and the *Berlin Townbook of Brandenburg Law*. It was attacked by theologians and some articles of it condemned in 1372, but long after this remained the basis of much legal literature and a major source of civil law right down to the adoption of the German Civil Code at the end of the nineteenth century.

Sacramentum. In the oldest kind of civil proceeding in Roman law (*legis actio sacramento*) a sum of money deposited by both the litigants as a stake, the loser in the action forfeiting his to the State. It was doubtless originally an oath, and the deposit of money originally an expiation of perjury. The judge had to decide which party's *sacramentum* was *iustum* and the amount depended on the value of the cause. Later, a deposit was not made but payment was guaranteed by sureties.

Sacrilege. Any profanation of sacred things. The Levitical code of Israel punished such acts and the Israelites had extensive rules to safeguard what was consecrated. In Greece sacrilege was connected with treason and the theft of temple property was criminal. In Christianity the term originally meant theft of sacred things, but by the fourth century was used more broadly. The Theodosian Code applied it to heresy, schism, apostasy, paganism, actions against the immunity of church or clergy, the desecration of sacraments, and other acts. In modern canon law it is a sin against the virtue of religion, or abuse or violation of a sacred person, place, or thing, dedicated to the service of God. Personal sacrilege is assault on, wrongful imprisonment or defamation of a member of a religious community. Local sacrilege is violation of a sacred place, such as a church or churchyard, as by robbing a grave. Real sacrilege is the misuse of sacred things. Sacrilege may be tortious or criminal, and if so is frequently viewed more seriously because of the affront to the feelings of persons who adhere to the religion in question.

In English law theft from a church was for a long time a capital offence, the last execution taking place in 1819.

Sadism. The form of sexual deviation which consists in achieving sexual gratification through the infliction of pain on another. It is so-called from the Marquis de Sade (1740–1814) who indulged in a wide variety of perversions. Sadism is not a distinct crime but may involve assault or various forms of sexual offences.

Sadler, Michael Thomas (1780–1835). Businessman, radical politician, and leader of the Factory Reform Movement, led the movement in Parliament for a limit of 10 hours daily on factory work by persons under 18 which resulted in the Factory Act of 1833 restricting hours for persons aged 13–17 to 12 hours daily. On his death, leadership of the Reform Movement passed to Shaftesbury (q.v.).

Safe-conduct. A written permission, given by a belligerent to enemy subjects or others, allowing them to proceed to a particular place for a defined object, or to return home, without molestation. The grantee is inviolable so long as he complies with the conditions imposed on him or necessitated by the circumstances of the case. Unless stated, a safe-conduct does not cover goods or luggage. They may be given also for ships and for goods. To be effective under international law the grant must have been arranged between belligerents.

St. Germain, Christopher (?1460–1540). Called to the Bar but confined himself to legal studies and literary work. He is chiefly remembered as author of *Doctor and Student* (in Latin, 1523, under the title *Dialogus de fundamentis legum et de conscientia*, in English 1530 and many later editions), two dialogues, the first in Latin, the second (and longer) in English, between a doctor of divinity and a student of the laws of England. There are Additions to the second dialogue comprising thirteen chapters on the power and jurisdiction of Parliament. Legal rules are criticized by religious and moral standards, and there are many enquiries about the law of reason and of nature, and the foundations of the common law. It put into popular form canonist learning as to the nature and objects of law and the different kinds of law, and facilitated the development of these principles on active lines. The book was very well known in the legal profession, frequently cited, and often reprinted, and it exercised great influence on the development of equity. It was answered by the *Replication of the Serjeant-at-law to the Doctor and Student* by an unknown serjeant, and this in turn provoked *A Little Treatise concerning Writs of Subpoena.*

St. Leonards, Lord. See SUGDEN, EDWARD.

St. Nicholas, Confraternity of. A guild of lawyers practising at the Parlement of Paris, established in 1342.

Saint-Pierre, Charles Irenee Castel de (1658–1743). French man of affairs, and philanthropist, author of a *Projet de paix universelle entre les potentats de l'Europe* (1713), proposing a kind of League of Nations for Europe.

St. Vigeans, Lord. See ANDERSON, DAVID.

Sale. The transfer by common consent of the ownership of a thing or an interest in land, or in incorporeal property, from one person to another in exchange for a price in money. The distinction between a sale and some other contracts such as hire-purchase, mortgage, and contract for work and labour may be a narrow one. For reasons partly historical, partly related to the nature of the thing, rules in many respects distinct have developed in common law and related systems for sales of chattels, sales of land, and sales of choses in action.

See SALE OF GOODS; SALE OF LAND.

Sale of goods. In the United Kingdom down to 1893 the law of England, Wales, and Ireland as to sale of goods was regulated almost entirely by common law, including the law merchant, and the law of Scotland by Scottish common law, in this respect following closely the Roman law. The Sale of Goods Act, 1893, largely codified the rules peculiar to sale but does not deal with matters common to sale and other contracts, such as capacity, or the effect of fraud, and common law principles continue to apply so far as not inconsistent with the Act. Apart from certain provisions applicable to England, Wales, and Ireland only, or to Scotland only, it applies to the whole U.K. and made revolutionary changes in the law of Scotland. The Act applies only to sales of goods, i.e. chattels personal other than things in action and money; the goods may be existing, or future goods—yet to be made or acquired; the price may be fixed by the contract, or be left to be fixed in manner thereby agreed, or be determined by the course of dealing between the parties.

Under the Act there is an implied condition that the seller has, or will have, a right to sell, and implied warranties that the buyer shall have quiet possession of the goods and that they will be free from any charge or encumbrance not declared or known to the buyer. Goods may be sold by description, in which case there is an implied condition that they shall correspond with the description; or by sample, in which case it is implied that the bulk shall correspond with the sample, or by sample and description, in which case it is implied that the bulk shall correspond with the sample in quality, that the buyer shall have a reasonable opportunity of comparing the bulk with the sample, and that the goods shall be free from any defect rendering them unmerchantable which would not be apparent on reasonable examination of the sample.

Apart from exceptions, there is under the Act no implied warranty or condition as to the quality or fitness for any particular purposes of goods supplied under a contract of sale, unless the buyer makes known to the seller any particular purpose for which the goods are required, in which case there is an implied condition of reasonable fitness for such purpose, unless the circumstances show that the buyer does not rely on the seller's skill and judgment. Where goods are bought by description from a seller who deals in goods of that description, there is an implied condition that the goods shall be of merchantable quality. Implied conditions and warranties may only be excluded or varied in non-consumer sales. If there is a breach of warranty, the buyer may set this up in diminution or extinction of the price or claim damages for breach of warranty. If there is a breach of condition, he may rescind the contract and claim damages for its total failure.

The property in the goods sold passes at the time when the parties intend it to pass, which is prima facie when the contract is made; the Act provides for other common cases. The owner's rights and liabilities, including the risk of loss, pass on the completion of the sale.

The Act makes provisions for the transfer of title of ownership by, in certain cases, a non-owner.

Delivery of the goods and payment of the price are concurrent conditions; a buyer is not deemed to have accepted goods until he has had a reasonable opportunity of examining them to ascertain whether they conform to the contract. Rejection is precluded unless made within a reasonable time.

An unpaid seller has a lien on the goods while still in his possession, a right of stopping the goods *in transitu* after he has parted with possession of them and, within limits, a right of resale; if the property in the goods has not passed, he has a right of withholding delivery. The seller may also bring an action for the price, or claim damages, for non-acceptance. If the seller defaults, the buyer may claim damages for non-delivery, or claim specific performance of the contract, or claim for the detention or conversion of the goods.

Sale of land. The sale of land is regulated mainly by common law and because of the differences in land law differs materially between England and Scotland. In England the main stages are a preliminary agreement, a formal contract and payment of a deposit, delivery by the vendor of an abstract of title and requisitions for further information on particular points, preparation by the purchaser and execution by the vendor of a conveyance, transfer of title deeds relating wholly to the property conveyed and payment of the purchase-money by the purchaser. In the conveyance the vendor usually enters into covenants for title in pursuance of his obligation to give a title clear of defects and encumbrances. Various covenants are implied by statute. Registered land is conveyed by the execution of an instrument of transfer followed by registration of the purchaser as proprietor of the land.

In Scotland the main stages are execution of a contract, usually by exchange of letters, called 'missives', delivery of the titles to and their examination by the buyer's solicitor, drafting by the buyer and execution by the seller of the disposition, the deed which transfers the seller's title to the buyer, and delivery thereof and of the title deeds to the buyer in return for payment of the purchase price. The disposition must be recorded in the General Register of Sasines.

Sale or return. A form of contract whereby goods are delivered to a party on terms that the property does not pass until the goods are approved or accepted and may be returned within a reasonable time if not wanted.

Saleilles, Raymond (1855–1912). French jurist, notable particularly in comparative law. His major writings include *Étude sur la théorie générale de l'obligation d'après le premier projet de code civil pour l'empire allemand* (1890), *De la declaration de volonté* (1901), *De la possession des meubles* (1907), and *De la personalité juridique* (1910) and he exercised great influence on the legal science of his time.

Salem witch trials (1692). Various girls in Salem, Mass. accused three Salem women of witchcraft. The magistrates established a special court and investigated various cases and as a result 19 were hanged, and many more persecuted before public opinion stopped and condemned the trials. The General Court of the colony adopted a resolution for repentance in 1696 and one of the judges publicly admitted the injustice of the trials.

Salford Hundred Court. This was one of the ancient Anglo-Saxon hundred courts, and its jurisdiction, limited in 1278, was extended in 1846 when it was made a court of record. The court was re-constituted in 1868 by the amalgamation of the ancient Salford Hundred Court and the Manchester Court of Record, which had been held by right of a charter of 1838 and had a jurisdictional area co-extensive with the areas of the Manchester and Salford county courts. Its jurisdiction was statutory but was similar to though less than that of a county court. The judge had to be a barrister of at least 10 years standing and was appointed by the Chancellor of the Duchy of Lancaster. Appeal lay to the Court of Appeal. The court was still active when it was abolished in 1971.

Salgado de Somoza, Francisco (?–1664). Spanish jurist, author of *Tractatus de supplicatione* (1639), *Labyrinthus creditorum concurrentium* (1653) on debtors and insolvency, and *Tractatus de regia protectione* (1654).

Salic Law. A primitive and very early Germanic code which seems to have been a compilation of ancient customs of the Salian Franks, made at different times, subsequently frequently amended, enlarged, and revised. It may have been first written down before A.D. 500 but revised versions extend down to about A.D. 800. The earliest version has neither pagan nor Christian elements, but Christian elements came into the later versions, as did capitularies of Merovingian kings.

The text is accompanied by glosses known as the Malbergian glosses (the term probably standing for malloberg, the assembly or popular court) which are practical instructions for applying the law to cases. Charlemagne had the law amended, the Malberg glosses eliminated, interpolations omitted, and other chapters added. It is predominantly a penal code containing a long list of fines, but contains some civil provisions, notably that on succession, providing that daughters cannot inherit land.

The Salic Code had great authority and wide influence by reason of the dominant position of the Frankish people in Western Europe. It became the foundation code for all the Frankish codes (Chamavian, Ripuarian, Thuringian). It influenced many later Germanic codes and was in force in Italy for Franks so long as the system of personal laws prevailed.

Hessels and Kern, *Lex Salica* (1880); Eckhardt, *Pactus Legis Salicae* (1954).

Salic law of succession. The principle, taking its name from the Salic Law or *Lex Salica* (q.v.), in certain sovereign dynasties, that persons descended from a previous sovereign through a female are excluded from the succession to the throne. The Salic Law seems first to have been adduced as a justification for the exclusion of descendants by a female from the French throne in the fifteenth century. The principle was adopted by statesmen in the sixteenth century as a fundamental law of the kingdom and from 1593 it was accepted as fundamental law in France. It was introduced into Spain in 1713. It applied in Hanover so that when Princess Victoria became Queen of Britain, the Crown of Hanover passed to her uncle and became separated from the succession to the British throne.

Salicetus, Bartholomaeus (?–1412), Italian jurist, a famous commentator, author of *Commentaria in Codicem* and *Commentaria in Digestum* (1574).

Salisbury, John of (Johannes Sarisberiensis) (*c.* 1115–80). English diplomat, secretary to Becket and bishop of Chartres, author of *The Policraticus,* sketching an ideal state like Plato's Republic, in which the prince can be deposed if he disobeys the law, the *Metalogicus,* fusing Aristotelian and Augustinian philosophy, an *Historia Pontificalis* and other works.

Salmond, Sir John William (1862–1924). Became a New Zealand barrister in 1887, Professor of Law at Adelaide, 1897–1906, and Wellington, N.Z., 1906–7. He was Solicitor-General for New Zealand, 1910–20, and later a judge of the Supreme Court of New Zealand. He was the author of *Essays in Jurisprudence and Legal History* (1891), *Jurisprudence* (1902 and later editions), *The Law of Torts* (1907 and later editions), and *The Law of Contracts* (completed by J. W. Williams). His *Jurisprudence* and *Torts* are classics and very influential.

Salman. In Germanic law a person charged with the conveyance of property to another. His office is sometimes thought to have been the origin of the concept of trustee.

Salmon, Cyril Barnet, Lord (1903–). Called to the Bar in 1925 he became a judge of the Queen's Bench Division in 1957, a Lord Justice of Appeal in 1964, and a Lord of Appeal in 1972.

Salusbury, Sir Thomas (eighteenth century). Judge of the Court of Admiralty, 1757–73, in which capacity he seems to have been of only ordinary quality.

Salvage. A service which saves or contributes to the safety of a vessel, her apparel, cargo, wreck, or to the lives of persons belonging to her when in real and substantial danger at sea, so long as rendered voluntarily and not under any legal, or official duty, or for self-preservation. Typical salvage services include towing out of danger, extinguishing fire, and saving cargo. Under maritime law such a service justifies a reward, also called salvage, which, to encourage such services and to reward skill and seamanship, is generous, or the service may be the subject of an express contract. The right to salvage reward for preserving life is statutory only. As a rule salvage is payable only if the service was successful, in at least contributing materially to ultimate safety. The principle of salvage has been extended to aircraft.

W. R. Kennedy on *Civil Salvage.*

Salvioli, Giuseppe (1857–1928). Italian jurist and economic historian, author of an important *Manuale di storia del diritto italiano* (1890) which emphasized the continuity of Roman influence.

Salvius Julianus (Lucius Octavius Cornelius Publius Salvius Julianus Aemilianus) (*c.* A.D. 100–*c.* A.D. 169). Roman jurist, disciple of Javolenus and last leader of the Sabinian school. He was commissioned by Hadrian to rearrange, revise, and put into final form the praetorian edict, a task which brought him a high reputation. Under Hadrian and Antoninus Pius he was a member of the imperial *consilium.* He wrote a *Digesta,* extensively quoted by later jurists, *Responsa* and commentaries on certain lesser jurists, and is generally regarded as the outstanding jurist with whom Roman legal science reached the height of its development.

Sanchez, Thomas (1550–1610). Spanish jurist, canonist, and theologian, author of a famous *Disputationes de sancto matrimonii sacramento* (1602) and of *Consilia seu opuscula moralia* (1634). The former brought him great renown and is still considered by Catholic authorities one of the classic works on marriage.

Sanction. Generally any reaction which indicates approval or disapproval of conduct and tends to induce conformity to a required standard of behaviour; more specifically, a provision for the enforcement of a rule of law by reward for observance, or punishment for its contravention. In practical terms every requirement, or provision, or prohibition of law requires a sanction; a rule without a sanction is valueless. Scholastic views regarded law as derived from reason and the power to apply sanctions as dependent on authority. Austin regarded a sanction as an essential of law and as operating through the subject's fear of evil. Kelsen agreed that sanction is of the essence of law, but how it influences the mind of the subject is irrelevant. Sociological jurisprudence makes coercion and force the foundation of law and sanctions. In primitive and undeveloped law sanctions included ridicule, ostracism, and outlawry. In more developed systems they include loss of property (fines), loss of liberty (imprisonment), bodily pain, and even death. But legal sanctions co-exist with social and other sanctions, such as fear of disgrace or dishonour, of loss of appointment, of social ostracism, and all operate as social controlling forces.

Sanctuary and Abjuration. Christian sanctuaries were first recognized by Roman law in the late fourth century and developed through recognition of the function of the bishops as intercessor, until they extended to substantial areas around churches. Justinian limited the privilege of sanctuary to persons not responsible for serious crimes. Canon law gave limited sanctuary to persons guilty of crimes of violence to enable compensation to be paid to the victim or his kinsmen.

In the Middle Ages—developed from ancient custom, Anglo-Saxon and Anglo-Norman law, modified by Christianity—it was accepted that a person who had committed a crime would flee to consecrated soil for refuge, frequently a local church or monastery. The coroner was then summoned, to whom he confessed his guilt. Then, on taking an oath to abjure the realm, he was allowed to go in safety to a designated port, where he had to embark within a stated time. It became settled that a man in sanctuary who refused to abjure could be starved into surrender. One who abjured, if he did not go to the port and take ship, might be hanged. The legal effect of abjuration was the same as condemnation to death; the criminal's goods were forfeited, his lands escheated, and his wife deemed a widow. Certain persons were not allowed to take sanctuary, and certain offences prevented it. Under Henry VIII the criminal was branded to facilitate identification if he ever returned. Legislation of 1540 restricted sanctuary and it was abolished in 1623–24. Certain so-called sanctuaries existed till the eighteenth century which gave practical immunity to fraudulent debtors and even to criminals. Into these places, such as the palatinates, and the liberties of St. Martin le Grand, of Westminster and elsewhere, even coroners might not go. The lords of these places registered persons who took sanctuary, exacted oaths of fealty and enforced discipline. Henry VIII abolished many of these and substituted eight cities of refuge, but these were abandoned under Edward VI. Nevertheless, some sanctuaries continued to exist without legal warrant and long thereafter, probably even to the present, certain areas exist into which the police hesitate to go.

In Scotland a person could escape imprisonment for debt by retiring to one of several sanctuaries and the Palace of Holyroodhouse was, and is possibly still, recognized as a sanctuary, but since the abolition of imprisonment for debt the question of sanctuary has not arisen. Under various statutes a person guilty of homicide committed in the heat of the moment had the benefit of sanctuary if he fled to a church or holy place. Sanctuary was abolished at the Reformation.

In Europe the right of sanctuary or asylum survived, though restricted in the sixteenth century, until the French Revolution.

The principle is the source of the rules of diplomatic immunity and of diplomatic asylum in embassies.

N. M. Trenholme, *The Right of Sanctuary in England* (1903); P. Halkerston, *History, Law and Privileges of the Palace and Sanctuary of Holyroodhouse* (1831); J. C. Cox, *Sanctuaries and Sanctuary Seekers of Mediaeval England* (1911).

Sande, Frederic à (seventeenth century). Dutch jurist, son of the next-named, author of *Commentaria in Gelriae et Zutphaniae consuetudines feudales* (1674).

Sande, Johannes à (Jan van den) (1568–1638). Dutch jurist, compiler of *Utriusque Hollandiae, Zelandiae, Frisiaeque Curiae Decisiones* (1615), an important source of Roman-Dutch law, and *De actionum cessione* (1623). He was also a noteworthy historian.

Sanderson, Sir Lancelot (1863–1944). Called to the Bar in 1886, he sat in Parliament and became Chief Justice of Bengal 1915–26, and thereafter a member of the Judicial Committee of the Privy Council.

Sanhedrin. The highest native governing body of the Jews during Roman times. It was probably composed of chief priests, leading laymen, and scribes or lawyers, presided over by the chief priest. By Roman practice it was allowed to regulate religious questions and, within limits, civil affairs, and was the supreme court of justice. It could probably pronounce a death sentence which had to be ratified and executed by the Roman procurator, as in the trial of Jesus. There were also local or provincial sanhedrins with lesser jurisdiction, under the supervision of the Great Sanhedrin.

Sankey, Sir John, 1st Viscount (1866–1948). Called to the Bar in 1892, practised at Cardiff, and made good progress, moving to London in 1907. In 1914 he became a puisne judge of the High Court. He was much concerned with the Church in Wales and he drafted the constitution of the Church in Wales. In 1919 he was made Chairman of the Commission on the Coal Industry, in the report of which he recommended nationalization of the mines. In the course of the hearings he had undergone a change of mind and sympathy and moved towards socialism. In 1928 he was promoted to the Court of Appeal. In 1929 when the Labour government was formed he was made Lord Chancellor and a peer. He went out of office in 1935.

During his period of office he secured the appointment in 1929 of the Committee on Ministers' Powers to examine the delegated legislative and judicial powers of ministers and departments, and in 1934 the appointment of a standing Law Revision Committee to consider the revision of various doctrines of law. He also sat as Chairman of the Inter-Imperial Relations Committee of the Imperial Conference, and then of the Federal Structure Committee of the Indian Round Table Conference. In consequence he sat less frequently as judge than is normal for a chancellor, but his judgments are sound though not usually distinguished. After his retirement he rarely sat.

Sasine. In Scots law, the acquisition of a real right in land by the transfer of symbols of the land itself at a public ceremony held thereon in pursuance of a charter or other deed containing a precept of sasine. The ceremony took place in the presence of at least two male witnesses and it was attested by an instrument of sasine prepared by a notary, detailing the ceremony, defining the nature and extent of the estate, and affording evidence of all the forms having been complied with.

Sasine, Instrument of. In Scottish conveyancing, a deed by a notary narrating that one person had given sasine of land to another by the appropriate symbols, and recorded in the Register of Sasines. Later the actual ceremony of giving sasine on the land was dispensed with and the instrument narrated that sasine had been given. After 1858 the deed of conveyance itself might be recorded and the instrument of sasine became superfluous.

Satisfaction. The discharge of an obligation by performance, payment, or some equivalent act. In equity the doctrine of satisfaction is applied in cases where the actings or language of testators or settlors is ambiguous. Equity recognizes satisfaction of legacies, portions, and debts, where there is a gift with the express or implied intention that it is to be taken as extinguishing a prior claim or demand. Thus, it may be held that a legacy has been satisfied by a portion, or a portion by a legacy, or a debt by a legacy.

Saturninus, Venuleius (first century A.D.). A Roman jurist, author of *de stipulationibus* (19 books), *de interdictis* (6 books), and other works.

Saunders, Edmund (?–1683). A man of obscure origins and self educated he got called to the Bar in 1664 and acquired a large practice. In 1683 he was suddenly made Chief Justice of the King's Bench, but died after only six months in office. He was deeply versed in the technicalities of the law and seems during his short tenure to have been a satisfactory judge. He was author of a famous volume of reports of King's Bench cases covering the years 1666–72, first published in law-French in 1686, which were highly esteemed, translated, and repeatedly re-edited, and on this his reputation rests. Later editions, especially those edited by Serjeant Williams (1799 and 1809) and by E. V. Williams (1824 and 1845), hence known as Williams' Saunders, included also extensive annotations. They were of especial value to special pleaders (q.v.).

Savigny, Friedrich Carl von (1779–1861). Distinguished German jurist, studied at Marburg and elsewhere. In 1800 he became a lecturer at Marburg, chiefly on Roman law, but in 1804–8 visited various libraries in quest of materials and in 1810 took up the chair of law at Berlin, where he lectured on Roman law and the Prussian *Landrecht*. He later served on the Prussian Council of State and as Counsellor to the Court of Revision and Cassation at Berlin. In 1842 he resigned his chair and was Prussian Minister of Justice till 1848, and as such effected some important reforms.

In 1803 he published his famous work on *Das Recht des Besitzes* (*Law of Possession*). In 1814 he published a pamphlet *The Vocation of our Age for Legislation and Jurisprudence* (revised 1828), an attack on the Code system imposed on much of Europe by Napoleon and attacking the advocacy by Thibaut of a German Code. He attacked the demand for a code on the grounds that it was impracticable, adequate juristic preparation being impossible and the German language being juridically underdeveloped, and that existing codes were unsuccessful. Over the years 1815–31 he published the great *Geschichte des römischen Rechts im Mittelalter (History of Roman Law in the Middle Ages)*, which deals with Roman law in the fifth century, then with the kingdoms carved from the ruin of the Roman Empire, the revival of the twelfth century, and on to the mediaeval jurists, and in 1840–49 his massive *System des heutigen römischen Rechts* (*System of Modern Roman Law*, 8 volumes) which demonstrated the great and living unity of much of the Roman law. The last volume deals with private international law. He also published in 1853 a treatise on contracts, *Das Obligationenrecht*.

Savigny was the leading exponent of the historical method, founded by Hugo (q.v.), believing that to trace the history of Roman law in Europe was the only method of arriving at the secret of the evolution of law generally. The content of law must accord with the *Volksgeist*, the Spirit of the people. Custom was the chief manifestation of law. This was a reaction against the hitherto-prevalent school of natural law. His followers and successors carried on his method triumphantly. His opposition to codification delayed progress in Germany for nearly a century but probably thereby benefited the German Code of 1900.

A. Stoll, *Friedrich Carl von Savigny* (1927–39); E. Wolf, *Grosse Rechtsdenker der deutschen Geistesgeschichte* (1963).

Saxons, Law of the. The first part of the *Lex Saxonum* contains ancient customary enactments of Saxony, represents pure Saxon law, without Frankish traces and consists of 20 chapters on penal matters. It is later than Charlemagne's conquest of the Saxons (772–804) and is preceded by two of Charlemagne's capitularies for Saxony, the *Capitulatio de partibus Saxoniae* (782) and the *Capitulare Saxonicum* (797). The second part, the *Lex Francorum*, is later and shows the influence of Frankish law. The whole compilation may date from about 800.

Scaevola, Publius Mucius (second century B.C.). A distinguished Roman jurist, said to have been a master of the *ius pontificium* and probably one of the first to write books on law containing not merely forms but independent discussion.

Scaevola, Quintus Mucius (first century B.C.). Son of P. M. Scaevola, a famous jurist, who published the first systematic treatise on the *ius civile* in 18 books, which was the basis of much later work (books *Ad Quintum Mucium*) and of a *Liber singularis Definitionum*, the earliest work cited in the *Digest*. His rather older cousin of the same name was also a great jurist and teacher of Cicero.

Scaevola, Quintus Cervidius (second century A.D.). A famous jurist, adviser of Marcus Aurelius and teacher of Paulus, Tryphoninus, and possibly of Papinian. He wrote *Responsa* (6 books), *Digesta* (40 books), *Notae* to the *Digesta* of Julian and Marcellus, *Regulae*, and *Quaestiones*.

Scaffold. A raised platform, particularly one on which an execution by beheading or guillotining took place.

Scandalum magnatum. An English statute of 1275 enacted that one who published false news or scandal tending to produce discord between king and people or magnates should be imprisoned until he produced in court the originator of the tale. It was re-enacted in 1378, 1388, 1554, and 1559. Magnates were peers, prelates, justices, and various high officials. In the late sixteenth century the common law courts applied the principle that an injured person could have a civil action for damages for breach of the statute, and thus developed an action for slander of a magnate, including words which would not have grounded an action at common law. The Star Chamber took up the criminal aspect of the statute, dealing with the slander of peers and seditious words, and, by adopting from Roman law the concept of the *libellus famosus*, extended the crime far beyond the scope of the statute. Libel was indeed primarily a political offence and this gave rise to modern criminal libel. The statutes were repealed in 1887.

Scandinavian law. In mediaeval times a branch of Germanic law which was found originally in Norway, Sweden, and Denmark and extended from there to Iceland, Greenland, the Faeroes, Finland, and to parts of Great Britain and Ireland and to Normandy. Originally there were a number of distinct systems of customary law which were approved and developed by the tribes themselves in their *things* or gatherings. Meetings of *things*, both local and for wider areas, were attended by law-men who stated law applicable to particular problems, laid down the law and in case of need stated new law with the consent of the *thing*.

Between the eleventh and thirteenth centuries the law of provinces began to be set down in writing, sometimes at the behest of kings, sometimes as private compilations. In Norway King Magnus V Lagaboter secured a common code for country districts in 1274 and for the towns in 1276. In Denmark three sets of provincial laws survived until King Christian V issued his Danish Law in 1683. These codes were generally reductions to written form of customary rules and included matters which would now be regarded as public, criminal, civil law, and procedure, but usually excluded ecclesiastical law. They showed little trace of external influence, but differed from most Germanic laws in providing for public measures against criminals rather than merely regulating blood-feud.

In Denmark extensive codification replaced the older codes in 1683 and King Christian's Norwegian Law did the same for Norway in 1687. In Sweden a revised version of the earlier code was made in the fifteenth century and a new code, which also applied to Finland was enacted in 1734. Since then all the Scandinavian countries have further developed their separate laws, but still influenced by their historical tradition.

See DANISH LAW; FINNISH LAW; NORWEGIAN LAW; SWEDISH LAW.

Scarlett, James, 1st Baron Abinger (1769–1844). Called to the Bar in 1791, and acquired a large jury practice in the King's Bench and on the

Northern circuit. He was one of the greatest advocates of the time, particularly before a jury. He became M.P. in 1818, and Attorney-General 1827–28 and 1829–30; he was knighted in 1827 and became Baron Abinger in 1835; and held office as Chief Baron of the Exchequer 1834–44; as a judge he was dictatorial on the bench and domineering towards juries but of an acute mind, and a sound but not distinguished judge.

P. C. Scarlett, *Memoir of Lord Abinger.*

Scarman, Leslie George, Baron Scarman of Quatt (1911–). Called to the Bar in 1936, he took silk in 1957, and became a judge of the Family Division in 1961, was Chairman of the Law Commission, 1965–73, and of the Council of Legal Education, 1973– , a Lord Justice of Appeal, 1973–77, and a Lord of Appeal in Ordinary, 1977–

Sceptre. An ornamented rod or staff borne by a ruler on ceremonial occasions as an emblem of authority. They are known to Greek, Roman, and Germanic rulers and were commonly used at coronations. In the British coronation the monarch is given two sceptres, one surmounted by a cross and one by a dove.

Schechter Poultry Corpn.* v. *U.S. (1935), 295 U.S. 495. The National Industrial Recovery Act of 1933, one of the measures enacted to restore business prosperity, provided for the establishment of codes of fair competition, worked out in detail by government officials and representatives of the industries affected. The *Schechter* case involved the regulation of the marketing in New York City of poultry, much of which had been brought from other states, and fixed hours and wages and working conditions of employees. The Supreme Court held the statute unconstitutional as involving a largely unrestrained delegation of legislative power, and as attempting to regulate interstate commerce before or after interstate shipment. This decision was among those which provoked President Roosevelt to attempt to alter the composition of the Supreme Court, to obtain justices more sympathetic to his recovery programme.

Schedule. An inventory, or a writing appendant to a deed, such as a list of fixtures in a lease. The term is applied particularly to the parts of statutes appended at the end of the enacted sections, stating matters of detail which cannot conveniently be included in the body of the Act, providing forms, lists of earlier Acts amended, Acts repealed, and similar provisions.

Scheme A document containing provisions for the distribution or management of property or the regulation of the conflicting rights of parties. Thus the Chancery Division may settle a scheme for the administration of a charitable trust. The conflicting claims of creditors of a company may be regulated by a Scheme of Arrangement approved by the Court.

Schenck* v. *United States (1919), 249 U.S. 47. This case raised the question whether the Espionage Act of 1917 contravened the First Amendment, which forbade Congress to make any law abridging freedom of speech. The question was whether the words prohibited by the Act were used in such circumstances and were of such a nature as to create a 'clear and present danger' of bringing about the substantive evil that Congress had a right to prevent, in which case they were not protected by the First Amendment.

Schilter, Johann (1632–1705). German jurist, author of *Institutiones juris ex principiis juris naturae, gentium et civilis ad usum fori hodierni accommodatae* (1685), *Ad jus feudale introductio* (1693), *Institutiones juris publici Romano-Germanici* (1696), *Praxis Juris Romani in foro Germanico* (1698), and many other works.

Schizophrenias. One of the main diagnostic groups of mental illnesses. The term does not mean 'split personality' but 'cut-off mind', and is applied to a wide variety of patients with very diverse symptoms. There may be several distinct forms of schizophrenias, such as the catatonic type, in which the patient will frequently assume and remain in a particular posture; the paranoid type, in which delusions of persecution are a main feature; the hebephrenic type, in which deterioration of reasoning is common; and the simple type in which the patient loses drive, will-power, and sense of responsibility, or the variety of symptoms may reflect the variety of ways in which personalities react to stress. Heredity seems to be a material causal factor.

Schmalz, Theodor Anton Heinrich von (1759–1831). German jurist, author of *Das Recht der Natur* (1795), *Encyclopädie des gemeines Rechts* (1804), *Das Europäische Völkerrecht* (1817), and other works.

Schlmelzing, Julius (nineteenth century). German jurist, author of *Systematischer Grundriss des praktischen europäischen Völkerrechts* (1818–20) and *Lehrbuch des europäischen Völkerrechts* (1821).

Schneidewen, Johan (1519–?). German jurist and humanist, author of *Epitome in usus feudorum* (1595) and *In quattuor Institutionum libros commentarii* (1720).

Schöffen. Mediaeval German magistrates, descended from the *scabini* appointed by Charlemagne in each district of his empire as permanent law-finders. They sat in groups of 12 or 14 and had the functions of law-finding and the proposing of judgments to the assemblies, held three times a year, at which all local inhabitants had to attend; their proposals were usually accepted but might be challenged. Some *Schöffen*, originally judges, in time became jurists who gave *responsa* and came to assume leadership in the development of German law. The main period of this activity was A.D. 1300–1500 and they administered purely German law with few traces of Roman influence. A few came to be described as *Oberhöfe*, having the prestige and influence of appellate courts. All were concerned with problem cases. The consultative function developed out of the strictly judicial work, but tended to produce specialization and a distinction between *Schöffen* courts for litigation and consultancy. The office was normally held by life-time appointment, made by local lords, city councils, or sometimes by co-optation. The office was one of honour and responsibility and the holders were usually men of high rank.

Particularly important was the practice of giving legal rulings to daughter cities and so transmitting to dependent communities the legal tradition of a mother-city. Thus Brandenburg was a daughter-city of Magdeburg, and in turn acquired Berlin as a daughter which in turn acquired Frankfort-am-Oder. Disputes could be referred to a mother or grandmother *Oberhöf* by inquiry from court officials or direct application by parties.

Many responses have been collected and published, and many towns and cities preserved records of responses addressed to themselves. Some collections were edited and seem to have been intended for use as guides, and some attained wide circulation. There are also known books of precedents as guides for internal use by *Oberhöfe* themselves.

Their work was of great importance in developing the customary law of German towns and districts before the Reception of Roman law in Germany in the 1500s.

The *Schöffen* were reintroduced in Germany in the nineteenth century as a counterbalance to juries in lesser criminal cases and, after eclipse during the Nazi era, were reintroduced in 1950. They are lay judges or assessors who sit along with professional judges, usually in twos, but in serious cases six sit. They are chosen by lot from local citizens and seem to be much influenced by the qualified judges.

Scholastic theories of law. Scholasticism, the predominant philosophy of the Christian schools in the Middle Ages, was founded on acquiring knowledge of much thought from earlier times, on systematic discussion of texts and problems (*quaestiones, disputationes*) and on attempts to reach comprehensive views of all attainable truth (*summae*), principally in the sphere of theology and philosophy. Scholasticism culminated in the work of St. Thomas Aquinas (q.v.) whose system included ideas on government and law, relating human law to divine law and natural law. Thereafter, it has never been obliterated and, since 1879, Thomist doctrine has been the approved philosophy of Catholicism.

In modern times the doctrines of the Catholic Church and many natural law theories are both founded on the scholastic basis. Thus Cathrein derives natural law from Christian revelation and presses scholastic doctrine to the point of declaring that all positive law contrary to natural law is void. Renard and Delos have linked Hauriou's theory of the institution with a Catholic conception of natural law, while Maritain, Dabin, and Rommen have restated the modern Catholic doctrine of natural law, opposing the omnipotence of the State and defending the liberties of groups and of rights of private property. All have had difficulty reconciling this view, with the observed fact in every developed society, of man-made obligatory rules of conduct.

Schools of jurisprudence. A term for the groups into which thinkers about law are commonly put. Such classification and grouping has value as labelling some of the major and most significant approaches to the main problems of legal theory, and as indicating similarities of interest and attitude among jurists, but clear distinguishing lines cannot be drawn and some thinkers are hard to classify. Bentham, for example, was both an analyst examining concepts, and a teleologist interested in the ends the law should pursue.

One group comprises the analytical jurists concerned primarily with examination of the structure and the linguistic analysis of concepts in systems of positive law, particularly Roman law and English law. Among these mention may be made of Bentham, Austin, Markby, Holland, and Salmond. Modern linguistic philosophers such as Hart may be put in this category also, as may Kelsen's pure theory of law.

A second group is the historical school which saw law as something which evolved from the instinctive sense of right of the community and developed in and by reason of particular social, economic, and other contexts. The leading name is Savigny, followed in England by Maine and Vinogradoff, and by legal historians such as Maitland and Holdsworth.

Again, there is the philosophic school, though this label includes jurists too numerous to mention, with the most diverse philosophic views, natural lawyers, scholastic and neo-scholastic jurists, idealists, utilitarians, and others. Kelsen's pure theory of law as a strictly logical construction may be included here, though seeking to purge law of all ethical content. It is not very helpful to group many jurists in this school because there are such divergences of view among them. Possibly the only common factor is that all are concerned with the philosophical bases of the institutions and doctrines of systems of law, and are seeking to discover the fundamental principles through philosophy.

Another is the so-called sociological or functional school, concerned with how law operates in society. The great names are Jhering and Pound, followed by Stone and others. A more radical branch of the same group are the realists whose work stems from the attitudes of Holmes, J., and were represented later by Frank, Llewellyn, and others.

The classification of jurists into schools should also not obscure the fact that there is some truth and value in all the approaches and methods, that they are complementary.

Schrorer, Willem (1717–1800). Flemish jurist, president of the court of Flanders, who edited and annotated Grotius' *Introduction to Roman-Dutch Law* (1767), a valuable older edition.

Schouler, James (1839–1902). American historian and jurist, author of books on *Domestic Relations* (1870), *Personal Property* (1872), *Bailments* (1880), *Husband and Wife* (1882), *Wills* (1887) and, his major work, a *History of the U.S.A. under the Constitution* (7 vols., 1880–1913).

Schröder, Richard (1838–1917). German jurist, author of two major historical works, *Geschichte des ehelichen Güterrechts in Deutschland* (1863–74) and *Lehrbuch des deutschen Rechtsgeschichte* (1889).

Schulte, Johann Friedrich von (1827–1914). German historian of canon law, author of *System des allgemeinen Katholischen Kirchenrechts* (1856), *Die Lehre von dem Quellen des Katholischen Kirchenrechts* (1860), and *Die Geschichte der Quellen und Literatur des Kanonischen Rechts* (1875–80).

Schultingh, Antonius (1659–1734). Dutch jurist, professor at Leyden, known as the Dutch Cujas. His works included *Jurisprudentia antejustiniana* (1717), *Notae ad veteres glossas verborum juris in Basilico, De Jurisprudentia Historica* (1724), and *Notae ad Digesta seu Pandectas* (1834).

Schwabenspiegel. The usual name for an anonymous mediaeval German law-book, the *Kaiserliches Land und Lehnrecht*, based partly on the *Sachsenspiegel* (q.v.), partly on the *Spiegel der deutschen Leute* (q.v.), and partly on a collection of older materials. It dates from about 1275 and was itself taken as the basis of some later books.

Schwarzenberg, Freiherr Johann von (1463–1528). German lay judge and penal reformer. As president of the Bishop of Bamberg's judicial tribunal he created the *Halsgerichtordnung* for the bishopric (*Constitutio criminalis bambergensis*), a synthesis of criminal law and procedure, reconciling German and foreign law which was very successful and influential in Germany, and became the model for the *Peinliche Gerichtsordnung Karls V* or *Carolina* (q.v.) which was really the German imperial criminal code until 1870.

Scialoja, Vittorio (1856–1933). Italian jurist and statesman, Senator and Minister of Justice, author of works on Roman law, legal history, *Istitutizioni di diritto Romano, Procedura civile romana,* and other works.

Science of law. A phrase sometimes used as a synonym for jurisprudence (q.v.) and sometimes for legal science (q.v.).

Scienter (knowingly). At common law a person was liable for harm done by a domestic animal if he knew that it was liable to do harm of that kind; proof of the knowledge was called proving the *scienter*.

Scintilla iuris. A mere spark of right, in particular a doctrine in the older law of real property that when a future interest was limited by way of an executory one, the whole seisin was carried over from the feofee to uses to the *cestui que use* who had a vested estate by the Statute of Uses, yet a possibility of seisin or a *scintilla juris* remained in him sufficient to give effect, when necessary, to contingent or executory uses.

Scire facias (that you cause him to know). A writ based on a judgment, letters patent, or other record, directing the sheriff to warn the person against whom it is directed to show cause why the person bringing it should not have the benefit of the judgment, etc. It was used to enforce forfeiture of an office held by letters patent. In some cases it was an original action, sometimes a continuation of a former action, and sometimes a process of execution. The writ is now in disuse. In civil proceedings it was abolished in 1947.

Sclopis di Salerano, Conte Federico (1798–1878). Italian historian and jurist, minister of justice of Piedmont and president of the *Alabama* (q.v.) Arbitration Commission. Among his many writings the most important is his *Storia della legislazione italiana* (1840–57).

Scobell, Henry (?–1660). Clerk of the House of Commons from 1648 and an authority on Parliamentary procedure during the Commonwealth. He published *Memorials of Method and Manner of Proceedings of Parliament in Passing Bills* (1656), an important treatise on the practice of the time, *Remembrances of some Methods, Orders and Proceedings of the House of Lords* (1657), and *Power of Lords and Commons in Parliament in Points of Judicature* (1680).

Scoble, Sir Andrew Richard (1831–1916). Called to the Bar in 1856, he was a member of the legislative council Bombay, 1872–77, and of the Council of the Governor-General of India, 1886–91, and a member of the Judicial Committee of the Privy Council from 1901.

Scot-and-Lot boroughs. Borough constituencies in the unreformed (pre-1832) British parliament where the vote was held by those who paid the local rates, 'Scot' being a customary tax or contribution and 'Lot' being a share.

Scotland, Secretary of State for. After the Union of 1707 the office of Secretary (or Secretaries) of State in Scotland seems to have lapsed. The Duke of Queensberry was added to the two U.K. Secretaries of State and given special responsibility for Scottish affairs and, from 1707 to 1725 and 1741 to 1745, there was a Secretary of State mainly responsible for Scotland. Between 1745 and 1827 the government business in Scotland was formally the concern of one of the two Secretaries of State, after 1782 the Secretary of State for the Home Department, but the Lord Advocate emerged as the principal Officer of State in Scotland and the real manager of government business in Scotland. From 1827 to 1885 the Home Secretary was responsible for Scottish affairs with the advice and assistance of the Lord Advocate. Over this period dissatisfaction increased over the mismanagement and neglect of Scottish affairs, and the ignorance of London departments of the needs of the country.

During the eighteenth and nineteenth centuries various boards were set up to administer certain functions, notably the Board of Manufactures, the Fishery Boards (amalgamated 1839–82), the Board of Supervision for the Relief of the Poor, replaced in 1894 by the Local Government Board for Scotland, the Scottish Insurance Commissioners, the Highlands and Islands (Medical Service) Boards—the last three being transferred to the Scottish Board of Health in 1919—the General Board of Commissioners in Lunacy (later the General Board of Control), the General Board of Directors (later the Prisons Commission for Scotland), the Scotch Education Department, the Crofters Commission, and the Board of Agriculture for Scotland.

In 1885 the office of Secretary for Scotland was created and there were transferred to him duties previously exercised by various authorities, including the Privy Council, the Treasury, and the Home Secretary.

Since 1892, except in wartime, the holder has always been a member of the Cabinet.

In 1926 the office became that of a Secretaryship of State, and in 1951 there was added a Minister of State, Scottish Office, and there are now three Under-Secretaries of State, offices created in 1919, 1940, and 1952. The Secretary of State is Keeper of the Great Seal of Scotland.

In 1939 nearly all the existing Boards were brought under the Secretary of State's wing and became his responsibility, being grouped in four major departments. Since then statute has added considerably to his responsibilities.

His main administrative offices are in Edinburgh and comprise five departments, Scottish Home and Health Department, Scottish Education Department, Scottish Development Department, Scottish Economic Planning Department and Department of Agriculture for Scotland, but to a larger extent the Secretary of State is Scotland's Minister and has powers and duties exercised in England by a wide range of Ministers. In legal matters, assisted by the Lord Advocate, he is responsible for seeing through Parliament all Scottish legislation. He has important supervisory functions in respect of the inferior courts and is responsible for the police and prison services and corrective establishments. He appoints justices of the peace in Scotland, recommends the appointment of Scottish judges and advises on the exercise of the royal prerogative in Scottish cases.

D. Milne, *The Scottish Office.*

Scots law. Though Scotland is a part of the United Kingdom for historical reasons there exists in Scotland a legal structure and a system of law very different from that existing in England and Wales. Its existence and continuance were guaranteed by the Treaty under which the Scottish and English Parliaments were replaced by the British Parliament in 1707. Among the major characteristics of difference may be mentioned a completely different history and course of development (it being remembered that Scotland had a distinct ruling house till 1603 and an independent Parliament till 1707), separate sets of courts and judges, a separate legal profession with distinct training and qualifications, different procedures and terminology, and on most topics distinct bodies of substantive law which have grown up differently from English law. Among other noteworthy features are the fact that there never existed in Scotland distinct jurisdictions in law and equity with consequential distinct bodies of principles, and the fact that Roman law was, until the nineteenth century, the most powerful single external influence. In many respects Scots law stands in a position intermediate between a civil law and a common law system.

Unlike English law, Scots law has long been a highly systematized body of principles. By 1681 it was possible for Stair in his *Institutions* (*vide infra*) to set out the whole law in a coherent, connected narrative, stating the law in systematic orderly form, in an order based on Justinian's *Institutes*, and this tradition has persisted to the present. Following from this, and consistently with the civilian tradition, the tendency of Scottish courts and judges is to seek the appropriate general principle and to reason from that to the decision of the case before them, rather than to seek the apt precedent.

A major difference is that although English law was spread all over the world, though Scots may have emigrated and settled, Scots law has remained restricted to its homeland. This had been due since 1707 to Cabinet and Parliamentary ignorance, neglect and exclusion of Scots law when dealing with British affairs, and the arbitrary rule that persons of British stock emigrating took with them English law with all its complications. It is always assumed that British law is English law. Similarly, the armed forces of the Crown are subject to bodies of law which are in fact pure English law, the only kind of law Parliament ever thought of enacting.

Historical development

Scotland owes its name to Scots who entered the country from Ireland and may have introduced some elements of Celtic culture. Though Roman forces traversed much of Scotland and established forts there, the country was not colonized by the Romans. A united kingdom of Alba was established by 850 and by the eleventh century the boundary between Scotland and England was settled along roughly its present line. Down to this time law was customary and local.

After the Norman Conquest of England, Anglo-Saxon and Anglo-Norman influences were felt and, in the reign of David I (1124–53), feudalism was largely accepted and became the prevailing form of government and land tenure. The institutions and officers of a feudal monarchy, such as chancellor, justiciar, and constable were introduced and a network of sheriffdoms began to spread over the country. Burghs began to be chartered, and a system of courts, of justiciars, barons, sheriffs, and burghs began to develop. The Church became powerful and much litigation took place before ecclesiastical tribunals, and through the Church knowledge of Roman and canon law entered Scotland.

In the thirteenth century there appeared *Regiam Majestatem*, a work of uncertain date and unknown authorship, based on Glanvill's *De Legibus et Consuetudinibus Angliae* with alterations and adaptations to Scottish conditions, and drawings on Roman and canon law. It attempts to cover the whole civil and criminal law from the standpoint of the royal courts. The *Quoniam Attachiamenta* of about a century later is a systematic manual of procedure in feudal courts based on practical experience.

The law at this period was mainly concerned with complaints against violence and wrongdoing, with tariffs of compensation for murder and injuries in lieu of blood feuds, and there was little law outside the field of wrong and crime, and the courts were manifestly unable to redress all the grievances brought to them. There was some borrowing from, and adaptation of, Anglo-Norman law and forms, but there was divergent rather than parallel development; the more technical developments from Glanvill to Bracton were not followed, but the Romano-canonical tradition was more influential.

A king's council had been in existence since the twelfth century and in the following century is sometimes described as sitting in exchequer, sometimes in colloquy, and sometimes in Parliament. By the end of the thirteenth century Parliament was a sitting of the council to declare the law, affirming it, applying it to a case, or sometimes altering it. There are traces of legislation from the twelfth century.

The orderly development of law and order was stopped by the repeated attempts of Edward I of England to subjugate Scotland between 1286 and his death, and, despite the Treaty of Northampton of 1328 by which England renounced all claims to superiority over Scotland, renewed by Edward III in the fourteenth century. But determined efforts were made in the fourteenth and fifteenth centuries to build an orderly society though those years were bedevilled by royal minorities and turbulent barons. Parliament was weak, and many expedients were resorted to to supply an administration of justice mainly through various committees of the King's Council and of Parliament, such as the Lords Auditors of Causes and the Session. From about 1425 there was a fairly regular royal court of justice.

Ecclesiastical courts continued to exercise important jurisdiction, mainly in matrimonial cases, and the existence of an alliance with France and common resort to French universities brought strong French and, through it, Roman influences to bear on Scottish civilization, and through this channel much Roman and feudal legal ideas trickled into Scotland. There was much legislation and several projects of revising the old laws. In 1426 it was enacted that all the King's subjects were to live by the King's laws and not by particular laws or privileges; there was to be a common law of Scotland. By this time Scots law was already developing on quite different lines from English law and diverging from it.

1532–1603. The College of Justice was founded in 1532, consisting of a clerical President and 14 other judges, half clerical and half lay, and this systematized the central administration of civil justice by establishing a permanent civil court, the Court of Session. The court sat as a collegiate body, but before 1600 it seems to have become the practice to have two or three lords sitting singly to deal with preliminary stages of cases and to take evidence. This was the origin of the modern Outer House.

In the sixteenth century statute was a major source of law, and there were no reports of cases, save notes by lawyers for their own use. There were several abortive projects for codification of the laws.

From the establishment of the Court of Session there developed a fresh kind of legal literature, Practicks, or practical observations, notes of judgments and points of law made, chiefly by judges, for private use, passed from hand to hand and subsequently copied. Two kinds of practicks developed, those containing notes of decisions, usually in chronological order, and those containing notes of cases, statutes, passages from *Regiam Majestatem,* and the like, usually digested under alphabetical subject-headings. From the first kind developed the later reports and from the second the later law dictionaries, digests, and encyclopaedias.

The Scottish Reformation was a widespread movement of religious opinion rather than a change by royal decree; in 1560 Parliament abolished the papal authority, the mass, and all practices contrary to the new creed, all papal jurisdiction and appeals to Rome. In 1563 there was established the Commissary Court of Edinburgh to determine all causes formerly competent to the Church courts, with jurisdiction in marriage, divorce, and bastardy, though the consistorial law of the old courts based on the canon law continued to be observed, except in so far as expressly altered by the reformers. Divorce for adultery was granted by the courts from this time onwards on the basis of common law, and divorce for desertion was introduced in 1573.

The ecclesiastical system which emerged in Scotland was presbyterian in form and Calvinist in doctrine, and the General Assembly became a powerful influence in the land.

Among the other important developments of the sixteenth century were the emergence of the office of King's Advocate, the introduction of the office of Justice of the Peace, the holding of justice-ayres and circuit courts of justiciary.

1603–1707. In 1603 King James VI of Scotland succeeded Queen Elizabeth, becoming King of England also, and projected an incorporating union of the two countries.

The ecclesiastical policy of James and his son, Charles I, seeking to impose episcopacy on Scotland, provoked opposition and resistance, and a Scottish army participated in the struggle between Parliament and king which culminated in the execution of the King. During this time the Parliament and the General Assembly of the Church came to play a more powerful part than ever before in the government of the country. After the execution of Charles I Scotland supported Charles II, but Cromwell defeated the Scottish forces and subjugated the country. Scotland and England were amalgamated into one commonwealth and Scottish representatives, in fact picked creatures, sat in the Commonwealth and Protectorate Parliaments. The Scottish Privy Council and the Court of Session ceased to function and seven commissioners were appointed to administer justice.

After the Restoration, the Court of Session was restored while the Privy Council was the effective government of Scotland; tyranny and persecution were widespread and the Scottish Parliament in 1689 offered the Scottish Crown also to William of Orange.

In 1672 the High Court of Justiciary was created as the supreme criminal court, replacing the mediaeval justiciars; it was to consist of the Justice-General (an hereditary office), the Justice-Clerk (an office which had commenced as that of merely clerk to the justiciar), and five judges of the Court of Session as Commissioners of Justiciary.

There flourished in these times a group of lawyers who all contributed to the literature of the law and its development. Craig, author of *Jus Feudale*, Skene, the first great legal antiquary, who published *Regiam Majestatem,* Hope, author of *Major Practicks* and *Minor Practicks*, Mackenzie, famed as a persecutor but author of many works, some still of great importance, but, above all, James Dalrymple, later Viscount Stair, President of the Court of Session, 1671–81 and 1689–95, who published two volumes of *Decisions* and the *Institutions of the Law of Scotland* (1681, revised 1693), still the greatest of the classic texts, a masterpiece of synthesis founded on wide learning and historical and comparative study, setting out principles in systematic form. To Stair, natural law and the Roman law were the basic law of Europe and the law of Scotland was one set of variations on this general theme. Stair not only systematized and rationalized the common law but undoubtedly introduced some new law from the Roman tradition.

Despite all the troubles the seventeenth century was a period of important development of private law and from this period date many statutes still in force, or still the foundation of the modern principles, such as on fraudulent alienations in bankruptcy, prescription, entails, execution of deeds, and the establishment of the Register of Sasines, the public register of deeds affecting land. In 1680 there was founded the great Advocates' Library, since 1925 the National Library of Scotland, and it is significant how rich the early library was in European, particularly French and Dutch, materials.

1707–89. In 1707, despite strong opposition, the Parliaments of Scotland and of England each passed Acts agreeing to Articles of Union proposed by a joint commission, establishing a single Parliament in place of the separate Parliaments, but providing that the Court of Session and Court of Justiciary should continue for all time, that no causes in Scotland should be cognizable by any court in Westminster Hall, that the separate systems of law should continue, save that laws which concern public right, policy, and civil government might be made the same throughout the United Kingdom but that no alteration be made in laws which concerned private right except for evident utility of the subjects within Scotland.

From the first the Treaty was flagrantly breached; the Scottish Privy Council was abolished, the ancient English law of treason applied to Scotland, and a commission of oyer and terminer superseded the Court of Justiciary in this field. During most of the eighteenth century there was profound dissatisfaction in Scotland with the Union. A Court of Exchequer was set up which used English forms. Appeals to the House of Lords were entertained in civil cases, though that had not been provided for by the Treaty. After both the Jacobite rebellions of 1715 and 1745, particularly the latter, harsh repressive measures were undertaken, prisoners were removed to England for trial, feudal rights and heritable jurisdictions were abolished, and much of Scotland was treated as a conquered and occupied province.

There was steady development of private law by the courts in the eighteenth century, and some very distinguished lawyers graced the bench. By the end of the century Scots law was a finished philosophical system, a systematic body of principles. The influence of Roman law declined and feudal conveyancing became elaborated; in mercantile spheres the influence of English law increased.

In legal literature, the period is notable for the *Institutes* of Bankton, the *Institute* of Erskine, the first great academic statement of the principles of the whole law and one second in standing only to that of Stair, and the valuable writings of Millar and Kames.

The eighteenth century saw the end of the practice which had prevailed since the fifteenth century of young Scots studying law in France, or, later, in Holland. In this way the civil law and the continental commentators thereon and teachers thereof had a long, consistent, and powerful influence on Scots lawyers, and much of developed civil law was absorbed into Scots law.

1787–1832. The period of the French Revolution and the Napoleonic Wars gave rise to movements for political reform, to repression and discontent, and treason trials. Precedent became increasingly important as a source of law and Roman law influence declined while that of England increased.

Some important changes were made in this time. In 1808 the Court of Session ceased to sit as a unitary court, and was divided into First and Second Divisions, while in 1825 the junior judges were detached from both Divisions and required to sit permanently in the Outer House to hear cases at first instance. In 1815, at the instance of English reformers, a separate Jury Court was established to try issues of fact remitted to it by the Court of Session or of Admiralty. It was not a great success and, in 1830, it was merged in the Court of Session, which continued the practice of trying some issues of fact with a jury. Also in 1830 the Scottish Court of Admiralty was merged in the Court of Session. The Scottish Court of Exchequer set up in 1707 was restricted in function in 1832, though not abolished till 1856.

The complication of feudal conveyancing continued while mercantile law steadily developed in bulk and importance.

In 1797 David Hume published his *Commentaries on the Law of Scotland respecting Crimes*, which has ever since been accepted as the supreme authority on Scottish criminal law, though the *Principles and Practice of the Criminal Law of Scotland* of Alison (1832–33) is also highly regarded. Hume's successor as Professor of Scots Law at Edinburgh, Bell, wrote two standard and authoritative books, the *Commentaries on the Law of Scotland* (1800), which did much to incorporate principles of mercantile law into Scots law, and *Principles of the Law of Scotland* (1829).

1832–1914. The latter nineteenth century was a time of reform. The House of Lords had, since 1707, been an utterly unsatisfactory court of ultimate appeal in civil cases from Scotland, in that Scottish appeals were heard by Lord Chancellors who knew nothing of Scots law and repeatedly imposed English law on Scotland. In 1866 Lord President McNeill was made Lord Colonsay to assist the House in handling Scottish appeals, and since the Appellate Jurisdiction Act, 1867, there has always been one, and latterly two, Lords of Appeal drawn from Scotland. But it is still mainly an appeal to foreign lawyers.

There were several reforms in the Court of Session, and in 1830 it absorbed the jurisdiction of the Admiralty Court (save the prize jurisdiction, which was transferred to the English Court of Admiralty), and in 1856 that of the Scottish Court of Exchequer. The Commissary Courts were abolished in 1836 and 1876.

In 1887 the High Court of Justiciary was enlarged to include all the judges, and indictments and procedure considerably simplified, while the jurisdiction of the sheriff courts was extended several times.

Social and economic conditions gave rise to many statutory developments, some general to the United Kingdom such as factories Acts, trade union Acts, companies Acts; some to cope with peculiarly Scottish problems, such as the series of conveyancing Acts, the entail Acts, the crofters' Acts, and some with different but parallel provision for Scotland and for England.

The development of private law was materially affected by the rising tide of legislation, frequently English in principle and terminology, and by the increasing tendency to look to English decisions for guidance, though not infrequently unsuitable or misunderstood precedents were prayed in aid. The House of Lords too frequently stated or assumed that Scots law must, or should, be the same as English, and imposed foreign ideas (such as common employment) on Scots law while some amending legislation (such as the Sale of Goods Act, 1893) made radical alterations of Scots law, not for the better.

Procedure was radically altered: from the sixteenth century pleading had mostly been by written papers, but these were abolished in 1825 and 1850 and briefer pleadings introduced, oral arguments becoming normal. Evidence was required to be heard before a Lord Ordinary and the use of shorthand writers to record the evidence authorized.

Legal scholarship did not, however, languish and some still important works were written such as Lord Fraser's *Husband and Wife*, Lord McLaren's *Wills and Succession*, and Rankine's *Landownership* and *Leases*, as well as many lesser works.

Since 1914. Since 1914 the main features have been the growing weight and complexity of statute law and subordinate legislation, increasing reliance on precedent, and growing pressure to assimilate Scots law to English, frequently resulting in an unhappy mixture, though there have been occasional instances of contrary influence.

Down to 1950 there was woeful lack of legal writing or scholarship, but since then there has been great improvement. In 1960 the Scottish Universities Law Institute was founded and it has promoted the writing of many new standard textbooks of quality and value, while others have appeared independently.

The future of Scots law is problematical; it is difficult for a system of law to survive within a country, when ultimate appeal lies to a predominantly English body (the House of Lords in London), and all legislation has for long been made by an overwhelmingly English legislature in which the Scottish members are the servants of London-based parties, when the head offices of most corporations and professional institutions are in England and think in English terms, and most decisions are taken ignoring or forgetting the existence of Scotland, or Scottish peculiarities and differences, which, if remembered, are treated as deviations from the norm and something abnormal and eccentric.

Scotland in everything is dragged at the heels of England, and neglected and ignored, and it is not surprising that some Scots look with envy at Ireland which threw off the English yoke; at Canada where, in Quebec, a civil law system survives in a federal state; and at South Africa where a system which has grown from some of the same roots as Scots law has independence to develop and to reject English principles and precedents as not necessarily the only or the best body of law.

Sources and literature

The major historical sources of Scots law are indigenous custom, canon law, Roman law, feudal law, the law merchant and maritime, and, since about 1800, English law.

The formal sources are legislation, judicial precedent, institutional writings, custom, equity.

Legislation comprises statutes passed by the Parliaments of Scotland down to 1707, of the Parliaments of Great Britain 1707–1800, and of the Parliaments of the United Kingdom since 1801 (all so far as applicable to Scotland, and not repealed or amended), and subordinate legislation made by authorized persons or bodies under authority of legislation.

Scotland suffers severely by not having a provincial legislature of its own, and legislation for Scotland by the United Kingdom Parliament may take the form of Acts applicable equally to Scotland and to England, frequently with a section dealing with 'application to Scotland' translating English technical terms used in the Act for Scotland; or Acts applicable to England with Parts, or sections, or odd phrases made applicable to Scotland by an 'application to Scotland' section; or Acts applicable to Scotland only, which contain the word '(Scotland)' in the title. Statutes passed by the U.K. Parliament are presumed to apply equally to the whole U.K., but this may be rebutted by internal evidence in the Act. Yet in too many cases it is uncertain whether a provision applies to Scotland and, if so, what it means in the context of Scots law and procedure, and much legislation is passed without understanding how it will affect Scots law and sometimes with peculiar side-effects. It is impossible to believe that it would not be beneficial both to the U.K. Parliament and to legislation affecting Scotland to create a Scottish provincial legislature.

On the interpretation of statutes there is one Scottish specialty; legislation of the pre-1707 Parliament of Scotland may be deemed abrogated by desuetude, and such legislation is liberally rather than literally interpreted, special weight being given to interpretations placed judicially on such statutes shortly after the legislation was passed. Apart from that the same general considerations and presumptions as to interpretation apply as in England, and in the interpretation of statutory provisions common to Scotland and to England decisions of the English courts are treated as authoritative.

Judicial precedents

Down to the nineteenth century judicial precedents were treated in Scotland more as examples of the courts' attitudes than precedents, and only a line of concurring decisions had binding force. During the nineteenth century, however, mainly under the influence of the House of Lords, a more rigid attitude crept in and the doctrine is now almost as rigid in Scotland as in England.

The principles now accepted are:

Decisions of the House of Lords in Scottish appeals are binding on the House itself, subject to the House's limited power to reconsider its own precedents, and on all lower Scottish courts. Decisions of the House in non-Scottish cases are highly persuasive, particularly if the point of law is one common to Scotland and England.

Decisions of either Division of the Inner House of the Court of Session are normally treated as binding on each Division, and on all inferior courts.

Decisions of the whole Court of Session or of a bench of judges larger than the normal size of either Division of the Inner House are binding on both Divisions and on lower courts. A bench larger than normal may sit to hear cases of difficulty or reconsider questioned precedents.

A Division may also disregard a precedent if by reason of change in social conditions it is obsolete.

Decisions of single judges in the Outer House of the Court of Session, of sheriffs-principal, and sheriffs are not binding, but are normally followed out of comity.

Decisions of English, Irish, Commonwealth, and U.S. courts are persuasive in degree related to the eminence of the court and the relevance of the principle involved to Scots law.

Decisions of the Privy Council are persuasive only, but in appropriate cases in high degree.

In the High Court of Justiciary a bench will normally follow the decisions of a previous bench, but a larger court may be convened to reconsider difficult precedents or settle difficult points. Decisions of the High Court are normally accepted as binding on single judges of the High Court and on judges of inferior criminal courts.

Institutional writings

Weight comparable to that of a decision of the Inner House of the Court of Session attaches to statements of principle contained in the works of the institutional writers, who all in their times wrote major works covering the whole of the law, or at least major branches of it.

The rank of institutional writing is enjoyed by Craig's *Jus Feudale* (1655), James Dalrymple, Viscount Stair's *Institutions of the Law of Scotland* (1681), Erskine's *Institute of the Law of Scotland* (1773), and Bell's *Commentaries on the Law of Scotland* (1800) and *Principles of the Law of Scotland* (1829), and is sometimes accorded to Bankton's *Institute of the Laws of Scotland* (1751–53) and Kames' *Principles of Equity*. Of these Stair, Erskine, and Bell are pre-eminent.

In criminal law Sir George Mackenzie's *Laws and Customs of Scotland in Matters Criminal* (1678), Hume's *Commentaries on the Law of Scotland respecting Crimes* (1797), and Alison's *Principles* (1832) and *Practice of the Criminal Law of Scotland* (1833) are held institutional.

Legal literature

As in England lesser weight attaches to statements in standard modern textbooks, but there is no convention against citing the work of a living writer. Among textbooks repeatedly cited have been Fraser on *Husband and Wife*, McLaren on *Wills and Succession*, and Gloag on *Contract*. In branches of law largely common to Scots law and to English, such as carriage by sea or company law, standard English texts are referred to in default of Scottish books.

Custom

Custom or usage underlies many practices long since recognized by institutional writers or cases, but custom, local or of trade, if proved in suitable cases, may still be held to supply a rule to be followed.

Equity

Equity in the English sense of a distinct body of principles, complementary to and sometimes conflicting with common law principles, has never been an element in Scots law. But principles, remedies, and judicial attitudes based on equity, natural justice, and fairness, have long been recognized by Scottish courts and jurists, and the Scottish courts have always administered an undifferentiated body of law combining legal and equitable principles, and most of the principles and remedies characteristic of equity in England can be paralleled by rules of Scottish common law. Apart from the ordinary equitable jurisdiction the Court of Session, and to a lesser extent, the High Court of Justiciary, has an extraordinary equitable jurisdiction, the *nobile officium*, to modify or abate the rigour of the law where justice requires it. In practice this power is usually exercised only in cases for which there is precedent or analogy.

Ideals of Justice

Failing guidance from any of the foregoing sources a judge may resort to his own view of what is fair, reasonable, and just in the circumstances.

Judicial organization

Superior courts. The supreme Scottish civil court is the House of Lords comprising the Lord Chancellor, the Lords of Appeal in Ordinary, and peers who hold or have held high judicial office. Customarily two of the Lords of Appeal are promoted from the Scottish bench or Bar, but Scottish appeals are normally heard by a bench of five or more Law Lords, a majority, and possibly all, of whom have no training in Scots law. No appeal lies to the House from Scottish criminal courts.

The superior Scottish civil court is the Court of Session, which is divided into the Inner House (itself sub-divided into the First Division, comprising the Lord President of the Court of Session and three Lords of Session, and the Second Division, comprising the Lord Justice-Clerk and three Lords of Session), which has a mainly appellate jurisdiction, and the Outer House, which consists of the remaining judges of the Court of Session sitting singly, sometimes with a jury of 12, with original jurisdiction only. The Court sits only in Edinburgh.

The superior criminal court is the High Court of Justiciary which consists of the Lord Justice-General of Scotland (who also holds the office of Lord President of the Court of Session), the Lord Justice-Clerk of Scotland, and Lords Commissioners of Justiciary (who also hold the offices of Lords of Session). Three or more judges sit as Court of Criminal Appeal, hearing appeals against conviction or sentence after trial on indictment, and as High Court of Justiciary hearing appeals against conviction or sentence after summary trial in the lower courts. Single Lords Commissioners, sitting with a jury of 15, try cases on indictment in Edinburgh or, on circuit, in major towns of the country.

Inferior courts. The chief inferior court is the sheriff court. Scotland is divided into six sheriffdoms, each with a full-time sheriff-principal, and a number of sheriffs, all legally qualified, resident and holding court in the main towns.

In civil cases the sheriff-principal is a judge of intermediate appeal, hearing appeals from decisions by the sheriffs. The latter have wide, and pecuniarily unlimited, jurisdiction, the only important cases excluded from their jurisdiction being actions affecting status. Appeal lies to the Inner House of the Court of Session, direct, or after intermediate appeal to the sheriff-principal.

In criminal cases sheriffs-principal and sheriffs have the same powers. They may, sitting with a jury of 15, try cases of moderate seriousness on indictment, or, sitting alone, try lesser cases summarily.

In burghs stipendiary magistrates (only in Glasgow) or lay justices sitting in district courts may deal summarily with minor offences. There are no coroners' courts.

Courts of special jurisdiction. Among those peculiar to Scotland mention must be made of the Scottish Land Court, which deals with questions between landlord and tenant, particularly in the Highlands and Islands; the Court of the Lord Lyon, dealing with claims to armorial bearings; the Lands Valuation Appeal Court, which consists of three judges of the Court of Session hearing appeals against the valuations set on property for rating purposes; the Teind Court, dealing with questions as to teinds or tithes.

The courts of the Church of Scotland (General Assembly, Synods, Presbyteries, and Kirk Sessions) are supreme and independent in their own spheres.

Other courts which may sit in Scotland are the Restrictive Practices Court, the Courts-Martial Appeal Court, and the Employment Appeal Tribunal.

Administrative tribunals. Scotland has a great number and wide range of administrative tribunals, many concerned with deciding questions arising out of the provision of social services, many of which are the counterparts of similar English tribunals.

Domestic tribunals. There is similarly a wide range and variety of domestic tribunals.

Public law

The highest executive of Scottish government is the United Kingdom Cabinet, with the qualification that some departments of state, such as the Treasury and the Ministry of Defence have functions in Scotland as in England, that others have no functions at all in Scotland, and the main executive functions in relation to Scotland are exercised by the Secretary of State for Scotland, operating through his five departments situated in Edinburgh (the Scottish Home and Health Department, Scottish Development Department, Scottish Education Department, Department of Agriculture for Scotland, and Scottish Economic Planning Department). Many government departments and branches have Scottish offices with some degree of delegated power.

The legislature for Scotland is the United Kingdom Parliament, but all purely Scottish legislation is dealt with, at second reading and committee stages in the Scottish Standing Committee (q.v.), and the Scottish Grand Committee (q.v.) has special opportunities to consider Scottish Estimates. But neither of these, nor any other body, has power to initiate Scottish legislation or secure for it an adequate place in an overcrowded annual legislative programme. Members are not necessarily Scots nor acquainted with Scottish conditions, and owe allegiance primarily to English-based parties, whose ideas are formulated in and for England only.

Though Scotland is subject to the same executive and legislature as England, the public law differs in many material respects and is too readily assumed to be the same as in England. Thus, there was no rule that 'the King can do no wrong' and the special procedures attaching to remedies against the Crown never applied in Scotland, although the Crown Proceedings Act, 1947, in parts affects Scots law also.

Public corporations in many cases operate on a United Kingdom basis, but in other cases, e.g. regional hospital boards, are organized differently and operate in Scotland only.

Local government authorities consist of popularly elected councils of regions, islands and districts within regions. They have generally similar responsibilities to their English counterparts but are governed by quite distinct bodies of law and practice.

In respect of judicial control the prerogative orders of English law and the procedure of application for judicial review do not apply, and the courts exercise control mainly through the ordinary common law remedies of declarator, interdict, reduction, and damages.

In respect of personal liberty, habeas corpus does not exist, but the same end can be attained by petition to the High Court of Justiciary. The common law protects liberty of speech, meeting, opinion, and religious belief by ordinary actions in generally the same manner and to the same extent as in England.

Criminal law and procedure

The criminal law is uncodified and, apart from offences expressly created by statute, is mainly contained in common law and only to a small extent in statute. Accordingly, the systematization of the law by text-writers has been particularly important. The High Court of Justiciary still claims the right to declare criminal conduct inherently harmful, but the power is used sparingly. The distinction between treasons, felonies, and misdemeanours has never been recognized, and in all crimes all participants, in whatever degree, are deemed equally liable (guilty 'art and part'). The basic principle is that conduct is criminal only if that kind of conduct, or conduct having prescribed consequences, brought about with the requisite state of mind, has been recognized by common law or statute as punishable.

The age of criminal responsibility is eight. Persons of unsound mind may be held free of criminal liability if irresponsible. The McNaghten Rules (q.v.) do not apply, but similar tests are applied by the courts.

Scottish judges in the nineteenth century admitted the concept of diminished responsibility, of mental weakness, not amounting to insanity, as a ground for reducing a charge of murder to one of culpable homicide, and this has been accepted ever since.

Among the major crimes treason is, since 1707, governed by the Treason Act, 1351. Homicide is distinguished into murder, culpable homicide, accidental homicide, and justifiable homicide, the last two not inferring criminal penalties. Theft, reset (i.e. receipt, receiving), assault in various degrees, rape, breach, and other common crimes are all regulated by common law, though Road Traffic Offences are the commonest single class of crimes.

Criminal procedure is largely statutory but based on common law. For procedural purposes crimes are divided into indictable crimes and summary offences. A person arrested on a charge of crime is brought before a judge when he may be released on bail, or remanded in custody, and may make a judicial declaration, i.e. statement to the judge. This is rarely done.

There is, and long has been, a system of public prosecution headed by the Lord Advocate. Apart from bodies and persons statutorily authorized to prosecute, private prosecution is almost unknown. All crimes of any consequence are reported to the Crown Office and the Lord Advocate and his staff direct whether and in what court the prosecution is to take place. There is no grand jury and no inquiry before an examining magistrate. A person is charged only if the Crown Office is satisfied that there is a case to answer. The mode of trial is that directed by the Lord Advocate, and neither prosecution nor accused can elect jury trial or summary trial. In the High Court the Crown is represented by one of the Law Officers or an advocate-depute. In sheriff courts the Crown is represented by the procurators-fiscal and their deputes.

Trial on indictment is in the High Court before a Lord of Justiciary (in the most serious cases) or in the sheriff court a sheriff-principal or sheriff (in less serious cases) and a jury of 15, who may convict by a bare majority. Summary trial is before a sheriff-principal, or sheriff, or stipendiary magistrate, or bench of justices of the peace. In the High Court the prosecution is conducted by the Lord Advocate, Solicitor-General, or one of advocates-depute appointed by the Lord Advocate; in the sheriff court it is by the procurator-fiscal; the police never appear as prosecutors.

Evidence depends mainly on common law, and there is a general requirement that evidence must be corroborated by independent testimony or by facts and circumstances.

In all courts there are three possible verdicts, Guilty, Not Proven, and Not Guilty. The latter two both result in the accused being liberated and free from future prosecution. Not Proven is appropriate where there is grave suspicion but evidence not quite sufficient or convincing for conviction.

There is a separate set of penal establishments in Scotland, and in many details the penalties competent are different from those in England.

Private law

The private law is uncodified and though in many branches affected, sometimes severely, by statutes, many of which were passed with English law and conditions chiefly (if not entirely) in contemplation, is in many respects still clearly based on the Roman law or canon law. The terminology is akin to continental systems and the institutional writers, and the best modern books exhibit strong civilian influence in scope, arrangement, and treatment.

In modern times much damage has been done to native principles by legislation, framed to suit English conditions, applied to Scotland in ignorance of or to override native rules so as 'to bring Scots law into line with English', as if that were the touchstone of excellence, and by incautious citation and following of English precedents, frequently misunderstood or unsuitable to Scottish conditions.

Law of persons and domestic relations

Persons are distinguished into pupils (up to age 14 if male or 12 if female), minors (from those ages to 18), and adults. A pupil has no legal capacity and is represented by a parent or other person as tutor; a minor has considerable legal capacity but should act with the consent and concurrence of his curator, if he has one. Adoption of children was introduced in 1930; the status of the illegitimate has been improved by many statutes.

Marriage may be constituted regularly, by a clergyman or before an authorized registrar, or irregularly: of the three modes of irregular marriage, marriage by bare exchange of consent, and marriage by promise followed by intercourse in reliance thereon, were abolished in 1940, but marriage by cohabitation and acquisition of the reputation of being married is still competent. Judicial separation is competent for adultery or cruelty (including habitual drunkenness). Marriages may be declared null for lack of consent or impotence of either spouse. Judicial divorce has been competent since the sixteenth century; the grounds now recognized are breakdown of the marriage (evidenced by adultery, intolerable behaviour, desertion, absence of cohabitation for two years and the defender consents, or for five years if there is no consent), while marriage may be dissolved if the death of the other spouse must be presumed. On divorce either party may be awarded a capital payment and/or periodical allowance from the guilty spouse. Provision is made for the custody of any children.

Contract

The law of contract has been heavily influenced by Roman law. A unilateral promise, if proved in writing, is obligatory without acceptance or consideration. A bilateral agreement needs no consideration to support it. The rules relative to constitution and proof of contract are complicated, some contracts being made orally, some requiring proof in writing, some requiring constitution in formal writing, and some, under statute, requiring special modes of constitution or proof. The grounds for setting aside a contract as induced by error, fraud, and similar vitiations is mainly Roman, but confused in some respects by incautious reference to English authority.

The remedies for breach of contract are broadly as in English law, though specific implement is a remedy claimable as of right and not one granted in the discretion of the court only. Penalty clauses could always be modified by the court, but English influence has introduced the concept of liquidated damages.

As in Roman law loan, deposit, and pledge are real contracts requiring actual transfer for the existence of the contract.

Sale was largely assimilated to the English law on the topic (and greatly confused) by the Sale of Goods Act, 1893; while hire-purchase is subject to statutory modifications on the same lines as in England. Hiring is purely common law, while carriage and employment have been subjected to much the same statutory disturbances as in England.

Quasi-contract. The principles of quasi-contract or restitution are purely Roman (*condictio indebiti, negotiorum gestio*, and the like), though the obligations to pay salvage and make general average contributions are derived from the general maritime law rather than directly from Roman law.

Delict

The law of delict has evolved from the later, generalized, developed *actio legis Aquiliae* and *actio injuriarum*, and in modern law there is a recognized general principle of liability; a person commits delict who, by act or omission, done intentionally, or without adequate regard for precautions, contravenes the duty to avoid conduct foreseeably likely in the circumstances to cause, and actually causing, unjustifiable harm to another in person, property, or right. In addition, in many cases statutes impose more strict, or even absolute, liability for harm done, as for industrial accidents.

The principle of vicarious liability is recognized, and all wrongdoers are liable for the whole harm done. One who pays has a right of relief against the one personally in fault.

To the multifarious ways in which delicts can be committed special names are frequently attached (such as assault, defamation, negligent injury, and the like), but, unlike the common law of tort, these names do not indicate particular wrongs or forms of action having their own specialties, but are only labels for delicts occurring in particular circumstances. Among points of distinction from the common law of torts there may be mentioned the absence of any distinction between libel and slander, both being comprehended under the wrong of defamation, itself a species of the wider wrong of verbal injury; the fact that negligence is not a nominate or distinct delict, but only a state of mind with which any kind of delict may be committed; the fact that strict liability under the rule in *Rylands* v. *Fletcher* is only doubtfully recognized; the fact that liability for wrongfully causing death has been long recognized at common law, before and without need for the Fatal Accidents Act; and the fact that malicious motive may make lawful conduct actionable.

The House of Lords was responsible for the imposition on Scots law of the rule of common employment, and the detailed duties owed to invitees, licensees, and trespassers, both of which had to be eliminated by statute.

Property

Property is distinguished into heritable property, which formerly descended to the heir-at-law, chiefly land and buildings, and moveable property which formerly descended to the next-of-kin, things and rights.

The land law, with exceptions in Orkney and Shetland, is still basically feudal; it is owned ultimately by the Crown which has granted pieces of land to tenants-in-chief who have sub-granted it, and so on. A person having actual possession of land holds it from a superior who holds from an over-superior and so on up to the Crown; each person in the chain has an estate in the land with rights and interests defined partly by law and partly by the terms of the grant under which the land is held. All deeds conferring rights in or relating to land must be recorded in the General Register of Sasines; proposals have been made for its conversion from a register of deeds into a register of titles to land. The rules as to servitudes and liferent draw heavily on the corresponding Roman law. The law of leases has been enormously complicated by legislation in the twentieth century, and principles have been almost submerged by statutory provisions dealing with particular difficulties.

Moveable property comprises moveable objects and incorporeal rights. In respect of moveable objects the leading rule is that no rights pass without physical delivery, though there are exceptions to this, introduced under English influence, in sale and hire-purchase. The rule of market overt has never been recognized. Transfer in security requires delivery, save that the concept of the floating charge has been introduced in the case of companies. The incorporeal rights of patents, trade-marks, copyrights, designs, and business goodwill are in substantials the same as in England and regulated by the same statutes. Other incorporeal rights, as to shares or rights under insurance policies, must be transferred by written assignation.

Trusts

Though a native development, and that without the assistance of a separate court of equity, the law of trusts has been largely influenced by English practice. In Scotland trust does not depend on the dualism of legal and equitable ownership, but on legal ownership vesting in the trustee, subject to enforceable rights in the beneficiaries to have the trust executed. Trusts are either private (enforceable by the beneficiaries), or public (enforceable by members of the public). The Lord Advocate has no function, comparable to that of the Attorney-General, to enforce public trusts. Public trusts include many which would be charitable in the Chancery understanding of that idea.

Statutes confer many general powers on trustees and make provision for the investment of trust funds. The courts scrutinize as closely as in England the actings of trustees and enforce a similar high standard of care. There is no Public Trustee.

The courts also have powers to appoint persons as judicial factors, with most of the powers of trustees, to hold property for persons under age, absent, incapacitated, or otherwise unable to act for themselves. Such persons are officers of court under the supervision of the Accountant of Court.

Bankruptcy

A person who is insolvent may be deemed notour (i.e. notorious) bankrupt, a state which has statutorily defined indicia and consequences. Such a person may be sequestrated on petition by himself or a creditor, in which case a trustee is appointed, and commissioners, to realize and ingather the assets and distribute them to the creditors after paying expenses and preferred claims. The procedure is almost entirely regulated by statute.

Succession

At common law spouses and children of a deceased (whether intestate or testate) have long been entitled to minimum proportions of his or her estate, their 'legal rights'; accordingly it has never been the case that a person could wholly disinherit his spouse or children. Legal rights now amount to, for a widow or widower, one-half of the deceased's moveables if he left no descendants, one-third if he did, and for descendants one-half of his moveables if he left no spouse, or one-third if he did.

Prior to 1964 heritable property devolved feudally, to the heir-at-law, in the choice of whom the rules of primogeniture and preference for males over females applied, and moveable property to the next of kin. Since 1964 all property devolves in the same way according to a statutory table, without preference for age or males, surviving spouses having preferential right to a house, furniture, and a sum of money.

Wills may be completely informal, and are valid if holograph (i.e. entirely or in all material respects in the testator's handwriting) and signed, though unwitnessed, or may be formal, signed and attested. Legacies not exceeding £100 Scots (£8.33 sterling) may be made orally (nuncupative legacies). A will is not revoked by marriage but may be by absence of provision for an after-born child (*conditio si testator sine liberis decesserit*). There is a large body of law on the interpretation of wills. By the principle *conditio si institutus sine liberis decesserit* a bequest to a legatee who predeceases may be held to belong to his descendants. Problems sometimes arise of election between accepting legal rights and provisions under a will.

On death all the deceased's estate now vests in one or more executors named expressly, or impliedly by him (executors-nominate), or in persons appointed by the court (executors-dative). Before dealing with the estate executors must obtain confirmation from the sheriff: confirmation of an executor-nominate is a testament–testamentar; that of an executor-dative is a testament-dative. Executors once confirmed have title to ingather the deceased's assets and transfer them to those entitled under the doctrines of legal rights and the deceased's will or the rules of intestacy.

Civil procedure

Civil procedure is regulated by a large number of statutes and by Rules of Court made by the Court of Session under authority thereof. Much importance is attached to the formulation of detailed written pleadings and cases have frequently to be debated, before evidence is heard, on such questions as jurisdiction, competency of the action, or relevancy of the pleadings (i.e. their sufficiency in point of law, even if the facts are admitted or proved, to justify the remedy of the defence). There are no Masters, summonses for directions, or the like, and the procedure is largely in the hands of the parties.

Trial is normally before a judge alone, but in cases enumerated by statute, particularly personal injuries and death cases, trial is by judge and a jury of 12. In the sheriff court jury trial is limited and uncommon.

Evidence is regulated by common law and various amending statutes. A fundamental rule, save in delict cases, is that each material piece of evidence must be corroborated by a further statement or by facts and circumstances, and uncorroborated evidence is inadequate.

Professional organization

The profession is sharply divided into advocates and solicitors. Advocates comprise the Scottish Bar and are members of the Faculty of Advocates, which corresponds to the Inns of Court in England. They have right of audience in all courts, unless excluded by statute, and also give opinions and advice on matters referred to them. The Bar is much smaller than in England and less specialized. The Bar is headed by the Dean, annually elected, assisted by a Vice-Dean and other officers. Its etiquette is similar to that of the English Bar. There is a roll of Queen's Counsel in Scotland, names being submitted to Her Majesty by the Lord Justice-General.

Solicitors, formerly called law agents, engage in general legal work, particularly conveyancing, executry, trust, and general business advice. They have right of audience in the sheriff courts and other inferior courts, and some practise largely in those courts. The profession is governed by the Law Society of Scotland. The Society of Writers to Her Majesty's Signet, the Society of Solicitors in the Supreme Courts, the Royal Faculty of Procurators in Glasgow, and the Society of Advocates in Aberdeen are much older societies of solicitors. The first of these formerly had certain exclusive privileges in Court of Session practice. The profession of notary was formerly a separate one, with the function of authenticating various kinds of formal acts and deeds, but admission as a notary is now confined to solicitors.

Legal education and training

Law has been studied in Scottish Universities since before the Reformation, initially civil and canon law only, but since the eighteenth century a wider range of subjects, and a full range of courses is now available in five universities. It is noteworthy that the municipal law of Scotland was being taught in Scottish universities long before Blackstone began to lecture on English law at Oxford. Scottish university law schools have traditionally been more closely related to the profession and its requirements than English law schools, and have long provided for the teaching of conveyancing, procedure, and other professional topics, now largely in post-graduate courses leading to a Diploma in Legal Practice.

Neither branch of the profession has ever conducted any law school and all who have sought instruction in law have had to seek it in the universities, not always, however, as graduating students. In consequence, the great majority of lawyers have for long been graduates. Each branch regulates its own requirements. A prospective advocate must pass the Faculty's examination, from which exemption, in whole or in part, may be obtained by university degree passes, and spend a period in practical training, partly in a solicitor's office, and partly as a pupil to a practising advocate. A prospective solicitor must serve a five years' apprenticeship with a practising solicitor (a period reduced to two years for graduates), pass the Law Society's examinations, from which exemption in whole or in part may be obtained by university degree passes, and obtain a university Diploma in Legal Practice.

History: (Stair Society) *Sources and Literature of Scots Law*; *Introduction to Scottish Legal History*; A. D. Gibb, *Law from Over the Border*. General: D. M. Walker, *The Scottish Legal System*; T. B. Smith, *British Justice—The Scottish Contribution*; *Doctrines of Judicial Precedent in Scots Law*. Public Law: J. D. B. Mitchell, *Constitutional Law*; J. Bennett Miller, *Outline of Administrative and Local Government Law*. Criminal Law: G. H. Gordon, *Criminal Law of Scotland*; R. W. Renton and H. H. Brown, *Criminal Procedure in Scotland*. Private Law: D. M. Walker, *Principles of Scottish Private Law*; *Law of Contracts in Scotland*; *Law of Delicts in Scotland*; *Law of Civil Remedies in Scotland*; J. G. Wilson and E. M. Clive, *Husband and Wife*; J. G. S. Cameron and G. Paton, *Landlord and Tenant*; J. Rankine, *Landownership*; W. A. Wilson and A. G. M. Duncan, *Trusts and Executors*; N. Walker, *Judicial Factors*; J. Bennett Miller, *Partnership*; A. E. Anton, *Private International Law*.

Scots money. Mention of money in Acts of the Parliament of Scotland and in public and judicial proceedings before the Union of 1707 means Scots money, the value of which at the Union was one-twelfth of that of sterling money. The relative values were as follows:

Scots	*Sterling*	*Decimal*
Doyt or penny	$\frac{1}{12}$d.	—
Bodle or twopence	$\frac{1}{6}$d.	—
Plack, groat or fourpence	$\frac{1}{3}$d.	—
Shilling	1d.	0.5p
Merk (13s. 4d. Scots)	1s. 1$\frac{1}{3}$d.	5.5p
Pound	1s. 8d.	8p

Scott, Austin Wakeman (1884–). U.S. jurist, joined the Harvard Law School Faculty in 1910 and became successively Story professor, 1919–38, and Dane professor, 1938–61. He was the author of *Scott on Trusts* and other works.

Scott, Dred* v. *Sandford (1857), 19 Howard 393 (U.S. Supreme Ct.). Dred Scott was a slave, held in Missouri; his owner took him to Illinois, where slavery was forbidden by statute, later into the upper portion of the Louisiana Purchase, where slavery was forbidden by the Missouri Compromise Act of Congress of 1820, and then back to Missouri. Later Scott sued for freedom, contending that he had become free by residence in free territory. The Supreme Court held that Congress had no power to make Scott free by virtue of residence in free territory, that even if free at some time he was now slave by virtue of Missouri law, that the Missouri Compromise was unconstitutional as contrary to the due process clause of the Fifth Amendment, and that, in any event, as a negro he could not be a citizen of the United States within the meaning of the Constitution and therefore could not sue for his freedom in a federal court. The decision stoked the fires of hatred between free and slave states, and the North could not accept the Supreme Court's limitation on Congress's power. It made the Civil War almost inevitable, though the actual result was nullified by the Fourteenth Amendment.

Scott, James Brown (1866–1943). From his time at law school specialized in international law and was for long director of the division of international law in the U.S. Department of State. He was also an advisor of or delegate to many international conferences, and congresses, and a member of various conciliation commissions. He established the Los Angeles Law School in 1896 and was dean of the School of Law of the University of Illinois at Urbana, 1899–1903, and taught at Columbia, 1903–6. Among many books he wrote *The Hague Peace Conferences of 1899 and 1907* (1909), *The Spanish Origin of International Law—Vitoria and his Law of Nations* (1934), *The Catholic Conception of International Law* (1934), *The Spanish Conception of International Law and of Sanctions* (1934), *Law, the State and the International Community* (1939). He founded the *American Journal of International Law* and he was managing editor, editor-in-chief, and honorary editor-in-chief from 1906 till his death. He was President of the American Institute of International Law, The American Society of International Law, and the Institute of International Law.

Scott, John, Earl of Clonmell (1739–98). Called to the Irish Bar in 1765, he became Solicitor-General in 1774 and Attorney-General in 1777 till 1782. In 1784 he became Chief Justice of the King's Bench in Ireland and Baron Earlsfort, in 1789 Viscount, and in 1793 Earl of Clonmell.

Scott, John, Lord Eldon (1751–1838). Brother of William Scott, later Lord Stowell (q.v.), eloped with an heiress to Scotland and married her in 1772, and was called to the Bar in 1776 and at first made slow progress, so much so that he contemplated leaving London, but was able to take silk in 1783, when he entered Parliament. In 1788 he became Solicitor-General, and Attorney-General in 1793, in which capacity he conducted the high treason trials in 1794.

He became Chief Justice of the Common Pleas and Lord Eldon in 1799 and in that office proved himself a very sound common lawyer before he became Chancellor in 1801. Save for the years 1806–7, when Erskine replaced him, he remained Chancellor till 1827. He was also, particularly during the Napoleonic wars, an important member of the government. He became an Earl in 1821. In politics he was conservative to the point of being reactionary and failed to see the need for reforms after 1815, introduced no material legal reforms and opposed many proposed reforms. Personally he was sociable, genial, and charitable. As Chancellor he was responsible for some poor appointments to the judicial bench, but as a judge he stands in the first rank. He was the most learned lawyer of his day, his judgments illuminating every branch of the civil law, and a complete master of equity, and his reputation rests on his development of equity. Many of his decisions are leading cases. His great contribution to equity consisted in seeking to settle the principles and to make the rules of equity nearly as fixed and ascertained as those of common law. His achievement of this, in very large measure, completed the work of Nottingham and Hardwicke and fixed the relations of equity to law. His desire, however, to investigate the authorities thoroughly and by a decision to settle the principle for the future, led him in too many cases to be very dilatory in judgment. The arrears of business to which his scrupulosity gave rise occasioned the creation in 1813 of the office of Vice-Chancellor and the appointment of a deputy-speaker of the House of Lords in 1824. Nevertheless the judicial duties of the Chancellor were never more carefully discharged than by Eldon, and never by a mind better stored with law or better capable to draw a fine distinction or appreciate a subtlety.

T. Twiss, *Life of Lord Chancellor Eldon*; V. E. Surtees, *Lives of Lord Eldon and Lord Stowell.*

Scott, William, Lord Stowell (1745–1836). Elder brother of John Scott, Lord Eldon (q.v.), became a Fellow of University College, Oxford, and was Camden reader in ancient history, 1774–85, where his lectures were as popular as Blackstone's. He was a close friend of Dr. Johnson. In 1779 he became a D.C.L. and a Fellow of Doctors' Commons and in 1780 was called to the Bar. He had immediate success as a civilian. In 1788 he was appointed King's advocate and Vicar-General of the province of Canterbury, in 1790 master of the faculties and in 1798 judge of the court of the Admiralty. He was also an M.P. from 1790 to 1821 when he became a peer; politically he was as conservative as was Eldon, and he was generally an opponent of legal reforms. He resigned from the court of Admiralty in 1828. He was sociable but gluttonous and parsimonious.

As a judge he stands in the very first rank. He had great intellectual and literary abilities and was given the opportunity, partly by the Napoleonic wars and partly by the absence of reports of cases in the admiralty and ecclesiastical courts, to create a system of prize law. In this he succeeded, and created a definite body of principles and rules of prize law. In this field many of his decisions are of outstanding importance. He also gave many decisions of great importance on maritime law, on marriage, and on ecclesiastical law. His work in its own spheres is comparable to that of Mansfield in common law and of Hardwicke and Eldon in equity, largely creative of the whole branch of law.

W. E. Surtees, *Lives of Lords Stowell and Eldon*; Townsend, *Twelve Eminent Judges*, II, 279; E. S. Roscoe, *Lord Stowell*; N. Bentwich, *Lord Stowell* in *Great Jurists of the World.*

Scott Dickson, Charles, Lord (1850–1922). Called to the Scottish Bar in 1877, having previously practised as a solicitor, he sat in Parliament, 1900–06 and 1909–15, became Solicitor-General for Scotland in 1896–1903, Lord Advocate in 1903–06, and was Dean of the Faculty of Advocates 1908–15. Thereafter he was Lord Justice-Clerk of Scotland 1915–22.

Scottish Grand Committee. A committee of the House of Commons consisting of all the members sitting for Scottish constituencies together with 10 to 15 other members appointed by the Committee of Selection to preserve the balance of the committee in proportion to the strengths of the parties in the House as a whole. It may, unless the members object to the Bill being so committed, consider in principle Bills relating exclusively to Scotland, and this has the effect of debate on second reading. It may also consider the Scottish Estimates on not more than six days each session, and report thereon to the House. Provision is also made, unless there is objection, for the reference to this Committee of other matters relating exclusively to Scotland.

Scottish Land Court. A court established in 1911 and assuming the judicial functions of the former Crofters' Commission. It consists of a legally experienced Chairman, who has the rank and dignity of a judge of the Court of Session, and four members familiar with agriculture, one of whom speaks Gaelic. It has all the powers of a court of law. Its jurisdiction, originally confined to the crofting counties, now extends over all Scotland, and it sits wherever business requires. It deals with questions arising under the Crofters Holdings Acts, Small

Landholders Acts, Agricultural Holdings and Agriculture Acts, all concerned with landholding and questions between landlords and tenants. Appeal lies by stated case to the Court of Session.

Scottish Standing Committee. One of the standing committees of the House of Commons, consisting of 30 members representing Scottish constituencies and not more than 20 other members. The members are nominated by the Committee of Selection for each Bill. If a Bill is certified by the Speaker as relating exclusively to Scotland, it is referred to this committee for discussion at committee stage.

Scrip (or scrip certificate). Probably an abbreviation of 'subscription certificate', an acknowledgment by a company or the authority issuing a loan that a person named therein, or the holder of the certificate, is entitled to a specified number of shares, debentures, bonds, etc. Such a certificate is commonly used where shares or bonds are payable by instalments. It is issued in exchange for the letter of allotment and replaced by the shares, debentures, or bonds which it represents. A scrip certificate is a negotiable instrument.

Scroggs, William (1623–83). Called to the Bar in 1653, became a bencher in 1669 and by influence and subservience to the Crown became a judge of the Common Pleas in 1676, and in 1678 Chief Justice of the King's Bench, where he showed partiality and zeal for the Crown interest, such that he had to be removed in 1681. He presided at many of the trials arising out of the Popish Plot. Scroggs was able and eloquent but coarse and unprincipled and not a consistent lawyer. His judicial conduct has been compared unfavourably even with that of Jeffreys. His conduct of the Popish Plot trials was a disgraceful exhibition of partiality, he was impeached in 1681 and, though it was not proceeded with, public opinion forced the King to dismiss him. He was undoubtedly one of the worst judges who ever disgraced the bench.

Scrope, John (?1662–1752). Called to the English Bar in 1692 and in 1708 appointed a Baron of the newly-created Court of Exchequer in Scotland, an office which he held till 1722 when he became an M.P. He retained his Scottish judgeship till 1724. He was also secretary to the Treasury from 1724 till his death. In 1710 he was one of the Commissioners of the Great Seal. He was author, with Clerk, of a *Historical View of the Forms and Powers of the Court of Exchequer in Scotland* (1820).

Scrutton, Sir Thomas Edward (1856–1934). He had an outstanding career at Cambridge and won the Yorke prize four times. He was called to the Bar in 1882 and, while a junior, also acted as professor of constitutional law and history at University College, London. His practice developed rapidly, especially in commercial cases, aided by the publication in 1886 of *The Contract of Affreightment as expressed in Charter parties and Bills of Lading*, which has ever since been a standard book. In the Commercial Court from 1895 he had many forensic tussles with J. A Hamilton (later Lord Sumner). In 1910 he became a judge of the K. B. Division, being promoted in 1916 to the Court of Appeal. There his abilities were shown to greater advantage and he proved himself a really great judge; in commercial law he has been called the greatest judge since Mansfield. In 1883 he published *The Laws of Copyright*, based on a Yorke essay. His other Yorke Prize essays were published as *The Influence of Roman Law on the Law of England* (1885), *Land in Fetters* (1886), and *Commons and Common Fields* (1887). He published other books on mercantile law.

Scutage or escuage (from Latin *scutum*, a shield). Shield-money, in mediaeval feudal law, a payment in lieu of military service, paid by a tenant-in-chief in respect of the service of knights which he owed to the Crown. His personal obligation to serve could not be discharged by scutage but only by fine. Payment of scutage, though known in France and Germany, was most highly developed in England where it became a general tax on knights' estates at rates which by the thirteenth century were standardized. King John demanded frequent and heavy scutages and Magna Carta forbade the levying of scutage without the consent of a general council. Scutage was divided between the King and the tenants-in-chief who gave personal service in the campaign. It became obsolete by the fourteenth century.

Scutage rolls. Contain the accounts of scutage covering 1215–1347.

Sea. The water covering a large part of the surface of the earth, but excluding rivers, entirely enclosed areas of water, i.e. lakes, and substantially landlocked areas of water, usually designated as inlets, gulfs or bays, which are regarded as internal waters of a state. The sea is distinguished legally into the territorial sea (q.v.), which is the belt of water adjacent to coastal states over which they claim to exercise sovereignty and jurisdiction, subject to a liberty of innocent passage for ships of all states, and the open sea or high seas, beyond those belts, which are open to all vessels for peaceful navigation. Coastal states also assert sovereignty over the seabed even beyond the limit of the territorial sea if it is part of the continental shelf (q.v.), so that the natural resources thereunder can be exploited, e.g. by drilling for oil. Coastal states commonly assert jurisdiction of various kinds and for various purposes, e.g. the prevention of smuggling or pollution of the sea, over contiguous zones (q.v.), belts wider than the territorial sea. States have also increasingly asserted exclusive fishing rights over coastal waters. In 1976 the U.K. declared its claim to a strip 200 miles from the coast in place of the former 12-mile zone.

H. A. Smith, *Law and Custom of the Sea* (1959); C. J. Colombos, *International Law of the Sea* (1962).

Sea laws. See CONSOLATO DEL MARE; OLERON.

Sealing. In mediaeval times, the mode of authenticating charters, deeds and other writings. A drop of molten wax was dropped on the document and pressed with a die containing the seal. In many cases the wax was dropped on a piece of ribbon hanging from the document and sealed on both sides. There developed, in the royal household, a variety of seals used for different purposes, notably the Great Seal, now used for letters patent, grants and charters conferring titles and dignities, for incorporating towns or other bodies. The Lord Chancellor is Keeper of the Great Seal and the Lord Privy Seal nominal keeper of the Privy or private Seal. The Secretary of State controls a signet and cachet Seals and, in certain cases, also a second secretarial seal. Hence, the more important posts in a government are conferred by the Crown by delivery of the seals of office and surrendered by delivery up of the seals.

Today having a seal is an essential part of an incorporated body and it must be used in the execution of formal deeds and instruments, e.g. share certificates, but its general use is less insisted on now than formerly.

In modern practice, a deed must be sealed and the party professing to be bound must do some act acknowledging the seal to be his. The seal may be of wax affixed to the deed, or attached to it by ribbon, or be a red wafer attached to the deed, or be simply impressed on the deed by a hand-press. The formal manner of sealing is for the grantor to place a finger on the seal and say 'I deliver this as my act and deed', but a signature on a document bearing wax or wafer or other indication of a seal with the intention of executing the document as a deed is sufficient recognition of the seal to amount to due execution as a deed. When an individual executes a deed he must sign it as well as sealing it.

In Scotland, sealing has not for long been a necessary part of executing a deed, save in the case of corporations. Since the Union of 1707 there have been, in Scotland, the Great Seal used to authenticate charters of lands and commissions to principal officers of the Crown, the Privy Seal, used to authenticate royal grants of assignable or personal rights, the Quarter Seal, and the Signet, used to authenticate summonses and other royal warrants connected with the administration of justice.

Seal of cause. In Scots law, a kind of charter or grant, the means of exercise of the power conferred on many royal burghs and superiors of burghs of barony in Scotland, of constituting subordinate corporations or crafts, and defining the privileges and powers to be enjoyed by the subordinate corporation. The power was exercised mostly to charter incorporations of merchants or tradesmen of the burgh.

Seamanship. Skill in handling a ship or boat afloat. The rules of good seamanship cover the whole conditions from getting a vessel under weigh until she returns to anchorage or moorings. The question of what is good seamanship is one of fact depending on the circumstances of the case, many practices being customary. Some rules, notably the Collision Regulations, are enacted by Order in Council under the Merchant Shipping Act, and these contain detailed provisions seeking to minimize collisions.

Search. (1) Examination by police officers, customs officers or other persons of the body, clothing, premises or other property pertaining to an individual to determine whether he should be arrested or to look for evidence relevant to a crime. The practice raises many issues as to the liberty of the individual. Most countries require some kind of warrant or authority before any search or seizure of anything found can lawfully be done, but the method of authorizing and the extent of authority which may be granted varies from one legal system to another.

In early common law, search warrants seem to have been unknown but they crept in, to permit searches for stolen property. The Star Chamber gave rise to the practice of issuing general warrants to search for libellous matter. The Wilkes cases (*Wilkes* v. *Wood* (1765), 19 St. Tr. 1153, *Leach* v. *Money* (1765), 19 St. Tr. 1001, and *Entick* v. *Carrington* (1765), 19 St. Tr. 1030), held general warrants illegal.

In England, the right of a constable to search a person arrested depends on the circumstances of the case; there is no general statutory power. A constable may be authorized by warrant to search premises for stolen goods in particular circumstances, and though no-one has been arrested or charged the police may take an article if they have reasonable grounds for believing that a serious offence has been committed and that the article was connected with the crime. The police have statutory powers, in order to effect an arrest, to enter and search any place where the person to be arrested is or is reasonably suspected to be.

Search warrants may be issued by judges or magistrates under many statutes, and the search need not always be ancillary to an arrest.

In the U.S. the Fourth Amendment protects persons from unreasonable searches and seizures of property, and unauthorized confiscation of guns and papers, interception of communications by electronic eavesdropping, and things discovered by unauthorized invasion of privacy have been held illegal exercises of search and seizure.

In international law, the right of search is the right of ships of war to visit and search merchant vessels to ascertain whether the vessel is liable to seizure as carrying goods of the enemy or goods to the enemy.

(2) In English law, an inquiry before contract is made for a purchase of land, made in the local land charges register, the register of agricultural charges, the registers kept by the registrar of companies, registers of applications for planning permission kept by local authorities and certain other registers for matters registrable therein. Registration, in many cases, is deemed actual notice to all persons and for all purposes connected with the land affected so that failure to search for and discover entries does not support a refusal to complete on the ground of ignorance.

In Scots law, a search is the examination of certain public registers to ascertain the owner of a particular piece of heritable property, any burdens affecting it, and any barriers to the owner voluntarily dealing with it. The term is also used for a certificate of search, i.e. a note by an official or professional searcher, of entries found in the register or registers in question affecting the property in question. Searches as to owners and burdens are made in the General Register of Sasines (q.v.) and as to personal bars to dealing with land in the Register of Inhibitions and Adjudications (q.v.).

Seashore or fore-shore. The portion of the land lying between high-water and low-water marks of ordinary tides. At common law it is vested in the Crown unless acquired by a subject by grant or by long possession. The public are entitled to pass over the foreshore in boats and to anchor there, and the owner of the foreshore must not do anything which interferes with the right of navigation. The public have, however, no right to pass over or along the foreshore, except along a right of way, or incidentally to the rights of navigation or fishing, nor to wander about on the foreshore or use it for recreation or cross it for bathing, nor to cross it to get to or from boats, nor to load or unload boats save in case of peril or necessity. But the inhabitants of a parish may have acquired customary rights to do many of these things. No person may remove sand, gravel or stone from the seashore save by right of profit *à prendre*. Extensive provision is made by statute against erosion of the land and encroachment by the sea.

Seavey, Warren Abner (1880–1966). Distinguished U.S. jurist, professor at Harvard from 1927, and author of *Cases on Torts* and other textbooks.

Seaworthiness. At common law a shipowner, when contracting to carry goods, impliedly warrants that his ship is seaworthy. Under statute his liability is only to have exercised due diligence to make her seaworthy. The term means not only being in a fit state as regards state of repair to encounter the ordinary perils of the sea, but includes provision of competent master and crew, engines, fuel and other equipment and everything necessary for the contemplated voyage. Whether a ship is or is not seaworthy at a given time is a question of fact. Seaworthiness means that the ship is in a fit state, as to repairs, equipment and crew, and in all other respects, to encounter the ordinary perils of her voyage.

Secession of the South. The separation of 11 states from the United States and their formation of the Confederate States of America. The basic differences between the southern and northern states were over the existence of slavery and the issue of states' rights. As the frontier moved westwards and new territories were settled and, in time, organized as states, the question of the extension of slavery arose. The Missouri Compromise of 1820 prohibited slavery in territories north of 36°30′. By 1837 there were 13 free and 13 slave states, but the prospect was that the expansion of the Union would be predominantly of free states.

Secession was averted for a time by the Compromise of 1850, proposed by Henry Clay, which provided for the admission of California to the Union as a free state, the organization of New Mexico and Utah territories without restriction on slavery, abolition of the slave trade in the District of Columbia, more stringent Fugitive Slave Laws and certain other matters. But the Compromise was nullified by the Kansas-Nebraska Act of 1854, which established the doctrine of popular sovereignty, i.e. of federal non-intervention in the territories.

The non-extension of slavery was a vital issue in the Presidential election of 1860 and Lincoln's election made southerners who wished to extend slavery realize that secession was the only way of attaining that object. In December 1860, South Carolina seceded, and was followed by six other southern states: Alabama, Florida, Mississippi, Georgia, Louisiana, Texas. A compromise, the Crittenden Compromise, was proposed but rejected in Congress. On its failure representatives from the seven seceded states met at the Montgomery

Convention in Montgomery, Alabama (Feb. 1861–Feb. 1862) and formed the Confederate States of America. A Peace Conference in February 1861 made a further attempt to compromise, Lincoln sought to relieve the beleaguered Fort Sumter and called for volunteers to maintain the Union. Four border slave states (Virginia, Arkansas, North Carolina, and Tennessee) also seceded, but four others (Maryland, Delaware, Missouri, and Kentucky) did not. Each side's forces, in fact, contained men from all states, as not all in seceding states accepted the state view. The secession was ended only by the military defeat of the Confederate States, and their ultimate readmission to the Union.

Seckel, Emil (1864–1924). German jurist, who particularly studied the work of Benedict Levita (q.v.), wrote a *Beiträge zur Geschichte beider Rechte in Mittelalter* (1898) on the Roman-canon law of the period of the Reception, and edited Heumann's *Handlexicon zu der Quellen des römischen Rechts* (1907) and Huschke's *Jurisprudentiae anteiustinianeae reliquae* (1908–11).

Secondary evidence. Evidence which is admitted only in default of primary evidence. Thus, hearsay evidence is admissible of what persons now deceased stated orally in their lifetimes.

Secret societies. Groups or organizations making use of secret initiations, rituals, oaths, signs of recognition, and the like. They were known in ancient Egypt, Greece, and Rome, particularly connected with particular religions, and are known in ancient and modern primitive societies. In Europe, mediaeval guilds were sometimes organized as secret societies and the most widespread modern secret society, freemasonry, developed from the guilds of the stonemasons. Such societies have been organized for many purposes—religious, criminal, revolutionary, subversive, and other. They have included the Fenians, the Carbonari, and the Ku Klux Klan. Whether particular secret societies can be tolerated by the general body of the community will depend on the aims of the society and how it operates, how far it is detrimental to the general safety, and how far it in fact truly controls the government of the community itself.

Secret trust. Where a testator gives property to a person on the express or implied promise by the legatee or devisee that he will hold the property in trust for another. Secret trusts are valid in English law if they would have been valid as an express trust.

Secretariat, Cabinet. A Secretary to the Cabinet was appointed only in 1917 and has existed ever since. The secretariat is headed by the Permanent Secretary to the Cabinet Office. Its functions are to circulate papers and minutes of the conclusions reached at meetings of the Cabinet and its committees, communicate decisions to those who have to act on them, and maintain records. The conclusions are the only official record of Cabinet meetings and are circulated only to the Queen and Cabinet Ministers.

R. K. Mosley, *The Story of the Cabinet Office*; S. S. Wilson, *The Cabinet Office to 1945*.

Secretariat, Commonwealth. This was established in 1965 and is based in London and headed by a Secretary-General. It is responsible for servicing Commonwealth conferences wherever they may be held, and works through a number of divisions concerned with development, aid and planning; international affairs; finance, trade and commodities; and legal matters; and has oversight of many forms of Commonwealth co-operation.

Secretariat (of U.N.). The office staff of the organization, headed by the Secretary-General, who is appointed by the General Assembly on the recommendation of the Security Council. The Secretary-General is the chief administrative officer and serves all the organs of the U.N. He submits an annual report on the work of the Organization and may bring to the attention of the Security Council any matter which, in his opinion, may threaten international peace and security.

Secretary. (1). Originally, one who writes for another, but in modern practice, used in many contexts. In the Civil Service, the senior officers in a Department are known as the Permanent Secretary (or Permanent Under-Secretary where the political head of the Department is entitled Secretary of State), Deputy Secretaries, and Assistant Secretaries. Every company must have a secretary who is responsible for giving effect to decisions of the board, meetings, records, and returns and the management of the office generally, and is charged with statutory duties. Societies and clubs have secretaries who execute decisions of the committee and conduct the business affairs of the body.

(2) One of the Officers of State of Scotland, known from the fourteenth century. He had a regular place in Parliament and the Privy Council from the fifteenth century. The office was held by important persons and, particularly after 1660, became one of some influence and patronage, and the task of the holder, trying to reconcile the policy of the king of London with that of the Estates in Edinburgh, was not easy. At about this time the title Secretary of State began to be used. The office seems to have disappeared at the Union of 1707.

R. K. Hannay, in *History of the Society of Writers to H. M. Signet*; D. Milne, *The Scottish Office*.

Secretary-General. The principal executive and administrative officer and the head of the secretariat of the former League of Nations and now of the United Nations Organization. A person is recommended by the Security Council and elected by simple majority by the General Assembly. The term of office is for five years which may be renewed or extended. Apart from specific functions imposed by certain articles of the Charter and by treaties, the Secretary-General as chief executive can convene the Security Council to deal with a threat to peace, and bring matters to the attention of U.N. organs, make investigations and exercise a general power to act on his own initiative, when necessary, in the interests of international peace and security, act as mediator and informal adviser to many governments, and generally exercise any functions entrusted to him by the U.N. organs. As chief administrator he must secure the efficient working of the U.N. organs and committees, co-ordinate the work of the various branches of the Secretariat, specialized agencies, and other inter-governmental organizations, have studies made and reports prepared, prepare the annual budget and annual report.

The holders of the office of the Secretary-General of the League of Nations were Sir Eric Drummond and M. Avenol.

The holders of the office of Secretary-General of the U.N. have been Trgyve Lie (1946–52), Dag Hammarskjold (1953–61), U Thant (1962–71) and Kurt Waldheim (1971–).

See also SECRETARIAT, COMMONWEALTH.

Secretary of State (U.K.). The King's Secretary is met with in England in the time of Henry III, and by the time of Henry VIII, he had a seat in Parliament and Council and ranked next to the greater officers of the Household. The office grew in precedence and importance and men of high ability were chosen to fill it, such as Cecil (Lord Burleigh), Walsingham, and Cecil (Lord Salisbury). After a decline under the Stuarts, the office rose again in importance from 1660.

From about 1689 to 1794, there were two Secretaries of State, and, from 1709 to 1746, there was a third for Scottish affairs and, from 1768 to 1782, a third for colonial affairs. While there were two, one was in charge of the Northern Department, dealing with the Northern European powers, the other in charge of the Southern Department, dealing with Southern European countries and also Home, Irish, and colonial affairs. In 1782 the Northern Department became the Foreign Office and the Southern the Home Office. In 1794 a Secretary of State for War was appointed, and in 1801 assumed the Colonial Office as well. In 1854 these offices were separated. The office of Secretary of State for the Colonies was, after 1925, sometimes combined with and sometimes separated from that for Dominion Affairs until abolished in 1966, being merged in that for Commonwealth Affairs. Secretaryships of State have also been created for India (1858–1937), India and Burma (1937–47), Air (1917–64), Scotland (1926), Dominion Affairs (1930) (later (1947) altered to Commonwealth Relations and (1966) to Commonwealth Affairs and merging into the Foreign and Commonwealth Office in 1968), Defence (1964) (replacing the offices of Secretaries of State for War and for Air), Education and Science (1964), Wales (1964), Employment (1968), Social Services (1968), the Environment (1970), Energy (1974), Industry (1974), Trade (1974), Prices and Consumer Protection (1974) and Northern Ireland (1975).

At various times, there have been Secretaries of State in charge of other departments of State. The modern tendency seems to be to mark the importance of ministries by designating the holder a Secretary of State.

It seems probable that this rank and title will continue to attach to the Ministers in charge of Home Affairs, Foreign and Commonwealth Affairs, Defence, Education and Science, and two or three others. Since 1962 an individual has sometimes been designated First Secretary of State, but this merely indicates primacy *inter se*.

Secretaries of State are appointed by the delivery to them of three seals engraved with the royal arms, and on recall of appointment the seals are delivered up. It is a relic of the origin of the office that constitutionally any one Secretary of State may act for any other; theoretically they are joint holders of one office, each having a special sphere of responsibility.

Secretary of State (U.S.). In the U.S., a Secretary of State has been appointed since 1789, his sphere of responsibility being originally home and foreign affairs. (There had been a Department of Foreign Affairs since 1781.) A separate Secretary of the Interior was appointed, first in 1849, since when the Secretary of State has been concerned with foreign affairs and diplomacy only.

G. Hunt, *The Department of State of the U.S.* (1914); G. H. Stuart, *The Department of State* (1949); S. F. Bemis, *The American Secretaries of State and their Diplomacy* (10 vols., 1927–29).

Securities. A term often used for shares, stock, debentures, and similar claims against government, local authorities, or companies, but which should be confined to claims in respect of which the holder, or someone on his behalf, such as a trustee for debenture holders, has a claim backed by an enforceable right in security. It should not be used of shares or stock.

Securities and Exchange Commission. An independent agency created by Congress to administer the Federal Securities Act of 1933 and the Securities Exchange Act of 1934. It is composed of five members appointed for five-year terms. It licenses stock exchanges, demands information about their organization and trading practices, has considerable quasi-judicial powers, and has developed a reputation for protecting investors.

Security. That which fortifies or makes more secure a person's undertaking, particularly to pay or repay money. There are two kinds of security, personal security, consisting in the guarantee by a third party (a surety or guarantor or, in Scotland, cautioner) of the debtor's obligation to pay, and real security, consisting in rights in or over some property belonging to the debtor, which are transferred by the debtor to his creditor to supplement the debtor's promise to pay and to make more secure the latter's claim for recovery of his debt by conferring claims beyond what the latter has equally with all other creditors.

In Roman law, the earliest real security was *fiducia*, in which ownership of the pledged object was transferred to the creditor by *mancipatio* or by *in iure cessio*, subject to the obligation to retransfer it to the debtor on payment of the debt. By the later form of *pignus*, the creditor acquired possession only and from it developed *hypothec*, in which the object pledged in security was left in the hands of the debtor, the creditor having neither ownership nor possession.

In modern English law, a security right of the latter kind may be constituted over goods by pawn or pledge, or lien, when possession is transferred to the creditor, by floating charge or hypothec when possession is not transferred, or by bill of sale (q.v.). Bills of sale are unknown in Scots law.

A security right of the latter kind may be created over real property in England by legal mortgage or by equitable mortgage, or over heritable property in Scotland by statutory standard security, and over choses in action or incorporeal property, such as shares in a company, by transfer in security or by deposit of a blank transfer along with the share certificate.

Security, Act of (1703). An Act of the Scottish Parliament providing for a Protestant succession to the throne of Scotland distinct from the English succession, unless Scottish government were freed from English influence.

Security Council (of U.N.). The Security Council comprises five permanent members, China, France, U.S.S.R., U.K., and U.S.A., and 10 non-permanent members elected for two years by the General Assembly. Its functions are to maintain international peace and security, to investigate any disputes or situation which might lead to international friction, to recommend methods of adjusting such disputes or terms of settlement, to formulate plans for the establishment of a system to regulate armaments, to determine the existence of a threat to peace or act of aggression and to recommend what action should be taken, to call on members to apply economic sanctions, to take military measures against an aggressor, to recommend the admission of new members, to exercise the trusteeship functions of U.N. in strategic areas, and to submit annual and special reports to the General Assembly. It is primarily responsible for maintaining peace and security. Voting requires the votes of nine members including the concurring votes of all five permanent members, but a member may not vote when a party to a dispute. It is able to function continuously and a representative of each member is always present at U.N. Headquarters. A non-member of the Council is invited to be present when its interests are particularly affected. It is advised by the Military Staff Committee and under it functions the Disarmament Commission.

Sedgwick, Theodore (1811–59). Was admitted to the Bar in 1833 and developed an extensive practice, which ill-health forced him to discontinue in 1850. He wrote extensively, however, on political and legal problems, most importantly *A Treatise on the Measure of Damages* (1847) and a *Treatise on Interpretation* (1857).

Sedition. The crime of doing acts, or speaking and publishing words, or publishing writings capable of being a libel, in each case with the intention to bring into hatred and contempt or excite disaffection against the Queen or the government and constitution of the United Kingdom, or either House of Parliament or the administration of justice, or to excite the Queen's subjects to attempt, otherwise than by lawful means, the alteration of any matter in Church or State by law established, or to incite persons to commit any crime in general disturbance of the peace, or to raise discontent and disaffection amongst Her Majesty's subjects or to promote feelings of ill will and hostility between different classes of those subjects. The seditious intention is essential, but the words may evidence the intention. Sedition is, accordingly, a crime against the Crown and government, but not so serious as treason.

Sederunt, Act of. See ACT OF SEDERUNT.

Seduction. Leading a woman astray whereby she is induced to have sexual relations and loses her virginity. In England, action could be brought by a parent or employer for enticing away the woman with the result that the plaintiff suffered a loss of her services (*per quod servitum amisit*), provided that the plaintiff had a legal right or interest in the woman's services. This action has now been abolished. In Scotland, the woman herself could and can sue but must establish that she was led astray and persuaded by guile to surrender her virtue.

Seebohm, Frederic (1833–1912). Historian, lawyer and banker, and author of notable works in early economic history, *The English Village Community* (1883), *the Tribal System in Wales* (1895) and *Tribal Custom in Anglo-Saxon England* (1902).

Segregation. The practice of restricting people, particularly those of a particular race or religion, to particular districts, or towns or facilities, such as transport or education. It has been practised many times in many places, as in confinement of Jews to ghettos, or of Bantu in South Africa to particular areas. In the southern states of the U.S., segregation in public facilities was legal from the late nineteenth century to the 1950s and was outlawed only by the Civil Rights Act, 1964.

Seignory. A manor or feudal lordship, and also the relationship between a feudal lord and one of his tenants or the land held by the tenant. Since, and by virtue of *Quia Emptores* (1289) no new seignory can be created. The only seignories now remaining are those of lords of manors.

Seisin. In feudal land law, possession of land by one who actually occupied and used it and whose right to do so strengthened with the passage of time.

In English common law, it was possession of freehold estates in land, the predominant concept in land law. All the real actions, the writ of right, and the possessory assizes which replaced it, determined who had right to seisin. The mediaeval law protected seisin, and the person seised could exercise all the rights of an owner and a person not seised could exercise none of these rights until he recovered seisin by action. Seisin was *prima facie* evidence of ownership and the best right to seisin was the only form of ownership recognized. The rules of law followed out logically the consequences of this and all rights attached to the person seised. Thus, two persons could not simultaneously exclusively possess the same land, though several persons could simultaneously be seised of different estates (q.v.) in the same land.

Seisin, livery of. The formal delivery of possession of a feudal estate in land. See also FEOFFMENT.

Selborne, Lord. See PALMER, ROUNDELL.

Selden, John (1584–1654). Was called to the Bar in 1612 and practised as a conveyancer with some success. Early, he became acquainted with Ben Jonson, Camden, and Cotton and turned more to legal and historical studies than to practice. By 1610, he had published several valuable works on legal history; then followed his *Titles of Honour* (1614), an edition of Fortescue's (q.v.) *De Laudibus Legum Angliae* (1616), *A Brief Discourse touching the Office of Lord Chancellor of England* (1617), the *History of Tythes* (1617), *Mare Clausum, seu de Dominio Maris* (1636) (a reply to Grotius' *Mare Liberum* of 1609), the *Privileges of the Baronage of England* (1641), an edition of *Fleta* (q.v.) (1647), and a posthumous *Of the Judicature in Parliaments.* His *Table Talk* (1689) composed by his secretary, Richard Milward, is a collection of notable utterances by him over a period of 20 years, covering a very wide range of topics.

In 1623 he entered Parliament but, even before that, utilized his vast knowledge of constitutional and legal history and records to assist the popular party in the Commons. He participated in the impeachment of Buckingham, defended Hampden, and was involved in all the great constitutional issues of the next few years. In the Long Parliament, he was a leader of the opposition to the Crown on many issues but took no part in public affairs after 1649. Thereafter, he continued the literary work which he had managed to continue despite his public duties.

Apart from his legal studies Selden conducted extensive researches in oriental learning, and he attained fame in this sphere by his *De Diis Syriis* (1617) and published many books of exposition of Jewish law. He is the eponymus of the Selden Society (q.v.).

Life by D. Wilkins, prefixed to the latter's edition of Selden's *Opera Omnia* (1726).

Selden Society. The principal society having the object of advancing the knowledge of the history of English law, founded by Maitland and others in 1886. It has published a long series of volumes, ably edited and with valuable introductions, falling mainly into three groups: editions or new editions of old texts important for legal history, selections of charters and documents, and editions of Year Books (q.v.) and of early reports of cases in various courts. The materials thereby made available have been invaluable for the understanding of mediaeval English legal history.

Select Committees. Both Houses of the U.K. Parliament appoint Select Committees to investigate and make recommendations on any matter referred to them. The House appointing defines the terms of reference and the committee is responsible to the House. A committee is normally set up on the motion of a Government Whip, composed of members selected by the House—the committee as a whole reflecting the strengths of the parties in the House. Committees are given the powers necessary for their enquiry, normally to summon witnesses and call for papers and records, but are not given executive power.

Select committees are of three kinds: Firstly, Sessional Committees, appointed each session by the House or under a Standing Order, and including, in the House of Commons, the Committee of Privileges, Select Committees on Public Petitions, Statutory Instruments (now a joint committee), Nationalized Industries, the Committee of Public Accounts, Public Expenditure (formerly Estimates) Committee, Committee of Selection, Standing Orders Committee and Business Committee, and in the House of Lords, the Committee of Selection, Standing Orders Committee, Special Orders Committee, Personal Bills Committee, Committee for Privileges, Select Committee on Procedure of the House, Offices Committee, and Committee for the Journals. Secondly, instead of referring a Bill to a Standing Committee it may be committed to a Select Committee before being further considered in a Committee of the Whole House. Thirdly, Select Committees may be appointed for any particular inquiry. In 1966 and after, Select Committees were appointed to consider policy in relation to specialized areas of government, such as Science and Technology.

Select Vestries Bill. A Bill which is given a first reading in the House of Lords each session before the debate on the Address in reply to the Queen's Speech. It is never proceeded with any further but merely asserts the rights of the House to deal with any business it thinks fit in priority to any business recommended in the Queen's Speech. It corresponds to the Clandestine Outlawries Bill (q.v.), in the House of Commons.

Selective employment tax. A tax, instituted in 1966, with the object of stimulating employment in productive industries and discouraging services and non-productive employment, and imposed on employers in respect of each employee employed for more than a stated minimum time per week. In certain industries, supposedly productive, the tax was refunded and, in certain industries in certain areas, the tax was refunded and a premium paid in addition. It was replaced by Value Added Tax (q.v.).

Self-defence. It is permissible to cause bodily harm or even death in order to defend oneself or another person from unlawful violence, provided that the person causing the harm or death did what he could to avoid the violence, as by retreating where possible, and inflicts no greater injury than he, in good faith and on reasonable grounds, believes to be necessary to protect himself or the other. If the defence is made out, the accused escapes liability entirely, the injury or death being justified; if not, he may be guilty of assault, or even murder. The defence is not confined to a person's defence of his own life, but extends to defence against rape, possibly against sodomy, and defence of another whom one reasonably should protect, such as a child.

Under international law, every state has an inherent right of self-defence against armed attack. The right is vague and very much open to abuse. Under the U.N. Charter, the right is preserved but subject to the review of the Security Council, which may find that self-defence was not properly invoked and which may act to enforce peace against the state invoking the doctrine. Measures taken in self-defence must be reported to the Security Council and self-defence must cease when the Security Council has taken measures to maintain international peace and security. Many states have entered into alliances or arrangements with other states for collective self-defence.

Self-determination. The claim of a group of people having some degree of national consciousness, to form their own state and govern themselves. During the First World War, the Allies accepted self-determination, particularly by races subject to the Austro-Hungarian and Ottoman empires, as a peace aim, and it was included in President Wilson's Fourteen Points. It was achieved to a substantial extent by the break-up of the Austro-Hungarian and Ottoman empires. After the Second World War, it became one of the objects of the U.N., the Charter of which applies the term to the claim of a state to choose its own social, political, economic, and other systems, and also the right of a people to constitute themselves a state or determine their relationship, if any, with another state. The latter aim has been largely attained as colonies have achieved independence from European powers. The real problem is always whether a group is sufficiently large and distinct to be justifiably allowed to form a state by itself.

Self-help. The term for those kinds of legal remedies which a person may use at his own hand, without need to seek any order of court. These include self-defence, distress damage feasant, abatement of nuisance, arrest of a criminal (qq.v.), and a few others.

Self-executing treaties. Treaties between states which do not require legislation or other action by the states parties thereto to make them operative internally and as part of their municipal law. Whether a treaty is intended to be self-executing or not is a question of interpretation, determined by the wording of the treaty. In the U.K., treaties are not self-executing and if a treaty is to have internal effect in English, Scots, or Northern Irish law it must be enacted or otherwise given effect to by statute. In the U.S., treaties which are self-executing and intended to be effective as domestic law without need for implementing legislation are part of the law of the land unless contrary to the constitution, and override prior congressional statutes and state constitutions and laws whenever enacted. Executive agreements made by the President are not, however, treaties for the purpose of constitutional law unless made under prior authority of Congress or ratified by it. In other countries, the constitution usually provides whether treaties require ratification or not before they have internal effect, assuming that they are interpreted as self-executing.

Self-incrimination, right against. The principle based on the maxim *nemo tenetur accusare ipsum* and the phrase in the Fifth Amendment of the U.S. Constitution: 'No person ... shall be compelled in any criminal case to be a witness against himself'. The Supreme Court extended the principle to any Federal proceeding and the Fourteenth Amendment was thought, by some, to have extended this guarantee to state courts, and this was accepted in *Malloy* v. *Hogan* (1964), 378 U.S. 1. A witness must expressly claim his immunity and for himself alone. The right protects personal but not corporate documents and the witness may have a reasonable fear of prosecution even though innocent. In English law, a defendant in a criminal case cannot be compelled to give evidence, nor to produce documents, or other evidence against himself, but his failure to give evidence may be commented on by the presiding judge to the jury and the jury may draw an inference from his failure to explain his conduct.

Self-preservation. In international law, a concept now largely discarded and replaced, if by anything, by the concept of necessity. This is adduced as justification of action in circumstances demanding immediate or extraordinary action and is rather wider than self-defence. Action may be justified under this principle where a state is not acting to protect itself from imminent danger of attack but to avert some other evil. How far self-preservation or necessity is a defence in international law is a matter of controversy. If justifiable the peril must be instant and overwhelming and the action taken must not be disproportionate to the harm threatened. The sinking of the French fleet at Oran in 1940, when France surrendered, may be justified by the need to ensure that the ships did not fall into enemy hands.

Selwyn, Sir Charles Jasper (1813–69). Was called to the Bar in 1840 and, though not brilliant, became distinguished as counsel in the Chancery Court. He took silk in 1856 and sat in Parliament, 1859–68. He became Solicitor-General in 1867 and a Lord Justice of Appeal in 1868 but was undistinguished as a judge.

Semantics. The study of meaning, of the relations between words as linguistic signs and what they refer to, of the formal relations of words to other words, and of words to those who utter or receive them. Since law is very largely a matter of words, in statutes, cases, contracts, wills, letters, speech, and other contexts, the study of semantics has importance for the study of law. Important are the facts that particular words, e.g. action, may refer to different kinds of things in different contexts, that different persons may use the same verbal form, e.g. democracy, right, to convey different ideas, and that some words are of very indeterminate reference, e.g. right, possession. Jurists influenced by linguistic philosophers have concentrated attention on how particular legal words are used in particular contexts. Words, both legal words and ordinary words, do not have 'proper' meanings, but only usual or customary meanings or applications, some of which are well-settled and some of which are unsettled.

Semble (it seems). A word used in reports of decisions and textbooks to preface a proposition of law which is not definitely or clearly stated but which appears to be implied or indicated by the terms of the opinion or statement.

Semiplena probatio. In Scots law, evidence inducing a reasonable belief but not complete evidence, formerly supplemented by the pursuer's oath in supplement. Since parties became admissible as witnesses the concept has disappeared.

Senate (Roman). At Rome, the assembly of the senior men, successor of the council of the kings. At first Senators were chosen by the kings, then by the consuls, and later by the censors. Membership depended largely on wealth and birth and consisted mainly of ex-magistrates, thereby depending indirectly on popular election. It existed to advise the magistrates and had great influence on internal and foreign policy, finance, and legislative proposals and it ratified acts of the *comitia*, though it was not itself either an executive or legislative body. In the last century of the Republic, it collapsed when military leaders destroyed the Senate's authority.

Under the Empire, the Senate became an hereditary order and most of the high offices in the State were reserved for senators. Augustus imposed a property qualification. Out of its former advisory capacity developed legislative power and *senatus consulta* had been fully recognized as legislation by A.D. 200 and judicial functions also developed. Though it lost independence, it remained nominally sovereign and conferred on the *princeps* his powers, and preserved a tradition of stability, good order, and regard for the public interest.

In the Later Empire the number of provincial senators increased and the Senate recovered much administrative authority. Proposals of the Emperor, passed as *senatus consulta*, were the normal means of legislation. It is last heard of in A.D. 580.

Constantine created another Senate in Constantinople which, in A.D. 359, was made equal to that at Rome.

Senate (U.S.). The Upper House of the Congress of the U.S., composed of two Senators from each state, elected for six-year terms by popular vote, one-third of the members retiring every two years. Senators must be 30, citizens of the U.S. for nine years, and resident in the State they represent. The Vice-President of the U.S. presides.

In the matter of legislation, it has coordinate power with the House of Representatives, but exclusive power to choose a Vice-President if there is a tie in the Electoral College, to ratify treaties (by two-thirds majority), to confirm or reject presidential appointments, including those of Cabinet members, ambassadors and Justices of the Supreme Court (by simple majority), and of acting as court of trial in impeachment cases, a two-thirds majority being required for conviction.

It operates mainly through standing committees, the chairmen of which enjoy great power, and each party has its floor leaders, whips, and steering committees, but senators have greater freedom of action than have representatives in the House and there is a less elaborate structure of party control.

Senate of the Inns of Court and the Bar. The Senate of the Four Inns of Court was established in 1966 as a single body, to act on behalf of the Inns collectively. In 1974, it and the General Council of the Bar were replaced by the Senate of the Inns of Court and the Bar to act on behalf of the Inns and the Bar collectively on matters of common interest and on which common policy was desirable. It is composed of *ex officio* members, representatives elected by the Benchers of the Inns, representatives of the hall members of the Inns, representatives elected by members of the Bar, and certain members appointed by the Senate. It operates through various standing committees. Within the framework of the Senate there is a new Bar Council concerned with the independence, honour, standards, services, and functions of the Bar. It has a close connection with the Council of Legal Education.

Senators of the College of Justice. The judges of the Court of Session, by virtue of the statute establishing that court. The advocates and Writers to the Signet are ordinary members of the College. See COLLEGE OF JUSTICE.

Senatus consultum. An expression of advice of the Roman senate to the magistrates. In republican times, it did not have legislative force, though, in fact, it was binding. During the Empire *senatus consulta* were implemented, at first, by a clause in the praetor's edict, but from the time of Hadrian they had immediate force of law. The jurists frequently named them after one of the consuls of the year, e.g. SC *Orfitianum*; or after the Emperor who proposed it, e.g. SC *Claudianum*; or after the event which occasioned the legislation, e.g. SC *Macedonianum*. The texts of *senatus consulta* were preserved in the *aerarium*. Some *senatus consulta* were translated into Greek and many are preserved in their Greek versions.

Senior (or senior counsel). A Queen's Counsel as distinct from a junior, or barrister. The terms have no reference to age. In the Republic of Ireland, the term Senior Counsel (s.c.) has replaced Queen's (or King's) Counsel.

Seniority. Priority by reason of greater age and/or length of service, a factor frequently relevant in appointments, promotions, and choice of persons to be made redundant.

Sennachie. The Celtic term for the official bard of the early Scottish kings, an office from which developed the modern Lord Lyon King of Arms (q.v.).

Sentence. The judgment of a criminal court stating the penalty imposed on a person who has pled guilty or been convicted. What sentences are competent depends in each legal system on the powers conferred on particular kinds of criminal courts, for particular kinds of defendants, and for particular kinds of crimes and offences. In many cases, the judge has discretion, within legally defined limits, to determine what kind of sentence to impose, e.g. imprisonment, probation, or fine, and what measures of sentence. In practice, certain kinds and measures of sentences are customary for the more ordinary offences but variations may be made in particular circumstances. Judicial conferences frequently discuss the kinds of sentences appropriate in particular cases. On a conviction for several offences, sentences for each must be consistent one with another, not, e.g., probation for one and imprisonment for another, and in the case of imprisonment may, in general, be ordered to run concurrently or consecutively. A right to appeal against sentence is commonly permitted and sentences may be modified by an appellate court. In ecclesiastical procedure, sentence is similar to judgment (q.v.) in a civil action.

Sentencing. The technique of determining in a criminal court what sentence is appropriate for the person convicted, having regard to the crime of which he has been convicted. In the United Kingdom, sentencing is a matter entirely for the court, not for the jury, nor may prosecuting counsel ask for or propose any sentence or kind of sentence. Counsel for the defence, in speaking in mitigation, may suggest one kind of sentence rather than another. The decision is governed by a number of factors, particularly the sentences prescribed as competent or permissible for the kind of court in question, and for the kind of offence in question, the offender's sex, age, mental and physical state, and criminal record, his apparent suitability for particular kinds of treatment and its apparent suitability for him, the nature of the offence and its circumstances, and special factors such as the need for more drastic sentences to stamp out particularly heinous or prevalent crime. Sentencing is also affected by particular judges' views of the purposes, value, and effect of particular kinds of penalties and by their willingness to take an unusual line in a particular case, such as to give an offender one final chance. In practice, for many crimes there are fairly well-settled ranges within which sentences vary and beyond which they only occasionally go. In some jurisdictions, judicial conferences are held periodically to discuss the factors which should be weighed in deciding on a sentence and to establish sentences regarded as appropriate for ordinary cases of different kinds.

Separate estate. At common law, a married woman could not own any property independently of her husband, but if property was settled on her to be held to her separate use and benefit, she was treated in equity in respect of that property as if she were unmarried. Separate property might also be acquired by pre-nuptial contract with her husband or by gift from any person. The position of the married woman with respect to property was almost completely abolished by statute in 1882 and now there are no specialties attaching to married women so that all property, however acquired, is held by them as separate estate.

Separate opinion. An opinion or judgment by a member of an appellate court who concurs with the majority in its disposal of the appeal but who has different reasons for doing so.

Separate property. The system of matrimonial property law under which each spouse continues, after marriage, to be the independent owner of his or her property, as distinct from systems under which the properties of each become one fund held in common.

Separate system. See PENNSYLVANIA SYSTEM.

Separation. In canon law *divortium a mensa et thoro* (as distinct from *divortium a vinculo matrimonii*), the action of spouses in ceasing to cohabit and in living apart. It may take place by consent, or be authorized by law at the request of one party on account of the other spouse's behaviour. If consensual, a separation agreement is frequently entered into providing for property, maintenance of children, and other matters. What circumstances justify a spouse in seeking legal separation from the other depend on the particular system. In England, separation may be granted where the petitioner has grounds for seeking divorce and orders may also be made for maintenance, settlement of a wife's property, and custody, maintenance, and education of children. It may be granted on numerous grounds in magistrates' courts, including adultery, cruelty, desertion and certain other grounds. Separation does not dissolve the marriage but merely modifies the parties' obligations to each other. Today, it is principally resorted to by persons whose religious views do not recognize divorce or as a preliminary to divorce.

Separation of powers. A doctrine, found originally in some ancient and mediaeval theories of government, contending that the processes of government should involve the different elements in society—the monarchical, aristocratic, and democratic elements. Locke argued that legislative power should be divided between king and parliament but the great modern formulation of the doctrine was that of Montesquieu, in *l'Esprit des Lois* (1748), who contended that liberty was most effectively safeguarded by the separation of powers, namely the division of the legislative, executive, and judicial functions of government between separate and independent persons or bodies. His view was founded on observation of the British constitution though his understanding of British politics was not wholly accurate. In fact, in the British constitution there was not then and is not, complete separation of powers; the Lord Chancellor is chairman of the House of Lords, an important Minister and head of the judiciary; the Cabinet and other Ministers who comprise the heads of the executive departments are also members of the legislature; the judiciary have delegated legislative powers and judges who are peers are members of the House of Lords, even in its capacity as a legislative chamber.

Montesquieu's work was widely read and very influential and profoundly influenced the framing of the U.S. Constitution. In modern constitutions and structures of government, the doctrine still has influence but is not so rigidly or dogmatically applied. In particular, the power of the executive has developed so much that it can influence, if not control, the legislature and the judiciary. Thus, in modern Britain, the political executive is composed of the leaders of the majority party in the House of Commons and they can generally lead the majority, which can impose its will on the minority. Also, the principle that an Act of Parliament is unchallengeable implies that any judicial decision unpleasing to the political executive can be overruled retrospectively by statute, and that any obnoxious rule of law can be altered in anticipation or retrospectively.

Septennial Act. An Act of 1716, superseding the Triennial Act of 1694, extending the duration of Parliaments of the U.K., existing and future, to a maximum of seven years, a period later reduced to five years by the Parliament Act, 1911. It was passed after the suppression of the Jacobite Rebellion of 1715 when the Ministry was apprehensive of Jacobite schemes and afraid lest a general election disturb the stability of the new Hanoverian regime. Many attempts were made in the eighteenth and nineteenth centuries to shorten the duration of Parliaments and, in practice, Parliament rarely lasted more than six years.

Sequestration. The removal, by judicial authority, of property from the person in possession of it, normally for its preservation until it is determined who properly has right to it. In Roman law, two persons disputing over property gave control of it to a *sequester* until the dispute could be settled. From this the concept has been incorporated into many civil and common law systems. In international law, it is the seizure of property by a government which uses it as its own property. In English law, it is a writ issued against a person in contempt for disobedience of the court by failing to satisfy a judgment or order for the recovery of property or the payment of money. Sequestrators are authorized to take possession of all the defendant's real and personal property until the contempt is cleared. In ecclesiastical law, sequestration is granted by the bishop to the churchwardens when a benefice is vacant, to manage the property of the benefice and provide for the care of souls.

In Scots law, it is the transfer of property by judicial order from private hands into the hands of an officer of court. A court takes such action in three main sets of circumstances, when property is for the time being ownerless, or in dispute, or danger, when a landlord is seeking to make effectual the security he has at common law over a tenant's goods in security of his rent (sequestration for rent), and when an insolvent's property is taken out of his hands in order that no creditor may secure an unfair preference, but so that the whole may be administered by a trustee for the benefit of the whole body of creditors (mercantile sequestration). This last form is what is commonly called bankruptcy.

Sequestration for rent. A remedy in Scots law which may be invoked by a landlord where rent is in arrears, exercised by petition to the sheriff praying for warrant to sequestrate the tenant's property subject to the landlord's hypothec, or right in security of his rent, and to sell it so far as necessary to satisfy his claim for rent. Its application has been much restricted by legislation.

Serfdom. A state, distinct from slavery, in which persons were attached to a hereditary plot of land and liable only for certain dues and services and protected by law from being maimed or killed. The serfs of mediaeval Europe were descended partly from the slaves of the late Roman Empire and the barbarian states. Most of them were, however, neither slaves nor freedmen but *coloni*, who surrendered their lands to a lord in return for protection or had taken up lands granted by him on terms; they were bound to the land, and though free were an inferior rank in society to free tenants.

In practice, serfs could frequently be bought and sold without land but depended on the plot of land for subsistence; they were dependent peasants and had to pay the lord a proportion of the produce of the plot and normally had, alternatively or in addition, to give a customary amount of work on the lord's own land. The essential mark of serfdom was not payment in money or in kind, or labour on the lord's land, but lack of freedom of movement of the family and inability to dispose freely of his property. There were usually also restrictions on marriage outwith the lord's lands. The status of serfdom was hereditary and could be ended only by enfranchisement or manumission. On the serf's death, the lord might allow the land to transmit to the serf's heirs, but if there were none, it fell to the lord.

By the end of the thirteenth century, the French kings began to manumit serfs on crown lands, and some Italian communes did the same, in both cases mainly for fiscal reasons. In England, the Black Death and all over Europe, peasant risings in the fourteenth and fifteenth centuries brought about abandonment of discipline by landlords and peasant servility and their replacement by free peasant tenure for money rent. In England, serfdom developed into villeinage and villein tenure into copyhold tenure. The free peasantry were not necessarily better off than the serfs but had free status rather than unfree.

In Eastern Europe, peasant status continued and their conditions deteriorated. Serfs were not freed in the Austrian Empire till the end of the eighteenth century nor in Russia, until 1861.

P. Vinogradoff, *Villeinage in England.*

Serjeant-at-Arms. In the House of Lords, the Serjeant-at-Arms attends the Lord Chancellor with the Mace; the office is combined with that of Yeoman Usher of the Black Rod. In the House of Commons, the Serjeant-at-Arms attends on the Speaker, is Keeper of the Mace, has charge of any person appearing at the Bar of the House as petitioner or offender, and maintains law and order throughout the precincts of the House. Both officers are appointed by the Crown.

Serjeants-at-Law (or sergeants counters). Formerly the highest order of counsel at the English Bar, and known as the Order of the Coif. In the Middle Ages, they were the leaders of the legal profession, developing from the mediaeval *narratores* who argued pleas before the justices. From the seventeenth century, their position was challenged by the rise of the King's counsel who came to take priority over them. The titles may have been introduced at the Conquest and *servientes Regis ad legem* acted as stewards of courts baron to manorial lords or appeared as suitors in courts. They were appointed by the Crown on the advice of the Lord Chancellor, who acted on the recommendation of the Chief Justice of the Common Pleas. Serjeants were created by writ of summons under the Great Seal and wore a gown and scarlet hood and ranked after knights bachelor. The constant element of their dress was the coif (q.v.), a small cap, a feature which gave the name to the whole order of serjeants. They had precedence over junior barristers but only serjeants with patents of precedence (q.v.) took precedence over Queen's Counsel. On appointment, serjeants gave their colleagues gold rings inscribed with mottoes. A few of the more distinguished serjeants were King's Serjeants, appointed by patent and summoned to Parliament. Until 1814, the two senior King's Serjeants took precedence of the Attorney-General and the Solicitor-General. There was also a Queen's Ancient Serjeant, the last holder of which office, James Manning, was appointed in 1849 and died in 1866, and there is mention of the rank of Queen's Premier Serjeant. The serjeants had their own Inns, one in Fleet Street held from 1443, the members of which, in 1758, united with the other in Chancery Lane, which had been held since 1416. On being made a serjeant, he left his Inn of Court and joined Serjeants' Inn. Serjeants had the privilege of immunity from being sued in any court other than the Common Pleas. In the nineteenth century, there were seldom more than about 40 serjeants, and frequently fewer. Till 1845, serjeants had a monopoly of practice in the Court of Common Pleas. From mediaeval times till the Judicature Act, 1873, the judges of the King's Bench and the Common Pleas were appointed from the serjeants, if necessary being made serjeants before elevation to the bench, and, for this reason, judges and serjeants addressed each other as 'Brother'. This rule applied to the Court of Exchequer only from 1579. After 1868 no person, except a judge designate, received the degree. In 1877, the order was dissolved and Serjeants Inn sold, but the dignity has never been formally abolished. The last surviving serjeant was Lord Lindley, successively judge, Lord Justice, Master of the Rolls, and Lord of Appeal. In Ireland, King's Serjeants were appointed until 1922 and the last survivor lived until 1959. The Common Serjeant is a judicial officer in the City of London.

J. Manning, *Serviens ad Legem* (1840); A. Pulling, *Order of the Coif* (1897); H. W. Woolrych, *Lives of Eminent Serjeants at Law* (1869).

Serjeanty, tenure by. A mode of tenure of land under feudalism differing from knight-service (q.v.) in that the vassal, instead of serving the King in his wars, was bound to do some personal service, as by carrying the King's banner or lance, or acting as butler at his coronation. The great officers of the royal household held their lands by serjeanty. Grants of land were also made to inferior persons to be held for lesser services, such as forester or falconer. Hence, a distinction was later drawn between grand and petty serjeanty. In 1660, serjeanty was converted into socage (q.v.), saving, however, the privilege of performing the honorary services peculiar to the higher ranks of serjeants.

Service, contract of. In Roman law *locatio operarum*, the contract traditionally called that of master and servant, and today that of employment; whereby the one undertakes to serve the other, and act subject to the direction and control of the other, in return for wages or salary. Today it is mainly regulated by statutes.

Service of heirs. The former process whereby, in Scots law, an heir judicially established his relationship, to the effect of being entitled to inherit heritable property vested in the ancestor. There were special service, connecting with the ancestor in specific subjects in which the latter died infeft, and general service, establishing identity as heir in general. Both services proceeded originally on brieves from Chancery directing officials to try the matter in hand, by the inquest of a jury. The brieve of mortancestry developed into the brieve of inquest. The brieve was retoured to Chancery. After 1847, service was obtained on petition to a sheriff. After 1964, the procedure became unnecessary.

Service of process. The means of bringing the contents of a writ to the knowledge of the person affected. Writs of summons and certain other judicial documents require direct service, in the case of persons by handing a copy of the writ to the defendant and, if requested, showing him the original, and in the case of ships by nailing the writ, and then leaving a copy of it nailed, to the mast, or substituted service, by serving the document on a person likely to bring it to the knowlege of the party concerned. Alternatively, the solicitor for the party affected may give a personal undertaking to enter appearance (q.v.) for him; this is called 'accepting service'. Other documents do not require direct or substituted service but are delivered to the party's address or to his solicitor. Summons to a magistrates' court may be sent by post.

Service out of the jurisdiction. With the leave of a judge a writ may, in certain circumstances, be allowed to be served on a person outside the jurisdiction of the English courts, e.g. where the contract sued on was entered into within the jurisdiction.

Services, contract for. In Roman law *locatio operis faciendi*, the contract whereby one party undertakes to render services, e.g. professional or technical services, to or for another, in the performance of which he is not subject to detailed direction and control but exercises professional or technical skill and uses his own knowledge and discretion. There are two major groups of such services, professional services of lawyers, accountants, surgeons and the like, and the technical services of building and engineering contractors, shipbuilders, garages, transport contractors, and many more.

Servitude. In Roman and Scots law, a burden affecting lands by which the proprietor is restrained from the full use of the lands in the interest of adjacent lands (negative servitude), or is obliged to allow another to do certain acts thereon which would otherwise be competent only to the owner (positive servitude), corresponding to easements and profits (q.v.) in common law. They are frequently divided into urban and rural servitudes. The main categories of servitudes are support of buildings, light and prospect, cutting peat or turf, taking water and right of way or passage. Servitudes may be created by express grant, in some cases by implied grant, or by possession for the prescriptive period and may be lost by abandonment, discharge, prescription and certain other modes.

A distinct class of rights over other persons' property consisted in personal servitudes, so designated only in post-classical law; these included *usufructus* (the right to use and draw produce from another's land), *usus* (the right to use but not to take profits), *habitatio* (the use of another's house or lodging), and *operae servorum vel animalium* (the labour of slaves or animals). Similarly, in Scots law, older writers call liferent (q.v.) a personal servitude.

The word servitude is also used meaning slavery or serfdom, and for apprenticeship leading to freedom of a borough or of a City Company.

Servitudes. In international law, international agreements impressing on territory a permanent status, such as those demilitarizing or neutralizing states, or creating rights over water, comparable to those imposed by servitude in municipal law. Particular kinds of servitudes recognized include freedom of transit, economic use of waters, customs-free zones, fishery rights, demilitarized and neutralized areas, access to enclaves in other states, and access by landlocked states to the sea.

Session of Parliament. The period, usually of about a year, between the opening of a Parliament, or a new session thereof, and prorogation (which ends the session) or dissolution (which ends that Parliament). A session may be interrupted by adjournments and recesses, e.g. over Christmas. It is usual to prorogue and commence a new session in the autumn of each year. On prorogation, all bills which have not completed their progress towards enactment lapse and have to be re-introduced and debated afresh in the new session, unless the House expressly resolves otherwise, as in the case of private bills.

Session, Court of. See COURT OF SESSION.

Session, Great, of Wales. A court, established by Henry VIII, with a jurisdiction similar to that of assizes in England. Latterly, it was exercised by two barristers, sitting for 18 days. In 1830, the court was abolished, the High Court exercising jurisdiction in Wales and Chester as in England.

Sessions of the peace. Sittings of justices of the peace in England and Wales for the exercise of their judicial powers. Petty sessions was a meeting of two or more justices for the county or petty sessional division, sitting to exercise summary jurisdiction; this is now properly called a magistrates' court. Special sessions are sittings of two or more justices on particular occasions for the exercise of a distinct branch of their authority, particularly for the grant or renewal of licences to sell excisable liquors, and similar functions. General or quarter sessions were meetings held quarterly by two or more justices to try indictable offences and to hear appeals from petty or special sessions. Borough quarter sessions were held by a Recorder as sole judge. Quarter sessions were abolished in 1971 and replaced by sittings of the Crown Court (q.v.).

Setenario. A comprehensive book intended to improve the law of Castile and Leon, started by King Ferdinand III about 1250 and continued by his son Alfonso X (the Wise). It never seems to have been completed and, indeed, does not contain only law, but history, philosophy, and philology. It is doubtful whether it was ever meant to be enacted as a code of positive law. It may have been intended as a manual for the education of Castilian princes or as a treatise for the instruction of judges and others.

Vanderford, *Alfonso el Sabio—Setenario* (1945)

Set-Off. The right to plead a compensating debt against a creditor and thereby to have his claim extinguished or diminished to that extent. It is limited to money claims and is a ground of defence, not a substantive claim. Cross claims in respect of damages, which are not yet fixed in amount, are counterclaims, not set-off.

Settle. To settle a document is to get its terms and language put into proper and satisfactory form. In cases of complexity, this may be done by counsel, or by a judge. In Chancery practice orders are settled by the registrar, unless the parties are agreed. To settle property is to limit it, or its income, to certain persons in succession.

Settled land. Land and any estate or interest therein which is the subject of a settlement (q.v.). Prior to 1856, settled land could be sold or leased only under the authority of a power in the settlement or of a private Act of Parliament. The Settled Land Acts, 1882 and 1890 enabled tenants for life to convert the whole or part of the settled land into money and to improve part of the settled land from money raised by the sale of part of it. The Settled Land Act, 1925, vests the fee simple of settled land in an owner, usually the tenant for life, who may transfer it to a purchaser for value free from the limitations and trusts of the settlement.

Settlement. (1) In English law, a disposition of property, real or personal, made by deed or will, or occasionally by statute, whereby trusts are constituted to regulate the enjoyment of the settled property by the persons or classes designated by the settlor. It includes marriage settlements, post-nuptial settlements, settlements for minors, strict settlements and resettlements of land, protective settlements, and settlements to minimize tax. For the purposes of the Settled Land Act, 1925, a settlement is one whereby any land stands for the time being limited in trust for any persons by way of succession, or limited in trust for any person in possession for one of certain interests, or limited in trust for any person for an estate in fee simple for a term of years absolute contingently on the happening of any event, of charged with the payment of any rentcharge. A settlement *inter vivos* for the purposes of that Act must be effected by a vesting deed and a trust instrument. A tenant for life or statutory owner under a settlement has statutory powers and may be given further powers.

(2) An agreement or compromise between litigants to settle the matters in dispute between them and conclude their litigation. The terms of settlement may be incorporated in a consent judgment to which the court grants authority, or in a contract between the parties.

Settlement, Act of (1701). An Act necessitated to regulate the succession to the English throne by reason of the death of Queen Anne's last surviving child. It was provided that the throne was to pass to the Electress Sophia of Hanover, granddaughter of King James I and daughter of his daughter Elizabeth of Bohemia, and the heirs of her body, being Protestants. It was intended to exclude from the throne James II (in exile in France since 1689) and his heirs, who were Catholics. Under these provisions, on the death of Queen Anne in 1714, Sophia having already died, the throne passed to her son King George I. The Act also limited the sovereign's authority in that future monarchs were forbidden to use English revenues for defence of foreign possessions without Parliament's authority, foreigners were excluded from both Houses of Parliament, the Privy Council and high offices, while judges were to hold office during good behaviour rather than at the Crown's pleasure. The effect of the Act was extended to Scotland by the Act of Union, 1707 and to Ireland by the Act of Union, 1800.

Seven Bishops, Trial of the. 12 St. Tr. 416. In 1688, James II reissued his Declaration of Indulgence, suspending the penal laws against Catholics and dissenters, and ordered the Anglican clergy to read it from their pulpits. The Archbishop of Canterbury (Sancroft) and six bishops petitioned against this order on the ground that the suspending power had been declared illegal by Parliament. They were then indicted in the King's Bench for seditious libel and, amid great public rejoicing, acquitted. The case confirmed the right of the subject to petition the Crown.

Sewell, Thomas (?–1784). He was called to the Bar in 1734, served in Parliament from 1754, but made no mark there, and held the office of Master of the Rolls, 1764 to 1784. He was a very efficient and capable judge though not distinguished.

Several (of claims and liabilities), separate or distinct, as opposed to 'joint'. In certain cases, as for tort, liability is joint and several, so that each defendant is liable along with the others and is also liable alone and individually. Property is said to belong to persons in severalty when the share of each is ascertained so that he can exclude the others from it, as distinct from joint property, common property and coparcenery, where the owners hold in undivided shares.

Sex. The distinction between the two types of human and other kinds of animal life, namely male and female, a differentiation marked by physiological, biochemical and psychological features essentially connected with their respective reproductive roles and characteristics. The distinction has great legal importance and gives rise to many legal problems, many overlapping into areas of medical, psychological, ethical, social, and religious concern.

The sexual instinct motivates much human conduct and strongly influences many aspects of social and cultural life. Rules regulating behaviour in relation to sex have been very common in legal systems from early times, being found in Near Eastern bodies of law in the second millennium B.C. These influenced Jewish thinking, are found in the Bible, notably in Leviticus, and thence influenced Christianity. In Western countries where and when the influence of the church was strong, the attitude of the law to everything involving sex was repressive, though social practice might be much more permissive. This has been connected with the Church concern for marriage, and the family, but Christianity has in many ways been anti-sex. The Church attitudes generally embody an attitude that everything connected with sex outside marriage is wrong and sinful.

Legal rules concerning sex and sexual conduct are of various kinds; there are protective rules, seeking to penalize sexual relations with females who do not consent or are deemed incapable of consenting, rules seeking to prevent offence to the public, including public sexual activity, exhibitionism, nudity and near-nudity, and rules seeking to maintain strict sexual morality, including extra-marital relations, incest, homosexuality, prostitution, transvestism, and the like. The trend in recent years has been to relax the rules on matters of private morality and some of the rules protecting public sensibilities, but maintaining others of the latter class.

In addition, there is in each society and in various strata thereof, a variety of customs, practices, taboos, and prohibitions which also control socio-sexual conduct, and may conflict with legal rules, though societies differ very much in what they accept as permissible or frown on as undesirable, in relation, e.g. to extra-marital coitus.

The sex to which an individual belongs is determined, in most cases decisively, by chromosomal distribution, humans having 22 pairs of non-sexual chromosomes plus in the case of females an XX pair and in the case of the males an XY pair. Abnormal distributions, however occur. But a child, when born, is normally assigned to one sex or the other on the basis of visual inspection and brought up in the clothes and social surroundings of that sex.

Some persons, however, under the influence of hormones, do not develop fully as persons of the sex to which they were originally thought to belong; particularly at puberty or adolescence when hormone production increases enormously they may develop characteristics of the sex to which they truly belong and the false characteristics of the other sex become suppressed. A person may, accordingly, have been brought up as a boy, or girl, and at puberty develop as a girl, or boy.

Intersexuality exists where a person has both male and female biological characteristics, such as the external genitals of one sex and the gonads of the other, or the internal and external features of both sexes, or any other mixture of male and female characteristics. Such biological abnormalities may lead to the person being brought up as of one sex but hormonal activity at puberty revealing that the child is truly of the other sex; they may lead to transsexualism.

Transsexualism is the condition in which a biologically normal person believes himself or herself to be of the other sex, and adopts the clothing and manners of the other sex. Sometimes the transsexual seeks, by hormone and surgical treatment, to be altered to what he or she believes himself or herself really to be. Such so-called 'sex change' operations do not, in fact, change sex, but assist a person to appear to be of the sex to which he or she believes he or she really belongs.

In normal and the great majority of cases, one of the consequences of sexual differentiation is that, at least from the ages of puberty or adolescence, there are strong instinctive urges for persons to interact sexually with one of the other sex. This instinctive desire gives rise to many problems, many of which have legal implications.

Another question is of what kinds of sexual conduct should be repressed, what tolerated and what legal consequences follow on particular kinds of sexual conduct. Most sexual behaviour in society is heterosexual, between one male and one female, and involves various stages and degrees of arousal, variously called kissing, cuddling, necking or petting, and finally coitus.

What is regarded as permissible and what is forbidden, and in what circumstances, depends on the accepted customs and practices of society and the stratum of society in which the parties live and may be modified or inhibited by example, warning, upbringing, ethical and religious beliefs, and other factors. The main distinctions are between necking and coitus and between pre-marital, marital and extra-marital coitus. Petting is generally legally irrelevant, save that if thrust on an unwilling partner it might be treated as assault.

In repressive societies, pre-marital coitus is frowned on, though rather less so if the couple intend marriage. In Britain and western countries, in recent years, there has been an increase in and increased tolerance of premarital coitus; in some societies it is almost normal. To protect girls against what is deemed premature indulgence, even with their consent, various offences of indecency and statutory rape exist. Marital coitus is accepted and indeed encouraged, and deemed such an important element of marriage that a union may be declared null if coitus has not taken place and cannot take place by reason of impotence. Extra-marital coitus, particularly by wives, is generally condemned but is probably being increasingly tolerated. In some societies, it has long been tolerated by husbands. These three varieties of coitus are legally distinguishable as fornication, legitimate marital relations, and adultery. Coitus with an unwilling female who does not consent is rape.

One variety of sexual relationship very commonly condemned is incest, sexual relations between persons deemed too closely related for such relations to be permitted. In Western society it has been much influenced by the prohibitions in the Bible (which also condemns fornication and adultery). It is criminal and, in earlier times, a capital crime.

Particular difficulties arise in cases of sexual deviations, which are forms of sexual gratification other than ordinary heterosexual relations.

Homosexuality is human sexual behaviour between two persons of the same sex and is the commonest deviation from sexual normality. Male homosexuals may prefer to act the male or the female role or alternate in both roles; some resort to female clothing. Female homosexuals may similarly be very masculine, alternately one and the other, or very feminine, and some will dress and act like a man.

The legal attitude to homosexuality has varied at different times and places; in the later twentieth century, it is being increasingly tolerated. Formerly, homosexual acts were punishable but in England, since 1967 there is an exception for acts in private between consenting adults over 21.

Bestiality is the deviation in which an animal is used for obtaining sexual gratification; it is rare.

Exhibitionism is, however, a common though lesser deviation; in it a person, normally a male, obtains sexual excitement by exposing his genitals publicly to persons of the other sex. The exhibitionist does not normally seek a sexual response, nor does he intend to hurt, frighten, or seduce.

Voyeurism or 'peeping Tom' activity is the deviation in which sexual excitement is obtained by watching females in the nude, or undressing, or watching couples making love. Such conduct may be treated as breach of the peace. A variant is viewing pornographic drawings, photographs, and films, the sale or exhibition of which may be treated as an attempt to corrupt public morals or as possessing obscene publications.

An important variant of this is the extent to which a society tolerates nudity or near-nudity in public, as on beaches and in cabaret shows. Western society has steadily grown more permissive, and only the genitals, and the breasts of females need be covered to avoid prosecution for indecent behaviour in public, while 'strip-tease' shows and nudity on the stage are now tolerated.

Minor sexual deviations are fetishism and transvestism. In fetishism a person obtains sexual pleasure from a body part, such as hair, or an object, such as clothing; the fetish is a substitute for a wanted live person. Legally, it is probably of no significance but it may motivate assault or theft.

Transvestism consists of obtaining sexual excitements by wearing clothes of the other sex and passing oneself off as of the other sex. Such persons are not necessary transsexuals nor homosexuals and, save when cross-dressing, act entirely normally. Transvestism is probably of no legal significance unless it is connected with some specific offence.

The body of legal rules relative to sex and persons' attitudes to them, are accordingly influenced by their social, moral, and religious attitudes and by increasing understanding of the biological and psychological factors underlying sexual customs, practices, and conduct generally.

Sex discrimination. Discrimination in the provision of services, employment, pay, education, and facilities generally between one sex and the other, particularly against women, are outlawed by the Equal Pay Act, 1970, the Sex Discrimination Act, 1975, and the Employment Protection Act, 1975. The Equal Opportunities Commission has been established to enforce non-discrimination save in the few circumstances where it is permitted.

Sext. In the canon law, the book of Decretals of Boniface VIII promulgated in 1298 as a sixth book (*liber sextus*) of Decretals, supplementing the five books of Decretals of Gregory IX of 1234.

Seydel, Max von (1846–1901). German jurist, chief proponent of states' rights in the German Empire, and author of a *Kommentar zur Verfassungsurkunde für das Deutsches Reich* (1873), *Grundzüge einer allgemeinen Staatslehre* (1873), *Bayerisches Staatsrecht* (1884–94), and other works.

Shadow Cabinet. The group of leaders of the main opposition party each assigned to watch a major department of government, to lead for the Opposition in debate on that topic and to watch, question, and challenge government policy in relation to that department. If there is a change of government, many of the Shadow Cabinet are likely to be appointed to the departments which they have shadowed while in opposition. Though the office of Leader of the Opposition is known to the U.K. constitution by being given a salary as such, the Shadow Cabinet is a development since the mid-nineteenth century and has no official standing.

Shadwell, Lancelot (1779–1850). Called to the Bar in 1803, he took silk in 1821. He became an M.P. in 1826 and, in 1827 succeeded as Vice-Chancellor, holding office till 1850. He was also twice a Commissioner of the Great Seal. He was a weak judge, though patient and courteous, and not without ability.

Shaftesbury, Anthony Ashley Cooper, 7th earl of (1801–85). A social and legal reformer. In 1845 he secured the passage of the first Lunacy Act to treat insane persons as mentally ill. He then became associated with the factory reform movement and secured the passage of the Mines Act, 1842, which excluded women and boys under 13 from employment in coal mines. He was also active in support of government housing, ragged schools, and missionary socities.

J. L. Hammond, *Lord Shaftesbury*.

Shand, Alexander Burns, Baron Shand of Woodhouse (1828–1904). Called to the Scottish Bar in 1853, he progressed rapidly and became a judge of Court of Session in 1872. He retired in 1890 but was created a Privy Councillor, and a peer (not a Lord of Appeal) in 1892. He sat regularly in the House of Lords and Privy Council till his death. He was a strong and independent judge, particularly in mercantile and commercial law, and sat on several Commissions of Inquiry.

Share. In company law, a fractional part of the ownership of the company. The share capital is the nominal total value of the company and is divided into a number of shares. There may be several kinds of shares with different rights, particularly to participate in profits, such as preference shares (carrying a preferential right to receive a dividend, normally of a fixed percentage), cumulative preference shares (where the preferential right, if not satisfied in one year, is added to the claim for the next year), ordinary shares, and sometimes other kinds. See also STOCK.

Shari'ah. The way to follow, or the law, the fundamental concept of Islam, which specifies how a Muslim must conduct himself in accordance with his religion. It was systematized during the eighth and ninth centuries A.D. and deals with the whole social, political, religious, and private life of persons who profess Islam. It differs from Western law in that it is, in theory, founded on divine revelation and is not man-made, though it has been much developed by interpretation. In many countries adhering to the Islamic faith it remains a part of, or has materially influenced, the modern secular law and, in some places, it remains the law of the land.

N. J. Coulson, *History of Islamic Law* (1964); J. Schacht, *Introduction to Islamic Law* (1964).

Shaw, Charles James Dalrymple, Lord Kilbrandon (1906–). Called to the Scottish Bar in 1932, he took silk in 1949, became Dean of the Faculty of Advocates in 1957, and a judge of the Court of Session in 1959. He was Chairman of the Scottish Law Commission, 1965–71, when he became a Lord of Appeal till 1976. He was also a member or chairman of many committees, particularly ones concerned with young persons, and of the Royal Commission on the Constitution, 1969–73.

Shaw, Lemuel (1781–1861). Admitted to the Bar in 1804, was Chief Justice of the Supreme Judicial Court of Massachusetts, 1830–60, and did much to develop the law of that state to accord with changing social and industrial conditions and to influence American legal opinion generally. He regularly interpreted public grants to railroads and other public utilities favourably to the community. In 1842, in *Commonwealth* v. *Hunt*, he abandoned the common-law rule that trade unions were illegal conspiracies in restraint of trade. He may, however, have originated the segregationist doctrine of separate but equal facilities in education, which was not disapproved till *Brown* v. *Board of Education* in 1954, and, in *Farwell* v. *Boston and Worcester Railroad* (1842), he formulated the doctrine of common employment and influenced the House of Lords to fasten it on Scottish and English law.

Chase: *Lemuel Shaw, C. J.*; Adlow, *The Genius of Lemuel Shaw*.

Shaw, Thomas, Lord, later Baron Craigmyle (1850–1937). Having passed advocate in 1875, he developed a good practice, but saw that the way to advance was through politics and became an M.P. in 1892. He became Solicitor-General for Scotland in 1894–95, and Lord Advocate in 1905. In 1909, he became Lord of Appeal in succession to Lord Robertson, having abandoned a case in which he was counsel and which was being heard, to hasten to London to ask the Prime Minister for the appointment. As a law Lord he was assiduous but made no contribution to the development of legal principles. He presided over a number of commissions and inquiries. He resigned in 1929 and received a barony, as Lord Craigmyle. He published a number of addresses and lectures and two verse plays and a sketch of Chief Justice John Marshall (q.v.) *John Marshall in Diplomacy and in Law* (1933).

Shelley's case, rule in (1581) 1 Co. Rep. 936. A rule of English law, older than but fully discussed in that case, applicable to deeds and wills, to the effect that if, in the same instrument, an estate of freehold was limited to one with remainder to that one's heirs, or to the heirs of his body, the remainder conferred the fee simple, or in the latter case the fee tail, on the named ancestor. The words 'his heirs' were deemed words of limitation and not words of purchase. The rule was abrogated in 1926.

Shepard's Citations. A series of legal indexes for the Supreme Court and the superior courts of each state of the Union, published in the U.S. since 1873, which facilitate the finding of cases, opinions, and articles in which a given case is cited as relevant, and tracing whether it has been subsequently approved, distinguished, overruled, or otherwise commented on by another court.

Shepherd, Sir Samuel (1760–1840). Was called to the English Bar in 1781 and acquired a considerable practice. He became Solicitor-General of England in 1813 and Attorney-General in 1817 and, though he could have become Chief Justice of the King's Bench or of the Common Pleas in 1818 declined, owing to deafness and a dislike of criminal work. In 1819, he became Chief Baron of Exchequer in Scotland, an office he held till 1830. He was a friend of Sir Walter Scott and popular in Scotland.

Sheppard, Jack (1702–24). Brought up in a workhouse and apprenticed as a carpenter, he achieved notoriety in 1723–24 as a robber, but incurred the enmity of Jonathan Wild and was arrested. He escaped twice from prison but was recaptured and hanged. He was made the hero of Harrison Ainsworth's novel, of 1839, bearing his name.

Sheppard, William (*fl.* 1660). He became a serjeant in 1656 but was deprived of his offices at the Restoration. He published various works, the most important being the *Touchstone of Common Assurances* (1641) said to be from a manuscript found in a library. He wrote a second part (1650) under the title *Law of Common Assurances.* He also published, in four parts, *A Grand Abridgment of the Common and Statute Law of England alphabetically digested under Proper Heads and Titles* (1675). It is in English but is otherwise inferior to Rolle's *Abridgment*, being little more than a recension and rearrangement of *Coke upon Littleton* and Coke's *Institutes* under aphabetical heads and brought up to date.

Shepway, Court of. A court formerly held before the Lord Warden of the Cinque Ports to hear proceedings by writ of error from the mayor and jurat of each port, with a further appeal to the King's Bench. It was abolished in 1855.

Sheriff. Originally shire reeve, a person who performs various administrative functions, in a county or district. The office existed before the Norman Conquest, being probably created as part of the reorganization of local government consequent on the reconquest of the Danelaw. He was a royal official and took over the functions, particularly collection of monies due to the King, formerly performed in burghal areas by the King's reeve. He became the King's representative in the shire and shire court and the King's main link with local administration. After William the Conqueror separated secular and ecclesiastical courts, the sheriff was the chief man in the county and presided in the county court. He executed writs and judged civil and criminal cases as well as calling out and leading the military levies of the shire. But from the mid-twelfth century, his jurisdiction was limited by the growth of the King's courts and he came to have the functions of conducting preliminary investigations of accused persons, trying minor offences and detaining greater offenders till the itinerent justices came. The development of the offices of coroner, local constable and, particularly of justices of the peace, all took away duties from the sheriffs and, from the time of the Tudors, the office was mainly ceremonial.

In England, a high sheriff is appointed annually for each county. To be appointed a person must have sufficient land to answer the Queen and her people and, in effect, must have considerable means. In November each year, persons fit to serve as high sheriffs are nominated, three for each county, except Cornwall and Lancaster, by any two or more of the Lord Chancellor, Chancellor of the Exchequer, Lord President and members of the Privy Council, and Lord Chief Justice, with the judges of the High Court or any two or more of them. The names are engrossed on parchment, presented to the Queen in Council and the Queen pricks with a bodkin one name for each county. In practice, the choice is pre-arranged with the nominees. The names are then notified in the *London Gazette* and warrants from the Privy Council sent to them. The sheriff of the county of Lancaster is appointed by the Queen in right of the Duchy, and the sheriff of the County of Cornwall by the Prince of Wales as Duke of

Cornwall. The office is held during the pleasure of the Crown and is an annual one. It is nominally a compulsory office of great antiquity and importance. He retains certain important ministerial functions.

The sheriff's former civil and criminal jurisdiction is now exercised by the county court and magistrates' courts, but he may have to summon a jury to assess compensation under the Lands Clauses Acts. As conservator of the Queen's peace he must suppress unlawful assemblies and riots and apprehend offenders and defend his county against invasion by the Queen's enemies and, in theory, he may still raise the *posse comitatus* to pursue and arrest felons within the county. He is returning officer at parliamentary elections and may have to make known royal proclamations, and the summoning of juries for the High Court, Crown Court, and Central Criminal Court.

He had, till 1971, to make arrangements for the reception and accommodation of assize judges on circuit and attend on them personally.

Every sheriff of a county must appoint an under-sheriff, who may be reappointed by the next high sheriff, and who normally performs all the ordinary duties of the high sheriff. In the City of London a similar position is occupied by the Secondary.

Every sheriff of a county, city, or town must appoint a deputy, who must reside or have an office within a mile from Inner Temple Hall, for the receipt of writs, grant of warrants, and similar purposes.

Bailiffs or sheriff's officers are appointed by sheriffs for the purpose of collecting fines and executing writs and processes. The sheriff is civilly liable for any fraud or wrongful act or omission by his under-sheriff, bailiff or officer in the course of his employment.

Nearly all writs of execution on judgments and orders of the Supreme Court are directed to the sheriff and he must have them executed. He is liable if anything is done in excess of the authority given by the writ. He is also the person required to effect arrests on civil process for non-payment of money, where this is still competent, and to levy all fines, penalties and forfeited recognizances payable to the Crown, and was formerly the person charged with the execution of the death penalty.

In Scotland, the sheriff and the sheriffdom were introduced in the twelfth century, under Anglo-Norman influence. The sheriffdom centred on a royal castle, and frequently associated with a burgh, was an instrument of royal government and the division of the county into sheriffdoms spread south-west, north, and later north-west. The sheriff had delegated royal powers but the earliest sheriffs were frequently barons who abrogated the powers to themselves. Early on, the office of sheriff became hereditary, and the central government was frequently impotent to control its sheriffs. From the thirteenth century, sheriffs appointed deputes, and the sheriff tended to become a figurehead, the deputes presiding at the main courts. The sheriff's powers were administrative, financial, military, and judicial. From the first he held a court at the royal castle or burgh tolbooth, head courts thrice a year, and intermediate courts for lesser matters more frequently. The court was composed of barons and freeholders owing suit of court for their lands, but the work of the court was frequently hampered by absence of suitors. The sheriff presided and directed on matters of law, but judgment was given by the suitors.

The court's civil jurisdiction was both at first instance and on appeal from baron courts, within the sheriffdom and extended to possession of lands and property and harm by one person to another. Its criminal jurisdiction was limited to homicide and theft when the slayer or thief was caught red-handed.

In 1617 James VI considered abolishing hereditary sheriffs but, after some temporary improvements during the Commonwealth, the hereditary system revived after the Restoration. In 1747 all heritable sheriffships were abolished, provision being made for sheriffs, who were never appointed, and sheriff-deputes, who had to be advocates of three years' standing but might continue in practice in the Court of Session. After 1828 sheriff deputes might be addressed as sheriff and the term 'sheriff principal' has been used since the early eighteenth century for that officer. They appointed substitutes, unqualified persons but resident in the sheriffdom, to deal with simple actions, relying on them to reserve difficult cases and appeals for the sheriff-depute.

In 1787 the salary of the substitutes became payable by the Crown and in 1825 they had to be advocates or solicitors of three years' (now 10 years') standing. They were given tenure *ad vitam aut culpam* in 1832 and in 1878 they became irremoveable save by the Lord President and the Lord Justice-Clerk and, in 1877, they were appointed by the Crown. They were full-time judges and had to be resident in the sheriffdom.

Since 1870, several statutes have effected amalgamations of sheriffdoms and this has reduced the number of sheriffs; but as business has increased, the number of sheriffs-substitute has also increased. Statutes have extended their jurisdiction considerably. In 1970 the sheriff was officially renamed 'sheriff-principal' and the sheriff-substitute renamed 'sheriff'.

In modern practice, the sheriff-principal occupies an office of great dignity, and has numerous administrative functions. He also supervises the judicial business of the sheriffdom, may participate in hearing some criminal cases, and hear appeals in civil cases from the sheriffs.

The sheriff-principal may appoint a few honorary sheriffs, some of whom are unqualified and appointed merely as an honour, while others are legally qualified and assist and deputize for the regular sheriffs. Other qualified persons may be appointed as temporary sheriffs.

In the U.S., the sheriff is usually an elected officer of the county, its chief executive officer, and an officer of the court. He appoints a deputy and both have police powers in the enforcement of criminal law. They may call out the *posse comitatus* (q.v.) to assist in maintaining order, and have civil functions such as the serving of process and execution of writs, particularly writs of judgment by sale and distress.

Sheriff court. The main inferior court in Scotland, which has had a continuous development since the twelfth century. The country is divided into a number of sheriffdoms with a sheriff-principal and a number of locally resident, full-time, sheriffs. In each sheriffdom courts are held regularly in all the major towns. The court corresponds to the English county court and middle tier of the Crown Court.

The civil jurisdiction is financially unlimited and extends to practically all kinds of cases, the main exception being actions relative to status (including divorce). Cases are heard normally by a sheriff alone, but sometimes by him with a jury of seven. Both advocates and solicitors have right of audience. Appeal lies to the sheriff principal and then to the Court of Session, or to the Court of Session direct, and, in certain circumstances, to the House of Lords. Minor pecuniary claims are heard in the Summary Court under a simplified procedure.

The criminal jurisdiction includes nearly all common law and statutory offences, tried summarily, or with a jury of 15, the powers of sentencing being greater in the latter case. In criminal jurisdiction, the powers of sheriff-principal and sheriff are the same. Appeal lies to the High Court of Justiciary. Practically all prosecutions are conducted in the public interest by the procurator-fiscal and his deputes.

Sheriff officers. In Scotland, persons appointed by the sheriff, after examination as to their knowledge and after their finding security for the due performance of their offices, to serve writs and carry through the steps involved in doing diligence (q.v.) in pursuance of a decree of court.

Sheriff of Middlesex's Case (1840), 11 Ad. & E. 273. Following the case of *Stockdale* v. *Hansard* (q.v.) Stockdale brought another action against Hansard for libel and was awarded damages. The Sheriff of Middlesex levied execution on the defendant's goods. When the House of Commons reassembled, it resolved that the levying of execution was a contempt of the House and resolved that the sheriff be taken into custody. On an application for habeas corpus the Serjeant-at-Arms made the return that he acted under the warrant of the Speaker of the House of Commons and gave no further specification of the facts. The Court of Queen's Bench held this a good answer as each House of Parliament, being a superior court, must have the power of protecting itself from interference. The sheriff was released on the passing of the Parliamentary Papers Act, 1840.

Sheriffs' Courts. In London there still nominally exist two ancient sheriffs' courts for the Poultry Compter and the Giltspur Street Compter respectively, of which the judges are the judges of the Mayor's and City of London Court.

Sheriffs, Inquest of. An investigation by Henry II of England, in 1170, into local administration, in consequence of which he dismissed a number of sheriffs found to have been lacking in diligence.

Sheriff's tourn. A court, which emerged in the thirteenth century from a combination of the sheriff's old jurisdiction at special sessions of the hundred court, to supervise the view of frankpledge (q.v.) and the new machinery of the jury of presentment introduced after the Norman Conquest and applied to criminal procedure by the Assize of Clarendon. With the decay of the frankpledge system, the important business of the tourn came to be the presentment of the jurors of the hundred to the sheriff. It was specially concerned with the maintenance of the peace and, because of the association of this with the King, it became a royal court and court of record, held by the sheriff as royal deputy. It was concerned with the punishment of petty officers and the supervision of the lower units of local government. In the Middle Ages, several statutes regulated procedure at the tourn, and an Act of 1401, which provided that sheriffs should no longer have power to arrest persons, but should transmit indictments to the justices of the peace, marked the end of the tourn. It was obsolete by the end of the sixteenth century but not abolished till much later, though, where the tourn was in private hands or vested in a borough, it remained active much longer under the name of court leet.

Sherman Act (1890) (or Anti-Trust Act). The main U.S. legislation directed against combinations in restraint of trade between the States or with foreign nations, passed in response to growing public dissatisfaction with the development of industrial monopolies and trusts. More than 10 years elapsed before it was ever used to break up any industrial monopoly. Its first use was against a striking union alleged to be acting in restraint of interstate commerce. From 1902, however, it was invoked by Roosevelt against railroad, oil, tobacco, and other monopolies and its use was upheld by the Supreme Court. In 1914 Wilson established the Federal Trade Commission with wide authority to prevent business practices, which would lead to monopoly, and the Clayton Anti-Trust Act supplemented the Sherman Act with provisions against interlocking directorates and other practices tending towards monopoly.

Shifting clause. In a settlement, a clause providing, in certain circumstances, for a devolution of property in a way other than that primarily prescribed.

Shifting use. A secondary or executory use which, if it operates, does so in derogation of a preceding estate.

Ship and shipping. Ships include every description of vessel used in navigation not propelled by oars. The law of shipping includes a vast volume of common law rules, legislation, and interpretative decisions, dealing with such subjects as property in ships, masters and crew, safety provisions, carriage of goods by sea under charterparties, and bills of lading, carriage of passengers, pilotage, towage, collisions, wreck and salvage, lighthouses, harbours, docks and piers, and related subjects. It is a major part of the general maritime law as applied in the Court of Admiralty.

Ship-broker. A person who acts as agent for a shipper seeking cargo space or for a shipowner seeking cargoes for shipment.

Ship-money, case of. See HAMPDEN'S CASE.

Shipping inquiry. An investigation into a shipping casualty by a magistrates' court or a wreck commissioner, with the assistance of assessors of maritime experience. It hears evidence and reports to the Board of Trade and has power to suspend or cancel the certificate of a master, mate, or engineer if it finds that the loss or abandonment of or damage to any ship was caused by his wrongful act or default.

Shire. A very ancient division of England, existing from at least the seventh century. Each shire comprised several hundreds (q.v.) and was administered by an ealdorman, who commanded the fyrd, administered justice, and was responsible for executing the laws of the shire. Under the Danish kings, the title of ealdorman was supplanted by that of earl. His deputy was the reeve, who latterly became the shire reeve or sheriff, and who later acquired all the administrative and judicial functions, the earl retaining only the command of the military forces.

The shire moot was the court of the shire and also the general assembly of the people of the shire. It was convened twice a year by the sheriff and attended by the ealdorman, the bishop and other public officers, all lords of lands, the representative reeve, and four men and the parish priest from each township. All these formed the judges of the court. Its jurisdiction extended to every kind of case, except those concerning a high officer of State or a King's thegn, but it could not be invoked until the hundred court had heard and denied the case, nor could appeal be taken to the King unless the shire moot had failed to give justice. The shire court lost most of its importance after the development of the itinerant justices by the Norman kings, but long continued to exercise an extensive civil jurisdiction, and remained the general assembly of the freeholders of the shire for county purposes.

Until the local government reforms of 1972, the term shire was generally synonymous with county and was the major rural local government area. Since then a shire is a geographical term only.

Shock (or mental injury). This is now recognized as an actionable harm as much as physical injury. For this purpose, shock is not merely a fright but identifiable physical and/or mental lesion brought about, not by physical impact but through the mind, by what has been seen, heard or otherwise experienced.

Shoplifting. The colloquial term for theft from a shop. It is not a distinct crime but punishable as theft.

Shop steward. A person elected by the trade union members working in a particular shop, department, or other division of an organization, to act as their representative in negotiating with management.

Short Parliament. A Parliament summoned by Charles I in 1640 which refused to grant supply for the war against the Scots and insisted on redress of grievances first. It was dissolved after only three weeks and, later in that year, Charles was forced to summon a Parliament, which became the Long Parliament.

Short title. A brief name, e.g. The Sale of Goods Act, 1893, which statutes, particularly since 1845, have been given by a section in each Act. An Act of 1845 enacted that, in citing statutes, it should be sufficient to cite them in this way, and the Short Titles Acts, 1892 and 1896 and certain other Acts have retrospectively given short titles to most previous Acts still in force and not already so equipped. See also COLLECTIVE TITLE; LONG TITLE.

Shorthand writers' notes. In trials in superior courts generally, a verbatim note of the evidence given, and of arguments heard incidentally thereto, is taken by a shorthand writer. In jury trials, a note is taken of the judge's summing up, and of all judgments delivered orally. Transcripts of such notes are made when required for consideration in case of appeal. Shorthand notes are also made of extempore judgments in the Court of Appeal and copies thereof are filed in the Bar Library.

Sichard, Johann. (1499–1552). A German humanist and author of works on the *Codex Theodosianus*, the *Breviarium Alaricianum*, and *Sichardus Redivivus: Dictata et praelectiones in Codicem* (1598).

Side-bar. A bar or partition, formerly existing in Westminster Hall from within which, each morning in term, attorneys moved the judges, on their way to their respective courts, for common rules known as side-bar rules. These included the rule to plead, to reply, and to rejoin. These motions were made *ex parte* and were granted of course. This practice was discontinued before 1825 and the common rules were obtainable in the rule office, though still known as side-bar rules.

In the Parliament House, Edinburgh, where the Court of Session sat, there was formerly a bar known as the side-bar at which the Lords Ordinary's rolls were called.

Siete Partidas, Codigo de las (Code of the Seven Parts). An important law-book of mediaeval Spain, compiled between 1256 and 1265, by order of King Alfonso X, the Wise, officially called the *Libro de las Leyes* and, from its division into seven parts, *Las Siete Partidas*. The seven parts dealt with religious life, powers and duties of administrators, legal justice, marriage, contracts, wills, and crime and punishment. It was in Castilian, based on the municipal charters and customs of Castile and Leon, the canon law, the Roman jurists contained in the Digest, and the works of the Italian commentators thereon, and its general character is of a systematic compendium of canon and Roman law. It seems to have been intended to be imposed as common law, but this was negatived by concession of new and confirmation of old local *fueros* even thereafter. The *Partidas* became a university text and a reference book and, in 1348, it was made obligatory in all matters not contradictory of the municipal *fueros* and certain other rules. It was taken overseas with Spanish colonists and entered into the law of most of central and south America, and of the states of the U.S. formerly part of Mexico, and even into the law of the states formed from the Louisiana Purchase.

Sign manual. The actual handwritten signature of the sovereign, as contrasted with documents authenticated by the signet. Warrants under the sign manual are necessary for the use of the Great Seal or Wafer Great Seal.

Signatura, Apostolic. The supreme tribunal of the Roman Church. It probably originated in the referendaries or reporters in civil chanceries in the Middle Ages who received petitions addressed to the sovereign, reported on them, and communicated his answers. In the Church, referendaries dealt with requests made to the popes. In the fifteenth century, they were called *Referendarii Signaturae* because they presented rescripts for papal signature or in some cases signed them. In the fifteenth century, the *Signatura referendariorum* was established with a president and office. Later the Signatura divided into the *Signatura Gratiae* and the *Signatura Justitiae*, and, in the sixteenth century, each Signatura had its own prefect. The former lost importance and was suppressed in 1899 but latter assumed the function of a tribunal. It was reformed several times and, from 1908, has been the supreme tribunal of the Holy See. It consists of a certain number of cardinals, one of whom is prefect, assisted by a college of voting referendaries, and a group of simple consultors. It deals mainly with procedure and due observance of the law and acts as court of cassation for the Rota. Its functions are revision of sentences of the Rota (q.v.) before official communication, civil jurisdiction as supreme tribunal of the Vatican state, delegated jurisdiction to rule on requests to the Pope for the committal of special cases to the Rota, and jurisdiction controlling the Rota in special cases. It does not give reasons for its decisions.

Signature. The name of a person written by himself, either in full or by the initials of the forename or names and the surname in full. In English law, a signature may be by initials only, by mark, rubber stamp, or proxy. In some contexts, it has been held enough to write a note on paper on which the party's name was printed. Signature is required to authenticate wills, deeds, and certain kinds of contracts. In Scotland, signature does not, save in cases statutorily authorized, include mark, rubber stamp, proxy, or a printed name on paper.

Signet. A seal commonly used for the sign manual (q.v.) of the sovereign. It is also a seal, the principal of the three delivered to a Secretary of State on his assuming office and redelivered when he ceases to act on behalf of the sovereign. In Scotland, a summons initiating an action in the Court of Session must 'pass the Signet', i.e. be sealed at the Signet Office of the Court, before being served.

Silent system. Sometimes known as the Auburn system, the silent system was a prison regime, developed in Boston, based on separate sleeping but with workshops for useful employment by day, but with silence strictly enforced. This system spread rapidly and many prisons were built on this model, including Sing Sing and Auburn. The silent system was adopted in England in the nineteenth century and again came into favour at the end of that century, though the privilege of talking might be earned by long-term prisoners in their later stages. The rule was extensively modified in 1921 and now the only restrictions are on excessive chatter during working hours.

Signing judgment. In the Queen's Bench Division, the party who has obtained judgment after trial prepares two forms of judgment in accordance with the certificate of the associate and takes them to the appropriate officer of court, who signs one form and files it, and stamps the other with the seal of the court and returns it to the party.

Silk. A Queen's Counsel (q.v.) wears a silk gown; hence 'to take silk' is to become a Q.C., and a 'silk' is a Q.C.

Simmler, Josias (1530–76). A Swiss jurist and theologian, author of *De Republica Helvetiorum Libri Duo* (1576), a book of wide influence and of continuing interest as depicting the public law of the old Swiss Confederation.

Simon, Jocelyn Edward Salis, Baron Simon of Glaisdale (1911–). Called to the Bar in 1934, he held political offices and became Solicitor-General, 1959–62, President of the Probate, Divorce and Admiralty Division, 1962–71, and a Lord of Appeal in Ordinary, 1971.

Simon, John Allsebrook Simon, 1st Viscount (1873–1954). Called to the Bar in 1899, he sat as a Liberal M.P. from 1906–18 and 1922–40. He was Solicitor-General (1910–13), Attorney-General (1913–15), refused the office of Lord Chancellor in 1915 to stay in politics, becoming Home Secretary (1915–16), and later Foreign Secretary, (1931–35), Home Secretary again (1935–37), Chancellor of the Exchequer (1937–40), and Lord Chancellor (1940–45). He was chairman of the statutory Commission on the Government of India, 1927–30. He made no mark as a legislator or law reformer but was a distinguished Chancellor.

J. Simon, *Retrospect*.

Simon de Montfort's Parliament. The first English Parliament to which, in 1265, two representatives of each city and borough were summoned, as well as earls, barons, bishops, and two knights from each shire. It was not, however, a truly representative assembly because the delegates were known supporters of de Montfort.

Simonds, Gavin Turnbull, Viscount (1881–1971). Was called to the Bar in 1906, took silk in 1924, became a judge of the Chancery Division in 1937, and was promoted a Lord of Appeal in 1944. He was also chairman of the National Arbitration Tribunal, 1940–44. He was Lord Chancellor 1951–54, though he had had no political background, becoming a hereditary peer in 1952 and a Viscount on his retirement, and a Lord of Appeal again 1954–62. A very distinguished and profound lawyer, he was editor-in-chief of the third edition of *Halsbury's Laws of England.*

Simony. The buying or selling of something connected with the spiritual or, more generally, a contract of any kind contrary to divine or ecclesiastical law, so-called from Simon Magus who tried to buy the powers of conferring the gifts of the Holy Spirit from the Apostles (Acts, 8:18). There developed, from the fifth century A.D., attempts to buy and sell offices and ranks in the Christian Church and there was much ecclesiastical legislation against it. The wrong was later extended to include all trafficking in benefices and consecrated objects. It was widespread in the ninth and tenth centuries but severely repressed and disappeared after the Reformation and Counter-Reformation. In England, it was made an offence by various statutes and a person about to be instituted into certain clerical offices must make a declaration that he had not been party to a simoniacal contract.

Simple contract. In English law, a contract made not under seal, but orally or in writing.

Sin. The concept of thought or conduct transgressing the standards set by a religion. In the Bible and Christian theology, sin is a falling short of the mark set by God for men and an offence against God, his nature, and the eternal law. Numerous individual kinds of sin are recognized, including evil thoughts, adultery, immorality, murder, theft, covetousness, wickedness, and deceit. Theologians have long distinguished between mortal and venial sins. Aquinas considered mortal sin to be *contra legem Dei* but venial sin merely *praeter legem*, and it has frequently been held that they differed in punishment due. From mediaeval times, there was general recognition of seven (or sometimes eight) capital or deadly sins; the list usually includes pride, greed, lust, gluttony, envy, anger, and sloth. These are not all necessarily mortal sins.

The Christian notion of sin, though quite distinct from legal crime and wrong, has powerfully influenced legal ideas in Western society and those sins which consisted in overt acts, words, and deeds, such as theft and murder, have, probably without exception, been treated as criminal or wrongful or both at common law. On the other hand, those sins which consisted only in thought, such as envy, hatred and anger, have been treated as religious and moral faults only and not as criminal or tortious, because, in general, the law cannot know or take account of a person's mental states but only of his conduct. Some kinds of conduct, e.g. adultery, at one time made criminal because sinful have now ceased to be punishable.

Single escheat. The forfeiture of all of a person's moveable goods to the Crown on his being declared rebel.

Singleton, Henry (1682–1759). Was called to the Irish Bar in 1707, sat in Parliament, and became Chief Justice of the Irish Common Pleas in 1740, surrendering that office in 1753. He became Master of the Rolls in Ireland in 1754.

Sinha, Satyendra Prasanno, Baron Sinha of Raipur (1864–1928). Studied at Calcutta and was called to the English Bar in 1886. Thereafter, he built up a large practice in Calcutta and in 1905, became Advocate-General of Bengal. In 1909, he became the first Indian to be legal member of the Governor-General's Council, a position which he filled with great success. He resigned in 1910 and took a leading part in the movement for Indian self-government. In 1914 he was knighted and, in 1919, he was raised to the peerage and made Parliamentary Under-Secretary for India, and later served as governor of Bihar and Orissa. In 1926 he was appointed to the Judicial Committee of the Privy Council where he gave evidence of great judicial ability. He was one of the first Indians summoned to share in the government of India and, indeed, of the British Empire, and received many honours.

Sinibaldo dei Fieschi (*c.* 1190–1254). A distinguished canon lawyer, author of an *Apparatus in quinque libros Decretalium*, and later cardinal and Pope Innocent IV.

Sirmondian constitutions. A small collection, named after its first editor, made about A.D. 430, probably in Gaul, of 16 Roman imperial constitutions dealing with ecclesiastical law.

Sist. In Scottish procedure, to stop the progress of a case for a time. Also to make a person a party to a case.

Sit-in. A mode of nonviolent disobedience or of industrial struggle. In either case, a group of protesters occupy a public place, building, or factory and remain there until forcibly evicted or satisfied that their grievances will be remedied. Legally, a sit-in is a trespass and, if persisted in, may be criminal. It has been a mode used by Indians during their struggle for independence, by American negroes in their civil rights struggles, and by trade unions protesting against closures of plants and redundancies.

Sittings. The sittings of common law and Chancery courts were formerly regulated by the terms (q.v.) and were distinguished as sittings in and after terms. Terms were abolished in 1873 and sittings substituted. The sittings are Hilary (January 11 to the Wednesday before Easter), Easter (the Tuesday after Easter week to the Friday before Whit Sunday), Trinity (the Tuesday after Whitsun week to July 31) and Michaelmas (October 1 to December 21). Sittings may be held at any place in England or Wales.

Sittings of each House of Parliament are meetings for public business. The House of Lords sits regularly for public business on Tuesdays, Wednesdays, and Thursdays and may sit on Mondays and Fridays. The House of Commons normally sits each weekday other than Saturday, occasionally on Saturdays and only in emergency on Sundays. Each occasion of meeting is a sitting and it continues till adjournment.

Sittings in banc. Sittings of the judges of the pre-1875 common law courts at Westminster to decide matters of law and other judicial business as distinct from sittings at *nisi prius* (q.v.), at which issues of fact were tried.

Six Acts. The collective term for six Acts passed in 1819 designed to check revolutionary propaganda by limiting freedom of speech and freedom of the press. They prohibited the training of persons in the use of arms, authorized magistrates to search for and seize arms, deprived defendants in cases of misdemeanour of the right of imparling, prohibited meetings of over 50 persons for the discussion of public grievances, permitted the seizure of copies of seditious libels and extended newspaper stamp duty to political pamphlets and periodicals.

Six articles, Statute of the (1539). When Henry VIII had established himself as supreme head of the Church of England, he determined to vindicate its doctrinal orthodoxy by imposing compulsory belief in the main doctrines of the Roman Church. The Statute of Six Articles declared that, in the sacrament of the Altar there is really present the body of Christ under the forms of bread and wine, that communion in both kinds is not necessary to salvation, that priests may not by the law of God marry, that vows of chastity should be observed, and that private masses and auricular confession should be retained. Stern penalties were provided for contravention. The Act was repealed in 1547 when more strongly reformist views prevailed.

Six clerks. Officials of the Old Court of Chancery, possibly descended from six *clerici praenotarii* of the Chancery. In the later sixteenth century, they were, in theory, the only attorneys of the court but, in fact, they were intermediaries between solicitors and the court. Their duties latterly corresponded to those later performed by the Clerks of Records and Writs to whom their remaining functions were transferred when their offices were abolished in 1842.

Sixpenny writ office. An office of the old Court of Chancery, the head of which had to attend to take an account of the writs sealed to receive a duty on them, to receive from the Purse Bearer the duty on all writs received at private seals, to pay various officials and pay the surplus to the Clerk of the Hanaper. It seems to have been abolished since 1852.

Sixty clerks. Assistants of the Six Clerks (q.v.) in the old Court of Chancery, remunerated by a percentage of the fees payable to the Six Clerks. Their offices were abolished in 1842.

Skene, Sir John, Lord Curriehill (?1543–1617). Studied abroad before being called to the Scottish Bar in 1575. Shortly thereafter, he was selected, with Sir James Balfour, to prepare a digest of Scots law. He served on embassies abroad and became ambassador to the States-General in 1591. He was joint Lord Advocate with David MacGill in 1589–94, and was zealous in the prosecution of witches. In 1592 he was again made one of a commission to examine the laws and acts of Parliament, to consider which should be printed, and was finally entrusted with the execution of the work. It was published in 1597 as *The Lawes and Actes of Parliament maid be King James the First and His successors, Kings of Scotland.* Also, in 1597, he published *De Verborum Significatione*, a dictionary of ancient Scottish legal terms. In 1594, he became Clerk Register and a judge of the Court of Session, as Lord Curriehill. He later served on various commissions, including one considering union with England in 1604. In 1609 he published, in English and in Latin, *Regiam Majestatem, the Ancient Laws and Constitutions of Scotland*, a book containing other ancient laws of Scotland, some of questionable authenticity. In 1611 he resigned the office of clerk-register, and retired from the Privy Council in 1616. He was the first great Scottish legal antiquary and defective though his work is, it rescued much that might otherwise have been lost. His son, James, became a judge in 1612 and Lord President in 1626–33.

Skinner* v. *East India Company (1666–69). The case which decided that the House of Lords had no original civil jurisdiction. Skinner presented a petition against the Company and the Lords, on a reference by the Crown, held the Company liable to pay Skinner £5,000. The Company presented a complaint to the Commons, which developed into a quarrel between the two Houses. Ultimately, the King recommended erasure from the journals of all that had passed, and thereafter the Lords abandoned all claim to original jurisdiction in civil cases. This did not, and does not, affect their appellate jurisdiction.

Skynner, Sir John (?1724–1805). Called to the Bar in 1748, became an M.P. in 1771, and was Chief Baron of Exchequer, 1777–87. He appears to have had considerable capacity.

Slains, letters of. In old Scots law, letters subscribed by the relations of a person who had been slain, declaring that they had received an assythment (q.v.) and concurring in an application to the Crown for a pardon for the offender.

Slander. A false statement of defamatory (q.v.) character concerning a person made by another person orally or in other non-permanent form. It is actionable *per se* if it imputes any unchastity or adultery to any woman or girl, or imputes to the plaintiff the commission of a crime punishable with imprisonment for a first offence, or the having of some contagious disorder which may exclude him from society, or is calculated to disparage him in his office, profession, calling, trade, or business. It is also actionable if it makes any other defamatory imputation and has thereby caused the plaintiff special, i.e. actual, damage. See also DEFAMATION; LIBEL.

Slander of goods. An action lies for slander of goods where a defendant has, orally or in writing, disparaged a person's goods, maliciously and in a manner calculated to, and actually producing, actual damage. The action is not one for slander, nor for libel, and the words need not be defamatory.

Slander of property. An action lies for slander of property where a defendant has, orally or in writing, disparaged a person's property, real or personal, maliciously and in a manner calculated to, and actually producing actual damage.

Slander of title. An action for slander of title lies where a defendant has orally, or in writing, disparaged the title of an owner of property, real or personal, and thereby caused him actual damage. The plaintiff must prove that the aspersion was false, publication thereof to the disparagement of the plaintiff's title, actual malice, and, save in certain cases, special damage to the plaintiff.

Slaughter-House cases (1873), 16 Wallace 36 (U.S. Supreme Ct.). A Louisiana statute created a monopoly of the business of slaughtering livestock in New Orleans, thereby driving the butchers out of business. Some of the latter challenged the statute on the ground that it deprived them of the privileges and immunities of citizens of the United States, contrary to the Fourteenth Amendment. The Supreme Court rejected this plea, distinguishing between the privileges inherent in State citizenship and those inherent in national citizenship, and holding that the Fourteenth Amendment protected the latter only. A minority emphasized the need for the protection of rights, and their view was the basis for later expansion of the meaning of the due process and equal protection clauses of the Fourteenth Amendment.

Slavery. A relationship in which one person has power enabling him to compel and direct the involuntary labour of another and to restrict the latter's freedom of action and movement. Slavery is distinguishable from serfdom, corvée or statute labour, contract labour, and contractual employment (qq.v.). A slave is owned and deprived of most rights and freedoms. He is not necessarily of low social status nor rightless, but does not have full rights of citizenship. In Rome, slaves might have responsibility, education, earnings, and be given, or buy, their freedom. Slavery is known from the pastoral stage of development of societies and is widespread in the agricultural stage, certainly when it goes beyond subsistence farming.

In most countries which have had slavery, there have been various sources of slaves, by capture in war, by kidnapping to meet a need for a supply of labour, by way of punishment for crime, by birth to slave parents, or by sale of a person under authority, e.g. a child, into slavery. The slave is, in general, the property of his master and can be bought and sold like an animal.

In the ancient Middle East, slavery was common, though there were also various semi-free classes and slaves might attain their freedom. In later Babylonian times, they had power to carry on businesses and to work as craftsmen and tradesmen. Babylonian and Jewish law contain many rules dealing with slavery. Slavery was an accepted institution in the Greek and Roman world. Slavery for debt existed at Athens till the sixth century and at Rome till the late fourth century B.C. but most slaves were acquired by war, piracy, kidnapping, import, and breeding.

In Greece, slaves had no legal personality and could be disposed of as property; they might belong to the State, to temples, or to private persons. Some had a measure of economic independence, some legal powers, and if managing estates or administering their master's businesses, could make legally binding declarations, could be parties to trials, and were partially free. Others had no independence and could not transact legally. There were also variations between different states, but generally they were protected against cruelty, could take refuge in an asylum and demand to be sold to another master.

Release from slavery could be effected by will, by notification before witnesses, in public places or before altars, by consecration or sale to a divinity. The slave then acquired freedom but not citizenship and remained dependent on his master.

At Rome, slaves originally could not have any legal rights but were objects only, but in time there was some recognition of their legal personality and they could acquire property for their master. Later by praetorian developments, they could, if engaged in managing a commercial undertaking, bind their masters also. There were also protective measures against bad treatment of slaves by their masters. Slaves might acquire money (*peculium*) for their personal use, and this might be deemed their own property and be employed in trade.

Slaves became free by manumission, which might be effected by will, or *vindicta*, originally a fictitious claim of liberty and later a solemn procedure before a magistrate, or by enrolment of the slave with his master's consent in the list of citizens; later informal declarations sufficed and the Christian state allowed manumission *in ecclesia*. The institution reached its highest level in Rome in the second and first centuries B.C. Slaves were extensively employed in cultivating large estates and in the mines, as domestics, in industry, and in brothels. Though excluded from political life, slaves were extensively employed in secretarial and managerial posts and in the Roman principate the Emperor's principal secretaries and officials were slaves or freedmen but politically influential.

Only later were these offices treated as public and opened to the equestrian order. Treatment of slaves varied greatly; some were, in practice, as well off and as free as freemen, others were subject to arbitrary cruelty.

From Aristotle, philosophers and jurists admitted that slavery was contrary to natural law but none argued for its abolition and a large body of law existed as to slavery. It was an accepted social institution and much of the time little difficulty arose between masters and slaves. There were large slave revolts in Sicily in 135–132 B.C. and again in 104–100 B.C. and the Sparticist rising in Italy in 73–71 B.C. and these were difficult to suppress. There were also great differences in the treatment of different classes of slaves.

In the Americas, slavery was not general until the arrival of the Europeans, though it existed in some areas. The Spanish conquerors initially used native peoples as serfs, but treated them as slaves in the mines. The harsh treatment was, to some extent, mitigated by Catholic missionaries but, from 1517, Africans were imported as slaves into the Caribbean islands, the adjacent mainland and Brazil. The first slaves arrived in Virginia in 1619 and slavery increased after the cultivation of tobacco developed. It accelerated after the introduction of cotton, and in all these areas the development of slavery was connected with the growth of the plantation economy in sugar, tobacco, and cotton.

Policies as to treatment of slaves varied, and it was generally better in Catholic than in Protestant countries; some of the Latin countries promulgated detailed laws as to the treatment of slaves, notably Louis XIV's *Code Noir* of 1685 and the Spanish Slave Code of 1789. The Barbados Code, copied in Virginia and elsewhere, was very harsh. Catholic clergy generally showed greater solicitude for slaves than did Protestant.

Slavery began in the U.S. when negroes were shipped to Virginia in 1619 and sold as indentured servants. It developed under the plantation system and spread from Virginia in the seventeenth century. Slaves became commercially important with the development of the tobacco industry and the extension of cotton. They were legally chattels and sold privately or by auction, but might be freed. The conditions in which they lived and worked probably varied greatly, but the majority were poorly or harshly treated.

As early as 1688, Pennsylvania Quakers opposed slavery, and an antislavery society was founded in Philadelphia in 1775 and, thereafter, the abolition movement made progress, except in the South. Jefferson condemned the slave trade in the first draft of the Declaration of Independence, though this did not pass. Rhode Island abolished slavery in 1774, Vermont in its constitution of 1777, and Pennsylvania in 1780. Other northern states did the same and an Ordinance of 1787 excluded slavery from the Northern Territory. All states had ended the legal importation of foreign slaves before 1803 though South Carolina removed restrictions in 1804. In 1807, Congress exercised its power under the Constitution to prohibit the slave trade from 31 December 1807, but illicit importation continued thereafter, and at about that time there were probably a million slaves in the U.S., amounting to about 40 per cent of the population in the southern states. By 1860 there were 4 million slaves in a southern population of 12 million.

The slave system caused little protest till the eighteenth century but, by the late eighteenth century, moral disapproval of the system was widespread. In 1807 the slave trade to British colonies was abolished, and prohibited by the U.S. Most of the Spanish territories in America abolished slavery when they secured their independence. In 1833 Britain abolished slavery in all British colonies and this encouraged the anti-slavery movement in the U.S. But under the Constitution, the question of slavery was a matter for the states and there were many constitutional difficulties and compromises before the southern states seceded from the Union in 1861. In 1863 Lincoln proclaimed emancipation of the slaves in areas in rebellion and, in 1865, the Thirteenth Amendment emancipated all slaves.

In modern times, the U.N. Universal Declaration of Human Rights and the European Convention on Human Rights have declared against the existence of slavery. But in some states labour camps, indistinguishable from slavery, exist.

I. Mendelsohn, *Legal Aspects of Slavery in Babylonia, Assyria and Palestine*; M. Finley, *Slavery in Classical Antiquity*; W. W. Brickland, *Roman Law of Slavery*; R. H. Barrow, *Slavery in The Roman Empire*; P. Vinogradoff, *Villeinage in England*; K. M. Stampp, *The Peculiar Institution: Slavery in the Ante-Bellum South.*

Slip. In negotiations for a contract of marine insurance, a note containing the terms of the proposed insurance, initialled by the underwriters who have agreed to underwrite the chance of loss to the extent stated thereon. It is a contract for the issue of a policy but not itself enforceable.

Slot-machine theory. A sarcastic term for the theory that judicial decisions are direct applications of existing, settled rules of law to new sets of facts or applications of rules strictly logically deduced from prescribed or previously settled rules to new sets of facts, with the consequence that decisions in undecided cases are, or should be, as predictable when the facts are put before the court as the reaction of the slot-machine when the appropriate coin is inserted. The theory has some appearance of truth in straightforward cases, such as of debt or undefended actions of divorce, but is seen to be oversimplified and to overrate certainty and logic in many cases, even where the relevant facts are admitted or clear. Judges may differ in opinion on whether certain facts amounted to reasonable care or fall short of it, on how discretion should be exercised, on penalties, and will sometimes decline to draw the strictly logical inference from previously established rules where they think that such a conclusion would be unreasonable.

Small claims courts. Informal tribunals in which claims of small amounts can be heard and decided quickly and cheaply. In England, a small claims procedure was introduced within the existing county court (q.v.) structure in 1971 to assist unrepresented claimants to make claims. In 1973 provision was made for either party to apply to a county court registrar for a matter to be referred to arbitration if the sum involved is less than £100; if more the parties must consent or the county court judge order it. A London Small Claims Court began operating in 1973; it is essentially a variant of arbitration, and deals with cases of contract, tort or landlord and tenant. An administrator prepares cases for hearing by part-time adjudicators. Acceptance of the jurisdiction is voluntary. Legal representation is not permitted and procedure is inquisitorial rather than adversary.

Small Debt Court. In Scotland, courts for the summary and cheap decision of claims for small amounts of money and certain other kinds of claims. There were two kinds. The Justice of the Peace Small Debt Court was established in 1799, having jurisdiction in debts not exceeding £5. This court was abolished in 1975. The Sheriff's Small Debt Court was established in 1826 and had jurisdiction in debts limited originally to £8. 6s. 8d. but extended by various statutes to £50. It was abolished in 1976 its cases being dealt with as summary cases in the Sheriff Court.

Smith, Adam (1723–90). British philosopher and economist, Professor of Logic (and later of Moral Philosophy) at Glasgow (1751–63) and subsequently tutor to the Duke of Buccleuch and a member of the board of customs in Scotland. His publications include *The Theory of Moral Sentiments* (1759), *An Inquiry into the Nature and Causes of the Wealth of Nations* (1776), a work of enormous importance in the history of economic thought and the basis of his fame, *Essays on Philosophical Subjects* (1795), and lecture notes published as *Lectures on Justice, Police, Revenue and Arms* (1896) and as *Lectures on Jurisprudence* (1977). Under moral philosophy he dealt with natural theology, ethics, jurisprudence, and political economy. His standpoint was that of natural law.

J. Rae, *Life of Adam Smith.*

Smith, Sir Archibald Levin (1836–1901). Called to the Bar in 1860, he quickly acquired a good practice and, in 1883, became a judge of the Queen's Bench Division. In 1888 he sat on the special commission inquiring into the allegations against Parnell. In 1892 he became a Lord Justice of Appeal and Master of the Rolls in 1900. He resigned in the following year and died shortly afterwards. He was an intelligent and industrious lawyer but not a great or learned one.

Smith, Edmund Munroe (1854–1926). U.S. jurist and political scientist, who studied in Europe and became Professor of Roman Law and Comparative Jurisprudence at Columbia in 1891 and Bryce Professor of European Legal History in 1922. He wrote little, the best being lectures published as *A General View of European Legal History* (1927) and *The Development of European Law* (1928), but was a profound scholar of wide interests who played a leading part in the development of Columbia.

Smith, Frederick Edwin, 1st Earl of Birkenhead (1872–1930). Won the Vinerian Scholarship at Oxford and was called to the Bar in 1899. At first he practised in Liverpool but moved to London in 1906, in which year he also entered Parliament and made a reputation with his maiden speech. He was engaged in the heaviest cases in the courts and in the political controversies of 1909–10. In World War I, he served in France but, in 1915, was appointed Attorney-General and as such prosecuted Sir Roger Casement for high treason. In 1918 he became Lord Chancellor, an unpopular appointment with Bench and Bar, and was made Lord Birkenhead. He became a viscount in 1921 and an earl in 1922.

From the first, he had a reputation for brilliant speaking, wit, irony, and crushing cross-examination but, as a chancellor and judge, he must be accorded a high place. Many judgments show learning and research, and are embodied in lively phraseology. He attempted to reform the circuit system and did much of the preliminary work leading up to the property legislation of 1925. In the Cabinet, his outstanding ability was of great value, and many of his speeches in the House of Lords were masterly, but he came to be disliked and distrusted in his own party for his support of the treaty establishing the Irish Free State. In 1924 he was made Secretary of State for India but resigned in 1928 and sought directorships in the City. In his latter years he was frequently involved in controversy, and wrote a number of unimportant books, including one on international law.

Birkenhead, *F.E.* (1959).

Smith, John (1657–1726). Was called to the English Bar in 1684, became a judge of the Irish Common Pleas in 1700, a baron of the English Exchequer in 1702, and, in 1708, was sent to Scotland as Lord Chief Baron of the Exchequer, though still allowed to retain his place in the English court, holding both offices till his death.

His voluminous private and diplomatic correspondence is an important historical source.

Strype, *Life of Sir Thomas Smith.*

Smith, Madeleine, trial of. In 1855, Madeleine Smith, the 21-year-old daughter of a Glasgow architect, became friendly with Pierre L'Angelier, a clerk, and they exchanged love letters and probably exchanged promises of marriage and had intercourse in reliance thereon—conduct which together, as Scots law then stood, constituted legal marriage. In 1857 a more eligible suitor for Madeleine appeared and she tried to break off her connection with L'Angelier and to obtain the return of her passionate love-letters. Shortly thereafter, L'Angelier had several attacks of severe intestinal pains and died. Madeleine was arrested and tried for murder, there being considerable evidence of her having purchased arsenic, of having given L'Angelier cups of cocoa, and of his having died of arsenic poisoning. Her trial before the High Court of Justiciary in Edinburgh caused intense public interest; the verdict was, by a majority, Not Proven (q.v.) and Madeleine went to America and lived till 1928. The case has given rise to controversy ever since.

Smith (ed.): *Trial of Madeleine Smith.*

Smith, Sir Michael (1740–1809). Was called to the Irish Bar in 1769, became a baron of Exchequer in Ireland, 1793, and Master of the Rolls in Ireland, 1801–06.

Smith, Sir Montagu Edward (1809–91). Was called to the Bar in 1835, took silk in 1853, and became an M.P. in 1859. He became a judge of the Common Pleas in 1865 and a paid member of the Judicial Committee of the Privy Council from 1871 to 1881, where he showed himself a sound lawyer and an accurate and painstaking judge.

Smith, Sir Thomas (1513–77). Became Regius Professor of Civil Law at Cambridge in 1544 and later served as a principal Secretary of State and on missions abroad. When the Protector Somerset fell in 1549, Smith was deprived of his offices and imprisoned in the Tower for a time. He re-entered public life in Elizabeth's reign, being a member of Parliament and later ambassador to France on several occasions. In 1572 he again became Secretary of State, and has generally been considered one of the most upright statesmen of his time. He was a noted classical scholar and published a great deal, but his principal work was his *De Republica Anglorum; the Manner of Government or Policie of the Realm of England* (1583), the most important description of the constitution of England in the Tudor period, which went through many editions and was translated into Latin, Dutch, and German.

Smith, Thomas Berry Cusack (1795–1866). Was called to the Irish Bar in 1819, became Solicitor-General and Attorney-General in 1842 and was Master of the Rolls in Ireland from 1846 till his death.

Smuggling. The clandestine movement of goods in a way intended to evade customs duties, import or export restrictions, prohibitions on drug-trafficking, and similar controls. Smuggling arises wherever prohibitions or substantial duties are imposed and profits can be made by evading them. It was at its height in the eighteenth century, when tea, spirits, tobacco, spices, and silk were smuggled into England in large quantities. Tobacco and salt were smuggled into France. In the East, opium was smuggled into China and many kinds of dutiable goods smuggled between one state and another in Africa. During the prohibition period in the U.S. (1920–33), liquor was extensively smuggled into the U.S. from Europe, the West Indies, and Canada. Today, the smuggling of drugs is an extensive industry. The modes of checking smuggling include customs posts at frontiers, with powers to search persons, clothes, baggage, vehicles, and the interception of boats by coastguards.

Smyth, Sir Edward (?1610–70). Called to the English Bar in 1635, he went to Ireland as a Commissioner of the Court of Claims in 1662 and was Chief Justice of the Common Pleas in Ireland, 1665–69.

Smyth* v. *Ames (1898), 169 U.S. 466. It had become the rule that the courts were the arbiters of the reasonableness of rates fixed by businesses affected by a public interest, and that to be constitutional the rate level must permit the earning of a fair return on the fair value of the property involved. *Smyth* v. *Ames* showed the indefiniteness of the criteria for measuring the value of the property, and led to much litigation with public utilities over changing rate structures.

Smythe, Sir Sidney Stafford (1705–78). Called to the Bar in 1728, became a Baron of Exchequer in 1750, Chief Baron, 1772–77, and was twice a Commissioner for the Great Seal. He was a sound lawyer and competent judge.

Soames, (Arthur) Christopher, Lord (1920–). Was an M.P., 1950–68, Secretary for War, 1958–60, Minister of Agriculture, 1960–64, Ambassador to France, 1968–72. and a Member and Vice-President of the Commission of the

European Communities, 1973–77. He became a life peer in 1978.

Socage. The mode of tenure of land by any certain and fixed but non-military service, such as to pay a rent in money, as opposed to tenure by knight-service (q.v.), which was deemed more honourable, but was more onerous. Latterly, all services were commuted into money. Socage was free from the incidents of scutage, wardship, and marriage which attached to knight-service. In the course of time, the money payments became not worth collecting. Socage was, accordingly the residuary tenure comprising all lands not held by knight-service, serjeanty, frankalmoin, or villeinage. Burgage and gavelkind were special varieties of socage. In 1925 socage became the only remaining form of tenure and all the incidents of tenure disappeared.

Social controls. The means and forces which maintain, regulate and limit the behaviour of individuals in a society or subgroup of society. The category accordingly includes law, morals, religion, custom, habit, etiquette, education, fashion, and other similar forces. These differ in their power in different societies and at different times, and in their sanctions. Accordingly, in any society or sub-group the rules of law are only one, and not necessarily the most important or powerful, of the forces or pressures influencing and controlling the conduct of individuals. This view was adumbrated by Montesquieu and developed by Ihering and more fully by modern sociological jurists. Legal and non-legal social controls are inter-dependent and inter-acting, and sometimes conflicting. The morals, customs, and habits of criminal sub-groups conflict with legal pressures applicable to them. Moreover, a legal proposition may be a rule significant for lawyers if and when it is applied or enforced; it is not law in a social sense until and unless social relations are actually being ordered thereby. The differentiation of law from other forces of social control is a feature of developed communities, and much less clear in undeveloped societies. In particular, differentiation accompanies the rise of distinct political authorities who take law as the instrument of their purposes, and as societies become more highly developed more and more matters shift or are shifted into the area of legal control. Thus, sex-equality and race relations, long left to common-sense, have been made fields of legal regulation.

Social defence. The principle that society is justified not only in punishing persons who have offended but in interfering to prevent harm where the harm is serious and the probability of its being done by an identifiable individual is considerable. To some extent this is done, as where mentally abnormal persons are compulsorily put into hospital for the protection of themselves, and of others, and where children are taken away from their parents and put into care when they are in danger of being subjected to bad influences, in the rules allowing arrest on suspicion and in the institution of preventive detention. But some urge that the principle should be applied more widely. Such an extension would have the difficulties of subjecting more persons to supervision or detention, would necessarily miss some cases and would be treated as an unwarranted interference with liberty.

Social engineering. A metaphor employed by some sociological jurists, notably Roscoe Pound, to illustrate their contention that the function of law in society is to achieve a proper balance between freedom and control, encouragement and prohibition, and to enable persons to interact with one another in society with the minimum of heat and friction, waste of energy and to use their energies to the maximum possible effect.

Social idealism. A term applied particularly to the legal thought of Stammler (q.v.), who founded on the Kantian idealist proposition that nothing should be deemed right which, if universalized, would destroy the fundamental idea of a legally organized society, but modified it by making it not the harmonizing of the conduct of each free-willing individual with all others but the conduct of each individual with the idea of society.

Social legislation. A general term for legislation with a predominantly social purpose, such as for education, provision of housing, control of rents, provision of health and welfare services, pensions, and other social security provisions. The earliest social legislation is probably the Old Poor Law but it began to be important and voluminous in the nineteenth century with Factories Acts, Health and Morals of Apprentices Acts, and Workmen's Compensation Legislation. See also SOCIAL SECURITY.

Social security. The compendious term for a set of interlocking bodies of legislation designed to protect individuals from the hardships which would otherwise result from old age, sickness, death, unemployment. The earliest legal, as distinct from voluntary and charitable, provision for these needs was the Old Poor Law, while in the nineteenth century, many forms of insurance and voluntary provision, such as friendly societies, developed. Between 1890 and 1950, most Western European countries adopted social security measures in different ways. In the U.K., landmarks have been the first Workmen's Compensation Act in 1897, the first Old Age Pensions Act in 1911 and the

Beveridge Report on Social Insurance in 1944, followed by the introduction of the National Health Service, National Insurance, National Insurance (Industrial Injuries), and National Assistance (now Supplementary Benefits) after 1945.

Social services. The sacred cows of modern Britain which constantly grow fatter and more expensive to keep, which it is deemed sacrilege even to touch, and which can never be killed off. Social services comprise public provision of education, housing, recreational and leisure facilities, public health and sanitation, health service and hospitals, roads and bridges, police, the maintenance of public order, the probation service, prisons and borstals, after-care, and other services provided by central, local, or statutory authorities at public expense. The term is sometimes used to include social security provisions (q.v.). There has been an enormous expansion of the provision of social services since the mid-nineteenth century with a concomitant enormous bureaucracy, a great volume of very complicated law and constantly growing expense.

Social solidarity. The major principle of the legal thought of Leon Duguit (q.v.). The basic fact of social life was social solidarity arising from similarity of needs and diversity of functions; the role of law is part of this observed fact of social solidarity. Law is valid not by virtue of any abstract authority of justice or natural law, nor by the command of any sovereign, but because acts done in violation of it cause social disorder and a reaction towards readjustment. The individual's attributes are not in terms of rights but of functions; the law would protect an individual so long as he performed a function which promoted social solidarity. Law exists to encourage conduct which furthers social solidarity, and the state is merely machinery for that end. The principle of solidarity was the content of the *règle de droit* or objective law which bound all members of society.

Socialist law. A general term for the law of Soviet Russia since 1917 and of the people's republics of Eastern Europe and of Asia with similar legal systems. The Soviet Russian system is based on a Romano-Germanic basis but has developed separately since 1917 because of the prevailing Marxist–Leninist social, political, and economic philosophy and the dictatorship of the Communist party. To a considerable extent, it has preserved the terminology and, to outward appearance, the structure of Romano-Germanic laws, but Soviet jurists assert the independence of their, and other related, legal systems from all Romano–Germanic ones because of the fundamental differences in underlying social, political, and economic structure. Moreover, terms and concepts such as trade union, contract, property, and liberty have quite different meanings in the Soviet and other socialist and Romano-Germanic or common law systems of law.

Société anonyme. In French law, an association, the name of which is not a personal name but indicates the association's object, in which the liability of all partners is limited. It requires a series of formalities for its formation, including registration of copies of the *statuts* at the registry of the Commercial Court.

Société en commandite. In French law, a partnership in which some of the partners are only lenders of money to the firm, and not liable for losses beyond the amount which they have contributed and not entitled to participate in the management of the firm, comparable to the British limited partnership. There are two forms, the *société en commandite* simple in which *commanditoires* are akin to sleeping partners, and the *société en commandite par actions,* in which the *commanditaires* are akin to shareholders rather than partners.

Society. (1) A term for the aggregate of persons living together in a more or less ordered community and for the whole complex of relations of man with his fellow-men.

In a sense, all human beings in the world form one society but normally societies are considered as comprising only those united by living within a determinate area, speaking the same language, and linked as members of a particular state, so that persons in the United Kingdom and the United States are deemed distinct societies. This sense of society is relevant for jurists concerned with law in society, or law as a social control or social force, with the social operation and effect of legal institutions, legal doctrines, and principles. Such jurists are more interested in the way law works in a society than in law as a logical body of principles.

(2) In a narrower sense, a society is a general term for a group of individuals associated for the furtherance of a particular end. Societies may be organized legally as unincorporated associations, as bodies incorporated under statute, or as chartered bodies. They are of many kinds. Among unincorporated bodies are friendly societies, loan societies, and many kinds of private clubs and societies; among incorporated bodies are building societies, industrial and provident societies (including co-operative societies) and some companies, and among chartered bodies are such major institutions as The Royal Society and The British Academy.

Literary and scientific institutions, whether incorporated or not, are regulated generally by the Literary and Scientific Institutions Acts, 1846 and 1854.

A society is unlawful if its objects are criminal,

seditious, or otherwise illegal, or, under statutes, if its members are obliged to take any oath or enter into any engagement prohibited by statute.

Socinus, Bartholomaeus (1436–1507). An Italian jurist of the commentator school, and author of *Consiliorum sive Responsorum Volumina* (1571) and commentaries on the Digest (1524).

Sociological jurisprudence. The general name for those approaches to the study of law, in general, which have more regard to the working of law in society than to its form or content. To a considerable extent, these draw on work on the sociology of law, criminology, and reports on the working of legal rules in society (e.g. in divorce and hire-purchase) and the need for alteration of them. Sociological jurists look on legal institutions, doctrines, and precepts functionally; the form of legal precepts is the means only. They have very divergent philosophical views.

Its initial stage was physical and mechanical, but this was succeeded by a biological stage influenced by Darwinian evolution, a psychological stage and finally an integrated or unified stage.

Montesquieu may be looked on as the forerunner of the school; he tried to trace the effect of social environment on law. The initial steps were taken by Jhering whose *Der Zweck im Recht* (*Purpose in Law*, translated as *Law as a Means to an End*) develops aspects of analytical positivism and combines them with some utilitarian ideas; the purpose was the protection of interests. In Germany the move for a sociological jurisprudence was led by Ehrlich and Kantorowicz and this led in Germany to the 'Jurisprudence of Interests' school of which the leader was Heck.

In the U.S., the forerunner was Mr. Justice Holmes, succeeded by Mr. Justice Cardozo and, among academics, Roscoe Pound, followed in Australia by Stone and Paton. Pound stated the programme of the sociological school as comprising: study of the actual social effects of legal institutions, precepts and doctrines; study in preparation for lawmaking; study of the means of making legal precepts effective in action; study of judicial method; sociological legal history; recognition of the importance of individualized application of legal precepts; a ministry of justice; and to make effort more effective in achieving the purpose of the legal order. Much attention is devoted to the interests which law seeks to protect, individual interests, social interests and public interests. This study does not, however, help to decide the crucial issues, how particular interests are to be weighed and compared one with another.

A divergent branch of sociological jurisprudence is realism, the approach of jurists principally in the U.S. who regard what the courts will, in fact, do in particular cases as being the law. Notable figures adopting this standpoint have been Gray, Mr. Justice Holmes, Frank, and Llewellyn.

Sociological jurisprudence has embraced so many groups of thinkers that it cannot be regarded as a single approach. The common factors are the broad concept of law, as including the legal system and the administration of law, and the insistence on observing the operations of law.

Sociological jurisprudence is not identical with sociology of law (q.v.), which is applied science, the study of the operation of law in particular contexts.

R. v. Jhering, *Law as a Means to an End*; E. Ehrlich, *Fundamental Principles of the Sociology of Law*; N. Timasheff, *Introduction to the Sociology of Law*; R. Pound, *Jurisprudence*; J. Stone, *Social Dimensions of Law and Justice*.

Sociology of law. A branch of sociology, which is concerned with legal institutions, rules, practices, procedures, and persons as elements in the totality which constitutes society, and which examines their functions, influences, and effects in particular societies. The emphasis of the study is on society and law is one of the phenomena in society. For the purpose of this study, law must be understood widely and not only in the narrow sense of rules laid down and enforced by a state authority. 'Sociology of law is that part of the sociology of the human spirit which studies the full social reality of law, beginning with its tangible and externally observable expressions, in effective collective behaviours and in the material basis. Sociology of law interprets these behaviours and material manifestations of law according to the internal meanings which, while inspiring and penetrating them, are at the same time in part transformed by them. It proceeds specially from jural symbolic patterns fixed in advance, such as organized law, procedures and sanctions, to jural symbols proper, such as flexible rules and spontaneous law. From the latter it proceeds to jural values and ideas which they express, and finally to the collective beliefs and intuitions which aspire to these values and grasp these ideas, and which manifest themselves in spontaneous 'normative facts', sources of the validity, that is to say, of the positivity of all law' (Gurvitch). It has been distinguished into systematic sociology of law, the study of the manifestation of law as a function of the forms of sociality and of the level of social reality; differential sociology of law, studying the manifestations of law as a function of real collective units whose solution is found in the jural typology of particular groups and inclusive societies; and genetic sociology of law, studying the regularities as tendencies and factors of the change, development and decay of the law within a particular type of society.

It is distinct from sociological jurisprudence (on

which see Jurisprudence) but casts light on jurisprudence. Sociological jurisprudence examines legal institutions, conceptions, rules and procedures from the standpoint of sociology, viewing law generally as a means of social control in society and in its social and functional setting.

G. Gurvitch, *Sociology of Law*; E. Ehrlich, *Fundamental Principles of the Sociology of Law* (1936); N. Timasheff, *Introduction to the Sociology of Law*.

Sodomy. Unnatural sexual relations by a male with another person, *per anum*, either between male homosexuals or between a man and a woman. Long a crime—'the abominable crime of buggery'—it is, in England since 1967, no longer a crime between males of full age (i.e. 21) in private if there is consent. See also BESTIALITY.

Sohm, Rudolf (1841–1917). German jurist, one of the most brilliant of his time, who worked on the commission which drafted the *Burgerliches Gesetzbuch*, and was author of *Institutionen: Geschichte und System des römischen privatrechts* (1883 and many later editions), *Die altdeutsche Reichs und Gerichtsverfassung* (1871); *Kirchenrecht* (1892), and many other writings.

Solatium. In Scots law, an award of money to an injured person as solace for hurt feelings. It arises chiefly in personal injury cases, as compensation for pain and suffering, loss of limbs, amenities of life and similar losses, in defamation and similar cases as solace for hurt feelings and damaged reputation, and in wrongful death cases (in this context now called 'loss of society award'), as compensation to surviving relatives for the non-financial loss caused by the deceased's premature death.

Solemn League and Covenant. An agreement made in 1643 between representatives of the nobles and commons of England, Scotland, and Ireland, to gain the assistance of the Scots for the English Parliamentarians in their struggle with Charles I. The parties pledged themselves to preserve the Church of Scotland, to reform religion in England and Ireland in doctrine, worship, discipline, and government according to the Word of God and the example of the best reformed churches, to bring the churches of the three kingdoms to the nearest possible uniformity in faith and government, and to extirpate popery and prelacy. Pursuant to this agreement, a Scottish Army entered England and co-operated with the Parliamentary army.

Solicitor (England). In the English courts of equity, the permanent clerical staff undertook the framing of pleadings and deeds, but the need for legal agents in other matters produced solicitors in the fifteenth century. Initially, they were general business agents rather than lawyers and inferior to attorneys but, in time, won recognition alongside the attorneys, particularly as practising in the courts of equity. Throughout the eighteenth century, attorneys and solicitors existed side by side, an Act of 1728 regulating admission to both professions, but by the next century the latter group had attained the ascendancy. Till the end of the eighteenth century, attorneys and solicitors were members of the Inns of Court but, in 1793, they were finally excluded therefrom. They had, however, monopolized the Inns of Chancery. Requirements for admission and for examination were imposed from 1729 but were, for long, a mere formality. In 1831 the Law Society was chartered and, by the mid-nineteenth century, regulations under statute required the passing of examinations for admission. It now has complete control over the qualifications for admission

By the Judicature Act all solicitors, attorneys, and proctors became solicitors of the Supreme Court and consequently officers of the court.

To practise as a solicitor today, a person must have qualified by passing the Law Society's examinations and serving a period under articles to a solicitor in practice, have his name on the roll of solicitors, and have in force an annual practising certificate. Solicitors must comply with stringent regulations as to accounts and clients' money.

The functions of the solicitor include general legal and business advice, negotiations for the sale and conveyance of real property, the drafting of wills, trusts and settlements, the administration of trusts, probate business, company and commercial work, and the instruction of counsel. Solicitors have right of audience in inferior courts and tribunals generally but not, in general, in the superior courts though the Lord Chancellor may direct that solicitors may appear in specified cases in the Crown Court. Their fees are regulated by scales prescribed under statutory authority.

The professional discipline of solicitors is entrusted to a disciplinary tribunal appointed by the Master of the Rolls, subject to an appeal to the High Court.

Solicitor (Scotland). Since 1933, the correct generic designation of legal practitioners formerly called law agents, writers, or procurators. It covers those who are members of the societies of Writers to the Signet, Solicitors in the Supreme Courts of Scotland, Royal Faculty of Procurators in Glasgow, Society of Advocates in Aberdeen and lesser local societies, as well as those who are enrolled though not members of any particular society. To practise as a solicitor a person must have undergone a period of training with a practising solicitor, passed certain examinations, been admitted, and have taken out an annual practising certificate. Solicitors are concerned with general legal and business advice, conveyancing, executry and trust business, commercial and company business, instruction of and preparation of cases for counsel, and, in some cases, representation of clients in sheriff and other inferior courts.

Solicitor-General. The second Law Officer of the Crown in England, in general the assistant to, colleague, and deputy of, and very frequently successor to, the Attorney-General. The office of King's Solicitor is known from about 1460 and appears under its modern title from 1515. It may originally have had special connection with Chancery practice. He is appointed by the Prime Minister, is a junior Minister, almost invariably an M.P., and largely concerned with legal advice to the Cabinet and Ministers and the conduct of Bills of a legal character through Parliament. He may act for the Attorney-General if the latter is ill or his office vacant. He is normally knighted, frequently made a Privy Councillor, but not necessarily at once nor automatically, but is not normally a Member of the Cabinet. Like the Attorney-General, he could till 1895, engage in private practice but since then he may not do so, and is remunerated by salary, not by fees. Frequently, though not always, a Solicitor-General succeeds to the office of Attorney-General on a vacancy, and frequently thereafter, or directly, to a judicial appointment, including in a number of cases advancement to be Lord Chancellor.

See also the list of holders of the office in the Appendix.

J. L. Edwards, *The Law Officers of the Crown.*

Solicitor-General for Scotland. A King's Solicitor is first mentioned in the seventeenth century and the office has certainly existed since the early eighteenth century. In the seventeenth and eighteenth centuries, there were frequently two Solicitors-General acting jointly but subsequently there has always been only one. He is a junior member of the Government, though he may hold the office on a non-political basis, and demits office on a change of government. In modern practice, he is the deputy and chief assistant of the Lord Advocate (q.v.) in all the latter's functions. In particular he appears for the Crown in Exchequer causes. He cannot prosecute *jure officii* but may do so if the Lord Advocate should die or be removed from office. In court he is entitled by custom to sit within the Bar on the left-hand side of the Bench. On a vacancy occurring, he is frequently promoted Lord Advocate and also frequently appointed to the Bench.

See also the list of holders of the office in the Appendix.

Solicitors-at-Law in Edinburgh, Society of. An association, formed in 1707, of lawyers practising before the Commissary Court of Edinburgh and later practising, to the exclusion of all others, in the Sheriff and Burgh courts of Edinburgh. They obtained a royal charter in 1780. The society vanished in the late nineteenth century.

Solicitors in the Supreme Courts of Scotland, Society of. A body of solicitors practising in the Court of Session in Scotland, chartered in 1797. Members are usually designated S.S.C.

Solitary confinement. A prison regime, carrying to its logical conclusion the suggestion of John Howard that physical and moral corruption of prisoners would be prevented by the provision of separate cells for sleeping, which insisted on solitary confinement of prisoners at all times 'to promote the calm contemplation which brings repentance'. This strict system was enforced in Pennsylvania from the late eighteenth century, and only after the expiry of the solitary stage were prisoners allowed to work in association, discipline being enforced by a return to solitary confinement. In the early nineteenth century, several prisons followed the Pennsylvania system of keeping the prisoners isolated in their cells during the whole prison term. Later, opinion changed to the silent system (q.v.) but allowing work in association. In modern thinking, separate confinement is reserved for untried prisoners, as a disciplinary or security measure, and for exceptional cases, but it is not regarded as a useful general regime.

Solon (*c.* 640–*c.* 560 B.C.). Athenian statesman and poet. About 594 B.C. he was given the task of terminating civil strife at Athens, which largely sprang from the monopoly of public offices, land, and wealth by the nobility, and achieved this by cancelling all existing debts and securities and forbidding any future borrowing on personal security. He thus ended serfdom at Athens. He also reformed the constitution; though he reserved the chief offices and the Areopagus (q.v.) for members of the highest of the four property-classes, this broke the monopoly of the old nobility of birth. He defined the rights of the Ecclesia and established the Boule to prepare business for the Ecclesia. He issued a new and more humane code of law, defining the powers and duties of the magistrates and replacing, except as to homicide, the harsh code of Draco (q.v.). Important changes attributed to him were allowing any citizen and not only an injured party to bring a suit, and allowing an appeal from the verdict of magistrates to the citizens at large. Despite some contention his reforms remained the basis of the Athenian state. He also encouraged industry and commerce and was a poet of some note.

C. Woodhouse, *Solon the Liberator*.

Someren, Johan van (1634–1706). President of the court of Utrecht and author of *Tractatus de jure novercarum* (1658) and *Tractatus de repraesentatione* (1676).

Somers, John, 1st Lord Somers (1651–1716). Called to the Bar in 1676, he made his name in defence of the Seven Bishops (1688), entered Parliament in 1689, and became chairman of the committee which drew up the Bills of Rights. He successively became Solicitor-General and Attorney-General (1692) under William III before becoming Lord Keeper in 1693 and Lord Chancellor in 1697. He was driven from office in 1700 and impeached, but this failed. Though out of office he promoted legal reforms, helped to pass the Regency Act of 1706, took an active part in carrying the Union of 1707 and was Lord President of the Council in 1708-10. He was highly regarded as a lawyer and his judgment in the *Bankers case* (1700), 14 S.T. 1, is a classic on the subject of the legal remedies available against the Crown, while in *Ashby* v. *White* (1700), 14 S.T. 695, he assisted the House of Lords to uphold Holt, C.J.'s dissent. He had learning, patience, industry and the qualities most needed in a Chancellor. His main contributions were as statesman and constitutional lawyer rather than as equity judge. He also wrote a number of pamphlets and his valuable library was the basis of the collection known as the Somers Tracts, subsequently published.

Somervell, Donald Bradley, Baron Somervell of Harrow (1889–1960). Entered Parliament in 1931, was Solicitor-General, 1933–36, Attorney-General, 1936–45, and Home Secretary, 1945. He became a Lord Justice of Appeal, 1946–54, and a Lord of Appeal in Ordinary, 1954–60.

Sommersett's case (1772), 20 St. Tr. 1. A writ of habeas corpus was brought by a negro slave, James Sommersett, who had been brought to England by his West Indian master, and granted—Lord Mansfield holding that a slave actually in England could not be sent back to a colony for punishment, and laying down incidentally that slavery was not recognized by the laws of England.

Somme rural. A treatise by Jehan Boutillier, stating the whole of French law about 1390, based on Roman and canon law, customs of the north of France, Flanders, Vermandois, and Normandy, decisions of tribunals in these areas, and earlier works. It was very influential for long afterwards.

Somner, William (1598–1669). Anglo-Saxon scholar, author of a Saxon-Latin-English Dictionary, *Observations on the Laws of King Henry I* (1644), and an important *Treatise on Gavelkind* (q.v.) (1660). He also translated Lambarde's Ancient Saxon Laws but this was never published.

Sosius (or Zoesius), Thomas (?–1598). Dutch jurist and judge of the Provincial Court of Utrecht.

Soto, Domingo de (1494–1560). Spanish jurist and theologian, a Dominican teacher at the University of Salamanca, and author of an important *De justitia et de jure* (1556), *Summulae in libros sententiarum commentarii* and other works. *Rationis ordinatio*, ordinance of reason, as the criterion of law was the foundation of his legal philosophy. He recognized the *ius gentium* as part of positive law.

Soup. It was the custom at certain courts, in cases not privately prosecuted, to allot the cases to counsel who were members of the bar-mess connected with that court and who regularly attended. Such briefs were known as 'soup' by analogy with soup doled out to poor persons at soup kitchens.

Sources of law. The phrase 'sources of law' is used in several different senses. In the first place, there are the historical sources of law, which are the acts and events in past time which have given rise to particular principles and rules of law. In the United Kingdom and Western European legal systems generally these include the Roman law, the canon law, the principles of feudalism and feudal customs, the law merchant, the general maritime law of Western Europe, and, in particular countries, particular acts and events, such as in England the clash between King and barons which produced Magna Carta, the constitutional disputes between King and Parliament in the seventeenth century, the dispute which provoked the Trades Disputes Act, 1906, and many more.

Secondly, the term 'sources' is sometimes applied to those theoretical or philosophical principles which have influenced law, motivated legislation or prompted change. Thus, the source of many principles of equity was the idea of natural justice, and that of much reforming legislation in the nineteenth century was utilitarianism, the philosophy of utility, that men should seek to promote the greatest good of the greatest number. In Eastern Europe, the philosophy of Marxism–Leninism has been the source of whole legal systems.

Thirdly, the term is used of the formal sources which, by reason of their accepted authority, confer validity and legal force on principles and rules drawn from them. They are the recognized law-creating and law-declaring agencies from which come valid rules of law. In this sense the formal sources of law in the United Kingdom are declarations by Parliament in the form of legislation, or by certain subordinate authorities to which legislative power has been delegated in the form of delegated legislation, statements of law by the superior courts contained in the *rationes decidendi* of cases decided by them, statements by authors of books of authority, customs, agreement of parties, and judicial ideas of justice, equity, morality, and reason.

Fourthly, the term is used of the documentary sources, the documents containing the authoritative statements of rules of law, where one finds the law authoritatively stated. In the United Kingdom, these are the volumes of statutes, statutory instruments, and of reports of case-law, and the writings of jurists which are recognized as books of authority, such as Coke and Blackstone.

Fifthly, the term is used of the literary sources, legal literature, the books to which one turns for information as to the law on any matter, where one finds the law unauthoritatively stated. These include encyclopaedias, treatises, text-books, and works of reference, which are based on the material sources, but have no authority in that a statement of law in any such book confers no validity on the rule so stated, and no judge is bound to accept the rules so stated as correctly stated. Every principle and rule of law has a source in each of the first four senses of the term. For study purposes the first and second are interesting and may cast light on the true scope and meaning of a principle. For practical purposes the third and fourth are important, and an alleged rule has no legal force or validity unless it proceeds from one of the material sources and can be found stated in one of the documentary sources. The literary sources are valuable for finding the law relevant to some point, for their synthesis of materials derived from various documentary and material sources relevant to a particular topic, and particularly for their evaluation and criticism of the trends in the law, its omissions, inconsistencies and defects.

Historically, custom is doubtless the original source of law and it becomes so in a particular community where customary practices are accepted as authoritative and binding and are treated as law. Custom itself is preceded by individual decisions and also will develop when decisions adopt consistent lines.

Every principle and rule has an historical origin, a philosophical basis and a formal source, but the principle or rule is legally valid, authoritative and enforceable as law only if drawn from a recognized formal source; the historical and philosophical sources confer no validity or authority. Thus, the strict liability of the common carrier is English law not because it originated in the Roman praetorian edict, nor because it was justifiable by expediency, but because the custom of the realm adopted it and this was applied and restated in many binding decisions of the superior courts.

Very early, however, in legal history there began the practice of legislating, of consciously making or restating or changing rules of law, of laying down the law on some point, initially a ruler's fiat, and increasingly by kings with the consent of their councils and their Parliament, developed. The earliest written records of law show legislation recognized.

The next stage historically, at least in the common law, was the development of the practice of judges treating the course of individual decisions, the practice of the courts in particular matters, as amounting to binding custom. Thus, case-law and the doctrine of following precedents, *stare decisis*, began to evolve as a distinct authoritative source. This development did not take place to anything like the same extent in Roman law and systems derived from it, in which systematic writings have always been more authoritative.

In modern times, importance increasingly attaches to treaties, international conventions and supra-national legislation and agreements, including particularly E.E.C. law, which are sometimes automatically part of the municipal law and sometimes become part of its law only if and when incorporated, as by Act of Parliament.

Legislation and judicial decisions may both in certain cases adopt principles or rules initially worked out in other legal systems, and reforming statutes may deliberately adopt a rule developed on the basis of comparative study of other legal systems.

As legal systems have evolved, each accordingly develops a recognized set of authoritative legal sources, which can be set out in a hierarchical order, according to which, in case of conflict, one takes priority over others. Different systems have different sets of sources and may rank them in different order. How this has come to be is a matter of the legal history of each system. What the legal sources are in a particular system may be determined by the Constitution, or the major Code, or the practice of the superior courts.

In Roman law, the sources changed substantially over the centuries. In the republican period (down to 31 B.C.) the sources were *leges*, enactments of the *comitia*, *plebiscita*, enactments of the *concilium plebis*, interpretation of these statutory materials by the *pontifices*, advice (*responsa*) by *jurisprudentes* to magistrates and parties, *edicta* issued by superior magistrates, notably the praetors, and custom.

In the principate, the classical period (31 B.C.–A.D. 235) legislation by the *comitia* and *concilium plebis* disappeared by the end of the first century A.D. *Senatus consulta* increasingly came to have authoritative force, and from the mid-second century *constitutiones principum* were regarded as having statutory force; these included *edicta, decreta, epistulae* and *mandata*. The praetorian edict was settled by Salvius Julianus and was thereafter unalterable by the praetors. *Responsa prudentium* continued to be an increasingly important source of law and from the time of Augustus some jurists had the *jus respondendi*, which gave their *responsa* higher authority.

In the dominate (*c.* A.D. 235–395) the Emperor was the principal legislator, expressing his will in various forms, letters, edicts, rescripts, pragmatic sanctions, and mandata; *responsa prudentium* had ceased to be a source of law.

At Constantinople under Justinian (A.D. 527–565) imperial constitutions were the major source of law, though sometimes still called *edicta, epistulae* or *decreta*; they are well exemplified in the Novels issued after A.D. 534; *responsa* were obsolete. Justinian enacted that imperial judgments, given in the presence of the parties, were to be valid as precedents in later similar cases, and also reserved power of interpretation of existing law.

In modern British law, the authoritative sources, in descending order of authority, are legislation of the European Community, decisions of the European Court, statutes (Acts of Parliament), statutory instruments and other subordinate delegated legislation, certain decisions of the superior courts, proven custom, statements in the books of authority, and extraneous sources, such as statements in textbooks, persuasive decisions, decisions from other jurisdictions, principles of morality, and values. The discovery from the authoritative sources of the principle or rule appropriate for the decision of a particular point in controversy is not, however, a simple or mechanical process, particularly as principles may have to be drawn from more than one source. In relation to statute it has to be determined what statute(s) and what section(s) thereof is or are relevant, and then to interpret the relevant sections to discover what the statutory provisions prescribe in the particular case.

In relation to case-law or judicial precedents (q.v.) there has to be discovered what judicial decisions appear to be in point, which of them are authoritative or binding and which merely persuasive, and in each case what the *ratio decidendi* or principle of law underlying the particular decision is. Only then is the court in a position to apply a principle extracted from the precedent cases to the case requiring decision.

In relation to custom (q.v.), the custom must be proved by evidence and satisfy the requirements for its recognition by the court; in particular it may not conflict with any rule of statute or common law.

Books of authority (q.v.) may provide authoritative statements on a matter in issue, but modern textbooks are not books of authority, though statements in them may be highly persuasive.

Extraneous sources include the decisions of the Privy Council, of superior courts in other English-speaking jurisdictions, such as the U.S. and Commonwealth countries, international law, textbooks and periodicals, and ultimately, the judge's view of the requirements of justice or of public policy, his moral standards, and other values.

In the U.S., the prime sources are the Constitution, treaties, federal statutes, federal executive orders and administrative regulations, state constitutions, state statutes, state administrative regulations and municipal ordinances, rules of procedure, judicial decisions, usages, treatises, and principles of morality and justice. As in British law decisions of superior courts are normally treated as authoritative and binding though there is rather more freedom to depart from precedents, or to elect which line to follow, because of the great number of co-ordinate jurisdictions.

In French law, in private and criminal law the sources are the codes and statutes, and subordinate legislation, such as presidential decrees and administrative ordinances, custom and usage, *jurisprudence*, i.e. decisions of the courts, *doctrine*, i.e. the body of opinion expressed in books and articles by legal scholars, textbooks and commentaries, and doctrines of equity, public policy and morals. In public, particularly administrative, law the sources are the statutes, executive orders, decrees and regulations, *jurisprudence* and *doctrine*.

In German law the sources are codes and statutes, and custom. Decisions of courts are binding on the parties only though they have some persuasive value. Writings of jurists are not authoritative but have persuasive authority.

The major differences between sources in common-law countries and sources in civil-law countries are that in the former, though the bulk and importance of statutory law is steadily increasing, principles of common law developed by courts remain of great importance; interpretation of statutes is literal and disregards *travaux préparatoires*; precedents are often absolutely binding; and the views of text-writers are very subsidiary. In the latter, code and statute law is the primary source; the only form of common law recognized is custom; interpretation is liberal and invokes *travaux préparatoires* and other extrinsic materials; precedents are never more than persuasive; and the views of text-writers may be material.

In the law of the European Community (q.v.), the sources of the law are the foundation treaties, and the supplementary conventions and treaties, conventions between member states, administrative acts (regulations, directives and decisions) of the chief organs of the Community, judicial decisions of the European Court (though these are persuasive only), and general principles of law common to the laws of the member states. Sometimes principles of international law have been referred to. The general approach to sources of law is the continental civil-law one rather than the common-law one.

In international law the Statute of the International Court of Justice provides for the application of international conventions, general or particular, establishing rules expressly recognized, international custom, as evidence of a general practice accepted as law, and the general principles of law recognized by civilized nations, and judicial decisions and the teaching of the most highly qualified publicists of the various nations, as subsidiary means for the determination of rules of law.

South African Law. The Dutch occupied the Cape of Good Hope in 1652 and brought with them the Roman-Dutch law (q.v.), an amalgam of mediaeval Dutch law of mainly Germanic origin and of the Roman law of Justinian, adopted at the time of the Reception. Thereafter till 1795, the law of Holland was, with variations, the law of the European settlements in South Africa. From 1795 to 1910, save for the years 1803–6, the Cape and other colonies in South Africa, into which white settlers had penetrated, were under British rule. By 1795 Roman-Dutch law had become firmly established, but thereafter legislation and judicial influences (including that of the Privy Council), and the fact that in Holland the native law had been replaced in 1809 by a Dutch copy of the Code Napoleon, in 1811 by the Code Napoleon itself and in 1838 by the *Burgerlijk Wetboek*, a mixture of Roman-Dutch and French law, caused it to be substantially overlaid and influenced by common law and equity principles derived from English law. In 1910 the Union of South Africa was created and in 1961 it left the Commonwealth and became a republic. Since 1910 and, even more so since 1961 there has been a strong nationalist revulsion against English principles and in favour of a return to Roman-Dutch principles, while independent legislation has produced many divergences from English law. Today Roman-Dutch law is the common law of South Africa, and also of Rhodesia, S.W. Africa, Lesotho, Botswana, and Swaziland.

First representative and then responsible government were established at the Cape (1853 and 1872 respectively) and in Natal (1856 and 1893).

In 1910 the Union of South Africa amalgamated the former Cape Colony, Natal, Transvaal, and Orange River Colony (renamed Orange Free State province) into a single state. The four colonies lost their legislatures but became provinces of the union, with provincial councils having subordinate delegated powers. The executive was headed by a governor-general, aided by an executive council of ministers. The legislature comprised a senate and a house of assembly. Provincial councils might make provincial ordinances on matters assigned to the provinces. A supreme court was created, the courts of the former colonies becoming provincial and local divisions of the supreme court.

As a whole, the legal system and the law stand in a position intermediate between a civil law system and a common law system, and the law is composite and has adopted elements from both Roman-Dutch and English law.

Sources

Among legislation, a few statutes of the States-General of the Netherlands still apply, as do some statutes of the old colonies and republics before Union. The main bodies of legislation are those of the South African Parliament since 1910. Unless otherwise provided legislation comes into force on publication in the *Gazette*. Pre-Union statutes may be held abrogated by disuse but statutes in force at the time of Union or promulgated later require express repeal.

The common law is the developed Roman-Dutch law. In elucidating Roman-Dutch law, courts rely heavily on the institutional writers, particularly Johannes Voet, Van der Keessel, Bynkershoek and Pauw, and much less on the collections of opinions that have come down, such as Naeranus' *Hollandsche Consultatien*. Roman law is still a living source, more attention being given to the *usus modernus Pandectarum* than to classical Roman law. In the years 1850–1910 many English principles overlaid Roman-Dutch rules and, in many cases, introduced confused doctrines; since 1910 and particularly since 1961 there has been a revulsion against such encrustations.

In modern practice, the doctrine of judicial precedent applies though rather less strict than that applied in England. Since the abolition of appeals to the Privy Council its decisions are no longer absolutely binding. Appellate Division decisions are binding on inferior courts and, in practice, on that Division itself, though exceptionally they may be departed from, but it is not bound by any pre-Union South African court decision, nor by decisions of any provincial or local division. In the divisions, courts are bound by decisions of superior or larger courts, and, normally, by their own previous decisions.

The great text-writers of South African law are the 'old Authorities', the major Roman-Dutch jurists, and these are still not infrequently cited and applied, as are decisions of the high courts of Holland of the seventeenth and eighteenth centuries. Modern text-writers' works are, at most, of persuasive force, but in some branches of law, textbooks have considerable reputation.

Custom is a relatively unimportant source, but may be relied on if proved to have been long established and uniformly observed, and if reasonable and certain. Trade usage may similarly be proved and applied.

In the last view the judges' view of what is fair and just may be applied.

Public law

The constitutional and administrative law is mainly founded on English lines.

The higher executive now consists of the State President and his Executive Council, or Cabinet. The government must have the support of the House of Assembly.

Since the Republic of South Africa Constitution Act, 1961, the South African Parliament is fully sovereign. It consists of the State President, a Senate, and House of Assembly, in the latter of which money bills must originate. The State President acts on the advice of the Executive Council, i.e. the Cabinet.

Each province has a provincial council, which may make ordinances within a limited compass; their powers do not extend to private law. The Transkeian Legislative Assembly has wider legislative power and the Legislative Assembly of South-West Africa wider powers again.

The courts cannot enquire into observance of the rules of legislation, but a division of the Supreme Court may declare a provincial ordinance invalid.

Many types of subordinate legislatures have delegated legislative capacity; powers are delegated to the State President, Ministers, public functionaries or bodies.

There is no distinct body of administrative law or hierarchy of administrative courts, but only an amorphous body of law drawn from various sources. There are departments of state, provincial administrations dealing with matters assigned to the provincial councils, and many kinds of local authorities and boards, and numerous statutory and public bodies conducting economic activities or public services for the community.

The government is, in general, civilly liable for breach of duty in the same way as a private individual. There are no guaranteed rights or civil liberties under the Constitution. The individual may act as he pleases, provided he does not transgress the law or the rights of others. The courts closely scrutinize legislation interfering with an individual's liberty. Unlawful interference with personal liberty can be terminated by invoking the *interdictum de homine libero exhibendo,* generally comparable to habeas corpus. The police force has long been organized nationally. It has extensive powers to invade private rights in the execution of its functions.

Racial laws have been in existence since 1809 and now form a complex body of restrictive regulation. A person's classification affects his liberty to marry, own and occupy land, to move within the country, his right to education, social security, and in other respects.

Courts and judges

The Supreme Court is composed of the Appellate Division, six provincial divisions (Cape, Transvaal, Natal, Orange Free State, Eastern Cape, South-West Africa) the judges of which also serve certain local divisions, and the Griqualand West Local Division, each having a Judge President and puisne judges. There are fixed quorums for certain kinds of business. The judges are appointed by the State President from senior counsel.

The principal inferior courts are magistrates' courts, with both civil and criminal jurisdiction. There are special water courts, courts for income tax appeals, the commissioner of patents, and courts of Bantu affairs commissioners.

The courts of the provincial divisions have both civil and criminal jurisdiction. Civil juries disappeared in 1927. In criminal cases, juries of nine males, seven at least to determine the verdict, existed, but various provisions exist permitting trial by a special court without a jury and, since 1958, trial is without jury unless the accused elect otherwise; in practice, it has almost completely disappeared. But trial by judge with two assessors is increasing. Prosecutions are in the hands of the Attorney-General in each province, assisted by staffs of professional assistants and public prosecutors.

Since 1927 native commissioners have had criminal jurisdiction, like magistrates' courts, in respect of offences committed by Bantu. Since then native commissioners' courts have had civil jurisdiction between natives only, and native chiefs or headmen have had jurisdiction in certain civil claims under native law and custom, with appeal to native commissioners' courts. There are now three native appeal courts, composed of senior native commissioners. These courts have judicial knowledge of native law and custom.

Criminal law and procedure

The criminal law is contained in Roman and Roman-Dutch law, written and unwritten, decisions of the courts, and in legislation. English law has been a powerful influence.

The general principles of criminal liability apply, that there must be proved both *actus reus* and intention, actual or legal (i.e. imputed by law). Negligence may justify a charge of culpable homicide. The M'Naghten rules are the basis of the law as to the defence of insanity. All participants in crime are equally liable. Under statute certain acts are criminal only if done by non-whites. There is no distinction between felonies and misdemeanours.

The rules of evidence in criminal trials are largely statutory, failing which English rules are followed.

Private law

The private law is strongly influenced by the Roman law, in arrangement, terminology, and actual rules.

Persons

The Roman-Dutch law has been very substantially maintained in the South African law of persons and family relations. Material differences in legal status exist between whites, natives, Asiatics, and coloured persons of mixed blood. There are some reservations in respect of native law and custom and thus, to some extent, a system of personal laws. Marriage is competent by civil or religious ceremony, in the presence of a marriage officer and two witnesses. Marriages of mixed races are prohibited. Adoption was introduced by statute in 1923. In the absence of ante-nuptial contract, marriage invests the husband with power over his wife's person and property and creates community of property, all their property becoming common property. The wife's legal position is analogous to that of a minor under guardianship. Judicial separation is competent on the ground of a matrimonial offence by the defendant. Divorce is competent on grounds of adultery, desertion, insanity, or imprisonment. On divorce, the innocent spouse may obtain an order that the other forfeits benefits from the marriage.

Contract

Though based on Roman-Dutch law, the South African law of contract has been heavily influenced by English law. Agreement must be reached by offer and acceptance. Every contract must have *justa causa,* which is not equivalent to English consideration. There are few requirements of formalities; the doctrine of part performance is not accepted. Contracts for the benefit of third persons are accepted, and in general contractual rights may be assigned.

In the event of breach, either specific performance or damages may be claimed, subject to the court's discretion. Breach of a non-essential term justifies damages only, but of an essential term gives a right to rescind and to claim damages for loss of the bargain.

In mercantile transactions, English influence has been more noticeable. The main contracts, such as sale, lease, suretyship are pure Roman-Dutch, modified in detail by statutes, but other such as agency, banking, insurance, and company law have been developed under English influence.

Quasi contract

The Roman law categories of unjust enrichment, such as *negotiorum gestio,* are fully accepted.

Delict

Delictual liability is almost entirely based on the *actio legis Aquiliae* and the *actio injuriarum,* affording remedies for wrongs to interests of substance and to interests of personality respectively. The latter covers assault, defamation, and the like. Vicarious liability has been adopted from English law. Contributory negligence has, since 1956, been a partial defence only, though the Act in question is not identical with the British legislation. Damages, as in English law, may be general or special. In Aquilian actions pecuniary loss only can be recovered, but in intentional wrongs actionable under the *actio injuriarum,* damages are awarded in *solatium* for injured feelings irrespective of pecuniary loss. By an extension of Aquilian liability the dependants of a deceased have an action for loss caused them by his death, but for pecuniary loss only, not for hurt to feelings. In a few instances, liability exists independently of fault; these include nuisance, liability for animals, and breach of statutory duty. The *Rylands* v. *Fletcher* rule (q.v.) is not accepted.

Property

In the law of property the Roman-Dutch law has been largely preserved. Property is divided into things corporeal and incorporeal, and the former into moveables and immoveables. In land ownership is absolute save as restricted by rules of public and private law. Deeds affecting land must be registered with the Registrar of Deeds. There is a large volume of law relating to mineral rights. Incorporeal moveable property is transferred by cession, i.e. by agreement.

Trusts

Roman-Dutch law knew no law of trusts, but *fideicommissa* ensured devolution within a family or provision for charitable purposes. Attempts to create trusts are construed as *fideicommissa*, or donations *sub modo.* But the courts are tending to acknowledge and develop a body of trust law. The administrators of a will may be directed to hold certain of the deceased's assets, and the testator may create temporary interests therein similar to *fideicommissa.*

Succession

English law has had considerable influence in succession, modifying the basic Roman-Dutch law. On a person's death his estate is wound up by an executor under the supervision of a Master of the Supreme Court. Intestacy is based on the mediaeval *Schependomsrecht* of Holland. Estate devolves successively on descendants, on parents and their descendants, on grandparents and their descendants, and so on. By statute a surviving spouse has certain rights of succession. The standard form of will is one executed under the testator's hand before two witnesses. Former restrictions on freedom of testation have been abolished. Life interests, such as the bequest of the use of property, may be created and *fideicommissa*, in the form of a bequest of a thing subject to a determining condition, are also common. Mutual wills of spouses are common.

Evidence

Statutory changes apart, the principles of evidence are generally those of English law.

Civil procedure

Civil proceedings are by way of action, with oral evidence, or application with evidence on affidavit.

Legal profession

The legal profession is strictly divided, in effect into three parts, the Bar (advocates), the Side-Bar (attorneys), and the Public Service (prosecutors and magistrates). There are Societies of Advocates, for each Division of the Supreme Court. Judges are appointed from the advocates, but magistrates from law graduates or persons who have been trained in the Department of Justice Training Centre. The Bar was formerly divided into Queen's Counsel and advocates and since 1961 into Senior Counsel and advocates.

Apart from the old Dutch authorities, South Africa has produced a considerable volume of legal literature.

J. W. Wessels: *History of the Roman-Dutch law*; H. L. Hahlo and E. Kahn: *The South African Legal System and its Background*; H. L. Hahlo and E. Kahn: *The Union of South Africa: The Development of its Laws and Constitution* (with *Supplement*); R. W. Lee: *Introduction to Roman-Dutch Law*; G. Wille: *Principles of South African Law*; A. F. Maasdorp: *Institutes of South African Law*; F. Gardner and C. Lansdown: *South African Criminal Law and Procedure; Annual Survey of South African Law.*

South Sea Bubble (1720). The South Sea Company was founded in 1711 to trade largely in slaves with Spanish America. Initially, the company was only moderately successful but George I became governor of the company in 1718 and this raised public confidence in it. In 1720 it made the proposal, which was accepted by Parliament, to take over a large part of the national debt, giving the holders of government stock the company's stock in exchange, and its shares rose sharply from 128.5 in January 1720 to 1,000 in August. Many investors were gulled by optimistic promoters of other companies floated at this time. By September the market collapsed and South Sea stock fell to 124, government and other stocks falling in sympathy. Many investors were ruined and a House of Commons inquiry uncovered the fact that several Ministers had taken bribes and had speculated. The fall brought disgrace on the directors though the company survived itself until 1853. The crisis gave rise to the Bubble Act which made it illegal to create a corporate body with transferable stock or shares without the authority of statute or Crown charter. This held up the development of companies on modern lines until the Act was repealed in 1825, but associations were formed under deeds of settlement.

Sovereign. The person who as King or Queen exercises the office of monarch or head of state under the British constitution. The old English Kingship was elective but, in the twelfth century, the kingship became more feudal in character and by the time of Richard I, election was becoming merely a recognition of hereditary right and from the accession of Edward II (1307), election disappeared and hereditary succession was the established rule. But Edward II and Richard II were deposed, under Henry IV the right of succession to the Crown was settled, altered and resettled four times, and under Henry VIII the succession was altered repeatedly by Parliament. On the death of Elizabeth, the Council proclaimed King James VI of Scotland to be also King James I of England but under Charles II, Parliament twice passed bills for the exclusion of the Duke of York from the succession. In 1688 Parliament offered the Crown to William of Orange and by the Act of Settlement (1701) entailed the succession on the heirs of Sophia, Electress of Hanover and under this provision, on the death of Queen Anne in 1714, the Crown passed to George of Hanover who became George I.

The title to the Crown now derives from the Act of Settlement, 1701, and this could probably be altered only by the common consent of those member nations of the Commonwealth which are monarchies. The inheritance of the Crown is regulated by feudal rules of descent, the Crown descending to the sons of the late sovereign according to seniority and failing sons, to daughters according to seniority (not as co-parceners). Failing children or more remote descendants the Crown passes to collaterals and their descendants. The powers of the Crown are the same whether the sovereign be a King or a Queen. By custom, the wife of the King is designated Queen (Consort) but no special rank or title attaches to the husband of a Queen-Regnant. Queen Victoria's husband, Prince Albert, was designated Prince Consort.

On the death of a sovereign, the new sovereign is at once proclaimed at an Accession Council to which all members of the Privy Council are summoned, and the Lords Spiritual and Temporal, the Lord Mayor of London and certain other prominent persons are present. The coronation follows the accession after about a year. It takes place in Westminster Abbey in the presence of representatives of the Lords, the Commons, leading personages from the United Kingdom and Commonwealth countries, and representatives from other countries, the form of service being much the same as centuries ago though frequently modified in detail.

The sovereign is, in law, head of the executive, an essential part of the legislature, head of the judiciary in each of the parts of the United Kingdom, Commander-in-Chief of the armed forces of the Crown and temporal governor of the Church of England, and the whole government of the U.K. is carried on in name of the sovereign by Her Majesty's Government.

Many acts of government require the sovereign's participation and the Queen confers honours, makes appointments to all important state offices, entrusts to a person the duty of forming a Cabinet and carrying on the government in her name, concludes treaties, makes war and peace, summons, prorogues and dissolves Parliament, and opens each session with a speech from the Throne, gives Royal Assent to Bills which have passed through all stages in both Houses of Parliament and pardons criminals. In all cases she acts on the recommendation or advice of her Ministers, though some honours are in the sovereign's personal gift and in exceptional cases she may have a choice of which person to invite to act as Prime Minister. But the Queen is always entitled to be informed and consulted and may enquire, encourage, or warn. The continuing importance of the sovereign is such that statutory provision has been made for a regency (q.v.) in the event of the sovereign being under the age of 18 on accession, or being incapacitated, while provision is made for discharge of the royal functions during illness or absence abroad by the appointment of Counsellors of State.

The sovereign has numerous ceremonial functions and takes precedence over all subjects, while royal presence, patronage or approbation is deemed a high honour.

The permanent home of the sovereign is in the United Kingdom, the Queen being represented in Northern Ireland by the Governor (an office now in suspense), in the Channel islands and the Isle of Man by Lieutenant-Governors, and in the other member-nations of the Commonwealth which continue to owe allegiance to the Crown by Governors-General, who are appointed by the Crown on the advice of the Ministers of the country concerned. These persons represent the Queen, not the government of the United Kingdom. In colonies and dependencies the Queen is represented by Governors, High Commissioners, Administrators or Residents who are appointed by the Crown and are responsible to the United Kingdom government for the administration of the territory in question.

The sovereign is Head of the Commonwealth, the living symbol of the continuing association of the group of independent countries which form the Commonwealth, and allegiance continues to be owed to the Crown by many of the member states. The Queen, that is, is also Queen of Canada, Australia, etc., but not of those countries, such as India, which remain in the Commonwealth, and recognize the Queen as Head thereof, but have become republics and do not owe allegiance to any monarch.

The relations of the sovereign with Ministers and the exercise of royal powers is discussed in royal and statesmen's biographies. See, e.g., Longford, *Victoria, R.I.*; Magnus, *King Edward VII*; Nicholson, *King George V*; Wheeler-Bennett, *King George VI*; Young, *A. J. Balfour*; Spender and Asquith, *Life of Lord Oxford and Asquith*; Middlemass, *Baldwin*; Churchill, *Winston S. Churchill.*

Sovereignty. The concept of supremacy or superiority in a state by virtue of which some person or body or group in that political society is supreme and can, in the last resort, impose his or its will on all other bodies and persons therein. That there must be some person, body, or institution in every political society which is sovereign is evident; otherwise there would be no end to disputes. Problems of sovereignty have long been discussed by political thinkers, notably by Bodin, Hobbes, Rousseau and, among jurists, by Austin. Austin resorted to the concept to distinguish positive law, which was in his view rules set by the sovereign, from other rules not so set, which were mere positive morality. Sovereignty was an attribute of a determinate person or body to whom the generality of the society gave habitual obedience and who or which was not itself habitually obedient to any other person or body. Within that society the sovereign had ultimate power to lay down the law.

It follows from this concept of sovereignty that a sovereign person or body cannot by law impose any legal limits on his or its future conduct or ability to change the law, even retrospectively. It follows also that a theory of law which depends on the concept of sovereignty cannot be reconciled with any view that a law of God or of nature or of other superior kind binds the state and regulates or limits its law-making and law-enforcing powers.

Sovereign and subject

It follows from this concept of sovereignty that the sovereign cannot be bound by legal duties to his subjects, nor can they have any legal rights as against the sovereign. Indeed, no legal relationship can exist between them, because there is no superior who could enforce law as between such parties. In practice, sovereigns frequently conceded quasi-rights to subjects, as under the British Crown Proceedings Act, 1947, but these are enforceable by grace only and always revocable, in that the Act could be repealed. In fact, what claims a subject has against the state he has by moral right and concession only. In law, the Queen in Parliament could refuse to pay the civil service and the armed forces and scrap the whole social security system.

Aspects of sovereignty

Sovereignty may be distinguished firstly into external and internal sovereignty. The first is concerned with whether a state in international relations has sovereignty and total control of its policies or whether it is ruled directly or indirectly by, or must obtain consent or concurrence of, another state, in which case it is not fully sovereign. Thus, colonies are normally not sovereign in external relations. The second is concerned with the exercise of sovereign powers within the state itself; who or what can in the last resort impose his or its will on every aspect of the government of that state?

Secondly, the exercise of sovereignty is distinguishable into various aspects which may be exercised by different persons or bodies. A despot might be legislator, chief executive and chief judge all in one, but in many states these functions are distributed. Thus, in the U.K., legislation is made by the Queen in Parliament, the execution of policy is effected by the Queen's Ministers, who by convention are all members of one or other House of Parliament and of the party which has a majority in the House of Commons, and the judicial functions are exercised by judges who are very largely independent of Parliament, though the final court of appeal for many disputes is the House of Lords, a chamber of Parliament, exercising the judicial powers of the Queen in Parliament. In other Western countries similarly, the powers which amount to sovereignty are divided in various ways.

Legal and practical sovereignty

A distinction must also be drawn between legal and political or practical sovereignty. Legal sovereignty may be vested in a person, such as a monarch, or a group, such as a Cabinet, and there may be no legal superior or legal control on what that person or body does. But if that person or body in fact defers to and acts in accordance with the wishes of some extra-governmental person or body, such as the party conference, or a group of trade union bosses, or the leaders of an international political movement living outside that country altogether, practical sovereignty belongs to that other person or body, and the legal sovereign is only nominal and not actual sovereign. Thus, in the United Kingdom legal sovereignty is vested in the Queen in Parliament in that that tripartite body (Queen, Lords, and Commons) can make or unmake any rule of law, make or overrule any judicial decision, or take any action it likes. But in practical terms the decision to do any of these things is determined by the Cabinet, and one must further inquire how far in doing anything a Cabinet is being dictated to by a few trade-union bosses backed by their power to bring economic life to a halt by calling a strike. Where the latter is the case, these bosses are in fact sovereign.

For historical reasons, because in the U.K. the monarch was originally in law and, in fact, sovereign and the person who actually exercised sovereign powers the monarch is still called 'the Sovereign' though for long now he or she has had only a share in the exercise of sovereign powers.

The location of sovereignty

Sovereignty in a state may be located in a monarch, or an oligarchy, an assembly, in certain groups or in the community, the people. Or it may be said to be located in the constitution, or the law, or in God. Theories of the latter kind founder in practice on the fact that if the constitution or the law or the Will of God is flouted how does that sovereign assert its authority? The constitution or the law cannot be sovereign if either can lawfully be modified by someone else, such as Parliament. In the U.K., sovereignty may generally be said to reside in the Queen in Parliament, or in the Cabinet which in practice largely controls Parliament. In the U.S., it cannot reside in the Congress, which is subject to the Constitution, nor in the Constitution because it can be amended, but to say that it resides in the bodies which collectively can pass a constitutional amendment is rather absurd, as that body never exists as a single body and has acted only a small number of times in two centuries. In any federation it is difficult to say where sovereignty resides; primarily, it is probably in the court which judges whether the constitution has been infringed by state or federal action; it may be ultimately in the electorate which can agree to amend the constitution. In a democracy it may be argued that sovereignty resides in the people, or rather in the electorate. There is a measure of truth in this but the electorate expresses its will only periodically and, in general terms, favouring for the greater part the X party rather than the Y, and the electorate has no control over the way in which the party elected to power actually exercises its power and no power by itself to take any action. In the cases of states which are members of the European Communities there is the further problem that sovereignty in certain matters in vested in the Council and Commission of the Communities but in other matters in its own governing authorities.

The limitations on sovereignty

The idea of limitations on sovereignty seems to involve a logical contradiction but, in practice, the idea is convenient and necessary. A person or body may be sovereign *de jure*, but *de facto* that sovereignty is always limited. Even the most absolute despot must at least retain the respect and obedience of his praetorian guard or janissaries or they will turn on him, and what he must do to retain their support is a practical limitation on his unfettered sovereignty. The Thirty Tyrants may be supreme only so long as they agree. In modern states, there are numerous practical limitations on the exercise of the sovereignty of the ruling group, moral, economic, political, and other, in that flouting generally held moral views, widespread economic exploitation or mismanagement, or cruel political subjection will sooner or later provoke uprising. Similarly, the leaders of the governing party in a democracy cannot act so as to antagonize too many of their supporters or they will lose the next election.

Increasingly, in modern times, the sovereignty of many states is in fact limited by their international agreements and by international law and morality. There are many things the doing of which by a state would flout agreements of one kind or another, or outrage other states, and the fear of sanctions or international disapproval will often restrain a state from doing what it may legally be fully entitled to do. In practice, accordingly, sovereignty is always limited and never absolute.

Sovereignty in international law

In international law legal persons are sovereign and independent states, and a state which is not sovereign is not a state in international law; hence states which are parts of a federal state, or are colonies or protectorates or dependencies of another state, are not sovereign and not full states. To be sovereign in international law a state must be able to exercise jurisdiction over a determinate tract of territory and the inhabitants therein, and have legally independent powers of government, administration and disposition over that territory. Where there has been a material change, e.g. after a revolution, problems of recognition (q.v.) by other states arise.

The corollaries of sovereignty and equality of states are exclusive jurisdiction over their several territories, a duty of non-intervention in the area of exclusive jurisdiction of other states and the dependence of obligations arising from customary law and treaties on the consent of the obligor.

Difficult cases arise of states which are nominally sovereign and independent but are in fact puppets or vassals of other states, with the exercise of their sovereign powers in fact controlled from outside.

M. Wilks, *Problem of Sovereignty in the Later Middle Ages*; C. E. Merriam, *History of the Theory of Sovereignty since Rousseau*; J. Austin, *Province of Jurisprudence Determined*; R. A. Eastwood and G. Keeton, *Austinian Theories of Law and Sovereignty*; H. Laski, *Studies in the Problem of Sovereignty*; A. Larson, *Sovereignty within the Law*; J. Mattern, *Concepts of State, Sovereignty and International Law*; H. Krabbe, *The Modern Idea of the State.*

Space law. Principles of law accepted by nations as binding on them and their nationals in engaging in activities in outer space, i.e. the area beyond the Earth's atmosphere, and in relation to celestial bodies. The subject has become one of practical importance since the beginnings of space flight in 1957. It was brought before the U.N. in 1958 and an *ad hoc* committee established. In 1959 it was transformed into a permanent Committee on the Peaceful Uses of Outer Space. It early recognized that international law applied to outer space and celestial bodies, and that they were free for exploration and use by all states in conformity with international law and not subject to national appropriation. In 1963 it framed a Declaration of Legal Principles Governing the Activities of States in the Exploration and Use of Outer Space which was adopted by the General Assembly, and in 1966 an Outer Space Treaty was signed by most states. It declares that space is free for exploration, that space and the celestial bodies are not subject to appropriation, that no military installations are to be established in space, and that states are responsible for their, and their nationals' activities in space. Many problems remain, notably as to communications satellites.

Spanish law. The earliest inhabitants of the peninsula were probably Iberians and Celts living under customary law, supplemented later by Cretans, Phoenicians, and Greeks. From about 200 B.C. to the fifth century A.D. Spain was part of the Roman Empire and in this period there was very substantial Roman influence, and records exist of imperial constitutions, statutes, and decrees. Recognition continued of provincial custom, which remained important.

On the collapse of Roman authority in the West the Visigoths under Euric and Alaric II established dominion in Spain, and their basically Germanic customary law in some ways resembled the earlier native law. Euric promulgated the first Visigothic code about A.D. 475. Spanish Romans continued to live under Roman law, and for them Alaric's *Breviary*, or the *Lex Romana Visigothorum*, was enacted in the early sixth century. Subsequently there were many statutes common to both communities and, starting in the mid-seventh century, with the first common code, then called the *Lex Visigothorum Reccesvindiana* or *Liber Judiciorum* and later known as the *Fuero Juzgo*, there was considerable fusion of Roman and Germanic elements and the *Lex Romana Visigothorum* was abrogated.

From the eighth to the thirteenth centuries most of Spain was under the dominion of the Arabs and this resulted in the application to them and those who adhered to their way of life of Muslim law, and to the others of a debased version of the Visigothic legislation. In the earlier period of the conquest law embraced various factors; the *Lex Romana Visigothorum* retained influence in some districts, elsewhere a mass of Germanic customs reappeared and elsewhere again the *Fuero Juzgo* was again invoked though interpolated and modified. In time diverse laws developed in the separate kingdoms of Castile, Aragon, Catalonia, Navarre, and the others, and within them there were variations in different localities as *fueros* proliferated. Count Ramon Berenguer I of Barcelona promulgated in the eleventh century the earliest laws used in the compilation of Catalan law known as the *usatges de Barcelona*.

In the thirteenth century many Spaniards studied in the Roman law schools of Italy and France and Italian jurists came to Spain. In Castile in the thirteenth century there was promulgated a model *fuero*, the *Fuero Real*, which perpetuated indigenous legislation, and other town *fueros* were granted or confirmed about the same time. There also developed royal ordinances, letters, and statutes of Cortes. The work of Justinian came to dominate legislation and was influential on the *Libro de las Leyes* (*c.* 1265) later commonly known as the *Codigo de las Siete Partidas* (q.v.) and received in the following century as a source of Castilian law. It influenced also the legislation and decision of the other Christian kingdoms.

The *Partidas* became a text and reference book, but thereafter there was a struggle between native law and Romanism. In the *Ordenamiento de Alcala* (1348) the King made the *Siete Partidas* authoritative where the *fueros* were silent, and hence it became valid law. In 1480 there was published a collection of the law enacted since the *Partidas*, the *Ordenanzas Reales* or *Ordenanzas de Montalvo*, in eight books. Then in 1505 the *Legas de Toro* were promulgated, laying down which of the older laws were authoritative. Developments in the other Spanish kingdoms were broadly similar. About 1370 a compilation of Catalan maritime law, the *Libre del consolat de mar* was prepared and widely accepted throughout Europe.

From the thirteenth century also there emerged Spanish jurists of importance, while Spaniards achieved fame in foreign universities and as canonists at Rome; among the latter was Raymond of Penaforte.

By the end of the fifteenth century the formation of native law had come to an end. While Spain was subject to the Hapsburgs, numerous orders and statutes continued to be issued and older and newer elements were combined in the official digest, known as the *Neuva Recopilacion* (1564) which did not however include all the law. There were also great masses of statutes relative to the colonies consolidated in the *Leyes de Indias* (1680), and to industry and commerce and attempts were made to codify many of the local ordinances.

In the Hapsburg period legal science was intensively cultivated in Spain and notable jurists included de Valderas, Guerra, Soto, Menchaca, de Cartagena, Covarrubias, Ayala and, above all, Vitoria, in international law; Suarez, Mariana, Molina, Montana, and others in public law; Soto, Vitoria, Molina, and others in criminal law; Antonio Agustin, Azpilcueta, Covarrubias, and others in canon law; Lopez, Gomez, Molina, and others in private law.

The Bourbons (1700–1808) added to the mass of statutes and issued revised and enlarged editions of the *Nueva Recopilacion*. They also annulled the special public laws enjoyed by Catalonia, Aragon, and Majorca and the special public and civil laws of Valencia. But diversity in legislation continued between the different kingdoms and even within Castile itself.

The digest known as the *Novisma Recopilacion* (1805) includes general provisions for the whole of Spain and law of Castile. It continued the validity of the old *fueros*, the *Fuero Juzgo* and the *Fuero Real* and of the *Partidas*.

Legal science and literature continued to flourish and in all branches of law jurists made valuable contributions to learning.

Spain is now a Catholic, social, and representative state and a kingdom. The constitution is set out in a series of Fundamental laws culminating in the Organic Law of the State in 1967. The head of state is the King who exercises supreme political and administrative power and a wide range of other functions. The highest central organ is the Council of Ministers and there is an office of president of the government to co-ordinate all the organs of government and administration. In each of the provinces civil governors represent the central government, while in municipalities mayors are the heads of local government; they represent the central government and have the function of maintaining public order. The Council of the Realm is a consultative body; it is composed of *ex officio* councillors and representatives of the Cortes. It is the highest administrative consultative body and acts through plenary sessions, divisions, and committees. Each province has a provincial House of Deputies and districts have municipal councils. The Cortes is the legislature; it comprises members representing provinces, municipalities, organizations of many kinds, and directly elected members.

The president of the Cortes is appointed by the head of State.

The judicial structure comprises a Supreme Court which may function in plenary session, as a disciplinary court, as a judicial council and, in six divisions as a court of justice. There are 15 territorial High Courts handling civil cases and 50 provincial High Courts trying criminal cases. Below these are numerous judicial districts with petty courts dealing with civil cases; and district courts, municipal courts, and petty courts dealing with lesser criminal cases. The judiciary is a professional career, as are members of the public prosecution service. There is a separate Court of Exchequer, labour magistracy, and juvenile court system.

A partial civil code was drafted in 1843–46, but opposition by provinces who wished to preserve their distinctive characteristics delayed the adoption of a civil code (Codigo civil) until 1889. Customary law continues to be important. Commercial codes were enacted in 1829 and 1885 and criminal codes in 1810 and later years.

F. de Sola Cañizares, *Bibliografia juridica español*; A. Garcia Gallo, *Manual de historia del derecho español*; M. Fernandez-Carvajal, *La Constitucion Española*; A. Quintano Ripolles, *Compendio de derecho penal*; D. Espin Canovas, *Manual de derecho civil español*; J. Manresa y Navarro, *Comentarios al código civil español*; J. Santamaria, *Comentarios al código civil*; E. Langle y Rubio, *Manual de derecho mercantil español.*

Speaker. The member chosen to preside in the House of Commons. He is elected by the members from among themselves at the commencement of each Parliament, and customarily re-elected if willing again to serve. In this context there is a convention that a Speaker does not issue an election address in his constituency, nor is he normally opposed there, at least officially. After election he submits himself, in the House of Lords, to the Royal Approbation which is signified by the Lord Chancellor and which has not been refused since 1678. He is the channel of communication between the Commons and the Sovereign, and between the Commons and the Lords.

Since 1841 it has been established that a Speaker should be re-elected and that, once elected, he dissociated himself from political controversy. He takes no part in debates and does not vote save when a division results in equality, in which case he votes for the *status quo.*

He presides at all times when the House is sitting, save when it goes into committee, regulates debates, rules on procedure and points of order, is the guardian of the privileges of the House, and has to decide such matters as whether a prima facie case has been made of breach of privilege. He has a particular duty to safeguard the rights of minorities.

The Speaker enjoys precedence immediately after the Lord President of the Council, has an official residence in the Palace of Westminster, and normally receives a peerage, formerly a viscountcy or barony, now a life peerage, on retirement. While on duty he wears a full-bottomed wig and black silk gown over Court dress, and, on special occasions, a gold-embroidered ceremonial gown.

The Lord Chancellor is Speaker of the House of Lords but is never spoken of by that title.

J. Dasent, *Speakers of the House of Commons*; M. MacDonagh, *Speaker of the House*; P. Laundy, *The Office of Speaker.*

Special Acts. Statutes not of general application and accordingly including only local, personal or private Acts. In the Clauses Acts (q.v.) particular Acts which may subsequently be passed concerning a particular undertaking but incorporating the relevant Clauses Act are frequently described as 'Special' Acts.

Special agent. An agent authorized to transact some particular business only for his principal, as distinct from a general agent.

Special case. (1) A mode of obtaining a decision on a statement of facts submitted to a court by the parties and raising one or more questions of law. Also a mode of raising any question of law arising in the course of an arbitration, or the award itself, or any part thereof for the decision of the court. Also the mode in which a person, aggrieved by a conviction or order in a court of summary jurisdiction, may have its rightness in point of law considered by the High Court.

(2) In Scottish procedure, a means of obtaining the opinion of the court on a matter of law where parties are agreed on the facts. The case is settled by their counsel, and sets out the facts and the questions of law on which the parties seek the opinion of the court. It is frequently utilized to obtain an opinion on the interpretation of a will or trust settlement.

Special commissioners (or Commissioners for the special purposes of the income tax). Consist of the Commissioners of Inland Revenue acting *ex officio*, together with other persons appointed to make certain assessments, directions, and apportionments and to determine appeals against certain assessments, with further appeal to the courts on a point of law.

Special constables. Persons not regular members of a police force appointed to assist regular constables in the performance of their duties.

Special damage. Such kinds of loss as will not be legally presumed to have followed from the defendant's wrongful act, but which must be specifically claimed in the pleadings and be proved by evidence to have been incurred, and that by reason of the defendant's breach of duty.

Special damages. Compensation for special damage which is not presumed to be a natural and probable consequence of the act or omission complained of, but which has in fact resulted. Special damages must be claimed specifically and proved strictly.

Special defence. In Scottish criminal procedure one of the defences of alibi, incrimination, self-defence, or insanity at the time of the crime; if one of these is relied on, it must be specially intimated to the Crown before the trial, otherwise evidence in support of it may not be led.

Special jury. A jury consisting of persons entitled to be designated esquire, banker or merchant, or occupier of premises of stated rateable value. After 1922 special juries could be had only in cases claiming compensation for land, and after 1949 special juries were abolished except City of London special juries in commercial causes. These also were abolished in 1971.

Special licence. A licence granted by the Archbishop of Canterbury authorizing a marriage to take place at any time or place.

Special Orders. The term used in the House of Lords for those statutory instruments which require affirmative resolutions of both Houses before having the force of Statute. In that House a Special Orders Committee exists to examine all such rules and orders and to report to the House any questions of principle, policy, and precedent raised by any such order. The required resolution may not be moved in the Lords until the Special Orders Committee has reported on the orders in question.

Special pleaders. Those members of the legal profession in England in the eighteenth and nineteenth century whose business it was to draw all the written proceedings in a suit at law, the declaration, plea, replication, rejoinder, etc. Special pleaders were not in general members of the Bar. It was common to spend a few years 'under the Bar' as a special pleader before being called, but some, such as Tidd, were special pleaders for many years. In view of the highly technical nature of the law at the time, when pleadings were very involved and lengthy, and the subtle and technical differences between different kinds of action, such as assumpsit, trespass, etc., special pleading was an involved and technical science in which wording and form were more important than substance. The reforms of law and procedure of the early nineteenth century abolished the science and the class of practitioners. The most famous of all special pleaders was William Tidd (1760–1847), author of *The Practice of the Court of King's Bench in Personal Actions* (1790–98) which was the leading authority for 50 years, *Practical Forms and Entries of Proceedings in the Courts of King's Bench, Common Pleas and Exchequer of Pleas* (1799), and other books on practice.

Special Procedure Orders. Orders made by a minister in respect of matters which would otherwise, and formerly, have required legislation by Private Bill. Such orders have largely, save in Scotland, superseded the system of Provisional Orders (q.v.). Under the Statutory Orders (Special Procedure) Acts, 1945 and 1965, special procedure applies if power to make or confirm orders is conferred on any authority and provision is made by the Act conferring the power that the order shall be subject to special parliamentary procedure. Provision is made for preliminary notices and enquiries, after these have been complied with, the order may be laid before Parliament. If a petition is presented against the order, it is referred for examination to the Lord Chairman of Committees and the Chairman of Ways and Means. If either House, on the Chairmen's report, so resolves, the order may be annulled, or the petition may be referred to a joint committee of both Houses. The committee may report the order as made, or with amendments, or that it should not be approved. If approved, the order comes into force when the report of the committee is laid before both Houses; if not approved, the order does not take effect unless confirmed by Act proceeding on a Public Bill.

Special traverse. A form of plea known at common law, consisting of two parts, the first or inducement consisting of affirmative matter, the second or *absque hoc* of denials or traverses. This form of plea was abolished in 1852.

Special verdict. In jury cases, findings by the jury on particular points of fact rather than a general finding in favour of either party. Argument is subsequently heard by the court on the effect of the findings and a decision is arrived at by the court on the proper interpretation of the facts thus found.

Specialized agencies. A large number of inter-governmental organizations, which operate under the general oversight of the U.N. and the activities of which are co-ordinated by the Economic and Social Council. Some of these long antedated the setting up of the U.N. The principal ones are the International Telecommunications Union, the Universal Postal Union, the International Labour Organization, the Food and Agricultural Organization, the International Monetary Fund, the International Bank for Reconstruction and Development, the International Finance Corporation, the United Nations Educational, Scientific, and Cultural Organization, the International Civil Aviation Organization, the World Health Organization, the World Meteorological Organization, and the Inter-Governmental Maritime Consultative Organization. Each has its own constitution, institutions and organs, and secretariat or bureau. All are organizations through which the members co-operate, but they are not supra-national in that their decisions do not *ipso facto* become law for their members though they can become law and create obligations for the members. The sanctions they can apply to their members vary.

Specialized committees of the House of Commons. From 1966 the House of Commons has appointed specialized committees to consider particular subjects or the working of particular departments of the executive. The committee on Science and Technology was subsequently re-appointed regularly, and others have been appointed, e.g. on Race Relations and Immigration, and on the Parliamentary Commissioner for Administration (q.v.), but the committees on departments were not continued. These committees have produced some valuable reports and have made M.P.s better informed about certain facets of government.

Speciality debts. Claims represented by bonds, mortgages, and debts secured by writing under seal which formerly ranked in priority to simple contract debts in the administration of the estate of a deceased.

Specific bequest or devise. A gift by will of a particular item of personal, or real, property as contrasted with a general bequest or devise.

Specific implement. In Scots law a remedy in a case of breach of contract whereby the defender is ordained specifically to implement his contract, and do what he undertook to do. While it is a general right, the court may decline to grant it to a pursuer if in the circumstances it would be unreasonable to do so.

Specific performance. An equitable remedy granted by a court to enforce against a defendant the duty of doing what he has undertaken by contract to do. The remedy was developed by the Chancery courts to deal with cases where damages recoverable at common law were not an adequate remedy. It is special and extraordinary in character and the court has discretion to grant it or to leave the parties to their rights at law, though the discretion is exercised on fixed principles in accordance with previous decisions. Jurisdiction to grant the remedy is now vested in all branches of the High Court, but as equity acts *in personam* a decree for specific performance will not normally be granted against a defendant outside the jurisdiction. It is appropriate to enforce a positive contract, where damages are not an adequate remedy, and to enforce an executory rather than an executed contract, and the court will not enforce the performance of contracts which involve continuing acts and would require the supervision of the court, nor for personal work or service, nor contracts not mutually enforceable, nor contracts merely ancillary to a principal contract itself unenforceable, nor of a contract to pay money, nor where the decree would be valueless or a payment of monetary damages would be adequate. To be enforceable in this way the contract must be certain in all material respects, not illegal, nor unequal, nor unequal or unfair. In certain circumstances a plaintiff may be entitled to specific performance with compensation for some immaterial inability exactly to perform the contract. Where specific performance could be ordered, the court may award damages in substitution for or in addition to specific performance.

Specificatio. In civil and Scots law, the mode of acquiring ownership where a new subject-matter has been created from materials belonging to another. If the new thing, e.g. a bronze statue, can be restored to its old form, the ownership remains with the owner of the materials, who must recompense the person who changed the form for his labour. If the new thing cannot be restored to its constituent elements, e.g. flour made into bread, the maker has property in the new species but must compensate the owner of the materials for having used them. See also CONFUSION OR INTERMIXTURE.

Specification. In relation to patents, a statement in writing describing in detail the nature of the invention. It may be provisional or complete, and will describe the invention, set out the claims of novelty made for it, and must give sufficient information to make the description intelligible.

Speech, freedom of. See Civil rights.

Speculator. A title given in mediaeval times to Durandus (q.v.).

Speculum et lumen juris canonici. A mediaeval title given to Joannes Andreae (q.v.).

Speculum juris. A title given in mediaeval times to Bartolus (q.v.).

Speculum Abbreviatum. A textbook of canon law based on the *Speculum Juris* of Durandus, attributed to Johannes de Stynna and written in Germany in the early fourteenth century. It deals with lawsuits in general, pleadings and legal documents, and legal maxims.

Spedalieri, Nicolas (1741–95). An Italian publicist, author of *De diritto del Uomo* (1791) and other writings on the rights of man.

Spelman, Sir Henry (?1564–1641). Lived as a country gentleman but developed antiquarian interests and became intimate with Camden, Cotton, and others. He entered Parliament in 1597 and was appointed by King James I a commissioner to determine unsettled titles to lands in Ireland. In 1613 he published *De non temerandis Ecclesiis,* an attack on lay expropriation of ecclesiastical property; his *History of Sacrilege* was not published till 1698. As a prelude to a great work on the bases of English law he prepared a dictionary of legal terms entitled *Archaeologus* (2 vols., 1626 and 1664). He then undertook his *Councils, Decrees, Laws and Constitutions of the English Church* (vol. I., (to 1066), 1639), the first attempt to deal systematically with the early documents relating to the Church and the basis of modern work on the subject. Vol. II was published posthumously, in 1644. He also founded an Anglo-Saxon lectureship at Cambridge. His last work was the *Original Growth, Propagation and Condition of Tenures by Knight Service* (1641). He also took part in public affairs, being a member of the council for New England and treasurer of the Guiana Company, a member of Parliament and member of several commissions of investigation.

Spencer, Herbert (1820–1903). English philosopher, author of *Principles of Psychology* (1855), *Principles of Sociology* (1876–96), *The Data of Ethics* (1879), *Man versus the State* (1884), and other works. He was an individualist of an extreme *laissez-faire* type, and an evolutionist. His *Ethics* may be his most valuable work. In his view government and law are completely incorporated in sociology, and his views on law are unimportant.

Spender, Sir Percy Claude (1897–). Australian statesman, judge of the International Court (1958–67), and President (1964–67).

Speransky, Mikhail Mikhailovich (1772–1839). Russian statesman who promoted administrative and financial reforms and proposed a new constitution in the period 1805–12. He prepared a *Complete Collection of the Laws of the Russian Empire* in 45 volumes (1830) and supervised a *Digest of the Laws* (15 vols., 1832–39).

Spiegel der deutschen Leute (or *Deutschenspiegel*). A mediaeval German law book of about 1250, based on the *Sachsenspiegel* (q.v.) but attempting to present Germanic common law, in particular south German law.

Spinning house. A prison at Cambridge used by the university authorities for the detention of women convicted of associating with undergraduates for immoral purposes. The power to detain them was abolished in 1849, but any such woman is punishable as an idle and disorderly person.

Spinoza, Baruch (or Benedict de) (1632–77). Dutch lens-maker and philosopher. Much influenced by Descartes he made a geometrical version of the latter's *Principia* and wrote an *Ethics, Tractatus Theologico-politicus* (1670) and other works. He was the father of modern metaphysics and moral and political philosophy. He made no formal study of law. He conceived of law as social control. Truly human life was possible only in an organized community or state, which depended on an agreement by citizens to obey the sovereign authority. But state sovereignty was never really absolute. A basic principle which he held was that Right was Might, though by might he understood human capacities and aptitudes, not merely force. The root idea of law was uniformity; it was a plan of living which served to render life and the State secure, prescribed by the community under sanction of a penalty. Spinoza's theory of positive law was influential in containing the seeds of Kant's system, and accordingly of most later metaphysical and sociological jurisprudence.

F. Pollock, *Spinoza, the Life and Philosophy*; R. Duff, *Spinoza's Ethical and Political Philosophy.*

Spiropoulos, Ioannes (1896–). Greek jurist, author of *Théorie générale du droit international* (1930) and *Traité théorique et pratique du droit international public* (1933).

Sponsalia per verba de praesenti (marriage by present consent to take each other for husband and wife). In canon law a mode of valid marriage, though not perfected until consummated by carnal copulation. It was consensual, consent having to be evidenced by words written or spoken before witnesses. It was distinct from betrothal, *Sponsalia per verba de futuro,* which was the usual preliminary to marriage, and if *copula* took place thereafter on the faith of the promise later to marry, the betrothal was converted into marriage thereby. Marriage by *de praesenti* consent and by promise *subsequente copula* were prohibited by the Council of Trent in 1563 but survived in Scotland till 1940.

Sponsio. In Roman law the oldest form of *stipulatio*, comprising a question and answer using the solemn verb *spondere*, viz. *spondes-ne? spondeo.* This form was always confined to Roman citizens, but in classical law question and answer could be expressed in other words, provided the answer agreed with the question.

Sponsio ludicra. In Scots law an agreement deemed made in sport, not seriously, and accordingly legally unenforceable. The category comprises principally betting and gaming transactions.

Spottiswoode (or **Spotteswood** or **Spotiswood**), **John** (1646–1728). Passed advocate in 1686. In view of the non-existence of a school of law in Edinburgh he established 'Spotiswood's College of Law' about 1703, himself becoming professor. For this college he composed a *Form of Process before the Lords of Council and Session* (1711) and *Introduction to the Knowledge of the Stile of Writs* (1707). He is sometimes credited with *A Compend or Abbreviat of the most important ordinary Securities* (1700), but this is sometimes also attributed to one Carruthers or to Sir Andrew Birnie, Lord Saline. He also published the *Decisions* of Lords President Gilmour and Falconer (1701). In 1706 he published the *Practicks of the Law of Scotland* of his grandfather, Sir Robert Spottiswood (q.v.). He died while putting through the Press his edition of Hope's (q.v.) *Minor Practicks* (1734) and left in manuscript two other works, a *Scots Law-Lexicon* and *Spotiswood's Practical Titles.* Most of his writings were reprinted several times and were very popular in their time.

Spottiswood, Sir Robert (1596–1646). Grandfather of John Spottiswood (q.v.), studied abroad and became first an extraordinary (1622) and later an ordinary (1626) Lord of the Court of Session, as Lord New Abbey. In 1633 he became Lord President. He supported King Charles I and acted as his Secretary of State in 1643. Taken prisoner when Montrose was defeated at Philiphaugh in 1645, he was executed at St. Andrews. He left a *Practicks of the Law of Scotland* published in 1706 by his grandson.

Springing use. A kind of limitation of an equitable estate in land recognized in equity, whereby a limitation to feoffees to the use of a named person on an event happening, e.g. when he married, operated to vest a legal estate in the person on the occurrence of the prescribed event. Since 1925 such an interest is an equitable interest only.

Spuilzie. In Scots law, the taking away of moveable goods from a person's possession, against his will and without order of law. Action lies for restoration of the goods and for all the profits which the owner could have made from the goods.

Spy. A person who does not wear military uniform or badges but seeks secretly to obtain information about a belligerent and communicate it to the other side. A spy, if caught, must be tried but may be punished by death or lesser penalty. If he rejoins his own forces but is subsequently captured by the enemy, he is entitled to be treated as a mere prisoner of war and not as a spy.

Squatter. A person who occupies buildings or other premises without title, such as tenancy, to do so, or whose title has terminated but who refuses to remove. Persons may squat in buildings by reason of inability to find other accommodation and may do so deliberately as a protest against shortage of housing in the area. A squatter is a trespasser and liable to criminal penalties if he forces entry against the opposition of the lawful occupier or if, having been warned, he fails to leave; he may also be arrested and removed. A squatter who, in England, encloses a piece of waste land and builds a hut on it and lives there may by prescription acquire a fee simple in the land enclosed.

Stadrecht. In mediaeval Europe, a compendium of the more important laws and regulations concerning a town and its citizens. Practically every town of any consequence had the right to draw up its own municipal *stadrecht* or body of law. Such might include charters, grants of privileges, constitutions of governing bodies, rules about the acquisition and loss of burgher's rights, regulations about fairs and markets, building restrictions, and even matters of more general application such as transfer of property. Early *Stadrechten* were usually unsystematic codifications. By the sixteenth century many were voluminous works, regularly revised and amended. Early *stadrechten* included those of Zutphen (1190), Middelburg (1217), Delft (1246), Leiden (1266), and Rotterdam (1299). One notable by reason of its quality was the *Rechtboek van Briel* by Jan Matthijsen, town-clerk of Briel, of the early fifteenth century. Many younger towns borrowed the customs of older towns, modifying them in accordance with their local requirements, so that families of *stadrechten* came into being; thus several towns borrowed from Leiden. When a new or difficult legal issue arose the daughter towns might send a delegate to the court of the mother town to obtain its opinion, but this practice disappeared when territorial appeal courts were established.

Stahl, Friedrich Julius (1802–61). German ecclesiastical lawyer, and statesman, professor at Wurzburg and Berlin. His major work *Die philsophie des Rechts nach geschichtlicher Ansicht* bases all law and political science on Christian revelation and maintains that a state Church must be confessional only. He sought to strengthen the Lutheran Church in Prussia.

Stahlberg, Kaarlo Juho (1865–1952). Finnish jurist and statesman, was professor of administrative law at Helsinki, 1908–18, and drafted the republican constitution of 1917 which was the basis of the constitution adopted in 1919. He was president of the republic, 1919–25, and was narrowly defeated for re-election in 1931 and 1937.

Stair Society. A Scottish society, taking its name from James Dalrymple, Viscount Stair (q.v.), founded in 1934 to encourage the study and advance the knowledge of the history of Scots law. It has published numerous hitherto unpublished works on Scots law and two valuable collaborative works, *The Sources and Literature of Scots Law* (1936) and *An Introduction to Scottish Legal History* (1958).

Stakeholder. A person with whom money or other property is deposited by the parties to a bet or wager pending the decision of the event on which the bet has been made.

Stallage. The liberty of pitching stalls in a fair or market; also the money payable for that right to the owner of the market.

Stallybrass, William Teulon Swan (1883–1948). Born Sonnenschein, he assumed the name Stallybrass in 1917. Called to the Bar in 1909 he devoted his life to teaching law at Oxford. He was not a great lecturer, but as editor of four editions of Salmond's (q.v.) *Law of Torts* won wide renown. He became Principal of Brasenose College in 1936.

Stammler, Rudolph (1856–1938). German jurist, professor at Leipzig, Giessen, Halle, and Berlin, heavily influenced by Kant, author of *Wirtschaft und Recht* (1896), *Lehre von dem richtigen Recht* (1926), *Theorie der Rechtswissenschaft* (1923), and *Lehrbuch der Rechtsphilosophie* (1928), and a very influential thinker. He set himself against the Marxist theories of economic determinism of law and justice and restated the problem of justice in terms of the social order. He set out to develop the idea of a right law in contrast to positivism, a kind of standard for distinguishing what among existing law is right law. Right law is positive law, the content of which possesses particular objective qualities. This exists in such general legal concepts as good faith, reasonable discretion, and the like; the method of judging is by critical reflection about the lawful end of the legal order. All right law is oriented towards the social ideal, the community of free-willing men. The philosophy is liberal, the community for which law exists having regard to individuals. The idea of right law is not equivalent to the idea of justice, though the two are closely related.

Stamp duties. Since the seventeenth century stamp duties have been levied on various kinds of instruments. The modern legislation imposes specified duties on specified kinds of instruments (not on the transactions effected or evidenced thereby, so that, if a contract can be and is entered into orally, it is not subject to stamp duty) under the sanction that, subject to certain qualifications, the instrument may not, except in criminal proceedings, be given in evidence or be available for any purpose, unless duly stamped. An instrument must be stamped according to its meaning and effect, not its name, and an instrument dealing with several matters is chargeable with duty separately in respect of each matter. Stamp duty must normally be paid by means of impressed stamps, but in certain cases adhesive stamps are permitted. In general instruments must be stamped before execution, but certain instruments may be after-stamped. The rates of duty on particular kinds of instruments are set out in the Stamp Act, 1891, as amended, and are in some cases fixed duties, e.g. on a deed not otherwise chargeable, and in other cases are *ad valorem,* related to the value of the transaction dealt with, e.g. conveyance on sale. In case of doubt a person may require the Commissioners of Inland Revenue to express the opinion whether any instrument is liable to any, and if so to what, duty, and in some cases this is required. Once stamped, or adjudged not liable to duty, it may be stamped with an adjudication stamp. An appeal lies to the High Court in case of dissatisfaction with commissioners' assessment.

Standards, Legal. Defined criteria by reference to which individual conduct may be judged. Standards may be either fixed, e.g. not more than X tons, or Y miles per hour, or Z persons; or be indefinite, fixed by a formula which permits individual judgment in the circumstances of the particular case, e.g. reasonable care, due diligence, the care of the reasonable man, the prudent trustee, the *bonus paterfamilias,* and so on. The statement of principles and rules of law frequently utilizes standards in their verbal formulations of rights and duties. The indefinite standard has been a great means of individualism of application of law and of taking account of circumstances, usual practice, and developing requirements. Indefinite standards are also largely used in legislation, e.g. fair rent, suitable alternative accommodation, unfair dismissal, and others. In each case it is for a court or judge to decide whether in the particular circumstances what is under consideration satisfied the standard or not.

Standing Civilian Courts. Courts consisting of an assistant judge advocate general, or in certain cases such an officer with two assessors, established in 1976 for the trial outside the United Kingdom of persons employed in the service of the armed forces or accompanying them but not subject to military law. They are comparable to magistrates' courts. They may try such civilians for offences for which a court-martial may try a civilian. A civilian convicted may appeal to a court-martial or petition a reviewing authority against finding, or sentence, or both.

Standing Committees. In the U.K. House of Commons the committee stage of Public Bills, other than money Bills, is taken in Standing Committees, unless the House orders otherwise. There is no limit to the number of such committees which may be established, but the greatest number that have sat at one time is nine. The selection of members for committees is made by the Committee of Selection and in each committee parties are represented in proportion to their strengths in the House as a whole. Each Committee, other than the Scottish and Welsh Standing Committees (q.v.) consists of 20 to 50 members appointed specifically for consideration of the Bill allocated to it; the average size is about 45. The composition of a Committee accordingly changes with each Bill. The debates are recorded verbatim, and the procedure and rules of debate of the whole House in committee apply. The Chairmen of Standing Committees are appointed by the Speaker from the Chairmen's Panel. They may select amendments for discussion and accept or reject a closure motion. The function of each Standing Committee is to discuss, clause by clause, the Bill committed to it, debate proposed amendments, and finally report it to the House.

Standing Orders. Both Houses of Parliament have separate bodies of Standing Orders, some formulating older practices and others based on the recommendations of various Select Committees reporting on procedure in various respects. Both Houses have Standing Orders Committees to determine whether Standing Orders may be dispensed with in particular cases. The procedure of the Lords more than of the Commons is regulated by Standing Orders. There are separate ones for Public and for Private Business.

Stannaries. In the thirteenth century English kings asserted rights over mines and metals, and in special districts the kings granted mineowners privileges in return for payment of dues. Thus, there developed franchise jurisdictions of which the most important were the Stannaries, districts in Devon and Cornwall where tin mining was carried on. In 1201 King John gave the Stannaries a charter and established a special Stannary jurisdiction, empowering the Lord Warden and his bailiffs to administer justice. Later charters gave the Stannary court jurisdiction over all causes arising in the Stannaries, except those affecting land, life, or limb. Courts of first instance were held by Stewards appointed by the Lord Warden, himself appointed by the Duke of Cornwall, at four places in each of Devon and Cornwall. They had criminal jurisdiction at great courts in spring and autumn, and main civil jurisdiction at three-weekly ordinary courts. Special customary courts were also held for the benefit of persons attending fairs. The Vice-Warden held a court to hear appeals from the Stewards and with an original jurisdiction in equity, and in the eighteenth century the Stewards' courts declined and it became common to commence actions in the Vice-Warden's court.

In 1836 and after statutes abolished the Stewards' courts and gave the Vice-Warden an original jurisdiction in law and equity, appeal lying to the Lord Warden, assisted by three or more members of the judicial committee of the Privy Council, and then to the House of Lords. The Court of the Lord Warden was merged in the Court of Appeal in 1873 and the court of the Vice-Warden abolished in 1898, the jurisdiction being transferred to the county courts.

The Stannaries had from the early sixteenth century Parliaments in Devon and Cornwall. The Cornish one consisted of 24 nominated by the councils of the four Stannary towns; the Devonshire one of 96 elected 24 from each of the four towns. These ceased to function in the eighteenth century but never seem to have been abolished.

Staple, Courts of the. In the Middle Ages the Crown had an interest to protect foreign merchants against the monopolistic and exclusive Gild merchants or associations of traders in and near particular towns. In the fourteenth century certain towns, Staple towns, were designated and only in them could trade be carried on in the more important articles of commerce; this system was consolidated by the Statute of the Staple, 1353. In each of the designated towns special courts were established for merchants resorting thereto, held by a mayor of the Staple and two constables, chosen annually. They had the assistance of two alien merchants. They were given jurisdiction in all kinds of pleas concerning debt, covenant, and trespass and the jurisdiction of the King's courts was excluded save in cases concerning freehold or felony, but were subject to the supervision of the Council. The law applied was the Law Merchant and not the common law. Cases involving aliens might be tried by a mixed jury, and special means were provided for recovering goods of which merchants had been robbed at sea, or which had been cast away and thrown up on the shore. These courts are now obsolete.

C. Gross and C. Hall, *Introduction to Select Cases concerning the Law Merchant* (Selden Socy.).

Staple Inn. One of the Inns of Chancery (q.v.).

Star Chamber, Court of. In the time of Edward III the royal council, sitting regularly for the despatch of business, is sometimes noted as having sat in the Star Chamber, one of the rooms in the Palace of Westminster. After the development of the Chancery with an independent jurisdiction, the Council seems usually to have sat in the Star Chamber for the disposal of other business.

The Tudor kings set themselves to master the lawlessness prevalent in England after 1485 and utilized the judicial powers of the Council. The Act *Pro Camera Stellata,* 1487, set out to establish a strong committee of the Council to deal with offences imperilling the safety of the State. It consisted of the Chancellor, Treasurer and Keeper of the Privy Seal, a bishop and a temporal lord from the Council, and the two Chief justices or two other judges. The Act empowered the summoning by subpoena and examination of persons accused of the offences set out in the Act. It did not prejudice the Council's wide jurisdiction and, being a committee thereof, the power of the Act and the mode of exercising them could be utilized by the Council.

In time the organization and work of the committee changed. In 1526 a distinction was drawn between the Council at court, which became the Privy Council, and the Council in the Star Chamber. The name is derived from the room at Westminster in which, formerly, starra or obligations of Jews were deposited. By the late sixteenth century the Star Chamber was a court distinct from but related to the Privy Council, with a distinct staff and procedure, and it absorbed some smaller statutory committees with limited judicial functions which had been created a century earlier. At that time it was an efficient and valuable court.

The Star Chamber jurisdiction was analogous to that of Chancery in principles and procedure, based on the inability of the common law courts to do full justice in criminal cases and cases of an extraordinary character. The civil jurisdiction comprised disputes between Englishmen and aliens, prize cases and other maritime matters, suits between corporations, and some testamentary actions. Most of these were gradually taken over by the ordinary courts. But the Star Chamber retained its criminal jurisdiction; it was a court of 'criminal equity' and dealt with riot, maintenance, fraud, forgery, perjury, libel, conspiracy, and generally with misdemeanours, particularly of a public character. It imposed heavy fines and imprisonment, and the pillory, whipping, branding, and mutilation.

Under the Stuarts the Star Chamber excited the jealousy of the common lawyers and was used as an instrument for the repression of the Puritans.

The court became very unpopular by reason of the tyrannical exercise and illegal extension of its powers, and it was abolished in 1641. It has subsequently, though not wholly justly, been a synonym for a tribunal of arbitrary tyranny.

There are several published volumes of cases in the Star Chamber.

C. Scofield, *Study of the Court of Star Chamber* (1900); I. S. Leadam, *Select Cases before the King's Council in The Star Chamber* (Selden Socy.) (1903–11).

Stare decisis. In Anglo-American jurisdictions there has developed the doctrine of *stare decisis* or of the binding force of precedent, to the effect that in deciding a case before him a judge not only has regard to precedents, i.e. to the principles applied by other judges in reaching the decisions in previously decided cases on the same or on closely related points, but that in certain circumstances he is *bound* to stand by the decided cases, and to accept and follow the principles of particular precedents, whether he personally approves them or not. This practice has been one of long growth and continuous development in English law. The main justifications are that it enables a judge to utilize the wisdom of his predecessors, that it makes for uniformity of application of law to similar cases, and that it makes the law predictable. It may, however, introduce an element of inflexibility into legal principles, and give rise to the drawing of fine distinctions between decisions on similar points.

In the application of this doctrine several questions arise, particularly: when is he bound and when free to decline to apply the precedent? by what element of the precedent is he bound? in what circumstances and by what techniques may he avoid holding himself bound?

In England there has developed a complicated body of rules to the general effect that the decisions of superior courts (House of Lords, Court of Appeal, Divisional Courts) bind for the future courts of the same rank and all courts of inferior rank to the court which decided the precedent, but that decisions of lesser superior courts (Divisional Courts, High Court (single judges)) do not bind later courts of the same rank but bind inferior courts (County courts). But the House of Lords has since 1966 asserted the power in exceptional cases to refuse to follow its own precedents, and there are certain exceptions in the Court of Appeal, where two precedents conflict, or a precedent is inconsistent with a later decision of the House of Lords, or the precedent was decided *per incuriam* in ignorance of a relevant statute or precedent. Decisions of the Privy Council are not binding on any British court but may be persuasive. The rule of *stare decisis* does not in general apply to special courts or tribunals, though they tend to follow their own prior decisions.

A precedent which is not binding on a future court under these rules may nevertheless be persuasive, i.e. influential, by virtue of the convincing quality of the judicial reasoning and the formulation and application of principle to circumstances. The persuasive quality of a precedent depends on such factors as the rank of the court which decided it, the personal eminence of the judges, the formulation and application of principles and similar factors. In particular, decisions of superior courts in other common-law jurisdictions may be persuasive on topics not regulated by any peculiar rule of the jurisdiction in which they are cited. Decisions of the Courts of Queen's Bench, Common Pleas, Exchequer, Chancery, and others which were superseded by the modern High Court may be only persuasive.

Where a precedent is deemed binding it is binding only in respect of its *ratio decidendi* (q.v.), but *obiter dicta,* dissenting judgments and other elements of the precedent may be persuasive. If the precedent is persuasive only, even the *ratio* is only persuasive.

When a subsequent court accepts the principle of a precedent it is said to *apply* it, where the same principle is utilized in relation to different facts, or to *follow* it, where a court accepts the principle of a precedent of a court superior to it, or utilizes the same principle in relation to a different class or kind of facts.

Notwithstanding the doctrine of *stare decisis* courts have some freedom in handling precedents which are binding under the principles already stated. If the later court is superior in the judicial hierarchy to the court which decided the precedent, it may *overrule* it, i.e. state expressly or impliedly that the precedent was wrongly decided or decided on a wrong principle. Thus the House of Lords can overrule any precedent; the Court of Appeal cannot overrule a precedent of the House of Lords but can overrule a decision of a single judge of the High Court. Less strong than overruling is *refusal to follow* the precedent; thus a single judge may decline to follow a prior single-judge ruling, leaving it to later courts to decide which conflicting precedent to follow. Regular refusal to follow a decision weakens its force. The later court may *distinguish* the precedent by pointing to some difference which it considers significant between the material facts underlying the decision of the precedent and the material facts of the instant case. If a precedent is repeatedly distinguished, it indicates that the principle of that precedent is not generally approved and the precedent tends to be regarded as confined more and more to its own special facts. Incidentally to refusing to follow, distinguishing, or overruling a precedent, a later court may *explain* a precedent, as by interpreting the *ratio* narrowly, confining it to its own facts, or otherwise putting a gloss on the decision. Distinguishing, refusing to follow, and explaining are the techniques which maintain some flexibility in the system and allow later judges to avoid mechanical application of the principles of precedents. It may be said that if a later judge approves a precedent and is willing to be bound by and to follow it, he treats it as binding; if he disapproves and wishes not to be bound nor to follow it, he seeks to explain and distinguish it. A judge is only bound if a precedent is indistinguishable, and much ingenuity has often been devoted to distinguishing precedents on very narrow grounds. Overruling is distinct from reversal of a decision on appeal. Overruling denotes the action of a later superior court condemning the principle of a decision as unsound; *reversal* is the action of an appellate court overturning the decision of the lower court in the same case, without necessarily disapproving the principle of law stated and applied, as a decision may be reversed on a matter of fact, but it may also involve disapproval of the principle stated and applied.

Overruling is retrospective in that a principle overruled is deemed never to have been the law, though this does not affect the actual case. This accordingly affects contracts, deeds, and transactions entered into prior to the overruling, and superior courts are accordingly reluctant to overrule a decision on the faith of which persons may have entered into commitments or regulated their affairs.

A case overruled may revive if the overruling decision is itself later reversed or overruled.

The weight and value of a precedent is accordingly determined by whether it has been applied, followed, and approved; or on the other hand not followed, disapproved, explained, distinguished, or overruled. Short of being overruled the authority of a decision may be strengthened or weakened by such factors as the unanimity or otherwise of the court and in the latter case the number and weight of the dissents, the eminence of the judges, subsequent judicial, professional, and academic approval of the decision, whether or not all relevant authorities were discussed, the quality of the report, and any specialties such as that a point was conceded, a party unrepresented, or the decision unconsidered.

None of these techniques need be utilized in relation to merely persuasive precedents, which can simply be ignored.

The principle of *stare decisis* has, largely by the influence of the House of Lords, been adopted in Scotland though greater freedom is there retained by the ability to refer difficult or conflicting precedents to a Court of Seven Judges or the Whole Court for decision.

In Commonwealth jurisdictions broadly the same principle of being bound by precedents of a higher court applies, persuasive value attaching to English decisions as the original home of the common law, which is the basis of the law of most Commonwealth jurisdictions.

In the United States the same general principle applies; in both federal and state courts a court is bound to follow a clearly applicable precedent of a court superior to it; but the Supreme Court may refuse to follow, or overrule, its own previous decisions.

State. A state or political society is an association of persons, living in a determinate part of the Earth's surface, legally organized and associated for their own government. The origins of states have probably to be sought in fighting between groups and the acquisition of territory, factors which give rise to cohesion and to the emergence of leaders and rulers. Ancient Greece knew the city state where each city and its inhabitants was a distinct entity, but particularly since the Reformation there have developed nation-states, where the territory and the population of each state is much greater.

The term 'the State' is applied to the community legally organized and personified. Thus personified and deemed a corporation, the State may own property, run industries, engage in commercial enterprises, provide services, prosecute offenders, and so on. Thus, in some countries prosecutions proceed at the instance of 'the State', 'the Commonwealth', or 'the people'.

In international law independent sovereign states are the subjects of the law, in which legal rights vest and which are subject to legal duties and responsibilities. In its own system of municipal law a state is normally a legal person or entity, frequently with rights, privileges, and immunities more extensive than enjoyed by any other legal person. In British law, for historical reasons, the rights and privileges are normally ascribed to 'the Crown' or 'the Queen' and the monarch as an incorporation rather than as an individual is treated as representing the community.

To be recognized as a state the group of persons must be substantial, usually millions, and have exclusive possession and control of a distinct portion of the Earth's surface, from which those people seek to exclude interference by others and within which they collectively seek to enforce their will. A nomadic people, however numerous and powerful, however well organized, would probably not normally be called a state. Generally the people have a common language and a substantial unanimity of customs, religious beliefs, and other characteristics, though there are exceptions to these general requisites. A state can comprise more than one nation, or group of people bound together by history, race, common traditions, and sentiments, and can comprise many substantial minority groups who are distinct in many cultural respects and who would frequently regard themselves as belonging to another nation. Thus, the United Kingdom comprises four national groups and many minority groups. Minority groups which remain for long in a state tend to identify with a predominant national group and to be assimilated to it, though they may long retain linguistic and cultural specialities.

In modern practice whether a community is deemed a state or not depends on general recognition by the existing group of other states. A new state may be formed by settlement of a distinct area of the Earth's surface and the adoption of legal and political organization, or, more commonly now, by the break-away from an existing state of a province or territory by the inhabitants thereof and their adoption of an independent organization, or by the grant of independence by one state to the inhabitants of a territory hitherto governed as a colony or dependency.

A state must have a relatively permanent legal organization, determining its structure and the relative powers of the major governing bodies, its constitution. This is frequently embodied in a fundamental legal instrument, the Constitution, round which develops a body of constitutional law interpreting the provisions of the Constitution and defining further the rights and duties of the main bodies of government and of citizens. The Constitution is itself pre-legal in that it does not have its source or basis in the law, but has its source and basis in facts, in usage, and on this basis creates a constitution and a body of constitutional law. This is most obvious in the case of a constitution such as that of the U.K. which has simply evolved and was never created by law, and that of the U.S.A. where the constituent states rejected the existing legal relationship with the U.K., asserted their independence, and created a constitution.

The most fundamental attribute in international law of the state of modern times is sovereignty, the possession of sufficient power to maintain independence externally, obedience and order internally, and the possession of supreme and independent authority to make, apply, and interpret a system of law within its territory.

A state may be unitary, when it has a single system of government applicable to all its parts, or federal when it has one system of government operating in certain respects in all its parts but also separate governments operating in other respects in distinct parts of the whole. The sub-divisions having separate governments are variously called states (as in U.S.A. and Australia), provinces (as in Canada), länder (in Germany), cantons (in Switzerland), or designated by other names. Both unitary and federal states have subordinate governments exercising delegated, local government functions in counties, districts and other sub-divisions.

States similarly may be classified according to their form of government into dictatorships or tyrannies, oligarchies, and democracies, and cross divided into communist, fascist, liberal, and socialist.

States may again be dependent or independent, the former being a constituent portion or dependency or non-contiguous portion of another (independent) state, the latter being wholly independent and sovereign by themselves. It is a question of terminology whether a dependent state is, or should be, called a 'state', or rather a colony, territory, or dependency. The countries of the British Commonwealth have generally evolved through various stages from colonies to dependent states and finally to independent states with only spiritual ties with the United Kingdom.

Membership of a state is constituted by citizenship, normally acquired by birth in the territory or, subject to conditions, by naturalization, and citizens normally enjoy rather greater rights than resident aliens, who are citizens of another state. Citizenship involves reciprocal rights and duties; the State owes protection to its members and they owe allegiance to it, implying duties of obedience, fidelity and services, and habitual submission to the will of the State.

Residence in the territory of a state involves certain rights and duties, but not full membership thereof.

The major functions of a state have traditionally been the defence of its territory and its inhabitants against external enemies, particularly other states, and the administration of justice within the territory, including legislation, but increasingly in modern times states have assumed numerous and other wide-ranging functions, particularly the provision of education, health, welfare, and social security functions, the control and management of natural resources and national assets, the provision of transport and communications services, the operation of certain industries deemed basic, and others. Down to the end of the nineteenth century the function of the State was largely negative, preventing disturbance, whereas in the twentieth century it has become increasingly positive, an agent actively seeking to organize a better life for its inhabitants. This increased and increasing extent of state intervention in the life of the community has necessitated and brought about a great increase in the volume of law and a swing towards greater importance of public law and the law of public administration. But it has also involved major interventions in private law, in such fields as landlord and tenant, employer and employee, manufacturer or seller and consumer.

A primary function of a state is the maintenance of law and order and the administration of justice within its borders including the exercise of legislative and judicial functions. A state may have one system of law and justice for its whole territory (apart from colonies or dependencies) or, as in the U.K., for historical reasons, several systems in district provinces or parts thereof; a federal state commonly has some differences in law or administration of justice in its component provinces, and some of the administration of justice is normally reserved to the provinces, but it could well have a single body of law administered federally.

The term 'the State' is also used as meaning the government of the State, as distinct from the subjects of the State, and in this sense there may be conflicts between the State and the citizen. In Britain the term 'the Crown' is frequently used for this sense of 'the State'. The State or the government consists of the officials who tax, coerce, and push the citizen around. In this sense the question of the rights of the individual against the State require consideration.

The State and the Constitution

The constitution is the formal structure of the State which has developed in a particular territory and community or which representatives of the inhabitants have created. The constitution determines the form of the State, monarchy, oligarchy, or democracy, the organs of the State and their respective powers, and their interaction. There must be a constitution though it may not be written but contained in a body of conventions, understandings, and practices. It must make provision for its development or amendment; if it does not, sooner or later a crisis will arise in which the constitution will be broken in a material respect. It depends on the constitution wherein in the particular state legal sovereignty lies, or, in short, who has the last word, the head of state, the assembly, the supreme court, the constitution itself, the body which may amend the constitution, or the citizens who elect the persons who decide to amend the constitution.

The State and the law

The State, if sovereign, cannot be bound by law whether externally created or created by the State itself but may make laws in whatever way and to whatever effect it chooses. An alternative view is that the State is bound by law, such as law of God or of nature; if this is so, the State is not fully sovereign. The difficulties about any such theory are of ascertaining what the law of God, or of nature, or other controlling force requires or forbids, and of enforcing it if the governors of the State choose to flout the prescriptions of God or nature. Kelsen has contended that law and the State are really the same; the State is merely a personified hierarchy of norms, and law is a more fundamental notion than the State, in that a legal order can take forms other than that of a state.

In international law a state is the relatively permanent population of a defined territory, politically organized and having an established government, and claiming as an entity to transact with other similar states. To rank as a state in international law such a community must be substantially sovereign and independent of external control; in modern conditions this can only mean independent of legal control by another state, but all states are in varying degrees subordinate to international and supranational organizations. Whether a particular community qualifies for acceptance internationally as a 'state' depends on facts and circumstances, general acceptance by already-recognized states, and the constitutions of international organizations. Thus membership of the L.N. and the U.N. is open only to fully sovereign states capable of fulfilling the obligations of membership, but a state not so qualified may be qualified for admission to other organizations and some organizations are open to dependent territories.

Two or more states linked by personal union and having the same person as head of State, such as England and Scotland, 1603–1707, or Great Britain and Hanover, 1714–1837, remain separate States in international law though the internal consequences may have international implications, e.g. in relation to nationality. Real unions, such as Austria–Hungary, 1867–1918, generally make a union a single State in international law. Federal States are generally single entities in international law, but sometimes constituent provinces have standing in international law for certain purposes.

The legal consequences of Statehood in international law are legal personality, independence, and equality.

Independence means absence of legal control or domination by any other State, though not necessarily absence of political or economic influence, and not absence of restriction by customary international law and treaty restrictions, or those imposed by membership of the international community and its principal organizations.

Equality means legal equality in respect of rights and duties, irrespective of size, military, political, and economic power and similar factors, and equal legal standing with other states in international law. But states may agree to endow certain other States with primacy or superior powers, as in relation to the permanent members of the Security Council, or accept the binding force of a majority decision, as under the U.N. Charter.

As corollaries thereto, each state has a right of jurisdiction over its territory and the population living there, a duty of non-intervention in the spheres of exclusive jurisdiction of all other states, and the dependence of obligations arising from customary international law and treaties on the consent of the party affected.

The right of jurisdiction involves legislative, administrative, and judicial powers; while jurisdiction is primarily territorial, in some respects it may be extra-territorial, extending to nationals of the State, wherever they may be. The right suffers exceptions in the cases of immunity, such as accorded to accredited diplomats, and of territorial privileges conceded to foreign armed forces in time of peace.

The duty of non-intervention involves that there are matters within the reserved domain or purely domestic jurisdiction of States, where the jurisdiction of the State is not bound by international law; the extent of this domain depends on international law and varies according to its development. It is commonly thought that no topic is irrevocably fixed within the reserved domain. The U.N. Charter provides that nothing therein authorizes the U.N. to intervene in matters which are essentially within the domestic jurisdiction of any State or require members to submit such matters to settlement under the Charter.

Philosophy of the State

Political philosophers from Machiavelli onwards have discussed the idea of the State. Machiavelli took for granted the power of the State as a single whole capable of being centrally controlled. Bodin saw sovereignty as the essential feature which distinguished the State from other organizations, but he made it subject to the constitutional law that determined succession. Hobbes sought to justify an absolutely powerful state, as being the only safeguard against the war of every man against every other. Locke, however, saw the State as resting on a social contract; a government was entrusted with power to protect life, liberty, and property. Rousseau insisted that the sovereign authority to legislate lay with the whole people. Hegel idealized the nation and venerated the State as the embodiment of national aspirations, and distinguished between the State and civil society; the State he glorified for its power, and this influenced later thinkers who exalted the nation-State, the heroic leader, military power, and war. Marx saw the State as part of the superstructure of society, but it was less real than the superstructure created by economic facts.

Kelsen has put forward the view that the State is merely a centralized legal order regulating the conduct of the inhabitants of the territory; it is the set of legal norms and the acts performed by natural persons in their capacities as organs of the State. He rejected the doctrine of State sovereignty and maintained that supreme authority must be given to international law.

E. Barker, *Principles of Social and Political Theory*; R. Maciver, *The Modern State*; H. Laski, *The State in Theory and Practice*; A. D. Lindsay, *The Modern Democratic State*, J. Mabbott, *The State and the Citizen*; W. E. Hocking, *Man and the State*; J. Maritain, *Man and the State*; G. Marshall, *Constitutional Theory*.

State, national or municipal law. The body of law of a distinct and independent nation-State, as distinct from international and supra-national law. In modern times principles and rules of law are predominantly state, or national, and applicable to persons present within a territorial State, rather than personal, attaching to an individual wherever he may be by virtue of his membership of a nation or tribe, or ethnic or religious group, or domicile or place of birth. French law, that is, is the law applicable by French courts in France, rather than the law applicable to Frenchmen wherever they may be, though there are circumstances in which the law of the country in which a person is domiciled continues to regulate his affairs though he is out of that country.

A country which is a single nation-State for the purposes of international relations may include within its frontiers several distinct systems of national law. This is the case in the U.K. and in the U.S.A., Canada, and other federal States. Conversely, it is theoretically possible for one system of law to be applicable in two or more independent States.

State or national law did not become a subject of study until nation-States had established themselves and less regard was paid to Roman and canon law which had till then been the sole subjects of legal study in universities. Swedish law was taught at Uppsala from 1620 and French law at the Sorbonne from 1679; the *Deutsches Recht* was first taught at Wittenberg in 1707, Spanish law in Spain in 1741, English law at Oxford in 1758 and at Cambridge in 1800, Scots law at Edinburgh in 1722.

While the bodies of law of different nation-States differ from one another in most cases, one system exhibits family resemblances to certain other systems (see FAMILIES OF LAWS) by reason of common historical origins, and a measure of community of thought on the part of jurists.

State, Department of, U.S. The first executive branch of the U.S. Government, established in 1789, and replacing the former Department of Foreign Affairs. Its responsibility is foreign affairs, and it is headed by the Secretary of State and a large staff.

See G. H. Stuart, *The Department of State*; S. F. Bemis, *The American Secretaries of State and their Diplomacy*.

State papers. The official documents of English and British government departments, dealing with all branches of government and administration.

State responsibility. The doctrine that a state as a legal person in international law may be liable for breach of engagement and illegal acts and be obliged to make restitution or reparation. In principle an act or omission which produces a result which is prima facie a breach of a legal obligation gives rise to a responsibility in international law, whether it rests on custom, treaty, or otherwise. Responsibility may be direct, for the action of the government and its executive officers, or vicarious, for the acts of their agents, or subjects, or even resident aliens, and even though unauthorized, such as the acts of rioters. Responsibility is in general determined objectively, for acts done by state organs or officials, despite absence of fault on their part. Grotius' view was, however, that the basis of state responsibility was *dolus* or *culpa*, and there is some support for this in cases. Acts in breach of legal obligation may be made by the legislature, judiciary, or executive organs or officials of the State in question. Defences include contributory negligence and assumption of risk; *force majeure* applies to acts of war; necessity is probably only exceptionally a defence.

Reparation may take the form of judicial declaration of the illegality of the act of the defendant state, satisfaction in the shape of acknowledgment of wrongdoing, or pecuniary statisfaction or compensation.

Questions of criminal responsibility may also arise, as for launching aggressive war or committing genocide.

States also have a responsibility to ensure the safety and freedom of aliens who enter their territory and at least permit them to obtain redress through the municipal legal system. Alternatively, or failing permitting a municipal law remedy, the State itself is liable for denial of justice. Similarly it must take steps to prosecute those offending criminally, on pain of being internationally liable for denial of justice.

State succession. The substitution of one sovereign State for another when territory is transferred from one State to another. This may take place by annexation, cession, emancipation, formation of a union or federation. Succession may be total, as where one State annexes another, or partial, as where one State acquires some territory from another, and the problems raised by the succession differ accordingly. In the nineteenth century the analogy of the death of a natural person was invoked, but this theory involved difficulties.

By virtue of its acquisition of sovereignty the successor State acquires all the property, rights, and claims of its predecessor in relation to the territory acquired, but not the private property of inhabitants. It may be a matter of dispute whether, e.g.

educational institutions are public property or private.

The legal system of the predecessor State continues in being until or unless modified, except for those parts of the constitutional and administrative law which are necessarily inconsistent with a change of sovereignty. The successor State may, however, legislate to change as much of the legal system as it likes.

The successor State may extend its nationality to persons living in the territory affected by the change and nationals of the predecessor State may lose nationality thereby; in each case it is a matter of the municipal law, and the result may be dual nationality or Statelessness. Treaties of cession sometimes give inhabitants an option.

Treaties which are personal and contractual in nature generally do not bind the successor State, whether the predecessor State is extinguished or merely diminished, though this principle has not always been applied where States have been formed by secession from a mother country. In some cases newly independent States have taken assignments of treaty rights and obligations. Treaties which impose a permanent status on a territory, e.g. dealing with rights of navigation of internal waterways, which are of the nature of the grant of servitudes, are generally succeeded to irrespective of the nature of the successor State. If a treaty has become part of customary law, it is binding on successor States. If a treaty is multilateral, the successor State may have to declare its attitude thereto. A successor State's treaties may extend automatically to the territory as extended by the succession, depending on the interpretation thereof.

Where a State succession takes place the successor State must respect the acquired rights of foreign nations, e.g. in real and personal property, but not in public offices or commercial monopolies. State contracts with individuals, such as economic concessions, lapse or are frustrated and the successor State is not bound to perform, but is subject to a liability to compensate.

The national debt of a State must be discharged by the successor State, in whole or in part, according as it has succeeded to the whole or part of the predecessor State; the matter has commonly been regulated by treaty. Local debts of autonomous provinces or cities pass to the successor State if it removes the fiscal autonomy of the area.

Problems similar to those of State succession arise in relation to succession of governments, particularly by revolution, and to succession of international organizations to one another.

State territory. The area of the surface of the world subject to the sovereignty of a particular State. The possession of some territory is an essential of an organized group of people being regarded as a State.

It falls to be distinguished not only from the territory of other States, but from *res nullius*, territory not yet placed under territorial sovereignty of any State, and *res communis*, such as the high seas and outer space, incapable of being subjected to sovereignty of any State. It is the area within which the State exercises its authority.

Territory includes the area of land, internal areas of water and waterways, the territorial sea, and the subsoil below, and the airspace above all of these. Warships, public vessels on the high seas, are treated as parts of the territory of the State and embassies in other countries are in many respects treated as if parts of the home State.

The boundaries of State territory are imaginary lines dividing the territory of one State from that of another, or from unappropriated territory, or from the open sea. These lines may be marked by physical marks or may run along natural features. Where the boundary line runs through water, the presumption is that it runs down the middle of a non-navigable river, or the middle of the Thalwag or mid-channel of a navigable river, or the mid-channel of a dividing strait, or the middle of land-locked lakes. In mountains the boundary is presumed to run along the watershed ridge. States sometimes assert that a particular line is their 'natural' frontier in the sense that by natural configuration it should be the boundary; but this is a political rather than legal concept.

Various legal rights may exist in respect of such territory: (1) sovereignty, or power of government, administration and disposal, (2) dominium or ownership of tracts of that territory, (3) jurisdiction or power of making and administering law for that territory and its inhabitants, (4) power of administration.

In Anglo-American law sovereignty and dominion are indistinguishable, but continental lawyers consider them distinguishable.

Sovereignty over territory may be divided between two or more States as in the case of *condominia*. or one State may have the significant powers in the government of another, as in cases of vassal States and protectorates. In a federal State sovereignty is divided and the territory of the member-states is collectively the territory of the federal State.

A State may acquire sovereignty over particular territory in various ways, by immemorial possession or prescription, by effective occupation or taking actual possession of territory hitherto *res nullius*, by cession effected by treaty, followed by effective occupation, by disposition by dominant powers (as after war), and probably by adjudication by an international court or arbitration. Prescription or long-continued possession, openly, peacefully and uninterruptedly exercised, may also confer title.

Discovery not followed by effective occupation confers no title, nor does merely symbolic annexation, though it may suffice in the case of *terra nullius*. Conquest is only a mode of acquisition if followed by express annexation, e.g. in proclamation or cession. By 'historic rights' a State may create title by creation of a specific custom coupled with absence of protest by the State.

Territory may be lost by cession, abandonment or dereliction, the operation of natural forces, subjugation by another State, annexation by another State, acquiescence in the claim of another or actual recognition of it, and revolt or secession from the State hitherto having sovereignty.

State territory may be subject of lease to another State or a right of occupation and use may be granted for a time or perpetuity. It may be conveyed in security and redeemed or the reversion extinguished. State servitudes may be created in the form of treaty restrictions on territorial supremacy whereby part of its territory is made permanently available to serve a certain purpose or interest of another State, but not merely in the form of grants to foreign corporations or individuals. Such servitudes may be positive or negative. Instances of servitudes are grants of fishery rights in the territorial sea of one State to another State and its subjects.

Rivers (q.v.) may be national, boundary, multinational or, if navigable from the sea, international, in which case vessels of all nations may navigate them. The flow of national rivers may be used by the local State, but in relation to other rivers no State may alter the natural conditions to the disadvantage of the territory of another State, as by diverting the flow. Lakes and inland seas may be national or multinational property, but only by treaty or by tolerance, international. Canals are national, though inter-oceanic canals are frequently declared open to vessels of all nations. Gulfs and bays, if small, are territorial, but if large, e.g. Hudson's Bay, may not be, and if bordered by the land of different territorial States, are not territorial. Straits may similarly be territorial, or boundary, or non-territorial.

Airspace above a State's territory is subject to its sovereignty and its jurisdiction may be exercised there, though aircraft other than on scheduled international air services may fly through a country's airspace and aircraft engaged on scheduled services may do so with permission. But sovereignty and jurisdiction do not extend indefinitely upwards, but probably only to the limit of the Earth's atmosphere, within which aircraft operate and beyond which satellites orbit. Outer space is, like the high seas, available for free passage by any State's satellites.

State Trials. Series of reports of important cases in British public and criminal law. The first edition was edited by Thomas Salmon in four volumes in 1719 and covers cases 1407–1710. Salmon also published an *Abridgment and Critical Review of the State Trials* (1720, 1731) and *A New Abridgment and Critical Review of the State Trials* (1737). The second edition by Sollom Emlyn in six volumes (1730) contains additional cases and comes down to 1727. There were later two supplementary volumes in 1735. There was a third edition in 1742, 9th and 10th volumes in 1766, and a fourth edition in 11 volumes by Francis Hargrave in 1775–81, covering the years 1387–1775. The fifth edition was in 33 volumes edited by T. B. Howell and T. J. Howell (1809–26), covers 1163–1820 and contains many additional cases. A new series, edited by Sir John Macdonald, in eight volumes (1885) includes cases 1820–58. The name is also given to some lesser series such as S. M. Phillips *State Trials* (2 vols., 1826).

Statelessness. The condition of a person who has no nationality, a situation where by a rule of municipal law, he loses one nationality but does not acquire another. It is possible to be stateless from birth, e.g. by birth in a State which bases nationality of the child on the *lex sanguinis* of the father and the father is a national of a State which bases it on the *lex soli*. The difficulty of a Stateless person is that he can call on no State to protect him against any other, and may find himself disentitled to enter or establish residence in any State at all. International attempts have been made to limit or exclude this condition, and a Convention of 1961 required parties thereto to grant nationality to persons born in their territories who would otherwise be stateless, subject to certain conditions. This applied to inhabitants of some former British dependent territories who, on the grant of independence, did not acquire citizenship of any other Commonwealth country.

Statement of claim. In a civil action in the High Court a written or printed statement by the plaintiff of the facts which he relies on to support his claim against the defendant, and the relief he claims. It replaces the former common law declaration and bill in Chancery. In a commercial cause a notification entitled 'points of claim' is the counterpart of a statement of claim.

Statement of defence. In a civil action in the High Court the defendant's pleading in reply to the plaintiff's statement of claim. It must deal with each allegation of fact made in the statement of claim, admitting or denying it, and, it may be, setting out further facts in explanation or avoidance of the facts stated by the plaintiff. It replaces the former pleas at common law and answers in Chancery. In a commercial cause a notification entitled 'points of defence' is the counterpart.

States General (*États généraux*). In pre-Revolutionary France the representative assembly of the three estates (nobles, clergy, and common people) summoned by the King to deal with major matters such as to approve taxation. It originated in the fourteenth century and was convened a number of times down to the early seventeenth century. During the Hundred Years' War (1337–1453) it was more important and asserted a claim to vote any new tax, demanded the removal of certain royal advisers and the establishment of a commission to share with the King in the government. But these moves came to naught. By the late fifteenth century the States had acquired by custom the right to be consulted on succession to the Crown, to assist in governing where the King was a minor, to give advice to the King and to consent to taxes imposed nationally. It became the practice for each estate to present a *cahier de doléance* (list of grievances). The monarchy was, however, able to govern without the States and the session of 1614 was the last till 1789. It revealed the impotence of the body in that the three estates could not co-operate. In 1789 in view of financial crisis the States were summoned again and the third estate established the National Assembly which began the Revolution.

States General. A body of delegates representing the provinces of the Dutch Republic, 1579–1795. It is today the name of the Netherlands Parliament.

States' rights. The political doctrine in the U.S. upholding the power of states against encroachment by federal authority. It is a development from colonial independence, and a doctrine which naturally commends itself to minorities and dissenters. The balance between federal rights and states' rights depends on the Constitution, legislative action by Congress and by the states, and judicial interpretation of these provisions. Initially the Constitution was rather strictly interpreted, but latterly the Supreme Court has viewed more favourably legislation relying on the powers impliedly given to Congress (Art. I, sec. 8), notwithstanding the powers reserved to states under the Tenth Amendment.

The issue of states' rights bulked largely in the discussions prior to the Civil War and was a major difference between the Union and the Confederate States.

The term is also given in other federal states such as Australia and Switzerland to the governmental powers retained by the individual states under the federal constitution.

Statham, Nicholas (or Stathum, John—*fl.* 1470—who may have been separate persons). He may have been a baron of exchequer about 1470. To Nicholas Statham is credited the authorship of the earliest *Abridgment*, printed about 1495, which contains under 258 titles (usually the names of the writs which commenced particular actions, in roughtly alphabetical order) about 3,750 notes on cases or on points of law. There is some attempt to arrange the notes chronologically and to bring together cases on the same point, but the arrangement is unscientific on the whole. The cases and points were probably drawn from manuscripts of the Year Books and abridgments of sections of the Year Books. Procedural points dominate and criminal law and trespass bulk largely. A translation of 1915, by M. C. Klingelsmith, is defective.

Stationers' Hall. The Hall of the Stationers' Company in London. Under the Copyright Act, 1842, registration of a published work at Stationers' Hall was necessary, not to secure copyright, but to confer the right to sue for an infringement. Accordingly 'Entered at Stationers' Hall' on the title page of books was a form of warning to pirates that the owner of the copyright could and might sue. This requirement disappeared with the Copyright Act, 1911.

Stationery Office, Her Majesty's. The department of government which supplies stationery and books for Parliament and executive departments. It is a very large publishing house and publishes and sells a great variety of governmental publications (Bills, Acts, Statutory Instruments, *Hansard*, White Papers, Blue Books, Royal Commission and other reports, maps, etc.). It secures the printing of all manner of materials required in government or for issue or sale to the public. The copyright in all government publications is vested in the Controller of H.M.S.O. He is also Queen's Printer of Acts of Parliament.

Status. In law, one of the attributes of legal personality (q.v.), particularly of the personality of natural persons, i.e. living human beings. Status is the particular person's standing in the law, which determines his rights and duties in particular contexts.

The status of a particular person is determined by the totality of his classifications under various heads, within each of which there are two or more mutually exclusive possibilities. The main heads are: (a) citizen and alien; (b) unborn, under full age, and of full age; (c) male and female; (d) legitimate, illegitimate, legitimated and adopted; (e) bodily and mentally capable, and bodily or mentally incapable; (f) single, married, divorced; (g) commoner and noble; (h) free and prisoner; (j) solvent and bankrupt; (k) lay and cleric; (l) civilian and member of the armed forces. Various other factors relevant to status in some legal systems at various times have been civil death, heresy, prodigality, serfdom or slavery, outlawry, race, colour, caste, whether *in potestate* or *sui iuris,* and others.

Most of the law is stated by reference to the standard individual who is a citizen, of full age, male, legitimate, and so on. The position of persons having different status is stated under each head as exceptions to those norms. To each aspect of status the law attaches certain rights and liabilities, capacities and incapacities, and the totality of a particular person's rights, capacities, and so forth is determined by the totality of his classifications.

A person's classification under most of these headings is determined by law, not by choice or agreement. In cases where a person can choose to change his status, e.g. by getting married, the consequences attaching to the changed status are still determined by law.

In relation to each legal issue which arises only some, or even only one, of the classifications, is relevant. Thus, capacity to make a contract may be affected by whether the person is minor or major, bankrupt or solvent, and mentally capable or otherwise, but not by sex or legitimacy. Some of the categories are important chiefly or only in public law.

Legal or juristic persons in general have the qualities of, and consequently equivalent status to, natural persons of full age, sound mind, and not subject to any incapacity, but the term status is not commonly used in relation to juristic persons.

The term status is also sometimes used of attributes which in particular respects give an individual special rights and duties, e.g. the status of an ambassador, a trustee, a registered medical practitioner, or a police constable. These are not true cases of status, but cases of special rights or duties attaching to persons holding particular offices, qualifications, or appointments, and only so long as they do so. These particular rights attach by virtue of office rather than by virtue of the person's legal status.

R. H. Graveson, *Status in the Common Law.*

Statuta Ecclesiae Antiqua. A compilation of canons made in Gaul probably about 480, possibly by one Gennadius of Marseilles. It followed the *Hispana Collectio* and is a major document for canon law and the liturgy of the period in Gaul. It includes a prologue, disciplinary canons, and ritual of ordinations and benedictions of persons of various ranks, and is of a reforming tendency.

Statuta gildae. A code of burgh laws, supplementary to the *Leges Quatuor Burgorum* (q.v.) which originated in Berwick in the thirteenth century.

Statute. Originally used of a decree of the sovereign, the word is now applied to Acts of Parliament (q.v.) both Public General and Local and Personal. There have been many collections and editions of the statutes. The first printed collection was the so-called *Nova Statuta,* covering 1327–1483; of 1484, followed by the *Antiqua Statuta* or *Vetera Statuta* from Magna Carta (1215) to 1327; of 1508, many times reprinted. The Second Part of the *Vetera Statuta* (1532) contains statutes prior to 1327 not included in the *Antiqua Statuta.* Another notable collection was William Rastell's *Collection in English of the Statutes from the beginning of Magna Carta now in force* (1557), of which many editions appeared down to 1621.

Of many later editions mention should be made of Keble's *Statutes at Large,* 1215–1676, continued later to 1733; Hawkins' *Statutes at Large* 1215–1734, continued to 1757; Cay's *Statutes at Large,* 1215–1757 continued to 1773; Ruffhead's *Statutes at Large,* 1215–1764, continued to 1800; Runnington's new edition of Ruffhead's *Statutes* to 1785, continued to 1801; Tomlins and Rathby's *Statutes at Large* 1215–1800, continued to 1869; and Pickering's *Statutes at Large,* 1215–1761, continued to 1806 and further continued to 1869. The *Statutes of the Realm,* 1225–1713, published under the aegis of the Record Commissioners is the fullest and most accurate edition.

The Statutes Revised were prepared under authority of the Statute Law Committee and omit repealed matter, and also local, personal, and private Acts. There are three editions, 1236–1878 in 18 volumes (1870–85); 1236–1886 in 16 volumes (1888–1901) with supplementary volumes to 1920; and 1236–1948 in 32 volumes (1951).

The *Statutes in Force* consists of pamphlets containing Acts so far as unrepealed, periodically replaced when further amended, and contained in loose-leaf binders. The pamphlets may be bound as the user chooses, though an arrangement is recommended.

See Sweet & Maxwell, *Guide to Law Reports and Statutes.*

Statute barred. The term applied to causes of action which are no longer legally enforceable by reason of the lapse of various periods of time, prescribed by Limitation Acts and Prescription Acts.

Statute Book. An inexact term for the whole body of statute law. In earlier times there are documents entered on the Statute Rolls (q.v.) but these are inaccurate. The early printed editions of the statutes were not authentic and included documents taken from the Parliament Rolls and other sources. In the nineteenth century the Record Commissioners had difficulty in deciding whether many documents were legislative or not, but their acceptance of some early materials and inclusion of them in the *Statutes of the Realm* (q.v.) has raised a presumption in favour of the authenticity of these materials as being legislative.

Today two prints on vellum are made of each Act of Parliament once it has been passed and received the Royal Assent, endorsed with the words of the Royal Assent and signed by the Clerk of the Parliaments, and preserved, one in the House of Lords and one in the Public Record Office. Copies are printed by the Queen's Printer and those passed in a particular calendar year (till 1940 those passed in a session of Parliament) are issued in one or more bound volumes. But there is no such thing as 'The Statute Book'.

Statute Labour. The requirement, formerly imposed by statute on the inhabitants of parishes, to do stated amounts of work each year on the highways repairable by the parish, or to pay a composition in lieu. The last Statute Labour Act was in 1773.

Statute law. The body of principles and rules of law laid down in statutes and statutory instruments as distinct from the common law, or the body of principles and rules of law developed and stated by judges in their opinions delivered on deciding particular disputes or claims, or derived inductively from examination of various opinions in various relevant cases. In theory statute law is a superior form of law in that it states general principles in an authoritative verbal formulation, whereas common law principles have to be distilled from judicial opinions and their verbal formulation is not authoritative. But in practice many statutes state law in very complicated and difficult language, which requires interpretation before the meaning of the legislative provision, as applicable to the facts of a particular case, can be ascertained. In authority statute law is certainly superior in that a statutory provision may amend, overrule, or abrogate any inconsistent principle or rule of law expressed in common law, authoritative book, textbook, or customary practice.

Statute Law Commissions. In 1833 a Royal Commission was appointed, one Commissioner being John Austin (q.v.), to digest into one statute all the statutes touching crimes and into another all the common law on the subject, to consider how far it might be expedient to combine both of these into one body of the criminal law, and how far it might be expedient to consolidate the other branches of the existing law for England. They made a general report in 1835 and seven reports on criminal law down to 1843. The Commission was dissolved in 1845.

In 1845 a second Commission was issued to Commissioners to consider the previous reports, complete the unfinished eighth report of the earlier Commission and prepare a Bill or Bills for codifying the whole criminal law. This Commission completed the eighth report and made five more of their own, appending a draft Bill containing an entire Digest on the Law of Crimes and Punishments. Their last work was a report on procedure in 1849. Several Bills were prepared based on these reports but did not result in legislation.

In 1853 Lord Chancellor Cranworth appointed a temporary Statute Law Board which presented three reports. In 1854 this Board was superseded by a Statute Law Commission, which comprised many distinguished judges and lawyers and presented four reports, 1854–59. The first Statute Law Revision Act, 1856, was based on their recommendations. There was much Parliamentary criticism of the work of these Commissions and a Register of Public General Acts, 1800–58, was published in 1859. On this basis another Statute Law Revision Act was passed in 1861 and seven criminal law consolidation Acts, based on the reports of the Commissioners of 1833 and 1859. A further Statute Law Revision Act was passed in 1863 and since then there have been numerous such Acts.

Statute Law Committee. In 1868 Lord Chancellor Cairns intimated his intention to have published an edition of the Statutes containing only the Statutes still in force after the first three Statute Law Revision Acts, and nominated a committee to superintend the execution of the work. This committee, under the name of the Statute Law Committee, has continued to the present to superintend Statute Law revision, and the publication of the *Chronological Table* and *Index to the Statutes in Force*. It has also initiated proposals for many consolidation Acts. The Committee dealt originally only with English, British, and United Kingdom legislation. In 1897 it was reinforced by Scottish members and effected measures for the revision of pre-1707 Scots Acts and the publication of editions of Scots Acts as revised. Since 1891 there has been a Statutory Publications Office responsible to the Committee for preparation of editions of the *Chronological Table of the Statutes, Index to the Statutes,* the *Statutes Revised*, the *Statutes in Force* and annual volumes of *Statutory Instruments.*

Statute Law Revision. The process of systematically going through the statute law and repealing, so far as not previously repealed, enactments which are spent, superseded, obsolete, or otherwise ineffective. The first Statute Law Revision Act, of 1856, was based on the recommendations of the first Statute Law Commission, 1854–59. Subsequent similar Acts of 1861, 1863, and various later years continued the process and it is now an essential instrument of law reform. The process is supervised by the Statute Law Committee (q.v.). Consolidation Acts (q.v.) effect a similar purpose by replacing many Acts of different dates by one modern measure.

Statute merchant. A bond acknowledged before the chief magistrate of a trading town pursuant to the statute *De Mercatoribus*, 1285. This was intended to encourage trade by giving a quick remedy for recovering debts. Each statute merchant had to be sealed with the seals of the debtor and of the King, and be enrolled, so that it became a bond of record. In the event of failure to pay by the debtor on the day assigned, execution might be awarded without mesne process to summon the debtor, or introduction of proofs. The debtor might be imprisoned, his goods seized, and his lands delivered to the creditor till the debt was satisfied from the rents and profits, during which time the creditor was known as tenant by statute merchant.

Statute of Frauds, 1677. An Act passed to minimize the frauds involved in contracts at a time when the law of evidence was undeveloped and the parties to a litigation not competent witnesses. The last of four attempts to prevent frauds and perjuries, it was framed by Lord Nottingham. It required certain conveyances of interests in land, wills of real estate, declarations or assignments of trusts, and certain classes of contracts, to be evidenced in prescribed written modes. The policy and drafting of the Act have been considerably criticized subsequently and a great deal of case-law grew up as to its interpretation. Much of the Statute has been repealed or replaced by later enactments.

Statute of Uses, 1535. From about the thirteenth century the practice developed in England of one man holding land to the use of another, and from about 1400 the Chancellors recognized and protected the rights of the beneficiary under a use, who had no legal right at common law, and thus developed the concept of his equitable rights in the land. The practice of conveying land to feofees to uses diminished the payment of feudal dues and by the Statute of Uses, 1535, Henry VIII attacked this diminution. The Act provided that when any person was seised of lands to the use of any other person the *cestui que use* (beneficiary) should be deemed to have lawful seisin of the land to the extent of his interest in the use. The Act was a serious grievance in increasing liability to the feudal incidents of land tenure and at least hindering the devise of land by will. The Act was circumvented by the device of conveying lands 'to A to the use of B to the use of C'. The Act passed the legal estate to B, but from 1634 (in *Sambach* v. *Dalston*, Tothill, 188) the courts held that such a conveyance gave C an equitable estate. Subsequently the second use was always designated a trust and the first feofee was omitted, the conveyance being made 'unto and to the use of B in trust for C'. By the time this had developed the feudal dues had been abolished and no person was concerned to preserve or enforce the purpose of the Statute. The Statute was repealed in 1925 as now spent.

Statute Roll. The rolls in the Public Record Office covering the years 1278–1431 and 1445–68 containing the statutes of the Parliament of England and also many documents which have never been regarded as of statutory force. The record was continued after 1468 by the *Inrollments of Acts of Parliament* (q.v.) from 1483 to the present or by the original Acts from 1497 to the present.

Statute staple. A bond of record acknowledged before the mayor of the staple in the presence at least of one of the constables of the staple and sealed with the seal of the staple, under the Statute of the Staple, 1353.

See also STATUTE MERCHANT.

Statutes at Large. A phrase apparently first applied to an edition of the statutes of England from 1215 to 1587 made in 1587 by one Christopher Barker, to distinguish his collection from prior collections. J. Rastell published two abridgements of some or all of the statutes and W. Rastell published an edition of the *Statutes at Large* in 1618. Later other private collections, such as those of Keble, Hawkins, Cay, Ruffhead, Runnington, Tomlins and Raithby, and Pickering, were also entitled *Statutes at Large*, though they were somewhat abridged. The term accordingly does not denote any particular collection or edition, nor imply total absence of abridgement, but only a degree of condensation less than in a collection expressed to be an abridgement.

Statutes, Editions of. Initially statutes were printed individually. Early collected editions distinguished between *Nova Statuta* (from Edward III (1327)) first printed by Machlinia about 1485, and *Vetera Statuta* (down to 1327) first printed by Pynson in 1508. In the sixteenth century there were numerous sessional publications, collections, and abridgements (see also ABRIDGEMENTS). The earliest attempt to cover the whole ground was Berthelet's two volume edition (1543) of all the statutes from Magna Carta to date. In 1557 William Rastell issued his *Collection of the Statutes* from Magna Carta now in force, which went through many editions. In 1618 Pulton issued a *Collection of Sundrie Statutes from 1225* but it was defective.

In 1667 the King's Printer issued an edition of

the statutes at large in force from 1641 to 1667 and Thames Manby issued a similar work, based on Pulton's edition. In 1673 Manby produced an abridgement under alphabetical heads of the statutes of Charles I and II, and in 1676 Joseph Keble issued an edition of the statutes from Magna Carta to date.

In 1735 serjeant Hawkins published an edition of the statutes from Henry III to 7 George II. It was followed by Cay's edition, down to 30 George II, in 1758, Ruffhead's edition of statutes down to 4 George III (1762–65), Danby Pickering's edition of statutes down to 1 George III (1762–66), Runnington's new edition of Ruffhead down to 1785 (1786), Tomlins and Raithby's Statutes at Large from Magna Carta to 1800 (1811). All these editions were continued by supplementary volumes.

In Scotland the first collection of statutes was the *Black Acts, 1424–1564* (1566) followed by Sir John Skene's *Lawes and Actes of Parliament, 1424–1597*, and Sir Thomas Murray of Glendook's *Laws and Acts of Parliament 1424–1681*. Sir James Stewart of Goodtrees published in 1702 an *Index or Abridgment of the Acts of Parliament*, Lord Kames also published an *Abridgment* in 1757, and John Swinton an *Abridgment of Acts relative to Scotland from 1707 to 1755* in the latter year.

In 1800 a select committee of the House of Commons recommended that a complete and authoritative edition be published, and the Record Commission prepared editions of the English statutes down to 1713 (9 vols. 1810–22) and the Scots statutes 1424–1707 (11 vols. 1814–24, plus Vol. I in 1844). Though better than earlier editions, this was in some respects defective. Sessional volumes of *Statutes at Large* were issued from 1801 to 1920. *Public General Statutes* have been issued annually since 1831. The *Statutes Revised* (1st edn. to 1878, 2nd edn. to 1886, extended to 1900, 3rd edn. to 1948) print the statutes as amended to the date of publication. The *Statutes in Force* is in loose-leaf format and continually being revised by the issue of replacement pages. There are also various unofficial collateral editions, *Chitty's Statutes of Practical Utility*, *Halsbury's Statutes of England*, and *Current Law Statutes*.

Statutes in Force. An official revised edition of the statutes of the United Kingdom and of England, Scotland, Northern Ireland, started to be published in 1972, in which each Act is printed as a separate booklet and the groups of Acts on particular subjects are contained in a binder. When Acts are heavily amended the booklet is replaced by a new one containing a print of the Act as amended. This edition will be constantly self-renewing and once complete will be substantially up-to-date. It supersedes the *Statutes Revised* (q.v.).

Statutes of the Realm. An edition in nine folio volumes, published by the Record Commission between 1810 and 1825, of the Statutes or Acts of the Parliament of England and of Great Britain from 1235 to 1713. It is based on manuscript collections and 61 printed editions, but contains incorrect evaluation of sources, mistakes in transcription, mistranslations, and inaccurate dates, particularly in respect of the *Statuta Incerti Temporis*, some of which are not truly legislation at all and which are grouped between the statutes of Edward II and Edward III. There is a gap between 1641 and 1660, covering the period of the Commonwealth and Protectorate.

The Interpretation Act, 1889, made the *Statutes of the Realm* authoritative as regards statutes passed before 1713, when cited in Acts subsequent to 1889.

Statutes of Westminster. At least four statutes of the English or U.K. Parliaments have been given this title.

The Statute of Westminster I (1275) in 51 chapters made many changes in procedure, some designed to protect subjects against the King's officers, and it may indeed be regarded as a supplement to Magna Carta. It protected the property of the Church, provided for freedom of popular elections by the freeholders in the county court, contained a declaration to enforce the provision in Magna Carta against excessive fines, corrected abuses of tenures, regulated the levying of tolls, and amended the criminal law.

The Statute of Westminster II (1285) extends to 50 chapters. Chapter 1, *De Donis*, provided that where a tenement of land was given on condition, the donee should not have power of alienating it, but it should descend to his issue or revert to the donor, if there were no issue; c. 18, established the writ of *elegit* whereby a judgment creditor could take the debtor's chattels and half of his land and the rent, an alternative to *fieri facias*; c. 24 provided that when in future a writ was found in one case but not another involving the same point of law, the Chancery clerks should frame a writ or adjourn the matter to the next Parliament to have a writ framed. This provided a regular procedure for the development of the common law by enlarging the available writs in defined circumstances. The Chancery clerks did not regard this as giving them wide powers; the only major extension of the law effected as a result of the Act was the writ of entry *in consimili casu*. Chapter 30 regulated the system of *nisi prius* justices, and c. 31 allowed challenge of the court's ruling by a bill of exceptions sealed by the trial judge. Other chapters were less, but still, important.

The Statute of Westminster III (1290), otherwise called *Quia Emptores*, provided that every freedman should be at liberty to sell his lands, the purchaser to hold them of the seller's lord, and prohibited subinfeudation, whereby the purchaser would have held of and under the seller, thereby defeating the superior lord's claims to the incidents of tenure. It made the future creation of seignories and manors impossible, and first authorized conveyances of feudal land.

The Statute of Westminster IV (1931) was passed to give effect to the changed relationships between the United Kingdom and the self-governing Dominions (Canada, Australia, New Zealand, South Africa, Eire, and Newfoundland) which had developed, particularly since 1914, and which was recognized, at the Imperial Conference in 1926, by the statement of their equality of status but union by their common allegiance to the Crown and membership of the British Commonwealth. The Statute removes various surviving rules indicative of subordination to the United Kingdom.

When British colonies were established and subsequently endowed by the United Kingdom Parliament with legislatures, those legislatures were subordinate and their powers depended on the statute, Order in Council, or Letters Patent granting them. Moreover, it was a rule of common law, and sometimes enacted in a statute creating a constitution for a colony, that a colonial statute was invalid if repugnant to fundamental principles of English law or inconsistent with U.K. legislation applying to the colony concerned or all colonies, but not invalid merely if altering non-fundamental rules of English law. The Colonial Laws Validity Act, 1865, provided that a colonial law repugnant to any U.K. statute extending to that colony or Order made thereunder should be invalid to the extent of the repugnancy but not otherwise, but not if merely repugnant to English Common law.

The Statute of Westminster, 1931, was passed to take account of the fact that certain colonies had attained representative and responsible self-government and had virtually attained independence of the U.K. Parliament. It was passed on the recommendation of the Imperial Conference, 1930, after the communication of resolutions of the Parliaments of Canada, Australia, New Zealand, Union of South Africa, Irish Free State, and Newfoundland.

The Act deals with legislative powers only. It provides that the Colonial Laws Validity Act should not apply to any law made after the 1931 Act by the Parliament of one of the six Dominions and that no Dominion legislation should be void by reason of repugnancy to the law of England, that the Parliament of a Dominion had power to make laws having extra-territorial application, and that no statute passed after the 1931 Act should be deemed to extend to a Dominion as part of the law thereof unless expressly declared in the Act that the Dominion had requested and consented to the enactment thereof.

The main provisions of the Act were not to apply to Australia, New Zealand, and Newfoundland until these Dominions, by their own legislation, adopted the specified sections: Australia did so in 1942, and New Zealand in 1947, while Newfoundland had not adopted the sections prior to becoming a Province of Canada. The Act ceased to apply to South Africa and Eire (later Republic of Ireland) when they left the Commonwealth.

K. C. Wheare, *The Statute of Westminster and Dominion Status.*

Statutes merchant and statutes staple. In mediaeval times merchants were much concerned to have machinery for quick and certain judgments for debt, and early adopted the system of insisting on judgment for repayment before actually paying money. In 1283 the Statute of Acton Burnell made provision for the enrolment of mercantile debts in principal towns, where the mayor was to keep a roll and the clerk was to draw up a deed and give it to the creditor. On default the mayor was to order sale of the debtor's chattels and burgage lands. In the Statute of Merchants 1285 the execution was improved and the process was to commence with imprisonment of the debtor. Recognizances entered into under this Act were termed statutes and a creditor in possession was said to be a tenant by statute merchant. This remained the principal form of security during the Middle Ages. Machinery was similarly established in every staple town in 1353, but the officials who took the recognizances were the staple officers and initially they were confined to operations in the staple commodities, such as wool and leather. Both statutes merchant and statutes staple came later to be used by non-merchants.

Statutes Revised. In 1868 Lord Chancellor Cairns, the ground having been cleared by the passage of the first three Statute Law Revision Acts, intimated his intention to have published an edition of the statutes containing only provisions in force, and appointed the Statute Law Committee to superintend the work. The first edition of the *Statutes Revised* was published in 15 volumes in 1870–78, including all statutes in force down to 1868. Three later volumes continued the edition to 1878 and substituted 18 volumes for 118. After other Statute Law Revision Acts a second edition of the *Statutes Revised* was published in 16 volumes from 1888 to 1901, containing the statutes as revised to 1886. Subsequent supplementary volumes brought the law down to 1920. The third edition of the *Statutes Revised* was published in 32 volumes in 1950, stating the law as amended to the end of 1948. It does not contain pre-1707 Scots Acts, of which a separate volume, *The Acts of the Parliaments of Scotland, 1424–1707* (1965) sets out the Acts still in force and as amended to 1965. In 1972, in place of a further edition of the *Statutes Revised*, publication was begun of the *Statutes in Force* (q.v.).

Statutory duty, Breach of. Where a statute imposes a duty to do or not to do something, it may itself specify the means of enforcement and the sanction for non-compliance, in which case this is presumed to be the only remedy, or the Statute may not so specify. In the latter case, if the statute affirms a liability which existed at common law and gives a special and peculiar remedy, this may co-exist with the common law remedy; if the statute creates a liability not previously existing at common law but provides no particular remedy, common law remedies may normally be resorted to for enforcement; and if the statute creates a liability not existing at common law and provides a special remedy, prima facie the obligation can be enforced in that manner only. In the absence of contrary provision, every breach of a statutory requirement is indictable at common law. In England the Attorney-General may at any time apply for an injunction against a breach of statutory duty or the infringement of a public right. Where a statute creates a public duty and provides no special remedy, the High Court in England may grant a mandamus that justice be done.

Statutory obligation. In many circumstances statutory provisions impose duties on persons and for the non-implement of the duty the prima facie sanction is criminal prosecution and imposition of a penalty. In some circumstances it has been held that a statutory duty is owed not only to the public but also to persons injured in consequence of the non-implement thereof, that it has consequently created a statutory obligation to those persons, who may in a civil action claim damages for the loss to them by reason of the non-implement. Whether a particular statutory provision creates such a statutory obligation, enforceable by any person to whom the duty was owed, depends entirely on the interpretation of the particular statute. Where such a statutory obligation exists it normally co-exists with an obligation arising from common-law tort, though the precise duties of care may not be identical.

Statutory instruments. The modern British term for the major class of subordinate legislation, formerly known as Statutory Rules and Orders. The statutory definition of statutory instruments is obscure. It comprises orders made under statutory authority by Her Majesty in Council as Orders in Council, and rules, regulations, and orders of a legislative character made by ministers, departments, and other persons or bodies authorized by statute to do so, where the parent Act provides that powers may be exercised in one of these ways, and orders by the Queen in Council or a minister confirming or approving subordinate legislation. It does not cover, on the one hand, statutes or Acts of Parliament, nor on the other hand provisional orders, special procedure orders, local authority by-laws, nor some kinds of sub-delegated legislation. Statutory instruments are published separately and in annual collected volumes under authority of the Statute Law Committee, but certain instruments are not required to be published.

Statutory Instruments Revised. A set of volumes containing all the statutory instruments in force at a stated date, grouped according to subject-matter, taking account of amendments, and omitting instruments revoked or spent.

Statutory interpretation. The process of determining what a statute or statutory instrument means in particular circumstances, and the body of guiding principles developed by the courts to assist in the process.

P. B. Maxwell, *Interpretation of Statutes.*

Statutory Rules and Orders. Formerly the main class of subordinate legislation, they were in 1948 renamed statutory instruments (q.v.). From 1890 they were published in a form corresponding to that of the annual statutes, and an index was periodically published, while rules from earlier years were collected as Statutory Rules and Orders Revised. The Rules Publication Act, 1893, required prior notice of intention to make rules in certain cases and publicity for draft rules, and provided for the printing, numbering, and sale of statutory rules.

Statutory trusts. Land held on the statutory trusts is held on trust to sell the same and stand possessed of the net proceeds of sale and of the net rents and profits until sale, upon such trusts and subject to such powers and provisions as may be requisite for giving effect to the rights of the persons interested in the land.

Statutory undertakers. Persons authorized by statute to construct and operate public utility undertakings such as the provision of gas, electricity, or water.

Statutum Walliae (Statute of Wales). By this Act of 1284 Wales and its inhabitants, hitherto feudally subject to the Kings of England, were wholly annexed and united to England. Material alterations were made in Welsh laws, but certain provincial immunities were retained, such as equal partition of lands on death among male issue. The Act established six counties according to the English model, with Justiciars of Chester and Snowdon each having responsibility for three. It established English criminal law for certain offences and introduced trial by jury. It provided for retaining Welsh procedure in real and personal actions. The statute did not indicate any intention to discriminate legally between Welsh and English, but Welshmen were excluded from the privileges of life in most boroughs as they were appendages of English castles and this gave rise to dislike and distrust, and, in the marcher lordships, flexibility of land tenure led to discrimination.

Staunford, Stamford or Stanford, Sir William (1509–58). Called to the Bar in 1536, sat in Parliament, and in 1551 was placed on a commission to resolve on the reformation of the canon laws. He became a serjeant in 1552 and a judge of the common pleas in 1554. He was a learned lawyer and author of *Les Plees del Coron* (1557), in three books, and heavily indebted to Bracton and Britton; it has a high reputation, was the first attempt to give a connected account of the criminal law, and influenced later writers on this branch of law. He also wrote *An Exposicion of the Kinge's Prerogative* (1567) based on matter from Fitzherbert's *Abridgment* and other writers, but showing independent thought, which was much relied on by later writers, and is said to have edited the earliest printed version of Glanville's *Tractatus*, published about 1555.

Stay of proceedings. A suspension of proceedings in an action, or sometimes a total discontinuance of it, as where an action has been compromised.

Steelyard. A depot established in London by Hanse merchants for their goods. The Hanse was deprived of trading privileges by Edward VI, but the Steelyard remained in its possession until it was expelled from England in 1597 and even thereafter the Hanseatic towns of Lubeck, Bremen, and Hamburg continued to own the steelyard until 1852.

Stefan Dusan (1308–55). The greatest king of Serbia (1331–55) who sought to forestall the Ottoman threat to Constantinople. He sought to establish the reign of law and in 1349 produced the famous *Zakonik Dusana,* modelled on and partly copied from the *Basilica.* It did much to produce the reception of Roman law in Serbia and remained in force in Serbia till 1830 and in Montenegro till 1888.

Stellionate. In Roman law, granting a second security without giving notice of the first one or alienating without disclosing charges. In English law it extends to the deceitful selling of a thing. In Scots law it covers any kind of fraud or deceit not having a specific name.

Stephen, Henry John (1787–1864). Called to the Bar in 1815 but made little headway professionally. He became well known, however, by his *Treatise on the Principles of Pleading in Civil Actions* (1824 and many later editions), a lucid exposition of an intricate subject. He became a member of the common law commission of 1828 and may have been offered a Judgeship. In 1834 he published a *Summary of the Criminal Law* and in 1841 his *New Commentaries on the Law of England*, partly founded on Blackstone but with extensive additions, and really an original work. Many other editions have been published by later editors, extensively altering the original work. Latterly he became a commissioner of bankruptcy at Bristol but never attained the professional position which his abilities merited.

Stephen, Sir James Fitzjames (1829–94). English judge and writer. Called to the Bar in 1854 he made slow headway and wrote extensively for the *Saturday Review* and the *Cornhill Magazine*, and later for the *Pall Mall Gazette*, many of his essays being reprinted later. In 1863 he published a *General View of the Criminal Law of England*, really an attempt to justify criminal law as a branch of social science, and in 1869 succeeded Maine as legal member of the Council in India. There he carried on the work of Macaulay and Maine on codification and, after his return to England in 1872, was much employed in attempts to carry out codification in England, preparing various measures none of which became law. The bases of these measures were published as a *Digest of the Law of Evidence* (1876), a *Digest of the Criminal Law* (1877), and *A Digest of the Law of Criminal Procedure in Indictable Offences* (1883). His *Liberty, Equality, Fraternity* (1873) was a vigorous protest against popular opinions, and in support of strong government, a feeling strengthened by his failures to have his codification measures approved. He became a judge of the Queen's Bench Division in 1879. Thereafter he wrote a *History of the Criminal Law* (3 vols., 1883) and an historical inquiry, *The Story of Nuncomar and Sir Elijah Impey* (1885).

He resigned due to ill health in 1891 and received a baronetcy. As a judge he lacked subtlety and disliked technicality, but was respected for his knowledge and fairness, particularly in criminal law.

Leslie Stephen, *Life of Sir J. F. Stephen.*

Sterilization. The process, by surgical treatment, of rendering a person incapable of begetting or of conceiving and giving birth to a child. It may be for contraceptive, therapeutic, eugenic, or punitive purposes. The operation may be carried through with the consent of the person being sterilized (though the consequences for any possible or existing marriage must be considered), or be advised on medical grounds. The operation involves interference with a fundamental human right and, if performed for non-therapeutic reasons and without the consent of the person, infringes that right. Consequently it is only exceptionally that it may be done on a minor or person of unsound mind. See *Re D.*, [1976] 1 All E.R. 326. In the U.S. compulsory eugenic sterilization is permitted in many states to safeguard the human race from degeneration, under statutes which are directed against the mentally ill and mentally deficient. Punitive sterilization of sex offenders is lawful in certain states.

Sterndale, Lord. See PICKFORD, W.

Steward of a manor. The business man of the lord of the manor, who kept the court rolls and transacted legal and other business. Nominally he was judge of the customary court baron and court leet and registrar of the freeholders' court baron qq.v.) Among his chief functions were to grant admittances to the copyhold lands of the manor, to receive surrenders and to keep the court roll.

Steward of Scotland. A steward was an officer appointed by the Scottish kings with jurisdiction over Crown lands and having the same jurisdiction as a lord of regality. The judicial office of steward was equivalent to that of sheriff. The Steward of Scotland administered the Crown revenues, superintended the affairs of the royal household, and had the privilege of standing in the army in battle second only to the King. The royal house of Stewart took its name from this officer succeeding to the throne when the Crown passed from Robert Bruce's son David to the son of his sister Marjory who had married Walter the Steward.

Stewards of the Household. A high office in the mediaeval households of the English kings. The office tended to become hereditary under the Norman kings and an early distinction appeared between the hereditary titular stewards and working household stewards. For a time several seem to have held office concurrently, but from the time of Edward II the office was normally held by one only. The duties of the office were mainly the business management of the royal household. The modern Lord Steward of the Queen's Household is in charge of the management of the royal household. He is a non-political officer and does not change on a change of government.

The office of Lord High Steward, originally hereditary, became merged in the Crown in the fifteenth century. The Lord High Steward presided in the House of Lords when a peer was being tried before his peers, prior to 1948 when the privilege was abolished. A peer was appointed Lord High Steward for the occasion.

Stewart, Sir James (1635–1715). Called to the Scottish Bar in 1661 he had to leave the country in 1670 in consequence of a political pamphlet he had published, *Jus Populi Vindicatum.* He returned to Scotland in 1679 but in 1681 again had to flee to Holland and was sentenced to death in his absence. He returned shortly before the Revolution.

In 1692 he became Lord Advocate and introduced many legal reforms, including particularly the Act for preventing delays in criminal trials (1701); he was a member of the Glencoe Commission and probably prepared the report. He resigned in 1709 but held office again in 1711–14.

His son, Sir James (1681–1727), was Solicitor-General 1709–14 and 1714–17, and his grandson, also Sir James Stewart, later Stewart Denham (1712–80), published a *Dissertation on the Doctrine and Principles of money* and an *Inquiry into the Principles of Political Economy* (1767) which was highly thought of until overshadowed by Adam Smith's *Wealth of Nations.*

Stewart, Alexander, 5th Earl of Moray (?1630–1700). Became a member of the Scottish Privy Council in 1661, and was made Lord Justice-General of Scotland in 1675, an extraordinary Lord of Session in 1680, and Secretary of State in the same year. He was active in suppressing the covenanters and at the Revolution was deprived of all his offices and retired into private life.

Stewartry. The authority of an officer appointed by the Scottish kings, the steward or stewart, to administer certain crown lands. The jurisdiction was generally equivalent to that of a lord of regality (q.v.), and was generally heritable. It was abolished in 1747. Certain districts, being administered by stewards, not sheriffs, were known as stewartries, not sheriffdoms, and this survives in that the Kirkcudbright district of Galloway is still known as The Stewartry.

Stewartry and bailiery courts. In mediaeval Scotland a stewart was a king's officer appointed over lands belonging to the King or lands created into a stewartry. Bailieries were lands with lesser jurisdiction attached to a stewartry. The Stewart and the Bailie each had rights to hold courts, the jurisdiction being at least as wide as that of regality or sheriff courts respectively, and their rights became hereditary. They were abolished in 1747.

Stifling a prosecution. Agreeing to abstain from prosecuting a person in return for some benefit. If the crime in question is a serious one, the stifling of a prosecution is invalid as an agreement and possibly criminal.

Stipendiary magistrate. A legally qualified and salaried person appointed under various statutory powers to sit in a magistrates' court and exercise the jurisdiction thereof in place of two or more justices. In London such officers are called Metropolitan Stipendiary Magistrates.

Stipulatio. In Roman law a formal contract made by a question by the creditor, e.g. "centum dare spondes?" and the reply by the debtor, "spondeo", whence it gained the alternative name of *sponsio*. It was one of the oldest institutions of Roman law and, originally restricted to a fixed sum of money, it became extended to any specific article and later to any legal transaction, and was allowed to be used by non-citizens. It came accordingly to be a general and the most usual mode of creating any kind of obligation. Though classical law required the personal presence of the parties, oral proceedings, and an immediate answer appropriate to the question, verbs other than *spondere*, such as *promittere, dare*, or *facere*, and even Greek equivalents were permitted. It came to be used also for accessory obligations guaranteeing the creditor against failure on the part of the principal debtor. The surety bound himself by a separate *sponsio* to undertake the same liability as the debtor did by his own *stipulatio*. In the case of non-citizens *fidepromissio* bound the surety. If the principal debt did not arise from *stipulatio*, suretyship had to be created by *fidejussio*. *Stipulatio* did not require witnesses but it early became customary to draw up a note (*instrumentum* or *cautio*) to secure evidence of its having been entered into and in post-classical law the note was sufficient, the *stipulatio* being deemed to have been entered into. Finally, an imperial rescript abolished the requirement of solemn words. In Justinian's law the only requirement remaining was the simultaneous presence of the parties and even this was assumed if the parties had been in the same place on the day mentioned in the document.

Stock. A fund or capital capable of being divided into and held in any amount. Thus government stocks are generally transferable in any multiple of one penny. Companies may convert their shares into stock, the only difference being that units of stock are not numbered but transferred as so many pounds' (nominal value) of stock. The origin of this was the fact that what is now called the capital of the company was originally called its "joint stock", being the fund contributed by the members jointly, and companies being called "joint stock companies".

Stock Exchange. A society of persons associated to buy and sell, on behalf of clients, Government stocks and stocks and shares in public companies. The principal stock exchange in the U.K. is the London Stock Exchange which is legally an unincorporated company composed of its proprietors and control of it is vested in its Council. The Council may make, amend, and revoke rules and regulations as to the admission and expulsion of members and the conditions on which business shall be transacted. Members, if engaged in active business, must be a broker or broker's clerk, or a jobber or jobber's clerk; no member may carry on business in both capacities; members may deal only with other members, and disputes between members must be referred to the arbitration of a member or members. Members may deal only in securities quoted on the Stock Exchange and subject to the rules of the Stock Exchange.

For his client a broker is an agent with fiduciary responsibilities; he must act in accordance with his instructions and execute each transaction to the best advantage according to his judgement at the time of dealing, and is entitled to indemnity from his client against any liability incurred by reason of having contracted on the latter's behalf. Contracts of sale or purchase of stock are made orally and noted by brokers in their books. Bargains are normally made for settlement during the next settlement period.

Stock option. A power to buy or sell stock in a designated company for a specified period at a fixed price, irrespective or movements in market price during that period. Speculators may purchase options in anticipation of changes in stock exchange prices. Stock options may be issued along with ordinary stock to make it more saleable and enable the buyer to increase his holding of stock in the company at will. Stock options are sometimes also given to senior executives as a form of incentive payment.

Stockdale* v. *Hansard (1839), 9 Ad. & E. 1; (1840), 11 Ad. & E. 253. Inspectors of prisons reported to the Secretary of State that improper books published by the plaintiff were permitted in Newgate prison. The report was published to the public by and under orders of the House of Commons. The plaintiff sued the publisher for defamation and the Court of Queen's Bench held for the plaintiff, holding that the extent of the privilege of the House of Commons was determined by law and could be examined by the courts. The House accepted the decision and allowed the damages to be paid, and has subsequently never refused to admit the jurisdiction of the courts in matters of privilege outside the walls of the House; though it has not relinquished the claim to decide whether a privilege exists. [The Parliamentary Papers Act, 1840, was then passed. It conferred absolute privilege on all reports, papers, votes, and proceedings published by order of either House.] In the second action the plaintiff sued *Hansard* for further publication of the alleged libel and recovered damages. The Sheriff of Middlesex levied execution but delayed to pay the damages. The House of Commons then resolved that the execution was a contempt of the House, and that the sheriff be taken into custody. The Queen's Bench ordered payment of the damages to the plaintiff.

See also SHERIFF OF MIDDLESEX'S CASE.

Stocks. An obsolete piece of machinery of punishment. Stocks consisted of two boards which, when fastened one on top of the other, left two holes just large enough to enclose the ankles of an individual sitting behind the boards. To have to sit in the stocks for a stated time was for long a common penalty for minor offences. It fell into disuse in the early nineteenth century.

See also PILLORY.

Stone, Harlan Fiske (1872–1946). Was admitted to the New York Bar in 1898, became a professor of law at Columbia (1903–05) and Dean (1910–23), carrying on simultaneously a heavy legal practice, before becoming Attorney-General of the U.S. (1924–25) in which capacity he did much to restore faith in the Department of Justice after the Teapot Dome scandal. He was an Associate Justice of the U.S. Supreme Court (1925–41) and Chief Justice (1941–46). In his early years he was a vigorous dissenter against the Courts' attitude to the New Deal.

A. T. Mason, *Harlan Fiske Stone: Pillar of the Law.*

Stone, Julius (1907–). Australian jurist, professor of international law and jurisprudence at Sydney, 1942–72, author of many works, including *The Province and Function of Law* (1946), later rewritten and expanded into *Legal System and Lawyer's Reasonings* (1964), *Human Law and Human Justice* (1965), *Social Dimensions of Law and Justice* (1966), *Legal Controls of International Conflict* (1954), and other works.

Stop list. A list maintained by a trade association of persons with whom members of the association are forbidden to trade, on the ground that they have offended the association, as by reselling goods at cut prices.

Stoppage *in transitu*. The right of an unpaid seller of goods to stop delivery of and resume possession of goods sold on credit if the buyer has become insolvent before the goods have got into his possession. It cannot be exercised after the goods have been delivered to the buyer or an agent for him.

Stormont. The suburb of Belfast in which is situated the Northern Ireland Parliament Building and Stormont House, formerly the official residence of the Prime Minister of Northern Ireland.

Story, Joseph (1779–1845). Graduated from Harvard, learned law on his own, and in 1801 began to practise law in Salem, Mass. He sat in the Massachusetts legislature (1805–8 and 1810–11) and headed a committee which recommended the creation of an equity court there. He sat in Congress 1808–9. In 1811 he was appointed an Associate Justice of the U.S. Supreme Court, the youngest ever appointed. On the bench his views had great weight with Marshall, and his judgments are impressive for breadth of learning. On circuit he did much to establish admiralty law. He published editions of Chitty's *Bills of Exchange* (1890), Abbott's *Merchant Shipping* (1810), *Public and General Statutes of the U.S., 1789–1827* (3 vols., 1828), and planned an American Digest similar to Comyn's *Digest.* In 1829 he also became the first Dane Professor at Harvard and over the next 15 years wrote numerous books of the highest authority and value, including *Bailments* (1832), *Commentaries on the Constitution* (1833), *Conflict of Laws* (1834), *Equity Jurisprudence* (1836), *Equity Pleading* (1838), *Agency* (1839), *Partnership* (1841), *Bills of Exchange* (1843), and *Promissory Notes* (1845), all of which have gone through many editions. Apart from James Kent no man has, by writing and teaching, had greater influence on American law and some of his books have been extensively referred to in England and elsewhere. Along with J. H. Ashmun, Royall Professor at Harvard, and Tapping Reeve and James Gould of the Litchfield Law School, he was a pioneer in the development of the law school, as contrasted with office, preparation for legal practice. The *Commentaries* were incredible achievements, and evidence immense industry and legal knowledge, and themselves entitled him to be ranked as a jurist of the first rank. His son, W. W. Story, wrote his life story, two standard textbooks on *Contracts* and *Sales*, but made his own name as a sculptor.

W. W. Story, *Life and Letters of Joseph Story*; C. Warren, *The Supreme Court in U.S. History.*

Stowell, Lord. See SCOTT, WILLIAM.

Straccha, Benvenuto (1509–78). Italian humanist jurist specializing in commercial law, one of the first to view commercial law as distinct from civil law, and to examine it from the practical standpoint, author of *De mercatura sive de mercatore* (1553), *De proxenetis atque proxeneticis* (1558), and *De Assecurationibus* (1569). Though cognizant of Roman law he disregarded scholastic theories and had regard to the practical problems of commerce.

He was the first to present a systematic exposition of commercial law.

Strand Inn, An Inn of Chancery (q.v.).

Strange, John (1696–1754). Called to the Bar in 1718, was one of the counsel who defended Macclesfield (q.v.) when impeached, became Solicitor-General in 1737 and recorder of London in 1739, but resigned these offices in 1742. From 1750 to 1754 he was Master of the Rolls and was deemed a very able lawyer and competent judge. He compiled a volume of reports which were published posthumously and which enjoy a good reputation.

Straw, Man of. A person of no means. The device is sometimes employed of transferring an onerous liability to a man of straw, which in many cases can have the effect of shedding liability.

Strict law. A phrase importing rigid adherence to literal requirements of law as distinct from more liberal compliance with the substance. In Roman law a distinction was drawn between procedure *stricti iuris* where the matter had to be decided according to the rules of law and procedure *bonae fidei* in which the judge was ordered to give a decision according to good faith and what was fair and just. In England a similar distinction has been drawn between the requirements of strict law and those of equity (q.v.).

Strict liability. The term in the law of tort for a standard of liability which is more stringent than the ordinary one of liability for failure to take reasonable care, yet not absolute, which is the standard sometimes set by statute, where liability arises if the harm to be prevented takes place, whatever care and precautions have been taken. Where strict liability exists there are certain, though limited, defences to liability—but having taken reasonable care is not among them. The principal instances of strict liability are liability for animals and liability in cases within the principle of *Rylands* v. *Fletcher* (q.v.) for harm caused by the escape of a danger from one's land. In criminal law strict or absolute liability is liability for having done something or having allowed it to happen, independently of intention, recklessness or negligence. It exists by virtue of various statutory provisions, but probably never at common law.

Strict settlement. A settlement of land intended to keep the land in the family descending in the eldest male line. The eldest son, in each generation, on the attainment of majority, is persuaded voluntarily to reduce his entailed interest in the land to a life interest and, by thus relinquishing his power of disentailment, to tie up the land for a further generation. It is normally effected by a disentailing assurance and a deed of resettlement. The process of resettlement is generally repeated once in every generation when the prospective heir attains majority, provisions being made for the younger sons and daughters by portions.

Strikes. A strike is the usual term for a simultaneous and concerted cessation of work by an employer's employees, or a substantial group of them, normally in pursuance of an industrial dispute. Strikes have developed from the nineteenth century concept of the ultimate reaction of employees to intolerable conditions to the twentieth century one of a weapon regularly and automatically threatened or resorted to, to enforce the union view on any matter on employers. Not infrequently a strike arises from an inter-union or demarcation dispute; frequently it is engineered by agitators. There is no legal distinction between an official and a unofficial strike; the former is one authorized or called by a trade union, whereas the latter is not so authorized or called. A strike is a breach of contract by each workman, unless due notice of intent to terminate employment be given by each, in which case it is not considered a breach of contract, but is not illegal or criminal unless it involves committing criminal acts, nor is it an actionable conspiracy so long as the predominant motive is the furtherance of a legitimate interest, such as improvement in the conditions of work. An employer may dismiss an employee for participating in a strike, but all employees must be treated equally. Courts are forbidden to require an employee to attend for work by order of specific performance of a contract of employment or by injunction restraining a breach or threatened breach of such a contract.

Trade union legislation gives numerous immunities from legal proceedings to unions and participants in strikes, notably immunity from tortious liability for inducing a person to break a contract of employment or threatening that a contract of employment will be broken, or for interfering with the trade, business or employment of a person, or for agreeing or combining to do or procure an act in contemplation or furtherance of a trade dispute if it would not have been actionable if done by one person alone. Moreover, peaceful picketing (q.v.) is lawful though picketing which amounts to intimidation may be criminal. The social and legal problems of the later twentieth century include the political strike, where the strike is really called to coerce the government. This has become possible by the growth of giant unions and sympathetic action by unions not themselves involved in the dispute, if indeed there is any dispute, beyond one engineered for the occasion. The legality of a political strike is very questionable.

There is also the problem of whether persons

engaged in public services such as the armed forces, the police, or the provision of essential services can be allowed to strike, because their withdrawal of their services causes harm to the whole community and to persons who have no part in any dispute there may be. The public badly lacks protection against strikes in which it has no concern.

Striking out. The action of a court in treating as deleted a pleading or part of a pleading, as where it discloses no possible ground of claim or defence, or is scandalous or merely frivolous.

Strube, David Georg (1694–1755). German jurist, author of a *Commentatio de jure villicorum* (1720) and *Vindiciae juris venandi nobilitatis germanicae* (1739).

Strupp, Karl (1886–1940). German jurist, author of *Grundzüge des positiven Völkerrechts* (1921) and other works on international law, and completer of Hatschek's *Wörterbuch des Völkerrechts und der Diplomatie* (1924–29).

Struve, Georg Adam (1619–93). A leading German jurist, active in presenting a practical law for Germany developed from native sources, rather than from received Roman law; author of *Juris feudalis syntagma, Syntagma jurisprudentiae secundum ordinem Pandectarum* (1658–83), *Jurisprudentia Romano-Germanica forensis* (1670), and *Centuria decisionum.*

Stryk, Samuel (1640–1710). A leading German jurist, active in developing German law from native sources rather than from received Roman law, creator of the phrase *usus modernus Pandectarum* in the work of that name, and author also of *Examen juris feudalis* (1675), *Dissertationum juridicarum Francofurtensium Volumina II* (1683), and many other works.

Stuart, John (1793–1876). Called to the Bar in 1819 and reported decisions in the Vice-Chancellor's court 1822–26. He became Q.C. in 1839 and a Vice-Chancellor in 1852, but was not a distinguished judge.

Stuff gown. The kind of robe worn by barristers who have not been made Queen's Counsel. Hence a "stuff gownsman" means a barrister or junior. See also SILK.

Sturgeon. One of the category of royal fish, which if cast ashore or caught near the coast is the property of the sovereign.

Stubbs, William (1825–1901). Regius professor of history at Oxford 1866–84 and then bishop of Chester 1884–88 and of Oxford 1888–1901. He edited many volumes of chronicles in the Rolls Series, published a standard *Select Charters and other illustrations of English Constitutional History from the Earliest Times to the Reign of Edward I* (1870), and wrote a magisterial *Constitutional History of England* (1874–78) which, though corrected and criticized has been extremely influential on later scholars and is still valuable. He also published *Lectures on the Study of Mediaeval and Modern History* (1887).

Style. The name or title of a person, as in the phrase "royal style and titles"; also, in Scotland, a model form or precedent of a deed or pleading.

Stylus Curiae Parliamenti (or *Style du Parlement*). The first treatise on civil procedure written in France, by William du Brevil, and advocate of the Parlement of Paris. It is a methodical statement of the practice of that Parlement in the early fourteenth century.

Suarez, Francisco (1548–1617). Theologian and jurist, one of the greatest thinkers of the Jesuit order and professor of philosophy at Segovia in Spain, at Rome, Alcala, Salamanca, and at Coimbra; he was almost the last eminent representative of scholasticism. Though founding on St. Thomas, he developed the latter's doctrines on political philosophy, the State, sovereignty, and related topics. He also wrote extensively on theology and philosophy, such as *De Virtute et Statu Religionis* (1608–09) and *Defensio Fidei Catholicae* (1613) in opposition to Anglican theologians who urged the divine right of kings philosophy of monarchy. As author of *Tractatus De Legibus et Deo legislatore* (1612) he was a forerunner of Grotius and Pufendorf in finding a natural law basis for a law of nations. In the *De Legibus* he declared the people to be the original holders of political authority, and the State the result of a social contract to which the people have consented. In his *De Bello et de Indis* he criticized the practices of the Spaniards in the Indies and argued for the natural rights of the individual to life, liberty, and property.

Mullaney, *Suarez on Human Freedom*; Wilenius, *Social and Political Theory of Francisco Suarez.*

Suarez, Karl Gottlief (1746–98). German jurist, entrusted in 1780 with preparing a general legal code, finally proclaimed in 1794 as the *Allgemeines Landrecht für die preussischen Staaten.* It was later taken as a model for legislation in Austria and was adopted in other German states, and remained fundamentally unaltered until replaced by the German Civil Code of 1900.

Subinfeudation. In feudal law the practice whereby a person, himself holding land of another as his vassal in return for services, granted to another a smaller feudal estate to be held of and under him. This process might be repeated several times so as to create a number of links in the feudal chain, each holding of the person above him, up to the Crown. In England subinfeudation was prohibited by *Quia Emptores* (q.v.) in 1290, but it continues possible in Scotland up to the present.

Subject of right. The person in whom legal rights are vested, the object being the person or thing against whom or in respect of which the right avails.

Subjection. The legal concept of being subject to the power of another in some respect, of being in the position of having one's legal rights altered by the exercise of another party of a legal power vested in him. Thus, a mortgagor is subject to the mortgagee's power to sell or foreclose, a judgment debtor subject to the creditor's power to issue execution against him, and so on. It is frequently equated with liability, the correlative concept in each case being power.

Subjugation. In international law, the destruction of the independent existence of a territory defeated in war and, normally, its annexation by the victor. The state of war may, at least nominally, continue though the government of the defeated state had been completely displaced. Subjugation may be temporary, in that the victor may disclaim any intention of annexation.

Submission. A deed by which parties submit a dispute to arbitration. The agreement to submit may be general, such as of all disputes arising between the parties, or particular, such as of a dispute which has arisen.

Subornation of perjury. The causing, persuading, or inducing of a person to commit perjury (q.v.). It is itself criminal.

Subpoena. A writ issued by a court or other authorized agency requiring a person to come before it at a stated place and time, subject to penalty (*sub poena*) if he does not comply. It takes two forms, *subpoena ad testificandum*, when the recipient is called to give evidence, and *subpoena duces tecum*, when he is required to bring documents or papers relevant to the controversy for examination by the court. The name was also formerly given to the writ whereby all persons other than peers were called to appear and answer to a bill in Chancery.

Subrogation. The principle that a person who discharges a liability incumbent primarily on another is put in the place of that other for the purpose of obtaining relief against any other party liable. Thus in insurance law, if an assured is compensated for his loss by the insurers, the latter are entitled to stand in the assured's place and exercise all rights competent to him to recover from the person who caused the loss.

Subsidy. In the fourteenth and fifteenth centuries the word generally meant the import and export duties on cloth, wool, leather and skins, and the tonnage and poundage granted to the Crown for special occasions. In the sixteenth century it usually meant a tax of 4*s.* in the pound on land and 2*s.*8*d.* on movables sometimes voted, but sometimes any tax imposed by Parliament. In the seventeenth century the word was extended to other taxes and the older duties were sometimes termed perpetual subsidies and other taxes temporary subsidies. In more modern usage it means any financial assistance, particularly given by the Crown to individuals, industries, local authorities, or others.

Substantive law. The major part of any legal system and of each branch thereof, the part which is concerned with the legal rights attributed to and legal duties imposed on particular legal persons in particular circumstances, as contrasted with adjective law (q.v.) which is concerned with the legal machinery whereby rights and duties may be declared and enforced. It is frequently difficult to say whether a particular matter is one of substantive or of adjective law.

The substantive law includes both primary or antecedent rights, such as to have performance of duties owed by another, or to have freedom from harm or interference by another, and also secondary or remedial rights, such as to remedies against a party in breach of duty or who has failed to satisfy the first party's primary right. It includes also both rights *in personam* and rights *in rem* (qq.v.).

Succession. The branch of law which deals with the devolution of a person's property to others on his death. It generally falls into three sub-branches, dealing with prior rights which some legal systems confer on certain relatives to specified shares in the estate of a deceased, with the mode of distribution prescribed by law in default of any provision made by the deceased (intestate succession or succession on intestacy), and the rules relative to the making, interpretation, and enforcement of wills or written instructions left by the deceased (testate succession). Closely allied to rules of succession are, on the one hand, inheritance taxes or death duties; and on the other, principles relating to the appointment and functioning of one or more executors or administrators to administer the deceased's estate, collect the deceased's assets, and distribute them according to law or his will.

Succession duty. A duty first imposed in 1853 on the gratuitous acquisition on death of property in respect of which no legacy duty (q.v.) was payable. The rate varied according to the propinquity of the successor. It was abolished in 1949.

Succession to the Crown. The succession to the Crown is determined by the Act of Settlement, 1700, extended to Scotland in 1707 and to Ireland in 1800, which entailed the Crown to the Princess Sophia, granddaughter of James I and Electress of Hanover and the heirs of her body being Protestants. Accordingly, the Crown descends in the same way as real property did at common law, males being preferred to females and the principle of primogeniture applying, save that where the succession opens to females co-parcenery does not apply and the eldest female succeeds. The only modification made was by His Majesty's Declaration of Abdication Act, 1936, which excluded Edward VIII and his issue (in fact there were none) from the throne and passed it to his next eldest brother as George VI. Even if that Act had not been passed Prince George's issue (Princess Elizabeth) would have succeeded Edward VIII.

Sue. To bring a civil action against a person. Contrast PROSECUTE.

Sue and labour clause. A clause in a policy of marine insurance designed to make clear that the assured and his servants or agents, particularly the master and crew, can take every step practicable to recover the insured property which is in peril, without loss of rights under the insurance, and will be repaid any expenditure incurred to avert the loss. It does not cover general average losses and contributions, nor salvage charges. The assured is obliged to take the steps authorized.

Suffrage. The legal right or privilege of voting at Parliamentary or local government elections. The extension of the suffrage is one of the themes of British constitutional history since the early nineteenth century.

Sugden, Edward Burtenshaw, 1st Lord St. Leonards (1781–1875). He began as a clerk to a firm of solicitors, then practised as a conveyancer under the Bar but was called in 1807 and later became leader of the Chancery Bar. In 1828 he entered Parliament, in the following year became Solicitor-General but lost office in 1830. In 1834 he became Lord Chancellor of Ireland but held office for only a few months. He held that office again in 1841–46, and in 1852 briefly held office as Chancellor, being created Lord St. Leonards. He refused the office again in 1858, but sat frequently in the House of Lords. He was generally critical of changes in the law, but as Chancellor promoted measures amending the law of wills, trusts, and procedure. As a judge he was highly regarded, his opinions being clear and authoritative. While still a student he published a book on real property and first published his famous book on *Vendors and Purchasers* in 1805, which was the foundation of his practice, and he occupied much of his time while out of office in revising that and other books on real property. His other main work was his *Practical Treatise on Powers* (1808 and later editions). Both in Ireland and in England he pronounced numerous important decisions which evidence his great legal learning and ability to expound the law lucidly. After his death his will was not discoverable, a fact which gave rise to a leading case on the proof of a lost will by secondary evidence.

Sui iuris. In Roman law a person, irrespective of age, who was not subject to the power of a father, husband, or master, as contrasted with one who was so subject (*alieni iuris*). In modern law the phrase is sometimes used of one who is not under the guardianship of a parent or other guardian, and not subject to any disability.

Suicide. Self-killing or the taking of one's own life. It is frequently a difficult question whether death results from accident, homicide, or suicide. In England suicide was formerly criminal and for long resulted in forfeiture of goods to the Crown and ignominious burial in the roadway with a stake driven through the body, but the latter was finally abolished in 1882. Suicide ceased to be criminal in 1961. Attempted suicide was similarly, but is not now, criminal. In respect of coroner's inquests (q.v.) suicide can only be found if there is adequate evidence of intent to commit suicide; it cannot be presumed.

In Scotland it was not treated as a crime, but formerly resulted in single escheat whereby the whole moveable estate of the deceased fell to the Crown. A suicide pact was an agreement between two persons that both would commit suicide or that one would kill the other and then himself. If one survived, he was formerly guilty of murder, but since 1957 of manslaughter only.

Suit. A civil legal proceeding or action, a term particularly used formerly of certain Chancery proceedings and now of proceedings for nullity or dissolution of marriage.

Suit of court. In the early Middle Ages attendance at the local county court was a burdensome duty which tended in time to be attached to pieces of land, the tenants of which owed the duty of suit of court. The suitors were thus a defined class, and they, under the presidency of the sheriff, were the judges of the county court. Similarly in the hundred courts, the duty of attending the court was incumbent on the holders of certain pieces of land.

Sullivan, Sir Edward (1822–?). Called to the Irish Bar in 1848. He became Solicitor-General 1865, Attorney-General 1868, Master of the Rolls in Ireland 1870, and Lord Chancellor of Ireland 1883–85.

Sullivan, Timothy (1874–1949). Called to the Irish Bar in 1895 and became successively President of the High Court, Irish Free State, 1924–36, President of the Supreme Court and Chief Justice of the Irish Free State, 1936–37, and of Eire, 1937–46.

Sulpicius Rufus, Servius (*c.* 106–43 B.C.), A famous Roman jurist, counsel in 51 B.C., friend and rival of Cicero, and author of nearly 180 books, including *Reprehensa Scaevolae capita*, i.e. corrections of Q. Mucius Scaevola's views, a book of dowries, and a commentary on the edict.

Summa. A name given to mediaeval compositions which present a concise, ordered statement of and commentary on the principal contents of a branch of the Roman law or of the *Decretum.* The term is also given to synopses of post-Justinianean legal writing. Initially merely summaries, *summae* became in time more systematic and further removed from the text. Many *summae* on the Decretum are known, such as Paucapalea (*c.* 1145), Rolando Bandinelli (*c.* 1148), Rufinus (*c.* 1160), Huguccio (*c.* 1190), and many mediaeval works, particularly 13th-century tracts on procedure, were called *Summae*, notably the works called briefly *Fet asaver*, *Hengham Parva* and *Hengham magna*, the *Mirror of Justices* (*qq.v.*), and various other.

Summary jurisdiction. The jurisdiction of a judge or magistrate to try accused persons without a jury. Since jury trial was, save in certain exceptional cases, the only mode of trial permitted by common law, summary jurisdiction is entirely statutory and dates from 1848–49. It is, however, limited to lesser offences though in certain circumstances some charges may be tried either summarily or on indictment, but all the more serious crimes must be tried on indictment before a jury and cannot be tried summarily. Summary trials are generally simpler and speedier than jury trials, and the number of cases now disposed of summarily is enormously greater than those tried by jury. In England certain civil proceedings, principally matrimonial, are also within the summary jurisdiction of magistrates' courts. In the U.S., notwithstanding guarantees of jury trial in federal and state constitutions, it is accepted that petty offences may be tried summarily.

Summing up. The summarizing of the main points in the evidence by a judge to a jury, together with his guidance on the form of verdict to be given, depending on the view the jury forms of the evidence.

Summons. A document calling on a person to attend before a judge or court for stated purposes. In the High Court a summons is a mode of application to a judge or master in chambers, e.g. to give directions as to procedure. In the Chancery Division summonses are either in a pending cause or matter, e.g. a summons to proceed; or originating summonses, e.g. for the determination of the rights of persons interested in an estate.

In Scottish civil procedure a summons is the pursuer's first pleading which initiates the claim, in the form of a writ in name of the sovereign, passing the signet, which calls on the defender to appear to answer it, sets out the remedy sought, and the condescendence (*q.v.*) or facts alleged to justify the grant of the remedy and pleas-in-law (*q.v.*) for the pursuer.

Sumner, Viscount. See HAMILTON, J.A.

Sumptuary laws. Formerly laws restricting extravagance in food, drink, dress, or furnishings, frequently imposed for moral or religious reasons, and more commonly today laws restricting the manufacture or consumption of various commodities imposed on grounds of morals, health, or social welfare. Regulation of this kind has been very common and was found in many European countries, and in the seventeenth century in some American colonies. Possibly the most striking attempt in recent times was the attempt in 1919–33 to impose prohibition in the U.S. and prohibit the manufacture, transport, or sale of alcoholic beverages. In modern times the effect can usually more readily be achieved by taxation than by direct prohibition.

Sunday legislation. Legislation proscribing various kinds of work, recreation, or other activities on Sundays. The Emperor Constantine in the fourth century condemned work on Sunday, and Sunday laws appeared in England in the early thirteenth century. An Act of 1677 prohibited worldly labour on Sundays and required all persons to attend church. Similar legislation appeared in American colonies from their foundation, but they vanished in the early nineteenth century. Some old statute law on Sunday Observance still survives. Many exceptions have been made to the prohibitions of Sunday work and the matter has frequently been before the Supreme Court. Several statutes of the Scottish Parliament of the sixteenth and seventeenth centuries imposed fines or other punishment for profanation of the Sabbath. In modern times restrictions on work on Sunday exist partly to preserve a certain degree of religious observance, and partly to ensure that Sunday is not normally a working day. But many kinds of recreations are permitted on Sundays.

Superficies. In Roman law, the alienation by an owner of rights necessary to build on the surface of the land subject to reservation of an annual payment.

Superior and vassal. In mediaeval feudal law, and still in Scots land law, the relationship whereby one person, the vassal, holds land of and under the other, the superior, on certain terms and conditions and in return for stipulated payment, nowadays always in money, known as feuduty. Each party simultaneously holds a distinct estate or interest in the lands, the superior holding the *dominium directum* or superiority, the vassal the *dominium utile* or fee, the latter carrying the rights of actual occupation, possession, and use of the land. The creation of fresh feus is permitted but feuduties cannot now be created and there is statutory provision for the redemption by lump sum payments of existing feuduties.

Superior courts. The higher courts which deal with more important cases, in which sit the higher grades of judges, and the decisions of which have weight as settling the law. In England the superior courts are the House of Lords, and the Supreme Court of Judicature comprising the Court of Appeal, the High Court, and the Crown Court. In Scotland the superior courts are the Court of Session and the High Court of Justiciary. In Northern Ireland the superior courts are the Supreme Court of Northern Ireland comprising the Court of Appeal, the High Court and the Crown Court. The Judicial Committee of the Privy Council is also a superior court. Among courts of special jurisdiction the Courts-Martial Appeal Court and the Employment Appeal Tribunal are superior courts.

Superior orders. Orders from a person higher up in the civil or, more commonly, military hierarchy, sometimes put forward as an excuse or justification for having committed crimes. Such orders are not an automatic defence but may be relevant as negativing *mens rea* or negligence, or showing that the accused had a claim of right to do what he did. It may not be a defence if the orders lead to a mistake of law, namely a mistaken impression that it was the accused's duty to do what he did.

Super-tax. A term used from 1918 for the additional tax on income above a total fixed each year, at rates also fixed each year. In 1927 it was replaced by surtax (*q.v.*), itself replaced in 1973.

Supervision order. An order making a person subject to the supervision of a probation officer, which is applied to young persons on release from a detention centre (*q.v.*) and to young short-term prisoners.

Supervisory jurisdiction. The ancient jurisdiction, vested formerly in the Court of King's Bench, and now in the Queen's Bench Division of the High Court in England, to supervise inferior courts, or tribunals, or persons exercising statutory powers of a judicial kind, e.g. commanding magistrates to do what their duty requires in cases where there is no special remedy and securing the liberty of the subject by summary interposition. The proceedings are known as proceedings on the Crown side of the Queen's Bench Division, and the jurisdiction is exercised on an application for judicial review, incorporating the former prerogative writ of habeas corpus, and the orders (formerly also prerogative writs) of certiorari, mandamus, and prohibition.

Supplementary benefits. Every person in Great Britain aged 16 or over whose resources are insufficient to meet his requirements is entitled to a supplementary pension or a supplementary allowance, depending on whether or not he has attained pensionable age and, where appropriate, benefit to meet medical requirements or an exceptional need. These benefits replace the scheme of National Assistance, which applied between 1948 and 1966. The administration is in the hands of the Supplementary Benefits Commission, which must exercise its functions in the way which best promotes the welfare of the persons affected. Whether a person is entitled to benefit and the amount of any benefit is determined by the Supplementary Benefits Commission, subject to appeal to an appeal tribunal, in accordance with regulations. The only qualifying condition is need. Persons are disqualified if aged under 16, employed, or unemployed by reason of a trade dispute (though this does not affect their dependants), and special conditions attach to persons refusing or neglecting to maintain themselves or their dependants.

Supply, Commissioners of. Bodies established in Scotland by statute in 1667 on a county basis, having originally the function of levying and collecting the land tax for the "supply" of the sovereign. They had prescribed qualifications by ownership of lands and heritages, but included also the sheriff of the county, his substitute, and representatives of the burghs in the county. As time went on other duties were given to them, including the police force, preparation of the valuation roll, registration of voters, and provision of court houses and prisons. Their powers were transferred to county councils in 1889.

Supply, Committee of. See COMMITTEE OF SUPPLY.

Support. The right of an owner of land to have it upheld in its natural position. This is presumed, and an owner of subjacent strata, of minerals, and of neighbouring land, is not entitled so to use his land so as to withdraw vertical or lateral support unless he has an easement (*q.v.*) entitling him to do so. The natural right to support does not, however, apply to land the weight of which has been increased by buildings and a right to support in such a case is not a natural right but an easement, which may be acquired by express grant or by prescription. Similarly the right to have one building supported by another is an easement. No right to support by underground percolating water is recognized, and such water may be withdrawn even if this causes letting down of the surface.

Supra-national organizations, institutions, and law. Supra-national organizations, institutions, and law differ from national ones in that they are not confined to the area of any one nation-state, and from international ones in that they do not purport or attempt to apply to all nation-states and as between any of them. They are, however, above and superior to groups of nation-states but yet not a federal superior to them in that they have a large measure of independence. Nor is a supra-national organization a mere linking relationship or alliance; in at least some respect it is superior to the members of the group and can exercise powers over them. The principal supra-national organization is the European Community with its institutions, Council, Commission, Parliament, Court of Justice, and its extensive body of law, some of which is directly applicable in member states. Such organizations as the Commonwealth and the Organization of American States are probably not properly described as supra-national in that the organization cannot direct or control the actings of a member but only co-ordinate the actings of members and facilitate co-operation.

Supremacy, Acts of (1534 and 1559). Statutes which gave effect to the Protestant Reformation of the Church in England. The Act of 1534 declared the sovereign (then Henry VIII) the supreme head or governor of the English Church.

After a Catholic reaction under Mary (Tudor) in 1553–58 the Act of 1559 passed by the first Parliament of Elizabeth I, substantially restated the Act of 1534.

Supremacy, Oath of. An oath which under the Bill of Rights had to be taken by various high officers of state; by it the swearer abjured the authority, ecclesiastical or spiritual, of any foreign person in England. It was abolished in 1868.

Supreme Court of Judicature. By the late 1860s the inconvenience of a multiplicity of courts in England had become very serious. There were conflicts of jurisdiction, different bodies of law, distinct procedures, and uncertainty as to which was the appropriate court in which to sue. After many Acts had modified and improved conditions in particular courts, the Judicature Commission in 1868 proposed the fusion of jurisdiction, law, and procedure. Accordingly the Judicature Act, 1873, which came into effect in 1875, consolidated into a single Supreme Court of Judicature nearly all the existing superior courts. The Supreme Court was divided into the High Court of Justice and the Court of Appeal.

The High Court was assigned the jurisdiction formerly exercised by: (1) the High Court of Chancery of England both as common law court and court of equity, including the jurisdiction of the Master of the Rolls; (2) the court of Queen's (or King's) Bench; (3) the court of Common Pleas; (4) the court of Exchequer; (5) the High Court of Admiralty; (6) the court of Probate; (7) the court for Divorce and Matrimonial Causes; (8) the London court of Bankruptcy; (9) the court of Common Pleas at Lancaster; (10) the court of Pleas at Durham; and (11) the court created by any of the Commissioners of Assize, of Oyer and Terminer, of Gaol Delivery. The jurisdiction of the Court of Crown Cases Reserved was vested in the judges of the High Court, or any five of them, the Lord Chief Justice always being one. This court was abolished in 1906 and replaced by the Court of Criminal Appeal.

The Courts of Chancery of the Counties Palatine of Lancaster and of Durham were merged in the High Court in 1971.

For convenience, in the distribution of business, the High Court was divided initially into five Divisions, namely, the Chancery Division, the Queen's Bench Division, the Common Pleas Division, the Exchequer Division, and the Probate, Divorce, and Admiralty Division. Matters generally of the kinds which had formerly been appropriate in the Court of Chancery were assigned to that

Division, and to the other Divisions were assigned the jurisdiction exercised formerly by the other courts. In 1881 the Queen's Bench, Common Pleas, and Exchequer Divisions were merged in the Queen's Bench Division. In 1969 the Probate, Divorce, and Admiralty Division was renamed the Family Division, and Admiralty work was transferred to the Queen's Bench Division. All jurisdiction vested in the High Court belongs to all the Divisions alike, and all the judges of the High Court have equal power and jurisdiction.

The jurisdiction of the High Court is both original and appellate. Original jurisdiction is normally exercised by a single judge of the High Court, but certain kinds of appellate business must be heard by divisional courts of two or three judges of one of the Divisions. In the Chancery Division divisional courts of two judges hear bankruptcy appeals from county courts. In the Queen's Bench Division divisional courts, usually of three judges, hear cases stated by magistrates' courts and certain other appeals.

In the P.D.A. (now Family) Division divisional courts of three judges hear appeals from magistrates' courts in relation to separation and maintenance and certain other classes of appeals. In each of the Divisions certain appeals lie under various statutes to a single judge.

The existing judges of the first seven courts mentioned were in 1875 constituted "Justices of the High Court". On appointment each judge is assigned to a Division, but may be transferred, and any judge may sit in any Division and exercise any of the jurisdictions of the High Court. The Divisions were to be presided over respectively by the Lord Chancellor, the Chief Justices of the Queen's Bench and Common Pleas, the Chief Baron of Exchequer, and the President of the P.D.A. Division. The title of Lord Chief Justice of England was conferred on the Chief Justice of the Queen's Bench in 1859, and the offices of Chief Justice of the Common Pleas and Chief Baron of Exchequer were abolished in 1881. In 1968 the Lord Chancellor was empowered to and did, appoint a Vice-Chancellor to deputize for him as head of the Chancery Division, and the President of the P.D.A. Division became President of the Family Division.

The Court of Appeal was to consist of the Lord Chancellor, the Master of the Rolls, the Lord Chief Justice of England, the President of the P.D.A. Division, and any ex-Lord Chancellor, if he consented to act, all *ex officiis*, and, originally five, judges of appeal, later styled Lords Justices of Appeal. In 1913 the Lords of Appeal in Ordinary, if qualified, were also made *ex officio* judges of the court. The number of Lords Justices has subsequently been increased. Any judge of the High Court may be required to sit as an additional judge of the Court of Appeal, and an ex-judge or ex-Lord Justice of Appeal may be invited to sit as additional judges.

The Court of Appeal was, in 1873, given the jurisdiction and powers of the Lord Chancellor and Court of Appeal in Chancery, both as court of equity and court of appeal in bankruptcy, of the court of Exchequer Chamber (*q.v.*), the jurisdiction of the Privy Council in lunacy appeals, and appeals from the instance jurisdiction of the Admiralty, the jurisdiction and powers of the court of Appeal in Chancery of the County Palatine of Lancaster and of the Chancellor of the Duchy and County Palatine of Lancaster, and the jurisdiction and powers of the court of the Lord Warden of the Stannaries. Its original jurisdiction is very limited, and its main jurisdiction is to hear appeals from the High Court, county courts, other courts of similar standing, in certain cases from divisional courts, and under many statutes.

The old procedure of appeal by writ of error, or by motion for a new trial, was superseded by appeal by way of rehearing of the case.

The Court of Appeal normally sits simultaneously in several divisions of three judges; the Master of the Rolls is the only one of the *ex officio* judges who sits, and he presides in one of the divisions. The Lord Chancellor may appoint a Lord Justice to be Vice-President of the court. Appeals may be heard by a full court of five judges, and interlocutory appeals are heard by two judges, and final appeals may, by consent of parties, be heard by two judges subject to rehearing before three judges if the two differ in opinion.

In 1966 the Court of Criminal Appeal (q.v.) was abolished and its jurisdiction transferred to the Court of Appeal which, since then, consists of the Civil Division, exercising the jurisdiction hitherto exercised by the Court of Appeal, and the Criminal Division, exercising the jurisdiction transferred from the Court of Criminal Appeal. At least three judges must sit in the Criminal Division. Any number of courts of either Civil or Criminal Division may sit simultaneously.

In 1971 the Crown Court was established and made part of the Supreme Court. The judges of the Crown Court are any judge of the High Court, any Circuit judge or Recorder, and justices of the peace sitting with a judge of the High Court, Circuit judge or Recorder. Judges of the Court of Appeal may be requested to sit as judges of the Crown Court. It has jurisdiction in criminal cases in trials on indictment, and in appeals from magistrates' courts. Appeal lies to the Court of Appeal (Criminal Division).

The Judicature Act abolished the jurisdiction of the House of Lords as final court of appeal, but the relevant provisions were suspended and in 1876 the Appellate Jurisdiction Act reformed the appellate jurisdiction of the House and provided for substantial assimilation of the judicial membership of the

House and the Judicial Committee of the Privy Council. The House of Lords as court of appeal is not part of the Supreme Court.

Supreme Court of the U.S. The Supreme Court was established by the Constitution as the third branch of government. It consists of a Chief Justice and such a number of Associate Justices as Congress may determine (5 in 1789, 4 in 1801, 5 in 1802, 6 in 1807, 8 in 1837, 9 in 1863, 6 in 1866, 8 in 1869). All justices are nominated by the President and must be confirmed by the Senate, which confirmation is not always granted. They may be removed only by impeachment, which has been utilized (unsuccessfully) only once (Justice Chase in 1804), but may resign if in danger of impeachment (Justice Fortas in 1969). A quorum is six and decisions are by a majority.

The jurisdiction of the court extends to all cases arising under the U.S. Constitution, statutes and treaties of the U.S., and such cases as lie outwith the States' jurisdiction. In the exercise of its power of judicial review it may have to examine federal or state legislation or executive acts to determine their constitutionality. In 1891 Congress created the circuit courts of appeal to deal with most cases outwith a state court's jurisdiction.

Cases come before it by way of appeal from the 11 U.S. courts of appeals (which hear appeals from U.S. district courts), from State courts, when cases involve a federal question, the U.S. Court of Claims, the U.S. Court of Customs and Patent Appeals (which hears appeals from the U.S. Customs Court) and it also reviews decisions of such administrative agencies as the Tax Court, the Federal Trade Commission, and the National Labor Relations Board. It has an original jurisdiction in actions between federal government and a state, or between two states.

By reason of its political power, the Court has inevitably from time to time been involved in political controversy. In the last quarter of the nineteenth century a conservative court usually adopted a *laissez-faire* attitude to business activities and an unsympathetic attitude to social legislations. This provoked the dissents not infrequently voiced by Brandeis, Harlan, and Holmes. Again in the 1930s the Court struck down a number of major items of Roosevelt's New Deal legislation and provoked Roosevelt to propose a plan to bring new blood to the Court. This was attacked as "packing the Court" and Congress did not in fact act, but the temper of the Court changed and large changes in personnel made it a much more liberal body. Since the Second World War it has done much to further equality of civil rights and liberties.

Nevertheless, the Court has established a high reputation and enjoys general confidence, respect and regard.

A list of the Chief and Associate Justices is in Appendix I. See further C. Warren, *The Supreme Court in U.S. History*; Carr, *The Supreme Court and Judicial Review*; C. Haines, *Role of the Supreme Court in American Government and Politics*, 2 vols.; C. Haines, *The American Doctrine of Judicial Supremacy*.

Surcharge. An overcharge; formerly also in local government a declaration by an auditor that a person has illegally expended public money and is personally liable to refund it.

Sûreté Nationale. A French police force, controlled by the Ministry of the Interior, operating throughout the whole country supervising the municipal police of provincial towns, each of which has a *commissaire de police*, and concerned with preventive police work and criminal investigation.

See also GENDARMERIE NATIONALE; PRÉFECTURE DE POLICE.

Surety. See GUARANTEE; PRINCIPAL AND SURETY.

Surrebutter. In old common law pleading, the pleading which followed the rebutter. It was the last kind of named pleading.

Surrejoinder. In old common law pleading, the pleading which followed the rejoinder and preceded the rebutter.

Surrender. In international law, cessation of resistance by a belligerent force, made without capitulation (q.v.) or special conditions. Intention to surrender is normally made by hoisting a white flag and stopping active resistance. The enemy must accept this provided it is clear that the white flag has been hoisted by or with the authority of the commander of the surrendering force.

In municipal law a surrender is the yielding up of some right, such as a lease before its expiration, a copyhold, a charter, or an insurance policy.

Surrogate. A person appointed in place of another, as by a bishop or chancellor, and in particular an officer appointed to dispense licences to marry without banns.

Surtax. An additional tax on incomes exceeding stated amounts. It replaced supertax in 1927 and disappeared in 1973, being replaced by higher rates of taxation levied on successive bands of income.

Survey, Courts of. A court of survey consists of a judicial member, either a county court judge, stipendiary magistrate, wreck commissioner, or other fit person, and two assessors, (persons of maritime, engineering, or other appropriate knowledge). It has jurisdiction to hear appeals by the owners of ships detained as unseaworthy, undermanned, or unsafe, or from a declaration of survey of a passenger steamer and in certain other cases.

Survivorship. The question whether one or more particular persons have continued in life after the death of one or more others. It is important in many questions of succession. In English law there is now a presumption, where persons have died in circumstances rendering it uncertain which survived longest, that such deaths are, for the purposes of title to property, presumed to have occurred in order of seniority. This may be excluded by contrary provision in a will, and as between an intestate and his spouse the estate is distributed as if the spouse had not survived. In Scots law there is now a presumption that the younger survived, but where they are husband and wife the presumption is that neither survived, and in other cases if the elder left property to the younger, whom failing, to a third party, the presumption for the purposes of that destination is that the elder survived.

Suspended sentence. A kind of judicial sentence whereby the judge may pronounce a sentence of imprisonment or fine but suspend its operation for a period. If during the period of suspension the offender does not offend again, the sentence is wiped out; if he does offend, the sentence is automatically enforced, quite apart from any penalty exigible for the second offence. The difficulty is that the offender's circumstances at the time it becomes enforceable may be so different as to make the sentence inappropriate. The suspended sentence of imprisonment was introduced in England in 1973 in the case of sentences of not more than two years' imprisonment, and the operation of the sentence is not automatic on reconviction but only if a court orders that the original sentence shall take effect, either as passed or as varied by the later court.

Suspending power. A royal power, utilized during the latter seventeenth century, to suspend temporarily the operation of any statute or group of statutes. In 1673 Charles II issued a Declaration of Indulgence, suspending the penal laws against nonconformists, really intended to benefit Catholics. It united churchmen and even dissenters against the king, in regarding the power as a despotic power capable of dangerous applications. The House of Commons petitioned the King to recall the declaration as being inconsistent with the legislative authority of the King and Parliament, which he eventually did, and Parliament then passed the Test Act. In 1687 James II issued a Declaration of Indulgence to a similar effect; a year later it was republished and ordered to be read in all churches, which provoked the petition of the Seven Bishops, who were then tried for seditious libel and acquitted amid great popular rejoicing. The Bill of Rights, 1689, declared that the pretended power of suspending of laws by royal authority without consent of Parliament was illegal.

See also DISPENSING POWER.

Suspension. In Scots law a remedy competent to stop the doing of diligence (q.v.) or, in limited circumstances, to review a decision of a court. It may be combined with interdict in a process of suspension and interdict. In Scottish criminal procedure a Bill of Suspension is a mode of having reviewed by the High Court of Justiciary an illegal warrant or irregular conviction or sentence. It may be combined, where the suspender has been imprisoned, with liberation in a Bill of Suspension and Liberation.

Swainson, William (1809–83). New Zealand lawyer, Attorney-General from its establishment as a separate colony in 1841 until the establishment of responsible government in 1856. He drafted the early laws of the colony; his legislation was logically arranged and lucidly expressed. He later served as Speaker of the legislative council and as a member thereof.

Swanimote, Court of. A court which, according to the Charter of the Forest, met thrice a year, for business connected with agistment, pannage, and fawning.

Swear. To take an oath, or to declare on oath. Swearing in the sense of cursing and using profane language in public is punishable under certain statutes and is frequently an element in minor breaches of the peace.

Sweatt* v. *Painter (1950), 339 U.S. 629. The Supreme Court held that a segregated law school for negroes did not provide them with equal educational opportunities with those enjoyed in a law school for whites, and therefore contravened the Fourteenth Amendment. The Court did not decide whether the doctrine of "separate but equal" facilities (see *Plessy* v. *Ferguson*) would still have applied, if the facilities had been unequal, but this was settled by *Brown* v. *Board of Education* (q.v.).

Swedish law. In Sweden early law was customary and there existed the institution of Law-men who recited the laws at folk assemblies and advised on its application to cases. In Sweden the Law-men soon became elected public officials and representatives enforcing justice in a Land, and having also political and administrative functions. The Law-men from various districts combined under Chieftains, who also recited the law as a kind of official oral code. The Land Things, or regional assemblies, acquired legislative authority.

Collections of written law are found from the thirteenth century, the oldest being the *Vestgota-lag*, but similar law books were prepared for other provinces in that century, and there were also municipal codes, but centralization and unification of Swedish law came later than in Denmark or Norway because of the independence of the Lands and Things. King Magnus Erikson in 1340–50 enacted a series of ordinances supplementing or changing older law, and appointed a commission to frame a code. It was formulated in 14 parts, but excluded church law. King Gustavus Adolphus II in the early seventeenth century effected reforms in the judicial system, creating appellate courts.

In 1686 the King at the request of the Riksdag appointed a commission to prepare a code; work continued from then till 1736 when it was adopted as Sveriges Rikes Lag. It is in nine parts, including parts on crime, punishment and court practice, and while influenced by Roman law is strongly based on the traditional law of the country. The commission continued to work till 1811 when a new commission undertook to prepare another civil code.

The code of 1736 was supplemented by later legislation. Fresh criminal codes were enacted between 1855 and 1864, and the judicial system was modernized in the nineteenth century.

Since 1809 there has been a written constitution, supplemented by constitutional legislation. The latest version is that of 1975. Legislation is supplemented by an extensive body of ordinances.

There is also a substantial volume of the legislation adopted in common by the Scandinavian countries, particularly maritime law adopted in 1891–93. Custom long continued important and is still utilized, especially in commercial matters, though less resorted to than in Denmark or Norway. The courts are not attached to the principle of *stare decisis*.

Legal science was flourishing from the sixteenth century and Loccenius produced the first methodical work on Swedish civil law in 1653.

The court system comprises a Supreme Court (Hogsta Domstolen) established in 1789, with a mainly appellate jurisdiction. It acts also as a Court of Impeachment. There are six intermediate courts of appeal and a large number of city or district courts, courts of first instance. All have both civil and criminal jurisdiction. Juries are very little used, but in many cases judges sit with lay assistants. There are a Supreme Administrative Court (1909), administrative courts of appeal, lower administrative courts, labour court, market court, social insurance court, and certain other special courts. The Attorney-General represents the Crown's interest in the correct administration of the law.

L. Orfield, *Growth of Scandinavian Law*; S. Juul, A. Malmström and J. Søndergaard, *Scandinavian Legal Bibliography*; N. Regner, *Svensk juridisk litteratur, 1865–1956* and Supp.; L. Frykholm and T. Byström, *Swedish Legal Publications* in *Scandinavian Studies in Law*, 1961. S. V. Thorelli. *The Constitution of Sweden. Svensk författningssamling*; *Sveriges Rikes Lag* (annual edition of entire statute law).

Swereford, Alexander de (?–1246). An ecclesiatic and civil servant employed in Exchequer and latterly a Baron of Exchequer. He probably compiled the *Black Book of the Exchequer* and the *Red Book of the Exchequer* (q.v.).

Swinfen-Eady, Charles, Baron Swinfen (1851–1919). First admitted a solicitor but called to the Bar in 1879 he developed a chancery practice. In 1901 he became a judge of the Chancery Division and in 1913 a Lord Justice of Appeal. In 1918 he became Master of the Rolls and in 1919 a peer but his health failed in 1919 and he resigned. As a judge he was learned and well regarded.

Swiss law. Until about 1500 Swiss legal history is a part of German legal history and Swiss law grew out of mediaeval German law. The Swiss Confederation developed out of the German Empire and was recognized internationally by the Peace of Westphalia in 1648. Germanic folk-laws percolated into Switzerland and continued to be referred to down to the tenth century.

From the thirteenth century various cities and rural districts developed as independent states, and there are distinguishable types of town law: Habsburg town law, Constance law, Zahringen law, Basel law, and Vaud-Savoy law, each of which was conferred by founders on various towns and carried from one town to another.

Between 1300 and 1800 there developed the Old Swiss Confederation which became the Confederation of the Thirteen Places in 1513. There were also Associated Places and territory held in common. The constitution of this Confederation was not in any single deed but deducible from alliances and the practice under them. There developed meetings of deputies, known as Diets.

There was a partial reception of Roman law in some parts of the country and elsewhere it was treated as of subsidiary force, supplementary to local law books, statutes, and agreements between lords and people.

An important source of knowledge is the *Eidgenossisches Stadt- und Landrecht* (Confederate Town and Rural Law) by Hans Jacob Leu of Zurich (1724–44).

Legal science was for long little developed in Switzerland, not least because the Roman law was not received, but after 1500 law was studied at Basel university. At Lausanne the first professor of law was Barbeyrac (q.v.) and at Geneva distinguished jurists were Hotman, Bonnefoi, Denis and Jacques Godefroi, and, in the eighteenth century, Burlamaqui and Vattel. Later jurists included von Keller and Bluntschli.

The Helvetian Republic was established in 1798 as a centralized state, being altered in 1803 to a Confederation, comprising 19 cantons, together with a confederate Diet, in 1815 to a Confederation of 22 cantons, and in 1848 to a federal state. The federal constitution of 1848 has several times been amended; the modern constitution is of 1874. Federal sovereignty is exercised by the Federal Assembly, the legislature, and the Federal Council, the chief administrative body.

Each of the 23 cantons has its own legislature, executive, and judiciary, usually at appeal and first instance levels. There is a Federal Supreme Court which has original jurisdiction in inter-canton and other federal matters, and appellate jurisdiction against cantonal authorities in matters of federal law, and a Federal Criminal Court. In the nineteenth century legislation began to oust customary law. In some cantons the French codes of 1804 were received when they were parts of France and were retained, in others they were imitated, while in others codes of private law were adopted which were developments of the Austrian Civil Code of 1811. In others, again, original codes were created, such as the Zurich code of 1853–55. The Swiss Civil Code, a uniform system applicable in all the cantons, was adopted in 1907 and came into force in 1912. It was later copied in Turkey.

The Swiss Civil Code (Code Civil Suisse or Schweizerisches Zivilgesetzbuch) was drafted by Eugen Huber (q.v.) and, though influenced by both French and German codes, includes much indigenous matter. It comprises a brief introductory section and four books dealing respectively with the law of persons and associations, family law, matrimonial regimes and guardianship, succession, and property. There is a separate federal code of commercial and personal obligation which became effective in 1881.

Criminal law was first codified in 1799 and again in 1937 and has been uniform since 1942, but criminal procedure is a matter for each canton.

C. A. Sladitz, *Guide to Foreign Legal Materials—France, Germany, Switzerland*; H. Planitz, *Quellenbuch der deutschen, österreichischen und schweizer Rechtsgeschichte*; E. Ruck, *Schweizerisches Staatsrecht*; O. A. Germann, *Schweizerisches Strafgesetzbuch*; A. B. Schwarz, *Das Schweizerisches Zivilgesetzbuch.*

Symond's Inn. An Inn of Chancery (q.v.).

Synallagmatic. Involving mutual obligations and duties.

Syndic. An agent, attorney, or person appointed by a corporation to act for it in a particular matter.

Syndicate. A body of persons associated temporarily for some piece of business or undertaking.

Synod, General, of the Church of England. A body established in 1970 renaming and reconstituting the National Assembly of the Church of England (the Church Assembly). It consists of the convocations of Canterbury and York combined in a House of Bishops (the Upper house), a House of Clergy (the Lower house), and a House of Laity elected in accordance with representation rules. It meets at least twice a year and its functions are to consider matters concerning the Church of England and to make provision therefor by Acts of Synod, orders or regulations, Canons made as previously by Convocations, and by Measures intended to be given the force and effect of Act of Parliament when approved by Parliament.

Syntagma Canonum et legum. A work compiled, in the Eastern Roman Empire in 1335 by Matthew Blastares and intended for the use of the clergy. It included material from the Nomocanon, the Procheiron, and the Basilica. It was translated into Slavonic in Serbia on the orders of Stephan Dusan and had authority there, in Romania, Bulgaria, and Russia.

Syntagma Canonum Antiochenum. A chronological collection of canons constituting an important source of Byzantine law. It comprises canons from the Council of Antioch (341) onwards down to about 800.

Syro-Roman law book. A work originally in Greek, but known only in versions in Oriental languages, dealing with family law, slaves, and succession, mainly in Roman law, and probably intended for academic purposes but used about the eighth century by ecclesiastics in the Middle East who were under Mohammedan rule but required a Christian law-book.

T

Tabard A short gown or coat, in particular the surcoat worn by a herald or pursuivant over his uniform, embroidered on front and back with the royal arms.

Table A, B, etc. Schedule I of the Companies Act, 1948, contains model forms of Memorandum and Articles of Association (q.v.) which may be adopted in whole or in part by various kinds of companies. Table A is the model form of articles of association of a company, not being a private company, limited by shares. Table B is the model form of Memorandum of Association of a company limited by shares. Table C is the model form of memorandum and articles of association of a company limited by guarantee and not having a share capital. Table D is the model form of memorandum and articles of association of a company limited by guarantee and having a share capital. Table E is the model form of memorandum and articles of association of an unlimited company having a share capital.

Tacit relocation. In Scots law, an implied renewal of an expired lease of premises on the same terms as the former lease for a year, if the previous lease had been for a year or longer, and if for a shorter duration, for the same period again. The same principle applies to some contracts of employment.

Taciturnity. Keeping silence in circumstances when a claim could have been made, giving rise, if prolonged, to an inference of abandonment of the claim.

Tack. A lease of land.

Tacking. In the law of mortgages the former principle that a mortgagee who made a further advance and took a further charge in ignorance of any mortgage made at a time between his first and subsequent advances was held to have his subsequent advance attached to the former one and ranked in priority to the intermediate mortgage. Tacking operated only where the mortgagee had obtained possession of the legal estate and had been ignorant of the intermediate mortgage.

The term is applied also in Parliamentary procedure to the addition of a non-financial matter to a 'money ' Bill, to prevent it being amended by the House of Lords. This practice was abused in the past and the Lords passed a resolution against it in 1702 as an infringement by the Commons of their privileges, and since 1807 respect for constitutional practice has prevented such subterfuges to get proposals through the Lords.

Taff Vale Railway Co.* v. *Amalgamated Society of Railway Servants [1901] A. C. 426. Two officials of the defendant union organized a strike on the plaintiff company's line and induced strike-breakers to decline, in breach of contact, to replace workmen on strike. It was held that the officials were acting in the course of their employment and that the union as an entity was liable for their wrongful conduct. This was a serious blow to the use of strikes by unions. The decision was overruled by the Trade Disputes Act, 1906.

Taft, William Howard (1857–1930). A graduate of Yale and Cincinnati Law School, Taft was admitted to the Ohio Bar in 1880 and became a judge of the Supreme Court of Ohio in 1887. From 1890 onwards Taft was a capable government executive and showed great administrative ability. In 1890 he became Solicitor-General of the U.S., judge of the U.S. Circuit Court of Appeals, 6th circuit, 1892–1900, and also Dean of the University of Cincinnati Law School, 1896–1900. After serving as President of the second Philippine Commission, 1900–1, he was Governor-General there, 1901–4, Secretary for War, 1904–8, and President of the United States, 1909–13. Thereafter he was Kent Professor of Law at Yale, 1913–21, and Chief Justice of the U.S. Supreme Court, 1921–30. He was unhappy as President and failed to win a second term. But he loved judicial work and was the last Chief Justice actually to sit in a circuit court. He was also the first Chief Justice to regard himself as the administrative head of the entire federal judicial system, and was mainly instrumental in securing the passage of the Judges Act in 1925 which greatly improved the judicial machinery and enabled the Court to catch up with its work, and to concentrate on constitutional cases and others of national significance. He was not, however, a leader of judicial thought, nor a jurist, though a sound, working judge.

H. S. Duffy, *W. H. Taft*; A McHale, *President and Chief Justice*; H. F. Pringle, *W. H. Taft*; F. C. Hicks, *W. H. Taft, Professor of Law*; A. E Regan, *Chief Justice Taft*; A. T. Mason, *W. H. Taft: Chief Justice*.

Taft-Hartley Act (1947). U.S. legislation prohibiting the closed shop and unions' contributions to political campaigns, establishing an 80-day 'cooling off' period before strikes threatening national health or safety, and requiring an affidavit of non-communism from union leaders.

Tail. (Fr. *tailler*, to cut off). An estate tail was formerly a freehold estate of inheritance in land which descended only to the specified descendants of the body of the devisee rather than to his heirs general, descendants, ascendants, or collaterals. It was the creation of the statute *De Donis*, 1285. Since 1925 an interest in tail (or tail male, or tail female, or tail special), may be created by way of trust in any property, real of personal. Entailed interests, unless barred, devolve according to the general law in force before 1926.

Taille. The main direct tax of pre-Revolutionary France. It was bitterly resented, not least because the clergy and nobles were exempt. It originated in the Middle Ages as a seigneurial aide, due by manorial tenants to their lord, and was extended from the royal domains to all France about 1400. Thereafter it became a permanent levy; in northern France it was levied on individuals; in the south it was levied on non-privileged land. Many exemptions had arisen by the eighteenth century, such as for inhabitants of large towns and holders of office which carried ennoblement and exemption. It was abolished in 1789.

Tailzie. (or tailye—pron. taillie). In Scots law, a line of succession, other than that prescribed by law, on whom a proprietor of lands, by deed of tailzie or entail, prescribed that the lands should devolve. Since 1914 it has been impossible to create new entails but older entails may still exist, unless the tenant in possession has effected a disentail.

Take-over. The acquisition by one company of the control of another by acquisition of a controlling interest in the share capital. The one company makes an offer to the shareholders of the other to buy their shares for cash, or for its own shares or debentures, or a mixture. The offer is normally conditional on acceptance by a stated proportion of shareholders. If the offer is accepted, the company acquired becomes a subsidiary of the acquiring company and the shareholders become shareholders in the acquiring company.

The directors of the company being acquired should always act in the best interests of the shareholders. In 1959 a working party in the City of London produced a City Code on Take-overs and Mergers, prescribing a code of good practice, administered by a Panel. The Stock Exchange imposes further safeguards.

Talbot, Charles, 1st Lord (1685–1737). After becoming a fellow of All Souls, Talbot took to law and was called to the Bar in 1711. He built up a practice, mainly in equity, and became M.P. in 1719 and Solicitor-General in 1726. In 1733 he became Lord Chancellor and a baron; he died in 1737. Opinion was that he was one of the ablest lawyers of his time with a clear, subtle mind, and some of his judgments are landmarks in the development of equity. His work as Chancery judge was a valuable prelude to the consolidation of equity doctrine effected by Hardwicke.

Talbot, Sir William (?–1691). Was secretary of the colony of Maryland, 1670–71, a commissioner of the revenue in Ireland, and Master of the Rolls in Ireland, 1689–90.

Tales (pron. Ta-lays—Lat. such men). If, when a jury has been summoned, a sufficient number do not appear, or by reason of exemptions or challenges a sufficient number is not left to make the proper number, either party may ask the court to make good the number. The judge may award a *tales de circumstantibus*, which is a command to the sheriff to return as many other qualified men as are present or can be found. Such jurors are called talesmen (pron. tailsmen), but a jury cannot be entirely so composed.

Talion. The principle that criminals should as punishment receive the same injuries and harms as they had inflicted on their victims, the principle of 'an eye for an eye and a tooth for a tooth'. Retributive punishment of this kind was known in Babylonian law, early Jewish law, and early Roman law. The Twelve Tables seem to have established the talio as the limit to which vengeance might be exercised, if the offender could not placate the victim in any other way. It had begun to disappear in Roman law by the fifth century B.C. The early Christian Church tended to regard the Mosaic law as divinely commanded and to this may be attributed much of the cruelty of later criminal law. The idea reappeared in South Germany in mediaeval times, and traces of it are found in Europe and in Scandinavia in the seventeenth and eighteenth centuries. It still seemed to Cocceji the proper principle underlying a theory of punishment, but was rejected by Pufendorf and later thinkers generally.

Tallage. In mediaeval Europe a tax imposed by a lord on his unfree tenants. It was originally in the lord's discretion both as to frequency and amount, but had become a fixed charge on many estates by the thirteenth century. In England from the twelfth century it had become established as a royal tax on estates in the King's possession and on boroughs. King John levied tallages frequently and the practice was attacked in Magna Carta. Thereafter parliamentary taxation of boroughs and of the King's estates began to be preferred and the last royal tallage was exacted in 1312.

Tally. A stick across one side of which notches were cut indicating payments. It was then split lengthwise and each party retained one half. From early times tallies were used in the Exchequer and this lasted until 1826. The burning of a large quantity of old tallies led to the burning down of the old Houses of Parliament.

Talmud. A compilation consisting of the Mishna, which is a collection of laws, originally oral, in supplement of scriptural laws, the Gemara, a commentary on and elaboration of the Mishna, and certain other materials. In Judaism it is of authority second only to the Bible. The term Talmud is frequently used of the Gemara only. There are two versions, the Palestinian or Jerusalem Talmud and the Babylonian Talmud, the Mishna of which are identical. The Midrash is a mode of biblical interpretation used in Talmudic literature and a body of commentaries on scripture using this method.

Tametsi (although). The first word of the opening sentence of a decree enacted by the Council of Trent in 1563 concerning the juridical form of marriage, requiring for the validity of marriage the presence of a priest and two or three witnesses. It also prescribed that banns were to be published before marriage was to take place and laid down a liturgical form for marriage. Clandestine marriages were previously considered valid. It was never made applicable throughout the whole Catholic Church and was never wholly successful in establishing a settled form of marriage law.

Tancred of Bologna (?–1236). A canonist, professor at Bologna, author of the *Ordo Judiciarius* (*c.* 1215) which superseded all the numerous earlier works on procedure in canon law, and a *Summa de sponsalibus et matrimonio* (1214).

Taney, Roger Brooke (1777–1864). A diligent student of law, was admitted to the Maryland Bar in 1799, elected to the House of Delegates of Maryland in the same year, and served as a State Senator, 1816–21, and Attorney-General of Maryland, 1827, before becoming Attorney-General of the U.S. in 1831. He was nominated as Secretary of the Treasury in 1833, but the Senate rejected the nomination. A year later he was appointed an Associate Justice and less than a year after that, and before the Senate had confirmed the appointment, Chief Justice of the U.S. Supreme Court, an office which he held from 1835 to 1864. As Chief Justice he reversed some of the trends which had characterized the attitude of Marshall, his predecessor. Thus he asserted that rights not specifically conferred by a corporate character could not be enforced, and that rights granted must be construed narrowly. This became a settled principle of U.S. law.

His most notorious decision was in *Dred Scott* v. *Sandford* (1857), a decision on the free or slave status of a negro, in which he based the majority decision on the invalidity of the Missouri Compromise, which resulted in personal attacks on him, and a serious drop in the authority of the court. The decision also precipitated the Civil War. He was an eminently competent Chief Justice, but latterly became out of touch and the *Dred Scott* decision ruined his reputation.

S. Tyler, *Memoir of R. B. Taney* (1876); B. C. Steiner, *Life of R. B. Taney*; C. B. Swisher, *Roger B. Taney* (1935).

Tanistry. A system of succession known in Ireland and also traced in the barbarian laws of Europe, whereby the eldest male member of the family, normally the deceased's eldest brother or a similar near relative, succeeded, in contrast to the feudal principle of succession by the eldest son.

Tarde, Gabriel (1843–1904). French magistrate, statistician, philosopher, sociologist, and criminologist, recognized the importance of imitation in crime and social life generally. He wrote *La Criminalité comparé* (1886), *La philosophie penale* (1890), attacking Lombroso's view, and was also author of *Les lois de l'imitation* (1890), *Les transformations du droit* (1893), *La criminalité professionelle* (1897), *Les lois sociales* (1898), and many other works.

Tariff. A table of kinds of merchandise stating the customs duties levied on each, as may be laid down by statute or settled by agreement between certain states.

Tartagnus, Alexander (1424–77). Italian jurist and commentator, author of *Lectura in librum III Decretalium* (1485), *Lectura in rubricam de fide* (1490), *Commentaria in Codicem*, and *in Digestum.*

Tax avoidance. The arranging of his affairs by a person in such a way as to avoid, in whole or in part, a tax which might otherwise be levied on his property, or the art of dodging tax without actually breaking the law. Such conduct is legal, but devices to avoid tax are sometimes hard to distinguish from tax evasion (q.v.).

Tax evasion. Action by a person to avoid, in whole or in part, a tax lawfully exigible from him, as by not declaring or under-declaring income or taxable transactions. Such conduct is criminal and subject to heavy penalties.

Tax law. The bodies of rules of law concerned with the determination of what receipts, or payments, or transactions, are taxable and at what rates, with the assessment of persons' liability in particular circumstances, and the computation of the amounts due from them, and with the enforcement of payment. Each tax has a distinct body of rules applicable to it. More than any other branch of municipal law tax law is open to the reproach of being utterly incomprehensible by the individuals affected, and even frequently by their legal advisers. The enormous complexity of the rules of law on each kind of tax gives rise to an enormous volume of dispute and argument, and a great deal of litigation by way of appeal from assessments made. Neither justice nor reason has any place in tax law, and many decisions of the superior courts are in plain conflict with all sense and reason.

Tax planning. The science of tax avoidance, or the planning of transactions, modes of holding property, modes and times of transferring property, and generally effecting legal results in the way which will minimize the tax payable.

Taxation. Traditionally the principal way in which the ruling classes in organized communities have oppressed, fleeced, and expropriated some of their subjects. It has been known from very early times, and from the earliest times the tax-gatherer has been an object of public fear, hatred, and execration. Taxation was originally a contribution levied from people generally to defray the major common expenses of the State, namely defence and the maintenance of law and order, but not only have the public purposes for which taxation is levied widened to include public health, education, housing, town planning, social services, subsidies to industries, and many other purposes, but taxation is now even in nominally liberal-democratic countries the major weapon of class-warfare, designed to rob some people of their earnings and property in the interest of 'redistribution of wealth'. Indeed, in effect some individuals are expected to work gratuitously, receiving a small percentage commission from the State for their efforts. The tax system is the greatest inhibitor of effort, ingenuity, and exercise of ability. There are no adequate rewards for ability, skill, ingenuity, and responsibility. Subsidiary purposes are to limit the expenditure on socially undesirable consumption goods, such as alcohol or tobacco, and to stimulate or inhibit economic activities. In practice all taxation does far more to inhibit than ever to stimulate economic activity or growth.

The taxing power and its exercise was frequently a matter of dispute between King and Commons and only from 1689 onwards has it been settled that taxation is controlled by Parliament and, within Parliament, by the House of Commons as nominal representive of the community.

The initiative in tax policy is vested in the Chancellor of the Exchequer and the Treasury, and is exercised mainly by the Chancellor's annual budget speech, which is given effect to by the annual Finance Act. There is statutory authority, by resolution of the House of Commons, to give provisional statutory effect to proposed new rates of the annual taxes. Supplementary budgets and other legislation to levy money may be introduced at any time during a parliamentary session. In practice most of the major taxes continue in general from year to year, but altered in detail and sometimes in substantial matters, every year.

In modern Britain taxes are levied on income and on expenditure, on earnings and on capital, during life and on death, by central government and by local government. They can be distinguished into direct taxes (income tax, corporation tax, capital gains tax, capital transfer tax, self-employed persons tax (called class 4 National Insurance contributions), payroll tax (called National Insurance surcharge)); and indirect taxes levied mostly in the form of customs and excise duties on petrol, wines, tobacco, cars, value added tax, and stamp duties on legal deeds. There are other more specialized taxes. Local rates are levied on the occupiers of land and buildings, without reference to capacity to pay.

In some cases one tax is levied on an object already taxed under another category. Thus V.A.T. is charged on petrol already inflated in price by excise duty.

A feature of modern taxation is its progressive nature, in that the larger the sum on which tax is levied the higher the rate. This has gone so far as to be practically expropriation in some cases.

The incidence and weight of taxation on any legal transaction is today a factor of major importance in considering whether, and how, to try to attain some desired result, and this frequently gives rise to involved legal devices seeking to avoid or minimize taxation.

Not the least evil features of the modern tax system are the army of unproductive civil servants concerned with the assessing and collecting of taxes, the enormous volume and constantly changing detail of the chaotic and largely incomprehensible body of verbiage called the law of taxation, the incomprehensible and frequently incorrect assessments, and the utterly irrational nature of the whole topic. In the law of taxation justice has no place at all.

J. Ramsay, *History of the Revenue of the Kings of England, 1066–1399* (1925); S. Dowell, *History of Taxation and Taxes in England* (1888); H. Hall, *History of Customs Revenue in England* (1892).

Taxation of costs. An order by a court for payment by one party to another of costs (q.v.) (in Scotland, expenses) normally directs that they be taxed, i.e. scrutinized by a taxing master (in Scotland, the auditor of court) when mistaken, excessive, or unnecessary expenditure is disallowed. Costs are taxed on one or other of several bases. The taxing master has wide discretion as to what he may allow or disallow. An appeal lies to the court against allowance or disallowance of items on taxation.

Taylor, John (1750–1824), U.S. lawyer and publicist, and the ablest theorist of the early state rights school. His writings include *An Inquiry into* *the Principles and Policy of the Government of the United States* (1814) and *New Views of the Constitution of the United States* (1823).

Taylor, Michael Angelo (1757–1834). Called to the Bar in 1774 and was an M.P. almost continuously from 1784 till 1834. From 1811 he was concerned to expose the abuses in the procedure of Chancery and the delays in House of Lords appeals, and also succeeded in carrying a useful Bill dealing with the lighting and paving of London streets.

Techniques of law. The practical skills of judges and lawyers, and their means of utilizing and applying their knowledge so as to decide disputes or achieve other desired results. Every branch of legal practice has a body of practical skills and methods. In deciding disputes the relevant techniques are those of framing pleadings, eliciting evidence, interpreting legislation, and handling precedents. In property law and conveyancing the techniques are of drafting, of minimizing taxation, of utilizing such legal concepts as trusts to achieve certain desired ends. In commercial law the techniques are of framing agreements and documents, ensuring payment, and facilitating quick and easy achievement of the desired practical results.

Teinds. In Scots law, the tenths of certain produce of the land appropriated to the maintenance of the Church and clergy. At the Reformation most of the property of the Church was acquired by the Crown, nobles, and landowners who appointed persons to vacant benefices, at first for life but later increasingly conferring perpetual heritable right. Such persons were termed Lords of Erection and titulars. Like the abbots and priors whom they succeeded they named vicars to serve in the Church and had to provide a stipend out of the profits of the lands they had acquired. The Privy Council provided in 1567 that a third of the revenues of lands should be applied to paying the clergy of the Reformed Church. Under various Acts in the seventeenth century teinds might be sold by titulars to proprietors of heritage, but under burden of paying stipend to the parish minister.

In 1925 the system was recast by statute and provision made for the standardization of stipend at a fixed value in money.

Teind Court. The common name for the Court of Session (q.v.) acting as Commission for the Plantation of Kirks and Valuation of Teinds. In this capacity the court inherited functions relating to churches, manses, teinds, and stipends conferred by statute on various commissions between 1617 and 1707. Since 1839 the eight Inner House judges and the second junior Lord Ordinary have been Lords Commissioners of Teinds, and since 1868 five have been a quorum. Since 1925 the functions of the court have been greatly diminished and business is now mostly non-contentious and uncontested.

Telephone tapping. The practice of listening to, and frequently also of recording, telephone conversations. Prima facie this is an invasion of personal privacy, but it is understood that the Home Secretary sometimes authorizes the police or the Security Service to tap telephone conversations for the purpose of obtaining evidence of the commission of crime. It is not clear what, if any, legal authority there is for the practice, but in the absence of interference with a recognized legally protected interest it is questionable if an aggrieved person has any civil remedy. Evidence thus obtained has been held admissible in criminal proceedings.

Templars. Officially the Poor Fellow Soldiers of Christ and the Temple of Solomon, one of the first of a number of religious military orders of knighthood established in the twelfth and thirteenth centuries, founded about 1119 to protect and guide pilgrims in the Holy Land. Baldwin, King of Jerusalem, handed over to them a part of his palace, believed to be the Temple of Solomon. By 1153 the order was established in many western countries. After the Christians had been expelled from the Holy Land in 1291 the Templars centred on Paris, and there and in London the Temple was the depository of their wealth. In 1312 Pope Clement V suppressed the order.

Temple. An area in London between the Strand and the river Thames originally belonging to the English branch of the Templars (q.v.) and still containing the mediaeval Temple Church. Not long after the Templars were suppressed lawyers are known to have settled in the area, which in time came to belong to the two societies of barristers known as the Inner Temple and the Middle Temple. Most of the area nearer the Strand has long been occupied by buildings containing barristers' chambers, but the area nearer the river is Temple Gardens.

J. B. Williamson, *History of the Temple* (1924); H. H. Bellot, *The Inner and Middle Temple* (1902).

Temple, Sir John (?1600–77). Was a member of Lincoln's Inn and obtained the office of Master of the Rolls of Ireland in 1641 and was prominent in the government of Ireland and negotiations with Cromwell and Charles II. He published a history of the Irish rebellion. He was father of Sir William Temple (q.v.) and grandfather of Viscount Palmerston.

Temple, Sir William (1628–99). Sat in the Irish Parliament, travelled abroad, promoted the Triple Alliance in 1668, and was envoy extraordinary at the Congress of Aix-la-Chapelle and ambassador at the Hague. He succeeded his father as Master of the Rolls in Ireland in 1677, was superseded in 1689, regained office in 1690 but resigned in 1696 in favour of his nephew. He also sat briefly in the English Parliament.

Courtenay, *Life of Sir William Temple.*

Temporalities. The secular possessions of an ecclesiastic, particularly the lands and revenues annexed to archiepiscopal and episcopal sees by kings and others. During a vacancy in the see the Crown has the custody of the temporalities.

Tenancy. The relation of a tenant to his land, the occupation of land belonging to another with his consent for a defined period and on defined terms, and also the legal interest which the tenant has. Tenancies are of many kinds. Joint tenancy and tenancy in common are both interests existing where several persons simultaneously have an interest as tenants in individual land, but in the case of joint tenancy there is one title only which accresces on death of any one tenant to the survivors until it comes to be vested in the last survivor, whereas tenants in common have each a distinct title to a distinct share which passes on each tenant's death to his own heirs. A tenancy at will has no certain duration but may be ended by either party at his pleasure. A tenancy for life is entitled for the life of himself or of another. A tenancy for years exists for a stated period of years. A tenancy from year to year exists for a year at a time from the time when it was first granted.

Tender. A quotation of price or offer to undertake a contract on stated terms. Also, an unconditional offer by a debtor to his creditor of the amount owed, and, in Scotland, an offer of a sum to settle a claim of damages.

Tender, Legal. The kinds of money in which only a debtor is entitled to make payment and which only a creditor is obliged to accept. Money of legal tender comprises: Bank of England notes to any amount; gold coins to any amount; cupro-nickel or silver coins of value more than 10p to amounts not exceeding £10; cupro-nickel or silver coins of values not exceeding 10p to amounts not exceeding £5; bronze coins to amount not exceeding 20p; and other coins, if made current by proclamation to any amount stated in the proclamation. Scottish bank notes are not legal tender.

Tenement. That which is the subject of tenure and may be held, particularly corporeal hereditaments, such as land, but also incorporeal hereditaments issuing as rents or other profits granted out of land. By transference a tenement means a house or a block containing a number of separate houses or flats.

Tenendas. In old charters the clause, commencing with the words *tenendas praedictas terras*, stating the tenure or terms on which the lands granted were to be held.

Tenendum. In old conveyances the clause commencing with the words *tenendum de me et haeredibus meis sibi et haeredibus suis*, stating by what tenure the grantee was to hold the lands of the granter.

Tenor. Properly speaking, the exact words of a document, but in common speech the general substance and purport of the document, though not its exact words.

See also PROVING THE TENOR.

Tenterden, Lord. See ABBOTT, CHARLES.

Tenure. In feudal landholding, the relationship between one person, the tenant or vassal, and another, the lord or superior, whereby the tenant holds certain lands not in full ownership but from and under the other in return for periodical rendering of services or payments in money. Under a system of tenure both lord and tenant simultaneously have an estate or interest in the land though the latter has the actual occupation and use of the land held. Various kinds of tenures (q.v.) were recognized having different rules, particularly as to the services or payments to be given to the lord.

Tenures. In feudal law and at common law in England and Scotland practically all land was held of and under the Crown, and from those who held land thus others below them might hold land. The relationship between each landholder in the feudal pyramid and the lord above him, and the terms on which he held that land, constituted the relationship of tenure. There were a number of recognized tenures based on differences in the services given in return for the land.

In English law the recognized tenures were: (a) free tenure, comprising (i) frankalmoin or spiritual tenure, the return being prayers and masses, (ii) lay tenure, of which there were three varieties, knight service, in which land was held for military services, tenure by serjeanty, in which it was held for personal service to the Crown, and socage tenure, in which the return was money or goods; and (b) non-free tenure, or villeinage, later called copyhold. The incidents of the free tenures were homage and fealty, relief and primer seisin, wardship and marriage, aids, escheat and forfeiture (qq.v.). In addition there were certain customary modes of landholding under which land was subject in various respects to certain abnormal incidents, notably gavelkind, borough English, and ancient demesne (qq.v.).

In Scotland the recognized tenures were: (a) mortification; (b) wardholding, or tenure by military service; (c) feuholding, or tenure for return in money or goods; (d) burgage, by which a burgh held its lands; and (e) blench, when lands were held for a blank or nominal return, such as a rose at midsummer.

Terce. In Scots law, the species of legal right (q.v.) formerly enjoyed by a widow, of a liferent of one-third of her late husband's heritable estate, and also claimable by a wife who divorced her husband. By statute of 1924 the right could be redeemed for a cash payment, and in 1964 it was abolished. Lesser terce was the right of a widow when her father-in-law's widow was still enjoying terce and was one-third of the remaining two-thirds.

Term of contract. A stipulation in a contract prescribing some matter of the agreement, e.g. quantity, quality, date of delivery, of the subject-matter of the contract. It is sometimes difficult to determine whether a statement is a representation inducing the other party to contract, or a term or stipulation of the contract. It is the terms which give content to the parties' agreement, and questions of whether a contract has been performed or breached commonly depend on ascertainment and interpretation of the terms agreed. Terms may be express, stated in the contract, or imported by reference to some other document where they are set out in full, or implied into the contract, by common law, by custom and usage, by statute, and in some other ways. Terms may be of varying importance, material terms, frequently called 'conditions' in English law, and non-material terms, frequently called 'warranties' in English law.

Term of years. In English law the right in land of a person having a lease for a determinate period. Term signifies both the interest which passes, and the duration for which it passes. Originally the termor had a personal claim against the lessor only, but from the Statute of Gloucester, 1278, the lessee could protect his term against third parties by the collusive use of a real action and, as against ejectors, he could later invoke a form of the action of trespass, *ejectio firmæ*, for damages, or by *quare ejecit* recover the land itself. *Ejectio firmæ* gave rise to the action of ejectment (q.v.). In modern law a term of years absolute is a term of years in possession or reversion, whether or not at a rent, and is a legal interest in land.

Termes de la Ley (or Certaine Difficult and Obscure Words and Termes of the Common Lawes of the Realme expounded). A legal dictionary in law French with a parallel English translation, by William Rastell, first published about 1620 and founded on Rastell's *Exposition of the Termes of the Lawes of England* (1563), itself a translation of his father, John Rastell's, *Expositiones Terminorum Legum Anglorum* (1527). It was later corrected, enlarged, and reprinted until as late as 1742.

Terminology, Legal. Law is constantly being dragged two ways; by the desire of jurists to develop an exact technical vocabulary comparable to those of medicine or engineering as an aid to clear thinking and reasoning and legal efficiency; and the other way by the ignorance of law and legal terms and their connotations by lay legislators, sometimes by their desire to make law understandable by the common man, and by the need in at least some branches of the law to have it stated in terms understandable by persons skilled in particular professions, business men and other laymen. To a small extent law utilizes technical terms, e.g. bailment, easement, negotiable instruments, which are not everyday words and which therefore have fairly precise connotations, but to a large extent it uses words of everyday speech, e.g. agreement, negligence, and wrong, and much of the confusion which arises stems from the fact that these words may not only have various meanings but may have usages, implications, and overtones in legal discourse other than those of everyday use. Moreover, legal terms and propositions cannot be translated into popular speech without some distortion or dilution of meanings. Much valuable analysis of legal terms has been done by analytical jurists from Austin to Hohfeld, and today, by jurists influenced by linguistic analysis, but much has also been done by the leading textbooks on various branches of law, which have exposed much confusion and fallacious reasoning.

Hexner, *Studies in Legal Terminology*.

Terms. From the thirteenth century the four portions of the year, designated Hilary, Easter, Trinity, and Michaelmas, during which only, according to court practice, judicial business could be transacted. The dates of the terms varied at different times. By the Judicature Acts, 1873–75, terms were abolished and the legal year divided into sittings and vacations, but terms are still maintained for computing the period required by the Inns of Court for call to the Bar and certain periods are 'dining terms' during which students must eat the required number of dinners.

Terms or Term days. In Scotland, the days at which rent, interest, and other yearly or half-yearly payments are normally due, and at which leases frequently begin and end. They are Whitsunday (15th May) and Martinmas (11th November).

Territorial law. Law is territorial if it is applicable to a determinate part of the Earth's surface and to persons and transactions within that territory, as distinct from a system of personal law whereby the law relative to a person and his transactions depends on the law of his tribe or race, or other group, wherever he may be. Territoriality was the basis of law in the Roman Empire and, after a period when law in Europe was personal, after the fall of the Roman Empire, territoriality became re-established as the ruling principle with the growth of feudalism and, in Italy, with the growth of the city-states. This trend was reinforced after the Reformation with the growth of modern nation-states, and the principle is now general, though not invariable, that a system of law is one developed and applied within a determinate territory by the courts of that territory and applicable, prima facie, to all persons and transactions within that territory, whatever their nationality, race, domicile, or other personal factors, and whatever the nature of the transaction. The major exception is the recognition, by each territory's rules of international private law that some matters, if having substantial connection with another territory and system of law, may have to be decided by that other territory's courts, or by reference to that other territory's system of law, or may have to be held decided by a judgment obtained in that other territory under its system of law. Thus the devolution of the estate of a foreigner dying in England is, by English law, to be decided by his *lex domicilii.*

Territorial sea. That part of the sea over which the authority of a state is deemed to extend. It forms a belt between the coastline and the open sea, and is sometimes called the maritime belt or marginal sea. Bartolus, and later Gentilis, contended for a 100 miles or two days' journey; Baldus and Bodin proposed 60 miles, or one day's journey. Philip II of Spain, in an Admiralty code issued for the Netherlands in 1563, prohibited acts of war against his subjects within sight of land or port, and Scottish practice claimed fishing rights within the area from which land could be seen from the mast-head. James VI and I defined this as 14 miles and the King of Denmark in 1691 adopted a similar view as regards fishing rights. This standard, the *portée de vue*, was adopted in treaties and proclamations of neutrality in the eighteenth century. Grotius suggested that a sovereign could assert authority over the sea so far as coercible from the land, and Bynkershoek's authority secured the adoption of the rule of *portée du canon*, the limit of cannon-shot from the shore. This was later defined as two leagues or six miles by Abren y Bertodano and Valin. De Martens set the limit as three leagues in 1789 and Azuni and Galiani as three miles. In time jurists tended to fix the *portée du canon* as one league or three miles. Lord Stowell's decisions introduced the three-mile limit into English law, and this view was adopted by the Territorial Waters Jurisdiction Act, 1878.

In many cases it was long uncertain whether a claim was to certain types of jurisdiction or was a general limit of sovereignty. There came, however, to be accepted a basic legal distinction between territorial sea as an extension of sovereignty and special jurisdiction areas, later called contiguous zones, over the high seas.

There is no general consensus as to the breadth of the territorial sea. The U.K. and the U.S. have supported the three-mile limit since the Napoleonic wars, but other states have claimed larger limits or declined to differentiate between territorial sea and jurisdictional zones, or claimed zones for particular purposes. A large number of states now claim more than three miles, or a limit plus a contiguous zone, though the view has been expressed that three miles is the minumum and 12 miles the maximum which can validly be claimed, but subject to historic rights and acquiescence by other states in greater claims. In many cases bilateral treaties regulate the matter in particular cases, but attempts to reach multilateral agreements have hitherto failed.

The territorial sea is measured from the low-water line round the coast or, in tideless seas, the average waterline of the coast. In cases of heavily indented and islet-fringed coasts a state may, at least if it has consistently claimed to do so, take as its base-lines straight lines on the map joining promontories, rather than lines following every indentation, though this principle may be limited by setting limits to the lengths of the straight lines. Title to bays is frequently claimed on the basis of historic title. Coastal islands are frequently treated as part of the mainland. Drilling platforms or other artificial constructions do not make their site part of the territorial sea.

The generally accepted view is that states have rights amounting to sovereignty over the territorial sea, but some jurists treat it as a qualified sovereignty only. It extends to the seabed and subsoil thereof, and to the airspace above. In English law the territorial sea is not at common law within the realm. The coastal state may reserve fishing for its own nationals and may exclude foreign vessels from navigation and trade along the coast. It may exercise jurisdiction in relation to sanitation and health, customs and taxation, and national security.

Custom recognizes the right of peaceful or innocent passage by foreign vessels through the territorial sea, including stopping and anchoring if incidental to normal navigation, or necessitated by weather or distress. It is doubtful whether there is right of passage for warships in peacetime, and prior authorization or notification is commonly required.

The coastal state may prevent passage which is not innocent. It may not divert or stop a ship passing through territorial sea to exercise civil jurisdiction over a person on board, not arrest the ship itself except in respect of obligations assumed by the ship itself in the course of or for the purpose of its voyage. It may exercise criminal jurisdiction over persons on board only if the crime disturbs the peace of the country or of the territorial sea, or its consequences extend to the coastal state, or the assistance of the authorities is requested by the ship. Warships and government ships are immune from local jurisdiction.

States also have jurisdictional rights, though not sovereignty, over parts of the high seas by the concept of the contiguous zone. Such zones, like the territorial sea, are contiguous to the land and share the same baselines, but do not attach automatically to a state and must be claimed. Not all states claim contiguous zones. They may be maintained for various purposes, notably for customs and fiscal regulation, immigration, prevention of pollution, sanitary regulation, security, neutrality, and fishing. Many states have a fishery limit more extensive than the territorial sea. In the prescribed zone or zones the coastal state may enforce compliance with its laws and exercise a police power.

Territorial sovereignty. The concept of the exercise by a state of actual power and control over parts of the surface of the world. It extends over tracts of land, the territorial sea appurtenant to the land, and the seabed and subsoil of the territorial sea. It does not extend to those parts of the world's surface, notably the high seas, which are common to all nations and not subject to the sovereignty of any state, nor to outer space, nor to any parts of the Earth's surface capable of being acquired by a state but not yet placed under territorial sovereignty. It is a characteristic of modern states that their sovereignty is territorial, limited to particular territories and waters. Some tracts of territory may, at least temporarily, be of indeterminate sovereignty, where there is ineffective occupation or competing titles.

In many cases territorial sovereignty has developed from the personal sovereignty over the territory of feudal or absolutist rulers, but in the last century the principle of national self-determination has emerged.

Territorial sovereignty may be acquired originally by occupation or by accretion, or derivatively by cession or other consent, by award of international conference, conquest, or prescription. To be legally effective occupation there must be actual possession and exercise of administration. Acquiescence or recognition by a competing claimant is evidence of the existence of sovereignty.

Territoriality of laws. As opposed to personality of laws, the principle that a body of law attaches to a territorial area and applies to all therein, irrespective of their origin or race. It developed first in the feudal period when, in each lordship, all men were vassals of the one lord and united by a common feudal law, while ecclesiastics all lived by the Church's law and townsmen by their burghal customs. The change took place about the year 1000, as in each region a particular system of law came to dominate. Thus, in northern France, Germanic law prevailed, in southern France a Romanistic system, in north Italy Lombard law, Roman law in central Italy, and so on. In many countries there were variations in territorial systems between different districts.

Territories of U.S. Divisions of the country not admitted to the right of statehood. The Northwest Ordinance of 1787 promised a territorial assembly, under a governor appointed by Congress, as soon as a district had a population of 5,000 free males. Statehood was promised as soon as a district attained 60,000 people, and a Bill of rights was granted as a bond between the existing states and people and states in the Northwest Territory. Both federal and state courts have held that the Northwest Ordinance is superior to the constitutions subsequently adopted by the states carved out of the Northwest Territory. The principles of this Ordinance have been followed down to the present. Unorganized territories are governed directly by federal officers. Organized territories are governed by elected representatives, but are dependent on Congress for their status and power. Their chief officers are appointed by the President and they may send one delegate to Congress, with a voice in territorial matters but no vote.

Most of the existing states of the Union passed through the stage of being territories before being admitted as states, being formed from parts of the unsettled areas then known as Louisiana Territory, Northwest Territory, and Oregon Territory. In 1959 Alaska and Hawaii were promoted from territories to states.

Tertullianus (second century A.D.). A Roman jurist, author of *De castrensi peculio liber singularis* and *Quaestiones* (8 books). He may be the same person as the theologian of that name.

Test Acts. Various statutes making eligibility for public office conditional on professing the established religion. In Scotland an Act of 1567 made profession of the reformed faith a condition of office.

The Declaration of Indulgence issued by Charles II in 1672, suspending the penal laws against Catholics and non-conformists, though cancelled, provoked the Commons to pass the Test Act, 1673 (25 Car. II, c. 2), as a measure of security against popish councillors and officials but it affected also most Protestant dissenters. It provided that all persons holding any office or place of trust, civil or military, or admitted to the King's or the Duke of York's household, should publicly receive the sacrament according to the rites of the Church of England, take the oath of supremacy, and subscribe a declaration against transubstantiation. The Act compelled the resignation of the Treasurer, Lord Clifford, and the Duke of York from the office of Lord High Admiral.

All the test acts were repealed in the nineteenth century.

Test case. An action brought to ascertain the law, one of a number of similar actions which will all be determined by the same principle.

Testa de Nevill sive libri Feudorum in Curia Scaccarii. A register in two volumes, made between 1190 and 1250 containing transcripts of inquisitions as to certain feudal holdings, so-called from either Jollan de Nevill, a justice in eyre who died in 1246, or Ralph de Nevill, an Accountant of the Exchequer about the same date. It was printed by the Record Commissioners in 1807 and again as the Book of Fees commonly called *Testa de Nevill* in 1921–31.

Testable. Able to test or having capacity to make a will. At common law a man was testable at 14. Since 1837 he must be of full age, with extensions for persons on actual service with the forces.

Testament. A will of personal property, usually found only in the formal phrase 'last will and testament': Hence testamentary, of or relating to a will.

Testate succession. The branch of the law of succession concerned with cases where the deceased has left a valid will regulating the succession to his property, as contrasted with intestate succession (q.v.).

Testatum. The part of an indenture which begins with the words 'Now this indenture witnesseth'.

Teste. The concluding part of a writ, so called from the words *teste meipso*, giving the date and place of issue. Writs are generally tested in the name of the Lord Chancellor.

Testimony. Particularly in American usage, evidence given orally in court.

Testing clause. In Scots law, the clause at the end of a solemnly executed deed narrating the names, addresses, and designations of the signatories and the witnesses to their signatures and other particulars relative to the execution of the deed.

Texas* v. *White (1869), 7 Wallace 700 (U.S. Supreme Court). After Texas seceded from the Union in 1861 its government authorized the sale of certain U.S. bonds to finance the war against the Union. After the war governors of Texas brought suits to recover the bonds, contending that the secession government had had no power to dispose of the bonds. The Supreme Court had to determine the status of Texas during the war. A majority held that Texas had never been out of the Union and that the acts of the secession government of that state had been illegal, and that it was still owner of and could recover the bonds.

Textbook. The general name for a book which seeks to state in systematic form the principles relative to a particular branch or topic of the law. Textbooks vary greatly in size and value from the large and detailed practitioners' book to the small, elementary, student's book. There is no clear distinction, but the term textbook is applicable particularly to a book for students and the term treatise particularly to a book for practitioners. According to British practice statements in textbooks (as distinct from books of authority (q.v.)) are not authoritative but may nevertheless be very valuable for their examination and evaluation of the authorities on the branch of law. Passages from textbooks of repute may be adopted by counsel and quoted as setting out their arguments, and they may be referred to by judges as evidence of general professional understanding or as supporting their views. An older view was that the book of a living author could not be cited as authority, but in modern practice some books by living authors have been cited in argument and judgment in ways hardly distinguishable from treating them as books of authority.

See also BOOKS OF AUTHORITY; INSTITUTIONAL WRITERS; TREATISES.

Textor, Johann Wolfgang (1638–1701). Director of the Hohenlohe Chancellery in Neuenstein and (1666) professor of the Institutes at Altdorf and later (1673) professor of jurisprudence at Heidelberg. He was author of an important *Tractatus juris publici de ratione status Germaniae modernae* (1667), *Disputatio de jure ecclesiastico* (1677), and *Synopsis juris gentium* (1680). From 1691 till his death he was syndic at Frankfurt where he wrote the *Tractatus de jure publico statuum imperii* (1701).

Thankerton, Lord. See WATSON, WILLIAM

Thant, U (1909—). Burmese teacher, civil servant, and delegate to the U.N. Later (1957) he became Burma's permanent U.N. representative and in 1959 Vice-President of the General Assembly. He served as Secretary-General of the U.N., 1962–71, and sought to maintain neutrality between East and West.

Thavies' Inn. An Inn of Chancery.

Thayer, Ezra Ripley (1866–1915). Son of J. B. Thayer (q.v.) served as Dean of the Harvard Law School, 1910–15, but wrote little.

Thayer, James Bradley (1831–1902). Admitted to the Boston Bar in 1856, and, having contributed to legal periodicals and worked on an edition of Kent's *Commentaries*, became a professor in the Harvard Law School in 1874 and collaborated with Langdell, Ames, and Gray in building the modern law school. His writings include *A Preliminary Treatise on Evidence at the Common Law* (1898), which contains a history of the jury, *Cases on Constitutional Law* (1895), and a brief biography of John Marshall.

Theft. In English law, the statutory offence of dishonestly appropriating property belonging to another with the intention of permanently depriving the other of it whether for gain of not. In 1968 it replaced larceny, embezzlement, and fraudulent conversion.

In Scots law theft is criminal at common law and may be constituted by unlawfully appropriating to oneself possession of goods belonging to another, or by finding and appropriating to oneself. Theft by housebreaking is an aggravated form of theft.

In both systems the thing must be capable of removal, and be owned by another. To catch a trout in a river is not theft, but to catch it in another's fishpond is. Akin to theft are blackmail, burglary or housebreaking, cheating, embezzlement, fraud, and robbery (qq.v.).

Theftbote. Taking a payment from a thief or receiving back some of the stolen goods from him to release him from criminal prosecution.

Thellusson Act. The Accumulations Act, 1800, setting limits on the times for which a settlor might direct that income of a trust might be accumulated, now replaced by modern legislation to the same general effect.

Theobald (*c.* 1090–1161). Archbishop of Canterbury from 1138, statesman and a patron of learning, including Roman law, brought Vacarius to Oxford to teach Roman law, and played an important role in setting Henry II on the throne.

Theodoric, Edict of. An enactment issued by Theodoric, king of the Ostrogoths in Italy, probably about A.D. 500, in 154 articles, based entirely on Roman law sources, chiefly Paul's *Sentences*, the Gregorian, Hermogenian, and Theodosian codes, and dealing mainly with criminal law and procedure. It was binding on Roman population and Gothic conquerors alike, and where it was silent each race was ruled by its own law. It was promulgated in Italy and also adjoining regions and influenced the civilization of Germanic tribes. Theodoric was also responsible for some lesser edicts.

Theodosian Code. See *Codex Theodosianus.*

Theophilus (sixth century). A professor of law at Constantinople who collaborated in Justinian's codifications and particularly in writing the *Institutes.* A Theophilus, probably the same, wrote a Greek paraphrase of the *Institutes*, which survives almost complete, dating from about 533 or 534. It was based on and followed closely a Greek paraphrase of Gaius' *Institutes.*

Theory of law. A phrase sometimes used as a synonym for jurisprudence, but more properly for a branch of that study, concerned with the nature of law.

Theresiana (or *Constitutio criminalis Theresiana*). An elaborate criminal code enacted for Austria in 1769, in many respects modern, as in considering deterrence and reform of the offender, but mediaeval in some of the kinds of conduct deemed criminal.

Thesiger, Frederick, 1st Lord Chelmsford (1794–1878). He worked in the West Indies and served in the navy before being called to the Bar in 1818. He entered Parliament in 1840 and became Solicitor-General in 1844 and Attorney-General in the next year, but lost office in 1846. He held the latter office again briefly in 1852 and was Chancellor in 1858–59 and again in 1866–68. In the interval he sat regularly in the House of Lords. He had been a great success at the Bar and in the House of Commons, but was not successful in the House of Lords. His judgments indicate that he was a skilled but not very profound lawyer, but some of them are certainly leading cases.

Thibaut, Anton Friedrich Justus (1772–1840). German jurist, professor at Kiel, Jena, and Heidelberg, author of *Theorie des logischen Anslegung des römischen Rechts* (1799) and an important *System des Pandektenrechts* (1803) substantially a codification of Roman law as applied in Germany.

Founding on natural law principles and patriotic sentiment in Germany he advocated the liberation of Germany from foreign oppression and foreign law and in *The Necessity for a Common Civil Law for Germany* (1814) called for a single civil code for all Germanic peoples, based on the principles of justice and right reason, adapted to Germanic custom and with foreign law eliminated and Roman law rejected. His pamphlet provoked Savigny's (q.v.) opposition.

Thing. In mediaeval Scandinavia provincial and local assemblies of freemen which met at fixed intervals and elected kings, legislated, and reached legal decisions. They were presided over by a local chief or a law speaker, who could expound the law orally, and though nominally democratic were dominated by the local chieftains. They lost power in the thirteenth and fourteenth centuries. In Iceland the *Althingi* was a national body which met at Thingvellir from 930 to 1798 and gave its name to the modern Icelandic Parliament.

Things. The objects of legal rights. They are distinguishable into things corporeal and things incorporeal, or into things real or immovable, which are lands, tenements, and other immovables; things personal or movable, which are goods and chattels; and things mixed, which partake of both of these categories. There are also recognized things common, such as air and running water, not capable of appropriation by anyone; things public, available to all persons; and things sacred, which are not objects of commercial dealing. In Roman law a common classification of rules of law was into law of persons and law of things; in this sense things comprise all proprietary rights, both those against other persons arising from obligation, such as from contract or tort, and those against persons generally arising from claims to objects.

Third party. Any person who is not a party to a relationship or transaction between any two others. The general rule is that a contract between two persons neither confers rights nor imposes liabilities on any third party, but to this there are certain exceptions. A party sued by a plaintiff may bring in a third party and claim contribution from or indemnity against him on the ground that the third party is alone, or at least partly, responsible for the default with which he is charged.

Thirlage. In old Scots law, a stipulation in ancient charters that the occupier of lands must have the grain grown on the land ground at a specified mill. It was a kind of trade monopoly of the same kind as the exclusive rights of trading in burghs. Under the obligation of thirlage payments in kind, known as multures, were due for grinding the grain.

Thirteen Colonies. The British colonies on the eastern seaboard of North America which declared independence in 1776, formed themselves into states, adopted a Constitution (1789) and became the original states of the U.S. They were Connecticut, Delaware, Georgia, Maryland, Massachusetts, New Hampshire, New Jersey, New York, North Carolina, Pennsylvania, Rhode Island, South Carolina, and Virginia.

Thirty-One Articles, Petition of. In 1406 when Henry IV's government was in confusion the Commons petitioned him, in effect, to conform to standards of good government; the King accepted the articles. The main provisions were: that the King should name 16 councillors to advise him until the next Parliament; that the council should determine nothing cognizable at common law, unless for reasonable cause and by the advice of the judges; that the council and chief officers of state were to be sworn in Parliament to observe the common law and the statutes and ordinances, for the king's household; and for the good government of the realm. Abuses of various kinds in council and the courts were listed and forbidden, and the administration of the household and the courts was to be regulated. The articles were to be in force only from the beginning of that Parliament until the next.

This day six months. The parliament term for the date to which further debate on second reading of a Bill is deferred, which in effect means never, the House rejecting the Bill.

Thöl, Johann Heinrich (1807–84). German jurist, first to put commercial law on a scientific basis in Germany and a member of the commission which framed the German commercial code. He wrote *Das Handelsrecht* (1841–80) and various other works.

Thole. In Scotland, to suffer, endure, put up with. To thole an assize is to suffer trial, which, if done, is a bar to any future trial on the same charge.

Thomas* v. *Sawkins [1935] 2 K.B. 249. Thomas and others organized in private premises a meeting which the public were invited to attend. Police officers, though refused admission, insisted on entering the hall as private individuals and staying during the meeting. It was held that the police were entitled to enter and remain on the premises if they had reasonable grounds for believing that, if they were not present, seditious speeches or a breach of the peace would take place, and were not guilty of trespass.

Thomas* v. *Sorrell (1674), Vaughan 330. The plaintiff claimed a sum as penalty for the defendant's having sold wine without a licence contrary to statute. The jury found that a royal patent had permitted the Vintners Company to sell wine without a licence, notwithstanding a statute to the contrary. The judge held that the King might dispense with an individual breach of a penal statute by which no man was injured, or with the continuous breach of a penal statute enacted for the King's benefit. The Bill of Rights, 1689, abolished the Crown's alleged dispensing power, save where this was authorized by Parliament.

Thomasius, Christian (1655–1728). German jurist, professor at Leipzig and Halle, who urged the teaching of German law and its establishment in practice in place of Roman law. He had great influence in promoting German law; he saw law as natural law, founded on divine decrees addressed by God to the hearts of men. He was author of *Institutiones iurisprudentiae divinae* (1688); *Fundamenta iuris naturae et gentium* (1705), *Usus modernus forensis ad Institutiones et Pandectas* (1713); *Dissertationes*, and many other works.

Thomson, Sir Alexander (1744–1817). Thomson was successively a practitioner, master in, and Accountant-General of, the Court of Chancery. He became a Baron of Exchequer in 1787 and Chief Baron in 1814. His reputation as a lawyer and as a judge was high.

Thomson, George Morgan Thomson, Lord, of Monifieth (1921–). Was an M.P., 1952–72, and Secretary for Commonwealth Affairs, 1967–68, Minister without Portfolio, 1968–69, and Chancellor of the Duchy of Lancaster and Minister for European Affairs, 1969–70, and then a Commissioner of the European Economic Community, 1972–76. He became a life peer in 1977.

Thomson, George Reid (1893–1961). Called to the Scottish Bar in 1922, he took silk in 1936, became Lord Advocate in 1945 and was Lord Justice-Clerk 1947–61.

Thomson, James Bruce (1810–73). Qualified in medicine and became resident surgeon at Perth Prison in 1858. In that capacity he made investigations which make him a pioneer of criminology, and sought to discover any connection between crime and mental and physical disease. He reached the conclusion that heredity was the prime factor of criminality, and his work gave a stimulus to criminal anthropology.

Thomson, Thomas (1768–1852). Passed advocate of the Scottish Bar in 1793 and in 1806 was appointed deputy Clerk Register, in charge of all the records of Scotland. As such he published Vols. II–XI of the Acts of the Parliaments of Scotland (Vol. I did not appear till 1844, edited by Cosmo Innes) and numerous other publications of the greatest importance for legal history and research. He succeeded Sir Walter Scott as President of the Bannatyne Club which published many other older works of historical importance, many edited by Thomson. In 1828 he became a principal clerk of session. In 1839 he was removed from his office of deputy Clerk-Register. His work is still of the highest importance for the history of Scots law.

Thonissen, Jean Joseph (1816–91). Belgian jurist and statesman, notable for his studies in criminal law and legal history. His works include *La constitution belge annotée* (1846), *Complement de code pénal* (1846), *Études sur l'histoire du droit criminel des peuples anciens* (1869), *Le droit penal de la république athenienne* (1875), and *L'organisation judiciare, le droit pénal et la procedure pénale de la loi salique* (1882). Later he played a major part in legal reform in Belgium.

Thornton, Gilbert de (*fl.* 1290). He became chief Justice of the King's Bench in 1289 in succession to Hengham and held that office till 1295. He composed a *Summa*, an epitome of Bracton's *De Legibus*, but omitting the cases and citations from the Roman law and modifying the arrangement.

Thorpe's case (1454), Rot. Parl. V, 239. Thorpe, Speaker of the Commons and a baron of Exchequer was imprisoned for non-payment of a fine for a trespass in seizing certain goods of the Duke of York. The Commons complained to the King and the Lords demanding release of the Speaker. The judges, consulted by the Lords, declined to declare the privilege of Parliament but thought that Thorpe was entitled to release, but the Lords declined to release him, and the Commons were ordered to elect a new Speaker. The case is an important stage in the establishment of the privilege of freedom from arrest.

Threats. Expressions of intention to cause hurt or harm or to do acts of a particular kind. Threats giving rise to reasonable fear of harmful consequences are an actionable tort, and threats of various kinds of harms, e.g. to kill, to destroy, or damage property, are criminal.

Three-mile limit. See TERRITORIAL WATERS.

Thring, Sir Henry, Lord (1818–1907). Called to the Bar in 1841 and became interested in statute law, and pamphlets by him brought him to notice as a draftsman. His first major work was the Merchant Shipping Act of 1854. In 1860 he was appointed home office counsel. In 1869 there was constituted the office of Parliamentary Counsel to the Treasury, the staff of which was to settle departmental Bills and draw all other Bills, except Scottish and Irish ones, as he might be required to draw. Thring held this office from 1868 to 1886, a period of great legislative activity. He introduced greatly improved style, embodied in his *Instructions to Draftsmen* and *Practical Legislation* (1902) and devoted his whole energies to improvement of the statute law. He was an original member of the statute law committee from 1868 and for many years chairman. He planned the *Chronological Table* of and *Index to the Statutes in Force* (both now published annually), planned numerous statute law revision Bills, and the publication of the living statute law in the *Statutes Revised*, which substituted 23 volumes for 195. His initiative also underlay the continuation of the *State Trials* from 1820 to 1858. He was made a peer in 1886 when he retired. He also supervised the compilation of the first edition of the War Office *Manual of Military Law*.

Thuringian code (or Code of the Angli and Werini). This is a small code of local customs, dating from about 800, dealing with penal and miscellaneous, mainly criminal, matters, and has affinities with the code of the Ripuarian Franks. Parts of it are based on the *Capitulare legi Ribuariae additum* of 803.

Thurlow, Edward Thurlow, 1st Lord (1731–1806). After studying in a solicitor's office he was called to the Bar in 1754; initially he had little practice but took silk (1762) and became M.P. in 1765. Thereafter his progress was rapid and he made his name by his speech for the appellant in the *Douglas Cause* (q.v.). He became Solicitor-General in 1770 and Attorney-General in the following year, in both of which capacities he appeared in important cases. He became Baron Thurlow and Chancellor in 1778 and with his commanding appearance and manner allied to legal learning and ability in advocacy became a leader in the government as well as in the courts. In 1783 he resigned on the formation of the Fox-North coalition, but later in the same year resumed office under Pitt and held office till 1792 when he had to resign after secret intrigues with the Prince of Wales came to light. He had a tremendous reputation as statesman and lawyer, despite his temper, his manners, and his delays in dealing with cases, but as a politician was self-seeking and unscrupulous. He never really mastered equity and relied on others for learning, but nevertheless delivered some judgments of lasting importance.

R. Gore-Browne, *Chancellor Thurlow*.

Tichborne case. The Tichbornes were a wealthy Hampshire family. Roger Charles Tichborne was born in 1829, served in the army 1849–52, sold out and went to South Africa. In 1854 he embarked at Rio de Janeiro in a ship which was lost and it was assumed that he had perished. His mother, however, believed that he was not dead, and in 1865 was informed that a man answering the description of her son had been found in Queensland. She persuaded this person to come to London in 1866, and professed to recognize him, though the rest of the family declared him an imposter. He succeeded, however, in persuading many persons of his identity with the lost son. In 1868 an issue was directed to be tried in the Court of Common Pleas as to whether the claimant was the heir of Sir James Tichborne (the real Roger's father, who had died in 1862). To stave off his creditors, his ejectment action against the Tichborne trustees came on before Bovill, C. J., and a special jury. The trial lasted 102 days (May 1871–March 1872), and the claimant was cross-examined for 22 days by Sir John Coleridge (later Lord Coleridge, C. J.). A large number of witnesses swore to the claimant's identity, but cross-examination revealed enormous ignorance of what the real Roger Tichborne must have known. The jury found for the defendants. The fact was that one Arthur Orton (1834–98) went to sea in 1849, deserted in South America, and in 1852 went to Australia. There he bore the name Tom Castro, adopted from persons who befriended him in South America, and worked as stockman, mail-rider, and probably bushranger. On being brought to England he had re-established contact with the Orton family in Wapping. He probably pressed his claim to try to get money to discharge the worst of his debts.

After the trial he was arrested for perjury and tried at Bar in 1873 before Cockburn, C. J., and Mellor and Lush, JJ. A great mass of evidence was called on both sides, but on the 188th day of the trial the jury found that the claimant was Arthur Orton, and he was sentenced to 14 years' penal servitude. Orton's counsel, Kenealy (q.v.) was disbarred for his gross breaches of professional etiquette at and in connection with the trial, but sought in Parliament to have the case reopened. Orton was released in 1884 and in 1895 published a signed confession. Some, however, continued to maintain that he was an illegitimate member of the Tichborne family.

J. B. Atlay, *Famous Trials*; F. H. Maugham, *The Tichborne Case (1936)*; D. Woodruff, *The Tichborne Claimant* (1957); D. N. B. Supp., s. v. Orton.

Ticket-of-leave. When penal servitude replaced transportation in 1853 persons serving such sentences were made eligible for tickets-of-leave after a certain time, in effect for release on licence. The 1853 Act made such release more or less automatic and little was done to ensure that the stipulations as to behaviour were complied with. This gave rise to public outcry and an 1864 Act made it more difficult to get a ticket, and supervision after release was made more stringent. Prisoners transported to Australia were also frequently granted tickets-of-leave and established themselves independently thereunder.

Tidd, William (1760–1847). Practised as a special pleader under the Bar for 30 years and was not called till 1813. He had a great reputation in the art of drawing special pleadings and many men, later distinguished in the law, were pupils of Tidd. His fame rests on his book *The Practice of the Court of King's Bench in Personal Actions* (1790–98) which was for 50 years nearly the only authority in matters of common law practice, and stood unequalled in its day as a statement of the rules of procedure. He also wrote *Practical Forms and Entries of Proceedings in the Courts of King's Bench, Common Pleas and Exchequer of Pleas* (1799) and certain less important books.

Tigni immittendi. In Roman law a servitude consisting in the right to insert a beam from one house into the wall of another making the latter take part of the weight of the beam.

Tillet, Jean du (?–1570). A French humanist jurist, collaborated with Cujas in editing the *Codex Theodosianus* (1550) and edited Ulpian's *Regulae*.

Time. The passage of time, marked by observed successions of days and nights, seasons and natural changes, birth, maturation, and death, and similar facts, has great legal importance. Statutes and contracts may provide for certain things being done for a certain time or within a certain time, and other rules of law provide for legal consequences following if something does or does not happen within a stated time. Time is important in computing birthdays and ages.

The passage of time has important consequences for legal rights. Two sets of rules are found in legal systems dealing with the effect of lapse of time, rules of prescription and rules of limitation. A rule of prescription is a rule of substantive law prescribing that after the lapse of a stated period of time a particular right is wholly extinguished; in some cases such a rule, such as the one which extinguishes claims by non-possessors to a piece of land, may have the effect of fortifying the possessor's title and rendering it indefeasible. A rule of limitation on the other hand is a procedural rule that, after the lapse of a stated time, legal action of a particular category, e.g. a claim of damages for personal injury, may not be brought.

Time is measured in days, weeks, months, and years. Days, months, and years are based respectively on the rotation of the Earth on its axis, the orbital motion of the moon about the Earth, and the orbital motion of the Earth about the Sun respectively; the week is not related to any astronomical cycle but is traditional, possibly originating in the intervals between market days and established among the Jews and thence in Christianity.

The calendar year has since 1752 begun on 1st January and consists of 365 days, or in years called leap years, being years the number of which is exactly divisible by four, 366 days. It is divided into 12 months four having 30, others 31 days, but the month of February having 28 days in common years and 29 days in leap years. Before 1582 the Julian Calendar applied throughout Christendom under which in every fourth year an extra day was intercalated between 24th and 25th February. Under this calendar the calendar year got out of step with the solar year and in 1582 Pope Gregory XIII ordained that 10 days be suppressed and the intercalated days reduced by three in every 400 years. The Gregorian Calendar was adopted throughout Christendom, including Scotland, but not in England or countries of Orthodox or Greek communion. The Calendar (New Style) Act, 1750, and the Calendar Act, 1751, assimilated the English to the Gregorian Calendar by omitting 11 days, modified the intercalation by omitting three leap years in every 400, and made the calendar year commence on 1st January, instead of 25th March. In Scotland the year had commenced on 1st January from 1600.

The method of determining Easter was left unaffected by the change from the Julian to the Gregorian calendar, and it is still the first Sunday after the first full moon which happens next after 21st March; if that happens on a Sunday, it is the next Sunday. The Easter Act, 1928, provides for a fixed Easter, but has never been brought into force.

The term 'year' in legal contexts may be a solar year of 365 days or a calendar year or any period of 365 days running from a date fixed. In statutes passed since 1889 in relation to taxation, a financial year is the period from 1st April to 31st March, but for income tax purposes the year is from 6th April to 5th April.

Particularly as between landlord and tenant reference is made to the quarter days, which divide the year into four; these are, in England, the four feast days, Lady Day (25th March), Midsummer Day (24th June), Michaelmas Day (29th September), and Christmas Day (25th December) and in Scotland, Candlemas Day (2nd Febuary), Whitsunday (15th May), Lammas Day (1st August), and Martinmas (11th November). The legal year in England was formerly divided into four terms, Hilary, Easter, Trinity, and Michaelmas, but is now divided into sittings and vacation, though the Inns of Court maintain the old terms, and the academic year is commonly divided into four terms under various names.

A month in statutes after 1850, in contracts,

deeds, and wills since 1925, and always by mercantile custom is a calendar month.

A month may run from any date in a month to that date minus one day in the next month.

A week is prima facie the period between 0001 hours on any Sunday to 2400 hours on the succeeding Saturday, but may mean any period of seven consecutive days, a calendar week.

A 'day' is prima facie the period between 0001 hours and 2400 hours immediately succeeding, but may mean any period of 24 hours, or the period between sunrise and sunset, or a similar period provided by statute.

The hour of the day is prima facie Greenwich mean time and not local time. Under the Summer Time Acts for certain periods each year, local time in the U.K. is one hour in advance of Greenwich mean time.

In England the almanac annexed to the Book of Common Prayer is part of the common law and recognized by statute. Certain feast days of the Christian Church are fixed by dates, e.g. Christmas, and certain are movable. The Calendar Act, 1750, contains Tables and Rules for the Feasts and Fast days of the Church. The Easter Act, 1928, to regulate the date of Easter has never been brought into force.

On Red Letter Days (holy days and saints' days indicated in early ecclesiastical calendars in red ink, the list of which was finally approved at the Council of Nicaea in A.D. 325, for which special services are appointed in the Book of Common Prayer, to which have been added certain dates such as that of the Queen's birthday), scarlet robes are worn by the judges of the Queen's Bench Division (as well as at sittings of the Crown court and on state occasions).

By common law Good Friday and Christmas Day are holidays in England and New Year's Day in Scotland. By the Banking and Financial Provisions Act, 1972, replacing Bank Holidays Act, 1871, certain days must be kept as holidays in all banks in England, Northern Ireland and Scotland respectively, and these were extended to customs and inland revenue offices by the Holidays Extension Act, 1875. Particular days may be appointed as holidays in special circumstances. The right to holidays in various business, trades, or industries is conferred by statute, or agreement, and shop assistants are entitled to a weekly half-holiday.

Ther are restrictions on doing various things on Sundays.

In computing a stated period of time, regard must always be had to the particular statute or contract, if any, which is in issue, but the general rule is that a period of time runs from the first moment of the next day, until the last moment of the final day of the period; a provision for 'clear days' excludes both the starting day and the final day from the period; thus, if an act required three clear days notice and notice is given on Monday, the notice does not expire until the first moment of Friday.

Expressions limiting time frequently have to be construed; principles of common law frequently use the expression 'a reasonable time'; this has to be construed according to the circumstances of the particular case and is accordingly entirely a question of fact.

Time immemorial or time of living memory is from a time whereof the memory of man is not to the contrary, such that no man alive has heard proof to the contrary. Time of legal memory runs from the commencement of the reign of Richard I (July 6 or September 3, 1189) because the Statute of Westminster I, 1275, fixed that period as the time of limitation for bringing certain real actions.

Time charter. A charter-party of a ship under which it is let to the charterer for a stated period of time, rather than for one or more voyages.

Tindal, Sir Nicholas Conyngham (1776–1846). Acquired a good practice as a special pleader before being called to the Bar in 1809. He made a reputation as a very learned lawyer and became Solicitor-General in 1826 and Chief Justice of the Common Pleas in 1829, which office he held till his death. He was the most eminent holder of that office for many years and a very great judge, dignified, learned, and lucid and accurate in exposition, and contributed greatly to the development of the common law.

Tinsel of the feu. In Scots law, forfeiture and termination of a feu-right which may be declared by a court in the case of the vassal's failure to pay feu-duty for two (now five) years, or in the case of incurring any ground of forfeiture stated in the feu-contract.

Tipstaff. An officer of the Supreme Court, akin to a constable. The functions of these officers are now confined to arresting persons guilty of contempt of court, but in the Queen's Bench Division the tipstaves have also in theory custody of any prisoner brought before the court or committed by the court when exercising criminal jurisdiction.

Tiraqueau, André (Tiraquellus, Andreas) (1480–1558). French humanist jurist, counsellor to the Parlement of Paris, author of *Tractatus de jure constituti possessorii*, *De legibus connubialibus* (1513), and *De nobilitite et jure primogenitorum* (1549).

Tithes. The tenth part of all fruits and profits due to God and the Church for the maintenance of the Church and clergy. Tithing goes back to Old Testament times, adopted by the early Christian Church and became obligatory all over Europe as

Christianity spread. It was enjoined by ecclesiastical law from the sixth century and enforced by secular law from the eighth. In England it became obligatory from the tenth century. At the Reformation tithes were continued for the benefit of the reformed Churches. Tithes were distinguished as praedial, if arising from crops, personal, if arising from the profits of industry and labour, and mixed, if arising from things nourished by the land, such as animals and poultry and their by-products. Certain lands were exempted from tithes.

The inconvenience of collecting tithe in kind early led to arrangements for commutation of tithe into corn rents or tithe rent-charges. The latter has now been extinguished, landowners paying instead to the Crown a redemption annuity for 60 years from 1936 at the end of which time lands will be free of tithes.

Under the name of teinds (q.v.) tithes were exacted in Scotland until modern times. They were abolished in Ireland in 1871 and have been abolished in most European countries, in France in 1789. Tithe was never a legal requirement in the United States, but members of some churches are required to, or do, pay contribution of comparable amounts to their churches.

Tithing. A Saxon subdivision of the hundred (q.v.) and the unit for local administration in some parts of England. Also a group of 10 men with their families associated in a group all of whom were bound to the King for their good and peaceable behaviour; the chief was the tithing-man, a name which came later to mean a kind of under-constable.

Titius, Gottlieb Gerhard (1661–1714). German jurist, author of *Specimen juris publici Romano-Germanici* (1698) and *Observationes in Pufendorfii libros de officiis hominis et civis* (1703).

Title. The legal connection between a person and a right constituted by some act or event having legal significance. Every right which a person has attaches to him by virtue of some title, and rights of the same character, e.g. ownership of goods, may be vested in different persons by virtue of different kinds of titles, by making, by gifts, by purchase, by inheritance, and so on.

Title may be acquired originally, where no person has previously had title to the right in question; by investitive facts, as by creation, e.g. title to copyright by writing the work, by occupation, by capture, by accession, where one's sheep have lambs, by injury done, vesting the injured person with a title to claim damages, by statute or judgment, by creation by the Crown; or derivatively, where the title is derived from another, as by gift, purchase or other contract, inheritance, or judgment of court.

Title may be lost by divestitive facts where it is transferred to another, as by gift, sale, or transmission on death, or where the subject-matter of the right is consumed, destroyed, or abandoned.

The word is most commonly used in connection with land. Various kinds of title to land may be recognized. The highest is the title of the absolute owner, in English law the owner of a fee simple absolute in possession. Lesser rights are conferred by the interest for life, the term of years, the mortgage, down to the title of occupancy or bare possession. The squatter has no title and may be ejected at any time.

The word 'title' also signifies a mode of address indicating an honour or dignity. Thus peers are known as 'Lord' and Knights as 'Sir' and certain persons are by courtesy addressed as 'The Honourable . . . ' or 'Lady'. Titles of honour are a kind of incorporeal hereditament.

In legal procedure every action, petition, or other proceeding has a title consisting of the name of the court, the names of the parties, and the reference to the record comprising the year, the initial of the first plaintiff's surname, and the serial number of the action. Every pleading commences with the title.

The title of a patent is a short description of the subject-matter of an invention and must be prefixed to every specification.

In land law 'titles' or 'title deeds' denote those deeds which evidence a person's title to particular land, which require to be examined by a prospective purchaser and which are transferred to him on completion of the purchase.

Title, Convenants for. In conveyances of real or personal property, certain covenants for title are implied. In a conveyance of real property, other than a gift or mortgage, there are covenants that the person conveying has the right to convey, that the person taking the conveyance shall enjoy the subject-matter without disturbance by the person conveying, and that he will execute all further assurances required for more perfectly assuring the subject-matter of the conveyance to the person to whom it is made. Generally similar covenants are implied in conveyances of leasehold, by way of mortgage, by way of settlement, and in certain other cases. In a conveyance for personal property there is an implied condition that the seller has a right to sell the goods or will have a right to sell when the property is to pass, an implied warranty that the goods will be free from any charge or encumbrance in favour of a third party, not declared or known to the buyer before or when the contract was made, and that the buyer will enjoy quiet possession of the goods except so far as it may be disturbed by the owner or other person entitled to the benefit of any charge or encumbrance so disclosed. If the seller is purporting to sell only a limited title, there is an implied warranty that all charges have been disclosed and that neither the seller nor a third party having a limited title will disturb the buyer's quiet possession of the goods.

Titles, Judicial. The titles attaching to various judicial offices are the product of history rather than of design. In the old English common law courts the men who were sent by the King to do justice in his name came to be called justiciars or, increasingly as time went on, Justices, whether of the King's Bench or the Common Bench (later Common Pleas) but those who sat in the Exchequer were Barons because the Exchequer evolved from a sitting of the King's barons and great officials to deal with Exchequer or financial business.

The Chancellor took his name from the *cancelli* or screen behind which the secretarial work of the royal household was carried on, he being originally the head of the office. The Master of the Rolls, originally a clerical officer, became the Chancellor's deputy in the Court of Chancery. In 1813 one, and later (1842) three, Vice-Chancellors were appointed to assist the Lord Chancellor to exercise his equity jurisdiction, and in 1851 the creation of the Court of Appeal in Chancery produced the rank of Lords Justices in Chancery. Down to 1831 there was a Chief Judge in Bankruptcy with three judges and six commissioners.

The judge of the Court of Admiralty also became Probate judge in 1857 and when the Court for Matrimonial Causes was established in 1857 he became judge of that court also, as the Judge Ordinary. His three jurisdictions became the Probate, Divorce, and Admiralty Division in 1875. In 1869 the London Court of Bankruptcy was established under a Chief Judge in Bankruptcy. In 1883 it became part of the High Court.

When the old common law courts were abolished in 1875 and the modern Supreme Court was created, the judges of all the Divisions became Judges of the High Court, and in 1877 Justices of the High Court. The judges of the Court of Appeal were designated Justices of Appeal in 1875 and in 1877 Lords Justices of Appeal.

The Lord Chief Justices of the Queen's Bench and of the Common Pleas lasted till 1881, as did the Lord Chief Baron of the Exchequer, though from 1859 the first of these had been entitled Lord Chief Justice of England. The office of President of the Probate, Divorce, and Admiralty (now Family) Division appeared in 1875.

The creation in 1876 of the salaried life peers to serve as judicial members of the House of Lords introduced the title of Lord of Appeal in Ordinary, but the term Law Lords was and is colloquially given to them and also to those peers who, having held high judicial office, might be and sometimes were and are asked to sit in the House to hear appeals.

In 1970 the title Vice-Chancellor was revived for the administrative head of the Chancery Division and from 1965 the presence of ladies on the High Court Bench has resulted in the title 'Mrs Justice X'.

It has long been traditional to confer a knighthood on persons appointed Justices of the High Court (D.B.E. for ladies), but a judge is nevertheless referred to as 'Mr. (or Mrs.) Justice Blank'. Lords Justices of Appeal are sworn of the Privy Council; thus they become Right Honourable, and this honour is sometimes conferred on an ordinary Justice after distinguished service.

According to Foss (VIII, 200), it was only during the eighteenth century that puisne judges began to be addressed as 'Your Lordship', a mode of address previously reserved to the Chancellor, Chief Justices, and Chief Baron, but today this is proper only when on the bench or on circuit. Vice-Chancellors and the Master of the Rolls were traditionally addressed as 'Your Honour' prior to 1875. Today all judges of the superior courts are addressed officially as 'Your Lordship'. Any judge sitting at the Central Criminal Court should now be addressed as 'Your Lordship'.

Judges of the county courts and Crown Court (circuit judges) are entitled Judge X and addressed as 'Your Honour'. The origin of the judicial title of Recorder is obscure; a Recorder is addressed as 'Sir'. There remain a few ancient titles such as the Vice-Chancellor of the County Palatine of Lancaster, the Common Serjeant, the Dean of the Arches. Magistrates are sometimes called 'Your Worship'.

Similar titles have spread with English law over most of the English-speaking world.

In Scotland the Court of Session, on its foundation, replaced committees of Council and of Session which had earlier sat to dispense justice, so that its judges became Lords of Council and Session as well as Senators of the College of Justice. Till 1887 some, and since then all, of the same persons sit also in the High Court of Justiciary (q.v.) as Lords Commissioners of Justiciary. The chief judge is both Lord President of the Court of Session and Lord Justice-General of Scotland, and the second chief judge is, in both courts, Lord Justice-Clerk. All are addressed judicially as 'Your Lordship'. They bear the courtesy title Lord X, Lord Y, and may keep their family name, or take a title from land, e.g. Lord Kilbrandon, but are not peers, nor even knighted, though a peerage or a knighthood may be conferred in particular cases. Sheriffs-principal and sheriffs hold a very ancient office and are also addressed in court as 'Your Lordship'.

Tocco, Carolus (thirteenth century A.D.). A professor at Bologna who prepared a critical apparatus of the entire gloss on the Lombard law which acquired an authority almost as great as that of the text.

Tocher. In Scotland, a marriage portion or dowry which a wife brings on her marriage.

Tolbooth. In Scottish burghs, a place where goods were weighed, a custom-house or prison.

Toler, John, Lord Norbury (1739–1831). Called to the Irish Bar in 1770, became Solicitor-General in 1789 and Attorney-General in 1798. In 1800 he became Chief Justice of the Common Pleas in Ireland and a baron. He retired in 1822, becoming Earl Norbury.

Toleration. Acknowledgment of the liberty of persons to hold beliefs or do things which were not permitted or are not acceptable to a large number of persons in the community. It is practical recognition of individual freedom of conscience and of belief. The term has frequently been used of permitting freedom of worship to non-Catholics in Catholic countries and conversely, but it extends more widely, to social and political views, to practices and conduct, e.g. nudity, homosexuality, and refusal to join a trade union. It is always a question of circumstances and degree in what respects and how far toleration should be permitted at a particular time by law. Restrictions on particular kinds of beliefs or conduct may be deemed desirable or necessary in the interest of national security, public health, general moral standards, or the health of the participants, as e.g. in drug-taking, and the protection of other members of society.

Toleration Act, 1689. Measure granting freedom of worship to non-conformist Protestants, subject to certain oaths of allegiance to the Crown. It did not apply to Catholics or Unitarians and continued to exclude them from public and political office. The Act was finally repealed in 1871.

Toleration Act, 1712. Measure following the appeal of Greenshields from the Court of Session to the House of Lords in 1711, which gave legal recognition to Scottish Episcopalians and thereby gave much offence in Scotland and seriously strained the newly-achieved Union of 1707.

Toll. A payment due for passing over or using a bridge, ferry, market, port, or other facility. The right to exact tolls may be an incident of a franchise (q.v.) or may be imposed by statute.

Toll and team. Words used in Norman times to grant jurisdiction. Toll meant the right to take tollage from one's villeins, and team probably the right to hold a court in which a stranger can be vouched as warrantor.

Tolpuddle martyrs. During the outburst of trade union activity in 1833–34 some Dorsetshire labourers established in Tolpuddle a branch of the Friendly Society for Agricultural Labourers. The government arrested six of them ostensibly for administering unlawful oaths, and they were convicted and sentenced in 1834 to seven years' transportation to Australia. The trial provoked considerable public reaction and Chartist demonstrations and the sentences were remitted in 1836.

Tomlin, Thomas James Chesshyre, Baron (1867–1935). Called to the Bar in 1891 and took silk in 1913. In 1923 he was made judge of the Chancery Division and in 1929 a Lord of Appeal in Ordinary. He was a learned and skilful judge and his judgments were lucid and well regarded.

Tonnage and poundage. Two subsidies granted by the English Parliament to the King from mediaeval times. Tonnage was a fixed duty on each tun of wine imported, and poundage was an *ad valorem* duty on all goods imported and exported. They were granted together from 1373 and from 1422 granted to successive kings for life. In the seventeenth century their grant became an issue between King and Parliament. In 1625 Parliament voted them to Charles I for one year only, and when the King sought to collect them without authority it passed resolutions forbidding their collection. In 1641 Parliament declared their levy to be illegal without its consent. In 1660 they were again granted for life and later were made perpetual. They were abolished in 1787.

Tontine. A kind of financial arrangement, invented by an Italian, Tonti, whereby subscribers to a loan each receive for life an annuity, which is increased as other subscribers die, until the last survivor receives the whole sum of the annuities. Such a scheme was promoted several times in the eighteenth century.

Tordesillas, Treaty of (1494). Disputes arose between Spain and Portugal over lands in America discovered by early explorers. In 1493 the Pope issued bulls demarcating zones of colonization by a north–south line 100 leagues west of the Cape Verde Islands, Spain being allocated the areas west and Portugal those east of the line. Disputes having continued, representatives of the two countries at Tordesillas affirmed the papal principle but fixed the line 370 leagues west. This enabled Portugal to claim most of Brazil, though exploration from there went beyond the line. A later Pope sanctioned the change in 1506.

Torelli, Lelius (1489–1576). An Italian humanist, who edited the Florentine text of the Pandects in three volumes in 1553, *Digestorum seu Pandectarum libri L ex Florentinis Pandectis repraesentati.*

Torrens, Robert Richard (1814–84). Held a post in the Customs Office in London before going in 1840 to Adelaide to become Collector of Customs. In 1853 he became Registrar-General of Deeds and an official member of the Legislative Council of South Australia. He was premier in 1857. In this office he worked out a system of the registration of title to land, with the object of simplifying conveyancing. In 1858 he carried in South Australia an Act establishing his system, which spread rapidly throughout Australia, having been adopted in all states by 1874 (though not in identical forms). The validity of title to land is guaranteed by the State and the title of a registered proprietor is paramount, subject to certain exceptions; the title deeds are replaced by a Certificate of Title. He returned to England in 1863 and was an M.P. 1868–74.

Tort (from *tortus*, twisted or wrong). The term in common law systems for a civilly actionable harm or wrong, and for the branch of law dealing with liability for such wrongs. Analytically the law of tort (or torts) is a branch of the law of obligations, where the legal obligations to refrain from harm to another and, if harm is done, to repair it or compensate for it, are imposed not by agreement, but independently of agreement by force of the general law. Socially the function of tort is to shift loss sustained by one to the person who is deemed to have caused it or been responsible for its happening, and in some measure to spread the loss over an enterprise or even the whole community. Historically there was no general principle of tortious liability, but the King's courts gave remedies for various forms of trespass, for direct injuries, and later allowed an action on the case (q.v.) for harm indirectly caused. Other forms of harm later became redressible, e.g. libel and slander, and distinct forms of action developed to redress paricular kinds of harm, so that the law of tort was concerned with a number of recognized kinds of wrong, each with distinct requirements and procedure. Statute added new entitlements to claim, e.g. in cases of fatal accidents, and new grounds of liability. Case-law has extended liability, e.g. from physical injuries to mental injuries, and from intentional harms to harms done negligently, i.e. by failure to show the standard of precautions deemed necessary in the circumstances. It remains the case, however, that the law of tort is a collection of circumstances in which the courts will give a remedy, normally by way of damages, for legally unjustified harm or injury done by one person to another rather than a general principle of liability applicable to manifold cases. It is potentially confusing to think of tort as connected with wrongs, as the wrongful element consists only in there having been a breach of legal duty, which may be purely technical and not involve any moral delinquency or criminality.

Tort and crime sprang from a common root but have diverged in many respects, but it is still true that many common law crimes are also actionable torts, e.g. assault, but not conversely.

Liability in general depends on the defendant having, by act or omission, acted in breach of a legal duty incumbent on him and infringed a recognized legal right vested in the plaintiff and thereby caused the plaintiff harm of a foreseeable kind. Not every harm is actionable; there is no liability for an inevitable accident, or an act of God; there are justifications such as statutory or common law authority. Till 1947 the Crown was in general not liable; since then it is in general liable in the same way as a private individual. The pecuniary consequences of liability may be shifted by liability insurance.

In tort law the principle of vicarious liability (q.v.) applies, and joint tortfeasors are all liable for the whole harm caused, with right of relief *inter se*. If the plaintiff was himself wholly or partly to blame for the damage, damages awarded may be reduced in proportion to the degree in which he was in fault.

The standard of care and precautions which imports liability for harm, is generally failure to take the care and precautions which were reasonable in the circumstances, but in certain cases strict liability (q.v.) applies, where the defendant is liable if he failed to avoid the evil consequences, unless he can establish one of certain limited defences and in cases of breach of statutory duty the liability may be absolute, i.e. there is liability if the prohibited harm happens at all, irrespective of precautions.

Torts may be classified into those involving intention, those involving negligence, and the wrongs of strict liability. They may also be classified into torts affecting the person (e.g. trespass, negligence), the family (wrongful death of a relative), reputation (libel and slander), property (e.g. trespass to land or goods, nuisance, conversion), economic rights (deceit, inducement of breach of contract, injurious falsehood), and certain miscellaneous torts such as conspiracy. There are certain kinds of conduct, such as infringement of privacy, which are not yet, but may come to be, recognized as actionable torts.

The normal remedy for a tort is an award of pecuniary damages in compensation for the harm done; in personal injury and death cases the computation of damages involves many complicated issues. In some circumstances, e.g. nuisance, an injunction is a competent remedy.

(Eng.) P. H. Winfield, *Province of the Law of Tort*; P. H. Winfield and J. A Jolowicz, *Textbook of the Law of Tort*; J. W. Salmond, *The Law of Torts*; H. Street, *Law of Torts*; J. G. Fleming, *Law of Tort.* (U.S.A.) W. Prosser on *Torts*; F. V. Harper and F. James, *Law of Torts.*

Torture. The infliction of physical or mental pain on a person to obtain from him a confession. At Rome, under the Republic and early Empire, citizens escaped torture but it was used on slaves and provincials, though a *Lex Julia Maiestatis* decreed that in treason cases all accused might be put to torture. But in the early Empire the custom began of applying it to citizens and it became of general application. When Roman law was revived in the twelfth century the practice was revived also and was substituted for ordeals which by this time were losing credence. By the fourteenth century it had become a general custom in Europe and one of the recognized institutions of inquisitorial criminal procedure.

Canon law under Roman influence recognized it, Gratian's *Decretum* repudiated it, but many of the commentators on the *Decretum* accepted it and such great commentators as Innocent IV and Durandus did not doubt it.

In the Germanic laws some admit torture, borrowed from Roman law, and the Visigothic law allowed torture of even a freeman in default of other proof. Torture, however, had no official place in feudal justice at least as between the higher grades of superiors and vassals. But French Ordinances recognized it from the thirteenth century, partly as part of energetic attempts to repress crime, partly because of the influence of the Roman law, and partly because in the procedure of that time, unless there was overwhelming evidence, the confession of the prisoner was indispensable on a capital charge. The prisoner had to ratify his confession after the torture had ceased, but if he retracted it he might be tortured again.

In the Middle Ages and early modern period torture was regularly employed in France, even where the accused confessed, and was frequently repeated several times. There were preparatory torture, used to extort confession of the crime charged, and preliminary torture, administered to condemned persons to compel them to disclose their accomplices. According to severity torture was divided into ordinary and extraordinary. Criminal ordinances regulated the circumstances under which recourse could be had to torture. In 1707 the Parlement of Paris issued a detailed memorandum with regard to torture which was adopted in many jurisdictions.

Modes of torture have been various. Torture in France in the seventeenth century was administered by water and by the boot; in Britanny it was done by squeezing the thumb or fingers or a leg with machines, or by roasting the sufferer's feet before a fire; in Besancon the man's arms were tied behind his back and he was raised into the air by a pulley attached to his bound arms, or sometimes a heavy weight was attached to the big toe of each foot which when he was lifted remained suspended from his feet; in Autun boiling oil was used.

In the seventeenth century torture was still accepted but protests began to be made and became stronger in the period of the Enlightenment. Montesquieu doubted it, Voltaire condemned it, and Beccaria attacked the practice but many jurists continued to accept it. By the end of the eighteenth century it was under attack from all sides and edict of 1788 abolished torture and the *cahiers* sent in by the three orders of the States General convoked for 1789 unanimously demanded that this be permanent.

In mediaeval German law torture was allowed even in addition to numerous presumptions of guilt. In the *Constitutio criminalis Carolina* of 1532, in the absence of confession, or several competent and credible witnesses, recourse was to be had to torture and confession obtained by torture could not be dispensed with. In Austria the position was the same but it was abolished by Joseph II's criminal code of 1787.

Torture was used in Spain even before the revived influence of Roman law. In the law of the *Siete Partidas* torture was extensively employed though it was not an essential feature of inquisitorial procedure.

In England the criminal procedure which developed in mediaeval times was accusatory not inquisitorial, but the Council in the sixteenth century adopted the practice of making a preliminary examination of accused persons and torture was used to discover facts whenever the Council deemed it necessary. But it was deemed contrary to the common law, though legal if inflicted under the authority of the Crown's extraordinary power to supersede the common law in emergency. Torture was justified as an extraordinary proceeding which the extraordinary power of the Crown could justify. Even Coke admitted the existence of this extraordinary power and saw no objection to the use of torture if thus authorized. His views changed later, however, and he condemned the practice as inconsistent with the liberty of the subject.

Torture never seems to have been practised so systematically in England as on the Continent, but it was possible to have recourse to it in the highest court, the King's Council. It disappeared about 1640.

In Scotland inferior courts used torture particularly in cases of witchcraft, but in theory required the consent of the Privy Council. In Scotland in the seventeenth century, in the religious persecutions of the covenanters, torture was extensively resorted to to extract renunciation of adherence to the Covenant. Common methods were the thumbscrews, and the boot (an iron boot enclosing both legs, wedges were then hammered in until the legs were crushed). Torture was made illegal in 1707.

Tosefta. A collection of oral traditions relative to Jewish law. Like the Mishna it is the work of Jewish scholars, mostly in Palestine, who collected and selected the main traditions from material developed from the time of Ezra (*c.* 450 B.C.) and was probably meant to complement the Mishna by preserving examples and explanations of oral law. The traditions included are not always self-consistent nor always consistent with the Mishna.

Totalitarianism. A term used first of the Communist régime of Russia, the Fascist régime of Italy, and the Nazi régime of Germany. It relates not so much to the political ideology of a government but to the extension of governmental control to all aspects of life and thought, social, intellectual, and sporting as much as political and economic, and to the direction of all efforts to a few related overriding objectives, particularly state power, imperialism, and aggression, without regard to individual rights or interests or to the rule of law. Totalitarian régimes are marked by extensive use of radio and other media to influence thought, by censorship and muzzling of information and opinion, suppression of dissent and enforced conformity as well as by the more traditional machinery of tyranny, police, armed forces, spying, terror, and the imprisonment or killing of nonconformists or opponents. The legal and judicial systems, and even thinking about law, are controlled and made subservient to the dominant political ideology. Law accordingly means something different from what it does in liberal-democratic societies. In public and criminal law every rule is distorted to serve political purposes; in civil and commercial law this is almost as true. In any event all legal justice is liable to be overruled by administrative action. The views of the political masters are the supreme law.

Tothill Fields. The Westminster Bridewell, known in the nineteenth century for a severe though not inhumane régime.

Tottel, Richard (d. 1594). Set up in business as a stationer and printer in London about 1550 and in 1553 he was granted a patent for seven years, later extended for his life, to print all duly authorized books on common law. For 40 years he worked in Fleet Street. He was an original member of and held all the high offices in the Stationers' Company. He published some literary books, but mainly law books, the last being *Dyer's Reports* in 1586.

Touchstone. See SHEPHERD, WILLIAM.

Toullier, Charles-Bonaventure-Marie (1752–1835). French jurist, author of an important *Le Droit civil français suivant l'ordre du Code Napoléon* (1811–31).

Tourtoulon, Pierre de (1867–1932). French jurist. His *Les Principes philosophiques de l'histoire du Droit* (1908) was translated as *Philosophy in the Development of Law*.

Towage. A contract to tow a vessel from one place to another. It is implied that the tug will be efficient for the purpose and her equipment equal to the task. Normally the tug is under control of the tow and must obey orders from those in charge of the tow, but in the absence of instructions it is the duty of the tug to direct the course and to look out both for herself and for the tow. A towage service is a service akin to salvage (q.v.) but one where the service consists solely in towing a vessel from one place to another and not in rescuing from danger. It gives a right to reward, but not on such a generous scale as salvage.

Town and country planning. A body of legislation developed since 1909 to secure the planned development and use of land. Local authorities are designated planning authorities and must prepare, and review every five years, development plans showing how land is (and is to be) used. Development is controlled by requiring every person intending to develop land to obtain planning permission. Powers exist also for controlling existing uses and for preservation of trees and buildings. Planning authorities are also given wide powers of compulsory purchase of land to secure its proper development.

Township. In English constitutional history a unit of local government, generally coinciding with a parish. Also a district of a town, tithing or vill, or an area of land containing a small town. In the U.S., particularly in the north-east, it is a subdivision of a county and a unit of local government. In some states a township meeting levies taxes and enacts by-laws, while the township board appoints officers and has administrative functions. Justices of the peace and constables are frequently elected from the township, though they are state officers. The tendency is for the township as a local government unit to decline.

Toxicology. The study of poisons and poisoning, a study allied to and drawing on biochemistry, pharmacology, pathology, and other sciences. It has important legal applications in connection with determining causes of injury or death, and is commonly studied in association with forensic medicine (q.v.).

Tracing. The equitable doctrine that where a trustee has handed over trust property wrongfully, the beneficiary may not only recover the property so long as not mixed with other property, but may recover it from anyone who has taken it with notice of the equitable interest or without having given value for it, or recover any specific asset into which the trust property has been converted, or have a charge on any fund into which the trust assets have been mixed, or any assets purchased with the trust funds. Tracing is not allowed if the result would be harsh and unconscionable on an innocent volunteer, i.e. one who has taken the trust property without notice of the trust but without having given value for it.

Tractatus tractatuum (or *Tractatus universi juris*). A large work, entitled in full *Tractatus tractatuum, sive Tractatus illustrium jurisconsultorum in utroque jure Caesareo et Pontificio*, published by Ziletti at Venice in 29 volumes, folio, in 1584–86. A list of the treatises printed in this collection is in George Draud's *Bibliotheca classica* (1611).

Trade Acts. A series of Acts passed by the British Parliament in the eighteenth century, designed to make the trade of the different parts of the British Empire complementary. British colonies were largely restricted to buying British manufactured goods, though export bounties on such goods kept the prices down. Colonial products were encouraged by favourable tariffs, though some Acts introduced restrictions to protect home industry. After 1763, however, the British Parliament tried to use these measures as a tax on the colonies, and played a part in precipitating the American Revolution.

See also NAVIGATION ACTS.

Trade associations. A term including manufacturers' or traders' associations, trade protection societies, and similar bodies. The constitutions and underlying agreements of such associations are subject to the common law rules that combinations of persons in unreasonable restraint of trade are void, and also to the law of conspiracy. Restrictive trading agreements, such as to maintain prices, limit output, share orders, and the like, are subject in most cases to the requirement of being registered and of being held void under the Restrictive Trade Practices Act, unless they can be shown not to be contrary to the public interest.

Trade Boards. Bodies established by statutes of 1909 and 1918 to fix minimum wages and replaced after 1945 by Wages Councils.

Trade combinations. A combination of persons in business is neither criminal nor tortious if its predominant purpose is to protect or further legitimate trade interests, such as to increase the combiners' share of the market, but it is tortious if illegal means are used, or if the predominant purpose were not the furtherance of legitimate trade interests but disinterested malevolence.

A combination of persons, whether employers or employees, is unenforceable and void at common law if in unreasonable restraint of trade. But associations for mutual protection, sickness and accident benefits, and the like are valid. Trade union legislation, however, has legalized the contracts of a trade union, with certain exceptions, notwithstanding that a trade union is not a lawful association at common law if, as is commonly the case, its main objects are in unreasonable restraint of trade. A combination in restraint of trade is not now criminal at common law, unless it involves agreement to effect any unlawful purpose, or to effect a lawful purpose by unlawful means.

Trade descriptions. The first general statute dealing with fraudulent marking of merchandise was passed in 1862 and was replaced by the Merchandise Marks Acts, 1887 to 1953, and that in turn by the Trade Descriptions Act, 1968. Under this Act it is an offence, in the course of trade or business, to apply a false or misleading trade description to any goods or to supply or offer to supply any goods to which a false or misleading trade description is applied. An oral statement may be a trade description. The information to be given in advertisements may be specified. False or misleading indications as to the price of goods are also offences. False representations as to prices, Royal approval, and the like are prohibited.

Trade dispute. Any dispute between employers and workers, or workers and workers, connected with the employment or non-employment, or the terms of employment or the conditions of labour of any workers. Acts done in pursuance of an agreement or combination by two or more persons, in contemplation or furtherance of a trade dispute, are not actionable unless they would be actionable without such agreement or combination. An act done by a person in contemplation or furtherance of a trade dispute is not actionable in tort on the ground only that it induces some other person to break a contract of employment, or only that it is an interference with the trade, business, or employment of some other person, or with the right of some other person to dispose of his capital or of his labour as he wills.

Where a trade dispute arises it may be referred by the Secretary of State for Employment to the Industrial Arbitration Board, or to one or more persons for arbitration, or to a board of arbitration, or a court of inquiry may be appointed to inquire into the causes and circumstances of a trade dispute and to report to the minister.

Trade Marks. Marks on goods indicative of their manufacturer were originally controlled by guilds of merchants and craftsmen. They became important after the Industrial Revolution as distinguishing one maker's goods from those of another. The courts were originally reluctant to protect traders against unauthorized use of their marks, but in 1824 the copying of a trade mark used by a manufacturer was held to give a cause of action where it had caused him loss. Before trade marks became registrable the user of a trade mark did not have to prove any length of use to give him a right of action. A register for certain trade marks was established in 1875 and the legislation was consolidated in 1938. Unregistered trade marks may still exist but no action for infringement lies save in the form of an action for passing-off.

The register of trade marks is maintained by the Comptroller-General of Patents, Design, and Trade Marks. It is divided into Parts A and B. To be registrable in Part A a mark must be distinctive and contain or consist of at least one of the name of a person or company represented in a special or particular manner, the signature of the applicant, an invented word or words, a word or words having no direct reference to the character or quality of the goods and not being a geographical name or surname, or any other distinctive mark. To be registrable in Part B a mark must be capable of distinguishing the goods for which it is registered having regard to both the inherent qualities of the mark and the extent to which user or other circumstances has rendered it so capable. Certain matters may not be registered and the Registrar always has a discretion to refuse registration. A trade mark must be registered in respect of particular goods or classes of goods and for this purpose goods are classified into 34 classes.

Registration gives the proprietor the exclusive right to the use of the trade mark in connection with the goods in respect of which it is registered, and any invasion of this, as by another using a mark the same or so similar as to be likely to confuse, is an infringement, actionable for an injunction, damages, or an account of profits.

Applications may be made to rectify the register, on such grounds as that the mark is not properly registrable, or is disused, or being misused.

A trade mark is assignable and transmissible with or without the goodwill of a business, the transfer being registered with the Registrar. A registered proprietor may permit use of the mark by registered users subject to the approval of the Registrar.

Trade names. No manufacturer or trader is entitled to represent his goods as those of somebody else, and any trader who uses a name, mark, or get-up which has become distinctive of his goods is considered to have a right of property in that name and can prevent others using the same or a similar name, mark, or get-up where that will deceive or is calculated to deceive members of the trade or public.

Similarly no person is entitled to represent that his business is that of or connected with another, save that the courts are very reluctant to prevent a man trading under his own name, even if he may thereby be thought to be another trader.

Trade unions. Trade unions developed in Britain as associations for self-protection of employees under modern industrial conditions. At common law such associations were deemed conspiracies in restraint of trade and consequently unlawful associations and from the thirteenth to the early nineteenth century various statutes were passed reinforcing the common law. But in the later eighteenth and early nineteenth centuries secret trade societies became more numerous, despite the Combination Acts of 1799–1800. In 1824 an Act repealed the previous legislation but the policy was reversed in 1825. Nevertheless, there was a considerable multiplication of trade unions in the years down to 1871 when the first modern Trade Union Act was passed. It provided that a trade union should not, merely because its purposes were in restraint of trade, be deemed unlawful nor render a member guilty of conspiracy or otherwise. It provided for registration of trade unions, and for their property. It was amended in 1876. In 1875 the Conspiracy and Protection of Property Act, 1875, provided that an agreement by two or more persons to do an act in contemplation or furtherance of a trade dispute should not be indictable as a conspiracy if such an act committed by one person would not be a crime, and legalized peaceful picketing. Under these conditions unionism developed steadily though unevenly. Employers' associations developed strongly from the 1890s.

In 1901 it was held (*Taff Vale Ry.* v. *Amalgamated Socy. of Railway Servants*, [1901] A.C. 426) that a union, as an entity, was liable in damages for loss caused by a strike it called. The unions and the Labour party resented this and, though a royal commission reported against change, the government passed in 1906 a Trade Disputes Act which exempted trade unions from liability for any tort, though officials and members might be personally liable, and gave persons acting in trade disputes immunity from liability for procuring breaches of contracts of employment. Then in 1910 it was held (*Osborne* v. *Amalgamated Socy, of Railway Servants*, [1910] A.C. 87) that provision under the union rules of payment of salaries or maintenance to M.P.s pledged to support the Labour Party, from contributions levied from members of the society was *ultra vires*. In 1913, at the instance of the Labour party the Trade Union Act, 1913, was passed, which permitted political activity, provided it obtained the assent of a majority of members, established a separate political fund, and allowed dissenting members to 'contract-out' from paying the political levy.

Trade union membership grew before and during the First World War; in 1926 the T.U.C. called a general strike in support of the Miners' Federation. The failure of the general strike was followed by the Trade Disputes and Trade Union Act, 1927, which made illegal a strike or lockout designed to coerce the government by inflicting hardship on the community or which had any object other than furthering a dispute within the trade or industry, and substituted 'contracting-in' for 'contracting-out' in the 1913 Act as to political funds. In 1946 the Labour government repealed the 1927 Act and restored 'contracting-out'.

In the boom years after the Second World War trade unions reached new heights of size and power and sought to dictate the government of the country, and won pay increases wholly disproportionate to increased output, thereby causing galloping inflation, which they then exploited by incessant and increasing wage demands and endless strikes.

In 1964 it was held (*Rookes* v. *Barnard*, [1964] A.C. 1129) where a threat of strike, in breach of an agreement not to do so, forced an employer to dismiss a non-union man, that this constituted intimidation and conspiracy to injure. The Trade Dispute Act, 1965, reversed the decision. There followed another Royal Commission on Trade Unions and Employers' Associations, and a Labour government's surrender in face of trade union objections to its proposed legislation to implement the Report thereof. The position had been reached where trade unions occupied a grossly privileged position under the law and would not countenance any judicial decision which in any way curbed their freedom to dictate, wreck, and damage with impunity.

In 1971 the Industrial Relations Act repealed nearly all the previous law, saving only the 1913 Act. It was repealed by the Trade Union and Labour Relations Act, 1974, which largely restated the pre-1971 law and conferred on trade unions a highly privileged position.

The position under the 1974 Act, as amended, is that a union is an organization of workers whose principal purposes include the regulation of relations between those workers and employers. Unions are not corporate bodies, but have contractual capacity. But collective agreements made by unions with employers are presumed not legally enforceable and can consequently be dishonoured with impunity. Their purposes, in so far as relating to the regulation of relations between workers and employers, are not unlawful merely by reason of being in restraint of trade. Acts done in comtemplation or furtherance of trade disputes are not actionable in tort on the ground only that they induce another person to break a contract of employment, or are an interference with the trade, business, or employment of another person or with the right of another person to dispose of his capital or labour as he wills. No action in tort lies in respect of any act alleged to have been done, or threatened, or intended to be done, against the union or its trustees, or any members, or officials on behalf of themselves and all other members, but saving criminal liability for conduct resulting in personal injury or breach of duty in connection with possession or use of property. So-called peaceful picketing is permitted.

The modern situation is that trade unions and their members have a legal position relative to the State very like that of the mediaeval Church and its clergy, having legal privileges and immunities from contractual or tortious liability, easily the most privileged groups in the community.

So far from existing to protect workmen from employers, unions now tyrannize over employers and dictate to them by strikes and threats of strikes, show brutal intolerance to workmen who will not join their ranks, and have no compunction over forcing employers to dismiss non-union employees. They are no more democratic or tolerant than the mediaeval Church at its worst. The common trade union insistence on the 'closed shop' is a fundamental infringement of the liberty of the individual, as vile as enforced conformity with any religion ever was, and contrary to the U. N. Universal Declaration of Human Rights. So-called peaceful picketing (q.v.) is commonly mob hooliganism and intimidation.

In public law unions are not only consulted by governments on every matter, but assert and dictate their views, arrogantly claim to represent the people of the country, and to make and unmake governments. In some parliamentary constituencies they determine the choice of member. They are dedicated to restrictive practices, laziness, selfishness, and insatiable greed and have no regard for anything but their own immediate advantage. The public interest frequently suffers enormously by strikes, inter-union demarcation disputes, or stoppages for the most obscure reasons, and from strikes obviously designed to coerce the elected government and the community generally. They represent the gravest threat to democracy, liberty, and economic progress and prosperity yet known and constantly call for the law to be kept out of industrial relations to enable anarchy to be promoted.

S. & B. Webb, *History of Trade Unionism*; G. D. H. Cole, *Short History of the British Working Class Movement*; H. Clegg, Fox and Thompson, *History of British Trade Unions*; N. Citrine, *Trade Union Law*; C. Grunfeld, *Modern Trade Union Law*.

Trades Union Congress. The national organization of British trade unions, founded in 1868, to which most trade unions, particularly of manual workers, are affiliated. It has developed a permanent staff and organization but has only limited control over affiliated unions. In World War II and thereafter it has increasingly been consulted by governments seeking to ensure workers' assent to and participation in governmental policies and it now behaves as if it were the real government of the U.K. It exercises powerful influence on the policies of the Labour party and accordingly at times on the government of the U.K.

Trading stamps. Printed stamps given by retail traders to customers in numbers proportional to the value of goods purchased. They are obtained by traders from one of several trading stamp companies and, if collected by customers, usually by way of being affixed in books of blank pages issued for the purpose, may be exchanged for money or goods at the depots of the trading stamp company. In effect the trader is giving the customer a discount. As a commercial practice such schemes were tried in the nineteenth century but did not become popular until the 1930s in the U.S. and the 1960s in the U.K.

Traditio. In Roman law and systems based thereon, the actual handing over or transfer of an article with intent to transfer a legal right of property, or at least of possession thereof.

Tradition. The aggregate of beliefs, customs, habits, and practice which develop in a particular culture and, by being continued, give it continuity. Legal institutions and at least some rules of law, particularly of customary and common law, are part of a country's traditions. Also in it are accepted attitudes, practices, and ways of thinking and working about law, which may have no other authority than that they have developed in time and are part of the tradition. Thus, legal dress, the respective functions of the different kinds of lawyers, the practice of *stare decisis* (q.v.) are all part of British tradition, in relation to law; they are acknowledged and accepted but sanctioned only by custom and usage, and it would not be illegal, e.g. to refuse to wear wig and gown, or to refuse to be bound by a House of Lords' decision.

Traffic commissioners. Persons appointed in areas of the country to regulate the grant of licences for public service vehicles and renamed in 1947 the licensing authority for public service vehicles. The chairman is licensing authority for goods vehicles.

Trailbaston, Commission of. A commission, dating from an ordinance of 1304, issued to persons directing enquiry as to persons who disturbed the peace or maintained malefactors or perverted the course of justice by ill-treating jurors, similar to but rather wider than a commission of oyer and terminer. The name seems to have been derived from a term for the offenders, as persons carrying sticks. They were obsolete by the fifteenth century.

Trainbands. The old term for the militia, or those men in the community trained in the use of arms.

Traitor's Gate. The gate of the Tower of London at the bank of the Thames by which traitors and state prisoners, brought by river, entered the Tower.

Trani, Laws of. A code of maritime law originating about 1063 at Trani on the Adriatic coast of Italy, entitled *Ordinamenta et Consuetudo Maris edita per consules Civitatis Trani*, and printed in Twiss' *Black Book of the Admiralty*.

Transaction. In Scots law, any agreement between two parties tending to the settlement of doubtful or controversial claims.

Transfer. The making over to another of rights in property, particularly voluntarily, and principally of shares in a company, and also the deed for effecting the change of ownership. An instrument of transfer is used to convey registered land, the transfer being entered on the register.

In procedure the term transfer is used of shifting an action from one court or division to another.

Transmission. The transfer of a right from one person to another, generally involuntarily, as on the occurrence of death or bankruptcy, as distinct from voluntary transfer on sale or gift.

Transport. The moving of persons and goods from place to place. The relevant law is generally called the law of carriage (q.v.) and distinct bodies of rules apply to carriage by road, by rail, by sea and by air, and to persons and to goods.

Transport Tribunal. A body which in 1947 replaced the former Railway Rates Tribunal and assumed also the jurisdiction of the Railway and Canal Commissioners. It is a court of record, consisting of a president, who is a lawyer, and four members, two experienced in transport, one in commercial affairs, and one in finance. It has a regulatory jurisdiction over land transport, as to the regulation of carriage of goods by road, operators' licences, transport managers' licences, rates and charges, and similar matters. Appeal lies to the Court of Appeal or Court of Session and, in one special case, to the House of Lords.

Transportation. The former mode of dealing with criminals, by removing them to a remote place overseas. It was never a common law punishment, but to some extent originated in the practice of abjuring the realm or being hanged, while later prisoners were sometimes given the alternative of being transported or hanged. In the seventeenth century prisoners of war, political offenders, and felons were on occasion sent to the plantations of America. In 1717 statute initiated systematic transportation to America, but this ended in 1776 and in 1784 transportation to Australia was authorized. In 1787 the first fleet sailed to Sydney. Governors Macquarie (1810–21) and Arthur (1824–36) were just and saw reformation as an object of policy. Ex-convicts were treated on a par with free settlers. A select committee in 1837, however, condemned transportation, but Van Diemen's Land remained a penal colony and Norfolk Island was a punitive settlement, save under the rule of Maconochie. From 1847 the whole term of prison confinement of prisoners was served in England, and only ticket-of-leave men were sent to Australia. An Act of 1857 prohibited sentences of transportation, but till 1867 sentences of penal servitude was partly carried out in Western Australia. Over 80 years about 160,000 convicts were sent from the U.K. to Australia. France also utilized transportation as a punishment, to Guiana and New Caledonia, and this lasted till 1938.

A. G. Shaw, *Convicts and the Colonies.*

Transsexualism. See Sex.

Transumpt. A copy of a legal document or writing.

Transvestism. See Sex.

Travaux préparatoires. Materials used in the preparation of, and having formative effect on, the ultimately adopted form of an agreement, or legislation, or an international treaty. Such materials include, in the domestic sphere, reports, proposals and technical advice, in the legislative sphere, Select Committee or Royal Commission or other reports, academic studies, Green Papers, White Papers, and the like, and in the international sphere reports of expert committees, discussions and proposals, drafts, and the like. The major question is whether such materials may be considered in interpreting and resolving doubts or ambiguities in the text of the finally agreed contract, statute, or treaty. In relation to contracts the accepted principle is that the contract is presumed to supersede entirely all preliminary discussions and to express the finally concluded view of the parties; examination of earlier negotiations is excluded and the contract must be interpreted by itself. In relation to legislation the British attitude is that reference to prior reports, Parliamentary debates, drafts, and the like is inadmissible, and that the meaning must be determined by examination of the words used only. But some legal systems take a different view. In particular, in France reference may be made to the texts of projects or proposals for laws and expositions of reasons for them, relative agreements, and debates in the Chamber, all of which are published in the *Journal Officiel.* In relation to international treaties resort to preparatory work is permissible and international courts may examine, e.g. minutes of conferences, correspondence, rejected drafts, and amendments rejected. The question is, however, of how much reliance should be placed on the words finally agreed and how much on the earlier materials. Even municipal courts in interpreting treaties have adopted the international approach and referred to *travaux préparatoires.*

Traverse. In a pleading or an affidavit, to deny an allegation of fact. Traverse of office or inquisition was an old mode of disputing an office or inquisition finding the Crown entitled to property obtained by the party traversing.

Treadmill (or treadwheel). A large wheel turned by persons stepping on the treads on the circumference, formerly utilized as a means of imposing hard labour on prisoners in prison. There were usually several wheels or a wheel with broad treads so that several men could work it simultaneously. The wheel was usually geared to grinding meal or pumping water. It was legalized as a prison occupation by the Penitentiary Act, 1779, but was not widely used until about 1818. Thereafter it was regarded as a godsend by penal reformers looking for a substitute for the death penalty and in the 1820s it was regarded as infallible in reforming efficacy, though from the start there were critics, notably those who wished more moral and religious education or industrial training. The treadmill was used in the nineteenth century but gradually fell into disfavour as being useless labour, and disappeared by the end of the century. When the Separate System (q.v.) replaced the Silent System (q.v.) in prisons, the treadmill was generally replaced by the crank-machine which could be set up in every individual cell. The treadmill was out of favour in the 1850s but made a reappearance after a Committee of 1863 commended it. It, and the crank machine, were abolished by the Prison Act, 1898.

Treason. The essence of treason is violation of the duty of allegiance which is owed to the sovereign and is due by all British subjects who are citizens of the U.K. and colonies, wherever they are, and by aliens under the protection of the Crown, so long as within the realm or still within its protection though having left the realm. In the United Kingdom it now depends on the Treason Act, 1351, passed to clarify the common law and extended later to Ireland and Scotland, and amended. It makes it punishable as treason to compass or imagine (intend) the death or bodily harm of the sovereign, her heirs and successors; and to utter such compassing or intending by publication or by any overt act or deed; to compass or imagine the death of the King's wife, or the wife of the sovereign's eldest son and heir, to violate the King's wife or his eldest daughter unmarried or the wife of his eldest son and heir; to levy war against the sovereign in her realm, to be adherent to the sovereign's enemies in the realm, giving them aid and comfort in the realm, or elsewhere; to slay the Chancellor, the Treasurer, or any of the judges of the High Court or Court of Session or Justiciary, when in their places and doing their offices; to endeavour to hinder any person next in succession from succeeding to the Crown, and maliciously by writing or printing to maintain that any persons have title to the Crown otherwise than according to the Bill of Rights, 1688, the Act of Settlement, 1700, and the Acts for the Union of England and Scotland. Many of the provisions have given rise to difficult questions of interpretation. All persons who incite, aid or abet treason, or receive or protect a traitor, are guilty of treason also.

In 1795 an expressed intention to compass the death of the King or to levy war was made treason. In 1848 the Treason Felony Act repealed the latter part of the 1795 Act and made the offences therein felonies only.

Prior to 1707 many kinds of offences had been declared by statute to be treasons in Scotland.

At common law the punishment for treason was drawing on a hurdle, hanging, and quartering, but these have been repealed though treason remains a capital crime. The Treason Act, 1945, simplified trials by applying to them the procedure of murder cases.

Petty treason consisted in a servant killing his master, a wife her husband, or an ecclesiastic his superior, such conduct being a violation of private allegiance. Petty treason was abolished in 1828, such killings being made murder only.

Misprision of treason is a common law offence committed by one who knows or has reasonable cause to believe that another has committed treason but fails to disclose this information to the proper authority within a reasonable time.

Treason (U.S.). Is defined in the Constitution as levying war against the U.S. or adhering to their enemies, but is provided as not effecting corruption of blood or forfeiture except during the life of the person attainted of treason. It is required that there be two witnesses to the same overt act of treason, or confession in a public court.

There have been few treason trials in the U.S., the most notable bring that of ex-Vice-President Aaron Burr in 1806 for an alleged conspiracy to dismember the Union. The charge failed. After the Civil War an amnesty was extended to those who had taken the Confederate side.

Treason felony. By the Treason Felony Act, 1848, this is the crime of compassing, imagining, devising, or intending to depose or deprive the sovereign from the Crown of the United Kingdom, or of levying war against the sovereign within the United Kingdom to compel her to change her measures or counsels, or to put force on, or intimidate, or overawe either House of Parliament, or to move any foreigner with force to invade the United Kingdom, and to express or declare such intentions by publishing any printing or writing, or by any overt act or deed.

Treasons, felonies and misdemeanours. At common law in England crimes were classified as treasons, felonies, or misdemeanours. Treasons were those crimes which amounted to treason (q.v.). Originally felonies were more serious than misdemeanours, but latterly this was not completely accurate. At common law felony, such as murder, arson, rape, robbery, involved forfeiture of the felon's land and goods and, before 1870, when forfeiture was abolished by statute, any offence involving forfeiture was a felony. After 1870 when felonies were created by statute, it was to signify that the offence was serious and to bring about the consequences which a felony entailed. All other offences were misdemeanours. The distinction between felony and misdemeanour had various consequences; powers of arrest were wider in the case of felony; it was misdemeanour to compound a felony, or to conceal a felony; in felonies a distinction was drawn between principals and accessories, whereas in treasons and misdemeanours all participants were charged as principals. The distinction was abolished in 1967, the law and practice applicable to misdemeanours being made applicable to both felonies and misdemeanours, but a new distinction between arrestable and non-arrestable offences was introduced.

Treasure trove. Gold or silver coin, plate, bullion, or other valuables hidden in the earth or another secret place and found, it being unknown and undiscoverable who the original owner was. It belongs by prerogative right to the Crown, but if the hider is known or discoverable it belongs to him. If property is lost or abandoned, the finder acquires a possessory right by finding. The finder of treasure trove must report it in England to the coroner for the district. The Crown may award things found or their value to the finder, and the Crown may grant a franchise of treasure trove to a subject.

Treasurer. One of the Officers of State for Scotland, first appointed about 1424, and having charge of the management of the King's revenue. The treasurer and the comptroller were at first subordinate to the chamberlain, but later superseded him. In 1685 the office was put into commission and since 1707 there has been one Treasury for Scotland and England.

Treasurer, Lord High. A great office of state in England, dating from Norman times. In the seventeenth century the office was several times entrusted to a Board and since 1714 it has always been in commission, and in 1667 the commission was made independent of the Privy Council. This resulted in the rise of the First Lord of the Treasury to the position of chief minister of the Crown and of the Chancellor of the Exchequer to the working headship of the Treasury.

Treasury. In the twelfth century the sitting of the *Curia Regis* to deal with financial business was known as the Exchequer from the chequered cloth on the table to aid calculation. The Exchequer branched into the Upper Exchequer which managed and audited the royal accounts, and developed into the Court of Exchequer which assumed the judicial business and the Lower Exchequer, concerned with the receipt of royal revenue, which developed into the Treasury. It emerged as a public department after 1688.

The modern functions of the Treasury are the imposition, in accordance with legislation made by Parliament, and the collection of revenue, tasks effected, in respect of direct taxes, by the Board of Inland Revenue and, in respect of indirect taxes, by the Board of Customs and Excise, the control of public expenditure through supervision of the estimates prepared by public departments, the provision of funds for the various public services and the relevant banking and borrowing. It increasingly exercises a measure of economic co-ordination. The Treasury is deemed the senior and chief department of state and to it are attached various sub-departments such as the Commissioners of Crown Lands, the office of Parliamentary Counsel to the Treasury, H.M. Stationery Office, and, formerly, the Civil Service Commission.

The office of Treasurer originated in the twelfth century but has been entrusted since 1714 to a board, the Lords Commissioners of H.M. Treasury, which consists of the First Lord of the Treasury (an office always now held by the Prime Minister), the Chancellor of the Exchequer (with whose office is now combined that of Under Treasurer), and five junior Lords of the Treasury (who along with the Parliamentary Secretary to the Secretary are the Government Whips in the House of Commons). The Chancellor of the Exchequer was originally appointed in the twelfth century to be a check on the Treasurer and to have charge of the seal of the Exchequer; he is now the working head of the Treasury and is assisted by a Chief Secretary and a Financial Secretary to the Treasury as well as by a large permanent staff.

Lord Bridges, *The Treasury*.

Treasury counsel. The common term for the group of counsel appointed by the Attorney-General to prosecute, on the directions of the Director of Public Prosecutions, on behalf of the Crown at the Central Criminal Court. They have in fact no connection with the Treasury, but the name is derived from the time when Crown counsel in that court were instructed by the Treasury Solicitor who also held the office of Director of Public Prosecutions in 1883–1908. They are designated first, second, third, fourth, etc., senior and first, second, third, fourth, etc. junior prosecution counsel, and must resign on becoming Queen's Counsel. They are quite distinct from the junior counsel to the Treasury who, on the instructions of the Treasury Solicitor, appear on behalf of the Crown in civil cases involving Treasury issues.

Treasury Solicitor. An office dating from about 1655, the Solicitor to the Treasury was in the nineteenth century responsible for preparing and conducting state prosecutions for the gravest offences, and was also responsible for the whole of the legal business of government, save for a few departments which had their own legal staffs. From 1883 to 1908 it was combined with the office of Director of Public Prosecutions. He now acts for the Crown in matters falling within the sphere of his department and other government departments which do not have their own solicitor. He is also H.M. Procurator-General, an office which does not now appear to have any functions.

Treaties of Rome, 1957. Two treaties between France, Germany, Italy, Belgium, the Netherlands, and Luxembourg. The first established the European Economic Community with a Council, Commission, parliamentary assembly, and court of justice. The second established the European Atomic Energy Commission (Euratom) with similar institutions. Simultaneously with the Rome treaties a Convention relating to certain Institutions common to the European Communities provided for a single parliamentary assembly and a single court of justice to serve all three communities. In 1965 the Brussels Treaty (the Merger treaty) established a single Council to replace the separate Councils of the three Communities and a single Commission to replace the High Authority of the E.C.S.C. and the Commissions of the E.E.C. and Euratom.

Treatises and textbooks, Legal. Legal treatises are books dealing with particular recognized branches and topics of law, for the use of students, or legal practitioners, or laymen professionally concerned with certain aspects of law. The term treatise is applied particularly to major books for practitioners, textbooks to books for students, but there is no fixed usage of the words and some books serve both purposes.

They may vary greatly in size and detail of discussion, from the small student's manual to the large and detailed practitioner's treatise, e.g. Russell on Crime, Williston on Contracts. Some cite few authorities, others many, but the function of the treatise is not primarily to collect and refer to all the cases on a point but rather those that the author considers significant.

They also vary greatly in scope of subject-matter from books on a small topic, e.g. The Hearsay Rule, to a wide and somewhat heterogeneous branch of law, e.g. Commercial Law, but most commonly deal with a distinct and recognized branch of law, e.g. Torts, Evidence, Insurance.

They also vary very much in respect of regard to the history and development of the branch of law in question, of regard to defects and inconsistencies in the rules and cases, and of regard to developments in other jurisdictions. Some treatises by their historical, comparative, and critical approach have exercised a definite formative influence on the law. Others concentrate on trying to state the existing law of the jurisdiction dealt with.

England

In England in earlier times books of practice, dealing with forms of actions, procedure, and pleading, predominated over treatises on law proper. Among these are *Brevia placitata, Fet asaver, Judicium essoniorum, Cum sit necessarium, Exceptiones ad cassandum brevia, Cadit Assisa*, Hengham's *Summa magna, Summa parva,* and *Casus placitorum* (qq.v.), all of the thirteenth century. There were also tracts on procedure in local courts. In the next three centuries there appeared the *Natura brevium, Novae narrationes, Articuli ad novas narrationes, Diversité de courts* and *Returna brevium* (qq.v.), while the *Registrum Brevium* was growing all the time from the twelfth century till it attained its final form in the fifteenth century.

Legal textbooks proper commence with Glanvill (twelfth century) and, later, Bracton's *De Legibus et Consuetudinibus Angliae* (thirteenth century), Littleton's *Tenures* (1481) and Coke's *Institutes* (1628–44), all of which deal with the whole, or large tracts, of the law of England as of their several dates. All have subsequently been regarded as of the highest authority. In the late thirteenth century two epitomes of Bracton appeared, *Britton* and *Fleta.*

In the late sixteenth and early seventeenth century Coke in his *Institutes* (1628–44) did much to adapt the medieval common law to the needs of early modern times, and in so doing covered most of the branches of the law.

In the eighteenth century Blackstone's *Commentaries* (1765) similarly surveyed the whole field; it was the most complete, orderly, and accurate book on the whole law of England which had yet appeared.

On land law and conveyancing, after Littleton and Coke, the earliest modern book on real property was that of Burton (1828) but there were many books of conveyancing precedents, notably Shepherd's *Touchstone of Common Assurances* (1641). In the medieval and early modern period there were numerous Books of Entries (q.v.).

In criminal law the earliest textbook was Staunford's *Plees del Coron* (1557), followed by Pulton's *De Pace Regis et Regni* (1609), Coke's *Institutes*, Part III (1644), Hawkins' *Pleas of the Crown* (1716), Hale's *Pleas of the Crown* (1678), Foster's *Crown Law* (1762), and Blackstone's *Commentaries*, Book IV (1769).

Books on evidence emerged only in the eighteenth century, the first being that of L. C. B. Gilbert of 1756, followed by the works of Peake (1801), Phillips (1814), Starkie (1824), Taylor (1848), and Best (1849).

The publication of such books has continued, including notably Chitty's *Practical Treatise on Pleading* (1809), his son, Joseph Chitty's *Precedents in Pleading* (1836), H. J. Stephen's *Treatise on the Principles of Pleading in Civil Actions* (1824), J. F. Archbold's *Digest of the Law Relative to Pleading and Evidence in Civil Actions* (1822), and his *Practice of the Court of King's Bench, Common Pleas and Exchequer* (1819 and many later editions), and Bullen and Leake's *Precedents of Pleadings* (1860). At the present time there is also the multi-volume *Encyclopaedia of Court Forms and Precedents* (1937).

The earliest work on the duties of the justices of the peace was Lambard's *Eirenarcha* (1581) which was followed by a series of compendious works down to the present time, notably those of Dalton, Burn, Archbold, and Stone.

In the nineteenth and twentieth centuries the

increasing bulk and complexity of the law, the greater number of cases to be cited, and the increasingly refined analysis of the cases and problems, have rendered the general treatise impossible, and resulted in the treatises on particular branches and topics of law. Of those published in the early part of the nineteenth century many have survived, though altered and sometimes remodelled by editors; such books as Abbott on *Shipping* (1802), Woodfall on *Landlord and Tenant* (1802), Russell on *Crimes* (1819), Archbold's *Criminal Pleading* (1822) are still standard. Later in the century Pollock on *Contract* (1876), and on *Torts* (1887), Anson on *Contracts* (1879) have had long periods as standard texts, but changes in the law have rendered Williams on *Real Property* (1845) largely obsolete and called forth Cheshire's *Modern Real Property*.

Scotland

The first textbook was *Regiam Majestatem* (thirteenth century) and was followed by volumes of *Practicks* (q.v.), a line of development which led partly to the modern Digest and partly to the modern Encyclopaedia.

The systematic legal textbook in Scotland was born full-grown with the publication of Stair's *Institutions* (1681), followed by Bankton's *Institute* (1751), Erskine's *Institute* (1773), Bell's *Commentaries* (1800), and Hume on *Crimes* (1797). Smaller general books, for students, appeared in Mackenzie's *Institutions* (1634), Erskine's *Principles* (1754), and Bell's *Principles* (1829), but since the early nineteenth century the tendency has been for books to be written on particular branches of the law. The latter part of the nineteenth century was an important period with Fraser's *Husband and Wife* (1846), *Parent and Child* (1846), and McLaren on *Wills* (1862). A very bleak period in the twentieth century was followed by a revival after 1960 when the works sponsored by the Scottish Universities' Law Institute began to appear.

The Commonwealth

Commonwealth countries prior to the mid-twentieth century produced little of note, but many jurisdictions have now valuable treatises of their own.

America

In the American colonies Blackstone's book was very widely read. Kent's *Commentaries* (1826) was modelled on Blackstone, and laid the foundations of American law. In the nineteenth century Story developed the treatise with his great series of *Commentaries* (1833–45) on various branches of the law. In the twentieth century there have been many outstanding publications such as Corbin on *Contracts*, Williston on *Contracts*, Wigmore on *Evidence*, Powell on *Real Property*, and Scott on *Trusts*.

Western Europe

The prevalence of codes in countries of continental Europe has made the treatise on the civil law, the commercial law, or the criminal law, each closely related to the prevalent code, the major form of legal literature. Thus, in France the multi-volume treatises of Aubry et Rau, Beudant, Planiol et Ripert are noteworthy, but increasingly treatises are being written on particular branches of law. In Germany and Italy the same is generally true.

Treatment tribunals. Proposed bodies to which would be transferred responsibility for deciding the disposal or treatment of persons convicted by criminal courts. The alleged merits of such decisions would be their social welfare approach, expertise, experience, and constant contact with treatment and training facilities. The idea is supported by the increasing importance attached to classification of prisoners and specialization of treatment. But such tribunals would have no personal knowledge of the circumstances of the crime, nor of the need for protection of the public. They would rely entirely on psychologists, welfare workers, and other experts. To some extent this approach is accepted in executive powers to transfer a prisoner from one kind of institution to another, to release on parole, or on licence. All indeterminate sentences are necessarily in the hands of the executive.

Treaty. An international agreement, normally in written form, passing under various titles (treaty, convention, protocol, covenant, charter, pact, statute, act, declaration, concordat, exchange of notes, agreed minute, memorandum of agreement) concluded between two or more states, or other subjects of international law, intended to create rights and obligations between them and governed by international law. Treaties perform many functions; some are akin to charters of incorporation, some to contracts, some to conveyances, some to legislation. Treaty-making power is confined to sovereign states and other international legal persons.

Negotiators other than heads of state, heads of government, and foreign ministers must normally give evidence of their authority to act on behalf of and to bind their states by exhibiting Full Powers (q.v.).

There are no particular requirements as to manner of negotiation, or reaching agreement, or the form of a treaty, but it normally results in the adoption of an agreed text setting out the mutual undertakings of parties. The text is authenticated by signature, or initialling, or incorporation in the final act of a conference. Signature entitles the signatory state to proceed to ratification or approval but does not bind the state to ratify, and binds it to refrain from conduct calculated to frustrate the objects of the treaty. If ratification is not necessary, signature evidences consent to be bound by the terms. In some multilateral treaties the text may be adopted or approved by the General Assembly of the U.N. and submitted to member states for their accession.

A treaty is probably void or voidable if induced by fraud, error, or coercion of a state or its representative, though the coercion inherent in a treaty between a victor and a vanquished is disregarded.

Treaties are unenforceable so far as inconsistent with obligations of members of the U.N. under the Charter, and probably if at variance with universally recognized principles of international law.

In many states by their constitutions the treaty must be submitted for ratification, as by the Crown in the U.K. or the Senate in the U.S. and, if ratified, instruments of ratification are formally exchanged or deposited. But some less formal agreements are specified not to require ratification and to be binding by signature only. If not ratified, the treaty falls. A treaty cannot be ratified in part or conditionally, nor amended at the ratification stage.

The treaty itself determines how and when the treaty comes into force; if not specified, it is presumed to come into force when ratified or accepted. Every treaty and international engagement entered into by members of the U.N. must be registered with the Secretariat and published by it; treaties not so registered may not be invoked before any organ of the U.N.

Where a state not a party to a treaty accepts its provisions and desires to become a party thereto, it does so by acceding thereto, which may be before or after the treaty comes into force. Alternatively, a convention may be approved by the General Assembly of the U.N. and be open for accession thereto by member states, in which case accession is the only means of becoming a party thereto.

A reservation is a unilateral statement by a state when signing, ratifying, or acceding to a treaty, varying or excluding the legal effect of some provisions of the treaty in its application to that state. The formerly accepted view was that reservations were valid only if the treaty permitted reservation and all other parties accepted the reservation, but in 1951 the International Court ruled that a state which has made a reservation objected to by one or more of the parties to the convention may be regarded as being a party to the convention if the reservation is compatible with the object and purpose of the convention.

In general a treaty can neither benefit nor damage other states not parties to it, but a treaty rule may come to bind other states if it becomes a part of international custom, and a treaty may provide for lawful sanctions for violations of the law to be imposed on an agressor state.

The interpretation of treaties is primarily a matter for the parties, but a treaty may itself confer jurisdiction to interpret on an *ad hoc* tribunal or on the International Court. On one view interpretation should be according to the intention of parties as expressed in the text of the treaty, on another it should be in accordance with the ordinary meaning of each term used, in the context of the treaty and in the light of the objects and purposes of the treaty. Preparatory work may be resorted to for interpreting controversial provisions. Subsequent practice may shed some light on parties' understanding of its interpretation.

Among means of securing performance of treaties, confirmation by solemn oaths, and the taking of hostages are obsolete, but the creation of a charge on the assets of a contracting state, occupation of part of its territory, and the guarantees of third-party states are competent. Other states may resort to economic and other sanctions under the U.N. Charter against a state in breach.

Treaties may expire when their purpose is achieved, or by lapse of time if entered into for a determinate period, or be cancelled by common consent, or a party may, after notice, withdraw, or they may be dissolved by material changes of circumstances, or inconsistency with subsequent international law. Violation by one party does not cancel a treaty, but it entitles the other party to cancel or denounce the treaty.

A treaty may specify a date of termination or provide for denunciation by the parties; if there is no such provision, the existence of a right of denunciation depends on the intention of parties. It may be terminated by agreement, or be affected when one state succeeds wholly or partly to the legal personality of another. War does not automatically terminate treaties between the parties to the conflict, and indeed some treaties are intended to regulate relations during war and the conduct of war. But many treaties are suspended during hostilities and may be terminated by fundamental change of circumstances. Material breach by one party entitles the other or others to suspend or terminate the treaty. Supervening impossibility of performance or fundamental change of circumstances may be invoked as grounds for terminating or withdrawing from the treaty.

Treaty (U.S.). By the U.S. Constitution, Art. II, sec. 2, treaties may be made by the President with the advice and consent of two-thirds of the Senate. Treaties are the supreme law of the land, taking precedence over state constitutions and legislation, and must be upheld by the courts.

The requirement of Senate consent has resulted in many treaties being defeated, most notably the rejection of President Wilson's proposal to take the U.S. into the League of Nations. An Executive

Agreement (q.v.) has the validity of a treaty, but does not require Senate consent, and for that reason is frequently resorted to. Secret treaties are not prohibited, but none has been ratified since 1800.

Treaty of Brussels, 1965 (the 'merger' treaty). A treaty signed in 1965 and effective in 1969, completing the merger of the superior institutions of the three European Communities, merging the High Authority of the E.C.S.C. and the Commissions of the E.E.C. and Euratom into a new Commission, and the separate Councils of the three Communities into a single Council. The establishment of a single Parliamentary Assembly and a single Court of Justice to serve all three Communities had already been achieved by a Convention of 1957.

Treaty of Brussels, 1972 (the 'Accession' treaty). A treaty signed in January 1972 between the European Communities, the original six member states and the four states which had completed negotiations for full membership (Denmark, Ireland, Norway, and the U.K.). It provided for their admission to the three Communities as new members, accepting all their rules, conditionally on the incorporation of the Community law into their municipal laws. (Norway later declined to ratify the treaty and did not proceed with entry.)

Treaty of Paris, 1951. The treaty between France, Germany, Italy, Belgium, the Netherlands, and Luxembourg which established the European Coal and Steel Community, with its High Authority, Consultative Committee, Special Council of Ministers, Assembly, and Court of Justice. These institutions were subsequently merged with the corresponding institutions of the E.E.C. and Euratom. It provided for abolition of duties and quantitative restrictions on trade in coal and steel between member states, and of discrimination and restrictive practices, and provided for the development of the coal and steel industries.

Treaty of Rome. See TREATIES OF ROME, 1957.

Trebatius Testa, Gaius (*fl. temp* Augusti). Roman jurist, legal adviser to Caesar, and author of a large treatise on pontifical law (*De religionibus*) and of a work *De iure civilis*.

Treby, Sir George (?1644–1700). Called to the Bar in 1671, he quickly came to the fore and was of the counsel who defended the Seven Bishops. He became Recorder of London (1680–3, 1688–92), Solicitor-General (1689), and Attorney-General (1689) and in 1692 Chief Justice of the Common Pleas. He was deemed a learned lawyer and an impartial judge, and frequently deputized for Lord Somers as speaker of the House of Lords. He edited a *Collection of Letters relating to the Popish Plot*.

Trendelenburg, Friedrich Adolf (1802–72). German philosopher, author of philosophical works, including *Naturrecht auf dem Grunde der Ethik* (1860).

Trent incident. In November, 1861, the U.S.S. *San Jacinto* stopped the British steamer *Trent* and seized two Confederate diplomatic agents, James M. Mason and John Slidell, on board en route to Britain. In Britain the incident was deeply resented and there was even a danger of war, but the matter was resolved by the release of the seized persons and the admission that the *Trent* with its personal contraband should have been brought into port for adjudication in prize.

Trespass. A form of action in medieval English law initiated by a writ calling on the defendant to show why he had done something to the plaintiff's damage and in breach of the King's peace. It became common in the fourteenth century, only when the *nisi prius* (q.v.) system had made it possible to try locally issues already settled in London. The origin is uncertain, possibly in the old appeal of larceny, possibly in the assize of novel disseisin. Apart from the action the courts dealt with indictable trespass, later called misdemeanours. Distinct varieties of trespass for personal harm, taking away chattels, and unlawfully entering on land became settled. The writ might allege that the conduct was done in breach of the King's peace, in which case only the King's court had jurisdiction, but this was not essential.

Later examples multiplied where the writ did not take one or other of these common forms but was specially framed with reference to the circumstances, such as breach of a statute or infringement of a custom of the realm, which had caused the plaintiff harm. These became known as actions on the case (q.v.) and trespass and case were clearly distinguished. Trespass lay only for injuries direct and immediate, but indirect or consequential injuries were remediable by case.

Another special form of trespass was trespass *de ejectione firmae*, available to a tenant against anyone who ejected him from the premises leased, and another that which alleged that the plaintiff had lost possession of something, that the defendant had found it and refused to restore it, but converted it to his own use. This gave rise to the action of trover, now known as conversion.

An important development of trespass arose when the narrative alleged that the defendant had undertaken (*assumpsit*, q.v.) to do something and had failed to do so. In this way an action based on trespass became the remedy for breaches of contract.

By the end of the seventeenth century much litigation was effected by using forms of actions which had developed from trespass, hence the epithet 'the fertile mother of actions'.

In modern law trespass is a voluntary wrongful act against the person of another or to the disturbance of his possession of property against his will. Certain trespasses, e.g. with firearms, or amounting to a breach of the peace, or in pursuit of game, are both actionable and criminal, but in general trespass is actionable only. Trespass is criminal only under particular statutes or where wilful and malicious damage is done.

There are three kinds of trespass, viz.: to the person, to goods, and to land. Trespass to the person comprises assault, assault and battery, and wrongful imprisonment. The act may be intentional or negligent but must be committed against the will of the plaintiff. Trespass to goods (*de bonis asportatis*) is an unlawful disturbance of the possession of goods by seizure or removal of them, or by a direct act causing damage to them. It is no defence to an action that a third party under whom the defendant does not claim, has a better right to possession than the plaintiff, unless the latter relies on a mere right of property. Trespass to land (*quare clausum fregit*) covers every unlawful entry by a person on land in the possession of another, even though no damage be done, and includes taking possession, pulling down or destroying anything permanently fixed to it, wrongfully abstracting minerals from it, discharging water or dumping rubbish on it, and the like. By statute passage of aircraft through airspace above land is, subject to conditions, not actionable. Any form of possession, so long as clear, exclusive, and exercised with intent to possess, will support an action of trespass. A person who enters on land with authority and while there abuses the authority by committing a wrongful act is deemed a trespasser *ab initio*. A trespasser may be requested to leave and, if need be, removed with no more force than is reasonably necessary. Damages are also competent, and an injunction against continuance or threatened repetition. It is a defence to prove that the defendant entered by permission, in the exercise of a legal right, or in circumstances of necessity.

In Scotland trespass is used in relation to entry on land only; as in England it is criminal only under statute, and it is not actionable unless some actual harm is done. In other respects it is generally comparable to trespass to land in English law.

Trespasser. A person who comes on to the land or premises of another not as a legitimate visitor and who has no business to be there. He may be requested to leave or be turned away, but while on the premises no harm may be deliberately done him and some care must be taken not negligently to cause him injury.

Treutler, Hieronymus (1565–1607). German humanist jurist, author of *Selectarum Disputationum ad jus civile Justinianeum* (1592) and *Consilia* (1625).

Trevelyan–Northcote Report (1853). Report of an inquiry by Sir Charles Trevelyan and Sir Stafford Northcote into the organization of the British civil service. They recommended the creation of a unified service with appointment on the basis of competitive examinations and promotion by ability and merit. The recommendations were put into force in 1870 and have remained fundamental elements ever since.

Trevethin, Lord. See LAWRENCE, A. T.

Trevor, Sir John (1637–1717). Called to the bar in 1661. A relative of Jeffreys, he sought to advance himself by ingratiating himself with that scoundrel and defended him in Parliament. He entered Parliament in 1673, became Speaker of the Commons in 1685, and in the same year Master of the Rolls, an office which he lost at the Revolution. He then became Speaker of the House but was found to have taken a bribe and expelled in 1695, but retained the office of Master of the Rolls to which he had been reappointed in 1693 and held that office till his death in 1717. Though coarse, avaricious, corrupt, and an unscrupulous politician, he was an able lawyer, upright as a judge, and some of his decisions were noteworthy.

Trevor, Sir Thomas (1658–1730). Called to the Bar in 1680, became Solicitor-General in 1692, Attorney-General in 1695, and Chief Justice of the Common Pleas in 1701. He became a peer in 1712 (the first Chief Justice of the Common Pleas to be ennobled while in office), but was dismissed in 1714, though he later became Lord Privy Seal (1726) and Lord President of the Council (1730). He is generally deemed an able and upright judge though politically he vacillated.

Trial. The general term for proceedings, civil or criminal, in a court of first instance, frequently involving the hearing of evidence, leading to the court's determination of the matter in issue. In various kinds of proceedings there are other terms for the same process, such as 'hearing' in the Chancery Division, where the trial is by affidavit, or 'proof' in the Scottish civil courts in cases without a jury.

At common law there were many modes of trial. The ancient Anglo-Saxon modes or forms of proof, by ordeal, by oath (compurgation, later called wager of law), by witnesses, and by production of charters continued till about the end of the twelfth century side by side with the Norman procedure of wager of battle. But then the principle of recognition by a sworn inquest or jury came to be applied to judicial inquiries. The use of a jury both for criminal presentment and for civil inquest is first mentioned in the Constitutions of Clarendon (1164) and in the Grand Assize it was applied to the decision of actions brought to try the right to land. By the Assize of Clarendon (1166) the principle of recognition by jury was extended to criminal cases. The petty or trial jury, originally to affirm or traverse the testimony of the jury of presentment, later became judges of fact, determining cases on the evidence of others, and down to the nineteenth century trial by jury was the standard mode. It has now declined because of, on the criminal side, the great number of offences which can be tried summarily and, on the civil side, the readiness of judges and parties to have cases heard before a judge alone.

In actions in the High Court there are now several modes of trial, particularly trial before a judge alone, trial before a judge and jury, before a judge with assessors, before a master, and before a master with assessors. In criminal cases a trial takes place in the Crown Court before a judge and jury, or summarily before a stipendiary magistrate or bench of lay justices.

Trial at Bar. The trial of an accused person before three judges in the Central Criminal Court, a mode of trial formerly allowed only on special application to the court, and on special grounds, but now incompetent. See also TICHBORNE CASE.

Tribonianus (*c.* A.D. 470–543). Byzantine jurist, esteemed particularly learned in earlier legal literature. He rose to high office, became a confidant of Justinian and was appointed to participate in the codification projects. He was a member of the commission which prepared the first *Codex*, possibly inspired and planned the *Digest*, and was certainly chairman of the commission, director of the work of excerpting the fragments and compiling the *Digest*, directed the writing of the *Institutiones* and, having been mainly responsible for the *Quinquaginta Decisiones* issued since the first *Codex*, was head of the commission which prepared, and was largely responsible for, the revised *Codex*. He remained quaestor or minister of justice till his death and was probably responsible for some of Justinian's earlier *Novellae*.

A. M. Honoré, *Tribonian* (1978).

Tribunal. In its most general sense, any person or body of persons having power to judge, adjudicate on, or determine claims or disputes. But in modern Britain the term means more particularly a person or body, formerly frequently called an administrative tribunal (q.v.), as distinguished from a court properly so-called with professional judges, formal procedure and decisions according to rules of law. Tribunals have greatly increased in number, variety, and jurisdiction since 1918, being frequently established to decide issues not wholly suitable for courts of law, to bring experience and expertise to the decision, to operate locally, speedily, and cheaply. Common features are their being composed largely or entirely of laymen, though the chairman is often legally qualified, simplicity and informality of procedure, and decisions based on discretion, impression, and experience rather than on the application of rules of law. Important examples are industrial tribunals, rent tribunals, and social security tribunals. Appeal to a court is commonly restricted or practically excluded, frequently confined to questions of law, but most tribunals are subject to the supervision of the Council on Tribunals and to the supervisory jurisdiction of the courts, whereby the courts will ensure that tribunals observe the rules of natural justice, do not overstep their jurisdiction, or otherwise act illegally.

Tribunal des Conflits. A French court with the function of determining whether a case falls within civil or administrative jurisdiction.

Tribunal of Inquiry. Under the Tribunals of Inquiry (Evidence) Act, 1921, the government may set up a public tribunal of inquiry, usually under a judicial chairman, to inquire into some matter of public importance. Evidence is adduced by or on behalf of the Attorney-General and parties involved may be represented by counsel.

Important instances have been into leakage of Budget proposals in 1936; the Lynskey Tribunal, 1948, which investigated charges of corruption against members of the government; and the Vassall Tribunal, 1963, enquiring into alleged disclosures of state secrets to foreign powers.

G. W. Keeton, *Trial by Tribunal.*

Tribunals, Council on. A council established by the Tribunals and Inquiries Act, 1958, not to be a court of appeal from administrative tribunals, but to keep under review the constitution and working of certain tribunals, to consider and report on such matters as may be referred to the Council under the Act with respect to tribunals other than the ordinary courts, and to consider and report on such matters as may be referred, or as the Council may determine to be of special importance, with respect to administrative procedures involving the holding by or on behalf of a Minister of a statutory inquiry. No power to make procedural rules for a tribunal may be exercised save after consultation with the Council. It consists of 10 to 15 members and has a Scottish Committee of two or three members together with three or four non-members. It makes an annual report.

Triennial Act. The Act of 1641 (16 Car. I, c. 1) passed following Charles I's period of personal rule 1628–40, to provide that Parliaments be held regularly. It enacted that if a fresh Parliament were not called by 10 September in the third year after the last sitting of the last Parliament, the old Parliament shall assemble again, on the second Monday in November, and that no Parliament thereafter should be prorogued or dissolved within 50 days of its meeting unless with its own consent.

Also an Act of 1694 (6 & 7 W. & M., c. 2) enacting that thereafter a new parliament be summoned within three years at most from the end of the last Parliament and that no parliament continue longer than three years. This latter period was first extended by the Septennial Act, 1715, to seven years and then reduced by the Parliament Act 1911 to five years.

Trimoda necessitas (erroneously called 'trinoda' by Selden). The triple burden attaching to land in pre-Norman England, comprising liability to military service and to contribute to the repair of fortresses and bridges (*fyrd-fare, burh-bot*, and *brycybot*). This was continued after feudal tenure was fully established, but all lands were subject also to additional obligations, due to the King as feudal lord, or to tenants-in-chief from mesne-tenants, or to mesne tenants from under-tenants, of which the chief was knight-service.

Trinity House (or the Master, Wardens, and Assistants of the Guild, Fraternity, or Brotherhood of the Most Glorious and Undivided Trinity and of St. Clement). Commonly called the Corporation of the Trinity House of Deptford Strond. The Masters are known as Elder Brethren and comprise Acting Masters and a number of other Masters, some of whom are honorary officers. The corporation was chartered in 1514 and has the responsibility for pilotage, lighthouses, beacons, and seamarks round the coast of England and Wales. The corresponding body in Scotland is the Commissioners of Northern Lighthouses. Acting Masters may sit with a judge in an Admiralty action to advise him on matters of seamanship or navigation, and may sit as assessors in prize courts.

Tronchet, François Denis (1726–1806). An experienced practising lawyer, chosen to defend Louis XVI, president of the Tribunal de cassation, one of the commission which drafted the French Civil Code and a champion of the tradition of the customary law therein.

Trover. A form of action in English law, derived from the action on the case (q.v.) taking its name from and based on the fiction that the defendant had found (trouvé) goods and then converted them to his own use. It is now called the action of conversion. It accordingly lies where a person has converted or wrongfully appropriated goods to his own use or the use of another, or wrongfully deprived the owner of the use or possession of them for a substantial time, or destroyed them. The plaintiff must have had a right of property and possession in the goods. The plaintiff may lawfully retake the goods, or recover damages for the conversion, or waive the tort and sue on an implied contract for money had and received.

Truce of God. An institution which appeared in Europe about 1025, as an attempt to prohibit fighting on certain days. Like the Peace of God, which aimed to place under permanent shelter from feudal wars all churches and clergy and the peaceful part of the population, it stemmed from the desire of the Church to reduce the risks of harm from war. The idea of Truce of God first appeared as a diocesan measure at the Synod of Elne in 1027, being at first limited to Sundays, but later to the period from Wednesday evening to Monday morning. It was taken up in other dioceses, and at the Council of Clermont in 1095 it was made general and this was confirmed by the Lateran Council in 1139. War was prohibited for, in effect, three-quarters of the year. The idea spread to Spain, Germany, and Italy but was not very effective since it depended on ecclesiastical censures. With the emergence of royal states in the twelfth century the idea of the King's peace tended to replace the peace of the Church.

Truck Acts. A series of Acts passed from 1831 onwards to combat the 'truck system' common on public works contracts in the nineteenth century whereby workmen were paid in the form of orders for goods redeemable only at the 'tommy shop' on the site, the owner of which was commonly in league with the employer and who commonly sold inferior goods at inflated prices. The Acts accordingly required payments of wages to be in cash and forbade deductions save under stringent conditions. In 1960 it was made permissible, notwithstanding the Truck Acts, at the employee's request, to pay by cheque, money order, or direct credit to a bank.

True bill. The finding of a grand jury (q.v.) endorsed on the bill of indictment, when, having heard the evidence, they were satisfied that there was a prima facie case to be answered.

Trumbull, Sir William (1639–1716). Was admitted to Doctors' Commons in 1668, served as Judge-Advocate of the Fleet, and as an ambassador to Constantinople and was later Secretary of State, 1695–97. He was both a learned civilian and man of letters.

Truro, Lord. See WILDE, THOMAS.

Trust. An arrangement for the holding and administration of property under which property or legal rights are vested by the owner of the property or rights (truster) in a person or persons (trustees) which they are then to hold or to exercise for or on behalf of another (*cestui que trust* or beneficiary), or others, or for the accomplishment of a particular purpose or purposes. The essence of the concept is the separation of legal and beneficial ownership, the property being legally vested in one or more trustees but in equity held for and belonging to others. The obligations involved are equitable and enforceable only in a court of equity. Trusts developed after uses were forbidden by the Statute of Uses (q.v.) and were recognized and enforced, and the doctrines and principles relevant thereto, elaborated and refined by the Court of Chancery thereafter. The concept of trust is the most important contribution of English equity to jurisprudence. It is not derived from Roman law, from such institutions as *fidei commissa*, and its origin is doubtful. Significance in its development attaches to the idea of conscience, which was treated as important by the early chancellors, who were ecclesiastics. Statutes have only to a small extent affected trust law, though the few statutes are important.

The trustees stand in a fiduciary relation to the *cestui que trust* and must hold the property or exercise the rights in a fiduciary capacity. In many relationships, e.g. partnership, a person stands towards another or others in a fiduciary relationship and may be deemed a trustee for some purposes.

A trust may be created expressly, by statute, by *inter vivos* declaration, or by will, or arise by operation of law, which includes constructive trusts and resulting trusts. It may be a private trust, for the benefit of individuals, or a public trust for charitable or public purposes. It may extend to any property or interest in property which a person can transfer, or assign, or dispose of. It may be in favour of any person, or for charitable or other lawful purposes. The objects of the trust must be declared with sufficient certainty to permit the court to enforce it.

Where a trust is created expressly there must be an adequate declaration of trust and certainty of intention, as to subject matter, and as to objects or persons to be benefited.

Among trusts arising by operation of law a constructive trust attaches to property which a person holding property in trust has acquired by means of his ownership of or dealings with that trust. A resulting trust arises where an intention to create a trust is indicated but the trust is not fully declared or fails in whole or in part, or where property is put in the possession of a person ostensibly for his own use but truly to effect a purpose which fails, or where property is put in the possession without intimation that he is to hold in trust but retention of the beneficial interest by the purchaser is presumed intended and is held to be equitable.

The Chancery courts have over the centuries greatly elaborated principles laying down the rights, powers, and duties of trustees. In particular, high standards have always been demanded of trustees in respect of care for the trust estate, careful investment, strict accounting, fair apportionment between income and capital, duty to pay the right beneficiaries, and absence of personal profit or self-interest and trustees have frequently been held liable to repay to the trust losses caused by lack of care or other breach of trust. Trust property can be recovered from a third party who has obtained it unless he obtained a legal title, for valuable consideration, and without notice that his acquisition was in breach of trust.

Equity has always shown great favour to charitable trusts, which comprise trusts for the relief of poverty, for the advancement of education, for the advancement of religion, and for other purposes beneficial to the community, and the court will not allow a charitable trust to fail because the purpose is uncertain but will give effect to the truster's intention by settling a scheme for administration of the trust to give effect as nearly as possible to the truster's intentions.

Persons qualified to be trustees include persons, corporate bodies having power to do so, judicial trustees appointed by a court and subject to its supervision, and the Public Trustee. Special kinds of trustees include trustees in bankruptcy, trustees for the purposes of the Settled Land Act, tenants for life, statutory owners, and trustees for sale.

A trustee must accept or disclaim office, and persons may be authorized to appoint new trustees, or the court may do so. A trustee may retire but remains liable for things done while he was a trustee, or may be removed by the court. A trustee must not make use of trust property for his private advantage and must account for profit made out of his trust. He is not entitled to any remuneration, may not acquire trust property, and dealings with beneficiaries are voidable.

The trustees' duties are to take possession of and preserve the trust property, be diligent and prudent in administering it, act personally, be impartial as between beneficiaries, keep accounts and give information to the beneficiaries as required, and invest the trust funds in the manner permitted by statute. If in doubt, trustees are entitled to obtain the opinion of the court as to the right course of action, and they may have specific questions determined by the court.

The court now has wide powers to approve

arrangements varying or revoking the trusts or enlarging the powers of the trustees.

Breach of trust is any act in contravention or excess of the duties imposed on the trustees by the trust, including neglects, omissions, and dishonesty, and trustees are liable in so far as loss has resulted to the trust estate. They may be relieved from liability by provision in the trust deed, or by the court under statutory power. Beneficiaries can obtain no relief against trustees if they concurred or acquiesced in the breach of trust.

In Scotland the law of trusts has developed separately; it is based not on the dualism of legal and equitable ownership nor on obligations enforceable in equity, there being no distinction of law and equity in Scotland, but on the principle of property being vested in trustees as legal owners subject to the burden of their holding and administering it for the trust purposes, for the benefit of persons who have claims on the trust estate. Different statutes apply; there is no Public Trustee. In respect of the general principles of the stringent duty of care and diligence incumbent on trustees, the avoidance of conflict between the interests of trustee and beneficiary, and the like similar principles to those of English law have been applied and English analogies have been influential.

The trust concept is practically unknown in civil law systems. It has analogies with but is distinct from the *fidei commissa* of such systems, which is really a different concept altogether.

T. A. Lewin on *Trusts*; A. Underhill on *Trusts*; (Scotland) W. A. Wilson and A. G. Duncan on *Trusts, Trustees and Executors*.

Trust (U.S.) In addition to the foregoing uses the term has commonly been used of large-scale business combinations and groups, frequently utilizing the device of the holding company, whereby one corporation holds enough stock in other companies to control and co-ordinate their operations. For long the U.S. Steel Corporation was the largest holding company in the U.S., but equally notorious were the oil trusts, e.g. Standard Oil Company. After 1890 the Sherman Anti-Trust Act outlawed many of their practices and the Supreme Court decreed dissolution of such trusts as the American Tobacco Company and the Standard Oil Company in 1911. See also ANTI-TRUST LAWS.

Trust for sale. A mode of dealing with land, frequently resorted to in settlements and wills, which has the effect of converting real property into personalty so that the proceeds pass to the beneficiaries as personalty unless they elect to take it unconverted, as realty. Sale may be postponed and when it is effected the proceeds must be paid to at least two trustees, or a trust corporation, or a sole personal representative. Trusts for sale may also be created by a conveyance of the land on trust for sale and an instrument declaring the trusts of the proceeds.

Trust territories. Territories consisting mainly of former colonies of defeated belligerents in the two World wars administered under the Trusteeship Council of the U.N. Since 1945 nearly all have become independent.

Trustee. A person having nominal title to some right or property which he holds not for his own sole beneficial interest but for the interest of another or others. Trustees may be persons, corporate bodies empowered by their constitutions to act as trustee, and, in England, the Public Trustee. Trustees may be appointed under express or implied trusts, both private and charitable, by the court and special categories of trustees include trustees in bankruptcy and trustees of the assets of friendly societies. Partners may hold land as trustees for their firm. The term is also used in a wider sense of persons who though not strictly trustees must have regard for the interests of others in their affairs; thus, directors of a company have been said to be trustees for the shareholders. A trustee is normally not entitled to any remuneration, but is entitled to be reimbursed his necessary outlays, must show a high standard of diligence and care, must not make profit from the office nor allow personal interest to conflict with the interest of those entitled under the trust, and must account strictly for his intromissions.

Trustee in bankruptcy. A person in whom the property of a person adjudged bankrupt is vested by law, for the benefit of the creditors. The trustee's duties are to investigate the circumstances of the bankruptcy, to discover, gather in, and convert into money all the bankrupt's assets, to investigate claims made by creditors and to admit or reject them according to the proofs tendered of them, and to distribute the available assets among the creditors according to their several rights and claims.

Trusteeship Council of U.N. An organ of U.N. composed of member states, administering trust territories, permanent members who do not do so, and certain other members elected for three-year terms by the General Assembly. The Council's functions are to formulate questionnaires on the political, economic, social, and educational advancement of the inhabitants of trust territories, make annual reports, to examine and discuss reports, to examine petitions in consultation with administering authorities, and to make periodic inspection visits. Since 1947 U.N. entered into a number of trusteeship agreements, many providing for transfer to trusteeship of territories formerly administered as mandates by the League of Nations, but most have subsequently terminated.

Trusteeship system. After World War II the United Nations established an international trusteeship system for the administration and supervision of trust territories. It succeeded and drew on the experience of the League of Nations system of mandates, and it took over such mandates as were still in force. The objects of the trusteeship system are to promote the political, economic, social, and educational advancement of the inhabitants of the trust territories. The General Assembly approved trusteeship agreements for various territories, some formerly mandated, and others newly taken into care. The normal function of supervision of the administration of trust territories is conferred on the Trusteeship Council and the administering authority must report annually to the General Assembly. While trusteeship agreements provide that the territories are to be administered as integral parts of the administering state, this does not confer sovereignty on that state, and the trusteeship may be revoked by the U.N.

Truth. The quality of agreement or consistency between what is believed, or alleged, or asserted and what in fact is or was. Truth arises particularly in relation to questions of fact, where the legal adviser or judge is seeking to discover the truth, or true state of some matter of fact, as the basis for application thereto of the relevant rule of law. Discovery of the truth, of what really is or was some state of fact, is commonly aided by regard to evidence of facts rendering it certain or probable that the truth of some fact being enquired into is to a particular effect.

The traditional legal view has regarded highly eye-witness evidence as likely to establish the truth, but modern psychological investigations have shown the great risk of erroneous observation and consequently mistaken (though honestly given) evidence, with consequent considerable danger of not establishing the truth. Many matters reported by eye-witnesses may be materially inaccurate and therefore untrue. Expert evidence explaining scientific examination, e.g. fingerprinting, blood-testing, is much more reliable and more likely to discover truth. Photographs and measurements are better than impressions and estimates. Good cross-examination can frequently disclose both innocent and deliberate untruth or inaccuracy.

Tubman and Postman. In the old English Court of Exchequer two barristers of seniority and eminence had these privileged positions conferred on them by the Lord Chief Baron. They were so called because of the places where they sat, the Postman at the left and the Tubman at the right end of the front row of the outer Bar, by the post anciently used as the measure of length in excise cases, and the tub used as the measure of capacity respectively. The positions were vacated when the holders took silk. The Postman had pre-audience over all other barristers, even the Law Officers, in common law matters, the Tubman in equity matters. Both offices ceased with the disappearance of the Court of Exchequer.

Tucker, Frederick James, Baron (1888–1975). Called to the Bar in 1914, he was a judge of the King's Bench Division, 1937–45, a Lord Justice of Appeal, 1945–50, and a Lord of Appeal in Ordinary, 1950–61.

Tugendhat, Christopher Samuel (1937–). Was a journalist and M.P., 1970–76, and then a member of the Commission of the European Communities from 1977.

Tulk* v. *Moxhay (1842), 2 Ph. 774. The plaintiff sold ground to Elms who covenanted to maintain the ground as a pleasure garden. It came into the hands of the defendant who knew of the covenant. The plaintiff obtained an injunction against the defendant's proposal to build on the garden ground. It was held that in equity the covenant ran with the land and was enforceable against any purchaser who bought with knowledge of the restriction.

Tuning, Gerard van (Tuningius) (1566–1610). Dutch jurist and humanist, author of *In institutiones commentaria* (1618), a work highly esteemed in its time.

Turbary. The liberty of digging turf on another's land. It may be created by grant or by prescription and be appurtenant to a house, to take turf as fuel for that house, or in gross.

Turkish law. The Ottoman Empire, which had ruled Turkey and much more widely from 1299, came to an end in 1922 and Turkey became an independent republic. Then followed a period of reform and Westernization. Under the constitution of 1961 Turkey is a democratic republic, a national, democratic, and secularist state based on human rights, the rule of law and social justice, with sovereignty vested in the nation. The constitution defines rights and liberties and their limits, and declares basic rights and liberties inviolable, but makes private property and private enterprise liable to be limited in the public interest. Legislative power is vested in the Grand National Assembly comprising the National Assembly of 165 and the Senate of the Republic of 450 members. Executive power is exercised by the President of the Republic and a Council of Ministers. The President is chosen by the Grand National Assembly from among its members for a seven-year term and cannot be re-elected. The Council of Ministers consists of a Prime Minister and ministers, the former chosen by the President from among members of the Assembly and the latter appointed by the President on the nomination of the Prime Minister. Government is organized into provinces, districts, and sub-districts.

The legal framework established by Suleiman the Magnificent in the sixteenth century survived in modified form until the nineteenth century. In 1869 a civil code, based on Hanafi canon law, was drawn up under the name of Mecelle (*Journal of Legal Provisions*) but it did not cover the whole range of personal relationships. Non-Muslim minorities lived by their personal law. In 1936 the Grand National Assembly adopted the Swiss Civil Code. In 1926 the Turkish Penal Code was adopted; it was inspired by the 1880 Italian Penal Code with the addition of many provisions of the old Turkish Penal Code, and was revised in 1971.

There is a Constitutional Court, established in 1962, designed to make it impossible for Parliament to pass laws contravening the Constitution, and which acts also as a High Court for the trial of senior officers of state and in legal proceedings against political parties. The Council of State founded in 1869 is supreme arbiter of administrative practice and it can also comment on Bills and examine rules and contracts. The Court of Cassation established in 1928 is the highest court and is also court of first instance in particular classes of cases. Below it are Civil Courts of First Instance, Commercial Courts dealing with commercial disputes, and Labour Courts dealing with industrial disputes. Civil Magistrates' Courts deal with cases of minor importance. Below the Court of Appeal in criminal matters there are High or Central Criminal Courts, Criminal Courts of the First Instance or Correctional Courts and Criminal Magistrates' Courts. The Criminal Courts' Procedure law of 1929 is based on the German Criminal Procedure Code of 1877 altered to suit Turkish conditions. There are also special courts of various kinds.

Courts of special jurisdiction include the Court of Accounts, the organ of financial control over state expenditure, and the Military Court of Appeal. There is also a Court for the Settlement of Conflicts of Jurisdiction, charged with settling in final form all disagreements and conflicts of authority and jurisdiction among courts of justice, administrative and military courts, and a High Council of Judges, an independent body to ensure the independence of the judiciary and its freedom from pressure.

Turner, Sir George James (1798–1867). Called to the Bar in 1821 and, with Russell, compiled a volume of Chancery reports in 1823–24. He became Q.C. in 1841 and M.P. in 1847. In 1851 he became a Vice-Chancellor and, in 1853, a Lord Justice of Appeal, and as such he was grave, steady, and sound. He was also a leading member of the commission on the Chancery Court of 1852, whose recommendations resulted in extensive reforms of Chancery procedures.

Turnor, or Turnour, Edward (1617–76). He was called to the Bar in 1640 and sat in Cromwell's Parliaments and in the Convention Parliament. He became Speaker of the House of Commons, 1661–71, Solicitor-General, 1670, and succeeded Hale as Chief Baron of Exchequer in 1671. He left no mark on the law.

Turnpikes. Roads the cost of which was defrayed by tolls levied on travellers at turnpikes or gates across the road, opened to allow the passage of those who had paid. Numerous local Acts in the eighteenth century established turnpike trusts to manage particular stretches of road and levy tolls for their maintenance. In the mid-nineteenth century there were about 1,000 such trusts. The number of such trusts declined in the nineteenth century as roads were increasingly taken over by local authorities, and by the end of that century they had disappeared. Turnpikes were introduced into the U.S. in 1785 when Virginia authorized the Little River Turnpike, and in 1794 a private corporation completed the Philadelphia Lancaster Turnpike in Pennsylvania. The Wilderness Road, created as a trail to link Virginia with Ohio, became a turnpike and a highway to the growing West. By 1819 about 300 turnpike corporations had been chartered and took the lead in developing a national road system. Turnpikes declined in 1820–40 and became obsolete with the growth of canals and railroads, though the name has been revived in the twentieth century, e.g. Pennsylvania Turnpike, for certain motorways on which tolls are payable.

Turpis causa. Base or disgraceful consideration for a promise, on which no action can be founded.

Tutors and curators. In Scots law, following Roman law, the two categories of guardians. Tutors represent, act for, and manage the affairs of pupils, i.e. boys under 14 and girls under 12, while curators act for persons of unsound mind or advise and consent to the acts of minors, i.e. young persons above pupillarity but not yet of full age.

Twelve Tables. The earliest code of laws at Rome and the start of the Roman system of law. According to tradition the Twelve Tables were framed by ten commissioners (*decemviri legibus scribundis*) in 451–450 B.C. There were originally ten, but two supplementary tables were added by a second set of commissioners. The tradition that the commissioners visited Greece may be true and is supported by flavour of Greek rules in the text. The drafts were enacted by the *comitia* and published in the Forum on tables of wood or bronze. They were intended to reduce to writing the most important rules of the customary law, knowledge of which had hitherto been confined to priests, and to reduce patrician and aristocratic privilege. There was accordingly also a measure of legislative reform and innovation on custom.

The Tables contained rules of public but mostly of private and criminal law and of procedure, but cannot have stated the whole rules accepted as law at that time. Only fragments of the text survive in later literature, including some passages in archaic language, difficult of interpretation. The provisions are brief and simple and framed as imperatives.

They recognize the patriarchal family and the prerogative of the patrician clan, the influence of religious custom, and enslavement for debt.

The Tables were highly regarded at Rome and were never abolished, but were superseded by later developments in Roman law, though a few fundamental rules remained in force until Justinian's time.

A. Berger, in *Pauly-Wissowa*; H. F. Jolowicz, *Historical Introduction to Roman Law*. Text in C. G. Bruns, *Fontes Juris Romani*, P. F. Girard, *Textes de Droit romain.*

Twiss, Sir Travers (1809–97). Called to the Bar in 1840 but was successively a fellow of University College, Oxford, and Drummond professor of political economy, 1842–47, professor of international law at King's College, London, 1852–55, and regius professor of civil law at Cambridge, 1855–70. He was a member of various royal commissions and also developed a practice in the ecclesiastical courts, and later in the Admiralty courts. He became Admiralty Advocate-General in 1862 and Queen's Advocate-General in 1867. In 1872 he resigned all his offices and ceased to practise, and devoted himself to scholarship. His publications included numerous pamphlets on current international and political issues, *The Law of Nations considered as Independent Political Communities* (2 vols., 1861–63), an edition of *The Black Book of the Admiralty* (1871–76) and an edition, generally condemned as inadequate, of Bracton (1878–83). Though a diligent, he was an inaccurate, scholar. He was active in the Institute of International Law.

Tyburn. A stream flowing into the Thames in London, now running in a culvert. The name is associated with the Middlesex Gallows which stood west of the stream near the modern Marble Arch. This was a place of public hanging of criminals from about 1300 till 1783. The gibbet was sometimes known as Tyburn tree.

Tynwald, Court of. The Governor, Legislative Council, and House of Keys (or legislative assembly) of the Isle of Man. Tynwald is the hill which was formerly the seat of the assembly and at which Manx legislation was formerly promulgated.

Tytler, Alexander Fraser, Lord Woodhouselee (1747–1813). Called to the Scottish Bar in 1770 and became in 1780 joint professor and in 1786 sole professor of universal history at Edinburgh. In 1790 he became Judge-Advocate of Scotland and in 1802 a judge of the Court of Session as Lord Woodhouselee. He published numerous works, including two supplementary volumes to Lord Kames' *Dictionary of Decisions* (1778), an *Essay on Military Law and the Practice of Courts Martial* (1800), and *Memoirs of the Life and Writings of Henry Home, Lord Kames* (2 vols., 1807). His son Patrick (1791–1849) wrote a still valuable history of Scotland, and a life of Sir Thomas Craig (q.v.).

U

U.K. The United Kingdom (q.v.).

U.N. The United Nations (q.v.).

U.N.E.S.C.O. United Nations Educational, Scientific, and Cultural Organization (q.v.).

U.N.O. United Nations Organization (q.v.).

U.S. United States, United States Reports.

U.S.A. United States of America (q.v.).

Uberrimae fidei. Of the utmost good faith, a term applied to a category of contracts and arrangements where each party must not only refrain from misrepresenting to the other but must voluntarily and positively disclose any factor which a reasonable person in the position of the other party might regard as material in determining whether or not to undertake the contract. This requirement applies to contracts of guarantee, insurance, partnership, family arrangements, and certain others, but not to such transactions as sale.

Ubi ius, ibi remedium (where there is a right, there is a remedy). A traditional maxim importing that where the common law confers a right, it gives also a remedy or right of action for interference with or infringement of that right. The leading example is *Ashby* v. *White* (q.v.), which held that where a man had a right to vote at a Parliamentary election, he could bring an action against the returning officer for refusing to admit his vote. It follows also from this principle that the fact that a claim is unprecedented is no defence; if a right has been infringed some claim to remedy is competent. The remedy is commonly though not always an action for damages.

Udal (or odal law). The system of law which came to Orkney and Shetland with the Norsemen in the ninth century, and now co-exists there with feudalism of later Scottish origin. The odal is the hereditary estate held in absolute property and not of any superior, nor for any homage or service, but for a payment called skat, and a personal obligation to appear at the host or Thing. It comprises the homestead, the common lands, and land let to a stranger. Since the fifteenth century, Scottish ideas and feudal tenure have crept in and much land has been feudalized by charter from the Crown or through the Earl or Bishop of Orkney.

Ugolini (twelfth–thirteenth century). Italian glossator and teacher of Roffredus, Jacobus di Ardizzone, and Odofredus. He made additions to Azo's *Summa* and wrote *Quaestiones, Dissensiones dominorum* and a *Summa super usibus feudorum*.

Ulpianus, Domitius (*c.* 160–228 A.D.). One of the last of the great classical Roman jurists. He held various official posts, being *magister libellorum* under Alexander Severus and *praefectus praetorio* from 222 till his death. Though an official colleague of Paulus they were rivals as jurists and scholars. He wrote extensively, notably a commentary on the praetorian edict (*Ad edictum*) in 83 books, a work *Ad Sabinum* on the *ius civile* in 51 books, with supplementary monographs on particular laws and topics of law, short text-books (*Institutiones Regulae*, seven books and *Liber singularis Regularum*) and general books for practitioners (*Responsa, Disputationes, Opiniones*). He was a compiler rather than an original thinker, but very learned and well-versed in the earlier literature. His work is superior to that of Paulus in clarity of exposition, but inferior in originality and acuteness of judgment. In later centuries, his authority was second only to Papinian. His writings were a principal source of Justinian's *Digest*, almost one-third of which consists of excerpts from Ulpian—2462 in number. The work known as the *Epitome of Ulpian* was formerly regarded as a late abridgment of the *Libri Singularis Regularum* but it is probably based on a late compilation based on classical works. The *Opiniones* may also be post-classical.

Ulster. One of the ancient provinces of Ireland. The name is often used interchangeably with Northern Ireland (q.v.), though the latter, which is politically a part of the United Kingdom, includes only six of the nine counties of Ulster.

Ultimatum. A written communication from one state to another formulating categorically and finally demands which must be fulfilled unless other measures, particularly war, are to be avoided. Theoretically, an ultimatum is not compulsion on the other state, but may have the same effect. It is a violation of international law to send an ultimatum without having previously tried to settle the difference by negotiation or other means of pacific settlement.

Ultimogeniture. The customary rule, also known as borough-English, recognized in certain places, that land devolved on the youngest son.

Ultimus haeres. The final heir, i.e. the Crown, which succeeds to a deceased's property failing all relatives.

Ultra vires (beyond the powers). A doctrine important in relation to the acts or contracts of public authorities, companies and others whose powers are limited by statute or constituting deed. Acts which are *ultra vires* are void and incapable of ratification. Thus, a company incorporated with objects A, B, and C, and powers necessary to attain these objects, cannot validly contract to do something classified as a class D object. Whether an act is *intra vires* or *ultra vires* is accordingly a question of interpretation of the authority or company's statute, charter, or Memorandum of Association. Other acts may be *ultra vires* of the directors if going beyond the powers delegated to them. Similarly, local authorities may act *ultra vires* if they act in bad faith, or make by-laws for purposes not authorized. The principle applies also to delegated and devolved legislative power so that, e.g. byelaws made by a local authority are challengeable if beyond its powers.

Umpire. A third person called in to decide if two arbitrators differ in opinion. Provision for an umpire is implied in a submission to arbitration if not expressed.

Unascertained goods. In the law of sale of goods (q.v.) goods which are not identified but are referred to by description only, e.g. two sheep from a flock, as contrasted with specific or ascertained goods, e.g. a car bearing a particular number. No right of property in unascertained goods can pass to the buyer until the goods being sold are ascertained.

Unborn child. Though not yet born and, accordingly, not a person in law, the existence of an unborn child is recognized for certain legal purposes. The killing of an unborn child is not murder or manslaughter but child destruction (q.v.) and the causing of an abortion (q.v.) is criminal save in special circumstances. A claim of damages for ante-natal injuries is probably competent provided the child is later born alive, but proof that the injuries were caused by the conduct alleged will be difficult. In the law of succession, it is deemed already born for the purposes of provisions beneficial to it, provided the child is later born alive.

Uncertainty. A ground of challenge of wills, charitable gifts, and similar provisions. The will, gift, or other deed is void for uncertainty if a court, construing the provisions, finds it impossible to say what the testator's or donor's intention was.

Unconscionable. A traditional term in equity, signifying what is contrary to good conscience. Courts of equity will, accordingly, generally decline to uphold an unconscionable bargain, such as bargains with heirs or reversioners, usurious loans to expectant heirs and other transactions not fraudulent but yet unfair.

Unde vi. In Roman law, the variety of interdict used to protect a person who had acquired possession by force used against the true owner. It provided the model for the canon law *actio spolii* and in turn for the assize of novel disseisin.

Under-lease. A grant by a lessee to another, the underlessee or sub-lessee, of part of his interest under the original lease, reserving the reversion, and differing from an assignment which transfers the lessee's whole interest to the assignee. Neither privity of estate nor privity of contract exists between the original lessor and the under-lessee.

Undertaking. A promise, frequently one given in the course of legal proceedings by one party or his counsel, and usually as a condition of some concession by the court or the other party. Thus, a court granting an interim or *ex parte* injunction may require an undertaking that the party obtaining it will pay damages if it turns out that the injunction was unjustified and causes the defendant loss. The term is also used in older Acts dealing with companies incorporated for the construction of public utilities, of the whole business, works and enterprise promoted by the company in question.

Underwriter. A person who, under a policy of marine insurance, undertakes to indemnify a person suffering loss by reason of an accident affecting sea adventure. Normally a number of underwriters each undertake the risk to the extent of a stated sum. The term is also used of persons who, before a public issue of shares by a company, undertake, in return for a fixed commission, to take all or a stated proportion of the shares offered and not taken up by the public.

Undischarged bankrupt. A person who has been judicially adjudicated bankrupt and not yet been discharged. Various kinds of conduct, such as obtaining credit without intimating that he is an undischarged bankrupt, are offences. To be an undischarged bankrupt is a disqualification for membership of the House of Commons.

Undue influence. In English law, the principle of equity, a branch of the doctrine of constructive fraud (q.v.) which provides that where one party obtains a benefit by contract or gift from another, the transaction may be set aside if the other party was prevented from exercising free and independent judgment as to the transaction. Where there is a confidential or fiduciary relationship between the parties, e.g. guardian and ward, doctor and patient, undue influence is presumed, but may be rebutted, and in any other case it may be proved to have existed.

In Scotland, a similar principle applies to benefits under contracts, gifts, or wills, but there is no presumption in any case that undue influence existed, but it may always be proved to have existed.

Unenforceable contract. A contract which is legal and valid but cannot be sued on or enforced legally, e.g. because it lacks written evidence in a case where such evidence is required.

Unfair dismissal. In practically every kind of employment, an employee has, since 1971, had the right not to be unfairly dismissed, with the remedy of complaint to an industrial tribunal. This is additional to, but has, in practice, largely superseded, the common law claim of damages for wrongful dismissal. The employer must show the reason for the dismissal and that it related to the capability or qualifications of the employee for performing work of the kind he was employed to do, or related to his conduct, or that the reason was that the employee was redundant or could not continue to work in the position he held without contravention of an enactment, or that there was some other substantial reason such as to justify dismissal. Dismissal for participation in trade union activities is unfair but dismissal for refusal to join a union is fair. Dismissal for participation in a strike is not unfair unless other employees were treated differently. Apart from these considerations the question is whether the employer can satisfy the tribunal that in the circumstances he acted reasonably in dismissing the employee. The result of this legislation is to make it extremely difficult for an employer, without incurring liability for compensation, to get rid of the incompetent, the lazy, and the troublemaker.

Unfitness to plead. The mental state in which a person accused of crime cannot be called on to plead to a charge or be tried thereon if he is so disordered mentally that it would not be just to proceed with a trial, whether or not he was mentally disordered at the time of the alleged crime. The first cases raising the problem seem to have been cases of deaf-mutes in whose cases it had to be determined whether they were mute of malice or by visitation of God. Only in the mid-eighteenth century did the insane prisoner have a real chance of being found unfit for trial. The Trial of Lunatics Act, 1800, provided that if a person was found insane on arraignment by a jury impanelled for the purpose, or later appeared insane to the trial jury, he should be kept in custody until His Majesty's pleasure be known. In modern practice, if a person has been committed for trial and is certified to be suffering from mental illness or severe subnormality he may be detained in hospital. If he is brought up for trial he may be found unfit to plead on arraignment. The issue must be tried by a jury impanelled for the purpose. If the accused is found unfit to plead he is admitted to hospital and may be detained indefinitely. The test of 'insanity' is not necessarily the same as that of insanity at the time of the crime, but is commonly equated with 'certifiability under the Mental Health Act'. In effect, in many cases, the decision is in the hands of the prison doctor.

In Scotland a broad view of 'insanity in bar of trial' has been reached so that it has been rare for a person to be considered fit to be tried and yet to have been found insane at the time of the crime.

Unfree tenure. In the older law of the tenure (q.v.) of land, one of the cases of villein tenures; these were pure villeinage, and the later copyhold and customary freehold tenure, and privileged villeinage or villein socage, from which developed tenure in ancient demesne.

Ungodly jumble. It is alleged, according to Maitland, that Oliver Cromwell once pronounced the English law of real property to be an ungodly jumble, and the phrase has stuck in relation to real property law. Whether the reforms of 1925 have rescued the law from that state is a matter of opinion.

Unger, Joseph (1828–1913). Austrian jurist and statesman, and, from 1881, president of the Austrian Imperial Court. His *System des österreichischen allgemeinen Privatrechts* (1856–64) did much to harmonize Austrian law with German legal science.

Unidroit. The International Institute for the Unification of Private Law, a body established in Rome in 1926, under the aegis of the League of Nations.

Unification of law. The elimination of national peculiarities in law on various topics. A movement towards this end developed in several spheres in the nineteenth century, particularly in those spheres where international transactions were common and the inconvenience of different bodies of law was most apparent. Thus, in 1851, proposals were made for an International Code of Commercial Law, and attempts were made to establish uniform laws throughout Europe on railway freights at conferences at Berne (1878, 1881, and 1886), on negotiable instruments at conferences at Bremen, Antwerp, and Frankfurt (1876–78) and elsewhere, and on the maritime law of general average at conferences at Glasgow, London, and York (1860–64) which resulted in the York rules, later revised and known as the York–Antwerp rules. More recently international conferences have given rise to conventions on collisions at sea, carriage of goods by sea, carriage by air, and by railway. Expert unions have also given rise to much unification of law. The Telegraphic Union was the first important administrative union; it embraced all Europe by 1852 and became universal in 1865. The Universal Postal Union was established in 1874 and others followed. Very important in securing internationally accepted minimum standards has been the International Labour Organization (1919). From 1893, countries have been moving towards unification of their rules of international private law through the work of The Hague Conferences from 1893 onwards.

In 1926 the League of Nations set up the International Institute for the Unification of Private Law but it has achieved little.

The United Nations has established a number of specialized agencies some of which have a unifying effect on the law of member states mostly in technical areas. The growth of regional organizations, such as the European Communities has, in some cases, a unifying effect in that approximation of the laws of member states on certain topics is a necessary aim of the organization.

Complete unification of the legal systems of the world would seem to be a Utopian dream. In every country the public law has been influenced heavily by its origins and historical development and is strongly influenced by political views and the felt needs of the times. Criminal law reflects history and social, moral, economic, and other views; it must be responsive to the views and needs of the particular community. Private law is shaped by many factors, by history, notably in the land law, by social, religious, and moral views, notably in the law of persons and the family, by economic and business considerations, notably in the law of contracts and commercial and industrial relations. The parts of each legal system which give most benefit from unification are those parts which frequently and naturally involve international transactions, such as carriage, bills and payment and shipping, and where international protection of rights is desirable, such as copyright and patents.

Uniform legislation. In federal states and between states closely linked by geography or commercial relations, there are great advantages in uniform legislation, particularly in commercial matters, in eliminating disputes. There is less need or justification for this in the law of persons, land law, or criminal law which all reflect historical and social differences more directly.

In the U.S., a movement for uniform state legislation began in 1889 with the appointment by the American Bar Association of a Committee on Uniform State Laws. Their annual conference is now known as the National Conference of Commissioners on Uniform State Laws. It has prepared draft legislation on many subjects the most successful and widely adopted being the Negotiable Instruments Act, the Warehouse Receipts Act, the Sale Act, the Bills of Lading Act, and the Stock Transfer Act. The American Law Institute has co-operated. A draft commercial code was framed in 1952 and later revised.

In Canada, a Conference on Uniformity of Legislation has met regularly since 1918 and, in Scandinavia, a number of identical laws have been adopted in the Scandinavian countries.

Uniformity, Acts of. English statutes seeking to enforce Protestantism on all persons in the country. The Act of 1549 enforced the exclusive use of Archbishop Cranmer's moderate Protestant Prayer Book but imposed only mild penalties on disobedient priests and none on dissenting laymen. It was repealed in 1553. The Act of 1552 enforced the use of Cranmer's revised (1552) Prayer Book, which was strongly Protestant, prescribed heavy penalties for non-use thereof, enforced attendance at the prescribed services and asserted that the Church's liturgy depended on parliamentary authority. It also was repealed in 1553. The Act of 1559 was passed to enforce the use of the 1559 Prayer Book which was a slightly modified version of that of 1552, and prescribed severe penalties for recalcitrant clergy. The Act of 1662 imposed the Book of Common Prayer and the Church of England liturgy on all ordained preachers, which drove a substantial proportion of the clergy out of the Church, and required a declaration from all in orders repudiating the Solemn League and Covenant, denying the right to take up arms against the King and undertaking to adopt the Church of England liturgy as established by law. It failed to secure uniformity and from that time Protestant nonconformity existed.

Unilateral act. An act in which the will of only one party is operative, as in making a will, as contrasted with a bilateral act, as where two parties make an agreement.

Unilateral obligation. An obligation under which only one party is bound, such as one constituted by promise or undertaking to another. In English law, such an obligation, lacking consideration, must be entered into under seal. In Scots law, such an obligation is valid without consideration but must be proved by the writ or the admission on oath of the alleged promisor. The other party to such an obligation, the promisee, is not bound to accept nor obliged in any way.

Unincorporated body. A group of persons associated for some common lawful purpose in a body, normally under such a name as association, club, or society, which body is, however, not incorporated and thereby made into a legal entity by itself. The consequences of not being incorporated are that the body has no existence apart from the members, that it cannot own property, which must accordingly be held by persons as trustees for all the members of the body, and generally cannot incur criminal liability. Common specimens of unincorporated bodies are partnerships, friendly societies, trade unions, and many kinds of associations, clubs, and societies.

Union. In the poor law, any parish or union of parishes having a separate board of guardians. Unions were effected under an Act of 1781 and, more particularly, the Poor Law Amendment Act, 1834. They were abolished in 1930–34. In ecclesiastical law, a union is two or more benefices united into one benefice.

Union, Acts of (1706). Acts of the English (The Union with Scotland Act, 1706) and Scottish (The Union with England Act, 1706) each ratifying a treaty negotiated by commissioners from each Parliament providing for the union of the two kingdoms, which since 1603 had been joined in a personal union only (apart from the period under the Commonwealth when they were united), the termination of both Parliaments, and the creation of a single Parliament of Great Britain, but for the continuing separate existence of the Churches of England and Scotland and of the systems of courts and of private and criminal law of each country. The succession on the Hanoverian line after Anne's death was settled.

Union, Act of (1800). An Act of the Parliament of Great Britain (The Union with Ireland Act, 1800) terminating the Parliament of Ireland and giving Ireland representation in the new Parliament of the United Kingdom. It was largely repealed in 1920–21 by the recognition of the Oireachtas or Parliament of the Irish Free State (now the Republic of Ireland) and the setting up of the Parliament of Northern Ireland (abolished 1973) under the Council of Ireland Act, 1920.

Union of the Crowns. The personal union between England and Scotland effected in 1603, when King James VI of Scotland succeeded Queen Elizabeth and became King James I of England. The union was personal only and each country retained its own Privy Council, Parliament, church, courts and legal system, taxation and foreign policy. An incorporating union, creating the Parliament of Great Britain, was effected in 1706 and came into force in 1707.

Union of Wales and England Act (1536). The Act of the English Parliament which completed the incorporation of Wales into England. It dissolved the marcher lordships and annexed them to existing counties, and divided the rest of Wales into five new counties. English law and legal administration were extended to Wales, which was to send 24 M.P.s to Parliament.

Union shop. An arrangement under which employees are required, within a specified time of entering employment there, to join a particular trade union. It differs from the closed shop (q.v.) in that non-union members may be recruited, though they must join as a condition of continuing to be employed. Union shop agreements are legal and common in many countries, including most states of the U.S.

Unit trust. A means of spreading risk by investing in a large range of stock exchange securities. A unit trust is normally constituted by a trust deed between managers, who select the securities in which money is invested and buy and sell them as they deem prudent, and trustees who are responsible for holding the securities and collecting and distributing the income therefrom. The managers issue units to investors, who acquire thereby an interest in an undivided trust fund in proportion to the number of units held as compared with the total number of units in existence. Unlike an investment trust, a unit trust has no fixed share capital and the managers may create and issue as many units as they like, at a price corresponding to the proportionate interest of a unit in the total trust fund. The income derived from the securities in the trust fund is normally distributed to unit holders twice yearly. The trust deed will define the kinds of securities in which the trust funds will be invested and many unit trusts specialize in, e.g. financial securities, or European securities, or mining shares. Unit trust schemes may be authorized by the Department of Trade if satisfied that various conditions are fulfilled.

Unitary state. A state in which the major institutions of government, legislature, executive, and judiciary, have power in all matters over the whole area and all persons in the territory of the state, as contrasted with a federal state where powers are distributed, some to the central or federal government and some to the state or provincial governments. A unitary state does not exclude the possibility of local or other governmental agencies having powers devolved or delegated to them by the main government, but power is delegated, not distributed, and, technically, all belongs to the main government.

United Kingdom of Great Britain and Northern Ireland. A state, commonly but incorrectly referred to as 'Great Britain', 'Britain', 'The British Isles', or even 'England'. It comprises (1) the kingdom of England; (2) the principality of Wales (q.v.) which was declared by the *Statutum Walliae*, passed in 1284 after Edward I had defeated Llewelyn ap Griffith, to be deemed incorporated into the kingdom of England (the introduction of English legal and political principles being completed by Henry VIII, particularly the Laws in Wales Act, 1535, though the judicial systems remained distinct till 1830); (3) the Kingdom of Scotland which became linked to England and Wales by a personal union in 1603, when King James VI of Scotland succeeded Queen Elizabeth I of England and became also King James I of England, and by the creation by treaty in 1707, ratified by the separate Parliaments, of the Parliament of Great Britain, replacing the separate Parliaments of the two countries, which then ceased to exist; and (4) Northern Ireland (q.v.), being that part of the kingdom of Ireland, united with Great Britain into the United Kingdom of Great Britain and Ireland by the Act of Union of 1800, which elected in 1920 to remain within the United Kingdom when the rest of Ireland asserted its independence (becoming successively the Irish Free State, Eire, and the Republic of Ireland). The Isle of Man (q.v.) and the Channel Islands (q.v.) though Crown dependencies are not parts of the United Kingdom, save for some statutory purposes.

Though the United Kingdom is the state for the purposes of international relations, it is not truly a unitary state in that, as well as the Parliament at Westminster, there was a Parliament in Northern Ireland at Belfast from 1921 to 1973, with substantial but not plenary legislative powers, an executive based on London, with another executive, subordinate in certain respects, in Northern Ireland, and with certain substantial powers devolved to Secretaries of State for Scotland and for Wales, based on Edinburgh and Cardiff respectively, distinct structures of local government in England and Wales, in Scotland, and in Northern Ireland respectively, separate hierarchies of civil and criminal courts for England and Wales, for Scotland, and for Northern Ireland, and in many and important respects quite distinct systems and bodies of law, civil, administrative and criminal, substantive and adjective, in England and Wales, in Scotland, and in Northern Ireland. There is, accordingly, no such thing as United Kingdom or British law, though there are some rules common to all parts of the U.K. and these terms may be used loosely for the law generally or particular branches of law as contrasted with, say, French or American law. In respects in which the law varies as between England, Scotland, and Northern Ireland, e.g. land law, the terms 'United Kingdom law' or 'British law' are meaningless. But in respect of taxation or social security where the same or substantially the same rules apply in each part of the U.K., such terms are meaningful.

Nor, however, is the United Kingdom a federal state in that there are no Parliaments of Scotland or of Wales and the Parliament of Northern Ireland (now abolished) had not only a defined legislative competence but was established expressly without prejudice to the overriding legislative supremacy of the United Kingdom Parliament, and any legislation of the Northern Ireland Parliament, even within its powers, was void if and in so far as repugnant to an Act of the United Kingdom Parliament and could be expressly overruled thereby. In no respect was the Northern Ireland Parliament a sovereign legislature.

As a state, the United Kingdom is, therefore, *sui generis* and in Scotland and Wales there are, in the 1970s, strong movements for federalism, or greater home rule, or even for independence. Moreover, in the four parts of the United Kingdom, there are very distinctive ethnic, linguistic, religious, and cultural differences, and strong national consciousness even when combined with consciousness of and regard for British institutions.

Constitution

There is no comprehensive constitutional document and the principles and rules as to the structure of government and its powers and operation have to be gathered from various statutes, decisions of the courts, books of authority, customs, and constitutional conventions, understandings, and practices. These are derived more from English than from Scots or Irish constitutional development. The constitution, moreover, is completely flexible in that no principle or rule is exempt from liability to be changed, and, apart from development in the course of time, the Queen in Parliament may, by statute, alter any rule as to the constitution as readily as it may alter any other part of the law. Parliament is accordingly supreme in that it may alter any rule of law whatever, and its legislation is not subject to any kind of judicial review. No person or court has power to declare legislation unconstitutional.

The constitution is monarchical, in that the Head of State is an hereditary monarch, but the monarchy is limited by customs and conventions which have developed over a long period to the effect that the monarch reigns but does not rule, and that the great majority of the legal powers vested in the Crown are exercised, and may only be exercised, on the advice of the Ministers of the Crown for the time being. It is parliamentary in that Parliament has supreme control over all branches of government, representative in that Parliament consists of representatives elected by adult suffrage, and responsible in that the Ministers of the Crown are responsible to Parliament, particularly to the House of Commons, and must retain the confidence of a majority in that House to continue in office and to secure approval of their proposals.

The separation of powers, of the legislative, executive, and judicial powers of government is only admitted to a modified extent. The Crown is an element in all three. The political executive, the Ministers of the Crown, must all be members of one or other House of Parliament, mainly of the House of Commons, though the permanent executive, the civil service, may not be members of either House. The Lord Chancellor, though head of the judiciary, is a member and chairman of the House of Lords and also a member of the political executive, in effect Minister of Justice. The House of Lords has legislative and judicial functions. Ministers of the Crown frequently have legislative and judicial powers delegated to them by Parliament. Parliament was originally a court of law and may, by virtue of its supremacy, do any act, executive or judical as much as legislative, by enacting a statute. But apart from the Lord Chancellor and the House of Lords the judiciary is independent of both executive and legislative.

Legislature

The legislature is the Queen in Parliament. Parliament (q.v.) consists of the Queen, the House of Lords (q.v.), and the House of Commons (q.v.). Its functions are to check, question, and criticize the executive, to debate its policy proposals, and to discuss and, if thought fit, to pass legislation proposed by the government or by private members or petitioned for by private interests. In practice, the majority party supplies the members of the Cabinet and junior ministers and it obediently votes through legislation proposed by the government to give effect to its policy. The royal right to veto legislation passed by Lords and Commons has not been exercised since the early eighteenth century and could not now be exercised without arousing a constitutional crisis. By long-settled convention it is now a pure formality. The once co-equal legislative power of the Lords has been reduced by the Parliament Acts of 1911 and 1949 to a power to delay public bills only, and excluding a bill to extend the maximum life of Parliament. A two-party political system has existed in Britain since the late seventeenth century and there is a perpetual conflict between Government and Opposition. Minor parties are only important when, as happened in 1974–79 they held the balance between closely matched Government and Opposition and were eventually instrumental in bringing the Government down.

Executive

The executive consists of the Queen, the Privy Council, (q.v.) and, much more importantly, the Cabinet (q.v.), which now consists of the leader of the majority party in the House of Commons, who is appointed Prime Minister, and those of his supporters who have been appointed on his recommendation to the headship of the main Departments of State. With the proliferation of ministries and departments not all heads of department are in the Cabinet, and some persons may be in the Cabinet though not at the head of a department. All Cabinet Ministers, if not already Privy Councillors, are made so. There are also a large number of junior ministers.

The functions of the Cabinet are to determine the policy to be submitted to Parliament, to control the national executive in accordance with that policy, and to coordinate the interests and work of all the departments. The Queen acts on the advice of the Cabinet, normally tendered through the Prime Minister. In effect, the Cabinet exercises the sovereignty of Parliament.

The execution of policy approved by Parliament and normally embodied in legislation made by it, and in greater detail in delegated legislation made by Ministers or bodies authorized by Parliament is effected by three main kinds of agencies, central government departments, public corporations, and local authorities.

Central government departments include the Treasury, the Foreign and Commonwealth Office, the Home Office, the Scottish Office, the Ministry of Defence, the Ministries of Health, Education and Science, and so on. Ministries are frequently created, enlarged, merged, renamed and dissolved. They are staffed by civil servants.

Public corporations (q.v.) include the Post Office, the British Railways Board, the National Coal Board, the British Electricity Authority, the British Broadcasting Corporation, and many others. They are staffed by their own employees.

Local authorities have different names, constitutions, powers and functions in England and Wales, Scotland and Northern Ireland. They are composed of elected members and have some discretion in the execution of their functions, which include housing, health, education, subject to constant direction, supervision and control by the relevant ministries of the central government. They are staffed by their own employees.

There is an extensive body of administrative law regulating the organization, functions, powers and duties and control of the main administrative authorities, the civil service, public corporations, and local authorities. Much of this differs as between England, Scotland, and Northern Ireland. But there is no *system* of administrative law, nor of administrative courts or tribunals, nor of Parliamentary or judicial control, only an enormous body of particular provisions.

Judiciary

Parliament was, and to a limited extent is, a High Court, but its judicial functions are limited to privileges, private bills, peerages, and impeachments.

The House of Lords in its judicial capacity may be a United Kingdom court, or an English, Scottish, or Northern Ireland court as the case may be, according to the jurisdication from which a case comes to it on appeal.

A few courts only are United Kingdom courts. The Courts-Martial Appeal Court (q.v.) is composed of judges from the three parts of the United Kingdom and may sit in or out of the United Kingdom. The Restrictive Practices Court (q.v.) is similarly composed and includes also laymen, and may sit in any part of the U.K.

There are distinct legal systems in England and Wales, in Scotland, and in Northern Ireland, each having their own structure of courts, superior and inferior, judges, and procedure, not to speak of their own history and traditions, etiquette, practices and customs. All ordinary judicial work belongs accordingly to the distinct legal systems of the U.K., rather than to the U.K. itself.

Substantive law

Some rules of substantive law, particularly rules established by statute, are common to two or more of the constituent parts of the United Kingdom, and some bodies of rules are largely the same, or very similar, but many bodies of principles and individual rules differ. In general, the law of England and Wales is the same, and that of Northern Ireland very similar thereto, all being developed from the medieval common law of England, which has Germanic roots and affinities. The law of Scotland is materially different, being developed from a distinct medieval common law which has Roman roots and continental affinities but has since 1750 been substantially influenced by imposition of English ideas and principles.

Legal professions

England and Wales, Scotland, and Northern Ireland have each separate professions, separately organized and with their own systems of education and training. Qualifications are not interchangeable though some persons are qualified in more than one of the three systems. In all three the profession is divided into barristers (advocates in Scotland) and solicitors (qq.v.).

See English Law; Northern Ireland Law; Scots Law; Welsh Law.

United Nations Educational, Scientific and Cultural Organization (U.N.E.S.C.O.). A body, established in 1946, with the purposes of contributing to peace and security by promoting collaboration among nations through education, science, and culture in order to further universal respect for justice, for the rule of law, and the human rights and fundamental freedoms which are affirmed by the Charter of the U.N. It operates by exchange of information and preparation of international conventions, and operational involvement in development projects. It is directed by a General Conference composed of representatives of member states, an Executive Board and a Director-General and staff. The headquarters are in Paris.

United Nations Forces. The U.N. may create military forces in the exercise of its responsibility for keeping the peace, and it has done so in several cases. They can only be deployed with the agreement of the state concerned. The U.N. established a force in the Middle East by resolution of the General Assembly, which was established as a subsidiary organ of the U.N. composed of international personnel. Its immunities and privileges were founded on the General Convention on the Privileges and Immunities of the U.N. The force in the Congo was established by resolution of the Security Council and a special agreement regulated its privileges and immunities. A force was also established in Cyprus.

United Nations Organization. During World War II, the establishment of a general organization of states to safeguard peace and promote international co-operation, replacing the League of Nations, (q.v.) was accepted as necessary. The name was devised by President F. D. Roosevelt and first used in 1942. The Moscow Declaration (U.K., U.S., Russia, and China) of 1943 recognized the need to establish a general international organization. The Organization was established by representatives of 50 countries who met at San Francisco in 1945, on the basis of principles formulated by representatives of the major powers at Dumbarton Oaks in 1944. It came into being officially when a majority of the signatories ratified the Charter in October 1945. Subsequently most other nations of the world have become members. The headquarters is in New York City and the official languages are Chinese, English, French, Russian, and Spanish, the working languages English and French. The aims and ideals of U.N.O. are set out in the Preamble to the Charter and include determination to prevent war, reaffirmation of faith in fundamental human rights, the maintenance of Treaty obligations and observance of international law.

The members are the original 50 states which participated in the San Francisco Conference and states later admitted, which are peace-loving, accept the obligations of the Charter and are deemed able and willing to carry out these obligations. States are admitted members on the recommendation of the Security Council and with a two-thirds vote of the General Assembly. Membership is not compulsory and a state may be expelled for persistent violation of the principles of the Charter, or, probably, withdraw.

The fundamental legal duties of members are to settle disputes by peaceful means, to refrain from the threat or use of force against the territorial integrity and political independence of any state, and the obligation to give the U.N. assistance in any action taken in accordance with the Charter and to refrain from assisting any state against which the U.N. is taking measures of prevention or enforcement. It is mandatory on the U.N. to ensure that non-member states act in accordance with the principles of the organization so far as necessary for the maintenance of internal peace and security. Nothing in the Charter authorizes the U.N. to intervene in matters essentially within the domestic jurisdiction of any state or requires the members to submit such matters to settlement under the Charter.

The U.N. is the legal organization of the international community and has legal personality distinct from that of its members, though a kind of legal personality distinct from that of a state, somewhat resembling that of a confederation of states.

The U.N. operates through six main organs, the

General Assembly, Security Council, Economic and Social Council, Trusteeship Council, International Court of Justice and Secretariat (qq.v.).

The General Assembly comprises representatives of all members and its functions are initiative, discussion, study, and recommendation, but not legislation. It is charged with initiating studies and making recommendations for promoting international co-operation in the political sphere and encouraging the progressive development and codification of international law, and furthering co-operation in the economic, social, cultural, and educational fields. It has functions under the trusteeship system (q.v.) and controls the Economic and Social Council and the Trusteeship Council. It meets in regular annual sessions and special sessions as required. Decisions are, in general, reached by simple majority, but certain important questions require a two-thirds majority.

The Security Council was composed of five permanent members (China, France, Russia, U.K., and U.S.A.) and six non-permanent members elected by the General Assembly for two years, all having one representative, and since 1965 the five permanent members and 10 non-permanent members. Its functions are primarily executive, almost entirely in relation to the maintenance of international peace and security. Members of the U.N. agree to accept and carry out the decisions of the Security Council reached in accordance with the Charter, but there is no legal obligation to treat recommendations of the Security Council as binding though they may provide authority for individual or collective action in pursuance of the Charter. The Security Council must be able to function continuously and each member is at all times represented at U.N. Headquarters. In addition to functions in relation to international peace, the Council has functions together with the General Assembly in relation to the working of the Organization as a whole. Decisions require the votes of nine members, including (except in matters of procedure) those of all the permanent members. In relation to the pacific settlement of disputes the parties must abstain from voting. This power of veto has frustrated many of the actions of the Security Council.

The Economic and Social Council is the main instrument of the Organization for the promotion of higher standards of living, economic and social progress and respect for the observance of human rights and fundamental freedoms. It consists of 27 (originally 18) members elected by the General Assembly and its functions are to make or initiate studies and reports on international economic, social, cultural, educational, health and related matters and to make recommendations thereon, to make recommendations for promoting respect for and observance of human rights and fundamental freedoms for all, to prepare draft conventions for the General Assembly and to call international conferences. It may consult with international and national organizations and the various specialized agencies of international administration, e.g. F.A.O., I.L.O., I.C.A.O., and U.N.E.S.C.O., are brought into relationship with the U.N. by agreements between the Economic and Social Council and them, though they retain internal autonomy.

The Trusteeship Council consists of states which administer trust territories, the permanent members of the Security Council if not administering such territories, and states elected by the General Assembly for three years. It is charged with supervising the administration of trust territories and may consider reports by the administering authority, accept petitions from the inhabitants of trust territories, arrange for periodic visits to trust territories, formulate questionnaires on the progress of these territories.

The International Court of Justice (q.v.) is the principal judicial organ of and an integral part of the U.N. The Secretariat is responsible for the administration of the principal organs of the U.N. other than the I.C.J. The Secretary General may bring to the attention of the Security Council any matter which, in his opinion, may threaten the maintenance of international peace. The Secretary-General is appointed by the General Assembly on the recommendation of the Security Council. The office has been held by Trygve Lie (1946–52) Dag Hammarskjold (1953–61), U Thant (1961–71), and Kurt Waldheim (1971–) (qq.v.). He is the chief administrative officer and also has important political functions, in particular, being charged with bringing before the Organization any matter which threatens international peace and security.

The U.N. is also a kind of superior to which a large number of specialized agencies are affiliated. Many of these were founded long before 1945. Among those with important legal functions are the International Labour Organization, the International Bank for Reconstruction and Development, and the International Monetary Fund (qq.v.). In 1957 the General Assembly established the International Law Commission (q.v.) to make recommendations for the progressive development and codification of international law.

The major purposes of the Organization are the maintenance of international peace and security, the development of friendly relations among nations and the achievement of international co-operation, and it has had some success in each of these fields. But the Security Council has frequently been frustrated by the failure of the five great powers (whose unanimity is required for substantive decisions) to agree. It has, on the whole, had more success under the heads of friendly relations and international co-operation.

E. A. Gross: *The United Nations: Structure for Peace*; M. Waters: *The United Nations: Institutional Organisation and Administration*; R. E. Asher: *The United Nations and the Promotion of General Welfare*; *United Nations Yearbook* (annually); *Annual Reports of the Secretary-General on the Work of the Organisation.*

United States Code. An official compilation of the public, general, and permanent laws of the United States in force, prepared under the supervision of a committee of the House of Representatives and published by the Government. The subject-matter is rearranged under 50 titles, many subdivided into parts and chapters, and numbered sections and, to a limited extent, the legislation is rewritten. Some titles have been re-enacted as positive law but, apart from these, the Code is a convenient restatement and the law remains the Revised Statutes or Statutes at Large from which the titles and sections of the Code have been derived. Nevertheless, for most practical purposes it has largely supplanted the statutes which are the real authority. The title and section numbers do not correspond to the official title and sections of the Act from which the material is derived. The Code first appeared in 1926 and new editions have appeared at intervals since then. There are several unofficial annotated editions, particularly the United States Code Annotated and the Federal Code Annotated, which contain additional matter not in the official edition.

United States of America law. Apart from federal constitutional law, there is no system of law common to all the states of the U.S. but the term is a convenient general one for the federal law and the bodies of law of those states. The federal constitutional law is common but, apart from that, each state has its own public, criminal and private law, both common law and statutory. Factors making for uniformity are the decisions of the Supreme Court and of the lower federal courts, particularly in the field of commercial law, the requirement of the Constitution that every state shall give full faith and credit to the public acts, records, and judicial proceedings of every other state, and the tendency for states to adopt Uniform Laws on particular matters, such as sales and commercial transactions. Despite the diversity there is a large measure of unity in broad principle, and both the general ideas of statutes and judicial doctrines commonly spread from one state to another.

The historical basis of most U.S. law is the common law of England as developed down to the eighteenth century. Systems based on civil law formerly prevailed in territory subject to France and Spain but little trace of this remains save in Louisiana and, in certain particulars, in certain other states.

The Colonial period—to 1776

It is an oversimplification to say that the law of the American colonies was essentially the common law of England, adopted to the extent appropriate to colonial conditions, and that it was uniform for all the colonies. From the first there were divergences from English law and differences between colonies, just as each tended to develop its own social and political system. At any particular time the law of a colony was made up of a set of traditions, some English, but some local.

The earliest white settlers on the eastern seaboard of North America carried with them some major principles of English common law and equity and the first lords proprietors of north American colonies had by their charters to govern by English law, and the settlers had the liberties of Englishmen. The Virginia Company ordered Governor Sir George Yeardley to introduce English common law and due process, to encourage private property, and to summon a representative assembly having power, with the appointed council of the colony, to pass local laws, subject to the Company's veto. In 1619 he did summon a legislative assembly, which seems to have been elected by manhood suffrage, and placed Virginia under the common law of England.

In the seventeenth century generally, the colonists adhered to the general notions of the common law, but it was generally the simpler, more popular and summary rules, with less regard for the more refined distinctions and artificial developments of the law of England. They claimed the general liberties of Englishmen but denied the binding character of the common law.

The first Court of Assistants of Massachusetts was held in 1630 and established rules of proceedings in civil and criminal matters. In the same year, the first General Court was held at Boston. Until 1639 it seems to have exercised the whole legislative and judicial power, and to have had jurisdiction in civil and criminal matters. Latterly, the President or Governor and Council also exercised a chancery jurisdiction.

In 1636 the general court of Massachusetts entreated their government to make a draft of laws agreeable to the Word of God to be the fundamental laws of the commonwealth, and in the meantime the magistrates were to determine all causes according to the laws then established and where there was no law as near the law of God as they can. The tendency was to develop a body of customary law. The general court pressed for a comprehensive body of law, but the magistrates were reluctant to fetter their discretion and resisted codification.

This agitation resulted in the passage of the

Massachusetts Body of Liberties in 1641, a code prepared by a minister with some legal training, Nathaniel Ward, revised by the General Court, and further revised after suggestions had been received from the towns, which was not enacted as law but which the General Court entreated those in authority to consider as laws. The *Body* evidences the theocratic nature of the colony but contains many provisions originated by the colonists in response to their special needs. In the actual conduct of cases the divine law, interpreted by the discretion of magistrates, was regarded as binding while the common law was at most referred to for illustration. Forms of contracts and deeds were, however, modelled on the common law. In 1646 one Dr Child attacked the government of the colony and called for a return to the 'Fundamental and wholesome Lawes of England'.

In 1648 the *Body of Liberties* was superseded by the *Massachusetts General Laws and Libertyes* which was the first modern legal code of the Western world, a collection of important rules arranged alphabetically rather than a systematic statement. It kept some elements of English procedure, adopted a system of freehold tenure of land, and established a public register of all deeds and mortgages. It established freedom of speech and travel, the right to bail and to appeal, the rule against double jeopardy and against self-incrimination, and secured the equal protection of the law for all.

In 1667 the Privy Council made specific objection to certain laws of Massachusetts as repugnant to the laws of England and the General Court made some amendments in 1681 but asserted the independence of the colony from English laws. After the Crown rescinded the colony's charter in 1684 there was a strong and continued effort to introduce the common law. In 1688 the governor and council were appointed a court of record to try civil and criminal cases, their proceedings to be agreeable to the laws and statutes of England, but the governor's conduct provoked a reaction and the colonists began to assert rights protected by the common law. From 1712, however, when the first professional lawyer, Lynde, became Chief Justice, English books and authors came to be frequently cited, though for long thereafter Massachusetts jurisprudence exhibited traces of its early informality.

In Connecticut the development was similar; in 1639 the main settlements adopted the *Fundamental Orders of Connecticut*, and this remained the basic law of the colony till it became a charter colony in 1662. The Connecticut code of 1642 was copied from that of Massachusetts. Connecticut and New Haven departed even further from common law in their system of popular courts, radical modification or absence of jury trial, discretion of the magistrates and, in the case of New Haven, the assertion of the binding force of divine law as a common law in all temporal matters and a guiding rule in civil and criminal jurisdictions. The various laws of the colony were revised and digested and remained the laws of the province from 1650 to 1686; they are known as the Blue Laws of Connecticut.

In New Hampshire, the General Assembly passed a body of general laws in 1679–80 in which they claimed the liberties belonging to free Englishmen but refused to admit the binding force of any code, law, or ordinance not made by the General Assembly and approved by the president and council. The code is simple but cites English statutes. The colonists were very impatient of regulation and had all matters of law and fact decided by juries. The early judges were all laymen, not till 1754 did a lawyer become Chief Justice, and even then no person considering himself bound by precedents or ruling of common law judges held judicial office in the eighteenth century. There is record of a court of chancery in 1705.

In Rhode Island, a Code of Civil and Criminal Law was adopted in 1647 but generally shows a very archaic concept of law, reminiscent of Anglo-Saxon codes, though in parts it is quite original.

In New York, the colony was under royal authority almost from the beginning and its rulers early accustomed it to the principles of common law, though Dutch speech and procedures lingered on. But, in 1665, an informal assembly promulgated what is known as the Duke of York's laws. The courts had much less of a popular element than in the Puritan colonies. The old court of burgomasters was changed to the Mayor's Court. In 1700 Attwood, an English professional lawyer, became chief justice and made it his purpose to introduce the common law and practice of English courts. The complete doctrine of the binding force of the common law was not declared before 1761.

In New Jersey, systems of popular courts were established, in East Jersey in 1675, in West Jersey in 1693. The early laws of East Jersey were founded on scriptural authority but, in 1698, the privileges of the English common law were assured to everyone. These colonies were reunited in 1697.

Pennsylvania had the most complete system of colonial laws, comprising the frame of government, the laws agreed upon in England in 1682, the Great Law or body of laws enacted at Chester in 61 chapters in 1682, the act of settlement passed at Philadelphia in 1683, laws in 80 chapters made at an assembly there in 1683, the frames of government of 1683 and 1696, and the laws of 1701. These contain many innovations; not least, in the laws of 1683 there is a separation of fundamental provisions, alterable only by the consent of six-sevenths of the council and assembly, from merely secondary provisions. Early court procedure was very informal, and legal and equitable principles were administered by the same courts.

In Maryland, full powers of government were given to the proprietor, who could establish courts and laws. The first legislative assembly passed a body of laws in 1635 which was rejected by the proprietor. Later, when there was deadlock the colonists claimed that they were governed by English common law, so far as applicable to their situation, but the controversy between him and the colonists went on till 1732. The colonists tended to insist more on the common law as subsidiary rules because of the wide powers of the governor and council and the limited legislative powers of this colony. In 1662 an Act declared that, failing provision in the laws of the colony, justice was to be administered according to the laws and statutes of England, which was the first definite recognition of the power of the courts in America to apply English common law to colonial conditions and to reject provisions deemed unsuitable. The struggle between proprietor and colonists concerning English law revived in 1722 over the applicability of English statutes and was settled in 1732. Maryland also had a complicated battery of courts on English models.

In Virginia, the instruction for the government of the colonies placed civil jurisdiction in the hands of governor and council sitting as a General Court. The first code, *Laws, Divine, Moral and Martial*, of 1612 was severe and in 1620 an attempt was made by the London company to compile a more humane code. The first legislative assembly, of 1619, passed various laws and there developed a system of county courts with a jurisdiction developed by custom. Governor and council and the house of burgesses made rules and decided cases. Governor and council sat as a Quarter Court and later as a General Court. County courts were established in 1623. The former laws were repealed and a new code enacted in 1661–62. No colony had greater hostility to professional lawyers than Virginia in the seventeenth century.

In the Carolinas, the proprietors attempted the enforcement of a very complete code, the *Fundamental Constitutions of Carolina* framed by John Locke and promulgated in 1668, and reissued in modified forms until finally abandoned in 1698. It was an elaborate scheme of government, mixing a romanticized feudalism with advanced ideas. A Grand Council of executive and judicial authority was to function along with a provincial assembly comprising Governor, hereditary nobles, and deputies. Land and rank were to be inseparable, and society was to comprise landgraves, caciques, lords of manors, freeholders, serfs and slaves. This system never became effective. The Carolinas were, however, among the earliest to adopt the English common law, this being done in South Carolina in 1712. In North Carolina in 1715 an Act provided that the common law should be in force so far as compatible with the colonists' ways of life and trade. Court proceedings seem to have been very informal, though trained lawyers were among the judges from an early date.

The Revolution Settlement of 1688 and following years in Britain provoked no disturbances in America, but is important because elements of that settlement, such as the Habeas Corpus Act of 1679, the Declaration of Rights of 1689, and the Act of Settlement of 1701, conferred benefits and established barriers to despotic government which extended to the colonists also.

Down to the end of the seventeenth century, accordingly, the general tendency of the colonists was towards law stated in code form, rather than acceptance of the complicated and frequently antiquated and unsuitable rules of the common law, ascertainable only from numerous books, reports, and statutes, mostly unavailable in the colonies. The codes frequently departed radically from the principles of the common law. The colonists knew of the ideas of reform and codification urged by Bacon and under the Commonwealth, and some of their codes were drafted by men in sympathy with these ideas. On the other hand, some colonies declared English common law subsidiary in cases not governed by colonial legislation, others established scriptural law as subsidiary. Everywhere there was great informality in procedure, particularly in the absence of a class of professional lawyers, but the basic institutions of jury, grand jury, writ, pleadings, and oral evidence were adopted from England. There appears to have been early a realization of the need to work out a new legal system. It is an overstatement at this stage to think of the reception of the common law. There were almost no law books even imported, and very few trained lawyers. Even chief justices were rarely lawyers; laymen administered a kind of natural equity. While ideas of liberty and public law were adopted, and general legal conceptions and terminology were derived from common law, the colonists did not by any means adopt it as a system, or with all its technicalities; they frequently repudiated it, denied its force, or deliberately departed from some of its best-settled principles.

Down to about 1774 most of what Britain did with regard to the American colonies was referable to commercial considerations. The mercantile system under which colonies existed mainly for the benefit of the mother country and must trade exclusively with her was expressed chiefly in the Acts of Trade and Navigation, the main principles of which were that all commerce between Britain and her colonies had to be conducted in vessels built, owned, and manned in Britain or the colonies, and that trade from the colonies to foreign countries had normally to be conducted through the mother country, and trade from foreign countries normally had to pass through a British port.

In the first three-quarters of the eighteenth

century all the proprietary colonies except those belonging to the Penns and the Calverts became royal provinces and of the originally corporate colonies only Rhode Island, Connecticut, and Hudson's Bay retained their charters. Royal governors were expected to enforce Acts of the British Parliament by executive means and by vetoing acts of colonial assemblies which conflicted with such Acts. Moreover Acts of colonial assemblies, though signed by the governor could, after a hearing, be disallowed by the British Privy Council, a right which, though not recklessly but judiciously exercised, provoked considerable discontent. In *Winthrop* v. *Lechmere* (1727–28) the Privy Council disallowed Connecticut intestacy laws. The judges of colonial superior courts held office at the King's pleasure, not as in Britain, during good behaviour, and appeal lay to the Privy Council in London, but the number of appeals was small. An Act of 1696 required all cases under the Acts of Trade and Navigation to be tried without juries, by royally appointed admiralty judges, but in some colonies the administration of justice was hindered or even reduced to impotence by local non-cooperation. Each colony continued to have a pyramid of courts. Prior to the Revolution, courts of chancery existed in some colonies and there were vice-admiralty courts dealing with maritime cases. The basic court was usually the county court.

By 1763 the colonies had grown in strength and confidence, and fundamental to them all was the idea of the rule of law. There was a very large measure of personal freedom, almost complete freedom of speech, press, and assembly, an absence of ancient restrictive customs, restrictive guilds, and privileged classes. The main grievances were the fact that the governor's instructions from the Crown were deemed mandatory, and that the tenure of judges was still during the King's pleasure. In all the colonies, except North Carolina, the Crown, after 1760, managed to force the colonies, despite rising resentment, to accept judges with that tenure. Nevertheless, it is important that the Revolution was not fought to obtain freedom from tyranny, but to preserve liberties and the rule of law which the colonists had long enjoyed.

In the eighteenth century, moreover, there was a more strict supervision of the colonies, the disallowance of colonial statutes for departure from settled law, more numerous appeals to England, the growth of a trained bench and bar, many of whom studied at the Inns of Court in London, the introduction of law books, and there followed a more general reception of the principles of English common law and the disappearance of many of the innovations though the development of a few others. A professional bar began to develop and a considerable number of young men went to the Inns of Court in London to study law. Twenty-five of the 56 signatories of the Declaration of Independence were lawyers as were 31 of the 55 members of the Constitutional Convention of 1787. Accordingly, when the Revolution came, English law ideas were dominant. What few law libraries there were contained only English law and when Blackstone's *Commentaries* were published in 1765–69 Americans were eager purchasers.

Colonial law was largely codified law, though every code had a history of its own. The *General Laws and Liberties of Massachusetts* were widely copied but there was no copying of case-law in the absence of printed reports. By the time of the Revolution, wholesale borrowing of codes had ended but the colonies were developing distinct bodies of statutory law partly their own and partly founded on English law.

It is difficult to seek to summarize the law of the period prior to independence. Regulation of business was limited but pervasive, and local; there was some regulation of public markets, the growth and sale of tobacco and inspection of goods. Criminal law was harsh. The *Laws and Liberties of Massachusetts* (1648) contained much criminal law, but there were fewer capital crimes in New England than in Old England. Benefit of clergy was recognized. Sexual offences seem to have been the commonest crimes. In theory, land law was governed by the principle of tenure from the King, not of any right of ownership of land, but the feudal system never took root, though there were instances of manors and manor-courts. The system of equitable estates was also recognized, but only in substance, not in detail. The colonies greatly modified the rules of succession. Some colonies adopted primogeniture but equal division became the norm. Wills and testaments on English lines appeared early in the colonies.

1776–1828

After the Declaration of Independence of 1776 every State betook itself to framing its constitution, in each case limiting the power of the representative government by a bill of rights, or statement of natural rights with which no government might interfere. The first, that of Virginia, was adopted by the Virginia convention on 12 June 1776, and much of it stems from English legal history, from Magna Carta of 1215, from the Petition of Right of 1628, and from the English Bill of Rights of 1689. Those of the other states, in general, followed the model of Virginia. The first Continental Congress of 1776 adopted a declaration of rights, declaring that the colonists were entitled to the common law of England and, in particular, the right of trial by jury. The idea of a bill of rights proved essential to ratification of the federal constitution.

In the Federal Convention of 1787, and in the Constitution of 1789, the influence of lawyers was great and the Judiciary Act of 1789 provided for a

Supreme Court of a Chief Justice and five associates, 13 district courts, and three circuit courts. The Supreme Court opened its first session on 2 February 1790. In the states, the division between legislative and judicial bodies was still frequently uncertain, and there was frequently no clear division between appellate and trial courts. Some states kept separate courts of equity.

New York took as its law the common and statute law of England as it stood at 19 April 1775 and the other original States and States admitted later to the Union did substantially the same, but in all the States the common law was adopted only so far as suited to American conditions. The period was essentially one of reception of English law. The first volume of American reports appeared in 1789 and only a few more volumes appeared before 1800. There were no American law books till Kent's *Commentaries.*

Unfortunately, after the Revolution, many of the better lawyers left the country and the practice of law was largely left in the hands of a lower type of less ability and education. There was public feeling against lawyers and the idea of a learned profession was repugnant to social conditions of the post-Revolution era.

Enormously important in the development of law and legal ideas were Blackstone's *Commentaries* which, first published at Oxford and Dublin in 1765–69, appeared in a Philadelphia edition in 1771–72, and new editions followed on both sides of the Atlantic repeatedly within the following century. It embodied and expressed natural law ideas, but also presented English law as a systematic body of rules. This was the most important influence on American law for long. Kent's *Commentaries on the American Law* (1826–30) were manifestly an American counterpart but made an enormous independent contribution to the growth of an independent system, a task greatly facilitated by the almost complete absence of precedents. At the end of the period Story's treatises on various branches of the law assisted greatly in the development of distinctively American bodies of principles, working out in different conditions fundamental principles formulated earlier in England. Other judges collected decisions, revised or digested the statutes of their states and wrote or revised treatises.

As the frontier was pushed westwards, the general principles of law recognized in the 13 colonies, founded mainly on the common law of England, were eventually carried forward to the Pacific. The Ordinance of 1787 prescribed basic law for what became the mid-West and the Ordinance of 1798 extended this to the south-west. In Michigan Territory, a lawyer Woodward compiled the Woodward Code, a collection adopted in 1805, and common-sense law expanded with settlement.

Only in Louisiana, the state organized as a territory in 1804 and admitted a state in 1812, did and does a different system prevail. Louisiana was originally a French crown colony, ceded to Spain in 1762, returned to France in 1801, and sold to the United States as part of the Louisiana Purchase in 1803. Its legal system is strongly influenced by the civil law; in 1803 it was a mixture of French and Spanish law; the Civil Code of 1825 (replacing a civil law code of 1808) was essentially a reproduction of the Code Napoleon and the French model exercised an equally strong influence on the Revised Civil Code of 1870, though, particularly in public law, commercial law, and procedure it has experienced the intrusion of common law institutions and methods. Spanish–Mexican law left some imprint on Texas but it has now long since been overlaid by common law, the constitution of 1836 requiring the introduction of English common law modified to suit local conditions.

1828–65

The spirit of frontier democracy was paralleled by boldness and originality in early nineteenth-century law. One of its manifestations was the system of popular election of judges, and the lowering of standards of admission to the Bar. Of great importance, however, were Joseph Story's textbooks, which had great effect in forming American law.

Reform by statute became more and more important, and there were great movements for social reform, the increased recognition of women's rights, the abolition of property qualifications, and legislation regulating banks and railroads. Many of the more technical and antiquated rules received from England vanished. The distinct administration of law and equity disappeared in many states.

An important feature of this period was the common adoption of codes of civil procedure, though frequently they have been harmed by legislative intereference and narrow and technical interpretation, and the general movement for codification.

The first half of the nineteenth century has been called the Golden Age of American Law but though there were great men on the bench and great jurists, the position of the Bar deteriorated. Election of judges began in 1832 and spread to many states. In many, moreover, the judge was reduced to a passive umpire, counsel having the only significant role in the trial of a case. The idea of a legal profession, with qualifications for entry, seemed undemocratic, and reversed the English tradition, which was becoming accepted before the Revolution.

But, by the first third of the nineteenth century, the reception of English common law as the basis of U.S. law had definitely been achieved, despite the general distrust of things English after independence, despite the attachment to France and the creation of the French civil code, and the rise of Jeffersonian and then Jacksonian democracy. There was some agitation for an American code, but by 1840 the common law had prevailed. The law merchant had been absorbed, largely by the work of Kent and Story, who also had played their part in the formation of equity.

In the years before the Civil War there was also much legislative reform of substantive law and procedure.

The movement for codification

In 1776 the Virginia legislature had entrusted to Jefferson the task of codifying and revising its criminal law and this was finally carried in 1796.

In Louisiana in 1805, Edward Livingston drafted a new code of procedure and saw it promptly adopted. In 1821 Livingston was entrusted with the preparation of a criminal code and, with collaborators, a civil code. He relied substantially on English and European jurists. But despite praise from jurists at home and abroad, the Louisiana legislature failed to adopt the code.

In the 1830s David Dudley Field began to agitate for legal reform in New York and, in 1847, he became a member of a commission which produced a code of civil (1848) and criminal (1865) procedure and, later, a penal code. The Code of Civil Procedure abolished the distinction between law and equity and all technicalities of pleading, substituting a single form of civil action. It was not till 1887 that New York adopted the penal code. All of Field's drafts were adopted in some of the Western states. Georgia adopted a civil code in 1860. Possibly most importantly, the New York Code of Civil Procedure was copied by similar codes in many states, so that 'code pleading' replaced common law pleading over large parts of the U.S. The New York Code was amended and enlarged in 1880.

Dakota, in 1805, adopted the civil code and penal code drafted by the New York commissioners, the first English territory to adopt a codified substantive law, and, since 1872, California has had a complete collection of four codes, political, civil, civil procedure, penal, and criminal procedure. In addition there have been many instances of partial codification, especially of procedure, some of them influenced by the nineteenth-century reforms in common law procedure in England.

Since 1865

The century since the Civil War has seen great developments, particularly in social and industrial legislation, related to the great and accelerating changes in society and in urbanization, industrialization, and technological change. But there remains great and justifiable discontent with procedural delays, with the influence of politics on legal processes, and with the frightful proliferation of statutes and reported decisions.

Reconstruction and the aftermath of the Civil War raised problems for the Supreme Court and lower courts alike. The Legal Tender cases, the Granger cases, cases under the Civil Rights Act of 1875. The rights of the negro were secured but only up to a point; *Plessy* v. *Ferguson* laid down that 'separate but equal' facilities were enough.

Towards the end of the nineteenth century, the Court had to rule on cases under the Interstate Commerce Act of 1887, the Sherman Anti-Trust Act of 1890, and the federal income tax law, which it invalidated. Its predominantly *laissez-faire* attitude changed towards the end of the century when proceedings under the Sherman Act against Northern Securities Company and Standard Oil Coy. were successful.

The movement for codification which began before the Civil War has taken the form since then of the adoption by many states of uniform laws on commercial matters promoted by the American Bar Association and framed by the Conference of Commissioners on Uniform State Laws. Almost equally important have been the series of *Restatements* published by the American Law Institute.

Higher educational and admission standards have steadily been demanded for the Bar and though election of judges is still common, they have increasingly come to enjoy longer and more secure tenure. Bar associations have developed; in 1870 the Bar of the City of New York was organized and there were 16 Bar Associations in existence in 1878 when the American Bar Association was formed. The improvement of legal education and the raising of the standards for admission to the Bar were and have remained a major policy objective of that body. About 1890 probably marked the nadir of standards of admission and control of the profession.

Before 1870 the Bar as a profession had sunk very low. Under Jacksonian democracy in many states only the most nominal qualifications were required to practise law.

In the latter half of the nineteenth century, many state and city Bar Associations were established to promote the interests of the profession and to raise professional ethical and intellectual standards. The American Bar Association was founded in 1878 and the National Bar Association in 1888, with its major goals, uniformity of state laws, improvement of the judicial system, and consideration of a code of international law. The latter, however, died out in 1893.

The twentieth century is also as notable for creative text-writing as the nineteenth was for the work of Kent, Story, Greenleaf, and a few others, in the form of such works as Wigmore's *Evidence*, Williston's *Contracts* and *Sales*, Beale's *Conflict of Laws*, and others.

See *Essays in Anglo-American Legal History*, II, 365 *et seq;* S. E. Baldwin (ed.), *Two Centuries' Growth of American Law*, 1701–1901; W. Hurst, *The Growth of American Law; The Law Makers*; C. Warren, *History of the American Bar*; C. Warren, *The Supreme Court in U.S. History*; C. G. Haines, *The Role of the Supreme Court in American Government and Politics*; A. Reppy (ed.), *Law: A Century of Progress*; R. Pound, *The Formative Era of American Law*.

The modern legal structure

The Constitution of 1788 establishes a federal system of government in which certain powers are delegated to the national government and all other powers fall to the states. The national (federal) government consists of legislative, executive, and judicial branches, interrelated and overlapping though distinct, and intended to check and balance each other. Each state has a constitution generally paralleling that of the federal government.

The legislature

The federal legislature is the Congress, comprising the Senate and the House of Representatives. The Senate is composed of two representatives from each state, elected for six years, one-third of the members retiring every two years. It operates mainly through a number of standing committees. The House is composed of representatives elected by direct vote for two-year terms, the number of representatives for each state being based on population. There are two major political parties, the Democrats being rather more conservative and the Republicans rather more liberal. The Congress has powers to levy taxes, borrow money, regulate inter-state commerce, declare war, and regulate its own procedure.

The executive

The executive is headed by the President who is Head of State, chief executive, and commander-in-chief of the armed forces. His responsibilities include the formulation of home and foreign policy, the proposing of legislation and leadership of his political party (which is not necessarily in the majority in either House of Congress). The members of the Cabinet are not members of Congress but head the different departments, State, Treasury, Defense, and other, but much power is also exercised by presidential aides not in the Cabinet.

The states

Each state has its own constitution, with its governor, legislature, executive, and judiciary. They have extensive functions with large administrative organizations.

The judicial system

The salient feature of the U.S. legal system is the dualism of federal courts and state courts.

The federal courts have jurisdiction in all cases arising under the Constitution, cases affecting ambassadors, admiralty and maritime cases, controversies to which the U.S. is a party, between two or more states, between a state and citizens of another state, and certain other cases. Cases of the latter-mentioned classes are within federal jurisdiction if the legislation so provides, and in some cases, the jurisdiction is concurrent with that of state courts.

The federal courts comprise the Supreme Court, which has original jurisdiction in cases affecting ambassadors and cases in which a state is a party, appellate jurisdiction, in certain cases, from the federal district courts and from all state courts in all cases involving a claimed federal right.

Below the Supreme Court are the U.S. Courts of Appeals with intermediate appellate jurisdiction; they exist in the District of Columbia and 10 other circuits, or groups of states, and have appellate jurisdiction from U.S. District Courts and, at first instance level, U.S. District Courts with jurisdiction in all federal civil proceedings and in criminal proceedings, there being at least one such court in each state.

There are also the Court of Military Appeals, the Court of Claims which determines claims against the U.S., and the Customs Court which reviews customs rulings.

The Constitution has been held by implication, since *Marbury* v. *Madison* in 1803, to have created jurisdiction to refuse to apply or enforce a statute of Congress which it deems unconstitutional and to enjoin any federal official to enforce the Act, and also to enjoin the enforcement of a state constitutional provision or statute repugnant to the Constitution, federal statute or treaty.

An important function of the courts, particularly of the Supreme Court, is to act as a check on federal or state legislation, by exercising the power of judicial review. A question of the disconformity of legislation to the federal or state constitution may be raised in a civil or criminal case by a party having a lawful interest to do so, and be decided as the, or a, reason for deciding the case one way or the other. The difficulty of the problems so raised arises from the vague and general language of many constitutional provisions, so that what is omitted or implied is as important as what is expressed. The power of the Supreme Court to refuse to enforce an act of Congress was first clearly enunciated in *Marbury* v. *Madison* (q.v.) in 1803 and not again exercised until *Dred Scott* v. *Sandford* in 1857. It exercised the power more frequently in respect of state legislation. The most controversial issues were the invalidation by the Court of major portions of Roosevelt's New Deal legislation in 1935–36, which gave rise to

Roosevelt's unsuccessful attempt to 'pack' the Court by enlarging it. It is difficult, in many of the issues raised in such cases, to distinguish law from politics and to eliminate social, economic, and political views from the interpretation of the Constitution.

The Courts also have an important function in acting as a check on the executive. This is exercised by considering the validity of administrative regulations and other delegated legislation, inquiring (as by habeas corpus proceedings) into the legality of arrest or detention, giving damages for wrongful arrest, or wrongful seizure of or damage to property by public officers, and by mandamus ordering official action in cases where it should have been taken.

Each state has such courts, exercising such jurisdiction as its own constitution and legislation has enacted. They also have jurisdiction over cases within the federal judicial power not withdrawn from them by Congress or expressly conferred on them by Congress. The court structure in each state has been determined largely by historical factors and later organized states have frequently borrowed from older states. Civil and criminal jurisdiction are commonly vested in the same courts. The structure commonly combines a state Supreme Court, circuit or district courts, county courts, and town or township courts, the last frequently called Justice of the Peace Courts.

There are also in many instances courts of special jurisdiction, particularly to deal with children.

Sources of law.

The sources comprise two groups of materials—legislation and case-law. Under the head of legislation there are, in order of importance, the Constitution of the U.S.; treaties made by the U.S.; federal statutes; federal executive orders made by the President and administrative rules and regulations made by federal administrative bodies empowered to make rules of a legislative character; state constitutions; state statutes; state administrative rules and regulations; and county and municipal ordinances, rules, and regulations.

Case-law comprises decisions of the Supreme Court, Court of Appeals, and other lower federal courts, and decisions of the supreme and lower courts of each state. In no case are decisions absolutely binding on the same court in a later case or on an inferior court but, for practical purposes, superior decisions are normally treated as binding. Decisions of courts of one state are only persuasive in any other state. Custom is unimportant as a source of law.

Subsidiary sources having, at most, persuasive authority include major treatises on branches of law, articles in legal periodicals, material contained in encyclopaedias, notably *American Jurisprudence, 2nd*, and *Corpus Juris Secundum*, or in the *Restatement of the Law* and legal dictionaries. Digests of case-law are not authoritative at all but merely guides to authority.

The major divisions of U.S. law

The distinction between law and equity was originally adopted from England but has greatly diminished and the procedural distinctions have disappeared in the federal and nearly all state courts. The division of substantive law into public and private (or public, criminal and private) is rarely of practical importance because there are no special courts for distinct public and private matters. Criminal law has a certain distinctness by reason of procedure. There is no division of private law into civil and commercial branches.

Public law. Constitutional law consists chiefly of the study of the decisions of the Supreme Court interpreting the federal Constitution; this includes problems of the preservation of individual rights to freedom of speech, assembly, press, and religious worship, secured mainly by the first 10 amendments (the Bill of Rights) and the protection afforded to the individual against the states, particularly by the fourteenth Amendment; Administrative law is concerned with the powers of procedure of agencies of government, other than legislatures or courts. The Interstate Commerce Commission (1887) was the first great regulatory commission. Trade regulation is chiefly effected by federal legislation, such as the Sherman Anti-trust Act (1890), and covers unfair competition. Labour law is largely covered by federal legislation. Tax law is chiefly federal income taxation and corporate taxation, but state and local authorities also levy income, sales, and property taxes. In the mid-twentieth century, a significant development has been the extension of civil rights to the coloured population, marked by *Brown* v. *Board of Education* (1954), which ended the legality of segregated schools.

Criminal law. The basis of criminal law is now mainly statutory; there are federal penal statutes enacted under the commerce, taxing, and postal powers but each state has its own criminal law which, in some cases, recognizes common law crimes, traditionally defined by case law. Only a few states have general codes of criminal law. The categories of felonies and misdemeanours are commonly recognized.

Private law. Private law is mainly state law and, accordingly, varies from one state to another. The law of contracts and torts has grown from the same roots as English law on these topics and exhibits many similarities. Property law, though rooted in English feudal land law, has moved far from its origins but varies more substantially from state to state. A material difference is that it is normal to have deeds affecting land recorded in public registers and, in some states, a system of registered title applies.

Family law is entirely state law and there is great variety of provisions; some states have achieved notoriety for the facility of divorce. Commercial law is largely a matter of state law but in this sphere Uniform Laws framed by the National Conference of Commissioners on Uniform State Laws have been adopted in many states, and the Uniform Commercial Code has been a massive contribution to uniformity. Business enterprises are almost invariably organized under state law. Tort law has been developed enormously, mostly by judges, since 1850. From 1867 to 1878 and since 1898 there has been a national Bankruptcy Act. Labour law has developed from the 1870s when big business and unions first confronted each other. In the 1890s, labour injunctions were regularly used against strikers and many statutes intended to protect employees were held unconstitutional, but since then protective legislation has been widely recognized.

Civil procedure

Though there is a fairly uniform basic pattern, frequently based on David Dudley Field's N.Y. Code of Civil Procedure of 1848, the details of procedure vary markedly from state to state. The process is served on the defendant, usually personally, and the petition or complaint answered by the defendant. The case may be disposed of on the pleadings or the issues of fact in dispute identified.

The trial of an issue of fact may be by judge or jury. Jury trial in some cases is a matter of right, or sometimes may be ordered by the court, or sometimes a particular issue of fact may be determined by a jury. Jury trial may be waived in some cases where a majority verdict is sufficient. The verdict may be a general one or a special one.

Criminal procedure

Criminal procedure operates within the framework of the requirement of 'due process of law' required by the Constitution and the specific limitations contained in the state constitution. There is also the important privilege against self-incrimination, that no person is to be compelled to be a witness against himself.

Investigation and prosecution is, in the case of federal crimes, the responsibility of the Attorney-General aided by the F.B.I. and, in the states, in the hands of the Attorney-General of the state and county, district, or state attorneys or prosecutors acting largely on the information of the police. Persons taken into custody are very thoroughly questioned and brought before a magistrate for preliminary examination. In some states, a grand jury must be summoned to determine whether there is evidence justifying the accused being tried; in other states the prosecutor may himself make no formal accusation. The importance of the grand jury is that it can compel the attendance of witnesses and compel answers to questions. If it finds a true bill, an indictment is served on the accused. If, at the trial, the accused does not plead guilty or guilty to a modified charge, trial takes place before a jury of 12 or by the court alone. The accused may waive his right to trial by jury. The prosecution makes an opening statement and leads evidence; the defence does likewise. Then usually follow closing statements by the defence and the prosecutor, after which the judge instructs or charges the jury and it returns its verdict; in most states the verdict must be unanimous. In most states the judge fixes the sentence, but in some the jury does so in some cases. There is substantial provision for obtaining review of verdicts.

Legal Profession

The legal profession is regulated independently by each state, each having its own requirements for admission to practice. A lawyer may practice only in the state in which he has been admitted. No distinction of function is recognized, as between barristers and solicitors; the terms attorney, counsellor and others have no special significance. But in practice, particularly in metropolitan districts, some lawyers are entirely or mainly court lawyers, others chamber lawyers and individuals specialize in particular fields of law. In such districts, practice in partnership is common, but over the U.S. as a whole individual practice is more common. A substantial number of lawyers are employed by large corporations and by federal, state, and municipal governments.

Judges are normally drawn from the practising bar and, occasionally, from government service or law teaching.

In state courts, judges may be appointed by the governor, elected by the legislature or popularly elected. There was a swing towards election in the later nineteenth century, and in the mid-twentieth century a swing towards appointment. Popular election is the predominant pattern. The Missouri Plan (1940) shifted responsibility for choice of a judge to a panel of lawyers and laymen; the man appointed was to run for election on his record, but is not opposed.

The profession has expanded enormously, both individuals and firms in private practice and lawyers on the payroll of corporations. Lawyers have frequently entered politics. Bar Associations have developed since the 1870s and everywhere the Bar is now organized on a statewide basis. The American Bar Association was formed in 1878 and adopted a canon of professional ethics in 1908.

Legal education

After the Revolution, most lawyers learned their law by apprenticeship in a practising lawyer's office. The first law schools developed from law offices which devoted serious attention to training. The earliest law school was opened in Litchfield, Connecticut by Judge Tapping Reeve in 1784. It acquired a wide reputation but closed in 1833. A few colleges taught law, the first American chair of law being established at William and Mary College in 1779–80 and held by George Wythe. This was followed by Virginia, Pennsylvania, and Maryland. Harvard established the Royall Chair in 1816 and Nathan Dane in 1829 established the Dane Chair, which was held by Story.

Legal education is now mainly provided at graduate level by university law schools, partly in 'national' schools which do not seek to prepare for practice in any particular state and partly in state law schools. It is strongly professionally orientated and, in many subjects, relies heavily on the case method (the analysis of decided cases) rather than on professorial exposition. Professional apprenticeship has disappeared.

Prior to 1850, educational requirements for admission to the Bar were frequently lax and, on occasion, non-existent. Since then standards have been steadily rising.

Legal literature

The earliest native, as distinct from imported, legal literature consisted of reports, beginning with Kirby's Connecticut Reports in 1789 and Dallas' Pennsylvania Cases in 1790. From the second volume of Dallas, Supreme Court cases were included and this begins the continuing series of Supreme Court Reports. Other volumes of reports of state courts followed. Publishers began to bring out American editions of textbooks, such as St. George Tucker's *Blackstone*, with appendices dealing with Virginia law, in 1803 and Reed's *Pennsylvania Blackstone* in 1831. From the first, American legal literature tended to be very practical and, from 1830, there was a steady flow, including Kent's *Commentaries on American Law* (1826–30), Story's *Commentaries* (1831–45), Wheaton's *International Law* (1836), Greenleaf on *Evidence* (1842), Sidgwick on *Damages* (1847), and others. Since these times, there has been an enormous growth in the volume of published case-law, of case-books for students and of treatises on every possible subject, some on major topics, some on highly specialized themes. Among the most notable have been Wigmore on *Evidence*, Williston on *Contracts* and on *Sales*, Corbin on *Contracts*, and Scott on *Trusts*.

The U.S. has contributed disappointingly little to the literature of legal history or jurisprudence; O. W. Holmes' *The Common Law*, J. C. Gray's *Nature and Sources of the Law* and later, the works of Roscoe Pound are notable exceptions. Notable enterprises have been the Law Reviews, edited by the senior students in the law schools; the best of these contain much very valuable material; and the *Restatement of the Law*, a series of unofficial but highly regarded codifications of the best accepted doctrines on the major branches of the law.

History: J. Willard Hurst, *The Growth of American Law* (1950); R. B. Morris, *Studies in the History of American Law* (1959); A. H. Kelly and W. A. Harbison, *The American Constitution: Its Origins and Development*; C. Warren, *The Supreme Court in U.S. History* (1922); L. Friedman and F. Israel, *The Justices of the U.S. Supreme Court* (1969).

Constitution and Sources: M. O. Price and H. Bitner, *Effective Legal Research*; B. Schwartz, *Commentary on the Constitution of the U.S.*; F. Frankfurter and J. M. Landis, *The Business of the Supreme Court*; *U.S. Government Organisation Manual.*

Public Law: C. H. Pritchett, *The American Constitution*; F. M. Riddick, *The U.S. Congress: Organisation and Procedure*; E. Corwin, *The Constitution and What It Means Today*; K. C. Davis, *Administrative Law Treatise.*

Criminal Law: R. M. Perkins, *Criminal Law*; W. L. Clark and T. Marshall, *Treatise on the Law of Crimes*; R. A. Anderson, *Wharton's Criminal law and Procedure.*

Private Law: A. Corbin on *Contracts*; S. Williston, *Treatise on the Law of Contracts*; W. L. Prosser, *Handbook of the Law of Torts*; F. V. Harper and F. James, *Law of Torts*; R. R. Powell, *Law of Real Property*; R. A. Brown, *Law of Personal Property*; A. W. Scott, *Law of Trusts*; J. H. Wigmore, *Evidence.*

Legal Profession: A. P. Blaustein and C. Porter, *The American Lawyer*; A. Chroust, *The Rise of the Legal Profession in America*; A. J. Harno, *Legal Education in the U.S.*

United States* v. *Wong Kim Ark (1898), 169 U.S. 649. An American-born Chinese, the son of parents ineligible for citizenship by naturalization, visited China. The federal government sought to prevent his return, but the Supreme Court held that he was a citizen, regardless of the ineligibility of his parents.

Universal Declaration of Human Rights. See HUMAN RIGHTS.

Universal Postal Union. The U.P.U. was founded as the General Union of Posts in 1874 and assumed its present form in 1878. It is based on the International Postal Convention which guaranteed to every member state the full use of postal services and agrees uniformity of classification and standards. Since 1968 it has been a specialized agency of the U.N.

Universal succession. The succession of one person, or several persons jointly, to the totality of the rights and duties of another on his death. In Roman law, universal succession might occur by adrogation, or by the acquisition of *manus* over a woman already *sui iuris* or, and most commonly, by succession on death. The universal successor on death was the *heres* who succeeded to the *universitas iuris*, or complex of rights and duties, of the deceased. In modern law, the universal successor is the executor or administrator, but he differs from the *heres* in that he is merely a personal representative appointed to carry out the testator's wishes or apply the rules of intestate succession, and does not assume the full legal personality of the deceased and in particular is not personally liable for the deceased's debts. The singular successor, on the other hand, succeeds to a person in respect of one item of property only, normally by gift or purchase.

Universities. Institutions, incorporated formerly by the Pope or other high ecclesiastical authority, and later by sovereigns or high lay authorities, each comprising a group of scholars and students, existing for the development and extension and the transmission of knowledge. The power to confer degrees, or titles indicative of having attained a distinct degree of level of skill in a branch of learning is an attribute of universities, but is sometimes conferred on other institutions; thus call to the Bar by an Inn of Court and admission to the Fellowship of the Royal College of Surgeons are deemed equivalent to degrees. There are three main grades of degrees, Bachelor or licentiate, Master, and Doctor. The word *universitas* was originally applicable to any corporate body but became appropriated to the corporations of learning. Mediaeval universities were called *studia generalia* because they welcomed scholars from all parts. The origins of universities must be sought in Plato's Academy and in the schools of later philosophers. In Byzantium, schools of advanced study were organized; a recognized one was founded at Constantinople in A.D. 425, and there was a great growth of foundations when Europe emerged from the Dark Ages after about A.D. 1000. In the mediaeval universities, the major studies were philosophy and the three higher faculties, theology, law, and medicine. The archetypal mediaeval universities were Salerno (famous for medicine), Bologna (notable for law), and Paris. In Pavia and Ravenna, there had been law-schools which antedated Bologna, but Bologna overshadowed them in size and fame. From Bologna, the scientific study of law in general, and Roman law in particular, spread all over Europe. In Britain, Oxford was founded about 1150, Cambridge about 1210, and St. Andrews in 1413. Subsequently, universities have been founded all over the world.

In many universities, from the earliest times, law has been recognized as a proper, and a major, subject of study. Originally civil (Roman) and canon law were the only subjects. In England down to the latter part of the nineteenth century, the study concentrated on civil (Roman) law and the more academic aspects of law, the more practical study of law being concentrated in the Inns of Court. Since then, university study of law in Britain has paid more attention to practical aspects and issues, but is still not conceived as solely or even primarily professional training. Probably the most important function of the law schools in modern British universities is research and writing, and since the mid-nineteenth century, a large proportion of the most important publications on law have come from them.

In the U.S. and some parts of the Commonwealth, legal studies in universities have a more professional purpose; they follow a general degree in Arts or Science and are conceived as strictly professional education. Here too the universities have made major contributions to legal literature.

In continental Europe, law is very widely studied in universities because it is the necessary or appropriate preparation for governmental and administrative as well as strictly legal careers.

H. Rashall, *Universities of Europe in The Middle Ages* (1936).

University courts. The older British universities were not only teaching institutions but jurisdictional areas.

At Oxford, through its chancellor, the university asserted the ecclesiastical jurisdiction over its members and won independence of bishop and archbishop, and he was also given jurisdiction in personal actions and crimes. The main court was that of the chancellor or his deputy, exercising civil, criminal, and ecclesiastical jurisdiction. It used the procedure of the canon, and later of the civil, law. Appeals lay to the Regent Masters, and from them to Convocation. A later charter created a court held by the Steward of the University for the trials of treason and felonies by members of the university, and the courts of the coroner and of the clerk of the market also existed. The civil jurisdiction, the court of the Steward and the University coroners, exists; the ecclesiastical jurisdiction is obsolete and the criminal is probably so.

At Cambridge, a chancellor's court also arose which could hear personal actions to any value and try nearly all criminal cases if one party was a member of the university. There was also a Steward's Court, the court of the clerk to the market and a court for Stourbridge fair. Statute in the nineteenth century restricted the powers of its courts. There are also disciplinary courts within the university. In both Oxford and Cambridge the jurisdictions of the Chancellors' courts, other than that existing in 1977, have been abolished.

The mediaeval Scottish universities, St. Andrews, Glasgow, and Aberdeen, had franchise jurisdictions combining the powers of secular and ecclesiastical courts. The Rector's court was the court of the university community, with appeal to the chancellor, and also heard more serious cases. In St. Andrews, the Rector's court became indistinguishable from the University Senate and discipline came to be exercised by it. In Glasgow, the bishop who founded the university, authorized the rector to judge in civil and less serious criminal cases, and his successor gave the rector full jurisdiction in all civil cases and all quarrels between students or between them and citizens. By virtue of these grants, the University long exercised jurisdiction and, in 1670, a student was tried for murder. All jurisdiction, other than purely internal disciplinary, is now obsolete.

Unjust enrichment. The general principle that a person who has obtained a benefit from another, not intended as a gift and not legally justifiable, must repay it or make restitution to or recompense the other party. In many cases, this is founded on an implied promise to repay or on the principle of money had and received. It is a matter of some doubt how far the principle of unjust enrichment has been adopted in English law. The term Restitution is increasingly being used to cover the categories where restitution can be claimed for the avoidance of unjust enrichment. In Roman law, this is the principle underlying the quasi-contractual obligations. In Scotland, the principle of unjust enrichment underlies the quasi-contractual obligations (q.v.) which oblige the party benefited to repay. The same principle is the fundamental justification for the maritime law obligations of salvage and general average (qq.v).

R. Goff and G. Jones, *Restitution.*

Unlawful. A term of vague and uncertain connotation. It may be used as a synonym of 'illegal', i.e. that which is contrary to law and criminally punishable or tortious. But sometimes it refers to promises or agreements which are merely legally ineffective though not illegal or criminal.

Unlawful Assembly. An assembly of three or more persons with intent to commit crime by open force or to carry out any common purpose, lawful or unlawful, in such a way as to give reasonably courageous persons reasonable grounds to apprehend a breach of the peace. All participants are guilty of an offence. They are guilty even if the parties change their minds and disperse without doing anything. An assembly originally lawful may become unlawful, if it acts on a proposal to do an act of violence, but a lawful assembly does not become unlawful because participants appreciate that others may oppose them and commit a breach of the peace. An unlawful assembly becomes a rout as soon as some act has been done towards the execution of the common purpose, and the rout becomes a riot when some act is done in part execution of this common purpose.

Unlawful Oaths. The crime of administering, assisting in, or being present at and consenting to the administration of an oath to engage in any mutinous or seditious purpose, or to disturb the public peace, or to be of any association formed for such a purpose, or to obey the orders of any body not lawfully constituted, or not to inform or give evidence against any associate, or not to reveal any illegal combination or act, or any illegal oath.

Unliquidated damages. Damages not fixed by prior agreement of parties or otherwise but to be determined by the court or jury, having regard to the circumstances of the case.

Unneutral service. The action of a neutral vessel or, probably, aircraft in carrying persons or dispatches for a belligerent. The persons must be members of the forces or such persons as might be made prisoners of war. The service may extend to transmission of radio messages. Such action is wrongful to and punishable by other belligerents, by capture and condemnation of the neutral vessel as prize.

Unsound mind, persons of. Persons of unsound mind or mentally ill have been variously classified and described at various times. The earlier distinction was between idiots, whose lack of mental capacity was from birth, and lunatics, who became insane after birth. The Lunacy Act, 1890, used the term 'lunatic' and was replaced by 'person of unsound mind' in the Mental Treatment Act, 1930.

The Mental Deficiency Act, 1913, defined mental deficiency and four sub-divisions (starting with the most severe), viz; idiocy, imbecility, moral imbecility and feeblemindedness. The Mental Health Act, 1959, replaced these with different categories and definitions; the genus was 'mental disorder', subdivided into severe subnormality, subnormality, psychopathic disorder, mental illness, and other disorders or disabilities of mind. These terminology and definitions are statutory definitions for the purposes of particular legislation and its administration. For the purposes of civil and criminal responsibility and certain other issues, such as capacity to make a contract or a will, insanity, as defined by various decisions, is the criterion.

Unsound mind, courts with jurisdiction in cases of. In England, the statute De Praerogativa Regis declared that the care and custody of persons of unsound mind and their estates was vested in the King. This power was delegated to the Lord Chancellor as Keeper of the Great Seal and was, from 1853, exercised by the Lords Justices of Appeal in Chancery. Later it was transferred to the Lord Chancellor alone or jointly with the judges of the Supreme Court entrusted with this jurisdiction, then to the Lord Chancellor, Master of the Rolls and Lords Justices of the Court of Appeal, and then to the Lord Chancellor and nominated judges of the Chancery Division.

A judge exercising the jurisdiction might direct an inquisition whether a person is of unsound mind and, if so, to make orders for the custody of such persons and the management of their property. From 1853, statutory powers supplemented the inherent jurisdiction and in modern times the jurisdiction is entirely statutory.

Apart from functions reserved to the Lord Chancellor or a nominated judge, the jurisdiction may be exercised by the Master or Deputy Master of the Court of Protection.

In Scotland, the Court of Session has jurisdiction to appoint a suitable person as *curator bonis* to look after the property of a person medically certified to be of unsound mind.

Untried prisoners. Persons charged with crime but not allowed bail and detained in prison pending trial. They must be kept strictly separate from the general body of prisoners in prison, as the latter have been convicted and sentenced, partly because they must be presumed innocent and partly to avoid conspiracy with or contamination by convicted prisoners (some of whom might be witnesses at an untried prisoner's subsequent trial). They are allowed greater privileges than convicted prisoners and facilities to see friends and legal advisers and prepare their defence.

Unwritten law. A term used with various connotations, sometimes as meaning natural law, moral law, or the law of God, sometimes for such practices as the conventions of the constitution (q.v.) and sometimes, rather inaccurately, as meaning case law. It is always used antithetically to written law (statute or enacted law).

Upjohn, Gerald Ritchie, Baron (1903–71). Was called to the Bar in 1929, became a judge of the Chancery Division in 1951, a Lord Justice of Appeal in 1960, and a Lord of Appeal in Ordinary, 1963–71. He was also a judge of the Restrictive Practices Court, 1956–60.

Upper Bench. The name of the King's Bench under the Commonwealth and Protectorate.

Upset Price. In auction sales, the amount set as the price at which the property will be sold to the first person bidding that price, unless there be a higher bid.

Urania Cottage. A house in Shepherd's Bush, London, established by Charles Dickens in 1846 as a home for prostitutes, seeking to redeem them, but extended to girls from prison, lost or starving, ill-treated, or otherwise requiring care and protection. Miss Angela Burdett-Coutts was associated with the project and paid the bills, but the planning and execution were chiefly Dickens'. The home took 13 girls and there were two superintendents. The system of management depended on marks for each inmate's daily attainments. Dickens was associated with the project until 1858.

Ure, Alexander, Baron Strathclyde (1853–1928). Called to the Scottish Bar in 1878, he attained early success and, entering Parliament in 1895, he became Solicitor-General for Scotland in 1905 and Lord Advocate in 1909. He attained great prominence and popularity as a politician, and was a most successful advocate. In 1913, he became Lord President of the Court of Session and in 1914 a peer. He proved himself a capable and efficient judge, but he was not a learned lawyer and his judgments added little to understanding of legal principles. Ill health forced him to resign in 1920.

Usage. Uniform and customary practice.

Usage of trade. General accepted practices and understandings of a particular kind of business, which may have the effect of implying terms into a commercial contract. To have this effect the usage must be consistent with the general law, well known and such that it must be taken to be accepted by both parties.

Usance. The time in which bills of exchange between this country and foreign countries were formerly payable. The term is now used for the periods for which bills on a foreign country are in practice usually drawn.

Uses (from *opus*, not *usus*). A concept of mediaeval English law whereby property could be held by one person to the use of, i.e. for the benefit of, another. Uses of land seem to have been quite common by the fourteenth century. The development seems to have been promoted by various factors, the holding of land for persons absent on the Crusades, for ecclesiastical and charitable purposes, and so on. The use enabled land to be disposed of after death by will, facilitated settlements and enabled ownership to be extended to unincorporated bodies. The beneficiary had initially no legal protection but, in the fifteenth century, Chancery began to intervene and by the end of that century a good deal of law had developed. Legislation was passed from 1377 to deal with applications of the use to facilitate fraud and, under the Tudors, there was much legislation with the general policy of treating the beneficiary as having the legal estate, so as to avoid feudal lords being deprived of the valuable incidents of tenure. This policy was continued by the Statute of Uses (q.v.).

Uses and trusts. The use was, in mediaeval English law, a doctrine whereby property was held by one to the use of, for the benefit of, another, whose interest in the property would be enforced. Though analogous to the usufruct or fideicommissum of Roman law, it is probably derived from the salman or treuhand of Germanic law, who developed into the executor. The common law courts declined to recognize the interest of the beneficiary but it was protected by the Chancery and this enabled many kinds of interests in land, impossible at law, to be recognized. Statute early intervened to prevent conveyance to uses from being used for improper purposes, such as to defraud creditors, or evade the statutes of mortmain. But uses could be used to avoid various burdensome exactions, notably the incidents of tenure of land, and this led to the Statute of Uses (q.v.) of 1535. The main principle of this Act was to transfer the legal estate from the feoffee to uses (the nominal owner) to the cestui que use (the beneficiary). The attempts to circumvent the Act gave rise to the modern equitable trust.

Uses, Statute of. In mediaeval English law, a use arose where one person held property not for himself but to the use of, or for the benefit of, another. Uses, however, defrauded feudal lords of the valuable incidents of feudal tenure, such as wardship, marriage, and relief. Such grievances on account of lost revenue led the Crown to have enacted the Statute of Uses in 1535. Landowners opposed the measure which deprived them of the power of making family settlements, and lawyers saw the danger of losing business. Under the Act, the 'use was executed', i.e. the beneficiary became the full legal owner for all purposes, and not merely the beneficial owner. There was no intention to abolish uses, nor did the Act apply to all uses. It subjected landowners to the liabilities of legal ownership and took away from them the power of devising their land, though this was largely restored by the Wills Act, 1540. The Chancery lawyers, however, subsequently circumvented the Act by recognizing trusts, circumstances where one person was under a moral duty to deal with property vested legally in him for the benefit of another, and from *Sambach* v. *Dalston*, in 1634, it was held that a devise to A to the use of B in trust for C, or, later and more briefly, unto and to the use of B in trust for C, gave B the legal estate but subject to a beneficial right in C.

Usher. A doorkeeper, who also maintains silence and order in a court. Black Rod (q.v.) is properly styled the Gentleman Usher of the Black Rod. The Usher of the White Rod or Principal Usher for Scotland is an hereditary office vested, under the Walker Trust Act, 1877, in trustees and exercised at a coronation by a deputy.

Usucapio. In Roman law, the acquisition of title to property by substantial, uninterrupted, and *bona fide* possession founded on a colourable title. According to the Twelve Tables the period was two years for land and one year for other things.

Usufruct. In Roman law, the right of using and enjoying the property of another, usually for life, without right to change the character of the property. The usufructuary had to maintain the property, pay ordinary outgoings and generally act like a responsible owner. The right is generally comparable to that of a tenant for life. In modern civil law systems, there are recognized perfect usufructs extending to those kinds of property which the usufructuary may use without changing their substance, and imperfect or quasi-usufruct, extending to kinds of property, such as money or agricultural produce, which can only be used by consuming, using or altering the substance of the property, but where there is liability to replace or compensate the estate for the value of what has been used up.

Usury. The charging of heavy rates of interest for the loan of money. The Christian Church condemned the practice, which helped to drive money-lending into the hands of Jews. Nevertheless, though the canon law condemned it and sought to suppress it, the practice was common in the Middle ages. In the sixteenth century, Calvin was prepared to allow interest on loans between business men. In both England and Scotland, many statutes were passed in fifteenth, sixteenth, and seventeenth centuries seeking to enforce a rate of interest and to penalize as usury contracts providing for interest at higher rates, and many devices were resorted to to evade the penalties and to allow interest at forbidden rates. Thus, a common device in England was to grant a lease in return for a lump sum for a term of years long enough to enable the lender to recover his advance and interest out of the revenues of the land. This device avoided the Church's prohibition of usury, and accordingly, the mortgage came to be effected by the grant of a term of years.

All usury laws were repealed in 1854 and it is now competent to stipulate for any rate of interest, but courts have wide powers to reopen extortionate transactions and reduce the rate of interest to what they deem reasonable in the circumstances.

J. B. Kelly, *History and Law of Usury* (1835).

Usus. In Roman law, a personal servitude whereby a person not the owner was entitled to the use of particular property but without wasting the capital or substance thereof. It differed from *usufructus* in that there was no right to *fructus*, i.e. to take the fruits of the subjects. In Roman law, *usus* was also a means of creating *manus*, i.e. of bringing the wife within the *patria potestas* of her husband. It was constituted by a year's uninterrupted cohabitation under a marriage not creative of *manus*.

Usus modernus Pandectarum. A term coined in the late seventeenth century by Samuel Stryk (q.v.) (in a four volume book of that title of 1690) for the Roman law, as received in northern Europe from about 1500 onwards, and modified and adapted for use in contemporary conditions, and as elaborately discussed and systematized by jurists. In the nineteenth century, this body of knowledge was developed to a very high pitch by the German jurists and reached its culmination in the works of such as Windscheid and Dernburg, thereby preparing the way and laying the foundations for the German Civil Code.

Uterine brother or sister. A half-brother or half-sister having the same mother, but a different father.

Uthwatt, Augustus Andrews, Baron (1879–1949). Born in Australia, he was called to the English Bar in 1904 and practised in Chancery. He became a judge in 1941 and a Lord of Appeal in 1946. His judicial career was cut short, but his judgments on many points are concise but illuminating.

Uti possidetis. In the Roman law, an interdict whereby the colourable possession of real property by a *bona fide* possessor was continued until the rights of parties were finally determined. The phrase is sometimes referred to as a principle under which property not expressly provided for in a treaty terminating hostilities is to remain in the hands of the party who happened to have possession of it when hostilities ended.

Utilitarianism. The philosophical principle, developed chiefly in and after the late eighteenth century, holding that the rightness of conduct and of law was to be judged by its tendency to promote the greatest happiness of the greatest number, or the reverse. According to Bentham, both morals and legislation rested on knowledge of pains and pleasures, and he tabulated these as affecting the individual. The legislator's power to control conduct lay in his use of pleasures and pains as sanctions. Benthamite utilitarianism was very influential in nineteenth-century Britain in promoting legal reform, notably in humanizing the criminal law and its administration, the removal of many personal incapacities and inequalities, and reforms of procedure and evidence. In Germany, Ihering criticized the individualistic utilitarianism of Bentham and his followers and urged a social utilitarianism, emphasizing social purposes and the valuation of individual purposes in terms of social purposes. Social mechanics was the means used by society for controlling individual purposes and shaping them more consistently with social purposes. These means included law, and the standard of good law was its fulfilment of human purposes or securing of interests of the individual, the state, the church, and society generally. Ihering's social utilitarianism is important as a link between nineteenth and twentieth century legal thought and it has been adopted by such jurists as Pound and Stone and goes far to explain much modern legislation.

Utter barrister. An outer barrister, that is, one who pleads 'without' the Bar. This includes all ordinary or junior barristers, but not Queen's Counsel or, formerly, serjeants-at-law, who have been called within the Bar.

Utrum, assize. An assize or jury inquisition introduced in the twelfth century to determine whether land was held in frankalmoin, i.e. by spiritual tenure, or by a lay tenure, a decision on which depended whether spiritual or lay courts had jurisdiction. As the assize could decide not only the actual points in issue but also collateral issues, it came to be treated as a jurata, known as the *jurata de utrum*, later corrupted into the *juris utrum*. Till 1571 it was chiefly used by clerics seeking to recover land which belonged to their churches.

Uttering. The crime of putting counterfeit coin into circulation. In Scotland, uttering is tendering a forged document with intent to defraud.

V

V.C. Vice-Chancellor (q.v.).

Vacant possession. Possession of land or buildings consisting not only in the legal right of possession but in the absence of any actual occupier, so that the legal possessor may himself actually occupy. The term is often used in relation to buildings for sale.

Vacarius (?1115–?1200). A civilian from Lombardy, possibly a pupil of Irnerius at Bologna, who came to England about 1143. He is recorded as lecturing on Justinian at Oxford in 1149 and as having prepared an abridgment in nine books of Justinian's Digest and Code, similar to Irnerius' *Summa Codicis*. It seems to have been popularly known, having been prepared for poorer scholars, as the *Summa Pauperum de Legibus* or the *Liber Pauperum*. About the end of his reign King Stephen destroyed all the books of Italian laws which he could lay hands on and this for a time stopped Vacarius' teaching. Thereafter he worked as an ecclesiastical lawyer and judge in the province of York, and as envoy to the papal court, and about 1170 composed two tracts *De Assumpto Homine* and *De Matrimonio*.

F. W. Maitland, 11 L.Q.R. 133, 270; F. de Zulueta, *Liber Pauperum* (Selden Socy.).

Vacations. The periods of the year during which courts are closed for ordinary business, though certain judges and officials attend periodically to deal with certain kinds of business and matters of urgency. In England there are four vacations, namely Easter, Whitsun, the Long Vacation (August and September), and Christmas (extending over the New Year).

Vagrancy. The state of a person who has no settled home but drifts from one place to another, normally having no regular means of support. Traditionally, a vagrant was an idle person living by begging and theft. At common law, a vagrant was liable to punishment varying from branding and whipping to transportation. Later statutes treated as vagrancy a variety of kinds of misconduct. In England, the Vagrancy Act, 1824, later amended, distinguishes idle and disorderly persons, rogues and vagabonds, and incorrigible rogues.

In the U.S. vagrancy has often been used as a concept justifying prosecution for many kinds of conduct from political demonstrations to loitering.

Valderas, Arias de (sixteenth century). A Spanish jurist and author of *De Bello et Ejus Justitia* (1533), a study of international law.

Valin, Réné Josué (1695–1765). French magistrate, author of *Commentaire sur la Coutume de la Rochelle* (1750), *Nouveau Commentaire sur l'Ordonnance de la Marine* (1681) (published 1760), and *Traité des Prises* (1762)

Valla, Lorenzo (1407–57). Italian humanist, writer on ethics, history, and theology, best remembered as the critic who, in 1440, exposed as a forgery the Donation of Constantine (q.v.) the document alleged to have been granted by Constantine giving Pope Sylvester I temporal supremacy over the whole Western World. He also engaged in many philosophical and theological controversies.

Valla, Nicholas (sixteenth century). French jurist and author of *De rebus dubiis et quaestionibus in jura controversis tractatus XX*.

Valor ecclesiasticus. A record of the whole property, spiritual and temporal, of the Church in England and Wales, its monasteries, priories and churches, made by a commission appointed under a statute of 1534 for the purpose of enabling the Crown to collect annates (q.v.) It has been called a Domesday Book of church property, and proved useful when monastic property was confiscated. An incomplete manuscript exists and is supplemented by a *Liber valorum*, and was edited and printed by the Record Commission in 1810–34.

Valuable consideration. See VALUE.

Valuation. The putting of a value in money on any kind of property. It has to be done for many purposes such as for assessment of compensation for compulsory purchase, as a basis for the levying of local rates, and in relation to administration of the estate of a deceased. Different bases of valuation may be prescribed for different purposes, but a common one is the price which would have been obtained by sale in the open market.

Value. An economic concept also frequently important in legal contexts. Value may consist in spiritual or aesthetic qualities, or in utility in use, or in the amount of money or other goods which could be obtained in exchange for the thing in question. In legal contexts, the last is the sense usually relevant, in such phrases as 'holder for value' or 'purchaser for value', in each of which 'value' means 'having paid a reasonable equivalent in money'. In English law, simple contracts require for their enforceability consideration in return for the promisor's promise; valuable consideration is some right, interest, profit, or benefit accruing to one party, or some forbearance, detriment, loss, or responsibility given, suffered, or undertaken by the other.

See also VALUES.

Value Added Tax. A tax based on the net value added to a taxable product by each person concerned with a distinct stage of manufacture, and chargeable also on services rendered. It was introduced in the U.K. in 1973, replacing Purchase Tax and Selective Employment Tax (q.v.), to harmonize the tax system with that of other E.E.C. countries. Certain goods and services are zero-rated and certain others exempt but other goods and services are charged at rates varied from time to time.

Valued policy. In marine insurance, a policy in which a specified value is put on the ship, goods or other effects insured, to avoid the trouble and necessity of proof of value in the event of total loss.

Values. Non-legal considerations which may be influential on conduct, including legislation, application of policy and judicial decisions. They are of the nature of ideals or general doctrines embodying assessments of worth or desirability which may, in doubtful cases, be effective in swinging judgment one way or the other. Among such factors are the national safety, the liberty of the subject, the general or public interest, the upholding of property rights, equality before the law, fairness, the maintenance of moral standards, and such lesser values as convenience, uniformity, practicability, and the like. The development of the law of negligence has been based largely on the value attached to the policy of making manufacturers liable to the consumers of their products. Sometimes two or more values are in conflict in that one can be invoked by each side; this raises questions of valuation and weighing of different values. Different values may carry different weights in war-time and peace-time, and with different judges, according to their personalities and their social, moral and other viewpoints. The weights of different values may be profoundly influenced over a period of time by changes in general social philosophy; thus, in the present century, the general or public interest has been given increasing weight and the upholding of property rights decreasing weight.

P. G. Stein and J. Shand, *Legal Values in Western Society* (1974).

Vangerow, Karl Adolf van (1808–70). German jurist, writer on Roman law, and the author of a standard *Lehrbuch der Pandekten* (1838–46).

Van Santvoord, George (1819–63). Was admitted to the Bar in 1844, published several books on practice and pleading, and *Sketches of the Lives and Judicial Services of the Chief Justices of the U.S.* (1854).

Vassal. In feudal society and law, a person holding a fief from a lord or superior in return for certain services. Those vassals who held fiefs directly from the Crown were the tenants-in-chief and formed the important feudal group, the barons, but most vassals held fiefs from barons or other lords who were themselves vassals of over-lords. Under the feudal contract, the lord provided the fief for his vassal and had to protect him and give him justice in his court, while the vassal owed to the lord feudal services, military, judicial and administrative, and feudal incidents, notably relief (q.v.). The vassal owed fealty to his lord and breach of this duty was felony (q.v.) a term which came to be widened to include most serious crimes. In time, vassals' rights to their fiefs were recognized as hereditary, and alienable, at first with the lord's consent but later, as of right, but on payment of an impost. So too in most countries, vassals were allowed to subinfeudate, i.e. to grant subordinate fiefs to vassals who held of and under them. If a vassal committed a felony or died without an heir, his fief escheated or reverted to his lord. In Scotland, the term vassal is still the technical term for a person holding land by feu-holding from a superior.

Vatican City. An independent state within the City of Rome created by the Lateran Pacts between the Pope and the government of Italy in 1929. The States of the Church were annexed to Italy in 1870 but the Holy See continued to regard itself as a separate entity in international law. The Lateran Pacts recognized the Holy See as possessing sovereignty in the international field and the Vatican City as a new state.

The Holy See and Vatican City are distinct entities in international law but united in the person of the Pope. The Holy See exercises sovereignty over Vatican City. Vatican City is a member of the international community and a member of many international organizations, but not of the U.N. It maintains permanent neutrality.

The constitution is contained in a complex of six laws issued by the Pope in 1929, comprising fundamental law, sources of law, and regulations on citizenships, residence, administration, public security and economic, commercial, and professional controls. Later, a judicial regulation and code of civil procedure was promulgated.

Sovereignty over Vatican City is exercised by the Pope as supreme head of the Catholic Church; when the papacy is vacant the college of cardinals assumes governmental power. There is no legislative body, the Pope having sole legislative, executive, and judicial powers. Foreign relations are conducted through the Papal Secretary of State. Administration is delegated to the Papal Commission for Vatican City, composed of three cardinals. There is a governor who has delegated power to issue ordinances and regulations. Justice is administered by a tribunal of first instance in a court of appeals and a court of cassation, and there is possible recourse to the Apostolic Signatura (q.v.).

Vattel, Emer or Emmerich de (1714–67). Swiss jurist, entered the Saxon diplomatic service, was minister at Berne and chief adviser of the Saxon government on foreign affairs and a Privy Councillor. He was influenced by Leibnitz and Wolff. His major work is *Le Droit des Gens ou Principes de la Loi Naturelle* (1758) but he also wrote a *Questions de Droit Naturel* (1762) and volumes of essays on international affairs. *Le Droit des Gens* modernized the whole of international law, bringing it down from the realms of speculation into the sphere of natural relations and problems, and became a classic, particularly in England and U.S.A.

Vaughan, Sir John (1603–74). Called to the Bar about 1630, he became a friend of Selden and Hyde (later Clarendon). A royalist, he retired from politics and the law during the Rebellion and the Commonwealth, but sat in the Restoration Parliament. He was made Chief Justice of the Common Pleas in 1668 in which capacity he displayed learning, discrimination, and sound judgment. He had a reputation for liberal opinions and was opposed to arbitrary authority. His fame today rests on the decision in *Bushell's case*, that jurors were not punishable for verdicts which the courts disapproved, while in *Thomas* v. *Sorrel* (1637), Vaughan 330, he sought to define the principles limiting the Crown's dispensing power. He also compiled reports which were published posthumously.

Vazquez de Menchaca, Fernando (1512–69). A distinguished Spanish jurist and writer on international law, author of *Libri Tres Controversarium* (1572), dealing with the laws of war as well as a *De Successionum creatione Tractatus* (1559) and a *De Successionibus libri IX* (1612).

Vendor. A seller, in English law particularly a seller of land, the body of law concerned with the sale of land being traditionally called the law of vendor and purchaser. It includes such topics as the conditions of sale, the contract, the abstract of title, requisitions, searches, and preparation and completion of the conveyance.

Venezuela boundary dispute. A controversy between Great Britain and Venezuela over the boundary between British Guiana (now Guyana) and Venezuela. It originated in 1814 and became important with the discovery of gold in the disputed area. In 1887, Venezuela broke off diplomatic relations with Britain and asked the U.S. to arbitrate; Britain refused to do so. In 1895, the U.S. asserted that British pressure on Venezuela would be declared a violation of the Monroe doctrine, but Britain took the view that the Doctrine did not apply to the case. In 1895, the U.S., with British assistance, created an independent commission, and a treaty in 1897 between Britain and Venezuela submitted the claim to arbitration. The tribunal sitting at Paris found substantially in favour of the British claim.

Venire de novo. In criminal practice, a writ issued by the Queen's Bench on a writ of error (q.v.) from the verdict in an inferior court, vacating the verdict and directing the sheriff to summon fresh jurors to try the issue afresh. It has been superseded by provisions for a new trial.

Venire facias. The name of a variety of writs all involving a summons to appear. They included a writ to summon a person to appear and be arraigned for an offence (*venire facias ad respondendum*), a writ to summon a new jury (*venire facias de novo*), a writ to summon a jury (*venire facias juratores*), and a writ to summon a jury of matrons (*venire facias tot matronas*).

Ventre inspiciendo, de. A writ which might be issued to summon a jury of matrons to discover whether or not a woman was pregnant, as where a landowner's widow alleged that she was so, and might pass off a child to defeat the claims of an heir-presumptive to the estate. The last reported case is believed to be in 1845 (*Re Blakemore*, 14 L.J.Ch. 336). At common law, two writs issued, one to ascertain whether the widow was pregnant and when she would be delivered, issues determined by 12 women in the presence of 12 knights, and the second, if she were so, directing her removal to a castle where the sheriff was to keep her safely and where some of the 12 were to see her daily and some to be present at the birth. By the eighteenth century, however, this had been modified to merely periodical inspection on reasonable notice by two women nominated by the heirs whose claims would be defeated by the birth of the child. Also, the issue of the writ might be suspended if, within a limited time, the widow submitted to examination by two midwives appointed by the applicants for the writ.

Venue. The locality in which a crime has been committed or that in which a court has jurisdiction to try the offence. The common law rule for criminal procedure was that the venue had to be laid (i.e. the indictment had to be preferred) in the county where the offence was committed, but statute now permits removal of the trial elsewhere in certain cases. In civil practice, the venue was the part of the declaration in an action which stated in which county the action was to be tried. Venue has been abolished and the place of trial fixed at the hearing of the summons for directions.

***Verba de praesenti* and *verba de futuro*.** In canon law, adopted into English and Scottish matrimonial law, the distinction was drawn between a contract to marry *per verba de praesenti*, which had the effect at common law of immediately making the couple man and wife, and a contract to marry *per verba de futuro*, which was a betrothal or engagement only.

Verbal injury. In Scots law, the generic term for harms done by words rather than by physical impact (real injuries), and comprising the three species of convicium, defamation, and malicious or injurious falsehood (qq.v). Some older cases confuse it with malicious falsehood.

Verbal obligation. In Roman law, *obligationes verbis*, constituted by the solemn form of question and answer known as *stipulatio*. In Scots law, the term is applied to those kinds of obligations which may be constituted orally and do not require to be constituted *literis*, by solemnly authenticated writings.

Verdict. The finding of a jury on the matter, or each of the matters, referred to it by the court. In English civil practice, the verdict is usually a general verdict, for the plaintiff or for the defendant, but a special verdict, finding certain facts proved and leaving to the court the application thereto of the law, is competent, but extremely rare.

In English criminal practice, the possible verdicts are Guilty and Not Guilty, and traditionally the jury had to be unanimous for a verdict of guilty but, since 1967, subject to certain conditions, verdicts of guilty by 10 of 12 or 11 jurors, or nine of ten jurors, are competent. A jury may find a defendant not guilty of the crime charged but guilty of another impliedly charged e.g. not guilty of theft but guilty of receiving.

Where a coroner's jury find the death of a person but do not find how he came to it, it is called an open verdict.

In Scottish civil practice, the verdict of the jury may be by a majority, and may be a general verdict, for the pursuer or the defender, or a special verdict, returning answers to specific questions of fact set by the trial judge, which is the basis for a subsequent interlocutor applying the verdict. In Scottish criminal practice, the verdict may be either Guilty, Not Proven, or Not Guilty (the latter two of which both have the effect of acquittal), and may be by a majority save that to convict a majority of the total votes must be for Guilty. Special verdicts in criminal practice are obsolete.

Verdross, Alfred (1890–). Austrian jurist, member of the International Law Commission and the Permanent Court of Arbitration at The Hague, President of the Institute of International Law, judge of the European Court of Human Rights, and author of *Die Verfassung der Völkerrechtsgemeinschaft* (1926), *Völkerrecht* (1937) and other works on international law.

Verge. The district within 12 miles of the royal court, an area in which the coroner of the county had no jurisdiction, being superseded by the Coroner of the Queen's Household or Coroner of the Verge. The latter's jurisdiction is now limited to the place where the sovereign is residing, and similarly the Lord High Steward of the Household has a domestic jurisdiction over the members of the royal household not within the verge but only within the palace or other residence.

***Veritas convicii*.** The truth of the imputation. In Scots law, proof of veritas is a defence in an action for defamation (q.v.), but not in an action for convicium (q.v.).

Verity, oath of. An oath required to be taken by a creditor petitioning for sequestration (q.v.) or claiming therein, as to the truth of his claim of debt.

Verney, John (1699–1741). Called to the Bar in 1721, he was an M.P. (1722–34), attorney-general to the Queen (1729) and Chief Justice of Chester (1734). In 1738, he became Master of the Rolls and held that office till 1741.

Versailles, Treaty of. Framed at the Versailles Peace Conference in 1919 to settle affairs after World War I. Important provisions included a statement of German guilt for the war, the demilitarization of the Rhineland, the loss of the Saar basin, abolition of the German general staff, limitation of the size of the German army and prohibition of German naval and air forces, the repayment by Germany to the allies for destroyed allied shipping, the stripping from Germany of her colonies and of some of her territory in Europe, the imposition on Germany of an indemnity, and the Covenant of the League of Nations.

The U.S. ultimately declined to ratify the Treaty and, in 1921, concluded separate peace treaties with Germany, Austria, and Hungary.

Vested remainder. An expectant interest limited or transmitted to a person capable of taking possession, if the particular interest should happen to determine.

Vesting. The concept of an interest destined to a person under a will or intestacy becoming the latter's property. The question of when an interest vests is frequently of importance and a matter of difficulty. This depends primarily on the testator's intention as disclosed by his will or settlement. An interest may vest in interest when the party has obtained legal right; or it may vest in possession when the party has obtained a right to actual present possession. Generally vesting means vested in interest. The presumption is that a gift vests on the testator's death but the terms of the will may make a different intention apparent. Vesting is not postponed merely because a life interest is interposed, nor because the date for actual payment is postponed. Vesting may be postponed by conditions, such as attainment of majority or marriage, and in some cases vesting may take place in a person but subject to divestiture or defeasance in the event of the happening of a subsequent occurrence such as the birth of a child.

Once an interest is indefeasibly vested in a person it is part of his property, notwithstanding that he may not yet have actual possession or enjoyment, and he may deal with and alienate the interest as his own. An interest not vested is contingent and defeasible.

Vesting declaration. A declaration in a deed of appointment of new trustees that any estate or interest in the trust property is to vest in the new trustees.

Vesting instrument. A deed, order of court, or assent constituting the evidence of the title of a tenant for life or statutory owner to the legal estate in settled property as estate owner and necessary for the settlement of a legal estate in land otherwise than by trust for sale.

Vesting order. An order which might be made by the Court of Chancery and may be made by the Chancery Division under various statutes passing the legal estate in property without other conveyance.

Vestitive facts. Facts which result in a person being vested with certain rights (investitive facts) or being divested of certain rights (divestitive facts). Investitive facts comprise original titles, which create a right which did not exist, as where a man catches fish and derivative titles, which transfer a title from another, as where he buys or inherits something, and divestitive facts comprise alienative facts, which transfer a title to another, as where a person sells or gives or bequeaths something, and extinctive facts which extinguish title completely, as where a person pays a debt or lapse of time extinguishes his duty to pay and the creditor's right to exact payment.

Vestry. A room or place adjacent to a church where the clergyman's vestments are kept. Also an assembly, usually convened in the vestry, to transact parochial business. Vestries formerly had non-ecclesiastical functions but these have been transferred to local authorities.

Vetera statuta (or *Antiqua Statuta*). The ancient statutes from Magna Carta down to Edward II, including some of uncertain date, as contrasted with the *Nova Statuta* (q.v.). They were printed by Pynson in 1508 and in many later editions. The distinction seems to have turned on the inaccurate assumption that the end of Edward II's reign (1327) saw the essentials of modern statutes settled, but may have followed from a merely accidental division in the manuscripts. A second part (*Secunda pars Veterum Statutorum*) containing statutes, down to 1327, not included in the original *Vetera Statuta* was published by Berthelet in 1532 and in several later editions.

Veterinary Surgeons, Royal College of. The body in the United Kingdom charged with the government of the profession of veterinary surgery, supervision of professional education, and with the maintenance of the register of veterinary surgeons. Through its disciplinary committee it regulates professional discipline and may remove from the register the name of a member who has been convicted of a criminal offence or been guilty of conduct disgraceful to him in a professional respect. An appeal lies to the Divisional Court of the Queen's Bench Division.

Veto. A prohibition or power to forbid. It applies particularly to the royal power (not exercised since 1708) to refuse assent to a Bill passed by both Houses of Parliament. The formula used was *le roy* (or *la reyne*) *s'avisera*.

In the United Nations, the term is used of the power enjoyed by the permanent members of the Security Council to block decisions by refusing to concur; the power is implied by the rule that all the permanent members must concur in certain decisions.

Veto power. The power of the President to prevent the enactment of legislation passed by Congress. If he refuses to sign it, he returns the Bill, which must be reconsidered and, if again approved by a two-thirds majority in each House, it becomes law notwithstanding his refusal to sign. If Congress is in Session, a measure not returned becomes law without the President's signature after 10 days. If Congress adjourns within the 10 days, the President's refusal to sign constitutes a Pocket Veto.

The grounds on which a President may refuse to sign are a matter of opinion. Presidents since Jackson have generally thought that the President may use his own judgment. Use of the veto is naturally most likely when President and Congress are in disagreement or of different parties.

State governors in the U.S. have a similar power and usually also power to veto particular items or parts of a measure.

Vexatious actions. Actions brought, not *bona fide*, but brought to annoy or embarrass the other party or not likely to lead to any practical result. The High Court has inherent power to stay any action brought merely to annoy or oppress. Statute also empowers the court if satisfied that a person has habitually and persistently instituted vexatious proceedings to forbid further proceedings without the leave of the court or a judge.

Vi et armis (by force of arms). Words inserted at common law in pleadings alleging a trespass.

Vicar. One who acts in place of another; in particular, in ecclesiastical law, one who is incumbent of an appropriated benefice, i.e. one perpetually annexed to a spiritual corporation, which is the patron of the living, as contrasted with one who is incumbent of a non-appropriated benefice, who is designated a rector.

Vicar-general. In ecclesiastical law, an officer appointed by a bishop to act under his direction in spiritual matters and in relation to the government of the Church. The modern vicar-general is a ministerial or judicial officer, not necessarily a cleric, who also holds the office of chancellor or official principal of the diocese.

Vicarious liability. The liability of one person for the conduct of another. In criminal law, vicarious liability is exceptional and probably never arises at common law but may exist under statute, though even then, in some cases, special limited defences exist, such as that a master used due diligence to enforce compliance with the statute, and the servant committed the offence without his consent.

In civil law, vicarious liability is recognized in tort or delict in certain cases. The principle is expressed but not explained or justified by the maxims *respondeat superior* and *qui facit per alium facit per se*. A principal is liable for wrong done by his agent if the latter were, at the time, acting within the scope of his authority. A master or employer is liable for wrong done by his servant or employee if the latter were, at the time, acting in the course of his employment, but not if the latter were, at the time, acting outwith the course of his employment. An employer is not, in general, liable for wrong done by an independent contractor employed by him. A parent is not vicariously liable for the wrong of his child, nor one spouse for the other.

Vice, inherent. Defect in goods carried in something incidental to the nature of the things themselves, which relieves a common carrier from liability for the safe delivery of the goods.

Vice, Succeeding in the. In old Scots law, an intrusion whereby one enters into possession in place of and by collusion with a tenant bound to remove. It rendered both tenant and successor liable as violent possessors.

Vice-admirals of the coast. The sea-coast of England and Wales is divided into 19 stretches, corresponding generally to coast counties, and for each a vice-admiral may be appointed as representing the Lord High Admiral or the Lords of the Admiralty. They have jurisdiction up to high-water mark and the first bridges on rivers. These courts are now obsolete.

Vice-admiralty courts. Despite the rise of the High Court of Admiralty in England, there remained in certain towns, till 1835, local courts of vice-admiralty. Vice-admiralty courts were established as Britain acquired colonies around the world, the earliest being in Jamaica in 1662. In 1890, the Colonial Courts of Admiralty Act abolished them all, with certain exceptions, and substituted colonial courts of admiralty.

Vice-Chancellor. A judicial office in England created in 1813 to assist the Lord Chancellor in the exercise of his first instance equity jurisdiction. The Vice-Chancellor could decide only cases specially delegated to him by the Chancellor, and appeal lay to the Chancellor and later to the Court of Appeal in Chancery. In 1842, two further Vice-Chancellors were appointed but the establishment of three did not become permanent till 1852. The office disappeared in the reorganization effected by the Judicature Act, 1873. A list of holders of the office is in the Appendix.

In 1970, the Lord Chancellor was authorized to appoint, from among the puisne judges of the Chancery Division of the High Court, a Vice-Chancellor who should be responsible to him for the organization and management of the business of the division and be the everyday head of the Division. For holders of the office see the Appendix.

In the Duchy of Lancaster, the Vice-Chancellor performed the active duties of the Chancellor. Since 1971 he is a circuit judge.

The title is also given to the chief officer and working head of British Universities, as deputy for the Chancellor or titular head of the university.

Vice-comes. The ancient term for a sheriff, being the deputy of the *comes* or earl of the county.

Vice-President of the U.S. The second highest executive office under the U.S. Constitution. The candidate for Vice-President tends to be chosen from a different part of the country and a different wing of the party from the candidate for President, and is all too frequently a person of lesser eminence and capacity than the Presidential candidate. He is the President's 'running mate' in the Presidential election. If no candidate gains a majority in the electoral college the choice is made by the Senate.

The Vice-President's sole official duty is to preside in the Senate, having a casting vote in case of a tie. In the event of the President dying in office, or resigning, he succeeds automatically and serves the rest of the late President's term. If he has served more than two years of the late President's term he may be elected President once only thereafter, but if he has served less, he may be re-elected twice (22nd Amendment). The Vice-President succeeds an Acting President if the President transmits to both Houses of Congress a written declaration that he is unable to discharge the duties of his office, or if the Vice-President and a majority of principal officers of the executive departments transmit to both Houses a written declaration that the President is unfit to act. If there is a vacancy in the office of Vice-President, the President must nominate a person to be Vice-President who shall take the office on confirmation by a majority vote of both Houses (25th Amendment). He is not a member of the Cabinet but, in modern practice, Presidents have permitted Vice-Presidents to participate in Cabinet discussions, and have invited them to undertake particular executive or administrative tasks.

Viceroy. A person who rules a county or province in name of and as representative of a monarch. It was the title given to the principal governors of Spain's colonies in America, notably the viceroyalties of New Spain (Mexico) and Peru, created in the sixteenth century, and of Granada and Rio de la Plata, created in the eighteenth century. The Portuguese captain-general in Brazil was styled viceroy from the seventeenth century. The term was sometimes applied to the English governors in Ireland from the fourteenth century and was the title of the British Governor-General of India from 1858 to 1947.

Vico, Giovanni Battista (or Giambattista) (1668–1744). Italian philosopher, historian, and jurist, became professor of rhetoric at Naples in 1697 but interested himself in jurisprudence and Roman law and attained distinction with his *De universi iuris uno principio* (1720) and *De Constantia Jurisprudentia* (1721). In 1725 he published his *Scienza Nuova* containing the idea of a natural law which could serve as a universal justice. He aimed to relate the history of law to that of the human mind and held that law emanated from the conscience of mankind. In 1735 he became historiographer royal at Naples. He was thoroughly acquainted with Roman law and with the historical development of law, and believed that Roman law developed in three stages, divine, heroic, and human. In his *De Uno principio*, he sought to deal with the origin, the course and the subsistence of jurisprudence by reference to a theological conception. Law is based on reason and authority, on philosophy and history. He sought to apply his theory of cycles or stages followed by all nations to government, jurisprudence, language, and literature.

Victimology. A branch of the study of criminology (q.v.) concerned with the effect on the victims of crime, physical and mental, of their experience. It supplies a valuable corrective to sole regard for the criminal.

Victims of violence. See CRIMINAL INJURIES COMPENSATION.

Vidocq, Francois-Eugene (1775–1857). French adventurer and detective who, having served spells in prison, created a new police department for Napoleon in 1809–12, the *police de sûreté*. After an unsuccessful period in business (1827–30), he again served in the detective department and, in 1832, created the first private detective agency, which was, however, suppressed by the authorities. He was a friend of men of letters and various volumes of memoirs are attributed to him.

View. An inspection by a jury before or during a trial of property in dispute or of a place where a crime was committed.

View of Frankpledge. See FRANKPLEDGE.

Vigilantes. Volunteer groups which were formed in parts of the U.S., particularly in the West before it was organized, and in areas where law enforcement was weak. When responsibly led, vigilantes maintained rough justice and were particularly effective in California during the gold rush and in parts of the West afflicted by cattle thefts, but in other circumstances, they were little better than mobs, and sometimes resorted to lynch law.

Vill. In mediaeval English law, a subdivision of the hundred and the smallest administrative unit, corresponding roughly to the later civil parish. It was usually identified with a township or village but extended beyond the urban area to the bounds of neighbouring vills.

Village. A small group of houses in separate occupation situated close together.

Villeinage. The tenure of land, ranking below free socage, by which agricultural labourers, free and unfree, held land in lieu of wages. The conditions of tenancy varied with the customs of different manors and the tenure was always more or less precarious. Later, a distinction was drawn between privileged villeinage, or villein socage (by which tenants on the King's demesnes held land on condition of performing base but certain services and who could not be removed so long as they were able and willing to perform the services due) and pure villeinage (by which tenants of mesne lords held land at the lord's will, being bound to do whatever work on his demesne was set them from day to day). In the fourteenth and fifteenth centuries, this labour service system gave way to a money payment system, and the rents remained stable despite the continuing fall in money values. When a tenant sold his land, the circumstances of the sale were recorded in the rolls of the lord's court and the buyer received a copy. As he held by a copy of the court roll, he came to be called a copyholder.

Down to 1925 the customs which constituted the local law governing the tenure varied from manor to manor and could not be discovered without examining the manorial records and was not uniform throughout England, while the mode of conveyance differed from that for socage. In 1925, copyhold was abolished and converted into land held by socage.

P. Vinogradoff, *Villeinage in England.*

Villeins. In mediaeval English law, a composite class of persons, comprising slaves as known in Anglo-Saxon law and free, yet dependent, cultivators of the soil whose tenure was defined by Norman lawyers as unfree because they owed duties to their lords indefinite in nature and extent. Some of the Roman rules of slavery were applied to them by Norman lawyers. They were things and the lord could dispose of them. The servile element early disappeared and the change from labour services to the lord to money rents destroyed villein tenure and reduced the importance of villein status. It survived, however, to the late sixteenth century before falling into complete disuse.

P. Vinogradoff, *Villeinage in England.*

Vindicatio. In Roman law, the action by which the owner of a thing could assert his title to it against anyone having possession (*rei vindicatio*) and, more generally, the actions by which one could assert title to a servitude (*vindicatio servitutis* or *usufructus*), or by which an *adsertor libertatis* could assert the freedom of a supposed slave (*vindicatio in libertatem*). In early Roman procedure, it signified the claim made by both parties. If the claimant were not *dominus ex jure Quiritium, rei vindicatio* was not competent, but if he were *in via usucapiendi*, the praetor would allow an equitable form (*actio utilis*) of the action (*actio Publiciana*) on the fiction that the time required for *usucapio* had expired. Under the formulary procedure, the plaintiff in a *rei vindicatio* could not compel the possessor to return the thing but could claim its value by *iusiurandum in litem* and by overstating the value exert pressure to return it.

Vindictive (or punitive) damages. A term sometimes applied to damages awarded, not only to compensate the plaintiff, but to punish a wilful wrongdoer.

Vinding, Rasmus or Erasmus (1615–84). Professor of Greek and later of History at Copenhagen, member and later referender in the Supreme Court, a member of several commissions to codify the law of Denmark, and author of a draft (*Codex Fredericius*) which was revised and adopted and proclaimed as *Kong Kristian V's Danske Lov* in 1683.

Viner, Charles (1678–1756). Viner had chambers in the Temple but was never called to the Bar. He spent many years in the compilation of *A General Abridgment of Law and Equity, Alphabetically digested under Proper Titles, with Notes and references to the Whole* (23 vols., 1742–53), based on Rolle's *Abridgment.* At first it was planned as a continuation of D'Anvers *Abridgment*, but later completed. An alphabetical index by Robert Kelham (1758) was incorporated in Viner's second edition (24 vols., 1791–94) and there was a supplement by various hands in six volumes, 1799–1806. It is the biggest of the *Abridgments* and generally very useful, and still of use despite defects in arrangement. Viner left the remaining copies of his *Abridgment* and the residue of his estate to Oxford University, effect being given to his bequests by the foundation of the Vinerian Chair, first occupied by Blackstone in 1758, the Vinerian scholarship and fellowships.

H. G. Hanbury, *Vinerian Chair and Legal Education.*

Vinnius, Arnoldus (Vinnen) (1588–1657). A leading Dutch jurist, with an international reputation, Professor of Law at Leiden, author of *Jurisprudentiae contractae libri IIII* (1624–31), *In IV libros Institutionum commentarius* (1642), a book long famous and used in law schools and later edited by Heineccius, and of *Selectarum Juris Quaestionum* (1653).

Vinogradoff, Sir Paul Gavrilovitch (1854–1925). Born in Russia, he early acquired a taste for the languages, literature, and philosophy of the West, and learned mediaeval history at Moscow University. Later at Berlin, he studied under Mommsen and Brunner. A visit to Italy imbued him with an interest in feudalism. In 1883 he visited England and, in the next year, found in the British Museum the manuscript collection of cases of the reign of Henry III made for Bracton, and later published by Maitland as *Bracton's Note-Book*. In 1887 he became Professor of History in Moscow, but left Russia in 1901. In 1903 he was appointed to succeed Pollock in the Corpus Chair of Jurisprudence at Oxford and, apart from visits to Russia and elsewhere spent the rest of his life in England. At Oxford, he instituted the first seminars on the continental model and trained in research methods, many who later became eminent as historians or lawyers. Among his many publications, the most noteworthy are *Villeinage in England* (1892); *The Growth of the Manor* (1905); *English Society in the Eleventh Century* (1908); *Roman Law in Mediaeval Europe* (1909); and *Outlines of Historical Jurisprudence* (2 vols., 1920–22). Two volumes of *Collected Papers* were published in 1928. He was first literary director of the Selden Society (1907–18) and edited volumes of the Society's *Year Books* series. He also prompted the British Academy volumes, *Records of English Economic and Social History* (to which he contributed the *Survey of the Honour of Denbigh*) and the Oxford Studies in Social and Legal History. Throughout his life he was an internationally known and respected scholar, multilingual and of enormous breadth and depth of erudition.

H. A. L. Fisher, *Memoir* in *Collected Papers*, I.

Vinson, Frederic Moore (1890–1953). Practised law in Kentucky and served in Congress, 1923–38, where he obtained the reputation of being a tax expert. Thereafter, he sat as a justice of the U.S. Court of Appeals for the District of Columbia (1938–43) and as a director of various federal agencies during World War II and was Secretary of the Treasury, 1945–46. He was Chief Justice of the U.S. Supreme Court, 1946–53, where probably his most notable opinion was his dissent in *Youngstown Sheet and Tube Co.* v. *Sawyer* (1952) in which the Court struck down a presidential order as executive lawmaking. In general, he interpreted broadly the powers of the federal government, which he upheld in opposition to claims of individual rights.

Violent profits. In Scots law, damages due from a tenant, for forcibly retaining possession after he ought to have removed, or by any other unlawful possessor *in mala fide*. They are a form of penal damages, comparable to English mesne profits. In rural tenements, they are the full profits the landlord could have made by possessing or letting the lands; in urban tenements, they are double the stipulated rent.

Viollet, Paul (1840–1914). French jurist, editor of the *Établissements de Saint Louis* (1881–86), and author of *Histoire du droit civil français* (1885) and *Droit public: Histoire des institutions politiques et administratives de la France* (1889–1903).

Virgate. In mediaeval English law, a measurement of land equivalent to a quarter of a carucate (q.v.) or to two bovates. Also known as a yardland, the area varied in different parts of the country, but was commonly about 20 acres.

Vis major. Overpowering force, whether of nature, e.g. a storm, or of men, e.g. a riotous mob. It is frequently equated with act of God (q.v.) and is an excuse for damage done or loss of property.

Viscount. The fourth grade in the U.K. peerage, ranking below dukes, marquesses, and earls but above barons. The title *vice-comes* was originally borne by the sheriff of a county long before the first viscount was created in 1440.

Visigoths, laws of the. The oldest of the *leges barbarorum*. The tribal law of the Visigoths was first set down by King Euric in the *Codex Eurici* (or *Euricianus*) about A.D. 475. Surviving fragments indicate that it was strongly influenced by Roman law. It was taken as a model by some later Germanic peoples. A code for the Roman subjects of the Visigothic kings was compiled by Euric's successor Alaric II and issued, after popular ratification, in A.D. 500 as the *Lex Romana Visigothorum*, or the *Breviarium Alaricianum* (q.v.) or the *Breviarium Aniani* (after its compiler). It was intended to summarise the most important rules of practice and remove the confusion resulting from the use of multiple texts of Roman law, and was founded on the Theodosian Code and other Roman materials.

The *Codex Eurici* remained in force, was revised under Leovigild (568–80), Recarred (586–601), and again under Recceswinth (649–72) to take account of the gradual merger of the Germanic and Roman populations, and to fuse the two systems of law. This code was applicable to Visigoths and Romans alike, and the materials were then arranged systematically in 12 books on the lines of Justinian's *Codex*. This was known as the *Lex Visigothorum* or *Liber Judiciorum*, and was to govern all subjects of the Visigothic kings in lower France and Spain. It appeared in 654. It was strongly Romanist in form and substance and contained large sections from the *Breviarium* which Recceswinth declared abolished as the code for Roman subjects in his dominions.

An enlarged and later compilation of Visigothic

law, attributed to King Ervig (680–687) is known as the *Lex Visigothorum Ervigiana* or *Lex Visigothorum renovata*. Though prolix and bombastic, it closely imitates Roman law and Church canons and was taken as a model and a source by other Germanic peoples. Despite the fall of the Visigothic Kingdom, it survived in regions where the Arab conquerors did not penetrate and, after their expulsion, was restored in places, a translation becoming the *Fuero* of Cordoba in the twelfth century.

P. W. King, *Law in the Visigothic Kingdom*.

Visigoths, Roman law of the. See *Breviarium Alaricianum*.

Visit and search. In international law, the right of belligerents to visit and search neutral ships to ascertain whether they are truly neutral and if so whether they are trying to break blockade (q.v.), or carry contraband (q.v.), or render unneutral service (q.v.). It may be exercised only during war, and not in the territorial sea of a neutral state, and not in respect of neutral warships, and possibly not in the case of neutral public ships such as mailboats. The right is exercised by stopping the ship and sending a boarding party; the vessel may be brought into port to be thoroughly searched. A ship which resists visit or search is at once captured and may be confiscated. A neutral ship sailing in an enemy convoy is equivalent to one resisting. Suspicion is raised by false ship's papers or by concealment or destruction of papers. Visit and search of foreign merchant ships in time of peace is contrary to international law, but has occasionally been resorted to as a kind of pacific blockade. Thus, in 1962 the U.S. intimated that vessels suspected of carrying missiles to Cuba would be stopped, visited and searched. This was enforced for only a short time.

Visitation. The power of certain persons to visit, inquire into, and correct any irregularities in the conduct of ecclesiastical and eleemosynary corporations. In ecclesiastical corporations, the Crown is visitor of the archbishops, the archbishops of the bishops within the province, and the bishops of subordinate corporations. In eleemosynary corporations and charities the founder and his heirs are visitors, or a person appointed by them, or the Crown. The judges of the High Court are visitors of the Inns of Court and may hear appeals against decisions as to professional misconduct of members. Some universities have a Visitor, frequently the Crown.

Visiting committees. Bodies of persons, replacing the pre-1877 system under which the local justices appointed Visiting Justices, from among magistrates, to be independent and non-official bodies with the right to enter any prison or Borstal at any time and to see any prisoner or inmate. Any prisoner or inmate may apply to see the committee, or any member of it, to make a complaint or application, and the institution is a valuable protection for prisoners against harshness, oppression or brutality by any member of the prison staff. A committee must meet at the prison monthly and at least one member is to visit once a week. The Visiting Committee has a general oversight of the management of the prison and makes an annual report on the way in which the statutes and rules are applied by the Governor and his staff, and is also the superior authority for the maintenance of discipline in prison. Serious offences against the rules, and in particular prison offences for which corporal punishment may be imposed, must be dealt with by the Visiting Committee, the Governor having power to deal only with minor offences. Similar bodies exist for Borstal institutions, where they are known as Boards of Visitors.

Visiting forces. Units of the armed forces of one state present on the territory of another state other than as part of an invading force, e.g. as part of an occupying force maintained after the conclusion of hostilities, or as part of a force stationed abroad under treaty. Generally, such forces are subject to the military law of their own state in respect of disciplinary offences but to the jurisdiction of local courts in other matters. But this principle is subject to the criminal jurisdiction of the state receiving them.

Visscher, Charles de (1884–) French jurist and author of *Théories et réalités en droit international public* (1953) and other writings on international law.

Vitious intromission. In Scots law, the act of any person, other than the executor confirmed by the court, who takes possession of, uses, or deals with any moveable property belonging to a deceased person. It matters not that the intromitter is a creditor or entitled to succeed to the deceased. Liability depends on unauthorized intermeddling and fraud need not be proved.

Vitium reale. Inherent vice, a defect in the title to property which transmits with it, such as the defect attaching to property which has been stolen, which cannot be purged even by sale for value to an innocent purchaser.

Vitoria (Victoria) Francisco de (*c.* 1483–1546). Spanish jurist and theologian, Professor of Theology at Salamanca from 1526, and very famous in that sphere, but also noteworthy legally as a founder of international law. He questioned the justice of the Spanish conquest of the New World and upheld the rights of the Indians. He was author of *De Indis et de iure belli relectiones*, a treatise of great importance, maintaining that looting and slavery were wrong, that non-combatants should be spared, and that pagans had rights to life, liberty, and property. He also wrote a commentary on Aquinas and was a leading scholar on Thomist thought.

C. B. Tralles, *Francisco de Vitoria, Fundador del Derecho Internacional moderno* (1927); J. B. Scott, *The Spanish Origin of International Law* (1933).

Voet, Johannes (1647–1713). Professor of Law at Utrecht and Leiden and son of Paulus Voet (1619–67) (q.v.). He was author of *Elementa Iuris* (1700), *Compendium Iuris* (1683), widely used as a student's book, and *Commentarius ad Pandectas* (1698–1704), dealing with the Roman law and the law of Voet's own day. It has always been of the highest authority in South Africa and has been translated into Italian, Dutch, and English.

Voet, Paulus (1619–67). Dutch jurist, professor at Utrecht, father of Johannes Voet (q.v.), and author of *De usu juris civilis et canonici in Belgio unito* (1657), *De statutis eorumque concursu liber singularis* (1661), a work on private international law, *De Mobilium et Immobilium natura* (1666), and *In Quatuor libros Institutionium imperialium Commentarius* (1668).

Void and voidable. An act, deed, or transaction is void when it has no legal effect and is a nullity, save that it may sometimes have some collateral legal consequences. It is a matter of decision in each branch of law, e.g. marriage, contract, wills, in what circumstances and for what reasons a purported act, deed or transaction is in fact void and legally ineffective. Where a vitiating factor renders an act or transaction void it normally does so *ab initio* notwithstanding that the vitiating defect is not observed at once, but in England a deed may, in certain circumstances, become void *ex post facto*. An act, deed, or transaction is voidable when it appears to be good and valid but, in fact, contains a defect entitling one or both parties to rescind it or have it cancelled by a court. But until or unless that is done, it has full force and effect and is legally valid. It is again a matter of law in each branch of law to determine in what circumstances an act, deed, or transaction is voidable. Rescission is competent only in certain circumstances and the right of rescission may be lost or waived. In some cases, it is disputed whether the effect of a defect is to render the act, deed, or transaction void or merely voidable.

Volenti non fit injuria (no injury is done to a willing person). The principle that an injured person cannot complain of harm done him if he knew of and voluntarily incurred the risk of that harm. Knowledge of risk alone is not sufficient, and a risk is not voluntarily undertaken merely because the injured person did not refuse, e.g. to do the work or seek other employment, and still less if he continued under protest only. The plea is practically excluded in employment cases but is still relevant to cases of participation in dangerous sports.

Volksgeist. The spirit of the people, or an innate consciousness. In F.C. von Savigny's (q.v.) philosophy of law, positive law should be consistent with the Volksgeist; it was part of the complex of a people's experience and character and, in time, the law reflects a people's general development, being founded in their social attitudes and ways. Accordingly, customary law was the principal and truest kind of law and codification should be simply the reduction of a people's law to writing and the giving to it of the stamp of state authority.

Volterra, Eduardo (1904–). Italian jurist and author of *Collatio Legum Mosaicarum et Romanarum* (1930) and *Istituzioni di Diritto Romano* (1961).

Voluntary. The word used of a gift, conveyance, settlement or similar transaction meaning that no valuable consideration (q.v.) was given in return for it. Equity regards voluntary transfers with suspicion and puts the onus on the party benefited to show that the donor appreciated what he was doing. If he cannot, the gift may be set aside at the instance of the donor or his representatives.

Voluntary conduct. Conduct which a man does in pursuance of an exercise of volition or will, as contrasted with involuntary conduct, such as sneezing or shivering. It is, in general, a prerequisite of incurring liability, civil or criminal, for doing anything that the conduct was voluntary. Accordingly, a man is not liable for harm caused when sleep-walking or overcome by illness.

Vote. An expression of view in favour of or against a particular motion put to a meeting, or for or against particular candidates standing for election to an office. In the management of companies, voting is in the first instance by show of hands but any member may demand a poll. At local authority and parliamentary elections voting by show of hands was abolished in 1872 and the poll is taken by secret ballot.

Vote, casting. A right to vote sometimes conferred on a chairman additional to his ordinary or deliberative vote, to enable him to decide an issue where the votes are equal. The right does not exist at common law and a person claiming a casting vote must justify it by the constitution of the body in question, by statute, or other authority. Where the chairman has a casting vote, it is proper, though not obligatory, to use it to maintain the *status quo*, since the contentions for change have not secured a majority.

Vote, The. The name in the House of Commons for a collection of papers issued daily during session to members. They include Votes and Proceedings (q.v.), Private Business, Notice Paper of Public Business (which is the agenda for the day), Supplement to the Votes (i.e. amendments to Public Bills to be taken in Committee of the Whole House and in Standing Committees), Minutes of Proceedings of Standing Committees on the previous day, and Division Lists.

Votes and Proceedings. A paper prepared by the Journal office from the minute books kept by the Clerks-at-the-Table and issued daily during sessions in the U.K. House of Commons, recording the decisions of the House on the previous day. It also lists documents presented to the House, such as Command Papers.

Vouch. To rely on or quote as authority. Hence, a voucher is a document which can be relied on as evidencing something, particularly payment. In the old law of real property, voucher to warranty arose where a person, sued for the recovery of land, called to court the person who had warranted the land to him, the vouchee, and required him to defend the title now challenged or to surrender to him land of equivalent value. In common recoveries (q.v.), the vouchee was usually the crier of the court, who was called the common vouchee.

Voyage. In carriage by sea, any passage by sea outside the confines of harbour or enclosed waters. Distinctions have to be drawn between voyage to a loading port, and the voyage from there to the port of unloading.

Voyage charter. A charter of a ship for one or more voyages, as distinct from a time-charter or a charter by demise (qq.v.).

Voyeurism (or scopophilia). The desire to obtain sexual stimulus by watching sexually stimulating scenes, normally by stealth. The stimulus and pleasure is obtained, e.g., by watching other persons undress or have sexual relations, and is not a preliminary to sexual advances. It is the reverse of exhibitionism.

Vulgar law. A term for the mixture of debased Roman law and local customs and practices which was, in fact, applied in the Roman provinces from about the fourth century onwards, and which can be seen exemplified in such legislation as the *Breviarium Alaricianum* (q.v.). The subtlety and theoretical refinement of the classical law was entirely lost and only simpler, more basic, forms and less complicated concepts were retained. Vulgar law was also found in the various codifications made in the Germanic kingdoms of the period of the great migrations, and it influenced the legislation of the Eastern Roman Empire also.

Vulteius, Hermann (1565–1634). German jurist, a humanist, professor at Marburg, and author of *Jurisprudentiae Romanae a Justiniano compositae libri II* (1590), *De Feudis eorundemque jure libri duo* (1595) and commentaries on the *Institutes.*

Vyshinsky, Andrei Yanuarevich (1883–1954). State prosecutor of Russia, 1935, and as such prosecuted Zinoviev and others in 1936–38. He was also Professor of Law at Moscow, head of the Institute of Law of the Academy of Sciences and represented the U.S.S.R. at the United Nations. He was considered a leading legal scholar, replaced Pashukanis as the spokesman of Soviet jurisprudence, edited *The Law of the Soviet State*, and wrote many books and articles.

W

W.S., Writer to the Signet (q.v.).

Wach, Adolf (1843–1926). German jurist, concerned chiefly with criminal procedure, and author of *Vorträge uber die Reichs-Zivilprozessordnung* (1879) and *Handbuch des deutschen Zivilprozessrechts* (1886).

Wadset. An ancient Scottish form of security over land, now obsolete, whereby lands were sold under right of reversion or re-sale. The lender received a conveyance of the lands with a clause of reversion to be exercised by the debtor after intimation. Under a proper wadset, the creditor entered into possession and drew the rents for his own benefit, not being bound to account for his intromissions. In improper wadset, the borrower remained in possession, the deed stipulating for lease of the premises back to him. Later, the transaction took the form of an absolute conveyance, qualified by a separate writing, a 'back-letter' or 'back-bond', stating the true nature of the transaction as a conveyance in security.

Waechter, Karl Georg von (1797–1880). German jurist and author of *Lehrbuch der deutschen Strafrechts* (1825), *Ueber die Collision der Privat-Gesetze* and other works.

Wager of law (or compurgation). A mediaeval mode of trial, found in the laws of many of the barbarian tribes which overran the Roman Empire, and adopted by the Church, in which a defendant who, on oath and in set words, had denied the charge against him and was called on to make his law, had to find a stated number of persons, fixed by the court according to the circumstances, and then to take an oath that he was innocent. His supporters then swore not to the facts of the case, but that the oath which he had taken was clean, that they believed him to be honourable in taking the oath. Increasingly detailed rules were formulated about it in England. It was not used in real actions, nor where the Crown was a party, but was allowed in the older personal actions, debt, detinue and account. It long persisted in ecclesiastical courts and, in London, even against accusations of felony, but declined in the sixteenth and seventeenth centuries, but was not abolished until 1833. The last case of it was *King* v. *Williams* (1824), 2 B. & C. 538.

Wagering. An arrangement whereby two persons, holding different views on the outcome of a future uncertain event, agree that, depending on the outcome, one will win and the other lose and the loser will pay to the winner a sum of money or other stake. It is a purely bipartite arrangement, thus differing from gaming (q.v.). A wager is void as a contract, but is not illegal. The term is in general synonymous with betting (q.v.).

Wages Councils. Bodies established under statute which make recommendations to the Department of Employment as to wages and conditions of employment.

Waifs and estrays. Waifs are things stolen and thrown away by the thief in flight, but not things left behind or hidden, and belong to the Crown by prerogative right if they have been seized on its behalf. The owner may reclaim the property if he can retake it before it has been seized for the Crown, and is entitled to restitution if he pursues the thief with due diligence, or later brings him to justice and obtains a conviction. Estrays are valuable tame animals, found wandering and ownerless. They belong to the Crown or, by virtue of grant or prescriptive right, to the lord of the manor, provided they have been proclaimed in the church and two adjacent market towns and remained unclaimed for a year and a day.

Waite, Morrison Remick (1816–88). Having graduated from Yale, was admitted in the Ohio Bar in 1839, served a term in the Ohio legislature, and declined appointment to the Ohio Supreme Court in 1863. He was of counsel in the Alabama arbitration, a delegate to and president of the Ohio State Constitutional Convention, 1873, and Chief Justice of theU.S. Supreme Court, 1874–88. During his period as Chief Justice the consequences of the Civil War were assimilated and there was developed the idea that a function of government was to protect the mass of individuals from the economic power of others. He founded the law of regulation of public utilities, and showed willingness to allow the states to assume regulatory power over corporate enterprises. He was a most competent Chief Justice, a good presiding judge, yet has not attained great public reputation as a judge, possibly because he was not a partisan and his career contained no spectacular case. During his time the court functioned very efficiently and competently.

C. Warren, *The Supreme Court in U.S. History*; B. R. Trimble, *Chief Justice Waite* (1938). L. Umbreit, *Our Eleven Chief Justices*, ch. 7.

Waiver. A renunciation or abandonment of a claim or remedy or immunity. It may be done expressly or impliedly, by taking no action to enforce the claim or remedy in the knowledge that enforcement is competent. In the case of waiver of immunity, e.g. diplomatic immunity, it must be done expressly and by the head of the mission concerned.

Wake. A gathering of persons, particularly at night, to watch over a dead body between death and burial, or a gathering for pleasure after a vigil connected with a local patron saint or with a religious purpose.

Waldheim, Kurt (1918–). Austrian diplomat, led the Austrian delegation to the U.N., served as Minister to Canada and ambassador to the U.N. (1964–68 and 1970–71), and Foreign Minister of Austria. In 1972 he succeeded U Thant as fourth Secretary-General of the U.N.

Waldock, Sir Claud Humphrey Meredith (1904–). Professor of Public International Law at Oxford, 1947–72, a member of the European Commission on Human Rights, 1954–61 (President 1955–61), Judge of the European Court of Human Rights, 1966–74 (Vice-President, 1968–71; President, 1971–74), and a judge of the International Court since 1973.

Wales, courts in. The Statute of Wales, 1284, annexed Wales to the English Crown and divided North and West Wales into six counties, with

Sheriffs, coroners, and bailiffs and subject to the justices of Snowdon, Chester, and South Wales, administering justice according to English rules and by English writs. The rest of Wales was governed by Lords Marchers (q.v.). Error lay to Parliament and, till the mid-sixteenth century, the courts of Wales were independent of the jurisdiction of the English common law courts. In 1536 the Act of Union between England and Wales made Welshmen subject to the same laws as Englishmen. Courts were to be held in English. In 1543 Wales was divided into 12 counties, and a Court of Great Sessions to be held twice a year in each group of three counties by the justices of Chester, North Wales and two persons named by the Crown. Sheriffs' tourns and hundred courts, county courts, coroners, justices of the peace, and quarter sessions were established. The jurisdiction of the President and Council of Wales and the Marches was defined. Writs of error lay from the Great Sessions to the King's Bench. In the seventeenth century, disputes arose over the powers of the courts at Westminster over Wales but, by the end of the eighteenth century, the courts at Westminster had established a right to exercise concurrent jurisdiction with the Courts of Great Sessions. In 1830 the courts of Great Sessions were abolished, the English circuit system extended to cover Wales and Chester, and jurisdiction in equity transferred to the Chancery and the Exchequer. Since 1942 Welsh may be used in legal proceedings in courts in Wales.

H. Owen, *Administration of English Law in Wales and the Marches* (1900); C. Abbot, *Jurisdiction and Practice of the Court of Great Sessions of Wales* (1795); W. R. Williams, *History of the Great Sessions in Wales 1542–1830* (1899).

Wales and the Marches, Council of. It grew out of the council which administered Edward IV's marcher lands, was revived by Henry VII, and later reconstituted by Wolsey. Special treatment was needed for Wales because it was then still only half-conquered. In 1534 it was given power to supervise justice in the franchises of the marcher lords. The Council was a large body with seven paid members. It had jurisdiction over the whole of Wales and the border-counties of Gloucester, Worcester, Hereford, Monmouth, and Shropshire, and could determine real and personal actions and reinstate persons violently ousted from their lands. It exercised conciliar jurisdiction, being a local Chancery and Star Chamber, and had commissions of the peace and oyer and terminer, and could even deal with treason. It had a criminal jurisdiction similar to that of the Star Chamber and a civil and equitable jurisdiction coordinate with that of the courts of Great Sessions.

After being in abeyance during the Civil War, it revived at the Restoration and was abolished only in 1689.

C. Skeel, *The Council in the Marches of Wales* (1904); R. Flenley, *Register of the Council in the Marches, 1569–91* (1916).

Wales, law of. Wales has been legally, administratively, and politically united with England since 1536 and since that date has had no independent system of law, though it had distinct courts till 1830 (See WALES, COURTS IN). Since the Wales and Berwick Act, 1746, mention of England in a statute is deemed to include Wales also. Though, in substantive and procedural law, its system is still identical with that of England there are elements of distinctiveness in public administration. The office of Secretary of State for Wales was created in 1964 and since then there has been a considerable growth in the powers of the Welsh Office at Cardiff, with responsibilities for administering education, health and social services, housing, and planning.

Andrews (ed.) *Welsh Studies in Public Law* (1970).

Wales, Prince of. A title dating from 1301 when Edward I of England, having conquered Wales and executed the last native Welsh prince, gave the title to his son, later Edward II. Since then, the title has been reserved for the eldest son of a monarch of England and later of the U.K. and been conferred on most but not all eldest sons. It is not inherited automatically but is conferred by the sovereign and the new prince is then invested at a ceremony in Wales. The title goes into abeyance when the son succeeds as king until regranted by him to his own son.

Wales, Statute of (1284). An Act of the English Parliament uniting Wales to England, following its subjugation by Edward I, and providing for its administration and legal system. It created six counties out of the Land of Snowdon, i.e. the territory previously held by Llewelyn and provided that the Justice of Snowdon should have jurisdiction within them. The Lord Marchers had jurisdiction over the rest of Wales.

Wales, Secretary of State for. A Minister for Welsh Affairs was appointed in 1951, and subsequently a Minister of State also. A Secretary of State was first appointed in 1964, with responsibility for local government, housing, town and country planning, and roads in Wales, and supervisory powers over the application in Wales of national policies. The Welsh Office is in Cardiff, with an office in London for liaison purposes.

Walker, Sir Samuel (1832–1911). Was called to the Irish Bar in 1855. He was Solicitor-General,

1883, Attorney-General, 1885 and 1886, and Lord Chancellor of Ireland, 1892–95 and 1905–11.

Walking possession. Where a writ of execution is issued to a sheriff, he is bound to take possession of goods belonging to the judgment debtor, but the bailiff may take 'walking possession' by agreement with the judgment debtor, in which case the goods are not actually removed from the debtor's custody nor a person left in charge of them, but the sheriff's officers are authorized to re-enter the premises and the debtor undertakes not to remove the goods, and to pay possession money each day.

Wallace, Sir Thomas, of Craigie (?–1680). Sat in Parliament, was made a baronet, became a Lord of Session in 1671, and Lord Justice-Clerk in 1675.

Wallis, Sir John Edward Power (1861–1946). Was called to the Bar in 1886 and was successively Advocate-General, 1900–7, puisne judge, 1907–14, and Chief Justice of Madras, 1914–21. He became a member of the Judicial Committee of the Privy Council in 1926.

Walsh, John Edward (1816–69). Was called to the Irish Bar in 1839 and was Master of the Rolls in Ireland 1866–69.

Walsingham, Lord. See DE GREY, WILLIAM.

Walter, Ferdinand (1794–1879). German jurist and politician and author of an important *Lehrbuch des Kirchenrechts* (1822). *Fontes iuris ecclesiastici antiqui et hodierni* (1862), *Corpus iuris Germanici antiqui* (1824), *Römische Rechtsgeschichte* (1836), *Deutsche Rechtsgeschichte* (1853), and other works.

Walter, Hubert (?–1205). A nephew of and brought up by Ranulf de Glanvill (q.v.), he early acted as a justiciar. He was Bishop of Salisbury, 1189–93, and Archbishop of Canterbury, 1193–1205. In 1190 he fought in Palestine with Richard I. He also acted as justiciar 1193–98, and, as such, governed England in the absence of Richard I. He was Chancellor under John, 1199–1205. He may have been the real author of the treatise ascribed to Glanvill (q.v.) and was one of the outstanding public servants of his time.

Walton, Frederick Parker (1858–1948). Was called to the Scottish Bar in 1886 and lectured on Roman law at Glasgow before being successively Dean of the Law Faculty at McGill (1897) and Director of the Khedivial School of Law at Cairo (1915–23). He was a pioneer in the study of comparative law and a learned civilian. His publications include a *Handbook of the Law of Husband and Wife according to the Law of Scotland* (1893), *Historical Introduction to the Roman Law* (1903), *The Egyptian Law of Obligations* (1920), *Introduction to French Law* (with Sir M. S. Amos, 1935), and Lord Hermand's *Consistorial Decisions* (1940).

Wambaugh, Eugene (1856–1940). Practised at the Cincinnati Bar and then taught law at Iowa State University and Harvard. He wrote *The Study of Cases* (1892) and various case-books.

Wapentake. The term in Yorkshire and East Midland counties for a division of the land corresponding to the hundred (q.v.). The origin of the name is obscure though it seems to have been based on an armed gathering of freemen.

War. A forcible contention between one state or group of states and another state or group of states through the application of armed force and other measures with the purpose of overpowering the other side and securing certain claims or demands. It is to be distinguished from unilateral use of force at least by a declaration that they are deemed to be acts of war. War proper is international and to be distinguished from force between individuals or groups within a state, repression of revolt, insurgency or piracy, and civil war, though if a revolting part of a state is itself a member of a federal state, or is recognized as a breakaway state, the revolt may become war proper.

War is a condition recognized and sought to be regulated by international law, even though it does involve breach of peaceful relations between the belligerent states; it is not a state of anarchy. Its existence brings into play a great body of legal rules attempting to define parties' rights and duties.

Early in the history of international law, a distinction was drawn between just and unjust wars.

Theologian-jurists were much concerned with the legitimacy of war, even in self-defence. They generally adopted the view that defensive war was justifiable and that, in certain cases, even offensive war might be justifiable; they further devoted attention to the criteria of a just war. Aquinas held that to be just, a war must have sufficient authority in the one who declares war, there must be a just cause of offence by a guilty party, and there must be a just intention to make war solely for the establishment of peace, the repression of the wicked or the assistance of the good. But these criteria were difficult to apply in practical cases.

In the sixteenth and seventeenth centuries, the theologians and canonists sought to cope with the difficulties of disputes where each party could claim some right. But in the seventeenth and eighteenth centuries, wars were fought on the old pagan bases that wars were a legitimate way of increasing colonial possessions, national wealth and trade. The times of Frederick the Great and Napoleon illustrate the belief that ambition and greed justified resort to force.

There was a recrudescence of this in the nineteenth and twentieth centuries when Germany's territorial ambitions and desire for aggrandisement led her into wars with Denmark, Austria, and France and finally into World War I. Desire for revenge and a revival of territorial ambition led her to make World War II. After it the desire of Russia and China to spread the doctrines of Communism led to the Korean and Vietnam wars.

The canonists allowed the state which waged a just war extensive rights over the persons and property of his enemies. The theory that the wager of a just war was executing justice against a criminal was held to justify seizing the object which was in dispute, razing the enemy's fortresses, exacting tribute, deposing the ruling prince, annexing territory, and occupying it with troops.

The just belligerent's troops might seize property as booty from the enemy; if an army succeeded in entering a besieged city by actual assault, not after surrender, pillage was allowed, and if the defence had been prolonged slaughter and sacking were normal, as in many cases from Rome in 1527 to Badajoz in 1812. Distinguished prisoners might be held to ransom, lesser men might be, and usually were, put to death.

In the eighteenth and nineteenth centuries, international law tended to recognize war both as a legitimate instrument of self-help against an international wrong, and as an act of national sovereignty for the purpose of altering existing rights.

All trading and commercial intercourse with alien enemies becomes illegal on the outbreak of war, unless specially permitted. Contracts made before, but still unperformed at the outbreak of war are discharged.

A belligerent may seize public enemy property on his territory at the outbreak of war, and enemy private property at sea in the belligerent ships, but not private property of enemies, though it may be taken into the custody of an official. Belligerent property in neutral states is immune from seizure by the other belligerents.

An outbreak of war formerly empowered a state to lay an embargo on enemy shipping in its harbours, to confiscate them, but since 1854 they have sometimes been allowed a reasonable time in which to depart. Enemy ships at sea are liable to capture and condemnation as prize, even if ignorant of the outbreak of war.

War may be terminated by mere cessation of hostilities, or by one belligerent subjugating his enemy, or by treaty of peace. Cessation of hostilities does not automatically bring the state of war to an end. Where cessation takes place each party is deemed to accept the situation, e.g. in regard to occupied territory, existing at the cessation.

In modern times, partly because wars have become so devastating and destructive, and partly because a war is recognized as a failure of international society to resolve disputes without resort to violence, much attention has been given to attempts to limit resort to war. A Hague Convention of 1907 prohibited, with certain exceptions, recourse to force as a legal remedy for enforcing obligations. The Covenant of the League of Nations deprived members of the right of war in certain cases. The General Treaty for the Renunciation of War of 1928 condemned recourse to war and the signatories renounced it as an instrument of national policy, though retaining it as a measure of permissible self-defence, as against a signatory who had broken the Treaty, as against non-signatories and as a means of collective action for the enforcement of international obligations by virtue of such instruments as the Charter of the United Nations. It is questionable whether the Treaty prohibited resort to force short of war.

Similarly, Geneva Conventions of 1949 provide that in non-international armed conflicts each party is bound to apply certain fundamental humanitarian provisions.

War, laws of. The rules of international law respecting warfare, its commencement, conduct and termination, based on the practices and usages of belligerents as modified by custom and treaties. In ancient and mediaeval times, war seems to have been almost entirely unrestrained though in the Middle Ages religion, chivalry, and the rise of humanist and rationalist sentiment gave rise to mitigating factors. The practice of warfare has been substantially modified since the seventeenth century by the recognition of principles of chivalry, mutual respect and fairness, and of humanity, moderating the kind and degrees of violence permitted. General treaties which have notably modified the customary laws of warfare include The Declaration of Paris of 1856 which abolished privateering, enacted that a blockade must be effective to be valid and that non-contraband neutral goods under an enemy flag and non-contraband enemy goods under a neutral flag must not be seized; the Geneva Conventions of 1864 and 1906 for the protection of wounded in the field, the Convention respecting the laws of war on land agreed at the Hague Conference of 1899 and revised in 1907 and various Hague Declarations and Conventions of 1907 as to particular topics, and the Geneva Conventions of 1949 dealing with prisoners, wounded and sick.

Usages of warfare, which by custom or treaty have become rules, are binding on belligerents under all conditions except in the case of reprisals as retaliation against a belligerent for illegitimate acts of warfare by his forces. Necessity does not justify departing from the rules.

Legitimate and illegitimate warfare

Warfare is legitimate in so far as belligerents conduct hostilities in accordance with international law; otherwise it is illegitimate. Belligerents may be directly responsible, or vicariously responsible for illegal acts of their forces.

Legitimate warfare may be secured by reprisals, punishment of war crimes by enemy soldiers and subjects, or the taking of hostages; or by complaints lodged with the enemy, or neutral states, or by good offices, mediation, or intervention by neutral states; or by claims for compensation.

Outbreak of war

Traditionally war had to be commenced by a declaration of war though at sea hostilities commenced without formalities. But there are many instances of commencement of hostilities without declaration and war may begin if one state resists acts of force used by another state by way of reprisals, or during intervention or pacific blockade.

Outbreak of war causes the cessation of diplomatic relations between belligerents, if indeed this has not already happened, and of consular services. Members of diplomatic and consular staff must be allowed to depart. The outbreak cancels all political treaties between the belligerents, but not treaties concluded with a view to war, nor political treaties intended to establish a permanent condition, nor non-political treaties, e.g. of commerce, though these are normally suspended. Multipartite treaties are not cancelled but their execution is suspended as between the belligerents. Belligerents' subjects in enemy territory must be allowed to return home though subjects who are actual or potential members of the armed forces may be detained as prisoners. Subjects who continue to reside in enemy territory may be placed under supervision or interned.

Land warfare

The objects of land warfare are defeat and destruction of the enemy's army so as to destroy his power and will to resist, and occupation of his territory.

Belligerents may prepare and carry on hostilities only in the region of war, a term more extensive than theatre of war, in which actual hostilities take place. The region of war comprises the whole of the territories and territorial seas of all belligerents. But a belligerent may renounce the right to treat certain territories as within the region of war. The territory of a neutral state may be within the region of war. Certain areas, e.g. Switzerland and the Vatican, may, however, be excluded from a region of war by neutralization, by permanent treaty or by special treaty between the belligerents. No part of the high seas is neutralized.

Belligerents' forces comprise primarily their regular military, naval, and air forces, including commando and raiding parties, and non-combatant and female personnel. Irregular forces are treated in the same way provided they are commanded by a person responsible for his subordinates, are distinguished by a recognizable emblem, carry arms openly, and conduct their operations in accordance with the laws and customs of war. Levies *en masse* of the population who spontaneously take up arms on an enemy's approach are also entitled to be treated as forces, but a levy *en masse* of the population of an occupied territory in order to free it is not entitled to be treated as part of the forces. Privateers (q.v.) were formerly a recognized part of belligerents' armed forces but privateering was abolished in 1856 by the Declaration of Paris.

In the conduct of operations of war belligerents are restricted in the means they may use for overpowering the enemy, partly by principles of chivalry and humanity and partly by law-making conventions. Legitimate modes of violence including killing or wounding combatants, if resisting, but not if sick, wounded or surrendering. Quarter must be given to troops who surrender or are made prisoners, save where they resist despite having surrendered or by way of reprisal for a refusal of quarter by the other side. The former rules that quarter might be refused to the defenders of a fortress carried by assault or a force which obstinately defended a place against overwhelming odds are obsolete. Law-making conventions prohibit various kinds of bullets, poison gas and chemical warfare.

Non-combatants may not be directly attacked but may be made prisoners. Chaplains and medical personnel may not be made prisoners.

Private enemy individuals may not be attacked, but are exposed to the risks of warfare if they stay in the theatre of operations, nor, in general, may they be made prisoners, but restrictions may be imposed on them.

By various conventions, provision is made for caring for the sick and wounded of both sides, and persons and places marked with the red cross are immune from attack. Dead must not be mutilated but buried or cremated and the identity of dead personnel recorded and communicated.

In former times, prisoners were frequently killed, mutilated, or enslaved, but by modern international law, they must be treated humanely and properly accommodated. Prisoners may be required to do work of various kinds, but not work directly contributing to the detaining power's war effort. It is legitimate to try to escape.

Bombardment, seige, and assault are legitimate but not against undefended localities. A besieging force is not obliged to allow non-combatants or aged, sick and others to leave the besieged locality, and if the defending commander expels such persons the besieging force may decline to allow them to pass.

Ruses, stratagems, and methods of deceit intended to mislead the enemy are all legitimate, but not the use of uniforms, badges or other insignia of the enemy since belligerent forces should be able to know who are friends and who enemies.

The use of spies to obtain information is lawful but such persons if caught may be punished. Similarly it is legitimate to seek to persuade or bribe persons treasonably to give information, though if detected their treason is punishable. It is legitimate to encourage enemy subjects to revolt against their own government and to encourage or assist underground movements in occupied territory to commit acts of sabotage or other war treason against an occupying power.

Occupation of part or the whole of enemy territory is an important aim of warfare, enabling the belligerent to use the enemy's resources for military purposes, deny them to the enemy, and hold the territory to obtain concessions from the enemy. Occupation is more permanent than mere raiding into a territory or invasion; it involves taking possession and control. Formerly a power which occupied enemy territory might treat it as its own and do what it liked with the territory. Later a distinction was drawn between occupation and acquisition of territory by conquest and subjugation. An occupying power acquires actual authority and must administer the territory not only for his own benefit but for the benefit of the inhabitants. It may not annex the territory, set it up as an independent state, nor sever it from the rest of the state to which it belonged. It is not bound by the constitution or laws of the territory occupied, but may not change these beyond what is necessitated by the maintenance and safety of the occupying forces, unless particular laws are in defiance of developed concepts of legality and justice. As the occupying power has military authority the inhabitants are subject to his martial law, but they owe no allegiance to the occupying power. The occupying power may levy ordinary taxes and dues but not general penalties on the inhabitants.

If a belligerent occupies enemy territory he may not sell public land or buildings of the enemy, though he may administer them and utilize their produce, and utilize them for necessary purposes. Public moveable property may be appropriated, but moveable property of municipalities and charitable and educational institutions cannot lawfully be seized, though in both World Wars I and II such property was frequently seized. Public enemy property captured on the battlefield may be appropriated by the victor.

Private land and buildings may not be appropriated or sold by an invading belligerent, but may be used for necessary purposes. Private moveable property usable for war purposes may be seized and used, but compensation must be paid, and they must be restored at the conclusion of peace; private moveable property not usable for war purposes may as a rule not be seized. But these rules were largely violated during both World Wars. Private enemy property captured on the battlefield may be seized if of military value, but not if purely private.

Belligerents may requisition all kinds of articles necessary for an army from the inhabitants of occupied territory, but they must be paid for in cash or at least a receipt given and payment made as soon thereafter as possible. Requisition may be made only by the commander of the troops in the locality. Similarly, troops may be quartered in the houses of private individuals, payment being made therefor. Belligerents may also levy contributions in money from municipalities or inhabitants for the needs of the army.

They may not deport inhabitants of occupied territory to do forced labour in the occupying power's own territory.

Wanton destruction of enemy property is not now permitted, unless imperatively demanded by the necessities of war, but damage or destruction for military purposes is justifiable, e.g. blowing a bridge.

Occupation terminates when an occupying power withdraws from or is driven from a territory.

Sea warfare

Similarly, in the case of war at sea, not all practices capable of injuring the enemy are deemed lawful.

The objects of sea warfare are defence of the home country against invasion, protection of the merchant fleet, defeat of the enemy fleet, capture or destruction of the enemy merchant fleet, cutting off of intercourse between the enemy and other countries, denial to the enemy of supplies, and support of military operations such as invasion of the enemy's territory.

It is legitimate to attack all enemy warships and public vessels, and enemy merchant ships if they refuse to submit to being visited; the latter may be armed and may defend themselves, but, privateering being abolished, may not act offensively. If a vessel hauls down her flag and surrenders, she must be given quarter. Public enemy ships once seized may be appropriated at once and sunk or taken into port, the crew being made prisoners of war. Private vessels once seized must be taken into port, considered by the Prize Court, and adjudicated as lawful prize, which alone makes appropriation by the capturing belligerent final. A captor may destroy a captured ship only if she cannot be taken into port for adjudication, or if she cannot spare a crew to take her into port. Vessels engaged in scientific missions, fishing boats, small boats employed in local trade, and hospital ships, may not be attacked or seized. If a ship is lawfully seized as prize and justifiably sunk, neutral owners of cargo have no claim for compensation. The sinking of merchant ships without warning is, accordingly, a war crime. A prize is lost if she is abandoned by the captor, escapes, or is recaptured.

If a ship is captured and condemned as lawful prize, the vessel and enemy goods fall to the capturing state. Goods on board are presumed enemy but if proved to be neutral and not contraband they cannot be confiscated.

It is legitimate to seek to wound or kill combatant personnel but not those who are sick, wounded, surrender or do not resist capture. Various kinds of projectiles which cause unnecessary injury are prohibited. Persons sick, wounded, or cast into the water by sinking ships or aircraft must be rescued if possible. Dead should be identified and buried. Hospital ships are immune from attack or capture, though these have sometimes been misused for naval purposes.

Ruses are legitimate including disguising the ship, using a neutral or enemy flag, save that before making an attack a vessel must display her national flag.

Defended places on the enemy coast may be bombarded by naval forces, but undefended places may not, except for depots, workshops, or military or naval establishments in such places.

Air warfare

The rules of air warfare are in a very underdeveloped state but analogies from principles of land and sea warfare give guidance.

It is clearly legitimate to use aircraft for observation, and in support of land or sea operations, and to attack enemy aircraft, guns, tanks, ships, etc.

Destruction by aerial bombing or machine-gunning of enemy personnel, equipment, military or naval installations is legitimate, but attacks on non-combatants, civilian population or targets of no military importance are not. But it is very difficult to distinguish permissible from impermissible targets.

Enemy civil aircraft may be captured and are subject to the same rules as enemy ships at sea.

Modern limitations on warfare

Since 1945, the use of force by one state against another is governed by the U.N. Charter, which limits the use of force to self-defence. It is open to argument whether this means defence against actual aggression or includes resistance to the threat of force. It is also questionable whether a state can still exercise other traditional rights of self-help, including reprisals and intervention (qq.v.).

L. Oppenheim, *International Law*, vol. 2 (1952); J. Stone, *Legal Controls of International Conflict* (1959); G. Schwarzenberger, *International Law*, vol. 2, *The Law of Armed Conflict* (1968); I. Brownlie, *International Law and the Use of Force by States* (1963); M. Greenspan, *The Modern Law of Land Warfare* (1959); R. W. Tucker, *The Law of War and Neutrality at Sea* (1955); J. M. Spaight, *Air Power and War Rights* (1947); F. Seyersted, *United Nations Forces in the Law of Peace and War* (1966).

War crimes. Conduct by governments, leaders, members of the forces, and civilians in wartime which goes beyond legitimate hostilities and which may be punished by the enemy on capture of the offenders. They include violations of recognized rules of warfare by members of the armed forces, e.g. killing wounded or prisoners, armed hostilities by persons not members of the armed forces, espionage and war treason, and all marauding acts. Violation of a rule of warfare is a war crime, notwithstanding that it has been ordered by a government or commander, and the plea of superior orders is no defence, certainly in the case of conduct which clearly violates general sentiments of humanity. Conversely, governments and commanders are responsible for illegalities committed by their troops unless they have genuinely tried to suppress these.

After World War II, three classes of offences against international law came to be regarded as war crimes, crimes against peace, as by planning or waging a war of aggression, conventional war crimes, or violations of the accepted laws or customs of warfare and crimes against humanity, including extermination, enslavement, deportation and other inhumane acts.

War criminals, punishment of. Since at least the later part of the nineteenth century, it has been recognized that international law imposes criminal responsibility on individuals for certain acts or omissions, and that they may be punished for such acts by international tribunals, or by national courts and military tribunals. After World War I an inter-allied Commission recommended war-crimes trials and there were demands to hang the Kaiser and to try other German leaders. In 1943 the allied powers set up a United Nations War Crimes Commission. In 1945 the victorious powers in World War II signed an agreement establishing an International Military Tribunal, declaring that individual responsibility attached to crimes against peace (namely planning, preparation, initiation, or waging of a war of aggression, or a war in violation of international agreements), war crimes (namely violations of the laws or customs of war) and crimes against humanity (namely murder, extermination, enslavement, deportation, or other inhumane acts committed against any civilian population before or during the war), or persecution on political, racial, or religious grounds. The tribunal sat at Nuremberg in 1945–46 and tried 24 former Nazi leaders and several groups and organizations such as the Gestapo. The majority of the defendants were convicted and sentenced to death or long terms of imprisonment. Other states later adhered to this Agreement and, in 1946, the U.N. General Assembly affirmed the principles of international law recognized by the Charter of the Nuremberg Tribunal and its judgment.

In 1946 the Supreme Commander in Japan established an International Military Tribunal for the Far East. It sat from 1946 to 1948 and tried 28 defendants, 25 of whom were sentenced to death or imprisonment. Apart from these major trials, thousands of Germans and Japanese were tried by various national courts for conventional war crimes of many kinds. All these trials have been attacked as merely vindictive and as making conduct criminal *ex post facto*. The doctrine of individual responsibility for war crimes was settled by law in 1945 and has been accepted ever since.

Since 1945, the Geneva Convention of 1949 made punishable offences against persons protected by the conventions and, in 1948, the U.N. approved a Convention on the Prevention and Punishment of Genocide, while in 1947, when establishing the International Law Commission, the U.N. requested that body to draft a code of offences against the peace and security of mankind, which it did in 1951. On several other occasions the U.N. has evidenced concern for the punishment of war crimes and crimes against humanity, and there have been numerous allegations of war crimes in the hostilities which have taken place since 1945, notably in Korea, Vietnam, and the Middle East. The question has also arisen whether the same principles apply in civil wars, wars of national liberation (genuine or so-called) and similar conflicts but this is disputed.

C. Mullins, *The Leipzig Trials* (1921); *History of the U.N. War Crimes Commission and the Development of the Laws of War* (1948); T. Taylor, *Nuremburg Trials, War Crimes and International Law* (1949); S. Horwitz, *The Tokio Trial* (1950); J. B. Keenan and B. F. Brown, *Crimes against International Law* (1950).

War treason. A term for various acts hostile to a belligerent committed by an inhabitant of occupied enemy territory and in conflict with the duty of such inhabitants to obey the law enforced by the occupying power. It includes the activities of members of an 'underground' or resistance movement (assuming they are not operating as organized bodies of guerillas).

Ward, Sir Edward (1638–1714). Was called to the Bar in 1670, became Attorney-General in 1693, and was Chief Baron of Exchequer, 1695–1714. He appears to have been an honest and intelligent judge.

Ward, Robert Plumer (1765–1846). Politician and author of an *Enquiry into the foundation and history of the law of nations in Europe* (1795), *A Treatise of the relative Rights and Duties of Belligerents and Neutral Powers in Maritime Affairs* (1801), *An Essay on Contraband* (1801), and various novels.

Ward. One of the casualties (q.v.) competent to the feudal tenure of ward-holding, by which the superior was entitled to the guardianship of a vassal's heir during his minority and the profits of his lands during that time. Ward might be taxed or restricted to a fixed sum to be paid annually by the minor heir. In its modern sense a ward is a young person under the care of a guardian. The term is also applied to divisions of boroughs existing for electoral purposes.

Ward-holding. In Scottish feudal land law, the tenure of land in return for military services, abolished in 1747.

Ward of court. In England, infants have always been regarded as specially under the protection of the Sovereign whose functions were delegated to the Lord Chancellor and are now exercised by the Chancery Division. An infant may be made a ward of court by order made on application therefor, in which case the court exercises special jurisdiction and control, parental and administrative, over his person and property.

The court will not allow a ward to be removed out of the jurisdiction, unless satisfied that the removal is for the ward's benefit, and restoration will be enforced. It may control the ward's education and must give consent to a ward's marriage. The wardship ends when the ward comes to be of full age.

Warden. A guardian, particularly an officer appointed to have charge of an institution. The Lord Warden of the Cinque Ports (q.v.) is an appointment of high honour, held by a succession of elder statesmen and, since 1978, by the Queen Mother. In the fourteenth and fifteenth centuries there were Lords Wardens of the Marches of England and Scotland, The head of All Souls College, Oxford, is designated The Warden, and The Master of a City Company is sometimes designated a warden.

Wardrobe. A department of the mediaeval English king's household which developed into an office of state and was important in the thirteenth and fourteenth centuries. The clerical staff became independent of those of the King's Chamber and, since they had custody of the King's ready money and jewels, readily acquired the power to keep accounts of the privy purse, make payments for household expenses, and receive money paid to the King as he moved around the country. The treasurer or keeper of the Wardrobe was the chief clerical officer of the household, kept the accounts and presented them at the Exchequer, while his subordinate, the controller, had custody of the Privy Seal. The power of the Wardrobe was attacked in the thirteenth century, and it was eclipsed in importance by the royal chamber and reverted to being merely a household department in the fifteenth century, when the raising of money began to depend more on grants authorized by Parliament.

T. F. Tout, *Chapters in the Administrative History of Mediaeval England*, vol. VI.

Wards and Liveries, Court of. A court established by statute in 1539 to look after the King's rights to wardships (q.v.) and other incidents of feudal tenure and, in 1542, becoming also surveyor of liveries. While it lasted, it generally exercised the jurisdiction delegated to the Chancellor of the Crown's powers and duties over persons of unsound mind. It was closed in 1646 and abolished in 1660.

H. E. Bell, *History and Records of the Court of Wards and Liveries* (1953); *Decrees and Cases in the Court of Wards and Liveries* 1553–81 (Selden Socy.).

Wardship and Marriage. In feudal law, the right of a lord to take back land held on tenure of knight service or grand serjeanty by a tenant on the latter's death if the heir were an infant or a woman, and to have the wardship of the heir or the right to determine the woman's marriage. In theory, these rights existed to protect the ward or the woman, but came to be rights valuable to the lord. In Glanvill's time, the wardships and marriage of both male and female wards were valuable rights, and Magna Carta asserted the lord's right in the matters boldly. The extent of the right was initially vague but was undoubtedly valuable and much law developed as to how they might be exercised.

Warehouse. A store, particularly one in which goods are stored, having been landed from ships, or while awaiting disposal. A bonded warehouse is one in respect of which the Commissioners of Customs and Excise have required the owner to give security by bond for the duty on goods deposited in warehouses without payment of duty or removed without payment of duty or in certain other cases. A Queen's warehouse is a place provided by the Crown or appointed by the Commissioners of Customs and Excise for the deposit of goods for security of them and of the duties chargeable on them. A warehouse-keeper's certificate that he holds certain goods on behalf of a named person is a document of title to those goods and used in business as representing those goods. It may be transferred and confers authority on the transferee to receive the goods referred to in the documents. In the case of mercantile agents in possession with the consent of the owner, sellers in possession after sale, and buyers in possession of the goods or documents of title thereto, a warehouse-keeper's certificate may pass by delivery, or endorsement and delivery, entitling the possessor to receive or transfer the goods represented thereby.

Warnkönig, Leopold August (1794–1866). German jurist and author of *Institutiones juris Romani* (1860), *Commentarii juris romani privati* (1835) and other works on Roman law, *Versuch einer Begrundung des Rechts durch eine Vernunftidee*, *Juristische Enzyclopädie* (1853), and *Philosophiae juris delineatio*.

Warrandice. In Scottish conveyancing, the obligation whereby a party conveying a subject or interest is bound to indemnify the grantee in the event of the latter's eviction, or of claims and burdens being made effectual against the subjects, arising out of obligations or transactions prior to the date of the conveyance. Warrandice might be personal or real; the latter is now abolished. Personal warrandice may be express or implied. Express warrandice may be simple, that the granter will do nothing inconsistent with the grant, from fact and deed, that the granter has done nothing and will do nothing contrary to his grant, or absolute, whereby the granter is liable for every defect in the right which he has granted. Unless qualified it extends to the full value of the subjects at the date of eviction.

Warrandice is implied to certain effects in certain kinds of grants; simple warrandice is implied in donations and absolute warrandice in onerous deeds. Warrandice is implied to stated effects by statute in particular circumstances.

Warrant. A written authority empowering a person to do some act, particularly to execute an arrest or a search; also a writ conferring a right or authority. A warrant could formerly only be executed in the borough or county in which it was issued. Since 1925, a warrant issued by a justice may be executed anywhere in England and Wales, but must be backed, i.e. indorsed by a justice of the other jurisdiction, if to be executed in Scotland, Northern Ireland or the Channel Islands, and conversely.

Warranty. Etymologically, and particularly in insurance law, something contractually guaranteed, and accordingly a material stipulation, breach of which is a repudiation of the contract. In sale of goods, the term was opposed to condition (q.v.) and, in England, in contracts generally, it denotes a term collateral to the main purpose of the contract, breach of which justifies a claim of damages but does not justify treating the contract as repudiated. In Scotland, on the other hand, warranty always denotes a material or fundamental term, breach of which does justify treating the contract as at an end.

In sale of goods, hire-purchase, and certain other contracts, certain warranties are implied by statute and can be excluded only to a limited extent.

In old English real property law, warranty was an obligation by the donor or feoffor of land to defend the donee or feoffee in the possession of the land and if he was evicted from it to give him other land of equal value.

Warren. Ground used for the breeding or keeping of hares or rabbits, and consequently the privilege of keeping and killing certain kinds of animals, particulary hares and rabbits on a piece of land. A free warren or warren in private hands was a franchise to have and keep beasts and fowl, or game, within the precincts of a manor or other known place. The owner of the warren has property in the beasts and fowls and a right to exclude all other persons from hunting and taking them. These franchises were abolished in 1971.

Warren, Charles (1868–1954). A Boston lawyer who became a leading authority on American constitutional law and Assistant Attorney-General of the U.S. His writings include *A History of the American Bar* (1911), *The Supreme Court in U.S. History* (revised, 1937), and *Congress, the Constitution and the Supreme Court* (1935).

Warren, Earl (1891–1974). Admitted to the California Bar in 1914, he was Attorney-General of California (1939–43), a candidate for Vice-President (1948), and Governor of California (1943–53) until appointed Chief Justice of the U.S. Supreme Court. He proved himself a liberal, notably in leading the Court in ruling in *Brown* v. *Board of Education* (1954) that segregation in schools was unconstitutional, and under him the Supreme Court set the U.S. on a new path in race relations, taking a stand against discrimination, and upholding law enforcement. After the assassination of President Kennedy, he headed the commission which investigated the murder. He retired in 1969.

L. Katcher, *Earl Warren* (1967); J. D. Weaver, *Warren: The Man, the Court, the Era* (1969); L. W. Levy, *The Supreme Court under Earl Warren* (1972).

Warrington, Thomas Rolls, Baron Warrington of Clyffe (1851–1937). Was called to the Bar in 1875 and acquired a good Chancery practice. In 1904 he became a judge of the Chancery Division and, in 1915, a Lord Justice of Appeal. He retired in 1926 but received a peerage and thereafter sat frequently in the House of Lords and the Privy Council. He was a sound, rather than a brilliant judge.

Warships. Warships are immune from the local jurisdiction of ports at which they call, though members of the crew are subject to the local jurisdiction if they go ashore but cannot be arrested.

Washburn, Emory (1800–77). Was admitted to the Bar in 1821, sat in the Massachusetts House of Representatives and was Governor of the state, 1853–54. In 1856 he became University Professor (later Bussey Professor) of Law at Harvard, and published *A Treatise on the American Law of Real Property* (2 vols., 1860–62), *A Treatise on the American Law of Easements and Servitudes* (1863), both very valuable works, and a *Sketches of the Judicial History of Massachusetts, 1630–1775.*

Washington, Treaty of (1871). An agreement between the U.S. and Great Britain providing for the submission to arbitration of boundary disputes, and the claims arising from the building and equipping of the *Alabama* and other Confederate raiders. It was significant for the expression of regret by the United Kingdom for permitting the escape of the *Alabama*, and the agreement on the rules of neutrality to be applied by the arbitration tribunal which was to sit to investigate the *Alabama* claims in the following year.

Wason, ex parte (1869), L.R. 4 Q.B. 573. After the failure of his action against *The Times* (*Wason* v. *Walter*, q.v.), Wason applied to a magistrate to prosecute certain peers for conspiracy to pervert the course of justice and to injure him by making untrue statements to prevent his petition to the House of Lords being granted. The magistrate, and on appeal, the Court of Queen's Bench, refused to do so, on the ground that no indictable offence was disclosed. Statements made by members of either House in the House could not be made the foundation of civil or criminal proceedings.

Wason* v. *Walter (1868), L.R. 4 Q.B. 73. The plaintiff, in a petition to the House of Lords, made a charge against the Chief Baron of Exchequer. In a debate, the plaintiff's charge was harshly criticized, and the criticism was published in a report in the *Times*, published by the defendant. The plaintiff sued for libel in respect of the report and of a leading article founded thereon. It was held that a fair and accurate report in a public newspaper of a debate in either House of Parliament was privileged and not actionable, and that, the matter being of public concern, comment thereon in an article was proper and privileged in the absence of malice.

Waste. In real property law, any land which has never been brought under cultivation as contrasted with arable or pasture land. The most important kind of waste was manorial waste, the part of the manorial lands subject to the tenants' right of common, and accordingly, waste is sometimes confused with land subject to tenants' rights of common.

In the law of torts, waste is any conduct doing permanent damage to the freehold or inheritance of land, or materially altering its nature or diminishing its value. It may be voluntary, when done deliberately, as by pulling down buildings, or permissive, when resulting from omission, as by allowing buildings to fall into disrepair. Waste may be checked by injunction or remedied by damages. Tenants, other than tenants in fee simple or in tail, are impeachable for waste, i.e. are liable for waste, unless the tenant's interest was granted him without impeachment for waste. A tenant from year to year is liable for voluntary waste only. Courts of equity always prevented so-called equitable waste which consisted in abuse of the privilege of not being impeachable for waste, as where a tenant for life without impeachment for waste committed wilful, malicious, or extravagant destruction, as by cutting timber not yet ripe.

Water. In tidal waters, the foreshore and the bed of the sea and of estuaries and rivers *prima facie* belong to the Crown but may have been granted to a subject. The water itself is *res nullius*.

In non-tidal waters, whether navigable or not, the ownership of the bed of the river is presumed to belong to the riparian proprietors, each *ad medium filum aquae*, but the water itself is not a subject of anyone's property.

The bed of pools and lakes entirely surrounded by land, and the water therein, belong entirely to the owner of the lands within which they are situated, or if on the boundary of two or more estates, to the owners of these lands in so far as their general lands abut on the shore.

Riparian owners are entitled to access to the water and to have the water reach their frontages in its natural quantity and state, and not to have it obstructed in its passage away from their lands. While the water is passing through, or along the boundary of his property each is entitled to take and use water for all ordinary and domestic purposes, such as washing, drinking and watering animals even to the extent, if necessary, of exhausting the stream entirely. He may also take water for extraordinary purposes, such as irrigation, a mill, bleach works or dye works, provided that it is taken at such times and in such manner that the flow of water is not perceptibly diminished and the water not unduly detained. But water may not be diverted from its channel for a non-riparian purpose, nor may a riparian owner dam up the stream or introduce obstructions to the flow of the water or obstruct the passage of fish.

Water percolating underground through the soil may be drawn off by any landowner, or diverted or appropriated; if one owner by draining his land dries up another's well or stream there is no actionable wrong.

An owner of land is not responsible for the consequences of the escape of water from his land by natural drainage, but if he has artificially gathered or impounded water on his land and it escapes he is strictly liable for the harm done under the principle of *Rylands* v. *Fletcher* (q.v.).

The management of water resources for human use is governed by a national policy effectuated by Regional Water Authorities and the Welsh National Water Development Authority. The pollution of water is an offence.

Water-bailiff. Formerly an officer in seaport towns whose duty was to search ships and, in London, an official who supervised fishing boats in the Thames, collected river tolls, and arrested persons on the river for debt or for crime. The title is now given to persons charged with enforcing fishing rights on rivers and trying to catch poachers.

Watkins* v. *U.S. (1957), 353 U.S. 178. Watkins testified as to his own former collaboration with the Communist party before the Committee of the House of Representatives on Un-American Activities but refused to disclose the names of other persons involved. He was punished for contempt. The Supreme Court held that abuse of the investigative process might lead to the abridgment of freedoms protected by the Constitution, and that such Committees had no power to expose for the sake of exposure. The decision goes some way to protect individuals from loss of reputation or punishment without conviction.

Watson, William, Baron Thankerton (1873–1948). Son of Lord Watson (q.v.), he was called to the Scottish Bar in 1899. He sat in Parliament in 1913–18 and 1924–29; in 1922 he was Solicitor-General and Lord Advocate in 1922–24 and 1924–29. In 1929 he became a Lord of Appeal where his work soon won respect and his judgments attained high respect.

Watson, William, Baron Watson of Thankerton, (1827–99) was called to the Scottish bar in 1851 and for long had no practice, but latterly was able to develop one. In 1874 he became Solicitor-General for Scotland, in the next year Dean of the Faculty of Advocates, and in the following year (1876), Lord Advocate and an M.P. In 1880 he became a Lord of Appeal as Baron

Watson of Thankerton. In that capacity he made a great reputation, not merely in Scottish, but in English and Privy Council cases, and has indeed been regarded as one of the greatest lawyers who ever sat on the British bench. His judgments are of very high authority, particularly on Scots law. His son followed in his footsteps as Lord Thankerton (q.v.).

Way. A right of passage, or the path along which that right is exercised. Ways are distinguishable into highways, (which are public rights of passage open to all subjects of the Crown along defined paths or roads, and consequently sometimes called the King's Highway) and private ways (which are rights of passage available only to particular persons or classes of persons). Private ways are incorporeal hereditaments. Ways are also distinguishable into footways, ways for riding or leading animals, and cartroads, a category which includes and implies the other two kinds. Private right of way may be claimed by grant, or by prescription and immemorial usage which presumes a grant, or by operation of law where a grant must be held implied. Such rights may be limited as to the extent of user authorized, e.g. footway, carriageway, etc., or as to the occasions on which the way may be used. A navigable river is held to be a kind of highway.

Wayleave. A right of way under, across or over land, as for running a pipeline under it, carrying goods across it, or suspending wires or cables above it. It is a kind of easement normally created by express grant or by reservation and a rent or charge may be exacted for the privilege.

Ways and Means business. The business of the House of Commons concerned with the provision of revenue, comprising resolutions for the imposition of new taxation, continuing an expiring tax, increasing an existing tax or extending the incidence of tax, and certain similar powers. The resolutions are given effect to by bills, of which the most important is the annual Finance Bill.

Ways and Means, Committee of. A committee of the whole House of Commons which, prior to 1966, considered the resolutions for the imposition of taxation and other measures to obtain revenue to meet national expenditure.

Weber, Max (1864–1920). German economist and sociologist, and professor at Berlin, Freiburg, and Munich. His *Roman Agrarian History* (1891) showed the interrelation of legal with social, political, and economic development. His *Economy and Society* (1922), itself part of a larger work, *Outline of Social Economics*, investigated the relations of the pursuit of wealth and law, and has been influential on many modern scholars. He also wrote *The Protestant Ethic and the Spirit of Capitalism* (1905) relating the rise of capitalism to the Protestant belief in the moral value of implementing worldly duties. A valuable selection is Rheinstein (ed.) *Max Weber on Law in Economy and Society* (1954).

Webster, Richard Everard, 1st Viscount Alverstone (1842–1915). Was called to the Bar in 1868 and became a Q.C. in 1878. He practised chiefly in commercial and patent cases. He was Attorney-General in 1885–86, 1886–92, and 1895–1900, and also an M.P. By this time he had a huge practice and appeared before such other tribunals as the Parnell Commission and the Behring Sea and Venezuela Arbitrations. In 1900 he became Baron Alverstone and Master of the Rolls but, later in the same year, became Lord Chief Justice, an office he held till his retiral in 1915. He was not a clever man, a learned lawyer, nor a good speaker, but succeeded by industry and forcible personality; as a judge he was undistinguished.

Wedderburn, Alexander, Lord Loughborough and Earl of Rosslyn (1733–1805). Son of a Scottish judge, Wedderburn became an advocate of the Scottish Bar in 1754 and began to attract attention in literature and debating. In 1757, however, he withdrew from the Scottish Bar and joined the English Bar. In 1760 he secured a seat in Parliament and gradually became recognized as one or the ablest counsel of the day. He kept his own advancement in view and changed sides when it suited him. In 1771 he became Solicitor-General and, in 1777, Attorney-General. In 1780 he became Chief Justice of the Common Pleas and Lord Loughborough. In 1793 he became Chancellor and, as such, supported all the government's repressive measures. In 1801 he was dismissed and made Earl of Rosslyn. Though a fine speaker and a sound judge, he subordinated everything to his own advancement and took a short-sighted view of many affairs of state, and turned his coat so often that no one trusted him. He is said to have lacked a real taste for legal learning. His judgments in equity are of more permanent value than those on matters of common law, though even of these, none made any striking developments in the principles of equity.

Weight of the evidence. The fact of the evidence for one party being of greater volume and cogency than that for the other. It is a fact relevant to the grant of a new trial that the jury's verdict is against the weight of the evidence.

Weights and measures. The weights and measures in use were originally customary and local, varying from one part of the country to another but, from an early date, the regulation of weights and measures has been held a prerogative of the Crown and effected by statute. A uniform system of weights and measures throughout the U.K. was effected by statute of 1824 which established Imperial measures, and after 1878 the use of local customary measures was illegal. The metric system of weights and measures was legalized in the U.K. in 1864 and increasingly enforced in the 1970s. The checking of weighing and measuring apparatus used in shops is a function of local authorities, which are now also responsible for enforcing trades description, consumer credit and other related legislation.

Weis, Philipp Friedrich (1766–1808). German jurist and author of *Historiae Novellarum literariae Particula* (1800).

Weiss, André (1858–1928). French jurist, president of the Institute of International Law, and judge and vice-president of the Permanent Court of International Justice. He wrote an important *Traité théorique et pratique de droit international privé* (1892–1905) and an elementary version (*Manuel*) for students, the first fundamental work in French on the subject.

Welfare state. The colloquial term for a society in which welfare and social security is extensively provided by the state through its departments and agencies. In early times, welfare was the responsibility of the group but this came to be increasingly assumed by the master or the feudal lord or, with the development of Christianity, by the Church. In England, a landmark was the Poor Law of 1601 which accepted the principles of individual moral responsibility and parochial administrative responsibility for the deserving poor, able and willing to work. The Poor Law Amendment Act 1834 affirmed the principle but substituted relief in kind (the poorhouse) with minimal cash payments to the destitute. In 1883–89, in Germany, provisions were introduced for sickness and maternity, injuries at work and old-age, invalidity and death.

In Britain, a patchwork of provisions developed, workmen's compensation in 1897, then old age pensions and other provisions and these were rationalized, largely on lines proposed by Sir William Beveridge in the years 1945–50, and subsequently substantially amended. Provision is now made for medical and hospital care, financial aid during unemployment, sickness and ill-health, financial provision for injury or illness resulting from work, for persons retired from employment, allowances for persons bringing up children, and supplementary benefits for those whose resources are inadequate. Related to welfare programmes are free education, school meals, subsidized housing, subsidized legal aid and advice schemes and other benefits. These schemes give rise to difficult social, political, economic and other problems.

This policy has given rise to a number of voluminous and very complicated bodies of law, mostly in the form of statutes and regulations but, in some cases, also of interpretative case-law.

In the U.S., the Social Security Act, 1935, later amended, covers the risks of old age, disability, and premature death. Unemployment insurance is a federal and state responsibility and workmen's compensation a state responsibility only. General assistance is administered at state and local level.

The details vary from country to country but very extensive welfare and security provision is now made by all Western countries and, indeed, in different ways and to different extents, by the great majority of organized states in the world.

Welsh Grand Committee. A standing committee of the House of Commons, including all the Members for constituencies in Wales together with not more than 25 other members chosen by the Committee of Selection chosen so as to make the Committee reflect the composition of the House as a whole. It considers such matters relating to Wales as may be referred to it by the House and reports thereon to the House.

Welsh law. West Wales was occupied by Irish-speaking people in the fifth century and, though frequently attacked, was not settled by the Norsemen. It developed into a land of petty princes and the only law was custom. Howel the Good (Hywel Dda, 910–950), styled King of all the Welsh, laid down a code or restatement of custom and this, amplified and re-edited by later generations, became the standard of tribal and personal relations throughout the country and was the greatest intellectual achievement of mediaeval Wales.

Thereafter, various principalities grew strong, sometimes one, sometimes another achieving primacy, but the same rules of Welsh law were applicable in any Welsh kingdom, though there were some local variations. But Wales was a legal unity. A Welsh law-book was suitable for use in any Welsh kingdom. The court of these law-books was peripatetic.

In 1282 Edward I of England defeated Llywellyn ap Gruffyd and conquered Wales. The *Statutum Walliae* of 1284 declared Wales to be annexed and incorporated into the Kingdom of England, and established six counties on the English model, three under the Justiciar of Snowdon and one under the Justiciar of Chester; no provision was made for the other two though the Justiciar of West Wales had authority there.

But the *Statutum Walliae* did not affect the parts of Wales outwith the Principality, namely the marcher lordships, the rulers of which claimed all the regalian rights of the Welsh rulers whom they had succeeded, and in 1354, it was provided that the Lords of the Marches of Wales should be attendant to the Crown of England, not to the Principality of Wales. In general, justice was administered in each separate jurisdiction and royal writs from Westminster did not run in Wales.

Even in areas subject to the *Statutum Walliae* the administration of law was not uniform, and Welsh law continued to be applied more in West Wales than in Gwynedd. The Statute applied English criminal law but retained Welsh procedure in real and personal actions, though the parties might opt for trial by jury.

In the lordships, the law varied from area to area; it was the custom of the march, a mixture, varying from place to place of Welsh and English law.

After the Glendower rising of 1400, numerous statutes were passed at Westminster limiting the power and privileges of Welshmen in England and otherwise reducing the Welsh to lower status. Henry VII, however, entered into agreements with Marcher Lords whereby they undertook to administer law impartially and granted charters of liberties to many Welsh districts. By 1520, moreover, many Welsh lordships had by inheritance or forfeiture got into the King's hands, and the development of the Council in the Marches of Wales gave rise to a body exercising from its headquarters at Ludlow a jurisdiction comparable to that of the Star Chamber.

The Laws in Wales Act, 1535 (truly 1536) united Wales with England and conferred on Welshmen all the laws, rights, and privileges of Englishmen and lands in Wales were to be inheritable by English law. The principality was divided into 12 counties for local government purposes, Monmouthshire becoming an English county, and the Welsh counties and boroughs were given seats in the English Parliament. Provision was made for administering justice according to the English pattern, and for the English language.

In 1542 the Laws in Wales Act placed the court of the President and Council of Wales, set up under Edward IV to suppress private feuds among the Lords Marchers and their retainers, on a legal footing, with very wide powers; it became an obsolete agency of oppression by the seventeenth century and was finally abolished in 1688. This Act also provided that courts of justice, under the title of 'the king's great sessions in Wales' should sit twice a year in every county in Wales, and four circuits, two for North, two for South, Wales were created, and lords lieutenant, sheriffs, coroners, and justices of the peace were appointed on the English model. All ancient Welsh laws and customs, at variance with English customs, were declared illegal; gavelkind was expressly abolished and replaced by primogeniture, and it was provided that all legal proceedings must be in English. But Wales remained distinct, in the retention of the name Principality, in the continuance of the Council in the Marches of Wales, and in the separate judiciary, while some statutes thereafter made special provision for Wales. Justice was administered through this system of great sessions until 1830 when they were abolished and replaced by the North and South Wales circuits (later the Wales and Chester circuit).

The Welsh language, however, continued and there were sharp religious contrasts with England, and from the later nineteenth century, special legislation for Wales began to appear. In 1919 a Welsh Board of Health was established—the first significant devolution of authority to Wales.

During and after World War II many government departments treated Wales as a unit and established offices there. In 1948 the Council for Wales and Monmouthshire was established as a consultative body; in 1951 a Minister for Welsh Affairs was appointed and, in 1964, a Secretaryship of State was created.

The Secretary of State for Wales, assisted by a Minister of State, has certain powers devolved to him and administered from Cardiff, and special provision is made for Wales in certain statutes, but unless otherwise stated, England in all legal contexts includes Wales, and there are no differences in private law, criminal law, or procedure. In Parliament since 1960, a Welsh Grand Committee has met to examine topics relating to Wales.

In 1942 the Welsh Courts Act authorised the use of the Welsh language in any court in Wales or Monmouthshire and the Welsh Language Act 1967 has provided that in any legal proceedings in Wales or Monmouthshire, the Welsh language may be spoken by any person who desires to use it. Provision is made for interpretation.

A. Owen (ed.) *Ancient Laws and Institutes of Wales* (1841); T. P. Ellis, *Welsh Tribal Law and Custom in the Middle Ages* (1925–6); H. Lewis, *Ancient Laws of Wales* (1892); A. W. Wade-Evans, *Welsh Mediaeval Law* (1909); I. Bowen, *The Statutes of Wales* (1908); C. A. Skeel, *The Council in the Marches of Wales* (1904); W. L. Williams, *The King's Court of Great Sessions in Wales* (1916); W. R. Williams, *History of the Great Sessions in Wales, 1542–1830* (1899); W. R. Williams, *Parliamentary History of the Principality of Wales, 1542–1830*; H. Owen, *The Administration of English Law in Wales and the Marches* (1900).

Welsher. A person who is entrusted with money to await the outcome of an event on which the money has been staked and who decamps with the money.

Welwood or Welwod, William (?–1622). Professor successively of Mathematics and of Law at St. Andrews, who published, in 1590, a treatise on *The Sea Law of Scotland* and, in 1594, a treatise *Juris Divini Judaeorum ac Juris Civilis Romanorum Parallela*, an early study in comparative jurisprudence. In 1597 he was removed for his views on ecclesiastical prerogatives and, despite royal intervention, it is doubtful if he was ever readmitted. In 1613 he published *An Abridgment of All Sea Lawes*, which included a comparison of the codes of Oleron and Wisby with the Roman law. This was reprinted, without his name, in 1686 in Malynes' *Consuetudo vel Lex Mercatoria*. In 1616 he republished part of it as *De Domino Maris*, upholding English claims to the supremacy of the Channel, a later edition of which provoked a reply from Graswinkel, *Maris liberi vindiciae* (1653).

Wensleydale, Lord. See PARKE, JAMES.

Wensleydale Peerage Case (1856), 5 H.L. Cas. 958. The Crown, to strengthen the judicial force in the House of Lords (there being then no law lords), sought by letters patent to create Sir James Parke (a Baron of the Court of Exchequer) (q.v.) a baron for life with a seat in the House of Lords, and issued a writ of summons to Parliament. The House of Lords Committee for Privileges held that neither act entitled Parke, B. to sit and vote in Parliament. The Queen then conferred on him an ordinary (hereditary) peerage. The creation of law lords having life peerages was authorized by the Appellate Jurisdiction Act, 1876.

Wergild. In mediaeval Germanic law, the amount of compensation payable by a person causing harm to another to that other or, if he caused death, to the victim's family. Wergild accordingly represents the first step away from the blood-feud, compensation in lieu of revenge. At first indeterminate, wergild early came to be fixed by law and mediaeval legal codes such as some of the *Leges Barbarorum* frequently contain extensive tariffs of wergild. In some countries, the wergild of a person was related to his status in society, that of a lord being much greater than that of an ordinary man.

In some cases, part of the wergild was payable to the King or to the lord as they had lost a subject or a vassal. Other claims were often related to wergild. One was *bot*, which was compensation for damage done and also extended to the repair of houses for those living on an estate. Another was *wite*, a fine payable to the King as penalty for the deed in addition to wergild, if the harm was intentional.

In time, as royal power increased, it became no longer permissible to expiate crimes by paying compensation and wrongdoers came to be punished by kings and lords, while victims and their relatives might claim damages as compensation for the loss to them.

Wernher, Jean Balthasar von (1675–1742). German jurist, professor at Wittenberg, counsellor at the imperial court at Vienna, and author of *Selectae observationes forenses* (1710) and *Compendium juris quo Germani hodie ac imprimis Saxonis in foro utuntur* (1728).

Wesel, Abraham à (1635–80). A Dutch jurist and author of *Commentarius ad novellas constitutiones Ultrajectinas* (1666), *Tractatus de connubiali bonorum societate et pactis dotalibus* (1674), and *Tractatus de remissione mercedis* (1678).

Wesenbeck, Matthaeus (1531–86). Low countries jurist and humanist and author of *Isagoge in libros IV Institutionum juris civilis*; *Commentarius in Institutiones*; *De Feudis*: *Tractatus Novus* (1583), and *Commentarius in Pandectas Juris Civilis et Codicis Justinianei libros VIII* (1589). His father, Petrus (1487–1562), and brothers Andreas (1527–1569) and Petrus (1546–1603) were also distinguished jurists; the last wrote an *Annotationes ad Pandectas*.

West, Richard (?–1726). Was called to the English Bar in 1714, was one of the managers of the impeachment of Lord Macclesfield and went to Ireland as Chancellor in 1725, but survived for only a year. He also wrote books on treason and bills of attainder, and on the manner of creating peers.

West Council of the. A council set up in 1537, modelled on the Council of the North (q.v.), with jurisdiction over Somerset, Dorset, Devon, and Cornwall. It administered law and gave civil remedies, but it disappeared about 1547.

Westbury, Lord. See BETHELL, RICHARD.

Western European Union. In 1954 the governments of seven Western European states created the Western European Union, superseding the Brussels Treaty Organization of 1948. Like N.A.T.O., W.E.U. is an organization for collective self-defence. It was directed to establish an agency for the control of armaments and given powers of decision by majority vote. The Council meets at ministerial or ambassadorial level. It reports to the Assembly of W.E.U., which is composed of the same representatives as those appointed by the member-states to the Consultative Assembly of the Council of Europe, and which may make recommendations or transmit opinions to the Council of W.E.U. on matters within its competence. It cannot overthrow the Council of Ministers. In 1959 its social and cultural work was transferred to the Council of Europe.

Westlake, John (1828–1913). Was called to the Bar in 1854, took silk in 1874 and, in 1888, succeeded Maine in the Whewell Chair of international law at Cambridge, a post he held till 1908. His *Treatise on Private International Law* (1858, rewritten 1880) was the first English book to systematize this branch of law and has been regarded with respect ever since. He also wrote *International Law, Peace* (1904) and *War* (1907) which has not survived so well. He was a member of The Hague International Court of Arbitration from 1900 to 1906. He was also a founder of the Institut de Droit International in 1873 and of the *Revue de Droit International et de Législation Comparée*, the first periodical on international law, in 1869, and an M.P. His *Collected Papers* were published in 1914.

Westminster Hall. A hall in the Palace of Westminster, originally the great hall of the palace, built by William Rufus, completed in 1099 and altered by Richard II, with a fine open timber roof. It is 238 feet long, 67 feet broad and 90 feet high. Simon de Montfort's Parliament and the Model Parliament met there. Many famous trials, such as of Wallace, Richard II, Sir Thomas More, the Gunpowder Plot conspirators, Charles I, Titus Oates, Warren Hastings, and Queen Caroline were held in the hall. The English common law courts sat in the hall until the Royal Courts of Justice in the Strand were opened in 1884. More recently, it has witnessed the lying-in-state of deceased kings of Britain and great statesmen, and been used for special parliamentary occasions.

Westminster, Statutes of. Three important English and one important U.K. legislative measures. The Statute of Westminster I (1275) protected church property from the King and nobles, provided for the freedom of popular elections of sheriff, coroner and conservators of the peace, contained a declaration to enforce the provision of Magna Carta (1215) against excessive fines, and amended criminal law and procedure.

Westminster II (1285) had as its first chapter *De Donis Conditionalibus* (q.v.).

Westminster III (1290) is usually known as *Quia Emptores* (q.v.) and forbade subinfeudation of land.

The Statute of Westminster (1931) fundamentally altered the relationship between the U.K. and the British Dominions of Canada, Australia, New Zealand, South Africa, Irish Free State, and Newfoundland, by providing that the Colonial Laws Validity Act, 1865, should not apply to legislation made by the Parliament of a Dominion, and that the U.K. Parliament would not legislate for a Dominion save at its request and with its consent.

Wharton, Francis (1820–89). Admitted to the Pennsylvania Bar in 1843, he was successful in practice, but later became a lay preacher, and then an Anglican clergyman and, latterly, a professor in the Episcopal Theological Seminary, Cambridge, Massachusetts. In his earlier years, he wrote several legal works, including a *Treatise on the Criminal Law of the United States* (1846), *State Trials of the United States* (1849), and *Treatise on the Law of Homicide* (1855). Later, as well as books on religious themes, he wrote a *Treatise on the Conflict of Laws* (1872), which made his reputation, *A Treatise on the Law of Negligence* (1874), *A Commentary on the Law of Evidence* (1877), *The Philosophy of Criminal Law* (1880), and a *Commentary on the Law of Contracts* (1882). In 1885 he became chief of the legal division in the Department of State and was entrusted with the compilation of *A Digest of the International Law of the United States* (3 vols., 1886), much of which was later incorporated by John Bassett Moore (q.v.) in his *Digest*, and the editing of *The Revolutionary Diplomatic Correspondence of the U.S.* (6 vols., 1889). Many of his books passed through several editions, and his *Conflict of Laws* is still of importance.

H. E. Wharton; *Francis Wharton, A Memoir.*

Wheatley, John, Lord, of Shettleston (1908–). Called to the Scottish Bar in 1932, he was Solicitor-General for Scotland in 1947, Lord Advocate, 1947–51, a judge of the Court of Session, 1954–72, and Lord Justice-Clerk from 1972. He was Chairman of the Royal Commission on Local Government in Scotland, 1966–69 and became a life peer in 1970.

Wheaton, Henry (1785–1848). Became a leader of the New York Bar. He reported the cases argued before the Supreme Court between 1816–27, reports which are noteworthy for their full notes, and served on the Commission to revise the laws of New York. He also published two *Digests of Decisions* and other books. He then served in diplomatic posts in Denmark (1827–35) and Prussia (1835–46), negotiated important treaties, and wrote a *History of the Northmen* (1831). He also wrote an *Elements of International Law* (1836), and a *History of the Law of Nations* (in French, 1841, in English, 1845), which have given him a place among the foremost American legal writers.

Whip. In fox-hunting, a whipper-in has the task of keeping hounds from straying from the pack. In the eighteenth century, the word was borrowed, abbreviated and applied to those members of both Government and Opposition parties whose duty is to ensure an attendance of their members for divisions and other necessary occasions, to have them furnished with information and to supply names of suitable members for service on Standing and Select Committees. The Government and Opposition Chief Whips arrange the business and negotiate when time is sought for the discussion of some special piece of business. The Government

Chief Whip holds the office of Parliamentary Secretary to the Treasury and other Whips the offices of junior Lords Commissioners of the Treasury, or of certain political offices in H.M. Household. There are also unpaid Assistant Whips. The Chief Whips do not usually take part in debates in the House, so as to minimize friction between them.

The term is also used of a document issued weekly by the two Chief Whips to their members stating the business for the following week. Sending the Whip to a member is a recognition that he is recognized as a member of the party in the House, and withdrawal signifies withdrawal of recognition or virtual expulsion from the party. Each item of business on the Whip is underlined, in modern practice from one to three times (though four and five underlinings were known in Victorian times). In general, a one-line whip indicates that no division is expected, a two-line that the business is of some importance, that a division may take place and that the member should attend, and a three-line that the business is very important, that a division is certain or very probable and that he must attend unless prevented.

Whipping. At common law whipping was a recognized part of the punishment for misdemeanours. It was inflicted in public, the victim being sometimes fastened to the tail of a moving cart. Until 1820 it could be inflicted on women as well as men. In the twentieth century, its use has been much curtailed and it was abolished in 1948.

White, Edward Douglass (1845–1921). Served in the Confederate Army, was admitted to the Bar in 1868, and practised in New Orleans. After service as a State Senator and an Associate Justice of the Supreme Court of Louisiana (1879–80) he was a U.S. Senator, 1891–94, before becoming, first, an Associate Justice (1894–1910) and then Chief Justice (1910–21) of the U.S. Supreme Court. He was the first justice to be promoted Chief Justice. White's appointment was based solely on merit. Scholarly and studious, he was a powerful advocate, successful in practice, and became a highly respected judge. His chief contribution to the jurisprudence of the Court was the principle that restraint of trade by a monopoly must be 'unreasonable' to be struck down by the Sherman anti-trust Act.

Klinkhamer, *Edward Douglass White* (1943); Umbreit, *Our Eleven Chief Justices*, Ch. 9.

White House. The official residence of the President of the U.S. in Washington, D.C. Built in the neo-classical style in 1792–1800 it was burned during the British invasion of 1814 and later rebuilt and enlarged. The name is derived from its being built in grey-white limestone or from having long been painted white. The name is often used as meaning the President, or the President and his Cabinet.

White Paper. The common term for a government memorandum, report, statement of intended policy, or the like issued in a white paper cover rather than the thicker blue paper cover of a Blue Book (q.v.).

White slave traffic. Traffic, particularly international, in prostitution, struck at by various international conventions.

White-collar crime. The term for crime committed by persons of good status, holding managerial or professional positions, in connection with their ordinary occupations. These include frauds, swindling, dishonest financial dealings, and the like. Such crimes have been estimated to involve more money than conventional crimes, such as theft and robbery.

Whitebonnet. A person who, at an auction sale, in collusion with the seller, makes bids with the intention of pushing up the price, not with the intention of buying.

Whitefriars. A place in London, between the Temple and Blackfriars, which was formerly a sanctuary (q.v.) and a place of privilege from arrest.

Whitehall. A street in Westminster, London, extending from Trafalgar Square to Parliament Square. Along it, or adjacent to it, are a large number of important government offices, and the name is accordingly a common label for the central government.

Whiteside, James (1806–76). Was called to the Irish Bar in 1830, became Solicitor-General in 1852, and Attorney-General in 1858–59, and in 1866. From 1866 to the time of his death he was Chief Justice of the Queen's Bench in Ireland.

Whitley councils. Bodies composed of representatives of management and of workers for the promotion of better industrial relations and, latterly, for wages negotiations. They were named after J. H. Whitley, chairman (1916–19) of the committee which recommended their creation.

Whitshed, William (1679–1727). Was called to the Irish Bar, sat in the Irish Parliament, and was Solicitor-General, 1709–11. He was Chief Justice of the King's Bench in Ireland, 1714–27, and then of the Common Pleas, 1727. He sought to have the

Drapier's Letters pronounced seditious and was lampooned by Swift.

Whitsunday. In Scotland, 15th May, one of the term days on which formerly many leases expired and many terms of employment ended.

Whole blood. In deducing relations, kinship derived from a pair of common ancestors, as contrasted with half-blood.

Wickens, Sir John (1815–73). Was called to the Bar in 1840 and made slow progress at first, but later became a leading equity practitioner. In 1871 he became a Vice-Chancellor but his health gave way shortly thereafter. His reputation was of sound knowledge, but slow and too close adherence to precedents.

Wicquefort, Abraham de (1598–1682). Dutch diplomat, long resident in Paris, and author of *L'ambassadeur et ses fonctions* (1681) and other works on history and diplomacy.

Widgery, John Passmore, Lord (1912–). Practised as a solicitor from 1933 to 1946 when he was called to the Bar. He took silk in 1958, became a judge of the High Court in 1961, a Lord Justice of Appeal in 1968, and Lord Chief Justice of England,.and a life peer, in 1971.

Wielant, Philip (?1441–1520). A Belgian jurist, counsellor to the Council of Flanders and member of the Council of Malines, and author of *Tractaet van den Leenrechten na da Hoven van Vlaenderen* (1491) *Practijcke civile* (1519), and *Practijcke crimineele* (1558); both the latter were widely used and taken by Damhouder (q.v.) as the basis of his works.

Wiener-Neustadt, Raymund von (fourteenth century). Author of *Summa Legum*, a textbook of private and criminal law composed about 1345 on the basis of Italian-Romanist literature but taking account of Germanic legal institutions, which was very influential on legal development in Austria, Hungary, and Poland.

Wig. An artificial head of hair, worn in various forms from very ancient times. In western Europe, wigs for men came into widespread use in the seventeenth century and certain professions adopted wigs as part of their official costume. The periwig was introduced into England from France in 1663 and was soon adopted in the courts as a part of fashionable dress. About 1705, white and grey powdered wigs became more popular in court as well as outside. After 1720 the fashion came to be for smaller wigs but judges kept the larger wigs as befitting their dignity. Accordingly, between 1720 and 1760, the judicial wig became a part of official dress as it had been retained after going out of fashion generally, and by this time it had become recognized as peculiar to judges and serjeants. A little later, however, judges took to wearing shorter wigs for everyday occasions and adopted what was then the fashion of a short wig with two queues hanging down behind. Barristers took to short wigs in 1730–40 and, about 1780, adopted the fashion of wearing a wig with two or three horizontal curls running round the back of the head and two queues hanging down behind.

In Scotland, white full-bottomed wigs seem to have been adopted by judges in the early eighteenth century, but both judges and advocates seem to have adopted short wigs from about 1750.

Irish legal dress never seems to have differed from English.

Today the wig is, along with the appropriate gown and neck-wear, a part of the uniform of judges and members of the Bar, both senior and junior, in all parts of the U.K. The full-bottomed wig (down to the shoulders) is worn by judges and Q.C.'s but not barristers on ceremonial occasions, the short wig being worn by all for everyday appearances. Wigs and gowns are worn also by the Speaker of the House of Commons and by some other legal officials as part of their official dress.

Contrary to popular belief, the wig is neither heavy nor uncomfortable and judges and counsel would feel strange without them. They are made of horsehair or, today, nylon.

See also GOWNS.

Wigmore, John Henry (1863—1943). As a student Wigmore was a founder of the *Harvard Law Review*, practised law in Boston, and later was Dean of the Law Faculty at Northwestern University, 1901–29. He was author of the massive *Treatise on Evidence* (1904, 3rd edn., 10 vols., 1940), other books on evidence, and a *Panorama of the World's Legal Systems* (3 vols. 1928). He also published other works, edited *Greenleaf on Evidence*, and was co-editor of various valuable series, such as the *Essays in Anglo-American Legal History*, *Evolution of Law* series, *Modern Legal Philosophy* series, and *Continental Legal History* series.

Wigram, James (1793–1866). Was called to the Bar in 1819 and practised in the Chancery court, latterly as a leader. He became a Vice-Chancellor in 1841 and a Privy Councillor in 1842, but had to resign from failing sight in 1850. He was well regarded as a judge, his judgments being lucid and well-founded. In 1831 he published a useful book on *Extrinsic Evidence in Aid of the Interpretation of Wills* and in 1836 a book on *Discovery*.

Wilberforce, Richard Orme, Baron (1907–). Called to the Bar in 1932, he was a judge of the Chancery Division, 1961–64, and a Lord of Appeal in Ordinary from 1964. He wrote the *Law of Restrictive Trade Practices* (1956).

Wild, Jonathan. A notable master-criminal in London in the early eighteenth century. He showed thieves how to manage their business more successfully and had a number of gangs under control, maintained discipline among them, and arranged the arrest of the recalcitrant. He became the receiver-in-chief of all stolen goods and set up an office for the recovery of missing property at which despoiled persons might recover their stolen property, at a price, if they asked no questions, and sought out and prosecuted offenders, who would not recognize his authority, for the sake of the reward which could then be obtained. Statute in 1717 made his conduct a capital offence. In 1725 he was arrested for helping a highwayman escape from custody and executed.

Wilde, James Plaisted, Lord Penzance (1816–99). A nephew of Lord Chancellor Truro (see next entry), was called to the Bar in 1839 and became a Baron of Exchequer in 1860. In 1863 he became judge of the Probate and Divorce Court. He became a peer in 1869 and retired in 1872 but sat thereafter as judge under the Public Worship Regulation Act. His best work was done in the Probate and Divorce Court but he supported some important law reform measures in the House of Lords.

Wilde, Sir Thomas, Lord Truro (1782–1855). Began life as an articled clerk to his father and later set himself up in a commercial business. He then became a special pleader and was called to the Bar in 1817. He had both ability and industry and entered Parliament in 1831. He was made Solicitor-General in 1839 and Attorney-General in 1841, and was reappointed in 1846, and shortly thereafter, became Chancellor as Lord Truro and held that office till 1852. As such, he initiated several measures for the reform of the Court of Chancery. As a judge he showed patience, impartiality, and acute intellect, and made some contributions to the development of the law.

Wilful. Intentional or deliberate, as contrasted with negligent or accidental.

Wilkes, John (1727–97). A journalist of profligate habits who entered Parliament for Aylesbury in 1757. In 1762 he began a periodical, the *North Briton*, which virulently attacked Bute, the King's favourite, and contributed to his downfall. In issue 45, in 1763, he fiercely attacked the King's Speech at the opening of Parliament. The King treated the article as a personal insult and a general warrant was issued for the arrest of the persons involved. Forty-eight persons were arrested as well as Wilkes. But Pratt C.J. held his arrest was breach of privilege and that general warrants were illegal.

He was then expelled from Parliament for publishing an obscene libel, the *Essay on Woman*, and lived in France till 1768 when he returned and was elected for Middlesex, but, having been imprisoned for the *Essay on Woman* and for No. 45, was thrice more re-elected and each time expelled by the House. This allowed him to pose as champion of the unenfranchised middle and lower classes and he organized support for a party of reform. After 1774 he was unmolested in Parliament and also became Lord Mayor of London. During the American Revolution he supported the cause of the colonists. In the Gordon Riots he took a leading part in crushing the disturbance. Thereafter his popularity declined; he did not seek re-election in 1790 and died insolvent in 1797. Despite his early indulgence in obscenity he was latterly conspicuously honest and honourable, and he secured legal principles of much importance, the illegality of general warrants, the freedom of choice of the electorate, and the freedom of the press.

H. W. Bleackley, *Life of John Wilkes* (1917); G. F. Rude, *Wilkes and Liberty: A Social Study of 1763 to 1774* (1962); I. R. Christie, *Wilkes, Wyvill and Reform: The Parliamentary Reform Movement in British Politics, 1760–1785* (1962).

Wilkins, David (1685–1745). Professor of Arabic at Cambridge and librarian to the Archbishop of Canterbury, he produced the definitive edition of the writings of Selden (*Johannis Seldeni, Jurisconsulti Opera Omnia*, 3 vols, 1725–26), while his *Concilia of the English Church* (4 vols, 1737) superseded Spelman's work on that theme and is itself not yet wholly superseded. He also produced an edition of the Anglo-Saxon laws *Leges Anglo-Saxonicae Ecclesiasticae et Civiles* (1721) incorporating Spelman's *Codex Legum Veterum* and including the collections of Anglo-Saxon customs made after the Conquest.

Will. The mental faculty or power or capacity by which a person may control his conduct, choose his course of action, and direct it towards the attainment of certain ends. Willing is a capacity distinct from knowing and from reasoning. As it implies capacity to choose one course of action rather than another, it cannot be attributed to young children or to persons of unsound mind. Persons of strong or weak will have strong or weak inclinations to determine their own course of action, free from persuasion, and even in face of opposition.

Willing is legally important in that, in general, responsibility and legal liability attach to a person only for conduct which he has willed and not for conduct which he did not will and over which he had no control, e.g. sneezing, acts done when sleepwalking. A person's will may also be overborne by duress, or be misdirected by reason of mistake or of fraud or undue influence practised on him. The will, moreover, must, to be legally effective, not be confined to choosing a course of action and formulating an intention, but be expressed or made manifest, sometimes informally, as by nodding to an auctioneer, sometimes formally as when making a will and some kinds of contracts. What a person has willed can indeed only be discovered from the external manifestations of his will.

Will or testament (synonymous terms). A declaration in a particular form by a person of his wishes as to certain matters, normally as to the disposal of his property, to take effect on or after his death. During his lifetime, it is an expression of intention only and may be revoked or altered in any way. To make a valid will a person must be of full age and sound mind, and the expression of intention must be in writing signed by or on behalf of the testator in the presence of, or acknowledged before, two witnesses at one time, who themselves sign as witnesses. A printed form may be used, or the blanks in such a form completed. Persons in active military service may make wills orally or informally. A will, unless made in contemplation of marriage, is revoked by marriage, and it may be revoked by physical destruction, by another later will or codicil, or duly executed writing showing an intention to revoke. A person may leave several wills, or a will and several codicils, and all those not revoked must be construed as one testamentary disposition.

A testator may, by will, dispose of any or all of his property to whomsoever he wishes, creating any such interests as the law allows in any part thereof, outright gifts, gifts subject to conditions, options, life or other terminable interests, successive and future interests, and so on. But the court may under statute make provision for the maintenance of a dependant not adequately provided for. If a donee accepts a conditional gift he must take it subject to the conditions and other burdens unless the condition be contrary to public policy and void. A donee may disclaim a bequest. A testamentary gift may be adeemed or taken away from the donee by a subsequent disposition of the gift by the testator in his lifetime to another, by a change in its ownership or nature as when it ceases to belong to the testator, or by a presumption that the testator did not intend to provide double portions for his children. A gift lapses if the legatee predeceases the testator.

A will normally disposes of the testator's property by way of legacies, special legacies of determinate subjects or general legacies of money or generic things, and provides in a residue clause for the disposal of all the testator's property not already disposed of.

A testator's property, real and personal, devolves on his personal representatives for administration, and it is customary to nominate one or more persons to act as executors, but failing nomination the court will grant persons letters of administration *cum testamento annexo* to administer the estate and give effect to the will.

Courts are frequently faced with the problems of the construction or interpretation of wills. The cardinal rule is that the testator's intention must be ascertained from the words and phrases of his testamentary writings, and given effect to, so far as consistent with law and with the circumstances which have happened. A will is construed as speaking from the moment of death, and evidence is admissible to enable the court to put itself in the testator's position as to persons and facts known to him when he made his will. In cases of doubt the court frequently relies on various presumptions as to intention and the meaning of words. If the court reaches the conclusion that there is no effective disposition of the estate or part of it, it falls into intestacy.

In Scots law, a nuncupative or oral bequest is valid up to £100 Scots (£8.33 sterling.) but otherwise a will must be in writing. A holograph will, entirely in the testator's handwriting and signed by him, is valid as is a will signed by the testator before two witnesses who also sign. A minor has testamentary capacity. The privilege of persons in active military service making wills informally does not apply.

A will is not revoked by marriage, but is by physical destruction, by another later will or codicil, a duly executed writing showing an intention to revoke, or by the subsequent birth of a child not provided for.

By virtue of the doctrine of legal rights (q.v.), a will may operate only on the part of the testator's estate not required to satisfy claims for legal rights. Persons entitled to legal rights may challenge the will so far as necessary to make good their claims.

T. Jarman on *Wills*; H. Theobald on *Wills*; J. McLaren on *Wills and Succession* (Scotland).

Will, At. A term for the duration of a partnership, tenancy, or other relationship, signifying that it is to endure so long as the parties will it, but that it can be terminated at any time by either or any party giving reasonable notice to the other of intention to terminate the relationship.

Willes, Edward (1702–68). Was called to the English Bar 1727 and went to Ireland as Chief Baron of Exchequer 1757 till his retiral in 1766.

Willes, Sir James Shaw (1814–72). Practised as a special pleader, was called to the Bar in 1840, and built up a leading practice in the court of Exchequer. He edited Smith's *Leading Cases* and helped to draft the Common Law Procedure Acts. Subsequently, he was a member of the Indian law commission (1861) and of the English and Irish law commission (1862). In 1855 he became a judge of the Court of Common Pleas and, in 1871, a Privy Counsellor, but overwork produced a breakdown and he shot himself. He was probably the most learned lawyer of his day, and also a fine scholar. Many of his judgments are classics, being clear, learned, and authoritative.

Manson, *Builders of our Law*, 127.

Willes, Sir John (1685–1761). He became a fellow of All Souls before being called to the Bar in 1713. He became Attorney-General in 1734 and was Chief Justice of the Common Pleas, 1737–61. Though of bad character, he aspired to be Chancellor and intrigued to that end; he was offered the Great Seal in 1756 but the King refused to ennoble him and the offer fell. He left a set of reports, which are among the best of the period and have always enjoyed a high reputation. He was a lawyer of high reputation in his day, though politically an unscrupulous intriguer.

William I, the Conqueror (1028–87). Succeeded as Duke of Normandy in 1035 and had a long fight to recover lost ducal rights and revenues and to reduce his duchy to order. In 1066, to enforce a bequest of the kingdom of England made to him by Edward the Confessor he invaded England and secured the crown. Then followed the Norman Conquest, the replacement of native lords and landowners by Normans, the introduction of the feudal system of government and land tenure, the increased power and influence of the Church, and the stronger enforcement of law and order. He did not subvert the existing law but allowed the native English to retain their courts and laws though in time most business went to feudal courts, and the King's law superseded local laws as the common law of England. In 1086 he ordered the economic and tenurial survey of the whole kingdom, the results of which are summarized in *Domesday Book*. He was a strong, vigorous and rough ruler, who did much to consolidate the kingdom. He was not a great legislator and the *Leis Willelme*, probably compiled in Mercia about 1120 by a private individual, containing some rules of the old English law as they were understood under the Norman kings, as well as some articles borrowed from Justinian and a translation of parts of the law of king Canute, does not indicate large-scale legislation. But the system of government introduced set English law off on a different tack and fundamentally affected its development.

D. C. Douglas, *William the Conqueror* (1964); F. Barlow, *William I and The Norman Conquest* (1965); F. Barlow, *The Feudal Kingdom of England 1042–1216* (1955).

Williams, John (1777–1846). A good classical scholar, was called to the Bar in 1804, sat in the House of Commons, 1822–32, became a Baron of Exchequer in 1834, and a judge of the King's Bench, 1834–46. When in Parliament, he constantly attacked the delays of the Court of Chancery and Lord Eldon's dilatory habits.

Willielmus de Cabriano (twelfth century). An Italian glossator and author of a *Summa Contrarietatum Casus Codicis.*

Williston, Samuel (1861–1963). Graduated from Harvard in 1888 and taught law there, 1890–1938. His major contributions to law are *The Law of Sales* (1909, 4 vols. 1924), and *The Law of Contracts* (1920, 9 vols. 1936–38), which have done much to promote uniform commercial practice in and between the States. He also played a leading part in drafting many of the Uniform Acts approved by the Commissioners of Uniform State Laws between 1905 and 1920.

Willoughby, Westel Woodbury (1867–1945). U.S. publicist, wrote extensively on political theory and jurisprudence, notably *The Constitutional Law of the United States* (1910), *The Fundamental Concepts of Public Law* (1924), and *The Ethical Basis of Political Authority* (1930).

Wilmot, Sir John Eardley (1709–92). He was called in 1732 and, though he acquired a good practice, was diffident about his abilities and resisted promotion. In 1755, however, he became a judge of the King's Bench, and became Chief Justice of the Common Pleas, 1766–71. Several times he declined the Great Seal. He was an outstanding lawyer and the ablest of the puisne judges under Mansfield. He left notes of cases heard by him, later published by his son.

Wilson, James (1742–98). Practised law in Philadelphia and supported the revolutionary cause. As a member of Congress (1782–83 and 1785–87) he laid the foundations for the Bank of the U.S., and he was, by reason of his grasp of law and economics, one of the most influential delegates to the Federal Constitutional Convention. He was later the main author of the Pennsylvanian constitution of 1790. In 1789 he was made an Associate Justice of the U.S. Supreme Court and, in 1790, the first Professor of Law at the College of Philadelphia. In his lectures, he departed from Blackstonian sovereignty and based law on the consent of the individual. He essayed, but did not complete a digest of the laws of Pennsylvania. He was greedy and ambitious. His major contribution to the founding of the Republic was his grasp of the idea of federalism.

Wilson, Woodrow (1856–1924). Practised law briefly before taking up academic work, becoming, in 1890, Professor of Jurisprudence and Political Economy at Princeton and, in 1902, president of that university. There he made great improvements in the organization of studies. In his academic years, he published *Congressional Government* (1885) and *The State* (1889). In 1910 he became Governor of New Jersey and, in 1912, President of the U.S. as a conservative reformer. He achieved many legislative reforms, not least the creation of the Federal Trade Commission and the passing of the Clayton Anti-Trust Act. During World War I he consistently strove for recognition of the rights of neutrals and explored possibilities of mediation. In 1917 events forced him to declare war on Germany, and he devoted all his energies to aid military operations. His own major contribution to victory was the formulation of war aims and, in January 1918, he formulated his famous Fourteen Points (q.v.), enunciating ideals for securing and preserving post-war peace, and it was to him that the Germans turned in October 1918; his negotiations saved Germany from invasion and unconditional surrender, but accelerated the end of hostilities. He was a leader at the Paris Peace Conference and forced acceptance of the Covenant of the League of Nations as an integral part of the treaty of peace. But later he had to agree to a series of compromises on many points. He failed, however, to secure Senate ratification of the Treaty and Covenant and suffered a complete breakdown. He was a man of great intellectual qualities, an idealist, a reformer, and an orator; he achieved much, though he failed to achieve all he might or wished.

R. S. Baker: *Woodrow Wilson: Life and Letters.*

Winding up. The term for bringing to an end the life and affairs of a company and certain analogous bodies. In the case of partnerships, winding up may be voluntary or by order of the court in an action for dissolution of the partnership. In the case of companies winding up may be done voluntarily, either by the members or the creditors, or by the court, or under the supervision of the court. In each case the steps involved are stopping the company's operations, realising its assets and collecting debts due, paying creditors and discharging other liabilities, adjusting the rights and liabilities of the members and dividing any surplus assets among the members. The actual operation of winding up is effected by a liquidator, a person elected for the purpose by the members and creditors or appointed by the court. The term is also used for the administration of a deceased's estate, bringing his affairs to a conclusion and distributing his estate.

Window tax. A tax first imposed in England in 1696 in substitution for the hearth tax and charged according to the number of windows in houses. It was substantially increased by the younger Pitt but was reduced in 1823 and abolished in 1851. It did not extend to Scotland.

Windscheid, Bernhard (1817–92). German jurist and civilian, professor at Heidelberg and Leipzig, and one of the last and greatest of the systematic writers on the modern Roman law used in Germany prior to the adoption of the Code in 1900. His *Lehrbuch des Pandektenrechts* (1862 and later editions) is a classic. He worked on the German Civil Code Commission and the Code as enacted owes much in substance and language to him.

Windsor. The family name of the British royal family since 1917. In that year, King George V, like all descendants of Queen Victoria and Prince Albert, a member of the German house of Saxe-Coburg-Gotha, declared by proclamation that all descendants of Queen Victoria in the male line who were also British subjects would adopt the surname Windsor, from Windsor Castle. Hence when Edward VIII abdicated in 1936 he was created Duke of Windsor. Queen Elizabeth II declared, shortly after her accession in 1952, that her issue in the male line would bear the name Windsor, not their father's adopted surname Mountbatten, but in 1960 modified this decision to the effect that issue other than those styled Royal Highness and prince or princess should bear the name Mountbatten-Windsor.

Winfield, Sir Percy Henry (1878–1958). Called to the Bar in 1903, he taught law in Cambridge from 1911, becoming, in 1928, Rouse Ball Professor of English Law. His *Chief Sources of English Legal History* is a bibliographical classic and his *Province of the Law of Tort* (1931) was very stimulating. His *Textbook of the Law of Tort* (1937) rapidly became a standard work. He also edited several editions of Pollock on *Contracts*, edited the *Cambridge Law Journal* and served on the Law Revision and other committees. He retired in 1943.

Winhoff, Melchior (1500–72). A Dutch lawyer, compiler of *Het Recht van Overissel* (1559) the most complete response of a province to an order by the Governor-General of the Netherlands to make a collection of the customary law of the province and towns. Though unofficial it is valuable for the study of the legal system of the period, and includes public, criminal and civil law.

Winkler, Karl Gottlieb von (1722–90). German jurist and author of *Institutiones jurisprudentiae naturalis in usum praelectionum.*

Wisby (or Visby), Laws of. A body of sea-law, named from the Hanse town of Wisby in Gotland in the Baltic. They appear to have been based on the code of Lubeck, on the Laws of Oleron (q.v.), and on the maritime ordinance of Amsterdam, and seem to have contained the accepted mercantile and maritime laws of Copenhagen and to have been generally accepted in the Baltic. They are printed in various collections, such as Cleirac's *Us et coustumes de la mer* (1647), Justices's *General Treatise of the Dominion of the Sea* (1705), and R. Peters' *Admiralty Decisions in Pennsylvania Districts* (1807).

***Wise* v. *Dunning*,** [1902] 1 K.B. 167. W., in addressing meetings, used language likely to insult and annoy Roman Catholics, and this was likely to and did provoke breaches of the peace. It was held that a magistrate had rightly bound over W. to be of good behaviour.

Wissenbach, Johann Jacob (1607–75). German jurist and author of *Disputationes ad Instituta Imperialia* (1648), and *In libros IV priores Codicis commentationes* (1660).

Witches and witchcraft. Witchcraft is an inherent mysterious power of certain weird and peculiar people to do harm to others. Belief in witchcraft is very common in ancient and modern societies and witches have frequently been blamed for events not readily explicable, such as bad harvests or sudden death. In ancient Greece and Rome, certain goddesses were associated with the performance of rituals at night with malevolent motives. Belief in and fear of witches was common among the Germanic people. From early in Christianity, legislation, both civil and ecclesiastical, was passed against witchcraft practices and beliefs but the Fathers of the Church and the canon law were comparatively moderate and lenient in measures against witches and witchcraft. Later, however, agitation against heretics led to a move to extirpate witchcraft and for three centuries there were regular outbursts of witch-hunting in various parts of Europe. The *Malleus Maleficarum* of two Dominicans, Kraemer and Sprenger, of 1486, contained a theory attributing witches' powers to links with the devil, so that witches became regarded in Europe as Satan's representatives on earth. Thereafter witches were regularly persecuted by Protestants and Catholics. Henry VIII promoted an act making witchcraft and sorcery felonies and this was confirmed by Acts of 1563 and 1603.

In Scotland witchcraft was made a capital offence in 1563. The Salem witch trials (q.v.) of 1692 were among the last outbursts of the witch craze. Prosecutions for witchcraft were prohibited in Britain in 1736 though burning had ended in England in 1712 and in Scotland in 1722.

J. C. Baroja, *The World of the Witches* (1964); H. C. Lea, *Materials Towards a History of Witchcraft* 3 vols. (1939); A. Macfarlane, *Witchcraft in Tudor and Stuart England* (1970); R. Robbins, *Encyclopaedia of Witchcraft and Demonology* (1959); G. L. Burr, *Narratives of the Witchcraft Cases, 1648–1706* (1959).

Wite. In old Germanic law, a penalty for murder or similar serious offence against a person, paid to the Crown, as contrasted with *wer* or *wergild* (q.v.) which was the sum payable to the relatives of the murdered man.

Witenagemot. In Anglo-Saxon England, the supreme council of the nation. It comprised the King, ealdormen, the King's thegns, bishops, abbots, and generally the leaders of the people. It could depose the King for misgovernment, elect the King, though preference seems to have been given to children born to the King after his accession and to a person recommended by the late King, and had a share in all government. Along with the King it enacted laws, levied taxes, made war and peace, appointed ealdormen, bishops and officers of state, and sometimes acted as a supreme court both in civil and criminal cases. After the Norman Conquest it was replaced by the Norman kings' *magnum concilium.*

Without prejudice. A phrase used in offers, intended to guard against any inference of waiver of right, and in correspondence seeking to negotiate a compromise, in which case the communication so made may not be put in evidence.

Witness. A person who, by affixing his signature to a document, affirms that he witnessed a party to the document signing it, and thus attests the veracity of the signature.

It also denotes a person called to give oral evidence at a trial or other court proceeding. Such a person is primarily, and was originally, a person who had witnessed with his own eyes some fact that is a subject of enquiry or relevant thereto, and can accordingly give direct evidence of it, but the word has been extended to all persons giving oral testimony, though that may be evidence not of anything seen but of opinion, based on observations, measurements, tests, and other investigations.

The general rule is that all persons who appear to have sufficient capacity to understand are competent as witnesses and also compellable. A witness who is served with a subpoena must attend in court.

Witnesses, apart from young children, are required to take the oath (q.v.) or to make a solemn affirmation to tell the truth.

A witness must answer all questions put to him by counsel or by the court, but may refuse to answer any question, the answer to which would have a tendency to expose the witness to criminal proceedings or to a forfeiture or penalty. The court should not allow counsel to put questions apparently irrelevant or apparently intended merely to annoy or humiliate the witness.

A hostile witness is one whose evidence is not merely unfavourable to the party calling him but whose conduct shows that he is hostile to that party and is not giving his evidence fairly. Counsel may in such a case, with leave of the judge, cross-examine such a witness.

Witnesses, trial by. In the twelfth century witnesses were persons produced by plaintiff or defendant to swear to a belief in his story; and the decision for plaintiff or defendant depended on counting their numbers, not on weighing their evidence. But in some cases in the thirteenth century, bands of witnesses were examined by the judges to see if their tales were consistent and credible. Trial by jury ousted trial by witnesses in the thirteenth and fourteenth centuries before the latter form of trial was fully developed. Trial by witnesses survived till 1834 in the one case of whether a husband was dead, so that his widow could claim dower, though in the seventeenth century, this came to be converted into trial by the justices upon proofs made before them.

Wogan, John de (?–1321). Royal justiciar of Ireland, 1295–1308 and 1309–12. He was sent out to reconcile the feuds of the nobles and to make the English lordship self-supporting and a treasury on which Edward I could draw for his Scottish and French wars. He made Ireland profitable to the English Crown and supplied men and provisions for Edward's Scottish wars.

His major achievement was the summoning, in 1297, of the first Irish Parliament which included knights from each shire. In 1310 he summoned a more representative assembly to Kilkenny, though it was representative of the Anglo-Irish and of the English part of Ireland only.

Wolfe, Arthur, Viscount Kilwarden (1739–1803). Called to the Irish Bar in 1766, he became Solicitor-General (1787), Attorney-General (1789), and then Chief Justice of the King's Bench in Ireland (1798–1803), being made Baron and, later, Viscount Kilwarden.

Wolff, Christian von (1679–1754). Mathematician and philosopher, professor at Halle, who wrote a nine-volume work on jurisprudence and international law, *Jus Naturae Methodo Scientifico pertractatum* (1710–9), and a smaller book setting out his views on international law, *Institutiones Juris Naturae et Gentium* (1754). Though treating international law as a department of the law of nature, which itself rested on an ethical basis, he recognized that agreement between states could also give rise to binding rules.

Wolsey, Thomas (*c.*1473–1530). English cleric, entered the royal service and rose rapidly under Henry VIII, becoming Archbishop of York in 1514, Cardinal in 1515, Chancellor in 1515, and papal legate *a latere* in 1518. He was a pluralist, and absentee bishop, greedy and unscrupulous, and thereby encouraged anti-clericalism. As Chancellor he greatly developed the Star Chamber, making it a court which administered quick and sure justice. As a diplomat he sought European peace. In 1529 he had lost the King's support for having failed to secure Henry's divorce and was found guilty on a praemunire (q.v.) charge of having misused his legatine powers, and was stripped of his secular offices. In 1530 he was arrested but died before being tried.

Women, legal status of. Biological and personality differences between men and women have given rise in every society to social and other differences, not least to differences in legal status. Some anthropologists have suggested that primitive societies were commonly matriarchal, and early Marxists accepted this as the ideal type of social organization. Later investigations distinguish between matriarchy, where females exercise authority, which is uncommon, and matrilineal kinship, where descent is traced through the female line, which is common in primitive societies. Patriarchal societies may be matrilineal. One anthropological conclusion is generally accepted, that in nomadic and hunting societies the status of women is low, but in agricultural societies it is substantially higher.

In the ancient civilizations, women enjoyed independence and high status in Babylonia, and in Egypt it was higher still; Egypt had many queens and women took full part in public life, owned property, and seem even to have been the dominant sex. In Greece, women in Sparta had equality and independence but in Athens they were secluded at home, had few rights and were treated mainly as mere housekeepers. In Rome, they were completely subordinated, first to their fathers or brothers and then to their husbands, had no political or public position, and yet had high social position and many are known to have been persons of character, ability, and influence.

Christianity treated women as servants behind the scenes and wholly excluded them from public life, yet on the other hand gave them a place in the Church through nunneries, whereby they frequently did have great power and influence. But they were largely excluded from education and public life save that at the top queens could exercise rule as legitimately as kings. Chivalry idealized womanhood but did not treat them as more than rather idealized objects of love.

The Renaissance and the Enlightenment caused regression in the position of women, though a few attained scholarly eminence, even becoming professors in Italian and Spanish universities. Puritanism and Calvinism also depressed women's status, and their status was low in the eighteenth century, but a movement for its elevation began with the intellectual ferment preceding the French Revolution. Demands for liberty, equality and human rights could not logically entirely exclude women. Condorcet wrote an essay in 1790 on the admission of women to full citizenship and Mary Wollstonecraft issued her *Vindication of the Rights of Women* in 1792.

In the nineteenth century, on the one hand, industrialization grossly degraded many women, forcing them to work under very bad conditions, whereas in the middle and upper classes, women were kept idle, apart from looking after homes and children. Education for women was neglected and only from about mid-century were serious efforts made to afford educational facilities at all comparable to those for boys while higher education and the professions long remained closed.

From the 1860s women began increasingly to be allowed to hold and manage property, a trend marked by a series of Married Women's Property Acts from 1880 onwards. Political equality was long denied, being first attained in New Zealand in 1893 followed by the Scandinavian countries; in most Western countries it was not conceded until after World War I. But the proportion of women entering Parliament has always continued small. In the mid-twentieth century, however, it was not exceptional to have women Ministers and women Prime Ministers. In Britain a woman became Prime Minister in 1979. Modern economies depend very largely on women workers of many kinds and with many different kinds of qualifications and abilities.

In Britain, notable developments have been the Sex Disqualification Removal Act, 1919, the Equal Pay Act, 1970, and the Sex Discrimination Act, 1975, but equally important have been growing acceptance of egalitarian philosophy and increasing evidence of the capacity of women for all kinds of work. Equality, coupled with freedom of the sexes, have made marriage and family life increasingly a partnership, and the older view of the husband as the master, or even the senior partner, has largely disappeared.

J. S. Mill, *The Subjection of Women* (1869); J. Langdon Davies, *A Short History of Women* (1927); R. H. Graveson and F. R. Crane, *A Century of Family Law* (1957); C. Rovar, *Love, Morals and the Feminists* (1970).

Wood, Thomas (1661–1722). Fellow of New College, Oxford, and barrister, though he eventually abandoned the law and entered the Church. He wrote first on Roman law and legal theory. He published a *New Institute of the Imperial or Civil Law* (1704) dealing with the influence of the Roman law and the differences between English and Roman and canon law, and *A Treatise of the First Principles of Laws in General* (1705), a translation of part of Domat's *Les Lois Civiles dans leur Ordre Naturel*, intended as an introduction to his *New Institute*. In 1708 he published *Some Thoughts concerning the Study of the Laws of England, particularly in the two Universities* and in 1720 *An Institute of the Laws of England, or Laws of England in their natural order according to common use*, founded on Finch's *Discourse*, to supply a methodical book on English law for the use of students at the Universities and the Inns of Court. It went through many editions down to 1772 and, in many respects, anticipated Blackstone, though it lacked Blackstone's literary qualities and proportion and was superseded by that work. He published various other, but unimportant, works.

Wooddeson, Richard (1745–1823). Barrister (1767), Vinerian Scholar and fellow, became Vinerian Professor of Law at Oxford, 1776–93 and published a *Systematical View of the Laws of England* (3 vols., 1793) consisting of the 60 lectures he gave as Vinerian Professor. Though clear, they lack the historical sense, the grasp of principle and the literary grace of Blackstone, but in some respects, particularly equity, usefully supplement Blackstone. He also published an *Elements of Jurisprudence* (1783), a scholarly work and a useful introduction to the study of English law. He was also author of a *Brief Vindication of the Rights of the British Legislature* (1799) and part author of Toller's book on *Tithes* (1808).

Woods and Forests, Commissioners of. A Government department which managed the Crown lands and collected the land revenues of the Crown. They were renamed Commissioners of Crown Lands in 1924 and Crown Estate Commissioners in 1956.

Woolsack. The large bale of wool covered in red cloth and with a small backrest projecting from the centre of it on which the Lord Chancellor sits when acting as Speaker of the House of Lords. The origin of this is that when an Act was passed in the time of Elizabeth I to prevent the exportation of wool from England, sacks of wool were placed in the House of Lords, on which the judges sat, as a constant reminder of this source of the national wealth. It is said to be the survivor of an original four woolsacks on which sat the judges, barons of exchequer, masters in chancery and serjeants at law. Technically the Woolsack is outside the precincts of the House, and if the Lord Chancellor acts as Speaker before his patent of creation as a peer has been made out, he may still occupy the Woolsack. If the Lord Chancellor wishes to speak in debate as a peer, as he may do, he leaves the Woolsack and moves to the left; but in committee he speaks from a place on the benches, though when he votes as a peer, he does so from the Woolsack.

Worde, Wynkyn de (*d.* ?1534). Correctly named Jan van Wynkyn (de Worde meaning from Worth in Alsace), he served an apprenticeship as a printer to Caxton in 1476–1491 and, on Caxton's death succeeded to his materials. He was the leading printer prior to 1600 and is credited with printing over 600 known books, including some legal ones.

Words. Words of the English language are of great importance in law because law is expressed in words, in statutes, judgments, and books of authority, and a great deal of legal work consists in trying to draw meaning from the words of statutes, decisions, books, contracts, wills, statements of parties, and other sources, and in trying to embody meaning in pleadings, contracts, conveyances, affidavits, evidence, and otherwise. The handling of words is a large part of legal work.

It is important to bear in mind that, apart from misuse and choice of inappropriate words, most words of the English language have a number of different senses and uses which have to be distinguished, and that many words have both everyday or common meanings and technical legal meanings, e.g. negligence, malice. Again, many words are commonly misused, and the misuse may be found in legal contexts, e.g. the confusion between 'common' and 'mutual'. Again, particular words may be defined in particular statutes, by the interpretation sections thereof, in terms narrower or wider than their everyday connotations, e.g. 'factory'. Some words again, e.g. 'bailment', are probably exclusively legal terms and little known to or used by the non-legal community. The law would be much more precise if it had a much larger stock of terms of purely technical, legal significance, as have medical, engineering, and other sciences.

In other legal contexts words are important. Words which a person considers unjustifiably cast an aspersion on his character or conduct may be made the basis of an action for libel or slander. Words used by parties contracting may have to be closely examined to discover what meaning other parties were reasonably entitled to draw from those words.

In trying to ascertain what parties intended to convey, or were reasonably entitled to think were conveyed, by particular words, a court will, in general, primarily consider what the word means in the ordinary everyday usage of the English language, of which standard dictionaries are good evidence; secondly, what technical legal meaning is conveyed by the particular word, particularly if used in a legal document and particularly one drafted professionally; and thirdly, what special or technical meaning is conveyed by the particular word when used in the context of a particular profession or trade or business practice.

G. Jacobs, *Law Dictionary* (1729, 2 vols. 1835); W. Byrne, *Dictionary of English Law* (1923); J. Wharton, *Law Lexicon* (1938); W. Jowitt, *Dictionary of English Law* (1959); W. Bell, *Dictionary and Digest of the Law of Scotland* (1870); J. Bouvier, *Law Dictionary* (U.S.) (1914).

Words of limitation, of procreation, and of purchase. In a conveyance or will, words of limitation are words which have the effect of marking the duration of the estate, e.g. to A and his heirs. Words of procreation have to be used in creating an entailed interest so as to confine the estate to the designated descendants of the first grantee, e.g. to A and the heirs of his body. Words of purchase are words which denote the person who is to take the estate, e.g. to A for life with remainder to his heirs. Words of limitation and words of purchase are accordingly opposed to each other, though there are circumstances in which words operate partly as words of purchase and partly as words of limitation.

Workhouse. In England, the Old Poor Law (1601) imposed responsibility for the poor on parishes which sometimes built workhouses as institutions for the infirm and paupers of the parish. They tended to be used to hold sick, orphans, lunatics, and criminals as well and were sometimes hard to distinguish from houses of correction. In the nineteenth century, groups of parishes combined to provide workhouses or poorhouses, and in the mid-twentieth century, the concept disappeared though some of the old buildings are still in use under other names.

Workmen's Compensation. A system, initiated in Germany in 1884 and in the United Kingdom in 1897, imposing liability on employers to compensate their employees for injuries caused 'by accident arising out of and in the course of the employment', irrespective of fault or negligence on the employer's part. The system did not prejudice claims for damages founded on common-law liability but, for such a claim to succeed, the workman had to prove fault and negligence on his employer's part. An injured workman might either claim compensation or damages at common law, but not both. The amount of compensation was fixed by reference to average weekly earnings and might be commuted for a lump sum. Disputed questions were settled by arbitration, normally by a county court judge or sheriff-substitute as arbitrator, with appeal to the Court of Appeal or Court of Session and to the House of Lords and the system gave rise to a huge volume of litigation interpreting the Acts. The system was extended and developed by successive Acts and in 1948 abolished and replaced by the National Insurance (Industrial Injuries) Scheme, now superseded by Social Security.

R. Willis, *Workmen's Compensation Acts.*

World Bank. See INTERNATIONAL BANK FOR RECONSTRUCTION AND DEVELOPMENT.

World Court. See INTERNATIONAL COURT OF JUSTICE.

World Health Organization. A specialized agency of the U.N., established in 1948, to further international co-operation for improvement of health. Its headquarters are in Geneva and it operates through an annual Assembly, an executive board, and a secretariat.

World Intellectual Property Organization. An organization, established in 1967, to further protection for intellectual property.

World Peace through Law Center. Founded in 1957 as a special committee of the American Bar Association, it became, in 1963, an independent association of the world's judges and legal profession devoted to the continued development of international law and legal institutions and to the strengthening of the world's legal system. It is based in Washington, D.C. Affiliated to the Center are four associations, the World Associations of Judges (1946), Lawyers (1975), Law Professors (1975), and Law Students (1976). The Center publishes *The World Jurist* and other publications, promotes research, and organizes world law conferences.

Woulfe, Stephen (1789–1840). Was called to the Irish Bar in 1814, became Solicitor-General in 1836, Attorney-General in 1837, and Chief Baron of Exchequer in Ireland, 1838–40.

Wound. Any kind of bodily lesion causing breach of the continuity of the whole skin. Even if a bone is broken there is no wounding if the skin is not broken. Unlawfully and maliciously to wound a person is criminal.

Wreck. Property, being a ship or her cargo or part thereof, or an aircraft, cast ashore within the ebb and flow of the tide following on shipwreck. Both under the royal prerogative and by statute, the Crown is entitled to all unclaimed wreck found in or on the seashores or in any tidal water, save that the right may have been granted to another; in Cornwall it belongs to the Duke of Cornwall, in Counties Palatine to the Earls Palatine, and in Wales to the Lords Marchers. Wreck includes flotsam, jetsam, and lagan, unless they are taken on the high seas when, unless the true owner be known, they belong to the finder. A person finding or taking possession of wreck must report the find to the local receiver of wreck.

Wreck commissioners. Persons appointed by the Lord Chancellor to hold investigations into shipping casualties.

Wrenbury, Lord. See BUCKLEY, H. B.

Wright, Sir Nathan (1654–1721). Called to the Bar in 1677, he was of counsel for the prosecution in the trial of the Seven Bishops. In default of anyone willing to take the post, he became Lord Keeper in 1700 and held the office till 1705, but he never wholly understood equity and chancery business fell heavily into arrears. Some of his decisions helped, however, to settle certain of the principles of equity. He was replaced in 1705.

Wright, (Philip) Quincy (1890–1970). American political scientist and international lawyer, professor at Chicago, and adviser to the U.S. State Department, U.N.E.S.C.O., and the Nuremberg tribunal. He wrote extensively, notably *The Enforcement of International Law through Municipal Law in the U.S.* (1916); *The Causes of War and the Conditions of Peace* (1935); *A Study of War* (1942); *Problems of Stability and Progress in International Relations* (1954); *The Study of International Relations* (1955); and *The Role of International Law in the Prevention of War* (1961).

Wright, Sir Robert (?–1689). Attained a good practice and became a Baron of Exchequer in 1684, a judge of the King's Bench in the following year, Chief Justice of the Common Pleas in 1687, and a few days later, Chief Justice of the King's Bench. He presided at the trial of the Seven Bishops and conducted the proceedings fairly and impartially, possibly being influenced by public opinion. A protégé of Jeffreys, he was utterly unfit to be a judge. On the Revolution occurring in 1688, he went into hiding, and died in Newgate Prison.

Wright, Robert Alderson, Lord, of Durley (1869–1964). Called to the Bar in 1900, he was a pupil of Scrutton (q.v.) and developed a large commercial practice. He took silk in 1917, became a judge in 1925, a Lord of Appeal in 1932, and served as such till 1947 save for the years 1935–37 when he acted as Master of the Rolls. He was Chairman of the Law Revision Committee and of the United Nations War Crimes Commission which collected material for the War Crimes trial at Nuremberg in 1945–46. He was highly regarded as a judge. He should not be confused with Sir Robert Samuel Wright (1839–1904), a judge of the Q.B.D. 1890–1904 and author (with Sir Frederick Pollock) of *An Essay on Possession in the Common Law* (1888).

Writ. (1) A written order or warrant. By the thirteenth century, there were three main recognized kinds of writs: charters, normally for grants of land or liberties in perpetuity; letters patent, for commissions to royal officials and grants of limited duration; and letters close, closed and sealed, conveying orders or information. The Anglo-Norman writs were substantially Anglo-Saxon writs turned into Latin and, like them, in many cases, contained executive orders. The Norman kings began to employ writs for judicial purposes, and standard forms were developed to meet the normal cases. The number grew rapidly in the thirteenth century. The most important were the original writs, necessary to commence actions in the courts, writs of entry, for the recovery of land by a person who had been dispossessed, and writs for assistance, for the transference of property. Other important writs were those of habeas corpus, mandamus, certiorari, and prohibition (qq.v.), known as the prerogative writs because issued by virtue of the royal prerogative at the discretion of the court on a *prima facie* case being made. The writ of habeas corpus (q.v.) remains, but the others have been altered to prerogative orders. Even in the thirteenth century, the power to issue new writs was checked and the number of writs obtainable became limited. This led to the growth of a definite Register of Writs. In Bracton's work, it is clear that the law depends on the writs; if a writ is obtainable there is a remedy by action, but not otherwise.

In modern practice, a writ is an order issued in name of the sovereign requiring the performance of some act. The commonest is the writ of summons (q.v.) used to initiate an action in the High Court. Others are interlocutory writs, such as writs of inquiry and writs for enforcing obedience to interlocutory orders by attachment, sequestration or otherwise; and writs of execution.

(2) A document issued by the Clerk of the Crown in Chancery, or formerly, the Governor of Northern Ireland, directing the returning officer of a parliamentary constituency to hold an election for a Member to serve in Parliament for that constituency. For a General Election, the Queen in Council orders the Lord Chancellor to issue the writs and for a bye-election, the Clerk of the Crown in Chancery issues the writ on a warrant issued by the Speaker, a motion that he do so being moved by the party which had previously held the Seat.

Writ of error. A procedure whereby, prior to the institution of the Court of Criminal Appeal in 1907, with the Attorney-General's consent, a criminal case could be taken to the Court of Appeal and thence to the House of Lords. It was competent only in a very few cases, where there was an error in law apparent on the face of the record, which excluded errors in the judge's charge to the jury and a wrong or even perverse verdict.

Writ of right. A writ obtainable as a matter of right, as contrasted with a prerogative writ, granted in the exercise of the royal prerogative as a matter of discretion only. In older real property law a writ of right was the real action which lay to enable a person to recover land in fee simple unjustly withheld from him. It was used where the person disseised had lost his right of entry or right to possession because, otherwise, a possessory action such as a writ of entry was more convenient.

Writ of subpoena. The first step in the commencement of a suit in equity after the initial filing of the bill addressed to the Lord Chancellor, stating the matters on which the plaintiff relied and praying the desired relief. The subpoena did not tell the defendant what the complaint against him was, nor mention any cause of action. Accordingly, it made equity a flexible system and, as compared with procedure of the common law, the procedure of equity was largely formless and very flexible. After 1852 the writ of subpoena was made unnecessary.

Writ of summons. The first step in an action in the High Court, a process issued at the request of the plaintiff giving the defendant notice of the claim made against him and requiring the defendant to appear and answer if he does not admit the claim. The writ consists of the body, the memoranda, and the indorsements. The body gives the title, i.e. reference number, name of the court and Division, and the names of the parties, the order to the defendant to appear, and the *teste*, i.e. the date and place of issue and the name of the Lord Chancellor. The memoranda state the time within which the writ must be served and the place where the defendant may enter appearance. The indorsements may be an indorsement of claim, which is a brief statement of the nature of the plaintiff's claim, or a special indorsement where the plaintiff's claim is for a debt or a liquidated claim, and also state the name and address of the plaintiff and his solicitor with an address for service if necessary, and the indorsement of service made by the person who serves the writ. Certain other indorsements are used in particular cases. A special form of writ of summons is used in Admiralty actions *in rem*. A writ remains in force for twelve months but this may be extended by a judge for a further six months.

Writ of Summons to Parliament. About the thirteenth century, it was settled in England that those who were summoned to counsel the King in Parliament were the holders of baronies. Accordingly, it followed that one who had received a writ of summons to the House and had taken his seat had acquired an hereditary right; baronies were recognized as created by writ of summons. Richard II introduced what is now the normal mode of creation, by letters patent under the Great Seal. Nevertheless, the members of the House of Lords, both spiritual and temporal, still receive writs of summons to a new Parliament, issued on the Queen's Behalf by the Clerk of the Crown in Chancery.

Writer. An old form of designation, formerly common and still affected by some solicitors in Scotland.

Writers to the Signet. The oldest and leading society of solicitors in Scotland. Originally persons of this designation seem to have been clerks in the office of the Scottish Secretary of State, who had charge of the King's seal. When the Court of Session was instituted in 1532 they were already an existing body and became members of the College of Justice. Writs initiating actions in the Court of Session had to pass the Signet and Writers to the Signet became engaged in the preliminary stages of actions and, in 1754, their title to do so was recognized. Prior to 1873, Writers of the Signet and Solicitors in the Supreme Courts had the exclusive right of practising as Solicitors in the Court of Session.

The titular head of the society is the Keeper of the Signet, appointed by the Crown, an office conjoined with that of Lord Clerk-Register, a Depute Keeper appointed by him and the actual head of the Society (D.K.S.) and other officials. Members are designated by the initials W.S. after their names. The society is recognized by immemorial custom as a corporation.

Firms of solicitors mostly in Edinburgh, at least one partner of which is a W.S. may append the letters W.S. to the firm name. In certain cases of long-established firms the designation used is C.S. (Clerks to the Signet)

R. K. Hannay, *History of the Society of Writers to Her Majesty's Signet* (1936).

Writs of assistance (q.v.). General search warrants issued by superior courts in the American colonies to assist the British government to enforce the trade and navigation laws by authorizing customs officers to search houses for smuggled goods, but not specifying either the house or the goods. They were first used in 1751 and aroused resentment when their use was repeated in 1761, but their legality was confirmed from England. They were again authorized by the Townshend Acts (1767) but were challenged repeatedly in the courts of the colonies and refused as illegal in most of the colonies. Their use was accordingly a major grievance of the colonists prior to the Revolution.

Writs, Register of. See REGISTER OF WRITS.

Writing. As a noun, writing means a document, written, printed or otherwise produced in permament form, as contrasted with oral communication. Certain kinds of contracts and transactions must be effected or evidenced in writing.

As a verb, writing generally means any process of representing words in visible form and includes typewriting, printing, etc., though, in some contexts, writing is distinguished from printing.

Written law. A term of indefinite connotation but usually meaning codified law and statute law as contrasted with customary law and judge-made law, though the latter of these is, and the former may be recorded in writing or print.

Wrongful dismissal. A dismissal of an employee by his employer before the expiry of the period of employment in circumstances which are legally unjustifiable. At common law, the employee has, in such a case, a claim of damages. At common law, there were no requirements of prior warning that the employee's conduct might lead to dismissal, nor to give him a hearing, nor to allow any kind of appeal against dismissal. The wrong and relative remedy have now been very largely superseded by the statutory claim for unfair dismissal (q.v.).

Wrongs. Instances of human conduct which are legally deemed wrongful. Wrongs are distinguishable into two overlapping categories: moral or natural wrongs, which are deemed wrong and reprehensible by a moral standard (including, e.g., envy, hatred, and overreaching) and legal wrongs, actings which are held to be wrong in accordance with the principles of a particular legal system. The categories overlap in many cases; thus assault, fraud, and malicious damage are both morally and legally wrongful; but by no means all morally wrongful conduct is legally wrongful; nor is all legally wrongful conduct also morally wrongful, unless one adopts the view that any infringement of the law is necessarily morally wrong. Legal wrongs are themselves distinguished into two overlapping categories, civil wrongs and criminal wrongs. Civil wrongs are those infringements of the rights of others or breaches of legal duty, normally causing personal or pecuniary loss, to others, which give rise to a civil remedy or claim for redress by the wronged individual, for injunction, damages, accounting or other civil remedy. They include, in particular, breaches of contract, torts, breaches of trust, and some breaches of statutory duty. Criminal wrongs are those infractions of the criminal code of a particular state which, by that code, are deemed for the sake of the general welfare to require repression and to justify punishment of the persons responsible. Criminal wrongs may again be contraventions of common law or of statute. Again the categories overlap; thus assault is both a tort and a crime, as are some kinds of contraventions of statute. But many kinds of conduct are either civil or criminal wrongs but not both, or are in both categories only subject to conditions. Thus possession of a knife may be criminal *per se*, but will be tortious (civilly wrongful) only if it causes harm.

Wyndham, Thomas, Lord (1681–1745). Was called to the English Bar in 1705, became Chief Justice of the Irish Common Pleas in 1724, and gained a high reputation. He became Chancellor of Ireland in 1726 and a peer in 1731. He retired in 1739.

Wynford, Lord. See BEST, WILLIAM.

Wythe, George (1726–1806). A Virginian lawyer, who was a delegate to the Continental Congress, and who, with Jefferson and Pendleton, revised the laws of Virginia. He was also a delegate to the Federal Constitutional Convention in 1787. He was a fine scholar, deemed the most distinguished lawyer of his generation, and occupied the chair of law at the College of William and Mary, the first in an American college, from 1779 to 1790. His view of law was influential on Jefferson, Marshall, Monroe, and Clay. His lectures, based on Blackstone, contrasted English and Virginian law, and virtually charted the way for American jurisprudence. After 1790 he formed a small law school of his own. He also served as a judge in the Virginia High Court of Chancery, 1778–88, and sole judge thereof, 1788–1801, being accordingly known as Chancellor Wythe. As a judge, he was erudite, logical, and impartial, and one of his decisions (*Commonwealth* v. *Caton* (1782), 4 Call, 5) is among the earliest enunciation of the doctrine of judicial review.

X

Xiphilinus, John (*c.*1010–1075). A jurist, theologian, and patriarch of Constantinople, he was one of the notable figures of the eleventh century, who became head of the law school when the university of Constantinople was reorganized in 1045. In 1054 he entered a monastery and later became patriarch of Constantinople.

Y

Year. The period of 365 consecutive days, or in leap years 366 days divided into 12 calendar months and 52 weeks plus one, or in leap years two, days. From 1752 the year has commenced in England on 1st January, having formerly begun on 25th March, but it had traditionally commenced on 1st January in Scotland and elsewhere. The government financial year runs from 1st April to 31st March, the income tax year from 6th April to 5th April, and the accounting year of companies runs from whatever date is thought most convenient. Leases frequently run for a year or a term of years.

Year and a day. A traditional period, probably to ensure that a full year had elapsed, relevant in various contexts. Probably the most important is the rule that if injury is done but death does not ensue until after a year and a day from the injury, the death is presumed not attributable to the injury so that an indictment does not lie for murder or manslaughter. Another is the rule that the owners of estrays must claim them within a year and a day, failing which they fall to the Crown. In feudal land law, the commission of a felony by a tenant caused his land to pass to his lord (escheat *proper delictum* *tenentis*) subject to being held by the Crown for a year and a day, being entitled during that time to waste (q.v.) the land. In practice, the lord usually took the land at once, compounding with the Exchequer for the Crown's right to 'year, day and waste'.

Year Books. A series of notes on debates on points of pleading in Norman French in mediaeval English cases, of the very first importance as sources of our knowledge of the mediaeval common law. They are accounts of the debates on the formulation by judge and counsel of the points in issue in cases rather than reports of decisions. The collection of cases by the regnal year seems to be a printer's device and manuscripts show cases arranged topically not chronologically.

It is not known when the first volumes were compiled, but some cases are believed to date from the 1270s. The series continued almost without interruption to 1535 and then unaccountably stopped.

The compilers are unknown. A now discredited story is that for most of the period of compilation they were produced by four official reporters paid by the Crown, and another that they were the unofficial work of four clerks of the Common Bench. Maitland considered that the earliest Year Books were notes taken by law students in court, later copied and multiplied. Whoever made them, they were made in court for the guidance of pleaders in later cases.

The style, content, and quality of the Year Book reports varies considerably. Many of them contain irrelevancies, notes of the répartee in court, and mistakes, but on the whole the contents are intensely practical. They record the procedural steps in actions, in a way frequently incomprehensible without reference to the Plea Rolls, and some of the cases in the Year Books are simply transcripts from the rolls or incorporate large portions of the roll.

After the Year Books ceased in 1535, they steadily became obsolete in language and substance and fell into professional disuse, and were replaced by other kinds of reports and books.

There are many manuscripts of Year Books extant and originally there must have been many copies of at least some of them.

Year Books seem first to have been printed about 1481–1482 by William de Machlinia, but the first systematic publisher of Year Books was Pynson who produced at least 50 editions. Thereafter Tottell printed some 225 editions, and also nearly all the other years which have been incorporated in subsequent editions, and reprinted all the years already published. He also introduced the practice of grouping years into one volume. The first collected edition was the quarto edition in 10 volumes, covering the years from Edward III to Henry VIII, published 1562–1640. Publishing of Year Books became less common in the seventeenth century, until, in 1679–80, there appeared the standard edition, known as Maynard's edition, in 11 folio parts (most of which had previously been printed) each, save one, covering the cases of a number of regnal years. All of these black-letter editions of the Year Books were defective in spelling, in Latin, in dating and in other respects.

The difficulty of finding materials in the Year Books led to the production of *Abridgments* (q.v.) and four of these, Statham, *The Abridgment of the Book of Assizes*, Fitzherbert and Brooke (qq.v.) are founded largely on the cases in the Year Books.

Since the latter part of the nineteenth century, many Year Books have been published, carefully edited, in the Rolls Series, edited successively by Horwood and Pike, and the Selden Society series, by various editors.

W. C. Bolland, *The Year Books* (1921); W. C. Bolland, *Manual of Year Book Studies* (1925); W. S. Holdsworth, *H.E.L.*, II, 525; P. H. Winfield, *Chief Sources*, 158: Introductions to volumes in Rolls Series and Selden Society series; J. H. Beale, *Bibliography of Early English Law Books.*

Year, day and waste. A royal prerogative right under which the Crown had, for a year and a day, the profits of the lands of those attainted of petty treason or felony with the right to waste the lands by demolishing the buildings, cutting the timber etc., unless the mesne lord undertook redemption of the waste, only thereafter restoring the lands to the lord. The right disappeared with the abolition of forfeiture.

Yellow-dog contract. In the U.S., an agreement connected with a contract of employment that the employee will not join a union during the course of his employment. Such contracts were common in the U.S. in the 1920s and enabled employers to take action against union organizers who sought to get workers to join unions. Since 1932 yellow-dog contracts have been unenforceable in federal courts.

Yelverton, Barry, Viscount Avonmore (1736–1805). Was called to the Irish Bar in 1764, supported Grattan in Parliament, became Attorney-General in 1782 and Chief Baron of Exchequer in Ireland in 1783–1805. He became a baron in 1795 and a Viscount in 1800.

Yelverton's Act (1782). This Act of the Irish Parliament extended the principle of Poynings' Act, 1492 (q.v.), to private estate Acts and shipping Acts.

Yeoman. A person owning free land of the value of 40 shillings yearly and consequently qualified to serve on juries, vote for knights of the shire, and do any other act for which a *probus et legalis homo* was required. It was a general term for lesser gentry, owners of small estates and gentlemen farmers.

Yeoman Usher of the Black Rod. The officer who is deputy to the Gentleman Usher of the Black Rod and also Serjeant-at-Arms in the House of Lords.

York, Chancery Court of. The provincial court of the Archbishop of York, having the same appellate jurisdiction as the Court of Arches (q.v.). Formerly the sole judge was the Auditor, who since 1874 has been the same person as is Dean of the Arches, appointed by the two Archbishops jointly with the approval of the Queen. Since 1963 the court is similarly composed to the Court of Arches, i.e. the Auditor, two persons in holy orders appointed by the Lower House of Convocation of the province and two laymen with judicial experience. Appeal lies in cases of faculty not involving doctrine, ritual, or ceremonial to the Judicial Committee of the Privy Council.

York-Antwerp rules. A series of rules, worked out first at international conferences at York, 1864, and Antwerp, 1877, and revised in 1890, 1924, 1950, and 1975, to secure uniformity of practice in the adjustment of claims for general average (q.v.), losses, and commonly incorporated by reference in bills of lading.

R. Lowndes and G. R. Rudolf, *General Average and the York-Antwerp Rules.*

Yorke, Charles (1722–70). Was second son of Philip Yorke, Lord Hardwicke (q.v.). Called to the Bar in 1746, he became an M.P. in 1748 and rapidly succeeded in both spheres, becoming Solicitor-General in 1756. He resigned in 1761 but in 1762, accepted the office of Attorney-General and acted in the proceedings against Wilkes. He resigned, probably under pressure, in 1763, but resumed the office in 1765. In 1770 he broke with all his political associates and friends to accept the Chancellorship, but the consciousness of having betrayed his friends and their condemnation of his defection brought on an illness from which he died three days later. He had never sat and the patent for his elevation as Lord Morden was never sealed. His career at the Bar had been brilliant and he had the capacity to be a great judge. He was not only a powerful advocate and a fine lawyer but a successful M.P., a man of letters and author of a tract on *Considerations on the Law of Forfeiture for High Treason* written in support of the Treason Act, 1744.

Yorke, Philip, 1st Earl of Hardwicke (1690–1764). After experience in an attorney's office, he was called to the Bar in 1715, and made rapid progress. He became an M.P. in 1719, Solicitor-General in 1720, and Attorney-General in 1725. As a law officer, he showed great abilities in constitutional and criminal law, but he also had a large equity practice. In 1733 he became Chief Justice of the King's Bench and a baron, and showed his great qualities of learning, diligence, and fairness in that office. He was influential in the House of Lords and became Chancellor in 1737. He became an earl in 1754 but, though he resigned the Seal in 1756, remained a member of the Cabinet till 1762. As a statesman he exercised a strong steadying and conciliating influence, as when he brought together Pitt and Newcastle. He promoted legal reforms but sought to maintain the constitution as it was. As lawyer and judge he stands in the first rank; he was courteous and fair, diligent and learned, and that over a wide range of law, not merely of England. His great achievement was the settlement of many of the principles of equity in their modern form, consolidating and systematizing the developments of the past century, since Nottingham, so that equity developed into a settled system of definite principles and rules, yet still having flexibility and a capacity for growth and adaptation. He also sought to arrest the growing confusion and delays in Chancery procedure. His abilities and work were of outstanding importance in the creation of modern equity. He wrote *A Discourse of the Judicial Authority belonging to the Office of Master of the Rolls in the High Court of Chancery* (1727).

P. C. Yorke, *Life and Correspondence of Lord Hardwicke.*

Yorke, William (1700–76). A kinsman of Lord Hardwicke, was called to the English Bar, and went to Ireland as a justice of the Common Pleas in 1743, becoming Chief Justice in 1753. In 1761 he surrendered the office and became Chancellor of the Exchequer, 1761–63, and a baronet.

Young offenders. In general, persons of under full age who have committed criminal offences, though statutory classifications have varied at different times. Down to the early nineteenth century, young offenders were not treated differently or separately from adults. In the early nineteenth century, reformatories were founded by private benevolence and, in 1854, the Reformatory Schools Act enabled courts to commit offenders under 16 to a reformatory for from two to five years. In 1857 local authorities were enabled to contribute to the establishment of reformatories and, from 1866, there was an Inspector of Reformatories. Industrial schools were established, from 1867, as training schools for children under 14 requiring care and protection, and some child offenders could be sent there too. Parkhurst Prison was used for young offenders from 1838 and to it were sent offenders under 18 sentenced to transportation; they might later be pardoned on going to a reformatory, or transported. Nevertheless, many young people were sent to prison, though from 1865, those under 16 were kept separate and given separate treatment. In 1908 the Borstal System (special institutions for the 16–21 age-group with the emphasis on discipline, training and hard work) was established. In 1933 the distinction between reformatory and industrial schools was abolished, and they were renamed approved schools, the age-range for committal being 10 to 17. Since 1948 a wide variety of alternatives to prison have been provided, namely, attendance centres, community homes (replacing approved schools and remand homes), detention centres and Borstal institutions; committal to each is hedged with qualifications and pre-requisites. Fines and probation are also extensively used for young offenders. Only very exceptionally are young adults (under 21) committed to prison and when this is done they are allocated to special prisons or parts of prisons reserved for this class of prisoner.

Young person. A person defined by reference to age variously in various statutes but generally meaning one between 14 and 17. Many statutes enact special protective measures for young persons.

Younger, Sir Robert, Baron Blanesburgh of Alloa (1861–1946). Was called to the Bar in 1884, became a Chancery judge in 1915, a Lord Justice of Appeal in 1919, and, in 1923, a Lord of Appeal. He retired in 1937. He was a quick and independent judge, who not infrequently dissented.

Youngstown Sheet & Tube Company* v. *Sawyer (1952), 343 U.S. 579. In pursuance of a dispute with the management, a union at the Youngstown plant give notice of a strike, which might have prejudiced the supply of equipment to American forces in Korea. The President directed the Secretary of the Interior to take over and operate the industry to avoid halting production. The Supreme Court held that the President had no power, unless authorized by statute, to take over the industry. The decision was a sharp check on the growth of Presidential power.

Z

Zabarella, Francesco (1360–1417). Italian canonist, ecclesiastical diplomat, and later cardinal. He was author of *Lectura super Clementinis* (1477); *Commentaria in quinque libros Decretalium* (1502), *Tractatus varii*, *Consilia* (1490), *Quaestiones*, and other works.

Zachariae von Lingenthal, Karl-Eduard (1812–94). German jurist, a leading authority on Byzantine law, editor of the *Procheiros Nomos* (1837), *Ecloga* and *Epanagoge* (1852), the *Ius Graeco-romanum* (1856–84) and the *Novellae* (1881). He described Byzantine law as a whole in *Historiae iuris graeco-romani delineatio* (1839) and *Geschichte des griechischen-römischen Rechts* (1856–64).

Zachariae von Lingenthal, Karl Salomo (1769–1843). German jurist, father of Karl-Eduard (q.v.), and author of many works, but most importantly of a *Handbuch des französischen Zivilrechts* (1808), the best systematic work on French law by a German.

Zafrullah Khan, Sir Mohammed (1893–) Pakistani statesman and judge, member of the Viceroy's Executive Council and judge of the Indian Federal Court (1941–47). He led the Pakistan delegation to the U.N., 1947–54, and became a judge of the International Court, 1954–61 (Vice-President, 1958–61) and again 1964–73 (President, 1970–73).

Zakonik. A code published by Stephen Dusan in Serbia in 1349 and reissued in enlarged form in 1354. It comprises a shortened version of Blastares' *Syntagma* (q.v.) and a collection called Justinian's Law, which is probably a code issued by Justinian II in the seventh or eighth century, and rules supplementary to these codes based on Serbian customary law or derived from Byzantine law.

Zaleucus (*c.* 650 B.C.). Lawgiver of Locri in Italy and compiler of the first written Greek codification, a body of rules famed for their severity and relying considerably on the *lex talionis* and prescriptions of exact penalties for each crime. His code was adopted in many cities of Italy and Sicily.

Zamora, The, [1916] 2 A.C. 77. The Zamora, a Swedish ship, was seized as prize by a British warship. The court held that the King in Council, or any branch of the executive, had no prerogative power to prescribe or alter the law, and also that the Prize Court administers not national or municipal law but international law, which could not be altered by Order in Council.

Zasius, Ulrich (1461–1536). A German jurist and leading humanist, one of the group which inaugurated a rational and scientific method for the study of Roman law. In 1500 he became professor at Freiburg. He was author of *Epitome in usibus feudorum* (1537), *Responsa sive Consilia* (1541), and *In titulos aliquot Pandectarum* (1543).

Zenzelinus de Cassanis (?–1334). French canonist, auditor of the Rota, author of a *Lectura* on the Sext, a commentary on the *Clementines*, and of glosses on some decretals of John XXII.

Ziletti, Francois (sixteenth century). Printer, publisher of the *Tractatus Tractatuum* (q.v.).

Zitelmann, Ernst (1852–1923). German jurist and author of *Die Rechtsgeschäfte im Entwurf eines bürgerlichen Gesetzbuchs für das deutsche Reich* (1889), *Das Recht des bürgerlichen Gesetzbuchs* (1900), and, particularly, an important *Internationales Privatrecht* (1897–1912).

Zoesius, Henricus (1571–1627). A Dutch humanist and author of *Commentarius ad Digestorum libros L* (1645), a work held in high regard, *Commentarius ad Decretales epistolas Gregorii IX* (1647), *Commentarius ad Institutionum libros* (1653), and *Praelectiones feudales* (1641).

Zonaras, John (?–*c.* 1160). Byzantine historian and canonist, author of an important *Epitome historiarum Libri*, a universal history to the year A.D. 1118, and of a massive commentary on the Apostolic Constitutions, the councils and synods, which is probably the greatest achievement of Byzantine canon law.

Zorn, Philipp (1850–1928). German jurist, notable in ecclesiastical and constitutional law, and author of *Lehrbuch des Kirchenrechts* (1888) and *Das Staatsrecht des Deutschen Reichs* (1880–83). Later he became interested in international law.

Zouche, Richard (1590–1661). Became an advocate of Doctors' Commons in 1617 and, in 1620, became Regius Professor of Civil Law at Oxford. Later he was an M.P. and Chancellor of the diocese of Oxford. In 1641 he became judge of the High Court of Admiralty but was replaced in 1643. In 1661 he recovered that office but held it only a month. His literary output was extensive and varied, including particularly *Elementa Jurisprudentiae* (1629), intended to lay down a generally applicable scheme of legal science, *Descriptio Juris et Judicii Feudalis* (1634), *Descriptio Juris et Judicii Temporalis* (1636), *Descriptio Juris et Judicii Ecclesiastici* (1636), *Descriptio Juris et Judicii Sacri* (1640), *Descriptio Juris et Judicii Militaris* (1640), *Descriptio Juris et Judicii Maritimi* (1640), and *Juris et Judicii Fecialis, sive Juris inter Gentes Explicatio* (1650), all of which deal with topics of his general logical scheme, and other works particularly *Solutio questionis veteris et novae, sive de legati delinquentis judice competente dissertatio* (1657) and *Quaestionum juris civilis centuria* (1600). His main importance rests on the *Juris et Judicii Fecialis* which is the first to set out the law of nations as an ordered system, relegating the law of warfare to a subordinate position therein. His approach is positivist and practical rather than doctrinal and he cites and founds on recorded instances. He preferred the title *jus inter gentes* to *jus gentium* for international law.

Zuichemus, Viglius d'Aytta de Zuichem (1507–77). Belgian jurist, successor of Alciatus at Bourges, president of the council of justice in the Netherlands and adviser to the regent (Margaret of Parma), first editor of the paraphrase of Theophilus, and author of *De Institutione jurisconsulti* (1530), *Commentaria in decem titulos Institutionum iuris civilis* (1534), *Praelectiones in titulum Pandectarum de rebus creditis* (1582), and other works.

Zypaeus (van den Zype) Franciscus (1578–1650). A Belgian jurist, author of *Jus pontificium novum* (1620), *Judex, magistratus, senator, libris IV exhibitus* (1633), and, most important, *Notitia iuris Belgici* (1635), the first attempt to systematise Belgian law. He also wrote a *De jurisdictione ecclesiastica et civili libri IV* (1649) and other works.

APPENDIX I

Lists of the holders of various offices since 1660

1. Kings and Queens Regnant of England, 975–1603

For earlier monarchs see *Handbook of British Chronology* (R. Hist. Soc.)

Edward the Martyr	975–979	
Ethelred (Unraed)	979–1016	
Swegn Forkbeard	1013–1014	
Edmund Ironside	1016	
Cnut	1016–1035	
Harold Harefoot	1035–1040	
Harthacnut	1040–1042	
Edward the Confessor	1042–1066	
Harold Godwinson	1066	
Norman kings		
William I (the Conqueror)	1066–1087	
William II (Rufus)	1087–1100	
Henry I	1100–1135	
Stephen	1135–1154	
Angevin kings		
Henry II	1154–1189	
Richard I (Lionheart)	1189–1199	
John	1199–1216	
Henry III	1216–1272	
Edward I	1272–1307	
Edward II	1307–1327	
Edward III	1327–1377	
Richard II	1377–1399	abdicated
House of Lancaster		
Henry IV	1399–1413	
Henry V	1413–1422	
Henry VI	1422–1461	deposed
House of York		
Edward IV	1461–1470	

102	*House of Lancaster*		
103	Henry VI	1470–1471	restored; deposed
104	*House of York*		
105	Edward IV	1471–1483	restored
106	Edward V	1483	deposed
107	Richard III	1483–1485	
108	*House of Tudor*		
109	Henry VII	1485–1509	
110	Henry VIII	1509–1547	
111	Edward VI	1547–1553	
112	Jane (Grey)	1553	deposed
113	Mary I (Tudor)	1553–1558	
114	Elizabeth I	1558–1603	

115

Kings and Queens Regnant of Scotland, 1005–1603

116	Malcolm II	1005–1034	
117	Duncan I	1034–1040	
118	Macbeth	1040–1057	
119	Lulach	1057–1058	
120	Malcolm III (Canmore)	1058–1093	
121	Donald Bane	1093–1097	
122	Duncan II	1094	
123	Edgar	1097–1107	
124	Alexander I	1107–1124	
125	David I	1124–1153	
126	Malcolm IV (the Maiden)	1153–1165	
127	William I (the Lion)	1165–1214	
128	Alexander II	1214–1249	
129	Alexander III	1249–1286	
130	Margaret (the Maid of Norway)	1286–1290	
131	Interregnum	1290–1292	
132	John (Balliol)	1292–1296	
133	Interregnum	1296–1306	
134	Robert I (the Bruce)	1306–1329	
135	David II	1329–1371	
136	*House of Stewart*		
137	Robert II	1371–1390	
138	Robert III	1390–1406	
139	James I	1406–1437	
140	James II	1437–1460	
141	James III	1460–1488	
142	James IV	1488–1513	
143	James V	1513–1542	
144	Mary, Queen of Scots	1542–1567	abdicated
145	James VI	1567–1625	
146	(from 1603 also James I of England)		

APPENDIX I

Kings and Queens Regnant of Great Britain and of the United Kingdom, 1603–

House of Stewart		
James I (also James VI of Scotland from 1567)	1603–1625	
Charles I	1625–1649	
Interregnum	1649–1653	
Oliver Cromwell, Lord Protector	1653–1658	
Richard Cromwell, Lord Protector	1658–1659	
Interregnum	1659–1660	
Charles II	1660–1685	(regnal years dated from 1649)
James II and VII	1685–1688	
Interregnum	1688–1689	
William III and Mary II (1689–94)	1689–1702	
Anne	1702–1714	
House of Hanover		
George I	1714–1727	
George II	1727–1760	
George III	1760–1820	regency, 1811–20
George IV	1820–1830	
William IV	1830–1837	
Victoria	1837–1901	
House of Saxe-Coburg and Gotha		
Edward VII	1901–1910	
House of Windsor		
George V	1910–1936	
Edward VIII	1936	abdicated
George VI	1936–1952	
Elizabeth II	1952–	

2. Lord Chancellors of England, 1660–1707

For Lord Chancellors prior to 1660 see *Handbook of British Chronology* (R. Hist. Soc., 1961), 80–87; Foss, Vols. 1–6; Campbell, Vols. 1–3.

Sir Edward Hyde	1658–67	Lord Hyde, 1660; E. of Clarendon, 1661
[Sir Orlando Bridgman, L.K.]	1667–72	
Anthony Ashley Cooper, E. of Shaftesbury	1672–73	
[Sir Heneage Finch, L.K.]	1673–75	Lord Finch, 1674
Lord Finch	1675–82	1st E. of Nottingham, 1681
[Sir Francis North, L.K.]	1682–85	1st L. Guildford, 1683
George Jeffreys, 1st L. Jeffreys	1685–89	Late C.J., K.B.
Great Seal in commission	1689–93	
[Sir John Somers, L.K.]	1693–97	
Sir John Somers	1697–1700	1st L. Somers, 1697
Great Seal in commission	1700	
[Sir Nathan Wright, L.K.]	1700–5	
[William Cowper, L.K.]	1705–7	1st L. Cowper, 1706

3. Lord Chancellors of Scotland, 1660–1707

For Lord Chancellors of Scotland prior to 1660 see *Handbook of British Chronology*, 173–77; *Introduction to Scottish Legal History* (Stair Soc. (1958), 459); Cowan, *The Lord Chancellors of Scotland*; Brunton and Haig, *Senators of the College of Justice.*

William Cunningham, 8th E. of Glencairn	1661–64	Late L.J.G.
[John Leslie, 7th E. of Rothes L.K.]	1664–67	
John Leslie, 7th E. of Rothes	1667–81	1st D. of Rothes, 1680
George Gordon of Haddo, 1st E. of Aberdeen	1682–84	
James Drummond, 4th E. of Perth	1684–88	
The Seal in commission	1689–92	
John Hay, 2nd E. of Tweeddale	1692–96	1st M. of Tweeddale, 1694
Patrick Hume, 1st Lord Polwarth	1696–1702	1st E. of Marchmont, 1697
James Ogilvy, 1st E. of Seafield	1702–4	See also 1705
John Hay, 2nd M. of Tweeddale	1704–5	
James Ogilvy, 1st E. of Seafield	1705–8	L.C.B. of Exchequer in Scotland, 1708; presided in the Court of Session as Chancellor, 1713

4. Lord Chancellors of Great Britain, 1707–

William, Lord Cowper	1707–08	1st E. Cowper, 1718; see 1714
The Seal in commission	1708–10	
[Sir Simon Harcourt, L. K.]	1710–13	1st L. Harcourt, 1711
Lord Harcourt	1713–14	1st Visc. Harcourt, 1721
Lord Cowper	1714–18	See 1707
The Seal in commission	1718	
Thomas Parker, 1st L. Macclesfield	1718–25	1st E. of Macclesfield, 1721
The Seal in commission	1725	
Peter King, 1st L. King	1725–33	Late C.J., C.P.
Charles Talbot, 1st L. Talbot	1733–37	
Philip Yorke, 1st L. Hardwicke	1737–56	Late C.J., K.B.; 1st E. of Hardwicke, 1754
The Seal in commission	1756–57	
[Sir Robert Henley, L.K.]	1757–61	1st L. Henley, 1760
Lord Henley	1761–66	1st E. of Northington, 1764
Charles Pratt, 1st L. Camden	1766–70	1st E. of Camden, 1786
Charles Yorke	1770	Died before created Lord Morden; never sat
The Seal in commission	1770–71	
Henry Bathurst, 1st L. Apsley	1771–78	2nd E. Bathurst, 1775
Edward Thurlow, 1st L. Thurlow	1778–83	See 1783
The Seal in commission	1783	
Lord Thurlow	1783–92	See 1778
The Seal in commission	1792–93	
Alexander Wedderburn, 1st L. Loughborough	1793–1801	1st E. of Rosslyn, 1801
John Scott, 1st L. Eldon	1801–6	See 1807
Thomas Erskine, 1st L. Erskine	1806–07	

Lord Eldon	1807–27	1st E. of Eldon, 1821
John Singleton Copley, 1st L. Lyndhurst	1827–30	See 1834, 1841
Henry Brougham, 1st L. Brougham and Vaux	1830–34	
Lord Lyndhurst	1834–35	See 1827, 1841
The Seal in commission	1835	
Charles Christopher Pepys, 1st L. Cottenham	1836–41	See 1846
Lord Lyndhurst	1841–46	See 1827, 1834
Lord Cottenham	1846–50	See 1836; 1st E. of Cottenham, 1850
The Seal in commission	1850	
Sir Thomas Wilde, 1st L. Truro	1850–52	
Sir Edward Burtenshaw Sugden, 1st L. St. Leonards	1852	
Robert Monsey Rolfe, 1st L. Cranworth	1852–58	See 1865
Sir Frederick Thesiger, 1st L. Chelmsford	1858–59	See 1866
John Campbell, 1st L. Campbell	1859–61	
Richard Bethell, 1st L. Westbury	1861–65	
Lord Cranworth	1865–66	See 1852
Lord Chelmsford	1866–68	See 1858
Hugh McCalmont Cairns, 1st L. Cairns	1868	See 1874
William Page Wood, 1st L. Hatherley	1868–72	
Roundell Palmer, 1st L. Selborne	1872–74	See 1880
Lord Cairns	1874–80	1st E. Cairns, 1878; see 1868
Lord Selborne	1880–85	1st E. of Selborne, 1882; see 1872
Hardinge Stanley Giffard, 1st L. Halsbury	1885–86	See 1886, 1895
Farrer Herschell, 1st L. Herschell	1886	See 1892
Lord Halsbury	1886–92	See 1885, 1895
Lord Herschell	1892–95	See 1886
Lord Halsbury	1895–1905	See 1885, 1886; 1st E. of Halsbury, 1898
Robert Threshie Reid, 1st L. Loreburn	1905–12	1st E. of Loreburn, 1911
Richard Burdon Haldane, 1st Visc. Haldane	1912–15	See 1924
Stanley Owen Buckmaster, 1st L. Buckmaster	1915–16	1st Visc. Buckmaster, 1932
Robert Bannatyne Finlay, 1st L. Finlay	1916–19	1st Visc. Finlay, 1919
Frederick Edwin Smith, 1st L. Birkenhead	1919–22	1st E. of Birkenhead, 1922
George Cave, 1st Visc. Cave	1922–24	See 1924
Viscount Haldane	1924	See 1912
Viscount Cave	1924–28	See 1922
Douglas McGarel Hogg, 1st L. Hailsham	1928–29	1st Visc. Hailsham, 1929
Sir John Sankey, 1st L. Sankey	1929–35	1st Visc. Sankey, 1935
Viscount Hailsham	1935–38	See 1928
Frederick Herbert Maugham, 1st L. Maugham	1938–39	Lord of Appeal, 1939

Thomas Walker Hobart Inskip, 1st Visc. Caldecote	1939–40	Later L.C.J.
John Allsebrook Simon, 1st Visc. Simon	1940–45	
William Allen Jowitt, 1st L. Jowitt	1945–51	1st Visc. Jowitt, 1947; 1st E. Jowitt, 1951
Gavin Turnbull Simonds, 1st L. Simonds	1951–54	1st Visc. Simonds, 1954
David Patrick Maxwell Fyfe, 1st Visc. Kilmuir	1954–62	
Reginald Edward Manningham-Buller, 1st Visc. Dilhorne	1962–64	Lord of Appeal, 1969
Gerald, Lord Gardiner	1964–70	Life Peer, 1964
Quintin McGarel Hogg, Lord Hailsham of St. Marylebone	1970–74	Renounced Viscountcy of Hailsham, 1963; Life peer, 1970
Frederick Elwyn Jones	1974–79	
Lord Hailsham of St. Marylebone	1979–	

5. Salaried Members of the Judicial Committee of the Privy Council, 1871–76

Appointed under the Judicial Committee Act, 1871

Sir James Colville	1871–1880	Late C.J., High Court, Bengal
Sir Montague Smith	1871–1881	Judge, C.P., 1865
Sir Robert Collier	1871–1886	Lord Monkswell, 1885
Sir Barnes Peacock	1872–1890	Late C.J., High Court, Bengal

6. Lords of Appeal in Ordinary, 1876–

Prior to the Appellate Jurisdiction Act, 1876, judges were occasionally ennobled to enable them to assist with the appellate work of the House of Lords. The 1876 Act authorized the appointment of two and later four Lords of Appeal in Ordinary. The number was raised in 1913 to six, in 1929 to seven, in 1947 to nine and in 1968 to eleven: Appellate Jurisdiction Acts, 1913, 1929, and 1947; Administration of Justice Act, 1968.

Peers who had held high judicial office, including judges given peerages during service or on retirement, have also sat in the House, but were not Lords of Appeal in Ordinary.

The numbers in brackets show the succession.

(1)	Colin Blackburn, Lord Blackburn	1876–87	Late judge, Q.B., 1859
(2)	Edward S. Gordon, Lord Gordon	1876–79	Late L.A. of Scotland
(2)	William Watson, Lord Watson	1880–99	Late L.A. of Scotland
(3)	John D. Fitzgerald	1882–89	Late judge, Q.B., Ireland
(1)	Edward Macnaghten	1887–1913	
(3)	Michael Morris	1889–1899	Late L.C.J. of Ireland; later Lord Killanin of Galway
(4)	James Hannen, Baron Hannen	1891–93	Late Pres., P.D.A. Div.
(4)	C.S.C. Bowen, Lord Bowen of Colwood	1893–94	
(4)	Charles Russell, Lord Russell of Killowen	1894	Later L.C.J.

(4)	Horace Davey, Lord Davey of Fernhurst	1894–1907	
(2)	James Patrick Bannerman Robertson, Lord Robertson	1899–1909	Late L.P., Ct. of Session
(3)	Nathaniel Lindley, Lord Lindley of East Carleton	1900–5	Late M.R.
(3)	John Atkinson, Lord Atkinson	1905–28	Late A.G. for Ireland
(4)	Richard Henn Collins, Lord Collins	1907–10	Late M.R.
(2)	Thomas Shaw, Lord Shaw	1909–29	Late L.A. of Scotland, Baron Craigmyle, 1929
(4)	William S. Robson, Lord Robson of Jesmond	1910–12	
(4)	John Fletcher Moulton, Lord Moulton of Bank	1912–21	
(1)	Robert John Parker, Lord Parker of Waddington	1913–18	Late judge, Ch. Div.
(5)	Andrew Graham Murray, Lord Dunedin	1913–32	Late L.P., Ct. of Session, Visc. Dunedin, 1926
(6)	James A. Hamilton, Lord Sumner of Ibstone	1913–30	Visc. 1927
(1)	George Cave, Visc. Cave	1919–22	Late L.C. Cave
(4)	Edward H. Carson, Lord Carson of Duncairn	1921–29	
(1)	Robert Younger, Lord Blanesburgh of Alloa	1923–37	
(3)	James R. Atkin, Lord Atkin of Aberdovey	1928–44	
(2)	William Watson, Lord Thankerton	1929–48	Late L.A. of Scotland
(7)	Thomas James Chesshyre Tomlin, Lord Tomlin	1929–35	
(4)	Rt. Hon. Frank Russell, Lord Russell of Killowen	1929–46	
(6)	Hugh Pattison Macmillan, Lord Macmillan of Aberfeldy	1930–39	Minister of Information 1939; see 1941
(5)	Robert A. Wright, Lord Wright of Durley	1932–35	Later M.R.; see 1937
(7)	Frederick H. Maugham, Lord Maugham	1935–38	Later L.C., Visc. 1939
(5)	Alexander A. Roche, Lord Roche	1935–38	
(1)	Lord Wright of Durley	1937–47	Late M.R.; see 1932
(5)	Mark Lemon Romer, Lord Romer of New Romney	1938–44	
(7)	Samuel Lowry Porter, Lord Porter of Longfield	1938–54	
(6)	Viscount Maugham	1939–41	
(6)	Lord Macmillan	1941–47	See 1930
(5)	Gavin Turnbull Simonds, Lord Simonds	1944–51	L.C. 1951–54; see 1954
(3)	Rayner Goddard, Lord Goddard	1944–46	Later L.C.J.
(4)	Augustus Andrewes Uthwatt, Lord Uthwatt	1946–49	Late judge, Ch. D.

(3) Herbert du Parcq, Lord du Parcq	1946–49	
(1) John Clarke MacDermott	1947–51	Judge, High Ct., N. Ireland; later L.C.J., N. Ireland
(8) Geoffrey Lawrence, Lord Oaksey	1947–57	
(9) Fergus Dunlop Morton, Lord Morton of Henryton	1947–59	
(6) Wilfrid Guild Normand, Lord Normand of Aberdour	1947–53	Late L.P., Ct. of Session
(2) James Scott Cumberland Reid, Lord Reid of Drem	1948–74	Late Dean of the Faculty of Advocates, Scotland
(4) Cyril John Radcliffe, Lord Radcliffe of Werneth	1949–64	Visc., 1962
(3) Wilfrid Arthur Greene, Lord Greene	1949–50	Late M.R.
(3) Frederick James Tucker, Lord Tucker	1950–61	
(5) Cyril Asquith, Lord Asquith of Bishopstone	1951–54	
(1) Lionel Leonard Cohen, Lord Cohen of Walmer	1951–60	
(6) James Keith, Lord Keith of Avonholm	1953–61	
(5) Donald Bradley Somervell, Lord Somervell of Harrow	1954–60	
(7) Gavin Turnbull Simonds, Viscount Simonds	1954–62	Late L.C.; see 1944
(8) Alfred Thompson Denning, Lord Denning	1957–62	Later M.R.
(9) David Llewellyn Jenkins, Lord Jenkins	1959–63	
(5) John William Morris, Lord Morris of Borth-y-Gest	1960–75	
(1) Francis Lord Chorlton Hodson, Lord Hodson	1960–71	
(6) Christopher William Graham Guest, Lord Guest of Graden	1961–71	
(3) Patrick Arthur Devlin, Lord Devlin	1961–64	
(7) Raymond Francis Evershed, Lord Evershed	1962–65	Late M.R.
(8) E. Holroyd Pearce, Lord Pearce	1962–69	
(9) Gerald Ritchie Upjohn, Lord Upjohn	1963–71	
(3) Terence Norbert Donovan, Lord Donovan	1964–71	
(4) Richard Orme Wilberforce, Lord Wilberforce	1964–	
(7) Colin Hargreaves Pearson, Lord Pearson	1965–74	
(10) William John Kenneth Diplock, Lord Diplock	1968–	
(8) Reginald Edward Manningham Buller, Viscount Dilhorne	1969–	Late L.C. 1962–64

(9) Geoffrey Cross, Lord Cross of Chelsea	1971–75	
(1) Joscelyn Edward Salis Simon, Lord Simon of Glaisdale	1971–78	Late Pres., P.D.A. Div.
(6) Charles James Dalrymple Shaw, Lord Kilbrandon	1971–76	
(3) Cyril Barnet Salmon, Lord Salmon	1972–	
(7) Herbert Edmund Davies, Lord Edmund-Davies	1974–	
(2) Walter Ian Reid Fraser, Lord Fraser of Tullybelton	1975–	
(9) Charles Ritchie Russell, Lord Russell of Killowen	1975–	
(6) Henry Shanks Keith, Lord Keith of Kinkel	1977–	
(1) Leslie George Scarman, Lord Scarman of Quatt	1978–	

7. Chief Justices, King's (or Queen's) Bench, 1660–1880

Sir Robert Foster	1660–63	
Sir Robert Hyde	1663–65	
Sir John Kelyng	1665–71	
Sir Matthew Hale	1671–76	Late C.B.
Sir Richard Rainsford	1676–78	
Sir William Scroggs	1678–81	
Sir Francis Pemberton	1681–83	Later C.J., C.P.
Sir Edmund Saunders	1683	
Sir George Jeffreys	1683–85	Later L.C.
Sir Edward Herbert	1685–87	Later C.J., C.P.
Sir Robert Wright	1687–89	
Sir John Holt	1689–1710	
Sir Thomas Parker	1710–18	Later Lord Parker of Macclesfield L.C. 1718; E. of Macclesfield, 1721
Sir John Pratt	1718–25	
Sir Robert Raymond	1725–33	Lord Raymond, 1731
Sir Philip Yorke, L. Hardwicke	1733–37	Later L.C. Hardwicke
Sir William Lee	1737–54	
Sir Dudley Ryder	1754–56	
Hon. William Murray, L. Mansfield	1756–88	Earl of Mansfield, 1776
Sir Lloyd Kenyon	1788–1802	Late M.R.
Sir Edward Law, L. Ellenborough	1802–18	
Sir Charles Abbott	1818–32	L. Tenterden, 1827
Sir Thomas Denman	1832–50	L. Denman, 1834
Sir John Campbell, L. Campbell	1850–59	Later L.C.
Sir Alexander Cockburn	1859–80	Late C.J., C.P. Designated Lord Chief Justice of England from 1859

8. Chief Justices, Court of Common Pleas, 1660–1880

Sir Orlando Bridgman	1660–67	Later Lord Keeper
Sir John Vaughan	1668–74	
Sir Francis North	1675–82	Later Lord Guildford
Sir Francis Pemberton	1683	Late C.J., K.B.
Sir Thomas Jones	1683–86	
Sir Henry Bedingfield	1686–87	
Sir Robert Wright	1687	Later C.J., K.B.
Sir Edward Herbert	1687–89	Late C.J., K.B.
Sir Henry Pollexfen	1689–91	
Sir George Treby	1692–1700	
Sir Thomas Trevor	1701–14	Lord Trevor, 1711
Sir Peter King	1714–25	Later Lord King and L.C.
Sir Robert Eyre	1725–36	Late C.B.
Sir Thomas Reeve	1736–37	
Sir John Willes	1737–61	
Sir Charles Pratt	1762–66	Lord Camden, 1765
Sir John Eardley Wilmot	1766–71	
Sir William de Grey	1771–80	Lord Walsingham, 1780
Sir Alexander Wedderburn	1780–93	Lord Loughborough, later L.C.
Sir James Eyre	1793–99	Late C.B.
Sir John Scott, Lord Eldon	1799–1801	Later L.C.
Sir Richard Pepper Arden, Lord Alvanley	1801–3	Later M.R.
Sir James Mansfield	1804–14	
Sir Vicary Gibbs	1814–18	Late C.B.
Sir Robert Dallas	1818–24	
Lord Gifford	1824	Later M.R.
Sir William Best	1824–29	Later Lord Wynford
Sir Nicolas Tindal	1829–46	
Sir Thomas Wilde	1846–50	Later L.C. Truro
Sir John Jervis	1850–56	
Sir Alexander Cockburn	1856–59	Later C.J., Q.B.
Sir William Erle	1859–66	
Sir William Bovill	1866–73	
Sir John Duke Coleridge	1873–80	Lord Coleridge, 1874

The office was abolished in 1880 and the Division merged with the Queen's Bench Division.

9. Chief Barons, Court of Exchequer, 1660–1880

Sir Orlando Bridgman	1660	Later C.J., C.P.
Sir Matthew Hale	1660–71	Later C.J., K.B.
Sir Edward Turnor	1671–76	
Hon. William Montagu	1676–86	
Sir Edward Atkyns	1686–89	
Sir Robert Atkyns	1689–94	
Sir Edward Ward	1695–1714	
Sir Samuel Dodd	1714–15	

Sir Thomas Bury	1716–22	
Sir James Montague	1722–23	
Sir Robert Eyre	1723–25	Later C.J., C.P.
Sir Jeffrey Gilbert	1725–26	
Sir Thomas Pengelly	1726–30	
Sir James Reynolds	1730–38	
Sir John Comyns	1738–40	
Sir Edward Probyn	1740–42	
Sir Thomas Parker	1742–72	
Sir Sydney Smythe	1772–77	
Sir John Skynner	1777–87	
Sir James Eyre	1787–93	Later C.J., C.P.
Sir Archibald Macdonald	1793–1813	
Sir Vicary Gibbs	1813	Later C.J., C.P.
Sir Alexander Thomson	1814–17	
Sir Richard Richards	1817–24	
Sir William Alexander	1824–31	
Lord Lyndhurst	1831–34	Late L.C., later L.C. again
Sir James Scarlett, Lord Abinger	1834–44	
Sir Jonathan Frederick Pollock	1844–66	
Sir Fitzroy Kelly	1866–80	

The office was abolished in 1880 and the Division merged with the Queen's Bench Division.

10. Lord Chief Justices of England since 1859

The office was created in 1859 by renaming the Chief Justice of the Queen's Bench, though the independent courts were not amalgamated into the Supreme Court of Judicature until 1875 and the offices of Lord Chief Justice of the Common Pleas and Lord Chief Baron of Exchequer were not abolished until 1880 when these Divisions were merged in the Q.B.D.

Sir Alexander Cockburn	1859–80	Late C.J., C.P.
Lord Coleridge	1880–94	Late C.J., C.P.
Lord Russell of Killowen	1894–1900	Late Lord of Appeal
Lord Alverstone	1900–13	Late M.R.
Sir Rufus Isaacs, later Lord Reading, Visc. Erleigh and E. Reading	1913–21	Late A.G.
Sir A. T. Lawrence, later Lord Trevithin	1921–22	Late puisne judge
Sir Gordon Hewart, Lord Hewart	1922–40	Late A.G.
Viscount Caldecote	1940–46	Late L.C.
Lord Goddard	1946–58	Late Lord of Appeal
Lord Parker of Waddington	1958–71	Late L.J.
Lord Widgery	1971–	Late L.J.

11. Masters of the Rolls from 1660

Sir Harbottle Grimston	1660–84	
Sir John Churchill	1685	
Sir John Trevor	1685–88	Later Speaker, H. of Commons
Sir Henry Powle	1689–92	

Sir John Trevor	1693–1717	Reinstated
Sir Joseph Jekyll	1717–38	
Hon. John Verney	1738–41	
William Fortescue	1741–49	
Sir John Strange	1750–54	
Sir Thomas Clarke	1754–64	
Sir Thomas Sewell	1764–84	
Sir Lloyd Kenyon	1784–88	Later C.J., K.B.
Sir Richard Pepper Arden	1788–1801	Later C.J., C.P.
Sir William Grant	1801–18	
Sir Thomas Plumer	1818–24	Late V.C.
Robert Gifford	1824–26	
John Singleton Copley	1826–27	Later L.C. Lyndhurst
Sir John Leach	1827–34	Late V.C.
Sir Charles Christopher Pepys	1834–36	Later L.C. Cottenham
Henry Bickersteth, L. Langdale	1836–51	
Sir John Romilly	1851–73	Baron, 1865
Sir George Jessel	1873–83	
Sir William Baliol Brett, L. Esher	1883–97	Later Visc. Esher, 1897
Sir Nathaniel Lindley	1897–1900	Later L. of Appeal
Sir R. E. Webster	1900	Later L.C.J. Alverstone
Sir A. L. Smith	1900–1	Late L.J.
Sir Richard Henn Collins	1901–7	Late L.J.; later L. of Appeal
Sir H. H. Cozens-Hardy	1907–18	Later Lord Cozens-Hardy of Letheringsett, 1914
Sir Charles Swinfen Eady	1918–19	Later L. Swinfen
Lord Sterndale	1919–23	Late Pres., P.D.A.
Sir Ernest Pollock	1923–35	Baron Hanworth, 1926; Visc., 1936
Lord Wright	1935–37	Late L. of Appeal; later L. of Appeal again
Sir Wilfrid Arthur Greene	1937–49	Later L. of Appeal
Sir Francis Raymond Evershed	1949–62	Later L. of Appeal
Lord Denning	1962–	Late L. of Appeal

12. Vice-Chancellors of England, 1813–75 and 1970–

Thomas Plumer	1813–18	Later M.R.
John Leach	1818–27	Later M.R.
Anthony Hart	1827	Later L.C. of Ireland
Lancelot Shadwell	1827–50	
* Sir J. L. Knight Bruce	1841–51	Later L.J.
* Sir James Wigram	1841–51	
Robert Monsey Rolfe	1850–51	Later L.J. and L.C. Cranworth
Sir J. Parker	1851–52	
Sir George J. Turner	1851–53	Later L.J.
R. T. Kindersley	1851–66	

* One Vice-Chancellor was appointed in 1813 and the number was raised to three in 1841.

John Stuart	1852–71	
W. Page Wood	1853–68	Later L.J. and L.C. Hatherley
Richard Malins	1866–81	
George M. Giffard	1868–69	Later L.J.
W. M. James	1869–70	Later L.J.
James Bacon	1870–86	Chief Judge in Bankruptcy, 1869–84
John Wickens	1871–73	
Charles Hall	1873–82	

The office of Vice-Chancellor was revived in 1970 and the holder became the working head of the Chancery Division

Sir John Pennycuick	1970–74	
Sir John Anthony Plowman	1974–76	
Sir Robert Edgar Megarry	1976–	

13. Lords Justices (Court of Appeal in Chancery, 1851–75)

Sir J. L. Knight Bruce	1851–66	Late V.C.
R. M. Rolfe	1851–52	Late V.C., later L.C. Cranworth
Sir G. J. Turner	1853–67	Late V.C.
H. M. Cairns	1866–68	Later L.C. Cairns
John Rolt	1867–68	
Charles J. Selwyn	1868–69	
W. Page Wood	1868	Late V.C., later L.C. Hatherley
G. M. Giffard	1869–70	Late V.C.
W. M. James	1870–81	Late V.C.
Geo. Mellish	1870–77	

In 1875 this court was merged in the new Court of Appeal.

14. Lords Justices of Appeal (called Justices of Appeal, 1875–77) Court of Appeal, 1875–

Sir Richard Baggallay	1875–85	
Sir Geo. W. W. Bramwell	1876–81	Later Lord Bramwell, 1882
Sir Wm. Baliol Brett	1876–83	Later M.R. and Lord Esher
Sir Richard P. Amphlett	1876–77	Late Baron of Exchequer, 1874–76
Sir Henry Cotton	1877–90	
Sir Alfred H. Thesiger	1877–80	
Sir Robert Lush	1880–81	
Sir Nathaniel Lindley	1881–97	Later M.R. and L. of Appeal
Sir John Holker	1882	
Sir Charles S. C. Bowen	1882–93	Later L. of Appeal

Sir Edward Fry	1883–92	
Sir Henry C. Lopes	1885–97	Later Lord Ludlow
Sir Edward E. Kay	1890–96	
Sir A. L. Smith	1892–1900	Later M.R.
Sir Horace Davey	1893–94	Later L. of Appeal
Sir John Rigby	1894–1901	
Sir Joseph W. Chitty	1897–99	
Sir Richard Henn Collins	1897–1901	Later M.R. and L. of Appeal
Sir Roland Vaughan Williams	1897–1914	
Sir Robert Romer	1899–1906	
Sir James Stirling	1900–1906	
Sir James C. Mathew	1901–1906	
Sir Herbert H. Cozens-Hardy	1901–1907	Later M.R.
Sir John Fletcher Moulton	1906–1912	Later L. of Appeal
Sir George Farwell	1906–1915	
Sir Henry B. Buckley	1906–15	Later Lord Wrenbury
Sir Wm. R. Kennedy	1907–15	
Sir J. A. Hamilton	1912–1913	Later L. of Appeal (Sumner)
Sir C. Swinfen Eady	1913–1918	Later M.R.
Sir W. G. F. Phillimore	1913–1916	Later Lord Phillimore
Sir Wm. Pickford	1914–1919	Later Lord Sterndale, M.R.
Sir John E. Bankes	1915–1927	
Sir T. R. Warrington	1915–1926	Later Lord Warrington of Clyffe, 1926
Sir T. E. Scrutton	1916–1934	
Sir H. E. Duke	1918–1919	Later Lord Merrivale, Pres. P.D.A.
Sir James R. Atkin	1919–1928	Later Lord Atkin, L. of Appeal
Sir R. T. Younger	1919–1923	Later Lord Blanesburgh, L. of Appeal
Sir C. Sargant	1923–1928	
Sir P. O. Lawrence	1926–1934	
Sir F. A. Greer	1927–1938	Later Lord Fairfield
Sir John Sankey	1928–29	Later L.C.
Frank Russell	1928–29	
Sir H. H. Slesser	1929–40	
Sir Mark L. Romer	1929–38	
Sir F. H. Maugham	1934–35	Later L. of Appeal and L.C.
Sir A. A. Roche	1934–35	Later L. of Appeal
Sir Wilfrid A. Greene	1935–37	Later M.R
Sir Leslie F. Scott	1935–48	
Sir F. D. Mackinnon	1937–46	
Sir A. C. Clauson	1938–42	Later Baron Clauson
William, 2nd Visc. Finlay	1938–45	
Sir A. F. C. C. Luxmoore	1938–44	
Sir Rayner Goddard	1938–44	Later L. of Appeal and L.C.J.
Sir Herbert du Parcq	1938–46	Later L. of Appeal
Sir Geoffrey Lawrence	1944–47	Later Lord Oaksey, L. of Appeal
Sir Fergus D. Morton	1944–47	Later Lord Morton of Henryton, L. of Appeal
Sir Fred. J. Tucker	1945–50	Later L. of Appeal
Sir Alfred T. Bucknill	1945–51	

Sir Donald B. Somervell	1946–54	Later L. of Appeal
Sir Lionel Leonard Cohen	1946–51	Later L. of Appeal
Sir Cyril Asquith	1946–51	Later L. of Appeal
Sir F. J. Wrottesley	1947–48	
Sir F. Raymond Evershed	1947–49	Later M.R.
Sir J. E. Singleton	1948–57	
Sir A. T. Denning	1948–57	Later L. of Appeal and M.R.
Sir D. L. Jenkins	1949–59	Later L. of Appeal
Sir W. Norman Birkett	1950–57	
Sir F. L. C. Hodson	1951–60	Later L. of Appeal
Sir John W. Morris	1951–60	Later L. of Appeal
Sir Charles R. R. Romer	1951–60	
Sir Hubert L. Parker	1954–58	Later L.C.J.
Sir F. A. Sellers	1957–68	
Sir Benjamin Ormerod	1957–63	
Sir E. Holroyd Pearce	1957–62	Later L. of Appeal
Sir H. Gordon Willmer	1959–69	
Sir Charles E. Harman	1959–70	
Sir Patrick A. Devlin	1960–61	Later L. of Appeal
Sir Gerald R. Upjohn	1960–65	Later L. of Appeal
Sir Terence Norbert Donovan	1960–63	Later L. of Appeal
Sir Harold Otto Danckwerts	1961–69	
Sir Colin Hargreaves Pearson	1961–65	Later L. of Appeal
Sir W. Arthian Davies	1961–74	
Sir William John Kenneth Diplock	1961–68	Later L. of Appeal
Sir Charles Ritchie Russell	1962–75	Later L. of Appeal
Sir Cyril Barnet Salmon	1964–72	Later L. of Appeal
Sir Charles Roger Noel Winn	1965–71	
Sir Eric Sachs	1966–73	
Sir Herbert Edmund Davies	1966–74	Later L. of Appeal
Sir John Passmore Widgery	1968–71	Later L.C.J.
Sir Fenton Atkinson	1968–71	
Sir Henry Josceline Phillimore	1968–74	
Sir Seymour Edward Karminski	1969–73	
Sir John Megaw	1969	
Sir Geoffrey Cross	1969–71	Later L. of Appeal
Sir Denys Burton Buckley	1970	
Sir David Arnold Scott Cairns	1970–77	
Sir Edward Blanshard Stamp	1971–78	
Sir John Frederick Eustace Stephenson	1971	
Sir Alan Stewart Orr	1971	
Sir Eustace Wentworth Roskill	1971	
Sir Frederick Horace Lawton	1972	
Sir Arthur Evan James	1973–76	
Sir Leslie George Scarman	1973–77	Later L. of Appeal
Sir Roger Fray Greenwood Ormrod	1974	
Sir Patrick Reginald Evelyn Browne	1974	
Sir Geoffrey Dawson Lane	1974	
Sir Reginald William Goff	1975	
Sir Nigel Cyprian Bridge	1975	
Sir Sebag Shaw	1975	

Sir George Stanley Waller	1976
Sir James Roualeyn Hovell-Thurlow Cumming-Bruce	1977
Sir Edward Walter Eveleigh	1977
Sir Henry Vivian Brandon	1978
Sir Sydney William Templeman	1978

15. Judges of the Court of Admiralty, 1660–1875

Richard Zouche	1641–43 and 1661	
Thomas Hyde	1660–61	
William Turner	1660	
John Exton	1661–68	
Sir Leoline Jenkins	1668–84	
Sir Richard Lloyd	1685–86	
Sir Thomas Exton	1686–89	
Sir Richard Raines	1686–89	
Charles Hedges	1689–1714	
Humphrey Henchman	1714	
Sir Henry Newton	1714–15	
Sir Henry Penrice	1715–51	
Sir Thomas Salusbury	1751–73	
Sir George Hay	1773–78	
Sir James Marriott	1778–98	
Sir William Scott, L. Stowell	1798–1828	
Sir Christopher Robinson	1828–33	
Sir John Nicholl	1833–38	
Stephen Lushington	1838–67	
Sir Robert Joseph Phillimore	1867–83	Sat in P.D.A. Div., 1875–83

16. Court of the Arches, 1660–

Giles Sweit	1660–72	
Sir Robert Wiseman	1672–84	
Sir Richard Lloyd	1684–86	
Sir Thomas Exton	1686–88	
George Oxendon	1689–1703	
Sir John Cooke	1703–10	
John Bettesworth	1710–51	
Sir George Lee	1751–58	
Sir Edward Simpson	1758–64	
Sir George Hay	1764–78	
Peter Calvert	1778–88	
Sir William Wynne	1788–1809	
John Nicholl	1809–34	
Herbert Jenner (later Jenner-Fust)	1834–52	
John Dodson	1852–57	
Stephen Lushington	1858–67	
Sir R. J. Phillimore	1867–75	Sat in P.D.A. till 1883
Lord Penzance	1875–98	

Sir Arthur Charles	1898–1903
Sir Lewis T. Dibdin	1903–34
Sir Philip W. Baker Wilbraham	1934–55
Sir Henry Willink	1955–71
Dr Walter Wigglesworth	1971–72
Sir Harold Kent	1972–76
Rev. Kenneth Elphinstone	1977–

17. Probate and Divorce Courts, 1857–75

Sir Cresswell Cresswell	1858–63	
Sir James Wilde	1863–72	Lord Penzance, 1869
Sir James Hannen	1872–91	Judge Q.B. 1868; Pres. P.D.&A. Div. 1875–91; later L. of Appeal

18. Presidents of the Probate, Divorce, and Admiralty Division, 1875–1971

Sir James Hannen	1875–91	L. of Appeal, 1891
Sir Charles Parker Butt	1891–92	
Sir Francis Henry Jeune	1892–1905	Later L. St Helier, 1905
Sir John Gorell Barnes	1905–9	Later Lord Gorell
Sir John Charles Bigham	1909–10	Later Lord Mersey of Toxteth; Viscount Mersey, 1916
Sir Samuel T. Evans	1910–18	
Sir William Pickford, Lord Sterndale	1918–19	Later M.R.
Sir Henry Edward Duke	1919–33	Later Lord Merrivale, 1925
Sir Frank Boyd Merriman	1933–62	Later Lord Merriman, 1941
Sir Jocelyn E. S. Simon	1962–71	Later Lord Simon of Glaisdale and L. of Appeal
Sir George Gillespie Baker	1971	Later President of the Family Division

19. Presidents of the Family Division, 1972–

Sir George Gillespie Baker	1971–	Late President, P.D.&A. Division

20. Presidents of the Restrictive Practices Court, 1956–

Sir Patrick Arthur Devlin	1956–60	Later L.J. and L. of Appeal
Sir Colin H. Pearson	1960–61	Later L.J. and L. of Appeal
Sir Kenneth Diplock	1961	Later L.J. and L. of Appeal
Sir Charles R. Russell	1961–62	Later L.J. and L. of Appeal
Sir John Megaw	1962–68	Later L.J.
Sir Denys B. Buckley	1968–70	Later L.J.
Sir Alan A. Mocatta	1970–	

21. Presidents of the National Industrial Relations Court, 1971–74

Sir John Donaldson	1971–74	

22. Presidents of the Employment Appeal Tribunal, 1976–

Sir John R. Phillips	1976–78	
Sir Gordon Slynn	1978–	

23. Chairmen of the Law Commission

Sir Leslie Scarman	1965–73	Seconded from P.D.A.D.; later L.J. and L. of Appeal, 1977
Sir Samuel Burgess Ridgway Cooke	1973–78	Seconded from Q.B.D.
Sir Michael R. E. Kerr	1978–	Seconded from Q.B.D.

24. Chairmen of the Scottish Law Commission

Charles James Dalrymple Shaw, Lord Kilbrandon	1965–71	Seconded from Court of Session. L. of Appeal, 1971
John Oswald Mair Hunter, Lord Hunter	1971–	Seconded from Court of Session

25. Attorneys-General of England from 1660

For earlier holders of the office see Foss, Vols. 1–6

Sir Geoffrey Palmer	1660–70	
Sir Heneage Finch	1670–73	Later L. Keeper
Sir Francis North	1673–75	Later C.J., C.P.
Sir William Jones	1675–79	
Sir Cresswell Levinz	1679–81	
Sir Robert Sawyer	1681–87	
Sir Thomas Powys	1687–88	
Sir Henry Pollexfen	1689	
Sir George Treby	1689–92	
Sir John Somers	1692–93	Later L. Keeper
Sir Edward Ward	1693–95	Later C.B.
Sir Thomas Trevor	1695–1701	Later C.J., C.P.
Sir Edward Northey	1701–7	
Sir Simon Harcourt	1707–8	Later L. Keeper
Sir James Montagu	1708–10	
Sir Edward Northey	1710–18	
Sir Nicholas Lechmere	1718–20	
Sir Robert Raymond	1720–24	
Sir Phillip Yorke	1724–34	Later C.J., K.B.

John Willes	1734–37	Later C.J., C.P.
Dudley Ryder	1737–54	Later C.J., K.B.
Hon. William Murray	1754–56	Later C.J., K.B.
Sir Robert Henley	1756–57	Later L. Keeper
Sir Charles Pratt	1757–62	Later C.J., C.P.
Hon. Charles Yorke	1762–63	See 1765
Hon. Fletcher Norton	1763–65	Speaker, House of Commons, 1770–80
Hon. Charles Yorke	1765–66	See 1762
Sir William de Grey	1766–71	Later C.J., C.P.
Sir Edward Thurlow	1771–78	Later L.C.
Sir Alexander Wedderburn	1778–80	Later C.J., C.P. and L.C.
Sir James Wallace	1780–82	
Sir Lloyd Kenyon	1782–83	
Sir James Wallace	1783	
Sir John Lee	1783	
Sir Lloyd Kenyon	1783–84	Later M.R.
Sir Richard Pepper Arden	1784–88	Later M.R.
Sir Archibald Macdonald	1788–93	Later C.B.
Sir John Scott	1793–99	Later C.J., CP.
Sir John Mitford	1799–1801	Later Speaker
Sir Edward Law	1801–2	Later C.J., K.B.
Sir Spencer Perceval	1802–6	
Sir Arthur Piggot	1806–7	
Sir Vicary Gibbs	1807–12	
Sir Thomas Plumer	1812–13	Later V.C.
Sir William Garrow	1813–17	
Sir Samuel Shepherd	1817–19	Later C.B. of Exch. in Scotland
Sir Robert Gifford	1819–24	Later C.J., C.P.
Sir John S. Copley	1824–26	Later M.R. and L.C. Lyndhurst
Sir Charles Wetherall	1826–27	
Sir James Scarlett	1827–28	
Sir Charles Wetherall	1828–29	
Sir James Scarlett	1829–30	
Sir Thomas Denman	1830–32	Later C.J., K.B.
Sir William Horne	1832–34	
Sir John Campbell	1834	
Sir Jonathan F. Pollock	1834–37	See 1841
Sir John Campbell	1837–41	Later L.C. of Ireland, 1841; L.C. 1859
Sir Thomas Wilde	1841	See 1846
Sir Jonathan F. Pollock	1841–44	Later C.B.
Sir William W. Follett	1844–45	
Sir Frederick Thesiger	1845–46	See 1852
Sir Thomas Wilde	1846	Later C.J., C.P. and L.C. Truro
Sir John Jervis	1846–50	Later C.J., C.P.
Sir John Romilly	1850–51	Later M.R.
Sir Alexander J. E. Cockburn	1851–52	
Sir Frederick Thesiger	1852	Later L.C. Chelmsford
Sir Alexander J. E. Cockburn	1852–56	Later C.J., C.P.
Sir Richard Bethell	1856–58	See 1859
Sir Fitzroy Kelly	1858–59	Later C.B.

Sir Richard Bethell	1859–61	Later L.C. Westbury
Sir William Atherton	1861–63	
Sir Roundell Palmer	1863–66	Later L.C. Selborne
Sir Hugh McCalmont Cairns	1866	Later L.C.
Sir John Rolt	1866–67	Later L.J.
Sir John B. Karslake	1867–68	See 1874
Sir Robert P. Collier	1868–71	Lord Monkswell, 1885
Sir John Duke Coleridge	1871–73	Later C.J., C.P. and C.J., Q.B.
Sir Henry James	1873	See 1880
Sir John B. Karslake	1874–75	
Sir Richard Baggallay	1874–75	Later L.J. 1875–85
Sir John Holker	1876–80	Later L.J.
Sir Henry James	1880–85	Later Lord James of Hereford
Sir Richard E. Webster	1885–86	See 1886
Sir Charles Russell	1886	See 1892
Sir Richard E. Webster	1886–92	See 1895
Sir Charles Russell	1892–94	Later L. of Appeal and L.C.J.
Sir John Rigby	1894	Later L.J.
Sir Robert Threshie Reid	1894–95	Later L.C. Loreburn
Sir Richard E. Webster	1895–1900	Later M.R. and L.C.J. Alverstone
Sir Robert B. Finlay	1900–5	Later L.C. Finlay
Sir John L. Walton	1905–8	
Sir William S. Robson	1908–10	Later L. of Appeal
Sir Rufus D. Isaacs	1910–13	Later L.C.J.
Sir John Simon	1913–15	Later L.C.
Sir Edward Carson	1915	Later L. of Appeal
Sir F. E. Smith	1915–19	Later L.C. Birkenhead
Sir Gordon Hewart	1919–22	Later L.C.J.
Sir Ernest M. Pollock	1922	Later M.R. and baronet
Sir Douglas M. Hogg	1922–24	See 1924
Sir Patrick Hastings	1924	
Sir Douglas M. Hogg	1924–28	Later L.C. Hailsham
Sir Thomas W. H. Inskip	1928–29	Later L.C. and L.C.J.
Sir William A. Jowitt	1929–32	Later L.C
Sir Thomas W. H. Inskip	1932–36	Later Minister for Defence, L.C. and L.C.J.
Sir Donald Bradley Somervell	1936–45	Later L.J. and L. of Appeal
Sir David P. Maxwell Fyfe	1945	Later L.C. Kilmuir
Sir Hartley W. Shawcross	1945–51	Later Lord Shawcross
Sir Frank Soskice	1951	Later Lord Stow Hill
Sir Lionel F. Heald	1951–54	
Sir Reginald Manningham Buller	1954–62	Later L.C. Dilhorne
Sir John Hobson	1962–64	
Sir Frederick Elwyn Jones	1964–70	Later L.C.
Sir Peter Rawlinson	1970–74	
Samuel Silkin	1974–79	
Sir Michael Havers	1979–	

26. Solicitors-General of England from 1660

For earlier holders of the office see Foss, Vols. 1–6

Sir Heneage Finch	1660–70	Later A.G.

Sir Edward Turnor	1670–71	Later C.B.
Sir Francis North	1671–73	Later A.G.
Sir William Jones	1673–74	Late A.G.
Sir Francis Winnington	1674–79	
Hon. Heneage Finch	1679–85	
Sir Thomas Powys	1686–87	
Sir William Williams	1687–89	
Sir George Treby	1689	
Sir John Somers	1689–92	
Sir Thomas Trevor	1693–95	Later A.G.
Sir John Hawles	1695–1702	
Sir Simon Harcourt	1702–7	Later A.G.
Sir James Montagu	1707–8	Later A.G.
Sir Robert Eyre	1708–10	Later J., Q.B.
Sir Robert Raymond	1701–14	
Nicholas Lechmere	1714–15	
John Fortescue-Aland	1715–17	
William Thomson	1717–20	
Sir Philip Yorke	1720–24	Later A.G.
Sir Clement Wearg	1724–26	
Sir Charles Talbot	1726–34	Later L.C.
Dudley Ryder	1734–37	Later A.G.
John Strange	1737–42	
Hon. William Murray	1742–54	Later A.G.
Sir Richard Lloyd	1754–76	
Hon. Charles Yorke	1756–62	Later A.G.
Fletcher Norton	1762–63	Later A.G.
William de Grey	1763–66	Later A.G.
Edward Willes	1766–68	
John Dunning	1768–70	
Alexander Wedderburn	1771–77	Later A.G.
James Wallace	1778–80	Later A.G.
James Mansfield	1780–82	
John Lee	1782	
Richard Pepper Arden	1782–83	
John Lee	1783	Later A.G.
James Mansfield	1783	
Richard Pepper Arden	1783–84	Later A.G.
Archibald Macdonald	1784–88	Later A.G.
John Scott	1788–93	Later A.G.
John Mitford	1793–99	Later A.G.
William Grant	1799–1801	Later M.R.
Spencer Perceval	1801–2	Later A.G.
Thomas Manners Sutton	1802–5	
Vicary Gibbs	1805–6	
Samuel Romilly	1806–7	
Thomas Plumer	1807–12	Later A.G.
William Garrow	1812–13	Later A.G.
Robert Dallas	1813	
Samuel Shepherd	1813–17	
Robert Gifford	1817–19	Later A.G.

John S. Copley	1819–24	Later A.G.
Charles Wetherall	1824–26	Later A.G.
Nicholas C. Tindal	1826–29	
Edward B. Sugden	1829–30	
William Horne	1830–32	Later A.G.
John Campbell	1832–34	Later A.G.
C. C. Pepys	1834	Later M.R.
R. M. Rolfe	1834	
William W. Follett	1834–35	See 1841
R. M. Rolfe	1835–39	
Thomas Wilde	1839–41	Later A.G.
William W. Follett	1841–44	Later A.G.
Frederick Thesiger	1844–45	Later A.G.
Fitzroy Kelly	1845–46	See 1852
John Jervis	1846	Later A.G.
David Dundas	1846–48	
John Romilly	1848–50	Later A.G.
Alexander J. E. Cockburn	1851	Later A.G.
William Page Wood	1851–52	
Fitzroy Kelly	1852	Later A.G.
Richard Bethell	1852–56	Later A.G.
Hon. J. S. Wortley	1856–57	
Henry S. Keating	1857–58	
Hugh M. Cairns	1858–59	
Henry S. Keating	1859–	Judge, C.P. 1859–75
William Atherton	1859–61	Later A.G.
Roundell Palmer	1861–63	Later A.G.
Robert P. Collier	1863–66	Later A.G.
William Bovill	1866	C.J., C.P. 1866
John B. Karslake	1866–67	Later A.G.
Charles J. Selwyn	1867	L.J. 1868
Richard Baggallay	1868	See 1874
William Baliol Brett	1868	Judge C.P. 1868, M.R. 1883
John Duke Coleridge	1868–71	Later A.G.
George Jessel	1871–73	M.R. 1873
Henry James	1873	Later A.G.
W. Vernon Harcourt	1873	
Richard Baggallay	1874	Later A.G.
John Holker	1874–75	
Hardinge Stanley Giffard	1875	
Farrer Herschell	1880–85	
John Eldon Gorst	1885	
Horace Davey	1886	L.J. 1893, L. of Appeal, 1894
Edward George Clarke	1886–92	
John Rigby	1892–94	Later A.G.
Robert Threshie Reid	1894	Later A.G.
Frank Lockwood	1894–95	
Robert B. Finlay	1895–1900	Later A.G.
Edward Carson	1900–5	Later A.G.
William S. Robson	1905–8	Later A.G.
Samuel T. Evans	1908–10	Later Pres. P.D.A.

Rufus Daniel Isaacs	1910	Later A.G.
John Allsebrook Simon	1910–13	Later A.G.
Stanley Owen Buckmaster	1913–15	
Frederick E. Smith	1915	Later A.G.
George Cave	1915–16	Later Home Secretary
Gordon Hewart	1916–19	Later A.G.
Ernest Murray Pollock	1919–22	Later A.G.
Leslie F. Scott	1922	Later L.J.
Thomas W. H. Inskip	1922–24	
Henry H. Slesser	1924	Later L.J.
Thomas W. H. Inskip	1924–28	Later A.G.
Frank Boyd Merriman	1928–29	
James B. Melville	1929–30	
Richard Stafford Cripps	1930–31	
Thomas W. H. Inskip	1931–32	Later A.G.
Frank Boyd Merriman	1932–33	Later Pres. P.D.A.
Donald Bradley Somervell	1933–36	Later A.G.
Terence J. O'Connor	1936–40	
William Allen Jowitt	1940–42	
David P. Maxwell Fyfe	1942–45	Later A.G.
Sir Walter Monckton	1945	Later Viscount Monckton
Frank Soskice	1945–51	Later A.G.
Arwyn Lynn Ungoed Thomas	1951	Later J.
Reginald E. Manningham Buller	1951–54	Later A.G.
Henry B. Hylton Foster	1954–59	Later Speaker of H. of Commons
Jocelyn E. S. Simon	1959–62	Later Pres. P.D.A.
John G. S. Hobson	1962–64	Later A.G.
Peter A. G. Rawlinson	1962–64	Later A.G.
Dingle Foot	1964–67	
Arthur Irvine	1966–70	
Geoffrey Howe	1970–72	
Michael Havers	1972–74	
Peter Kingsley	1974–79	
Ian Percival	1979–	

27. King's (or Queen's) Advocates, 1660–1872 (often called Advocate-General)

John Godolphin	1660–78	
Thomas Exton	1679–86	
Sir Thomas Pinfold	1687	
Thomas Exton	1687–88	
Sir John Cooke	1701–10	
Sir Nathaniel Lloyd	1715–27	
George Paul	1727–55	
Sir G. Hay	1755–64	
James Marriott	1764–78	Judge of Admiralty Ct. 1778
Dr (later Sir W.) Wynne	1778–88	
William Scott	1788–98	Later Judge of Admiralty, 1798

John Nicholl	1798–1809	Judge of Admiralty, 1833
Christopher Robinson	1809–28	Judge of Admiralty, 1828
Herbert Jenner (later Jenner-Fust)	1828–34	
John Dodson	1834–52	Dean of the Arches, 1852
John D. Harding	1852–62	
Robert Joseph Phillimore	1862–67	
Travers Twiss	1867–72	

On the retirement of Sir Travers Twiss the office came to an end. Sir James Parker Deane was legal adviser to the Foreign Office 1872–86, but was never Queen's Advocate.

28. Admiralty Advocates, 1660–1870 (sometimes called Admiralty Advocate-General)

T. Hyde	1660	
W. Turner	1661	
Sir W. Walker	*c.* 1670	
R. Lloyd	1674	
T. Pinfold	1685	
W. Oldys	1686	
F. Littleton	1693	
H. Newton	1697	
H. Penrice	1714	
R. Fuller	1715	
E. Sayer	1727	
E. Isham	1731	
W. Strahan	1742	
T. Salusbury	1748	
C. Pinfold	1751	
J. Bettesworth	1756	
G. Harris	1764	
William Scott	1782	King's Advocate 1788; Judge of the Admiralty Court 1798 (Lord Stowell)
T. Bever	1788	
W. Battine	1791	
John Dodson	1829	King's Advocate 1834
Joseph Phillimore	1834	
Robert Joseph Phillimore	1855	Queen's Advocate 1862; Judge of the Admiralty Court 1867
Travers Twiss	1862	
James Parker Deane	1868	Legal Adviser to the Foreign Office 1872–86

29. Directors of Public Prosecutions

Sir John B. Maule	1879–83	
Sir Augustus Stephenson	1883–94	
Sir Hamilton Cuffe	1894–1908	Later Lord Desart
Sir Charles Mathews	1908–20	

Sir Archibald Bodkin	1920–30
Sir Edward Tindal Atkinson	1930–44
Sir Theobald Mathew	1944–64
Sir Norman Skelhorn	1964–77
Thomas Chalmers Hetherington	1977–

30. Lords Justice-General of Scotland from 1660

For earlier holders see *Handbook of British Chronology* (R. Hist. Soc.), 194

John Murray, 2nd E. of Atholl	1661–75	1st M. of Atholl, 1676
Alexander Stewart, 5th E. of Moray	1675–76	
Sir Archibald Primrose of Carrington	1676–78	
Sir George Mackenzie of Tarbat	1678–80	Earl of Cromarty 1703. See 1705
William Douglas, 3rd E. of Queensberry	1680–82	1st D. of Queensberry, 1684
James Drummond, 1st E. of Perth	1682–84	
George Livingston, 3rd E. of Linlithgow	1684–88	
Robert Ker, 4th E. of Lothian	1689–1703	1st M. of Lothian 1701
George Mackenzie, 1st E. of Cromarty	1705–10	See 1678
Archibald Campbell, Earl of Ilay	1710–61	3rd D. of Argyll, 1743
John Hay, 4th M. of Tweeddale	1761–62	
Charles Douglas, 3rd D. of Queensberry	1763–78	
David Murray, 7th Visc. Stormont	1778–94	2nd E. of Mansfield, 1792
James Graham, 3rd D. of Montrose	1795–1836	Died 30 Dec. 1836

By the Court of Session Act, 1830, s. 18, the office of Justice-General was, on the death of the then holder, to be held thereafter by the holder of the office of Lord President of the Court of Session. Under this Lord President Hope was installed as Lord Justice-General on 23rd January 1837.

31. Lords President of the Court of Session from 1660

For earlier holders of the office see *Handbook of British Chronology* (R. Hist. Soc.) 192 and *Introduction to Scottish Legal History* (Stair Soc.), 459–60.

Sir John Gilmour of Craigmillar	1661–71	
Sir James Dalrymple	1671–81	See 1689
Sir George Gordon of Haddo, 1st E. of Aberdeen	1681–82	L. Ch. of Sc. 1682
Sir David Falconer of Newton	1682–85	
Sir George Lockhart of Carnwath	1685–89	
Sir James Dalrymple	1689–95	1st Visc. Stair, 1690
Sir Hew Dalrymple of North Berwick	1698–1737	
Duncan Forbes of Culloden	1737–48	
Robert Dundas of Arniston	1748–54	
Robert Craigie of Glendoick	1754–60	
Robert Dundas of Arniston	1760–87	
Sir Thomas Millar of Barskimming and Glenlee	1787–89	
Sir Ilay Campbell of Succoth	1789–1808	
Robert Blair of Avontoun	1808–11	
Charles Hope of Granton	1811–41	
David Boyle of Shewalton	1841–52	

Duncan McNeill of Colonsay	1852–67	An ordinary L. of Session, 1851–52, 1st L. Colonsay, 1867
John Inglis of Glencorse	1867–91	
James Patrick Bannerman Robertson	1891–99	L. of Appeal, 1899
John Blair Balfour	1899–1905	1st L. Kinross, 1902
Andrew Graham Murray, 1st B. Dunedin	1905–13	L. of Appeal, 1913
Alexander Ure	1913–20	1st B. Strathclyde, 1914
James Avon Clyde	1920–35	
Wilfrid Guild Normand	1935–47	L. of Appeal and 1st B. Normand of Aberdour, 1947
Thomas Mackay Cooper	1947–54	1st B. Cooper of Culross 1954
James Latham McDiarmid Clyde	1954–72	
George Carlyle Emslie	1972	An ordinary L. of Session, 1970–72

32. Lords Justice-Clerk of Scotland from 1660

For earlier holders of the office see *Introduction to Scottish Legal History* (Stair Soc.), 461–2; 27 J.R. 342, 375, 47 J.R. 311; Brunton and Haig.

Sir Robert Murray	1651–63	
Sir John Home of Renton, Lord Renton	1663–71	
Sir James Lockhart of Lee, Lord Lee	1671–74	
Sir William Lockhart of Lee	1674–75	
Sir Thomas Wallace of Craigie, Lord Craigie	1675–80	
Richard, Lord Maitland	1680–84	4th E. of Lauderdale
Sir James Foulis of Colinton, Lord Colinton	1684–88	
Sir John Dalrymple, Earl of Stair	1688–90	
Sir Alexander (? or George) Campbell of Cessnock, Lord Cessnock	1690–92	
Sir Adam Cockburn of Ormiston, Lord Ormiston	1692–99	See 1705, 1714
Sir John Maxwell of Pollok, Lord Pollok	1699–1702	
Roderick Mackenzie of Prestonhall, Lord Prestonhall	1702–4	
Sir William Hamilton of Whitelaw, Lord Whitelaw	1704–5	
Sir Adam Cockburn of Ormiston, Lord Ormiston	1705–10	See 1714
Sir James Erskine of Grange, Lord Grange	1710–14	
Adam Cockburn of Ormiston, Lord Ormiston	1714–35	
Andrew Fletcher, Lord Milton	1735–48	
Charles Erskine, Lord Tinwald	1748–63	
Sir Gilbert Eliot of Minto, Lord Minto	1763–66	
Sir Thomas Miller of Barskimming, and Glenlee, Bt., Lord Glenlee	1766–87	Later L. Pres. 1787–89

Robert Macqueen of Braxfield, Lord Braxfield	1787–99	
Sir David Rae of Eskgrove, Lord Eskgrove	1799–1804	
Charles Hope of Granton	1804–11	Later L. Pres. 1811–41
David Boyle of Shewalton	1811–41	Later L. Pres. 1841–52
John Hope	1841–58	
John Inglis of Glencorse, Lord Glencorse	1858–67	Later L. Pres. 1867–91
George Patton, Lord Glenalmond	1867–69	
James Moncrieff of Tullibole	1869–88	
Sir John Hay Athole Macdonald, Lord Kingsburgh	1888–1915	
Charles Scott Dickson	1915–22	
Robert Munro, Lord Alness	1922–33	
Craigie Mason Aitchison	1933–41	
Thomas Mackay Cooper	1941–47	Later L. Pres. 1947–54
Alexander Moncrieff	1947	Ordinary Lord of Session 1926–47
George Reid Thomson	1947–62	
William Grant	1962–72	
John Wheatley	1972–	Ordinary Lord of Session, 1954–72

33. Chief Barons of the Court of Exchequer in Scotland, 1707–1832

James Ogilvy, 1st E. of Seafield	1707–8	
John Smith	1709–26	
Matthew Lant	1726–41	
John Idle	1741–55	
Robert Ord	1755–75	
Sir James Montgomery of Stanhope	1775–1801	Later Lord Advocate
Robert Dundas	1801–19	Late Dean of Faculty
Sir Samuel Shepherd	1819–30	
James Abercromby	1830–32	Later Baron Dunfermline

The office was abolished in 1832 and the Court was merged with the Court of Session in 1856.

34. Lords Commissioners of the Jury Court in Scotland, 1815–30

L. Ch. Commr. William Adam	1815–30	Judge, C. of Session, 1830–33
David Monypenny, Lord Pitmilly	1815–30	Judge, C. of Session, 1813–30
Allan Maconochie, Lord Meadowbank	1815–16	Judge, C. of Session, 1796–1816
Adam Gillies, Lord Gillies	1816–30	Judge, C. of Session, 1811–37; later B. of Exchequer, 1837–42
Joshua Henry Mackenzie, Lord Mackenzie	1825–30	Judge, C. of Session, 1822–51
James Wolfe Murray, Lord Cringletie	1825–30	Judge, C. of Session, 1825–34

The Jury Court was merged with the Court of Session in 1830.

35. Chairmen, Scottish Land Court, 1912–

* Neil J. D. Kennedy, Lord Kennedy	1912–18	
David Anderson, Lord St. Vigeans	1918–34	
R. Macgregor Mitchell, Lord Macgregor Mitchell	1934–38	
T. D. King Murray, Lord Murray	1938–41	Later Judge, C. of Session (Lord Birnam)
Robert Gibson, Lord Gibson	1941–65	
Harald Leslie, Lord Birsay	1965–78	
Walter Archibald Elliott, Lord Elliott	1978–	

* Chairman of the Crofters Commission, 1908–1912. He succeeded Sir David Brand, who was Chairman, 1886–1908.

36. Judges of the Scottish Court of Admiralty, 1661–1830

James Robertson of Bedlay	1661–64	
?	1664–68	
John Stewart of Kettleston	*c.* 1668–72	
Walter Pringle of Graycrook	1672–74	
Patrick Lyon of Carse	1675–83	
Sir David Thoirs of Inverkeithing	1684–89	
Walter Dundas of Kincavil	1689–(?)95	
Sir Archibald Sinclair	1695–99	(*fl.* 1664–1719)
Sir Robert Forbes of Learney	1699–1701 and 1704–10 jointly	
James Graham of Airth	1702–46	
James Philip of Greenlaw	1746–82	
Archibald Cockburn of Cockpen	1782–90	
Sir George Buchan Hepburn	1790–1800	Baron of Exchequer, 1800–14
Robert Hodgson Cay	1800–10	
John Burnett	1810	
James Wolfe Murray of Cringletie	1811–16	Judge, C. of Session, 1816
Sir John Connell	1816–30	

37. Lords Lyon King of Arms

For earlier holders of the office and biographical notes see Grant (ed.) *Court of the Lord Lyon, 1318–1945* (Scottish Record Socy.)

Sir Alexander Durham of Largs	1660
Sir Charles Erskine of Cambo, Bart.	1663
Sir Alexander Erskine of Cambo, Bart.	1672
Sir Alex. Erskine of Cambo, younger, Bart. (jointly)	1701
Alexander Brodie of that Ilk	1727
John Hooke Campbell of Bangeston	1754
Robert Boswell (interim)	1795
Robert Auriol, Earl of Kinnoull	1796
Thomas Robert, Earl of Kinnoull	1804
George Burnett, advocate	1866

Sir James Balfour Paul	1890
Captain George Sitwell Campbell Swinton	1927
Sir Francis James Grant	1929
Sir Thomas Innes of Learney	1945
Sir James Monteith Grant	1970

38. Lord Advocates of Scotland from 1660

For earlier holders see Omond, *Lord Advocates of Scotland*

Sir John Fletcher	1661–64	
Sir John Nisbet	1664–77	Lord Dirleton
Sir Geo. Mackenzie of Rosehaugh	1677–87	See 1688
Sir John Dalrymple	1687–88	See 1689
Sir George Mackenzie of Rosehaugh	1688–89	
Sir John Dalrymple	1689–92	Master of Stair, later 1st E. of Stair
Sir James Steuart, Bt., of Goodtrees	1692–1709	See 1711
Sir David Dalrymple of Hailes	1709–11	See 1714
Sir James Steuart, Bt. of Goodtrees	1711–13	
Thomas Kennedy of Dunure	1714	Later Baron of Exchequer, 1721
Sir David Dalrymple of Hailes, Bt.	1714–20	
Robert Dundas, Yr., of Arniston	1720–25	Later Lord Pres. Arniston
Duncan Forbes of Culloden	1725–37	Later Lord Pres. Forbes
Charles Erskine of Barjarg	1737–42	Later L. J. C. Tinwald
Robert Craigie of Glendoick	1742–46	Later L. P. Craigie
William Grant of Prestongrange	1746–52	Later Lord Prestongrange
Robert Dundas of Arniston	1754–60	Later Lord Pres. Arniston
Thomas Miller of Barskimming and Glenlee, Bt.	1760–66	Later L.J.C. and L.P. Glenlee
James Montgomery of Stanhope	1766–75	Later C.B. Exch.
Rt. Hon. Henry Dundas	1775–83	Later Viscount Melville
The Hon. Henry Erskine	1783	See 1806
Ilay Campbell of Succoth	1783–89	Later L.P. Succoth
Robert Dundas of Arniston	1789–1801	Later Chief Baron Dundas
Charles Hope of Granton	1801–4	Later L.J.C. and L.P. Hope
Sir James Montgomery, Bt.	1804–6	
The Hon. Henry Erskine	1806–7	
Archibald Campbell (later Campbell Colquhoun)	1807–16	Later Lord Clerk Register
Alexander Maconochie	1816–19	Later Lord Meadowbank
Rt. Hon. Sir William Rae, Bt.	1819–30	See 1834
Francis Jeffrey	1830–34	Later Lord Jeffrey
John Archibald Murray	1834–	See 1835
Rt. Hon. Sir William Rae, Bt.	1834–35	See 1841
John Archibald Murray	1835–39	Later Lord Murray
Andrew Rutherfurd	1839–41	See 1846
Rt. Hon. Sir William Rae, Bt.	1841–42	
Duncan McNeill	1842–46	Later L.P. McNeill and Baron Colonsay
Andrew Rutherfurd	1846–51	Later Lord Rutherfurd
James Moncrieff	1851–52	See 1852
Adam Anderson	1852	Later Lord Anderson

John Inglis	1852	See 1858
James Moncrieff	1852–58	See 1859
John Inglis	1858	Later L.J.C. and L.P. Inglis
Charles Baillie	1858–59	Later Lord Jerviswoode
David Mure	1859	Later Lord Mure
James Moncrieff	1859–66	See 1868
George Patton	1866–67	Later L.J.C. Patton
Edward Strathearn Gordon	1867–68	See 1874
James Moncrieff	1868–69	Later L.J.C. Moncrieff
George Young	1869–74	Later Lord Young
Edward S. Gordon	1874–76	Later Baron Gordon
William Watson	1876–80	Later Baron Watson
John McLaren	1880–81	Later Lord McLaren
J. B. Balfour	1881–85	See 1886
J. H. A. Macdonald. Q.C.	1885–86	See 1886
J. B. Balfour, Q.C.	1886–	See 1892
J. H. A. Macdonald, Q.C.	1886–88	Later L.J.C. Macdonald
J. P. B. Robertson, Q.C.	1888–91	Later L.P. Robertson and Baron Robertson
Sir Charles J. Pearson, Q.C.	1891–92	See 1895
J. B. Balfour, Q.C.	1892–95	Later L.P. Kinross
Sir Charles J. Pearson, Q.C.	1895–96	Later Lord Pearson
A. Graham Murray, Q.C.	1896–1903	Later L.P. Dunedin
Charles Scott Dickson, Q.C.	1903–5	Later L.J.C. Scott Dickson
Thomas Shaw, Q.C.	1905–9	Later Baron Shaw and Baron Craigmyle
Alexander Ure, K.C.	1909–13	Later L.P. Strathclyde
Robert Munro, K.C.	1913–16	Later L.J.C. Alness
James A. Clyde, K.C.	1916–20	Later L.P. Clyde
T. B. Morison, K.C.	1920–22	Later Lord Morison
C. D. Murray, K.C.	1922	Later Lord Murray
William Watson, K.C.	1922–24	See 1924
H. P. Macmillan, K.C.	1924–	Later Baron Macmillan
William Watson, K.C.	1924–29	Later Baron Thankerton
A. M. MacRobert, K.C.	1929	
Craigie M. Aitchison, K.C.	1929–33	Later L.J.C. Aitchison
W. G. Normand, K.C.	1933–35	Later L.P. Normand and (1947) Baron Normand
T. M. Cooper, K.C.	1935–41	Later L.J.C. and L.P. Cooper
J. S. C. Reid, K.C.	1941–45	Later Baron Reid of Drem
G. R. Thomson, K.C.	1945–47	Later L.J.C. Thomson
John Wheatley, K.C.	1947–51	Later Lord Wheatley
J. L. M. Clyde, K.C.	1951–55	Later L.P. Clyde
W. R. Milligan, Q.C.	1955–60	Later Lord Milligan
William Grant, Q.C.	1960–62	Later L.J.C. Grant
Ian H. Shearer, Q.C.	1962–64	Later Lord Avonside
G. Gordon Stott, Q.C.	1964–67	Later Lord Stott
H. S. Wilson, Q.C.	1967–70	Later Baron Wilson of Langside (life peer, 1969), Director, Scottish Courts Administration and Sheriff of Glasgow

Norman R. Wylie, Q.C.	1970–74	Later Lord Wylie
Ronald King Murray, Q.C.	1974–79	Later Lord Murray
James P. H. Mackay, Q.C.	1979–	Later Lord Mackay of Clashfern (life peer, 1979)

39. Solicitors-General for Scotland, 1660–

For further information see 54 Juridical Review, 67, 125.

Robert Dalgleish	1647–62	
Sir William Purves	1662–84	
John Purves	1678–84	
Sir William Purves	1683–84	
George Bannerman		Appointment ineffective
George Bannerman	1684–87	
Robert Colt	1684–87	
James Graham	1687–89	
Sir William Lockhart	1689–93	
Sir James Ogilvy	1693–96	Later Secretary of State and (1702) Lord Chancellor
Sir Patrick Hume	1696–1700	
Sir David Dalrymple of Hailes	1701–6	Later L.A.
William Carmichael	1701–9	
Thomas Kennedy	1709–14	Later L.A.
Sir James Steuart, Bt.	1709–14	
John Carnegie of Boyseck	1714–16	
Sir James Steuart, Bt.	1714–17	
Robert Dundas of Arniston	1717–20	Later L.A.
Walter Stewart	1720–21	
John Sinclair	1721–33	Later Lord Murkle
Charles Binning	1721–25	
Charles Erskine of Tinwald	1725–37	Later L.A., Lord Tinwald, 1742; L.J.C. 1748
William Grant of Prestongrange	1737–42	Later L.A. and Lord of Session
Robert Dundas of Arniston	1742–46	Later L.A.
Patrick Haldane of Gleneagles	1746–55	
Alexander Hume	1746–55	Later Principal Clerk of Session
Andrew Pringle of Alemore	1755–59	Later Lord Alemore
Thomas Miller of Barskimming and Glenlee	1759–60	Later L.A., L.J.C. and L.P.
James Montgomery	1760–64	
Francis Garden	1760–64	Later Lord Gardenstone
James Montgomery	1764–66	Later L.A.
Henry Dundas	1766–75	Later L.A.
Alexander Murray of Henderland	1775–83	Later Lord Henderland
Ilay Campbell of Succoth	1783	Later L.A. and L.P.
Alexander Wight	1783–84	
Robert Dundas of Arniston	1784–89	
Robert Blair of Avontoun	1789–1806	Later L.A. and L.P.
John Clerk of Eldin	1806–7	Later Lord Eldin
David Boyle of Shewalton	1807–11	Later L.J.C. Boyle

David Monypenny of Pitmilly	1811–13	Later Lord Pitmilly
Alexander Maconochie	1813–16	Later L.A. and Lord Meadowbank
James Wedderburn	1816–22	
John Hope	1822–30	Later L.J.C. Hope
Henry Thomas Cockburn	1830–34	Later Lord Cockburn
Andrew Skene	1834–35	
Duncan McNeill	1835	See 1841
John Cunninghame	1835–37	Later Lord Cunninghame
Andrew Rutherfurd	1837–39	Later L.A.
James Ivory	1839–40	Later Lord Ivory
Thomas Maitland of Dundrennan	1840–41	See 1846
Duncan McNeill	1841–42	Later L.A. and L.P.
Adam Anderson	1842–46	Later L.A.
Thomas Maitland of Dundrennan	1846–50	Later Lord Dundrennan
James Moncrieff	1850–51	Later L.A.
John Cowan	1851	Later Lord Cowan
George Deas	1851–52	Later Lord Deas
John Inglis of Glencorse	1852	Later L.A.
Charles Neaves	1852–53	Later Lord Neaves
Robert Handyside	1853	Later Lord Handyside
James Craufurd	1853–55	Later Lord Ardmillan
Thomas Mackenzie	1855	Later Lord Mackenzie
Edward Francis Maitland	1855–58	See 1859
Charles Baillie	1858	Later L.A.
David Mure	1858–59	Later L.A.
George Patton	1859	Later L.A.
Edward Francis Maitland	1859–62	Later Lord Barcaple
George Young	1862–66	See 1868
Edward Strathearn Gordon	1866–67	Later L.A.
John Millar	1867–68	See 1874
George Young	1868–69	Later L.A.
Andrew Rutherfurd Clark	1869–74	Later Lord Rutherfurd Clark
John Millar	1874–76	Later Lord Craighill
William Watson	1874–76	Later L.A.
John Hay Athole Macdonald	1876–80	Later L.A.
John Blair Balfour	1880–81	Later L.A.
Alexander Asher	1881–85	See 1886, 1892
James Patrick Bannerman Robertson	1885–86	See 1886
Alexander Asher	1886	See 1892
J. P. B. Robertson	1886–88	Later L.A.
Moir Tod Stormonth Darling	1888–90	Later Lord Stormonth Darling
Sir Charles John Pearson	1890	Later L.A.
Andrew Graham Murray	1891–92	See 1895
Alexander Asher	1892–94	Later Dean of Faculty
Thomas Shaw	1894–95	Later L.A.
Andrew Graham Murray	1895–96	Later L.A.
Charles Scott Dickson	1896–1903	Later L.A.
David Dundas	1903–5	Later Lord Dundas
Edward Theodore Salvesen	1905	Later Lord Salvesen
James Avon Clyde	1905	Later L.A.
Alexander Ure	1905–9	Later L.A.

Arthur Dewar	1909–10	Later Lord Dewar
William Hunter	1910–11	Later Lord Hunter
Andrew Macbeth Anderson	1911–13	Later Lord Anderson
Thomas Brash Morison	1913–20	Later L.A.
Charles David Murray	1920–22	Later L.A.
Andrew Henderson Briggs Constable	1922	Later Lord Constable
Hon. William Watson	1922	Later L.A.
David Pinkerton Fleming	1922–23	See 1924
Frederick Charles Thomson	1923–24	
John Charles Fenton	1924	Later Sheriff of the Lothians
David Pinkerton Fleming	1924–26	Later Lord Fleming
Alexander Munro MacRobert	1926–29	Later L.A.
Wilfrid Guild Normand	1929	See 1931
John Charles Watson	1929–31	Later Sheriff of Caithness
Wilfrid Guild Normand	1931–33	Later L.A. and L.P.
Douglas Jamieson	1933–35	Later L.A.
Thomas Mackay Cooper	1935	Later L.A., L.J.C. and L.P.
Albert Russell	1935	Later Lord Russell
James Scott Cumberland Reid	1936–41	Later L.A. and Lord of Appeal
Sir David King Murray	1941–45	Later Lord Birnam
Daniel Patterson Blades	1945–47	Later Lord Blades
John Wheatley	1947	Later L.A., Lord Wheatley and L.J.C. (also Life Peer)
Douglas Harold Johnston	1947–51	Later Lord Johnston
William Rankine Milligan	1951–55	Later L.A.
William Grant	1955–60	Later L.A. and L.J.C.
David Colvill Anderson	1960–64	Later Permanent Reporter of Inquiries to the Secretary of State
Norman Russell Wylie	1964	Later L.A. and Lord Wylie
James Graham Leechman	1964–65	Later Lord Leechman
Henry Stephen Wilson	1965–67	Later L.A., Director of Scottish Courts Administration and Sheriff of Glasgow
Ewan George Francis Stewart	1967–70	Later Lord Stewart
David William Robert Brand	1970–72	Later Lord Brand
William Ian Stewart	1972–74	Later Lord Allanbridge
John H. McCluskey	1974–79	Later Lord McCluskey of Churchill (Life Peer)
Nicholas Fairbairn	1979–	

40. Lord Chancellors of Ireland

For earlier holders of the office see Ball, *The Judges in Ireland*

Sir Maurice Eustace	1660
Archbp. Michael Boyle	1665
Sir Charles Porter	1686
Alexander Fitton	1687
Sir Charles Porter	1690
John Methven	1697

Sir Richard Cox	1703	
Richard Freeman	1707	
Sir Constantine Phipps	1710	
Alan, Viscount Brodrick	1714	
Richard West	1725	
Thomas, Lord Wyndham	1726	
Robert, Viscount Jocelyn	1739	
John, Lord Bowes	1757	
James, Viscount Hewitt	1767	
John, Lord Fitzgibbon	1789	
John Mitford	1802	
George Ponsonby	1806	
Thomas Maurice Sutton	1807	
Anthony Hart	1827	
William Conyngham Plunket	1830	See 1835
Edward Burtenshaw Sugden	1834	See 1841
William Conyngham Plunket	1835	See 1830
John Campbell	1841	Later L.C. of G.B.
Edward Burtenshaw Sugden	1841	Later L.C. of G.B. (St. Leonards)
Maziere Brady	1846	See 1852, 1859
Francis Blackburne	1852	See 1866
Maziere Brady	1852	See 1846, 1859
Joseph Napier	1858	
Maziere Brady	1859	See 1846, 1852
Francis Blackburne	1866	See 1852
Abraham Brewster	1867	
Thomas O'Hagan	1868	See 1880
The seal in commission	1874	
John Thomas Ball	1875	
Thomas O'Hagan	1880	See 1868
Hugh Law	1881	
Edward Sullivan	1883	
John Naish	1885	See 1881
Edward Gibson, Lord Ashbourne	1885	See 1886, 1896
John Naish	1886	See 1885
Edward Gibson, Lord Ashbourne	1886	See 1885, 1890
Samuel Walker	1892	
Edward Gibson, Lord Ashbourne	1895	See 1885, 1886
Samuel Walker	1905	
Redmond John Barry	1911	
Ignatius John O'Brien	1913	
James Henry Mussen Cambell	1918	Later Lord Glenavy
John Ross	1921	

41. Masters of the Rolls in Ireland

For earlier holders of the office see Ball, *The Judges in Ireland*

Sir John Temple	1641	
Sir William Temple	1677	See 1690
Sir William Talbot	1689	

Sir William Temple	1690	
William, Lord Berkeley	1696	
Thomas Carter	1731	
Henry Singleton	1754	
Richard Rigby	1759	
William Robert Fitzgerald, Duke of Leinster	1788	
John Crosbie, Earl of Glandore and John Joshua Proby, Earl of Carysfort	1789	
Michael Smith	1801	
John Philpot Curran	1806	
William MacMahon	1814	
Michael O'Loghlen	1837	
Francis Blackburne	1842	Later L.C.J., Q.B.I. and L.C.I.
Thomas Berry Cusack Smith	1846	
John Edward Walsh	1866	
Edward Sullivan	1870	Later L.C.I.
Andrew Marshall Porter	1883	
Richard Edmund Meredith	1906	
Charles Andrew O'Connor	1912–24	Later judge of the Supreme Court of the Irish Free State, 1924–25

42. Vice-Chancellor of Ireland

Hedges Eyre Chatterton	1867

43. Chief Justices of the Upper Bench or King's (or Queen's) Bench in Ireland, 1660–1924

For earler holders of the office see Ball, *The Judges in Ireland*

James Barry	1660	
Sir John Povey	1673	
Sir Robert Booth	1679	
Sir William Davys	1681	
Thomas Nugent	1687	
Sir Richard Reynell	1690	
Sir Richard Pyne	1695	
Alan Brodrick	1709	
Sir Richard Cox	1711	
William Whitshed	1714	Later C.J., C.P.I.
John Rogerson	1727	
Thomas Marlay	1741	
St. George Caufeild	1751	
Warden Flood	1760	

John Gore, Lord Annaly	1764	
John Scott, Earl of Clonmell	1784	
Arthur Wolfe, Visc. Kilwarden	1798	
William Downes	1803	
Charles Kendal Bushe	1822	
Edward Pennefather	1841	
Francis Blackburne	1846	Later M.R.I.; later L.C.I.
Thomas Langlois Lefroy	1852	
James Whiteside	1866	
George Augustus Chichester May	1877	
Michael Morris	1887	Chief Justice of Ireland, Lord of Appeal 1889
Peter O'Brien	1889	
Richard Robert Cherry	1913	
James Henry Mussen Campbell	1916	Later L.C.I.
Thomas Francis Molony	1918–24	

44. Chief Justices of the Common Bench or Common Pleas in Ireland, 1660–1887

For earlier holders of the office see Ball, *The Judges in Ireland*

Sir James Donnellan	1660	
Sir Edward Smythe	1665	
Sir Robert Booth	1670	Later C.J., K.B.I.
John Keatinge	1679	
Sir Richard Pyne	1691	Later C.J., K.B.I.
Sir John Hely	1695	
Sir Richard Cox	1701	Later L.C.I. and C.J., K.B.I.
Robert Doyne	1703	
John Forster	1714	
Sir Richard Levinge	1720	
Thomas Wyndham	1724	Later L.C.I.
William Whitshed	1727	
James Reynolds	1727	
Henry Singleton	1740	Later M.R.I.
William Yorke	1753	
Richard Aston	1761	Later Judge, English K.B.
Richard Clayton	1765	
Marcus Paterson	1770	
Hugh Carleton, Visc. Carleton	1787	
John Toler	1800	
William Conyngham Plunket	1827	Later L.C.I.
John Doherty	1830	
James Henry Monahan	1850	
Michael Morris	1876	Later C.J., Q.B.I. and C.J. of Ireland 1887

Court merged with Court of Queen's Bench, 1887.

45. Chief Barons of the Exchequer in Ireland, 1660–1898

For earlier holders of the office see Ball, *The Judges in Ireland*

John Bysse	1660	
Henry Henn	1680	
Sir Stephen Rice	1687	
Sir John Hely	1690	Later C.J., C.P.I.
Robert Doyne	1695	Later C.J., C.P.I.
Nehemiah Donnellan	1703	
Richard Freeman	1706	Later L.C.I.
Robert Rochfort	1707	
Joseph Deane	1714	
Jeffrey Gilbert	1715	Later C.B. Exch. in England
Bernard Hale	1722	
Thomas Dalton	1725	
Thomas Marlay	1730	Later C.J., K.B.I.
John Bowes	1741	Later L.C.I.
Edward Willes	1757	
Anthony Foster	1766	
James Dennis, Lord Tracton	1777	
Walter Hussey Burgh	1782	
Barry Yelverton, Visc. Avonmore	1783	
Standish O'Grady	1805	
Henry Joy	1831	
Stephen Woulfe	1838	
Maziere Brady	1840	Later L.C.I.
David Richard Pigot	1846	
Christopher Palles	1874	

Court merged in the Queen's Bench Division, 1898.

46. Lord Chief Justices of Ireland, 1887–1924

Sir Michael Morris	1887–89	Later Lord of Appeal
Peter O'Brien	1889–1913	
Richard Robert Cherry	1913–16	
James Henry Mussen Campbell	1916–18	Later L.C.I.
Thomas Francis Molony	1918–24	

The office ended when Ireland was divided into Irish Free State and Northern Ireland.

47. Chief Justices of Irish Free State, later Eire, later Republic of Ireland, 1924–

Hugh Kennedy	1924–36
Timothy Sullivan	1936–46
Conar Alexander Maguire	1946–61

Cearbhall O'Dalaigh	1961–72	Later judge of the European Court and President of the Republic of Ireland
William O. B. Fitzgerald	1973–74	
Thomas F. O'Higgins	1974–	

48. Lord Chief Justices of Northern Ireland, 1922–

Sir Denis Stanislaus Henry	1922–25	
Sir William Moore	1925–37	Baronet, 1932
Sir James Andrews	1937–51	
Lord MacDermott	1951–71	Lord of Appeal, 1947–51
Robert Lynd Erskine Lowry	1971–	Judge of High Court, Northern Ireland, 1964–71

49. Speakers of the House of Commons

House of Commons of Parliament of England, 1660–1707

Sir Harbottle Grimston	1660–
Sir Edward Turnour	1661–71
Sir Job Charlton	1673
Sir Edward Seymour	1673–78
Sir William Gregory	1679
Sir William Williams	1680–81
Sir John Trevor	1685–87
Henry Powle	1689
Sir John Trevor	1690–95
Paul Foley	1695–98
Sir Thomas Littleton	1698–1700
Robert Harley	1701–5
John Smith	1705–8

House of Commons of Parliament of Great Britain, 1707–1801

Sir Richard Onslow	1708–10	Baron Onslow
William Bromley	1710–13	
Sir Thomas Hanmer	1714–15	
Sir Spencer Compton	1715–27	Earl of Wilmington
Arthur Onslow	1728–61	
Sir John Cust	1761–70	
Sir Fletcher Norton	1770–80	Baron Grantley
Charles Cornwall	1780–89	
William Wyndham Grenville	1789	Baron Grenville
Henry Addington	1789–1801	Visc. Sidmouth

House of Commons of the Parliaments of the United Kingdom

Sir J. Mitford	1801–2	Later Baron Redesdale
C. Abbot	1802–17	Later Baron Colchester
C. Manners-Sutton	1817–34	Later Visc. Canterbury
J. Abercromby	1835–39	Later Baron Dunfermline
G. Shaw Lefevre	1839–57	Later Visc. Eversley
J. E. Denison	1857–72	Later Visc. Ossington
H. B. W. Brand	1872–84	Later Baron Dacre and Visc. Hampden
A. W. Peel	1884–95	Later Visc. Peel
W. C. Gully	1895–1905	Later Visc. Selby
J. W. Lowther	1905–21	Later Visc. Ullswater
J. H. Whitley	1921–28	
E. A. Fitzroy	1928–43	Widow made Viscs. Daventry
Douglas Clifton-Brown	1943–51	Later Visc. Ruffside
W. S. Morrison	1951–59	Later Visc. Dunrossil
Sir Harry B. Hylton-Foster	1959–65	Widow made Baroness Hylton-Foster
Dr. Horace King	1965–70	Later Baron Maybray-King
Selwyn Loyd	1971–76	Later Baron Selwyn Lloyd
George Thomas	1976–	

50. Chief Justices and Associate Justices of the Supreme Court of the U.S.A., 1789–

(Chief Justices in bold type)

Numbers show succession, each judge succeeding the previous justice bearing the same number. Superior letters following names refer to Presidential nominations; see list at end of section. For biographical information see Friedman and Israel, *The Justices of the U.S. Supreme Court*, 1789–1969, 4 vols.

1.	**John Jay (N.Y.)**[a]	1789–95	Later Governor of N.Y.
2.	John Rutledge (S.C.)[a]	1789–91	Resigned; never sat
3.	William Cushing (Mass.)[a]	1789–1810	
4.	James Wilson (Pa.)[a]	1789–98	
5.	John Blair (Va.)[a]	1789–96	
	Robert H. Harrison (Md.)[a]	1789	Never served
6.	James Iredell (N.C.)[a]	1790–98	
2.	Thomas Johnson (Md.)[a]	1791–93	
2.	William Paterson (N.J.)[a]	1793–1806	
1.	**John Rutledge (S.C.)**[a]	1795	Appointment not confirmed
5.	Samuel Chase (Md.)[a]	1796–1811	
1.	**Oliver Ellsworth (Conn.)**[a]	1796–99	Later U.S. Commissioner to France
4.	Bushrod Washington (Va.)[b]	1798–1829	
6.	Alfred Moore (N.C.)[b]	1799–1804	
1.	**John Marshall (Va.)**[b]	1801–35	
6.	William Johnson (S.C.)[c]	1804–34	
2.	Henry Brockholst Livingston (N.Y.)[c]	1806–23	
7.*	Thomas Todd (Ky.)[c]	1807–26	
3.	Joseph Story (Mass.)[d]	1811–45	

5.	Gabriel Duvall (Md.)[d]	1812–35	
2.	Smith Thomson (N.Y.)[e]	1823–43	
7.	Robert Trimble (Ky.)[f]	1826–28	
7.	John McLean (Ohio)[g]	1829–61	
4.	Henry Baldwin (Pa.)[g]	1830–44	
6.	James M. Wayne (Ga.)[g]	1835–67	
1.	**Roger B. Taney (Md.)**[g]	1836–64	
5.	Philip B. Barbour (Va.)[g]	1836–41	
8.†	John Catron (Tenn.)[g]	1837–65	
9.†	John McKinley (Ala.)[h]	1837–52	
5.	Peter V. Daniel (Va.)[h]	1841–60	
2.	Samuel Nelson (N.Y.)[j]	1845–72	
3.	Levi Woodbury (N.H.)[k]	1845–51	
4.	Robert C. Grier (Pa.)[k]	1846–70	
3.	Benjamin R. Curtis (Mass.)[l]	1851–57	
9.	John A. Campbell (Ala.)[m]	1853–61	
3.	Nathan Clifford (Me.)[n]	1858–81	
7.	Noah H. Swayne (Ohio)[o]	1862–81	
5.	Samuel F. Miller (Iowa)[o]	1862–90	
9.	David Davis (Ill.)[o]	1862–77	
10.‡	Stephen J. Field (Calif.)[o]	1863–97	
1.	**Salmon P. Chase** (Ohio)[o]	1864–73	
4.	William Strong (Pa.)[p]	1870–80	
6.	Joseph P. Bradley (N.J.)[p]	1870–92	
2.	Ward Hunt (N.Y.)[p]	1873–82	
1.	**Morrison R. Waite (Ohio)**[p]	1874–88	
9.	John M. Harlan (Ky.)[q]	1877–1911	
4.	William B. Woods (Ga.)[q]	1881–87	
7.	Stanley Matthews (Ohio)[r]	1881–89	
3.	Horace Gray (Mass.)[s]	1882–1902	
2.	Samuel Blatchford (N.Y.)[s]	1882–93	
4.	Lucius Q.C. Lamar (Miss.)[t]	1888–93	
1.	**Melville W. Fuller (Ill.)**[t]	1888–1910	
7.	David J. Brewer (Kan.)[u]	1889–1910	
5.	Henry B. Brown (Mich.)[u]	1891–1906	
6.	George Shiras, Jr. (Pa.)[u]	1892–1903	
4.	Howell E. Jackson (Tenn.)[u]	1893–1905	
2.	Edward D. White (La.)[v]	1894–1910	Later C.J., 1910–21
4.	Rufus W. Peckham (N.Y.)[v]	1895–1909	
8.	Joseph McKenna (Calif.)[w]	1898–1925	
3.	Oliver Wendell Holmes (Mass.)[x]	1902–32	
6.	William R. Day (Ohio)[x]	1903–22	
5.	William H. Moody (Mass.)[x]	1906–10	
4.	Horace H. Lurton (Tenn.)[y]	1910–14	
1.	**Edward D. White (La.)**[y]	1910–21	Assoc. Justice, 1894–1910
7.	Charles E. Hughes (N.Y.)[y]	1910–16	Secretary of State, 1921–25; Judge of P.C.I.J., 1928–30; C.J., 1930
2.	Willis van Devanter (Wyo.)[y]	1911–37	
5.	Joseph R. Lamar (Ga.)[y]	1911–16	
9.	Mathlon Pitney (N.J.)[y]	1912–22	

4.	James C. McReynolds (Tenn.)[z]	1914–41	
5.	Louis D. Brandeis (Mass.)[z]	1916–39	
7.	John H. Clarke (Ohio)[z]	1916–22	
1.	**William H. Taft (Conn.)[aa]**	1921–30	Formerly President of U.S., 1909–13
7.	George Sutherland (Utah)[aa]	1922–38	
6.	Pierce Butler (Minn.)[aa]	1922–39	
9.	Edward T. Sanford (Tenn.)[aa]	1923–30	
8.	Harlan F. Stone (N.Y.)[bb]	1925–41	Later C.J., 1941–46
1.	**Charles E. Hughes (N.Y.)[cc]**	1930–41	
9.	Owen J. Roberts (Pa.)[cc]	1930–45	
3.	Benjamin N. Cardozo (N.Y.)[cc]	1932–38	
2.	Hugo L. Black (Ala.)[dd]	1937–71	
7.	Stanley F. Reed (Ky.)[dd]	1938–57	
3.	Felix Frankfurter (Mass.)[dd]	1939–62	
5.	William O. Douglas (Conn.)[dd]	1939–75	Longest service on Court
6.	Frank Murphy (Mich.)[dd]	1940–49	
1.	**Harlan F. Stone (N.Y.)[dd]**	1941–46	Assoc. Justice, 1925–41
4.	James F. Byrnes (S.C.)[dd]	1941–42	
8.	Robert H. Jackson (N.Y.)[dd]	1941–54	
4.	Wiley B. Rutledge (Iowa)[dd]	1943–49	
9.	Harold H. Burton (Ohio)[ee]	1945–58	
1.	**Fred M. Vinson (Ky.)[ee]**	1946–53	
6.	Tom C. Clark (Tex.)[ee]	1949–67	
4.	Sherman Minton (Ind.)[ee]	1949–56	
1.	**Earl Warren (Calif.)[ff]**	1953–69	
8.	John M. Harlan (N.Y.)[ff]	1955–71	
4.	William J. Brennan, Jr. (Ind.)[ff]	1956	
7.	Charles E. Whittaker (Mo.)[ff]	1957–62	
9.	Potter Stewart (Ohio)[ff]	1958	
7.	Byron R. White (Colo.)[gg]	1962	
3.	Arthur J. Goldberg (Ill.)[gg]	1962–65	Later U.S. Ambassador to U.N.
3.	Abe Fortas (Tenn.)[hh]	1965–69	Nomination as C.J. not confirmed; resigned
6.	Thurgood Marshall (N.Y.)[hh]	1967	
1.	**Warren Earl Burger (Minn.)[jj]**	1969	
3.	Harry A. Blackmun (Minn.)[jj]	1970	
2.	Lewis F. Powell (Virg.)[jj]	1971	
8.	William H. Rehnquist (Ariz.)[jj]	1971	
5.	John Paul Stevens (Ill.)[kk]	1976	

* The Court was increased from six to seven in 1807.

† The Court was increased from seven to nine in 1837.

‡ The Court was increased from nine to ten in 1863 but when John Catron died in 1865 he was not replaced and seat 10 merged with seat 8. Since 1869 the Court has been fixed at nine.

(*a*)	Washington	(*k*)	Polk	(*t*)	Cleveland	(*cc*)	Hoover
(*b*)	Adams	(*l*)	Fillmore	(*u*)	B. Harrison	(*dd*)	F. D. Roosevelt
(*c*)	Jefferson	(*m*)	Pierce	(*v*)	Cleveland	(*ee*)	Truman
(*d*)	Madison	(*n*)	Buchanan	(*w*)	McKinley	(*ff*)	Eisenhower
(*e*)	Monroe	(*o*)	Lincoln	(*x*)	T. Roosevelt	(*gg*)	Kennedy
(*f*)	J. Q. Adams	(*p*)	Grant	(*y*)	Taft	(*hh*)	L. B. Johnson
(*g*)	Jackson	(*q*)	Hayes	(*z*)	Wilson	(*jj*)	Nixon
(*h*)	Van Buren	(*r*)	Garfield	(*aa*)	Harding	(*kk*)	Ford
(*j*)	Tyler	(*s*)	Arthur	(*bb*)	Coolidge		

51. Attorney-Generals of the U.S.A., 1789–

Pres. Washington	Edmund Randolph (Va.)	1789
	William Bradford (Va.)	1794
	Charles Lee (Va.)	1795
Pres. John Adams	Charles Lee (Va.)	1797
	Theophilus Parsons (Mass.)	1801
Pres. Jefferson	Levi Lincoln (Mass.)	1801
	Robert Smith (Md.)	1805
	John Breckenridge (Ky.)	1805
	Caesar A. Rodney (Del.)	1807
Pres. Madison	Caesar A. Rodney (Del.)	1809
	William Pinkney (Md.)	1811
	Richard Rush (Pa.)	1814
Pres. Monroe	Richard Rush (Pa.)	1817
	William Wirt (Va.)	1817
Pres. J. Q. Adams	William Wirt (Va.)	1817
Pres. Jackson	John McP. Berrien (Ga.)	1829
	Roger B. Taney (Md.)	1831
	Benjamin F. Butler (N.Y.)	1833
Pres. Van Buren	Benjamin F. Butler (N.Y.)	1837
	Felix Grundy (Tenn.)	1838
	Henry D. Gilpin (Pa.)	1840
Pres. W. H. Harrison	John J. Crittenden (Ky.)	1841
Pres. Tyler	John J. Crittenden (Ky.)	1841
	Hugh S. Legaré (S.C.)	1841
	John Nelson (Md.)	1843
Pres. Polk	John Y. Mason (Va.)	1845
	Nathan Clifford (Me.)	1846
	Issac Toncay (Conn.)	1848
Pres. Taylor	Reverdy Johnson (Md.)	1849
Pres. Fillmore	John J. Crittenden (Ky.)	1850
Pres. Pierce	Caleb Cushing (Mass.)	1853
Pres. Buchanan	Jeremiah S. Black (Pa.)	1857
	Edwin M. Stanton (Ohio)	1860
Pres. Lincoln	Edward Bates (Mo.)	1861
	Titian J. Coffey (Pa.)	1863
	James Speed (Ky.)	1864
Pres. Johnson	James Speed (Ky.)	1865
	Henry Stanbury (Ohio)	1866
	William M. Evarts (N.Y.)	1868
Pres. Grant	Ebenezer R. Hoar (Mass.)	1869
	Amos T. Akerman (Ga.)	1870
	George M. Williams (Ore.)	1870
	Edwards Pierrepont (N.Y.)	1875
	Alphonso Taft (Ohio)	1876
Pres. Hayes	Charles Devens (Mass.)	1877
Pres. Garfield	Wayne MacVeagh (Pa.)	1881
Pres. Arthur	Benjamin H. Brewster (Pa.)	1881
Pres. Cleveland	Augustus Garland (Ark.)	1885
Pres. B. Harrison	William H. H. Miller (Ind.)	1889

Pres. Cleveland	Richard Olney (Mass.)	1893
	Judson Harmon (Ohio)	1895
Pres. McKinley	Joseph McKenna (Cal.)	1897
	John W. Griggs (N.J.)	1898
	Philander C. Knox (Pa.)	1901
Pres. T. Roosevelt	Philander C. Knox (Pa.)	1901
	William H. Moody (Mass.)	1904
	Charles J. Bonaparte (Md.)	1906
Pres. Taft	George W. Wickersham (N.Y.)	1909
Pres. Wilson	James C. McReynolds (Tenn.)	1913
	Thomas W. Gregory (Tex.)	1914
	A. Mitchell Palmer (Pa.)	1919
Pres. Harding	Harry M. Daugherty (Ohio)	1921
Pres. Coolidge	Harry M. Daugherty (Ohio)	1923
	Harlan F. Stone (N.Y.)	1924
	John S. Sargent (Vt.)	1925
Pres. Hoover	William D. Mitchell (Minn.)	1929
Pres. F. D. Roosevelt	Homer S. Cummings (Conn.)	1933
	Frank Murphy (Mich.)	1939
	Robert H. Jackson (N.Y.)	1940
	Francis Biddle (Pa.)	1941
Pres. Truman	Tom C. Clark (Tex.)	1945
	J. Howard McGrath (R.I.)	1949
	J. P. McGranery (Pa.)	1952
Pres. Eisenhower	Herbert Brownell, Jr. (N.Y.)	1953
	William P. Rodgers (N.Y.)	1957
Pres. Kennedy	Robert F. Kennedy (Va.)	1961
Pres. L. B. Johnson	Robert F. Kennedy (Va.)	1963
	Nicholas de B. Katzenbach (Pa.)	1964
Pres. Nixon	John Mitchell ( )	1969
	Richard G. Kleindienst ( )	1972
	Elliot Richardson ( )	1973
	William S. Saxbe (Ohio)	1973
Pres. Ford	Edward H. Levi ( )	1975
Pres. Carter	Griffin Bell (Georgia)	1977

52. Judges of the Permanent Court of International Justice, 1921–30

Judges

M.	Altamira	(Spain)	
	Anzilotti	(Italy)	President, 1928–30
	Barboza	(Brazil)	
	de Bustamente	(Cuba)	
Lord	Finlay	(U.K.)	
M.	Huber	(Switzerland)	President, 1925–27, Vice-President, 1928–30
	Loder	(Netherlands)	President, 1921–24
	J. B. Moore	(U.S.A.)	
	Nyholm	(Denmark)	
	Oda	(Japan)	
	Weiss	(France)	Vice-President, 1921–28

Deputy-Judges

M.	Beichmann	(Norway)
	Negulesco	(Roumania)
	Wang	(China)
	Yovanovitch	(Jugoslavia)

Changes in the Court 1921–30

M. da Silva Pessoa (Brazil) succeeded M. Barboza, 1923.
Charles Evans Hughes (U.S.A.) succeeded J. B. Moore, 1928
M. Fromageot (France) succeeded M. Weiss, 1929.
Sir Cecil Hurst (U.K.) succeeded Lord Finlay, 1929.
Frank B. Kellogg (U.S.A.) succeeded Charles Evans Hughes, 1930.

1930–39

(Biographical notices in Seventh Annual Report of P.C.I.J.)

Judges

M. Adatci	(Japan) President, 1931–34
R. Altamira y Crevea	(Spain)
D. Anzilotti	(Italy)
A. S. de Bustamente y Sirvan	(Cuba)
H. Fromageot	(France)
Jonkheer van Eysinga	(Netherlands)
J. G. Guerrero	(Salvador) Vice-President, 1931–36, President, 1936–39
Sir Cecil Hurst	(U.K.) President, 1934–36, Vice-President, 1936–39
Baron Rolin Jaequemyns	(Belgium)
Frank B. Kellogg	(U.S.A.)
D. Negulesco	(Roumania)
Count M. Rostworowski	(Poland)
W. Schucking	(Germany)
F. J. Urrutia	(Colombia)
Wang Chung-Hui	(China)

Changes in the Court, 1930–39

(Biographical notices of new judges in Annual Reports of P.C.I.J. for year of election)

Harnkazu Nagaoka (Japan) succeeded M. Adatci, 1935.
M. A. Hammarskjold (Sweden) succeeded W. Schucking, 1936.
Manley O. Hudson (U.S.A.) succeeded F. B. Kellogg, 1936
Cheng Tien-Hsi (China) succeeded Wang Chung Hiu, 1936
Charles de Visscher (Belgium) succeeded Baron Rolin Jaequemyns, 1937
Rafael Waldemar Erich (Finland) succeeded M. Hammarskjold, 1938

1939–

The election of a new court, due to take place in 1939, was never held. The court did not sit from 1940 but continued in being until superseded by the International Court of Justice in 1946.

53. Judges of the International Court of Justice, 1946–

(Regular elections for five seats held triennially in and after 1945, for triennia commencing 1946, 1949, etc.)

For biographical information see *Year Book of the International Court of Justice* and *International Year Book and Statesmen's Who's Who.*

A. Alvarez	(Chile)	1946–55	
J. Azevedo	(Brazil)	1946–51	Died 1951
A. H. Badawi	(Egypt)	1946–67	V.P. 1955–58
J. Basdevant	(France)	1946–64	V.P. 1946–49; President, 1949–52
Ch. de Visscher	(Belgium)	1946–52	
I. Fabela	(Mexico)	1946–52	
J. G. Guerrero	(Salvador)	1946–60	President, 1946–49; V.P. 1949–55
G. H. Hackworth	(U.S.A.)	1946–61	President, 1955–58
Hsu Mo	(China)	1946–56	Died 1956
H. Klaestad	(Norway)	1946–61	President, 1958–61
S. B. Krylov	(U.S.S.R.)	1946–52	
Sir Arnold McNair	(U.K.)	1946–55	President, 1952–55
J. E. Read	(Canada)	1946–58	
B. Winiarski	(Poland)	1946–67	President, 1961–64
M. Zoricic	(Jugoslavia)	1946–58	
Levi Carneiro	(Brazil)	1951–55	Vice Azevedo, died
Sir B. N. Rau	(India)	1952–53	Died 1953
Sir M. Zafrulla Khan	(India)	1953–61 and 1964–73	Vice Rau, died; V.P., 1958–61; President, 1970–73
E. C. Armand Ugon	(Uruguay)	1952–61	
S. A. Golunsky	(U.S.S.R.)	1952–53	Resigned 1953
F. I. Kojevnikov	(U.S.S.R.)	1953–61	Vice Golunsky, resigned
Hersch Lauterpacht	(U.K.)	1955–60	Died 1960
L. M. Moreno Quintana	(Argentine)	1955–64	
R. Cordova	(Mexico)	1955–64	
Dr. V. K. Wellington Koo	(China)	1956–67	Vice Hsu Mo, died; V.P. 1964–67
Sir Percy C. Spender	(Australia)	1958–67	President, 1964–67
Jean Spiropoulos	(Greece)	1958–67	
Ricardo J. Alfaro	(Panama)	1960–64	Vice Guerrero, died; V.P. 1961–64
Sir Gerald Fitzmaurice	(U.K.)	1960–73	Vice Lauterpacht, died
J. L. Bustamente y Rivero	(Peru)	1961–70	President 1967–70
Philip C. Jessup	(U.S.A.)	1961–70	
Vladimir M. Koretsky	(U.S.S.R.)	1961–70	V.P. 1967–70
Gaetano Morelli	(Italy)	1961–70	
Kotara Tanaka	(Japan)	1961–70	
Issac Forster	(Senegal)	1964–82	
André Gros	(France)	1964–82	
Luis Padilla Nervo	(Mexico)	1964–73	
Fouad Ammoun	(Lebanon)	1967–76	V.P. 1970–76
Cesar Bengzon	(Philippines)	1967–76	
Sture Petren	(Sweden)	1967–76	

Manfred Lachs	(Poland)	1967–85	President, 1973–76
Charles D. Onyeama	(Nigeria)	1967–76	
Hardy C. Dillard	(U.S.A)	1970–79	
Louis Ignacio-Pinto	(Dahomey)	1970–79	
Federico de Castro	(Spain)	1970–79	
Platon D. Morozov	(U.S.S.R.)	1970–79	
Eduardo Jimenez de Arechaga	(Uruguay)	1970–79	President, 1976–79
Sir Humphrey Waldock	(U.K.)	1973–82	
Nagendra Singh	(India)	1973–82	V.P., 1976–79
Jose Maria Ruda	(Argentina)	1973–82	
Herman Mosler	(Germany)	1976–85	
Taslim O. Elias	(Nigeria)	1976–85	
Salah El Dine Tarazi	(Syria)	1976–85	
Shigeru Ode	(Japan)	1976–85	

54. Judges of the Court of Justice of the European Coal and Steel Community 1954–58, and of the European Communities, 1958–

M. Pilotti	(Italy)	1954–59	President, 1954–59
P. J. S. Serrarens	(Netherlands)	1954–58	
Ch. L. Hammes	(Belgium)	1954–68	President, 1965–68
O. Riese	(Germany)	1954–63	
L. Delvaux	(Belgium)	1954–65	
J. Rueff	(France)	1954–62	
A. van Kleffens	(Netherlands)	1954–59	
A. M. Donner	(Netherlands)	1958–	President, 1959–65
T. Rossi	(Italy)	1959–68	
N. Catalano	(Italy)	1959–62	
A. Trabucchi	(Italy)	1962–73	
R. Lecourt	(France)	1962–77	President, 1968–77
W. Strauss	(Germany)	1963–70	
R. Monaco	(Italy)	1965–76	
J. Mertens de Wilmars	(Belgium)	1968–	
P. Pescatore	(Luxembourg)	1968–	
H. Kutscher	(Germany)	1971–	President, 1977–
Lord Mackenzie Stuart	(U.K.)	1973–	
C. O'Dalaigh	(Ireland)	1973–74	
M. Sørensen	(Denmark	1973–	
Aindrias O'Caoimh (A. O'Keeffe)	(Ireland)	1974–	
Giacinto Bosco	(Italy)	1976–	
Adolphe Touffait	(France)	1977–	

APPENDIX II
Bibliographical Note

A bibliography of the literature of and about Western law would itself be voluminous, but it may be useful to append this note as a pointer to the most important guides to further information. It does not include common standard reference books and general encyclopaedias.

1. Bibliographies

Annual Legal Bibliography, 1961– (Harvard Law School).

Beardsley, A. S., and Orman, O. C., *Legal Bibliography and the Use of Law Books*, 1947.

Besterman, T. *Law and International Law: A bibliography of bibliographies*, 1971.

Bibliografia giuridica internazionale, 1932– .

Caes, L. and Henrion, R., *Collectio Bibliographica Operum as Jus Romanum pertinentium*, 21 vols. and Supps.

Friend, W. L., *Anglo-American legal bibliographies*, 1944.

Grandin, A., *Bibliographie Générale des Sciences Juridiques*, 3 vols. and Supps.

Hicks, F. C., *Materials and Methods of Legal Research*, 1942.

Howell, M. A., *Bibliography of Bibliographies of Legal Material*, 1969.

International Association of Legal Science: *Cahiers de Bibliographie juridique.*

— *Catalogue des Sources de Documentation Juridique dans le Monde*, 1957.

Karlsruher juristische Bibliographie, 1965– .

Manual of Legal Citations, 2 vols., 1959–60 (Inst. of Adv. Legal Studies).

O'Higgins, P., *Bibliography of Periodical Literature relating to Irish Law*, 1966.

Price, M. O., and Bitner, H., *Effective Legal Research*, 1953.

Repertorium Bibliographicum Institutionum et Sodalitatum Juris Historiae, 1969 and Supp.

Revue Bibliographique des Ouvrages du Droit, 1894– .

Stollreither, K., *Internationale Bibliographie der juristischen Nachschlagwerke*, 1955.

Sweet & Maxwell, *Legal Bibliography of the British Commonwealth of Nations*, 1955– .

U.N.E.S.C.O., *Register of Legal Documentation in the World*, 1957.

2. Indexes to Periodical Literature

Blaustein, A. P., *Manual on Foreign Legal Periodicals and their Index*, 1962.

British Humanities Index, 1962– .

Index to Foreign Legal Periodicals, 1960– (Inst. of Adv. Legal Studies).

Index to Legal Periodicals, 1908– (Am. Assoc. of Law Librarians).

Index to Periodical Articles related to Law, 1958–

International Index, 1955–65.

International Index to Periodicals, 1922–55.

Jones–Chipman, *Index to Legal Periodical Literature*, 6 vols., 1888–1939.

Poole, W. F., *Index to Periodical Literature*, 1853, and Supps to 1907.

Readers' Guide to Periodical Literature, Supplement, 2 vols., 1907–19.

Social Sciences and Humanities Index, 1965–

Subject-Index to Periodicals, 1915–60.

3. Legal Biography and Profession

Allgemeine deutsche biographie, 1875–1910.

Atlay, J. B., *The Victorian Chancellors*, 1906.

Biographie nationale de Belgique, 1866–1938.

Biographie universelle, ancienne et moderne, ed. L. G. Michaud, 1843–65.

Bland, D. E., *Bibliography of the Inns of Court and Chancery.*

Campbell, J., *Lives of the Lord Chancellors*, 1868.

— *Lives of the Chief Justices*, 1849.

Chevalier, C. U. J., *Repertoire des sources historiques du moyen age*, 1903–7.

Dictionary of American Biography, 20 vols. and Supps.

Dictionary of National Biography, 63 vols. and Supps.

Dictionnaire biographique du Canada, 1966– .

Dictionnaire de biographie française, 1933– .

Die Grossen Deutschen, 1956–57.

Dizionario Biographico degli Italiani, 1960– .

Foss, E., *Biographical Dictionary of the Judges of England, 1066–1870.*

Friedman, L., and Israel, F. L., *The Justices of the U.S. Supreme Court, 1789–1969*, 1969.
Heuston, R. F. V., *Lives of the Lord Chancellors, 1885–1940*, 1964.
Macdonnell, J., *Great Jurists of the World*, 1913.
Manson, E., *Builders of Our Law*, 1904.
National Biografisch Woordenboek, 1964–74.
Neue deutsche Biographie, 1953– .
Nieuw Nederlandsch Biografisch woordenboek, 1911–37.
Nouvelle biographie generale, ed. J. C. F. Hofer, 1854–66.
Serle, P., *Dictionary of Australian Biography*, 1949.
Tompkins, D. C., *The Supreme Court of the U.S. – a bibliography*, 1959.
Townsend, W., *Lives of Twelve Eminent Judges*, 1846.
Warren, C., *History of the American Bar*, 1912.

4. Legal History

Beale, J. H., *Bibliography of Early English Law Books*, 1926.
Berger, A., *Encyclopedic Dictionary of Roman Law*, 1953.
Brunner, H., *Deutsche Rechtsgeschichte*, 1906–28.
Buckland, W. W., *Textbook of Roman Law*, 1963.
Calasso, F., *Medioevo del diritto*, 1954.
Calhoun, G. M. *A Working Bibliography of Greek Law*, 1927.
Continental Legal History Series, Vol. I–*General Survey of Continental Legal History*, 1912.
Fournier, P., and Le Bras, G., *Histoire des Collections canoniques en Occident*, 1931–32.
Garcia Gallo, A., *Manual de Historia del Derecho español*, 1967.
Gavet, G., *Sources de l'histoire des institutions et du droit français*, 1899.
Gilissen, J., *Introduction bibliographique à l'histoire du droit et a l'ethnologie juridique*, 1965– .
Holdsworth, Sir W., *History of English Law*, 17 vols., 1966.
Ius Romanum Medii Aevi, 1961– .
Jolowicz, H. F., and Nicholas, J. K. B., *Historical Introduction to the Study of Roman Law*, 1972.
Lepointe, G., *Histoire des Institutions et des faits sociaux*, 1956.
Naz, R., *Dictionnaire du droit canonique*, 1935.
Olivier-Martin, F., *Histoire du droit français*, 1961.
Pauly-Wissowa, *Real-Encyclopädie der classischen Altertumswissenschaft*, 1893– .
Plucknett, T. F. T., *Early English Legal Literature*, 1958.
Pollock, Sir F., and Maitland, F. W., *History of English Law before the time of Edward I*, 1898.
Recueils de la Société Jean Bodin pour l'histoire comparative des institutions.
Savigny, F. C. von, *Geschichte des römischen Rechts im Mittelalter* 1834–51.
Schiller, A. A., *Bibliography of Roman Law*, 1966.
Schroder, R., and von Kunssberg, E., *Lehrbuch der deutschen Rechtsgeschichte*, 1932.
Schulte, J. F. von, *Geschichte der Quellen und Literatur des canonischen Rechts*, 1875–77.
Schulz, F., *History of Roman Legal Science*, 1946.
— *Classical Roman Law*, 1951.
Selden Society, publications of.
Solmi, A., *La Storia del diritto italiano*, 1922.
Stair Society, *Sources and Literature of Scots Law*, 1936.
— *Introduction to Scottish Legal History*, 1955.
Vortrage und Forschungen.
Villien, A., and Magnin, E., *Dictionnaire de droit canonique*, 1924.
Winfield, P. H., *Chief Sources of English Legal History*, 1925.
Zachariae von Lingenthal, C. E., *Geschichte des griechisch-romischen Rechts*, 1892.

5. Jurisprudence

Berolzheimer, F., *The World's Legal Philosophies*, 1912.
Cairns, H., *Legal Philosophy from Plato to Hegel*, 1951.
Chambliss, W. J., and Seidman, R. B., *Sociology of the Law*, 1970.
Dias, R. W. M., *Bibliography of Jurisprudence*, 1970.
Friedrich, C. J., *Philosophy of Law in Historical Perspective*, 1958.
Jones, J. W., *Historical Introduction to the Theory of Law*, 1940.
Pound, R., *Introduction to the Philosophy of Law*, 1954.
— *Jurisprudence*, 5 vols., 1959.
— *Outlines of Lectures on Jurisprudence*, 1943.

6. Comparative Legal Studies

Arminjon, P., Nolde, B., and Wolff, M., *Droit comparé*, 1950–52.
Juris-classeur de droit comparé.
Schlegelberger, F., *Rechtsvergleichendes Handwörterbuch*, 1929–39.
Schnitzer, A., *Vergleichende Rechtslehre*, 1961.
Szladits, C., *Bibliography on Foreign and Comparative Law*, 1955 and Supps.
Zweigert, K. (ed.), *Encyclopaedia of Comparative Law*, 1969– .

7. International law and institutions

Aufricht, H., *Guide to League of Nations Publications, 1920–47*, 1951.
Bibliography of the International Court of Justice, 1946.
Borchard, E. M., *Bibliography of International law and continental law*, 1912.

Delupis, I., *Bibliography of International Law*, 1975.
Douma, J., *Bibliography of the International Court, 1918–64*, 1966.
Francesakis, P., *Répertoire de droit international*, 1968–69 and Supp.
Harvard Law School Library, *Catalogue of International Law and Relations*, 1965–67.
Heere, W. P., *International Bibliography of Air Law, 1900–71*, 1972.
Holtzendorff, F. von, *Handbuch des Völkerrechts*, 1885–89.
International Court of Justice, *Yearbook*, 1946– .
Johnson, H. S., *International Organization: A Classified Bibliography*, 1969.
La Pradelle, A. de, and Niboyet, J. B., *Répertoire de droit international*, 11 vols., 1929–34.
O'Connell, D. P., *International Law*, 1965.
Olivart, M. de, *Bibliographie du droit international*, 1905.
Oppenheim, L., *International Law*, 1952–55.
Robinson, J., *International Law and Organisation*, 1967.
Speeckaert, G. P., *Bibliographie sélective sur l'organisation internationale, 1885–1964*, 1965.
Strupp, K., *Worterbuch des Völkerrechts*, 1960–62.
United Nations Yearbook, 1946– .
Walker, T. A., *History of the Law of Nations*, 1899.
Winton, H. N. M., *Publications of the U.N. System*, 1972.
Yearbook of International Organisations, 1948– .

8. European Communities Law

The Atlantic Community: An introductory bibliography, 1962.
Campbell, A., *Common Market Law*, 1969 and Supps.
European Yearbook, 1955– .
Ganshof van der Meersch, W. J., *Droit des communautés europeenes*, 1969.
Megret, J., and others, *Le droit de la Communauté économique europeene*, 1970.
N.A.T.O., *Bibliography*, 1964.
Padelford, N.J., *Selected bibliography on regionalism and regional arrangements*, 1956.
Parry, A., and Hardy, S., *E.E.C. Law*, 1973.
Pehrsson, H., and Wulf, H., *The European Bibliography*, 1965.
Robertson, A. H., *European Institutions*, 1973.
Selected Bibliography of European Law (Council of Europe, 1971).
Wohlfarth, E., and others, *Die Europaische Wirtschafigemeinschaft*, 1960.

9. Commonwealth Law

Roberts-Wray, K., *Commonwealth and Colonial Law*, 1966.

10. Particular Countries

Austria

Langer, O., *Bibliographia juridica austriaca*, 1959.
Maultasche, F., *Rechtslexikon, Handbuch des österreichischen Rechts*, 1956.

Australia

Paton, G. W., *The Commonwealth of Australia*, 1952.
Sweet & Maxwell's *Legal Bibliography of the British Commonwealth*, Vol. 6, 1958.

Belgium

Bosly, H., *Répertoire bibliographique du droit belge*, 1919–45 and Supp.
Brunet, E., *Répertoire pratique du droit belge*, 1949.
Dekkers, R., *Bibliotheca Belgica Juridica*, 1951.
Graulich, P., *Guide to Foreign Legal Materials–Belgium, Luxembourg, Netherlands*, 1968.

Canada

Boult, R., *Bibliography of Canadian Law*, 1960.
Gall, G.R., *The Canadian Legal System*, 1977.
Sweet & Maxwell's *Legal Bibliography of the British Commonwealth*, Vol. 3, 1955.

Denmark

Sondergard, J., *Danish legal Publications*, in Scand. Studies in Law, 7.
See also SCANDINAVIA.

England and Wales

Chloros, A. G., *Bibliographical Guide to U.K. Law*, 1953.
Holdsworth, W. S., *Sources and Literature of English Law*, 1925.
Sweet & Maxwell's *Legal Bibliography of the British Commonwealth*, Vols 1 and 2, 1954–57.
Walker, R. J., and Walker, M. G., *The English Legal System*, 1972

Finland

Reinikainen, V., *Litterature sur le droit finlandais*, 1957

France

Amos, M. S., and Walton, F. P.: *Introduction to French Law*, 1967.
David, R., and de Vries, H. P.: *The French Legal System*, 1958.
David, R., *Bibliographie du droit français*, 1945–60.
— *Le droit français*, 1962.
Grandin, A., *Bibliographie generale des sciences juridiques, politiques, economiques et sociales*, 1926 and Supps.
Szladits, C., *Guide to Foreign Legal Materials, French, German, Swiss*, 1959.

Germany

Borchard, E. M., *Guide to the Law and Legal Literature of Germany*, 1912.
Gesellschaft fur Rechtsvergleichung (F. Baur, ed.). *Bibliography of German Law*, 1964.

H.M.S.O., *Manual of German Law*, 1968.
Szladits, C. See under FRANCE.

Greece

Zepos, P. J., *Greek Law*, 1949.

Ireland

Sweet & Maxwell's *Legal Bibliography of the British Commonwealth*, Vol. 4, 1955.

Italy

Cappelletti, M., *The Italian Legal System*, 1967.
Enciclopedia del diritto, 1958– .
Grisoli, A., *Guide to Foreign Legal Materials–Italian*, 1965.

Luxembourg

Frieden, F., *Bibliographie luxembourgeoise*, 1946. See also under BELGIUM.

Netherlands

Van der Berg, J. D. *Klapper op de rechtspraak en de rechtsliteratuur*, 1952– .
Fokkema, D. C., *Introduction to Dutch Law*, 1975. See also under BELGIUM.

New Zealand

Robson, J. L.: *New Zealand*, 1967.
Sweet & Maxwell's *Legal Bibliography of the British Commonwealth*, Vol. 6, 1955.

Northern Ireland

Chloros, A. G., *Bibliographical Guide to U.K. Law*, 1973.
Sweet & Maxwell's *Legal Bibliography of the British Commonwealth*, Vol. 4, 1955.

Norway

See SCANDINAVIA.

Scandinavia

Iuul, S.: *Scandinavian Legal Bibliography*, 1961.

Scotland

Chloros, A. G., *Bibliographical Guide to U.K. Law*, 1973.
Stair Society, *Sources and Literature of Scots Law*, 1936.
Sweet & Maxwell's *Legal Bibliography of the British Commonwealth*, Vol. 5, 1957.
Walker, D. M., *The Scottish Legal System*, 1976.

South Africa

Hahlo, H. R., and Kahm, E., *The South African Legal System*, 1968.
Roberts, A. A., *South African Legal Bibliography*, 1942.

Spain

Fernandez de Villavicencio, and de Sola Canizares, F., *Bibliografia Juridica Espanola*, 1954.
Palmer, T. W., *Guide to the Law of Spain*, 1915.

Sweden

Frykholm, L., *Swedish legal publications*, 1935–60, in Scand. Studies in Law, 5.
Regner, N., *Svensk juridisk litteratur*, 1865–1956 and Supps.

Switzerland

Christen, H., *Schweizer Rechtsbibliographie*.
Szladits, C. See under FRANCE.

Turkey

Ansay, T., and Wallace, D., *Introduction to Turkish Law*, 1966.

U.S.A.

Andrews, J. L., *The Law in the U.S.A.*, 1965.
Farnsworth, E. A., *Introduction to the Legal System of the U.S.A.*, 1963.
Harvard Law Library, *Annual Legal Bibliography*, 1960– .
Hicks, F. C., *Materials and Methods of Legal Research*, 1942.
Mayers, A., *The American Legal System*, 1955.
Price, M. O., and Bitner, H., *Effective Legal Research*, 1953.